COMPUTATIONAL FLUID DYNAMICS

Second Edition

This revised second edition of *Computational Fluid Dynamics* represents a significant improvement from the first edition. However, the original idea of including all computational fluid dynamics methods (FDM, FEM, FVM); all mesh generation schemes; and physical applications to turbulence, combustion, acoustics, radiative heat transfer, multiphase flow, electromagnetic flow, and general relativity is maintained. This unique approach sets this book apart from its competitors and allows the instructor to adopt this book as a text and choose only those subject areas of his or her interest.

The second edition includes new sections on finite element EBE-GMRES and a complete revision of the section on the flowfield-dependent variation (FDV) method, which demonstrates more detailed computational processes and includes additional example problems. For those instructors desiring a textbook that contains homework assignments, a variety of problems for FDM, FEM, and FVM are included in an appendix. To facilitate students and practitioners intending to develop a large-scale computer code, an example of FORTRAN code capable of solving compressible, incompressible, viscous, inviscid, 1-D, 2-D, and 3-D for all speed regimes using the flowfield-dependent variation method is available at http://www.uah.edu/cfd.

T. J. Chung is distinguished professor emeritus of mechanical and aerospace engineering at the University of Alabama in Huntsville. He has also authored *General Continuum Mechanics* and *Applied Continuum Mechanics*, both published by Cambridge University Press.

To my family

COMPUTATIONAL FLUID DYNAMICS

Second Edition

T. J. CHUNG
University of Alabama in Huntsville

CAMBRIDGE
UNIVERSITY PRESS

CAMBRIDGE
UNIVERSITY PRESS

32 Avenue of the Americas, New York NY 10013-2473, USA

Cambridge University Press is part of the University of Cambridge.

It furthers the University's mission by disseminating knowledge in the pursuit of education, learning and research at the highest international levels of excellence.

www.cambridge.org
Information on this title: www.cambridge.org/9781107425255

First edition published 2002
Second edition published 2010
Reprinted 2012, 2013 (twice)
First paperback edition 2014

A catalogue record for this publication is available from the British Library

Library of Congress Cataloguing in Publication data
Chung, T. J., 1929–
Computational fluid dynamics / T. J. Chung. – 2nd ed.
p. cm.
Includes bibliographical references and index.
ISBN 978-0-521-76969-3
1. Fluid dynamics – Data processing. I. Title.
QA911 .C476 2010
532′.050285 – dc22 2010029493

ISBN 978-0-521-76969-3 Hardback
ISBN 978-1-107-42525-5 Paperback

Additional resources for this publication at www.cambridge.org/9781107425255

Contents

APPENDIXES

Preface to the First Edition

This book is intended for the beginner as well as for the practitioner in computational fluid dynamics (CFD). It includes two major computational methods, namely, finite difference methods (FDM) and finite element methods (FEM) as applied to the numerical solution of fluid dynamics and heat transfer problems. An equal emphasis on both methods is attempted. Such an effort responds to the need that advantages and disadvantages of these two major computational methods be documented and consolidated into a single volume. This is important for a balanced education in the university and for the researcher in industrial applications.

Finite volume methods (FVM), which have been used extensively in recent years, can be formulated from either FDM or FEM. FDM is basically designed for structured grids in general, but is applicable also to unstructured grids by means of FVM. New ideas on formulations and strategies for CFD in terms of FDM, FEM, and FVM continue to emerge, as evidenced in recent journal publications. The reader will find the new developments interesting and beneficial to his or her area of applications. However, the subject material is often inaccessible due to barriers caused by different training backgrounds. Therefore, in this book, the relationship among all currently available computational methods is clarified and brought to a proper perspective.

To the uninitiated beginner, this book will serve as a convenient guide toward the desired destination. To the practitioner, however, preferences and biases built over the years can be relaxed and redeveloped toward other possible options. Having studied all methods available, the reader may then be able to pursue the most reasonable directions to follow, depending on the specific physical problems of each reader's own field of interest. It is toward this flexibility that the present volume is addressed.

The book begins with Part One, Preliminaries, in which the basic principles of FDM, FEM, and FVM are illustrated by means of a simple differential equation, each leading to the identical exact solution. Most importantly, through these examples with step-by-step hand calculations, the concepts of FDM, FEM, and FVM can be easily understood in terms of their analogies and differences. The introduction (Chapter 1) is followed by the general forms of governing equations, boundary conditions, and initial conditions encountered in CFD (Chapter 2), prior to embarking on details of CFD methods.

Parts Two and Three cover FDM and FEM, respectively, including both historical developments and recent contributions. FDM formulations and solutions of various types of partial differential equations are discussed in Chapters 3 and 4, whereas the counterparts for FEM are covered in Chapters 8 through 11. Incompressible and compressible flows are treated in Chapters 5 and 6 for FDM and in Chapters 12 through 14 for FEM, respectively. FVM is included in both Part Two (Chapter 7) and Part Three (Chapter 15) in accordance with its original point of departure. Historical developments are important for the beginner, whereas the recent contributions are included as they are required for advanced applications given in Part Five. Chapter 16, the last chapter in Part Three, discusses the detailed comparison between FDM and FEM and other methods in CFD.

Full-scale complex CFD projects cannot be successfully accomplished without automatic grid generation strategies. Both structured and unstructured grids are included. Adaptive methods, computing techniques, and parallel processing are also important aspects of the industrial CFD activities. These and other subjects are discussed in Part Four (Chapters 17 through 20).

Finally, Part Five (Chapters 21 through 27) covers various applications including turbulence, reacting flows and combustion, acoustics, combined mode radiative heat transfer, multiphase flows, electromagnetic fields, and relativistic astrophysical flows.

It is intended that as many methods of CFD as possible be included in this text. Subjects that are not available in other textbooks are given full coverage. Due to a limitation of space, however, details of some topics are reduced to a minimum by making a reference, for further elaboration, to the original sources.

This text has been classroom tested for many years at the University of Alabama in Huntsville. It is considered adequate for four semester courses with three credit hours each: CFD I (Chapters 1 through 4 and 8 through 11), CFD II (Chapters 5 through 7 and 12 through 16), CFD III (Chapters 17 through 20), and CFD IV (Chapters 21 through 27). In this way, the elementary topics for both FDM and FEM can be covered in CFD I with advanced materials for both FDM and FEM in CFD II. FVM via FDM and FVM via FEM are included in CFD I and CFD II, respectively. CFD III deals with grid generation and advanced computing techniques covered in Part IV. Finally, the various applications covered in Part V constitute CFD IV. Since it is difficult to study all subject areas in detail, each student may be given an option to choose one or two chapters for special term projects, more likely dictated by the expertise of the instructor, perhaps toward thesis or dissertation topics.

Instead of providing homework assignments at the end of each chapter, some selected problems are shown in Appendix E. An emphasis is placed on comparisons between FDM, FEM, and FVM. Through these exercises, it is hoped that the reader will gain appreciation for studying all available methods such that, in the end, advantages and disadvantages of each method may be identified toward making decisions on the most suitable choices for the problems at hand. Associated with Appendix E is a Web site http://www.uah.edu/cfd that provides code (FORTRAN 90) for solutions of some of the homework problems. The student may use this as a guide for programming with other languages such as C++ for the class assignments.

More than three decades have elapsed since the author's earlier book on FEM in CFD was published [McGraw-Hill, 1978]. Recent years have witnessed great progress in FEM, parallel with significant achievements in FDM. The author has personally experienced the advantage of studying both methods on an equal footing. The purpose of this book is, therefore, to share the author's personal opinion with the reader, wishing that this idea may lead to further advancements in CFD in the future. It is hoped that all students in the university will be given an unbiased education in all areas of CFD. It is also hoped that the practitioners in industry will benefit from many alternatives that may impact their new directions of future research in CFD applications.

In completing this text, the author recalls with sincere gratitude a countless number of colleagues and students, both past and present. They have contributed to this book in many different ways.

My association with Tinsley Oden has been an inspiration, particularly during the early days of finite element research. Among many colleagues are S. T. Wu and Gerald Karr, who have shared useful discussions in CFD research over the past three decades.

I express my sincere appreciation to Kader Frendi, who contributed to Sections 23.2 (pressure mode acoustics) and 23.3 (vorticity mode acoustics) and to Vladimir Kolobov for Section 26.3.2 (semiconductor plasma processing).

My thanks are due to J. Y. Kim, L. R. Utreja, P. K. Kim, J. L. Sohn, S. K. Lee, Y. M. Kim, O. Y. Park, C. S. Yoon, W. S. Yoon, P. J. Dionne, S. Warsi, L. Kania, G. R. Schmidt, A. M. Elshabka, K. T. Yoon, S. A. Garcia, S. Y. Moon, L. W. Spradley, G. W. Heard, R. G. Schunk, J. E. Nielsen, F. Canabal, G. A. Richardson, L. E. Amborski, E. K. Lee, and G. H. Bowers, among others. They assisted either during the course of development of earlier versions of my CFD manuscript or at the final stages of completion of this book.

I would like to thank the reviewers for suggestions for improvement. I owe a debt of gratitude to Lawrence Spradley, who read the entire manuscript, brought to my attention numerous errors, and offered constructive suggestions. I am grateful to Francis Wessling, Chairman of the Department of Mechanical & Aerospace Engineering, UAH, who provided administrative support, and to S. A. Garcia and Z. Q. Hou, who assisted in typing and computer graphics. Without the assistance of Z. Q. Hou, this text could not have been completed in time. My thanks are also due to Florence Padgett, Engineering Editor at Cambridge University Press, who has most effectively managed the publication process of this book.

T. J. Chung

Preface to the Revised Second Edition

This revised second edition of *Computational Fluid Dynamics* represents a significant improvement from the first edition. However, the original idea of including all computational fluid dynamics methods (FDM, FEM, FVM); all mesh generation schemes; and physical applications to turbulence, combustion, acoustics, radiative heat transfer, multiphase flow, electromagnetic flow, and general relativity is maintained. This unique approach sets this book apart from its competitors and allows the instructor to adopt this book as a text and choose only those subject areas of his or her interest.

The second edition includes new sections on finite element EBE-GMRES and a complete revision of the section on the flowfield-dependent variation (FDV) method, which demonstrates more detailed computational processes and includes additional example problems. For those instructors desiring a textbook that contains homework assignments, a variety of problems for FDM, FEM, and FVM are included in an appendix. To facilitate students and practitioners intending to develop a large-scale computer code, an example of FORTRAN code capable of solving compressible, incompressible, viscous, inviscid, 1-D, 2-D, and 3-D for all speed regimes using the flowfield-dependent variation method is available at http://www.uah.edu/cfd.

PART ONE

PRELIMINARIES

The dawn of the twentieth century marked the beginning of the numerical solution of differential equations in mathematical physics and engineering. Numerical solutions were carried out by hand and using desk calculators for the first half of the twentieth century, then by digital computers for the later half of the century. In Section 1.1, a brief summary of the history of computational fluid dynamics (CFD) will be given, along with the organization of text.

Before we proceed with details of CFD, simple examples are presented for the beginner, demonstrating how to solve a simple differential equation numerically by hand calculations (Sections 1.2 through 1.7). Basic concepts of finite difference methods (FDM), finite element methods (FEM), and finite volume methods (FVM) are easily understood by these examples, laying a foundation or providing a motivation for further explorations. Even the undergraduate student may be brought to an adequate preparation for advanced studies toward CFD. This is the main purpose of Preliminaries.

Furthermore, in Preliminaries, we review the basic forms of partial differential equations and some of the governing equations in fluid dynamics (Sections 2.1 and 2.2). These include nonconservation and conservation forms of the Navier-Stokes system of equations as derived from the first law of thermodynamics and are expressed in terms of the control volume/surface integral equations, which represent various physical phenomena such as inviscid/viscous, compressible/incompressible, subsonic/supersonic flows, and so on.

Typical boundary conditions are briefly summarized, with reference to hyperbolic, parabolic, and elliptic equations (Section 2.3). Examples of Dirichlet, Neumann, and Cauchy (Robin) boundary conditions are also examined, with additional and more detailed boundary conditions to be discussed later in the book.

CHAPTER ONE

Introduction

1.1 GENERAL

1.1.1 HISTORICAL BACKGROUND

The development of modern computational fluid dynamics (CFD) began with the advent of the digital computer in the early 1950s. Finite difference methods (FDM) and finite element methods (FEM), which are the basic tools used in the solution of partial differential equations in general and CFD in particular, have different origins. In 1910, at the Royal Society of London, Richardson presented a paper on the first FDM solution for the stress analysis of a masonry dam. In contrast, the first FEM work was published in the *Aeronautical Science Journal* by Turner, Clough, Martin, and Topp for applications to aircraft stress analysis in 1956. Since then, both methods have been developed extensively in fluid dynamics, heat transfer, and related areas.

Earlier applications of FDM in CFD include Courant, Friedrichs, and Lewy [1928], Evans and Harlow [1957], Godunov [1959], Lax and Wendroff [1960], MacCormack [1969], Briley and McDonald [1973], van Leer [1974], Beam and Warming [1978], Harten [1978, 1983], Roe [1981, 1984], Jameson [1982], among many others. The literature on FDM in CFD is adequately documented in many text books such as Roache [1972, 1999], Patankar [1980], Peyret and Taylor [1983], Anderson, Tannehill, and Pletcher [1984, 1997], Hoffman [1989], Hirsch [1988, 1990], Fletcher [1988], Anderson [1995], and Ferziger and Peric [1999], among others.

Earlier applications of FEM in CFD include Zienkiewicz and Cheung [1965], Oden [1972, 1988], Chung [1978], Hughes et al. [1982], Baker [1983], Zienkiewicz and Taylor [1991], Carey and Oden [1986], Pironneau [1989], Pepper and Heinrich [1992]. Other contributions of FEM in CFD for the past two decades include generalized Petrov-Galerkin methods [Heinrich et al., 1977; Hughes, Franca, and Mallett, 1986; Johnson, 1987], Taylor-Galerkin methods [Donea, 1984; Löhner, Morgan, and Zienkiewicz, 1985], adaptive methods [Oden et al., 1989], characteristic Galerkin methods [Zienkiewicz et al., 1995], discontinuous Galerkin methods [Oden, Babuska, and Baumann, 1998], and incompressible flows [Gresho and Sani, 1999], among others.

There is a growing evidence of benefits accruing from the combined knowledge of both FDM and FEM. Finite volume methods (FVM), because of their simple data structure, have become increasingly popular in recent years, their formulations being related to both FDM and FEM. The flowfield-dependent variation (FDV) methods [Chung, 1999] also point to close relationships between FDM and FEM. Therefore, in this book we are seeking to recognize such views and to pursue the advantage of studying FDM and FEM together on an equal footing.

Historically, FDMs have dominated the CFD community. Simplicity in formulations and computations contributed to this trend. FEMs, on the other hand, are known to be more complicated in formulations and more time-consuming in computations. However, this is no longer the case in many of the recent developments in FEM applications. Many examples of superior performance of FEM have been demonstrated. Our ultimate goal is to be aware of all advantages and disadvantages of all available methods so that if and when supercomputers grow manyfold in speed and memory storage, this knowledge will be an asset in determining the computational scheme capable of rendering the most accurate results, and not be limited by computer capacity. In the meantime, one may always be able to adjust his or her needs in choosing between suitable computational schemes and available computing resources. It is toward this flexibility and desire that this text is geared.

1.1.2 ORGANIZATION OF TEXT

This book covers the basic concepts, procedures, and applications of computational methods in fluids and heat transfer, known as computational fluid dynamics (CFD). Specifically, the fundamentals of finite difference methods (FDM) and finite element methods (FEM) are included in Parts Two and Three, respectively. Finite volume methods (FVM) are placed under both FDM and FEM as appropriate. This is because FVM can be formulated using either FDM or FEM. Grid generation, adaptive methods, and computational techniques are covered in Part Four. Applications to various physical problems in fluids and heat transfer are included in Part Five.

The unique feature of this volume, which is addressed to the beginner and the practitioner alike, is an equal emphasis of these two major computational methods, FDM and FEM. Such a view stems from the fact that, in many cases, one method appears to thrive on merits of other methods. For example, some of the recent developments in finite elements are based on the Taylor series expansion of conservation variables advanced earlier in finite difference methods. On the other hand, unstructured grids and the implementation of Neumann boundary conditions so well adapted in finite elements are utilized in finite differences through finite volume methods. Either finite differences or finite elements are used in finite volume methods in which in some cases better accuracy and efficiency can be achieved. The classical spectral methods may be formulated in terms of FDM or they can be combined into finite elements to generate spectral element methods (SEM), the process of which demonstrates usefulness in direct numerical simulation for turbulent flows. With access to these methods, readers are given the direction that will enable them to achieve accuracy and efficiency from their own judgments and decisions, depending upon specific individual needs. This volume addresses the importance and significance of the in-depth knowledge of both FDM and FEM toward an ultimate unification of computational fluid dynamics strategies in general. A thorough study of all available methods without bias will lead to this goal.

Preliminaries begin in Chapter 1 with an introduction of the basic concepts of all CFD methods (FDM, FEM, and FVM). These concepts are applied to solve simple one-dimensional problems. It is shown that all methods lead to identical results. In this process, it is intended that the beginner can follow every step of the solution with simple hand calculations. Being aware that the basic principles are straightforward, the reader may be adequately prepared and encouraged to explore further developments in the rest of the book for more complicated problems.

Chapter 2 examines the governing equations with boundary and initial conditions which are encountered in general. Specific forms of governing equations and boundary and initial conditions for various fluid dynamics problems will be discussed later in appropriate chapters.

Part Two covers FDM, beginning with Chapter 3 for derivations of finite difference equations. Simple methods are followed by general methods for higher order derivatives and other special cases.

Finite difference schemes and solution methods for elliptic, parabolic, and hyperbolic equations, and the Burgers' equation are discussed in Chapter 4. Most of the basic finite difference strategies are covered through simple applications.

Chapter 5 presents finite difference solutions of incompressible flows. Artificial compressibility methods (ACM), SIMPLE, PISO, MAC, vortex methods, and coordinate transformations for arbitrary geometries are elaborated in this chapter.

In Chapter 6, various solution schemes for compressible flows are presented. Potential equations, Euler equations, and the Navier-Stokes system of equations are included. Central schemes, first order and second order upwind schemes, the total variation diminishing (TVD) methods, preconditioning process for all speed flows, and the flowfield-dependent variation (FDV) methods are discussed in this chapter.

Finite volume methods (FVM) using finite difference schemes are presented in Chapter 7. Node-centered and cell-centered schemes are elaborated, and applications using FDV methods are also included.

Part Three begins with Chapter 8, in which basic concepts for the finite element theory are reviewed, including the definitions of errors as used in the finite element analysis. Chapter 9 provides discussion of finite element interpolation functions.

Applications to linear and nonlinear problems are presented in Chapter 10 and Chapter 11, respectively. Standard Galerkin methods (SGM), generalized Galerkin methods (GGM), Taylor-Galerkin methods (TGM), and generalized Petrov-Galerkin (GPG) methods are discussed in these chapters.

Finite element formulations for incompressible and compressible flows are treated in Chapter 12 and Chapter 13, respectively. Although there are considerable differences between FDM and FEM in dealing with incompressible and compresible flows, it is shown that the new concept of flowfield-dependent variation (FDV) methods is capable of relating both FDM and FEM closely together.

In Chapter 14, we discuss computational methods other than the Galerkin methods. Spectral element methods (SEM), least squares methods (LSM), and finite point methods (FPM, also known as meshless methods or element-free Galerkin), are presented in this chapter. Chapter 15 discusses finite volume methods with finite elements used as a basic structure.

Finally, the overall comparison between FDM and FEM is presented in Chapter 16, wherein analogies and differences between the two methods are detailed. Furthermore, a general formulation of CFD schemes by means of the flowfield-dependent variation (FDV) algorithm is shown to lead to most all existing computational schemes in FDM and FEM as special cases. Brief descriptions of available methods other than FDM, FEM, and FVM such as boundary element methods (BEM), particle-in-cell (PIC) methods, Monte Carlo methods (MCM) are also given in this chapter.

Part Four begins with structured grid generation in Chapter 17, followed by unstructured grid generation in Chapter 18. Subsequently, adaptive methods with structured grids and unstructured grids are treated in Chapter 19. Various computing techniques, including domain decomposition, multigrid methods, and parallel processing, are given in Chapter 20.

Applications of numerical schemes suitable for various physical phenomena are discussed in Part Five (Chapters 21 through 27). They include turbulence, chemically reacting flows and combustion, acoustics, combined mode radiative heat transfer, multiphase flows, electromagnetic flows, and relativistic astrophysical flows.

1.2 ONE-DIMENSIONAL COMPUTATIONS BY FINITE DIFFERENCE METHODS

In this and the following sections of this chapter, the beginner is invited to examine the simplest version of the introduction of FDM, FEM, FVM via FDM, and FVM via FEM, with hands-on exercise problems. Hopefully, this will be a sufficient motivation to continue with the rest of this book.

In finite difference methods (FDM), derivatives in the governing equations are written in finite difference forms. To illustrate, let us consider the second-order, one-dimensional linear differential equation,

$$\frac{d^2u}{dx^2} - 2 = 0 \quad 0 < x < 1 \tag{1.2.1a}$$

with the Dirichlet boundary conditions (values of the variable u specified at the boundaries),

$$\begin{cases} u = 0 & \text{at } x = 0 \\ u = 0 & \text{at } x = 1 \end{cases} \tag{1.2.1b}$$

for which the exact solution is $u = x^2 - x$.

It should be noted that a simple differential equation in one-dimensional space with simple boundary conditions such as in this case possesses a smooth analytical solution. Then, all numerical methods (FDM, FEM, and FVM) will lead to the exact solution even with a coarse mesh. We shall examine that this is true for this example problem.

The finite difference equations for du/dx and d^2u/dx^2 are written as (Figure 1.2.1)

$$\left(\frac{du}{dx}\right)_i \approx \frac{u_{i+1} - u_i}{\Delta x} \quad \textit{forward difference} \tag{1.2.2a}$$

$$\left(\frac{du}{dx}\right)_i \approx \frac{u_i - u_{i-1}}{\Delta x} \quad \textit{backward difference} \tag{1.2.2b}$$

$$\left(\frac{du}{dx}\right)_i \approx \frac{u_{i+1} - u_{i-1}}{2\Delta x} \quad \textit{central difference} \tag{1.2.2c}$$

$$\frac{d^2u}{dx^2} = \frac{d}{dx}\left(\frac{du}{dx}\right) \cong \frac{1}{\Delta x}\left[\left(\frac{du}{dx}\right)_{i+1} - \left(\frac{du}{dx}\right)_i\right] = \frac{1}{\Delta x}\left(\frac{u_{i+1} - u_i}{\Delta x} - \frac{u_i - u_{i-1}}{\Delta x}\right) \tag{1.2.3}$$

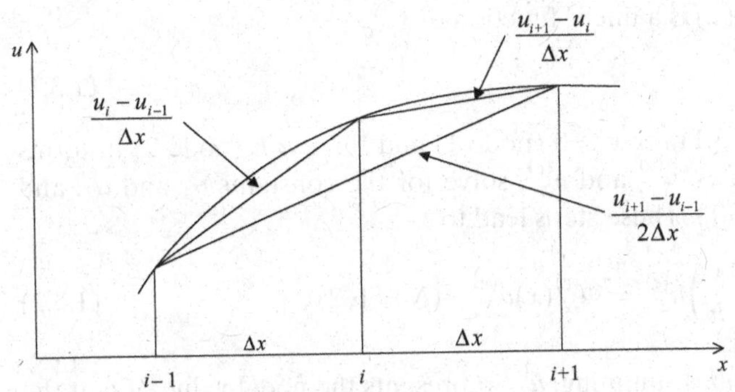

Figure 1.2.1 Finite difference approximations.

Substitute (1.2.3) into (1.2.1a) and use three grid points to obtain

$$\frac{u_{i+1} - 2u_i + u_{i-1}}{\Delta x^2} = 2 \tag{1.2.4}$$

With $u_{i-1} = 0$, $u_{i+1} = 0$, as specified by the given boundary conditions, the solution at $x = 1/2$ with $\Delta x = 1/2$ becomes $u_i = -1/4$. This is the same as the exact solution given by

$$u_i = (x^2 - x)_{x=\frac{1}{2}} = -\frac{1}{4} \tag{1.2.5}$$

In what follows, we shall demonstrate that the same exact solution is obtained, using other methods: FEM and FVM.

1.3 ONE-DIMENSIONAL COMPUTATIONS BY FINITE ELEMENT METHODS

For illustration, let us consider a one-dimensional domain as depicted in Figure 1.3.1a. Let the domain be divided into subdomains; say two local elements ($e = 1, 2$) in this example as shown in Figure 1.3.1b,c. The end points of elements are called nodes.

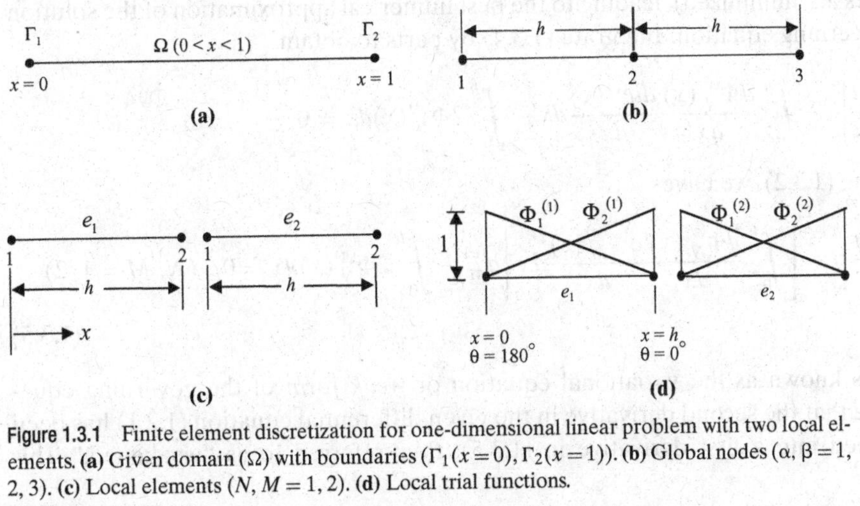

Figure 1.3.1 Finite element discretization for one-dimensional linear problem with two local elements. **(a)** Given domain (Ω) with boundaries ($\Gamma_1(x=0)$, $\Gamma_2(x=1)$). **(b)** Global nodes ($\alpha, \beta = 1, 2, 3$). **(c)** Local elements ($N, M = 1, 2$). **(d)** Local trial functions.

Assume that the variable $u^{(e)}(x)$ is a linear function of x

$$u^{(e)}(x) = \alpha_1 + \alpha_2 x \tag{1.3.1}$$

Write two equations from (1.3.1) for $x = 0$ (node 1) and for $x = h$ (node 2) in terms of the nodal values of variables, $u_1^{(e)}$ and $u_2^{(e)}$, solve for the constants α_1 and α_2, and substitute them back into (1.3.1). These steps lead to

$$u^{(e)}(x) = \left(1 - \frac{x}{h}\right)u_1^{(e)} + \left(\frac{x}{h}\right)u_2^{(e)} = \Phi_N^{(e)}(x)u_N^{(e)} \quad (N = 1, 2) \tag{1.3.2}$$

where the repeated index implies summing, $u_N^{(e)}$ represents the nodal value of u at the local node N for the element (e), and $\Phi_N^{(e)}(x)$ are called the local domain (element) *trial functions* (alternatively known as interpolation functions, shape functions, or basis functions),

$$\Phi_1^{(e)}(x) = 1 - \frac{x}{h}, \qquad \Phi_2^{(e)}(x) = \frac{x}{h} \tag{1.3.3a}$$

$$0 \leq \Phi_N^{(e)}(x) \leq 1 \tag{1.3.3b}$$

These functions are shown in Figure 1.3.1d, indicating that trial functions assume the value of one at the node under consideration and zero at the other node, linearly varying in between.

There are many different ways to formulate finite element equations (as detailed in Part Three). One of the simplest approaches is known as the Galerkin method. The basic idea is to construct an inner product of the residual $R^{(e)}$ of the local form of the governing equation (1.2.1a) with the *test functions* chosen the same as the trial functions given by (1.3.3) and in (1.3.2):

$$(\Phi_N^{(e)}(x), R^{(e)}) = \int_0^h \Phi_N^{(e)}(x)\left(\frac{d^2u^{(e)}(x)}{dx^2} - 2\right)dx = 0 \tag{1.3.4}$$

This represents an orthogonal projection of the residual error onto the subspace spanned by the test functions summed over the domain, which is then set equal to zero (implying that errors are minimized), leading to the best numerical approximation of the solution to the governing equation. Integrate (1.3.4) by parts to obtain

$$\overset{*}{\Phi}{}_N^{(e)} \frac{du}{dx}\bigg|_0^h - \int_0^h \frac{d\Phi_N^{(e)}(x)}{dx}\frac{du^{(e)}(x)}{dx}dx - \int_0^h 2\Phi_N^{(e)}(x)dx = 0$$

or by using (1.3.2), we have

$$\overset{*}{\Phi}{}_N^{(e)} \frac{du}{dx}\bigg|_0^h - \left[\int_0^h \frac{d\Phi_N^{(e)}(x)}{dx}\frac{d\Phi_M^{(e)}(x)}{dx}dx\right]u_M^{(e)} - \int_0^h 2\Phi_N^{(e)}(x)dx = 0 \quad (N, M = 1, 2) \tag{1.3.5}$$

This is known as the variational equation or *weak form* of the governing equation. Note that the second derivative in the given differential equation (1.2.1) has been transformed into a first derivative in (1.3.5), thus referred to as "weakened." This implies that, instead of solving the second order differential equation directly, we are to solve the first order (weakened) integro-differential equation as given by (1.3.5), thus leading to a *weak solution*, as opposed to a *strong solution* that represents the analytical solution of (1.2.1). The derivative du/dx in the first term is no longer the variable within the domain, but it is the Neumann boundary condition (constant) to be specified at $x = 0$ or $x = h$ if so required. Likewise, the test function is no longer the function of x, thus given a special notation $\overset{*}{\Phi}{}_N^{(e)}$, called the Neumann boundary test function, as opposed to the domain test function $\Phi_N^{(e)}(x)$. The Neumann boundary test function assumes the value of 1 if the Neumann boundary condition is applied at node N, and 0 otherwise, similar to a Dirac delta function. This represents one of the limit values given by (1.3.3b) at $x = 0$ or $x = h$, indicating that it is no longer the function of x within the domain. Furthermore, appropriate direction cosines must be assigned, reduced from two-dimensional configurations (Figure 8.2.3). Depending on the Neumann boundary condition being applied on either the left-hand side ($x = 0$) or the right-hand side ($x = h$), we obtain

$$\left.\frac{du}{dx}\right|_{x=0} = \frac{du}{dx}\cos\theta\bigg|_{\theta=180^\circ} = -\frac{du}{dx}, \qquad \left.\frac{du}{dx}\right|_{x=h} = \frac{du}{dx}\cos\theta\bigg|_{\theta=0^\circ} = \frac{du}{dx} \tag{1.3.6a}$$

To prove (1.3.6a), we must first refer to the 2-D geometry as shown in Figure 8.2.3, and integration by parts is carried out as follows:

$$\iint \Phi_N^{(e)}(x)\frac{d^2u}{dx^2}dxdy \Rightarrow \int \overset{*}{\Phi}{}_N^{(e)}\frac{du}{dx}dy = \int \overset{*}{\Phi}{}_N^{(e)}\frac{du}{dx}\cos\theta\, d\Gamma = \overset{*}{\Phi}{}_N^{(e)}\frac{du}{dx}\cos\theta$$

$$= \overset{*}{\Phi}{}_N^{(e)}\frac{du}{dx}\bigg|_{x=0,\theta=180^\circ}^{x=h,\theta=0^\circ} \tag{1.3.6b}$$

in which only the integrated term is shown (omitting the differentiated term) and the direction cosines for 1-D are applied at both ends of an element ($\theta = 0^\circ$ for $x = h$, $\theta = 180^\circ$ for $x = 0$). This represents the simplification of 2-D geometry into a 1-D problem.

Using a compact notation, we rewrite (1.3.5) as

$$K_{NM}^{(e)}\, u_M^{(e)} = F_N^{(e)} + G_N^{(e)} \quad (N, M = 1, 2) \tag{1.3.7}$$

This leads to a system of local algebraic finite element equations, consisting of the following quantities [henceforth the functional representation (x) in the domain trial and test functions will be omitted for simplicity unless confusion is likely to occur]:

Stiffness (Diffusion *or* Viscosity) Matrix (associated with the physics arising from the second derivative term)

$$K_{NM}^{(e)} = \int_0^h \frac{d\Phi_N^{(e)}}{dx}\frac{d\Phi_M^{(e)}}{dx}dx = \begin{bmatrix} \int_0^h \frac{d\Phi_1^{(e)}}{dx}\frac{d\Phi_1^{(e)}}{dx}dx & \int_0^h \frac{d\Phi_1^{(e)}}{dx}\frac{d\Phi_2^{(e)}}{dx}dx \\ \int_0^h \frac{d\Phi_2^{(e)}}{dx}\frac{d\Phi_1^{(e)}}{dx}dx & \int_0^h \frac{d\Phi_2^{(e)}}{dx}\frac{d\Phi_2^{(e)}}{dx}dx \end{bmatrix}$$

$$= \begin{bmatrix} K_{11}^{(e)} & K_{12}^{(e)} \\ K_{21}^{(e)} & K_{22}^{(e)} \end{bmatrix} = \frac{1}{h}\begin{bmatrix} 1 & -1 \\ -1 & 1 \end{bmatrix}$$

Source Vector

$$F_N^{(e)} = -\int_0^h 2\Phi_N^{(e)} dx = -h\begin{bmatrix}1\\1\end{bmatrix}$$

Neumann Boundary Vector

$$G_N^{(e)} = \overset{*}{\Phi}{}_N^{(e)} \frac{du}{dx}\bigg|_0^h = \overset{*}{\Phi}{}_N^{(e)} \frac{du}{dx}\cos\theta$$

Contributions of local elements calculated above ($e = 1, 2$) can be assembled into global nodes ($\alpha, \beta = 1, 2, 3$) simply by summing the adjacent elemental contributions to the global node shared by both elements. In this example, global node 2 is shared by local node 2 of element 1 and local node 1 of element 2.

$$K_{\alpha\beta} = \begin{bmatrix} K_{11} & K_{12} & K_{13} \\ K_{21} & K_{22} & K_{23} \\ K_{31} & K_{32} & K_{33} \end{bmatrix} = \begin{bmatrix} K_{11}^{(1)} & K_{12}^{(1)} & 0 \\ K_{21}^{(1)} & K_{22}^{(1)} + K_{11}^{(2)} & K_{12}^{(2)} \\ 0 & K_{21}^{(2)} & K_{22}^{(2)} \end{bmatrix} = \frac{1}{h}\begin{bmatrix} 1 & -1 & 0 \\ -1 & 2 & -1 \\ 0 & -1 & 1 \end{bmatrix} \tag{1.3.8}$$

$$F_\alpha = \begin{bmatrix} F_1 \\ F_2 \\ F_3 \end{bmatrix} = \begin{bmatrix} F_1^{(1)} \\ F_2^{(1)} + F_1^{(2)} \\ F_2^{(2)} \end{bmatrix} = -h\begin{bmatrix} 1 \\ 2 \\ 1 \end{bmatrix} \tag{1.3.9}$$

$$G_\alpha = \begin{bmatrix} G_1 \\ G_2 \\ G_3 \end{bmatrix} = \begin{bmatrix} G_1^{(1)} \\ G_2^{(1)} + G_1^{(2)} \\ G_2^{(2)} \end{bmatrix} = \begin{bmatrix} \overset{*}{\Phi}_1 \\ \overset{*}{\Phi}_2 \\ \overset{*}{\Phi}_3 \end{bmatrix} \frac{du}{dx}\cos\theta = \begin{bmatrix} \overset{*}{\Phi}{}_1^{(1)} \\ \overset{*}{\Phi}{}_2^{(1)} + \overset{*}{\Phi}{}_1^{(2)} \\ \overset{*}{\Phi}{}_2^{(2)} \end{bmatrix} \frac{du}{dx}\cos\theta$$

$$= \begin{bmatrix} 0 \\ 0 \\ 0 \end{bmatrix} \frac{du}{dx}\cos\theta \tag{1.3.10}$$

with $\overset{*}{\Phi}_1 = \overset{*}{\Phi}_2 = \overset{*}{\Phi}_3 = 0$ indicating that the Neumann boundary conditions are not to be applied to any of the global nodes for the solution of (1.2.1a,b). This implies that, if the Neumann boundary conditions are not applied, then the Neumann boundary vector is zero even if the gradient du/dx is not zero. If the Neumann boundary conditions are to be applied, then the boundary test function $\overset{*}{\Phi}{}_N^{(e)}$ assumes the value of one and the du/dx as given is simply imposed at the node under consideration. This is a part of the FEM formulation that makes the process more complicated than in FDM, but it is a distinct advantage when the Neumann boundary conditions are to be specified exactly.

Notice that the 2×2 local stiffness matrices for element 1 and element 2 are overlapped (superimposed) at the global node 2 with the contributions algebraically summed together,

$$K_{22} = K_{22}^{(1)} + K_{11}^{(2)}$$

and similarly,

$$F_2 = F_2^{(1)} + F_1^{(2)}, \qquad G_2 = G_2^{(1)} + G_1^{(2)}$$

In view of the above, we obtain the final global algebraic equations in the form

$$\begin{bmatrix} 1 & -1 & 0 \\ -1 & 2 & -1 \\ 0 & -1 & 1 \end{bmatrix} \begin{bmatrix} u_1 \\ u_2 \\ u_3 \end{bmatrix} = -h^2 \begin{bmatrix} 1 \\ 2 \\ 1 \end{bmatrix} \tag{1.3.11}$$

It will be shown in Chapter 8 that the global finite element equations (1.3.11) may be obtained directly from the global form of (1.3.4),

$$(\Phi_\alpha, R) = \int_0^1 \Phi_\alpha \left(\frac{d^2u}{dx^2} - 2 \right) dx = 0 \tag{1.3.12}$$

which will lead to (1.3.11), or

$$K_{\alpha\beta} u_\beta = F_\alpha + G_\alpha \quad (\alpha, \beta = 1, 2, 3) \tag{1.3.13}$$

Expanding (1.3.11) at the global node 2 yields

$$-u_1 + 2u_2 - u_3 = -2h^2 \quad (h = \Delta x) \tag{1.3.14}$$

or

$$\frac{u_{i+1} - 2u_i + u_{i-1}}{\Delta x^2} = 2 \tag{1.3.15}$$

This result is identical to the FDM formulation (1.2.4).

The Galerkin finite element method described here is called the standard Galerkin method (SGM). It works well for linear differential equations, but is not adequate for nonlinear problems in fluid mechanics. In this case, the test functions must be of the form different from the trial functions. This will be one of the topics to be discussed in Part Three.

1.4 ONE-DIMENSIONAL COMPUTATIONS BY FINITE VOLUME METHODS

Finite volume methods (FVM) utilize the control volumes and control surfaces as depicted in Figure 1.4.1. The control volume for node i covers $\Delta x/2$ to the right and left of node i with the control surface being located at $i - 1/2$ and $i + 1/2$. Finite volume formulations can be obtained either by a finite difference basis or a finite element basis. The results are identical for one-dimensional problems.

1.4.1 FVM VIA FDM

The basic idea for the formulation of FVM is similar to the finite element method (1.3.12) with the test function being set equal to unity, as applied to the differential equation (1.2.1a),

$$(\Phi_\alpha, R) = (1, R) = \int_0^1 (1) \left(\frac{d^2u}{dx^2} - 2 \right) dx = 0, \quad 0 < x < 1 \tag{1.4.1}$$

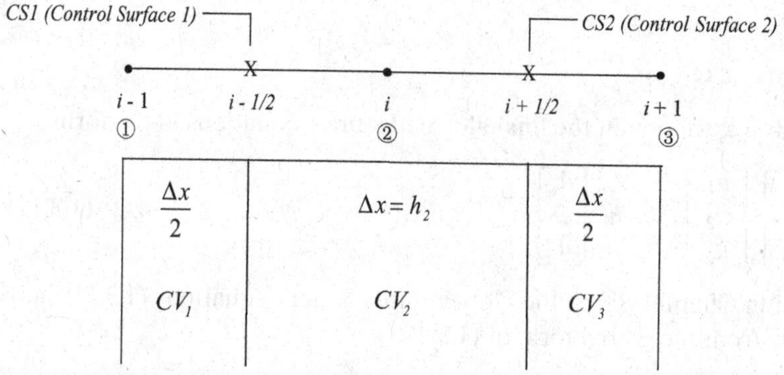

Figure 1.4.1 Finite volume approximations.

Integrating (1.4.1) yields

$$\left.\frac{du}{dx}\right|_0^1 - \int_0^1 2dx = 0 \tag{1.4.2a}$$

or

$$\sum_{CS1,2} \frac{\Delta u}{\Delta x} - \sum_{CV2} 2\Delta x = 0 \tag{1.4.2b}$$

The integration limits of 0 and 1 are now replaced by discrete control surfaces ($CS1$ and $CS2$) between $i - 1/2$ and $i + 1/2$, and the source term is to be evaluated for the control volume ($CV2$), with reference to Figure 1.4.1. This implies that du/dx in (1.4.2a) is to be evaluated at the control surfaces and that the diffusion flux du/dx is conserved between $i - 1$ and i through the control surface $i - 1/2$ or $CS1$ and between i and $i + 1$ through the control surface $i + 1/2$ or $CS2$. This is accomplished when the control surface equations are assembled at $i - 1, i$, and $i + 1$. This conservation property is the most significant aspect of the finite volume methods.

To complete the illustrative process, (1.4.2) can be written using finite difference representation for the control surfaces between $i - 1/2$ and $i + 1/2$ as

$$\underset{(CS2)}{\frac{u_{i+1} - u_i}{\Delta x}} - \underset{(CS1)}{\frac{u_i - u_{i-1}}{\Delta x}} = \underset{(CV2)}{2\Delta x} \tag{1.4.3}$$

Dividing (1.4.3) by Δx, we obtain

$$\frac{u_{i+1} - 2u_i + u_{i-1}}{\Delta x^2} = 2 \tag{1.4.4}$$

which is identical to (1.2.4) for the finite difference method. Note that $CV1$ and $CV3$ do not contribute to this process since nodes $i - 1$ and $i + 1$ are the boundaries whose influence is contained in (1.4.3) through control surfaces $CS1$ and $CS2$.

1.4.2 FVM VIA FEM

In order to demonstrate that FVM can also be formulated by FEM, we evaluate du/dx analytically from the trial functions (1.3.2), (Figure 1.3.1d), for the finite volume representation of (1.4.2a),

$$u^{(e)} = \Phi_N^{(e)} u_N^{(e)} = \left(1 - \frac{x}{h}\right) u_1^{(e)} + \frac{x}{h} u_2^{(e)}$$

or

$$\frac{du^{(e)}}{dx} = \frac{u_2^{(e)} - u_1^{(e)}}{h}$$

so that, from (1.3.6), we obtain

$$\left.\frac{du^{(1)}}{dx}\right|_{CS1} = \frac{u_2^{(1)} - u_1^{(1)}}{h} \cos\theta \Big|_{\theta=180^\circ} = \frac{u_2^{(1)} - u_1^{(1)}}{h}(-1) \tag{1.4.5}$$

$$\left.\frac{du^{(2)}}{dx}\right|_{CS2} = \frac{u_2^{(2)} - u_1^{(2)}}{h} \cos\theta \Big|_{\theta=0^\circ} = \frac{u_2^{(2)} - u_1^{(2)}}{h}(1) \tag{1.4.6}$$

Here, $CS1$ provides the direction cosine, $\cos\theta = \cos 180^\circ = -1$, whereas $CS2$ gives $\cos\theta = \cos 0^\circ = 1$, with reference to Figure 1.4.1.

Summing the fluxes through $CS1$ and $CS2$ at the control volume center (node 2) in terms of the global nodes

$$\sum_{CS1,2} \frac{du}{dx} = \frac{u_2 - u_1}{h}(-1) + \frac{u_3 - u_2}{h}(1) \tag{1.4.7}$$

Note that, using (1.4.7), the finite volume representation (1.4.2) is given by

$$\frac{u_3 - 2u_2 + u_1}{\Delta x^2} = 2 \tag{1.4.8}$$

Once again, the result is the same as all other previous analyses.

1.5 NEUMANN BOUNDARY CONDITIONS

So far, we have dealt with only the Dirichlet boundary conditions for numerical examples. However, it has been seen that the Neumann boundary condition, du/dx, arises automatically from the finite element or finite volume formulations through integration by parts. This information, if given as an input, may be implemented at the boundary nodes under consideration. This is not the case for finite difference methods.

To demonstrate this point, let us return to the differential equation examined in Section 1.2.

$$\frac{d^2u}{dx^2} - 2 = 0 \quad 0 < x < 1 \tag{1.5.1}$$

with the following boundary conditions:

$$u(0) = 0 \qquad \text{(Dirichlet)} \quad \text{at } x = 0 \tag{1.5.2}$$

$$\frac{du}{dx}(1) = 1 \qquad \text{(Neumann)} \quad \text{at } x = 1 \tag{1.5.3}$$

where it is reminded that the given differential equation (1.5.1) is described only within the domain, $0 < x < 1$, not including the boundaries, $x = 0$ and $x = 1$, which are reserved for the specification of boundary conditions, either Dirichlet or Neumann. Only when the governing equation is integrated are the boundary points ($x = 0, x = 1$) needed and used.

In the following subsections, implementations of the Neumann boundary conditions will be demonstrated.

1.5.1 FDM

One way to implement the Neumann boundary condition of the type (1.5.3) is to install a phantom (ghost, imaginary, fictitious) node 4 as shown in Figure 1.5.1. Writing the finite difference equation and the Neumann boundary condition (slope) at the boundary node 3, we have

$$u_4 - 2u_3 + u_2 = 2\Delta x^2 \tag{1.5.4}$$

$$\frac{u_4 - u_2}{2\Delta x} = 1 \tag{1.5.5}$$

Substitute (1.5.5) into (1.5.4),

$$2\Delta x + u_2 - 2u_3 + u_2 = 2\Delta x^2 \tag{1.5.6}$$

Writing the finite difference equation at node 2, we have

$$u_3 - 2u_2 + u_1 = 2\Delta x^2 \tag{1.5.7}$$

Solve (1.5.6) and (1.5.7) simultaneously to obtain

$$u_2 = -1/4, \quad \text{with } u_3 = 0$$

which is the exact solution. This is because the approximation given by (1.5.5) is reasonable with respect to the exact solution. The phantom node method may give a large

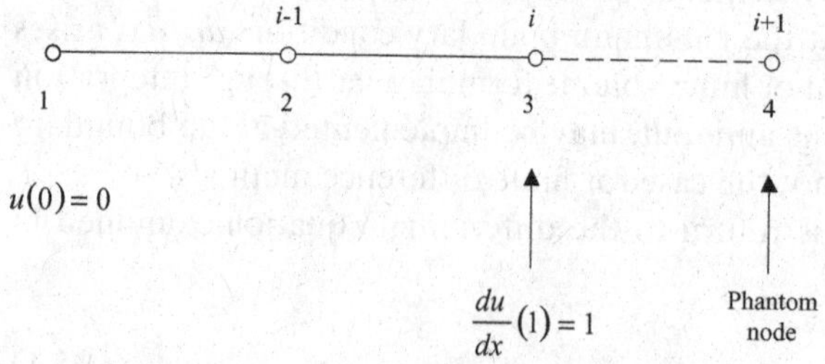

Figure 1.5.1 Installation of phantom node for Neumann boundary condition in finite difference method.

error if this is not the case, or if the solution is unsymmetric with respect to the interior and phantom node.

Instead of using a phantom node, we may utilize the higher order finite difference equation at the Neumann boundary node. For example, we use the second order accurate finite difference formula for du/dx at node 3 (see Chapter 3 for derivation),

$$\left(\frac{du}{dx}\right)_3 = \frac{3u_3 - 4u_2 + u_1}{2\Delta x} = 1 \tag{1.5.8}$$

Solve u_3 from the above and substitute the result into (1.5.7) and obtain once again the exact solution $u_2 = -1/4$, $u_3 = 0$.

1.5.2 FEM

It follows from (1.3.6b) that, at the Neumann boundary node 3,

$$G_N^{(e)} = \overset{*}{\Phi}{}_N^{(e)} \frac{du}{dx}\bigg|_0^h, \quad with\ \overset{*}{\Phi}{}_3^{(e)} = 1 \tag{1.5.9a}$$

Thus

$$G_3 = (1)\frac{du}{dx}\bigg|_{x=h} = (1)\frac{du}{dx}\cos 0^\circ = 1 \tag{1.5.9b}$$

It follows from (1.3.11) that, having applied the Dirichlet boundary condition at node 1 ($u(0) = 0$), the global finite element equation becomes

$$\begin{bmatrix} 2 & -1 \\ -1 & 1 \end{bmatrix}\begin{bmatrix} u_2 \\ u_3 \end{bmatrix} = -h^2\begin{bmatrix} 2 \\ 1 \end{bmatrix} + h\begin{bmatrix} 0 \\ 1 \end{bmatrix} \tag{1.5.10}$$

from which we obtain the exact solution $u_2 = -1/4$ and $u_3 = 0$. Notice that FEM accommodates the Neumann boundary conditions exactly within the formulation itself, not through those approximations required in FDM.

At this point it is important to realize that, if the Neumann boundary condition $du/dx = -1$ is specified on the left end, then we have

$$G_1 = \frac{du}{dx}\bigg|_{x=0} = \frac{du}{dx}\cos 180^\circ = (-1)(-1) = 1$$

Thus, we have

$$\begin{bmatrix} 1 & -1 \\ -1 & 2 \end{bmatrix}\begin{bmatrix} u_1 \\ u_2 \end{bmatrix} = -h^2\begin{bmatrix} 1 \\ 2 \end{bmatrix} + h\begin{bmatrix} 1 \\ 0 \end{bmatrix}$$

This will once again give the exact solution, $u_1 = 0$ and $u_2 = -1/4$.

1.5.3 FVM VIA FDM

The finite volume equation is given by Figure 1.4.1,

$$\frac{du}{dx}\bigg|_{i-\frac{1}{2}}^{i+\frac{1}{2}} - \int_{i-\frac{1}{2}}^{i+\frac{1}{2}} 2dx = 0$$

or in terms of finite differences at node 2,

$$\frac{u_3 - u_2}{\Delta x} - \frac{u_2 - u_1}{\Delta x} - 2\Delta x = 0 \tag{1.5.11}$$

at node 3,

$$\left.\frac{du}{dx}\right|_3 - \left.\frac{du}{dx}\right|_{i+\frac{1}{2}} - 2\frac{\Delta x}{2} = 0 \quad \text{or} \quad 1 - \frac{u_3 - u_2}{\Delta x} - 2\frac{\Delta x}{2} = 0 \tag{1.5.12}$$

Combining (1.5.11) and (1.5.12), we obtain

$$\begin{bmatrix} 2 & -1 \\ -1 & 1 \end{bmatrix} \begin{bmatrix} u_2 \\ u_3 \end{bmatrix} = -h^2 \begin{bmatrix} 2 \\ 1 \end{bmatrix} + h \begin{bmatrix} 0 \\ 1 \end{bmatrix}$$

It is interesting to note that this is identical to the FEM formulation (1.5.10). Solving, we have the exact solution ($u_2 = -1/4$, $u_3 = 0$). In this manner, FVM via FDM is capable of implementing the Neumann boundary conditions exactly, unlike FDM.

1.5.4 FVM VIA FEM

We return to (1.4.2a),

$$\left.\frac{du}{dx}\right|_0^1 - \int_0^1 2dx = 0 \tag{1.5.13}$$

where at node 2 we have, from (1.4.5) and (1.4.6),

$$\left.\frac{du}{dx}\right|_0^1 - \int_0^1 2dx = \frac{du}{dx}\cos 180^\circ + \frac{du}{dx}\cos 0^\circ - 2h$$

or

$$\frac{u_2 - u_1}{h}(-1) + \frac{u_3 - u_2}{h}(1) - 2h = 0 \tag{1.5.14}$$

at node 3,

$$\left.\frac{du}{dx}\right|_{2\frac{1}{2}} + \left.\frac{du}{dx}\right|_3 - 2\frac{h}{2} = 0$$

or

$$\frac{u_3 - u_2}{h}(-1) + 1 - h = 0 \tag{1.5.15}$$

Combining (1.5.14) and (1.5.15), we have

$$\begin{bmatrix} 2 & -1 \\ -1 & 1 \end{bmatrix} \begin{bmatrix} u_2 \\ u_3 \end{bmatrix} = \begin{bmatrix} -2h^2 \\ h - h^2 \end{bmatrix}$$

This gives the exact solution, $u_2 = -1/4$ and $u_3 = 0$. Once again in FVM via FEM the treatment of the Neumann boundary condition is precise.

1.6 EXAMPLE PROBLEMS

Here we provide additional examples, illustrating further applications of boundary conditions and including treatment of source terms.

1.6.1 DIRICHLET BOUNDARY CONDITIONS

Consider the three-element system as shown in Figure 1.6.1a to solve the differential equation with the source term $f(x)$,

$$\frac{d^2u}{dx^2} - 2u = f(x) \quad 0 < x < 1 \tag{1.6.1}$$

$$f(x) = 4x^2 - 2x - 4$$

subject to the Dirichlet boundary conditions:

$$u = 0 \quad \text{at } x = 0$$
$$u = -1 \quad \text{at } x = 1$$

whose exact solution is given by $u = -2x^2 + x$.

FDM

Write FDE at nodes 2 and 3.

Node 2

$$\frac{u_3 - 2u_2 + u_1}{\Delta x^2} - 2u_2 = f_2$$

$$\frac{u_3 - 2u_2 + 0}{(1/3)^2} - 2u_2 = 4\left(\frac{1}{3}\right)^2 - 2\left(\frac{1}{3}\right) - 4 = \frac{-38}{9}$$

$$9(u_3 - 2u_2) - 2u_2 = \frac{-38}{9}$$

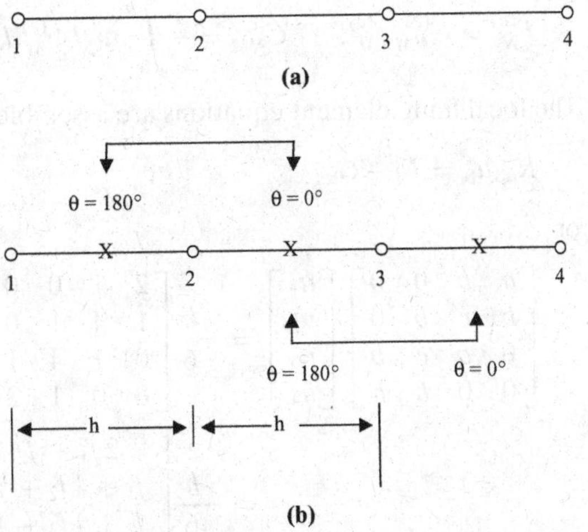

Figure 1.6.1 Example problem, Dirichlet and Neumann boundary conditions. (**a**) Three elements, four nodes for FDM and FEM. (**b**) Direction cosines at control surfaces as a result of integration by parts for FVM.

Node 3

$$\frac{u_4 - 2u_3 + u_2}{(1/3)^2} - 2u_3 = 4\left(\frac{2}{3}\right)^2 - 2\left(\frac{2}{3}\right) - 4 = \frac{-32}{9}$$

$$9(-1 - 2u_3 + u_2) - 2u_3 = \frac{-32}{9}$$

Combining, we have

$$\begin{bmatrix} -20 & 9 \\ 9 & -20 \end{bmatrix}\begin{bmatrix} u_2 \\ u_3 \end{bmatrix} = \begin{bmatrix} -\frac{38}{9} \\ \frac{49}{9} \end{bmatrix}$$

$$\begin{bmatrix} u_2 \\ u_3 \end{bmatrix} = \begin{bmatrix} 0.111 \\ -0.222 \end{bmatrix}$$

These values represent the exact solution.

FEM

The local Galerkin finite element analog is given by

$$\int_0^h \Phi_N^{(e)}\left(\frac{d^2u}{dx^2} - 2u - f(x)\right)dx = 0$$

where the source term $f(x)$ may be linearly approximated in the form

$$f(x) = \Phi_N^{(e)}(x) f_N^{(e)}$$

Integrating by part, the local algebraic equations are written as

$$K_{NM}^{(e)} u_M^{(e)} = F_N^{(e)} + G_N^{(e)}$$

where

$$K_{NM}^{(e)} = \int_0^h \left(\frac{d\Phi_N^{(e)}}{dx}\frac{d\Phi_M^{(e)}}{dx} + 2\Phi_N^{(e)}\Phi_M^{(e)}\right)dx = \frac{1}{h}\begin{bmatrix} 1 & -1 \\ -1 & 1 \end{bmatrix} + \frac{2h}{6}\begin{bmatrix} 2 & 1 \\ 1 & 2 \end{bmatrix}$$

$$F_N^{(e)} = C_{NM}^{(e)} f_M^{(e)}, \qquad C_{NM}^{(e)} = -\int_0^h \Phi_N^{(e)}\Phi_M^{(e)} dx = -\frac{h}{6}\begin{bmatrix} 2 & 1 \\ 1 & 2 \end{bmatrix}, \qquad G_{NM}^{(e)} = \begin{bmatrix} 0 \\ 0 \end{bmatrix}$$

The local finite element equations are assembled into the global form,

$$K_{\alpha\beta} u_\beta = F_\alpha + G_\alpha$$

or

$$\begin{bmatrix} a & b & 0 & 0 \\ b & c & b & 0 \\ 0 & b & c & b \\ 0 & 0 & b & a \end{bmatrix}\begin{bmatrix} u_1 \\ u_2 \\ u_3 \\ u_4 \end{bmatrix} = -\frac{h}{6}\begin{bmatrix} 2 & 1 & 0 & 0 \\ 1 & 4 & 1 & 0 \\ 0 & 1 & 4 & 1 \\ 0 & 0 & 1 & 2 \end{bmatrix}\begin{bmatrix} f_1 \\ f_2 \\ f_3 \\ f_4 \end{bmatrix} + \frac{du}{dx}\begin{bmatrix} 0 \\ 0 \\ 0 \\ 0 \end{bmatrix}$$

$$= -\frac{h}{6}\begin{bmatrix} 2f_1 + f_2 \\ f_1 + 4f_2 + f_3 \\ f_2 + 4f_3 + f_4 \\ f_3 + 2f_4 \end{bmatrix} = \frac{h}{54}\begin{bmatrix} 110 \\ 220 \\ 184 \\ 68 \end{bmatrix}$$

with

$$a = (1/h) + (2h/3) = 29/9$$
$$b = -(1/h) + (2h/6) = -26/9,$$
$$c = 58/9$$

$$f_\beta = \begin{bmatrix} f_1 \\ f_2 \\ f_3 \\ f_4 \end{bmatrix} = \begin{bmatrix} -4 \\ -\frac{38}{9} \\ -\frac{32}{9} \\ -2 \end{bmatrix}$$

The first and last equations are replaced by the Dirichlet boundary conditions $u(0) = 0$ and $u(1) = -1$, and the rest of the equations are modified as follows:

$$u_1 = 0$$
$$cu_2 + bu_3 = F_2$$
$$bu_2 + cu_3 + b(-1) = F_3$$
$$u_4 = -1$$

Rewriting the above in matrix form,

$$\begin{bmatrix} 1 & 0 & 0 & 0 \\ 0 & c & b & 0 \\ 0 & b & c & 0 \\ 0 & 0 & 0 & 1 \end{bmatrix} \begin{bmatrix} u_1 \\ u_2 \\ u_3 \\ u_4 \end{bmatrix} = \begin{bmatrix} 0 \\ F_2 \\ F_3 \\ -1 \end{bmatrix} - \begin{bmatrix} 0 \\ 0 \\ -b \\ 0 \end{bmatrix}$$

The solution of the above equations again results in the exact solution,

$$\begin{bmatrix} u_1 \\ u_2 \\ u_3 \\ u_4 \end{bmatrix} = \begin{bmatrix} 0 \\ 0.111 \\ -0.222 \\ -1 \end{bmatrix}$$

Notice that the first and last equations may be deleted and only the second and third equations solved to once again arrive at the exact solution.

FVM via FDM

Finite volume methods require the use of control volumes and control surfaces centered around a node. The governing differential equation is integrated similarly as in finite element formulations, but with the test functions set equal to unity at a node under consideration and zero elsewhere. At node 2 for control volume 1, we have

$$\int_{x_i - \frac{1}{2} = 1\frac{1}{2}}^{x_i + \frac{1}{2} = 2\frac{1}{2}} (1)\left[\frac{d^2u}{dx^2} - 2u - f(x)\right] dx = 0$$

$$\left.\frac{du}{dx}\right|_{1\frac{1}{2}}^{2\frac{1}{2}} - \int_{1\frac{1}{2}}^{2\frac{1}{2}} 2u dx = \int_{1\frac{1}{2}}^{2\frac{1}{2}} f(x) dx, \qquad \frac{u_3 - u_2}{\Delta x} - \frac{u_2 - u_1}{\Delta x} - 2u_2\Delta x = f_2\Delta x$$

Similarly, at node 3 for control volume 2

$$\frac{u_4 - u_3}{\Delta x} - \frac{u_3 - u_2}{\Delta x} - 2u_3\Delta x = f_3\Delta x$$

These equations are identical to FDM, giving the exact solution.

FVM via FEM

$$\left.\frac{du}{dx}\right|_0^1 - \int_0^1 2u dx = \int_0^1 f(x)dx$$

For control volume 1 with $CS1$ and $CS2$ involved, we have

$$\sum_{CS1,2} \frac{du}{dx} = \frac{u_2 - u_1}{h}(\cos 180°) + \frac{u_3 - u_2}{h}(\cos 0°) = \frac{u_2 - u_1}{h}(-1) + \frac{u_3 - u_2}{h}(1)$$

$$\frac{u_3 - 2u_2 + u_1}{h} - 2u_2 h = f_2 h$$

or

$$\frac{u_3 - 2u_2 + u_1}{\Delta x^2} - 2u_2 = f_2$$

Similarly, for control volume 2 with $CS1$ and $CS2$ involved,

$$\frac{u_4 - 2u_3 + u_2}{\Delta x^2} - 2u_3 = f_3$$

It is seen that the result is identical to FVM via FDM.

1.6.2 NEUMANN BOUNDARY CONDITIONS

Here we demonstrate methods for treating the Neumann boundary conditions depending on the side of the boundary to which they are applied.

Neumann Boundary Condition Specified at Right End Node. Given the same differential equation as in (1.6.1), Figure 1.6.1b:

$$\frac{d^2u}{dx^2} - 2u = f(x) \quad 0 < x < 1$$

$$f(x) = 4x^2 - 2x - 4$$

subject to boundary conditions:

$$u = 0 \qquad \text{at } x = 0$$

$$\frac{du}{dx} = -3 \quad \text{at } x = 1$$

which has the exact solution:

$$u = -2x^2 + x$$

FDM

From the given Neumann boundary conditions without using the phantom node, we have

$$\frac{u_4 - u_3}{(1/3)} = -3, \qquad u_4 = u_3 - 1$$

with FDM equations at nodes 2 and 3 given by

$$\frac{u_3 - 2u_2 + u_1}{(1/3)^2} - 2u_2 = f_2$$

$$\frac{u_4 - 2u_3 + u_2}{(1/3)^2} - 2u_3 = f_3$$

Thus we obtain

$$9(u_3 - 2u_2 + 0) - 2u_2 = -\frac{38}{9}$$

$$9(u_3 - 1 - 2u_3 + u_2) - 2u_3 = -\frac{32}{9}$$

or

$$\begin{bmatrix} -20 & 9 \\ 9 & -11 \end{bmatrix} \begin{bmatrix} u_2 \\ u_3 \end{bmatrix} = \begin{bmatrix} -\frac{38}{9} \\ -\frac{32}{9} \end{bmatrix} + \begin{bmatrix} 0 \\ 9 \end{bmatrix} = \begin{bmatrix} -\frac{38}{9} \\ \frac{49}{9} \end{bmatrix}$$

or

$$\begin{bmatrix} u_2 \\ u_3 \end{bmatrix} = \begin{bmatrix} -0.018 \\ -0.51 \end{bmatrix}$$

$$u_4 = -1 - 0.51 = -1.51, \quad 50\% \text{ error}$$

In order to improve the solution, we may use a three-element system with the phantom node 5,

$$\left.\frac{du}{dx}\right|_{x=1} = -3 = \frac{u_5 - u_3}{2\Delta x}, \qquad u_5 = u_3 - 2$$

$$9u_3 - 20u_4 + 9u_5 = f_4$$

$$9u_3 - 20u_4 + 9(u_3 - 2) = -2$$

$$\begin{bmatrix} -20 & 9 & 0 \\ 9 & -20 & 9 \\ 0 & 18 & -20 \end{bmatrix} \begin{bmatrix} u_2 \\ u_3 \\ u_4 \end{bmatrix} = \begin{bmatrix} -\frac{38}{9} \\ -\frac{32}{9} \\ 16 \end{bmatrix}$$

This gives the exact solution

$$u_1 = 0, \qquad u_2 = 1/9, \qquad u_3 = -2/9, \qquad u_4 = -1$$

Another method is to use the second order accurate formula for du/dx [(3.2.5) or (3.2.20) in Chapter 3] written at node 4,

$$\frac{3u_4 - 4u_3 + u_2}{2\Delta x} = -3$$

or with d^2u/dx^2 written at node 4 as

$$\frac{\left(\frac{du}{dx}\right)_4 - \left(\frac{du}{dx}\right)_{4-\frac{1}{2}}}{\Delta x/2} = \frac{2}{\Delta x}\left(-3 - \frac{u_4 - u_3}{\Delta x}\right)$$

and combining with FDM equations written at nodes 2 and 3, we again obtain the exact solution. The reader may verify that the solution deteriorates significantly if only two elements are used. This is because the implementation of Neumann boundary conditions is difficult in FDM, contrary to FEM, as shown in the next example.

FEM

The Neumann boundary conditions at $x = 1$ are written as

$$G_N^{(e)} = \overset{*}{\Phi}_N \frac{du}{dx}\bigg|_0^h, \qquad G_2^{(2)} = (1)\left(\frac{du}{dx}\right)_2 = -3$$

with $\overset{*}{\Phi}_N = 0$ everywhere except at the Neumann boundary node. Assembly of all contributions of elements for the global stiffness matrix and the load vector for a two-element system results in the following:

$$\begin{bmatrix} c & b \\ b & a \end{bmatrix}\begin{bmatrix} u_2 \\ u_3 \end{bmatrix} = \frac{-h}{6}\begin{bmatrix} f_1 + 4f_2 + f_3 \\ f_2 + 2f_3 \end{bmatrix} = -\frac{1}{12}\begin{bmatrix} -22 \\ -8 \end{bmatrix}$$

with

$$\begin{aligned} a &= (1/h) + (2h/3) = 2 + 1/3 = 7/3 \\ b &= -(1/h) + (h/3) = -2 + 1/6 = -11/6 \\ c &= 14/3 \end{aligned}$$

so that the final algebraic equations together with the Neumann boundary vector are written as

$$\begin{bmatrix} \frac{14}{3} & -\frac{11}{6} \\ -\frac{11}{6} & \frac{7}{3} \end{bmatrix}\begin{bmatrix} u_2 \\ u_3 \end{bmatrix} = \frac{1}{12}\begin{bmatrix} 22 \\ 8 \end{bmatrix} + \begin{bmatrix} 0 \\ -3 \end{bmatrix}$$

or

$$\begin{bmatrix} u_2 \\ u_3 \end{bmatrix} = \begin{bmatrix} 0 \\ -1 \end{bmatrix}$$

Once again, the exact solution has been obtained with only two elements.

FVM via FEM and FDM (two elements)

For node 3 via FEM, we have

$$\frac{du}{dx}\bigg|_{2\frac{1}{2}} + \frac{du}{dx}\bigg|_3 - 2u_3\frac{h}{2} = f_3\frac{h}{2}$$

$$\frac{u_3 - u_2}{h}(-1) - 3 - 2u_3\frac{h}{2} = -2\left(\frac{h}{2}\right)$$

Similarly, for node 3 via FDM, we obtain

$$\left.\frac{du}{dx}\right|_{2\frac{1}{2}}^{3} - 2u_3\frac{h}{2} = f_3\frac{h}{2}$$

$$-3 - \frac{u_3 - u_2}{h} - 2u_3\frac{h}{2} = -2\left(\frac{h}{2}\right)$$

Thus, for both methods, we have

$$\begin{bmatrix} -\frac{5}{2} & 1 \\ 1 & -\frac{5}{4} \end{bmatrix}\begin{bmatrix} u_2 \\ u_3 \end{bmatrix} = \begin{bmatrix} -1 \\ \frac{5}{4} \end{bmatrix}$$

or

$$\begin{bmatrix} u_2 \\ u_3 \end{bmatrix} = \begin{bmatrix} 0 \\ -1 \end{bmatrix}$$

It is seen that both methods give the same results.

Neumann Boundary Condition Specified at Left End Node. To demonstrate treatment of the Neumann boundary condition if given at the left end node, we consider the following data:

$$\frac{du}{dx} = 1 \quad at\ x = 0$$
$$u = -1 \quad at\ x = 1$$

FDM

(1) Phantom node method (phantom node created, corresponding to u_0)

$$\frac{u_2 - u_0}{2\Delta x} = 1$$

(2) Second order accurate formula for du/dx at node 1

$$\frac{-3u_1 + 4u_2 - u_3}{2\Delta x} = 1$$

(3) d^2u/dx^2 written at node 1 as

$$-\frac{\left(\frac{du}{dx}\right)_1 - \left(\frac{du}{dx}\right)_{1\frac{1}{2}}}{\frac{\Delta x}{2}} = \frac{-2}{\Delta x}\left(1 - \frac{u_2 - u_1}{\Delta x}\right)$$

With either one of these three methods, we obtain the exact solution. The reader should carry out the calculations for verification of the above results.

FEM

$$\begin{bmatrix} a & b & 0 \\ b & c & b \\ 0 & b & a \end{bmatrix}\begin{bmatrix} u_1 \\ u_2 \\ u_3 \end{bmatrix} = \begin{bmatrix} F_1 \\ F_2 \\ F_3 \end{bmatrix} + \begin{bmatrix} \overset{*}{\Phi}_1 \left.\frac{du}{dx}\right|_{x=0} \\ 0 \\ 0 \end{bmatrix}$$

with

$$\overset{*}{\Phi}_1 \left.\frac{du}{dx}\right|_{x=0} = (1)\frac{du}{dx}\cos(180^\circ) = (1)(1)(-1)$$

$$\begin{bmatrix} 2.333 & -1.888 & 0 \\ -1.888 & 4.666 & -1.833 \\ 0 & -1.833 & 2.333 \end{bmatrix}\begin{bmatrix} u_1 \\ u_2 \\ u_3 \end{bmatrix} = \begin{bmatrix} 1 \\ 1.833 \\ 0.666 \end{bmatrix} + \begin{bmatrix} -1 \\ 0 \\ 0 \end{bmatrix}$$

$$\begin{bmatrix} u_1 \\ u_2 \\ u_3 \end{bmatrix} = \begin{bmatrix} 0 \\ 0 \\ -1 \end{bmatrix}$$

Note that, although $\frac{du}{dx}(0) = 1$ at the left end node, we obtain $G_1 = -1$ because of the direction cosine, $\cos 180^\circ = -1$. The reader is reminded that it is important to recognize the role of direction cosines as depicted in Figure 8.2.3.

FVM via FEM

$$\text{Node 1:} \quad \left.\frac{du}{dx}\right|_1 + \left.\frac{du}{dx}\right|_{1\frac{1}{2}} - 2u_1\frac{h}{2} = f_1\frac{h}{2}$$

$$\text{Node 2:} \quad \left.\frac{du}{dx}\right|_{1\frac{1}{2}} + \left.\frac{du}{dx}\right|_{2\frac{1}{2}} - 2u_2 h = f_2 h$$

Specifying the Neumann boundary data with correct direction cosine (−1), we obtain

$$\text{Node 1:} \quad (1)(-1) + \frac{u_2 - u_1}{h}(1) - 2u_1\left(\frac{h}{2}\right) = -4\left(\frac{h}{2}\right)$$

$$\text{Node 2:} \quad \frac{u_2 - u_1}{h}(-1) + \frac{u_3 - u_2}{h}(1) - 2u_2 h = -4h$$

$$\begin{bmatrix} -\frac{5}{4} & 1 \\ 1 & -\frac{5}{2} \end{bmatrix}\begin{bmatrix} u_1 \\ u_2 \end{bmatrix} = \begin{bmatrix} 0 \\ 0 \end{bmatrix}$$

from which, again, we obtain the same results.

FVM via FDM

$$\text{Node 1:} \quad \left.\frac{du}{dx}\right|_1^{1\frac{1}{2}} = \frac{u_2 - u_1}{h} - 1$$

$$\text{Node 2:} \quad \left.\frac{du}{dx}\right|_{1\frac{1}{2}}^{2\frac{1}{2}} = \frac{u_3 - u_2}{h} - \frac{u_2 - u_1}{h}$$

The formulation and results here are the same as in FVM via FEM.

1.7 SUMMARY

The purpose of this chapter was to acquaint the reader with all available computational methods through very simple one-dimensional linear second order differential equations. For one-dimensional problems presented in this chapter, it is seen that all methods, finite differences, finite elements, and finite volumes provide the final forms of algebraic equations identical to each other, giving the same results for Dirichlet problems. Neumann boundary conditions are approximated in FDM, but they are implemented exactly in FEM and FVM. They "naturally" arise in due course of the formulation. For this reason, Neumann boundary condition is often called "natural" boundary condition. This is not the case for FDM, although exact solutions were obtained for simple examples.

The formulation of FDM equations in one dimension is simple, whereas the concept of algebra involved in FEM is complex. This complicated algebra, however, will be quite useful in multidimensional, arbitrary geometries, and boundary conditions.

Although we have shown only one-dimensional problems in this chapter, we may be able to predict what will happen in multidimensional problems. Mesh configurations for FDM must be *structured* for multidimensional problems as shown in Figure 1.7.1a. Inclined or curved mesh lines can be transformed into orthogonal coordinates so that finite difference equations can be written in orthogonal directions for 2-D or 3-D. This can not be done for FDM if the mesh configuration is *unstructured* as in Figure 1.7.1b. In this case, FEM and FVM can still be accommodated to arbitrary geometries and arbitrary mesh configurations (triangular or quadrilateral elements for 2-D, tetrahedral or hexahedral elements for 3-D).

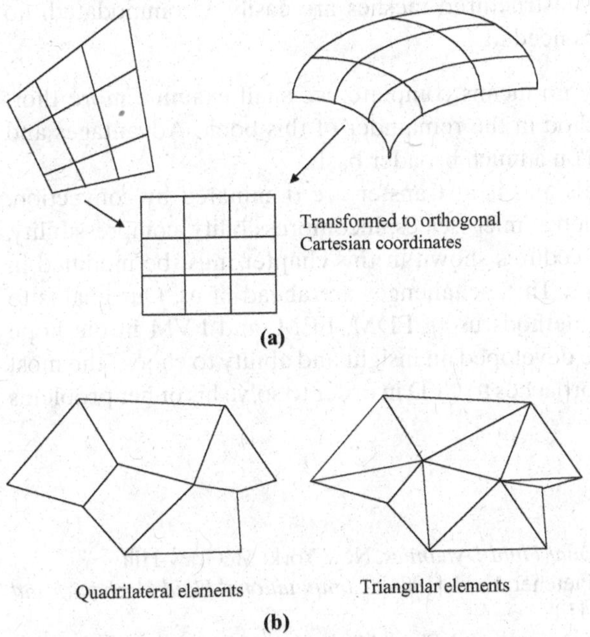

Figure 1.7.1 Geometric mesh configurations in two dimensions. (a) Structured grids for finite difference, mesh lines intersecting two ways (2-D) and three ways (3-D). They must be transformed into orthogonal cartesian coordinates. (b) Unstructured grids for finite elements or finite volumes. No coordinate transformations are required.

There are differences and analogies (similarities) among all methods, irrespective of geometric dimensions. Some of the relatively well known properties are listed below.

FDM

1. Easy to formulate.
2. For multidimensional problems, meshes must be structured in either two or three dimensions. Curved meshes must be transformed into orthogonal cartesian coordinates so that finite difference equations can be written on structured cartesian meshes.
3. Neumann boundary conditions can only be approximated, not exactly enforced.

FEM

1. Underlying principles and formulations require a mathematical rigor.
2. Complex geometries and unstructured meshes are easily accommodated, no coordinate transformations needed.
3. Neumann boundary conditions are enforced exactly.

FVM

1. Formulations can be based on either FDM or FEM.
2. Surface integrals of normal fluxes guarantee the conservation properties throughout the domain.
3. Complex geometries and unstructured meshes are easily accommodated, no coordinate transformations needed.

The above assessments are by no means complete; we shall examine more thoroughly all the details of each method in the remainder of this book. Advantages and disadvantages are to be evaluated on a much broader basis.

Many of the problems in fluids and heat transfer are dominated by convection, shock wave discontinuities, turbulence microscales, incompressibility, compressibility, viscosity, etc. Thus the simple procedures shown in this chapter must be modified in accordance with physical situations. These challenges are ahead of us. Our goal is to explore all major computational methods using FDM, FEM, and FVM in the hope that in the end the reader will have developed an insight and ability to choose the most accurate, efficient, and suitable approaches to CFD in order to solve his or her problems of interest.

REFERENCES

Anderson, J. D., Jr. [1995]. *Computational Fluid Dynamics*. New York: McGraw-Hill.

Anderson, D. A., Tannehill, J. C., and Pletcher, R. H. [1984]. *Computational Fluid Mechanics and Heat Transfer*. New York: McGraw-Hill.

Baker, A. J. [1983]. *Finite Element Computational Fluid Mechanics*. New York: Hemisphere, McGraw-Hill.

Beam, R. M. and Warming, R. F. [1978]. An implicit factored scheme for the compressible Navier-Stokes equations. *AIAA J.*, 16, 393–401.

Briley, W. R. and McDonald, H. [1973]. Solution of the Three-dimensional Compressible Navier-Stokes Equations by an Impucit Technique. Proc. Fourth Int. Conf. Num. Methods Fluid

Dyn., Boulder, Colorado. *Lecture Notes in Physics*. Vol. 35. New York: Springer-Verlag, pp. 105–110.

Carey, G. and Oden, J. T. [1986]. *Finite Elements, Fluid Mechanics*. Vol. 6. Englewood Cliffs, NJ: Prentice Hall.

Chung, T. J. [1978]. *Finite Element Analysis in Fluid Dynamics,* McGraw-Hill.

———. [1999]. Transitions and interactions of inviscid/viscous, compressible/incompressible and laminar/turbulent flows. *Int. J. Num. Methods in Fluids,* 31, 223–46.

Courant, R., Friedrichs, K. O., and Lewy, H. [1928]. Uber die partiellen differenz-gleichungen der mathematischen Physik. *Mathematische Annalen*, 100, 32–74. *IBM J.* [1967], 215–34. [English translation].

Donea, J. [1984]. A Taylor-Galerkin method for convective transport problems. *Int. J. Num. Methods Eng.*, 20, 101–19.

Evans, M. E. and Harlow, F. H. [1957]. The Particle-in-Cell Method for Hydrodynamic Calculations, Los Alamos Scientific Laboratory Report LA-2139. Los Alamos: New Mexico.

Ferziger, J. H. and Peric, M. [1999]. *Computational Methods for Fluid Dynamics*. Berlin: Springer-Verlag.

Fletcher, C. A. [1988]. *Computational Techniques for Fluid Dynamics*. Vol. 1: Fundamental and General Techniques. Berlin: Springer-Verlag.

———. [1988]. *Computational Techniques for Fluid Dynamics*. Vol. 2: Specific Techniques for Different Flow Categories. Berlin: Springer-Verlag.

Godunov, S. K. [1959]. A difference scheme for numerical computation of discontinuous solutions of hydrodynamic equations (in Russian). *Math. Sibornik*, 47, 271–306. United States Joint Publications Research Service, JPRS 7226 [1960]. [English translation].

Gresho, P. M. and Sani, R. L. [1999]. *Incompressible Flows and Finite Element Method*. New York: Wiley.

Harten, A. [1978]. The artificial compression method for computation of shocks and contact discontinuities. III. Self-adjusting hybrid schemes. *Math. Comput*, 32, 363–89.

———. [1983]. High resolution schemes for hyperbolic conservation laws. *J. Comp. Physics*, 49, 357–93.

Heinrich, J. C., Huyakorn, P. S., Zienkiewicz, O. C., and Mitchell, A. R. [1977]. An upwind finite element scheme for two-dimensional convective transport equation. *Int. J. Num. Meth. Eng.*, 11, no. 1, 131–44.

Hirsch, C. [1988]. *Numerical Computation of Internal and External Flows*, Vol. 1: Fundamentals of Numerical Discretization, New York: Wiley.

———. [1990]. *Numerical Computation of Internal and External Flows*, Vol. 2: Computational Methods for Inviscid and Viscous Flows, New York: Wiley.

Hoffmann, K. A. [1989]. *Computational Fluid Dynamics for Engineers*, Engineering Education System, Austin, TX.

Hughes, T. J. R. and Brooks, A. N. [1982]. A theoretical framework for Petrov-Galerkin methods with discontinuous weighting functions: application to the streamline upwind procedure. In R. H. Gallagher, et al. (eds). *Finite Elements in Fluids*, London: Wiley.

Hughes, T., and Mallet, M. [1986]. A new finite element formulation for computational fluid dynamics: IV. A discontinuity capturing operator for multidimensional advective-diffusive systems. *Comp. Meth. Appl. Mech. Eng.*, 58, 329–36.

Hughes, T., Franca, L., and Mallet, M. [1986]. A new finite element formulation for computational fluid dynamics: I. Symmetric forms of the compressible Euler and Navier-Stokes equations and the second law of thermodynamics. *Comp. Meth. Appl. Mech. Eng.*, 54, 223–34.

Hughes, T., Mallet, M., and Mizukami, A. [1986]. A new finite element formulation for computational fluid dynamics: It. Beyond SUPG, *Comp. Meth. Appl. Mech. Eng.*, 54, 341–55.

Jameson, A. [1982]. Transonic aerofoil calculations using the Euler equations. In P. L. Roe (ed.) *Numerical Methods in Aeronautical Fluid Dynamics*, New York: Academic Press.

Johnson, C. [1987]. *Numerical Solution of Partial Differential Equation's on the Finite Element Method*, Lund, Sweden: Student Litteratur.

Lax, P. D. and Wendroff, B. [1960]. Systems of conservation laws. *Comm. Pure and Appl. Math.*, 13, 217–37.

Löhner, R., Morgan, K., and Zienkiewicz, O. C. [1985]. An adaptive finite element procedure for compressible high speed flows. *Comp. Meth. Appl. Mech. Eng.*, 51, 441–65.

MacCormack, R. W. [1969]. The effect of viscosity in hypervelocity impact cratering. AIAA Paper 69–354, Cincinnati, Ohio.

Oden, J. T. [1972]. *Finite Elements of Non Linear Continua*. New York: McGraw-Hill.

———. [1988]. *Adaptive FEM in Complex Flow Problems. The Mathematics of Finite Elements with Applications*. Edited by J. R. Whiteman, London Academic Press, Vol. 6, 1–29.

Oden, J. T., Babuska, I., and Baumann, C. E. [1998]. A discontinuous *hp* finite element method for diffusion problems. *J. Comp. Phys.*, 146, 491–519.

Oden, J. T. and Demkowicz, L. [1991]. h-p adaptive finite element methods in computational fluid dynamics. *Comp. Meth. Appl. Mech. Eng.*, 89 (1–3): 1140.

Oden, J. T., Demkowicz, L., Rachowicz, W., and Westerman, T. A. [1989]. Toward a universal *h-p* adaptive finite element strategy, Part 2: A posteriori error estimation, *Comp. Meth. Appl. Mech. Eng.* 77, 113–80.

Oden, J. T. and Wellford, L. C. Jr. [1972]. Analysis of viscous flow by the finite element method. *AIAA J*, 10, 1590–9.

Patankar, S. V. [1980]. *Numerical Heat Transfer and Fluid Flow*. New York: Hemisphere/McGraw-Hill.

Pepper, D. W. and Heinrich, J. [1992]. *The Finite Element Method: Basic Concepts and Applications*. UK: Taylor & Francis.

Peyret, R. and Taylor, T. D. [1983]. *Computational Methods for Fluid Flow*. New York: Springer-Verlag.

Pironneaau, O. [1989]. *Finite Element Methods for Flows*. New York: Wiley.

Richardson, L. F. [1910]. The approximate arithmetical solution by finite differences of physical problems involving differential equations with an application to the stresses in masonry dam. *Trans. Roy. Soc. Lond.*, Ser. A, 210, 307–57.

Roe, P. L. [1981]. Approximate Riemann solvers, parameter vectors and difference schemes. *J. Comp. Phys.*, 43, 357–72.

———. [1984]. Generalized formulation of TVD Lax-Wendroff schemes. ICASE Report 84-53. NASA CR-172478, NASA Langley Research Center.

Roache, P. J. [1972]. *Computational Fluid Dynamics*, Albuquerque, NM: Hermosa Publications.

———. [1999]. *Fundamentals of Computational Fluid Dynamics*. Albuquerque, NM: Hermosa Publications.

Tannehill, J. C., Anderson, D. A., and Pletcher, R. H. [1997]. *Computational Fluid Mechanics and Heat Transfer*. 2nd ed., New York: McGraw-Hill.

Turner, M. J., Clough, R. W., Martin, H. C., and Topp, L. P. [1956]. Stiffness and deflection analysis of complex structures. *J. Aeron. Soc.*, 23, 805–23.

Van Leer, B. [1974]. Towards the ultimate conservative difference scheme. II. Monotonicity and conservation combined in a second order scheme. *J. Comp. Phys.*, 14, 361–70.

Zienkiewicz, O. C. and Cheung, Y. K. [1965]. Finite elements in the solution of field problems. *The Engineer*, 507–10.

Zienkiewicz, O. C., and Codina, R. [1995]. A general algorithm for compressible and incompressible flow – Part I. Characteristic-based scheme. *Int. J. Num. Methods in Fluids*, 20, 869–85.

Zienkiewicz, O. C., and Taylor, R. L. [1991]. The *Finite Element Method*, Vol. 2. New York: McGraw-Hill.

CHAPTER TWO

Governing Equations

2.1 CLASSIFICATION OF PARTIAL DIFFERENTIAL EQUATIONS

Partial differential equations (PDEs) in general, or the governing equations in fluid dynamics in particular, are classified into three categories: (1) elliptic, (2) parabolic, and (3) hyperbolic. The physical situations these types of equations represent can be illustrated by the flow velocity relative to the speed of sound as shown in Figure 2.1.1. Consider that the flow velocity u is the velocity of a body moving in the quiescent fluid. The movement of this body disturbs the fluid particles ahead of the body, setting off the propagation velocity equal to the speed of sound a. The ratio of these two competing speeds is defined as Mach number

$$M = \frac{u}{a}$$

For subsonic speed, $M < 1$, as time t increases, the body moves a distance, ut, which is always shorter than the distance at of the sound wave (Figure 2.1.1a). The sound wave reaches the observer, prior to the arrival of the body, thus warning the observer that an object is approaching. The zones outside and inside of the circles are known as the zone of silence and zone of action, respectively.

If, on the other hand, the body travels at the speed of sound, $M = 1$, then the observer does not hear the body approaching him prior to the arrival of the body, as these two actions are simultaneous (Figure 2.1.1b). All circles representing the distance traveled by the sound wave are tangent to the vertical line at the position of the observer. For supersonic speed, $M > 1$, the velocity of the body is faster than the speed of sound (Figure 2.1.1c). The line tangent to the circles of the speed of sound, known as a Mach wave, forms the boundary between the zones of silence (outside) and action (inside). Only after the body has passed by does the observer become aware of it.

The governing equations for subsonic flow, transonic flow, and supersonic flow are classified as elliptic, parabolic, and hyperbolic, respectively. We shall elaborate on these equations below. Most of the governing equations in fluid dynamics are second order partial differential equations. For generality, let us consider the partial differential equation of the form [Sneddon, 1957] in a two-dimensional domain

$$A\frac{\partial^2 u}{\partial x^2} + B\frac{\partial^2 u}{\partial x \partial y} + C\frac{\partial^2 u}{\partial y^2} + D\frac{\partial u}{\partial x} + E\frac{\partial u}{\partial y} + Fu + G = 0 \qquad (2.1.1)$$

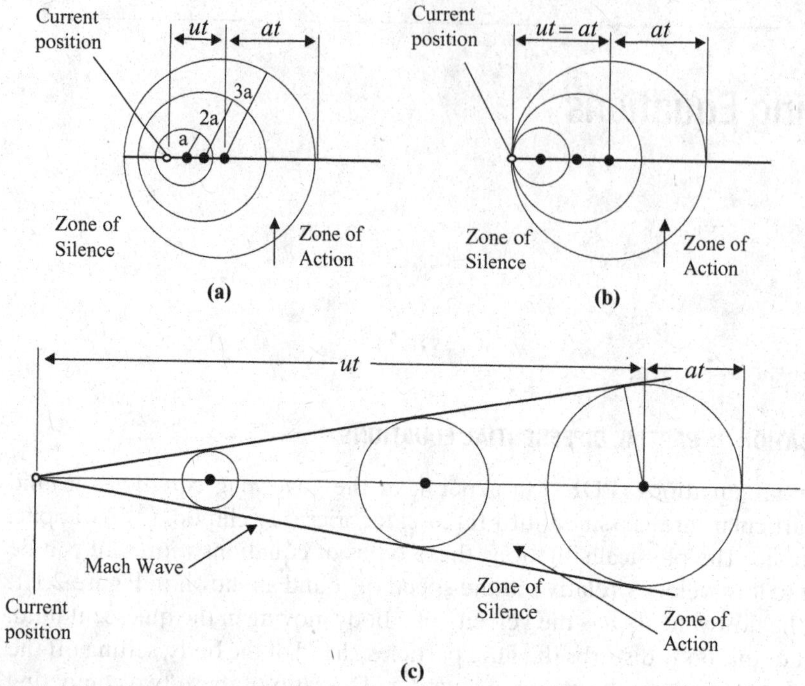

Figure 2.1.1 Subsonic, sonic, and supersonic flows. **(a)** Subsonic ($u < a,\ M < 1$). **(b)** Sonic ($u = a,\ M = 1$). **(c)** Supersonic ($u > a,\ M > 1$).

where the coefficients A, B, C, D, E, and F are constants or may be functions of both independent and/or dependent variables. To assure the continuity of the first derivative of u, $u_x \equiv \partial u/\partial x$ and $u_y \equiv \partial u/\partial y$, we write

$$du_x = \frac{\partial u_x}{\partial x}dx + \frac{\partial u_x}{\partial y}dy = \frac{\partial^2 u}{\partial x^2}dx + \frac{\partial^2 u}{\partial x \partial y}dy \tag{2.1.2a}$$

$$du_y = \frac{\partial u_y}{\partial x}dx + \frac{\partial u_y}{\partial y}dy = \frac{\partial^2 u}{\partial x \partial y}dx + \frac{\partial^2 u}{\partial y^2}dy \tag{2.1.2b}$$

Here u forms a solution surface above or below the $x - y$ plane and the slope dy/dx representing the solution surface is defined as the characteristic curve.

Equations (2.1.1), (2.1.2a), and (2.1.2b) can be combined to form a matrix equation

$$\begin{bmatrix} A & B & C \\ dx & dy & 0 \\ 0 & dx & dy \end{bmatrix} \begin{bmatrix} u_{xx} \\ u_{xy} \\ u_{yy} \end{bmatrix} = \begin{bmatrix} H \\ du_x \\ du_y \end{bmatrix} \tag{2.1.3}$$

where

$$H = -\left(D\frac{\partial u}{\partial x} + E\frac{\partial u}{\partial y} + Fu + G \right) \tag{2.1.4}$$

Since it is possible to have discontinuities in the second order derivatives of the dependent variable along the characteristics, these derivatives are indeterminate. This

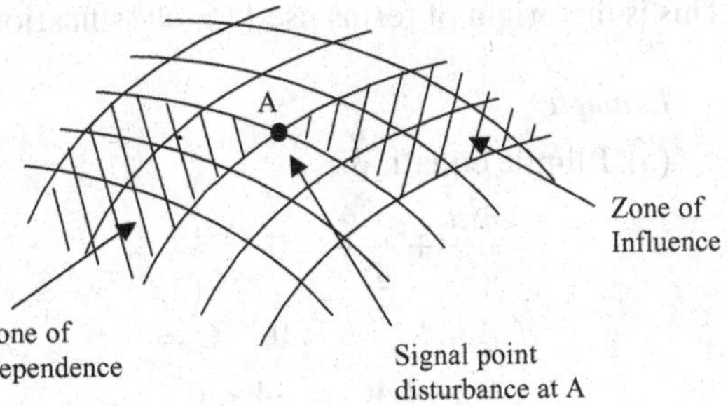

Figure 2.1.2 Propagation of disturbance and characteristics.

happens when the determinant of the coefficient matrix in (2.1.3) is equal to zero.

$$\begin{vmatrix} A & B & C \\ dx & dy & 0 \\ 0 & dx & dy \end{vmatrix} = 0 \tag{2.1.5}$$

which yields

$$A\left(\frac{dy}{dx}\right)^2 - B\left(\frac{dy}{dx}\right) + C = 0 \tag{2.1.6}$$

Solving this quadratic equation yields the equation of the characteristics in physical space,

$$\frac{dy}{dx} = \frac{-B \pm \sqrt{B^2 - 4AC}}{2A} \tag{2.1.7}$$

Depending on the value of $B^2 - 4AC$, characteristic curves can be real or imaginary. For problems in which real characteristics exist, a disturbance propagates only over a finite region (Figure 2.1.2). The downstream region affected by this disturbance at point A is called the zone of influence. A signal at point A will be felt only if it originates from a finite region called the zone of dependence of point A.

The second order PDE is classified according to the sign of the expression $(B^2 - 4AC)$.

(a) Elliptic if $B^2 - 4AC < 0$
In this case, the characteristics do not exist.
(b) Parabolic if $B^2 - 4AC = 0$
In this case, one set of characteristics exists.
(c) Hyperbolic if $B^2 - 4AC > 0$
In this case, two sets of characteristics exist.

Note that (2.1.1) resembles the general expression of a conic section,

$$AX^2 + BXY + CY^2 + DX + EY + F = 0 \tag{2.1.8}$$

in which one can identify the following geometrical properties:

$B^2 - 4AC < 0$	ellipse
$B^2 - 4AC = 0$	parabola
$B^2 - 4AC > 0$	hyperbola

This is the origin of terms used for classification of partial differential equations.

Examples

(a) Elliptic equation

$$\frac{\partial^2 u}{\partial x^2} + \frac{\partial^2 u}{\partial y^2} = 0 \tag{2.1.9}$$

$A = 1, \quad B = 0, \quad C = 1$

$B^2 - 4AC = -4 < 0$

(b) Parabolic equation

$$\frac{\partial u}{\partial t} - \alpha \frac{\partial^2 u}{\partial x^2} = 0 \quad (\alpha > 0) \tag{2.1.10}$$

$A = -\alpha, \quad B = 0, \quad C = 0$

$B^2 - 4AC = 0$

(c) Hyperbolic equation

1-D First Order Wave Equation

$$\frac{\partial u}{\partial t} + a \frac{\partial u}{\partial x} = 0 \quad (a > 0) \tag{2.1.11}$$

1-D Second Order Wave Equation

Differentiating (2.1.11) with respect to x and t,

$$\frac{\partial^2 u}{\partial t \partial x} + a \frac{\partial^2 u}{\partial x^2} = 0 \tag{2.1.12a}$$

$$\frac{\partial^2 u}{\partial t^2} + a \frac{\partial^2 u}{\partial t \partial x} = 0 \tag{2.1.12b}$$

Combining (2.1.12a) and (2.1.12b) yields

$$\frac{\partial^2 u}{\partial t^2} - a^2 \frac{\partial^2 u}{\partial x^2} = 0 \tag{2.1.13}$$

where

$A = 1, \quad B = 0, \quad C = -a^2$

$B^2 - 4AC = 4a^2 > 0$

(d) Tricomi equation

$$y \frac{\partial^2 u}{\partial x^2} + \frac{\partial^2 u}{\partial y^2} = 0 \tag{2.1.14}$$

$A = y, \quad B = 0, \quad C = 1$

$B^2 - 4AC = -4y$

elliptic $\quad y > 0$

parabolic $\quad y = 0$

hyperbolic $\quad y < 0$

(e) 2-D small disturbance potential equation

$$(1 - M^2)\frac{\partial^2 \phi}{\partial x^2} + \frac{\partial^2 \phi}{\partial y^2} = 0 \tag{2.1.15}$$

$$A = 1 - M^2, \quad B = 0, \quad C = 1$$

$$B^2 - 4AC = -4(1 - M^2)$$

elliptic $M < 1$

parabolic $M = 1$

hyperbolic $M > 1$

In CFD applications, computational schemes and specification of boundary conditions depend on the types of PDEs. In many cases, the governing equations in fluids and heat transfer are of mixed types. For this reason, selections of computational schemes and methods to apply boundary conditions are important subjects in CFD. We shall examine them in detail for the remainder of this book.

2.2 NAVIER-STOKES SYSTEM OF EQUATIONS

Physics of fluids and heat transfer as a part of continuum mechanics has now been well established. The nonconservation form of the governing equations in fluids can be derived from the first law of thermodynamics, written as [Truesdell and Toupin, 1960; Chung, 1996]

$$\frac{DK}{Dt} + \frac{DU}{Dt} = M + Q \tag{2.2.1}$$

where K, U, M, and Q denote the kinetic energy, internal energy, mechanical power, and heat energy, respectively,

$$K = \int_\Omega \frac{1}{2}\rho \mathrm{v}_i \mathrm{v}_i d\Omega \tag{2.2.2}$$

$$U = \int_\Omega \rho \varepsilon d\Omega \tag{2.2.3}$$

$$M = \int_\Omega \rho F_i \mathrm{v}_i d\Omega + \int_\Gamma \sigma_{ij} \mathrm{v}_j n_i d\Gamma \tag{2.2.4}$$

$$Q = \int_\Omega \rho r d\Omega \pm \int_\Gamma q_i n_i d\Gamma \tag{2.2.5}$$

with

$$\varepsilon = c_p T - \frac{p}{\rho} \tag{2.2.6a}$$

$$\sigma_{ij} = -p\delta_{ij} + \tau_{ij} \tag{2.2.6b}$$

$$\tau_{ij} = \mu(\mathrm{v}_{i,j} + \mathrm{v}_{j,i}) - \frac{2}{3}\mu \mathrm{v}_{k,k}\delta_{ij} \tag{2.2.6c}$$

$$q_i = \pm kT_{,i} \tag{2.2.6d}$$

where the repeated indices imply summing and the comma denotes partial derivatives with respect to the independent variables x_i, Ω represents the domain of the flowfield with n_i being the components of a vector normal to the boundary surface Γ, with ρ = density per unit mass, v_i = components of the velocity vector, ε = internal energy per unit mass, F_i = components of body force vector, c_p = specific heat at constant pressure, σ_{ij} = total stress tensor, τ_{ij} = viscous stress tensor, μ = coefficient of dynamic viscosity, p = pressure, q_i = heat flux, T = temperature, k = coefficient of thermal conductivity, and r = heat supply per unit mass. Note that δ_{ij} denotes the Kronecker delta with $\delta_{ij} = 1$ for $i = j$ and $\delta_{ij} = 0$ for $i \neq j$.

The dynamic viscosity and thermal conductivity coefficients are functions of temperature as given by Sutherland's law,

$$\mu = \frac{C_1 T^{3/2}}{T + C_2} \tag{2.2.7}$$

$$k = \frac{C_3 T^{3/2}}{T + C_4} \tag{2.2.8}$$

with C_1, C_2, C_3, and C_4 being the constants for a given gas. For air at moderate temperatures, we may use $C_1 = 1.458 \times 10^{-6}$ kg/(m s $\mathrm{K}^{1/2}$), $C_2 = 110.4$ K, $C_3 = 2.495 \times 10^{-3}$ kg m/(s^3 $\mathrm{K}^{3/2}$), and $C_4 = 194$ K.

Substituting (2.2.2) through (2.2.5) into (2.2.1) and using the Green-Gauss theorem, we obtain the governing equations of continuity, momentum, and energy,

Continuity

$$\frac{\partial \rho}{\partial t} + (\rho \mathrm{v}_i)_{,i} = 0 \tag{2.2.9a}$$

Momentum

$$\rho \frac{\partial \mathrm{v}_j}{\partial t} + \rho \mathrm{v}_{j,i} \mathrm{v}_i + p_{,j} - \tau_{ij,i} - \rho F_j = 0 \tag{2.2.9b}$$

Energy

$$\rho \frac{\partial \varepsilon}{\partial t} + \rho \varepsilon_{,i} \mathrm{v}_i + p \mathrm{v}_{i,i} - \tau_{ij} \mathrm{v}_{j,i} + q_{i,i} - \rho r = 0 \tag{2.2.9c}$$

with the equation of state

$$p = \rho R T \tag{2.2.10}$$

where R is the specific gas constant. Note that equations (2.2.9a) through (2.2.9c) are known as the nonconservation form of the Navier-Stokes system of equations for compressible viscous flows.

The above equations may be recast in the so-called conservation form of the Navier-Stokes system of equations,

$$\frac{\partial \mathbf{U}}{\partial t} + \frac{\partial \mathbf{F}_i}{\partial x_i} + \frac{\partial \mathbf{G}_i}{\partial x_i} = \mathbf{B} \tag{2.2.11}$$

where $\mathbf{U}$, $\mathbf{F}_i$, $\mathbf{G}_i$, and $\mathbf{B}$ are the conservation flow variables, convection flux variables, diffusion flux variables, and source terms, respectively

$$\mathbf{U} = \begin{bmatrix} \rho \\ \rho v_j \\ \rho E \end{bmatrix}, \quad \mathbf{F}_i = \begin{bmatrix} \rho v_i \\ \rho v_i v_j + p\delta_{ij} \\ \rho E v_i + p v_i \end{bmatrix}, \quad \mathbf{G}_i = \begin{bmatrix} 0 \\ -\tau_{ij} \\ -\tau_{ij} v_j + q_i \end{bmatrix}, \quad \mathbf{B} = \begin{bmatrix} 0 \\ \rho F_j \\ \rho r + \rho F_j v_j \end{bmatrix}$$

with E being the total (stagnation) energy,

$$E = \varepsilon + \frac{1}{2} v_j v_j \tag{2.2.12a}$$

which is related by the pressure and temperature as

$$p = (\gamma - 1)\rho\left(E - \frac{1}{2} v_j v_j\right) \tag{2.2.12b}$$

$$T = \frac{1}{c_v}\left(E - \frac{1}{2} v_j v_j\right) \tag{2.2.12c}$$

with c_v being the specific heat at constant volume. The Navier-Stokes system of equations is simplified to the Euler equations if the diffusion flux variables $\mathbf{G}_i$ are neglected.

It should be noted that, upon differentiation as implied in (2.2.11), we recover the nonconservation form of the Navier-Stokes system of equations given by (2.2.9).

On the other hand, integrating (2.2.11) spatially over the volume of the domain,

$$\int_\Omega \left(\frac{\partial \mathbf{U}}{\partial t} + \frac{\partial \mathbf{F}_i}{\partial x_i} + \frac{\partial \mathbf{G}_i}{\partial x_i} - \mathbf{B}\right) d\Omega = 0 \tag{2.2.13}$$

we obtain another form of governing equations,

$$\int_\Omega \left(\frac{\partial \mathbf{U}}{\partial t} - \mathbf{B}\right) d\Omega + \int_\Gamma (\mathbf{F}_i + \mathbf{G}_i) n_i d\Gamma = 0 \tag{2.2.14}$$

Note that the surface integral in (2.2.14) represents the convection and diffusion fluxes through the control surfaces, which are in balance with $\partial\mathbf{U}/\partial t$ and $\mathbf{B}$ inside the control volume. The surface integral in (2.2.14) has two important roles. First, it lays the foundation for the finite volume methods (FVM). Second, it provides appropriate numerical treatments for high gradient flows or discontinuities such as shock waves. Conservation properties across the discrete element boundary surfaces are satisfied if the surface integral components in (2.2.14) are properly implemented in the numerical solution.

Various types of fluid flows emerge from the Navier-Stokes system of equations in nonconservation and conservation forms. In general, computational schemes are dictated from the physics of flows characterized by special forms of the governing equations.

We have written the governing equations in fluid dynamics in three different ways. Equations (2.2.9a) through (2.2.9c) derived from the **F**irst **L**aw of **T**hermodynamics (FLT) are the nonconservation form of the Navier-Stokes system of equations in terms of the primitive variables ρ, v_i, p, T, whereas the **C**onservation form of **N**avier-**S**tokes system (CNS) of (2.2.11) are written in terms of the conservation variables $\mathbf{U}$, $\mathbf{F}_i$, and $\mathbf{G}_i$. In contrast, the **C**ontrol **V**olume-**S**urface (CVS) equations (2.2.14) are expressed in volume and surface integral forms, but still in terms of the conservation variables $\mathbf{U}$, $\mathbf{F}_i$, and $\mathbf{G}_i$. All of these three different forms of the governing equations represent certain types of numerical schemes to be developed, each playing special roles in CFD.

The FLT equations are convenient when the primitive variables ρ, v_i, p, T are to be solved directly, whereas this is not possible if CNS or CVS equations are used. It is seen that the conservation variables must be solved first with primitive variables extracted indirectly. Despite this inconvenience, the CNS or CVS equations are preferred in many CFD problems. For example, when the solution of density ρ is discontinuous, such as in shock waves, the solution through FLT is difficult. On the other hand, the mass flow ρv_i is a smooth function and so are all other conservation variables, whereby the solution of CNS or CVS equations makes it possible to obtain discontinuous solution of primitive variables (indirectly). So, the conclusion here is that we can use FLT if the solution does not contain discontinuities such as in incompressible flows (no shock waves). This is known as the *pressure-based* formulation. Otherwise, CNS or CVS equations can be chosen, in which satisfactory results are assured in general, when the solution may contain discontinuities such as in compressible flows. This is known as the *density-based* formulation.

The Navier-Stokes system of equations as given by (2.2.11) may be simplified by disregarding one or more equations and/or some of the terms of each equation. For example, the momentum equations (2.2.9b) alone are often called the Navier-Stokes equations, thus distinguished from the Navier-Stokes *system* of equations which includes all equations (2.2.9a) through (2.2.9c). If all viscous terms are eliminated from the Navier-Stokes system of equations, then the resulting equations are known as Euler equations. The momentum equations without the pressure gradients are called the Burgers' equation. The Burgers' equation can be inviscid linear (no viscosity terms with convection terms being linearized), inviscid nonlinear, linear viscous, and nonlinear viscous. Simpler forms of these equations will be treated in Chapter 4. The governing equations for incompressible and compressible flows are discussed in Chapters 5 and 6 for FDM and Chapters 12 and 13 for FEM. More complicated governing equations are the subjects of Chapters 21 through 27.

The Navier-Stokes system of equations can be modified into various different forms, corresponding to particular physical phenomena, with the following subject areas included: compressible viscous flow (Navier-Stokes system of equations), compressible inviscid flow (diffusion terms are neglected), incompressible viscous flow (temporal and spatial variations of density are neglected), incompressible inviscid flow (both diffusion and density variations are neglected), vortex flow in terms of vorticity and stream function, compressible inviscid flow in terms of velocity potential function, turbulence, chemically reacting flows and combustion, acoustics, combined mode radiative heat transfer, and two-phase flows, as summarized in Table 2.2.1.

The governing equations in fluids and heat transfer in general are of mixed types: elliptic, parabolic, and hyperbolic partial differential equations. The presence or absence of each of the terms in these equations will determine their specific classifications. It will be shown throughout the book that numerical schemes depend on the types of partial

Table 2.2.1 Various types of flows

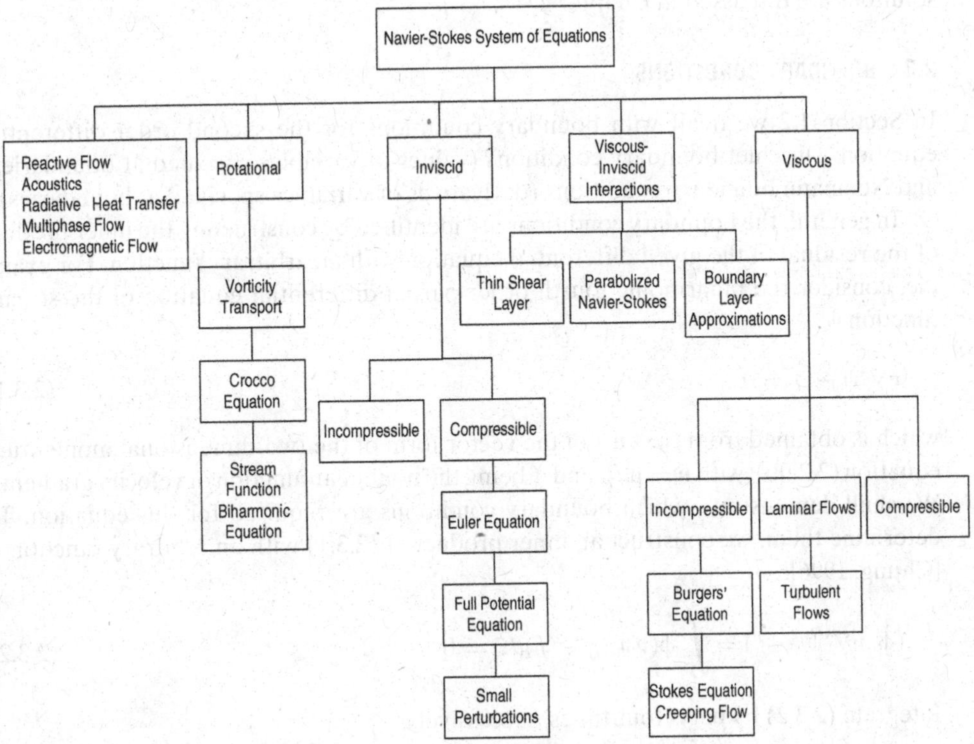

differential equations. In general, physical phenomena dictate the types of equations to be used, which are then accommodated by appropriate numerical schemes for solutions of the equations.

The Navier-Stokes system of equations presented above is cast in the Eulerian coordinates in which the current flowfield is fixed at the reference coordinates. In dealing with multiphase flows, however, it is convenient to work with the Lagrangian coordinates in which displacements of fluid or solid particles are tracked relative to the initial reference coordinates. Both Eulerian and Lagrangian coordinates may be coupled in dealing with certain physical phenomena. These and other topics of coordinate systems are discussed in Section 16.4 and Chapter 25. Detailed mathematics of Eulerian and Lagrangian coordinates are given in Chung [1996].

For flows coupled with magnetic and electric forces, it is necessary to solve the Maxwell's equations together with the modified Navier-Stokes system of equations. Applications of these equations to coronal mass ejection and semiconductor plasma processing are presented in Chapter 26.

The Navier-Stokes system of equations discussed in this section is based on the macroscopic nonrelativistic continuum view. In dealing with extremely high velocities such as occur in supernova explosions, the cosmic expansion, and cosmic singularity, however, the relativity principles based on the microscopic kinetic theory must be used. The governing equations for the relativistic astrophysical flows and their numerical solutions are discussed in Chapter 27.

2.3 BOUNDARY CONDITIONS

In Section 1.2 we dealt with boundary conditions for the second order differential equation: Dirichlet boundary conditions (values of variables specified at boundaries) and Neumann boundary conditions (derivatives of variables specified at boundaries).

In general, the boundary conditions are identified by constructing the inner product of the residual of the given differential equation with an arbitrary function. For example, consider the biharmonic fourth order partial differential equation of the stream function ψ

$$\nu\nabla^4\psi - f = 0 \tag{2.3.1}$$

which is obtained from the curl of the vector form of the two-dimensional momentum equation (2.2.9b), with $\nu = \mu/\rho$ and f being the nonlinear function of velocity gradients. We shall demonstrate which boundary conditions are required for this equation. To determine them, we construct an inner product of (2.3.1) with an arbitrary function ϕ [Chung, 1996]:

$$(\phi, \nu\nabla^4\psi - f) = \int_\Omega \phi(\nu\psi_{,iijj} - f)d\Omega = 0 \tag{2.3.2}$$

Integrate (2.3.2) by parts four times, successively,

$$\int_\Gamma \phi\nu\psi_{,iij}n_j d\Gamma - \int_\Omega \phi_{,j}\nu\psi_{,iij}d\Omega - \int_\Omega \phi f d\Omega = 0$$

$$\int_\Gamma \phi\nu\psi_{,iij}n_j d\Gamma - \int_\Gamma \phi_{,j}\nu\psi_{,ii}n_j d\Gamma + \int_\Omega \phi_{,jj}\nu\psi_{,ii}d\Omega - \int_\Omega \phi f d\Omega = 0$$

$$\int_\Gamma \phi\nu\psi_{,iij}n_j d\Gamma - \int_\Gamma \phi_{,j}\nu\psi_{,ii}n_j d\Gamma + \int_\Gamma \phi_{,jj}\nu\psi_{,i}n_i d\Gamma - \int_\Omega \phi_{,jji}\nu\psi_{,i}d\Omega - \int_\Omega \phi f d\Omega = 0$$

Finally,

$$\int_\Gamma (\phi\nu\psi_{,iij}n_j - \phi_{,j}\nu\psi_{,ii}n_j + \phi_{,jj}\nu\psi_{,i}n_i - \phi_{,jji}\nu\psi n_i)d\Gamma + \int_\Omega \phi_{,jjii}\nu\psi\, d\Omega - \int_\Omega \phi f d\Omega = 0 \tag{2.3.3}$$

where the boundary conditions consist of two Neumann and two Dirichlet conditions:

Neumann Boundary Conditions

$\psi_{,iij}n_j$ *normal stress gradient*

$\psi_{,ii}n_j$ *normal velocity gradient* (2.3.4a)

Dirichlet Boundary Conditions

$\psi_{,i} n_i$ *normal velocity* (2.3.4b)
ψ *stream function*

It is seen that, for the $2m$th order differential equation, the Neumann boundary conditions are of the order $2m-1, 2m-2, \ldots m$ and the Dirichlet boundary conditions are of the order $m-1, m-2, \ldots 0$. These boundary conditions are to be prescribed on the boundary surfaces. Similarly, for the second order equation ($\nabla^2\psi = 0$), there is one Neumann boundary condition ($\psi_{,i}\, n_i$) and one Dirichlet boundary condition (ψ). It was seen in Chapter 1 that the implementation of the Neumann boundary conditions "naturally" arises in the formulation process of FEM, whereas in FDM they must be carried out "manually" with appropriate forms of the difference equations.

Often, mixed Dirichlet and Neumann conditions (called Cauchy or Robin conditions) are used. For example, for the second order differential equation such as in combined conductive and convective heat transfer boundary conditions, we may write

$$\alpha T + \beta \frac{\partial T}{\partial n} = \gamma \tag{2.3.5}$$

with

$$\frac{\partial T}{\partial n} = (\boldsymbol{n} \cdot \nabla) T = T_{,i} n_i = \frac{\partial T}{\partial x} n_1 + \frac{\partial T}{\partial y} n_2 + \frac{\partial T}{\partial z} n_3 \tag{2.3.6}$$

$\beta = 0$ Dirichlet

$\alpha = 0$ Neumann

$\alpha \neq 0, \beta \neq 0$ Cauchy/Robin

Note that the notation $\partial T/\partial n$ is misleading since n in this derivative is neither the unit normal vector $\boldsymbol{n}$, nor its components n_i. However, this unfortunate notation has been generally accepted in the literature.

For time dependent problems, we must provide initial conditions as well as boundary conditions. Let us consider the case of hyperbolic, parabolic, and elliptic equations as shown in Figure 2.3.1.

(1) Hyperbolic equations associated with Cauchy conditions in an open region (Figure 2.3.1a).

Second Order Equation

$$\frac{\partial^2 u}{\partial t^2} - a^2 \frac{\partial^2 u}{\partial x^2} = 0 \quad 0 < x < 1 \tag{2.3.7}$$

Two initial conditions given $\left\{ u(x,0) \text{ and } \frac{\partial u}{\partial t}(x,0) \right.$

Two boundary conditions given $\left\{ u(0,t) \text{ or } \frac{\partial u}{\partial x}(0,t) \right.$

$\left\{ u(1,t) \text{ or } \frac{\partial u}{\partial x}(1,t) \right.$

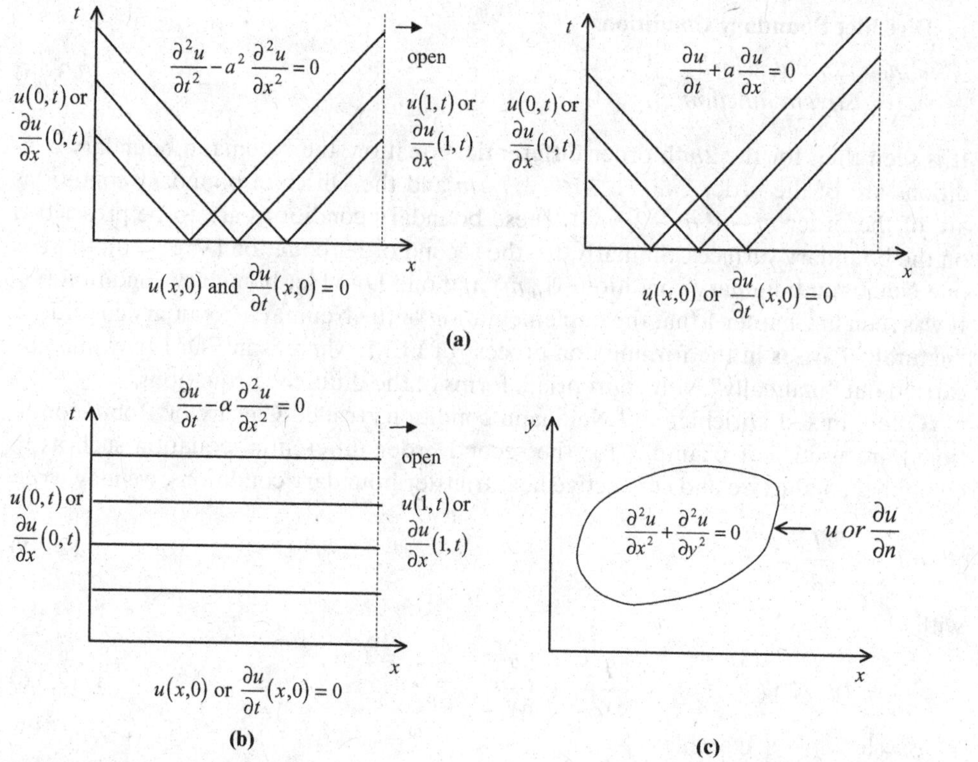

Figure 2.3.1 Initial and boundary conditions for hyperbolic, parabolic, and elliptic equations. **(a)** Hyperbolic equations (two sets of characteristics), Cauchy conditions in open region for second order equation. **(b)** Parabolic equations (one set of characteristics), Dirichlet or Neumann boundary conditions in an open region. **(c)** Elliptic equations (no real characteristics), Dirichlet or Neumann boundary conditions in closed region.

First Order Equation

$$\frac{\partial u}{\partial t} + a\frac{\partial u}{\partial x} = 0 \quad 0 < x < 1 \tag{2.3.8}$$

One initial condition given $\left\{ u(x,0) \text{ or } \frac{\partial u}{\partial t}(x,0) \right.$

One boundary condition given at $x = 0$ $\left\{ u(0,t) \text{ or } \frac{\partial u}{\partial x}(0,t) \right.$

(2) Parabolic equations associated with Dirichlet or Neumann conditions in an open region (Figure 2.3.1b).

$$\frac{\partial u}{\partial t} - \nu\frac{\partial^2 u}{\partial x^2} = 0 \quad 0 < x < 1 \tag{2.3.9}$$

One initial condition given $\left\{ u(x,0) \text{ or } \frac{\partial u}{\partial t}(x,0) \right.$

Two boundary conditions given $\begin{cases} u(0,t) & \text{or} \quad \dfrac{\partial u}{\partial x}(0,t) \\ u(1,t) & \text{or} \quad \dfrac{\partial u}{\partial x}(1,t) \end{cases}$

(3) Elliptic equations associated with Dirichlet or Neumann conditions in a closed region (Figure 2.3.1c).

$$\frac{\partial^2 u}{\partial x^2} + \frac{\partial^2 u}{\partial y^2} = 0 \quad \text{in} \quad \Omega \tag{2.3.10}$$

Two boundary conditions given

$$u \quad \text{on} \quad \Gamma_D$$
$$\frac{\partial u}{\partial n} \quad \text{on} \quad \Gamma_N$$

where Γ_D and Γ_N denote the Dirichlet and Neumann boundaries, respectively.

In general, more complicated boundary and initial conditions are required for CFD. Discussions on detailed boundary conditions for the Euler equations and the Navier-Stokes system of equations in FDM will be presented in Section 6.7, various aspects of boundary conditions associated with FEM in Sections 10.1.2, 11.1, and 13.6.6, and special boundary conditions for multiphase flows in Section 22.2.6.

2.4 SUMMARY

The basic properties of partial differential equations have been described and classified as elliptic, parabolic, and hyperbolic equations. The Navier-Stokes system of equations which represents mixed elliptic, parabolic, and hyperbolic partial differential equations can be written in three different forms: first law of thermodynamics (FLT) nonconservation form, conservation form of Navier-Stokes system (CNS), and control volume-surface integral form (CVS). The nonconservation form of the Navier-Stokes system of equations is derived from the first law of thermodynamics (FLT) written in terms of primitive variables, suitable for low-speed incompressible flows in which the solution surfaces are relatively smooth and not discontinuous. The conservation form of the Euler equations or Navier-Stokes (CNS) system of equations, on the other hand, is convenient for discontinuities such as in shock waves, thus suitable for high-speed compressible flows. Another conservation form is the control volume-surface (CVS) integral equations, applicable for the finite volume methods in which conservation requirements through discrete interior boundary surfaces as well as the exterior boundary surfaces are self-enforced. Relationships of these three forms of the Navier-Stokes system of equations have been mathematically linked together, traced back to the first law of thermodynamics [Chung, 1996].

The governing equations presented in this chapter are based on the Eulerian coordinates, which are fixed on the reference coordinates in which velocity components of fluid particles are calculated at any fixed point rather than tracing the particles downstream. In some problems, however, it is convenient to use the Lagrangian coordinates where the coordinate points are allowed to move together with fluid particles such as in multiphase flows. This subject will be discussed in Section 16.4.2 and Chapter 25.

In this chapter, we also discussed the boundary conditions for simple geometries and simple physics. The general method of identifying the existence of Neumann and Dirichlet boundary conditions of higher order partial differential equations was demonstrated. However, in reality, determination of boundary conditions is a difficult task in multidimensional, complex geometrical configurations with complex physical phenomena. Applications of boundary conditions will be the subject of discussion throughout the remainder of this book.

REFERENCES

Chung, T. J. [1996]. *Applied Continuum Mechanics*. New York: Cambridge University Press.

Sneddon, I. N. [1957]. *Elements of Partial Differential Equations*. New York: McGraw-Hill.

Truessdell, C. and Toupin, R. A. [1960]. *The Classical Field Theories in Encyclopedia of Physics*. S. Flugge (ed.). Berlin: Springer-Verlag.

PART TWO

FINITE DIFFERENCE METHODS

Part Two presents the finite difference methods (FDM) and topics related to finite difference approximations. The subjects to be covered here include basic concepts of finite difference theory, various formulation strategies, and applications to incompressible and compressible flows. Finite volume methods (FVM) via FDM are also presented.

Although FDM as applied to CFD is widespread and many textbooks are available, the purpose of Part Two is to make detailed comparisons with other methods such as finite element methods (FEM) to be presented in Part Three (particularly in Chapter 16) for the benefit of the beginner and the practitioner alike. Historical developments, traditional treatments of finite difference methods, and some recent advancements are presented for this reason.

Chapter 3 discusses derivations of finite difference equations, followed in Chapter 4 by various finite difference schemes for solutions of elliptic, parabolic, hyperbolic, and Burgers' equations. General fluid dynamics problems of incompressible and compressible flows are presented in Chapters 5 and 6, respectively. Finally, finite volume methods (FVM) via FDM are discussed in Chapter 7.

CHAPTER THREE

Derivation of Finite Difference Equations

The basic idea of finite difference methods is simple: derivatives in differential equations are written in terms of discrete quantities of dependent and independent variables, resulting in simultaneous algebraic equations with all unknowns prescribed at discrete mesh points for the entire domain.

In fluid dynamics applications, appropriate types of differencing schemes and suitable methods of solution are chosen, depending on the particular physics of the flows, which may include inviscid, viscous, incompressible, compressible, irrotational, rotational, laminar, turbulent, subsonic, transonic, supersonic, or hypersonic flows. Different forms of the finite difference equations are written to conform to these different physical phenomena encountered in fluid dynamics.

In this chapter, we present various methods for deriving finite difference equations of low and high orders of accuracy. Truncation errors, as related to the orders of accuracy involved in the approximations, will also be discussed.

3.1 SIMPLE METHODS

Consider a function $u(x)$ and its derivative at point x,

$$\frac{\partial u(x)}{\partial x} = \lim_{\Delta x \to 0} \frac{u(x+\Delta x) - u(x)}{\Delta x} \tag{3.1.1}$$

If $u(x+\Delta x)$ is expanded in Taylor series about $u(x)$, we obtain

$$u(x+\Delta x) = u(x) + \Delta x \frac{\partial u(x)}{\partial x} + \frac{(\Delta x)^2}{2}\frac{\partial^2 u(x)}{\partial x^2} + \frac{(\Delta x)^3}{3!}\frac{\partial^3 u(x)}{\partial x^3} + \cdots \tag{3.1.2}$$

Substituting (3.1.2) into (3.1.1) yields

$$\frac{\partial u(x)}{\partial x} = \lim_{\Delta x \to 0}\left(\frac{\partial u(x)}{\partial x} + \frac{\Delta x}{2}\frac{\partial^2 u(x)}{\partial x^2} + \cdots\right) \tag{3.1.3}$$

Or it is seen from (3.1.2) that

$$\frac{u(x+\Delta x) - u(x)}{\Delta x} = \frac{\partial u(x)}{\partial x} + \frac{\Delta x}{2}\frac{\partial^2 u(x)}{\partial x^2} + \cdots = \frac{\partial u(x)}{\partial x} + O(\Delta x) \tag{3.1.4}$$

The derivative $\frac{\partial u(x)}{\partial x}$ in (3.1.4) is of first order in Δx, indicating that the truncation error $O(\Delta x)$ goes to zero like the first power in Δx. The finite difference form given by (3.1.1), (3.1.3), and (3.1.4) is said to be of the first order accuracy.

Referring to Figure 1.2.1, we may write u in Taylor series at $i+1$ and $i-1$,

$$u_{i+1} = u_i + \Delta x\left(\frac{\partial u}{\partial x}\right)_i + \frac{\Delta x^2}{2}\left(\frac{\partial^2 u}{\partial x^2}\right)_i + \frac{\Delta x^3}{3!}\left(\frac{\partial^3 u}{\partial x^3}\right)_i + \frac{\Delta x^4}{4!}\left(\frac{\partial^4 u}{\partial x^4}\right)_i + \cdots \quad (3.1.5)$$

$$u_{i-1} = u_i - \Delta x\left(\frac{\partial u}{\partial x}\right)_i + \frac{\Delta x^2}{2}\left(\frac{\partial^2 u}{\partial x^2}\right)_i - \frac{\Delta x^3}{3!}\left(\frac{\partial^3 u}{\partial x^3}\right)_i + \frac{\Delta x^4}{4!}\left(\frac{\partial^4 u}{\partial x^4}\right)_i + \cdots \quad (3.1.6)$$

Rearranging (3.1.5), we arrive at the forward difference:

$$\left(\frac{\partial u}{\partial x}\right)_i = \frac{u_{i+1} - u_i}{\Delta x} + O(\Delta x) \quad (3.1.7)$$

Likewise, from (3.1.6), we have the backward difference:

$$\left(\frac{\partial u}{\partial x}\right)_i = \frac{u_i - u_{i-1}}{\Delta x} + O(\Delta x) \quad (3.1.8)$$

A central difference is obtained by subtracting (3.1.6) from (3.1.5):

$$\left(\frac{\partial u}{\partial x}\right)_i = \frac{u_{i+1} - u_{i-1}}{2\Delta x} + O(\Delta x^2) \quad (3.1.9)$$

It is seen that the truncation errors for the forward and backward differences are first order, whereas the central difference yields a second order truncation error.

Finally, by adding (3.1.5) and (3.1.6), we have

$$\frac{u_{i+1} - 2u_i + u_{i-1}}{\Delta x^2} = \left(\frac{\partial^2 u}{\partial x^2}\right)_i + \frac{(\Delta x)^2}{12}\left(\frac{\partial^4 u}{\partial x^4}\right)_i + \cdots \quad (3.1.10)$$

This leads to the finite difference formula for the second derivative with second order accuracy,

$$\left(\frac{\partial^2 u}{\partial x^2}\right)_i = \frac{u_{i+1} - 2u_i + u_{i-1}}{\Delta x^2} + O(\Delta x^2) \quad (3.1.11)$$

Note that these results were intuitively obtained in Section 1.2 by approximations of slopes of a curve, without the notion of truncation errors.

3.2 GENERAL METHODS

In general, finite difference equations may be generated for any order derivative with any number of points involved (any order accuracy). For example, let us consider a first derivative associated with three points such that

$$\left(\frac{\partial u}{\partial x}\right)_i = \frac{au_i + bu_{i-1} + cu_{i-2}}{\Delta x} \quad (3.2.1)$$

The coefficients a, b, c may be determined from a Taylor series expansion of upstream nodes u_{i-1} and u_{i-2} about u_i (one-sided upstream or backward difference)

$$u_{i-1} = u_i + (-\Delta x)\left(\frac{\partial u}{\partial x}\right)_i + \frac{(-\Delta x)^2}{2}\left(\frac{\partial^2 u}{\partial x^2}\right)_i + \frac{(-\Delta x)^3}{3!}\left(\frac{\partial^3 u}{\partial x^3}\right)_i + \cdots \tag{3.2.2a}$$

$$u_{i-2} = u_i + (-2\Delta x)\left(\frac{\partial u}{\partial x}\right)_i + \frac{(-2\Delta x)^2}{2}\left(\frac{\partial^2 u}{\partial x^2}\right)_i + \frac{(-2\Delta x)^3}{3!}\left(\frac{\partial^3 u}{\partial x^3}\right)_i + \cdots \tag{3.2.2b}$$

from which we obtain

$$au_i + bu_{i-1} + cu_{i-2} = (a+b+c)u_i - \Delta x(b+2c)\left(\frac{\partial u}{\partial x}\right)_i + \frac{\Delta x^2}{2}(b+4c)\left(\frac{\partial^2 u}{\partial x^2}\right)_i + O(\Delta x^3) \tag{3.2.3}$$

It follows from (3.2.1) and (3.2.3) that the following three conditions must be satisfied:

$$a+b+c=0 \tag{3.2.4a}$$

$$b+2c=-1 \tag{3.2.4b}$$

$$b+4c=0 \tag{3.2.4c}$$

The solution of (3.2.4) yields $a = 3/2$, $b = -2$, and $c = 1/2$. Thus, from (3.2.1) we obtain

$$\left(\frac{\partial u}{\partial x}\right)_i = \frac{3u_i - 4u_{i-1} + u_{i-2}}{2\Delta x} + O(\Delta x^2) \tag{3.2.5}$$

If the downstream nodes u_{i+1} and u_{i+2} are used (one-sided downstream or forward difference), then we have

$$\left(\frac{\partial u}{\partial x}\right)_i = \frac{-3u_i + 4u_{i+1} - u_{i+2}}{2\Delta x} + O(\Delta x^2) \tag{3.2.6}$$

A similar approach may be used to determine the finite difference formula for a second derivative. In view of (3.2.3) and setting

$$a+b+c=0 \tag{3.2.7a}$$

$$b+2c=0 \tag{3.2.7b}$$

$$b+4c=2 \tag{3.2.7c}$$

we obtain

$$\left(\frac{\partial^2 u}{\partial x^2}\right)_i = \frac{u_i - 2u_{i-1} + u_{i-2}}{\Delta x^2} + \Delta x\frac{\partial^3 u}{\partial x^3} + \cdots \tag{3.2.8}$$

This implies that the one-sided formula provides only the first order accuracy in contrast to the two-sided formula, which gives the second order accuracy as seen in (3.1.11).

The foregoing procedure may be transformed into a systematic form in terms of "displacement" and "difference" operators so that difference formulas may be obtained with a preselected order of accuracy [Hildebrand, 1956; Kopal, 1961; Collatz, 1966], among others. These results are summarized next.

Forward Difference Formulas

The Taylor series expansion (3.1.2) may be written in terms of the displacement operator E and the derivative operator D,

$$Eu(x) = [1 + \Delta x D + (\Delta x D)^2/2! + (\Delta x D)^3/3! + \cdots] u(x) \tag{3.2.9}$$

with $Du = \frac{\partial u}{\partial x}$, $E = e^{\Delta x D}$, and $D = \frac{1}{\Delta x}\ln E$. These definitions lead to the first derivative of u at i in the form

$$\left(\frac{\partial u}{\partial x}\right)_i = \frac{1}{\Delta x}\ln(1+\delta^+)u_i = \frac{1}{\Delta x}\left(\delta^+ - \frac{\delta^{+2}}{2} + \frac{\delta^{+3}}{3} - \frac{\delta^{+4}}{4} + \cdots\right)u_i \tag{3.2.10}$$

where δ^+ is the forward difference operator,

$$\delta^+ = E - 1, \qquad \delta^+ u_i = u_{i+1} - u_i \tag{3.2.11}$$

with E being defined such that

$$Eu_i = u_{i+1}, \qquad E^n u_i = u_{i+n} \tag{3.2.12}$$

It is now obvious that the order of accuracy increases with the number of terms kept on the right-hand side of (3.2.10) given by

$$\left(\frac{\partial u}{\partial x}\right)_i = \frac{1}{\Delta x}\left((E-1) - \frac{(E-1)^2}{2} + \frac{(E-1)^3}{3} - \frac{(E-1)^4}{4} + \cdots\right)u_i \tag{3.2.13}$$

which leads to

First Order Accuracy

$$\left(\frac{\partial u}{\partial x}\right)_i = \frac{u_{i+1} - u_i}{\Delta x} - \frac{\Delta x}{2}\frac{\partial^2 u}{\partial x^2} \tag{3.2.14}$$

Second Order Accuracy

$$\left(\frac{\partial u}{\partial x}\right)_i = \frac{-3u_i + 4u_{i+1} - u_{i+2}}{2\Delta x} + \frac{\Delta x^2}{3}\frac{\partial^3 u}{\partial x^3} \tag{3.2.15}$$

Backward Difference Formulas

A backward difference formula can be derived similarly in the form

$$\begin{aligned}\left(\frac{\partial u}{\partial x}\right)_i &= \frac{-1}{\Delta x}\ln(1-\delta^-)u_i = \frac{1}{\Delta x}\left(\delta^- + \frac{\delta^{-2}}{2} + \frac{\delta^{-3}}{3} + \frac{\delta^{-4}}{4} + \cdots\right)u_i \\ &= \frac{1}{\Delta x}\left[(1-E^{-1}) + \frac{(1-E^{-1})^2}{2} + \frac{(1-E^{-1})^3}{3} + \frac{(1-E^{-1})^4}{4} + \cdots\right]u_i\end{aligned} \tag{3.2.16}$$

where δ^- is the backward difference operator,

$$\delta^- = 1 - E^{-1}, \qquad \delta^- u_i = u_i - u_{i-1} \tag{3.2.17}$$

with

$$E^{-1}u_i = u_{i-1} \tag{3.2.18}$$

These definitions lead to the following schemes:

First Order Accuracy

$$\left(\frac{\partial u}{\partial x}\right)_i = \frac{u_i - u_{i-1}}{\Delta x} + \frac{\Delta x}{2}\frac{\partial^2 u}{\partial x^2} \tag{3.2.19}$$

Second Order Accuracy

$$\left(\frac{\partial u}{\partial x}\right)_i = \frac{3u_i - 4u_{i-1} + u_{i-2}}{2\Delta x} + \frac{\Delta x^2}{3}\frac{\partial^3 u}{\partial x^3} \tag{3.2.20}$$

Central Difference Formulas

The central difference formulas are derived using the following definitions:

$$\delta u_i = u_{i+1/2} - u_{i-1/2} = \left(E^{1/2} - E^{-1/2}\right)u_i \tag{3.2.21}$$

with

$$\delta = e^{\Delta x D/2} - e^{-\Delta x D/2} = 2\sinh(\Delta x D/2) \tag{3.2.22}$$

which leads to the first derivative of u at i in the form

$$\left(\frac{\partial u}{\partial x}\right)_i = \frac{1}{\Delta x}\left(2\sinh^{-1}\frac{\delta}{2}\right)u_i = \frac{1}{\Delta x}\left(\delta - \frac{\delta^3}{24} + \frac{3\delta^5}{640} - \frac{5\delta^7}{7168} + \cdots\right)u_i \tag{3.2.23}$$

With these definitions, we obtain

Second Order Accuracy (with the first term)

$$\left(\frac{\partial u}{\partial x}\right)_i = \frac{u_{i+\frac{1}{2}} - u_{i-\frac{1}{2}}}{\Delta x} - \frac{\Delta x^2}{24}\frac{\partial^3 u}{\partial x^3} \tag{3.2.24}$$

Fourth Order Accuracy (with the first two terms)

$$\left(\frac{\partial u}{\partial x}\right)_i = \frac{1}{24\Delta x}\left(-u_{i+\frac{3}{2}} + 27u_{i+\frac{1}{2}} - 27u_{i-\frac{1}{2}} + u_{i-\frac{3}{2}}\right) + \frac{3}{640}\Delta x^4\frac{\partial^5 u}{\partial x^5} \tag{3.2.25}$$

The half-integer mesh points may be avoided by choosing

$$\left(\frac{\partial u}{\partial x}\right)_i = \frac{1}{\Delta x}\left(\bar{\delta} - \frac{\bar{\delta}^3}{3!} + \frac{3^2}{5!}\bar{\delta}^5 + \cdots\right)u_i \tag{3.2.26}$$

where $\bar{\delta}$ is the alternative central difference operator such that

$$\bar{\delta}u_i = \frac{1}{2}\left(E - E^{-1}\right)u_i = \frac{1}{2}(u_{i+1} - u_{i-1}) \tag{3.2.27}$$

These definitions provide

Second Order Accuracy

$$\left(\frac{\partial u}{\partial x}\right)_i = \frac{u_{i+1} - u_{i-1}}{2\Delta x} - \frac{(\Delta x)^2}{6}\frac{\partial^3 u}{\partial x^3} \tag{3.2.28}$$

Fourth Order Accuracy

$$\left(\frac{\partial u}{\partial x}\right)_i = \frac{-u_{i+2} + 8u_{i+1} - 8u_{i-1} + u_{i-2}}{12\Delta x} + \frac{\Delta x^4}{30}\frac{\partial^5 u}{\partial x^5} \tag{3.2.29}$$

3.3 HIGHER ORDER DERIVATIVES

Finite difference formulas for higher-order derivatives may be derived using the operator technique similarly to the one employed for the first order derivative. Let us consider the forward difference relation given by (3.2.10) and extend it to higher order derivatives as

$$\left(\frac{\partial^n u}{\partial x^n}\right)_i = \frac{1}{\Delta x^n}[\ln(1+\delta^+)]^n u_i$$
$$= \frac{1}{\Delta x^n}\left[\delta^{+n} - \frac{n}{2}\delta^{+(n+1)} + \frac{n(3n+5)}{24}\delta^{+(n+2)} - \frac{n(n+2)(n+3)}{48}\delta^{+(n+3)} + \cdots\right]u_i \tag{3.3.1}$$

Similarly for the backward difference, we write

$$\left(\frac{\partial^n u}{\partial x^n}\right)_i = \frac{-1}{\Delta x^n}[\ln(1-\delta^-)]^n u_i$$
$$= \frac{1}{\Delta x^n}\left(\delta^- + \frac{\delta^{-2}}{2} + \frac{\delta^{-3}}{3} + \cdots\right)^n u_i$$
$$= \frac{1}{\Delta x^n}\left[\delta^{-n} + \frac{n}{2}\delta^{-(n+1)} + \frac{n(3n+5)}{24}\delta^{-(n+2)} + \frac{n(n+2)(n+3)}{48}\delta^{-(n+3)} + \cdots\right]u_i \tag{3.3.2}$$

The central difference formulas are in the form

$$\left(\frac{\partial^n u}{\partial x^n}\right)_i = \left(\frac{2}{\Delta x}\sinh^{-1}\frac{\delta}{2}\right)^n u_i$$
$$= \frac{1}{\Delta x^n}\left[\delta - \frac{\delta^3}{24} + \frac{3\delta^5}{640} - \frac{5\delta^7}{7168} + \cdots\right]^n u_i$$
$$= \frac{1}{\Delta x^n}\delta^n\left[1 - \frac{n}{24}\delta^2 + \frac{n}{64}\left(\frac{22+5n}{90}\right)\delta^4 - \frac{n}{4^5}\left(\frac{5}{7} + \frac{n-1}{5} + \frac{(n-1)(n-2)}{3^5}\right)\delta^6 + \cdots\right]u_i \tag{3.3.3}$$

If n is even, the difference formulas are obtained at the integer mesh points. If n is uneven, however, the difference formulas involve half-integer mesh points. In order to maintain the integer mesh points, we may use

$$\left(\frac{\partial^n u}{\partial x^n}\right)_i = \frac{\mu}{\left(1+\frac{\delta^2}{4}\right)^{\frac{1}{2}}}\left(\frac{2}{\Delta x}\sinh^{-1}\frac{\delta}{2}\right)^n u_i$$
$$= \mu\frac{\delta^n}{\Delta x^n}\left[1 - \frac{n+3}{24}\delta^2 + \frac{5n^2+52n+135}{5760}\delta^4 + \cdots\right]u_i \tag{3.3.4}$$

where

$$\mu = \left(1 + \frac{\delta^2}{4}\right)^{\frac{1}{2}}$$

Based on these formulas, we summarize the second, third, and fourth order derivatives below.

Second Order Derivative ($n = 2$)

$$\left(\frac{\partial^2 u}{\partial x^2}\right)_i = \frac{1}{\Delta x^2}\left(\delta^{+2} - \delta^{+3} + \frac{11}{12}\delta^{+4} - \frac{5}{6}\delta^{+5} + \cdots\right)u_i, \quad \text{from (3.3.1)} \tag{3.3.5a}$$

$$\left(\frac{\partial^2 u}{\partial x^2}\right)_i = \frac{1}{\Delta x^2}\left(\delta^{-2} + \delta^{-3} + \frac{11}{12}\delta^{-4} + \frac{5}{6}\delta^{-5} + \cdots\right)u_i, \quad \text{from (3.3.2)} \tag{3.3.5b}$$

$$\left(\frac{\partial^2 u}{\partial x^2}\right)_i = \frac{1}{\Delta x^2}\left(\delta^2 - \frac{\delta^4}{12} + \frac{\delta^6}{90} - \frac{\delta^8}{560} + \cdots\right)u_i, \quad \text{from (3.3.3)} \tag{3.3.5c}$$

$$\left(\frac{\partial^2 u}{\partial x^2}\right)_i = \frac{\mu}{\Delta x^2}\left(\delta^2 - \frac{5\delta^4}{24} + \frac{259}{5760}\delta^6 + O(\Delta x^8)\right)u_i, \quad \text{from (3.3.4)} \tag{3.3.5d}$$

Forward Difference

First Order Accuracy

$$\left(\frac{\partial^2 u}{\partial x^2}\right)_i = \frac{1}{\Delta x^2}(u_{i+2} - 2u_{i+1} + u_i) - \Delta x \frac{\partial^3 u}{\partial x^3} \tag{3.3.6}$$

Second Order Accuracy

$$\left(\frac{\partial^2 u}{\partial x^2}\right)_i = \frac{1}{\Delta x^2}(2u_i - 5u_{i+1} + 4u_{i+2} - u_{i+3}) + \frac{11}{12}\Delta x^2 \frac{\partial^4 u}{\partial x^4} \tag{3.3.7}$$

Backward Difference

First Order Accuracy

$$\left(\frac{\partial^2 u}{\partial x^2}\right)_i = \frac{1}{\Delta x^2}(u_i - 2u_{i-1} + u_{i-2}) + \Delta x \frac{\partial^3 u}{\partial x^3} \tag{3.3.8}$$

Second Order Accuracy

$$\left(\frac{\partial^2 u}{\partial x^2}\right)_i = \frac{1}{\Delta x^2}(2u_i - 5u_{i-1} + 4u_{i-2} - u_{i-3}) - \frac{11}{12}\Delta x^2 \frac{\partial^4 u}{\partial x^4} \tag{3.3.9}$$

Central Difference

Second Order Accuracy

$$\left(\frac{\partial^2 u}{\partial x^2}\right)_i = \frac{1}{\Delta x^2}(u_{i+1} - 2u_i + u_{i-1}) - \frac{\Delta x^2}{12}\frac{\partial^4 u}{\partial x^4} \tag{3.3.10}$$

Fourth Order Accuracy

$$\left(\frac{\partial^2 u}{\partial x^2}\right)_i = \frac{1}{12\Delta x^2}(-u_{i+2} + 16u_{i+1} - 30u_i + 16u_{i-1} - u_{i-2}) + \frac{\Delta x^4}{90}\frac{\partial^6 u}{\partial x^6} \tag{3.3.11}$$

Central Difference – Half Integer Points

Second Order Accuracy

$$\left(\frac{\partial^2 u}{\partial x^2}\right)_i = \frac{1}{2\Delta x^2}\left(u_{i+\frac{3}{2}} - u_{i+\frac{1}{2}} - u_{i-\frac{1}{2}} + u_{i-\frac{3}{2}}\right) - \frac{5}{24}\Delta x^2 \frac{\partial^4 u}{\partial x^4} \tag{3.3.12}$$

Fourth Order Accuracy

$$\left(\frac{\partial^2 u}{\partial x^2}\right)_i = \frac{1}{48\Delta x^2}\left(-5u_{i+\frac{5}{2}} + 39u_{i+\frac{3}{2}} - 34u_{i+\frac{1}{2}} - 34u_{i-\frac{1}{2}} + 39u_{i-\frac{3}{2}} - 5u_{i-\frac{5}{2}}\right) + \frac{259}{5760}\Delta x^4 \frac{\partial^6 u}{\partial x^6} \tag{3.3.13}$$

Note that the last scheme requires six mesh points to achieve the fourth order accuracy, whereas for the same accuracy, the scheme given by (3.3.11) requires only five mesh points.

Third Order Derivative ($n = 3$)

Forward Difference

First Order Accuracy

$$\left(\frac{\partial^3 u}{\partial x^3}\right)_i = \frac{1}{\Delta x^3}(u_{i+3} - 3u_{i+2} + 3u_{i+1} - u_i) - \frac{\Delta x}{2}\frac{\partial^4 u}{\partial x^4} \tag{3.3.14}$$

Second Order Accuracy

$$\left(\frac{\partial^3 u}{\partial x^3}\right)_i = \frac{1}{2\Delta x^3}(-3u_{i+4} + 14u_{i+3} - 24u_{i+2} + 18u_{i+1} - 5u_i) + \frac{21}{12}\Delta x^2 \frac{\partial^5 u}{\partial x^5} \tag{3.3.15}$$

Backward Difference

First Order Accuracy

$$\left(\frac{\partial^3 u}{\partial x^3}\right)_i = \frac{1}{\Delta x^3}(u_i - 3u_{i-1} + 3u_{i-2} - u_{i-3}) + \frac{\Delta x}{2}\frac{\partial^4 u}{\partial x^4} \tag{3.3.16}$$

Second Order Accuracy

$$\left(\frac{\partial^3 u}{\partial x^3}\right)_i = \frac{1}{2\Delta x^3}(5u_i - 18u_{i-1} + 24u_{i-2} - 14u_{i-3} + 3u_{i-4}) - \frac{21}{12}\Delta x^2 \frac{\partial^5 u}{\partial x^5} \tag{3.3.17}$$

Central Difference

Second Order Accuracy

$$\left(\frac{\partial^3 u}{\partial x^3}\right)_i = \frac{1}{2\Delta x^3}(u_{i+2} - 2u_{i+1} + 2u_{i-1} - u_{i-2}) - \frac{1}{4}\Delta x^2 \frac{\partial^5 u}{\partial x^5} \tag{3.3.18}$$

Fourth Order Accuracy

$$\left(\frac{\partial^3 u}{\partial x^3}\right)_i = \frac{1}{8\Delta x^3}(-u_{i+3} + 8u_{i+2} - 13u_{i+1} - 13u_{i-1} - 8u_{i-2} + u_{i-3}) + \frac{7}{120}\Delta x^4 \frac{\partial^7 u}{\partial x^7} \tag{3.3.19}$$

Central Difference – Half Integer Points

Second Order Accuracy

$$\left(\frac{\partial^3 u}{\partial x^3}\right)_i = \frac{1}{\Delta x^3}\left(u_{i+\frac{3}{2}} - 3u_{i+\frac{1}{2}} + 3u_{i-\frac{1}{2}} - u_{i-\frac{3}{2}}\right) - \frac{\Delta x^2}{8}\frac{\partial^5 u}{\partial x^5} \tag{3.3.20a}$$

Fourth Order Accuracy

$$\left(\frac{\partial^3 u}{\partial x^3}\right)_i = \frac{1}{8\Delta x^3}\left(-u_{i+\frac{5}{2}} + 13u_{i+\frac{3}{2}} - 34u_{i+\frac{1}{2}} + 34u_{i-\frac{1}{2}} - 13u_{i-\frac{3}{2}} + u_{i-\frac{5}{2}}\right)$$
$$+\frac{37}{1920}\Delta x^4\frac{\partial^7 u}{\partial x^7} \tag{3.3.20b}$$

Fourth Order Derivative

Forward Difference (first order accuracy)

$$\left(\frac{\partial^4 u}{\partial x^4}\right)_i = \frac{1}{\Delta x^4}(u_{i+4} - 4u_{i+3} + 6u_{i+2} - 4u_{i+1} + u_i) - 2\Delta x\frac{\partial^5 u}{\partial x^5} \tag{3.2.21}$$

Backward Difference (first order accuracy)

$$\left(\frac{\partial^4 u}{\partial x^4}\right)_i = \frac{1}{\Delta x^4}(u_i - 4u_{i-1} + 6u_{i-2} - 4u_{i-3} + u_{i-4}) + 2\Delta x\frac{\partial^5 u}{\partial x^5} \tag{3.2.22}$$

Central Difference (second order accuracy)

$$\left(\frac{\partial^4 u}{\partial x^4}\right)_i = \frac{1}{\Delta x^4}(u_{i+2} - 4u_{i+1} + 6u_i - 4u_{i-1} + u_{i-2}) - \frac{\Delta x^2}{6}\frac{\partial^6 u}{\partial x^6} \tag{3.2.23}$$

Various order finite difference formulas up to fourth order derivatives are summarized in Table 3.3.1.

3.4 MULTIDIMENSIONAL FINITE DIFFERENCE FORMULAS

Multidimensional finite difference formulas can be derived using the results of one-dimensional formulas. For two-dimensions, we consider

$$x_i = x_0 + i\Delta x$$
$$y_j = y_0 + j\Delta y$$

as defined in Figure 3.4.1. The forward and backward operators are now given by $\delta_x^{\pm}$ and $\delta_y^{\pm}$ for x- and y-directions, respectively. The first partial derivatives in the x- and y-directions are

$$\left(\frac{\partial u}{\partial x}\right)_{ij} = \frac{1}{\Delta x}\delta_x^+ u_{ij} + O(\Delta x) = \frac{u_{i+1,j} - u_{i,j}}{\Delta x} + O(\Delta x) \tag{3.4.1}$$

$$\left(\frac{\partial u}{\partial y}\right)_{ij} = \frac{1}{\Delta y}\delta_y^+ u_{ij} + O(\Delta x) = \frac{u_{i,j+1} - u_{i,j}}{\Delta y} + O(\Delta y) \tag{3.4.2}$$

Table 3.3.1 Various Order Finite Difference Formulas

(a) Forward Difference, O(Δx)

	u_i	u_{i+1}	u_{i+2}	u_{i+3}	u_{i+4}
$\Delta x \frac{\partial u}{\partial x}$	–1	1			
$\Delta x^2 \frac{\partial^2 u}{\partial x^2}$	1	–2	1		
$\Delta x^3 \frac{\partial^3 u}{\partial x^3}$	–1	3	–3	1	
$\Delta x^4 \frac{\partial^4 u}{\partial x^4}$	1	–4	6	–4	1

(b) Backward Difference, O(Δx)

	u_{i-4}	u_{i-3}	u_{i-2}	u_{i-1}	u_i
$\Delta x \frac{\partial u}{\partial x}$				–1	1
$\Delta x^2 \frac{\partial^2 u}{\partial x^2}$			1	–2	1
$\Delta x^3 \frac{\partial^3 u}{\partial x^3}$		–1	3	–3	1
$\Delta x^4 \frac{\partial^4 u}{\partial x^4}$	1	–4	6	–4	1

(c) Central Difference, O(Δx^2)

	u_{i-2}	u_{i-1}	u_i	u_{i+1}	u_{i+2}
$2\Delta x \frac{\partial u}{\partial x}$		–1	0	1	
$\Delta x^2 \frac{\partial^2 u}{\partial x^2}$		1	–2	1	
$2\Delta x^3 \frac{\partial^3 u}{\partial x^3}$	–1	2	0	2	1
$\Delta x^4 \frac{\partial^4 u}{\partial x^4}$	1	–4	6	–4	1

(d) Forward Difference, O(Δx^2)

	u_i	u_{i+1}	u_{i+2}	u_{i+3}	u_{i+4}	u_{i+5}
$2\Delta x \frac{\partial u}{\partial x}$	–3	4	–1			
$\Delta x^2 \frac{\partial^2 u}{\partial x^2}$	2	–5	4	–1		
$2\Delta x^3 \frac{\partial^3 u}{\partial x^3}$	–5	18	–24	14	–3	
$\Delta x^4 \frac{\partial^4 u}{\partial x^4}$	3	–14	26	–24	11	–2

(e) Backward Difference, O(Δx^2)

	u_{i-5}	u_{i-4}	u_{i-3}	u_{i-2}	u_{i-1}	u_i
$2\Delta x \frac{\partial u}{\partial x}$				1	–4	3
$\Delta x^2 \frac{\partial^2 u}{\partial x^2}$			–1	4	–5	2
$2\Delta x^3 \frac{\partial^3 u}{\partial x^3}$		3	–14	24	–18	5
$\Delta x^4 \frac{\partial^4 u}{\partial x^4}$	–2	11	–24	26	–14	3

(f) Central Difference, O(Δx^4)

	u_{i-3}	u_{i-2}	u_{i-1}	u_i	u_{i+1}	u_{i+2}	u_{i+3}
$12\Delta x \frac{\partial u}{\partial x}$		1	–8	0	8	–1	
$12\Delta x^2 \frac{\partial^2 u}{\partial x^2}$		–1	16	–30	16	–1	
$8\Delta x^3 \frac{\partial^3 u}{\partial x^3}$	1	–8	13	0	–13	8	–1
$6\Delta x^4 \frac{\partial^4 u}{\partial x^4}$	–1	12	–39	56	–39	12	–1

Similarly, the second order central difference formulas for the second order derivatives are of the form

$$\left(\frac{\partial^2 u}{\partial x^2}\right)_{ij} = \frac{u_{i+1,j} - 2u_{i,j} + u_{i-1,j}}{\Delta x^2} - \frac{\Delta x^2}{12}\frac{\partial^4 u}{\partial x^4} \tag{3.4.3}$$

$$\left(\frac{\partial^2 u}{\partial y^2}\right)_{ij} = \frac{u_{i,j+1} - 2u_{i,j} + u_{i,j-1}}{\Delta y^2} - \frac{\Delta y^2}{12}\frac{\partial^4 u}{\partial x^4} \tag{3.4.4}$$

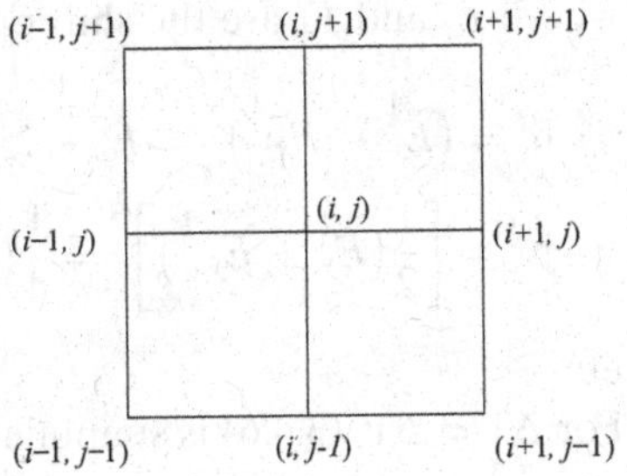

Figure 3.4.1 Two-dimensional mesh.

Let us now consider the Laplace equation

$$\nabla^2 u = \frac{\partial^2 u}{\partial x^2} + \frac{\partial^2 u}{\partial y^2} = 0$$

whose finite difference formula is obtained as the sum of (3.4.3) and (3.4.4), resulting in a five-point scheme

$$\Delta u_{ij} = \left(\frac{\delta_x^2}{\Delta x^2} + \frac{\delta_y^2}{\Delta y^2}\right) u_{ij} = \frac{u_{i-1,j} - 2u_{i,j} + u_{i+1,j}}{\Delta x^2} + \frac{u_{i,j-1} - 2u_{i,j} + u_{i,j+1}}{\Delta y^2} + O(\Delta x^2, \Delta y^2) \tag{3.4.5a}$$

For $\Delta x = \Delta y$

$$\Delta^{(1)} u_{ij} = \frac{u_{i+1,j} + u_{i-1,j} + u_{i,j-1} + u_{i,j+1} - 4u_{i,j}}{\Delta x^2} - \frac{\Delta x^2}{12}\left(\frac{\partial^4 u}{\partial x^4} + \frac{\partial^4 u}{\partial y^4}\right) \tag{3.4.5b}$$

as graphically shown in Figure 3.4.2a.

An alternative representation of (3.4.5a) is given by

$$\begin{aligned}\Delta^{(2)} u_{ij} &= \left[\left(\frac{1}{\Delta x}\mu_y \delta_x\right)^2 + \left(\frac{1}{\Delta y}\mu_x \delta_y\right)^2\right] u_{ij} \\ &= \left[\frac{1}{4\Delta x^2}(E_y + 2 + E_y^{-1})(E_x - 2 + E_x^{-1}) \right. \\ &\quad \left. + \frac{1}{4\Delta y^2}(E_x + 2 + E_x^{-1})(E_y - 2 + E_y^{-1})\right] u_{ij}\end{aligned} \tag{3.4.6}$$

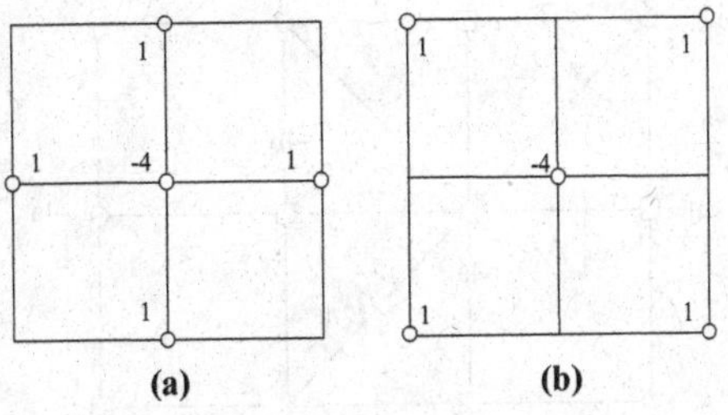

Figure 3.4.2 Five-point finite difference mesh. **(a)** Regular operator. **(b)** Shift operator.

where E_x and E_y are the shift operators resulting from

$$\delta_x^2 = \left(E_x^{\frac{1}{2}} - E_x^{-\frac{1}{2}}\right)^2 = E_x - 2 + E_x^{-1}$$

$$\mu_y^2 = \left[\frac{1}{2}\left(E_y^{\frac{1}{2}} + E_y^{-\frac{1}{2}}\right)\right]^2 = \frac{1}{4}\left(E_y + 2 + E_y^{-1}\right)$$

etc.

For $\Delta x = \Delta y$, (3.4.6) is simplified as (Figure 3.4.2b)

$$\Delta^{(2)} u_{ij} = \frac{1}{4\Delta x^2}(u_{i+1,j+1} + u_{i+1,j-1} + u_{i-1,j-1} + u_{i-1,j+1} - 4u_{i,j}) \tag{3.4.7}$$

For higher order terms, we may write

$$\begin{aligned}\Delta^{(2)} u_{ij} &= \left(1 + \frac{\Delta y^2}{4}\frac{\partial^2}{\partial y^2}\right)\left(\frac{\partial^2 u}{\partial x^2} + \frac{\Delta x^2}{12}\frac{\partial^4 u}{\partial x^4}\right)_{ij} + \left(1 + \frac{\Delta x^2}{4}\frac{\partial^2}{\partial x^2}\right)\left(\frac{\partial^2 u}{\partial y^2} + \frac{\Delta y^2}{12}\frac{\partial^4 u}{\partial y^4}\right)_{ij} \\ &= \Delta u_{ij} + \frac{1}{12}\Delta x^2 \frac{\partial^4 u}{\partial x^4} + \frac{1}{12}\Delta y^2 \frac{\partial^4 u}{\partial y^4} + \left(\frac{\Delta x^2 + \Delta y^2}{4}\right)\frac{\partial^4 u}{\partial x^4 \partial y^4} + \cdots\end{aligned} \tag{3.4.8}$$

with the truncation error being $O(\Delta x^2, \Delta y^2)$. Note that this scheme involves the odd-numbered nodes detached from the even-numbered nodes (Figure 3.4.3). Note that point (i, j) is coupled to the points marked by a square, while there is no connection to the even-numbered points marked by a circle. Thus, the solution oscillates between the two values a and b when passing from an even to odd-numbered point, satisfying the difference equation $\Delta^{(2)} u_{ij} = 0$. However, it will not satisfy the difference equation (3.4.5).

The well-known nine-point formula can be derived by combining (3.4.8) with $\Delta^{(1)} u_{ij}$.

$$\begin{aligned}\Delta^{(3)} u_{ij} &= \left(a\Delta^{(1)} + b\Delta^{(2)}\right) u_{ij} \\ &= \frac{1}{\Delta x^2}\left[(\delta x^2 + \delta y^2) + \frac{b}{2}\delta x^2 \delta y^2\right] u_{ij} = \Delta^{(1)} u_{ij} + \frac{b}{2}\delta x^2 \delta y^2 u_{ij} \\ &= \Delta u_{ij} + \frac{\Delta x^2}{12}\left[\frac{\partial^4 u}{\partial x^4} + \frac{\partial^4 u}{\partial y^4} + 6b\frac{\partial^4 u}{\partial x^4 \partial y^4}\right]\end{aligned} \tag{3.4.9}$$

where $a + b = 1$. For $b = 2/3$, we arrive at the scheme depicted in Figure 3.4.4a, which can also be obtained from finite elements. For $b = 1/3$, the Dahlquist and Bjorck scheme

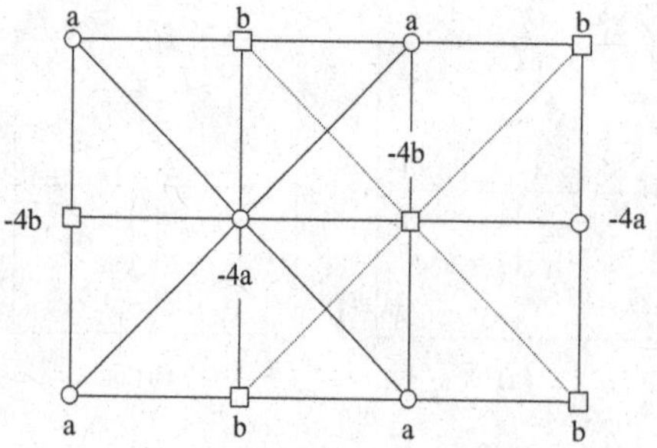

Figure 3.4.3 Odd-even oscillations of the five-point scheme.

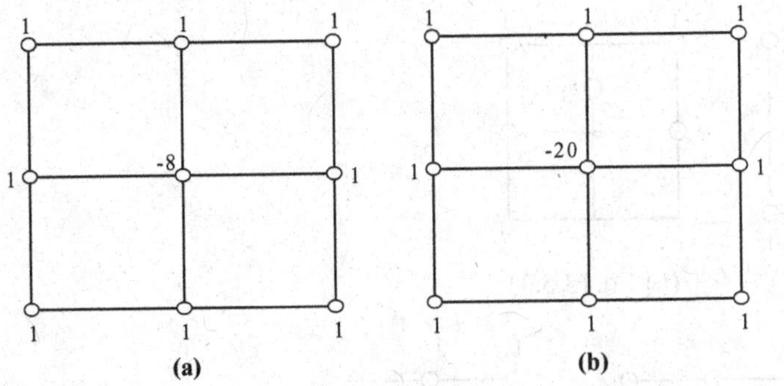

Figure 3.4.4 Nine-point molecule. (a) Nine-point formula with $b = 2/3$. (b) Nine-point formula with $b = 1/3$.

[1974] arises as shown in Figure 3.4.4b, providing the truncation error

$$-\frac{\Delta x^2}{12}\left(\frac{\partial^2}{\partial x^2}+\frac{\partial^2}{\partial y^2}\right)^2 u = -\frac{\Delta x^2}{12}\Delta^2 u$$

For $\Delta u = \lambda u$, the nine-point operator with $\Delta^{(3)} = \frac{2}{3}\Delta^{(1)} + \frac{1}{3}\Delta^{(2)}$ gives a truncation error

$$-\lambda^2 \frac{\Delta x^2}{12} u$$

Therefore, the corrected difference scheme

$$\Delta^{(3)} u_{i,j} = \left(\lambda + \lambda^2 \frac{\Delta x^2}{12}\right) u$$

has a fourth order truncation error.

An extension to three-dimensional geometries is straightforward. Some applications to 3-D problems will be discussed in Chapter 7.

3.5 MIXED DERIVATIVES

The simplest, second order central formula for the mixed derivative is obtained from the application of (3.2.3) in both directions x and y.

$$\left(\frac{\partial^2 u}{\partial x \partial y}\right)_{ij} = \frac{1}{\Delta x \Delta y} \mu_x \delta_x \left[\left(1 - \frac{\delta x^2}{6} + O(\Delta x^4)\right)\right] \mu_y \delta_y \left[\left(1 - \frac{\delta y^2}{6} + O(\Delta y^4)\right)\right] u_{i,j} \tag{3.5.1}$$

This leads to a second order accuracy (Figure 3.5.1a),

$$\begin{aligned}\left(\frac{\partial^2 u}{\partial x \partial y}\right)_{ij} &= \frac{1}{\Delta x \Delta y}(\mu_x \delta_x \mu_y \delta_y)\, u_{i,j} + O(\Delta x^2, \Delta y^2) \\ &= \frac{u_{i+1,j+1} - u_{i+1,j-1} - u_{i-1,j+1} + u_{i-1,j-1}}{4\Delta x \Delta y} + O(\Delta x^2, \Delta y^2)\end{aligned} \tag{3.5.2}$$

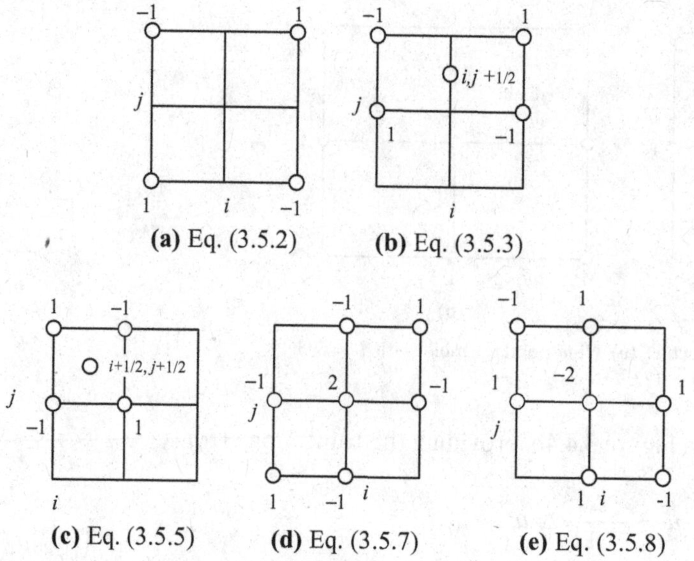

Figure 3.5.1 Mixed derivatives.

An alternative approach as shown in Figure 3.5.1b is given by

$$\left(\frac{\partial^2 u}{\partial x \partial y}\right)_{ij} = \frac{1}{\Delta x \Delta y}(\mu_x \delta_x \delta_y^+)\, u_{ij} + O(\Delta x^2, \Delta y)$$

$$= \frac{1}{2\Delta x \Delta y}(u_{i+1,j+1} - u_{i-1,j+1} - u_{i+1,j} + u_{i-1,j}) + O(\Delta x^2, \Delta y) \qquad (3.5.3)$$

A similar form can be obtained for the truncation error of $O(\Delta x, \Delta y^2)$. A first order in both x and y is derived in the form

$$\left(\frac{\partial^2 u}{\partial x \partial y}\right)_i = \frac{1}{\Delta x \Delta y}\delta_x^+ \delta_y^+ u_{ij} + O(\Delta x, \Delta y)$$

$$= \frac{1}{\Delta x \Delta y}(u_{i+1,j+1} - u_{i+1,j} - u_{i,j+1} + u_{i,j}) + O(\Delta x, \Delta y) \qquad (3.5.4)$$

This scheme can be altered to give a second order accuracy at $i + \frac{1}{2}, j + \frac{1}{2}$,

$$\left(\frac{\partial^2 u}{\partial x \partial y}\right)_{i+\frac{1}{2},j+\frac{1}{2}} = \frac{1}{\Delta x \Delta y}\delta_x \delta_y u_{i+\frac{1}{2},j+\frac{1}{2}} + O(\Delta x^2, \Delta y^2)$$

$$= \frac{1}{\Delta x \Delta y}(u_{i+1,j+1} - u_{i+1,j} - u_{i,j+1} + u_{i,j}) + O(\Delta x^2, \Delta y^2) \qquad (3.5.5)$$

as shown in Figure (3.5.1c).

Applying backward differences in both directions, we obtain

$$\left(\frac{\partial^2 u}{\partial x \partial y}\right)_i = \frac{1}{\Delta x \Delta y}\delta_x^- \delta_y^- u_{ij} + O(\Delta x, \Delta y)$$

$$= \frac{1}{\Delta x \Delta y}(u_{i-1,j-1} - u_{i-1,j} - u_{i,j-1} + u_{i,j}) + O(\Delta x, \Delta y)$$

$$= \frac{1}{\Delta x \Delta y}\delta_x \delta_y u_{i-\frac{1}{2},j-\frac{1}{2}} + O(\Delta x, \Delta y) \qquad (3.5.6)$$

Summing (3.5.4) and (3.5.6), we obtain a second order formula,

$$\left(\frac{\partial^2 u}{\partial x \partial y}\right)_i = \frac{1}{2\Delta x \Delta y}[\delta_x^+\delta_y^+ + \delta_x^-\delta_y^-]u_{ij} + O(\Delta x^2, \Delta y^2)$$
$$= \frac{1}{2\Delta x \Delta y}[u_{i+1,j+1} - u_{i+1,j} - u_{i,j+1} + u_{i-1,j-1} - u_{i-1,j} - u_{i,j-1} + 2u_{ij}]$$
$$+ O(\Delta x^2, \Delta y^2) \tag{3.5.7}$$

This is shown in Figure 3.5.1d. Another form can be obtained by combining forward and backward differences as (Figure 3.5.1e)

$$\left(\frac{\partial^2 u}{\partial x \partial y}\right)_i = \frac{1}{2\Delta x \Delta y}[\delta_x^+\delta_y^- + \delta_x^-\delta_y^+]u_{ij} + O(\Delta x^2, \Delta y^2)$$
$$= \frac{1}{2\Delta x \Delta y}[u_{i+1,j} - u_{i+1,j-1} + u_{i,j+1} + u_{i,j-1} - u_{i-1,j+1} + u_{i-1,j} - 2u_{ij}]$$
$$= \frac{1}{2\Delta x \Delta y}(\delta_x\delta_y u_{i+\frac{1}{2},j-\frac{1}{2}} + \delta_x\delta_y u_{i-\frac{1}{2},j+\frac{1}{2}}) + O(\Delta x^2 \Delta y^2) \tag{3.5.8}$$

Combining (3.5.7) and (3.5.8), we recover the fully central second order approximation (3.5.2). Therefore, the most general second order mixed derivative approximation can be obtained by an arbitrary linear combination of (3.5.7) and (3.5.8) [Mitchell and Griffiths, 1980].

$$\left(\frac{\partial^2 u}{\partial x \partial y}\right)_i = \frac{1}{2\Delta x \Delta y}\delta_x\delta_y(au_{i+\frac{1}{2},j+\frac{1}{2}} + au_{i-\frac{1}{2},j-\frac{1}{2}} + bu_{i+\frac{1}{2},j-\frac{1}{2}} + bu_{i-\frac{1}{2},j+\frac{1}{2}})$$
$$+ O(\Delta x^2, \Delta y^2) \tag{3.5.9}$$

with $a + b = 1$.

3.6 NONUNIFORM MESH

The standard Taylor series expansion may be applied to nonuniform meshes. The first derivative one-sided first order formula takes the form

$$\left(\frac{\partial u}{\partial x}\right)_i = \frac{u_{i+1} - u_i}{\Delta x_{i+1}} - \frac{\Delta x_{i+1}}{2}\frac{\partial^2 u}{\partial x^2} \tag{3.6.1a}$$

The backward formula becomes

$$\left(\frac{\partial u}{\partial x}\right)_i = \frac{u_i - u_{i-1}}{\Delta x_i} + \frac{\Delta x_i}{2}\frac{\partial^2 u}{\partial x^2} \tag{3.6.1b}$$

where $\Delta x_i = x_i - x_{i-1}$, etc.

The central difference is obtained by combining (3.6.1a) and (3.6.1b), which will lead to the second order formula

$$\left(\frac{\partial u}{\partial x}\right)_i = \frac{1}{\Delta x_i + \Delta x_{i+1}}\left[\frac{\Delta x_i}{\Delta x_{i+1}}(u_{i+1} - u_i) + \frac{\Delta x_{i+1}}{\Delta x_i}(u_i - u_{i-1})\right] - \frac{\Delta x_i \Delta x_{i+1}}{6}\frac{\partial^3 u}{\partial x^3} \tag{3.6.2}$$

It can also be shown that Taylor expansion leads to a forward or backward scheme. For example, for a forward scheme, we obtain

$$\left(\frac{\partial u}{\partial x}\right)_i = \left(\frac{\Delta x_{i+1} + \Delta x_{i+2}}{\Delta x_{i+2}}\frac{u_{i+1} - u_i}{\Delta x_{i+1}} - \frac{\Delta x_{i+1}}{\Delta x_{i+2}}\frac{u_{i+2} - u_i}{\Delta x_{i+1} + \Delta x_{i+2}}\right) + \frac{\Delta x_{i+1}(\Delta x_{i+1} + \Delta x_{i+2})}{6}\frac{\partial^3 u}{\partial x^3} \tag{3.6.3}$$

The three-point central difference formula for the second derivative is of the form

$$\left(\frac{\partial^2 u}{\partial x^2}\right)_i = \left(\frac{u_{i+1} - u_i}{\Delta x_{i+1}} - \frac{u_i - u_{i-1}}{\Delta x_i}\right)\frac{2}{\Delta x_{i+1} + \Delta x_i} + \frac{1}{3}(\Delta x_{i+1} - \Delta x_i)\frac{\partial^3 u}{\partial x^3} - \frac{\Delta x_{i+1}^3 + \Delta x_i^3}{12(\Delta x_{i+1} + \Delta x_i)}\frac{\partial^4 u}{\partial x^4} \tag{3.6.4}$$

Note that a loss of accuracy in nonuniform meshes is expected to occur and abrupt changes in mesh size in (3.6.4) result in the first order accuracy. For example, the third order accuracy of (3.6.4) is reduced to the second order for $\Delta x_{i+1} = \Delta x_i$.

3.7 HIGHER ORDER ACCURACY SCHEMES

For many applications in fluid dynamics with discontinuities and/or high gradients such as in shock waves and turbulence, it is necessary that higher order accuracy be provided in constructing difference equations for the first order, second order, and higher order derivatives. Lele [1992] presents various finite difference schemes which are generalization of the Padé scheme [Hildebrand, 1956; Kopal, 1961; Collatz, 1966]. These generalizations for the first order derivatives are given by

$$\beta u'_{i-2} + \alpha u'_{i-1} + u'_i + \alpha u'_{i+1} + \beta u'_{i+2} = a\frac{u_{i+1} - u_{i-1}}{2\Delta x} + b\frac{u_{i+2} - u_{i-2}}{4\Delta x} + c\frac{u_{i+3} - u_{i-3}}{6\Delta x} \tag{3.7.1}$$

with $u' = du/dx$. The relations between the coefficients a, b, c and α and β are derived by matching the Taylor series coefficients of various orders. Similarly, the generalizations for the second order derivatives are given by

$$\beta u''_{i-2} + \alpha u''_{i-1} + u''_i + \alpha u''_{i+1} + \beta u''_{i+2} = a\frac{u_{i+1} - 2u_i + u_{i-1}}{\Delta x^2} + b\frac{u_{i+2} - 2u_i + u_{i-2}}{4\Delta x^2} + c\frac{u_{i+3} - 2u_i + u_{i-3}}{9\Delta x^2} \tag{3.7.2}$$

with $u'' = d^2u/dx^2$. Again, the relations between the coefficients a, b, c and α and β are derived by matching the Taylor series coeffcients of various orders.

Higher Order Accuracy for the First Order Derivatives

Fourth Order Accuracy. Note that, for $\alpha = \beta = 0$ and $a = 4/3, b = -1/3$, and $c = 0$ inserted in (3.7.1), the first order derivative in (3.7.1) leads to the well-known fourth order central difference scheme.

$$u'_i = \frac{du_i}{dx} = \frac{1}{12\Delta x}(u_{i-2} - 8u_{i-1} + 8u_{i+1} - u_{i+2}) \tag{3.7.3}$$

Other higher order accuracy schemes for the first order derivative are obtained from (3.7.1) as follows:

Sixth Order Accuracy

$\alpha = 1/3, \quad \beta = 0, \quad a = 14/9, \quad b = 1/9, \quad c = 0$

Eighth Order Accuracy

$\alpha = 4/9, \quad \beta = 1/36, \quad a = 40/27, \quad b = 25/54, \quad c = 0$

Higher Order Accuracy for the Second Order Derivatives

Fourth Order Accuracy. The fourth order accuracy for the second order derivative arises from (3.7.2) by inserting the same constants as in the first order derivative.

$$u_i'' = d^2u_i/dx^2 = \frac{1}{12\Delta x^2}(-u_{i-2} + 16u_{i-1} - 30u_i + 16u_{i+1} - u_{i+2}) \tag{3.7.4}$$

Higher order accuracy schemes for the second order derivative are obtained by inserting the following constants in (3.7.2):

Sixth Order Accuracy

$\alpha = 2/11, \quad \beta = 0, \quad a = 12/11, \quad b = 3/11, \quad c = 0$

Eighth Order Accuracy

$$\alpha = 344/1179, \qquad \beta = \frac{38\alpha - 9}{214},$$

$$a = \frac{696 - 1191\alpha}{428}, \qquad b = \frac{2454\alpha - 294}{535}, \qquad c = \frac{1179\alpha - 344}{2140}$$

These higher order accuracy derivatives have been used extensively in the analysis of shock waves and turbulence, as will be discussed in Part Five, Applications.

3.8 ACCURACY OF FINITE DIFFERENCE SOLUTIONS

The finite difference formulas and their subsequent use in boundary value problems must assure accuracy in portraying the physical aspect of the problem that has been modeled. The accuracy depends on consistency, stability, and convergence as defined below:

(a) *Consistency* A finite difference equation is consistent if it becomes the corresponding partial differential equation as the grid size and time step approach zero, or truncation errors are zero. This is usually the case if finite difference formulas are derived from the Taylor series.

(b) *Stability* A numerical scheme used for the solution of finite difference equations is stable if the error remains bounded. Certain criteria must be satisfied in order to achieve stability. This subject will be elaborated upon in Sections 4.2 and 4.3.

(c) *Convergence* A finite difference scheme is convergent if its solution approaches that of the partial differential equation as the grid size approaches zero. Both consistency and stability are prerequisite to convergence.

The ultimate goal of any numerical scheme is a convergence to the exact solution as the mesh size is reduced. Discrete time step sizes are chosen adequately as related to the mesh sizes so that the solution process is stable. The finite difference formulas studied in this chapter will be used for developing such numerical schemes. Here, the stability and convergence are important factors for the success in CFD projects and will be addressed continuously for the rest of this book.

3.9 SUMMARY

In this chapter, we have demonstrated that finite difference equations can be derived in many different ways. Simple methods and more rigorous general methods by means of finite difference operator, derivative operator, forward difference operator, and backward difference operator are introduced. Applications to various order derivatives in multidimensions are presented.

We have also shown how to obtain finite difference equations for higher order accuracy. They are particularly useful for complex physical phenomena such as in shock waves and turbulence, as will be shown in Part Five, Applications.

Our ultimate goal is the accuracy of the solution of differential equations. In order to achieve this accuracy, it is necessary that difference equations satisfy three criteria: consistency, stability, and convergence. Among these, the properties of consistency and stability reside in the realm of the development of finite difference equations. Convergence prevails if the requirements of consistency and stability are satisfied. The consequence of satisfaction of these criteria leads to the assurance of accuracy in CFD.

REFERENCES

Adam, Y. [1975]. A Hermitian finite difference method for the solution of parabolic equations. *Comp. Math. Appl.*, 1, 393–406.
Collatz, L. [1966]. *The Numerical Treatment of Differential Equations*. New York: Springer-Verlag.
Dahlquist, G. and Bjork, A. [1974]. *Numerical Methods*. Englewood Cliffs, NJ: Prentice-Hall.
Hildebrand, F. B. [1956]. *Introduction to Numerical Analysis*. New York: McGraw-Hill.
Kopal, Z. [1961]. *Numerical Analysis*. New York: Wiley.
Lely, S. K. [1992]. Compact finite difference schemes with spectral-like resolution. *J. Comp. Phys.*, 183, 16–42.
Mitchell, A. R. and Griffiths, D. F. [1980]. *The Finite Difference Method in Partial Differential Equations*. New York: Wiley.

CHAPTER FOUR

Solution Methods of Finite Difference Equations

In this chapter, solution methods for elliptic, parabolic, hyperbolic equations, and Burgers' equations are presented. These equations do not represent actual fluid dynamics problems, but the methods discussed in this chapter will form the basis for solving incompressible and compressible flow problems which are presented in Chapters 5 and 6, respectively. Although the computational schemes for these equations have been in existence for many years and are well documented in other text books, they are summarized here merely for the sake of completeness and for references in later chapters.

4.1 ELLIPTIC EQUATIONS

Elliptic equations represent one of the fundamental building blocks in fluid mechanics. Steady heat conduction, diffusion processes in viscous, turbulent, and boundary layer flows, as well as chemically reacting flows are characterized by the elliptic nature of the governing equations. Various difference schemes for the elliptic equations and some solution methods are also presented in this chapter.

4.1.1 FINITE DIFFERENCE FORMULATIONS

Consider the Laplace equation which is one of the typical elliptic equations,

$$\frac{\partial^2 u}{\partial x^2} + \frac{\partial^2 u}{\partial y^2} = 0 \tag{4.1.1}$$

The five-point and nine-point finite differences for the Laplace equation are, respectively,

$$\frac{u_{i+1,j} - 2u_{i,j} + u_{i-1,j}}{\Delta x^2} + \frac{u_{i,j+1} - 2u_{i,j} + u_{i,j-1}}{\Delta y^2} = 0 \tag{4.1.2}$$

$$\frac{-u_{i-2,j} + 16u_{i-1,j} - 30u_{i,j} + 16u_{i+1,j} - u_{i+2,j}}{12\Delta x^2} + \frac{-u_{i,j-2} + 16u_{i,j-1} - 30u_{i,j} + 16u_{i,j+1} - u_{i,j+2}}{12\Delta y^2} = 0 \tag{4.1.3}$$

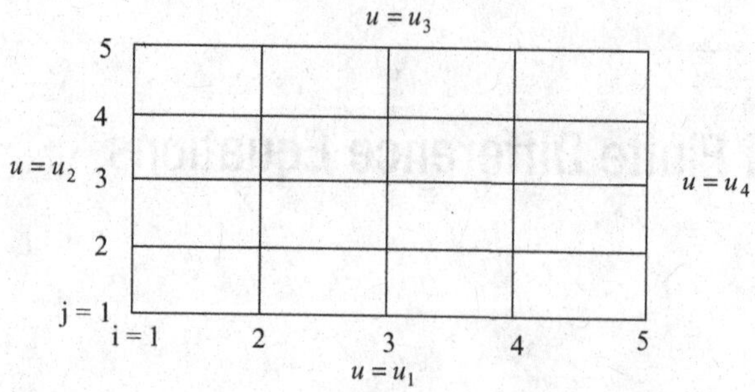

Figure 4.1.1 Finite difference grids with Dirichlet boundary conditions specified at all boundary nodes.

as discussed in Chapter 3. For illustration, let us consider the five-point scheme (4.1.2) for the geometry given in Figure 4.1.1.

$$u_{i+1,j} + u_{i-1,j} + \beta^2 u_{i,j+1} + \beta^2 u_{i,j-1} - 2(1+\beta^2)u_{i,j} = 0 \tag{4.1.4}$$

where β is defined as $\beta = \Delta x/\Delta y$. For Dirichlet boundary conditions, the values of u at all boundary nodes are given. Thus, writing (4.1.4) at all interior nodes and setting

$$\gamma = -2(1+\beta^2)$$

we obtain for the discretization as shown in Figure 4.1.1,

$$\begin{bmatrix} \gamma & 1 & 0 & \beta^2 & 0 & 0 & 0 & 0 & 0 \\ 1 & \gamma & 1 & 0 & \beta^2 & 0 & 0 & 0 & 0 \\ 0 & 1 & \gamma & 0 & 0 & \beta^2 & 0 & 0 & 0 \\ \beta^2 & 0 & 0 & \gamma & 1 & 0 & \beta^2 & 0 & 0 \\ 0 & \beta^2 & 0 & 1 & \gamma & 1 & 0 & \beta^2 & 0 \\ 0 & 0 & \beta^2 & 0 & 1 & \gamma & 0 & 0 & \beta^2 \\ 0 & 0 & 0 & \beta^2 & 0 & 0 & \gamma & 1 & 0 \\ 0 & 0 & 0 & 0 & \beta^2 & 0 & 1 & \gamma & 1 \\ 0 & 0 & 0 & 0 & 0 & \beta^2 & 0 & 1 & \gamma \end{bmatrix} \begin{bmatrix} u_{2,2} \\ u_{3,2} \\ u_{4,2} \\ u_{2,3} \\ u_{3,3} \\ u_{4,3} \\ u_{2,4} \\ u_{3,4} \\ u_{4,4} \end{bmatrix} = \begin{bmatrix} -u_{1,2} - \beta^2 u_{2,1} \\ -\beta^2 u_{3,1} \\ -u_{5,2} - \beta^2 u_{4,1} \\ -u_{1,3} \\ 0 \\ -u_{5,3} \\ -u_{1,4} - \beta^2 u_{2,5} \\ -\beta^2 u_{3,5} \\ -u_{5,4} - \beta^2 u_{4,5} \end{bmatrix} \tag{4.1.5}$$

Notice that the matrix on the left-hand side is always pentadiagonalized for the five-point scheme. The nine-point schemes given by (4.1.3), although more complicated, can be written similarly as in (4.1.5).

There are two types of solution methods for the linear algebraic equations of the form (4.1.5). The first kind includes the direct methods such as Gauss elimination, Thomas algorithm, Chelosky method, etc. The second kind includes the iterative methods such as Jacobi iteration, point Gauss-Seidel iteration, line Gauss-Seidel iteration, point-successive over-relaxation (PSOR), line successive over-relaxation (LSOR), alternating direction implicit (ADI), and so on.

The disadvantage of the direct methods is that they are more time consuming than iterative methods. Additionally, direct methods are susceptible to round-off errors which, in large systems of equations, can be catastrophic. In contrast, errors in each step of an iterative method are corrected in the subsequent step, thus round-off errors are usually not a concern. We elaborate on some of the iterative methods in Section 4.1.2, and a direct method of Gaussian elimination in Section 4.1.3. Other methods will be presented in later chapters, including conjugate gradient methods (CGM) (Section 10.3.1) and generalized minimal residual (GMRES) algorithm (Section 11.5.3).

4.1.2 ITERATIVE SOLUTION METHODS

Jacobi Iteration Method

In this method, the unknown u at each grid point is solved in terms of the initial guess values or previously computed values. Thus, from (4.1.4), we compute a new value of $u_{i,j}$ at the new iteration $k+1$ level as

$$u_{i,j}^{k+1} = \frac{1}{2(1+\beta^2)}[u_{i+1,j}^k + u_{i-1,j}^k + \beta^2(u_{i,j+1}^k + u_{i,j-1}^k)] \tag{4.1.6}$$

where k represents the previously computed values or the initial guesses for the first round of computations. The computation is carried out until a specified convergence criterion is achieved.

We may use the newly computed values of the dependent variables to compute the neighboring points when available. This process leads to efficient schemes such as the Gauss-Seidel method.

Point Gauss-Seidel Iteration Method

In this method, the current values of the dependent variables are used to compute neighboring points as soon as they are available. This will increase the convergence rate. The solution for the independent variables is obtained as

$$u_{i,j}^{k+1} = \frac{1}{2(1+\beta^2)}[u_{i+1,j}^k + u_{i-1,j}^{k+1} + \beta^2(u_{i,j+1}^k + u_{i,j-1}^{k+1})] \tag{4.1.7}$$

The $k+1$ level on the right-hand side of (4.1.7) indicates that the solution process takes advantage of the values at $i-1$ and $j-1$ which have just been calculated in the previous step.

Line Gauss-Seidel Iteration Method

Equation (4.1.5) may be solved for the three unknowns at $(i-1, j)$, (i, j), $(i+1, j)$, as follows:

$$u_{i-1,j}^{k+1} - 2(1+\beta^2)u_{i,j}^{k+1} + u_{i+1,j}^{k+1} = -\beta^2(u_{i,j+1}^k + u_{i,j-1}^{k+1}) \tag{4.1.8}$$

which leads to a tridiagonal matrix. Note that $u_{i,j-1}^{k+1}$ is known at the $k+1$ level, whereas $u_{i,j+1}^k$ was determined at the kth level. This method converges faster than the point Gauss-Seidel method, but it takes more computer time per iteration. The line iteration technique is useful when the variable changes more rapidly in the direction of the iteration because of the use of the updated values.

Point Successive Over-Relaxation Method (PSOR)

Convergence of the point Gauss-Seidel method can be accelerated by rearranging (4.1.7),

$$u_{i,j}^{k+1} = u_{i,j}^{k} + \frac{1}{2(1+\beta^2)}[u_{i+1,j}^{k} + u_{i-1,j}^{k+1} + \beta^2(u_{i,j+1}^{k} + u_{i,j-1}^{k+1}) - 2(1+\beta^2)u_{i,j}^{k}] \tag{4.1.9}$$

The idea is to make $u_{i,j}^{k}$ approach $u_{i,j}^{k+1}$ faster. To this end, we introduce the relaxation parameter, ω, to be multiplied to the terms with brackets on the right-hand side of (4.1.9),

$$u_{i,j}^{k+1} = u_{i,j}^{k} + \frac{\omega}{2(1+\beta^2)}[u_{i+1,j}^{k} + u_{i-1,j}^{k+1} + \beta^2(u_{i,j+1}^{k} + u_{i,j-1}^{k+1}) - 2(1+\beta^2)u_{i,j}^{k}]$$

or

$$u_{i,j}^{k+1} = (1-\omega)u_{i,j}^{k} + \frac{\omega}{2(1+\beta^2)}[u_{i+1,j}^{k} + u_{i-1,j}^{k+1} + \beta^2(u_{i,j+1}^{k} + u_{i,j-1}^{k+1})] \tag{4.1.10}$$

where we choose $1 < \omega < 2$ for convergence. This is known as the point successive over-relaxation procedure. For certain problems, however, a better convergence may be achieved by under-relaxation, where the relaxation parameter is chosen as $0 < \omega < 1$. Note that for $\omega = 1$ we recover the Gauss-Seidel iteration method.

For a rectangular domain subjected to Dirichlet boundary conditions with constant step size, we obtain the optimum relaxation parameter

$$\omega_{opt} = \frac{2 - \sqrt{1-a}}{a} \tag{4.1.11}$$

with

$$a = \left[\frac{\cos\left(\dfrac{\pi}{IM-1}\right) + \beta^2\cos\left(\dfrac{\pi}{JM-1}\right)}{1+\beta^2}\right]^2 \tag{4.1.12}$$

where IM and JM refer to the maximum numbers of i and j, respectively. Further details are found in Wachspress [1966] and Hageman and Young [1981].

Line Successive Over-Relaxation Method (LSOR)

The idea of relaxation may also be applied to the line Gauss-Seidel method,

$$\omega u_{i-1,j}^{k+1} - 2(1+\beta^2)u_{i,j}^{k+1} + \omega u_{i+1,j}^{k+1} = -(1-\omega)\lfloor 2(1+\beta^2)\rfloor u_{i,j}^{k} - \omega\beta^2(u_{i,j+1}^{k} + u_{i,j-1}^{k+1}) \tag{4.1.13}$$

where an optimum relaxation parameter ω can be determined experimentally, or by (4.1.11).

Alternating Direction Implicit (ADI) Method

In this method, a tridiagonal system is solved for rows first and then followed by columns, or vice versa. Toward this end, we recast (4.1.8) into two parts:

$$u_{i-1,j}^{k+\frac{1}{2}} - 2(1+\beta^2)u_{i,j}^{k+\frac{1}{2}} + u_{i+1,j}^{k+\frac{1}{2}} = -\beta^2\left(u_{i,j+1}^{k} + u_{i,j-1}^{k+\frac{1}{2}}\right) \tag{4.1.14a}$$

and

$$\beta^2 u_{i,j-1}^{k+1} - 2(1+\beta^2)u_{i,j}^{k+1} + \beta^2 u_{i,j+1}^{k+1} = -\left(u_{i+1,j}^{k+\frac{1}{2}} + u_{i-1,j}^{k+1}\right) \tag{4.1.14b}$$

Here (4.1.14a) and (4.1.14b) are solved implicitly in the x-direction and y-direction, respectively. The relaxation parameter ω may be introduced to accelerate the convergence,

$$\omega u_{i-1,j}^{k+\frac{1}{2}} - 2(1+\beta^2)u_{i,j}^{k+\frac{1}{2}} + \omega u_{i+1,j}^{k+\frac{1}{2}} = -(1-\omega)[2(1+\beta^2)]u_{i,j}^{k} - \omega\beta^2\left(u_{i,j+1}^{k} + u_{i,j-1}^{k+\frac{1}{2}}\right) \tag{4.1.15a}$$

and

$$\omega\beta^2 u_{i,j-1}^{k+1} - 2(1+\beta^2)u_{i,j}^{k+1} + \omega\beta^2 u_{i,j+1}^{k+1} = -(1-\omega)[2(1+\beta^2)]u_{i,j}^{k+\frac{1}{2}} - \omega\left(u_{i+1,j}^{k+\frac{1}{2}} + u_{i-1,j}^{k+1}\right) \tag{4.1.15b}$$

with the optimum ω being determined experimentally as appropriate for different physical problems.

4.1.3 DIRECT METHOD WITH GAUSSIAN ELIMINATION

Consider the simultaneous equations resulting from the finite difference approximation of (4.1.2) in the form

$$\begin{aligned} k_{11}u_1 + k_{12}u_2 + \cdots &= g_1 \\ k_{21}u_1 + k_{22}u_2 + \cdots &= g_2 \\ \vdots \\ k_{n1}u_n \cdot \quad \cdots &= g_n \end{aligned} \tag{4.1.16}$$

Here, our objective is to transform the system into an upper triangular array. To this end, we choose the first row as the "pivot" equation and eliminate the u_1 term from each equation below it. To eliminate u_1 from the second equation, we multiply the first equation by k_{21}/k_{11} and subtract it from the second equation. We continue similarly until u_1 is eliminated from all equations. We then eliminate $u_2, u_3, \ldots$ in the same manner until we achieve the upper triangular form,

$$\begin{bmatrix} k_{11} & k_{12} & \cdot & \cdot \\ & k'_{22} & \cdot & \cdot \\ & & \cdot & \cdot \\ & & & k'_{nn} \end{bmatrix} \begin{bmatrix} u_1 \\ u_2 \\ \cdot \\ u_n \end{bmatrix} = \begin{bmatrix} g_1 \\ g'_2 \\ \cdot \\ g'_n \end{bmatrix} \tag{4.1.17}$$

It is seen that backsubstitution will determine all unknowns.

An example for the solution of a typical elliptical equation is shown in Section 4.7.1.

4.2 PARABOLIC EQUATIONS

The governing equations for some problems in fluid dynamics, such as unsteady heat conduction or boundary layer flows, are parabolic. The finite difference representation

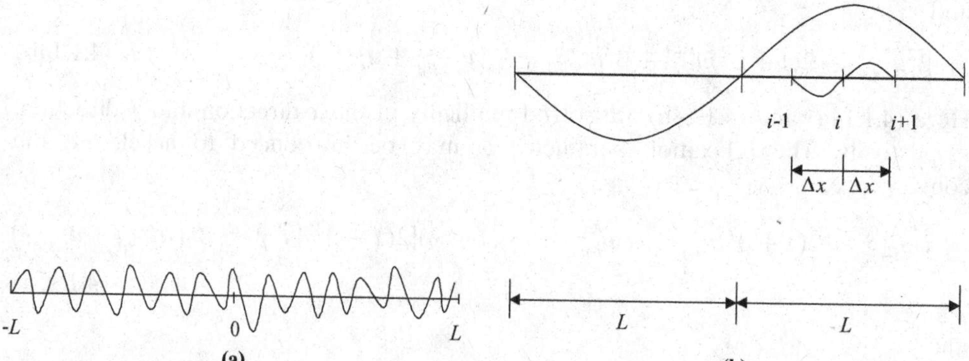

Figure 4.2.1 Fourier representation of the error on interval $(-L, L)$. **(a)** Error distribution. **(b)** Maximum and minimum wavelength.

of these equations may be represented in either explicit or implicit schemes, as illustrated below.

4.2.1 EXPLICIT SCHEMES AND VON NEUMANN STABILITY ANALYSIS

Forward-Time/Central-Space (FTCS) Method

A typical parabolic equation is the unsteady diffusion problem characterized by

$$\frac{\partial u}{\partial t} - \alpha \frac{\partial^2 u}{\partial x^2} = 0 \tag{4.2.1}$$

An explicit finite difference equation scheme for (4.2.1) may be written in the forward difference in time and central difference in space (FTCS) as (see Figure 4.2.1a)

$$\frac{u_i^{n+1} - u_i^n}{\Delta t} = \frac{\alpha\left(u_{i+1}^n - 2u_i^n + u_{i-1}^n\right)}{\Delta x^2} + O(\Delta t, \Delta x^2) \tag{4.2.2a}$$

or

$$u_i^{n+1} = u_i^n + d(u_{i+1}^n - 2u_i^n + u_{i-1}^n) \tag{4.2.2b}$$

where d is the diffusion number

$$d = \frac{\alpha \Delta t}{\Delta x^2} \tag{4.2.3}$$

By definition, (4.2.2) is explicit because u_i^{n+1} at time step $n+1$ can be solved *explicitly* in terms of the known quantities at the previous time step n, thus called an *explicit scheme.*

In order to determine the stability of the solution of finite difference equations, it is convenient to expand the difference equation in a Fourier series. Decay or growth of an amplification factor indicates whether or not the numerical algorithm is stable. This is known as the von Neumann stability analysis [Ortega and Rheinbolt, 1970]. Assuming that at any time step n, the computed solution u_i^n is the sum of the exact solution $\bar{u}_i^n$ and error ε_i^n

$$u_i^n = \bar{u}_i^n + \varepsilon_i^n \tag{4.2.4}$$

and substituting (4.2.4) into (4.2.2a), we obtain

$$\frac{\bar{u}_i^{n+1} - \bar{u}_i^n}{\Delta t} + \frac{\varepsilon_i^{n+1} - \varepsilon_i^n}{\Delta t} = \frac{\alpha}{(\Delta x)^2}\left(\bar{u}_{i+1}^n - 2\bar{u}_i^n + \bar{u}_{i-1}^n\right) + \frac{\alpha}{(\Delta x)^2}\left(\varepsilon_{i+1}^n - 2\varepsilon_i^n + \varepsilon_{i-1}^n\right) \tag{4.2.5}$$

or

$$\frac{\varepsilon_i^{n+1} - \varepsilon_i^n}{\Delta t} = \frac{\alpha}{(\Delta x)^2}\left(\varepsilon_{i+1}^n - 2\varepsilon_i^n + \varepsilon_{i-1}^n\right) \tag{4.2.6}$$

Writing (4.2.4) – (4.2.6) for the entire domain leads to

$$\mathbf{U}^n = \bar{\mathbf{U}}^n + \boldsymbol{\varepsilon}^n \tag{4.2.7}$$

with

$$\boldsymbol{\varepsilon}^n = \begin{bmatrix} \cdot \\ \varepsilon_{i-1}^n \\ \varepsilon_i^n \\ \varepsilon_{i+1}^n \\ \cdot \end{bmatrix} \tag{4.2.8}$$

$$\bar{\mathbf{U}}^{n+1} + \boldsymbol{\varepsilon}^{n+1} = \mathbf{C}(\bar{\mathbf{U}}^n + \boldsymbol{\varepsilon}^n) \tag{4.2.9}$$

$$\boldsymbol{\varepsilon}^{n+1} = \mathbf{C}\boldsymbol{\varepsilon}^n \tag{4.2.10}$$

with

$$C = 1 + d(E - 2 + E^{-1}) = \begin{bmatrix} \cdot & \cdot & \cdot & \cdot & \cdot \\ d & (1-2d) & d & 0 & 0 \\ \cdot & d & (1-2d) & d & 0 \\ \cdot & 0 & d & (1-2d) & d \\ \cdot & \cdot & \cdot & \cdot & \cdot \end{bmatrix} \tag{4.2.11}$$

If the boundary conditions are considered as periodic, the error ε^n can be decomposed into a Fourier series in space at each time level n. The fundamental frequency in a one-dimensional domain between $-L$ and L (Figure 4.2.1) corresponds to the maximum wave length of $\lambda_{max} = 2L$. The wave number $k = 2\pi/\lambda$ becomes minimum as $k_{min} = \pi/L$, whereas the maximum wave number k_{max} is associated with the shortest wavelength λ on a mesh with spacing Δx corresponding to $\lambda_{min} = 2\Delta x$, leading to $k_{max} = \pi/\Delta x$. Thus, the harmonics on a finite mesh are

$$k_j = jk_{min} = j\pi/L = j\pi/(N\Delta x), \quad j = 0, 1, \ldots N \tag{4.2.12}$$

with $\Delta x = L/N$. The highest value of j is equal to the number of mesh intervals N. Any finite mesh function, such as ε_i^n or the full solution u_i^n, can be decomposed into a Fourier series

$$\varepsilon_i^n = \sum_{j=-N}^{N} \bar{\varepsilon}_j^n e^{Ik_j(i\Delta x)} = \sum_{j=-N}^{N} \bar{\varepsilon}_j^n e^{Iji\pi/N} \tag{4.2.13}$$

with $I = \sqrt{-1}$, $\bar{\varepsilon}_j^n$ being the amplitude of the jth harmonic, and the spatial phase angle ϕ is given as

$$\phi = k_j \Delta x = j\pi/N \tag{4.2.14}$$

with $\phi = \pi$ corresponding to the highest frequency resolvable on the mesh, namely the frequency of the wavelength $2\Delta x$. Thus

$$\varepsilon_i^n = \sum_{j=-N}^{N} \bar{\varepsilon}_j^n e^{Ii\phi} \tag{4.2.15}$$

Substituting (4.2.15) into (4.2.6) yields

$$\frac{\bar{\varepsilon}^{n+1} - \bar{\varepsilon}^n}{\Delta t} e^{Ii\phi} = \frac{\alpha}{\Delta x^2}\left(\bar{\varepsilon}^n e^{I(i+1)\phi} - 2\bar{\varepsilon}^n e^{Ii\phi} + \bar{\varepsilon}^n e^{I(i-1)\phi}\right)$$

or

$$\bar{\varepsilon}^{n+1} - \bar{\varepsilon}^n - d\bar{\varepsilon}^n(e^{I\phi} - 2 + e^{-I\phi}) = 0 \tag{4.2.16}$$

The computational scheme is said to be stable if the amplitude of any error harmonic $\bar{\varepsilon}^n$ does not grow in time, that is, if the following ratio holds:

$$|g| = \left|\frac{\bar{\varepsilon}^{n+1}}{\bar{\varepsilon}^n}\right| \leq 1 \quad \text{for all } \phi \tag{4.2.17}$$

where $g = \bar{\varepsilon}^{n+1}/\bar{\varepsilon}^n$ is the amplification factor, and is a function of time step Δt, frequency, and the mesh size Δx. It follows from (4.2.16) that

$$g = 1 + d(e^{I\phi} - 2 + e^{-I\phi}) \tag{4.2.18a}$$

or

$$g = 1 - 2d(1 - \cos\phi) \tag{4.2.18b}$$

Thus, the stability condition is

$$g \leq 1 \tag{4.2.19}$$

or

$$1 - 2d(1 - \cos\phi) \geq -1 \tag{4.2.20}$$

Since the maximum of $1 - \cos\phi$ is 2, we arrive at, for stability,

$$0 \leq d \leq 1/2 \tag{4.2.21}$$

The von Neumann stability analysis shown above can be used to determine the computational stability properties of other finite difference schemes to be discussed subsequently.

OTHER EXPLICIT SCHEMES

Richardson Method

If the diffusion equation (4.2.1) is modeled by the form

$$\frac{u_i^{n+1} - u_i^{n-1}}{2\Delta t} = \frac{\alpha\left(u_{i+1}^n - 2u_i^n + u_{i-1}^n\right)}{\Delta x^2}, \quad O(\Delta t^2, \Delta x^2) \tag{4.2.22}$$

This is known as the Richardson method and is unconditionally unstable.

Dufort-Frankel Method

The finite difference equation for this method is given by

$$\frac{u_i^{n+1} - u_i^{n-1}}{2\Delta t} = \frac{\alpha\left(u_{i+1}^n - 2\dfrac{u_i^{n+1} + u_i^{n-1}}{2} + u_{i-1}^n\right)}{\Delta x^2} \tag{4.2.23a}$$

or

$$(1 + 2d)u_i^{n+1} = (1 - 2d)u_i^{n-1} + 2d\left(u_{i+1}^n + u_{i-1}^n\right), \quad O(\Delta t^2, \Delta x^2, (\Delta t/\Delta x)^2) \tag{4.2.23b}$$

This scheme can be shown to be unconditionally stable by the von Neumann stability analysis.

4.2.2 IMPLICIT SCHEMES

Laasonen Method

Contrary to the explicit schemes, the solution for *implicit schemes* involves the variables at more than one nodal point for the time step $(n + 1)$. For example, we may write the difference equation for (4.2.1a) in the form

$$\frac{u_i^{n+1} - u_i^n}{\Delta t} = \frac{\alpha\left(u_{i+1}^{n+1} - 2u_i^{n+1} + u_{i-1}^{n+1}\right)}{\Delta x^2}, \quad O(\Delta t, \Delta x^2) \tag{4.2.24}$$

This equation is written for all grid points at $n + 1$ time step, leading to a tridiagonal form. The scheme given by (4.2.24) is known as the Laasonen method. This is unconditionally stable.

Crank-Nicolson Method

An alternative scheme of (4.2.24) is to replace the diffusion term by an average between n and $n + 1$,

$$\frac{u_i^{n+1} - u_i^n}{\Delta t} = \frac{\alpha}{2}\left[\frac{u_{i+1}^{n+1} - 2u_i^{n+1} + u_{i-1}^{n+1}}{\Delta x^2} + \frac{u_{i+1}^n - 2u_i^n + u_{i-1}^n}{\Delta x^2}\right], \quad O(\Delta t^2, \Delta x^2) \tag{4.2.25}$$

This may be rewritten as

$$A + B = C + D \tag{4.2.26}$$

where

$$A = \frac{u_i^{n+\frac{1}{2}} - u_i^n}{\Delta t/2}, \qquad B = \frac{u_i^{n+1} - u_i^{n+\frac{1}{2}}}{\Delta t/2}, \qquad C = \frac{\alpha\left(u_{i-1}^n - 2u_i^n + u_{i+1}^n\right)}{(\Delta x)^2},$$

$$D = \frac{\alpha\left(u_{i-1}^{n+1} - 2u_i^{n+1} + u_{i+1}^{n+1}\right)}{(\Delta x)^2}$$

Note that $A = C$ and $B = D$ represent explicit and implicit scheme, respectively. This scheme is known as the Crank-Nicolson method. It is seen that $A = C$ is solved explicitly for the time step $n + 1/2$ and the result is substituted into $B = D$. The scheme is unconditionally stable.

β-Method

A general form of the finite difference equation for (4.2.1) may be written as

$$\frac{u_i^{n+1} - u_i^n}{\Delta t} = \alpha\left[\frac{\beta\left(u_{i+1}^{n+1} - 2u_i^{n+1} + u_{i-1}^{n+1}\right)}{(\Delta x)^2} + \frac{(1-\beta)\left(u_{i+1}^n - 2u_i^n + u_{i-1}^n\right)}{(\Delta x)^2}\right] \tag{4.2.27}$$

This is known as the β-method. For $1/2 \leq \beta \leq 1$, the method is unconditionally stable. For $\beta = 1/2$, equation (4.2.27) reduces to the Crank-Nicolson scheme, whereas $\beta = 0$ leads to the FTCS method.

A numerical example for the solution of a typical parabolic equation characterized by Couette flow is presented in Section 4.7.2.

4.2.3 ALTERNATING DIRECTION IMPLICIT (ADI) SCHEMES

Let us now examine the solution of the two-dimensional diffusion equation,

$$\frac{\partial u}{\partial t} - \alpha\left(\frac{\partial^2 u}{\partial x^2} + \frac{\partial^2 u}{\partial y^2}\right) = 0 \tag{4.2.28}$$

with the forward difference in time and the central difference in space (FTCS). We write an explicit scheme in the form

$$\frac{u_{i,j}^{n+1} - u_{i,j}^n}{\Delta t} = \alpha\left(\frac{u_{i+1,j}^n - 2u_{i,j}^n + u_{i-1,j}^n}{\Delta x^2} + \frac{u_{i,j+1}^n - 2u_{i,j}^n + u_{i,j-1}^n}{\Delta y^2}\right), \quad O(\Delta t, \Delta x^2, \Delta y^2) \tag{4.2.29}$$

It can be shown that the system is stable if

$$d_x + d_y \leq \frac{1}{2} \tag{4.2.30}$$

Here, diffusion numbers d_x and d_y are defined as

$$d_x = \frac{\alpha \Delta t}{\Delta x^2}, \qquad d_y = \frac{\alpha \Delta t}{\Delta y^2} \tag{4.2.31}$$

For simplicity, let $d_x = d_y = d$ for $\Delta x = \Delta y$. This will give $d \leq 1/4$ for stability, which is twice as restrictive. To avoid this restriction, consider an implicit scheme

$$\frac{u_{i,j}^{n+1} - u_{i,j}^{n}}{\Delta t} = \alpha\left(\frac{u_{i+1,j}^{n+1} - 2u_{i,j}^{n+1} + u_{i-1,j}^{n+1}}{\Delta x^2} + \frac{u_{i,j+1}^{n+1} - 2u_{i,j}^{n+1} + u_{i,j-1}^{n+1}}{\Delta y^2}\right) \tag{4.2.32}$$

or

$$d_x u_{i+1,j}^{n+1} + d_x u_{i-1,j}^{n+1} - (2d_x + 2d_y + 1)u_{i,j}^{n+1} + d_y u_{i,j+1}^{n+1} + d_y u_{i,j-1}^{n+1} = -u_{i,j}^{n} \tag{4.2.33}$$

This leads to a pentadiagonal system.

An alternative is to use the alternating direction implicit scheme, by splitting (4.2.25) into two equations:

$$\frac{u_{i,j}^{n+\frac{1}{2}} - u_{i,j}^{n}}{\Delta t/2} = \alpha\left(\frac{u_{i+1,j}^{n+\frac{1}{2}} - 2u_{i,j}^{n+\frac{1}{2}} + u_{i-1,j}^{n+\frac{1}{2}}}{\Delta x^2} + \frac{u_{i,j+1}^{n} - 2u_{i,j}^{n} + u_{i,j-1}^{n}}{\Delta y^2}\right) \tag{4.2.34a}$$

and

$$\frac{u_{i,j}^{n+1} - u_{i,j}^{n+\frac{1}{2}}}{\Delta t/2} = \alpha\left(\frac{u_{i+1,j}^{n+\frac{1}{2}} - 2u_{i,j}^{n+\frac{1}{2}} + u_{i-1,j}^{n+\frac{1}{2}}}{\Delta x^2} + \frac{u_{i,j+1}^{n+1} - 2u_{i,j}^{n+1} + u_{i,j-1}^{n+1}}{\Delta y^2}\right) \tag{4.2.34b}$$

This scheme is unconditionally stable. These two equations can be written in a tridiagonal form as follows:

$$\underbrace{-d_1 u_{i+1,j}^{n+\frac{1}{2}} + (1+2d_1)u_{i,j}^{n+\frac{1}{2}} - d_1 u_{i-1,j}^{n+\frac{1}{2}}}_{\text{implicit in } x\text{-direction}} = \underbrace{d_2 u_{i,j+1}^{n} + (1-2d_2)u_{i,j}^{n} + d_2 u_{i,j-1}^{n}}_{\text{explicit in } y\text{-direction}} \tag{4.2.35a}$$

$$\underbrace{-d_2 u_{i,j+1}^{n+1} + (1+2d_2)u_{i,j}^{n+1} - d_2 u_{i,j-1}^{n+1}}_{\text{unknown}} = \underbrace{d_1 u_{i+1,j}^{n+\frac{1}{2}} + (1-2d_1)u_{i,j}^{n+\frac{1}{2}} + d_1 u_{i-1,j}^{n+\frac{1}{2}}}_{\text{known}} \tag{4.2.35b}$$

where

$$d_1 = \frac{1}{2}d_x = \frac{1}{2}\frac{\alpha\Delta t}{\Delta x^2}$$

$$d_2 = \frac{1}{2}d_y = \frac{1}{2}\frac{\alpha\Delta t}{\Delta y^2}$$

Note that (4.2.35a) is implicit in the x-direction and explicit in the y-direction, known as the x-sweep. The solution of (4.2.35a) provides the data for (4.2.35b) so that the y-sweep can be carried out in which the solution is implicit in the y-direction and explicit in the x-direction.

4.2.4 APPROXIMATE FACTORIZATION

The ADI formulation can be shown to be an approximate factorization of the Crank-Nicolson scheme. To this end, let us write the Crank-Nicolson scheme for (4.2.25) in the form

$$\frac{u_{i,j}^{n+1}-u_{i,j}^{n}}{\Delta t}=\frac{\alpha}{2}\left[\begin{array}{l}\dfrac{u_{i+1,j}^{n+1}-2u_{i,j}^{n+1}+u_{i-1,j}^{n+1}}{\Delta x^2}+\dfrac{u_{i+1,j}^{n}-2u_{i,j}^{n}+u_{i-1,j}^{n}}{\Delta x^2}\\ +\dfrac{u_{i,j+1}^{n+1}-2u_{i,j}^{n+1}+u_{i,j-1}^{n+1}}{\Delta y^2}+\dfrac{u_{i,j+1}^{n}-2u_{i,j}^{n}+u_{i,j-1}^{n}}{\Delta y^2}\end{array}\right],$$

$$O(\Delta t^2, \Delta x^2, \Delta y^2) \quad (4.2.36)$$

Introducing a compact notation,

$$\delta_x^2 u_{i,j} = u_{i+1,j} - 2u_{i,j} + u_{i-1,j}$$

$$\delta_y^2 u_{i,j} = u_{i,j+1} - 2u_{i,j} + u_{i,j-1}$$

we may rewrite (4.2.36) as

$$\left[1-\frac{1}{2}(d_x\delta_x^2+d_y\delta_y^2)\right]u_{i,j}^{n+1}=\left[1+\frac{1}{2}(d_x\delta_x^2+d_y\delta_y^2)\right]u_{i,j}^{n} \quad (4.2.37)$$

To compare (4.2.37) with the ADI formulation, we use (4.2.36) to rewrite the ADI equations as

$$\frac{u_{i,j}^{n+\frac{1}{2}}-u_{i,j}^{n}}{\frac{\Delta t}{2}}=\alpha\left(\frac{\delta_x^2 u_{i,j}^{n+\frac{1}{2}}}{\Delta x^2}+\frac{\delta_y^2 u_{i,j}^{n}}{\Delta y^2}\right) \quad (4.2.38a)$$

$$\frac{u_{i,j}^{n+1}-u_{i,j}^{n+\frac{1}{2}}}{\frac{\Delta t}{2}}=\alpha\left(\frac{\delta_x^2 u_{i,j}^{n+\frac{1}{2}}}{\Delta x^2}+\frac{\delta_y^2 u_{i,j}^{n+1}}{\Delta y^2}\right) \quad (4.2.38b)$$

Rearranging (4.2.38a,b)

$$\left(1-\frac{1}{2}d_x\delta_x^2\right)u_{i,j}^{n+\frac{1}{2}}=\left(1+\frac{1}{2}d_y\delta_y^2\right)u_{i,j}^{n} \quad (4.2.39a)$$

$$\left(1-\frac{1}{2}d_y\delta_y^2\right)u_{i,j}^{n+1}=\left(1+\frac{1}{2}d_x\delta_x^2\right)u_{i,j}^{n+\frac{1}{2}} \quad (4.2.39b)$$

and eliminating $u_{i,j}^{n+\frac{1}{2}}$ between (4.2.39a) and (4.2.39b),

$$\left(1-\frac{1}{2}d_x\delta_x^2\right)\left(1-\frac{1}{2}d_y\delta_y^2\right)u_{i,j}^{n+1}=\left(1+\frac{1}{2}d_x\delta_x^2\right)\left(1+\frac{1}{2}d_y\delta_y^2\right)u_{i,j}^{n} \quad (4.2.40)$$

or

$$\left[1-\frac{1}{2}(d_x\delta_x^2+d_y\delta_y^2)+\frac{1}{4}d_xd_y\delta_x^2\delta_y^2\right]u_{i,j}^{n+1}=\left[1+\frac{1}{2}(d_x\delta_x^2+d_y\delta_y^2)+\frac{1}{4}d_xd_y\delta_x^2\delta_y^2\right]u_{i,j}^{n} \quad (4.2.41)$$

We note that, compared to (4.2.37), the additional term in (4.2.41)

$$\frac{1}{4}d_xd_y\delta_x^2\delta_y^2(u_{i,j}^{n+1}-u_{i,j}^{n})$$

is smaller than the truncation error of (4.2.37). Thus, it is seen that the ADI formulation is an approximate factorization of the Crank-Nicolson scheme.

4.2.5 FRACTIONAL STEP METHODS

An approximation of multidimensional problems similar to ADI or approximate factorization schemes is also known as the method of fractional steps. This method splits the multidimensional equations into a series of one-dimensional equations and solves them sequentially. For example, consider a two-dimensional equation

$$\frac{\partial u}{\partial t} = \alpha\left(\frac{\partial^2 u}{\partial x^2} + \frac{\partial^2 u}{\partial y^2}\right) \tag{4.2.42}$$

The Crank-Nicolson scheme for (4.2.36) can be written in two steps:

$$\frac{u_{i,j}^{n+\frac{1}{2}} - u_{i,j}^n}{\frac{\Delta t}{2}} = \frac{\alpha}{2}\left[\frac{u_{i+1,j}^{n+\frac{1}{2}} - 2u_{i,j}^{n+\frac{1}{2}} + u_{i-1,j}^{n+\frac{1}{2}}}{\Delta x^2} + \frac{u_{i+1,j}^n - 2u_{i,j}^n + u_{i-1,j}^n}{\Delta x^2}\right] \tag{4.2.43a}$$

$$\frac{u_{i,j}^{n+1} - u_{i,j}^{n+\frac{1}{2}}}{\frac{\Delta t}{2}} = \frac{\alpha}{2}\left[\frac{u_{i,j+1}^{n+1} - 2u_{i,j}^{n+1} + u_{i,j-1}^{n+1}}{\Delta y^2} + \frac{u_{i,j+1}^{n+\frac{1}{2}} - 2u_{i,j}^{n+\frac{1}{2}} + u_{i,j-1}^{n+\frac{1}{2}}}{\Delta y^2}\right]$$
$$+ O(\Delta t^2, \Delta x^2, \Delta y^2) \tag{4.2.43b}$$

This scheme is unconditionally stable.

4.2.6 THREE DIMENSIONS

The ADI method can be extended to three-space dimensions for the time intervals $n, n+1/3, n+2/3$, and $n+1$. Consider the unsteady diffusion problem,

$$\frac{\partial u}{\partial t} = \alpha\left(\frac{\partial^2 u}{\partial x^2} + \frac{\partial^2 u}{\partial y^2} + \frac{\partial^2 u}{\partial z^2}\right) \tag{4.2.44}$$

The three-step FDM equations are written as

$$\frac{u_{i,j,k}^{n+\frac{1}{3}} - u_{i,j,k}^n}{\Delta t/3} = \alpha\left(\frac{\delta_x^2 u_{i,j,k}^{n+\frac{1}{3}}}{\Delta x^2} + \frac{\delta_y^2 u_{i,j,k}^n}{\Delta y^2} + \frac{\delta_z^2 u_{i,j,k}^n}{\Delta z^2}\right) \tag{4.2.45a}$$

$$\frac{u_{i,j,k}^{n+\frac{2}{3}} - u_{i,j,k}^{n+\frac{1}{3}}}{\Delta t/3} = \alpha\left(\frac{\delta_x^2 u_{i,j,k}^{n+\frac{1}{3}}}{\Delta x^2} + \frac{\delta_y^2 u_{i,j,k}^{n+\frac{2}{3}}}{\Delta y^2} + \frac{\delta_z^2 u_{i,j,k}^{n+\frac{1}{3}}}{\Delta z^2}\right) \tag{4.2.45b}$$

$$\frac{u_{i,j,k}^{n+1} - u_{i,j,k}^{n+\frac{2}{3}}}{\Delta t/3} = \alpha\left(\frac{\delta_x^2 u_{i,j,k}^{n+\frac{2}{3}}}{\Delta x^2} + \frac{\delta_y^2 u_{i,j,k}^{n+\frac{2}{3}}}{\Delta y^2} + \frac{\delta_z^2 u_{i,j,k}^{n+1}}{\Delta z^2}\right), \quad O(\Delta t, \Delta x^2, \Delta y^2, \Delta z^2) \tag{4.2.45c}$$

This method is conditionally stable with $(d_x + d_y + d_z) \leq 3/2$. A more efficient method may be derived using the Crank-Nicolson scheme.

$$\frac{u^*_{i,j,k} - u^n_{i,j,k}}{\Delta t} = \alpha\left[\frac{1}{2}\frac{\delta_x^2 u^*_{i,j,k} + \delta_x^2 u^n_{i,j,k}}{\Delta x^2} + \frac{\delta_y^2 u^n_{i,j,k}}{\Delta y^2} + \frac{\delta_z^2 u^n_{i,j,k}}{\Delta z^2}\right]$$

$$\frac{u^{**}_{i,j,k} - u^n_{i,j,k}}{\Delta t} = \alpha\left[\frac{1}{2}\frac{\delta_x^2 u^*_{i,j,k} + \delta_x^2 u^n_{i,j,k}}{\Delta x^2} + \frac{1}{2}\frac{\delta_y^2 u^{**}_{i,j,k} + \delta_y^2 u^n_{i,j,k}}{\Delta y^2} + \frac{\delta_z^2 u^n_{i,j,k}}{\Delta z^2}\right]$$

$$\frac{u^{n+1}_{i,j,k} - u^n_{i,j,k}}{\Delta t} = \alpha\left[\frac{1}{2}\frac{\delta_x^2 u^*_{i,j,k} + \delta_x^2 u^n_{i,j,k}}{\Delta x^2} + \frac{1}{2}\frac{\delta_y^2 u^{**}_{i,j,k} + \delta_y^2 u^n_{i,j,k}}{\Delta y^2} + \frac{1}{2}\frac{\delta_z^2 u^{n+1}_{i,j,k} + \delta_z^2 u^n_{i,j,k}}{\Delta z^2}\right] \tag{4.2.46}$$

In this scheme, the final solution $u^{n+1}_{i,j,k}$ is obtained in terms of the intermediate steps $u^*_{i,j,k}$ and $u^{**}_{i,j,k}$.

4.2.7 DIRECT METHOD WITH TRIDIAGONAL MATRIX ALGORITHM

Consider the implicit FDM discretization for the transient heat conduction equation in the form,

$$\frac{T_i^{n+1} - T_i^n}{\Delta t} = \frac{\alpha}{\Delta x^2}\left(T_{i+1}^{n+1} - 2T_i^{n+1} + T_{i-1}^{n+1}\right) \tag{4.2.47}$$

This may be rewritten as

$$a_i T_{i-1}^{n+1} + b_i T_i^{n+1} + c_i T_{i+1}^{n+1} = g_i \tag{4.2.48}$$

with

$$a_i = c_i = -\frac{\alpha\Delta t}{\Delta x^2}, \qquad b_i = 1 + \frac{2\alpha\Delta t}{\Delta x^2}, \qquad g_i = T_i^n \tag{4.2.49}$$

If Dirichlet boundary conditions are applied to this problem, we obtain the following tridiagonal form, known as tridiagonal matrix algorithm (TDMA) or Thomas algorithm [Thomas, 1949]:

$$\begin{bmatrix} b_1 & c_1 & 0 & \cdot & \cdot & \cdot & \cdot \\ a_2 & b_2 & c_2 & 0 & \cdot & \cdot & \cdot \\ 0 & a_3 & b_3 & c_3 & 0 & \cdot & \cdot \\ \cdot & \cdot & * & * & * & \cdot & \cdot \\ \cdot & \cdot & \cdot & * & * & * & \cdot \\ \cdot & \cdot & \cdot & \cdot & * & * & c_{NI-1} \\ 0 & \cdot & \cdot & \cdot & \cdot & a_{NI} & b_{NI} \end{bmatrix} \begin{bmatrix} T_1^{n+1} \\ T_2^{n+1} \\ T_3^{n+1} \\ * \\ * \\ * \\ T_{NI}^{n+1} \end{bmatrix} = \begin{bmatrix} g_1 \\ g_2 \\ g_3 \\ * \\ * \\ * \\ g_{NI} \end{bmatrix} \tag{4.2.50}$$

An upper triangular form of the tridiagonal matrix may be obtained as follows:

$$b_i = b_i - \frac{a_i}{b_{i-1}} c_{i-1} \quad i = 2, 3, \ldots NI$$

$$g_i = g_i - \frac{a_i}{b_{i-1}} g_{i-1} \quad i = 2, 3, \ldots NI$$

$$T_{NI} = \frac{g_{NI}}{b_{NI}}$$

$$T_j = \frac{g_j - c_j T_{j+1}}{b_j} \qquad j = NI-1, \quad NI-2, \ldots, 1$$

It should be noted that Neumann boundary conditions can also be accommodated into this algorithm with the tridiagonal form still maintained.

4.3 HYPERBOLIC EQUATIONS

Hyperbolic equations, in general, represent wave propagation. They are given by either first order or second order differential equations, which may be approximated in either explicit or implicit forms of finite difference equations. Various computational schemes are examined below.

4.3.1 EXPLICIT SCHEMES AND VON NEUMANN STABILITY ANALYSIS

Euler's Forward Time and Forward Space (FTFS) Approximations

Consider the first order wave equation (Euler equation) of the form

$$\frac{\partial u}{\partial t} + a\frac{\partial u}{\partial x} = 0, \quad a > 0 \tag{4.3.1}$$

The Euler's forward time and forward space approximation of (4.3.1) is written in the FTFS scheme as

$$\frac{u_i^{n+1} - u_i^n}{\Delta t} = -a\frac{u_{i+1}^n - u_i^n}{\Delta x} \tag{4.3.2}$$

It follows from (4.2.15) and (4.3.2) that the amplification factor assumes the form

$$g = 1 - C(e^{I\phi} - 1) = 1 - C(\cos\phi - 1) - IC\sin\phi = 1 + 2C\sin^2\frac{\phi}{2} - IC\sin\phi \tag{4.3.3}$$

with C being the Courant number or CFL number [Courant, Friedrichs, and Lewy, 1967],

$$C = \frac{a\Delta t}{\Delta x}$$

and

$$|g|^2 = g\,g^* = \left(1 + 2C\sin^2\frac{\phi}{2}\right)^2 + C^2\sin^2\phi = 1 + 4C(1+C)\sin^2\frac{\phi}{2} \geq 1 \tag{4.3.4}$$

where g^* is the complex conjugate of g. Note that the criterion $|g| \leq 1$ for all values of ϕ can not be satisfied ($|g|$ lies outside the unit circle for all values of ϕ, Figure 4.3.1). Therefore, the explicit Euler scheme with FTFS is unconditionally unstable.

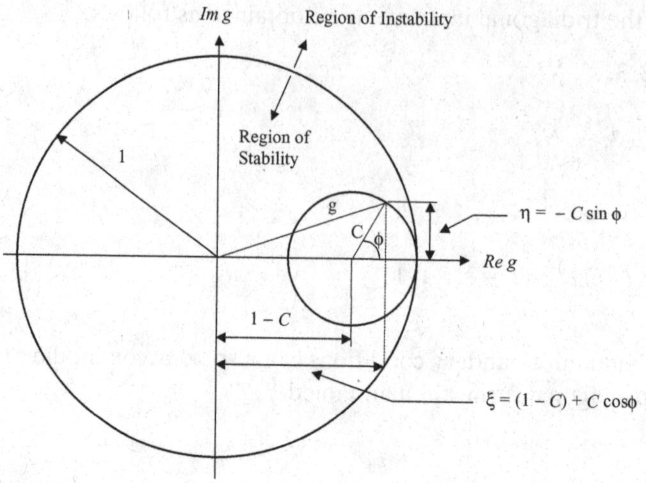

Figure 4.3.1 Complex g plane for upwind scheme with unit circle representing the stability region.

Euler's Forward Time and Central Space (FTCS) Approximations

In this method, Euler's forward time and central space approximation of (4.3.1) is used:

$$\frac{u_i^{n+1} - u_i^n}{\Delta t} = -a\frac{(u_{i+1}^n - u_{i-1}^n)}{2\Delta x}, \quad O(\Delta t, \Delta x) \tag{4.3.5}$$

The von Neumann analysis shows that this is also unconditionally unstable.

Euler's Forward Time and Backward Space (FTBS) Approximations – First Order Upwind Scheme

The Euler's forward time and backward space approximations (also known as upwind method) is given by

$$\frac{u_i^{n+1} - u_i^n}{\Delta t} = -a\frac{u_i^n - u_{i-1}^n}{\Delta x}, \quad O(\Delta t, \Delta x) \tag{4.3.6}$$

The amplification factor takes the form

$$\begin{aligned} g &= 1 - C(1 - e^{-I\phi}) = 1 - C(1 - \cos\phi) - IC\sin\phi \\ &= 1 - 2C\sin^2\frac{\phi}{2} - IC\sin\phi \end{aligned} \tag{4.3.7}$$

or

$$g = \xi + I\eta, \qquad |g| = \left[1 - 4C(1-C)\sin^2\frac{\phi}{2}\right]^{1/2} \tag{4.3.8a,b}$$

with

$$\xi = 1 - 2C\sin^2\frac{\phi}{2} = (1 - C) + C\cos\phi$$

$$\eta = -C\sin\phi$$

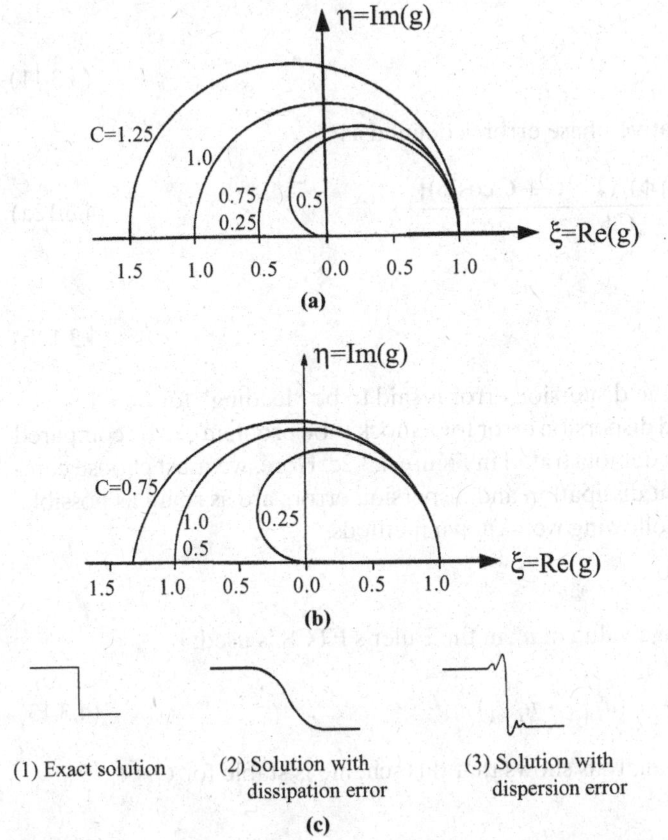

Figure 4.3.2 Dissipation and dispersion errors compared to exact solution. (a) Dissipation error (amplification factor modulus $|g|$). (b) Dispersion error (relative phase error, $\Phi/\tilde{\Phi}$). (c) Comparison of exact solution with dissipation error and dispersion error for shock tube problem.

which represents the parametric equation of a unit circle centered on the real axis ξ at $(1 - C)$ with radius C (Figure 4.3.1), whereas the modulus of the amplification factor, $|g|$, for various values of C are shown in Figure 4.3.2a.

In this complex plane of g, the stability condition (4.3.7) states that the curve representing g for all values of $\phi = k\Delta x$ should remain within the unit circle. It is seen that the scheme is stable for

$$0 < g < 1 \tag{4.3.9}$$

Hence, the scheme (4.3.6) is conditionally stable. Equation (4.3.9) is known as the Courant-Friedrich-Lewy (CFL) condition.

We have so far discussed the amplification factor g which represents dissipation error (Figure 4.3.2a). In numerical solutions of finite difference equations, we are also concerned with dispersion (phase) error as shown in Figure 4.3.2b. The phase Φ as determined by the adopted numerical scheme is given by the arctangent of the ratio of imaginary and real parts of g,

$$\Phi = \tan^{-1}\frac{\text{Im}(g)}{\text{Re}(g)} = \tan^{-1}\frac{\eta}{\xi} = \tan^{-1}\frac{-C\sin\phi}{1 - C + C\cos\phi} \tag{4.3.10}$$

The phase angle $\tilde{\Phi}$ is

$$\tilde{\Phi} = ka\Delta t = C\phi \tag{4.3.11}$$

The dispersion error or relative phase error is defined as

$$\varepsilon_\phi = \frac{\Phi}{\tilde{\Phi}} = \frac{\tan^{-1}[(-C\sin\phi)/(1-C+C\cos\phi)]}{C\phi} \tag{4.3.12a}$$

or

$$\varepsilon_\phi \approx 1 - \frac{1}{6}(2C^2 - 3C + 1)\,\phi^2 \tag{4.3.12b}$$

As shown in Figure 4.3.2b, the dispersion error is said to be "leading" for $\varepsilon_\phi > 1$.

The dissipation error and dispersion error for a shock tube problem can be compared to the exact solution. This is demonstrated in Figure 4.3.2c. Here, we must choose computational schemes such that dissipation and dispersion errors are as small as possible. To this end, we review the following well-known methods.

Lax Method

In this method, an average value of u_i^n in the Euler's FTCS is used:

$$u_i^{n+1} = \frac{1}{2}(u_{i+1}^n + u_{i-1}^n) - \frac{C}{2}(u_{i+1}^n - u_{i-1}^n) \tag{4.3.13}$$

The von Neumann stability analysis shows that this scheme is stable for $C \le 1$.

Midpoint Leapfrog Method

Central differences for both time and spaces are used in this method:

$$\frac{u_i^{n+1} - u_i^{n-1}}{2\Delta t} = -\frac{a(u_{i+1}^n - u_{i-1}^n)}{2\Delta x}, \quad O(\Delta t^2, \Delta x^2) \tag{4.3.14}$$

This scheme is stable for $C \le 1$. It has a second order accuracy, but requires two sets of initial values when the starter solution can provide only one set of initial data. This may lead to two independent solutions which are inaccurate.

Lax-Wendroff Method

In this method, we utilize the finite difference equation derived from Taylor series,

$$u(x, t+\Delta t) = u(x,t) + \frac{\partial u}{\partial t}\Delta t + \frac{1}{2!}\frac{\partial^2 u}{\partial t^2}\Delta t^2 + O(\Delta t^3) \tag{4.3.15a}$$

or

$$u_i^{n+1} = u_i^n + \frac{\partial u}{\partial t}\Delta t + \frac{1}{2!}\frac{\partial^2 u}{\partial t^2}\Delta t^2 + O(\Delta t^3) \tag{4.3.15b}$$

Differentiating (4.3.1) with respect to time yields

$$\frac{\partial^2 u}{\partial t^2} = -a\frac{\partial}{\partial x}\left(\frac{\partial u}{\partial t}\right) = a^2\frac{\partial^2 u}{\partial x^2} \tag{4.3.16}$$

Substituting (4.3.1) and (4.3.16) into (4.3.15b) leads to

$$u_i^{n+1} = u_i^n + \Delta t\left(-a\frac{\partial u}{\partial x}\right) + \frac{\Delta t^2}{2}\left(a^2\frac{\partial^2 u}{\partial x^2}\right) \tag{4.3.17}$$

Using central differencing of the second order for the spatial derivative, we obtain

$$u_i^{n+1} = u_i^n - a\,\Delta t\left(\frac{u_{i+1}^n - u_{i-1}^n}{2\Delta x}\right) + \frac{1}{2}(a\,\Delta t)^2\left(\frac{u_{i+1}^n - 2u_i^n + u_{i-1}^n}{(\Delta x)^2}\right), \quad O(\Delta t^2, \Delta x^2) \tag{4.3.18}$$

This method is stable for $C \leq 1$.

4.3.2 IMPLICIT SCHEMES

Implicit schemes for approximating (4.3.1) are unconditionally stable. Two representative implicit schemes are Euler's FTCS method and the Crank-Nicolson method.

Euler's FTCS Method

$$\frac{u_i^{n+1} - u_i^n}{\Delta t} = \frac{-a}{2\Delta x}\left(u_{i+1}^{n+1} - u_{i-1}^{n+1}\right), \quad O(\Delta t, \Delta x^2) \tag{4.3.19}$$

or

$$\frac{C}{2}u_{i-1}^{n+1} - u_i^{n+1} - \frac{C}{2}u_{i+1}^{n+1} = -u_i^n \tag{4.3.20}$$

Crank-Nicolson Method

$$\frac{u_i^{n+1} - u_i^n}{\Delta t} = -\frac{a}{2}\left[\frac{u_{i+1}^{n+1} - u_{i-1}^{n+1}}{2\Delta x} + \frac{u_{i+1}^n - u_{i-1}^n}{2\Delta x}\right], \quad O(\Delta t^2, \Delta x^2) \tag{4.3.21}$$

or

$$\frac{C}{4}u_{i-1}^{n+1} - u_i^{n+1} - \frac{C}{4}u_{i+1}^{n+1} = -\frac{C}{4}u_{i-1}^n - u_i^n + \frac{C}{4}u_{i+1}^n \tag{4.3.22}$$

Examples of the numerical solution procedure for a typical first order hyperbolic equation using the explicit and implicit schemes are shown in Section 4.7.3.

4.3.3 MULTISTEP (SPLITTING, PREDICTOR-CORRECTOR) METHODS

Computational stability, convergence, and accuracy may be improved using multistep (intermediate step between n and $n+1$) schemes, such as Richtmyer, Lax-Wendroff, and McCormack methods. The two-step schemes for these methods are shown below.

Richtmyer Multistep Scheme

Step 1

$$\frac{u_i^{n+\frac{1}{2}} - \frac{1}{2}\left(u_{i+1}^n + u_{i-1}^n\right)}{\Delta t/2} = -a\frac{\left(u_{i+1}^n - u_{i-1}^n\right)}{2\Delta x} \tag{4.3.23a}$$

Step 2

$$\frac{u_i^{n+1} - u_i^n}{\Delta t} = -a \frac{\left(u_{i+1}^{n+\frac{1}{2}} - u_{i-1}^{n+\frac{1}{2}}\right)}{2\Delta x} \tag{4.3.23b}$$

These equations can be rearranged in the form

Step 1

$$u_i^{n+\frac{1}{2}} = \frac{1}{2}(u_{i+1}^n + u_{i-1}^n) - \frac{C}{4}(u_{i+1}^n - u_{i-1}^n) \tag{4.3.24a}$$

Step 2

$$u_i^{n+1} = u_i^n - \frac{C}{2}\left(u_{i+1}^{n+\frac{1}{2}} - u_{i-1}^{n+\frac{1}{2}}\right), \quad O(\Delta t^2, \Delta x^2) \tag{4.3.24b}$$

This scheme is stable for $C \leq 2$.

Lax-Wendroff Multistep Scheme

Step 1

$$u_{i+\frac{1}{2}}^{n+\frac{1}{2}} = \frac{1}{2}(u_{i+1}^n + u_i^n) - \frac{C}{2}(u_{i+1}^n - u_i^n), \quad O(\Delta t^2, \Delta x^2) \tag{4.3.25a}$$

Step 2

$$u_i^{n+1} = u_i^n - C\left(u_{i+\frac{1}{2}}^{n+\frac{1}{2}} - u_{i-\frac{1}{2}}^{n+\frac{1}{2}}\right), \quad O(\Delta t^2, \Delta x^2) \tag{4.3.25b}$$

The stability condition is $C \leq 1$. Note that substitution of (4.3.25a) into (4.3.25b) recovers the original Lax-Wendroff equation (4.3.18). The same result is obtained with (4.3.24a) and (4.3.24b).

MacCormack Multistep Scheme

Here we consider an intermediate step u_i^* which is related to $u_i^{n+\frac{1}{2}}$:

$$u_i^{n+\frac{1}{2}} = \frac{1}{2}(u_i^n + u_i^*) \tag{4.3.26}$$

Step 1

$$\frac{u_i^* - u_i^n}{\Delta t} = -a\frac{(u_{i+1}^n - u_i^n)}{\Delta x} \tag{4.3.27a}$$

Step 2

$$\frac{u_i^{n+1} - u_i^{n+\frac{1}{2}}}{\Delta t/2} = -a\frac{(u_i^* - u_{i-1}^*)}{\Delta x} \tag{4.3.27b}$$

Substituting (4.3.26) into (4.3.27b) yields

Predictor

$$u_i^* = u_i^n - C(u_{i+1}^n - u_i^n) \tag{4.3.28a}$$

Corrector

$$u_i^{n+1} = \frac{1}{2}[(u_i^n + u_i^*) - C(u_i^* - u_{i-1}^*)], \quad O(\Delta t^2, \Delta x^2) \tag{4.3.28b}$$

with the stability criterion of $C \leq 1$.

The MacCormack multistep method is well suited for nonlinear problems. It becomes equivalent to the Lax-Wendroff method for linear problems.

4.3.4 NONLINEAR PROBLEMS

A classical nonlinear first order hyperbolic equation is the Euler's equation

$$\frac{\partial u}{\partial t} = -u\frac{\partial u}{\partial x} \tag{4.3.29}$$

which in conservation form may be written as

$$\frac{\partial u}{\partial t} = -\frac{\partial}{\partial x}\left(\frac{u^2}{2}\right) \tag{4.3.30a}$$

or

$$\frac{\partial u}{\partial t} = -\frac{\partial F}{\partial x} \quad \text{with } F = \left(\frac{u^2}{2}\right) \tag{4.3.30b}$$

The solution of (4.3.30b) may be obtained by several methods: Lax method, Lax-Wendroff method, MacCormack method, and Beam-Warming implicit method. These are described below.

Lax Method

In this method, the FTCS differencing scheme is used.

$$\frac{u_i^{n+1} - u_i^n}{\Delta t} = -\frac{F_{i+1}^n - F_{i-1}^n}{2\Delta x}, \quad O(\Delta t, \Delta x^2) \tag{4.3.31}$$

To maintain stability, we replace u_i^n by its average,

$$u_i^{n+1} = \frac{1}{2}(u_{i+1}^n + u_{i-1}^n) - \frac{\Delta t}{2\Delta x}(F_{i+1}^n - F_{i-1}^n) \tag{4.3.32}$$

or

$$u_i^{n+1} = \frac{1}{2}(u_{i+1}^n + u_{i-1}^n) - \frac{\Delta t}{4\Delta x}[(u_{i+1}^n)^2 - (u_{i-1}^n)^2] \tag{4.3.33}$$

The solution will be stable if

$$\left|\frac{\Delta t}{\Delta x} u_{\max}\right| \leq 1 \tag{4.3.34}$$

Lax-Wendroff Method

In this method, the finite difference equation is derived from the Taylor series expansion,

$$u_i^{n+1} = u_i^n + \frac{\partial u}{\partial t}\Delta t + \frac{1}{2!}\frac{\partial^2 u}{\partial t^2}\Delta t^2 + \cdots \tag{4.3.35}$$

Using (4.3.30b) we have

$$\frac{\partial^2 u}{\partial t^2} = -\frac{\partial}{\partial t}\left(\frac{\partial F}{\partial x}\right) = -\frac{\partial}{\partial x}\left(\frac{\partial F}{\partial t}\right) \tag{4.3.36}$$

where

$$\frac{\partial F}{\partial t} = \frac{\partial F}{\partial u}\frac{\partial u}{\partial t} = \frac{\partial F}{\partial u}\left(-\frac{\partial F}{\partial x}\right) = -A\frac{\partial F}{\partial x} \tag{4.3.37}$$

with A being the Jacobian.

$$A = \frac{\partial F}{\partial u} = \frac{\partial}{\partial u}\left(\frac{u^2}{2}\right) = u \tag{4.3.38}$$

Thus

$$\frac{\partial^2 u}{\partial t^2} = -\frac{\partial}{\partial x}\left(-A\frac{\partial F}{\partial x}\right) = \frac{\partial}{\partial x}\left(A\frac{\partial F}{\partial x}\right) \tag{4.3.39}$$

Substituting (4.3.39) and (4.3.30b) into (4.3.35) yields

$$u_i^{n+1} = u_i^n + \left(-\frac{\partial F}{\partial x}\right)\Delta t + \frac{\partial}{\partial x}\left(A\frac{\partial F}{\partial x}\right)\frac{\Delta t^2}{2} + O(\Delta t^3)$$

or

$$\frac{u_i^{n+1} - u_i^n}{\Delta t} = -\frac{\partial F}{\partial x} + \frac{\partial}{\partial x}\left(A\frac{\partial F}{\partial x}\right)\frac{\Delta t}{2} + O(\Delta t^2)$$

Approximating the spatial derivatives by central differencing of order 2,

$$\frac{u_i^{n+1} - u_i^n}{\Delta t} = -\frac{F_{i+1}^n - F_{i-1}^n}{2\Delta x} + \frac{\Delta t}{2\Delta x}\left[\left(A\frac{\partial F}{\partial x}\right)_{i+\frac{1}{2}}^n - \left(A\frac{\partial F}{\partial x}\right)_{i-\frac{1}{2}}^n\right] \tag{4.3.40}$$

The last term above is approximated as

$$\begin{aligned}\frac{\left(A\frac{\partial F}{\partial x}\right)_{i+\frac{1}{2}}^n - \left(A\frac{\partial F}{\partial x}\right)_{i-\frac{1}{2}}^n}{\Delta x} &= \frac{A_{i+\frac{1}{2}}^n\dfrac{F_{i+1}^n - F_i^n}{\Delta x} - A_{i-\frac{1}{2}}^n\dfrac{F_i^n - F_{i-1}^n}{\Delta x}}{\Delta x}\\ &= \frac{\dfrac{1}{2\Delta x}(A_{i+1}^n + A_i^n)(F_{i+1}^n - F_i^n) - \dfrac{1}{2\Delta x}(A_i^n + A_{i-1}^n)(F_i^n - F_{i-1}^n)}{\Delta x}\end{aligned} \tag{4.3.41}$$

For $A = u$, we obtain

$$\begin{aligned}u_i^{n+1} &= u_i^n - \frac{\Delta t}{2\Delta x}(F_{i+1}^n - F_{i-1}^n)\\ &\quad + \frac{1}{4}\frac{\Delta t^2}{\Delta x^2}\left[(u_{i+1}^n + u_i^n)(F_{i+1}^n - F_i^n) - (u_i^n + u_{i-1}^n)(F_i^n - F_{i-1}^n)\right]\end{aligned} \tag{4.3.42}$$

This is second order accurate with the stability requirement,

$$\left|\frac{\Delta t}{\Delta x}u_{\max}\right| \le 1$$

MacCormack Method

In this method, the multilevel scheme is used as given by

$$u_i^* = u_i^n - \frac{\Delta t}{\Delta x}(F_{i+1}^n - F_i^n) \tag{4.3.43a}$$

$$u_i^{n+1} = \frac{1}{2}\left[u_i^n + u_i^* - \frac{\Delta t}{\Delta x}(F_i^* - F_{i-1}^*)\right] \tag{4.3.43b}$$

Because of the two-level splitting, the solution performs better than the Lax method or the Lax-Wendroff method. One of the most widely used implicit schemes is the Beam-Warming method, discussed below.

Beam-Warming Implicit Method

Let us consider the Taylor series expansion,

$$u(x,t+\Delta t) = u(x,t) + \left.\frac{\partial u}{\partial t}\right|_{x,t}\Delta t + \left.\frac{\partial^2 u}{\partial t^2}\right|_{x,t}\frac{\Delta t^2}{2} + O(\Delta t^3) \tag{4.3.44}$$

and

$$u(x,t) = u(x,t+\Delta t) - \left.\frac{\partial u}{\partial t}\right|_{x,t+\Delta t}\Delta t + \left.\frac{\partial^2 u}{\partial t^2}\right|_{x,t+\Delta t}\frac{\Delta t^2}{2!} + O(\Delta t^3) \tag{4.3.45}$$

Subtracting (4.3.45) from (4.3.44)

$$\begin{aligned}2u(x,t+\Delta t) &= 2u(x,t) + \left.\frac{\partial u}{\partial t}\right|_{x,t}\Delta t + \left.\frac{\partial u}{\partial t}\right|_{x,t+\Delta t}\Delta t \\ &\quad + \left.\frac{\partial^2 u}{\partial t^2}\right|_{x,t}\frac{\Delta t^2}{2!} - \left.\frac{\partial^2 u}{\partial t^2}\right|_{x,t+\Delta t}\frac{\Delta t^2}{2!} + O(\Delta t^3)\end{aligned}$$

or

$$u_i^{n+1} = u_i^n + \frac{1}{2}\left[\left(\frac{\partial u}{\partial t}\right)_i^n + \left(\frac{\partial u}{\partial t}\right)_i^{n+1}\right]\Delta t + \frac{1}{2}\left[\left(\frac{\partial^2 u}{\partial t^2}\right)_i^n - \left(\frac{\partial^2 u}{\partial t^2}\right)_i^{n+1}\right]\frac{\Delta t^2}{2!} + O(\Delta t^3)$$

where

$$\left(\frac{\partial^2 u}{\partial t^2}\right)_i^{n+1} = \left(\frac{\partial^2 u}{\partial t^2}\right)_i^n + \frac{\partial}{\partial t}\left(\frac{\partial^2 u}{\partial t^2}\right)_i^n \Delta t + O(\Delta t^2)$$

Thus, we arrive at

$$u_i^{n+1} = u_i^n + \frac{1}{2}\left[\left(\frac{\partial u}{\partial t}\right)_i^n + \left(\frac{\partial u}{\partial t}\right)_i^{n+1}\right]\Delta t + O(\Delta t^3) \tag{4.3.46}$$

For the model equation

$$\frac{\partial u}{\partial t} = -\frac{\partial F}{\partial x} \tag{4.3.47}$$

Using (4.3.46) in (4.3.47), we obtain

$$\frac{u_i^{n+1} - u_i^n}{\Delta t} = -\frac{1}{2}\left[\left(\frac{\partial F}{\partial x}\right)_i^n + \left(\frac{\partial F}{\partial x}\right)_i^{n+1}\right] + O(\Delta t^2) \tag{4.3.48}$$

This indicates that (4.3.48) leads to the second order accuracy.

Recall that the nonlinear term $F = u^2/2$ was applied at the known time level n, and the resulting FDE in explicit form was linear. The resulting FDE in implicit formulation is nonlinear, and therefore a procedure is used to linearize the FDE. To this end, we write a Taylor series for $F(t+\Delta t)$ in the form

$$\begin{aligned} F(t+\Delta t) &= F(t) + \frac{\partial F}{\partial t}\Delta t + O(\Delta t^2) \\ &= F(t) + \frac{\partial F}{\partial u}\frac{\partial u}{\partial t}\Delta t + O(\Delta t^2) \end{aligned}$$

or

$$F^{n+1} = F^n + \frac{\partial F}{\partial u}\left(\frac{u^{n+1} - u^n}{\Delta t}\right)\Delta t + O(\Delta t^2) \tag{4.3.49}$$

Taking a partial derivative of (4.3.49) yields

$$\left(\frac{\partial F}{\partial x}\right)^{n+1} = \left(\frac{\partial F}{\partial x}\right)^n + \frac{\partial}{\partial x}[A(u^{n+1} - u^n)] \tag{4.3.50}$$

Combining (4.3.48) and (4.3.50) gives

$$\frac{u_i^{n+1} - u_i^n}{\Delta t} = -\frac{1}{2}\left\{\left(\frac{\partial F}{\partial x}\right)_i^n + \left(\frac{\partial F}{\partial x}\right)_i^n + \frac{\partial}{\partial x}[A(u_i^{n+1} - u_i^n)]\right\}$$

or

$$u_i^{n+1} = u_i^n - \frac{1}{2}\Delta t\left\{2\left(\frac{\partial F}{\partial x}\right)_i^n + \frac{\partial}{\partial x}[A(u_i^{n+1} - u_i^n)]\right\} \tag{4.3.51}$$

Using a second order central differencing for the terms with A on the right-hand side of (4.3.51) and linearizing, we obtain

$$\begin{aligned} u_i^{n+1} &= u_i^n - \frac{1}{2}\Delta t\left[\frac{2(F_{i+1}^n - F_{i-1}^n)}{2\Delta x} + \frac{A_{i+1}^n u_{i+1}^{n+1} - A_{i-1}^n u_{i-1}^{n+1}}{2\Delta x}\right. \\ &\quad \left. - \frac{A_{i+1}^n u_{i+1}^n - A_{i-1}^n u_{i-1}^n}{2\Delta x}\right] \end{aligned} \tag{4.3.52}$$

Modifying (4.3.52) to a tridiagonal form

$$\begin{aligned} &-\frac{\Delta t}{4\Delta x}A_{i-1}^n u_{i-1}^{n+1} + u_i^{n+1} + \frac{\Delta t}{4\Delta x}A_{i+1}^n u_{i+1}^{n+1} \\ &= u_i^n - \frac{1}{2}\frac{\Delta t}{\Delta x}(F_{i+1}^n - F_{i-1}^n) + \frac{\Delta t}{4\Delta x}A_{i+1}^n u_{i+1}^n - \frac{\Delta t}{4\Delta x}A_{i-1}^n u_{i-1}^n + D \end{aligned} \tag{4.3.53}$$

This scheme is second order accurate, unconditionally stable, but dispersion errors may arise. To prevent this, a fourth order smoothing (damping) term is explicitly added:

$$D = -\frac{\omega}{8}(u_{i+2}^n - 4u_{i+1}^n + 6u_i^n - 4u_{i-1}^n + u_{i-2}^n),$$

with $0 < \omega < 1$. Since the added damping term is of fourth order, it does not affect the second order accuracy of the method.

4.3.5 SECOND ORDER ONE-DIMENSIONAL WAVE EQUATIONS

Let us consider the second order one-dimensional wave equation,

$$\frac{\partial^2 u}{\partial t^2} = a^2 \frac{\partial^2 u}{\partial x^2} \tag{4.3.54}$$

Here we require two sets of initial conditions,

$$u(x, 0) = f(x)$$
$$\frac{\partial u}{\partial t}(x, 0) = g(x)$$

and two sets of boundary conditions,

$$u(0, t) = h_1(t)$$
$$u(L, t) = h_2(t)$$

We may use the midpoint leapfrog method for this problem,

$$u_i^{n+1} = 2u_i^n - u_i^{n-1} + C^2\left(u_{i-1}^n - 2u_i^n + u_{i+1}^n\right) \tag{4.3.55}$$

If we choose $\dfrac{\partial u(x, 0)}{\partial t} = 0$, then

$$\frac{u_i^{n+1} - u_i^{n-1}}{2\Delta t} = 0$$

or

$$u_i^{n+1} = u_i^{n-1}$$

Thus, from (4.3.55), we obtain

$$u_i^{n+1} = u_i^n + \frac{1}{2}C^2\left(u_{i-1}^n - 2u_i^n + u_{i+1}^n\right) \tag{4.3.56}$$

This is called the midpoint leapfrog method. An example problem for the second order hyperbolic equation is demonstrated in Section 4.7.4.

4.4 BURGERS' EQUATION

The Burgers' equation is a special form of the momentum equation for irrotational, incompressible flows in which pressure gradients are neglected. It is informative to study this equation in the one-dimensional case before we launch upon full-scale CFD problems.

Consider the Burgers' equation written in various forms:

$$\frac{\partial u}{\partial t} + a\frac{\partial u}{\partial x} = \nu\frac{\partial^2 u}{\partial x^2} \tag{4.4.1}$$

$$\frac{\partial u}{\partial t} + u\frac{\partial u}{\partial x} = \nu\frac{\partial^2 u}{\partial x^2} \tag{4.4.2}$$

$$\frac{\partial u}{\partial t} + \frac{\partial F}{\partial x} = \nu\frac{\partial^2 u}{\partial x^2} \tag{4.4.3}$$

with $F = 1/2\, u^2$. These equations are mixed hyperbolic, elliptic, and parabolic types. If steady state is considered, then they become mixed hyperbolic and elliptic equations. Because of these special properties, various solution schemes have been tested extensively for the Burgers' equations. In what follows, we shall examine some of the well-known numerical schemes.

4.4.1 EXPLICIT AND IMPLICIT SCHEMES

FTCS Explicit Scheme

In this scheme (FTCS), approximations of forward differences in time and central differences in space are used:

$$\frac{u_i^{n+1} - u_i^n}{\Delta t} + a\frac{u_{i+1}^n - u_{i-1}^n}{2\Delta x} = \nu\frac{u_{i+1}^n - 2u_i^n + u_{i-1}^n}{\Delta x^2} \tag{4.4.4}$$

where the truncation error is $O(\Delta t, \Delta x^2)$. The central difference for the convective term tends to introduce significant damping.

FTBS Explicit Scheme

This is the same as in FTCS except that backward differences are used for the convective term,

$$\frac{u_i^{n+1} - u_i^n}{\Delta t} + a\frac{u_i^n - u_{i-1}^n}{\Delta x} = \nu\frac{u_{i+1}^n - 2u_i^n + u_{i-1}^n}{\Delta x^2} \tag{4.4.5}$$

Here the first order approximation of the convective term may introduce an excessive dissipation error. A compromise is to use higher order schemes such as (3.2.20) for the second order. With (3.2.1) modified for four points, the third order scheme may be written as

$$\frac{u_i^{n+1} - u_i^n}{\Delta t} + a\left(\frac{11u_i - 18u_{i-1} + 9u_{i-2} - 2u_{i-3}}{6\Delta x}\right) = \nu\frac{u_{i+1}^n - 2u_i^n + u_{i-1}^n}{\Delta x^2} \tag{4.4.6}$$

DuFort-Frankel Explicit Scheme

In this scheme, we use second order central differences for all derivatives,

$$\frac{u_i^{n+1} - u_i^{n-1}}{2\Delta t} + a\frac{u_{i+1}^n - u_{i-1}^n}{2\Delta x} = \nu\frac{u_{i+1}^n - (u_i^{n-1} + u_i^{n+1}) + u_{i-1}^n}{\Delta x^2},$$

$$O\left(\Delta t^2, \Delta x^2, \left(\frac{\Delta t}{\Delta x}\right)^2\right) \tag{4.4.7a}$$

or

$$u_i^{n+1} = \left(\frac{1-2d}{1+2d}\right)u_i^{n-1} + \left(\frac{C+2d}{1+2d}\right)u_{i-1}^n - \left(\frac{C-2d}{1+2d}\right)u_{i+1}^n \tag{4.4.7b}$$

This is stable for $C \leq 1$.

MacCormack Explicit Scheme

The two-step or predictor-corrector scheme is written as

Step 1

$$u_i^* = u_i^n - a\frac{\Delta t}{\Delta x}(u_{i+1}^n - u_i^n) + \nu\frac{\Delta t}{\Delta x^2}(u_{i+1}^n - 2u_i^n + u_{i-1}^n) \tag{4.4.8a}$$

Step 2

$$u_i^{n+1} = \frac{1}{2}\left[u_i^n + u_i^* - a\frac{\Delta t}{\Delta x}(u_i^* - u_{i-1}^*) + \nu\frac{\Delta t}{\Delta x^2}(u_{i+1}^* - 2u_i^* + u_{i-1}^*)\right] \tag{4.4.8b}$$

This method is second order accurate with the stability requirement

$$\Delta t \leq \frac{1}{\dfrac{a}{\Delta x} + \dfrac{2\nu}{\Delta x^2}} \tag{4.4.9}$$

The following alternate form may be used:

Step 1

$$\Delta u_i^n = -a\frac{\Delta t}{\Delta x}(u_{i+1}^n - u_i^n) + \frac{\nu\Delta t}{\Delta x^2}(u_{i+1}^n - 2u_i^n + u_{i-1}^n)$$

$$u_i^* = u_i^n + \Delta u_i^n \tag{4.4.10a}$$

Step 2

$$\Delta u_i^* = -a\frac{\Delta t}{\Delta x}(u_i^* - u_{i-1}^*) + \frac{\nu\Delta t}{\Delta x^2}(u_{i+1}^* - 2u_i^* + u_{i-1}^*)$$

$$u_i^{n+1} = \frac{1}{2}(u_i^n + u_i^* + \Delta u_i^*) \tag{4.4.10b}$$

MacCormack Implicit Scheme

One of the most frequently used implicit schemes is the MacCormack scheme.

Step 1

$$\left(1 + \lambda\frac{\Delta t}{\Delta x}\right)\delta u_i^* = \Delta u_i^n + \lambda\frac{\Delta t}{\Delta x}\delta u_{i+1}^*$$

$$u_i^* = u_i^n + \delta u_i^* \tag{4.4.11a}$$

Step 2

$$\left(1 + \lambda\frac{\Delta t}{\Delta x}\right)\delta u_i^{n+1} = \Delta u_i^* + \lambda\frac{\Delta t}{\Delta x}\delta u_{i-1}^{n+1}$$

$$u_i^{n+1} = \frac{1}{2}(u_i^n + u_i^* + \delta u_i^{n+1}) \tag{4.4.11b}$$

where

$$\lambda \geq \max\left[\frac{1}{2}\left(|a| + \frac{2\nu}{\Delta x} - \frac{\Delta x}{\Delta t}\right), 0\right] \tag{4.4.12}$$

Note that equations (4.4.11a,b) form a tridiagonal system. The method is unconditionally stable and second order accurate as long as the diffusion number, $d = \nu\Delta t/\Delta x^2$, is bounded for the limiting process for which Δt and Δx approach zero.

4.4.2 RUNGE-KUTTA METHOD

The transient nonlinear inviscid Burgers' equation can be written as

$$\frac{\partial u}{\partial t} + u\frac{\partial u}{\partial x} = 0$$

or

$$\frac{\partial u}{\partial t} + \frac{\partial F}{\partial x} = 0, \quad F = \frac{u^2}{2}$$

For nonlinear transient problems, the Runge-Kutta method is known to be efficient and has been used extensively. This method is briefly introduced below.

Let us consider an equation of the type

$$\frac{\partial u}{\partial t} = R(u) \tag{4.4.13}$$

One of the popular approaches is the fourth order Runge-Kutta scheme written as

Step 1

$$u^{(1)} = u^n + \frac{\Delta t}{2}R^n$$

Step 2

$$u^{(2)} = u^n + \frac{\Delta t}{2}R^{(1)}$$

Step 3

$$u^{(3)} = u^n + \Delta t R^{(2)}$$

Step 4

$$u^{n+1} = u^n + \frac{\Delta t}{6}\left(R^n + 2R^{(1)} + 2R^{(2)} + R^{(3)}\right) \tag{4.4.14}$$

with

$$R^{(1)} = R\left(t^{n+1/2}, u^{(1)}\right)$$
$$R^{(2)} = R\left(t^{n+1/2}, u^{(2)}\right)$$
$$R^{(3)} = R\left(t^{n}, u^{(3)}\right)$$

It is seen that higher order Runge-Kutta schemes require more steps for the evaluation of $R(u)$, resulting in additional computer time requirements.

An example of the solution procedure for the nonlinear Burgers' equation is presented in Section 4.7.5.

4.5 ALGEBRAIC EQUATION SOLVERS AND SOURCES OF ERRORS

4.5.1 SOLUTION METHODS

As a result of FDM formulations, we obtain linear or nonlinear simultaneous algebraic equations which must be solved. As we discussed in previous sections, either direct methods or iterative methods may be used. Recall that, as direct methods, we examined the Gaussian elimination in Section 4.1.3 and the Thomas algorithm (tridiagonal matrix algorithm, TDMA) in Section 4.2.7. We also discussed the Runge-Kutta method in Section 4.4.2 for the nonlinear time dependent equations.

In general, the number of arithmetic operations of a direct method can be very high particularly for a large system of equations – much larger than the total number of operations in an iterative method. Therefore, for fluid mechanics problems with nonlinear sparse matrices, it is more convenient, and often necessary, to work with iterative methods.

There are many iterative methods other than those already introduced in the earlier sections of this chapter. They include conjugate gradient method, generalized minimum residual (GMRES) algorithm, and multigrid method. These methods are well documented in the literature. Among them are Varga [1962], Wachspress [1966], Dahlquist and Bjork [1974], and Saad [1996].

Some of these advanced iterative methods will be presented in Parts Three and Four. Conjugate gradient method, generalized minimum residual method, and multigrid method are presented in Sections 10.3.1, 11.5.2, and 20.2, respectively. This is because of the convenience of presentation as appropriate to the topical arrangements of this book. Namely, the iterative solution methods are included in Part Three since the element-by element method of FEM assembly requires special treatments of iterative solution procedures, whereas the multigrid method is included in Part Four as it is related to other topics including automatic grid generation. Newton-Raphson methods for nonlinear algebraic equations are discussed in Section 11.5.1. Thus, the reader may find it useful in visiting these sections as needed for his/her studies in FDM, Part Two.

4.5.2 EVALUATION OF SOURCES OF ERRORS

Recall that computational errors were discussed in terms of an amplification factor g in Sections 4.2 and 4.3. For $g < 1$, the result is numerical diffusion (sometimes known as numerical damping or numerical dissipation). On the other hand, for $g > 1$, the result is numerical instability. Both of these cases lead to amplitude errors as shown in Figure 4.5.1, which may be equivalent to the severely damped shock wave as depicted in Figure 4.3.2c(2).

If waves of different wavelengths travel in a medium, such a phenomenon is known as dispersion. The dispersion arises from discrete spatial approximations and results in a numerical error, called the numerical dispersion or phase error as shown in Figure 4.5.1b or Figure 4.3.2c(3). The dispersion error occurs in convection or wave equations, but not in diffusion equations.

In numerical simulations, the so-called Gibb's phenomenon occurs due to discretization of the domain by a limited number of nodal points (Figure 4.5.1c). They appear as overshoots and undershoots near the steep gradients, similar to the diffusion errors.

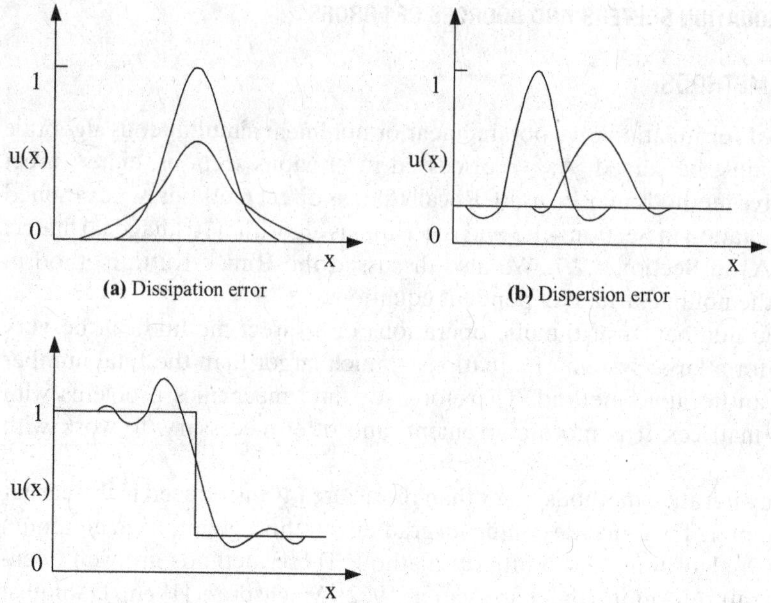

(a) Dissipation error
(b) Dispersion error
(c) Gibb's error

Figure 4.5.1 Various numerical errors.

Next we shall discuss these errors, which are associated with the diffusion transport and convection transport equations.

Diffusive Transport

Parabolic equations represent the diffusion process associated with both spatial and temporal variations. A general form of (4.2.2a) may be written in the form

$$\frac{u^{n+1}{}_i - u_i^n}{\Delta t} = \frac{\alpha\theta}{\Delta x^2}\left(u_{i+1}^{n+1} - 2u_i^{n+1} + u_{i-1}^{n+1}\right) + \frac{\alpha(1-\theta)}{\Delta x^2}\left(u_{i+1}^n - 2u_i^n + u_{i-1}^n\right) \tag{4.5.1}$$

with $0 \le \theta \le 1$.

The method is fully explicit for $\theta = 0$ and partially implicit for $0 < \theta < 1$, with $\theta = 1$ being fully implicit. The scheme with $\theta = 1/2$, known as the centered scheme, provides reasonably stable and accurate solutions in general.

Using the definitions given in (4.2.12–4.2.15), the analytical solution of the diffusion equation (4.2.1) may be written in the form

$$u(x,t) = u(t)e^{Ikx} \tag{4.5.2}$$

or

$$u(x,t) = u_0 e^{-\alpha k^2 t} e^{Ikx} \tag{4.5.3}$$

Substituting (4.5.2) into (4.5.1) and using the definition of the amplification factor (4.2.17), we obtain the amplification factor for various values of θ,

$$|g|_\theta = \frac{[1 - d(1-\theta)(1-\cos(k\Delta x))]^{1/d}}{[1 + d\theta(1-\cos(k\Delta x))]^{1/d}} \tag{4.5.4}$$

The amplification factors for explicit scheme (E), centered scheme (C), and fully implicit scheme (I) for $\theta = 0, \theta = 1/2, \theta = 1$ are shown in Figure 4.5.2. It is seen that the explicit and centered schemes behave irregularly for high values of diffusion number, whereas the fully implicit scheme is stable.

For multidimensional problems, the implicit method requires the inversion of large and sparse matrix equations and is computationally expensive. Although the solution may be stable with large time steps, numerical diffusion becomes excessive, resulting in inaccuracy. On the other hand, the explicit scheme is less expensive, but small time steps are necessary in order to achieve accuracy. The amplitude errors are significant in the diffusive transport equations, whereas dispersion errors and Gibb's errors dominate in the convective transport equations.

Convective Transport

Hyperbolic equations represent convection and wave phenomena. A typical convection equation may be written in the finite difference form

$$\frac{u_i^{n+1}-u_i^n}{\Delta t}=-a\left[\theta\frac{(u_{i+1}^{n+1}-u_{i-1}^{n+1})}{2\Delta x}+\left((1-\theta)\frac{(u_{i+1}^n-u_{i-1}^n)}{2\Delta x}\right)\right] \tag{4.5.5}$$

or in terms of the Courant number $C = a\Delta t/\Delta x$.

$$u_i^{n+1}+\frac{\theta C}{2}(u_{i+1}^{n+1}-u_{i-1}^{n+1})=u_i^n-\frac{(1-\theta)C}{2}(u_{i+1}^n-u_{i-1}^n) \tag{4.5.6}$$

Note that the values of u at $n+1$ for $\theta > 0$ (implicit scheme) are calculated in terms of the values at n, but are involved in three different spatial locations, resulting in a

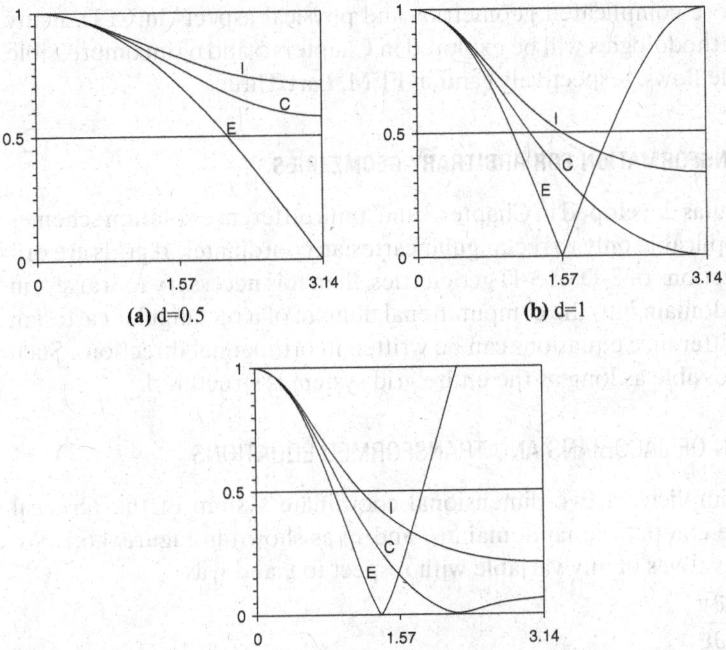

Figure 4.5.2 Amplification factors for diffusion equation, E = explicit, C = centered, I = fully implicit.

tridiagonal matrix. Although the explicit scheme ($\theta = 0$) reduces to a simple algebraic equation, computational difficulties in stability and accuracy are likely to occur.

In view of (4.2.15), (4.3.3), and (4.5.6), the amplification factors for $\theta = 0$, $\theta = 1/2$, and $\theta = 1$ can be written in the form, respectively,

$$|g|_E = |1 - IC\sin(k\Delta x)| \tag{4.5.7}$$

$$|g|_C = \left|\frac{1 - I\frac{C}{2}\sin(k\Delta x)}{1 + I\frac{C}{2}\sin(k\Delta x)}\right| \tag{4.5.8}$$

$$|g|_I = \left|\frac{1}{1 + IC\sin(k\Delta x)}\right| \tag{4.5.9}$$

Similarly, using (4.3.18), the amplification factor for the Lax-Wendroff scheme is derived in the form,

$$|g|_L = |1 - C^2(1 - \cos(k\Delta x)) - IC\sin(k\Delta x)| \tag{4.5.10}$$

These results (Figure 4.5.3) show that the explicit scheme performs poorly in the region $k\Delta x = \pi/2$, whereas the Lax-Wendroff scheme behaves quite satisfactorily in the high wave number region.

As seen in other schemes studied in Section 3, computational errors including amplitude errors, dispersion errors, and Gibb's errors must be carefully examined, particularly in multidimensional problems. Some of the schemes used in one-dimensional problems may be extended to multidimensional problems, although the conclusions reached for one-dimensional problems discussed here are by no means universally applicable. In order to deal with more complicated geometries and physical aspects in CFD, many other schemes and methodologies will be explored in Chapters 5 and 6 (incompressible flows and compressible flows, respectively) and in FEM, Part Three.

4.6 COORDINATE TRANSFORMATION FOR ARBITRARY GEOMETRIES

Finite difference formulas developed in Chapter 3 and finite difference solution schemes discussed so far are applicable only to rectangular cartesian coordinates. If grids are oriented in arbitrary directions of 2-D or 3-D geometries, then it is necessary to transform the arbitrary physical domain into the computational domain of a rectangular cartesian system so that finite difference equations can be written in orthogonal directions. Such transformations are possible as long as the entire grid system is structured.

4.6.1 DETERMINATION OF JACOBIANS AND TRANSFORMED EQUATIONS

Let us consider for simplicity a two-dimensional coordinate system of the physical domain (x, y), and the computational domain (ξ and η) as shown in Figure 4.6.1. We begin with spatial derivatives of any variable with respect to ξ and η as

$$\frac{\partial}{\partial \xi} = \frac{\partial}{\partial x}\frac{\partial x}{\partial \xi} + \frac{\partial}{\partial y}\frac{\partial y}{\partial \xi}$$

$$\frac{\partial}{\partial \eta} = \frac{\partial}{\partial x}\frac{\partial x}{\partial \eta} + \frac{\partial}{\partial y}\frac{\partial y}{\partial \eta}$$

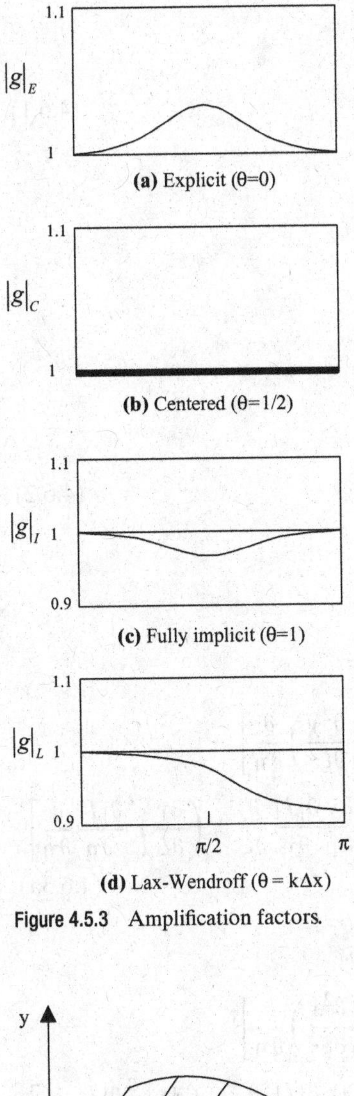

$$|g|_E = |1 + IC\sin(k\Delta x)|$$

(a) Explicit (θ=0)

$$|g|_C = \left|\frac{1 - I\dfrac{C}{2}\sin(k\Delta x)}{1 + I\dfrac{C}{2}\sin(k\Delta x)}\right|$$

(b) Centered (θ=1/2)

$$|g|_I = \left|\frac{1}{1 + IC\sin(k\Delta x)}\right|$$

(c) Fully implicit (θ=1)

$$|g|_L = |1 - C^2(1 - \cos(k\Delta x)) - IC\sin(k\Delta x)|$$

(d) Lax-Wendroff (θ = kΔx)

Figure 4.5.3 Amplification factors.

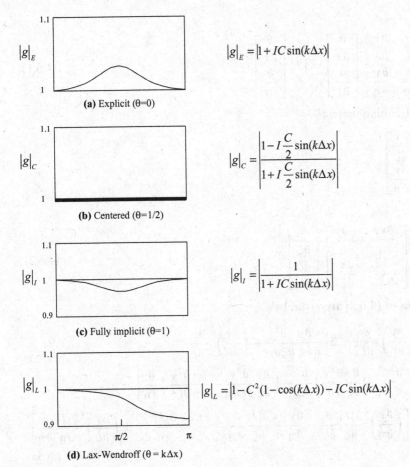

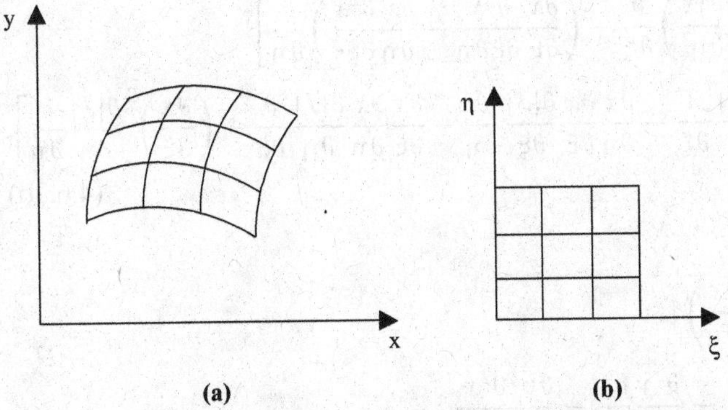

Figure 4.6.1 Transformation from curvilinear grid system into rectangular grid system. **(a)** Original curvilinear grid. **(b)** Transformed cartesian grid.

or

$$\begin{bmatrix} \frac{\partial}{\partial \xi} \\ \frac{\partial}{\partial \eta} \end{bmatrix} = \begin{bmatrix} \frac{\partial x}{\partial \xi} & \frac{\partial y}{\partial \xi} \\ \frac{\partial x}{\partial \eta} & \frac{\partial y}{\partial \eta} \end{bmatrix} \begin{bmatrix} \frac{\partial}{\partial x} \\ \frac{\partial}{\partial y} \end{bmatrix} = [J] \begin{bmatrix} \frac{\partial}{\partial x} \\ \frac{\partial}{\partial y} \end{bmatrix} \tag{4.6.1}$$

where $[J]$ is the Jacobian matrix

$$[J] = \begin{bmatrix} \frac{\partial x}{\partial \xi} & \frac{\partial y}{\partial \xi} \\ \frac{\partial x}{\partial \eta} & \frac{\partial y}{\partial \eta} \end{bmatrix}$$

Thus

$$\begin{bmatrix} \frac{\partial}{\partial x} \\ \frac{\partial}{\partial y} \end{bmatrix} = \frac{1}{|J|} \begin{bmatrix} \frac{\partial y}{\partial \eta} & \frac{-\partial y}{\partial \xi} \\ \frac{-\partial x}{\partial \eta} & \frac{\partial x}{\partial \xi} \end{bmatrix} \begin{bmatrix} \frac{\partial}{\partial \xi} \\ \frac{\partial}{\partial \eta} \end{bmatrix} \tag{4.6.2}$$

Second derivatives of (4.6.2) are given by

$$\begin{aligned} \frac{\partial^2}{\partial \mathrm{x}^2} = \frac{1}{|J|^2}\Bigg[&\left(\frac{\partial y}{\partial \eta}\right)^2 \frac{\partial^2}{\partial \xi^2} - 2\frac{\partial y}{\partial \eta}\frac{\partial y}{\partial \xi}\frac{\partial^2}{\partial \xi \partial \eta} + \left(\frac{\partial y}{\partial \xi}\right)^2 \frac{\partial^2}{\partial \eta^2} \\ &+ \left(\frac{\partial y}{\partial \eta}\frac{\partial^2 y}{\partial \xi \partial \eta} - \frac{\partial y}{\partial \xi}\frac{\partial^2 y}{\partial \eta^2}\right)\frac{\partial}{\partial \xi} + \left(\frac{\partial y}{\partial \xi}\frac{\partial^2 y}{\partial \xi \partial \eta} - \frac{\partial y}{\partial \eta}\frac{\partial^2 y}{\partial \xi^2}\right)\frac{\partial}{\partial \eta}\Bigg] \\ &- \frac{1}{|J|^3}\left[\left(\frac{\partial y}{\partial \eta}\right)^2 \frac{\partial |J|}{\partial \xi}\frac{\partial}{\partial \xi} - \frac{\partial y}{\partial \eta}\frac{\partial y}{\partial \xi}\frac{\partial |J|}{\partial \xi}\frac{\partial}{\partial \eta} - \frac{\partial y}{\partial \xi}\frac{\partial y}{\partial \eta}\frac{\partial |J|}{\partial \eta}\frac{\partial}{\partial \xi} + \left(\frac{\partial y}{\partial \xi}\right)^2 \frac{\partial |J|}{\partial \eta}\frac{\partial}{\partial \eta}\right] \end{aligned} \tag{4.6.3a}$$

$$\begin{aligned} \frac{\partial^2}{\partial y^2} = \frac{1}{|J|^2}\Bigg[&\left(\frac{\partial x}{\partial \eta}\right)^2 \frac{\partial^2}{\partial \xi^2} - 2\frac{\partial x}{\partial \eta}\frac{\partial x}{\partial \xi}\frac{\partial^2}{\partial \xi \partial \eta} + \left(\frac{\partial x}{\partial \xi}\right)^2 \frac{\partial^2}{\partial \eta^2} \\ &+ \left(\frac{\partial x}{\partial \eta}\frac{\partial^2 x}{\partial \xi \partial \eta} - \frac{\partial x}{\partial \xi}\frac{\partial^2 x}{\partial \eta^2}\right)\frac{\partial}{\partial \xi} + \left(\frac{\partial x}{\partial \xi}\frac{\partial^2 x}{\partial \xi \partial \eta} - \frac{\partial x}{\partial \eta}\frac{\partial^2 x}{\partial \xi^2}\right)\frac{\partial}{\partial \eta}\Bigg] \\ &- \frac{1}{|J|^3}\left[\left(\frac{\partial x}{\partial \eta}\right)^2 \frac{\partial |J|}{\partial \xi}\frac{\partial}{\partial \xi} - \frac{\partial x}{\partial \eta}\frac{\partial x}{\partial \xi}\frac{\partial |J|}{\partial \xi}\frac{\partial}{\partial \eta} - \frac{\partial x}{\partial \xi}\frac{\partial x}{\partial \eta}\frac{\partial |J|}{\partial \eta}\frac{\partial}{\partial \xi} + \left(\frac{\partial x}{\partial \xi}\right)^2 \frac{\partial |J|}{\partial \eta}\frac{\partial}{\partial \eta}\right] \end{aligned} \tag{4.6.3b}$$

where

$$\begin{aligned} \frac{\partial |J|}{\partial \xi} &= \frac{\partial}{\partial \xi}\left(\frac{\partial x}{\partial \xi}\frac{\partial y}{\partial \eta} - \frac{\partial y}{\partial \xi}\frac{\partial x}{\partial \eta}\right) \\ &= \frac{\partial^2 x}{\partial \xi^2}\frac{\partial y}{\partial \eta} + \frac{\partial x}{\partial \xi}\frac{\partial^2 y}{\partial \xi \partial \eta} - \frac{\partial^2 y}{\partial \xi^2}\frac{\partial x}{\partial \eta} - \frac{\partial y}{\partial \xi}\frac{\partial^2 x}{\partial \xi \partial \eta} \\ \frac{\partial |J|}{\partial \eta} &= \frac{\partial}{\partial \eta}\left(\frac{\partial x}{\partial \xi}\frac{\partial y}{\partial \eta} - \frac{\partial y}{\partial \xi}\frac{\partial x}{\partial \eta}\right) \\ &= \frac{\partial^2 x}{\partial \xi \partial \eta}\frac{\partial y}{\partial \eta} + \frac{\partial x}{\partial \xi}\frac{\partial^2 y}{\partial \eta^2} - \frac{\partial^2 y}{\partial \eta \partial \xi}\frac{\partial x}{\partial \eta} - \frac{\partial y}{\partial \xi}\frac{\partial^2 x}{\partial \eta^2} \end{aligned}$$

Consider the governing equations in the form

$$\frac{\partial \mathbf{U}}{\partial t} + u\frac{\partial \mathbf{U}}{\partial x} + \mathrm{v}\frac{\partial \mathbf{U}}{\partial y} - \nu\left(\frac{\partial^2 \mathbf{U}}{\partial x^2} + \frac{\partial^2 \mathbf{U}}{\partial y^2}\right) - \mathbf{f} = 0 \tag{4.6.4}$$

with

$$\mathbf{U} = \begin{bmatrix} u \\ \mathrm{v} \end{bmatrix}, \qquad \mathbf{f} = \begin{bmatrix} f_x \\ f_y \end{bmatrix}$$

Applying (4.6.3) to (4.6.4) yields

$$\frac{\partial \mathbf{U}}{\partial t} + \bar{u}\frac{\partial \mathbf{U}}{\partial \xi} + \bar{\mathrm{v}}\frac{\partial \mathbf{U}}{\partial \eta} - \nu\left[\frac{1}{|J|^2}\left(a\frac{\partial^2 \mathbf{U}}{\partial \xi^2} - 2b\frac{\partial^2 \mathbf{U}}{\partial \xi \partial \eta} + c\frac{\partial^2 \mathbf{U}}{\partial \eta^2}\right) + p\frac{\partial \mathbf{U}}{\partial \xi} + q\frac{\partial \mathbf{U}}{\partial \eta}\right] - f = 0 \tag{4.6.5}$$

where

$$\bar{u} = \frac{1}{|J|}\left(u\frac{\partial y}{\partial \eta} - \mathrm{v}\frac{\partial x}{\partial \eta}\right)$$

$$\bar{\mathrm{v}} = \frac{1}{|J|}\left(\mathrm{v}\frac{\partial x}{\partial \xi} - u\frac{\partial y}{\partial \xi}\right)$$

$$p = \frac{1}{|J|^3}\left[-\frac{\partial y}{\partial \eta}\left(a\frac{\partial^2 x}{\partial \xi^2} - 2b\frac{\partial^2 x}{\partial \xi \partial \eta} + c\frac{\partial^2 x}{\partial \eta^2}\right) + \frac{\partial x}{\partial \eta}\left(a\frac{\partial^2 y}{\partial \xi^2} - 2b\frac{\partial^2 y}{\partial \xi \partial \eta} + c\frac{\partial^2 y}{\partial \eta^2}\right)\right]$$

$$q = \frac{1}{|J|^3}\left[\frac{\partial y}{\partial \xi}\left(a\frac{\partial^2 x}{\partial \xi^2} - 2b\frac{\partial^2 x}{\partial \xi \partial \eta} + c\frac{\partial^2 x}{\partial \eta^2}\right) - \frac{\partial x}{\partial \xi}\left(a\frac{\partial^2 y}{\partial \xi^2} - 2b\frac{\partial^2 y}{\partial \xi \partial \eta} + c\frac{\partial^2 y}{\partial \eta^2}\right)\right]$$

$$a = \left(\frac{\partial x}{\partial \eta}\right)^2 + \left(\frac{\partial y}{\partial \eta}\right)^2$$

$$b = \frac{\partial x}{\partial \xi}\frac{\partial x}{\partial \eta} + \frac{\partial y}{\partial \xi}\frac{\partial y}{\partial \eta}$$

$$c = \left(\frac{\partial x}{\partial \xi}\right)^2 + \left(\frac{\partial y}{\partial \xi}\right)^2$$

4.6.2 APPLICATION OF NEUMANN BOUNDARY CONDITIONS

Neumann boundary conditions are applied in the transformed coordinates based on the same procedure described above. For example, let us consider the gradient of $\mathbf{U}$ with respect to η.

$$\frac{\partial \mathbf{U}}{\partial \eta} = \frac{\partial \mathbf{U}}{\partial x}\frac{\partial x}{\partial \eta} + \frac{\partial \mathbf{U}}{\partial y}\frac{\partial y}{\partial \eta} \tag{4.6.6}$$

Using a first order backward difference for $\frac{\partial \mathbf{U}}{\partial \eta}$, $\frac{\partial x}{\partial \eta}$ and $\frac{\partial y}{\partial \eta}$, we have

$$\frac{\mathbf{U}_{i,j} - \mathbf{U}_{i,j-1}}{\Delta \eta} = \frac{\partial \mathbf{U}}{\partial x}\frac{x_{i,j} - x_{i,j-1}}{\Delta \eta} + \frac{\partial \mathbf{U}}{\partial y}\frac{y_{i,j} - y_{i,j-1}}{\Delta \eta}$$

or

$$\mathbf{U}_{i,j} = \mathbf{U}_{i,j-1} + \frac{\partial \mathbf{U}}{\partial x}\Delta x + \frac{\partial \mathbf{U}}{\partial y}\Delta y \tag{4.6.7}$$

4.6.3 SOLUTION BY MACCORMACK METHOD

The transformed governing equations (4.6.5) may be solved using the MacCormack method as follows:

Predictor

$$\mathbf{U}^*_{i,j} = \mathbf{U}^n_{i,j} + \Delta t\left\{-\left(\bar{u}\frac{\partial \mathbf{U}}{\partial \xi} + \bar{v}\frac{\partial \mathbf{U}}{\partial \eta}\right)^n_{i,j} + \nu\Delta t\left[\frac{1}{J^2}\left(a\frac{\partial^2 \mathbf{U}}{\partial \xi^2} - 2b\frac{\partial^2 \mathbf{U}}{\partial \xi \partial \eta} + c\frac{\partial^2 \mathbf{U}}{\partial \eta^2}\right) + p\frac{\partial \mathbf{U}}{\partial \xi} + q\frac{\partial \mathbf{U}}{\partial \eta}\right]^n_{i,j} + \mathbf{f}^n_{i,j}\right\} \tag{4.6.8a}$$

Corrector

$$\mathbf{U}^{n+1}_{i,j} = \frac{1}{2}(\mathbf{U}^*_{i,j} + \mathbf{U}^n_{i,j}) + \frac{\Delta t}{2}\left[-\left(\bar{u}\frac{\partial \mathbf{U}}{\partial \xi} + \bar{v}\frac{\partial \mathbf{U}}{\partial \eta}\right)^*_{i,j}\right] + \frac{\nu\Delta t}{2}\left[\frac{1}{J^2}\left(a\frac{\partial^2 \mathbf{U}}{\partial \xi^2} - 2b\frac{\partial^2 \mathbf{U}}{\partial \xi \partial \eta} + c\frac{\partial^2 \mathbf{U}}{\partial \eta^2}\right) + p\frac{\partial \mathbf{U}}{\partial \xi} + q\frac{\partial \mathbf{U}}{\partial \eta}\right]^*_{i,j} + \frac{\Delta t}{2}\mathbf{f}^{n+1}_{i,j} \tag{4.6.8b}$$

It is now clear that the solution of the governing equation (4.6.4) is replaced by the solution of transformed equation (4.6.5) in which finite difference formulas of Chapter 3 can be used using the grid system of Figure 4.6.1b. This cumbersome procedure can be avoided if finite volume methods (Chapter 7) or finite element methods (Part Three) are used.

4.7 EXAMPLE PROBLEMS

The purpose of this chapter was to list or summarize the existing numerical schemes for later references in forthcoming chapters. Thus, examples shown in this section are limited to simple problems for the benefit of the uninitiated reader.

4.7.1 ELLIPTIC EQUATION (HEAT CONDUCTION)

In this example, we demonstrate the solution of steady state heat conduction,

$$\frac{\partial^2 T}{\partial x^2} + \frac{\partial^2 T}{\partial y^2} = 0$$

with the geometry and boundary conditions as shown in Figure 4.7.1.1a. The analytical

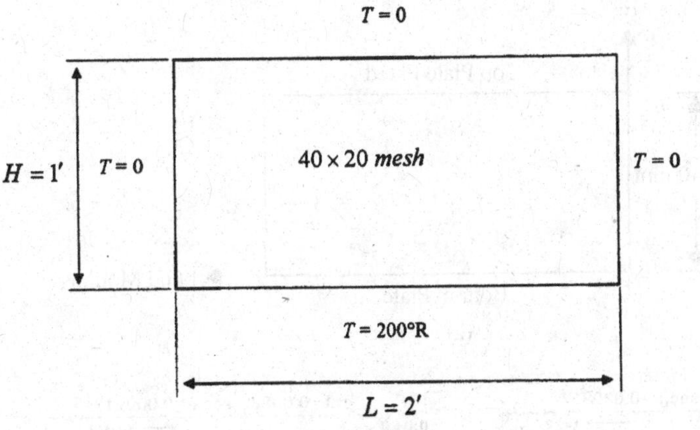

(a)

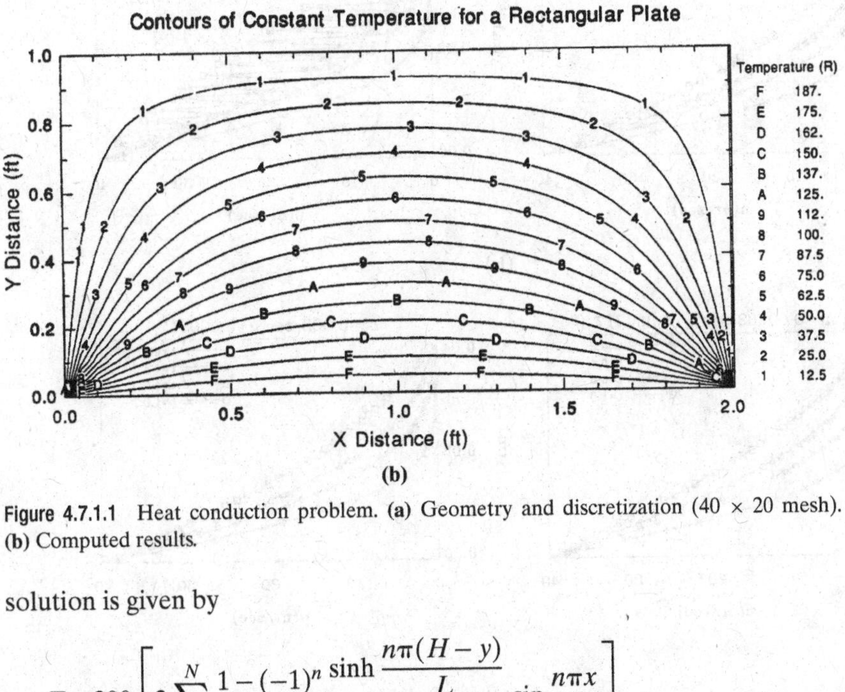

(b)

Figure 4.7.1.1 Heat conduction problem. **(a)** Geometry and discretization (40 × 20 mesh). **(b)** Computed results.

solution is given by

$$T = 200\left[2\sum_{n=1}^{N}\frac{1-(-1)^n}{n\pi}\frac{\sinh\dfrac{n\pi(H-y)}{L}}{\sinh\dfrac{n\pi H}{L}}\sin\frac{n\pi x}{L}\right]$$

Required: Solve using the point successive over-relaxation (PSOR).

Solution: The results for 40×20 mesh are shown in Figure 4.7.1.1b. The optimum relaxation parameter in this case is $\omega = 1.7$. The average error is approximately 0.5% as compared with the analytical solution ($N = 100$).

Remarks: For this simple problem, all methods introduced in this section will provide similar results.

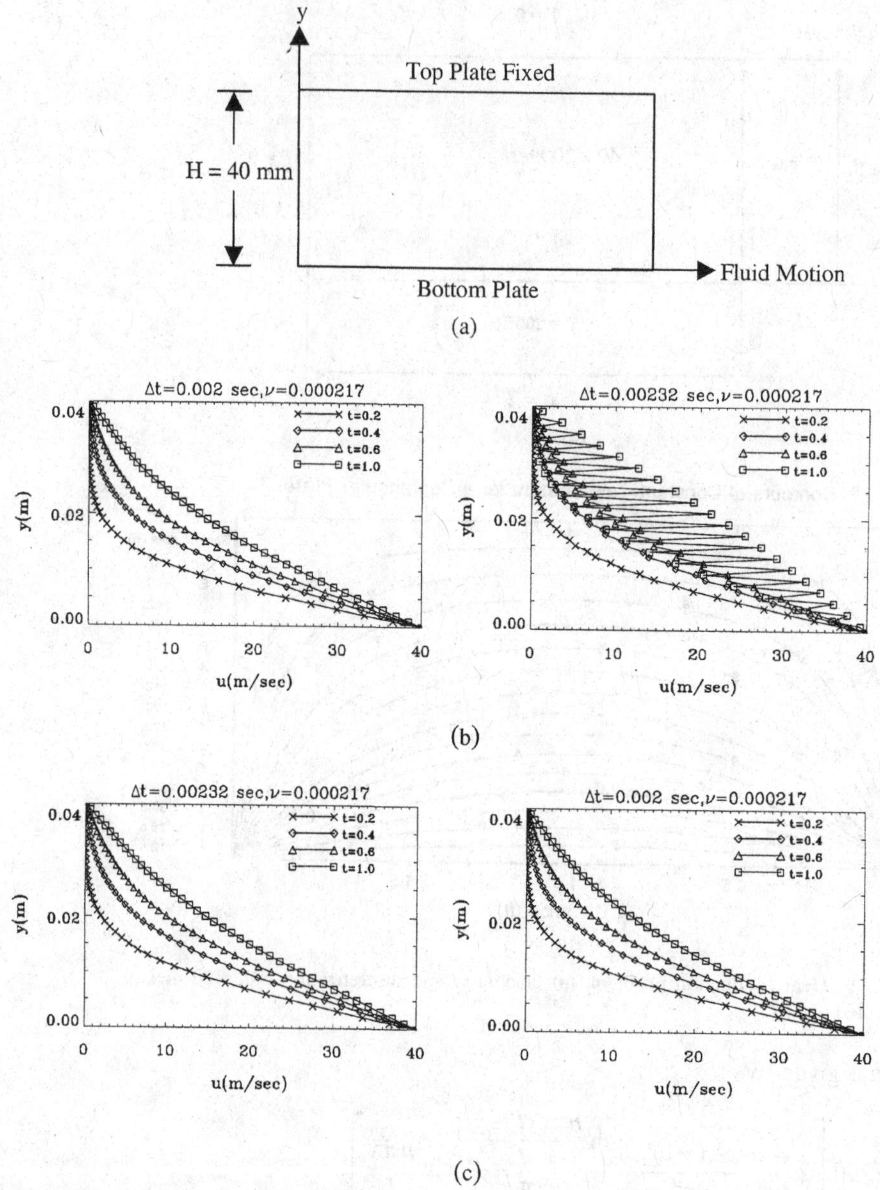

Figure 4.7.2.1 Couette flow. **(a)** Couette flow geometry. **(b)** Velocity profiles for FTCS explicit method (40 elements). **(c)** Velocity profiles for Crank-Nicolson method (40 elements).

4.7.2 PARABOLIC EQUATION (COUETTE FLOW)

Consider the Couette flow characterized by the parabolic equation,

$$\frac{\partial u}{\partial t} - \nu \frac{\partial^2 u}{\partial y^2} = 0, \quad \nu = 0.000217\,\text{m}^2/\text{s}$$

with the geometry given in Figure 4.7.2.1a and

Initial conditions at $t = 0$ $\begin{cases} u = u_0 = 40\,\text{m/s}, & y=0 \\ u=0, & 0<y\leq h \end{cases}$

Boundary conditions at $t > 0$ $\begin{cases} u = u_o = 40\,\text{m/s}, & y=0 \\ u=0, & y=h \end{cases}$

Required: Solve by FTCS and Crank-Nicolson methods with the initial and boundary conditions as shown below.

Solution: The results are shown in Figure 4.7.2.1b. As expected, FTCS for $d = .5034 > 1/2$ is unstable whereas Crank-Nicolson gives stable results regardless of diffusion number ranges.

4.7.3 HYPERBOLIC EQUATION (FIRST ORDER WAVE EQUATION)

The governing equation is given by

$$\frac{\partial u}{\partial t} + a\frac{\partial u}{\partial x} = 0, \qquad a = 300\frac{m}{s}$$

with

Initial conditions at $t=0$ $\begin{cases} u(x)=0 & 0\leq x\leq 50 \\ u(x)=100\sin\pi\dfrac{(x-50)}{60} & 50\leq x\leq 110 \\ u(x)=0 & 110\leq x\leq 300 \end{cases}$

Boundary conditions at $t>0$ $\begin{cases} u(x)=0 & x=0 \\ u(x)=0 & x=L \end{cases}$

Explicit Schemes

Required: Solve by explicit schemes, (a) first order upwind scheme (FTBS), (b) Lax-Wendroff scheme, and FTCS implicit scheme.

$\Delta x = 5$, $\Delta t = 0.01666$ $(C = 0.9996)$ (CFL number)
$\Delta x = 5$, $\Delta t = 0.015$ $(C = 0.9)$
$\Delta x = 5$ $\Delta t = 0.0075$ $(C = 0.45)$

Solution: The results are as shown in Figure 4.7.3.1. Note that the exact solution is obtained for both methods for $C = 1$. However, as C decreases, FTBS becomes dissipative, whereas the Lax-Wendroff scheme (second order accurate) becomes dispersive.

Implicit Schemes

Required: Solve by implicit scheme (FTCS).

Solution: The results are shown in Figure 4.7.3.2. This scheme is very dissipative at high C values. Although unconditionally stable, the results are poor, particularly with large time steps (large Courant number).

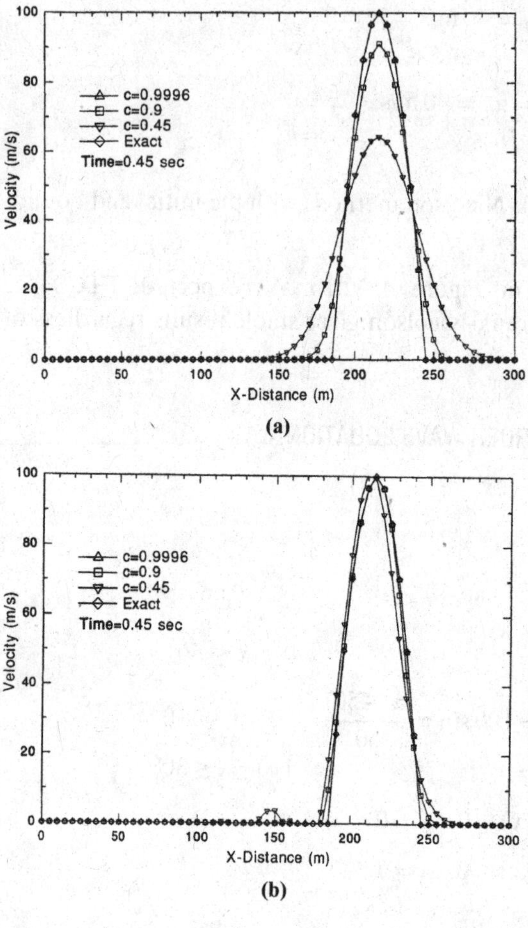

Figure 4.7.3.1 Solutions of first order wave equation by FTBS and Lax-Wendroff schemes, 60 nodes. **(a)** First order upwind (FTBS). **(b)** Lax-Wendroff scheme.

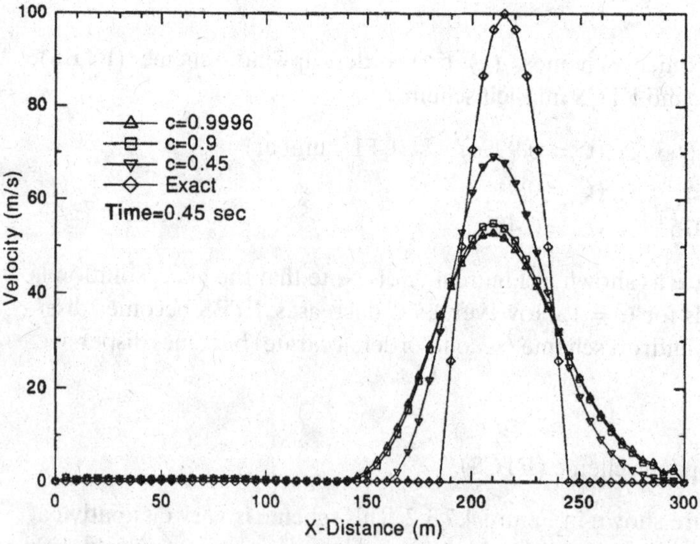

Figure 4.7.3.2 Solution of first order wave equation by FTCS implicit scheme.

4.7.4 HYPERBOLIC EQUATION (SECOND ORDER WAVE EQUATION)

The second order wave equation is considered in this example.

$$\frac{\partial^2 u}{\partial t^2} = a^2 \frac{\partial^2 u}{\partial x^2}$$

Two sets of initial conditions are required:

Initial conditions

$$\text{(a) at } t = 0 \quad \begin{cases} u(x) = 0 & 0 \leq x \leq 100 \\ u(x) = 100 \sin\left[\dfrac{\pi(x - 100)}{120}\right] & 100 \leq x \leq 220 \\ u(x) = 0 & 220 \leq x \leq 300 \end{cases}$$

$$\text{(b) at } t = 0 \quad \frac{\partial u(x)}{\partial t} = 0$$

Boundary conditions

$$t = 0 \quad \begin{cases} u(x) = 0 & x = 0 \\ u(x) = 0 & x = L \end{cases}$$

Required: Solve by the midpoint leapfrog scheme.

Solution: The results (Figure 4.7.4.1) are obtained at $t = 0.28$ seconds. The best solution occurs for $C = 1$. Note that dispersion errors occur for C less than 1.

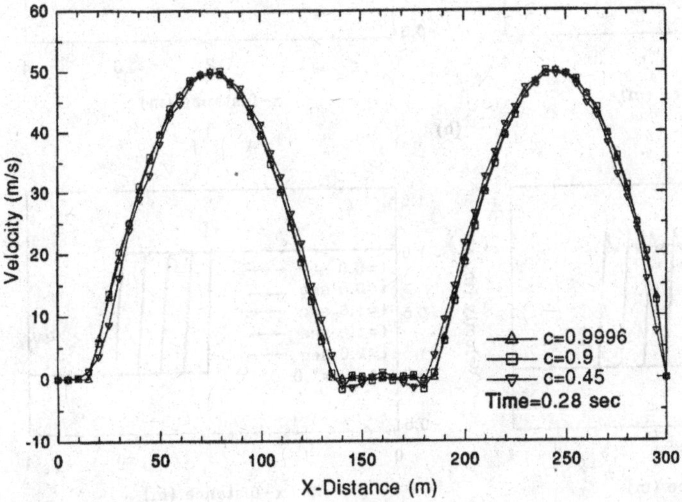

Figure 4.7.4.1 Solution of second order wave equation by midpoint leapfrog scheme, 60 nodes.

4.7.5 NONLINEAR WAVE EQUATION

Consider the nonlinear wave equation in the form

$$\frac{\partial u}{\partial t} + u\frac{\partial u}{\partial x} = 0$$

or

$$\frac{\partial u}{\partial t} + \frac{\partial F}{\partial x} = 0 \quad \text{with } F = \frac{1}{2}u^2$$

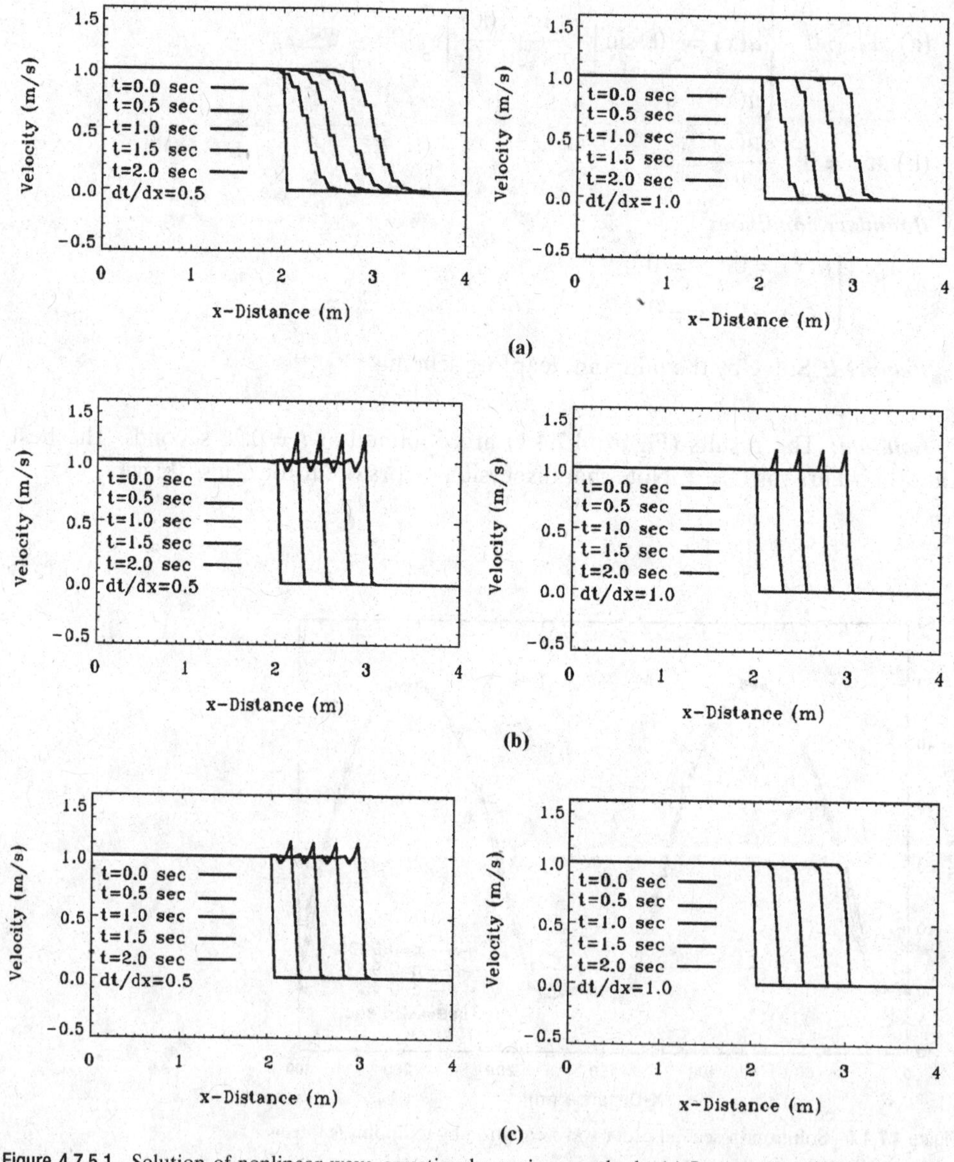

Figure 4.7.5.1 Solution of nonlinear wave equation by various methods. **(a)** Lax method (80 elements). **(b)** Lax-Wendroff method (80 elements). **(c)** MacCormack method (80 elements).

The following initial and boundary conditions are to be used:

$$u(x,0) = 1 \quad 0 \leq x \leq 2$$

$$u(x,0) = 0 \quad 2 \leq x \leq 4$$

Required: Solve by (a) Lax method, (b) Lax-Wendroff method, and (c) MacCormack method.

Solution: The results are obtained with $\Delta t/\Delta x = 1$ and $\Delta t/\Delta x = 0.5$. Referring to Figure 4.7.5.1, the Lax method is dissipative, whereas the Lax-Wendroff method is dispersive. This trend is worse when the Courant number is smaller. The MacCormack method gives better results particularly with Courant number near 1. It is still dispersive at lower Courant number, but better than the Lax-Wendroff scheme.

4.8 SUMMARY

In this chapter, FDM schemes for typical elliptic, parabolic, and hyperbolic partial differential equations and Burgers' equation have been presented. These equations do not represent complete fluid dynamics phenomena, but the computational schemes described herein do constitute the basis for computations involved in incompressible and compressible flows. Concepts of explicit and implicit schemes with von Neumann stability analyses are expected to play significant roles in all aspects of computational methods in fluid dynamics and heat transfer.

Although most of the computational schemes for FDM presented in this chapter are in terms of one-dimensional applications, their extensions to multidimensions including noncartesian orientations of physical domain can be accomplished by transformation into the cartesian computational domain.

In practical applications, most physical phenomena in fluid mechanics and heat transfer are multidimensional. Thus, significant modifications and improvements over the simple approaches introduced in this chapter are required in dealing with incompressible and compressible flows, which are the subjects of the subsequent chapters.

REFERENCES

Courant, R., Friedrichs, K. O., and Lewy, H. [1962]. On the partial differential equations of mathematical physics. *IBM J. Res. Dev.* 11, 215–24.

Demel, J. W. [1997]. *Applied Numerical Linear Algebra*. Philadelphia, PA: SIAM.

Dahlquist, G. and Bjork, A. [1974]. *Numerical Methods*. Englewood Cliffs, NJ: Prentice-Hall.

Hageman, P. and Young, D. M. [1981]. *Applied Iterative Methods*. New York: Academic Press.

Ortega, J. M. and Rheinboldt, W. C. [1970]. *Iterative Solutions of Non-Linear Equations in Several Variables*. London: Wiley.

Saad, Y. [1996]. *Iterative Methods for Sparse Linear Systems*. Boston: PWS Publishing.

Thomas, L. H. [1949]. Elliptic problems in linear differential equations over a network. Watson Sci. Comp. Lab., Columbia University, NY.

Varga, R. S. [1962]. *Matrix Iterative Analysis*. Englewood Cliffs, NJ: Prentice-Hall.

Wachspress, E. L. [1966]. *Iterative Solution of Elliptic Systems*. Englewood Cliffs, NJ: Prentice-Hall.

CHAPTER FIVE

Incompressible Viscous Flows via Finite Difference Methods

5.1 GENERAL

The basic concepts in FDM and applications to simple partial differential equations have been presented in the previous chapters. This chapter will focus on incompressible viscous flows in which the physical property of the fluid, *incompressibility*, requires substantial modifications of computational schemes discussed in Chapter 4.

In general, a flow becomes incompressible for low speeds, that is, $M < 0.3$ for air, and compressible for higher speeds, that is, $M \geq 0.3$, although the effect of compressibility may appear at the Mach number as low as 0.1, depending on pressure and density changes relative to the local speed of sound. Computational schemes are then dictated by various physical conditions: viscosity, incompressibility, and compressibility of the flow. The so-called pressure-based formulation is used for incompressible flows to keep the pressure field from oscillating, which may arise due to difficulties in preserving the conservation of mass or *incompressibility condition* as the sound speed becomes so much higher than convection velocity components. The pressure-based formulation for incompressible flows uses the primitive variables (p, v_i, T), whereas the density-based formulation applicable for compressible flows utilizes the conservation variables (ρ, ρv_i, ρE).

Incompressible viscous flows are usually computed by means of the continuity and momentum equations. If temperature changes in natural and/or forced convection heat transfer are considered, then the energy equation is also added. For simplicity in demonstrating the computational strategies for incompressible flows in general, we shall consider only the isothermal case in this chapter. In Chapter 6, it will be shown that computational schemes for incompressible flows can also be developed from preconditioning processes of the density-based formulation which is originally intended for compressible flows. This process leads to implementations of an algorithm applicable for both compressible and incompressible flows [Merkle et al., 1998].

In dealing with incompressible flows, there are two approaches: primitive variable methods and vortex methods. The primitive variable approach includes the artificial compressibility method (ACM) [Chorin, 1967], and the pressure correction methods (PCM) including the marker and cell (MAC) method [Harlow and Welch, 1965], the semi-implicit method for pressure linked equations (SIMPLE) [Patankar and Spalding, 1972], and the pressure implicit with splitting of operators (PISO) [Issa, 1985]. The main difficulty in incompressible flows is the accurate solution for pressure. Thus, the purpose of the vortex methods is to remove the pressure terms from the momentum equations, which can be achieved by solving the vorticity transport equation(s) (one scalar equation for 2-D and three vector component equations for 3-D).

In view of the fact that the transition between incompressible and compressible flows involves a complex process of interactions between inviscid and viscous properties, it is reasonable to seek a unified approach in which both incompressible and compressible flows can be accommodated. This subject will be discussed in Section 6.4, Preconditioning Process for Compressible Flows and Viscous Flows, and in Section 6.5 on the flowfield-dependent variation (FDV) methods. For this reason, treatments of incompressible flows in this chapter will be brief.

5.2 ARTIFICIAL COMPRESSIBILITY METHOD

The governing equations for incompressible viscous flows, known as the incompressible Navier-Stokes system of equations, are written in nondimensionalized form as

Continuity

$$v_{i,i} = 0 \tag{5.2.1}$$

Momentum

$$\frac{\partial v_i}{\partial t} + v_{i,j}v_j = -p_{,i} + \frac{1}{Re}v_{i,jj} \tag{5.2.2}$$

where the following nondimensional quantities are used:

$$v_i = \frac{v_i^*}{v_\infty}, \quad x_i = \frac{x_i^*}{L}, \quad p = \frac{p^*}{\rho v_\infty^2}, \quad t = \frac{t^* v_\infty}{L}, \quad Re = \frac{v_\infty L}{\nu}$$

with asterisks implying the physical variable and *Re* being the Reynolds number.

In the artificial compressibility method (ACM), the continuity equation is modified to include an artificial compressibility term which vanishes when the steady state is reached [Chorin, 1967]:

$$\frac{\partial \tilde{\rho}}{\partial \tilde{t}} + v_{i,i} = 0 \tag{5.2.3}$$

where $\tilde{\rho}$ is an artificial density, equated to the product of artificial compressibility factor β and pressure,

$$\tilde{\rho} = \beta^{-1} p \tag{5.2.4}$$

Here $\frac{\partial \tilde{\rho}}{\partial \tilde{t}} \to 0$ at the steady state and $\tilde{t}$ is a fictitious time.

With these definitions and combining (5.2.1–5.2.4), we may write the incompressible Navier-Stokes system of equations in the form

$$\frac{\partial \mathbf{W}}{\partial t} + \mathbf{A}_i \frac{\partial \mathbf{W}}{\partial x_i} = \frac{1}{Re}\frac{\partial}{\partial x_i}\left(\mathbf{B}_{ij}\frac{\partial \mathbf{W}}{\partial x_j}\right) \tag{5.2.5}$$

with

$$\mathbf{W} = \begin{bmatrix} p \\ \mathrm{v}_j \end{bmatrix}, \qquad \mathbf{A}_i = \frac{\partial \mathbf{D}_i}{\partial \mathbf{W}}, \qquad \mathbf{D}_i = \begin{bmatrix} \beta \mathrm{v}_i \\ \mathrm{v}_i \mathrm{v}_j + p\delta_{ij} \end{bmatrix}, \qquad \mathbf{B}_{ij} = \begin{bmatrix} 0 & \\ & \delta_{ij} \end{bmatrix}$$

$$\mathbf{A}_1 = \frac{\partial \mathbf{D}_1}{\partial \mathbf{W}} = \begin{bmatrix} 0 & \beta & 0 & 0 \\ 1 & 2u & 0 & 0 \\ 0 & \mathrm{v} & u & 0 \\ 0 & w & 0 & u \end{bmatrix} \qquad \mathbf{A}_2 = \frac{\partial \mathbf{D}_2}{\partial \mathbf{W}} = \begin{bmatrix} 0 & 0 & \beta & 0 \\ 0 & \mathrm{v} & u & 0 \\ 1 & 0 & 2\mathrm{v} & 0 \\ 0 & 0 & w & \mathrm{v} \end{bmatrix}$$

$$\mathbf{A}_3 = \frac{\partial \mathbf{D}_3}{\partial \mathbf{W}} = \begin{bmatrix} 0 & 0 & 0 & \beta \\ 0 & w & 0 & u \\ 0 & 0 & w & \mathrm{v} \\ 1 & 0 & 0 & 2w \end{bmatrix}$$

Let us now investigate the eigenvalues of $\mathbf{A}_i$,

$$|\mathbf{A}_i - \lambda_i \mathbf{I}| = 0$$

where the eigenvalues of $\mathbf{A}_i$ ($i = 1, 2, 3$) are, respectively,

$$(u, u, u \pm \sqrt{u^2 + \beta}), \qquad (\mathrm{v}, \mathrm{v}, \mathrm{v} \pm \sqrt{\mathrm{v}^2 + \beta}), \qquad (w, w, w \pm \sqrt{w^2 + \beta}) \tag{5.2.6}$$

in which $\sqrt{\beta}$ is the artificial speed of sound (often called the artificial compressibility factor) with β being chosen adequately (between 0.1 and 10 as suggested by Kwak et al. [1986]). The idea is to maintain low enough β (close to the convective velocity) to overcome stiffness associated with a disparity in the magnitudes of the eigenvalues, but high enough such that pressure waves (moving with infinite speed at incompressible limit) be allowed to travel far enough to balance viscous effects. As a result, the conservation of mass or incompressibility condition is assured by means of an artificial compressibility. In this process, it is possible to obtain the correct pressure distributions. The solution of (5.2.5) is usually obtained by the Crank-Nicolson method.

From the point of view of linear algebra, the finite difference algebraic equations resulting from (5.2.5) are well conditioned (with a proper choice of β), as compared to the original equations (5.2.1) and (5.2.2). This is due to the well-conditioned eigenvalues given by (5.2.6). All other solution schemes for incompressible flows without using the artificial compressibility must employ special approaches as discussed below.

5.3 PRESSURE CORRECTION METHODS

5.3.1 SEMI-IMPLICIT METHOD FOR PRESSURE-LINKED EQUATIONS (SIMPLE)

It is well known that, if the finite difference equation is written in control volume grids (Section 1.4) for continuity $\mathrm{v}_{i,i} = 0$, this will lead to nonphysical, checkerboard-type oscillations of velocity in each one-dimensional direction (same values repeated at every other node, assuming that the velocity distribution between the adjacent nodes is linear). As a consequence, the mass is not conserved, thus causing the pressure to undergo similar oscillations. This is particularly true when pressure becomes constant ($p_{,i} = 0$) for the same reason as $\mathrm{v}_{i,i} = 0$. These difficulties can be shown to be remedied by using staggered grids [velocity nodes staggered with respect to pressure nodes (Figure 5.3.1)]

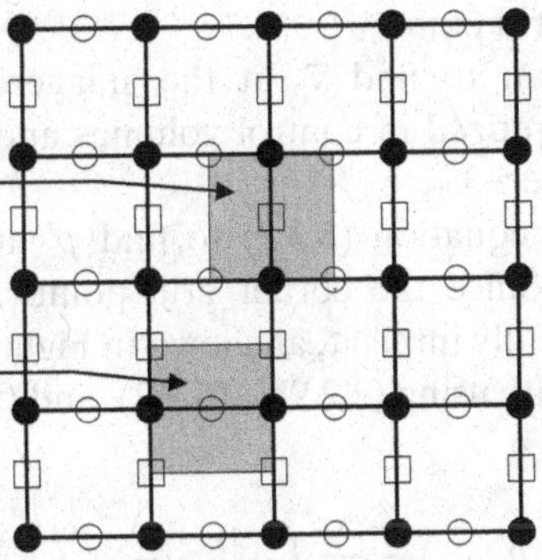

Figure 5.3.1 Computational domain for staggered grid.

in the algorithm known as SIMPLE [Patankar and Spalding, 1972]. In this method, the predictor-corrector procedure with successive pressure correction steps is used:

$$p = \overline{p} + p' \tag{5.3.1}$$

where p is the actual pressure, $\overline{p}$ is the estimated pressure, and p' is the pressure correction. Likewise, the actual velocity components in two-dimensions are

$$u = \overline{u} + u' \tag{5.3.2a}$$

$$\mathrm{v} = \overline{\mathrm{v}} + \mathrm{v}' \tag{5.3.2b}$$

The pressure corrections are related to the velocity corrections by approximate momentum equations,

$$\rho \frac{\partial u'}{\partial t} = -\frac{\partial p'}{\partial x} \tag{5.3.3a}$$

$$\rho \frac{\partial \mathrm{v}'}{\partial t} = -\frac{\partial p'}{\partial y} \tag{5.3.3b}$$

or

$$u' = -\frac{\Delta t}{\rho}\frac{\partial p'}{\partial x} \tag{5.3.4a}$$

$$\mathrm{v}' = -\frac{\Delta t}{\rho}\frac{\partial p'}{\partial y} \tag{5.3.4b}$$

Combining (5.3.2) and (5.3.4) and substituting the result into the continuity equation, we obtain the so-called pressure-correction Poisson equation of an elliptic form,

$$p'_{,ii} = -\frac{\rho}{\Delta t}\left(\frac{\partial \mathrm{v}_i}{\partial x_i} - \frac{\partial \overline{\mathrm{v}}_i}{\partial x_i}\right) = \frac{\rho}{\Delta t}\frac{\partial \overline{\mathrm{v}}_i}{\partial x_i} \qquad (i = 1, 2) \tag{5.3.5}$$

where we set $\frac{\rho}{\Delta t}\frac{\partial \mathrm{v}_i}{\partial x_i} = 0$ to enforce the mass conservation at the current iteration step. An iterative procedure is used to obtain a solution as follows [Raithby and Schneider, 1979].

(a) Guess the pressure $\overline{p}$ at each grid point.
(b) Solve the momentum equation to find $\overline{\mathrm{v}}_i$ at the staggered grid ($i+1/2$, $i-1/2$, $j+1/2$, $j-1/2$), discretized in control volumes and control surfaces (Section 1.4) as shown in Figure 5.3.1.
(c) Solve the pressure correction equation (5.3.5) to find p' at $(i, j), (i, j-1)$, $(i, j+1), (i-1, j), (i+1, j)$. Since the corner grid points are avoided, the scheme is "semi-implicit," not fully implicit, as shown in Figure 5.3.1.
(d) Correct the pressure and velocity using (2.2.9b), (5.3.2), and (5.3.4).

$$p = \overline{p} + p'$$

$$u = \overline{u} - \frac{\Delta t}{2\rho\,\Delta x}(p'_{i+1,j} - p'_{i-1,j}) - \frac{\Delta t}{\rho}\left(A^{(1)}_{i+\frac{1}{2},j} - A^{(1)}_{i-\frac{1}{2},j}\right) \tag{5.3.6}$$

$$\mathrm{v} = \overline{\mathrm{v}} - \frac{\Delta t}{2\rho\,\Delta y}(p'_{i,j+1} - p'_{i,j-1}) - \frac{\Delta t}{\rho}\left(A^{(2)}_{i,j+\frac{1}{2}} - A^{(2)}_{i,j-\frac{1}{2}}\right)$$

where

$$A^{(1)} = (\rho \mathrm{v}'_k \mathrm{v}'_1)_{,k} - \mu\left(\mathrm{v}'_{1,kk} + \frac{1}{3}\mathrm{v}'_{k,k1}\right) \quad (k = 1, 2)$$

$$A^{(2)} = (\rho \mathrm{v}'_k \mathrm{v}'_2)_{,k} - \mu\left(\mathrm{v}'_{2,kk} + \frac{1}{3}\mathrm{v}'_{k,k2}\right) \quad (k = 1, 2)$$

with μ being the dynamic viscosity.
(e) Replace the previous intermediate values of pressure and velocity $(\overline{p}, \overline{\mathrm{v}}_i)$ with the new corrector values (p, v_i) and return to (b).
(f) Repeat Steps (b) through (e) until convergence.

Often the convergence of the above process is not satisfactory because of the tendency for overestimation of p'. A remedy to this difficulty may be found by the use of under-relaxation parameter α,

$$p = \overline{p} + \alpha p' \tag{5.3.7}$$

However, in many cases a proper choice of α is not easy ($\alpha \cong 0.8$ is often used). Thus, a further corrective measure is to use SIMPLER (SIMPLE revised) in which a complete Poisson equation is used for pressure corrections.

$$\nabla^2 p = -\rho(\mathrm{v}_{i,j}\mathrm{v}_j)_{,i} \tag{5.3.8a}$$

or

$$\nabla^2 p = 2\rho\left(\frac{\partial u}{\partial x}\frac{\partial \mathrm{v}}{\partial y} - \frac{\partial \mathrm{v}}{\partial x}\frac{\partial u}{\partial y}\right) \tag{5.3.8b}$$

Here u and v will be replaced by (5.3.2) and subsequently (5.3.5) replaced by (5.3.8).

Instead of using the time-dependent formulation described above, it is convenient to use a steady state approach with finite volume discretizations as shown in Figure 5.3.2.

$$a_p\phi_p = \alpha\left(\sum a_{nb}\phi_{nb} + b\right) + (1-\alpha)a_p\phi_p^0 \tag{5.3.9}$$

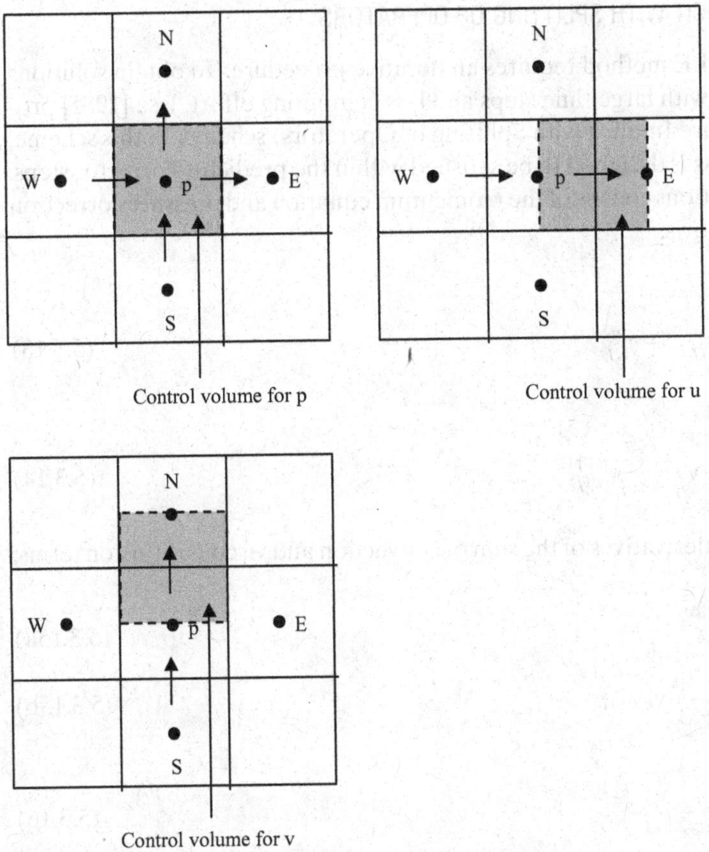

Figure 5.3.2 Computational domain for steady-state problems.

where ϕ is any conservation variable and α is the under-relaxation parameter with the subscripts p and nb denoting the node under consideration and neighbor contributions, respectively.

Convergence of (5.3.9) may be improved using SIMPLEC proposed by Van Doormaal and Raithby [1984] in which more "consistent" approximation of (5.3.9) is implemented:

$$\left(a_e - \sum a_{nb}\right) u'_e = A_e(p'_p - p'_E) \tag{5.3.10}$$

with

$$u_e = u'_e + d_e(p'_p - p'_E) \tag{5.3.11}$$

$$u'_e = \frac{\sum a_{nb}u_{nb} + b}{a_e}, \qquad d_e = \frac{A_e}{a_e - \sum a_{nb}} \tag{5.3.12}$$

Examples of computations reported by Van Doormaal and Raithby [1984] show that SIMPLEC is most effective, followed by SIMPLER and SIMPLE.

5.3.2 PRESSURE IMPLICIT WITH SPLITTING OF OPERATORS

We note that the SIMPLE method requires an iterative procedure. To obtain solutions without iterations, and with large time steps and less computing effort, Issa [1985] proposed the PISO (Pressure Implicit with Splitting of Operators) scheme. In this scheme, the conservation of mass is designed to be satisfied within the predictor-corrector steps.

The governing equations consist of the momentum equation and pressure correction equation written as follows:

Momentum

$$\frac{\rho}{\Delta t}(v_j^{n+1} - v_j^n) = -s_{ij,i}^{n+1} - p_{,j}^{n+1} \tag{5.3.13}$$

Pressure Corrector

$$p_{,jj}^{n+1} = -\frac{\rho}{\Delta t}(v_{j,j}^{n+1} - v_{j,j}^n) - s_{ij,ij}^{n+1} \tag{5.3.14}$$

where $s_{ij,ij}$ refers to the derivatives of the sum of convection and viscous diffusion terms, $s_{ij,i}$.

$$s_{ij,i} = (\rho v_i v_j)_{,i} - \tau_{ij,i} \tag{5.3.15a}$$

$$\tau_{ij} = \mu(v_{i,j} + v_{j,i}) - \frac{2\mu}{3} v_{k,k}\delta_{ij} \tag{5.3.15b}$$

(a) Predictor

$$\frac{\rho}{\Delta t}(v_j^* - v_j^n) = -s_{ij,i}^* - p_{,j}^n \tag{5.3.16}$$

(b) Corrector I

$$p_{,jj}^* = -\frac{\rho}{\Delta t}(v_{j,j}^* - v_{j,j}^n) - s_{ij,ij}^* = \frac{\rho}{\Delta t} v_{j,j}^n - s_{ij,ij}^* \tag{5.3.17}$$

$$\frac{\rho}{\Delta t}(v_j^{**} - v_j^n) = -s_{ij,i}^* - p_{,j}^* \tag{5.3.18}$$

with $v_{j,j}^*$ set equal to zero in (5.3.17) in order to enforce the conservation of mass.

(c) Corrector II

$$p_{,jj}^{**} = \frac{\rho}{\Delta t} v_{j,j}^n - s_{ij,ij}^{**} \tag{5.3.19}$$

$$\frac{\rho}{\Delta t}(v_j^{***} - v_j^n) = -s_{ij,i}^{**} - p_{,j}^{**} \tag{5.3.20}$$

with $v_{j,j}^{**} = 0$ being once again enforced in (5.3.19). Thus, in the above process, there are no iterative steps involved.

In order to increase stability and accuracy, we may split $s_{ij,i}$ into diagonal and non-diagonal terms.

$$s_{ij,i} = s_{ij,i}^{(D)} + s_{ij,i}^{(N)} = A_{ji}^{(D)} v_i + s_{ij,i}^{(N)} \tag{5.3.21}$$

To illustrate this splitting of diagonal term, consider a one-dimensional case

$$s_{ij,i} \Rightarrow \frac{\partial}{\partial x}\left(\rho u\phi - k\frac{\partial \phi}{\partial x}\right) \tag{5.3.22}$$

or

$$s_{ij,i} \Rightarrow \frac{(\rho u\phi)_{i+1} - (\rho u\phi)_{i-1}}{2\Delta x} - \frac{(k\phi)_{i+1} - 2(k\phi)_i + (k\phi)_{i-1}}{\Delta x^2}$$

$$\Rightarrow \frac{1}{\Delta x_i}\left[(\rho u\phi)_{i+\frac{1}{2}} - (\rho u\phi)_{i-\frac{1}{2}} - \frac{k_{i+\frac{1}{2}}}{\Delta x_{i+\frac{1}{2}}}(\phi_{i+1} - \phi_i) + \frac{k_{i-\frac{1}{2}}}{\Delta x_{i-\frac{1}{2}}}(\phi_i - \phi_{i-1})\right]$$

Construct an upwind scheme to get

$$(\rho u\phi)_{i+\frac{1}{2}} - (\rho u\phi)_{i-\frac{1}{2}} \Rightarrow \begin{cases} (\rho u)_{i+\frac{1}{2}}\phi_i - (\rho u)_{i-\frac{1}{2}}\phi_{i-1} & \text{for } (+u) \\ (\rho u)_{i+\frac{1}{2}}\phi_{i+1} - (\rho u)_{i-\frac{1}{2}}\phi_i & \text{for } (-u) \end{cases}$$

Then we arrive at

$$\left.\begin{aligned} (\rho u\phi)_{i+\frac{1}{2}} &= (\rho u)^+_{i+\frac{1}{2}}\phi_i + (\rho u)^-_{i+\frac{1}{2}}\phi_{i+1} \\ (\rho u\phi)_{i-\frac{1}{2}} &= (\rho u)^+_{i-\frac{1}{2}}\phi_{i-1} + (\rho u)^-_{i-\frac{1}{2}}\phi_i \end{aligned}\right\} \quad \text{with } (\rho u)^{\pm} = \frac{1}{2}(\rho u \pm |\rho u|)$$

Thus $s_{ij,i}$ can be written as

$$s_{ij,i} \Rightarrow \frac{1}{\Delta x_i}(\alpha\phi_{i+1} + \beta\phi_i + \gamma\phi_{i-1}) \tag{5.3.23}$$

where

$$\alpha = (\rho u)^-_{i+\frac{1}{2}} - \frac{k_{i+\frac{1}{2}}}{\Delta x_{i+\frac{1}{2}}}, \qquad \beta = (\rho u)^+_{i+\frac{1}{2}} - (\rho u)^-_{i-\frac{1}{2}} + \frac{k_{i+\frac{1}{2}}}{\Delta x_{i+\frac{1}{2}}} + \frac{k_{i-\frac{1}{2}}}{\Delta x_{i-\frac{1}{2}}},$$

$$\gamma = -(\rho u)^+_{i-\frac{1}{2}} - \frac{k_{i-\frac{1}{2}}}{\Delta x_{i-\frac{1}{2}}}$$

Rewriting (5.3.23), we have

$$\begin{bmatrix} \beta & \alpha & & \\ \gamma & \beta & \alpha & \\ & \gamma & \beta & \alpha \\ & & \gamma & \beta \end{bmatrix}\begin{bmatrix} \phi_1 \\ \phi_2 \\ \phi_3 \\ \phi_4 \end{bmatrix} = \begin{bmatrix} \beta & 0 & 0 & 0 \\ 0 & \beta & 0 & 0 \\ 0 & 0 & \beta & 0 \\ 0 & 0 & 0 & \beta \end{bmatrix}\begin{bmatrix} \phi_1 \\ \phi_2 \\ \phi_3 \\ \phi_4 \end{bmatrix} + \begin{bmatrix} 0 & \alpha & 0 & 0 \\ \gamma & 0 & \alpha & 0 \\ 0 & \gamma & 0 & \alpha \\ 0 & 0 & \gamma & 0 \end{bmatrix}\begin{bmatrix} \phi_1 \\ \phi_2 \\ \phi_3 \\ \phi_4 \end{bmatrix} \tag{5.3.24}$$

or for multidimensions, we write (5.3.24) as

$$s_{ij,i} = A^{(D)}_{ji}\mathrm{v}^*_i + s^{(N)}_{ij,i} \tag{5.3.25}$$

Note that $s_{ij,i}$ is diagonally dominant for low Mach number flows,

$$(\rho u)_{i+\frac{1}{2}} > (\rho u)_{i-\frac{1}{2}}$$

or

$$\beta > |\alpha| + |\gamma|$$

If Mach number increases (high speed or compressible flow), then $(\rho u)_{i-\frac{1}{2}} > (\rho u)_{i+\frac{1}{2}} > 0$, or

$$\beta < |\alpha| + |\gamma|.$$

This implies that the diagonal dominance diminishes at high speed or compressible flows. We discuss a remedy for this problem in Section 6.3.3 on the PISO scheme for compressible flows.

With the splitting of $s_{ij,i}$ into the diagonal and nondiagonal parts, we proceed as follows:

(a) Predictor

$$\left(\frac{\rho}{\Delta t}\delta_{ij} + A_{ji}^{(D)}\right)\mathrm{v}_i^* = -s_{ij,i}^{*(N)} - p_{,j}^n + \frac{\rho}{\Delta t}\mathrm{v}_j^n \tag{5.3.26}$$

(b) Corrector I

$$\left(\frac{\rho}{\Delta t}\delta_{ij} + A_{ji}^{(D)}\right)(\mathrm{v}_i^{**} - \mathrm{v}_i^*) = -(p_{,j}^* - p_{,j}^n) \tag{5.3.27}$$

$$\left[\left(\frac{\rho}{\Delta t}\delta_{ij} + A_{ji}^{(D)}\right)^{-1}(p^* - p^n)_{,j}\right]_{,i} = \mathrm{v}_{i,i}^* \tag{5.3.28}$$

Solve $(p^* - p^n)$ and insert the result into (5.3.26) to obtain new v_i^{**}.

(c) Corrector II

$$\left(\frac{\rho}{\Delta t}\delta_{ij} + A_{ji}^{(D)}\right)\mathrm{v}_i^{**} - \frac{\rho}{\Delta t}\mathrm{v}_i^n = -s_{ij,i}^{*(N)} - p_{,j}^* \tag{5.3.29}$$

$$\left(\frac{\rho}{\Delta t}\delta_{ij} + A_{ji}^{(D)}\right)\mathrm{v}_i^{***} - \frac{\rho}{\Delta t}\mathrm{v}_j^n = -s_{ij,i}^{**(N)} - p_{,j}^{**} \tag{5.3.30}$$

Subtracting (5.3.29) from (5.3.30), we obtain

$$\left(\frac{\rho}{\Delta t}\delta_{ij} + A_{ji}^{(D)}\right)(\mathrm{v}_i^{***} - \mathrm{v}_i^{**}) = -\left(s_{ij,i}^{**(N)} - s_{ij,i}^{*(N)}\right) - (p^{**} - p^*)_{,j} \tag{5.3.31}$$

For $\mathrm{v}_{i,i}^{***} = 0$, we must have

$$\left[\left(\frac{\rho}{\Delta t}\delta_{ij} + A_{ji}^{(D)}\right)^{-1}(p^{**} - p^*)_{,i}\right]_{,j} = -\left(\frac{\rho}{\Delta t}\delta_{ik} + A_{ki}^{(D)}\right)^{-1}\left(s_{ij,k}^{**(N)} - s_{ij,k}^{*(N)}\right)_{,j} + \mathrm{v}_{i,i}^{**} \tag{5.3.32}$$

Solution of (5.3.32) leads to

$$\mathrm{v}_i^{***} = \mathrm{v}_i^{n+1} \tag{5.3.33}$$

$$p^{**} = p^{n+1} \tag{5.3.34}$$

This completes the splitting process in which the v_i^{***} and p^{**} fields imply the exact solution v_i^{n+1} and p^{n+1}. For additional information on this procedure, see Issa, Gosman, and Watkins [1986].

5.3.3 MARKER-AND-CELL (MAC) METHOD

This is one of the earliest methods developed for the solution of incompressible flows, although its use in the original form is no longer pursued, but it has been altered to other more efficient schemes. The basic idea of MAC as originally introduced by Harlow and Welch [1965] is one of the pressure correction schemes developed on a staggered mesh, seeking to trace the paths of fictitious massless marker particles introduced on the free surface. The solution is advanced in time by solving the momentum equations for velocity components using the current estimates of the pressure distributions. The pressure is improved by numerically solving the Poisson equation,

$$p_{,ii} = f \tag{5.3.35}$$

with

$$f = S - \frac{\partial D}{\partial t} \tag{5.3.36}$$

$$S = \lfloor -(\rho v_i v_j)_{,i} + \mu v_{j,ii} \rfloor_{,j} \tag{5.3.37}$$

$$D = v_{i,i} \tag{5.3.38}$$

Here, the correction in pressure is required to compensate for the nonzero dilatation D (5.3.38) at the current iteration level. The Poisson equation is then solved for the revised pressure field. The improved pressure may then be used in the momentum equations for a better solution at the present time step. If D does not vanish, cyclic process of solving the momentum equations and the Poisson equation is repeated until the velocity field is divergent free.

The original MAC method was based on an explicit time-marching scheme. Subsequently, implicit schemes have been implemented by various authors [Briley, 1974; Ghia, Hankey, and Hodge, 1979].

5.4 VORTEX METHODS

Two-Dimensional Vorticity Transport Equation

In the previous sections, we dealt with primitive variables, v_i and p. An alternative approach is to use the vortex methods in which we utilize the vorticity and stream functions as variables.

$$\boldsymbol{\omega} = \nabla \times \mathbf{v} \tag{5.4.1}$$

$$\mathbf{v} = \varepsilon_{ij}\psi_{,j}\mathbf{i}_i \tag{5.4.2}$$

where $\boldsymbol{\omega}$ is the vorticity vector, ε_{ij} is the second order tensor of the permutation symbol for 2-D,

$$\varepsilon_{ij} = \begin{cases} 1 & \text{for } \varepsilon_{12} \\ -1 & \text{for } \varepsilon_{21} \\ 0 & \text{otherwise} \end{cases}$$

and ψ is the stream function.

For incompressible two-dimensional flows, the scalar vorticity transport equation is written as

$$\frac{\partial \omega}{\partial t} + \omega_{,i} \mathrm{v}_i = \nu \omega_{,ii} \tag{5.4.3}$$

where $\omega = \omega_3$ is the component of the vorticity vector $\boldsymbol{\omega}$ in the direction normal to the x-y plane. Auxiliary equations required are

$$\nabla^2 \psi = -\omega \tag{5.4.4}$$

$$\nabla^2 p = 2\rho \left(\frac{\partial u}{\partial x} \frac{\partial \mathrm{v}}{\partial y} - \frac{\partial u}{\partial y} \frac{\partial \mathrm{v}}{\partial x} \right) \tag{5.4.5}$$

or

$$\nabla^2 p = 2\rho \left[\frac{\partial^2 \psi}{\partial x^2} \frac{\partial^2 \psi}{\partial y^2} - \left(\frac{\partial^2 \psi}{\partial x \partial y} \right)^2 \right] \tag{5.4.6}$$

It is seen that the variables v_i, p, ψ, and ω may be computed using equations (5.4.1) through (5.4.6).

For simplicity, let us consider a wall located at $y = 0$. Referring to Figure 5.4.1, we have

$$\left(\frac{\partial p}{\partial x} \right)_{\text{wall}} = -\mu \left(\frac{\partial \omega}{\partial y} \right)_{\text{wall}} \tag{5.4.7}$$

or

$$\frac{p_{i+1,1} - p_{i-1,1}}{2\Delta x} = -\mu \frac{-3\omega_{i,1} + 4\omega_{i,2} - \omega_{i,3}}{2\Delta y} \tag{5.4.8}$$

Here, the pressure must be specified on the wall surface. The pressure at the adjacent point can be determined with a first order, one-sided difference expression for $\partial p/\partial x$ in (5.4.7). Thereafter, (5.4.8) can be used to determine the pressure at all other wall points.

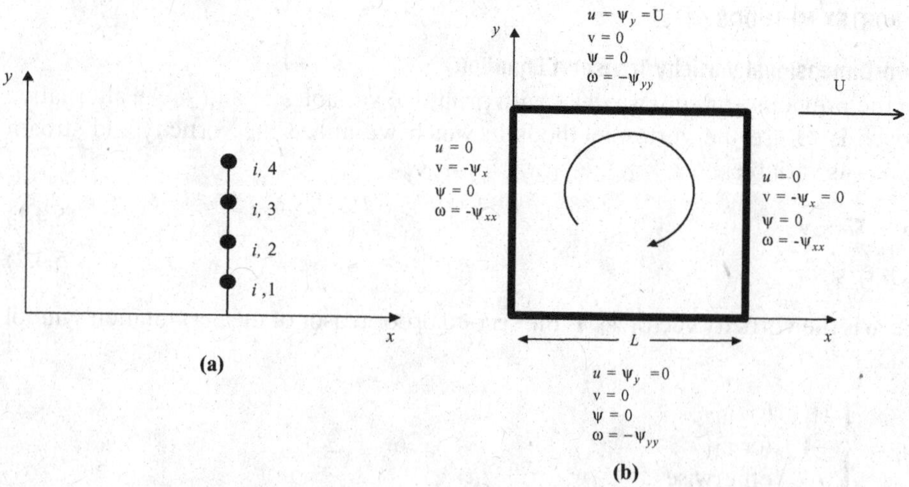

Figure 5.4.1 Illustration of vortex methods. **(a)** Grid point normal to a flat plate. **(b)** Driven cavity problem.

Notice that for the simultaneous solutions of (5.4.2), (5.4.3), (5.4.4), and (5.4.6), we may use the finite difference schemes presented in Chapter 3. For example, the nonlinear terms on the right-hand side of (5.4.6) may be represented as

$$\nabla^2 p = 2\rho_{i,j}\left[\left(\frac{\psi_{i+1,j} - 2\psi_{i,j} + \psi_{i-1,j}}{(\Delta x)^2}\right)\left(\frac{\psi_{i,j+1} - 2\psi_{i,j} - \psi_{i,j-1}}{(\Delta y)^2}\right)\right.$$
$$\left. - \left(\frac{\psi_{i+1,j+1} - \psi_{i+1,j-1} - \psi_{i-1,j+1} + \psi_{i-1,j-1}}{4\Delta x \Delta y}\right)^2\right] \tag{5.4.9}$$

where the alternative mixed derivative may be chosen as shown in Section 3.5.

For a steady state problem, the Poisson equation for pressure is solved once, that is, after the steady-state values of ω and ψ have been computed.

For time dependent problems, the solution of the vorticity transport equation and the Poisson equation requires that boundary conditions for ψ and ω be specified. At the wall, ψ is a constant and may be set equal to a reference value, that is, $\psi = 0$. To find ω at the wall surface, we write ψ in terms of Taylor series about the wall point $(i, 1)$,

$$\psi_{i,2} = \psi_{i,1} + \left.\frac{\partial \psi}{\partial y}\right|_{i,1} \Delta y + \frac{1}{2}\left.\frac{\partial^2 \psi}{\partial y^2}\right|_{i,1} (\Delta y)^2 + \cdots \tag{5.4.10}$$

where

$$\left.\frac{\partial \psi}{\partial y}\right|_{i,1} = u_{i,1} = 0 \tag{5.4.11a}$$

$$\left.\frac{\partial^2 \psi}{\partial y^2}\right|_{i,1} = \left.\frac{\partial u}{\partial y}\right|_{i,1} \tag{5.4.11b}$$

$$\omega_{i,1} = \left.\frac{\partial v}{\partial x}\right|_{i,1} - \left.\frac{\partial u}{\partial y}\right|_{i,1} = -\left.\frac{\partial^2 \psi}{\partial y^2}\right|_{i,1} \tag{5.4.11c}$$

thus, rewriting (5.4.10) as

$$\psi_{i,2} = \psi_{i,1} - \frac{1}{2}\omega_{i,1}\Delta y^2 + O(\Delta y^3) \tag{5.4.12}$$

$$\omega_{i,1} = \frac{2(\psi_{i,1} - \psi_{i,2})}{\Delta y^2} + O(\Delta y)$$

$$u_{i,1} = \left.\frac{\partial \psi}{\partial y}\right|_{i,2} = \frac{-3\psi_{i,1} + 4\psi_{,2} + \psi_{i,3}}{4\Delta y} \tag{5.4.13}$$

Three-Dimensional Vorticity Transport Equations

For three-dimensional problems, the vorticity transport equations are of the form [Chung, 1996]:

$$\frac{\partial \boldsymbol{\omega}}{\partial t} + (\mathbf{v} \cdot \nabla)\boldsymbol{\omega} - (\boldsymbol{\omega} \cdot \nabla)\mathbf{v} = \nu \nabla^2 \boldsymbol{\omega} \tag{5.4.14}$$

$$\mathbf{v} = \nabla \psi \times \hat{\mathbf{n}} = \nabla \times \boldsymbol{\Psi} \tag{5.4.15}$$

with

$$v_i = \varepsilon_{ijk}\Psi_{k,j} \tag{5.4.16}$$

$$\begin{aligned} i = 1 \quad & v_1 = \Psi_{3,2} - \Psi_{2,3} \\ i = 2 \quad & v_2 = \Psi_{1,3} - \Psi_{3,1} \\ i = 3 \quad & v_3 = \Psi_{2,1} - \Psi_{1,2} \end{aligned} \tag{5.4.17}$$

$$\Psi_k = \hat{n}_k \psi$$

and

$$\boldsymbol{\omega} = -\nabla^2 \boldsymbol{\Psi} \tag{5.4.18}$$

Note that $\nabla\psi$ is perpendicular to the velocity vector $\mathbf{v}$ and $\hat{\mathbf{n}}$ is perpendicular to the plane $\nabla\psi$ and $\mathbf{v}$, whereas $\boldsymbol{\Psi}$ is known as the three-dimensional stream function vector. The geometric properties of the stream function vector are presented in Section 12.2.

Another approach is to use the fourth order stream function vector equation of the form

$$\frac{\partial}{\partial t}\nabla^2 \boldsymbol{\Psi} + (\nabla \times \boldsymbol{\Psi} \cdot \nabla)\nabla^2 \boldsymbol{\Psi} - (\nabla^2 \boldsymbol{\Psi} \cdot \nabla)(\nabla \times \boldsymbol{\Psi}) = \nu \nabla^4 \boldsymbol{\Psi} \tag{5.4.19}$$

with the boundary conditions extended to three-dimensional geometries.

Solutions may be obtained from either (5.4.14) or (5.4.19) using the definitions given by (5.4.15) and (5.4.18). These and other subjects on applications in three-dimensional stream function vector components are further detailed in Section 12.2.

The Curl of Vorticity Transport Equations

We have noted that the advantage of the vorticity transport equation(s) is the numerical stability accrued from removing pressure gradient terms from the solution process. However, the velocity must be calculated from solving simultaneously (5.4.14) through (5.4.18) or from (5.4.19). These steps can be eliminated if we take a curl of the vorticity transport equation (5.4.14), in which the velocity is the only variable. This subject will be discussed in Section 12.2.1.

5.5 SUMMARY

The incompressible flow analysis based on the artificial compressibility method and the pressure-based formulation using SIMPLE, SIMPLER, SIMPLEC, and PISO have been presented. It was shown that these methods are devised in order to ensure the conservation of mass so that pressure oscillations can be prevented. Vortex methods in which pressure terms are absent are preferred in dealing with rotational incompressible flows as they are computationally efficient. Accurate physics of fluids can be obtained without difficulties which may arise from inaccurate pressure calculations in other methods.

The current trend appears to be in favor of preconditioning of the time-dependent term of the density-based formulation so that both compressible and incompressible flows can be treated. This is because, in many practical situations, high- and low-speed regions are coupled particularly in high-speed boundary layer flows and the analysis capable of handling both compressible and incompressible flows is frequently in demand. Details of the preconditioning process for the combined density- and pressure-based formulations for the incompressible flow analysis are presented in Section 6.4.

Since the solution of incompressible flows can be obtained as a part of the compressible flow formulation, it appears that more attention is given to the compressible flow analysis. This leads to a motivation toward attempting to develop a general purpose program, anticipating that the results of incompressible flows arise automatically when the flow velocity decreases at low Mach number. This topic is addressed in Section 6.5.

The theoretical basis for three-dimensional vorticity transport equations is examined. Numerical examples for the three-dimensional vortex methods based on the three-dimensional stream function vector components will be discussed in Section 12.2.

Although not presented in this chapter, other methods have been used in the past. One of the significant developments in the late 1950s was the particle-in-cell (PIC) method [Evans and Harlow, 1957, 1959], particularly efficient in the flows with large distortions (see Section 16.4.3). Recent developments dealing with multiphase incompressible flows will be presented in Chapter 25.

REFERENCES

Briley, W. R. [1974]. Numerical method for predicting three-dimensional steady viscous flow in ducts. *J. Comp. Phys.*, 14, 8–28.

Chorin, A. J. [1967]: A numerical method for solving incompressible viscous flow problems. *J. Comp. Phys.*, 2, 12–26.

Chung, T. J. [1996]. *Applied Continuum Mechanics*. New York: Cambridge University Press.

Evans, M. W. and Harlow, F. H. [1957]. The particle-in-cell method for hydrodynamic calculations. Los Alamos Scientific Laboratory Report No. LA-2139.

———. [1959]. Calculation of unsteady supersonic flow past a circular cylinder. *ARS Journal*, 29, 46–51.

Ghia, K. N., Hankey Jr., W. L., and Hodge, J. K. [1979]. Use of primitive variables in the solution of incompressible Navier-Stokes equations. *AIAA J.*, 17, 298–301.

Harlow, F. H. and Welch, J. E. [1965]. Numerical calculation of time-dependent viscous incompressible flow of fluid with free surface. *Phys. Fluids*, 8, 2182–89.

Issa, R. [1985]. Solution of the implicitly discretized fluid flow equations by operator splitting. *J. Comp. Phys.*, 62, 40–65.

Issa, R. I., Gosman, A. D., and Watkins, A. P. [1986]. The computation of compressible and incompressible recirculating flows by a non-iterative implicit scheme. *J. Comp. Phys.*, 62, 66–82.

Kwak, D. C., Chang, J. L. C., Shanks, S. P., and Chakravarthy, S. K. [1986]. A three-dimensional incompressible Navier-Stokes solver using primitive variables. *AIAA Journal.*, 24, 390–96.

Merkle, C. L., Sullivan, J. Y., Buelow, P. E. O., and Ventateswaran, S. [1998]. Computation of flows with arbitrary equations of state. *AIAA J.*, 36, 4, 515–21.

Patankar, S. V. and Spalding, D. B. [1972]: A calculation procedure for heat, mass and momentum transfer in three-dimensional parabolic flows. *Int. J. Heat Mass Transfer*, 15, 1787–1806.

Raithby, G. D. and Schneider, G. E. [1979]. Numerical solution of problems in incompressible fluid flow: Treatment of the velocity-pressure coupling. *Num. Heat Transfer*, 2, 417–40.

Van Doormaal, J. P. and Raithby, G. D. [1984]. Enhancements of the SIMPLE methods for predicting incompressible fluid flows. *Num. Heat Transfer.*, 7, 147–63.

CHAPTER SIX

Compressible Flows via Finite Difference Methods

In general, the physical behavior of compressible flows is more complicated than in incompressible flows. Compressible flows may be viscous or inviscid, depending on flow velocities. Compressible inviscid flows are analyzed using the potential or Euler equations, whereas compressible viscous flows are solved from the Navier-Stokes system of equations. Shock waves may occur in compressible flows and require special attention as to the solution methods. Furthermore, shock wave turbulent boundary layer interactions in compressible viscous flows constitute one of the most important physical phenomena in computational fluid dynamics. Let us consider air flows at speeds greater than 100 m/s, which corresponds to a Mach number of approximately 0.3, but less than 1700 m/s, or approximately Mach 5. Air flows in this range ($0.3 \leq M \leq 5$) may be considered as compressible and inviscid. This range is usually subdivided into regions identified as subsonic ($0.3 < M < 0.8$), transonic ($0.8 \leq M \leq 1.2$), and supersonic ($1.2 < M \leq 5$). For $M > 5$, the flow is referred to as hypersonic. Hypersonic flows around a solid body are usually coupled with viscous boundary layers. Effects of dilatational dissipation due to compressibility, high temperature gradients, vortical motions within the secondary boundary layers, radiative heat transfer, vibrational and electronic energies, and chemical reactions are examples of some of the complex physical phenomena associated with hypersonic flows.

In order to take into account the compressibility and variations of density in high-speed flows, we utilize the conservation form of the governing equations, using the density-based formulation. This is in contrast to the pressure-based formulation for incompressible flows discussed in Chapter 5. For compressible flows, we encounter some regions of the flow domain (close to the wall, for example) in which low Mach numbers or incompressible flows prevail. In this case, the density-based formulations become ineffective, with the solution convergence being extremely slow. To resolve such problems, various schemes have been developed. Among them are the preconditioning process for the time-dependent term toward improving the stiff convection eigenvalues and the flowfield-dependent variation (FDV) methods allowing the transitions and interactions of various flow properties as well as all speed flows.

For simple cases of compressible inviscid flows (irrotational, isentropic, isothermal), the potential equation can be used, whereas the Euler equations are preferred for more general compressible inviscid flows. For compressible viscous flows, various approximate governing equations such as boundary layer equations or parabolized

Navier-Stokes system of equations are utilized. However, the most general and complete analysis is to invoke the full Navier-Stokes system of equations, which is the emphasis in this book.

FDM formulations and solution procedures for the potential equation are presented in Section 6.1, with Euler equations and the Navier-Stokes system of equations in Sections 6.2 and 6.3, respectively. The solution of the Navier-Stokes system of equations for compressible and incompressible flows using the preconditioning process will be presented in Section 6.4, followed by the flowfield-dependent variation (FDV) methods in Section 6.5 and various other methods in Section 6.6. Finally, the boundary conditions for compressible flows in general are discussed in Section 6.7.

6.1 POTENTIAL EQUATION

6.1.1 GOVERNING EQUATIONS

The governing equation for steady-state compressible inviscid flows may be represented by the potential equation of the form (2-D),

$$\left[1-\left(\frac{u}{a}\right)^2\right]\frac{\partial u}{\partial x}+\left[1-\left(\frac{\mathrm{v}}{a}\right)^2\right]\frac{\partial \mathrm{v}}{\partial y}-\frac{u\mathrm{v}}{a^2}\left(\frac{\partial u}{\partial y}+\frac{\partial \mathrm{v}}{\partial x}\right)=0 \tag{6.1.1a}$$

or

$$\left[1-\left(\frac{u}{a}\right)^2\right]\frac{\partial u}{\partial x}+\left[1-\left(\frac{\mathrm{v}}{a}\right)^2\right]\frac{\partial \mathrm{v}}{\partial y}-\frac{2}{a^2}u\mathrm{v}\frac{\partial u}{\partial y}=f \tag{6.1.1b}$$

with

$$f=\frac{1}{a^2}u\mathrm{v}\left(\frac{\partial \mathrm{v}}{\partial x}-\frac{\partial u}{\partial y}\right) \tag{6.1.2}$$

and $f=0$ for irrotational flow. In terms of the velocity potential function ϕ, (6.1.1) may be written as

$$\phi_{,ii}-\frac{1}{a^2}\phi_{,i}\phi_{,j}\phi_{,ij}=0$$

or

$$(1-M_x^2)\frac{\partial^2\phi}{\partial x^2}+(1-M_y^2)\frac{\partial^2\phi}{\partial y^2}-\frac{2}{a^2}\frac{\partial\phi}{\partial x}\frac{\partial\phi}{\partial y}\frac{\partial^2\phi}{\partial x\partial y}=0 \tag{6.1.3}$$

with $u=\partial\phi/\partial x$, $\mathrm{v}=\partial\phi/\partial y$, $M_x=u/a$, and $M_y=\mathrm{v}/a$.

For small perturbation approximations in irrotational flow, we obtain

$$(1-M_\infty^2)\frac{\partial^2\phi}{\partial x^2}+\frac{\partial^2\phi}{\partial y^2}=M_\infty^2\left(\frac{1+\gamma}{U_\infty}\right)\frac{\partial\phi}{\partial x}\frac{\partial^2\phi}{\partial x^2} \tag{6.1.4}$$

For unsteady flows, using the first and second laws of thermodynamics for isentropic and irrotational flows, (6.1.1) is modified to

$$\phi_{,ii}-\frac{1}{a^2}\phi_{,i}\phi_{,j}\phi_{,ij}-\frac{1}{a^2}\left[\frac{\partial^2\phi}{\partial t^2}+\frac{\partial}{\partial t}(\phi_{,i}\phi_{,i})\right]=0 \tag{6.1.5}$$

where

$$a^2 = \frac{\partial p}{\partial \rho} = \frac{\gamma p}{\rho} = (\gamma - 1)H = (\gamma - 1)\left[H_0 - \frac{1}{2}\phi_{,i}\phi_{,i} - \frac{\partial \phi}{\partial t}\right]$$

In the case of isentropic flows with stagnation density ρ_0 and stagnation enthalpy H_0, we have

$$\frac{\rho}{\rho_0} = \left[1 - \frac{1}{2H_0}\phi_{,i}\phi_{,i} - \frac{1}{H_0}\frac{\partial \phi}{\partial t}\right]^{\frac{1}{\gamma-1}} \tag{6.1.6}$$

with

$$H_0 = H + \frac{1}{2}\mathbf{v}\cdot\mathbf{v} \tag{6.1.7}$$

For steady flows, (6.1.6) takes the form

$$\frac{\rho}{\rho_0} = \left[1 - \frac{1}{2H_0}\phi_{,i}\phi_{,i}\right]^{\frac{1}{\gamma-1}} \tag{6.1.8}$$

or

$$\frac{\rho}{\rho_0} = \left[1 - \frac{\gamma - 1}{2a_0^2}\phi_{,i}\phi_{,i}\right]^{\frac{1}{\gamma-1}} = \left[1 - \frac{\gamma - 1}{2}M^2\right]^{\frac{1}{\gamma-1}} \tag{6.1.9}$$

If a nonisentropic process with rotational flows is considered, the momentum equation is written as

$$T\nabla S + \mathbf{v} \times \boldsymbol{\omega} - \nabla H_0 = 0 \tag{6.1.10}$$

where S is the entropy per unit mass. Combining (6.1.10) and (6.1.2), we obtain for two dimensions

$$f = -\frac{1}{V^*}\left(H_{0,i}n_i - \frac{a^2}{\gamma R}S_{,i}n_i\right)\frac{uv}{a^2} \tag{6.1.11}$$

with

$$V^* = vn_1 - un_2$$

It follows from (6.1.10) and (6.1.9) that

$$\frac{\rho}{\rho_0} = \left[\exp\left(\frac{-\Delta S}{c_v}\right)\left(1 - \frac{1}{2H_0}\phi_{,i}\phi_{,i}\right)^{\frac{1}{\gamma-1}}\right] \tag{6.1.12}$$

where ΔS is the entropy increase over the shock. This is equivalent to a modification of the stagnation density ρ_0

$$\frac{\rho}{\rho_{02}} = \left(1 - \frac{1}{2H_0}\phi_{,i}\phi_{,i}\right)^{\frac{1}{\gamma-1}} \tag{6.1.13}$$

with

$$\rho_{02} = \rho_{01}\left(\frac{p_{02}}{p_{01}}\right)^{\frac{1}{\gamma}}$$

where the subscripts 1 and 2 denote upstream and downstream of the shock, respectively.

6.1.2 SUBSONIC POTENTIAL FLOWS

For irrotational flow [$f = 0$ in (6.1.3)] with $\Delta x = \Delta y = 1$, the finite difference scheme may be written as

$$\left(1 - M_x^2\right)_{i,j}(\phi_{i+1,j} - 2\phi_{i,j} + \phi_{i-1,j}) + \left(1 - M_y^2\right)_{i,j}(\phi_{i,j+1} - 2\phi_{i,j} + \phi_{i,j-1})$$
$$- \frac{1}{2}(M_x M_y)_{i,j}(\phi_{i+1,j+1} - \phi_{i+1,j-1} - \phi_{i-1,j+1} + \phi_{i-1,j-1}) = 0 \qquad (6.1.14)$$

It is interesting to note that (6.1.14) is diagonally dominant for subsonic flows, while this is not true for transonic and supersonic flows. This implies that the elliptic nature of (6.1.14) changes to parabolic and hyperbolic forms.

Another scheme is to use the continuity equation,

$$\nabla \cdot (\rho \mathbf{v}) = \nabla \cdot (\rho \nabla \phi) = (\rho \phi_{,i})_{,i} = 0 \qquad (6.1.15)$$

Thus, the finite difference form of (6.1.15) may be written as

$$\rho_{i+\frac{1}{2},j}(\phi_{i+1,j} - \phi_{i,j}) - \rho_{i-\frac{1}{2},j}(\phi_{i,j} - \phi_{i-1,j}) + \rho_{i,j+\frac{1}{2}}(\phi_{i,j+1} - \phi_{i,j})$$
$$- \rho_{i,j-\frac{1}{2}}(\phi_{i,j} - \phi_{i,j-1}) = 0 \qquad (6.1.16)$$

To solve (6.1.16), the so-called Taylor linearization [Murman and Cole, 1971] may be used:

$$\nabla \cdot (\rho^n \nabla \phi^{n+1}) = 0 \qquad (6.1.17)$$

or

$$\nabla \cdot \left[\rho\left(|\nabla \phi^n|^2\right) \nabla \phi^{n+1}\right] = 0 \qquad (6.1.18)$$

where density is calculated from the known values of the velocities obtained at the previous iteration step n.

6.1.3 TRANSONIC POTENTIAL FLOWS

As the Mach number approaches unity, the potential equation tends toward parabolic, leading to instability or nonconvergence of the numerical scheme. To cope with this difficulty, a number of numerical methods have been developed. They include artificial viscosity, artificial compressibility, artificial flux or upwinding, and iterations with over-relaxation, among others.

(a) Artificial Viscosity with Nonconservative Equation

In order to resolve shock discontinuities, we consider two forms of finite differences:

Central Differences

$$\left.\frac{\partial^2 \phi}{\partial x^2}\right|_{i,j}^{(c)} = \frac{1}{(\Delta x)^2}(\phi_{i+1,j} - 2\phi_{i,j} + \phi_{i-1,j}) \qquad (6.1.19)$$

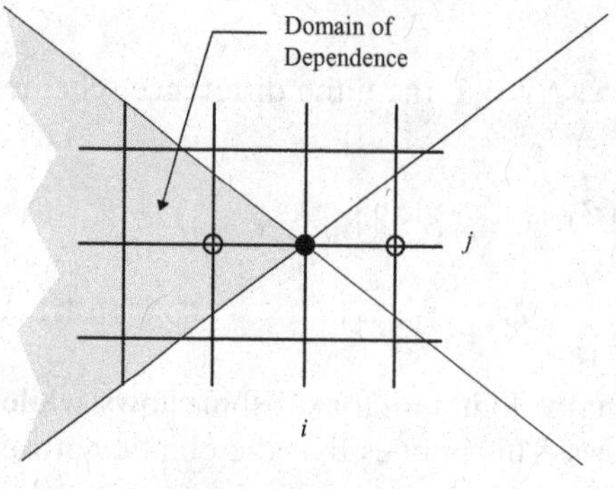

Figure 6.1.1 Region of dependence.

Backward (upwind) Differences

$$\left.\frac{\partial^2\phi}{\partial x^2}\right|_{i,j}^{(B)} = \frac{1}{(\Delta x)^2}(\phi_{i-2,j} - 2\phi_{i-1,j} + \phi_{i,j}) \tag{6.1.20}$$

Obviously, the central difference is in opposition to the physical properties of supersonic flows since only the points located within the region of dependence (Figure 6.1.1) can have an effect on the flow properties at the point under consideration ($\phi_{i,j}$). Subtracting (6.1.19) from (6.1.20), we get

$$\left.\frac{\partial^2\phi}{\partial x^2}\right|_{i,j}^{(B)} - \left.\frac{\partial^2\phi}{\partial x^2}\right|_{i,j}^{(C)} = -\frac{1}{(\Delta x)^2}(\phi_{i+1,j} - 3\phi_{i,j} + 3\phi_{i-1,j} - \phi_{i-2,j})$$

or

$$\left.\frac{\partial^2\phi}{\partial x^2}\right|_{i,j}^{(B)} = \left.\frac{\partial^2\phi}{\partial x^2}\right|_{i,j}^{(C)} - \Delta x\frac{\partial^3\phi}{\partial x^3} \tag{6.1.21}$$

Thus, it is seen that the backward difference amounts to adding an artificial viscosity, $\Delta x\frac{\partial^3\phi}{\partial x^3}$, to the central difference. Therefore, the upwind differencing automatically adds an entropy condition in the form of artificial dissipation terms which are proportional to the mesh size.

For applications to the small perturbation equation, we have the following options [Murman and Cole, 1971]:

$$(1 - M^2)\phi_{xx}^{(B)} + \phi_{yy}^{(C)} = 0 \qquad \text{for } M > 1 \tag{6.1.22}$$

$$(1 - M^2)\phi_{xx}^{(C)} + \phi_{yy}^{(C)} = -\Delta x(M^2 - 1)\phi_{xxx} \qquad \text{for } M > 1 \tag{6.1.23}$$

$$(1 - M^2)\phi_{xx}^{(C)} + \phi_{yy}^{(C)} = 0 \qquad \text{for } M < 1 \tag{6.1.24}$$

To apply these conditions to the full potential equation, we must take into account the local flow direction (Figure 6.1.2). Jameson [1974] introduced "rotational difference scheme" for this purpose.

Choosing the local streamline coordinates as (ℓ, n),

$$(1 - M^2)\phi_{\ell\ell} - \phi_{nn} = 0 \tag{6.1.25}$$

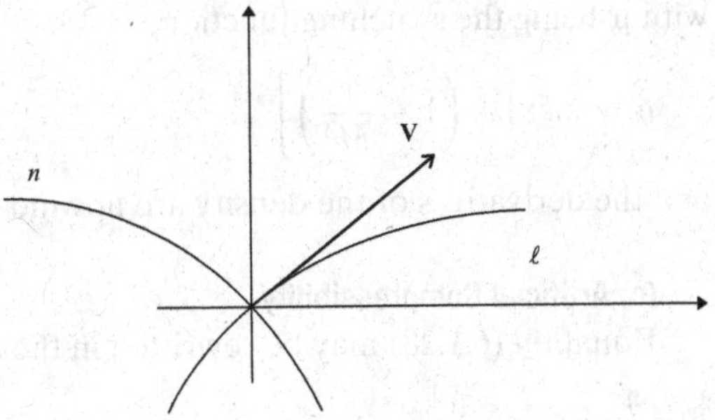

Figure 6.1.2 Local streamlines.

where

$$\phi_{\ell\ell} = \frac{1}{q^2}(u^2\phi_{xx} + 2uv\phi_{xy} + v^2\phi_{yy}) \tag{6.1.26a}$$

$$\phi_{nn} = \frac{1}{q^2}(v^2\phi_{xx} - 2uv\phi_{xy} + u^2\phi_{yy}) \tag{6.1.26b}$$

with $q = (u^2 + v^2)^{\frac{1}{2}}$ and

$$\phi_{xy}^{(C)} = \frac{1}{4\Delta x \Delta y}(\phi_{i+1,j+1} - \phi_{i+1,j-1} - \phi_{i-1,j+1} + \phi_{i-1,j-1})$$

$$\phi_{xy}^{(B)} = \frac{1}{4\Delta x \Delta y}(\phi_{i,j} - \phi_{i-1,j} - \phi_{i,j-1} + \phi_{i-1,j-1})$$

$$= \phi_{xy}^{(C)} - \frac{\Delta x}{2}\phi_{xxy} - \frac{\Delta y}{2}\phi_{xyy}$$

$$\phi_{\ell\ell}^{(B)} = \phi_{\ell\ell}^{(C)} - \frac{u^2}{q}\Delta x\phi_{xxx} - \frac{v^2}{q}\Delta y\phi_{yyy} - \frac{uv}{q}(\Delta x\phi_{xxy} + \Delta y\phi_{xyy})$$

Thus, the rotational difference scheme takes the form

$$(1 - M^2)\phi_{\ell\ell}^{(C)} - \phi_{nn}^{(C)} = g \tag{6.1.27}$$

$$g = \frac{1}{q^2}(1 - M^2)[\Delta x(u^2\phi_{xxx} + uv\phi_{xxy}) + \Delta y(v^2\phi_{yyy} + uv\phi_{xyy})]$$

(b) Artificial Viscosity with Conservative Equation

The potential equation in conservation form with artificial viscosity is written as

$$\boldsymbol{\nabla} \cdot (\rho\,\boldsymbol{\nabla}\phi + \mathbf{A}) = 0 \tag{6.1.28}$$

where $\boldsymbol{A}$ is the artificial viscosity vector,

$$\boldsymbol{A} = -\mu(u\rho_x\Delta x\mathbf{i}_x + v\rho_y\Delta y\mathbf{i}_y) \tag{6.1.29}$$

$$\rho_x = -\frac{\rho}{a^2}\mathbf{v} \cdot \frac{\partial \mathbf{v}}{\partial x} = -\frac{\rho}{a^2}\left(u\frac{\partial u}{\partial x} + v\frac{\partial v}{\partial x}\right)$$

$$\rho_y = -\frac{\rho}{a^2}\mathbf{v} \cdot \frac{\partial \mathbf{v}}{\partial y} = -\frac{\rho}{a^2}\left(u\frac{\partial u}{\partial y} + v\frac{\partial v}{\partial y}\right)$$

with μ being the switching function,

$$\mu = \max\left[0, \left(1 - \frac{1}{M^2}\right)\right] \tag{6.1.30}$$

and the derivatives of the density are upwind differenced.

(c) Artificial Compressibility

Equation (6.1.28) may be rewritten in the form [Holst and Ballhaus, 1979],

$$\frac{\partial}{\partial x}(\overline{\rho}\phi_x) + \frac{\partial}{\partial y}(\overline{\overline{\rho}}\phi_y) = 0 \tag{6.1.31}$$

where

$$\overline{\rho} = \rho - \mu\rho_x\Delta x \tag{6.1.32a}$$

$$\overline{\overline{\rho}} = \rho - \mu\rho_y\Delta y \tag{6.1.32b}$$

The artificial densities are prescribed at the midpoints $(i \pm \frac{1}{2}, j)$ and $(i, j \pm \frac{1}{2})$.
For $u_{i+\frac{1}{2},j} > 0$

$$\overline{\rho}_{i+\frac{1}{2},j} = \rho_{i+\frac{1}{2},j} - \mu_{ij}\left(\rho_{i+\frac{1}{2},j} - \rho_{i-\frac{1}{2},j}\right) \tag{6.1.33a}$$

For $u_{i+\frac{1}{2},j} < 0$

$$\overline{\rho}_{i+\frac{1}{2},j} = \rho_{i+\frac{1}{2},j} + \mu_{i+1,j}\left(\rho_{i+\frac{1}{2},j} - \rho_{i+\frac{3}{2},j}\right) \tag{6.1.33b}$$

For $\mathrm{v}_{i,j+\frac{1}{2}} > 0$

$$\overline{\overline{\rho}}_{i,j+\frac{1}{2}} = \rho_{i,j+\frac{1}{2}} - \mu_{ij}\left(\rho_{i.j+\frac{1}{2}} - \rho_{i,j-\frac{1}{2}}\right) \tag{6.1.34a}$$

For $\mathrm{v}_{i,j+\frac{1}{2}} < 0$

$$\overline{\overline{\rho}}_{i,j+\frac{1}{2}} = \rho_{i,j+\frac{1}{2}} + \mu_{i,j+1}\left(\rho_{i,j+\frac{1}{2}} - \rho_{i,j+\frac{3}{2}}\right) \tag{6.1.34b}$$

An alternative form for artificial compressibility may be given as

$$\nabla\cdot(\tilde{\rho}\nabla\phi) = 0 \tag{6.1.35}$$

where

$$\tilde{\rho} = \rho - \mu\frac{\partial\rho}{\partial\ell}\Delta\ell = \rho - \mu\Delta\ell\left(\frac{u}{q}\rho_x + \frac{\mathrm{v}}{q}\rho_y\right) \tag{6.1.36a}$$

or

$$\tilde{\rho} = \rho - \mu\left(\frac{u}{q}\rho_x\Delta x + \frac{\mathrm{v}}{q}\rho_y\Delta y\right) \tag{6.1.36b}$$

Various switching functions have been suggested for stability, such as

$$\mu = \max\left[0, \left(1 - \frac{M_c^2}{M^2}\right)CM^2\right] \tag{6.1.37}$$

where M_c is a cutoff Mach number of the order of $M \cong 0.95$, $1 \leq C \leq 2$. The cutoff

Mach number M_c activates the switching function in the small subsonic region $M_c \leq M \leq 1$ close to the sonic lines.

(d) Artificial Flux or Flux Upwinding

For switching at sonic points to avoid unwanted expansion peaks, we may utilize controlled monotone schemes such as those used in Euler equations [Engquist and Osher, 1980; Osher, Hafez, and Whitlow, 1985]. To this end, we may write the continuity equation in the form

$$\frac{\partial \rho}{\partial \ell} = -\frac{\rho q}{a^2}\frac{\partial q}{\partial \ell} \tag{6.1.38}$$

and

$$\begin{aligned}\frac{\partial \rho q}{\partial \ell} &= -\frac{\rho}{a^2}q^2\frac{\partial q}{\partial \ell} + \rho\frac{\partial q}{\partial \ell} = \rho(1 - M^2)\frac{\partial q}{\partial \ell} \\ &= q\left(1 - \frac{1}{M^2}\right)\frac{\partial \rho}{\partial \ell}\end{aligned} \tag{6.1.39}$$

The corrected upwinded flux $\tilde{\rho} q$ can be written in supersonic regions as

$$\tilde{\rho} q = \rho q - q\left(1 - \frac{1}{M^2}\right)\frac{\partial \rho}{\partial \ell}\Delta \ell \tag{6.1.40}$$

or

$$\tilde{\rho} q = \rho q - \frac{\partial}{\partial \ell}(\rho q)\Delta \ell \tag{6.1.41}$$

A modification of (6.1.41) results in

$$\overline{\rho q} = \rho q - \frac{\partial}{\partial \ell}[\mu(\rho q - \rho^* q^*)]\Delta \ell \tag{6.1.42}$$

where $\rho^* q^*$ denotes the sonic flux, $\mu = 0$ for subsonic flow ($M \leq 1, q \leq q^*, \rho \geq \rho^*$) and $\mu = 1$ for supersonic flows ($M > 1, q > q^*, \rho < \rho^*$) (see Figure 6.1.3). The discrete form

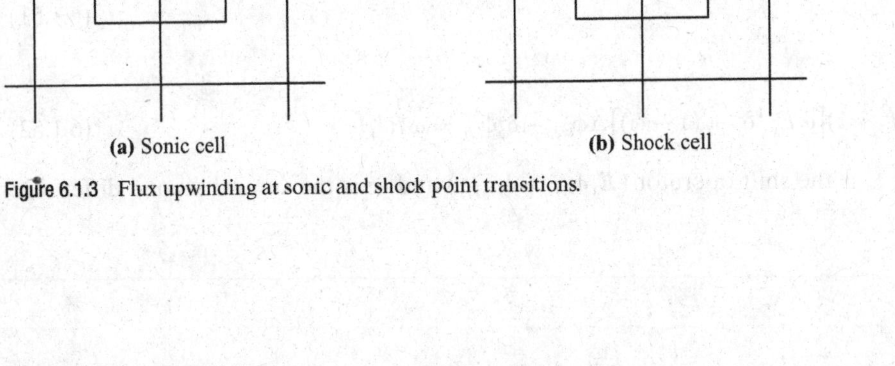

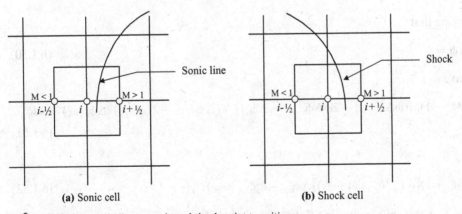

Figure 6.1.3 Flux upwinding at sonic and shock point transitions.

of (6.1.42) becomes

$$(\overline{\rho q})_{i+\frac{1}{2},j} = (\rho q)_{i+\frac{1}{2},j} - \mu_{i+\frac{1}{2},j}(\rho q - \rho^* q^*)_{i+\frac{1}{2},j} + \mu_{i-\frac{1}{2},j}(\rho q - \rho^* q^*)_{i-\frac{1}{2},j} \qquad (6.1.43)$$

It is similar for other points. Thus, we have
For $M < 1$

$$(\overline{\rho q})_{i+\frac{1}{2},j} = (\rho q)_{i+\frac{1}{2},j} \qquad (6.1.44)$$

For $M > 1$

$$(\overline{\rho q})_{i+\frac{1}{2},j} = (\rho q)_{i-\frac{1}{2},j} \qquad (6.1.45)$$

For $M = 1$

$$(\overline{\rho q})_{i+\frac{1}{2},j} = (\rho^* q^*) \qquad (6.1.46)$$

Notice that this scheme guarantees that expansion shocks will not occur when $(\rho q)_{i+\frac{1}{2},j} < (\rho^* q^*)$.

At a shock transition, we obtain

$$(\overline{\rho q})_{i+\frac{1}{2},j} = (\rho q)_{i+\frac{1}{2},j} + (\rho q - \rho^* q^*)_{i-\frac{1}{2},j} \qquad (6.1.47)$$

At shock points, the switching ensures that there is only one mesh point in the shock region since the corresponding cell is treated as fully supersonic or fully subsonic as soon as the shock cell is left. This results in a very sharp shock.

(e) Over-Relaxation Scheme

To solve (6.1.22) in the supersonic region, we write

$$(1 - M^2)(\phi^{n+1}_{i-2,j} - 2\phi^{n+1}_{i-1,j} + \overline{\phi^{n+1}_{i,j}}) + (\overline{\phi^{n+1}_{i,j+1}} - 2\overline{\phi^{n+1}_{i,j}} + \phi^{n+1}_{i,j-1}) = 0 \qquad (6.1.48)$$

where

$$\phi^{n+1} = \phi^n + \omega\left(\overline{\phi^{n+1}} - \phi^n\right) \qquad (6.1.49)$$

Denoting that

$$\Delta\phi = \phi^{n+1} - \phi^n \qquad (6.1.50)$$

we have

$$(M^2 - 1)_{i,j}(\omega\Delta\phi_{i-2,j} - 2\omega\Delta\phi_{i-1,j} + \Delta\phi_{i,j}) - (\Delta\phi_{i,j+1} - 2\Delta\phi_{i,j} + \Delta\phi_{i,j-1}) = \omega R^n_{i,j} \qquad (6.1.51)$$

or

$$(M^2 - 1)[\omega E_x^{-1}\delta_x^2 + (1 - \omega)]\Delta\phi_{i,j} - \delta_y^2\phi_{i,j} = \omega R^n_{i,j} \qquad (6.1.52)$$

where E is the shift operator ($E_x\phi_{i,j} = \phi_{i+1,j}$) and δ^2 is the central second difference operator. The equivalent artificial time dependent formulation is

$$(M^2 - 1)[\omega\phi_{xxt} + (1-\omega)\phi_t] - \phi_{yyt} = \frac{\omega}{\tau}R \tag{6.1.53}$$

where

$$\phi_t \cong \frac{\Delta\phi}{\tau} \tag{6.1.54}$$

with τ being a fictitious time step and where ϕ_{xxt} is backward differenced and R is the differential potential equation.

In (6.1.53), ϕ_{yy} is represented by

$$\phi_{yy} = \delta_y^2\phi^{n+1} + (\omega - 1)\delta_y^2\phi^n \tag{6.1.55}$$

but the appropriate procedure in the supersonic region is to march in the flow direction, such that $\phi_{i,j}^{n+1}$ can be determined only as a function of the new values $\phi_{i-2,j}^{n+1}$ and $\phi_{i-1,j}^{n+1}$ determined on the previous columns. This implies that ϕ_{yy} should be represented by $\delta_y^2\phi^{n+1}$ in the supersonic region. Note that the scheme (6.1.48) satisfies this requirement for $\omega = 1$. For a general relaxation procedure, this condition can be satisfied by taking the y-derivative terms at the new level $n+1$, instead of the intermediate level, introducing a factor ω in front of the y second difference operator of (6.1 52).

The analysis using the potential equation has been well established, but important physical phenomena such as rotational, nonisentropic, or nonisothermal effects are not taken into account in the governing equation. For this reason, the most general approach to the analysis of compressible inviscid flows must resort to the Euler equations. This is the subject of the next section.

6.2 EULER EQUATIONS

Compressible inviscid flows including rotational, nonisentropic, and nonisothermal effects require simultaneous solutions of continuity, momentum, and energy equations. In this approach, however, specialization for small perturbation or linearization outside of transonic flow as done in the potential equation can not be allowed. Thus, the difficulty encountered in transonic flows with shock discontinuities must be resolved with special computational schemes.

The most basic requirement for the solution of the Euler equations is to assure that solution schemes provide an adequate amount of artificial viscosity required for rapid convergence toward an exact solution. Furthermore, eigenvalues and compatibility relations associated with convection terms are important factors in the resolution of shock and expansion waves.

Solution schemes for the Euler equations may be grouped into three major categories: (1) central schemes, (2) first order upwind schemes, and (3) second order upwind schemes and essentially nonoscillatory schemes. These schemes are tabulated in Table 6.2.1 and elaborated in the following subsections.

Table 6.2.1 Various Computational Schemes for Euler Equations

Central Schemes	First Order Upwind Schemes	Second Order Upwind Schemes
1. Combined Space-Time Integration	**1. Flux Vector Splitting**	**1. Extrapolation**
(a) Explicit Schemes	Courant, Isaacson, and Reeves (1952)	**(a) Variable Extrapolation (MUSCL)**
Lax-Friendrichs – First order (1954)	Moretti (1979)	Van Leer (1979)
Lax-Wendroff – Second order (1960)	Steger and Warming (1981)	**(b) Flux Extrapolation**
(b) Two-Step Explicit Schemes	VanLeer (1982)	Van Leer (1979)
Richtmyer and Morton (1967)	**2. Godunov Methods-Riemann Solvers**	**2. Explicit TVD Upwind**
MacCormack (1969)	**(a) Exact Riemann Solvers**	VanLeer (1974)
LeRat and Peyret (1974)	Godunov (1959) – First order	Harten (1983)
(c) Implicit Schemes	VanLeer (1979) – Second order	Osher (1984)
MacCormack (1981)	Woodward and Colella (1984)	Osher and Chakravarthy (1984)
Casier, Deconinck, Hirsch (1983)	Ben-Artzi and Falcovitz (1984)	**3. Implicit TVD Upwind**
LeRat (1979, 1983)	**(b) Approximate Riemann Solvers**	Yee (1986)
2. Separate Space-Time Integration	Roe (1981)	**4. Central TVD Implicit or Explicit**
(a) Implicit Schemes	Enquist and Osher (1980)	Davis (1984)
Briley and McDonald (1975)	Osher (1982)	Roe (1985)
Beam and Warming (1976)	Harten, Lax, Van Leer (1983)	Yee (1985)
(b) Explicit Schemes (Multistage Runge-Kutta)		**5. Essentially Nonoscillatory Scheme**
Jameson, Schmidt, Turkel (1981)		Harten and Osher (1987)
		6. Flux Corrected Transport
		Boris and Book (1973)

6.2.1 MATHEMATICAL PROPERTIES OF EULER EQUATIONS

6.2.1.1 Quasilinearization of Euler Equations

The Euler equations may be linearized in terms of conservation variables or primitive (nonconservation) variables. Consider the conservation form of the Euler equations,

$$\frac{\partial \mathbf{U}}{\partial t} + \frac{\partial \mathbf{F}_i}{\partial x_i} = 0, \quad \text{or} \quad \frac{\partial \mathbf{U}}{\partial t} + \mathbf{a}_i \frac{\partial \mathbf{U}}{\partial x_i} = 0 \quad (i = 1, 2, 3) \tag{6.2.1}$$

with

$$\mathbf{U} = \begin{bmatrix} \rho \\ \rho \mathrm{v}_j \\ \rho E \end{bmatrix}, \qquad \mathbf{F}_i = \begin{bmatrix} \rho \mathrm{v}_i \\ \rho \mathrm{v}_i \mathrm{v}_j + p\delta_{ij} \\ \rho E \mathrm{v}_i + p\mathrm{v}_i \end{bmatrix}, \qquad \mathbf{a}_i = \frac{\partial \mathbf{F}_i}{\partial \mathbf{U}}, \tag{6.2.2}$$

For two dimensions, components of the convection Jacobian $\mathbf{a}_i$ $(i = 1, 2)$ are given by

$$\mathbf{a}_1 = \begin{bmatrix} 0 & 1 & 0 & 0 \\ \frac{(\gamma-3)u^2}{2} + \frac{(\gamma-1)v^2}{2} & (3-\gamma)u & -(\gamma-1)v & \gamma-1 \\ -uv & v & u & 0 \\ -\gamma uE + (\gamma-1)uq^2 & \gamma E - \frac{\gamma-1}{2}(v^2+3u^2) & -(\gamma-1)uv & \gamma u \end{bmatrix} \tag{6.2.3a}$$

$$\mathbf{a}_2 = \begin{bmatrix} 0 & 0 & 1 & 0 \\ -uv & v & u & 0 \\ \frac{(\gamma-3)v^2}{2} + \frac{(\gamma-1)u^2}{2} & -(\gamma-1)u & (3-\gamma)v & \gamma-1 \\ -\gamma vE + (\gamma-1)vq^2 & -(\gamma-1)uv & \gamma E - \frac{\gamma-1}{2}(u^2+3v^2) & \gamma v \end{bmatrix} \tag{6.2.3b}$$

Alternatively, the Euler equations may be written in nonconservation form for isentropic flow in terms of the primitive variable $\mathbf{V}$ as

$$\frac{\partial \mathbf{V}}{\partial t} + \mathbf{A}_i \frac{\partial \mathbf{V}}{\partial x_i} = 0 \tag{6.2.4}$$

with

$$\mathbf{V} = \begin{bmatrix} \rho \\ u \\ v \\ p \end{bmatrix} = \begin{bmatrix} \rho \\ u \\ v \\ (\gamma-1)\left(\rho E - \rho \frac{(u^2+v^2)}{2}\right) \end{bmatrix} \tag{6.2.5}$$

$$\mathbf{A}_1 = \begin{bmatrix} u & \rho & 0 & 0 \\ 0 & u & 0 & \frac{1}{\rho} \\ 0 & 0 & u & 0 \\ 0 & \rho a^2 & 0 & u \end{bmatrix}, \quad \mathbf{A}_2 = \begin{bmatrix} v & 0 & \rho & 0 \\ 0 & v & 0 & 0 \\ 0 & 0 & v & \frac{1}{\rho} \\ 0 & 0 & \rho a^2 & v \end{bmatrix} \tag{6.2.6}$$

Introducing a transformation between the conservation and nonconservation variables,

$$\mathbf{M} = \frac{\partial \mathbf{U}}{\partial \mathbf{V}} \tag{6.2.7}$$

or

$$\mathbf{M} = \begin{bmatrix} 1 & 0 & 0 & 0 \\ u & \rho & 0 & 0 \\ v & 0 & \rho & 0 \\ \frac{q^2}{2} & \rho u & \rho v & \frac{1}{\gamma-1} \end{bmatrix} \mathbf{M}^{-1} = \begin{bmatrix} 1 & 0 & 0 & 0 \\ \frac{-u}{\rho} & \frac{1}{\rho} & 0 & 0 \\ \frac{-v}{\rho} & 0 & \frac{1}{\rho} & 0 \\ \frac{\gamma-1}{2}q^2 & -(\gamma-1)u & -(\gamma-1)v & \gamma-1 \end{bmatrix} \tag{6.2.8}$$

and combining (6.2.1), (6.2.4), and (6.2.7), we obtain

$$\mathbf{M}\frac{\partial \mathbf{V}}{\partial t} + \mathbf{a}_i \mathbf{M}\frac{\partial \mathbf{V}}{\partial x_i} = 0 \tag{6.2.9}$$

Multiplying (6.2.9) by $\mathbf{M}^{-1}$, we obtain the form given by (6.2.4):

$$\frac{\partial \mathbf{V}}{\partial t} + \mathbf{A}_i \frac{\partial \mathbf{V}}{\partial x_i} = 0 \tag{6.2.10}$$

with

$$\mathbf{A}_i = \mathbf{M}^{-1}\mathbf{a}_i\mathbf{M}, \qquad \mathbf{a}_i = \mathbf{M}\mathbf{A}_i\mathbf{M}^{-1} \tag{6.2.11}$$

Note that **M** represents the transformation matrix between the conservation variables **U** and the primitive variables **V**.

6.2.1.2 Eigenvalues and Compatibility Relations

In order to examine the oscillatory behavior of the equations such as (6.2.4), we write **V** in the form

$$\mathbf{V} = \overline{\mathbf{V}} e^{\mathrm{I}(\boldsymbol{\kappa}\cdot\mathbf{x}-\omega t)} = \overline{\mathbf{V}} e^{\mathrm{I}(\kappa_i x_i - \omega t)} \tag{6.2.12}$$

Substituting (6.2.12) into (6.2.10) leads to

$$(-\omega + \mathbf{A}_i \kappa_i)\overline{\mathbf{V}} = 0 \tag{6.2.13}$$

or

$$|\mathbf{K} - \lambda \mathbf{I}| = 0 \tag{6.2.14}$$

with

$$\omega = \lambda \mathbf{I}, \quad \mathbf{A}_i \kappa_i = \mathbf{K}$$

For one dimension, (6.2.14) becomes

$$\begin{vmatrix} u-\lambda & \rho & 0 \\ 0 & u-\lambda & \dfrac{1}{\rho} \\ 0 & \rho a^2 & u-\lambda \end{vmatrix} = 0 \tag{6.2.15}$$

where $\lambda_1 = u$, $\lambda_2 = u + a$, $\lambda_3 = u - a$, constitute eigenvalues.

Return to (6.2.10) for one-dimensional case and write

$$\mathbf{L}^{-1}\frac{\partial \mathbf{V}}{\partial t} + \mathbf{L}^{-1}\mathbf{A}\frac{\partial \mathbf{V}}{\partial x} = 0, \quad \text{with} \quad \mathbf{L} = \frac{\partial \mathbf{V}}{\partial \mathbf{W}} \tag{6.2.16}$$

where $\mathbf{L}^{-1}$ is the matrix which will diagonalize the matrix $\mathbf{K} = \mathbf{A}_i\kappa_i$. Using (6.2.7) in (6.2.16) leads to

$$\mathbf{L}^{-1}\mathbf{M}^{-1}\left(\frac{\partial \mathbf{U}}{\partial t} + \mathbf{A}\frac{\partial \mathbf{U}}{\partial x}\right) = 0 \tag{6.2.17}$$

Similarly, we may define the variable $\mathbf{P}$ in the form

$$\mathbf{P} = \frac{\partial \mathbf{U}}{\partial \mathbf{W}} = \frac{\partial \mathbf{U}}{\partial \mathbf{V}}\frac{\partial \mathbf{V}}{\partial \mathbf{W}} = \mathbf{ML}, \qquad \mathbf{P}^{-1} = \mathbf{L}^{-1}\mathbf{M}^{-1}$$

so that the diagonalized eigenvalue matrix becomes

$$\boldsymbol{\Lambda} = \mathbf{L}^{-1}\mathbf{M}^{-1}\mathbf{KML} = \mathbf{P}^{-1}\mathbf{KP} \tag{6.2.18}$$

where $\mathbf{P}^{-1}$ and $\mathbf{P}$ denote the left eigenvector and right eigenvector, respectively.

Let us now postulate an existence of the characteristic variables $\mathbf{W}$ such that

$$\delta\mathbf{W} = \mathbf{L}^{-1}\delta\mathbf{V} \tag{6.2.19}$$

Substituting (6.2.19) into (6.2.16) yields

$$\frac{\partial \mathbf{W}}{\partial t} + \mathbf{L}^{-1}\mathbf{AL}\frac{\partial \mathbf{W}}{\partial x} = 0 \tag{6.2.20}$$

which is known as the compatibility equation.

The characteristic variables are also related by

$$\delta\mathbf{W} = \mathbf{P}^{-1}\delta\mathbf{U}, \quad \text{or} \quad \delta\mathbf{U} = \mathbf{P}\delta\mathbf{W} \tag{6.2.21}$$

Thus, the relations between the three sets of variables (Equations 6.2.7, 6.2.19, and 6.2.21) may be summarized as shown in Figure 6.2.1.

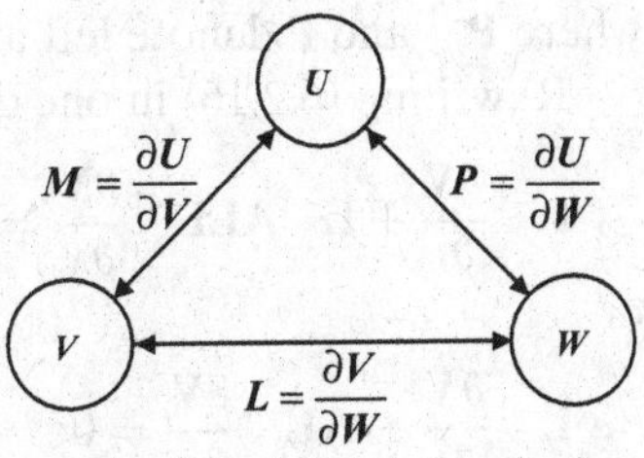

Figure 6.2.1 Relation between conservation variable **U**, primitive variable **V**, and characteristic variables **W**.

6.2.1.3 Characteristic Variables

The existence of characteristic variables postulated in (6.2.19) and (6.2.21) may now be examined for one-dimensional flow.

For the eigenvalues determined from (6.2.15), the three left eigenvectors of **K** are given by

$$\begin{bmatrix} \ell^{(1)} \\ \ell^{(2)} \\ \ell^{(3)} \end{bmatrix} = \begin{bmatrix} \alpha & 0 & -\dfrac{\alpha}{a^2} \\ 0 & \beta & \dfrac{\beta}{\rho a} \\ 0 & \delta & \dfrac{-\delta}{\rho a} \end{bmatrix} \tag{6.2.22}$$

where α, β, δ are the three normalization coefficients for the eigenvalues, $\lambda_1 = u$, $\lambda_2 = u + a$, $\lambda_3 = u - a$. With $\alpha = \beta = \delta = 1$, the diagonalization matrices are

$$\mathbf{L}^{-1} = \begin{bmatrix} 1 & 0 & -\dfrac{1}{a^2} \\ 0 & 1 & \dfrac{1}{\rho a} \\ 0 & 1 & \dfrac{-1}{\rho a} \end{bmatrix}, \quad \mathbf{L} = \begin{bmatrix} 1 & \dfrac{\rho}{2a} & -\dfrac{\rho}{2a} \\ 0 & \dfrac{1}{2} & \dfrac{1}{2} \\ 0 & \dfrac{\rho a}{2} & -\dfrac{\rho a}{2} \end{bmatrix} \tag{6.2.23}$$

where $\mathbf{L}^{-1}$ and $\mathbf{L}$ denote left and right eigenvector of **K**, respectively.

Similarly, transformation matrices $\mathbf{P}^{-1}$ and $\mathbf{P}$ can be derived.

$$\mathbf{P}^{-1} = \mathbf{L}^{-1}\mathbf{M}^{-1} = \begin{bmatrix} 1 - \dfrac{\gamma-1}{2}\dfrac{u^2}{a^2} & (\gamma-1)\dfrac{u}{a} & \dfrac{-(\gamma-1)}{a^2} \\ \left(\dfrac{\gamma-1}{2}u^2 - ua\right)\dfrac{1}{\rho a} & \dfrac{1}{\rho a}[a - (\gamma-1)u] & \dfrac{-(\gamma-1)}{\rho a} \\ -\left(\dfrac{\gamma-1}{2}u^2 + ua\right)\dfrac{1}{\rho a} & \dfrac{1}{\rho a}[a + (\gamma-1)u] & \dfrac{-(\gamma-1)}{\rho a} \end{bmatrix} \tag{6.2.24a}$$

$$\mathbf{P} = \mathbf{M}\mathbf{L} = \begin{bmatrix} 1 & \dfrac{\rho}{2a} & -\dfrac{\rho}{2a} \\ u & \dfrac{\rho}{2a}(u+a) & -\dfrac{\rho}{2a}(u-a) \\ \dfrac{u^2}{2} & \dfrac{\rho}{2a}\left(\dfrac{u^2}{2} + ua + \dfrac{a^2}{\gamma-1}\right) & -\dfrac{\rho}{2a}\left(\dfrac{u^2}{2} - ua + \dfrac{a^2}{\gamma-1}\right) \end{bmatrix} \tag{6.2.24b}$$

where $\mathbf{P}^{-1}$ and $\mathbf{P}$ denote left and right eigenvectors of $\mathbf{a}_i$, respectively.

Rewriting (6.2.16) in one dimension,

$$\mathbf{L}^{-1}\frac{\partial \mathbf{V}}{\partial t} + \mathbf{L}^{-1}\mathbf{A}\mathbf{L}\mathbf{L}^{-1}\frac{\partial \mathbf{V}}{\partial x} = 0 \tag{6.2.25a}$$

or

$$\mathbf{L}^{-1}\frac{\partial \mathbf{V}}{\partial t} + \mathbf{\Lambda}\mathbf{L}^{-1}\frac{\partial \mathbf{V}}{\partial x} = 0 \tag{6.2.25b}$$

where

$$\Lambda = \begin{bmatrix} u & & \\ & u+a & \\ & & u-a \end{bmatrix} \tag{6.2.26}$$

Expanding (6.2.25b) results in the continuity and momentum equations written in the form

$$\frac{\partial \rho}{\partial t} - \frac{1}{a^2}\frac{\partial p}{\partial t} + u\frac{\partial \rho}{\partial x} - \frac{u}{a^2}\frac{\partial p}{\partial x} = 0 \tag{6.2.27a}$$

$$\frac{\partial u}{\partial t} + \frac{1}{\rho a}\frac{\partial p}{\partial t} + (u+a)\left(\frac{\partial u}{\partial x} + \frac{1}{\rho a}\frac{\partial p}{\partial x}\right) = 0 \tag{6.2.27b}$$

$$\frac{\partial u}{\partial t} - \frac{1}{\rho a}\frac{\partial p}{\partial t} + (u-a)\left(\frac{\partial u}{\partial x} - \frac{1}{\rho a}\frac{\partial p}{\partial x}\right) = 0 \tag{6.2.27c}$$

which are known as compatibility equations. It follows from (6.2.19) or (6.2.27), by introducing an arbitrary variation δ, that

$$\delta W_1 = \delta\rho - \frac{1}{a^2}\delta p = 0 \tag{6.2.28a}$$

$$\delta W_2 = \delta u + \frac{1}{\rho a}\delta p = 0 \tag{6.2.28b}$$

$$\delta W_3 = \delta u - \frac{1}{\rho a}\delta p = 0 \tag{6.2.28c}$$

and subsequently from (6.2.20) or (6.2.28) that

$$\frac{\partial}{\partial t}\begin{bmatrix} W_1 \\ W_2 \\ W_3 \end{bmatrix} + \begin{bmatrix} u & & \\ & u+a & \\ & & u-a \end{bmatrix}\frac{\partial}{\partial x}\begin{bmatrix} W_1 \\ W_2 \\ W_3 \end{bmatrix} = 0 \tag{6.2.29}$$

If the characteristic variables W_1, W_2, W_3 remain constant, they are known as Riemann variables, Riemann invariants, or Riemann solution, defined as follows:

$$W_1 = \rho - \int \frac{dp}{a^2} = \text{constant along the } C_0 \text{ characteristic, stream line}$$

$$W_2 = u + \int \frac{dp}{\rho a} = \text{constant along the } C_+ \text{ characteristic}$$

$$W_3 = u - \int \frac{dp}{\rho a} = \text{constant along the } C_- \text{ characteristic}$$

These characteristic lines are schematically shown in Figure 6.2.2, and propagation of flow lines associated with characteristic lines are shown in Figure 6.2.3.

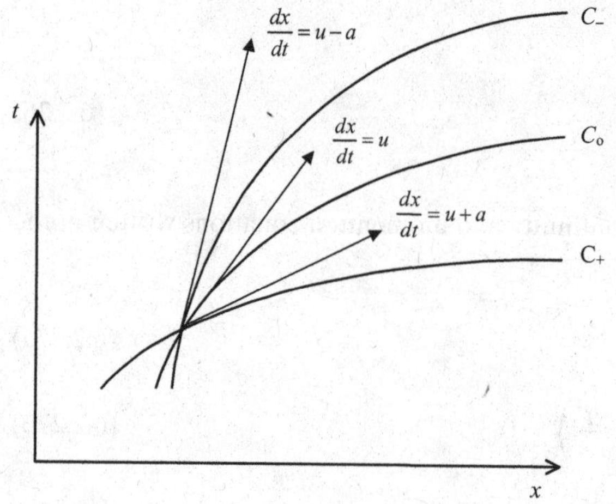

Figure 6.2.2 Characteristic lines for one-dimensional flow.

For isentropic flow it can be shown that

$$W_1 = \frac{p}{\rho^\gamma} \tag{6.2.30a}$$

$$W_2 = u + \frac{2a}{\gamma - 1} = J_+ \tag{6.2.30b}$$

$$W_3 = u - \frac{2a}{\gamma - 1} = J_- \tag{6.2.30c}$$

or

$$a = \frac{\gamma - 1}{4}(J_+ - J_-) \tag{6.2.31a}$$

$$u = \frac{1}{2}(J_+ J_-) \tag{6.2.31b}$$

If the values of J_+ and J_- are known at a given point in the x–t plane , then (6.2.31a,b) immediately give the local values of u and a at that point.

The propagation of flow lines associated with characteristic lines as related to expansion wave and shock waves are shown in Figure 6.2.3. The number of boundary conditions to be specified at inflow and outflow boundaries is determined by the eigenvalue spectrum of the Jacobian matrices (6.2.6) in terms of the primitive variables associated with the normal to the boundaries. Details for boundary conditions will be presented in Section 6.7.

6.2.2 CENTRAL SCHEMES WITH COMBINED SPACE-TIME DISCRETIZATION

Central finite differences may be formulated using combined space-time discretization (often known as Lax-Wendroff scheme). An alternative is to use independent space-time discretization which is discussed in Section 6.2.3. In this section, we shall examine the combined space-time discretization schemes.

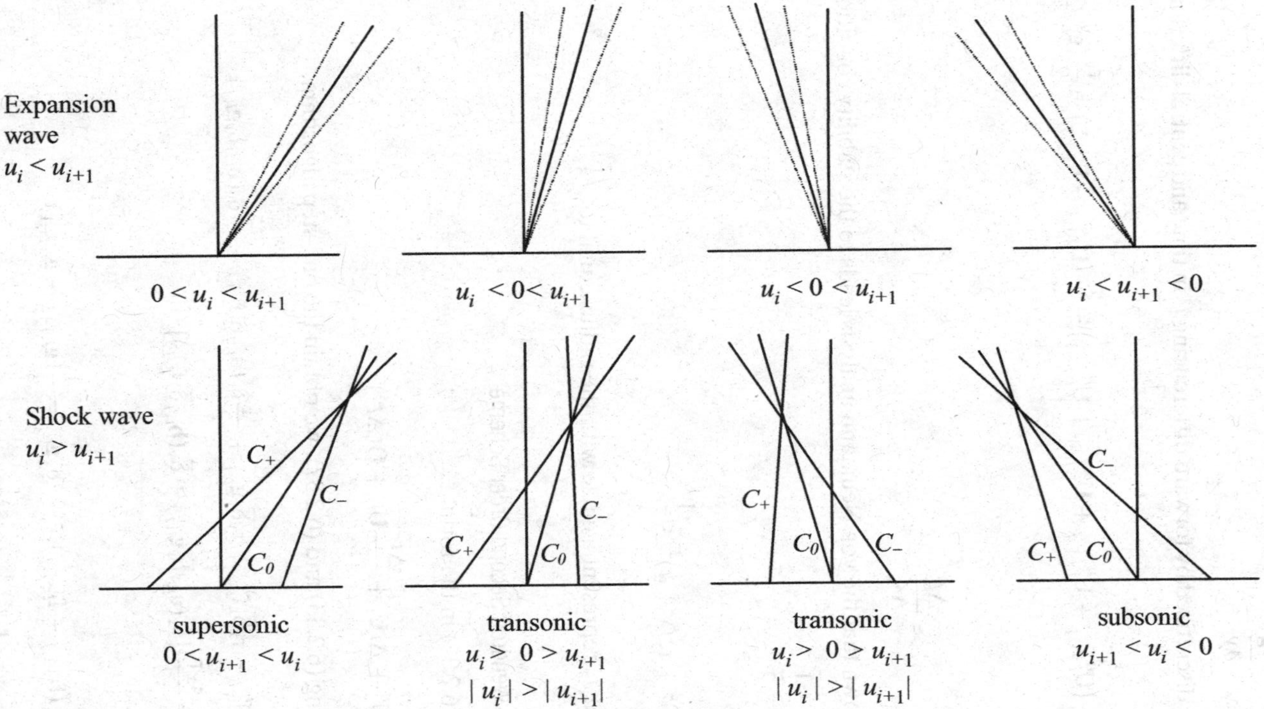

Figure 6.2.3 Propagation of flow quantities.

6.2.2.1 Lax-Friedrichs First Order Scheme

Consider the two-dimensional system of Euler equations in the form

$$\frac{\partial \mathbf{U}}{\partial t} + \frac{\partial \mathbf{f}}{\partial x} + \frac{\partial \mathbf{g}}{\partial y} = 0 \tag{6.2.32}$$

This may be discretized by forward differencing **U** in time and central differencing **f** and **g** in space.

$$\mathbf{U}_{i,j}^{n+1} = \frac{1}{4}\left(\mathbf{U}_{i+1,j}^{n} + \mathbf{U}_{i-1,j}^{n} + \mathbf{U}_{i,j+1}^{n} + \mathbf{U}_{i,j-1}^{n}\right) - \frac{\tau_x}{2}\left(\mathbf{f}_{i+1,j}^{n} - \mathbf{f}_{i-1,j}^{n}\right) - \frac{\tau_y}{2}\left(\mathbf{g}_{i,j+1}^{n} - \mathbf{g}_{i,j-1}^{n}\right) \tag{6.2.33}$$

with

$$\tau_x = \frac{\Delta t}{\Delta x}, \quad \tau_y = \frac{\Delta t}{\Delta y} \tag{6.2.34}$$

It can be shown that the von Neumann analysis leads to the stability condition,

$$J_x^2 + J_y^2 \leq \frac{1}{2} \tag{6.2.35}$$

or

$$\tau_x^2(u+a)^2 + \tau_y^2(\mathrm{v}+a)^2 \leq \frac{1}{2} \tag{6.2.36}$$

where J_x and J_y represent a circle with the radius equal to $\sqrt{\frac{1}{2}}$.

6.2.2.2 Lax-Wendroff Second Order Scheme

We rewrite (6.2.32) in the form

$$\mathbf{U}^{n+1} = \mathbf{U}^{n} + \Delta t \mathbf{U}_t + \frac{\Delta t^2}{2}\mathbf{U}_{tt} + \mathbf{O}(\Delta t^3) \tag{6.2.37}$$

and combining (6.2.33) and (6.2.37), we obtain the one-step algorithm,

$$\begin{aligned}\mathbf{U}_{i,j}^{n+1} = {} & \mathbf{U}_{i,j}^{n} - \tau_x \bar{\delta}_x \mathbf{f}_{i,j}^{n} - \tau_y \bar{\delta}_y \mathbf{g}_{i,j}^{n} + \frac{\tau_x^2}{2}\delta_x(\mathbf{a}_{i,j}\delta_x \mathbf{f}_{i,j}) + \frac{\tau_y^2}{2}\delta_y(\mathbf{b}_{i,j}\delta_y \mathbf{g}_{i,j}) \\ & + \frac{\tau_x \tau_y}{2}[\bar{\delta}_x(\mathbf{a}_{i,j}\bar{\delta}_y \mathbf{g}_{i,j}) + \bar{\delta}_y(\mathbf{b}_{i,j}\bar{\delta}_x \mathbf{f}_{i,j})]\end{aligned} \tag{6.2.38}$$

where

$$\bar{\delta}_x \mathbf{f}_{i,j} = \frac{1}{2}(\mathbf{f}_{i+1,j} - \mathbf{f}_{i-1,j}), \qquad \bar{\delta}_y \mathbf{g}_{i,j} = \frac{1}{2}(\mathbf{g}_{i,j+1} - \mathbf{g}_{i,j-1}) \tag{6.2.39}$$

$$\delta_x \mathbf{f}_{i,j} = \mathbf{f}_{i+1,j} - \mathbf{f}_{i-1,j}, \qquad \delta_y \mathbf{g}_{i,j} = \mathbf{g}_{i,j+1} - \mathbf{g}_{i,j-1} \tag{6.2.40}$$

etc.

Determination of Jacobians in (6.2.38) is cumbersome. To avoid this, it is preferable to employ a two-step scheme [Lerat and Peyret, 1974].

Step 1

$$\mathbf{U}_{i,j}^{n+\frac{1}{2}} = \frac{1}{4}\left(\mathbf{U}_{i+1,j}^n + \mathbf{U}_{i-1,j}^n + \mathbf{U}_{i,j+1}^n + \mathbf{U}_{i,j-1}^n\right) - \frac{\tau_x}{2}\left(\mathbf{f}_{i+1,j}^n - \mathbf{f}_{i-1,j}^n\right) - \frac{\tau_y}{2}\left(\mathbf{g}_{i,j+1}^n - \mathbf{g}_{i,j-1}^n\right) \tag{6.2.41}$$

Step 2

$$\mathbf{U}_{i,j}^{n+1} = \mathbf{U}_{i,j}^n - \tau_x\left(\mathbf{f}_{i+1,j}^{n+\frac{1}{2}} - \mathbf{f}_{i-1,j}^{n+\frac{1}{2}}\right) - \tau_y\left(\mathbf{g}_{i,j+1}^{n+\frac{1}{2}} - \mathbf{g}_{i,j-1}^{n+\frac{1}{2}}\right) \tag{6.2.42}$$

The stability condition is shown to be, for $\Delta x = \Delta y$

$$\frac{\Delta t}{\Delta x}(|\mathbf{v}| + a) \leq \frac{1}{\sqrt{2}}$$

The following two-step scheme was introduced by MacCormack and Paullay [1972]:

$$\overline{\mathbf{U}}_{i,j} = \mathbf{U}_{i,j}^n - \tau_x\left(\mathbf{f}_{i+1,j}^n - \mathbf{f}_{i,j}^n\right) - \tau_y\left(\mathbf{g}_{i,j+1}^n - \mathbf{g}_{i,j}^n\right) \tag{6.2.43}$$

$$\overline{\overline{\mathbf{U}}}_{i,j} = \mathbf{U}_{i,j}^n - \tau_x\left(\overline{\mathbf{f}}_{i,j} - \overline{\mathbf{f}}_{i-1,j}\right) - \tau_y\left(\overline{\mathbf{g}}_{i,j} - \overline{\mathbf{g}}_{i,j-1}\right) \tag{6.2.44}$$

$$\mathbf{U}^{n+1} = \frac{1}{2}\left(\overline{\mathbf{U}}_{i,j} + \overline{\overline{\mathbf{U}}}_{i,j}\right) \tag{6.2.45}$$

The corresponding stability condition is

$$\Delta t \leq \left[\frac{|\lambda(A)|\max}{\Delta x} + \frac{|\lambda(B)|\max}{\Delta y}\right]^{-1} \tag{6.2.46}$$

or

$$\Delta t \leq \frac{1}{(|u| + a)/\Delta x + (|\mathbf{v}| + a)/\Delta y} \leq \frac{\Delta x \Delta y}{|u|\Delta y + |\mathbf{v}|\Delta x + a\sqrt{(\Delta x)^2 + (\Delta y)^2}} \tag{6.2.47}$$

6.2.2.3 Lax-Wendroff Method with Artificial Viscosity

The Lax-Wendroff approaches with three-point central schemes lead to oscillations around sharp discontinuities. For one dimension,

$$\mathbf{U}_i^{n+1} - \mathbf{U}_i^n = -\tau\left(\mathbf{f}_{i+\frac{1}{2}} - \mathbf{f}_{i-\frac{1}{2}}\right) \tag{6.2.48}$$

where $\mathbf{f}_{i+\frac{1}{2}}$ and $\mathbf{f}_{i-\frac{1}{2}}$, called the numerical flux, are equal at steady state

$$\mathbf{f}_{i+\frac{1}{2}} = \frac{\mathbf{f}_{i+1} + \mathbf{f}_i}{2} - \frac{\tau}{2}a_{i+\frac{1}{2}}(\mathbf{f}_{i+1} - \mathbf{f}_i) \tag{6.2.49a}$$

$$\mathbf{f}_{i-\frac{1}{2}} = \frac{\mathbf{f}_i + \mathbf{f}_{i-1}}{2} - \frac{\tau}{2}a_{i-\frac{1}{2}}(\mathbf{f}_i - \mathbf{f}_{i-1}) \tag{6.2.49b}$$

Now with the artificial viscosity added to (6.2.49), we write

$$\mathbf{f}_{i+\frac{1}{2}} = \frac{\mathbf{f}_{i+1} + \mathbf{f}_i}{2} - \frac{1}{2}\tau a_{i+\frac{1}{2}}(\mathbf{f}_{i+1} - \mathbf{f}_i) + \mathbf{D}_{i+\frac{1}{2}}(\mathbf{U}_{i+1} - \mathbf{U}_i) \tag{6.2.50}$$

where $\mathbf{D}$ is any positive function of $\mathbf{U}_{i+1} - \mathbf{U}_i$ which vanishes at least linearly with $\mathbf{U}_{i+1} - \mathbf{U}_i$.

Substituting (6.2.50) into (6.2.48) leads to

$$\mathbf{U}_i^{n+1} - \mathbf{U}_i^n = -\tau \left(\mathbf{f}_{i+\frac{1}{2}} - \mathbf{f}_{i-\frac{1}{2}}\right)_{LW} + \tau \left[\mathbf{D}_{i+\frac{1}{2}}(\mathbf{U}_{i+1} - \mathbf{U}_i) - \mathbf{D}_{i-\frac{1}{2}}(\mathbf{U}_i - \mathbf{U}_{i-1})\right]$$

where it is seen that the artificial viscosity terms are those discretized as

$$\Delta x \frac{\partial}{\partial x}\left(\mathbf{D}\frac{\partial \mathbf{U}}{\partial x}\right)$$

Hence, the addition of an artificial viscosity term can be seen as

$$\mathbf{f}^{(AV)} = \mathbf{f} - \Delta x \mathbf{D}\frac{\partial \mathbf{U}}{\partial x}$$

or

$$\mathbf{f}_{i+\frac{1}{2}}^{(AV)} = \mathbf{f}_{i+\frac{1}{2}}^{(LW)} - \mathbf{D}_{i+\frac{1}{2}}(\mathbf{U}_{i+1} - \mathbf{U}_i)$$

6.2.2.4 Explicit MacCormack Method

Let us consider a quasi–one-dimensional problem such as occurs in a nozzle with a variable cross-sectional area S.

$$\frac{\partial S\mathbf{U}}{\partial t} + \frac{\partial S\mathbf{F}}{\partial x} - \frac{dS}{dx}\mathbf{B} = 0, \quad S = \hat{S}(x) \tag{6.2.51}$$

with

$$\mathbf{U} = \begin{bmatrix} \rho \\ \rho u \\ \rho E \end{bmatrix}, \quad \mathbf{F} = \begin{bmatrix} \rho u \\ \rho u^2 + p \\ (\rho E + p)u \end{bmatrix}, \quad \mathbf{B} = \begin{bmatrix} 0 \\ p \\ 0 \end{bmatrix}$$

An explicit MacCormack predictor-corrector scheme is formulated as follows:

Predictor

$$\frac{S\mathbf{U}_i^* - S\mathbf{U}_i^n}{\Delta t} + \frac{S\mathbf{F}_{i+1}^n - S\mathbf{F}_i^n}{\Delta x} - \left.\frac{dS}{dx}\right|_i \mathbf{B}_i^n = 0 \tag{6.2.52}$$

or

$$S\mathbf{U}_i^* = S\mathbf{U}_i^n - \frac{\Delta t}{\Delta x}(S\mathbf{F}_{i+1}^n - S\mathbf{F}_i^n) + \Delta t\left(\frac{dS}{dx}\right)_i \mathbf{B}_i^n + \mathbf{D}_i(\mathbf{U}) \tag{6.2.53}$$

where the artificial viscosity term $\mathbf{D}_j$ may be given by

$$\mathbf{D}_i(U) = \frac{\omega}{8}\left[S\mathbf{U}_{i+2}^n - 4S\mathbf{U}_{i+1}^n + 6S\mathbf{U}_i^n - 4S\mathbf{U}_{i-1} + S\mathbf{U}_{i-2}^n\right] \tag{6.2.54}$$

with $0 \leq \omega \leq 2$, $\Delta t = C\Delta x/u + a$.

Corrector

$$\mathbf{SU}_i^{n+1} = \frac{1}{2}\left\{\mathbf{SU}_i^n + \mathbf{SU}_i^* - \frac{\Delta t}{\Delta x}(\mathbf{SF}_i^* - \mathbf{SF}_{i-1}^*) + \Delta t\left(\frac{dS}{dx}\right)_i \mathbf{B}_i^* + \mathbf{D}_i^*(\mathbf{U})\right\} \tag{6.2.55}$$

Numerical applications for this case are demonstrated in Section 6.8.1.

6.2.3 CENTRAL SCHEMES WITH INDEPENDENT SPACE-TIME DISCRETIZATION

Instead of using combined space-time discretization, we may employ independent time discretization while maintaining central differences for space [Briley and McDonald, 1975; Beam and Warming, 1976, 1978]. We begin with the general form

$$\frac{d\mathbf{U}_{i,j}}{dt} = -\frac{1}{2}\left[\frac{\mathbf{f}_{i+1,j} - \mathbf{f}_{i-1,j}}{\Delta x} + \frac{\mathbf{g}_{i,j+1} - \mathbf{g}_{i,j-1}}{\Delta y}\right]$$

where various finite difference schemes of the time derivative term may be applied. The two level time integration of (6.2.51) leads to

$$(1+\xi)\Delta\mathbf{U}^{n+1} - \xi\Delta\mathbf{U}^n = \Delta t\theta\left[\left(\frac{\partial\mathbf{f}}{\partial x} + \frac{\partial\mathbf{g}}{\partial y}\right)^{n+1} - \left(\frac{\partial\mathbf{f}}{\partial x} + \frac{\partial\mathbf{g}}{\partial y}\right)^n\right]$$

with $\xi > -1/2$, $\theta \geq 1/2(\xi + 1)$ for linear stability.

The two-level integration scheme takes the form

$$(1+\xi)\Delta\mathbf{U}^{n+1} + \Delta t\theta\left(\frac{\partial\mathbf{f}}{\partial x} + \frac{\partial\mathbf{g}}{\partial y}\right)^{n+1} = -\Delta t\left(\frac{\partial\mathbf{f}}{\partial x} + \frac{\partial\mathbf{g}}{\partial y}\right)^n + \xi\Delta\mathbf{U}^n$$

or

$$\left[(1+\xi) + \Delta t\theta\left(\frac{\partial\mathbf{a}}{\partial x} + \frac{\partial\mathbf{b}}{\partial y}\right)\right]\Delta\mathbf{U}^{n+1} = -\Delta t\left(\frac{\partial\mathbf{f}}{\partial x} + \frac{\partial\mathbf{g}}{\partial y}\right)^n + \xi\Delta\mathbf{U}^n$$

Introducing a central discretization, we obtain

$$[(1+\xi) + \theta(\tau_x\overline{\delta}_x\mathbf{a} + \tau_y\overline{\delta}_y\mathbf{b})]\Delta\mathbf{U}_{i,j}^{n+1} = -\tau_x\overline{\delta}_x\mathbf{f}_{i,j}^n + \tau_y\overline{\delta}_y\mathbf{g}_{i,j}^n + \xi\Delta\mathbf{U}_{i,j}^n \tag{6.2.56}$$

or

$$\begin{aligned}
&(1+\xi)\Delta\mathbf{U}_{i,j}^{n+1} + \theta\frac{\Delta t}{2\Delta x}(\mathbf{a}_{i+1,j}\Delta\mathbf{U}_{i+1,j} - \mathbf{a}_{i-1,j}\Delta\mathbf{U}_{i-1,j})^{n+1} \\
&\quad + \theta\frac{\Delta t}{2\Delta y}(\mathbf{b}_{i,j+1}\Delta\mathbf{U}_{i,j+1} - \mathbf{b}_{i,j-1}\Delta\mathbf{U}_{i,j-1})^{n+1} \\
&\quad = -\Delta t\left(\frac{\mathbf{f}_{i+1,j}^n - \mathbf{f}_{i-1,j}^n}{2\Delta x} + \frac{\mathbf{g}_{i,j+1}^n - \mathbf{g}_{i,j-1}^n}{2\Delta y}\right) + \xi\Delta\mathbf{U}_{i,j}^n
\end{aligned} \tag{6.2.57}$$

and with an ADI factorization, for $\xi = 0$

$$(1 + \theta\tau_x\overline{\delta}_x)(1 + \theta\tau_y\overline{\delta}_y\mathbf{b}^n)\Delta\overline{\mathbf{U}}_{i,j} = -(\tau_x\overline{\delta}_x\mathbf{f}_{i,j}^n + \tau_y\overline{\delta}_y\mathbf{g}_{i,j}^n) \tag{6.2.58a}$$

$$(1 + \theta\tau_y\overline{\delta}_y\mathbf{b}^n)\Delta\mathbf{U}_{i,j}^{n+1} = \Delta\overline{\mathbf{U}}_{i,j} \tag{6.2.58b}$$

Notice that each step is a tridiagonal system along the x lines for $\Delta\overline{\mathbf{U}}$ and along the y lines for $\Delta\mathbf{U}^{n+1}$.

6.2.4 FIRST ORDER UPWIND SCHEMES

In general, the central schemes tend to provide excessive damping with shock discontinuities not well resolveld. To compensate for this trend, first order upwind schemes can be used. However, overshoots and undershoots may occur at discontinuities. A remedy for this difficulty can be provided by low- or high-resolution second order upwind schemes. In this section, we discuss the first order upwind schemes, followed by the second order upwind schemes in Section 6.2.5. High-resolution second order upwind schemes will be discussed in Section 6.2.6. The first order upwind schemes are divided into two groups: flux vector splitting schemes and Godunov schemes. These and other topics are presented below.

6.2.4.1 Flux Vector Splitting Method

The basic strategy here is to split the flux and eigenvalues into positive and negative components and apply the one-dimensional splitting to each flux component separately according to the sign of the associated eigenvalues. This method is known as the flux vector splitting method.

Consider the two-dimensional flow in the form

$$\frac{\partial \mathbf{U}}{\partial t} + \frac{\partial \mathbf{f}}{\partial x} + \frac{\partial \mathbf{g}}{\partial y} = 0 \tag{6.2.59a}$$

or

$$\frac{\partial \mathbf{U}}{\partial t} + \mathbf{A}\frac{\partial \mathbf{U}}{\partial x} + \mathbf{B}\frac{\partial \mathbf{U}}{\partial y} = 0 \tag{6.2.59b}$$

with the convection Jacobians **A** and **B** as related by diagonalized eigenvalue matrices,

$$\Lambda_1 = \mathbf{P}_1^{-1}\mathbf{A}\mathbf{P}_1 = \begin{bmatrix} u & & & \\ & u & & \\ & & u+a & \\ & & & u-a \end{bmatrix} \tag{6.2.60a}$$

$$\Lambda_2 = \mathbf{P}_2^{-1}\mathbf{B}\mathbf{P}_2 = \begin{bmatrix} \mathrm{v} & & & \\ & \mathrm{v} & & \\ & & \mathrm{v}+a & \\ & & & \mathrm{v}-a \end{bmatrix} \tag{6.2.60b}$$

For the one-dimensional problem, we have

$$\mathbf{f}^+ = \mathbf{f}, \quad \mathbf{f}^- = \mathbf{0} \quad \text{for supersonic flow}$$
$$\mathbf{f}^+ = \mathbf{0}, \quad \mathbf{f}^- = \mathbf{f} \quad \text{for subsonic flow}$$

and

$$\Lambda_1^+ = \begin{bmatrix} u & & & \\ & u & & \\ & & u+a & \\ & & & 0 \end{bmatrix}, \quad \Lambda_1^- = \begin{bmatrix} 0 & & & \\ & 0 & & \\ & & 0 & \\ & & & u-a \end{bmatrix}$$

with

$$\mathbf{A}^+ = \mathbf{P}_1\Lambda_1^+\mathbf{P}_1^{-1}, \quad \mathbf{A}^- = \mathbf{P}_1\Lambda_1^-\mathbf{P}_1^{-1}$$

and similarly for **B**. The split fluxes are defined by

$$\mathbf{f}^\pm = \mathbf{A}^\pm\mathbf{U} \quad \mathbf{g}^\pm = \mathbf{B}^\pm\mathbf{U}$$

The general eigenvalue matrix may written as given by Steger and Warming [1980]

$$\Lambda = \begin{bmatrix} \lambda_1 & & & \\ & \lambda_2 & & \\ & & \lambda_3 & \\ & & & \lambda_4 \end{bmatrix} \tag{6.2.61}$$

which will allow the split flux components to be written as follows:

$$f = \frac{\rho}{2\gamma}\begin{bmatrix} \eta \\ \eta u + a(\lambda_2 - \lambda_3) \\ \eta v \\ \eta\frac{u^2+v^2}{2} + ua(\lambda_2 - \lambda_3) + a^2\frac{\lambda_2+\lambda_3}{\gamma-1} \end{bmatrix} \tag{6.2.62a}$$

$$g = \frac{\rho}{2\gamma}\begin{bmatrix} \eta \\ \eta u \\ \eta v + a(\lambda_2 - \lambda_3) \\ \eta\frac{u^2+v^2}{2} + va(\lambda_2 - \lambda_3) + a^2\frac{\lambda_2+\lambda_3}{\gamma-1} \end{bmatrix} \tag{6.2.62b}$$

with

$$\eta = 2(\gamma - 1)\lambda_1 + \lambda_2 + \lambda_3$$

Rewriting (6.2.59a) in a discrete form for a variable cross section $S(x)$:

$$\mathbf{U}_{i,j}^{n+1} - \mathbf{U}_{i,j}^n = -\frac{\Delta t}{\Delta x}\left(\mathbf{f}^*_{i+\frac{1}{2},j} - \mathbf{f}^*_{i-\frac{1}{2},j}\right) - \frac{\Delta t}{\Delta y}\left(\mathbf{g}^*_{i,j+\frac{1}{2}} - \mathbf{g}^*_{i,j-\frac{1}{2}}\right) \tag{6.2.63}$$

with

$$\mathbf{f}^*_{i+\frac{1}{2},j} = \mathbf{f}^-_{i+1,j} + \mathbf{f}^+_{i,j}, \qquad \mathbf{g}^*_{i,j+\frac{1}{2}} = \mathbf{g}^-_{i,j+1} + \mathbf{g}^+_{i,j} \tag{6.2.64}$$

For quasi–one-dimensional problems such as a nozzle with variable cross-section area considered in (6.2.51), the solution procedure using the flux vector splitting is presented below.

$$\frac{\partial S\mathbf{U}}{\partial t} + \frac{\partial S\mathbf{F}}{\partial x} - \frac{dS}{dx}\mathbf{B} = 0 \tag{6.2.65}$$

Linearizing the above,

$$\mathbf{F}^{n+1} = \mathbf{F}^n + \frac{\partial \mathbf{F}^n}{\partial t}\Delta t = \mathbf{F}^n + \frac{\partial \mathbf{F}^n}{\partial \mathbf{U}}\Delta\mathbf{U} = \mathbf{F}^n + \mathbf{a}\Delta\mathbf{U}$$

$$\mathbf{B}^{n+1} = \mathbf{B}^n + \frac{\partial \mathbf{B}^n}{\partial t}\Delta t = \mathbf{B}^n + \frac{\partial \mathbf{B}^n}{\partial \mathbf{U}}\Delta\mathbf{U} = \mathbf{B}^n + \mathbf{b}\Delta\mathbf{U}$$

where

$$F_1 = \rho u = U_2$$

$$F_2 = \rho u^2 + p = \frac{\rho^2 u^2}{\rho} + (\gamma - 1)\left(\rho E - \frac{\rho^2 u^2}{2\rho}\right) = \frac{U_2^2}{U_2} + (\gamma - 1)\left(U_3 - \frac{U_2^2}{2U_1}\right)$$

$$= \frac{3-\gamma}{2}\left(\frac{U_2^2}{U_1}\right) + (\gamma - 1)U_3$$

$$F_3 = (\rho E + p)u = \frac{\gamma U_3 U_2}{U_1} - (\gamma - 1)\frac{U_2^2}{2U_1^2}$$

The flux vector **F** can be split into subvectors such that each is associated with either positive or negative eigenvalues of **a**.

$$\mathbf{a} = \mathbf{a}^+ + \mathbf{a}^- = \mathbf{P\Lambda^+P^{-1}} + \mathbf{P\Lambda^-P^{-1}}$$

$$\Lambda = \Lambda^+ + \Lambda^- = \begin{bmatrix} u & 0 & 0 \\ 0 & u+a & 0 \\ 0 & 0 & 0 \end{bmatrix} + \begin{bmatrix} 0 & 0 & 0 \\ 0 & 0 & 0 \\ 0 & 0 & u-a \end{bmatrix}$$

$$\mathbf{F}^+ = \mathbf{a}^+\mathbf{U}, \quad \mathbf{F}^- = \mathbf{a}^-\mathbf{U}, \quad \mathbf{a}^+ = \mathbf{P\Lambda^+P^{-1}}, \quad \mathbf{a}^- = \mathbf{P\Lambda^-P^{-1}}$$

For $M < 1\ (u < a)$

$$\left.\begin{matrix} u > 0 \\ u + a > 0 \\ u - a < 0 \end{matrix}\right\} \quad a^+ = 0, \quad a^- = a$$

For $M > 1\ (u > a)$

$$\left.\begin{matrix} u > 0 \\ u + a > 0 \\ u - a > 0 \end{matrix}\right\}, \quad a^+ = a, \quad a^- = 0$$

The above criteria require that backward differencing (upwinding) be used for terms associated with positive eigenvalues, whereas forward differencing should be used for terms involved in negative eigenvalues. Furthermore, the number of boundary conditions to apply are also dictated by the eigenvalues, compatibility relations, and characteristic variables as discussed in Sections 6.2.1 and 6.7.

In order to write the finite difference equations, we return to the governing equation (6.2.65) and obtain the discretized form

$$S\frac{\Delta \mathbf{U}}{\Delta t} + \frac{\partial}{\partial x}\left[S\left(\mathbf{F} + \frac{\partial \mathbf{F}}{\partial \mathbf{U}}\Delta \mathbf{U}\right)\right] - \frac{dS}{dx}\left[\mathbf{B} + \frac{\partial \mathbf{B}}{\partial \mathbf{U}}\Delta \mathbf{U}\right] = 0$$

In terms of the Jacobians,

$$S\frac{\Delta \mathbf{U}}{\Delta t} + \frac{\partial}{\partial x}(S\mathbf{a}\Delta \mathbf{U}) - \frac{dS}{dx}(\mathbf{b}\Delta \mathbf{U}) = -\frac{\partial S\mathbf{F}}{\partial x} + \frac{dS}{dx}\mathbf{B}$$

or

$$\left[S\mathbf{I} + \Delta t\frac{\partial}{\partial x}S\mathbf{a} - \Delta t\frac{dS}{dx}\mathbf{b}\right]\Delta \mathbf{U} = -\Delta t\left[\frac{\partial S\mathbf{F}}{\partial x} - \frac{dS}{dx}\mathbf{B}\right]$$

Introducing the flux vector splitting, we write

$$\left\{S\mathbf{I} + \Delta t\left[\frac{\partial}{\partial x}(S\mathbf{a}^+ + S\mathbf{a}^-) - \frac{dS}{dx}\mathbf{b}\right]\right\}\Delta\mathbf{U} = -\Delta t\left[\frac{\partial}{\partial x}(S\mathbf{F}^+ + S\mathbf{F}^-) - \frac{dS}{dx}\mathbf{B}\right]$$

A backward (upwind) differencing is used for $\mathbf{a}^+$ and $\mathbf{F}^+$ as follows:

$$\left\{S\mathbf{I} + \frac{\Delta t}{\Delta x}[(S\mathbf{a}_j^+ - S\mathbf{a}_{j-1}^+) + (S\mathbf{a}_{j+1}^- - S\mathbf{a}_j^-)] - \Delta t\frac{dS}{dx}\mathbf{b}_j\right\}\Delta\mathbf{U}$$
$$= -\frac{\Delta t}{\Delta x}[(S\mathbf{F}_j^+ - S\mathbf{F}_{j-1}^+) + (S\mathbf{F}_{j+1}^- - S\mathbf{F}_j^-)] + \Delta t\frac{dS}{dx}\mathbf{B}_j$$

The above results may be rearranged in the form:

$$\alpha\Delta\mathbf{U}_{j+1} + \beta\Delta\mathbf{U}_j + \gamma\Delta\mathbf{U}_{j-1} = \delta \tag{6.2.66}$$

where

$$\alpha = \frac{\Delta t}{\Delta x}S_{j+1}\mathbf{a}_{j+1}^-$$

$$\beta = S\mathbf{I} + \frac{\Delta t}{\Delta x}(S_j\mathbf{a}_j^+ - S_j\mathbf{a}_j^-) - \Delta t\left.\frac{dS}{dx}\right|_j \mathbf{b}_j$$

$$\gamma = -\frac{\Delta t}{\Delta x}S_{j-1}\mathbf{a}_{j-1}^+$$

$$\delta = -\frac{\Delta t}{\Delta x}(S_j\mathbf{F}_j^+ - S_{j-1}\mathbf{F}_{j-1}^+ + S_{j+1}\mathbf{F}_{j+1}^- - S_j\mathbf{F}_j^-) + \Delta t\left.\frac{dS}{dx}\right|_j \mathbf{B}_j$$

with

$$\Delta\mathbf{U} = \mathbf{U}^{n+1} - \mathbf{U}^n$$

Here it is seen that for supersonic flow, $\mathbf{a}^- = 0$, making the scheme upwinded with the diagonal term β being maximum. For subsonic flow, the diagonal term β is still large with the eigenvalue Λ^- or $\mathbf{a}^-$ being negative. The scheme provides a stable solution. A numerical example for this problem is demonstrated in Section 6.8.1.

The flux vector splitting has been applied to many first and second order upwind schemes such as in MUSCL (monotone upstream centered schemes for conservation laws) and TVD (total variation diminishing). For example, in the construction of the second order schemes, MUSCL approach has been used to extrapolate the primitive variables to the cell interface rather than the fluxes. Similar approaches and various other versions of flux vector splitting were used for TVD. These and other topics on the flux vector splitting will be discussed in Section 6.2.5.

6.2.4.2 Godunov Methods

The basic idea of a Godunov scheme is to use the finite volume structure of spatial discretization (Figure 6.2.4a) and a piecewise constant distribution of the variable u with the shock discontinuities occurring at each cell interface in order to obtain the

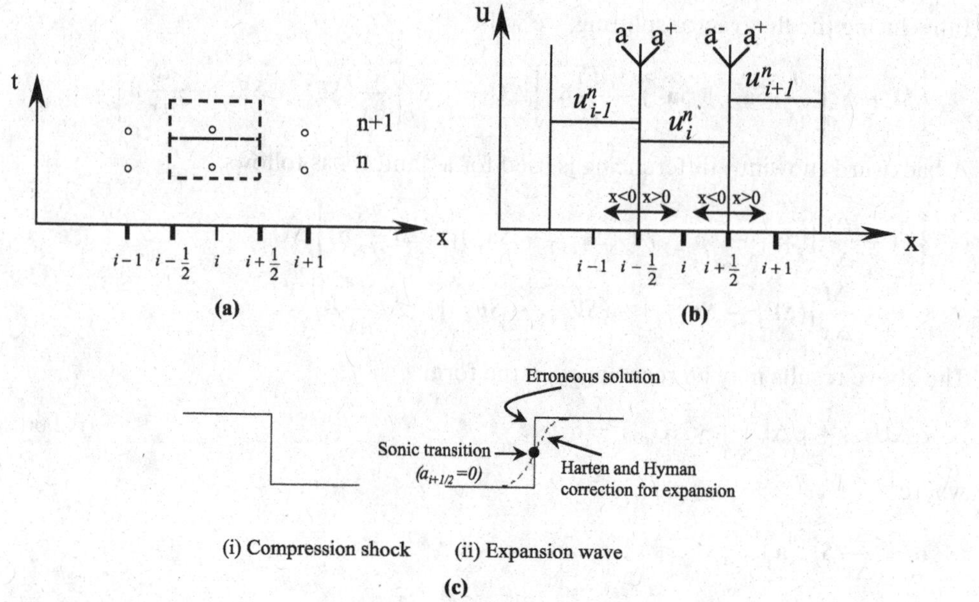

Figure 6.2.4 Control volume and piecewise constant distribution of u. **(a)** Control volume for Godunov method. **(b)** Piecewise constant distribution of u at $t = n\Delta t$. **(c)** Compression shock and expansion wave.

exact Riemann solution (Figure 6.2.4b). Here, the dependent variable u may be written as

$$u_i^{n+1} = u_i^n - \frac{\Delta t}{\Delta x}[f(u_{i+1/2}) - f(u_{i-1/2})] \tag{6.2.67a}$$

with the value of u over the volume element given by the average value u_i,

$$u_i = \frac{1}{\Delta x}\int_{x-\Delta x/2}^{x+\Delta x/2} u(x,t)\,dx$$

and the flux time-averaged at the control volume surface:

$$f = \frac{1}{\Delta t}\int_t^{t+\Delta t} f dt$$

As an exact Riemann solution, we note that (6.2.67a) can be reduced to

$$u_i^{n+1} = u_i^n - \frac{a\Delta t}{\Delta x}[u_i^n - u_{i-1}^n] \quad \text{for } a > 0 \tag{6.2.67b}$$

$$u_i^{n+1} = u_i^n - \frac{a\Delta t}{\Delta x}[u_{i+1}^n - u_i^n] \quad \text{for } a < 0 \tag{6.2.67c}$$

with $|a_{\max}|\Delta t/\Delta x \leq 1/2$ for stability. This is because the wave can travel at almost half the cell.

Godunov's idea has been extended for improvements by various investigators including Roe [1981] and Enquist and Osher [1980] among others. One of the most widely used schemes is the Roe's approximate Riemann solver. This scheme is described below.

Roe's Approximate Riemann Solver

The original Godunov scheme (6.2.67) may be *approximated* by splitting the Jacobian a into positive and negative components as

$$a_{i-1/2} = a^+_{i-1/2} - a^-_{1-1/2}, \qquad a_{i+1/2} = a^+_{i+1/2} - a^-_{1+1/2}$$

For $a > 0$, the upwinding scheme is given by

$$u_i^{n+1} = u_i^n - \frac{\Delta t}{\Delta x}(f_i - f_{i-1})$$

with the flux terms split in terms of positive and negative Jacobians,

$$\begin{aligned} f_{i-1/2} - f_{i-1} &= a^-_{i-1/2}(u_i - u_{i-1}) \\ f_i - f_{i-1/2} &= a^+_{i-1/2}(u_i - u_{i-1}) \end{aligned} \tag{6.2.68a,b}$$

Subtracting (6.2.68b) from (6.2.68a) leads to

$$f_{i-1/2} = \frac{f_i + f_{i-1}}{2} - \frac{1}{2}|a_{i-1/2}|(u_i - u_{i-1}) = f^*_{i-1/2} \tag{6.2.69a}$$

with the symbol * representing the first order upwind numerical flux at $i - 1/2$.

Similarly, for $a < 0$, the upwinding scheme is written as

$$u_i^{n+1} = u_i^n - \frac{\Delta t}{\Delta x}(f_{i+1} - f_i)$$

with

$$f_{i+1/2} - f_i = a^-_{i+1/2}(u_{i+1} - u_i)$$

$$f_{i+1} - f_{i+1/2} = a^+_{i+1/2}(u_{i+1} - u_i)$$

from which we obtain

$$f_{i+1/2} = \frac{f_{i+1} + f_i}{2} - \frac{1}{2}|a_{i+1/2}|(u_{i+1} - u_i) = f^*_{i+1/2} \tag{6.2.69b}$$

with * denoting the first order upwind numerical flux at $i + 1/2$.

Unfortunately, the scheme given by (6.2.68b) above does not recognize the possible occurrence of expansion wave at a sonic transition identified by

$$|a_{i+1/2}| = a^+_{i+1/2} - a^-_{i+1/2} = 0$$

at which the scheme computes as a shock discontinuity that represents a nonphysical behavior, violating the entropy condition. A remedy for this situation can be found in Harten and Hyman [1983] in which the following modification is made to (6.2.69b):

$$|a_{i+1/2}| = \begin{cases} |a_{i+1/2}| & \text{for } |a_{i+1/2}| \geq \varepsilon \\ \varepsilon & \text{for } |a_{i+1/2}| < \varepsilon \end{cases}, \qquad \varepsilon = \max\left(0, \left|\frac{a_{i+1} - a_i}{2}\right|\right) \tag{6.2.70}$$

Note that for expansion we have $\varepsilon = (u_{i+1} - u_i)/2$ and this requires a modification. On the other hand, for compression ($\varepsilon = 0$), no modification is needed. This accommodation will allow a correct expansion wave to develop instead of the nonphysical discontinuity as shown in Figure 6.2.4c(ii).

The expressions given by (6.2.69a,b) can be substituted into (6.2.67a) in the form of finite volume discretization. This process can easily be formulated in terms of diagonalized Jacobians with eigenvalues and eignevectors for multidimensional Euler equations as shown in Section 6.2.1. Thus, the first order upwind scheme for the solution of Euler equations is of the form:

$$u_i^{n+1} = u_i^n - \frac{\Delta t}{\Delta x}\left[f_{i+1/2}^* - f_{i-1/2}^*\right] \tag{6.2.71}$$

with the numerical fluxes determined from (6.2.69a,b).

6.2.5 SECOND ORDER UPWIND SCHEMES WITH LOW RESOLUTION

There are two approaches for the second order upwind schemes with low resolution: (1) variable extrapolation and (2) flux extrapolation. In each of these approaches, an additional predictor step may or may not be included. In these schemes it is intended that the second order upwind approaches lead to greater accuracy.

(1) Variable Extrapolation – MUSCL Approach

In this approach, known as Monotone Upstream-Centered Schemes for Conservation Laws (MUSCL) [Van Leer, 1979], the variables are extrapolated instead of the flux terms.

$$\mathbf{f}_{i+\frac{1}{2}}^{**} = \frac{1}{2}\left[\mathbf{f}\left(\mathbf{U}_{i+\frac{1}{2}}^L\right) + \mathbf{f}\left(\mathbf{U}_{i+\frac{1}{2}}^R\right) - |\mathbf{a}|_{i+\frac{1}{2}}\left(\mathbf{U}_{1+\frac{1}{2}}^R - \mathbf{U}_{i+\frac{1}{2}}^L\right)\right] \tag{6.2.72}$$

with ** representing the second order scheme and

$$\mathbf{U}_{i+\frac{1}{2}}^L = \mathbf{U}_i + \frac{1}{4}[(1-\kappa)(\mathbf{U}_i - \mathbf{U}_{i-1}) + (1+\kappa)(\mathbf{U}_{i+1} - \mathbf{U}_i)]$$

$$\mathbf{U}_{i+\frac{1}{2}}^R = \mathbf{U}_{i+1} - \frac{1}{4}[(1+\kappa)(\mathbf{U}_{i+1} - \mathbf{U}_i) + (1-\kappa)(\mathbf{U}_{i+2} - \mathbf{U}_{i+1})]$$

where the superscripts L and R refer to the left and right sides at the considered boundary and κ denotes a weight ($\kappa = -1, 0, 1$) leading to various extrapolation schemes Figure 6.2.5a,b).

The final solution is obtained as

$$\mathbf{U}_i^{n+1} = \mathbf{U}_i^n - \tau\left(\mathbf{f}_{i+\frac{1}{2}}^{**} - \mathbf{f}_{i-\frac{1}{2}}^{**}\right) \tag{6.2.73}$$

Second order upwind schemes in space and time are obtained with an additional predictor step

$$\mathbf{U}_{i+\frac{1}{2}}^{L^*} = \overline{\mathbf{U}}_i + \frac{1}{4}[(1-\kappa)(\mathbf{U}_i - \mathbf{U}_{i-1}) + (1+\kappa)(\mathbf{U}_{i+1} - \mathbf{U}_i)] \tag{6.2.74}$$

$$\overline{\mathbf{U}}_i = \mathbf{U}_i^n - \frac{\Delta t}{2\Delta x}\left(\mathbf{f}_{i+\frac{1}{2}}^* - \mathbf{f}_{i-\frac{1}{2}}^*\right) \tag{6.2.75a}$$

$$\mathbf{U}_{i+\frac{1}{2}}^{R^*} = \overline{\mathbf{U}}_{i+1} - \frac{1}{4}[(1+\kappa)(\mathbf{U}_{i+1} - \mathbf{U}_i) + (1-\kappa)(\mathbf{U}_{i+2} - \mathbf{U}_{i+1})] \tag{6.2.75b}$$

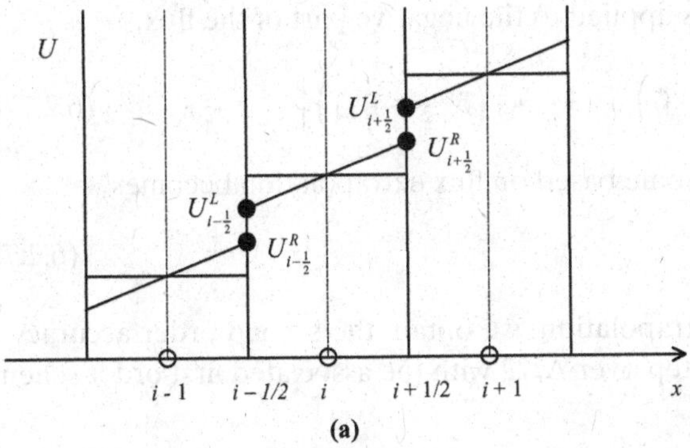

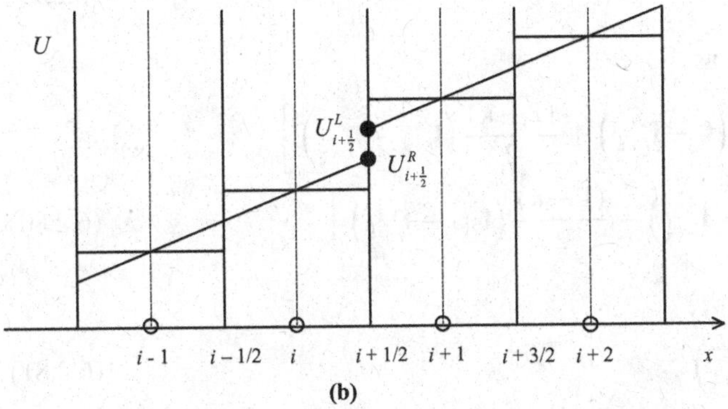

Figure 6.2.5 Variable extrapolation. **(a)** Piecewise linear representation within cells. **(b)** Linear one-sided extrapolation of interface values for $\kappa = -1$.

with * indicating the first order approximation. Thus, (6.2.72) may be replaced by

$$\overline{\mathbf{f}^{**}_{i+\frac{1}{2}}} = \frac{1}{2}\left[\mathbf{f}^*\left(\mathbf{U}^{L^*}_{i+\frac{1}{2}}\right) + \mathbf{f}^*\left(\mathbf{U}^{R^*}_{i+\frac{1}{2}}\right) - |\mathbf{a}|_{i+\frac{1}{2}}\left(\mathbf{U}^{R^*}_{i+\frac{1}{2}} - \mathbf{U}^{L^*}_{i+\frac{1}{2}}\right)\right]$$

Finally, we obtain

$$\mathbf{U}^{n+1}_i = \mathbf{U}^n_i - \tau\left(\overline{\mathbf{f}^{**}_{i+\frac{1}{2}}} - \overline{\mathbf{f}^{**}_{i-\frac{1}{2}}}\right) \tag{6.2.76}$$

This is one of the most widely used schemes for capturing discontinuities in capturing shock discontinuities in compressible flows.

(2) Flux Extrapolation Approach

In the previous approach, the state variables are directly extrapolated to the cell interfaces. The fluxes at the cell boundaries are then calculated from these values. In the flux extrapolation approach, the fluxes in the cell are directly extrapolated to the boundaries.

The extrapolation formulas for the fluxes are the same as the formulas applied to the variables. A general backward extrapolation of the positive flux is given by

$$\mathbf{f}^{+b}_{i+\frac{1}{2}} = \mathbf{f}^+_i + \frac{1}{4}\left[(1-\kappa)\left(\mathbf{f}_i - \mathbf{f}^*_{i-\frac{1}{2}}\right) + (1+\kappa)\left(\mathbf{f}_{i+1} - \mathbf{f}^*_{i+\frac{1}{2}}\right)\right] \tag{6.2.77a}$$

whereas a forward extrapolation is applied to the negative part of the flux,

$$\mathbf{f}^{-f}_{i+\frac{1}{2}} = \mathbf{f}^{-}_{i+1} - \frac{1}{4}\left[(1+\kappa)\left(\mathbf{f}^{*}_{i+\frac{1}{2}} - \mathbf{f}_i\right) + (1-\kappa)\left(\mathbf{f}^{*}_{i+\frac{3}{2}} - \mathbf{f}_{i+1}\right)\right] \tag{6.2.77b}$$

Thus, the second order upwind scheme based on flux extrapolation becomes

$$\mathbf{f}^{**}_{i+\frac{1}{2}} = \mathbf{f}^{+b}_{i+\frac{1}{2}} + \mathbf{f}^{-f}_{i+\frac{1}{2}} \tag{6.2.78}$$

Similarly, as in the variable extrapolation, we obtain the second order accuracy in time by adding a first integration step over $\Delta t/2$ with the associated first order scheme (6.2.74). Defining

$$\overline{\mathbf{f}^{*}_{i+\frac{1}{2}}} = \mathbf{f}^{*}(\overline{\mathbf{U}}_i, \overline{\mathbf{U}}_{i+1}) \tag{6.2.79}$$

we obtain the numerical flux as

$$\begin{aligned}\overline{\mathbf{f}^{**}_{i+\frac{1}{2}}} = \overline{\mathbf{f}^{*}_{i+\frac{1}{2}}} &+ \frac{1}{2}\left[\frac{(1-\kappa)}{2}\left(\mathbf{f}_i - \mathbf{f}^{*}_{i-\frac{1}{2}}\right) + \frac{(1+\kappa)}{2}\left(\mathbf{f}_{i+1} - \mathbf{f}^{*}_{i+\frac{1}{2}}\right)\right] \\ &+ \frac{1}{2}\left[\frac{(1+\kappa)}{2}\left(\mathbf{f}_i - \mathbf{f}^{*}_{i+\frac{1}{2}}\right) + \frac{(1-\kappa)}{2}\left(\mathbf{f}_{i+1} - \mathbf{f}^{*}_{i+\frac{1}{2}}\right)\right]\end{aligned} \tag{6.2.80}$$

Finally,

$$\mathbf{U}^{n+1}_i = \mathbf{U}^{n}_i - \tau\left(\overline{\mathbf{f}^{**}_{i+\frac{1}{2}}} - \overline{\mathbf{f}^{**}_{i-\frac{1}{2}}}\right) \tag{6.2.81}$$

Unfortunately, the above schemes have had some difficulty; they are unable to control overshoots and undershoots at shock discontinuities. A remedy is found in second order upwind schemes with high resolution, discussed in the following subsection.

6.2.6 SECOND ORDER UPWIND SCHEMES WITH HIGH RESOLUTION (TVD SCHEMES)

The most important development in computational fluid dynamics may be the second order upwind schemes with high resolution, known as the total variation diminishing (TVD) schemes, pioneered by Godunov [1959], VanLeer [1973, 1979], Harten and Lax [1981], Harten [1983, 1984], Osher [1984], Osher and Chakravarthy [1984] as reviewed by Hirsch [1990], which are based on the following physical properties:

- *Entropy condition* – A decrease of entropy associated with expansion shocks must not be admitted.
- *Monotonicity condition* – This condition must be enforced to prevent oscillatory behavior in the numerical scheme.
- *Total Variation Diminishing (TVD)* – The total variation of any physically admissible solution must not be allowed to increase in time.

In general, undesirable gradients, undershoots, overshoots which may occur in second order upwind schemes with low resolution may be controlled by providing nonlinear corrections, called limiters, which satisfy the above properties: entropy condition, monotonicity condition, and total variation diminishing.

(1) Definition of High Resolution Schemes

(a) Entropy Condition. The solution of the Euler equations may contain discontinuities of variable gradients involving an entropy increase (or compression shocks) and, unfortunately, an unrealistic entropy decrease (or expansion shocks) which violate the second law of thermodynamics. In order to eliminate such undesirable (numerically generated) entropy decrease or expansion shocks, we must guarantee that

$$a_R < a < a_L \tag{6.2.82}$$

where a is the speed of propagation of the discontinuity satisfying the Rankine-Hugoniot relations and

$$a_R = \frac{df_R}{du}, \quad a_L = \frac{df_L}{du} \tag{6.2.83}$$

with R and L being the right and left sides of the discontinuity. The requirement (6.2.82) is schematically shown in Figure 6.2.6, implying that a_R and a_L must intersect along the surface of discontinuity, resulting in a compression shock. In terms of eigenvalues, the entropy condition is given by

$$\lambda_k(U_R) < a < \lambda_k(U_L) \tag{6.2.84}$$

Another treatment of the entropy condition may be given by the smooth and positive entropy function $S(u_i)$ such that

$$\frac{\partial^2 S}{\partial u_i \partial u_j} > 0 \tag{6.2.85}$$

as proposed by Lax [1973]. This is equivalent to the existence of a system of equations with an artificial viscosity ν such that

$$U_t + aU_x = \nu U_{xx} \tag{6.2.86}$$

the solution of which confirms the entropy condition given by (6.2.85) in the limit as the

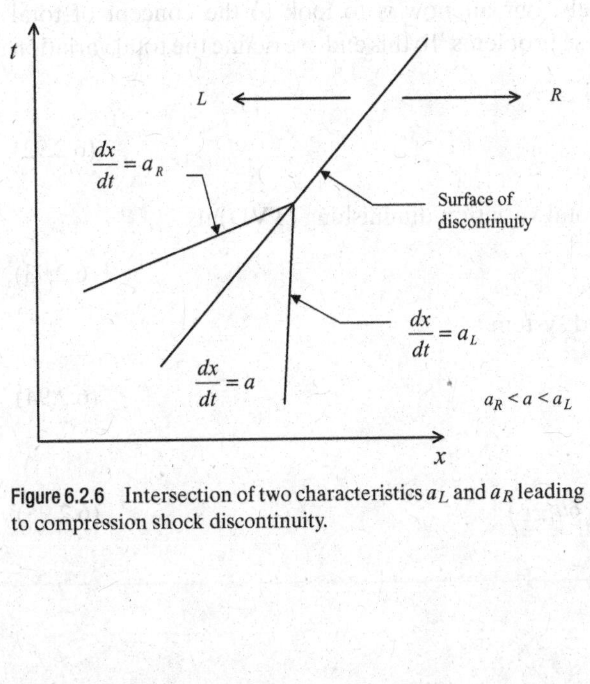

Figure 6.2.6 Intersection of two characteristics a_L and a_R leading to compression shock discontinuity.

artificial viscosity vanishes. In general, however, the satisfaction of entropy conditions alone may lead to oscillatory motions (overshoots and undershoots) along the discontinuities. Remedies can be found in the concepts of monotonicity and total variation diminishing, which are described below.

(b) Monotonicity Condition. A monotonicity condition refers to the nonoscillatory behavior of the numerical solution. Consider the solution of Euler equation to be in the form

$$u_i^{n+1} = H\left(u_{i-k}^n, u_{i-k+1}^n, \ldots, u_{i+k}^n\right) \tag{6.2.87}$$

This scheme is monotone if H is a monotonically increasing function such that

$$\frac{\partial H}{\partial u_j}(u_{i-k}, u_{i-k+1}, \ldots, u_{i+k}) \geq 0 \tag{6.2.88}$$

for all $i - k \leq j \leq i + k$, with

$$u_i^{n+1} = u_i^n - \tau\left(f_{i+\frac{1}{2}}^* - f_{i-\frac{1}{2}}^*\right) \tag{6.2.89}$$

$$f_{i+\frac{1}{2}}^* = f^*\left(u_{i-k+1}^n, u_{i-k+2}^n, \ldots, u_{i+k}^n\right) \tag{6.2.90}$$

The condition for monotonicity is given by

$$\frac{\partial f_{i+\frac{1}{2}}^*}{\partial u_{i-k+1}} \geq 0, \quad \frac{\partial f_{i+\frac{1}{2}}^*}{\partial u_{i+k}} \leq 0 \tag{6.2.91}$$

This represents a severe limitation, resulting in a scheme that is too diffusive. A compromise is the total variation diminishing concept, described next.

(c) Total Variation Diminishing (TVD) Schemes. As we have seen that the satisfaction of entropy and monotonicity conditions may still be restricted with oscillatory motions and excessive damping, respectively, our air now is to look to the concept of total variation diminishing to resolve these problems. To this end, we define the total variation [Lax, 1973] as

$$TV = \int \left|\frac{\partial u}{\partial x}\right| dx \tag{6.2.92}$$

A numerical scheme is said to be total variation diminishing (TVD) if

$$TV(u^{n+1}) \leq TV(u^n) \tag{6.2.93}$$

Let us consider the semi-discretized system

$$\frac{du_i}{dt} = -\frac{1}{\Delta x}\left(f_{i+\frac{1}{2}}^* - f_{i-\frac{1}{2}}^*\right) \tag{6.2.94}$$

or

$$\frac{du_i}{dt} = -\frac{1}{\Delta x}\left(C_{i+\frac{1}{2}}^- \delta u_{i+\frac{1}{2}} + C_{i-\frac{1}{2}}^+ \delta u_{i-\frac{1}{2}}\right) \tag{6.2.95}$$

with

$$C^-_{i+\frac{1}{2}}\delta u_{i+\frac{1}{2}} = f^*_{i+\frac{1}{2}} - f_i = a^-_{i+\frac{1}{2}}(u_{i+1} - u_i) \tag{6.2.96a}$$

$$C^+_{i-\frac{1}{2}}\delta u_{i-\frac{1}{2}} = f_i - f^*_{i-\frac{1}{2}} = a^+_{i-\frac{1}{2}}(u_i - u_{i-1}) \tag{6.2.96b}$$

and

$$C^+_{i+\frac{1}{2}} + C^-_{i+\frac{1}{2}} = \frac{f_{i+1} - f_i}{u_{i+1} - u_i} = \frac{\delta f_{i+\frac{1}{2}}}{\delta u_{i+\frac{1}{2}}} \equiv a_{i+\frac{1}{2}} \tag{6.2.97}$$

To compare the results above with the central scheme, we consider

$$f^*_{i+\frac{1}{2}} = \frac{1}{2}(f_i + f_{i+1}) - \frac{1}{2}\overline{D}_{i+\frac{1}{2}}\delta u_{i+\frac{1}{2}} \tag{6.2.98}$$

where $\overline{D}$ denotes the numerical viscosity coefficient. Combining (6.2.96) and (6.2.98), we obtain the wave speeds $C^-_{i+\frac{1}{2}}$ and $C^+_{i+\frac{1}{2}}$ in the form

$$C^-_{i+\frac{1}{2}} = \frac{1}{2}(a_{i+\frac{1}{2}} - \overline{D}_{i+\frac{1}{2}}) \tag{6.2.99}$$

$$C^+_{i+\frac{1}{2}} = \frac{1}{2}(a_{i+\frac{1}{2}} + \overline{D}_{i+\frac{1}{2}}) \tag{6.2.100}$$

from which the numerical viscosity coefficient becomes

$$\overline{D}_{i+\frac{1}{2}} = C^+_{i+\frac{1}{2}} - C^-_{i+\frac{1}{2}} \tag{6.2.101}$$

Thus the viscosity is expected to be proportional to the difference between the positive and negative wave speeds.

It follows from (6.2.101) that the semi-discrete system (6.2.95) is TVD if and only if

$$C^-_{i+\frac{1}{2}} \geq 0 \quad \text{and} \quad C^+_{i+\frac{1}{2}} \leq 0 \tag{6.2.102}$$

Once again from (6.2.95) we obtain, using the sign function $s_{i+\frac{1}{2}} = \text{sign}(\delta u_{i-\frac{1}{2}})$,

$$\begin{aligned}\frac{d}{dt}[TV(u)] &= \sum_i s_{i+\frac{1}{2}}\frac{d}{dt}(u_{i+1} - u_i) = \frac{1}{\Delta x}\sum_i s_{i+\frac{1}{2}}\Big[(C^- - C^+)_{i+\frac{1}{2}}\delta u_{i+\frac{1}{2}} \\ &\quad - C^-_{i+\frac{3}{2}}\delta u_{i+\frac{3}{2}} + C^+_{i-\frac{1}{2}}\delta u_{i-\frac{1}{2}}\Big] \\ &= \frac{1}{\Delta x}\sum_i [s_{i+\frac{1}{2}}(C^-_{i+\frac{1}{2}} - C^+_{i+\frac{1}{2}}) - s_{i-\frac{1}{2}}C^-_{i+\frac{1}{2}} + s_{i+\frac{3}{2}}C^+_{i+\frac{1}{2}}]\delta u_{i+\frac{1}{2}}\end{aligned} \tag{6.2.103}$$

The TVD condition requires that the right-hand side of (6.2.103) be nonpositive to ensure (6.2.102). This condition is satisfied for $\delta u_{i+\frac{1}{2}} = 1$ and $\delta u_{i+\frac{3}{2}} = \delta u_{i-\frac{1}{2}} = 0$. Thus, from (6.2.95) with an explicit Euler method

$$u_i^{n+1} = u_i^n - \frac{\Delta t}{\Delta x}\left(C^-_{i+\frac{1}{2}}\delta u_{i+\frac{1}{2}} + C^-_{i-\frac{1}{2}}\delta u_{i-\frac{1}{2}}\right)^n \tag{6.2.104}$$

with the CFL-like condition

$$\tau\left(C^+_{i+\frac{1}{2}} - C^-_{i+\frac{1}{2}}\right) \leq 1 \tag{6.2.105}$$

Integrating (6.2.95) and (6.2.103) and combining the results, together with (6.2.102), we obtain [Harten, 1983],

$$TV(u^{n+1}) \leq \sum_i \left\{ [1 - \tau(C^+ - C^-)_{i+\frac{1}{2}}] |\delta u_{i+\frac{1}{2}}| - \tau C^-_{i+\frac{1}{2}} |\delta u_{i+\frac{1}{2}}| + \tau C^+_{i+\frac{1}{2}} |\delta u_{i+\frac{1}{2}}| \right\}$$
$$= \sum_i |\delta u_{i+\frac{1}{2}}| = TV(u^n) \tag{6.2.106}$$

This is the basic requirement for the total variation diminishing.

Note that the second order upwind scheme (6.2.77a,b) with $\kappa = -1$ can be written as

$$\frac{du_i}{dt} = -\frac{1}{2\Delta x}(3f_i^+ - 4f_{i-1}^+ + f_{i-2}^+) - \frac{1}{2\Delta x}(-3f_i^- + 4f_{i+1}^- - f_{i+2}^-)$$
$$= -\frac{a^+}{2\Delta x}[3(u_i - u_{i-1}) - (u_{i-1} - u_{i-2})] - \frac{a^-}{2\Delta x}[3(u_{i+1} - u_i) - (u_{i+2} - u_{i+1})] \tag{6.2.107}$$

in which oscillations along discontinuities may still prevail. In what follows, we shall discuss the TVD schemes with limiters to achieve accuracy and stability based on (6.2.107).

(2) TVD Schemes with Limiters

The TVD scheme described above may have over- and under-shoots which can be treated with the concept of limiters [Roe, 1984; Sweby, 1984]. To this end, rewrite (6.2.107) in the form,

$$\frac{du_i}{dt} = -\frac{a^+}{\Delta x}\left[(u_i - u_{i-1}) + \frac{1}{2}(u_i - u_{i-1}) - \frac{1}{2}(u_{i-1} - u_{i-2})\right]$$
$$- \frac{a^-}{\Delta x}\left[(u_{i+1} - u_i) + \frac{1}{2}(u_{i+1} - u_i) - \frac{1}{2}(u_{i+2} - u_{i+1})\right] \tag{6.2.108}$$

Here, the variations in the second and third terms within the square brackets will be limited as follows:

$$\frac{du_i}{dt} = -\frac{a^+}{\Delta x}\left[(u_i - u_{i-1}) + \frac{1}{2}\Psi^+_{i-\frac{1}{2}}(u_i - u_{i-1}) - \frac{1}{2}\Psi^+_{i-\frac{3}{2}}(u_{i-1} - u_{i-2})\right]$$
$$- \frac{a^-}{\Delta x}\left[(u_{i+1} - u_i) + \frac{1}{2}\Psi^-_{i+\frac{1}{2}}(u_{i+1} - u_i) - \frac{1}{2}\Psi^-_{i+\frac{3}{2}}(u_{i+2} - u_{i+1})\right] \tag{6.2.109}$$

Now the TVD conditions are obtained by rewriting (6.2.109) in the form similar to (6.2.95),

$$\frac{du_i}{dt} = -\frac{a^+}{\Delta x}\left[1 + \frac{1}{2}\Psi^+_{i-\frac{1}{2}} - \frac{1}{2}\frac{\Psi_{i-\frac{3}{2}}}{r^+_{i-\frac{3}{2}}}\right](u_i - u_{i-1})$$
$$- \frac{a^-}{\Delta x}\left[1 + \frac{1}{2}\Psi^-_{i+\frac{1}{2}} - \frac{1}{2}\frac{\Psi^-_{i+\frac{3}{2}}}{r^-_{i+\frac{3}{2}}}\right](u_{i+1} - u_i) \tag{6.2.110}$$

with

$$
\begin{aligned}
r^+_{i+\frac{1}{2}} &= \frac{u_{i+2}-u_{i+1}}{u_{i+1}-u_i}, \quad & r^-_{i+\frac{1}{2}} &= \frac{u_i-u_{i-1}}{u_{i+1}-u_i} \\
r^+_{i-\frac{1}{2}} &= \frac{u_{i+1}-u_i}{u_i-u_{i-1}}, \quad & r^-_{i-\frac{1}{2}} &= \frac{u_{i-1}-u_{i-2}}{u_i-u_{i-1}} \\
r^+_{i-\frac{3}{2}} &= \frac{u_i-u_{i-1}}{u_{i-1}-u_{i-2}}, \quad & r^-_{i-\frac{3}{2}} &= \frac{u_{i-2}-u_{i-3}}{u_{i-1}-u_{i-2}} \\
r^-_{i+\frac{3}{2}} &= \frac{u_{i+3}-u_{i+2}}{u_{i+2}-u_{i+1}}, \quad & r^-_{i+\frac{3}{2}} &= \frac{u_{i+1}-u_i}{u_{i+2}-u_{i+1}}
\end{aligned}
\tag{6.2.111}
$$

$$
\Psi^+_{i-\frac{1}{2}} = \Psi\left(r^+_{i-\frac{1}{2}}, r^+_{i+\frac{3}{2}}\right), \qquad \Psi^-_{i+\frac{1}{2}} = \Psi\left(r^-_{i+\frac{1}{2}}, r^-_{i+\frac{3}{2}}\right) \tag{6.2.112}
$$

Thus, the TVD conditions are

$$
\Psi^+ = 1 + \frac{1}{2}\Psi^+_{i-\frac{1}{2}} - \frac{1}{2}\frac{\Psi^+_{i-\frac{3}{2}}}{r^+_{i-\frac{3}{2}}} \geq 0 \tag{6.2.113a}
$$

$$
\Psi^- = 1 + \frac{1}{2}\Psi^-_{i+\frac{1}{2}} - \frac{1}{2}\frac{\Psi^-_{i+\frac{3}{2}}}{r^-_{i+\frac{3}{2}}} \geq 0 \tag{6.2.113b}
$$

It is interesting to note that the basic Godunov's scheme (6.2.67b) is recovered for $\Psi^+ = \Psi^- = 1$. With more restricted definitions for the limiter,

$$
\begin{aligned}
\Psi^+_{i-\frac{1}{2}} &= \Psi\left(r^+_{i-\frac{1}{2}}\right), \quad & \Psi^+_{i-\frac{3}{2}} &= \Psi\left(r^+_{i-\frac{3}{2}}\right) \\
\Psi^-_{i+\frac{1}{2}} &= \Psi\left(r^-_{i+\frac{1}{2}}\right), \quad & \Psi^-_{i+\frac{3}{2}} &= \Psi\left(r^-_{i+\frac{3}{2}}\right)
\end{aligned}
\tag{6.2.114}
$$

the TVD conditions (6.2.113a,b) may be written in the form [Roe, 1984, 1985; Swevy, 1984] as

$$
\frac{\Psi\left(r^+_{i-\frac{3}{2}}\right)}{r^+_{i-\frac{3}{2}}} - \Psi\left(r^+_{i-\frac{1}{2}}\right) \leq 2 \tag{6.2.115a}
$$

$$
\frac{\Psi\left(r^-_{i+\frac{3}{2}}\right)}{r^-_{i+\frac{3}{2}}} - \Psi\left(r^-_{i+\frac{1}{2}}\right) \leq 2 \tag{6.2.115b}
$$

which may be generalized in the following form for all values of r and s:

$$
\frac{\Psi(r)}{r} - \Psi(s) \leq 2 \tag{6.2.116}
$$

with the following constraints:

$$
\Psi(r) \geq 0 \quad \text{for } r \geq 0 \tag{6.2.117a}
$$

$$
\Psi(r) = 0 \quad \text{for } r < 0 \tag{6.2.117b}
$$

where (6.2.117b) is designed to avoid nonmonotone behavior. Thus, the sufficient condition becomes

$$0 \leq \Psi(r) \leq 2r \tag{6.2.118}$$

Let us examine the above condition with another scheme such as the explicit second order Warming and Beam scheme:

$$u_i^{n+1} - u_i^n = -\sigma(u_i - u_{i-1})^n - \frac{\sigma}{2}(1-\sigma)(u_i - 2u_{i-1} + u_{i-2})^n \tag{6.2.119}$$

which may be rewritten as

$$u_i^{n+1} = u_i^n - \sigma(u_i - u_{i-1})^n - \frac{\sigma}{2}(1-\sigma)\delta^-\left[\Psi\left(r_{i-\frac{1}{2}}^+\right)(u_i - u_{i-1})^n\right] \tag{6.2.120}$$

Using the conditions (6.2.107) and (6.2.120), we obtain

$$0 \leq \tau\, C_{i-\frac{1}{2}}^+ = \sigma\left\{1 + \frac{1}{2}(1-\sigma)\left[\Psi\left(r_{i-\frac{1}{2}}^+\right) - \frac{1}{2}\frac{\Psi\left(r_{i-\frac{3}{2}}^+\right)}{r_{i-\frac{3}{2}}^+}\right]\right\} \leq 1 \tag{6.2.121a}$$

$$C_{i-1/2}^- = 0 \tag{6.2.121b}$$

This requires, for arbitrary values of r and s,

$$\frac{\Psi(r)}{r} - \Psi(s) \leq \frac{2}{1-\sigma} \tag{6.2.122a}$$

and

$$\Psi(s) - \frac{\Psi(r)}{r} \leq \frac{2}{\sigma} \tag{6.2.122b}$$

Combining (6.2.118) and (6.2.122), the second order upwind scheme is TVD for

$$0 \leq \Psi(r) \leq \min(2r, 2) \tag{6.2.123}$$

with $\Psi = 1$ for the Warming and Beam Scheme and $\Psi(r) = r$ for the Lax-Wendroff scheme.

Various limiters for second order schemes are summarized below:

(a) TVD regions for $\Psi(r)$ in general

(b) Van Leer's limiter $\Psi = \dfrac{r + |r|}{1 + r}$ (6.2.124)

(c) Minimum modulus (minmod) $\Psi(r) = \begin{cases} \min(r, 1) & \text{if } r > 0 \\ 0 & \text{if } r \leq 0 \end{cases}$ (6.2.125a)

$$\text{minmod}(x, y) = \begin{cases} x & \text{if } |x| < |y|, \quad xy > 0 \\ y & \text{if } |x| > |y|, \quad xy > 0 \\ 0 & \text{if } xy < 0 \end{cases} \tag{6.2.125b}$$

(d) Roe's Superbee limiter $\Psi(r) = \max[0, \min(2r, 1), \min(r, 2)]$ (6.2.126)

(e) General β-limiters $\Psi = \max[0, \min(\beta r, 1), \min(r, \beta)],\ 1 \leq \beta \leq 2$ (6.2.127)

(f) Chakravarthy and Osher limiter $\Psi(r) = \max[0, \min(r, \beta)],\ 1 \leq \beta \leq 2$ (6.2.128)

In these limiters, we observe the following features:

(i) For $r < 1$ or

$$\frac{u_{i+1} - u_i}{\Delta x} < \frac{u_i - u_{i-1}}{\Delta x}$$

Then set $\Psi(r) = r$ and the contribution $u_i^n - u_{i-1}^n$ to u_i^{n+1} is replaced by the smaller quantity $(u_{i+1}^n - u_i^n)$.

(ii) If $r > 1$, the contribution $(u_i - u_{i-1})$ remains unchanged.

(iii) If the slopes of consecutive intervals change sign, then the updated point i receives no contribution from the upstream interval.

The limiters as defined above may be applied to numerical fluxes in the form

$$\frac{du_i}{dt} = -\frac{1}{\Delta x}\left[1 + \frac{1}{2}\Psi\left(r^+_{i-\frac{1}{2}}\right) - \frac{1}{2}\frac{\Psi\left(r^+_{i-\frac{3}{2}}\right)}{r^+_{i-\frac{3}{2}}}\right](f_i - f^*_{i-\frac{1}{2}})$$
$$-\frac{1}{\Delta x}\left[1 + \frac{1}{2}\Psi\left(r^-_{i+\frac{1}{2}}\right) - \frac{1}{2}\frac{\Psi\left(r^-_{i+\frac{3}{2}}\right)}{r^-_{i+\frac{3}{2}}}\right](f^*_{i+\frac{1}{2}} - f_i) \quad (6.2.129)$$

with

$$r^+_{i-\frac{3}{2}} = \frac{f_i - f^*_{i-\frac{1}{2}}}{f_i^+ - f^*_{i-1}}, \quad r^-_{i+\frac{3}{2}} = \frac{f^*_{i+\frac{1}{2}} - f_i}{f^-_{i-1} - f^-_i} \quad (6.2.130)$$

and equivalently,

$$\frac{du_i}{dt} = -\frac{1}{\Delta x}\left[1 + \frac{1}{2}\Psi\left(r^+_{i-\frac{1}{2}}\right) - \frac{1}{2}\frac{\Psi\left(r^+_{i-\frac{3}{2}}\right)}{r^+_{i-\frac{3}{2}}}\right]a^+_{i-\frac{1}{2}}(u_i - u_{i-1})$$
$$-\frac{1}{\Delta x}\left[1 + \frac{1}{2}\Psi\left(r^-_{i+\frac{1}{2}}\right) - \frac{1}{2}\frac{\Psi\left(r^-_{i+\frac{3}{2}}\right)}{r^-_{i+\frac{3}{2}}}\right]a^-_{i+\frac{1}{2}}(u_{i+1} - u_i) \quad (6.2.131)$$

Here it is seen that with redefinition of slope ratios (6.2.130), the limiters are generalized to nonlinear scalar conservation equations from (6.2.113).

(3) Time Integration Methods for TVD Schemes

So far we have been concerned with the second order space-accurate TVD schemes only. We are now prepared to discuss integration of the time dependent term.

Recall that there are two types of time integration methods: (1) the combined space time methods (Section 6.2.2) and (2) separate space-time methods (Section 6.2.3). The former is more suitable for time dependent problems (time accurate), whereas the latter are more suitable for steady-state problems (not time accurate).

(a) Explicit TVD Schemes of First Order Accuracy in Time. Consider the first order time integration of (6.2.128) in the form

$$\frac{du_i}{dt} = -\frac{1}{\Delta x}\left[1+\frac{1}{2}\Psi\left(r^+_{i-\frac{1}{2}}\right)-\frac{1}{2}\frac{\Psi\left(r^+_{i-\frac{3}{2}}\right)}{r^+_{i-\frac{3}{2}}}\right]\left(f_i - f^*_{i-\frac{1}{2}}\right)^n$$
$$-\frac{1}{\Delta x}\left[1+\frac{1}{2}\Psi\left(r^-_{i+\frac{1}{2}}\right)-\frac{1}{2}\frac{\Psi\left(r^-_{i+\frac{3}{2}}\right)}{r^-_{i+\frac{3}{2}}}\right]\left(f^*_{i+\frac{1}{2}} - f_i\right)^n \tag{6.2.132}$$

This scheme without the limiter ($\Psi = 1$) is unstable, whereas the nonlinear TVD version with $\Psi > 1$ is conditionally stable, as seen from (6.2.107).

Define the local, positive, and negative CFL numbers,

$$\sigma^+_{i+\frac{1}{2}} = \tau\frac{f_{i+1} - f^*_{i+\frac{1}{2}}}{u_{i+1} - u_i} = \tau\frac{f^+_{i+1} - f^+_i}{u_{i+1} - u_i} \tag{6.2.133a}$$

$$\sigma^-_{i+\frac{1}{2}} = \tau\frac{f^*_{i+\frac{1}{2}} - f_i}{u_{i+1} - u_i} = \tau\frac{f^-_{i+1} - f^-_i}{u_{i+1} - u_i} \tag{6.2.133b}$$

with

$$\sigma_{i+\frac{1}{2}} = \sigma^+_{i+\frac{1}{2}} + \sigma^-_{i+\frac{1}{2}} = \tau\frac{f_{i+1} - f_i}{u_{i+1} - u_i} = \tau a_{i+\frac{1}{2}} \tag{6.2.134}$$

$$|\sigma|_{i+\frac{1}{2}} = \sigma^+_{i+\frac{1}{2}} - \sigma^-_{i+\frac{1}{2}} = \tau\left|\frac{f_{i+1} - f_i}{u_{i+1} - u_i}\right| = \tau|a|_{i+\frac{1}{2}} \tag{6.2.135}$$

and

$$\tau C^+_{i-\frac{1}{2}} = \sigma^+_{i-\frac{1}{2}}\left[1+\frac{1}{2}\Psi\left(r^+_{i-\frac{1}{2}}\right)-\frac{1}{2}\frac{\Psi\left(r^+_{i-\frac{3}{2}}\right)}{r^+_{i-\frac{3}{2}}}\right] \tag{6.2.136a}$$

$$\tau C^-_{i+\frac{1}{2}} = \sigma^-_{i+\frac{1}{2}}\left[1+\frac{1}{2}\Psi\left(r^-_{i+\frac{1}{2}}\right)-\frac{1}{2}\frac{\Psi\left(r^-_{i+\frac{3}{2}}\right)}{r^-_{i+\frac{3}{2}}}\right] \tag{6.2.136b}$$

Thus, the TVD condition (6.2.107) with (6.2.115) is given by

$$\tau\left(C^+_{i+\frac{1}{2}} - C^-_{i+\frac{1}{2}}\right) \le \tau|a|_{i+\frac{1}{2}}\left(\frac{1+\alpha}{2}\right) \le 1 \tag{6.2.137}$$

where

$$\left|\Psi(s) - \frac{\Psi(r)}{r}\right| \le \alpha$$

with $0 < \alpha \le 2$. The CFL condition for this case is

$$|\sigma| \le \frac{2}{2+\alpha} \tag{6.2.138}$$

The stability conditions for various limiters are

minmod limiter: $|\sigma| < \frac{2}{3}$

superbee limiter: $|\sigma| < \frac{1}{2}$

and so on.

(b) Implicit TVD Schemes. An implicit multistep method for the second order TVD scheme may be written as

$$\Delta u_i^n + \tau\theta\left(f_{i+\frac{1}{2}}^{*n+1} - f_{i-\frac{1}{2}}^{*n+1}\right) = -\tau(1-\theta)\left(f_{i+\frac{1}{2}}^{*n} - f_{i-\frac{1}{2}}^{*n}\right) \tag{6.2.139}$$

Using (6.2.104), we may rewrite (6.2.139) as

$$\left[1 + \tau\theta\left(C_{i+\frac{1}{2}}^{-}\delta^{+} + C_{i-\frac{1}{2}}^{+}\delta^{-}\right)\right]^n \Delta u_i^n = -\tau\left(f_{i+\frac{1}{2}}^{*} - f_{i-\frac{1}{2}}^{*}\right)^n \tag{6.2.140}$$

or

$$[1 + \tau\theta(C^+ - C^-)]\Delta u_i + \tau\theta C^- \Delta u_{i+1} - \tau\theta C^+ = -\tau\left(f_{i+\frac{1}{2}}^{*} - f_{1-\frac{1}{2}}^{*}\right) \tag{6.2.141}$$

It is now seen from (6.2.105) that the left-hand side of (6.2.141) is diagonally dominant. The CFL-like condition is given as

$$\tau(1-\theta)\left(C_{i+\frac{1}{2}}^{-} - C_{i-\frac{1}{2}}^{+}\right) \leq 1 \tag{6.2.142}$$

(c) Explicit Second Order TVD Schemes. Consider (6.2.74) with $\kappa = -1$ in (6.2.80),

$$\overline{u}_i = u_i^n - \frac{\tau}{2}\delta^- f_{i+\frac{1}{2}}^{*} \tag{6.2.143a}$$

$$\overline{f}_{i+\frac{1}{2}}^{*} = f^*(\overline{u}_i, \overline{u}_{i+1}) \tag{6.2.143b}$$

$$u_i^{n+1} = u_i^n - \tau\delta^{-1}\left[\overline{f}_{i+\frac{1}{2}}^{*} + \frac{1}{2}\Psi_{i-\frac{1}{2}}^{+}\left(f_i^n - f_{i-\frac{1}{2}}^{*}\right) + \Psi_{i+\frac{3}{2}}^{-}\left(f_{i+1}^n - f_{i+\frac{3}{2}}^{*}\right)\right] \tag{6.2.143c}$$

Applying (6.2.143c) to the linear convection equation, we obtain

$$u_i^{n+1} = u_i^n - \sigma\left[1 + \frac{1}{2}\left(\Psi_{i-\frac{1}{2}}^{+} - \sigma\right) - \frac{1}{2}\frac{\left(\Psi_{i-\frac{3}{2}}^{+} - \sigma\right)}{r_{i-\frac{3}{2}}^{+}}\right]\left(u_i^n - u_{i-1}^n\right) \tag{6.2.144}$$

with the TVD conditions,

$$0 \leq \sigma\left[2 + \Psi(s) - \sigma - \frac{\Psi(r) - \sigma}{r}\right] \leq 2 \tag{6.2.145}$$

where

$$s = r_{i-\frac{1}{2}}^{+}, \quad r = r_{i-\frac{3}{2}}^{+}$$

and

$$0 \leq \Psi(r) \leq (2-\sigma)r + \sigma \tag{6.2.146a}$$

$$0 \leq \Psi(r) \leq \frac{2}{\sigma} \tag{6.2.146b}$$

These conditions lead to, for $0 < \sigma \leq 1$,

$$\Psi(r) \leq \min(2, 2r) \tag{6.2.147}$$

It is interesting to recognize that the limited terms in (6.2.143c) represent the difference (prior to limiting) between the second and the first order numerical fluxes, $f^{**} - f^{*}$, and that this is the antidiffusive flux of the Flux Corrected Transport (FCT) [Boris and Book, 1973].

Various second order TVD schemes are identified as follows:

(i) Explicit Second-Order Schemes with Variable Extrapolation (MUSCL) Approach. Once again, from (6.2.74), (6.2.80), and (6.2.72a,b), we obtain

$$u_i = u_i^n - \frac{\Delta t}{2\Delta x}\left(f^*_{i+\frac{1}{2}} - f^*_{i-\frac{1}{2}}\right) \tag{6.2.148a}$$

$$\tilde{u}^L_{i+\frac{1}{2}} = \overline{u}_i + \frac{1}{2}\hat{\Psi}^L(u_i - u_{i-1}) \tag{6.2.148b}$$

$$\tilde{u}^R_{i+\frac{1}{2}} = \overline{u}_{i+1} - \frac{1}{2}\hat{\Psi}^R(U_{i+1} - u_i) \tag{6.2.148c}$$

with the tilde indicating monotonicity conditions, leading to

$$u_i^{n+1} - u_i^n = -\tau\left(\overline{f}^{**}_{i+\frac{1}{2}} - \overline{f}^{**}_{i-\frac{1}{2}}\right) \tag{6.2.149}$$

where

$$\overline{f}^{**}_{i+\frac{1}{2}} = f^*\left(\tilde{u}^L_{i+\frac{1}{2}}, \tilde{u}^R_{i+\frac{1}{2}}\right) \tag{6.2.150}$$

(ii) Lax-Wendroff TVD Scheme. This is an application of TVD to the Lax-Wendroff Scheme [Davis, 1984; Roe, 1984]. Here, the Lax-Wendroff numerical flux,

$$f_{i+\frac{1}{2}} = \frac{1}{2}(f_i + f_{i+1})$$

is transformed into an equivalent flux split form by decomposing the fluxes and the Jacobians into their positive and negative parts,

$$f^{*(LW)}_{i+\frac{1}{2}}\Big|_{TVD} = f_i^+ + f_{i+1}^- + \frac{1}{2}\left(1 + \tau A^+_{i+\frac{1}{2}}\right)\delta f^+_{i+\frac{1}{2}} - \frac{1}{2}\left(1 + \tau A^-_{i+\frac{1}{2}}\right)\delta f^-_{i+\frac{1}{2}} \tag{6.2.151}$$

Thus, the TVD Lax-Wendroff scheme becomes

$$f^{*(LW)}_{i+\frac{1}{2}}\Big|_{TVD} = f_i^+ + f_{i+1}^- + \frac{1}{2}\Psi\left(\frac{1}{r^+_{i-\frac{1}{2}}}\right)\left(1-\sigma^+_{i+\frac{1}{2}}\right)\delta f^+_{i+\frac{1}{2}}$$

$$-\frac{1}{2}\Psi\left(\frac{1}{r^-_{i+\frac{3}{2}}}\right)\left(1+\sigma^-_{i+\frac{1}{2}}\right)\delta f^-_{i+\frac{1}{2}} \qquad (6.2.152)$$

where the symmetry property

$$\frac{\Psi(r)}{r} = \Psi\left(\frac{1}{r}\right)$$

is utilized. Note that the presence of the functional dependence on $r_{i+3/2}$ is required, leading to a five-point scheme to satisfy the TVD and second order accuracy conditions.

(iii) Harten's Modified Flux Method. The first order upwind scheme has a truncation error h_x such that

$$u_t + f_x + h_x = 0 \qquad (6.2.153)$$

with

$$h = \Delta t \beta(u) u_x \qquad (6.2.154)$$

Equation (6.2.153) represents a second order approximation to $u_t + f_x = 0$. For the first order upwind scheme,

$$f^*_{i+\frac{1}{2}} = \frac{1}{2}(f_i + f_{i+1}) - \frac{1}{2}|a|_{i+\frac{1}{2}}(u_{i+1} - u_i) \qquad (6.2.155)$$

the truncation error becomes

$$h = \frac{\Delta x}{2}|a|(1-\tau|a|)u_x + O(\Delta x^2) = \frac{\Delta x}{2\tau}|\sigma|(1-|\sigma|)u_x + O(\Delta x^2) \qquad (6.2.156)$$

Thus, the numerical flux for (6.2.153) assumes the second order form,

$$f^{**}_{i+\frac{1}{2}} = \frac{1}{2}(f_i + f_{i+1}) + \frac{1}{2}(h_i + h_{i+1}) - \frac{1}{2}|a+b|_{i+\frac{1}{2}}(u_{i+1} - u_i) \qquad (6.2.157)$$

with

$$h_{i+\frac{1}{2}} = |a|_{i+\frac{1}{2}}\left(1 - |\sigma|_{i+\frac{1}{2}}\right)\frac{u_{i+1} - u_i}{2} = \frac{h_{i+1} + h_i}{2} \qquad (6.2.158a)$$

$$b_{i+\frac{1}{2}} = \frac{h_{i+1} - h_i}{u_{i+1} - u_i} \qquad (6.2.158b)$$

This scheme is TVD with

$$\tau|a+b|_{i+\frac{1}{2}} \le 1 \qquad (6.2.159)$$

and

$$h_i = \min\text{mod}\left(h_{i-\frac{1}{2}}, h_{i+\frac{1}{2}}\right) \qquad (6.2.160)$$

(d) Artificial Dissipation and TVD Schemes. Let us rearrange (6.2.152) in the form

$$f^{*(LW)}_{i+\frac{1}{2}}\Big|_{TVD} = \frac{1}{2}(f_i + f_{i+1}) - \frac{1}{2}|a|_{i+\frac{1}{2}}(u_{i+1} - u_i) + \frac{1}{2}[\Psi^+(1-\sigma^+)a^+ - \Psi^-(1-\sigma^-)a^-]_{i+\frac{1}{2}}(u_{i+1} - u_i) \tag{6.2.161}$$

where

$$\begin{aligned} &\Psi^+ a^- = 0 \quad \text{or} \quad \Psi^+ f^- = 0 \\ &\Psi^- a^+ = 0 \quad \text{or} \quad \Psi^- f^+ = 0 \end{aligned} \tag{6.2.162}$$

Thus

$$f^{*(LW)}_{i+\frac{1}{2}}\Big|_{TVD} = \frac{1}{2}(f_i + f_{i+1}) - \frac{1}{2}|a|_{i+\frac{1}{2}}(u_{i+1} - u_i) + \frac{1}{2}(\Psi^+ + \Psi^-)[(1-\sigma^+)a^+ - (1+\sigma^-)a^-]_{i+\frac{1}{2}}(u_{i+1} - u_i) \tag{6.2.163a}$$

or

$$f^{*(LW)}_{i+\frac{1}{2}}\Big|_{TVD} = \frac{1}{2}(f_i + f_{i+1}) - \frac{1}{2}|a|_{i+\frac{1}{2}}(u_{i+1} - u_i) + \frac{1}{2}(\Psi^+ + \Psi^-)[|a|(1-|\sigma|)]_{i+\frac{1}{2}}(u_{i+1} - u_i) \tag{6.2.163b}$$

Written alternatively,

$$f^{*(LW)}_{i+\frac{1}{2}}\Big|_{TVD} = \frac{1}{2}(f_i + f_{i+1}) - \frac{1}{2}\tau a^2_{i+\frac{1}{2}}(u_{i+1} - u_i) + \frac{1}{2}(\Psi^+ + \Psi^- - 1)[|a|(1-|\sigma|)]_{i+\frac{1}{2}}(u_{i+1} - u_i) \tag{6.2.164}$$

Comparing with (6.2.50), $\Psi^\pm$ are identified as

$$D_{i+\frac{1}{2}} = \frac{1}{2}(1 - \Psi^+ - \Psi^-)[|a|(1-|\sigma|)]_{i+\frac{1}{2}} \tag{6.2.165}$$

Similarly, a TVD MacCormack scheme is given by

$$\begin{aligned} &\overline{u}_i = u_i^n - \tau(f_{i+1} - f_i)^n \\ &\overline{\overline{u}} = u_i^n - \tau(\overline{f}_i - \overline{f}_{i-1}) \\ &u_i^{n+1} = \frac{1}{2}(\overline{u}_i + \overline{\overline{u}}_i) + \tau\left[D_{i+\frac{1}{2}}(u_{i+1} - u_i) - D_{i-\frac{1}{2}}(u_i - u_{i-1})\right] \end{aligned} \tag{6.2.166}$$

Although the artificial viscosity of the central schemes is analogous to the TVD schemes, accuracy and efficiency of the TVD schemes have proven to be superior.

In summary, the TVD schemes, although capable of resolving shock waves, are not uniformly high order accurate. They are reduced to first order accurate at local extrema of the solutions, while maintaining second order accuracy in other smooth regions. To circumvent this difficulty, the essentially nonoscillatory (ENO) schemes have been introduced. This is the subject of the next section.

6.2.7 ESSENTIALLY NONOSCILLATORY SCHEME

In the previous sections, we have studied low- and high-resolution schemes of Godunov, MUSCL, and TVD. In this section we examine a generalization and extension of these schemes, leading to a uniformly high order accurate essentially nonoscillatory scheme (ENO) as advanced by Harten and Osher [1987], and subsequently by Shu and Osher [1988, 1989], among others.

In the ENO scheme, high-order accuracy is obtained, whenever the solution is smoothed by means of a piecewise polynomial reconstruction procedure, yielding high order pointwise information from the cell averages of the solution. When applied to piecewise smooth initial data, this reconstruction enables a flux computation which is of high order accuracy, whenever the function is smooth, and avoids nonconvergence.

Initially, ENO schemes were developed in terms of cell averages conducive to FVM applications, followed by numerical fluxes for FDM applications with TVD Runge-Kutta discretization. These two types of ENO schemes were compared and evaluated by Casper, Shu, and Atkins [1994]. Recently, the ENO scheme has been extended to the Navier-Stokes system of equations [Zhong, 1994] and to unstructured triangular grids [Abgrall,1994; Suresh and Jorgenson, 1995; Stanescu and Habashi, 1998], among others. The basic theory of ENO is briefly summarized below.

The purpose of ENO is to achieve uniformly high order accuracy by avoiding the growth of spurious oscillations at shock discontinuities known as Gibb's phenomena. To this end, we employ piecewise polynomial reconstruction in the numerical solution based on an adaptive stencil. Such stencil is chosen according to the local smoothness of the flow variable.

Although ENO schemes have been applied to multidimensional Euler and Navier-Stokes system of equations, we illustrate the procedure using one-dimensional hyperbolic conservation law,

$$\frac{\partial \mathbf{U}}{\partial t} + \frac{\partial \mathbf{F}}{\partial x} = 0 \tag{6.2.167}$$

For simplicity, we consider a one-dimensional scalar function and reconstruct the point values $u(x)$ of a piecewise smooth function u from its known values of cell average $\overline{u}_i$.

$$\overline{u}_i = \int_{x_i-1/2}^{x_i+1/2} u(\xi)d\xi \tag{6.2.168}$$

with $h_i = x_{i+1/2} - x_{i-1/2}$. Let us now reconstruct $u(x)$ from $\overline{u}_i$ by interpolating the primitive function $U(x)$,

$$U(x) = \int_{x_0}^{x} u(\xi)d\xi \tag{6.2.169}$$

The point value of the primitive function at $x = x_{i+1/2}$ is given by

$$U_{i+1/2} = \sum_{i=i_0}^{i} \overline{u}_i h_i \tag{6.2.170}$$

Table 6.2.2 Illustration of Divided Difference

	x^k	Δ^0	Δ^1 (1st divided difference)	Δ^2 (2nd divided difference)	Δ^3 (3rd divided difference)
p_0	x_0	U_0			
			$\frac{U_1 - U_0}{x_1 - x_0} = \Delta U_0$		
p_1	x_1	U_1		$\frac{\Delta U_1 - \Delta U_0}{x_2 - x_0} = \Delta^2 U_0$	
			$\frac{U_2 - U_1}{x_2 - x_1} = \Delta U_1$		$\frac{\Delta^2 U_1 - \Delta^2 U_0}{x_3 - x_0} = \Delta^3 U_0$
p_2	x_2	U_2		$\frac{\Delta U_2 - \Delta U_1}{x_3 - x_1} = \Delta^2 U_1$	
			$\frac{U_3 - U_2}{x_3 - x_2} = \Delta U_2$		
p_3	x_3	U_3			

Since we have

$$u(x) = \frac{d}{dx} U(x) \tag{6.2.171}$$

it is now possible to obtain a piecewise polynomial interpolation function $H_m(x, U)$ of degree m by interpolating the point values of $U_{i+1/2}$ from (6.2.170) and arrive at the reconstruction polynomial of the form

$$R(x, \overline{u}) = \frac{d}{dx} H_m(x, U) \tag{6.2.172}$$

where, for cell $x_{i-1/2}$ and $x_{i+1/2}$, $H_m(x, U)$ represents the mth degree polynomial that interpolates the values of $U_{i+1/2}$ at $m+1$ successive points $x_{j+1/2}(j_m \leq j \leq j_m + m)$ including $x_{i-1/2}$ and $x_{i+1/2}$. Thus, our objective is to choose a stencil with $H_m(x, U)$ being the smoothest. This can be extracted from a table of divided differences of $U(x)$ such as shown in Table 6.2.2.

The one-dimensional ENO reconstruction described above has been extended to two dimensions via primitive function [Casper, 1992]. The cell average can be carried out as follows:

$$U_{ij} = \frac{1}{\Delta x_i} \int_{x_i - 1/2}^{x_i + 1/2} \overline{U}_j(x) dx \tag{6.2.173}$$

with

$$\overline{U}_j(x) = \frac{1}{\Delta y_j} \int_{y_i - 1/2}^{y_i + 1/2} \overline{U}(x, y) dy \tag{6.2.174}$$

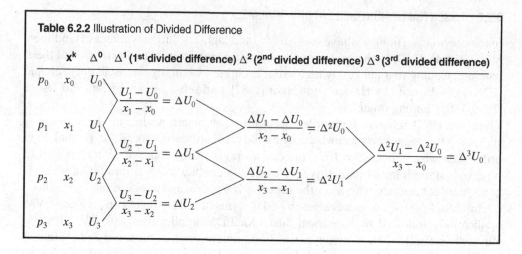

Recently, applications of ENO to the Euler equations in unstructured triangular grids have been reported by Abgrall [1994], Suresh and Jorgenson [1995], and Stanescu and Habashi [1998]. The reconstruction via extrapolation allows the selection in one step of all of the cells in the required stencil for each cell. For an rth order-of-accuracy, the approximation polynomials of degree $m = r - 1$ are written as

$$R_i[U,r] = \sum_{p=0}^{m} \sum_{j+k=p} a_{jk} X^j Y^k \tag{6.2.175}$$

Here, we have $M = (m+1)(m+2)/2$ unknowns, requiring a stencil of M cells to build the interpolation polynomial.

The ENO reconstruction such as given by (6.2.175) can be used to compute fluxes so that the solution procedure of any scheme presented in the previous subsections will be followed.

6.2.8 FLUX-CORRECTED TRANSPORT SCHEMES

The flux-corrected transport (FCT) scheme was originally developed by Boris and Book [1973] and subsequently generalized by Zalesak [1979] in which monotonicity is assured in multidimensional problems. The basic idea is to combine a high order scheme with a low order scheme in such a way that the high order scheme is employed in smooth regions of the flow, whereas the low order scheme is used near discontinuities in an attempt to obtain a monotonic solution. The following six steps are used for the solution.

(1) Compute $F^L_{i+1/2}$, the transportive flux given by some low order method guaranteed to give monotonic results.
(2) Compute $F^H_{i+1/2}$, the transportive flux given by some high order method. This flux is mathematically more accurate, but can lead to physically unacceptable ripples in the solution.
(3) Compute the updated low order, transported and diffused solution,

$$U_i^{TD} = U_i^o - \frac{\Delta t}{\Delta x_i}\left(F^L_{i+1/2} - F^L_{i-1/2}\right) \tag{6.2.176}$$

(4) Define the antidiffusive flux which becomes the amount of the monotone transportive flux that we would like to limit before correcting the transported and diffused conservation variables of step (3).

$$F^{AD}_{i+1/2} = F^H_{i+1/2} - F^L_{i+1/2} \tag{6.2.177}$$

Limit the antidiffusive fluxes $F^A_{i+1/2}$ so that U^n as computed in step (4) is free of the overshoots and undershoots which also do not appear in U_i^{TD}.

$$F^C_{i+1/2} = C_{i+1/2} F^{AD}_{i+1/2} \quad 0 \le C_{i+1/2} \le 1 \tag{6.2.178}$$

Apply the limited antidiffusive fluxes to get the new values U_i^n,

$$U_i^n = U_i^{TD} - \frac{\Delta t}{\Delta x}\left(F^C_{i+1/2} - F^C_{i-1/2}\right) \tag{6.2.179}$$

Note that if $F^C_{i+1/2} = F^{AD}_{i+1/2}$ for $C_{i+1/2} = 1$, the U_i^n reduces to the time-advanced higher order method without the required monotonicity correction.

The procedure described above can be generalized to the two-dimensional case,

$$U^n_{i,j} = U^o_{i,j} - \frac{\Delta t}{A_{i,j}}[(F_x)_{i+1/2,j} - (F_x)_{i-1/2,j} + (F_y)_{i,j+1/2} - (F_y)_{i,j-1/2}] \tag{6.2.180}$$

where $A_{i,j}$ is the two-dimensional area element centered on grid point (i, j). Here, two sets of transportive fluxes F_x and F_y are treated as follows:

a. Compute $(F_x)^L_{i+1/2,j}$ and $(F_y)^L_{i,j+1/2}$ by a lower order method.
b. Compute $(F_x)^H_{i+1/2,j}$ and $(F_y)^H_{i,j+1/2}$ by a higher order method.
c. Compute the previously updated low order, transported and diffused solution.

$$U^{TD}_{i,j} = U^n_{i,j} - \frac{\Delta t}{A_{i,j}}\left[(F_x)^L_{i+1/2,j} - (F_x)^L_{i-1/2,j} + (F_y)^L_{i,j+1/2} - (F_y)^L_{i,j-1/2}\right] \tag{6.2.181}$$

d. Define the vector components of the antidiffusive fluxes

$$F^{AD}_{i+1/2,j} = (F_x)^H_{i+1/2,j} - F^L_{i+1/2,j}$$
$$F^{AD}_{i,j+1/2,} = (F_y)^H_{i+1/2,j} - F^L_{i,j+1/2} \tag{6.2.182}$$

e. Limit the antidiffusive fluxes so that there are no overshoots or undershoots in $U^n_{i,j}$ of step (f) below that do not appear in $U^n_{i,j}$ of step (c).

$$F^C_{i+1/2,j} = C_{i+1/2,j} F^{AT}_{i+1/2,j} \quad 0 \le C_{i+1/2,j} \le 1$$
$$F^C_{i,j+1/2,} = C_{i+1/2,j} F^{AT}_{i,j+1/2} \quad 0 \le C_{i,j+1/2,j} \le 1 \tag{6.2.183}$$

f. Apply the limited antidiffusive fluxes to get the new values $U^n_{i,j}$

$$U^n_{i,j} = U^{TD}_{i,j} - \frac{\Delta t}{A_{i,j}}\left[F^C_{i+1/2,j} - F^C_{i-1/2,j} + F^C_{i,j+1/2} - F^C_{i,j-1/2}\right] \tag{6.2.184}$$

Here, it is important to limit the antidiffusive fluxes $F^{AT}_{i+1/2,j}$ and $F^{AT}_{i,j+1/2}$ by choosing the cell-interface flux-correcting factors $C_{i+1,j}$ and $C_{i,j+1}$ such that the combination of four fluxes acting together, through (6.2.184), does not allow $U^n_{i,j}$ to exceed some maximum value $U^{max}_{i,j}$ or to fall below some minimum value $U^{min}_{i,j}$. It should be noted that determination of suitable values of flux-correcting factors $C_{i+1,j}$ and $C_{i,j+1}$ is analogous to the TVD limiters. There are many possible ways to determine these limiters, as suggested in Zalesak [1979].

6.3 NAVIER-STOKES SYSTEM OF EQUATIONS

Diffusion processes due to viscosity and thermal conductivity are characterized by the Navier-Stokes system of equations. As the Reynolds number increases, boundary layers are formed and the laminar flow undergoes a transition toward turbulence. In high Reynolds number and high Mach number flows, shock waves and turbulent boundary layer interactions are most likely to occur. Furthermore, diffusivity due to chemical reactions also adds to the complexity of governing equations and computations. In general, such physical properties make the length and time scales of the variables widely disparate, thus causing the resulting algebraic finite difference equations to become "stiff." The subjects of turbulence and chemical reactions will not be discussed until Part Five.

Although implicit schemes are used predominantly in dealing with stiff equations for compressible viscous flows, explicit schemes have also been used in relatively low

Reynolds number flows. In this section, some of the prominent explicit and implicit schemes are discussed, followed by PISO.

6.3.1 EXPLICIT SCHEMES

The compressible viscous flow in its most general form was presented in Chapter 2. An expanded form in 3-D is shown below, but without source terms.

$$\frac{\partial \mathbf{U}}{\partial t} + \frac{\partial \mathbf{A}}{\partial x} + \frac{\partial \mathbf{B}}{\partial y} + \frac{\partial \mathbf{C}}{\partial z} = 0 \tag{6.3.1}$$

with

$$\mathbf{U} = \begin{bmatrix} \rho \\ \rho u \\ \rho \mathrm{v} \\ \rho w \\ \rho E \end{bmatrix} \qquad \mathbf{A} = \begin{bmatrix} \rho u \\ \rho u^2 + p - \tau_{xx} \\ \rho u \mathrm{v} - \tau_{xy} \\ \rho u w - \tau_{xz} \\ (\rho E + p)u - u\tau_{xx} - \mathrm{v}\tau_{xy} - w\tau_{xz} + q_x \end{bmatrix}$$

$$\mathbf{B} = \begin{bmatrix} \rho \mathrm{v} \\ \rho \mathrm{v} u - \tau_{yx} \\ \rho \mathrm{v}^2 + p - \tau_{yy} \\ \rho \mathrm{v} w - \tau_{yz} \\ (\rho E + p)\mathrm{v} - u\tau_{yx} - \mathrm{v}\tau_{yy} - w\tau_{yz} + q_y \end{bmatrix}$$

$$\mathbf{C} = \begin{bmatrix} \rho w \\ \rho w u - \tau_{zx} \\ \rho \mathrm{v} w - \tau_{zy} \\ \rho w^2 + p - \tau_{zz} \\ (\rho E + p)w - u\tau_{zx} - \mathrm{v}\tau_{zy} - w\tau_{zz} + q_z \end{bmatrix}$$

$$\tau_{xx} = \frac{2}{3}\mu\left(2\frac{\partial u}{\partial x} - \frac{\partial \mathrm{v}}{\partial y} - \frac{\partial w}{\partial z}\right), \qquad \tau_{yy} = \frac{2}{3}\mu\left(2\frac{\partial \mathrm{v}}{\partial y} - \frac{\partial u}{\partial x} - \frac{\partial w}{\partial z}\right),$$

$$\tau_{zz} = \frac{2}{3}\mu\left(2\frac{\partial w}{\partial z} - \frac{\partial u}{\partial x} - \frac{\partial \mathrm{v}}{\partial y}\right)$$

$$\tau_{xy} = \mu\left(\frac{\partial u}{\partial y} + \frac{\partial \mathrm{v}}{\partial x}\right) = \tau_{yx}, \qquad \tau_{xz} = \mu\left(\frac{\partial u}{\partial z} + \frac{\partial w}{\partial x}\right) = \tau_{zx}, \qquad \tau_{yz} = \mu\left(\frac{\partial w}{\partial y} + \frac{\partial \mathrm{v}}{\partial z}\right) = \tau_{zy}$$

$$q_x = -k\frac{\partial T}{\partial x}, \qquad q_y = -k\frac{\partial T}{\partial y}, \qquad q_z = -k\frac{\partial T}{\partial z}$$

In terms of a curvilinear coordinate system (ξ, η, ζ) (see Section 4.6), equations (6.3.1) are transformed to

$$\frac{\partial}{\partial t}\left(\frac{\mathbf{U}}{J}\right) + \frac{\partial}{\partial \xi}\left[\frac{1}{J}(\xi_x\mathbf{A} + \xi_y\mathbf{B} + \xi_z\mathbf{C})\right] + \frac{\partial}{\partial \eta}\left[\frac{1}{J}(\eta_x\mathbf{A} + \eta_y\mathbf{B} + \eta_z\mathbf{C})\right]$$
$$+ \frac{\partial}{\partial \zeta}\left[\frac{1}{J}(\zeta_x\mathbf{A} + \zeta_y\mathbf{B} + \zeta_z\mathbf{C})\right] = 0 \tag{6.3.2}$$

with

$$J = [x_\xi(y_\eta z_\zeta - y_\zeta z_\eta) - x_\eta(y_\xi z_\zeta - y_\zeta z_\xi) - x_\zeta(y_\xi z_\eta - y_\eta z_\xi)]^{-1}$$

$$\xi_x = J(y_\eta z_\zeta - y_\zeta z_\eta), \qquad \xi_y = -J(x_\eta z_\zeta - x_\zeta z_\eta), \qquad \xi_z = J(x_\eta y_\zeta - x_\zeta y_\eta),$$

$$\eta_x = -J(y_\xi z_\zeta - y_\zeta z_\xi), \qquad \eta_y = J(x_\xi z_\zeta - x_\zeta z_\xi), \qquad \eta_z = -J(x_\xi y_\zeta - x_\zeta y_\xi),$$

$$\zeta_x = J(y_\xi z_\eta - y_\eta z_\xi), \qquad \zeta_y = -J(x_\xi z_\eta - x_\eta z_\xi), \qquad \zeta_z = J(x_\xi y_\eta - x_\eta y_\xi)$$

$$\tau_{xx} = \frac{2}{3}\mu[2(\xi_x u_\xi + \eta_x u_\eta + \zeta_x u_\zeta) - (\xi_y v_\xi + \eta_y v_\eta + \zeta_y v_\zeta) - (\xi_z w_\xi + \eta_z w_\eta + \zeta_z w_\zeta)]$$

$$\tau_{yy} = \frac{2}{3}\mu[2(\xi_y v_\xi + \eta_y v_\eta + \zeta_y v_\zeta) - (\xi_x u_\xi + \eta_x u_\eta + \zeta_x u_\zeta) - (\xi_z w_\xi + \eta_z w_\eta + \zeta_z w_\zeta)]$$

$$\tau_{zz} = \frac{2}{3}\mu[2(\xi_z w_\xi + \eta_z w_\eta + \zeta_z w_\zeta) - (\xi_x u_\xi + \eta_x u_\eta + \zeta_x u_\zeta) - (\xi_y v_\xi + \eta_y v_\eta + \zeta_y v_\zeta)]$$

$$\tau_{xy} = \mu(\xi_x u_\xi + \eta_y u_\eta + \zeta_y u_\zeta + \xi_x v_\xi + \eta_x v_\eta + \zeta_x v_\zeta)$$

$$\tau_{xz} = \mu(\xi_z u_\xi + \eta_z u_\eta + \zeta_z u_\zeta + \xi_x w_\xi + \eta_x w_\eta + \zeta_x w_\zeta)$$

$$\tau_{yz} = \mu(\xi_z v_\xi + \eta_y v_\eta + \zeta_y v_\zeta + \xi_y w_\xi + \eta_y w_\eta + \zeta_y w_\zeta)$$

$$q_x = -k(\xi_x T_\xi + \eta_x T_\eta + \zeta_x T_\zeta), \qquad q_x = -k(\xi_x T_\xi + \eta_x T_\eta + \zeta_x T_\zeta),$$

$$q_x = -k(\xi_x T_\xi + \eta_x T_\eta + \zeta_x T_\zeta)$$

Equations (6.3.1) and (6.3.2) are mixed sets of hyperbolic and parabolic equations in time. If the unsteady terms are dropped, then a mixed set of hyperbolic-elliptic system results. As a consequence, the compressible Navier-Stokes system of equations are normally solved in their unsteady form using the time dependent approach, in which the equations are integrated forward in time until either the desired time is reached or a steady-state solution is obtained asymptotically after a sufficient number of time steps. If only the steady-state solution is desired, an implicit finite difference scheme can be used, where fewer iterations are necessary. If time accuracy is required, then a second order accurate explicit scheme may be used with small time increments.

Explicit schemes include the leapfrog/DuFort-Frankel method, Lax-Wendroff method, Runge-Kutta method, MacCormack method, among others. Highlights of the explicit MacCormack scheme [MacCormack, 1969] with predictor and corrector steps are presented below.

Predictor

$$\overline{U^{n+1}_{i,j,k}} = \mathbf{U}^n_{i,j,k} - \frac{\Delta t}{\Delta x}(\mathbf{A}^n_{i+1,j,k} - \mathbf{A}^n_{i,j,k}) - \frac{\Delta t}{\Delta y}(\mathbf{B}^n_{i,j+1,k} - \mathbf{B}^n_{i,j,k}) - \frac{\Delta t}{\Delta z}(\mathbf{C}^n_{i,j,k+1} - \mathbf{C}^n_{i,j,k}) \tag{6.3.3}$$

Corrector

$$\mathbf{U}^{n+1}_{i,j,k} = \frac{1}{2}\Big[\mathbf{U}^n_{i,j,k} + \overline{U^{n+1}_{i,j,k}} - \frac{\Delta t}{\Delta x}(\overline{\mathbf{A}^{n+1}_{i+1,j,k}} - \overline{\mathbf{A}^{n+1}_{i,j,k}}) - \frac{\Delta t}{\Delta y}(\overline{\mathbf{B}^{n+1}_{i,j+1,k}} - \overline{\mathbf{B}^{n+1}_{i,j,k}})$$
$$- \frac{\Delta t}{\Delta z}(\overline{\mathbf{C}^{n+1}_{i,j,k+1}} - \overline{\mathbf{C}^{n+1}_{i,j,k}})\Big] \tag{6.3.4}$$

with $x = i\Delta x$, $y = j\Delta y$, $z = k\Delta z$. This explicit scheme is second order accurate in both space and time, and useful for time accurate calculations or problems with low to moderate Reynolds numbers. Although forward differences are used for all spatial derivatives in the predictor step while backward differences are used in the correction step, the forward and backward differencing can be alternated between predictor and corrector steps as well as between the three spatial derivatives in order to eliminate any bias.

Unfortunately, no analytical stability analysis is available to determine limiting time step requirements because of the nonlinear nature of the governing equations, but the following empirical formula [Tannehill, Hoist, and Rakich, 1975] is proposed.

$$\Delta t \leq \frac{\sigma(\Delta t)_{\mathrm{CFL}}}{1 + 2/\mathrm{Re}_\Delta} \tag{6.3.5}$$

with $\sigma \cong 0.7 - 0.9$

$$(\Delta t)_{\mathrm{CFL}} \leq \left[\frac{|u|}{\Delta x} + \frac{|v|}{\Delta y} + \frac{|w|}{\Delta z} + a\sqrt{\frac{1}{\Delta x^2} + \frac{1}{\Delta y^2} + \frac{1}{\Delta z^2}}\right]^{-1}$$

$$\mathrm{Re}_\Delta = \min(\mathrm{Re}_{\Delta x}, \mathrm{Re}_{\Delta y}, \mathrm{Re}_{\Delta z}) \geq 0$$

$$\mathrm{Re}_{\Delta x} = \frac{\rho|u|\Delta x}{\mu}, \qquad \mathrm{Re}_{\Delta y} = \frac{\rho|v|\Delta y}{\mu}, \qquad \mathrm{Re}_{\Delta z} = \frac{\rho|w|\Delta z}{\mu}$$

It is often necessary to add artificial viscosity using the fourth order derivatives of the form,

$$-\varepsilon(\Delta x_i \Delta x_j \Delta x_k \Delta x_m)\frac{\partial^4 \mathbf{U}}{\partial x_i \partial x_j \partial x_k \partial x_m} \tag{6.3.6}$$

where ε is an experimentally determined parameter.

For high Reynolds number flows (thin viscous layers), the mesh must be refined (small time steps), leading to long computer times. To circumvent this difficulty, implicit methods may be used. We discuss this subject in the following section.

6.3.2 IMPLICIT SCHEMES

Earlier developments of implicit schemes for the Navier-Stokes system of equations include Briley and McDonald [1975], Beam and Warming [1978], and MacCormack [1981], among others. First, let us consider the Navier-Stokes system of equations in the general form

$$\frac{\partial \mathbf{U}}{\partial t} + \frac{\partial \mathbf{F}_i}{\partial x_i} + \frac{\partial \mathbf{G}_i}{\partial x_i} = 0 \tag{6.3.7}$$

Here, it is assumed that the convection and diffusion fluxes are functions of the conservation flow variables $\mathbf{U}$. In addition, the diffusion flux is assumed to be a function of the gradient of conservation flow variables. These functional relations are characterized by the convection Jacobian $\mathbf{a}_i$, diffusion Jacobian $\mathbf{b}_i$, and diffusion gradient Jacobian $\mathbf{c}_{ij}$.

$$\mathbf{a}_i = \frac{\partial \mathbf{F}_i}{\partial \mathbf{U}}, \qquad \mathbf{b}_i = \frac{\partial \mathbf{G}_i}{\partial \mathbf{U}}, \qquad \mathbf{c}_{ij} = \frac{\partial \mathbf{G}_i}{\partial \mathbf{U}_{,j}} \tag{6.3.8}$$

To evaluate the Jacobians, we set new variables $\ell = \rho u$, $m = \rho \mathrm{v}$, $e = \rho E$, and $\mu_R = \lambda + 2\mu$ for two-dimensional flows,

$$\mathbf{U} = \begin{bmatrix} U_1 \\ U_2 \\ U_3 \\ U_4 \end{bmatrix} = \begin{bmatrix} \rho \\ \rho u \\ \rho \mathrm{v} \\ \rho E \end{bmatrix} = \begin{bmatrix} \rho \\ \ell \\ m \\ e \end{bmatrix}$$

$$\mathbf{F}_1 = \begin{bmatrix} F_1^1 \\ F_1^2 \\ F_1^3 \\ F_1^4 \end{bmatrix} = \begin{bmatrix} \rho u \\ p + \rho u^2 \\ \rho u \mathrm{v} \\ \rho E u + p u \end{bmatrix} = \begin{bmatrix} \ell \\ p + \ell^2/\rho \\ \ell m/\rho \\ (p+e)\dfrac{\ell}{\rho} \end{bmatrix}$$

$$\mathbf{F}_2 = \begin{bmatrix} F_2^1 \\ F_2^2 \\ F_2^3 \\ F_2^4 \end{bmatrix} = \begin{bmatrix} \rho \mathrm{v} \\ \rho \mathrm{v} u \\ p + \rho \mathrm{v}^2 \\ \rho E \mathrm{v} + p \mathrm{v} \end{bmatrix} = \begin{bmatrix} m \\ \ell m/\rho \\ p + m^2/\rho \\ (p+e)\dfrac{m}{\rho} \end{bmatrix}$$

where

$$p = (\gamma - 1)\rho\left(E - \frac{1}{2}\mathrm{v}_j \mathrm{v}_j\right) = (\gamma - 1)\rho\left[E - \frac{1}{2}(u^2 + \mathrm{v}^2)\right] = (\gamma - 1)\left[e - \frac{1}{2\rho}(\ell^2 + m^2)\right]$$

$$T = \frac{1}{c_v}\left(E - \frac{1}{2}\mathrm{v}_j \mathrm{v}_j\right) = \frac{1}{c_v}\left[E - \frac{1}{2}(u^2 + \mathrm{v}^2)\right] = \frac{1}{\rho c_v}\left[e - \frac{1}{2\rho}(\ell^2 + m^2)\right]$$

The convective Jacobian $\mathbf{a}_i$ can be evaluated as

$$\mathbf{a}_1 = \frac{\partial \mathbf{F}_1}{\partial \mathbf{U}} = \begin{bmatrix} \dfrac{\partial F_1^1}{\partial U_1} & \dfrac{\partial F_1^1}{\partial U_2} & \dfrac{\partial F_1^1}{\partial U_3} & \dfrac{\partial F_1^1}{\partial U_4} \\ \dfrac{\partial F_1^2}{\partial U_1} & \dfrac{\partial F_1^2}{\partial U_2} & \dfrac{\partial F_1^2}{\partial U_3} & \dfrac{\partial F_1^2}{\partial U_4} \\ \dfrac{\partial F_1^3}{\partial U_1} & \dfrac{\partial F_1^3}{\partial U_2} & \dfrac{\partial F_1^3}{\partial U_3} & \dfrac{\partial F_1^3}{\partial U_4} \\ \dfrac{\partial F_1^4}{\partial U_1} & \dfrac{\partial F_1^4}{\partial U_2} & \dfrac{\partial F_1^4}{\partial U_3} & \dfrac{\partial F_1^4}{\partial U_4} \end{bmatrix}$$

$$\mathbf{a}_1 = \frac{\partial \mathbf{F}_1}{\partial \mathbf{U}}$$

$$= \begin{bmatrix} 0 & 1 & 0 & 0 \\ \dfrac{\gamma - 3}{2}u^2 + \dfrac{\gamma - 1}{2}\mathrm{v}^2 & (3-\gamma)u & -(\gamma - 1)\mathrm{v} & \gamma - 1 \\ -u\mathrm{v} & \mathrm{v} & u & 0 \\ -\dfrac{\gamma e u}{\rho} + (\gamma - 1)u(u^2 + \mathrm{v}^2) & \dfrac{\gamma e}{\rho} + \dfrac{1-\gamma}{2}(3u^2 + \mathrm{v}^2) & (1-\gamma)u\mathrm{v} & \gamma u \end{bmatrix}$$

$$\mathbf{a}_2 = \frac{\partial \mathbf{F}_2}{\partial \mathbf{U}}$$

$$= \begin{bmatrix} 0 & 0 & 1 & 0 \\ -uv & v & u & 0 \\ \frac{\gamma-3}{2}v^2 + \frac{\gamma-1}{2}u^2 & -(\gamma-1)u & (3-\gamma)v & \gamma-1 \\ -\frac{\gamma e v}{\rho} + (\gamma-1)v(u^2+v^2) & (1-\gamma)uv & \frac{\gamma e}{\rho} + \frac{1-\gamma}{2}(3v^2+u^2) & \gamma v \end{bmatrix} \tag{6.3.9a}$$

Similarly, the diffusion terms with their Jacobians are of the form

$$\mathbf{G}_1 = \begin{bmatrix} G_1^1 \\ G_1^2 \\ G_1^3 \\ G_1^4 \end{bmatrix} = -\begin{bmatrix} 0 \\ \tau_{11} \\ \tau_{12} \\ \tau_{11}u + \tau_{12}v - q_1 \end{bmatrix} \quad \mathbf{G}_2 = \begin{bmatrix} G_2^1 \\ G_2^2 \\ G_2^3 \\ G_2^4 \end{bmatrix} = -\begin{bmatrix} 0 \\ \tau_{21} \\ \tau_{22} \\ \tau_{21}u + \tau_{22}v - q_2 \end{bmatrix}$$

$$\mathbf{b}^1 = \frac{\partial \mathbf{G}_1}{\partial \mathbf{U}} = \begin{bmatrix} 0 & 0 & 0 & 0 \\ b_{21}^1 & b_{22}^1 & b_{23}^1 & 0 \\ b_{31}^1 & b_{32}^1 & b_{33}^1 & 0 \\ b_{41}^1 & b_{42}^1 & b_{43}^1 & b_{44}^1 \end{bmatrix} \quad \mathbf{b}_2 = \frac{\partial \mathbf{G}_2}{\partial \mathbf{U}} = \begin{bmatrix} 0 & 0 & 0 & 0 \\ b_{21}^2 & b_{22}^2 & b_{23}^2 & 0 \\ b_{31}^2 & b_{32}^2 & b_{33}^2 & 0 \\ b_{41}^2 & b_{42}^2 & b_{43}^2 & b_{44}^2 \end{bmatrix} \tag{6.3.9b}$$

with $m_1 = \rho u$, $m_2 = \rho v$

$$b_{21}^1 = -\frac{1}{\rho^2}\left(-\mu_R m_{1,1} - \lambda m_{2,2} + 2\mu_R \frac{m_1 \rho_{,1}}{\rho} + 2\lambda \frac{m_2 \rho_{,2}}{\rho}\right) \quad b_{22}^1 = \frac{\mu_R}{\rho^2}\rho_{,1} \quad b_{23}^1 = \frac{\lambda}{\rho^2}\rho_{,2}$$

$$b_{31}^1 = -\frac{\mu}{\rho^2}\left(-m_{2,1} - m_{1,2} + 2\frac{m_1 \rho_{,2}}{\rho} + 2\frac{m_2 \rho_{,1}}{\rho}\right) \qquad b_{32}^1 = \frac{\mu}{\rho^2}\rho_{,2}. \; b_{33}^1 = \frac{\mu}{\rho^2}\rho_{,1}$$

$$b_{41}^1 = u b_{21}^1 + v b_{31}^1 - \frac{1}{\rho^2}(m_1\tau_{11} + m_2\tau_{12}) + \frac{k}{\rho^2 c_v}[-(\rho E)_{,1} + 2u m_{1,1}$$
$$+ 2v m_{2,1} + (2E - 3u^2 - 3v^2)\rho_{,1}]$$

$$b_{42}^1 = -\frac{\tau_{11}}{\rho} + u b_{22}^1 + v b_{32}^1 + \frac{k}{\rho^2 c_v}[-m_{1,1} + 2u\rho_{,1}]$$

$$b_{43}^1 = -\frac{\tau_{12}}{\rho} + u b_{23}^1 + v b_{33}^1 - \frac{k}{\rho^2 c_v}[-m_{2,1} + 2v\rho_{,1}] b_{44}^1 - \frac{k}{\rho^2 c_v}\rho_{,1}$$

$$b_{21}^2 = b_{31}^1, \quad b_{22}^2 = b_{32}^1, \quad b_{23}^2 = b_{33}^1, \quad b_{31}^2 = \frac{1}{\rho^2}(\lambda m_{1,1} + \mu_R m_{2,2} + \mu_R u \rho_{,1} - \mu_R v \rho_{,2}),$$

$$b_{32}^2 = \frac{\lambda}{\rho^2}\rho_{,1}, \quad b_{33}^2 = \frac{\mu_R}{\rho^2}\rho_{,2}$$

$$b_{41}^2 = u b_{21}^2 + v b_{31}^2 + \frac{1}{\rho^2}(m_1\tau_{12} + m_2\tau_{22}) - \frac{k}{\rho^2 c_v}[-(\rho E)_{,2} + 2u m_{1,2}$$
$$+ 2v m_{2,2} + (2E - 3u^2 - 3v^2)\rho_{,2}],$$

$$b_{42}^2 = -\frac{\tau_{12}}{\rho} + ub_{22}^2 - \mathrm{v}b_{32}^2 + \frac{k}{\rho^2 c_v}[-m_{1,2} + 2u\rho_{,2}],$$

$$b_{43}^2 = -\frac{\tau_{22}}{\rho} + ub_{23}^2 + \mathrm{v}b_{33}^2 - \frac{k}{\rho^2 c_v}[-m_{2,2} + 2\mathrm{v}\rho_{,2}], \quad b_{44}^2 = \frac{k}{\rho^2 c_v}\rho_{,2}$$

The diffusion gradient Jacobians are evaluated as

$$\mathbf{c}_{11} = \frac{\partial \mathbf{G}_1}{\partial \mathbf{U}_{,1}} = -\begin{bmatrix} 0 & 0 & 0 & 0 \\ c_{21}^{11} & c_{22}^{11} & 0 & 0 \\ c_{31}^{11} & 0 & c_{33}^{11} & 0 \\ c_{41}^{11} & c_{42}^{11} & c_{43}^{11} & c_{44}^{11} \end{bmatrix} \qquad \mathbf{c}_{12} = \frac{\partial \mathbf{G}_1}{\partial \mathbf{U}_{,2}} = -\begin{bmatrix} 0 & 0 & 0 & 0 \\ c_{21}^{12} & 0 & c_{23}^{12} & 0 \\ c_{31}^{13} & c_{32}^{12} & 0 & 0 \\ c_{41}^{12} & c_{42}^{12} & c_{43}^{12} & 0 \end{bmatrix}$$

$$\mathbf{c}_{21} = \frac{\partial \mathbf{G}_2}{\partial \mathbf{U}_{,1}} = -\begin{bmatrix} 0 & 0 & 0 & 0 \\ c_{21}^{21} & 0 & c_{23}^{21} & 0 \\ c_{31}^{21} & c_{32}^{21} & 0 & 0 \\ c_{41}^{21} & c_{42}^{21} & c_{43}^{21} & 0 \end{bmatrix} \qquad \mathbf{c}_{22} = \frac{\partial \mathbf{G}_2}{\partial \mathbf{U}_{,2}} = -\begin{bmatrix} 0 & 0 & 0 & 0 \\ c_{22}^{22} & c_{22}^{22} & 0 & 0 \\ c_{31}^{22} & 0 & c_{33}^{22} & 0 \\ c_{41}^{22} & c_{42}^{22} & c_{43}^{22} & c_{44}^{22} \end{bmatrix} \tag{6.3.9c}$$

with

$$c_{21}^{11} = -(2\mu + \lambda)\frac{m_1}{\rho^2}, \quad c_{22}^{11} = (2\mu + \lambda)\frac{1}{\rho}, \quad c_{31}^{11} = -\mu\frac{m_2}{\rho^2}, \quad c_{33}^{11} = \frac{\mu}{\rho},$$

$$c_{41}^{11} = -(2\mu + \lambda)\frac{m_1^2}{\rho^3} - \mu\frac{m_2^2}{\rho^3} + \frac{k}{c_v}\left(-\frac{e}{\rho^2} + \frac{m_1^2 + m_2^2}{\rho^3}\right),$$

$$c_{42}^{11} = \left(2\mu + \lambda - \frac{k}{c_v}\right)\frac{m_1}{\rho^2}, \quad c_{43}^{11} = \left(\mu - \frac{k}{c_v}\right)\frac{m_2}{\rho^2}, \quad c_{44}^{11} = \frac{k}{c_v}\frac{1}{\rho},$$

$$c_{21}^{12} = -\lambda\frac{m_2}{\rho^2}, \quad c_{23}^{12} = \frac{\lambda}{\rho}, \quad c_{31}^{12} = -\mu\frac{m_1}{\rho^2}, \quad c_{32}^{12} = \frac{\mu}{\rho},$$

$$c_{41}^{12} = -(\mu + \lambda)\frac{m_1 m_2}{\rho^3}, \quad c_{42}^{12} = \mu\frac{m_2}{\rho^2}, \quad c_{43}^{12} = \lambda\frac{m_1}{\rho^2},$$

$$c_{21}^{21} = -\mu\frac{m_2}{\rho^2}, \quad c_{23}^{21} = \frac{\mu}{\rho}, \quad c_{31}^{21} = -\lambda\frac{m_1}{\rho^2}, \quad c_{32}^{21} = \frac{\lambda}{\rho}$$

$$c_{41}^{21} = -(\mu + \lambda)\frac{m_1 m_2}{\rho^3}, \quad c_{42}^{21} = \lambda\frac{m_2}{\rho^2}, \quad c_{43}^{21} = \mu\frac{m_1}{\rho^2},$$

$$c_{21}^{22} = -\mu\frac{m_1}{\rho^2}, \quad c_{22}^{22} = \frac{\mu}{\rho}, \quad c_{31}^{22} = -(2\mu + \lambda)\frac{m_2}{\rho^2}, \quad c_{33}^{22} = (2\mu + \lambda)\frac{1}{\rho},$$

$$c_{41}^{22} = -(2\mu + \lambda)\frac{m_2^2}{\rho^3} - \mu\frac{m_1^2}{\rho^3} + \frac{k}{c_v}\left(-\frac{e}{\rho^2} + \frac{m_1^2 + m_2^2}{\rho^3}\right),$$

$$c_{42}^{22} = \left(\mu - \frac{k}{c_v}\right)\frac{m_1}{\rho^2}, \quad c_{43}^{22} = \left(2\mu + \lambda - \frac{k}{c_v}\right)\frac{m_2}{\rho^2}, \quad c_{44}^{22} = \frac{k}{c_v}\frac{1}{\rho}$$

An extension to three-dimensional flux Jacobians follows the similar procedure. The 3-D convection, diffusion, and diffusion gradient flux Jacobians are presented in Appendix A.

A typical implicit scheme may be constructed by linearizing the convection flux, diffusion flux, and diffusion gradient as follows:

$$\mathbf{F}_i^{n+1} = \mathbf{F}_i^n + \frac{\partial \mathbf{F}_i^n}{\partial \mathbf{U}} \Delta \mathbf{U}^{n+1} = \mathbf{F}_i^n + \mathbf{a}_i^n \Delta \mathbf{U}^{n+1} \tag{6.3.10a}$$

$$\mathbf{G}_i^{n+1} = \mathbf{G}_i^n + \frac{\partial \mathbf{G}_i^n}{\partial \mathbf{U}} \Delta \mathbf{U}^{n+1} + \frac{\partial \mathbf{G}_i^n}{\partial \mathbf{U}_{,j}} \Delta \mathbf{U}_{,j}^{n+1} = \mathbf{G}_i^n + \mathbf{b}_i^n \Delta \mathbf{U}^{n+1} + \mathbf{c}_{ij}^n \Delta \mathbf{U}_{,j}^{n+1} \tag{6.3.10b}$$

An unsteady implicit scheme for (6.3.7) can be represented as an average of the flowfield between the current and previous time steps,

$$\frac{\Delta \mathbf{U}^{n+1}}{\Delta t} = -\frac{1}{2}\left[\left(\frac{\partial \mathbf{F}_i}{\partial x_i} + \frac{\partial \mathbf{G}_i}{\partial x_i}\right)^n + \left(\frac{\partial \mathbf{F}_i}{\partial x_i} + \frac{\partial \mathbf{G}_i}{\partial x_i}\right)^{n+1}\right] \tag{6.3.11}$$

Substituting (6.3.10) into (6.3.11) and using the relation,

$$\mathbf{c}_{ij} \Delta \mathbf{U}_{,j} = (\mathbf{c}_{ij} \Delta \mathbf{U})_{,j} - \mathbf{c}_{ij,j} \Delta \mathbf{U} \tag{6.3.12}$$

it follows that

$$\left\{\mathbf{I} + \frac{\Delta t}{2}\left[\frac{\partial}{\partial x_i}(\mathbf{a}_i + \mathbf{b}_i - \mathbf{c}_{ij,j}) + \frac{\partial^2 \mathbf{c}_{ij}}{\partial x_i \partial x_j}\right]^n\right\} \Delta \mathbf{U}^{n+1} = -\frac{\Delta t}{2}\left(\frac{\partial \mathbf{F}_i}{\partial x_i} + \frac{\partial \mathbf{G}_i}{\partial x_i}\right)^n \tag{6.3.13}$$

Although (6.3.13) can be used for general applications, it may be modified specifically for ADI procedure, leading to the so-called Beam-Warming scheme [Beam and Warming, 1978], described below.

For simplicity of notation, let the Navier-Stokes system of equations be written as

$$\frac{\partial \mathbf{U}}{\partial t} = \mathbf{W}, \qquad \mathbf{W} = -\frac{\partial \mathbf{F}_i}{\partial x_i} - \frac{\partial \mathbf{G}_i}{\partial x_i}$$

The Beam-Warming implicit method begins with an introduction of implicitness parameters ξ and θ such that

$$\frac{1}{\Delta t}[(1+\xi)\Delta \mathbf{U}^{n+1} - \xi \Delta \mathbf{U}^n] = \theta \mathbf{W}^{n+1} + (1-\theta)\mathbf{W}^n \tag{6.3.14a}$$

with $0 \le (\xi, \theta) \le 1$, $\Delta \mathbf{U}^{n+1} = \mathbf{U}^{n+1} - \mathbf{U}^n$, and $\Delta \mathbf{U}^n = \mathbf{U}^n - \mathbf{U}^{n-1}$. Equivalently, we may write (6.3.14a) in the form,

$$\Delta \mathbf{U}^{n+1} = \frac{\Delta t}{1+\xi}\left[\frac{\partial}{\partial t}(\theta \Delta \mathbf{U}^{n+1} + \mathbf{U}^n) + \xi \frac{\Delta \mathbf{U}^n}{\Delta t}\right] \tag{6.3.14b}$$

Using the linearization procedure of (6.3.10) in (6.3.14), we obtain

$$\frac{1}{\Delta t}[(1+\xi)\Delta \mathbf{U}^{n+1} - \xi \Delta \mathbf{U}^n] = -\theta\left[\frac{\partial}{\partial x_i}(\mathbf{a}_i \Delta \mathbf{U} + \mathbf{b}_i \Delta \mathbf{U} + \mathbf{c}_{ij} \Delta \mathbf{U}_{,j})\right]^{n+1} - \left[\frac{\partial}{\partial x_i}(\mathbf{F}_i + \mathbf{G}_i)\right]^n$$

or

$$\left\{\mathbf{I} + \frac{\theta \Delta t}{1+\xi}\left[\frac{\partial}{\partial x_i}(\mathbf{a}_i + \mathbf{b}_i - \mathbf{c}_{ij,j}) + \frac{\partial^2 \mathbf{c}_{ij}}{\partial x_i \partial x_j}\right]^n\right\} \Delta \mathbf{U}^{n+1} = \frac{\xi}{1+\xi} \Delta \mathbf{U}^n - \frac{\Delta t}{1+\xi}\left(\frac{\partial \mathbf{F}_i}{\partial x_i} + \frac{\partial \mathbf{G}_i}{\partial x_i}\right)^n \tag{6.3.15}$$

At this point, we anticipate difficulties handling the cross derivatives of the viscosity terms in the ADI procedure. Therefore, the diffusion flux terms are separated into two parts: normal derivatives and cross derivatives so that the differentiation of the diffusion gradient Jacobians is performed only for normal derivatives, whereas the cross derivative Jacobians on the left-hand side are excluded from the $(n+1)$th step in (6.3.15). Furthermore, the θ terms on the right-hand side are to retain only the cross derivatives (shear stresses) and allowed to lag to the n-1 step explicitly. With these arrangements, (6.3.15) is rewritten as

$$\left\{\mathbf{I}+\frac{\theta\Delta t}{1+\xi}\left[\frac{\partial}{\partial x_i}(\mathbf{a}_i+\mathbf{b}_i-\mathbf{c}_{ij,j})+\frac{\partial^2\mathbf{c}_{ij}}{\partial x_i\partial x_j}\right]^n\right\}\Delta\mathbf{U}^{n+1}$$
$$=\frac{\xi}{1+\xi}\Delta\mathbf{U}^n-\frac{\Delta t}{1+\xi}\left(\frac{\partial\mathbf{F}_i}{\partial x_i}+\frac{\partial\mathbf{G}_i}{\partial x_i}\right)^n+\frac{\Delta t\theta}{1+\xi}\frac{\partial\mathbf{G}_{(i)}^{n-1}}{\partial x_{(j)}} \tag{6.3.16}$$

with $(i)\neq(j)$. Here, it should be noted that in Beam and Warming [1978] the cross derivative terms alone become associated with the implicitness parameter θ at the $(n-1)$ step. This will allow (6.3.16) to be solved in two steps in the spirit of ADI with a block tridiagonal form. In step 1, set $i=1$ and $j=1,2$ in the x-direction with only the normal derivative Jacobians ($\mathbf{c}_{11}$ of $\mathbf{c}_{ij}$) retained on the left-hand side. Step 2 is to set $i=2$ and $j=1,2$ with only $\mathbf{c}_{22}$ being involved in the y-direction on the left-hand side and place the solution obtained in step 1 on the right-hand side to determine the final solution. In this process, the diffusion gradient Jacobian components, $\mathbf{c}_{12}$ and $\mathbf{c}_{21}$ are never used, contrary to the general case of (6.3.14). Expansion of (6.3.16) as described above leads to the following expressions.

$$\left\{\mathbf{I}+\frac{\theta\Delta t}{1+\xi}\left[\frac{\partial}{\partial x}(\mathbf{a}_1+\mathbf{b}_1-\mathbf{c}_{11.1})+\frac{\partial^2\mathbf{c}_{11}}{\partial x^2}+\frac{\partial}{\partial y}(\mathbf{a}_2+\mathbf{b}_2-\mathbf{c}_{22.2})+\frac{\partial^2\mathbf{c}_{22}}{\partial y^2}\right]^n\right\}\Delta\mathbf{U}^{n+1}$$
$$=\text{RHS} \tag{6.3.17}$$

$$\text{RHS}=\frac{\xi}{1+\xi}\Delta\mathbf{U}^n-\frac{\Delta t}{1+\xi}\mathbf{W}^n+\frac{\theta\Delta t}{1+\xi}\left[\frac{\partial\tau_{xy}}{\partial x}+\frac{\partial\tau_{yx}}{\partial y}\right]^n$$
$$+O\left[\left(\theta-\frac{1}{2}-\xi\right)\Delta t^2,\Delta t^3\right]$$

with

$$\mathbf{b}_1-\mathbf{c}_{11.1}=\frac{1}{\rho}\begin{bmatrix}0&0&0&0\\-u\left(\frac{4}{3}\mu\right)_x&\left(\frac{4}{3}\mu\right)_x&0&0\\-v\mu_x&0&\mu_x&0\\-u^2\left(\frac{4}{3}\mu\right)_x-v^2\mu_x&u\left(\frac{4}{3}\mu\right)_x&v\mu_x&0\end{bmatrix} \tag{6.3.18a}$$

$$\mathbf{b}_2-\mathbf{c}_{22.2}=\frac{1}{\rho}\begin{bmatrix}0&0&0&0\\-u\mu_x&\mu_y&0&0\\-v\left(\frac{4}{3}\mu\right)_y&0&\left(\frac{4}{3}\mu\right)_y&0\\-v^2\left(\frac{4}{3}\mu\right)_y-u^2\mu_y&u\mu_y&v\frac{4}{3}\mu_y&0\end{bmatrix} \tag{6.3.18b}$$

The solution of (6.3.17) is carried out in the manner of an ADI scheme as follows:

Step 1

$$\left\{\mathbf{I} + \frac{\theta\Delta t}{1+\xi}\left[\frac{\partial}{\partial x}(\mathbf{a}_1 + \mathbf{b}_1 - \mathbf{c}_{11,1}) + \frac{\partial^2 \mathbf{c}_{11}}{\partial x^2}\right]^n\right\}\Delta\mathbf{U}^* = \text{RHS} \tag{6.3.19a}$$

Step 2

$$\left\{\mathbf{I} + \frac{\theta\Delta t}{1+\xi}\left[\frac{\partial}{\partial y}(\mathbf{a}_2 + \mathbf{b}_2 - \mathbf{c}_{22,2}) + \frac{\partial^2 \mathbf{c}_{22}}{\partial y^2}\right]^n\right\}\Delta\mathbf{U}^{n+1} = \Delta\mathbf{U}^* \tag{6.3.19b}$$

where it should be noted that the substitution of (6.3.19b) into (6.3.19a) is equivalent to (6.3.17), but with additional higher order terms which may be neglected. This approach is known as the approximate factorization [Beam and Warming, 1978].

For assurance of convergence, an explicit artificial viscosity of fourth order derivatives (6.3.6) may be added to the right-hand side of (6.3.19a). Furthermore, implicit second order derivative artificial viscosities may be added to the left-hand side of both (6.3.19a) and (6.3.19b) in the x- and y-directions, respectively. The stability analysis by Beam and Warming [1978] shows that $\xi \geq 0.385$ and $\theta = 1/2 + \xi$, leading to $0.639 \leq \frac{\theta}{1+\xi} \leq 0.75$.

The Beam-Warming scheme has been used successfully and many improvements have been reported for the last two decades. An important question still remains. That is, dominance of implicitness or excessive artificial dissipation enforced uniformly everywhere in the flow domain must be adjusted according to the actual local flow physics such as inviscid-viscous interactions, transition to turbulence, shock wave boundary layer interactions, etc. This subject will be presented in the flowfield-dependent variation (FDV) methods in Section 6.5.

6.3.3 PISO SCHEME FOR COMPRESSIBLE FLOWS

Recall that in Section 5.3.2 we discussed the PISO scheme for incompressible flows. We demonstrate here that a similar procedure may be followed for compressible flows except that an additional corrector stage must be incorporated because the coupling between the momentum, energy, and pressure (continuity) equations involves the density and temperature [Issa, Gosman, and Watkins, 1986].

We begin with the continuity, momentum, and energy equations using the notations given in Section 5.3.2.

$$\frac{1}{\Delta t}(\rho^{n+1} - \rho^n) + (\rho v_i)^{n+1}_{,i} = 0 \tag{6.3.20a}$$

$$\frac{1}{\Delta t}[(\rho v_j)^{n+1} - (\rho v_j)^n] = -S^{n+1}_{ij,i} - p^{n+1}_{,j} \tag{6.3.20b}$$

$$\frac{1}{\Delta t}[(\rho E)^{n+1} - (\rho E)^n] + (\rho E v_i)^{n+1}_{,i} = -(p v_i)^{n+1}_{,i} - (\tau_{ij} v_i)^{n+1}_{,j} \tag{6.3.20c}$$

The predictor and corrector steps are as follows:

(a) Momentum Predictor

$$\left(\frac{\delta_{ij}}{\Delta t} + \frac{A_{ji}^{(D)}}{\rho^n}\right)(\rho^n \mathrm{v}_i^*) = -S_{ij,i}^{*(N)} - p_{,j}^n + \frac{\rho^n \mathrm{v}_j^n}{\Delta t} \tag{6.3.21}$$

(b) Momentum Corrector I

$$\left(\frac{\delta_{ij}}{\Delta t} + \frac{A_{ji}^{(D)}}{\rho^n}\right)(\rho^* \mathrm{v}_i^{**}) = -S_{ij,i}^{*(N)} - p_{,j}^* + \frac{\rho^n \mathrm{v}_j^n}{\Delta t} \tag{6.3.22}$$

Subtracting (6.3.22) from (6.3.21) gives

$$\rho^* \mathrm{v}_j^{**} - \rho^n \mathrm{v}_j^* = -\left(\frac{\delta_{ij}}{\Delta t} + \frac{A_{ji}^{(D)}}{\rho^n}\right)^{-1} (p^* - p^n)_{,i} \tag{6.3.23}$$

Writing (6.3.20a) in the form

$$(\rho^* \mathrm{v}_i^{**})_{,i} = -\frac{1}{\Delta t}(\rho^* - \rho^n) \tag{6.3.24}$$

Differentiating (6.3.23) and using (6.3.24) we obtain

$$\left[\left(\frac{\delta_{ij}}{\Delta t} + \frac{A_{ji}^{(D)}}{\rho^n}\right)^{-1} (p^* - p^n)_{,i}\right]_{,j} = (\rho^n \mathrm{v}_j^*)_{,j} + \frac{1}{\Delta t}(\rho^* - \rho^n) \tag{6.3.25}$$

Introducing the equation of state in the form

$$\rho^* = p^* \phi(p^n, T^n) \tag{6.3.26}$$

it is seen that (6.3.25) combined with (6.3.24) leads to

$$\left[\left(\frac{\delta_{ij}}{\Delta t} + \frac{A_{ji}^{(D)}}{\rho^n}\right)^{-1} (p^* - p^n)_{,i}\right]_{,j} - \frac{\phi(p^n, T^n)}{\Delta t}(p^* - p^n) = (\rho^n \mathrm{v}_j^*)_{,j} \tag{6.3.27}$$

from which p^*, ρ^*, and v_j^{**} can be solved.

(c) Energy Predictor

$$\left(\frac{1}{\Delta t} + \frac{B^{(D)}}{\rho^*}\right)(\rho^* E^*) = -(\rho E \mathrm{v}_i)_{,i}^{*(N)} - (p^* \mathrm{v}_i^{**})_{,i} + (\tau_{ij} \mathrm{v}_i^{**})_{i} + \frac{\rho^n E^n}{\Delta t} \tag{6.3.28}$$

with $B^{(D)}$ being the diagonal components of the convective terms.

(d) Momentum Corrector II

$$\left(\frac{\delta_{ij}}{\Delta t} + \frac{A_{ji}^{(D)}}{\rho^*}\right)(\rho^{**} \mathrm{v}_i^{***}) = -S_{ij,i}^{**(N)} - p_{,j}^{**} + \frac{\rho^n \mathrm{v}_j^n}{\Delta t} \tag{6.3.29}$$

$$(\rho^{**} \mathrm{v}_i^{***})_{,i} = -\frac{1}{\Delta t}(\rho^{**} - \rho^n) \tag{6.3.30}$$

$$\left[\left(\frac{\delta_{ij}}{\Delta t}+\frac{A_{ji}^{(D)}}{\rho^*}\right)^{-1}(p^{**}-p^*)_{,i}\right]_{,j}-\frac{\phi(p^*,T^*)}{\Delta t}(\rho^{**},\mathrm{v}_i^{***})(p^{**}-p^*)$$

$$=\left[\left(\frac{\delta_{ij}}{\Delta t}+\frac{A_{ji}^{(D)}}{\rho^*}\right)^{-1}\left(S_{ki,k}^{**(N)}-S_{ki,k}^{*(N)}\right)-A_{ji}\left(\frac{\rho^*-\rho^n}{\rho^n}\right)\mathrm{v}_i^{**}\right]_{,j}$$

$$+\frac{p^*}{\Delta t}[\phi(p^n,T^n)-\phi(p^n,T^n)] \tag{6.3.31}$$

with

$$\rho^{**}=p^{**}\phi(p^*,T^*) \tag{6.3.32}$$

Now (6.3.31) can be solved for p^{**}, whereas ρ^{**} and v_i^{***} are calculated from Eqs. (6.3.32) and (6.3.30), respectively,

(e) Energy Corrector

The energy equation is updated in the form

$$\left(\frac{1}{\Delta t}+\frac{B^{(D)}}{\rho^{**}}\right)(\rho^{**}E^{**})=-(\rho E\mathrm{v}_i)_{,i}^{**(N)}-(p^{**}\mathrm{v}_i^{***})_{,i}+(\tau_{ij}\mathrm{v}_j^{***})_{,i}+\frac{\rho^n E^n}{\Delta t} \tag{6.3.33}$$

from which E^{**}, v_i^{***}, and T^{**} are evaluated.

(f) Momentum Corrector III

This is the final step for all variables:

Momentum Equation

$$\left(\frac{\delta_{ij}}{\Delta t}-\frac{A_{ji}^{(D)}}{\rho^{**}}\right)(\rho^{***}\mathrm{v}_i^{****})=-S_{ij,i}^{***(N)}-p_{,j}^{***}+\frac{\rho^n\mathrm{v}_j^n}{\Delta t} \tag{6.3.34}$$

Continuity Equation

$$(\rho^{***}\mathrm{v}_i^{****})=-\frac{1}{\Delta t}(\rho^{***}-\rho^n) \tag{6.3.35}$$

Pressure Equation

$$\left[\left(\frac{\delta_{ij}}{\Delta t}+\frac{A_{ji}^{(D)}}{\rho^{**}}\right)^{-1}(p^{***}-p^{**})_{,i}\right]_{,j}-\frac{\phi(p^{**},T^{**})}{\Delta t}(p^{***}-p^{**})$$

$$=\left[\left(\frac{\delta_{ij}}{\Delta t}+\frac{A_{ji}^{(D)}}{\rho^{**}}\right)^{-1}\left(S_{ki,k}^{***(N)}-S_{ki,k}^{**(N)}\right)-A_{ji}\left(\frac{\rho^{**}-\rho^n}{\rho^n}\right)\mathrm{v}_i^{***}\right]_{,j}$$

$$+\frac{p^{**}}{\Delta t}[\phi(p^{**},T^{**})-\phi(p^n,T^n)] \tag{6.3.36}$$

with

$$\rho^{***}=p^{***}\phi(p^{**},T^{**}) \tag{6.3.37}$$

It is seen that p^{***} can be solved from (6.3.36) with ρ^{***} and v_i^{***} determined from (6.3.37) and (6.3.34).

6.4 PRECONDITIONING PROCESS FOR COMPRESSIBLE AND INCOMPRESSIBLE FLOWS

6.4.1 GENERAL

For the analysis of compressible flows, it is possible that some regions of the flow domain such as in the boundary layers have low speeds and thus are incompressible. As a result, the density-based formulations in terms of conservation variables may suffer extremely slow or nonconvergence of the solution. This is due to an ill-conditioned system of algebraic equations contributed by the stiff eigenvalues of convection terms. The reason for this is that the acoustic speed is so much higher than the flow velocity in incompressible flows. This phenomenon then appears to be numerical, but it is important to realize that actually physical aspects of the fluid flows precipitate such numerical disorder. For example, transitions and interactions between inviscid/viscous flows induce physical disturbances or instabilities, which may then contribute to transitions and interactions between laminar and turbulent flows and/or compressible and incompressible flows. We address the subjects of transitions and interactions between different properties of fluid flows in Section 6.5 on flowfield-dependent variation (FDV) methods. In this section, our discussion will be limited strictly to the numerical aspect of the transition from the compressible flow to incompressible flow or vice versa. Our objective is to begin with the density-based formulation and subsequently by providing the preconditioning matrix to the time-dependent terms we improve the convection eigenvalues for low Mach number or incompressible flows.

The numerical difficulties of the density-based formulation dealing with low Mach number flows or incompressible flows have been addressed by a number of investigators [Peyret and Vivian, 1985; Choi and Merkle, 1993; Pletcher and Chen, 1993; Merkle et al., 1998], among others. In this vein, we construct Jacobian matrices transforming the conservation variables into the primitive variables such that

$$\frac{\partial \mathbf{U}}{\partial t} + \frac{\partial \mathbf{F}_i}{\partial x_i} + \frac{\partial \mathbf{G}_i}{\partial x_i} = 0 \tag{6.4.1}$$

$$\mathbf{A}\frac{\partial \mathbf{Q}}{\partial t} + \mathbf{B}_i\frac{\partial \mathbf{Q}}{\partial x_i} + \mathbf{C}_i\frac{\partial \mathbf{Q}}{\partial x_i} + \mathbf{C}_{ij}\frac{\partial^2 \mathbf{Q}}{\partial x_i \partial x_j} = 0 \tag{6.4.2}$$

where $\mathbf{Q}$ is the primitive variables, $\mathbf{Q} = [\rho, u, v, w, T]^T$, and $\mathbf{A}$ is the time Jacobian.

$$\mathbf{A} = \frac{\partial \mathbf{U}}{\partial \mathbf{Q}} = \begin{bmatrix} \rho\beta_T & 0 & 0 & 0 & -\rho\alpha_p \\ \rho\beta_T u & \rho & 0 & 0 & -\rho\alpha_p u \\ \rho\beta_T v & 0 & \rho & 0 & -\rho\alpha_p v \\ \rho\beta_T w & 0 & 0 & \rho & -\rho\alpha_p w \\ e_1^p & \rho u & \rho v & \rho w & e_4^p \end{bmatrix} \tag{6.4.3}$$

with

$$\beta_T = \frac{1}{\rho}\left(\frac{\partial \rho}{\partial p}\right)_T, \qquad \alpha_p = -\frac{1}{\rho}\left(\frac{\partial \rho}{\partial T}\right)_p \tag{6.4.4}$$

$$e_1^p = \rho\beta_T E_1 - \alpha_p T, \qquad e_4^p = -\rho\alpha_p E_1 + \rho c_p \tag{6.4.5}$$

$$E_1 = H + K = c_p T + \frac{1}{2}v_i v_i \tag{6.4.6}$$

Here, the convection eigenvalues can be examined from

$$|\mathbf{A}^{-1}\mathbf{B}_i - \lambda_i \mathbf{I}| = 0 \tag{6.4.7}$$

However, for incompressible limits, the eigenvalues become stiff as the algebraic equations resulting from (6.4.2) are ill-conditioned, with the acoustic speed being infinite.

6.4.2 PRECONDITIONING MATRIX

To improve the eigenvalues of (6.4.7), let us examine the quantities of the first column of the time Jacobian which contain the derivative of density with respect to pressure at the constant temperature.

$$\left(\frac{\partial\rho}{\partial p}\right)_T = \frac{1}{RT} = \frac{\gamma}{a^2} \tag{6.4.8}$$

Note that this derivative vanishes for incompressible flows ($a = \infty$), leading to the stiff eigenvalues in (6.4.7). To circumvent this problem, we may adjust (6.4.8) in the form

$$\rho\beta_T = \left(\frac{\partial\rho}{\partial p}\right)_T = \frac{1}{RT} = \frac{1}{\gamma RT} + \frac{1}{c_p T} = \frac{1}{a^2} + \frac{1}{c_p T} = \frac{1}{V_r^2} + \frac{\alpha_p}{c_p} = \frac{1}{V_r^2} - \frac{1}{c_p \rho}\left(\frac{\partial\rho}{\partial T}\right)_p \tag{6.4.9}$$

where V_r is a reference velocity which may be defined differently for compressible and incompressible flows. A logical choice would be that $V_r = a$ for compressible flows and $V_r = (\mathrm{v}_i \mathrm{v}_i)^{1/2}$ for incompressible flows. Thus, the time Jacobian matrix $\mathbf{A}$ is adjusted to $\hat{\mathbf{A}}$ implying the preconditioning matrix with $\rho\beta_T$ in (6.4.3) given by (6.4.9). For highly viscous flow such as in the boundary layers, it is necessary to choose the reference velocity to be governed by the diffusion velocity such that

$$V_r = \max(V_r, \nu/\Delta x)$$

The adjusted eigenvalues of the preconditioned system are determined from

$$|\hat{\mathbf{A}}^{-1}\mathbf{B}_i - \lambda_i \mathbf{I}| = 0 \tag{6.4.10}$$

$$\Lambda = diag(u, u, u, u^* + a^*, u^* - a^*) \tag{6.4.11}$$

with

$$u^* = \frac{1}{2}u\left[1 - \left(\rho\beta_T - \frac{\alpha_p}{c_p}\right)V_r^2\right] \tag{6.4.12a}$$

$$a^* = \frac{1}{2}\left\{\left[1 - \left(\rho\beta_T - \frac{\alpha_p}{c_p}\right)V_r^2\right]u^2 + V_r^2\right\}^{1/2} \tag{6.4.12b}$$

Here, it is seen that, for $V_r \geq a$, the eigenvalues in (6.4.12) become $u \pm a$, whereas if $V_r \cong 0$, all eigenvalues are of the same order as u. This shows that the eigenvalues of the preconditioned system remain well conditioned at all speeds.

To provide efficiency and time-accurate solutions, one may utilize a dual time stepping by introducing a pseudo-time derivative term into (6.4.2) in linearized iteration steps:

$$\hat{\mathbf{A}}\frac{\partial\Delta\mathbf{Q}}{\partial\tau} + \mathbf{A}\frac{\partial\Delta\mathbf{Q}}{\partial t} + \mathbf{B}_i\frac{\partial\Delta\mathbf{Q}}{\partial x_i} + \mathbf{C}_i\frac{\partial\Delta\mathbf{Q}}{\partial x_i} + \mathbf{C}_{ij}\frac{\partial^2\Delta\mathbf{Q}}{\partial x_i \partial x_j} = -\mathbf{H} \tag{6.4.13}$$

with **H** given by (6.4.1). As the pseudo time τ approaches infinity, the pseudo time term vanishes and we recover (6.4.1) at steady state.

Pletcher and Chen [1993] constructs the time Jacobian matrix in nondimensional quantities with the first column and last row of (6.4.3) in terms of Mach number to obtain the pseudo-time preconditioning matrix by dividing by the Mach number so that the fatal ill-conditioning can be eliminated. Some examples have been presented by Merkle et al. [1998] using the pseudo-time preconditioning of the type given by (6.4.3) with (6.4.9).

6.5 FLOWFIELD-DEPENDENT VARIATION METHODS

So far, the major portions of the historical developments in FDM have been covered and various computational schemes for the various flow properties have also been discussed. In this section we explore a general approach which leads to most of the currently available computational schemes as special cases. Such an approach, called the flowfield-dependent variation (FDV) methods, is examined in this section.

6.5.1 BASIC THEORY

The original idea of FDV methods began from the need to address the physics involved in shock wave turbulent boundary layer interactions [Chung, 1999; Schunk et al., 1999]. In this situation, transitions and interactions of inviscid/viscous, compressible/incompressible, and laminar/turbulent flows constitute not only the physical complexities but also computational difficulties. This is where the very low velocity in the vicinity of the wall ($M \cong 0$, $\mathrm{Re} \cong 0$) and very high velocity far away from the wall (e.g., $M \cong 20$, $\mathrm{Re} \cong 10^9$) coexist within a domain of study. Transitions from one type of flow to another and interactions between two distinctly different flows have been studied for many years, both experimentally and numerically. Incompressible flows were analyzed using the pressure-based formulation with the primitive variables for the implicit solution of the Navier-Stokes system of equations. The precondition process for the time-dependent term intended for all speed flows was also discussed. Compressible flows were analyzed using the density-based formulation with the conservation variables for the solution of the Navier-Stokes system of equations. However, in dealing with the domain which contains all speed flows with various physical properties where the equations of state for compressible and incompressible flows are different, and where the transitions between laminar and turbulent flows are involved in dilatational dissipation due to compressibility, we must provide very special and powerful numerical treatments. The FDV scheme has been devised toward resolving these issues.

For the purpose of the discussion, we shall consider the conservation form of the Navier-Stokes system (2.2.11) without the source terms (see Section 13.6 for the source terms not equal to zero in FEM).

$$\frac{\partial \mathbf{U}}{\partial t} + \frac{\partial \mathbf{F}_i}{\partial x_i} + \frac{\partial \mathbf{G}_i}{\partial x_i} = 0 \tag{6.5.1}$$

In expanding $\mathbf{U}^{n+1}$ in a special form of Taylor series about $\mathbf{U}^n$, we introduce the parameters s_a and s_b for the first and second order derivatives of $\mathbf{U}$ with respect to time, respectively,

$$\mathbf{U}^{n+1} = \mathbf{U}^n + \Delta t \frac{\partial \mathbf{U}^{n+s_a}}{\partial t} + \frac{\Delta t^2}{2} \frac{\partial^2 \mathbf{U}^{n+s_b}}{\partial t^2} + O(\Delta t^3) \tag{6.5.2}$$

where

$$\frac{\partial \mathbf{U}^{n+s_a}}{\partial t} = \frac{\partial \mathbf{U}^n}{\partial t} + s_a \frac{\partial \Delta \mathbf{U}^{n+1}}{\partial t} \qquad 0 \le s_a \le 1 \tag{6.5.3a}$$

$$\frac{\partial^2 \mathbf{U}^{n+s_b}}{\partial t^2} = \frac{\partial^2 \mathbf{U}^n}{\partial t^2} + s_b \frac{\partial^2 \Delta \mathbf{U}^{n+1}}{\partial t^2} \qquad 0 \le s_b \le 1 \tag{6.5.3b}$$

with $\Delta \mathbf{U}^{n+1} = \mathbf{U}^{n+1} - \mathbf{U}^n$. Substituting (6.5.3) into (6.5.2) yields

$$\mathbf{U}^{n+1} = \mathbf{U}^n + \Delta t \left(\frac{\partial \mathbf{U}^n}{\partial t} + s_a \frac{\partial \Delta \mathbf{U}^{n+1}}{\partial t} \right) + \frac{\Delta t^2}{2} \left(\frac{\partial^2 \mathbf{U}^n}{\partial t^2} + s_b \frac{\partial^2 \Delta \mathbf{U}^{n+1}}{\partial t^2} \right) + O(\Delta t^3) \tag{6.5.4}$$

Introducing the Jacobians of convection, diffusion, and diffusion gradients, we write the first and second derivatives of the conservation variables in the form,

$$\frac{\partial \mathbf{U}}{\partial t} = -\frac{\partial \mathbf{F}_i}{\partial x_i} - \frac{\partial \mathbf{G}_i}{\partial x_i} \tag{6.5.5}$$

$$\frac{\partial^2 \mathbf{U}}{\partial t^2} = -\frac{\partial}{\partial x_i} \left(\mathbf{a}_i \frac{\partial \mathbf{U}}{\partial t} \right) - \frac{\partial}{\partial x_i} \left(\mathbf{b}_i \frac{\partial \mathbf{U}}{\partial t} \right) - \frac{\partial^2}{\partial x_i \partial x_j} \left(\mathbf{c}_{ij} \frac{\partial \mathbf{U}}{\partial t} \right) \tag{6.5.6a}$$

in which the convection Jacobian $\mathbf{a}_i$, the diffusion Jacobian $\mathbf{b}_i$, and the diffusion gradient Jacobian $\mathbf{c}_{ij}$ are defined as in (6.3.9) for 2-D and Appendix A for 3-D. Combining (6.5.5) and (6.5.6a) leads to

$$\frac{\partial^2 \mathbf{U}}{\partial t^2} = \frac{\partial}{\partial x_i} (\mathbf{a}_i + \mathbf{b}_i) \left(\frac{\partial \mathbf{F}_j}{\partial x_j} + \frac{\partial \mathbf{G}_j}{\partial x_j} \right) + \frac{\partial^2}{\partial x_i \partial x_k} \mathbf{c}_{ik} \left(\frac{\partial \mathbf{F}_j}{\partial x_j} + \frac{\partial \mathbf{G}_j}{\partial x_j} \right) \tag{6.5.6b}$$

Substituting (6.5.5) and (6.5.6b) into (6.5.4), and assuming the product of the diffusion gradient Jacobian with third order spatial derivatives to be negligible, we have

$$\begin{aligned} \Delta \mathbf{U}^{n+1} = \Delta t & \left[-\frac{\partial \mathbf{F}_i^n}{\partial x_i} - \frac{\partial \mathbf{G}_i^n}{\partial x_i} + s_a \left(-\frac{\partial \Delta \mathbf{F}_i^{n+1}}{\partial x_i} - \frac{\partial \Delta \mathbf{G}_i^{n+1}}{\partial x_i} \right) \right] \\ & + \frac{\Delta t^2}{2} \left\{ \frac{\partial}{\partial x_i} (\mathbf{a}_i + \mathbf{b}_i) \left(\frac{\partial \mathbf{F}_j^n}{\partial x_j} + \frac{\partial \mathbf{G}_j^n}{\partial x_j} \right) \right. \\ & \left. + s_b \left[\frac{\partial}{\partial x_i} (\mathbf{a}_i + \mathbf{b}_i) \left(\frac{\partial \Delta \mathbf{F}_j^{n+1}}{\partial x_j} + \frac{\partial \Delta \mathbf{G}_j^{n+1}}{\partial x_j} \right) \right] \right\} + O(\Delta t^3) \end{aligned} \tag{6.5.7}$$

The parameters s_a and s_b which appear in (6.5.7) above may be given appropriate physical roles by calculating them from the flowfield-dependent quantities. For example, if s_a is associated with the temporal changes (fluctuations) of convection, it may be calculated from the changes of Mach number between adjacent nodal points so that $s_a = 0$ would imply no changes in convection fluctuations. The functional dependency

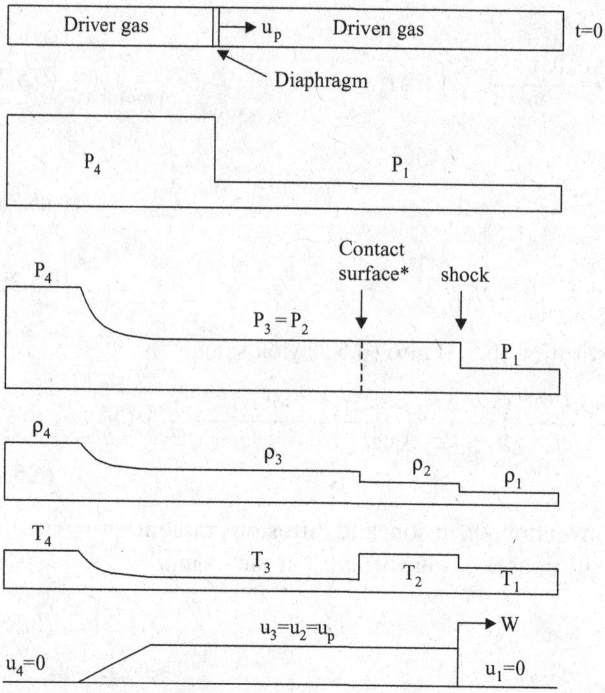

*Analogous to sliplines ($P_3=P_2$, $T_3 \neq T_2$, $\rho_3 \neq \rho_2$, $u_3=u_2$) leading to entropy discontinuity at the contact surface.

Figure 6.5.1 Mechanism of shock wave discontinuities as related to s_1 in terms of the changes of Mach number with respect to the velocity and square root of pressure, density, or temperature, $s_a = f(u/\sqrt{p/\rho}) = f(u/\sqrt{RT}) = f(M)$.

of s_a on Mach number is illustrated from the shock tube physics as shown in Figure 6.5.1. Here it is seen that discontinuities of pressure, density, and temperature are related as a function of Mach number,

$$s_a = f(u/\sqrt{p/\rho}) = f(u/\sqrt{RT}) = f(M)$$

Similarly, if s_a is associated with the changes (fluctuations) of diffusion, such as in boundary layers, then it may be calculated from the changes of Reynolds number or Peclet number between adjacent nodal points such that $s_a = 0$ would signify no changes in diffusion fluctuations. Therefore, the role of s_a for diffusion is different from that of convection. For example, we may define the fluctuation quantities associated with s_a as

$$s_a\left(\frac{\partial \Delta \mathbf{F}_i^{n+1}}{\partial x_i} + \frac{\partial \Delta \mathbf{G}_i^{n+1}}{\partial x_i}\right) \Rightarrow s_1 \frac{\partial \Delta \mathbf{F}_i^{n+1}}{\partial x_i} + s_3 \frac{\partial \Delta \mathbf{G}_i^{n+1}}{\partial x_i}$$
$$= \frac{\sqrt{M_{\max}^2 - M_{\min}^2}}{M_{\min}} \frac{\partial \Delta \mathbf{F}_i^{n+1}}{\partial x_i} + \frac{\sqrt{Re_{\max}^2 - Re_{\min}^2}}{Re_{\min}} \frac{\partial \Delta \mathbf{G}_i^{n+1}}{\partial x_i} \tag{6.5.8}$$

where it is seen that the parameter s_a originally adopted as a single mathematical or numerical parameter has now turned into multiple physical parameters such as the changes of Mach numbers and Reynolds numbers (or Peclet numbers) between adjacent nodal points. The magnitudes of fluctuations of convection, diffusion, and source terms are dictated by the current flowfield situations in space and time. Similar assessments can be applied to the parameter s_b as associated with its corresponding fluctuation terms of convection and diffusion. Thus, in order to provide variations to the changes of convection and diffusion differently in accordance with the current flowfield situations, we reassign s_a and s_b associated with convection and diffusion as follows:

$$s_a \Delta \mathbf{F}_i \Rightarrow s_1 \Delta \mathbf{F}_i, \quad s_a \Delta \mathbf{G}_i \Rightarrow s_3 \Delta \mathbf{G}_i$$
$$s_b \Delta \mathbf{F}_i \Rightarrow s_2 \Delta \mathbf{F}_i, \quad s_b \Delta \mathbf{G}_i \Rightarrow s_4 \Delta \mathbf{G}_i$$

with the various parameters, called the flowfield-dependent variation (FDV) parameters or simply variation parameters, defined as follows:

s_1 = first order convection FDV parameter

s_2 = second order convection FDV parameter

s_3 = first order diffusion FDV parameter

s_4 = second order diffusion FDV parameter

The first order FDV parameters s_1 and s_3 are flowfield-dependent, whereas the second order FDV parameters s_2 and s_4 are exponentially proportional to the first order FDV parameters, and mainly act as artificial viscosity. Details of these FDV parameters are given below.

6.5.2 FLOWFIELD-DEPENDENT VARIATION PARAMETERS

As has been pointed out, the success of FDV methods depends on accurate calculations of the flowfield-dependent variation parameters. Specifically, the convection FDV parameters s_1 and s_2 and diffusion FDV parameters s_3 and s_4 are dependent on Mach numbers and Reynolds numbers or Peclet numbers, respectively. The first order FDV parameters s_1 and s_3 dictate the flowfield solution accuracy, whereas the second order FDV parameters s_2 and s_4 maintain the solution stability.

Convection FDV Parameters

$$s_1 = \begin{cases} \min(r, 1), & r > \alpha \\ 0 & r < \alpha, \; M_{\min} \neq 0 \\ 1 & M_{\min} = 0 \end{cases} \tag{6.5.9a}$$

$$s_2 = \frac{1}{2}(1 + s_1^{\eta}), \quad 0.05 < \eta < 0.2 \tag{6.5.9b}$$

with

$$r = \sqrt{M_{\max}^2 - M_{\min}^2} / M_{\min} \tag{6.5.10}$$

where the maximum and minimum Mach numbers are calculated between the local

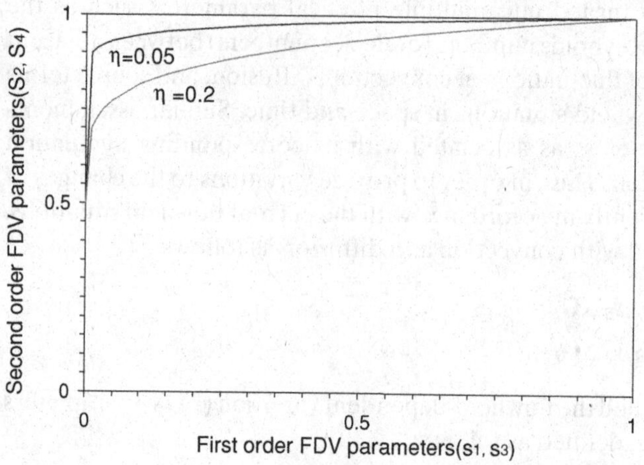

Figure 6.5.2 Relationship between the first and second order variation parameters $s_2 = (1 + s_1^{\eta})/2$, $s_4 = (1 + s_3^{\eta})/2$, with $0.05 < \eta \leq 0.2$.

adjacent nodal points with α being the user-specified small number ($\alpha \cong 0.01$). The ranges of the second order FDV parameter exponent η are given, exponentially proportional to the first order FDV parameter, as shown in Figure 6.5.2. It appears that the range in $0.05 \leq \eta \leq 0.2$ is adequate in most of the examples that have been tested.

Diffusion FDV Parameters

$$s_3 = \begin{cases} \min(r, 1), & r > \alpha, \quad \alpha \cong 0.01 \\ 0 & r < \alpha, \quad \mathrm{Re}_{\min} \neq 0, \quad \text{or } Pe_{\min} \neq 0 \\ 1 & \mathrm{Re}_{\min} = 0, \quad \text{or } Pe_{\min} = 0 \end{cases} \tag{6.5.11a}$$

$$s_4 = \frac{1}{2}(1 + s_3^{\eta}), \quad 0.05 < \eta < 0.2 \tag{6.5.11b}$$

with

$$r = \sqrt{\mathrm{Re}_{\max}^2 - \mathrm{Re}_{\min}^2} \Big/ \mathrm{Re}_{\min} \quad \text{or } r = \sqrt{Pe_{\max}^2 - Pe_{\min}^2} \Big/ Pe_{\min} \tag{6.5.12a,b}$$

where the maximum and minimum Reynolds numbers or maximum and minimum Peclet numbers are calculated between the local adjacent nodal points, and α is a user-specified small number ($\alpha \cong 0.01$). If temperature gradients are large, it is possible that Peclet numbers instead of Reynolds numbers may dictate the diffusion FDV parameters. The larger value of s_3 is to be chosen, as obtained either from (6.5.12a) or (6.5.12b). Adequate ranges of η for the second order FDV variation parameter are the same as for the case of convection.

Relationships between the first and second order FDV parameters are graphically shown in Figure 6.5.3a. The ranges of these convection and diffusion FDV parameters for a typical compression corner high-speed flow are illustrated in Figure 6.5.3b. They represent the trend of an exhaustive numerical experimentation for various physical situations.

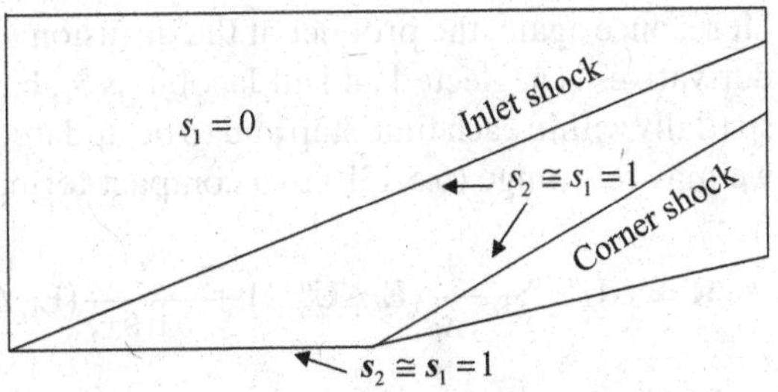

(a) Ranges of convection variation parameters

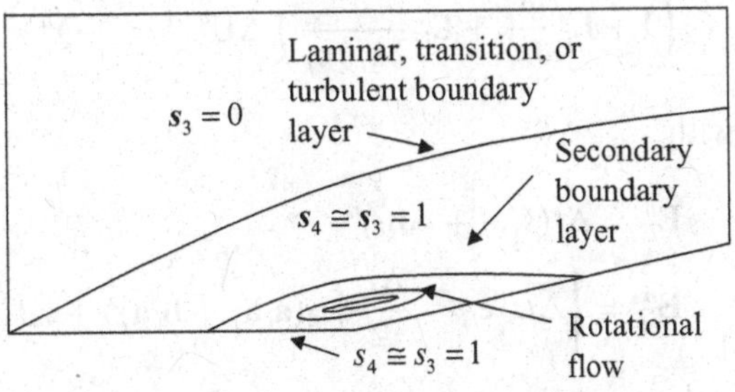

(b) Ranges of diffusion variation parameters

Figure 6.5.3 Typical ranges of first order (convection) and second order (diffusion) variation parameters for the compression corner high-speed flow.

6.5.3 FDV EQUATIONS

The final form of the FDV equations can be obtained by substituting the FDV parameters as defined in (6.5.8) through (6.5.12) into (6.5.7), leading to the residual of the form,

$$\mathbf{R} = \Delta\mathbf{U}^{n+1} - \Delta t\left[-\frac{\partial \mathbf{F}_i^n}{\partial x_i} - \frac{\partial \mathbf{G}_i^n}{\partial x_i} - s_1\frac{\partial \Delta\mathbf{F}_i^{n+1}}{\partial x_i} - s_3\frac{\partial \Delta\mathbf{G}_i^{n+1}}{\partial x_i}\right]$$
$$-\frac{\Delta t^2}{2}\left[\frac{\partial}{\partial x_i}(\mathbf{a}_i + \mathbf{b}_i)\left(\frac{\partial \mathbf{F}_j^n}{\partial x_j} + \frac{\partial \mathbf{G}_j^n}{\partial x_j}\right)\right] - \frac{\Delta t^2}{2}\left\{s_2\left[\frac{\partial}{\partial x_i}(\mathbf{a}_i + \mathbf{b}_i)\left(\frac{\partial \Delta\mathbf{F}_j^{n+1}}{\partial x_j}\right)\right]\right.$$
$$\left. + \frac{\partial}{\partial x_i}(\mathbf{a}_i + \mathbf{b}_i)\left(s_4\frac{\partial \Delta\mathbf{G}_j^{n+1}}{\partial x_j}\right)\right\} + O(\Delta t^3) \qquad (6.5.13a)$$

Now, rearranging and expressing the remaining terms associated with the FDV parameters in terms of the Jacobians, we have

$$\Delta\mathbf{U}^{n+1} + \Delta t\left[s_1\left(\frac{\partial \mathbf{a}_i\Delta\mathbf{U}^{n+1}}{\partial x_i}\right) + s_3\left(\frac{\partial \mathbf{b}_i\Delta\mathbf{U}^{n+1}}{\partial x_i} + \frac{\partial^2 \mathbf{c}_{ij}\Delta\mathbf{U}^{n+1}}{\partial x_i\partial x_j}\right)\right]$$
$$-\frac{\Delta t^2}{2}\left\{s_2\left[\frac{\partial^2(\mathbf{a}_i\mathbf{a}_j + \mathbf{b}_i\mathbf{a}_j)\Delta\mathbf{U}^{n+1}}{\partial x_i\partial x_j}\right] + s_4\left[\frac{\partial^2(\mathbf{a}_i\mathbf{b}_j + \mathbf{b}_i\mathbf{b}_j)\Delta\mathbf{U}^{n+1}}{\partial x_i\partial x_j}\right]\right\}$$
$$+\Delta t\left(\frac{\partial \mathbf{F}_i^n}{\partial x_i} + \frac{\partial \mathbf{G}_i^n}{\partial x_i}\right) - \frac{\Delta t^2}{2}\left[\frac{\partial}{\partial x_i}(\mathbf{a}_i + \mathbf{b}_i)\left(\frac{\partial \mathbf{F}_j^n}{\partial x_j} + \frac{\partial \mathbf{G}_j^n}{\partial x_j}\right)\right] + O(\Delta t^3) = 0$$
$$(6.5.13b)$$

Here, once again, the product of the diffusion gradient Jacobian with third order spatial derivatives is neglected and all Jacobians $\mathbf{a}_i$, $\mathbf{b}_i$, and $\mathbf{c}_{ij}$ are assumed to remain constant spatially within each time step and to be updated at subsequent time steps. For simplicity, we may rearrange (6.5.13b) in a compact form,

$$\mathbf{R} = \Delta\mathbf{U}^{n+1} + \frac{\partial}{\partial x_i}(\mathbf{E}_i \Delta\mathbf{U}^{n+1}) + \frac{\partial^2}{\partial x_i \partial x_j}(\mathbf{E}_{ij}\Delta\mathbf{U}^{n+1}) + \mathbf{Q}^n + O(\Delta t^3), \tag{6.5.13c}$$

or, lagging $\mathbf{E}_i$ and $\mathbf{E}_{ij}$ one time step behind,

$$\left(\mathbf{I} + \mathbf{E}_i^n \frac{\partial}{\partial x_i} + \mathbf{E}_{ij}^n \frac{\partial^2}{\partial x_i \partial x_j}\right)\Delta\mathbf{U}^{n+1} = -\mathbf{Q}^n \tag{6.5.14}$$

with

$$\mathbf{E}_i^n = \Delta t(s_1 \mathbf{a}_i + s_3 \mathbf{b}_i)^n \tag{6.5.15a}$$

$$\mathbf{E}_{ij}^n = \left\{\Delta t s_3 \mathbf{c}_{ij} - \frac{\Delta t^2}{2}[s_2(\mathbf{a}_i\mathbf{a}_j + \mathbf{b}_i\mathbf{a}_j) + s_4(\mathbf{a}_i\mathbf{b}_j + \mathbf{b}_i\mathbf{b}_j)]\right\}^n \tag{6.5.15b}$$

$$\mathbf{Q}^n = \frac{\partial}{\partial x_i}[\Delta t(\mathbf{F}_i^n + \mathbf{G}_i^n)] - \frac{\partial^2}{\partial x_i \partial x_j}\left[\frac{\Delta t^2}{2}(\mathbf{a}_i + \mathbf{b}_i)(\mathbf{F}_j^n + \mathbf{G}_j^n)\right] \tag{6.5.15c}$$

Note that the Beam-Warming scheme [1978] discussed in Section 6.3.2 can be written in the form similar to (6.5.14) with the following definitions of $\mathbf{E}_i$, $\mathbf{E}_{ij}$, and $\mathbf{Q}^n$:

$$\mathbf{E}_i = m\Delta t(\mathbf{a}_i + \mathbf{b}_i), \quad \text{with } m = \theta/(1+\xi) \tag{6.5.16a}$$

$$\mathbf{E}_{ij} = m\Delta t \mathbf{c}_{ij} \tag{6.5.16b}$$

$$\mathbf{Q}^n = \frac{\Delta t}{1+\xi}\left(\frac{\partial \mathbf{F}_i^n}{\partial x_i} + \frac{\partial \mathbf{G}_i^n}{\partial x_i}\right) + \frac{\xi}{1+\xi}\Delta\mathbf{U}^n \tag{6.5.16c}$$

where the cross-derivative terms appearing in $\mathbf{Q}^n$ for the Beam-Warming scheme are included in the second derivative terms on the left-hand side. The Beam-Warming scheme is seen to be a special case of the FDV equations if we set $s_1 = s_3 = m$, $s_2 = s_4 = 0$, in (6.5.14), with adjustments of $\mathbf{Q}^n$ on the right-hand side as in (6.5.16c). The stability analysis of the Beam-Warming scheme requires $\xi \geq 0.385$ and $\theta = 1/2 + \xi$. This will fix the FDV parameter m to be $0.639 \leq m \leq 0.75$.

We realize that all physical phenomena are dictated by the FDV parameters in the FDV equations (6.5.14). Either FDM, FEM, or FVM approximations can be applied to (6.5.14). However, their roles are merely to provide different options of discretization, with physics governed by the FDV theory itself. Furthermore, the FDV equations are capable of producing many existing FDM and FEM schemes as special cases, as demonstrated in Chapter 16.

For FDM applications, the first derivative for $\mathbf{E}_i\Delta\mathbf{U}^{n+1}$ and the second derivative for $\mathbf{E}_{ij}\Delta\mathbf{U}^{n+1}$ in (6.5.14) may be approximated by many options of finite difference equations including high order accuracy schemes introduced in Section 3.7 or using the flux vector splitting for the term involved in $\mathbf{a}_i$ for $\mathbf{E}_i$ in (6.5.14). However, the physical aspects accommodated in the FDV theory through the various FDV parameters are unique and they play important roles, as elaborated next.

6.5.4 INTERPRETATION OF FLOWFIELD-DEPENDENT VARIATION PARAMETERS

The flowfield-dependent variation (FDV) parameters as defined earlier are capable of allowing various numerical schemes to be automatically generated. They are summarized as follows:

(1) *First order FDV parameters.* The first order FDV parameters s_1 and s_3 control all high gradient phenomena such as shock waves and turbulence. These parameters as calculated from the changes of local Mach numbers and Reynolds (or Peclet) numbers within each element and are indicative of the actual local element flowfields. The contours of these parameters closely resemble the flowfields themselves, with both s_1 and s_3 being large (close to unity) in regions of high gradients, but small (close to zero) in regions where the gradients are small. The fact that the contours of s_1 and s_3 resemble the flowfield (Mach number or density contours) is demonstrated in Figure 13.7.3.2. The basic role of s_1 and s_3 is to provide computational accuracy.

(2) *Second order FDV parameters.* The second order FDV parameters s_2 and s_4 are also flowfield dependent, exponentially proportional to the first order FDV parameters. However, their primary role is to provide adequate computational stability (artificial viscosity) as they were originally introduced into the second order time derivative term of the Taylor series expansion of the conservation flow variables $\mathbf{U}^{n+1}$.

(3) *Parabolic/elliptic* ($s_1 = 0$). The s_1 terms represent convection. This implies that if $s_1 \cong 0$ then the effect of convection is small. The computational scheme is automatically altered to take this effect into account, with the governing equations being predominantly parabolic-elliptic.

(4) *Hyperbolic* ($s_3 = 0$). The s_3 terms are associated with diffusion. Thus, with $s_3 \cong 0$, the effect of viscosity or diffusion is small and the computational scheme is automatically switched to that of Euler equations where the governing equations are predominantly hyperbolic.

(5) *Mixed elliptic/parabolic/hyperbolic* ($s_1 \neq 0, s_3 \neq 0$). If the first order FDV parameters s_1 and s_3 are nonzero, this indicates a typical situation for the mixed hyperbolic, parabolic, and elliptic nature of the Navier-Stokes system of equations, with convection and diffusion being equally important. This is the case for incompressible flows at low speeds. The unique property of the FDV scheme is its capability to control pressure oscillations adequately without resorting to the separate hyperbolic elliptic pressure Poisson equation for pressure corrections. The capability of the FDV scheme to handle incompressible flows is achieved by a delicate balance between s_1 and s_3 as determined by the local Mach numbers and Reynolds (or Peclet) numbers. If the flow is completely incompressible ($M = 0$), the criteria given by (6.5.9) leads to $s_1 = 1$, whereas the variation parameter s_3 is to be determined according to the criteria given in (6.5.11). Make a note of the presence of convection-diffusion interaction terms given by the product of $\mathbf{b}_i\mathbf{a}_j$ in the s_2 terms and $\mathbf{a}_i\mathbf{b}_j$ in the s_4 terms. These terms allow interactions between convection and diffusion in the viscous incompressible and/or viscous compressible flows.

(6) *High temperature gradient flow.* If temperature gradients rather than velocity gradients dominate the flowfield, then s_3 is governed by the Peclet number rather

than by the Reynolds number. Such cases arise in high-speed, high-temperature compressible flows close to the wall.

(7) *Transition to turbulence.* The transition to turbulence is a natural flow process as the Reynolds number increases, causing the gradients of any or all flow variables to increase. This phenomenon is physical instability and is detected by the increase of s_3 if the flow is incompressible, but by both s_3 and s_1 if the flow is compressible. Such physical instability is likely to trigger the numerical instability, but will be countered by the second order FDV parameters s_2 and/or s_4 to ensure numerical stability automatically. In this process, these flowfield dependent variation parameters are capable of capturing relaminarization, compressibility effect or dilatational turbulent energy dissipation, and turbulent unsteady fluctuations. They are characterized by the product of s_3 and the fluctuations of stress tensor ($s_3 \Delta\tau_{ij}$) in which the stresses consist of mean and fluctuation parts. As a consequence, some regions of the flow domain such as in boundary layers may always be unsteady ($\Delta\tau_{ij} \neq 0$), even though the steady state may have been reached away from the wall. However, in order for these fluctuation parts to be correctly determined, it is necessary that Kolmogorov scales be resolved in sufficiently refined grids such as in the direct numerical simulation (DNS). Thus, for a coarse mesh, the advantage of FDV process cannot be expected.

6.5.5 SHOCK-CAPTURING MECHANISM

The shock-capturing mechanism is built into the FDV equations of continuity, momentum, and energy. For example, let us examine (6.5.7) or (6.5.13) and write the momentum equations, with all diffusion terms neglected.

$$\begin{aligned}
&\Delta(\rho v_j)^{n+1} + \Delta t[(\rho v_i v_j)_{,i} + p_{,j}]^n \\
&= -s_1 \Delta t(\Delta\rho v_i v_j + \Delta p \delta_{i,j})^{n+1}_{,i} + s_2 \frac{\Delta t^2}{2}\left(a_k^{(m)} + b_k^{(m)}\right)[\Delta(\rho v_i v_j)_{,i} + \Delta p_{,j}]^{n+1}_{,k} \\
&\quad + \frac{\Delta t^2}{2}\left(a_k^{(m)} + b_k^{(m)}\right)[(\rho v_i v_j)_{,i} + p_{,j}]^n_{,k}
\end{aligned} \tag{6.5.17}$$

where $a_k^{(m)}$ and $b_k^{(m)}$ denote the convection and diffusion Jacobians, respectively. To identify the shock capturing mechanism in the FDV formulation as compared to the TVD finite difference scheme, let us rewrite (6.5.17) for the 1-D momentum equation, retaining only the convection flux without the pressure gradients.

$$\Delta u^{n+1} = -\Delta t s_1 \frac{\partial a \Delta u^{n+1}}{\partial x} + \frac{\Delta t^2}{2} s_2 a^2 \frac{\partial^2 \Delta u^{n+1}}{\partial x^2} - \Delta t \frac{\partial f^n}{\partial x} + \frac{\Delta t^2}{2} a \frac{\partial^2 f^n}{\partial x^2} \tag{6.5.18a}$$

or

$$\begin{aligned}
\Delta u^{n+1} &= -\Delta t \frac{\sqrt{M_{\max}^2 - M_{\min}^2}}{M_{\min}} \frac{\partial a \Delta u^{n+1}}{\partial x} + \frac{\Delta t^2}{2}\left(\frac{\sqrt{M_{\max}^2 - M_{\min}^2}}{M_{\min}}\right)^{\eta} a^2 \frac{\partial^2 \Delta u^{n+1}}{\partial x^2} \\
&\quad - \Delta t \frac{\partial f^n}{\partial x} + \frac{\Delta t^2}{2} a \frac{\partial^2 f^n}{\partial x^2}
\end{aligned} \tag{6.5.18b}$$

where f is the convection flux and a is the 1-D convection Jacobian or speed of sound.

The FDM analog of (6.5.18) at node i becomes

$$\frac{\Delta u_i^{n+1}}{\Delta t} = -s_1 a \frac{1}{\Delta x}\left(\Delta u_i^{n+1} - \Delta u_{i-1}^{n+1}\right) + s_2 a^2 \Delta t \frac{1}{2\Delta x^2}\left(\Delta u_i^{n+1} - 2\Delta u_{i-1}^{n+1} + \Delta u_{i-2}^{n+1}\right)$$
$$-\frac{1}{\Delta x}\left(f_i^n - f_{i-1}^n\right) + a\Delta t \frac{1}{2\Delta x^2}\left(f_i^n - 2f_{i-1}^n + f_{i-2}^n\right) \tag{6.5.19}$$

The second order TVD semi-discretized scheme (6.2.110) with limiter functions (6.2.111) is written at node i as

$$\frac{du_i}{dt} = -\frac{a^+}{\Delta x}\left[1 + \Psi_{i-1/2}^+ - \frac{1}{2}\frac{\Psi_{i-3/2}^+}{r_{i-3/2}^+}\right](u_i - u_{i-1})$$
$$-\frac{a^-}{\Delta x}\left[1 + \Psi_{i+1/2}^- - \frac{1}{2}\frac{\Psi_{i+3/2}^-}{r_{i+3/2}^-}\right](u_{i+1} - u_i) \tag{6.5.20}$$

where Ψ and r denote the limiter function and velocity ratio, respectively,

$$r_{i-3/2}^+ = \frac{u_i - u_{i-1}}{u_{i-1} - u_{i-2}}, \qquad r_{i+3/2}^- = \frac{u_{i+1} - u_i}{u_{i+2} - u_{i+1}} \tag{6.5.21}$$

Inserting (6.5.21) into (6.5.20) yields

$$\frac{du_i}{dt} = -\frac{a^+}{\Delta x}\left[(u_i - u_{i-1}) + \frac{1}{2}\Psi_{i-1/2}^+(u_i - u_{i-1}) - \Psi_{i-3/2}^+(u_{i-1} - u_{i-2})\right]$$
$$-\frac{a^-}{\Delta x}\left[(u_{i+1} - u_i) + \frac{1}{2}\Psi_{i+1/2}^-(u_{i+1} - u_i) - \Psi_{i+3/2}^-(u_{i+2} - u_{i+1})\right] \tag{6.5.22}$$

Let us assume that, for positive-going waves,

$$u_i = u_i^n + s\Delta u_i^{n+1}, \qquad a^- = 0, \qquad a^+ = a, \qquad \Psi_{i-1/2}^+ = 2\Psi_{i-3/2}^+ = -\Psi$$

Substituting the above into (6.5.22), the TVD equation may be expressed as

$$\frac{\Delta u_i^{n+1}}{\Delta t} = -sa\frac{1}{\Delta x}\left(\Delta u_i^{n+1} - \Delta u_{i-1}^{n+1}\right) + \frac{\Psi\Delta x}{2\Delta x^2}\left(\Delta u_i^{n+1} - 2\Delta u_{i-1}^{n+1} + \Delta u_{i-2}^{n+1}\right)$$
$$-\frac{1}{\Delta x}\left(f_i^n - f_{i-1}^n\right) + \frac{\Psi\Delta x}{2\Delta x^2}\left(f_i^n - 2f_{i-1}^n + f_{i-2}^n\right) \tag{6.5.23}$$

If we set

$$s_1 = s, \quad s_2 = \frac{s\Delta x\Psi}{a\Delta t}, \quad \Psi = \frac{a\Delta t}{\Delta x}, \quad s_2 = s_1$$

it is seen that the FDV equation (6.5.19) becomes identical to the TVD equation (6.5.23). Note that in TVD either a^+ or a^- must be chosen from the flowfield and the FDV parameters s_1 and s_2 in FDV are automatically calculated. Of course, the precise shock-capturing mechanism of both methods is not exactly the same, because all the assumptions made above are not true in general. However, it is interesting to note that the first order convection FDV parameter s_1 is related to the TVD limiter function Ψ as

$$s_1 = \frac{s\Delta x}{a\Delta t}\Psi$$

in which it is shown that the convection FDV parameters (s_1, s_2) are proportional or equivalent to the TVD limiter functions. A similar process can be shown also for negative-going waves $(a^- = a, a^+ = 0)$.

Considering that the motivations and procedures of derivation are completely different, the analogy between the TVD scheme and FDV formulation as demonstrated above is remarkable. Notice that, beyond this analogy, the FDV formulation is to couple the convection variation parameters (s_1, s_2) with all other variation parameters (s_3, s_4) so that shock wave interactions with all other physical properties can be resolved. They are involved also in transitions and interactions of compressible/incompressible, inviscid/viscous, and laminar/turbulent flows.

In the TVD methods, the resulting Euler equations are based on positive and negative eigenvalues or Jacobians, either $a^- = 0$ or $a^+ = 0$, which will switch the scheme to either backward differenceing for positive waves or forward differencing for negative waves in one dimension, respectively.

To illuminate the consequence of the FDV theory, it is infomative to write (6.5.18a) in the form,

$$u_i^{n+1} = u_i^n - s_1 a^{(+,-)} \frac{\Delta t}{\Delta x}\left(\delta u_i^{(+,-)n+1} - \delta u_i^{(+,-)n}\right) - s_2 a^2 \frac{\Delta t^2}{2\Delta x^2}$$
$$\times \left[\left(\delta^2 u_i^{(+,-)}\right)^{n+1} - \left(\delta^2 u_i^{(+,-)}\right)^n\right] - \frac{\Delta t}{2\Delta x}(\delta f_i^{(+,-)})^n + \frac{a\Delta t^2}{2\Delta x^2}(\delta^2 f_i^{(+,-)})^n \tag{6.5.24a}$$

where the flux vector splitting scheme is used with $a = a^+ + a^-$ and the following definitions:

For $M > 1,\quad a^+ = a,\quad a^- = 0,\quad \delta u_i^+ = u_i - u_{i-1},\quad \delta f_i^+ = f_i - f_{i-1},$
$\delta^2 u_i^+ = u_i - 2u_{i-1} + u_{i-2},\quad \delta^2 f_i^+ = f_i - 2f_{i-1} + f_{i-2},$
For $M < 1,\quad a^+ = 0,\quad a^- = a,\quad \delta u_i^- = u_{i+1} - u_i,\quad \delta f_i^- = f_{i+1} - f_i,$
$\delta^2 u_i^- = u_{i+2} - 2u_{i+1} + u_i,\quad \delta^2 f_i^- = f_{i+2} - 2f_{i+1} + f_i,$

Thus, the finite-differenced FDV equation takes the form

$$u_i^{n+1} + s_1 a^{(+,-)} \frac{\Delta t}{\Delta x}\delta u_i^{(+,-)n+1} + s_2 a^2 \frac{\Delta t^2}{2\Delta x^2}\left(\delta^2 u_i^{(+,-)}\right)^{n+1}$$
$$= u_i^n + s_1 a^{(+,-)} \frac{\Delta t}{\Delta x}\delta u_i^{(+,-)n} + s_2 a^2 \frac{\Delta t^2}{2\Delta x^2}\left(\delta^2 u_i^{(+,-)}\right)^n$$
$$- \frac{\Delta t}{2\Delta x}\left(\delta f_i^{(+,-)}\right)^n + \frac{a\Delta t^2}{2\Delta x^2}\left(\delta^2 f_i^{(+,-)}\right)^n \tag{6.5.24b}$$

The main difference between the finite-differenced FDV theory and the TVD schemes lies in the fact that in FDV methods variation parameters control the shock capturing mechanism and play the role similar to the limiters in TVD.

In the finite-differenced FDV methods, calculated variation parameters affect the convection and diffusion Jacobians associated with E_1^n and E_{ij}^n in (6.5.15a,b) based on the Mach number and Reynolds number changes between adjacent nodes in multidimensions. Thus, for high values of the variation parameters indicative of high gradients of variables, characterize the discontinuous physical behavior of the variables. The contours of these variation parameters closely resemble the flowfield itself (see Figure 13.7.3.2). An example for a triple shock wave boundary layer interaction problem using FDV-FDM is shown in Figures 6.8.21 through 6.8.24. Other examples of the FDV methods are demonstrated in Sections 13.7, 15.3, and 27.3.

6.5.6 TRANSITIONS AND INTERACTIONS BETWEEN COMPRESSIBLE AND INCOMPRESSIBLE FLOWS

One of the most significant aspects of the FDV scheme is that, for low Mach numbers (incompressible flow), the scheme will automatically adjust itself to prevent pressure oscillations by ensuring the conservation of mass. This can be evidenced by the presence of the second derivatives of pressure arising in the equations of momentum, continuity, and energy. We note that the FDV momentum equations given by (6.5.17) may be rearranged in the form,

$$\frac{\partial}{\partial t}(\rho v_j)^{n+1} + (\rho v_i v_j + p\delta_{ij} - \tau_{ij})^n_{,i} = S_j(m) \tag{6.5.25}$$

with

$$\begin{aligned} S_j(m) = &-\lfloor s_1(\Delta\rho v_i v_j + \Delta p\delta_{ij}) - s_3\Delta\tau_{ij}\rfloor^{n+1}_{,i} \\ &+ \frac{\Delta t}{2}\left[\left(a_k^{(m)} + b_k^{(m)}\right)((\rho v_i v_j)_{,i} + p_{,j} - \tau_{ij,i})\right]^n_{,k} \\ &+ \frac{\Delta t}{2}\left[\left(a_k^{(m)} + b_k^{(m)}\right)(s_2(\Delta(\rho v_i v_j)_{,i} + \Delta p_{,j}) - s_4\Delta\tau_{ij,i})\right]^{n+1}_{,k} \end{aligned} \tag{6.5.26}$$

Similarly, the FDV equation for continuity becomes

$$\Delta\rho^{n+1} = \Delta t[-(\rho v_i)^n_{,i} - s_1\Delta(\rho v_j)^{n+1}_{,j}] + \frac{\Delta t^2}{2}\left[\left(a_i^{(c)}(\rho v_j)_{,j}\right)_{,i} + s_2\left(a_i^{(c)}\Delta(\rho v_j)_{,j}\right)_{,i}\right]^{n+1} \tag{6.5.27}$$

with $a_i^{(c)}$ being the convection Jacobian for the continuity equation. Substituting (6.5.17) into (6.5.27) and rearranging the differential equation of continuity,

$$\frac{\partial\rho^{n+1}}{\partial t} + (\rho v_i)^n_{,i} = S(c) \tag{6.5.28}$$

with

$$\begin{aligned} S(c) = &\Delta t s_1[(\rho v_i v_j)_{,i} + p_{,j} - \tau_{ij,i}]^n_{,j} - \Delta t s_1[s_1(\Delta(\rho v_i v_j)_{,i} + \Delta p_{,j}) - s_3\Delta\tau_{ij,i}]^{n+1}_{,j} \\ &- \frac{\Delta t^2}{2}s_1[(a_k^{(m)} + b_k^{(m)})((\rho v_i v_j)_{,i} + p_{,j} - \tau_{ij,i})]^n_{,kj} \\ &- \frac{\Delta t^2}{2}s_1[(a_k^{(m)} + b_k^{(m)})(s_2(\Delta(\rho v_i v_j)_{,i} + \Delta p_{,j}) - s_4\Delta\tau_{ij,i})]^{n+1}_{,kj} \\ &+ \frac{\Delta t}{2}(a_i^{(c)}(\rho v_j)^n_{,j})_{,i} \end{aligned} \tag{6.5.29}$$

where the third derivative associated with s_2 is neglected. A glance at (6.5.26) and (6.5.29) reveals that the right-hand side terms $S(m)$ for momentum and $S(c)$ for continuity are the additional terms of higher order derivatives arising from the process of derivations of the FDV equations.

The FDV equation for energy is of the form,

$$\begin{aligned}\Delta(\rho E)^{n+1} = {} & \Delta t\lfloor -(\rho E\mathrm{v}_i + p\mathrm{v}_i)_{,i} + (\tau_{ij}\mathrm{v}_j)_{,i} + kT_{,ii}\rfloor^n \\ & - \Delta t\{s_1[\Delta(\rho E\mathrm{v}_i) + p\mathrm{v}_i]_{,i} - s_3\lfloor\Delta(\tau_{ij}\mathrm{v}_j) + kT_{,i}\rfloor_i\}^{n+1} \\ & + \frac{\Delta t^2}{2}\{(a_k^{(e)} + b_k^e)[(\rho E\mathrm{v}_i + p\mathrm{v}_i)_{,i} - (\tau_{ij}\mathrm{v}_j)_{,i} - kT_{,ii}]\}_k^n \\ & + \frac{\Delta t^2}{2}\{(a_k^{(e)} + b_k^e)[s_2\Delta(\rho E\mathrm{v}_i + p\mathrm{v}_i)_{,i} - s_4\Delta(\tau_{ij}\mathrm{v}_j + kT_{,i})_{,i}]\}_{,k}^{n+1}\end{aligned} \tag{6.5.30}$$

which leads to the reconstructed equation of energy,

$$\frac{\partial(\rho E)^{n+1}}{\partial t} + [-(\rho E\mathrm{v}_i + p\mathrm{v}_i)_{,i} + (\tau_{ij}\mathrm{v}_j)_{,i} + kT_{,ii}]^n = S(e) \tag{6.5.31}$$

with

$$\begin{aligned}S(e) = {} & -\{s_1[\Delta(\rho E\mathrm{v}_i) + p\mathrm{v}_i]_i - s_3[\Delta(\tau_{ij}\mathrm{v}_j) + kT_{,i}]_{,i}\}^{n+1} \\ & + \frac{\Delta t}{2}\{(a_k^{(e)} + b_k^e)[(\rho E\mathrm{v}_i + p\mathrm{v}_i)_{,i} - (\tau_{ij}\mathrm{v}_j)_{,i} - kT_{,ii}]\}_{,k}^n \\ & + \frac{\Delta t}{2}\{(a_k^{(e)} + b_k^{(e)})[s_2\Delta(\rho E\mathrm{v}_i + p\mathrm{v}_i)_{,i} - s_4\Delta(\tau_{ij}\mathrm{v}_j + kT_{,i})_{,i}]\}_{,k}^{n+1}\end{aligned} \tag{6.5.32}$$

The physical implications of the right-hand side terms for all equations are quite complex. There exist not only the second derivatives of pressure for the terms having no variation parameters at the temporal station n, but also the inviscid/viscous interactions contributed by the s_2 and s_4 terms at the temporal station $n+1$. Thus, the transitions and interactions between compressible and incompressible flows are contributed by inviscid/viscous interactions or convection/diffusion interactions.

The most crucial aspect of the transition between compressible and incompressible flows is the relationship of the equation of state shared by both compressible and incompressible flows. To this end, consider that initially the fluid is a perfect gas and that the total energy is given by

$$E = c_p T - \frac{p}{\rho} + \frac{1}{2}\mathrm{v}_i\mathrm{v}_i \tag{6.5.33}$$

The momentum equation for steady-state incompressible rotational flow may be integrated to give

$$\int \left(p + \frac{1}{2}\rho\mathrm{v}_j\mathrm{v}_j\right)_{,i} dx_i = \int (\mu\mathrm{v}_{i,jj} + \rho\varepsilon_{ijk}\mathrm{v}_j\omega_k)dx_i \tag{6.5.34}$$

$$p + \frac{1}{2}\rho\mathrm{v}_j\mathrm{v}_j = p_0 + Q \tag{6.5.35}$$

with

$$Q = \frac{1}{n}\int (\mu v_{i,jj} + \rho\varepsilon_{ijk} v_j \omega_k) dx_i$$

where p_0 is the constant of integration, and n is the spatial dimension.

Substituting (6.5.33) into (6.5.35) leads to the following relationship:

$$p_0 = \rho(c_p T + \frac{1}{2} v_i v_i - E) - Q \tag{6.5.36}$$

If p_0 as given by (6.5.36) remains a constant, equivalent to a stagnation (total) pressure, then the compressible flow as assumed in the conservation form of the Navier-Stokes system of equations has now been turned into an incompressible flow, which is expected to occur when the flow velocity is sufficiently reduced (approximately $0.1 \leq M < 0.3$ for air). Thus, (6.5.36) serves as an equivalent equation of state for an incompressible flow. This can be identified nodal point by nodal point or element by element for the entire domain. Figure 13.7.4e,f shows that both density and stagnation pressure begin to vary in the cavity flow problem for $M = 0.1$, whereas they remain constant for $M = 0.01$.

We may begin with the condition given by (6.5.35) for compressible flows. If computations are involved in low-speed flows, then the governing equations and computational schemes initially intended for high-speed compressible flows are automatically switched to those for low-speed incompressible flows with p_0 remaining constant for all low Mach number flows (approximately $0.1 \leq M \leq 0.3$) based on the flowfield-dependent variation parameters. If the flow reverses to compressible, then the stagnation pressure becomes variable, allowing the density to change.

An advantage of the FDV scheme is to avoid the so-called pressure correction process, preconditioning approach, or the implementation of a separate hyperbolic-elliptic equation as is the case with other computational schemes designed to accommodate flows of all speed regimes. In the case of the FDV formulation, a computational scheme similar to pressure correction (keeping pressure from oscillating) automatically arises by means of the Mach number and Reynolds number-dependent variation parameters. This approach is particularly useful for the inviscid-viscous interaction regions and boundary layers close to the wall such as in hypersonic aircraft or shock wave turbulent boundary layer interactions in general.

6.5.7 TRANSITIONS AND INTERACTIONS BETWEEN LAMINAR AND TURBULENT FLOWS

When inviscid flow becomes viscous, we may expect that the flow may become laminar or turbulent through inviscid/viscous interactions across the boundary layer. Below the laminar boundary layer, if viscous actions are significant, then the fluid particles are unstable, causing the changes of Mach number and Reynolds number between adjacent nodal points (assuming they are closely spaced) to be irregular, the phenomenon known as transition instability prior to the state of full turbulence. How can these processes be modeled in FDV formulation?

Fluctuations due to turbulence are characterized by the presence of the terms in the equation of momentum, continuity, and energy such as

$$s_3 \Delta\tau_{ij} = \frac{\sqrt{\mathrm{Re}_{max}^2 - \mathrm{Re}_{min}^2}}{\mathrm{Re}_{min}} \Delta\tau_{ij} \tag{6.5.37}$$

Physically, the above quantity represents the fluctuations of total stresses (physical viscous stresses plus Reynolds stresses) controlled by the Reynolds number changes between the local adjacent nodal points. Thus, the FDV solution contains the sum of the mean flow variables and the fluctuation parts of the variables.

Once the solution of the Navier-Stokes system of equations is carried out and all flow variables are determined, then we compute the fluctuation part, f' of any variable f,

$$f' = f - \overline{f} \tag{6.5.38}$$

where f and $\overline{f}$ denote the Navier-Stokes solution and its time or mass average, respectively. This process may be replaced by the fast Fourier transform of the Navier-Stokes solution. Unsteady turbulence statistics (turbulent kinetic energy, Reynolds stresses, and various energy spectra) can be calculated once the fluctuation quantities of all variables are determined.

Although the solutions of the Navier-Stokes system of equations using FDV are assumed to contain the fluctuation parts as well as the mean quantities, it will be unlikely that such information is reliable when the Reynolds number is very high and if mesh refinements are not adequate to resolve the Kolmogorov microscales. In this case, it is necessary to invoke the level of mesh refinements as required for the direct numerical simulation (DNS). It is expected that FDV methods lead to accurate solutions at high Mach number and high Reynolds number flows if the mesh refinements required for DNS are used.

It is important to recognize that unsteadiness in turbulent fluctuations may prevail in the vicinity of the wall, although a steady state may have been reached far away from the wall. This situation can easily be verified by noting that $\Delta\mathbf{U}^{n+1}$ will vanish only in the region far away from the wall, but remain fluctuating in the vicinity of the wall, as dictated by the changes of Reynolds number in the variation parameter s_3 between the nodal points and fluctuations of the stresses due to both physical and turbulent viscosities in $\Delta\tau_{ij}$ characterized by (6.5.37).

CONCLUDING REMARKS

Transitions and interactions between inviscid/viscous, compressible/incompressible, and laminar/turbulent flows can be resolved by the FDV theory. It is shown that variation parameters initially introduced in the Taylor series expansion of the conservation variables of the Navier-Stokes system of equations are translated into flowfield-dependent physical parameters responsible for the characterization of fluid flows. In particular, the convection FDV parameters (s_1, s_2) are identified as equivalent to the TVD limiter functions in a specialized case. The FDV equations are shown to contain the terms of fluctuation variables automatically generated in the course of developments, varying in time and space, but following the current physical phenomena. In addition, adequate numerical controls (artificial viscosity) to address both nonfluctuating and fluctuating parts of variables are automatically activated according to the current flowfield. It has been shown that some existing numerical schemes in FDM are the special cases of the FDV theory.

An example of three-dimensional triple shock wave boundary layer interactions is demonstrated in Section 6.8.2. Some simple problems of FDV methods for supersonic compression corner and driven cavity using FEM are shown in Section 13.7. Applications of FDV theory using FVM-FEM are demonstrated in Section 15.3. Finally, applications of FDV-FEM methods to relativistic astrophysical flows are presented in Section 27.3.

6.6 OTHER METHODS

6.6.1 ARTIFICIAL VISCOSITY FLUX LIMITERS

The convection flux vector may be written in the form [Jameson et al., 1981],

$$F_{j+1} = F\left(\frac{U_{j+1} + U_j}{2}\right) - d_{j+2} \tag{6.6.1}$$

with

$$d_{j+2} = \varepsilon^{(2)}_{j+1/2}(U_{j+1} - U_j) - \varepsilon^{(4)}_{j+1/2}(U_{j+2} - 3U_{j+1} + 3U_j - U_{j-1})$$

$$\varepsilon^{(2)}_{j+1/2} = k^{(2)} R_{j+1/2} \Psi_{j+1/2}$$

$$\varepsilon^{(4)}_{j+1/2} = \max\left(0, k^{(2)} R_{j+1/2} - \varepsilon^{(2)}_{j+1/2}\right)$$

where $k^{(2)}$ and $k^{(4)}$ are real numbers fixing the amount of diffusion brought up by the second and fourth order dissipative operators. $R_{j+1/2}$ is the spectral radius of the Jacobian $\partial\mathbf{F}/\partial\mathbf{U}$ at the cell face $j+1$. $\Psi_{j+1/2}$ is a limiter based on

$$\Psi_j = \frac{|\overline{p}_{j+1} - 2\overline{p}_j + \overline{p}_{j-1}|}{|\overline{p}_{j+1} + 2\overline{p}_j + \overline{p}_{j-1}|}$$

$$\Psi_{j+1/2} = \max\left(\Psi_j, \Psi_{j+1/2}\right) \tag{6.6.2}$$

Thus, the flux vectors may be written in terms of limiters in the form,

$$F_{j+1} = \frac{1}{2}(U_j\Psi_j + U_{j+1}\Psi_{j+1}) = \frac{1}{2}(U_j + U_{j+1})$$

$$F_{j-1} = \frac{1}{2}(U_{j-1}\Psi_{j-1} + U_j\Psi_j) = \frac{1}{2}(U_{j-1} + U_j) \tag{6.6.3}$$

Using the flux of the mean value, we obtain

$$F_{j+1/2} = \frac{1}{4}(U_j + U_{j+1})(\Psi_j + \Psi_{j-1})$$

$$F_{j-1/2} = \frac{1}{4}(U_{j-1} + U_j)(\Psi_{j-1} + \Psi_j) \tag{6.6.4}$$

which represents a semi-discrete equation using a skew-symmetric form of second order. It is designed to reduce the aliasing errors that are crucial in low order nondissipative schemes useful in problems such as large eddy simulations of turbulence (see Section 21.7.3).

6.6.2 FULLY IMPLICIT HIGH ORDER ACCURATE SCHEMES

The Navier-Stokes system of equations in terms of the primitive flow variables,

$$\mathbf{Q} = [\rho \quad v_i \quad p]^T$$

may be written as

$$\frac{\partial \mathbf{Q}}{\partial t} + \mathbf{A}_i \frac{\partial \mathbf{Q}}{\partial x_i} + \frac{\partial \mathbf{G}_i}{\partial x_i} = 0 \tag{6.6.5}$$

with

$$\mathbf{A}_i \frac{\partial \mathbf{Q}}{\partial x_i} = \mathbf{A}_i^+ \mathbf{Q}_{,i}^- + \mathbf{A}_i^- \mathbf{Q}_{,i}^+$$
$$\mathbf{A}_i^{\pm} = \mathbf{P}\,\Lambda_i^{\pm}\,\mathbf{P}^{-1} \tag{6.6.6}$$

where $\mathbf{A}_i$ is the convection flux Jacobian matrix.

The fully implicit finite difference approximations of (6.6.5) may be written as

$$\frac{3\mathbf{Q}^{n+1} - 4\mathbf{Q}^n + \mathbf{Q}^{n-1}}{2\Delta t} + (\mathbf{A}_i^+ \mathbf{Q}_{,i}^- + \mathbf{A}_i^- \mathbf{Q}_{,i}^+)^{n+1} + \left(\frac{\partial \mathbf{G}_i}{\partial x_i}\right)^{n+1} = 0 \tag{6.6.7}$$

The Newton-Raphson solution of (6.6.7) may be written in the form

$$\left(\mathbf{I} + \Delta t \frac{\partial \mathbf{H}}{\partial \mathbf{Q}}\right) \Delta \mathbf{Q}^{m+1} = -\Delta \mathbf{Q}^m + \Delta t \mathbf{H}^m \tag{6.6.8}$$

with

$$\Delta \mathbf{Q}^{m+1} = \left(\frac{3}{2}\mathbf{Q}^{n+1} - 2\mathbf{Q}^n + \frac{1}{2}\mathbf{Q}^{n-1}\right)^{m+1}$$
$$\Delta \mathbf{Q}^{m} = \left(\frac{3}{2}\mathbf{Q}^{n+1} - 2\mathbf{Q}^n + \frac{1}{2}\mathbf{Q}^{n-1}\right)^{m}$$
$$\mathbf{H}^m = \left[-(\mathbf{A}_i^+ \mathbf{Q}_{,i}^- + \mathbf{A}_i^- \mathbf{Q}_{,i}^+) + \frac{\partial \mathbf{G}_i}{\partial x_i}\right]^m$$

where the superscript m represents the m-th iteration step, with $\mathbf{Q}^+$ and $\mathbf{Q}^-$ indicating the forward and backward finite differences. Rai and Moin [1993] used fifth order accurate finite differences for large eddy simulation calculations in compressible flows with a seven-point stencil,

$$\mathbf{Q}_x^- = \frac{-6\mathbf{Q}_{i+2} + 60\mathbf{Q}_{i+1} + 40\mathbf{Q}_i - 120\mathbf{Q}_{i-1} + 30\mathbf{Q}_{i-2} - 4\mathbf{Q}_{i-3}}{120\Delta \mathbf{x}}$$
$$\mathbf{Q}_x^+ = \frac{4\mathbf{Q}_{i+3} - 30\mathbf{Q}_{i+2} + 120\mathbf{Q}_{i-2} - 40\mathbf{Q}_i - 60\mathbf{Q}_{i-1} + 6\mathbf{Q}_{i-2}}{120\Delta \mathbf{x}} \tag{6.6.9}$$

on a grid that is equidistanced in the x-direction. The remaining convective terms are evaluated in a similar manner. The above scheme is used in Section 21.7.3.

6.6.3 POINT IMPLICIT METHODS

In order to circumvent stiff equations due to widely disparate time scales in source terms (such as occur in chemically reactive flows), it is advantageous to use the point implicit scheme in which the source terms are provided implicitly. Thus, the Navier-Stokes system of equations are written as

$$\frac{\Delta \mathbf{U}^{n+1}}{\Delta t} + \left(\frac{\partial \mathbf{F}_i}{\partial x_i} + \frac{\partial \mathbf{G}_i}{\partial x_i}\right)^n - \left(\mathbf{B}^n + \frac{\partial \mathbf{B}}{\partial \mathbf{U}}\Delta \mathbf{U}^{n+1}\right) = 0 \tag{6.6.10}$$

Rearranging, we obtain

$$\left(\mathbf{I} - \Delta t\frac{\partial \mathbf{B}}{\partial \mathbf{U}}\right)^{n+1} \Delta \mathbf{U}^{n+1} = -\Delta t\left(\frac{\partial \mathbf{F}_i}{\partial x_i} + \frac{\partial \mathbf{G}_i}{\partial x_i} - \mathbf{B}\right)^n \tag{6.6.11}$$

where the source term Jacobian is evaluated implicitly. Note that derivatives of the convection and diffusion terms may be discretized with the fourth order accuracy finite difference scheme as used in Section 22.6.2.

6.7 BOUNDARY CONDITIONS

Mathematical theories of boundary conditions have been reported extensively in the literature. They include Kreiss [1970], Rudy and Strikwerda [1980], Gustafsson [1982], Dutt [1988], Oliger and Sundström [1978], and Nordström [1989], among others. Incorrect specifications of boundary conditions result in solution instability, nonconvergence of solutions, and/or convergence to inaccurate results. Boundary conditions must be correctly specified in accordance with speed regimes at inlet and outlet, viscous interactions on solid walls, one-dimensional or multidimensional geometries, reflecting and nonreflecting boundaries, and farfield boundaries.

Recall that derivations of Neumann boundary conditions and specification of boundary conditions in general for hyperbolic, parabolic, and elliptic equations were presented in Section 2.3. Discussions on boundary conditions associated with FEM will be included in Sections 10.1.2, 11.1, and 13.6.6. Multiphase flow boundary conditions are also presented in Section 22.2.6. In what follows, various boundary conditions involved in FDM are described.

6.7.1 EULER EQUATIONS

6.7.1.1 One-Dimensional Boundary Conditions

As mentioned in Section 6.2.1.3, the number of boundary conditions to be specified at inflow and outflow boundaries is determined by the eigenvalue spectrum of the Jacobian matrices (6.2.6) in terms of the primitive variables associated with boundary conditions normal to the surface. They are the characteristic variables or Riemann invariants $\mathbf{W}_1, \mathbf{W}_2$, and $\mathbf{W}_3$ in one dimension as given by (6.2.15) and (6.2.30). The general rule is that the number of Dirichlet boundary conditions for primitive variables is equal to the number of positive eigenvalues of the Jacobian matrix, which are prescribed as *physical boundary conditions*. In contrast, the negative eigenvalues represent the *numerical boundary conditions* which must be extrapolated from the flowfield.

Propagation of flow quantities in a one-dimensional flow are shown for expansion and shock waves in Figure 6.2.3. Note that C_- wave is negative for subsonic flow whereas it is positive for supersonic flow in the domain of dependence. A summary of general boundary conditions is shown in Figure 6.7.1. At an inlet point, the characteristics C_0 and C_+ have slopes u and $u+a$, which are always positive for a flow in the positive x- direction. Thus, they will carry information from the boundaries toward the inside domain. The third characteristic C_- has a slope whose sign depends on the inlet Mach number. For the supersonic inlet, C_- has a positive sign, whereas it has a negative sign for subsonic flow. Therefore, no boundary conditions associated with C_- for the subsonic inlet can be specified. Similar considerations can be made for the outlet. Namely, no boundary conditions are to be specified for C_+ and C_0. As to C_-, however, we must provide boundary conditions for subsonic outlet, but not for supersonic outlet.

Note that each characteristic variable transports a given information and the quantities transported from the inside of the domain toward the boundary will dictate the situation along this boundary. Thus, only variables transported from the boundaries toward the interior are identified as physical boundary conditions. The remaining variables transported outside of the domain depend on the computed flow situations or part of the solution. This additional information, known as the numerical boundary conditions, can be linearly or quadratically extrapolated from the downstream (inflow) or upstream (outflow) flowfield information. These physical and numerical boundary conditions are summarized in Table 6.7.1.

Characteristic Boundary Conditions

If the full information on the incoming and outgoing characteristics is recovered from the imposed combinations of conservation variables **U** and primitive variables **V**, then the problem is said to be well posed. Let us consider the subsonic outlet in which one physical boundary condition is allowed, say pressure p. From the relations (6.2.19) and (6.2.23) together with (6.2.28) we may write

$$\Delta\mathbf{W} = \begin{bmatrix} \Delta\mathbf{W}_a \\ \Delta\mathbf{W}_b \end{bmatrix} = \begin{bmatrix} \mathbf{L}_{aa}^{-1} & \mathbf{L}_{ab}^{-1} \\ \mathbf{L}_{ba}^{-1} & \mathbf{L}_{bb}^{-1} \end{bmatrix} \begin{bmatrix} \Delta\mathbf{V}_a \\ \Delta\mathbf{V}_b \end{bmatrix} = \begin{bmatrix} -1/\rho a & 0 & 1 \\ -1/a^2 & 1 & 0 \\ 1/\rho a & 0 & 1 \end{bmatrix}_0 \begin{bmatrix} \Delta p \\ \Delta\rho \\ \Delta u \end{bmatrix} \tag{6.7.1}$$

where the subscript 0 denotes end conditions, with a and b indicating the physical

Table 6.7.1 Physical and Numerical Boundary Conditions

		Subsonic	Supersonic
Inlet	Physical	W_1, W_2	W_1, W_2, W_3
	Numerical	W_3	None
Outlet	Physical	W_3	None
	Numerical	W_1, W_2	W_1, W_2, W_3

Supersonic Inflow

Supersonic Outflow

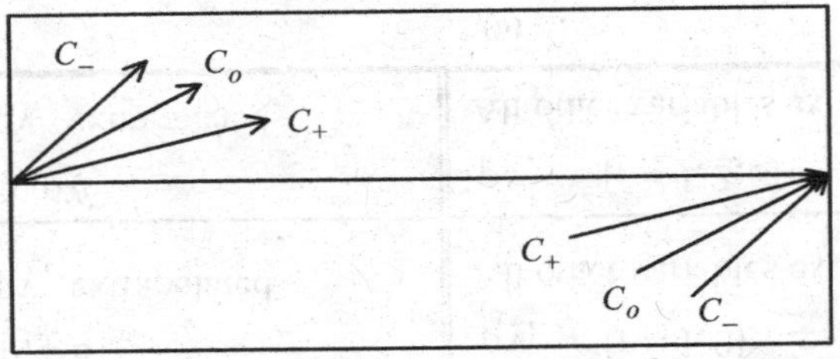

Three characteristics going into the domain. (Three physical boundary conditions required)

Three characteristics going out of the domain. (No physical boundary conditions. Instead, three numerical boundary conditions required)

	INFLOW		OUTFLOW	
	Euler	Navier-Stokes	Euler	Navier-Stokes
1-D	$\rho, \rho v_i, \rho E \quad (i = 1)$		Dirichlet, none (Set free) All variables extrapolated	
2-D	$\rho, \rho v_i, \rho E \quad (i = 1, 2)$	$\rho, \rho v_i, \rho E \quad (i = 1, 2)$	Dirichlet, none (Set free) $\rho E_{,i} n_i = 0$	Dirichlet, none (Set free) $\tau_{ij} n_j = 0 \quad (i = 1, 2)$ $\rho E_{,i} n_i = 0$
3-D	$\rho, \rho v_i, \rho E \quad (i = 1, 2, 3)$	$\rho, \rho v_i, \rho E \quad (i = 1, 2, 3)$	Dirichlet, none (Set free) $\rho E_{,i} n_i = 0$	Dirichlet none (Set free) $\tau_{ij} n_j = 0 \quad (i = 1, 2, 3)$ $\rho E_{,i} n_i = 0$

(a)

Figure 6.7.1 Summary of boundary conditions for compressible flows. **(a)** Supersonic flow boundary conditions. **(b)** Subsonic flow boundary conditions.

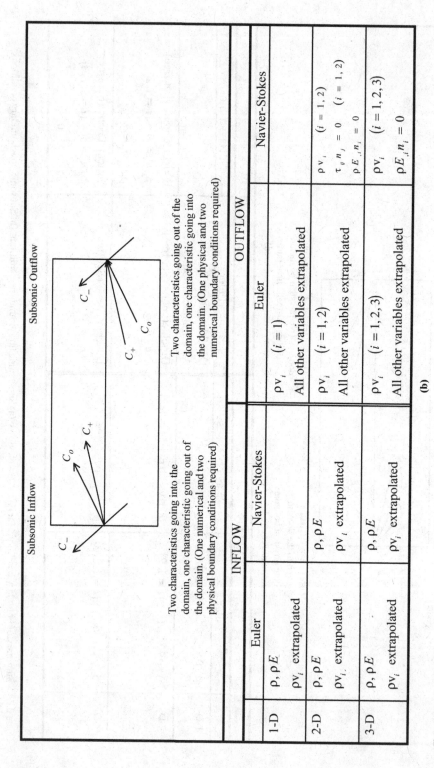

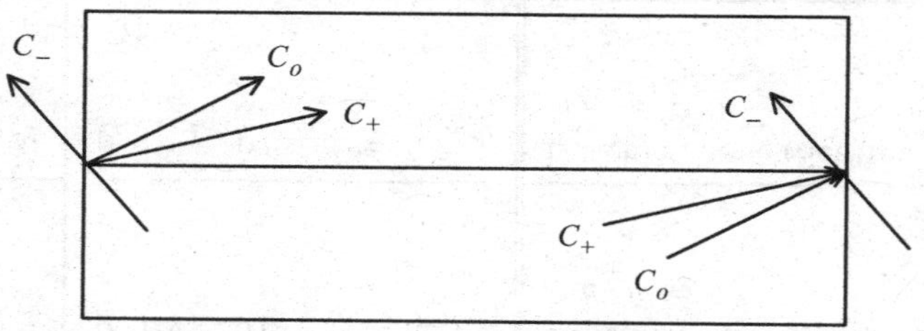

Two characteristics going into the domain, one characteristic going out of the domain. (One numerical and two physical boundary conditions required)

Two characteristics going out of the domain, one characteristic going into the domain. (One physical and two numerical boundary conditions required)

	INFLOW		OUTFLOW	
	Euler	Navier-Stokes	Euler	Navier-Stokes
1-D	$\rho, \rho E$ ρv_i extrapolated		$\rho v_i \quad (i = 1)$ All other variables extrapolated	
2-D	$\rho, \rho E$ ρv_i extrapolated	$\rho, \rho E$ ρv_i extrapolated	$\rho v_i \quad (i = 1, 2)$ All other variables extrapolated	$\rho v_i \quad (i = 1, 2)$ $\tau_{ij} n_j = 0 \quad (i = 1, 2)$ $\rho E_{,i} n_i = 0$
3-D	$\rho, \rho E$ ρv_i extrapolated	$\rho, \rho E$ ρv_i extrapolated	$\rho v_i \quad (i = 1, 2, 3)$ All other variables extrapolated	$\rho v_i \quad (i = 1, 2, 3)$ $\rho E_{,i} n_i = 0$

(b)

Figure 6.7.1 (*continued*)

(imposed variable) and numerical (free variable) boundary conditions, respectively.

$$\mathbf{W}_a = \mathbf{W}_3, \qquad \mathbf{W}_b = \begin{bmatrix} \mathbf{W}_1 \\ \mathbf{W}_2 \end{bmatrix}, \qquad \mathbf{V}_a = p, \qquad \mathbf{V}_b = \begin{bmatrix} \rho \\ u \end{bmatrix}$$

$$\mathbf{L}_{aa}^{-1} = -1/\rho a, \qquad \mathbf{L}_{ab}^{-1} = \begin{bmatrix} 0 & 1 \end{bmatrix}, \qquad \mathbf{L}_{ba}^{-1} = \begin{bmatrix} -1/a^2 \\ 1/\rho a \end{bmatrix}, \qquad \mathbf{L}_{bb}^{-1} = \begin{bmatrix} 1 & 0 \\ 0 & 1 \end{bmatrix}$$

Solving $\Delta\mathbf{V}_b$ from (6.7.1) yields

$$\Delta\mathbf{V}_b = \left(\mathbf{L}_{bb}^{-1}\right)^{-1}\left[\Delta\mathbf{W}_b - \mathbf{L}_{ba}^{-1}\Delta\mathbf{V}_a\right] \tag{6.7.2}$$

Obviously, the nonsingularity of $\mathbf{L}_{bb}^{-1}$ in (6.7.2) constitutes the condition for *well-posedness*. Thus, we require

$$\left|\mathbf{L}_{bb}^{-1}\right| \neq 0 \tag{6.7.3}$$

This can be applied for the various combinations of primitive variables at the boundaries. At a subsonic outlet it is shown by (6.7.1) that any of three variables ρ, u, p can be chosen as a physical boundary condition. This is because the first column of the transformation matrix in (6.7.1) contains all nonzero terms and thus none of the submatrices defining $\mathbf{W}_b$ is zero. For a subsonic inlet the physical boundary conditions $\mathbf{W}_a$ consist of ρ and u, with $\mathbf{W}_b = p$. This leads to

$$\Delta\mathbf{W} = \begin{bmatrix} \Delta\mathbf{W}_a \\ \Delta\mathbf{W}_b \end{bmatrix} = \begin{bmatrix} 1 & 0 & -1/a^2 \\ 0 & 1 & 1/\rho a \\ 0 & 1 & -1/\rho a \end{bmatrix}_0 \begin{bmatrix} \Delta\rho \\ \Delta u \\ \Delta p \end{bmatrix} \tag{6.7.4}$$

where it is seen that the bottom row of the transformation matrix has one zero term corresponding to the density ρ so that

$$\Delta\mathbf{W}_b = \Delta u - \frac{\Delta p}{(\rho a)_0} \tag{6.7.5}$$

which indicates that it is not possible to define $\Delta\rho$ at the boundary and the choice of u and p as a physical boundary condition is not well-posed. However, for any other combination involving ρ as a physical condition, one can determine the remaining free variable using (6.7.5).

Extrapolation Methods

For simplicity, let us use the variable Q to denote either the conservation variables (U), primitive variables (V), and characteristic variables (W) or any other combination, with the conditions for an inlet boundary designated as $i = 1, 2, 3\ldots$ and outlet boundary as $i = p, p-1, p-2, \ldots$ The common practice is to use a linear (first order) extrapolations as follows:

Space extrapolation:

$$\mathbf{Q}_p^{n+1} = 2\mathbf{Q}_{p-1}^{n+1} - \mathbf{Q}_{p-2}^{n+1}$$
$$\Delta\mathbf{Q}_p^n = 2\Delta\mathbf{Q}_{p-1}^n - \Delta\mathbf{Q}_{p-2}^n \tag{6.7.6a,b}$$

Space-time extrapolation:

$$\mathbf{Q}_p^{n+1} = 2\mathbf{Q}_{p-1}^n - \mathbf{Q}_{p-2}^n$$
$$\Delta\mathbf{Q}_p^n = 2\Delta\mathbf{Q}_{p-1}^{n-1} - \Delta\mathbf{Q}_{p-2}^{n-1} \tag{6.7.7a,b}$$

Time extrapolation:

$$\mathbf{Q}_p^{n+1} = 2\mathbf{Q}_p^n - \mathbf{Q}_p^{n-1}$$
$$\Delta\mathbf{Q}_p^n = \Delta\mathbf{Q}_p^{n-1} \tag{6.7.8a,b}$$

These schemes were studied by Griffin and Anderson [1977] and Gottlieb and Turkel [1978] for applications to the two-step Lax-Wendroff schemes. They show that the space-extrapolation methods do not stabilize these schemes nor reduce the stability limits.

Another approach is to discretize the equations at the boundary points in a one-sided manner such as (3.2.5) and to add this equation to the interior scheme. For instance, one could add a first order appropriate upwind equation in the Lax-Wendroff scheme and provide the missing information. Some of the examples for boundary conditions are shown in Figure 6.7.2.

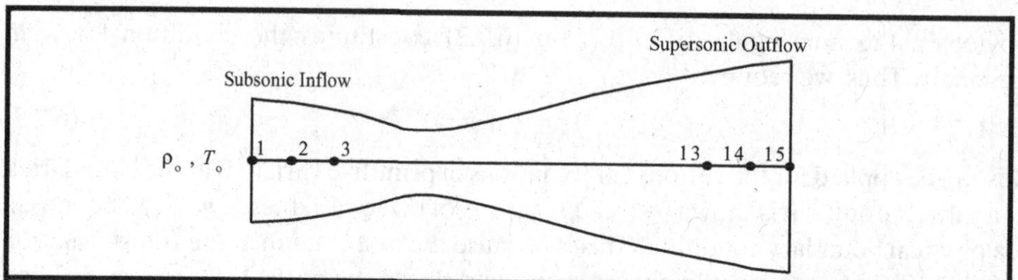

INFLOW	OUTFLOW
Physical boundary conditions, (ρ_1, T_1)	No physical boundary conditions
Numerical boundary conditions Extrapolate: $u_1 = 2u_2 - u_3$ Compute: $p = \rho RT$	Numerical boundary conditions Extrapolate: $u_{15} = 2u_{14} - u_{13}$, $\rho_{15} = 2\rho_{14} - \rho_{13}$, $T_{15} = 2T_{14} - T_{13}$
Initial conditions, $\rho(x,0) = T(x,0) = u(x,0) = 0$; If outflow is subsonic, specify p_{15} and set the others free.	

(a)

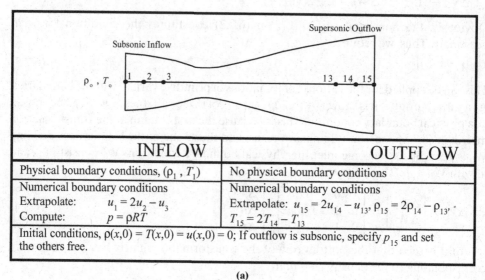

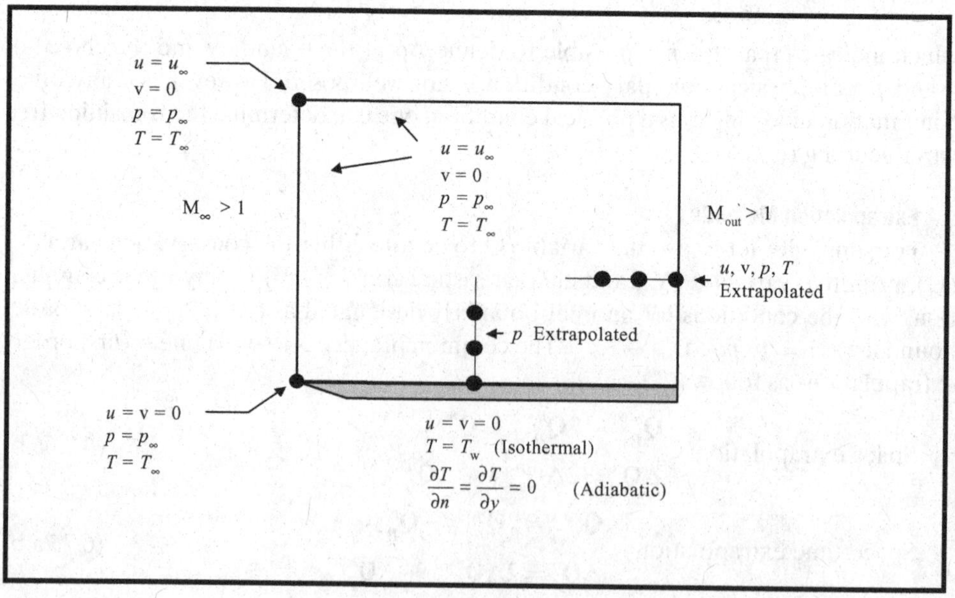

(b)

Figure 6.7.2 Examples for boundary conditions **(a)** 1-D boundary conditions for variable cross sections. **(b)** 2-D boundary conditions for a flat plate.

Characteristic Extrapolation Methods

This is an alternative method to the one-sided discretization of the compatibility equations corresponding to the outgoing characteristics [Yee, Beam, and Warming, 1982]. It follows from (6.7.2) that the numerical characteristic variables $\Delta \mathbf{W}_b$ are defined by an extrapolation such as in (6.7.6b):

$$\Delta \mathbf{W}_{b,p} = 2\Delta \mathbf{W}_{b,p-1} - \Delta \mathbf{W}_{b,p-2} \tag{6.7.9}$$

The values at $i = p-1$ and $i = p-2$ are obtained from the primitive variables by an explicit evaluation of (6.7.2):

$$\Delta \mathbf{W}_b = L_{ba}^{-1} \Delta \mathbf{V}_a + L_{bb}^{-1} \Delta \mathbf{V}_b \quad \text{for } i = p-1, p-2 \tag{6.7.10}$$

where the matrix elements are evaluated at time level n. By setting $\Delta \mathbf{V}_a = 0$ in (6.7.2) we obtain

$$\Delta \mathbf{V}_{b,p} = \left(L_{bb,p}^{-1}\right)^{-1} \Delta \mathbf{W}_{b,p} \tag{6.7.11}$$

where for time dependent problems $\Delta \mathbf{V}_a \neq 0$. This will be determined by the imposed time variation. The free variables $\mathbf{V}_{b,p}$ are transformed to the conservation variables through (6.2.7).

$$\Delta \mathbf{U}_p = \mathbf{M}_p \begin{bmatrix} \Delta \mathbf{V}_a \\ \Delta \mathbf{V}_{b,p} \end{bmatrix} = \mathbf{M}_p \begin{bmatrix} 0 \\ \Delta \mathbf{V}_{b,p} \end{bmatrix} \tag{6.7.12}$$

For subsonic outflow boundary with pressure imposed, we observe that

$$\Delta \mathbf{W}_{b,i} = \begin{bmatrix} \Delta w_1 \\ \Delta w_2 \end{bmatrix}_i = \begin{bmatrix} -1/a^2 \\ 1/\rho a \end{bmatrix} \Delta p_i + \begin{bmatrix} 1 & 0 \\ 0 & 1 \end{bmatrix} \Delta \begin{bmatrix} \rho \\ u \end{bmatrix}_i \tag{6.7.13}$$

$$\Delta \mathbf{W}_{b,p} = 2\begin{bmatrix} -\Delta p/a^2 + \Delta\rho \\ \Delta p/\rho a + \Delta u \end{bmatrix}_{p-1} - \begin{bmatrix} -\Delta p/a^2 + \Delta\rho \\ \Delta p/\rho a + \Delta u \end{bmatrix}_{p-2} = \begin{bmatrix} \Delta w_1 \\ \Delta w_2 \end{bmatrix}_p \tag{6.7.14}$$

with $i = p-1, i = p-2$. It follows from (6.7.11) and (6.7.14) that

$$\Delta \mathbf{V}_{b,p} = \begin{bmatrix} \Delta\rho \\ \Delta u \end{bmatrix}_p = \begin{bmatrix} 1 & 0 \\ 0 & 1 \end{bmatrix} \begin{bmatrix} \Delta w_1 \\ \Delta w_2 \end{bmatrix}_p \tag{6.7.15}$$

Similarly, the corresponding conservation variables are given by (6.7.12),

$$\Delta \mathbf{U}_p = \mathbf{M} \begin{bmatrix} \Delta\rho \\ \Delta u \\ 0 \end{bmatrix}_p = \begin{bmatrix} \Delta\rho \\ \Delta(\rho u) \\ \Delta(\rho E) \end{bmatrix}_p \tag{6.7.16}$$

Thus the boundary condition equation for $(\Delta\rho)_p$ may be written as

$$(\Delta\rho)_p + 2\left(\frac{\Delta p}{a^2} - \Delta\rho\right)_{p-1} - \left(\frac{\Delta p}{a^2} - \Delta\rho\right)_{p-2} = 0 \tag{6.7.17}$$

This should be added to the interior point $p-1$. Similar equations can be written for $\Delta(\rho u)$ and $\Delta(\rho E)$.

6.7.1.2 Multi-Dimensional Boundary Conditions

Evaluation of multidimensional boundary conditions may be carried out similarly as in one dimension. The number of physical boundary conditions to be imposed at a boundary with the normal vector $\mathbf{n}$ pointing toward the flow domain is determined by the signs of the eigenvalues of the matrix $\mathbf{K}$ in terms of the primitive variable Jacobian $\mathbf{A}_i$ or the conservation variable Jacobian $\mathbf{a}_i$.

$$\begin{aligned} \mathbf{K} &= \mathbf{A}_i \kappa_i \qquad (i = 1, 2, 3) \\ \mathbf{K}^* &= \mathbf{a}_i \kappa_i \qquad (i = 1, 2, 3) \end{aligned} \tag{6.7.18a,b}$$

The eigenvalues Λ of both matrices $\mathbf{K}$ and $\mathbf{K}^*$ are equal,

$$\Lambda = \begin{bmatrix} \mathrm{v}_i\kappa_i & 0 & 0 & 0 & 0 \\ 0 & \mathrm{v}_i\kappa_i & 0 & 0 & 0 \\ 0 & 0 & \mathrm{v}_i\kappa_i & 0 & 0 \\ 0 & 0 & 0 & \mathrm{v}_i\kappa_i + a & 0 \\ 0 & 0 & 0 & 0 & \mathrm{v}_i\kappa_i - a \end{bmatrix} \tag{6.7.19}$$

in which the normal velocities $\mathrm{v}_i n_i = \mathrm{v}_i \kappa_i$ determine the signs of the eigenvalues.

Note that, for the inflow and outflow boundaries, if an eigenvalue λ is positive, the information carried by the corresponding characteristics propagates toward the interior domain and a physical boundary condition is to be imposed. If λ is negative, then the numerical boundary condition must be imposed. For example, at the subsonic inlet, two thermodynamic variables (temperature and pressure) and two velocity components are available as physical boundary conditions and one velocity component can be used as a numerical boundary condition.

For a solid wall, a single physical boundary condition is required as only one characteristic enters the flow boundaries. This is equivalent to

$$\begin{aligned} \mathrm{v}_i n_i &= 0 \\ p n_i &\neq 0 \end{aligned} \tag{6.7.20a,b}$$

Here, the wall pressure is numerically extrapolated from adjacent points.

Two-dimensional compatibility or characteristic relations are written as an extension of (6.2.28) as

$$\delta \mathbf{W} = \begin{bmatrix} \delta w_1 \\ \delta w_2 \\ \delta w_3 \\ \delta w_4 \end{bmatrix} = \begin{bmatrix} \delta\rho - \dfrac{\delta p}{a^2} \\ \kappa_y \delta u - \kappa_x \delta \mathrm{v} \\ n_i \delta \mathrm{v}_i + \dfrac{\delta p}{\rho a} \\ -n_i \delta \mathrm{v}_i + \dfrac{\delta p}{\rho a} \end{bmatrix} \tag{6.7.21}$$

which may be recast into (6.2.20) and (6.2.29). Thus, if the pressure and the velocity are uniform in the boundary surface, it is seen that we recover the one-dimensional condition given by (6.7.20).

6.7.1.3 Nonreflecting Boundary Conditions

Physical boundary conditions may be replaced by specification of nonreflecting boundary conditions. Let a constant pressure be imposed at a subsonic exit section as

$$\Delta p = p^{n+1} - p^n = 0 \tag{6.7.22}$$

This is equivalent to allowing perturbation waves to be reflected at the boundaries. Since the amplitude of the local perturbation wave carried by the incoming characteristic is $\Delta w_3 = \Delta u - \Delta p/\rho a$, imposing $\Delta p = 0$ amounts to the generation of an incoming wave of intensity $\Delta w_3 = \Delta u$ reflected from the exit boundary.

Engquist and Majda [1979] and Hedstrom [1979] proposed that the nonreflecting boundary conditions be implemented by making the local perturbations propagated along incoming characteristics vanish.

$$\frac{\partial w_k}{\partial t} = 0 \tag{6.7.23}$$

This will require that, for subsonic flows, we have

Inlet boundary conditions

$$\Delta w_1 = \Delta\rho - \frac{\Delta p}{a^{2n}} = 0 \tag{6.7.24a}$$

$$\Delta w_2 = \Delta u + \frac{\Delta p}{\rho^n a^n} = 0 \tag{6.7.24b}$$

Outlet boundary condition

$$\Delta w_3 = \Delta u - \frac{\Delta p}{\rho^n a^n} \tag{6.7.25}$$

These characteristic variables are not constant across a shock wave and will result in a reflection wave if a shock passes through a boundary.

Rudy and Strickwerda [1980] observed that the nonreflecting condition (6.7.23) does not ensure (6.7.22) or $p = p^*$ and that an ad hoc treatment may be to replace (6.7.23) for the incoming characteristic by, at the exit boundary,

$$\frac{\partial u}{\partial t} - \frac{1}{\rho a}\frac{\partial p}{\partial t} - \frac{\alpha}{\rho a}(p - p^*) = 0 \tag{6.7.26}$$

for $\alpha > 0$. The parameter α is problem dependent. For example, it has been suggested that we may choose $0.1 \leq \alpha \leq 0.2$ for $M = 0.8$ and $\alpha \cong 1$ for $M = 0.4$.

6.7.2 NAVIER-STOKES SYSTEM OF EQUATIONS

The Navier-Stokes system of equations may be considered as mixed hyperbolic, parabolic, and elliptic equations, or refered to as incompletely parabolic equations [Strikwerda, 1976; Gustafsson and Sundström, 1978]. Let us consider the Navier-Stokes system of equations in the form

$$\frac{\partial \mathbf{U}}{\partial t} + \mathbf{a}_i\frac{\partial \mathbf{U}}{\partial x_i} + \mathbf{b}_i\frac{\partial \mathbf{U}}{\partial x_i} + \mathbf{c}_{ij}\frac{\partial^2 \mathbf{U}}{\partial x_i \partial x_j} = 0 \tag{6.7.27}$$

which is obtained from (6.3.7) by inserting the convection Jacobian $\mathbf{a}_i$, diffusion Jacobian $\mathbf{b}_i$, and diffusion gradient Jacobian $\mathbf{c}_{ij}$. To determine the number of boundary condition, we must convert the conservation variables, $\mathbf{U} = [\rho \;\; \rho v_i \;\; \rho E]^T$ into nonconservation variables (primitive variables), $\mathbf{V} = [\rho \;\; v_i \;\; p]^T$ such that

$$\frac{\partial \mathbf{V}}{\partial t} + \mathbf{A}_i\frac{\partial \mathbf{V}}{\partial x_i} + \mathbf{B}_i\frac{\partial \mathbf{V}}{\partial x_i} + \mathbf{C}_{ij}\frac{\partial^2 \mathbf{V}}{\partial x_i \partial x_j} = 0 \tag{6.7.28}$$

Here, procedures similar to those employed for the case of the Euler equations in Section 6.2.1 may be followed to obtain the eigenvalues for the diffusion Jacobian $\mathbf{B}_i$ and the diffusion gradient Jacobian $\mathbf{C}_{ij}$ through the transformation matrix of the form (6.2.7)

$$\mathbf{M} = \frac{\partial \mathbf{U}}{\partial \mathbf{V}}$$

so that

$$\mathbf{A}_i = \mathbf{M}^{-1}\mathbf{a}_i\mathbf{M}, \quad \mathbf{B}_i = \mathbf{M}^{-1}\mathbf{b}_i\mathbf{M}, \quad \mathbf{C}_{ij} = \mathbf{M}^{-1}\mathbf{c}_{ij}\mathbf{M}$$

Introduce the oscillatory behavior in (6.7.28) with the wave number κ_i and frequency ω in the form,

$$\mathbf{V} = \overline{\mathbf{V}}e^{I(\kappa_i x_i - \omega t)} \tag{6.7.29}$$

leading to

$$(-\omega + \mathbf{A}_i\kappa_i + \mathbf{B}_i\kappa_i + \mathbf{C}_{ij}\kappa_i\kappa_j)\overline{\mathbf{V}} = 0 \tag{6.7.30}$$

which has a nontrivial solution if and only if

$$|\mathbf{K} - \lambda\mathbf{I}| = 0 \tag{6.7.31}$$

with

$$\lambda\mathbf{I} = \omega \tag{6.7.32}$$

$$\mathbf{K} = \mathbf{A}_i\kappa_i + \mathbf{B}_i\kappa_i + \mathbf{C}_{ij}\kappa_i\kappa_j \tag{6.7.33}$$

The eigenvalue problem similar to (6.7.31) was obtained by Nordström [1989], neglecting $\mathbf{B}_i\kappa_i$.

For multidimensional problems, the extra boundary conditions for the Navier-Stokes system of equations are obtained by

$$\int_\Omega \tau_{ij,i}d\Omega = \int_\Gamma \tau_{ij}n_i d\Gamma \tag{6.7.34}$$

with

$$\tau_{ij}n_i = \mu\left[(v_{i,j} + v_{j,i})n_i - \tfrac{2}{3}v_{i,i}n_j\right] \tag{6.7.35}$$

where the velocity gradients are taken in the flow directions.

Unlike Euler equations, the Navier-Stokes system of equations require the no-slip boundary conditions at solid walls, resulting in the relative velocity between the fluid and the solid wall being zero.

For an adiabatic wall, we have

$$q_w = -kT_{,i}n_i = 0 \tag{6.7.36}$$

The wall temperature $T = T_w$ may also be fixed. The second thermodynamic variable at the solid wall can be obtained either by extrapolation from the inside or by applying the normal pressure equation

$$\frac{\partial p}{\partial n} = \tau_{ij,i}n_j \tag{6.7.37}$$

which vanishes for thin shear layers. A summary of boundary conditions for the Navier-Stokes system of equations is shown in Figure 6.7.2.

Since the exact form of eigenvalues of K in (6.7.33) depends on many different physical and geometrical conditions, the number of physical boundary conditions (positive eigenvalues) and the number of numerical boundary conditions (negative eigenvalues) cannot be determined exactly for all arbitrary physical and geometrical situations.

As mentioned earlier, the accuracy and convergence of numerical solution of the Navier-Stokes system of equations depend on correct applications of boundary conditions. Rudy and Strickwerder [1980] and Nordström [1989] examine various options of boundary conditions and evaluate the rates of solution convergence (well-posedness) associated with appropriate choices of boundary conditions. Other theoretical studies of boundary conditions include Kreiss [1970], Strickwerder [1976, 1977], Gustaffson and Sundström [1978], and Engquist and Gustaffson [1987], among others.

6.8 EXAMPLE PROBLEMS

Since benchmark problems using the central schemes, low and high order upwinding schemes including MUSCL, TVD, FCT, and ENO have been amply demonstrated in the literature, no attempt is made to include them here except for a simplest example for the benefit of the beginner. FDM applications of the FDV theory for high-speed flows have not appeared in the literature, and so they are illustrated in this section. Some incompressible and compressible flow problems using the FDV theory via FEM are presented in Section 13.7.

6.8.1 SOLUTION OF EULER EQUATIONS

In this example, solutions of Euler equations are given in a quasi–one-dimensional nozzle with variable cross section, NACA 1135, using McCormack explicit scheme and flux vector splitting method.

Given:

$$S(x) = 1.398 + 0.347\tanh(0.8x - 4)\text{ft}^2 \text{ (NACA 1135)}$$

$$\gamma = 1.4$$

$$R = 1716\frac{\text{ft}^2}{\text{sec}^2 R}$$

Case 1

Supersonic inflow – supersonic outflow.

Boundary Conditions

Inflow

$M = 1.5$
$p = 1000$ psf
$\rho = 0.00237$ slug/ft^3
$\rho u = 2.7323$ slug/ft^3 sec
$\rho E = 4075$ slug/ft sec^2

Outflow. Full extrapolation of **U** is required since all eigenvalues are positive.

Initial Conditions

$$\begin{aligned}\rho &= 0.00237\\ \rho u &= 2.7323 \qquad 0 \le x \le 10\\ \rho E &= 4075\end{aligned}$$

Case 2

Supersonic inflow – subsonic outflow.

Boundary Conditions

Inflow – same as before
Outflow – $u = 390.75$ ft/sec.

Other quantities are extrapolated since two eigenvalues are positive.

Initial Conditions

$$\text{for } x \le 2.8 \begin{cases}\rho = 0.00237\\ \rho u = 2.7323\\ \rho E = 4075\end{cases}, \quad \text{for } x > 2.8 \begin{cases}\rho = 0.00237\\ \rho u = 0.92608\\ \rho E = 2680.93\end{cases}$$

Results: The computational results for both the McCormack and flux vector splitting methods are shown in Figure 6.8.1.1a for Case 1 and Figure 6.8.1.1b for Case 2. The solution for both methods was obtained using a total of eighty grid points.

Case 1

Both schemes demonstrate a good level of accuracy, with the flux vector splitting scheme converging faster than the McCormack explicit scheme.

Case 2

Here again we find that the flux vector splitting scheme converges faster than the McCormack explicit scheme, but the level of accuracy is not as good as in the first case (supersonic outflow). In this case, the solution exhibits dispersion errors at the shock.

6.8.2 TRIPLE SHOCK WAVE BOUNDARY LAYER INTERACTIONS USING FDV THEORY

The FDV theory is utilized to analyze the flowfield produced from a triple shock/boundary layer interaction using 3-D FDM discretization [Schunk et al., 1999]. Flowfields of this nature are often encountered in the inlets of high-speed vehicles such as the scramjet engine of NASA's Hyper-X research vehicle. For this analysis, the FDV numerical results are compared to the experimental measurements and FDM calculations via $k-\varepsilon$ turbulent model reported by Garrison et al. [1994]. As indicated earlier, the FDV theory is expected to simulate turbulent flow accurately if DNS mesh refinements are provided. However, such mesh refinements are not available at the present time due to limited computer resources. No turbulence modeling is used in the present analysis.

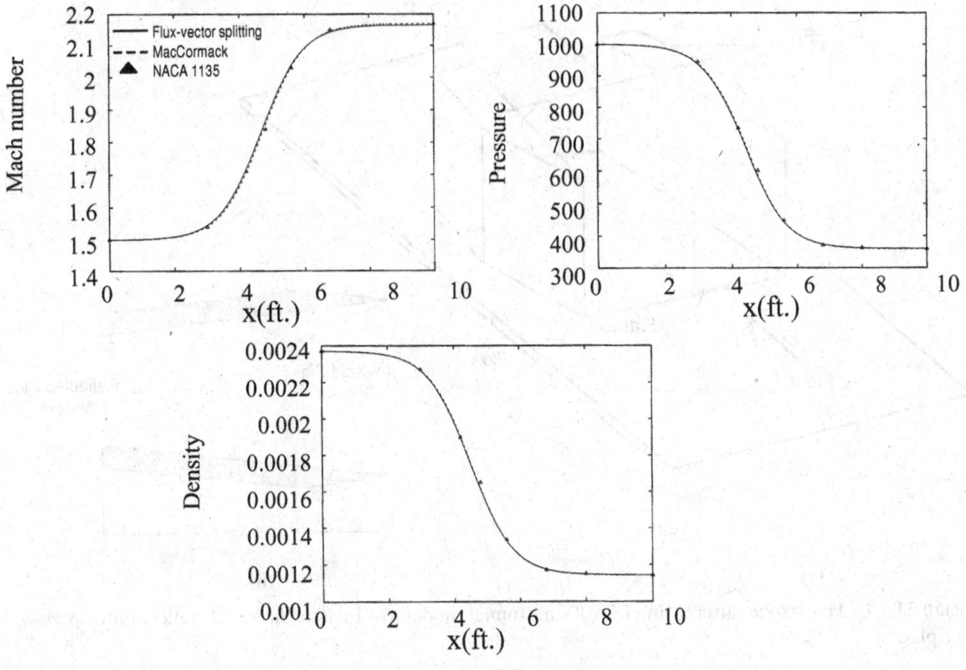

(a)

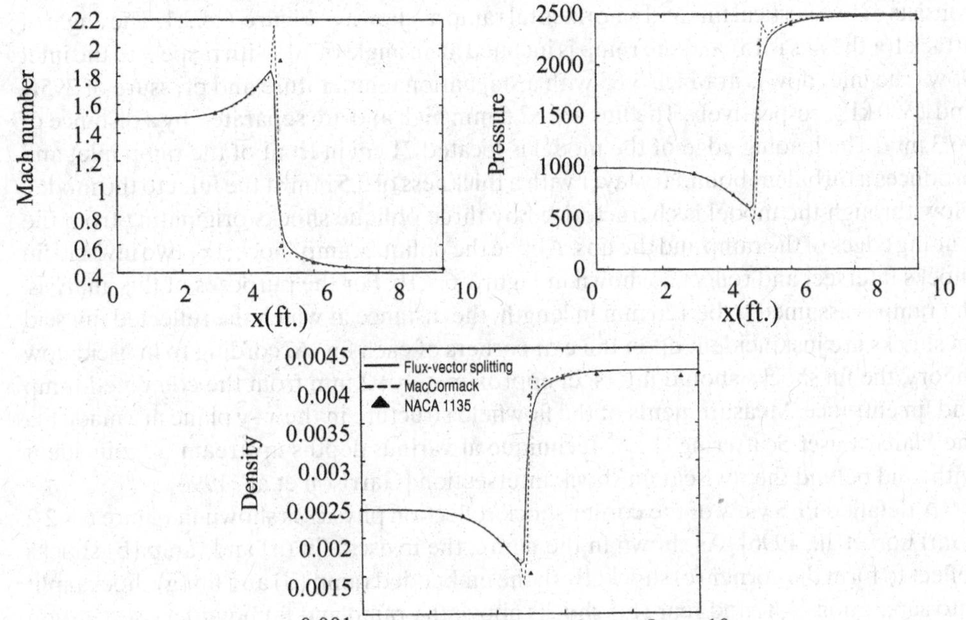

(b)

Figure 6.8.1.1 Quasi–one-dimensional supersonic nozzle flow. **(a)** Case 1 supersonic inflow-supersonic outflow. **(b)** Case 2 supersonic inflow-subsonic outflow.

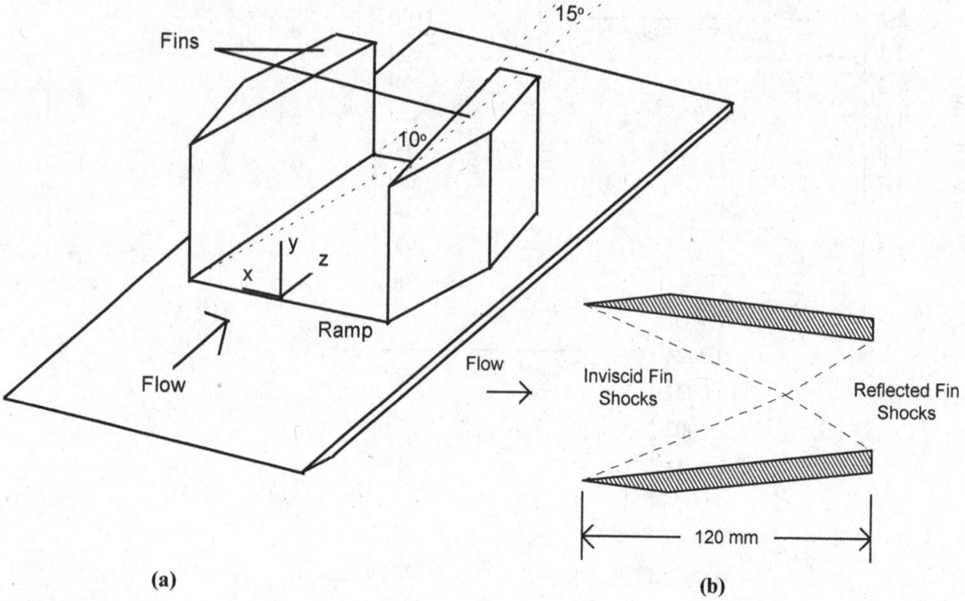

Figure 6.8.2.1 Hypersonic aircraft inlet. **(a)** Wind tunnel model. **(b)** Inviscid fin shock reflection (top view, x–z plane).

The wind tunnel model used to produce the triple shock/boundary layer interaction consists of two vertical fins and a horizontal ramp as shown in Figure 6.8.2.1. The angle of attack for the fins is 15° and the ramp is inclined at an angle of 10° with respect to the inlet flow. The inlet flow is at Mach 3.85 with a stagnation temperature and pressure of 295K and 1500 kPa, respectively. The fins are 82.5 mm high and are separated by a distance of 96.3 mm. The leading edge of the model is located 21 cm in front of the ramp inlet and produces a turbulent boundary layer with a thickness of 3.5 mm at the inlet to the model. Flow through the model is characterized by three oblique shocks originating from the leading edges of the ramp and the fins. Above the oblique ramp shock, the two inviscid fin shocks intersect and reflect as shown in Figure 6.7.1b. For the purposes of this analysis, the ramp is assumed to be 120 mm in length, the distance at which the reflected inviscid fin shocks are just incident upon the exit corners of each fin. According to inviscid flow theory, the fin shocks should intersect approximately 92 mm from the combined ramp and fin entrance. Measurements of the flowfield structure in the x–y plane are made via the Planar Laser Scattering (PLS) technique at various depths upstream of, coincident with, and behind the inviscid fin shock intersection [Garrison et al., 1996].

A detailed PLS view of the corner shock reflection physics is shown in Figure 6.8.2.2 [Garrison et al., 1996]. As shown in the figure, the inviscid fin (a) and ramp (b) shocks reflect to form the corner (c) shock. Both the embedded ramp (d) and fin (g) shocks split into separation (e,h) and rear (f,i) shocks above the ramp and fin boundary/separation layers. The ramp separated region (j) and the slip lines (k) dividing the different velocity regions as induced by the shock structure are also visible in the image.

Since the two fins are symmetric about the centerline, only half of the wind tunnel model is included in the computational model. Two finite difference computational grids, varying in resolution, are developed for the FDV analysis. The coarse grid model,

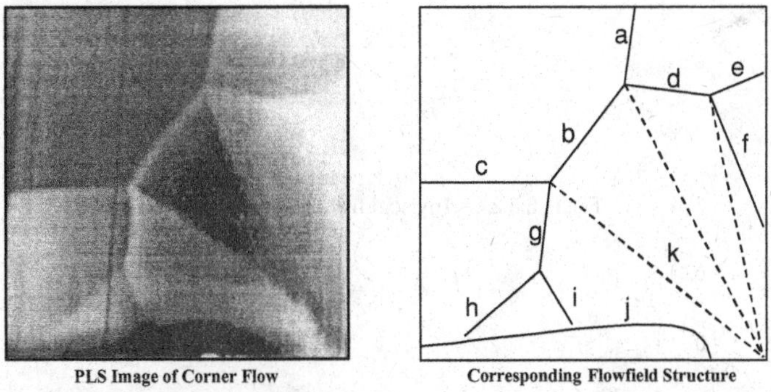

Figure 6.8.2.2 Fin/ramp shock structure in the x–y plane [Garrison et al., 1996], **a**) inviscid fin shock, **b**) corner shock, **c**) inviscid ramp shock, **d**) embedded ramp shock, **e**) ramp separation shock, **f**) ramp rear shock, **g**) embedded fin shock, **h**) separation fin shock, **i**) rear fin shock, **j**) separated region, **k**) sliplines.

consisting of a nonuniform nodal resolution of $31 \times 41 \times 55$ (in the x, y, and z directions) is shown in Figure 6.8.2.3. The viscous grid is clustered close to the fin and ramp surfaces. Results from the coarse grid analysis are used as the starting condition for the fine grid model. The fine grid model is obtained by interpolating the flow variables against the coarse mesh. Doubling the number of grid points in each direction produces a fine grid with over 538,000 nodal points ($61 \times 81 \times 109$). Recall that the most important aspect of the FDV theory is that the shock capturing mechanism and the transition and interaction between compressible/incompressible, viscous/inviscid, and laminar/turbulent flows are incorporated into the FDV formulation. No special treatments are required to simulate these physical phenomena. Thus, the finite difference discretization requires no special schemes. Simple central differences can be used to discretize the FDV equations given by (6.5.14).

The inlet conditions to the model are fixed with the freestream conditions described above ($M = 3.85$, $P_0 = 1500$ kPa, and $T_0 = 295$ K) and include a superimposed boundary layer 3.5 mm in height. At the fin and ramp surfaces, no-slip velocity boundary conditions are imposed and the normal pressure and temperature gradients are set to zero. In the symmetry plane and for the bounding surface on top (x-z plane), all of the

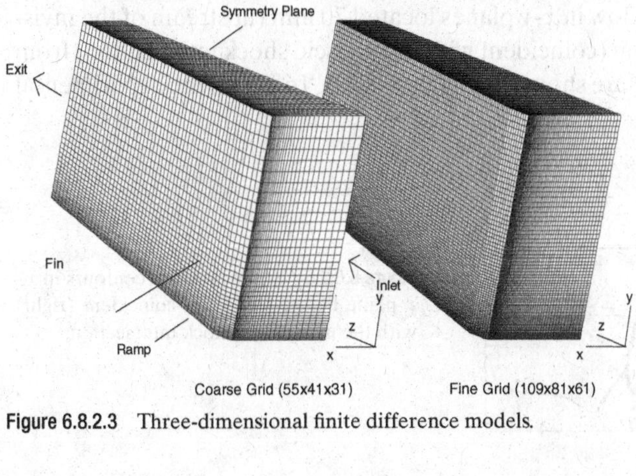

Figure 6.8.2.3 Three-dimensional finite difference models.

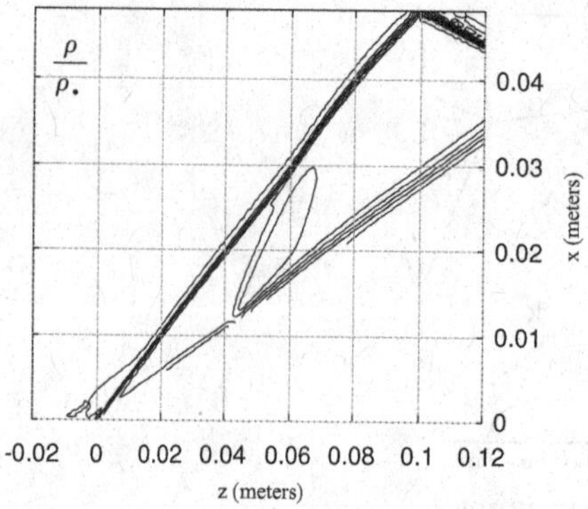

Figure 6.8.2.4 Inviscid fin shock interaction.

flow variables are computed such that the normal gradients vanish except for the normal flux, which is explicitly set to zero. At the exit, all of the flow variables are extrapolated from interior grid points.

In order to test the hypothesis that the FDV equations contain the necessary terms to model turbulence, no turbulence model is included in the analysis. It is theorized that turbulent fluctuations result from the interaction of the convective and diffusion Jacobians present in the second order terms of the FDV equations as dictated by the FDV parameters (s_1, s_2, s_3, s_4). The conclusions drawn from this study will be limited to predictions of the boundary layer separation height since the experimental results contain no measurements of turbulent statistics such as spectral energy density versus wave numbers, etc., since no Kolmogrov microscales are resolved in this analysis.

Density contours for the inviscid shock interaction (x-z plane, as viewed from above the wind tunnel model) are shown in Figure 6.8.2.4. The 15° fins produce inviscid shocks that are predicted to intersect and reflect approximately 97 mm from the ramp entrance (as opposed to 92 mm as predicted by inviscid flow theory). The reflected shock does not intersect with the exit corner of the ramp as expected. The discrepancy between the numerical prediction and inviscid flow theory could be due to the secondary oblique shock that is formed behind the fin shock (approximately 45 mm from the entrance). This is apparently an anomalous condition and could be due to the formation of a nonphysical boundary layer on the fin, possibly due to the discretization of the flowfield close to the fin wall.

Static pressure contours for flow in x-y planes located 70 mm (upstream of the inviscid shock intersection) and 97 mm (coincident with the inviscid shock intersection) from the combined fin/ramp entrance are shown in Figure 6.8.2.5. To match the experimental

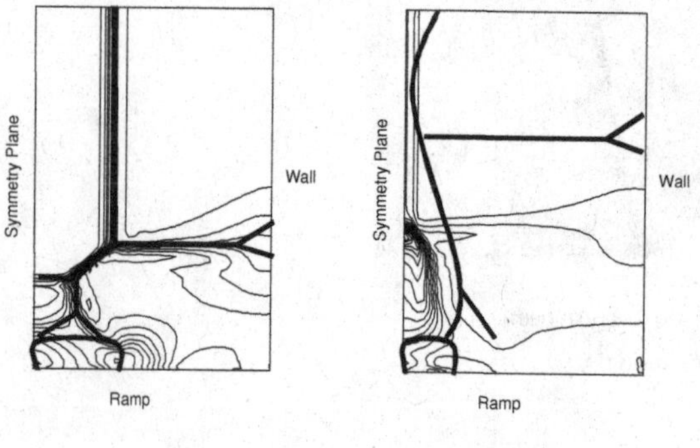

Figure 6.8.2.5 Static pressure contours in x–y plane before (left) and coincident (right) with the inviscid fin shock intersection.

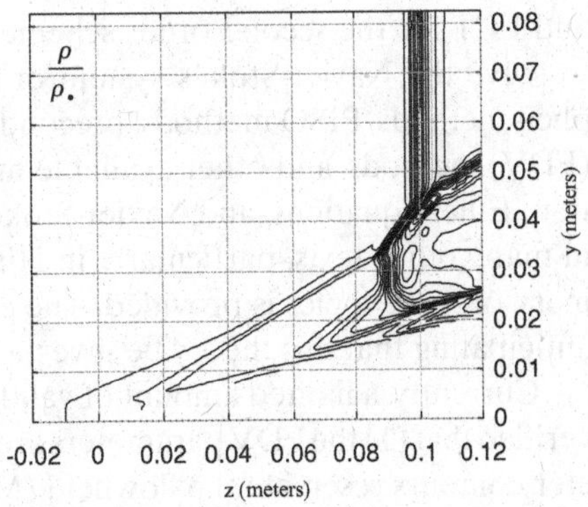

Figure 6.8.2.6 Boundary layer separation on the ramp.

images, the z locations of the x-y planes are scaled relative to the predicted inviscid fin shock intersection. The numerical predictions upstream of the fin shock intersection (left) correlate well with the experimental PLS images. Evident in the upstream figure are the inviscid ramp and fin shocks as well as the corner reflection. The flow separation from the ramp is also visible, appearing as concentric isobaric rings. Although not well resolved, it appears that both the embedded fin and ramp shocks split into separation and rear shocks above the respective surface separation/boundary layers. Coincident with the shock intersection (see right), the inviscid fin shocks merge together in the symmetry plane. No curvature of inviscid fin shock is observed in the numerical predictions as in the experimental results. The reflection of the corner shock about the symmetry plane is observed, but the ramp embedded shock is much lower relative to the height of the fin than in the experimental results. The ramp boundary layer separation is not strongly resolved in the static pressure contours. It is important to note that these results from the FDV theory qualitatively reveal the boundary layer separation predicted by Garrison et al. [1996] using a k-ε turbulence model.

An alternative view (Figure 6.8.2.6) of the flowfield in the symmetry plane (y–z plane, $x = 0$) shows the boundary layer separation and the reflection of the fin intersection shocks through the weaker ramp shock. No experimental imagery is available to compare to this figure, but it is nonetheless informative. Boundary layer separation appears to be approximately 5 mm at the exit.

More fundamental studies for validation of the FDV theory are presented in Chapter 13 using FEM. Contour plots of the FDV parameters are shown to resemble the actual flowfields of the supersonic compression corner flow. Transition between compressible and incompressible flows is also demonstrated for the driven cavity problems. Thus, these fundamental examples are not duplicated in this chapter. The reader is invited to examine Examples (3) and (4), Section 13.7, for details.

6.9 SUMMARY

History of compressible flow computations using potential equations, Euler equations, and the Navier-Stokes system of equations is long, and so is this chapter. Our focus was to study how to capture shocks in both inviscid flows and viscous flows. In compressible inviscid flows using Euler equations, we studied central schemes, first order upwind schemes, and second order upwind schemes. Specifically, we examined the flux vector splitting and Godunov method for the first order scheme and MUSCL, TVD, ENO, and FCT for the second order scheme. For compressible viscous flows, it is necessary to solve the Navier-Stokes system of equations. We examined explicit methods, implicit methods, PISO methods, preconditioning methods, flowfield-dependent variation (FDV) methods, and other available methods. Exhaustive coverage of potential equation, Euler equations, and Navier-Stokes system of equations has been made available in many other texts, particularly in Hirsch [1990]. Thus, in this text, only a brief summary of these topics is provided. The emphasis has been placed on the FDV methods, anticipating that this theory be investigated more thoroughly in the future.

Currently, a limited amount of validation of the FDV theory is available. It has been verified that (1) the FDV parameters are equivalent to the TVD limiters, (2) FDV parameter contours resemble the flowfield (Mach number or density contours), and (3) transitions and interactions between inviscid/viscous flows, compressible/incompressible flows, and laminar/turbulent flows are characterized by the FDV process. Examples demonstrating these fundamental properties are presented in Section 13.7. An extensive and rigorous future research on FDV theory will be required not only for its own theoretical foundation, but also for closer examinations as to the relationships with other methods.

REFERENCES

Abgrall, R. [1994]. On essentially non-oscillatory schemes on unstructured meshes analysis and implementation. *J. Comp. Phys.*, 114, 45–58.

Beam, R. M. and Warming, R. F. [1976]. An implicit finite-difference algorithm for hyperbolic systems in conservation law form. *J. Comp. Phys.*, 22, 87–110.

———. [1978]. An implicit factored scheme for the compressible Navier-Stokes equations. *AIAA J.*, 16, 393–401.

Ben-Artzi, M. and Falcovitz, J. [1984]. A second order Godunov-type scheme for compressible fluid dynamics. *J. Comp. Phys.*, 55, 1–32.

Boris, J. P. and Book, D. L. [1973]. Flux corrected transport 1, SHASTA, a fluid transport algorithm that works., *J. Comp. Phys.*, 11, 38–69.

Briley, W. R. and McDonald, H. [1975]. Solution of the three-dimensional compressible Navier-Stokes equations by an implicit technique. Proc. Fourth International Conference on Numerical Methods in Fluid Dynamics, Lecture Notes in Physics, Vol. 35, Berlin: Springer-Verlag.

Casier, F., Deconinck, H. and Hirsch C. [1983]. A class of central bidiagonal schemes with implicit boundary conditions for the solution of Eulers equations. AIAA Paper 83-0126, AIAA 21st Aerospace Sciences Meeting. See also *AIAA J.*, 22, 1556–63.

Casper, J. [1992]. Finite volume implementation of high-order essentially nonoscillatory schemes in two dimensions. *AIAA J.* 30, 12, 2829–35.

Casper, J., Shu, C. W., and Atkins, H. [1994]. Comparison of two formulations for high-order accurate essentially nonoscillatory schemes. *AIAA J.*, 32, 10, 1970–77.

Choi, D. and Merkle, C. L. [1993]. The application of preconditioning for viscous flows. *J. Comp. Phys.*, 105, 203–23.

Chung, T. J. [1999]. Transitions and interactions of inviscid/viscous, compressible/incompressible, and laminar/turbulent flows. *Int. J. for Num. Meth. in Fl.*, 31, 223–46.

Courant, R., Isaacson, E., and Reeves, M. [1952]. On the solution of nonlinear hyperbolic differential equations by finite differences. *Comm. Pure Appl. Math.*, 5, 243–55.

Davis, S. F. [1984]. TVD finite difference schemes and artificial viscosity. ICASE Report 84-20, NASA CR-172373, NASA Langley Research Center.

Dutt, P. [1988]. Stable boundary conditions and difference schemes for Navier-Stokes equations. *SIAM J. Num. Anal.*, 25, 245–67.

Engquist, B. and Majda, A. [1979]. Radiation boundary conditions for acoustic and elastic wave calculations. *Comm. Pure Appl. Math.*, 32, 629–51.

Engquist, B. and Osher, S. [1980]. Stable and entropy satisfying approximations for transonic flow calculations. *Math. of Comp.*, 34, 45–75.

Garrison, T. J., Settles, G. S., Narayanswami, N., Knight, D. D., Horstman, C. C. [1996]. Flowfield surveys and computations of a crossing-shock wave/boundary-layer interaction. *AIAA J.*, 34, no. 1, 57–64.

Godunov, S. K. [1959]. Finite-difference method for numerical computation of discontinuous solutions of the equations of fluid dynamics., *Mat. Sb.*, 47, 271–306.

Godunov, S., Zabrodine, A., Ivanov, M., Kraiko, A. and Prokopov, G. [1979]. Resolution Numerique des Problems Multidimensionnels de la Dynamique des Gaz. Moscow, USSR: Editions MIR.

Gottlieb, D. and Turkel, E. [1978]. Boundary conditions for multistep finite difference methods for time-dependent equations. *J. Comp. Phys.*, 26, 181–96.

Griffin, M. D. and Anderson, J. D. [1977]. On the application of boundary conditions to time-dependent computations for quasi–one-dimensional fluid flows. *Comp. and Fl.*, 5, 127–37.

Gustafsson, B. [1982]. The choice of numerical boundary conditions for hyperbolic systems. *J. Comp. Phys.*, 48, 270–83.

Gustafsson, B. and Sundström, A. [1978]. Incompletely parabolic problems in fluid dynamics. *SIAM J. Appl. Math.*, 35, 343–57.

Harten, A. [1983]. High resolution schemes for hyperbolic conservation laws. *J. Comp. Phys.*, 9, 357–393.

———. [1984]. On a class of high resolution total variation stable finite difference schemes. *SIAM J. Num. Anal.*, 21, 1–23.

Harten, A. and Lax, P. D. [1981]. A random choice finite difference scheme for hyperbolic conservation laws. *SIAM J. Num. Anal.*, 18, 289–315.

Harten, A., Lax, P. D., and Van Leer, B. [1983]. On upstream differencing and Godunov-type schemes for hyperbolic conservation laws. SIAM Review, 25, 35–61.

Harten, A. and Osher, S. [1987]. Uniformity high-order accurate nonoscillatory schemes I. *SIAM J. Num. Anal.*, 24, 279–309.

Hedstrom, G. W. [1979]. Non-reflecting boundary conditions for non-linear hyperbolic systems. *J. Comp. Phys.*, 30, 222–37.

Hirsch, C. [1990]. *Numerical Computation of Internal and External Flows*. Vol. 2. New York: Wiley.

Holst, T. L. and Ballhaus, W. F. [1979]. Fast, conservative schemes for the full potential equation applied to transonic flows *AIAA J.*, 17, 145–52.

Issa, R. I., Gosman, A. D., and Watkins, A. P. [1986]. The computation of compressible and incompressible recirculating flows by a non-iterative schmes. *J. Comp. Phys.*, 62, 66–82.

Jameson, A. [1974]. Iterative solutions of transonic flows over airfoils and wings, including flows at Mach 1. *Comm. Pure Appl. Math.*, 27, 283–309.

Jameson, A., Schmidt, W., and Turkel, E. [1981]. Numerical simulation of the Euler equations by finite volume methods using Runge-Kutta time stepping schemes. AIAA Paper 81-1259, AIAA 5th Computational Fluid Dynamics Conference.

Kreiss, H. O. [1970]. Initial boundary value problem for hyperbolic systems. *Comm. Pure Appl. Math.*, 23, 273–98.

Lax, P. D. [1954]. Weak solutions of nonlinear hyperbolic equations and their numerical computation. *Comm. Pure Appl. Math.*, 7, 159–93.

Lax, P. D. [1973]. Hyperbolic systems of conservation laws and mathematical theory of shock waves. *Society for Industrial and Applied Mathematics*. Philadelphia, PA.

Lax, P. D. and Wendroff, B. [1960]. Systems of conservation laws. *Comm. Pure Appl. Math.*, 15, 363.

Lerat, A. [1979]. Numerical shock structure and nonlinear corrections for difference schemes in conservation form. Lecture Notes in Physics, 20, 345–51, New York: Springer-Verlag.

Lerat, A. [1983]. Implicit methods of second order accuracy for the Euler equations. AIAA Paper, 83–1925.

Lerat, A. and Peyret, R. [1974]. Noncentered schemes and shock propagation problems. *Comp. Fl.*, 2, 35–52.

MacCormack, R. W. [1969]. The effect of viscosity in hypervelocity impact cratering. AIAA Paper, 66–354.

MacCormack, R. W. and Paullay, A. J. [1972]. Computational efficiency achieved by time splitting of finite difference operators. AIAA Paper 72–154.

MacCormack, R. W. [1981]. A numerical method for solving the equations of compressible viscous flow, AIAA Paper, 81–0110.

Merkle, C. L., Sullivan, J. Y., Buelow, P. E. O., and Venkateswaran, S. [1998]. Computation of flows with arbitrary equations of state. *AIAA J.*, 36, 4, 515–21.

Moretti, G. [1979]. Conformal Mappings for the Computation of Steady Three-dimensional Supersonic Flows, Numerical/Laboratory Computer Methods in Fluid Mechanics. A. A. Pouring and V. I. Shah, eds. New York: ASME, 13–28.

Murman, E. M. and Cole, J. D. [1971]. Calculation of plane steady transonic flows. *AIAA J.*, 9, 114–21.

Nordström, J. [1989]. The influence of open boundary conditions on the convergence to steady state for the Navier-Stokes equations. *J. Comp. Phys.*, 85, 210–44.

Oliger, J. and Sundsröm, A. [1978]. Theoretical and practical aspects of some initial boundary value problems in fluid dynamics. *SIAM J. Appl. Math.*, 35, 419–46.

Osher, S. [1982]. Shock modelling in aeronautics. In K. W. Morton and M. J. Baines (eds.) *Numerical Methods for Fluid Dynamics*, London: Academic Press, 179–218.

———. [1984]. Riemann solvers, the entropy condition and difference approximations. *SIAM J. Num. Anal.*, 21, 217–35.

Osher, S. and Chakravarthy, S. R. [1984]. High resolution schemes and the entropy condition. *SIAM J. Num. Anal.*, 21, 955–84.

Osher, S., Hafez, M., and Whitlow, W. Jr. [1985]. Entropy condition satisfying approximations for the full potential equation of transonic flow. *Math. Comp.*, 44, 1–29.

Peyret, R. and Viviand, H. [1985]. Pseudo-unsteady methods for inviscid or viscous flow computations. *Recent Advances in the Aerospace Sciences, C. Casi (ed). Plenum, New York.*

Pletcher, R. H. and Chen, K. H. [1993]. On solving the compressible Navier-Stokes equations for unsteady flows at very low Mach numbers. AIAA Paper, 93-3368.

Rai, M. M. and Moin, P. [1993]. Direct numerical simulation of transitional and turbulence in a spatially evolving boundary layer. *J. Comp. Phys.*, 109, 169–92.

Raithby, G. D. [1976]. Skew upstream differencing schemes for problems involving fluid flow. *Comput. Meth. Appl. Mech. Eng.*, 9, 153–64.

Richtmyer, R. D. and Morton, K. W. [1967]. *Difference Methods for Initial-Value Problems*, 2nd ed., New York: Wiley, Interscience Publishers.

Roe, P. L. [1981]. Approximate Riemann solvers, parameter vectors and difference schemes. *J. Comp. Phys.*, 43, 357–372.

———. [1984]. Generalized formulation of TVD Lax-Wendroff schemes. ICASE Report 84-53. NASA CR-172478, NASA Langley Research Center.

———. [1985]. Upwind schemes using various formulations of the Euler equations. In F. Angrand et al. (eds.) *Numerical Methods for the Euler Equations of Fluid Dynamics*, Philadelphia, PA: SIAM Publications.

Rudy, D. H. and Strickwerda, J. C. [1980]. A non-reflecting outflow boundary condition for subsonic Navier-Stokes calculations. *J. Comp. Phys.*, 36, 55–70.

Schunk, R. G., Canabal, Heard, G. W., and Chung, T. J. [1999]. Unified CFD methods via flowfield-dependent variation theory. AIAA Paper, 99-3715.

Shu, C. W. and Osher, S. [1988]. Efficient implementation of essentially non-oscillatory shock-capturing schemes. *J. Comp. Phys.*, 77, 439–71.

Shu, C. W. and Osher, S. [1989]. Efficient implementation of essentially non-oscillatory shock-capturing schemes, II. *J. Comp. Phys.*, 83, 32–78.

Suresh, A. and Jorgenson, P. C. E. [1995]. Essentially nonoscillatory reconstructions via extrapolation. *AIAA Paper*, 95-0467.

Steger, J. L. and Warming, R. F. [1981]. Flux vector splitting of the inviscid gas-dynamic equations with applications to finite difference methods. *J. Comp. Phys.*, 40, 263–93.

Ştanescu, D. and Habashi, G. [1998]. Essentially nonoscillatory Euler solutions on unstructured meshes using extrapolation. *AIAA J.*, 36, 8, 1413–16.

Strikwerda, J. C. [1976]. Initial boundary value problems for incompletely parabolic systems. Ph. D. dissertation, Standford University.

Strikwerda, J. C. [1977]. Initial boundary value problems for incompletely parabolic systems. *Comm. Pure Appl. Math.*, 30, 797–822.

Sweby, P. K. [1984]. High resolution schemes using flux limiters for hyperbolic conservation laws. *SIAM J. Num. Anal.*, 21, 995–1011.

Tannehill, J. C., Hoist, T. L., and Rakich, J. V. [1975]. Numerical computation of two-dimensional viscous blunt body flows with an impinging shock, AIAA Paper 75-154, Pasadena, California.

Van Leer, B. [1973]. Towards the ultimate conservative difference scheme. I. The quest of monotonicity. Lecture Notes in Physics, Vol. 18, 163–68. Berlin: Springer Verlag.

———. [1974]. Towards the ultimate conservative difference scheme. II. Monotonicity and conservation combined in a second order scheme. *J. Comp. Phys.*, 14, 361–70.

———. [1979]. Towards the ultimate conservative difference scheme. V. A second order sequel to Godunovs method. *J. Comp. Phys.*, 32, 101–36.

———. [1982]. Flux vector splitting for the Euler equations. Proc. 8th International Conference on Numerical Methods in Fluid Dynamics, Berlin: Springer-Verlag.

Woodward, P. R. and Colella, P. [1984]. The numerical simulation of two-dimensional fluid flow with strong shocks. *J. Comp. Phys.*, 54, 115–73.

Yanenko, N. N. [1979]. *The Method of Fractional Steps*. New York: Springer-Verlag. In J. R. Bunch and D. J. Rose (eds.), *Space Matrix Computations*, New York: Academic Press.

Yee, H. C. [1985]. On symmetric and upwind TVD schemes. Proc. 6th GAMM Conference on Numerical Methods in Fluid Mechanics, 399-407, Braunschweig: Vieweg.

Yee, H. C. [1986]. Linearized form of implicit TVD schemes for multidimensional Euler and Navier-Stokes equations. *Comp. Math. Appl.*, 12A, 413–32.

Yee, H. C., Beam, R. M., and Warming, R. F. [1982]. Boundary approximation for implicit schemes for one-dimensional inviscid equations of gas dynamics. *AIAA J.*, 20, 1203–11.

Zalesak, S. T. [1979]. Fully multidimensional flux corrected transport algorithm for fluids. *J. Comp. Phys.*, 31, 335–62.

Zhong, X. [1994]. Application of essentially nonoscillatory schemes to unsteady hypersonic shock-shock interference heating problems. *AIAA J.*, 32, 8, 1606–16.

CHAPTER SEVEN

Finite Volume Methods via Finite Difference Methods

7.1 GENERAL

Finite volume methods (FVM), often called control volume methods, are formulated from the inner product of the governing partial differential equations with a unit function, **I**. This process results in the spatial integration of the governing equations. The integrated terms are approximated by either finite differences or finite elements, discretely summed over the entire domain. Recall that we briefly discussed this subject in Section 1.4 for one-dimensional problems.

One of the most important features of FVM is their flexibility for unstructured grids. The traditional curvilinear coordinate transformation required for FDM is no longer needed. Designation of the components of a vector normal to boundary surfaces in FVM accommodates the unstructured grid configuration with each boundary surface integral constructed between nodal points.

For illustration, consider the conservation form of the Navier-Stokes system of equations

$$\mathbf{R} = \frac{\partial \mathbf{U}}{\partial t} + \frac{\partial \mathbf{F}_i}{\partial x_i} + \frac{\partial \mathbf{G}_i}{\partial x_i} - \mathbf{B} \tag{7.1.1}$$

The finite volume equations are obtained as

$$(\mathbf{I}, \mathbf{R}) = \int_\Omega \mathbf{R}\, d\Omega = \int_\Omega \left(\frac{\partial \mathbf{U}}{\partial t} + \frac{\partial \mathbf{F}_i}{\partial x_i} + \frac{\partial \mathbf{G}_i}{\partial x_i} - \mathbf{B} \right) d\Omega = 0 \tag{7.1.2}$$

or

$$\int_\Omega \left(\frac{\partial \mathbf{U}}{\partial t} - \mathbf{B} \right) d\Omega + \int_\Gamma (\mathbf{F}_i + \mathbf{G}_i) n_i \, d\Gamma = 0 \tag{7.1.3}$$

where n_i denotes the component of a unit vector normal to the boundary surface. Discretizing (7.1.3) and summing over all discrete nodes or cells (elements) throughout the control volumes (CV) and control surfaces (CS), we obtain

$$\sum_{CV} \left(\frac{\Delta \mathbf{U}}{\Delta t} - \mathbf{B} \right) \Delta\Omega + \sum_{CS} (\mathbf{F}_i + \mathbf{G}_i) n_i \Delta\Gamma = 0 \tag{7.1.4}$$

or

$$\sum_{CV}(\Delta\mathbf{U} - \Delta t\mathbf{B})\Delta\Omega + \sum_{CS}\Delta t(\mathbf{F}_i + \mathbf{G}_i)n_i\Delta\Gamma = 0 \tag{7.1.5}$$

The basic idea of FVM is to obtain a system of algebraic equations for the discretized control volume and control surfaces written such as in (7.1.5). In this process, the conservation of all variables is enforced across the control surfaces. Thus, when a specific quantity of a conserved variable is transported out of one control volume, the same quantity is transported into the adjacent control volumes. As a result there is no artificial creation or destruction of conserved variable. Inaccuracies that arise in coarse meshes, therefore, are not the result of a failure of any variable, but rather are due to approximation errors. Another advantage of FVM is that the discretized governing equations retain their physical interpretation, rather than possibly distorting the physics due to numerical discretization of each derivative term.

The finite volume methods are cost effective, because the calculation of flows at the surface of the adjoining control volumes need be performed only once since the expression is the same for both control volumes, differing only in sign. This gives rise to both cost reduction and algorithmic simplicity. In this chapter, finite volume methods via FDM are presented. Finite volume methods via FEM will be discussed in Chapter 15.

7.2 TWO-DIMENSIONAL PROBLEMS

There are two types of control volume formulations: the node-centered control volume and the cell (element)-centered control volume. These topics are discussed below.

7.2.1 NODE-CENTERED CONTROL VOLUME

For illustration, let us consider the two-dimensional configuration as shown in Figure 7.2.1a. Node 1 is connected to adjacent nodes 5, 7, 9, 11, and 2. The quadrilaterals A, B, C, D, and E are subdivided by connecting midpoints of lines between nodes with quadrants associated with node 1, forming the control volume for node 1 consisting of subcontrol volumes CV_1A, CV_1B, CV_1C, CV_1D, and CV_1E. Directions normal to two control surfaces of each element are identified by the arrows pointing outward, with angles $\theta^{(a)}$ and $\theta^{(b)}$ in a subcontrol volume (Figure 7.2.1b).

Let us examine the FVM formulation for the Poisson equation,

$$u_{,ii} - f = 0 \quad (i = 1, 2) \tag{7.2.1}$$

The finite volume equation becomes

$$\int_\Gamma u_{,i}n_i d\Gamma = \int_\Omega f d\Omega \tag{7.2.2}$$

or

$$\sum_{CS}^{A,B,C,D,E}\left(\frac{\Delta u}{\Delta x}n_1 + \frac{\Delta u}{\Delta y}n_2\right)\Delta\Gamma = \sum_{CV} f\Delta\Omega \tag{7.2.3}$$

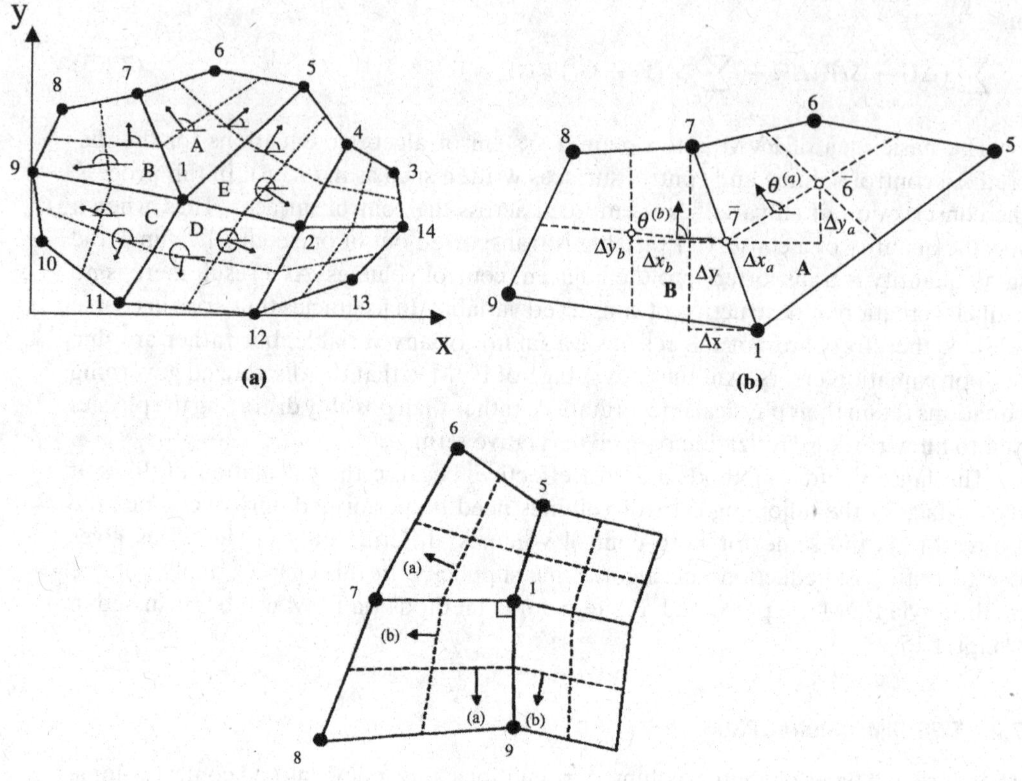

Figure 7.2.1 Control volumes and control surfaces. **(a)** Control volume at node 1 in unstructured grid system. **(b)** Control surfaces between nodes 1 and 7. **(c)** Modifications required for points 7-1 on horizontal (x) line and 9-1 on vertical (y) line.

The FDM discretization of (7.2.3) yields

$$(u_7 - u_1)S_{7,1} + (u_9 - u_1)S_{9,1} + (u_{11} - u_1)S_{11,1} + (u_2 - u_1)S_{2,1} + (u_5 - u_1)S_{5,1} = f_1\Omega_1 \tag{7.2.4}$$

with Ω_1 being the sum of the control volume areas surrounding node 1,

$$\Omega_1 = CV_1A + CV_1B + CV_1C + CV_1D + CV_1E$$

and $S_{7,1}$, $S_{9,1}$, etc. represent the surface parameters determined from the direction cosines. For example, the surface parameter $S_{7,1}$ associated with $u_7 - u_1$ is given by

$$S_{7,1} = \left(\cos\theta\frac{\Delta\Gamma}{\Delta x} + \sin\theta\frac{\Delta\Gamma}{\Delta y}\right)^{(a)}_{7,1} + \left(\cos\theta\frac{\Delta\Gamma}{\Delta x} + \sin\theta\frac{\Delta\Gamma}{\Delta y}\right)^{(b)}_{7,1} \tag{7.2.5}$$

where (a) and (b) refer to the adjacent control surfaces in the counterclockwise direction. Note also that

$$\Delta y_{(a)} = (\cos\theta\Delta\Gamma)^{(a)}, \quad \Delta x_{(a)} = (\sin\theta\Delta\Gamma)^{(a)}$$

refer to, respectively, the y and x components of $\Delta\Gamma$ on the control surface for the control volume A (see Figure 7.2.1b). Orientations of these surfaces are determined by the angle

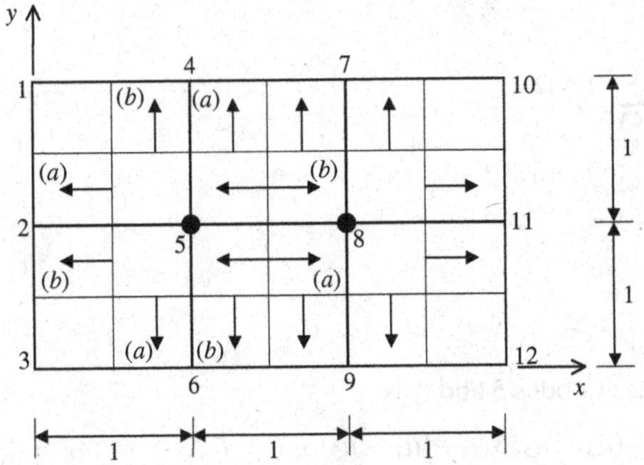

Figure E7.2.1 FVM solution of the Poisson equation.

θ of the direction cosines always measured counterclockwise from the x-axis as defined in Figure 7.2.1a. Note also that $\Delta x_{7,1} = x_7 - x_1$, $\Delta y_{7,1} = y_7 - y_1$, etc. It should be cautioned that, if two points are horizontal or vertical ($\Delta y = 0$ or $\Delta x = 0$), as may be the case in Figure 7.2.1c, then we set

$$\sin\theta \frac{\Delta\Gamma}{\Delta y} = 0 \quad \text{for } \Delta y = 0, \qquad \cos\theta \frac{\Delta\Gamma}{\Delta x} = 0 \quad \text{for } \Delta x = 0 \tag{7.2.6a}$$

This is to avoid division by zero ($\Delta y_{7,1} = 0$, $\Delta x_{9,1} = 0$). For a node in a rectangular geometry such as node 5 of Figure E7.2.1, the direction cosine is zero so that the division by zero is avoided by setting

$$\sin 180^\circ \Delta\Gamma = 0, \qquad \cos 270^\circ \Delta\Gamma = 0 \tag{7.2.6b}$$

This restriction allows the FVM formulation to yield the result identical to the FDM methods for rectangular grids. For all nonrectangular arbitrary geometries, the definitions given in (7.2.5) should be used.

Detailed computational steps for a simple geometry (Figure E7.2.1) are demonstrated in Example 7.2.1.

Example 7.2.1

Given: $\nabla^2 u = f(x, y)$, with the exact solution ($u = 2x^2y^2$), Dirichlet boundary conditions.

Required: Solve using the finite volume method via finite differences (3×2 unit square mesh, Figure E7.2.1). Dirichlet boundary data for all exterior boundaries and the source term are calculated from the exact solution.

$$u_1, u_2, u_3, u_6, u_9, u_{12} = 0$$

$$u_4 = 8, \quad u_7 = 32, \quad u_{10} = 72, \quad u_{11} = 18$$

$$f_5 = 8, \quad f_8 = 20$$

Solution:

$$\sum_{CS}\left(\frac{\Delta u}{\Delta x}n_1 + \frac{\Delta u}{\Delta y}n_2\right)\Delta\Gamma = \sum_{CV} f\Delta\Omega$$

$$\sum_{CS}\Delta u S = \sum_{CV} f\Delta\Omega$$

where

$$S = \frac{n_1\,\Delta\Gamma}{\Delta x} + \frac{n_2\,\Delta\Gamma}{\Delta y}$$

We write the finite difference analogs at nodes 5 and 8 as

$$(u_2 - u_5)S_{2,5} + (u_6 - u_5)S_{6,5} + (u_8 - u_5)S_{8.5} + (u_4 - u_5)S_{4,5} = f_5 A_5$$
$$(u_5 - u_8)S_{5,8} + (u_9 - u_8)S_{9,8} + (u_{11} - u_8)S_{11,8} + (u_7 - u_8)S_{7,8} = f_8 A_8$$

with

$$f_5 A_5 = 8, \quad f_8 A_8 = 20$$

$$S_{2,5} = \left[\left(S_{2,5}^{(1)} + S_{2,5}^{(2)}\right)^{(a)} + \left(S_{2,5}^{(1)} + S_{2,5}^{(2)}\right)^{(b)}\right]$$
$$= \left[\frac{\Delta y_{(a)}}{\Delta x_{2,5}} + 0\right] + \left[\frac{\Delta y_{(b)}}{\Delta x_{2,5}} + 0\right] = \left[\frac{-1/2}{-1} + 0\right] + \left[\frac{-1/2}{-1} + 0\right] = 1$$

and

$$S_{6,5} = \left[0 + \frac{\Delta x_{(a)}}{\Delta y_{6,5}}\right] + \left[0 + \frac{\Delta x_{(b)}}{\Delta y_{6,5}}\right] = \left[0 + \frac{-1/2}{-1}\right] + \left[0 + \frac{-1/2}{-1}\right] = 1$$

$$S_{8,5} = S_{4,5} = 1, \quad \text{etc.}$$

Solving the above two equations for nodes 5 and 8 with the boundary conditions imposed, we obtain

$$\begin{bmatrix} -4 & 1 \\ 1 & -4 \end{bmatrix}\begin{bmatrix} u_5 \\ u_8 \end{bmatrix} = \begin{bmatrix} 8 - 8 \\ 20 - 50 \end{bmatrix} = \begin{bmatrix} 0 \\ -30 \end{bmatrix}$$

$$\begin{bmatrix} u_5 \\ u_8 \end{bmatrix} = \begin{bmatrix} 2 \\ 8 \end{bmatrix}$$

which is the exact solution. For the structured orthogonal grids, the process is the same as in FDM.

Example 7.2.2

Given: Same as Example 7.2.1 with Neumann data:

$$\left(\frac{\partial u}{\partial x}\right)_4 = 16, \qquad \left(\frac{\partial u}{\partial y}\right)_4 = 8$$

Solution: The additional equation required at node 4 becomes

$$\left.\frac{\partial u}{\partial x}\right|_4 + \left.\frac{\partial u}{\partial y}\right|_4 + \left.\frac{\Delta u}{\Delta y}\right|_{5,4} + \left.\frac{\Delta u}{\Delta x}\right|_{1,4} + \left.\frac{\Delta u}{\Delta x}\right|_{7,4} = f_4 A_4$$

$$16 + 8 + \frac{u_{5/2} - u_4}{\Delta y/2} + \frac{u_{1/2} - u_4}{\Delta x/2} + \frac{u_{7/2} - u_4}{\Delta x/2} = 20\left(\frac{1}{2}\right)$$

Combining equations written at nodes 5 and 8 from Example 7.2.1, we obtain

$$\begin{bmatrix} -4 & 1 & 1 \\ 1 & -4 & 0 \\ 1 & 0 & -6 \end{bmatrix} \begin{bmatrix} u_5 \\ u_8 \\ u_4 \end{bmatrix} = \begin{bmatrix} 8 \\ -30 \\ -46 \end{bmatrix}$$

Thus

$$\begin{bmatrix} u_5 \\ u_8 \\ u_4 \end{bmatrix} = -\frac{1}{86} \begin{bmatrix} 24 & 6 & 4 \\ 6 & 23 & 1 \\ 4 & 1 & 15 \end{bmatrix} \begin{bmatrix} 8 \\ -30 \\ -46 \end{bmatrix} = \begin{bmatrix} 2 \\ 8 \\ 8 \end{bmatrix}$$

This is the exact solution. Note that for unstructured grids with sloped boundaries, specification of the Neumann boundary conditions must be adjusted for direction cosines.

7.2.2 CELL-CENTERED CONTROL VOLUME

In the previous section, we dealt with the case in which nodes are identified with the surrounding subcontrol volumes (node-centered control volume). Instead of subcontrol or tributary control volumes surrounding the node, it is possible to consider control volumes constructed by adjacent nodes as shown in Figure 7.2.2a,b,c. Here, control surfaces are identified between adjacent nodes for a structured grid system, leading to the cell-centered control volume. However, this requirement lacks the generality prevailing in the unstructured grid system.

For illustration, let us consider the cell-centered FVM scheme as shown in Figure 7.2.2 for the solution of the Poisson equation examined in Section 7.2.1. The corresponding FVM equation is given by (7.2.2).

$$\sum_{CS}\left(\frac{\Delta u}{\Delta x}n_1 + \frac{\Delta u}{\Delta y}n_2\right)\Delta\Gamma = \sum_{CV} f\Delta\Omega \tag{7.2.7}$$

This can be written for the cell-centered scheme in the form,

$$\begin{aligned}
&\left(\frac{\Delta u}{\Delta x}\right)_{i,j-1/2} n_1\Delta\Gamma_{AB} + \left(\frac{\Delta u}{\Delta y}\right)_{i,j-1/2} n_2\Delta\Gamma_{AB} + \left(\frac{\Delta u}{\Delta x}\right)_{i+1/2,j} n_1\Delta\Gamma_{BC} \\
&+ \left(\frac{\Delta u}{\Delta y}\right)_{i+1/2,j} n_2\Delta\Gamma_{BC} + \left(\frac{\Delta u}{\Delta x}\right)_{i,j+1/2} n_1\Delta\Gamma_{CD} + \left(\frac{\Delta u}{\Delta y}\right)_{i,j+1/2} n_2\Delta\Gamma_{CD} \\
&+ \left(\frac{\Delta u}{\Delta x}\right)_{i-1/2,j} n_1\Delta\Gamma_{DA} + \left(\frac{\Delta u}{\Delta y}\right)_{i-1/2,j} n_2\Delta\Gamma_{DA} = (f\Delta\Omega)_{i,j}
\end{aligned} \tag{7.2.8}$$

where $\Delta u/\Delta x$ and $\Delta u/\Delta y$ may be approximated by using tributary areas and

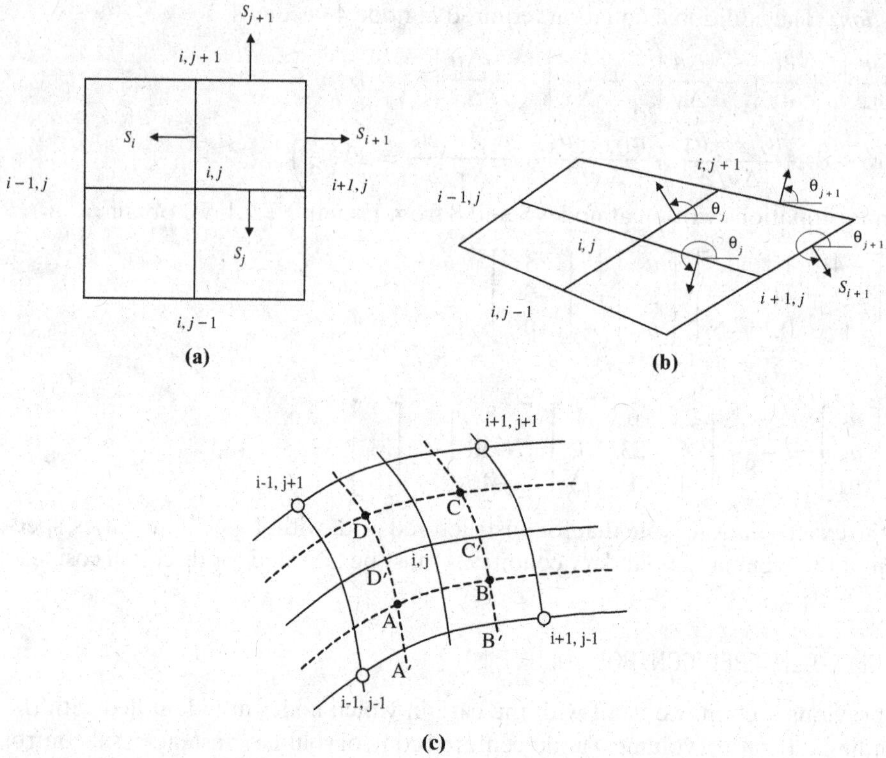

Figure 7.2.2 Cell-centered control volume. **(a)** Square or rectangular grids. **(b)** Skewed grids. **(c)** Curvilinear system.

corresponding boundary surface areas. For example, we have (Figure 7.2.2c)

$$\left(\frac{\Delta u}{\Delta x}\right)_{i,j-1/2} = \frac{1}{\Delta\Omega_{i,j-1/2}} \sum_{A'B'C'D'} u_{i,j-1/2}\Delta y_{i,j-1/2}$$
$$= (u_{i,j-1}\Delta y_{A'B'} + u_B \Delta y_{B'C'} + u_{i,j}\Delta y_{C'D'} + u_A \Delta y_{D'A'})/\Delta\Omega_{i,j-1/2}$$

with

$$u_A = \frac{1}{4}(u_{i,j} + u_{i-1,j} + u_{i-1,j-1} + u_{i,j-1})$$

$$u_B = \frac{1}{4}(u_{i,j} + u_{i+1,j} + u_{i+1,j-1} + u_{i,j-1})$$

Other quantities in (7.2.8) are calculated similarly. It can be shown that the above procedure gives the identical results for the problem in Example 7.2.1. Note that the cell-centered FVM presented here can not be applied to the unstructured grid shown in Figure 7.2.1.

The cell-centered FVM scheme for Euler equations takes the form

$$\sum_{CV}\left(\frac{\Delta \mathbf{U}}{\Delta t}\right)\Delta\Omega = -\sum_{CS}(\mathbf{F}_i S_i), \qquad S_i = n_i \Delta\Gamma \tag{7.2.9}$$

This may be solved using the operator splitting scheme or the fractional step scheme:

(1) Operator Splitting Scheme

Step 1

$$\overline{\mathbf{U}}_{i,j}^{n+1} = \mathbf{U}_{i,j}^{n} - \frac{\Delta t}{\Delta\Omega_{i,j}}\left(\mathbf{F}_{i,j}^{n} S_{j+1} + \mathbf{F}_{i,j-1}^{n} S_j + \mathbf{F}_{i,j}^{n} S_{i+1} + \mathbf{F}_{i-1,j}^{n} S_i\right) \tag{7.2.10a}$$

Step 2

$$\mathbf{U}_{i,j}^{n+1} = \frac{1}{2}\left[\mathbf{U}_{i,j}^{n} + \overline{\mathbf{U}}_{i,j}^{n+1} - \frac{\Delta t}{\Delta\Omega_{i,j}}\left(\overline{\mathbf{F}}_{i,j+1}^{n+1} S_{j+1} + \overline{\mathbf{F}}_{i,j}^{n+1} S_j + \overline{\mathbf{F}}_{i+1,j}^{n+1} S_{i+1} + \overline{\mathbf{F}}_{i,j}^{n+1} S_i\right)\right] \tag{7.2.10b}$$

These steps are repeated until steady-state is reached.

(2) Fractional Step Scheme

In this scheme, a half-time step is introduced in order to increase accuracy.

Step 1

$$\overline{\mathbf{U}}_{i,j}^{n+\frac{1}{2}} = \mathbf{U}_{i,j}^{n} - \frac{\Delta t}{\Delta\Omega_{i,j}}\left(\mathbf{F}_{i,j}^{n} S_{j+1} + \mathbf{F}_{i,j-1}^{n} S_j\right) \tag{7.2.11a}$$

$$\mathbf{U}_{i,j}^{n+\frac{1}{2}} = \frac{1}{2}\left[\mathbf{U}_{i,j}^{n} + \overline{\mathbf{U}}_{i,j}^{n+\frac{1}{2}} - \frac{\Delta t}{\Delta\Omega_{i,j}}\left(\overline{\mathbf{F}}_{i,j+1}^{n+\frac{1}{2}} S_{j+1} + \overline{\mathbf{F}}_{i,j}^{n+\frac{1}{2}} S_j\right)\right] \tag{7.2.11b}$$

Step 2

$$\overline{\mathbf{U}}_{i,j}^{n+1} = \mathbf{U}_{i,j}^{n+\frac{1}{2}} - \frac{\Delta t}{\Delta\Omega_{i,j}}\left(\mathbf{F}_{i,j}^{n+\frac{1}{2}} S_{i+1} + \mathbf{F}_{i-1,j}^{n+\frac{1}{2}} S_i\right) \tag{7.2.11c}$$

$$\mathbf{U}_{i,j}^{n+1} = \frac{1}{2}\left[\mathbf{U}_{i,j}^{n+\frac{1}{2}} + \overline{\mathbf{U}}_{i,j}^{n+1} - \frac{\Delta t}{\Delta\Omega_{i,j}}\left(\overline{\mathbf{F}}_{i+1,j}^{n+1} S_{i+1} + \overline{\mathbf{F}}_{i,j}^{n+1} S_i\right)\right] \tag{7.2.11d}$$

Here, S_i, S_{i+1}, S_j, S_{j+1} are the control surfaces as oriented by the direction cosine components in the structured grid system.

7.2.3 CELL-CENTERED AVERAGE SCHEME

The cell-centered average scheme was proposed by Ni [1982]. To illustrate, we consider the Euler equation written in the form

$$\frac{\partial \mathbf{U}}{\partial t} = -\frac{\partial \mathbf{F}}{\partial x} - \frac{\partial \mathbf{G}}{\partial y}$$

where

$$\mathbf{U} = \begin{bmatrix} \rho \\ \rho u \\ \rho v \\ \rho E \end{bmatrix} \qquad \mathbf{F} = \begin{bmatrix} \rho u \\ p + \rho u^2 \\ \rho u v \\ \rho E u + p u \end{bmatrix} \qquad \mathbf{G} = \begin{bmatrix} \rho v \\ \rho v u \\ p + \rho v^2 \\ \rho E v + p v \end{bmatrix}$$

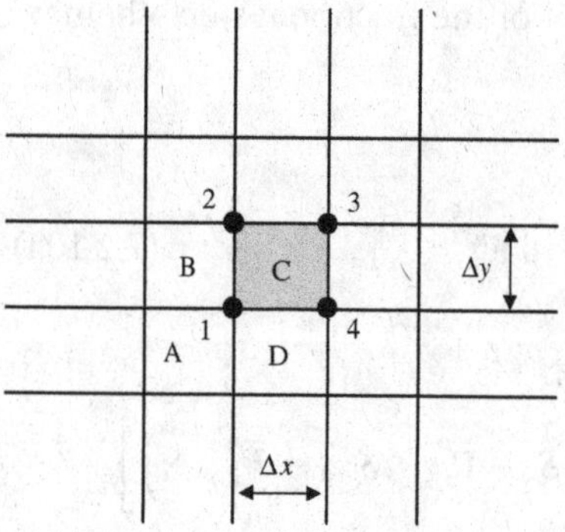

Figure 7.2.3 Cell-centered average scheme.

The control volumes for two-dimensional problems are shown in Figure 7.2.3. The change of flow variables $\mathbf{U}_c$ for the control volume C is given by

$$\Delta\mathbf{U}_c = \frac{\Delta t}{2\Delta x\Delta y}[(F_1+F_2)\Delta y - (F_3+F_4)\Delta y + (G_1+G_4)\Delta x - (G_2+G_3)\Delta x] \tag{7.2.12}$$

The corrections to the grid points associated with the control volume C (distribution formula) are determined by

$$\begin{aligned}
(\delta\mathbf{U}_1)_c &= \frac{1}{4}\left[\Delta\mathbf{U}_c - \frac{\Delta t}{\Delta x}\Delta\mathbf{F}_c - \frac{\Delta t}{\Delta y}\Delta\mathbf{G}_c\right]\\
(\delta\mathbf{U}_2)_c &= \frac{1}{4}\left[\Delta\mathbf{U}_c - \frac{\Delta t}{\Delta x}\Delta\mathbf{F}_c + \frac{\Delta t}{\Delta y}\Delta\mathbf{G}_c\right]\\
(\delta\mathbf{U}_3)_c &= \frac{1}{4}\left[\Delta\mathbf{U}_c + \frac{\Delta t}{\Delta x}\Delta\mathbf{F}_c + \frac{\Delta t}{\Delta y}\Delta\mathbf{G}_c\right]\\
(\delta\mathbf{U}_4)_c &= \frac{1}{4}\left[\Delta\mathbf{U}_c + \frac{\Delta t}{\Delta x}\Delta\mathbf{F}_c - \frac{\Delta t}{\Delta y}\Delta\mathbf{G}_c\right]
\end{aligned} \tag{7.2.13}$$

where

$$\Delta\mathbf{F}_c = \left(\frac{\partial\mathbf{F}}{\partial\mathbf{U}}\right)_c\Delta\mathbf{U}_c, \qquad \Delta\mathbf{G}_c = \left(\frac{\partial\mathbf{G}}{\partial\mathbf{U}}\right)_c\Delta\mathbf{U}_c \tag{7.2.14}$$

and

For arbitrary curvilinear coordinates, the change of flow variables for the control volume C takes the form

$$\begin{aligned}
\Delta U_c = \frac{\Delta t}{\Delta\Omega}\Bigg\{&\left[\frac{F_1+F_2}{2}(y_2-y_1) - \frac{G_1+G_2}{2}(x_2-x_1)\right]\\
&-\left[\frac{F_3+F_4}{2}(y_3-y_4) - \frac{G_3+G_4}{2}(x_3-x_4)\right]\\
&+\left[\frac{G_1+G_4}{2}(x_4-x_1) - \frac{F_1+F_4}{2}(y_4-y_1)\right]\\
&-\left[\frac{G_2+G_3}{2}(x_3-x_2) - \frac{F_2+F_3}{2}(y_3-y_2)\right]\Bigg\}
\end{aligned} \tag{7.2.15}$$

with

$$\Delta\Omega = -\frac{1}{2}[(x_3 - x_1)(y_4 - y_2) - (x_4 - x_2)(y_3 - y_1)] \tag{7.2.16}$$

The flow variables at point 1 are updated as

$$\mathbf{U}_1^{n+1} = \mathbf{U}_1^n + \delta\mathbf{U}_1 \tag{7.2.17}$$

with

$$\delta\mathbf{U}_1 = (\delta\mathbf{U}_1)_A + (\delta\mathbf{U}_1)_B + (\delta\mathbf{U}_1)_C + (\delta\mathbf{U}_1)_D \tag{7.2.18}$$

where A through D refer to control volumes surrounding the grid point 1. Here the CFL condition is given by

$$\Delta t \leq \min\left(\frac{\Delta x}{|u| + a}, \frac{\Delta y}{|v| + a}\right) \tag{7.2.19}$$

It should be noted that, for transonic and supersonic flows, an artificial viscosity must be added for stability. For example,

$$(\delta\mathbf{U}_1)_c = \frac{1}{4}\left[\Delta\mathbf{U}_c - \frac{\Delta t}{\Delta x}\Delta\mathbf{F}_c - \frac{\Delta t}{\Delta y}\Delta\mathbf{G}_c + \mu(\overline{\mathbf{U}} - \mathbf{U}_1)\right] \tag{7.2.20}$$

with

$$\overline{\mathbf{U}} = \frac{1}{4}(\mathbf{U}_1 + \mathbf{U}_2 + \mathbf{U}_3 + \mathbf{U}_4) \tag{7.2.21}$$

$$\mu = \sigma\left(\frac{\Delta t}{\Delta x} + \frac{\Delta t}{\Delta y}\right) \tag{7.2.22}$$

where σ is an artificial damping factor usually taken as $0 < \sigma < 0.1$.

It is seen that the corrections defined in (7.2.13) together with (7.2.20) guarantee the proper domain of dependence regardless of local flow direction and wave speed, leading to a stable second order solution.

7.3 THREE-DIMENSIONAL PROBLEMS

7.3.1 3-D GEOMETRY DATA STRUCTURE

For three-dimensional problems dealing with arbitrary unstructured meshes, an efficient algorithm for data structure will be important. For illustration, consider the geometry shown in Figure 7.3.1, where all nodes are on the exterior global boundaries except two interior nodes, 10 and 11. The control volume for node 10 and its control surfaces are represented in Figure 7.3.2. Let us examine any inclined control surface arbitrarily located in three-dimensional reference coordinates (x, y, z) as shown in Figure 7.3.3. Note that local cartesian coordinates (x', y', z') are constructed such that the $x' - y'$ plane coincide with the control surface. The origin is located at node 1 with the x' axis lying on the line connecting nodes 1 and 2. The z' axis is in the direction of the unit vector $\mathbf{n}$ normal to the control surface. The y' axis can be determined once the unit normal vector is known. The origin of natural or isoparametric coordinates (ξ, η)

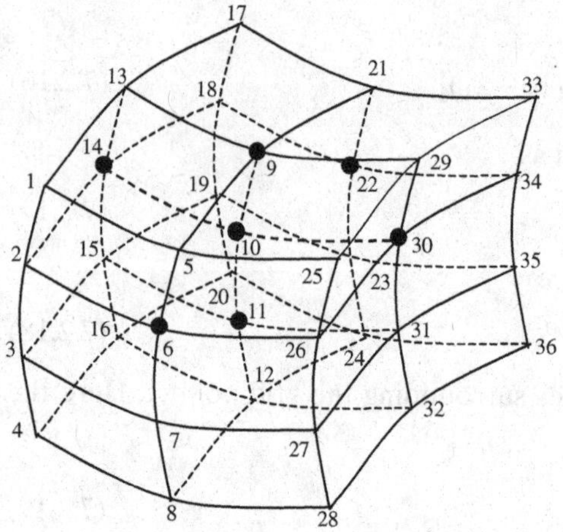

Figure 7.3.1 Illustration of 3-D finite volume discretization, 2 interior nodes 10 and 11; all other nodes are on the boundaries.

may be embedded at the controid of the quadrilateral so that the surface area can be calculated easily.

The unit vector normal to the surface is found by establishing the unit vectors $\mathbf{e}_{12}$ and $\mathbf{e}_{14}$ along the lines 1-2 and 1-4, respectively, as follows. Between nodes 1 and 2, we have

$$\mathbf{e}_{12} = \lambda_i \mathbf{i}_i \tag{7.3.1}$$

where

$$\lambda_1 = \frac{x_{12}}{L_{12}}, \quad \lambda_2 = \frac{y_{12}}{L_{12}}, \quad \lambda_3 = \frac{z_{12}}{L_{12}}$$
$$L_{12} = \left(x_{12}^2 + y_{12}^2 + z_{12}^2\right)^{1/2}$$

with $x_{12} = x_1 - x_2$, etc. Similarly, for the unit vector along the line on nodes 1 and 4, we have

$$\mathbf{e}_{14} = \mu_i \mathbf{i}_i \tag{7.3.2}$$

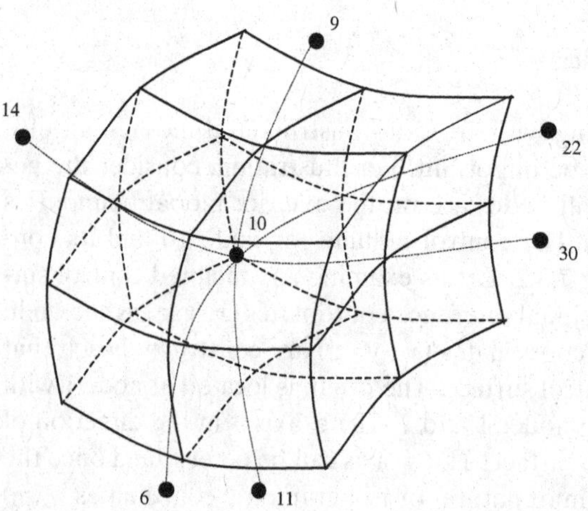

Figure 7.3.2 Control volumes and control surfaces for the interior node 10; connected neighboring nodes are 6, 14, 22, 30, 9, and 11.

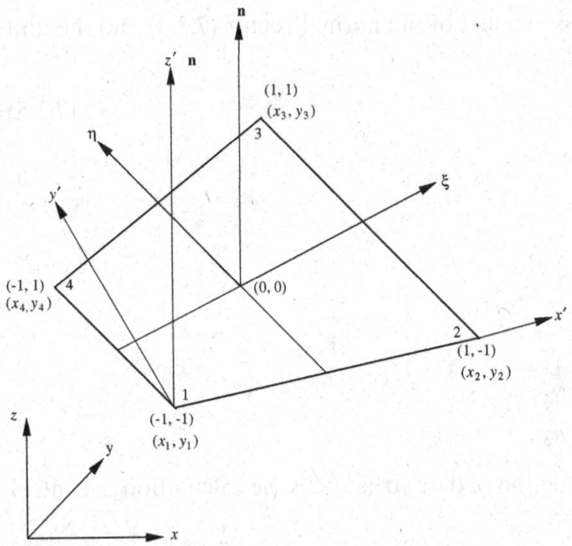

Figure 7.3.3 Control surface on the natural (isoparameteric) coordinates (ξ, η), oriented in terms of the local cartesian coordinates (x', y', z'), the unit normal vector coinciding with the z' axis.

where

$$\mu_1 = \frac{x_{14}}{L_{14}}, \qquad \mu_2 = \frac{y_{14}}{L_{14}}, \qquad \mu_3 = \frac{z_{14}}{L_{14}}$$

$$L_{14} = \left(x_{14}^2 + y_{14}^2 + z_{14}^2\right)^{1/2}$$

with $x_{14} = x_1 - x_4$, etc.

The unit vector normal to the surface is given by the cross product of these two unit vectors along the lines 1-2 and 1-4.

$$\mathbf{n} = \mathbf{e}_{12} \times \mathbf{e}_{14} = \varepsilon_{ijk}\lambda_i\mu_j\mathbf{i}_k = n_k\mathbf{i}_k \tag{7.3.3}$$

with

$$n_1 = \lambda_2\mu_3 - \lambda_3\mu_2$$
$$n_2 = \lambda_3\mu_1 - \lambda_1\mu_3$$
$$n_3 = \lambda_1\mu_2 - \lambda_2\mu_1$$

To calculate the control surface areas surrounding the control volume such as in Figure 7.3.3, it is necessary to carry out the coordinate transformation between the local coordinates (x', y', z') and the global reference coordinates (x, y, z), since the control surface plane is located arbitrarily in the three-dimensional configurations.

$$x_i' = a_{ij}x_j \tag{7.3.4}$$

where a_{ij} is the transformation matrix. The components of a_{ij} corresponding to the x_1' are the same as those for the unit vector $\mathbf{e}_{12}$,

$$a_{11} = \lambda_1, \quad a_{12} = \lambda_2, \quad a_{13} = \lambda_3$$

To determine the rest of the direction cosines, we must find the unit vector along the

y' axis. This can be done by the cross product of the normal vector (7.3.3) and the unit vector along the 1-2 direction,

$$\mathbf{e}_y = \mathbf{n} \times \mathbf{e}_{12} = \varepsilon_{ijk} n_i \lambda_j \mathbf{i}_k = \gamma_k \mathbf{i}_k \tag{7.3.5}$$

with

$$\gamma_1 = n_2\lambda_3 - n_3\lambda_2$$
$$\gamma_2 = n_3\lambda_1 - n_1\lambda_3$$
$$\gamma_3 = n_1\lambda_2 - n_2\lambda_1$$

Thus, we have

$$a_{21} = \gamma_1, \qquad a_{22} = \gamma_2, \qquad a_{23} = \gamma_3$$
$$a_{31} = n_1, \qquad a_{32} = n_2, \qquad a_{33} = n_3$$

The remaining task for the construction of data structure is the calculation of control surface areas and control volumes.

Control Surface Area

$$A = \iint dx' dy' = \int_{-1}^{1}\int_{-1}^{1} |J'| d\xi d\eta \tag{7.3.6}$$

with $|J'|$ being the determinant of the control surface Jacobian (see Section 9.3.3 for derivation),

$$|J'| = \begin{vmatrix} \dfrac{\partial x'}{\partial \xi} & \dfrac{\partial y'}{\partial \xi} \\ \dfrac{\partial x'}{\partial \eta} & \dfrac{\partial y'}{\partial \eta} \end{vmatrix} \tag{7.3.7}$$

$$x' = a_{11}x + a_{12}y + a_{13}z \tag{7.3.8}$$
$$y' = a_{21}x + a_{22}y + a_{23}z$$

$$x = \Phi_N(\xi, \eta) x_N \quad (N = 1, 2, 3, 4) \tag{7.3.9}$$

$$\Phi_1 = \frac{1}{4}(1-\xi)(1-\eta), \qquad \Phi_2 = \frac{1}{4}(1+\xi)(1-\eta), \qquad \Phi_3 = \frac{1}{4}(1+\xi)(1+\eta),$$
$$\Phi_4 = \frac{1}{4}(1-\xi)(1+\eta) \tag{7.3.10}$$

with Φ_N being the interpolation functions derived in Section 9.3.3. Substituting (7.3.10) into (7.3.9), (7.3.8), and (7.3.7), we obtain

$$\begin{aligned}
\frac{\partial x'}{\partial \xi} &= a_{11}\frac{\partial \Phi_N}{\partial \xi}x_N + a_{12}\frac{\partial \Phi_N}{\partial \xi}y_N + a_{13}\frac{\partial \Phi_N}{\partial \xi}z_N \\
\frac{\partial y'}{\partial \xi} &= a_{21}\frac{\partial \Phi_N}{\partial \xi}x_N + a_{22}\frac{\partial \Phi_N}{\partial \xi}y_N + a_{23}\frac{\partial \Phi_N}{\partial \xi}z_N \\
\frac{\partial x'}{\partial \eta} &= a_{11}\frac{\partial \Phi_N}{\partial \eta}x_N + a_{12}\frac{\partial \Phi_N}{\partial \eta}y_N + a_{13}\frac{\partial \Phi_N}{\partial \eta}z_N \\
\frac{\partial y'}{\partial \eta} &= a_{21}\frac{\partial \Phi_N}{\partial \eta}x_N + a_{22}\frac{\partial \Phi_N}{\partial \eta}y_N + a_{23}\frac{\partial \Phi_N}{\partial \eta}z_N
\end{aligned} \tag{7.3.11}$$

Integration of (7.3.6) can be carried out most accurately by using the Gaussian quadrature, which is detailed in Section 9.3.3.

Control Volume

$$V = \iiint dxdydz = \int_{-1}^{1}\int_{-1}^{1}\int_{-1}^{1} |J|d\xi d\eta d\zeta \tag{7.3.12}$$

$$|J| = \begin{vmatrix} \frac{\partial x}{\partial \xi} & \frac{\partial y}{\partial \xi} & \frac{\partial z}{\partial \xi} \\ \frac{\partial x}{\partial \eta} & \frac{\partial y}{\partial \eta} & \frac{\partial z}{\partial \eta} \\ \frac{\partial x}{\partial \zeta} & \frac{\partial y}{\partial \zeta} & \frac{\partial z}{\partial \zeta} \end{vmatrix} \tag{7.3.13}$$

with $|J|$ being the determinant of the control volume Jacobian in terms of the natural or isoparameteric coordinates (ξ, η, ζ) with reference to the global cartesian coordinates (x, y, z) as shown in Figure 7.3.4. See Section 9.4.3 for derivation and details of integration using the Gaussian quadrature.

The control surface and control volume for a three-dimensional geometry may be calculated alternatively as follows. Referring to Figure 7.3.5, the surface area A_{1234} is equal to one-half of the absolute value of the cross product between the diagonal unit vectors times their corresponding physical lengths.

$$A_{1234} = |\mathbf{A}_{1234}| = \frac{1}{2}|\mathbf{e}_{13}L_{13} \times \mathbf{e}_{24}L_{24}| \tag{7.3.14}$$

Here, the calculation of the components of the unit vectors follow the same procedure as in (7.3.1) and (7.3.2). These surface areas should be oriented by the unit normal vector calculated from (7.3.3).

Similarly, the control volume is equal to one third of the dot product of the sum of any three adjacent surface area vectors and the unit vector times its physical length,

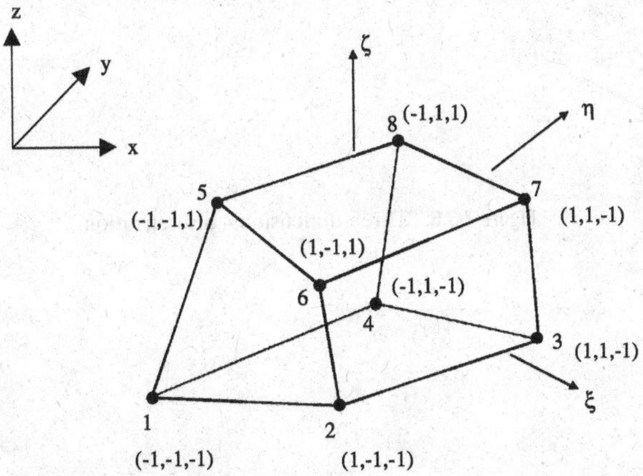

Figure 7.3.4 Three-dimensional control volume with hexahedral isoparameteric coordinates.

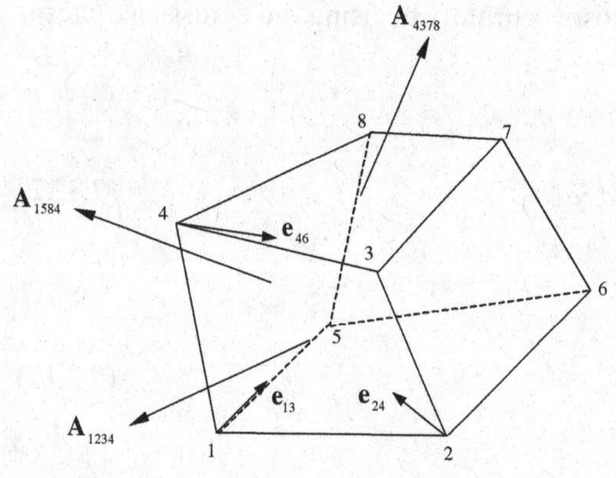

Figure 7.3.5 Alternative method for calculations of control surface areas and control volume.

connecting two nodes diagonally with one of them containing the three surfaces under consideration.

$$
\begin{aligned}
V &= \frac{1}{3}(\mathbf{A}_{1234} + \mathbf{A}_{4378} + \mathbf{A}_{1584}) \cdot \mathbf{e}_{46} L_{46} \\
&= \frac{1}{3}(\mathbf{e}_{13} L_{13} \times \mathbf{e}_{24} L_{24} + \mathbf{e}_{47} L_{47} \times \mathbf{e}_{38} L_{38} + \mathbf{e}_{18} L_{18} \times e_{54} L_{54}) \cdot \mathbf{e}_{46} L_{46} \qquad (7.3.15)
\end{aligned}
$$

in which node 4 is common to the three surfaces and node 6 is in the diagonal direction constituting the unit vector $\mathbf{e}_{46}$, with all unit vectors calculated similarly as in (7.3.1).

7.3.2 THREE-DIMENSIONAL FVM EQUATIONS

Three-dimensional FVM via FDM can be formulated as a direct extension of the two-dimensional case discussed in Section 7.2. A typical control volume element configuration is shown in Figure 7.3.6. The cell-centered control volume procedure for the Euler

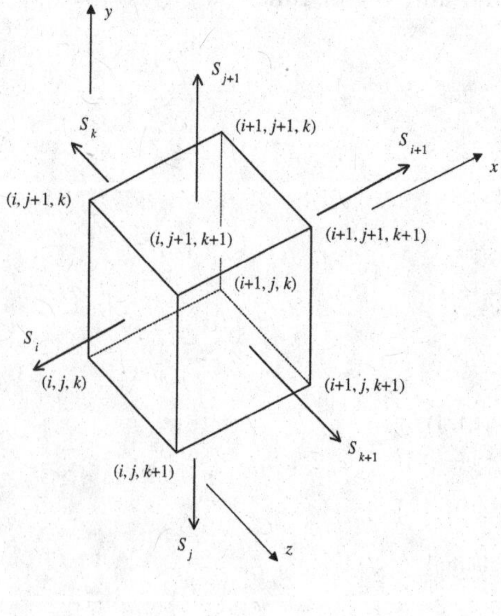

Figure 7.3.6 Three-dimensional discretization.

equation using the operator splitting and fractional step scheme is described below [Rizzi and Inouye, 1973].

(1) Operator Splitting Scheme

Step 1

$$\overline{\mathbf{U}}^{n+1}_{i,j,k} = \mathbf{U}^n_{i,j,k} - \frac{\Delta t}{\Delta\Omega_{i,j,k}}(\mathbf{F}^n_{i,j,k}S_{j+1} + \mathbf{F}^n_{i,j-1,k}S_j + \mathbf{F}^n_{i,j,k}S_{i+1} + \mathbf{F}^n_{i-1,j,k}S_i + \mathbf{F}^n_{i,j,k}S_{k+1} + \mathbf{F}^n_{i,j,k-1}S_k) \tag{7.3.16a}$$

Step 2

$$\mathbf{U}^{n+1}_{i,j,k} = \frac{1}{2}\left[\mathbf{U}^n_{i,j,k} + \overline{\mathbf{U}}^{n+1}_{i,j,k} - \frac{\Delta t}{\Delta\Omega_{i,j,k}}(\overline{\mathbf{F}}^{n+1}_{i,j+1,k}S_{j+1} + \overline{\mathbf{F}}^{n+1}_{i,j,k}S_j + \overline{\mathbf{F}}^{n+1}_{i+1,j,k}S_{i+1} + \overline{\mathbf{F}}^{n+1}_{i,j,k}S_i + \overline{\mathbf{F}}^{n+1}_{i,j,k+1}S_{k+1} + \overline{\mathbf{F}}^{n+1}_{i,j,k}S_k)\right] \tag{7.3.16b}$$

(2) Fractional Step Scheme

Step 1

$$\overline{\mathbf{U}}^{n+\frac{1}{3}}_{i,j,k} = \mathbf{U}^n_{i,j,k} - \frac{\Delta t}{\Delta\Omega_{i,j,k}}(\mathbf{F}^n_{i,j,k}S_{j+1} + \mathbf{F}^n_{i,j-1,k}S_j) \tag{7.3.17a}$$

$$\mathbf{U}^{n+\frac{1}{3}}_{i,j,k} = \frac{1}{2}\left[\mathbf{U}^n_{i,j,k} + \overline{\mathbf{U}}^{n+\frac{1}{3}}_{i,j,k} - \frac{\Delta t}{\Delta\Omega_{i,j,k}}\left(\overline{\mathbf{F}}^{n+\frac{1}{3}}_{i,j+1,k}S_{j+1} + \overline{\mathbf{F}}^{n+\frac{1}{3}}_{i,j,k}S_j\right)\right] \tag{7.3.17b}$$

Step 2

$$\overline{\mathbf{U}}^{n+\frac{2}{3}}_{i,j,k} = \mathbf{U}^{n+\frac{1}{3}}_{i,j,k} - \frac{\Delta t}{\Delta\Omega_{i,j,k}}\left(\mathbf{F}^{n+\frac{1}{3}}_{i,j,k}S_{i+1} + \mathbf{F}^{n+\frac{1}{3}}_{i-1,j,k}S_i\right) \tag{7.3.18a}$$

$$\mathbf{U}^{n+\frac{2}{3}}_{i,j,k} = \frac{1}{2}\left[\mathbf{U}^{n+\frac{1}{3}}_{i,j,k} + \overline{\mathbf{U}}^{n+\frac{2}{3}}_{i,j,k} - \frac{\Delta t}{\Delta\Omega_{i,j,k}}\left(\overline{\mathbf{F}}^{n+\frac{2}{3}}_{i+1,j,k}S_{i+1} + \overline{\mathbf{F}}^{n+\frac{2}{3}}_{i,j,k}S_i\right)\right] \tag{7.3.18b}$$

Step 3

$$\overline{\mathbf{U}}^{n+1}_{i,j,k} = \mathbf{U}^{n+\frac{2}{3}}_{i,j,k} - \frac{\Delta t}{\Delta\Omega_{i,j,k}}\left(\mathbf{F}^{n+\frac{2}{3}}_{i,j,k}S_{k+1} + \mathbf{F}^{n+\frac{2}{3}}_{i,j,k-1}S_k\right) \tag{7.3.19a}$$

$$\mathbf{U}^{n+1}_{i,j,k} = \frac{1}{2}\left[\mathbf{U}^{n+\frac{2}{3}}_{i,j,k} + \overline{\mathbf{U}}^{n+1}_{i,j,k} - \frac{\Delta t}{\Delta\Omega_{i,j,k}}\overline{\mathbf{F}}^{n+1}_{i,j,k+1}S_{k+1} + \overline{\mathbf{F}}^{n+1}_{i,j,k}S_k\right] \tag{7.3.19b}$$

Stability conditions may be given as

$$\Delta t \le \min(\Delta t_x, \Delta t_y, \Delta t_z) \tag{7.3.20}$$

with

$$\Delta t_x \le \min_{i,j,k}\left[\frac{\Delta\Omega_{i,j,k}}{(|\mathbf{q}\cdot\mathbf{S}_i| + aS_i)_{i,j,k}}\right] \tag{7.3.21a}$$

$$\Delta t_y \leq \min_{i,j,k}\left[\frac{\Delta\Omega_{i,j,k}}{(|\mathbf{q}\cdot\mathbf{S}_j| + a S_j)_{i,j,k}}\right] \tag{7.3.21b}$$

$$\Delta t_z \leq \min_{i,j,k}\left[\frac{\Delta\Omega_{i,j,k}}{(|\mathbf{q}\cdot\mathbf{S}_k| + a S_k)_{i,j,k}}\right] \tag{7.3.21c}$$

where $\mathbf{q}$ and a are the resultant velocity vector and speed of sound, respectively.

The node-centered control volume approach as demonstrated for two dimensions may also be used for three dimensions. We discuss this subject for the FDV equations in the following section.

7.4 FVM-FDV FORMULATION

The FDV concept introduced in Section 6.5 can be used for the FVM formulation. To this end, we begin with the FDV governing equations given by (6.5.14)

$$\mathbf{R} = \left(\mathbf{I} + \mathbf{E}_i^n\frac{\partial}{\partial x_i} + \mathbf{E}_{ij}^n\frac{\partial^2}{\partial x_i\partial x_j}\right)\Delta\mathbf{U}^{n+1} + \mathbf{Q}^n \tag{7.4.1}$$

The FVM integration equation is of the form

$$\int_\Omega \mathbf{R}\,d\Omega = \int_\Omega\left[\left(\mathbf{I} + \mathbf{E}_i^n\frac{\partial}{\partial x_i} + \mathbf{E}_{ij}^n\frac{\partial^2}{\partial x_i\partial x_j}\right)\Delta\mathbf{U}^{n+1} + \mathbf{Q}^n\right]d\Omega = 0 \tag{7.4.2}$$

Integrating (7.4.2) with respect to the spatial coordinates, we obtain

$$\int_\Omega \Delta\mathbf{U}^{n+1}\,d\Omega + \int_\Gamma\left(\mathbf{E}_i\Delta\mathbf{U}^{n+1} + \mathbf{E}_{ij}\Delta\mathbf{U}_{,j}^{n+1}\right)n_i\,d\Gamma = -\int_\Omega \mathbf{Q}^n\,d\Omega \tag{7.4.3}$$

or

$$\sum_{CV}\Delta\mathbf{U}^{n+1}\,\Delta\Omega + \sum_{CS}\left(\mathbf{E}_i\Delta\mathbf{U}^{n+1} + \mathbf{E}_{ij}\Delta\mathbf{U}_{,j}^{n+1}\right)n_i\,\Delta\Gamma = -\int_\Omega \mathbf{Q}^n\,d\Omega \tag{7.4.4}$$

where

$$\int_\Omega \mathbf{Q}^n\,d\Omega = \int_\Gamma\left(\mathbf{H}_i^n + \mathbf{H}_{ij,j}^n\right)n_i\,d\Gamma = \sum_{CS}\left(\mathbf{H}_i^n + \mathbf{H}_{ij,j}^n\right)n_i\Delta\Gamma \tag{7.4.5}$$

with

$$\mathbf{H}_i^n = \Delta t\left(\mathbf{F}_i^n + \mathbf{G}_i^n\right),\qquad \mathbf{H}_{ij}^n = \frac{\Delta t^2}{2}(\mathbf{a}_i + \mathbf{b}_i)\left(\mathbf{F}_j^n + \mathbf{G}_j^n\right) \tag{7.4.6a,b}$$

Let us now illustrate the solution procedure (7.4.4) based on the node-centered control volume as shown in Figure 7.2.1 and Example 7.2.1. The control surface computations on the left-hand side of (7.4.4) include terms with $\mathbf{E}_i$ without derivative and those with $\mathbf{E}_{ij}$ with the first order derivatives and similarly for $\mathbf{H}_i^n$ and $\mathbf{H}_{ij}^n$ on the right-hand side of (7.4.4). Thus, the FVM equation at node 1 for Figure 7.2.1 becomes (with $\Delta \mathbf{U}^{n+1} = \boldsymbol{\Psi}$)

$$\begin{aligned}
&\boldsymbol{\Psi}_1 \Delta\Omega_1 + \frac{1}{2}(\boldsymbol{\Psi}_7 + \boldsymbol{\Psi}_1) R_{7,1} + (\boldsymbol{\Psi}_7 - \boldsymbol{\Psi}_1) S_{7,1} + \frac{1}{2}(\boldsymbol{\Psi}_9 + \boldsymbol{\Psi}_1) R_{9,1} + (\boldsymbol{\Psi}_9 - \boldsymbol{\Psi}_1) S_{9,1} \\
&+ \frac{1}{2}(\boldsymbol{\Psi}_{11} + \boldsymbol{\Psi}_1) R_{11,1} + (\boldsymbol{\Psi}_{11} - \boldsymbol{\Psi}_1) S_{11,1} + \frac{1}{2}(\boldsymbol{\Psi}_2 + \boldsymbol{\Psi}_1) R_{2,1} + (\boldsymbol{\Psi}_2 - \boldsymbol{\Psi}_1) S_{2,1} \\
&+ \frac{1}{2}(\boldsymbol{\Psi}_5 + \boldsymbol{\Psi}_1) R_{5,1} + (\boldsymbol{\Psi}_5 - \boldsymbol{\Psi}_1) S_{5,1} = -\mathbf{Q}_1
\end{aligned} \tag{7.4.7}$$

where

$$\mathbf{R}_{7,1} = \left\lfloor (\mathbf{E}_1 \mathbf{n}_1 \Delta\Gamma + \mathbf{E}_2 \mathbf{n}_2 \Delta\Gamma)^{(a)} + (\mathbf{E}_1 \mathbf{n}_1 \Delta\Gamma + \mathbf{E}_2 \mathbf{n}_2 \Delta\Gamma)^{(b)} \right\rfloor_{7,1} \tag{7.4.8}$$

$$\mathbf{S}_{7,1} = \left\{ \begin{aligned} &\left[(\mathbf{E}_{11}\mathbf{n}_1 + \mathbf{E}_{21}\mathbf{n}_2) \frac{\Delta\Gamma}{\Delta \mathbf{x}} + (\mathbf{E}_{12}\mathbf{n}_1 + \mathbf{E}_{22}\mathbf{n}_2) \frac{\Delta\Gamma}{\Delta \mathbf{y}} \right]^{(a)} \\ &+ \left[(\mathbf{E}_{11}\mathbf{n}_1 + \mathbf{E}_{21}\mathbf{n}_2) \frac{\Delta\Gamma}{\Delta \mathbf{x}} + (\mathbf{E}_{12}\mathbf{n}_1 + \mathbf{E}_{22}\mathbf{n}_2) \frac{\Delta\Gamma}{\Delta \mathbf{y}} \right]^{(b)} \end{aligned} \right\}_{7,1} \tag{7.4.9}$$

with $\mathbf{E}_i$ and $\mathbf{E}_{ij}$ given by (6.3.31a) and (6.3.31b), respectively, etc., and $\frac{1}{\Delta x}$ and $\frac{1}{\Delta y}$ calculated similarly as in (7.2.5).

The right-hand side terms of $\mathbf{H}$ are obtained in a manner similar to the left-hand side.

$$\mathbf{Q}_1 = \overline{\mathbf{R}}_{7,1}^n + \overline{\mathbf{S}}_{7,1}^n + \overline{\mathbf{R}}_{9,1}^n + \overline{\mathbf{S}}_{9,1}^n + \overline{\mathbf{R}}_{11,1}^n + \overline{\mathbf{S}}_{11,1}^n + \overline{\mathbf{R}}_{2,1}^n + \overline{\mathbf{S}}_{2,1}^n + \overline{\mathbf{R}}_{5,1}^n + \overline{\mathbf{S}}_{5,1}^n \tag{7.4.10}$$

$$\begin{aligned}
\overline{\mathbf{R}}_{7,1}^n = &\frac{1}{2}\{[(\mathbf{H}_1\mathbf{n}_1 + \mathbf{H}_2\mathbf{n}_2)\Delta\Gamma]_7 + [(\mathbf{H}_1\mathbf{n}_1 + \mathbf{H}_2\mathbf{n}_2)\Delta\Gamma]_1\}^{(a)} \\
&+ \frac{1}{2}\{[(\mathbf{H}_1\mathbf{n}_1 + \mathbf{H}_2\mathbf{n}_2)\Delta\Gamma]_7 + [(\mathbf{H}_1\mathbf{n}_1 + \mathbf{H}_2\mathbf{n}_2)\Delta\Gamma]_1\}^{(b)}
\end{aligned}$$

$$\begin{aligned}
\overline{\mathbf{S}}_{7,1}^n = &\left\{ [(\mathbf{H}_{11})_7 - (\mathbf{H}_{11})_1]\mathbf{n}_1 \frac{\Delta\Gamma}{\Delta \mathbf{x}} + [(\mathbf{H}_{12})_7 - (\mathbf{H}_{12})_1]\mathbf{n}_1 \frac{\Delta\Gamma}{\Delta \mathbf{y}} \right. \\
&\left. + [(\mathbf{H}_{21})_7 - (\mathbf{H}_{21})_1]\mathbf{n}_2 \frac{\Delta\Gamma}{\Delta \mathbf{x}} + [(\mathbf{H}_{22})_7 - (\mathbf{H}_{22})_1]\mathbf{n}_2 \frac{\Delta\Gamma}{\Delta \mathbf{y}} \right\}_{7,1}^{(a)} \\
&+ \left\{ [(\mathbf{H}_{11})_7 - (\mathbf{H}_{11})_1]\mathbf{n}_1 \frac{\Delta\Gamma}{\Delta \mathbf{x}} + [(\mathbf{H}_{12})_7 - (\mathbf{H}_{12})_1]\mathbf{n}_1 \frac{\Delta\Gamma}{\Delta \mathbf{y}} \right. \\
&\left. + [(\mathbf{H}_{21})_7 - (\mathbf{H}_{21})_1]\mathbf{n}_2 \frac{\Delta\Gamma}{\Delta \mathbf{x}} + [(\mathbf{H}_{22})_7 - (\mathbf{H}_{22})_1]\mathbf{n}_2 \frac{\Delta\Gamma}{\Delta \mathbf{y}} \right\}_{7,1}^{(b)}
\end{aligned} \tag{7.4.11}$$

with $\mathbf{H}_i$ and $\mathbf{H}_{ij}$ given by (7.4.6a,b), respectively.

The FVM equation at node 2 is written similarly and the solution for $\mathbf{U}^{n+1}$ for nodes 1 and 2 can be obtained with appropriate boundary conditions applied similarly, as demonstrated in Examples 7.2.1 and 7.2.2. If all Dirichlet data are provided, then we have

$$\begin{bmatrix} \mathbf{K}_{11} & \mathbf{K}_{12} \\ \mathbf{K}_{21} & \mathbf{K}_{22} \end{bmatrix} \begin{bmatrix} \Delta\mathbf{U}_1 \\ \Delta\mathbf{U}_2 \end{bmatrix}^{n+1} = - \begin{bmatrix} \mathbf{Q}_1 \\ \mathbf{Q}_2 \end{bmatrix}^n + \begin{bmatrix} \mathbf{D}_1 \\ \mathbf{D}_2 \end{bmatrix} \tag{7.4.12}$$

where $\mathbf{D}_1$ and $\mathbf{D}_2$ represent the source vector as a result of the Dirichlet boundary conditions. Note that $\mathbf{K}_{11}$ and $\mathbf{K}_{22}$ denote the collective sum of contributions for nodes 1 and 2, respectively, whereas $\mathbf{K}_{12}$ and $\mathbf{K}_{21}$ are the interactions between node 1 and node 2, respectively,

$$\mathbf{K}_{12} = \frac{1}{2}\mathbf{R}_{2,1} - \mathbf{S}_{2,1}$$
$$\mathbf{K}_{21} = \frac{1}{2}\mathbf{R}_{1,2} - \mathbf{S}_{1,2} \tag{7.4.13a,b}$$

Implementation of Neumann boundary conditions is carried out similarly as in Example 7.2.2. If the Neumann boundary condition is prescribed at node 7, then the FDV equations (7.4.12) will be modified to include ΔU_7 as one of the unknowns with the Neumann data directly imposed on the right-hand side of (7.4.12).

For three-dimensional applications such as in Figure 7.3.1, FDV equations in terms of FVM are written similarly as in 2-D, following the procedure of (7.4.7) through (7.4.13). For example, at node 10 (Figure 7.3.1), the adjacent nodes connected to node 10 are as shown in Figure 7.3.2. Direction cosines of the normal vector are calculated (Figure 7.3.3), with control surface areas and control volumes determined as described in Section 7.3. Let us examine the FDV finite volume equations at node 10 (Figures 7.3.1 and 7.3.2).

$$\begin{aligned}
&\boldsymbol{\Psi}_{10}\Delta\Omega_{10} + \frac{1}{2}(\boldsymbol{\Psi}_6 + \boldsymbol{\Psi}_{10})R_{6,10} + (\boldsymbol{\Psi}_6 - \boldsymbol{\Psi}_{10})S_{6,10} + \frac{1}{2}(\boldsymbol{\Psi}_{14} + \boldsymbol{\Psi}_{10})R_{14,10}\\
&\quad + (\boldsymbol{\Psi}_{14} - \boldsymbol{\Psi}_{10})S_{14,10} + \frac{1}{2}(\boldsymbol{\Psi}_{22} + \boldsymbol{\Psi}_{10})R_{22,10} + (\boldsymbol{\Psi}_{22} - \boldsymbol{\Psi}_{10})S_{22,10}\\
&\quad + \frac{1}{2}(\boldsymbol{\Psi}_{14} + \boldsymbol{\Psi}_{10})R_{14,10} + (\boldsymbol{\Psi}_{14} - \boldsymbol{\Psi}_{10})S_{14,10} + \frac{1}{2}(\boldsymbol{\Psi}_9 + \boldsymbol{\Psi}_{10})R_{9,10}\\
&\quad + (\boldsymbol{\Psi}_9 - \boldsymbol{\Psi}_{10})S_{9,10} + \frac{1}{2}(\boldsymbol{\Psi}_{11} + \boldsymbol{\Psi}_{10})R_{11,10} + (\boldsymbol{\Psi}_{11} - \boldsymbol{\Psi}_{10})S_{11,10} = -\mathbf{Q}_{10}
\end{aligned} \tag{7.4.14}$$

with

$$R_{6,10} = \begin{bmatrix} (E_1 n_1 \Delta\Gamma + E_2 n_2 \Delta\Gamma + E_3 n_3 \Delta\Gamma)^{(a)} + (E_1 n_1 \Delta\Gamma + E_2 n_2 \Delta\Gamma + E_3 n_3 \Delta\Gamma)^{(b)} \\ + (E_1 n_1 \Delta\Gamma + E_2 n_2 \Delta\Gamma + E_3 n_3 \Delta\Gamma)^{(c)} + (E_1 n_1 \Delta\Gamma + E_2 n_2 \Delta\Gamma + E_3 n_3 \Delta\Gamma)^{(d)} \end{bmatrix}_{6,10} \tag{7.4.15}$$

$$S_{6,10} = \left\{ \begin{aligned}
&\left[(E_{11}n_1 + E_{21}n_2 + E_{31}n_3)\frac{\Delta\Gamma}{\Delta x} + (E_{12}n_1 + E_{22}n_2 + E_{32}n_3)\frac{\Delta\Gamma}{\Delta y} + (E_{13}n_1 + E_{23}n_2 + E_{33}n_3)\frac{\Delta\Gamma}{\Delta z}\right]^{(a)}\\
&+\left[(E_{11}n_1 + E_{21}n_2 + E_{31}n_3)\frac{\Delta\Gamma}{\Delta x} + (E_{12}n_1 + E_{22}n_2 + E_{32}n_3)\frac{\Delta\Gamma}{\Delta y} + (E_{13}n_1 + E_{23}n_2 + E_{33}n_3)\frac{\Delta\Gamma}{\Delta z}\right]^{(b)}\\
&+\left[(E_{11}n_1 + E_{21}n_2 + E_{31}n_3)\frac{\Delta\Gamma}{\Delta x} + (E_{12}n_1 + E_{22}n_2 + E_{32}n_3)\frac{\Delta\Gamma}{\Delta y} + (E_{13}n_1 + E_{23}n_2 + E_{33}n_3)\frac{\Delta\Gamma}{\Delta z}\right]^{(c)}\\
&+\left[(E_{11}n_1 + E_{21}n_2 + E_{31}n_3)\frac{\Delta\Gamma}{\Delta x} + (E_{12}n_1 + E_{22}n_2 + E_{32}n_3)\frac{\Delta\Gamma}{\Delta y} + (E_{13}n_1 + E_{23}n_2 + E_{33}n_3)\frac{\Delta\Gamma}{\Delta z}\right]^{(d)}
\end{aligned} \right\}_{6,10} \tag{7.4.16}$$

$$\left(\frac{1}{\Delta x}\right)_{6,10} = \left(\frac{\Delta y \Delta z}{\Delta\Omega}\right)_{6,10}, \quad \left(\frac{1}{\Delta y}\right)_{6,10} = \left(\frac{\Delta z \Delta x}{\Delta\Omega}\right)_{6,10}, \quad \left(\frac{1}{\Delta z}\right)_{6,10} = \left(\frac{\Delta x \Delta y}{\Delta\Omega}\right)_{6,10} \tag{7.4.17}$$

where $(\Delta x)_{6,10} = |x_6 - x_{10}|$, etc., $\Delta\Gamma$ being the subcontrol surface areas corresponding

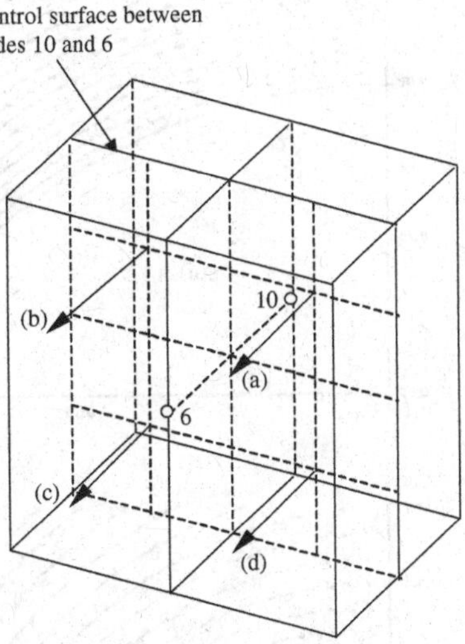

Figure 7.4.1 Control surface [normal vectors (a), (b), (c), (d)] between nodes 10 and 6 for the control volume containing node 10 of Figure 7.3.1.

to the normal vector components designated by (a), (b), (c), (d), and $(\Delta\Omega)_{6,10}$ represents the subcontrol volume for node 10 toward node 6 (Figure 7.3.2). As indicated in Section 7.2.1, if a cell is coincident with the x-, y-, or z-coordinate, then the left-hand side quantities in (7.4.17) must be used. However, the right-hand side quantities are used for the directions in which the coordinate components are zero to avoid singularities.

Similarly, we compute the right-hand side of (7.4.14) as in (7.4.10). The final form of the FDV/FVM equations is similar to (7.4.12) corresponding to the two interior nodes 10 and 11 (Figures 7.3.1 and 7.4.1).

It should be noted that the node-centered scheme described above is capable of accommodating any arbitrary unstructured grid system. Recall that in FDV equations, all physical aspects of the flow for all speed regimes have been accommodated as detailed in Section 6.5. Some applications of FDV methods via FVM/FEM are shown in Section 15.3.

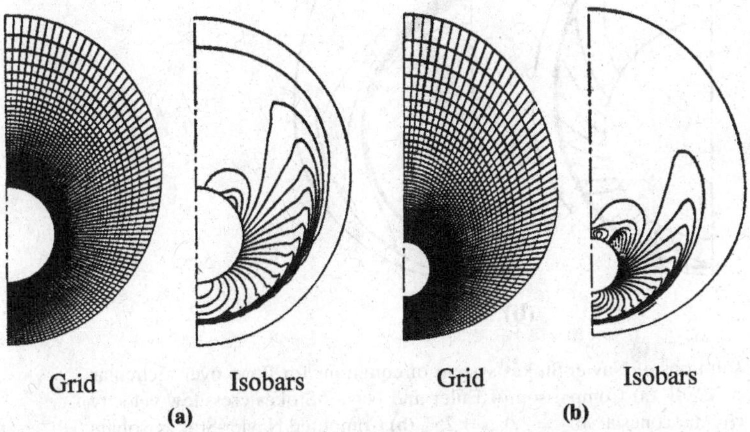

Figure 7.5.1 FVM/FDM solutions of Euler equations for flows over a circular cone [Siclari and Jameson, 1989]. **(a)** Euler grid and computed isobars for a 20° circular cone at $M_\infty = 2.0$, $\alpha = 25°$. **(b)** Euler grid and computed isobars for a 10° circular cone at $M_\infty = 2.0$, $\alpha = 25°$.

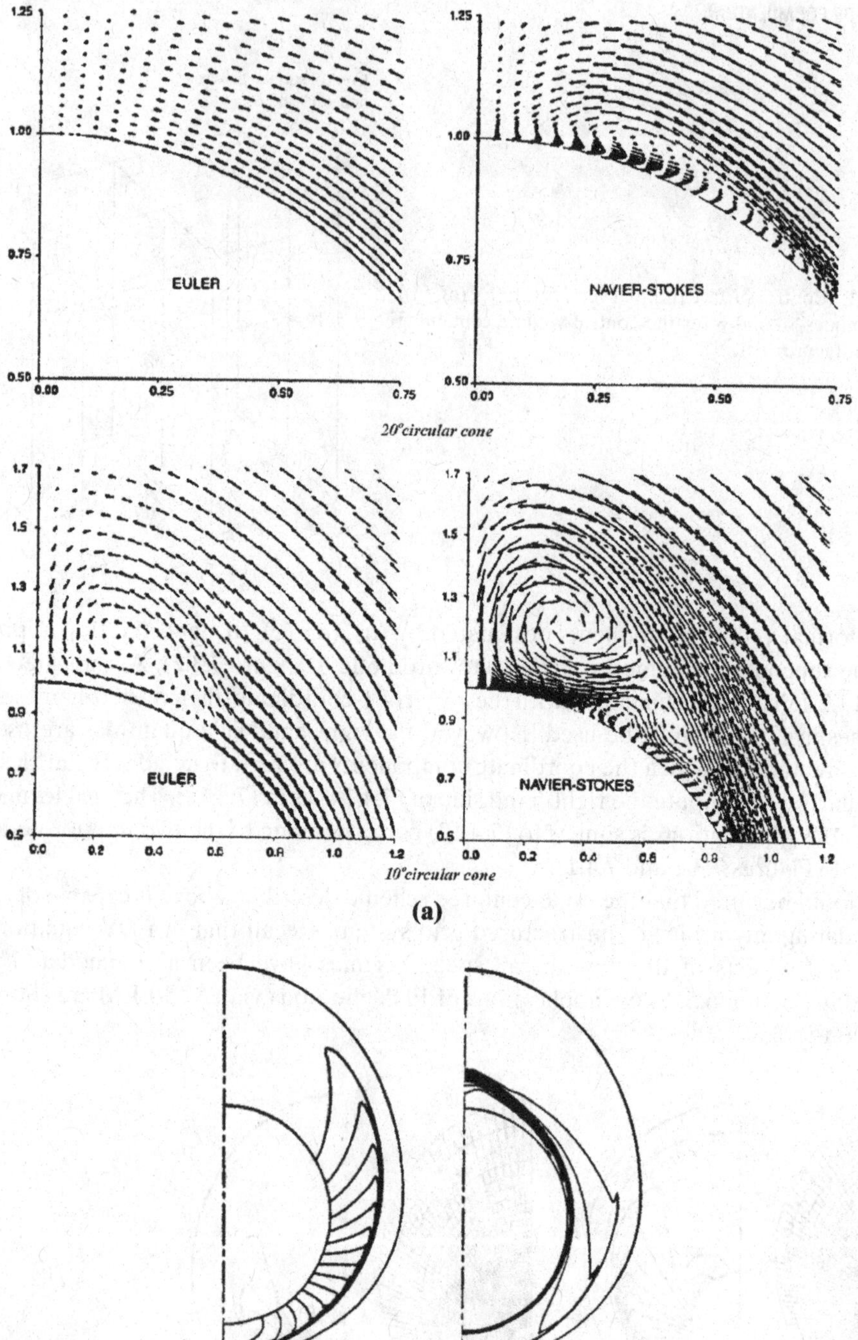

Figure 7.5.2 FVM/FDM solutions of Navier-Stokes system of equations for flows over a circular cone [Siclari and Jameson, 1989]. **(a)** Comparison of Euler and Navier-Stokes crossflow velocity vectors for a 10° and 20° circular cones at $M_\infty = 2.0$, $\alpha = 25°$. **(b)** Computed Navier-Stokes isobars and entropy contours for a 10° circular cone at $M_\infty = 7.95$, $\alpha = 12°$, $Re = 3.6 \times 10^6$.

7.5 EXAMPLE PROBLEMS

(1) Solution of Euler Equation Using FVM/FDM

The work presented here is reported by Siclari and Jameson [1989] on a node centered, finite volume, central difference scheme to solve the Euler equations. High-speed flows over a circular cone using spherical coordinates are investigated with FVM/FDM. To expedite the solution convergence, they used multigrid methods, which will be discussed in Section 20.2.

Figure 7.5.1a shows the (81×50) grid for a 20° circular cone with $M_\infty = 2.0$ and the cone angle of $\alpha = 25°$. The resulting isobar solution shows that a weak crossflow shock occurs on the lee side of the cone with attached flow.

The geometry and discretization (81×50) for a 10° circular cone $(M_\infty = 2.0, \alpha = 25°)$ and the resulting isobar distributions are shown in Figure 7.5.1b. In this example, a strong crossflow shock develops on the lee side, resulting in shock-induced separation.

(2) Solution of Navier-Stokes System of Equations Using FVM/FDM

Siclari and Jameson [1989] solved the same problem above for the case of viscous flows. This requires additional attention, providing refined discretization, and higher order artificial dissipation as discussed in Section 6.3.

With the grid (81×68), the computed results are displayed in Figure 7.5.2a, compared with the case of inviscid flow. For the 20° cone, the Euler solution shows attached flow, whereas the Navier-Stokes solution shows a small separation. The Euler solution for the 10° cone shows a shock vorticity induced separation. The Navier-Stokes solution shows a more complex separated flow pattern including primary, secondary, and tertiary vortices.

Figure 7.5.2b shows the computed isobars and entropy contours. The leeside boundary layer separates at this incidence as indicated by the entropy contours.

7.6 SUMMARY

In this chapter, it has been shown that any finite difference schemes can be implemented in FVM with either structured or unstructured grids. There are two advantageous features in FVM: (1) Physically, the conservation of mass, momentum, and energy is assured in the formulation itself; and (2) Numerically, unstructured grids and arbitrary geometries are accommodated without coordinate transformation.

The conclusion appears to be that FVM is preferred to FDM for arbitrary geometries. For structured grids, however, such conclusion is premature. Personal preferences may persist for many years to come. The final outcome may be determined by convenience in applications associated with computing techniques from the viewpoints of data structure managements, which will be discussed in Chapter 20.

REFERENCES

Ni, R. N. [1982]. A multiple grid scheme for solving the Euler equations. *AIAA J.*, 20, 11, 1565–71.

Rizzi, A. W. and Inouye, M. [1973]. Time split finite volume method for three-dimensional blunt-body flows. *AIAA J.*, 11, 11, 1478–85.

Siclari, M. J. and Jameson, A. [1989]. A multigrid finite volume method for solving the Euler and Navier-Stokes equations for high speed flows. AIAA paper, AIAA-89-0283.

PART THREE

FINITE ELEMENT METHODS

Finite element methods (FEM) and topics related to finite element applications are presented in Part Three. We have seen in Chapter 1 that the finite element methods based on the standard Galerkin integral lead to results identical to those of the finite difference methods (FDM) for the examples of simple linear problems. In dealing with nonlinear or convection-dominated flow problems in fluid dynamics, however, the standard Galerkin methods are no longer adequate. Various special strategies must be designed to assure stability and convergence, as we noted also in FDM. Dissipation and dispersion errors can be minimized with a high level of accuracy achieved in much the same way as in FDM. In this vein, the reader will see that finite element methods are analogous to finite difference methods in dealing with all aspects of the physics of fluids. Developments of both approaches in close alliance are shown to be complementary to each other. It is with this expectation that our journey begins.

CHAPTER EIGHT

Introduction to Finite Element Methods

8.1 GENERAL

The finite element theory as applied to one-dimensional problems was discussed in Part One, Preliminaries. In general, finite element methods (FEM) are versatile in applications to multidimensional complex irregular geometries. Initial applications of FEM began with structural analysis in the late 1950s and primarily were based on variational principles. During the early days of the development of FEM, applications were made for simple flow problems, beginning with Zienkiewicz and Cheung [1965], followed by Oden and Wellford [1972], Chung [1978], and Baker [1983], among others. Significant contributions in CFD began with the streamline upwind Petrov-Galerkin (SUPG) methods [Heinrich, Huyakorn, Zienkiewicz, and Mitchell, 1977; Hughes and Brooks, 1982; Hughes, Mallet, and Mizukami, 1986] or streamline diffusion methods (SDM) [Johnson, 1987], Taylor-Galerkin methods (TGM) [Donea, 1984; Löhner, Morgan, and Zienkiewicz, 1985], and *hp* adaptive methods [Oden and Demkowicz, 1991], among many other related works.

New approaches and various alternative methodologies are preponderant in the literature. Efforts are made in this book to simplify and unify some of the terminologies. For example, the original approaches of SUPG or SDM for convection-dominated flows have grown into GLS (Galerkin/least squares) when some changes in the formulation are introduced. It is suggested that all methods related to numerical diffusion test functions be called the generalized Petrov-Galerkin (GPG) methods. Hughes and his co-workers have contributed significantly in the past two decades to the GPG methodologies associated with the problems of convection-dominated flows and shock discontinuities.

Another example is the algorithm arising from the Taylor series expansion such as TGM. Zienkiewicz and his co-workers [Zienkiewicz and Codina, 1995] have applied for the past decade the concept of characteristic Galerkin methods (CGM) which produce results similar to TGM in dealing with convection-dominated problems for both compressible and incompressible flows.

The idea of treating discontinuities developed in the finite difference methods (FDM) flux vector splitting, TVD, and ENO associated with the first and second order upwinding can be utilized in the discontinuous Galerkin methods (DGM) as demonstrated by Oden and his co-workers [Oden, Babuska, and Baumann, 1998]. Clearly, this represents the merit of studying FDM and FEM closely together.

Recall that in FDM we explored solutions for all-speed flows. Among them was the concept of flowfield-dependent variation (FDV) methods [Chung, 1999] as detailed in Section 6.5. This was an attempt to resolve transitions and interactions of various physical properties such as inviscid/viscous, compressible/incompressible, and laminar/turbulent flows. The same approach can be applied to FEM. It can be shown that FDV methods are capable of generating most of the existing computational schemes in both FDM and FEM.

Although the various forms of Galerkin methods constitute the finite element methods in which the test functions are the same as the trial functions, there are other methods where the test functions are different from the trial functions, generally known as the weighted residual methods. Some examples include spectral element methods (SEM), least square methods (LSM), and finite point methods (FPM).

The finite element literature is enriched with mathematical error analysis. Mathematical proofs of convergence, stability, and accuracy are important in the so-called *hp* adaptive methods in which accuracy improves as the mesh is refined and the approximating polynomial degrees are increased in accordance with the flowfield gradients. This subject was developed by Babuska and his co-workers and Oden and his co-workers for the last two decades.

In this chapter, the FEM formulation presented in Chapter 1 will be repeated with more rigorous mathematical notations and expanded into multidimensional problems. Definitions used in error estimates and convergence properties are also introduced in this chapter.

The finite element analysis begins with the interpolation functions of the variables for one-dimensional, two-dimensional, and three-dimensional elements of various geometries with linear and high order approximations, presented in Chapter 9. This will be followed by linear steady and unsteady problems in Chapter 10 and nonlinear problems with convection-dominated flows in Chapter 11.

In Chapters 12 and 13, we present FEM formulations for incompressible flows and compressible flows, respectively. The major issues in CFD as observed in Part Two for FDM are as follows: (1) Difficulties of satisfying the conservation of mass in incompressible flows (incompressibility condition), resulting in checkerboard type pressure oscillations; (2) shock discontinuities in compressible flows; and (3) convection-dominated flows in both incompressible and compressible flows. Mixed methods, penalty methods, and pressure correction methods were developed to cope with the incompressibility condition. On the other hand, the Taylor-Galerkin methods (TGM) and generalized Petrov-Galerkin (GPG) methods have been successful in dealing with shock discontinuities and convection-dominated flows. Recent developments include computational methods capable of analyzing both compressible and incompressible flows by a single formulation and a single computer code using the various schemes extended from TGM, GPG, and FDV (Chapter 13), leading to "all speed flows."

Weighted residual methods including spectral element methods (SEM) and least square methods (LSM) are presented in Chapter 14. Finite point methods (FPM) using only the nodal points without element meshes (meshless methods) are also discussed in this chapter. The finite volume methods (FVM) via FEM are elaborated in Chapter 15.

Finally, in Chapter 16, we examine some of the significant analogies between FDM and FEM. Most of the existing computational schemes in both FDM and FEM are shown to be special cases of the flowfield-dependent variation (FDV) methods. There are many numerical methods other than the FDM, FEM, and FVM which are based on the standard Eulerian coordinates. They include boundary element methods (BEM), coupled-Eulerian-Lagrangian (CEL) methods, particle-in-cell (PIC) methods, and Monte Carlo methods (MCM). For the sake of completeness, these methods are briefly discussed in Section 16.4.

8.2 FINITE ELEMENT FORMULATIONS

The basic concept of finite element formulations was presented in Chapter 1, for simple one-dimensional problems, using the Galerkin methods. In the Galerkin methods, the variable of the partial differential equation is approximated as a linear combination of the trial (interpolation, shape, or basis) functions. It was shown that local properties were assembled into a global form by superposition intuitively. In this chapter, we demonstrate this process directly from the global form using Boolean algebra, with the local properties then arising indirectly as a consequence.

For simplicity of illustration, let us consider a one-dimensional domain as depicted in Figure 8.2.1. Let the domain be divided into subdomains; say two local elements in this example. The end points of elements are called nodes. The finite element model $\overline{\Omega}$

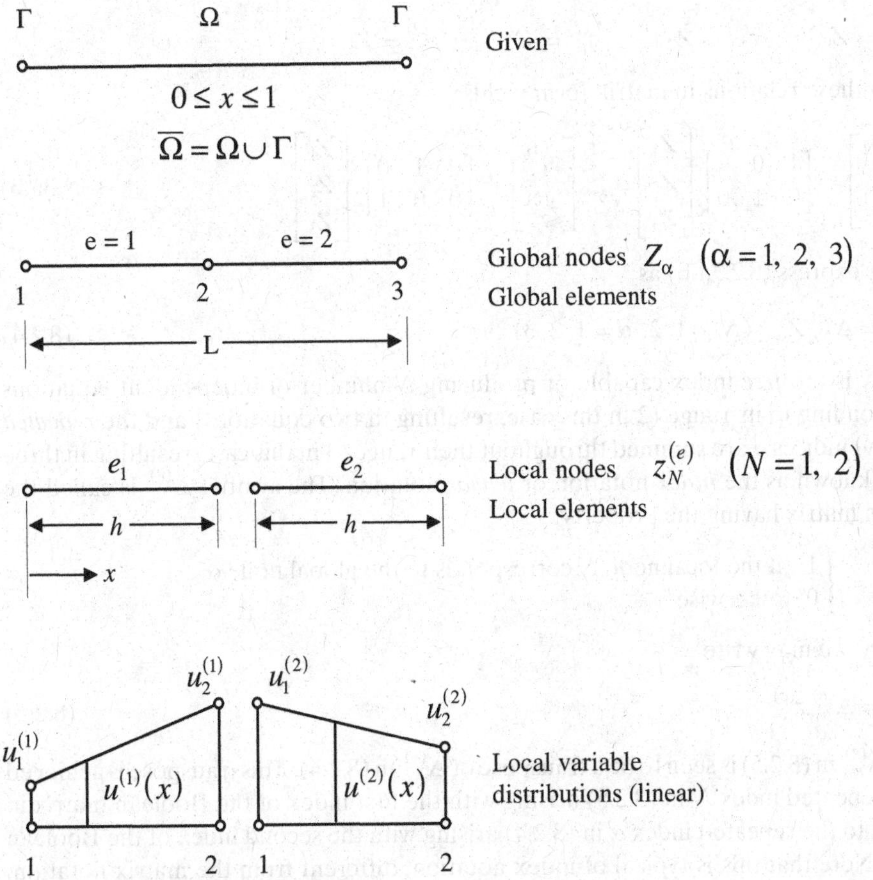

Figure 8.2.1 Finite element approximations.

is expressed as the union of the domain Ω and its boundaries Γ,

$$\overline{\Omega} = \Omega \bigcup \Gamma \tag{8.2.1}$$

We now isolate all elements from the global domain. Each local element $\overline{\Omega}_e$ is identified as

$$\overline{\Omega}_e = \Omega_e \bigcup \Gamma_e$$

The boundaries of this element and the neighboring element are the intersection

$$\Gamma_e \bigcap \Gamma_f = \Gamma_{ef}$$

Thus, the connected finite element model (8.2.1) is the union of all elements

$$\overline{\Omega} = \bigcup_{e=1}^{E} \overline{\Omega}_e \tag{8.2.2}$$

where E is the total number of elements. The global nodes of the connected model $\overline{\Omega}$ and the local nodes of isolated elements are identified by Z_α ($\alpha = 1, 2, 3$, being the number of global nodes) and $z_N^{(e)}$ ($N = 1, 2$, being the number of local nodes) with $e = 1, 2$, being the number of local elements, respectively. They are related as follows:

$$z_1^{(1)} = Z_1, \qquad z_2^{(1)} = Z_2, \qquad z_1^{(2)} = Z_2, \qquad z_2^{(2)} = Z_3,$$

Writing these relations in matrix form yields

$$\begin{bmatrix} z_1^{(1)} \\ z_2^{(1)} \end{bmatrix} = \begin{bmatrix} 1 & 0 & 0 \\ 0 & 1 & 0 \end{bmatrix} \begin{bmatrix} Z_1 \\ Z_2 \\ Z_3 \end{bmatrix}, \quad \begin{bmatrix} z_1^{(2)} \\ z_2^{(2)} \end{bmatrix} = \begin{bmatrix} 0 & 1 & 0 \\ 0 & 0 & 1 \end{bmatrix} \begin{bmatrix} Z_1 \\ Z_2 \\ Z_3 \end{bmatrix} \tag{8.2.3a,b}$$

We may express (8.2.3a,b) as

$$z_N^{(e)} = \Delta_{N\alpha}^{(e)} Z_\alpha \quad (N = 1, 2, \ \alpha = 1, 2, 3) \tag{8.2.4}$$

where N is the *free* index capable of producing N number of independent equations corresponding to its range (2 in this case, resulting in two equations) and the *repeated (dummy)* indices α are summed throughout their range (3 in this case, resulting in three terms), known as the *index* notation or *tensor* notation. The symbol $\Delta_{N\alpha}^{(e)}$ is called the Boolean matrix having the property:

$$\Delta_{N\alpha}^{(e)} = \begin{cases} 1 & \text{if the local node } N \text{ corresponds to the global node } \alpha \\ 0 & \text{otherwise} \end{cases}$$

Similarly, we may write

$$Z_\alpha = \Delta_{N\alpha}^{(e)} z_N^{(e)} \tag{8.2.5}$$

where $\Delta_{N\alpha}^{(e)}$ in (8.2.5) is seen to be a transpose of $\Delta_{N\alpha}^{(e)}$ in (8.2.4). This transpose is achieved by the repeated index N in (8.2.5) arising with the first index of the Boolean matrix in contrast to the repeated index α in (8.2.4) arising with the second index of the Boolean matrix. Note that this is typical of index notation, different from the matrix notation.

Inserting (8.2.4) into (8.2.5) yields

$$Z_\alpha = \Delta^{(e)}_{N\alpha}\Delta^{(e)}_{N\beta} Z_\beta \tag{8.2.6}$$

from which we obtain the relation

$$\Delta^{(e)}_{N\alpha}\Delta^{(e)}_{N\beta} = \delta_{\alpha\beta} \tag{8.2.7}$$

where $\delta_{\alpha\beta}$ is the Kronecker delta,

$$\delta_{\alpha\beta} = \begin{cases} 1 & \text{if } \alpha = \beta \\ 0 & \text{if } \alpha \neq \beta \end{cases}$$

Likewise, substituting (8.2.5) into (8.2.4) gives

$$z^{(e)}_N = \Delta^{(e)}_{N\alpha}\Delta^{(e)}_{M\alpha} z^{(e)}_M \tag{8.2.8}$$

Once again, we obtain

$$\Delta^{(e)}_{N\alpha}\Delta^{(e)}_{M\alpha} = \delta_{NM} \tag{8.2.9}$$

In matrix notation, the above relation shows that

$$\begin{bmatrix} 1 & 0 & 0 \\ 0 & 1 & 0 \end{bmatrix}\begin{bmatrix} 1 & 0 \\ 0 & 1 \\ 0 & 0 \end{bmatrix} = \begin{bmatrix} 1 & 0 \\ 0 & 1 \end{bmatrix} \tag{8.2.10}$$

The use of Boolean matrix $\Delta^{(e)}_{N\alpha}$ will prove to be convenient in derivations of finite element equations, relating the properties between the local and global systems. However, in actual executions of finite element computations, these Boolean matrices will never be constructed but instead are replaced by computer programs based on local and global node number correspondence.

It should be noted, at this point, that we make use of tensor notation in which a free single index implies the components of a column vector whereas free double indices denote a matrix with its size determined by the ranges of the indices. The free index must match at both sides of the equality sign within an equation. The advantage of using tensor notation in FEM will become obvious as we develop finite element equations more extensively in later chapters.

To obtain the finite element equations, the concept of classical variational or weighted residual methods is used. Toward this end, we require suitable functions for the variable to be approximated *locally* within an element or subdomain. This is in contrast to the classical variational methods or weighted residual methods where the *global* approximating functions are used, in which the satisfaction of boundary conditions is difficult, if not impossible, for complex geometries.

Suppose that the variable u may be approximated linearly within a local element e, $(0 \le x \le h)$, as shown in Figure 1.3.1, Figure 8.2.1, Figure 8.2.2:

$$u^{(e)}(x) = \Phi^{(e)}_N(x) u^{(e)}_N \tag{8.2.11}$$

where $\Phi^{(e)}_N(x)$ are called the local element trial functions [interpolation functions, shape functions, or basis functions as shown in (1.3.3)]. For simplicity, the argument (x) will be omitted in what follows unless confusion is likely to occur. They have the

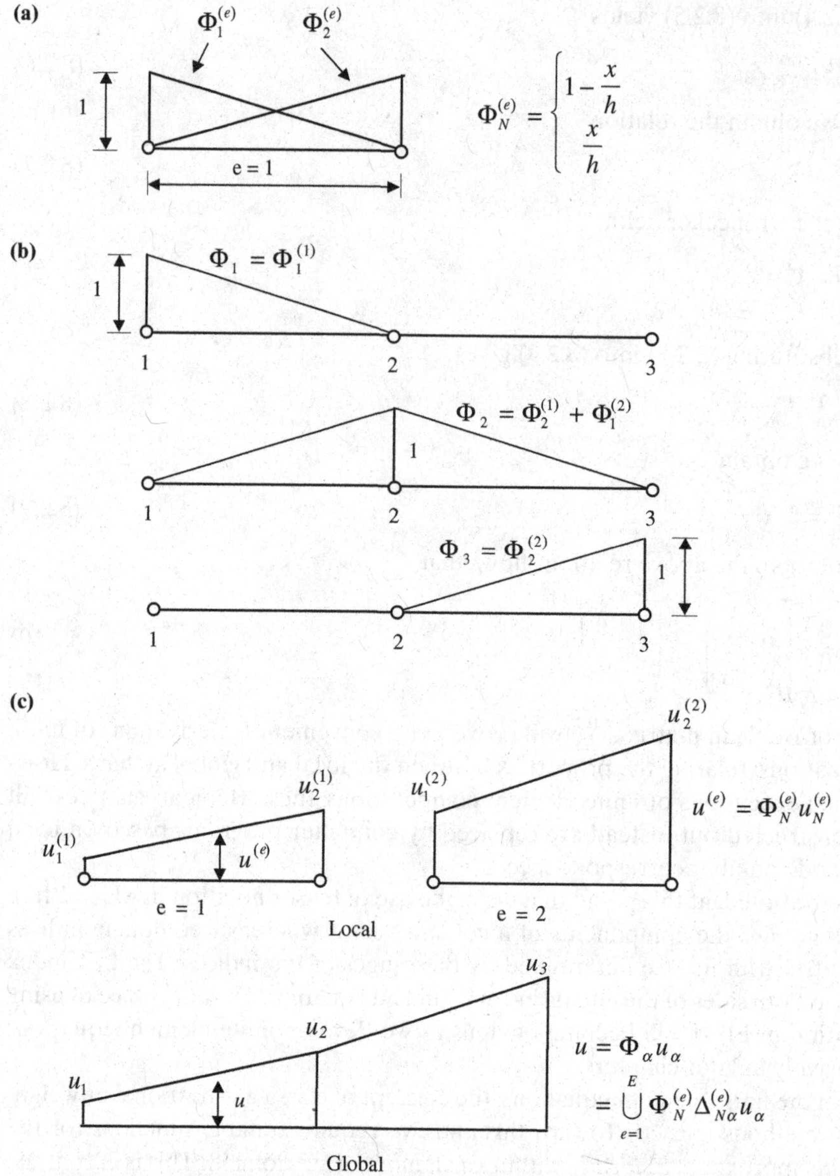

Figure 8.2.2 Local and global interpolation (trial) functions. **(a)** Local interpolation functions. **(b)** Global interpolation functions $\phi_\alpha = \bigcup_{e=1}^{E} \phi_N^{(e)} \Delta_{N\alpha}^{(e)}$. **(c)** Local and global values $(u_N^{(e)}, u_\alpha)$.

properties

$$0 \leq \Phi_N^{(e)} \leq 1, \qquad \sum_{N=1}^{2} \Phi_N^{(e)} = 1, \qquad \Phi_N^{(e)}\left(z_M^{(e)}\right) = \delta_{NM} \tag{8.2.12}$$

The local nodal values can be related to the global nodal values in a manner similar to (8.2.4):

$$u_N^{(e)} = \Delta_{N\alpha}^{(e)} u_\alpha \tag{8.2.13}$$

Thus, for the total number of elements, E, the global function can be written as the union of all local element contributions:

$$u = \bigcup_{e=1}^{E} u^{(e)} = \bigcup_{e=1}^{E} \Phi_N^{(e)} u_N^{(e)} = \bigcup_{e=1}^{E} \Phi_N^{(e)} \Delta_{N\alpha}^{(e)} u_\alpha \tag{8.2.14a}$$

or

$$u = \Phi_\alpha u_\alpha \tag{8.2.14b}$$

where Φ_α is called the global trial (interpolation, shape, or basis) function,

$$\Phi_\alpha = \bigcup_{e=1}^{E} \Phi_N^{(e)} \Delta_{N\alpha}^{(e)} \tag{8.2.15}$$

with

$$\Phi_\alpha(Z_\beta) = \delta_{\alpha\beta} \tag{8.2.16}$$

It follows from (8.2.15) that the expanded form of (8.2.14) appears as shown in Figure 8.2.2b,c. Note that the union operation in (8.2.14) and (8.2.15) is subject to the constraint (8.2.16). Thus, (8.2.14) through (8.2.16) lead to $u = u_1$ at node 1, $u = u_2$ at node 2, and $u = u_3$ at node 3. The union operation implies a Boolean summing rather than algebraic summing in this process.

With these preliminaries, we are now prepared to revisit the differential equation (1.2.1) for a more formal approach to the finite element solution process. There are two options for the formulation of finite element equations: (a) variational methods and (b) weighted residual methods. In the variational methods, we minimize the variational principle for the governing differential equation, which is a common practice in structural mechanics. Unfortunately, however, variational principles are not available in exact forms for nonlinear fluid mechanics equations in general. Thus, it is logical to seek the weighted residual methods in fluid mechanics where the variational principles are not required. The basic idea of the weighted residual methods is to construct a mathematical process in which the error or the residual of the governing differential equation(s), R (for example, $R = \nabla^2 u$), is minimized to zero. This can be done by forming a subspace spanned by test functions or weighting functions, W_α, and projecting the residual R orthogonally onto this subspace. This process is known as the *inner product* of the test function and the residual, which can be expressed as follows:

$$(W_\alpha, R) = \int_0^1 W_\alpha R dx = 0, \quad 0 < x < 1 \tag{8.2.17}$$

where the test functions W_α are known also as weighting functions. The integral given by (8.2.17) implies that the error at each point in the domain orthogonally projected onto a functional space spanned by the weighting function summed over the entire domain is set equal to zero. This process will provide necessary algebraic equations from which unknowns can be calculated. Thus, the finite element method is sometimes called the projection method.

If the test functions W_α are replaced by the trial functions Φ_α, then the scheme is known as the Galerkin method,

$$(\Phi_\alpha, R) = \int_0^1 \Phi_\alpha R dx = 0 \quad 0 < x < 1 \tag{8.2.18}$$

Note that Φ_α act as trial functions in (8.2.14) but are treated as test functions in (8.2.18). Formulations with test functions different from trial functions such as in the generalized Petrov-Galerkin methods will be discussed in Chapter 11 for nonlinear or convection-dominated flows.

For the purpose of illustration, let the residual be given by the differential equation (1.2.1a). We then obtain the so-called global Galerkin integral,

$$(\Phi_\alpha, R) = \int_0^1 \Phi_\alpha\left(\frac{d^2u}{dx^2} - 2\right)dx = 0 \tag{8.2.19}$$

This is in contrast to the local Galerkin integral (1.3.4). Integrate (8.2.19) by parts to arrive at the form known as the *variational equation*,

$$\overset{*}{\Phi}_\alpha \frac{du}{dx}\bigg|_0^1 - \int_0^1 \frac{d\Phi_\alpha}{dx}\frac{du}{dx}dx - \int_0^1 2\Phi_\alpha dx = 0 \tag{8.2.20}$$

where $du/dx|_0^1$ is the global Neumann boundary condition to be specified either at $x = 0$ or at $x = 1$ if required. Note also that $\overset{*}{\Phi}_\alpha$ indicates the global boundary test function defined only at $x = 0$ or $x = 1$, which is no longer a continuous function of x [Chung, 1978]. This is because the role of Φ_α is no longer the same at boundaries as in the domain. It is important to realize that $du/dx|_0^1$ in (8.2.20) arises from the one-dimensional assumption of the two-dimensional problem (Figure 8.2.3),

$$\iint \frac{d^2u}{dx^2}dxdy = \int \frac{du}{dx}dy = \int \frac{du}{dx}\cos\theta d\Gamma = \frac{du}{dx}\cos\theta = \frac{du}{dx}\bigg|_0^1$$

where the integral $\int_\Gamma d\Gamma$ is unity in one dimension, and

$$\frac{du}{dx}\bigg|_{x=1} = \frac{du}{dx}\cos 0^\circ = \frac{du}{dx}(1) \tag{8.2.21a}$$

$$\frac{du}{dx}\bigg|_{x=0} = \frac{du}{dx}\cos 180^\circ = \frac{du}{dx}(-1) \tag{8.2.21b}$$

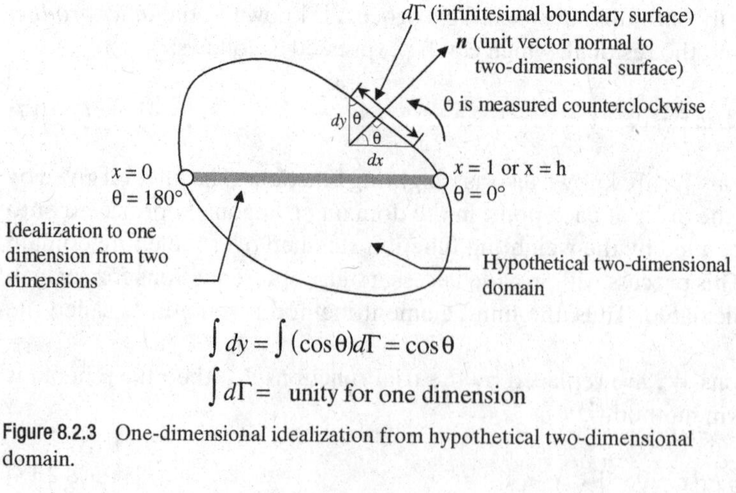

Figure 8.2.3 One-dimensional idealization from hypothetical two-dimensional domain.

The above result is due to the one-dimensional idealization of two-dimensional problems presented in Chapter 10. An important application of the above development is demonstrated for implementation of Neumann boundary conditions at the right or left end nodes in Section 1.5. Substituting (8.2.14b) into (8.2.20) gives

$$\left[\int_0^1 \frac{d\Phi_\alpha}{dx}\frac{d\Phi_\beta}{dx}dx\right]u_\beta = -\int_0^1 2\Phi_\alpha dx + \overset{*}{\Phi}_\alpha \frac{du}{dx}\bigg|_0^1 \tag{8.2.22}$$

We recognize that the left-hand side integral of (8.2.22) represents the first order derivative, known as a *weak form*, reduced (weakened) from the original second order derivative of the governing equation. The solution obtained from this weak form is known as the *weak solution*.

At this time, it is informative to point out that the result similar to (8.2.22) can be obtained using the variational principle approach [Chung, 1978]. The variational principle for the governing differential equation (1.2.1a) is of the form

$$I = \int_0^1 \left[\frac{1}{2}\left(\frac{\partial u}{\partial x}\right)^2 + 2u\right]dx \tag{8.2.23}$$

In the variational methods, the above integral is minimized with respect to the nodal value of the variable.

$$\delta I = \frac{\partial I}{\partial u_\alpha}\delta u_\alpha = 0$$

Since δu_α is arbitrary, we require

$$\frac{\partial I}{\partial u_\alpha} = 0$$

It can easily be verified that the minimization (differentiation) of (8.2.23) with respect to the nodal values of u_α as indicated above results in (8.2.22) except that the Neumann boundary condition must be manually added. This analogy does not exist in nonlinear fluid mechanics equations, because the integration of the nonlinear convection term by parts can not be carried out in an exact form.

With compact notation, we rewrite (8.2.22) in the form

$$K_{\alpha\beta}u_\beta = F_\alpha + G_\alpha \tag{8.2.24}$$

where $K_{\alpha\beta}$ is the global stiffness, diffusion, or viscosity matrix, F_α is the global load or source vector, and G_α is the global Neumann boundary vector, as deduced from (8.2.22):

$$K_{\alpha\beta} = \int_0^1 \frac{d\Phi_\alpha}{dx}\frac{d\Phi_\beta}{dx}dx = \bigcup_{e=1}^{E}\int_0^h \frac{d\Phi_N^{(e)}}{dx}\frac{d\Phi_M^{(e)}}{dx}dx\Delta_{N\alpha}^{(e)}\Delta_{M\beta}^{(e)} = \bigcup_{e=1}^{E} K_{NM}^{(e)}\Delta_{N\alpha}^{(e)}\Delta_{M\beta}^{(e)} \tag{8.2.25}$$

$$F_\alpha = -\int_0^1 2\Phi_\alpha dx = -\bigcup_{e=1}^{E}\int_0^h 2\Phi_N^{(e)}dx\Delta_{N\alpha}^{(e)} = \bigcup_{e=1}^{E} F_N^{(e)}\Delta_{N\alpha}^{(e)} \tag{8.2.26a}$$

$$G_\alpha = \overset{*}{\Phi}_\alpha \frac{du}{dx}\bigg|_0^1 = \bigcup_{e=1}^{E}\overset{*}{\Phi}{}_N^{(e)}\Delta_{N\alpha}^{(e)}\frac{du}{dx}\bigg|_0^h = \bigcup_{e=1}^{E} G_N^{(e)}\Delta_{N\alpha}^{(e)} \tag{8.2.26b}$$

where $\overset{*}{\Phi}{}_N^{(e)}$ is the local Neumann boundary interpolation function,

$$\overset{*}{\Phi}{}_N^{(e)} = \delta(\overset{*}{z}_N - z_N), \qquad \overset{*}{\Phi}{}_N^{(e)}(z_M^{(e)}) = \delta_{NM} \tag{8.2.27}$$

indicating that $\overset{*}{\Phi}{}_N^{(e)}$ is the Dirac delta function at $x = 0$ or $x = h$, being unity at $\overset{*}{z}_N = z_N$ where du/dx is prescribed at node $\overset{*}{z}_N$ and zero elsewhere. This implies that, if the Neumann boundary condition is to be applied to a node, then we set $\overset{*}{\Phi}{}_N^{(e)} = 1$ for that node. Otherwise, we set $\overset{*}{\Phi}{}_N^{(e)} = 0$. Assembled in a global system, we obtain

$$\overset{*}{\Phi}_\alpha = \delta(\overset{*}{Z}_\alpha - Z_\alpha), \qquad \overset{*}{\Phi}_\alpha(Z_\beta) = \delta_{\alpha\beta} \tag{8.2.28}$$

In this process, the Neumann boundary conditions are actually enforced between the adjacent elements, with positive and negative gradients cancelled throughout the domain (thus establishing the "energy balance" across the adjacent local element interfaces) until the end point is reached. This is where the actual Neumann boundary conditions are to be physically applied. This process is explicitly demonstrated by having constructed the FEM equations (8.2.24) in a global form instead of beginning with the local form and assembling the element stiffness matrices to a global form afterward. This is contrary to the traditional FEM formulations shown in other textbooks.

The global stiffness matrix (1.3.8), source vector (1.3.9), and Neumann boundary vector (1.3.10) are now assembled from the local element properties as

$$
\begin{aligned}
K_{\alpha\beta} &= \begin{bmatrix} K_{11} & K_{12} & K_{13} \\ K_{21} & K_{22} & K_{23} \\ K_{31} & K_{32} & K_{33} \end{bmatrix} = \begin{bmatrix} K_{11}^{(1)} & K_{12}^{(1)} & 0 \\ K_{21}^{(1)} & K_{22}^{(1)} + K_{11}^{(2)} & K_{12}^{(2)} \\ 0 & K_{21}^{(2)} & K_{22}^{(2)} \end{bmatrix} \\
&= \begin{bmatrix} 1 & 0 \\ 0 & 1 \\ 0 & 0 \end{bmatrix} \frac{1}{h} \begin{bmatrix} 1 & -1 \\ -1 & 1 \end{bmatrix} \begin{bmatrix} 1 & 0 & 0 \\ 0 & 1 & 0 \end{bmatrix} + \begin{bmatrix} 0 & 0 \\ 1 & 0 \\ 0 & 1 \end{bmatrix} \frac{1}{h} \begin{bmatrix} 1 & -1 \\ -1 & 1 \end{bmatrix} \begin{bmatrix} 0 & 1 & 0 \\ 0 & 0 & 1 \end{bmatrix} \\
&= \frac{1}{h} \begin{bmatrix} 1 & -1 & 0 \\ -1 & 2 & -1 \\ 0 & -1 & 1 \end{bmatrix}
\end{aligned} \tag{8.2.29}
$$

$$
F_\alpha = \begin{bmatrix} F_1 \\ F_2 \\ F_3 \end{bmatrix} = \begin{bmatrix} F_1^{(1)} \\ F_2^{(1)} + F_1^{(2)} \\ F_2^{(2)} \end{bmatrix} = -\begin{bmatrix} 1 & 0 \\ 0 & 1 \\ 0 & 0 \end{bmatrix} h \begin{bmatrix} 1 \\ 1 \end{bmatrix} - \begin{bmatrix} 0 & 0 \\ 1 & 0 \\ 0 & 1 \end{bmatrix} h \begin{bmatrix} 1 \\ 1 \end{bmatrix} = -h \begin{bmatrix} 1 \\ 2 \\ 1 \end{bmatrix} \tag{8.2.30}
$$

$$
\begin{aligned}
G_\alpha &= \begin{bmatrix} G_1 \\ G_2 \\ G_3 \end{bmatrix} = \begin{bmatrix} G_1^{(1)} \\ G_2^{(1)} + G_1^{(2)} \\ G_2^{(2)} \end{bmatrix} = \left\{ \begin{bmatrix} 1 & 0 \\ 0 & 1 \\ 0 & 0 \end{bmatrix} \begin{bmatrix} \overset{*}{\Phi}{}_1^{(1)} \\ \overset{*}{\Phi}{}_2^{(1)} \end{bmatrix} + \begin{bmatrix} 0 & 0 \\ 1 & 0 \\ 0 & 1 \end{bmatrix} \begin{bmatrix} \overset{*}{\Phi}{}_1^{(2)} \\ \overset{*}{\Phi}{}_2^{(2)} \end{bmatrix} \right\} \frac{du}{dx} \cos\theta \\
&= \left\{ \begin{bmatrix} 1 & 0 \\ 0 & 1 \\ 0 & 0 \end{bmatrix} \begin{bmatrix} 0 \\ 0 \end{bmatrix} + \begin{bmatrix} 0 & 0 \\ 1 & 0 \\ 0 & 1 \end{bmatrix} \begin{bmatrix} 0 \\ 0 \end{bmatrix} \right\} \frac{du}{dx} \cos\theta = \begin{bmatrix} \overset{*}{\Phi}_1 \\ \overset{*}{\Phi}_2 \\ \overset{*}{\Phi}_3 \end{bmatrix} \frac{du}{dx} \cos\theta = \begin{bmatrix} 0 \\ 0 \\ 0 \end{bmatrix} \frac{du}{dx} \cos\theta
\end{aligned} \tag{8.2.31}
$$

with $\overset{*}{\Phi}_1 = \overset{*}{\Phi}_2 = \overset{*}{\Phi}_3 = 0$ indicating that the Neumann boundary conditions are not to be applied to any of the global nodes for the solution of (1.2.1a,b). This implies that, if the Neumann boundary conditions are not applied, then the Neumann boundary vector is zero even if the gradient du/dx is not zero. Recall that in Section 1.3 the assembly of local properties into a global form was achieved intuitively. This has now been verified with a mathematical rigor of Boolean matrices. In practice, however, these Boolean matrices are never constructed, but they are replaced by computer codes based on the nodal correspondence between global and local nodes as detailed in (10.1.15c,d).

For multidimensional problems, the formulation of the finite element equations is carried out similarly as in one-dimensional problems. For example, let us examine the Poisson equation,

$$R = \nabla^2 u - f = 0 \tag{8.2.32}$$

The corresponding finite element equation takes the form

$$\int_\Omega \Phi_\alpha (u_{,ii} - f) d\Omega = 0 \tag{8.2.33}$$

$$\int_\Gamma \overset{*}{\Phi}_\alpha u_{,i} n_i d\Gamma - \int_\Omega \Phi_{\alpha,i} u_{,i} d\Omega - \int_\Omega \Phi_\alpha f d\Omega = 0 \tag{8.2.34}$$

or

$$K_{\alpha\beta} u_\beta = F_\alpha + G_\alpha \tag{8.2.35}$$

with

$$K_{\alpha\beta} = \int_\Omega \Phi_{\alpha,i} \Phi_{\beta,i} d\Omega = \bigcup_{e=1}^{E} \int_\Omega \Phi_{N,i}^{(e)} \Phi_{M,i}^{(e)} d\Omega \Delta_{N\alpha}^{(e)} \Delta_{M\beta}^{(e)} \tag{8.2.36}$$

$$F_\alpha = \int_\Omega \Phi_\alpha f d\Omega = \bigcup_{e=1}^{E} \int_\Omega \Phi_N^{(e)} f d\Omega \Delta_{N\alpha}^{(e)} \tag{8.2.37}$$

$$G_\alpha = \int_\Gamma \overset{*}{\Phi}_\alpha u_{,i} n_i d\Gamma = \bigcup_{e=1}^{E} \int_\Gamma \overset{*}{\Phi}{}_N^{(e)} u_{,i} n_i d\Gamma \Delta_{N\alpha}^{(e)} \tag{8.2.38}$$

For two-dimensional problems, trial and test functions, $\Phi_N^{(e)}$, are functions of x and y and thus the Neumann boundary test functions, $\overset{*}{\Phi}{}_N^{(e)}$, are functions of one dimension around the boundary contour. This will require the numerical integration around the boundaries. Step-by-step details of assembly for applications to multidimensional geometries will be presented in Chapter 10.

Before we proceed further, we must recognize the special mathematical and physical implications of the expression given by (8.2.34). This is the variational equation or the *weak* form of the original governing equation (8.2.32), which is the two-dimensional form of (8.2.20). Physically, if the residual (8.2.32) represents the force, then the integral given by (8.2.33) implies the energy contained in the domain Ω. Once integrated by parts as in (8.2.34), the consequence implies the energy balance between the domain Ω and the boundary surface Γ containing the Neumann boundary conditions (normal gradients of u). Thus, the physical consequence of the variational equation (or energy) allows us to add any number of physical constraints in variational forms. These constraints can be those terms playing a role of numerical diffusion (viscosity) as necessary. Many of the recent developments of FEM take advantage of this variational concept, which we shall discuss in greater detail in later chapters.

Unfortunately, the Galerkin methods described in (8.2.33–8.2.35) lead to unstable and inaccurate solutions in fluid dynamics equations in which the flow is convection-dominated. In this case, we must use the methods of weighted residual (MWR) with test functions W_α chosen differently from the trial functions Φ_α such that

$$(W_\alpha, R) = \int_\Omega W_\alpha R d\Omega = 0 \tag{8.2.39}$$

Thus, the determination of the most suitable test functions W_α remains the crucial task in order to be successful in dealing with convection-dominated flows. The most commonly used test functions are the Galerkin test functions Φ_α plus the numerical diffusion test functions Ψ_α. In this case, the finite element equations are of the form,

$$((\Phi_\alpha + \Psi_\alpha), R) = \int_\Omega (\Phi_\alpha + \Psi_\alpha) R d\Omega = 0 \tag{8.2.40}$$

Here, the numerical diffusion test functions Ψ_α play a role of numerical viscosities, equivalent to those used in FDM formulations. Some specific applications include streamline upwind Petrov-Galerkin (SUPG) methods, Taylor-Galerkin methods (TGM), generalized Petrov-Galerkin (GPG) methods, characteristic Galerkin methods (CGM), discontinuous Galerkin methods (DGM), and flowfield-dependent variation (FDV) methods, discussed in Chapters 11 through 13.

For multidimensional time-dependent problems, $R_j = \frac{\partial \mathrm{v}_j}{\partial t} + \mathrm{v}_{j,i}\mathrm{v}_i - \mathrm{v}_{j,ii} - f_j$, the general approach is to construct a double inner product of space and time in the form,

$$(\hat{W}(\xi), ((\Phi_\alpha + \Psi_\alpha), R_j)) = \int_\xi \hat{W}(\xi) \int_\Omega (\Phi_\alpha + \Psi_\alpha) R_j d\Omega d\xi = 0 \tag{8.2.41}$$

where $\hat{W}(\xi)$ is the temporal test function approximating the temporal variation between the discrete time steps with ξ being the nondimensional time variable. Note that the temporal approximation used here is independent of and discontinuous from the spatial approximations. Details on transient time-dependent problems with and without convection will be presented in Chapters 10 through 14 for linear and nonlinear cases.

8.3 DEFINITIONS OF ERRORS

Definitions of errors and error estimates for finite element methods have been well developed since the early 1970s. Finite element computational errors are defined in various norms. The most frequently used error norms are the pointwise error, L_2 norm error, and energy norm error. These error norms are the special cases of the more rigorous and general norm, called the Sobolev space norm, which can then be simplified into more meaningful and practical error definitions.

Sobolev Space (W_p^m) Norm Error

Let us define the global node error e_α as

$$e_\alpha = u_\alpha - \hat{u}_\alpha \tag{8.3.1}$$

where u_α and $\hat{u}_\alpha$ denotes the finite element approximate solution and exact solution, respectively. Then, the Sobolev space norm error is defined as

$$\|e\|_{W_p^m} = \left\{ \int \left[e^p + \left(\frac{de}{dx}\right)^p + \left(\frac{d^2e}{dx^2}\right)^p + \cdots + \left(\frac{d^me}{dx^m}\right)^p \right] dx \right\}^{\frac{1}{p}} \tag{8.3.2}$$

where m denotes the highest order of the weak derivatives of the $2m$th governing equation and p represents the power to which the derivatives are raised. Here, weak derivatives refer to the order $m, m-1, \ldots 0$. The Sobolev space (W_p^m) is defined as the functional space which includes all weak derivatives with p integrable functions, $0 \leq p \leq \infty$.

Hilbert Space (H^m) Norm Error

The Hilbert space (H^m) is the Sobolev space (W_p^m) with p equal to 2, $H^m = W_2^m$. Thus

$$\|e\|_{H^m} = \|e\|_{W_2^m} = \left\{ \int \left[e^2 + \left(\frac{de}{dx}\right)^2 + \left(\frac{d^2e}{dx^2}\right)^2 + \cdots + \left(\frac{d^me}{dx^m}\right)^2 \right] dx \right\}^{\frac{1}{2}} \tag{8.3.3}$$

It is seen that the Hilbert space is the square integrable function ($p = 2$) complete in the inner product space.

Energy Norm Error

The energy norm error, $\|e\|_E$ is a special case of the Hilbert space norm error H^m in the $2m$th order differential equation. Thus, for the fourth order equation ($m = 2$), we have

$$\|e\|_E = \|e\|_{H^2} = \|e\|_{W_2^2} = \left\{ \int \left[e^2 + \left(\frac{de}{dx}\right)^2 + \left(\frac{d^2e}{dx^2}\right)^2 \right] dx \right\}^{\frac{1}{2}} \tag{8.3.4}$$

Notice that, for the second order differential equation ($m = 1$), we write

$$\|e\|_E = \|e\|_{H^1} = \|e\|_{W_2^1} = \left\{ \int \left[e^2 + \left(\frac{de}{dx}\right)^2 \right] dx \right\}^{\frac{1}{2}} \tag{8.3.5}$$

which can be written in terms of nodal errors e_α with $e = \Phi_\alpha e_\alpha$,

$$\|e\|_E = \left\{ \int \left[\Phi_\alpha \Phi_\beta + \frac{d\Phi_\alpha}{dx}\frac{d\Phi_\beta}{dx} \right] dx e_\alpha e_\beta \right\}^{\frac{1}{2}} \tag{8.3.6}$$

Here, as usual, the global interpolation functions are obtained by means of assembly of the local interpolation functions $\Phi_N^{(e)}$.

L_2 Space Norm Error

The L_2 space arises from the Banach space (L_p) with $p = 2$, equivalent to the Hilbert space (H^m) with $m = 0$. Thus

$$\|e\|_{L_2} = \|e\|_{H^0} = \|e\|_{W_2^0} = \left(\int e^2 dx\right)^{\frac{1}{2}} \tag{8.3.7}$$

in which no rates of change of errors are involved.

p-Norm (Banach Space Norm) Error

The Banach space (L_p) is defined as the complete normed linear space such that

$$\|e\|_{L_p} = \left(\int e^p dx\right)^{\frac{1}{p}}$$

For $p = 1$ and $p = \infty$, we obtain L_1 and L_∞ norms, respectively,

$$\|e\|_{L_1} = \int e dx = \sum_{j=1}^{n} (|e_1| + |e_2| + \cdots + |e_n|) \tag{8.3.8}$$

$$\|e\|_{L_\infty} = \max_j |e_j| \tag{8.3.9}$$

It should be noted that the L_2 norm is a special case of the Banach space norm ($p = 2$), and is one of the most widely used error norm. Other norms of Banach space (other than $p = 1, 2, \infty$) are seldom used in practice.

Pointwise Error or Root Mean Square (RMS) Error

This is the simplest form of an error definition given by

$$\|e\|_{\text{RMS}} = \left(\sum e^2\right)^{\frac{1}{2}} = (e_\alpha e_\alpha)^{\frac{1}{2}} \tag{8.3.10}$$

Here the percent error may be defined as

$$\|e\|_{\%} = \frac{\|e\|_{\text{RMS}}}{\left(\sum u^2\right)^{\frac{1}{2}}} = \left(\frac{e_\alpha e_\alpha}{u_\beta u_\beta}\right)^{\frac{1}{2}} \tag{8.3.11}$$

Note that there is no integral involved in this approach, thus it is called the pointwise error, or often known as the root mean square (RMS) error.

Matrix Norms

Matrix norms are an important concept in determining the computational stability of the finite element equations such as in (8.2.35) in terms of the so-called *condition number*. To demonstrate this concept, we write (8.2.35) in the matrix form

$$\mathbf{Ku} = \mathbf{F} \tag{8.3.12}$$

If $\mathbf{K}$ is an $n \times n$ matrix and $\mathbf{u}$ any vector with n components, then there exists a constant c such that

$$\|\mathbf{Ku}\| \leq c\|\mathbf{F}\| \tag{8.3.13}$$

where $\mathbf{u} \neq \mathbf{0}, \|\mathbf{u}\| > 0$, and the constant c is given by

$$c \geq \frac{\|\mathbf{Ku}\|}{\|\mathbf{u}\|} \tag{8.3.14}$$

The smallest c is known as the matrix norm of **K,** denoted by $\|\mathbf{K}\|$.

$$\|\mathbf{K}\| \leq \max \frac{\|\mathbf{Ku}\|}{\|\mathbf{u}\|} \tag{8.3.15}$$

with the matrix norm being calculated from

$$\|\mathbf{K}\|_{L_1} = \max \sum_{\beta} |\mathbf{K}_{\alpha\beta}|, \qquad \|\mathbf{K}\|_{L_2} = (K_{\alpha\beta}K_{\beta\alpha})^{1/2}, \qquad \|\mathbf{K}\|_{L_\infty} = \max \sum_{\alpha} |\mathbf{K}_{\alpha\beta}|$$

Combining (8.3.13) and (8.3.15), we obtain

$$\|\mathbf{Ku}\| \leq \|\mathbf{K}\|\|\mathbf{u}\| \tag{8.3.16}$$

If we define the condition number N as

$$N(\mathbf{K}) = \|\mathbf{K}\|\|\mathbf{K}^{-1}\| \tag{8.3.17}$$

the following theorem can be established.

Theorem: A linear system of equations given by (8.3.12) is said to be *well-conditioned* if the condition number as defined in (8.3.17) is small.

Proof: It follows from (8.3.12) and (8.3.16) that $\|\mathbf{F}\| \leq \|\mathbf{K}\|\|\mathbf{u}\|$. Let $\mathbf{F} \neq \mathbf{0}, \mathbf{u} \neq \mathbf{0}$. Then, we have

$$\frac{1}{\|\mathbf{u}\|} \leq \frac{\|\mathbf{K}\|}{\|\mathbf{F}\|} \tag{8.3.18}$$

Let the residual be given by

$$\mathbf{R} = \mathbf{K}(\mathbf{u} - \hat{\mathbf{u}}) \tag{8.3.19}$$

Combining (8.3.16) and (8.3.19) leads to

$$\|\mathbf{u} - \hat{\mathbf{u}}\| = \|\mathbf{K}^{-1}\mathbf{R}\| \leq \|\mathbf{K}^{-1}\|\|\mathbf{R}\| \tag{8.3.20}$$

From (8.3.18) and (8.3.20) we obtain

$$\frac{\|\mathbf{u} - \hat{\mathbf{u}}\|}{\|\mathbf{u}\|} \leq \frac{1}{\|\mathbf{u}\|}\|\mathbf{K}^{-1}\|\|\mathbf{R}\| \leq \frac{\|\mathbf{K}\|}{\|\mathbf{F}\|}\|\mathbf{K}^{-1}\|\|\mathbf{R}\| = N(\mathbf{K})\frac{\|\mathbf{R}\|}{\|\mathbf{F}\|} \tag{8.3.21}$$

This proves that a small relative error results from the small condition number with the system being well-conditioned. Otherwise, the system is ill-conditioned.

Example 8.3.1

Given:

$$\mathbf{e} = \begin{bmatrix} 1 \\ -2 \\ -3 \\ 2 \end{bmatrix}$$

Required: Find the vector norms in L_1, L_2, L_∞.

Solution: $\|\mathbf{e}\|_{L_1} = 8$; $\|\mathbf{e}\|_{L_2} = \sqrt{18}$; $\|\mathbf{e}\|_{L\infty} = 3$

Example 8.3.2

Given:

$$\mathbf{K} = \begin{bmatrix} 0 & 0 & 10 & 0 \\ 1 & 1 & 5 & 1 \\ 0 & 1 & 5 & 1 \\ 0 & 0 & 5 & 1 \end{bmatrix}$$

Required: Find the matrix norms in L_1, L_2, L_∞.

Solution: $\|\mathbf{K}\|_{L_1} = \max\{1, 2, 25, 3\} = 25$; $\|\mathbf{K}\|_{L_2} = \sqrt{181}$; $\|\mathbf{K}\|_{L_\infty} = \max\{10, 8, 7, 6\} = 10$

Typical convergence properties are shown in Figure 8.3.1. It is seen in Figure 8.3.1a that convergence is achieved at the point N and that further refinements or the increase of polynomial degrees do not affect the exact solution. The convergence to the exact solution depends on the so-called *mesh parameter*. The mesh parameter h is defined as "diameter" of the largest element in a given domain. For one-dimensional problems, it is simply the length h of the domain with $0 < h < 1$. Let e_1 and e_2 be the errors for the mesh parameters h_1 and h_2, respectively. Assume that reduction of mesh parameters results in the increase of the order p of the rate of convergence. This relation may be written in the form (Figure 8.3.1b)

$$\frac{\|e_1\|}{\|e_2\|} = \left(\frac{h_1}{h_2}\right)^p \tag{8.3.22}$$

Taking the natural logarithm on both sides, we obtain

$$p = \frac{\ln\|e_1\| - \ln\|e_2\|}{\ln h_1 - \ln h_2} \tag{8.3.23}$$

where the magnitude of p is indicative of the rate of convergence of the finite element solution to the exact solution. In plotting the computed results to examine the convergence, one may choose at least three different mesh parameters. They should be chosen in the range where convergence to the exact solution has not been achieved as illustrated in points 1, 2, and 3 of Figure 8.3.1a,b. The slope p is seen to be a straight line with accuracy increasing with a steeper slope. If the mesh parameter is chosen too small beyond convergence, the slope p will become horizontal ($p = 0$), such as points 4, 5, and 6 in Figure 8.3.1a. If computational round-off errors are accumulated due to the

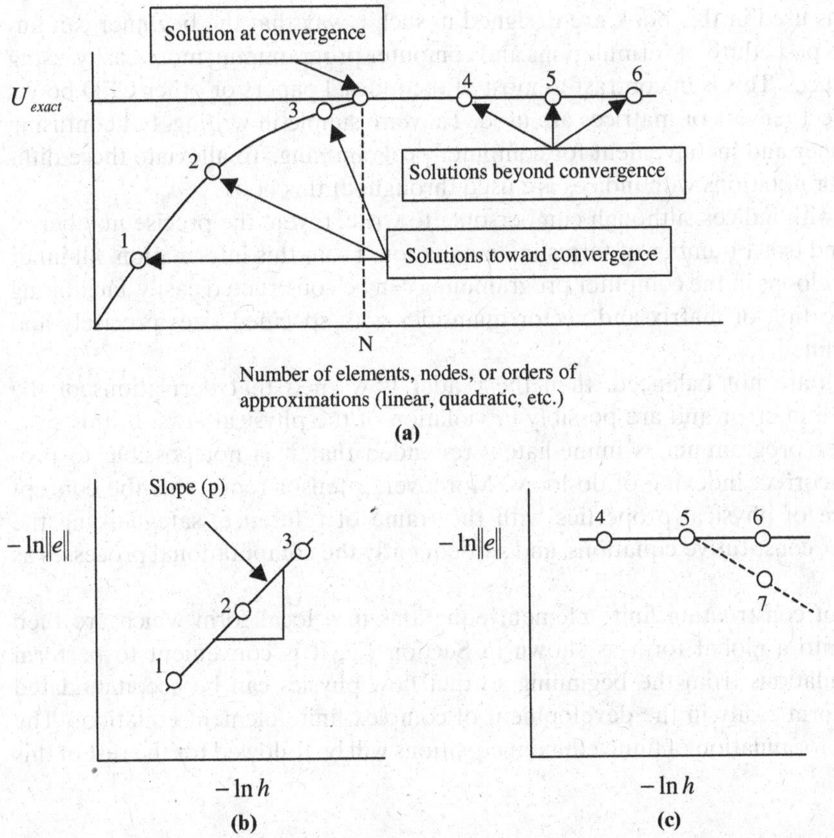

Figure 8.3.1 Convergence toward and beyond exact solutions. Notice that the order (p) of rate of convergence becomes horizontal ($p = 0$) for the solutions beyond convergence but may turn negative ($p < 0$) due to round-off errors. **(a)** Solutions vs. refinements. **(b)** Solutions toward convergence. **(c)** Solutions beyond convergence.

limitation of the computer through no fault of the computational methodology itself, then the slope may tend to deviate from the horizontal line (point 7 in Figure 8.3.1b). This is not an indication of deterioration of accuracy or rate of convergence, but rather it is meaningless to show the rate of convergence beyond the point at which convergence has already been achieved.

In recent years, error estimates particularly in the adaptive *h-p* methods have been studied extensively by Babuska and his co-workers [Babuska and Guo, 1988] and Oden and his co-workers [Oden and Demkowicz, 1991], among others. Some discussions on this topic will be presented in Chapter 19.

8.4 SUMMARY

In this chapter, we revisited Chapter 1 and reintroduced the finite element theory with more rigorous mathematical foundations as applied to multidimensional problems. Definitions of errors in terms of various functional norms and convergence vs. errors have also been presented.

Notations used in this book are designed in such a way that the beginner can understand the procedure of formulations and computer programming more easily, using tensorial indices. This is in contrast to most of the journal papers or other CFD books in which direct tensors or matrices are used. They are simple in writing, but confusing to the beginner and inconvenient for computer programming. To alleviate these difficulties, tensor notations with indices are used throughout this book.

Tensors with indices, although cumbersome to write, reveal the precise number of equations and exact number of terms in an equation. From this information, all inner and outer do-loops in the computer programming can be constructed easily, facilitating the multiplication of matrix and vector quantities with specified sizes precisely and explicitly defined.

If indices are not balanced, then the reader is warned that derivations of the equations are in error and are possibly in violation of the physical laws. In this case, the computer programmer is immediately reminded that it is not possible to proceed with incorrect indexing of do-loops. Moreover, a tensor represents the concept of invariance of physical properties with the frame of reference, safeguarding the physical laws, constitutive equations, and subsequently the computational processes as well.

Instead of constructing finite element equations in a local form which are then assembled into a global form as shown in Section 1.3, it is convenient to perform global formulations from the beginning so that flow physics can be accommodated in a global form easily in the development of complex finite element equations. The direct global formulation of finite element equations will be followed for the rest of this book.

REFERENCES

Babuska, I. and Guo, B. Q. [1988]. The *h-p* version of the finite element method for domains with curved boundaries. *SIAM J. Num. Anal.*, 25, 4, 837–61.

Babuska, I., Szabo, B. A., and Katz, I. N. [1981]. The p-version of the finite element method. *SIAM J. Num. Anal.*, 18, 512–45.

Baker, A. J. [1983]. *Finite Element Computational Fluid Mechanics*. New York: Hemisphere, McGraw-Hill.

Chung, T. J. [1978]. *Finite Element Analysis in Fluid Dynamics*. New York: McGraw-Hill.

———. [1999]. Transitions and interactions of inviscid/viscous, compressible/incompressible and laminar/turbulent flows. *Int. J. Num. Meth. Fl.*, 31, 223–46.

Donea, J. [1984]. A Taylor-Galerkin method for convective transport problems. *Int. J. Num. Meth. Eng.*, 20, 101–19.

Heinrich, J. C., Huyakorn, P. S., Zienkiewicz, O. C., and Mitchell, A. R. [1977]. An upwind finite element scheme for two-dimensional convective transport equation. *Int. J. Num. Meth. Eng.*, 11, 1, 131–44.

Hughes, T. J. R. and Brooks, A. N. [1982]. A theoretical framework for Petrov-Galerkin methods with discontinuous weighting functions: application to the streamline upwind procedure. In R. H. Gallagher et al. (eds). *Finite Elements in Fluids*, London: Wiley.

Hughes, T., Mallet, M. and Mizukami, A. [1986]. A new finite element formulation for computational fluid dynamics I. Beyond SUPG, *Comp. Meth. Appl. Mech. Eng.*, 54, 341–55.

Johnson, C. [1987]. Numerical Solution of Partial Differential Equations on the Element Method. student litteratur, Lund, Sweden.

Löhner, R., Morgan, K., and Zienkiewicz, O. C. [1985]. An adaptive finite element procedure for compressible high speed flows. *Comp. Meth. Appl. Mech. Eng.*, 51, 441–65.

Oden, J. T., Babuska, I., and Baumann, C. E. [1998]. A discontinuous hp finite element methods for diffusion problems. *J. Comp. Phy.*, 146, 491–519.

Oden, J. T. and Demkowicz, L. [1991]. h-p adaptive finite element methods in computational fluid dynamics. *Comp. Meth. Appl. Mech. Eng.*, 89 (1–3): 1140.

Oden, J. T. and L. C. Wellford, Jr. [1972]. Analysis of viscous flow by the finite element method. *AIAA J.*, 10, 1590–99.

Zienkiewicz, O. C. and Cheung, Y. K. [1965]. Finite elements in the solution of field problems. *The Engineer*, 507–10.

Zienkiewicz, O. C. and Codina, R. [1995]. A general algorithm for compressible and incompressible flow–Part I. Characteristic-based scheme. *Int. J. Num. Meth. Fl.*, 20, 869–85.

CHAPTER NINE

Finite Element Interpolation Functions

9.1 GENERAL

We saw in Section 1.3 that finite element equations are obtained by the classical approximation theories such as variational or weighted residual methods. However, there are some basic differences in philosophy between the classical approximation theories and finite element methods. In the finite element methods, the global functional representations of a variable consist of an assembly of local functional representations so that the global boundary conditions can be implemented in local elements by modification of the assembled algebraic equations. The local interpolation (shape, basis, or trial) functions are chosen in such a manner that continuity between adjacent elements is maintained.

The finite element interpolations are characterized by the shape of the finite element and the order of the approximations. In general, the choice of a finite element depends on the geometry of the global domain, the degree of accuracy desired in the solution, the ease of integration over the domain, etc.

In Figure 9.1.1, a two-dimensional domain is discretized by a series of triangular elements and quadrilateral elements. It is seen that the global domain consists of many subdomains (the finite elements). The global domain may be one-, two-, or three-dimensional. The corresponding geometries of the finite elements are shown in Figure 9.1.2. A one-dimensional element (as we have studied in Chapters 1 and 8) is simply a straight line, a two-dimensional element may be triangular, rectangular, or quadrilateral, and a three-dimensional element can be a tetrahedron, a regular hexahedron, an irregular hexahedron, etc. The three-dimensional domain with axisymmetric geometry and axisymmetric physical behavior can be represented by a two-dimensional element generated into a three-dimensional ring by integration around the circumference. In general, the interpolation functions are the polynomials of various degrees, but often they may be given by transcendental or special functions. If polynomial expansions are used, the linear variation of a variable within an element can be expressed by the data provided at the corner nodes. For quadratic variations, we add a side node located midway between the corner nodes (Figure 9.1.3). Cubic variations of a variable are represented by two side nodes in addition to the corner nodes. Sometimes a complete expansion of certain degree polynomials may require installation of nodes at various points within the element (interior nodes). Thus, there are three different types of nodes: vertex nodes in which only corner nodes are installed at vertices, side nodes

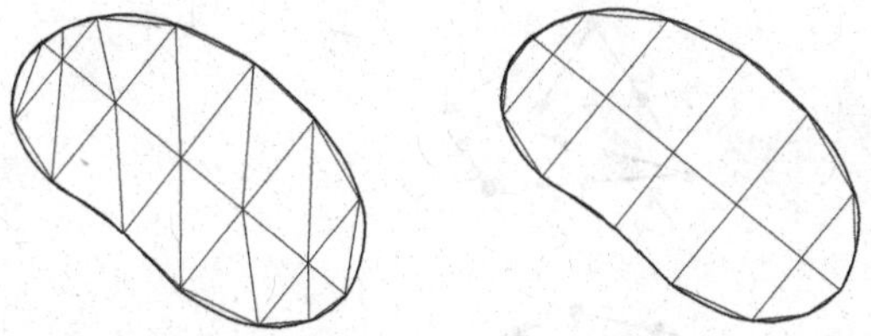

(a) Discretization by triangular elements

(b) Discretization by quardrilateral elements

Figure 9.1.1 Finite element discretization of a two-dimensional domain.

in which one or more nodes are installed along the element sides, and internal nodes in which one or more interior nodes are provided inside of an element.

Nodal configurations and corresponding polynomials may be selected from the so-called Pascal triangle, Pascal tetrahedron, two-dimensional hypercube, or three-dimensional hypercube, as shown in Figure 9.1.4. Various combinations between the number of nodes and degrees of polynomials for two-dimensional geometries can be selected as illustrated in Figures 9.1.5 and 9.1.6. Similar approaches may be used for three-dimensional geometries. In choosing a suitable element, the number of nodes

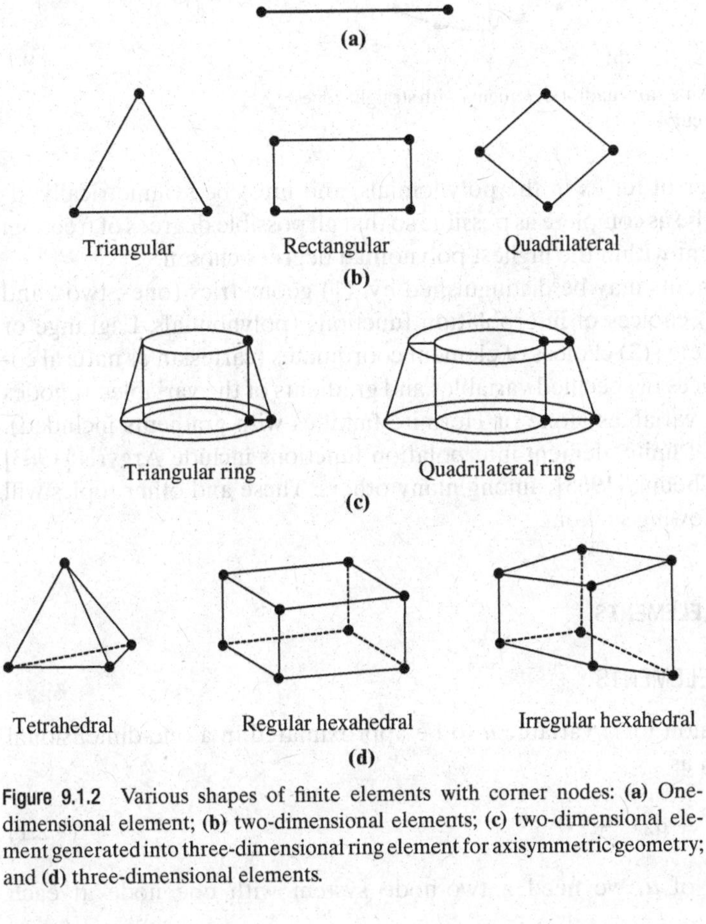

Figure 9.1.2 Various shapes of finite elements with corner nodes: (a) One-dimensional element; (b) two-dimensional elements; (c) two-dimensional element generated into three-dimensional ring element for axisymmetric geometry; and (d) three-dimensional elements.

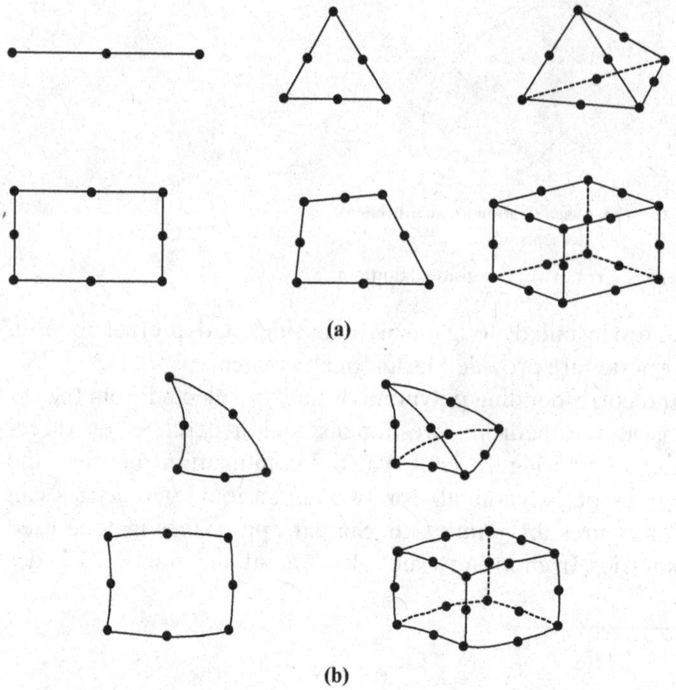

Figure 9.1.3 Quadratic elements. **(a)** Quadratic elements with straight edges. **(b)** Quadratic elements with curved edges.

must match the number of terms in the polynomials, and must be symmetrically arranged. They must also be as complete as possible so that all possible degrees of freedom are allowed to be present within the highest polynomial degrees chosen.

Types of finite elements may be distinguished by: (1) geometries (one-, two-, and three dimensional); (2) choices of interpolation functions (polynomials, Lagrange or Hermite polynomials), etc.; (3) choices of element coordinates (cartesian or natural coordinates); and (4) choices of specified variables and gradients of the variables at nodes (Lagrange family with variables alone or Hermite families with gradients included). Earlier developments of finite element interpolation functions include Argyris [1963] and Zienkiewicz and Cheung [1965], among many others. These and other topics will be presented in the following sections.

9.2 ONE-DIMENSIONAL ELEMENTS

9.2.1 CONVENTIONAL ELEMENTS

The polynomial expansion for a variable u to be approximated in a one-dimensional element may be written as

$$u = \alpha_1 + \alpha_2 x + \alpha_3 x^2 + \alpha_4 x^3 + \cdots \tag{9.2.1}$$

For a linear variation of u, we need a two-node system with one node at each

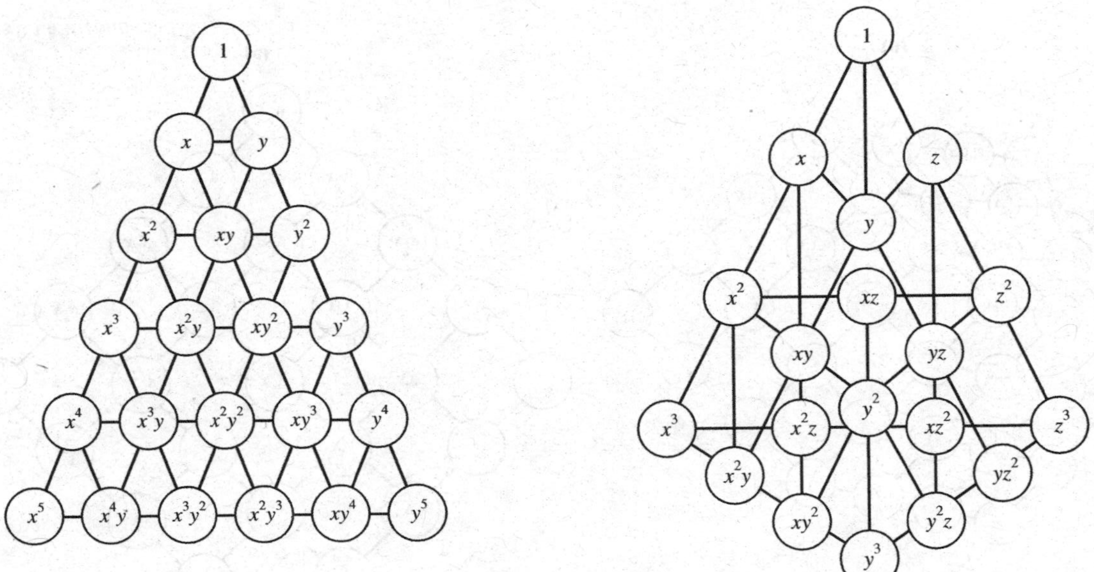

(a) (b)

Figure 9.1.4 Polynomial expansions in finite elements. **(a)** Pascal triangle. **(b)** Pascal tetrahedron. **(c)** Two-dimensional hypercube. **(d)** Three-dimensional hypercube.

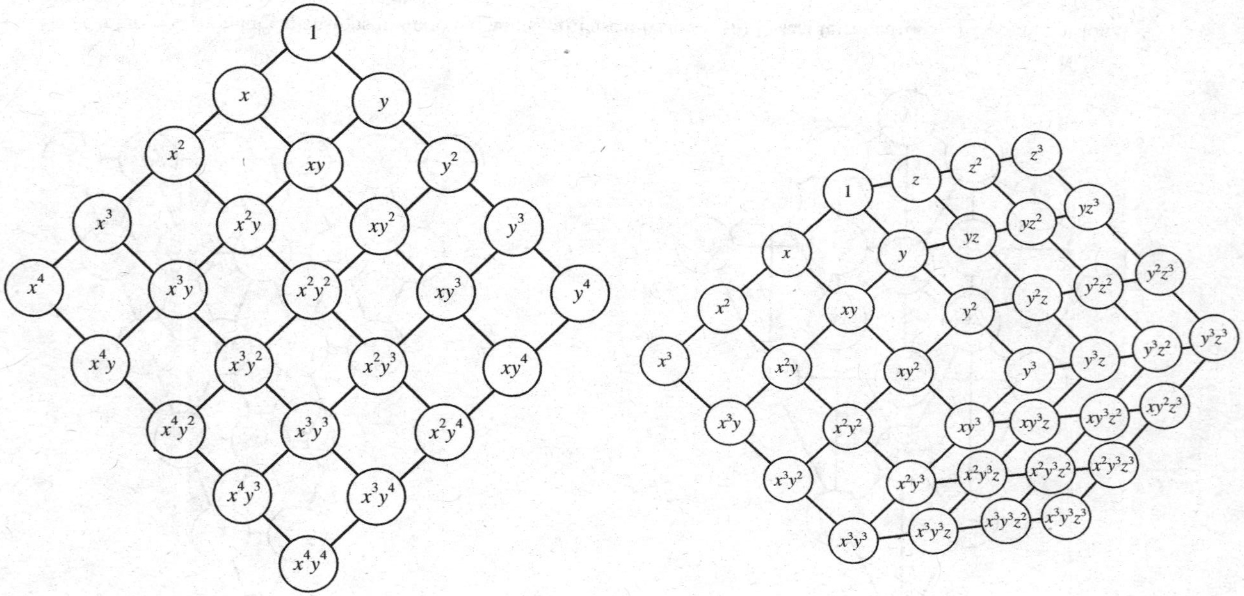

(c) (d)

Figure 9.1.4 *(continued)*

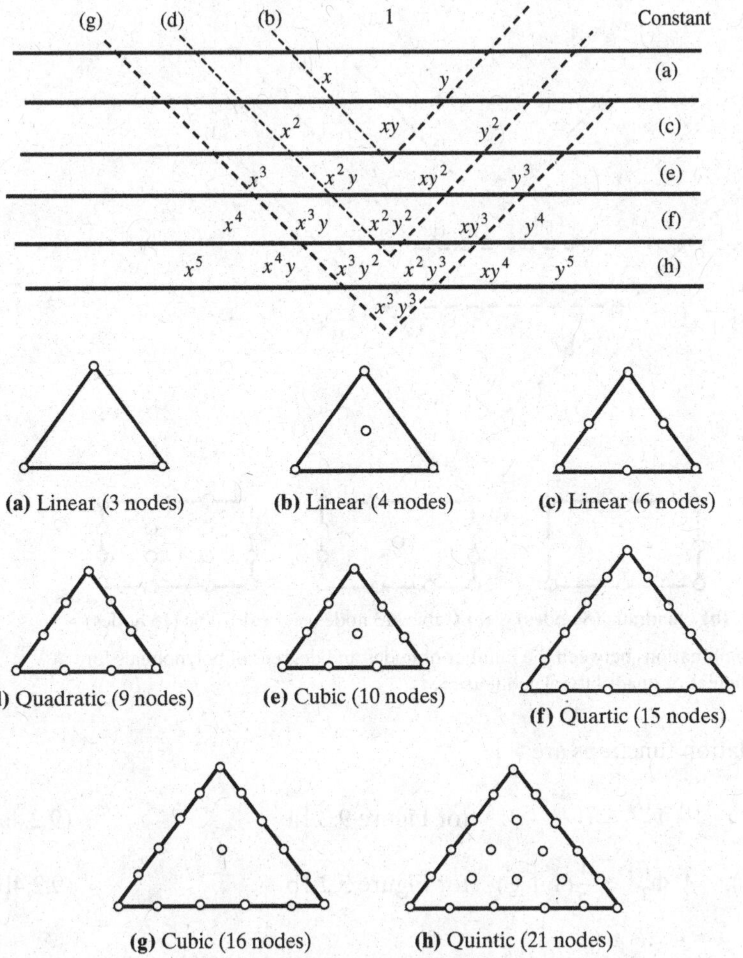

Figure 9.1.5 Various combinations between the number of nodes and degrees of polynomials for two-dimensional triangular geometries.

end. The interpolation functions for this case were derived in Section 1.3, based on Figure 1.3.1. An alternative method, perhaps the more general approach, is to use the natural (nondimensional) coordinate, ξ, with the origin set as in Figure 9.2.1a ($\xi = x/h$) or Figure 9.2.1b ($\xi = x/(h/2)$). Then (9.2.1) becomes

$$u = \alpha_1 + \alpha_2\xi + \alpha_3\xi^2 + \alpha_4\xi^3 + \cdots \tag{9.2.2}$$

For a linear element (two-node system), we have

$$u = \alpha_1 + \alpha_2\xi \tag{9.2.3}$$

Writing (9.2.3) at each node, solving for the constants, and substituting them into (9.2.3) for an element, we obtain $u^{(e)}$ [u in element (e)]:

$$u^{(e)} = \Phi_1^{(e)}u_1^{(e)} + \Phi_2^{(e)}u_2^{(e)} = \Phi_N^{(e)}u_N^{(e)}, \quad (N = 1, 2)$$

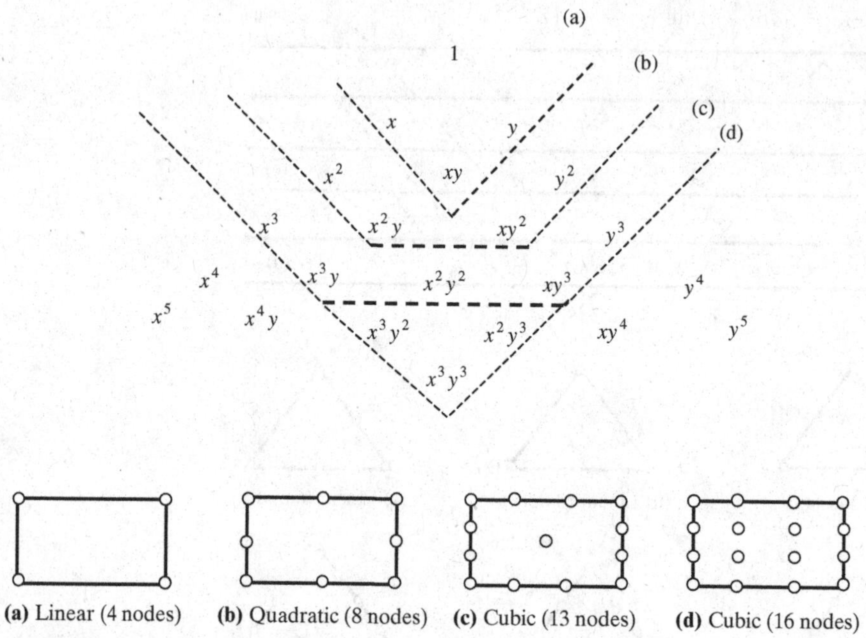

Figure 9.1.6 Various combinations between the number of nodes and degrees of polynomials for two-dimensional rectangular or quadrilateral geometries.

where the interpolation functions are

$$\Phi_1^{(e)} = 1 - \xi, \qquad \Phi_2^{(e)} = \xi \qquad \text{for Figure 9.2.1a} \tag{9.2.4a}$$

$$\Phi_1^{(e)} = \frac{1}{2}(1 - \xi), \qquad \Phi_2^{(e)} = \frac{1}{2}(1 + \xi) \quad \text{for Figure 9.2.1b} \tag{9.2.4b}$$

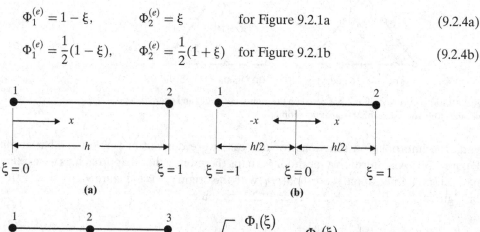

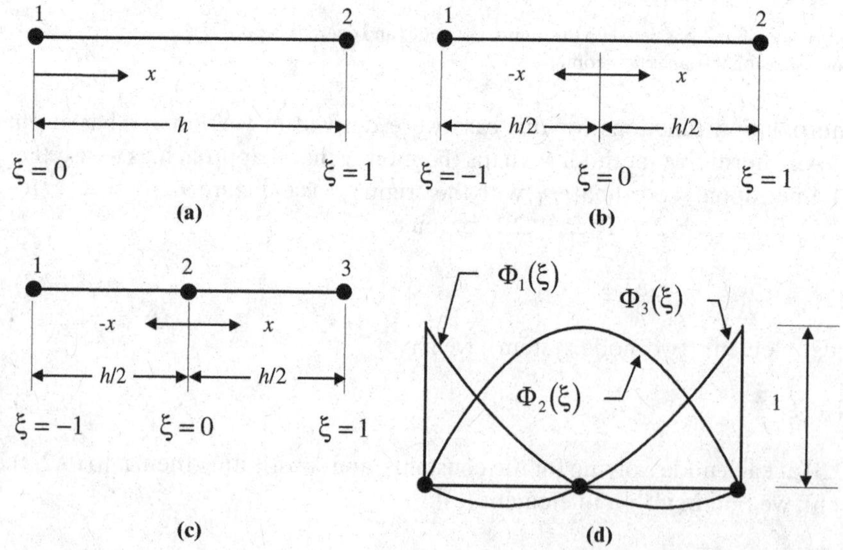

Figure 9.2.1 One-dimensional element. **(a)** Origin at end node (linear variation). **(b)** Origin at center (linear variation). **(c)** Origin at center node (quadratic variation). **(d)** Quadratic variation.

Likewise, for quadratic approximations in which we require an additional node, preferably at the midside (Figure 9.2.1c), we have

$$u = \alpha_1 + \alpha_2\xi + \alpha_3\xi^2 \tag{9.2.5}$$

and writing (9.2.5) at each node yields

$$u_1 = \alpha_1 - \alpha_2 + \alpha_3, \qquad u_2 = \alpha_1, \qquad u_3 = \alpha_1 + \alpha_2 + \alpha_3 \tag{9.2.6}$$

Evaluating the constants, we obtain

$$u^{(e)} = \Phi_1^{(e)} u_1^{(e)} + \Phi_2^{(e)} u_2^{(e)} + \Phi_3^{(e)} u_3^{(e)} = \Phi_N^{(e)} u_N^{(e)}, \quad (N = 1, 2, 3) \tag{9.2.7}$$

where the interpolation functions are (see Figure 9.2.1d)

$$\Phi_1^{(e)} = \frac{1}{2}\xi(\xi - 1), \qquad \Phi_2^{(e)} = 1 - \xi^2, \qquad \Phi_3^{(e)} = \frac{1}{2}\xi(\xi + 1) \tag{9.2.8}$$

It is easily seen that the limits of integration of the interpolation functions should be changed such that

$$\int_{-h/2}^{h/2} f(x)dx = \int_{-1}^{1} f(\xi)\frac{\partial x}{\partial \xi}d\xi = \frac{h}{2}\int_{-1}^{1} f(\xi)d\xi \tag{9.2.9}$$

where $x = (h/2)\xi$. If the interpolation functions are derived in terms of nondimensionalized spatial variables, then such a normalized system is called a natural coordinate. Note that the basic properties of interpolation functions as given by (8.2.12) are satisfied for both (9.2.4) and (9.2.8).

9.2.2 LAGRANGE POLYNOMIAL ELEMENTS

To avoid the inversion of the coefficient matrix for higher order approximations, we may use the Lagrange interpolation function L_N, which can be obtained as follows. Let $u(x)$ be given by (Figure 9.2.2)

$$u(x) = L_1(x)u_1 + L_2(x)u_2 + \cdots L_n(x)u_n$$

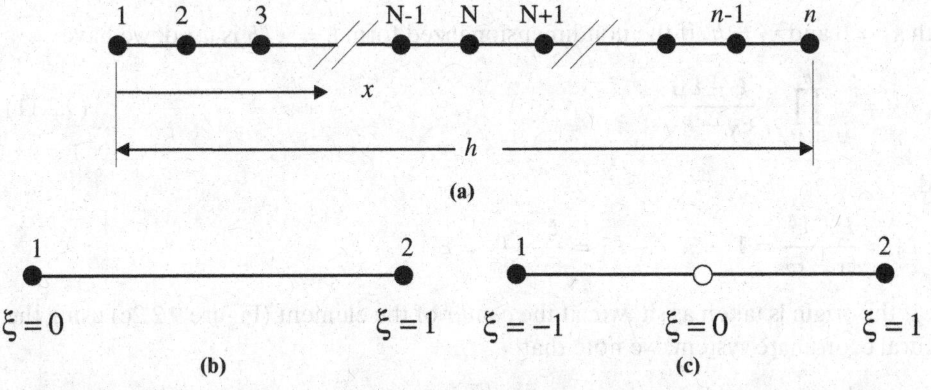

Figure 9.2.2 Lagrange element with natural coordinates. **(a)** Lagrange element of the n-1th degree approximation. **(b)** Linear approximation with origin at the left node. **(c)** Linear variation with origin at the center.

where $L_N(x)$ is chosen such that

$$L_N(x_M) = \delta_{NM}$$

$L_N(x)$ may be expanded in the form

$$L_N(x) = c_N(x - x_1)(x - x_2)\cdots(x - x_{N-1})(x - x_{N+1})\cdots(x - x_n)$$

where

$$L_N(x_M) = \begin{cases} 0 & M \neq N \\ 1 = c_N \prod\limits_{M=1,M\neq N}^{n} (x_N - x_M) & M = N \end{cases}$$

Solving for the coefficient c_N and substituting it to the expression for $L_N(x)$, we obtain

$$\Phi_N^{(e)}(x) = L_N(x) = \prod_{M=1,M\neq N}^{n} \frac{x - x_M}{x_N - x_M} \tag{9.2.10}$$

$$= \frac{(x - x_1)(x - x_2)\cdots(x - x_{N-1})(x - x_{N+1})\cdots(x - x_n)}{(x_N - x_1)(x_N - x_2)\cdots(x_N - x_{N-1})(x_N - x_{N+1})\cdots(x_N - x_n)}$$

with the symbol $\prod$ denoting a product of binomials over the range $M = 1, 2, \ldots, n$ (see Figure 9.2.2). Here the element is divided into equal length segments by the $n = m+1$ nodes, with m and n equal to the order of approximations and the number of nodes in an element, respectively. Let us consider a first order approximation of a variable u such that

$$u^{(e)} = L_N u_N^{(e)} \quad (N = 1, 2)$$

with

$$L_1 = \frac{x - x_2}{x_1 - x_2} = \frac{x - h}{-h} = 1 - \frac{x}{h}$$

$$L_2 = \frac{x - x_1}{x_2 - x_1} = \frac{x}{h}$$

with $x_1 = 0$ and $x_2 = h$. If the nondimensionalized form $\xi = x/h$ is used, we have

$$L_N = \prod_{M=1,M\neq N}^{n} \frac{\xi - \xi_M}{\xi_N - \xi_M} \tag{9.2.11}$$

and

$$L_1 = \frac{\xi - \xi_2}{\xi_1 - \xi_2} = 1 - \xi, \qquad L_2 = \frac{\xi - \xi_1}{\xi_2 - \xi_1} = \xi$$

If the origin is taken as shown, at the center of the element (Figure 9.2.2c) using the natural coordinate system, we note that

$$L_1 = \frac{1}{2}(1 - \xi), \qquad L_2 = \frac{1}{2}(1 + \xi)$$

These functions are the same as in (9.2.4b).

For quadratic approximations, we have $n = m+1 = 3$ and

$$L_1 = \frac{(\xi-\xi_2)(\xi-\xi_3)}{(\xi_1-\xi_2)(\xi_1-\xi_3)} = 2\left(\xi-\frac{1}{2}\right)(\xi-1)$$

$$L_2 = \frac{(\xi-\xi_1)(\xi-\xi_3)}{(\xi_2-\xi_1)(\xi_2-\xi_3)} = -4\xi(\xi-1)$$

$$L_3 = \frac{(\xi-\xi_1)(\xi-\xi_2)}{(\xi_3-\xi_1)(\xi_3-\xi_2)} = 2\xi\left(\xi-\frac{1}{2}\right)$$

For the natural coordinate system with the origin at the center, we obtain

$$L_1 = \frac{1}{2}\xi(\xi-1), \qquad L_2 = 1-\xi^2, \qquad L_3 = \frac{1}{2}\xi(\xi+1)$$

which are identical to (9.2.8), the results one would expect to obtain.

The interpolation functions derived using the natural coordinates are convenient to generate multidimensional element interpolation functions by means of tensor products as shown in Section 9.3.2.

9.2.3 HERMITE POLYNOMIAL ELEMENTS

If continuity of the derivative of a variable at common nodes is desired, one efficient way of assuring this continuity is to use the Hermite polynomials. For a one-dimensional element with two end nodes, the development of Hermite polynomials for a variable u begins with

$$u = \alpha_1 + \alpha_2\xi + \alpha_3\xi^2 + \alpha_4\xi^3$$

We write the nodal equations for $u(\xi)$ and $du(\xi)/d\xi$ at two end nodes and evaluate the constants to obtain

$$u^{(e)}(\xi) = H_N^0(\xi)u_N^{(e)} + H_N^1(\xi)\left(\frac{\partial u}{\partial \xi}\right)_N^{(e)} \quad (N = 1, 2) \tag{9.2.12a}$$

or

$$u^{(e)}(\xi) = \Phi_r^{(e)} Q_r \quad (r = 1, 2, 3, 4) \tag{9.2.12b}$$

where the Hermite polynomials have the properties [see Hildebrand, 1956]

$$H_N^0(\xi_M) = \delta_{NM}, \qquad \frac{d}{d\xi}H_N^1(\xi_M) = \delta_{NM}$$

Here $H_N^0(\xi)$ and $H_N^1(\xi)$, which are now used as the finite element interpolation functions,

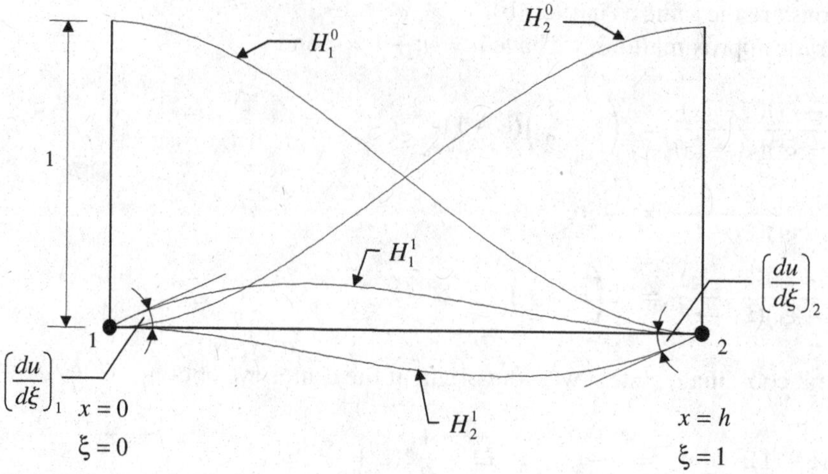

Figure 9.2.3 Hermite interpolation functions.

are the cubic polynomials of the form (see Figure 9.2.3)

$$\begin{aligned}
\Phi_1^{(e)} &= H_1^0 = 1 - 3\xi^2 + 2\xi^3 & Q_1 &= u_1^{(e)} \\
\Phi_2^{(e)} &= H_2^0 = 3\xi^2 - 2\xi^3 & Q_2 &= u_2^{(e)} \\
\Phi_3^{(e)} &= H_1^1 = \xi - 2\xi^2 + \xi^3 & Q_3 &= \left(\frac{\partial u}{\partial \xi}\right)_1^{(e)} \\
\Phi_4^{(e)} &= H_2^1 = \xi^3 - \xi^2 & Q_4 &= \left(\frac{\partial u}{\partial \xi}\right)_2^{(e)}
\end{aligned} \tag{9.2.13}$$

with $\xi = x/h$, h being the length of the element. Note that $\Phi_3^{(e)}$ and $\Phi_4^{(e)}$ must be multiplied by h, if the nodal values of derivatives are given by $\partial u/\partial x$.

If the second as well as the first derivative is to be specified at the end nodes, we require a fifth degree Hermite polynomial such that

$$u^{(e)}(\xi) = H_N^0(\xi)u_N^{(e)} + H_N^1(\xi)\left(\frac{\partial u}{\partial \xi}\right)_N^{(e)} + H_N^2(\xi)\left(\frac{\partial^2 u}{\partial \xi^2}\right)_N^{(e)} \tag{9.2.14}$$

with

$$\begin{aligned}
\Phi_1^{(e)} &= H_1^0 = 1 - 10\xi^3 + 15\xi^4 - 6\xi^5 & Q_1 &= u_1^{(e)} \\
\Phi_2^{(e)} &= H_2^0 = 10\xi^3 - 15\xi^4 + 6\xi^5 & Q_2 &= u_2^{(e)} \\
\Phi_3^{(e)} &= H_1^1 = \xi - 6\xi^3 + 8\xi^4 - 3\xi^5 & Q_3 &= \left(\frac{\partial u}{\partial \xi}\right)_1^{(e)} \\
\Phi_4^{(e)} &= H_2^1 = -4\xi^3 + 7\xi^4 - 3\xi^5 & Q_4 &= \left(\frac{\partial u}{\partial \xi}\right)_2^{(e)}
\end{aligned}$$

$$\Phi_5^{(e)} = H_2^2 = \frac{1}{2}(\xi^2 - 3\xi^3 + 3\xi^4 - \xi^5) \qquad Q_5 = \left(\frac{\partial^2 u}{\partial \xi^2}\right)_1^{(e)}$$

$$\Phi_6^{(e)} = H_2^2 = \frac{1}{2}(\xi^3 - 2\xi^4 + \xi^5) \qquad Q_6 = \left(\frac{\partial^2 u}{\partial \xi^2}\right)_2^{(e)}$$

Note that $\Phi_3^{(e)}$, $\Phi_4^{(e)}$, and $\Phi_5^{(e)}$, $\Phi_6^{(e)}$ must be multiplied by h and h^2, respectively, if the nodal values of derivatives are given by $\partial u/\partial x$, and $\partial^2 u/\partial x^2$.

Additional discussions of Hermite polynomials can be found in Birkoff, Schultz, and Varga [1968].

9.3 TWO-DIMENSIONAL ELEMENTS

Among the two-dimensional elements, the triangular element was the first investigated in the early days of development. In recent years, however, the four-sided isoparametric element has become equally popular, or more convenient in some applications. Various features of these elements are described below.

9.3.1 TRIANGULAR ELEMENTS

As noted in the one-dimensional element, we may use the standard rectangular cartesian coordinates or the natural coordinates (nondimensionalized) to obtain the interpolation functions. It will be seen that the choice of a particular coordinate system influences the amount of algebra required in the formulation of finite element equations. For higher order approximations (with higher order polynomials), an evaluation of constants is particularly easy if natural coordinates are used.

Cartesian Coordinate Triangular Elements

In this element, the properties of the element are determined in terms of the local rectangular cartesian coordinates (x_i) with their origin at the centroid of the triangle (Figure 9.3.1).

$$x_1 + x_2 + x_3 = 0 \quad \text{and} \quad y_1 + y_2 + y_3 = 0$$

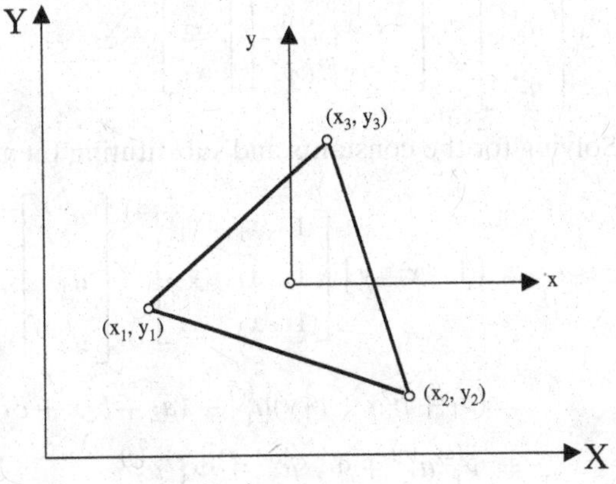

Figure 9.3.1 Cartesian coordinate triangular element.

or

$$\sum_{N=1}^{3} x_{Ni} = 0 \quad (N = 1, 2, 3, i = 1, 2)$$

with $x_{N1} = x_N$ and $x_{N2} = y_N$. If this triangle is identified from the global rectangular cartesian coordinates (X_i) with their origin outside the triangle, we note that the following relationships hold:

$$x_1 = X_1 - \frac{1}{3}(X_1 + X_2 + X_3)$$

$$x_2 = X_2 - \frac{1}{3}(X_1 + X_2 + X_3)$$

$$\vdots$$

$$y_3 = Y_3 - \frac{1}{3}(Y_1 + Y_2 + Y_3)$$

Or, combining these equations, we write

$$x_{Ni} = X_{Ni} - \frac{1}{3}\sum_{N=1}^{3} X_{Ni} \quad (N = 1, 2, 3,\ i = 1, 2) \tag{9.3.1}$$

Now consider the polynomial expansion of a variable $u^{(e)}$ in the form

$$u^{(e)} = \alpha_1 + \alpha_2 x + \alpha_3 y \tag{9.3.2}$$

This represents a linear variation of u in both x and y directions within the triangular element. To evaluate the three constants α_1, α_2, and α_3, we must provide three equations in terms of the known values of u, x, and y at each of the three nodes.

$$u_1^{(e)} = \alpha_1 + \alpha_2 x_1 + \alpha_3 y_1$$

$$u_2^{(e)} = \alpha_1 + \alpha_2 x_2 + \alpha_3 y_2$$

$$u_3^{(e)} = \alpha_1 + \alpha_2 x_3 + \alpha_3 y_3$$

Writing in a matrix form, we obtain

$$\begin{bmatrix} u_1^{(e)} \\ u_2^{(e)} \\ u_3^{(e)} \end{bmatrix} = \begin{bmatrix} 1 & x_1 & y_1 \\ 1 & x_2 & y_2 \\ 1 & x_3 & y_3 \end{bmatrix} \begin{bmatrix} \alpha_1 \\ \alpha_2 \\ \alpha_3 \end{bmatrix} \tag{9.3.3}$$

Solving for the constants and substituting them into (9.3.2) gives

$$u^{(e)} = \begin{bmatrix} 1 & x & y \end{bmatrix} \begin{bmatrix} 1 & x_1 & y_1 \\ 1 & x_2 & y_2 \\ 1 & x_3 & y_3 \end{bmatrix}^{-1} \begin{bmatrix} u_1^{(e)} \\ u_2^{(e)} \\ u_3^{(e)} \end{bmatrix}$$

$$= (a_1 + b_1 x + c_1 y)u_1^{(e)} + (a_2 + b_2 x + c_2 y)u_2^{(e)} + (a_3 + b_3 x + c_3 y)u_3^{(e)}$$

$$= \Phi_1^{(e)} u_1^{(e)} + \Phi_2^{(e)} u_2^{(e)} + \Phi_3^{(e)} u_3^{(e)}$$

or

$$u^{(e)} = \Phi_N^{(e)} u_N^{(e)} \quad (N = 1, 2, 3)$$

where the interpolation function $\Phi_N^{(e)}$ is given by

$$\Phi_N^{(e)} = a_N + b_N x + c_N y \tag{9.3.4}$$

$$a_1 = \frac{1}{|D|}(x_2 y_3 - x_3 y_2) \quad a_2 = \frac{1}{|D|}(x_3 y_1 - x_1 y_3) \quad a_3 = \frac{1}{|D|}(x_1 y_2 - x_2 y_1) \tag{9.3.4a}$$

$$b_1 = \frac{1}{|D|}(y_2 - y_3) \quad b_2 = \frac{1}{|D|}(y_3 - y_1) \quad b_3 = \frac{1}{|D|}(y_1 - y_2) \tag{9.3.4b}$$

$$c_1 = \frac{1}{|D|}(x_3 - x_2) \quad c_2 = \frac{1}{|D|}(x_1 - x_3) \quad c_3 = \frac{1}{|D|}(x_2 - x_1) \tag{9.3.4c}$$

with

$$|D| = \det \begin{bmatrix} 1 & x_1 & y_1 \\ 1 & x_2 & y_2 \\ 1 & x_3 & y_3 \end{bmatrix} = 2A$$

where A denotes the area of triangle.

Note that the node numbers 1, 2, 3, are assigned counterclockwise in Figure 9.3.1. If assigned clockwise, however, it is seen that the determinant |D| yields $-2A$, twice the negative area. Observe that the fundamental requirements of the interpolation functions for one dimension,

$$\sum_{N=1}^{3} \Phi_N^{(e)} = 1, \qquad 0 \le \Phi_N^{(e)} \le 1, \qquad \Phi_N^{(e)}(z_M) = \delta_{NM}$$

are also established in this case in two dimensions.

In view of (9.3.1) and (9.3.4a), we note that

$$\begin{aligned} a_1 &= \frac{1}{2A}(x_2 y_3 - x_3 y_2) \\ &= \frac{1}{2A}\left\{\left(X_2 - \frac{1}{3}\sum_{N=1}^{3} X_N\right)\left(Y_3 - \frac{1}{3}\sum_{N=1}^{3} Y_N\right) - \left(X_3 - \frac{1}{3}\sum_{N=1}^{3} X_N\right)\right. \\ &\quad \left. \times \left(Y_2 - \frac{1}{3}\sum_{N=1}^{3} Y_N\right)\right\} \\ &= \left(\frac{1}{2A}\right)\frac{1}{3}\begin{vmatrix} 1 & X_1 & Y_1 \\ 1 & X_2 & Y_2 \\ 1 & X_3 & Y_3 \end{vmatrix} = \left(\frac{1}{2A}\right)\frac{2A}{3} = \frac{1}{3} \end{aligned}$$

Similarly, we may prove that $a_1 = a_2 = a_3 = 1/3$.

If the variable u is assumed to vary quadratically or cubically, then we require additional nodes along the sides and possibly at the interior. The evaluation of constants would require an inversion of a matrix of the size corresponding to the total number of

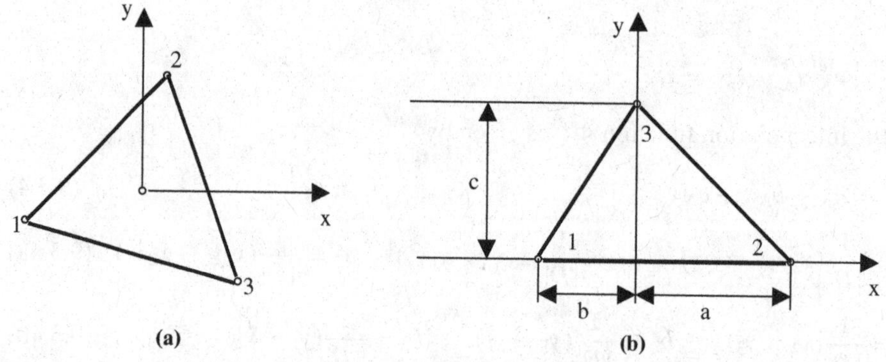

Figure 9.3.2 Integration over the triangular element using cartesian coordinates. **(a)** Integration with origin at centroid; **(b)** Integration with x-axis along one side of triangle.

nodes. An explicit inversion of a large size matrix in terms of nodal coordinate values is difficult, but such complications are avoided if natural coordinates are used.

With the interpolation functions constructed for various degrees of approximations, one generally encounters integration over the spatial domain of the form

$$\iint f\left(\Phi_N^{(e)}\right) dxdy = \iint f(x, y) dxdy$$

If the functions $f(x, y)$ are of higher order, the explicit integration becomes extremely cumbersome. Let us consider an integral

$$P_{rs} = \iint x^r y^s dxdy$$

The limits of this integral must be calculated from the slope of each side of the triangle oriented from the reference cartesian coordinates. The final form of the integral consists of the sum of the integrals performed along all three sides of the triangle. With the origin of the cartesian coordinates at the centroid (Figure 9.3.2a), the following results are obtained:

$$\begin{aligned}
& n = r + s \\
& n = 1 \quad P_{rs} = \iint x dxdy = \iint y dxdy = 0 \\
& n = 2 \quad P_{rs} = \frac{A}{12}\left(x_1^r y_1^s + x_2^r y_2^s + x_3^r y_3^s\right) \\
& n = 3 \quad P_{rs} = \frac{A}{30}\left(x_1^r y_1^s + x_2^r y_2^s + x_3^r y_3^s\right) \qquad (9.3.5) \\
& n = 4 \quad P_{rs} = \frac{A}{60}\left(x_1^r y_1^s + x_2^r y_2^s + x_3^r y_3^s\right) \\
& n = 5 \quad P_{rs} = \frac{2A}{105}\left(x_1^r y_1^s + x_2^r y_2^s + x_3^r y_3^s\right)
\end{aligned}$$

Triangular Element with Origin on One Side

Integration formulas for $r + s > 5$ are difficult to obtain for the triangular element with the origin at the centroid. An easier, more compact integration formula can be derived from a triangle with the origin on the side between nodes 1 and 2 designated as the x-axis with the y-axis passing through node 3 as shown in Figure 9.3.2b. In this triangle, we obtain the integration formula as follows:

$$\begin{aligned}
\iint x^r y^s\,dxdy &= \int_0^c \int_{-\frac{b}{c}(c-y)}^{\frac{a}{c}(c-y)} x^r y^s\,dxdy \\
&= \int_0^c \frac{1}{r+1}\left[x^{r+1}\right]_{-\frac{b}{c}(c-y)}^{\frac{a}{c}(c-y)} y^s\,dy \\
&= \frac{1}{r+1}\frac{a^{r+1}-(-b)^{r+1}}{c^{r+1}}\int_0^c (c-y)^{r+1}y^s\,dy \\
&\vdots \\
&= \frac{r!s!}{(s+r+2)!}\left[a^{r+1}-(-b)^{r+1}\right]c^{s+1}
\end{aligned} \tag{9.3.6}$$

The triangular element characterized by (9.3.6) is effective in the solution of fourth order differential equations [Cowper, et al., 1969].

Example 9.3.1 Local Element Stiffness Matrix

Given: Consider the local element stiffness matrix which arises from the two-dimensional Laplace equation $\nabla^2 u = 0$ in the form

$$K_{NM}^{(e)} = \iint \left(\frac{\partial \Phi_N^{(e)}}{\partial x}\frac{\partial \Phi_M^{(e)}}{\partial x} + \frac{\partial \Phi_N^{(e)}}{\partial y}\frac{\partial \Phi_M^{(e)}}{\partial y}\right) dxdy$$

Required: Determine the explicit form of the above expression in a linear triangular element using the interpolation functions given by (9.3.4).

Solution: Using the formula given by (9.3.3), we obtain

$$\frac{\partial \Phi_N^{(e)}}{\partial x}\frac{\partial \Phi_M^{(e)}}{\partial x} = b_N b_M, \qquad \frac{\partial \Phi_N^{(e)}}{\partial y}\frac{\partial \Phi_M^{(e)}}{\partial y} = c_N c_M$$

Since the area of the triangle is given by

$$\iint dxdy = A$$

the local element stiffness matrix becomes

$$K_{NM}^{(e)} = A(b_N b_M + c_N c_M) = A\begin{bmatrix} b_1^2 + c_1^2 & b_1 b_2 + c_1 c_2 & b_1 b_3 + c_1 c_3 \\ b_2 b_1 + c_2 c_1 & b_2^2 + c_2^2 & b_2 b_3 + c_2 c_3 \\ b_3 b_1 + c_3 c_1 & b_3 b_2 + c_3 c_2 & b_3^2 + c_3^2 \end{bmatrix}$$

where b_N and c_N are explicitly shown by (9.3.4b) and (9.3.4c), respectively. The cartesian coordinate triangular element is simple to use as long as the interpolation function is linear. It is cumbersome for nonlinear interpolation functions with $n = r + s > 5$ in (9.3.5). Notice that the element characterized by the integration formula (9.3.6) is free from this restriction.

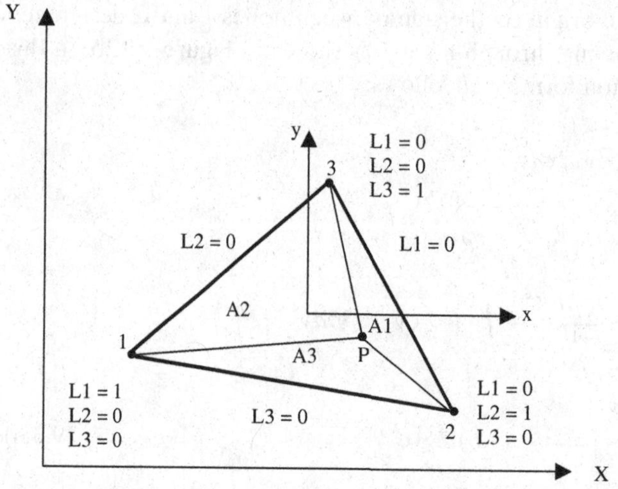

Figure 9.3.3 Natural coordinate triangular element (linear variation).

Natural Coordinate Triangular Element

Consider a triangle with the natural coordinates L_N whose values are zero along the sides and unity on the vertices with a linear variation in between, as shown in Figure 9.3.3. These coordinates are defined as

$$L_i = \frac{A_i}{A}$$

Here $L_1 = A_1/A$, $L_2 = A_2/A$, and $L_3 = A_3/A$ with A_1, A_2, and A_3 being the areas obtained by connecting the three vertices from any point within the triangle such that the total area A is

$$A = A_1 + A_2 + A_3 \tag{9.3.7}$$

$$1 = L_1 + L_2 + L_3 \tag{9.3.8}$$

with A_1 = area (P23), A_2 = area (P31), and A_3 = area (P12). It is now possible to establish a relationship between the cartesian coordinates x and the natural coordinates L_N in the form

$$x = L_1 x_1 + L_2 x_2 + L_3 x_3 \tag{9,3.9a}$$

$$y = L_1 y_1 + L_2 y_2 + L_3 y_3 \tag{9.3.9b}$$

Writing (9.3.8) and (9.3.9) in matrix form, we obtain

$$\begin{bmatrix} 1 \\ x \\ y \end{bmatrix} = \begin{bmatrix} 1 & 1 & 1 \\ x_1 & x_2 & x_3 \\ y_1 & y_2 & y_3 \end{bmatrix} \begin{bmatrix} L_1 \\ L_2 \\ L_3 \end{bmatrix} \tag{9.3.10}$$

We note that the 3×3 matrix on the right-hand side of (9.3.10) is the transpose of the matrix appearing on the right-hand side of (9.3.3). Solving for the natural coordinates, we obtain the interesting result

$$L_1 = \Phi_1^{(e)} \qquad L_2 = \Phi_2^{(e)} \qquad L_3 = \Phi_3^{(e)} \text{ or } L_N = \Phi_N^{(e)} \tag{9.3.11}$$

with $\Phi_N^{(e)}$ identical to (9.3.4).

The natural coordinates as used here for the triangular element are often called the area coordinates or triangular coordinates. Any variable u may now be written as

$$u^{(e)} = L_N u_N^{(e)}$$

It is possible to write (9.3.9a) in the form

$$x = \alpha_1 L_1 + \alpha_2 L_2 + \alpha_3 L_3 \tag{9.3.12}$$

Writing for each node, we obtain

$$x_1 = \alpha_1, \qquad x_2 = \alpha_2, \qquad x_3 = \alpha_3$$

Substituting these into (9.3.12) yields the same expression as (9.3.9a).

Advantages of the natural coordinates can be demonstrated for higher order elements. Notice that, if cartesian coordinates are used for quadratic elements, we require an inversion of the 6×6 matrix corresponding to the three corner nodes plus three side nodes. This difficulty can be avoided in the natural coordinate system. For example, for quadratic approximations, we may write

$$x = \alpha_1 L_1 + \alpha_2 L_2 + \alpha_3 L_3 + \alpha_4 L_1 L_2 + \alpha_5 L_2 L_3 + \alpha_6 L_3 L_1 \tag{9.3.13}$$

Referring to Figure 9.3.4 with three additional nodes installed at midsides of the triangle, we may write (9.3.13) at each corner and midside node,

$$x_1 = \alpha_1 \qquad x_2 = \alpha_2 \qquad x_3 = \alpha_3$$

$$x_4 = \frac{1}{2}\alpha_1 + \frac{1}{2}\alpha_2 + \frac{1}{4}\alpha_4 \qquad x_5 = \frac{1}{2}\alpha_2 + \frac{1}{2}\alpha_3 + \frac{1}{4}\alpha_5 \qquad x_6 = \frac{1}{2}\alpha_1 + \frac{1}{2}\alpha_3 + \frac{1}{4}\alpha_6$$

Solving for the constants and substituting them into (9.3.13) yields

$$x = \Phi_r^{(e)} x_r \quad (r = 1, 2, \ldots, 6) \tag{9.3.14}$$

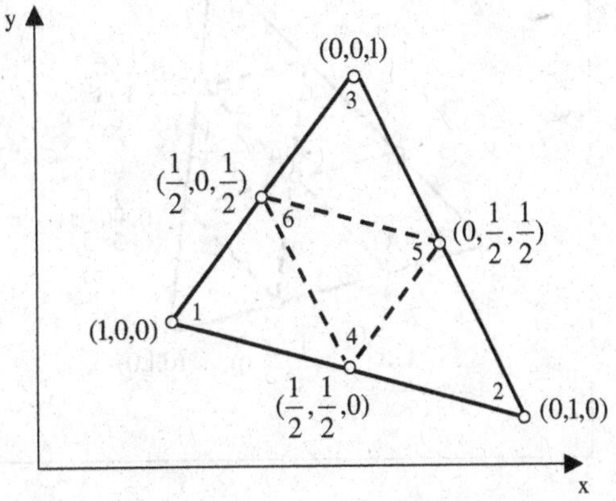

Figure 9.3.4 Natural coordinate triangular element (quadratic variation).

with

$$\Phi_1^{(e)} = (2L_1 - 1)L_1 \qquad \Phi_2^{(e)} = (2L_2 - 1)L_2 \qquad \Phi_3^{(e)} = (2L_3 - 1)L_3$$
$$\Phi_4^{(e)} = 4L_1L_2 \qquad \Phi_5^{(e)} = 4L_2L_3 \qquad \Phi_6^{(e)} = 4L_3L_1 \tag{9.3.15}$$

Similarly, we write

$$y = \Phi_r^{(e)} y_r$$

and consequently, for any variable u

$$u^{(e)} = \Phi_r^{(e)} u_r^{(e)}$$

Using the index notations for a cubic variation, we may proceed similarly as follows (see Figure 9.3.5):

$$x = a_N L_N + b_{NM} L_N L_M + c_{NMQ} L_N L_M L_Q \tag{9.3.16}$$

with $N, M, Q = 1, 2, 3$ and $b_{NM} = 0$ for $N = M$ and $c_{NMQ} = 0$ for $N = M = Q$. Writing (9.3.16) for the three corner nodes, six side nodes (equally spaced), and the interior node, we evaluate the ten constants. Returning to (9.3.16) with these constants, we can now write

$$x = \Phi_r^{(e)} x_r \quad (r = 1, 2, \ldots, 10) \tag{9.3.17}$$

Here, for corner nodes:

$$\Phi_N^{(e)} = \frac{1}{2}(3L_N - 1)(3L_N - 2)L_N \quad (N = 1, 2, 3)$$

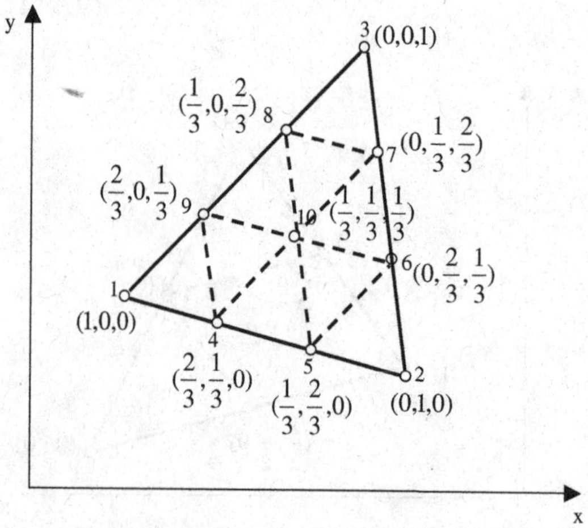

Figure 9.3.5 Natural coordinate triangular element (cubic variation).

for side nodes:

$$\Phi_4^{(e)} = \frac{9}{2} L_1 L_2 (3L_1 - 1) \qquad \Phi_7^{(e)} = \frac{9}{2} L_2 L_3 (3L_3 - 1)$$

$$\Phi_5^{(e)} = \frac{9}{2} L_1 L_2 (3L_2 - 1) \qquad \Phi_8^{(e)} = \frac{9}{2} L_3 L_1 (3L_3 - 1)$$

$$\Phi_6^{(e)} = \frac{9}{2} L_2 L_3 (3L_2 - 1) \qquad \Phi_9^{(e)} = \frac{9}{2} L_3 L_1 (3L_1 - 1)$$

for interior node:

$$\Phi_{10}^{(e)} = 27 L_1 L_2 L_3 \tag{9.3.18}$$

It has been shown that the determination of the interpolation functions for the natural coordinate triangular element can be accomplished quite easily by noting the special geometrical features that make it possible to avoid the inversion.

An additional feature, which should be noted, is the fact that the Lagrange interpolation formula can be used to generalize the procedure. Consider the higher order elements as depicted in Figure 9.3.6. The Lagrange interpolation formula may be

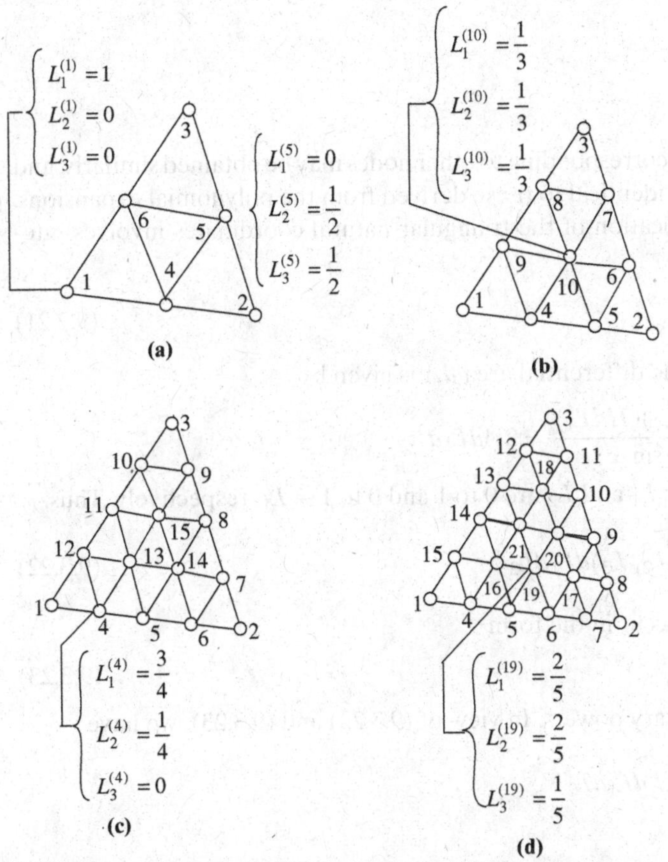

Figure 9.3.6 High order natural coordinate elements. **(a)** Quadratic (m = 2); **(b)** cubic (m = 3); **(c)** quartric (m = 4); **(d)** quintic (m = 5).

transformed to natural coordinates by

$$B^{(r)}(L_N) = \begin{cases} \prod_{s=1}^{s=d} \frac{1}{s}(mL_N - s + 1) & \text{for } d \geq 1 \\ 1 & \text{for } d = 0 \end{cases} \tag{9.3.19}$$

with $d = mL_N^{(r)}$. Here m denotes the degree of approximations and $L_N^{(r)}$ $(N = 1, 2, 3, r = 1, 2, \ldots, n, n = \text{total number of nodes})$ represents the values of area coordinates at each node. The interpolation functions are given by

$$\Phi_r^{(e)} = B^{(r)}(L_1)B^{(r)}(L_2)B^{(r)}(L_3) \tag{9.3.20}$$

To determine $\Phi_1^{(e)}$, we write (for $m = 2$)

$$\Phi_1^{(e)} = B^{(1)}(L_1)B^{(1)}(L_2)B^{(1)}(L_3)$$

$$B^{(1)}(L_1) = (2L_1 - 1 + 1)\frac{1}{2}(2L_1 - 2 + 1)$$

$$B^{(1)}(L_2) = 1$$

$$B^{(1)}(L_3) = 1$$

Thus,

$$\Phi_1^{(e)} = L_1(2L_1 - 1)$$

The interpolation functions corresponding to other nodes may be obtained similarly, and we note that the results are identical to those derived from the polynomial expansions.

The finite element application of the triangular natural coordinates involves integration of a typical form

$$I = \int_A f(L_1, L_2, L_3)dA \tag{9.3.21}$$

Referring to Figure 9.3.7, the differential area dA is given by

$$dA = \frac{(dh)(dH)}{\sin\alpha} = \frac{(hdL_2)(HdL_1)}{\sin\alpha} = 2AdL_1dL_2$$

The limits of integration for L_1 and L_2 are 0 to 1 and 0 to $1 - L_1$, respectively. Thus,

$$I = 2A\int_0^1 \int_0^{1-L_1} f(L_1, L_2, L_3)dL_1dL_2 \tag{9.3.22}$$

where the function f may occur in the form

$$f(L_1, L_2, L_3) = L_1^m L_2^n L_3^p \tag{9.3.23}$$

with m, n, p being the arbitrary powers. In view of (9.3.22) and (9.3.23), we have

$$I = 2A\int_0^1 \int_0^{1-L_1} L_1^m L_2^n L_3^p dL_1dL_2$$

or

$$I = 2A\int_0^1 JL_1^m dL_1 \tag{9.3.24}$$

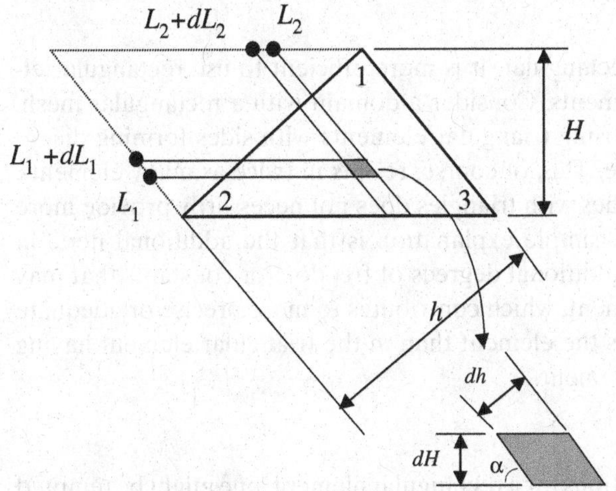

Figure 9.3.7 Geometry for area coordinate integration.

where

$$J = \int_0^{1-L_1} L_2^n L_3^p dL_2 = \int_0^{1-L_1} L_2^n (1 - L_1 - L_2)^p dL_2 \tag{9.3.25}$$

Integrating (9.3.25) by parts gives

$$J = \left[\frac{L_2^{n+1}}{n+1} (1 - L_1 - L_2)^p \right]_0^{1-L_1} + \int_0^{1-L_1} \frac{pL_2^{n+1}}{n+1} (1 - L_1 - L_2)^{p-1} dL_2$$

$$= \frac{p}{n+1} \int_0^{1-L_1} L_2^{n+1} (1 - L_1 - L_2)^{p-1} dL_2$$

$$= \frac{p(p-1)}{(n+1)(n+2)} \int_0^{1-L_1} L_2^{n+2} (1 - L_1 - L_2)^{p-2} dL_2$$

or

$$J = \frac{p!n!}{(n+p)!} \int_0^{1-L_1} L_2^{n+p} dL_2 = \frac{p!n!(1 - L_1)^{n+p+1}}{(n+p+1)!} \tag{9.3.26}$$

Substituting (9.3.26) into (9.3.24) and integrating by parts again, we obtain

$$I = \frac{2Am!p!n!}{(n+m+p+2)!} \tag{9.3.27}$$

For example, if $m = 2$, $n = 0$, and $p = 3$, we obtain

$$\int_A L_1^2 L_3^3 dxdy = \frac{2A(2!)(0!)(3!)}{(2+0+3+2)!} = \frac{A}{210}$$

It is clear that the advantage of this element is that higher order elements are generated easily and a simple integration formula is available without limitation to the order of polynomial degrees.

9.3.2 RECTANGULAR ELEMENTS

If the entire domain of study is rectangular, it is more efficient to use rectangular elements rather than triangular elements. Consider a domain with a rectangular mesh. The mesh can also be generated using triangular elements with sides forming diagonals passed through each rectangle. This, of course, results in twice as many elements. That such a system of refined meshes with triangles does not necessarily provide more accurate results is well known. A simple explanation is that the additional node in the rectangular element leads to additional degrees of freedom or constants that may be specified at all nodes of an element, which contributes to more precise or adequate representation of a variable across the element than in the triangular element having an area equal to the rectangular element.

Cartesian Coordinate Elements

To construct interpolation functions for a rectangular element, one might be tempted to use a polynomial expansion in terms of the standard cartesian coordinates.

$$u^{(e)} = \alpha_1 + \alpha_2 x + \alpha_3 y + \alpha_4 xy + \ldots \tag{9.3.28}$$

The necessary terms of polynomials corresponding to the side and interior nodes, as well as the corner nodes as related to the degrees of approximations of a variable, must be chosen wisely. Polynomials are often incomplete for the desired inclusion of side and interior nodes. Furthermore, the inverses of coefficient matrices may not exist in some cases. The natural coordinates, on the other hand, usually provide an efficient means of obtaining acceptable forms of the interpolation functions. Lagrange and Hermite polynomials, as discussed in the one-dimensional case, are also frequently used for the rectangular elements. A special element popularly known as an isoparametric element is perhaps the most widely adopted. Among the many desirable features of the isoparametric element is the fact that it may be used not only for the rectangular geometry but also for irregular quadrilateral geometries.

Lagrange and Hermite Elements

The advantage of using Lagrange or Hermite elements for a rectangular element is that desired interpolation functions are constructed simply by a tensor product of the one-dimensional counterparts for the x and y directions, respectively.

Consider the Lagrange interpolations in two dimensions, as shown in Figure 9.3.8. For a linear variation of u (Figure 9.3.8a), we write

$$u^{(e)} = \Phi_N^{(e)} u_N^{(e)} \quad (N = 1, 2, 3, 4) \tag{9.3.29}$$

with

$$\Phi_1^{(e)} = L_1^{(x)} L_1^{(y)}, \qquad \Phi_2^{(e)} = L_2^{(x)} L_1^{(y)}, \qquad \Phi_3^{(e)} = L_2^{(x)} L_2^{(y)} \quad \text{and} \quad \Phi_4^{(e)} = L_1^{(x)} L_2^{(y)}$$

where

$$L_1^{(x)} = \frac{1}{2}(1-\xi), \qquad L_2^{(x)} = \frac{1}{2}(1+\xi), \qquad L_1^{(y)} = \frac{1}{2}(1-\eta),$$

$$L_2^{(y)} = \frac{1}{2}(1+\eta), \qquad \xi = \frac{2x}{a}, \qquad \eta = \frac{2y}{b}$$

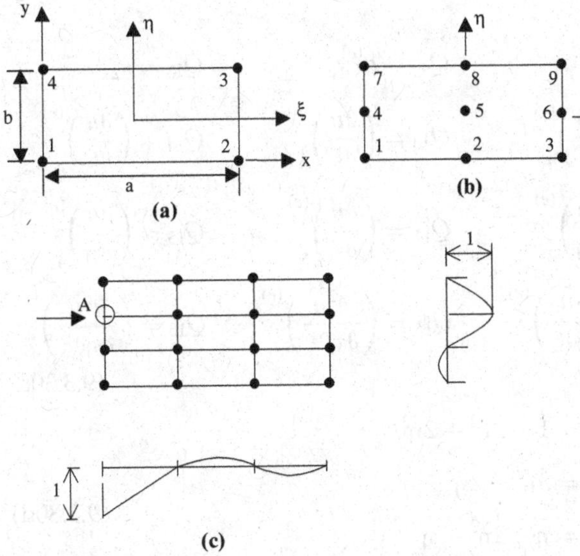

Figure 9.3.8 Lagrange interpolation functions: **(a)** linear, **(b)** quadratic, **(c)** cubic (variations of functions along the line through A).

Interpolations of quadratic and cubic variations can be constructed in the same way (see Figure 9.3.8 b,c).

The Hermite polynomials may be applied similarly to the rectangular element as the Lagrange polynomials. For bicubic Hermite polynomials, we have (Figure 9.3.9):

$$u^{(e)} = \Phi_r^{(e)} Q_r \quad (r = 1, 2, \ldots 16) \tag{9.3.30a}$$

with

$$
\begin{aligned}
\Phi_1^{(e)} &= H_{1(x)}^0 H_{1(y)}^0 & \Phi_5^{(e)} &= H_{2(x)}^0 H_{1(y)}^0 & \Phi_9^{(e)} &= H_{1(x)}^0 H_{2(y)}^0 & \Phi_{13}^{(e)} &= H_{2(x)}^0 H_{2(y)}^0 \\
\Phi_2^{(e)} &= H_{1(x)}^1 H_{1(y)}^0 & \Phi_6^{(e)} &= H_{2(x)}^1 H_{1(y)}^0 & \Phi_{10}^{(e)} &= H_{1(x)}^1 H_{2(y)}^0 & \Phi_{14}^{(e)} &= H_{2(x)}^1 H_{2(y)}^0 \\
\Phi_3^{(e)} &= H_{1(x)}^0 H_{1(y)}^1 & \Phi_7^{(e)} &= H_{2(x)}^0 H_{1(y)}^1 & \Phi_{11}^{(e)} &= H_{1(x)}^0 H_{2(y)}^1 & \Phi_{15}^{(e)} &= H_{2(x)}^0 H_{2(y)}^1 \\
\Phi_4^{(e)} &= H_{1(x)}^1 H_{1(y)}^1 & \Phi_8^{(e)} &= H_{2(x)}^1 H_{1(y)}^1 & \Phi_{12}^{(e)} &= H_{1(x)}^1 H_{2(y)}^1 & \Phi_{16}^{(e)} &= H_{2(x)}^1 H_{2(y)}^1
\end{aligned}
\tag{9.3.30b}
$$

$$
\begin{cases}
Q_1 = u_1^{(e)} \\
Q_2 = (\partial u/\partial \xi)_1^{(e)} \\
Q_3 = (\partial u/\partial \eta)_1^{(e)} \\
Q_4 = (\partial^2 u/\partial \xi \partial \eta)_1^{(e)}
\end{cases}
$$

$$
\begin{cases}
Q_{13} = u_4^{(e)} \\
Q_{14} = (\partial u/\partial \xi)_4^{(e)} \\
Q_{15} = (\partial u/\partial \eta)_4^{(e)} \\
Q_{16} = (\partial^2 u/\partial \xi \partial \eta)_4^{(e)}
\end{cases}
$$

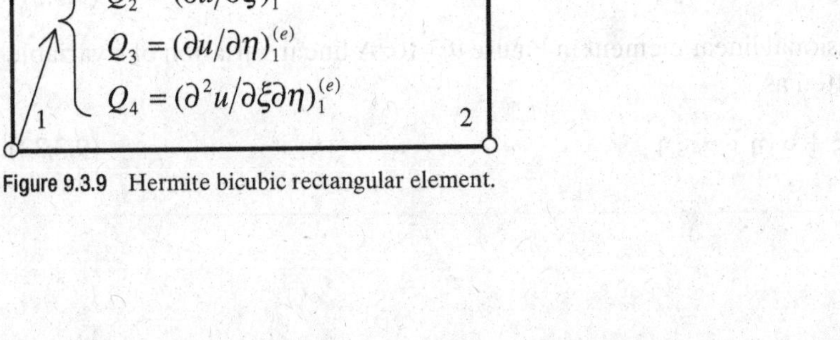

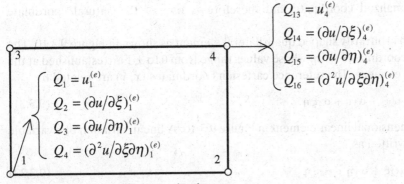

Figure 9.3.9 Hermite bicubic rectangular element.

and

$$
\begin{array}{llll}
Q_1 = u_1^{(e)} & Q_5 = u_2^{(e)} & Q_9 = u_3^{(e)} & Q_{13} = u_4^{(e)} \\
Q_2 = \left(\dfrac{\partial u}{\partial \xi}\right)_1^{(e)} & Q_6 = \left(\dfrac{\partial u}{\partial \xi}\right)_2^{(e)} & Q_{10} = \left(\dfrac{\partial u}{\partial \xi}\right)_3^{(e)} & Q_{14} = \left(\dfrac{\partial u}{\partial \xi}\right)_4^{(e)} \\
Q_3 = \left(\dfrac{\partial u}{\partial \eta}\right)_1^{(e)} & Q_7 = \left(\dfrac{\partial u}{\partial \eta}\right)_2^{(e)} & Q_{11} = \left(\dfrac{\partial u}{\partial \eta}\right)_3^{(e)} & Q_{15} = \left(\dfrac{\partial u}{\partial \eta}\right)_4^{(e)} \\
Q_4 = \left(\dfrac{\partial^2 u}{\partial \eta \partial \xi}\right)_1^{(e)} & Q_8 = \left(\dfrac{\partial^2 u}{\partial \eta \partial \xi}\right)_2^{(e)} & Q_{12} = \left(\dfrac{\partial^2 u}{\partial \eta \partial \xi}\right)_3^{(e)} & Q_{16} = \left(\dfrac{\partial^2 u}{\partial \eta \partial \xi}\right)_4^{(e)}
\end{array}
\tag{9.3.30c}
$$

$$
\begin{array}{ll}
H_{1(x)}^0 = 1 - 3\xi^2 + 2\xi^3 & H_{1(y)}^0 = 1 - 3\eta^2 + 2\eta^3 \\
H_{2(x)}^0 = 3\xi^2 - 2\xi^3 & H_{2(y)}^0 = 3\eta^2 - 2\eta^3 \\
H_{1(x)}^1 = \xi - 2\xi^2 + \xi^3 & H_{1(y)}^1 = \eta - 2\eta^2 + \eta^3 \\
H_{2(x)}^1 = \xi^3 - \xi^2 & H_{2(y)}^1 = \eta^3 - \eta^2
\end{array}
\tag{9.3.30d}
$$

Note that, because of the combinations of the Hermite polynomials for both x and y directions, the mixed second derivatives must be included as nodal generalized coordinates. Higher order Hermite polynomials may be constructed similarly using (9.2.14).

A similar approach can be used to generate three-dimensional elements $\Phi_1^{(e)} = L_1^{(x)} L_2^{(y)} L_3^{(z)}$, etc. for Lagrange elements and similarly for Hermite elements. However, it should be noted that for nonorthogonal elements (arbitrary quadrilateral and hexahedral), appropriate coordinate transformation (geometrical Jacobian) will be required as discussed in the following section.

9.3.3 QUADRILATERAL ISOPARAMETRIC ELEMENTS

The isoparametric element was first studied by Zienkiewicz and his associates [see Zienkiewicz, 1971]. The name "isoparametric" derives from the fact that the "same" parametric function which describes the geometry may be used for interpolating spatial variations of a variable within an element. The isoparametric element utilizes a nondimensionalized coordinate and therefore is one of the natural coordinate elements.

Consider an arbitrarily shaped quadrilateral element as shown in Figure 9.3.10. The isoparametric coordinates (ξ, η) whose values range from 0 to ± 1 are established at the centroid of the element. The reference cartesian coordinates (x, y) are related to

$$x, y = \alpha_1 + \alpha_2\xi + \alpha_3\eta + \alpha_4\xi\eta \tag{9.3.31}$$

for the two-dimensional linear element in Figure 9.3.10. A linear variation of a variable u may also be written as

$$u^{(e)} = \alpha_1 + \alpha_2\xi + \alpha_3\eta + \alpha_4\xi\eta \tag{9.3.32}$$

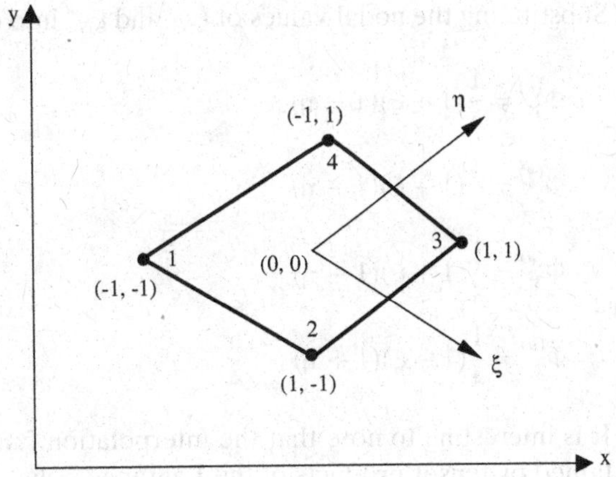

Figure 9.3.10 Quadrilateral isoparametric element (linear variation).

Writing (9.3.31) in terms of nodal values yields

$$
\begin{aligned}
x_1 &= \alpha_1 + \alpha_2(-1) + \alpha_3(-1) + \alpha_4(-1)(-1) \\
x_2 &= \alpha_1 + \alpha_2(1) + \alpha_3(-1) + \alpha_4(1)(-1) \\
x_3 &= \alpha_1 + \alpha_2(1) + \alpha_3(1) + \alpha_4(1)(1) \\
x_4 &= \alpha_1 + \alpha_2(-1) + \alpha_3(1) + \alpha_4(-1)(1)
\end{aligned}
\tag{9.3.33a}
$$

In a matrix form, we may rewrite (9.3.33a) as

$$[x] = [C][\alpha] \tag{9.3.33b}$$

Here the coefficient matrix [C] is given by

$$[C] = \begin{bmatrix} 1 & -1 & -1 & 1 \\ 1 & 1 & -1 & -1 \\ 1 & 1 & 1 & 1 \\ 1 & -1 & 1 & -1 \end{bmatrix}$$

Thus,

$$[\alpha] = [C]^{-1}[x] \tag{9.3.34}$$

with

$$[C]^{-1} = \frac{1}{4}\begin{bmatrix} 1 & 1 & 1 & 1 \\ -1 & 1 & 1 & -1 \\ -1 & -1 & 1 & 1 \\ 1 & -1 & 1 & -1 \end{bmatrix}$$

Substituting (9.3.34) into (9.3.31) yields

$$x_i = \Phi_N^{(e)} x_{Ni} \tag{9.3.35}$$

Here $\Phi_N^{(e)}$ is called the isoparametric function and has the form

$$\Phi_N^{(e)} = \frac{1}{4}(1 + \xi_{N1}\xi_1)(1 + \xi_{N2}\xi_2) \tag{9.3.36}$$

Substituting the nodal values of ξ_{N1} and ξ_{N2} into (9.3.36) gives, with $\xi_1 = \xi$, $\xi_2 = \eta$,

$$\begin{aligned}
\Phi_1^{(e)} &= \frac{1}{4}(1-\xi)(1-\eta) \\
\Phi_2^{(e)} &= \frac{1}{4}(1+\xi)(1-\eta) \\
\Phi_3^{(e)} &= \frac{1}{4}(1+\xi)(1+\eta) \\
\Phi_4^{(e)} &= \frac{1}{4}(1-\xi)(1+\eta)
\end{aligned} \tag{9.3.37}$$

It is interesting to note that the interpolation functions derived in (9.3.37) can be obtained by tensor products of the Lagrange polynomials with the origin at the centroid from (9.3.29) for the case in Figure (9.3.8a).

The quadratic element requires midside nodes as shown in Figure 9.3.11. Thus, we may approximate x or y in the form (see Figure 9.1.6)

$$x, y = \alpha_1 + \alpha_2\xi + \alpha_3\eta + \alpha_4\xi\eta + \alpha_5\xi^2 + \alpha_6\eta^2 + \alpha_7\xi^2\eta + \alpha_8\xi\eta^2 \tag{9.3.38}$$

A similar procedure as in the linear element may be used to determine $[C]$ and $[C]^{-1}$, and we obtain

at corner nodes:

$$\Phi_N^{(e)}(\xi_i) = \frac{1}{4}(1+\xi_{N1}\xi_1)(1+\xi_{N2}\xi_2)(\xi_{N1}\xi_1 + \xi_{N2}\xi_2 - 1) \tag{9.3.39}$$

at midside nodes:

$$\begin{aligned}
\Phi_N^{(e)}(\xi_i) &= \frac{1}{2}(1-\xi_1^2)(1+\xi_{N2}\xi_2) \quad \text{for } \xi_{N1} = 0 \\
\Phi_N^{(e)}(\xi_i) &= \frac{1}{2}(1+\xi_{N1}\xi_1)(1-\xi_2^2) \quad \text{for } \xi_{N2} = 0
\end{aligned} \tag{9.3.40}$$

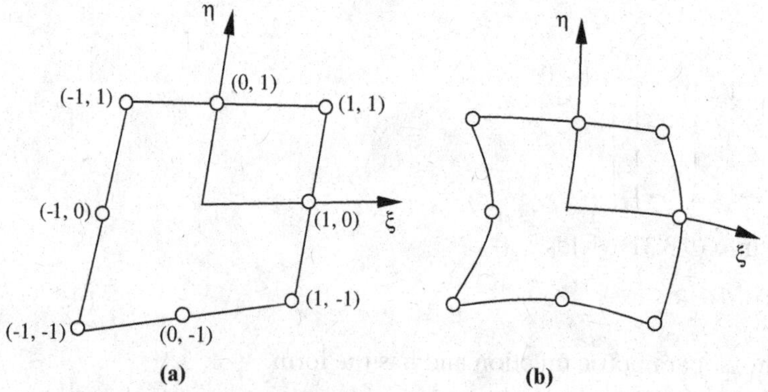

Figure 9.3.11 Quadrilateral isoparametric element (quadratic variation): **(a)** straight edges, **(b)** curved edges.

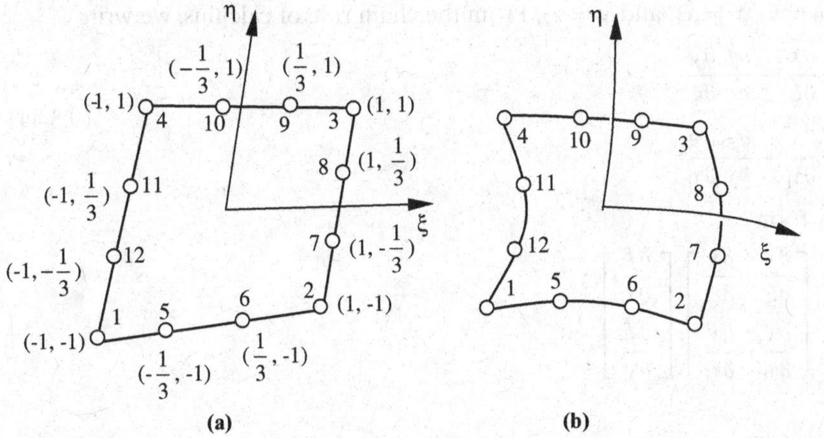

Figure 9.3.12 Quadrilateral isoparametric element (cubic variation): **(a)** straight edges; **(b)** curved edges.

For a cubic element as shown in Figure 9.3.12, we have (see Figure 9.1.6)

$$x, y = \alpha_1 + \alpha_2\xi + \alpha_3\eta + \alpha_4\xi\eta + \alpha_5\xi^2 + \alpha_6\eta^2 + \alpha_7\xi^2\eta + \alpha_8\xi\eta^2 + \alpha_9\xi^3 \\ + \alpha_{10}\eta^3 + \alpha_{11}\xi^3\eta + \alpha_{12}\xi\eta^3 \tag{9.3.41}$$

where we notice that the $x^2y^2(\xi^2\eta^2$ here) term is omitted from the complete cubic expansion (Figure 9.1.6) in order to match the number of nodes chosen here (12 terms instead of 13 terms)

at corner nodes:

$$\Phi_N^{(e)}(\xi_i) = \frac{1}{32}(1 + \xi_{N1}\xi_1)(1 + \xi_{N2}\xi_2)[9(\xi_1^2 + \xi_2^2) - 10] \tag{9.3.42a}$$

at side nodes:

$$\Phi_N^{(e)}(\xi_i) = \frac{9}{32}(1 + \xi_{N1}\xi_1)(1 - \xi_2^2)(1 + 9\xi_{N2}\xi_2) \quad \text{for } \xi_{N1} = \pm 1 \text{ and } \xi_{N2} = \pm\tfrac{1}{3}$$
$$\Phi_N^{(e)}(\xi_i) = \frac{9}{32}(1 + \xi_{N2}\xi_2)(1 - \xi_1^2)(1 + 9\xi_{N1}\xi_1) \quad \text{for } \xi_{N2} = \pm 1 \text{ and } \xi_{N1} = \pm\tfrac{1}{3} \tag{9.3.42b}$$

It should be remarked that for higher order isoparametric elements, Lagrange polynomials can still be used without interior nodes but with side constraints imposed.

In engineering applications, we are concerned with a derivative and the integration of quantity associated with a variable with respect to the cartesian reference coordinates. Since the variable is represented in terms of the nondimensionalized isoparametric coordinates, we require a transformation between the two coordinate systems. Consider a quantity given by

$$\iint \frac{\partial}{\partial x_i} f(\xi, \eta)\, dx\, dy \tag{9.3.43}$$

with $\xi = \xi_1, \eta = \xi_2, x = x_1$, and $y = x_2$. From the chain rule of calculus, we write

$$\begin{aligned} \frac{\partial f}{\partial \xi} &= \frac{\partial f}{\partial x}\frac{\partial x}{\partial \xi} + \frac{\partial f}{\partial y}\frac{\partial y}{\partial \xi} \\ \frac{\partial f}{\partial \eta} &= \frac{\partial f}{\partial x}\frac{\partial x}{\partial \eta} + \frac{\partial f}{\partial y}\frac{\partial y}{\partial \eta} \end{aligned} \tag{9.3.44}$$

or in a matrix form

$$\begin{bmatrix} \dfrac{\partial f}{\partial \xi} \\ \dfrac{\partial f}{\partial \eta} \end{bmatrix} = \begin{bmatrix} \dfrac{\partial x}{\partial \xi} & \dfrac{\partial y}{\partial \xi} \\ \dfrac{\partial x}{\partial \eta} & \dfrac{\partial y}{\partial \eta} \end{bmatrix} \begin{bmatrix} \dfrac{\partial f}{\partial x} \\ \dfrac{\partial f}{\partial y} \end{bmatrix}$$

Thus,

$$\begin{bmatrix} \dfrac{\partial f}{\partial x} \\ \dfrac{\partial f}{\partial y} \end{bmatrix} = [J]^{-1} \begin{bmatrix} \dfrac{\partial f}{\partial \xi} \\ \dfrac{\partial f}{\partial \eta} \end{bmatrix} \tag{9.3.45}$$

where J is called the Jacobian given by

$$[J] = \begin{bmatrix} \dfrac{\partial x}{\partial \xi} & \dfrac{\partial y}{\partial \xi} \\ \dfrac{\partial x}{\partial \eta} & \dfrac{\partial y}{\partial \eta} \end{bmatrix} \tag{9.3.46}$$

Here the derivatives $\partial f/\partial x$ or $\partial f/\partial y$ are determined from the inverse of the Jacobian and the derivatives $\partial f/\partial \xi$ and $\partial f/\partial \eta$. The integration over the domain referenced to the cartesian coordinates must be changed to the domain now referenced to the isoparametric coordinates

$$\iint dxdy = \int_{-1}^{1}\int_{-1}^{1} |J| d\xi d\eta \tag{9.3.47}$$

To prove (9.3.47), we consider the two coordinate systems shown in Figure 9.3.13. The directions of the cartesian coordinates and the arbitrary nonorthogonal (possibly curvilinear) isoparametric coordinates are given by the unit vectors $\mathbf{i}_1, \mathbf{i}_2$, and the tangent vectors $\mathbf{g}_1, \mathbf{g}_2$, respectively, related by

$$\mathbf{g}_1 = \frac{\partial x}{\partial \xi}\mathbf{i}_1 + \frac{\partial y}{\partial \xi}\mathbf{i}_2$$

$$\mathbf{g}_2 = \frac{\partial x}{\partial \eta}\mathbf{i}_1 + \frac{\partial y}{\partial \eta}\mathbf{i}_2$$

The differential area (shaded) is

$$dx\,\mathbf{i}_1 \times dy\,\mathbf{i}_2 = dxdy\,\mathbf{i}_3 = \mathbf{g}_1 d\xi \times \mathbf{g}_2 d\eta = \begin{vmatrix} \mathbf{i}_1 & \mathbf{i}_2 & \mathbf{i}_3 \\ \dfrac{\partial x}{\partial \xi} & \dfrac{\partial y}{\partial \xi} & 0 \\ \dfrac{\partial x}{\partial \eta} & \dfrac{\partial y}{\partial \eta} & 0 \end{vmatrix} d\xi d\eta$$

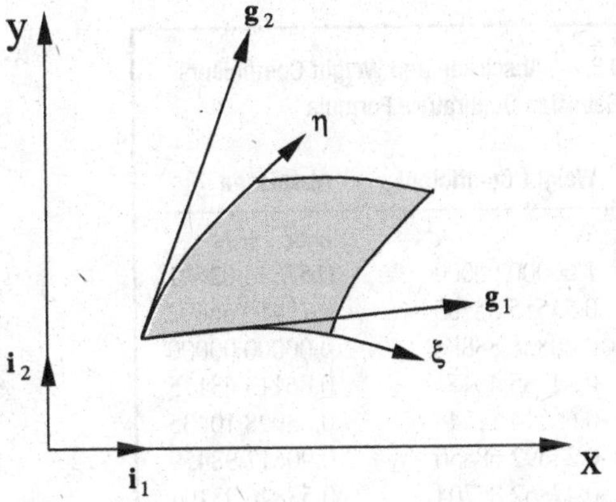

Figure 9.3.13 Coordinate transformation.

or

$$dx dy\, \mathbf{i}_3 = |J| d\xi d\eta\, \mathbf{i}_3$$

with

$$|J| = \begin{vmatrix} \dfrac{\partial x}{\partial \xi} & \dfrac{\partial y}{\partial \xi} \\ \dfrac{\partial x}{\partial \eta} & \dfrac{\partial y}{\partial \eta} \end{vmatrix}$$

Thus, we obtain the relations

$$dx dy = |J| d\xi d\eta \tag{9.3.48}$$

and

$$\iint \frac{\partial f}{\partial x} dx dy = \int_{-1}^{1}\int_{-1}^{1}\left(\overline{J}_{11}\frac{\partial f}{\partial \xi} + \overline{J}_{12}\frac{\partial f}{\partial \eta}\right)|J| d\xi d\eta = \int_{-1}^{1}\int_{-1}^{1} g_x(\xi,\eta) d\xi d\eta \tag{9.3.49}$$

$$\iint \frac{\partial f}{\partial y} dx dy = \int_{-1}^{1}\int_{-1}^{1}\left(\overline{J}_{21}\frac{\partial f}{\partial \xi} + \overline{J}_{22}\frac{\partial f}{\partial \eta}\right)|J| d\xi d\eta = \int_{-1}^{1}\int_{-1}^{1} g_y(\xi,\eta) d\xi d\eta \tag{9.3.50}$$

where $\overline{J}_{11}, \overline{J}_{12}, \overline{J}_{21}$, and $\overline{J}_{22}$ are the components of the inverted Jacobian matrix (9.3.46).

The integration (9.3.49) may be performed most efficiently by means of the Gaussian quadrature [see Hildebrand, 1956]. For a one-dimensional case, we may write

$$\int_{-1}^{1} f(\xi) d\xi = \sum_{j=1}^{n} w_j f(\xi_j)$$

or, when extended to a tensor product in two dimensions, we write

$$\int_{-1}^{1}\int_{-1}^{1} f(\xi,\eta) d\xi d\eta = \sum_{j=1}^{n}\sum_{k=1}^{n} w_j w_k f(\xi_j,\eta_k)$$

where w_j and w_k are the weight coefficients, and $f(\xi_k)$ and $f(\xi_j,\eta_k)$ denote the abscissae representing the values of the functions $f(\xi)$ and $f(\xi,\eta)$ corresponding to the n Gaussian points. The weight coefficients and abscissae for the first ten Gaussian points

Table 9.3.1 Abscissae and Weight Coefficients of the Gaussian Quadrature Formula

	Weight Coefficient	Abscissae
N	W_k	$\pm\xi_k, \pm\eta_k$
2	1.00000 00000	0.57735 02691
3	0.55555 55555	0.77459 66692
	0.88888 88888	0.00000 00000
4	0.34785 48451	0.86113 63115
	0.65214 51548	0.33998 10435
5	0.23692 68850	0.90617 98459
	0.47862 86704	0.53846 93101
	0.56888 88888	0.00000 00000
6	0.17132 44923	0.93246 95142
	0.36076 15730	0.66120 93864
	0.46791 39345	0.23861 91860
7	0.12948 49661	0.94910 79123
	0.27970 53914	0.74153 11855
	0.38183 00505	0.40584 51513
	0.41795 91836	0.00000 00000
8	0.10122 85362	0.96028 98564
	0.22238 10344	0.79666 64774
	0.31370 66458	0.52553 24099
	0.36268 37833	0.18343 46424
9	0.08127 43883	0.96816 02395
	0.18064 81606	0.83603 11073
	0.26061 06964	0.61336 14327
	0.31234 70770	0.32425 34234
	0.33023 93550	0.00000 00000
10	0.06667 13443	0.97390 65285
	0.14945 13491	0.86506 33666
	0.21908 63625	0.67940 95682
	0.26926 67193	0.43339 53941
	0.29552 42247	0.14887 43389

are shown in Table 9.3.1. In general, accuracy of integration increases with an increase of Gaussian points, but it can be shown that only a very few Gaussian points may lead to an acceptable accuracy. The basic idea of Gaussian quadrature is shown in Appendix B.

The Gaussian quadrature numerical integration may be easily extended to the three-dimensional element. Extension of the Gaussian quadrature integration to the triangular or tetrahedral elements are also possible with some modification of the procedure.

Example 9.3.2 Stiffness Matrix of an Isoparametric Element

Given:

$$K_{NM}^{(e)} = \iint \left(\frac{\partial \Phi_N^{(e)}}{\partial x} \frac{\partial \Phi_M^{(e)}}{\partial x} + \frac{\partial \Phi_N^{(e)}}{\partial y} \frac{\partial \Phi_M^{(e)}}{\partial y} \right) dxdy$$

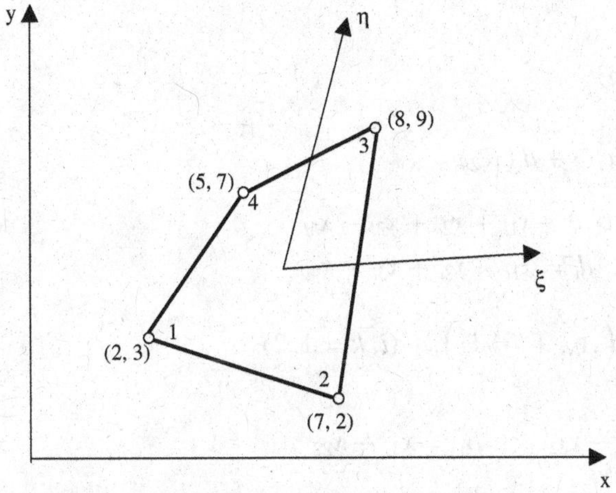

Figure E9.3.2 Geometry for Example 9.3.2.

Required: Work out the detailed algebra necessary for computer integration using the Gaussian quadrature. Compute the integral with 3, 4, 5 Gaussian points (Figure E9.3.2).

Solution:

$$K_{NM}^{(e)} = \iint \left(\frac{\partial \Phi_N^{(e)}}{\partial x} \frac{\partial \Phi_M^{(e)}}{\partial x} + \frac{\partial \Phi_N^{(e)}}{\partial y} \frac{\partial \Phi_M^{(e)}}{\partial y} \right) dxdy$$

$$= \int_{-1}^{1} \int_{-1}^{1} \left(\frac{\partial \Phi_N^{(e)}}{\partial x} \frac{\partial \Phi_M^{(e)}}{\partial x} + \frac{\partial \Phi_N^{(e)}}{\partial y} \frac{\partial \Phi_M^{(e)}}{\partial y} \right) |J| d\xi d\eta$$

$$= \int_{-1}^{1} \int_{-1}^{1} k_{NM}(\xi, \eta) d\xi d\eta$$

$$= \sum_{j=1}^{n} \sum_{k=1}^{n} w_j w_k k_{NM}(\xi_j, \eta_k)$$

where

$$k_{NM}(\xi, \eta) = \left[\left(\overline{J}_{11} \frac{\partial \Phi_N^{(e)}}{\partial \xi} + \overline{J}_{12} \frac{\partial \Phi_N^{(e)}}{\partial \eta} \right) \left(\overline{J}_{11} \frac{\partial \Phi_M^{(e)}}{\partial \xi} + \overline{J}_{12} \frac{\partial \Phi_M^{(e)}}{\partial \eta} \right) + \left(\overline{J}_{21} \frac{\partial \Phi_N^{(e)}}{\partial \xi} + \overline{J}_{22} \frac{\partial \Phi_N^{(e)}}{\partial \eta} \right) \left(\overline{J}_{21} \frac{\partial \Phi_M^{(e)}}{\partial \xi} + \overline{J}_{22} \frac{\partial \Phi_M^{(e)}}{\partial \eta} \right) \right] |J|$$

with

$$\overline{J}_{11} = \frac{1}{|J|} \frac{\partial y}{\partial \eta}, \qquad \overline{J}_{12} = -\frac{1}{|J|} \frac{\partial y}{\partial \xi}, \qquad \overline{J}_{21} = -\frac{1}{|J|} \frac{\partial x}{\partial \eta}, \qquad \overline{J}_{22} = \frac{1}{|J|} \frac{\partial x}{\partial \xi}$$

$$|J| = \frac{\partial x}{\partial \xi} \frac{\partial y}{\partial \eta} - \frac{\partial y}{\partial \xi} \frac{\partial x}{\partial \eta}$$

with

$$\Phi_N^{(e)} = \frac{1}{4}(1 + \xi_{N1}\xi_1)(1 + \xi_{N2}\xi_2)$$

$$x_i = \Phi_N^{(e)} x_{Ni} = \frac{1}{4}(a_i + b_i\xi_1 + c_i\xi_2 + d_i\xi_1\xi_2)$$

$$a_i = x_{1i} + x_{2i} + x_{3i} + x_{4i}, \qquad b_i = -x_{1i} + x_{2i} + x_{3i} - x_{4i}$$

$$c_i = -x_{1i} - x_{2i} + x_{3i} + x_{4i}, \qquad d_i = x_{1i} - x_{2i} + x_{3i} - x_{4i}$$

$$\frac{\partial \Phi_N^{(e)}}{\partial x_i} = (J_{ik})^{-1}\frac{\partial \Phi_N^{(e)}}{\partial \xi_k} = \frac{1}{8|J|}\left(A_{Ni} + B_{Ni}^k \xi_k\right), \quad (i, k = 1, 2)$$

with

$$\begin{aligned}
A_{11} &= x_{22} - x_{42}, & B_{11}^1 &= x_{42} - x_{32}, & B_{11}^2 &= x_{32} - x_{22} \\
A_{21} &= x_{32} - x_{12}, & B_{21}^1 &= x_{32} - x_{42}, & B_{21}^2 &= x_{12} - x_{42} \\
A_{31} &= x_{42} - x_{22}, & B_{31}^1 &= x_{12} - x_{22}, & B_{31}^2 &= x_{42} - x_{12} \\
A_{41} &= x_{12} - x_{32}, & B_{41}^1 &= x_{22} - x_{12}, & B_{41}^2 &= x_{22} - x_{32} \\
A_{12} &= x_{41} - x_{21}, & B_{12}^1 &= x_{31} - x_{41}, & B_{12}^2 &= x_{21} - x_{31} \\
A_{22} &= x_{11} - x_{31}, & B_{22}^1 &= x_{41} - x_{31}, & B_{22}^2 &= x_{41} - x_{11} \\
A_{32} &= x_{21} - x_{41}, & B_{32}^1 &= x_{21} - x_{11}, & B_{32}^2 &= x_{11} - x_{41} \\
A_{42} &= x_{31} - x_{11}, & B_{42}^1 &= x_{11} - x_{21}, & B_{42}^2 &= x_{31} - x_{21}
\end{aligned}$$

$$|J| = \frac{\partial x_1}{\partial \xi_1}\frac{\partial x_2}{\partial \xi_2} - \frac{\partial x_2}{\partial \xi_1}\frac{\partial x_1}{\partial \xi_2} = \frac{1}{8}(\alpha_0 + \alpha_1\xi_1 + \alpha_2\xi_2)$$

$$\alpha_0 = (x_{41} - x_{21})(x_{12} - x_{32}) - (x_{11} - x_{31})(x_{42} - x_{22})$$

$$\alpha_1 = (x_{31} - x_{41})(x_{12} - x_{22}) - (x_{11} - x_{21})(x_{32} - x_{42})$$

$$\alpha_2 = (x_{41} - x_{11})(x_{22} - x_{32}) - (x_{21} - x_{31})(x_{42} - x_{12})$$

where

$$x_{22} - x_{42} = y_2 - y_4, \qquad x_{11} - x_{31} = x_1 - x_3, \text{ etc.}$$

$$\frac{\partial \Phi_N^{(e)}}{\partial x_1} = \frac{1}{8|J|}(A_{N1} + B_{N1}^k \xi_k) = C_{N1}, \qquad \frac{\partial \Phi_N^{(e)}}{\partial x_2} = \frac{1}{8|J|}(A_{N2} + B_{N2}^k \xi_k) = C_{N2}$$

If we chose $n = 3$, then from Table 9.3.1 we have

$$w_1 = 0.55555555, \qquad w_2 = 0.88888888, \quad w_3 = 0.55555555$$

$$(\xi_1, \eta_1) = -0.77459666, \qquad (\xi_2, \eta_2) = 0.0, \qquad (\xi_3, \eta_3) = 0.77459666$$

We are now prepared to calculate

$$K_{NM}^{(e)} = \sum_{i=1}^{n}\sum_{j=1}^{n} w_i w_j k_{NM}(\xi_i, \eta_j)$$

where

$$k_{NM}(\xi_i, \eta_j) = (C_{N1}C_{M1} + C_{N2}C_{M2})|J|$$

Thus,

$$K_{NM}^{(e)} = \begin{bmatrix} 0.5449 & -0.2773 & -0.1035 & -0.1640 \\ -0.2773 & 0.8771 & 0.1380 & -0.7377 \\ -0.1035 & 0.1380 & 0.6378 & -0.6723 \\ -0.1640 & -0.7377 & -0.6723 & 1.5740 \end{bmatrix}$$

Similarly,
for $n = 4$

$$K_{NM}^{(e)} = \begin{bmatrix} 0.5457 & -0.2776 & -0.1026 & -0.1655 \\ -0.2776 & 0.8771 & 0.1377 & -0.7372 \\ -0.1026 & 0.1377 & 0.6390 & -0.6741 \\ -0.1655 & -0.7372 & -0.6741 & 1.5768 \end{bmatrix}$$

for $n = 5$

$$K_{NM}^{(e)} = \begin{bmatrix} 0.5457 & -0.2776 & -0.1025 & -0.1656 \\ -0.2776 & 0.8771 & 0.1376 & -0.7372 \\ -0.1025 & 0.1376 & 0.6391 & -0.6742 \\ -0.1656 & -0.7372 & -0.6742 & 1.5770 \end{bmatrix}$$

We notice that an asymptotic convergence is evident as the Gaussian integration point n increases from 3 to 5.

Example 9.3.3 Transition from Linear to Quadratic Element

Figure E9.3.3 presents irregular elements with transition from a linear element to a quadratic element. In this case, side (1-5-2) is quadratic for the element ($e = 1$). Element 2 is fully quadratic, whereas element 1 is partially linear and partially quadratic. Interpolation functions for element 1 can be derived by constructing tensor products as follows:

$$\Phi_1^{(e)} = L_1^{(2)}(\xi)L_1^{(1)}(\eta) = \frac{1}{4}\xi(\xi - 1)(1 - \eta)$$

$$\Phi_2^{(e)} = L_3^{(2)}(\xi)L_1^{(1)}(\eta) = \frac{1}{4}\xi(\xi + 1)(1 - \eta)$$

$$\Phi_3^{(e)} = L_2^{(1)}(\xi)L_2^{(1)}(\eta) = \frac{1}{4}(1 + \xi)(1 + \eta)$$

$$\Phi_4^{(e)} = L_1^{(1)}(\xi)L_2^{(1)}(\eta) = \frac{1}{4}(1 - \xi)(1 + \eta)$$

$$\Phi_5^{(e)} = L_2^{(2)}(\xi)L_1^{(1)}(\eta) = \frac{1}{2}(1 - \xi^2)(1 - \eta)$$

where the superscripts (1) and (2) for Lagrange polynomials denote linear and quadratic functions, respectively.

Example 9.3.4 Irregular Elements with an Irregular Node

Consider the irregular elements that may occur in the process of refinements as seen in Figure E9.3.4. All elements are to be approximated linearly. Interpolation functions

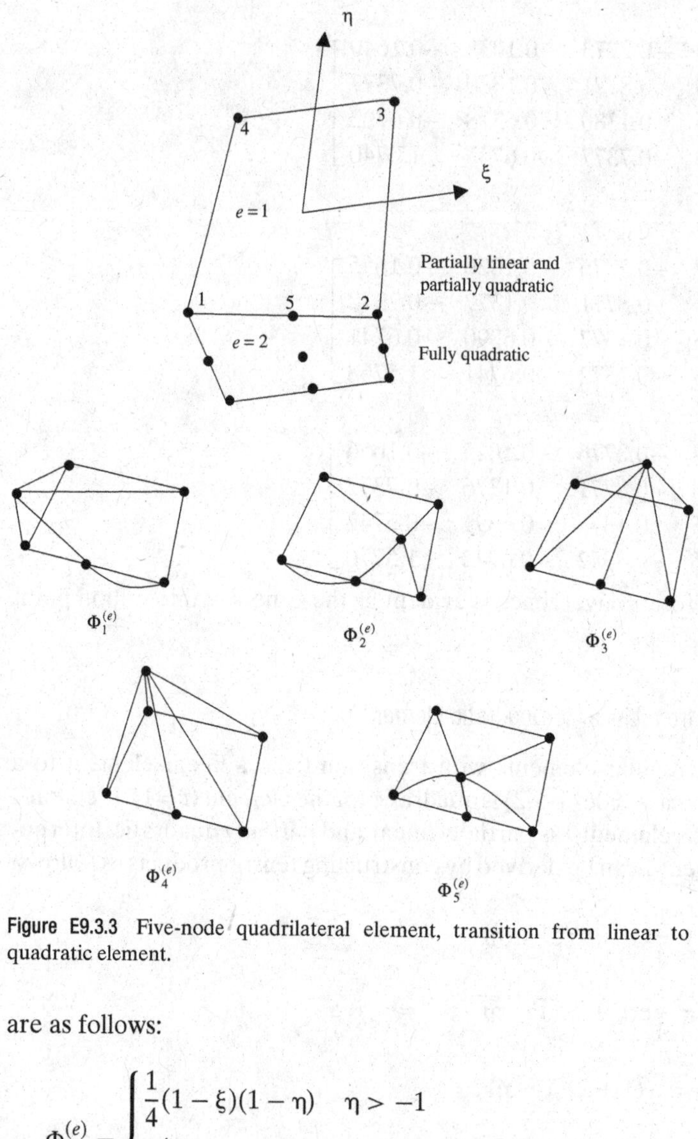

Figure E9.3.3 Five-node quadrilateral element, transition from linear to quadratic element.

are as follows:

$$\Phi_1^{(e)} = \begin{cases} \frac{1}{4}(1-\xi)(1-\eta) & \eta > -1 \\ -\xi & \eta = -1, \quad -1 \le \xi \le 0 \\ 0 & \eta = -1, \quad 0 \le \xi \le 1 \end{cases}$$

$$\Phi_2^{(e)} = \begin{cases} \frac{1}{4}(1+\xi)(1-\eta) & \eta > -1 \\ \xi & \eta = -1, \quad 0 \le \xi \le 1 \\ 0 & \eta = -1, \quad -1 \le \xi \le 0 \end{cases}$$

$$\Phi_3^{(e)} = \frac{1}{4}(1+\xi)(1+\eta)$$

$$\Phi_4^{(e)} = \frac{1}{4}(1-\xi)(1+\eta)$$

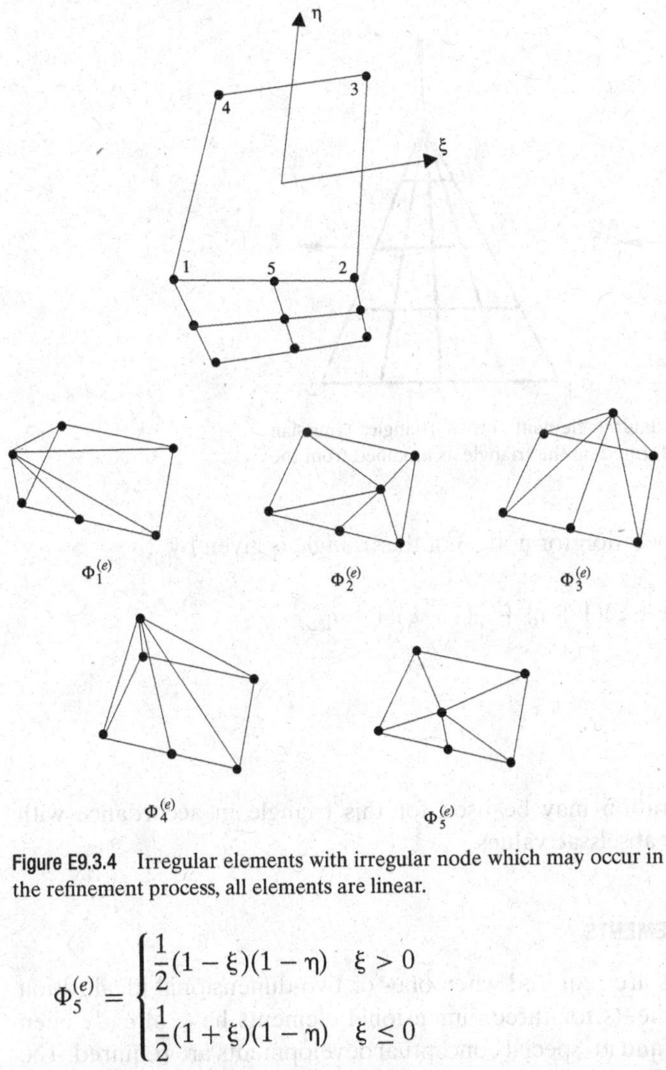

Figure E9.3.4 Irregular elements with irregular node which may occur in the refinement process, all elements are linear.

$$\Phi_5^{(e)} = \begin{cases} \dfrac{1}{2}(1-\xi)(1-\eta) & \xi > 0 \\ \dfrac{1}{2}(1+\xi)(1-\eta) & \xi \leq 0 \end{cases}$$

Here $\Phi_5^{(e)}$ for the midside node (hanging node) may be eliminated by readjusting the corner node functions, as is usually the case in adaptive mesh refinement methods (see Chapter 19).

Example 9.3.5 Collapse of Quadrilateral to Triangle

A quadrilateral element may be collapsed into a triangle by combining two of the quadrilateral nodes into one (Figure E9.3.5), as follows:

$$u^{(e)} = \Phi_1^{(e)} u_1^{(e)} + \Phi_2^{(e)} u_2^{(e)} + \Phi_3^{(e)} u_3^{(e)} + \Phi_4^{(e)} u_4^{(e)}$$

Equating $u_4^{(e)} = u_3^{(e)}$ we have for the triangle

$$u^{(e)} = \Phi_1^{(e)} u_1^{(e)} + \Phi_2^{(e)} u_2^{(e)} + (\Phi_3^{(e)} + \Phi_4^{(e)}) u_3^{(e)} = \Phi_1^{(e)} u_1^{(e)} + \Phi_2^{(e)} u_2^{(e)} + \overline{\Phi}_3^{(e)} u_3^{(e)}$$

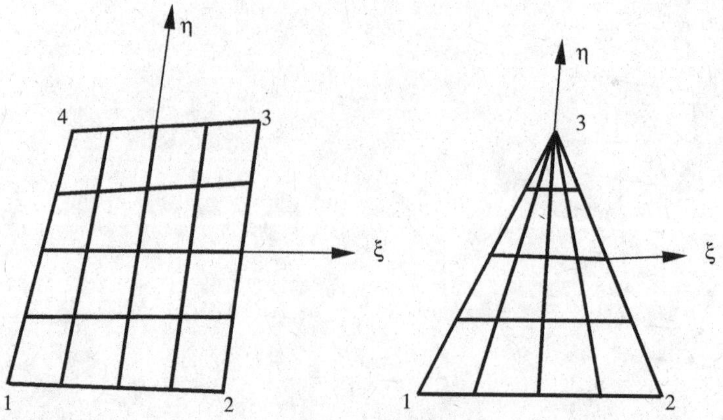

Figure E9.3.5 Collapsing a quadrilateral element into a triangle. Gaussian quadrature integration can be performed on the triangle as modified from the quadrilateral.

in which the modified interpolation for node 3 of the triangle is given by

$$\overline{\Phi}_3^{(e)} = \Phi_3^{(e)} + \Phi_4^{(e)} = \frac{1}{4}(1+\xi)(1+\eta) + \frac{1}{4}(1-\xi)(1+\eta)$$

Thus

$$\overline{\Phi}_3^{(e)} = \frac{1}{2}(1+\eta)$$

Gaussian quadrature integration may be used for this triangle in accordance with Table 9.3.1 with appropriate abscissae values.

9.4 THREE-DIMENSIONAL ELEMENTS

Three-dimensional elements are required when one- or two-dimensional idealization is not possible. Basic ingredients for three-dimensional elements have already been presented in earlier sections and no special conceptual developments are required. The three-dimensional elements may be constructed quite easily by direct extension of the ideas used for two-dimensional elements.

9.4.1 TETRAHEDRAL ELEMENTS

Consider the tetrahedral elements as shown in Figure 9.4.1. For linear variation of a variable (Figure 9.4.1a), we write

$$u^{(e)} = \alpha_0 + \alpha_1 x + \alpha_2 y + \alpha_3 z \tag{9.4.1}$$

It is a simple matter to write the above equation at each node, which yields a total of four equations. Evaluating the constants from these equations, we obtain

$$u^{(e)} = \Phi_N^{(e)} u_N^{(e)} \quad (N = 1, 2, 3, 4) \tag{9.4.2}$$

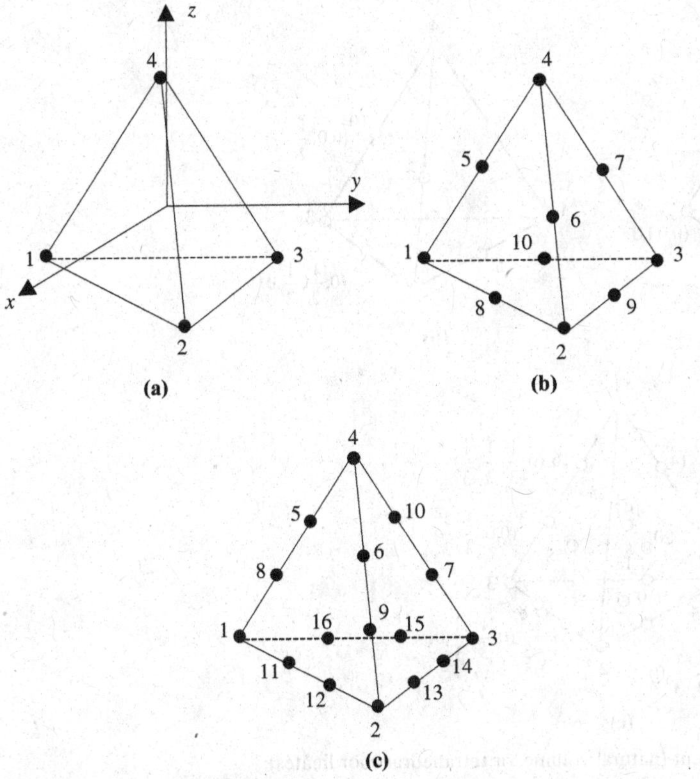

Figure 9.4.1 Tetrahedral element (cartesian coordinate): (a) linear variation, (b) quadratic variation, (c) cubic variation.

where

$$\Phi_N^{(e)} = a_N + b_N x + c_N y + d_N z \tag{9.4.3}$$

For $N = 1$, the coefficients a_1, b_1, c_1, d_1 are of the form

$$a_1 = \begin{vmatrix} x_2 & y_2 & z_2 \\ x_3 & y_3 & z_3 \\ x_4 & y_4 & z_4 \end{vmatrix} \frac{1}{|D|}, \qquad b_1 = -\begin{vmatrix} 1 & y_2 & z_2 \\ 1 & y_3 & z_3 \\ 1 & y_4 & z_4 \end{vmatrix} \frac{1}{|D|}$$

$$c_1 = \begin{vmatrix} 1 & x_2 & z_2 \\ 1 & x_3 & z_3 \\ 1 & x_4 & z_4 \end{vmatrix} \frac{1}{|D|}, \qquad d_1 = -\begin{vmatrix} 1 & x_2 & y_2 \\ 1 & x_3 & y_3 \\ 1 & x_4 & y_4 \end{vmatrix} \frac{1}{|D|} \tag{9.4.4}$$

$$|D| = \begin{vmatrix} 1 & x_{11} & x_{12} & x_{13} \\ 1 & x_{21} & x_{22} & x_{23} \\ 1 & x_{31} & x_{32} & x_{33} \\ 1 & x_{41} & x_{42} & x_{43} \end{vmatrix} = \begin{vmatrix} 1 & x_1 & y_1 & z_1 \\ 1 & x_2 & y_2 & z_2 \\ 1 & x_3 & y_3 & z_3 \\ 1 & x_4 & y_4 & z_4 \end{vmatrix} = 6V \tag{9.4.5}$$

where V is the volume of the tetrahedron. The rest of the coefficients can be determined similarly.

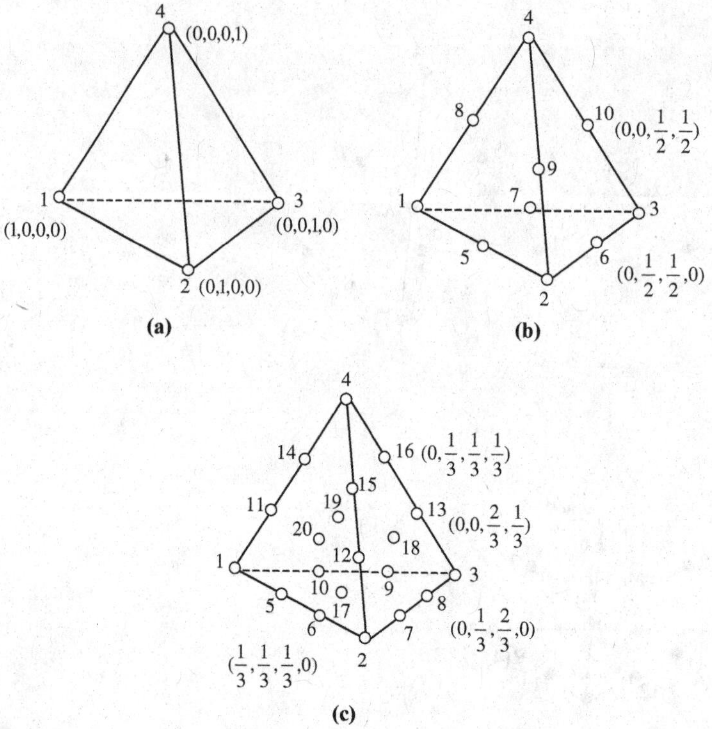

Figure 9.4.2 Tetrahedral element (natural volume, or tetrahedral coordinates): **(a)** linear variation; **(b)** quadratic variation; **(c)** cubic variation.

For higher order approximations, the coefficient matrix becomes very large in size and a resort to natural coordinates is inevitable. The most suitable choice is the volume coordinate system extended from the area coordinates for a two-dimensional triangular element.

If the three-dimensional natural coordinates (tetrahedral or volume coordinates) are used, a node having the coordinate of one decreases to zero as it moves to the opposite triangular surface formed by the rest of the nodes (Figure 9.4.2). For the linear element (Figure 9.4.2a), the interpolation functions are

$$\Phi_N^{(e)} = L_N \quad (N = 1, 2, 3, 4) \tag{9.4.6}$$

For higher order interpolations (Figure 9.4.2b,c), we invoke a formula similar to (9.3.20),

$$\Phi_r^{(e)} = B^{(r)}(L_1)B^{(r)}(L_2)B^{(r)}(L_3)B^{(r)}(L_4) \tag{9.4.7}$$

where $B^{(r)}(L_N)$ is given by (9.3.19). This provides the following results:
For quadratic variation (Figure 9.4.2b):
at corner nodes:

$$\Phi_N^{(e)} = (2L_N - 1)L_N$$

at midside nodes:

$$\Phi_6^{(e)} = 4L_2L_3, \qquad \Phi_{10}^{(e)} = 4L_3L_4, \text{ etc.}$$

For cubic variation (Figure 9.4.2c):
at corner nodes:

$$\Phi_N^{(e)} = \frac{1}{2}(3L_N - 1)(3L_N - 2)L_N$$

at side nodes:

$$\Phi_8^{(e)} = \frac{9}{2}L_2L_3(3L_3 - 1), \qquad \Phi_{13}^{(e)} = \frac{9}{2}L_3L_4(3L_3 - 1), \text{ etc.}$$

at midside nodes:

$$\Phi_{17}^{(e)} = 27L_1L_2L_3, \qquad \Phi_{18}^{(e)} = 27L_2L_3L_4, \text{ etc.}$$

The spatial integration of the tetrahedral coordinates may be derived similarly as in the triangular coordinates. This results in

$$I = \int_V L_1^m L_2^n L_3^p L_4^q dv$$

or

$$I = \frac{6Vm!n!p!q!}{(m+n+p+q+3)!} \tag{9.4.8}$$

We may use a hexahedral element to generate five tetrahedral elements as shown in Figure 9.4.3. This approach is desirable in some applications where both hexahedral

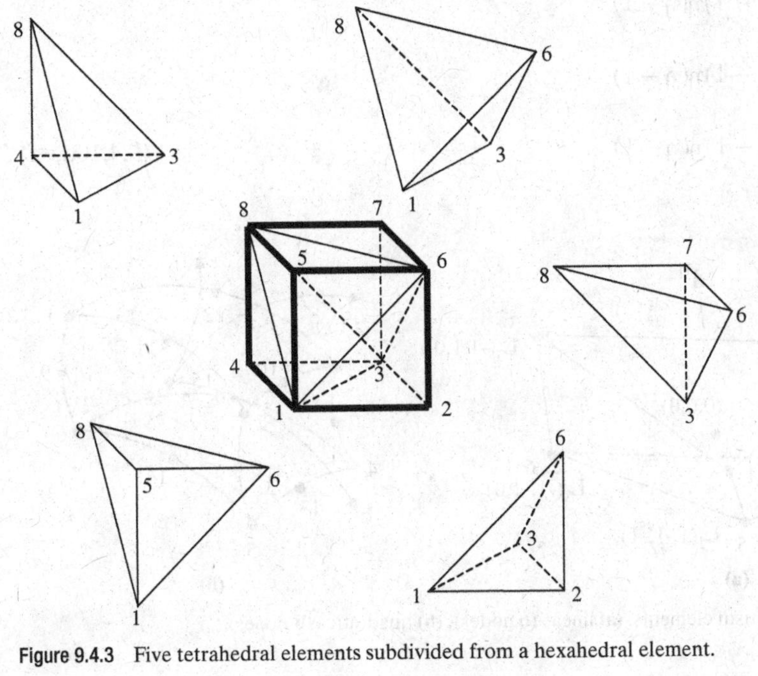

Figure 9.4.3 Five tetrahedral elements subdivided from a hexahedral element.

and tetrahedral elements are used. It is also convenient for the structured automatic grid generation.

9.4.2 TRIANGULAR PRISM ELEMENTS

It is possible to extend the tetrahedral element into triangular prism elements as shown in Figure 9.4.4. Note that triangular shapes may be completely arbitrary with the curvilinear coordinates ξ, η, ζ being distorted. Interpolation functions for linear and quadratic approximations are given as follows:

Linear (6 nodes)

$$\Phi_1^{(e)} = \frac{L_1(1+\eta)}{2}, \quad \Phi_2^{(e)} = \frac{L_2(1+\eta)}{2}, \quad \Phi_3^{(e)} = \frac{L_3(1+\eta)}{2} \tag{9.4.9a,b,c}$$

$$\Phi_4^{(e)} = \frac{L_1(1-\eta)}{2}, \quad \Phi_5^{(e)} = \frac{L_2(1-\eta)}{2}, \quad \Phi_6^{(e)} = \frac{L_3(1-\eta)}{2} \tag{9.4.9d,e,f}$$

Quadratic (15 nodes)

Corner nodes

$$\begin{aligned}
\Phi_1^{(e)} &= \frac{1}{2}L_1(2L_1-1)\eta(\eta+1)\\
\Phi_2^{(e)} &= \frac{1}{2}L_2(2L_2-1)\eta(\eta+1)\\
\Phi_3^{(e)} &= \frac{1}{2}L_3(2L_3-1)\eta(\eta+1)
\end{aligned} \tag{9.4.10a,b,c}$$

$$\begin{aligned}
\Phi_4^{(e)} &= \frac{1}{2}L_1(2L_1-1)\eta(\eta-1)\\
\Phi_5^{(e)} &= \frac{1}{2}L_2(2L_2-1)\eta(\eta-1)\\
\Phi_6^{(e)} &= \frac{1}{2}L_3(2L_3-1)\eta(\eta-1)
\end{aligned} \tag{9.4.10d,e,f}$$

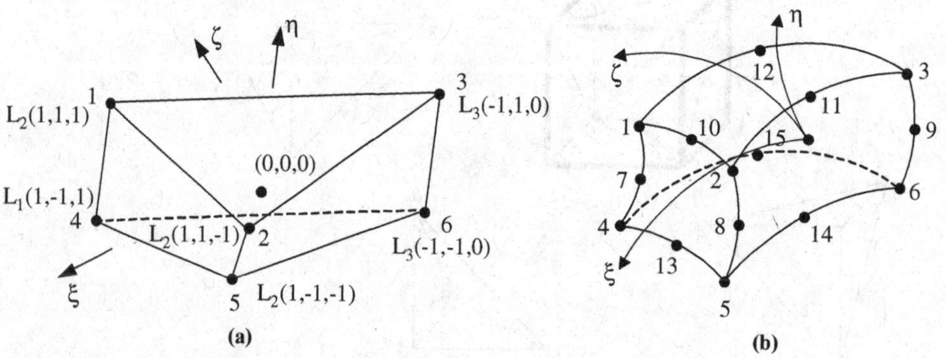

Figure 9.4.4 Triangular prism elements: **(a)** linear (6 nodes), **(b)** quadratic (15 nodes).

Midsides of Triangle

$$\Phi_{10}^{(e)} = 2L_1L_2\eta(\eta+1),\quad \Phi_{11}^{(e)} = 2L_2L_3\eta(\eta+1),\quad \Phi_{12}^{(e)} = 2L_1L_3\eta(\eta+1) \qquad (9.4.11\text{a,b,c})$$

$$\Phi_{13}^{(e)} = 2L_1L_2\eta(\eta-1),\quad \Phi_{14}^{(e)} = 2L_2L_3\eta(\eta-1),\quad \Phi_{15}^{(e)} = 2L_1L_3\eta(\eta-1) \qquad (9.4.11\text{d,e,f})$$

Midsides of Quadrilateral

$$\Phi_7^{(e)} = L_1(1-\eta^2),\qquad \Phi_8^{(e)} = L_2(1-\eta^2),\qquad \Phi_9^{(e)} = L_3(1-\eta^2) \qquad (9.4.12\text{a,b,c})$$

9.4.3 HEXAHEDRAL ISOPARAMETRIC ELEMENTS

The four-sided two-dimensional elements may be extended to three-dimensional elements (Figure 9.4.5). The rectangular and arbitrary quadrilateral elements are developed into a regular hexahedron (brick) and irregular hexahedron. For a regular hexahedron, we may use either the Lagrange or Hermite element, but this becomes cumbersome as higher order approximations must include interior and surface nodes as well as corner and side nodes. Besides, neither may be applicable for irregular hexahedrons. An element which is free from these disadvantages is the isoparametric element.

In the isoparametric element for a linear variation of the geometry and variable, we write (see Figure 9.4.5a)

$$x, y, z = \alpha_1 + \alpha_2\xi + \alpha_3\eta + \alpha_4\zeta + \alpha_5\xi\eta\zeta + \alpha_6\xi\eta + \alpha_7\eta\zeta + \alpha_8\xi\zeta \qquad (9.4.13)$$

Using the same procedure as in the two-dimensional element, we obtain

$$\Phi_N^{(e)} = \frac{1}{8}(1+\xi_{N1}\xi_1)(1+\xi_{N2}\xi_2)(1+\xi_{N3}\xi_3) \qquad (9.4.14)$$

For a quadratic variation (Figure 9.4.4b), we have

$$\begin{aligned} x, y, z = {} & \alpha_1 + \alpha_2\xi + \alpha_3\eta + \alpha_4\zeta + \alpha_5\xi\eta\zeta + \alpha_6\xi\eta + \alpha_7\eta\zeta + \alpha_8\xi\zeta \\ & + \alpha_9\xi^2 + \alpha_{10}\eta^2 + \alpha_{11}\zeta^2 + \alpha_{12}\xi^2\eta + \alpha_{13}\xi\eta^2 + \alpha_{14}\eta^2\zeta + \alpha_{15}\eta\zeta^2 \\ & + \alpha_{16}\xi^2\zeta + \alpha_{17}\xi\zeta^2 + \alpha_{18}\xi^2\eta\varsigma + \alpha_{19}\xi\eta^2\varsigma + \alpha_{20}\xi\eta\varsigma^2 \end{aligned} \qquad (9.4.15)$$

The interpolation functions are:
at corner nodes :

$$\Phi_N^{(e)} = \frac{1}{8}(1+\xi_{N1}\xi_1)(1+\xi_{N2}\xi_2)(1+\xi_{N3}\xi_3)(\xi_{N1}\xi_1+\xi_{N2}\xi_2+\xi_{N3}\xi_3-2) \qquad (9.4.16\text{a})$$

at midside nodes :

$$\Phi_N^{(e)} = \frac{1}{4}(1-\xi_1^2)(1+\xi_{N2}\xi_2)(1+\xi_{N3}\xi_3) \qquad (9.4.16\text{b})$$

for

$$\xi_{N1} = 0,\qquad \xi_{N2} = \pm 1,\qquad \xi_{N3} = \pm 1, \text{ etc.}$$

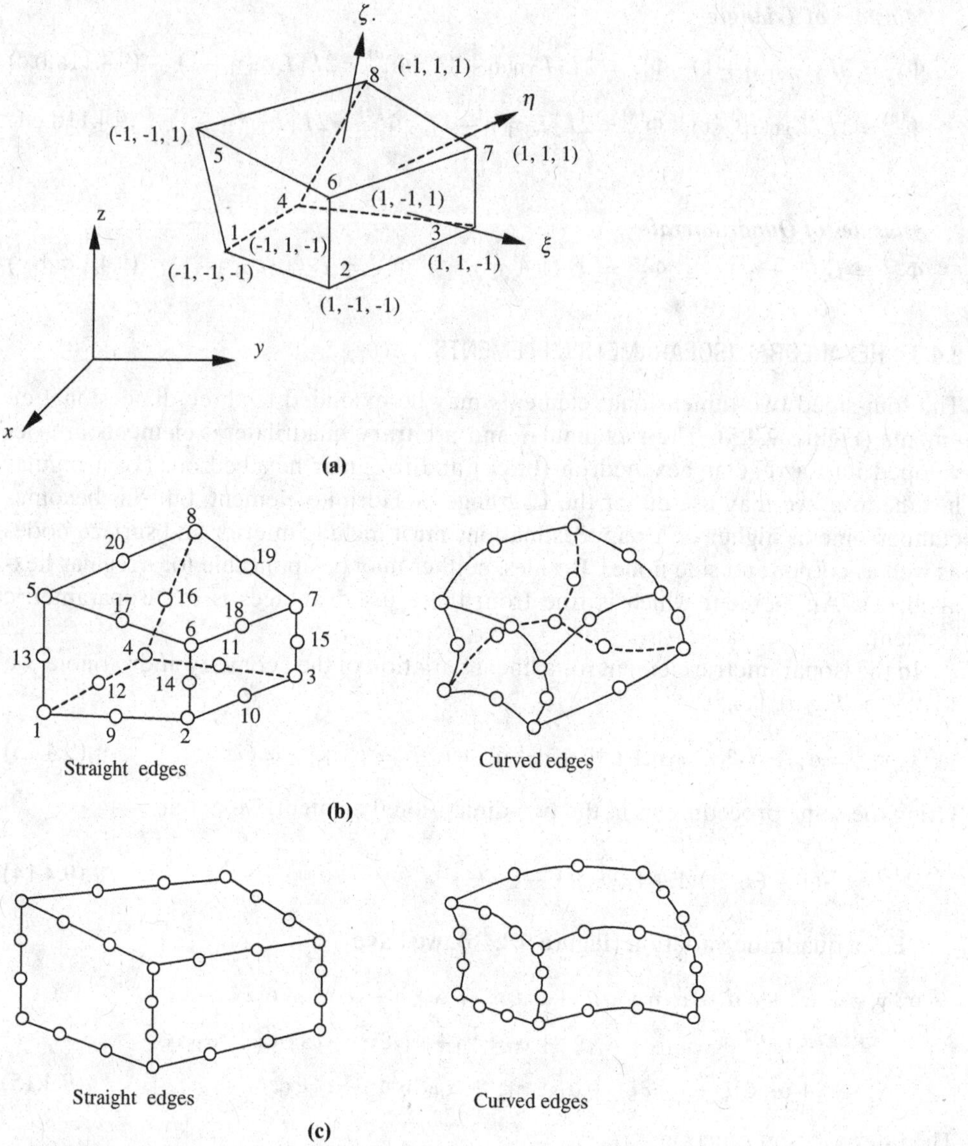

Figure 9.4.5 Hexahedral isoparametric element: **(a)** linear variation (8 nodes); **(b)** quadratic variation (20 nodes); **(c)** cubic variation (32 nodes).

Once again, Lagrange polynomials may be used to determine three-dimensional interpolation functions without interior nodes, but with side node constraint conditions. We now require integration of the form

$$\iiint \frac{\partial}{\partial x} f(\xi, \eta, \zeta) dx dy dz \tag{9.4.17}$$

with $\xi = \xi_1$, $\eta = \xi_2$, $\zeta = \xi_3$, $x = x_1$, $y = x_2$, and $z = x_3$. Proceeding similarly as in the two-dimensional case, we obtain

$$\iiint \frac{\partial f}{\partial x}dxdydz = \int_{-1}^{1}\int_{-1}^{1}\int_{-1}^{1}\left(\overline{J}_{11}\frac{\partial f}{\partial \xi} + \overline{J}_{12}\frac{\partial f}{\partial \eta} + \overline{J}_{13}\frac{\partial f}{\partial \zeta}\right)|J|d\xi d\eta d\zeta$$
$$= \int_{-1}^{1}\int_{-1}^{1}\int_{-1}^{1} g(\xi,\eta,\zeta)d\xi d\eta d\zeta \tag{9.4.18}$$

where $\overline{J}_{11}, \overline{J}_{12},$ and $\overline{J}_{13}$ are the first row of the 3×3 inverted Jacobian matrix

$$[J] = \begin{bmatrix} \frac{\partial x}{\partial \xi} & \frac{\partial y}{\partial \xi} & \frac{\partial z}{\partial \xi} \\ \frac{\partial x}{\partial \eta} & \frac{\partial y}{\partial \eta} & \frac{\partial z}{\partial \eta} \\ \frac{\partial x}{\partial \zeta} & \frac{\partial y}{\partial \zeta} & \frac{\partial z}{\partial \zeta} \end{bmatrix}$$

We may carry out differentiations of f with respect to y and z similarly, and write the general form of integration as follows:

$$\int_{-1}^{1}\int_{-1}^{1}\int_{-1}^{1} g(\xi,\eta,\zeta)d\xi d\eta d\zeta = \sum_{i=1}^{n}\sum_{j=1}^{n}\sum_{k=1}^{n} w_i w_j w_k g(\xi_i,\eta_j,\zeta_k) \tag{9.4.19}$$

The weight coefficients w_i, w_j, w_k, and the abscissae $g(\xi_i, \eta_j, \zeta_k)$ are obtained from Table 9.3.1 as a tensor product in three directions. A procedure similar to Example 9.3.1 may be followed for three dimensions to perform Gaussian quadrature integrations.

9.5 AXISYMMETRIC RING ELEMENTS

If the three-dimensional domain of study is axisymmetric, then any two-dimensional element may be used with the spatial integral replaced by

$$\iiint f(x,y,z)dxdydz = \int_0^{2\pi}\iint f(r,z)rd\theta drdz \tag{9.5.1}$$

where $dx = dr$, $dy = rd\theta$, and $dz = dz$ (see Figure 9.5.1). For quadrilateral isoparametric elements, we have

$$\int_0^{2\pi}\int_{-1}^{1}\int_{-1}^{1} f(\xi,\eta)rd\theta|J|d\xi d\eta = 2\pi\int_{-1}^{1}\int_{-1}^{1} f(\xi,\eta)r(\xi,\eta)|J|d\xi d\eta$$

or

$$2\pi\int_{-1}^{1}\int_{-1}^{1} g(\xi,\eta)d\xi d\eta = 2\pi\sum_{j=1}^{n}\sum_{k=1}^{n} w_j w_k g(\xi_j,\eta_k) \tag{9.5.2}$$

This represents a three-dimensional ring element generated by a two-dimensional element.

Note that the applications arise in the flowfields of missiles and rockets at zero angle of attack. For a nonzero angle of attack, the flowfields become asymmetric. In this case, the axisymmetric ring element can no longer be used and three-dimensional elements must be invoked instead. Another alternative is to keep the ring element and use Fourier

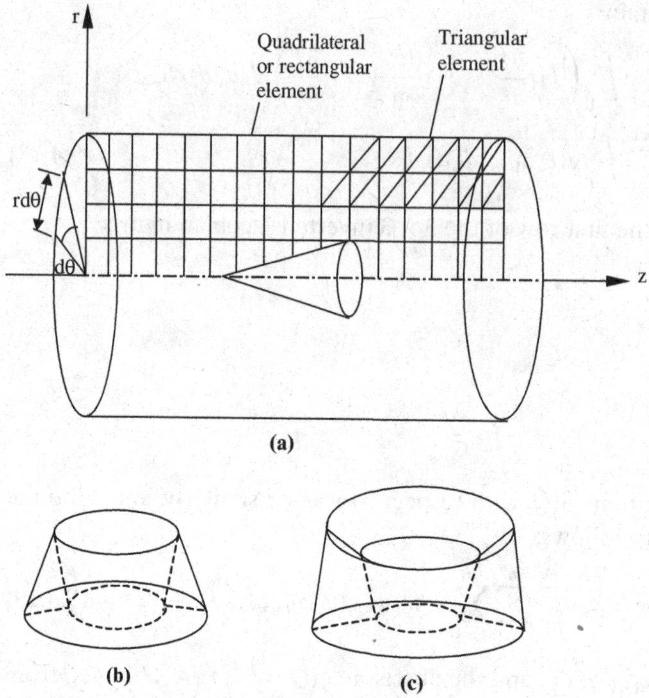

Figure 9.5.1 Axisymmetric ring elements. (a) Discretized geometry of a cylinder. (b) Triangular ring. (c) Quadrilateral ring.

series expansions around the circumference in order to accommodate nonaxisymmetric flowfields at every few degrees apart. This may result in a process just as expensive as three-dimensional elements.

9.6 LAGRANGE AND HERMITE FAMILIES AND CONVERGENCE CRITERIA

All finite elements, regardless of their geometrical shapes, may be grouped into two categories: Lagrange and Hermite families. The Lagrange family consists of finite elements in which the values of a variable are specified at nodes, whereas the Hermite family includes derivatives of the variable as well as its values defined at nodes.

Both Lagrange and Hermite families may be represented by polynomials derived from the Pascal triangle (Figure 9.1.4a) and Pascal tetrahedron (Figure 9.1.4b), or from two-dimensional hypercube (Figure 9.1.4c) and three-dimensional hypercube (Figure 9.1.4d).

In the Lagrange family, the polynomial terms contained in the circles represent the corresponding number of nodes required. However, in general, interior nodes lead to cumbersome bookkeeping and subsequent removal of some of the polynomial terms, resulting in an incomplete polynomial.

For a Hermite family, the number of nodes and polynomial terms required increases since derivatives in addition to the variable itself are to be specified. However, a reasonable compromise can be met by eliminating the values of a variable and specifying only the normal derivatives at side nodes. Let us consider twenty-one terms of a quintic

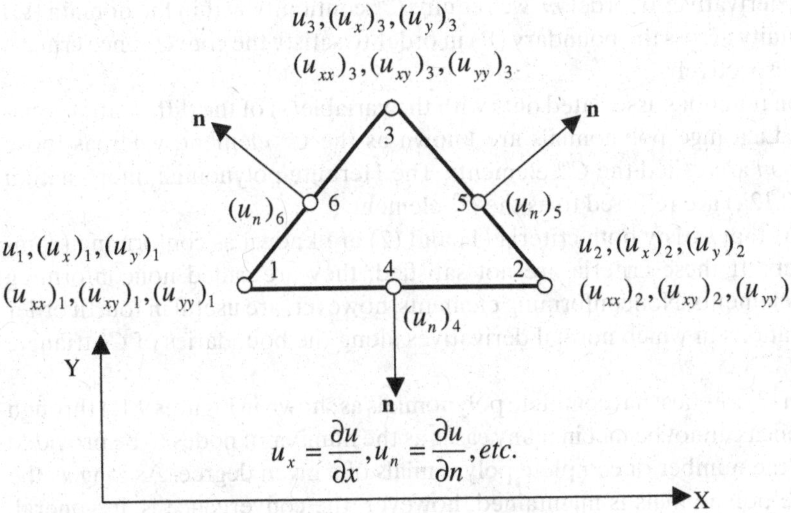

Figure 9.6.1 Hermite triangle, 21 constants to be determined.

polynomial for a triangular element given by

$$
\begin{aligned}
u = &\alpha_1 \\
&+ \alpha_2 x + \alpha_3 y \\
&+ \alpha_4 x^2 + \alpha_5 xy + \alpha_6 y^2 \\
&+ \alpha_7 x^3 + \alpha_8 x^2 y + \alpha_9 xy^2 + \alpha_{10} y^3 \\
&+ \alpha_{11} x^4 + \alpha_{12} x^3 y + \alpha_{13} x^2 y^2 + \alpha_{14} xy^3 + \alpha_{15} xy^4 \\
&+ \alpha_{15} x^5 + \alpha_{17} x^4 y + \alpha_{18} x^3 y^2 + \alpha_{19} x^2 y^3 + \alpha_{20} xy^4 + \alpha_{21} y^5
\end{aligned}
\tag{9.6.1}
$$

The nodal values of u and derivatives of u to be specified are shown in Figure 9.6.1. Notice that we can write only eighteen equations with six equations at each node. We require three more equations which are furnished by writing normal derivatives at midsides. In this way, all twenty-one constants can be evaluated.

The $2m$th order differential equations associated with many of the engineering problems are in the form

$$\nabla^2 u = f \quad (m = 1) \tag{9.6.2}$$

$$\nabla^4 u = f \quad (m = 2) \tag{9.6.3}$$

Thus, the weak derivatives that appear in the Galerkin finite element formulations have $m = 1$ for (9.6.2) and $m = 2$ for (9.6.3). The choice of interpolation functions must ensure the convergence of solutions of the given differential equations. Toward this end, the following criteria should be satisfied.

(1) Smooth within the interior domain
(2) Continuity across each element
(3) Completeness

To satisfy (1), the degree of polynomial, p, should be $p \geq m$ so that the integrand of the finite element equation does not vanish (remaining at least a constant). For the stiffness integrands with derivatives of order m, we require C^m continuity within the domain (Ω) and C^{m-1} continuity across the boundary (Γ) in order to satisfy the convergence criteria of (1) and (2), respectively.

Interpolation functions associated only with the variable(s) of the differential equation such as in Lagrange polynomials are known as the C° element, whereas those with derivatives m are called the C^m elements. The Hermite polynomial interpolation functions of (9.2.12a) are referred to as the C^1 element.

The elements that satisfy both criteria (1) and (2) are known as conforming (compatible) elements. If these criteria are not satisfied, they are called nonconforming (incompatible) elements. Nonconforming elements, however, are useful in fourth order differential equations in which normal derivatives along the boundaries of C^1 triangle are specified.

The criterion (3) implies that complete polynomials as shown in Figures 9.1.4 through 9.1.6 be used, which cannot be met in many cases as the number of nodes to be provided does not match the number of complete polynomials of a given degree. As long as the symmetry of the polynomials is maintained, however, the convergence is, in general, not affected.

9.7 SUMMARY

Although the standard textbooks on finite elements provide information presented in this chapter, it was intended that a complete summary of finite element interpolation functions serve as a counterpart of Chapter 3, Derivation of Finite Difference Equations, as well as this text being self-contained and adequately balanced between FEM and FDM.

It is clear now that, instead of writing finite difference approximations using as many nodal points as necessary for desired order accuracy in FDM, we achieve similar objectives in FEM through interpolation functions. Instead of Taylor series expansions or Pade approximations used in finite difference equations, we resort to polynomial expansion in finite element interpolation functions. Although not covered in this chapter, special functions such as Chebyshev polynomials, Legendre polynomials, or Laguerre polynomials have been used in association with spectral elements. This subject will be discussed in Section 14.1.

REFERENCES

Argyris, J. H. [1963]. *Recent Advances in Matrix Methods of Structural Analysis by Finite Elements*. Elmsford, NY: Pergamon Press.

Birkhoff, G., Schultz, M. H., and Varga, R. [1968]. Piecewise Hermite interpolation in one and two variables with applications to partial differential equations. *Num. Math.*, 11, 232–56.

Cowper, G., Kosko, E., Lindberg, G., and Olson, M. [1969]. Static and dynamic applications of a high precision triangular plate bending element, *AIAA J.*, 7, 1957–65.

Hildebrand, F. B. [1956]. *Introduction to Numerical Analysis*, New York: McGraw-Hill.

Zienkiewicz, O. C. [1971]. *The Finite Element Method in Engineering Science*, 2nd. ed. New York: McGraw-Hill.

Zienkiewicz, O. C. and Cheung, Y. K. [1965]. *The Finite Element Method in Engineering Science*. New York: McGraw-Hill.

CHAPTER TEN

Linear Problems

In this chapter, we discuss procedures for obtaining finite element equations and their solutions in linear two-dimensional boundary value problems. Implementations of boundary conditions are detailed and example problems for steady and unsteady cases are presented. Multivariable simultaneous partial differential equations and simple Stokes flow problems are also included.

10.1 STEADY-STATE PROBLEMS – STANDARD GALERKIN METHODS

10.1.1 TWO-DIMENSIONAL ELLIPTIC EQUATIONS

We have illustrated procedures for constructing finite element equations for one-dimensional problems in Chapters 1 and 8. Extension to two-dimensional cases follows the same general guidelines. The only difference is the appropriate interpolation functions for two-dimensional geometries, specification of Neumann boundary conditions, integration over the domain, and directional variables.

Consider the second order elliptic partial differential equation of the form,

$$R = \nabla^2 u + f(x, y) = 0 \quad \text{in } \Omega \tag{10.1.1}$$

As shown in Chapters 1 and 8, the Standard Galerkin Method (SGM) for (10.1.1) is the inner product of the residual with the test function Φ_α

$$(\Phi_\alpha, R) = \int_\Omega \Phi_\alpha [u_{,ii} + f(x, y)] d\Omega = 0 \tag{10.1.2}$$

Assuming that the variable u is approximated in the form

$$u = \Phi_\alpha u_\alpha \tag{10.1.3}$$

and integrating (10.1.2) by parts we obtain

$$\int_\Gamma \overset{*}{\Phi}_\alpha u_{,i} n_i d\Gamma - \left(\int_\Omega \Phi_{\alpha,i} \Phi_{\beta,i} d\Omega \right) u_\beta + \int_\Omega \Phi_\alpha f(x, y) d\Omega = 0$$

or

$$K_{\alpha\beta} u_\beta = F_\alpha + G_\alpha \tag{10.1.4}$$

where

Stiffness matrix $$K_{\alpha\beta} = \int_\Omega \Phi_{\alpha,i}\Phi_{\beta,i}d\Omega \tag{10.1.5a}$$

Source vector $$F_\alpha = \int_\Omega \Phi_\alpha f(x, y)d\Omega \tag{10.1.5b}$$

Neumann boundary vector $$G_\alpha = \int_\Gamma \overset{*}{\Phi}_\alpha u_{,i}n_i d\Gamma \tag{10.1.5c}$$

As we noted in the one-dimensional problem, the interpolation function originally defined in the domain is now a function of boundary coordinate Γ in the boundary integral G_α, with $\overset{*}{\Phi}_\alpha$ indicating the dependency on Γ, not on Ω. It represents the interpolation function describing the way the Neumann data $u_{,i}n_i$ varies along the boundaries. Thus, a suitable form for $\overset{*}{\Phi}_\alpha(\Gamma)$ would be the one-dimensional linear interpolation function.

The global forms (10.1.5) can be obtained by the assembly of local forms similarly as in the one-dimensional problems,

$$K_{\alpha\beta} = \bigcup_{e=1}^{E} K_{NM}^{(e)}\Delta_{N\alpha}^{(e)}\Delta_{M\beta}^{(e)} \tag{10.1.6a}$$

$$F_\alpha = \bigcup_{e=1}^{E} F_N^{(e)}\Delta_{N\alpha}^{(e)} \tag{10.1.6b}$$

$$G_\alpha = \bigcup_{e=1}^{E} G_N^{(e)}\Delta_{N\alpha}^{(e)} \tag{10.1.6c}$$

where

$$K_{NM}^{(e)} = \int_\Omega \Phi_{N,i}^{(e)}\Phi_{M,i}^{(e)}d\Omega \tag{10.1.7a}$$

$$F_N^{(e)} = \int_\Omega \Phi_N^{(e)} f(x, y)d\Omega \tag{10.1.7b}$$

$$G_N^{(e)} = \int_\Gamma \overset{*}{\Phi}{}_N^{(e)} u_{,i}n_i d\Gamma \tag{10.1.7c}$$

The source term $f(x, y)$ and the Neumann data $g(\Gamma) = u_{,i}n_i$ can be interpolated as follows:

$$f(x, y) = \Phi_\alpha(x, y)f_\alpha, \qquad f_\alpha = [f(x, y)]_\alpha \tag{10.1.8a}$$

$$g(\Gamma) = \overset{*}{\Phi}_\alpha(\Gamma)g_\alpha, \qquad g_\alpha = (u_{,i}n_i)_\alpha \tag{10.1.8b}$$

These approximations allow the corresponding source term $f(x, y)$ and the Neumann data $u_{,i}n_i$ to be entered directly to the particular node under consideration. Substituting (10.1.8a) and (10.1.8b) into (10.1.5b) and (10.1.5c), respectively, we obtain

$$\begin{aligned} F_\alpha &= \left(\int_\Omega \Phi_\alpha\Phi_\beta d\Omega\right) f_\beta = C_{\alpha\beta} f_\beta = \bigcup_{e=1}^{E} C_{NM}^{(e)}\Delta_{N\alpha}^{(e)}\Delta_{M\beta}^{(e)} f_p^{(e)}\Delta_{p\beta}^{(e)} \\ &= \bigcup_{e=1}^{E} C_{NM}^{(e)} f_M^{(e)}\Delta_{N\alpha}^{(e)} = \bigcup_{e=1}^{E} F_N^{(e)}\Delta_{N\alpha}^{(e)} \end{aligned} \tag{10.1.9}$$

and similarly,

$$G_\alpha = \left(\int_\Gamma \overset{*}{\Phi}_\alpha \overset{*}{\Phi}_\beta d\Gamma\right) g_\beta = \overset{*}{C}_{\alpha\beta}\, g_\beta = \bigcup_{e=1}^{E} G_N^{(e)} \Delta_{N\alpha}^{(e)} \tag{10.1.10}$$

where

$$F_N^{(e)} = C_{NM}^{(e)} f_M^{(e)} \tag{10.1.11}$$

$$G_N^{(e)} = \overset{*}{C}{}_{NM}^{(e)} g_M^{(e)} \tag{10.1.12}$$

with

$$C_{NM}^{(e)} = \int_\Omega \Phi_N^{(e)} \Phi_M^{(e)} d\Omega \tag{10.1.13a}$$

$$\overset{*}{C}{}_{NM}^{(e)} = \int_\Gamma \overset{*}{\Phi}{}_N^{(e)} \overset{*}{\Phi}{}_M^{(e)} d\Gamma \tag{10.1.13b}$$

For linear variations of $u_{,i} n_i$ for a boundary element of length l, $\overset{*}{\Phi}{}_N^{(e)} = (1 - \Gamma/l, \Gamma/l)$, the integration of (10.1.13b) gives the result,

$$\overset{*}{C}{}_{NM}^{(e)} = \frac{l}{6}\begin{bmatrix} 2 & 1 \\ 1 & 2 \end{bmatrix}$$

It is clear that, regardless of the choice of the local finite elements for the domain, whether triangular or quadrilateral, the boundary integral (10.1.13b) can remain independent.

As shown in Section 8.2, the Neumann boundary data interpolation functions $\overset{*}{\Phi}{}_N^{(e)}$ and $\overset{*}{\Phi}_\alpha$ are given by

$$\begin{aligned} &\overset{*}{\Phi}{}_N^{(e)} = \delta(\overset{*}{z}{}_N^{(e)} - z_N^{(e)}), \qquad \overset{*}{\Phi}{}_N^{(e)}(z_M^{(e)}) = \delta_{NM} \\ &\overset{*}{\Phi}_\alpha = \delta(\overset{*}{Z}_\alpha - Z_\alpha), \qquad \overset{*}{\Phi}_\alpha(Z_\beta) = \delta_{\alpha\beta} \end{aligned} \tag{10.1.14}$$

implying that $\overset{*}{\Phi}{}_N^{(e)} = 1$ if the Neumann boundary condition is applied at the boundary node N and zero, otherwise. This applies also to $\overset{*}{\Phi}_\alpha$.

The significance and importance of (10.1.14) cannot be overemphasized. Reexamine (10.1.5c), (10.1.6c), (10.1.7c), and (10.1.8b) in conjuction with (10.1.14). The process through these relations indicates that the local Neumann data are passed along across the local adjacent elements normal to the boundary surfaces to ensure the continuity of gradients or "energy balance" (incoming and outgoing normal gradients are cancelled at element boundaries) until the domain edge boundaries are reached, where the Neumann boundary conditions are applied and where the Neumann boundary condition interpolation functions $\overset{*}{\Phi}{}_N^{(e)}$ and $\overset{*}{\Phi}_\alpha$ assume the value of unity if *applied*, zero otherwise. Notice that this logic is established easily and clearly by having constructed the finite element equations in a global form from the beginning, called the "*global approach*," and by seeking the local element contributions in terms of the Boolean matrix algebra afterward. This is contrary to the traditional approach to the finite element formulations, from local to global, called the "*local approach*," in which the passage of Neumann data through element boundary surfaces cannot be defined

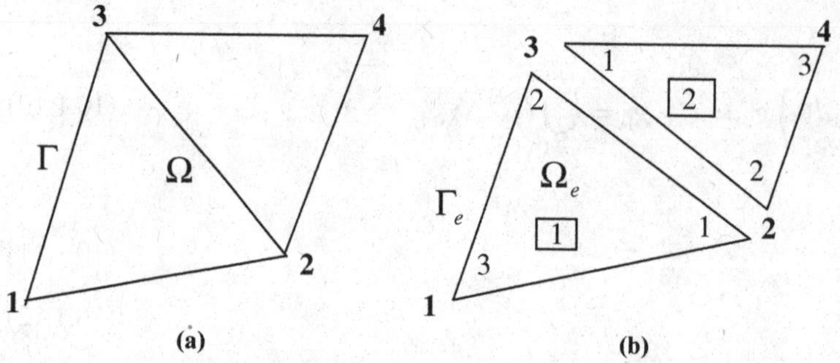

Figure 10.1.1 Finite element discretization. (a) Global nodes; (b) Local nodes.

easily and automatically. The global approach presented here is in contrast to the finite volume methods in which algebraic equations are generated by physically enforcing the normal gradients across the local element boundary surfaces. The consequences of operations involved in both FEM and FVM, however, are analogous, with the conservation properties maintained in both methods.

The assembly of local elements into a global form follows the same procedure as in the one-dimensional case. To obtain the global matrices $K_{\alpha\beta}$ and F_α, let us consider the two triangular elements in Figure 10.1.1. Although the expansion (10.1.6a) can be performed by summing the repeated indices, we may show such operations by matrix multiplications as follows:

First, we prepare the nodal correspondence table (Table 10.1.1) which indicates the correspondence of the local node with the global node for all elements.

$$
\begin{aligned}
K_{\alpha\beta} &= \bigcup_{e=1}^{E} K_{NM}^{(e)} \Delta_{N\alpha}^{(e)} \Delta_{M\beta}^{(e)} \\
&= \begin{bmatrix} 0 & 0 & 1 \\ 1 & 0 & 0 \\ 0 & 1 & 0 \\ 0 & 0 & 0 \end{bmatrix}
\begin{bmatrix} K_{11}^{(1)} & K_{12}^{(1)} & K_{13}^{(1)} \\ K_{21}^{(1)} & K_{22}^{(1)} & K_{23}^{(1)} \\ K_{31}^{(1)} & K_{32}^{(1)} & K_{33}^{(1)} \end{bmatrix}
\begin{bmatrix} 0 & 1 & 0 & 0 \\ 0 & 0 & 1 & 0 \\ 1 & 0 & 0 & 0 \end{bmatrix} \\
&\quad + \begin{bmatrix} 0 & 0 & 0 \\ 0 & 1 & 0 \\ 1 & 0 & 0 \\ 0 & 0 & 1 \end{bmatrix}
\begin{bmatrix} K_{11}^{(2)} & K_{12}^{(2)} & K_{13}^{(2)} \\ K_{21}^{(2)} & K_{22}^{(2)} & K_{23}^{(2)} \\ K_{31}^{(2)} & K_{32}^{(2)} & K_{33}^{(2)} \end{bmatrix}
\begin{bmatrix} 0 & 0 & 1 & 0 \\ 0 & 1 & 0 & 0 \\ 0 & 0 & 0 & 1 \end{bmatrix}
\end{aligned}
$$

or

$$
K_{\alpha\beta} = \begin{bmatrix} K_{11} & K_{12} & K_{13} & K_{14} \\ K_{21} & K_{22} & K_{23} & K_{24} \\ K_{31} & K_{32} & K_{33} & K_{34} \\ K_{41} & K_{42} & K_{43} & K_{44} \end{bmatrix}
= \begin{bmatrix} K_{33}^{(1)} & K_{31}^{(1)} & K_{32}^{(1)} & 0 \\ K_{13}^{(1)} & K_{11}^{(1)} + K_{22}^{(2)} & K_{12}^{(1)} + K_{21}^{(2)} & K_{23}^{(2)} \\ K_{23}^{(1)} & K_{21}^{(1)} + K_{12}^{(2)} & K_{22}^{(1)} + K_{11}^{(2)} & K_{13}^{(2)} \\ 0 & K_{32}^{(2)} & K_{31}^{(2)} & K_{33}^{(2)} \end{bmatrix}
\tag{10.1.15a}
$$

Table 10.1.1 Nodal Correspondence Table

e ⇒ N ⇓	1	2
1	2	3
2	3	2
3	1	4

* Entries indicate global node numbers corresponding to the local nodes (see Figure 10.1.1)

Similarly,

$$F_\alpha = \bigcup_{e=1}^{E} F_N^{(e)} \Delta_{N\alpha}^{(e)}$$

or

$$F_\alpha = \begin{bmatrix} F_1 \\ F_2 \\ F_3 \\ F_4 \end{bmatrix} = \begin{bmatrix} F_3^{(1)} \\ F_1^{(1)} + F_2^{(2)} \\ F_2^{(1)} + F_1^{(2)} \\ F_3^{(2)} \end{bmatrix} \tag{10.1.15b}$$

The procedure of assembly implied here requiring determination of Boolean matrices for all elements is quite cumbersome. They are useful and convenient in deriving finite element equations, but are useless in actual performance of assembly operations. Thus, we should avoid Boolean matrices and implement a scheme that can handle complex geometries with a simple algorithm. An intuitive and more convenient approach is schematically shown below.

$$K_{NM}^{(1)} = \begin{matrix} & \text{②} & \text{③} & \text{①} \\ & 1 & 2 & 3 \\ \text{②}\ 1 & K_{11}^{(1)} & K_{12}^{(1)} & K_{13}^{(1)} \\ \text{③}\ 2 & K_{21}^{(1)} & K_{22}^{(1)} & K_{23}^{(1)} \\ \text{①}\ 3 & K_{31}^{(1)} & K_{32}^{(1)} & K_{33}^{(1)} \end{matrix} \qquad K_{NM}^{(2)} = \begin{matrix} & \text{③} & \text{②} & \text{④} \\ & 1 & 2 & 3 \\ \text{③}\ 1 & K_{11}^{(2)} & K_{12}^{(2)} & K_{13}^{(2)} \\ \text{②}\ 2 & K_{21}^{(2)} & K_{22}^{(2)} & K_{23}^{(2)} \\ \text{④}\ 3 & K_{31}^{(2)} & K_{32}^{(2)} & K_{33}^{(2)} \end{matrix}$$

$$K_{\alpha\beta} = \begin{matrix} & \text{①} & \text{②} & \text{③} & \text{④} \\ \text{①} & K_{33}^{(1)} & K_{31}^{(1)} & K_{32}^{(1)} & 0 \\ \text{②} & K_{13}^{(1)} & K_{11}^{(1)} + K_{22}^{(2)} & K_{12}^{(1)} + K_{21}^{(2)} & K_{23}^{(2)} \\ \text{③} & K_{23}^{(1)} & K_{21}^{(1)} + K_{12}^{(2)} & K_{22}^{(1)} + K_{11}^{(2)} & K_{13}^{(2)} \\ \text{④} & 0 & K_{32}^{(2)} & K_{31}^{(2)} & K_{33}^{(2)} \end{matrix} \tag{10.1.15c}$$

Similarly,

$$F_N^{(1)} = \begin{bmatrix} F_1^{(1)} \\ F_2^{(1)} \\ F_3^{(1)} \end{bmatrix} \begin{matrix} 1 \; ② \\ 2 \; ③ \\ 3 \; ① \end{matrix} \qquad F_N^{(2)} = \begin{bmatrix} F_1^{(2)} \\ F_2^{(2)} \\ F_3^{(2)} \end{bmatrix} \begin{matrix} 1 \; ③ \\ 2 \; ② \\ 3 \; ④ \end{matrix}$$

$$F_\alpha = \begin{bmatrix} F_3^{(1)} \\ F_1^{(1)} + F_2^{(2)} \\ F_2^{(1)} + F_1^{(2)} \\ F_3^{(2)} \end{bmatrix} \begin{matrix} ① \\ ② \\ ③ \\ ④ \end{matrix} \tag{10.1.15d}$$

Here, the node number with a circle indicates global node. It is seen that the assembled global matrix is obtained by finding the appropriate entries from the local matrices with the local node numbers replaced by the corresponding incident global node numbers. For example, $K_{11}^{(1)}$ of the first element goes to the second row and second column in the global matrix because the local node 1 is incident with the global node 2. Similarly, $K_{12}^{(1)}$ enters in the second row and third column of the global matrix since the global node number 2 is incident with the global node 3. All entries in the same rows and columns are algebraically added together as we move to the second element. The same procedure applies in order to obtain F_α. In this way, we avoid the need to construct the Boolean matrices, and the entire assembly procedure can be programmed very efficiently.

The global load vector may be obtained more conveniently in the form

$$F_\alpha = C_{\alpha\beta} f_\beta$$

in which only $C_{\alpha\beta}$ is assembled from the local contributions with f_β evaluated at global nodes. This will be shown in Example 10.1.2. The assembly of the Neumann boundary data G_α and the method of implementation will be discussed in Section 10.1.2.

Example 10.1.1 Assembly of Two Triangular Elements

Given:

$$K_{NM}^{(e)} = \iint \left(\frac{\partial \Phi_N^{(e)}}{\partial x} \frac{\partial \Phi_M^{(e)}}{\partial x} + \frac{\partial \Phi_N^{(e)}}{\partial y} \frac{\partial \Phi_M^{(e)}}{\partial y} \right) dxdy$$

Required: Calculate $K_{\alpha\beta} = \bigcup_{e=1}^{E} K_{NM}^{(e)} \Delta_{N\alpha}^{(e)} \Delta_{M\beta}^{(e)}$ by assembling two local linear triangular elements (Figure E10.1.1) to a global form and compare the results with a single isoparametric element of Example 9.3.2. for $n = 4$ and $n = 5$.

Solution:

$$K_{NM}^{(e)} = \iint \left(\frac{\partial \Phi_N^{(e)}}{\partial x} \frac{\partial \Phi_M^{(e)}}{\partial x} + \frac{\partial \Phi_N^{(e)}}{\partial y} \frac{\partial \Phi_M^{(e)}}{\partial y} \right) dxdy = A(b_N b_M + c_N c_M)$$

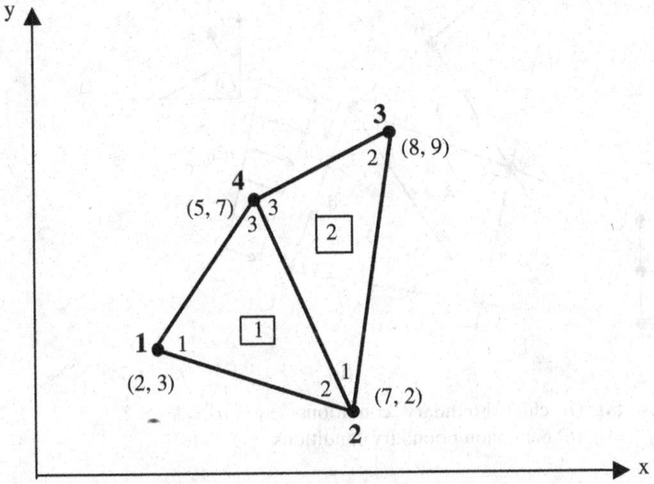

Figure E10.1.1 Assembly of two triangular elements.

where b_N and c_M are given in Example 9.3.1 and A is the triangle area.

$$K_{\alpha\beta} = \begin{bmatrix} K_{11}^{(1)} & K_{12}^{(1)} & 0 & K_{13}^{(1)} \\ & K_{22}^{(1)} + K_{11}^{(2)} & K_{12}^{(2)} & K_{23}^{(1)} + K_{13}^{(2)} \\ & & K_{22}^{(2)} & K_{23}^{(2)} \\ \text{sym} & & & K_{33}^{(1)} + K_{33}^{(2)} \end{bmatrix}$$

$$= \begin{bmatrix} 0.6304 & -0.3043 & 0 & -0.3261 \\ & 0.8856 & 0.1053 & -0.6865 \\ & & 0.7632 & -0.8684 \\ \text{sym} & & & 1.8810 \end{bmatrix}$$

This compares somewhat differently with the single isoparametric element of Example 9.3.2, although diagonal and off-diagonal values are, respectively, about 10% larger and 20% smaller than those for the single quadrilateral isoparametric element (Example 9.3.2). This may influence undoubtedly the solution of differential equations (see Example 10.2.1). It will be shown that the larger diagonal and smaller off-diagonal values in the stiffness matrix result in smaller responses to given boundary input data in general. It is concluded that the triangular element is "stiffer" than the quadrilateral element, as demonstrated in Example 10.2.1. The reason for this behavior is that triangular elements possess only three data points, whereas quadrilateral elements provide four data points allowing an additional degree of freedom and "flexibility."

10.1.2 BOUNDARY CONDITIONS IN TWO DIMENSIONS

(a) Standard Approach

Dirichlet Boundary Conditions. Dirichlet boundary conditions for multidimensional problems can be treated exactly the same as in the case of one-dimensional problems.

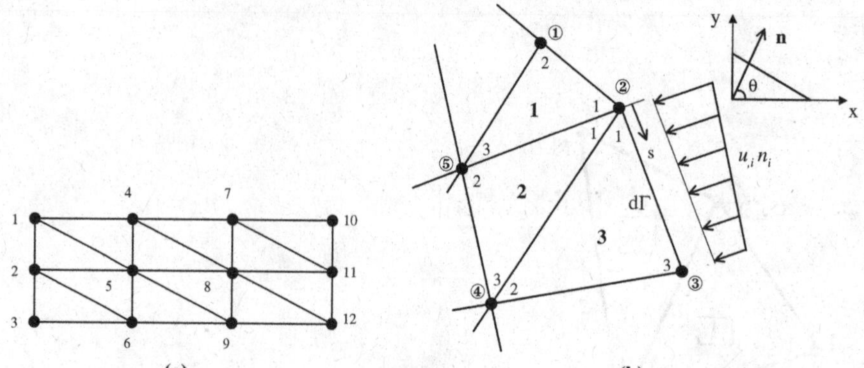

Figure 10.1.2 Boundary conditions. **(a)** Dirichlet boundary conditions ($u_1 = u_2 = u_3 = 2$, $u_4 = u_6 = u_7 = u_9 = u_{10} = u_{11} = u_{12} = 0$). **(b)** Neumann boundary conditions.

That is, the global finite element equations are modified, reflecting the specified Dirichlet data. For example, let us consider that the global finite element equations using either triangular elements or quadrilateral elements have been obtained in the form

$$K_{\alpha\beta}u_\beta = F_\alpha + G_\alpha \tag{10.1.16}$$

where we set $G_\alpha = 0$ because Neumann boundary conditions are not to be specified in this case. Only Dirichlet data are furnished as shown in Figure 10.1.2a. We begin with the assembled global equations,

$$\begin{bmatrix} K_{11} & K_{12} & \cdot & \cdot & \cdot & K_{1\,12} \\ K_{21} & K_{22} & \cdot & \cdot & \cdot & K_{2\,12} \\ \vdots & \vdots & \vdots & \vdots & \vdots & \vdots \\ K_{12\,1} & K_{12\,2} & \cdot & \cdot & \cdot & K_{12\,12} \end{bmatrix} \begin{bmatrix} u_1 \\ u_2 \\ \vdots \\ u_{12} \end{bmatrix} = \begin{bmatrix} F_1 \\ F_2 \\ \vdots \\ F_{12} \end{bmatrix} \tag{10.1.17a}$$

Now, if we apply the Dirichlet boundary conditions in (10.1.17a) as given in Figure 10.1.2a, we obtain

$$\begin{bmatrix}
1 & 0 & 0 & 0 & 0 & 0 & 0 & 0 & 0 & 0 & 0 & 0 \\
0 & 1 & 0 & 0 & 0 & 0 & 0 & 0 & 0 & 0 & 0 & 0 \\
0 & 0 & 1 & 0 & 0 & 0 & 0 & 0 & 0 & 0 & 0 & 0 \\
0 & 0 & 0 & 1 & 0 & 0 & 0 & 0 & 0 & 0 & 0 & 0 \\
0 & 0 & 0 & 0 & K_{55} & 0 & 0 & K_{58} & 0 & 0 & 0 & 0 \\
0 & 0 & 0 & 0 & 0 & 1 & 0 & 0 & 0 & 0 & 0 & 0 \\
0 & 0 & 0 & 0 & 0 & 0 & 1 & 0 & 0 & 0 & 0 & 0 \\
0 & 0 & 0 & 0 & K_{85} & 0 & 0 & K_{88} & 0 & 0 & 0 & 0 \\
0 & 0 & 0 & 0 & 0 & 0 & 0 & 0 & 1 & 0 & 0 & 0 \\
0 & 0 & 0 & 0 & 0 & 0 & 0 & 0 & 0 & 1 & 0 & 0 \\
0 & 0 & 0 & 0 & 0 & 0 & 0 & 0 & 0 & 0 & 1 & 0 \\
0 & 0 & 0 & 0 & 0 & 0 & 0 & 0 & 0 & 0 & 0 & 1
\end{bmatrix}
\begin{bmatrix} u_1 \\ u_2 \\ u_3 \\ u_4 \\ u_5 \\ u_6 \\ u_7 \\ u_8 \\ u_9 \\ u_{10} \\ u_{11} \\ u_{12} \end{bmatrix}
=
\begin{bmatrix} 0 \\ 0 \\ 0 \\ 0 \\ F_5 \\ 0 \\ 0 \\ F_8 \\ 0 \\ 0 \\ 0 \\ 0 \end{bmatrix}
+
\begin{bmatrix} 2 \\ 2 \\ 2 \\ 0 \\ -D_5 \\ 0 \\ 0 \\ -D_8 \\ 0 \\ 0 \\ 0 \\ 0 \end{bmatrix} \tag{10.1.17b}$$

with $D_5 = K_{51}(2) + K_{52}(2) + K_{53}(2)$ and $D_8 = K_{81}(2) + K_{82}(2) + K_{83}(2)$. It is seen that the rows and columns corresponding to the Dirichlet nodes are zero with unity at the diagonal position. The influence of Dirichlet boundary conditions, as imposed here, is reflected in the Dirichlet boundary vector D_α, so that

$$\overline{K}_{\alpha\beta}u_\beta = F_\alpha + D_\alpha \tag{10.1.18}$$

where D_α is given by the second column on the right-hand side with $\overline{K}_{\alpha\beta}$ as modified in (10.1.17) from the given Dirichlet boundary conditions. It is obvious that, if there are so many Dirichlet boundary nodes, then it is convenient to modify the above matrix equations in the form

$$\begin{bmatrix} K_{55} & K_{58} \\ K_{85} & K_{88} \end{bmatrix} \begin{bmatrix} u_5 \\ u_8 \end{bmatrix} = \begin{bmatrix} F_5 \\ F_8 \end{bmatrix} + \begin{bmatrix} -D_5 \\ -D_8 \end{bmatrix} \tag{10.1.19}$$

in which all rows and columns corresponding to Dirichlet boundary nodes are eliminated.

Neumann Boundary Conditions. Neumann boundary conditions are implemented using the integral form of (10.1.5c) with the local contributions coming from adjacent elements to the node at which Neumann data $g_M^{(e)}$ are prescribed in the form (10.1.8b),

$$g_M^{(e)} = (u_{,i}n_i)_M = \left(\frac{\partial u}{\partial x}\cos\theta + \frac{\partial u}{\partial y}\sin\theta\right)_M \tag{10.1.20}$$

as shown in Figure 10.1.2b with the normal angle θ measured counterclockwise from the axis.

Often in boundary value problems, there are instances in which the Dirichlet and Neumann boundary conditions are combined at the same location. For example, consider a heat conduction equation

$$k\nabla^2 T = 0$$

Here, for a resistance layer on the boundary, we specify

$$kT_{,i}n_i + \overline{\alpha}(T - T') = -q \tag{10.1.21}$$

where T, T', $\overline{\alpha}$, and q denote the surface temperature, ambient temperature, heat transfer coefficient, and surface heat flux, respectively. This is referred to as the Cauchy or Robin boundary condition and can be handled by substitution:

$$kT_{,i}n_i = -Q - \overline{\alpha}T$$

with

$$Q = q - \overline{\alpha}T'$$

Thus, we write

$$G_\alpha = \hat{G}_\alpha - \overset{*}{C}_{\alpha\beta}T_\beta \tag{10.1.22}$$

with

$$\hat{G}_\alpha = -\int_\Gamma \overset{*}{\Phi}_\alpha Q d\Gamma = \bigcup_{e=1}^{E} \hat{G}_N^{(e)} \Delta_{N\alpha}^{(e)}, \quad \hat{G}_N^{(e)} = -\int_\Gamma \overset{*}{\Phi}_N^{(e)} Q d\Gamma$$

$$\overset{*}{C}_{\alpha\beta} = \int_\Gamma \overline{\alpha}\, \overset{*}{\Phi}_\alpha \overset{*}{\Phi}_\beta d\Gamma = \bigcup_{e=1}^{E} \overset{*}{C}{}_{NM}^{(e)} \Delta_{N\alpha}^{(e)} \Delta_{M\beta}^{(e)}, \quad \overset{*}{C}{}_{NM}^{(e)} = \int_\Gamma \overline{\alpha}\, \overset{*}{\Phi}{}_N^{(e)} \overset{*}{\Phi}{}_M^{(e)} d\Gamma$$

This process then modifies (10.1.4) in the form

$$(K_{\alpha\beta} + \overset{*}{C}_{\alpha\beta}) T_\beta = F_\alpha + \hat{G}_\alpha \tag{10.1.23}$$

It should be noted that $\overset{*}{C}_{\alpha\beta}$ is activated only if the convection or Cauchy boundary conditions are present. That is, if a global node does not coincide with the boundary node at which the Neumann boundary conditions are prescribed, then $\overset{*}{C}_{\alpha\beta}$ is empty from the definition, $\overset{*}{\Phi}_\alpha(Z_\beta) = \delta_{\alpha\beta}$. It is cautioned that the local boundary surface matrix is (2×2), which is simply added to the local triangular element stiffness matrix (3×3) in correspondence with the nodal incidence along the boundaries.

(b) Lagrange Multipliers Approach

Any boundary condition prescribed at a boundary node may be imposed through Lagrange multipliers. Consider the boundary conditions of the form

$$u_1 = 0 \tag{10.1.24a}$$

$$u_2 = a \tag{10.1.24b}$$

$$u_3 - u_4 = b \tag{10.1.24c}$$

Obviously, if $b = 0$, then the second expression implies $u_3 = u_4$. Otherwise, it represents Neumann boundary conditions $(du/dx)\cos\theta$ or $(du/dy)\sin\theta$, prescribed at the global node Z_3 connected to the adjacent boundary node Z_4. For example, if $du/dx = c$ at Z_3 and the boundary line of length l between Z_3 and Z_4 is inclined an angle of θ from the x axis, then we write

$$\frac{du}{dx} = \frac{u_3 - u_4}{l\cos\theta} = c \tag{10.1.25}$$

or

$$u_3 - u_4 = b \quad \text{with } b = cl\cos\theta$$

Equation (10.1.24) can be written in the form

$$\begin{bmatrix} 1 & 0 & 0 & 0 & 0 & \cdots \\ 0 & 1 & 0 & 0 & 0 & \cdots \\ 0 & 0 & 1 & -1 & 0 & \cdots \end{bmatrix} \begin{bmatrix} u_1 \\ u_2 \\ u_3 \\ u_4 \\ \vdots \\ u_n \end{bmatrix} = \begin{bmatrix} 0 \\ a \\ b \end{bmatrix} \tag{10.1.26}$$

which may be rearranged as

$$q_{r\alpha}u_{\alpha} = E_r \tag{10.1.27}$$

with $r = 1, \ldots, m$ (total number of boundary conditions, $m = 3$ in this case) and $\alpha = 1, \ldots, n$ (total number of global nodes). Here, $q_{r\alpha}$ is called the boundary condition matrix. Let us now introduce quantities λ_r, referred to as Lagrange multipliers, and regarded as constraints or forces required to maintain the boundary conditions. Then, the product of (10.1.27) with the Lagrange multiplier λ_r

$$\lambda_r(q_{r\alpha}u_{\alpha} - E_r) = 0 \tag{10.1.28}$$

may be considered as an invariant or energy required to maintain such boundary conditions.

At this point, we transform the global finite element equation (10.1.16) into a variational energy,

$$\delta I = (K_{\alpha\beta}u_{\beta} - H_{\alpha})\delta u_{\alpha} = 0 \tag{10.1.29}$$

or

$$\delta I = \delta\left(\frac{1}{2}K_{\alpha\beta}u_{\alpha}u_{\beta} - H_{\alpha}u_{\alpha}\right) = 0 \tag{10.1.30}$$

for which the stationary condition is given by

$$I = \frac{1}{2}K_{\alpha\beta}u_{\alpha}u_{\beta} - H_{\alpha}u_{\alpha} \tag{10.1.31}$$

This may be considered as the actual energy contained in the domain. To this we may add (10.1.28),

$$I = \frac{1}{2}K_{\alpha\beta}u_{\alpha}u_{\beta} - H_{\alpha}u_{\alpha} + \lambda_r(q_{r\alpha}u_{\alpha} - E_r) \tag{10.1.32}$$

The expression (10.1.32) refers to the total variational energy in equilibrium with the imposed boundary conditions. The variation of (10.1.32) with respect to every u_{α} and λ_r will lead to the stationary condition

$$\delta I = \frac{\partial I}{\partial u_{\alpha}}\delta u_{\alpha} + \frac{\partial I}{\partial \lambda_r}\delta\lambda_r = 0 \tag{10.1.33}$$

Since u_{α} and λ_r are arbitrary, it is necessary that $\partial I/\partial u_{\alpha}$ and $\partial I/\partial\lambda_r$ vanish. These conditions yield

$$K_{\alpha\beta}u_{\alpha} + \lambda_r q_{r\alpha} = H_{\alpha}$$

$$q_{r\alpha}u_{\alpha} = E_r$$

Writing these two equations in matrix form, we obtain

$$\begin{bmatrix} K_{\alpha\beta} & q_{r\alpha} \\ q_{r\beta} & 0 \end{bmatrix}\begin{bmatrix} u_{\beta} \\ \lambda_r \end{bmatrix} = \begin{bmatrix} H_{\alpha} \\ E_r \end{bmatrix} \tag{10.1.34}$$

which may be expanded with the boundary conditions of (10.1.26) in the form

$$\begin{bmatrix} K_{11} & K_{12} & \cdot & \cdot & \cdot & \cdot & K_{1n} & 1 & 0 & 0 \\ K_{21} & K_{22} & \cdot & \cdot & \cdot & \cdot & K_{2n} & 0 & 1 & 0 \\ \cdot & \cdot & \cdot & \cdot & \cdot & \cdot & \cdot & 0 & 0 & 1 \\ \cdot & \cdot & \cdot & \cdot & \cdot & \cdot & \cdot & 0 & 0 & -1 \\ \cdot & \cdot & \cdot & \cdot & \cdot & \cdot & \cdot & 0 & 0 & 0 \\ \cdot & \cdot & \cdot & \cdot & \cdot & \cdot & \cdot & \cdot & \cdot & \cdot \\ K_{n1} & K_{n2} & \cdot & \cdot & \cdot & \cdot & K_{nn} & 0 & 0 & 0 \\ 1 & 0 & 0 & 0 & 0 & \cdot & 0 & 0 & 0 & 0 \\ 0 & 1 & 0 & 0 & 0 & \cdot & 0 & 0 & 0 & 0 \\ 0 & 0 & 1 & -1 & 0 & \cdot & 0 & 0 & 0 & 0 \end{bmatrix} \begin{bmatrix} u_1 \\ u_2 \\ u_3 \\ u_4 \\ \cdot \\ u_n \\ \lambda_1 \\ \lambda_2 \\ \lambda_3 \end{bmatrix} = \begin{bmatrix} H_1 \\ H_2 \\ \cdot \\ \cdot \\ H_n \\ E_1 \\ E_2 \\ E_3 \end{bmatrix} \tag{10.1.35}$$

The solution to these equations provides the values of Lagrange multipliers λ_r as well as the unknowns u_α. Here λ_r, interpreted as the boundary forces, assisted in imposing the boundary conditions. Note that the left-hand side matrix (10.1.35) is still symmetric, but matrix rearrangements are required to avoid zeros on the diagonal before a standard equation solver is applied.

Remarks: The Lagrange multiplier approach for implementing boundary conditions is useful if the finite element formulations are performed by means of methods of least squares, moments, or collocation in which the Neumann boundary conditions do not arise naturally since integration by parts is not involved in these methods.

10.1.3 SOLUTION PROCEDURE

In order to illustrate the solution procedure and implementation of both Dirichlet and Neumann boundary conditions, we present the following examples.

Example 10.1.2 Solution of Poisson Equation by Triangular Elements

Given:

$$u_{,ii} = f \quad (i = 1, 2)$$

with $f = 4(x^2 + y^2)$, exact solution: $u = 2x^2y^2$.
Consider the geometry (Figure E10.1.2) with Dirichlet boundary conditions:

(1) $u_2 = u_3 = u_6 = u_9 = u_{12} = 0$
(2) $u_{11} = 1,458$
(3) $u_1 = 0, \quad u_4 = 450, \quad u_7 = 3,528, \quad u_{10} = 5,832$

Neumann boundary conditions along nodes 1, 4, 7, and 10:

(4) $\left(\frac{\partial u}{\partial x}\right)_1 = 0, \quad \left(\frac{\partial u}{\partial x}\right)_4 = 300, \quad \left(\frac{\partial u}{\partial x}\right)_7 = 1{,}176, \quad \left(\frac{\partial u}{\partial x}\right)_{10} = 1{,}296,$

$\left(\frac{\partial u}{\partial y}\right)_1 = 0 \quad \left(\frac{\partial u}{\partial y}\right)_4 = 180 \quad \left(\frac{\partial u}{\partial y}\right)_7 = 1008 \quad \left(\frac{\partial u}{\partial y}\right)_{10} = 1{,}944$

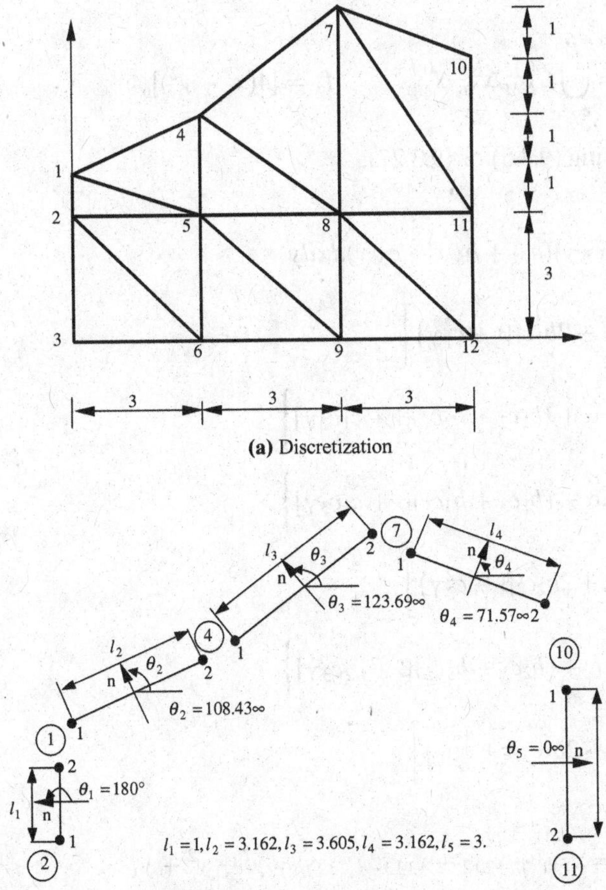

(a) Discretization

(b) Neumann boundary elements

Figure E10.1.2 Two-dimensional problem with linear triangular elements.

Required: Using the linear triangular elements, solve the differential equation with the boundary conditions:

(a) – (1), (2), and (3)
(b) – (1), (2), and (4)

Solution:

$$K_{\alpha\beta}u_\beta = F_\alpha + G_\alpha$$

$$K_{\alpha\beta} = \bigcup_{e=1}^{E} K_{NM}^{(e)}\Delta_{N\alpha}^{(e)}\Delta_{M\beta}^{(e)}$$

$$K_{NM}^{(e)} = \int_\Omega \Phi_{N,i}^{(e)}\Phi_{M,i}^{(e)}d\Omega = A(b_N b_M + c_N c_M)$$

$$F_\alpha = -\int_\Omega \Phi_\alpha f d\Omega = -\int_\Omega \Phi_\alpha \Phi_\beta d\Omega f_\beta = -C_{\alpha\beta} f_\beta = -\bigcup_{e=1}^{E} C_{NM}^{(e)}\Delta_{N\alpha}^{(e)}\Delta_{M\beta}^{(e)} f_\beta$$

or

$$F_\alpha = -C_{\alpha\beta} f_\beta, \qquad C_{\alpha\beta} = \bigcup_{e=1}^{E} C_{NM}^{(e)} \Delta_{N\alpha}^{(e)} \Delta_{M\beta}^{(e)}, \qquad f_\alpha = [4(x^2 + y^2)]_\alpha$$

The $C_{NM}^{(e)}$ may be determined using (9.3.5) or (9.3.27).
From (9.3.5), we have

$$C_{NM}^{(e)} = \iint (a_N + b_N x + c_N y)(a_M + b_M x + c_M y)dxdy$$

$$C_{11}^{(e)} = A^{(e)}\left[\frac{1}{9} + \frac{1}{12}\left(b_1^2\alpha + 2b_1c_1\beta + c_1^2\gamma\right)\right]$$

$$C_{12}^{(e)} = A^{(e)}\left\{\frac{1}{9} + \frac{1}{12}[b_1b_2\alpha + (b_1c_2 + b_2c_1)\beta + c_1c_2\gamma]\right\}$$

$$C_{13}^{(e)} = A^{(e)}\left\{\frac{1}{9} + \frac{1}{12}[b_1b_3\alpha + (b_1c_3 + b_3c_1)\beta + c_1c_3\gamma]\right\}$$

$$C_{22}^{(e)} = A^{(e)}\left[\frac{1}{9} + \frac{1}{12}\left(b_2^2\alpha + 2b_2c_2\beta + c_2^2\gamma\right)\right]$$

$$C_{23}^{(e)} = A^{(e)}\left\{\frac{1}{9} + \frac{1}{12}[b_2b_3\alpha + (b_2c_3 + b_3c_2)\beta + c_2c_3\gamma]\right\}$$

$$C_{33}^{(e)} = A^{(e)}\left[\frac{1}{9} + \frac{1}{12}\left(b_3^2\alpha + 2b_3c_3\beta + c_3^2\gamma\right)\right]$$

with

$$\alpha = x_1^2 + x_2^2 + x_3^2, \qquad \beta = x_1y_1 + x_2y_2 + x_3y_3, \qquad \gamma = y_1^2 + y_2^2 + y_3^2$$

After some algebra, it can be shown that

$$C_{NM}^{(e)} = \frac{A^{(e)}}{12}\begin{bmatrix} 2 & 1 & 1 \\ 1 & 2 & 1 \\ 1 & 1 & 2 \end{bmatrix}$$

This result can be obtained easily from (9.3.11 and 9.3.27) using the natural coordinate triangular element.

$$C_{NM}^{(e)} = \int_\Omega \Phi_N^{(e)} \Phi_M^{(e)} d\Omega = \iint \begin{bmatrix} L_1L_1 & L_1L_2 & L_1L_3 \\ L_2L_1 & L_2L_2 & L_2L_3 \\ L_3L_1 & L_3L_2 & L_3L_3 \end{bmatrix} dxdy = \frac{A^{(e)}}{12}\begin{bmatrix} 2 & 1 & 1 \\ 1 & 2 & 1 \\ 1 & 1 & 2 \end{bmatrix}$$

Thus, the global load vector is calculated from the assembly of $C_{NM}^{(e)}$ matrices for each element into a global form $C_{\alpha\beta}$ to be multiplied by the global nonhomogeneous data f_β determined at each global node.

The Neumann boundary vector G_α can be calculated as follows:

$$G_\alpha = \int_\Gamma \overset{*}{\Phi}_\alpha u_{,i} n_i d\Gamma = \int_\Gamma \overset{*}{\Phi}_\alpha \overset{*}{\Phi}_\beta d\Gamma g_\beta = \bigcup_{e=1}^{E} \overset{*}{C}{}_{NM}^{(e)} \Delta_{N\alpha}^{(e)} g_M^{(e)} = \bigcup_{e=1}^{E} G_N^{(e)} \Delta_{N\alpha}^{(e)}$$

where

$$\overset{*}{C}{}_{NM}^{(e)} = \int_0^l \overset{*}{\Phi}{}_N^{(e)} \overset{*}{\Phi}{}_M^{(e)} d\Gamma = \frac{l}{6}\begin{bmatrix} 2 & 1 \\ 1 & 2 \end{bmatrix}$$

Thus

$$G_N^{(e)} = \frac{l}{6}\begin{bmatrix} 2 & 1 \\ 1 & 2 \end{bmatrix}\begin{bmatrix} g_1^{(e)} \\ g_2^{(e)} \end{bmatrix} = \frac{l}{6}\begin{bmatrix} 2g_1^{(e)} + g_2^{(e)} \\ g_1^{(e)} + 2g_2^{(e)} \end{bmatrix}$$

where $\overset{*}{\Phi}{}_M^{(e)}$ vanishes everywhere except at Neumann boundary nodes. Recall that $\overset{*}{\Phi}{}_M^{(e)}(z_M) = \delta_{NM}$ and thus, $\overset{*}{\Phi}{}_M^{(e)} = 0$ if the boundary node N does not have the Neumann data prescribed, and $\overset{*}{\Phi}{}_M^{(e)} = 1$ if the boundary node N has the Neumann boundary data prescribed.

$$G_N^{(1)} = \frac{l_1}{6}\begin{bmatrix} 0 & 0 \\ 0 & 2 \end{bmatrix}\begin{bmatrix} g_1^{(1)} \\ g_2^{(1)} \end{bmatrix} = \frac{l_1}{6}\begin{bmatrix} 0 \\ 2g_2^{(1)} \end{bmatrix}$$

with $\overset{*}{\Phi}{}_N^{(1)} = 0$, because the Neumann data are not prescribed at the local node 1 for the boundary element 1.

$$G_N^{(2)} = \frac{l_2}{6}\begin{bmatrix} 2 & 1 \\ 1 & 2 \end{bmatrix}\begin{bmatrix} g_1^{(2)} \\ g_2^{(2)} \end{bmatrix} = \frac{l_2}{6}\begin{bmatrix} 2g_1^{(2)} + g_2^{(2)} \\ g_1^{(2)} + 2g_2^{(2)} \end{bmatrix}$$

$$G_N^{(3)} = \frac{l_3}{6}\begin{bmatrix} 2 & 1 \\ 1 & 2 \end{bmatrix}\begin{bmatrix} g_1^{(3)} \\ g_2^{(3)} \end{bmatrix} = \frac{l_3}{6}\begin{bmatrix} 2g_1^{(3)} + g_2^{(3)} \\ g_1^{(3)} + 2g_2^{(3)} \end{bmatrix}$$

$$G_N^{(4)} = \frac{l_4}{6}\begin{bmatrix} 2 & 1 \\ 1 & 2 \end{bmatrix}\begin{bmatrix} g_1^{(4)} \\ g_2^{(4)} \end{bmatrix} = \frac{l_4}{6}\begin{bmatrix} 2g_1^{(4)} + g_2^{(4)} \\ g_1^{(4)} + 2g_2^{(4)} \end{bmatrix}$$

$$G_N^{(5)} = \frac{l_5}{6}\begin{bmatrix} 2 & 0 \\ 0 & 0 \end{bmatrix}\begin{bmatrix} g_1^{(5)} \\ g_2^{(5)} \end{bmatrix} = \frac{l_5}{6}\begin{bmatrix} 2g_1^{(5)} \\ 0 \end{bmatrix}$$

with $\overset{*}{\Phi}{}_2^{(5)} = 0$ and

$$g_2^{(1)} = \left(\frac{\partial u}{\partial x}\cos\theta + \frac{\partial u}{\partial y}\sin\theta\right)_2^{(1)} = \left(\frac{\partial u}{\partial x}\right)_2^{(1)}(-1) = 0$$

$$g_1^{(2)} = \left(\frac{\partial u}{\partial x}\right)_1^{(2)}(-0.316) + \left(\frac{\partial u}{\partial y}\right)_1^{(2)}(0.948) = 0$$

$$g_2^{(2)} = \left(\frac{\partial u}{\partial x}\right)_2^{(2)}(-0.316) + \left(\frac{\partial u}{\partial y}\right)_2^{(2)}(0.948)$$
$$= (300)(-0.316) + (180)(0.948) = 75.84$$

Similarly,

$$g_1^{(3)} = -16.74, \quad g_2^{(3)} = 185.97, \quad g_1^{(4)} = 1,328.2, \quad g_2^{(4)} = 2,254.4,$$

$$g_1^{(5)} = \left(\frac{\partial u}{\partial x}\right)_1^{(5)} (1) = 1,296$$

$$G_\alpha = \begin{bmatrix} G_1 \\ G_4 \\ G_7 \\ G_{10} \end{bmatrix} = \frac{1}{6}\begin{bmatrix} \ell_1 2g_2^{(1)} + \ell_2(2g_1^{(2)} + g_2^{(2)}) \\ \ell_2(g_1^{(2)} + 2g_2^{(2)}) + \ell_3(2g_1^{(3)} + g_2^{(3)}) \\ \ell_3(g_1^{(3)} + 2g_2^{(3)}) + \ell_4(2g_1^{(4)} + g_2^{(4)}) \\ \ell_4(g_1^{(4)} + 2g_2^{(4)}) + \ell_5(2g_1^{(5)}) \end{bmatrix} = \begin{bmatrix} 40.00 \\ 172.03 \\ 2,802.05 \\ 4,372.02 \end{bmatrix}$$

with $G_\alpha = 0$ elsewhere. The sum of $F_\alpha + G_\alpha$ is given by

$$F_\alpha + G_\alpha = -\begin{bmatrix} 113.50 \\ 134.00 \\ 27.00 \\ 629.00 \\ 609.50 \\ 216.00 \\ 1673.50 \\ 2008.00 \\ 648.00 \\ 613.50 \\ 1652.00 \\ 810.00 \end{bmatrix} + \begin{bmatrix} 40.00 \\ 0.00 \\ 0.00 \\ 172.03 \\ 0.00 \\ 0.00 \\ 2802.05 \\ 0.00 \\ 0.00 \\ 4372.02 \\ 0.00 \\ 0.00 \end{bmatrix}$$

Note that G_α is obtained by an assembly of local data $g_M^{(e)}$. However for F_α, it is preferable to construct the $C_{\alpha\beta}$ matrix independent of local data $f_M^{(e)}$ and use the global data f_α instead.

The solution is carried out, and the results are shown in Table E10.1.1. It is seen that the solution for the Neumann data is less accurate than for the Dirichlet data. It can be shown that accuracy improves with mesh refinements. This is demonstrated in Section 10.4.1 for isoparametric elements.

10.1.4 STOKES FLOW PROBLEMS

Stokes flows or creeping flows occur in highly viscous, slowly moving fluids and are characterized by the conservation of mass and momentum. For a steady state, the governing equations take the form

$$\nabla \cdot \mathbf{v} = 0 \tag{10.1.36a}$$

$$-\mu \nabla^2 \mathbf{v} + \nabla p - \rho \mathbf{F} = 0 \tag{10.1.36b}$$

Although these equations are still linear (note that convective terms are absent), their solutions may not be easy to obtain because the enforcement of incompressibility

Table E10.1.1 Computed Results for Example 10.1.2

(a) Dirichlet Problem with the Boundary Conditions (1), (2), and (3)

Node	Exact Solution	FEM Solution	% Error
1	0.00	0.00	0.00
2	0.00	0.00	0.00
3	0.00	0.00	0.00
4	450.00	450.00	0.00
5	162.00	110.72	−31.66
6	0.00	0.00	0.00
7	3528.00	3528.00	0.00
8	648.00	508.92	−21.46
9	0.00	0.00	0.00
10	5832.00	5832.00	0.00
11	1458.00	1458.00	0.00
12	0.00	0.00	0.00

(b) Neumann Problem with the Boundary Conditions (1), (2), and (4)

Node	Exact Solution	FEM Solution	% Error
1	0.00	0.00	0.00
2	0.00	0.00	0.00
3	0.00	0.00	0.00
4	450.00	392.33	−12.82
5	162.00	79.57	−50.88
6	0.00	0.00	0.00
7	3528.00	3264.54	−7.47
8	648.00	458.15	−29.30
9	0.00	0.00	0.00
10	5832.00	5031.26	−13.73
11	1458.00	1458.00	0.00
12	0.00	0.00	0.00

conditions (conservation of mass) is difficult. As a result, the computed pressure, p, may be spurious and oscillatory, known as checkerboard type oscillations.

To cope with these difficulties, many methods have been reported in the literature [Carey and Oden, 1986; Zienkiewicz and Taylor, 1991]. Among them are the mixed methods and penalty methods, which are presented below.

Mixed Methods

The momentum equation has the second derivative of velocity ($\mathbf{v} \,\varepsilon\, H^2$) and first derivative of pressure ($p \,\varepsilon\, H^1$). In order to enforce the mass conservation (incompressibility condition) we must use an appropriate function for the pressure consistent with the functional space for the velocity. This is known as the "consistency condition" or "LBB condition" after Ladyzhenskaya [1969], Babuska [1973], and Brezzi [1974]. This condition requires that the trial function for pressure in the momentum equation and

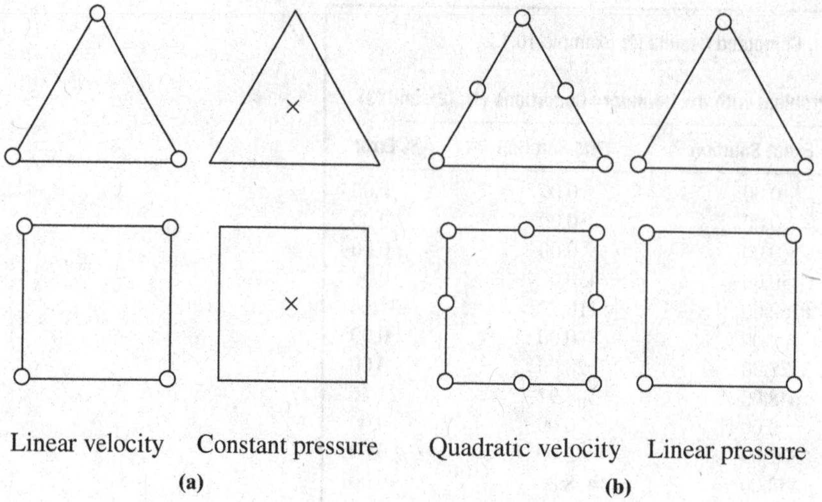

Figure 10.1.3 Mixed methods with triangles and quadrilaterals. **(a)** Mixed interpolation with constant pressure. **(b)** Mixed interpolation with linear pressure.

the test function for the continuity equation be chosen one order lower than the test function for the momentum equation and trial function for the velocity in the continuity equation, respectively.

Based on these requirements, the SGM equations of (10.1.36a,b) are of the form

$$\begin{bmatrix} A_{\alpha\beta ik} & B_{\alpha\beta i} \\ B_{\alpha\beta k} & 0 \end{bmatrix} \begin{bmatrix} v_{\beta k} \\ p_\beta \end{bmatrix} = \begin{bmatrix} F_{\alpha i} \\ 0 \end{bmatrix} + \begin{bmatrix} G_{\alpha i} \\ 0 \end{bmatrix} \tag{10.1.37}$$

If pressure is interpolated as constant (pressure node at the center of an element) and velocity as a linear function (velocity defined at corner nodes), then such element becomes over-constrained (known as "locking element") (Figure 10.1.3a). To avoid this situation, we may use linear pressure and quadratic velocity interpolations (Figure 10.1.3b). However, experience has shown that further improvements are needed in order to expedite convergence toward acceptable solutions. This subject will be elaborated in Chapter 12.

Penalty Methods

Penalty methods are designed such that the continuity equation which actually represents a constraint condition can be eliminated from the solution process. This is achieved by setting

$$p = -\lambda \nabla \cdot \mathbf{v} \tag{10.1.38}$$

where λ is the penalty parameter, equivalent to the Lagrange multiplier. The idea is to set λ equal to a large number ($\lambda \to \infty$) in the hope that $\nabla \cdot \mathrm{v} \approx 0$ as seen from

$$\nabla \cdot \mathbf{v} + \frac{p}{\lambda} \cong 0 \tag{10.1.39}$$

Substituting (10.1.38) into (10.1.36b), we obtain

$$-\mu \nabla^2 \mathrm{v} - \lambda \nabla(\nabla \cdot \mathrm{v}) - \rho \mathbf{F} = 0 \tag{10.1.40}$$

Here λ is seen to act as dilatational viscosity. It is now clear that pressure is eliminated from the solution of (10.1.40) in which the mass conservation is enforced through (10.1.39). Once the velocity components are calculated from (10.1.40), then pressure is calculated by means of (10.3.38).

Unfortunately, however, the solution of (10.1.40) is difficult because the penalty term dominates as λ becomes large, which is analogous to the over-constraint in the mixed methods. In other words, the consistency condition is violated. To cope with this difficulty, the finite element equation integral term involving the penalty function (pressure term) is given a special treatment by means of "reduced" Gaussian quadrature numerical integration. Specifically, we under-integrate the penalty term one point less than the shear viscosity term. For example, one point Gaussian quadrature rule for the penalty term is performed against the two-point rule for the shear viscosity term of a linear element. Similarly, a two-point rule for the penalty term against a three-point rule for the shear viscosity term of a quadratic element is recommended, and so on.

Once again, the mixed methods and penalty methods represent relatively earlier developments. They are being replaced by more efficient and advanced techniques to be discussed in Chapter 12 for incompressible viscous flows.

10.2 TRANSIENT PROBLEMS – GENERALIZED GALERKIN METHODS

10.2.1 PARABOLIC EQUATIONS

To describe the time-dependent behavior, we may use either the continuous space-time (CST) method or the discontinuous space-time (DST) method. In the CST method, continuous interpolation functions in both space and time are used so that

$$u(\mathbf{x}, t) = \Phi_\alpha(\mathbf{x}, t)u_\alpha \tag{10.2.1}$$

Alternatively, the DST method allows separation of variables between the spatial and temporal domains,

$$u(\mathbf{x}, t) = \Phi_\alpha(\mathbf{x})u_\alpha(t) \tag{10.2.2}$$

This requires interpolations of $\Phi_\alpha(\mathbf{x})$ in the spatial domain and the nodal values $u_\alpha(t)$ for the temporal domain.

The disadvantage of the CST method is the increase in computational dimension requiring the finite element in time. For this reason, our discussions in the sequel will be limited to the DST method, in which a time marching procedure is followed.

Consider a parabolic equation or the time-dependent differential equation in the form

$$R = \frac{\partial u(\mathbf{x}, t)}{\partial t} - \nabla^2 u(\mathbf{x}, t) - f(\mathbf{x}, t) = 0 \tag{10.2.3}$$

Let the nondimensional temporal variable be given by

$$\xi = t/\Delta t \tag{10.2.4}$$

where t and Δt denote time and a small time step, respectively.

In the past, the so-called semidiscrete method was used, in which the SGM equation for (10.2.3) is written as

$$(\Phi_\alpha, R) = \int_\Omega \Phi_\alpha\left(\frac{\partial u}{\partial t} - u_{,ii} - f\right) d\Omega = 0$$

where the time derivative of u is approximated by finite differences. Instead, our approach in DST is to seek a temporal test function independently and discontinuously from the spatial test function.

The DST method consists of first constructing the inner product of the residual (10.2.3) with the spatial test function $\Phi_\alpha(x)$ over the spatial domain and, subsequently, constructing another inner product of the resulting residual with the temporal weighting function or test function $\hat{W}(\xi)$ over the temporal domain. These steps lead to

$$(\hat{W}(\xi), (\Phi_\alpha, R)) = \int_0^1 \hat{W}(\xi)\left[\int_\Omega \Phi_\alpha\left(\frac{\partial u}{\partial t} - u_{,ii} - f\right) d\Omega\right] d\xi = 0 \tag{10.2.5}$$

which represents the SGM with DST approximations. The double projections of the residual onto the subspaces spanned by spatial and temporal test functions are referred to as the generalized Galerkin Method (GGM) as opposed to SGM. As noted in (8.2.41), the temporal weighting function $\hat{W}(\xi)$ is independent of and discontinuous from the spatial approximations.

Substituting (10.2.2) into (10.2.5) yields

$$\int_0^1 \hat{W}(\xi)\left[A_{\alpha\beta}\frac{\partial u_\beta(t)}{\partial t} + K_{\alpha\beta}u_\beta(t) - H_\alpha\right] d\xi = 0 \tag{10.2.6}$$

where we may define

Mass Matrix

$$A_{\alpha\beta} = \int_\Omega \Phi_\alpha \Phi_\beta d\Omega \tag{10.2.7}$$

Stiffness Matrix

$$K_{\alpha\beta} = \int_\Omega \Phi_{\alpha,i}\Phi_{\beta,i} d\Omega \tag{10.2.8}$$

$$H_\alpha = F_\alpha + G_\alpha \tag{10.2.9}$$

with

Source Vector $\qquad F_\alpha = \int_\Omega \Phi_\alpha f d\Omega$

Neumann Boundary Vector $\qquad G_\alpha = \int_\Gamma \overset{*}{\Phi}_\alpha u_{,i} n_i d\Gamma.$

If linear variations of $u_\alpha(t)$ are assumed within a small time step, we may write

$$u_\alpha(t) = \hat{\Phi}_m(\xi)u_\alpha^m \quad (m = 1, 2) \tag{10.2.10}$$

where the temporal trial functions may be derived from the standard one-dimensional configuration,

$$\hat{\Phi}_1 = 1 - \xi, \qquad \hat{\Phi}_2 = \xi$$

Thus,

$$u_\alpha(t) = (1-\xi)u_\alpha^n + \xi u_\alpha^{n+1} \tag{10.2.11}$$

in which $m = 1$ and $m = 2$ are replaced by the time steps n and $n+1$, respectively.

Differentiating (10.2.11) with respect to time, we obtain

$$\frac{\partial u_\alpha(t)}{\partial t} = \frac{\partial u_\alpha(\xi)}{\partial \xi}\frac{\partial \xi}{\partial t} = \frac{1}{\Delta t}(u_\alpha^{n+1} - u_\alpha^n) \tag{10.2.12}$$

which is identical to the forward finite difference of $\partial u(t)/\partial t$. Substituting (10.2.12) into (10.2.6) yields

$$[A_{\alpha\beta} + \eta\Delta t K_{\alpha\beta}]\, u_\beta^{n+1} = [A_{\alpha\beta} - (1-\eta)\Delta t K_{\alpha\beta}]\, u_\beta^n + \Delta t H_\alpha \tag{10.2.13}$$

where H_α may be regarded as the forcing function. If H_α is time dependent, then it may be expanded in a manner similar to u_α given in (10.2.11).

$$H_\alpha = (1-\xi)H_\alpha^n + \xi H_\alpha^{n+1}$$

Temporal Parameter

We define η as the temporal parameter,

$$\eta = \frac{\int_0^1 \hat{W}(\xi)\xi d\xi}{\int_0^1 \hat{W}(\xi) d\xi} \tag{10.2.14}$$

Evaluation of the temporal parameter requires an explicit form for the temporal test function $\hat{W}(\xi)$ as introduced in Zienkiewicz and Taylor [1991]. Some of the examples for $\hat{W}(\xi)$ and the corresponding temporal parameters are shown in Table 10.2.1. A glance at the temporal parameters suggested above reveals that they remain in the range

$$0 \le \eta \le 1$$

Equation (10.2.13) may be written in the form

$$D_{\alpha\beta}u_\beta^{n+1} = Q_\alpha^n \tag{10.2.15}$$

Table 10.2.1 Temporal Parameters for Parabolic Equations

$\hat{W}(\xi)$	η
$1-\xi$	1/3
ξ	2/3
1	1/2
$\delta(\xi-0)$	0
$\delta(\xi-1/2)$	1/2
$\delta(\xi-1)$	1

with

$$D_{\alpha\beta} = A_{\alpha\beta} + \eta\Delta t K_{\alpha\beta}$$
$$Q_\alpha^n = [A_{\alpha\beta} - (1-\eta)\Delta t K_{\alpha\beta}]u_\beta^n + \Delta t H_\alpha$$

Notice that, to solve (10.2.15), we must first apply the boundary conditions in a manner similar to that used in the steady-state problems. Initial conditions can be specified in Q_α^n. Initially, $n = 0$, and $u_\alpha^{(1)}$ for the first step is calculated from $Q_\alpha^{(0)}$. Then $u_\beta^{(2)}$ for the second time step will be calculated from $u_\alpha^{(1)}$ substituted into $Q_\alpha^{(1)}$, thus continuously marching in time until the desired time has been reached. An adequate choice of the temporal parameter η and the time step Δt is regarded as crucial to the success of the analysis. To this end, we examine the two cases in which $\eta = 0$ and $\eta \neq 0$, corresponding to the *explicit* scheme and the *implicit* scheme, respectively. Notice that $\eta = 1/2$ corresponds to the so-called Crank-Nicolson scheme (Section 4.3.2).

Explicit Scheme

The explicit scheme refers to the case $\eta = 0$. Rewrite (10.2.13) in the form

$$u_\alpha^{n+1} = A_{\alpha\gamma}^{-1}[(A_{\gamma\beta} - \Delta t K_{\gamma\beta})u_\beta^n + \Delta t H_\gamma] \tag{10.2.16}$$

and assume that errors are generated each time step, giving ε_α^n and ε_α^{n+1} corresponding to u_α^n and u_α^{n+1}, respectively, such that

$$u_\alpha^{n+1} + \varepsilon_\alpha^{n+1} = A_{\alpha\gamma}^{-1}[(A_{\gamma\beta} - \Delta t K_{\gamma\beta})(u_\beta^n + \varepsilon_\beta^n) + \Delta t H_\gamma] \tag{10.2.17}$$

Subtracting (10.2.16) from (10.2.17) yields

$$\varepsilon_\alpha^{n+1} = g_{\alpha\gamma}\varepsilon_\gamma^n \tag{10.2.18}$$

where $g_{\alpha\gamma}$ is the amplification matrix

$$g_{\alpha\gamma} = \delta_{\alpha\gamma} - A_{\alpha\delta}^{-1}K_{\delta\gamma}\Delta t \tag{10.2.19}$$

For stable solutions, we must assure that errors at the nth step do not grow toward the $(n+1)$th step; that is,

$$|\varepsilon_\alpha^{n+1}| \leq |\varepsilon_\alpha^n|$$

This requirement can be met when

$$|g_{\alpha\gamma}| = |\delta_{\alpha\gamma} - A_{\alpha\delta}^{-1}K_{\gamma\delta}\Delta t| \leq |\delta_{\alpha\gamma}| = 1 \tag{10.2.20}$$

Thus, in view of (10.2.19) and (10.2.20), and setting

$$\varepsilon_\alpha^{n+1} = \lambda\varepsilon_\alpha^n \tag{10.2.21}$$

we write

$$(g_{\alpha\gamma} - \lambda\delta_{\alpha\gamma})\varepsilon_\gamma^n = 0 \tag{10.2.22}$$

The stability of the solution of (10.2.16) can be assured if each and every eigenvalue λ_α of the amplification matrix $g_{\alpha\gamma}$ is made smaller than unity,

$$|\lambda_\alpha| \leq 1$$

The largest eigenvalue, called the *spectral radius*, governs the stability. Since there exists a bound for Δt outside of which stability can no longer be maintained, the explicit scheme is said to be *conditionally stable.*

Implicit Scheme

The implicit scheme arises for $\eta \neq 0$ in (10.2.13). Solving for u_α^{n+1}, we obtain

$$u_\alpha^{n+1} = (A_{\alpha\gamma} + \eta\Delta t K_{\alpha\gamma})^{-1}\{[A_{\gamma\beta} - (1-\eta)\Delta t K_{\gamma\beta}]u_\beta^n + \Delta t H_\gamma\} \tag{10.2.23}$$

The amplification matrix becomes

$$g_{\alpha\beta} = E_{\alpha\gamma}^{-1} D_{\gamma\beta}$$

with

$$E_{\alpha\gamma} = A_{\alpha\gamma} + \eta\Delta t K_{\alpha\gamma}$$
$$D_{\gamma\beta} = A_{\gamma\beta} - (1-\eta)\Delta t K_{\gamma\beta}$$

For all values of Δt, it is seen that we have $g_{\alpha\beta} \le \delta_{\alpha\beta}$, and the implicit scheme is *unconditionally stable.*

To study the stability behavior of (10.2.23) let us examine one-dimensional linear finite element approximation of (10.2.23) with three nodes,

$$\begin{aligned}&\frac{1}{6}(\Delta u_{j-1}^{n+1} + 4\Delta u_j^{n+1} + \Delta u_{j+1}^{n+1}) + \eta D(-u_{j-1}^{n+1} + 2u_j^{n+1} - u_{j+1}^{n+1})\\ &\quad = -D(-u_{j-1}^n + 2u_j^n - u_{j+1}^n)\end{aligned} \tag{10.2.24}$$

with $\Delta u_j^{n+1} = u_j^{n+1} - u_j^n$, $h = \Delta x$, and D being the nondimensional convergence parameter.

$$D = \nu\frac{\Delta t}{\Delta x^2}$$

The combined spatial and temporal response of the amplitude u^n may be written as

$$u_j^n = e^{ikx}e^{\omega t} = e^{ikj\Delta x}e^{ckn\Delta t} = e^{ikj\Delta x}g^n \tag{10.2.25}$$

where $g = e^{ck\Delta t}$ is the amplification factor, with k and c being the wave number and wave velocity, respectively. Thus,

$$\Delta u_j^{n+1} = e^{ikj\Delta x}(g-1)g^n \tag{10.2.26}$$

Substituting (10.2.25) and (10.2.26) into (10.2.24) leads to

$$e^{ikj\Delta x}g^n\left\{(g-1)\left[\frac{1}{6}(e^{-i\theta} + 4 + e^{i\theta}) + \eta D(-e^{-i\theta} + 2 - e^{i\theta})\right] + D(-e^{-i\theta} + 2 - e^{i\theta})\right\} = 0$$

with

$$\theta = k\Delta x$$

or

$$g = 1 + \frac{2D\sin^2\left(\dfrac{\theta}{2}\right)}{-\frac{1}{3} - \frac{1}{6}\cos\theta + \eta D(\cos\theta - 1)}$$

For $\theta \to 0$, the amplification factor takes the form

$$g = 1 - D\theta^2$$

It is seen that stability is maintained for $g < 1$ or

$$D\theta^2 > 0$$

which shows that the stability is proportional to the square of the phase angle.

10.2.2 HYPERBOLIC EQUATIONS

Consider the hyperbolic equation in the form

$$R = \frac{\partial^2 u}{\partial t^2} - u_{,ii} - f(x, y) = 0 \tag{10.2.27}$$

in which the time dependent term is of the second order. Proceeding in a manner similar to the parabolic equation, we write the DST/GGM equations as

$$(\hat{W}(\xi), (\Phi_\alpha, R)) = \int \hat{W}(\xi)(A_{\alpha\beta}\ddot{u}_\beta + K_{\alpha\beta}u_\beta - H_\alpha)d\xi = 0 \tag{10.2.28}$$

In order to handle the second order derivative of u with respect to time, we must provide at least quadratic trial functions for u_α,

$$u_\alpha = \hat{\Phi}_m u_\alpha^m \quad (m = 1, 2, 3)$$

Here, $\hat{\Phi}_m$ may be defined in $0 < \xi < 1$ or $-1 < \xi < 1$ as follows:

For $0 < \xi < 1$	For $-1 < \xi < 1$
$\hat{\Phi}_1 = 2\left(\xi - \frac{1}{2}\right)(\xi - 1)$	$\hat{\Phi}_1 = \frac{1}{2}\xi(\xi - 1)$
$\hat{\Phi}_2 = -4\xi(\xi - 1)$	$\hat{\Phi}_2 = 1 - \xi^2$
$\hat{\Phi}_3 = 2\xi\left(\xi - \frac{1}{2}\right)$	$\hat{\Phi}_3 = \frac{1}{2}\xi(\xi + 1)$

Using the interval $-1 < \xi < 1$, since this interval is more convenient for integration, we obtain

$$\ddot{u}_\alpha = \frac{\partial}{\partial t}\dot{u}_\alpha = \frac{\partial \dot{u}_\alpha}{\partial \xi}\frac{\partial \xi}{\partial t} = \frac{\partial}{\partial \xi}\frac{\partial u_\alpha}{\partial \xi}\left(\frac{\partial \xi}{\partial t}\right)^2 = \frac{1}{\Delta t^2}(u_\alpha^{n-1} - 2u_\alpha^n + u_\alpha^{n+1}) \tag{10.2.29}$$

which is identical to the finite difference form for the second derivative of u_α.

Defining the temporal parameters η and ζ in the form

$$\eta = \frac{\frac{1}{2}\int_{-1}^{1}\hat{W}\xi(1+\xi)d\xi}{\int_{-1}^{1}\hat{W}d\xi}, \qquad \zeta = \frac{\int_{-1}^{1}\hat{W}\left(\xi + \frac{1}{2}\right)d\xi}{\int_{-1}^{1}\hat{W}d\xi} \tag{10.2.30}$$

the recursive finite element equation takes the form

$$\begin{aligned}(A_{\alpha\beta} + \eta\Delta t^2 K_{\alpha\beta})u_\beta^{n+1} &= \left[2A_{\alpha\beta} - \left(\frac{1}{2} - 2\eta + \zeta\right)\Delta t^2 K_{\alpha\beta}\right]u_\beta^n \\ &\quad - \left[A_{\alpha\beta} + \left(\frac{1}{2} + \eta - \zeta\right)\Delta t^2 K_{\alpha\beta}\right]u_\beta^{n-1} + \Delta t^2 H_\alpha \end{aligned} \tag{10.2.31}$$

Table 10.2.2 Temporal Parameters for Hyperbolic Equations

$\hat{W}(\xi)$	η	ζ
$\delta(\xi+1)$	0	1/2
$\delta(\xi-0)$	0	1/2
$\delta(\xi-1)$	1	3/2
$1, 0 \le \xi \le 1$	1/6	1/2
$1+\xi, -1 \le \xi \le 0$	4/5	3/2
$1-\xi, -1 \le \xi \le 0$	1/12	1/2
$-\xi, 0 \le \xi \le 1$	1/4	1/2
ξ	1/4	1/2
$1-\xi^2$	1/10	1/2
$(1/2)\xi(1+\xi)$	4/5	3/2

Once again, $\eta = 0$ and $\eta = 1$ lead to the explicit and implicit schemes, respectively. Various values for $\hat{W}$, and the corresponding temporal parameters η and ζ, are presented in Table 10.2.2.

For highly oscillatory motions, quadratic approximations may be inadequate and cubic approximations are required for acceptable accuracy. Cubic variations can be formulated using the Lagrange polynomials for $-1 \le \xi \le 1$ so that u_α and $\ddot{u}_\alpha$ take the forms

$$u_\alpha = -\frac{9}{16}\left(\xi+\frac{1}{3}\right)\left(\xi-\frac{1}{3}\right)(\xi-1)u_\alpha^{n-2} + \frac{27}{16}(\xi+1)\left(\xi-\frac{1}{3}\right)(\xi-1)u_\alpha^{n-1} - \frac{27}{16}(\xi+1)\left(\xi+\frac{1}{3}\right)(\xi-1)u_\alpha^{n} + \frac{9}{16}(\xi+1)\left(\xi+\frac{1}{3}\right)\left(\xi-\frac{1}{3}\right)u_\alpha^{n+1}$$

and

$$\ddot{u}_\alpha = \frac{1}{\Delta t^2}\left[-\frac{9}{16}(6\xi-2)u_\alpha^{n-2} + \frac{27}{16}\left(6\xi-\frac{2}{3}\right)u_\alpha^{n-1} - \frac{27}{16}\left(6\xi+\frac{2}{3}\right)u_\alpha^{n} + \frac{9}{16}(6\xi+2)u_\alpha^{n+1}\right]$$

Substituting the above into (10.2.28), we arrive at

$$\begin{aligned}
&\left[A_{\alpha\beta}\frac{9}{16}(6\gamma+2) + \Delta t^2\frac{9}{16}\left(\eta+\zeta-\frac{1}{9}\gamma-\frac{1}{9}\right)K_{\alpha\beta}\right]u_\beta^{n+1} \\
&+\left[A_{\alpha\beta}\frac{27}{16}\left(-6\gamma-\frac{2}{3}\right) + \Delta t^2\frac{27}{16}\left(-\eta-\frac{1}{3}\zeta+\gamma+\frac{1}{3}\right)K_{\alpha\beta}\right]u_\beta^{n} \\
&+\left[A_{\alpha\beta}\frac{27}{16}\left(6\gamma-\frac{2}{3}\right) + \Delta t^2\frac{27}{16}\left(\eta-\frac{1}{3}\zeta-\gamma+\frac{1}{3}\right)K_{\alpha\beta}\right]u_\beta^{n-1} \\
&+\left[A_{\alpha\beta}\frac{9}{16}(-6\gamma+2) + \Delta t^2\frac{9}{16}\left(-\eta+\zeta+\frac{1}{9}\gamma-\frac{1}{9}\right)K_{\alpha\beta}\right]u_\beta^{n-2} \\
&-\Delta t^2(F_\alpha+G_\alpha) = 0
\end{aligned} \tag{10.2.32}$$

with

$$\eta = \frac{\int_{-1}^{1} \hat{W}(\xi)\xi^3 d\xi}{\int_{-1}^{1} \hat{W} d\xi}, \qquad \zeta = \frac{\int_{-1}^{1} \hat{W}(\xi)\xi^2 d\xi}{\int_{-1}^{1} \hat{W} d\xi}, \qquad \gamma = \frac{\int_{-1}^{1} \hat{W}(\xi)\xi d\xi}{\int_{-1}^{1} \hat{W} d\xi}$$

Appropriate choices of $\hat{W}(\xi)$ will lead to a variety of integration formulas.

Using the Newton backward difference (Chung, 1975), it can be shown that the cubic approximations may also be given as

$$\begin{aligned}&[11A_{\alpha\beta} + 6\Delta t(1-\theta)K_{\alpha\beta}]\mathrm{v}_\beta^{n+1} + [-18A_{\alpha\beta} + 6\Delta t\theta K_{\alpha\beta}]\mathrm{v}_\beta^n \\ &\quad + A_{\alpha\beta}\left(9\mathrm{v}_\beta^{n-1} - 2\mathrm{v}_\beta^{n-2}\right) - 6\Delta t H_\beta = 0\end{aligned} \tag{10.2.33}$$

where $0 \le \theta \le 1$.

10.2.3 MULTIVARIABLE PROBLEMS

The finite element formulation of multivariable problems which occur in two- or three-dimensional problems may be best handled using tensors. Let us consider a differential equation of the form

$$\frac{\partial \mathbf{v}}{\partial t} - \nabla^2\mathbf{v} - \nabla(\nabla\cdot\mathbf{v}) - \mathbf{f} = 0 \tag{10.2.34a}$$

or

$$R_i = \frac{\partial \mathrm{v}_i}{\partial t} - \mathrm{v}_{i,jj} - \mathrm{v}_{j,ji} - f_i = 0 \tag{10.2.34b}$$

where the variables v_i may be approximated spatially as

$$\mathrm{v}_i = \Phi_\alpha \mathrm{v}_{\alpha i} \quad (i = 1, 2) \text{ for 2-D} \tag{10.2.35}$$

Note that $\mathrm{v}_{\alpha i}$ implies v_i at the global node α. The GGM equations for (10.2.34b) become

$$(\hat{W}(\xi), (\Phi_\alpha, R_i)) = \int_\xi \hat{W}(\xi)\left[\int_\Omega \Phi_\alpha\left(\frac{\partial \mathrm{v}_i}{\partial t} - \mathrm{v}_{i,jj} - \mathrm{v}_{j,ji} - f_i\right)d\Omega\right]d\xi = 0 \tag{10.2.36}$$

which yields

$$\int_\xi \hat{W}(\xi)[A_{\alpha\beta}\delta_{ik}\dot{\mathrm{v}}_{\beta k} + (K^{(1)}_{\alpha i\beta k} + K^{(2)}_{\alpha j\beta j}\delta_{ik})\mathrm{v}_{\beta k} - F_{\alpha i} - G_{\alpha i}]d\xi = 0$$

where

$$A_{\alpha\beta} = \int_\Omega \Phi_\alpha\Phi_\beta d\Omega = \bigcup_{e=1}^{E}\int_\Omega \Phi_N^{(e)}\Phi_M^{(e)} d\Omega \Delta_{N\alpha}^{(e)}\Delta_{M\beta}^{(e)} = \bigcup_{e=1}^{E} A_{NM}^{(e)}\Delta_{N\alpha}^{(e)}\Delta_{M\beta}^{(e)}$$

$$K^{(1)}_{\alpha i\beta k} = \int_\Omega \Phi_{\alpha,i}\Phi_{\beta,k} d\Omega = \bigcup_{e=1}^{E}\int_\Omega \Phi_{N,i}^{(e)}\Phi_{M,k}^{(e)} d\Omega \Delta_{N\alpha}^{(e)}\Delta_{M\beta}^{(e)} = \bigcup_{e=1}^{E} K_{NiMk}^{(1)(e)}\Delta_{N\alpha}^{(e)}\Delta_{M\beta}^{(e)}$$

$$K^{(2)}_{\alpha j\beta j} = \int_\Omega \Phi_{\alpha,j}\Phi_{\beta,j} d\Omega = \bigcup_{e=1}^{E}\int_\Omega \Phi_{N,j}^{(e)}\Phi_{M,j}^{(e)} d\Omega \Delta_{N\alpha}^{(e)}\Delta_{M\beta}^{(e)} = \bigcup_{e=1}^{E} K_{NjMj}^{(2)(e)}\Delta_{N\alpha}^{(e)}\Delta_{M\beta}^{(e)}$$

$$F_{\alpha i} = \int_\Omega \Phi_\alpha \Phi_\beta d\Omega \delta_{ik} f_{\beta k} = C_{\alpha\beta}\delta_{ik} f_{\beta k} = \bigcup_{e=1}^{E} C_{NM}^{(e)} \Delta_{N\alpha}^{(e)} \Delta_{M\beta}^{(e)} \delta_{ik} f_{\beta k}$$

$$C_{NM}^{(e)} = \int_\Omega \Phi_N^{(e)} \Phi_M^{(e)} d\Omega$$

$$G_{\alpha i} = \int_\Gamma \overset{*}{\Phi}(\mathrm{v}_{i,j} n_j + \mathrm{v}_{j,j} n_i) d\Gamma = \bigcup_{e=1}^{E} G_{Ni}^{(e)} \Delta_{N\alpha}^{(e)}$$

For the case of Figure E10.1.2, we have

$$G_{\alpha i}^{(e)} = \bigcup_{e=1}^{E} \int_\Gamma \overset{*}{\Phi}{}_N^{(e)} \overset{*}{\Phi}{}_M^{(e)} d\Gamma \delta_{ik} g_{Mk}^{(e)} \Delta_{N\alpha}^{(e)} = \bigcup_{e=1}^{E} \overset{*}{C}{}_{NM}^{(e)} \delta_{ik} g_{Mk}^{(e)} \Delta_{N\alpha}^{(e)}$$

$$= \frac{l}{6} \begin{bmatrix} 2 & 0 & 1 & 0 \\ 0 & 2 & 0 & 1 \\ 1 & 0 & 2 & 0 \\ 0 & 1 & 0 & 2 \end{bmatrix} \begin{bmatrix} g_{11}^{(e)} \\ g_{12}^{(e)} \\ g_{21}^{(e)} \\ g_{22}^{(e)} \end{bmatrix} = \frac{l}{6} \begin{bmatrix} 2g_{11}^{(e)} + g_{21}^{(e)} \\ 2g_{12}^{(e)} + g_{22}^{(e)} \\ g_{11}^{(e)} + 2g_{21}^{(e)} \\ g_{12}^{(e)} + 2g_{22}^{(e)} \end{bmatrix}$$

where

$$g_{M1}^{(e)} = (2\mathrm{v}_{1,1} + \mathrm{v}_{2,2}) n_1 + \mathrm{v}_{1,2} n_2$$

$$g_{M2}^{(e)} = \mathrm{v}_{2,1} n_1 + (\mathrm{v}_{1,1} + 2\mathrm{v}_{2,2}) n_2$$

With linear temporal approximations, the global finite element equations take the form

$$\left[A_{\alpha\beta}\delta_{ik} + \eta\Delta t\left(K_{\alpha i\beta k}^{(1)} + K_{\alpha j\beta j}^{(2)}\delta_{ik}\right)\right]\mathrm{v}_{\beta k}^{n+1} = \left[A_{\alpha\beta}\delta_{ik} - (1-\eta)\Delta t\left(K_{\alpha i\beta k}^{(1)} + K_{\alpha j\beta j}^{(2)}\delta_{ik}\right)\right]\mathrm{v}_{\beta k}^{n} + \Delta t(F_{\alpha i} + G_{\alpha i}) \tag{10.2.37}$$

The solution of (10.2.37) will proceed similarly as a single variable problem except that the multivariables $\mathrm{v}_{\beta k}$ are to be solved simultaneously.

10.2.4 AXISYMMETRIC TRANSIENT HEAT CONDUCTION

Consider the transient heat conduction, without convection, in an axisymmetric geometry,

$$\rho c_p \frac{\partial T}{\partial t} - k\left(\frac{\partial^2 T}{\partial r^2} + \frac{\partial^2 T}{\partial z^2} + \frac{1}{r}\frac{\partial T}{\partial r}\right) = 0 \tag{10.2.38}$$

where ρ, c_p, T, k, and r are the density, specific heat at constant pressure, temperature, coefficient of thermal conductivity, and radius of a cylindrical geometry, respectively.

The generalized Galerkin finite element formulation of (10.2.38) leads to

$$\int_0^1 \hat{W}(\xi)\left\{\int_0^{2\pi}\iint \Phi_\alpha\left[\rho c_p \frac{\partial T}{\partial t} - k\left(\frac{\partial^2 T}{\partial r^2} + \frac{\partial^2 T}{\partial z^2} + \frac{1}{r}\frac{\partial T}{\partial r}\right)\right] r\,d\theta\,dr\,dz\right\} d\xi = 0 \tag{10.2.39}$$

Here, the partial integration of the term containing $\partial^2 T/\partial r^2$ in (10.2.39) becomes

$$\int_0^{2\pi}\iint \Phi_\alpha \frac{\partial^2 T}{\partial r^2} r d\theta dr dz = 2\pi\left(\int \overset{*}{\Phi}_\alpha \frac{\partial T}{\partial r} r dz - \iint \frac{\partial \Phi_\alpha}{\partial r}\frac{\partial T}{\partial r} r dr dz - \iint \Phi_\alpha \frac{\partial T}{\partial r} dr dz\right)$$

Thus, after canceling out the $\partial T/\partial r$ terms, we have

$$\int_0^1 \hat{W}(\xi)(A_{\alpha\beta}\dot{T}_\beta + K_{\alpha\beta}T_\beta - G_\alpha)d\xi = 0 \tag{10.2.40}$$

where, for isoparametric quadrilateral elements, with $r = \Phi_\gamma r_\gamma$, we obtain

$$A_{\alpha\beta} = 2\pi \int_{-1}^1\int_{-1}^1 \rho c_p \Phi_\alpha \Phi_\beta \Phi_\gamma r_\gamma |J|\, d\xi d\eta$$

Here, $d\xi$ refers to the isoparametric coordinates rather than the nondimensional time,

$$K_{\alpha\beta} = 2\pi\left\{\int_{-1}^1\int_{-1}^1 k\left(\frac{\partial \Phi_\alpha}{\partial r}\frac{\partial \Phi_\beta}{\partial r} + \frac{\partial \Phi_\alpha}{\partial z}\frac{\partial \Phi_\beta}{\partial z}\right)\Phi_\gamma r_\gamma |J| d\xi d\eta\right\}$$

$$G_\alpha = 2\pi \int_\Gamma \overset{*}{\Phi}_\alpha k T_{,i} n_i r d\Gamma = 2\pi \int_\Gamma - \overset{*}{\Phi}_\alpha \overline{\alpha}(T - T')r d\Gamma$$

$$= 2\pi\left[\int_\Gamma -\overline{\alpha}\,\overset{*}{\Phi}_\alpha\,\overset{*}{\Phi}_\beta\, r d\Gamma T_\beta + \int_\Gamma \overline{\alpha}\,\overset{*}{\Phi}_\alpha T' r d\Gamma\right] = \overset{*}{K}_{\alpha\beta}T_\beta + \overset{*}{G}_\alpha$$

where we set

$$-kT_{,i}n_i = \overline{\alpha}(T - T')$$

with $\overline{\alpha}$ and T' being defined as the heat transfer coefficient and ambient temperature, respectively. Here, $\overset{*}{K}_{\alpha\beta}$ is the convection boundary stiffness matrix representing the contribution of ambient temperature toward the boundary surface:

$$\overset{*}{K}_{\alpha\beta} = 2\pi \int_\Gamma \overset{*}{\Phi}_\alpha\,\overset{*}{\Phi}_\beta\,\overset{*}{\Phi}_\gamma\, r_\gamma d\Gamma$$

$$\overset{*}{G}_\alpha = 2\pi \int_\Gamma T'\overline{\alpha}\,\overset{*}{\Phi}_\alpha\,\overset{*}{\Phi}_\gamma\, r_\gamma d\Gamma$$

where $\overset{*}{K}_{\alpha\beta}$ should be combined with $K_{\alpha\beta}$ but its contribution is restricted only to the convection boundary nodes along the surface of convection boundaries as shown in (10.1.23). Thus,

$$\int_0^1 \hat{W}(\xi)(A_{\alpha\beta}\dot{T}_\beta + (K_{\alpha\beta} + \overset{*}{K}_{\alpha\beta})T_\beta - \overset{*}{G}_\alpha)d\xi = 0 \tag{10.2.41}$$

This ordinary differential equation will then be integrated over the temporal domain as in Section 10.2.1.

10.3 SOLUTIONS OF FINITE ELEMENT EQUATIONS

Solutions of simultaneous algebraic equations are carried out by using either direct or iterative methods. The direct methods yield answers in a finite number of operations (Section 4.2.7). They include Gauss elimination, Thomas algorithm, etc., which are suitable for linear equations. The iterative methods [Saad, 1996] include Gauss-Seidel methods, relaxation methods, conjugate gradient methods (CGM), and generalized minimal residual (GMRES) methods, among others. Here, solutions are obtained through a number of iterative steps, accuracy being increased with an increase of iterations. These methods are suitable for nonlinear as well as linear equations.

For a large system of equations, it is expected that the assembly of element stiffness matrices into a global form would take a prohibitive amount of computer time. This can be avoided by the so-called element-by-element (EBE) solution scheme [Hughes, Levit, and Winget, 1983; Carey and Jiang, 1986; Wathen, 1989, etc.]. In this approach, we replace the matrix assembly process by vector operations. This will be presented in Section 10.3.2.

The coverage of solution methods for algebraic equations in general is beyond the scope of this book. However, we select the conjugate gradient method (CGM) as one of the most popular schemes in CFD and present its brief description, followed by the EBE approach for finite element equations.

10.3.1 CONJUGATE GRADIENT METHODS (CGM)

Let us consider the global system of finite element equations in the form

$$K_{\alpha\beta} U_\beta = F_\alpha \tag{10.3.1}$$

The iterative solution by the conjugate gradient methods (CGM) can be obtained, using the following steps:

(1) Assume initial values $U_\alpha^{(r)}$

(2) Determine the residual $E_\alpha^{(r)}$

$$E_\alpha^{(r)} = F_\alpha - K_{\alpha\beta} U_\beta^{(r)} \tag{10.3.2}$$

(3) Define the auxiliary variables $P_\alpha^{(r)}$

$$P_\alpha^{(r)} = E_\alpha^{(r)}$$

(4) Compute rth iteration residual

$$\overline{E}_\alpha^{(r)} = K_{\alpha\beta} P_\beta^{(r)} \tag{10.3.3}$$

(5) Compute the correction factor $a^{(r)}$

$$a^{(r)} = \frac{E_\alpha^{(r)} P_\alpha^{(r)}}{\overline{E}_\beta^{(r)} P_\beta^{(r)}} \tag{10.3.4}$$

(6) Compute the solution $U_\alpha^{(r+1)}$

$$U_\alpha^{(r+1)} = U_\alpha^{(r)} + a^{(r)} P_\alpha^{(r)} \tag{10.3.5}$$

(7) Compute the residual $E_\alpha^{(r+1)}$

$$E_\alpha^{(r+1)} = E_\alpha^{(r)} - a^{(r)}\overline{E}_\alpha^{(r)} \tag{10.3.6}$$

(8) Compute the correction factor $b^{(r+1)}$

$$b^{(r+1)} = \frac{E_\alpha^{(r+1)} E_\alpha^{(r+1)}}{E_\beta^{(r)} E_\beta^{(r)}} \tag{10.3.7}$$

(9) Define the auxiliary variables $P_\alpha^{(r+1)}$

$$P_\alpha^{(r+1)} = E_\alpha^{(r+1)} + b^{(r+1)} P_\alpha^{(r)} \tag{10.3.8}$$

(10) Return to Step 4 and repeat the process until convergence.

If the matrix $K_{\alpha\beta}$ is nonsymmetric, then it is possible to symmetrize $K_{\alpha\beta}$ by multiplying the transpose of the stiffness matrix in (10.3.1) as follows:

$$[K]^T[K][U] = [K]^T[F]$$

or

$$K_{\gamma\alpha} K_{\gamma\beta} U_\beta = K_{\gamma\alpha} F_\gamma$$

This can be written in the form

$$A_{\alpha\beta} U_\beta = \overline{F}_\alpha \tag{10.3.9}$$

with

$$A_{\alpha\beta} = K_{\gamma\alpha} K_{\gamma\beta}, \qquad \overline{F}_\alpha = K_{\gamma\alpha} F_\gamma$$

The same procedure as given in Steps 1 through 10 above can be applied to (10.3.9). However, this will require extremely large operations and we may avoid them by constructing the product of the transpose of the stiffness matrix and the auxiliary variables as follows:

(1) Start with the initial guess $U_\alpha^{(o)}$

(2) $E_\alpha^{(o)} = K_{\gamma\alpha}(F_\gamma - K_{\gamma\beta} U_\beta)$

(3) $P_\alpha^{(r)} = E_\alpha^{(r)}$

(4) $\overline{E}_\alpha^{(r)} = K_{\gamma\alpha} K_{\gamma\beta} P_\beta^{(r)}$

(5) $a^{(r)} = \dfrac{E_\alpha^{(r)} P_\alpha^{(r)}}{\overline{E}_\beta^{(r)} P_\beta^{(r)}}$

(6) $U_\alpha^{(r+1)} = U_\alpha^{(r)} + a^{(r)} P_\alpha^{(r)}$

(7) $E_\alpha^{(r+1)} = E_\alpha^{(r)} - a^{(r)} \overline{E}_\alpha^{(r)}$

(8) $b^{(r+1)} = \dfrac{E_\alpha^{(r+1)} E_\alpha^{(r+1)}}{E_\beta^{(r)} E_\beta^{(r)}}$

(9) $P_\alpha^{(r+1)} = E_\alpha^{(r+1)} + b^{(r)} P_\alpha^{(r)}$

(10) Return to step (4) and repeat the process until convergence.

Example 10.3.1

Given: Consider a system of algebraic equations of the form,

$$\begin{bmatrix} 1 & -1 & 0 \\ -1 & 2 & -2 \\ 0 & -2 & 1 \end{bmatrix} \begin{bmatrix} U_1 \\ U_2 \\ U_3 \end{bmatrix} = \begin{bmatrix} 0 \\ -1 \\ -1 \end{bmatrix}$$

Required: Solve using the CGM algorithm and compare with the exact solution: $U_1 = 1, U_2 = 1, U_3 = 1$.

Solution:

(1) Assume $U_\alpha^{(o)} = \begin{bmatrix} 0 \\ 0 \\ 0 \end{bmatrix}$

(2) $E_\alpha^{(o)} = \begin{bmatrix} 0 \\ -1 \\ -1 \end{bmatrix} - \begin{bmatrix} 1 & -1 & 0 \\ -1 & 2 & -2 \\ 0 & -2 & 1 \end{bmatrix} \begin{bmatrix} 0 \\ 0 \\ 0 \end{bmatrix} = \begin{bmatrix} 0 \\ -1 \\ -1 \end{bmatrix}$

(3) $P_\alpha^{(o)} = \begin{bmatrix} 0 \\ -1 \\ -1 \end{bmatrix}$

(4) $\overline{E}_\alpha^{(o)} = \begin{bmatrix} 1 & -1 & 0 \\ -1 & 2 & -2 \\ 0 & -2 & 1 \end{bmatrix} \begin{bmatrix} 0 \\ -1 \\ -1 \end{bmatrix} = \begin{bmatrix} 1 \\ 0 \\ 1 \end{bmatrix}$

(5) $a^{(o)} = \dfrac{0+1+1}{0+0-1} = -2$

(6) $U_\alpha^{(1)} = \begin{bmatrix} 0 \\ 0 \\ 0 \end{bmatrix} + (-2) \begin{bmatrix} 0 \\ -1 \\ -1 \end{bmatrix} = \begin{bmatrix} 0 \\ 2 \\ 2 \end{bmatrix}$

(7) $E_\alpha^{(1)} = \begin{bmatrix} 0 \\ -1 \\ -1 \end{bmatrix} - (-2) \begin{bmatrix} 1 \\ 0 \\ 1 \end{bmatrix} = \begin{bmatrix} 2 \\ -1 \\ 1 \end{bmatrix}$

(8) $b^{(1)} = \dfrac{4+1+1}{0+1+1} = 3$

(9) $P_\alpha^{(1)} = \begin{bmatrix} 2 \\ -1 \\ 1 \end{bmatrix} + (3) \begin{bmatrix} 0 \\ -1 \\ -1 \end{bmatrix} = \begin{bmatrix} 2 \\ -4 \\ -2 \end{bmatrix}$

(10) $\overline{E}_\alpha^{(1)} = \begin{bmatrix} 1 & -1 & 0 \\ -1 & 2 & -2 \\ 0 & -2 & 1 \end{bmatrix} \begin{bmatrix} 2 \\ -4 \\ -2 \end{bmatrix} = \begin{bmatrix} 6 \\ -6 \\ 6 \end{bmatrix}$

(11) $a^{(1)} = \dfrac{4+4-2}{12+24-12} = \dfrac{6}{24} = 0.25$

(12) $U_\alpha^{(2)} = \begin{bmatrix} 0 \\ 2 \\ 2 \end{bmatrix} + (0.25) \begin{bmatrix} 2 \\ -4 \\ -2 \end{bmatrix} = \begin{bmatrix} 0.5 \\ 1 \\ 1.5 \end{bmatrix}$

Repeating another cycle of iteration, we obtain

$$U_\alpha^{(3)} = \begin{bmatrix} 1.0002 \\ 1 \\ 0.9998 \end{bmatrix}$$

The next step (7) shows the residual $E_\alpha^{(3)}$ to be zero and the exact answers, $U_1 = U_2 = U_3 = 1$, are obtained.

If the stiffness matrix $K_{\alpha\beta}$ is nonsymmetric or nonlinear, then the procedure for (10.3.9) can be used. It is expected that convergence toward the exact solution will be much slower. The GMRES methods suitable for CFD equations will be covered in Section 11.5.2.

10.3.2 ELEMENT-BY-ELEMENT (EBE) SOLUTIONS OF FEM EQUATIONS

A large system of equations is encountered when the number of finite element nodes increases in order to improve accuracy. The assembly of element stiffness matrices into a global form and solutions may occupy a large portion of computing time. To avoid this inconvenience, we shall examine the so-called element-by-element (EBE) approach [Hughes et al., 1983; Carey and Jiang, 1986; Wathen, 1989, etc.], in which the assembly of entire stiffness matrices is eliminated. The EBE methods using the Jacobi-iteration and conjugate gradient methods are described below.

Let us consider the global finite element equations of the form,

$$K_{\alpha\beta} U_\beta = F_\alpha \tag{10.3.10}$$

The global stiffness matrix $K_{\alpha\beta}$ can be split into the diagonal components $D_{\alpha\beta}$ and the off-diagonal matrix $N_{\alpha\beta}$ as follows:

$$K_{\alpha\beta} = D_{\alpha\beta} + N_{\alpha\beta} \tag{10.3.11}$$

leading to

$$(D_{\alpha\beta} + N_{\alpha\beta}) U_\beta = F_\alpha \tag{10.3.12}$$

or

$$D_{\alpha\beta} U_\beta^{(r+1)} \cong F_\alpha^{(r)} - N_{\alpha\beta} U_\beta^{(r)} \tag{10.3.13}$$

where the diagonal matrix and off-diagonal matrix are allowed to be associated with the iteration steps of U_β at $(r+1)$ and (r), respectively. Subtracting $D_{\alpha\beta} U_\beta^{(r)}$ from the left- and right-hand sides of (10.3.13), we obtain

$$D_{\alpha\beta}\left(U_\beta^{(r+1)} - U_\beta^{(r)}\right) = F_\alpha^{(r)} - (N_{\alpha\beta} + D_{\alpha\beta}) U_\beta^{(r)} \tag{10.3.14}$$

or

$$U_\alpha^{(r+1)} = U_\alpha^{(r)} - D_{\alpha\beta}^{-1}\left(\overline{F}_\beta^{(r)} - F_\beta^{(r)}\right) \tag{10.3.15}$$

with the diagonal matrix playing the role of the preconditioning matrix and

$$\overline{F}_\alpha^{(r)} = (N_{\alpha\beta} + D_{\alpha\beta})U_\beta^{(r)} = K_{\alpha\beta}U_\beta^{(r)} = \bigcup_{e=1}^{E} \overline{F}_N^{(e)} \Delta_{N\alpha}^{(e)}$$

$$\overline{F}_N^{(e)} = K_{NM}^{(e)} U_M^{(e)} \tag{10.3.16}$$

It is clearly seen that the assembly of the stiffness matrix has been replaced by the element-by-element basis as a column vector, identical to the assembly of the source vector $F_\alpha^{(r)}$ such as in (10.1.15b). Thus, the solution of (10.3.10) is obtained as

$$\begin{bmatrix} U_1 \\ U_2 \\ \vdots \end{bmatrix}^{(r+1)} = \begin{bmatrix} U_1 \\ U_2 \\ \vdots \end{bmatrix}^{(r)} - \begin{bmatrix} (\overline{F}_1 - F_1)/D_{11} \\ (\overline{F}_2 - F_2)/D_{22} \\ \vdots \end{bmatrix}^{(r)} \tag{10.3.17}$$

In order to increase convergence and accuracy, it is necessary to implement a standard relaxation process in the form

$$U = \xi U^{(r+1)} + (1 - \xi)U^{(r)}$$

with $0 < \xi < 1$ or preferably $\xi = 0.8$. The procedure described above resembles the Jacobi iteration method and, thus, this scheme is called the EBE Jacobi method [Hughes et al., 1983].

The EBE scheme may be incorporated into any high-accuracy iterative equation solver. For example, let us consider the conjugate gradient method. Here, we may adopt the following step-by-step procedure.

(1) Assume initial values $U_\alpha^{(r)}$.

(2) Compute the residual $E_\alpha^{(r)}$

$$E_\alpha^{(r)} = F_\alpha - K_{\alpha\beta}U_\beta = F_\alpha - \overline{F}_\alpha \tag{10.3.18}$$

with

$$\overline{F}_\alpha = \bigcup_{e=1}^{E} \overline{F}_N^{(e)} \Delta_{N\alpha}^{(e)}$$

$$\overline{F}_N^{(e)} = K_{NM}^{(e)} U_M^{(e)}$$

(3) Set the residual as the auxiliary variables $P_\alpha^{(r)}$

$$P_\alpha^{(r)} = E_\alpha^{(r)} \tag{10.3.19}$$

(4) Determine the rth iteration residual $\overline{E}_\alpha^{(r)}$ as

$$\overline{E}_\alpha^{(r)} = K_{\alpha\beta} P_\beta^{(r)} = \bigcup_{e=1}^{E} H_N^{(e)} \Delta_{N\alpha}^{(e)} \tag{10.3.20}$$

with

$$H_N^{(e)} = K_{NM}^{(e)} P_M^{(r)}$$

(5) Determine the correction factor $a^{(r)}$

$$a^{(r)} = \frac{E_\alpha^{(r)} P_\alpha^{(r)}}{\overline{E}_\beta^{(r)} P_\beta^{(r)}} \tag{10.3.21}$$

(6) Solve $U_\alpha^{(r+1)}$

$$U_\alpha^{(r+1)} = U_\alpha^{(r)} + a^{(r)} P_\alpha^{(r)} \tag{10.3.22}$$

(7) Determine the residual $E_\alpha^{(r+1)}$

$$E_\alpha^{(r+1)} = E_\alpha^{(r)} - a^{(r)} \overline{E}_\alpha^{(r)} \tag{10.3.23}$$

(8) Compute the correction factor $b^{(r+1)}$

$$b^{(r+1)} = \frac{E_\alpha^{(r+1)} E_\alpha^{(r+1)}}{E_\beta^{(r)} E_\beta^{(r)}} \tag{10.3.24}$$

(9) Determine the auxiliary variables $P_\alpha^{(r+1)}$

$$P_\alpha^{(r+1)} = E_\alpha^{(r+1)} + b^{(r+1)} P_\alpha^{(r)} \tag{10.3.25}$$

(10) Return to (4) and repeat until convergence.

For time-dependent and nonlinear problems, procedures similar to those above can be used. In order to expedite the convergence, however, appropriate preconditioning processes are important. These and other topics on the equation solvers such as GMRES and the EBE algorithms will be presented in Section 11.5.

10.4 EXAMPLE PROBLEMS

10.4.1 SOLUTION OF POISSON EQUATION WITH ISOPARAMETRIC ELEMENTS

In this example, we repeat Example 10.1.2 using 6 and 24 bilinear (4 node) isoparametric elements by removing the diagonals (Figure 10.4.1.1). Use the three-point Gaussian

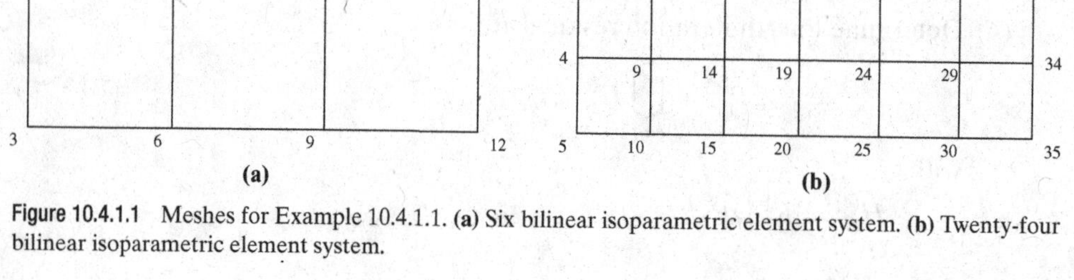

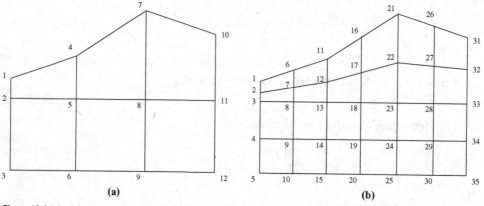

Figure 10.4.1.1 Meshes for Example 10.4.1.1. **(a)** Six bilinear isoparametric element system. **(b)** Twenty-four bilinear isoparametric element system.

quadrature integration. The solution procedure is as follows:

$$K_{\alpha\beta} = \bigcup_{e=1}^{E} K_{NM}^{(e)} \Delta_{N\alpha}^{(e)} \Delta_{M\beta}^{(e)}$$

$$K_{NM}^{(e)} = \int_\Omega \Phi_{N,i}^{(e)} \Phi_{M,i}^{(e)} d\Omega = \sum_{p=1}^{n} \sum_{q=1}^{n} w_p w_q k_{NM}(\xi_p, \eta_q)$$

$$F_\alpha = C_{\alpha\beta} f_\beta = \bigcup_{e=1}^{E} C_{NM}^{(e)} \Delta_{N\alpha}^{(e)} \Delta_{M\beta}^{(e)} f_\beta = \bigcup_{e=1}^{E} \left[\sum_{p=1}^{n} \sum_{q=1}^{n} w_p w_q c_{NM}(\xi_p, \eta_q) \right] \Delta_{N\alpha}^{(e)} \Delta_{M\beta}^{(e)} f_\beta$$

It is obvious that no local evaluation of the load vector is necessary and it is convenient to leave $f_\beta = [4(x^2 + y^2)]_\beta$ in the global form, unlike the Neumann boundary vector which was evaluated in the local level and assembled into a global form.

The Neumann boundary vector remains the same as in the case of triangular elements, and is independent of the Gaussian quadrature integration. If desired, however, the Neumann boundary vector may be rederived from the one-dimensional isoparametric (natural) coordinate. The results would be the same.

The Neumann boundary vector G_α for the six-element problem is the same as in Example 10.1.2, although the load vector F_α is different due to the different integration scheme. The summary of results is given in Table E10.4.1.1.

The following conclusions are drawn from Examples 10.1.2 and 10.1.3.

(1) The six isoparametric elements provide higher accuracy than twelve triangular elements. At interior nodes (5 and 8), triangular elements give answers smaller than the exact solutions, whereas the isoparametric elements lead to larger values, indicating that triangular elements are stiffer than the isoparametric elements as seen in Examples 10.1.2 and 10.1.3.

(2) In the coarse grid system, the Neumann problem is not as accurate as in the Dirichlet problem.

10.4.2 PARABOLIC PARTIAL DIFFERENTIAL EQUATION IN TWO DIMENSIONS

Consider the two-dimensional linear partial differential equation of the form

$$\frac{\partial u}{\partial t} - \nu \left(\frac{\partial^2 u}{\partial x^2} + \frac{\partial^2 u}{\partial y^2} \right) - f_x = 0$$

$$\frac{\partial v}{\partial t} - \nu \left(\frac{\partial^2 v}{\partial x^2} + \frac{\partial^2 v}{\partial y^2} \right) - f_y = 0$$

with

$$f_x = -\frac{1}{(1+t)^2} - 2\nu y, \qquad f_y = -\frac{1}{(1+t)^2} - 2\nu x$$

Table E10.4.1.1 Computed Results for Example 10.4.1.1

(a) Dirichlet Data (6 elements)

Node	Exact Solution	FEM Solution	% Error
1	0.00	0.00	0.00
2	0.00	0.00	0.00
3	0.00	0.00	0.00
4	450.00	450.00	0.00
5	162.00	197.05	21.64
6	0.00	0.00	0.00
7	3528.00	3528.00	0.00
8	648.00	667.45	3.00
9	0.00	0.00	0.00
10	5832.00	5832.00	0.00
11	1458.00	1458.00	0.00
12	0.00	0.00	0.00

(b) Neumann Data (6 elements)

Node	Exact Solution	FEM Solution	% Error
1	0.00	−28.99	0.00
2	0.00	0.00	0.00
3	0.00	0.00	0.00
4	450.00	339.18	24.63
5	162.00	130.63	19.36
6	0.00	0.00	0.00
7	3528.00	3221.45	8.69
8	648.00	601.47	7.18
9	0.00	0.00	0.00
10	5832.00	5697.71	2.30
11	1458.00	1458.00	0.00
12	0.00	0.00	0.00

(c) Dirichlet Data (24 elements)

Node	Exact Solution	FEM Solution	% Error
1	0.00	0.00	0.00
2	0.00	0.00	0.00
3	0.00	0.00	0.00
4	0.00	0.00	0.00
5	0.00	0.00	0.00
6	91.13	91.13	0.00
7	63.28	65.68	3.79
8	40.50	44.09	8.86
9	10.13	12.10	19.47
10	0.00	0.00	0.00
11	450.00	450.00	0.00
12	288.00	287.79	.07
13	162.00	170.18	5.05
14	40.50	44.51	9.90
15	0.00	0.00	0.00
16	1458.00	1458.00	0.00
17	820.13	830.87	1.31
18	364.50	379.19	4.03
19	91.13	94.76	3.99
20	0.00	0.00	0.00
21	3528.00	3528.00	0.00
22	1800.00	1812.86	.71
23	648.00	648.80	.12
24	162.00	163.41	.87
25	0.00	0.00	0.00
26	4753.13	4753.13	0.00
27	2538.28	2530.26	.32
28	1012.50	1005.26	.71
29	253.13	252.50	.25
30	0.00	0.00	0.00
31	5832.00	5832.00	0.00
32	3280.50	3280.50	0.00
33	1458.00	1458.00	0.00
34	364.50	364.50	0.00
35	0.00	0.00	0.00

(d) Neumann Data (24 elements)

Node	Exact Solution	FEM Solution	% Error
1	0.00	−3.69	
2	0.00	0.00	0.00
3	0.00	0.00	0.00
4	0.00	0.00	0.00
5	0.00	0.00	0.00
6	91.12	70.60	22.52
7	63.28	49.13	22.36
8	40.50	31.82	21.43
9	10.12	6.51	35.67
10	0.00	0.00	0.00
11	450.00	409.87	8.92
12	288.00	257.40	10.63
13	162.00	148.10	8.58
14	40.50	34.16	15.65
15	0.00	0.00	0.00
16	1458.00	1392.81	4.47
17	820.12	781.97	4.65
18	364.50	349.83	4.02
19	91.12	81.24	10.84
20	0.00	0.00	0.00
21	3528.00	3381.90	4.14
22	1800.00	1746.66	2.96
23	648.00	615.65	4.99
24	162.00	150.18	7.30
25	0.00	0.00	0.00
26	4753.12	4659.78	1.96
27	2538.28	2449.15	3.51
28	1012.50	983.92	2.82
29	253.12	244.03	3.59
30	0.00	0.00	0.00
31	5832.00	5586.42	4.21
32	3280.50	3280.50	0.00
33	1458.00	1458.00	0.00
34	364.50	364.50	0.00
35	0.00	0.00	0.00

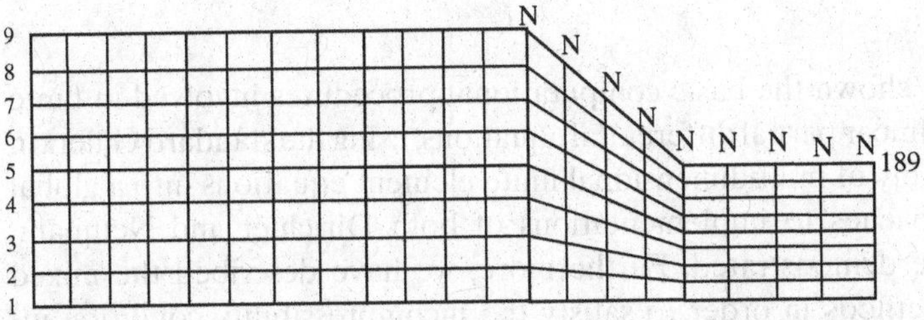

Figure 10.4.2.1 Geometry and discretization for Section 10.4.2 with N representing the Neumann boundary conditions. Dirichlet and Neumann boundary conditions are prescribed from the exact solution.

Exact Solution:

$$u = \frac{1}{1+t} + x^2 y, \quad \mathrm{v} = \frac{1}{1+t} + xy^2$$

Required: Solve the above partial differential equations using GGM for the coarse, intermediate, and fine meshes with the Dirichlet and Neumann boundary data as shown in Figure 10.4.2.1. Set $\nu = 1$, $\Delta t = 10^{-4}$, $\eta = 1/2$ Set $u = \mathrm{v} = 0$ initially at all interior nodes and observe convergence behavior.

Solution: The steady state is reached at $t \cong 0.25$ and 0.4 for u and v, respectively, at $x = 4.5$ and $y = 0.75$ to the almost exact steady-state values as shown in Figure 10.4.2.2. In Section 11.6.4, the results with nonlinear convection terms will be presented, demonstrating the solution convergence as a function of grid refinements.

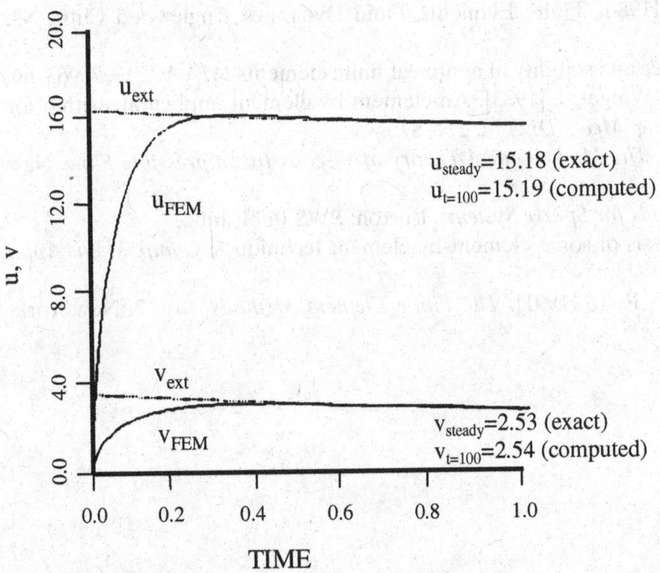

Figure 10.4.2.2 Convergence history of u and v($\nu = 1.0$, $\Delta t = 0.01$, $x = 4.5$ and $y = 0.75$).

10.5 SUMMARY

In this chapter, we have shown the basic computational procedures involved in finite element calculations for linear partial differential equations, using the standard Galerkin methods (SGM). Assembly of multidimensional finite element equations into a global form and various approaches to implementations of both Dirichlet and Neumann boundary conditions are demonstrated. Furthermore, we have described the mixed methods and penalty methods in order to satisfy the incompressibility condition involved in the Stokes flow.

In dealing with time-dependent problems, formulations with the generalized Galerkin methods (GGM) for parabolic and hyperbolic partial differential equations are presented. In particular, it was shown that temporal approximations can be provided independently and discontinuously from spatial approximations.

Solution procedures of finite element equations in general and solution approaches using element-by-element assembly techniques in particular are also elaborated. It is shown that, by means of the element-by-element (EBE) vector operations, the formulation of entire stiffness matrix array can be avoided.

Note that convective or nonlinear terms are not included in this chapter, which constitute one of the most important aspects of fluid dynamics, both physically and numerically. This is the subject of the next chapter.

REFERENCES

Babuska, I. [1973]. The finite element method with Lagrange multipliers. *Num. Math.*, 20, 179–92.

Brezzi, F. [1974]. On the existence, uniqueness and approximation of saddle point problems arising from Lagrangian multiplier, RAIRO, Ser. *Rouge Anal. Numer.*, R-2, 129–51.

Carey, G. F. and Jiang, B. [1986]. Element-by-element linear and nonlinear solution schemes. *Comm. Appl. Num. Meth.*, 2, 103–53.

Carey, G. F. and Oden, J. T. [1986]. Finite Elements, Fluid Dynamics. Englewood Cliffs, NJ: Prentice Hall.

Chung, T. J. [1975]. Convergence and stability of nonlinear finite elements. *AIAA J.*, 13, 7, 963–66.

Hughes, T. J. R., Levit, I., and Winget, J. [1983]. An element-by-element implicit algorithm for heat conduction. *ASCE J. Eng. Mech. Div.*, 74, 271–87.

Ladyszhenskaya, O. A. [1969]. *The Mathematical Theory of Viscous Incompressible Flow*. New York: Gordon and Breach.

Saad, Y. [1996]. *Iterative Methods for Sparse Systems*. Boston: PWS Publishing.

Wathen, A. J. [1989]. An analysis of some element-by-element techniques. *Comp. Meth. Appl. Mech. Eng.*, 74, 271–87.

Zienkiewicz, O. C. and Taylor, R. L. [1991]. *The Finite Element Methods*, Vol. 2. New York: McGraw-Hill.

CHAPTER ELEVEN

Nonlinear Problems/Convection-Dominated Flows

For fluid dynamics associated with nonlinearity and discontinuity, there have been significant developments in the last two decades both in finite difference methods (FDM) and finite element methods (FEM). Concurrent with upwind schemes in space and Taylor series expansion of variables in time for FDM formulations with various orders of accuracy, numerous achievements have been made in FEM applications since the publication of an earlier text [Chung, 1978]. These new developments include generalized Galerkin methods (GGM), Taylor-Galerkin methods (TGM) [Donea, 1984], and the streamline upwind Petrov-Galerkin (SUPG) methods [Heinrich et al., 1977; Hughes and Brooks, 1982], alternatively referred to as the streamline diffusion method (SDM) [Johnson, 1987], and Galerkin/least squares (GLS) methods [Hughes and his co-workers, 1988–1998]. In the sections that follow, it will be shown that computational strategies such as SUPG or SDM and other similar methods can be grouped under the heading of generalized Petrov-Galerkin (GPG) methods. Recent developments include unstructured adaptive methods [Oden et al., 1986; Löhner, Morgan, and Zienkiewicz, 1985], characteristic Galerkin methods (CGM) [Zienkiewicz and his co-workers, 1994–1998], discontinuous Galerkin methods (DGM) [Oden and his co-workers, 1996–1998], and flowfield-dependent variation (FDV) methods [Chung and his coworkers, 1995–1999], among others. On the other hand, the concepts of FDM and FEM have been utilized in developing finite volume methods in conjunction with unstructured grids [Jameson, Baker, and Weatherill, 1986]. It appears that FDM and FEM continue to co-exist and develop into a mature technology, mutually benefitting from each other.

We begin in this chapter with the general discussion of boundary conditions for the nonlinear momentum equations, followed by Taylor-Galerkin methods (TGM) and generalized Petrov-Galerkin (GPG) methods as applied to Burgers' equations. Some special topics such as Newton-Raphson methods and artificial viscosity are also discussed in this chapter. Applications to the Navier-Stokes system of equations characterizing incompressible and compressible flows are presented in Chapters 12 and 13, respectively.

11.1 BOUNDARY AND INITIAL CONDITIONS

Detailed treatments of boundary conditions with reference to FDM were presented in Section 6.7. In FEM formulations, Neumann boundary conditions arise from the partial integration of the inner product governing equations. This is an important aspect unique and advantageous in FEM, not available in FDM.

In general, precise definitions and implementations of boundary and initial conditions play decisive roles in obtaining acceptable and accurate solutions in fluid mechanics and heat transfer. As seen in Chapters 1 and 2, Neumann boundary conditions are derived from the inner product of the partial differential equation with test functions and by means of partial integrations of this inner product down to the mth order from the $2m$th order derivatives of the governing partial differential equations. Neumann boundary conditions arise "naturally" in this process with derivatives of order $2m-1, 2m-2, \ldots m$ (weak derivatives). Derivatives of order below m $(m-1, m-2, \ldots 0)$ are referred to as Dirichlet boundary conditions. These definitions as given in Chapters 1 and 2 for linear problems are applied to the nonlinear convective flows in this section.

Specification of boundary conditions depends on the types of partial differential equations (elliptic, parabolic, or hyperbolic) and types of flows (incompressible, compressible, vortical, irrotational, laminar, turbulent, chemically reacting, thermal radiation, surface tension, etc.). We shall limit our discussions of boundary and initial conditions to simpler and general topics of incompressible and compressible flows in this section. More complicated and specific subjects will be treated in their respective chapters and sections, Part Five, Applications.

11.1.1 INCOMPRESSIBLE FLOWS

For simplicity, let us first examine the steady-state incompressible flow governed by the conservation of mass and momentum. In order to obtain the correct forms for the boundary conditions, the governing equations must be written in conservation form. This is because the conservation form allows the partial integration to be carried out correctly. Thus, we write

Continuity

$$\mathrm{v}_{i,i} = 0 \tag{11.1.1a}$$

Momentum

$$\frac{\partial}{\partial x_i}(\rho \mathrm{v}_i \mathrm{v}_j - \sigma_{ij}) - \rho F_j = 0 \tag{11.1.1b}$$

where σ_{ij} is the total stress tensor,

$$\sigma_{ij} = -p\delta_{ij} + \tau_{ij} = -p\delta_{ij} + \mu(\mathrm{v}_{i,j} + \mathrm{v}_{j,i})$$

To determine the existence of Neumann (natural) boundary conditions, we construct an inner product of the residual of the governing partial differential equation with an appropriate variable which leads to a weak form. Since the primary variable is the velocity for the momentum equation, we write the energy due to the momentum as

$$J = (\mathrm{v}_j, R_j) = \int_\Omega \mathrm{v}_j \left[\frac{\partial}{\partial x_i}(\rho \mathrm{v}_i \mathrm{v}_j + p\delta_{ij} - \tau_{ij}) - \rho F_j\right] d\Omega \tag{11.1.2a}$$

Integrating (11.1.2a) by parts, we obtain

$$J = \int_\Gamma v_j(\rho v_i v_j + p\delta_{ij} - \tau_{ij}) n_i d\Gamma - \int_\Omega [\rho v_{j,i} v_i v_j + v_{j,j} p - \mu(v_{j,i} v_{i,j} + v_{j,i} v_{j,i}) + \rho F_j v_j] d\Omega \tag{11.1.2b}$$

Mathematically, the boundary integral in (11.1.2b) denotes the Neumann boundary conditions. Physically, it represents the energy required on the boundaries to be in balance with that available in the domain. Here, the boundary forces per unit normal surface area are identified as

$$\rho v_i v_j n_i + p\delta_{ij} n_i - \tau_{ij} n_i = S_j^{(1)} - S_j^{(2)} \quad \text{on } \Gamma_N \tag{11.1.3}$$

where $S_j^{(1)}$ and $S_j^{(2)}$ indicate the normal surface convective stress and normal surface viscous stress (traction), respectively.

$$S_j^{(1)} = \rho v_i v_j n_i + p\delta_{ij} n_i = \begin{cases} \rho u(un_1 + vn_2) + pn_1 \\ \rho v(un_1 + vn_2) + pn_2 \end{cases} \quad (i, j = 1, 2) \tag{11.1.4}$$

$$S_j^{(2)} = \tau_{ij} n_i = \mu(v_{i,j} + v_{j,i}) n_i = \begin{cases} \mu\left(2\dfrac{\partial u}{\partial x} n_1 + \left(\dfrac{\partial u}{\partial y} + \dfrac{\partial v}{\partial x}\right) n_2\right) \\ \mu\left(\left(\dfrac{\partial v}{\partial x} + \dfrac{\partial u}{\partial y}\right) n_1 + 2\dfrac{\partial v}{\partial y} n_2\right) \end{cases} \quad (i, j = 1, 2) \tag{11.1.5}$$

with $v_1 = u$, $v_2 = v$. A glance at (11.1.4) indicates that $S_j^{(1)} = pn_j$ for a solid wall ($v_i n_i = 0$), as defined in Figure 11.1.1 where the pressure is to be specified as a Neumann boundary condition. The normal surface traction $S_j^{(2)}$ is contributed by viscous stress normal to the surface. For example, consider the vertical surface where $\theta = 0°$ on the right side (outlet) and $\theta = 180°$ on the left side (inlet). Let us examine the outlet face where $n_1 = \cos(0°) = 1$ and $n_2 = \sin(0°) = 0$. For horizontal (x-axis) and vertical

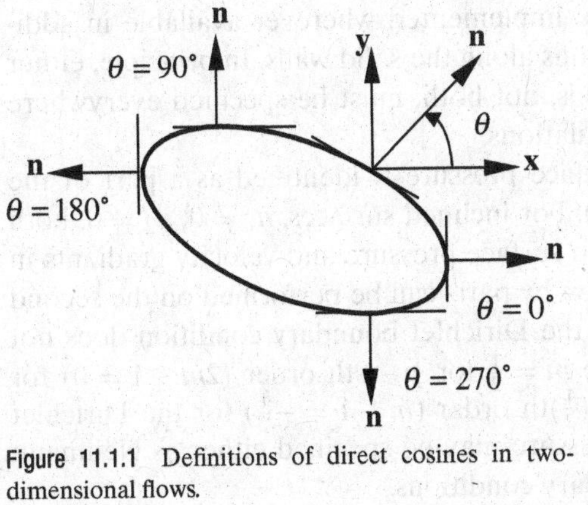

Figure 11.1.1 Definitions of direct cosines in two-dimensional flows.

(y-axis) directions, we have, respectively,

$$S_j^{(2)} = \tau_{ij} n_i = \begin{cases} 2\mu\left(\dfrac{\partial u}{\partial x}\right) \\ \mu\left(\dfrac{\partial \mathrm{v}}{\partial x} + \dfrac{\partial u}{\partial y}\right) \end{cases} \quad (i, j = 1, 2) \tag{11.1.6}$$

Alternatively, along the left side (inlet), $n_1 = \cos(180°) = -1$ and $n_2 = \sin(180°) = 0$,

$$S_j^{(2)} = \begin{cases} -2\mu\left(\dfrac{\partial u}{\partial x}\right) \\ -\mu\left(\dfrac{\partial \mathrm{v}}{\partial x} + \dfrac{\partial u}{\partial y}\right) \end{cases} \quad (i, j = 1, 2) \tag{11.1.7}$$

Similarly, for the top and bottom horizontal surfaces, respectively, with $\theta = 90°$ and $\theta = 270°$

$$S_j^{(2)} = \begin{cases} \mu\left(\dfrac{\partial u}{\partial y} + \dfrac{\partial \mathrm{v}}{\partial x}\right) \\ 2\mu\left(\dfrac{\partial \mathrm{v}}{\partial y}\right) \end{cases} \quad (i, j = 1, 2) \tag{11.1.8}$$

and

$$S_j^{(2)} = \begin{cases} -\mu\left(\dfrac{\partial u}{\partial y} + \dfrac{\partial \mathrm{v}}{\partial x}\right) \\ -2\mu\left(\dfrac{\partial \mathrm{v}}{\partial y}\right) \end{cases} \quad (i, j = 1, 2) \tag{11.1.9}$$

This completes the discussion of Neumann boundary conditions for the momentum equation. The continuity equation (11.1.1a) is a constraint condition for incompressibility or conservation of mass and is incapable of producing the Neumann boundary conditions. The Dirichlet (essential) boundary conditions arise from further integration by parts of the domain integral terms of (11.1.2b). Intuitively, we identify them as

$$\mathrm{v}_i = \overline{\mathrm{v}}_i \quad \text{on } \Gamma_D \tag{11.1.10}$$

Dirichlet boundary conditions may be implemented wherever available in addition to commonly assumed no-slip conditions along the solid walls. In principle, either Dirichlet or Neumann boundary conditions, not both, must be specified everywhere along the boundary surfaces for elliptic equations.

It is important to realize that the surface pressure is identified as a part of the Neumann boundary conditions in (11.1.4). For inclined surfaces, $n_1 \neq 0$, $n_2 \neq 0$, both components S_1 and S_2 contain the nonzero surface pressure and velocity gradients in both directions. Since no further integration by parts can be performed on the second term of the domain integral in (11.1.2b), the Dirichlet boundary condition does not arise. The reason for this is that we have $m = \frac{1}{2}$ for $p_{,i}$, 0th order ($2m - 1 = 0$) for the Neumann boundary condition and $-(\frac{1}{2})$th order ($m - 1 = -\frac{1}{2}$) for the Dirichlet boundary condition, implying that the pressure may be specified either as Neumann boundary conditions or as Dirichlet boundary conditions.

In view of these basic rules, any deviation arbitrarily chosen by practitioners may lead to incorrect solutions. Moreover, it is cautioned that any boundary nodes without specification of either Dirichlet or Neumann data are automatically construed as having enforced $S_i^{(1)} = S_i^{(2)} = 0$, because the finite element analog of the Neumann boundary vector in (11.1.2b) vanishes if either Dirichlet or Neumann data are not provided.

The numerical analysis involved in incompressible flows often requires the solution of Poisson equation for pressure in order to maintain the mass conservation and obtain accurate solutions of momentum equations. The pressure Poisson equation is obtained by constructing the divergence of the momentum equation. For incompressible flows, this operation leads to

$$p_{,ii} + (\rho v_{i,j} v_j)_{,i} = 0 \tag{11.1.11}$$

The inner product of (11.1.11) with p becomes

$$J = \int_\Omega p[p_{,ii} + (\rho v_{i,j} v_j)_{,i}]d\Omega = 0$$

or

$$J = \int_\Gamma p\,(p_{,i}n_i + \rho v_{i,j} v_j n_i)\,d\Gamma - \int_\Omega (p_{,i}p_{,i} + \rho p_{,i} v_{i,j} v_j)\,d\Omega \tag{11.1.12}$$

It follows that Neumann boundary conditions are

$$S^{(1)} = p_{,i}n_i = \frac{\partial p}{\partial x}\cos\theta + \frac{\partial p}{\partial y}\sin\theta \tag{11.1.13a}$$

$$S^{(2)} = \rho(v_i n_i)_{,j} v_j = \rho\left(\frac{\partial u}{\partial x}\cos\theta + \frac{\partial v}{\partial x}\sin\theta\right)u + \rho\left(\frac{\partial u}{\partial y}\cos\theta + \frac{\partial v}{\partial y}\sin\theta\right)v \tag{11.1.13b}$$

Here $S^{(1)}$ represents the normal surface pressure gradients. These data should be provided along the boundaries wherever the Dirichlet boundary conditions are not available. Notice that $S^{(2)}$ vanishes if $v_i n_i = 0$ along the boundary nodes. In this case, of course, the pressure must be specified as Dirichlet boundary conditions alone, contrary to the case in the momentum equation, where pressure is treated as Neumann boundary conditions.

For transient problems, the momentum equation is written as

$$\rho\frac{\partial v_j}{\partial t} + \frac{\partial}{\partial x_i}(\rho v_i v_j - \sigma_{ij}) - \rho F_j = 0 \tag{11.1.14}$$

In this case, the initial conditions consist of the initial data at $t = 0$ along the boundaries and the domain. For the velocity-pressure solutions of (11.1.1), the required initial conditions are

$$v_i(x_i, 0) = v_i^0 \quad \text{in } \overline{\Omega} = \Omega \cup \Gamma \tag{11.1.15a}$$

$$v_i n_i(x_i, 0) = v_i^0 n_i \quad \text{on } \Gamma \tag{11.1.15b}$$

In addition to these initial data, the Neumann boundary conditions of (11.1.4) and (11.1.5) at $t = 0$ should also be satisfied. Incompressibility conditions, $v_{i,i}^0(x_i, 0) = 0$ in

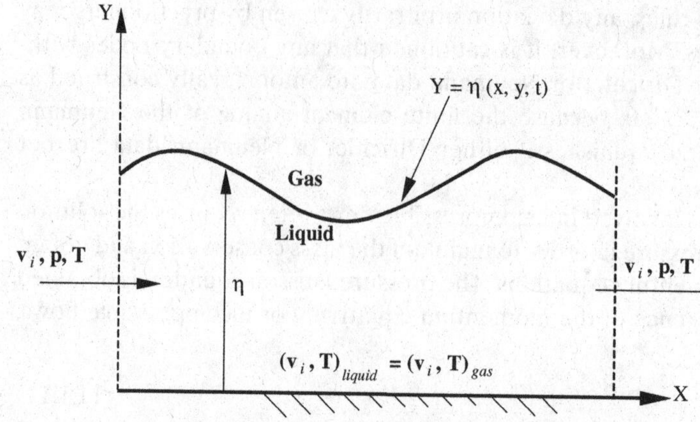

Figure 11.1.2 Free surface flow boundary conditions with variable temperatures as well as velocity and pressure between liquid and gas.

Ω are difficult to enforce at $t = 0$. We require that the mass conservation be satisfied for $t > 0$ through an adequate numerical control such as the pressure Poisson equation to be discussed in Chapter 12.

The momentum equation (11.1.14), which is characterized as an elliptic-parabolic equation, tends toward parabolic if time dependency dominates and toward elliptic if spatial dependency dominates. Boundary conditions must be specified everywhere along the boundaries for elliptic equations, but not specified at the outlet surface for parabolic equations, as dicussed in Section 2.4.

Boundary conditions involved in two-phase flows or free surfaces of variable densities require additional information. As shown in Figure 11.1.2, the free surface is the boundary between liquid and gas,

$$Y = \eta(x, y, t)$$
$$p_{(liquid)} = p_{(gas)} - \sigma\left(R_x^{-1} + R_y^{-1}\right)$$
$$\mathrm{v}_{(liquid)} = \mathrm{v}_{(gas)} = \frac{D\eta}{Dt} = \frac{\partial\eta}{\partial t} + u\frac{\partial\eta}{\partial x} + \mathrm{v}\frac{\partial\eta}{\partial y}$$
$$\sigma_{ij(liquid)} = \sigma_{ij(gas)}$$
$$q_{y(liquid)} = q_{y(gas)}, \quad q_y = -k\frac{\partial T}{\partial y}$$

where σ is the surface tension, R_x and R_y are the curvatures,

$$R_x^{-1} = \frac{\partial}{\partial x}\left\{\frac{\partial\eta}{\partial x}\left[1 + \left(\frac{\partial\eta}{\partial x}\right)^2 + \left(\frac{\partial\eta}{\partial y}\right)^2\right]^{-\frac{1}{2}}\right\}$$

$$R_y^{-1} = \frac{\partial}{\partial y}\left\{\frac{\partial\eta}{\partial y}\left[1 + \left(\frac{\partial\eta}{\partial x}\right)^2 + \left(\frac{\partial\eta}{\partial y}\right)^2\right]^{-\frac{1}{2}}\right\}$$

For simplified free-surface conditions between liquid and air, we may assume that

$$p_{(liquid)} \cong p_{(gas)} - \sigma\left(\frac{\partial^2\eta}{\partial x^2} + \frac{\partial^2\eta}{\partial y^2}\right)$$

$$\mathrm{v}_{(liquid)} \cong \frac{\partial\eta}{\partial t}$$

$$\frac{\partial \mathrm{v}}{\partial y}_{(liquid)} \cong 0, \qquad \frac{\partial T}{\partial y}_{(liquid)} \cong 0$$

$$p_{(liquid)} \cong p_{(atm)}$$

In addition, we specify the velocity, pressure, and temperature at the inlet and outlet as well as the no-slip condition (v = 0) at the wall. More detailed treatments of boundary conditions associated with surface tension will be given in Chapter 25, Multiphase Flows.

11.1.2 COMPRESSIBLE FLOWS

Compressible flows are characterized by additional terms for dilatation in the stress tensor and temporal and spatial variations of density.

$$\frac{\partial}{\partial t}(\rho \mathrm{v}_j) + \frac{\partial}{\partial x_i}(\rho \mathrm{v}_i \mathrm{v}_j + p\delta_{ij} - \tau_{ij}) - \rho F_j = 0 \qquad (11.1.16a)$$

$$\frac{\partial \rho}{\partial t} + (\rho \mathrm{v}_i)_{,i} = 0 \qquad (11.1.16b)$$

with

$$\tau_{ij} = \mu(\mathrm{v}_{i,j} + \mathrm{v}_{j,i}) - \frac{2}{3}\mu \mathrm{v}_{k,k}\delta_{ij}$$

For compressible flows, the normal surface convective stress, $S_j^{(1)}$, remains the same as in (11.1.4), but the normal surface traction, $S_j^{(2)}$, is modified as

$$S_j^{(2)} = \begin{cases} \mu\left[\dfrac{\partial u}{\partial x}n_1 + \dfrac{\partial u}{\partial y}n_2 + \dfrac{\partial u}{\partial x}n_1 + \dfrac{\partial \mathrm{v}}{\partial x}n_2 - \dfrac{2}{3}\left(\dfrac{\partial u}{\partial x} + \dfrac{\partial \mathrm{v}}{\partial y}\right)n_1\right] \\ \mu\left[\dfrac{\partial \mathrm{v}}{\partial x}n_1 + \dfrac{\partial \mathrm{v}}{\partial y}n_2 + \dfrac{\partial u}{\partial y}n_1 + \dfrac{\partial \mathrm{v}}{\partial y}n_2 - \dfrac{2}{3}\left(\dfrac{\partial u}{\partial x} + \dfrac{\partial \mathrm{v}}{\partial y}\right)n_2\right] \end{cases} \quad (j = 1, 2) \qquad (11.1.17)$$

Thus, equations (11.1.6)–(11.1.9) are written as follows:
For $\theta = 0°$

$$S_j^{(2)} = \begin{cases} \mu\left(\dfrac{4}{3}\dfrac{\partial u}{\partial x} - \dfrac{2}{3}\dfrac{\partial \mathrm{v}}{\partial y}\right) \\ \mu\left(\dfrac{\partial \mathrm{v}}{\partial x} + \dfrac{\partial u}{\partial y}\right) \end{cases} \quad (j = 1, 2) \qquad (11.1.18)$$

For $\theta = 180^\circ$

$$S_j^{(2)} = \begin{cases} -\mu\left(\dfrac{4}{3}\dfrac{\partial u}{\partial x} - \dfrac{2}{3}\dfrac{\partial \mathrm{v}}{\partial y}\right) \\ -\mu\left(\dfrac{\partial \mathrm{v}}{\partial x} + \dfrac{\partial u}{\partial y}\right) \end{cases} \quad (j = 1, 2) \tag{11.1.19}$$

For $\theta = 90^\circ$

$$S_j^{(2)} = \begin{cases} \mu\left(\dfrac{\partial u}{\partial y} + \dfrac{\partial \mathrm{v}}{\partial x}\right) \\ \mu\left(\dfrac{4}{3}\dfrac{\partial \mathrm{v}}{\partial y} - \dfrac{2}{3}\dfrac{\partial u}{\partial x}\right) \end{cases} \quad (j = 1, 2) \tag{11.1.20}$$

For $\theta = 270^\circ$

$$S_j^{(2)} = \begin{cases} -\mu\left(\dfrac{\partial u}{\partial y} + \dfrac{\partial \mathrm{v}}{\partial x}\right) \\ -\mu\left(\dfrac{4}{3}\dfrac{\partial \mathrm{v}}{\partial y} - \dfrac{2}{3}\dfrac{\partial u}{\partial x}\right) \end{cases} \quad (j = 1, 2) \tag{11.1.21}$$

For compressible flows, combined solutions of the pressure Poisson equation are not required as the enforcement of the incompressibility condition is not necessary. Thus, the pressure will not be used as Dirichlet boundary conditions. It is still a part of the Neumann boundary conditions as specified in (11.1.4).

Dirichlet boundary conditions and initial conditions for compressible flows are the same as the incompressible flows. Enforcement of incompressibility conditions as initial conditions, however, is no longer necessary.

The elliptic-parabolic nature of (11.1.14) tends toward a hyperbolic type in high-speed flows if the viscosity effect is negligible, resulting in the Euler equation. In this case, the outflow boundary conditions are not to be specified but, rather, should be determined by the calculated upstream flows since the downstream effect toward upstream is not allowed. Details were discussed in Section 6.7 and will be covered also in Section 13.6.6 for compressible flows.

CONCLUDING REMARKS

In identifying the Neumann boundary conditions, the conservation form of the momentum equations is used, in general, where convective terms as well as diffusion terms are integrated by parts. If the convective terms are not written in conservation form, however, no integration by parts is performed for the convective terms. In this case, the Neumann boundary conditions do not arise from the convective terms. This is the case for incompressible flows. In contrast, the conservation form is more convenient for compressible flows, and integration by parts for the convective term is carried out, resulting in the Neumann boundary conditions for compressible flows. This rule does not apply if a special test function (i.e., numerical diffusion test function) is used to induce artificial dissipation for the convective term as discussed in Section 11.3.

Specification of boundary conditions required for the Navier-Stokes system of equations is considerably more complicated, and will be discussed in Chapter 13.

11.2 GENERALIZED GALERKIN METHODS AND TAYLOR-GALERKIN METHODS

11.2.1 LINEARIZED BURGERS' EQUATIONS

To demonstrate the basic concept of generalized Galerkin methods (GGM), we consider the linearized Burgers' equations in the form,

$$R_i = \frac{\partial v_i}{\partial t} + \overline{v}_j v_{i,j} - \nu v_{i,jj} - f_i = 0 \quad (i = 1, 2, 3) \tag{11.2.1}$$

where $\overline{v}_j$ is temporarily held constant in the time-marching steps and/or iteration cycles but updated in the following steps and/or iteration cycles. The standard finite element formulation of (11.2.1) with DST approximations was introduced as the GGM in Section 10.2. This requires the successive inner products of the form

$$(\hat{W}(\xi), E_i) = (\hat{W}(\xi), [W_\alpha(x), R_i]) = \int_\xi \hat{W}(\xi)\left[\int_\Omega W_\alpha(x) R_i \, d\Omega\right] d\xi = 0 \tag{11.2.2}$$

in which $W_\alpha(x)$ and $\hat{W}(\xi)$ denote the spatial and temporal test functions, respectively. Furthermore, the trial functions for nodal values of variables are related as follows:

$$v_i = \Phi_\alpha(x_i) v_{\alpha i} \tag{11.2.3}$$

$$v_{\alpha i} = \hat{\Phi}_m(\xi) v_{\alpha i}^m \tag{11.2.4}$$

where $\Phi_\alpha(x)$ and $\hat{\Phi}_m(\xi)$ denote spatial and temporal trial functions, respectively, $\xi = t/\Delta t$, α = global spatial nodes, and m = local temporal station ($n+1, n, n-1$, etc.).

Setting the spatial test function W_α equal to the spatial trial function Φ_α and integrating (11.2.2) by parts in the spatial domain, we obtain

$$\int_\xi \hat{W}(\xi)[A_{\alpha\beta}\dot{v}_{\beta i} + (B_{\alpha\beta} + K_{\alpha\beta}) v_{\beta i} - F_{\alpha i} - G_{\alpha i}] d\xi = 0 \tag{11.2.5}$$

with

$$A_{\alpha\beta} = \int_\Omega \Phi_\alpha \Phi_\beta d\Omega, \qquad B_{\alpha\beta} = \int_\Omega \Phi_\alpha \Phi_{\beta,j} \overline{v}_j \, d\Omega$$

$$K_{\alpha\beta} = \int_\Omega \nu \Phi_{\alpha,j} \Phi_{\beta,j} \, d\Omega \qquad G_{\alpha i} = \int_\Gamma \nu \overset{*}{\Phi}_\alpha \overset{*}{\Phi}_\beta d\Gamma g_{\beta i} \qquad F_{\alpha i} = \int_\Omega \Phi_\alpha \Phi_\beta d\Omega f_{\beta i}$$

Notice here that all matrices are the same as in Chapter 10 except for $B_{\alpha\beta}$, which is called the convection matrix. Choosing a linear variation of a variable in the temporal domain

$$v_{\alpha i} = (1 - \xi) v_{\alpha i}^n + \xi v_{\alpha i}^{n+1}$$

we obtain from (11.2.5)

$$[A_{\alpha\beta} + \eta \Delta t (B_{\alpha\beta} + K_{\alpha\beta})] v_{\beta i}^{n+1} = [A_{\alpha\beta} - \Delta t (1 - \eta)(B_{\alpha\beta} + K_{\alpha\beta})] v_{\beta i}^n + \Delta t (F_{\alpha i} + G_{\alpha i}) \tag{11.2.6}$$

where the temporal parameter η is defined as

$$\eta = \frac{\int_0^1 \hat{W}(\xi)\,\xi\,d\xi}{\int_0^1 \hat{W}(\xi)\,d\xi}$$

For $\hat{W}(\xi) = \delta(\xi - 1/2)$ or $\hat{W}(\xi) = 1$ with $0 \le \xi \le 1$, the temporal parameter becomes $\eta = 1/2$. Thus,

$$\left[A_{\alpha\beta} + \frac{\Delta t}{2}(B_{\alpha\beta} + K_{\alpha\beta})\right] \mathrm{v}_{\beta i}^{n+1} = \left[A_{\alpha\beta} - \frac{\Delta t}{2}(B_{\alpha\beta} + K_{\alpha\beta})\right] \mathrm{v}_{\beta i}^{n} + \Delta t(F_{\alpha i} + G_{\alpha i}) \tag{11.2.7}$$

We may rearrange (11.2.7) in the form

$$\left[A_{\alpha\beta} + \frac{\Delta t}{2}(B_{\alpha\beta} + K_{\alpha\beta})\right] \frac{(\mathrm{v}_{\beta i}^{n+1} - \mathrm{v}_{\beta i}^{n})}{\Delta t} = -(B_{\alpha\beta} + K_{\alpha\beta})\,\mathrm{v}_{\beta i}^{n} + F_{\alpha i} + G_{\alpha i} \tag{11.2.8}$$

This is identical to the special case of the Taylor-Galerkin Methods (TGM) reported by Donea [1984]. If $\bar{\mathrm{v}}_j$ in (11.2.1) is no longer held constant, then the temporal trial functions $\hat{\Phi}_\alpha(\xi)$ or temporal test functions $\hat{W}(\xi)$, or both, may be chosen as higher order polynomials, which would introduce additional temporal stations as shown in Section 10.2. Note that the scheme as given by (11.2.8) is implicit and resembles the Crank-Nicholson scheme. In contrast to (11.2.7) in which $\eta = 1/2$ is fixed, we may choose $0 \le \eta \le 1$. Such choice is general and the expression given by (11.2.6) is known as the generalized Galerkin method (GGM) for the linearized convection-diffusion equation.

To prove that (11.2.8) is the same as the TGM of Donea [1984], we proceed as follows: Expanding v_i^{n+1} in Taylor series about v_i^n, we write

$$\mathrm{v}_i^{n+1} = \mathrm{v}_i^n + \Delta t \frac{\partial \mathrm{v}_i^n}{\partial t} + \frac{\Delta t^2}{2}\frac{\partial^2 \mathrm{v}_i^n}{\partial t^2} + \frac{\Delta t^3}{6}\frac{\partial^3 \mathrm{v}_i^n}{\partial t^3} + O(\Delta t^4) \tag{11.2.9}$$

Taking a time derivative of (11.2.1) for the time step n and substituting the result into the above leads to

$$\begin{aligned}\frac{\mathrm{v}_i^{n+1} - \mathrm{v}_i^n}{\Delta t} &= \left(-\bar{\mathrm{v}}_j \frac{\partial}{\partial x_j} + \nu \frac{\partial^2}{\partial x_j \partial x_j}\right)\mathrm{v}_i^n + \frac{\Delta t}{2}\left(-\bar{\mathrm{v}}_j \frac{\partial}{\partial x_j} + \nu \frac{\partial^2}{\partial x_j \partial x_j}\right)\frac{\partial \mathrm{v}_i^n}{\partial t} \\ &+ \frac{\Delta t^2}{6}\left(\bar{\mathrm{v}}_j \bar{\mathrm{v}}_k \frac{\partial^2}{\partial x_j \partial x_k} - 2\nu\bar{\mathrm{v}}_j \frac{\partial^3}{\partial x_j \partial x_k \partial x_k} + \nu^2 \frac{\partial^4}{\partial x_j \partial x_j \partial x_k \partial x_k}\right)\frac{\partial \mathrm{v}_i^n}{\partial t} + f_i\end{aligned} \tag{11.2.10a}$$

with

$$\frac{\partial \mathrm{v}_i^n}{\partial t} = \frac{\mathrm{v}_i^{n+1} - \mathrm{v}_i^n}{\Delta t}$$

Although the third order time derivative in (11.2.9) may be useful for the convection dominated flows without the viscous terms, we shall choose to neglect it for our purpose here to establish the analogy between GGM and TGM. Rearranging (11.2.10a) leads to

$$\left[1-\frac{\Delta t}{2}\left(-\overline{\mathrm{v}}_j\frac{\partial}{\partial x_j}+\nu\frac{\partial^2}{\partial x_j\partial x_j}\right)\right]\frac{\mathrm{v}_i^{n+1}-\mathrm{v}_i^n}{\Delta t}=\left(-\overline{\mathrm{v}}_j\frac{\partial}{\partial x_j}+\nu\frac{\partial^2}{\partial x_j\partial x_j}\right)\mathrm{v}_i^n+f_i \tag{11.2.10b}$$

The Galerkin finite element analog for (11.2.10b) yields

$$\int_\Omega \Phi_\alpha\left\{\left[1-\frac{\Delta t}{2}\left(-\overline{\mathrm{v}}_j\frac{\partial}{\partial x_j}+\nu\frac{\partial^2}{\partial x_j\partial x_j}\right)\right]\frac{\Phi_\beta\left(\mathrm{v}_{\beta i}^{n+1}-\mathrm{v}_{\beta i}^n\right)}{\Delta t}\right.$$
$$\left.+\left(\overline{\mathrm{v}}_j\frac{\partial}{\partial x_j}-\nu\frac{\partial^2}{\partial x_j\partial x_j}\right)\Phi_\beta\mathrm{v}_{\beta i}^n-\Phi_\beta f_{\beta i}\right\}d\Omega=0 \tag{11.2.10c}$$

Integrating the above equation by parts, we obtain the result identical to (11.2.8):

$$\left[A_{\alpha\beta}+\frac{\Delta t}{2}(B_{\alpha\beta}+K_{\alpha\beta})\right]\frac{\left(\mathrm{v}_{\beta i}^{n+1}-\mathrm{v}_{\beta i}^n\right)}{\Delta t}=-(B_{\alpha\beta}+K_{\alpha\beta})\,\mathrm{v}_{\beta i}^n+F_{\alpha i}+G_{\alpha i} \tag{11.2.11a}$$

which can then be rearranged in the form shown in (11.2.7),

$$\left[A_{\alpha\beta}+\frac{\Delta t}{2}(B_{\alpha\beta}+K_{\alpha\beta})\right]\mathrm{v}_{\beta i}^{n+1}=\left[A_{\alpha\beta}-\frac{\Delta t}{2}(B_{\alpha\beta}+K_{\alpha\beta})\right]\mathrm{v}_{\beta i}^n+\Delta t(F_{\alpha i}+G_{\alpha i}) \tag{11.2.11b}$$

It has been shown that the GGM approach with the temporal test function given by $\hat{W}(\xi)=\delta(\xi-1/2)$ or $\hat{W}(\xi)=1$ is identical to TGM proposed by Donea [1984] without the effect of the third order time derivative in the Taylor series expansion. This analogy of GGM to TGM does not hold true for the nonlinear Burgers' equations ($\mathrm{v}_j\neq\overline{\mathrm{v}}_j$) as will be demonstrated in Section 11.2.5 in which an explicit numerical diffusion arises in TGM, contributing to both stability and accuracy for the solution of nonlinear equations in general. The presence of the third order time derivative in the Taylor series expansion as originally proposed by Donea [1984] will be discussed in Section 11.2.3 in relation with the Euler method, leap-frog method, and Crank-Nicolson method.

Numerical Diffusion

In general, for convection dominated flows, numerical diffusion is required to stabilize the solution process. To see whether the algorithm of GGM or TGM as given by (11.2.8) or (11.2.11a) does provide such a numerical diffusion, we may trace from (11.2.11b) back to (11.2.10a) with Δt^2 terms neglected.

$$\int_\Omega \Phi_\alpha\left(\frac{\partial \mathrm{v}_i}{\partial t}+\overline{\mathrm{v}}_j\mathrm{v}_{i,j}-\nu\mathrm{v}_{i,jj}-f_i\right)d\Omega=-\int_\Omega\frac{\Delta t}{2}\overline{\mathrm{v}}_j\Phi_{\alpha,j}(\overline{\mathrm{v}}_k\mathrm{v}_{i,k}-\nu\mathrm{v}_{i,kk}-f_i)d\Omega$$

in which the difference equation has been converted to the differential equation, with boundary integrals neglected upon integration by parts in the right-hand side. Note also that integration by parts was performed only for the convective terms. The viscous terms and body forces on the right-hand side may be neglected. The GGM formulation can then be applied to the left-hand side. It is clear that the first term on the right-hand side,

$$C_{\alpha\beta} = \int_\Omega \frac{\Delta t}{2} \overline{v}_k \overline{v}_j \Phi_{\alpha,k} \Phi_{\beta,j} d\Omega = \int_\Omega \overline{\nu}_{kj} \Phi_{\alpha,k} \Phi_{\beta,j} d\Omega \qquad (11.2.12a)$$

represents the numerical diffusion matrix with $\overline{\nu}_{kj} = \frac{\Delta t}{2} \overline{v}_k \overline{v}_j$ being the artificial viscosity for convection. The numerical diffusion matrix $C_{\alpha\beta}$ should be added to the convection matrix $B_{\alpha\beta}$ in (11.2.8) particularly for high-speed convection-dominated flows.

$$B_{\alpha\beta} = \int_\Omega \Phi_\alpha \Phi_{\beta,j} \overline{v}_j d\Omega + \int_\Omega \overline{\nu}_{kj} \Phi_{\alpha,k} \Phi_{\beta,j} d\Omega \qquad (11.2.12b)$$

We shall further discuss this issue for the nonlinear Burgers' equations in Section 11.2.5. Note that a variety of approximations in GGM for the temporal test function $\hat{W}(\xi)$ and the temporal trial functions in (11.2.4) may lead to different forms of numerical diffusion. Similar consequences arise for TGM if the third order time derivative in the Taylor series expansion in (11.2.9) is retained.

Remarks: In general, we may consider TGM to be a special case of GGM with $\eta = 1/2$ being chosen in (11.2.6). This is not true in some special cases of TGM as derived by Donea [1984].

11.2.2 TWO-STEP EXPLICIT SCHEME

Nonlinear problems can be solved explicitly by splitting the equation into two parts within a time step. Equation (11.2.7) or (11.2.8) may be rewritten in the form

Step 1

$$A_{\alpha\beta} X_{\beta i}^{(1)} = -(B_{\alpha\beta} + K_{\alpha\beta}) v_{\beta i}^n + F_{\alpha i} + G_{\alpha i} \qquad (11.2.13a)$$

Step 2

$$A_{\alpha\beta} X_{\beta i}^{(2)} = -\frac{\Delta t}{2} (B_{\alpha\beta} + K_{\alpha\beta}) X_{\beta i}^{(1)} \qquad (11.2.13b)$$

where

$$X_{\beta i}^{(1)} = \frac{\Delta v_{\beta i}^{(1)}}{\Delta t}, \qquad X_{\beta i}^{(2)} = \frac{\Delta v_{\beta i}^{(2)} - \Delta v_{\beta i}^{(1)}}{\Delta t} \qquad (11.2.14a,b)$$

Note that substitution of (11.2.14) into (11.2.13b) recovers (11.2.11) if the following assumption is made upon convergence:

$$\Delta v_{\beta i}^{(2)} - \Delta v_{\beta i}^{(1)} = v_{\beta i}^{n+1} - v_{\beta i}^n \qquad (11.2.15)$$

A glance at (11.2.13a) and (11.2.13b) suggests that the solution of (11.2.13a) for $X_{\beta i}^{(1)}$ (Step 1) can be substituted into the right-hand side of (11.2.13b) to determine $X_{\beta i}^{(2)}$ (Step 2). At convergence, it is seen that

$$\frac{\Delta v_{\beta i}^{(2)}}{\Delta t} \to \frac{\Delta v_{\beta i}^{(1)}}{\Delta t} \to \frac{\Delta v_{\beta i}^{n+1}}{\Delta t} = \frac{v_{\beta i}^{n+1} - v_{\beta i}^n}{\Delta t} \cong 0$$

and that (11.2.11b) arises by combining (11.2.13a) with (11.2.13b). This process is known as the two-step scheme, similar to the Lax-Wendroff scheme, contributing to an increase in accuracy and/or convergence.

It follows from (11.2.14) and (11.2.15) that the unknowns $v_{\beta i}^{n+1}$ can be computed from

$$v_{\beta i}^{n+1} = v_{\beta i}^{n} + \Delta t\left(X_{\beta i}^{(1)} + X_{\beta i}^{(2)}\right) \tag{11.2.16}$$

which will then be substituted back into Step 1 (11.2.13a) for the next time step, thus continuously marching in time until steady-state is reached.

In (11.2.13a) and (11.2.13b) the inverse of the mass matrix $A_{\alpha\beta}$ would be simple if we chose to use the so-called lumped mass matrix as follows: Let $A_{\alpha\beta}^{(L)}$ be the lumped mass matrix, $A_{\alpha\beta}^{(C)}$ the consistent mass matrix as defined by $A_{\alpha\beta}$ in (11.2.13).

The lumped mass matrix is diagonal with entries from the tributary areas (sum of the row contributions). For example, the lumped mass matrix, $A_{NM}^{(L)}$, for a triangular element may be obtained from the consistent mass matrix, $A_{NM}^{(C)}$, as follows:

$$A_{NM}^{(C)} = \frac{A}{12}\begin{bmatrix} 2 & 1 & 1 \\ 1 & 2 & 1 \\ 1 & 1 & 2 \end{bmatrix}$$

$$A_{NM}^{(L)} = \sum_{p=1}^{3} A_{(N)p}^{(C)} \delta_{NM} = A_{(NN)}^{(L)} = \begin{bmatrix} A_{(11)}^{(L)} & 0 & 0 \\ 0 & A_{(22)}^{(L)} & 0 \\ 0 & 0 & A_{(33)}^{(L)} \end{bmatrix} \tag{11.2.17}$$

with

$$A_{(11)}^{(L)} = A_{(11)}^{(C)} + A_{(12)}^{(C)} + A_{(13)}^{(C)} = \frac{4A}{12}$$

$$A_{(22)}^{(L)} = A_{(21)}^{(C)} + A_{(22)}^{(C)} + A_{(23)}^{(C)} = \frac{4A}{12}$$

$$A_{(33)}^{(L)} = A_{(31)}^{(C)} + A_{(32)}^{(C)} + A_{(33)}^{(C)} = \frac{4A}{12}$$

Notice here that the index within the parentheses is not associated with summing. Thus we obtain

$$A_{(NM)}^{(L)} = \frac{A}{3}\begin{bmatrix} 1 & 0 & 0 \\ 0 & 1 & 0 \\ 0 & 0 & 1 \end{bmatrix}$$

Write (11.2.13a) or (11.2.13b) in the form

$$A_{\alpha\beta}^{(C)} Y_{\beta i} = W_{\alpha i}$$

or

$$\left(A_{\alpha\beta}^{(C)} + A_{\alpha\beta}^{(L)} - A_{\alpha\beta}^{(L)}\right) Y_{\beta i} = W_{\alpha i}$$

which may be rewritten as

$$A^{(L)}_{\alpha\beta} Y_{\beta i} = W_{\alpha i} - A^{(C)}_{\alpha\beta} Y_{\beta i} + A^{(L)}_{\alpha\beta} Y_{\beta i}$$

Let the left-hand side and the right-hand side be the $r+1$ iterative cycle and the r iterative cycle, respectively:

$$A^{(L)}_{\alpha\beta} \Delta Y^{r+1}_{\beta i} = W^r_{\alpha i} - A^{(C)}_{\alpha\beta} Y^r_{\beta i} \tag{11.2.18}$$

where

$$\Delta Y^{r+1}_{\beta i} = Y^{r+1}_{\beta i} - Y^r_{\beta i}$$

The iterations implied by (11.2.18) may be applied to Step 1 (11.2.13a) and then to Step 2 (11.2.13b) until each step acquires a satisfactory convergence. It has been shown that, in many instances, the lumped mass approach often leads to excellent results.

For two-dimensional problems, the $A^{(e)}_{NM}$ matrix must be expanded so that both x- and y-direction components of v_i can be accommodated. As noted earlier, this may be achieved by means of the Kronecker delta. This will expand (11.2.18) into a 6×6 matrix for triangular elements and an 8×8 matrix for quadrilateral elements when coupled with $A_{\alpha\beta}$.

To transform the generalized finite element equations given by (11.2.7) to the two-step solution scheme, we may establish the following procedure. Consider the matrix form of (11.2.7) written as

$$D\mathrm{v}^{n+1} = E\mathrm{v}^n + \Delta t H \tag{11.2.19}$$

where

$$D = A + B + C, \qquad E = A - B - C \tag{11.2.20}$$

(a) Rearrange (11.2.19) in the form

$$D\frac{\mathrm{v}^{n+1} - \mathrm{v}^n}{\Delta t} = F\frac{\mathrm{v}^n}{\Delta t} + H \tag{11.2.21}$$

with $F = E - D$

(b) Define

$$\Delta\mathrm{v}^{(2)} - \Delta\mathrm{v}^{(1)} = \mathrm{v}^{n+1} - \mathrm{v}^n \tag{11.2.22}$$

$$X^{(1)} = \frac{\Delta\mathrm{v}^{(1)}}{\Delta t} \tag{11.2.23a}$$

$$X^{(2)} = \frac{\Delta\mathrm{v}^{(2)} - \Delta\mathrm{v}^{(1)}}{\Delta t} \tag{11.2.23b}$$

(c) Write Step 1

$$AX^{(1)} = F\frac{\mathrm{v}^n}{\Delta t} + H \tag{11.2.24}$$

(d) Write Step 2

$$AX^{(2)} = (A - D)X^{(1)} \tag{11.2.25}$$

It can be shown that substitution of (11.2.24) into (11.2.25) together with (11.2.22) and (11.2.23) recovers (11.2.21) and subsequently (11.2.19).

If quadratic approximations are used for the temporal domain, then we write

$$Dv^{n+1} = Ev^n + Gv^{n-1} + \Delta t H \tag{11.2.26}$$

The two-step scheme becomes

Step 1

$$AX^{(1)} = F\frac{v^n}{\Delta t} + \frac{Gv^{n-1}}{\Delta t} + H \tag{11.2.27}$$

Step 2

$$AX^{(2)} = (A - D)X^{(1)} \tag{11.2.28}$$

The data for Gv^{n-1} are saved from the previous time station and used as additional source terms. A similar approach can be used for all higher approximations which will contain the terms of v^{n-2}, v^{n-3}, etc.

If f_i is time dependent, and if $\overline{v}_j$ in (11.2.1) is treated as a variable, and not held constant even during the discrete time step, then the second derivative in the Taylor series expansion would carry additional terms. In this case, $\overline{v}_j$ on the left-hand side of (11.2.10b) becomes v_j^n, and $\overline{v}_j$ on the right-hand side of (11.2.10b) takes the form with a fractional step (i.e., $n+1/2$),

$$\overline{v}_j - v_j^{n+\frac{1}{2}} = v_j^n + \frac{\Delta t}{2}\frac{\partial v_j}{\partial t} \tag{11.2.29}$$

and

$$f_j - f_j^{n+\frac{1}{2}} = f_j^n + \frac{\Delta t}{2}\frac{\partial f_j}{\partial t} \tag{11.2.30}$$

which would require the three-step solution scheme.

Step 1

$$A_{\alpha\beta}X_{\beta i}^{(0)} = -\frac{1}{2}(B_{\alpha\beta} + K_{\alpha\beta})v_{\beta i}^n + F_{\alpha i} + G_{\alpha i} \tag{11.2.31}$$

with

$$X_{\beta i}^{(0)} = \frac{v_{\beta i}^{n+\frac{1}{2}} - v_{\beta i}^n}{\Delta t}$$

Step 2

$$A_{\alpha\beta}X_{\beta i}^{(1)} = -B_{\alpha\beta}v_{\beta i}^{n+\frac{1}{2}} - K_{\alpha\beta}v_{\beta i}^n + F_{\alpha i}^{n+\frac{1}{2}} + G_{\alpha i} \tag{11.2.32}$$

Step3

$$A_{\alpha\beta}X_{\beta i}^{(2)} = -\frac{1}{2}(B_{\alpha\beta} + K_{\alpha\beta})\Delta t X_{\beta i}^{(1)} \tag{11.2.33}$$

The GGM analog for the three-step scheme requires the use of quadratic functions in the temporal trial functions $\hat{\Phi}_m$, which will involve Δt^2 and three time steps, including a fractional time step.

11.2.3 RELATIONSHIP BETWEEN FEM AND FDM

It is interesting to note that the GGM formulations lead to finite difference results such as Euler Method, Leapfrog Method, Crank-Nicolson Method, etc. We will examine these results below.

Euler Method

Consider the convection equation

$$\frac{\partial \mathrm{v}_i}{\partial t} + \overline{\mathrm{v}}_j \mathrm{v}_{i,j} = 0 \tag{11.2.34}$$

Taking a time derivative of (11.2.34) gives

$$\frac{\partial^2 \mathrm{v}_i}{\partial t^2} + \overline{\mathrm{v}}_j \left(\frac{\partial \mathrm{v}_i}{\partial t} \right)_{,j} \Rightarrow \frac{\partial^2 \mathrm{v}_i}{\partial t^2} - \overline{\mathrm{v}}_j \overline{\mathrm{v}}_k \mathrm{v}_{i,kj} = 0 \tag{11.2.35}$$

A further differentiation of (11.2.35) yields

$$\frac{\partial^3 \mathrm{v}_i}{\partial t^3} - \overline{\mathrm{v}}_j \overline{\mathrm{v}}_k \left(\frac{\partial \mathrm{v}_i}{\partial t} \right)_{,kj} = 0 \tag{11.2.36}$$

Expanding v_i^{n+1} in Taylor series about v_i^n to the third order derivative, we obtain

$$\mathrm{v}_i^{n+1} = \mathrm{v}_i^n + \Delta t \frac{\partial \mathrm{v}_i^n}{\partial t} + \frac{\Delta t^2}{2!} \frac{\partial^2 \mathrm{v}_i^n}{\partial t^2} + \frac{\Delta t^3}{3!} \frac{\partial^3 \mathrm{v}_i^n}{\partial t^3} \tag{11.2.37}$$

Rearranging (11.2.37) to determine the first derivative of v_i^n gives

$$\frac{\mathrm{v}_i^{n+1} - \mathrm{v}_i^n}{\Delta t} = \frac{\partial \mathrm{v}_i^n}{\partial t} + \frac{\Delta t}{2} \frac{\partial^2 \mathrm{v}_i^n}{\partial t^2} + \frac{\Delta t^2}{6} \frac{\partial^3 \mathrm{v}_i^n}{\partial t^3} \tag{11.2.38}$$

Substituting (11.2.34) through (11.2.36) into (11.2.38) leads to

$$\frac{\mathrm{v}_i^{n+1} - \mathrm{v}_i^n}{\Delta t} = -\overline{\mathrm{v}}_j \mathrm{v}_{i,j}^n + \frac{\Delta t}{2} \overline{\mathrm{v}}_j \overline{\mathrm{v}}_k \mathrm{v}_{i,kj}^n + \frac{\Delta t^2}{6} \overline{\mathrm{v}}_j \overline{\mathrm{v}}_k \left(\frac{\partial \mathrm{v}_i^n}{\partial t} \right)_{,kj} \tag{11.2.39}$$

with

$$\frac{\partial \mathrm{v}_i^n}{\partial t} = \frac{\mathrm{v}_i^{n+1} - \mathrm{v}_i^n}{\Delta t}$$

Equation (11.2.39) may be written as

$$\left(1 - \frac{\Delta t^2}{6} \overline{\mathrm{v}}_j \overline{\mathrm{v}}_k \frac{\partial^2}{\partial x_j \partial x_k} \right) \frac{\Delta \mathrm{v}_i^{n+1}}{\Delta t} = -\overline{\mathrm{v}}_j \mathrm{v}_{i,j}^n + \frac{\Delta t}{2} \overline{\mathrm{v}}_j \overline{\mathrm{v}}_k \mathrm{v}_{i,kj}^n \tag{11.2.40}$$

where $\Delta \mathrm{v}_i^{n+1} = \mathrm{v}_i^{n+1} - \mathrm{v}_i^n$.

We construct the Galerkin finite element integral for (11.2.40) in the form

$$\left(A_{\alpha\beta} + \frac{\Delta t^2}{6} K_{\alpha\beta}\right)\frac{\Delta \mathrm{v}_{\beta i}^{n+1}}{\Delta t} = -\left(B_{\alpha\beta} + \frac{\Delta t}{2} K_{\alpha\beta}\right)\mathrm{v}_{\beta i}^{n} + \frac{\Delta t}{2} G_{\alpha i} \tag{11.2.41}$$

where

$$A_{\alpha\beta} = \int_\Omega \Phi_\alpha \Phi_\beta \, d\Omega, \qquad B_{\alpha\beta} = \int_\Omega \Phi_\alpha \Phi_{\beta,j} \bar{\mathrm{v}}_j \, d\Omega,$$

$$K_{\alpha\beta} = \int_\Omega \bar{\mathrm{v}}_j \bar{\mathrm{v}}_k \Phi_{\alpha,j} \Phi_{\beta,k} \, d\Omega,$$

$$G_{\alpha i} = \int_\Gamma \overset{*}{\Phi}_\alpha \overset{*}{\Phi}_\beta \, d\Gamma g_{\beta i}, \quad g_{\beta i} = (\bar{\mathrm{v}}_j \bar{\mathrm{v}}_k \mathrm{v}_{i,k} n_j)_\beta$$

It should be noted that (11.2.41) is equivalent to the Generalized Galerkin finite element equations,

$$\left(A_{\alpha\beta} + \frac{\Delta t^2}{6} K_{\alpha\beta}\right)\mathrm{v}_{\beta i}^{n+1} = \left(A_{\alpha\beta} - \Delta t \, B_{\alpha\beta} - \frac{\Delta t^2}{3} K_{\alpha\beta}\right)\mathrm{v}_{\beta i}^{n} + \frac{\Delta t^2}{2} G_{\alpha i} \tag{11.2.42}$$

The two-step solution scheme for (11.2.41) becomes

$$A_{\alpha\beta} X_{\beta i}^{(1)} = -\left(B_{\alpha\beta} + \frac{\Delta t^2}{2} K_{\alpha\beta}\right)\mathrm{v}_{\beta i}^{n} + \frac{\Delta t}{2} G_{\alpha i} \tag{11.2.43}$$

$$A_{\alpha\beta} X_{\beta i}^{(2)} = -\frac{\Delta t^2}{6} K_{\alpha\beta} X_{\beta i}^{(1)} \tag{11.2.44}$$

with $X_{\beta i}^{(1)}$ and $X_{\beta i}^{(2)}$ defined as in (11.2.14). Notice that, in dealing with the advection equation with diffusion, we have included the third order time derivative [see (11.2.37)] which resulted in the numerical (artificial) diffusion characterized by the second order spatial derivative in (11.2.40) or the matrix $K_{\alpha\beta}$ in (11.2.41). The presence of these terms is responsible for the stability of numerical solution.

Leapfrog Method

The leapfrog method is obtained by writing the Taylor series of v_i^{n-1} about v_i^n to the third order,

$$\mathrm{v}_i^{n-1} = \mathrm{v}_i^n - \Delta t \frac{\partial \mathrm{v}_i^n}{\partial t} + \frac{\Delta t^2}{2!}\frac{\partial^2 \mathrm{v}_i^n}{\partial t^2} - \frac{\Delta t^3}{3!}\frac{\partial^3 \mathrm{v}_i^n}{\partial t^3} \tag{11.2.45}$$

Subtracting (11.2.45) from (11.2.37) and rearranging, we obtain

$$\left(1 - \frac{\Delta t^2}{6}\bar{\mathrm{v}}_j \bar{\mathrm{v}}_k \frac{\partial^2}{\partial x_j \partial x_k}\right)\frac{\Delta \mathrm{v}_i^{n+1}}{2\Delta t} = -\bar{\mathrm{v}}_j \mathrm{v}_{i,j}^n \tag{11.2.46}$$

with $\Delta \mathrm{v}_i^{n+1} = \mathrm{v}_i^{n+1} - \mathrm{v}_i^{n-1}$. The finite element analog of (11.2.46) becomes

$$\left(A_{\alpha\beta} + \frac{\Delta t^2}{6} K_{\alpha\beta}\right)\frac{\Delta \mathrm{v}_{\beta i}^{n+1}}{2\Delta t} = -B_{\alpha\beta}\mathrm{v}_{\beta i}^{n} \tag{11.2.47}$$

The corresponding Generalized Galerkin finite element equations, neglecting the

Neumann boundary conditions, are given by

$$\left(A_{\alpha\beta} + \frac{\Delta t^2}{6} K_{\alpha\beta}\right) v_{\beta i}^{n+1} = -2\Delta t\, B_{\alpha\beta} v_{\beta i}^{n} + \left(A_{\alpha\beta} + \frac{\Delta t^2}{6} K_{\alpha\beta}\right) v_{\beta i}^{n-1} \tag{11.2.48}$$

The two-step solution scheme consists of

$$\frac{1}{2} A_{\alpha\beta} X_{\beta i}^{(1)} = -\Delta t\, B_{\alpha\beta} v_{\beta i}^{n} \tag{11.2.49}$$

$$A_{\alpha\beta} X_{\beta i}^{(2)} = -\frac{\Delta t^2}{6} K_{\alpha\beta} X_{\beta i}^{(1)} \tag{11.2.50}$$

By definition for the leapfrog method, the variables $v_{\beta i}^{n+1}$ are calculated as

$$v_{\beta i}^{n+1} = v_{\beta i}^{n-1} + 2\Delta t\left(X_{\beta i}^{(1)} + X_{\beta i}^{(2)}\right) \tag{11.2.51}$$

Thus, initially both $v_{\alpha i}^{n}$ and $v_{\alpha i}^{n-1}$ are assumed to be known and, for the next time step, $v_{\alpha i}^{n}$ becomes $v_{\alpha i}^{n-1}$.

The leapfrog scheme may be revised to involve $v_{\alpha i}^{n}$ instead of $v_{\alpha i}^{n-1}$ (11.2.51) in the incremental form. This will alter the process as follows:

$$\left(A_{\alpha\beta} + \frac{\Delta t^2}{6} K_{\alpha\beta}\right) \frac{\Delta v_{\beta i}^{n+1}}{\Delta t} = \frac{1}{\Delta t}\left[\left(-2\Delta t\, B_{\alpha\beta} - A_{\alpha\beta} - \frac{\Delta t^2}{6} K_{\alpha\beta}\right) v_{\beta i}^{n} + \left(A_{\alpha\beta} + \frac{\Delta t^2}{6} K_{\alpha\beta}\right) v_{\beta i}^{n-1}\right] \tag{11.2.52}$$

The two-step solution scheme is now in the form

$$A_{\alpha\beta} X_{\beta i}^{(1)} = \frac{1}{\Delta t}\left[\left(-2\Delta t\, B_{\alpha\beta} - A_{\alpha\beta} - \frac{\Delta t^2}{6} K_{\alpha\beta}\right) v_{\beta i}^{n} + \left(A_{\alpha\beta} + \frac{\Delta t^2}{6} K_{\alpha\beta}\right) v_{\beta i}^{n-1}\right] \tag{11.2.53}$$

$$A_{\alpha\beta} X_{\beta i}^{(2)} = -\frac{\Delta t^2}{6} K_{\alpha\beta} X_{\beta i}^{(1)} \tag{11.2.54}$$

This will then allow the variables $v_{\alpha i}^{n+1}$ to be calculated as

$$v_{\alpha i}^{n+1} = v_{\alpha i}^{n} + \Delta t\left(X_{\alpha i}^{(1)} + X_{\alpha i}^{(2)}\right) \tag{11.2.55}$$

Crank-Nicolson Method

The Crank-Nicolson method is obtained by writing the Taylor series of v_i^n about v_i^{n+1} to the third order:

$$v_i^n = v_i^{n+1} - \Delta t \frac{\partial v_i^{n+1}}{\partial t} + \frac{\Delta t^2}{2!}\frac{\partial^2 v_i^{n+1}}{\partial t^2} - \frac{\Delta t^3}{3!}\frac{\partial^3 v_i^{n+1}}{\partial t^3} \tag{11.2.56}$$

Making use of the relation

$$\frac{1}{2}\left(\frac{\partial v_i^{n+1}}{\partial t} + \frac{\partial v_i^{n}}{\partial t}\right) = \frac{v_i^{n+1} - v_i^{n}}{\Delta t} \tag{11.2.57}$$

and in view of (11.2.35) and (11.2.36), and subtracting (11.2.56) from (11.2.37), we arrive at

$$\left(1 - \frac{\Delta t^2}{6}\overline{\mathrm{v}}_j\overline{\mathrm{v}}_k\frac{\partial^2}{\partial x_j \partial x_k}\right)\frac{\Delta \mathrm{v}_i^{n+1}}{\Delta t} = -\frac{\overline{\mathrm{v}}_j}{2}\left(\frac{\partial \mathrm{v}_i^n}{\partial x_j} + \frac{\partial \mathrm{v}_i^{n+1}}{\partial x_j}\right) + \frac{\Delta t}{4}\overline{\mathrm{v}}_j\overline{\mathrm{v}}_k\left(\frac{\partial^2 \mathrm{v}_i^n}{\partial x_j \partial x_k} - \frac{\partial^2 \mathrm{v}_i^{n+1}}{\partial x_j \partial x_k}\right) \quad (11.2.58)$$

$$\left(A_{\alpha\beta} - \frac{\Delta t^2}{12}K_{\alpha\beta} + \frac{\Delta t}{2}B_{\alpha\beta}\right)\mathrm{v}_{\beta i}^{n+1} = \left(A_{\alpha\beta} + \frac{\Delta t^2}{12}K_{\alpha\beta} - \frac{\Delta t}{2}B_{\alpha\beta}\right)\mathrm{v}_{\beta i}^n \quad (11.2.59)$$

This is the implicit Crank-Nicolson scheme. However, we may convert (11.2.59) into a two-step explicit scheme as follows:

(a) Rewrite the finite element equation in the time-step difference form

$$\left(A_{\alpha\beta} - \frac{\Delta t^2}{12}K_{\alpha\beta} + \frac{\Delta t}{2}B_{\alpha\beta}\right)\frac{\Delta \mathrm{v}_{\beta i}^{n+1}}{\Delta t} = -B_{\alpha\beta}\mathrm{v}_{\beta i}^n \quad (11.2.60)$$

(b) The two-step explicit form is written using the procedure described earlier,

$$A_{\alpha\beta}X_{\beta i}^{(1)} = -B_{\alpha\beta}\mathrm{v}_{\beta i}^n \quad (11.2.61)$$

$$A_{\alpha\beta}X_{\beta i}^{(2)} = \left(\frac{\Delta t^2}{12}K_{\alpha\beta} - \frac{\Delta t}{2}B_{\alpha\beta}\right)X_{\beta i}^{(1)} \quad (11.2.62)$$

Remarks: Appropriate choices of the finite element test functions for W_α, Φ_α, $\hat{\Phi}_m$, and $W(\xi)$ enable the finite element analogs of Euler (11.2.42), leapfrog (11.2.48), and Crank-Nicolson (11.2.59) to be generated without the Taylor series expansion. Other forms of finite difference schemes may be generated by adding discontinuous functions to W_α, which we shall elaborate in Section 11.3.

11.2.4 CONVERSION OF IMPLICIT SCHEME INTO EXPLICIT SCHEME

It follows from the approaches discussed in previous sections for the explicit schemes that it is possible to convert all implicit schemes into explicit schemes. Consider the generalized temporal-spatial finite element equations written in matrix form.

$$(A+B)\mathrm{v}^{n+1} = (A+C)\mathrm{v}^n + (A+D)\mathrm{v}^{n-1} + (A+E)\mathrm{v}^{n-2} + \cdots - \Delta t H \quad (11.2.63)$$

where $B = B_1 + B_2 + \cdots$, $C = C_1 + C_2 + \cdots$, $D = D_1 + D_2 + \cdots$, $E = E_1 + E_2 + \cdots$, etc. Note that various forms of (11.2.63) result from unlimited choices of functions in Φ_α, $\hat{\Phi}_m$, and $\hat{W}(\xi)$ in Section 11.2.

The conversion process consists of the following steps:

(a) Write (11.2.63) in an incremental form,

$$(A+B)\frac{\Delta \mathrm{v}^{n+1}}{\Delta t} = [(A+C) - (A+B)]\frac{\mathrm{v}^n}{\Delta t} + (A+D)\frac{\mathrm{v}^{n-1}}{\Delta t} + (A+E)\frac{\mathrm{v}^{n-2}}{\Delta t} + \cdots - H \quad (11.2.64)$$

where

$$\Delta v^{n+1} = v^{n+1} - v^n \tag{11.2.65}$$

(b) Step 1 is constructed by rewriting (11.2.64) with all terms other than the mass matrix A removed from the left-hand side of (11.2.64) and designating Δv^{n+1} as $\Delta v^{(1)}$, called the first increment,

$$AX^{(1)} = [(A+C) - (A+B)]\frac{v^n}{\Delta t} + (A+D)\frac{v^{n-1}}{\Delta t} + \cdots - H \tag{11.2.66}$$

where

$$X^{(1)} = \frac{\Delta v^{(1)}}{\Delta t}$$

(c) Step 2 is constructed by setting the product of the mass matrix and the second increment $X^{(2)}$, which is equated to the variant of the first increment,

$$AX^{(2)} = [A - (A+B)]X^{(1)} \tag{11.2.67}$$

where

$$X^{(2)} = \frac{\Delta v^{n+1} - \Delta v^{(1)}}{\Delta t} \tag{11.2.68}$$

(d) The variable v^{n+1} is given by

$$v^{n+1} = v^n + \Delta t\left(X^{(1)} + X^{(2)}\right) \tag{11.2.69}$$

A glance at (11.2.69) reveals that, for a steady-state condition, $t \to \infty$, and $v = v^{n+1} = v^n = v^{n-1} = v^{n-2} = \cdots$, we obtain

$$(B + C + D + E + \cdots)v = H \tag{11.2.70}$$

Thus, it is expected that a steady-state solution would result as recursive calculations are carried out consecutively.

11.2.5 TAYLOR-GALERKIN METHODS FOR NONLINEAR BURGERS' EQUATIONS

Let us consider the nonlinear Burgers' equations of the form

$$\frac{\partial v_i}{\partial t} + v_j v_{i,j} - \nu v_{i,jj} = f_i \tag{11.2.71}$$

The Taylor series expansion of (11.2.71) as given in (11.2.9) without the third order derivative term becomes

$$\Delta v_i^{n+1} = -\Delta t(v_j v_{i,j} - \nu v_{i,jj} - f_i)^n + \frac{\Delta t^2}{2}\left[v_k \frac{\partial}{\partial x_k}(v_j v_{i,j} - \nu v_{i,jj} - f_i)\right.$$
$$\left. + v_{i,j}(v_k v_{j,k} - \nu v_{j,kk} - f_j) - \nu \frac{\partial^2}{\partial x_j \partial x_j}(v_k v_{i,k} - \nu v_{i,kk} - f_i) + \frac{\partial f_i}{\partial t}\right]^n \tag{11.2.72}$$

from which the original differential equation can be recovered in the form,

$$\frac{\partial v_i}{\partial t} + v_j v_{i,j} - \nu v_{i,jj} - f_i = S_i \tag{11.2.73}$$

where

$$S_i = \frac{\Delta t}{2}\left[\mathrm{v}_k \frac{\partial}{\partial \mathrm{x}_k}(\mathrm{v}_j \mathrm{v}_{i,j} - \nu \mathrm{v}_{i,jj} - f_i)\right] \tag{11.2.74}$$

with higher order derivative terms and products of the gradients in (11.2.72) being neglected. It is clear that the right-hand side of (11.2.74) appears as numerical diffusion.

Applying the Galerkin integral to the right-hand side of (11.2.74) and integrating by parts, we obtain

$$\int_\Omega \Phi_\alpha S_i \, d\Omega = -\frac{\Delta t}{2}\int_\Omega \mathrm{v}_k \mathrm{v}_j \Phi_{\alpha,k}\Phi_{\beta,j} d\Omega \mathrm{v}_{\beta i} \tag{11.2.75}$$

where all terms other than the convective terms are negligible in practical applications. Thus, the numerical diffusion matrix is identified as

$$C_{\alpha\beta} = \int_\Omega \overline{\nu}_{kj}\Phi_{\alpha,k}\Phi_{\beta,j} d\Omega \tag{11.2.76}$$

with the numerical viscosity,

$$\overline{\nu}_{kj} = \frac{\Delta t}{2}\mathrm{v}_k \mathrm{v}_j \tag{11.2.77}$$

It is interesting to note that, using an entirely different approach, the numerical diffusion similar to (11.2.76) and (11.2.77) arises in the generalized Petrov-Galerkin (GPG) methods to be presented in Sections 11.3 and 11.4. More general treatments of TGM will be covered in Section 13.2.

11.3 NUMERICAL DIFFUSION TEST FUNCTIONS

In GGM described in Section 11.2, various degrees of polynomials (linear, quadratic, cubic, etc.) may be adopted for desired accuracy of solution. However, in convection-dominated problems, an adequate amount of numerical diffusion or artificial viscosity is required for numerical stability. To this end, the so-called streamline-upwind Petrov-Galerkin (SUPG) method [Heinrich et al., 1977; Hughes and Brooks, 1982] has been successfully used. In this case, the local finite element test functions consist of standard Galerkin test functions plus numerical diffusion test functions. There are many forms of numerical diffusion test functions as reported by Hughes and his co-workers during the 1980s. A similar approach is referred to as the streamline diffusion method (SDM) by Johnson [1987].

Computational stability is provided effectively through various forms of SUPG, SDM, or other similar strategies. All of these approaches are nonstandard Galerkin methods and, for simplicity, they may be combined into a single name "Generalized Petrov-Galerkin (GPG) methods. The concept of GPG for the one-dimensional Burgers' equation will be introduced first in order to identify a one-dimensional numerical diffusion test function which provides the numerical stability, followed by multi-dimensional numerical diffusion test functions representing the streamline diffusion and discontinuity-capturing schemes.

11.3.1 DERIVATION OF NUMERICAL DIFFUSION TEST FUNCTIONS

The concept of streamline diffusion began with the backward (often called upwinding) finite difference scheme for the convection-diffusion equation first given by Spalding [1972]. The idea is to introduce the numerical diffusion in the direction of flow or along the streamline parallel to the velocity in order to obtain stable solutions. In the following, we use the convection-diffusion equation to demonstrate the concept of streamline upwinding or streamline diffusion. Our objective here is to prove that numerical stability can be achieved by test functions written in the form,

$$W_N^{(e)} = \Phi_N^{(e)} + \Psi_N^{(e)} \tag{11.3.1}$$

where $W_N^{(e)}$ represents the generalized Petrov-Galerkin test functions which are the sum of the standard Galerkin test function $\Phi_N^{(e)}$ and the numerical diffusion test function $\Psi_N^{(e)}$. The numerical diffusion test function $\Psi_N^{(e)}$ in (11.3.1) is intended for adding numerical diffusion practiced in the finite difference literature. However, in the sequel, it will be shown that the derivation of numerical diffusion test functions involves significant physical aspects of convection-dominated flows.

To elucidate the argument involved in this approach, we look at the unsteady convection equation of the form

$$\frac{\partial u}{\partial t} + a\frac{\partial u}{\partial x} = 0 \tag{11.3.2}$$

Substituting (11.3.2) into Taylor series of the type (11.2.9), we obtain

$$u^{n+1} = u^n + \Delta t\left(-a\frac{\partial u^n}{\partial x}\right) + \frac{\Delta t^2}{2}\left(a^2\frac{\partial^2 u^n}{\partial x^2}\right) \tag{11.3.3}$$

If $u^{n+1} = u^n$ at steady-state, we may set $a\Delta t = C\Delta x$, where C is the nondimensional artificial viscosity (equal to Courant number for $a = u$, or $C = u\Delta t/\Delta x$), and rewrite (11.3.3) in the form

$$a\left(\frac{\partial u}{\partial x} - \frac{C\Delta x}{2}\frac{\partial^2 u}{\partial x^2}\right) = 0 \tag{11.3.4}$$

in which the second term of the left-hand side of (11.3.4) represents the numerical diffusion, equivalent to the artificial viscosity. Denoting $\alpha = C/2$ and $h = \Delta x$ as the nondimensional numerical diffusion parameter and the mesh parameter, respectively, we may construct the following inner product:

$$\int \Phi_N^{(e)} a\left(\frac{\partial u}{\partial x} - \alpha h\frac{\partial^2 u}{\partial x^2}\right) dx = 0 \tag{11.3.5}$$

Integrating (11.3.5) by parts, we obtain

$$\int\left(\Phi_N^{(e)} + \alpha h\frac{\partial \Phi_N^{(e)}}{\partial x}\right) a\frac{\partial u}{\partial x} dx = \overset{*}{\Phi}_N a\alpha h\frac{\partial u}{\partial x}$$

where the integral on the left-hand side is known as the Petrov-Galerkin integral. For

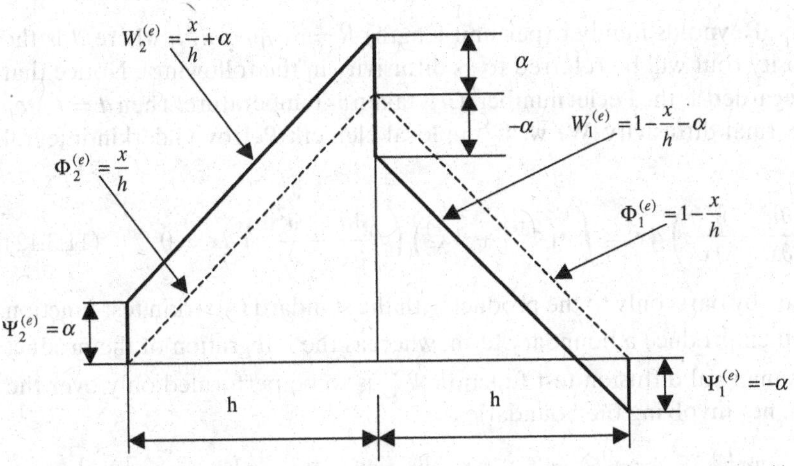

Figure 11.3.1 Linear generalized test functions for one-dimensional element $W_N^{(e)}$ with $\Psi_N^{(e)} = \alpha h(\partial\Phi_N^{(e)}/\partial x)$ constant within the element, discontinuous at boundaries.

Dirichlet problems $\overset{*}{\Phi}_N = 0$, we have

$$\int W_N^{(e)}\left(a\frac{\partial u}{\partial x}\right) dx = 0 \tag{11.3.6}$$

with

$$W_N^{(e)} = \Phi_N^{(e)} + \Psi_N^{(e)} \tag{11.3.7}$$

$$\Phi_N^{(e)} = \left(1 - \frac{x}{h}, \frac{x}{h}\right) \tag{11.3.8}$$

$$\Psi_N^{(e)} = \alpha h \frac{\partial \Phi_N^{(e)}}{\partial x} \tag{11.3.9}$$

The foregoing process indicates that the numerical diffusion may be applied to the convective term in (11.3.2) through (11.3.6), with the explicit form given by (11.3.9) representing the numerical diffusion test function. This is a variational weak form as constructed by the inner product of the numerical diffusion test function and the convection term.

Substituting (11.3.8) into (11.3.9), we arrive at

$$\Psi_N^{(e)} = [-\alpha, \quad \alpha] \tag{11.3.10}$$

which indicates that the numerical test function is constant within an element, equal to one-half of the Courant number, but discontinuous at boundaries, as shown in Figure 11.3.1. It will be shown that one-dimensional numerical diffusion test functions given in (11.3.9) arise from the two-dimensional numerical diffusion test functions to be discussed in Section 11.3.2.

11.3.2 STABILITY AND ACCURACY OF NUMERICAL DIFFUSION TEST FUNCTIONS

Let us examine the convection-diffusion equation,

$$\hat{R}\frac{\partial u}{\partial x} - \frac{\partial^2 u}{\partial x^2} = 0 \tag{11.3.11}$$

with $\hat{R}$ being the Reynolds number (per unit length) $\hat{R} = \rho u/\mu = u/d$, where d is the kinematic viscosity (but will be referred to as diffusivity in the following). Notice that $\hat{R} = \rho c_p u/k$ is regarded as the Peclet number if u is taken as temperature. Then $d = k/\rho c_p$ becomes the thermal diffusivity. We write the local element Petrov-Galerkin integral for (11.3.11) as

$$\int_0^h W_N^{(e)} \left(\hat{R}\frac{\partial u}{\partial x} - \frac{\partial^2 u}{\partial x^2} \right) dx = \int_0^h \left(\Phi_N^{(e)} + \Psi_N^{(e)} \right) \left(\hat{R}\frac{\partial u}{\partial x} - \frac{\partial^2 u}{\partial x^2} \right) dx = 0 \tag{11.3.12}$$

Apply integration by parts only to the product with the standard Galerkin test function $\Phi_N^{(e)}$ which will then produce a boundary term, whereas the integration of the product term with the numerical diffusion test function $\Psi_N^{(e)}$ is to be performed only over the interior domain, not involving the boundaries.

$$\left\{ \int_0^h \left[\hat{R}\left(\Phi_N^{(e)} \frac{\partial \Phi_M^{(e)}}{\partial x} + \alpha h \frac{\partial \Phi_N^{(e)}}{\partial x}\frac{\partial \Phi_M^{(e)}}{\partial x} \right) + \frac{\partial \Phi_N^{(e)}}{\partial x}\frac{\partial \Phi_M^{(e)}}{\partial x} - \alpha h \frac{\partial \Phi_N^{(e)}}{\partial x}\frac{\partial^2 \Phi_M^{(e)}}{\partial x^2} \right] dx \right\} u_M^{(e)}$$
$$= \overset{*}{\Phi}{}_N^{(e)} \frac{\partial u}{\partial x}\bigg|_0^h \tag{11.3.13a}$$

If linear trial functions are used, then the second derivative term vanishes, so that we have

$$\left(B_{NM}^{(e)} + C_{NM}^{(e)} \right) u_M^{(e)} + K_{NM}^{(e)} u_M^{(e)} = G_N^{(e)} \tag{11.3.13b}$$

where

$$B_{NM}^{(e)} = \int_0^h \hat{R}\Phi_N^{(e)} \frac{\partial \Phi_M^{(e)}}{\partial x} dx$$

is the standard convection matrix and

$$C_{NM}^{(e)} = \int_0^h \hat{R}\alpha h \frac{\partial \Phi_N^{(e)}}{\partial x}\frac{\partial \Phi_M^{(e)}}{\partial x} dx$$

represents the numerical diffusion matrix implying the numerical diffusion arising from the convection term. The last integral term $K_{NM}^{(e)}$ is identified as the physical diffusion matrix.

$$K_{NM}^{(e)} = \int_0^h \frac{\partial \Phi_N^{(e)}}{\partial x}\frac{\partial \Phi_M^{(e)}}{\partial x} dx$$

Evaluating these integrals, we obtain

$$B_{NM}^{(e)} + C_{NM}^{(e)} = \frac{\hat{R}}{2}\begin{bmatrix} -1+2\alpha & 1-2\alpha \\ -1-2\alpha & 1+2\alpha \end{bmatrix}$$

$$K_{NM}^{(e)} = \frac{1}{h}\begin{bmatrix} 1 & -1 \\ -1 & 1 \end{bmatrix}$$

Consider a two-element system with nodes at $i-1$, i, and $i+1$ and the global form of (11.3.13). Expanding the global equation corresponding to the node at i and assuming that the Neumann boundary conditions are unspecified ($\overset{*}{\Phi}{}_N^{(e)} = 0$), we obtain

$$\left[1+\frac{R}{2}(2\alpha+1)\right]u_{i-1}-(2+2R\alpha)\,u_i+\left[1+\frac{R}{2}(2\alpha-1)\right]u_{i+1}=0 \tag{11.3.14}$$

where R is the local Reynolds number, $R=\hat{R}h$. Equation (11.3.14) represents the forward, central, and backward finite difference equations for $\alpha=-1/2$, $\alpha=0$, and $\alpha=1/2$, respectively. The backward difference form ($\alpha=1/2$) given by

$$\hat{R}\frac{u_i-u_{i-1}}{h}-\frac{u_{i+1}-2u_i+u_{i-1}}{h^2}=0 \tag{11.3.15a}$$

can be modified by transforming the convection term into the central difference form to identify the numerical diffusion with the coefficient $\hat{R}h/2$,

$$\hat{R}\left(\frac{u_{i+1}-u_{i-1}}{2h}\right)-\left(\frac{\hat{R}h}{2}+1\right)\left(\frac{u_{i+1}-2u_i+u_{i-1}}{h^2}\right)=0 \tag{11.3.15b}$$

This is equivalent to the differential equation

$$\hat{R}\frac{\partial u}{\partial x}-\hat{\alpha}\frac{\partial^2 u}{\partial x^2}-\frac{\partial^2 u}{\partial x^2}=0 \tag{11.3.16}$$

with $\hat{\alpha}=\hat{R}h/2$ being the coefficient of numerical viscosity and $\hat{\alpha}(\partial^2 u/\partial x^2)$ representing the effect of numerical diffusion. We say that the effect of numerical diffusion is built into this equation if the backward difference is used. We may consider $\hat{\alpha}$ as being equivalent to the artificial viscosity.

To obtain the condition for stability (11.3.14), we proceed as follows: Let $G=1+R\alpha$ and $H=R/2$. Rewrite (11.3.14) in the form

$$(G-H)u_{i+1}-2Gu_i+(G+H)u_{i-1}=0 \tag{11.3.17}$$

where we assume the relations at the nodes $i+1$, i, and $i-1$ as

$$u_i=c\phi^i,\qquad u_{i+1}=c\phi^{i+1},\qquad u_{i-1}=c\phi^{i-1} \tag{11.3.18a,b,c}$$

Substituting the above into (11.3.17) yields

$$(G-H)\,\phi^{i+1}-2G\phi^i+(G+H)\,\phi^{i-1}=0$$

For $i=1$, we obtain the quadratic equation

$$(G-H)\phi^2-2G\phi+(G+H)=0$$

Solving for ϕ, we arrive at two values of ϕ

$$\phi=\begin{cases}1\\ \dfrac{G+H}{G-H}\end{cases}$$

These results call for two constants in (11.3.18).

Now we revise the relation in (11.3.18a) in the form

$$u_i=c_1+c_2\left(\frac{G+H}{G-H}\right)^i=c_1+c_2\left[\frac{1+\frac{R}{2}(2\alpha+1)}{1+\frac{R}{2}(2\alpha-1)}\right]^i \tag{11.3.19}$$

For stability, the denominator of the c_2 term must be positive,

$$G - H > 0$$

or

$$1 + \frac{R}{2}(2\alpha - 1) > 0 \tag{11.3.20a}$$

which provides the stability criteria

$$\begin{cases} \alpha = 0 & \text{if } R < 2 \\ \alpha \geq \dfrac{1}{2} - \dfrac{1}{R} & \text{if } R \geq 2 \end{cases} \tag{11.3.20b,c}$$

It is clear that the forward difference with $\alpha = -1/2$ (11.3.20a) becomes unconditionally unstable for $R > 1$, whereas the central difference ($\alpha = 0$) is conditionally stable and the backward difference ($\alpha = 1/2$) provides an unconditional stability. For accuracy, we set the exact solution as

$$u = c_1 + c_2 e^{\hat{R}x}$$

which, for $x = hi$, becomes

$$u_i = c_1 + c_2 e^{Ri} \tag{11.3.21}$$

Setting (11.3.21) equal to (11.3.19), we obtain the relationship

$$\left[\frac{1 + \frac{R}{2}(2\alpha + 1)}{1 + \frac{R}{2}(2\alpha - 1)}\right]^i = e^{Ri}$$

Taking a natural logarithm of the above leads to

$$\ln\left(\frac{G+H}{G-H}\right) = 2\coth^{-1}\left(\frac{G}{H}\right) = 2\coth^{-1}\left(\frac{1 + R\alpha}{R/2}\right) = R$$

from which we obtain

$$2\alpha = \coth\left(\frac{R}{2}\right) - \frac{2}{R} \tag{11.3.22}$$

with

$$\alpha = \frac{1}{2}C = \frac{1}{2}\overline{\alpha} \tag{11.3.23}$$

This is the criterion for accuracy. Here, the one-dimensional numerical diffusion parameter α, which assures the accuracy, is found to be a function of the local Reynolds number. It should be noted that the value of α is one-half of that in Heinrich et al. [1977], and $\overline{\alpha} = C$, called the effective numerical diffusion parameter, is indeed the Courant number.

Substituting (11.3.23) into (11.3.22) leads to

$$\overline{\alpha} = \coth H - \frac{1}{H} \tag{11.3.24}$$

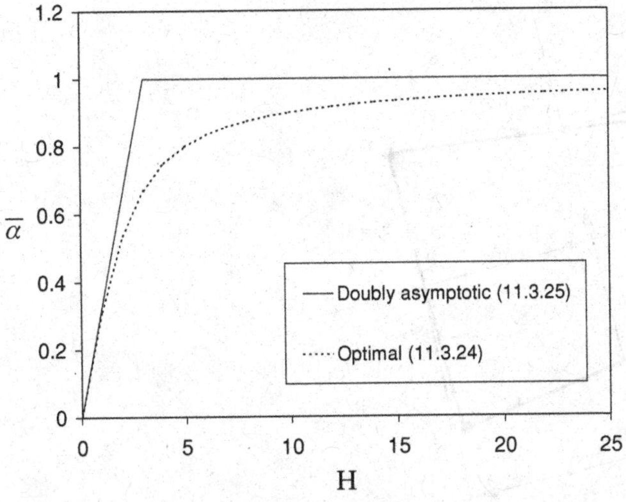

Figure 11.3.2 Effective numerical diffusivity $\overline{\alpha}$.

It can be shown that, expanding $\coth H$ in infinite series and retaining terms of fourth order accuracy in H (doubly asymptotic approximation) results in

$$\overline{\alpha} = H/3, \quad \text{if } -3 \leq H \leq 3 \tag{11.3.25a}$$

$$\overline{\alpha} = \operatorname{sgn} H, \quad \text{if } |H| > 3 \tag{11.3.25b}$$

The values of $\overline{\alpha}$ determined by (11.3.20), (11.3.24), and (11.3.25) are referred to as the critical value, optimal value, and higher order value, respectively (Figure 11.3.2) [Heinrich et al., 1977; Brooks and Hughes, 1982]. It is seen that the doubly asymptotic approximation (11.3.25) is the simpler and practical approach.

It follows from these observations that, for two-dimensional isoparametric elements, the numerical diffusion parameters α_ξ and α_η are defined as (Figure 11.3.3)

$$\alpha_\xi = \frac{1}{2}\overline{\alpha}_\xi \tag{11.3.26a}$$

$$\alpha_\eta = \frac{1}{2}\overline{\alpha}_\eta \tag{11.3.26b}$$

with the two-dimensional effective numerical diffusion parameters, $\overline{\alpha}_\xi$ and $\overline{\alpha}_\eta$, defined as

$$\overline{\alpha}_\xi = \coth\left(\frac{R_\xi}{2}\right) - \frac{2}{R_\xi} \tag{11.3.27a}$$

$$\overline{\alpha}_\eta = \coth\left(\frac{R_\eta}{2}\right) - \frac{2}{R_\eta} \tag{11.3.27b}$$

where the local Reynolds numbers in the ξ and η directions are of the form

$$R_\xi = \frac{v_\xi h_\xi}{d}, \qquad R_\eta = \frac{v_\eta h_\eta}{d}$$

For multidimensional convection-dominated problems, the directional properties of velocity are expected to play a key role. The numerical diffusion must be provided in the direction of flow or along the streamlines parallel to the velocity in both steady and

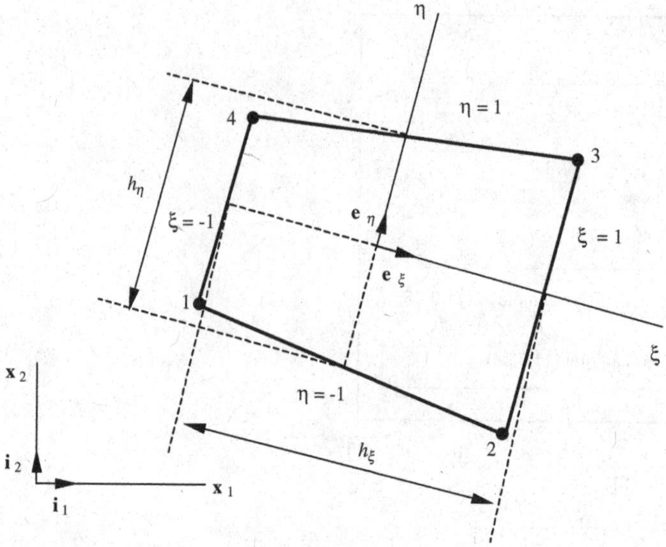

Figure 11.3.3 Four node isoparametric element; $\mathbf{e}_\xi$ and $\mathbf{e}_\eta$ are the unit vectors along the ξ and η axes, respectively.

time-dependent problems as well. First of all, the gradient of the standard test function in (11.3.9) is of a vector quantity. This will require that the dot product of the gradient of the standard test function $\Phi_N^{(e)}$ with another vector quantity such as the velocity vector along the streamlines be formed such that a scalar function arises at each local node N,

$$\mathbf{v} \cdot \nabla \Phi_N^{(e)} = \mathrm{v}_i \Phi_{N,i}^{(e)} \quad (i = 1, 2) \tag{11.3.28}$$

Thus, the quantity αh in (11.3.9) for the one-dimensional case must be altered to accommodate the appropriate dimensional properties in (11.3.28) which call for a scalar, say τ, such that

$$\Psi_N^{(e)} = \tau \mathrm{v}_i \Phi_{N,i}^{(e)} \tag{11.3.29}$$

with τ being the numerical diffusion factor having dimensions of time (often called the intrinsic time scale),

$$\tau = \frac{1}{2}(\alpha_\xi h_\xi \mathrm{v}_\xi + \alpha_\eta h_\eta \mathrm{v}_\eta)/S = \frac{1}{4}(\overline{\alpha}_\xi h_\xi \mathrm{v}_\xi + \overline{\alpha}_\eta h_\eta \mathrm{v}_\eta)/S = \frac{1}{\sqrt{16}}(\overline{\alpha}_\xi h_\xi \mathrm{v}_\xi + \overline{\alpha}_\eta h_\eta \mathrm{v}_\eta)/S \tag{11.3.30}$$

where $S = \mathbf{v} \cdot \mathbf{v}$, and $\overline{\alpha}_\xi = \coth(R_\xi/2) - 2/R_\xi$, etc. The coefficient $1/\sqrt{16}$ for the numerical diffusion factor τ in (11.3.30) disagrees with an arbitrary value of $1/\sqrt{15}$ adopted by Raymond and Garder [1976], and subsequently Brooks and Hughes [1982] as determined from the numerical experimentation for unsteady flows. The derivation demonstrated here, however, is based on the definition of Courant number and the criterion for accuracy which leads to $1/\sqrt{16}$ instead of $1/\sqrt{15}$. For the purpose of generality, the solution schemes employing the numerical diffusion test function given by (11.3.29) are termed "generalized Petrov-Galerkin (GPG)" instead of SUPG. The unfortunate choice of the term "SUPG" for various reasons was discussed in Hughes [1987]. The SUPG methods as referred to today imply far beyond the classical upwind methods [Spalding, 1972] or classical Petrov-Galerkin methods [Mikhlin, 1964] so that more general identification appears to be in order. Thus, it is suggested that the term "generalized Petrov-Galerkin (GPG)" may be a reasonable compromise.

For two-dimensional elements with isoparametric coordinates (Figure 11.3.3), we express the velocity components as

$$\mathrm{v}_\xi = \mathbf{v} \cdot \mathbf{e}_\xi, \qquad \mathrm{v}_\eta = \mathbf{v} \cdot \mathbf{e}_\eta$$

where the isoparametric unit vectors $\mathbf{e}_\xi$ and $\mathbf{e}_\eta$ are given by

$$\mathbf{e}_\xi = \frac{1}{\sqrt{J_\xi}} \frac{\partial x_i}{\partial \xi} \mathbf{i}_i, \quad \mathbf{e}_\eta = \frac{1}{\sqrt{J_\eta}} \frac{\partial x_i}{\partial \eta} \mathbf{i}_i, \quad J_\xi = \left(\frac{\partial x}{\partial \xi}\right)^2 + \left(\frac{\partial y}{\partial \xi}\right)^2, \ J_\eta = \left(\frac{\partial x}{\partial \eta}\right)^2 + \left(\frac{\partial y}{\partial \eta}\right)^2$$

It follows from (11.3.27) that the two-dimensional numerical diffusion test function reduces to that of one dimension given by (11.3.9):

$$\dot{\Psi}_N^{(e)} = \tau \mathrm{v}_1 \Phi_{N,1}^{(e)} = \left(\frac{\alpha h u}{u^2}\right) u \frac{\partial \Phi_N^{(e)}}{\partial x} = \alpha h \frac{\partial \Phi_N^{(e)}}{\partial x} \tag{11.3.31}$$

which establishes the complete link between the one- and two-dimensional aspects of the numerical diffusion test functions. It is interesting to note that, in due course of derivation of the one-dimensional numerical diffusion test function (11.3.9), the notion of time scale for the numerical diffusion factor τ did not arise, but is now taken into account as the numerical diffusion must be applied in the direction of flow with velocity specified in multidimensional cases.

Due to the fact that the gradient $\nabla \Phi_N^{(e)}$ is included in $\Psi_N^{(e)}$, it is clear that the use of the generalized test functions (11.3.1) brings the numerical diffusion automatically into the formulation. This is equivalent to the retention of artificial viscosity terms in FDM.

Using the similar procedure, the test functions for 3-D problems (with isoparametric coordinates ξ, η, and ζ) can be obtained. The three-dimensional test function may still be written in the general form (11.3.29).

$$\Psi_N^{(e)} = \tau \mathrm{v}_i \Phi_{N,i}^{(e)}, \quad (i = 1, 2, 3) \tag{11.3.32}$$

where

$$\tau = \frac{1}{6}(\overline{\alpha}_\xi h_\xi \mathrm{v}_\xi + \overline{\alpha}_\eta h_\eta \mathrm{v}_\eta + \overline{\alpha}_\zeta h_\zeta \mathrm{v}_\zeta)/S \tag{11.3.33}$$

$$\overline{\alpha}_\zeta = \coth\left(\frac{R_\zeta}{2}\right) - \frac{2}{R_\zeta}, \qquad R_\zeta = \frac{\mathrm{v}_\zeta h_\zeta}{d}, \qquad S = u^2 + \mathrm{v}^2 + w^2$$

Thus

$$\Psi_N^{(e)} = \tau \left(u \frac{\partial \Phi_N^{(e)}}{\partial x} + \mathrm{v} \frac{\partial \Phi_N^{(e)}}{\partial y} + w \frac{\partial \Phi_N^{(e)}}{\partial z} \right)$$

Once again, it should be emphasized that the numerical diffusion is activated along the stream line direction, which provides numerical stability. However, it has been observed that, as the convection domination becomes significant, it is not possible to eliminate entirely some numerical oscillations. We require additional measures in order to resolve numerical stability, known as the discontinuity-capturing scheme, which is discussed next.

11.3.3 DISCONTINUITY-CAPTURING SCHEME

In the presence of very high gradients of a variable such as in shock waves, one may apply numerical diffusion parallel to the direction of velocity gradients in addition to the streamline direction. Hughes et al. [1986] and Johnson [1987] investigated the so-called "discontinuity-capturing scheme" (DCS) and demonstrated improvements over the case of a numerical diffusion function applied only along the streamline direction. To be consistent with the present notations adopted in this book, a slightly modified version of DCS is presented below.

The basic idea is that the numerical diffusion is applied not only in the direction of velocity along the streamline, $v^i = v_i^{(a)}$, but also along the direction, $v_i = v_i^{(b)}$, parallel to the velocity gradients directed toward acceleration as shown in Figure 11.3.4. Note that $v_i^{(b)}$ is the projection of $v_i^{(a)}$ and thus the effect of the discontinuity-capturing will be significant if the angle θ becomes small, which represents very sharp surface discontinuities such as in shock waves.

To implement this scheme, we consider that the numerical diffusion test functions consist of the sum of the streamline numerical diffusion test function, $\Psi_N^{(a)}$, and the gradient numerical diffusion test function, $\Psi_N^{(b)}$,

$$\Psi_N^{(e)} = \Psi_N^{(a)} + \Psi_N^{(b)} \tag{11.3.34}$$

with

$$\Psi_N^{(a)} = \tau v_i^{(a)} \Phi_{N,i}^{(e)} = \tau v_i \Phi_{N,i}^{(e)} \quad \text{for streamline diffusion} \tag{11.3.35a}$$

$$\Psi_N^{(b)} = \tau v_i^{(b)} \Phi_{N,i}^{(e)} = \tau^{(b)} v_i \Phi_{N,i}^{(e)} \quad \text{for discontinuity-capturing} \tag{11.3.35b}$$

where $v_i^{(a)} = v_i$, and $v_i^{(b)}$ is the projection of $v_i^{(a)}$ parallel to velocity gradients directed toward acceleration per unit mass, A_j,

$$A_j = v_{j,k} v_k \tag{11.3.36}$$

$$v_i^{(b)} = v_i^{(a)} \cos\theta = v_i v_j A_j / \gamma \tag{11.3.37}$$

$$\tau^{(b)} = \tau v_j A_j / \gamma, \quad \text{discontinuity-capturing factor} \tag{11.3.38}$$

with $\gamma = |v_j||A_j|$. Thus, it is seen that the discontinuity-capturing scheme is simply to add an extra numerical diffusion test function applied parallel to the velocity gradients directed toward acceleration. For distorted elements, we may encounter $\tau^{(b)} - \tau$ to be

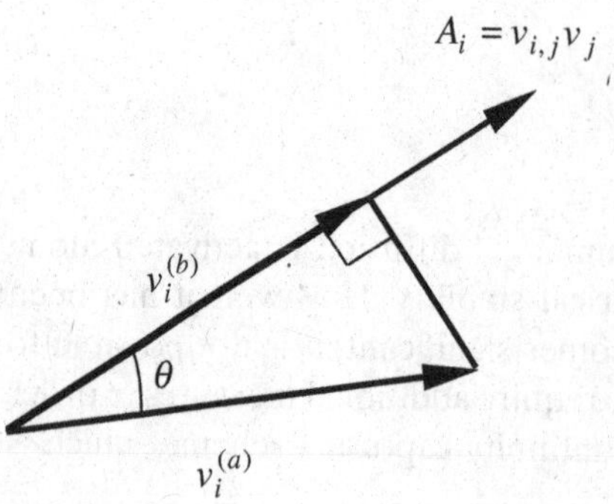

Figure 11.3.4 Discontinuity capturing scheme, $v_i^{(a)} = v_i$, $v_i^{(b)} = v_i^{(a)} \cos\theta = v_i v_j A_j / |v_j||A_j|$

negative. In this case we choose

$$\tau^{(b)} - \tau = \max\left(0, \quad \tau^{(b)} - \tau\right) \tag{11.3.39}$$

so that $\tau^{(b)} - \tau$ always remains positive. Further details are found in Hughes et al. [1986].

11.4 GENERALIZED PETROV-GALERKIN (GPG) METHODS

11.4.1 GENERALIZED PETROV-GALERKIN METHODS FOR UNSTEADY PROBLEMS

For illustration, let us consider the Burgers' equation in the form,

$$R_i = \frac{\partial v_i}{\partial t} + v_{i,j}v_j - \nu v_{i,jj} - f_i = 0$$

The finite element formulation of the generalized Petrov-Galerkin (GPG) methods using the numerical diffusion test functions projected on the discontinuous temporal test function or DST as given in (8.2.41) or (10.2.5) is written in the form.

$$\int_\xi \hat{W}(\xi) \int_\Omega W_\alpha R_i d\Omega d\xi = 0 \tag{11.4.1}$$

Here, the temporal test functions $\hat{W}(\xi)$ were discussed in Section 10.2.1, whereas the Petrov-Galerkin test functions W_α are the global form of the local test functions as the sum of the standard Galerkin test functions and the numerical diffusion test function for streamline diffusion.

$$W_\alpha = \Phi_\alpha + \Psi_\alpha^{(a)} \tag{11.4.2}$$

If the discontinuity-capturing scheme is desired, this can be added to (11.4.1) by constructing the product of $\Psi_\alpha^{(b)}$ and the convection term of the residual, leading to the GPG equations of the form,

$$\int_\xi \hat{W}(\xi) \int_\Omega \left[\left(\Phi_\alpha + \Psi_\alpha^{(a)}\right)\left(\frac{\partial v_i}{\partial t} + v_{i,j}v_j - \nu v_{i,jj} - f_i\right) + \Psi_\alpha^{(b)} v_j v_{i,j}\right] d\Omega d\xi = 0 \tag{11.4.3}$$

Note that the integration by parts is to be performed only with respect to the Galerkin test functions, which will lead to the Neumann boundary conditions, whereas those terms of the residual associated with numerical diffusion test functions will not be integrated by parts since they should be contained within the elements as a measure of numerical diffusion. Thus, the GPG integral takes the form, known as the *variational equation*,

$$\begin{aligned}
&\int_\xi \hat{W}(\xi)\left[\int_\Omega \left(\Phi_\alpha \Phi_\beta \frac{\partial v_{\beta i}}{\partial t} + v_j \Phi_\alpha \Phi_{\beta,j} v_{\beta i} + \nu \Phi_{\alpha,j}\Phi_{\beta,j} v_{\beta i} - \Phi_\alpha f_i\right) d\Omega\right.\\
&\left. - \int_\Gamma \Phi_\alpha^* \nu v_{i,j} n_j d\Gamma\right] d\xi + \int_\xi \hat{W}(\xi) \int_\Omega \tau v_k \Phi_{\alpha,k}\left(\frac{\partial \Phi_\beta v_{\beta i}}{\partial t} + v_j \Phi_{\beta,j} v_{\beta i}\right.\\
&\left. - \nu \Phi_{\beta,jj} v_{\beta i} - f_i\right) d\Omega d\xi + \int_\xi \hat{W}(\xi) \int_\Omega \tau^{(b)} v_k v_j \Phi_{\alpha,k} \Phi_{\beta,j} v_{\beta i} d\Omega d\xi = 0
\end{aligned} \tag{11.4.4}$$

The first integral indicates the Galerkin integral, with the second representing the streamline diffusion, and the third integral indicates the discontinuity-capturing.

Assume that the trial function is linear, independent of time, with the numerical diffusion due to the source term being negligible. Furthermore, if the temporal test function, $W(\xi) = \delta(\xi - 1/2)$ or $W(\xi) = 1$ is used and the variation of nodal values of the variables v_i is linear, then we obtain [see (10.2.13) or (11.2.6)]

$$[A_{\alpha\beta} + \eta\Delta t(B_{\alpha\beta} + C_{\alpha\beta} + K_{\alpha\beta})]\, v_{\beta i}^{n+1} = [A_{\alpha\beta} - (1-\eta)\Delta t(B_{\alpha\beta} + C_{\alpha\beta} + K_{\alpha\beta})]\, v_{\beta i}^{n} + \Delta t(F_{\alpha i} + G_{\alpha i}) \tag{11.4.5}$$

where the definitions of all terms are shown in Section 11.2 except that various forms of the numerical diffusion matrix, $C_{\alpha\beta}$, are given below.

$$C_{\alpha\beta} = \int_\Omega \tau v_k v_j \Phi_{\alpha,k}\Phi_{\beta,j}\,d\Omega \tag{11.4.6a}$$

for streamline diffusion, and

$$C_{\alpha\beta} = \int_\Omega \left(\tau + \tau^{(b)}\right) v_k v_j \Phi_{\alpha,k}\Phi_{\beta,j}\,d\Omega \tag{11.4.6b}$$

for combined streamline diffusion and discontinuity-capturing. It is seen that the numerical difffusion factor τ or $\tau + \tau^{(b)}$ in GPG corresponds to $\Delta t/2$ in (11.2.76) for TGM, but is much more complicated and actually flowfield-dependent. Note also that effects of numerical diffusion associated with terms other than convection are neglected in (11.4.5). The complexity of the numerical diffusion factor increases significantly for the case of the Navier-Stokes system of equations as discussed in Section 13.3.

Various options for temporal approximations or higher order accuracy may be selected as discussed in Section 10.2. For the case of streamline diffusion (11.4.6a) with the temporal parameter, $\eta = 1$, and linear trial and test functions of finite elements, the expression given by (11.4.5) is identical to equation 25 of Shakib and Hughes [1991] for the constant-in-time approximations of the space-time Galerkin/least squares (GLS) in one-dimensional problems. The GLS formulation will be described in the following section.

11.4.2 SPACE-TIME GALERKIN/LEAST SQUARES METHODS

The formal discussion of the least squares methods (LSM) of obtaining the FEM equations will be presented in the later chapters. However, in order to understand the Galerkin/least squares (GLS) methods reported by Hughes and his co-workers, we examine briefly a basic procedure for the least squares formulation. First, let us introduce the least squares variational function,

$$\Pi = \int_\Omega \frac{1}{2} R_j R_j\,d\Omega$$

which is then to be minimized with respect to the nodal variables $v_{\alpha i}$. In this process, we multiply Π by the numerical diffusion factor, τ.

$$\delta\Pi = \frac{\partial \tau\Pi}{\partial v_{\alpha i}}\delta v_{\alpha i} = 0 \tag{11.4.7}$$

or

$$\frac{\partial \tau\Pi}{\partial v_{\alpha i}} = \tau\int_\Omega \frac{\partial R_j}{\partial v_{\alpha i}} R_j\,d\Omega = 0 \tag{11.4.8}$$

Performing the differentiation in (11.4.8) and applying the temporal approximations, we obtain

$$\int_{\xi} \hat{W}(\xi) \int_{\Omega} \tau \left(\frac{\partial}{\partial t} + \mathrm{v}_k \frac{\partial}{\partial x_k} - \nu \frac{\partial^2}{\partial x_k \partial x_k} \right) \Phi_\alpha \left(\frac{\partial \mathrm{v}_i}{\partial t} + \mathrm{v}_j \mathrm{v}_{i,j} - \nu \mathrm{v}_{i,jj} - f_i \right) d\Omega d\xi = 0 \tag{11.4.9}$$

which may be written as

$$\int_{\xi} \hat{W}(\xi) \int_{\Omega} \tau (L\Phi_\alpha)(L\mathrm{v}_i - f_i)\, d\Omega d\xi = 0 \tag{11.4.10}$$

where L is the differential operator,

$$L = \frac{\partial}{\partial t} + \mathrm{v}_k \frac{\partial}{\partial x_k} - \nu \frac{\partial^2}{\partial x_k \partial x_k} \tag{11.4.11}$$

At this point, we add the least squares integral (11.4.10) and the discontinuity-capturing term as developed in Section 11.3.3 to the standard Galerkin integral. If we choose only the convective term in (11.4.11), then, these steps lead to the form identical to the generalized Petrov-Galerkin scheme given by (11.4.4). The sum of the standard Galerkin integral, the discontinuity capturing term, and the least squares integral represented by (11.4.10) is referred to as the space-time Galerkin/least squares (GLS) methods [Hauke and Hughes, 1998]. Note that the contributions from additional terms other than the convective terms in (11.4.11) are negligible.

The space-time GLS formulation is another form of generalized Petrov-Galerkin (GPG) methods in which the only difference from the GPG methods of Section 11.4.1 is the numerical diffusion test functions for streamline diffusion,

$$\Psi_\alpha^{(a)} = \tau L\Phi_\alpha \tag{11.4.12}$$

where the numerical diffusion factor τ can be constructed by introducing the local curvilinear coordinate contravariant metric tensor [Shakib and Hughes, 1991],

$$g^{ij} = \left(\frac{\partial x_k}{\partial \xi_i} \frac{\partial x_k}{\partial \xi_j} \right)^{-1} \tag{11.4.13}$$

With some algebra, it can be shown that one possible option for τ is of the form

$$\tau = \left[\left(\frac{2}{\Delta t} \right)^2 + \left(\frac{2|\mathrm{v}_i|}{|h_i|} \right)^2 + 9 \left(\frac{4\nu}{|h_i|^2} \right)^2 \right]^{-\frac{1}{2}} \tag{11.4.14}$$

where h_i denotes the average element size in local coordinates. Note that if only the convective term is chosen in (11.4.9), then the GLS formulation becomes identical to the GPG formulation given by (11.4.4). The standard least squares methods will be discussed in Section 12.1.8 for incompressible flows and in Section 14.2 for compressible flows.

Applications of GPG to the Navier-Stokes system of equations require some modifications for the numerical diffusion test functions in which entropy variables can be employed to advantage. This subject will be discussed in Section 13.4.

Remarks: The temporal integral with the temporal test function $\hat{W}(\xi)$ first introduced in (10.2.5) plays the role identical to the process referred to as the discontinuous space-time integral [Shakib and Hughes, 1991; Tezduyar, 1997]. Many possible options of this temporal test function can be chosen (Tables 10.2.1 and 10.2.2). Explicit forms of integrals (11.4.4) plus the least squares integrals (11.4.9) as applied to the Navier-Stokes system of equations are shown in (13.3.19).

11.5 SOLUTIONS OF NONLINEAR AND TIME-DEPENDENT EQUATIONS AND ELEMENT-BY-ELEMENT APPROACH

As was shown in Section 10.3.2, the global assembly of local stiffness matrices can be avoided via the element-by-element (EBE) scheme. In dealing with nonlinear and time-dependent equations, however, some modifications are required. We discuss in this section the Newton-Raphson methods of solving nonlinear time-dependent equations, followed by the generalized minimal residual (GMRES) equation solver and EBE scheme.

11.5.1 NEWTON-RAPHSON METHODS

Recall that in Section 11.2.1 we held $\overline{\mathrm{v}}_j$ constant in $\overline{\mathrm{v}}_j \mathrm{v}_{i,j}$, which was meant to be updated in each step of calculations. Otherwise, GGM or GPG, methods described in the previous sections, must be modified in order to solve nonlinear equations. For example, we may write (11.2.6) of the GGM formulation in the form where $\overline{\mathrm{v}}_j$ is no longer held constant.

$$E_{\alpha i} = A_{\alpha\beta}\mathrm{v}_{\beta i}^{n+1} + \eta\Delta t\left(B_{\alpha\beta j\gamma}\mathrm{v}_{\gamma j}^{n+1}\mathrm{v}_{\beta i}^{n+1} + K_{\alpha\beta}\mathrm{v}_{\beta i}^{n+1}\right) - A_{\alpha\beta}\mathrm{v}_{\beta i}^{n} + (1-\eta)\Delta t\left(B_{\alpha\beta j\gamma}\mathrm{v}_{\gamma j}^{n}\mathrm{v}_{\beta i}^{n} + K_{\alpha\beta}\mathrm{v}_{\beta i}^{n}\right) - \Delta t(F_{\alpha i} + G_{\alpha i}) = 0 \tag{11.5.1}$$

with

$$B_{\alpha\beta j\gamma} = \int_\Omega \Phi_\alpha \Phi_{\beta,j}\Phi_\gamma d\Omega \tag{11.5.2}$$

This form is based on the assumption that the squares and products of velocity components vary linearly within the time step as in (11.2.6),

$$\mathrm{v}_{\gamma j}^{n+1}\mathrm{v}_{\beta i}^{n+1} = (1-\xi)\mathrm{v}_{\gamma j}^{n}\mathrm{v}_{\beta i}^{n} + \xi\mathrm{v}_{\gamma j}^{n+1}\mathrm{v}_{\beta i}^{n+1} \tag{11.5.3}$$

One of the most efficient approaches to solve nonlinear equations is the Newton-Raphson method developed from the Taylor series expansion of the residual of the type in (11.5.1).

$$E_{\alpha i}^{n+1,r+1} = E_{\alpha i}^{n+1,r} + \frac{\partial E_{\alpha i}^{n+1,r}}{\partial \mathrm{v}_{\beta j}^{n+1,r}}\Delta\mathrm{v}_{\beta j}^{n+1,r+1} + \cdots = 0 \tag{11.5.4}$$

which implies that the residual at a given time station $n+1$ as incremented to the $r+1$ iteration cycle from the previous cycle r should vanish if (11.5.1) is to be satisfied. Retaining only up to and including the first order term in (11.5.4), we obtain

$$J_{\alpha\beta ij}^{n+1,r}\Delta\mathrm{v}_{\beta j}^{n+1,r+1} = -E_{\alpha i}^{n+1,r} \tag{11.5.5}$$

where

$$\Delta\mathrm{v}_{\beta j}^{n+1,r+1} = \mathrm{v}_{\beta j}^{n+1,r+1} - \mathrm{v}_{\beta j}^{n+1,r} \tag{11.5.6}$$

and $J_{\alpha\beta ij}^{n+1,r}$ is the Jacobian,

$$J_{\alpha\beta ij}^{n+1,r} = \frac{\partial E_{\alpha i}^{n+1,r}}{\partial v_{\beta j}^{n+1,r}}$$

or

$$\begin{aligned}
J_{\alpha\beta ij}^{n+1,r} &= A_{\alpha\eta}\frac{\partial v_{\eta i}^{n+1,r}}{\partial v_{\beta j}^{n+1,r}} + \eta\Delta t\left[B_{\alpha\eta k\gamma}\left(\frac{\partial v_{\gamma k}^{n+1,r}}{\partial v_{\beta j}^{n+1,r}}v_{\eta i}^{n+1,r} + v_{\gamma k}^{n+1,r}\frac{\partial v_{\eta i}^{n+1,r}}{\partial v_{\beta j}^{n+1,r}}\right) + K_{\alpha\eta}\frac{\partial v_{\eta i}^{n+1,r}}{\partial v_{\beta j}^{n+1,r}}\right]\\
&= A_{\alpha\eta}\delta_{\eta\beta}\delta_{ij} + \eta\Delta t\left[B_{\alpha\eta k\gamma}\left(\delta_{\gamma\beta}\delta_{kj}v_{\eta i}^{n+1,r} + v_{\gamma k}^{n+1,r}\delta_{\eta\beta}\delta_{ij}\right) + K_{\alpha\eta}\delta_{\eta\beta}\delta_{ij}\right]\\
&= A_{\alpha\beta}\delta_{ij} + \eta\Delta t\left[B_{\alpha\gamma j\beta}v_{\gamma i}^{n+1,r} + B_{\alpha\beta k\gamma}\delta_{ij}v_{\gamma k}^{n+1,r} + K_{\alpha\beta}\delta_{ij}\right] \qquad (11.5.7)
\end{aligned}$$

with

$$B_{\alpha\beta j\gamma} = \int_\Omega \Phi_\alpha\Phi_{\gamma,j}\Phi_\beta\, d\Omega, \quad B_{\alpha\beta k\gamma} = \int_\Omega \Phi_\alpha\Phi_{\beta,k}\Phi_\gamma\, d\Omega$$

The Newton-Raphson procedure described above may be simplified by revising the Jacobian matrix and the right-hand side residual as follows:

$$J_{\alpha\beta ij}^{n+1,r} = A_{\alpha\beta}\delta_{ij} + \frac{\Delta t}{2}(B_{\alpha\beta ij} + K_{\alpha\beta ij})$$

with

$$B_{\alpha\beta} = \int_\Omega \Phi_\alpha\Phi_{\beta,j}\overline{v}_j d\Omega$$

and (11.5.1) being replaced by

$$\begin{aligned}
E_{\alpha i}^{n+1,r} &= A_{\alpha\beta}v_{\beta i}^{n+1} + \frac{\Delta t}{2}(B_{\alpha\beta} + K_{\alpha\beta})v_{\beta i}^{n+1} - A_{\alpha\beta}v_{\beta i}^{n} + \frac{\Delta t}{2}(B_{\alpha\beta} + K_{\alpha\beta})v_{\beta i}^{n}\\
&\quad - \Delta t(F_{\alpha i} + G_{\alpha i}) = A_{\alpha\beta}\Delta v_{\beta i}^{n+1,r} + \frac{\Delta t}{2}(B_{\alpha\beta} + K_{\alpha\beta})\Delta v_{\beta i}^{n+1,r} - \Delta t(F_{\alpha i} + G_{\alpha i})
\end{aligned}$$

The Newton-Raphson iterations are performed using (11.5.5) within each time step until convergence which requires that $\Delta v_{\beta j}^{n+1,r+1} \cong 0$ in (11.5.5) before proceeding to the next time step in (11.5.7).

11.5.2 ELEMENT-BY-ELEMENT SOLUTION SCHEME FOR NONLINEAR TIME DEPENDENT FEM EQUATIONS

The linear and nonlinear simultaneous algebraic equations arising from the entire assembled global system of FEM formulations may be solved using direct or iterative methods. For a very large system, iterative methods are preferable to direct methods. Furthermore, it is often necessary to devise special techniques such as the frontal methods [Irons, 1970; Hood, 1976] or element-by-element (EBE) solution methods [Fox and Stanton, 1968; Irons, 1970]. In these methods, the standard assembly process of local stiffness matrices is not necessary. Instead, the product of a matrix by a vector can be obtained by assembling the product of local element matrices and the corresponding part of the vector, thus reducing the cost of computer time and storage. Initial contributions of the EBE concept to a large system of equations include Ortiz, Pinsky, and Taylor

[1983], Hughes, Frencz, and Hallquist [1987], Nour-Omid [1984], and Nour-Omid and Parlett [1985], among others.

Recall that we discussed the EBE algorithm for the linear equations in Section 10.3. For nonlinear stiffness matrices and time dependent problems, the procedure for EBE must be modified. These topics are elaborated below.

If the system of equations is nonlinear, then we may replace the preconditioner $D_{\alpha\beta}$ (see Section 10.3.2) by the Newton-Raphson Jacobian matrix. The global FEM nodal error can be written as

$$E_\alpha = K_{\alpha\beta} U_\beta - F_\alpha \tag{11.5.8}$$

Applying the Newton-Raphson scheme as shown in Section 11.5.1, we may rewrite (10.3.15) in the form

$$U_\alpha^{r+1} = U_\alpha^r - J_{\alpha\beta}^{-1}(\overline{F}_\beta - F_\beta)^r \tag{11.5.9}$$

where the EBE scheme is applied to the stiffness matrix as presented in Section 10.3.2 and the Jacobian matrix $J_{\alpha\beta}$ is given by

$$J_{\alpha\beta} = \frac{\partial E_\alpha}{\partial U_\beta} \tag{11.5.10}$$

which is considered as the preconditioning matrix. Here, as shown in (10.3.17), we may replace $J_{\alpha\beta}$ in (11.5.9) by the main diagonal of $J_{\alpha\beta}$ so that

$$U_\alpha^{r+1} = U_\alpha^r - J_{(\alpha\alpha)}^{-1}(\overline{F}_\alpha - F_\alpha)^r \tag{11.5.11}$$

The solution is obtained similarly as in (10.3.17) except that $J_{(\alpha\alpha)}$ and $\overline{F}_\alpha$ are nonlinear and must be updated at each iteration. Note that $\overline{F}_\alpha$ is converted from the EBE-based stiffness matrices.

In order to improve the solution accuracy, we may use the preconditioned conjugate gradient (PCG) method or the method known as the Lanczos/ORTHORES solver [Jea and Young, 1983]. In this method, begin with a starting value U_α^o and compute

$$U_\alpha^{r+1} = a^{r+1}(b^{r+1} D_\alpha^r + U_\alpha^r) + (1 - a^{r+1}) U_\alpha^r \tag{11.5.12}$$

with

$$b^{r+1} = \frac{(D_\alpha^r E_\alpha^r)}{(D_\delta^r K_{\delta\beta} D_\beta^r)} \qquad r = 0, 1, \ldots$$

$$a^{r+1} = \begin{cases} 1 & r = 0 \\ \left[1 - \dfrac{b^{r+1}}{b^r} \dfrac{(D_\alpha^r E_\alpha^r)}{(D_\delta^{r-1} E_\delta^{r-1})} \dfrac{1}{a^r}\right]^{-1} & r \geq 1 \end{cases}$$

$$E_\alpha^r = \begin{cases} F_\alpha - K_{\alpha\beta} U_\beta^o & r = 0 \\ a^r(-b^r K_{\alpha\beta} D_\beta + E_\alpha^{r-1}) + (1 - a^r) E_\alpha^{r-2} & r = 1, 2, \ldots \end{cases}$$

$$D_\alpha^r = Q_{\alpha\beta}^{-1} E_\beta^r \quad r \geq 0 \tag{11.5.13}$$

where $Q_{\alpha\beta}$ is the Jacobi preconditioner,

$$Q_{\alpha\beta} = dia(K_{\alpha\beta}) \tag{11.5.14}$$

Thus, the inverse of $Q_{\alpha\beta}$ is the reciprocal of the diagonal of $K_{\alpha\beta}$ which can be partitioned for EBE computations.

The preconditioner may be constructed from the square root of the main diagonal of the stiffness matrix. To this end, we write (11.5.11) in the form

$$E_\alpha = \overline{F}_\alpha - \overline{K}_{\alpha\beta}\overline{U}_\beta \tag{11.5.15}$$

with

$$\overline{K}_{\alpha\beta} = W_{\alpha\gamma}^{-\frac{1}{2}} K_{\gamma\eta} W_{\eta\beta}^{-\frac{1}{2}} \qquad \overline{U}_\beta = W_{\beta\gamma}^{-\frac{1}{2}} U_\gamma$$

$$\overline{F}_\alpha = \bigcup_{e=1}^{E} \overline{F}_N^{(e)} \Delta_{N\alpha}^{(e)} \qquad \overline{F}_N^{(e)} = W_{NR}^{(e)\frac{1}{2}} F_R^{(e)}$$

$$W_{NR}^{(e)} = dia\left(K_{NR}^{(e)}\right)$$

For known initial solution vector U_β^o, compute

$$E_\alpha^o = \overline{F}_\alpha - \overline{K}_{\alpha\beta} U_\beta^o \tag{11.5.16}$$

Subsequent steps are the same as in (11.5.15). The final solution is obtained as

$$U_\alpha = W_{\alpha\beta}^{-\frac{1}{2}} \overline{U}_\beta = dia(K_{\alpha\beta})^{-\frac{1}{2}} \overline{U}_\beta \tag{11.5.17}$$

The Lanczos/ORTHOMIN solver [Jea and Young, 1983] may be used. In this scheme, the preconditioning processes (11.5.15) through (11.5.16) are used together with the following steps:

Step 1

$$E_\alpha^o = \overline{F}_\alpha - \overline{K}_{\alpha\beta} U_\beta^o$$

$$P_\alpha^o = E_\alpha^o$$

$$D_\alpha^o = \tilde{P}_\alpha^o = E_\alpha^o$$

$$b^o = \frac{(D_\alpha^o E_\alpha^o)}{(D_\delta^o \overline{K}_{\delta\beta} D_\beta^o)}$$

$$U_\alpha^1 = U_\alpha^o + b^o P_\alpha^o$$

Step 2

$$b^r = \frac{(D_\alpha^r E_\alpha^r)}{(D_\delta^r \overline{K}_{\delta\beta} D_\beta^r)}$$

$$P_\alpha^r = E_\alpha^r + \overline{b}^r P_\alpha^{r-1}$$

$$\tilde{P}_\alpha^r = D_\alpha^r + \overline{b}^r \tilde{P}_\alpha^{r-1}$$

$$\overline{b}^r = \frac{(D_\alpha^r E_\alpha^r)}{(D_\delta^{r-1} E_\delta^{r-1})}$$

$$E_\alpha^{r+1} = E_\alpha^r - \overline{b}^r \overline{K}_{\alpha\beta} P_\beta^r$$

$$D_\alpha^{r+1} = D_\alpha^r - b^r \overline{K}_{\alpha\beta} \tilde{P}_\beta^r$$

$$U_\alpha^{r+1} = U_\alpha^r + b^r P_\beta^r$$

Iterative solutions through the above steps lead to the final converged solution as

$$U_\alpha = W_{\alpha\beta}^{-\frac{1}{2}} \overline{U}_\beta = dia(K_{\alpha\beta})^{-\frac{1}{2}} \overline{U}_\beta \tag{11.5.18}$$

For time-dependent problems, we may consider the main diagonal of the mass matrix as the preconditioner. For example, the matrix equation

$$(M_{\alpha\beta} + \theta \Delta t K_{\alpha\beta}) U_\beta^{n+1} = [M_{\alpha\beta} + (1-\theta)\Delta t K_{\alpha\beta}] U_\beta^n + \Delta t F_\alpha \tag{11.5.19}$$

can be written as

$$\left(\delta_{\alpha\beta} + \theta \Delta t M_{\alpha\gamma}^{-\frac{1}{2}} K_{\gamma\eta} M_{\eta\beta}^{-\frac{1}{2}}\right) \overline{U}_\beta^{n+1} = \left[\delta_{\alpha\beta} - (1-\theta)\Delta t M_{\alpha\gamma}^{-\frac{1}{2}} K_{\gamma\eta} M_{\eta\beta}^{-\frac{1}{2}}\right] \overline{U}_\beta^n + \Delta t \overline{F}_\alpha \tag{11.5.20}$$

where $\overline{U}_\alpha^n = M_{\alpha\beta}^{\frac{1}{2}} U_\beta$ and $\overline{F}_\alpha = M_{\alpha\beta}^{-\frac{1}{2}} F_\beta$. Note that the eigenvalues of (11.5.22) are the same as those of (11.5.20) such that

$$|\delta_{\alpha\beta} + \theta \Delta t\ M_{\alpha\gamma} K_{\gamma\beta}| = \left| M_{\alpha\gamma}^{-\frac{1}{2}} \left(\delta_{\gamma\eta} + \theta \Delta t\ M_{\gamma\zeta}^{-\frac{1}{2}} K_{\zeta\delta} M_{\delta\eta}\right) M_{\eta\beta}^{-\frac{1}{2}} \right| \tag{11.5.21}$$

Rewriting (11.5.15) in the form

$$E_\alpha = A_{\alpha\beta} \overline{U}_\beta^{n+1} - B_{\alpha\beta} \overline{U}_\beta^n - \Delta t \overline{F}_\alpha \tag{11.5.22}$$

it is now possible to apply steps 1 and 2 of the steady-state case with initial and boundary conditions applied to (11.5.22).

11.5.3 GENERALIZED MINIMAL RESIDUAL ALGORITHM

The conjugate gradient method discussed in Section 10.3.1 is accurate and efficient for linear symmetric matrix equations. However, for problems in CFD where nonsymmetric nonlinear, indefinite matrices are involved, the Generalized Minimal Residual (GMRES) algorithm has been proved to be efficient [Saad and Schultz, 1986; Saad, 1996]. This method is based on the property of minimizing the norm of the residual vector over a Krylov space. The Krylov space is a general concept based on the simple observation that in any sequence of iterates there will be a smallest set of consecutive iterates which are linearly dependent, and that the coefficients of a vanishing combination are the coefficients of a divisor to the characteristic polynomial. See Householder [1964] for a detailed discussion of the Krylov space.

For the purpose of our discussion, let us consider the global form of the finite element equations in the form,

$$K_{\alpha\beta} U_\beta = F_\alpha \tag{11.5.23}$$

in which preconditioning through the EBE scheme is to be implemented as in Section 11.5.2.

One of the most effective iteration methods for solving large sparse asymmetric linear and nonlinear systems of equations is a combination of the CGM with preconditions in minimizing the norm of residual vector over a Krylov space

$$K^{(r)} = span\left[U_0, KU_0, K^2 U_0 \ldots, K^{(r-1)} U_0\right] \tag{11.5.24}$$

This algorithm is a generalization of the MINRES [Paige and Saunders, 1975] for solving nonsymmetric linear systems and Arnoldi process [Arnoldi, 1951] which is an analogue of the Lanczos algorithm for nonsymmetric matrices [Lanczos, 1950]. In the GMRES scheme, we determine $U_\beta^{(o)} + \overline{U}_\beta$ where $U_\beta^{(o)}$ is the initial guess and $\overline{U}_\beta$ is a member of the Krylov space K of dimension r such that the L_2 norm error

$$\| E_\alpha \| = \left\| F_\alpha - K_{\alpha\beta}\left(U_\beta^{(o)} + \overline{U}_\beta\right)\right\| \tag{11.5.25}$$

is minimized. Here, we use a smaller value for r and restarting the algorithm after every r step; thereby, the amount of storage required can be minimized.

The step-by-step GMRES scheme is as follows:

First, let us define:

$E_\alpha^{(r)}$ = total error vector
$\overline{E}_\alpha^{(i)}$ = error coefficient vector
$\| E_\alpha^{(j)} \|$ = normed error
$\tilde{E}_\alpha^{(j)}$ = adjusted error
$\| \tilde{E}_\alpha^{(j)} \|$ = normed adjusted error
$a^{(i,j)}$ = normed error coefficient
$y^{(j)}$ = minimizer error vector

(1) Choose $U_\beta^{(o)}$ and compute

$$E_\alpha^{(o)} = F_\alpha - K_{\alpha\beta} U_\beta^{(0)} = F_\alpha - \overline{F}_\alpha^{(0)}, \quad \overline{F}_\alpha = \bigcup_{e=1}^{E} \overline{F}_N^{(0)(e)} \Delta_{N\alpha}^{(e)},$$

$$\overline{F}_N^{(0)(e)} = K_{NM}^{(e)} U_M^{(0)(e)}$$

$$\overline{E}_\alpha^{(1)} = E_\alpha^{(o)} / \| E_\alpha^{(o)} \| \quad \text{(Gram-Schmidt orthogonalization)}$$

(2) Iterate for $i = 1, 2, \ldots r$

$$a^{(i+1,j)} = \tilde{E}_\alpha^{(i+1)} \overline{E}_\alpha^{(j)} = K_{\alpha\beta} \overline{E}_\alpha^{(i)} \overline{E}_\beta^{(j)}, \quad j = 1, 2, \ldots, i$$

$$\tilde{E}_\alpha^{(i)} = K_{\alpha\beta} \overline{E}_\beta^{(i)} - \sum_{j=1}^{i} a^{(i+1,j)} \overline{E}_\alpha^{(j)}$$

$$\overline{E}_\alpha^{(i+1)} = \tilde{E}_\alpha^{(i)} / \| \tilde{E}_\alpha^{(i)} \|$$

(3) Approximate solution:

Let us consider a matrix consisting of the columns of residuals in the form

$$B_{\beta\xi}^{(r)} = \begin{bmatrix} \overline{E}_1^{(1)} & \overline{E}_1^{(2)} & \cdots & \overline{E}_1^{(r)} \\ \overline{E}_2^{(1)} & \overline{E}_2^{(2)} & \cdots & \overline{E}_2^{(r)} \\ \vdots & \vdots & & \vdots \\ \overline{E}_n^{(1)} & \overline{E}_n^{(2)} & \cdots & \overline{E}_n^{(r)} \end{bmatrix} \tag{11.5.26}$$

Then, it can be shown that

$$K_{\alpha\beta} B_{\beta\xi}^{(r,r)} = B_{\alpha\eta}^{(r,r+1)} H_{\eta\xi}^{(r+1,r)} \tag{11.5.27}$$

where $H_{\eta\xi}^{(r+1,r)}$ is the upper Hessenberg matrix of the form

$$H_{\eta\xi}^{(r+1,1)} = \begin{bmatrix} a^{(1,1)} & a^{(2,1)} & \cdot\ \cdot & a^{(r,1)} \\ \| \tilde{E}_\alpha^{(1)} \| & a^{(2,2)} & \cdot\ \cdot & a^{(r,2)} \\ 0 & \| \tilde{E}_\alpha^{(2)} \| & \cdot\ \cdot & a^{(r,3)} \\ \vdots & \vdots & \vdots & \vdots \\ 0 & 0 & \cdot\ \cdot & \| \tilde{E}^{(r)} \| \end{bmatrix} \tag{11.5.28}$$

Here, the idea is to find a vector y_ξ which will minimize the residual error as follows:

$$\begin{aligned} \min \| F_\alpha - K_{\alpha\beta}(U_\beta^{(0)} + E_\beta) \| &= \min \| E_\alpha^{(0)} - K_{\alpha\beta} B_{\beta\xi}^{(r,r)} y_\xi \| \\ &= \| B_{\alpha\eta}^{(r+1)} (e_\eta - H_{\eta\xi}^{(r+1)} y_\xi) \| \\ &= \| \overline{e}_\alpha - \overline{H}_{\alpha\xi}^{(r+1)} y_\xi \| \cong 0 \end{aligned} \tag{11.5.29}$$

with

$$\overline{e}_\alpha = \{ \| E_\alpha^{(1)} \|, 0, \ldots 0 \}^T \tag{11.5.30}$$

$$y_\xi = \overline{H}_{\alpha\xi}^{-1} \overline{e}_\alpha \tag{11.5.31}$$

The minimization process above does not provide the approximate solution explicitly at each step. Thus, it is difficult to determine when to stop. This may be simplified using the so-called Q-R algorithm as suggested by Saad and Schultz [1986]. In this approach, we utilize the Givens-Householder rotation matrix, $R_{\alpha\eta}$, such that

$$\overline{H}_{\alpha\xi} = R_{\alpha\eta} H_{\eta\xi} \tag{11.5.32}$$

where

$$R_{\alpha\xi} = R_r R_{r-1} \ldots . R_1$$

$$R_{\alpha\eta} = \begin{bmatrix} 1 & & & & & & & \\ & \cdot & & & & & & \\ & & \cdot & & & & & \\ & & & 1 & & & & \\ & & & & c_r & s_r & & \\ & & & & -s_r & c_r & & \\ & & & & & & 1 & \\ & & & & & & & \cdot \\ & & & & & & & & \cdot \\ & & & & & & & & & 1 \end{bmatrix} \tag{11.5.33}$$

with $c_r^2 + s_r^2 = 1$ and the size of the matrix being $(m+1) \times (m \times 1)$ for m steps of the GMRES iterations. The scalars c_r and s_r of the rth rotation R_r, being orthogonal, are defined as

$$c_r = \frac{H_{rr}}{\sqrt{(H_{rr})^2 + H_{r+1,r}^2}}, \qquad s_r = \frac{H_{r+1,r}}{\sqrt{((H_{rr})^2 + H_{r+1,r}^2)}} \tag{11.5.34}$$

For example, let us assume r steps of the GMRES iterations so that (11.5.28) is written as

$$\|e_\alpha - H_{\alpha\xi}^{r+1} y_\xi\| = \|R_{\alpha\xi}(\overline{e}_\alpha - \overline{H}_{\alpha\xi}^{r+1} y_\xi)\| = \|\overline{e}_\alpha - \overline{H}_{\alpha\xi}^{r+1} y_\xi\| \tag{11.5.35}$$

leading to the minimization,

$$\min\|\overline{e}_\alpha - \overline{H}_{\alpha\xi}^{r+1} y_\xi\| = |\overline{e}_\alpha^{r+1}| \tag{11.5.36}$$

and y_ξ satisfies

$$\begin{bmatrix} \overline{H}_{1,1} & \cdot & \cdot & \overline{H}_{1,r-1} & \overline{H}_{1,r} \\ 0 & \cdot & \cdot & \cdot & \cdot \\ 0 & 0 & \cdot & \cdot & \cdot \\ 0 & 0 & 0 & \overline{H}_{r-1,r-1} & \overline{H}_{r-1,r} \\ 0 & 0 & 0 & 0 & \overline{H}_{r,r} \end{bmatrix} \begin{bmatrix} y_1 \\ \vdots \\ \\ y_{r-1} \\ y_r \end{bmatrix} = \begin{bmatrix} \overline{e}_1 \\ \cdot \\ \cdot \\ \overline{e}_{r-1} \\ \overline{e}_r \end{bmatrix} \tag{11.5.37}$$

in which the back substitution provides the inverse required in (11.5.31).

To obtain the Hessenberg matrix in (11.5.37), we proceed as follows. If $m = 5$, then we have

$$\overline{H}_5 = \begin{bmatrix} h_{11} & h_{12} & h_{13} & h_{14} & h_{15} \\ h_{21} & h_{22} & h_{23} & h_{24} & h_{25} \\ & h_{32} & h_{33} & h_{34} & h_{35} \\ & & h_{43} & h_{44} & h_{45} \\ & & & h_{54} & h_{55} \\ & & & & h_{65} \end{bmatrix} \tag{11.5.38}$$

$$h^{(1)} = \begin{bmatrix} h_{11} \\ h_{21} \\ 0 \\ 0 \\ 0 \\ 0 \end{bmatrix} = \begin{bmatrix} a^{(1,1)} \\ \|\tilde{E}_\alpha\| \\ 0 \\ 0 \\ 0 \\ 0 \end{bmatrix} \tag{11.5.39}$$

$$r_1 = \left(h_{11}^2 + h_{21}^2\right)^{1/2}, \qquad c_1 = h_{11}/r_1, \qquad s_1 = h_{21}/r_1$$

The first column of $\overline{H}_5$ becomes

$$\overline{h}^{(1)} = R_1 h^{(1)} = \begin{bmatrix} r_1 \\ 0 \\ 0 \\ 0 \\ 0 \\ 0 \end{bmatrix}, \qquad \overline{h}^{(m)} = R_m R_{m-1} \cdots R_2 \overline{h}^{(1)}$$

Similarly,

$$\overline{e}_\alpha^{(1)} = R_1 \overline{e}_\alpha^{(0)}, \qquad \overline{e}^{(m)} = R_m R_{m-1} \cdots R_2 \overline{e}_\alpha^{(0)}$$

This process leads to the tridiagonalized form,

$$\overline{H}^{(5)} = \begin{bmatrix} h_{11}^{(5)} & h_{12}^{(5)} & h_{13}^{(5)} & h_{14}^{(5)} & h_{15}^{(5)} \\ & h_{22}^{(5)} & h_{23}^{(5)} & h_{24}^{(5)} & h_{25}^{(5)} \\ & & h_{33}^{(5)} & h_{34}^{(5)} & h_{35}^{(5)} \\ & & & h_{44}^{(5)} & h_{45}^{(5)} \\ & & & & h_{55}^{(5)} \\ & & & & 0 \end{bmatrix} \tag{11.5.40}$$

which is then inserted in (11.5.37) to determine y_ξ, required in (11.5.31).

(4) Calculate the error residuals $\overline{U}_\alpha^{(r)}$,

$$E_\alpha^{(r)} = E_{\alpha r} - y_r$$

(5) The converged solution is obtained as

$$U_\alpha = U_\alpha^o + E_\alpha^{(r)}$$

Example 11.5.1

Solve the following equations with an unsymmetric stiffness matrix using the GMRES algorithm. Compare with the exact solution: $U_1 = 1$, $U_2 = 2$, $U_3 = 3$.

$$\begin{bmatrix} 3 & 2 & -2 \\ -4 & -1 & 1 \\ 5 & -2 & -1 \end{bmatrix} \begin{bmatrix} U_1 \\ U_2 \\ U_3 \end{bmatrix} = \begin{bmatrix} 1 \\ -3 \\ -2 \end{bmatrix}$$

Solution:

Note that the EBE process is omitted here for simplicity. (The global matrix equation is used instead of the EBE column vector.) The EBE process must be used for a large system of equations. See Section 11.5.4 for EBE implementations.

1. Choose $U_\beta^{(0)} = \begin{bmatrix} 3 \\ 2 \\ 1 \end{bmatrix}$ (This is a deliberate choice to be much different from the exact solution.)
2. Compute

$$E_\alpha^{(0)} = F_\alpha - K_{\alpha\beta} U_\beta^{(0)} = \begin{bmatrix} -10 \\ 10 \\ -12 \end{bmatrix} \qquad \| E_\alpha^{(0)} \| = \sqrt{344} = 18.5472$$

$$\overline{E}_\alpha^{(1)} = \frac{E_\alpha^{(0)}}{\| E_\alpha^{(0)} \|} = \begin{bmatrix} -0.5392 \\ 0.5392 \\ -0.6470 \end{bmatrix}$$

3. Iterate for $i = 1, 2, \ldots, r$
   (a) $i = 1$:

$$\tilde{E}_\alpha^{(1)} = K_{\alpha\beta} \overline{E}_\beta^{(1)} = \begin{bmatrix} 0.7543 \\ 0.9705 \\ -3.1272 \end{bmatrix} \qquad \text{For } j = 1, \ldots, i:$$

$$a^{(1,1)} = \tilde{E}_\alpha^{(1)}\overline{E}_\alpha^{(1)} = 2.1395$$

$$\tilde{E}_\alpha^{(1)} = \tilde{E}_\alpha^{(1)} - a^{(1,1)}\overline{E}_\alpha^{(1)} = \begin{bmatrix} 1.9084 \\ -0.1831 \\ -1.7429 \end{bmatrix}$$

$$\| \tilde{E}_\alpha^{(1)} \| = 2.5910$$

$$\overline{E}_\alpha^{(2)} = \frac{\tilde{E}_\alpha^{(1)}}{\| \tilde{E}_\alpha^{(1)} \|} = \begin{bmatrix} 0.7366 \\ -0.0707 \\ -0.6727 \end{bmatrix}$$

(b) $i = 2$:

$$\tilde{E}_\alpha^{(2)} = K_{\alpha\beta}\overline{E}_\beta^{(2)} = \begin{bmatrix} 3.4137 \\ -3.5482 \\ 4.4968 \end{bmatrix}$$

For $j = 1, 2$ Do

$j = 1$:

$$a^{(2,1)} = \tilde{E}_\alpha^{(2)}\overline{E}_\alpha^{(1)} = -6.6630$$

$$\tilde{E}_\alpha^{(2)} = \tilde{E}_\alpha^{(2)} - a^{(2,1)}\overline{E}_\alpha^{(1)} = \begin{bmatrix} -0.1788 \\ 0.0442 \\ 0.1858 \end{bmatrix}$$

$j = 2$:

$$a^{(2,2)} = \tilde{E}_\alpha^{(2)}\overline{E}_\alpha^{(2)} = -0.2598$$

$$\tilde{E}_\alpha^{(2)} = \tilde{E}_\alpha^{(2)} - a^{(2,2)}\overline{E}_\alpha^{(2)} = \begin{bmatrix} 0.0126 \\ 0.0259 \\ 0.0111 \end{bmatrix}$$

$$\| \tilde{E}_\alpha^{(2)} \| = 0.0308$$

$$\overline{E}_\alpha^{(3)} = \frac{\tilde{E}_\alpha^{(2)}}{\| \tilde{E}_\alpha^{(2)} \|} = \begin{bmatrix} 0.4084 \\ 0.8392 \\ 0.3590 \end{bmatrix}$$

(c) $i = 3$:

$$\tilde{E}_\alpha^{(3)} = K_{\alpha\beta}\overline{E}_\beta^{(3)} = \begin{bmatrix} 2.1856 \\ -2.1138 \\ 0.0045 \end{bmatrix}$$

For $j = 1, \ldots 3$ Do

$j = 1$:

$$a^{(3,1)} = \tilde{E}_\alpha^{(3)}\overline{E}_\alpha^{(1)} = -2.3209$$

$$\tilde{E}_\alpha^{(3)} = \tilde{E}_\alpha^{(3)} - a^{(3,1)}\overline{E}_\alpha^{(1)} = \begin{bmatrix} 0.9342 \\ -0.8624 \\ -1.4972 \end{bmatrix}$$

$j = 2$:

$$a^{(3,2)} = \tilde{E}_\alpha^{(3)} \overline{E}_\alpha^{(2)} = -1.7561$$

$$\tilde{E}_\alpha^{(3)} = \tilde{E}_\alpha^{(3)} - a^{(3,2)} \overline{E}_\alpha^{(2)} = \begin{bmatrix} -0.3593 \\ -0.7383 \\ -0.3159 \end{bmatrix}$$

$j = 3$:

$$a^{(3,3)} = \tilde{E}_\alpha^{(3)} \overline{E}_\alpha^{(3)} = -0.8798$$

$$\tilde{E}_\alpha^{(3)} = \tilde{E}_\alpha^{(3)} - a^{(3,3)} \overline{E}_\alpha^{(3)} \approx \begin{bmatrix} 0 \\ 0 \\ 0 \end{bmatrix}$$

$$\| \tilde{E}_\alpha^{(3)} \| = 0$$

$$\overline{E}_\alpha^{(4)} = \frac{\tilde{E}_\alpha^{(3)}}{\| \tilde{E}_\alpha^{(3)} \|} = 0.$$

4. Construct Hessenberg matrix

$$\begin{bmatrix} a^{(1,1)} & a^{(2,1)} & a^{(3,1)} \\ \| \tilde{E}_\alpha^{(1)} \| & a^{(2,2)} & a^{3,2} \\ 0 & \| \tilde{E}_\alpha^{(2)} \| & a^{3,3} \\ 0 & 0 & \| \tilde{E}_\alpha^{(3)} \| \end{bmatrix} \begin{bmatrix} y^{(1)} \\ y^{(2)} \\ y^{(3)} \end{bmatrix} = \begin{bmatrix} \| E_\alpha^{(0)} \| \\ 0 \\ 0 \end{bmatrix}$$

$$\begin{bmatrix} 2.1395 & -6.6630 & -2.3210 \\ 2.5910 & -0.2598 & 1.7561 \\ 0 & 0.0308 & -0.8798 \end{bmatrix} \begin{bmatrix} y^{(1)} \\ y^{(2)} \\ y^{(3)} \end{bmatrix} = \begin{bmatrix} 18.5472 \\ 0 \\ 0 \end{bmatrix}$$

5. Apply Givens rotation to reduce matrix for tridiagonalization.
   (a) First rotation:

$$c_j = \frac{h_{jj}}{r_j}, \qquad s_j = \frac{h_{j+1,j}}{r_j}, \qquad r_j = \sqrt{h_{jj}^2 + h_{j+1,j}^2}$$

$$c_1 = \frac{a^{(1,1)}}{\sqrt{\left(a^{(1,1)}\right)^2 + \left(\| \tilde{E}_\alpha^{(1)} \|\right)^2}} = 0.6367 \; s_1 = \frac{\tilde{E}_\alpha^{(1)}}{\sqrt{\left(a^{(1,1)}\right)^2 + \left(\| \tilde{E}_\alpha^{(1)} \|\right)^2}} = 0.7711$$

$$\begin{bmatrix} c & s & \\ -s & c & \\ & & 1 \end{bmatrix} \begin{bmatrix} a^{(1,1)} & a^{(2,1)} & a^{(3,1)} \\ \| \tilde{E}_\alpha^{(1)} \| & a^{(2,2)} & a^{(3,2)} \\ & \| \tilde{E}_\alpha^{(2)} \| & a^{(3,3)} \\ & & \| \tilde{E}_\alpha^{(3)} \| \end{bmatrix} \begin{bmatrix} y_1 \\ y_2 \\ y_3 \end{bmatrix} = \begin{bmatrix} c & s & 0 \\ -s & c & 0 \\ 0 & 0 & 1 \end{bmatrix} \begin{bmatrix} \| E_\alpha^{(0)} \| \\ 0 \\ 0 \end{bmatrix}$$

$$\begin{bmatrix} 3.3602 & -4.4429 & -0.1237 \\ 0 & 4.9723 & 2.9079 \\ 0 & 0.0308 & -0.8798 \end{bmatrix} \begin{bmatrix} y_1 \\ y_2 \\ y_3 \end{bmatrix} = \begin{bmatrix} 11.8097 \\ -14.3014 \\ 0 \end{bmatrix}$$

   (b) Second rotation:

$$\begin{bmatrix} 1 & 0 & 0 \\ 0 & c & s \\ 0 & -s & c \end{bmatrix} \begin{bmatrix} 3.3602 & -4.4429 & -0.1237 \\ 0 & 4.9723 & 2.9079 \\ 0 & 0.0308 & -0.8798 \end{bmatrix} \begin{bmatrix} y_1 \\ y_2 \\ y_3 \end{bmatrix}$$

$$= \begin{bmatrix} 1 & 0 & 0 \\ 0 & c & s \\ 0 & -s & c \end{bmatrix} \begin{bmatrix} 11.8097 \\ -14.3014 \\ 0 \end{bmatrix}$$

$$c_2 = 0.9999 \qquad s_2 = 0.0062$$

$$\begin{bmatrix} 3.3602 & -4.4429 & -0.1237 \\ 0 & 4.9724 & 2.9024 \\ 0 & 0 & -0.8798 \end{bmatrix} \begin{bmatrix} y_1 \\ y_2 \\ y_3 \end{bmatrix} = \begin{bmatrix} 11.8097 \\ -14.3012 \\ 0.0886 \end{bmatrix}$$

$$\begin{bmatrix} y^{(1)} \\ y^{(2)} \\ y^{(3)} \end{bmatrix} = \begin{bmatrix} -0.2157 \\ -2.8185 \\ -0.0988 \end{bmatrix}$$

6. Compute residual

$$\begin{bmatrix} \overline{E}_1^{(1)} & \overline{E}_1^{(2)} & \overline{E}_1^{(3)} \\ \overline{E}_2^{(1)} & \overline{E}_2^{(2)} & \overline{E}_2^{(3)} \\ \overline{E}_3^{(1)} & \overline{E}_3^{(2)} & \overline{E}_3^{(3)} \end{bmatrix} \begin{bmatrix} y_1 \\ y_2 \\ y_3 \end{bmatrix} = \begin{bmatrix} E_1^{(r)} \\ E_2^{(r)} \\ E_3^{(r)} \end{bmatrix}$$

$$\begin{bmatrix} E_1^{(r)} \\ E_2^{(r)} \\ E_3^{(r)} \end{bmatrix} = \begin{bmatrix} -0.5392 & 0.7366 & 0.4084 \\ 0.5392 & -0.0707 & 0.8392 \\ -0.6470 & -0.6727 & 0.3590 \end{bmatrix} \begin{bmatrix} -0.2157 \\ -2.8185 \\ -0.0987 \end{bmatrix} = \begin{bmatrix} -2 \\ 0 \\ 2 \end{bmatrix}$$

7. Update U_β

$$\begin{bmatrix} U_1 \\ U_2 \\ U_3 \end{bmatrix} = \begin{bmatrix} U_1^{(0)} \\ U_2^{(0)} \\ U_3^{(0)} \end{bmatrix} + \begin{bmatrix} E_1^{(r)} \\ E_2^{(r)} \\ E_3^{(r)} \end{bmatrix} = \begin{bmatrix} 1 \\ 2 \\ 3 \end{bmatrix}$$

Note that the exact solution has been obtained.

11.5.4 COMBINED GPG-EBE-GMRES PROCESS

We consider the solution by generalized Petrov-Galerkin (GPG) method using EBE-GMRES solver. The global GPG equation (11.4.5) may be written in a local form.

$$\left[A_{NM}^{(e)} + \eta\Delta t\left(B_{NM}^{(e)} + C_{NM}^{(e)} + K_{NM}^{(e)}\right)\right] v_{Mi}^{(e)n+1}$$
$$= \left[A_{NM}^{(e)} - (1-\eta)\Delta t\left(B_{NM}^{(e)} + C_{NM}^{(e)} + K_{NM}^{(e)}\right)\right] v_{Mi}^{(e)n} + \Delta t\left(F_{Mi}^{(e)n} + G_{Mi}^{(e)n}\right) \quad (11.5.41)$$

or

$$R_{NM}^{(e)} v_{Mi}^{(e)n+1} = Q_{Ni}^{(e)n} \quad (11.5.42)$$

For illustration, let us consider the global and local configurations as given in Figure 11.5.4.1.

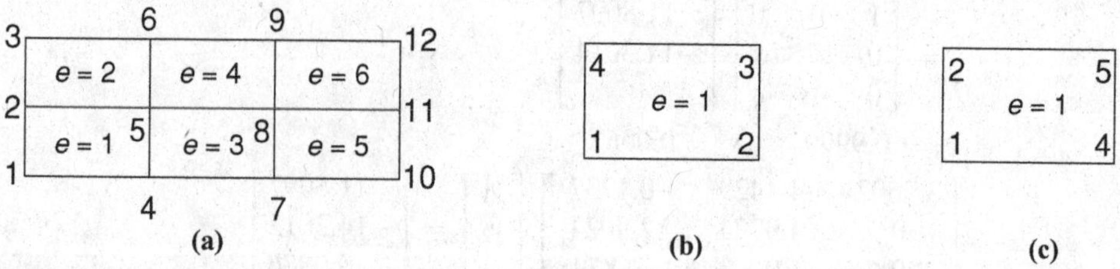

Figure 11.5.4.1 Global and local configurations. (**a**) Global system. (**b**) Local. (**c**) Global.

Using the four-node isoparametric element on the left-hand side of (11.5.42) for $e = 1$, we have

$$D_{Ni}^{(1)(n+1)} = R_{NM}^{(1)}\, v_{Mi}^{(1)(n+1)} \tag{11.5.43}$$

or

$$\begin{bmatrix} D_{11}^{(1)} \\ D_{12}^{(1)} \\ D_{41}^{(1)} \\ D_{42}^{(1)} \\ D_{51}^{(1)} \\ D_{52}^{(1)} \\ D_{21}^{(1)} \\ D_{22}^{(1)} \end{bmatrix}^{(n+1)} = \begin{bmatrix} R_{11}^{(1)} & 0 & R_{14}^{(1)} & 0 & R_{15}^{(1)} & 0 & R_{12}^{(1)} & 0 \\ 0 & R_{11}^{(1)} & 0 & R_{11}^{(1)} & 0 & R_{11}^{(1)} & 0 & R_{11}^{(1)} \\ R_{41}^{(1)} & 0 & R_{44}^{(1)} & 0 & R_{45}^{(1)} & 0 & R_{42}^{(1)} & 0 \\ 0 & R_{41}^{(1)} & 0 & R_{44}^{(1)} & 0 & R_{45}^{(1)} & 0 & R_{42}^{(1)} \\ R_{51}^{(1)} & 0 & R_{54}^{(1)} & 0 & R_{55}^{(1)} & 0 & R_{52}^{(1)} & 0 \\ 0 & R_{51}^{(1)} & 0 & R_{54}^{(1)} & 0 & R_{55}^{(1)} & 0 & R_{52}^{(1)} \\ R_{21}^{(1)} & 0 & R_{24}^{(1)} & 0 & R_{25}^{(1)} & 0 & R_{22}^{(1)} & 0 \\ 0 & R_{21}^{(1)} & 0 & R_{24}^{(1)} & 0 & R_{25}^{(1)} & 0 & R_{22}^{(1)} \end{bmatrix} \begin{bmatrix} v_{11}^{(1)} \\ v_{12}^{(1)} \\ v_{41}^{(1)} \\ v_{42}^{(1)} \\ v_{51}^{(1)} \\ v_{52}^{(1)} \\ v_{21}^{(1)} \\ v_{22}^{(1)} \end{bmatrix}^{(n+1)}$$

with the local element node numbers being replaced by the global node numbers for global assembly.

The assembled column vector $D_{\alpha i}$ takes the form

$$D_{\alpha i} = \bigcup_{e=1}^{E} D_{Ni}^{(e)} \Delta_{N\alpha}^{(e)} = \bigcup_{e=1}^{E} R_{NM}^{(e)}\, v_{Mi}^{(e)}\, \Delta_{N\alpha}^{(e)} \tag{11.5.44}$$

This operation is identical to the summing process, as shown in Table 11.5.1. with

$$D_{11}^{(1)} = R_{11}^{(1)} v_{11}^{(1)} + R_{14}^{(1)} v_{41}^{(1)} + R_{15}^{(1)} v_{51}^{(1)} + R_{12}^{(1)} v_{21}^{(1)}$$
$$D_{12}^{(1)} = R_{11}^{(1)} v_{12}^{(1)} + R_{14}^{(1)} v_{42}^{(1)} + R_{15}^{(1)} v_{52}^{(1)} + R_{12}^{(1)} v_{22}^{(1)}$$

etc.

For illustration let us consider the geometry given in Figure 11.5.4.1c. It represents $189 \times 2 = 378$ equations given by the column vector $D_{\alpha i}$, which is assembled from 8×8 local stiffness matrices multiplied by the 8×1 local variable unknown column vectors.

Table 11.5.1 Global Summing Procedure

Node	$e=1$	$e=2$	$e=3$	$e=4$	$e=5$	$e=6$	$D_{\alpha i}$ (sum)
1	$D_{11}^{(1)}$						$D_{11} = D_{11}^{(1)}$
	$D_{12}^{(1)}$						$D_{12} = D_{12}^{(1)}$
2	$D_{21}^{(1)}$	$D_{21}^{(2)}$					$D_{21} = D_{21}^{(1)} + D_{21}^{(2)}$
	$D_{22}^{(1)}$	$D_{22}^{(2)}$					$D_{22} = D_{22}^{(1)} + D_{21}^{(2)}$
3		$D_{31}^{(2)}$					$D_{31} = D_{31}^{(2)}$
		$D_{32}^{(2)}$					$D_{32} = D_{32}^{(2)}$
4	$D_{41}^{(1)}$		$D_{41}^{(3)}$				$D_{41} = D_{41}^{(1)} + D_{41}^{(2)}$
	$D_{42}^{(1)}$		$D_{42}^{(3)}$				$D_{42} = D_{42}^{(1)} + D_{42}^{(2)}$
5	$D_{51}^{(1)}$	$D_{51}^{(2)}$	$D_{51}^{(3)}$	$D_{51}^{(4)}$			$D_{51} = D_{51}^{(1)} + D_{51}^{(2)} + D_{51}^{(3)} + D_{51}^{(4)}$
	$D_{52}^{(1)}$	$D_{52}^{(2)}$	$D_{52}^{(3)}$	$D_{52}^{(4)}$			$D_{52} = D_{52}^{(1)} + D_{52}^{(2)} + D_{52}^{(3)} + D_{52}^{(4)}$
6		$D_{61}^{(2)}$		$D_{61}^{(4)}$			$D_{61} = D_{61}^{(2)} + D_{61}^{(4)}$
		$D_{62}^{(2)}$		$D_{62}^{(4)}$			$D_{62} = D_{62}^{(2)} + D_{62}^{(4)}$
7			$D_{71}^{(3)}$		$D_{71}^{(5)}$		$D_{71} = D_{71}^{(3)} + D_{71}^{(5)}$
			$D_{72}^{(3)}$		$D_{72}^{(5)}$		$D_{72} = D_{72}^{(3)} + D_{72}^{(5)}$
8			$D_{81}^{(3)}$	$D_{81}^{(4)}$	$D_{81}^{(5)}$	$D_{81}^{(6)}$	$D_{81} = D_{81}^{(3)} + D_{81}^{(4)} + D_{81}^{(5)} + D_{81}^{(6)}$
			$D_{82}^{(3)}$	$D_{82}^{(4)}$	$D_{82}^{(5)}$	$D_{82}^{(6)}$	$D_{82} = D_{82}^{(3)} + D_{82}^{(4)} + D_{82}^{(5)} + D_{82}^{(6)}$
9				$D_{91}^{(4)}$		$D_{91}^{(6)}$	$D_{91} = D_{91}^{(4)} + D_{91}^{(6)}$
				$D_{92}^{(4)}$		$D_{92}^{(6)}$	$D_{92} = D_{92}^{(4)} + D_{92}^{(6)}$
10					$D_{10,1}^{(5)}$		$D_{10,1} = D_{10,1}^{(5)}$
					$D_{10,2}^{(5)}$		$D_{10,2} = D_{10,2}^{(5)}$
11					$D_{11,1}^{(5)}$	$D_{11,1}^{(6)}$	$D_{11,1} = D_{11,1}^{(5)} + D_{11,1}^{(6)}$
					$D_{11,2}^{(5)}$	$D_{11,2}^{(6)}$	$D_{11,2} = D_{11,2}^{(5)} + D_{11,2}^{(6)}$
12						$D_{12,1}^{(6)}$	$D_{12,1} = D_{12,1}^{(6)}$
						$D_{12,2}^{(6)}$	$D_{12,2} = D_{12,2}^{(6)}$

We follow the procedure similar to the one given in Example 11.5.1 except that we use the EBE process here. Thus, instead of global matrix $K_{\alpha\beta}$ (378×378) we now have a column vector $D_{\alpha i}(378 \times 1)$.

1. Specify initial and boundary conditions on all boundary nodes and assume values for all interior nodes ($v_{Mi}^{(e)} = 0$, for example)
2. Compute the error coefficient vector $\overline{E}_{\alpha i}^{(1)}$

$$E_{\alpha i}^{(0)} = Q_{\alpha i} - D_{\alpha i},$$

with $Q_{\alpha i} = \cup_{e=1}^{E} Q_{Ni}^{(e)} \Delta_{N\alpha}^{(e)}$ and $D_{\alpha i}$ as determined from (11.5.44).

$$\overline{E}_{\alpha i}^{(1)} = \frac{E_{\alpha i}^{(0)}}{\| E_{\alpha i}^{(0)} \|} \quad \text{(Gram-Schmidt process)}$$

3. Iterate for $i = 1, 2, 3, \ldots r$, say $r = 4$

   For this example calculate the adjusted error vector $\tilde{E}^{(1)}_{\alpha i}$, the normed error coefficient $a^{(1,1)}$, and a new error coefficient vector $\overline{E}^{(2)}_{\alpha i}$.

   (a) $i = 1$:

$$\tilde{E}^{(1)}_{\alpha i} = \bigcup_{e=1}^{E} E^{(e)(1)}_{Ni} \Delta^{(e)}_{N\alpha} = \bigcup_{e=1}^{E} R^{(e)}_{NM} \overline{E}^{(e)}_{Mi} \Delta^{(e)}_{N\alpha}$$

   $j = 1$:

$$a^{(1,1)} = \tilde{E}^{(1)}_{\alpha i} \overline{E}^{(1)}_{\alpha i}$$

$$\tilde{E}^{(1)}_{\alpha i} = \tilde{E}^{(1)}_{\alpha i} - a^{(1,1)} \overline{E}^{(1)}_{\alpha i}$$

$$\overline{E}^{(2)}_{\alpha i} = \frac{\tilde{E}^{(1)}_{\alpha i}}{\| \tilde{E}^{(1)}_{\alpha i} \|}$$

   (b) $i = 2$: (Calculate, similarly, new adjusted error vector, normed error coefficients, and error coefficient vector.)

$$\tilde{E}^{(2)}_{\alpha i} = \bigcup_{e=1}^{E} E^{(e)(2)}_{Ni} \Delta^{(e)}_{N\alpha} = \bigcup_{e=1}^{E} R^{(e)}_{NM} \overline{E}^{(e)(2)}_{Mi} \Delta^{(e)}_{N\alpha}$$

   $j = 1$:

$$a^{(2,1)} = \tilde{E}^{(2)}_{\alpha i} \overline{E}^{(1)}_{\alpha i}$$

$$\tilde{E}^{(2)}_{\alpha i} = \tilde{E}^{(2)}_{\alpha i} - a^{(2,1)} \overline{E}^{(1)}_{\alpha i}$$

   $j = 2$:

$$a^{(2,2)} = \tilde{E}^{(2)}_{\alpha i} \overline{E}^{(2)}_{\alpha i}$$

$$\tilde{E}^{(2)}_{\alpha i} = \tilde{E}^{(2)}_{\alpha i} - a^{(2,2)} \overline{E}^{(2)}_{\alpha i}$$

$$\overline{E}^{(3)}_{\alpha i} = \frac{\tilde{E}^{(2)}_{\alpha i}}{\| \tilde{E}^{(2)}_{\alpha i} \|}$$

   (c) $i = 3$, similarly,

$$\tilde{E}^{(3)}_{\alpha i} = \bigcup_{e=1}^{E} E^{(e)(3)}_{Ni} \Delta^{(e)}_{N\alpha} = \bigcup_{e=1}^{E} R^{(e)}_{NM} \overline{E}^{(e)(3)}_{Mi} \Delta^{(e)}_{N\alpha}$$

   $j = 1$:

$$a^{(3,1)} = \tilde{E}^{(3)}_{\alpha i} \overline{E}^{(1)}_{\alpha i}$$

$$\tilde{E}^{(3)}_{\alpha i} = \tilde{E}^{(3)}_{\alpha i} - a^{(3,1)} \overline{E}^{(1)}_{\alpha i}$$

   $j = 2$:

$$a^{(3,2)} = \tilde{E}^{(3)}_{\alpha i} \overline{E}^{(2)}_{\alpha i}$$

$$\tilde{E}^{(3)}_{\alpha i} = \tilde{E}^{(3)}_{\alpha i} - a^{(3,2)} \overline{E}^{(2)}_{\alpha i}$$

   $j = 3$:

$$a^{(3,3)} = \tilde{E}^{(3)}_{\alpha i} \overline{E}^{(3)}_{\alpha i}$$

$$\tilde{E}^{(3)}_{\alpha i} = \tilde{E}^{(3)}_{\alpha i} - a^{(3,3)} \overline{E}^{(3)}_{\alpha i}$$

$$\overline{E}^{(4)}_{\alpha i} = \frac{\tilde{E}^{(3)}_{\alpha i}}{\| \tilde{E}^{(3)}_{\alpha i} \|}$$

(d) $i = 4$: Again similarly,

$$\tilde{E}^{(4)}_{\alpha i} = \bigcup_{e=1}^{E} E^{(e)(4)}_{Ni} \Delta^{(e)}_{N\alpha} = \bigcup_{e=1}^{E} R^{(e)}_{NM} \overline{E}^{(e)(4)}_{Mi} \Delta^{(e)}_{N\alpha}$$

$j = 1$:

$$a^{(4,1)} = \tilde{E}^{(4)}_{\alpha i} \overline{E}^{(1)}_{\alpha i}$$

$$\tilde{E}^{(4)}_{\alpha i} = \tilde{E}^{(4)}_{\alpha i} - a^{(4,1)} \overline{E}^{(1)}_{\alpha i}$$

$j = 2$:

$$a^{(4,2)} = \tilde{E}^{(4)}_{\alpha i} \overline{E}^{(2)}_{\alpha i}$$

$$\tilde{E}^{(4)}_{\alpha i} = \tilde{E}^{(4)}_{\alpha i} - a^{(4,2)} \overline{E}^{(2)}_{\alpha i}$$

$j = 3$:

$$a^{(4,3)} = \tilde{E}^{(4)}_{\alpha i} \overline{E}^{(3)}_{\alpha i}$$

$$\tilde{E}^{(4)}_{\alpha i} = \tilde{E}^{(4)}_{\alpha i} - a^{(4,3)} \overline{E}^{(3)}_{\alpha i}$$

$j = 4$:

$$a^{(4,4)} = \tilde{E}^{(4)}_{\alpha i} \overline{E}^{(4)}_{\alpha i}$$

$$\tilde{E}^{(4)}_{\alpha i} = \tilde{E}^{(4)}_{\alpha i} - a^{(4,4)} \overline{E}^{(3)}_{\alpha i} \approx 0$$

$$\overline{E}^{(5)}_{\alpha i} = \frac{\tilde{E}^{(4)}_{\alpha i}}{\| \tilde{E}^{(4)}_{\alpha i} \|} \approx 0$$

4. Construct Hessenberg matrix to calculate the minimizer vector y_r ($r = 4$ in this case)

$$\begin{bmatrix} a^{(1,1)} & a^{(2,1)} & a^{(3,1)} & a^{(4,1)} \\ \| \tilde{E}^{(1)}_{\alpha i} \| & a^{(2,2)} & a^{(3,2)} & a^{(4,2)} \\ & \| \tilde{E}^{(2)}_{\alpha i} \| & a^{(3,3)} & a^{(4,3)} \\ & & \| \tilde{E}^{(3)}_{\alpha i} \| & a^{(4,4)} \end{bmatrix} \begin{bmatrix} y_1 \\ y_2 \\ y_3 \\ y_4 \end{bmatrix} = \begin{bmatrix} \| E^{(0)}_{\alpha i} \| \\ 0 \\ 0 \\ 0 \end{bmatrix}$$

where $\| \tilde{E}^{(4)}_{\alpha i} \| \cong 0$ is assumed.

5. Apply Givens rotations to reduce Hessenberg matrix to an upper triangular form in order to find the minimizer error vector y, as shown in step 5 of Example 11.5.1
6. Compute residuals (for the case of Figure 11.6.3.1a)

$$\underset{(378 \times 4)}{\begin{bmatrix} \overline{E}^{(1)}_1 & \overline{E}^{(2)}_1 & \overline{E}^{(3)}_1 & \overline{E}^{(4)}_1 \\ \overline{E}^{(1)}_2 & \overline{E}^{(2)}_2 & \overline{E}^{(3)}_2 & \overline{E}^{(4)}_2 \\ \cdot & \cdot & \cdot & \cdot \\ \cdot & \cdot & \cdot & \cdot \\ \cdot & \cdot & \cdot & \cdot \\ \cdot & \cdot & \cdot & \cdot \\ \overline{E}^{(1)}_{378} & \overline{E}^{(2)}_{378} & \overline{E}^{(3)}_{378} & \overline{E}^{(4)}_{378} \end{bmatrix}} \underset{(4 \times 1)}{\begin{bmatrix} y_1 \\ y_2 \\ y_3 \\ y_4 \end{bmatrix}} = \underset{(378 \times 1)}{\begin{bmatrix} E^{(r)}_1 \\ E^{(r)}_2 \\ \cdot \\ \cdot \\ \cdot \\ \cdot \\ E^{(r)}_{378} \end{bmatrix}}$$

7. Update $U_{\beta i}$

$$\begin{bmatrix} v_1 \\ v_2 \\ \cdot \\ \cdot \\ \cdot \\ \cdot \\ v_{378} \end{bmatrix} = \begin{bmatrix} v_1^{(0)} \\ v_2^{(0)} \\ \cdot \\ \cdot \\ \cdot \\ \cdot \\ v_{378}^{(0)} \end{bmatrix} + \begin{bmatrix} E_1^{(r)} \\ E_2^{(r)} \\ \cdot \\ \cdot \\ \cdot \\ \cdot \\ E_{378}^{(r)} \end{bmatrix}$$

If the adjusted error vector $\tilde{E}_{\alpha i}^{(4)}$ and the error coefficient vector $\overline{E}_{\alpha i}^{(5)}$ are not approximately zero, then further iterations will be required.

11.5.5 PRECONDITIONING FOR EBE-GMRES

Although Krylov subspace methods such as the GMRES method are well founded theoretically, they are likely to suffer from slow convergence for fluid dynamics applications, especially in the problems involving high Mach numbers and high Reynolds numbers. Preconditioning is a key ingredient in the success of Krylov subspace methods in these applications. In creating a preconditioner for the EBE equations, the first step is to normalize each element matrix using a scaling transformation that can be viewed as an initial level of preconditioning, often called "pre-preconditioning" [Saad, 1996; Shakib et al. 1991]. Typically, a diagonal, or a block diagonal, scaling is first applied to the element matrices to obtain scaled element matrices.

Step 1: Pre-preconditioning

Consider the local finite element equations given by

$$R_{NMrs}^{(e)}\,\Delta U_{Ms}^{(e)} = Q_{Nr}^{(e)} \tag{11.5.45}$$

The left-hand side may be written as

$$C_{Nr}^{(e)} = R_{NMrs}^{(e)}\,\Delta U_{Ms}^{(e)} \tag{11.5.46}$$

The EBE process provides

$$C_{\alpha r}^{n+1} = \bigcup_{e=1}^{E} C_{Nr}^{(e)}\,\Delta_{N\alpha}^{(e)} \tag{11.5.47}$$

with

$$Q_{\alpha r}^{n+1} = \bigcup_{e=1}^{E} Q_{Nr}^{(e)}\,\Delta_{N\alpha}^{(e)} \tag{11.5.48}$$

Construct the diagonal scaling matrix $D_{\alpha\beta rs}$ in the form

$$D_{\alpha\beta rs} = \bigcup_{e=1}^{E} R_{\alpha prs}^{(e)}\,\delta_{pM}\Delta_{M\beta}^{(e)}$$

Note that since the off-diagonal terms of $D_{\alpha\beta rs}$ are zero, $D_{\alpha\beta rs}$ can be stored as a vector.

Performing the preconditioning operations on the unassembled element equations requires three steps:

(1) Gather, or localize, the components of the global diagonal vector into local element vectors. Let $D^{(e)}_{NMrs}$ denote the local diagonal matrix for element (e).

(2) Perform the preconditioning operations on the element level. Equation (11.5.48) is transformed into

$$\tilde{R}^{(e)}_{NMrs}\,\Delta\tilde{U}^{(e)}_{Ms} = \tilde{Q}^{(e)}_{Nr} \tag{11.5.49}$$

where

$$\tilde{R}^{(e)}_{NMrs} = (\tilde{D}^{(e)}_{Np})^{-\frac{1}{2}} R_{pqrs} (\tilde{D}^{(e)}_{qM})^{-\frac{1}{2}}$$
$$\Delta\tilde{U}^{(e)}_{Mr} = (D^{(e)}_{Mp})^{-\frac{1}{2}} \Delta U^{(e)}_{pr}$$
$$Q^{(e)}_{Nr} = (D^{(e)}_{Np})^{-\frac{1}{2}} Q^{(e)}_{pr}$$

with

$$\tilde{C}^{(e)}_{Nr} = \tilde{R}^{(e)}_{NMrs}\Delta U^{(e)}_{Ms}$$

(3) Scatter, or globalize, the components of the local element vectors into the global vectors as follows:

$$\tilde{C}^{(e)}_{\alpha r} = \bigcup_{e=1}^{E} \tilde{C}^{(e)}_{Nr}\,\Delta^{(e)}_{N\alpha}, \quad \tilde{Q}^{(e)}_{\alpha r} = \bigcup_{e=1}^{E} \tilde{Q}^{(e)}_{Nr}\,\Delta^{(e)}_{N\alpha} \tag{11.5.50}$$

Step 2: Main preconditioning by upper and lower triangular matrices

The second step in defining an EBE preconditioner is to regularize the transformed element matrices from step 1. Using Winget regularization, the diagonal of each coefficient matrix is forced to be the identity matrix. In other words, the regularized matrix is defined as

$$\bar{R}^{(e)}_{NMrs} = \tilde{R}^{(e)}_{NMrs} - diag(\tilde{R}^{(e)}_{NMrs}) + I_{NMrs} \tag{11.5.51}$$

Finally, the factorization must be chosen for the preconditioning matrix. We choose the LU factorization for the regularized matrix $\bar{R}^{(e)}_{NMrs}$ to produce the preconditioning matrix $G^{(e)}_{NMrs}$ of the form

$$G^{(e)}_{NMrs} = L^{(e)}_{Nprt}\, U^{(e)}_{pMts} \tag{11.5.52}$$

where $L^{(e)}_{Nprt}$ and $U^{(e)}_{pMts}$ are obtained by factoring the regularized matrix $\bar{R}^{(e)}_{NMrs}$ into a unit lower and an upper triangular matrix. In other words,

$$G^{(e)} = \begin{bmatrix} 1 & 0 & 0 & 0 & \cdots & 0 \\ L_{21} & 1 & 0 & 0 & \cdots & \vdots \\ L_{31} & L_{32} & 1 & 0 & \cdots & \vdots \\ L_{41} & L_{42} & \cdots & \ddots & 0 & \\ \vdots & \cdots & \cdots & \cdots & \ddots & 0 \\ L_{M1} & L_{M2} & \cdots & \cdots & \cdots & 1 \end{bmatrix} \begin{bmatrix} U_{11} & U_{12} & U_{13} & \cdots & \cdots & U_{1M} \\ 0 & U_{22} & U_{23} & U_{24} & \cdots & U_{2M} \\ 0 & 0 & U_{33} & U_{34} & \cdots & U_{3M} \\ 0 & 0 & 0 & U_{44} & \cdots & \vdots \\ \vdots & \vdots & \vdots & 0 & \ddots & \vdots \\ 0 & 0 & \cdots & \cdots & 0 & U_{MM} \end{bmatrix}$$

where the indices rt and ts are omitted for simplicity.

Notice that in practice, $L^{(e)}_{Nprt}$ and $U^{(e)}_{PMts}$ can be stored together.

We premultiply the left-hand and right-hand sides of (11.5.49) by the inverse of the preconditioned local element matrices as follows:

$$G^{(e)-1}_{pNrt}\, R^{(e)}_{NMts}\, \Delta U^{(e)}_{Ms} = G^{(e)-1}_{pNrs}\, Q_{Ns} \tag{11.5.53}$$

However, in practice we do not actually calculate the inverse of the preconditioning matrix. Instead, consider writing the right-hand side of (11.5.53) as

$$\hat{Q}^{(e)}_{Nr} = L^{(e)-1}_{NMrt}\, U^{(e)-1}_{Mpts}\, \tilde{Q}^{(e)}_{ps}, \quad \text{or} \quad L^{(e)}_{NMrt}\, U^{(e)}_{Mpts}\, \hat{Q}^{(e)}_{ps} = \tilde{Q}^{(e)}_{Nr} \tag{11.5.54}$$

Consider rewriting (11.5.54) as

$$L^{(e)}_{NMrs}\, Z^{(e)}_{Ms} = \tilde{Q}^{(e)}_{Nr} \tag{11.5.55}$$

where $Z^{(e)}_{Mr} = U^{(e)}_{MNrs}\, \hat{Q}^{(e)}_{Ns}$. Since $L^{(e)}_{NMrs}$ is lower triangular, Equation (11.5.55) can be solved for $Z^{(e)}_{Mr}$ using forward reduction. Then, the equation $U^{(e)}_{MNrs}\, \hat{Q}^{(e)}_{Ns} = Z^{(e)}_{Mr}$ can be solved for $\hat{Q}^{(e)}_{Ns}$, which is the right-hand side of (11.5.53), by back substitution. A similar operation is performed to evaluate the left-hand side of (11.5.53). The element values are then mapped to the global column vector as shown below.

$$\hat{C}^{(e)}_{Nr} = G^{(e)-1}_{pNrt}\, R^{(e)}_{pMts}\, \Delta \tilde{U}^{(e)}_{Ms}, \quad \hat{C}_{\alpha r} = \bigcup_{e=1}^{E} \hat{C}^{(e)}_{Nr}\, \Delta^{(e)}_{N\alpha}$$

$$\hat{Q}^{(e)}_{Nr} = G^{(e)-1}_{NMrs}\, \hat{Q}^{(e)}_{Ms}, \quad \hat{Q}_{\alpha r} = \bigcup_{e=1}^{E} Q^{(e)}_{Nr}\, \Delta^{(e)}_{N\alpha}$$

The pre-conditioned GMRES process begins with

$$E^{(0)}_{\alpha r} = \hat{Q}^{(0)}_{\alpha r} - \hat{C}^{(0)}_{\alpha r}$$

and

$$\overline{E}^{(1)}_{\alpha i} = \frac{E^{(0)}_{\alpha i}}{\left\| E^{(0)}_{\alpha i} \right\|}$$

Step 2 of the GMRES procedure described in Section 11.5.3 is rewritten as follows: GMRES iteration: For $i = 1, 2, 3, \ldots, r$ Do

$$\overline{E}^{(i+1)}_{\alpha r} = G^{-1}_{\alpha\gamma rt}\, R_{\gamma\beta ts}\, \overline{E}^{(i)}_{\beta s} = \bigcup_{e=1}^{E} G^{(e)-1}_{NMrt}\, R^{(e)}_{Mpts}\, \overline{E}^{(e)}_{ps}\, \Delta^{(e)}_{N\alpha}$$

The rest follows identically as in step 2 through step 6.

11.6 EXAMPLE PROBLEMS

11.6.1 NONLINEAR WAVE EQUATION (CONVECTION EQUATION)

Consider the first order nonlinear wave equation of the form used in Section 4.7.5.

$$\frac{\partial u}{\partial t} + u\frac{\partial u}{\partial x} = 0, \quad 0 \le x \le 4$$
$$u(x,0) = 1 \quad 0 \le x \le 2$$
$$u(x,0) = 0 \quad 2 \le x \le 4$$

Required: Solve with GPG using the numerical diffusion given by (11.3.32).

Solution: The GPG formulation begins with

$$\int_0^L W(\xi)\left[\int \Phi_\alpha\left(\frac{\partial u}{\partial t} + u\frac{\partial u}{\partial x}\right)dx + \int \Psi_\alpha u\frac{\partial u}{\partial x}dx\right]d\xi = 0$$

with

$$\Psi_N^{(e)} = \tau u\frac{\partial \Phi_N^{(e)}}{\partial x}$$

where τ is the numerical diffusion factor (intrinsic time scale),

$$\tau = \frac{C}{2}\frac{h}{u}$$

with C being the CFL number,

$$C = \overline{\alpha} = \coth H - \frac{1}{H}$$

which is characterized by the numerical diffusion as shown in Figure 11.3.2 defining the accuracy and stability for the solution of the nonlinear convection equation.

As a result, it is seen that dispersion or dissipation errors decrease with mesh refinements, as shown in Figure 11.6.1. Accuracies deteriorate significantly with inadequate numerical diffusivity constants outside the stability and accuracy criteria.

11.6.2 PURE CONVECTION IN TWO DIMENSIONS

The two-dimensional pure convection equation for a concentration cone placed in a rotating velocity field, as shown in Figure 11.6.2a is given by

$$\frac{\partial u}{\partial t} + A_i\frac{\partial u}{\partial x_i} = 0$$

where

$$A_i = (a\cos\theta, a\sin\theta) \quad \text{with } a = 1/2$$

Initial Data:

$$u_0 = \begin{cases} \frac{1}{2}(1+\cos 4\rho\pi) & \rho \le \frac{1}{4} \\ 0 & \text{otherwise} \end{cases}$$

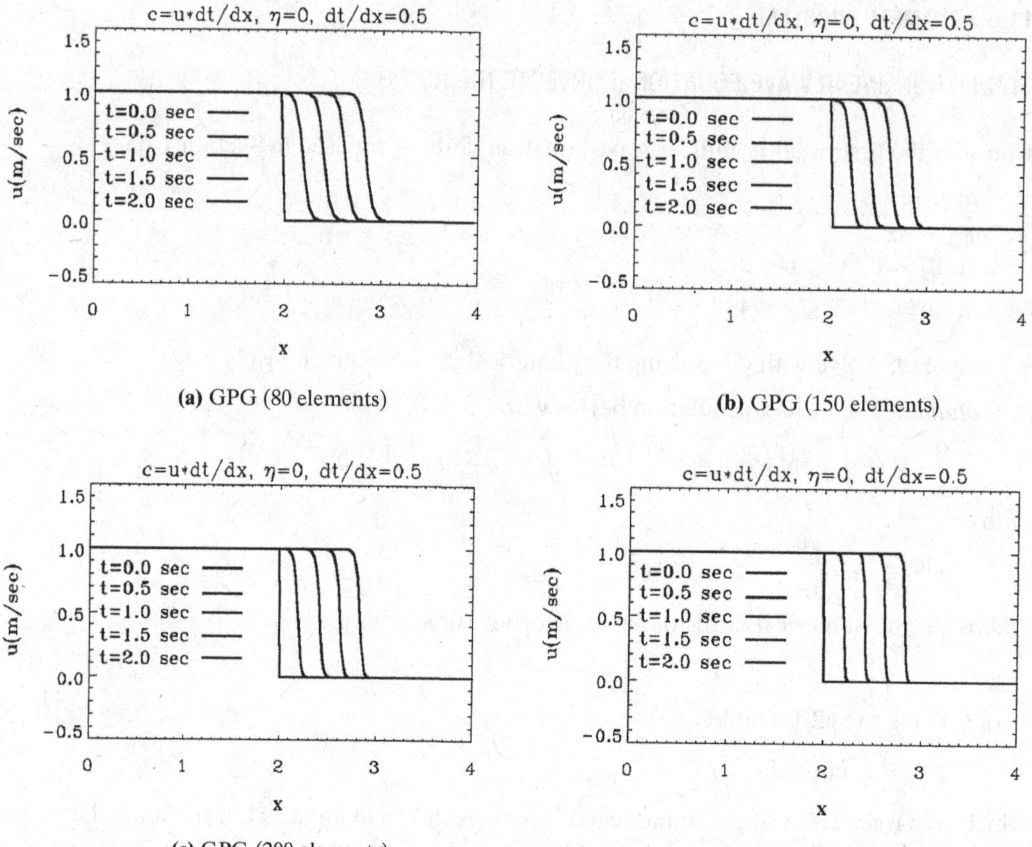

(a) GPG (80 elements)
(b) GPG (150 elements)
(c) GPG (200 elements)
(d) GPG (400 elements)

Figure 11.6.1 GPG solutions for nonlinear convection shock wave propagation (lumped mass matrix).

where

$$\rho^2 = (x - 0)^2 + (y + 0.5)^2$$

Required: Solve using the GTG and GPG methods with lumped and consistent mass matrices. Carry out until 1 revolution is reached.

Solution: For the computation, a 32×32 grid mesh in a 2.0×2.0 domain is chosen, and initial cosine hill with unit magnitude is centered at (0.0, −0.5) whose base radius spans eight elements in Figure 11.6.2b. Use a constant time step, $\Delta t = 2\pi/400$. The total number of nodes is 1089, and all boundary conditions are Dirichlet type, $u = 0$, a complete rotation is accomplished in 400 time steps. The Courant number at the peak of the cone is approximately 1/4.

For the GTG method with the lumped mass, the solution with one iteration (Figure 11.6.2c) has wiggles and reduced cone height more than those with three iterations (Figure 11.6.2d); an improved solution is obtained for the case of consistent mass (Figure 11.6.2e) for $t = \pi/4$ as compared with that for lumped mass. The results of the GPG method at $t = \pi/4$ are shown in (Figure 11.6.2f) (1), (2), (3), and (4) corresponding to the numerical diffusivity of $\overline{\alpha} = 10^{-4}, 10^{-2}, 1$, and 10^2, respectively. In Figure 11.6.2g,

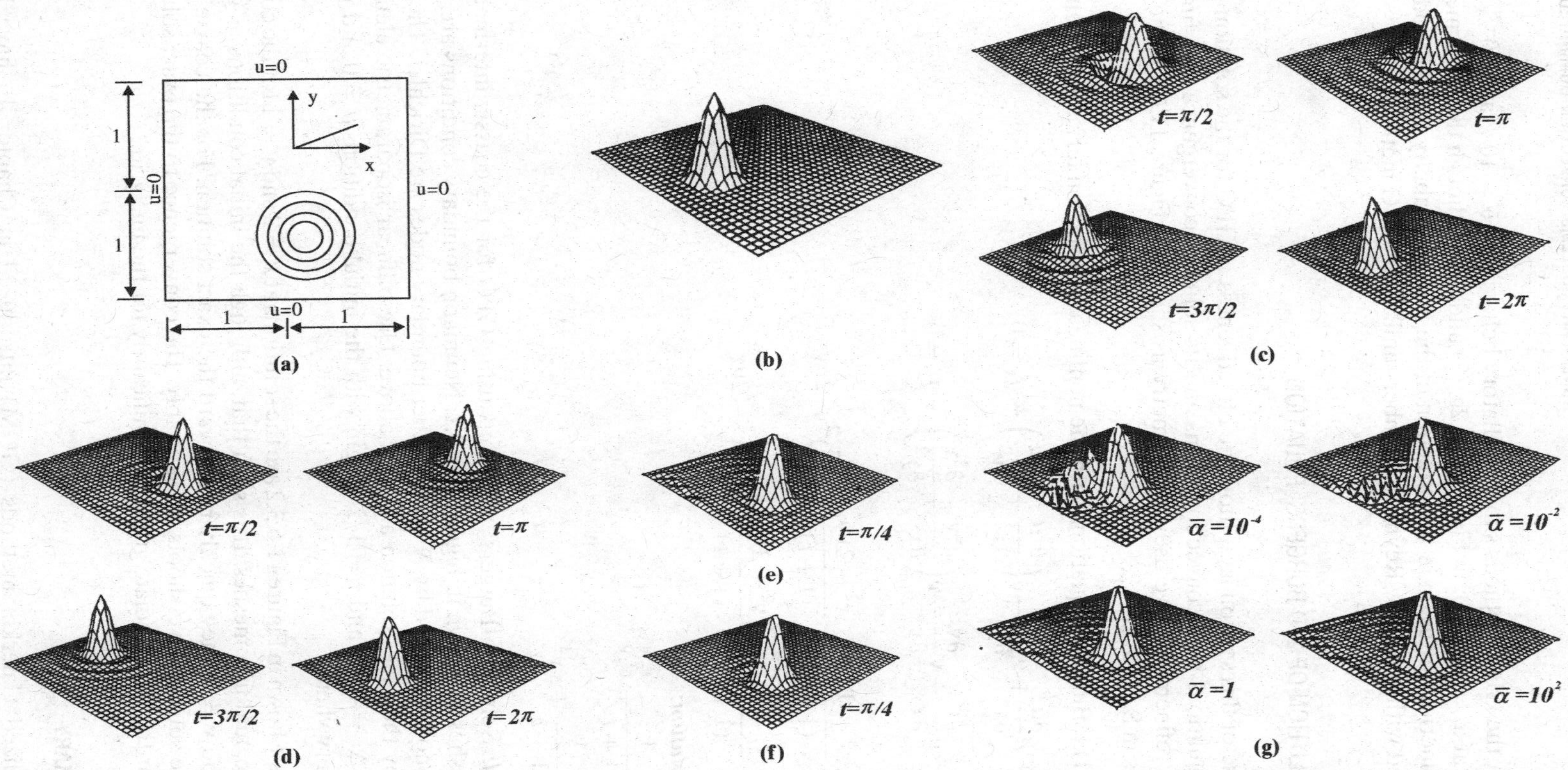

Figure 11.6.2 Rotating cone with cosine hill. **(a)** Geometry, rotating cone. **(b)** Unit initial cosine hill at $x = 0, y = -0.5, t = 0$. **(c)** Lumped mass, one iteration. **(d)** Lumped mass, three iterations. **(e)** Consistent mass, GTG. **(f)** Lumped mass, GTG. **(g)** Consistent mass, GPG.

(1) and (2), the GPG methods show oscillatory behavior at $\overline{\alpha} = 10^{-4}$ and 10^{-2}, which disappears at $\overline{\alpha} = 1$ and 10^2 in Figure 11.6.2g, (3) and (4). Although the GPG methods provide numerical diffusion in the direction of the flow for stability, the methods may be restricted within the low Reynolds numbers unlike the GTG methods.

11.6.3 SOLUTION OF 2-D BURGERS' EQUATION

The purpose of this section is to show the effectiveness of GPG for the solution of the Burgers' equations with convection terms and its solution convergence as a function of the grid refinements. We use the geometry as shown in Figure 11.6.3.1, the same geometry as in Section 10.4.2.

Given: The Burgers' equations with the nonlinear convection terms are given by

$$\frac{\partial u}{\partial t} + u\frac{\partial u}{\partial x} + v\frac{\partial u}{\partial y} - \nu\left(\frac{\partial^2 u}{\partial x^2} + \frac{\partial^2 u}{\partial y^2}\right) - f_x = 0$$

$$\frac{\partial v}{\partial t} + u\frac{\partial v}{\partial x} + v\frac{\partial v}{\partial y} - \nu\left(\frac{\partial^2 v}{\partial x^2} + \frac{\partial^2 v}{\partial y^2}\right) - f_y = 0$$

with

$$f_x = -\frac{1}{(1+t)^2} + \frac{x^2 + 2xy}{(1+t)} + 3x^3y^2 - 2\nu y$$

$$f_y = -\frac{1}{(1+t)^2} + \frac{y^2 + 2xy}{(1+t)} + 3y^3x^2 - 2\nu x$$

Exact Solution:

$$u = \frac{1}{1+t} + x^2 y$$

$$v = \frac{1}{1+t} + xy^2$$

Required: Solve the Burgers' equations using GPG for the coarse, intermediate, and fine meshes as shown in Figure 11.6.3.1. Neumann boundary conditions are to be specified at nodes marked by N and all other boundary nodes are Dirichlet. They are computed by the exact solution as given above. Use bilinear isoparametric elements with $\nu = 1$, $\Delta t = 10^{-4}$, and $\eta = 1/2$. Begin with the initial conditions $u = 0$ and $v = 0$ specified everywhere.

Solution: Shown in Figure 11.6.3.2 are the solutions at $x = 2$ and $y = 1$ for the coarse, intermediate, and fine meshes. It is seen that, although the initial conditions as given are $u = 0$ and $v = 0$, they quickly rise toward the exact solution. For the coarse grid, however, the solution overshoots considerably. The convergence to the exact solution is evident for the intermediate grid and significantly for the fine grid.

11.7 SUMMARY

The generalized Galerkin methods (GGM) introduced in Chapter 10 have been extended to the Taylor Galerkin methods (TGM) and to the generalized Petrov-Galerkin

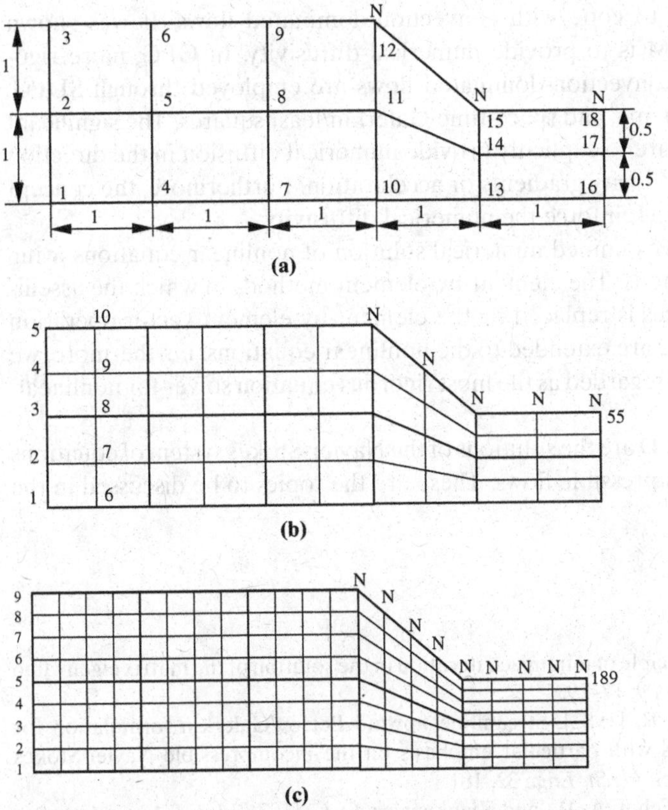

Figure 11.6.3.1 Geometries for Section 11.6.3, N representing Neumann boundary conditions, Dirichlet and Neumann boundary conditions are prescribed from the exact solution. **(a)** Coarse grid. **(b)** Intermediate grid (halved from the coarse grid). **(c)** Fine grid (halved from the intermediate grid).

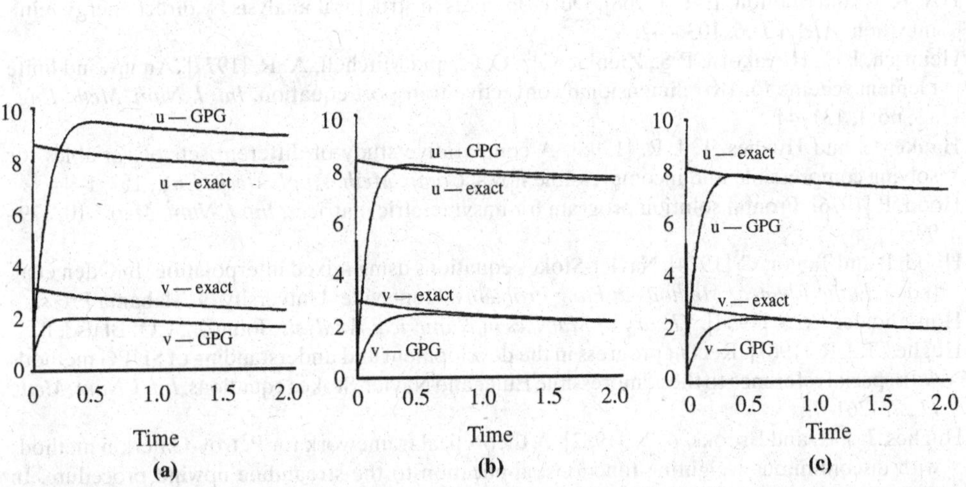

Figure 11.6.3.2 Solution of 2-D Burgers' equations, $x = 2$, $y = 1$ ($\nu = 1$). **(a)** Coarse grid. **(b)** Intermediate grid. **(c)** Fine grid.

(GPG) methods in order to cope with convection-dominated flows. It was shown that the basic idea of TGM is to provide numerical diffusivity. In GPG, more rigorous approaches to treat convection-dominated flows are employed through SUPG, discontinuity-capturing scheme, and space-time Galerkin/least squares. The significant features available in GPG are to explicitly provide numerical diffusion in the direction of streamline and toward velocity gradients or acceleration. Furthermore, the concept of least squares is applied to reinforce the numerical diffusivity.

In this chapter, we also examined numerical solution of nonlinear equations using the Newton-Raphson methods. The element-by-element methods in which the assembly of total stiffness matrices is replaced by the element-by-element vector operation introduced in Section 10.3.2 are extended to the nonlinear equations. Furthermore, we reviewed GMRES which is regarded as the most rigorous equation solver for nonlinear, nonsymmetric matrices.

Major applications in CFD are the solutions of the Navier-Stokes system of equations for incompressible and compressible flows. These are the topics to be discussed in the next two chapters.

REFERENCES

Arnoldi, W. A. [1951]. The principle of minimized iteration in the solution of the matrix eigenvalue problem. *Quart. Appl. Math.*, 9, 17–29.

Brooks, A. and Hughes, T. J. R. [1982]. Streamline upwind Petrov/Galerkin formulation for convection dominated flows with particular emphasis on the incompressible Navier-Stokes equations. *Comp. Meth. Appl. Mech. Eng.*, 32, 181.

Christie, I., Griffiths, D. F., Mitchel, A. R., and Zienkiewicz, O. C. [1976]. *Int. J. Num. Eng.*, 10, 1389–96.

Chung, T. J. [1978]. *Finite Element Analysis in Fluid Dynamics*. New York: McGraw-Hill.

———. [1999]. Transitions and interactions of inviscid/viscous, compressible/incompressible and laminar/turbulent flows. *Int. J. Num. Meth. Fl.*, 31, 223–46.

Donea, J. [1984]. A Taylor-Galerkin method for convective transport problems. *Int. J. Num. Meth. Eng.*, 20, 101–19.

Fox, R. L. and Stanton, E. L. [1968]. Developments in structural analysis by direct energy minimization. *AIAA J.*, 6, 1036–42.

Heinrich, J. C., Huyakorn, P. S., Zienkiewicz, O. C., and Mitchell, A. R. [1977]. An upwind finite element scheme for two-dimensional convective transport equation. *Int. J. Num. Meth. Eng.*, 11, no. 1, 131–44.

Hauke, G. and Hughes. T. J. R. [1998]. A comparative study of different sets of variables for solving compressible and incompressible flows. *Comp. Meth. Appl. Mech. Eng.*, 153, 1–44.

Hood, P. [1976]. Frontal solution program for unsymmetric matrices. *Int. J. Num. Meth.*, 10, 379–99.

Hood, P. and Taylor, C. [1974]. Navier-Stokes equations using mixed interpolation. In Oden et al. (eds.), *Finite Element Methods in Flow Problems*, Huntsville: University of Alabama Press.

Householder, A. S. [1964]. *Theory of Matrices in Numerical Analysis*. Johnson, CO: Blaisdell.

Hughes, T. J. R. [1987]. Recent progress in the development and understanding of SUPG methods with special reference to the compressible Euler and Navier-Stokes equations. *Int. J. Num. Meth. Fl.*, 7, 1261–75.

Hughes, T. J. R. and Brooks, A. N. [1982]. A theoretical framework for Petrov-Galerkin methods with discontinuous weighting functions: application to the streamline upwind procedure. In R. H. Gallagher et al. (eds.), *Finite Elements in Fluids*, London: Wiley.

Hughes, T. J. R, Franca L. P., and Hulbert, G. M. [1986]. A new finite element formulation for computational fluid dynainics: IV. A discontinuity-capturing operator for multidimensional advective-diffusive systems. *Comp. Meth. Appl. Mech. Eng.*, 58, 329–36.

Hughes, T. J. R., Frencz, R. M., and Hallquist, J. O. [1987]. Large scale vectorized implicit calculations in solid mehanics on a Cray–MP/48 utilizing EBE preconditioned conjugate gradients. *Comp. Meth. Appl. Mech. Eng.*, 61, 215–48.

Hughes, T. J. R., Levit, I., and Winget, J. [1983]. An element-by-element implicit algorithm for heat conduction, *ASCE J. Eng. Mech. Div.*, 109, 576–85.

Hughes, T. J. R. and Mallet, M. [1986]. A new finite element formulation for computational fluid dynamics: III. The generalized streamline operator for multi-dimensional advective-diffusive systems. *Comp. Meth. Appl. Mech. Eng.*, 58, 305–28.

Hughes, T., Mallet, M., and Mizukami, A. [1986]. A new finite element formulation for computational fluid dynamics: II. Beyond SUPG. *Comp. Meth. Appl. Mech. Eng.*, 54, 341–55.

Hughes, T. J. R. and Tezduyar, T. E. [1984]. Finite element methods for first order hyperbolic systems with particular emphasis on the compressible Euler equations. *Comp. Meth. Appl. Mech. Eng.*, 45, 217–84.

Irons, B. M. [1970]. A frontal solution program for finite element analysis. *Int. J. Num. Meth. Eng.*, 2, 5–32.

Jameson, A., Baker, T. J., and Weatherill, N. P. [1986]. Calculation of inviscid transonic flow over a complete aircraft. AIAA-86-0103.

Jea, K. C. and Young, D. M. [1983]. On the simplification of generalized conjugate-gradient methods for nonsymmetrizable linear systems. *Linear Algebra Appl.*, 52, 399–417.

Johnson, C. [1987]. Numerical Solution of Partial Differential Equations on the Element Method Student litteratur, Lund, Sweden.

Lanczos, C. [1950]. An iteration method for the solution of the eigenvalue problem of linear differential and integral operators. *J. Res. Nat. Bur. Stand.*, 45, 255–82.

Löhner, R., Morgan, K., and Zienkiewicz, O. C. [1985]. An adaptive finite element procedure for compressible high speed flows. *Comp. Meth. Appl. Mech. Eng.*, 51, 441–65.

Mikhlin, S. G. [1964]. *Variational Methods in Mathematical Physics*. Oxford, UK: Pergamon Press.

Nour-Omid, B. [1984]. A preconditioned conjugate gradient method for finite element equations. In W. K. Liu et al. (eds.), *Innovative Methods for Nonlinear Problems*, England: Swansea.

Nour-Omid, B. and Parlett, B. N. [1985]. Element preconditioning using splitting techniques. *SIAM J. Sci. Comp.*, 6, 761–70.

Oden, J. T., Babuska, I., and Baumann, C. E. [1998]. A discontinuous *hp* finite element methods for diffusion problems. *J. Comp. Phy.*, 146, 491–519.

Oden, J. T. and Demkowicz, L. [1991]. h-p adaptive finite element methods in computational fluid dynamics. *Comp. Meth. Appl. Mech. Eng.*, 89, (1–3): 1140.

Oden, J. T., Demkowicz, L., Strouboulis, T., and Devloo, P. [1986]. Adaptive finite element methods for the analysis of inviscid compressible flow: I. Fast refinement/unrefinement and moving mesh methods for unstructured meshes. *Comp. Meth. Appl. Mech. Eng.*, 59, 327–62.

Ortiz, M., Pinsky, P. M., Taylor, R. L. [1983]. Unconditionally stable element-by-element algorithm for synamic problems. *Comp. Meth. Appl. Mech. Eng.*, 36, 223–39.

Paige, C. C. and Saunders, M. A. [1975]. Solution of sparse indefinite systems of linear equations. *SIAM J. Num. Anal.*, 12, 617–24.

Raymond, W. H. and Garder, A. [1976]. Selective damping in a Galerkin method for solving wave problems with variable grids. *Mon. Weather Rev.* 104, 1583–90.

Saad, Y. [1996]. *Iterative Methods for Sparse Linear System*. Boston: PWS Publishing.

Saad, Y. and Schultz, M. H. [1986]. GMRES: a generalized minimal residual algorithm for solving nonsymmetric linear systems. *SIAM J. Sci. Stat. Comp.*, 7, 856–69.

Shakib, F. and Hughes, T. J. R. [1991]. A new finite element formulation for computational fluid dynamics: IX. Fourier analysis of space-time Galerkin/least squares algorithms. *Comp. Meth. Appl. Mech. Eng.*, 87, 35–58.

Spalding, D. B. [1972]. A novel finite-difference formulation for differential expressions involving both first and second derivatives. *Int. J. Num. Meth. Eng.*, 4, 551–59.
Tezduyar, T. [1997]. Advanced Flow Simulation and Modeling. AHPCRC 97-050, Minneapolis: University of Minnesota.
Zienkiewicz, O. C. and Codina, R. [1995]. A general algorithm for compressible and incompressible flow – Part I. Characteristic-based scheme. *Int. J. Num. Meth. Fl.*, 20, 869–85.

CHAPTER TWELVE

Incompressible Viscous Flows via Finite Element Methods

As noted in Chapter 5, the condition of incompressibility for incompressible flows is difficult to satisfy. The consequence of this difficulty results in a checkerboard type pressure oscillation which occurs when the primitive variables (pressure and velocity) are calculated directly from the governing equations of continuity and momentum. Various methods are used to overcome this difficulty. Among them are: mixed methods, penalty methods, pressure correction methods, generalized Petrov-Galerkin (GPG) methods, operator splitting (fractional) methods, and semi-implicit pressure correction methods. Another approach is to use the vortex methods in which stream functions and vorticity are calculated, thus avoiding the pressure term. Some of the earlier and recent contributions to the finite element analyses of incompressible flows are found in [Hughes, Liu, and Brooks, 1979; Carey and Oden, 1986; Zienkiewicz and Taylor, 1991; Gunzburger and Nicholaides, 1993; Gresho and Sani, 1999], among many others.

Instead of being limited to incompressible flows, we may begin with the conservation form of the Navier-Stokes system of equations for compressible flows, in which special steps can be devised to obtain solutions near incompressible limits ($M_\infty \cong 0$). This allows us to use a single formulation to handle both compressible and incompressible flows. We shall address this subject in Section 13.6. For this reason, treatments of incompressible flows in this chapter will be brief.

12.1 PRIMITIVE VARIABLE METHODS

12.1.1 MIXED METHODS

Consider the governing equations of continuity and momentum for incompressible flow in the form:

Continuity

$$v_{i,i} = 0 \tag{12.1.1a}$$

Momentum

$$\rho v_{i,j} v_j + p_{,i} - \mu v_{i,jj} = 0 \tag{12.1.1b}$$

It is well known that the standard Galerkin formulation of the simultaneous system of equations for continuity and momentum (12.1.1a,b) becomes ill-conditioned, known as the LBB condition [Ladyszhenskaya, 1969; Babuska, 1973; Brezzi, 1974] as pointed out in Section 10.1.4. In order to circumvent the numerical instability, trial functions for pressure are chosen one order lower than those for the velocity, defined as shown in Figure 10.1.3. We may write the standard Galerkin integrals in nondimensional form as follows:

$$\int_\Omega \Phi_\alpha \left(v_{i,j} v_j + p_{,i} - \frac{1}{\text{Re}} v_{i,jj} \right) d\Omega = 0 \tag{12.1.2a}$$

$$\int_\Omega \hat{\Phi}_\alpha v_{i,i} d\Omega = 0 \tag{12.1.2b}$$

where the pressure approximation is of one order lower than the velocity approximation so that the incompressibility condition may be satisfied as discussed in Section 10.1.4. Combining (12.1.2a,b) yields

$$\begin{bmatrix} D_{\alpha\beta ij} & C_{\alpha\beta i} \\ C_{\alpha\beta j} & 0 \end{bmatrix} \begin{bmatrix} v_{\beta j} \\ p_\beta \end{bmatrix} = \begin{bmatrix} G_{\alpha i} \\ 0 \end{bmatrix} \tag{12.1.3}$$

with

$$D_{\alpha\beta ij} = \int_\Omega \left(\Phi_\alpha \Phi_{\beta,k} v_k \delta_{ij} + \frac{1}{\text{Re}} \Phi_{\alpha,k} \Phi_{\beta,k} \delta_{ij} \right) d\Omega$$

$$C_{\alpha\beta i} = \int_\Omega \Phi_\alpha \hat{\Phi}_{\beta,j} \delta_{ij} d\Omega, \qquad C_{\alpha\beta j} = \int_\Omega \hat{\Phi}_\alpha \Phi_{\beta,j} d\Omega,$$

$$G_{\alpha i} = \int_\Gamma \frac{1}{\text{Re}} \overset{*}{\Phi}_\alpha v_{i,j} n_j d\Gamma$$

where the test function $\hat{\Phi}_\alpha$ for continuity is the same as the pressure trial function.

As mentioned in Section 10.1.4, if pressure is interpolated as constant (pressure node at the center of an element) and velocity as a linear function (velocity defined at corner node, Figure 10.1.3a), then such element becomes overconstrained (known as locking element). This situation can be alleviated by using linear pressure and quadratic velocity approximations (Figure 10.1.3b). In this process of unequal order approximations for pressure, we seek to achieve the computational stability. Many other available options are discussed below.

12.1.2 PENALTY METHODS

As seen in Section 10.1.4, the incompressibility condition can be satisfied by means of the penalty function λ such that

$$p = -\lambda v_{i,i} \tag{12.1.4a}$$

$$p_{,i} = -\lambda v_{j,ji} \tag{12.1.4b}$$

which is designed to replace the pressure gradient term in (12.1.2a). The reduced Gaussian quadrature integration for the penalty term is still required to avoid being over-constrained, as discussed in Section 10.1.4. In this way, we obtain the solution of

(12.1.2a) without (12.1.2b), but the mass conservation is maintained through the penalty constraint.

Another approach is to combine the penalty formulation with the mixed method of (12.1.2a,b). This can be achieved by replacing the continuity equation with the Galerkin integral of (12.1.4a),

$$\int_\Omega \Phi_\alpha\left(v_{i,i} + \frac{p}{\lambda}\right) d\Omega = 0 \tag{12.1.5}$$

This will then revise (12.1.3) in the form

$$\begin{bmatrix} D_{\alpha\beta ij} & C_{\alpha\beta i} \\ C_{\alpha\beta j} & E_{\alpha\beta} \end{bmatrix} \begin{bmatrix} v_{\beta j} \\ p_\beta \end{bmatrix} = \begin{bmatrix} G_{\alpha i} \\ 0 \end{bmatrix} \tag{12.1.6}$$

with

$$E_{\alpha\beta} = \int_\Omega \frac{1}{\lambda} \Phi_\alpha \Phi_\beta d\Omega$$

which provides an additional computational stability in comparison with (12.1.3).

12.1.3 PRESSURE CORRECTION METHODS

The basic idea of the pressure correction methods is to split the pressure and velocity in the form [Patankar and Spalding, 1972]

$$p^{n+1} = p^n + p' \tag{12.1.7a}$$

$$v_i^{n+1} = v_i^* + v_i' \tag{12.1.7b}$$

where v_i^* denotes the intermediate step velocity. Using (12.1.7) in (12.1.1b) we obtain, for the case of unsteady flow,

$$\left(\frac{\partial v_i}{\partial t}\right)^* + \left(\frac{\partial v_i}{\partial t}\right)' \cong \left(\frac{1}{\text{Re}} v_{i,jj}^* - v_{i,j}^* v_j^n - (p_{,i})^n - (p_{,i})'\right)$$

which may be split into

$$\left(\frac{\partial v_i}{\partial t}\right)^* = \frac{1}{\text{Re}} v_{i,jj}^* - v_{i,j}^* v_j^n - (p_{,i})^n \tag{12.1.8a}$$

$$\left(\frac{\partial v_i}{\partial t}\right)' = -(p_{,i})' \tag{12.1.8b}$$

where the asterisk and prime indicate intermediate and correction values. The solution of (12.1.8a) is not expected, in general, to satisfy the conservation of mass. In order to rectify this situation, we take a divergence of (12.1.8b) and write

$$p'_{,ii} = -\frac{\partial}{\partial t}(v_{i,i})' \tag{12.1.9a}$$

which may be recast in a difference form

$$p'_{,ii} \cong -\frac{1}{\Delta t}(v_{i,i}^{n+1} - v_{i,i}^*) \tag{12.1.9b}$$

Here we intend to force $v_{i,i}^{n+1}$ to vanish for mass conservation so that

$$p'_{,ii} = \frac{1}{\Delta t}(v^*_{i,i}) \tag{12.1.10}$$

Thus, the solution procedure consists of

(1) Solve (12.1.8a) for v_i^* with initial and boundary conditions and assumed pressure.
(2) Solve (12.1.10) for pressure corrections, p', with the boundary conditions $p' = 0$ on Γ_D and $p'_{,i}n_i$ on Γ_N.
(3) Determine v'_i from (12.1.8b).
(4) Determine

$$p^{n+1} = p^n + p'$$

$$v_i^{n+1} = v_i^* + v'_i$$

(5) Repeat steps (1) through (4) until convergence has been achieved.

The generalized Galerkin formulations may be used for (12.1.8a), (12.1.10), and (12.1.8b). Mixed interpolations (between velocity and pressure) are not required. Although the mass conservation is achieved through the pressure correction methods, the convective terms may still contribute to nonconvergence if convection dominates the flowfield. Toward this end, the generalized Galerkin formulation can be replaced by GPG methods.

12.1.4 GENERALIZED PETROV-GALERKIN METHODS

The mixed method may be modified so that both pressure and velocity can be interpolated in a same order. The convection and pressure gradient terms are treated with generalized Petrov-Galerkin (GPG), and the pressure is updated using the standard pressure Poisson equation.

$$\int_0^1 \hat{W}(\xi)\left\{\int_\Omega \left[\Phi_\alpha\left(\frac{\partial v_i}{\partial t} + v_{i,j}v_j - \frac{1}{Re}v_{i,jj}\right) + \Psi_\alpha(v_{i,j}v_j + p_{,i})\right]d\Omega\right\}d\xi = 0 \tag{12.1.11}$$

$$\int_\Omega \Phi_\alpha[p_{,ii} + (v_{i,j}v_j)_{,i}]d\Omega = 0 \tag{12.1.12}$$

Integrating (12.1.11) by parts leads to

$$\left[A_{\alpha\beta} + \frac{\Delta t}{2}(B_{\alpha\beta} + C_{\alpha\beta} + K_{\alpha\beta})\right]v_{\beta i}^{n+1} = \left[A_{\alpha\beta} - \frac{\Delta t}{2}(B_{\alpha\beta} + C_{\alpha\beta} + K_{\alpha\beta})\right]v_{\beta i}^{n} + \Delta t\,(F_{\alpha i} + G_{\alpha i}) \tag{12.1.13}$$

where

$$F_{\alpha i} = -\int_\Omega \tau v_k \Phi_{\alpha,k}\Phi_{\beta,i}d\Omega p_\beta \tag{12.1.14}$$

with all other quantities being the same as in (11.4.5) except for the Reynolds number.

The nodal pressure p_β will be updated from (12.1.12), which assumes the form

$$E_{\alpha\beta}p_\beta = H_\alpha + Q_\alpha \tag{12.1.15}$$

with

$$E_{\alpha\beta} = \int_\Omega \Phi_{\alpha,i}\Phi_{\beta,i}d\Omega$$

$$H_\alpha = \int_\Omega \Phi_\alpha(\overline{v}_{i,j}\overline{v}_j)_{,i}d\Omega$$

$$Q_\alpha = \int_\Gamma \overset{*}{\Phi}_\alpha p_{,i}n_i d\Gamma$$

Note that pressure oscillations are suppressed not only from (12.1.15) but also the damping effect built into (12.1.14).

Remarks: We note that GPG methods can be applied to the incompressible Navier-Stokes system of equations in which the special treatment for pressure is no longer required. In this case, the conservation form of the Navier-Stokes system of equations can be utilized and it is possible to formulate various schemes which can handle both compressible and incompressible flows. Furthermore, the conservation variables can be transformed into primitive variables in order to accommodate the incompressible nature of the flow. In this case, details of derivations of GPG schemes for incompressible flows are the same as in the case of compressible flows, which will be presented in Section 13.3.

12.1.5 OPERATOR SPLITTING METHODS

The pressure correction methods may be solved with fractional steps, often called operator splitting methods or fractional step methods [Yanenko, 1971], such that equations of hyperbolic, parabolic, and elliptic types are solved separately [Chorin, 1967]. To this end, we consider the standard Galerkin finite element equations of momentum and continuity in the form

$$A_{\alpha\beta}\dot{v}_{\beta i} + E_{\alpha\beta j\gamma}v_{\beta i}v_{\gamma j} - C_{\alpha\beta i}p_\beta + K_{\alpha j\beta j}v_{\beta i} - G_{\alpha i} = 0 \tag{12.1.16}$$

$$C_{\alpha\beta}v_{\beta i} = 0 \tag{12.1.17}$$

(1) Hyperbolic Fractional Step Operator for Convective Terms

$$A_{\alpha\beta}\dot{v}_{\beta i} = -E_{\alpha\beta j\gamma}v_{\beta i}v_{\gamma j} + G_{\alpha i} \tag{12.1.18}$$

where

$$E_{\alpha\beta j\gamma} = B_{\alpha\beta j\gamma} + C_{\alpha\beta j\gamma}$$

with $C_{\alpha\beta j\gamma}$ indicating the term constructed from the numerical diffusion test functions.

The solution of (12.1.18) is obtained from the GPG formulation,

$$\left(A_{\alpha\beta} + \frac{\Delta t}{2} E_{\alpha\beta j\gamma} v_{\gamma j}^{n}\right) \hat{v}_{\beta i}^{n+1} = A_{\alpha\beta} v_{\beta i}^{n} - \frac{\Delta t}{2} E_{\alpha\beta j\gamma} v_{\beta i}^{n} v_{\gamma j}^{n} + \Delta t\, G_{\alpha i} \tag{12.1.19}$$

(2) Parabolic Fractional Step Operator for Dissipation Term

$$A_{\alpha\beta} \dot{v}_{\beta i} = -K_{\alpha j\beta j} v_{\beta i} \tag{12.1.20}$$

$$\begin{cases} v_i = \bar{v}_i & \text{on } \Gamma_D \\ v_{i,j} n_j = g_i & \text{on } \Gamma_N \end{cases}$$

We solve (12.1.20) with TGM formulation so that

$$\left(A_{\alpha\beta} + \frac{\Delta t}{2} E_{\alpha\beta ij}\right) \tilde{v}_{\beta i}^{n+1} = A_{\alpha\beta} \hat{v}_{\beta i}^{n+1} - \frac{\Delta t}{2} K_{\alpha j\beta j} \hat{v}_{\beta i}^{n+1} + \Delta t\, G_{\alpha i} \tag{12.1.21}$$

(3) Elliptic Fractional Step Operator for Pressure Term

$$A_{\alpha\beta} \frac{v_{\beta i}^{n+1} - \tilde{v}_{\beta i}^{n+1}}{\Delta t} = C_{\alpha\beta i}\, p_{\beta}^{n+1} \tag{12.1.22}$$

$$C_{\alpha\beta} v_{\beta i}^{n+1} = 0 \tag{12.1.23}$$

$$\begin{cases} p = p_0 & \text{on } \Gamma_D \\ p_{,i} n_i = g_i & \text{on } \Gamma_N \end{cases}$$

Here the enforcement of incompressibility is achieved by substituting the first term on the right-hand side of (12.1.22) by (12.1.23).

$$D_{\alpha\beta i}\, p_{\beta}^{n+1} = -\frac{1}{\Delta t} C_{\alpha\beta} v_{\beta i}^{n+1} \tag{12.1.24}$$

where

$$D_{\alpha\beta i} = C_{\alpha\gamma} A_{\gamma\delta}^{-1} C_{\delta\beta i} \tag{12.1.25}$$

We calculate p_{β}^{n+1} from (12.1.25) and determine the final velocity from (12.1.22),

$$v_{\alpha i}^{n+1} = \tilde{v}_{\alpha i}^{n+1} + \Delta t\, A_{\alpha\gamma}^{-1} C_{\gamma\beta i}\, p_{\beta}^{n+1} \tag{12.1.26}$$

Note that the fractional step methods are similar to the pressure correction methods, although there are two distinctly different aspects:

(1) The solution involved in (12.1.8a) is split into two steps: hyperbolic step and parabolic step.
(2) The processes (12.1.8b) and (12.1.10) of pressure correction methods are combined into an elliptic step of the fractional step methods. The pressure Poisson equation is not used here.

It should be noted that (12.1.22) may be differentiated spatially to obtain the pressure Poisson equation as in the pressure correction method, expediting convergence to a certain extent.

12.1.6 SEMI-IMPLICIT PRESSURE CORRECTION

In this scheme, the GPG method is used for convection dominated flows, but we resort to the pressure correction method to maintain conservation of mass and to suppress pressure oscillations.

With the continuity equation written in the form

$$\frac{1}{c^2}\frac{\partial p}{\partial t} + (\rho \mathrm{v}_i)_{,i} = 0 \tag{12.1.27}$$

we obtain the finite element equations as follows:

Continuity

$$D_{\alpha\beta}\dot{p}_\beta + C_{\alpha\beta i}\mathrm{v}_{\beta i} = 0 \tag{12.1.28}$$

Momentum

$$A_{\alpha\beta}\dot{\mathrm{v}}_{\beta i} + (B_{\alpha\beta jj} + K_{\alpha j\beta j})\mathrm{v}_{\beta i} + C_{\alpha\beta i}p_\beta = 0 \tag{12.1.29}$$

where $B_{\alpha\beta jj}$ contains the GPG terms.

Denote the following:

$$\Delta p^n_\alpha = p^{n+1}_\alpha - p^n_\alpha \tag{12.1.30}$$

$$\Delta \mathrm{v}^n_{\alpha i} = \mathrm{v}^{n+1}_{\alpha i} - \mathrm{v}^n_{\alpha i} = \Delta \mathrm{v}^{n(1)}_{\alpha i} - \Delta \mathrm{v}^{n(2)}_{\alpha i} \tag{12.1.31}$$

and

$$\begin{aligned} p &= (1-\theta)p^n + \theta p^{n+1} \\ &= \theta(p^{n+1} - p^n) + p^n \\ &= \theta(\Delta p^n) + p^n \end{aligned} \tag{12.1.32}$$

$$\begin{aligned} \mathrm{v}_i &= (1-\theta)\mathrm{v}^n_i + \theta \mathrm{v}^{n+1}_i \\ &= \theta\left(\Delta \mathrm{v}^{n(1)}_i - \Delta \mathrm{v}^{n(2)}_i\right) + \mathrm{v}^n_i \\ &= \theta\left(\Delta \mathrm{v}^n_i\right) + \mathrm{v}^n_i \end{aligned} \tag{12.1.33}$$

Substituting (12.1.32) into (12.1.29) and taking a temporal approximation, we obtain

$$\Delta \mathrm{v}^n_{\beta i} = -\Delta t\left[(B_{\alpha\beta jj} + K_{\alpha j\beta j})\left(\theta \Delta \mathrm{v}^n_{\beta i} + \mathrm{v}^n_{\beta i}\right) + C_{\alpha\beta i}\left(\theta \Delta p^n_\beta + p^n_\beta\right)\right] \tag{12.1.34}$$

Combining (12.1.32) into (12.1.34) and separating the resulting equation into two parts leads to

$$[\Delta t\theta(B_{\alpha\beta jj} + K_{\alpha j\beta j})]\Delta \mathrm{v}^{(1)}_{\alpha i} = \Delta t\left[(B_{\gamma\beta jj} + K_{\gamma j\beta j})\mathrm{v}^n_{\beta i} + C_{\gamma\beta i}p^n_\beta\right] \tag{12.1.35a}$$

$$[A_{\alpha\beta} + \Delta t\theta(B_{\alpha\beta jj} + K_{\alpha j\beta j})]\Delta \mathrm{v}^{(2)}_{\alpha i} = \Delta t\theta C_{\gamma\beta i}\Delta p^n_\beta \tag{12.1.35b}$$

Substituting (12.1.32) into (12.1.28) and using (12.1.33) and (12.1.35), we obtain

$$\left(C_{\alpha\gamma i}Q^{-1}_{\gamma\delta}C_{\delta\beta i}\right)\Delta p^n_\beta = -C_{\alpha\beta i}\Delta t\left(\mathrm{v}^n_{\beta i} + \theta\Delta \mathrm{v}^{n(1)}_{\beta i}\right) \tag{12.1.36}$$

where

$$D_{\alpha\beta} = \int_\Omega \frac{1}{\alpha^2}\Phi_\alpha \Phi_\beta d\Omega = \int_\Omega \frac{M^2}{q^2}\Phi_\alpha \Phi_\beta d\Omega, \qquad Q_{\alpha\beta} = A_{\alpha\beta} + \Delta t\theta(B_{\alpha\beta jj} + K_{\alpha j\beta j}) \tag{12.1.37}$$

For incompressible flows, we have $D_{\alpha\beta} = 0$. This gives

$$\Delta t^2\theta^2 C_{\alpha\gamma i} Q^{-1}_{\gamma\delta} C_{\delta\beta i}\Delta p^n_\beta = C_{\alpha\beta i}\Delta t(\mathrm{v}^n_{\beta i} + \theta\Delta \mathrm{v}^{n(1)}_{\beta i}) \tag{12.1.38}$$

The von Neumann analysis shows that, for stable solutions, Δt must be limited by

$$\Delta t \le \frac{h}{|\mathrm{v}|}\sqrt{\frac{1}{\mathrm{Re}} + 1} - \frac{1}{\mathrm{Re}} \tag{12.1.39}$$

Upon solution of the pressure equation (12.1.38), we return to (12.1.34) for the corrected velocity components.

A simplified version of the previous approach arises in the absence of viscosity terms:

$$\frac{1}{a^2}\frac{\partial p}{\partial t} + \mathrm{v}_{i,i} = 0 \tag{12.1.40}$$

$$\frac{\partial \mathrm{v}_i}{\partial t} + p_{,i} = 0 \tag{12.1.41}$$

Rewriting (12.1.40) and (12.1.41) yields

$$\frac{1}{a^2}\Delta p^n + \Delta t(\mathrm{v}^n_i + \theta\Delta \mathrm{v}^{n(2)}_i - \theta\Delta \mathrm{v}^{n(1)}_i)_{,i} = 0 \tag{12.1.42}$$

$$\Delta \mathrm{v}^{n(2)}_i + \theta\Delta t\Delta p^n_{,i} = 0 \tag{12.1.43}$$

Substituting (12.1.43) into (12.1.42), we obtain

$$\frac{1}{a^2}\Delta p^n + \Delta t\,(\mathrm{v}^n_i + \theta\Delta \mathrm{v}^{n(1)}_i)_{,i} - (\theta\Delta t)^2\Delta p^n_{,ii} = 0 \tag{12.1.44}$$

With the finite element approximation,

$$\mathrm{v}_i = \Phi_\alpha \mathrm{v}_{\alpha i}, \qquad p = \hat{\Phi}_\alpha p_\alpha$$

we have

$$(D_{\alpha\beta} - \Delta t^2\theta^2 E_{\alpha i\beta i})\Delta p^n_\beta = -G_{\alpha\beta i}\Delta t\,(\Delta \mathrm{v}^n_{\beta i} + \theta\Delta \mathrm{v}^{n(1)}_{\beta i}) \tag{12.1.45}$$

The pressure correction as obtained from (12.1.45) can be used to solve (12.1.44) in which the viscosity term is now restored.

12.2 VORTEX METHODS

Recall that the vortex methods as examined in Section 5.4 utilize the vortex transport equation in which the terms with pressure gradients vanish upon satisfaction of the conservation of mass. Thus, the pressure oscillation is not expected to occur in the solution of the vortex transport equation.

In many engineering problems, it is not feasible to make two-dimensional simplifications because the flowfield is physically three-dimensional, such as in high-speed rotational flows and high-Reynolds number turbulent flows. Thus, we begin with three-dimensional formulations and demonstrate that the two-dimensional analysis can be derived easily as a simplification of the three-dimensional process if permitted by the special physical situations.

12.2.1 THREE-DIMENSIONAL ANALYSIS

Three-Dimensional Vorticity Transport Equations

The system of three-dimensional vorticity transport equation takes the form

$$\frac{\partial \boldsymbol{\omega}}{\partial t} + (\mathbf{v}\cdot\nabla)\,\boldsymbol{\omega} - (\boldsymbol{\omega}\cdot\nabla)\,\mathbf{v} = \nu\nabla^2\boldsymbol{\omega} \tag{12.2.1}$$

with

$$\boldsymbol{\omega} = \nabla\times\mathbf{v} \tag{12.2.2}$$

$$\nabla^2 p = \rho\nabla\cdot[(\mathbf{v}\cdot\nabla)\,\mathbf{v}] \tag{12.2.3}$$

The above system provides seven unknowns (ω, $\mathbf{v}$, p) and seven equations in three dimensions. We may use GGM , TGM, or GPG to solve the system of equations (12.2.1–12.2.3).

Three-Dimensional Biharmonic Equation with Stream Function

It is also possible to write (12.2.1) in terms of the stream function vector Ψ as defined in (5.4.15),

$$\frac{\partial}{\partial t}(\nabla^2\Psi) + (\nabla\times\Psi\cdot\nabla)\nabla^2\Psi - (\nabla^2\Psi\cdot\nabla)(\nabla\times\Psi) = \nu\nabla^4\Psi \tag{12.2.4a}$$

or

$$\frac{\partial}{\partial t}(\Psi_{i,jj}) + \varepsilon_{rjk}\Psi_{k,j}\Psi_{i,mmr} - \varepsilon_{isk}\Psi_{r,jj}\Psi_{k,sr} = \nu\Psi_{i,jjkk} \tag{12.2.4b}$$

with

$$\omega = \nabla(\nabla\cdot\Psi) - \nabla^2\Psi = -\nabla^2\Psi$$

To obtain the TGM equation for (12.2.4b), we proceed as follows:

$$\int_0^1 \hat{W}(\xi)\int_\Omega \Phi_\alpha\left(\frac{\partial}{\partial t}(\Psi_{i,jj}) + \varepsilon_{rjk}\Psi_{k,j}\Psi_{i,mmr} - \varepsilon_{isk}\Psi_{r,jj}\Psi_{k,sr} - \nu\Psi_{i,jjkk}\right)d\Omega d\xi = 0 \tag{12.2.5}$$

Integrate (12.2.5) twice to obtain

$$A_{\alpha\beta}\delta_{ij}\frac{\partial\Psi_{\beta j}}{\partial t} - B_{\alpha\beta\gamma k}\Psi_{\beta k}\Psi_{\gamma i} + C_{\alpha\beta\gamma imk}\Psi_{\beta m}\Psi_{\gamma k} + K_{\alpha\beta}\delta_{ij}\Psi_{\beta j} = -G_{\alpha i} \tag{12.2.6}$$

with

$$A_{\alpha\beta} = \int_\Omega \Phi_{\alpha,k}\Phi_{\beta,k}\, d\Omega$$

$$B_{\alpha\beta\gamma k} = \int_\Omega \varepsilon_{njk}\Phi_\alpha \Phi_{\beta,j}\Phi_{\gamma,mmn}\, d\Omega$$

$$C_{\alpha\beta\gamma imk} = \int_\Omega \varepsilon_{isk}\Phi_\alpha \Phi_{\beta,jj}\Phi_{\gamma,sm}\, d\Omega$$

$$K_{\alpha\beta} = \int_\Omega \nu\Phi_{\alpha,kk}\Phi_{\beta,mm}\, d\Omega$$

$$G_{\alpha i} = \int_\Gamma \nu\overset{*}{\Phi}_\alpha \Psi_{i,jjk} n_k\, d\Gamma - \int_\Gamma \nu\overset{*}{\Phi}_{\alpha,k}\Psi_{i,jj} n_k\, d\Gamma$$

Here, there are nine variables (Ψ_1, Ψ_2, Ψ_3, $\Psi_{1,2}$, $\Psi_{1,3}$, $\Psi_{2,1}$, $\Psi_{2,3}$, $\Psi_{3,1}$, and $\Psi_{3,2}$) to be specified and calculated at each of the eight nodes of the 3-D cubic isoparametric element. To this end, we require thirty-two constants to be determined with four of them (Ψ_i, $\Psi_{i,1}$, $\Psi_{i,2}$, $\Psi_{i,3}$) at each of the eight nodes as follows (Figure 12.2.1):

$$1, \xi, \eta, \zeta, \xi\eta, \xi\zeta, \eta\zeta, \xi\eta\zeta, \xi^2, \eta^2, \zeta^2, \xi^2\eta, \xi^2\zeta, \eta^2\xi, \eta^2\zeta, \zeta^2\xi, \zeta^2\eta, \xi^2\eta\zeta, \xi\eta^2\xi, \xi\eta\zeta^2, \xi^3,$$
$$\eta^3, \zeta^3, \xi^3\eta, \xi^3\zeta, \eta^3\xi, \eta^3\zeta, \zeta^3\xi, \zeta^3\eta, \xi^3\zeta\eta, \xi\eta^3\zeta, \xi\eta\zeta^3$$

The TGM Newton-Raphson formulation of (12.2.6) takes the form

$$J^{(r)}_{\alpha\beta ij}\Delta\Psi^{(n+1)(r+1)}_{\beta j} = -R^{(n+1)(r)}_{\alpha i} \tag{12.2.7}$$

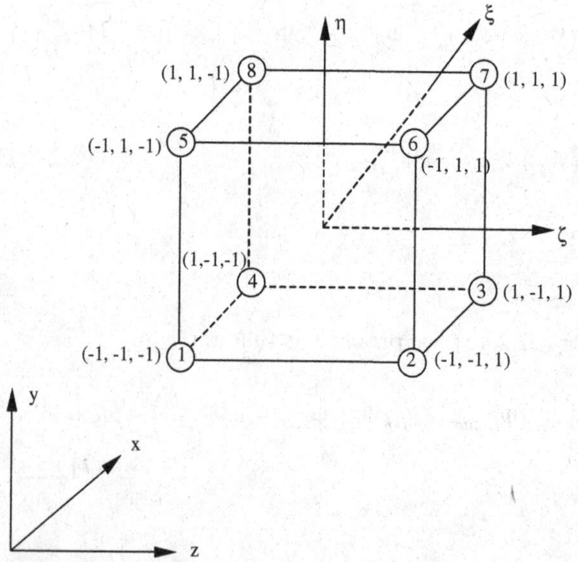

Figure 12.2.1 Hexahedral element for 3-D vortex transport analysis.

where

$$
\begin{aligned}
R_{\alpha i}^{(n+1)(r)} &= A_{\alpha\beta}\delta_{ij}\Psi_{\beta j}^{(n+1)} + \frac{\Delta t}{2}\Big(B_{\alpha\beta\gamma k}\Psi_{\beta k}^{(n+1)}\Psi_{\gamma i}^{(n+1)} - C_{\alpha\beta\gamma imk}\Psi_{\beta m}^{(n+1)}\Psi_{\gamma k}^{(n+1)} \\
&\quad - K_{\alpha\beta}\delta_{ij}\Psi_{\beta j}^{(n+1)}\Big) - A_{\alpha\beta}\delta_{ij}\Psi_{\beta j}^{(n)} + \frac{\Delta t}{2}\Big(B_{\alpha\beta\gamma k}\Psi_{\beta k}^{(n)}\Psi_{\gamma i}^{(n)} - C_{\alpha\beta\gamma imk}\Psi_{\beta m}^{(n)}\Psi_{\gamma k}^{(n)} \\
&\quad - K_{\alpha\beta}\delta_{ij}\Psi_{\beta j}^{(n)}\Big) - \Delta t G_{\alpha i}
\end{aligned}
\tag{12.2.8}
$$

$$
J_{\alpha\beta ij}^{(r)} = \frac{\partial R_{\alpha i}^{(n+1)(r)}}{\partial \Psi_{\beta j}^{(n+1)(r)}} = A_{\alpha\beta}\delta_{ij} + \frac{\Delta t}{2}(B_{\alpha\beta\gamma j}\Psi_{\gamma i} + B_{\alpha\beta\gamma k}\delta_{ij}\Psi_{\gamma k} - 2C_{\alpha\beta\gamma ijk}\Psi_{\gamma k} - K_{\alpha\beta}\delta_{ij})
\tag{12.2.9}
$$

First of all, the local element interpolation functions must be polynomials of at least third degree which will allow the stream function to be linear. The total number of element unknowns are thirty-two with four at each node (Figure 12.2.1). Explicit interpolation functions have been described in Elshabka and Chung [1999].

Typical Neumann and Dirichlet boundary conditions associated with the 3-D stream function vector components are shown in Figure 12.2.2. The Newton-Raphson solution of (12.2.7) is expected to be free of numerical oscillations because of the Jacobian matrix which is well-conditioned.

Computations of (12.2.7) based on the definition of the three-dimensional stream function vector components as given in (5.4.15) have been carried out in Elshabka [1995]. Some of the highlights are given in Section 12.3.

The Curl of Three-Dimensional Vorticity Transport Equations

The vorticity transport equations (12.2.1) are derived by taking a curl of the momentum equations. In this process, the pressure gradient terms of the momentum equations are eliminated, resulting in computationally more stable formulations. However, both vorticity and velocity are coupled together in the vorticity transport equations. The vorticity transport equations are written in a modified form,

$$
\frac{\partial \omega_i}{\partial t} + \varepsilon_{ijk}S_{k,j} - \nu\omega_{i,jj} = 0 \tag{12.2.10}
$$

with

$$
S_i = (\mathrm{v}_i\mathrm{v}_j)_{,j}
$$

To arrive at a single variable, say velocity alone, we take a curl of (12.2.10) and obtain

$$
\frac{\partial}{\partial t}(\mathrm{v}_{i,jj}) + S_{i,jj} - (S_j)_{,ji} - \nu \mathrm{v}_{i,jjkk} = 0 \tag{12.2.11}
$$

or

$$
\frac{\partial}{\partial t}(\mathrm{v}_{i,jj}) + (\mathrm{v}_i\mathrm{v}_k)_{,kjj} - (\mathrm{v}_j\mathrm{v}_k)_{,kji} - \nu \mathrm{v}_{i,jjkk} = 0 \tag{12.2.12}
$$

This will allow calculations of velocity by solving (12.2.12) alone. Other options include solving (12.2.10) and (12.2.11) simultaneously with $\omega = \nabla \times \mathbf{v}$.

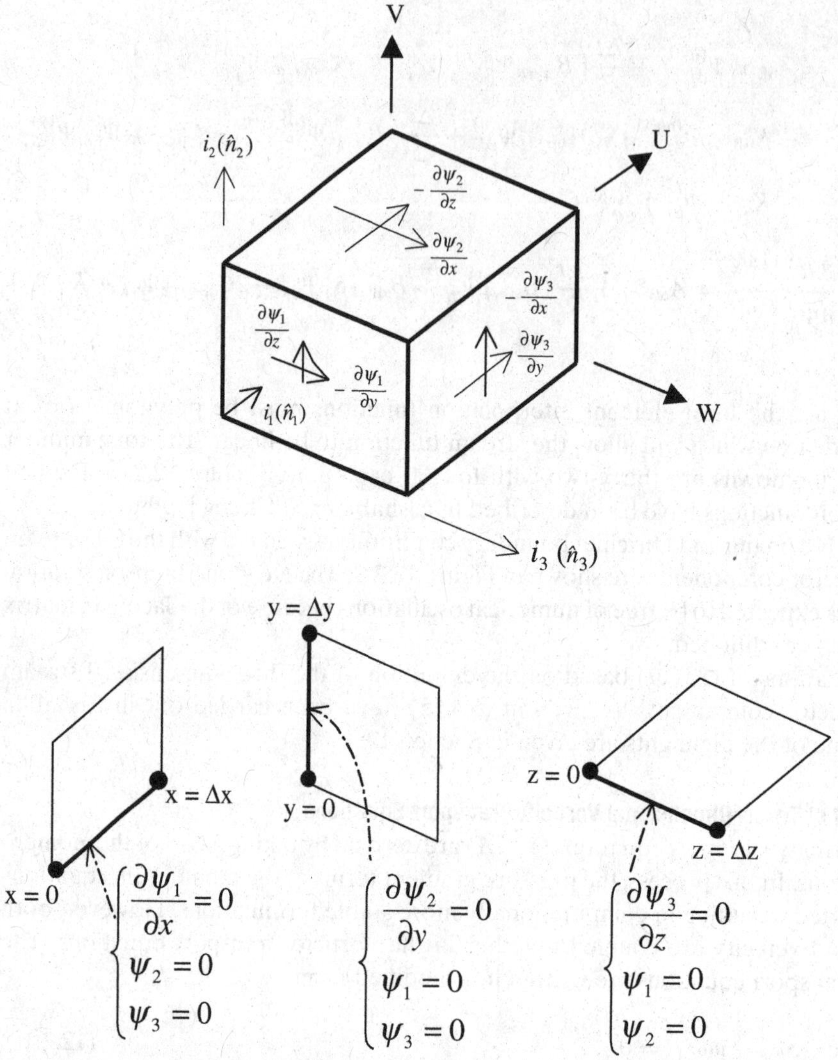

Figure 12.2.2 Three-dimensional boundary conditions.

12.2.2 TWO-DIMENSIONAL ANALYSIS

The two-dimensional vorticity transport equation is simplified to

$$\frac{\partial \omega}{\partial t} + (\mathbf{v} \cdot \nabla)\,\omega = \nu \nabla^2 \omega \tag{12.2.13a}$$

with

$$\omega = \frac{\partial \mathrm{v}}{\partial x} - \frac{\partial u}{\partial y} \tag{12.2.13b}$$

$$\frac{\partial \mathrm{v}}{\partial y} + \frac{\partial u}{\partial x} = 0 \tag{12.2.13c}$$

Here, there are three unknowns (u, v, ω) in the system of three equations (12.2.13a,b,c). The pressure is then calculated from the Poisson equation.

$$\nabla^2 p = 2\rho\left(\frac{\partial u}{\partial x}\frac{\partial \mathrm{v}}{\partial y} - \frac{\partial \mathrm{v}}{\partial x}\frac{\partial u}{\partial y}\right) \tag{12.2.14}$$

We may rewrite (12.2.13a) in terms of a scalar stream function, ψ,

$$\frac{\partial}{\partial t}(\psi_{,jj}) + \varepsilon_{ik}\psi_{,k}\psi_{,jji} - \nu\psi_{,iijj} = 0 \tag{12.2.15}$$

The TGM equation for (12.2.15) becomes

$$A_{\alpha\beta}\frac{\partial \psi_\beta}{\partial t} + B_{\alpha\beta\gamma}\psi_\beta\psi_\gamma - K_{\alpha\beta}\psi_\beta = G_\alpha \tag{12.2.16}$$

where

$$A_{\alpha\beta} = \int_\Omega \Phi_{\alpha,i}\Phi_{\beta,i}d\Omega$$

$$B_{\alpha\beta\gamma} = \int_\Omega \varepsilon_{ik}\Phi_\alpha\Phi_{\beta,k}\Phi_{\gamma,jji}d\Omega$$

$$K_{\alpha\beta} = \int_\Omega \nu\Phi_{\alpha,ii}\Phi_{\beta,jj}d\Omega$$

$$G_{\alpha i} = \int_\Gamma \nu\overset{*}{\Phi}_\alpha\psi_{,iij}n_j d\Gamma - \int_\Gamma \nu\overset{*}{\Phi}_{\alpha,j}\psi_{,ii}n_j d\Gamma$$

Here, there are three variables (Ψ, $\Psi_{,1}$, $\Psi_{,2}$) which are to be specified and calculated at each of the four nodes of the 2-D isoparametric element. To this end, we require twelve constants to be determined, with three of them (Ψ, $\Psi_{,1}$, $\Psi_{,2}$) at each of the four nodes:

$$1, \xi, \eta, \xi\eta, \xi^2, \eta^2, \xi^2\eta, \eta^2\xi, \xi^3, \eta^3, \xi\eta^3, \xi^3\eta$$

The 2-D TGM Newton-Raphson formulation of (12.2.16) can be constructed similarly as in (12.2.7) for the 3-D case with the boundary conditions reduced to the two-dimensional geometry from Figure 12.2.2 and Table 12.2.1.

12.2.3 PHYSICAL INSTABILITY IN TWO-DIMENSIONAL INCOMPRESSIBLE FLOWS

Unstable motions occur during the transition from laminar to turbulent flows. To examine such motions, the so-called Orr-Sommerfeld equation is solved. Here we may begin with the 2-D velocity and vorticity as a sum of the mean and fluctuation components,

$$\mathrm{v}_i = \overline{\mathrm{v}}_i + \mathrm{v}_i^* \quad (i = 1, 2) \tag{12.2.17a}$$

$$\omega_i = \overline{\omega}_i + \omega_i^* \quad (i = 3) \tag{12.2.17b}$$

where ($^-$) and (*) denote mean and fluctuation quantities, respectively.

Table 12.2.1 Boundary Conditions (3-D cavity)

At $x = 0, 1$	$\psi_{1,1} = \psi_2 = \psi_{2,2} = \psi_{2,3} = \psi_3 = \psi_{3,2} = \psi_{3,3}$ $\psi_{1,3} - \psi_{3,1} = 0$ $\psi_{2,1} - \psi_{1,2} = 0$
At $y = 0$	$\psi_1 = \psi_{1,1} = \psi_{1,3} = \psi_{2,2} = \psi_3 = \psi_{3,1} = \psi_{3,3}$ $\psi_{3,2} - \psi_{2,3} = 0$ $\psi_{2,1} - \psi_{1,2} = 0$
At $y = 1$	$\psi_1 = \psi_{1,1} = \psi_{1,3} = \psi_{2,2} = \psi_3 = \psi_{3,1} = \psi_{3,3}$ $\psi_{3,2} - \psi_{2,3} = U_{max}$ $\psi_{2,1} - \psi_{1,2} = 0$
At $z = 0, 1$	$\psi_1 = \psi_{1,1} = \psi_{1,2} = \psi_2 = \psi_{2,1} = \psi_{2,2} = \psi_{3,3}$ $\psi_{3,2} - \psi_{2,3} = 0$ $\psi_{1,3} - \psi_{3,1} = 0$
At $z = 0.5$	$\psi_1 = \psi_{1,1} = \psi_{1,2} = \psi_2 = \psi_{2,1} = \psi_{2,2} = \psi_{3,3}$

For two-dimensional flows with v_i $(i = 1, 2)$, ω_i $(i = 3)$, the vorticity transport equation takes the form

$$\frac{\partial}{\partial t}(-\psi^*_{,ii}) + \varepsilon_{ik}\overline{v}_{k,ij}\overline{v}_j + \varepsilon_{ik}\overline{v}_{k,ij}\varepsilon_{jr}\psi^*_{,r} - \psi^*_{,iij}\overline{v}_j - \psi^*_{,iij}\varepsilon_{jr}\psi^*_{,r} - \frac{1}{\text{Re}}(\varepsilon_{ik}\overline{v}_{k,ijj} - \psi^*_{,iijj}) = 0 \tag{12.2.18}$$

where we have used the following relationship:

$$\overline{\omega} = \varepsilon_{ik}\overline{v}_{k,i}$$

$$\omega^* = \varepsilon_{ik}v^*_{k,i} = \varepsilon_{ik}\varepsilon_{kr}\psi^*_{,ri} = -\psi^*_{,ii}$$

Denote

$$\psi^*(x, y, t) = q(x, y)e^{-i\beta t} = Q(y)e^{ikx}e^{-i\beta t} \tag{12.2.19a}$$

$$\beta = \beta^{(R)} + i\beta^{(I)} \tag{12.2.19b}$$

where $\beta^{(R)}$ is the circular frequency and $\beta^{(I)}$ is the amplification factor, related as

$$\beta = kc, \quad c = c^{(R)} + ic^{(I)} \tag{12.2.20}$$

with k = wave number and c is the velocity of propagation, (R) and (I) indicating the real and imaginary parts, respectively. In view of (12.2.18) and (12.2.19) and neglecting higher order terms ($\varepsilon_{ik}\overline{v}_{k,ij}\overline{v}_j$, $\psi^*_{,iij}\varepsilon_{jr}\psi^*_{,r}$, and $\varepsilon_{ik}\overline{v}_{k,ijj}$), we obtain

$$-i\beta q_{,ii} + \varepsilon_{ik}\overline{v}_{k,ij}\varepsilon_{jr}q_{,r} + q_{,iij}\overline{v}_j - \frac{1}{\text{Re}}q_{,iijj} = 0 \tag{12.2.21}$$

We further denote that

$$\overline{v}_1 = \mathbf{U}(y) \quad \text{and} \quad \overline{v}_2 = 0 \tag{12.2.22a}$$

and

$$q(x, y) = Q(y)e^{ikx} \tag{12.2.22b}$$

Combine (12.2.22) with (12.2.21) to obtain

$$-i\beta Q(ik)^2 - i\beta Q_{,22} + U_{,22}Q(ik) + UQ(ik)^3 + UQ_{,22}(ik)$$
$$-\frac{1}{\text{Re}}[Q_{,2222} + Q(ik)^4 - 2Q_{,22}(ik)^2] = 0 \qquad (12.2.23)$$

Dividing (12.2.23) by ik, we arrive at the Orr-Sommerfeld equation

$$c(k^2Q - Q_{,22}) - QU_{,22} - k^2QU + UQ_{,22} + \frac{i}{k\text{Re}}(Q_{,2222} + k^4Q - 2k^2Q_{,22}) = 0 \qquad (12.2.24)$$

or

$$(U-c)\left(\frac{d^2Q}{dy^2} - k^2Q\right) - Q\frac{d^2U}{dy^2} = -\frac{i}{k\text{Re}}\left(\frac{d^4Q}{dy^4} - 2k^2\frac{d^2Q}{dy^2} + k^4Q\right) = 0 \qquad (12.2.25)$$

Since (12.2.25) represents variations only in the lateral direction y, the trial functions are constructed in one dimension. The finite element formulations of (12.2.25) can be carried out in a standard manner, resulting in the form,

$$(K_{\alpha\beta} - cM_{\alpha\beta})Q_\beta = 0 \qquad (12.2.26)$$

with the boundary conditions

$$Q_\alpha = 0 \quad \text{and} \quad \partial Q_\alpha/\partial y = 0 \qquad (12.2.27)$$

The expression (12.2.26) is a standard eigenvalue problem,

$$|K_{\alpha\beta} - cM_{\alpha\beta}|Q_\beta = 0 \qquad (12.2.28)$$

Eigenvalues are the phase velocity (c) with real and imaginary parts as defined in (16.6.20),

$$c^{(I)} < 0 \quad \text{stable} \qquad (12.2.29a)$$

$$c^{(I)} = 0 \quad \text{neutral stability} \qquad (12.2.29b)$$

$$c^{(I)} > 0 \quad \text{unstable} \qquad (12.2.29c)$$

Eigenvectors Q_β represent fluctuation parts of stream function, which provide fluctuation parts of velocity $v_i^* = \varepsilon_{ij}\psi^*_{,j}$. The eigenvalue problem involved in a complex number may be solved using the so-called QR algorithm [Wilkinson, 1965].

12.3 EXAMPLE PROBLEMS

Three-Dimensional Vorticity Transport Equations

A convenient benchmark problem is the lid-driven cubic cavity flow as shown in Figure 12.3.1. The corresponding boundary conditions are shown in Table 12.2.1.

In Figure 12.3.2, we show comparisons between the TGM solution of the 3-D vorticity transport equations (12.2.4) and the results of other approaches reported by Takami and Kuwahara [1974] with the 20 × 10 × 20 FDM velocity-pressure formulation, Goda [1979] with the 20 × 10 × 20 FDM velocity-pressure formulation, and Mahallati and

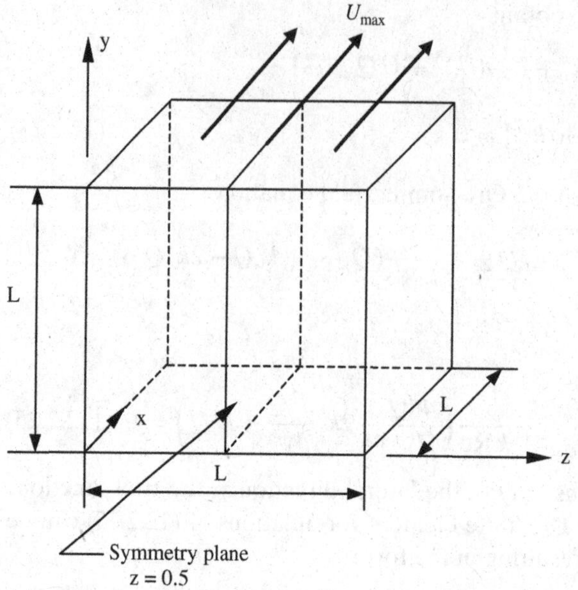

Figure 12.3.1 Geometry of the cubic cavity flow.

Militzer [1993] with the 21 × 11 × 21 vorticity vector potential formulation using the finite analytic method (FAM).

In Figure 12.3.3, the x-velocity profiles for Re = 100 at different x-planes are shown. Note that the effect of boundary layers is clearly evident in the y–z-planes, indicating that the velocity profiles in planes closer to the wall are less developed due to the boundary layer effects than in the symmetry planes.

The 3-D cavity streamline distributions for Re = 10 and Re = 100 at different planes are as shown in Figure 12.3.4. It is noted that, for higher Reynolds number (Re = 100), the vortex center moves toward downstream.

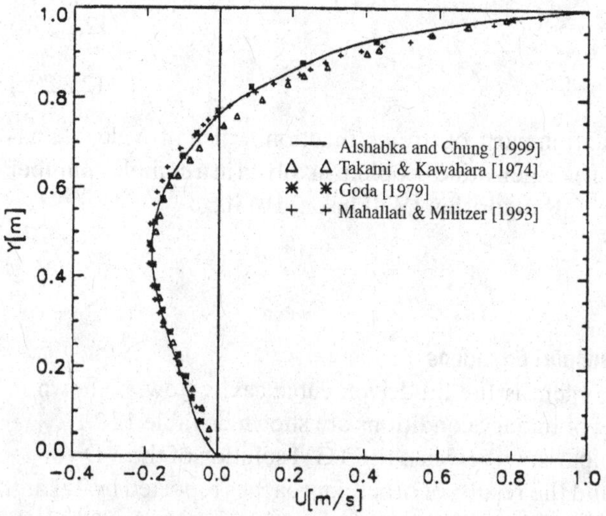

Figure 12.3.2 Velocity profiles on vertical centerline of the 3-D cavity for Re = 100.

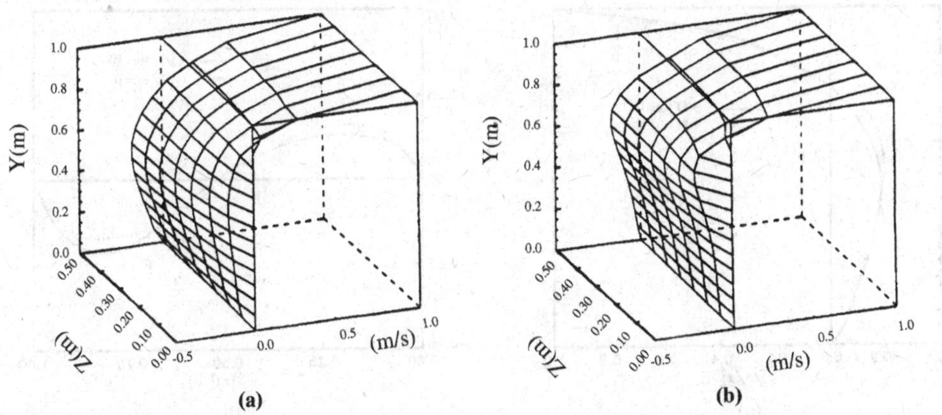

Figure 12.3.3 Profiles of the x-component of the velocity of the 3-D cavity flow at Re = 100. (**a**) The $X = 0.5$ plane. (**b**) The $x = 0.786$ plane.

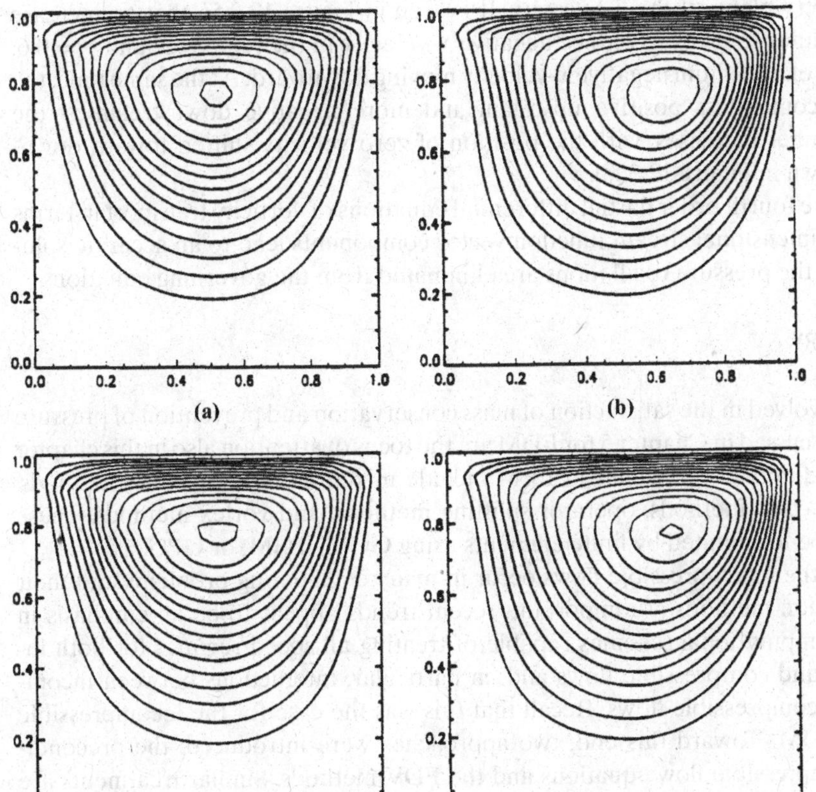

Figure 12.3.4 The 3-D cavity streamlines (ψ_3). (**a**) The symmetry plane ($z = 0.5$) for Re = 10. (**b**) The symmetry plane ($z = 0.5$) for Re = 100. (**c**) The $Z = 0.2$ plane for Re = 10. (**d**) The $Z = 0.2$ plane for Re = 100.

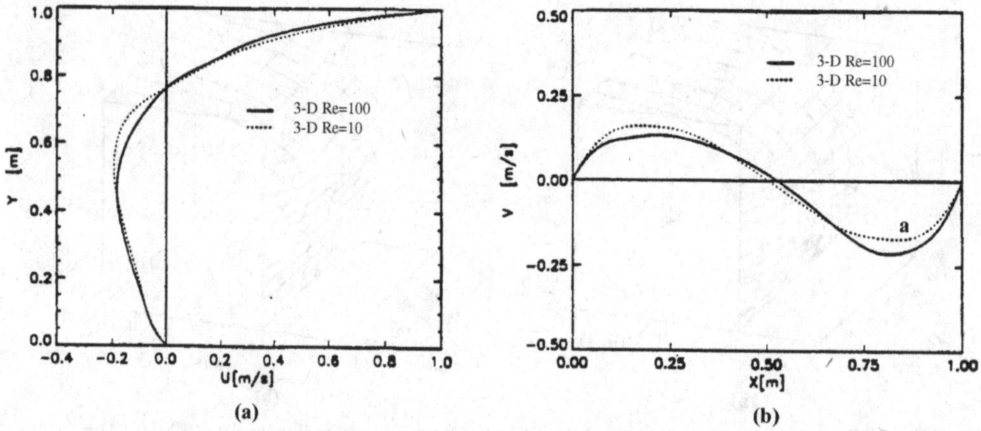

Figure 12.3.5 Velocity profile on the 3-D cavity. (a) Vertical centerline. (b) X-horizontal centerline.

Figure 12.3.5 shows the velocity profiles along the vertical and horizontal centerlines of the symmetry plane of the 3-D cavity. It is seen in Figure 12.3.5a that an increase in Reynolds number tends to reduce negative x-velocity in the region around $y = 0.6$, with the point of maximum negative x-velocity moving downward. At the same time, the y-velocity becomes less positive upstream and more negative downstream as the Reynolds number increases, with the position of zero velocity shifted toward downstream as shown in Figure 12.3.5b.

Overall, the fourth order partial differential equations of vorticity transport in terms of the three dimensional stream function vector components lead to an accurate solution, in which the pressure oscillations are eliminated from the governing equations.

12.4 SUMMARY

Difficulties involved in the satisfaction of mass conservation and prevention of pressure oscillations discussed in Chapter 5 for FDM are the focus of attention also in this chapter for FEM. Traditional approaches in FEM include mixed methods, penalty methods, pressure correction methods, operator splitting methods, and vortex methods. These methods can be formulated by finite elements using GGM, TGM, or GPG.

Although the incompressible flows occur in many engineering problems and their accurate solution methods are important, recent trends appear to be an emphasis in developing computational schemes capable of treating all speed regimes for both incompressible and compressible flows and, in particular, interactions between incompressible and compressible flows. Recall that this was the case for the incompressible flows using FDM. Toward this end, two approaches were introduced: the preconditioning of compressible flow equations and the FDV methods. Similar treatments are available for FEM applications. These and other topics will be discussed in the next chapter on compressible flows.

REFERENCES

Babuska, I. [1973]. The finite element method with Lagrange multipliers. *Num. Math.*, 20, 179–92.

Brezzi, F. [1974]. On the existence, uniqueness and approximation of saddle point problems arising from Lagrangian multiplier. RAIRO, series Rouge Analy. Numer., R-2, 129–51.

Elshabka, A. M. [1995]. Existence of three-dimensional stream function vector components and their applications in three-dimensional flow. Ph.D. disseration, The University of Alabama.

Elshabka, A. M. and Chung, T. J. [1999]. Numerical solution of three-dimensional stream function vector components of vorticity transport equations. *Comp. Meth. Appl. Mech. Eng.*, 170, 131–53.

Carey, G. F. and Oden, J. T. [1986]. Finite Elements: Fluid Mechanics. Englewood Cliffs, NJ: Prentice-Hall.

Chorin, A. J. [1967]. A numerical method for solving incompressible viscous flow problems. *J. Comp. Phys.*, 2, 12–26.

Francis, J. G. F. [1962]. The QR transformation. *Comp. J.*, 4, 265–71.

Gresho, P. M. and Sani. R. L. [1999]. *Incompressible Flows and Finite Element Method.* New York: Wiley.

Goda, K. [1979]. A multistep technique with implicit difference schemes for calculating two- or three-dimensional cavity flows. *J. Comp. Phys.*, 30, 76–95.

Gunzburger, M. D. and Nicholaides, R. A. [1993]. *Incompressible Computational Fluid Dynamics Trends and Advances*. UK: Cambridge University Press.

Hughes, T. J. R., Liu, W. K., and Brooks, A. N. [1979]. Finite element analysis of incompressible viscous flows by the penalty function formulation. *J. Comp. Phys.*, 30, 1–60.

Ladyszhenskaya, O. A. [1969]. *The Mathematical Theory of Viscous Incompressible Flow*. New York: Gordon and Breach.

Mahallati, A. and Militzer, J. [1993]. Application of the piecewise parabolic finite analytic methods to the three-dimensional cavity flow. *Num. Heat Trans.*, 24, Part B, 337–51.

Patankar, S. V. and Spalding, D. B. [1972]. A calculation procedure for heat, mass and momentum transfer in three-dimensional parabolic flows. *Int. J. Heat Mass Trans.*, 15, 1787–1806.

Takami, H. and Kuwahara, K. [1974]. Numerical study of three-dimensional flow within a cavity. *J. Phys. Soc. Japan.*, 73, 6, 1695–98.

Wilkinsom, J. H. [1965]. The algebraic eigenvalue problem. London: Clarendon Press.

Yanenko, N. N. [1971]. *The Method of Fractional Steps.* New York: Springer-Verlag.

Zienkiewicz, O. C. and Taylor, R. L. [1991]. *The Finite Element Method*, Vol. 2. UK: McGraw-Hill.

CHAPTER THIRTEEN

Compressible Flows via Finite Element Methods

In this chapter, finite element analyses of both inviscid and viscous compressible flows are examined. Traditionally, computational schemes for compressible inviscid flow are developed separately from compressible viscous flows, governed by Euler equations and Navier-Stokes system of equations, respectively. However, it is our desire in this chapter to study numerical schemes capable of treating a compressible flow with or without the effect of viscosity or diffusion. Furthermore, it would be desirable to develop a scheme that can handle all speed regimes – not only the compressible flow, but the incompressible flow as well. To accomplish this goal, the most suitable governing equations to use are the Navier-Stokes system of equations written in conservation form in terms of conservation variables. Advantages of transforming the conservation variables into entropy variables and primitive variables will be explored. One of the most prominent features in compressible flow calculations is the ability of numerical schemes to resolve shock waves or discontinuities in high-speed flows. Furthermore, compressible viscous flows at high Mach numbers and high Reynolds numbers lead to significant numerical difficulties. We shall address these and other issues in this chapter.

To this end, we begin with the general description of the governing equations in Section 13.1, followed by the Taylor-Galerkin methods (TGM), generalized Galerkin methods (GGM), generalized Petrov-Galerkin (GPG) methods, characteristic Galerkin methods (CGM), and discontinuous Galerkin methods (DGM) in Sections 13.2 through 13.4. Finally, the flowfield-dependent variation (FDV) methods introduced in FDM and discussed earlier in Section 6.5 will be presented for FEM applications (Section 13.6). This subject will be treated again in Chapter 16, where many of the methods in both FDM and FEM can be shown to be the special cases of FDV methods.

13.1 GOVERNING EQUATIONS

So far in the previous chapters, we have dealt with Stokes flows (no convection terms, Section 10.1.4), Burgers' equations (with convective terms but without pressure gradients, Chapter 11), and incompressible flows (with continuity and momentum equations, Chapter 12). More general types of flows include compressibility or density variations as a function of space and time and in nonisothermal environments, which are characterized by the Navier-Stokes system of equations for conservation of mass, momentum, and energy. Although we discussed these equations in Chapters 2 and 6, we shall repeat them here for convenience.

Continuity Equation

$$\frac{\partial \rho}{\partial t} + (\rho \mathbf{v}_i)_{,i} = 0 \tag{13.1.1a}$$

Momentum Equation

$$\rho \frac{\partial \mathbf{v}_j}{\partial t} + \rho \mathbf{v}_{j,i} \mathbf{v}_i + p_{,j} - \tau_{ij,i} - \rho F_j = 0 \tag{13.1.1b}$$

Energy Equation

$$\rho \frac{\partial \varepsilon}{\partial t} + \rho \varepsilon_{,i} \mathbf{v}_i + p \mathbf{v}_{i,i} - \tau_{ij} \mathbf{v}_{j,i} + q_{i,i} = 0 \tag{13.1.1c}$$

where τ_{ij}, ε, and q_i denote viscous stress tensor, internal energy, and heat flux, respectively.

Stress Tensor

$$\tau_{ij} = \mu \left(\mathrm{v}_{i,j} + \mathrm{v}_{j,i} - \frac{2}{3} \mathrm{v}_{k,k} \delta_{ij} \right)$$

Internal Energy

$$\varepsilon = c_p T - \frac{p}{\rho} = c_v T$$

Heat Flux

$$q_i = -k T_{,i}$$

where the dynamic viscosity μ and thermal conductivity k are given by Sutherland's law [(2.2.7) and (2.2.8)], respectively; and c_p and c_v represent specific heats at constant pressure and volume, respectively.

These equations may be combined into a conservation form

$$\frac{\partial \mathbf{U}}{\partial t} + \frac{\partial \mathbf{F}_i}{\partial x_i} + \frac{\partial \mathbf{G}_i}{\partial x_i} = \mathbf{B} \tag{13.1.2}$$

where $\mathbf{U}$, $\mathbf{F}_i$, $\mathbf{G}_i$, and $\mathbf{B}$ are the conservation variables, convection flux, diffusion flux, and body force vector, respectively.

$$\mathbf{U} = \begin{bmatrix} \rho \\ \rho \mathrm{v}_j \\ \rho E \end{bmatrix}, \quad \mathbf{F}_i = \begin{bmatrix} \rho \mathrm{v}_i \\ \rho \mathrm{v}_i \mathrm{v}_j + p\delta_{ij} \\ \rho E \mathrm{v}_i + p \mathrm{v}_i \end{bmatrix}, \quad \mathbf{G}_i = \begin{bmatrix} 0 \\ -\tau_{ij} \\ -\tau_{ij} \mathrm{v}_j + q_i \end{bmatrix}, \quad \mathbf{B} = \begin{bmatrix} 0 \\ \rho F_j \\ \rho F_j \mathrm{v}_j \end{bmatrix}$$

with E being the total energy,

$$E = \varepsilon + \frac{1}{2} \mathrm{v}_j \mathrm{v}_j \tag{13.1.3}$$

and the pressure p given by the equation of state,

$$p = \rho RT \tag{13.1.4a}$$

$$p = (\gamma - 1)\,\rho\left(E - \frac{1}{2}\mathrm{v}_i\mathrm{v}_i\right) \tag{13.1.4b}$$

$$T = \frac{1}{c_v}\left(E - \frac{1}{2}\mathrm{v}_i\mathrm{v}_i\right) \tag{13.1.4c}$$

where R is the specific gas constant which may be related to the specific heats as follows:

$$R = \frac{c_p(\gamma - 1)}{\gamma}, \qquad \gamma = \frac{c_p}{c_v}$$

The equation of state plays the role of a constraint for the Navier-Stokes system of equations.

For the purpose of generality, we shall keep the source terms **B** so that numerical formulations can be accommodated for the reacting flows as discussed in Chapter 22.

Nondimensional Form of Navier-Stokes System of Equations

The numerical solution of the Navier-Stokes system of equations in dimensional form typically involves operations between terms that vary by several orders of magnitude. This leads to a situation in which the numerical solution fails or becomes unstable as the computer floating point limits are exceeded. For this reason, the governing equations are often put into nondimensional form. Placing the flow variables in dimensionless form insures that variations are maintained within certain prescribed limits between 0 and 1. Additionally, writing the Navier-Stokes system of equations in dimensionless form facilitates generalization to embody a large range of problems. Also, the dimensionless form has the advantage that characteristic parameters such as Mach number, Reynolds number. Prandtl number, etc., can be regulated independently. Toward this end, we introduce the nondimensional variables

$$\begin{aligned} &x_i^* = \frac{x_i}{L}, \quad t^* = \frac{t}{L/\mathrm{v}_\infty}, \quad \rho^* = \frac{\rho}{\rho_\infty}, \quad \mathrm{v}_i^* = \frac{\mathrm{v}_i}{\mathrm{v}_\infty}, \quad E^* = \frac{E}{\mathrm{v}_\infty^2} \\ &p^* = \frac{p}{\rho_\infty \mathrm{v}_\infty^2}, \quad T^* = \frac{T}{T_\infty}, \quad \mu^* = \frac{\mu}{\mu_\infty}, \quad F_i^* = \frac{F_i}{\mathrm{v}_\infty^2/L} \end{aligned} \tag{13.1.5}$$

where an asterisk denotes nondimensional variables, infinity represents freestream conditions, and L is the reference length used in the Reynolds number

$$\mathrm{Re} = \frac{\rho_\infty \mathrm{v}_\infty L}{\mu_\infty} \tag{13.1.6}$$

With the nondimensional variables above, the dimensionless form of Navier-Stokes system of equations in conservation form (13.1.2) becomes

$$\frac{\partial \mathbf{U}^*}{\partial t^*} + \frac{\partial \mathbf{F}_i^*}{\partial x_i^*} + \frac{\partial \mathbf{G}_i^*}{\partial x_i^*} = \mathbf{B}^* \tag{13.1.7}$$

where the conservation flow variable vector, the convection flux vector, the diffusion flux vector, and the source vector in nondimensional form are defined by

$$\mathbf{U}^* = \begin{bmatrix} \rho^* \\ \rho^* v_j^* \\ \rho^* E^* \end{bmatrix}, \quad \mathbf{F}_i^* = \begin{bmatrix} \rho^* v_i^* \\ \rho^* v_i^* v_j^* + p^* \delta_{ij} \\ \rho^* E^* v_i^* + p^* v_i^* \end{bmatrix}, \quad \mathbf{G}_i^* = \frac{1}{\mathrm{Re}} \begin{bmatrix} 0 \\ -\tau_{ij}^* \\ -\tau_{ij}^* v_j^* + q_i^* \end{bmatrix},$$

$$\mathbf{B}^* = \begin{bmatrix} 0 \\ \rho^* F_j^* \\ \rho^* F_j^* v_j^* \end{bmatrix}, \quad \mathbf{b}_i = \frac{1}{\mathrm{Re}} \frac{\partial \mathbf{G}_i}{\partial \mathbf{U}}, \quad \mathbf{c}_{ij} = \frac{1}{\mathrm{Re}} \frac{\partial \mathbf{G}_i}{\partial \mathbf{U}_{,j}}$$

Here the nondimensional stagnation energy, the viscous stress tensor, and the thermal conductivity are

$$E^* = \frac{p^*}{(\gamma - 1)\rho^*} + \frac{1}{2} v_j^* v_j^* \tag{13.1.8}$$

$$\tau_{ij}^* = \mu^* \left(v_{i,j}^* + v_{j,i}^* - \frac{2}{3} v_{k,k}^* \delta_{ij} \right) \tag{13.1.9}$$

$$k^* = \frac{\mu^*}{(\gamma - 1) M_\infty^2 \,\mathrm{Pr}} = \frac{k}{\mu_\infty V_\infty^2 / T} \tag{13.1.10}$$

with Sutherland's law in the nondimensional form,

$$\mu^* = \frac{1 + S_o/T_\infty}{T^* + S_o/T_\infty} (T^*)^{\frac{3}{2}} \tag{13.1.11}$$

and the freestream Mach number,

$$M_\infty = \frac{V_\infty}{\sqrt{\gamma(\gamma - 1) c_v T_\infty}} \tag{13.1.12}$$

The nondimensional equations of state (13.1.4b,c) become

$$p^* = (\gamma - 1)\rho^* \left(E^* - \frac{1}{2} v_j^* v_j^* \right), \quad E^* = c_v^* T^* = \frac{T^*}{\gamma(\gamma - 1) M_\infty^2} \tag{13.1.13}$$

or

$$T^* = \frac{1}{c_v^*} \left(E^* - \frac{1}{2} v_j^* v_j^* \right) \tag{13.1.14}$$

where the nondimensional specific heat at constant volume,

$$c_v^* = \frac{1}{\gamma(\gamma - 1) M_\infty^2}, \quad c_p^* = \frac{c_p}{v_\infty^2 / T_\infty} = \frac{1}{(\gamma - 1) M_\infty^2} \tag{13.1.15}$$

An alternative form of the nondimensional state equations is expressed by

$$p^* = \rho^* R^* T^* \tag{13.1.16}$$

with

$$R^* = \frac{1}{\gamma M_\infty^2} \tag{13.1.17}$$

For convenience, the asterisks will now be omitted, but we continue to work with the dimensionless form of the governing equations in all or the following discussions.

13.2 TAYLOR-GALERKIN METHODS AND GENERALIZED GALERKIN METHODS

In Chapter 11, the Taylor Galerkin methods (TGM) were formulated by expanding the variables into Taylor series. It was also shown that similar results can be obtained from the generalized Galerkin methods (GGM) using the double projections of the residual onto the spatial and temporal test functions for the linearized Burgers' equations in which the numerical diffusion is absent. For the nonlinear Burgers' equations (Section 11.2.5), however, it was shown that TGM was capable of explicitly providing the numerical diffusion. In this section, we examine TGM as applied to the Navier-Stokes system of equations with the convection and diffusion fluxes transformed to the conservation variables through Jacobians. It will be shown that the numerical diffusion arises in much more complicated form than it does for the nonlinear Burgers' equations. We then discuss GGM, which is simpler but not as effective as TGM associated with convection-dominated flows or discontinuities. In Chapter 11 dealing with the Burgers equations, TGM was identified as a special case of GGM. This is no longer the case in this chapter working with the Navier-Stokes system of equations. This is because many different forms of TGM result from various approximations in Taylor series expansion of the conservation variables. We elaborate these and other topics below.

13.2.1 TAYLOR-GALERKIN METHODS

One of the well-known schemes in FEM as introduced in Chapter 11 is the Taylor-Galerkin methods (TGM) as applied to the Navier-Stokes system of equations. In dealing with the Navier-Stokes system of equations, unlike the Burgers' equations discussed in Chapter 11, it is convenient to work with conservation variables transformed from the convection and diffusion fluxes as follows [Hassan, Morgan, and Peraire, 1991]:

$$\frac{\partial \mathbf{F}_i}{\partial t} = \mathbf{a}_i \frac{\partial \mathbf{U}}{\partial t} \tag{13.2.1}$$

$$\frac{\partial \mathbf{G}_i}{\partial t} = \mathbf{b}_i \frac{\partial \mathbf{U}}{\partial t} + \mathbf{c}_{ij} \frac{\partial \mathbf{U}_{,j}}{\partial t} \tag{13.2.2}$$

with the convection Jacobian $\mathbf{a}_i$, diffusion Jacobian $\mathbf{b}_i$, and diffusion gradient Jacobian $\mathbf{c}_{ij}$ being defined as in (6.3.8).

Let us consider the Taylor series expansion of $\mathbf{U}^{n+1}$ in the form,

$$\mathbf{U}^{n+1} = \mathbf{U}^n + \Delta t \frac{\partial \mathbf{U}^n}{\partial t} + \frac{\Delta t^2}{2} \frac{\partial^2 \mathbf{U}^{n+1}}{\partial t^2} + O(\Delta t^3) \tag{13.2.3}$$

in which the second derivative is set at the implicit form $(n+1)$. Substituting (13.1.2) into (13.2.3) gives

$$\Delta \mathbf{U}^{n+1} = \Delta t \left(-\frac{\partial \mathbf{F}_i}{\partial x_i} - \frac{\partial \mathbf{G}_i}{\partial x_i} + \mathbf{B} \right)^n + \frac{\Delta t^2}{2} \frac{\partial}{\partial t} \left(-\frac{\partial \mathbf{F}_i}{\partial x_i} - \frac{\partial \mathbf{G}_i}{\partial x_i} + \mathbf{B} \right)^{n+1} + O(\Delta t^3) \tag{13.2.4}$$

Using the definitions of convection, diffusion, and diffusion gradient Jacobians, the temporal rates of change of convection and diffusion variables may be written as follows:

$$\frac{\partial \mathbf{F}_i^n}{\partial t} = \left(\mathbf{a}_i \frac{\partial \mathbf{U}}{\partial t}\right)^n = \left[\mathbf{a}_i\left(-\frac{\partial \mathbf{F}_j}{\partial x_j} - \frac{\partial \mathbf{G}_j}{\partial x_j} + \mathbf{B}\right)\right]^n$$

$$\frac{\partial \mathbf{F}_i^{n+1}}{\partial t} = \mathbf{a}_i\left(-\frac{\partial \mathbf{F}_j^{n+1}}{\partial x_j} - \frac{\partial \mathbf{G}_j^{n+1}}{\partial x_j} + \mathbf{B}^{n+1}\right)$$

$$= \mathbf{a}_i\left[\left(-\mathbf{a}_j \frac{\partial}{\partial x_j}(\mathbf{U}^{n+1} - \mathbf{U}^n) - \frac{\partial \mathbf{F}_j^n}{\partial x_j} - \frac{\partial \mathbf{G}_j^{n+1}}{\partial x_j} + \mathbf{B}^{n+1}\right)\right] \tag{13.2.5}$$

$$\frac{\partial \mathbf{G}_i^{n+1}}{\partial t} = \left(\mathbf{b}_i \frac{\partial \mathbf{U}}{\partial t}\right)^{n+1} + \left[\mathbf{c}_{ij} \frac{\partial}{\partial t}\left(\frac{\partial \mathbf{U}}{\partial x_j}\right)\right]^{n+1}$$

or

$$\frac{\partial \mathbf{G}_i^{n+1}}{\partial t} = \left(\mathbf{b}_i - \frac{\partial \mathbf{c}_{ij}}{\partial x_j}\right)\frac{\Delta \mathbf{U}^{n+1}}{\Delta t} + \frac{\partial}{\partial x_j}\left(\mathbf{c}_{ij}\frac{\Delta \mathbf{U}}{\Delta t}\right)^{n+1} \tag{13.2.6}$$

Substituting (13.2.5) and (13.2.6) into (13.2.4) yields

$$\begin{aligned}\Delta \mathbf{U}^{n+1} &= \Delta t\left(-\frac{\partial \mathbf{F}_i}{\partial x_i} - \frac{\partial \mathbf{G}_i}{\partial x_i} + \mathbf{B}\right)^n \\ &+ \frac{\Delta t^2}{2}\left\{\frac{\partial}{\partial x_i}\left[-\mathbf{a}_i\left(-\mathbf{a}_j\frac{\partial \Delta \mathbf{U}^{n+1}}{\partial x_j} - \frac{\partial \mathbf{F}_j^n}{\partial x_j} - \frac{\partial \mathbf{G}_j^{n+1}}{\partial x_j} + \mathbf{B}^{n+1}\right)\right.\right. \\ &\left.\left. - \left(\mathbf{e}_i + \frac{\partial \mathbf{c}_{ij}}{\partial x_j}\right)\frac{\Delta \mathbf{U}^{n+1}}{\Delta t}\right] + \frac{\partial \mathbf{B}^{n+1}}{\partial t}\right\}\end{aligned} \tag{13.2.7}$$

with

$$\mathbf{e}_i = \mathbf{b}_i - \frac{\partial \mathbf{c}_{ij}}{\partial x_j}$$

Neglecting the spatial and temporal derivatives of **B**, we rewrite (13.2.7) in the form

$$\left\{1 + \frac{\Delta t}{2}\frac{\partial \mathbf{e}_i}{\partial x_i} - \frac{\Delta t^2}{2}\frac{\partial}{\partial x_i}\left(\mathbf{a}_i\mathbf{a}_j - \frac{\mathbf{c}_{ij}}{\Delta t}\right)\frac{\partial}{\partial x_j}\right\}\Delta \mathbf{U}^{n+1}$$

$$= \Delta t\left(-\frac{\partial \mathbf{F}_i}{\partial x_i} - \frac{\partial \mathbf{G}_i}{\partial x_i} + \mathbf{B}\right)^n + \frac{\Delta t^2}{2}\frac{\partial}{\partial x_i}\left(\mathbf{a}_i\frac{\partial \mathbf{F}_j}{\partial x_j}\right)^n$$

Here the second derivatives of $\mathbf{G}_i$ are neglected and all Jacobians are assumed to remain constant within an incremental time step, but updated at subsequent time steps.

We now introduce the trial functions for the various variables in the form,

$$\mathbf{U} = \Phi_\alpha \mathbf{U}_\alpha, \qquad \mathbf{F}_i = \Phi_\alpha \mathbf{F}_{\alpha i}, \qquad \mathbf{G}_i = \Phi_\alpha \mathbf{G}_{\alpha i}, \qquad \mathbf{B} = \Phi_\alpha \mathbf{B}_\alpha$$

Substituting the above into (13.2.8) leads to an implicit scheme,

$$(A_{\alpha\beta}\delta_{rs} + B_{\alpha\beta rs})\Delta U_{\beta s}^{n+1} = H_{\alpha r}^n + N_{\alpha r}^{n+1} + \overline{N}_{\alpha r}^n \tag{13.2.8}$$

where

$$A_{\alpha\beta} = \int_\Omega \Phi_\alpha \Phi_\beta d\Omega$$

$$B_{\alpha\beta rs} = \frac{\Delta t}{2}\int_{\Omega} e_{irs}\Phi_{\alpha}\Phi_{\beta,i}d\Omega + \frac{\Delta t^2}{2}\int_{\Omega}\left[\left(a_{irq}a_{jqs} - \frac{c_{ijrs}}{\Delta t}\right)\Phi_{\alpha,i}\Phi_{\beta,j}\right]d\Omega$$

$$H^n_{\alpha r} = \Delta t\int_{\Omega}\left[\Phi_{\alpha,i}\Phi_{\beta}\left(F^n_{\beta ir} + G^n_{\beta ir}\right) + \Phi_{\alpha}\Phi_{\beta}B^n_{\beta r} - \frac{\Delta t}{2}a_{irs}\Phi_{\alpha,i}\Phi_{\beta,j}F^n_{\beta js}\right]d\Omega$$

$$N^{n+1}_{\alpha r} = \frac{\Delta t^2}{2}\int_{\Gamma}\left(a_{irq}a_{jqs} - \frac{c_{ijrs}}{\Delta t}\right)\overset{*}{\Phi}_{\alpha}\Delta U^{n+1}_{s,j}n_i d\Gamma$$

$$\overline{N}^n_{\alpha r} = -\int_{\Gamma}\left[\Delta t\overset{*}{\Phi}_{\alpha}\left(F^n_{ir} + G^n_{ir}\right) - \frac{\Delta t^2}{2}a_{irs}\overset{*}{\Phi}_{\alpha}F^n_{js,j}\right]n_i d\Gamma$$

where the indices α, β denote the global node, r, s represent the equation number listed in (13.1.2), and i, j indicate spatial coordinates. Note also that all quantities with r, s are lightfaced, indicating that they are no longer vector quantities.

It should be recognized that the integral

$$\frac{\Delta t^2}{2}\int_{\Omega}\mathbf{a}_{irq}\mathbf{a}_{jsq}\Phi_{\alpha,i}\Phi_{\beta,j}d\Omega = \int_{\Omega}\nu_{ijrs}\Phi_{\alpha,i}\Phi_{\beta,j}d\Omega \tag{13.2.9}$$

contained in $B_{\alpha\beta rs}$ represents the numerical diffusion, corresponding to that given in (11.2.76) for the Burgers' equations. We note that the velocity components for the Burgers' equations are simply replaced by the convection Jacobian components for the Navier-Stokes system of equations.

Instead of simulating the second order time derivatives implicitly, we may leave them in an explicit form so that the standard Taylor series can be used.

$$\mathbf{U}^{n+1} = \mathbf{U}^n + \Delta t\frac{\partial\mathbf{U}^n}{\partial t} + \frac{\Delta t^2}{2}\frac{\partial^2\mathbf{U}^n}{\partial t^2} + O(\Delta t^3) \tag{13.2.10}$$

where

$$\frac{\partial\mathbf{U}}{\partial t} = -\frac{\partial\mathbf{F}_i}{\partial x_i} - \frac{\partial\mathbf{G}_i}{\partial x_i} + \mathbf{B} = -\mathbf{a}_i\frac{\partial\mathbf{U}}{\partial x_i} - \frac{\partial\mathbf{G}_i}{\partial x_i} + \mathbf{B} \tag{13.2.11}$$

$$\frac{\partial^2\mathbf{U}}{\partial t^2} = -\frac{\partial}{\partial t}\left(\mathbf{a}_i\frac{\partial\mathbf{U}}{\partial x_i} + \frac{\partial\mathbf{G}_i}{\partial x_i} - \mathbf{B}\right)$$

or

$$\frac{\partial^2\mathbf{U}}{\partial t^2} = \frac{\partial}{\partial x_j}\left(\mathbf{a}_i\mathbf{a}_j\frac{\partial\mathbf{U}}{\partial x_i}\right) + \frac{\partial}{\partial x_i}\left(\mathbf{a}_i\frac{\partial\mathbf{G}_j}{\partial x_j}\right) - \frac{\partial}{\partial x_i}(\mathbf{a}_i\mathbf{B}) + \frac{\partial\mathbf{B}}{\partial t} \tag{13.2.12}$$

Substituting (13.2.11) and (13.2.12) into (13.2.10), we obtain

$$\Delta\mathbf{U}^{n+1} = \Delta t\left\{-\frac{\partial\mathbf{F}_i}{\partial x_i} - \frac{\partial\mathbf{G}_i}{\partial x_i} + \mathbf{B} + \frac{\Delta t}{2}\left[\frac{\partial}{\partial x_j}\left(\mathbf{a}_i\mathbf{a}_j\frac{\partial\mathbf{U}}{\partial x_i}\right) + \frac{\partial^2(\mathbf{a}_i\mathbf{G}_j)}{\partial x_i\partial x_j} - \frac{\partial}{\partial x_i}(\mathbf{a}_i\mathbf{B}) + \frac{\partial\mathbf{B}}{\partial t}\right]\right\}^n \tag{13.2.13}$$

or

$$\Delta\mathbf{U}^{n+1} = \Delta t\left(-\frac{\partial\mathbf{F}_i}{\partial x_i} - \frac{\partial\mathbf{G}_i}{\partial x_i} + \mathbf{B}\right)^n + \frac{\Delta t^2}{2}\left\{\frac{\partial}{\partial x_i}\left(\mathbf{a}_i\mathbf{a}_j\frac{\partial\Delta\mathbf{U}^{n+1}}{\partial x_j} + \mathbf{a}_i\frac{\partial\mathbf{F}^n_j}{\partial x_j}\right) + \frac{\partial^2(\mathbf{a}_i\mathbf{G}_j)^{n+1}}{\partial x_i\partial x_j} + \frac{\partial}{\partial x_i}(\mathbf{a}_i\mathbf{B})^{n+1} + \frac{\partial\mathbf{B}^{n+1}}{\partial t}\right\} \tag{13.2.14}$$

Rearranging (13.2.14) gives

$$\left[1 - \frac{\Delta t^2}{2}\frac{\partial}{\partial x_i}\left(\mathbf{a}_i\mathbf{a}_j - \frac{\mathbf{c}_{ij}}{\Delta t}\right)\frac{\partial}{\partial x_j}\right]\Delta\mathbf{U}^{n+1}$$
$$= \Delta t\left(-\frac{\partial \mathbf{F}_i}{\partial x_i} - \frac{\partial \mathbf{G}_i}{\partial x_i} + \mathbf{B}\right)^n + \frac{\Delta t^2}{2}\frac{\partial}{\partial x_i}\left(\mathbf{a}_i\frac{\partial \mathbf{F}_j}{\partial x_j}\right)^n \tag{13.2.15}$$

where the second derivatives of $\mathbf{G}_i$ are assumed to be negligible and $\mathbf{B}$ is constant in space and time. We then arrive at an implicit finite element scheme,

$$(A_{\alpha\beta}\delta_{rs} + B_{\alpha\beta rs})\,\Delta U_{\beta s}^{n+1} = H_{\alpha r}^n + N_{\alpha r}^{n+1} + \overline{N}_{\alpha r}^n \tag{13.2.16}$$

where

$$A_{\alpha\beta} = \int_\Omega \Phi_\alpha\Phi_\beta\, d\Omega$$

$$B_{\alpha\beta rs} = \frac{\Delta t^2}{2}\int_\Omega\left[\left(a_{irq}a_{jqs} - \frac{c_{ijrs}}{\Delta t}\right)\Phi_{\alpha,i}\Phi_{\beta,j}\right]d\Omega$$

$$H_{\alpha r}^n = \Delta t\int_\Omega\left[\Phi_{\alpha,i}\Phi_\beta\left(F_{\beta ir}^n + G_{\beta ir}^n\right) + \Phi_\alpha\Phi_\beta B_{\beta r}^n - \frac{\Delta t}{2}a_{irs}\Phi_{\alpha,i}\Phi_{\beta,j}F_{\beta js}^n\right]d\Omega$$

$$N_{\alpha r}^{n+1} = \frac{\Delta t^2}{2}\int_\Gamma\left(a_{irq}a_{jqs} - \frac{c_{ijrs}}{\Delta t}\right)\overset{*}{\Phi}_\alpha\Delta U_{s,j}^{n+1}n_i\, d\Gamma$$

$$\overline{N}_{\alpha r}^n = -\int_\Gamma\left[\Delta t\,\overset{*}{\Phi}_\alpha\left(F_{ir}^n + G_{ir}^n\right) - \frac{\Delta t^2}{2}a_{irs}\overset{*}{\Phi}_\alpha F_{js,j}^n\right]n_i\, d\Gamma$$

It is interesting to note that both (13.2.8) and (13.2.16) are identical if the first integral of $B_{\alpha\beta rs}$ in (13.2.8) is negligible or $\mathbf{e}_i = \mathbf{b}_i - \frac{\partial \mathbf{c}_{ij}}{\partial x_j} \cong 0$, in which the role of the diffusion Jacobian $\mathbf{b}_i$ no longer exists. However, in other formulations such as in FDV (see Section 6.5 and Section 13.6), the diffusion Jacobian is shown to be important in modeling convection-diffusion interactions.

13.2.2 TAYLOR-GALERKIN METHODS WITH OPERATOR SPLITTING

If the source term $\mathbf{B}$ contains time scales widely disparate in comparison with fluid convection time scales such as occur in chemical reactions, then it is advantageous to split the Navier-Stokes system of equations into two parts so that the flow can be treated explicitly whereas the source terms are accommodated implicitly, a scheme known as the point implicit method. To this end, we may split the governing equations (13.1.7) into two parts:

$$\frac{\partial \mathbf{U}}{\partial t} + \frac{\partial \mathbf{F}_i}{\partial x_i} + \frac{\partial \mathbf{G}_i}{\partial x_i} = \mathbf{0}$$
$$\frac{\partial \mathbf{U}}{\partial t} = \mathbf{B} \tag{13.2.17a,b}$$

where (13.2.17) is identified as the fluid operator written in two-step Taylor-Galerkin method.

Step 1

$$\Delta \mathbf{U}^{n+1/2} = \mathbf{U}^{n+1/2} - \mathbf{U}^n = -\frac{\Delta t}{2}\left(\frac{\partial \mathbf{F}_i}{\partial x_i} + \frac{\partial \mathbf{G}_i}{\partial x_i}\right)^n, \quad A_{\alpha\beta}\delta_{rs}U_{\beta s}^{n+1/2} = Q_{\alpha r}^n \quad (13.2.18a)$$

Step 2

$$\Delta \mathbf{U}^{n+1} = -\Delta t\left(\frac{\partial \mathbf{F}_i}{\partial x_i} + \frac{\partial \mathbf{G}_i}{\partial x_i}\right)^{n+1/2}, \qquad A_{\alpha\beta}\delta_{rs}U_{\beta s}^{n+1} = Q_{\alpha r}^{n+1/2} \quad (13.2.18b)$$

with the right-hand side of (13.2.18a,b) consisting of domain and boundary integrals as usual.

The source term operator is provided with the intermediate iterative increment $m+1$ and m between $n+1$ and n so that

$$\frac{\partial \mathbf{U}^{m+1}}{\partial t} = \mathbf{B}^{m+1} \quad (13.2.19)$$

where

$$\frac{\partial \mathbf{U}^{m+1}}{\partial t} = \frac{\mathbf{U}^{m+1} - \mathbf{U}^n}{\Delta t} = \frac{\Delta \mathbf{U}^{m+1}}{\Delta t} + \frac{\Delta \mathbf{U}^m}{\Delta t} \quad (13.2.20a)$$

$$\mathbf{B}^{m+1} = \mathbf{B}^m + \frac{\partial \mathbf{B}}{\partial \mathbf{U}}\Delta \mathbf{U}^{m+1} \quad (13.2.20b)$$

with

$$\Delta \mathbf{U}^{m+1} = \mathbf{U}^{m+1} - \mathbf{U}^m, \qquad \Delta \mathbf{U}^m = \mathbf{U}^m - \mathbf{U}^n$$

Substituting (13.2.20a,b) into (13.2.19) yields

Step 3

$$\left(\mathbf{I} - \Delta t\frac{\partial \mathbf{B}}{\partial \mathbf{U}}\right)\Delta \mathbf{U}^{m+1} = -\Delta \mathbf{U}^m + \Delta t \mathbf{B}^m \quad (13.2.21)$$

To implement these three steps, we must first obtain the finite element analogs (13.2.18a,b) using the standard approach. The Galerkin finite element formulation of (13.2.21) gives

$$(A_{\alpha\beta}\delta_{rs} - \Delta t\, B_{\alpha\beta rs})\,\Delta \mathbf{U}_{\beta s}^{m+1} = -A_{\alpha\beta}\delta_{rs}\Delta \mathbf{U}_{\beta s}^m + \Delta t\, A_{\alpha\beta}\delta_{rs}\mathbf{B}_{\beta s}^m \quad (13.2.22)$$

with

$$A_{\alpha\beta} = \int_\Omega \Phi_\alpha \Phi_\beta \, d\Omega \quad (13.2.23a)$$

$$B_{\alpha\beta rs} = \int_\Omega f_{rs}\Phi_\alpha \Phi_\beta \, d\Omega \quad (13.2.23b)$$

$$f_{rs} = \frac{\partial \mathbf{B}_{(r)}}{\partial \mathbf{U}_{(s)}} \quad (13.2.23c)$$

Here, ΔU^m is set equal to ΔU^{n+1} with the final solution being ΔU^{m+1}. The solution will begin with the initial and boundary conditions, followed by steps 1, 2, and 3 being repeated until convergence. Applications of this scheme are demonstrated in Section 22.6.1.

13.2.3 GENERALIZED GALERKIN METHODS

Recall that, in Section 11.2, TGM is shown to be a special case of generalized Galerkin methods (GGM) in dealing with the linearized Burgers' equations. Such is not the case for the Navier-Stokes system of equations, as demonstrated by the nonlinear Burgers' equations in Section 11.2.5.

Constructing the double projections of the residual of the Navier-Stokes system of equations in terms of Jacobians onto the spatial and temporal test functions, we obtain

$$(\hat{W}(\xi), (\Phi_\alpha, \mathbf{R})) = \int_\xi \hat{W}(\xi) \int_\Omega \Phi_\alpha \left(\frac{\partial \mathbf{U}}{\partial t} + \mathbf{a}_i \frac{\partial \mathbf{U}}{\partial x_i} + \mathbf{b}_i \frac{\partial \mathbf{U}}{\partial x_i} + \mathbf{c}_{ij} \frac{\partial^2 \mathbf{U}}{\partial x_i \partial x_j} - \mathbf{B} \right) d\Omega d\xi = 0 \tag{13.2.24}$$

or without the Jacobians,

$$(\hat{W}(\xi), (\Phi_\alpha, \mathbf{R})) = \int_\xi \hat{W}(\xi) \int_\Omega \Phi_\alpha \left(\frac{\partial \mathbf{U}}{\partial t} + \frac{\partial \mathbf{F}_i}{\partial x_i} + \frac{\partial \mathbf{G}_i}{\partial x_i} - \mathbf{B} \right) d\Omega d\xi = 0 \tag{13.2.25}$$

Using the various forms of the temporal test functions $W(\xi)$ and temporal parameters η as given in Chapter 10, we obtain numerous options for the finite element equations from (13.2.24) or from (13.2.25).

For simplicity, let us examine (13.2.24), using the temporal test function, $W(\xi) = \delta(\xi - \frac{1}{2})$ or $W(\xi) = 1$ with linear variations of nodal values of the conservation variables. The generalized Galerkin finite element equations are of the form

$$\left(A_{\alpha\beta}\delta_{rs} + \frac{\Delta t}{2}(B_{\alpha\beta rs} + K_{\alpha\beta rs}) \right) \Delta U^{n+1}_{\beta s} = H^n_{\alpha r} + N^n_{\alpha r} \tag{13.2.26}$$

where

$$A_{\alpha\beta} = \int_\Omega \Phi_\alpha \Phi_\beta d\Omega \qquad B_{\alpha\beta rs} = -\int_\Omega (a_{irs} + b_{irs}) \Phi_{\alpha,i} \Phi_\beta d\Omega$$

$$K_{\alpha\beta rs} = \int_\Omega c_{ijrs} \Phi_{\alpha,i} \Phi_{\beta,j} d\Omega \qquad H^n_{\alpha r} = \Delta t \int_\Omega \Phi_\alpha \Phi_\beta B_{\beta r} d\Omega$$

$$N^n_{\alpha r} = \Delta t \int_\Gamma \overset{*}{\Phi}_\alpha \left(F^n_{ir} + G^n_{ir} \right) n_i d\Gamma$$

Similarly, for (13.2.25), we obtain

$$A_{\alpha\beta}\delta_{rs}\Delta U^{n+1}_{\beta s} = \frac{\Delta t}{2}\left[E_{\alpha\beta i}\left(F^n_{\beta ir} + G^n_{\beta ir} \right) \right] + \Delta t \left(H^n_{\alpha r} + N^n_{\alpha r} \right) \tag{13.2.27}$$

where

$$E_{\alpha\beta i} = \int_\Omega \Phi_{\alpha,i} \Phi_\beta \, d\Omega$$

with all other notations being the same as in (13.2.26).

For the solution of (13.2.27), we may begin with the fractional step $n + 1/2$ in an explicit scheme, which is updated in the following step, $n + 1$.

Step 1

$$A_{\alpha\beta}\delta_{rs}\Delta U_{\beta s}^{n+1/2} = \frac{\Delta t}{2}\left[E_{\alpha\beta i}\left(F_{\beta ir}^{n} + G_{\beta ir}^{n}\right)\right] + 2\left(H_{\alpha r}^{n} + N_{\alpha r}^{n}\right) \tag{13.2.28}$$

Step 2

$$A_{\alpha\beta}\delta_{rs}\Delta U_{\beta s}^{n+1} = \frac{\Delta t}{2}\left[E_{\alpha\beta i}\left(F_{\beta ir}^{n+\frac{1}{2}} + G_{\beta ir}^{n+\frac{1}{2}}\right)\right] + 2\left(H_{\alpha r}^{n+\frac{1}{2}} + N_{\alpha r}^{n}\right) \tag{13.2.29}$$

The nodal values, $F_{\beta ir}^{n+1/2}$, $G_{\beta ir}^{n+1/2}$, and $H_{\alpha r}^{n+1/2}$ at step 1, are estimated or determined from the boundary conditions, and $F_{\beta ir}^{n+1}$, $G_{\beta i}^{n+1}$, and $H_{\alpha r}^{n+1}$ at step 2 are calculated from $U_{\beta s}^{n+\frac{1}{2}}$ of step 1.

As was demonstrated in (11.2.12), it is necessary to add the numerical diffusion integral (13.2.9) to the convection matrix in (13.2.26) for high-speed convection-dominated flows.

13.3 GENERALIZED PETROV-GALERKIN METHODS

13.3.1 NAVIER-STOKES SYSTEM OF EQUATIONS IN VARIOUS VARIABLE FORMS

In Chapter 11, we studied the generalized Petrov-Galerkin (GPG) methods, also known as the streamline upwind Petrov-Galerkin (SUPG) methods, streamline diffusion methods (SDM), or Galerkin/least squares (GLS) as discussed in Sections 11.2 and 11.3. They were originally developed for incompressible flows, and subsequently extended to compressible flows governed by the Navier-Stokes system of equations. These methods were explored extensively by Hughes and others and are now considered as some of the most robust computational schemes that deal with discontinuities such as in shock waves. In Sections 11.3 and 11.4, it was suggested that SUPG, SDM, and GLS be called GPG for the sake of uniformity and convenience. This is because all of these methods provide numerical diffusion test functions of various forms in addition to the standard Galerkin test functions, leading to the Petrov-Galerkin methods. The concept of space/time approximations suggests and lends itself to the generalized Petrov-Galerkin (GPG) methods.

As demonstrated in Sections 11.3 and 11.4, the basic idea is to apply numerical diffusion in the direction of the streamline parallel to the velocity as in (11.3.29). Sharp discontinuities require additional numerical diffusion parallel to the velocity gradients directed toward the acceleration as in (11.3.35b), known as the discontinuity-capturing scheme. These treatments were developed for Burgers' equations where the velocity can be identified as a single variable.

In dealing with multivariables such as in the Navier-Stokes system of equations, however, numerical diffusion test functions are modified accordingly. To this end, let us consider the conservation form of the Navier-Stokes system of equations,

$$\mathbf{R} = \frac{\partial \mathbf{U}}{\partial t} + \frac{\partial \mathbf{F}_i}{\partial x_i} + \frac{\partial \mathbf{G}_i}{\partial x_i} - \mathbf{B} = 0 \tag{13.3.1a}$$

or

$$\mathbf{R} = \frac{\partial \mathbf{U}}{\partial t} + (\mathbf{a}_i + \mathbf{b}_i)\frac{\partial \mathbf{U}}{\partial x_i} + \mathbf{c}_{ij}\frac{\partial^2 \mathbf{U}}{\partial x_i \partial x_j} - \mathbf{B} = 0 \tag{13.3.1b}$$

where $\mathbf{a}_i$, $\mathbf{b}_i$, and $\mathbf{c}_{ij}$ denote the Jacobians of convection, diffusion, and diffusion gradients, respectively, as shown in Section 13.2. It should be noted that, in some applications in CFD, the diffusion Jacobian $\mathbf{b}_i$ is neglected, but it is important where inviscid-viscous interactions are taken into account such as in FDV to be discussed in Section 13.6.

Although the governing equations given by either (13.3.1a) or (13.3.1b) may be solved using the GPG methods, it is possible that improved solutions are obtained if the conservation variables are transformed into entropy variables in which the Clausius-Duhem inequality is satisfied, contributing to numerical stability [Harten, 1983; Tadmor, 1984; Hughes, Franca, and Mallet, 1986; Hauke and Hughes, 1998].

The relationship between conservation variables U and entropy variables V can be established using the following definitions:

Conservation Variables

$$\mathbf{U} = \begin{bmatrix} U_1 \\ U_2 \\ U_3 \\ U_4 \\ U_5 \end{bmatrix} = \begin{bmatrix} \rho \\ \rho v_1 \\ \rho v_2 \\ \rho v_3 \\ \rho E \end{bmatrix} = \rho\varepsilon \begin{bmatrix} -V_5 \\ V_2 \\ V_3 \\ V_4 \\ 1 - \dfrac{V_2^2 + V_3^2 + V_4^2}{2V_5} \end{bmatrix} \tag{13.3.2}$$

Entropy Variables

$$\mathbf{V} = \begin{bmatrix} V_1 \\ V_2 \\ V_3 \\ V_4 \\ V_5 \end{bmatrix} = \frac{1}{\rho\varepsilon}\begin{bmatrix} -U_5 + \rho\varepsilon(\gamma + 1 - s) \\ U_2 \\ U_3 \\ U_4 \\ -U_1 \end{bmatrix} = \frac{1}{c_v T}\begin{bmatrix} H - c_v T s - \frac{1}{2}v_i v_i \\ v_1 \\ v_2 \\ v_3 \\ -1 \end{bmatrix} \tag{13.3.3}$$

where H is the enthalpy and s is the dimensionless entropy

$$s = \gamma - V_1 + (V_2^2 + V_3^2 + V_4^2)/2V_5 \tag{13.3.4a}$$

with

$$\rho\varepsilon = U_5 - (U_2^2 + U_3^2 + U_4^2)/2U_1 \tag{13.3.4b}$$

Substituting (13.3.2) and (13.3.3) into (13.3.1) leads to

$$\mathbf{R} = \mathbf{C}\frac{\partial \mathbf{V}}{\partial t} + \mathbf{C}_i\frac{\partial \mathbf{V}}{\partial x_i} + \mathbf{C}_{ij}\frac{\partial^2 \mathbf{V}}{\partial x_i \partial x_j} - \mathbf{B} = 0 \tag{13.3.5}$$

where the entropy variable Jacobians are defined as

$$\mathbf{C} = \frac{\partial \mathbf{U}}{\partial \mathbf{V}}, \qquad \mathbf{C}_i = (\mathbf{a}_i + \mathbf{b}_i)\mathbf{C}, \qquad \mathbf{C}_{ij} = \mathbf{c}_{ij}\mathbf{C} \tag{13.3.6a,b,c}$$

with the explicit form of the entropy variable Jacobian $\mathbf{C}$ being given in terms of the entropy variables as follows [Shakib, Hughes, and Johan, 1991]:

$$\mathbf{C} = \frac{\rho\varepsilon}{\overline{\gamma} V_5}\begin{bmatrix} -V_5^2 & e_1 & e_2 & e_3 & V_5(1-k_1) \\ & c_1 & d_1 & d_2 & V_2 k_2 \\ & & c_2 & d_3 & V_3 k_2 \\ & & & c_3 & V_4 k_2 \\ symm. & & & & -k_3 \end{bmatrix} \tag{13.3.7}$$

with

$$\overline{\gamma} = \gamma - 1 \qquad c_1 = \overline{\gamma} V_5 - V_2^2 \qquad e_1 = V_2 V_5$$
$$k_1 = \frac{1}{2}\left(V_2^2 + V_3^2 + V_4^2\right)/V_5 \qquad c_2 = \overline{\gamma} V_5 - V_3^2 \qquad e_2 = V_3 V_5$$
$$k_2 = k_1 - \gamma \qquad c_3 = \overline{\gamma} V_5 - V_4^2 \qquad e_3 = V_4 V_5$$
$$k_3 = k_1^2 - 2\gamma k_1 + \gamma \qquad d_1 = -V_2 V_3$$
$$k_4 = k_2 - \gamma \qquad d_2 = -V_2 V_4$$
$$k_5 = k_2^2 - \overline{\gamma}(k_1 + k_2) \qquad d_3 = -V_3 V_4$$

It should be noted that all coefficient matrices, $\mathbf{C}$, $\mathbf{C}_i$, and $\mathbf{C}_{ij}$ are symmetric, and the eigenvalues associated with the convective terms are well conditioned.

Primitive Variables

For calculations involving both compressible and incompressible flows, the formulations based on conservation variables may lead to difficulties when the incompressible limit ($M_\infty = 0$) is approached. In this case, convergence toward a steady state can be very slow. To circumvent such difficulties, the concept of preconditioning is introduced as in FDM [Choi and Merkle, 1993] and also as in FEM [Hauke and Hughes, 1998] by means of the primitive variable Jacobian,

$$\mathbf{D} = \frac{\partial \mathbf{U}}{\partial \mathbf{W}} \tag{13.3.8}$$

where $\mathbf{W}$ represents the primitive variables,

$$\mathbf{W} = \begin{bmatrix} \rho \\ v_1 \\ v_2 \\ v_3 \\ T \end{bmatrix} \tag{13.3.9}$$

Introducing (13.3.8) and (13.3.9) into (13.3.1), we obtain

$$\mathbf{R} = \mathbf{D}\frac{\partial \mathbf{W}}{\partial t} + \mathbf{D}_i \frac{\partial \mathbf{W}}{\partial x_i} + \mathbf{D}_{ij}\frac{\partial^2 \mathbf{W}}{\partial x_i \partial x_j} - \mathbf{B} = 0 \tag{13.3.10}$$

with

$$\mathbf{D}_i = (\mathbf{a}_i + \mathbf{b}_i)\mathbf{D} \tag{13.3.11}$$

$$\mathbf{D}_{ij} = \mathbf{c}_{ij}\mathbf{D} \tag{13.3.12}$$

where the explicit form of the primitive variable Jacobian D is given below,

$$\mathbf{D} = \begin{bmatrix} 1 & 0 & 0 & 0 & 0 \\ \mathrm{v}_1 & \rho & 0 & 0 & 0 \\ \mathrm{v}_2 & 0 & \rho & 0 & 0 \\ \mathrm{v}_3 & 0 & 0 & \rho & 0 \\ \hat{\varepsilon} & \rho \mathrm{v}_1 & \rho \mathrm{v}_2 & \rho \mathrm{v}_3 & \rho c_v \end{bmatrix} \tag{13.3.13}$$

with

$$\hat{\varepsilon} = c_p T + \frac{1}{2}\mathrm{v}_i\mathrm{v}_i - c_v T(\gamma - 1)$$

The governing equations given by (13.3.10) are well behaved as the eigenvalues of the convective terms are well conditioned even when the incompressible limit is reached.

13.3.2 THE GPG WITH CONSERVATION VARIABLES

Following the procedure presented in Section 11.4, let us now consider the GPG formulations of the Navier-Stokes system of equations in terms of conservation variables given by (13.3.1).

$$\int_\xi \hat{W}(\xi) \int_\Omega \left[(\Phi_\alpha + \Psi_\alpha^{(a)}) \left(\frac{\partial \mathbf{U}}{\partial t} + (\mathbf{a}_i + \mathbf{b}_i)\frac{\partial \mathbf{U}}{\partial x_i} + \mathbf{c}_{ij}\frac{\partial^2 \mathbf{U}}{\partial x_i \partial x_j} - \mathbf{B} \right) + \Psi^{(b)}\mathbf{a}_i \frac{\partial \mathbf{U}}{\partial x_i} \right] d\Omega d\xi = 0 \tag{13.3.14}$$

As shown earlier in Section 11.4, the integration by parts is to be performed only to those terms associated with the Galerkin test function Φ_α. With assumptions made similarly as in the case of the Burgers equation for those terms associated with the numerical diffusion test function for streamline diffusion, we obtain

$$\int_\xi \hat{W}(\xi) \left[\int_\Omega \left\{ \Phi_\alpha \left(\frac{\partial \mathbf{U}}{\partial t} - \mathbf{B} \right) - (\Phi_{\alpha,i}(\mathbf{a}_i + \mathbf{b}_i)\mathbf{U} + \Phi_{\alpha,i}\mathbf{c}_{ij}\mathbf{U}_{,j}) \right\} d\Omega + \int_\Gamma \overset{*}{\Phi}_\alpha (\mathbf{F}_i + \mathbf{G}_i) n_i d\Gamma \right] d\xi + \int_\xi \hat{W}(\xi) \int_\Omega [\Psi_\alpha^{(a)}(\mathbf{a}_i \mathbf{U}_{,i} + \mathbf{c}_{ij}\mathbf{U}_{,ji}) + \Psi_\alpha^{(b)}\mathbf{a}_i \mathbf{U}_{,i}] d\Omega d\xi = 0 \tag{13.3.15}$$

where the numerical diffusion test functions are given by

$$\Psi_\alpha^{(a)} = \tau \mathbf{a}_i \Phi_{\alpha,i}, \quad \text{streamline diffusion in GPG} \tag{13.3.16a}$$

$$\Psi_\alpha^{(a)} = \tau \mathbf{L} \Phi_\alpha, \quad \text{streamline diffusion in GLS} \tag{13.3.16b}$$

$$\Psi_\alpha^{(b)} = \tau^{(b)} \mathbf{a}_i \Phi_{\alpha,i}, \quad \text{discontinuity-capturing} \tag{13.3.17}$$

The differential operator L in (13.3.16b) is written as

$$\mathbf{L} = \frac{\partial}{\partial t} + \mathbf{a}_i \frac{\partial}{\partial x_i} + \mathbf{c}_{ij} \frac{\partial^2}{\partial x_i \partial x_j} \tag{13.3.18}$$

With the trial functions applied to the conservation variables, together with linear temporal test functions (Section 10.2), we arrive at

$$\begin{aligned}&[A_{\alpha\beta rs} + \eta\Delta t(C_{\alpha\beta rs} + D_{\alpha\beta rs} - B_{\alpha\beta rs} - K_{\alpha\beta rs})]U^{n+1}_{\beta s}\Delta t\\ &\quad = [A_{\alpha\beta rs} - (1-\eta)(C_{\alpha\beta rs} + D_{\alpha\beta rs} - B_{\alpha\beta rs} - K_{\alpha\beta rs})]U^n_{\beta s}\\ &\quad + \Delta t\left(H^n_{\alpha r} + N^n_{\alpha r}\right)\end{aligned} \tag{13.3.19}$$

with

$$A_{\alpha\beta rs} = \int_\Omega \delta_{rs}\Phi_\alpha\Phi_\beta d\Omega \tag{13.3.20a}$$

$$C_{\alpha\beta rs} = \int_\Omega (\overline{\tau} a_{irt} a_{jts} + \overline{\upsilon}\delta_{ij}\delta_{rs})\Phi_{\alpha,i}\Phi_{\beta,j} d\Omega \tag{13.3.20b}$$

$$K_{\alpha\beta rs} = \int_\Omega c_{ijrs}\Phi_{\alpha,i}\Phi_{\beta,j} d\Omega \tag{13.3.20c}$$

$$B_{\alpha\beta rs} = \int_\Omega (a_{irs} + b_{irs})\Phi_{\alpha,i}\Phi_\beta d\Omega \tag{13.3.20d}$$

$$D_{\alpha\beta rs} = \int_\Omega \overline{\tau} a_k c_{ijrs}\Phi_{\alpha,k}\Phi_{\beta,ji} d\Omega \tag{13.3.20e}$$

$$H^n_{\alpha r} = \int_\Omega \Phi_\alpha B_r d\Omega \tag{13.3.20f}$$

$$N^n_{\alpha r} = \int_\Gamma \overset{*}{\Phi}_\alpha (F_{ir} + G_{ir}) n_i d\Gamma \tag{13.3.20g}$$

where the intrinsic time scale $\overline{\tau}$ and the discontinuity-capturing factor $\overline{\upsilon}$ constitute the equivalent artificial diffusivity,

$$\overline{\tau} = \left(g^{ij} a_{irt} a_{jst} C^{-1}_{rs}\right)^{-\frac{1}{2}} \tag{13.3.21}$$

$$\overline{\upsilon} = \max(0, \overline{\upsilon}_d - \overline{\upsilon}_s) \tag{13.3.22}$$

with

$$\overline{\upsilon}_d = \left(\frac{C^{-1}_{rs} a_{itr} a_{jus} U_{t,i} U_{u,j}}{C_{vw} g^{mn} U_{v,m} U_{w,n}}\right)^{\frac{1}{2}}$$

$$\overline{\upsilon}_s = \frac{\overline{\tau} a_{irt} a_{jst} U_{r,i} U_{s,j}}{C_{uv} U_{u,k} U_{v,k}}$$

where C_{rs} is the entropy variable Jacobian (13.3.5) and g^{mn} is the contravariant metric tensor in the curvilinear isoparametric coordinates (Figure 11.3.3),

$$g^{mn} = \frac{\partial \xi_m}{\partial x_p}\frac{\partial \xi_n}{\partial x_p}$$

Here, the indicies i, j, k, m, n, p refer to the spatial coordinates (1,2,3) and r, s, t, v, v, w

denote the equation number (1,2,3,4,5) in the Navier-Stokes system of equations. It should be noted that the criterion used in (13.3.21) is motivated by the fact that the gradients of all variables are involved in determining the dimensionally equivalent artificial diffusivity rather than artificial time scale associated with only the velocity and velocity gradients. This is in contrast to the case of the numerical diffusion test functions developed for the Burgers' equations as given by (11.3.35b) and (11.3.38). Note also that another criterion in (13.3.22) is to ensure positive numerical diffusion for highly distorted elements.

There are other versions of numerical diffusion factors, as proposed in Hauke and Hughes [1998], Aliabadi and Tezduyar [1993], and other related references for the past decade. The basic idea is to apply the numerical diffusion in the direction of velocity for streamline diffusion and in the direction of gradients for discontinuity-capturing, as described in Section 11.3.

Instead of using the linear temporal variations, we may enhance temporal approximations with a second order accuracy of the form

$$\frac{\partial \mathbf{U}}{\partial t} = \frac{3\mathbf{U}^{n+1} - 4\mathbf{U}^n + \mathbf{U}^{n-1}}{2\Delta t} \tag{13.3.23}$$

together with quadratic variations of $\mathbf{U}_\beta$ between nodes,

$$\mathbf{U}_\beta = \frac{5}{8}\mathbf{U}_\beta^{n+1} + \frac{3}{4}\mathbf{U}_\beta^n - \frac{3}{8}\mathbf{U}_\beta^{n-1} \tag{13.3.24}$$

These approximations lead to

$$\begin{aligned}\left[3A_{\alpha\beta rs} + \frac{5}{4}\Delta t(C_{\alpha\beta rs} - K_{\alpha\beta rs})\right]\mathbf{U}_{\beta s}^{n+1} &= \left[4A_{\alpha\beta rs} - \frac{3}{2}\Delta t(B_{\alpha\beta rs} - D_{\alpha\beta rs})\right]\mathbf{U}_{\beta s}^n \\ &\quad - \left[A_{\alpha\beta rs} - \frac{3}{4}\Delta t(B_{\alpha\beta rs} - D_{\alpha\beta rs})\right]\mathbf{U}_{\beta s}^{n-1} \\ &\quad + \Delta t\left(H_{\alpha r}^n + N_{\alpha r}^n\right)\end{aligned} \tag{13.3.25}$$

Other possibilities for temporal approximations such as discussed in Section 10.2 may be considered for applications to various physical problems as required for higher order accuracy.

13.3.3 THE GPG WITH ENTROPY VARIABLES

The GPG formulations in terms of entropy variables can be carried out similarly as in (13.3.14) using (13.3.5),

$$\begin{aligned}&\int_\Omega \hat{W}(\xi)\int_\Omega \left[\left(\Phi_\alpha + \Psi_\alpha^{(a)}\right)\left(\mathbf{C}\frac{\partial \mathbf{V}}{\partial t} + \mathbf{C}_i\frac{\partial \mathbf{V}}{\partial x_i} + \mathbf{C}_{ij}\frac{\partial^2 \mathbf{V}}{\partial x_i \partial x_j} - \mathbf{B}\right)\right. \\ &\quad \left. + \Psi_\alpha^{(b)}\mathbf{a}_i\mathbf{C}\frac{\partial \mathbf{V}}{\partial x_i}\right] d\Omega d\xi = 0\end{aligned} \tag{13.3.26}$$

which leads to

$$\begin{aligned}&[A_{\alpha\beta rs} + \eta\Delta t(C_{\alpha\beta rs} + D_{\alpha\beta rs} - B_{\alpha\beta rs} - K_{\alpha\beta rs})]V_{\beta s}^{n+1} \\ &= [A_{\alpha\beta rs} - (1-\eta)\Delta t(C_{\alpha\beta rs} + D_{\alpha\beta rs} - B_{\alpha\beta rs} - K_{\alpha\beta rs})]V_{\beta s}^n \\ &\quad + \Delta t\left(H_{\alpha r}^n + N_{\alpha r}^n\right)\end{aligned} \tag{13.3.27}$$

where

$$A_{\alpha\beta rs} = \int_\Omega C_{rs}\Phi_\alpha \Phi_\beta d\Omega$$

$$C_{\alpha\beta rs} = \int_\Omega (\overline{\tau} a_{irt} a_{jst} + \overline{\upsilon}_{rs}\delta_{ij})\Phi_{\alpha,i}\Phi_{\beta,j} d\Omega$$

$$K_{\alpha\beta rs} = \int_\Omega c_{ijrt} C_{st}\Phi_{\alpha,i}\Phi_{\beta,j} d\Omega$$

$$B_{\alpha\beta rs} = \int_\Omega a_{irt} C_{st}\Phi_{\alpha,i}\Phi_\beta d\Omega$$

$$D_{\alpha\beta rs} = \int_\Omega \overline{\tau} a_k c_{ijrt} C_{st}\Phi_{\alpha,k}\Phi_{\beta,ij} d\Omega$$

$$H^n_{\alpha r} = \int_\Omega \Phi_\alpha B_r d\Omega$$

$$N^n_{\alpha r} = -\int_\Gamma \overset{*}{\Phi}_\alpha (F_{ir} + G_{ir}) n_i d\Gamma$$

with

$$\overline{\tau} = (g^{ij} C_{irt} C_{jst} C^{-1}_{rs})^{-\frac{1}{2}} \tag{13.3.28}$$

$$\overline{\upsilon}_{rs} = \max(0, \overline{\upsilon}_d - \overline{\upsilon}_s) C_{rs} \tag{13.3.29}$$

$$\overline{\upsilon}_d = \left(\frac{C^{-1}_{rs} C_{itr} C_{jus} V_{t,i} V_{u,j}}{C_{vw} g^{mn} V_{v,m} V_{w,n}} \right)^{\frac{1}{2}}$$

$$\overline{\upsilon}_s = \frac{\overline{\tau} C_{irt} C_{jst} V_{r,i} V_{s,j}}{C_{rs} V_{r,k} V_{s,k}}$$

The criterion given in (13.3.29) is to ensure that the discontinuity-capturing diffusivity is larger than the streamline diffusivity, which may not be true for highly distorted elements. As in the case of conservation variables, temporal approximations may be enhanced with a second order accuracy as in (13.3.22). Further details of the GPG with entropy variables are found in Hughes et al. [1986], Shakib et al. [1991], and Hauke and Hughes [1998].

13.3.4 THE GPG WITH PRIMITIVE VARIABLES

The projections of the residuals of the governing equations in terms of primitive variables (13.3.10) onto the various test functions are given by

$$\int_\xi \hat{W}(\xi) \int_\Omega \left[(\Phi_\alpha + \Psi^{(a)}_\alpha) \left(\mathbf{D}\frac{\partial W}{\partial t} + \mathbf{D}_i \frac{\partial W}{\partial x_i} + \mathbf{D}_{ij}\frac{\partial^2 W}{\partial x_i \partial x_j} - \mathbf{B} \right) \right.$$

$$\left. + \Psi^{(b)}_\alpha \mathbf{a}_i \mathbf{D}\frac{\partial \mathbf{W}}{\partial x_i} \right] d\Omega d\xi = 0 \tag{13.3.30}$$

The resulting algebraic equations are of the form

$$[A_{\alpha\beta rs} + \eta\Delta t(C_{\alpha\beta rs} - K_{\alpha\beta rs})]W^{n+1}_{\beta s} = [A_{\alpha\beta rs} - (1-\eta)\Delta t(B_{\alpha\beta rs} + D_{\alpha\beta rs})]W^n_{\beta s} + \Delta t\left(H^n_{\alpha r} + N^n_{\alpha r}\right) \tag{13.3.31}$$

where

$$A_{\alpha\beta rs} = \int_\Omega D_{rs}\Phi_\alpha\Phi_\beta d\Omega$$

$$C_{\alpha\beta rs} = \int_\Omega (\overline{\tau} a_{irt} a_{jts} + \overline{\upsilon}_{rs}\delta_{ij})\Phi_{\alpha,i}\Phi_{\beta,j} d\Omega$$

$$K_{\alpha\beta rs} = \int_\Omega c_{ijrt} D_{ts}\Phi_{\alpha,i}\Phi_{\beta,i} d\Omega$$

$$B_{\alpha\beta rs} = \int_\Omega a_{irt} D_{ts}\Phi_{\alpha,i}\Phi_\beta d\Omega$$

$$D_{\alpha\beta rs} = \int_\Omega \overline{\tau} a_{krt} c_{ijtu} D_{us}\Phi_{\alpha,k}\Phi_{\beta,ji} d\Omega$$

$$H^n_{\alpha r} = \int_\Omega \Phi_\alpha B_r d\Omega$$

$$N^n_{\alpha r} = -\int_\Gamma \overset{*}{\Phi}_\alpha (F_{ir} + G_{ir}) n_i d\Gamma$$

with

$$\overline{\tau} = \left(g^{ij} D_{irt} D_{jts} D^{-1}_{rs}\right)^{-\frac{1}{2}} \tag{13.3.32}$$

$$\overline{\upsilon}_{rs} = \max(0, \overline{\upsilon}_d - \overline{\upsilon}_s) D_{rs} \tag{13.3.33}$$

$$\overline{\upsilon}_d = \left(\frac{D^{-1}_{rs} D_{itr} D_{jus} W_{t,i} W_{u,j}}{D_{vw} g^{mn} W_{v,m} W_{w,n}}\right)^{\frac{1}{2}}$$

$$\overline{\upsilon}_s = \frac{\overline{\tau} D_{irt} D_{jts} W_{r,i} W_{s,j}}{D_{uv} W_{u,k} W_{v,k}}$$

Once again, the transformation of the conservation variable into primitive variables results in appropriate modifications of the parameters involved in the numerical diffusion test functions.

13.4 CHARACTERISTIC GALERKIN METHODS

The characteristic Galerkin methods (CGM) are based on the concept of trajectories or characteristics [Zienkiewicz and Codina, 1995; Zienkiewicz et al., 1998; Codina, Vazquez, and Zienkiewicz, 1998] with

$$x^n_i = x^{n+1}_i - \Delta t \mathrm{v}^n_i \tag{13.4.1}$$

Differentiating (13.4.1) with respect to time, we have

$$\mathrm{v}^n_i = \mathrm{v}^{n+1}_i - \Delta t \mathrm{v}^n_j \frac{\partial \mathrm{v}^n_i}{\partial x_j} \tag{13.4.2}$$

Combining (13.4.1) and (13.4.2) leads to

$$x_i^{n+1} - x_i^n = \Delta t \mathrm{v}_i^n - \frac{\Delta t^2}{2} \mathrm{v}_j^n \frac{\partial \mathrm{v}_i^n}{\partial x_j} \tag{13.4.3}$$

The main idea of CGM is to write the governing equations along the characteristics so that the Navier-Stokes system of equations may be recast in the form similar to (13.4.3).

$$\Delta \mathbf{U}^{n+1} = \Delta t \mathbf{R}^n - \frac{\Delta t^2}{2} \mathbf{a}_j^n \frac{\partial \mathbf{R}^n}{\partial x_j} \tag{13.4.4}$$

where $\mathbf{a}_j^n$ is the convection Jacobian, with $\mathbf{R}^n$ is the residual defined as

$$\mathbf{R}^n = -\left(\frac{\partial \mathbf{F}_i{}^n}{\partial x_i} + \frac{\partial \mathbf{G}_i^n}{\partial x_i} - \mathbf{B}^n \right)$$

Instead of solving (13.4.4) directly, the fractional step approach may be used for convenience. Here, the momentum equations are solved first without pressure, followed by the continuity equation to compute the pressure. With these results, we return to the momentum equations again to update the flowfield, before the energy equation is solved.

Momentum (initially):

$$\Delta \overline{\rho \mathrm{v}_i^n} = \Delta t R_i^n - \frac{\Delta t^2}{2} \mathrm{v}_k \frac{\partial \hat{R}_i^n}{\partial x_k} \tag{13.4.5}$$

with

$$R_i^n = -\frac{\partial}{\partial x_j}(\rho \mathrm{v}_i \mathrm{v}_j - \tau_{ij}) + \rho g_i$$

$$\hat{R}_i^n = R_i^n - \frac{\partial p^n}{\partial x_i}$$

Continuity:

$$\Delta \rho^n = -\Delta t \frac{\partial}{\partial x_i}\left(\rho \mathrm{v}_i^n + \theta_1 \Delta \overline{\rho \mathrm{v}_i^n}\right) + \theta_1 \Delta t^2 \frac{\partial^2 p^{n+\theta_2}}{\partial x_i \partial x_i} \tag{13.4.6}$$

with

$$0 \le \theta_1, \theta_2 \le 1$$

Momentum (updated):

$$\Delta \rho \mathrm{v}_i^n = \Delta \overline{\rho \mathrm{v}_i^n} - \Delta t \frac{\partial p^{n+\theta_2}}{\partial x_i} \tag{13.4.7}$$

Energy:

$$\Delta \rho E^n = \Delta t R^n - \frac{\Delta t^2}{2} \mathrm{v}_k \frac{\partial R^n}{\partial x_k} \tag{13.4.8}$$

with

$$R^n = -\frac{\partial}{\partial x_i}\left[(\rho E + p)\mathrm{v}_i - k\frac{\partial T}{\partial x_i} - \tau_{ij}\mathrm{v}_j\right]$$

The standard Galerkin approximations can now be applied to these equations separately and the solution proceeds as follows:

(1) Solve the momentum equations (13.4.5).
(2) Solve the continuity equation (13.4.6), using the mass flux obtained from step 1 to calculate the pressure.
(3) Update the mass flux with (13.4.7), using the pressure from step 2.
(4) Solve the energy equation (13.4.8) to obtain the total energy or temperature using the results obtained from step 3.
(5) Repeat the steps 1 through 4 until the steady state is reached.

To explore the physical significance of the CGM procedure, let us substitute (13.4.5) into (13.4.7) to obtain

$$\frac{\partial}{\partial t}(\rho \mathrm{v}_i) + (\rho \mathrm{v}_i \mathrm{v}_j)_{,j} + p_{,i} - \tau_{ij,j} - \rho f_i = S_i(m) \tag{13.4.9}$$

with

$$S_i(m) = \frac{\Delta t}{2}\{\mathrm{v}_k[(\rho \mathrm{v}_i \mathrm{v}_j)_{,j} + p_{,i} - \tau_{ij,j} - \rho f_i]\}_{,k} \tag{13.4.10}$$

Similarly, the continuity equation (13.4.6) and energy equation (13.4.8) are rewritten, respectively, as

$$\frac{\partial \rho}{\partial t} + (\rho \mathrm{v}_i)_{,i} = S(c) \tag{13.4.11}$$

with

$$S(c) = \frac{\Delta t}{2}[(\rho \mathrm{v}_i \mathrm{v}_j - \tau_{ij})_{,ji} + p_{,ii} - (\rho f_i)_{,i}] \tag{13.4.12}$$

by setting $\theta_1 = 1/2$ and $\theta_2 = 0$ in (13.4.6), and

$$\frac{\partial \rho E}{\partial t} + [(\rho E + p)\mathrm{v}_i - kT_{,i} - \tau_{ij}\mathrm{v}_j]_{,i} = S(e) \tag{13.4.13}$$

with

$$S(e) = \frac{\Delta t}{2}\mathrm{v}_j\,[(\rho E\mathrm{v}_i + p\mathrm{v}_i - kT_{,i} - \tau_{ik}\mathrm{v}_k)_{,i}]_{,j} \tag{13.4.14}$$

The consequence of the CGM process is that additional terms $S(m)$, $S(c)$, and $S(e)$ on the right-hand side of momentum, continuity, and energy equations, respectively, have been generated as numerical diffusion. It is remarkable that the combination of all equations, (13.4.5) through (13.4.8), which represents (13.4.4) can be identified in the TGM equations. The similar results arise in TGM with the right-hand side of (13.2.14) revised by substituting

$$a_j \frac{\partial \Delta U^{n+1}}{\partial x_j} = \frac{\partial \Delta F_j^{n+1}}{\partial x_j}$$

The advantage of the fractional step approach is the fact that the continuity equation can be written in the form given by (13.4.11) in which the spatial second derivatives of pressure arise explicitly, acting as numerical diffusion. Of course, this effect is present, implicitly embedded, when the entire equations are solved simultaneously in TGM. An important conclusion here is that the CGM concept is found to be identical to TGM. It will be shown in Section 13.6.3 that these results arise as a special case of the flowfield-dependent variation methods.

Direct assessments of the fractional step approach can be made by applying the Galerkin formulation of (13.4.9) and (13.4.11) separately and combining the results in a matrix form:

$$\begin{bmatrix} K_{\alpha\beta ij} & C_{\alpha\beta i} \\ D_{\alpha\beta j} & B_{\alpha\beta} \end{bmatrix} \begin{bmatrix} v_{\beta j} \\ p_\beta \end{bmatrix} = \begin{bmatrix} E_{\alpha i} \\ F_\alpha \end{bmatrix} \tag{13.4.15}$$

where it can be shown that the presence of $B_{\alpha\beta}$ is due to the numerical diffusion terms characterized by $S_i(m)$ and $S(c)$ in (13.4.9) and (13.4.11), respectively. Otherwise, $B_{\alpha\beta}$ would have been zero, resulting in numerical instability. In this case, the so-called LBB restriction requires a special treatment in incompressible flow as discussed in Chapter 12. It is reminded that the simultaneous solution of all equations in terms of the conservation variables have the advantage of versatility and simplicity with all numerical diffusion terms appearing on the left-hand side rather than on the right-hand side.

13.5 DISCONTINUOUS GALERKIN METHODS OR COMBINED FEM/FDM/FVM METHODS

The basic idea of discontinuous Galerkin methods (DGM) is to combine FDM schemes with upwind finite differences into the FEM formulation such as standard Galerkin methods or Taylor-Galerkin methods. In this process, integration by parts in the FEM equations provides the boundary terms in which the convection numerical flux terms are discretized using the upwind FDM schemes via finite volume approximations. Thus, in DGM, all currently available CFD schemes are combined together, alternatively referred to as the combined FEM/FDM/FVM methods. Various authors have contributed to DGM. Among them are La Saint and Raviart [1974], Johnson and Pitkäranta [1986], Cockburn, Hou, and Shu [1990, 1997], and Oden, Babuska, and Baumann [1998].

In the DGM approach, we begin with the standard Galerkin integral,

$$\int_\Omega \Phi_\alpha \left(\frac{\partial \mathbf{U}}{\partial t} + \mathbf{F}_{i,i} + \mathbf{G}_{i,i} - \mathbf{B} \right) d\Omega = 0 \tag{13.5.1}$$

or

$$\int_\Omega \Phi_\alpha \left[\frac{\partial \mathbf{U}}{\partial t} + (\mathbf{a}_i \mathbf{U})_{,i} + (\mathbf{b}_i \mathbf{U} + \mathbf{c}_{ij} \mathbf{U}_{,j})_{,i} - \mathbf{B} \right] d\Omega = 0 \tag{13.5.2}$$

Integrating (13.5.1) or (13.5.2) by parts, we obtain

$$\int_\Omega \Phi_\alpha \frac{\partial \mathbf{U}}{\partial t} d\Omega - \int_\Omega \Phi_{\alpha,i}(\mathbf{F}_i + \mathbf{G}_i) d\Omega - \int_\Omega \Phi_\alpha \mathbf{B} d\Omega + \int_\Gamma \overset{*}{\Phi}_\alpha \mathbf{F}_i n_i d\Gamma + \int_\Gamma \overset{*}{\Phi}_\alpha (\mathbf{G}_i + \hat{\mathbf{G}}_i) n_i d\Gamma = 0 \tag{13.5.3}$$

with

$$\mathbf{F}_i = \mathbf{a}_i \mathbf{U}, \quad \mathbf{G}_i = \mathbf{b}_i \mathbf{U}, \quad \hat{\mathbf{G}}_i = \mathbf{c}_{ij} \mathbf{U}_{,i} \tag{13.5.4}$$

In a compact notation, we write (13.5.3) in the form

$$(A_{\alpha\beta} + B_{\alpha\beta}) \Delta \mathbf{U}_\beta^{n+1} = \mathbf{F}_\alpha + \mathbf{G}_\alpha + \mathbf{H}_\alpha \tag{13.5.5}$$

with

$$A_{\alpha\beta} = \int_\Omega \Phi_\alpha \Phi_\beta d\Omega \tag{13.5.6a}$$

$$B_{\alpha\beta} = \Delta t \int_\Omega ((a_i + b_i)(\Phi_{\alpha,i} \Phi_\beta - c_{ij} \Phi_{\alpha,i} \Phi_{\beta,j})) d\Omega \tag{13.5.6b}$$

$$\mathbf{F}_\alpha = \Delta t \int_\Omega \Phi_\alpha \mathbf{B} d\Omega \tag{13.5.6c}$$

$$\mathbf{G}_\alpha = -\Delta t \int_\Gamma \overset{*}{\Phi}_\alpha \mathbf{F}_i n_i d\Gamma \tag{13.5.6d}$$

$$\mathbf{H}_\alpha = -\Delta t \int_\Gamma \overset{*}{\Phi}_\alpha (\mathbf{G}_i + \hat{\mathbf{G}}_i) n_i d\Gamma \tag{13.5.6e}$$

Instead of using the standard Galerkin formulation of (13.5.1–13.5.3), we may utilize the Taylor-Galerkin methods (TGM) as described in Section 13.2. In this case, the expression given by (13.2.15) is used instead of (13.5.3).

$$\int_\Omega \Phi_\alpha \left\{ 1 - \frac{\Delta t^2}{2} \frac{\partial}{\partial x_i} \left(\mathbf{a}_i \mathbf{a}_j - \frac{\mathbf{c}_{ij}}{\Delta t} \right) \frac{\partial}{\partial x_j} \right\} \Delta \mathbf{U}^{n+1} d\Omega = \int_\Omega \Phi_\alpha \left\{ \Delta t \left(-\frac{\partial \mathbf{F}_i}{\partial x_i} - \frac{\partial \mathbf{G}_i}{\partial x_i} + \mathbf{B} \right)^n + \frac{\Delta t^2}{2} \frac{\partial}{\partial x_i} \left(\mathbf{a}_i \frac{\partial \mathbf{F}_j}{\partial x_j} \right)^n \right\} d\Omega \tag{13.5.7}$$

Note that the first integral on the right-hand side of (13.5.5), upon integration by parts, becomes identical to the form given in (13.5.3), resulting in the same boundary integrals. All quantities resulting from (13.5.5) are identical to those given in (13.5.6) except for (13.5.6b,c),

$$B_{\alpha\beta} = \frac{\Delta t^2}{2} \int_\Omega \left(\left(a_i a_j - \frac{c_{ij}}{\Delta t} \right) \Phi_{\alpha,i} \Phi_{\beta,j} \right) d\Omega \tag{13.5.8a}$$

$$\mathbf{F}_\alpha = \Delta t \int_\Omega \Phi_\alpha \mathbf{B} d\Omega - \Delta t \int_\Omega \Phi_{\alpha,i} (\mathbf{F}_i + \mathbf{G}_i) d\Omega \tag{13.5.8b}$$

Here, the boundary integrals (13.5.6d) for convection represent possible discontinuities characterized by the eigenvalues and eigenvectors of the convection Jacobian $\mathbf{a}_i$ in the spirit of flux vector splitting. Similarly, the flux variables F_i may be reconstructed using the various FDM second order upwind schemes [Godunov, 1959; Harten, 1984; Roe, 1984; Osher, 1984; van Leer, 1979; etc.] or flux-corrected transport (FCT) [Boris and Book, 1976; Zalesak, 1979]. Recall that FDM schemes were presented in Chapter 6. The idea of DGM is to combine FDM into FEM. Some examples of various FDM schemes which may be combined to DGM are the flux vector splitting for the convection Jacobian and various second order upwind schemes as detailed in Section 6.2.

Some numerical applications of (13.5.5) have been reported by Baumann and Oden [1999] for the *hp* adaptive first order upwind scheme and by Atkins and Shu [1998] for the second order TVD upwind scheme, among others.

13.6 FLOWFIELD-DEPENDENT VARIATION METHODS

Recall that the flowfield-dependent variation (FDV) theory was developed in Section 6.5, in which the FDV equations were solved using FDM. The basic theory of FDV will not be repeated here. Thus, the reader should review the process of development presented in Section 6.5 thoroughly. In this section, some additional items of interest such as the source terms of gravity, surface tension, and chemical species reaction rate are included. These and other aspects of the FDV theory to be emphasized are presented next.

13.6.1 BASIC FORMULATION

As stated in Section 6.5, the FDV theory was devised in response to the need to characterize the complex physics of shock wave turbulent boundary layers in which transitions between, and interactions of, inviscid/viscous, incompressible/compressible, and laminar/turbulent flows constitute the most complex physical phenomena in fluid dynamics [Chung and his co-workers, 1996–1999]. The complexities of physics, in general, lead directly to computational difficulties. This is where the very low velocity in the vicinity of the wall and very high velocity far away from the wall coexist within a domain of study. Transitions from one type of flow to another and interactions between two distinctly different flows have been studied for many years, both experimentally and numerically. Incompressible flows were analyzed using the pressure-based formulation with the primitive variables for the implicit solution of the Navier-Stokes system of equations together with the pressure Poisson equation. On the other hand, compressible flows were analyzed using the density-based formulation with the conservation variables for the explicit solution of the Navier-Stokes system of equations.

In a given domain, however, dealing with all speed flows of various physical properties, we encounter different equations of state for compressible and incompressible flows, transitions between laminar and turbulent flows, dilatational dissipation due to compressibility as well as difficulties of satisfying the mass conservation or incompressibility condition. To cope with this situation, we must provide very special and powerful numerical treatments. The FDV scheme has been devised toward resolving these issues.

For most of the CFD methods, the numerical formulation begins with a particular physical phenomenon. Thus, if the physics is changed, then the numerics must be accordingly changed. Our goal in FDV, instead, is to derive a scheme in which all possible physical aspects are already taken into account in the final form of the governing equations so that FDM or FEM is reduced to an option of how to discretize between nodal points or elements. Thus, the formulation of FDV procedure in terms of FEM is identical to that of FDM.

To this end, we shall consider the most general form of Navier-Stokes system of equations in conservation form, including the chemically reacting species equations and source terms for the body force, surface tension, and chemical reaction rates, which will be useful for applications of FDV to problems in Part Five.

$$\frac{\partial \mathbf{U}}{\partial t} + \frac{\partial \mathbf{F}_i}{\partial \mathbf{x}_i} + \frac{\partial \mathbf{G}_i}{\partial \mathbf{x}_i} = \mathbf{B} \tag{13.6.1}$$

where $\mathbf{U}, \mathbf{F}_i, \mathbf{G}_i$, and $\mathbf{B}$ denote the conservation flow variables, convection flux variables, diffusion flux variables, and source terms, respectively,

$$\mathbf{U} = \begin{bmatrix} \rho \\ \rho v_j \\ \rho E \\ \rho Y_k \end{bmatrix}, \quad \mathbf{F}_i = \begin{bmatrix} \rho v_i \\ \rho v_i v_j + p\delta_{ij} \\ \rho E v_i + p v_i \\ \rho Y_k v_i \end{bmatrix}, \quad \mathbf{G}_i = \begin{bmatrix} 0 \\ -\tau_{ij} \\ -\tau_{ij} v_j - kT_{,i} - \sum \rho c_{pk} T D_{km} Y_{k,i} \\ -\rho D_{km} Y_{k,i} \end{bmatrix},$$

$$\mathrm{B} = \begin{bmatrix} 0 \\ \rho f_j \\ -\sum H_k^0 \omega_k + \rho f_j v_j \\ \omega_k \end{bmatrix}$$

where $f_j = \sum_{k=1}^N Y_k f_{kj}$ is the body force, Y_k is the chemical species, H_k^o is the zero-point enthalpy, ω_k is the reaction rate, and D_{km} is the binary diffusivity. Additional equations for vibrational and electronic energies may be included in (13.6.1) for hypersonics (see Section 22.5).

Using the Taylor series expansion of $\mathbf{U}^{n+1}$ in terms of the FDV parameters, following the process given by (6.5.2) through (6.5.13a,b) together with the source terms, the residual of the Navier-Stokes system of equations can be written as

$$\begin{aligned}
\mathbf{R} = \Delta \mathbf{U}^{n+1} &- \Delta t\left[-\frac{\partial \mathbf{F}_i^n}{\partial x_i} - \frac{\partial \mathbf{G}_i^n}{\partial x_i} + \mathbf{B}^n - s_1 \frac{\partial \Delta \mathbf{F}_i^{n+1}}{\partial x_i} - s_3 \frac{\partial \Delta \mathbf{G}_i^{n+1}}{\partial x_i} + s_5\, \Delta \mathbf{B}^{n+1}\right] \\
&- \frac{\Delta t^2}{2}\left\{\left[\frac{\partial}{\partial x_i}(\mathbf{a}_i + \mathbf{b}_i)\left(\frac{\partial \mathbf{F}_j^n}{\partial x_j} + \frac{\partial \mathbf{G}_j^n}{\partial x_j} - \mathbf{B}^n\right) - \mathbf{d}\left(\frac{\partial \mathbf{F}_i^n}{\partial x_i} + \frac{\partial \mathbf{G}_i^n}{\partial x_i} - \mathbf{B}^n\right)\right]\right. \\
&+ s_2\left[\frac{\partial}{\partial x_i}(\mathbf{a}_i + \mathbf{b}_i)\left(\frac{\partial \Delta \mathbf{F}_j^{n+1}}{\partial x_j}\right) - \mathbf{d}\frac{\partial \Delta \mathbf{F}_i^{n+1}}{\partial x_i}\right] + \left[\frac{\partial}{\partial x_i}(\mathbf{a}_i + \mathbf{b}_i)\right. \\
&\left.\left.\times\left(s_4 \frac{\partial \Delta \mathbf{G}_j^{n+1}}{\partial x_j} . - s_6\, \Delta \mathbf{B}^{n+1}\right) - \mathbf{d}\left(s_4 \frac{\partial \Delta \mathbf{G}_i^{n+1}}{\partial x_i} . - s_6 \Delta \mathbf{B}^{n+1}\right)\right]\right\} + \mathrm{O}(\Delta \mathrm{t}^3)
\end{aligned} \tag{13.6.2a}$$

with the convection, diffusion, and diffusion gradient Jacobians (a_i, b_i, c_{ik}) being defined in (6.3.9) for 2-D and Appendix A for 3-D. The source term Jacobian is given by

$$\mathbf{d} = \frac{\partial \mathbf{B}}{\partial \mathbf{U}}$$

Now, rearranging and expressing the remaining terms associated with the variation parameters in terms of the Jacobians, we have

$$\Delta\mathbf{U}^{n+1} + \Delta t\left[s_1\left(\frac{\partial \mathbf{a}_i\Delta\mathbf{U}^{n+1}}{\partial x_i}\right) + s_3\left(\frac{\partial \mathbf{b}_i\Delta\mathbf{U}^{n+1}}{\partial x_i} + \frac{\partial^2 \mathbf{c}_{ij}\Delta\mathbf{U}^{n+1}}{\partial x_i\partial x_j}\right) - s_5\,\mathbf{d}\Delta U^{n+1}\right]$$
$$-\frac{\Delta t^2}{2}\left\{s_2\left[\frac{\partial^2(\mathbf{a}_i\mathbf{a}_j + \mathbf{b}_i\mathbf{a}_j)\Delta\mathbf{U}^{n+1}}{\partial x_i\partial x_j} - \mathbf{d}\frac{\partial \mathbf{a}_i\Delta\mathbf{U}^{n+1}}{\partial x_i}\right] + s_4\left[\left(\frac{\partial^2(\mathbf{a}_i\mathbf{b}_j + \mathbf{b}_i\mathbf{b}_j)\Delta\mathbf{U}^{n+1}}{\partial x_i\partial x_j}\right)\right.\right.$$
$$\left.\left.-\mathbf{d}\left(\frac{\partial \mathbf{b}_i\Delta\mathbf{U}^{n+1}}{\partial x_i} + \frac{\partial^2 \mathbf{c}_{ij}\Delta\mathbf{U}^{n+1}}{\partial x_i\partial x_j}\right)\right] - s_6\left[\mathbf{d}\frac{\partial(\mathbf{a}_i + \mathbf{b}_i)\Delta\mathbf{U}^{n+1}}{\partial x_i} - \mathbf{d}^2\,\Delta\mathrm{U}^{n+1}\right]\right\}$$
$$+\Delta t\left(\frac{\partial \mathbf{F}_i^n}{\partial x_i} + \frac{\partial \mathbf{G}_i^n}{\partial x_i} - \mathbf{B}^n\right) - \frac{\Delta t^2}{2}\left[\frac{\partial}{\partial x_i}(\mathbf{a}_i + \mathbf{b}_i)\left(\frac{\partial \mathbf{F}_j^n}{\partial x_j} + \frac{\partial \mathbf{G}_j^n}{\partial x_j} - \mathbf{B}^n\right)\right.$$
$$\left.-\mathbf{d}\left(\frac{\partial \mathbf{F}_i^n}{\partial x_i} + \frac{\partial \mathbf{G}_i^n}{\partial x_i} - \mathbf{B}^n\right)\right] + O(\Delta t^3) = 0 \tag{13.6.2b}$$

with

$$\Delta\mathbf{B}^{n+1} = \frac{\partial\mathbf{B}}{\partial\mathbf{U}}\Delta\mathbf{U}^{n+1} = \mathbf{d}\Delta\mathbf{U}^{n+1} \tag{13.6.3}$$

Here, the product of the diffusion gradient Jacobian with third order spatial derivatives is neglected and all Jacobians a_i, b_i, c_{ij}, and d are assumed to remain constant spatially within each time step and to be updated at subsequent time steps. The FDV parameters s_1, s_2, s_3, s_4 are defined in Section 6.5.1 and Figures 6.5.1 through 6.5.3. Additional parameters for source terms s_5, s_6 are defined in a similar manner:

$$\begin{aligned} s_a\Delta\mathbf{B} &\Rightarrow s_5\Delta\mathbf{B} \\ s_b\Delta\mathbf{B} &\Rightarrow s_6\Delta\mathbf{B} \end{aligned} \tag{13.6.4}$$

where the source term FDV parameters s_5 (first order source term FDV parameter) and s_6 second order source term FDV parameter) are evaluated as

$$s_5 = \begin{cases} \min(r, 1) & r > \alpha, \quad \alpha \cong 0.01 \\ 0 & r < \alpha, \quad Da_{\min} \neq 0 \\ 1 & \qquad\quad Da_{\min} = 0 \end{cases} \tag{13.6.5a}$$

$$s_6 = \frac{1}{2}(1 + s_5^{\eta}), \quad 0.05 < \eta < 0.2 \tag{13.6.5b}$$

with

$$r = \sqrt{Da_{\max}^2 - Da_{\min}^2}\Big/ Da_{\min} \tag{13.6.5c}$$

where the Damköhler number Da can be defined in five different ways as shown in Table 22.2.1.

For simplicity, we may rearrange (13.6.2b) in a compact form,

$$\mathbf{R} = \mathbf{A}\Delta\mathbf{U}^{n+1} + \frac{\partial}{\partial x_i}(\mathbf{E}_i\,\Delta\mathbf{U}^{n+1}) + \frac{\partial^2}{\partial x_i\partial x_j}(\mathbf{E}_{ij}\,\Delta\mathbf{U}^{n+1}) + \mathbf{Q}^n + O(\Delta t^3), \tag{13.6.6}$$

or, lagging E_i and E_{ij} one time step behind,

$$\left(\mathbf{A} + \mathbf{E}_i^n \frac{\partial}{\partial x_i} + \mathbf{E}_{ij}^n \frac{\partial^2}{\partial x_i \partial x_j}\right) \Delta \mathbf{U}^{n+1} = -\mathbf{Q}^n \tag{13.6.7}$$

with

$$\mathbf{A} = \mathbf{I} - \Delta t\, s_5\, \mathbf{d} - \frac{\Delta t^2}{2} s_6\, \mathbf{d}^2 \tag{13.6.8a}$$

$$\mathbf{E}_i^n = \left\{\Delta t(s_1 \mathbf{a}_i + s_3 \mathbf{b}_i) + \frac{\Delta t^2}{2}[s_6\, \mathbf{d}(\mathbf{a}_i + \mathbf{b}_i) + s_2\, \mathbf{d}\mathbf{a}_i + s_4 \mathbf{d}\mathbf{b}_i]\right\}^n \tag{13.6.8b}$$

$$\mathbf{E}_{ij}^n = \left\{\Delta t\, s_3 \mathbf{c}_{ij} - \frac{\Delta t^2}{2}[s_2(\mathbf{a}_i \mathbf{a}_j + \mathbf{b}_i \mathbf{a}_j) + s_4(\mathbf{a}_i \mathbf{b}_j + \mathbf{b}_i \mathbf{b}_j - \mathbf{d}\mathbf{c}_{ij})]\right\}^n \tag{13.6.8c}$$

$$\mathbf{Q}^n = \frac{\partial}{\partial x_i}\left[\left(\Delta t + \frac{\Delta t^2}{2}\mathbf{d}\right)(\mathbf{F}_i^n + \mathbf{G}_i^n) + \frac{\Delta t^2}{2}(\mathbf{a}_i + \mathbf{b}_i)\mathbf{B}^n\right]$$
$$- \frac{\partial^2}{\partial x_i \partial x_j}\left[\frac{\Delta t^2}{2}(\mathbf{a}_i + \mathbf{b}_i)(\mathbf{F}_j^n + \mathbf{G}_j^n)\right] - \left(\Delta t + \frac{\Delta t^2}{2}\mathbf{d}\right)\mathbf{B}^n \tag{13.6.8d}$$

An alternative scheme is to allow the source term in the left-hand side of (13.6.7) to lag from $n+1$ to n so that (13.6.7) may be written as

$$\left(\mathbf{I} + \mathbf{E}_i^n \frac{\partial}{\partial x_i} + \mathbf{E}_{ij}^n \frac{\partial^2}{\partial x_i \partial x_j}\right) \Delta \mathbf{U}^{n+1} = -\mathbf{Q}^n \tag{13.6.9}$$

$$\mathbf{Q}^n = \frac{\partial}{\partial x_i}\left[\left(\Delta t + \frac{\Delta t^2}{2}\mathbf{d}\right)(\mathbf{F}_i^n + \mathbf{G}_i^n) + \frac{\Delta t^2}{2}(\mathbf{a}_i + \mathbf{b}_i)\mathbf{B}^n\right] - \frac{\partial^2}{\partial x_i \partial x_j}$$
$$\times \left[\frac{\Delta t^2}{2}(\mathbf{a}_i + \mathbf{b}_i)(\mathbf{F}_j^n + \mathbf{G}_j^n)\right] - \left(\Delta t\, s_5 + \frac{\Delta t^2}{2} s_6 \mathbf{d}\right)\mathbf{d}\Delta \mathbf{U}^n - \left(\Delta t + \frac{\Delta t^2}{2}\mathbf{d}\right)\mathbf{B}^n \tag{13.6.10}$$

13.6.2 INTERPRETATION OF FDV PARAMETERS ASSOCIATED WITH JACOBIANS

The flowfield-dependent FDV parameters as defined earlier are capable of allowing various numerical schemes to be automatically generated as summarized in Section 6.5.4. For the purpose of completeness and emphasis, they are repeated here along with additional features associated with FEM and the source terms.

The first order FDV parameters s_1 and s_3 control all high-gradient phenomena such as shock waves and turbulence. These parameters as calculated from the changes of local Mach numbers, and Reynolds (or Peclet) numbers between adjacent nodes are indicative of the actual local element flowfields. The contours of these parameters closely resemble the flowfields themselves, with both s_1 and s_3 being large (close to unity) in regions of high gradients, but small (close to zero) in regions where the gradients are small (see Figures 6.5.1 through 6.5.3).

The second order FDV parameters s_2 and s_4 are also flowfield dependent, exponentially proportional to the first order FDV parameters. However, their primary role is to provide adequate computational stability (artificial viscosity) as they were originally introduced into the second order time derivative term of the Taylor series expansion of the conservation flow variables $\mathbf{U}^{n+1}$.

The s_1 terms represent convection. This implies that if $s_1 \cong 0$, then the effect of convection is small. The computational scheme is automatically altered to take this effect into account, with the governing equations being predominantly parabolic-elliptic. The s_3 terms are associated with diffusion. Thus, with $s_3 \cong 0$, the effect of viscosity or diffusion is small and the computational scheme is automatically switched to that of Euler equations where the governing equations are predominantly hyperbolic. If the first order variation parameters s_1 and s_3 are nonzero, this indicates a typical situation for the mixed hyperbolic, parabolic, and elliptic nature of the Navier-Stokes system of equations, with convection and diffusion being equally important. This is the case for incompressible flows at low speeds.

The unique property of the FDV scheme is its capability to control pressure oscillations adequately without resorting to the separate hyperbolic-elliptic pressure equation for pressure corrections. The capability of the FDV scheme to handle incompressible flows is achieved by a delicate balance between s_1 and s_3 as determined by the local Mach numbers and Reynolds (or Peclet) numbers. If the flow is completely incompressible ($M = 0$), the criteria given by (13.6.9) leads to $s_1 = 1$, whereas the FDV parameter s_3 is to be determined according to the criteria given in (13.6.11). Make a note of the presence of the convection-diffusion interaction terms given by the product of $b_i a_j$ in the s_2 terms and $a_i b_j$ in the s_4 terms. These terms allow interactions between convection and diffusion in the viscous incompressible and/or viscous compressible flows.

If temperature gradients rather than velocity gradients dominate the flowfield, then s_3 is governed by the Peclet number rather than by the Reynolds number. Such cases arise in high-speed, high-temperature compressible flows close to the wall.

The transition to turbulence is a natural flow process as the Reynolds number increases, causing the gradients of any or all flow variables to increase. This phenomenon is a physical instability and is detected by the increase of s_3 if the flow is incompressible, but by both s_3 and s_1 if the flow is compressible. Such physical instability is likely to trigger the numerical instability, but will be countered by the second order variation parameters s_2 and/or s_4 to ensure numerical stability automatically. In this process, these flowfield dependent variation parameters are capable of capturing relaminarization, compressibility effect or dilatational turbulent energy dissipation, and turbulent unsteady fluctuations. These physical phenomena are originated from transitions and interactions between inviscid and viscous flows. They are characterized by the product of s_3 and the fluctuation stress tensor ($s_3\tau_{ij}$) in which the stresses consist of mean and fluctuation parts. As a consequence, $\Delta\mathbf{U}^{n+1}$ in (13.6.3) or (13.6.5) may not uniformly vanish, indicating that some regions of the domain (such as in the boundary layers) remain unsteady if the flow is turbulent. However, if turbulent microscales (Kolmogorov microscale) are to be resolved, then we must allow mesh refinements normally required for the direct numerical simulation (DNS).

A unique feature in finite element applications of the FDV theory is the FDV parameters, which can be used as error indicators for adaptive meshing. The source terms such as those contributing to the finite rate chemistry were not included in Section 6.5. These topics are elaborated next.

FDV Parameters Used as Error Indicators for Adaptive Mesh. An important contribution of the first and second order FDV parameters is the fact that they can be used as error indicators for adaptive mesh generations (see Figure 19.2.5, Section 19.2.1). That is, the larger the FDV parameters, the higher the gradients of any flow variables. Whichever governs (largest first or second order variation parameters) will indicate the need for mesh refinements. In this case, all variables (density, velocity, pressure, temperature, species mass fraction) participate in resolving the adaptive mesh, contrary to the conventional definitions of the error indicators.

Finite Rate Chemistry. In the case of reacting flows, the source term **B** contains the reaction rates which are functions of the flowfield variables. With widely disparate time and length scales involved in the fast and slow chemical reaction rates of various chemical species as characterized by Damköhler numbers, the first order source term variation parameter s_5 is instrumental in dealing with the stiffness of the resulting equations to obtain convergence to accurate solutions. On the other hand, the second order source term FDV parameter s_6 contributes to the stability of solutions. It is seen that the criteria given by (13.6.5) will adjust the reaction rate terms in accordance with the ratio of the diffusion time to the reaction time in finite rate chemistry so as to assure the accurate solutions in dealing with stiffness and computational stability.

Influence of FDV Parameters on Jacobians. Physically, the FDV parameters will influence the magnitudes of Jacobians. The diffusion variation parameters s_3 and s_4 as calculated from Reynolds number and Peclet number can be applied to the Jacobians (a_i, b_i, c_{ij}), corresponding to the momentum equations and energy equation, respectively. Furthermore, two different definitions of Peclet number (Pe_I, Pe_{II}) (see Table 22.2.1) would require the s_3 and s_4 as calculated from the energy and species equations to be applied to the corresponding terms of the Jacobians. Similar applications for the source term variation parameters s_5 and s_6 should be followed for the source term Jacobian d, based on the various definitions of Damköhler number (Da_I, Da_{II}, Da_{III}, Da_{IV}, Da_V) as shown in Table 22.2.1. In this way, high temperature gradients arising from the momentum and energy equations and the finite rate chemistry governed by the energy and species equations can be resolved accordingly.

13.6.3 NUMERICAL DIFFUSION

Note that the numerical diffusion is implicitly embedded in the FDV equations. This can be demonstrated by writing (13.6.2a) separately for the equations of momentum, continuity, and energy. Combining the momentum and continuity equations and reconstructing the original differential equations, we identify the numerical diffusion terms which are produced for all governing equations as a consequence of FDV formulations. We summarize the reconstructed equations of momentum, continuity, and energy without the source terms from (6.5.25), (6.5.28), (6.5.31). It is interesting to note that if we neglect all incremental (fluctuation) terms, we arrive at the results identical or analogous to many of the recent developments in FEM for the treatment of convection dominated flows, including the generalized Petrov-Galerkin (GPG) methods, characteristic Galerkin methods (CGM), etc., presented in the previous chapters. To demonstrate this analogy, let us neglect all incremental and higher order terms, but retain only the second order derivative terms, with $s_1 = 1/2$, so that we may arrive at the form more easily recognizable. Here, all components of convection and diffusion Jacobians can be shown to be the velocity components, $a_i^{(m)} = a_i^{(c)} = a_i^{(e)} = \mathrm{v}_i$. These arrangements lead to

Momentum

$$\frac{\partial}{\partial t}(\rho \mathrm{v}_j) + (\rho \mathrm{v}_i \mathrm{v}_j)_{,i} + p_{,j} - \tau_{ij,i} = S_j(m) \tag{13.6.11}$$

with

$$S_j(m) = \frac{\Delta t}{2}[\mathrm{v}_k\,(\rho \mathrm{v}_i \mathrm{v}_j + p\delta_{ij} - \tau_{ij})_{,i}]_{,k} \tag{13.6.12}$$

Continuity

$$\frac{\partial \rho}{\partial t} + (\rho \mathrm{v}_i)_{,i} = S(c) \tag{13.6.13}$$

with

$$S(c) = \frac{\Delta t}{2}[(\rho \mathrm{v}_i \mathrm{v}_j)_{,ij} + p_{,jj} - \tau_{ij,ij} + (\mathrm{v}_i(\rho \mathrm{v}_j)_{,j})_{,i}] \tag{13.6.14}$$

Energy

$$\frac{\partial}{\partial t}(\rho E) + [(\rho E + p)\mathrm{v}_i - kT_{,i} - \tau_{ij}\mathrm{v}_j]_{,i} = S(e) \tag{13.6.15}$$

with

$$S(e) = \frac{\Delta t}{2}\{\mathrm{v}_k[((\rho E + p)\mathrm{v}_i)_{,i} - kT_{,ii} - (\tau_{ij}\mathrm{v}_j)_{,i}]\}_{,k} \tag{13.6.16}$$

Examining the right-hand side terms for all equations, they are identified as numerical diffusions which arise from GPG or CGM formulations. It is seen that second derivatives of pressure arise on the right-hand side explicitly. Direct comparisons can be made with reference to CGM through (13.4.9) through (13.4.14).

13.6.4 TRANSITIONS AND INTERACTIONS BETWEEN COMPRESSIBLE AND INCOMPRESSIBLE FLOWS AND BETWEEN LAMINAR AND TURBULENT FLOWS

In order to understand how the FDV scheme handles computations involving both compressible and incompressible flows, fundamental definitions of pressure as involved in compressible and incompressible flows must be recognized, as pointed out in Section 6.5.6. In view of (6.5.33) through (6.5.36), we note that, if p_o as given by (6.5.36) remains a constant, equivalent to a stagnation (total) pressure, then the compressible flow as assumed in the conservation form of the Navier-Stokes system of equations has now been turned into an incompressible flow, which is expected to occur when the flow velocity is sufficiently reduced (approximately $0.1 \leq M < 0.3$ for air). Thus, (6.6.36) serves as an equivalent equation of state for an incompressible flow. This can be identified nodal point by nodal point or element by element for the entire domain.

When inviscid flow becomes viscous, we may expect that the flow may become laminar or turbulent through inviscid/viscous interactions across the boundary layer. Below the laminar boundary layer, if viscous actions are significant, then the fluid particles are unstable, causing the changes of Mach number and Reynolds number between adjacent nodal points (assuming they are closely spaced) to be irregular, the phenomenon known as transition instability prior to the state of full turbulence. Fluctuations due to turbulence are characterized by the presence of the terms such as in (6.5.37). Physically, this quantity represents the fluctuations of total stresses (physical viscous stresses plus Reynolds stresses) controlled by the Reynolds number changes between the local adjacent nodal points. Thus, the FDV solution contains the sum of the mean flow variables and the fluctuation parts of the variables. Once the solution of the Navier-Stokes system of equations is carried out and all flow variables are determined, then we compute the fluctuation part, f' of any variable f, as given in (6.5.38). Unsteady turbulence statistics (turbulent kinetic energy, Reynolds stresses, and various energy spectra) can be calculated once the fluctuation quantities of all variables are determined. Although the solutions of the Navier-Stokes system of equations using FDV are assumed to contain the fluctuation parts as well as the mean quantities, it will be unlikely that such information is reliable when the Reynolds number is very high and if mesh refinements are not adequate to resolve Kolmogorov microscales. In this case, it is necessary to invoke the level of mesh refinements as required for DNS.

Unsteadiness in turbulent fluctuations may prevail in the vicinity of the wall, although a steady-state may have been reached far away from the wall. This situation can easily be verified by noting that $\Delta\mathbf{U}^{n+1}$ will vanish only in the region far away from the wall, but remain fluctuating in the vicinity of the wall, as dictated by the changes of Mach number in the variation parameter s_3 between the nodal points and fluctuations of the stresses due to both physical and turbulent viscosities in $\Delta\tau_{ij}$ characterized by (6.5.37).

13.6.5 FINITE ELEMENT FORMULATION OF FDV EQUATIONS

We recall that all the provisions and numerical aspects for the physical phenomena such as discontinuities and fluctuations of flow variables have already been incorporated in the FDV equations. The standard Galerkin integral formulations of the FDV equations are all that will be necessary. Thus, we begin by expressing the conservation and flux variables and source terms as a linear combination of trial functions Φ_α with the nodal values of these variables in the form,

$$\mathbf{U}(\mathbf{x},t) = \Phi_\alpha(\mathbf{x})\mathbf{U}_\alpha(t), \qquad \mathbf{F}_i(\mathbf{x},t) = \Phi_\alpha(\mathbf{x})\mathbf{F}_{\alpha i}(t)$$
$$\mathbf{G}_i(\mathbf{x},t) = \Phi_\alpha(\mathbf{x})\mathbf{G}_{\alpha i}(t), \qquad \mathbf{B}(\mathbf{x},t) = \Phi_\alpha(\mathbf{x})\mathbf{B}_\alpha(t)$$

Applying the standard Galerkin approximations to (13.6.7), we obtain

$$\int_\Omega \Phi_\alpha \mathbf{R}(\mathbf{U},\mathbf{F}_i,\mathbf{G}_i,\mathbf{B})\,d\Omega = 0 \tag{13.6.17}$$

or

$$(A_{\alpha\beta}\eta_{rs} + B_{\alpha\beta rs})\,\Delta U_{\beta s}^{n+1} = H_{\alpha r}^n + N_{\alpha r}^n \tag{13.6.18}$$

where

$$A_{\alpha\beta} = \int_\Omega \Phi_\alpha \Phi_\beta \, d\Omega, \quad \eta_{rs} = \delta_{rs} + \Delta t \, s_5 \, d_{rs} + \frac{\Delta t^2}{2} s_6 \, d_{rm} \, d_{ms} \tag{13.6.19}$$

$$\begin{aligned} B_{\alpha\beta rs} = & \int_\Omega \Bigg[- \left\{ \Delta t (s_1 a_{irs} + s_3 \, b_{irs}) + \frac{\Delta t^2}{2} [s_2 d_{rt} a_{its} + s_6 d_{rt} \, (a_{its} + b_{its}) + s_4 d_{rt} b_{its}] \right\} \\ & \times \Phi_{\alpha,i} \Phi_\beta - \Bigg\{ \Delta t \, s_3 c_{ijrs} - \frac{\Delta t^2}{2} [s_2 (a_{irt} a_{jts} + b_{irt} a_{jts}) + s_4 (a_{irt} b_{jts} + b_{irt} b_{jts} \\ & - d_{rt} c_{ijts})] \Bigg\} \Phi_{\alpha,i} \Phi_{\beta,j} \Bigg] d\Omega + \int_\Gamma \Bigg[\Bigg\{ \Delta t (s_1 a_{irs} + s_3 \, b_{irs}) + \frac{\Delta t^2}{2} [s_2 d_{rt} a_{its} \\ & + s_6 d_{rt} \, (a_{its} + b_{its}) + s_4 d_{rt} b_{its}] \Bigg\} \overset{*}{\Phi}_\alpha \overset{*}{\Phi}_\beta + \Bigg\{ \Delta t \, s_3 c_{ijrs} - \frac{\Delta t^2}{2} [s_2 (a_{irt} a_{jts} \\ & + b_{irt} a_{jts}) + s_4 (a_{irt} b_{jts} + b_{irt} b_{jts} - d_{rt} c_{ijts})] \Bigg\} \overset{*}{\Phi}_\alpha \overset{*}{\Phi}_{\beta,j} \Bigg] n_i \, d\Gamma \end{aligned} \tag{13.6.20}$$

$$\begin{aligned} H^n_{\alpha r} = & \int_\Omega \Bigg\{ \left[\Delta t (F^n_{\beta ir} + G^n_{\beta ir}) + \frac{\Delta t^2}{2} d_{rs} (F^n_{\beta is} + G^n_{\beta is}) + \frac{\Delta t^2}{2} (a_{irs} + b_{irs}) B^n_{\beta s} \right] \Phi_{\alpha,i} \Phi_\beta \\ & - \frac{\Delta t^2}{2} (a_{irs} + b_{irs}) (F^n_{\beta js} + G^n_{\beta js}) \Phi_{\alpha,i} \Phi_{\beta,j} + \left[\Delta t \, B^n_{\beta r} + \frac{\Delta t^2}{2} d_{rs} \, B^n_{\beta s} \right] \Phi_\alpha \Phi_\beta \Bigg\} d\Omega \end{aligned} \tag{13.6.21}$$

$$\begin{aligned} N^n_{\alpha r} = & \int_\Gamma \Bigg\{ \left[- \Delta t (F^n_{\beta ir} + G^n_{\beta ir}) - \frac{\Delta t^2}{2} d_{rs} (F^n_{\beta is} + G^n_{\beta is}) - \frac{\Delta t^2}{2} (a_{irs} + b_{irs}) B^n_{\beta s} \right] \overset{*}{\Phi}_\alpha \overset{*}{\Phi}_\beta \\ & + \frac{\Delta t^2}{2} (a_{irs} + b_{irs}) (F^n_{\beta js} + G^n_{\beta js}) \overset{*}{\Phi}_\alpha \overset{*}{\Phi}_{\beta,j} \Bigg\} n_i \, d\Gamma \end{aligned} \tag{13.6.22}$$

Here all Jacobians must be updated at each iteration step, $\overset{*}{\Phi}_\alpha$ represents the Neumann boundary trial and test functions, with α, β denoting the global node number and r, s providing the number of conservation variables at each node. For three dimensions, $i, j = 1, 2, 3$ associated with the Jacobians imply directional identification of each Jacobian matrix $(a_1, a_2, a_3, b_1, b_2, b_3, c_{11}, c_{12}, c_{13}, c_{21}, c_{22}, c_{23}, c_{31}, c_{32}, c_{33})$ with $r, s = 1, 2, 3, 4, 5$ denoting entries of each of the 5×5 Jacobian matrices. These indices can be reduced similarly for 2-D.

Evaluation of integrals in (13.6.19)–(13.6.22) must begin with local elements of the form

$$\left(A^{(e)}_{NM} \delta_{rs} + B^{(e)}_{NMrs} \right) \Delta U^{(e)}_{Ms} = H^{n(e)}_{Nr} + N^{n(e)}_{Nr}$$

We shall describe the procedure for two-dimensional isoparametric elements using Gaussian quadrature integrations with an EBE process for assembly into a global form as shown in Section 10.3.2. The local FDV finite element equation given above represents a system of 16 equations with $N, M = 1, 2, 3, 4$ and $r, s = 1, 2, 3, 4$. These matrix equations are constructed by summing terms with repeated indices. A simple computer algorithm can be developed to achieve this process. For example, $A_{NM}^{(e)}\delta_{rs}\Delta U_{Ms}^{(e)}$ takes the form

$$
\begin{bmatrix}
A_{11} & & & & A_{12} & & & & A_{13} & & & & A_{14} & & & \\
& A_{11} & & & & A_{12} & & & & A_{13} & & & & A_{14} & & \\
& & A_{11} & & & & A_{12} & & & & A_{13} & & & & A_{14} & \\
& & & A_{11} & & & & A_{12} & & & & A_{13} & & & & A_{14} \\
A_{21} & & & & A_{22} & & & & A_{23} & & & & A_{24} & & & \\
& A_{21} & & & & A_{22} & & & & A_{23} & & & & A_{24} & & \\
& & A_{21} & & & & A_{22} & & & & A_{23} & & & & A_{24} & \\
& & & A_{21} & & & & A_{22} & & & & A_{23} & & & & A_{24} \\
A_{31} & & & & A_{32} & & & & A_{33} & & & & A_{34} & & & \\
& A_{31} & & & & A_{32} & & & & A_{33} & & & & A_{34} & & \\
& & A_{31} & & & & A_{32} & & & & A_{33} & & & & A_{34} & \\
& & & A_{31} & & & & A_{32} & & & & A_{33} & & & & A_{34} \\
A_{41} & & & & A_{42} & & & & A_{43} & & & & A_{44} & & & \\
& A_{41} & & & & A_{42} & & & & A_{43} & & & & A_{44} & & \\
& & A_{41} & & & & A_{42} & & & & A_{43} & & & & A_{44} & \\
& & & A_{41} & & & & A_{42} & & & & A_{43} & & & & A_{44}
\end{bmatrix}
\begin{bmatrix}
\Delta U_{11} \\ \Delta U_{12} \\ \Delta U_{13} \\ \Delta U_{14} \\ \Delta U_{21} \\ \Delta U_{22} \\ \Delta U_{23} \\ \Delta U_{24} \\ \Delta U_{31} \\ \Delta U_{32} \\ \Delta U_{33} \\ \Delta U_{34} \\ \Delta U_{41} \\ \Delta U_{42} \\ \Delta U_{43} \\ \Delta U_{44}
\end{bmatrix}
$$

Similarly, $B_{NMrs}^{(e)}\Delta U_{Ms}^{(e)}$ is of the form

$$
\begin{bmatrix}
B_{1111} & B_{1112} & B_{1113} & B_{1114} & B_{1211} & B_{1212} & B_{1213} & B_{1214} & B_{1311} & B_{1312} & B_{1313} & B_{1314} & B_{1411} & B_{1412} & B_{1413} & B_{1414} \\
B_{1121} & B_{1122} & B_{1123} & B_{1124} & B_{1221} & B_{1222} & B_{1223} & B_{1224} & B_{1321} & B_{1322} & B_{1323} & B_{1324} & B_{1421} & B_{1422} & B_{1423} & B_{1424} \\
B_{1131} & B_{1132} & B_{1133} & B_{1134} & B_{1231} & B_{1232} & B_{1233} & B_{1234} & B_{1331} & B_{1332} & B_{1333} & B_{1334} & B_{1431} & B_{1432} & B_{1433} & B_{1434} \\
B_{1141} & B_{1142} & B_{1143} & B_{1144} & B_{1241} & B_{1242} & B_{1243} & B_{1244} & B_{1341} & B_{1342} & B_{1343} & B_{1344} & B_{1441} & B_{1442} & B_{1443} & B_{1444} \\
B_{2111} & B_{2112} & B_{2113} & B_{2114} & B_{2211} & B_{2212} & B_{2213} & B_{2214} & B_{2311} & B_{2312} & B_{2313} & B_{2314} & B_{2411} & B_{2412} & B_{2413} & B_{2414} \\
B_{2121} & B_{2122} & B_{2123} & B_{2124} & B_{2221} & B_{2222} & B_{2223} & B_{2224} & B_{2321} & B_{2322} & B_{2323} & B_{2324} & B_{2421} & B_{2422} & B_{2423} & B_{2424} \\
B_{2131} & B_{2132} & B_{2133} & B_{2134} & B_{2231} & B_{2232} & B_{2233} & B_{2234} & B_{2331} & B_{2332} & B_{2333} & B_{2334} & B_{2431} & B_{2432} & B_{2433} & B_{2434} \\
B_{2141} & B_{2142} & B_{2143} & B_{2144} & B_{2241} & B_{2242} & B_{2243} & B_{2244} & B_{2341} & B_{2342} & B_{2343} & B_{2344} & B_{2441} & B_{2442} & B_{2443} & B_{2444} \\
B_{3111} & B_{3112} & B_{3113} & B_{3114} & B_{3211} & B_{3212} & B_{3213} & B_{3214} & B_{3311} & B_{3312} & B_{3313} & B_{3314} & B_{3411} & B_{3412} & B_{3413} & B_{3414} \\
B_{3121} & B_{3122} & B_{3123} & B_{3124} & B_{3221} & B_{3222} & B_{3223} & B_{3224} & B_{3321} & B_{3322} & B_{3323} & B_{3324} & B_{3421} & B_{3422} & B_{3423} & B_{3424} \\
B_{3131} & B_{3132} & B_{3133} & B_{3134} & B_{3231} & B_{3232} & B_{3233} & B_{3234} & B_{3331} & B_{3332} & B_{3333} & B_{3334} & B_{3431} & B_{3432} & B_{3433} & B_{3434} \\
B_{3141} & B_{3142} & B_{3143} & B_{3144} & B_{3241} & B_{3242} & B_{3243} & B_{3244} & B_{3341} & B_{3342} & B_{3343} & B_{3344} & B_{3441} & B_{3442} & B_{3443} & B_{3444} \\
B_{4111} & B_{4112} & B_{4113} & B_{4114} & B_{4211} & B_{4212} & B_{4213} & B_{4214} & B_{4311} & B_{4312} & B_{4313} & B_{4314} & B_{4411} & B_{4412} & B_{4413} & B_{4414} \\
B_{4121} & B_{4122} & B_{4123} & B_{4124} & B_{4221} & B_{4222} & B_{4223} & B_{4224} & B_{4321} & B_{4322} & B_{4323} & B_{4324} & B_{4421} & B_{4422} & B_{4423} & B_{4424} \\
B_{4131} & B_{4132} & B_{4133} & B_{4134} & B_{4231} & B_{4232} & B_{4233} & B_{4234} & B_{4331} & B_{4332} & B_{4333} & B_{4334} & B_{4431} & B_{4432} & B_{4433} & B_{4434} \\
B_{4141} & B_{4142} & B_{4143} & B_{4144} & B_{4241} & B_{4242} & B_{4243} & B_{4244} & B_{4341} & B_{4342} & B_{4343} & B_{4344} & B_{4441} & B_{4442} & B_{4443} & B_{4444}
\end{bmatrix}
\begin{bmatrix}
\Delta U_{11} \\ \Delta U_{12} \\ \Delta U_{13} \\ \Delta U_{14} \\ \Delta U_{21} \\ \Delta U_{22} \\ \Delta U_{23} \\ \Delta U_{24} \\ \Delta U_{31} \\ \Delta U_{32} \\ \Delta U_{33} \\ \Delta U_{34} \\ \Delta U_{41} \\ \Delta U_{42} \\ \Delta U_{43} \\ \Delta U_{44}
\end{bmatrix}
$$

For example, let us examine one of the terms in B_{1214},

$$
B_{1214} = \frac{\Delta t^2}{2}\int_\Omega s_2 a_{i1t} a_{jt4} \Phi_{1,i}\Phi_{2,j} d\Omega + \cdots \quad \text{with } i, j = 1, 2, \quad t = 1, 2, 3, 4
$$

All integrals are to be integrated using Gaussian quadrature.

The domain integrals on the right-hand side are evaluated similarly. However, they will result in a column vector compatible with left-hand side.

The evaluation of boundary integrals that appear in both left-hand side and right-hand side are discussed in the next section.

13.6.6 BOUNDARY CONDITIONS

Treatment of boundary conditions in finite element methods is simple and straightforward as discussed in Section 10.1.2. Particularly, in FDV formulations where all regimes of velocity are to be accommodated in multidimensions, implementations of boundary conditions are self-explanatory. Neumann boundary conditions in FDV occur in both left-hand side and right-hand side. The left-hand side Neumann boundary integrals are evaluated and summed into the corresponding domain integrals as first discussed in Section 10.2.4, whereas the right-hand side Neumann boundary conditions appear as a column vector as shown in Section 10.1.3.

The Neumann boundary conditions

To illustrate, let us consider one of the boundary integrals multiplied by the conservation variable vector on the left-hand side.

(1) $N = 1, r = 1, M = 1, 2, s = 1, 2, 3, 4,\ i, j = 1, 2$

$$\int_\Gamma a_{irs} \overset{*}{\Phi}_N \overset{*}{\Phi}_M n_i d\Gamma \Delta U_{Ms} = \int_\Gamma \{(a_{111}n_1 + a_{211}n_2)\Delta U_{11} + (a_{112}n_1 + a_{212}n_2)\Delta U_{12}$$
$$+(a_{113}n_1 + a_{213}n_2)\Delta U_{13} + (a_{114}n_1 + a_{214}n_2)\Delta U_{14}\} \overset{*}{\Phi}_1 \overset{*}{\Phi}_1 d\Gamma$$
$$+ \int_\Gamma \{(a_{111}n_1 + a_{211}n_2)\Delta U_{21} + (a_{112}n_1 + a_{212}n_2)\Delta U_{22}$$
$$+(a_{113}n_1 + a_{214}n_2)\Delta U_{23} + (a_{114}n_1 + a_{214}n_2)\Delta U_{24}\} \overset{*}{\Phi}_1 \overset{*}{\Phi}_2 d\Gamma$$

(2) $N = 1, r = 2, M = 1, 2, s = 1, 2, 3, 4,\ i, j = 1, 2$
(3) $N = 1, r = 3, M = 1, 2, s = 1, 2, 3, 4,\ i, j = 1, 2$
(4) $N = 1, r = 4, M = 1, 2, s = 1, 2, 3, 4,\ i, j = 1, 2$
(5) $N = 2, r = 1, M = 1, 2, s = 1, 2, 3, 4,\ i, j = 1, 2$
(6) $N = 2, r = 2, M = 1, 2, s = 1, 2, 3, 4,\ i, j = 1, 2$
(7) $N = 2, r = 3, M = 1, 2, s = 1, 2, 3, 4,\ i, j = 1, 2$
(8) $N = 2, r = 4, M = 1, 2, s = 1, 2, 3, 4,\ i, j = 1, 2$

Note that the terms with repeated indices will be summed for the free indices $N = 1, 2$ and $r = 1, 2, 3, 4$ for the two-node boundary line elements, resulting in the 8×8 square matrix corresponding to the $8 \times 1 \Delta U_{Ms}$ (see Figures 10.1.2 and 13.6.1). If two nodes, node 1 and node 2, of the boundary line element coincide with node 1 and node 2 of the local element adjoining the boundary line shown in Figure 13.6.1, then the 8×8 boundary line element matrix is algebraically added to the corresponding 16×16 local element matrix. This is the influence of the boundary conditions affecting the domain at the current time step $n + 1$.

The situation is different for the case of the right-hand side boundary integrals at the time step n. They simply result in a column vector as is the case for the regular time-dependent finite element equations. Note also that various Jacobians are

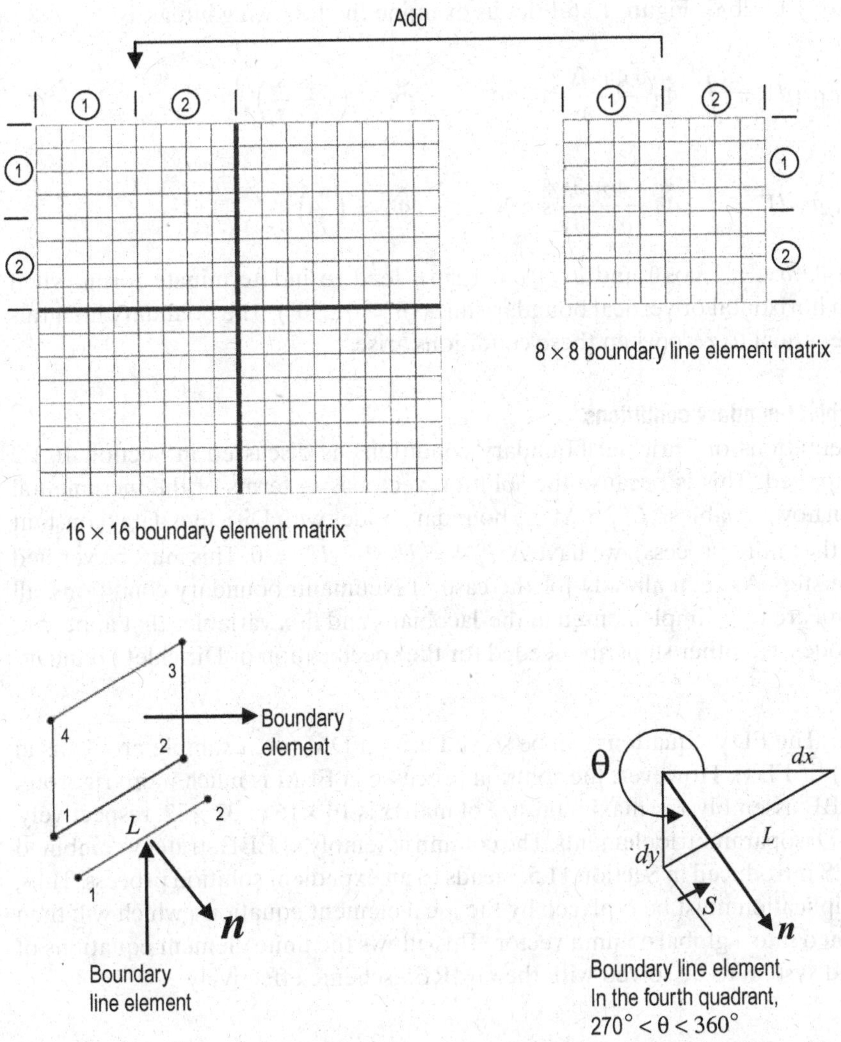

Figure 13.6.1 The 8 × 8 boundary line element matrix added to the adjoining 16 × 16 boundary element matrix for the left-hand side boundary integral terms.

multiplied by convection flux variables and diffusion flux variables, indicating that not only the boundary values within the Jacobians but also the flux variable as specified at the boundary nodes are to influence the outcome of the CFD solution.

It should be noted that one of the advantages of the FDV approach is the ability to specify all the values that occur in the Jacobians (a_i, b_i, c_{ij}) and flux variables (F_i, G_i) at the boundary nodes. For example, primitive variables (Dirichlet boundary conditions) and their gradients can be specified at boundary nodes directly through the Jacobians and flux variables.

To illustrate the boundary integrals with derivatives of trial functions, we consider a typical case as in Figure 13.6.1.

$$\int_\Gamma a_{irs} F^n_{Mjs} \overset{*}{\Phi}_N \overset{*}{\Phi}_{M,j} n_i d\Gamma$$

Using Figure 10.1.2b or Figure 13.6.1, let us examine the following integrals:

$$\int_\Gamma \overset{*}{\Phi}_1 \overset{*}{\Phi}_{1,1} n_1 d\Gamma = \int_0^L \overset{*}{\Phi}_1 \frac{\partial \overset{*}{\Phi}_1}{\partial s} \frac{\partial s}{\partial x} \cos\theta ds, \qquad \overset{*}{\Phi}_1 = \left(1 - \frac{s}{L}\right)$$

$$\int_\Gamma \overset{*}{\Phi}_1 \overset{*}{\Phi}_{1,2} n_2 d\Gamma = \int_0^L \overset{*}{\Phi}_1 \frac{\partial \overset{*}{\Phi}_{21}}{\partial s} \frac{\partial s}{\partial y} \sin\theta ds, \qquad \overset{*}{\Phi}_2 = \left(\frac{s}{L}\right)$$

Notice that $\partial s/\partial x = 1/\cos\theta$ and $\partial s/\partial y = 1/\sin\theta$ lead to indeterminate forms when dealing with horizontal or vertical boundary lines ($\theta = 0°, 90°$). The boundary integrals should be set equal to zero when these conditions arise.

The Dirichlet boundary conditions

Implementations of Dirichlet boundary conditions as discussed in Section 10.1.2 cannot be applied. This is because the solution vector is in terms of the incremental conservation flow variables $\Delta U_{\beta s}^{n+1}$. At the boundary nodes with Dirichlet data (constant throughout the entire process), we have $\Delta U^{n+1} = U^{n+1} - U^n = 0$. This must be verified at each time step. As seen already for the case of Neumann boundary conditions, all Dirichlet data are to be implemented in the Jacobians and flux variables that appear at boundary nodes. No other steps are needed for the specification of Dirichlet boundary conditions.

Remarks: The FDV equations can be solved using FDM (see example problems in Figure 6.8.2) or FEM. However, the solution process via FEM is much more rigorous. Using the EBE assembly, the maximum size of matrix is 16 ×16 or 32 ×32, respectively, for 2-D or 3-D isoparametric elements. The column assembly of EBE strategy combined with GMRES introduced in Section 11.5.3 leads to an expedient solution process. Thus, matrix multiplication must be replaced by the local element equations, which will then be transformed into a global column vector. This allows the finite element equations of the large grid system to be solved with the GMRES scheme effectively.

13.7 EXAMPLE PROBLEMS

(1) Quasi–1-D Supersonic Flows (Euler Equations) with Two-Step GPG

Given: Quasi–one-dimensional rocket nozzle given in Section 6.8.1.

Solution: This problem was solved using 500 linear finite elements with two-step GPG. The computed results are shown to be in good agreement with the analytical solution in Figure 13.7.1.

(2) Two-dimensional Supersonic Flows (Euler Equations) with Two-Step TGM

Given: Geometry and initial and boundary conditions are as shown in Figure 13.7.2a.

Solution: The results of calculations using TGM are shown in Figure 13.7.2b-e. The L_2 norm error convergence history of all variables is shown in Figure 13.7.2f.

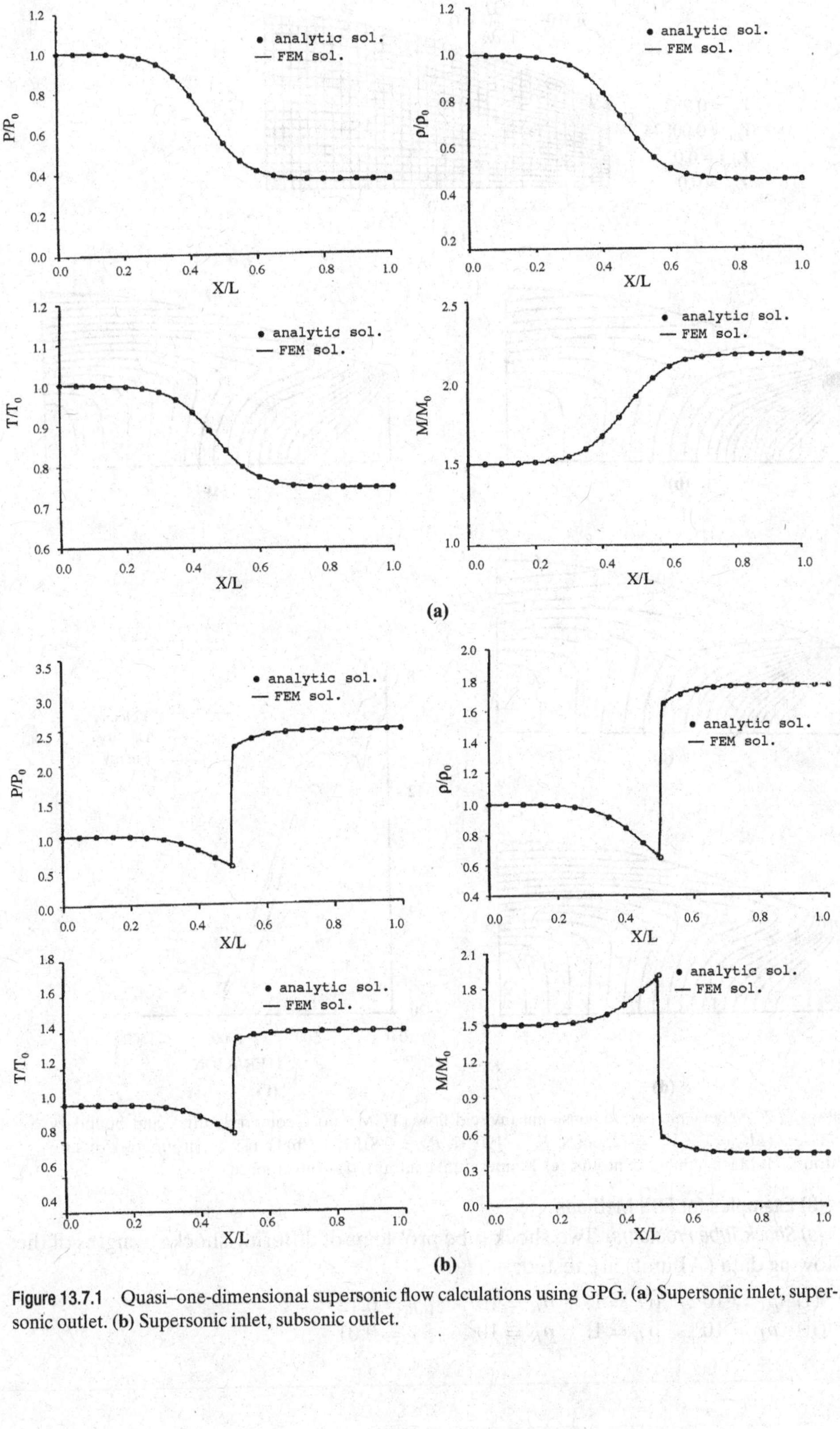

Figure 13.7.1 Quasi–one-dimensional supersonic flow calculations using GPG. (**a**) Supersonic inlet, supersonic outlet. (**b**) Supersonic inlet, subsonic outlet.

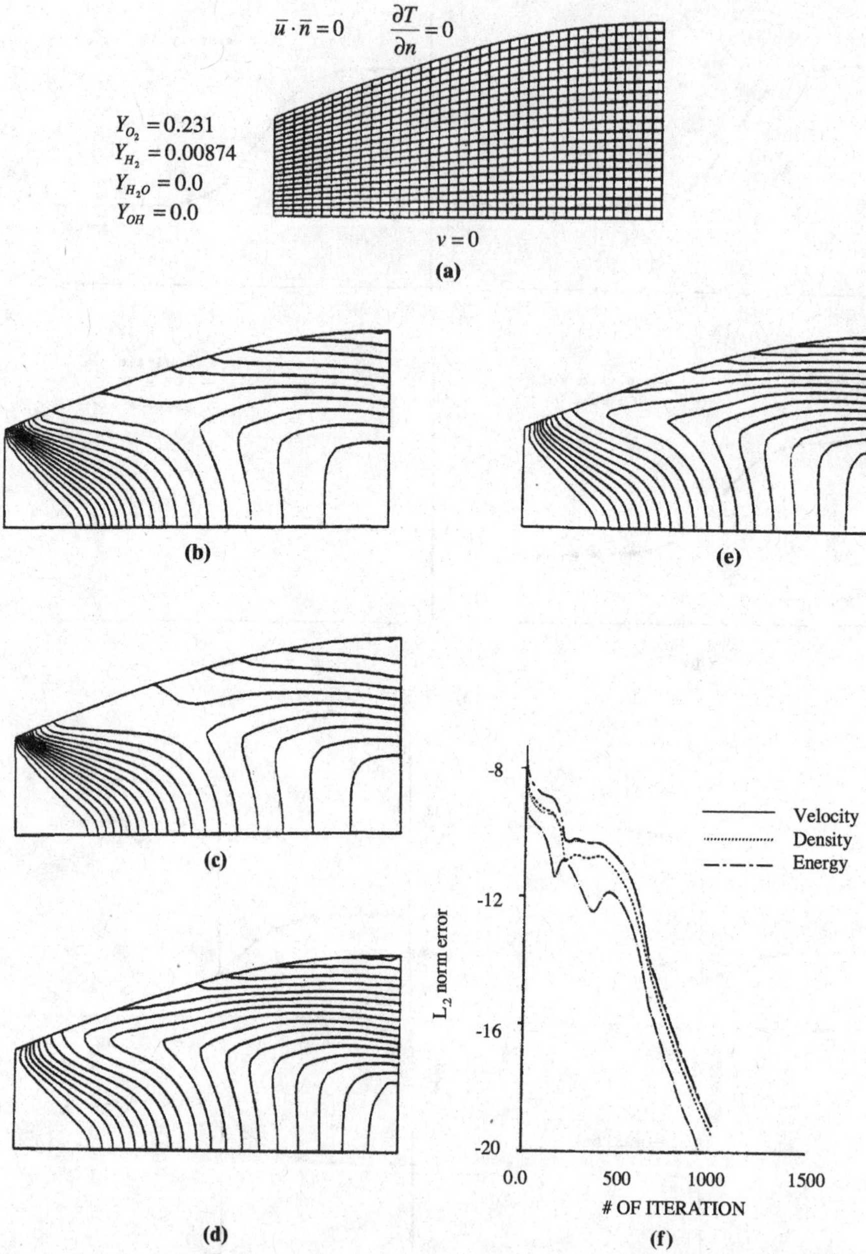

Figure 13.7.2 Supersonic two-dimensional inviscid flow (TGM). **(a)** Geometry, initial, and boundary conditions ($M_\infty = 1.4$, $V_\infty = 1230\text{m/s}$, $T_\infty = 1900\text{K}$, $P_\infty = 0.81\text{MPa}$). **(b)** Density contours. **(c)** Pressure contours. **(d)** Mach number contours. **(e)** Temperature contours. **(f)** Convergence.

(3) Examples for FDV Methods

(a) Shock Tube Problems. Two shock tube problems of differing shock strengths of the following data (AI unit) are tested:

(i) $p_L = 10^5, \quad \rho_L = 1, \quad p_R = 10^4, \quad \rho_R = 0.125$

(ii) $p_L = 10^5, \quad \rho_L = 1, \quad p_R = 10^3, \quad \rho_R = 0.01$

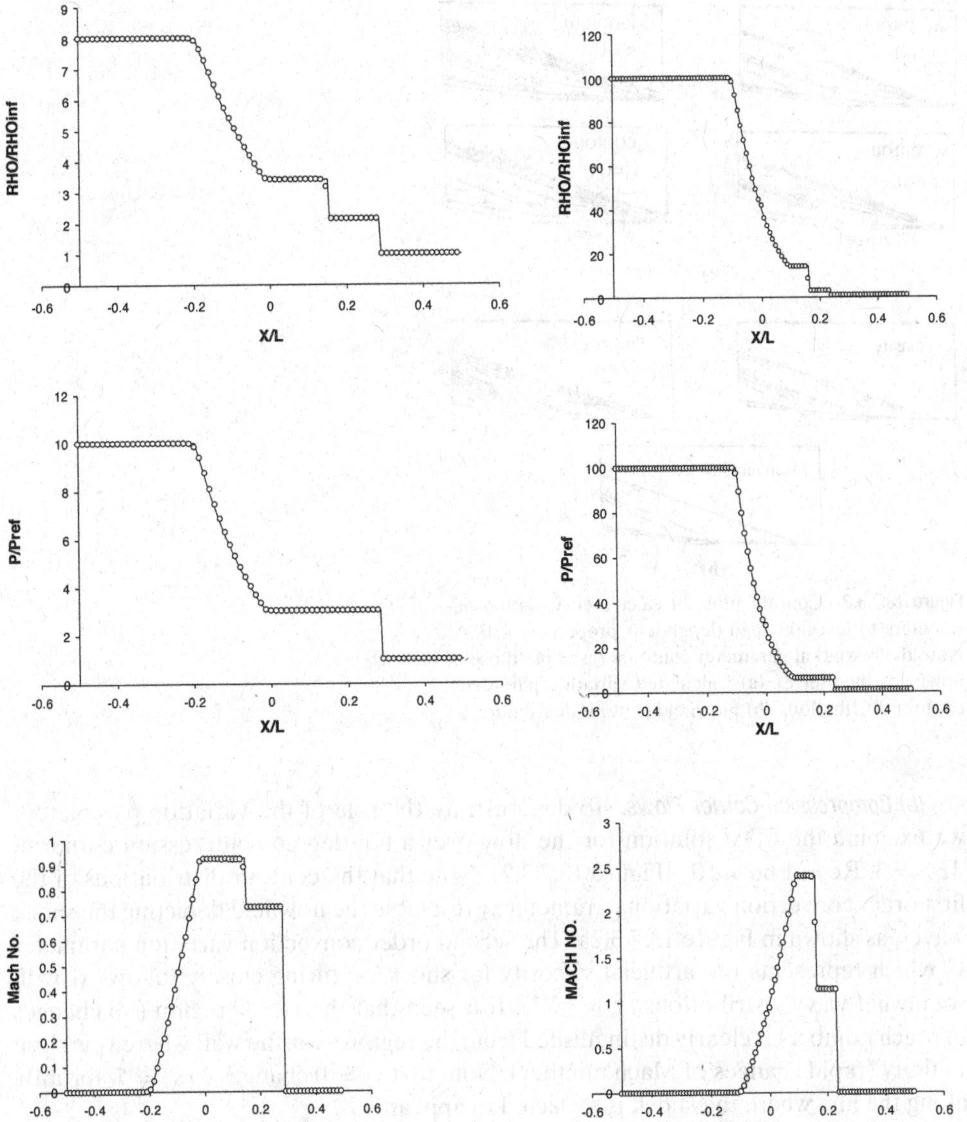

(a) $P_L = 10^5, \rho_L = 1, P_R = 10^4, \rho_R = 0.125,$
$t = 5.2$ ms

(b) $P_L = 10, \rho_L = 1, P_R = 10, \rho_R = 0.01,$
$t = 2.65$ ms

Figure 13.7.3.1 Shock tube calculations (1,200 elements) using the FDV theory, solid lines and symbols indicating analytical solutions and numerical results, respectively.

The FDV solutions for the above shock tube cases indicate perfect agreements with the analytical solutions as shown in Figure 13.7.3.1. The advantage of the FDV theory is an automatic switch from the Navier-Stokes system of equations to Euler equations with the calculated diffusion variation parameters (s_3, s_4) being zero everywhere in the domain. Only the convection variation parameters (s_1, s_2) remain nonzero.

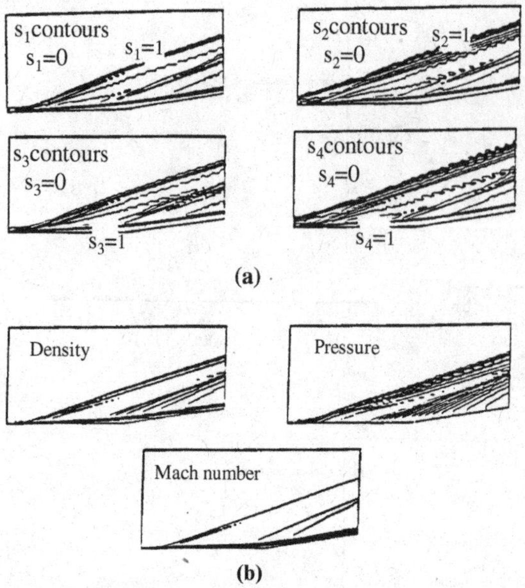

Figure 13.7.3.2 Contour plots of calculated variation parameters to test flow field-dependent properties in FDV. Note that variation parameter contours resemble those of flowfields themselves. **(a)** Calculated variation parameter contour distributions. **(b)** Flowfield contour distributions.

(b) Compression Corner Flows. To demonstrate the role of the variation parameters, we examine the FDV solution for the flow over a ten-degree compression corner at $M_\infty = 3$, Re $= 1.68 \times 10^4$ (Figure 13.7.3.2). Note that the contour distributions of the first order convection variation parameter s_1 resemble the flowfield depicting the shock waves, as shown in Figure 13.7.3.2a. The second order convection variation parameter s_2 which represents the artificial viscosity for shock capturing closely follows s_1 with somewhat wavy distributions ($s_2 = s_1^{1/4}$). It is seen that the $s_1 = 0$ region (no changes in Mach number) is clearly distinguished from the region near the wall where s_1 is close to unity (rapid changes of Mach number). Note that $s_1 = 0$ changes to $s_1 = 1$ abruptly along the line where the shock is expected to appear.

It is seen that the contour distributions of the first order diffusion variation parameter s_3 resemble the boundary layer formation in the vicinity of the wall with thickening of contours toward the wall as compared to the first order convection variation parameter s_1. The second order diffusion variation parameter s_4 whose role is to provide numerical diffusion for stability for the calculation of fluctuations of turbulent motions follows the trend of s_3 with wavy distributions ($s_4 = s_3^{1/4}$). No change in Reynolds number is indicated by $s_3 = 0$ in the upper upstream region, which coincides with $s_1 = 0$ for convection as expected.

The actual flowfield calculations based on these variation parameters are shown in Figure 13.7.3.2b. As the FDV theory dictates, the first order variation parameters (s_1, s_3) control the physics and accuracy, whereas the second order variation parameters (s_2, s_4) address numerical diffusion for stability. These variation parameters are updated throughout the computational process until the steady-state is reached, with their contours continuously resembling the actual flowfield.

It should be noted that the physical interactions between inviscid/viscous, compressible/incompressible, and laminar/turbulent flows are simultaneously controlled by the first and second order convection/diffusion variation parameters. These assessments will be verified from additional example problems presented below.

(4) Driven Cavity Flow Problems to Test Compressibility/Incompressibility Characteristics

This example is to demonstrate that the FDV scheme is capable of reaching the incompressible limit at low speeds as well as the shock capturing capability at high speeds. The cavity flow problem [Ghia et al., 1982; Yoon et al., 1998] is examined here for two different Mach numbers ($M = 0.01$ and $M = 0.1$). Streamline and vorticity contours shown in Figure 13.7.4a–d are in good agreement with FDM results of Ghia et al. [1982]. Density distributions (Figure 13.7.4e) for $M = 0.01$ are constant throughout the domain, whereas at $M = 0.1$ we note that variations begin to occur near the downstream upper region. The most significant feature is the distribution

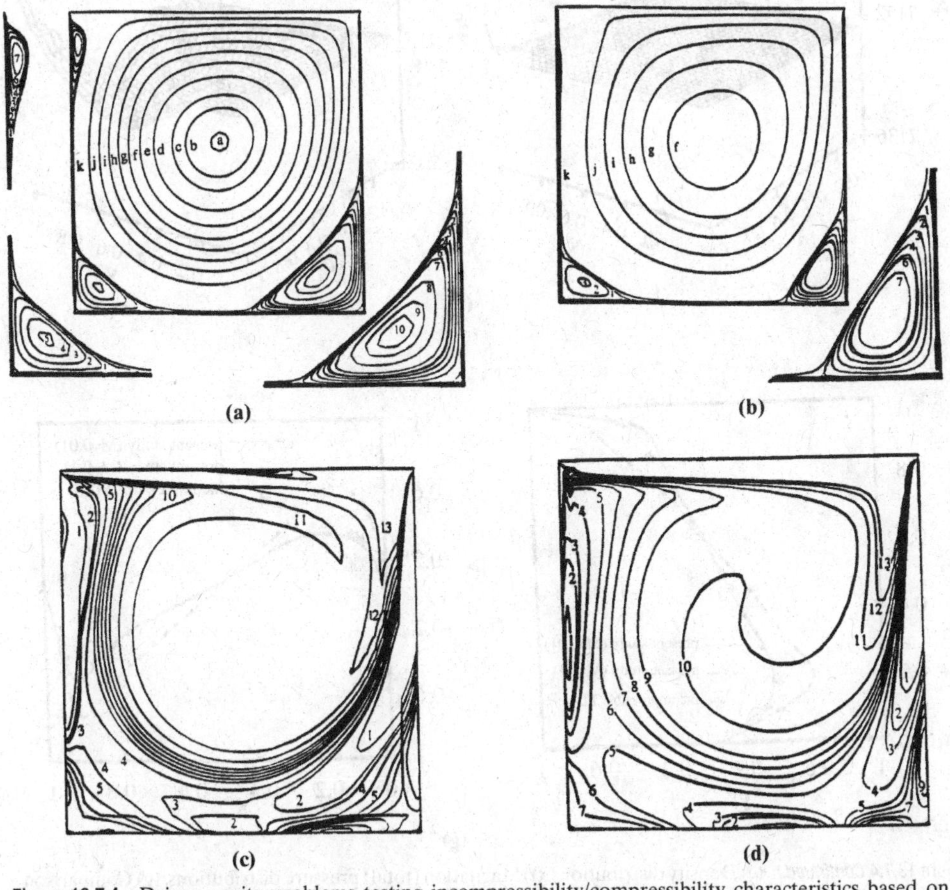

Figure 13.7.4 Driven cavity problems testing incompressibility/compressibility characteristics based on FDV theory. **(a)** Streamlines for $M = 0.01$. **(b)** Streamlines for $M = 0.1$. **(c)** Vorticity contours for $M = 0.1$. **(d)** Vorticity contours for $M = 0.01$.

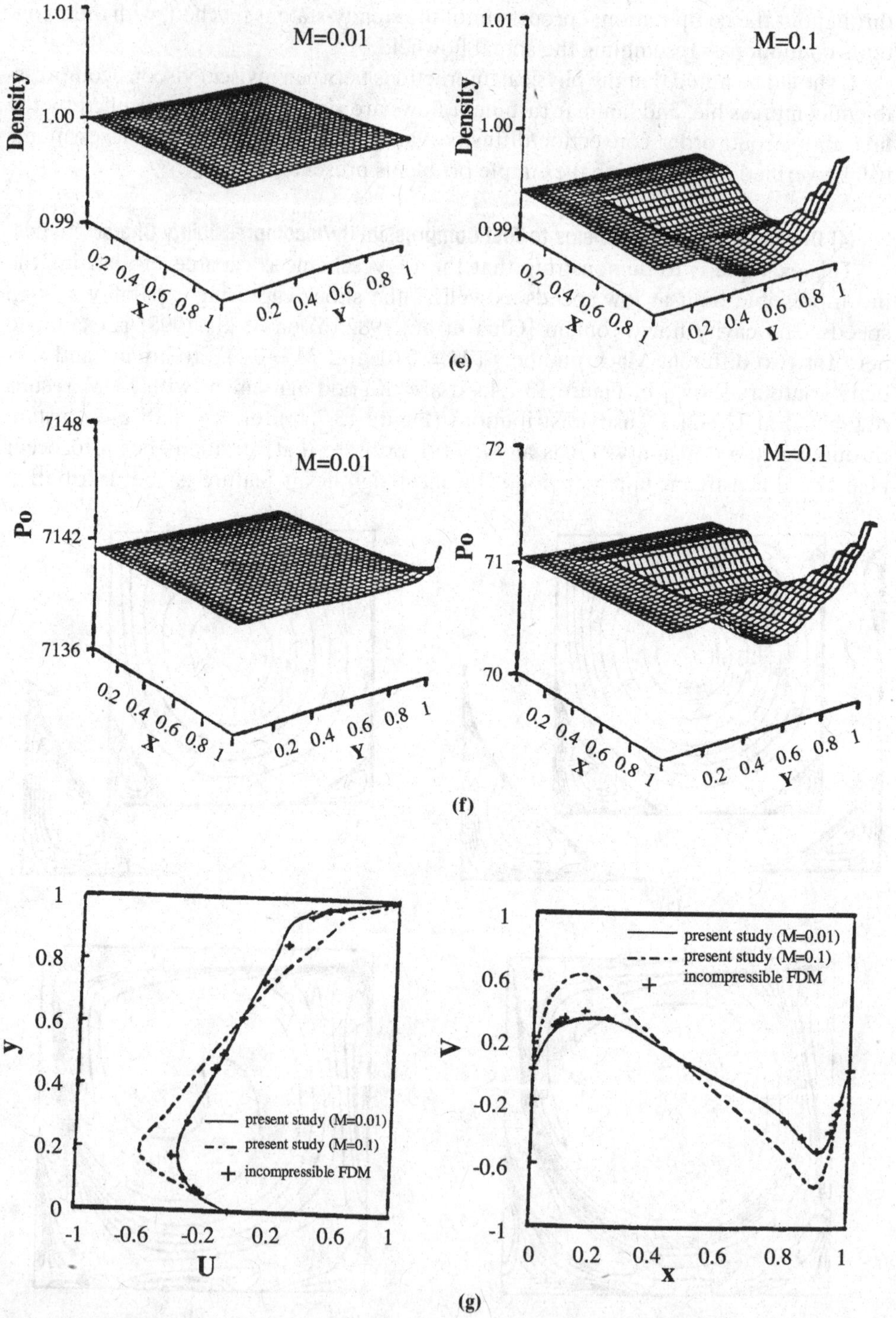

Figure 13.7.4 *Continued.* (e) Density distributions. (f) Stagnation (total) pressure distributions. (g) Comparison of velocity distributions with experiments.

of the stagnation (total) pressure (Figure 13.7.4f) as calculated from (6.5.36), indicating that the stagnation pressure is constant at $M = 0.01$ and it begins to vary at $M = 0.1$, almost exactly the same way as density. This proves that (6.5.36) acts as the equation of state encompassing the incompressible and compressible flows. Comparisons of the FDV solutions for the velocity distributions at the centerlines (Figure 13.7.4g) confirm the trend disclosed in Figure 13.7.4e,f. The velocity distributions for $M = 0.01$ are identical to the results of the experimental data for incompressible flow, whereas the solution for $M = 0.1$ (compressible effect present) deviates from the incompressible case. The evidence is overwhelming that the FDV scheme is capable of treating the transition automatically between the incompressible and compressible limit.

(5) Hypersonic Flow Solutions by the FDV Method, $M = 20$, Re = 300,000, with Impinging Shock Wave on a Flat Inlet Combustion Chamber

This example uses the impinging shock wave angle of 12.7° corresponding to the deflection angle of 10°. The solution clearly shows the advantage of the FDV method, with

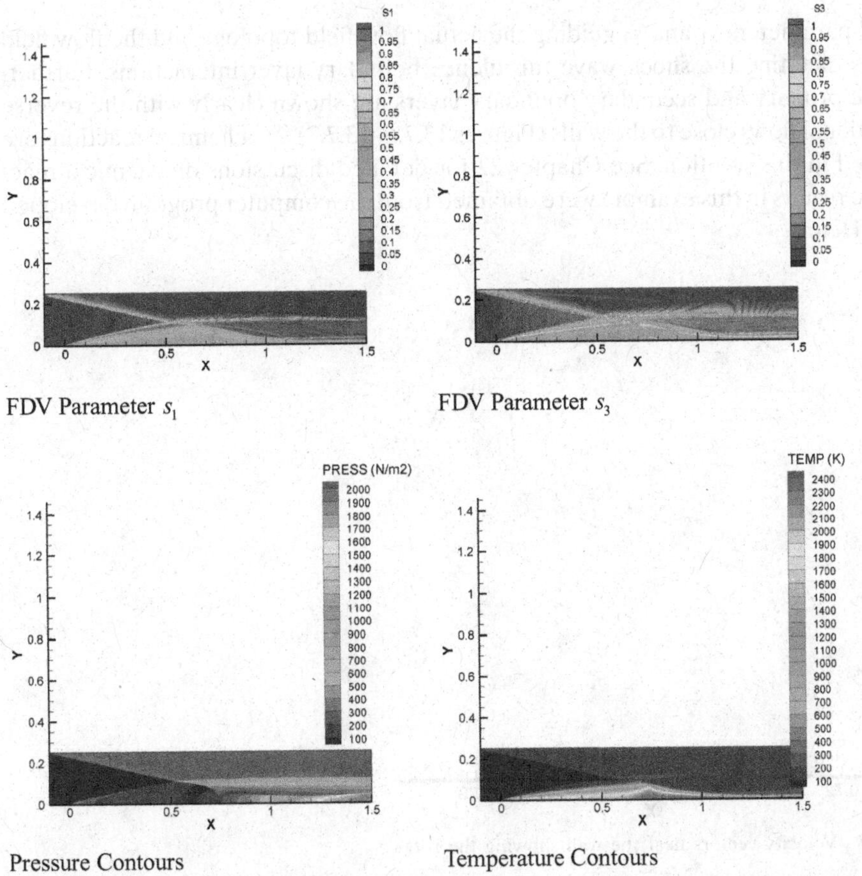

Figure 13.7.5 FDV parameters s_1 and s_2 as calculated from the local Mach numbers and Reynolds numbers resembling the flow field itself.

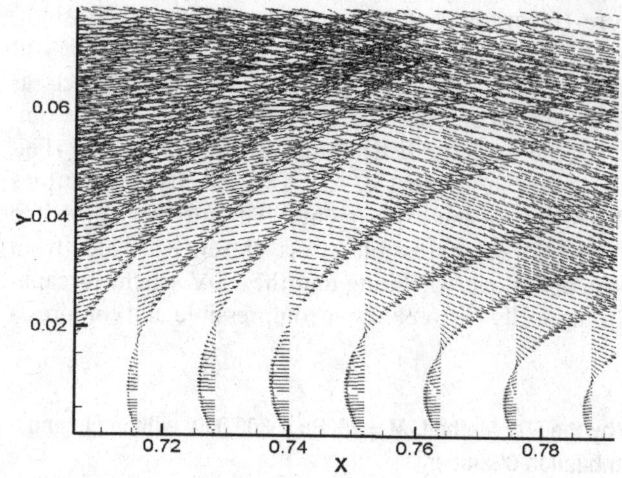

Figure 13.7.6 Velocity vectors near the wall showing the primary and secondary boundary layers and reversed flows.

the FDV parameters s_1 and s_3 guiding the actual flow field topology and the flow field Jacobians dictating the shock wave turbulence boundary layer interactions. Furthermore, the primary and secondary boundary layers are shown clearly with the reverse and rotational flows close to the walls (Figures 13.7.5–13.7.7). No chemical reactions are considered in this solution. See Chapter 22 for detailed discussions on chemical reactions. The results in this example were obtained from the computer program developed by Gary Heard.

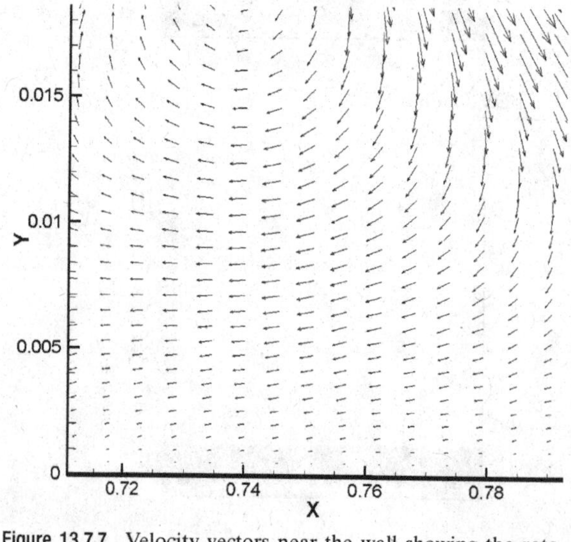

Figure 13.7.7 Velocity vectors near the wall showing the rotational flow.

13.8 SUMMARY

In this chapter, most of the currently available compressible flow analyses using FEM have been presented. They include GGM (generalized Galerkin methods), TGM (Taylor-Galerkin methods), GPG (generalized Petrov-Galerkin methods), CGM (characteristic Galerkin methods), DGM (discontinuous Galerkin methods), and FDV (flowfield-dependent variation methods). Exhaustive numerical results on TGM and GPG are available in the literature, and no attempt is made to introduce them here. Only a few selective examples are shown in Section 13.7 for illustration.

Transitions and interactions between inviscid/viscous, compressible/incompressible, and laminar/turbulent flows can be resolved by the FDV theory. It is shown that the FDV parameters initially introduced in the Taylor series expansion of the conservation variables of the Navier-Stokes system of equations are translated into flowfield-dependent physical parameters responsible for the characterization of fluid flows. In particular, the convection FDV parameters (s_1, s_2) are identified as equivalent to the TVD limiter functions. The FDV equations are shown to contain the terms of fluctuation variables automatically generated in due course of developments, varying in time and space, but following the current physical phenomena. In addition, adequate numerical controls (artificial viscosity) to address both nonfluctuating and fluctuating parts of variables are automatically activated according to the current flowfield. Just as important are the Jacobians providing interactions of any one variable with all other variables in the conservation form of the governing equations. It has been shown that practically all existing numerical schemes in FDM and FEM are the special cases of the FDV theory.

Some simple example problems have demonstrated most of the features available in the FDV theory. It was shown that the calculated FDV parameters resemble the flowfield itself. The program originally designed for the solution of the supersonic flows is used to resolve incompressible flows of driven cavity problems, with the transition from incompressibility to compressibility automatically realized.

There are other methods related to FEM which are not introduced in this chapter. They include spectral element methods, least square methods, and finite point methods. These are the subjects of the next chapter.

REFERENCES

Aliabadi, S. K. and Tezduyar, T. E. [1993]. Space-time finite element computation of compressible flows involving moving boundaries. *Comp. Meth. Appl. Mech. Eng.*, 107, 209–23.

Atkins, H. L. and Shu, C. W. [1998]. Quadrature-free implementation of discontinuous Galerkin method for hyperbolic equations. *AIAA J.*, 36, 5, 775–82.

Baumann, C. E. and Oden, J. T. [1999]. A discontinuous *hp* finite element methods for the Euler and Navier-Stokes equations. *Int. J. Num. Meth. Fl.*, 31, 79–95.

Boris, J. P. and Book, D. L. [1976]. Solution of the continuity equation by the method of flux corrected transport. *J. Comp. Phys.*, 16, 85–129.

Choi, D. and Merkle, C. L. [1993]. The application of preconditioning for viscous flows. *J. Comp. Phys.*, 105, 203–23.

Chung, T. J. [1999]. Transitions and interactions of inviscid/viscous, compressible/incompressible and laminar/turbulent flows. *Int. J. Num. Meth. Fl.*, 31, 223–46.

Cockburn, S., Hou, S., and Shu, C. W. [1990]. TVD Runge-Kutta local projection discontinuities Galerkin finite element for conservation laws, IV. The multidimensional case. *Math. Comp.* 54–65.

——— [1997]. The Runge-Kutta discontinuous Galerkin method for conservation laws. V. Multidimensional systems. ICASE Report 97–43.

Codina, R., Vazquez, M., and Zienkiewicz, O. C. [1998]. A general algorithm for compressible and incompressible flows. Part III: The semi-implicit form. *Int. J. Num. Meth. Fl.*, 27, 13–32.

Ghia, U., Ghia, K. N., and Shin, C. T. [1982]. High-Reynolds number solutions for incompressible flow using the Navier-Stokes equations and Multigrid method. *J. Comp. Phys.*, 48, 387–411.

Godunov, S. K. [1959]. A difference scheme for numerical computation of discontinuous solution of hydrodynamic equations. *Math. Sbornik*, 47, 271–306.

Harten, A. [1983]. On the symmetric form of systems of conservation laws with entropy. *J. Comp. Phys.*, 49, 151–64.

——— [1984]. On a class of high resolution total variation stable finite difference schemes. *SIAM J. Num. Anal.*, 21, 1–23.

Hassan, O., Morgan, K., and Peraire, J. [1991]. An implicit explicit element method for high-speed flows. *Int. J. Num. Meth. Eng.*, 32(1): 183.

Hauke, G. and Hughes. T. J. R. [1998]. A comparative study of different sets of variables for solving compressible and incompressible flows. *Comp. Meth. Appl. Mech. Eng.*, 153, 1–44.

Hughes, T., Franca, L., and Mallet, M. [1986]. A new finite element formulation for computational fluid dynamics: I. Symmetric forms of the compressible Euler and Navier-Stokes equations and the second law of thermodynamics. *Comp. Meth. Appl. Mech. Eng.*, 54, 223–34.

Johnson, C. and Pitkäranta, J. [1986]. An analysis of the discontinuous Galerkin method for a scalar hyperbolic equation. *Math. Comp.*, 46, 173, 1–26.

LaSaint, P. and Raviart, P. A. [1974]. On a finite element method for solving the neutron transport equations. In C. deBoor (ed.) *Mathematical Aspects of Finite Elements in Partial Differential Equations*. New York: Academic Press.

Oden, J. T., Babuska, I., and Baumann, C. [1998]. A discontinuous *hp* finite element method for diffusion problems. *J. Comp. Phys.*, 146, 491–519.

Osher, S. [1984]. Rieman solvers, the entropy condition and difference approximations. *SIAM J. Num Anal.*, 21, 217–35.

Richardson, G. A., Cassibly, J. T., Chung, T. J., and Wu, S. T. [2010]. Finite element form of FDV for widely varying flowfilds. *J. of Com. Physics*, 229, 149–167.

Roe, P. L. [1984]. Generalized formulation of TVD Lax–Wendroff schemes. ICASE Report 84–53. NASA CR-172478, NASA Langley Research Center.

Schunk, R. G., Canabal, F., Heard, G. A., and Chung, T. J. [1999]. Unified CFD methods via flowfield-dependent variation theory, AIAA paper, 99–3715.

Shakib, F., Hughes, T., and Johan, Z . [1991]. A new finite element formulation for computational fluid dynamics: X. The compressible Euler and Navier-Stokes equations. *Comp. Meth. Appl. Mech. Eng.*, 89, (1–3): 141–220.

Tadmor, E. [1984]. The large time behavior of the scalar, genuinely nonlinear Lax-Friedrichs scheme. *Math. Comp.*, 43, 353–68.

Van Leer, B. [1979]. Towards the ultimate conservative difference scheme. V. A second order sequel to Godunov's method. *J. Comp. Phys.*, 32, 101–36.

Yoon, K. T. and Chung, T. J. [1996]. Three-dimensional mixed explicit-implicit generalized Galerkin spectral element methods for high-speed turbulent compressible flows. *Comp. Meth. Appl. Mech. Eng.*, 135, 343–67.

Yoon, K. T., Moon, S. Y., Garcia, S. A., Heard, G. W., and Chung, T. J. [1998]. Flowfield-dependent mixed-implicit methods for high and low speed and compressible and incompressible flows. *Comp. Meth. Appl. Mech. Eng.*, 151, 75–104.

Zalesak, S. T. [1979]. Fully multidimensional flux corrected transport algorithm for fluids. *J. Comp. Phys.*, 31, 335–62.

Zienkiewicz, O. C. and Codina, R. [1995]. A general algorithm for compressible and incompressible flows – Part I. The split characteristic based scheme. *Int. J. Num. Meth. Fl.*, 20, 869–85.

Zienkiewicz, O. C., Satya Sai, B. V. K., Morgan, K., and Codina, R. [1998]. Split, characteristic based demi-implicit algorithm for laminar/turbulent incompressible flows. *Int. J. Num. Meth. Fl.*, 23, 787–809.

CHAPTER FOURTEEN

Miscellaneous Weighted Residual Methods

In the previous chapters, with an exception of GPG, the finite element formulations are based on the Galerkin methods in which test functions are chosen to be the same as the trial functions. This is not required in the weighted residual methods.

Weighted residual methods other than the Galerkin methods include spectral element methods (SEM), least square methods (LSM), moment methods, or collocation methods, in which the test functions or weighting functions are not necessarily the same as the trial functions. In spectral element methods (SEM), polynomials in terms of nodal values of the variables are combined with special functions such as Chebyshev or Legendre polynomials. For least square methods, the test functions are constructed by the derivative of the residual with respect to the nodal values of the variables. Some arbitrary functions are chosen as test functions for the moment and collocation methods. Recently, the weighted residual concept has been used in meshless configurations, known as the finite point method (FPM), partition of unity method, meshless cloud method, or element-free method.

In the following sections, we shall describe a certain type of spectral element methods, least square methods, optimal control methods (OCM), and finite point methods (FPM). They are selected here for discussion because of their possible future potential for further developments.

14.1 SPECTRAL ELEMENT METHODS

The term "spectral" as used here implies a special function. Examples of such functions may be Chebyshev, Legendre, or Laguerre polynomials. These functions are expected to portray physical phenomena more realistically and precisely than other functions that have been discussed previously, leading to a greater solution accuracy. However, their applications are limited to simple geometries and simple boundary conditions.

The spectral element methods (SEM) represent a recent development as a combination of the classical spectral methods and finite element methods, thus the term "spectral element." The classical spectral methods resemble the classical method of weighted residuals.

In the classical spectral methods, trial and test functions are chosen such that they satisfy global boundary conditions. In the spectral element method, the trial and test functions are local and combined with isoparametric finite element functions as first proposed by Patera [1984]. Applications of the spectral element methods to triangular finite elements were reported by Sherwin and Karniadakis [1995]. The basic idea, however, was employed earlier in the so-called *p*-version finite elements [Babuska, 1958]. Later extensions can be seen in the *h-p* methods [Oden et al., 1989] and the flowfield-dependent variation spectral element methods (FDV-SEM) [Yoon and Chung, 1996]. The classical spectral methods are well documented in the book by Canuto et al. [1987]. Here, in this section, we utilize the concept of the classical spectral methods and apply it to the finite element method in such a way that the accuracy and efficiency are realized with a reasonable compromise. The most important aspect of SEM as applied to the FDV scheme is to portray turbulent behavior in direct numerical simulation (DNS) calculations. This will allow direct numerical simulation to be more efficient in which turbulence models are no longer required, as indicated in Section 13.6.

In SEM formulations, we may use either Chebyshev polynomials or Legendre polynomials. Patera [1984] demonstrated the SEM formulation using Chebyshev polynomials. We illustrate the use of Legendre polynomials [Szabo and Babuska, 1991] as test functions in the following subsection.

14.1.1 SPECTRAL FUNCTIONS

In the traditional spectral methods, we use spectral functions that are normally provided by Chebyshev polynomials or Legendre polynomials. Either one of these polynomials can be used in the spectral element methods. Before we proceed to SEM, we briefly summarize the basic properties involved in the Chebyshev polynomials and Legendre polynomials.

Chebyshev Polynomials

The basic concept of the least squares approximations is used to derive the Chebyshev polynomials in which orthogonality properties are preserved. To this end, consider a polynomial $\Phi_r(x)$ of degree r in x such that

$$\int_{-1}^{1} W(x)\Phi_r(x)q_{r-1}(x)dx = 0 \tag{14.1.1}$$

where $W(x)$ is the weighting function

$$W(x) = \frac{1}{\sqrt{1-x^2}} \tag{14.1.2}$$

and $q_{r-1}(x)$ is an arbitrary polynomial of degree $r-1$ or less in x.

Let us now introduce the change in variables

$$x = \cos\theta \tag{14.1.3}$$

Substituting (14.1.3) into (14.1.2) and (14.1.1) yields

$$\int_{0}^{\pi} \Phi_r(\cos\theta)q_{r-1}(\cos\theta)d\theta = 0 \tag{14.1.4}$$

which is satisfied by

$$\int_0^{\pi} \Phi_r(\cos\theta)\cos k\theta d\theta = 0 \quad (k = 0, 1, \ldots, r-1) \tag{14.1.5}$$

with

$$\Phi_r(\cos\theta) = C_r \cos r\theta \tag{14.1.6}$$

It follows from (14.1.3) that

$$\Phi_r(x) = C_r \cos(r\cos^{-1} x) \tag{14.1.7}$$

are the required orthogonal polynomials with $C_r = 1$. These polynomials are known as Chebyshev polynomials, which possess the orthogonality property

$$\int_{-1}^{1} (1-x^2)^{-\frac{1}{2}} T_r(x) T_s(x) dx = 0 \quad (r \neq s) \tag{14.1.8}$$

$$T_{r+1}(x) = 2x T_r(x) - T_{r-1}(x) \tag{14.1.9}$$

$$T_o(x) = 1, \qquad T_1(x) = x \tag{14.1.10}$$

The orthogonal square factor γ_r is given by

$$\gamma_r = \int_{-1}^{1} (1-x^2)^{-\frac{1}{2}} T_r^2(x) dx = 0 \tag{14.1.11}$$

Since $x = \cos\theta$, $T_r(x) = \cos r\theta$, we have

$$\gamma_r = \int_0^{\pi} \cos^2 r\theta d\theta = \begin{cases} \pi, & r = 0 \\ \frac{\pi}{2}, & r \neq 0 \end{cases} \tag{14.1.12}$$

Thus, the nth degree least squares polynomial approximation to $f(x)$ in $(-1, 1)$, relevant to the weighting function $W(x) = (1-x^2)^{-\frac{1}{2}}$, is defined as

$$y(x) = \sum_{r=0}^{n} a_r T_r(x) \quad (-1 \leq x \leq 1) \tag{14.1.13}$$

The least squares approximations require that

$$\int_{-1}^{1} W(x)[f(x) - y(x)]^2 dx = \min \tag{14.1.14}$$

or

$$\frac{\partial}{\partial a_r} \int_{-1}^{1} W(x) \left[f(x) - \sum_{r=0}^{n} a_r T_r(x) \right]^2 dx = 0 \tag{14.1.15}$$

$$a_r \left[\int_{-1}^{1} W(x) T_r^2 dx \right] - \int_{-1}^{1} W(x) f(x) T_r(x) dx = 0 \tag{14.1.16}$$

with

$$a_k = \frac{\int_{-1}^{1} W(x) f(x) T_r(x) dx}{\int_{-1}^{1} W(x) T_r^2 dx}$$

$$a_0 = \frac{1}{\pi} \int_{-1}^{1} (1 - x^2)^{-\frac{1}{2}} f(x) dx$$

$$a_r = \frac{2}{\pi} \int_{-1}^{1} (1 - x^2)^{-\frac{1}{2}} f(x) Tr(x) dx \tag{14.1.17a}$$

or in general

$$a_r = \frac{2}{NC_r} \sum_{j=0}^{N} \frac{1}{c_j} f(x_j) T_r(x_j) \qquad \begin{cases} x_j = \cos \frac{j\pi}{N} & j = 0, \quad 1 \ldots \\ C_0 = C_N = 2, & C_r = 1 \end{cases}$$

which has all polynomials of degree n or less, the integrated weighted square error

$$\int_{-1}^{1} (1 - x^2)^{-\frac{1}{2}} [f(x) - y_n(x)]^2 dx \tag{14.1.17b}$$

is the least when $y_n(x)$ is identified with the right-hand side of (14.1.13).

In terms of the nondimensional variable $\xi = x/\Delta x$, the Chebyshev polynomials are summarized as follows:

$$T_n(\xi) = \cos n\theta, \quad \theta = \cos^{-1} \xi \quad -1 \leq \xi \leq 1$$

$$T_0(\xi) = \cos 0 = 1$$

$$T_1(\xi) = \cos(\cos^{-1} \xi) = \xi$$

$$\xi T_n(\xi) = \cos n\theta \cos \theta = \frac{1}{2}[\cos(n-1)\theta + \cos(n+1)\theta]$$

or

$$\xi T_n(\xi) = \frac{1}{2}[T_{n-1}(\xi) + T_{n+1}(\xi)]$$

thus, the general formula is given by

$$T_{n+1}(\xi) = 2\xi T_n(\xi) - T_{n-1}(\xi) \tag{14.1.18}$$

$$T_0(\xi) = 1$$

$$T_1(\xi) = \xi$$

$$T_2(\xi) = 2\xi^2 - 1$$

$$T_3(\xi) = 4\xi^3 - 3\xi$$

$$T_4(\xi) = 8\xi^4 - 8\xi^2 + 1$$

$$T_5(\xi) = 16\xi^5 - 20\xi^3 - 5\xi$$

$$\vdots$$

Similar developments are applied to other directions for 2-D and 3-D geometries, which will then be utilized through tensor products for applications to multidimensional problems. Applications of the Chebyshev polynomials to a spectral element method will be shown in Section 22.6.4.

Legendre Polynomials

The Legendre polynomials are based on the orthogonal properties of the least square concept. To this end, we require a polynomial $\Phi_r(x)$ of degree r in x such that

$$\int_a^b W(x)\Phi_r(x)q_{r-1}(x)dx = 0 \tag{14.1.19}$$

where $W(x) = 1$ is used for the Legendre polynomial. Consider the notation

$$W(x)\Phi_r(x) = \frac{d^r u_r(x)}{dx^r} \tag{14.1.20}$$

Thus, it follows from (14.1.19) and (14.1.20) that

$$\int_a^b u_r^{(r)}(x)q_{r-1}(x)dx = 0 \tag{14.1.21}$$

Integrating by parts

$$[u_r^{(r-1)}q_{r-1} - u_r^{(r-2)}q'_{r-1} + \cdots + (-1)^{r-1}u_r q_{r-1}^{(r-1)}]_a^b = 0 \tag{14.1.22}$$

The requirement for the function Φ_r (x) defined by (14.1.20)

$$\Phi_r(x) = \frac{1}{W(x)}\frac{d^r u_r(x)}{dx^r} \tag{14.1.23}$$

be a polynomial of degree r implies that $u_r(x)$ must satisfy the differential equation

$$\frac{d^{r+1}}{dx^{r+1}}\left[\frac{1}{W(x)}\frac{d^r u_r(x)}{dx^r}\right] = 0 \tag{14.1.24}$$

in [a, b] with the $2r$ boundary conditions

$$\begin{aligned} u_r(a) = u'_r(a) = u''_r(a) = \cdots = u_r^{(r-1)}(a) = 0 \\ u_r(b) = u'_r(b) = u''_r(b) = \cdots = u_r^{(r-1)}(b) = 0 \end{aligned} \tag{14.1.25}$$

For the least squares approximation over an interval of finite length, it is convenient to suppose that a linear change in variables has transformed that interval into the interval $[-1, 1]$. With $W(x) = 1$, we obtain

$$\frac{d^{2r+1}u_r}{dx^{2r+1}} = 0 \tag{14.1.26}$$

Using the boundary conditions (14.1.15) for $(-1, 1)$

$$u_r = \lambda_r(x^2 - 1)^r \tag{14.1.27}$$

where λ_r is an arbitrary constant. Hence, from (14.1.23) it follows that the rth relevant orthogonal polynomial is of the form

$$\Phi_r(x) = \lambda_r \frac{d^r}{dx^r}(x^2 - 1)^r \tag{14.1.28}$$

with

$$\lambda_r = \frac{1}{2^r p!} \tag{14.1.29}$$

The polynomial obtained in this manner is the rth Legendre polynomial

$$L_r(x) = \frac{1}{2^r r!}\frac{d^r}{dx^r}(x^2-1)^r \tag{14.1.30}$$

From the orthogonal property it follows that

$$\int_{-1}^{1} L_r(x) L_s(x) dx = 0 \quad r \neq s \tag{14.1.31}$$

The value assigned to λ_r is such that $L_r(x) = 1$ and it is true that $|L_r(x)| \leq 1$ when $|x| \leq 1$. With the nondimensional variable, this gives

$$L_0(\xi) = 1$$

$$L_1(\xi) = \xi$$

$$L_2(\xi) = \frac{1}{2}(3\xi^2 - 1)$$

$$L_3(\xi) = \frac{1}{2}(5\xi^3 - 3\xi)$$

$$L_4(\xi) = \frac{1}{8}(35\xi^4 - 30\xi^2 + 3)$$

$$L_5(\xi) = \frac{1}{8}(63\xi^5 - 70\xi^3 - 15\xi)$$

$$L_6(\xi) = \frac{1}{16}(231\xi^6 - 315\xi^4 + 105\xi^2 - 5)$$

$$L_7(\xi) = \frac{1}{16}(429\xi^7 - 693\xi^5 + 315\xi^3 - 35\xi)$$

$$\vdots$$

The recurrence formula is given by

$$L_r(\xi) = \frac{1}{2^r r!}\frac{d^r}{d\xi^r}(\xi^2-1)^r$$

$$L_{r+1}(\xi) = \frac{2r+1}{r+1}\xi L_r(\xi) - \frac{r}{r+1}L_{r-1}(\xi) \tag{14.1.32}$$

Applications of the Legendre polynomials to a spectral element method will be shown in the next section.

14.1.2 SPECTRAL ELEMENT FORMULATIONS BY LEGENDRE POLYNOMIALS

The most efficient approach toward multidimensional applications of the spectral element methods is to utilize the isoparametric elements (quadrilaterals for 2-D and hexahedrals for 3-D). Using a linear element with only corner nodes, but accepting as high a spectral degree of freedom as desired for the side and interior modes for 2-D

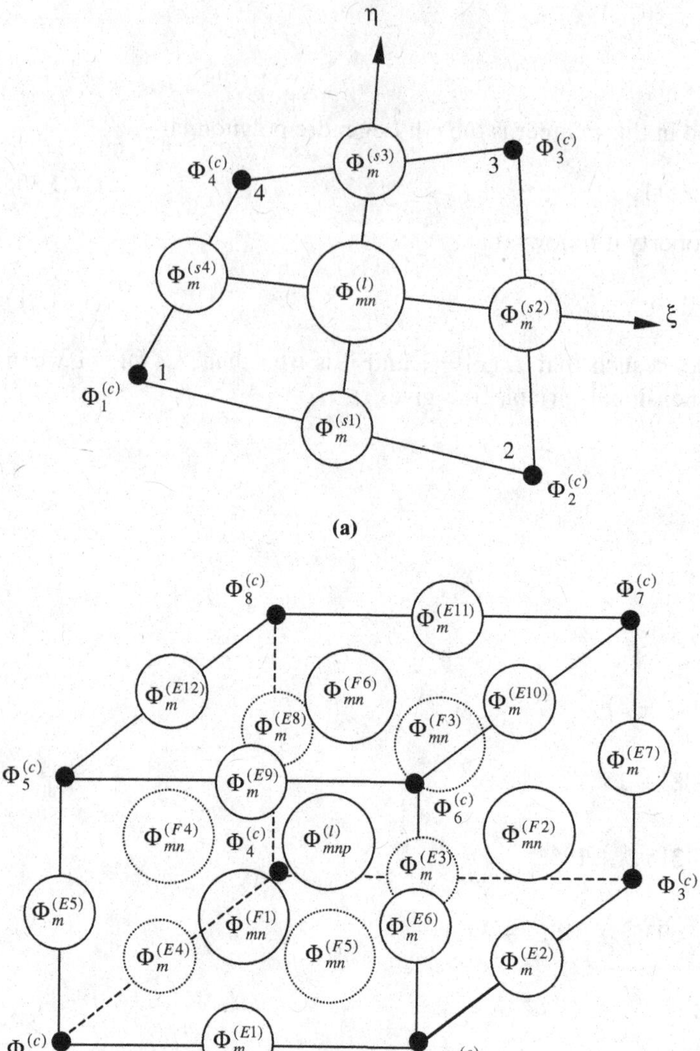

Figure 14.1.1 Spectral element configuration. **(a)** Two-dimensional spectral element. **(b)** Three-dimensional spectral element.

and the edge, face, and interior modes for 3-D, it is possible to construct Galerkin spectral element integrals. By means of static condensation, all spectral mode degrees of freedom are eliminated, resulting in the final algebraic equations only in terms of the corner nodes.

The spectral functions of the Legendre polynomials (14.1.32) are zero at the corner nodes but exhibit high order approximations elsewhere. In this process, any variable is to change nonlinearly everywhere but it is calculated only at the element corner nodes. To illustrate this point, let us consider the approximation of the variable **U**:

For two dimensions (Figure 14.1.1a) we have

Corner nodes: linear isoparametric function, $\Phi_N^{(c)}$

Side modes: Legendre spectral mode functions, $\Phi_m^{(s)}$
Interior modes: Legendre spectral mode functions, $\Phi_{mn}^{(I)}$

$$\mathbf{U} = \Phi_\alpha \mathbf{U}_\alpha + \Phi_m^{(s)} \hat{\mathbf{U}}_m + \Phi_{mn}^{(I)} \hat{\mathbf{U}}_{mn} \tag{14.1.33}$$

For three dimensions (Figure 14.1.1b) we have

Corner nodes: linear isoparametric function, $\Phi_N^{(c)}$
Edge modes: Legendre spectral mode functions, $\Phi_m^{(E)}$
Face modes: Legendre spectral mode functions, $\Phi_{mn}^{(F)}$
Interior modes: Legendre spectral mode functions, $\Phi_{mnp}^{(I)}$

$$\mathbf{U} = \Phi_\alpha \mathbf{U}_\alpha + \Phi_m^{(E)} \hat{\mathbf{U}}_m + \Phi_{mn}^{(F)} \hat{\mathbf{U}}_{mn} + \Phi_{mnp}^{(I)} \hat{\mathbf{U}}_{mnp} \tag{14.1.34}$$

where $\mathbf{U}_\alpha$ are the variables to be calculated at the corner nodes and $\hat{\mathbf{U}}_m$, $\hat{\mathbf{U}}_{mn}$, and $\hat{\mathbf{U}}_{mnp}$ denote spectral degrees of freedom.

The global trial functions Φ_α are assembled from the corner node linear isoparametric functions $\Phi_N^{(C)}$. The Legendre functions for the side modes $\Phi_m^{(s)}$ and the interior modes $\Phi_{mn}^{(I)}$ for two dimensions, and edge modes $\Phi_m^{(E)}$, face modes $\Phi_{mn}^{(F)}$, and interior modes $\Phi_{mnp}^{(I)}$ for three dimensions are given as follows:

For Two Dimensions

Side modes:

$$\begin{aligned}
\Phi_m^{(S1)} &= \frac{1}{2}(1-\eta)G_m(\xi)\\
\Phi_m^{(S2)} &= \frac{1}{2}(1+\xi)G_m(\eta)\\
\Phi_m^{(S3)} &= \frac{1}{2}(1+\eta)G_m(\xi)\\
\Phi_m^{(S4)} &= \frac{1}{2}(1-\xi)G_m(\eta)
\end{aligned} \tag{14.1.35}$$

with $m = 2, \ldots q$; $N^{(S)} = 4(q-1)$; $q \geq 2$

Interior modes:

$$\Phi_{mn}^{(I)} = G_m(\xi)G_n(\eta) \tag{14.1.36}$$

with $m, n = 2, \ldots, q-2$; $(m+n) = 2, \ldots, q$; $N^{(I)} = \frac{1}{2}[(q-2)(q-3)]$, $q \geq 4$

where $N^{(S)}$ and $N^{(I)}$ denote, respectively, the total number of functional modes available for sides (1, 2, 3, 4) and interior. The highest polynomial order chosen is denoted by q, and G_m refers to the Legendre polynomials defined as

$$G_m(\xi) = \frac{1}{\sqrt{2(2m-1)}}[L_m(\xi) - L_{m-2}(\xi)] \tag{14.1.37}$$

with the recursive formula given by

$$L_{m+1}(\xi) = \frac{2m+1}{m+1}\xi L_m(\xi) - \frac{m}{m+1}L_{m-1}(\xi) \tag{14.1.38}$$

Similar results are obtained for the η-direction. For illustration, variable orders of Legendre polynomials specified in different elements are shown in Figure 14.1.2. At

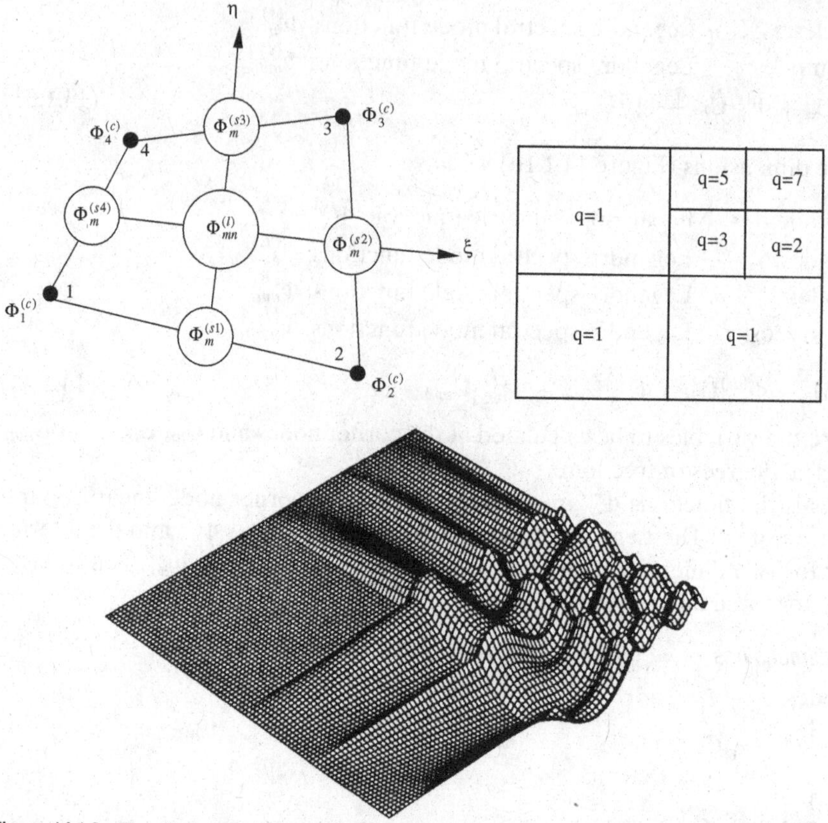

Figure 14.1.2 Two-D interpolation functions constructed by Legendre polynominal, $\Phi_N^{(C)}$ (corner nodes), $\Phi_M^{(S)}$ (side nodes), $\Phi_{mn}^{(1)}$ (interior nodes).

boundaries, higher order functions prevail over the lower order functions. In addition to the above polynomial space, (called $S1$) we may use another option of the space (called $S2$) in which $(q-1)^2$ interior modes are applied.

For Three Dimensions

$$\begin{aligned}
\text{Edge mode:} \quad \Phi_m^{(E1)} &= \tfrac{1}{4}(1-\eta)(1-\zeta)G_m(\xi) \\
\Phi_m^{(E2)} &= \tfrac{1}{4}(1+\xi)(1-\zeta)G_m(\eta) \\
&\text{etc.}
\end{aligned} \tag{14.1.39}$$

with $m = 2, \ldots, q; \quad N^{(E)} = 12(q-1); \quad q \geq 2$

$$\begin{aligned}
\text{Face mode:} \quad \Phi_{mn}^{(F1)} &= \tfrac{1}{2}(1-\eta)G_m(\xi)G_n(\eta) \\
\Phi_{mn}^{(F2)} &= \tfrac{1}{2}(1+\xi)G_m(\eta)G_n(\zeta) \\
&\text{etc.}
\end{aligned} \tag{14.1.40}$$

with $m, n = 2, \ldots, q-2; \quad (m+n) = 4, \ldots, q;$
$N^{(F)} = 3(q-2)(q-3); \quad q \geq 4$

$$\text{Interior mode:} \quad \Phi_{mnp}^{(I)} = G_m(\xi)G_n(\eta)G_p(\zeta) \tag{14.1.41}$$

$$\text{with } m, n, p = 2, \ldots, q-4; \quad (m+n+p) = 6, \ldots, q;$$
$$N^{(I)} = (q-3)(q-4)(q-5)/6; \quad q \geq 6$$

In addition to the above polynomials ($S1$), we may use an optional space ($S2$) in which $(q-1)^2$ face modes and $(q-1)^3$ interior modes ($q \geq 2$) are applied.

14.1.3 TWO-DIMENSIONAL PROBLEMS

Spectral element methods may be implemented through the generalized Galerkin scheme. A more rigorous approach such as the FDV-FEM technique introduced in Chapter 13 can be combined with the spectral functions. This is particularly useful for dealing with high-speed flows where shock wave/turbulent boundary layer interactions occur.

In general, the spectral element formulation begins with the Galerkin integral expressed in the following form:

For Corner Nodes

$$\int_\Omega \Phi_\alpha \mathbf{R}(\Delta\mathbf{U})d\Omega = 0 \tag{14.1.42a}$$

For Side Modes

$$\int_\Omega \Phi_m^{(S)} \mathbf{R}(\Delta\mathbf{U})d\Omega = 0 \tag{14.1.42b}$$

For Interior Modes

$$\int_\Omega \Phi_{mn}^{(I)} \mathbf{R}(\Delta\mathbf{U})d\Omega = 0 \tag{14.1.42c}$$

where the conservation variables $\mathbf{U}$ in the residual $\mathbf{R}(\Delta\mathbf{U})$ of the Navier-Stokes system of equations are approximated by the trial functions, and the source terms are assumed to be zero.

Substituting (14.1.33) into (14.1.42) yields the matrix equations,

$$\begin{bmatrix} A_{\alpha\beta}\delta_{rs} + B_{\alpha\beta rs} & A_{\alpha n}^{\delta}\delta_{rs} + B_{\alpha nrs}^{\delta} & A_{\alpha np}\delta_{rs} + B_{\alpha nprs} \\ A_{m\beta}^{\gamma}\delta_{rs} + B_{m\beta rs}^{\gamma} & A_{mn}^{\gamma\delta}\delta_{rs} + C_{mnrs}^{\gamma\delta} & A_{mnp}^{\gamma}\delta_{rs} + C_{mnprs}^{\gamma} \\ A_{mk\beta}\delta_{rs} + B_{mk\beta rs} & A_{mkn}^{\delta}\delta_{rs} + C_{mknrs}^{\delta} & A_{mknp}\delta_{rs} + D_{mknprs} \end{bmatrix} \begin{bmatrix} \Delta U_{\beta s} \\ \Delta \hat{U}_{ns}^{\delta} \\ \Delta \hat{U}_{nps} \end{bmatrix}^{n+1}$$
$$= \begin{bmatrix} W_{\alpha r} \\ \hat{W}_{mr}^{\gamma} \\ \hat{W}_{mkr} \end{bmatrix}^{n} \tag{14.1.43}$$

where α, β denote the product of the global corner node number times the total number of physical variables, whereas m, n, p, and q refer to degrees of freedom from the side and internal modes of Legendre polynomials with $\gamma, \delta = 1, 4$ and r, s denoting the number of conservation variables (4 in two dimensions and 5 in three dimensions).

If the residual $\mathbf{R}(\Delta\mathbf{U})$ is chosen to be the same as (13.1.2) for the FDV-FEM scheme without source terms, we obtain the matrix entries of (14.1.43) as follows:

$$A_{\alpha\beta} = \int_\Omega \Phi_\alpha \Phi_\beta \, d\Omega \qquad A^{\delta}_{\alpha n} = \int_\Omega \Phi_\alpha \hat{\Phi}^{\delta}_{n} \, d\Omega \qquad A_{\alpha np} = \int_\Omega \Phi_\alpha \hat{\Phi}_{np} \, d\Omega$$

$$A^{\gamma}_{m\beta} = \int_\Omega \hat{\Phi}^{\gamma}_{m} \Phi_\beta \, d\Omega \qquad A^{\gamma\delta}_{mn} = \int_\Omega \hat{\Phi}^{\gamma}_{m} \hat{\Phi}^{\delta}_{n} \, d\Omega \qquad A^{\gamma}_{mnp} = \int_\Omega \hat{\Phi}^{\gamma}_{m} \hat{\Phi}_{np} \, d\Omega$$

$$A_{mk\beta} = \int_\Omega \hat{\Phi}_{mk} \Phi_\beta \, d\Omega \quad A^{\delta}_{mkn} = \int_\Omega \hat{\Phi}_{mk} \hat{\Phi}^{\delta}_{n} \, d\Omega \quad A_{mknp} = \int_\Omega \hat{\Phi}_{mk} \hat{\Phi}_{np} \, d\Omega \tag{14.1.44}$$

$$\begin{aligned} B_{\alpha\beta rs} &= \int_\Omega \Big\{ \Delta t \left[-s_1 a_{irs} \Phi_{\alpha,i} \Phi_\beta - s_3 (b_{irs} \Phi_{\alpha,i} \Phi_\beta + c_{ijrs} \Phi_{\alpha,i} \Phi_{\beta,j}) \right] \\ &\quad + \frac{\Delta t^2}{2} ([s_2 (a_{irt} a_{jts} + b_{irt} a_{jts}) \Phi_{\alpha,i} \Phi_{\beta,j} - d_{rt} a_{its} \Phi_{\alpha,i} \Phi_\beta] \\ &\quad + s_4 [(a_{irt} b_{jts} + b_{irt} b_{jts} - d_{rt} c_{ijts}) \Phi_{\alpha,i} \Phi_{\beta,j} - d_{rt} b_{its} \Phi_{\alpha,i} \Phi_\beta]) \Big\} d\Omega \end{aligned}$$

$$\begin{aligned} B^{\delta}_{\alpha nrs} &= \int_\Omega \Big\{ -\Delta t [(s_1 a_{irs} + s_3 b_{irs}) \Phi_{\alpha,i} \hat{\Phi}^{\delta}_{n} + s_3 c_{ijrs} \Phi_{\alpha,i} \hat{\Phi}_{n,j}] \\ &\quad + \frac{\Delta t^2}{2} (s_2 d_{ijrs} + s_4 e_{ijrs}) \Phi_{\alpha,i} \hat{\Phi}^{\delta}_{n,j} \Big\} d\Omega \end{aligned}$$

with

$$\begin{aligned} d_{ijrs} &= a_{irt} a_{jts} + b_{irt} a_{jts} \\ e_{ijrs} &= a_{irt} b_{jts} + b_{irt} b_{jts} \\ B_{\alpha nprs} &= \int_\Omega \Big\{ -\Delta t \left[(s_1 a_{irs} + s_3 b_{irs}) \Phi_{\alpha,i} \hat{\Phi}_{np} + s_3 c_{ijrs} \Phi_{\alpha,i} \hat{\Phi}_{np,j} \right] \\ &\quad + \frac{\Delta t^2}{2} (s_2 d_{ijrs} + s_4 e_{ijrs}) \Phi_{\alpha,i} \hat{\Phi}_{np,j} \Big\} d\Omega \\ B^{\gamma}_{m\beta rs} &= \int_\Omega \Big\{ -\Delta t \left[(s_1 a_{irs} + s_3 b_{irs}) \hat{\Phi}^{\gamma}_{m,i} \Phi_\beta + s_3 c_{ijrs} \hat{\Phi}^{\gamma}_{m,i} \Phi_{\beta,j} \right] \\ &\quad + \frac{\Delta t^2}{2} (s_2 d_{ijrs} + s_4 e_{ijrs}) \hat{\Phi}^{\gamma}_{m,i} \Phi_{\beta,j} \Big\} d\Omega \\ C^{\gamma\delta}_{mnrs} &= \int_\Omega \Big\{ -\Delta t \left[(s_1 a_{irs} + s_3 b_{irs}) \hat{\Phi}^{\gamma}_{m,i} \hat{\Phi}^{\delta}_{n} + s_3 c_{ijrs} \hat{\Phi}^{\gamma}_{m,i} \hat{\Phi}^{\delta}_{n,j} \right] \\ &\quad + \frac{\Delta t^2}{2} (s_2 d_{ijrs} + s_4 e_{ijrs}) \hat{\Phi}^{\gamma}_{m,i} \hat{\Phi}^{\delta}_{n,j} \Big\} d\Omega \\ C^{\gamma}_{mnprs} &= \int_\Omega \Big\{ -\Delta t \left[(s_1 a_{irs} + s_3 b_{irs}) \hat{\Phi}^{\gamma}_{m,i} \hat{\Phi}_{np} + s_3 c_{ijrs} \hat{\Phi}^{\gamma}_{m,i} \hat{\Phi}_{np,j} \right] \\ &\quad + \frac{\Delta t^2}{2} (s_2 d_{ijrs} + s_4 e_{ijrs}) \hat{\Phi}^{\gamma}_{m,i} \hat{\Phi}_{np,j} \Big\} d\Omega \end{aligned}$$

$$
\begin{aligned}
B_{mk\beta rs} &= \int_\Omega \Big\{ -\Delta t \left[(s_1 a_{irs} + s_3 b_{irs}) \hat{\Phi}_{mk,i} \Phi_\beta + s_3 c_{ijrs} \hat{\Phi}_{mk,i} \Phi_{\beta,j} \right] \\
&\quad + \frac{\Delta t^2}{2} (s_2 d_{ijrs} + s_4 e_{ijrs}) \hat{\Phi}_{mk,i} \Phi_{\beta,j} \Big\} d\Omega \\
C^\delta_{mknrs} &= \int_\Omega \Big\{ -\Delta t \left[(s_1 a_{irs} + s_3 b_{irs}) \hat{\Phi}_{mk,i} \hat{\Phi}^\delta_n + s_3 c_{ijrs} \hat{\Phi}_{mk,i} \hat{\Phi}^\delta_{n,j} \right] \\
&\quad + \frac{\Delta t^2}{2} (s_2 d_{ijrs} + s_4 e_{ijrs}) \hat{\Phi}_{mk,i} \hat{\Phi}^\delta_{n,j} \Big\} d\Omega \\
D_{mknprs} &= \int_\Omega \Big\{ -\Delta t \left[(s_1 a_{irs} + s_3 b_{irs}) \hat{\Phi}_{mk,i} \hat{\Phi}_{np} + s_3 c_{ijrs} \hat{\Phi}_{mk,i} \hat{\Phi}_{np,j} \right] \\
&\quad + \frac{\Delta t^2}{2} (s_2 d_{ijrs} + s_4 e_{ijrs}) \hat{\Phi}_{mk,i} \hat{\Phi}_{np,j} \Big\} d\Omega
\end{aligned}
\tag{14.1.45}
$$

$$
W_{\alpha r} = H^n_{\alpha r} + N^n_{\alpha r} + \overline{N}^{n+1}_{\alpha r} \tag{14.1.46}
$$

with

$$
\begin{aligned}
H^n_{\alpha r} &= \int_\Omega \Big\{ \Delta t\, \Phi_{\alpha,i} \Phi_\beta (F^n_{\beta ir} + G^n_{\beta ir}) - \frac{\Delta t^2}{2} (a_{irs} + b_{irs}) \Phi_{\alpha,i} \Phi_{\beta,j} (F^n_{\beta js} + G^n_{\beta js}) \Big\} d\Omega \\
N^n_{\alpha r} &= \int_\Gamma \overset{*}{\Phi}_\alpha \Big[-\Delta t\, (F^n_{ir} + G^n_{ir}) + \frac{\Delta t^2}{2} (a_{irs} + b_{irs}) (F^n_{js,j} + G^n_{js,j}) \Big] n_i \, d\Gamma \\
\overline{N}^{n+1}_{\alpha r} &= \int_\Gamma \overset{*}{\Phi}_\alpha \Big\{ -\Delta t \left[(s_1 a_{irs} + s_3 b_{irs}) \Delta U^{n+1}_s + s_3 c_{ijrs} \Delta U^{n+1}_{s,j} \right] \\
&\quad + \frac{\Delta t^2}{2} (s_2 d_{ijrs} + s_4 e_{ijrs}) \Delta U^{n+1}_{s,j} \Big\} n_i \, d\Gamma \\
\hat{W}^\gamma_{mr} &= \int_\Omega \Big\{ \Delta t\, \hat{\Phi}^\gamma_{m,i} \Phi_\beta (F^n_{\beta ir} + G^n_{\beta ir}) - \frac{\Delta t^2}{2} (a_{irs} + b_{irs}) \hat{\Phi}^\gamma_{m,i} \Phi_{\beta,j} (F^n_{\beta js} + G^n_{\beta js}) \Big\} d\Omega \\
&\quad + \int_\Gamma \overset{*}{\hat{\Phi}}{}^\gamma_m \Big\{ \Delta t \left[-s_1 \left(a_{irs} \Delta U^{n+1}_s \right) - s_3 \left(b_{irs} \Delta U^{n+1}_s + c_{ijrs} \Delta U^{n+1}_{s,j} \right) \right] \\
&\quad + \frac{\Delta t^2}{2} \left[s_2 \left(a_{irt} a_{jts} + b_{irt} a_{jts} \right) \Delta U^{n+1}_{s,j} + s_4 \left(a_{irt} b_{jts} + b_{irt} b_{jts} \right) \Delta U^{n+1}_{s,j} \right] \Big\} n_i \, d\Gamma \\
&\quad + \int_\Gamma \overset{*}{\hat{\Phi}}{}^\gamma_m \Big\{ -\Delta t\, (F^n_{ir} + G^n_{ir}) + \frac{\Delta t^2}{2} (a_{irs} + b_{irs}) (F^n_{js,j} + G^n_{js,j} - B^n_s) \Big\} n_i \, d\Gamma \\
\hat{W}_{mkr} &= \int_\Omega \Big\{ \Delta t\, \hat{\Phi}_{mk,i} \Phi_\beta (F^n_{\beta ir} + G^n_{\beta ir}) \\
&\quad - \frac{\Delta t^2}{2} (a_{irs} + b_{irs}) \hat{\Phi}_{mk,i} \Phi_{\beta,j} (F^n_{\beta js} + G^n_{\beta js}) \Big\} d\Omega
\end{aligned}
\tag{14.1.47}
$$

If the Neumann boundary conditions for spectral modes are not specified, then, by definition, $\hat{\Phi}^*_m = \hat{\Phi}^*_{mn} = 0$ and only the corner nodes are subjected to the Neumann boundary conditions. However, these spectral Neumann boundary conditions may be computed and added after the initial corner node computation, resulting in possible improvements for the final solution.

The orthogonal properties of the Legendre polynomials give rise to sparse local matrices. For example, the following orthogonal properties arise for diffusion terms:

$$\int_\Omega \Phi_{N,i}\hat{\Phi}^\delta_{n,i}\, d\Omega \neq 0 \qquad \text{if and only if } n = 2, \text{ or } 3, \text{ zero otherwise}$$

$$\int_\Omega \Phi_{N,i}\hat{\Phi}_{np,i}\, d\Omega \equiv 0 \qquad \text{always}$$

$$\int_\Omega \hat{\Phi}^\gamma_{m,i}\Phi^\delta_{n,i}\, d\Omega \neq 0 \qquad \text{if and only if } \gamma - \delta = \text{even and } m = n \text{ or } m = n \pm 2, \text{ zero otherwise}$$

$$\int_\Omega \hat{\Phi}^\delta_{m,i}\hat{\Phi}_{np,i}\, d\Omega \neq 0 \qquad \text{if and only if } m = p \text{ and } n = 2 \text{ or } 3, \text{ with } \gamma = 1 \text{ or } 3;\ m = n \text{ and } p = 2 \text{ or } 3, \text{ with } \gamma = 2 \text{ or } 4; \text{ zero otherwise}$$

$$\int_\Omega \hat{\Phi}_{mk,i}\,\hat{\Phi}_{np,i}\, d\Omega \neq 0 \qquad \text{if and only if } m = n \text{ or } m = n \pm 2 \text{ and } k = p;\ k = p \text{ or } k = p \pm 2, \text{ and } m = n; \text{ zero otherwise}$$

It should be noted that these results are also obtained by using the Gaussian quadrature routine for integration.

Although the direct solution of (14.1.43) can be obtained, a number of other options are available. For example, we may initially consider only the corner node equations,

$$(A_{\alpha\beta}\delta_{rs} + B_{\alpha\beta rs})\,\Delta U^{n+1}_{\beta s} = W_{\alpha r} \tag{14.1.48}$$

The solution of (14.1.48) can be subsequently applied to the side-mode and edge-mode equations of (14.1.43) to solve

$$\begin{bmatrix} A^{\gamma\delta}_{mn}\delta_{rs} + C^{\gamma\delta}_{mnrs} & A^{\gamma}_{mnp}\delta_{rs} + C^{\gamma}_{mnprs} \\ A^{\delta}_{mkn}\delta_{rs} + C^{\delta}_{mkmrs} & A_{mknp}\,\delta_{rs} + D_{mknprs} \end{bmatrix} \begin{bmatrix} \Delta\hat{U}^\delta_{ns} \\ \Delta\hat{U}_{nps} \end{bmatrix} = \begin{bmatrix} \hat{W}^\gamma_{mr} - X^\gamma_{mr} \\ \hat{W}_{mkr} - X_{mkr} \end{bmatrix} \tag{14.1.49}$$

where

$$X^\gamma_{mr} = \left(A^\gamma_{m\beta}\delta_{rs} + B^\gamma_{m\beta rs}\right)\Delta U_{\beta s}$$

$$X_{mkr} = \left(A_{mk\beta}\delta_{rs} + B_{mk\beta rs}\right)\Delta U_{\beta s}$$

This allows (14.1.48) to be revised as

$$(A_{\alpha\beta}\delta_{rs} + B_{\alpha\beta rs})\Delta U^{n+1}_{\beta s} = W_{\alpha r} - \left(A^\delta_{\alpha n}\delta_{rs} + B^\delta_{\alpha nrs}\right)\Delta\hat{U}^\delta_{ns} - (A_{\alpha np}\delta_{rs} + B_{\alpha nprs})\Delta\hat{U}_{nps} \tag{14.1.50}$$

This approach resembles the so-called static condensation performed in reverse order. Thus, the solutions between (14.1.50) and (14.1.49) may be repeated until the desired convergence is obtained.

Notice that one advantage of this formulation is that, although the corner node isoparametric finite element function remains linear, the side and interior mode spectral orders can vary from element to element (Figure 14.1.2) as high as desired in order to simulate particular physical phenomena such as turbulence. Furthermore, the corner node linear isoparametric functions allow the computation of variables only at the corner nodes, irrespective of high order spectral functions chosen for side and interior modes.

Remark: It has been demonstrated that the SEM is effective for nonlinear problems, particularly for problems with singularities such as in shock waves and with high gradients such as in turbulence. For linear partial differential equations with smooth exact solutions, the numerical analysis by SEM may produce results which are worse than those of linear FEM (corner nodes only). This is an important observation in that the imposition of the higher order functions (Legendre polynomials) upon the linear solution surface may distort the numerical solution. This distortion may be drastic in some cases. Therefore, SEM is not recommended for linear problems. To illustrate, consider the results shown in the example below of the SEM solutions of a Laplace equation in comparison with the FEM solutions.

14.1.4 THREE-DIMENSIONAL PROBLEMS

For three-dimensional problems, the Galerkin integral is expressed in the following form:

For Corner Nodes

$$\int_\Omega \Phi_\alpha \mathbf{R}(\Delta\mathbf{U})d\Omega = 0 \tag{14.1.51a}$$

For Edge Modes

$$\int_\Omega \Phi_m^{(E)} \mathbf{R}(\Delta\mathbf{U})d\Omega = 0 \tag{14.1.51b}$$

For Face Nodes

$$\int_\Omega \Phi_{mn}^{(F)} \mathbf{R}(\Delta\mathbf{U})d\Omega = 0 \tag{14.1.51c}$$

For Interior Nodes

$$\int_\Omega \Phi_{mnp}^{(I)} \mathbf{R}(\Delta\mathbf{U})d\Omega = 0 \tag{14.1.51d}$$

Substituting (14.1.16) into (14.1.1) gives

$$\begin{bmatrix} A_{\alpha\beta}\delta_{rs} + B_{\alpha\beta rs} & A^\delta_{\alpha n}\delta_{rs} + B^\delta_{\alpha nrs} & A^\eta_{\alpha np}\delta_{rs} + B^\eta_{\alpha nprs} & A_{\alpha npq}\delta_{rs} + B_{\alpha npqrs} \\ A^\gamma_{m\beta}\delta_{rs} + B^\gamma_{m\beta rs} & A^{\gamma\delta}_{mn}\delta_{rs} + C^{\gamma\delta}_{mnrs} & A^{\gamma\eta}_{mnp}\delta_{rs} + C^{\gamma\eta}_{mnprs} & A^\gamma_{mnpq}\delta_{rs} + C^\gamma_{mnpqrs} \\ A^\xi_{mk\beta}\delta_{rs} + B^\xi_{mk\beta rs} & A^{\xi\delta}_{mkn}\delta_{rs} + C^{\xi\delta}_{mknrs} & A^{\xi\eta}_{mknp}\delta_{rs} + D^{\xi\eta}_{mknprs} & A^\xi_{mknpq}\delta_{rs} + D^\xi_{mknpqrs} \\ A_{mku\beta}\delta_{rs} + B_{mku\beta rs} & A^\delta_{mkun}\delta_{rs} + C^\delta_{mkunrs} & A^\eta_{mkunp}\delta_{rs} + D^\eta_{mkunprs} & A_{mkunpq}\delta_{rs} + E_{mkunpqrs} \end{bmatrix}$$

$$\times \begin{bmatrix} \Delta U_{\beta s} \\ \Delta\hat{U}^\delta_{ns} \\ \Delta\hat{U}^\eta_{nps} \\ \Delta\hat{U}_{npqs} \end{bmatrix}^{n+1} = \begin{bmatrix} W_{\alpha r} \\ \hat{W}^\gamma_{mr} \\ \hat{W}^\xi_{mkr} \\ \hat{W}_{mkur} \end{bmatrix}^n \tag{14.1.52}$$

with $\gamma, \delta = 1 \to 12; \xi, \eta = 1 \to 8$; $m, k, n, p, q,$ = degrees of freedom from edge, face, and interior modes; α, β = corner node variables; r, s = conservation variable degrees of freedom.

Note that all matrix entries are identical to the two-dimensional case with the following exception:

$$A_{\alpha npq} = \int_\Omega \Phi_\alpha \hat{\Phi}_{npq}\, d\Omega \quad A_{mku\beta} = \int_\Omega \hat{\Phi}_{mku} \Phi_\beta\, d\Omega \quad A^\gamma_{mnpq} = \int_\Omega \hat{\Phi}^\gamma_m \hat{\Phi}_{npq}\, d\Omega$$
$$A^\delta_{mkun} = \int_\Omega \hat{\Phi}_{mku} \hat{\Phi}^\delta_n\, d\Omega \quad A^\xi_{mknpq} = \int_\Omega \hat{\Phi}^\xi_{mk} \hat{\Phi}_{npq}\, d\Omega \quad A^\eta_{mkunp} = \int_\Omega \hat{\Phi}_{mku} \hat{\Phi}^\eta_{np}\, d\Omega$$
$$A_{mkunpq} = \int_\Omega \hat{\Phi}_{mku} \hat{\Phi}_{npq}\, d\Omega \tag{14.1.53}$$

$$B_{\alpha npqrs} = \int_\Omega \left\{ -\Delta t[(s_1 a_{irs} + s_3 b_{irs})\Phi_{\alpha,i} \hat{\Phi}_{npq} + s_3 c_{ijrs} \Phi_{\alpha,i} \hat{\Phi}_{npq,j}] + \frac{\Delta t^2}{2}(s_2 d_{ijrs} + s_4 e_{ijrs})\Phi_{\alpha,i} \hat{\Phi}_{npq,j} \right\} d\Omega$$

$$B_{mku\beta rs} = \int_\Omega \left\{ -\Delta t\,[(s_1 a_{irs} + s_3 b_{irs})\hat{\Phi}_{mku,i} \Phi_\beta + s_3 c_{ijrs} \hat{\Phi}_{mku,i} \Phi_{\beta,j}] + \frac{\Delta t^2}{2}(s_2 d_{ijrs} + s_4 e_{ijrs})\hat{\Phi}_{mku,i} \Phi_{\beta,j} \right\} d\Omega$$

$$C^\delta_{mkunrs} = \int_\Omega \left\{ -\Delta t\,[(s_1 a_{irs} + s_3 b_{irs})\hat{\Phi}_{mku,i} \hat{\Phi}^\delta_n + s_3 c_{ijrs} \hat{\Phi}_{mku,i} \hat{\Phi}^\delta_{n,j}] + \frac{\Delta t^2}{2}(s_2 d_{ijrs} + s_4 e_{ijrs})\hat{\Phi}_{mku,i} \hat{\Phi}^\delta_{n,j} \right\} d\Omega$$

$$C^\gamma_{mnpqrs} = \int_\Omega \left\{ -\Delta t\,[(s_1 a_{irs} + s_3 b_{irs})\hat{\Phi}^\gamma_{m,i} \hat{\Phi}_{npq} + s_3 c_{ijrs} \hat{\Phi}^\gamma_{m,i} \hat{\Phi}_{npq,j}] + \frac{\Delta t^2}{2}(s_2 d_{ijrs} + s_4 e_{ijrs})\hat{\Phi}^\gamma_{m,i} \hat{\Phi}_{npq,j} \right\} d\Omega$$

$$D^\xi_{mknpqrs} = \int_\Omega \left\{ -\Delta t\,[(s_1 a_{irs} + s_3 b_{irs})\hat{\Phi}^\xi_{mk,i} \hat{\Phi}_{npq} + s_3 c_{ijrs} \hat{\Phi}^\xi_{mk,i} \hat{\Phi}_{npq,j}] + \frac{\Delta t^2}{2}(s_2 d_{ijrs} + s_4 e_{ijrs})\hat{\Phi}^\xi_{mk,i} \hat{\Phi}_{npq,j} \right\} d\Omega$$

$$D^\eta_{mkunprs} = \int_\Omega \left\{ -\Delta t\,[(s_1 a_{irs} + s_3 b_{irs})\hat{\Phi}_{mku,i} \hat{\Phi}^\eta_{np} + s_3 c_{ijrs} \hat{\Phi}_{mku,i} \hat{\Phi}^\eta_{np,j}] + \frac{\Delta t^2}{2}(s_2 d_{ijrs} + s_4 e_{ijrs})\hat{\Phi}_{mku,i} \hat{\Phi}^\eta_{np,j} \right\} d\Omega$$

$$E_{mkunpqrs} = \int_\Omega \left\{ -\Delta t\,[(s_1 a_{irs} + s_3 b_{irs})\hat{\Phi}_{mku,i} \hat{\Phi}_{npq} + s_3 c_{ijrs} \hat{\Phi}_{mku,i} \hat{\Phi}_{npq,j}] + \frac{\Delta t^2}{2}(s_2 d_{ijrs} + s_4 e_{ijrs})\hat{\Phi}_{mku,i} \hat{\Phi}_{npq,j} \right\} d\Omega \tag{14.1.54}$$

$$\hat{W}^{\gamma}_{mr} = \int_{\Omega}\left\{\Delta t\,\hat{\Phi}^{\gamma}_{m,i}\Phi_{\beta}(F^n_{\beta ir}+G^n_{\beta ir}) - \frac{\Delta t^2}{2}(a_{irs}+b_{irs})\hat{\Phi}^{\gamma}_{m,i}\Phi_{\beta,j}(F^n_{\beta js}+G^n_{\beta js})\right\}d\Omega$$
$$+\int_{\Gamma}\overset{*}{\hat{\Phi}}{}^{\gamma}_{m}\left\{\Delta t\left[-s_1\left(a_{irs}\Delta U^{n+1}_s\right) - s_3\left(b_{irs}\Delta U^{n+1}_s + c_{ijrs}\Delta U^{n+1}_{s,j}\right)\right]\right.$$
$$\left.+\frac{\Delta t^2}{2}\left[s_2(a_{irt}a_{jts}+b_{irt}a_{jts})\Delta U^{n+1}_{s,j} + s_4(a_{irt}b_{jts}+b_{irt}b_{jts})\Delta U^{n+1}_{s,j}\right]\right\}n_i\,d\Gamma$$
$$+\int_{\Gamma}\overset{*}{\hat{\Phi}}{}^{\gamma}_{m}\left\{-\Delta t\,(F^n_{ir}+G^n_{ir}) + \frac{\Delta t^2}{2}(a_{irs}+b_{irs})(F^n_{js,j}+G^n_{js,j}-B^n_s)\right\}n_i\,d\Gamma$$

$$\hat{W}^{\xi}_{mkr} = \int_{\Omega}\left\{\Delta t\,\hat{\Phi}^{\xi}_{mk,i}\Phi_{\beta}(F^n_{\beta ir}+G^n_{\beta ir}) - \frac{\Delta t^2}{2}(a_{irs}+b_{irs})\hat{\Phi}^{\xi}_{mk,i}\Phi_{\beta,j}(F^n_{\beta js}+G^n_{\beta js})\right\}d\Omega$$
$$+\int_{\Gamma}\overset{*}{\hat{\Phi}}{}^{\gamma}_{mk}\left\{\Delta t\left[-s_1\left(a_{irs}\Delta U^{n+1}_s\right) - s_3\left(b_{irs}\Delta U^{n+1}_s + c_{ijrs}\Delta U^{n+1}_{s,j}\right)\right]\right.$$
$$\left.+\frac{\Delta t^2}{2}\left[s_2(a_{irt}a_{jts}+b_{irt}a_{jts})\Delta U^{n+1}_{s,j} + s_4(a_{irt}b_{jts}+b_{irt}b_{jts})\Delta U^{n+1}_{s,j}\right]\right\}n_i\,d\Gamma$$
$$+\int_{\Gamma}\overset{*}{\hat{\Phi}}{}^{\gamma}_{mk}\left\{-\Delta t\,(F^n_{ir}+G^n_{ir}) + \frac{\Delta t^2}{2}(a_{irs}+b_{irs})(F^n_{js,j}+G^n_{js,j}-B^n_s)\right\}n_i\,d\Gamma$$

$$\hat{W}_{mkur} = \int_{\Omega}\left\{\Delta t\,\hat{\Phi}_{mku,i}\Phi_{\beta}(F^n_{\beta ir}+G^n_{\beta ir})\right.$$
$$\left.-\frac{\Delta t^2}{2}(a_{irs}+b_{irs})\hat{\Phi}_{mku,i}\Phi_{\beta,j}(F^n_{\beta js}+G^n_{\beta js})\right\}d\Omega \tag{14.1.55}$$

As mentioned earlier for the case of two dimensions, the Neumann boundary conditions involved in all spectral degrees of freedom do not exist and are not applied, initially. However, they may be computed and added after the initial corner node computation. As in 2-D, we begin with

$$(A_{\alpha\beta}\delta_{rs} + B_{\alpha\beta rs})\Delta U^{n+1}_{\beta s} = W^n_{\alpha r} \tag{14.1.56}$$

In this process, the FDV-FEM computations are carried out with h-adaptivity until all shock waves are resolved. The next step is to resolve turbulent microscales using the spectral portion of the computations

$$\begin{bmatrix} A^{\gamma\delta}_{mn}\delta_{rs}+C^{\gamma\delta}_{mnrs} & A^{\gamma\eta}_{mnp}\delta_{rs}+C^{\gamma\eta}_{mnprs} & A^{\gamma}_{mnpq}\delta_{rs}+C^{\gamma}_{mnpqrs} \\ A^{\xi\delta}_{mkn}\delta_{rs}+C^{\xi\delta}_{mknrs} & A^{\xi\eta}_{mknp}\delta_{rs}+D^{\xi\eta}_{mknprs} & A^{\xi}_{mknpq}\delta_{rs}+D^{\xi}_{mknpqrs} \\ A^{\delta}_{mkun}\delta_{rs}+C^{\delta}_{mkunrs} & A^{\eta}_{mkunp}\delta_{rs}+D^{\eta}_{mkunprs} & A_{mkunpq}\delta_{rs}+E_{mkunpqrs} \end{bmatrix}\begin{bmatrix}\Delta\hat{U}^{\delta}_{ns}\\ \Delta\hat{U}^{\eta}_{nps}\\ \Delta\hat{U}_{npqs}\end{bmatrix}$$
$$=\begin{bmatrix}\hat{W}^{\gamma}_{mr}\\ \hat{W}^{\xi}_{mkr}\\ \hat{W}_{mkur}\end{bmatrix} - \begin{bmatrix}X^{\gamma}_{mr}\\ X^{\xi}_{mkr}\\ X_{mkur}\end{bmatrix} \tag{14.1.57}$$

where

$$X^{\gamma}_{mr} = (A^{\gamma}_{m\beta}\delta_{rs} + B^{\gamma}_{m\beta rs})\Delta U_{\beta s}$$
$$X^{\xi}_{mkr} = (A^{\xi}_{mk\beta}\delta_{rs} + B^{\xi}_{mk\beta rs})\Delta U_{\beta s}$$
$$X_{npqs} = (A_{mku\beta}\delta_{rs} + B_{mku\beta rs})\Delta U_{\beta s}$$

which act as source terms or coupling effect of the corner nodes upon spectral behavior through side, face, and interior modes. The final step is to combine (14.1.56) and (14.1.57) by

$$(A_{\alpha\beta}\delta_{rs} + B_{\alpha\beta rs})\,\Delta U_{\beta s}^{n+1} = W_{\alpha r}^{n} + Y_{\alpha r} \tag{14.1.58}$$

with

$$\begin{aligned} Y_{\alpha r} &= \left(A_{\alpha n}^{\delta}\delta_{rs} + B_{\alpha\beta rs}^{\delta}\right)\Delta\hat{U}_{ns}^{\delta} + \left(A_{\alpha np}^{\eta}\delta_{rs} + B_{\alpha nprs}^{\eta}\right)\Delta\hat{U}_{nps}^{\eta} \\ &\quad + (A_{\alpha npq}\delta_{rs} + B_{\alpha npqrs})\Delta\hat{U}_{npqs} \end{aligned}$$

Thus, the convergence toward shock wave turbulent boundary layer interactions can be achieved through iterations between (14.1.57) and (14.1.58). Note that in this process, the convection implicitness parameters s_1 and s_2 are held constant, whereas the diffusion implicitness parameters s_3 and s_4 are updated through Reynolds numbers. Some examples are shown in Section 14.4.

14.2 LEAST SQUARES METHODS

The least squares methods (LSM) have been used in FEM by a number of authors such as Lynn [1974], Bramble and Shatz [1970], Fix and Gunzburger [1978], Carey and Jiang [1987], among others. In LSM, the inner products of the governing equations are constructed, which are then differentiated (minimized) with respect to the nodal values of the variables. The integration by parts which is normally required in the standard Galerkin method is not involved. As a consequence, higher order derivatives remain, which will then require higher order trial functions. The basic formulation strategies are described next.

14.2.1 LSM FORMULATION FOR THE NAVIER-STOKES SYSTEM OF EQUATIONS

To illustrate the procedure, let us consider the Navier-Stokes system of equations,

$$\mathbf{R} = \frac{\partial \mathbf{U}}{\partial t} + \mathbf{a}_i\frac{\partial \mathbf{U}}{\partial x_i} + \mathbf{b}_i\frac{\partial \mathbf{U}}{\partial x_i} + \mathbf{c}_{ij}\frac{\partial^2 \mathbf{U}}{\partial x_i \partial x_j} - \mathbf{B} \tag{14.2.1}$$

where

$$\mathbf{U} = \Phi_\alpha \mathbf{U}_\alpha \tag{14.2.2}$$

The least squares formulation of (14.2.1) leads to

$$\frac{\partial}{\partial \mathbf{U}_\alpha}\frac{1}{2}(\mathbf{R},\mathbf{R}) = \frac{\partial}{\partial \mathbf{U}_\alpha}\int_\Omega \frac{1}{2}\mathbf{R}^2 d\Omega = 0$$

This leads to

$$\int_\Omega W_\alpha \mathbf{R}\, d\Omega = 0 \tag{14.2.3}$$

with the test function W_α given by

$$W_\alpha = \frac{\partial \mathbf{R}}{\partial \mathbf{U}_\alpha} \tag{14.2.4}$$

or

$$W_\alpha = \frac{\partial \Phi_\alpha}{\partial t} + \mathbf{a}_i \frac{\partial \Phi_\alpha}{\partial x_i} + \mathbf{b}_i \frac{\partial \Phi_\alpha}{\partial x_i} + \mathbf{c}_{ij} \frac{\partial^2 \Phi_\alpha}{\partial x_i \partial x_j} \tag{14.2.5}$$

It is seen that the trial function Φ_α is not a function of time and the first term in (14.2.5) must vanish. To avoid this situation, we rewrite (14.2.1) in the form

$$\mathbf{R} = \mathbf{U}^{n+1} - \mathbf{U}^n + \Delta t \left(\mathbf{a}_i \frac{\partial \mathbf{U}}{\partial x_i} + \mathbf{b}_i \frac{\partial \mathbf{U}}{\partial x_i} + \mathbf{c}_{ij} \frac{\partial^2 \mathbf{U}}{\partial x_i \partial x_j} - \mathbf{B} \right) \tag{14.2.6}$$

This will allow the test function W_α to be written as

$$W_\alpha = \frac{\partial \mathbf{R}}{\partial \mathbf{U}_\alpha^{n+1}} = \Phi_\alpha + \frac{\Delta t}{2} (\mathbf{a}_i \Phi_{\alpha,i} + \mathbf{b}_i \Phi_{\alpha,i} + \mathbf{c}_{ij} \Phi_{\alpha,ij}) \tag{14.2.7}$$

with $U = (U^{n+1} + U^n)/2$. Thus, (14.2.3) takes the form

$$K_{\alpha\beta} \mathbf{U}_\beta^{n+1} = \mathbf{F}_\alpha^n \tag{14.2.8}$$

where the stiffness matrix $K_{\alpha\beta}$ is of the form

$$K_{\alpha\beta} = \int_\Omega \left[\Phi_\alpha + \frac{\Delta t}{2} (\mathbf{a}_i \Phi_{\alpha,i} + \mathbf{b}_i \Phi_{\alpha,i} + \mathbf{c}_{ij} \Phi_{\alpha,ij}) \right] \times \left[\Phi_\beta + \frac{\Delta t}{2} (\mathbf{a}_k \Phi_{\beta,k} + \mathbf{b}_k \Phi_{\beta,k} + \mathbf{c}_{km} \Phi_{\beta,km}) \right] d\Omega$$

and

$$\mathbf{F}_\alpha^n = \int_\Omega \left[\Phi_\alpha + \frac{\Delta t}{2} (\mathbf{a}_i \Phi_{\alpha,i} + \mathbf{b}_i \Phi_{\alpha,i} + \mathbf{c}_{ij} \Phi_{\alpha,ij}) \right] \times \left[\Phi_\beta - \frac{\Delta t}{2} (\mathbf{a}_k \Phi_{\beta,k} + \mathbf{b}_k \Phi_{\beta,k} + \mathbf{c}_{km} \Phi_{\beta,km}) \right] d\Omega \mathbf{U}_\beta^n + \int_\Omega \left[\Phi_\alpha + \frac{\Delta t}{2} (\mathbf{a}_i \Phi_{\alpha,i} + \mathbf{b}_i \Phi_{\alpha,i} + \mathbf{c}_{ij} \Phi_{\alpha,ij}) \right] \mathbf{B}^n \, d\Omega \tag{14.2.9}$$

As noted from (14.2.7), the test function arising from the LSM formulation resembles the GPG methods discussed in Section 13.5. The functions W_α are flowfield-dependent through the Jacobians $\mathbf{a}_i$, $\mathbf{b}_i$, and $\mathbf{c}_{ij}$. Various simplifications are available [Carey and Jiang, 1987 and others].

14.2.2 FDV-LSM FORMULATION

It is possible to use the FDV scheme for applications to LSM formulation. The advantage of FDV-LSM is to contain the time dependent terms for transient analysis. We begin with the FDV equations of the form (13.6.6):

$$\mathbf{R} = \Delta \mathbf{U}^{n+1} + \mathbf{E}_i \frac{\partial \Delta \mathbf{U}^{n+1}}{\partial x_i} + \mathbf{E}_{ij} \frac{\partial^2 \Delta \mathbf{U}^{n+1}}{\partial x_i \partial x_j} + \mathbf{Q}^n \tag{14.2.10}$$

or

$$\mathbf{R} = \left(\Phi_\alpha + \mathbf{E}_i \frac{\partial \Phi_\alpha}{\partial x_i} + \mathbf{E}_{ij} \frac{\partial^2 \Phi_\alpha}{\partial x_i \partial x_j} \right) \Delta \mathbf{U}_\alpha^{n+1} + \mathbf{Q}^n$$

The test function for the LSM scheme is

$$W_\alpha = \frac{\partial \mathbf{R}}{\partial \mathbf{U}_\alpha^{n+1}} = \Phi_\alpha + \mathbf{E}_i \Phi_{\alpha,i} + \mathbf{E}_{ij} \Phi_{\alpha,ij} \tag{14.2.11}$$

Substituting (14.2.10) and (14.2.11) into (14.2.3) leads to (14.2.6)

$$K_{\alpha\beta} \Delta \mathbf{U}_\beta^{n+1} = \mathbf{F}_\alpha^n$$

where

$$\begin{aligned} K_{\alpha\beta} = \int_\Omega &(\Phi_\alpha \Phi_\beta + \mathbf{E}_k \Phi_\alpha \Phi_{\beta,k} + \mathbf{E}_{km} \Phi_\alpha \Phi_{\beta,km} \\ &+ \mathbf{E}_i \Phi_{\alpha,i} \Phi_\beta + \mathbf{E}_i \mathbf{E}_k \Phi_{\alpha,i} \Phi_{\beta,k} + \mathbf{E}_i \mathbf{E}_{km} \Phi_{\alpha,i} \Phi_{\beta,km} \\ &+ \mathbf{E}_{ij} \Phi_{\alpha,ij} \Phi_\beta + \mathbf{E}_{ij} \mathbf{E}_k \Phi_{\alpha,ij} \Phi_{\beta,k} + \mathbf{E}_{ij} \mathbf{E}_{km} \Phi_{\alpha,ij} \Phi_{\beta,km}) \, d\Omega \end{aligned} \tag{14.2.12}$$

and

$$\mathbf{F}_\alpha^n = \int_\Omega (\Phi_\alpha + \mathbf{E}_i \Phi_{\alpha,i} + \mathbf{E}_{ij} \Phi_{\alpha,ij}) \mathbf{Q}^n \, d\Omega \tag{14.2.13}$$

Once again, the computational requirements for the FDV-LSM formulation are significantly greater than those of the FDV Galerkin method.

14.2.3 OPTIMAL CONTROL METHOD

The optimal control method (OCM) was applied to a highly nonlinear integrodifferential equation such as in combined mode radiative heat transfer problems [Chung and Kim, 1984; Utreja and Chung, 1989]. It resembles the standard LSM except that penalty functions are used to provide constraints.

The basic idea is to construct a cost function in the form

$$J = \frac{1}{2} \int_\Omega (R_n R_n + \lambda_{(m)} S_m S_m) \, d\Omega \tag{14.2.14}$$

where R_n represents the residual of any governing equation and $S_m^{(i)}$ denotes a constraint function which will convert a first derivative into a second derivative with λ_m being the penalty parameter (see Section 12.1.2). For example, consider a steady-state two-dimensional Burgers equation of the form

$$
\begin{aligned}
R_1 &= u\frac{\partial u}{\partial x} + \mathrm{v}\frac{\partial u}{\partial x} - \nu\left(\frac{\partial \overline{S}_1}{\partial x} + \frac{\partial \overline{S}_2}{\partial y}\right) = 0 \\
R_2 &= u\frac{\partial \mathrm{v}}{\partial x} + \mathrm{v}\frac{\partial \mathrm{v}}{\partial x} - \nu\left(\frac{\partial \overline{S}_3}{\partial x} + \frac{\partial \overline{S}_4}{\partial y}\right) = 0
\end{aligned}
\tag{14.2.15}
$$

with

$$
\begin{aligned}
S_1 &= \overline{S}_1 - \frac{\partial u}{\partial x} = 0 \\
S_2 &= \overline{S}_2 - \frac{\partial u}{\partial y} = 0 \\
S_3 &= \overline{S}_3 - \frac{\partial \mathrm{v}}{\partial x} = 0 \\
S_4 &= \overline{S}_4 - \frac{\partial \mathrm{v}}{\partial y} = 0
\end{aligned}
\tag{14.2.16}
$$

Substituting (14.2.15) and (14.2.16) into (14.2.14) and minimizing the cost function J, we obtain

$$
\delta J = \frac{\partial J}{\partial u_\alpha}\delta u_\alpha + \frac{\partial J}{\partial \mathrm{v}_\alpha}\delta \mathrm{v}_\alpha + \lambda_{(m)}\frac{\partial J}{\partial S_m}\delta S_m = 0 \tag{14.2.17}
$$

Since δu_α, $\delta \mathrm{v}_\alpha$, and δS_m are arbitrary, it follows from (14.2.17) that

$$
\begin{aligned}
&\int_\Omega \left(R_n\frac{\partial R_n}{\partial u_\alpha} + \lambda_m\frac{\partial S_m}{\partial u_\alpha}\right)d\Omega = 0 \\
&\int_\Omega \left(R_n\frac{\partial R_n}{\partial \mathrm{v}_\alpha} + \lambda_m\frac{\partial S_m}{\partial v_\alpha}\right)d\Omega = 0 \quad (n = 1, 2,\ m, r = 1, 4) \\
&\int_\Omega \left(R_n\frac{\partial R_n}{\partial \overline{S}_{m\alpha}} + \lambda_r\frac{\partial S_r}{\partial \overline{S}_{m\alpha}}\right)d\Omega = 0
\end{aligned}
\tag{14.2.18}
$$

For other problems such as in combined mode radiative heat transfer where radiation source terms are to be separately calculated iteratively, the concept of penalty functions is particularly useful. Although simultaneous solutions of these equations are costly, they are quite useful for highly nonlinear problems. Applications of the OCM are demonstrated in Sections 24.3 and 24.4.

14.3 FINITE POINT METHOD (FPM)

Mesh configurations including local elements and nodal points are required for all computational methods discussed so far. In recent years, various methods which depend on finite number of points rather than meshes (meshless methods) have been developed. The so-called smooth particle hydrodynamics (SPH) [Lucy, 1977; Monaghan, 1988] has been used for the analysis of exploding stars and dust clouds using finite number of points with a functional representation of the variable $u(x)$ as

$$u(x) = \int_\Omega w(x - x_i)u(x_i)\, d\Omega = \Phi_i u_i \tag{14.3.1}$$

where $w(x - x_i)$ is the kernel, wavelets, or weight function and Φ_i is the SPH interpolation function, with the kernel being approximated by exponential, cubic spline, or quartic spline.

The concept of SPH can be extended to a meshless approach in terms of element-free Galerkin method (EFG) [Belytschko et al., 1996] or fixed least squares (FLS) and moving least square (MLS) procedures [Lancaster and Salkauskas, 1981; Onate et al., 1996]. In the FLS and MLS methods, we replace the integral (14.3.1) of the variable $u(x)$ by

$$u(x) = P_i(x)a_i(x) \tag{14.3.2}$$

where $P_i(x)$ are the monomial basis functions and $a_i(x)$ are their coefficients.

$$P_i = (1, x, x^2 \ldots) \qquad \text{1D} \tag{14.3.3a}$$

$$P_i = (1, x, y, x^2, xy, y^2, \ldots.) \qquad \text{2D} \tag{14.3.3b}$$

Expanding (14.3.2) to cover nodal points, we rewrite (14.3.2) as

$$u_i = P_{ik}a_k \tag{14.3.4}$$

where

$$P_{ik} = \begin{bmatrix} P_1(x_1) & P_2(x_1) & \cdots & P_m(x_1) \\ P_1(x_2) & P_2(x_2) & \cdots & P_m(x_2) \\ \vdots & \vdots & \vdots & \vdots \\ P_1(x_n) & P_2(x_n) & \cdots & P_m(x_n) \end{bmatrix} \tag{14.3.5}$$

In order to determine the unknown coefficients a_i,, we introduce in (14.3.4) the weighted least squares operation in the form,

$$\frac{\partial J}{\partial a_i} = 0 \tag{14.3.6}$$

where J is the weighted least squares function,

$$J = W_{ij}(P_{ik}a_k - u_i)(P_{jm}a_m - u_j) \tag{14.3.7}$$

with W_{ij} being the second order tensor weight functions,

$$W_{ij} = \begin{bmatrix} W(x - x_1) & 0 & \cdots & 0 \\ 0 & W(x - x_2) & \cdots & 0 \\ \vdots & \vdots & \vdots & \vdots \\ 0 & 0 & \cdots & W(x - x_n) \end{bmatrix} \tag{14.3.8}$$

Performing the differentiation in (14.3.6) leads to

$$a_i = (W_{nj}P_{nk}P_{jm})^{-1}W_{km}P_{ir}u_r \tag{14.3.9}$$

Substituting (14.3.9) into (14.3.2), we obtain

$$u(x) = \Phi_i u_i \tag{14.3.10}$$

where Φ_i is the finite point interpolation function,

$$\Phi_i = P_s(W_{nj}P_{nk}P_{jm})^{-1}W_{km}P_{si} \tag{14.3.11}$$

with

$$\Phi_i(x_j) = \delta_{ij} \tag{14.3.12}$$

and the diagonal component of the weighting functions may be chosen as a Gaussian function

$$W_{ij} = \frac{\exp[-(x/c)^2] - \exp[-(x_m/c)^2]}{1 - \exp[-(x_m/c)^2]} \tag{14.3.13}$$

where x_m is the half size of the support and c is a parameter determining the geometrical shape.

Another meshless (finite point) method, known as the partition of unity (PUM) or *h-p* cloud method, was advanced by Duarte and Oden [1996] and Melenk and Babuska [1996], which is suitable for an unstructured adaptive method (Chapter 19). In this method, the variable $u(x)$ is expressed as

$$u(x) = \Phi_i u_{i(mnp)} \tag{14.3.14}$$

where Φ_i is the MLS function of (14.3.11) and $u_{i(mnp)}$ is the spectral function consisting of either Lagrange or Legendre polynomials with m, n, p representing orders of polynomials similarly as in (14.1.16).

The functional representation of SPH, MLS, and PUM is based on the meshless approach. Lumping them all together, these meshless methods may be called the finite point methods (FPM), as suggested by Onate et al. [1996]. The advantage of FPM is obviously the elimination of the need for grid generation, which is itself a major task.

14.4 EXAMPLE PROBLEMS

In this section, we present some example problems of FDV spectral element methods using the Legendre polynomials [Yoon and Chung, 1996]. Spectral elements of Legendre polynomial degree 2 ($q2$) in space 2 ($S2$) are applied in the spatially evolving three-dimensional boundary layers with shock wave boundary layer interactions in a single and double sharp leading edged fins.

14.4.1 SHARP FIN INDUCED SHOCK WAVE BOUNDARY LAYER INTERACTIONS

To investigate the interaction of a shock wave with a boundary layer in three dimensions, a sharp leading edged fin is adopted as a model problem. Figure 14.4.1.1a shows the physical domain for a 3-D sharp fin ($\alpha = 20°$) with a general flowfield structure

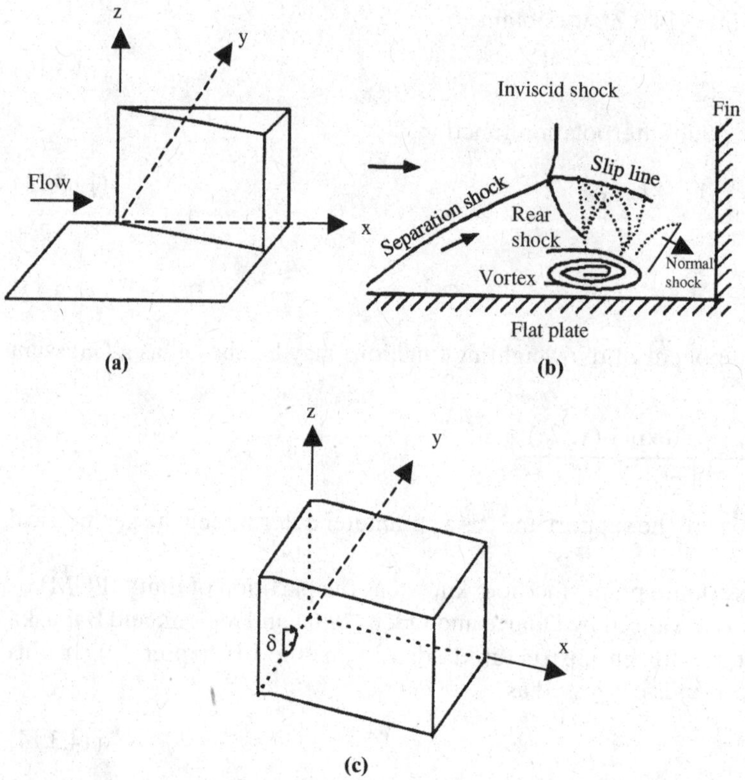

Figure 14.4.1.1 Computational domain for a 3-D 20° fin and flowfield structure with $M_\infty = 2.93$, $P_\infty = 20.57$ kPa, $T_\infty = 92.39$ K, $Re_\infty = 7 \times 10^8$/m. The inlet boundary conditions are obtained from the boundary layer analysis. On the solid surface, no-slip and adiabatic wall boundary conditions are applied. **(a)** 3-D 20° fin. **(b)** 20° fin interaction flowfield structures. **(c)** Computational domain.

(Figure 14.4.1.1b) [Settles and Dolling, 1990]. The inlet boundary conditions and the corresponding flowfield structure are the same as in Knight et al. [Settles and Dolling, 1990]. Here, the freestream Mach number and temperature are $M_\infty = 2.93$ and $T_\infty =$ 92.39 K, corresponding to the chamber pressure and temperature of 680 kPa and 251 K, respectively, with the Reynolds number of 7×10^8/m. The boundary layer thickness δ_o at the apex of the fin is 1.4 cm, yielding a Reynolds number $Re_{\delta o} = 9.8 \times 10^5$. In order to match the boundary conditions as used for the experiments [Settles and Dolling, 1990], the flowfield behind the fin is calculated as a flat plate boundary layer such that the computed boundary layer thickness δ_o is set equal to the experimental value of 1.4 cm. On the solid surfaces, no-slip and adiabatic wall boundary conditions are applied. On the upper, lateral, and downstream exit boundaries, the flow variables are set free. Adaptive spaced grid points are 33, 41, and 31 in the streamwise, spanwise, and vertical directions, respectively. Spectral elements of Legendre polynomial degree 2 in space 2 are applied in the boundary layer.

Figure 14.4.1.2 shows the background flowfield based on the geometric configurations and boundary conditions described in Figure 14.4.1.1, as observed from the front

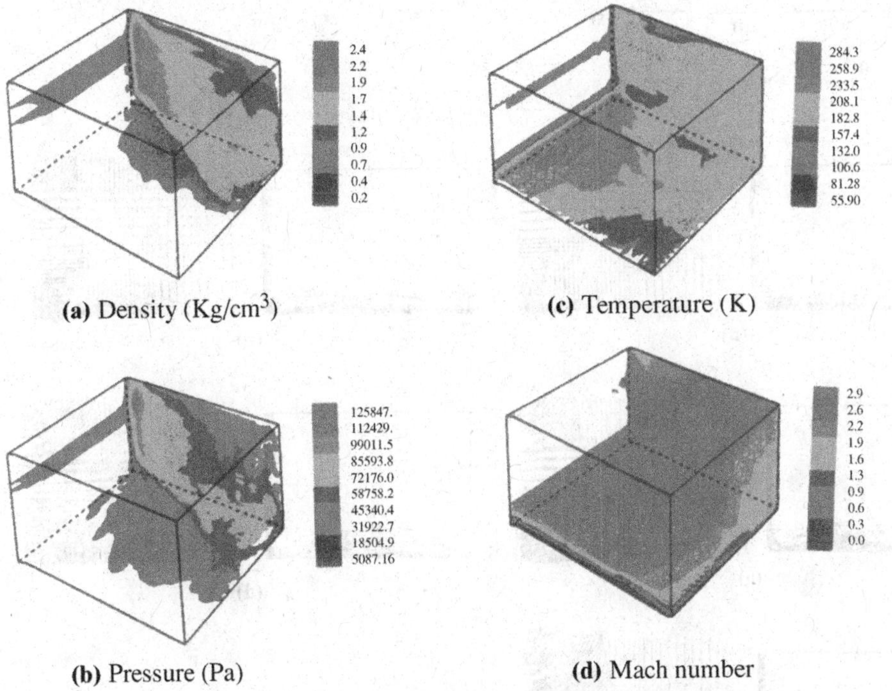

Figure 14.4.1.2 Background flowfield as observed from the front (x-z plane and y-z plane).

(x-z and y-z faces). As such, no details of the hidden portion are shown. It is noticed that the trend is in reasonable agreement with the results of Narayanswami, Hortzman, and Knight [1993], with density and pressure increasing drastically along the shock waves, the temperature rise being distributed along the flat plate, and Mach number sharply decreasing through the shock waves toward the flat plate boundary.

Vorticity variations at different planes are shown in Figures 14.4.1.3a through 14.4.1.3e. The contours of vorticity component in the streamwise planes (y-z planes) in the x-direction with each plane identified as a, b, c, d, e are shown. The corresponding velocity vectors are plotted on the right-hand side. Clearly, the vortex stretching occurs toward downstream with the evidence of separation shocks, slip lines, and vortex centers close to the wall. These physical phenomena become more significant toward downstream in agreement with the schematics shown in Figure 14.4.1.2.

Figure 14.4.1.4a shows the contours of vorticity component in the spanwise vertical planes ($x \cos\alpha$-z planes) in the $y \cos\alpha$-direction, with each plane identified as a, b, c, d. The vortex stretching occurs again toward downstream and moving upward away from the shock. The growth of vorticity is concentrated within the boundary layer close to the wall.

In Figure 14.4.1.4b, the spanwise horizontal plane vorticity contours are presented at various locations (a:$2\delta_o$, b:$2\delta_o$, c:$20.5\delta_o$) where δ_o is the boundary layer thickness. It is seen that vorticity increases toward the wall, with its intensity increasing toward downstream as expected.

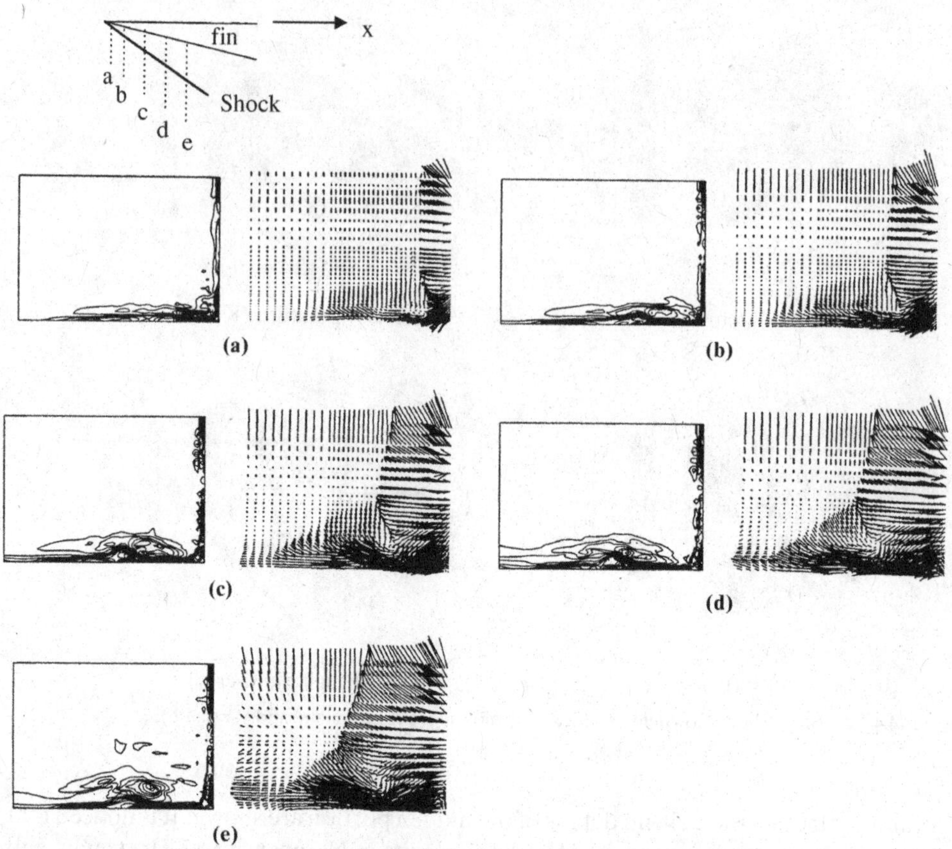

Figure 14.4.1.3 Streamwise vorticity contours and the corresponding velocity vectors ($t = 0.3965$ ms). The vortex stretching occurs toward downstream with the evidence of separation shocks, slip lines, and vortex centers close to the wall.

14.4.2 ASYMMETRIC DOUBLE FIN INDUCED SHOCK WAVE BOUNDARY LAYER INTERACTION

Complex three-dimensional shock wave boundary layer interactions occur on asymmetric double fins. Schematic representation of an asymmetric crossing shock wave turbulent boundary layer interaction is shown in Figure 14.4.2.1a. The dimensions and freestream conditions employed in the experiment by Knight et al. [1995] are shown in Figure 14.4.2.1b. The same dimensions and freestream conditions are used in the present investigation.

Figures 14.4.2.2a and 14.4.2.2b display density and pressure contours, respectively. Existence of crossing shock waves and expansion waves in the asymmetric double fins is clearly evident in these figures. Figure 14.4.2.3 shows velocity vectors at different streamwise planes (y-z planes) in the x-direction. It is evident that vortices are generated near the surface toward downstream.

The present result is compared with experimental data [Knight et al., 1995] for wall pressure. The comparisons on the throat middle line and at streamwise location

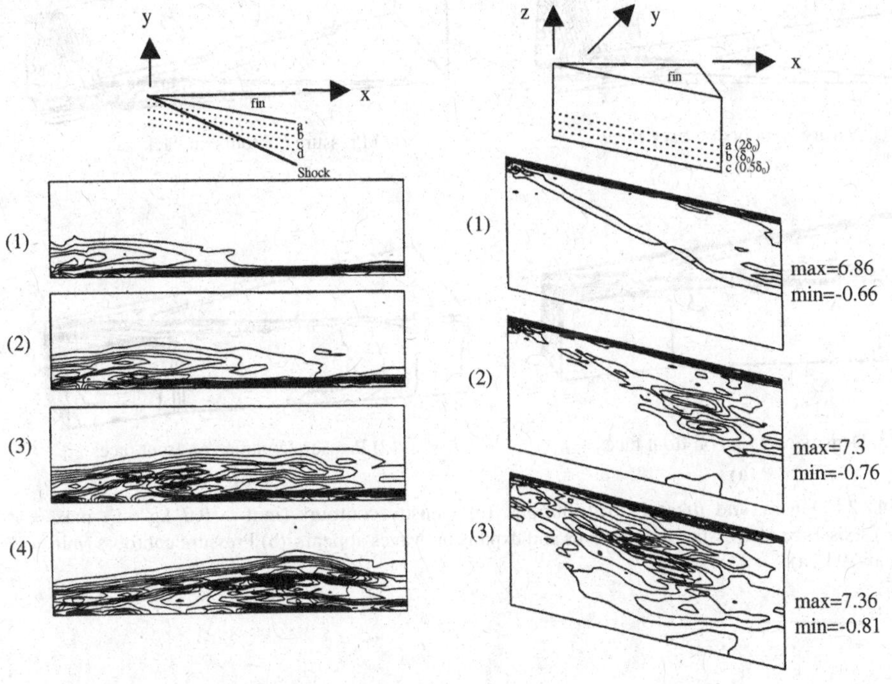

(a) (b)

Figure 14.4.1.4 Spanwise vertical and horizontal plane vorticity contours. **(a)** Spanwise vertical plane ($x \cos \alpha$-z plane) vorticity contours at various locations ($t = 0.3965$ ms, $0 \leq z/\delta_0 \leq 2.5$). The vortex stretching occurs again toward downstream and moving upward away from the shock. The growth of vorticity is concentrated within the boundary layer close to the wall. **(b)** Spanwise horizontal plane ($x \cos \alpha$-y plane) vorticity contours at various locations ($t = 0.3965$ ms). Is is seen that vorticity increases toward the wall with its intensity increasing toward downstream as expected.

$x = 46$ mm are displayed in Figure 14.4.2.4a and Figure 14.4.2.4b, respectively. The present and experimental surface pressure on throat middle line are in general agreement at upstream, but deviate toward downstream. At $x = 46$ mm, the present and experimental surface pressures show a close agreement.

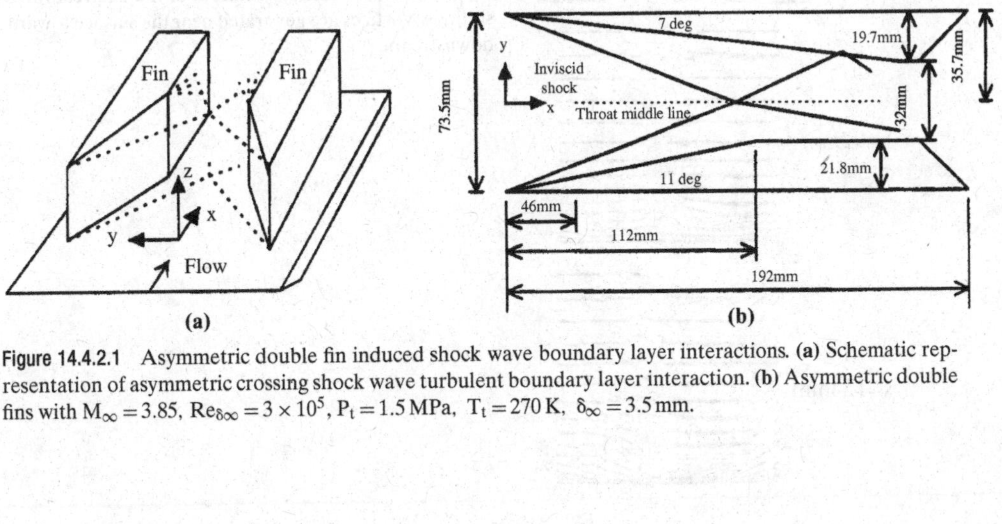

(a) (b)

Figure 14.4.2.1 Asymmetric double fin induced shock wave boundary layer interactions. **(a)** Schematic representation of asymmetric crossing shock wave turbulent boundary layer interaction. **(b)** Asymmetric double fins with $M_\infty = 3.85$, $Re_{\delta\infty} = 3 \times 10^5$, $P_t = 1.5$ MPa, $T_t = 270$ K, $\delta_\infty = 3.5$ mm.

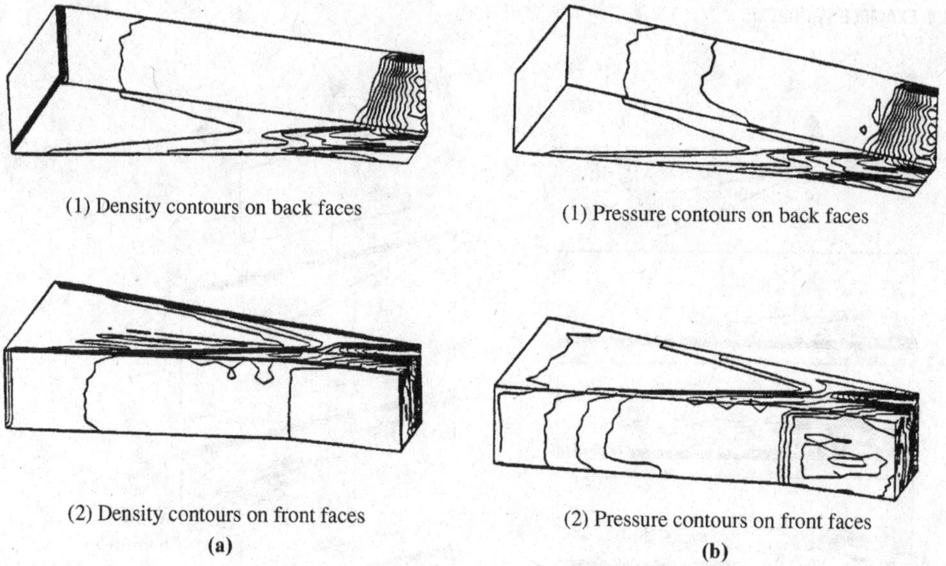

Figure 14.4.2.2 Density and pressure distributions. **(a)** Density contours (min = 0.6 kg /m^3, max = 2.3 kg/ m^3), existence of crossing shock waves and expansion waves appears. **(b)** Pressure contours (min = 11 kPa, max-79 kPa).

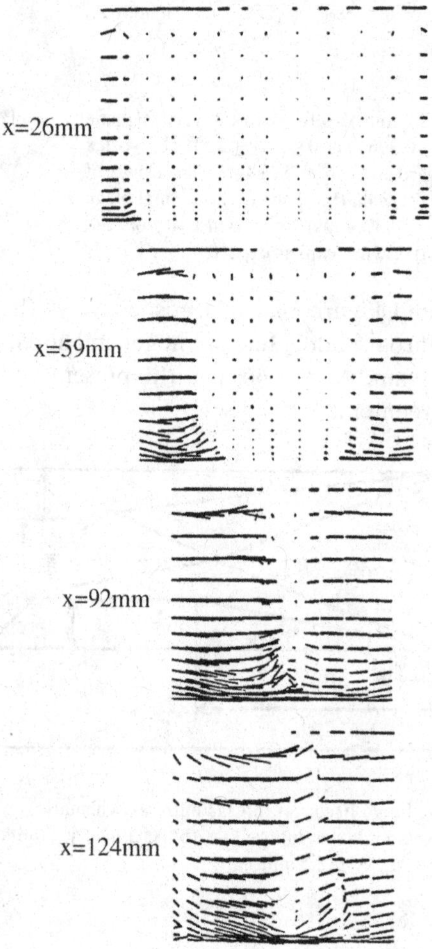

Figure 14.4.2.3 Velocity vectors at different streamwise stations. Vortices are generated near the surface toward downstream.

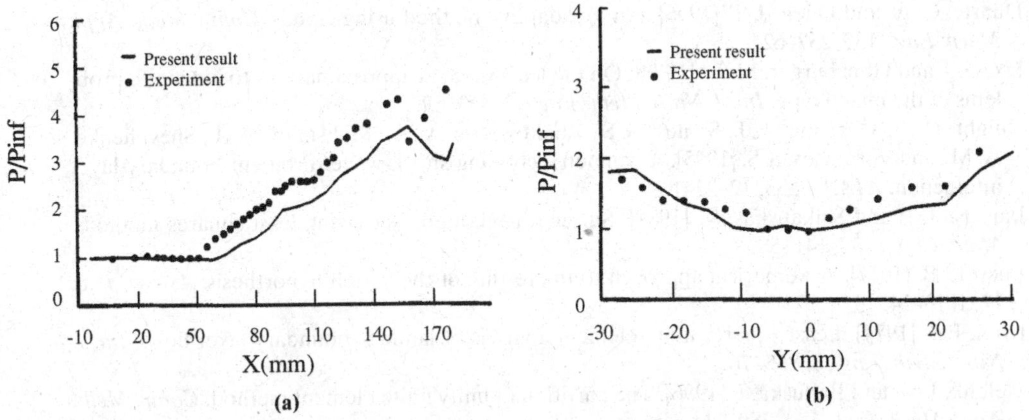

Figure 14.4.2.4 Comparison of pressure distributions with experimental data. (**a**) Comparison between the present result and experimental data of wall pressure on throat middle line. The present and experimental surface pressures on throat middle line are in general agreement at upstream, but deviate toward downstream. (**b**) Comparison of wall pressure at $x = 46$ mm for the present result and experimental data. At $x = 46$ mm, the present and experimental surface pressure show close agreement.

14.5 SUMMARY

In this chapter, we reviewed various methods that are related to FEM or weighted residual methods. Although the spectral element methods (SEM) are accurate for simple geometries and simple boundary conditions, the SEM applications to complex multidimensional problems are not practical. The least squares methods (LSM) can be applied to complicated geometries, but computations involved are quite time-consuming. The research in meshless methods or finite point methods (FPM) has begun recently. Active research in FPM in the future appears to be promising.

As we come to the end of finite element applications, we recall that, in Part Two, the finite volume methods (FVM) can be formulated using FDM as shown in Chapter 7. Thus, a similar treatment of FVM using FEM is the subject of the next chapter.

REFERENCES

Babuska, I. [1958]. The *p* and *h-p* versions of the finite element method. The state of the art. In D. L. Dwoyer, M. Y. Hussaini, and R. G. Voigt (eds.). *Finite Elements Theory and Application*, 199–239, New York: Springer-Verlag.

Belytschko, T., Krongauz, Y., Organ, D., Fleming, M., and Krysl, P. [1996]. Meshless methods: An overview and recent developments. *Comp. Meth. Appl. Mech. Eng.*, 139, 3–47.

Bramble, J. H. and Shatz, A. H. [1970]. On the numerical solution of elliptic boundary-value problems by least-squares approximation of the data. In B. Hubbered (ed.). *Numerical Solution of PDE, Vol. 2*, New York: Academic Press.

Canuto, C., Hussani, M. Y., Quarteroni, A., and Zang, T. A. [1987]. *Spectral Methods in Fluid Dynamics*. New York: Springer-Verlag.

Carey, G. F. and Jiang, B. N. [1987]. Least squares finite element method and preconditioned conjugate gradient solution. *Int. J. Num. Meth. Eng.*, 24, 1283–96.

Chung, T. J. and Kim, J. Y. [1984]. Two-dimensional, combined-mode heat transfer by conduction, convection and radiation in emitting, absorbing and scattering media – solution by finite elements. *J. Heat Trans.*, 106, 448–52.

Duarte, C. A. and Oden, J. T. [1996]. An *hp* adaptive method using clouds. *Comp. Meth. Appl. Mech. Eng.*, 139, 237–62.

Fix, G. J. and Gunzburger, M. D. [1978]. On the least squares approximations to indefinite problems of the mixed type. *Int. J. Num. Meth. Eng.* 12, 453–69.

Knight, D. D., Garrison, T. J., Senles, G. S., Zheltovodov, A. A., Maksimov, A. I., Shevehenko, A. M., and Vorontsov, S. S. [1995]. Asymmetric crossing-shock-wave/turbulent-boundary-layer interaction. *AIAA J.*, 33, 12, 2241.

Lancaster, P. and Salkauskas, K. [1981]. Surfaces generated by moving least squares methods. *Math. Comp.*, 37, 141–58.

Lucy, L. B. [1977]. A numerical approach to the testing of the fission hyporthesis. *Astron. J.*, 8, 12, 1013–24.

Lynn, P. P. [1974]. Least squares finite element analysis of laminar boundary layer flows, *Int. J. Num. Meth. Eng.*, 8, 865–76.

Melenk, J. M. and Babuska, I. [1996]. The partition of unity finite element method. *Comp. Meth. Appl. Mech. Eng.*, 139, 289–314.

Monaghan, J. J. [1988]. An introduction to SPH. *Comp. Phys. Comm.*, 48, 89–96.

Narayanswami, N., Hortzman, C. C., and Knight, D. D. [1993]. Computation of crossing shock/turbulence layer interaction at Mach 8.3. *AIAA J.*, 31, 1369–76.

Oden, J., Demkowicz, L., Rachowicz, W., and Westermann, T. A. [1989]. Toward a universal *h-p* adaptive finite element strategy: Part II. A posteriori error estimation. *Comp. Meth. Appl. Mech. Eng.*, 77, 113–80.

Onate, E., Idelsohn, S., Zienkiewicz, O. C., Taylor, R. L., and Sacco, C. [1996]. A stabilized finite point method for analysis of fluid mechanics problem. *Comp. Meth. Appl. Mech. Eng.*, 139, 315–46.

Patera, A. T. [1984]. A spectral method for fluid dynamics, laminar flow in a channel expansion. *J. Comp. Phys.*, 54, 468–88.

Settles, G. S. and Dolling, D. S. [1990]. Swept shock/boundary-layer interactions: Tutorial and update. *AIAA* 90–0375.

Sherwin, S. J. and Karniadakis, G. E. [1995]. A triangular spectral element methods; applications to the incompressible Navier-Stokes equations. *Comp. Meth. Appl. Mech. Eng.*, 123, 189–229.

Szabo, B. A. and Babuska, I. [1991]. *Finite Element Analysis*. New York: Wiley.

Utreja, L. R. and Chung, T. J. [1989]. Combined convection-conduction-radiation boundary layer flows using optimal control penalty finite elements. *J. Heat Trans.*, 111, 433–37.

Yoon, K. T. and Chung, T. J. [1996]. Three-dimensional mixed explicit-inplicit generalized Galerkin spectral element methods for high-speed turbulent compressible flows. *Comp. Meth. Appl. Mech. Eng.*, 135, 343–67.

CHAPTER FIFTEEN

Finite Volume Methods via Finite Element Methods

15.1 GENERAL

The finite volume methods (FVM) via FDM discussed in Chapter 7 may also be formulated using finite element methods (FEM). Schneider and Raw [1987], Masson, Saabas, and Baliga [1994], and Darbandi and Schneider [1999], among many others, contributed to the earlier and recent developments of FVM via FEM.

The FVM equations via finite elements are the same as those given in (7.1.4) for the case of the Navier-Stokes system of equations using finite differences,

$$\sum_{CV}\left(\frac{\Delta \mathbf{U}}{\Delta t} - \mathbf{B}\right)\Delta\Omega + \sum_{CS}(\mathbf{F}_i + \mathbf{G}_i)n_i\Delta\Gamma = 0 \tag{15.1.1a}$$

or

$$\sum_{CV}(\Delta \mathbf{U} - \Delta t\mathbf{B})\Delta\Omega + \Delta t\sum_{CS}(\mathbf{F}_i + \mathbf{G}_i)n_i\Delta\Gamma = 0 \tag{15.1.1b}$$

It is seen that quantities to be evaluated are involved in control volumes $\Delta\Omega$ and control surfaces $\Delta\Gamma$. We shall demonstrate how they are evaluated using finite elements in this chapter.

Consider the two-dimensional geometry as shown in Figure 15.1.1a. Note that global node 1 is surrounded by five elements, with each element divided into quadrilateral isoparametric elements (Figure 15.1.1b). A quadrant of each element is connected to node 1, forming five subcontrol volumes (CV1-A, CV1-B, CV1-C, CV1-D, and CV1-E). Each subcontrol volume has two control surfaces with outward normal directions with angles θ measured counterclockwise from the global reference cartesian x-coordinate.

It is reasonable to approximate $\Delta \mathbf{U}$ in control volumes with quadratic trial functions whereas the fluxes ($\mathbf{F}_i$ and $\mathbf{G}_i$) in control surfaces may be approximated by linear trial functions. Fluxes evaluated for all control volumes along the control surfaces plus the control volume quantities ($\Delta \mathbf{U}$ and $\mathbf{B}$) are to be assembled into each global node (control volume center), resulting in simultaneous algebraic equations for the entire system.

Note that the fluxes along the control surfaces are equal with opposite signs between neighboring control surfaces. This process renders all fluxes completely conserved – a distinctive advantage of FVM.

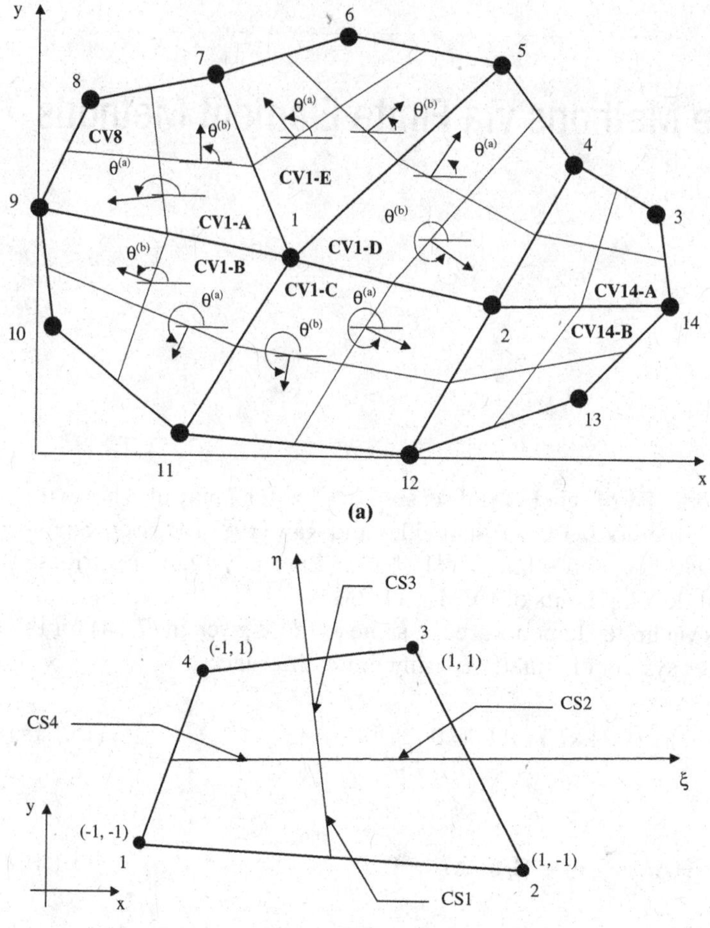

Figure 15.1.1 Unstructured grids for finite elements – node-centered control volume. **(a)** Subcontrol volumes CV1-A, B, C, D, E surrounding Node 1 with components of vectors normal to all control surfaces, subcontrol volume for node 8 (CV 8), subcontrol volumes for node 14 (CV14-A, B). **(b)** Control surfaces CS1, 2, 3, 4 with integration points along $\eta = 0$, $\xi = 0$ axes at centers of control surfaces in isoparametric element with corner nodes 1, 2, 3, 4.

Implementation of the finite element approximations toward FVM for two- and three-dimensional problems will be presented in the following subsections.

15.2 FORMULATIONS OF FINITE VOLUME EQUATIONS

15.2.1 BURGERS' EQUATIONS

To compare the formulation and solution procedure of FVM with FEM, let us consider the two-dimensional Burgers' equation in the form

$$\frac{\partial \mathbf{U}}{\partial t} + u\frac{\partial \mathbf{U}}{\partial x} + \mathrm{v}\frac{\partial \mathbf{U}}{\partial y} - \nu\left(\frac{\partial^2 \mathbf{U}}{\partial x^2} + \frac{\partial^2 \mathbf{U}}{\partial y^2}\right) - \mathbf{F} = 0 \tag{15.2.1}$$

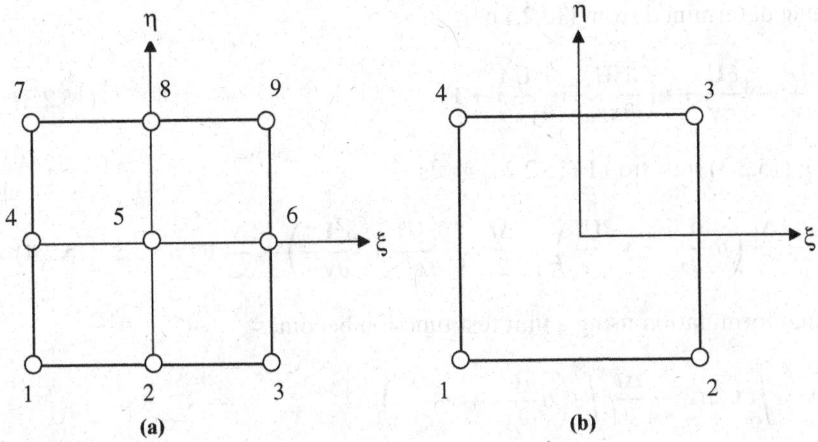

Figure 15.2.1 Isoparametric elements. (a) Quadratic approximation for control volumes. (b) Linear approximation for control surface.

where

$$\mathbf{U} = \begin{bmatrix} u \\ \mathrm{v} \end{bmatrix}, \qquad \mathbf{F} = \begin{bmatrix} f_x \\ f_y \end{bmatrix}$$

$$f_x = -\frac{1}{(1+t)^2} + \frac{x^2 + 2xy}{(1+t)} + 3x^3y^2 - 2vy$$

$$f_y = -\frac{1}{(1+t)^2} + \frac{y^2 + 2xy}{(1+t)} + 3y^3x^2 - 2vx$$

with the exact solution

$$u = \frac{1}{1+t} + x^2y, \qquad \mathrm{v} = \frac{1}{1+t} + xy^2$$

To illustrate the implementation of both the Dirichlet and Neumann boundary conditions on inclined surfaces, we consider the discretized geometries as shown in Figure 15.2.1 on which basic FVM equations will be written in terms of isoparametric finite elements.

Finite volume equations may be constructed within the framework of a two-step Taylor-Galerkin formulation. Toward this end, we begin with

$$\mathbf{U}^{n+1} = \mathbf{U}^n + \Delta t \frac{\partial \mathbf{U}^n}{\partial t} + O(\Delta t^2)$$

This may be split into two steps:

Step 1

$$\mathbf{U}^{n+\frac{1}{2}} = \mathbf{U}^n + \frac{\Delta t}{2}\frac{\partial \mathbf{U}^n}{\partial t} \tag{15.2.2a}$$

Step 2

$$\mathbf{U}^{n+1} = \mathbf{U}^n + \Delta t \frac{\partial \mathbf{U}^{n+\frac{1}{2}}}{\partial t} \tag{15.2.2b}$$

with $\partial \mathbf{U}/\partial t$ being determined from (15.2.1):

$$\frac{\partial \mathbf{U}}{\partial t} = -u\frac{\partial \mathbf{U}}{\partial x} - \mathrm{v}\frac{\partial \mathbf{U}}{\partial y} + \nu\left(\frac{\partial^2 \mathbf{U}}{\partial x^2} + \frac{\partial^2 \mathbf{U}}{\partial y^2}\right) + \mathbf{F} \tag{15.2.3}$$

Substituting (15.2.3) into step 1 (15.2.2a) gives

$$\mathbf{U}^{n+\frac{1}{2}} = \mathbf{U}^n - \frac{\Delta t}{2}\left(u\frac{\partial \mathbf{U}^n}{\partial x} + \mathrm{v}\frac{\partial \mathbf{U}^n}{\partial y}\right) + \frac{\Delta t}{2}\nu\left(\frac{\partial^2 \mathbf{U}^n}{\partial x^2} + \frac{\partial^2 \mathbf{U}^n}{\partial y^2}\right) + \frac{\Delta t}{2}\mathbf{F}^n \tag{15.2.4}$$

Finite volume formulation using a unit test function becomes

$$\int_\Omega \mathbf{U}^{n+\frac{1}{2}} d\Omega = \int_\Omega \mathbf{U}^n d\Omega - \frac{\Delta t}{2}\int_\Omega \left(u\frac{\partial \mathbf{U}^n}{\partial x} + \mathrm{v}\frac{\partial \mathbf{U}^n}{\partial y}\right) d\Omega$$
$$+ \frac{\Delta t}{2}\nu \int_\Omega \left(\frac{\partial^2 \mathbf{U}^n}{\partial x^2} + \frac{\partial^2 \mathbf{U}^n}{\partial y^2}\right) d\Omega + \frac{\Delta t}{2}\int_\Omega \mathbf{F}^n d\Omega \tag{15.2.5}$$

Integrating by parts, we have

$$\int_\Omega \mathbf{U}^{n+\frac{1}{2}} d\Omega = \int_\Omega \mathbf{U}^n d\Omega - \frac{\Delta t}{2}\int_\Omega \left(u^n\frac{\partial \mathbf{U}^n}{\partial x} + \mathrm{v}^n\frac{\partial \mathbf{U}^n}{\partial y}\right) d\Omega$$
$$+ \frac{\Delta t}{2}\nu \int_\Gamma \left(\frac{\partial \mathbf{U}^n}{\partial x}n_1 + \frac{\partial \mathbf{U}^n}{\partial y}n_2\right) d\Gamma + \frac{\Delta t}{2}\int_\Omega \mathbf{F}^n d\Omega \tag{15.2.6}$$

Rewriting the integral as summations,

$$\sum_{CV} \mathbf{U}^{n+\frac{1}{2}} \Delta\Omega = \sum_{CV}\left[\mathbf{U}^n - \frac{\Delta t}{2}\left(u^n\frac{\partial \mathbf{U}^n}{\partial x} + \mathrm{v}^n\frac{\partial \mathbf{U}^n}{\partial y}\right) + \frac{\Delta t}{2}\mathbf{F}^n\right]\Delta\Omega$$
$$+ \frac{\Delta t}{2}\nu \sum_{CS}\left(\frac{\partial \mathbf{U}^n}{\partial x}n_1 + \frac{\partial \mathbf{U}^n}{\partial y}n_2\right)\Delta\Gamma \tag{15.2.7}$$

Similarly for step 2 (15.2.2b), we have

$$\sum_{CV} \mathbf{U}^{n+1} \Delta\Omega = \sum_{CV}\left[\mathbf{U}^n - \frac{\Delta t}{2}\left(u^{n+\frac{1}{2}}\frac{\partial \mathbf{U}^{n+\frac{1}{2}}}{\partial x} + \mathrm{v}^{n+\frac{1}{2}}\frac{\partial \mathbf{U}^{n+\frac{1}{2}}}{\partial y}\right) + \frac{\Delta t}{2}\mathbf{F}^{n+\frac{1}{2}}\right]\Delta\Omega$$
$$+ \frac{\Delta t}{2}\nu \sum_{CS}\left(\frac{\partial \mathbf{U}^{n+\frac{1}{2}}}{\partial x}n_1 + \frac{\partial \mathbf{U}^{n+\frac{1}{2}}}{\partial y}n_2\right)\Delta\Gamma \tag{15.2.8}$$

Note that in these two-step solutions, (15.2.7) and (15.2.8), derivatives of $\mathbf{U}(d\mathbf{U}/dx$ and $d\mathbf{U}/dy)$ are involved within the control volumes and along the control surfaces. Quadratic and linear isoparametric finite element approximations are used, respectively, for control volumes and control surfaces, as shown in Figure 15.2.2. Derivatives of $\mathbf{U}$ involve the transformation between the isoparametric and cartesian coordinates as shown in Chapter 9. Derivatives involved in control volumes and control surfaces are carried out as follows:

For Control Volumes (quadratic approximation)

$$\frac{\partial \mathbf{U}}{\partial x_i} = \sum_{N=1}^{9} \left(\frac{\partial \Phi_N^{(e)}}{\partial x_i} \mathbf{U}_N \right)_{\xi=0,\,\eta=0} = \begin{bmatrix} \frac{1}{4|J|}[(\mathbf{U}_2 - \mathbf{U}_8)(y_6 - y_4) - (\mathbf{U}_6 - \mathbf{U}_4)(y_2 - y_8)] \\ \frac{1}{4|J|}[(\mathbf{U}_2 - \mathbf{U}_8)(x_6 - x_4) - (\mathbf{U}_6 - \mathbf{U}_4)(x_2 - x_8)] \end{bmatrix} \tag{15.2.9}$$

with

$$|J| = \frac{1}{4}[(x_2 - x_8)(y_6 - y_4) - (x_6 - x_4)(y_2 - y_8)] \tag{15.2.10}$$

For Control Surfaces (linear approximation)

$$\begin{aligned}\sum_{CS} \left(\frac{\partial \mathbf{U}}{\partial x} n_1 + \frac{\partial \mathbf{U}}{\partial y} n_2 \right) \Delta\Gamma = &\sum_{CS2,3} \left(\frac{\partial \mathbf{U}}{\partial x} n_1 + \frac{\partial \mathbf{U}}{\partial y} n_2 \right) \Delta\Gamma + \sum_{CS4,3} \left(\frac{\partial \mathbf{U}}{\partial x} n_1 + \frac{\partial \mathbf{U}}{\partial y} n_2 \right) \Delta\Gamma \\ &+ \sum_{CS1,4} \left(\frac{\partial \mathbf{U}}{\partial x} n_1 + \frac{\partial \mathbf{U}}{\partial y} n_2 \right) \Delta\Gamma + \sum_{CS1,2} \left(\frac{\partial \mathbf{U}}{\partial x} n_1 + \frac{\partial \mathbf{U}}{\partial y} n_2 \right) \Delta\Gamma\end{aligned} \tag{15.2.11}$$

with

$$\frac{\partial \mathbf{U}}{\partial x} = \frac{1}{|J|} \left(\frac{\partial \mathbf{U}}{\partial \xi} \frac{\partial y}{\partial \eta} - \frac{\partial \mathbf{U}}{\partial \eta} \frac{\partial y}{\partial \xi} \right) \tag{15.2.12a}$$

$$\frac{\partial \mathbf{U}}{\partial y} = \frac{1}{|J|} \left(-\frac{\partial \mathbf{U}}{\partial \xi} \frac{\partial x}{\partial \eta} + \frac{\partial \mathbf{U}}{\partial \eta} \frac{\partial x}{\partial \xi} \right) \tag{15.2.12b}$$

$$|J| = \frac{\partial x}{\partial \xi} \frac{\partial y}{\partial \eta} - \frac{\partial y}{\partial \xi} \frac{\partial x}{\partial \eta} \tag{15.2.13}$$

The above quantities are to be evaluated for each of the subcontrol volumes A, B, C, and D, corresponding to control surfaces (see Figure 15.2.2):

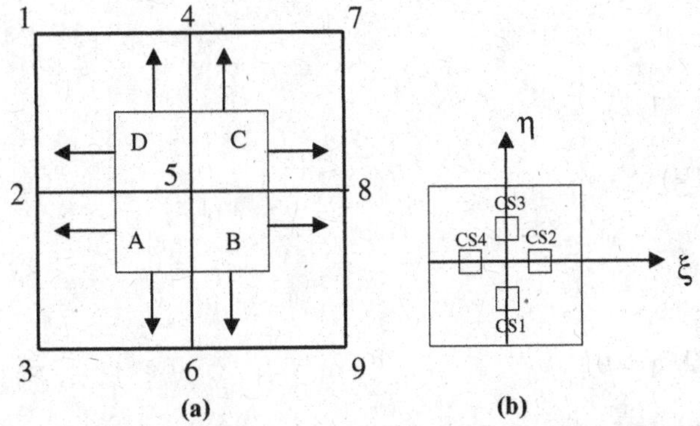

Figure 15.2.2 Control surfaces and their contributions to control volume at node 5 consisting of subcontrol volumes A, B, C, and D. **(a)** Control surfaces contributing to control volume. **(b)** Control surfaces evaluated at midpoints for each subcontrol volume.

CS2 and CS3 for A
CS3 and CS4 for B
CS4 and CS1 for C
CS1 and CS2 for D

Subcontrol Volume A

Control Surface CS2 ($\xi = 1/2, \eta = 0$)

$$\frac{\partial \mathbf{U}}{\partial \xi} = \frac{1}{4}(-\mathbf{U}_3 + \mathbf{U}_6 + \mathbf{U}_5 - \mathbf{U}_2)$$

$$\frac{\partial \mathbf{U}}{\partial \eta} = \frac{1}{8}(-\mathbf{U}_3 - 3\mathbf{U}_6 + 3\mathbf{U}_5 + \mathbf{U}_2)$$

Control Surface CS3 ($\xi = 0, \eta = 1/2$)

$$\frac{\partial \mathbf{U}}{\partial \xi} = \frac{1}{8}(-\mathbf{U}_3 + \mathbf{U}_6 + 3\mathbf{U}_5 - 3\mathbf{U}_2)$$

$$\frac{\partial \mathbf{U}}{\partial \eta} = \frac{1}{4}(-\mathbf{U}_3 - \mathbf{U}_6 + \mathbf{U}_5 + \mathbf{U}_2)$$

Sum the Control Surfaces CS2 and CS3

$$\sum_{CS2,3}^{A}\left(\frac{\partial \mathbf{U}}{\partial x}n_1 + \frac{\partial \mathbf{U}}{\partial y}n_2\right)\Delta\Gamma = \left(\frac{\partial \mathbf{U}}{\partial x}\cos\theta_2 + \frac{\partial \mathbf{U}}{\partial y}\sin\theta_2\right)\Delta\Gamma + \left(\frac{\partial \mathbf{U}}{\partial x}\cos\theta_3 + \frac{\partial \mathbf{U}}{\partial y}\sin\theta_3\right)\Delta\Gamma$$

with

$$|J| = \frac{1}{32}[(-x_3 + x_6 + 3x_5 - 3x_2)(-y_3 - y_6 + y_5 + y_2) - (-y_3 + y_6 + 3y_5 - 3y_2)(-x_3 - x_6 + x_5 + x_2)]$$

Subcontrol Volume B

Control Surface CS3 ($\xi = 0, \eta = 1/2$)

$$\frac{\partial \mathbf{U}}{\partial \xi} = \frac{1}{8}(-\mathbf{U}_6 + \mathbf{U}_9 + 3\mathbf{U}_8 - 3\mathbf{U}_5)$$

$$\frac{\partial \mathbf{U}}{\partial \eta} = \frac{1}{4}(-\mathbf{U}_6 - \mathbf{U}_9 + \mathbf{U}_8 + \mathbf{U}_5)$$

Control Surface CS4 ($\xi = -1/2, \eta = 0$)

$$\frac{\partial \mathbf{U}}{\partial \xi} = \frac{1}{4}(-\mathbf{U}_6 + \mathbf{U}_9 + \mathbf{U}_8 - \mathbf{U}_5)$$

$$\frac{\partial \mathbf{U}}{\partial \eta} = \frac{1}{8}(-\mathbf{U}_6 - 3\mathbf{U}_9 + 3\mathbf{U}_8 + \mathbf{U}_5)$$

Subcontrol Volume C

Control Surface CS4 ($\xi = -1/2, \eta = 0$)

$$\frac{\partial U}{\partial \xi} = \frac{1}{4}(-U_5 + U_8 + U_7 - U_4)$$

$$\frac{\partial U}{\partial \eta} = \frac{1}{8}(-3U_5 - U_8 + U_7 + 3U_4)$$

Control Surface CS1 ($\xi = 0, \eta = -1/2$)

$$\frac{\partial U}{\partial \xi} = \frac{1}{8}(-3U_5 + 3U_8 + U_7 - U_4)$$

$$\frac{\partial U}{\partial \eta} = \frac{1}{4}(-U_5 - U_8 + U_7 + U_4)$$

Subcontrol Volume D

Control Surface CS1 ($\xi = 0, \eta = -1/2$)

$$\frac{\partial U}{\partial \xi} = \frac{1}{8}(-3U_2 + 3U_5 + U_4 - U_1)$$

$$\frac{\partial U}{\partial \eta} = \frac{1}{4}(-U_2 - U_5 + U_4 + U_1)$$

Control Surface CS2 ($\xi = 1/2, \eta = 0$)

$$\frac{\partial U}{\partial \xi} = \frac{1}{4}(-U_2 + U_5 + U_4 - U_1)$$

$$\frac{\partial U}{\partial \eta} = \frac{1}{8}(-U_2 - 3U_5 + 3U_4 + U_1)$$

Assembly of the entire system is achieved by collecting contributions to an element from surrounding nodes in the first step and contributions to a node from surrounding elements in the second step, as shown in Figure 15.2.3.

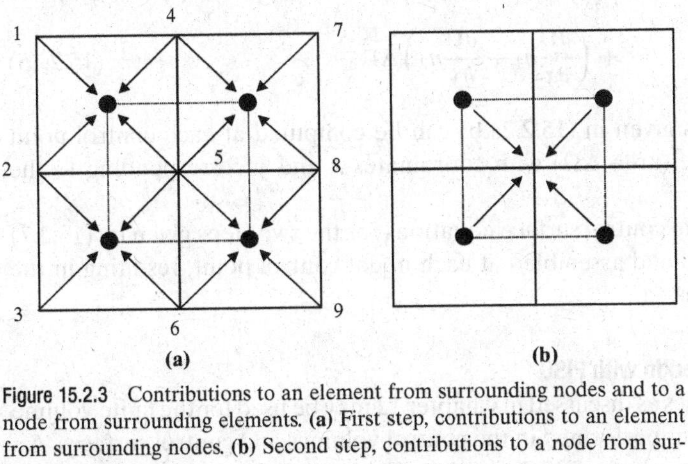

Figure 15.2.3 Contributions to an element from surrounding nodes and to a node from surrounding elements. **(a)** First step, contributions to an element from surrounding nodes. **(b)** Second step, contributions to a node from surrounding elements.

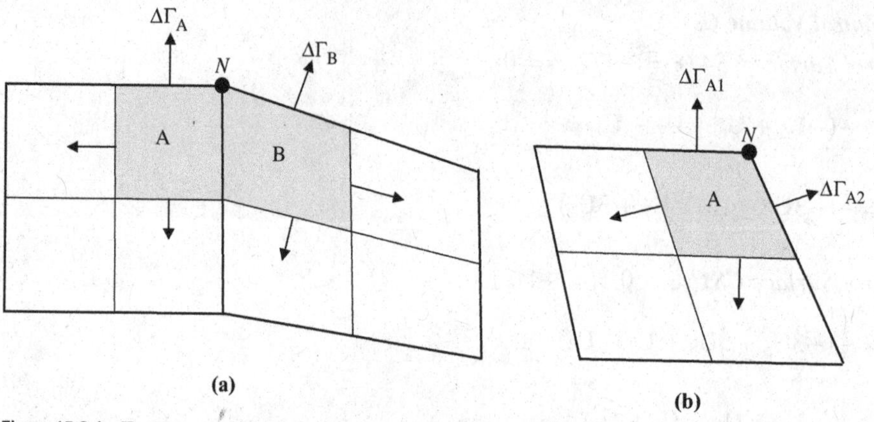

Figure 15.2.4 Treatment of Neumann boundary conditions. **(a)** Neumann boundary condition given at N joined by two inclined surfaces. **(b)** Neumann boundary condition given at corner node N.

Implementation of Neumann boundary conditions at the node with elements of inclined surfaces can be carried out easily. Consider the Neumann node N with two elements connected as shown in Figure 15.2.4a. The control surface equations for the Neumann node N become

$$\sum_{CS}\left(\frac{\partial \mathbf{U}}{\partial x}n_1+\frac{\partial \mathbf{U}}{\partial y}n_2\right)\Delta\Gamma=\sum_{CS3,2}^{A}\left(\frac{\partial \mathbf{U}}{\partial x}n_1+\frac{\partial \mathbf{U}}{\partial y}n_2\right)\Delta\Gamma+\sum_{CS4,3}^{B}\left(\frac{\partial \mathbf{U}}{\partial x}n_1+\frac{\partial \mathbf{U}}{\partial y}n_2\right)\Delta\Gamma$$
$$+\left(\frac{\partial \mathbf{U}}{\partial x}n_1+\frac{\partial \mathbf{U}}{\partial y}n_2\right)_A\Delta\Gamma_A+\left(\frac{\partial \mathbf{U}}{\partial x}n_1+\frac{\partial \mathbf{U}}{\partial y}n_2\right)_B\Delta\Gamma_B \tag{15.2.14}$$

The Neumann boundary conditions given at a corner with only one element (Figure 15.2.4b) can be implemented as

$$\sum_{CS}\left(\frac{\partial \mathbf{U}}{\partial x}n_1+\frac{\partial \mathbf{U}}{\partial y}n_2\right)\Delta\Gamma=\sum_{CS3,2}^{A}\left(\frac{\partial \mathbf{U}}{\partial x}n_1+\frac{\partial \mathbf{U}}{\partial y}n_2\right)\Delta\Gamma+\left(\frac{\partial \mathbf{U}}{\partial x}n_1+\frac{\partial \mathbf{U}}{\partial y}n_2\right)\Delta\Gamma_{A1}$$
$$+\left(\frac{\partial \mathbf{U}}{\partial x}n_1+\frac{\partial \mathbf{U}}{\partial y}n_2\right)\Delta\Gamma_{A2} \tag{15.2.15}$$

The source term $\mathbf{F}$ as given in (15.2.2a,b) can be computed at each control point using the control volume (area $\Delta\Omega$) with coordinates x and y corresponding to the control point.

All control volume and control surface equations for the two steps given by (15.2.7) and (15.2.8) are collected and assembled at each nodal control point, resulting in the global algebraic equations.

Generalized Petrov-Galerkin with PISO

Computational schemes as discussed in Chapter 12 may be used for the finite volume equations with integrations performed in the control volumes and control surfaces. An alternative technique is to use the PISO concept (Section 5.3.2), which is particularly conducive to FVM formulation. In this approach, the predictor corrector steps are constructed as follows.

Step 1. Predictor. Integrating the momentum equations and writing them in control volumes and control surfaces,

$$\sum_{CV} \rho\left(v_j^* - v_j^n\right)\frac{\Delta\Omega}{\Delta t} = -\sum_{CS} \rho\left(\bar{v}_i v_j - \mu v_{j,i} + p\delta_{ij}\right)n_i \Delta\Gamma \tag{15.2.16}$$

with

$\bar{v}_i =$ the old time step value

$p = \Phi_N p_N$

$$v_j = \begin{cases} W_N v_{Nj} & \text{in convective term} \\ \Phi_N v_{Nj} & \text{otherwise} \end{cases}$$

$W_N = \Phi_N + \Psi_N = \Phi_N + \beta g_k \Phi_{N,k}$

We may recast (15.2.16) in the form

$$K_N v_{Nj}^* = R_j \tag{15.2.17}$$

$$K_N = \sum_{CV} \rho \frac{\Delta\Omega}{\Delta t}\Phi_N + \sum_{CS} \left(\rho v_i^* W_N - \mu \Phi_{N,i}\right)n_i$$

$$R_j = \sum_{CV} \rho \frac{\Delta\Omega}{\Delta t} v_{Nj}^* \Phi_N - \sum_{CS} \Phi_N n_j p_N$$

Here we solve v_{Nj}^* implicitly:

Step 2. (Corrector I). The momentum control volume and the control surface equations are corrected as

$$\sum_{CV} \rho v_j^{**}\frac{\Delta\Omega}{\Delta t} = -\sum_{CS} \left(\rho v_i^* v_j^* - \mu v_{i,j}^* + p\delta_{ij}\right)n_i \Delta\Gamma + \sum_{CV} \rho v_j^n \frac{\Delta\Omega}{\Delta t} \tag{15.2.18}$$

To obtain the pressure correction equation, we differentiate spatially the momentum equation and integrate over the control volume in which we apply $v_{i,j}^{**} = 0$. The resulting control surface equations become

$$\sum_{CS} p_{,i}^* n_i \Delta\Gamma = -\sum_{CS} \left(\rho v_j^n n_j \frac{\Delta\Gamma}{\Delta t} - \rho v_{i,j}^* v_j^* + \rho v_{j,i}^* v_j^*\right) n_i \Delta\Gamma \tag{15.2.19}$$

where $v_{j,ji}^* = 0$ with linear variation of Φ_α. In this step we compute p^* from (15.2.19) and v_j^{**} from (15.2.18) explicitly.

Step 3. (Corrector II). This is exactly the same as step 2 with (*) replaced by (**) and (**) replaced by (***). We solve for pressure p^{**} using

$$\sum_{CS} \Phi_{N,i} n_i \Delta\Gamma p_N^{**} = -\sum_{CS} \left(\rho v_i^n n_i \frac{\Delta\Gamma}{\Delta t} - \rho v_{i,j}^{**} v_j^{**} + \rho v_{j,i}^{**} v_j^{**}\right) n_i \Delta\Gamma \tag{15.2.20}$$

and solve for velocity v_j^{***} explicitly using

$$\sum_{CV} \rho v_j^{***} \frac{\Delta\Omega}{\Delta t} = -\sum_{CS} (\rho v_i^{**} v_j^{**} - \mu v_{j,i}^{**} + p^{**}\delta_{ij}) n_i \Delta\Gamma + \sum_{CV} \rho v_j^n \frac{\Delta\Omega}{\Delta t} \tag{15.2.21}$$

The three steps are to be repeated until convergence is obtained.

15.2.2 INCOMPRESSIBLE AND COMPRESSIBLE FLOWS

(1) FVM with Two-step GTG Scheme

For the Burgers' equations considered in the previous sections, we evaluated derivatives along the control surfaces. If the Navier-Stokes system of equations is solved from the FVM equations of the type given by (15.1.1b), then we must evaluate the convection and diffusion fluxes ($\mathbf{F}_i$ and $\mathbf{G}_i$) directly along the boundary surfaces.

The FEM approximations for $\mathbf{U}$, $\mathbf{F}_i$, and $\mathbf{G}_i$ are given by

$$\begin{aligned} \Delta\mathbf{U} &= \Phi_N^{(e)} \Delta\mathbf{U}_N \\ \Delta\mathbf{F}_i &= \Phi_N^{(e)} \Delta\mathbf{F}_{Ni} \\ \Delta\mathbf{G}_i &= \Phi_N^{(e)} \Delta\mathbf{G}_{Ni} \end{aligned} \tag{15.2.22}$$

The two-step GTG scheme is the same as in (15.2.3):

Step 1

$$\sum_{CV} \mathbf{U}^{n+\frac{1}{2}} \Delta\Omega = \sum_{CV} (\mathbf{U}^n + \mathbf{B}^n)\Delta\Omega - \frac{\Delta t}{2} \sum_{CS} (\mathbf{F}_i^n + \mathbf{G}_i^n) n_i \Delta\Gamma \tag{15.2.23}$$

Step 2

$$\sum_{CV} \mathbf{U}^{n+1} \Delta\Omega = \sum_{CV} (\mathbf{U}^n + \mathbf{B}^n)\Delta\Omega - \frac{\Delta t}{2} \sum_{CS} (\mathbf{F}_i^{n+\frac{1}{2}} + \mathbf{G}_i^{n+\frac{1}{2}}) n_i \Delta\Gamma \tag{15.2.24}$$

The evaluation of $\mathbf{F}_i$, and $\mathbf{G}_i$ is carried out along the control surfaces, using (15.2.23 and 15.2.24) at the midpoints similarly as in the case of Burgers' equations presented in Section (15.2.1).

(2) FVM with PISO Approach

The FVM via FEM PISO approach can be extended to compressible flows similarly as in incompressible flows. This begins with integrating the momentum equations and writing them in control volumes and control surfaces,

$$\sum_{CV} \rho^n (v_j^* - v_j^n) \frac{\Delta\Omega}{\Delta t} = -\sum_{CS} [\rho^n \bar{v}_i v_j - \mu(v_{i,j} - v_{j,j}) + p\delta_{ij}] n_i \Delta\Gamma \tag{15.2.25}$$

The rest of the formulation follows the steps given in Section 6.3.4 by converting them into control volumes and control surfaces as shown in Section 15.2.2 for incompressible flows.

(3) FVM with Upwind Finite Elements

$$\frac{\partial \mathbf{U}}{\partial t} + \frac{\partial \mathbf{F}_i}{\partial x_i} + \frac{\partial \mathbf{G}_i}{\partial x_i} = 0$$

$$\int_\Omega \frac{\partial \mathbf{U}}{\partial t} d\Omega + \int_\Omega \left(\frac{\partial \mathbf{F}_i}{\partial x_i} + \frac{\partial \mathbf{G}_i}{\partial x_i} \right) d\Omega = \int_\Omega \frac{\partial \mathbf{U}}{\partial t} d\Omega + \int_\Gamma (\mathbf{F}_i + \mathbf{G}_i) n_i d\Gamma = 0 \quad (15.2.26)$$

(a) Inviscid Algorithm. Consider a typical flux change on the side r, s,

$$\Delta \mathbf{F}_i = \mathbf{F}_{ir} - \mathbf{F}_{is} = \mathbf{a}_i \Delta \mathbf{U} = \mathbf{a}_i (\mathbf{U}_r - \mathbf{U}_s) \quad \text{with } \mathbf{a}_i = \frac{\partial \mathbf{F}_i}{\partial \mathbf{U}} \quad (15.2.27)$$

in which we may use the Roe's average,

$$\mathbf{F}_i = \frac{1}{2}[\mathbf{F}_{ir} + \mathbf{F}_{is} - |\mathbf{a}_i|(\mathbf{U}_r - \mathbf{U}_s)] \quad (15.2.28)$$

as given by (6.2.67).

Implicit time stepping is constructed as

$$\Delta \mathbf{U}^{n+1} = \frac{\Delta t}{\Delta \Omega} \sum \frac{1}{2} [\mathbf{F}^n_{ir} + \mathbf{F}^n_{is} - |\mathbf{a}_i|(\mathbf{U}^n_r - \mathbf{U}_s)] n_i \Delta\Gamma \quad (15.2.29)$$

Linearizing, we get

$$\left(\mathbf{I} + \frac{\Delta t}{2\Delta\Omega} \sum |\mathbf{a}^*_i| n_i \Delta\Gamma \right) \Delta \mathbf{U}^{n+1} = \frac{\Delta t}{2\Delta\Omega} \sum [\mathbf{F}^n_{ir} + \mathbf{F}^*_{is} - |\mathbf{a}^*_i|(\mathbf{U}^*_r - \mathbf{U}^n_s)] n_i \Delta\Gamma \quad (15.2.30)$$

Here the linearization is performed with an iterative solution in mind, and the asterisk indicates that the term is evaluated using the latest available solution in an adjacent element. Then the iterative procedure may be regarded as a point Gauss-Seidel method requiring the inversion of a 4×4 matrix for each element in the computational grid.

(b) Viscous Contributions. The inviscid equation (15.2.31) may be modified to include the viscous contributions. Noting that

$$\int_\Gamma \mathbf{G}^{n+1}_i n_i d\Gamma = \int_\Gamma (\mathbf{G}^n_i + \Delta \mathbf{G}_i) n_i d\Gamma$$

or

$$\sum \mathbf{G}^{n+1}_i n_i d\Gamma = \sum (\mathbf{G}^n_i + \mathbf{b}_i \Delta \mathbf{U}) n_i d\Gamma \quad (15.2.31)$$

Substituting (15.2.32) into (15.2.26) through (15.2.31) we obtain

$$\left[\mathbf{I} + \frac{\Delta t}{\Delta\Omega} \sum \left(\frac{1}{2} |\mathbf{a}^*_i| - \mathbf{b}_i \right) n_i \Delta\Gamma \right] \Delta \mathbf{U}^{n+1} = -\frac{\Delta t}{\Delta\Omega} \sum \left\{ \frac{1}{2} [\mathbf{F}^n_{ir} + \mathbf{F}^*_{is} - |\mathbf{a}^*_i|(\mathbf{U}^*_r - \mathbf{U}^n_s)] - \mathbf{G}^*_i \right\} n_i \Delta\Gamma \quad (15.2.32)$$

The Galerkin approximation of (15.2.30 or 15.2.32) with the upwinded finite element equations in the finite volume formulation leads to

$$(A_{\alpha\beta}\delta_{rs} + B_{\alpha\beta rs})\Delta \mathbf{U}_{\beta s} = W_{\alpha r} \tag{15.2.33}$$

Here the diffusion terms are calculated along the control surfaces similarly as the convection terms.

(4) FVM with FDV

The FDV concept introduced in Sections 6.5 and 13.6 can be used for FVM formulations. To this end, we begin with the FDV governing equations,

$$\mathbf{R} = \left(\mathbf{I} + \mathbf{E}_i^n \frac{\partial}{\partial x_i} + \mathbf{E}_{ij}^n \frac{\partial^2}{\partial x_i \partial x_j}\right)\Delta\mathbf{U}^{n+1} + \mathbf{Q}^n \tag{15.2.34}$$

The FVM integration equation is of the form

$$\int_\Omega \mathbf{R}\, d\Omega = \int_\Omega \left[\left(\mathbf{I} + \mathbf{E}_i^n \frac{\partial}{\partial x_i} + \mathbf{E}_{ij}^n \frac{\partial^2}{\partial x_i \partial x_j}\right)\Delta\mathbf{U}^{n+1} + \mathbf{Q}^n\right] d\Omega = 0 \tag{15.2.35}$$

Integrating (15.2.35) with respect to the spatial coordinates, we obtain

$$\int_\Omega \Delta\mathbf{U}^{n+1}\, d\Omega + \int_\Gamma \left(\mathbf{E}_i \Delta\mathbf{U}^{n+1} + \mathbf{E}_{ij}\Delta\mathbf{U}_{,j}^{n+1}\right) n_i\, d\Gamma = -\int_\Omega \mathbf{Q}^n\, d\Omega \tag{15.2.36}$$

or

$$\sum_{CV} \Delta\mathbf{U}^{n+1}\, \Delta\Omega + \sum_{CS} \left(\mathbf{E}_i \Delta\mathbf{U}^{n+1} + \mathbf{E}_{ij}\Delta\mathbf{U}_{,j}^{n+1}\right) n_i\, \Delta\Gamma = -\int_\Omega \mathbf{Q}^n\, d\Omega \tag{15.2.37}$$

where

$$\int_\Omega \mathbf{Q}^n\, d\Omega = \int_\Gamma \left(\mathbf{H}_i^n + \mathbf{H}_{ij,j}^n\right) n_i\, d\Gamma = \sum_{CS} \left(\mathbf{H}_i^n + \mathbf{H}_{ij,j}^n\right) n_i\, \Delta\Gamma \tag{15.2.38}$$

with

$$\mathbf{H}_i^n = \Delta t\left(\mathbf{F}_i^n + \mathbf{G}_i^n\right), \quad \mathbf{H}_{ij}^n = \frac{\Delta t^2}{2}(\mathbf{a}_i + \mathbf{b}_i)\left(\mathbf{F}_j^n + \mathbf{G}_j^n\right) \tag{15.2.39a,b}$$

15.2.3 THREE-DIMENSIONAL PROBLEMS

Three-dimensional geometries may be discretized using hexahedral elements or tetrahedral elements. Determination of direction cosines for the subcontrol surfaces, subcontrol surface areas, and subcontrol volumes follows the same procedures for FVM via FDM. Formulations and solutions of FVM equations via FEM for three-dimensional problems are carried out similarly as in the two-dimensional case which has been detailed in Section 15.2.

Although hexahedral elements are easy for implementation in general, we may use tetrahedrals with each volume subdivided internally into four volumes corresponding to each vertexs, as shown in Figure 15.2.5a. Within a single tetrahedral, each node shares a common face with each of the neighboring nodes within the tetrahedral. The Green-Gauss theorem is applied to the sub-volume surrounding each vertex to equate the change in mass, momentum, and energy to the convective and diffusive fluxes passing

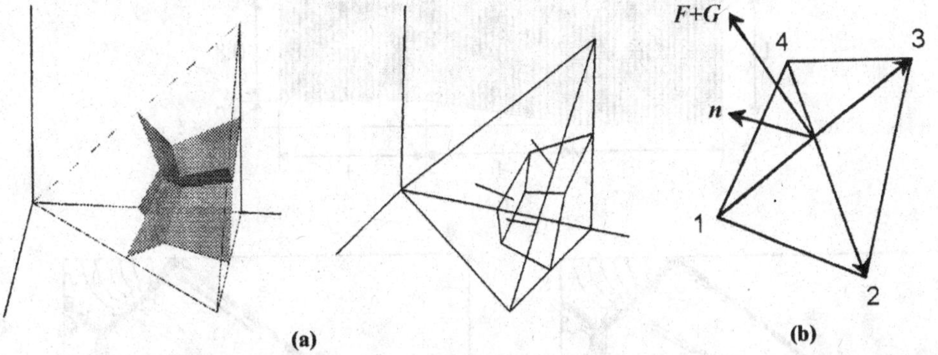

Figure 15.2.5 Tetrahedral element discretization and control volume representation (a) Tetrahedral element discretization (b) Flux through tetrahedral control volume.

through the control volume faces. Surface normals for each face are obtained via a cross-product as shown in Figure 15.2.5b. Finite element shape functions are used to interpolate the convective and diffusive fluxes at the center of each face.

An overall balance is obtained for a given nodal point by summing the contributions from all of the tetrahedral subvolumes within the mesh that happen to contain the given nodal point. (The nodal control volume is the sum of all of the subvolumes from the tetrahedrals that contain the node.) Note that the fluxes between adjacent tetrahedral volumes cancel since the flux is contained within a single nodal control volume, while identical fluxes through tetrahedral surfaces exposed on the external boundary do not.

15.3 EXAMPLE PROBLEMS

(1) Two-Dimensional Euler Equations, Scramjet Flame Holder Problem

Given:

$$\frac{\partial \mathbf{U}}{\partial t} + \frac{\partial \mathbf{F}_i}{\partial x_i} = 0$$

Inlet Boundary Conditions:

$$\gamma = 1.4, \qquad R = 1716\frac{\text{ft}^2}{\text{s}^2\,{}^\circ\text{R}} = 287\frac{\text{m}^2}{\text{s}^2\,{}^\circ\text{K}}$$

$$M = 2, \qquad \rho = 0.002378\frac{\text{slugs}}{\text{ft}^3} = 1.2215\frac{\text{kg}}{\text{m}^3}$$

$$\text{v} = 0, \qquad p = 2116\frac{\text{lbf}}{\text{ft}^2} = 101314.08\,\text{Pa}$$

Outlet Boundary Conditions. Supersonic outflow

Initial Conditions. Use inlet boundary conditions as initial conditions for all nodes.

Required: Use FVM via FEM using two step TGM.

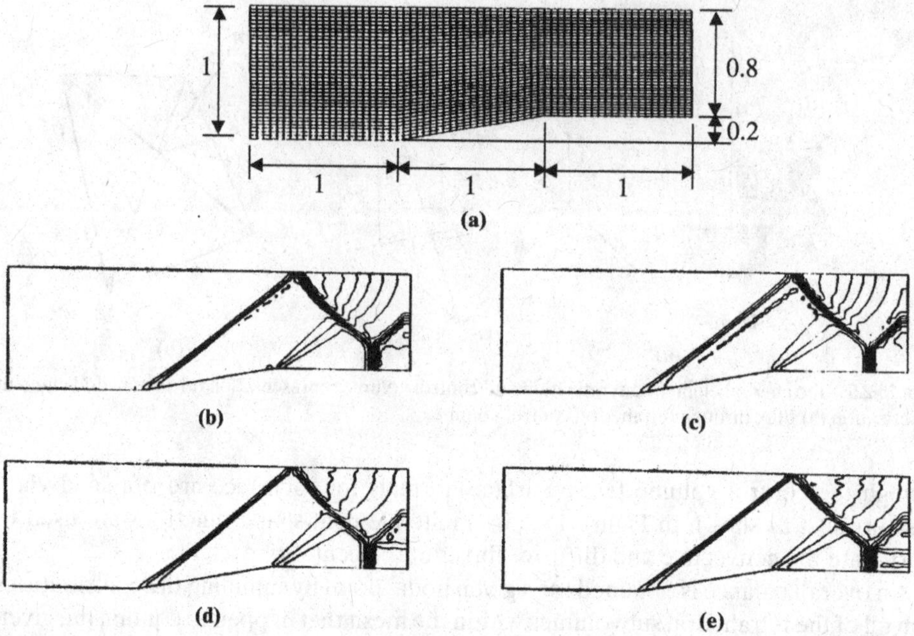

Figure 15.3.1 Solution of Euler equation by FVM-FEM. (a) Geometry and discretization. (b) Density contours. (c) Pressure. (d) Temperature contours. (e) Mach number contours.

Solution Procedure: The two steps given by (15.2.23) and 15.2.24) will be followed. Here the diffusion terms are zero and the details of the evaluation of convection terms along the control surfaces are calculated as follows:

Step 1

$$\mathbf{U}^{n+\frac{1}{2}} = \mathbf{U}^n - \frac{\Delta t}{2}\left(\frac{\partial \mathbf{F}_x^n}{\partial x} + \frac{\partial \mathbf{F}_y^n}{\partial y}\right)$$

or

$$\begin{aligned}
\mathbf{U}_e^{n+\frac{1}{2}} &= \mathbf{U}_e^n - \frac{\Delta t}{2}\sum_{CS}\left(\mathbf{F}_x^n n_1 + \mathbf{F}_y^n n_2\right)\frac{\Delta\Gamma}{\Delta\Omega_e} \\
&= \frac{1}{4}\left(\mathbf{U}_1^n + \mathbf{U}_2^n + \mathbf{U}_3^n + \mathbf{U}_4^n\right) \\
&\quad - \frac{\Delta t}{4}\{[(\mathbf{F}_{x1}^n + \mathbf{F}_{x2}^n)n_1 + (\mathbf{F}_{y1}^n + \mathbf{F}_{y2}^n)n_2]\Delta\Gamma_1 \\
&\quad + [(\mathbf{F}_{x2}^n + \mathbf{F}_{x3}^n)n_1 + (\mathbf{F}_{y2}^n + \mathbf{F}_{y3}^n)n_2]\Delta\Gamma_2 \\
&\quad + [(\mathbf{F}_{x3}^n + \mathbf{F}_{x4}^n)n_1 + (\mathbf{F}_{y3}^n + \mathbf{F}_{y4}^n)n_2]\Delta\Gamma_3 \\
&\quad + [(\mathbf{F}_{x4}^n + \mathbf{F}_{x1}^n)n_1 + (\mathbf{F}_{y4}^n + \mathbf{F}_{y1}^n)n_2]\Delta\Gamma_4\}/\Delta\Omega_e
\end{aligned}$$

Step 2

$$\mathbf{U}^{n+1} = \mathbf{U}^n - \frac{\Delta t}{2}\left(\frac{\partial \mathbf{F}_x^{n+\frac{1}{2}}}{\partial x} + \frac{\partial \mathbf{F}_y^{n+\frac{1}{2}}}{\partial y}\right)$$

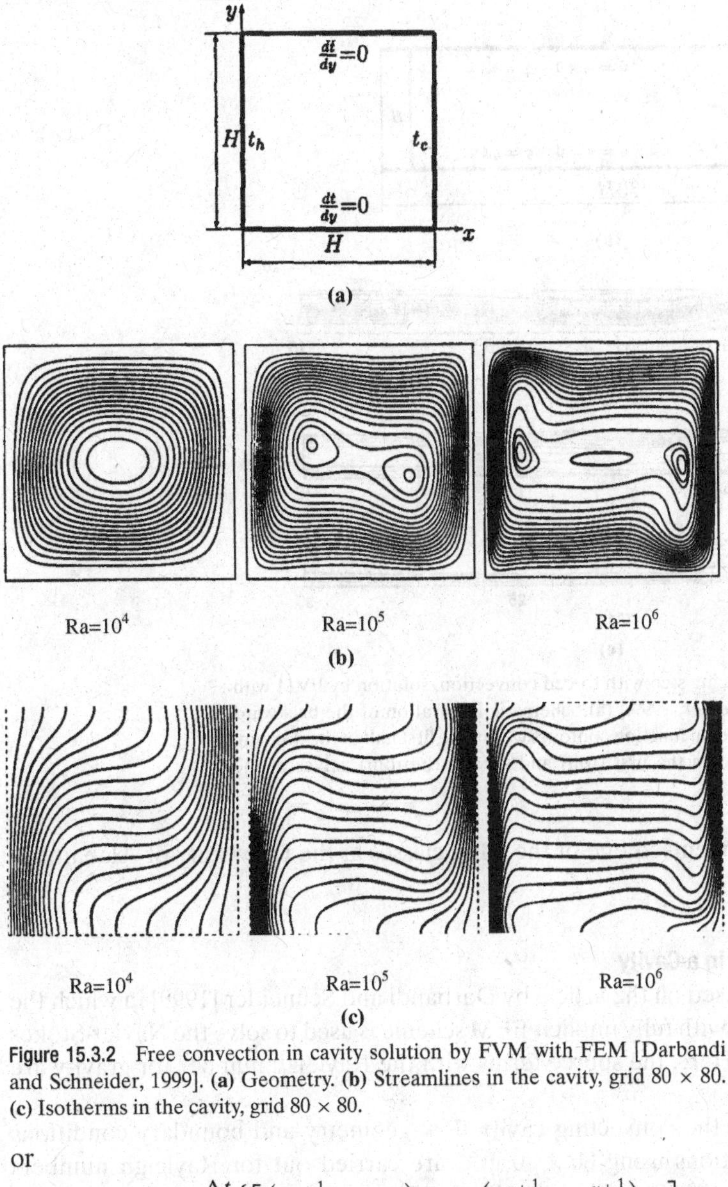

Figure 15.3.2 Free convection in cavity solution by FVM with FEM [Darbandi and Schneider, 1999]. **(a)** Geometry. **(b)** Streamlines in the cavity, grid 80 × 80. **(c)** Isotherms in the cavity, grid 80 × 80.

or

$$\begin{aligned}\mathbf{U}_e^{n+1} = \mathbf{U}_e^n - \frac{\Delta t}{2}\Big\{&\Big[\Big(\mathbf{F}_{xe1}^{n+\frac{1}{2}} + \mathbf{F}_{xe2}^{n}\Big)n_1 + \Big(\mathbf{F}_{ye1}^{n+\frac{1}{2}} + \mathbf{F}_{ye2}^{n+\frac{1}{2}}\Big)n_2\Big]\Delta\Gamma_1 \\ &+ \Big[\Big(\mathbf{F}_{xe2}^{n+\frac{1}{2}} + \mathbf{F}_{xe3}^{n+\frac{1}{2}}\Big)n_1 + \Big(\mathbf{F}_{ye2}^{n+\frac{1}{2}} + \mathbf{F}_{ye3}^{n+\frac{1}{2}}\Big)n_2\Big]\Delta\Gamma_2 \\ &+ \Big[\Big(\mathbf{F}_{xe3}^{n+\frac{1}{2}} + \mathbf{F}_{xe4}^{n+\frac{1}{2}}\Big)n_1 + \Big(\mathbf{F}_{ye3}^{n+\frac{1}{2}} + \mathbf{F}_{ye4}^{n+\frac{1}{2}}\Big)n_2\Big]\Delta\Gamma_3 \\ &+ \Big[\Big(\mathbf{F}_{xe4}^{n+\frac{1}{2}} + \mathbf{F}_{xe1}^{n+\frac{1}{2}}\Big)n_1 + \Big(\mathbf{F}_{ye4}^{n+\frac{1}{2}} + \mathbf{F}_{ye1}^{n+\frac{1}{2}}\Big)n_2\Big]\Delta\Gamma_4\Big\}\Big/\Delta\Omega_e\end{aligned}$$

The above procedure was carried out, using the geometry and discretization (2479 nodes) as shown in Figure 15.3.1a. It is seen that shock waves develop at the compression corner and expansion waves at the expansion corner as expected. This work is a part of

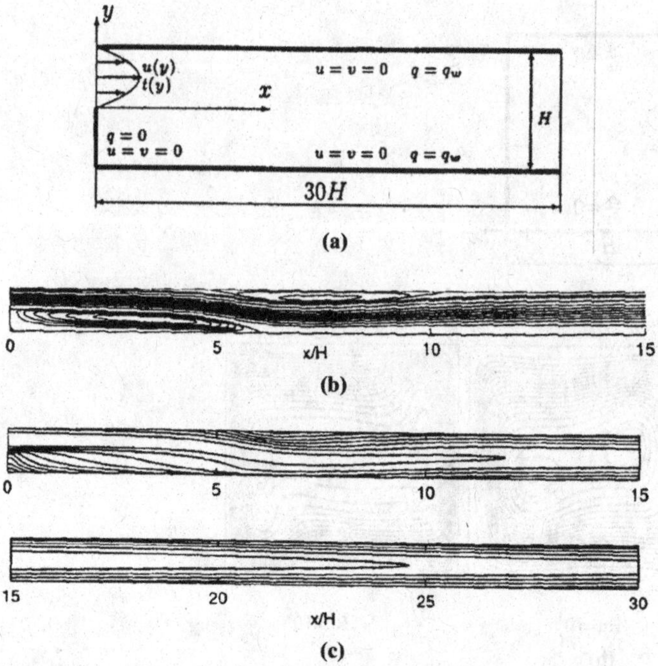

Figure 15.3.3 Backward facing step with forced convection, solution by FVM with FEM [Darbandi and Schneider, 1999.] **(a)** Schematic illustration of the backward facing step problem. **(b)** Stream function contours within the first half of the domain, grid 80 × 20. **(c)** Isotherms in the first (top) and second (bottom) halves of the domain.

the homework assignments in one of the CFD classes at the University of Alabama in Huntsville.

(2) Free Convection in a Cavity

This example is based on the article by Darbandi and Schneider [1999] in which the finite volume method with fully implicit FEM scheme is used to solve the Navier-Stokes system of equations. Here, the source terms with the Rayleigh number for gravity are also included.

In Figure 15.3.2a, the convecting cavity flow geometry and boundary conditions are shown. Computations using 80 × 80 grid are carried out for Rayleigh numbers of $Ra = 10^4, 10^5$, and 10^6. The corresponding results are shown in Figure 15.3.2b and 15.3.2c for the isotherms and streamlines, respectively. Effects of Rayleigh numbers are clearly shown, with distorted distributions being more prominent for higher Rayleigh numbers. Further details are found in Darbandi and Schneider [1999].

(3) Backward Facing Step with Forced Convection

Another example reported by the same authors above is the backward facing step with forced convection (Figure 15.3.3a). Solutions using 80 × 20 grid show stream function contours and isotherms in Figures 15.3.3b and 15.3.3c, respectively. The advantages of using FVM with FEM have been demonstrated in this work with further details found in Darbandi and Schneider [1999].

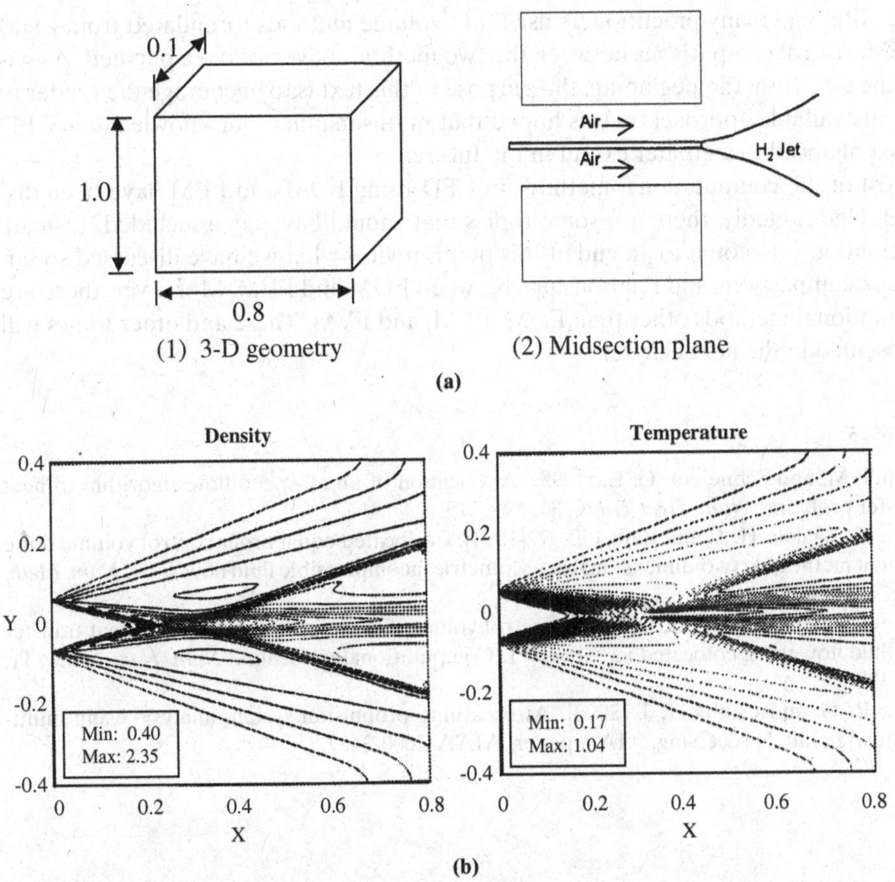

Figure 15.3.4 Density and temperature distributions, supersonic hydrogen-air injection flow analysis using finite volume tetrahedral elements (nonreacting case) with FVM-FDM-FDV [Schunk and Chung, 2000]. **(a)** Analysis by FVM with tetrahedral elements of Figure 15.2.5. **(b)** Density and temperature contours for nonreacting flowfield.

(4) Three-Dimensional Supersonic Propulsion Injection Flows

This is an example to demonstrate the use of three-dimensional tetrahedral elements with FVM-FE-FDV as shown in Figure 15.2.5 [Schunk and Chung, 2000]. Pure hydrogen is injected into a Mach 1.9 airstream at 1495 K (Figure 15.3.4a). The hydrogen is injected at Mach 2.0 and 251 K. Hydrogen is preburned in the air stream to produce a flow that contains 28% water along with 48% hydrogen and 24% oxygen. The static pressure of both the jet and the airstream is 1 atmosphere.

Steady-state density and temperature contours are shown in Figure 15.3.4b for the nonreacting flow case. It is shown that expansion waves are formed as the air flow is turned into and mixes with the hydrogen jet. Downstream, oblique shocks are formed as the main flow is turned back parallel with the free stream.

15.4 SUMMARY

In this chapter, we have shown that the finite volume methods can be formulated using FEM. This is the counterpart of Chapter 7 where the FDM was used to formulate

FVM. Although many practitioners use finite volume methods formulated from FDM or FEM, critical comparisons between the two methods have not been pursued. As has been the case from the beginning, the purpose of this text is to encourage the reader to learn all available approaches. It is hoped that in this manner, our knowledge in CFD will be enhanced to a greater extent in the future.

Most of the computational methods in CFD using FDM and FEM have been discussed. Undoubtedly, there are some topics that should have been included. Instead, our intention is to come to an end at this point, review what we have discussed so far, and seek comparisons and relationships between FDM and FEM. Moreover, there are computational methods other than FDM, FEM, and FVM. These and other topics will be presented in the next chapter.

REFERENCES

Darbandi, M. and Schneider, G. E. [1999]. Application of an all-speed flow algorithm to heat transfer problems. *Num. Heat Trans.*, 35, 695–715.

Masson, C., Saabas, H. J., and Baliga, B. R. [1994]. Co-located equal order control volume finite element method for two-dimensional axisymmetric incompressible fluid flow. *Int. J. Num. Meth. Eng.*, 18, 12–26.

Schneider, G. E. and Raw, M. J. [1987]. Control volume finite element method for heat transfer and fluid flow using colocated variables – 1. Computational procedure. *Num. Heat Trans.*, 11, 363–399.

Schunk, R. G. and Chung, T. J. [2000]. Airbreathing propulsion system analysis using multithreaded parallel processing. AIAA paper, AIAA-2000-3467.

CHAPTER SIXTEEN

Relationships Between Finite Differences and Finite Elements and Other Methods

Our explorations on the methods of finite differences and finite elements have come to an end. In Chapter 1, it was intended that the reader recognize the analogy between these two methods in one dimension. In fact, such an analogy exists for linear problems in all multidimensional geometries as long as the grid configurations are structured. In structured grids, with adjustments of the temporal parameters in generalized Galerkin methods and both temporal and convection diffusion parameters in generalized Petrov-Galerkin methods, the analogy between finite difference methods (FDM) and finite element methods (FEM) can be shown to exist also.

Traditionally, FEM equations are developed in unstructured grids as well as in structured grids. The FEM equations written in unstructured grids have global nodes irregularly connected around the entire domain, thus resulting in a large sparse matrix system, but the data management can be handled efficiently by using the element-by-element (EBS) assembly as discussed in Sections 10.3.2 and 11.5. FDM equations cannot be written in unstructured grids unless through FVM formulations. Thus, the FDM equations written only in structured grids cannot be directly compared with FEM equations written in general unstructured grids. Thus, the notion of FEM being more complicated, requiring more computer time than FDM, is an unfortunate comparison. For fair comparisons, FEM equations must be written in structured grids as in FDM.

In unstructured adaptive methods (Chapter 19), our assessments as to the merits and demerits of FDM versus FEM will be faced with a new challenge. This is because adaptive methods are instrumental in resolving many problems of numerical difficulties such as in shock waves and turbulence, making the fair comparison between FDM and FEM difficult.

Additionally, there are special numerical schemes in which both FDM and FEM are involved such as in DGM (discontinuous Galerkin methods, Section 13.5), FVM via FDM (Chapter 7), and FVM via FEM (Chapter 15). The most logical and simple comparison between FDM and FEM can be made in the flowfield-dependent variation (FDV) methods in which FDM (Section 6.5) and FEM (Section 13.6) contribute only through their unique discretization schemes, because all the physics required are already contained in the FDV equations. Indeed, it was demonstrated in Sections 6.8 and 13.7 that the choice between FDM and FEM is inconsequential if FDV equations are used.

Although the analogy between FDM and FEM is well understood, we must recognize some differences. One of the most significant differences between these two methodologies is the variational (or weak) formulation employed in FEM, not only for the governing equations but also for all constraint conditions particularly useful for solution stability and accuracy. Any number of variational constraint conditions can be introduced and simply added to the variational forms of the governing equations. This subject was covered in Chapters 11 through 14.

Thus, in this chapter, we are first concerned with analogies between FDM and FEM, with finite element equations written only in structured grids. We begin with simple elliptic, parabolic, and hyperbolic equations, followed by non-linear, multidimensional, and unstructured grid systems.

Historically, many methods other than FDM, FEM, and FVM have been developed, which are efficient for certain types of problems in physics and engineering. They include boundary element methods (BEM), coupled Eulerian-Lagrangian (CEL) methods, particle-in-cell (PIC) methods, and Monte Carlo methods (MCM), among others. For the sake of completeness, these methods will be briefly discussed in this chapter.

16.1 SIMPLE COMPARISONS BETWEEN FDM AND FEM

(1) Elliptic Equations

Consider an elliptic equation of the form

$$\frac{\partial^2 u}{\partial x^2} + \frac{\partial^2 u}{\partial y^2} = 0 \tag{16.1.1}$$

Using the four linear triangular elements, arranged in structured grids as shown Figure 16.1.1a, the assembled 5 × 5 finite element equations via SGM (Section 10.1) provide the global equation at nodes corresponding to (16.1.1) as follows:

$$\frac{u_4 - 2u_5 + u_2}{\Delta x^2} + \frac{u_1 - 2u_5 + u_3}{\Delta y^2} = 0 \tag{16.1.2}$$

This is identical to the five-point FDM equation written for the case of Figure 16.1.1b.

Similarly, it can be shown that the finite element equation for either eight linear triangular elements or four linear rectangular elements written at node 5 (Figure 16.1.1c) is identical to the nine-point FDM formula (Figure 16.1.1d) as follows:

$$u_1 + u_3 + u_7 + u_9 - \frac{2(\Delta x^2 - 5\Delta y^2)}{\Delta x^2 + \Delta y^2}(u_4 + u_6) + \frac{2(5\Delta x^2 - \Delta y^2)}{\Delta x^2 + \Delta y^2}(u_2 - u_8) - 20u_5 = 0 \tag{16.1.3}$$

The solution of these equations may be carried out using the procedure of FDM such as Jacobi iteration method, point Gauss-Seidel iteration, line Gauss-Seidel iteration, point successive over-relaxation, line successive relaxation, or alternating direction implicit (ADI) method, as discussed in Chapter 4.

(2) Parabolic Equations

A typical parabolic equation is given by

$$\frac{\partial u}{\partial t} - \alpha\frac{\partial^2 u}{\partial x^2} = 0 \tag{16.1.4}$$

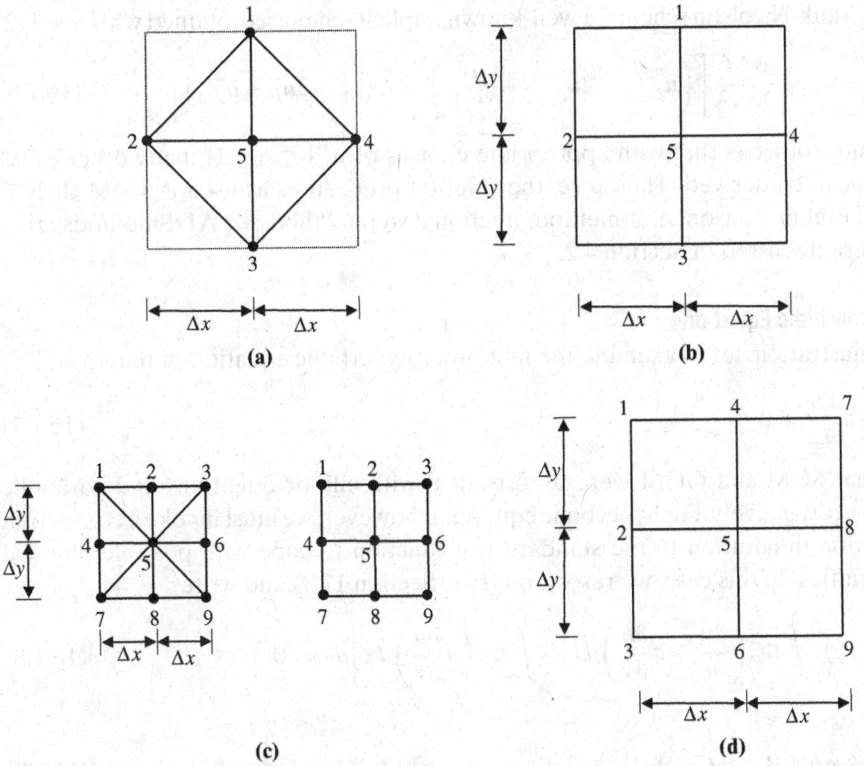

Figure 16.1.1 Analogy between FEM and FDM. **(a)** 4 × 4 finite element equations. **(b)** 5-point finite difference equations. **(c)** 9 × 9 finite element equations. **(d)** 9-point finite difference equations.

The finite element equations using GGM (Section 10.2) with linear approximations are of the form

$$(A_{\alpha\beta} + \eta \Delta t K_{\alpha\beta}) u_\beta^{n+1} = [A_{\alpha\beta} + (1-\eta)\Delta t K_{\alpha\beta}] u_\beta^n$$

where the Neumann boundary conditions are assumed to vanish. The local element stiffness matrix and lumped mass matrix are, respectively,

$$K_{NM}^{(e)} = \frac{1}{\Delta x}\begin{bmatrix} 1 & -1 \\ -1 & 1 \end{bmatrix}$$

$$A_{NM}^{(e)} = \frac{\Delta x}{2}\begin{bmatrix} 1 & 0 \\ 0 & 1 \end{bmatrix}$$

Here, the lumped mass matrix is used instead of the consistent mass matrix in order to arrive at the results identical to the finite difference equations.

Assembly of two equal elements with three nodes leads to the global finite element equation for the center node i in terms of the end nodes $i-1$ and $i+1$, with $\eta = 0$:

$$u_i^{n+1} = u_i^n + d(u_{i+1}^n - 2u_i^n + u_{i-1}^n) \tag{16.1.5}$$

This is an explicit scheme known as FTCS finite difference formula.

The Crank-Nicolson scheme, a well-known implicit scheme is obtained with $\eta = 1/2$,

$$u_i^{n+1} = u_i^n + \frac{\alpha\Delta t}{2\Delta x^2}\left[\left(u_{i+1}^{n+\frac{1}{2}} - 2u_i^{n+\frac{1}{2}} + u_{i-1}^{n+\frac{1}{2}}\right) + \left(u_{i+1}^n - 2u_i^n + u_{i-1}^n\right)\right] \tag{16.1.6}$$

It is now obvious that with appropriate choices of $\eta(0 \le \eta \le 1)$ many other FDM formulas can be derived. Therefore, the solution procedures as used in FDM such as DuFort-Frankel, Laasonen, β-method, fractional step methods, or ADI methods arise, which were discussed in Section 4.2.

(3) Hyperbolic Equations

For illustration, let us examine the first order hyperbolic equation of the form

$$\frac{\partial u}{\partial t} + a\frac{\partial u}{\partial x} = 0 \tag{16.1.7}$$

Recall that SGM and GGM were used to deal with elliptic equations and parabolic equations, respectively. For hyperbolic equations, however, we must invoke a convection test function in addition to the standard test function to cope with possible physical discontinuities. In this case, we resort to GPG (Section 11.3) and write

$$\int_0^1 W(\xi)\left[\int \Phi_\alpha\left(\frac{\partial u}{\partial t} + a\frac{\partial u}{\partial x}\right)dx + \int \Psi_\alpha\left(a\frac{\partial u}{\partial x}\right)dx\right]d\xi = 0 \tag{16.1.8}$$

or

$$[A_{\alpha\beta} + \eta\Delta t(B_{\alpha\beta} + C_{\alpha\beta})]u_\beta^{n+1} = [A_{\alpha\beta} - (1-\eta)\Delta t(B_{\alpha\beta} + C_{\alpha\beta})]u_\beta^n \tag{16.1.9}$$

For two elements with three nodes with lumped mass, we obtain

$$\left\{\frac{\Delta x}{2}\begin{bmatrix}1 & 0 & 0\\ 0 & 2 & 0\\ 0 & 0 & 1\end{bmatrix} + \frac{\eta\Delta t a}{2}\begin{bmatrix}-1 & 1 & 0\\ -1 & 0 & 1\\ 0 & -1 & 1\end{bmatrix} + \eta\Delta t a\alpha\begin{bmatrix}1 & -1 & 0\\ -1 & 2 & -1\\ 0 & -1 & 1\end{bmatrix}\right\}\begin{bmatrix}u_1\\ u_2\\ u_3\end{bmatrix}^{n+1}$$

$$= \left\{\frac{\Delta x}{2}\begin{bmatrix}1 & 0 & 0\\ 0 & 2 & 0\\ 0 & 0 & 1\end{bmatrix} - (1-\eta)\frac{\Delta t a}{2}\begin{bmatrix}-1 & 1 & 0\\ -1 & 0 & 1\\ 0 & -1 & 1\end{bmatrix}\right.$$

$$\left. + (1-\eta)\Delta t a\alpha\begin{bmatrix}1 & -1 & 0\\ -1 & 2 & -1\\ 0 & -1 & 1\end{bmatrix}\right\}\begin{bmatrix}u_1\\ u_2\\ u_3\end{bmatrix}^n$$

Expanding at node 2 or i in terms of $i-1$ and $i+1$ nodes, we have

$$u_i^{n+1} + \eta\frac{\Delta t a}{\Delta x}\left[\left(\frac{1}{2} - \alpha\right)u_{i+1} + 2\alpha u_i - \left(\frac{1}{2} + \alpha\right)u_{i-1}\right]^{n+1}$$

$$= u_i^n - (1-\eta)\frac{\Delta t a}{\Delta x}\left[\left(\frac{1}{2} - \alpha\right)u_{i+1} + 2\alpha u_i - \left(\frac{1}{2} + \alpha\right)u_{i-1}\right]^n \tag{16.1.10}$$

With appropriate choices of the temporal parameter $\eta(0 \le \eta \le 1)$ and the convection parameter $\alpha(a \le \alpha \le b)$ with a and b satisfying both the stability and accuracy criteria (11.3.20, 11.3.22), we arrive at various finite difference schemes.

With $\eta = 0$ and $\alpha = 1/2$ we obtain the FTBS scheme,

$$\frac{u_i^{n+1} - u_i^n}{\Delta t} = -a\frac{(u_i^n - u_{i-1}^n)}{\Delta x} \tag{16.1.11}$$

To demonstrate that the Lax-Wendroff scheme can be derived, we begin with the Taylor Series expansion of (16.1.7) in the form

$$u_i^{n+1} = u_i^n - a\Delta t\frac{\partial u}{\partial x} + \frac{(a\Delta t)^2}{2}\frac{\partial^2 u}{\partial x^2} \tag{16.1.12}$$

or the equivalent partial differential equation,

$$\frac{\partial u}{\partial t} = -a\frac{\partial u}{\partial x} + \frac{a^2\Delta t}{2}\frac{\partial^2 u}{\partial x^2} \tag{16.1.13}$$

The GPG formulation of (16.1.13) leads to

$$\int \Phi_\alpha\left(\frac{\partial u}{\partial t} + a\frac{\partial u}{\partial x} - \frac{a^2\Delta t}{2}\frac{\partial^2 u}{\partial x^2}\right)dx + \int \Psi_\alpha\left(a\frac{\partial u}{\partial x}\right)dx = 0 \tag{16.1.14}$$

Integrating by parts and rearranging, we obtain

$$\begin{aligned} &u_i^{n+1} + \eta\frac{\Delta t a}{\Delta x}\left[\left(\frac{1}{2} - \alpha\right)u_{i+1} + 2\alpha u_i - \left(\frac{1}{2} + \alpha\right)u_{i-1}\right]^{n+1} \\ &- \eta a^2\frac{\Delta t}{2\Delta x^2}(u_{i+1} - 2\alpha u_i + u_{i-1})^{n+1} = u_i^n - (1-\eta)\frac{\Delta t a}{\Delta x}\left[\left(\frac{1}{2} - \alpha\right)u_{i+1}\right. \\ &\left. + 2\alpha u_i - \left(\frac{1}{2} + \alpha\right)u_{i-1}\right]^n + a^2\frac{\Delta t}{2\Delta x^2}(u_{i+1} - 2\alpha u_i + u_{i-1})^n \end{aligned} \tag{16.1.15}$$

For $\eta = 0$ and $\alpha = 0$, (16.1.15) becomes

$$u_i^{n+1} = u_i^n - \frac{a\Delta t}{2\Delta x}(u_{i+1}^n - u_{i-1}^n) + \frac{(a\Delta t)^2}{2\Delta x^2}(u_{i+1}^n - 2u_i^n - u_{i-1}^n) \tag{16.1.16}$$

This is identical to the explicit Lax-Wendroff scheme presented in (4.3.15).

Implicit schemes such as Euler FTCS and Crank-Nicolson are generated as follows:

Euler FTCS ($\eta = 1$ *and* $\alpha = 0$)

$$\frac{u_i^{n+1} - u_i^n}{\Delta t} = \frac{a(u_{i+1}^{n+1} - u_{i-1}^{n+1})}{2\Delta x} \tag{16.1.17}$$

Crank-Nicolson ($\eta = 1/2$ *and* $\alpha = 0$)

$$\frac{u_i^{n+1} - u_i^n}{\Delta t} = -\frac{a}{2}\left[\frac{(u_{i+1}^{n+1} - u_{i-1}^{n+1})}{2\Delta x} + \frac{(u_{i+1}^n - u_{i-1}^n)}{2\Delta x}\right] \tag{16.1.18}$$

Obviously, many other difference schemes can be derived using the unlimited rages of η and α through the GPG formulations. Once the finite element equations are obtained in the form analogous to finite difference equations, then the FDM solution procedure can be followed as long as structured grid configurations are used.

16.2 RELATIONSHIPS BETWEEN FDM AND FDV

It was suggested in Section 6.5 that almost all existing FDM schemes can arise from the FDV scheme. We examine the analogies of FDV to some of the FDM schemes in this section.

Referring to (6.5.13 or 13.6.2) with the source terms neglected, we write

$$\left\{\mathbf{I}+\Delta t\,(s_1\mathbf{a}_i+s_3\mathbf{b}_i)\frac{\partial}{\partial x_i}+\left[\Delta t s_3\mathbf{c}_{ij}-\frac{\Delta t^2}{2}s_2(\mathbf{a}_i\mathbf{a}_j+\mathbf{b}_i\mathbf{a}_j)\right.\right.$$
$$\left.\left.-\frac{\Delta t^2}{2}s_4(\mathbf{a}_i\mathbf{b}_j+\mathbf{b}_i\mathbf{b}_j)\right]\frac{\partial^2}{\partial x_i\partial x_j}\right\}\Delta\mathbf{U}^{n+1}=-\frac{\Delta t}{2}\left(\frac{\partial\mathbf{F}_i^n}{\partial x_i}+\frac{\partial\mathbf{G}_i^n}{\partial x_i}\right)$$
$$+\frac{\Delta t^2}{2}(\mathbf{a}_i+\mathbf{b}_i)\frac{\partial}{\partial x_i}\left(\frac{\partial\mathbf{F}_j^n}{\partial x_j}+\frac{\partial\mathbf{G}_j^n}{\partial x_j}\right)\qquad(16.2.1)$$

where the Jacobians $\mathbf{a}_i$, $\mathbf{b}_i$, $\mathbf{c}_{ij}$, are flowfield dependent, but held constant within a discrete numerical integration time and updated for each successive time step. Here, (16.2.1) is regarded as the most general form which may be reduced to other CFD schemes in FDM and FEM.

(1) Beam-Warming Scheme

To show that a simplified special case of (16.2.1) resembles one of the most popular FDM schemes, let us express the Beam-Warming [1978] method using the notation of FDV,

$$\left\{\mathbf{I}+\frac{\theta\Delta t}{1+\xi}\left[\frac{\partial}{\partial x_i}(\mathbf{a}_i+\mathbf{b}_i)+\frac{\partial^2\mathbf{c}_{ij}}{\partial x_i\partial x_j}\right]\right\}\Delta\mathbf{U}^{n+1}$$
$$=\frac{\Delta t}{1+\xi}\left(\frac{\partial\mathbf{F}_i^n}{\partial x_i}+\frac{\partial\mathbf{G}_i^n}{\partial x_i}\right)+\frac{\theta\Delta t}{1+\xi}\frac{\partial\mathbf{G}_i^n}{\partial x_i}+\frac{\xi}{1+\xi}\Delta\mathbf{U}^n\qquad(16.2.2)$$

with $0\le(\theta,\xi)\le 1$. It is seen that the analogy of FDV to the Beam-Warming scheme is readily evident, although the main difference is that the parameters θ and ξ are chosen arbitrarily instead of being flowfield-dependent.

In general, the FDV scheme can be written in the form (6.5.14 or 13.6.9),

$$\left(\mathbf{I}+\mathbf{E}_i^n\frac{\partial}{\partial x_i}+\mathbf{E}_{ij}^n\frac{\partial^2}{\partial x_i\partial x_j}\right)\Delta\mathbf{U}^{n+1}=-\mathbf{Q}^n\qquad(16.2.3)$$

The Beam-Warming scheme and other related schemes such as Euler explicit, Euler implicit, three-point implicit, trapezoidal implicit, and leapfrog explicit schemes are summarized in Table 16.2.1.

Other schemes of FDM are compared with FDV as follows:

(2) Lax-Wendroff Scheme

The Lax-Wendroff scheme without artificial viscosity takes the form

$$\Delta\mathbf{U}_i^{n+1}=-\frac{\Delta t}{\Delta x}(\mathbf{F}_{i+\frac{1}{2}}-\mathbf{F}_{i-\frac{1}{2}})-\frac{\Delta t^2}{2\Delta x^2}\left[\mathbf{a}_{i+\frac{1}{2}}\mathbf{F}_{i+1}-(\mathbf{a}_{i+\frac{1}{2}}-\mathbf{a}_{i-\frac{1}{2}})\mathbf{F}_i+\mathbf{a}_{i-\frac{1}{2}}\mathbf{F}_{i-1}\right]\qquad(16.2.4)$$

Table 16.2.1 Comparison of FDV with Beam-Warming and Related Schemes

	s_1	s_3	E_i	E_{ij}	Q^n	Truncation Error
Beam-Warming [1]	$\frac{\theta}{1+\xi}$	$\frac{\theta}{1+\xi}$	$\frac{\theta\Delta t}{1+\xi}(\mathbf{a}_i+\mathbf{b}_i)$	$\frac{\theta\Delta t}{1+\xi}\mathbf{c}_{ij}$	$\frac{\Delta t}{1+\xi}\mathbf{W}^n+\frac{\xi}{1+\xi}\Delta\mathbf{U}^n$	$O[(\theta-\frac{1}{2}-\xi)\,\Delta t^2, \Delta t^3]$
Euler explicit	0	0	*	*	*	$O(\Delta t^2)$
Euler implicit	1	1	*	*	*	$O(\Delta t^2)$
Three-point implicit	2/3	2/3	*	*	*	$O(\Delta t^3)$
Trapezoidal implicit	1/2	1/2	*	*	*	$O(\Delta t^3)$
Leap frog explicit	0	0	*	*	*	$O(\Delta t^3)$

* Not applicable

This scheme arises if we set in FDV,

$$\mathbf{a}_{i+\frac{1}{2}} = \mathbf{a}_{i-\frac{1}{2}} = \mathbf{a}, \qquad s_1 = 0, \qquad s_2 = 0, \qquad s_3 = 0, \qquad s_4 = 0$$

(3) Lax-Wendroff Scheme with Viscosity

The Lax-Wendroff scheme with artificial viscosity is given by

$$\Delta\mathbf{U}_i^{n+1} = -\frac{\Delta t}{\Delta x}\left(\mathbf{F}_{i+\frac{1}{2}} - \mathbf{F}_{i-\frac{1}{2}}\right) \tag{16.2.5}$$

with

$$\mathbf{F}_{i+\frac{1}{2}} = \frac{\mathbf{F}_{i+1}+\mathbf{F}_i}{2} - \frac{\Delta t}{2\Delta x}\mathbf{a}_{i+\frac{1}{2}}(\mathbf{F}_{i+1}-\mathbf{F}_i) + D_{i+\frac{1}{2}}(\mathbf{U}_{i+1}-\mathbf{U}_i)$$

$$\mathbf{F}_{i-\frac{1}{2}} = \frac{\mathbf{F}_i+\mathbf{F}_{i-1}}{2} - \frac{\Delta t}{2\Delta x}\mathbf{a}_{i-\frac{1}{2}}(\mathbf{F}_i-\mathbf{F}_{i-1}) + D_{i-\frac{1}{2}}(\mathbf{U}_i-\mathbf{U}_{i-1})$$

This scheme arises if we set

$$D_{i+\frac{1}{2}} = D_{i-\frac{1}{2}} = \mathbf{a}s_1, \qquad s_2 = 0, \qquad s_3 = 0, \qquad s_4 = 0$$

This implies that the artificial viscosity is proportional to the FDV parameter s_1, but here it is manually implemented in the Lax-Wendroff scheme.

(4) Explicit MacCormack Scheme

Combining the predictor corrector steps of the MacCormack scheme, we write

$$\begin{aligned}\Delta\mathbf{U}_i^{n+1} &= -\frac{\Delta t}{\Delta x}(\mathbf{F}_{i+1}^n - \mathbf{F}_i^n) - \frac{\Delta t}{\Delta x}(\mathbf{F}_i^* - \mathbf{F}_{i-1}^*) + D_i \\ &= -\frac{\Delta t}{\Delta x}(\mathbf{F}_{i+1}^n - \mathbf{F}_i^n) - \frac{\Delta t}{\Delta x}\left(\mathbf{F}_{i+\frac{1}{2}} - \mathbf{F}_{i-\frac{1}{2}}\right) \\ &\quad - \frac{\Delta t^2}{\Delta x^2}\left[\mathbf{a}_{i+\frac{1}{2}}\mathbf{F}_{i+1} - (\mathbf{a}_{i+\frac{1}{2}}+\mathbf{a}_{i-\frac{1}{2}})\mathbf{F}_i + \mathbf{a}_{i-\frac{1}{2}}\mathbf{F}_{i-1}\right] + D_i\end{aligned} \tag{16.2.6}$$

The FDV becomes identical to this scheme with the following adjustments:

$$\mathbf{a}_{i+\frac{1}{2}} = \mathbf{a}_{i-\frac{1}{2}} = \mathbf{a}$$

$$\mathbf{F}_i^n - \mathbf{F}_{i-1}^n = \mathbf{F}_{i+1}^n - \mathbf{F}_i^n + \mathbf{F}_{i+\frac{1}{2}} - \mathbf{F}_{i-\frac{1}{2}}$$

$$s_1 = 0, \qquad s_2 = 0, \qquad s_3 = 0, \qquad s_4 = 0$$

and the s_2 term in the FDV method is equivalent to

$$D_i = \frac{\omega}{8}\left(\mathbf{U}_{i+\frac{1}{2}}^n - 4\mathbf{U}_{i+1}^n + 6\mathbf{U}_i^n - 4\mathbf{U}_{i-1}^n + \mathbf{U}_{i-2}^n\right)$$

This again is a manifestation that shows the equivalent of the s_2 terms is manually supplied in the MacCormack method.

(5) First Order Upwind Scheme

This scheme is written as

$$\begin{aligned}\Delta\mathbf{U}_i^{n+1} &= -\frac{\Delta t}{\Delta x}\left(\mathbf{F}_{i+\frac{1}{2}}^* - \mathbf{F}_{i-\frac{1}{2}}^*\right)\\ &= -\frac{\Delta t}{\Delta x}\left\{\left[\frac{1}{2}\left(\mathbf{F}_i^n + \mathbf{F}_{i+1}^n\right) - \frac{1}{2}|\mathbf{a}|\left(\mathbf{U}_{i+1}^n - \mathbf{U}_i^n\right)\right]\right.\\ &\quad\left. - \left[\frac{1}{2}\left(\mathbf{F}_i^n + \mathbf{F}_{i-1}^n\right) - \frac{1}{2}|\mathbf{a}|\left(\mathbf{U}_i^n - \mathbf{U}_{i-1}^n\right)\right]\right\}\end{aligned} \tag{16.2.7}$$

The FDM analogy is obtained by setting

$$\mathbf{F}_i^n = \frac{1}{2}\mathbf{F}_{i+1}^n, \qquad \mathbf{F}_{i-1}^n = \frac{1}{2}\mathbf{F}_{i-1}^n$$

$$s_2\mathbf{a}C\left(\Delta\mathbf{U}_i^{n+1} - 2\Delta\mathbf{U}_{i-1}^{n+1} + \Delta\mathbf{U}_{i-2}^{n+1}\right) = |\mathbf{a}|\left(\mathbf{U}_{i+1}^n - \mathbf{U}_{i-1}^n\right)$$

where C is the Courant number.

(6) Implicit MacCormack Scheme

With all second order derivatives removed from (16.2.1), we obtain the implicit MacCormack scheme by setting $s_1 = 1, s_2 = 0, s_3 = 0, s_4 = 0$. However, it is necessary to divide the process into the predictor and corrector steps. Once again the flowfield-dependent variation parameters for FDV will allow the computation to be performed in a single step.

(7) TVD Scheme

Another example is the analogy of FDV-FDM to the FDM-TVD scheme. To see this, we write (6.5.13) in one dimension using linear trial and test functions with all Neumann boundary conditions neglected.

$$\begin{aligned}&\frac{1}{6\Delta t}\left(\Delta\mathbf{U}_{i+1}^{n+1} + 4\Delta\mathbf{U}_i^{n+1} + \Delta\mathbf{U}_{i-1}^{n+1}\right) = \frac{1}{2\Delta x}(s_1\mathbf{a} + s_3\mathbf{b})\left(\Delta\mathbf{U}_{i+1}^{n+1} - \Delta\mathbf{U}_{i-1}^{n+1}\right)\\ &\quad + \frac{1}{2\Delta x^2}\{2s_3\mathbf{c} - \Delta t[s_2(\mathbf{a}^2 + \mathbf{ab}) + s_4(\mathbf{ba} + \mathbf{b}^2)]\}\left(\Delta\mathbf{U}_{i+1}^{n+1} - 2\Delta\mathbf{U}_i^{n+1} + \Delta\mathbf{U}_{i-1}^{n+1}\right)\\ &\quad + \frac{1}{2\Delta x}\left(\mathbf{F}_{i+1}^n - \mathbf{F}_{i-1}^n + \mathbf{G}_{i+1}^n - \mathbf{G}_{i-1}^n\right) - \frac{\Delta t}{2\Delta x^2}(\mathbf{a} + \mathbf{b})\\ &\quad \times\left(\mathbf{F}_{i+1}^n - 2\mathbf{F}_i^n + \mathbf{F}_{i-1}^n + \mathbf{G}_{i+1}^n - 2\mathbf{G}_i^n + \mathbf{G}_{i-1}^n\right)\end{aligned} \tag{16.2.8}$$

Neglecting all diffusion terms, adopting a lumped mass system, and moving one nodal point upstream, we have

$$\frac{\Delta \mathbf{U}_i^{n+1}}{\Delta t} = \frac{s_1 \mathbf{a}}{\Delta x}\left(\Delta \mathbf{U}_i^{n+1} - \Delta \mathbf{U}_{i-1}^{n+1}\right) - \frac{s_2 \mathbf{a}^2 \Delta t}{2\Delta x^2}\left(\Delta \mathbf{U}_i^{n+1} - 2\Delta \mathbf{U}_{i-1}^{n+1} + \Delta \mathbf{U}_{i-2}^{n+1}\right)$$
$$+ \frac{1}{\Delta x}\left(\mathbf{F}_i^n - \mathbf{F}_{i-1}^n\right) - \frac{\mathbf{a}\Delta t}{2\Delta x^2}\left(\mathbf{F}_i^n - 2\mathbf{F}_{i-1}^n + \mathbf{F}_{i-2}^n\right) \quad (16.2.9)$$

The FDM-TVD for the 1-D Euler equation is written as

$$\frac{d\mathbf{U}_i}{dt} = -\frac{\mathbf{a}^+}{\Delta x}\left[(\mathbf{U}_i - \mathbf{U}_{i-1}) + \frac{1}{2}\Psi^+_{i-\frac{1}{2}}(\mathbf{U}_i - \mathbf{U}_{i-1}) - \frac{1}{2}\Psi^+_{i-\frac{3}{2}}(\mathbf{U}_{i-1} - \mathbf{U}_{i-2})\right]$$
$$- \frac{\mathbf{a}^-}{\Delta x}\left[(\mathbf{U}_{i+1} - \mathbf{U}_i) + \frac{1}{2}\Psi^-_{i+\frac{1}{2}}(\mathbf{U}_{i+1} - \mathbf{U}_i) - \frac{1}{2}\Psi^-_{i+\frac{3}{2}}(\mathbf{U}_{i+2} - \mathbf{U}_{i+1})\right] \quad (16.2.10)$$

with

$$\mathbf{a}^+ = \max(0,\mathbf{a}) = \frac{1}{2}(\mathbf{a} + |\mathbf{a}|)$$
$$\mathbf{a}^- = \min(0,\mathbf{a}) = \frac{1}{2}(\mathbf{a} - |\mathbf{a}|)$$

Introducing variation parameter s for the time derivative on the right-hand side of (16.2.10) the form

$$\mathbf{U}_i = \mathbf{U}_i^n + s\Delta \mathbf{U}_i^{n+1} \quad (16.2.11)$$

Substituting (16.2.11) into (16.2.10) and assuming that

$$\mathbf{a}^- = 0, \qquad \mathbf{a}^+ = \mathbf{a}, \qquad \Psi^+_{i-\frac{1}{2}} = \Psi^+_{i-\frac{3}{2}} = \Psi$$

we obtain

$$\frac{\Delta \mathbf{U}_i^{n+1}}{\Delta t} = \frac{s\mathbf{a}}{2\Delta x}\left(\Delta \mathbf{U}_i^{n+1} - \Delta \mathbf{U}_{i-1}^{n+1}\right) - \frac{s\Psi \mathbf{a}\Delta x}{2\Delta x^2}\left(\Delta \mathbf{U}_i^{n+1} - 2\Delta \mathbf{U}_{i-1}^{n+1} + \Delta \mathbf{U}_{i-2}^{n+1}\right)$$
$$- \frac{1}{\Delta x}\left(\mathbf{F}_i^n - \mathbf{F}_{i-1}^n\right) - \frac{\Psi \Delta x}{2\Delta x^2}\left(\mathbf{F}_i^n - 2\mathbf{F}_{i-1}^n + \mathbf{F}_{i-2}^n\right) \quad (16.2.12)$$

Comparing (16.2.9) and (16.2.12) reveals that, with

$$s_1 = -\frac{s}{2}, \qquad s_2 = \frac{s\Delta x \Psi}{\mathbf{a}\Delta t}$$

and -1 for the coefficient of $(\mathbf{F}_i^n - \mathbf{F}_{i-1}^n)$ term, we note that the FDV-FDM formulation and FDM-TVD scheme are analogous; in fact, they are identical under the assumptions made above. The variation parameters s_1 and s_2 in the FDV-FEM scheme play the role of TVD limiters, Ψ. However, the implicitness parameters s_3 and s_4, beyond the concept of TVD scheme, together with s_1 and s_2, are expected to govern complex physical phenomena such as turbulent boundary layer interactions with shock waves, finite rate chemistry [with s_5 and s_6 (13.6.5a,b)], widely disparate length and time scales, compressibility effects in high Mach number flows, etc.

(8) PISO and SIMPLE

The basic idea of PISO and SIMPLE is analogous to FDV-FEM in that the pressure correction process is a separate step in PISO or SIMPLE, whereas the concept of pressure correction is implicitly embedded in FDV-FEM by updating the variation parameters based on the upstream and downstream Mach numbers and Reynolds numbers within an element.

The elliptic nature of the pressure Poisson equation in the pressure correction process resembles the terms embedded in the $B_{\alpha\beta rs}$ terms in (13.6.22). Specifically, examine the s_2 terms involving $a_{irq}a_{jsq}$ and $b_{irq}a_{jsq}$ and s_4 term involving $a_{irq}b_{jsq}$. All of these terms are multiplied by $\Phi_{\alpha,i}\Phi_{\beta,j}$ which provide dissipation against any pressure oscillations. Question: Exactly when is such dissipation action needed? This is where the importance of FDV variation parameters based on flowfield parameters comes in. As the Mach number becomes very small (incompressibility effects dominate) the variation parameters s_2 and s_4 calculated from the current flowfield will be indicative of pressure correction required. Notice that a delicate balance between Mach number (s_2 is Mach number dependent) and Reynolds number or Peclet number (s_4 is Reynolds number or Peclet number dependent) is a crucial factor in achieving convergent and stable solutions. Of course, on the other hand, high Mach number flows are also dependent on these variation parameters. In this case all variation parameters, s_1, s_2, s_3, s_4 will play important roles.

16.3 RELATIONSHIPS BETWEEN FEM AND FDV

(1) Taylor-Galerkin Methods (TGM) with Convection and Diffusion Jacobians

Earlier developments for the solution of Navier-Stokes system of equations were based on TGM without using the variation parameters. They can be shown to be special cases of FDV-FEM.

In terms of the both the diffusion Jacobian and the diffusion gradient Jacobian, we write

$$\frac{\partial \mathbf{G}_i}{\partial t} = \mathbf{b}_i \frac{\partial \mathbf{U}}{\partial t} + \mathbf{c}_{ij}\frac{\partial \mathbf{V}_j}{\partial t}$$

with

$$\mathbf{b}_i = \frac{\partial \mathbf{G}_i}{\partial \mathbf{U}}, \qquad \mathbf{c}_{ij} = \frac{\partial \mathbf{G}_i}{\partial \mathbf{V}_j}, \qquad \mathbf{V}_j = \frac{\partial \mathbf{U}}{\partial x_j}$$

Thus, it follows from (13.6.2) with $s_1 = s_3 = s_4 = s_5 = s_6 = 0$ and $s_2 = 1$ that

$$\Delta \mathbf{U}^{n+1} = \Delta t\left(-\frac{\partial \mathbf{F}_i}{\partial x_i} - \frac{\partial \mathbf{G}_i}{\partial x_i} + \mathbf{B}\right)^n + \frac{\Delta t^2}{2}\frac{\partial}{\partial t}\left(-\frac{\partial \mathbf{F}_i}{\partial x_i} - \frac{\partial \mathbf{G}_i}{\partial x_i} + \mathbf{B}\right)^{n+1} + O(\Delta t^3) \tag{16.3.1}$$

Using the definitions of convection, diffusion, and diffusion rate Jacobians discussed in Section 13.6, the temporal rates of change of the convection and diffusion variables may be written as follows:

$$
\begin{aligned}
\frac{\partial \mathbf{F}_i^n}{\partial t} &= \left(\mathbf{a}_i \frac{\partial \mathbf{U}}{\partial t}\right)^n = \left[\mathbf{a}_i\left(-\frac{\partial \mathbf{F}_j}{\partial x_j} - \frac{\partial \mathbf{G}_j}{\partial x_j} + \mathbf{B}\right)\right]^n \\
\frac{\partial \mathbf{F}_i^{n+1}}{\partial t} &= \mathbf{a}_i\left[\left(-\mathbf{a}_j \frac{\partial}{\partial x_j}(\mathbf{U}^{n+1} - \mathbf{U}^n) - \frac{\partial \mathbf{F}_j^n}{\partial x_j} - \frac{\partial \mathbf{G}_j^{n+1}}{\partial x_j} + \mathbf{B}^{n+1}\right)\right] \\
\frac{\partial \mathbf{G}_i^{n+1}}{\partial t} &= \left(\mathbf{b}_i \frac{\partial \mathbf{U}}{\partial t}\right)^{n+1} + \left[\mathbf{c}_{ij}\frac{\partial}{\partial t}\left(\frac{\partial \mathbf{U}}{\partial x_j}\right)\right]^{n+1}
\end{aligned}
\tag{16.3.2}
$$

or

$$
\frac{\partial \mathbf{G}_i^{n+1}}{\partial t} = \left(\mathbf{b}_i - \frac{\partial \mathbf{c}_{ij}}{\partial x_j}\right)\frac{\Delta \mathbf{U}}{\Delta t}^{n+1} + \frac{\partial}{\partial x_j}\left(\mathbf{c}_{ij}\frac{\Delta \mathbf{U}}{\Delta t}\right)^{n+1}
\tag{16.3.3}
$$

Substituting (16.3.2) and (16.3.3) into (16.3.1) yields

$$
\begin{aligned}
\Delta \mathbf{U}^{n+1} &= \Delta t\left(-\frac{\partial \mathbf{F}_i}{\partial x_i} - \frac{\partial \mathbf{G}_i}{\partial x_i} + \mathbf{B}\right)^n \\
&\quad + \frac{\Delta t^2}{2}\left\{\frac{\partial}{\partial x_i}\left[-\mathbf{a}_i\left(-\mathbf{a}_j\frac{\partial \Delta \mathbf{U}^{n+1}}{\partial x_j} - \frac{\partial \mathbf{F}_j^n}{\partial x_j} - \frac{\partial \mathbf{G}_j^{n+1}}{\partial x_j} + \mathbf{B}^{n+1}\right)\right.\right. \\
&\quad \left.\left. + \left(\mathbf{e}_i + \frac{\partial \mathbf{c}_{ij}}{\partial x_j}\right)\frac{\Delta \mathbf{U}}{\Delta t}^{n+1} + \frac{\partial \mathbf{B}^{n+1}}{\partial t}\right]\right\}
\end{aligned}
\tag{16.3.4}
$$

Assuming that

$$
\mathbf{e}_i = \mathbf{b}_i - \frac{\partial \mathbf{c}_{ij}}{\partial x_j} \cong 0
$$

and neglecting the spatial and temporal derivatives of **B**, we rewrite (16.3.4) in the form

$$
\begin{aligned}
&\left\{1 - \frac{\Delta t^2}{2}\frac{\partial}{\partial x_i}\left(\mathbf{a}_i\mathbf{a}_j - \frac{\mathbf{c}_{ij}}{\Delta t}\right)\frac{\partial}{\partial x_j}\right\}\Delta \mathbf{U}^{n+1} = \mathbf{H}^n \\
&\mathbf{H}^n = \Delta t\left(-\frac{\partial \mathbf{F}_i}{\partial x_i} - \frac{\partial \mathbf{G}_i}{\partial x_i} + \mathbf{B}\right)^n + \frac{\Delta t^2}{2}\frac{\partial}{\partial x_i}\left(\mathbf{a}_i\frac{\partial \mathbf{F}_j}{\partial x_j}\right)^n
\end{aligned}
\tag{16.3.5}
$$

Here the second derivatives of $\mathbf{G}_i$ are neglected and all Jacobians are assumed to remain constant within an incremental time step but updated at subsequent time steps.

Applying the Galerkin finite element formulation, we have an implicit scheme,

$$
(\mathbf{A}_{\alpha\beta}\delta_{rs} + \mathbf{B}_{\alpha\beta rs})\,\Delta \mathbf{U}_{\beta s}^{n+1} = \mathbf{H}_{\alpha r}^n + \mathbf{N}_{\alpha r}^{n+1} + \mathbf{N}_{\alpha r}^n
\tag{16.3.6}
$$

where

$$
\begin{aligned}
\mathbf{B}_{\alpha\beta rs} &= \frac{\Delta t^2}{2}\int_\Omega \left[\left(\mathbf{a}_{irq}\mathbf{a}_{jsq} - \frac{\mathbf{c}_{ijrs}}{\Delta t}\right)\Phi_{\alpha,i}\Phi_{\beta,j}\right] d\Omega \\
\mathbf{H}_{\alpha r}^n &= \Delta t \int_\Omega \left[\Phi_{\alpha,i}\Phi_\beta(\mathbf{F}_{\beta ir}^n + \mathbf{G}_{\beta ir}^n) + \Phi_\alpha\Phi_\beta\mathbf{B}_{\beta r}^n - \frac{\Delta t}{2}\mathbf{a}_{irs}\Phi_{\alpha,i}\Phi_{\beta,j}\mathbf{F}_{\beta js}^n\right] d\Omega \\
\mathbf{N}_{\alpha r}^{n+1} &= \frac{\Delta t^2}{2}\int_\Gamma \left(\mathbf{a}_{irq}\mathbf{a}_{jsq} - \frac{\mathbf{c}_{ijrs}}{\Delta t}\right)\overset{*}{\Phi}_\alpha \Delta \mathbf{U}_{s,j}^{n+1} n_i\, d\Gamma \\
\mathbf{N}_{\alpha r}^n &= -\int_\Gamma \left[\Delta t\,\overset{*}{\Phi}_\alpha(\mathbf{F}_{ir}^n + \mathbf{G}_{ir}^n) - \frac{\Delta t^2}{2}\mathbf{a}_{irs}\overset{*}{\Phi}_\alpha \mathbf{F}_{js,j}^n\right] n_i\, d\Gamma
\end{aligned}
$$

Here we note that the algorithm given by (16.3.6) results from (13.6.20) in FDV by setting $s_1 = s_3 = s_4 = 0$, $s_2 = 1$, $b_{irq}a_{jsq} = c_{ijrs}/\Delta t$, and neglecting the terms with b_{jrs} and derivatives of $\mathbf{G}_i$ and $\mathbf{B}$, the form identical to that introduced in Section 13.2.1.

(2) Taylor Galerkin Methods (TGM) with Convection Jacobians

Diffusion Jacobians may be neglected if their influence is negligible. In this case the Taylor-Galerkin finite element analog may be derived using only the convective Jacobian from the Taylor series expansion,

$$\mathbf{U}^{n+1} = \mathbf{U}^n + \Delta t \frac{\partial \mathbf{U}^n}{\partial t} + \frac{\Delta t^2}{2}\frac{\partial^2 \mathbf{U}^n}{\partial t^2} + O(\Delta t^3) \tag{16.3.7}$$

where

$$\frac{\partial \mathbf{U}}{\partial t} = -\frac{\partial \mathbf{F}_i}{\partial x_i} - \frac{\partial \mathbf{G}_i}{\partial x_i} + \mathbf{B} = -\mathbf{a}_i \frac{\partial \mathbf{U}}{\partial x_i} - \frac{\partial \mathbf{G}_i}{\partial x_i} + \mathbf{B} \tag{16.3.8}$$

$$\frac{\partial^2 \mathbf{U}}{\partial t^2} = -\frac{\partial}{\partial t}\left(\mathbf{a}_i \frac{\partial \mathbf{U}}{\partial x_i} + \frac{\partial \mathbf{G}_i}{\partial x_i} - \mathbf{B}\right)$$

or

$$\frac{\partial^2 \mathbf{U}}{\partial t^2} = \frac{\partial}{\partial x_j}\left(\mathbf{a}_i\mathbf{a}_j \frac{\partial \mathbf{U}}{\partial x_i}\right) + \frac{\partial}{\partial x_i}\left(\mathbf{a}_i \frac{\partial \mathbf{G}_j}{\partial x_j}\right) - \frac{\partial}{\partial x_i}(\mathbf{a}_i\mathbf{B}) + \frac{\partial \mathbf{B}}{\partial t} \tag{16.3.9}$$

Substituting (16.3.8) and (16.3.9) into (16.3.7), we obtain

$$\Delta \mathbf{U}^{n+1} = \Delta t\left\{-\frac{\partial \mathbf{F}_i}{\partial x_i} - \frac{\partial \mathbf{G}_i}{\partial x_i} + \mathbf{B} + \frac{\Delta t}{2}\left[\frac{\partial}{\partial x_j}\left(\mathbf{a}_i\mathbf{a}_j \frac{\partial \mathbf{U}}{\partial x_i}\right) + \frac{\partial^2(\mathbf{a}_i\mathbf{G}_j)}{\partial x_i \partial x_j} + \frac{\partial}{\partial x_i}(\mathbf{a}_i\mathbf{B}) + \frac{\partial \mathbf{B}}{\partial t}\right]\right\}^n \tag{16.3.10a}$$

Expanding $\partial \mathbf{F}_j/\partial t$ at $(n+1)$ time step

$$\frac{\partial \mathbf{F}_i^{n+1}}{\partial t} = \left[\mathbf{a}_i\left(-\frac{\partial \mathbf{F}_j}{\partial x_j} - \frac{\partial \mathbf{G}_j}{\partial x_j} + \mathbf{B}\right)\right]^{n+1} = \mathbf{a}_i^{n+1}\left[-\mathbf{a}_j \frac{\partial \Delta \mathbf{U}^{n+1}}{\partial x_j} - \frac{\partial \mathbf{F}_j^n}{\partial x_j} - \frac{\partial \mathbf{G}_j^{n+1}}{\partial x_j} + \mathbf{B}^{n+1}\right]$$

and substituting the above into (16.3.7–16.3.9), we arrive at $\Delta \mathbf{U}^{n+1}$ in a form different from (16.3.10a):

$$\Delta \mathbf{U}^{n+1} = \Delta t\left(-\frac{\partial \mathbf{F}_i}{\partial x_i} - \frac{\partial \mathbf{G}_i}{\partial x_i} + \mathbf{B}\right)^n + \frac{\Delta t^2}{2}\left\{\frac{\partial}{\partial x_i}\left(\mathbf{a}_i\mathbf{a}_j \frac{\partial \Delta \mathbf{U}^{n+1}}{\partial x_j} + \mathbf{a}_i \frac{\partial \mathbf{F}_j^n}{\partial x_j}\right) + \frac{\partial^2(\mathbf{a}_i\mathbf{G}_j)^{n+1}}{\partial x_i \partial x_j} + \frac{\partial}{\partial x_i}(\mathbf{a}_i\mathbf{B})^{n+1} + \frac{\partial \mathbf{B}^{n+1}}{\partial t}\right\} \tag{16.3.10b}$$

$$\mathbf{H}^n = \left[1 - \frac{\Delta t^2}{2}\frac{\partial}{\partial x_i}\left(\mathbf{a}_i\mathbf{a}_j - \frac{\mathbf{c}_{ij}}{\Delta t}\right)\frac{\partial}{\partial x_j}\right]\Delta \mathbf{U}^{n+1} \tag{16.3.10c}$$

$$\mathbf{H}^n = \Delta t\left(-\frac{\partial \mathbf{F}_i}{\partial x_i} - \frac{\partial \mathbf{G}_i}{\partial x_i} + \mathbf{B}\right)^n + \frac{\Delta t^2}{2}\frac{\partial}{\partial x_i}\left(\mathbf{a}_i \frac{\partial \mathbf{F}_j}{\partial x_j}\right)^n$$

where second derivatives of $\mathbf{G}_i$ are assumed to be negligible and $\mathbf{B}$ is constant in space and time, arriving at an implicit finite element scheme,

$$(\mathbf{A}_{\alpha\beta}\delta_{rs} + \mathbf{B}_{\alpha\beta rs})\,\Delta\mathbf{U}^{n+1}_{\beta s} = \mathbf{H}^{n}_{\alpha r} + \mathbf{N}^{n+1}_{\alpha r} + \mathbf{N}^{n}_{\alpha r} \tag{16.3.11}$$

where

$$\mathbf{A}_{\alpha\beta} = \int_\Omega \Phi_\alpha \Phi_\beta d\Omega$$

$$\mathbf{B}_{\alpha\beta rs} = \frac{\Delta t^2}{2}\int_\Omega \left[\left(\mathbf{a}_{irq}\mathbf{a}_{jsq} - \frac{\mathbf{c}_{ijrs}}{\Delta t}\right)\Phi_{\alpha,i}\Phi_{\beta,j}\right] d\Omega$$

$$\mathbf{H}^{n}_{\alpha r} = \Delta t \int_\Omega \left[\Phi_{\alpha,i}\Phi_\beta(\mathbf{F}^{n}_{\beta ir} + \mathbf{G}^{n}_{\beta ir}) - \Phi_\alpha\Phi_\beta\mathbf{B}^{n}_{\beta r} - \frac{\Delta t^2}{2}\mathbf{a}_{irs}\Phi_{\alpha,i}\Phi_{\beta,j}\mathbf{F}^{n}_{\beta js}\right] d\Omega$$

$$\mathbf{N}^{n+1}_{\alpha r} = \frac{\Delta t^2}{2}\int_\Gamma \left(\mathbf{a}_{irq}\mathbf{a}_{jsq} - \frac{\mathbf{c}_{ijrs}}{\Delta t}\right)\overset{*}{\Phi}_\alpha \Delta\mathbf{U}^{n+1}_{s,j} n_i d\Gamma$$

$$\mathbf{N}^{n}_{\alpha r} = -\int_\Gamma \left[\Delta t\,\overset{*}{\Phi}_\alpha(\mathbf{F}^{n}_{ir} + \mathbf{G}^{n}_{ir}) - \frac{\Delta t^2}{2}\mathbf{a}_{irs}\overset{*}{\Phi}_\alpha \mathbf{F}^{n}_{js,j}\right] n_i d\Gamma$$

It should be noted that the form (16.3.10c) arises from (13.6.20) in FDV with $s_1 = s_3 = s_4 = b_j = 0$ and $s_2 = 1$, an algorithm similar to TGM introduced in Section 13.2.1.

(3) Generalized Petrov-Galerkin

The Generalized Petrov-Galerkin (GPG) method can be identified in FDV by setting $s_1 = s_2 = 1$, $s_3 = s_4 = 0$, $\mathbf{b}_i = \mathbf{c}_{ij} = \mathbf{d} = 0$, $\mathbf{Q}^n = 0$, $\mathbf{E}_i = \mathbf{a}_i$, and $\mathbf{E}_{ij} = \frac{1}{2}\Delta t^2 \mathbf{a}_i\mathbf{a}_j$, so that (13.6.20) takes the form

$$\frac{\Delta\mathbf{U}}{\Delta t} + \mathbf{a}_i\frac{\partial\Delta\mathbf{U}}{\partial x_i} - \frac{\Delta t}{2}\mathbf{a}_i\mathbf{a}_j\frac{\partial^2\Delta\mathbf{U}}{\partial x_i\partial x_j} = 0 \tag{16.3.12}$$

For the steady-state nonincremental form in 1-D, we write (16.3.12) in the form

$$a\frac{\partial u}{\partial x} - \Delta t\frac{a^2}{2}\frac{\partial^2 u}{\partial x^2} = 0 \tag{16.3.13}$$

Taking the Galerkin integral of (16.3.13) leads to

$$\int \Phi^{(e)}_N\left(a\frac{\partial u}{\partial x} - \Delta t\frac{a^2}{2}\frac{\partial^2 u}{\partial x^2}\right)dx = 0, \qquad \int W^{(e)}_N a\frac{\partial u}{\partial x}dx = 0 \tag{16.3.14}$$

for vanishing Neumann boundaries. Here $W^{(e)}_N$ is the Petrov-Galerkin test function,

$$W^{(e)}_N = \Phi^{(e)}_N + \alpha h\frac{\partial\Phi^{(e)}_N}{\partial x} \tag{16.3.15}$$

with $\alpha = C/2$ and $C = a\Delta t/\Delta x$ being the Courant number.

For isoparametric coordinates in two dimensions, the Petrov-Galerkin test function assumes the form

$$W^{(e)}_N = \Phi^{(e)}_N + \beta g_i\frac{\partial\Phi^{(e)}_N}{\partial x} \tag{16.3.16}$$

with

$$\beta = \frac{1}{4}(\overline{\alpha}_\xi h_\xi + \overline{\alpha}_\eta h_\eta)$$

$$\overline{\alpha}_\xi = \coth\left(\frac{R_\xi}{2}\right) - \frac{2}{R_\xi}, \qquad \overline{\alpha}_\eta = \coth\left(\frac{R_\eta}{2}\right) - \frac{2}{R_\eta}$$

$$g_i = \frac{v_i}{\sqrt{v_j v_j}}$$

where R_ξ is the Reynolds number or Peclet number in the direction of isoparametric coordinates (ξ, η). Note that the GPG process given by (16.3.12)–(16.3.16) leads to the streamline upwinding Petrov-Galerkin (SUPG) scheme as a special case, thus leading to the analogy between FDV and GPG.

16.4 OTHER METHODS

We have examined in the previous chapters most of the currently available CFD methods. Throughout this text, it was intended that the reader be given adequate information so that he/she could make a final decision to choose the most suitable method for the problem at hand. Though biases or preferences in choosing CFD methods are often common among practitioners, this text may still serve as a guide and possibly toward re-orientation. It was shown that FVM can be formulated from either FDM or FEM. The FDV methods discussed in Chapters 6 and 13 as well as other methods are expected to meet these challenges. In particular, the ability of FDV methods to generate other prominent CFD schemes has been demonstrated. In the past, numerical methods other than those presented in the previous chapters have been used also. Among them are the boundary element methods (BEM), coupled Eulerian-Lagrangian (ECL) methods, particle-in-cell (PIC) methods, and Monte Carlo methods (MCM). The detailed coverage of these topics is beyond the scope of this book; but, for the sake of historical perspectives, we shall briefly review them next.

16.4.1 BOUNDARY ELEMENT METHODS

The boundary element methods (BEM) are based on boundary integral equations in which only the boundaries of a region are used to obtain apparoximate solutions. Interpolation functions for the surface behavior are coupled with the solutions to the governing equations which apply over the domain. The resulting equations are solved numerically for values on the boundary alone, and values at interior points are calculated subsequently from the surface data.

It is thus clear that fewer equations are involved in the solution by the BEM. On the other hand, it is required that the governing equations be linear but this can be overcome by linearization through Kirchhoff transformation [Brebbia, 1978; Brebbia, Telles, and Wrobel, 1983].

Green's Function and Boundary Integral Equation

To illustrate, let us consider the Laplace equation,

$$\nabla^2\phi = 0 \tag{16.4.1}$$

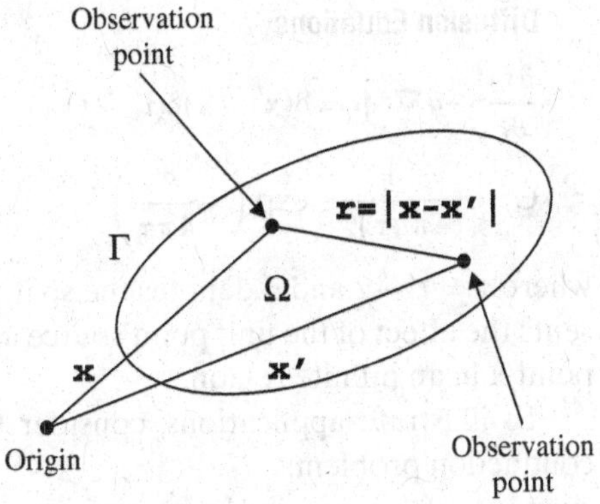

Figure 16.4.1 Location of source and field points.

Assume a weighting function ψ and the weighted residual integral of (16.4.1) such that

$$\int_\Omega \psi \nabla^2 \phi d\Omega = 0 \tag{16.4.2}$$

Integrating this by parts twice,

$$\int_\Omega (\psi \nabla^2 \phi - \phi \nabla^2 \psi) d\Omega = \int_\Gamma [\psi (\mathbf{n} \cdot \nabla \phi) - \phi (\mathbf{n} \cdot \nabla \psi)] d\Gamma \tag{16.4.3}$$

It follows from (16.4.1) and (16.4.2) that

$$\int_\Omega \phi \nabla^2 \psi \, d\Omega = \int_\Gamma \left(\phi \frac{\partial \psi}{\partial n} - \psi \frac{\partial \phi}{\partial n} \right) d\Gamma = 0 \tag{16.4.4}$$

which is known as the Green's identity. Here, the weighting function ψ is denoted as the Green's function, $G(\mathbf{x}'|\mathbf{x})$, which is assumed to be the solution of

$$\nabla^2 G(\mathbf{x}'|\mathbf{x}) = \delta(\mathbf{x}' - \mathbf{x}) \tag{16.4.5}$$

where $\delta(\mathbf{x}' - \mathbf{x})$ is the Dirac delta function with $\mathbf{x}$ and $\mathbf{x}'$ being the source point and the observation point, respectively, such that (Figure 16.4.1)

$$\int_\Omega \phi(\mathbf{x}) \delta(\mathbf{x}' - \mathbf{x}) d\Omega = \phi(\mathbf{x}') \tag{16.4.6}$$

For a polar coordinate system (r, θ), it can easily be shown that the solution of (16.4.5) is of the form

$$G = \frac{1}{2\pi} \ln r \tag{16.4.7}$$

or, for a three-dimensional domain,

$$G = \frac{1}{4\pi r} \tag{16.4.8}$$

The fundamental solutions for other types of partial differential equations are as follows:

Helmholtz Equations

$$\nabla^2 G + k^2 G = \delta(\mathbf{x}' - \mathbf{x}) \tag{16.4.9}$$

$$G = \frac{1}{4\pi} \frac{e^{ikr}}{r} \quad \text{for 3-}D$$

Diffusion Equations

$$\frac{\partial G}{\partial t} - a\nabla^2\phi = \delta(\mathbf{x}' - \mathbf{x})\delta(t' - t) \tag{16.4.10}$$

$$G = \frac{1}{(4\pi a\tau)^{d/2}} \exp\left(-\frac{r^2}{4\pi\tau}\right)$$

where $\tau = t' - t$ and d denotes the spatial dimension. The fundamental solution represents the effect of the unit point source applied at the observation point $\mathbf{x}'$ on the source point $\mathbf{x}$ in an infinite region.

To illustrate applications, consider the governing equation for an unsteady heat conduction problem:

$$\frac{\partial T}{\partial t} - aT_{,ii} - \frac{Q}{\rho c} = 0 \tag{16.4.11}$$

subject to boundary conditions

$$\begin{aligned} T &= T_1 && \text{on } \Gamma_1 \\ -kT_{,i}n_i &= q_2 && \text{on } \Gamma_2 \\ -kT_{,i}n_i &= \overline{\alpha}(T_3 - T') && \text{on } \Gamma_3 \end{aligned}$$

Recast (16.4.11) in terms of Green's identity and integrate with respect to time,

$$\beta T = \int_0^{t'}\int_\Gamma (aT_{,i}n_i - aTG_{,i}n_i)d\Gamma dt + \int_0^{t'}\int_\Omega \frac{Q}{\rho c} G d\Omega dt + \int_\Omega TG\bigg|_{t=0} \tag{16.4.12}$$

Introducing the interpolation functions in the form,

$$T = \Phi_\alpha T_\alpha$$
$$q = \Phi_\alpha q_\alpha$$

and rewriting (16.4.12) using the above approximations, we obtain

$$A_{\alpha\beta}^{(n+1)} T_\beta^{(n+1)} = F_\alpha^{(n)} \tag{16.4.13}$$

where

$$F_\alpha^n = B_{\alpha\beta}^{(n+1)} q_\beta^{(n+1)} + A_{\alpha\beta}^{(n)} T_\beta^{(n)} + B_{\alpha\beta}^{(n)} q_\beta^{(n)} + C_\alpha^{(\Gamma)} + C_\alpha^\Omega$$

$$A_{\alpha\beta}^{(n+1)} = \frac{1}{2}\delta_{\alpha\beta} - A_{\alpha\beta}^* \quad \text{for smooth boundary}$$

$$A_{\alpha\beta}^* = -\int_0^{t'}\left[\int_{\Gamma_2} a(G_{,i}n_i)_\alpha \Phi_\beta d\Gamma - \int_{\Gamma_3} a(G_{,i}n_i)_\alpha \Phi_\beta d\Gamma\right]dt$$

$$B_{\alpha\beta} = \int_0^t \int_{\Gamma_1} a(G)_\alpha \Phi_\beta d\Omega$$

$$C_\alpha^{(\Gamma)} = \int_0^{t'}\left[\int_{\Gamma_2} \frac{-q_{\Gamma_2}}{\rho c}(G)_\alpha d\Gamma - \int_{\Gamma_3} \frac{\overline{\alpha}}{\rho c}(T_{\Gamma_3} - T')d\Gamma\right]dt + \int_0^{t'}\int_\Omega (G)_\alpha \frac{1}{\rho c}\Phi_\beta Q_\beta d\Omega dt$$

$$C_\alpha^{(\Omega)} = \int_\Omega (G)_\alpha \Phi_\beta T_\beta d\Omega|_{t=0}$$

Since the algebraic equations given by (16.4.13) are linear, the solution involves a simple marching in time until desired time is reached.

16.4.2 COUPLED EULERIAN-LAGRANGIAN METHODS

It should be pointed out that all methods introduced in the previous chapters are based on the Eulerian coordinates in which computational nodes are fixed in space and all variables are calculated at these fixed nodes. In some instances in reality, however, it is of interest to compute variables in the Lagrangian coordinates where the mesh points are allowed to move along with the fluid particles. Furthermore, it is often convenient to have both Eulerian and Lagrangian coordinates coupled, known as the coupled Eulerian-Lagrangian (CEL) methods, useful in highly distorted flows or multiphase flows. Precise mathematical representations and treatments of Eulerian and Lagrangian coordinates are presented in Chung [1996].

The CEL methods were first developed by Noh [1964]. The basic idea is that the boundary Γ of the region Ω given by

$$\Omega = \bigcup_{i=1}^{n} \Omega_i$$

and the curves D_i which separate the subregions Ω_i are to be approximated by time-dependent Lagrangian lines $L_i(t)$. A subregion R_i which is approximated by the time-independent Eulerian mesh E will consequently have its boundary Γ_i prescribed by the Lagrangian calculations. Thus, the Eulerian calculation reduces to a calculation on a fixed mesh having a prescribed moving boundary and therefore contributes one of the central calculations in the CEL methods. The calculations that are made at each time step are divided into three main parts: Lagrange calculations, Eulerian calculations, and a calculation that couples the Eulerian and Lagrangian regions by defining that part of the Eulerian mesh which is active and by determining the pressures from the Eulerian region which act on the Lagrangian boundaries.

Physically, the local sound speed (and fluid velocity) can vary considerably in different regions of the fluid, and the mesh size in general will also be a function of the region being approximated. It is therefore to be expected that the different subregions will have different stability requirements. Thus, it is desirable to allow these different regions their characteristic time interval in hydrodynamic calculations. Approximations for difference equations for Eulerian coordinates (Figure 16.4.2a) and Lagrangian coordinates (Figure 16.4.2b) are given below.

Eulerian Difference Equations

The differential equations for Eulerian coordinates are the same as given in Chapter 2. To obtain finite difference equations for the above equations, we first introduce the following definitions:

$$\text{(i)}\quad u_{k+1,l+1}^{n+1} = \frac{1}{2}\left(u_{k+1,l}^{n+1/2} + u_{k+1,l+1}^{n+1/2}\right) \tag{16.4.14a}$$

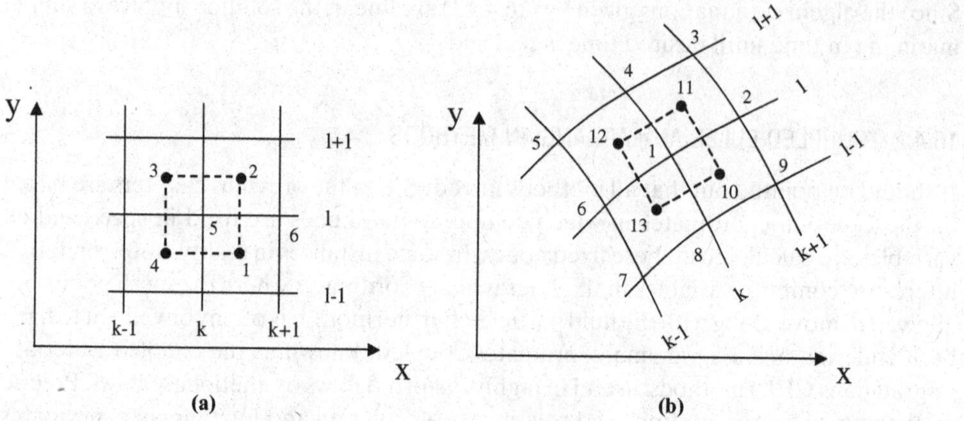

Figure 16.4.2 Eulerian and Lagrangian descriptions. **(a)** Eulerian description. **(b)** Lagrangian description.

$$\text{(ii)}\ v^{n+1}_{k+1,l+1} = \frac{1}{2}\left(u^{n+1/2}_{k+1,l+1} + u^{n+1/2}_{k+1,l+1}\right) \tag{16.4.14b}$$

$$\text{(iii)}\ (fu\Delta y)^{n+1/2}_{k+1,l+1} = \begin{cases} f^n_{k+1,l+1/2}u^{n+1/2}_{k+1,l+1/2}(y_{l+1} - y_l) & \text{if } u^{n+1/2}_{k+1,l+1/2} \geq 0 \\ f^n_{k+1,l+3/2}u^{n+1/2}_{k+1,l+1/2}(y_{l+1} - y_l) & \text{if } u^{n+1/2}_{k+1,l+1/2} \leq 0 \end{cases} \tag{16.4.14c}$$

$$\text{(iv)}\ (fv\Delta x)^{n+1/2}_{k+1,l+1} = \begin{cases} f^n_{k+1,l+1/2}u^{n+1/2}_{k+1,l+1/2}(x_{k+1} - x_k) & \text{if } v^{n+1/2}_{k+1,l+1/2} \geq 0 \\ f^n_{k+1,l+3/2}u^{n+1/2}_{k+1,l+1/2}(x_{k+1} - y_k) & \text{if } v^{n+1}_{k+1,l+1} \leq 0 \end{cases} \tag{16.4.14d}$$

$$\text{(v)}\ (\nabla \cdot f\mathbf{U})^{n+1}_{k+1,l+1} = \frac{(fu\Delta y)^{n+1/2}_{k+1,l+1/2} + (fv\Delta x)^{n+1/2}_{k+1/2,l+1} - (fu\Delta y)^{n+1/2}_{k,l+1/2} + (fv\Delta x)^{n+1/2}_{k+1/2,l}}{(x_{k+1} - x_k)(y_{l+1} - y_l)} \tag{16.4.14e}$$

$$\text{(vi)}\ \left(\frac{\Delta\overline{p}}{\Delta x}\right)^n_{k,l} = \frac{1}{2}\frac{\overline{p}^n_1 + \overline{p}^n_2 - \overline{p}^n_3 - \overline{p}^n_4}{x_1 - x_4} \tag{16.4.15a}$$

$$\text{(vii)}\ \left(\frac{\Delta\overline{p}}{\Delta y}\right)^n_{k,l} = \frac{1}{2}\frac{\overline{p}^n_2 + \overline{p}^n_3 - \overline{p}^n_1 - \overline{p}^n_4}{y_2 - y_1} \tag{16.4.15b}$$

$$\text{(viii)}\ (fu\Delta y)^{n-1}_{k+1/2,l} = (y_2 - y_1)\frac{(u^{n-1/2}_5 + u^{n-1/2}_6)}{2}\begin{cases} f^{n-1}_5 & \text{if } (u^{n-1/2}_5 u^{n-1/2}_6) \geq 0 \\ f^{n-1}_6 & \text{if } (u^{n-1/2}_5 u^{n-1/2}_6) \leq 0 \end{cases} \tag{16.4.15c}$$

$$\text{(ix)}\ (fv\Delta x)^{n-1/2}_{k/2,l+1} = (x_2 - x_3)\frac{(v^{n-1/2}_5 + v^{n-1/2}_7)}{2}\begin{cases} f^{n-1}_5 & \text{if } (v^{n-1/2}_5 u^{n-1/2}_7) \geq 0 \\ f^{n-1}_7 & \text{if } (v^{n-1/2}_5 v^{n-1/2}_7) \leq 0 \end{cases} \tag{16.4.15d}$$

$$\text{(x)}\quad (\nabla\cdot f\mathbf{U})_{k,l}^{n-1} = \frac{(fu\Delta y)_{k+1,l}^{n-1/2} + (fv\Delta x)_{k,l+1/2}^{n-1/2} - (fu\Delta y)_{k-1/2,l}^{n-1/2} + (fv\Delta x)_{k/2,l-1/2}^{n-1/2}}{(x_1 - x_4)(y_2 - y_1)} \tag{16.4.15e}$$

with $\overline{p} = p + q, \qquad q = \frac{1}{2}\rho v_i v_i.$

Based on the above definitions, the finite difference equations for inviscid flows are of the form:

Continuity

$$\rho_{k+1/2,l+1/2}^{n+1} = \rho_{k+1/2,l+1/2}^{n} - \Delta t(\nabla\cdot\rho\mathbf{U})_{k+1/2,l+1/2}^{n+1/2} \tag{16.4.16}$$

Momentum

$$M_{k,l}^{n+1/2} = M_{k,l}^{n-1/2} - \Delta t\left[(\nabla\cdot M\mathbf{U})_{k,l}^{n-1/2} + \left(\frac{\Delta\overline{p}}{\Delta x}\right)_{k,l}^{n}\right], \quad M = \rho u \tag{16.4.17a}$$

$$N_{k,l}^{n+1/2} = M_{k,l}^{n-1/2} - \Delta t\left[(\nabla\cdot M\mathbf{U})_{k,l}^{n-1/2} + \left(\frac{\Delta\overline{p}}{\Delta y}\right)_{k,l}^{n}\right], \quad N = \rho v \tag{16.4.17b}$$

Energy

$$\varepsilon_{k+1/2,l+1/2}^{n+1} = \varepsilon_{k+1/2,l+1/2}^{n} - \Delta t\Big[(\nabla\cdot\varepsilon U)_{k+1/2,l+1/2}^{n+1/2} + (\overline{p})_{k+1/2,l+1/2}^{n+1/2} + q_{k+1/2,l+1/2}^{n+1/2}(\nabla\cdot\mathbf{U})_{k+1/2,l+1/2}^{n+1/2}\Big] \tag{16.4.18}$$

Lagrangian Difference Equations

The differential equations in Lagrangian coordinates are given by

$$\frac{\partial u}{\partial t} = -\frac{1}{\rho}\frac{\partial p}{\partial x}, \quad \frac{\partial v}{\partial t} = -\frac{1}{\rho}\frac{\partial p}{\partial y} \tag{16.4.19}$$

$$u = \frac{\partial x}{\partial t}, \quad v = \frac{\partial y}{\partial t} \tag{16.4.20}$$

$$\rho J = const., \qquad \frac{\partial\varepsilon}{\partial t} = \frac{p}{\rho^2}\frac{\partial\rho}{\partial t}, \qquad p = p(\varepsilon,\rho) \tag{16.4.21}$$

with J being the Jacobian between the cartesian and curvilinear coordinates (Figure 16.4.2b).

The Lagrangian difference equations corresponding to (16.4.19–21) are written as follows.

$$u_{k,l}^{n+1} = u_{k,l}^{n-1/2} - \Delta t\frac{(\overline{p},y)_{k,l}^{n}}{(\rho J)_{k,l}} \tag{16.4.22a}$$

$$v_{k,l}^{n+1} = v_{k,l}^{n-1/2} - \Delta t\frac{(\overline{p},y)_{k,l}^{n}}{(\rho J)_{k,l}} \tag{16.4.22b}$$

$$x_{k.l}^{n+1} = x_{k.l}^{n} + \Delta t u_{k.l}^{n+1} \tag{16.4.23a}$$

$$y_{k.l}^{n+1} = y_{k.l}^{n} + \Delta t u_{k.l}^{n+1} \tag{16.4.23b}$$

$$\rho^{n+1}_{k+1,l+1} = \rho^{n}_{k+1/2,l+1/2} \frac{J^{n}_{k+1/2,l+1/2}}{J^{n+1/2}_{k+1/2,l+1/2}} \tag{16.4.24}$$

$$\varepsilon^{n+1}_{k+1,l+1} = \varepsilon^{n}_{k+1/2,l+1/2} + \overline{p}^{n+1/2}_{k+1/2,l+1/2} \frac{(\rho^{n+1} - \rho^{n})_{k+1/2,l+1/2}}{\rho^{n+1}\rho^{n}_{k+1/2,l+1/2}} \tag{16.4.25}$$

The velocity equations (16.4.23a,b) must be modified for the points of the lattice which define the boundaries of the Lagrangian region, but the remaining equations hold for all points of the mesh.

Finite elements have been used in CEL methods as applied to multiphase flows. Surface tension on the interfaces between different fluids can also be taken into account. These and other topics using CEL are discussed in Chapter 25.

16.4.3 PARTICLE-IN-CELL (PIC) METHOD

This is one of the early methods developed in the Los Alamos Scientific Laboratory in dealing with highly distorted flows with slippages or colliding interfaces [Evans and Harlow, 1957; Harlow, 1964]. In this method, Eulerian mesh is used and the cell is filled with particles of the same kind or a mixture of different kinds. The calculation of changes in the fluid configuration proceeds through a series of time steps or cycles. Each cell is characterized by a set of variables describing the mean components of velocity, the internal energy, the density, and the pressure in the cell. In the Eulerian part of the calculations, only the cellwise quantities are changed and the fluid is assumed to be momentarily completely at rest. In order to accomplish the particle motion, it is convenient to prepare as a first step for the possibility of particles moving across cell boundaries. For this purpose, the specific quantities in each of the cells are transformed to cellwise totals.

The results of a calculation applied to the formation of a crater by an explosion in an atmosphere above a dense material are shown in Figure 16.4.3 [Harlow, 1964]. The initial one for time $t = 0$ shows cold ground above which is a small and intensely heated sphere in an otherwise cold atmosphere. The second frame, two time units later, is shown in order to demonstrate the intense packing of particles in the initially heated sphere. The third frame shows a strong shock in the ambient atmosphere, together with considerable depression of the ground. The final frame shows, at time sixty units, the configuration just before the particles began to fall off the computation regions.

16.4.4 MONTE CARLO METHODS (MCM)

Monte Carlo methods have been successfully used in many problems in physics and engineering where stochastic or statistical approaches can describe the physical phenomena more realistically [Hammersley and Handscomb, 1964; Binder, 1984]. They have been extensively applied to electron distributions, neutron diffusion, radiative heat transfer, probability density functions for turbulent microscale eddies, etc.

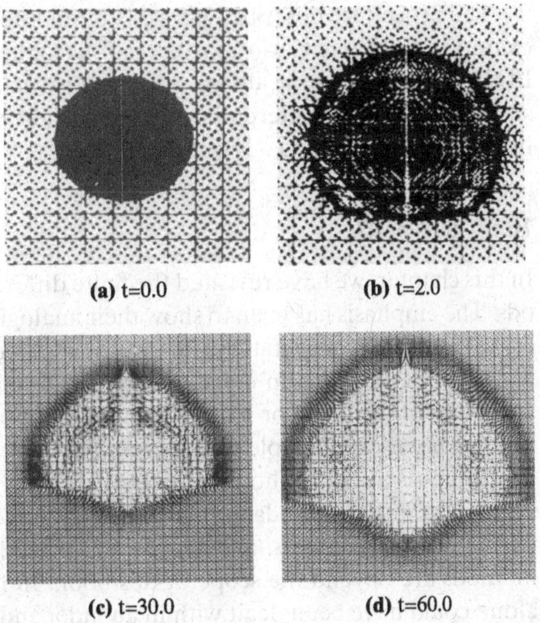

Figure 16.4.3 Configurations of particles at four times in the crater formulation problem; grid lines show every other cell boundary [Harlow, 1964].

In general, the Monte Carlo method is a statistical approach to the solution of multiple integrals of the type

$$I(\xi_1, \xi_2, \ldots\ldots, \xi_k) = \int_0^1 \int_0^1 w(\xi_1, \xi_2, \ldots\ldots, \xi_k) dP_1(\xi_1) dP_2(\xi_2) \ldots\ldots dP_k(\xi_k) \tag{16.4.26}$$

Monte Carlo becomes indispensable whenever multiple integrals have variables and can not be evaluated efficiently by standard numerical techniques.

As an example, let us consider the heat conduction equation,

$$\frac{\partial^2 T}{\partial x^2} + \frac{\partial^2 T}{\partial y^2} = 0$$

The integral (16.4.26) corresponding to heat conduction may be written as

$$I(\xi) = \int_0^1 w(\xi_1) dP_1(\xi_1) \tag{16.4.27}$$

In terms of the finite difference discretization, the integral (16.4.27) represents a finite difference equation written for the temperature at nodes (i, j) as

$$T_{i,j} = P_{x^+} T_{i+1,j} + P_{y^+} T_{i,j+1} + P_{x^-} T_{i-1,j} + P_{y^-} T_{i,j-1} \tag{16.4.28}$$

with

$$P_{x^+} = P_{x^-} = \frac{\Delta y/\Delta x}{2(\Delta y/\Delta x + \Delta x/\Delta y)} \tag{16.4.29a}$$

$$P_{y^+} = P_{y^-} = \frac{\Delta x/\Delta y}{2(\Delta y/\Delta x + \Delta x/\Delta y)} \tag{16.4.29b}$$

The procedure described above is often known as the random walk. In this simple example, the Monte Carlo approximations for heat conduction resembles the four-point FDM. In conduction, an abstraction using particles or random walks is used to simulate a solution of a partial differential equation, whereas in radiation a physical phenomenon – the transfer of photons – is simulated.

16.5 SUMMARY

In this chapter, we have revisited the finite difference methods and finite element methods. The emphasis has been to show their analogies. In this process, differences between these two major computational methods have been recognized. The advantage of studying both methods on an equal footing has been stressed. The finite volume methods based on either FDM or FEM are increasingly popular in applications to many engineering projects. Example problems in Part Five will demonstrate these trends.

Computational methods other than FDM, FEM, and FVM have been briefly reviewed, including boundary element methods, coupled Eulerian-Lagrangian methods, particle-in-cell methods, and Monte Carlo methods. Detailed presentations of these methods are beyond the scope of this book. In fact, the topics covered in this chapter alone could have been dealt with in an independent part.

As we look back on the chapters in Part Two and Part Three, our focus has been to introduce to the reader what has been accomplished in CFD for the past century. It was not possible to cover all minute details of every method that was introduced. Pertinent references are provided at the end of each chapter. Obviously, the reader should consult these references for further guidance.

This chapter marks the end of Parts Two and Three, including FDM, FEM, and FVM, but we have not discussed other important subjects: automatic grid generation, adaptive methods, and computing techniques. We shall examine them in the next several chapters, Part Four.

REFERENCES

Binder, K. [1984]. *Applications of the Monte Carlo Method in Statistical Physics*. Berlin: Springer-Verlag.

Brebbia, C. A. [1978]. *The Boundary Element Method for Engineers*. London: Pentech Press.

Brebbia, C. A., Telles, J., and Wrobel, L. [1983]. *Boundary Element Methods – Theory and Applications*. New York: Springer-Verlag.

Chung, T. J. [1996]. *Applied Continuum Mechanics*. London: Cambridge University Press.

Evans, M. W. and Harlow, F. H. [1957]. The particle-in-cell method for hydrodynamic calculations. Los Alamos Scientific Laboratory Report No. LA-2139.

Hammersley, J. M. and Handscomb, D. C. [1964]. *Monte Carlo Methods*. London: Methuen.

Harlow, F. H. [1964]. The particle-in-cell computing method for fluid dynamics. In F. H. Harlow, (ed.). *Methods in Computational Physics*. New York: Academic Press.

Noh, W. F. [1964]. CEL: A time-dependent, two-space-dimensional, coupled Eulerian-Lagrange code. In F. H. Harlow (ed.). *Methods in Computational Physics*. New York: Academic Press.

PART FOUR

AUTOMATIC GRID GENERATION, ADAPTIVE METHODS, AND COMPUTING TECHNIQUES

Automatic grid generation techniques have contributed significantly toward the application of computational fluid dynamics in large-scale industrial problems. Without such techniques the most accurate numerical schemes may fail to prove their full potential or effectiveness. Automatic grid generation in complicated geometries such as those of a complete aircraft is now considered a routine exercise and an important part of CFD projects.

There are two types of grid generation: structured and unstructured. In structured grids, all grid lines are oriented regularly in either two or three directions so that coordinate transformations of curvilinear lines result in a square or cube for two-dimensional or three-dimensional problems, respectively. In unstructured grids, however, there are no such restrictions, but at the expense of more complicated computer programming. Once the automatic grid generation is completed, a challenging task still remains – an adaptive mesh in which the most suitable mesh distributions are achieved to obtain the most accurate solution. This can be made possible by placing finer meshes in regions where gradients of variables are high. Furthermore, computing techniques including domain decomposition, multigrid methods, and parallel processing, among others, play an important role for the success of CFD projects.

We shall examine these and other subjects in Part Four. Structured grid generation is discussed in Chapter 17, unstructured grids in Chapter 18, adaptive methods for structured and unstructured grids in Chapter 19, and computing techniques in Chapter 20.

CHAPTER SEVENTEEN

Structured Grid Generation

Structured grids are generated in two- or three-dimensional geometries (with plane or curved surfaces). In general, two types of structured grid generation are in use: algebraic methods and partial differential equation (PDE) mapping methods. For more complex geometries, it is preferable to construct multiblocks initially, with refined grids filled in for each of the multiblocks subsequently. Detailed procedures are presented in the following sections.

17.1 ALGEBRAIC METHODS

In algebraic methods, geometric data of the cartesian coordinates in the interior of a domain are generated from the values specified at boundaries through interpolations or specific functions of the curvilinear coordinates. Toward this end, we begin first with the unidirectional interpolations of various functional representations, followed by multidirectional interpolations.

17.1.1 UNIDIRECTIONAL INTERPOLATION

Unidirectional interpolation refers to the functional representation in only one direction. Among the most widely used are Lagrange polynomials, Hermite polynomials, and cubic spline functions. These polynomials, some of which were discussed in Chapter 9, are briefly reviewed below.

(a) Lagrange Polynomials

The Lagrange polynomials, as used in FEM for interpolations of a variable (Section 9.2.2), may be used for grid generation in interpolation between cartesian and curvilinear coordinates (Figure 17.1.1).

$$x = \Phi_N(\xi)x_N, \qquad \Phi_N(\xi_M) = \delta_{NM} \tag{17.1.1}$$

with $\Phi_N(\xi)$ being the Lagrange polynomials

$$\Phi_N = \prod_{M=1,\, M\neq N}^{n} \frac{\xi - \xi_M}{\xi_N - \xi_M}, \qquad \xi = \frac{x}{h} \tag{17.1.2}$$

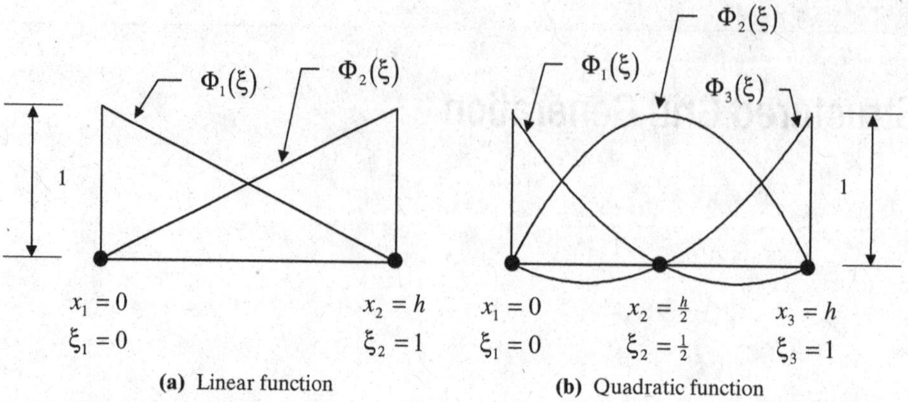

Figure 17.1.1 Lagrange polynomials.

This formula provides:

Linear Interpolation

$$\Phi_1 = 1 - \xi, \qquad \Phi_2 = \xi \tag{17.1.3}$$

Quadratic Interpolation

$$\Phi_1 = 2\left(\xi - \frac{1}{2}\right)(\xi - 1), \qquad \Phi_2 = -4\xi(\xi - 1), \qquad \Phi_3 = 2\xi\left(\xi - \frac{1}{2}\right) \tag{17.1.4}$$

(b) Hermite Polynomials

Hermite polynomials as used in FEM provide functional representation not only of coordinate values, but also of gradients of coordinate values. For example, the functional representation including the zeroth order and first order derivatives of coordinate values are given by cubic functions (Figure 17.1.2),

$$x = H_N^0(\xi)x_N + H_N^1(\xi)\,\theta_N, \qquad \theta_N = \frac{\partial x_N}{\partial \xi} \tag{17.1.5a}$$

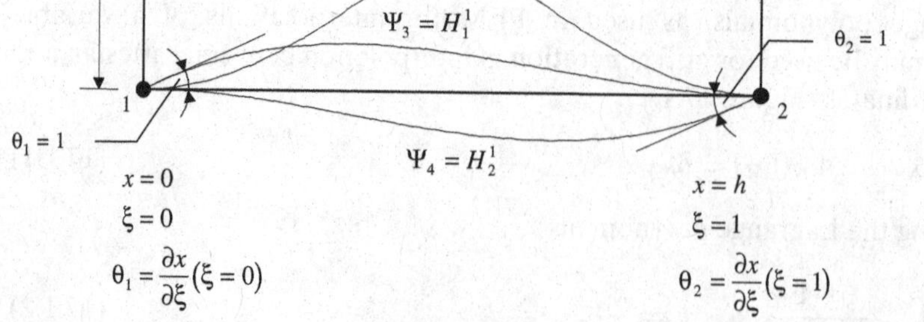

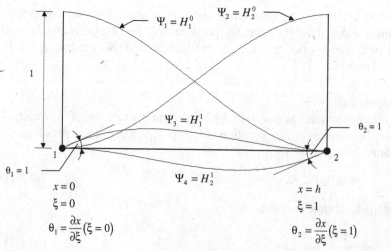

Figure 17.1.2 Hermite polynomial.

or

$$x = H_1^0 x_1 + H_2^0 x_2 + H_1^1 \theta_1 + H_2^1 \theta_2 \tag{17.1.5b}$$

with $H_N^0(\xi_M) = \delta_{NM}$, $H_N^1(\xi_M) = \delta_{NM}$
Thus

$$x = \Psi_r Q_r, \quad (r = 1, 2, 3, 4) \tag{17.1.5c}$$

with

$$\Psi_1 = H_1^0(\xi) = 1 - 3\xi^2 + 2\xi^3 \tag{17.1.6a}$$

$$\Psi_2 = H_2^0(\xi) = 3\xi^2 - 2\xi^3 \tag{17.1.6b}$$

$$\Psi_3 = H_1^1(\xi) = \xi - 2\xi^2 + \xi^3 \tag{17.1.6c}$$

$$\Psi_4 = H_2^1(\xi) = \xi^3 - \xi^2 \tag{17.1.6d}$$

These functions match the two boundary values x_1, and x_2 and the first derivatives, $(\partial x/\partial\xi)_1$, and $(\partial x/\partial\xi)_2$ at the two boundaries.

The advantage of specifying $(\partial x/\partial\xi)$ as well as x can be used to make the grid orthogonal at the boundary. This will be useful in multidirectional grid generation.

(c) Cubic Spline Functions

One of the difficulties with conventional polynomial interpolations, particularly if the polynomials are of high order, is the oscillatory character. To remedy this disadvantage, the cubic spline functions can be used to achieve smoother curves.

Consider two arbitrary adjacent points x_i and x_{i+1}. We wish to fit a cubic to these two points and use this cubic as the interpolation function between them.

$$F_i(x) = a_0 + a_1 x + a_2 x^2 + a_3 x^3, \quad (x_i \le x \le x_{i+1}) \tag{17.1.7}$$

Note that two constants in (17.1.7) may be determined by end conditions and two others by the slope (first derivative) and curvature (second derivative). Here the second derivative of a cubic line is a straight line (Figure 17.1.3) so that

$$g''(x) = g''(x_i) + \frac{x - x_i}{x_{i+1} - x_i}[g''(x_{i+1}) - g''(x_i)] \tag{17.1.8}$$

Integrating (17.1.8) twice, we obtain

$$\begin{aligned} g(x) = F_i(x) = {} & \frac{g''(x_i)}{6}\left[\frac{(x_{i+1} - x)^3}{\Delta x_i} - \Delta x_i(x_{i+1} - x)\right] \\ & + \frac{g''(x_{i+1})}{6}\left[\frac{(x - x_i)^3}{\Delta x_i} - \Delta x_i(x - x_i)\right] \\ & + f(x_i)\left(\frac{x_{i+1} - x}{\Delta x_i}\right) + f(x_{i+1})\left(\frac{x - x_i}{\Delta x_i}\right) \end{aligned} \tag{17.1.9}$$

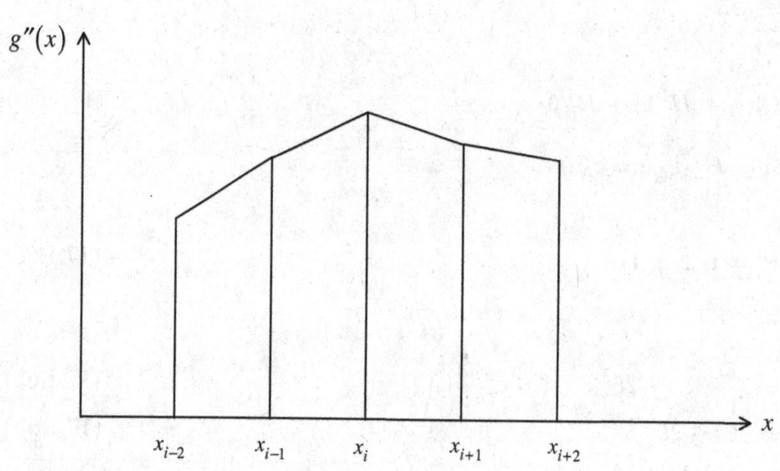

Figure 17.1.3 Cubic spline representation.

with $\Delta x_i = x_{i+1} - x_i, i = 0, 1, \ldots n-1, g(x_i) = f(x_i)$ and $g(x_{i+1}) = f(x_{i+1})$. Since the second derivatives $g''(x_i)$ $(i = 0, 1, \ldots n)$ are still unknown, these must be evaluated as follows:

$$F_i'(x_i) = F_{i-1}'(x_i) \tag{17.1.10a}$$

$$F_i''(x_i) = F_{i-1}''(x_i) \tag{17.1.10b}$$

Evaluation of (17.1.10a) leads to a set of simultaneous linear equations of the form

$$\begin{aligned} &\frac{\Delta x_{i-1}}{\Delta x_i} g''(x_{i-1}) + \frac{2(x_{i+1} - x_{i-1})}{\Delta x_i} g''(x_i) + g''(x_{i+1}) \\ &= 6\left[\frac{f(x_{i+1}) - f(x_i)}{(\Delta x_i)^2} - \frac{f(x_i) - f(x_{i-1})}{(\Delta x_i)(\Delta x_{i-1})}\right] \end{aligned} \tag{17.1.11}$$

This represents $n-1$ equations in the $n+1$ unknowns $g''(x_0), g''(x_1), \ldots, g''(x_n)$. The two necessary additional equations are

$$g''(x_0) = 0 \tag{17.1.12a}$$

$$g''(x_n) = 0 \tag{17.1.12b}$$

The resulting $g(x)$ is called a natural cubic spline.

In terms of nondimensional coordinates, (17.1.11) and (17.1.12) are written as

$$(\xi_i - \xi_{i-1})x_{i-1}'' + 2(\xi_{i+1} - \xi_{i-1})x_i'' + (\xi_{i+1} - \xi_i)x_{i+1}'' = 6\left(\frac{x_{i+1} - x_i}{\xi_{i+1} - \xi_i} - \frac{x_i - x_{i-1}}{\xi_i - \xi_{i-1}}\right) \tag{17.1.13}$$

with

$$x_1'' = 0 \tag{17.1.14a}$$

$$x_n'' = 0 \tag{17.1.14b}$$

The solution x'' is substituted into

$$x = \frac{(\xi_{i+1}-\xi)^3}{6(\xi_{i+1}-\xi_i)}x''_i + \frac{(\xi-\xi_i)^3}{6(\xi_{i+1}-\xi_i)}x''_{i+1} + \left[\frac{x_i}{\xi_{i+1}-\xi_i} - \frac{\xi_{i+1}-\xi_i}{6}x''_i\right](\xi_{i+1}-\xi)$$
$$+ \left[\frac{x_{i+1}}{\xi_{i+1}-\xi_i} - \frac{\xi_{i+1}-\xi_i}{6}x''_{i+1}\right](\xi-\xi_i) \qquad (17.1.15)$$

It is seen that (17.1.15) may be written in the form similar to (17.1.1) as a linear combination of interpolation functions and nodal values of the first and second derivatives of x at nodal points i and $i+1$.

Additional interpolation functions useful for surface grid generations are available. These functions will be discussed in Section 17.3.

17.1.2 MULTIDIRECTIONAL INTERPOLATION

There are two multidirectional interpolation methods available: domain vertex methods developed from FEM interpolation functions and transfinite interpolation methods predominantly used in FDM, constructed by means of tensor products of unidirectional functional representation in multidimensions.

17.1.2.1 Domain Vertex Method

Domain vertex methods utilize tensor products of unidirectional interpolation functions for two or three dimensions. Let us consider a two-dimensional domain with physical coordinates (x, y) and transformed computational domain (ξ, η) as shown in Figure 17.1.4a, related by

$$x_i = \hat{\Phi}_N(\xi)\hat{\Phi}_M(\eta)x_{iNM}, \quad (i = 1, 2, \ N, M = 1, 2) \qquad (17.1.16a)$$

or

$$x_i = \Phi_N(\xi, \eta)x_{iN}, \quad (i = 1, 2, \ N = 1, 2, 3, 4) \qquad (17.1.16b)$$

where i denotes the physical coordinate directions and N and M represent node numbers in the direction of the coordinate $\hat{\Phi}_N(\xi)$, and $\hat{\Phi}_M(\eta)$ are the unidirectional functions whereas $\Phi_N(\xi, \eta)$ indicates the tensor product.

$$\Phi_N(\xi, \eta) = \begin{cases} \Phi_1 = (1-\xi)(1-\eta) & = \hat{\Phi}_1(\xi)\hat{\Phi}_1(\eta) \\ \Phi_2 = \xi(1-\eta) & = \hat{\Phi}_2(\xi)\hat{\Phi}_1(\eta) \\ \Phi_3 = \xi\eta & = \hat{\Phi}_2(\xi)\hat{\Phi}_2(\eta) \\ \Phi_4 = (1-\xi)\eta & = \hat{\Phi}_1(\xi)\hat{\Phi}_2(\eta) \end{cases} \qquad (17.1.17)$$

which are known as "blending functions."

Similarly for three dimensions (Figure 17.1.4b), we obtain

$$x_i = \hat{\Phi}_N(\xi)\hat{\Phi}_M(\eta)\hat{\Phi}_P(\zeta)x_{iNMP}, \quad (i = 1, 2, 3, \ N, M, P = 1, 2) \qquad (17.1.18a)$$

or

$$x_i = \Phi_N(\xi, \eta, \zeta)x_{iN}, \quad (i = 1, 2, 3, \ N = 1, \ldots, 8) \qquad (17.1.18b)$$

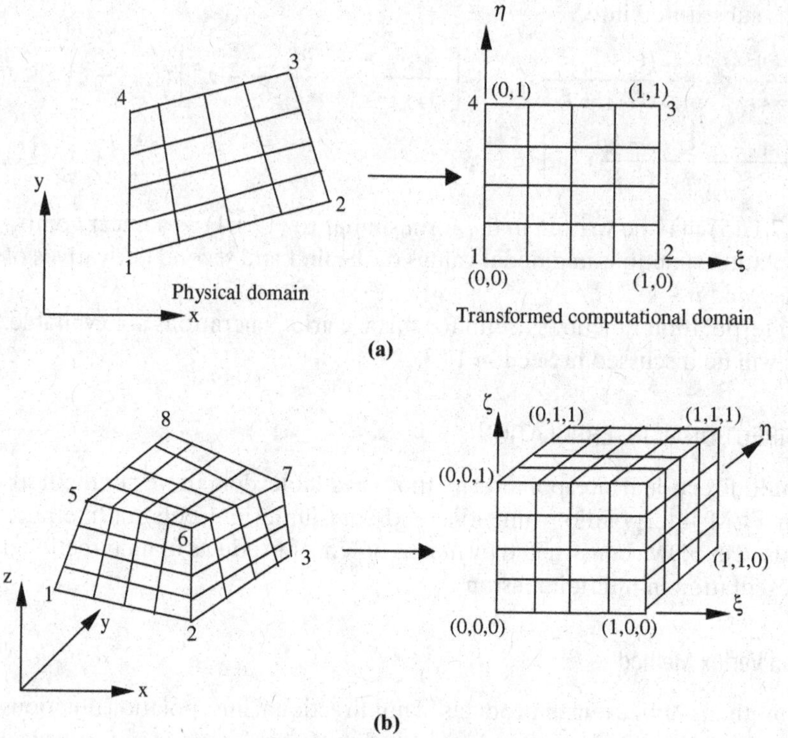

Figure 17.1.4 Multidimensional interpolation, all interior lines (as many as desired) are generated from (17.1.16) with the corner node coordinates and the interior values of ξ and η. **(a)** Two-dimensional domain. **(b)** Three-dimensional domain.

with

$$\Phi_N(\xi,\eta,\zeta) = \begin{cases} \Phi_1 = (1-\xi)(1-\eta)(1-\zeta) \\ \Phi_2 = \xi(1-\eta)(1-\zeta) \\ \Phi_3 = \xi\eta(1-\zeta) \\ \Phi_4 = (1-\xi)\eta(1-\zeta) \\ \Phi_5 = (1-\xi)(1-\eta)\zeta \\ \Phi_6 = \xi(1-\eta)\zeta \\ \Phi_7 = \xi\eta\zeta \\ \Phi_8 = (1-\xi)\eta\zeta \end{cases} \tag{17.1.19}$$

Extensions of the above processes can be made to accommodate higher order interpolations by providing interior nodes along each side (see Figure 17.1.5 for quadratic mapping). Furthermore, triangular elements and tetrahedral elements can also be constructed, following the FEM geometries discussed in Chapter 9.

Example 17.1.1 Trapezoidal Geometry

Given: Four points $A(0,0)$, $B(L,0)$, $C(L,H_2)$, and $D(0,H_1)$. Generate a mesh corresponding to ξ,η at 0.2 apart. Assume $L = 20$, $H_1 = 5$, $H_2 = 10$.

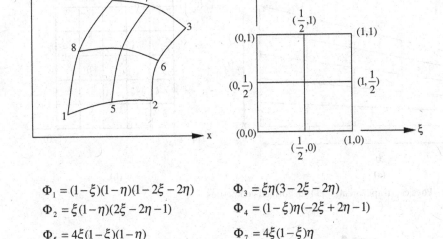

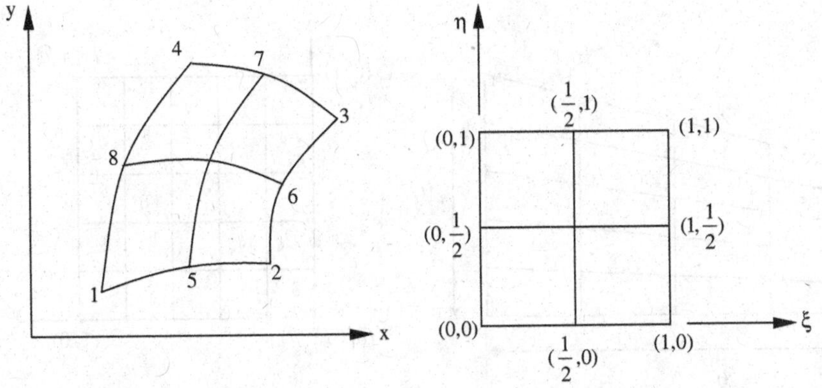

$$\Phi_1 = (1-\xi)(1-\eta)(1-2\xi-2\eta) \qquad \Phi_3 = \xi\eta(3-2\xi-2\eta)$$
$$\Phi_2 = \xi(1-\eta)(2\xi-2\eta-1) \qquad \Phi_4 = (1-\xi)\eta(-2\xi+2\eta-1)$$
$$\Phi_5 = 4\xi(1-\xi)(1-\eta) \qquad \Phi_7 = 4\xi(1-\xi)\eta$$
$$\Phi_6 = 4\xi\eta(1-\eta) \qquad \Phi_8 = 4(1-\xi)\eta(1-\eta)$$

Figure 17.1.5 Quadratic interpolation by inserting any values of ξ and η, interior coordinates are generated from the above functions (as many as desired).

Solution:

$$\begin{aligned} x &= (1-\xi)(1-\eta)x_1 + \xi(1-\eta)x_2 + \xi\eta x_3 + (1-\xi)\eta x_4 \\ &= [\xi(1-\eta)+\xi\eta]\,L \\ &= 20\,\xi \end{aligned}$$
with $x_1 = 0, \quad x_2 = L, \quad x_3 = L, \quad x_4 = 0, \quad L = 20$
$$\begin{aligned} y &= (1-\xi)(1-\eta)y_1 + \xi(1-\eta)y_2 + \xi\eta y_3 + (1-\xi)\eta y_4 \\ &= \xi\eta H_2 + (1-\xi)\eta H_1 \\ &= 10\,\xi\eta + 5(1-\xi)\eta \end{aligned}$$
with $y_1 = 0, \quad y_2 = 0, \quad y_3 = H_2 = 10, \quad y_4 = H_1 = 5$

The grid points or lines x, y can now be generated, and the results are shown in Figure E17.1.1.

Example 17.1.2

Consider a quarter circular disk as shown in Figure E17.1.2a. Using quadratic Lagrange polynomials develop a program to generate a 7×16 mesh:

Solution: The quadratic Lagrange interpolation functions (9 node) are given by

$$\Phi_N(\xi,\eta) = \prod_{M=1,\, N\neq M}^{n} \frac{\xi-\xi_M}{\xi_N-\xi_M}\frac{\eta-\eta_M}{\eta_N-\eta_M}$$

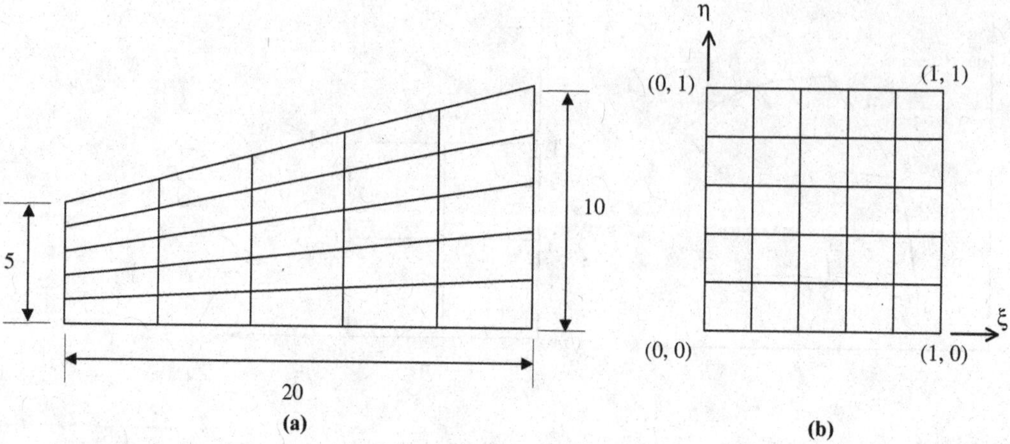

Figure E17.1.1 Physical (trapezoidal) and transformed geometries.

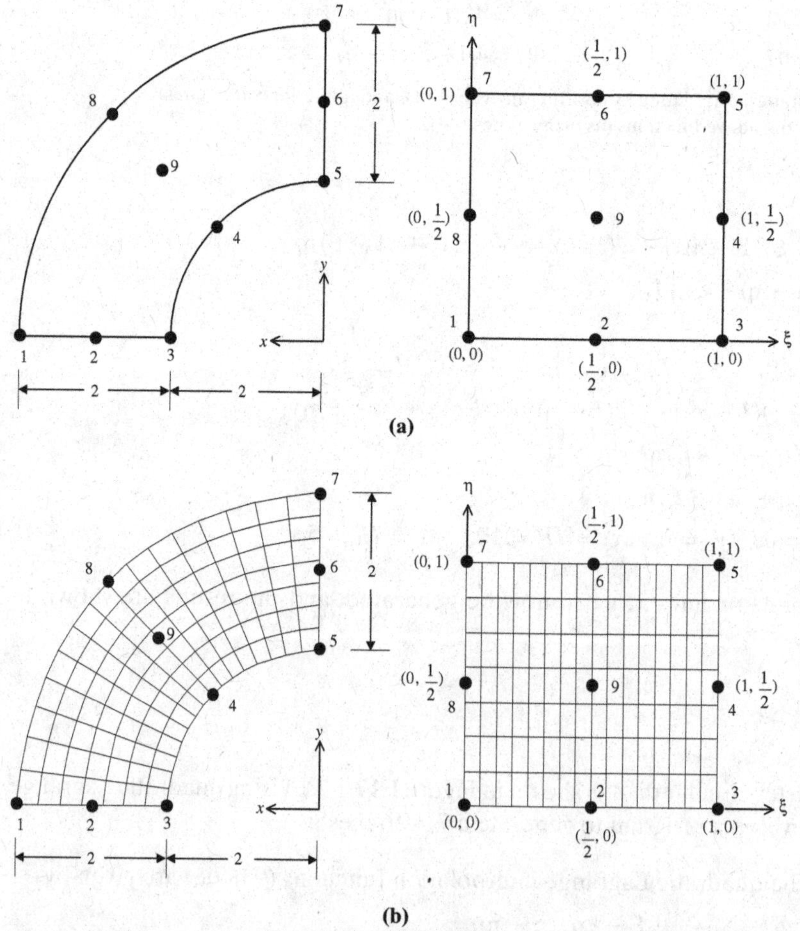

Figure E17.1.2 Quadratic Lagrange polynomials. **(a)** Quarter circle disk. **(b)** Mesh generated for a quarter circular disk using quadratic Lagrange polynomials.

$$
\Phi_N(\xi,\eta) = \begin{cases}
\Phi_1 = 4\left(\xi - \frac{1}{2}\right)(\xi - 1)\left(\eta - \frac{1}{2}\right)(\eta - 1) \\
\Phi_2 = -8\xi(\xi - 1)\left(\eta - \frac{1}{2}\right)(\eta - 1) \\
\Phi_3 = 4\xi\left(\xi - \frac{1}{2}\right)\left(\eta - \frac{1}{2}\right)(\eta - 1) \\
\Phi_4 = -8\xi\left(\xi - \frac{1}{2}\right)\eta(\eta - 1) \\
\Phi_5 = 4\xi\left(\xi - \frac{1}{2}\right)\eta\left(\eta - \frac{1}{2}\right) \\
\Phi_6 = -8\xi(\xi - 1)\eta\left(\eta - \frac{1}{2}\right) \\
\Phi_7 = 4\left(\xi - \frac{1}{2}\right)(\xi - 1)\eta\left(\eta - \frac{1}{2}\right) \\
\Phi_8 = -8\left(\xi - \frac{1}{2}\right)(\xi - 1)\eta(\eta - 1) \\
\Phi_9 = 16\xi(\xi - 1)\eta(\eta - 1)
\end{cases}
$$

$$
x_i(\xi,\eta) = \Phi_N(\xi,\eta)x_{iN}
$$

The results are shown in Figure E17.1.2.

Example 17.1.3

Consider a three-dimensional geometry as shown in Figure E17.1.3. Develop a computer program to generate an $11 \times 11 \times 11$ mesh (1331 points) using the domain vertex method. First derive the Lagrange polynomial interpolation functions with the data (x, y, z) as given in Table E17.1.3. Use a cubic between nodes 5–8 and 8–11.

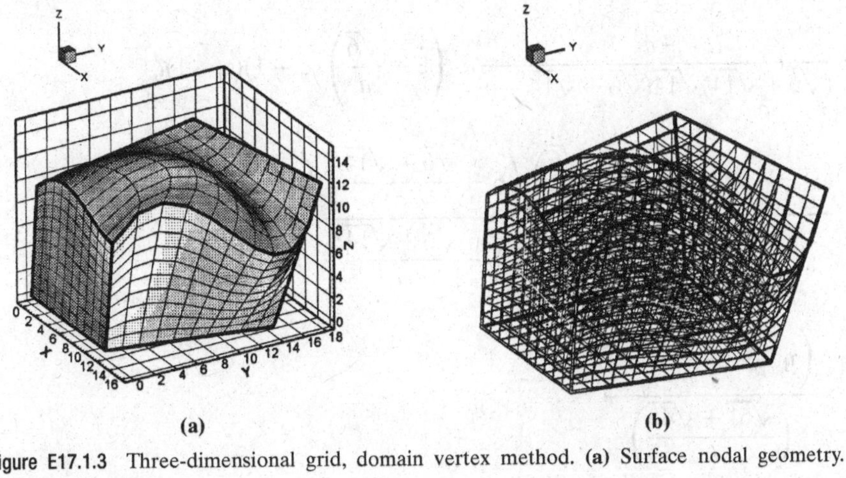

Figure E17.1.3 Three-dimensional grid, domain vertex method. (a) Surface nodal geometry. (b) Interior nodes.

Table E17.1.3 Interpolation Function Data

Point	(x, y, z)	Point	(x, y, z)	Point	(x, y, z)	Point	(x, y, z)
1	(0, 0, 0)	2	(12, 0, 0)	3	(15, 14, 0)	4	(0, 16, 0)
5	(1, 0, 10)	6	(3, 1, 11)	7	(6, 2, 9)	8	(8, 3, 7)
9	(12, 9, 10)	10	(14, 14, 6)	11	(16, 18, 12)	12	(4, 13, 14)

Physical domain coordinates are given:

$$a = \text{Length between 5 and 8} = \sqrt{6} + \sqrt{14} + 3$$

$$b = \text{Length between 8 and 11} = \sqrt{61} + \sqrt{45} + \sqrt{56}$$

$$\Phi_N(\xi,\eta,\zeta) = \prod_{M=1,\ N\neq M}^{n} \frac{\xi - \xi_M}{\xi_N - \xi_M}\frac{\eta - \eta_M}{\eta_N - \eta_M}\frac{\zeta - \zeta_M}{\zeta_N - \zeta_M}$$

with $n = \hat{n} + 1$, $\hat{n}$ being the total number of inside edge nodes in each direction (ξ, η, ζ).

$$\Phi_1 = \frac{(\xi - \xi_2)}{(\xi_1 - \xi_2)}\frac{(\eta - \eta_2)}{(\eta_1 - \eta_2)}\frac{(\zeta - \zeta_2)}{(\zeta_1 - \zeta_2)} = \frac{(\xi - 1)}{(0 - 1)}\frac{(\eta - 1)}{(0 - 1)}\frac{(\zeta - 1)}{(0 - 1)}$$

$$= -(\xi - 1)(\eta - 1)(\zeta - 1)$$

$$\Phi_2 = \xi(\eta - 1)(\zeta - 1) \qquad \Phi_3 = -\xi\,\eta(\zeta - 1) \qquad \Phi_4 = (\xi - 1)\,\eta(\zeta - 1)$$

$$\Phi_5 = \frac{a^2}{\sqrt{6}(\sqrt{6} + \sqrt{14})}\left(\xi - \frac{\sqrt{6}}{a}\right)\left(\xi - \frac{\sqrt{6} + \sqrt{14}}{a}\right)(\xi - 1)(\eta - 1)\zeta$$

$$\Phi_6 = \frac{a^3}{\sqrt{6}\sqrt{14}(\sqrt{6} - a)}\xi\left(\xi - \frac{\sqrt{6} + \sqrt{14}}{a}\right)(\xi - 1)(\eta - 1)\zeta$$

$$\Phi_7 = \frac{-a^3}{(\sqrt{6} + \sqrt{14})\sqrt{14}(\sqrt{6} + \sqrt{14} - a)}\xi\left(\xi - \frac{\sqrt{6}}{a}\right)(\xi - 1)(\eta - 1)\zeta$$

$$\Phi_8 = \frac{\xi(\xi - 1)(\eta - 1)\zeta\left(\xi - \frac{\sqrt{6}}{a}\right)}{\left(1 - \frac{\sqrt{6}}{a}\right)}\frac{\left(\xi - \frac{\sqrt{6} + \sqrt{14}}{a}\right)}{\left(1 - \frac{\sqrt{6} + \sqrt{14}}{a}\right)}\frac{\left(\eta - \frac{\sqrt{61}}{b}\right)}{\left(\frac{\sqrt{61}}{b}\right)}$$

$$\times\frac{\left(\eta - \frac{\sqrt{61} + \sqrt{45}}{b}\right)}{\left(\frac{\sqrt{61} + \sqrt{45}}{b}\right)}$$

$$\Phi_9 = \frac{\xi\eta\zeta}{\left(\frac{\sqrt{61}}{b}\right)} \frac{\left(\eta - \frac{\sqrt{61}+\sqrt{45}}{b}\right)}{\left(\frac{\sqrt{61}}{b} - \frac{\sqrt{61}+\sqrt{45}}{b}\right)} \frac{(\eta - 1)}{\left(\frac{\sqrt{61}}{b} - 1\right)}$$

$$\Phi_{10} = \frac{\xi\eta\zeta}{\left(\frac{\sqrt{61}+\sqrt{45}}{b}\right)} \frac{\left(\eta - \frac{\sqrt{61}}{b}\right)}{\left(\frac{\sqrt{61}+\sqrt{45}}{b} - \frac{\sqrt{61}}{b}\right)} \frac{(\eta - 1)}{\left(\frac{\sqrt{61}+\sqrt{45}}{b} - 1\right)}$$

$$\Phi_{11} = \xi\eta\zeta \frac{\left(\eta - \frac{\sqrt{61}}{b}\right)\left(\eta - \frac{\sqrt{61}+\sqrt{45}}{b}\right)}{\left(1 - \frac{\sqrt{61}}{b}\right)\left(1 - \frac{\sqrt{61}+\sqrt{45}}{b}\right)} \qquad \Phi_{12} = -(\xi - 1)\eta\zeta$$

Example 17.1.4

Clustering of boundary layers at the wall or interior domain may be achieved using exponential relations between the physical domain and transformed domain (Figure E17.1.4).

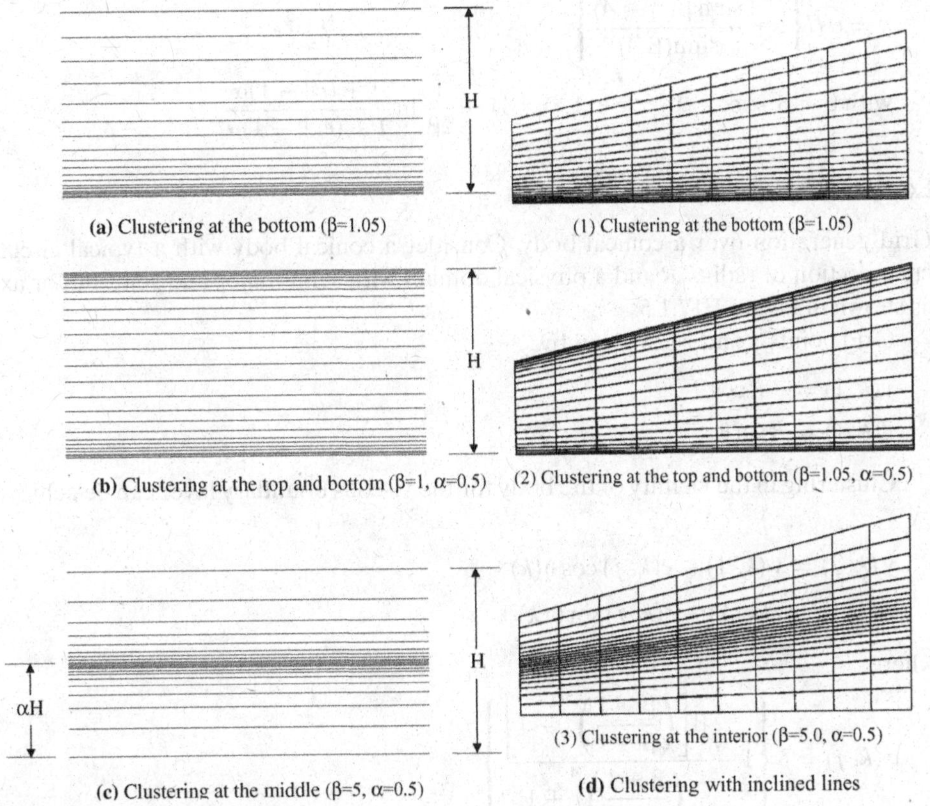

Figure E17.1.4 Clustering of mesh lines.

(a) Clustering at the Bottom Wall

$$x = \xi$$

$$y = H\frac{(\beta+1)-(\beta-1)\left(\frac{\beta+1}{\beta-1}\right)^{1-\eta}}{\left(\frac{\beta+1}{\beta-1}\right)^{1-\eta}+1}$$

with $1 < \beta < \infty$

(b) Clustering at Top and Bottom Walls

$$x = \xi$$

$$y = H\frac{(2\alpha+\beta)\left(\frac{\beta+1}{\beta-1}\right)^{\frac{\eta-\alpha}{1-\alpha}}+2\alpha-\beta}{(2\alpha+1)\left[\left(\frac{\beta+1}{\beta-1}\right)^{\frac{\eta-\alpha}{1-\alpha}}+1\right]}$$

with $0 < \alpha, \beta < \infty$

(c) Clustering at Interior Domain

$$x = \xi$$

$$y = \alpha H\left\{1+\frac{\sinh\beta(\eta-A)}{\sinh(\beta A)}\right\}$$

with $0 < \beta < \infty, \quad 0 < \alpha < 1, \qquad A = \frac{1}{2\beta}\ln\frac{1+(e^{\beta}-1)\alpha}{1+(e^{-\beta}-1)\alpha}$

Example 17.1.5

Grid generation over a conical body. Consider a conical body with a typical circular cross section of radius R and a physical domain with semi-major and semi-minor axes as shown in Figure E17.1.5.

Grid points y and z are given by

$$y(k,1) = -R\cos\theta$$
$$z(k,1) = R\sin\theta$$

Clustering in the vicinity of the body for the viscous boundary layer can be achieved by

$$y(k,j) = y(k,1) - c(k,j)\cos\theta(k)$$
$$z(k,j) = z(k,1) + c(k,j)\sin\theta(k)$$

where

$$c(k,j) = \delta\left\{1-\frac{\beta\left[\left(\frac{\beta+1}{\beta-1}\right)^{\eta}-1\right]}{\left(\frac{\beta+1}{\beta-1}\right)^{\eta}+1}\right\}$$

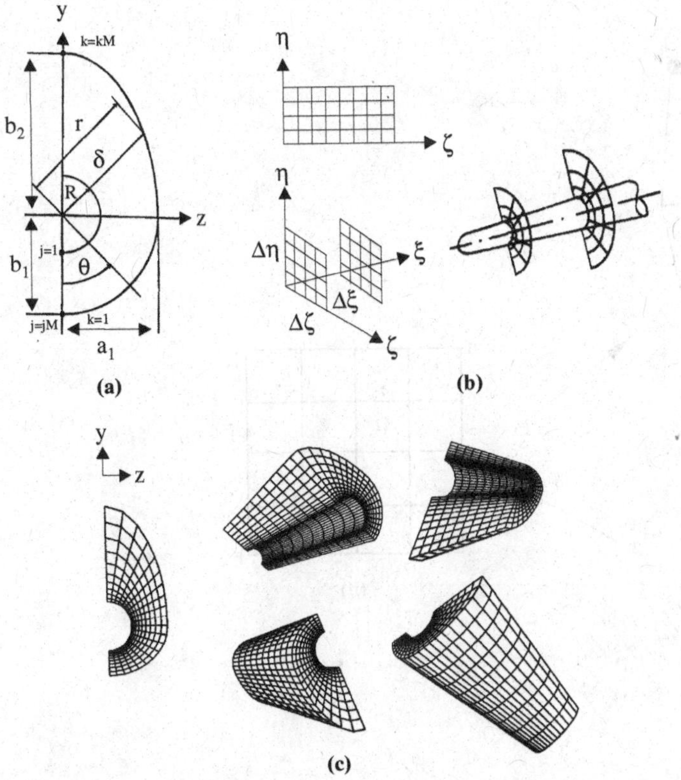

Figure E17.1.5 Algebraic grid generation of conical body. **(a)** Given data. **(b)** Transformed cross sections. **(c)** Finalized mesh.

with

$$\delta(k) = r(k) - R(k), \qquad r = \left[\left(\frac{\sin\theta}{a}\right)^2 + \left(\frac{\cos\theta}{b}\right)^2\right]^{-1/2}$$

The grid generated in Figure E17.1.5(c) will then be repeated in the (x, ξ) direction for the entire three-dimensional domain. Here, $R = 1$, $a_1 = 2.5$, $b_2 = 4$ were chosen in Figure E17.1.5(c).

17.1.2.2 Transfinite Interpolation Methods (TFI)

An alternative approach to the domain vertex methods is to use the unidirectional interpolation functions introduced in Section 17.1.1 and form tensor products in two or three directions as in the domain vertex methods, but with all sides of the boundaries interpolated and matched as well as the corner nodes. To this end, Lagrange polynomials, Hermite polynomials, or spline functions may be used. A transformed computational domain mapped into various arbitrary physical domains is shown in Figure 17.1.6.

Consider a region $\overline{\Omega}$, $[0, 1] \times [0, 1]$ and postulate the existence of a function **F** (vector valued) which maps $\overline{\Omega}$ into Ω such that **F**: F $\rightarrow \overline{\Gamma}$. Our objective is to construct a univalent (one-to-one) function **U**: $\overline{\Omega} \rightarrow \Omega$ which matches **F** on the boundary

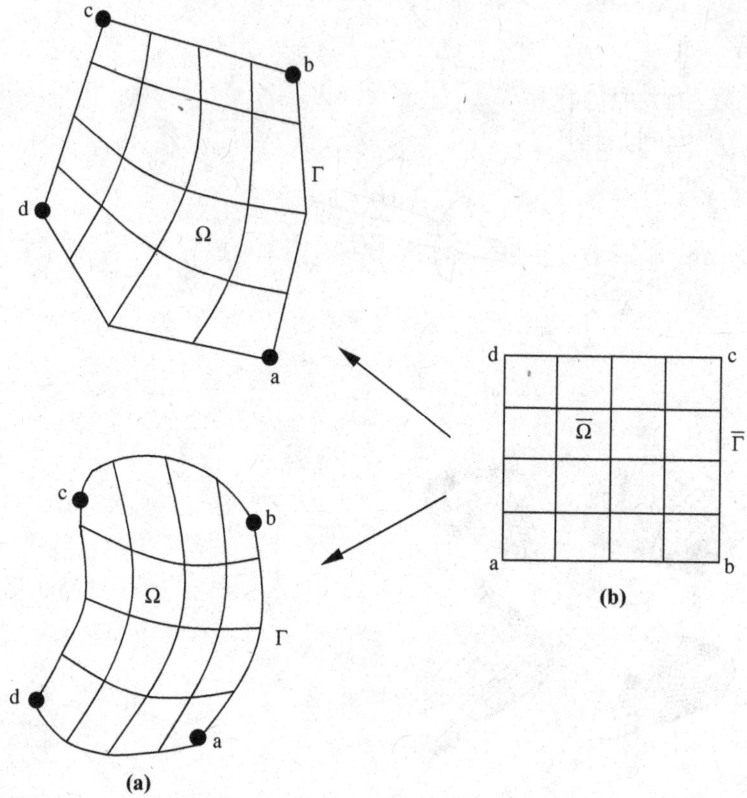

Figure 17.1.6 Physical domain and transformed computational domain for transfinite interpolation. (a) Physical domain. (b) Transformed computational domain.

of $\overline{\Gamma}$, that is,

$$\mathbf{U}(0,\eta) = \mathbf{F}(0,\eta), \quad \mathbf{U}(\xi,0) = \mathbf{F}(\xi,0) \tag{17.1.20a}$$

$$\mathbf{U}(1,\eta) = \mathbf{F}(1,\eta), \quad \mathbf{U}(\xi,1) = \mathbf{F}(\xi,1) \tag{17.1.20b}$$

A function **U** which interpolates to **F** at a finite set of points is defined as the transfinite interpolant of **F**. The isoparametric interpolation scheme is a special case of the transfinite interpolation schemes.

Consider now a linear operator known as a projector $\wp$, such that $\mathbf{U} \to \wp[\mathbf{F}]$ is a univalent map of $\overline{\Omega} \to \Omega$ satisfying the desired interpolatory properties [Gordon and Hall, 1973].

$$\wp(\xi)\,[\mathbf{F}(\eta)] = \Phi_1(\xi)\mathbf{F}_1(0,\eta) + \Phi_2(\xi)\mathbf{F}_2(1,\eta) \tag{17.1.21a}$$

$$\wp(\eta)[\mathbf{F}(\xi)] = \Phi_1(\eta)\mathbf{F}_1(\xi,0) + \Phi_2(\eta)\mathbf{F}_2(\xi,1) \tag{17.1.21b}$$

Then the tensor product projection

$$\wp(\xi)\ \wp(\eta)\,[\mathbf{F}] = \Phi_N(\xi)\Phi_M(\eta)\mathbf{F}_{NM} \tag{17.1.22}$$

interpolates to **F** at four corners of $[0,1] \times [0,1]$. Here $\mathbf{F}_{NM}$ matches the function at the four corners, but it may not match the function on all the boundaries as illustrated in Figure 17.1.7. Similar effects occur on all other boundaries. These discrepancies can be

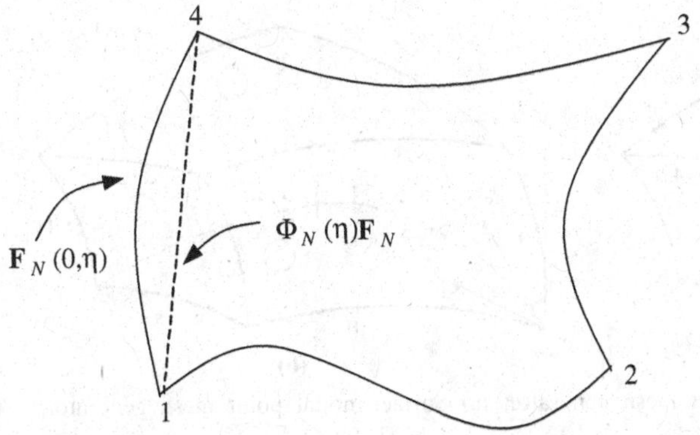

Figure 17.1.7 $\mathbf{F}_{NM}$ match the function at the four corners but not on all boundaries.

removed by subtracting from the sum of (17.1.21a,b) a function formed by interpolating the discrepancies (17.1.22), which represents the Boolean sum projection [Coons, 1967].

$$\mathbf{U} = [\wp(\xi) \oplus \wp(\eta)]\,[\mathbf{F}] = \wp(\xi)\,\mathbf{F}(\eta) + \wp(\eta)\,\mathbf{F}(\xi) - \wp(\xi)\,\wp(\eta)\,[\mathbf{F}] \tag{17.1.23}$$

where the symbol $\oplus$ implies the tensor product and $\mathbf{F}(\eta)$ and $\mathbf{F}(\xi)$ are the parameterization of the sides of the domain and $[\mathbf{F}]$ represents the corresponding vertices. This matches the function not only at the corners but also at all boundaries. Here $\mathbf{U}$ is a transfinite interpolant to $\mathbf{F}$. The functions $\Phi_N(\xi)$ and $\Phi_M(\eta)$ given in (17.1.21) and (17.1.22) are referred to as blending functions. The most commonly used blending functions are of the Lagrange polynomial type

$$\begin{aligned}\mathbf{U} = {} & (1-\xi)\mathbf{F}(0,\eta) + \xi\mathbf{F}(1,\eta) + (1-\eta)\mathbf{F}(\xi,0) + \eta\mathbf{F}(\xi,1) \\ & - [(1-\xi)(1-\eta)\mathbf{F}(0,0) + (1-\xi)\,\eta\mathbf{F}(0,1) + \xi\,(1-\eta)\mathbf{F}(1,0) + \xi\eta\,\mathbf{F}(1,1)]\end{aligned} \tag{17.1.24}$$

For a quadratic variation of boundaries, the blending function $\wp(\xi)$ and $\wp(\eta)$ can simply be replaced by the quadratic Lagrange polynomials.

The following rules are applied in choosing the transfinite interpolation functions:

(1) Pick four points on Γ and identify these as being the images of the four corners of $\overline{\Gamma}$.
(2) These four points separate $\overline{\Gamma}$ into four curve segments which we identify as being the graphs of the four vector valued functions $\mathbf{F}(0,\eta)$, $\mathbf{F}(1,\eta)$, $\mathbf{F}(\xi,0)$, and $\mathbf{F}(\xi,1)$, that is, the four segments of the boundary of Ω are defined to be the images of the four sides of $\overline{\Omega}$.
(3) Use the formulas of $\mathbf{F}(0,\eta)$, $\mathbf{F}(1,\eta)$, $\mathbf{F}(\xi,0)$, $\mathbf{F}(\xi,1)$ in (17.1.24) to define a bilinearly blended transfinite function $\mathbf{U}(\xi,\eta)$, and recall that $\mathbf{U} = \mathbf{F}$ for points (ξ,η) on the perimeter of $\overline{\Omega}$; that is, $\mathbf{U}$ maps the boundary of $\overline{\Omega}$ onto the boundary of Ω.
(4) Test to see if the univalency criteria are satisfied, that is, Jacobian is nonsingular.
(5) Higher order transfinite interpolation functions should be used if necessary (irregular boundaries).

Grid generation for three-dimensional geometries using transfinite interpolation functions was studied by Coons [1967] and extended by Cook [1974]. The procedure includes the surface nodal point mesh generator and volume nodal point mesh generator.

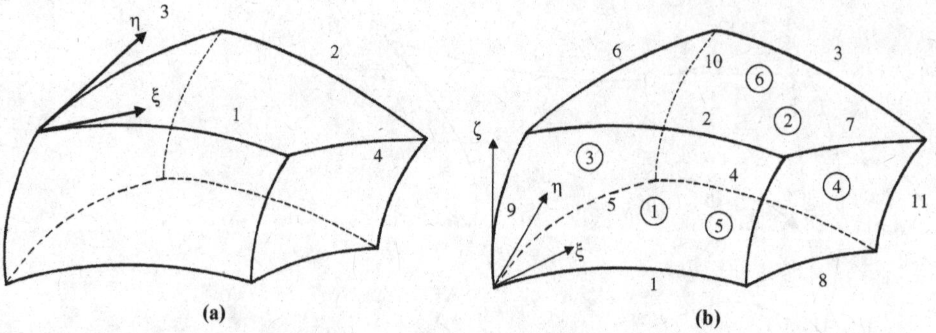

Figure 17.1.8 Surface and volume point mesh generator. **(a)** Surface nodal point mesh generator. **(b)** Volume nodal point mesh generator.

The transfinite interpolation formulas for three-dimensional problems are of the form

$$
\begin{aligned}
\mathbf{U} &= [\wp(\xi) \oplus \wp(\eta) \oplus \wp(\zeta)]\,[\mathbf{F}] \\
&= \wp(\xi)\,[\mathbf{F}(\eta, \zeta)] + \wp(\eta)\,[\mathbf{F}(\xi, \zeta)] + \wp(\zeta)\,[\mathbf{F}(\xi, \eta)] \\
&\quad - [\wp(\xi)\wp(\eta)[\mathbf{F}] + \wp(\eta)\wp(\zeta)\,[\mathbf{F}] + \wp(\zeta)\wp(\xi)[\mathbf{F}] + \wp(\xi)\,\wp(\eta)\wp(\zeta)[\mathbf{F}]]
\end{aligned} \tag{17.1.25}
$$

Consider the coordinate system as shown in Figure 17.1.8 in which the following relations can be established.

Boundary 1: $\eta = 0, \quad x = f_1(\xi), \quad y = g_1(\xi), \quad z = h_1(\xi)$
Boundary 2: $\eta = 1, \quad x = f_2(\xi), \quad y = g_2(\xi), \quad z = h_2(\xi)$
Boundary 3: $\xi = 0, \quad x = f_3(\eta), \quad y = g_3(\eta), \quad z = h_3(\eta)$
Boundary 4: $\xi = 1, \quad x = f_4(\eta), \quad y = g_4(\eta), \quad z = h_4(\eta)$

These definitions lead to the surface nodal point coordinates (Figure 17.1.8a):

$$
\begin{aligned}
x(\xi, \eta) &= (1-\eta) f_1(\xi) + \eta f_2(\xi) + (1-\xi) f_3(\eta) + \xi\, f_4(\eta) \\
&\quad - x\,(0, 0)\,(1-\xi)(1-\eta) - x\,(1, 0)\,\xi(1-\eta) \\
&\quad - x\,(0, 1)\,(1-\xi)\eta - x\,(1, 1)\,\xi\eta
\end{aligned} \tag{17.1.26}
$$

Similarly for $y(\xi, \eta)$ and $z(\xi, \eta)$.

For the volume nodal point mesh generator, we utilize the ξ, η, ζ coordinates normalized as follows (Figure 17.1.8b):

Boundary edge 1: $\eta = 0, \quad \zeta = 0, \quad x = f_1(\xi), \quad y = g_1(\xi), \quad z = h_1(\xi)$
Boundary edge 2: $\eta = 0, \quad \zeta = 1, \quad x = f_2(\xi), \quad y = g_2(\xi), \quad z = h_2(\xi)$
⋮
Boundary edge 12: $\eta = 0, \quad \xi = 1, \quad x = f_{12}(\zeta), \quad y = g_{12}(\zeta), \quad z = h_{12}(\zeta)$

With these boundary edge functions, the linearly blended interpolation functions are

$$
\begin{aligned}
x(\xi, \eta, \zeta) &= (1-\eta)(1-\zeta)\, f_1(\xi) + (1-\eta)\zeta f_2(\xi) + \eta\zeta f_3(\xi) + \eta(1-\zeta)\, f_4(\xi) \\
&\quad + (1-\xi)(1-\zeta)\, f_5(\eta) + (1-\xi)\zeta\, f_6(\eta) + \xi\,\zeta\, f_7(\eta) + \xi\,(1-\zeta)\, f_8(\eta) \\
&\quad + (1-\xi)(1-\eta)\, f_9(\zeta) + (1-\xi)\eta\, f_{10}(\zeta) + \xi\,\eta\, f_{11}(\zeta) \\
&\quad + \xi\,(1-\eta)\, f_{12}(\zeta) + c\,(\xi, \eta, \zeta)
\end{aligned} \tag{17.1.27}
$$

where

$$
\begin{aligned}
c(\xi, \eta, \zeta) = -3[&(1-\xi)(1-\eta)(1-\zeta)\,x(0,0,0) + (1-\xi)(1-\eta)\zeta\,x(0,0,1) \\
&+ (1-\xi)\eta(1-\zeta)\,x(0,1,0) + (1-\xi)\eta\zeta\,x(0,1,1) \\
&+ \xi(1-\eta)(1-\zeta)\,x(1,0,0) + \xi(1-\eta)\zeta\,x(1,0,1) \\
&+ \xi\eta(1-\zeta)\,x(1,1,0) + \xi\eta\zeta\,x(1,1,1)]
\end{aligned} \tag{17.1.28}
$$

Similarly for $y(\xi, \eta, \zeta)$ and $z(\xi, \eta, \zeta)$.

It is desirable to write (17.1.27) in terms of boundary surfaces:

$$
\begin{array}{llllll}
\text{Boundary surface} & 1: & \eta = 0, & x = \bar{f}_1(\xi, \zeta), & y = \bar{g}_1(\xi, \zeta), & z = \bar{h}_1(\xi, \zeta) \\
 & 2: & \eta = 1, & x = \bar{f}_2(\xi, \zeta), & y = \bar{g}_2(\xi, \zeta), & z = \bar{h}_2(\xi, \zeta) \\
 & 3: & \xi = 0, & x = \bar{f}_3(\eta, \zeta), & y = \bar{g}_3(\eta, \zeta), & z = \bar{h}_3(\eta, \zeta) \\
 & 4: & \xi = 1, & x = \bar{f}_4(\eta, \zeta), & y = \bar{g}_4(\eta, \zeta), & z = \bar{h}_4(\eta, \zeta) \\
 & 5: & \zeta = 0, & x = \bar{f}_5(\xi, \eta), & y = \bar{g}_5(\xi, \eta), & z = \bar{h}_5(\xi, \eta) \\
 & 6: & \zeta = 1, & x = \bar{f}_6(\xi, \eta), & y = \bar{g}_6(\xi, \eta), & z = \bar{h}_6(\xi, \eta)
\end{array}
$$

Thus, the boundary surface functions may be written in terms of boundary edge functions:

$$
\begin{aligned}
x(\xi, \eta, \zeta) = \frac{1}{2}\{&(1-\eta)\bar{f}_1(\xi, \zeta) + \eta\bar{f}_2(\xi, \zeta) + (1-\xi)\bar{f}_3(\eta, \zeta) \\
&+ \xi\bar{f}_4(\eta, \zeta) + (1-\zeta)\bar{f}_5(\xi, \eta) + \zeta\bar{f}_6(\xi, \eta) + 2c(\xi, \eta, \zeta)\}
\end{aligned} \tag{17.1.29}
$$

where

$$
\begin{aligned}
\bar{f}_1(\xi, \zeta) = {} & (1-\zeta) f_1(\xi) + \zeta f_2(\xi) + (1-\xi) f_9(\zeta) + \xi f_{12}(\zeta) \\
&- (1-\xi)(1-\zeta)\,x(0,0,0) - (1-\xi)\zeta\,x(0,0,1) \\
&- \xi(1-\zeta)\,x(1,0,0) - \xi\zeta\,x(1,0,1)
\end{aligned} \tag{17.1.30}
$$

etc.

With these coordinate transformation equations, the interior nodal point may be calculated if the interior nodal point can be described in terms of the ξ, η, ζ coordinate system and if the boundary surface functions are known [Cook, 1974].

Example 17.1.6

Repeat Example 17.1.2 using the transfinite interpolation functions (Figure E17.1.6).

The quadratic blending functions are

$$
\Phi_N(\xi) = \begin{cases} 2\left(\xi - \frac{1}{2}\right)(\xi - 1) \\ -4\xi(\xi - 1) \\ 2\xi\left(\xi - \frac{1}{2}\right) \end{cases}, \quad \Phi_N(\eta) = \begin{cases} 2\left(\eta - \frac{1}{2}\right)(\eta - 1) \\ -4\eta(\eta - 1) \\ 2\eta\left(\eta - \frac{1}{2}\right) \end{cases}
$$

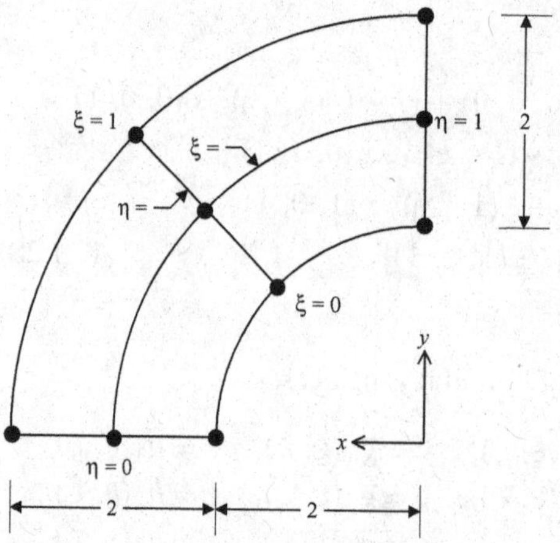

$$x(\xi, \eta) = -(2+2\xi)\cos\frac{\pi}{2}\eta$$
$$y(\xi, \eta) = (2+2\xi)\sin\frac{\pi}{2}\eta$$

$$\mathbf{F}(\xi, \eta) = \begin{bmatrix} x(\xi, \eta) \\ y(\xi, \eta) \end{bmatrix}$$

Figure E17.1.6 Quarter-circle disk with TIF method.

with the projections

$$\wp_\xi[\mathbf{F}] = \sum_{N=1}^{3} \Phi_N(\xi)\,\mathbf{F}(\xi_N, \eta)$$

$$\wp_\eta[\mathbf{F}] = \sum_{N=1}^{3} \Phi_N(\eta)\,\mathbf{F}(\xi, \eta_N)$$

and the product projections

$$\wp_\xi\,\wp_\eta[\mathbf{F}] = \sum_{N=1}^{3}\sum_{M=1}^{3} \Phi_N(\xi)\,\Phi_M(\eta)\,\mathbf{F}(\xi_N, \eta_M)$$

Thus, the transfinite interpolation functions are

$$\begin{aligned}\mathbf{U}(\xi, \eta) &= \wp_\xi \oplus \wp_\eta[\mathbf{F}] = \wp_\xi[\mathbf{F}] + \wp_\eta[\mathbf{F}] - \wp_\xi\,\wp_\eta[\mathbf{F}] \\ &= \sum_{N=1}^{3} \Phi_N(\xi)\,\mathbf{F}(\xi_N, \eta) + \sum_{M=1}^{3} \Phi_M(\eta)\,\mathbf{F}(\xi, \eta_M) \\ &\quad - \sum_{N=1}^{3}\sum_{M=1}^{3} \Phi_N(\xi)\,\Phi_M(\eta)\,\mathbf{F}(\xi_N, \eta_M)\end{aligned}$$

Thus,

$$\begin{bmatrix} x(\xi, \eta) \\ y(\xi, \eta) \end{bmatrix} = \sum_{N=1}^{3} \Phi_N(\xi) \begin{bmatrix} x(\xi_N, \eta) \\ y(\xi_N, \eta) \end{bmatrix} + \sum_{M=1}^{3} \Phi_M(\eta) \begin{bmatrix} x(\xi, \eta_M) \\ y(\xi, \eta_M) \end{bmatrix}$$
$$- \sum_{N=1}^{3}\sum_{M=1}^{3} \Phi_N(\xi)\,\Phi_M(\eta) \begin{bmatrix} x(\xi_N, \eta_M) \\ y(\xi_N, \eta_M) \end{bmatrix}$$

The primitive function $\mathbf{F}(\xi, \eta)$ is

$$\mathbf{F}(\xi, \eta) = \begin{bmatrix} x(\xi, \eta) \\ y(\xi, \eta) \end{bmatrix} = \begin{bmatrix} -(2+2\xi)\cos\frac{\pi}{2}\eta \\ (2+2\xi)\sin\frac{\pi}{2}\eta \end{bmatrix}$$

Thus,

$$\mathbf{U}(\xi, \eta) = \begin{bmatrix} x(\xi, \eta) \\ y(\xi, \eta) \end{bmatrix} = \sum_{N=1}^{3} \Phi_N(\xi) \begin{bmatrix} -(2+2\xi_N)\cos\frac{\pi}{2}\eta \\ (2+2\xi_N)\sin\frac{\pi}{2}\eta \end{bmatrix}$$

$$+\sum_{M=1}^{3} \Phi_M(\eta) \begin{bmatrix} -(2+2\xi)\cos\frac{\pi}{2}\eta_M \\ (2+2\xi)\sin\frac{\pi}{2}\eta_M \end{bmatrix}$$

$$-\sum_{N=1}^{3}\sum_{M=1}^{3} \Phi_N(\xi)\,\Phi_M(\eta) \begin{bmatrix} -(2+2\xi_N)\cos\frac{\pi}{2}\eta_M \\ (2+2\xi_N)\sin\frac{\pi}{2}\eta_M \end{bmatrix}$$

The results are identical to those for the domain vertex method in Example 17.1.2.

Additional discussions on algebraic methods will be presented for surface grid generation in Section 17.3.2. Although algebraic methods are convenient if the geometry can be represented by simple analytical expressions, severe limitations would occur when the computational domain is complicated and suitable functional representation of the geometry is unavailable.

17.2 PDE MAPPING METHODS

Grid generation can be achieved by solving partial differential equations with the dependent and independent variables being the physical domain coordinates and transformed computational domain coordinates, respectively. These PDEs may be of elliptic, hyperbolic, or parabolic form. In general, PDE mapping methods are more complicated than algebraic methods, but provide a smoother grid generation [Thompson, Warsi, and Mastin, 1985].

In the following sections, we shall discuss the basic concepts of elliptic, hyperbolic, and parabolic grid generators, including their advantages and disadvantages.

17.2.1 ELLIPTIC GRID GENERATOR

17.2.1.1 Derivation of Governing Equations

Let us consider a simply connected physical domain and transformed computational domain as shown in Figure 17.2.1. The basic idea stems from the fact that the grid generation in two dimensions is analogous to the solution of Laplace equations for stream function (ψ) and velocity potential function (ϕ).

$$\nabla^2\psi = 0 \tag{17.2.1a}$$

$$\nabla^2\phi = 0 \tag{17.2.1b}$$

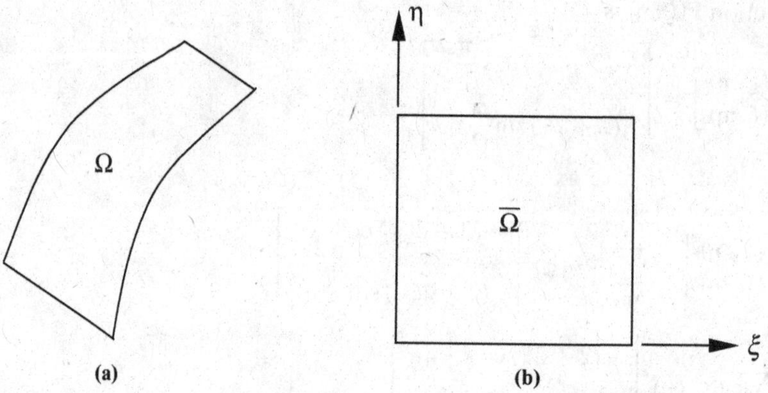

Figure 17.2.1 Simply connected domain. (a) Physical domain. (b) Transformed computational domain.

Solutions of these equations are shown in Figure 17.2.2. Thus, we may now consider $\psi \to x$ and $\phi \to y$ so that we intend to solve

$$\nabla^2 x = \frac{\partial^2 x}{\partial \xi^2} + \frac{\partial^2 x}{\partial \eta^2} = 0 \tag{17.2.2a}$$

$$\nabla^2 y = \frac{\partial^2 y}{\partial \xi^2} + \frac{\partial^2 y}{\partial \eta^2} = 0 \tag{17.2.2b}$$

Toward this end, we employ the curvilinear coordinates for the equation

$$\nabla^2 \mathbf{r} = 0 \tag{17.2.3}$$

where $\mathbf{r} = (x, y, z)^T$ and ∇ is the curvilinear differential operator defined as

$$\nabla = \mathbf{g}^i \frac{\partial}{\partial \xi_i} \quad (i = 1, 2, 3) \tag{17.2.4}$$

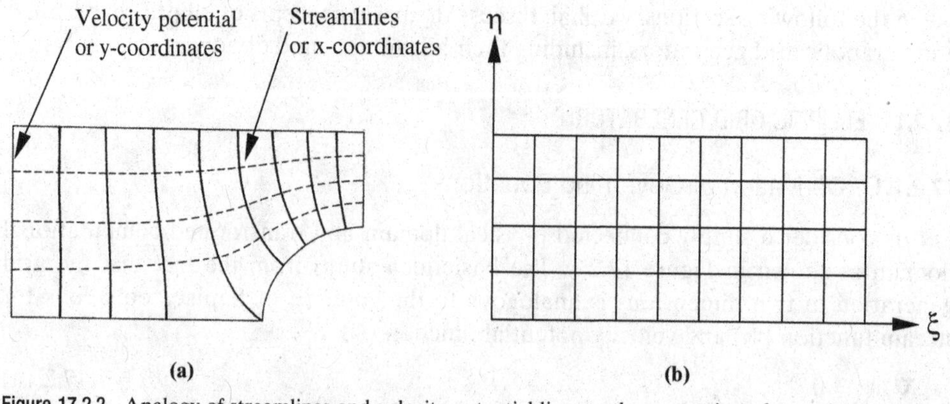

Figure 17.2.2 Analogy of streamlines and velocity potential lines to the generation of x- and y-coordinate grid lines. (a) Physical domain. (b) Computational domain.

with $\mathbf{g}^i$ being the contravariant tangent vector,

$$\mathbf{g}^i = \frac{\partial \xi_i}{\partial x_m}\mathbf{i}_m \tag{17.2.5}$$

Here x_m refers to the cartesian spatial coordinates and ξ_i denotes the curvilinear coordinates.

Using the standard tensor analysis, we obtain [Chung, 1988]

$$\begin{aligned}\nabla^2\mathbf{r} &= \left(\mathbf{g}^j\frac{\partial}{\partial \xi_j}\cdot\mathbf{g}^i\frac{\partial}{\partial \xi_i}\right)\mathbf{r}\\ &= \mathbf{g}^j\cdot\mathbf{g}^i_{,j}\mathbf{r}_{,i} + \mathbf{g}^j\cdot\mathbf{g}^i\mathbf{r}_{,ij}\\ &= \mathbf{g}^j\cdot(g^{ik}\mathbf{g}_k)_{,j}\mathbf{r}_{,i} + g^{ij}\mathbf{r}_{,ij} = 0\\ &= g^{ij}_{,i}\mathbf{r}_{,j} + g^{ij}\Gamma^k_{ki}\mathbf{r}_{,j} + g^{ij}\mathbf{r}_{,ij}\end{aligned} \tag{17.2.6a}$$

or

$$\nabla^2\mathbf{r} = \frac{1}{\sqrt{g}}\left(\sqrt{g}g^{ij}\mathbf{r}_{,j}\right)_{,i} = 0 \tag{17.2.6b}$$

where the comma denotes partial derivatives with respect to the curvilinear coordinates, Γ^r_{st} represents the Christoffel symbol of the second kind, g^{ij} is the contravariant metric tensor, and g is the determinant of the covariant metric tensor g_{ij}.

Equation (17.2.6) may be recast in the form

$$\nabla^2\mathbf{r} = g^{ij}\mathbf{r}_{,ij} + P^j\mathbf{r}_{,j} = 0 \tag{17.2.7}$$

where P^j is known as the control function

$$\begin{aligned}P^j &= g^{ij}_{,i} + g^{ij}\Gamma^k_{ki} = g^{ij}_{,i} + \frac{1}{g}\frac{\partial g}{\partial g_{ij}}\Gamma^k_{ki}\\ &= \frac{\partial}{\partial \xi_i}\left(\frac{\partial \xi_i}{\partial x_m}\frac{\partial \xi_j}{\partial x_m}\right) + \frac{\partial \xi_i}{\partial x_m}\frac{\partial \xi_j}{\partial x_m}\frac{\partial^2 x_p}{\partial \xi_i\partial \xi_k}\frac{\partial \xi_k}{\partial x_p}\end{aligned} \tag{17.2.8}$$

with $g^{ij}_{,i} = 0$.

Physically, the derivative of the contravariant metric tensor and the product of covariant metric tensor and the Christoffel symbol of the second kind represent the deformation process between the physical domain and the transformed computational domain.

In particular, P^j represents control functions capable of inducing two lines or two points to be pulled (attraction, tension) or pushed away (repelled, compression) as effected by the first derivatives and to be bent or twisted as dictated by second derivatives. This behavior is analogous to the differential equations corresponding to normal and shear strains and flexural (bending and torsion) strains in elasticity.

Notice that in this process of "deformation" or geometric transformation, the Laplace equation (17.2.3) has been changed into a Poisson equation (17.2.7).

For three dimensions, (17.2.7) is expanded as

$$\begin{aligned}&g^{11}\mathbf{r}_{,11} + g^{22}\mathbf{r}_{,22} + g^{33}\mathbf{r}_{,33} + 2g^{12}\mathbf{r}_{,12} + 2g^{23}\mathbf{r}_{,23} + 2g^{31}\mathbf{r}_{,31}\\ &\quad + P^1\mathbf{r}_{,1} + P^2\mathbf{r}_{,2} + P^3\mathbf{r}_{,3} = 0\end{aligned} \tag{17.2.9}$$

with

$$g^{ij} = \frac{1}{g}\frac{\partial g}{\partial g_{ij}} = \frac{1}{g}\frac{\partial}{\partial g_{ij}}\begin{vmatrix} g_{11} & g_{12} & g_{13} \\ g_{21} & g_{22} & g_{23} \\ g_{31} & g_{32} & g_{33} \end{vmatrix}$$

$$g^{11} = \frac{1}{g}(g_{22}g_{33} - g_{23}g_{32}), \quad g^{22} = \frac{1}{g}(g_{33}g_{11} - g_{31}g_{13}), \quad g^{33} = \frac{1}{g}(g_{11}g_{22} - g_{12}g_{21})$$

$$g^{12} = \frac{1}{g}(g_{23}g_{31} - g_{21}g_{33}), \quad g^{13} = \frac{1}{g}(g_{32}g_{21} - g_{31}g_{22}), \quad g^{23} = \frac{1}{g}(g_{31}g_{12} - g_{32}g_{11})$$

Similarly for two dimensions,

$$g^{11}\mathbf{r}_{,11} + g^{22}\mathbf{r}_{,22} + 2g^{12}\mathbf{r}_{,12} + P^1\mathbf{r}_{,1} + P^2\mathbf{r}_{,2} = 0 \tag{17.2.10a}$$

or

$$\frac{1}{g}(g_{22}\mathbf{r}_{,11} + g_{11}\mathbf{r}_{,22} - 2g_{12}\mathbf{r}_{,12}) + P^1\mathbf{r}_{,1} + P^2\mathbf{r}_{,2} = 0 \tag{17.2.10b}$$

with

$$g = |g_{ij}| = \begin{vmatrix} g_{11} & g_{12} \\ g_{21} & g_{22} \end{vmatrix} = J^2$$

$$g_{11} = \left(\frac{\partial x}{\partial \xi}\right)^2 + \left(\frac{\partial y}{\partial \xi}\right)^2, \quad g_{22} = \left(\frac{\partial x}{\partial \eta}\right)^2 + \left(\frac{\partial y}{\partial \eta}\right)^2, \quad g_{12} = \frac{\partial x}{\partial \xi}\frac{\partial x}{\partial \eta} + \frac{\partial y}{\partial \xi}\frac{\partial y}{\partial \eta}$$

where the Jacobian J is given by

$$J = \begin{vmatrix} \dfrac{\partial x}{\partial \xi} & \dfrac{\partial y}{\partial \xi} \\ \dfrac{\partial x}{\partial \eta} & \dfrac{\partial y}{\partial \eta} \end{vmatrix}$$

Note that the contravariant component P^i is the same as the physical component P_i since the control function is a scalar to be prescribed.

Finally, we obtain from (17.2.10b) two equations, using the notation $x_\xi = \partial x/\partial \xi$, etc:

$$(x_\eta^2 + y_\eta^2)x_{\xi\xi} + (x_\xi^2 + y_\xi^2)x_{\eta\eta} - 2(x_\xi x_\eta + y_\xi y_\eta)x_{\xi\eta} = -J^2(Px_\xi + Qx_\eta) \tag{17.2.11a}$$

and

$$(x_\eta^2 + y_\eta^2)y_{\xi\xi} + (x_\xi^2 + y_\xi^2)y_{\eta\eta} - 2(x_\xi x_\eta + y_\xi y_\eta)y_{\xi\eta} = -J^2(Py_\xi + Qy_\eta) \tag{17.2.11b}$$

with $P_1 = P$ and $P_2 = Q$.

Note that these equations are nonlinear and must be solved iteratively to determine the grid coordinate values (x, y). Geometries for this purpose are assumed to be amenable to one-to-one transformation (mapping) between physical domain and computational domain whether simply connected, doubly connected, or multiply connected. A typical doubly connected domain and a multiply connected domain are shown in Figure 17.2.3. Note that transformed computational domain is obtained by introducing the process of unwrapping of the doubly or multiply connected domain. In this way,

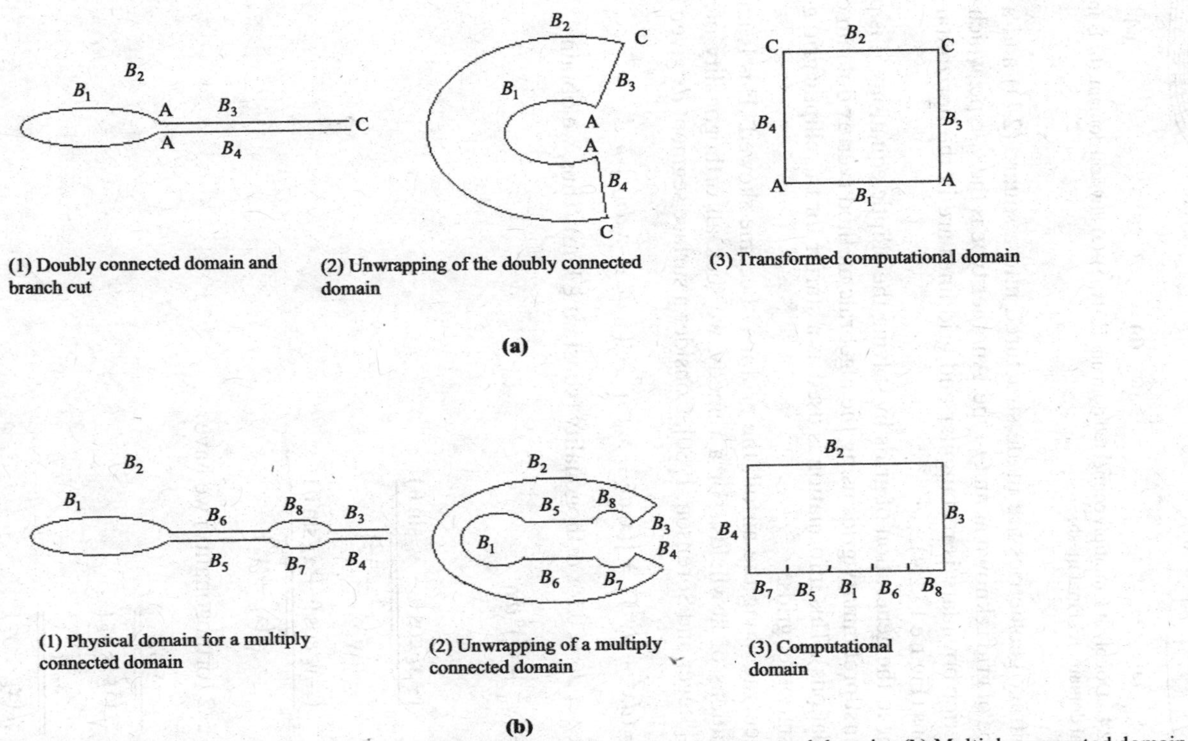

Figure 17.2.3 Doubly and multiply connected domains (O-type). **(a)** Doubly connected domain. **(b)** Multiply connected domain.

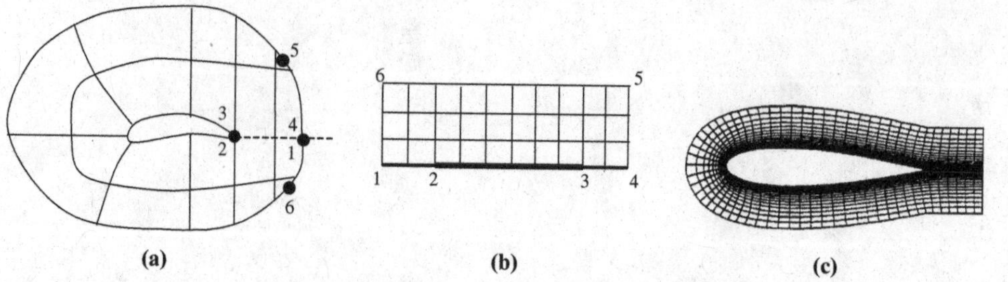

Figure 17.2.4 Doubly and multiply connected domains (C-type). **(a)** Physical domain. **(b)** Transformed computational domain. **(c)** Final mesh.

the arbitrary geometries are made structured into a square (2-D) and a cube (3-D). This type of grid is known as an O-type. Another type is the C-type, which occurs when the exterior boundary is rectangular and grid lines are in the horizontal and vertical directions (Figure 17.2.4).

Prior to the generation of grids by solving the elliptic equations, it is preferable to obtain first preliminary grids using the algebraic method (vertex domain or transfinite interpolations). This information is used as a starter for the elliptic grid generation to obtain smoother grids.

In general, along the airfoil, the grid may become skewed. This is undesirable in computations of flowfields. For a remedy, we seek an orthogonality of grids at the surface [Steger and Sorenson, 1980]. Consider a surface segment ds given by

$$ds = (dx^2 + dy^2)^{\frac{1}{2}} = [(x_\xi d\xi + x_\eta d\eta)^2 + (y_\xi d\xi + y_\eta d\eta)^2]^{\frac{1}{2}} \tag{17.2.12}$$

Here we set $d\xi = 0$ for orthogonality (constant ξ line). Thus, we obtain

$$ds = \left(x_\eta^2 + y_\eta^2\right)^{\frac{1}{2}} d\eta \tag{17.2.13}$$

where

$$x_\eta = s_\eta \frac{(-x_\xi \cos\theta - y_\xi \sin\theta)}{\sqrt{(x_\xi^2 + y_\xi^2)}}$$

$$y_\eta = s_\eta \frac{(-y_\xi \cos\theta + x_\xi \sin\theta)}{\sqrt{(x_\xi^2 + y_\xi^2)}}$$

For $\theta = \pi/2$ (orthogonality) we have

$$x_\eta = \frac{-s_\eta y_\xi}{\sqrt{(x_\xi^2 + y_\xi^2)}}$$

$$y_\eta = \frac{s_\eta x_\xi}{\sqrt{(x_\xi^2 + y_\xi^2)}}$$

$$s_\eta = \left.\frac{\partial s}{\partial \eta}\right|_{\xi = const} = \frac{\Delta s}{\Delta \eta}$$

The finite difference representation for the second derivatives of x and y with respect to η may be written as [Steger and Sorenson, 1980],

$$x_{\eta\eta}(i,\,1) = \frac{-7x_{i,1} + 8x_{i,2} - x_{i,3}}{2\Delta\eta^2} - \frac{3x_\eta(i,\,1)}{\Delta\eta}$$

$$y_{\eta\eta}(i,\,1) = \frac{-7y_{i,1} + 8y_{i,2} - y_{i,3}}{2\Delta\eta^2} - \frac{3y_\eta(i,\,1)}{\Delta\eta}$$

Solutions of elliptic equations (17.2.11) will proceed with central differences for the left-hand side terms (second order derivatives). The first order terms on the right-hand side may be forward-differenced for $P > 0$ and backward-differenced for $Q < 0$.

The control functions, P and Q, are to be used for clustering of grids and are discussed in the following section.

17.2.1.2 Control Functions

In view of the governing equations (17.2.7) or (17.2.11a,b), we may seek to determine the control functions, P and Q, in the form

$$\begin{bmatrix} x_\xi & x_\eta \\ y_\xi & y_\eta \end{bmatrix} \begin{bmatrix} P \\ Q \end{bmatrix} = \begin{bmatrix} R \\ S \end{bmatrix} \tag{17.2.14}$$

where

$$R = -\frac{1}{J^2}\left[(x_\eta^2 + y_\eta^2)x_{\xi\xi} + (x_\xi^2 + y_\xi^2)x_{\eta\eta} - 2(x_\xi x_\eta + y_\xi y_\eta)x_{\xi\eta}\right]$$

$$S = -\frac{1}{J^2}\left[(x_\eta^2 + y_\eta^2)y_{\xi\xi} + (x_\xi^2 + y_\xi^2)y_{\eta\eta} - 2(x_\xi x_\eta + y_\xi y_\eta)y_{\xi\eta}\right]$$

Solving for the control functions P and Q,

$$P = \frac{1}{J}(y_\eta R - x_\eta S) \tag{17.2.15a}$$

$$Q = \frac{1}{J}(x_\xi S - y_\xi R) \tag{17.2.15b}$$

The one-dimensional case of (17.2.15) can be shown to be in the form

$$P = -\frac{\partial^2 x}{\partial \xi^2}\bigg/\frac{\partial x}{\partial \xi}$$

which physically corresponds to (17.2.8), representing the deformation process between the physical domain and transformed computational domain, in which the first and second derivatives imply compression or tension and bending or twisting, respectively. Thus, the control functions may be assumed to be of the form

$$P = \hat{P}\left[\alpha(\xi)e^{-\beta(\xi,\,\eta)}\right] \tag{17.2.16a}$$

$$Q = \hat{Q}\left[\alpha(\eta)e^{-\gamma(\xi,\,\eta)}\right] \tag{17.2.16b}$$

Accordingly, we may adopt a form [Thompson et al., 1985]

$$
\begin{aligned}
P(\xi, \eta) &= -\sum_{i=1}^{n} a_i \, |\xi - \xi_i| \exp[-c_i \, |\xi - \xi_i|] \\
&\quad - \sum_{i=1}^{m} b_i \, |\xi - \xi_i| \exp[-d_i \sqrt{(\xi - \xi_i)^2 + (\eta - \eta_i)^2}] \\
Q(\xi, \eta) &= -\sum_{i=1}^{n} a_i \, |\eta - \eta_i| \exp[-c_i \, |\eta - \eta_i|] \\
&\quad - \sum_{i=1}^{m} b_i \, |\eta - \eta_i| \exp[-d_i \sqrt{(\xi - \xi_i)^2 + (\eta - \eta_i)^2}]
\end{aligned}
\qquad (17.2.17)
$$

where n and m denote the number of lines of ξ and η of the grid, respectively, with a_i and b_i being the amplification factors, and c_i and d_i being the decay factors.

(1) Amplification factors (a_i, b_i):

$a_i > 0$ lines ξ are attracted to lines, ξ_i

$b_i > 0$ lines ξ are attracted to points (ξ_i, η_i)

Similarly for η coordinates.

(2) Decay factors (c_i, d_i):

These decay factors are to modulate the amplifications from a_i and b_i.

For $a_i < 0$ and $b_i < 0$ the attraction is transformed into a repulsion. Obviously $P = Q = 0$ removes these effects.

In summary, advantages and disadvantages of the elliptic grid generators are as follows:

Advantages

(1) Smooth grid point distribution is achieved. Boundary point discontinuities are smoothed out in the interior domain.
(2) Orthogonality at boundaries can be maintained.

Disadvantages

(1) Computer time is large.
(2) Control functions are often difficult to determine.

Example 17.2.1 Elliptic Grid Generation and Comparison with TFI Method

The results are shown in Figure E17.2.1, with all of them using the 51×31 O-type grid.

17.2.2 HYPERBOLIC GRID GENERATOR

In dealing with an open domain, the hyperbolic grid generator is well suited and efficient. This is because the solution of a hyperbolic differential equation utilizes a marching scheme, which is computationally efficient. There are two methods commonly used to develop a hyperbolic grid generator: one is the cell area (Jacobian) method, and the second is an arc-length method [Steger and Sorenson, 1980].

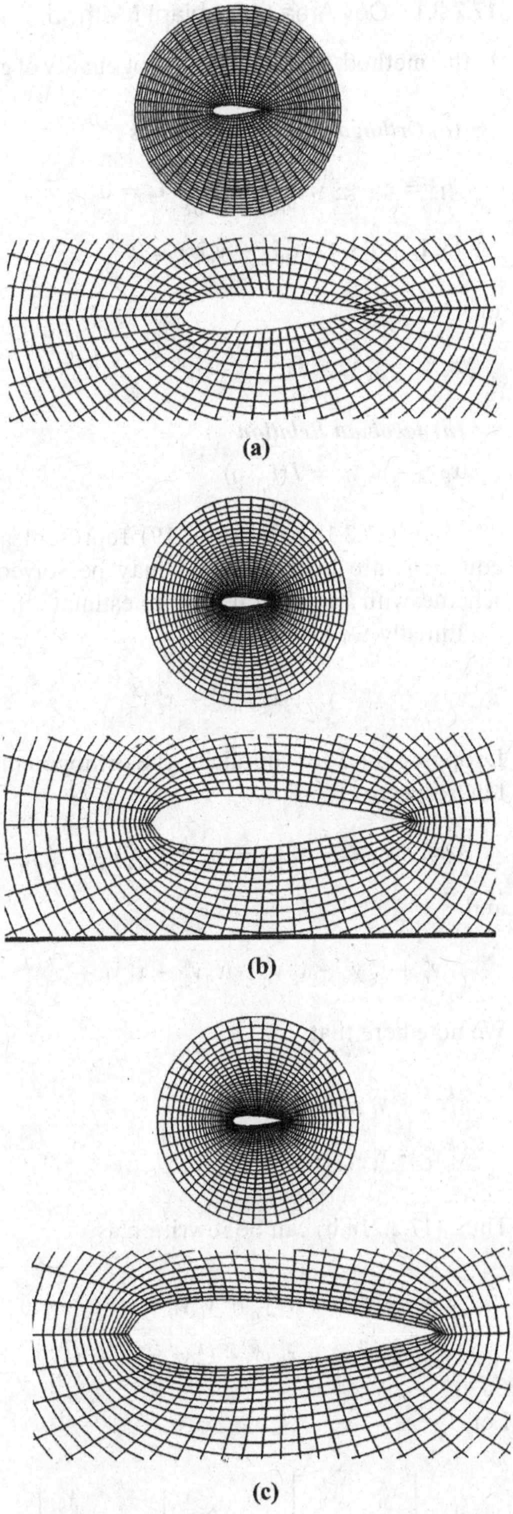

Figure E17.2.1 Elliptic grid generation compared with TFI method. **(a)** Elliptic grid generation without control function. **(b)** Elliptic grid generation with control function. **(c)** Transfinite interpolation approach.

17.2.2.1 Cell Area (Jacobian) Method

In this method, we establish orthogonality of grid lines and a Jacobian relation as follows:

(a) Orthogonality of Grid Lines

$$g_{12} = \mathbf{g}_1 \cdot \mathbf{g}_2 = \frac{\partial x_m}{\partial \xi_1}\mathbf{i}_m \cdot \frac{\partial x_n}{\partial \xi_2}\mathbf{i}_n = 0$$

$$(x_\xi \mathbf{i}_1 + y_\xi \mathbf{i}_2) \cdot (x_\eta \mathbf{i}_1 + y_\eta \mathbf{i}_2) = 0$$

or

$$x_\xi x_\eta + y_\xi y_\eta = 0 \tag{17.2.18}$$

(b) Jacobian Relation

$$x_\xi y_\eta - x_\eta y_\xi = J(\xi, \eta) \tag{17.2.19}$$

Here (17.2.18) and (17.2.19) represent a system of hyperbolic equations. These equations are nonlinear and may be solved using the standard Newton's iterative scheme, with an algebraic grid to estimate the Jacobian.

Initially, we assume that

$$x_\xi y_\eta = x_\xi^{k+1} y_\eta^k + x_\xi^k y_\eta^{k+1} - x_\xi^k y_\eta^k \tag{17.2.20}$$

Dropping $k+1$ for simplicity, the orthogonality and the Jacobian relation may be written, respectively, as

$$x_\xi x_\eta^k + x_\xi^k x_\eta - x_\xi^k x_\eta^k + y_\xi y_\eta^k + y_\xi^k y_\eta - y_\xi^k y_\eta^k = 0 \tag{17.2.21a}$$

and

$$x_\xi y_\eta^k + x_\xi^k y_\eta - x_\xi^k y_\eta^k - x_\eta y_\xi^k - x_\eta^k y_\xi + x_\eta^k y_\xi^k = J \tag{17.2.21b}$$

We note here that

$$x_\xi^k x_\eta^k + y_\xi^k y_\eta^k = 0 \tag{17.2.22a}$$

$$x_\eta^k y_\xi^k - x_\xi^k y_\eta^k = -J^k \tag{17.2.22b}$$

Thus, (17.2.21a,b) can be rewritten as

$$x_\xi x_\eta^k + x_\xi^k x_\eta + y_\xi y_\eta^k + y_\xi^k y_\eta = 0 \tag{17.2.23a}$$

$$x_\xi y_\eta^k + x_\xi^k y_\eta - x_\eta y_\xi^k - x_\eta^k y_\xi = J + J^k \tag{17.2.23b}$$

Let

$$A = \begin{bmatrix} x_\eta^k & y_\eta^k \\ y_\eta^k & -x_\eta^k \end{bmatrix}, \quad B = \begin{bmatrix} x_\xi^k & y_\xi^k \\ -y_\xi^k & x_\xi^k \end{bmatrix}, \quad R = \begin{bmatrix} x \\ y \end{bmatrix}, \quad H = \begin{bmatrix} 0 \\ J + J^k \end{bmatrix}$$

Then

$$AR_\xi + BR_\eta = H \tag{17.2.24a}$$

or

$$CR_\xi + R_\eta = B^{-1}H \tag{17.2.24b}$$

with

$$C = B^{-1}A = \frac{1}{D}\begin{bmatrix} x_\xi^k x_\eta^k - y_\xi^k y_\eta^k & x_\xi^k y_\eta^k + x_\eta^k y_\xi^k \\ x_\xi^k y_\eta^k + x_\eta^k y_\xi^k & -(x_\xi^k x_\eta^k - y_\xi^k y_\eta^k) \end{bmatrix}$$

$$D = \left(x_\xi^k\right)^2 + \left(y_\xi^k\right)^2$$

Thus, (17.2.24b) becomes hyperbolic if the eigenvalues of C

$$\lambda = \pm\left[\frac{\left(x_\eta^k\right)^2 + \left(y_\eta^k\right)^2}{D}\right]^{\frac{1}{2}}$$

are real. For real eigenvalues, we must assure that

$$\left(x_\xi^k\right)^2 + \left(y_\xi^k\right)^2 \neq 0$$

Now the solution of (17.2.24a) can be obtained with the use of central differences for ξ-derivatives and first order backward differences for η-derivatives. This will result in a block diagonal system, marching in the η-direction with an initial distribution of grid points on the surface and boundary lines given. At the boundaries either forward or backward differences may be employed, with the orthogonality conditions enforced. Further details are found in Steger and Sorenson [1980].

17.2.2.2 Arc-Length Method

In this method, the Jacobian equation (17.2.19) is replaced by the relation defining the tangent line

$$\mathbf{g}_i \cdot \mathbf{g}_i = g_{ii} = g_{11} + g_{22} = F(\xi, \eta) \tag{17.2.25a}$$

or

$$F(\xi, \eta) = (x_\xi \mathbf{i}_1 + y_\xi \mathbf{i}_2)\cdot(x_\xi \mathbf{i}_1 + y_\xi \mathbf{i}_2) + (x_\eta \mathbf{i}_1 + y_\eta \mathbf{i}_2)\cdot(x_\eta \mathbf{i}_1 + y_\eta \mathbf{i}_2)$$

$$= x_\xi^2 + y_\xi^2 + x_\eta^2 + y_\eta^2 \tag{17.2.25b}$$

This relation may also be obtained by

$$ds^2 = dx^2 + dy^2 \tag{17.2.26a}$$

which represents an arc-length

$$ds^2 = (x_\xi d\xi + x_\eta d\eta)^2 + (y_\xi d\xi + y_\eta d\eta)^2 \tag{17.2.26b}$$

Setting $\Delta\xi = \Delta\eta = 1$, we obtain

$$\Delta s^2 = x_\xi^2 + y_\xi^2 + x_\eta^2 + y_\eta^2 \tag{17.2.27}$$

Equating (17.2.25b) and (17.2.27) leads to

$$F(\xi, \eta) = \Delta s^2 \tag{17.2.28}$$

The arc-length Δs may be specified by the user. For a constant ξ-line, we obtain

$$\Delta s^2 = x_\eta^2 + y_\eta^2 \tag{17.2.29}$$

Linearization and finite difference approximations for (17.2.28) and (17.2.29) can be carried out similarly as in the cell area method.

In summary, it is seen that the hyperbolic grid generation system is less general, although it is much faster than the elliptic generation system. The specification of the cell volume distribution avoids the possible grid line overlapping that otherwise can occur with concave boundaries. Disadvantages include boundary slope discontinuities being propagated into the field, with shocklike solutions possibly resulting in an unsmooth grid generation.

17.2.3 PARABOLIC GRID GENERATOR

The parabolic system provides a compromise between the elliptic and hyperbolic systems:

(a) *Diffusiveness:* Propagation of boundary discontinuities are prevented similarly as in the elliptic system.

(b) *Marching scheme:* Solutions are fast, similar to the hyperbolic systems.

The governing equations are modified from the Poisson equations as [Nakamura, 1991]

$$x_\eta - Ax_{\xi\xi} = S_x \tag{17.2.30a}$$

$$y_\eta - Ay_{\xi\xi} = S_y \tag{17.2.30b}$$

where A = constant and S_x, S_y = source terms.

Here, the source terms act as control functions. Implementations of (17.2.30) are not as convenient as in the case of elliptic and hyperbolic systems, but the solution of a tridiagonal system for (17.2.30a,b) is much faster than the elliptic grid generator. However, orthogonality is not achieved as directly as in the hyperbolic system. Implementation of control functions through the source terms S_x and S_y remains undeveloped.

17.3 SURFACE GRID GENERATION

A surface mesh is a prerequisite for three-dimensional grid generation. Although the surface grid generation is considered a part of the unstructured three-dimensional mesh generation, it is often convenient to obtain the surface grid in a structured configuration using algebraic methods [De Boor, 1972; Bezier, 1986; Farin, 1988] or elliptic PDE methods [Warsi and Koomullil, 1991; Arina and Casella, 1991; Nakamura et al., 1991]. It is possible to combine the algebraic or elliptic PDE approaches in a structured fashion close to the surface with unstructured grids elsewhere away from the surface. Such a scheme is particularly useful in boundary layer flows.

17.3.1 ELLIPTIC PDE METHODS

The elliptic PDE methods for surface grid generation require derivations of governing equations based on the theory of surfaces or differential geometries. A brief review of the theory of differential geometry applicable to surface grid generation is given below [Chung, 1988, p. 229]:

17.3.1.1 Differential Geometry

Consider a reference surface characterized by a curvilinear coordinate system ($\xi^1, \xi^2, \xi^3 = 0$) with an origin located at P by a position vector $\mathbf{r}_o$, as shown in Figure 17.3.1a. Here, the usual practice of writing the curvilinear coordinates in terms of contravariant component ξ^i with indices placed as superscripts will be followed unlike in the previous sections. Let ξ^3 be the distance along the normal to the reference surface ($\xi^3 = 0$) and $\hat{\mathbf{n}}_3 = \mathbf{n}$ be the unit normal vector. An arbitrary point Q on the ξ^3 coordinate is defined by a position vector $\mathbf{r} = x_i \mathbf{i}_i$ where x_i's are the cartesian coordinate ($i = 1, 2, 3$):

$$\mathbf{r} = x_i \mathbf{i}_i = \mathbf{r}_o + \xi^3 \hat{\mathbf{a}}_3 = \mathbf{r}_o + \xi^3 \mathbf{n} \tag{17.3.1}$$

The tangent base vectors along the curvilinear coordinates $\xi^\alpha (\alpha = 1, 2)$ on the reference surface, often called the middle surface, are represented by the partial derivatives of $\mathbf{r}_o$ with respect to ξ^α:

$$\frac{\partial \mathbf{r}_o}{\partial \xi^\alpha} = \mathbf{r}_{o,\alpha} = \mathbf{a}_\alpha \tag{17.3.2}$$

Here, $\mathbf{a}_\alpha$ is the covariant surface tangent vector. Likewise, the tangent vectors along ξ^α on the arbitrary surface at $\mathbf{r}$ are

$$\frac{\partial \mathbf{r}}{\partial \xi^\alpha} = \mathbf{r}_{,\alpha} = \mathbf{r}_{o,\alpha} + \xi^3 \mathbf{n}_{,\alpha} = \mathbf{g}_\alpha \tag{17.3.3}$$

or

$$\mathbf{g}_\alpha = \mathbf{a}_\alpha + \xi^3 \mathbf{n}_{,\alpha} \tag{17.3.4a}$$

and

$$\mathbf{g}_3 = \mathbf{a}_3 = \hat{\mathbf{a}}_3 = \mathbf{n} \tag{17.3.4b}$$

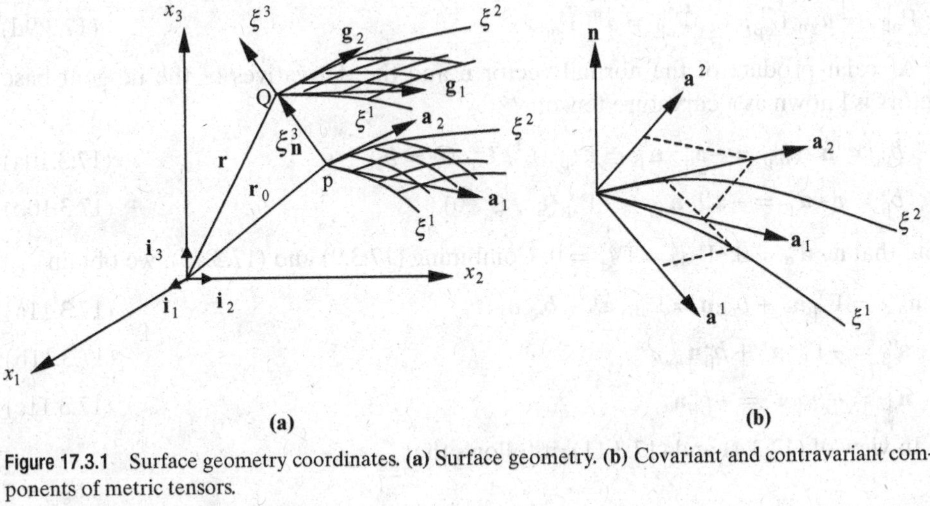

Figure 17.3.1 Surface geometry coordinates. (a) Surface geometry. (b) Covariant and contravariant components of metric tensors.

The reciprocal base vector or the contravariant component of the tangent vector $\mathbf{a}^i$ has the property (Figure 17.3.1b),

$$\mathbf{a}^i \cdot \mathbf{a}_j = \delta^i_j \tag{17.3.5}$$

and

$$\mathbf{a}^\alpha = a^{\alpha\beta}\mathbf{a}_\beta, \qquad \mathbf{a}_\alpha = a_{\alpha\beta}\mathbf{a}^\beta \tag{17.3.6}$$

in which $a_{\alpha\beta} = \mathbf{a}_\alpha \cdot \mathbf{a}_\beta$, $a^{\alpha\beta} = \mathbf{a}^\alpha \cdot \mathbf{a}^\beta$ are the covariant and contravariant components of the metric tensor, respectively. Note that $\mathbf{a}^\alpha$ is the contravariant surface tangent vector normal to the ξ^α surface. It also follows that

$$
\begin{aligned}
&a^{\alpha\beta} = g^{\alpha\beta}(\xi^1, \xi^2, 0), \qquad a_{\alpha\beta} = g_{\alpha\beta}(\xi^1, \xi^2, 0)\\
&a^{\alpha\beta}a_{\beta\gamma} = \delta^\alpha_\gamma\\
&|a_{\alpha\beta}| = a = g\,(\xi^1, \xi^2, 0)\\
&|a^{\alpha\beta}| = \frac{1}{a}\\
&a^{11} = \frac{a_{22}}{a}, \qquad a^{22} = \frac{a_{11}}{a}, \qquad a^{12} = -\frac{a_{12}}{a}
\end{aligned}
\tag{17.3.7}
$$

An elemental volume bound by the coordinate surface is given by

$$d\Omega = \mathbf{g}_1 d\xi^1 \times \mathbf{g}_2 d\xi^2 \cdot \mathbf{g}_3 d\xi^3 = \sqrt{g_{123}}\,\mathbf{g}^3 \cdot \mathbf{g}_3 d\xi^1 d\xi^2 d\xi^3 = \sqrt{g}\,d\xi^1 d\xi^2 d\xi^3 \tag{17.3.8}$$

The curvatures of a surface are defined through scalar products of the base vectors and the derivatives of the base vectors through the Christoffel symbols of the first kind ($\Gamma_{\alpha\beta\gamma}$) and the second kind $\Gamma^\gamma_{\alpha\beta}$:

$$\Gamma_{\alpha\beta\gamma}(\xi^1, \xi^2, 0) = \mathbf{a}_\gamma \cdot \mathbf{a}_{\alpha,\beta} = \Gamma_{\alpha\beta\gamma} \tag{17.3.9a}$$

$$\Gamma^\gamma_{\alpha\beta}(\xi^1, \xi^2, 0) = \mathbf{a}^\gamma \cdot \mathbf{a}_{\alpha,\beta} = -\mathbf{a}_\alpha \cdot \mathbf{a}^\gamma_{,\beta} = \Gamma^\gamma_{\alpha\beta} \tag{17.3.9b}$$

$$\Gamma_{\alpha\beta\gamma} = \frac{1}{2}(a_{\alpha\gamma,\beta} + a_{\beta\gamma,\alpha} - a_{\alpha\beta,\gamma}) \tag{17.3.9c}$$

$$\Gamma_{\alpha\beta\gamma} = a_{\gamma\eta}\,\Gamma^\eta_{\alpha\beta}, \qquad \Gamma^\eta_{\alpha\beta} = a^{\eta\gamma}\,\Gamma_{\alpha\beta\gamma} \tag{17.3.9d}$$

A scalar product of the normal vector $\mathbf{n}$ and the derivatives of the tangent base vectors is known as a curvature tensor:

$$b_{\alpha\beta} = \mathbf{n} \cdot \mathbf{a}_{\alpha,\beta} = -\mathbf{a}_\alpha \cdot \mathbf{n}_{,\beta} = \Gamma_{\alpha\beta3}(\xi^1, \xi^2, 0) = b_{\beta\alpha} \tag{17.3.10a}$$

$$b^\alpha_\beta = \mathbf{n} \cdot \mathbf{a}^\alpha_{,\beta} = -\mathbf{a}^\alpha \cdot \mathbf{n}_{,\beta} = -\Gamma^\alpha_{3\beta}(\xi^1, \xi^2, 0) \tag{17.3.10b}$$

Note that $\mathbf{n} \cdot \mathbf{n}_{,\alpha} = 0$, $\Gamma_{3\alpha3} = \Gamma^3_{3\alpha} = 0$. Combining (17.3.9) and (17.3.10), we obtain

$$\mathbf{a}_{\alpha,\beta} = \Gamma^\gamma_{\alpha\beta}\mathbf{a}_\gamma + b_{\alpha\beta}\mathbf{n} = \Gamma_{\alpha\beta\gamma}\mathbf{a}^\gamma + b_{\alpha\beta}\mathbf{n} \tag{17.3.11a}$$

$$\mathbf{a}^\alpha_{,\beta} = -\Gamma^\alpha_{\beta\gamma}\mathbf{a}^\gamma + b^\alpha_\beta\mathbf{n} \tag{17.3.11b}$$

$$\mathbf{n}_{,\beta} = -b_{\alpha\beta}\mathbf{a}^\alpha = -b^\alpha_\beta\mathbf{a}_\alpha \tag{17.3.11c}$$

In view of (17.3.4) and (17.3.11), it follows that

$$\mathbf{g}_\alpha = \mathbf{a}_\alpha - \xi^3 b_\alpha^\beta \mathbf{a}_\beta = \mathbf{a}_\alpha - \xi^3 b_{\alpha\beta}\mathbf{a}^\beta \tag{17.3.12}$$

$$g_{\alpha\beta} = a_{\alpha\beta} - 2\xi^3 b_{\alpha\beta} + (\xi^3)^2 b_{\alpha\gamma} b_\beta^\gamma \tag{17.3.13a}$$

$$g_{\alpha 3} = 0, \qquad g_{33} = 1 \tag{17.3.13b}$$

The changes in the position vector and the normal vector are given by

$$d\mathbf{r}_o = \mathbf{r}_{o,\alpha} d\xi^\alpha = \mathbf{a}_\alpha d\xi^\alpha \tag{17.3.14a}$$

$$d\mathbf{n} = \mathbf{n}_{,\alpha} d\xi^\alpha = -b_{\alpha\beta}\mathbf{a}^\beta d\xi^\alpha \tag{17.3.14b}$$

The scalar products of (17.3.14) are

$$d\mathbf{r}_o \cdot d\mathbf{r}_o = ds_o^2 = \mathbf{r}_{o,\alpha} d\xi^\alpha \cdot \mathbf{r}_{o,\beta} d\xi^\beta = a_{\alpha\beta} d\xi^\alpha d\xi^\beta \tag{17.3.15a}$$

$$d\mathbf{r}_o \cdot d\mathbf{n} = \mathbf{a}_\alpha d\xi^\alpha \cdot \mathbf{n}_{,\beta} d\xi^\beta = -b_{\alpha\beta} d\xi^\alpha d\xi^\beta \tag{17.3.15b}$$

$$\begin{aligned} d\mathbf{n} \cdot d\mathbf{n} &= \mathbf{n}_{,\alpha} d\xi^\alpha \cdot \mathbf{n}_{,\beta} d\xi^\beta = (-b_\alpha^\gamma \mathbf{a}_\gamma) \cdot (-b_\beta^\mu \mathbf{a}_\mu)\, d\xi^\alpha d\xi^\beta \\ &= b_\alpha^\gamma b_\beta^\mu a_{\gamma\mu} d\xi^\alpha d\xi^\beta = b_\alpha^\gamma b_{\beta\gamma} d\xi^\alpha d\xi^\beta = c_{\alpha\beta} d\xi^\alpha d\xi^\beta \end{aligned} \tag{17.3.15c}$$

Here, $a_{\alpha\beta}, b_{\alpha\beta}$, and $c_{\alpha\beta}$ are called the first, second, and third fundamental tensors, respectively.

It can be shown that the second order covariant derivative of any covariant component of a first order tensor is of the form

$$\begin{aligned} A_{i\,|\,jk} &= (A_{i\,|\,j})_{,k} - \Gamma_{ik}^r A_{r\,|\,j} - \Gamma_{jk}^r A_{i\,|\,r} \\ &= (A_{i,j} - \Gamma_{ij}^r A_r)_{,k} - \Gamma_{ik}^r (A_{r,j} - \Gamma_{rj}^s A_s) - \Gamma_{jk}^r (A_{i,r} - \Gamma_{ir}^s A_s) \\ &= A_{i,jk} - (\Gamma_{ij}^r)_{,k} A_r - \Gamma_{ij}^r A_{r,k} - \Gamma_{ik}^r A_{r,j} + \Gamma_{ik}^r \Gamma_{rj}^s A_s - \Gamma_{jk}^r A_{i,r} + \Gamma_{jk}^r \Gamma_{ir}^s A_s \end{aligned} \tag{17.3.16a}$$

Similarly,

$$A_{i\,|\,kj} = A_{i,kj} - (\Gamma_{ik}^r)_{,j} A_r - \Gamma_{ik}^r A_{r,j} - \Gamma_{ij}^r A_{r,k} + \Gamma_{ij}^r \Gamma_{rk}^s A_s - \Gamma_{kj}^r A_{i,r} + \Gamma_{kj}^r \Gamma_{ir}^t A_t \tag{17.3.16b}$$

Subtracting (17.3.16b) from (17.3.16a) yields

$$\begin{aligned} A_{i\,|\,jk} - A_{i\,|\,kj} &= \Gamma_{ik}^r \Gamma_{rj}^s A_s - (\Gamma_{ij}^r)_{,k} A_r - \Gamma_{ij}^r \Gamma_{rk}^s A_s + (\Gamma_{ik}^r)_{,j} A_r \\ &= \left[(\Gamma_{ik}^r)_{,j} - (\Gamma_{ij}^r)_{,k} + \Gamma_{ik}^s \Gamma_{sj}^r - \Gamma_{ij}^s \Gamma_{sk}^r\right] A_r \\ &= R_{ijk}^r A_r \end{aligned} \tag{17.3.17}$$

where R_{ijk}^r is a mixed tensor of order four, known as the Riemann-Christoffel tensor of the second kind. Since the left-hand side of (17.3.17) is zero, it follows that

$$R_{ijk}^r = 0 \tag{17.3.18}$$

The associated tensor

$$R_{ijkl} = g_{ir} R_{jkl}^r \tag{17.3.19}$$

is the Riemann-Christoffel tensor of the first kind, which may be written in the form

$$R_{ijkl} = \frac{1}{2}(g_{il,jk} + g_{jk,il} - g_{ik,jl} - g_{jl,ik}) + g^{mn}(\Gamma_{jkm}\Gamma_{iln} - \Gamma_{jlm}\Gamma_{ikn}) \tag{17.3.20}$$

$$R_{ijkl} = -R_{jikl} = -R_{ijlk} = R_{klij} \tag{17.3.21}$$

which implies that R_{ijkl} is skew-symmetric in ij and kl. We also note that there are six different components of R_{ijkl}, namely,

$$R_{3131}, \qquad R_{3232}, \qquad R_{1212}, \qquad R_{3132}, \qquad R_{3212}, \qquad R_{3112}$$

The Riemann-Christoffel tensors for the reference surface with $\xi^3 = 0$ are often of the form

$$R^{\lambda}_{\alpha\beta\gamma} = \Gamma^{\lambda}_{\alpha\gamma,\beta} - \Gamma^{\lambda}_{\alpha\beta,\gamma} + \Gamma^{\mu}_{\alpha\rho}\Gamma^{\lambda}_{\mu\beta} - \Gamma^{\mu}_{\alpha\beta}\Gamma^{\lambda}_{\mu\gamma} + \Gamma^{3}_{\alpha\gamma}\Gamma^{\lambda}_{3\beta} - \Gamma^{3}_{\alpha\beta}\Gamma^{\lambda}_{3\gamma} \tag{17.3.22}$$

or

$$R^{\lambda}_{\alpha\beta\gamma} = \overline{R}^{\lambda}_{\alpha\beta\gamma} + \Gamma^{3}_{\alpha\gamma}\Gamma^{\lambda}_{3\beta} - \Gamma^{3}_{\alpha\beta}\Gamma^{\lambda}_{3\gamma} = 0$$

$$R^{3}_{\alpha\beta\gamma} = \Gamma^{3}_{\alpha\gamma,\beta} - \Gamma^{3}_{\alpha\beta,\gamma} + \Gamma^{m}_{\alpha,\gamma}\Gamma^{3}_{m\beta} - \Gamma^{m}_{\alpha\beta}\Gamma^{3}_{m\gamma} = 0$$

$$\overline{R}^{\lambda}_{\alpha\beta\gamma} = \Gamma^{3}_{\alpha\beta}\Gamma^{\lambda}_{3\gamma} - \Gamma^{3}_{\alpha\gamma}\Gamma^{\lambda}_{3\beta} = b_{\alpha\beta}(-b^{\lambda}_{\gamma}) - b_{\alpha\gamma}(-b^{\lambda}_{\beta})$$

$$\overline{R}_{\lambda\alpha\beta\gamma} = a_{\lambda\mu}\overline{R}^{\mu}_{\alpha\beta\gamma} = b_{\alpha\gamma}b_{\lambda\beta} - b_{\alpha\beta}b_{\lambda\gamma}$$

From the symmetry of $\Gamma^{\gamma}_{\alpha\beta}$ and $b_{\alpha\beta}$, we obtain

$$\overline{R}_{\alpha\gamma\beta\gamma} = \overline{R}_{\alpha\beta\gamma\gamma} \quad (\alpha, \gamma \text{ are not summed})$$

and

$$\overline{R}_{1212} = \overline{R}_{2121} = -\overline{R}_{2112} = -\overline{R}_{1221}$$

Hence, every nonzero component of $R_{\alpha\beta\gamma\delta}$ is equal to $\overline{R}_{1212}$ or to $-\overline{R}_{1212}$, and it follows that

$$\overline{R}_{1212} = |b_{\alpha\beta}| = b_{11}b_{22} - b_{12}^2 \tag{17.3.23}$$

We introduce an invariant K, called the Gaussian curvature:

$$K = \frac{\overline{R}_{1212}}{a} = \frac{|b_{\alpha\beta}|}{|a_{\alpha\beta}|} = |b^{\alpha}_{\beta}| = b^1_1 b^2_2 - b^1_2 b^2_1 \tag{17.3.24}$$

Another important invariant, H, called the mean curvature of the surface, is of the form

$$H = \frac{1}{2}a^{\alpha\beta}b_{\alpha\beta} = \frac{1}{2}b^{\alpha}_{\alpha} = \frac{1}{2}(b^1_1 + b^2_2) \tag{17.3.25}$$

Since $R^3_{\alpha\beta\gamma}$ also vanishes, we obtain from (17.3.22) that

$$\begin{aligned} R^3_{\alpha\beta\gamma} &= \Gamma^{3}_{\alpha\gamma,\beta} - \Gamma^{3}_{\alpha\beta\gamma} + \Gamma^{m}_{\alpha\gamma}\Gamma^{3}_{m\beta} + \Gamma^{m}_{\alpha\beta}\Gamma^{3}_{m\gamma} \\ &= b_{\alpha\gamma,\beta} - b_{\alpha\beta,\gamma} + \Gamma^{m}_{\alpha\gamma}b_{mb} - \Gamma^{m}_{\alpha\beta}b_{m\gamma} \end{aligned}$$

Defining $b_{\alpha\gamma|\beta} = b_{\alpha\gamma,\beta} - b_{\alpha\lambda}\Gamma^{\lambda}_{\gamma\beta} - b_{\lambda\gamma}\Gamma^{\lambda}_{\alpha\beta}$, we have

$$b_{\alpha\gamma|\beta} = b_{\alpha\beta|\gamma} \tag{17.3.26}$$

which represents either of two equations, namely,

$$b_{11|2} = b_{12|1} \quad \text{or} \quad b_{21|2} = b_{22|1} \tag{17.3.27}$$

These equations, (17.3.27), are called the Codazzi equations of the surface and are useful in establishing compatibility of deformations.

17.3.1.2 Surface Grid Generation

Returning to (17.3.11a), we write the derivative of the surface tangent base vector as

$$\mathbf{a}_{\alpha,\beta} = \mathbf{r}_{o,\alpha\beta} = \Gamma^{\gamma}_{\alpha\beta}\mathbf{r}_{o,\gamma} + b_{\alpha\beta}\mathbf{n} \tag{17.3.28}$$

where $\mathbf{r}_o$, the position vector to the surface, implies the cartesian coordinate values of the surface grid. Multiplying (17.3.28) by $a^{\alpha\beta}$, we obtain

$$\begin{aligned} a^{\alpha\beta}\mathbf{r}_{o,\alpha\beta} &= a^{\alpha\beta}\Gamma^{\gamma}_{\alpha\beta}\mathbf{r}_{o,\gamma} + a^{\alpha\beta}b_{\alpha\beta}\mathbf{n} \\ &= P^{\gamma}\mathbf{r}_{o,\gamma} + a^{\alpha\beta}b_{\alpha\beta}\mathbf{n} \end{aligned} \tag{17.3.29}$$

where

$$P^{\gamma} = a^{\alpha\beta}\Gamma^{\gamma}_{\alpha\beta} \tag{17.3.30}$$

is the control function. Note also that

$$g^{\alpha\beta}b_{\alpha\beta}|_{surface} = a^{\alpha\beta}b_{\alpha\beta} = b^{\alpha}_{\alpha} \tag{17.3.31}$$

This is known as the principal curvature, which is twice the mean curvature (17.3.26).

It is seen that if the surface is degenerated into a plane, then $b^{\alpha}_{\alpha} = 0$ (zero mean curvature), and (17.3.30) becomes identical to that of a two-dimensional plane geometry as given in (17.2.7).

The governing equation for the surface grid generation takes the form

$$a^{11}\mathbf{r}_{o,11} + a^{22}\mathbf{r}_{o,22} + 2a^{12}\mathbf{r}_{o,12} = P^{1}\mathbf{r}_{o,1} + P^{2}\mathbf{r}_{o,2} + \left(b^{1}_{1} + b^{2}_{2}\right)\mathbf{n} \tag{17.3.32a}$$

or

$$\frac{1}{a}\left(a_{22}\mathbf{r}_{o,11} + a_{11}\mathbf{r}_{o,22} - 2a_{12}\mathbf{r}_{o,12}\right) - P_1\mathbf{r}_{o,1} - P_2\mathbf{r}_{o,2} = \left(b^{1}_{1} + b^{2}_{2}\right)\mathbf{n} \tag{17.3.32b}$$

with

$$\mathbf{r}_{o,11} = \begin{bmatrix} x_{\xi\xi} \\ y_{\xi\xi} \\ z_{\xi\xi} \end{bmatrix}, \quad \mathbf{r}_{o,22} = \begin{bmatrix} x_{\eta\eta} \\ y_{\eta\eta} \\ z_{\eta\eta} \end{bmatrix}, \quad \mathbf{r}_{o,12} = \begin{bmatrix} x_{\xi\eta} \\ y_{\xi\eta} \\ z_{\xi\eta} \end{bmatrix}, \quad \sqrt{a} = \begin{vmatrix} x_{\xi} & x_{\eta} & 0 \\ y_{\xi} & y_{\eta} & 0 \\ z_{\xi} & z_{\eta} & 1 \end{vmatrix}$$

$$a_{11} = x_{\xi}^2 + y_{\xi}^2 + z_{\xi}^2$$

$$a_{22} = x_{\eta}^2 + y_{\eta}^2 + z_{\eta}^2$$

$$a_{12} = x_{\xi}x_{\eta} + y_{\xi}y_{\eta} + z_{\xi}z_{\eta}$$

Principal curvatures are given by

$$a^{\alpha\beta}b_{\alpha\beta} = b^{\alpha}_{\alpha} = \mathbf{n}\cdot\mathbf{a}^{\alpha}_{,\alpha} = -\mathbf{a}^{\alpha}\cdot\mathbf{n}_{,\alpha} = -\Gamma^{\alpha}_{\alpha 3}(\xi^1,\xi^2,0) = -a^{\alpha\gamma}\Gamma_{\alpha 3\gamma}$$
$$= -(a^{11}\Gamma_{131} + a^{12}\Gamma_{132} + a^{21}\Gamma_{231} + a^{22}\Gamma_{232}) \qquad (17.3.33)$$

with

$$a^{11} = \frac{a_{22}}{a}, \qquad a^{22} = \frac{a_{11}}{a}, \qquad a^{12} = -\frac{a_{12}}{a}$$
$$\Gamma_{\alpha 3\beta} = \frac{\partial^2 x_m}{\partial\xi^{\alpha}\partial\xi^3}\frac{\partial x_m}{\partial\xi^{\beta}} = \frac{\partial^2 x_1}{\partial\xi^{\alpha}\partial\xi^3}\frac{\partial x_1}{\partial\xi^{\beta}} + \frac{\partial^2 x_2}{\partial\xi^{\alpha}\partial\xi^3}\frac{\partial x_2}{\partial\xi^{\beta}} + \frac{\partial^2 x_3}{\partial\xi^{\alpha}\partial\xi^3}\frac{\partial x_3}{\partial\xi^{\beta}} \qquad (17.3.34)$$

Example 17.3.1

Consider surface coordinates (x, y, z) given as

$$z = f(x, y), \quad \text{e.g.,} \quad z = h\sin\frac{\pi x}{A}\sin\frac{\pi y}{B}$$

(a) Surface Area

$$dA = \sqrt{1 + z_x^2 + z_y^2}\,dx\,dy$$

(b) Surface Unit Normal Vector

$$\mathbf{n} = n_i\,\mathbf{i}_i, \qquad \mathbf{n} = \frac{\mathbf{a}_1\times\mathbf{a}_2}{\sqrt{a}}$$

$$n_1 = \frac{-z_x}{\sqrt{1 + z_x^2 + z_y^2}}, \qquad n_2 = \frac{-z_y}{\sqrt{1 + z_x^2 + z_y^2}}, \qquad n_3 = \frac{1}{\sqrt{1 + z_x^2 + z_y^2}}$$

(c) Surface Length Element

$$ds = \sqrt{(1 + z_x^2)dx^2 + 2z_x z_y dx dy + (1 + z_y^2)dy^2}$$

(d) Principal Curvatures

$$b^{\alpha}_{\alpha} = \frac{(1 + z_y^2)z_{xx} - 2z_x z_y z_{xy} + (1 + z_x^2)z_{yy}}{(1 + z_x^2 + z_y^2)^{\frac{3}{2}}}$$

Example 17.3.2

Prolate ellipsoid defined by

$$x = a\cos\zeta, \qquad y = b\sin\zeta\cos\zeta, \qquad z = b\sin\zeta\sin\zeta$$

From (17.3.33) and (17.3.34) we obtain the curvature tensor as

$$b^{\alpha}_{\alpha} = \frac{-a[a^2\sin^2\zeta + b^2(1 + \cos^2\zeta)]}{b(a^2\sin^2\zeta + b^2\cos^2\zeta)^{\frac{3}{2}}}$$

The governing equations (17.3.32) may be solved using finite differences or finite elements. Control functions can be selected similarly as discussed in Section 17.2. These functions are set before the solution algorithm begins, either directly through input or by calculation from the boundary point distributions.

17.3.2 ALGEBRAIC METHODS

In algebraic methods, we are not concerned with differential equations, but rather involved in points, curves, elementary surfaces, and the global surface. Earlier works on this subject include Coons [1967], DeBoor [1972], Bezier [1986], Farin [1987, 1988], and George [1991], among others.

17.3.2.1 Points and Curves

Control points which are used in defining some higher order entities (curves and surfaces), points of the curves and surfaces, and the points created by the mesh generator are to be addressed in the algebraic methods.

A point is given either explicitly or is the result of a computation (intersection of two curves). Furthermore, the points given can be present in the surface approximation or else merely serve as supports for information. In this case, they will not exist in this approximation but are used to define the set of points to be created on the surface.

The curves are created from points and relatively complex functions to ensure certain continuity properties (in particular at the junction of two curves). Three types of construction can be established:

(a) The curve is defined by points and passes through them.
(b) The curve is defined by points but does not necessarily pass through them.
(c) The curve is defined by points and additional constraints such as directional derivatives.

We are now confronted with the problem of constructing a piecewise polynomial function of s, of degree n and of class C^{r_i-1} in s_i with $0 \le r_i \le n$, such that

$$C(s) = SMQ \tag{17.3.35}$$

where M is the matrix of coefficients of dimension $(n+1) \times (n+1)$, with S and Q being the basis polynomials of the representation (a line vector) and the control (column) vector so that

$$S = [s^n,\ s^{n-1}, \ldots, s, 1] \tag{17.3.36a}$$

$$Q = \left[q_o, q_1, \ldots, q_{\frac{n+1}{2}}, \dot{q}_o, \dot{q}_1, \ldots, \dot{q}_{\frac{n+1}{2}}\right] \tag{17.3.36b}$$

To illustrate, we shall examine the Lagrange polynomial, Hermite polynomial, and Bezier curve.

(a) Lagrange Polynomial
The Lagrange polynomials in the context of (17.3.35) are written as

$$C(s) = \sum_{i=0}^{n} \phi_i(s) Q_i \tag{17.3.37}$$

with

$$\phi_i(s) = \prod_{\substack{r=0\\ r\neq i}}^{n} \frac{s - s_r}{s_i - s_r} \tag{17.3.38}$$

in which $n+1$ specified points are involved and

$$\phi_i(s_j) = \delta_{ij}$$
$$C(s_i) = Q_i$$

With these definitions, the recurrence formula for (17.3.37) becomes

$$C_i^m(s) = \frac{s_{i+m} - s}{s_{i+m} - s_i} C_i^{m-1}(s) + \frac{s - s_i}{s_{i+m} - s_i} C_{i+1}^{m-1}(s) = 0 \tag{17.3.39}$$

with $i = 0, \ldots n - m$, $m = 1, \ldots n$, which is known as the Aitken's algorithm.

Notice that (17.3.35) and (17.3.39) are identical. To see this, let us consider $n = 1$. Then, (17.3.35) becomes

$$C(s) = [s \quad 1] \begin{bmatrix} \dfrac{1}{s_o - s_1} & \dfrac{1}{s_1 - s_0} \\ \dfrac{-s_1}{s_o - s_1} & \dfrac{-s_o}{s_1 - s_0} \end{bmatrix} \begin{bmatrix} Q_1 \\ Q_2 \end{bmatrix} = [s \quad 1] \begin{bmatrix} -1 & 1 \\ 1 & 0 \end{bmatrix} \begin{bmatrix} Q_1 \\ Q_2 \end{bmatrix}$$
$$= (1 - s)Q_1 + sQ_2$$

The same result arises from (17.3.38).

Similarly, for $n = 2$, we obtain

$$C(s) = [s^2 \quad s \quad 1] \begin{bmatrix} \dfrac{1}{(s_o - s_1)(s_o - s_2)} & \dfrac{1}{(s_1 - s_0)(s_1 - s_2)} & \dfrac{1}{(s_2 - s_0)(s_2 - s_1)} \\ \dfrac{-s_1 - s_2}{(s_o - s_1)(s_o - s_2)} & \dfrac{-s_o - s_2}{(s_1 - s_0)(s_1 - s_2)} & \dfrac{-s_o - s_1}{(s_2 - s_0)(s_2 - s_1)} \\ \dfrac{s_1 s_2}{(s_o - s_1)(s_o - s_2)} & \dfrac{s_o s_2}{(s_1 - s_0)(s_1 - s_2)} & \dfrac{s_o s_1}{(s_2 - s_0)(s_2 - s_1)} \end{bmatrix} \begin{bmatrix} Q_1 \\ Q_2 \\ Q_3 \end{bmatrix}$$

or

$$C(s) = [s^2 \quad s \quad 1] \begin{bmatrix} 2 & -4 & 2 \\ -3 & 4 & -1 \\ 1 & 0 & 0 \end{bmatrix} \begin{bmatrix} Q_1 \\ Q_2 \\ Q_3 \end{bmatrix}$$
$$= (2s^2 - 3s + 1)Q_1 + (-4s^2 + 4s)Q_2 + (2s^2 - s)Q_3$$

It is seen that this is the second order Lagrange polynomial representation.

(b) Hermite Polynomial

Proceeding similarly as in the Lagrange polynomial, but with derivatives of Q, we write for $n = 3$,

$$S = [s^3 \quad s^2 \quad s \quad 1], \qquad Q = [q_o \quad q_1 \quad \dot{q}_o \quad \dot{q}_1], \qquad M = \begin{bmatrix} 2 & -2 & 1 & 1 \\ -3 & 3 & -2 & -1 \\ 0 & 0 & 1 & 0 \\ 1 & 0 & 0 & 0 \end{bmatrix}$$

representing the cubic Hermite polynomials.

(c) Bezier Curve

An algebraic form of this approximation uses the Bernstein polynomials of the form

$$C(s) = \sum_{i=0}^{n} c_i^n s^i (1-s)^{n-i} Q_i \tag{17.3.40}$$

with

$$c_i^n = \frac{n!}{(n-i)!\, i!} \tag{17.3.41}$$

for which the matrix of coefficient takes the form

$$M = \begin{bmatrix} -1 & 3 & -3 & 1 \\ 3 & -6 & 3 & 0 \\ -3 & 3 & 0 & 0 \\ 1 & 0 & 0 & 0 \end{bmatrix} \quad \text{with} \quad S = [s^3 \quad s^2 \quad s \quad 1], \quad Q = [q_o \quad q_1 \quad q_2 \quad q_3] \tag{17.3.42}$$

These polynomials can be shown to be identical to the cubic Hermite polynomials if we consider a third degree polynomial satisfying the following four constraints:

$$C_i(0) = Q_i, \qquad C_i(1) = Q_{i+1}$$
$$\dot{C}_i(0) = \dot{Q}_i, \qquad \dot{C}_i(1) = \dot{Q}_{i+1}$$

To this end, we set

$$C_i(s) = a_i + b_i s + c_i s^2 + d_i s^3$$

and obtain

$$Q_i = a_i$$

$$Q_{i+1} = a_i + b_i + c_i + d_i$$

$$\dot{Q}_i = b_i$$

$$\dot{Q}_{i+1} = b_i + 2c_i + 3d_i$$

This gives

$$S = [s^3 \quad s^2 \quad s \quad 1], \quad Q = [Q_i \quad Q_{i+1} \quad \dot{Q}_i \quad \dot{Q}_{i+1}], \quad M = \begin{bmatrix} 2 & -2 & 1 & 1 \\ -3 & 3 & -2 & -1 \\ 0 & 0 & 1 & 0 \\ 1 & 0 & 0 & 0 \end{bmatrix} \tag{17.3.43}$$

Here, $C_i(s) = SMQ$ represents the cubic Hermite polynomial.

Another example is given for the case involving four consecutive points. (Q_{i-1}, Q_i, Q_{i+1}, Q_{i+2}) with a cubic polynomial mapped between [0, 1] and the curve passing through Q_i and Q_{i+1} and its tangent at these points being fixed to the value $\dot{Q}_i = \frac{1}{2}(Q_{i+2} - Q_i)$. These conditions lead to

$$Q_i = a_i$$

$$Q_{i+1} = a_i + b_i + c_i + d_i$$

$$Q_{i+1} - Q_{i-1} = 2b_i$$

$$Q_{i+2} - Q_i = 2b_i + 4c_i + 6d_i$$

and

$$S = [s^3 \quad s^2 \quad s \quad 1], \quad Q = [q_{i-1} \quad q_i \quad q_{i+1} \quad q_{i+2}], \quad M = \frac{1}{2}\begin{bmatrix} -1 & 3 & -3 & 1 \\ 2 & -5 & 4 & -1 \\ -1 & 0 & 1 & 0 \\ 0 & 2 & 0 & 0 \end{bmatrix} \tag{17.3.44}$$

This is known as the Catmull-Rom form.

A general form of (17.3.44), called the cardinal spline basis, is given as

$$M = \begin{bmatrix} -\alpha & 2-\alpha & \alpha-2 & \alpha \\ 2\alpha & \alpha-3 & 3-2\alpha & -\alpha \\ -\alpha & 0 & \alpha & 0 \\ 0 & 1 & 0 & 0 \end{bmatrix} \tag{17.3.45}$$

where $\alpha = 1$ leads to the Catmull-Rom form.

Similarly, the coefficient matrices for B-spline and Beta spline forms are given as follows:

B-Spline

$$M = \frac{1}{6}\begin{bmatrix} -1 & 3 & -3 & 1 \\ 3 & -6 & 3 & 0 \\ -3 & 0 & 3 & 0 \\ 1 & 4 & 1 & 1 \end{bmatrix} \tag{17.3.46}$$

Beta Spline

$$M = \frac{1}{\Delta}\begin{bmatrix} -2\beta_1^3 & 2(\beta_2 + \beta_1^3 + \beta_1^2 + \beta_1) & -2(\beta_2 + \beta_1^2 + \beta_1 + 1) & 2 \\ 6\beta_1^3 & -3(\beta_2 + 2\beta_1^3 + 2\beta_1^2) & 3(\beta_2 + 2\beta_1^2) & 0 \\ -6\beta_1^3 & 6(\beta_1^3 - \beta_1) & 6\beta_1 & 0 \\ 2\beta_1^3 & \beta_2 + 4(\beta_1^2 + \beta_1) & 2 & 0 \end{bmatrix} \tag{17.3.47}$$

with $\Delta = \beta_2 + 2\beta_1^3 + 4\beta_1^2 + 4\beta_1 + 2$. For $\beta_1 = 1$, $\beta_2 = 0$ the classic B-spline form is found there, β_1 (the bias) and β_2 (the tension) are introduced in B-spline form in order to control the curve by moving it toward the control points.

17.3.2.2 Elementary and Global Surfaces

The different methods to construct a curve can be extended to a surface by using tensor product in two or three directions.

$$C(s,u) = SMQ(u) \tag{17.3.48}$$

with

$$Q(u) = UMQ_{(ij)} \tag{17.3.49}$$

where i denotes the dependence with respect to parameter s and j that with respect to parameter u, U is the equivalent in u to S (i.e., the associated basis polynomial), and $Q_{(ij)}$, is a $(n+1) \times (n+1)$ matrix constructed on control points. Substituting (17.3.49) into (17.3.48) yields

$$\overline{C}(s,u) = SMQ_{(ij)}^T M^T U^T \tag{17.3.50a}$$

or

$$\overline{C}(s,u) = \sum_{i=0}^{n}\sum_{j=0}^{m} b_{ij} s_i u_j \tag{17.3.50b}$$

where b_{ij} depends on the method selected (n and m being arbitrary). In case of the Bezier form, $\overline{C}(s,u)$ can be expressed in terms of the Bernstein polynomials:

$$B_i^n(s) = C_i^n s^i (1-s)^{n-i} \tag{17.3.51}$$

so that

$$\overline{C}(s,u) = \sum_{i=0}^{n}\sum_{j=0}^{m} B_i^n(s) B_j^m(s)\, Q_{(ij)} \tag{17.3.52}$$

This represents the surface by Bezier patches leading to quadrilateral elements (Figure 17.3.2a). To produce triangular patches (Figure 17.3.2b), we use the polynomials

$$B_{ijk}^n(r,s,t) = \frac{n!}{i!\,j!\,k!} r^i s^j t^k, \quad i+j+k=n \tag{17.3.53}$$

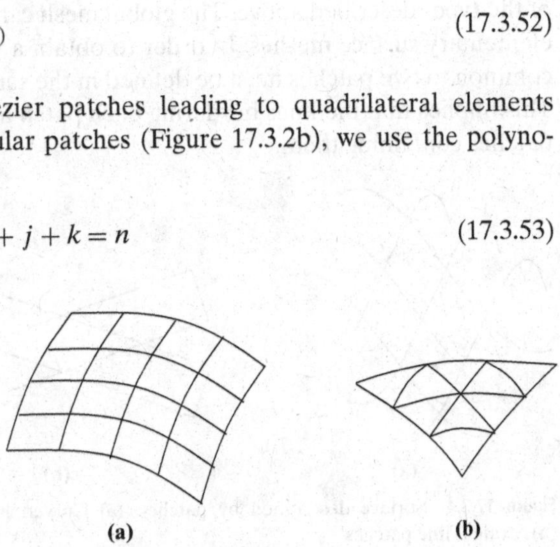

Figure 17.3.2 Quadrilateral and triangular element patches. (a) Quadrilateral element. (b) Triangular element.

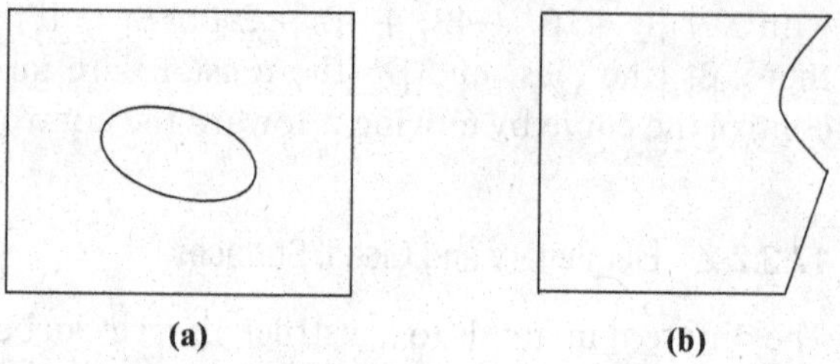

Figure 17.3.3 Basis patches. (a) Hole inside. (b) Deformed corner.

where r, s, t denote the barycentric coordinates, Q_{ijk}, representing a series of points representing a triangular network acting as control points.

A global surface is defined as a series of elementary surfaces. Let us consider two cases: (a) a patch with a hole (Figure 17.3.3a) and (b) a patch deformed at one of its corners (Figure 17.3.3b) for which the region of interest is the zone remaining or its complementary.

Let us consider the Catmull-Rom form of the third degree. The surface is described by a series of patches which are constructed using the control point s and the representation chosen. Each patch is joined to its neighbors with the properties present in this representation.

The surface is defined by a coarse grid, derived from the control points used in the case of the description for the Catmull-Rom method. Thus, to define a grid on the surface which includes n_1 divisions in one direction (s) and n_2 in the other (u), one has to provide $n_1 + 3$ series of $n_2 + 3$ control points. The end points only serve to define the shape of the surface by its boundary. Each patch is then determined by its four points and by the points of its neighbors (Figure 17.3.4), passing through the four points which define it and sharing a continuous junction with its neighboring patches as the points of the latter are taken into account in its definition.

17.3.2.3 Surface Mesh Generation

Assume that the surface under consideration is known by a series of elementary surfaces of the types described above. The global mesh can be seen as the union of the different elementary surface meshes. In order to obtain a valid mesh, we ensure that any point common to two patches must be defined in the same way in each patch that contains it. This implies that the lines bordering each patch are meshed in the same way in all the patches containing them.

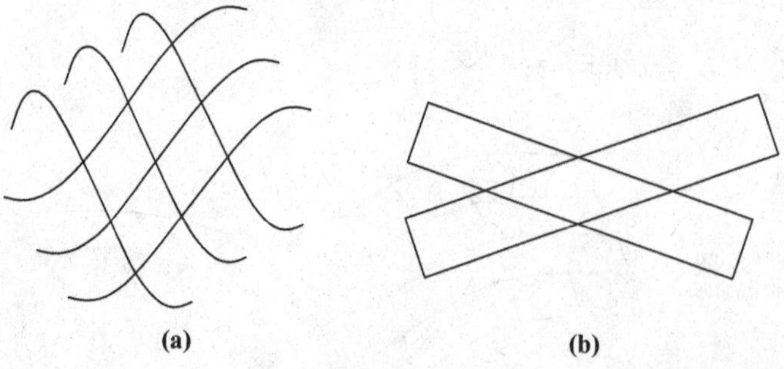

Figure 17.3.4 Surface description by patches. (a) Curved line patches. (b) Straight line patches.

Toward this end for each boundary line when considering a patch processed previously, we now perform discretization, compatible with the previously meshed lines. The global surface is obtained using, for example, the Catmull-Rom form of the third degree.

Example 17.3.3

Describe in detail the implementation of a Bezier curve for surface grid generation.

(1) Initial Step
A global surface is obtained from the union of elementary surfaces or patches. For the Catmull-Rom method, the surface is defined by a coarse grid of patches derived from user-specified control points. To define a grid on the surface which has n divisions in the s-direction and m divisions in the u-direction requires $(n+2) \times (m+2)$ control points. The extra end points serve to define the shape of the surface at its boundary. Each Bezier patch is then determined from its four points (the vertices of the quadrilateral element) and the points in its corresponding neighbors.

(2) Valid Mesh
In order to obtain a valid mesh, we must ensure that any point which is common to two patches is defined in the same way for each patch that contains it. This implies that the lines bordering each patch are meshed the same way in all patches containing them.

(3) Creation of Mesh
When all the lines forming the boundaries of the patches have been discretized, the mesh of all the patches is created as follows:

(3-1) If the patch is quadrilateral or triangular and if none of its boundary lines contains intermediary points, then it is considered an element of the mesh.

(3-2) If the patch is quadrilateral or triangular and if all of its boundary lines contain a given number of intermediary points compatible with a regular partitioning, then it is meshed by a suitable method. For example, use the Catmull-Rom method as follows:

Step 1

do for $i = 0$ and $i = N$
do for $j = 0$ to M, do
- Consider the location in R^3 of node (i, j) (located on a boundary line previously meshed)
- Compute values of the associated parameters

end do for $j = 0$ to M;
end do for $i = 0$ and $i = N$
for $j = 0$ and $j = M$, do
for $i = 0$ to N, do
- Consider the location in R^3 of node (i, j)
- Compute values of the associated parameters

end do for $i = 0$ to N

end do for $i = 0$ and $i = M$;
end do for Step 1:

Step 2

Create the mesh in space (t, u) of the unit square $[0, 1] \times [0, 1]$ as a function of its boundary discretazation
end do for Step 2

Step 3

do, for $i = 1$ to $N - 1$, do
for $j = 1$ to $M - 1$, do

- *Definition of Connectivity*: the vertices of the element created have the following couples as vertex numbers: $(i, j), (i+1, j), (i+1, j+1)$ and $(i, j+1)$, each of which will have a global number associated with it
- *Compute Vertex Location*: evaluate t and u corresponding to i and j and find the location using $C(t, u) = \sum_{i=0}^{n} \sum_{j=0}^{m} b_{ij} t^i u^j$ with P_{ij} the matrix of control points

end do for $j = 1$ to $M - 1$;
end do for $i = 1$ to $N - 1$;

(3-2) Any two-dimensional method can be implemented in (t, u) space, the problem being to know if the mapping in R^3 of mesh points in the space of parameters is valid, close to the surface, and good quality.

Example 17.3.4

This example is based on the surface grid generation via Bezier curve polynomials [Warsi, 1992]. Figure E17.3.4a shows a generic forebody surface grid of an aircraft, with the number of points increased in the canopy region (Figure E17.3.4b). Discontinuities in a surface may be handled easily by selecting appropriate patches so that spline constructions do not occur at the discontinuities.

Figure E17.3.4c shows a set of curves generated for a generic re-entry vehicle as an example of curve generation and editing facilities. Since actual surface definition data are not available, each of the curves shown is generated with the curve segment generator in the program. The majority of the curves are generated using the Bezier generator, and the complex curves at the trailing edge of the wing are generated by appending multiple Bezier curves, elliptical, circular and straight line segments.

Figure E17.3.4d shows the initial surface grid generated for the generic re-entry vehicle using the previously designed curves shown in Figure E17.3.4c, and the surface generation facilities of splining cross-sectional data and transfinite interpolation with specified edge curves. The final surface grid for the generic pre-entry vehicle after using the surface editing facilities is shown in Figure E17.3.4e. Notice that grid distributions are now much smoother and point resolution in areas of interest is better, while the original surface geometry is maintained.

A sample far-field boundary and blocking arrangement for the entry vehicle after performing a domain decomposition is shown in Figure E17.3.4f, with the mesh on selected block faces around the re-entry vehicle shown in Figure E17.3.4g. Figure E17.3.4h shows a global view of the surface grids generated for a win/pylon/lead configuration.

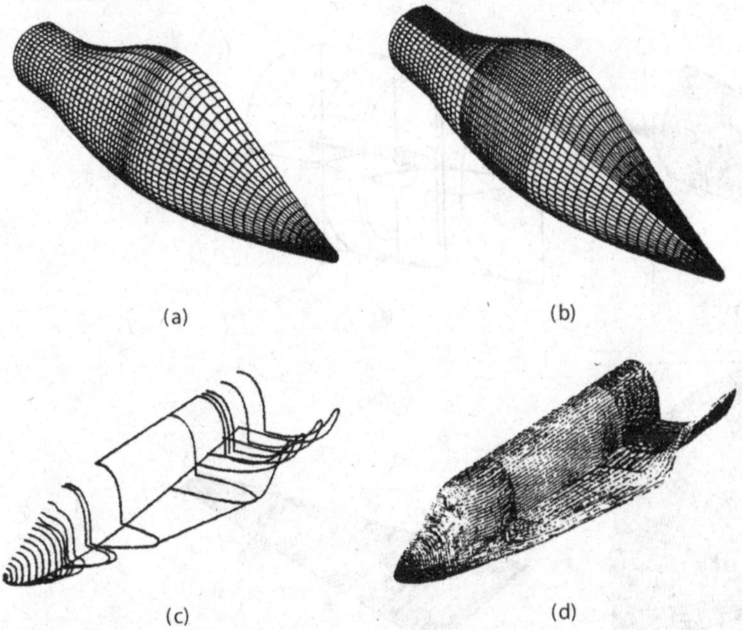

Figure E17.3.4 Surface grid generation via Bezier curve polynomials [Warsi, 1992]. **(a)** Generic forebody surface grid. **(b)** Enrichment of grid points in canopy region of forebody surface. **(c)** Surface definition curves for generic re-entry vehicle. **(d)** Initial surface grid for generic re-entry vehicle.

Figure E17.3.4i shows some details of the surface grids in the wing/pylon interaction region.

17.4 MULTIBLOCK STRUCTURED GRID GENERATION

An efficient approach to the grid generation in complex domain, particularly in three-dimensional geometries, is to establish block configurations initially, construct the grid with increasing details, and make modifications on an existing grid with minimum restrictions. Such a sequential procedure is known as multiblock grid generation, which is conducive to parallel processing to be discussed in Section 20.4. Ecer, Spyropoulos, and Maul [1985] presented the multiblock structured finite element grid generation. Brief descriptions of this approach are given below.

A convenient way of generating the finite element multigrid system is to use isoparametric elements in 2-D or 3-D. Linear, quadratic, or cubic interpolation functions may be used to divide the domain roughly by a desired number of blocks, each of which will then be subdivided into as many elements as required for computation. For geometries with a pointed nose or leading and trailing edges of an airfoil, it is necessary to use wedge type elements such as a triangle collapsed from a quadrilateral element for 2-D (see Example 9.3.5) or the counterpart for 3-D with a tetrahedron collapsed from a hexahedron.

Consider the modeling of a complete aircraft geometry as an example. The geometric modeling package provides information in three steps as shown in Figure 17.4.1a

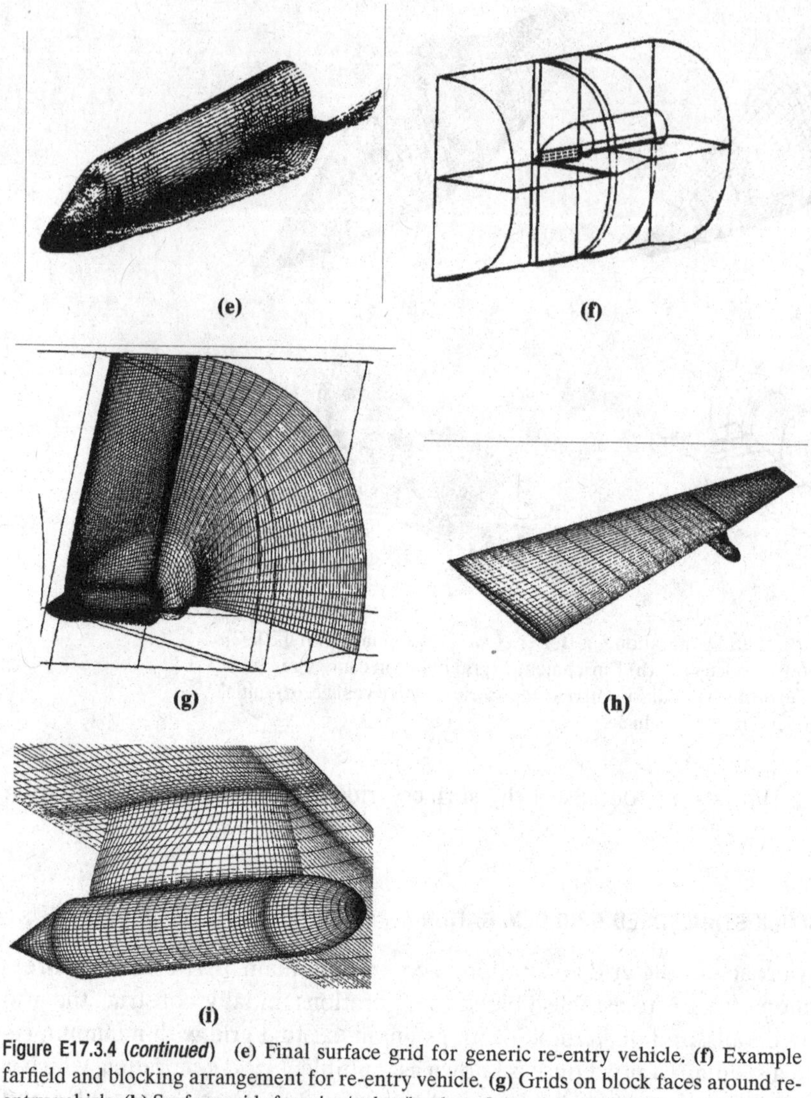

Figure E17.3.4 (*continued*) (e) Final surface grid for generic re-entry vehicle. (f) Example farfield and blocking arrangement for re-entry vehicle. (g) Grids on block faces around re-entry vehicle. (h) Surface grids for wing/pylon/load configuration. (i) Detail of surface grids in the wing/pylon intersection region.

by digitizing points on several sections [Figure 17.4.1a(1)]. These digitized points are connected by a series of B-splines and through Boolean operations [Figure 17.4.1a(2,3)].

The next step is to describe the blocks surrounding the aircraft. It is convenient to define four regions with two of them [I and II of Figure 17.4.1b(1)] for radial directions and another set of two regions [III and IV of Figure 17.4.1b(2)] for two pairs of wings along the aircraft. As a result, the block structured for the developed aircraft grid is obtained as shown in Figure 17.4.1b(3) with the coordinate system of radial, tangential, and longitudinal directions. It is shown that there are 336 blocks with the inner two subvolumes having smaller number of blocks than the outer subvolumes. Figure 17.4.1c shows the resulting grid generated around the nose.

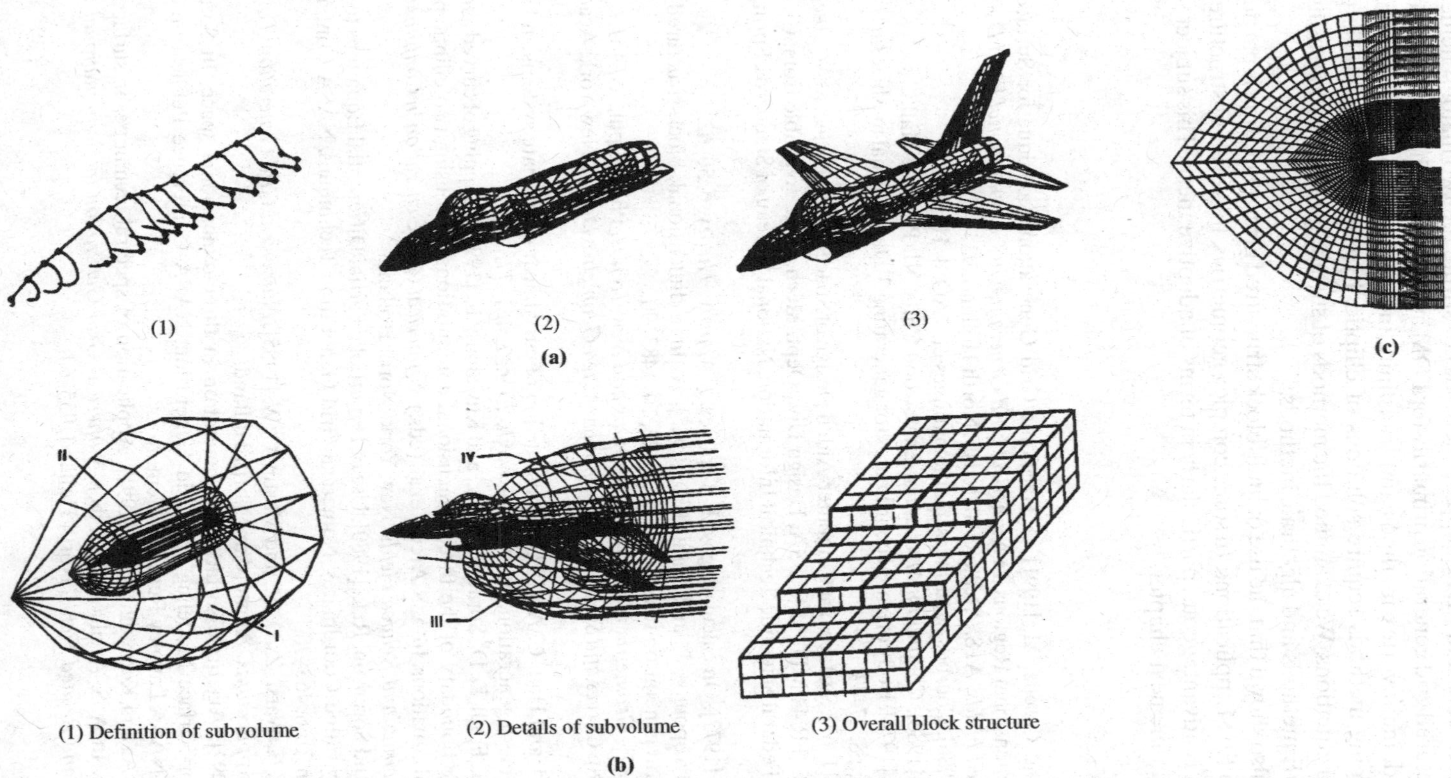

Figure 17.4.1 Multiblock structured grid generation [Ecer, 1986]. **(a)** Procedure of describing the aircraft geometry. **(b)** Geometric description of block structured around the aircraft. **(c)** Across section of the final grid for part of the aircraft geometry.

17.5 SUMMARY

Algebraic methods and PDE mapping methods constitute the two major schemes used in the structured grid generation primarily for FDM applications. The algebraic methods consist of domain vertex methods and transfinite interpolation methods, whereas the PDE mapping methods require solutions of elliptic, hyperbolic, or parabolic partial differential equations. We examined the methods of surface grid generation, using both elliptic PDE methods and algebraic methods.

It was also shown that the use of multiblock structured grid generation is particularly effective in FEM applications. In some complex geometries, however, unstructured grid generation is advantageous, particularly in terms of adaptive mesh. This subject will be presented in the next chapter.

REFERENCES

Arina, R. and Casella, M. [1991]. A Harmonic Grid Generation Technique for Surfaces and Three-Dimensional Regions. In *Numerical Grid Generation in Computational Fluid Dynamics and Related Fields*. A. S. Arcilla et al. (eds.). North Holland, 935–46.

Bezier, P. [1986]. Courbes et surfaces, mathematiques et CAO. 4, Hermes.

Chung, T. J. [1988]. *Continuum Mechanics*. Englewood Cliffs, NJ: Prentice-Hall.

Cook, W. A. [1974]. Body oriented coordinates for generating 3-Dimensional meshes. *Int. J. Num. Meth. Eng.*, 8, 27–43.

Coons, S. A. [1967]. Surfaces for Computer-Aided Design of Space Forms. Project MAC, Technical Rep. MAC-TR 44 MIT, MA, USA, Design Div., Dept. Mech. Eng., Available from: Clearinghouse for Federal Scientific-Technical Information, National Bureau of Standards, Springfield, VA, USA.

De Boor, C. [1972]. On calculating with B-splines. *J. Approx. Theory*, 6, 50–62.

Ecer, A., Spyropoulos, J., and Maul, J. D. [1985]. A three-dimensional, block-structured finite element grid generation scheme. *AIAA J.*, 23, 10, 1483–90.

Farin, G. [1987]. *Geometric Modeling: Algorithms and New Trends*. Philadelphia: *SIAM*.

———. [1988]. *Curves and Surfaces for Computer Aided Geometric Design*. New York: Academic Press.

Gordon, W. J. and Hall, C. A. [1973]. Construction of curvilinear coordinate systems and applications to mesh generation. *Int. J. Num. Meth. Eng.*, 7, 461–77.

Nakamura, S., Fradl, D. D., Spradling M. L., and Kuwahara, K. [1991]. Mapping of curved surfaces onto a side boundary of the three-dimensional computational grid using two elliptic partial differential equations. In A. S. Arcilla et al. (eds.). *Numerical Grid Generation in Computational Fluid Dynamics and Related Fields*. New York: North Holland.

Steger, J. L. and Sorenson, R. L. [1980]. Use of Hyperbolic Partial Differential Equations to Generate Body Fitted Coordinates, Numerical Grid Generation Techniques. NASA Conference Publication 2166, 463–78.

Thompson, J. F., Warsi, Z. U. A., and Mastin, C. W. [1985]. *Numerical Grid Generation: Foundations and Applications*. Amsterdam: North-Holland.

Warsi, S. [1992]. Algebraic surface grid generation in three-dimensional space. In Software Systems for Surface Modeling and Grid Generation, NASA Conference Publication 3143, Hampton: NASA Langley Research Center.

Warsi, Z. U. A. and Koomullil, G. P. [1991]. Application of spectral techniques in surface grid generation. In A. S. Arcilla et al. (eds.). *Numerical Grid Generation in Computational Fluid Dynamics and Related Fields*. North Holland, 955–64.

CHAPTER EIGHTEEN

Unstructured Grid Generation

The structured grid generation presented in Chapter 17 is restricted to those cases where the physical domain can be transformed into a computational domain through one-to-one mapping. For irregular geometries, however, such mapping processes may become either inconvenient or impossible to apply. In these cases, the structured grid generation methods are abandoned and we turn to unstructured grids where transformation into the computational domain from the physical domain is not required. Even for the regular geometries, an unstructured grid generation may be preferred for the purpose of adaptive meshing in which the structured grids initially constructed become unstructured as adaptive refinements are carried out.

Finite volume and finite element methods can be applied to unstructured grids. This is because the governing equations in these methods are written in integral form and numerical integration can be carried out directly on the unstructured grid domain in which no coordinate transformation is required. This is contrary to the finite difference methods in which structured grids must be used.

There are two major unstructured grid generation methods: Delaunay-Voronoi methods (DVM) and advancing front methods (AFM) for triangles (2-D) and tetrahedrals (3-D). Numerous other methods for quadrilaterals (2-D) and hexahedrals (3-D) are available (tree methods, paving methods, etc.). We shall discuss these and other topics in this chapter.

18.1 DELAUNAY-VORONOI METHODS

A two-dimensional domain may be triangulated as shown in Figure 18.1.1a (light lines). Each side line of the triangles can be bisected in a perpendicular direction such that these three bisectors join a point within the triangle (heavy lines in Figure 18.1.1a), forming a polygon surrounding the vertex of each triangle, known as the Voronoi polygon (diagram) [Voronoi, 1908]. A collection of Voronoi polygons is known as the Dirichlet tessellation [Dirichlet, 1850], and the resulting triangles as Delaunay triangulation [Delaunay, 1934].

Any three points in the plane may be connected by a circle, called the circumcircle (Figure 18.1.1b). The center of this circle, called circumcenter, may (triangle ABC) or may not (triangle DEF) remain within the triangles, although perpendicular bisectors

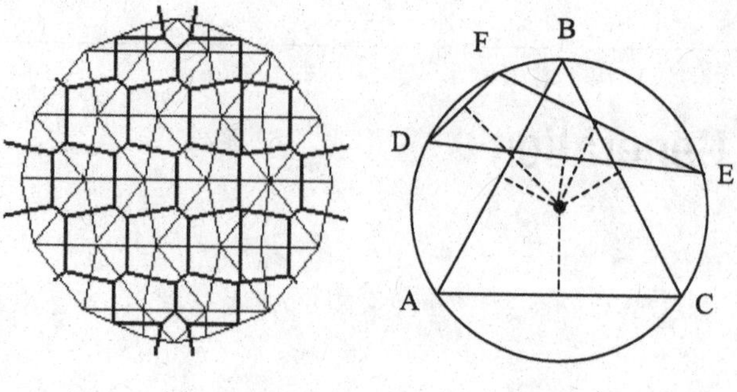

Figure 18.1.1 Delaunay triangulation with Voronoi polygon, and triangle's circumcircle. (a) Delaunay triangulation and Voronoi polygon. (b) Triangle and its circumcircle.

of sides of both triangles meet at the circumcenter. Here triangle ABC is accepted whereas triangle DEF is rejected, an obvious preference toward a triangle as close to an equilateral triangle as possible versus distorted triangular shapes.

The above geometrical requirements call for distinct points in the plane, P_1, $P_2, \ldots, P_N$ with the sets V_i $(i = 1, 2, \ldots, N)$ such that

$$V_i = \{x : \|x - P_i\| < \|x - P_j\|, \forall j \neq i\} \tag{18.1.1}$$

Thus, V_i represents a region of the plane whose points are nearer to node P_i than any other node and is an open convex polygon (a Voronoi polygon) whose boundaries are portions of the perpendicular bisector lines joining node P_i and P_j when V_i and V_j are contiguous. Connecting the node P of adjacent polygons forms a triangle T_k. The set of triangles $\{T_k\}$ constitutes the Delaunay triangulation.

A triangulation must satisfy the in-circle criterion that no point of the set P_i is interior to the circumcircle of any triangle $T(P_i)$. This criterion may be demonstrated in Figure 18.1.2 in which triangles A-B-C and A-C-D are avoided, but instead triangles A-B-D and D-B-C are chosen.

In implementation of Delaunay-Voronoi methods, various algorithms have been developed. Among them are the Watson algorithm [Watson, 1981; Cavendish, Field, and Frey, 1985] and Bowyer algorithm [Bowyer, 1981; Weatherill, 1992]. These algorithms are presented below for 2-D geometries. Extensions to 3-D geometries are also included.

18.1.1 WATSON ALGORITHM

Consider that three given nodes will form a Delaunay triangulation if and only if the circumdisk (interior of the circumcircle) defined by these nodes contains no other node points in its interior. In effect, Watson's algorithm is to reject, from the set of all possible triangles which might be formed, those with nonempty associated circumdisks. Then, those triangles not rejected form the Delaunay triangulation.

In practice, the procedure begins with inserting nodes, one by one, re-triangularizing upon insertion of each node. The detailed procedure is as follows:

(1) Initialize the algorithm by calculating the x and y coordinates of boundary points surrounding all the node points to be inserted (Figure 18.1.5). Also calculate the circumradius and x and y coordinates of the circumcircle.

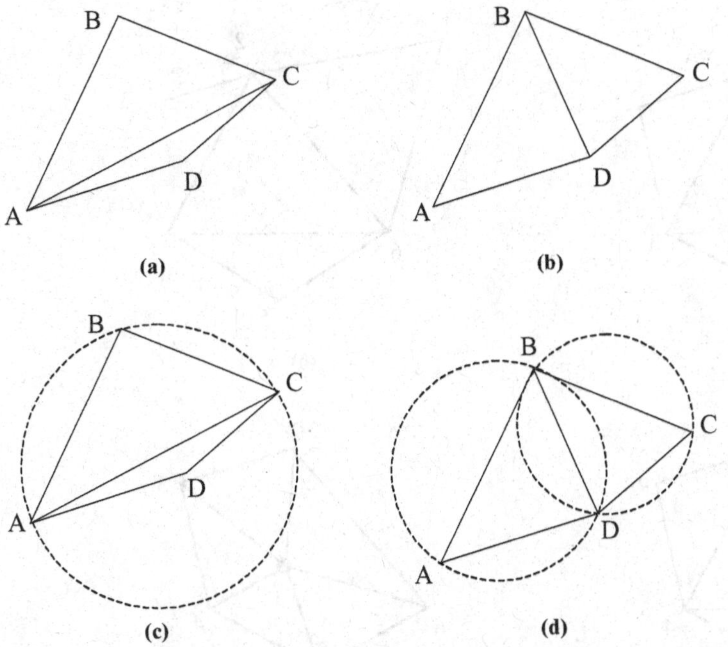

Figure 18.1.2 A triangulation must satisfy the in-circle criterion that no point of the set P_i is interior to the circumcircle of any triangle $T(P_i)$. **(a)** Undesirable triangle, maximum-minimum criterion is not satisfied. **(b)** Desirable triangulation maximum-minimum criterion is satisfied. **(c)** Unacceptable because the circumcircle ABC includes point D interior to the circumcircle. Similarly, if circumcircle ACD is drawn, then B will be interior to it. **(d)** Acceptable because no point is interior to the circumcircles (ABD or BCD).

(2) Introduce a new point.
(3) Conduct a search of all the current triangles to identify those whose circumdisks contain the new point. For each such disk, the associated triangle is flagged for removal.
(4) With the union of all such triangles, an insertion polygon is formed. Here no previously inserted node is contained in the interior of the polygon. Also, each boundary node of the polygon may be connected to the new node by a straight line lying entirely within the polygon. These lines form a new triangulation of the region, which can be shown to be a new Delaunay triangulation.
(5) Repeat Steps 2 through 4 until all nodes have been inserted.

To illustrate the procedure described above, consider triangle 2-4-6 and neighboring triangles 1-2-6, 2-3-4, and 4-5-6 as shown in Figure 18.1.3a. Introduce a new point inside the triangle 2-4-6 (denoted by 7). Each triangle has a circumdisk as defined by the circles containing all three vertices. By default, a new point lies on the circumdisk of the new triangle upon which it was introduced. Check to see if the new point lies within the circumdisk of the neighboring triangles by comparing the distance between the new point and the circumcenter to the radius for each triangle. Point 7 lies within the circumdisks of neighboring triangles 2-3-4 and 4-5-6, but not triangle 1-2-6 as shown in Figure 18.1.3b. Flag those triangles for removal that have circumdisks which contain the new point. In the example, triangles 2-3-4, 4-5-6, and 2-4-6 are flagged for

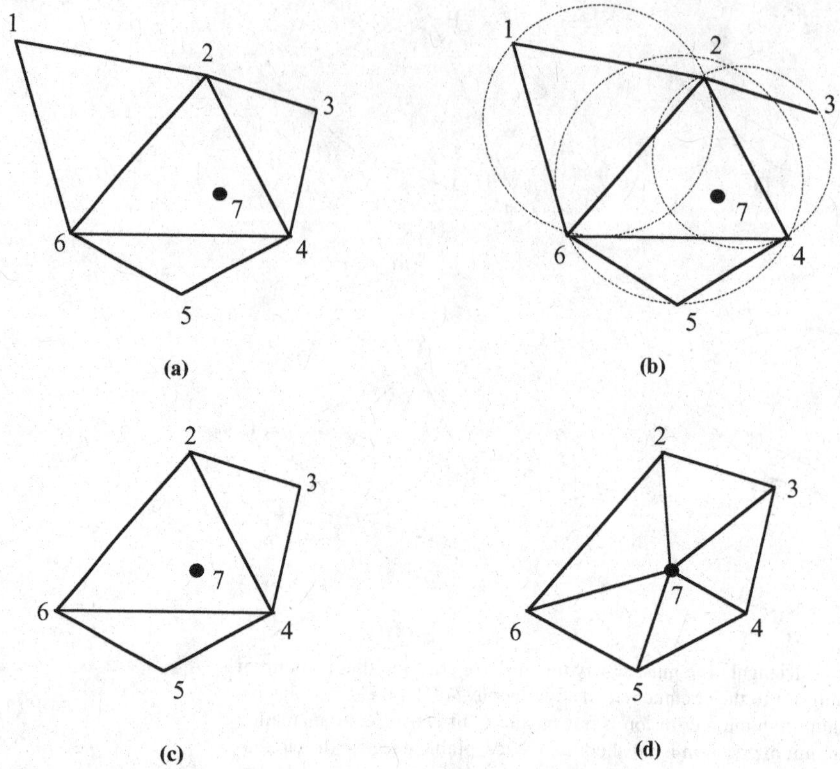

Figure 18.1.3 Watson algorithm of Delaunay-Voronoi method. **(a)** Initial stage. **(b)** New point 7 and circumcircles for all triangles. **(c)** Flag triangles for removal. **(d)** Triangulated insertion polygon and new triangles.

removal. Find the insertion polygon, the polygon remaining after the flagged triangles have been removed. The insertion polygon, which contains the flagged triangles, is shown in Figure 18.1.3c. When the insertion polygon is triangulated, the number of triangles is increased by the number of sides of the polygon. The sides of each triangle are the sides of the polygon plus the straight lines from the end points of the sides to the new point as shown in Figure 18.1.3d.

For 3-D geometries, we begin with a tetrahedron containing all the points to be inserted. New internal tetrahedra are formed as points are inserted one-by-one. This is done by testing a new point to determine which circumballs (interior of circumsphere) of existing tetrahedra contain the point. These tetrahedra are removed, leaving an insertion polyhedron which contains the new point. New tetrahedra are created by forming edges connecting the new point to all triangular faces of the insertion polyhedron. There exist some difficulties with the Watson algorithm which must be addressed: These are the problems of degenerate cases and slivers [Cavendish et al., 1985].

Degenerate cases occur in practice when a newly inserted node appears to lie on the surface of the circumsphere associated with an existing tetrahedron. The problem becomes apparent when the distance from a newly entered nodal point to an existing circumsphere is less than the accumulated truncation error. When this happens, there is a danger of making an incorrect or inconsistent decision regarding acceptance or rejection of a given tetrahedron. This in turn produces structural inconsistencies in

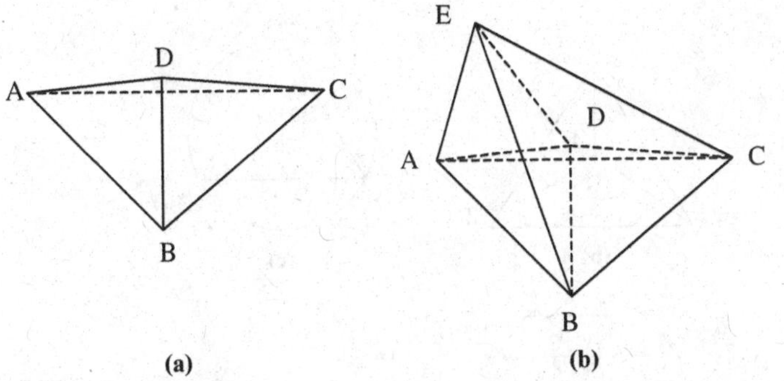

Figure 18.1.4 Treatment of undesirable of elements: (**a**) silver (badly distorted, D being slightly out of the plane of A-B-C) (**b**) Share a common vertex at E.

the mesh (overlapping tetrahedra or gaps in the mesh). A solution to this problem is to slightly perturb the coordinates of a newly entered point whenever that point is found to lie ambiguously on a circumsphere. At the completion of the triangulation, all perturbed nodes are restored to their original positions.

A sliver is a thin, badly distorted tetrahedron whose faces are well-proportioned triangles but whose volume can be made arbitrarily small (Figure 18.1.4a). In practice these are identified when the ratio

$$a = \frac{\textit{radius of inscribed sphere}}{\textit{radius of circumsphere}}$$

becomes "small" (less than 0.01). Slivers are removed in one of two ways, depending on how the tetrahedron fits into the mesh. Consider a tetrahedron $ABCD$ which is determined to be a sliver (Figure 18.1.4b). First we must determine the four tetrahedra that neighbor $ABCD$. If two of these share a common vertex, say node E, the sliver is removed from the collection of tetrahedra, and elements $\{ABDE, BCDE\}$ are replaced by elements $\{ABCE, ACDE\}$. When no two of the surrounding tetrahedra share a common vertex, the node point D is arbitrarily moved to improve the aspect ratio of the sliver.

Finally, we must post-process the mesh to obtain the final mesh over the given geometry. The above described process leads to a triangulation of the original tetrahedron. The tetrahedra associated with interior element nodes are distinguished because they have none of the four initial points as vertex. Of these interior tetrahedra, we remove the ones that lie outside of the geometry to be meshed. These are the ones whose centroids lie outside of the boundary surface.

For illustration, let us consider triangulation of a circle. The step-by-step procedure is described as follows:

(1) First of all, we define the convex hull within which all points will lie. Specify required points as shown in Figure 18.1.5a.
(2) Introduce a new point. Check to see if the new point lies on the circumdisk and if the distance from the new point to the circumcenter is less than the circumradius. Flag those triangles that contain the new point. Find the insertion polygon, the polygon remaining after the flagged triangles have been removed. First, identify the flagged triangles. Then, for each side of the triangles, check on the neighbor

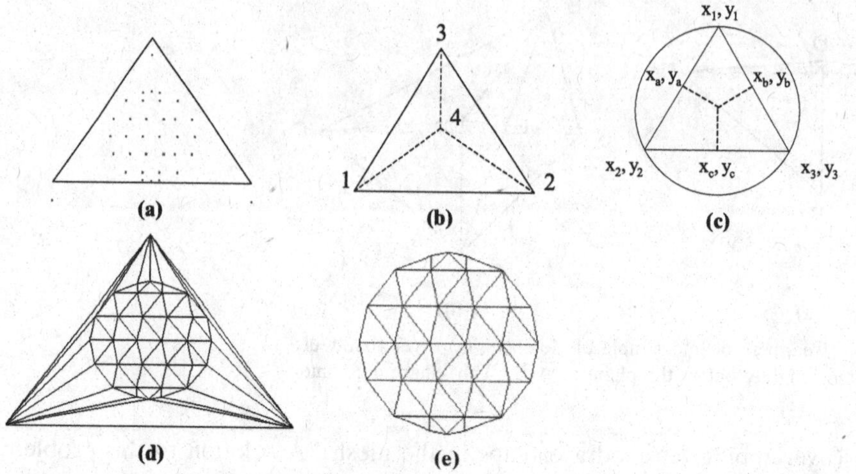

Figure 18.1.5 Illustration of the triangulation of a circle. **(a)** Convex hull. **(b)** Introduce a new point. **(c)** Circumradius and circumcenter. **(d)** Delaunay triangles with convex hull. **(e)** Final grids.

triangle to that side. If that triangle is to be removed also, then the side is not part of the polygon. If that triangle is not to be removed, or is a boundary, then that side is part of the polygon. When the polygon is triangulated, the number of triangles is increased by the number of sides of the polygon. The sides of each triangle are the side of the polygon plus the straight lines from the end points of the side to the new point.

(3) The insertion polygon resulting from adding point 4 is shown as solid lines and the triangulation results in the dashed lines as shown in Figure 18.1.5b.

(4) Calculate the circumradius and circumcenter, as seen in Figure 18.1.5c, the slope of the perpendicular bisector being the negative reciprocal of the slope of the triangle sides.

$$m_a = \frac{x_1 - x_2}{y_2 - y_1} = \frac{y_a - y_{center}}{x_a - x_{center}}$$

$$m_b = \frac{x_2 - x_3}{y_3 - y_2} = \frac{y_b - y_{center}}{x_b - x_{center}}$$

$$m_c = \frac{x_1 - x_3}{y_3 - y_1}$$

Thus, we obtain

$$y_{center} = y_a - m_a(x_a - x_{center})$$

$$m_b = \frac{y_b - y_a + m_a(x_a - x_{center})}{x_b - x_{center}}$$

$$m_b(x_b - x_{center}) = y_b - y_a + m_a(x_a - x_{center})$$

$$x_{center}(m_a - m_b) = y_b - y_a + m_a x_a - m_b x_b$$

$$x_{center} = \frac{y_b - y_a + m_a x_a - m_b x_b}{m_a - m_b}$$

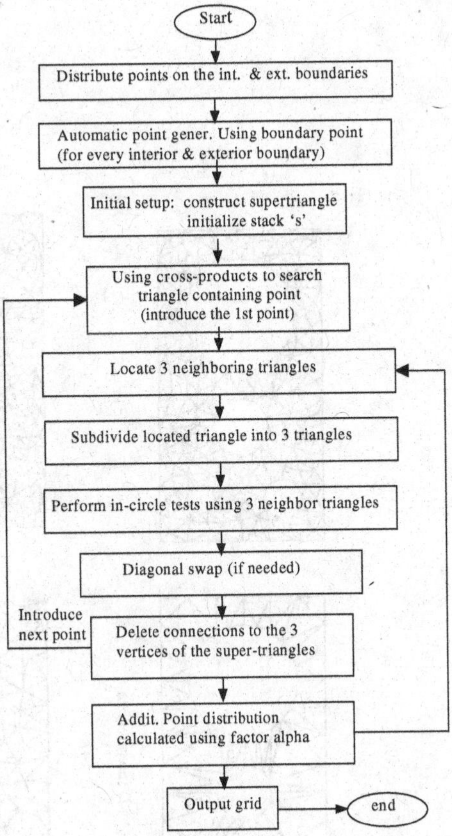

Figure 18.1.6 Delaunay-Voronoi-Watson algorithm flow chart for airfoil grid generation.

This gives the circumradius

$$r = \sqrt{(x_1 - x_{center})^2 + (y_1 - y_{center})^2}$$

(5) Degenerate case. This occurs when a newly inserted node appears to lie on the surface of a circumcircle/circumsphere. This can be resolved by slightly perturbing the coordinates of the newly entered point.

(6) The procedure described above leads to the results shown in Figure 18.1.5d,e.

The computer code flow chart and examples for mesh generation of a circle using the Delaunay-Voronoi method with Watson algorithm are shown in Figure 18.1.6 and Figure 18.1.7, respectively.

18.1.2 BOWYER ALGORITHM

In this algorithm, we utilize the forming points (points which define a Delaunay triangle and Voronoi vertex (vertex of a Voronoi polygon) as shown in Figure 18.1.8.

We recognize that it is possible to completely describe the structure of the Voronoi diagram and Delaunay triangulation by constructing two lists for each Voronoi vertex. These are a list of forming points for the vertex, and a list of the neighboring Voronoi vertices.

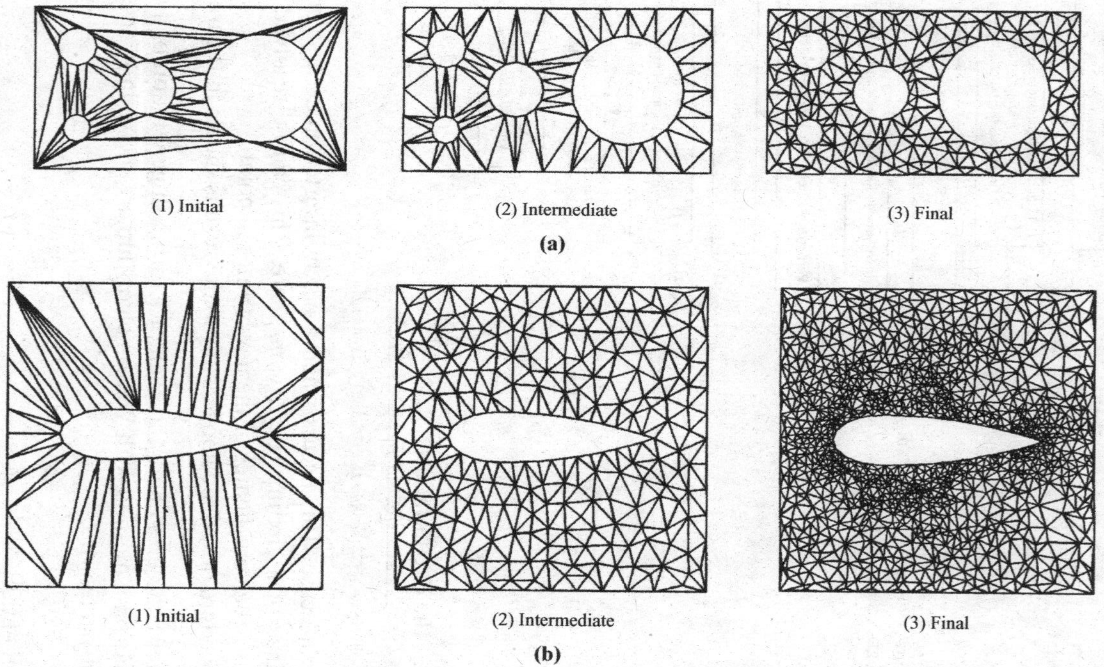

Figure 18.1.7 Delaunay-Voronoi triangulation with Watson algorithm. **(a)** Triangulation around circles. **(b)** Triangulation around airfoil.

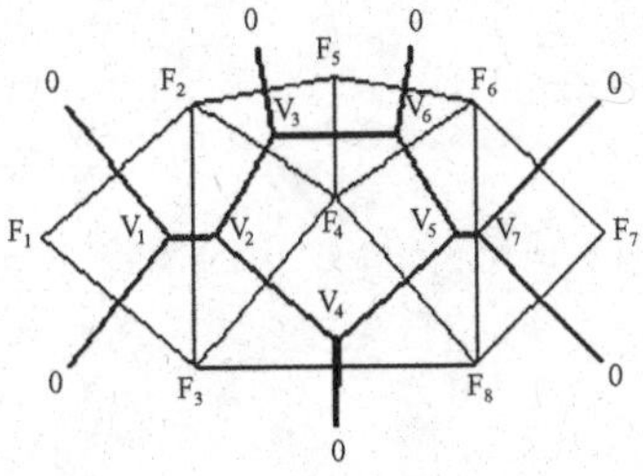

Vertex	Forming Point	Neighboring Vertices
V_1	$F_1F_2F_3$	V_2 0 0
V_2	$F_2F_3F_4$	$V_1V_3V_4$
V_3	$F_2F_4F_5$	V_2V_6 0
V_4	$F_3F_4F_8$	V_2V_5 0
V_5	$F_4F_6F_8$	$V_4V_6V_7$
V_6	$F_4F_5F_6$	V_3V_5 0
V_7	$F_6F_7F_8$	V_5 0 0

Figure 18.1.8 Forming points (F_1-F_8) and Voronoi vertices (V_1-V_7).

Similar to the previously described Watson algorithm, this is a sequential process. Each new point is introduced into the structure, one at a time, and the structure is reformulated onto a new Delaunay triangulation. The steps are as follows:

(1) Define a convex hull within which all points will lie. Specify four points with the associated Voronoi diagram.
(2) Introduce a new point.
(3) Determine all vertices of the Voronoi diagram to be deleted. A vertex to be deleted is one whose circumcircle (defined by three forming points) contains the new point. This is similar to step 3 in Watson's algorithm.
(4) Find the forming points of deleted Voronoi vertices, which are contiguous points to the new point. This is similar to step 4 of Watson's algorithm in which the new point is connected to the insertion polygon by straight lines.
(5) Determine the neighboring Voronoi vertices to the deleted vertices which have not been themselves deleted. These data provide the necessary information to enable valid combinations of contiguous points to be constructed.
(6) Determine the forming points of the Voronoi vertices. These must include the new point together with two other points which are contiguous to the new point, and form an edge of the neighboring triangle.
(7) Determine the neighboring Voronoi vertices to the new Voronoi vertices. From step 6, the forming points of all new vertices have been computed. For each new vertex, conduct a search through the forming points of the neighboring vertices found in step 5 to identify common pairs of forming points. When a common combination occurs, then the two associated vertices are neighbors of the Voronoi diagram.
(8) Reorder the Voronoi diagram data structure overwriting the entries of deleted vertices.
(9) Return to step 2 until all points have been inserted.

This process will generate regions that are both interior and exterior to the domain. For grid generation purposes, it is necessary that such triangles which are not within the domain of interest be removed before the next step of the procedure. To do this, in the initial generation of the list of points defining the physical domain, the outer domain boundary points should be listed in a counterclockwise fashion while any and all interior boundaries be listed in clockwise fashion. With this method, the sign of the cross-product of the face tangent vector with a vector to the cell centroid can be used to determine if a triangle lies either to the interior or exterior of the boundary and then

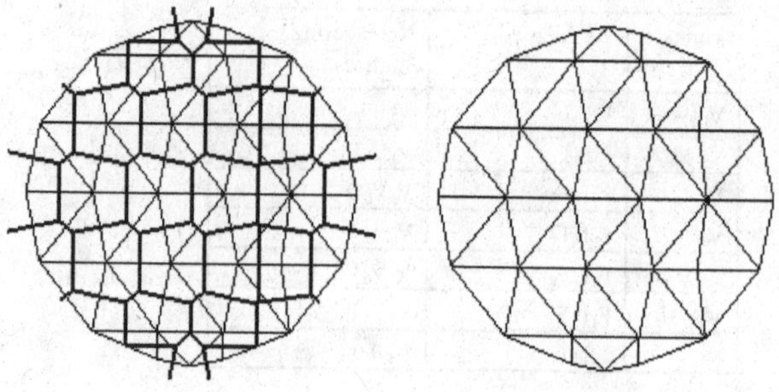

(a) (b)

Figure 18.1.9 Bowyer algorithm for triangulating a circle. (a) Voronoi polygon. (b) Delaunay triangle.

can be easily removed (by defining the triangle connectivities) if it should lie outside the desired domain. Once the initial triangulation of the domain has been performed, all triangles that have a node associated with the initial user-defined superstructure are removed. Following this process, the Voronoi polygons and the final triangulation are shown in Figure 18.1.9.

In summary, the Watson and Bowyer algorithms are quite similar. Each algorithm starts with an initial grid surrounding the geometry to be discretized. New points are introduced one at a time, and triangles whose circumdisk contain the new point are deleted. The region is then re-triangularized by connecting points on the deleted triangles to the new point. The basic difference between the Watson and Bowyer algorithms, however, is in the initial superstructure and the data structures. Note that the Bowyer algorithm maintains essentially a list of only Voronoi polygons and can then form the triangle lists from the Voronoi diagram, whereas the Watson algorithm chooses simply to maintain a list of the triplets of node numbers which represent the completed triangles, in which a running list of circumcircle center and circumradius for each formed triangle is kept.

18.1.3 AUTOMATIC POINT GENERATION SCHEME

In both the Watson and Bowyer algorithms, "a new point is introduced." The method for producing the points, however, has not been addressed. An algorithm for automatic generation of points can be developed as follows [Weatherill, 1992]:

(1) Compute the point distribution function for each boundary point x_i, y_i:

$$dP_i = \frac{1}{2}\left[\sqrt{(x_{i+1} - x_i)^2 + (y_{i+1} - y_i)^2} + \sqrt{(x_i - x_{i-1})^2 + (y_i - y_{i-1})^2}\right]$$

where the points $i + 1$ and $i - 1$ are contiguous to i.

(2) Generate the Delaunay triangulation of the boundary points.

(3) For all triangles within the domain:

(a) Define a prospective point to be at the centroid of the triangle.

(b) Derive the point distribution, dP_m, for the prospective point by interpolating the point distribution from the nodes of the triangle.

(c) Compute the distances, d_m $(m = 1, 2, 3)$ from the prospective node to each of the triangles. Then,

If $d_m < \alpha d P_m$, then reject the point and return to step 3a.
If $d_m > \alpha d P_m$, then insert the point using the Delaunay triangulation algorithm where the coefficient α is the parameter which controls the grid point density.

(d) Assign the interpolated value of the point distribution function to the new node.

(e) Move on to the next triangle.

18.2 ADVANCING FRONT METHODS

In contrast to the Delaunay-Voronoi methods (DVM), the advancing front methods (AFM) seek to achieve internal nodal formation and triangulation by marching techniques that advance front cell faces from the domain boundary, with or without background grid configurations. Various schemes of AFM have been reported [Lo, 1985, 1989; Peraire et al., 1987; Lohner, 1988] for both two dimensions (triangular elements) and three dimensions (tetrahedral elements). The AFM concept may be extended to a generation of quadrilateral elements [Zhu et al., 1991; Blacker and Stephenson, 1991]. We shall examine these and other topics in this section.

The simplest description of AFM begins with specification of boundaries, as shown in Figure 18.2.1 where the exterior boundaries move counterclockwise and interior boundaries (if they exist, i.e., multiply connected domain) move clockwise. For example, for the case of a simply connected domain (Figure 18.2.2a), exterior boundaries (nodes 1 through 6, Figure 18.2.2b) are used as initial active front faces. Node 7 is created to form a triangle 1-2-7 and then side 1-2 is deleted so that we now have two new front faces 1-7 and 2-7 (Figure 18.2.2c). Choose a new interior node 8 (Figure 18.2.2d) which will then allow side 2-3 to be deleted. The process continues (Figure 18.2.2e through Figure 18.2.2j) until all front faces are deleted. Deleted sides then represent the generated mesh.

The unstructured mesh generation by AFM described above may be controlled with node spacing more favorably maintained (node space control method). This method begins by constructing a coarse background grid of triangular elements which completely covers the domain of interest (Figure 18.2.3a). For the elements to be generated (Figure 18.2.3b), it is convenient to define a node spacing δ, the value of a stretching parameter s, and a direction of stretching α. Then the generated elements will have typical length $s\delta$ in the direction parallel to α and a typical length δ normal to α as shown in Figure 18.2.3b.

At each node on the background grid, nodal values of δ, s, α must be specified. During grid generation, local values will be obtained from interpolation of the nodal values on the background mesh. Note that if δ is required to be initially uniform and

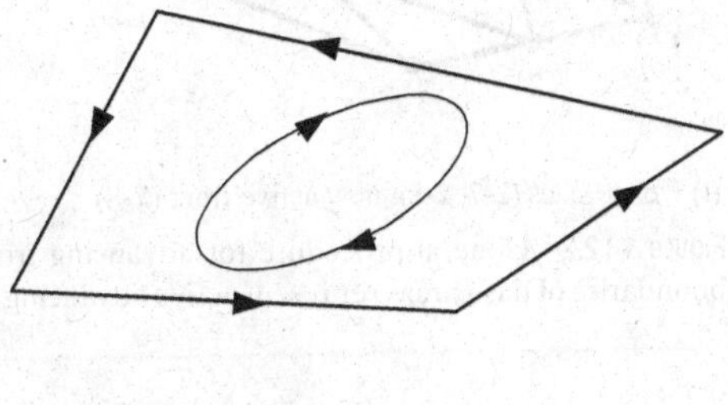

Figure 18.2.1 Multiply connected domain, counterclockwise advancing for outer boundaries, clockwise advancing for inner boundaries.

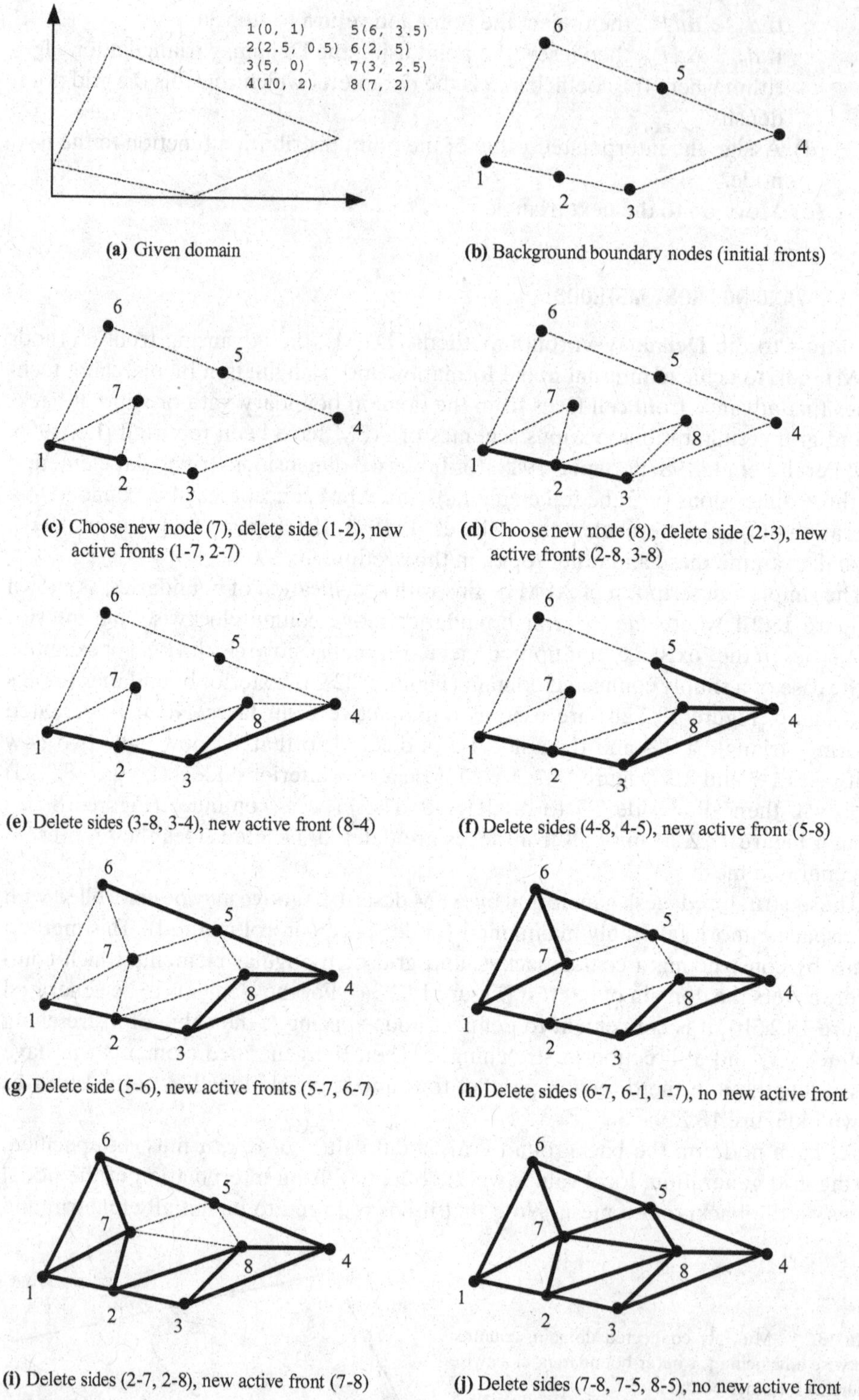

(a) Given domain

(b) Background boundary nodes (initial fronts)

(c) Choose new node (7), delete side (1-2), new active fronts (1-7, 2-7)

(d) Choose new node (8), delete side (2-3), new active fronts (2-8, 3-8)

(e) Delete sides (3-8, 3-4), new active front (8-4)

(f) Delete sides (4-8, 4-5), new active front (5-8)

(g) Delete side (5-6), new active fronts (5-7, 6-7)

(h) Delete sides (6-7, 6-1, 1-7), no new active front

(i) Delete sides (2-7, 2-8), new active front (7-8)

(j) Delete sides (7-8, 7-5, 8-5), no new active front

Figure 18.2.2 General procedure for advancing front methods (AFM) for simply connected domain, boundaries of dash areas represent active advancing fronts. Deleted lines constitute the generated mesh.

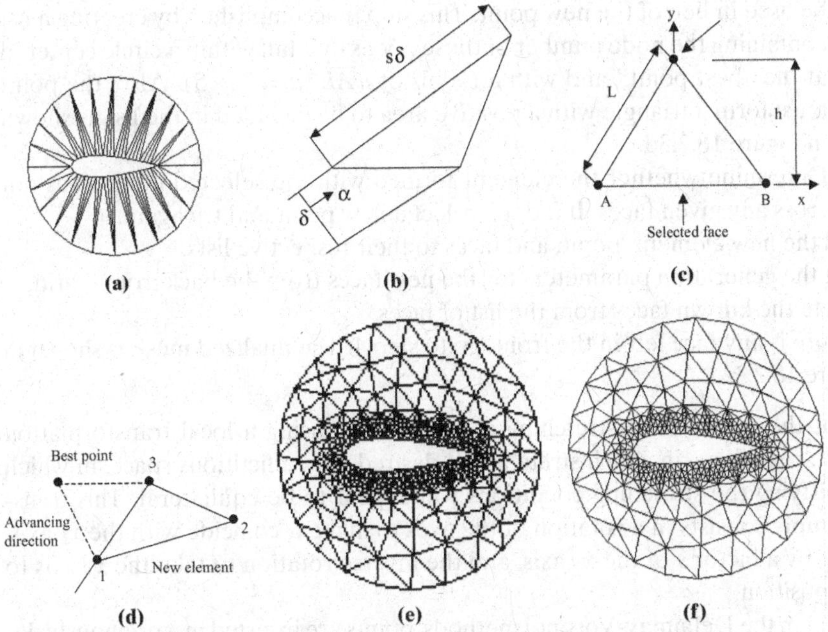

Figure 18.2.3 AFM procedure. (a) Background mesh. (b) Determination of mesh parameter. (c) Search for best point. (d) Undesirable element. (e) Finalized mesh. (f) Close-up view.

no stretching is to be specified, then the background grid need be only one triangle covering the entire domain.

Nodes are placed on the boundaries first, and the exterior boundary nodes are numbered counterclockwise, while any interior boundaries run clockwise. Thus, as the boundaries are traversed, the region to be triangulated always lies to the left.

At the start of the process, the front consists of the sequence of straight-line segments which connect consecutive boundary points. During the generation process, any straight-line segment that is available to form an element side is termed active, whereas any segment that is no longer active is removed from the front.

The following steps are involved in the process of generating new triangles in the mesh.

(1) Set up a background grid to define the spatial variation of the size, the stretching, and the stretching direction of the element to be generated (Figure 18.2.3b).

(2) Define the boundaries of the domain to be gridded, using the algebraic equations for each boundary.

(3) Using the information from Step 2, set up the initial front of faces. These faces are defined as segments between two consecutive points along the boundaries.

(4) Select the next face to be deleted from the front. In order to avoid large elements crossing over regions of small elements, the face forming the smallest new element is selected as the next face to be deleted from the list of faces.

(5) The following procedure is used for face deletion:

(a) The "best point" is calculated as shown in Figure 18.2.3c (equilateral).

(b) Determine whether a point exists in the already generated grid that should be used in lieu of the new point. This step is accomplished by creating a list containing the node number of those nodes that fall within a circle centered at the "best point" and with a radius of $n\overline{AB}$ ($n = 3 \sim 5$). Also, the point must form a triangle with a positive area to be included in the list as shown in Figure 18.2.3d.

(c) Determine whether the element formed with the selected point does not cross any given faces. If it does, select a new point and try again.

(6) Add the new element, point, and faces to their respective lists.
(7) Find the generation parameters for the new faces from the background grid.
(8) Delete the known faces from the list of faces.
(9) If there is any face left in the front, go to step 4. The finalized mesh is shown in Figure 18.2.3e,f.

Note that the inclusion of stretching is achieved by using a local transformation that maps the real plane, in which stretching is desired, into a fictitious space, in which triangles satisfying the stretching conditions will appear to be equilateral. This transformation simply consists of a rotation of the axes to make α coincide with the x_1 axis, and a scaling by a factor s of the x_1 axis, and the inverse rotation to take the x_1 axis to the original position.

Recall that in the Delaunay-Voronoi methods, points are inserted in a previously determined manner, and then the entire mesh is re-triangulated. In contrast, the advancing front methods determine where to put the points directly from the space control scheme.

Mesh Smoothing

Practical implementations of either advancing front or Delaunay-Voronoi grid generators indicate that in certain regions of the mesh, abrupt variations in element shape or size may be present. These variations appear even when trying to generate perfectly uniform grids. The best way to circumvent this problem is to improve the uniformity of the mesh by smoothing.

The so-called Laplacian smoother or the "spring-analogy" smoother may be used. In this method, the sides of the element are assumed to represent springs. These springs are then relaxed in time using explicit time stepping, until an equilibrium of spring forces has been established [Spradley, 1999].

In each subdomain, the standard Laplacian smoother is employed. Each side of the element can be visualized to represent a spring. Thus, the force acting on each point is given by

$$f_i = c\sum_{j=1}^{ns_i}(x_j - x_i)$$

where c denotes the spring constant, x_i the coordinates or the point, and the sum extends over all the points, ns_i, surrounding the point i. The spring constant is set in the computation software, based on tests of the method.

The time advancement for the coordinates is accomplished as follows:

$$\Delta x_i = \Delta t \frac{1}{ns_i} f_i$$

At the boundary of the subdomain, the points are allowed to "slide" along the boundaries, but not to "leave" the boundary.

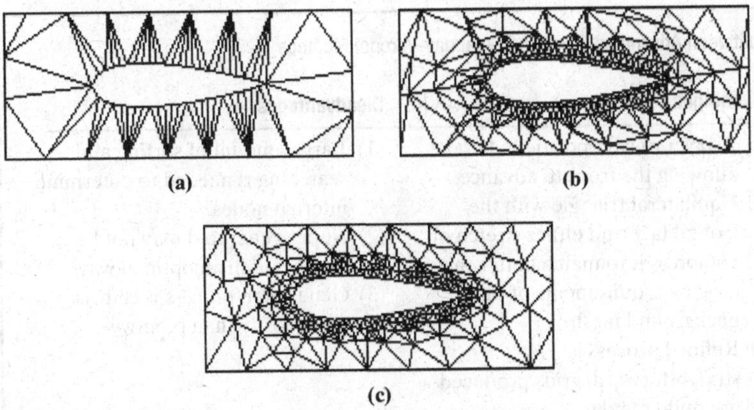

Figure 18.2.4 Mesh smoothing process, AFM. (a) Background mesh. (b) Finalized mesh without mesh smoothing. (c) After mesh smoothing.

The time step is also set in the code based on experience with using it. Usually, 5–10 time steps or passes over the mesh will smooth it sufficiently. The final results using the advancing front method without mesh smoothing and with mesh smoothing are shown in Figure 18.2.4. A sample program using C^{++} is listed in Figure 18.2.5.

```
//*****************************************************************
// Module Name: Mesh Smoothing, Advancing Front Method
//*****************************************************************
void Mesh_SmoothingMethod::meshSmoothing(int times)
                    // the parameter is the times of mesh smoothing, usually 10 is enough.
{
        int i, k;
        double deltaX, deltaY, deltaXY;
        int step[10]={10,9,8,7,6,5,4,3,2,1};

        numPoints=0;
        numTriangle=1;
        numEdge=0;

        readMeshFromFile(); // read triangle mesh from file
        formAllEdgeFromTriangleMesh(); // find all edges of triangle mesh
        findAllEdgeIndexForPoints(); // find point index for all edges

        for(k=0; k<times; k++)
        {
                calculateForceForPoints();  // calculate the force of points
                                    // f_i = c \sum_{j=1}^{ns_i} (x_j - x_i) , X and Y direction
                                    // \Delta x_i = \Delta t \frac{1}{ns_i} f_i
                for(i=0; i<numPoints; i++)
                {
                        if(pointSetData[i].type==SDC_INTERIOR)
                                // if the type of point is interior, then deform the position based
                                // on the force
                        {
                                deltaX=pointSetData[i].deltaX/step[k];
                                deltaY=pointSetData[i].deltaY/step[k];;
                                deltaXY=pointSetData[i].deltaXY/step[k];;
                                pointSetData[i].x+=deltaX;
                                pointSetData[i].y+=deltaY;
                        }
                }
        }
}
```

Figure 18.2.5 Mesh smoothing computer program using C^{++} [Z. Q. Hou, UAH].

Table 18.2.1 Comparison between Advancing Front and Delaunay-Voronoi Methods

	Advantages	Disadvantages
Advancing Front Methods	(1) A layer of well-positioned nodes allowing the front to advance. (2) Equilateral triangle with the frontal face and either stretch or compress it to match better the spacing requirements of the background mesh. (3) Refined process is straightforward, grids produced are quite regular. (4) High node distribution quality.	(1) Large amount of sorting and searching is needed to determine internal nodes. (2) Nodes generated may not be connected in an optimal way. (3) Generation process is cellwise, more costly than pointwise.
Delaunay-Voronoi Methods	(1) Each node is surrounded by its Voronoi region that compresses that part of the plane which is closer to this node than to any other node. (2) A unique triangulation is obtained by connecting the nodes whose Voronoi regions share a common boundary, forming a triangle with the three nodes that are closest to each other. (3) Generation process is pointwise, less costly than cellwise. (4) Optimal connectivity.	(1) Refined process is much more random; grids produced are not as regular. (2) Searching for the largest cell for a skewed cell with the largest circumcircle for each new node is very costly. (3) The skewness criterion is expensive (three square roots involved in the circle ratio).

The choice between DVM and AFM depends on the personal preference and the requirements for a given problem. This decision may be made upon the overall review of advantages and disadvantages presented in Table 18.2.1.

18.3 COMBINED DVM AND AFM

Having studied both DVM and AFM, it appears to be a logical approach to combine both methods in order to make use of advantages and discard disadvantages of both methods [Müller et al, 1993]. In this approach, we use a background grid to interpolate local mesh size parameters that is taken from the triangulation and create a set of nodes by means of AFM, and this set is subsequently introduced into the existing mesh, thus providing an updated DVM triangulation. The procedure is repeated until further improvements can not be obtained by inserting new nodes.

The results of a grid around a three element airfoil [Müller, Roer, and Deconinck, 1993] are shown in Figure 18.3.1. Here, the high node distribution quality of the AFM with the optimal connectivity of the DVM triangulation is demonstrated. Precise control of node spacing is achieved by the use of the initial triangulation of the boundary nodes as background mesh with no additional effort of the user. The node generation does not

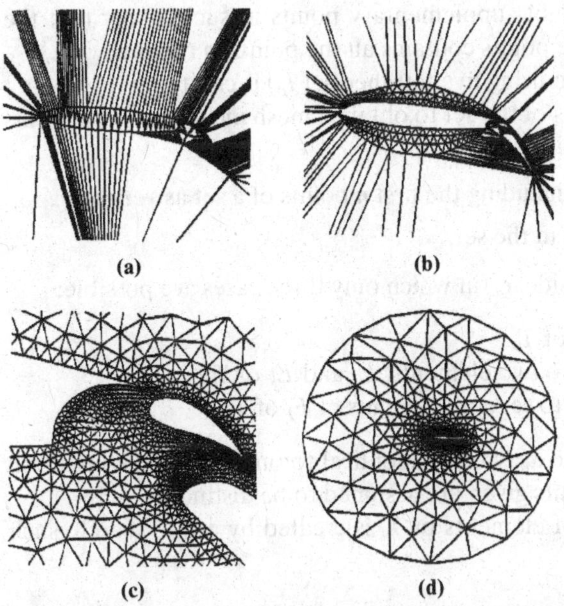

Figure 18.3.1 Three-element airfoil with combined DVM and AFM [Muller et al., 1993]. (a) Background grid. (b) After three rows of nodes inserted. (c) Details of grid around three-element airfoil. (d) Final grid of three-element airfoil.

require explicit tracking of the front and is independent of the order in which triangles are listed.

The resulting grids are very smooth and exhibit a high degree of regularity in cell shape and node distribution. This regularity is retained at singular points like corners or trailing edges. The use of a background grid that is derived from the initial triangulation of the boundary nodes results in a smooth variation in cell size of many orders of magnitude (about 10^5).

All features of this concept can be extended to three dimensions, where the optimal operation count and the simplicity of the front tracking and node construction of the method become even more attractive.

18.4 THREE-DIMENSIONAL APPLICATIONS

The basic concepts used in 2-D grid generations by means of DVM or AFM can be extended to 3-D geometries. Some of the earlier contributions include Baker [1989] for DVM and Löhner and Parikh [1988] for AFM. A brief review of these works is presented below.

18.4.1 DVM IN 3-D

Initially, the boundary surface grid generation using any one of the methods discussed in Section 18.3 is performed.

(1) The boundary points of the domain are created.

(2) Calculate the location of eight supplementary points in such a way that the hexahedron formed by these points contains all the points in the set.
(3) The mesh of this hexahedron using five tetrahedra (T_o) is created.
(4) Insert, one by one, the points of the set to obtain a mesh including these points as its element vertices. To this end, define:

T_n = a triangulation including the first n points of a set as vertices.

x_{n+1} = the next point in the set.

According to step 3, point x_{n+1} is inside T_n, in which only three cases are possible:

(a) x_{n+1} is inside an element E_i of T_n.
(b) x_{n+1} is on the face common to two elements, E_j and E_k of T_n.
(c) x_{n+1} is on the edge common to several elements of E_i of T_n.

The fourth possibility corresponds to x_{n+1} being identical to one of the existing mesh points; but this is rejected as the points given are assumed to be distinct.

Using element(s) E_i, the set S of elements of T_n is created by a tree search, such that

(i) x_{n+1} is identical to the circumsphere associated with the elements of S. Triangulation of T_{n+1} is constructed in the same way as in (a) above:
(ii) Include the elements of T_n , not included in S, in T_{n+1}.
(iii) Remove the elements of S and remesh this set by joining point x_{n+1} to the external faces of S.

Once all points of the initial set have been introduced, the initial hexahedron is constructed from these tetrahedra.

18.4.2 AFM IN 3-D

The AFM in 3-D geometries follows basically the same procedure as in AFM for 2-D except that triangles are replaced by tetrahedra:

(1) Set up a background grid to define the spatial variation of the size, the stretching, and the stretching direction of the elements to be generated. The background grid consists of tetrahedra. At the nodes, we define the desired element size, element stretching, and stretching direction.
(2) Define the boundary surfaces of the domain to be gridded.
(3) Using the information stored on the background grid, set up faces on all these boundaries. This yields the initial front of triangular faces. At the same time, find the generation parameters (element size, element stretching, and stretching direction) for these faces from the background grid.
(4) Select the next face to be deleted from the front. In order to avoid large elements crossing over regions of small elements, the face forming the smallest new element is selected as the next face to be deleted from the list of faces.
(5) For the face to be deleted:
    (a) Select a "best point" position for the introduction of a new point.

(b) Determine whether a point exists in the already generated grid that should be used in lieu of the new point. If there is such a point, set this point as a new point and continue searching.

(c) Determine whether the element formed with the selected point does not cross any given faces. If it does, select a new point and try again.

(6) Add the new element, point, and faces to their respective lists.
(7) Find the generation parameters of the new faces from the background grid.
(8) Delete the known faces from the list of faces.
(9) If there are any faces left in the front, go to step 4.

18.4.3 CURVED SURFACE GRID GENERATION

Two approaches for surface grid generation may be considered: (1) Boolean operations on solids and (2) Boolean operations on surfaces.

(1) Boolean operations on solids – the domain to be gridded from primitives (box, sphere, cylinder, etc.). The user combines these primitives through Boolean operations (union, intersection, exclusion, etc.) to define the domain to be gridded. The surface is then obtained in a post-processing operation.
(2) Boolean operations on surfaces – here only the surface of the domain to be gridded is defined in terms of independent surface patches. The surface patches are then combined using Boolean operations to yield the final surface of the domain to be gridded.

Boolean operations include points, lines, and surfaces which are obtained similarly as in the case of structured grids of Chapter 17.

18.4.4 EXAMPLE PROBLEMS

Examples of unstructured AFM three-dimensional mesh generation using tetrahedral elements are demonstrated in Figures 18.4.1 through 18.4.3.

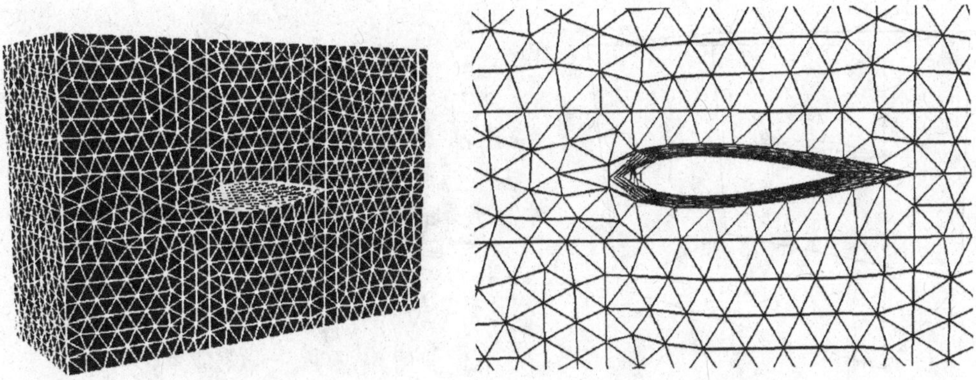

(a) (b)

Figure 18.4.1 Unstructured tetrahedral AFM mesh for NACA0012 airfoil [Spradley, 1999]. (a) Surface mesh. (b) Close-up view.

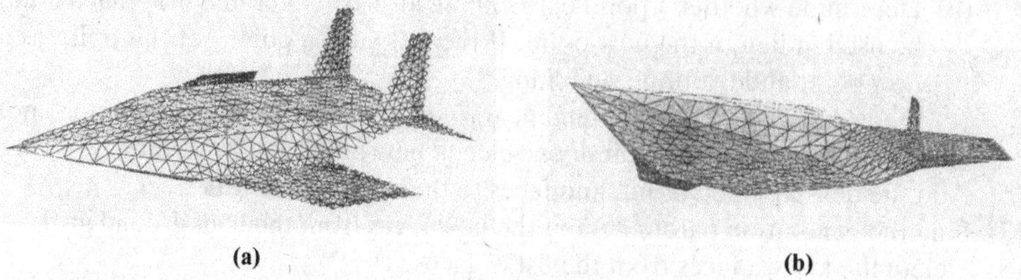

Figure 18.4.2 Unstructured AFM mesh for generic hypersonic plane [Spradley, 1999]. (a) Surface mesh. (b) Bottom view.

(1) NACA0012 Airfoil

Figure 18.4.1a shows the mesh generation of NACA airfoil geometry [Spradley, 1999]. A tetrahedral mesh is generated with the element sizes and clustering kept simple for illustrating. The surface mesh shown in Figure 18.4.1a is convenient to view the unstructured mesh. The interior looks very much the same as the surface. A close-up view near the airfoil is shown in Figure 18.4.1b.

(2) Hypersonic Airplane

Figure 18.4.2 shows two views of the surface mesh of a hypersonic airplane with a relatively coarse mesh for illustration only [Spradley, 1999]. The three-dimensional CFD computational domain with tetrahedral mesh is demonstrated in Figure 18.4.3 [Spradley, 1999].

18.5 OTHER APPROACHES

There are many other approaches in generating unstructured grids. Among them are Zhu et al. [1991] for quadrilateral grids by means of modified AFM, Blacker and Stephenson [1991] also for quadrilateral grids through paving technique, Yerry and Shephard [1984] for 2-D and 3-D mesh generations using quadtree and octree methods, respectively.

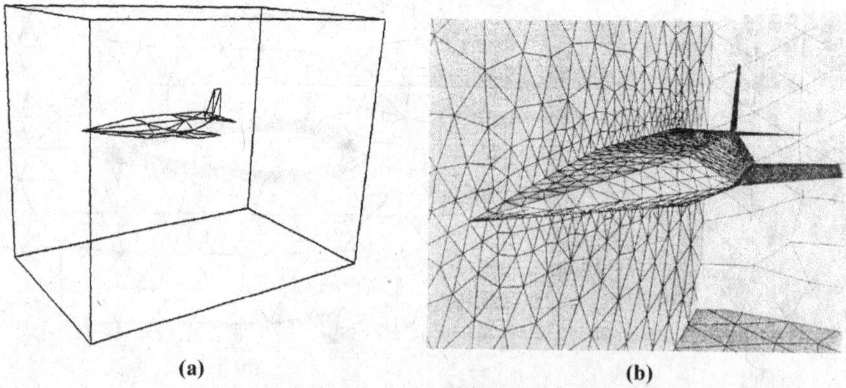

Figure 18.4.3 Unstructured AFM mesh for generic hypersonic plane with CFD domain [Spradley, 1999]. (a) CAD geometry for generic aircraft. (b) Unstructured AFM mesh.

In what follows, we shall briefly review these methods for the purpose of comparison. The reader should consult the original sources for details.

18.5.1 AFM MODIFIED FOR QUADRILATERALS

In this method, an even number of boundary nodes is first generated, according to the mesh size specification, on the boundary of the domain. The curvilinear boundary of the domain is then transformed into a polygon with an even number of sides. The main feature of the quadrilateral generation process is that first two triangles which have a common side are generated and then the triangles are combined to form a quadrilateral. This process continues until the domain is fully covered by nonoverlapping quadrilaterals.

The advantage of forming a quadrilateral by two triangles is that quadrilaterals with internal angles equal to and greater than 180° are allowed to be generated. These types of elements play an important role in the generation of transitions which are often necessary in a strongly graded mesh.

These steps can be readily achieved by any existing triangular mesh generator with some modifications; Zhu et al. [1991] utilized the AFM for this purpose. Note that the distinctive features of the AFM for triangular mesh generation are (1) that elements and nodes are generated simultaneously, and (2) that directional distribution of the elements can be achieved by introducing stretching in certain described directions so that the generated element size can be varied throughout the domain.

The following procedure is used for the quadrilateral element generation:

(1) Select an active side with nodes A and B in the generation front as a base to generate the first triangle (Figure 18.5.1a).
(2) Generate a triangle with nodes A, B, and C following the triangle generation process of AFM. The node D is determined according to the specified nodal spacing and it is either an active node in the current generation front or a newly constructed node in the region to be gridded.
(3) Set up the generation subfronts SF1 and SF2 (Figure 18.5.1b) by dividing the current generation front into two, if node D is an active node in the current generation front. Both SF1 and SF2 contain a closed loop of active sides and active nodes. No subfront is formed if D is a new node.

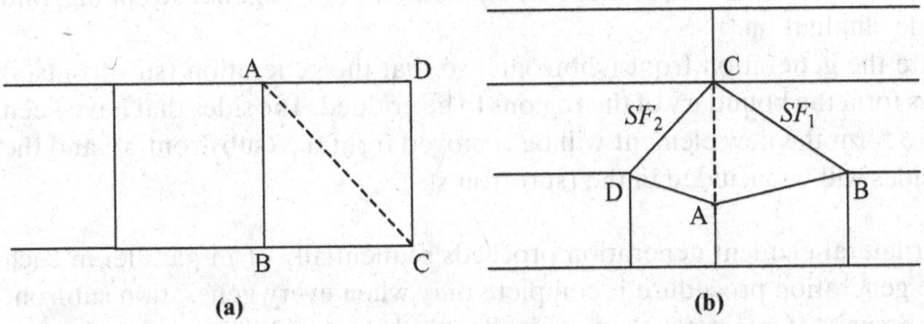

Figure 18.5.1 Generation of a quadrilateral element ABCD from AFM triangles. (a) No subfront is introduced after the generation of element ABCD. (b) Subfronts SF_1 and SF_2 are found.

(4) Examine the sides of triangle ABC. A side is flagged as active if it can be used as a side to form a new triangle. It is flagged as inactive otherwise.
(5) Select an active side which is also a side of triangle ABC as the base to generate the second triangle. Three possibilities may be encountered:
   (a) If no subfront has been formed in the generation of the first triangle ABC, in other words, node D is a new node being generated, then any side which is active in triangle ABC can be used to generate the second triangle.
   (b) If subfronts SF1 and SF2 are formed in the generation of triangle ABC, but one of the subfronts is empty, that is, all the sides in the subfront are flagged as inactive, then any active side in triangle ABC can again be chosen as the base for the second triangle generation.
   (c) Both subfronts SF1 and SF2 contain active sides. In this case one subfront, say SF1, contains an even number of active sides, the other subfront, say SF2, contains an odd number of active sides. The active side, which is in subfront SF2, in triangle ABC will be chosen as the base to generate the second triangle.
(6) Suppose the chosen active side for the second triangle has nodes A and C. Then the second triangle, say ACD, is generated by following the node and triangle generation process of AFM, with the requirement that each of the possible two new subfronts, formed as a result of the generation of the second triangle ACD, must contain an even number of active sides. Here steps (5) and (6) ensure that, after a quadrilateral is generated, every subfront formed in the element generation process contains a closed loop with an even number of active sides.
(7) The mesh parameters such as nodal spacing and element orientation for the new nodes are interpolated from the background mesh.
(8) Form quadrilateral element ABCD (Figure 18.5.1a,b). At this stage of the element generation, no consideration is given to the shape of the quadrilateral being generated. Indeed, there is no restriction on the shape of the quadrilaterals and any type of combination of two common side sharing triangles is allowed to form a quadrilateral. Such flexibility makes the generation of quadrilateral elements almost as easy as the generation of triangular elements. The enhancement of the element shape and therefore the enhancement of the quality of the mesh is an important part of this method. This can be achieved by means of (a) node elimination, (b) element elimination, (c) diagonal swapping, and (d) side elimination.
(9) Update the generation front (subfronts) so that the generation (sub)fronts(s) always form the boundary of the regions to be gridded. The sides that have been used to form the new element will be removed from the (sub)front(s), and the new sides will be included in the (sub)fronts(s).

The quadrilateral element generation proceeds sequentially or in parallel in each subfront. The generation procedure is complete only when every generation subfront is empty. The domain of interest is then covered entirely by quadrilateral elements.

In this method, no consideration has been given to the element orientation in the forming of the quadrilaterals. This is a subject of future research. Some typical results are shown in Figure 18.5.2 [Zhu et al., 1991].

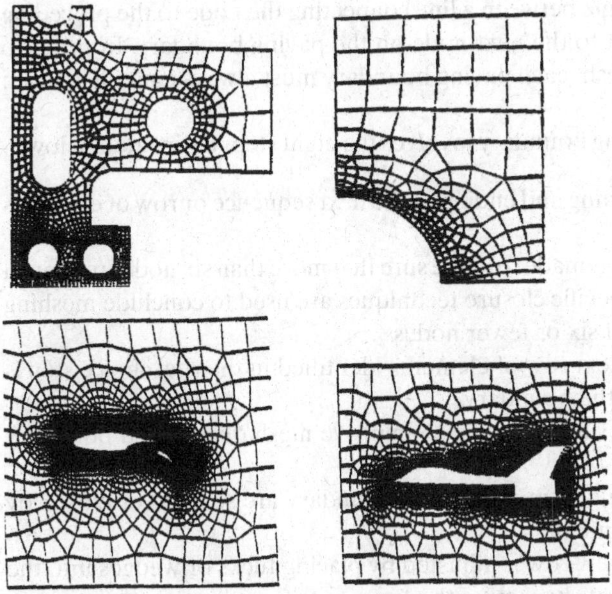

Figure 18.5.2 Examples of AFM modified for quadrilaterals [Zhu et al., 1991].

18.5.2 ITERATIVE PAVING METHOD

The paving technique meshes arbitrary 2-D geometries with quadrilaterals by iteratively layering or paving rows of elements to the interior of a region's boundary(ies) in a fashion similar to AFM. This technique was first proposed by Blacker and Stephenson [1991]. Paving allows varying element size distribution on the boundary as well as interior of a region. Similar to the AFM, the exterior boundary is ordered counterclockwise, and the interior boundary is ordered clockwise.

During mesh generation, the paving technique operates on boundaries of connected nodes referred to as paving boundaries. Initially, each paving boundary is identical to a permanent (fixed) boundary. As with permanent boundaries, paving boundaries are categorized as either exterior or interior boundaries, with exterior paving done counterclockwise and interior done clockwise.

The nodes are characterized into three types: (1) A fixed node is on a permanent boundary. (2) A floating node is any node not on a permanent boundary. (3) A paving node is any node on a paving boundary. Each paving node has an interior angle

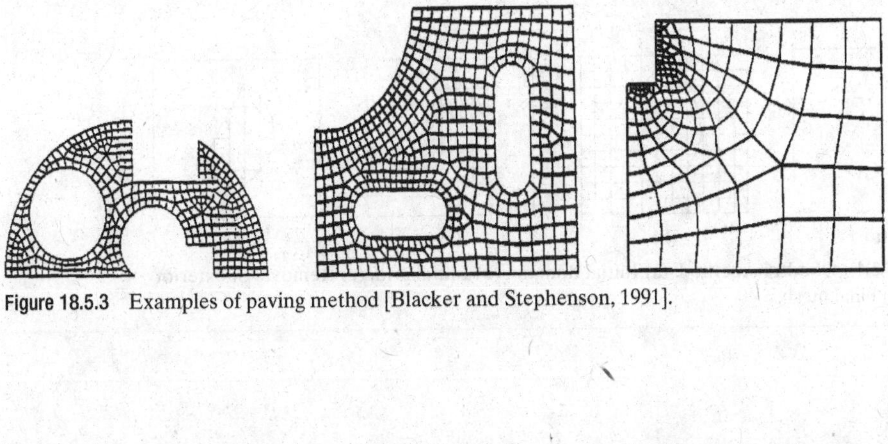

Figure 18.5.3 Examples of paving method [Blacker and Stephenson, 1991].

associated with it. This is the angle between a line connecting the node to the preceding node, and the line connecting it to the next node on the paving boundary. In order to generate an all quadrilateral mesh, each paving boundary must always contain an even number of nodes.

The propagation of the paving boundary involves the eight steps delineated below:

(1) Row choice – The beginning and ending of the next sequence or row of elements to be added is found.
(2) Closure check – A check is made to make sure that more than six nodes remain in the paving boundary. Specific closure techniques are used to conclude meshing for paving boundaries of six or fewer nodes.
(3) Row generation – The next row of elements identified in the row choice step is incrementally added to the boundary.
(4) Smooth – Floating nodes are adjusted to improve mesh quality and boundary smoothness.
(5) Seam – Small interior angles in the paving boundary are seamed or closed by connecting opposing elements.
(6) Row adjustment – The new row is adjusted by placing tucks or wedges into the row to correct for elements becoming too large or too small.
(7) Intersection – The paving boundary is checked for intersections with itself or with other paving boundaries. Intersections are connected to form new, often separate, paving boundaries.
(8) Cleanup – The completed mesh is adjusted where element deletion and/or addition improves the overall quality.

Some typical results using this technique are shown in Figure 18.5.3. Extension of the paving technique to 3-D geometries (hexahedral grid) can also be made.

18.5.3 QUADTREE AND OCTREE METHOD

Quadtree and octree methods were developed from the concept of superposition-deformation [Yerry and Shephard, 1984]. These methods construct a mesh of the domain under consideration, essentially from the data of points on its contour. A regular grid, or a grid based on a quadtree construction in three dimensions, is defined in such a way as to contain the domain. It is composed of squares and cubic boxes for quadtree and octree grids, respectively. This partitioning may then be deformed to resemble the real geometry. This process is shown in Figure 18.5.4 for a 2-D domain.

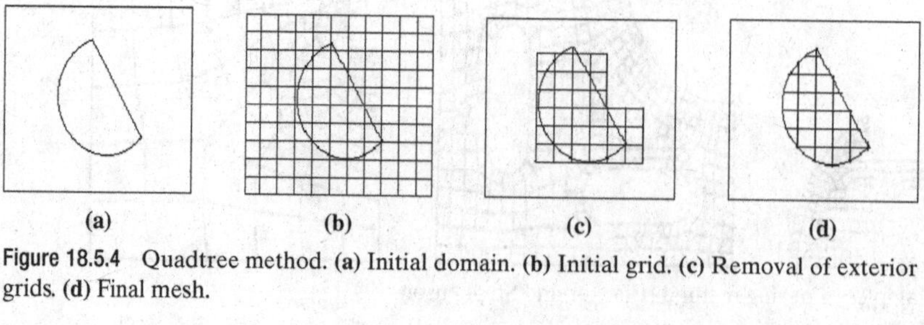

Figure 18.5.4 Quadtree method. (a) Initial domain. (b) Initial grid. (c) Removal of exterior grids. (d) Final mesh.

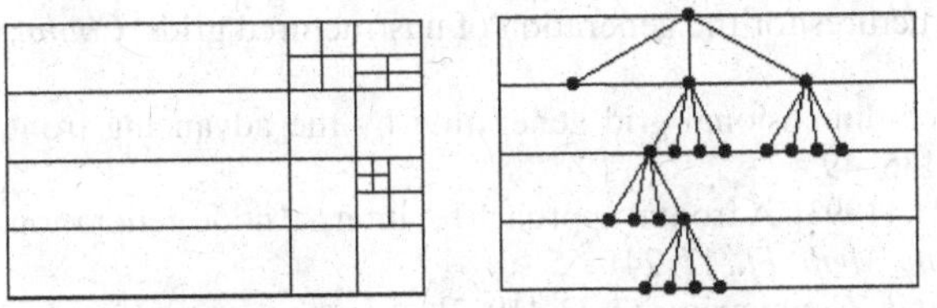

Figure 18.5.5 Quadtree data structure. (a) Recursive subdividing. (b) Parent-son-grandson relation.

Let us consider the quadtree shown in Figure 18.5.5a in which the mesh is refined in four levels. The parent is split into sons and grandsons as depicted in Figure 18.5.5b. For 3-D geometries, the method begins with a cube which is recursively split into sons and grandsons, that represents an octree structure. Thus, the final graded mesh can be constructed using this process.

18.6 SUMMARY

It was shown in this chapter that the two most popular methods of unstructured mesh generation are the Delaunay-Voronoi methods (DVM) and the advancing front methods (AFM). We examined the Watson and Bowyer algorithms for DVM, followed by AFM. Mesh smoothing in AFM was also discussed.

We presented the merits and demerits of DVM and AFM and showed that it is possible to combine both DVM and AFM taking advantage of the merits of both methods. We further examined three-dimensional applications for both DVM and AFM, followed by the surface grid generation. Other approaches such as AFM with quadrilaterals, iterative paving methods, and quadtree/octree methods were also examined. Unstructured grid generation is particularly useful in applications to adaptive methods. These and other subjects will be discussed in the next chapter.

REFERENCES

Baker, T. J. [1989]. Automatic mesh generation for complex three-dimensional regions using a constrained Delaunay triangulation. *Eng. Comm.*, 5, 161–75.

Blacker, T. D. [1991]. Analysis automation with paving: a new quadrilateral meshing technique. *Adv. Eng. Soft.*, 13: 5/6.

Blacker, T. D. and Stephenson, M. B. [1991]. Paving: a new approach to automated quadrilateral mesh generation. *Int. J. Num. Meth. Eng.*, 32, 811–47.

Bowyer, A. [1981]. Computing Dirichlet tessellations. *Comp. J.*, 24, 2, 162–66.

Cavendish, J. C., D. A., Field, and W. H. Frey [1985]. An approach to automatic three-dimensional finite element mesh generation. *Int. J. Num. Meth. Eng.*, 21, 329–47.

Delaunay, B. [1934]. Sur la sphere vide. Izvestiva Akademii Navk SSSR, *Math. Nat. Sci. Div.*, 6, 793–800.

Dirichlet, G. L. [1850]. Uber die Reduction der positiven Quadratschen Formen mit drei undestimmten ganzen Zahlen. *Z. Reine Angew. Math.*, 40, 3, 209–27.

Lo, S. H. [1985]. A new mesh generation scheme for arbitrary planar domains. *Int. J. Num. Meth. Eng.*, 21, 1403–26.

———. [1989]. Delaunay triangulation of non-convex planar domains. *Int. J. Num. Meth. Eng.*, 28, 2695–2707.

Löhner, R. [1988]. Some useful data structures for the generation of unstructured grids. *Comm. Appl. Num. Meth.*, 4, 123–35.

Löhner, R. and Parikh, P. [1988]. Three-dimensional grid generation by the advancing front method. *Int. J. Num. Meth. Eng.*, 8, 1135–49.

Müller, J. D., Roe, P. L., and Deconinck, H. [1993]. A frontal approach for internal node generation in Delaunay triangulations. *Int. J. Num. Meth. Fl.*, 17, 241–55.

Peraire, J., Vahdati, M., Morgan, K., and Zienkiewicz, O. C. [1987]. Adaptive remeshing for compressible flow computations. *J. Comp. Phys.*, 72, 2, 449–66.

Spradley, L. W. [1999]. A generalized meshing environment for adaptive computational fluid dynamics. Ph.D. dissertation, The University of Alabama, Huntsville.

Voronoi, G. [1908]. Nouvelles applications des parametres contilnus a la theorie des formes quadratiques. Rescherches sur les paralelloedres primitifs. *J. Reine angew, Math.* 134.

Watson, D. F. [1981]. Computing the N-dimensional Delaunay tesselation with application to Voronoi polytopes. *Comp. J.*, 24, 2, 167–72.

Weatherill, N. P. [1992]. Delaunay triangulation in computational fluid dynamics. *Comp. Math. Appl.*, 24, 5, 129–50.

Yerry, M. A. and Shephard, M. S. [1984]. Automatic 3D mesh generation by the modified-octree technique. *Int. J. Num. Meth. Eng.*, 20, 1965–90.

Zhu, J. Z., Zienkiewicz, O. C., Hinton, E., and Wu, J. [1991]. A new approach to the development of automatic quadrilateral mesh generation, *Int. J. Num. Meth. Eng.*, 32, 849–66.

CHAPTER NINETEEN

Adaptive Methods

The ultimate goal in computational fluid dynamics is to obtain desired solutions as accurately as possible while minimizing the requirement for computational resources. Thus, we ask: how do we achieve both "accuracy" and "efficiency" at the same time? Often we exercise a compromise where we may choose to sacrifice some accuracy for the sake of expediting a solution, or vice versa. Does an acceptable compromise exist? These are the types of questions that typically enter the minds of the CFD practitioner before undertaking a major project.

Given a fixed computational method and limited computer resources, one is confronted with a decision as to which direction to follow. The most feasible approach under these restricted circumstances will be to seek the best computational grid arrangement which will lend itself to the best possible accuracy and maximum efficiency. Adaptive methods are designed to achieve both accuracy and efficiency, with mesh refinements provided selectively only where needed.

The basic concept for adaptive methods consists of providing mesh refinements for efficiency (cost reduction) as dictated by predetermined criteria. The criteria for mesh refinements and unrefinements (coarsening) are determined by error indicators. The error indicators are usually represented by gradients of a suitable variable – the larger the gradient, the finer the mesh required.

In line with the two different grid generation schemes, structured and unstructured, two different adaptive methods are available, one for structured grids and another for unstructured grids. The structured adaptive methods are presented in Section 19.1, with the unstructured adaptive methods in Section 19.2.

19.1 STRUCTURED ADAPTIVE METHODS

Structured adaptive meshes may be constructed either by a control function approach or by a variational function approach. We shall discuss both of these methods next.

19.1.1 CONTROL FUNCTION METHODS

19.1.1.1 Basic Theory

In this method, grid points are moved in accordance with weights or control functions reflecting the gradients of the variables, the process known as redistribution. Adaptive redistribution of the points is based on the principle of equal distribution of error by which a point distribution is set so as to make the product of the spacing and a weight function W constant over the points. This idea is represented by

$$Wdx = \text{constant} \tag{19.1.1}$$

With the point distribution defined by a function $x(\xi)$, where ξ varies by a unit increment between points, the equal distribution principle can be expressed as

$$Wx_\xi d\xi = Wx_\xi = \text{constant}, \quad d\xi = 1 \tag{19.1.2}$$

This one-dimensional equation can be applied in each direction in an alternating fashion (in the spirit of ADI). However, a direct extension to multiple dimensions can be made in either of two ways: control function approach, or variational approach. In the control function methods, we combine the elliptic grid generation system with the equal distribution principle given by (19.1.2).

$$g^{ij}\mathbf{r}_{,ij} + P^i\mathbf{r}_{,i} = 0 \tag{19.1.3}$$

where g^{ij} are the elements of the contravariant metric tensor [Chung, 1996]:

$$g^{ij} = \frac{\partial \xi_i}{\partial x_m}\frac{\partial \xi_j}{\partial x_m} \tag{19.1.4}$$

These elements are more conveniently expressed computationally in terms of the elements of the covariant metric tensor g_{ij}:

$$g_{ij} = \frac{\partial x_m}{\partial \xi_i}\frac{\partial x_m}{\partial \xi_j} \tag{19.1.5}$$

which can be calculated directly. Thus

$$g^{ij} = \frac{1}{g}\frac{\partial g}{\partial g_{ij}} \tag{19.1.6}$$

where

$$g = |g_{ij}| \tag{19.1.7a}$$

or

$$g = \left|\frac{\partial x_m}{\partial \xi_i}\frac{\partial x_m}{\partial \xi_j}\right| \tag{19.1.7b}$$

with $\mathbf{r}$ being the cartesian coordinates and ξ_i the curvilinear coordinates. The $P^i = P_i$ denotes the control function which controls the spacing and orientation of the grid lines in the field.

The one-dimensional form of this system is

$$\frac{\partial^2 x}{\partial \xi^2} + P\frac{\partial x}{\partial \xi} = 0 \tag{19.1.8}$$

Differentiation of (19.1.2) yields

$$W\frac{\partial^2 x}{\partial \xi^2} + \frac{\partial W}{\partial \xi}\frac{\partial x}{\partial \xi} = 0 \tag{19.1.9}$$

It follows from (19.1.8) and (19.1.9) that

$$-P = \frac{\partial^2 x/\partial \xi^2}{\partial x/\partial \xi} = -\frac{\partial W/\partial \xi}{W} \tag{19.1.10}$$

from which the control function P can be taken as

$$P = \frac{1}{W}\frac{\partial W}{\partial \xi} \tag{19.1.11}$$

This may be extended to three-dimensional geometries as

$$P_i = \frac{1}{W}\frac{\partial W}{\partial \xi_i} \quad (i = 1, 2, 3) \tag{19.1.12a}$$

or

$$P_i = \frac{1}{W_{(i)}}\frac{g^{ij}}{g^{(ii)}}\frac{\partial W_{(i)}}{\partial \xi_j} \tag{19.1.12b}$$

where the latter version (19.1.12b) requires the weight functions to be specified in all three directions [Eiseman, 1987].

19.1.1.2 Weight Functions in One Dimension

As seen in (19.1.12), the effect of the weight function W is to reduce the point spacing function x_ξ if W is large. Therefore, the weight function should be set as some measure of the solution error or the solution variation. The simplest choice in one-dimensional problems is the solution gradient, that is,

$$W = u_x \tag{19.1.13}$$

Substituting (19.1.13) into (19.1.2) yields

$$u_x x_\xi = constant$$

or

$$u_\xi = constant$$

With the solution gradient used as a weight function, the point distribution can be adjusted in such a manner that the same change in the solution occurs over each grid, as illustrated in Figure 19.1.1a. This choice for the weight function has the disadvantage of making the spacing infinitely large when the solution is constant.

In contrast, consider the solution gradient in the form

$$W = \sqrt{1 + u_x^2} \tag{19.1.14}$$

An increment of arc length, ds, on the solution curve $u(x)$ is given by

$$ds^2 = dx^2 + du^2 = \left(1 + u_x^2\right) dx^2$$

so that this form of the weight function may be written as

$$W = s_x$$

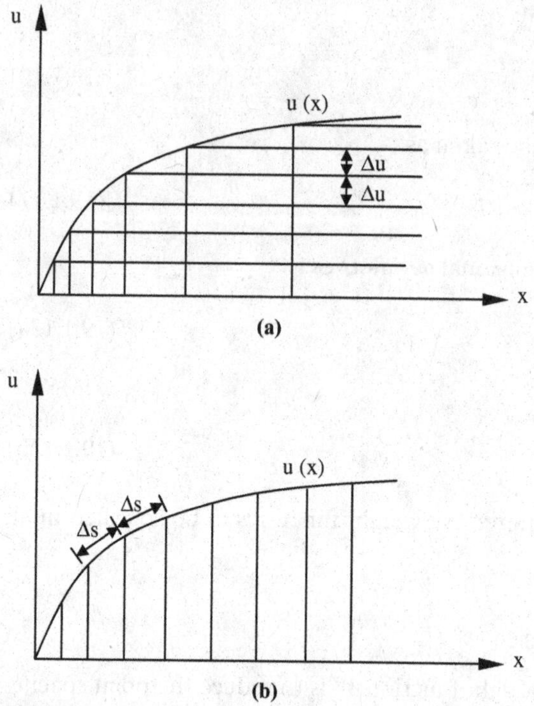

Figure 19.1.1 Relation between grid spacing and weight functions. (a) Constant solution gradient with $w = u_x$. (b) Constant solution gradient with $w = (1 + u_x^2)^{\frac{1}{2}}$.

which gives

$$s_x x_\xi = constant \tag{19.1.15a}$$

or

$$s_\xi = constant \tag{19.1.15b}$$

Thus, with the weight function defined by (19.1.14), the grid point distribution is such that the same increment in arc length on the solution curve occurs over each grid interval (Figure 19.1.1b).

Unlike the previous choice, this weight function gives uniform spacing when the solution is constant. The concentration of points in the high-gradient region, however, is not as great.

In order to maintain desirable concentration of nodes at high gradient regions and peak solutions, the following weight function has been suggested [Eiseman, 1985]:

$$W = (1 + \alpha^2 u_x^2)^{\frac{1}{2}}(1 + \beta^2 |k|) \tag{19.1.16}$$

where α and β are user specified parameters and k is the curvature defined as

$$k = \frac{u_{xx}}{(1 + u_x^2)^{\frac{3}{2}}} \tag{19.1.17}$$

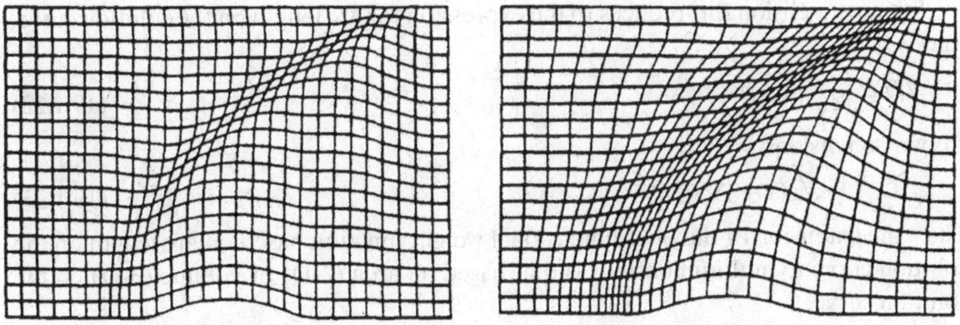

(a) Low α and β values (b) High α and β values

Figure 19.1.2 Effect of inclusion of curvature as well as gradient [Eiseman, 1985].

Here, large values of α and β contribute to closer nodal spacing, respectively, near the high solution gradient regions and solution extrema regions.

An alternative to (19.1.16) is to use

$$W = 1 + \alpha\,|u_x| + \beta\,|u_{xx}| \tag{19.1.18}$$

where α and β are non-negative parameters to be specified. An example based on (19.1.16) is shown in Figure 19.1.2a for low values of α and β and in Figure 19.1.2b for high values of α and β.

19.1.1.3 Weight Function in Multidimensions

The one-dimensional weight factors (19.1.13) based on the arc length on the solution curve can be generated to higher dimensions. Consider the position vector characterized by both the geometrical space **r** and solution space **u** such that

$$\mathbf{R} = \mathbf{r} + u\mathbf{e} = x_i\,\mathbf{i}_i + u\mathbf{e} \tag{19.1.19}$$

The covariant metric tensor is then given by

$$G_{ij} = \frac{\partial \mathbf{R}}{\partial \xi_i}\cdot\frac{\partial \mathbf{R}}{\partial \xi_j} = (\mathbf{r}_{\xi_i} + u_{\xi_i}\mathbf{e})\cdot(\mathbf{r}_{\xi_j} + u_{\xi_j}\mathbf{e})$$

or

$$G_{ij} = g_{ij} + u_{\xi_i}u_{\xi_j} \tag{19.1.20a}$$

where g_{ij} is the metric tensor in the physical space. Since

$$u_{\xi_i} = \nabla u\cdot\mathbf{r}_{\xi_i}$$

we obtain

$$G_{ij} = g_{ij} + \left(\nabla u\cdot\mathbf{r}_{\xi_i}\right)\left(\nabla u\cdot\mathbf{r}_{\xi_j}\right) \tag{19.1.20b}$$

and

$$|G_{ij}| = (1 + |\nabla u|^2)|g_{ij}| \tag{19.1.21}$$

In one dimension this reduces to the expression for arc length on the solution curve, that is,

$$\sqrt{G} = x_\xi \sqrt{1 + u_x^2} \tag{19.1.22}$$

In two dimensions, we have

$$\sqrt{|G_{ij}|} = \sqrt{g}(1 + |\nabla u|^2)^{\frac{1}{2}} \tag{19.1.23}$$

Thus, the extension of the one-dimensional weight function based on arc length on the solution curve to multidimensions is that based on area (2-D) or volume (3-D) on the solution curve,

$$W = (1 + |\nabla u|^2)^{\frac{1}{2}} \tag{19.1.24}$$

The weight functions as defined above can then be applied to the expressions for the control function given in (19.1.12a,b).

In multiple dimensions, adaptation should, in general, occur in all directions in a mutually dependent manner. If the solution varies only in one direction (say x) predominantly, then the adaptation may be carried out in that direction alone, using the one-dimensional weight function, with x replaced by the arc length along this line.

Examples of applications of the above schemes include Dwyer, Smook, and Kee [1982], Gnoffo [1980], and Nakamura [1982], among others.

19.1.2 VARIATIONAL METHODS

The classical theory of calculus of variations can be applied to problems requiring optimization or achieving the maximum degree of equal distribution of error [Brackbill and Saltzman, 1982]. With this in mind, we will examine the basic theory associated with equal distribution by means of variational methods.

19.1.2.1 Variational Formulation

The computational error can be reduced by distributing the grid points in such a way that the same positive weight function, $W(x)$, is equally distributed over the field as shown in Section 19.1.1. The nonuniform point distribution can be considered to be a transformation, $x(\xi)$, from a uniform grid in ξ-space, with the coordinate ξ serving to identify the grid points.

Let us now invoke a spring analogy so that, if the weight function $W(\xi)$ is a spring constant and x_ξ is the extension of the spring at ξ, then the energy stored in the spring is of the form

$$I = \frac{1}{2}\int_0^1 W(\xi)\, x_\xi^2 \, d\xi \tag{19.1.25}$$

It follows from the theory of calculus of variations that the integrand in (19.1.25) constitutes the "variational functional," F,

$$F(\xi, x, x_\xi) = \frac{1}{2} W(\xi)\, x_\xi^2 \tag{19.1.26}$$

and the energy stored in the spring, I, is known as the "variational principle."

There are two ways to obtain an optimum grid spacing:

(1) Minimization of the variational principle given by (19.1.25).
(2) Solving the differential equation(s) resulting from the so-called Euler-Lagrange equation,

$$\frac{\partial}{\partial \xi}\left[\frac{\partial F}{\partial(\partial x/\partial \xi)}\right] - \frac{\partial F}{\partial x} = 0 \tag{19.1.27}$$

Substituting (19.1.26) into (19.1.27) yields

$$\frac{\partial W}{\partial \xi}\frac{\partial x}{\partial \xi} + W\frac{\partial^2 x}{\partial \xi^2} = 0 \tag{19.1.28}$$

It is interesting to note that (19.1.28) is identical to (19.1.9), which originated from the general elliptic PDE representation (19.1.3).

The above process confirms the standard variational approach in which, given the differential equation [in this case the Poisson equation (19.1.28)], the corresponding variational functional (19.1.26) when substituted into the corresponding Euler-Lagrange equation (19.1.27) recovers the original differential equation (19.1.28).

This argument implies that, instead of using the PDE of the form (19.1.3), the variational approach suggests that, if there are means of obtaining many different forms of the variational functional, there will be a host of differential equations arising from this process, other than those of the standard form such as (19.1.3). The differential equations obtained in this manner are characterized by a variety of weight functions and subsequently the control functions leading to desired forms of adaptive procedures.

19.1.2.2 Smoothness Orthogonality and Concentration

To achieve adaptation with a maximum degree of smoothness, orthogonality, and desired concentration, our focus is to construct desirable forms of variational functionals [Brackbill and Saltzmann, 1982]. Toward this end, the following forms of variational principles are suggested:

(1) Smoothness

$$I_s = \int g^{ii}\, d\mathbf{x} \tag{19.1.29a}$$

$$I_s = \int \frac{1}{\sqrt{g}}(g_{ii}g_{jj} - g_{ij}g_{ij})\, d\xi \tag{19.1.29b}$$

(1) Orthogonality

$$I_o = \int g^{\frac{3}{2}} g^{ij} g^{ij}\, d\mathbf{x} \tag{19.1.30a}$$

$$I_o = \int (g_{ij}g_{ik} - g_{ii}g_{jk})(g_{mj}g_{mk} - g_{mn}g_{jk})\, d\xi \tag{19.1.30b}$$

(1) Concentration

$$I_W = \int W^2(\mathbf{x})\sqrt{g}d\mathbf{x} \tag{19.1.31a}$$

$$I_W = \int W^2(x)\, g\, d\xi \tag{19.1.31b}$$

The above formulation may be generalized using the directional control as follows [Brackbill, 1982]:

$$I(\xi) = \int_{\mathbf{x}} F(\xi)\, d\mathbf{x} \tag{19.1.32}$$

with

$$F(\xi) = \frac{1}{2} g^{ij}(\mathbf{x}) g_{\alpha\beta}(\xi) \frac{\partial \xi_\alpha}{\partial x_i} \frac{\partial \xi_\beta}{\partial x_j} \tag{19.1.33}$$

The Euler-Lagrange equation in general curvilinear coordinates takes the form

$$\left[g^{ij} \frac{\partial F}{\partial(\partial \xi_\alpha / \partial x_j)} \right]_{|i} - (g^{\alpha\lambda} F)_{|\alpha} = 0 \tag{19.1.34}$$

where the stroke " | " denotes the covariant derivative. Performing the covariant differentiation on (19.1.34) leads to

$$\frac{1}{\sqrt{g}} \frac{\partial}{\partial x_i} \left(g^{ij} \frac{\partial \xi_\lambda}{\partial x_j} \sqrt{g} \right) + g^{ij} \Gamma^\lambda_{\alpha\beta} \frac{\partial \xi_\lambda}{\partial x_i} \frac{\partial \xi_\beta}{\partial x_j} = 0 \tag{19.1.35}$$

where

$$\Gamma^\lambda_{\alpha\beta} = \frac{1}{2} g^{\lambda\gamma} [g_{\gamma\alpha,\beta} + g_{\gamma\beta,\alpha} - g_{\alpha\beta,\gamma}]$$

It can be shown that the second term in (19.1.35) vanishes. Taking a variational derivative of (19.1.35) gives

$$\delta I = \int_{\mathbf{x}} \left[\frac{1}{\sqrt{g}} \frac{\partial}{\partial x_i} \left(g^{ij} \frac{\partial \xi_\lambda}{\partial x_j} \sqrt{g} \right) \right] \delta\xi\, d\mathbf{x} = 0 \tag{19.1.36}$$

Integrating (19.1.36) by parts,

$$\delta I = \int_\Gamma g^{ij} \frac{\partial \xi_\lambda}{\partial x_j} n_i \delta\xi_\lambda\, d\Gamma - \int_{\mathbf{x}} g^{ij} \frac{\partial \xi_\lambda}{\partial x_j} \delta\xi_{\lambda,i}\, d\mathbf{x} = 0$$

or

$$\delta I = \delta \left(\int_{\mathbf{x}} \frac{1}{2} g^{ij} \frac{\partial \xi_\lambda}{\partial x_j} \frac{\partial \xi_\lambda}{\partial x_i}\, d\mathbf{x} - \int_\Gamma g^{ij} \frac{\partial \xi_\lambda}{\partial x_j} n_i \delta\xi_\lambda\, d\Gamma \right) = 0 \tag{19.1.37}$$

This provides the variational principle to be given in the form

$$I = \int_{\mathbf{x}} \frac{1}{2} g^{ij} \frac{\partial \xi_\lambda}{\partial x_j} \frac{\partial \xi_\lambda}{\partial x_i}\, d\mathbf{x} \tag{19.1.38}$$

with the Neumann boundary condition

$$S_\lambda = \frac{\partial \xi_\lambda}{\partial x_i} n_i \tag{19.1.39}$$

Thus, it follows that

$$I = \int_{\mathbf{x}} \frac{1}{W} \nabla \xi_i \cdot \nabla \xi_i \, d\mathbf{x}$$

$$\nabla \cdot \left(\frac{1}{W} \nabla \xi_i \right) = 0$$

$$\nabla^2 \xi_i = \frac{1}{W} \frac{\partial W}{\partial \xi_j} \frac{\partial \xi_j}{\partial x_k} \frac{\partial \xi_i}{\partial x_k} = \frac{1}{W} \frac{\partial W}{\partial \xi_j} g^{ij}$$

$$\nabla^2 \xi_k = -g^{ij} \mathbf{r}_{,ij} \cdot \nabla \xi_k = \frac{1}{W} g^{ik} \frac{\partial W}{\partial \xi_i}$$

$$\nabla \xi^i \cdot \mathbf{r}_{,j} = \delta_{ij}$$

and

$$g^{ij} \frac{\partial}{\partial \xi_i} \left(W \frac{\partial \mathbf{r}}{\partial \xi_j} \right) \cdot \nabla \xi_k = 0$$

Finally,

$$g^{ij} \frac{\partial}{\partial \xi_i} \left(W \frac{\partial \mathbf{r}}{\partial \xi_j} \right) = 0 \tag{19.1.40}$$

An adaptive grid with directional control can be constructed and the mesh alignment control variational principle takes the form

$$I_d = \int_{\mathbf{x}} \frac{1}{W} [(\mathbf{A} \times \nabla \xi_1)^2 + (\mathbf{B} \times \nabla \xi_2)^2 + (\mathbf{C} \times \nabla \xi_3)^2] \, d\mathbf{x} \tag{19.1.41}$$

where

$$(\mathbf{A} \times \nabla \xi_1)^2 = (\mathbf{A} \times \nabla \xi_1) \cdot (\mathbf{A} \times \nabla \xi_1), \quad \text{etc.}$$

Let

$$W = \frac{|\nabla U|}{|U|} \tag{19.1.42}$$

where U is the variable under consideration and choose $\mathbf{A}, \mathbf{B}, \mathbf{C}$ to be mutually orthogonal. This implies that

$$\mathbf{B} \times \mathbf{C} \parallel \frac{\partial \mathbf{r}}{\partial \xi_1}, \quad \mathbf{C} \times \mathbf{A} \parallel \frac{\partial \mathbf{r}}{\partial \xi_2}, \quad \mathbf{A} \times \mathbf{B} \parallel \frac{\partial \mathbf{r}}{\partial \xi_3}$$

Using the vector identity

$$\sqrt{g} g^{ij} = \frac{1}{W} (V_k V_k \delta^{ij} - V^i V^j)$$

we obtain the variational functional in the form

$$F(\xi) = \sqrt{g_{(A)}} g^{ij}_{(A)} \frac{\partial \xi_1}{\partial x_i} \frac{\partial \xi_1}{\partial x_j} + \sqrt{g_{(B)}} g^{ij}_{(B)} \frac{\partial \xi_2}{\partial x_i} \frac{\partial \xi_2}{\partial x_j} + \sqrt{g_{(C)}} g^{ij}_{(C)} \frac{\partial \xi_3}{\partial x_i} \frac{\partial \xi_3}{\partial x_j} \tag{19.1.43}$$

Substituting (19.1.43) into the Euler-Lagrange equation yields

$$\left\{ g^{ij}_{(A)} \left[\frac{\partial}{\partial \xi_i} \left(W \frac{\partial \mathbf{r}}{\partial \xi_j} \right) - W[\nabla(\mathbf{A} \cdot \mathbf{A}) - (\nabla \cdot \mathbf{A})\mathbf{A}] \right] \right\} \cdot \nabla \xi_1 = 0 \tag{19.1.44a}$$

$$\left\{ g^{ij}_{(B)} \left[\frac{\partial}{\partial \xi_i} \left(W \frac{\partial \mathbf{r}}{\partial \xi_j} \right) - W[\nabla(\mathbf{B} \cdot \mathbf{B}) - (\nabla \cdot \mathbf{B})\mathbf{B}] \right] \right\} \cdot \nabla \xi_2 = 0 \tag{19.1.44b}$$

$$\left\{ g^{ij}_{(C)} \left[\frac{\partial}{\partial \xi_i} \left(W \frac{\partial \mathbf{r}}{\partial \xi_j} \right) - W[\nabla(\mathbf{C} \cdot \mathbf{C}) - (\nabla \cdot \mathbf{C})\mathbf{C}] \right] \right\} \cdot \nabla \xi_3 = 0 \tag{19.1.44c}$$

with

$$g^{ij}_{(A)} = \mathbf{A} \cdot \mathbf{A} g^{ij} - (\mathbf{A} \cdot \nabla \xi_i)(\mathbf{A} \cdot \nabla \xi_j), \quad \text{etc.}$$

Here, if $\nabla \xi_1 \parallel \mathbf{A}$, the $\mathbf{g}(A) = 0$ and the variational functional $\mathbf{F}$ has no extremum with respect to ξ_1. Thus (19.1.44a) has no solution.

The above difficulty may be resolved by introducing "regularization." To this end, we divide each reference vector by its maximum magnitude and regulate the variational principle in the form

$$I = (1 - \lambda) I_s + \lambda I_d \tag{19.1.45}$$

with $0 \leq \lambda \leq 1$, λ being the user specified constant. It is seen that an increase of λ leads to an increase of the alignment of the reference vectors.

The solution of (19.1.44) has been carried out for applications to the Kelvin-Helmholtz instability problem in a driven magnetic flowfield (Figure 19.1.3) [Brackbill, 1982]. The complexity of formulation of the governing equations is considered a major difficulty if one chooses to use the variational functional of the form given in (19.1.43).

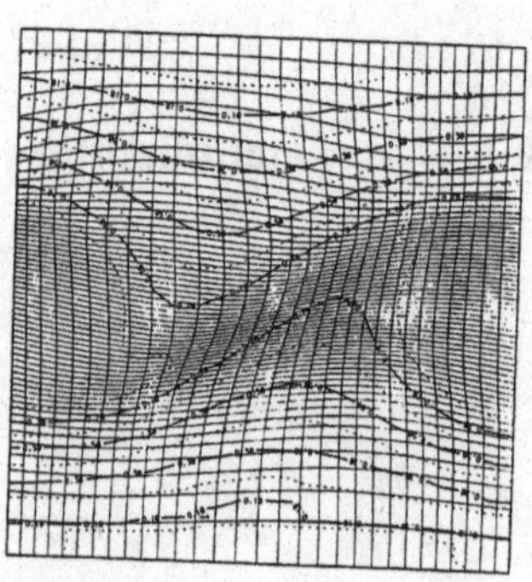

Figure 19.1.3 Magnetic flowfields based on direction control adaptive computational mesh [Brackbill, 1982].

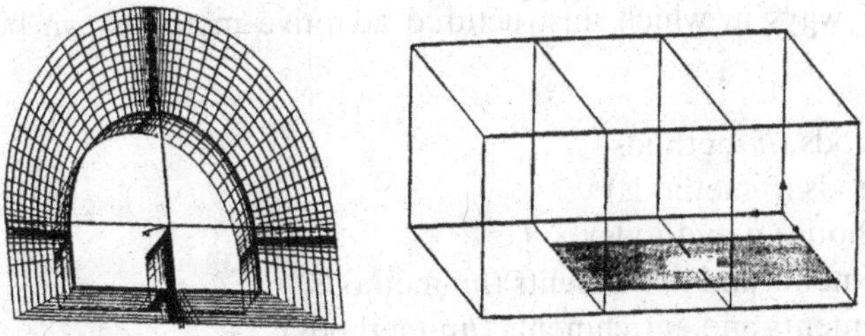

Figure 19.1.4 ONENA M6 wing multiblock [Kim and Thompson, 1990].

19.1.3 MULTIBLOCK ADAPTIVE STRUCTURED GRID GENERATION

An adaptive structured grid mesh can be constructed in a multiblock environment, such as an aircraft wing shown in Figure 19.1.4 [Kim and Thompson, 1990]. Any block can be linked to any other block or to itself, with complete or lesser continuity across the block interfaces as specified by input. The adaptive control function with either an elliptic or a variational method is then formed at each point in the field by combining the interpolated components. For complex multiblock systems, the evaluation from the algebraic grid can be used.

In Figure 19.1.4, the three-initial grid is generated by the elliptic system with the $97 \times 17 \times 17$ grid divided into three blocks in the spanwise direction. The final grid adapted to the pressure gradients at the 50% span location and over the upper wing surface is shown in Figure 19.1.5a. The leading edge and upper surface details are shown in Figures 19.1.5b and 19.1.5c, respectively.

19.2 UNSTRUCTURED ADAPTIVE METHODS

For nearly two decades, unstructured adaptive methods have been extensively developed by Oden and his co-workers and Babuska and his co-workers. Some of their later works are summarized in Babuska et al. [1986] and Oden et al. [1986].

In the previous section, we presented structured adaptive methods in which an initially structured grid remains structured even after the adaptive process. In unstructured adaptive methods, however, we may begin with either a structured or unstructured grid system, but the final grid system, upon adaptation, becomes subsequently unstructured or remains unstructured, respectively.

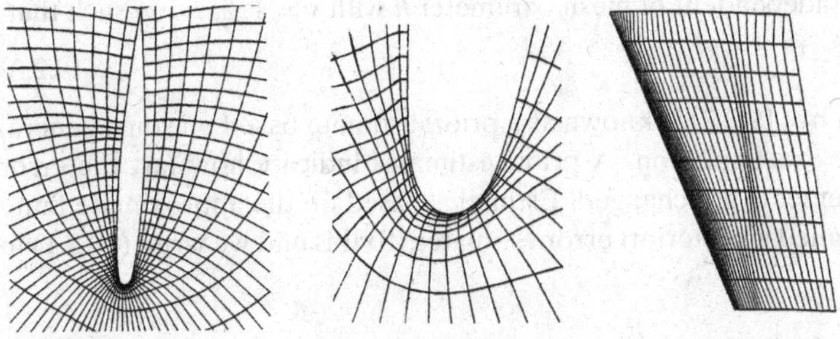

(a) 50% span (b) Leading-edge detail (c) Upper surface

Figure 19.1.5 Final adaptive grid for Figure 23.3.1.

There are many different ways in which unstructured adaptive methods can be developed:

(1) mesh refinement methods (h-methods)
(2) mesh movement methods (r-methods)
(3) mesh enrichment methods (p-methods)
(4) combined mesh refinements and movements (hr-methods)
(5) combined mesh refinements and enrichments (hp-methods)

These methods will be elaborated in the next section.

19.2.1 MESH REFINEMENT METHODS (h-METHODS)

The basic concept of mesh refinement methods is to refine the element in which a posteriori error indicator (measure of error based on solution gradients) is larger than the preset criterion. This procedure is ideally applied to the finite element methods.

19.2.1.1 Error Indicators

In general, the solution error is not available a priori. Even after the solution has been completed "a posteriori error" must be evaluated from the so-called error indicator as the exact solution is not available. The a posteriori error indicator may be predicted from the solution gradients of variables: density, velocity, pressure, or temperature. For inviscid flows with shock waves, we may consider density gradients to be a best measure of error, whereas velocity gradients may play a key role in the case of compressible viscous flows. Pressure or temperature gradients can also be considered an important factor in determining the error indicator.

Let $\hat{u}$ and u be the exact solution and approximate solution, respectively. We then specify an error e in elliptic problems as

$$\|e\|_E = \|\hat{u} - u\|_E \leq \alpha \tag{19.2.1}$$

where $\|e\|_E$ represents the energy norm and α denotes a specific tolerance. We may rewrite (19.2.1) in the form

$$\|e\|_E \leq Csh \tag{19.2.2}$$

where C is a constant independent of mesh parameter h with $s = 1, 2, \ldots, n$ such that

$$(s-1)h \leq x \leq sh \tag{19.2.3}$$

The estimate given by (19.2.2) is known as a priori estimate based on some general information about the exact solution. A priori estimates indicate how fast the error changes as the h-refinements are changed. Estimates based on the approximate finite element solution are called a posteriori error estimates. To this end we write (19.2.1) in an alternate form.

$$\|e\|_E \leq \frac{k}{N^p} \tag{19.2.4}$$

where k and p are constants and N represents the number of degrees of freedom.

If p is the degree of polynomials for interpolations and (19.2.4) is extended to the h-adaptivity, then we have

$$\|e\|_E \leq \frac{k}{\exp(\beta N^\gamma)} \tag{19.2.5}$$

where k, β, and γ are positive constants. Taking a logarithm in (19.2.5) yields

$$\log \|e\|_E \leq \log k - \beta N^\gamma \log e \tag{19.2.6}$$

This represents the rate of convergence to be exponential [Babuska and Suri, 1990; Oden, Wu, and Legat, 1995]. If there are singular points in the domain, then the rate of convergence is algebraic.

To achieve the error estimate described above, the h-adaptivity proceeds with an error indicator, a dimensionless quantity, given in terms of ratios of gradients or rates of changes of gradients of appropriate variables. These variables may be density, velocity, pressure, or temperature. For example, density and velocity are usually chosen for the shock wave turbulent boundary layer flows.

The nondimensional error indicator is defined as

$$\theta = f(h, H^m), \quad m = 0, 1, 2 \tag{19.2.7}$$

where h is the mesh parameter and H^m is the Hilbert space (Section 8.3). For density and velocity as governing variables for determining the error indicator, we define, in terms of various semi-norms.

for density:

$$\theta = h\,|\rho|_{H^1}/|\rho|_{H^0}, \qquad \theta = h|\rho|_{H^2}/|\rho|_{H^1} \tag{19.2.8a}$$

for velocity:

$$\theta = h|\mathrm{v}_i|_{H^1}/|\mathrm{v}_i|_{H^0}, \qquad \theta = h|\mathrm{v}_i|_{H^2}/|\mathrm{v}_i|_{H^1} \tag{19.2.8b}$$

with

$$|\rho|_{H^0} = \left[\int_{\Omega_e} \rho^2\, d\Omega\right]^{\frac{1}{2}}, \qquad |\rho|_{H^1} = \left[\int_{\Omega_e} \frac{\partial\rho}{\partial x_i}\frac{\partial\rho}{\partial x_i}\, d\Omega\right]^{\frac{1}{2}},$$

$$|\rho|_{H^2} = \left[\int_{\Omega_e} \frac{\partial^2\rho}{\partial x_i\partial x_j}\frac{\partial^2\rho}{\partial x_i\partial x_j}\, d\Omega\right]^{\frac{1}{2}} \tag{19.2.9a}$$

$$|\mathrm{v}_i|_{H^0} = \left[\int_{\Omega_e} \mathrm{v}_i\mathrm{v}_i\, d\Omega\right]^{\frac{1}{2}}, \qquad |\mathrm{v}_i|_{H^1} = \left[\int_{\Omega_e} \frac{\partial \mathrm{v}_i}{\partial x_j}\frac{\partial \mathrm{v}_i}{\partial x_j}\, d\Omega\right]^{\frac{1}{2}},$$

$$|\mathrm{v}_i|_{H^2} = \left[\int_{\Omega_e} \frac{\partial^2 \mathrm{v}_i}{\partial x_j\partial x_j}\frac{\partial^2 \mathrm{v}_i}{\partial x_k\partial x_k}\, d\Omega\right]^{\frac{1}{2}} \tag{19.2.9b}$$

Here the mesh parameter h is determined for one-, two-, and three-dimensional elements as follows:

h = length of 1-D element

h = diameter of the circumcircle containing a 2-D element

h = diameter of the circumsphere containing a 3-D element

The order of the Hilbert space m may be increased to four if the fourth order biharmonic equations are to be solved.

It is often useful to rearrange the error indicator in terms of the energy norm E for any variable ϕ,

$$\theta = \left[\frac{h^2|\nabla^2\phi|}{h|\nabla\phi| + \varepsilon|\phi|} \right]_E \tag{19.2.10}$$

where ε is the computational noise control parameter [Löhner and Baum, 1990]. This criterion is particularly effective in shock wave discontinuities.

The FDV parameters discussed in Sections 6.5 and 13.6 can be used as the error indicator. Since the FDV parameters are calculated by changes in Mach number, Reynolds number, Peclet number, and Damköhler number, they represent more precise variations of the gradients of whichever variable(s) are dominant. Comparisons of the h refinements by various error indicators are shown in the following subsections.

19.2.1.2 Two-Dimensional Quadrilateral Element

In Figure 19.2.1a, a simple example for two-level adaptation (refinements) is shown. Note that irregular nodes or hanging nodes arise in the process of refinements. For convenience, we shall permit only one irregular node along the side of unrefined element (nodes c and d, Figure 19.2.1b). For discretizations as shown in Figure 19.2.1c, however, the unrefined element D and B are subdivided even if not required by the error indicator.

In order that elements D and B remain linear in Figure 19.2.1b, we must eliminate nodes c and d as follows:

$$u_c = \frac{1}{2}(u_Q + u_R), \qquad u_d = \frac{1}{2}(u_R + u_S) \tag{19.2.11}$$

For the transition element, T1 we have

$$\begin{aligned} u^{(T1)} &= \Phi_1 u_1 + \Phi_2 u_2 + \Phi_3 u_3 + \Phi_4 u_4 \\ &= \Phi_1\left(\frac{u_4 + u_R}{2}\right) + \Phi_2 u_2 + \Phi_3 u_3 + \Phi_4 u_4 \\ &= [\Phi_1 \quad \Phi_2 \quad \Phi_3 \quad \Phi_4] \begin{bmatrix} \frac{1}{2} & 0 & 0 & \frac{1}{2} \\ 0 & 1 & 0 & 0 \\ 0 & 0 & 1 & 0 \\ 0 & 0 & 0 & 1 \end{bmatrix} \begin{bmatrix} u_R \\ u_2 \\ u_3 \\ u_4 \end{bmatrix} \end{aligned}$$

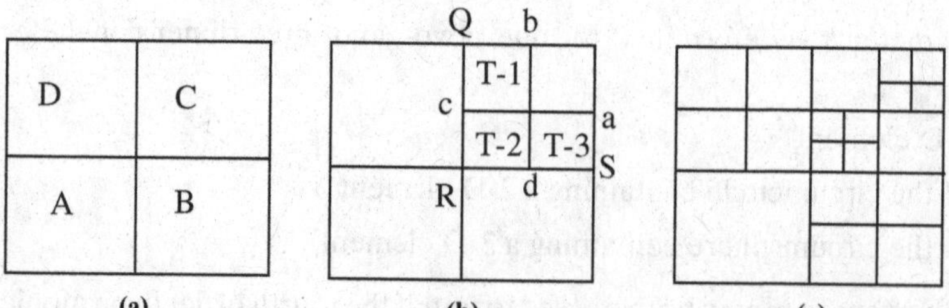

Figure 19.2.1 Simple example for mesh refinements. (a) Inital mesh. (b) One-level refinements. (c) Two-level refinements.

or

$$u^{(T1)} = \Phi_N H_{NM}^{(T1)} u_M^{(T1)} = \Phi_M^{(T1)} u_M^{(T1)} \tag{19.2.12}$$

with $\Phi_M^{(T1)} = \Phi_N H_{NM}^{(T1)}$ and $H_{NM}^{(T1)}$ being the interpolations and auxiliary matrix for the irregular element, respectively,

$$H_{NM}^{(T1)} = \begin{bmatrix} \frac{1}{2} & 0 & 0 & \frac{1}{2} \\ 0 & 1 & 0 & 0 \\ 0 & 0 & 1 & 0 \\ 0 & 0 & 0 & 1 \end{bmatrix}, \qquad \Phi_M^{(T1)} = [\Phi_1/2 \quad \Phi_2 \quad \Phi_3 \quad \Phi_1/2 + \Phi_4],$$

$$u_M^{(T1)} = [u_R \quad u_2 \quad u_3 \quad u_4]^T$$

Similarly,

$$u^{(T2)} = \Phi_N H_{NM}^{(T2)} u_M^{(T2)} = \Phi_M^{(T2)} u_M^{(T2)} \tag{19.2.13}$$

$$u^{(T3)} = \Phi_N H_{NM}^{(T3)} u_M^{(T3)} = \Phi_M^{(T3)} u_M^{(T3)} \tag{19.2.14}$$

with

$$H_{NM}^{(T2)} = \begin{bmatrix} 1 & 0 & 0 & 0 \\ \frac{1}{2} & \frac{1}{2} & 0 & 0 \\ 0 & 0 & 1 & 0 \\ \frac{1}{2} & 0 & 0 & \frac{1}{2} \end{bmatrix}, \qquad u_M^{(T2)} = [u_1 \quad u_S \quad u_3 \quad u_Q]^T$$

$$H_{NM}^{(T3)} = \begin{bmatrix} \frac{1}{2} & \frac{1}{2} & 0 & 0 \\ 0 & 1 & 0 & 0 \\ 0 & 0 & 1 & 0 \\ 0 & 0 & 0 & 1 \end{bmatrix}, \qquad u_M^{(T3)} = [u_R \quad u_2 \quad u_3 \quad u_4]^T$$

In this manner the irregular nodes c and d in the first level refinements (Figure 19.2.1b) are eliminated and replaced by the global nodes Q, R, and S. Note that for two or higher level refinements, no new types of auxiliary matrix arise. This is because only one hanging node (irregular node) is to be allowed for any refinement process.

The procedure for h-refinement is as follows:

Step 1

A coarse finite element mesh is constructed, which contains only a small number of elements, sufficient to model basic geometrical features and flow characteristics. Obtain the preliminary flow solution on this initial mesh.

Step 2

Compute the error indicator for each element.

Step 3

If $\theta \geq \alpha E$, refine by subdividing the quadrilateral through midpoints. If $\theta \leq \beta E$, unrefine by reversing the refining process. Typically set $\alpha = 0.2$ and $\beta = 0.5$ with E being the user-specified tolerance.

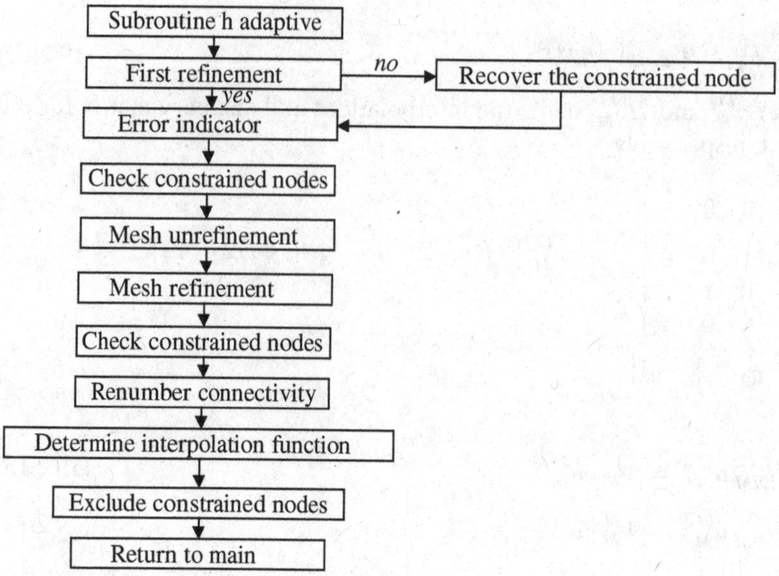

Figure 19.2.2 Flow diagram for mesh refinements.

The overall flow diagram is shown in Figure 19.2.2. Some applications to two-dimensional problems with triangular elements and quadrilateral elements using the error indicator (19.2.10) are shown in Figure 19.2.3 and Figure 19.2.4, respectively [Yoon and Chung, 1991].

Instead of using the primitive variable error indicators, we may take advantage of the FDV parameters as discussed in Section 13.6. To this end, we examine a compression

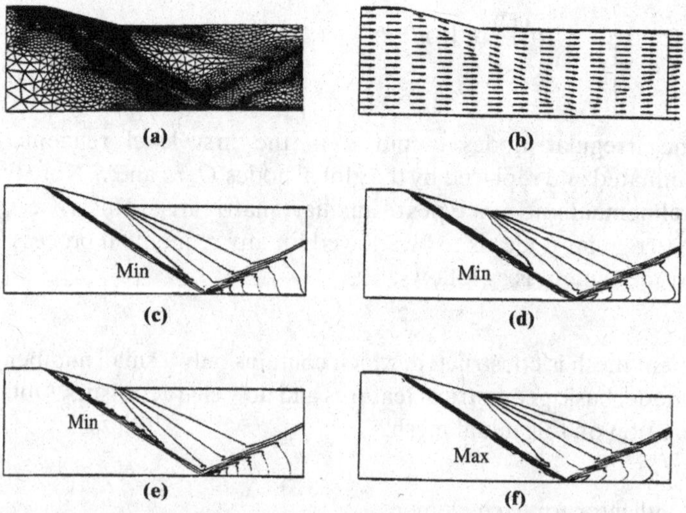

Figure 19.2.3 Mesh refinements (*h*-method), triangular elements. **(a)** Adapted mesh configuration (5004 elements, 2600 nodes, t = 0.059 sec). **(b)** Velocity field (Inlet vel. = 1506 m/sec). **(c)** Density contours (Max = 1.648, Min = 0.418, $\Delta = 0.123$ kg/m^3. **(d)** Temperature contours (Max = 1588, Min = 863, $\Delta = 73$K. **(e)** Pressure contours (Max = 0.725, Min = 0.115, $\Delta = 0.061$ MPa. **(f)** Mach number contours (Max = 2.603, Min = 1.152, $\Delta = 0.145$.

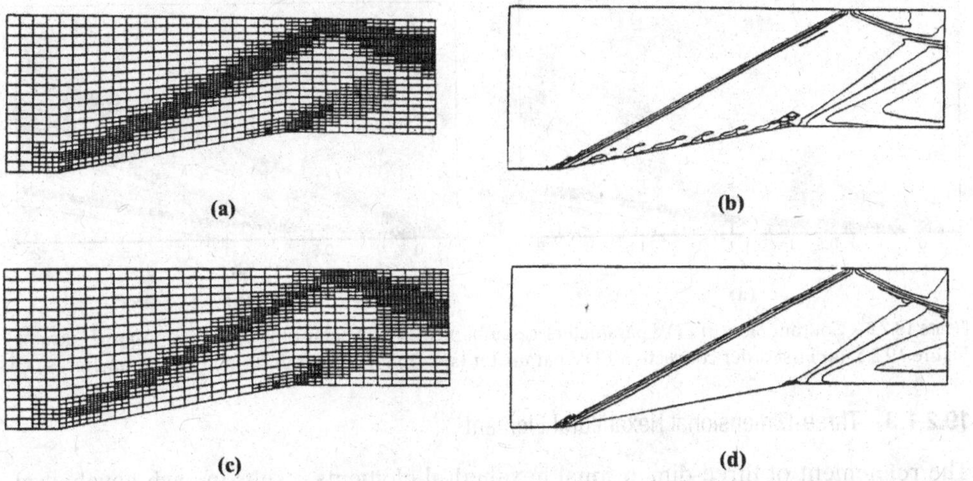

Figure 19.2.4 Mesh refinements (h-method), quadrilateral elements. (a) Adaptive mesh at t = 12 sec (2078 elements, 2317 nodes). (b) Mach number contours at t = 12 sec (max = 2.9, min = 1.4, $\Delta = 0.15$). (c) Adaptive mesh at t = 16 sec (1847 elements, 2078 nodes). (d) Mach number contours at t = 16 sec (max = 2.9, min = 1.3, $\Delta = 0.16$).

corner supersonic flow as shown in Figure 19.2.5a with initial grid of 4,600 elements (Figure 19.2.5b) [Heard and Chung, 2000]. Adaptative refinement is made for s_1 or s_3 greater than 0.45. Unrefinement is to be applied if the FDV parameters are less than 0.2. The contours of the first order FDV parameters (s_1, s_3) are shown in Figure 19.2.6. As demonstrated elsewhere (Section 6.6), these FDV parameters resemble closely the flowfield itself. Two level adaptation refinements have been carried out as shown in Figure 19.2.7. For the purpose of comparison, the results of computations based on the standard primitive variable error indicators are displayed. It is shown that the FDV results provide lesser number of adapted elements for both adaptative refinement levels. In addition, the refined regions are narrower for the FDV calculations. This is influenced by the FDV parameters being sensitive to the current flowfield physics dictating the decision for either refinement or unrefinement. These trends result in lesser computer time for the FDV-based error indicator. An additional advantage for the FDV approach is that the FDV parameters are already available in the formulation. The Mach number contours using the FDV error indicators and the primitive variable error indicators as shown in Figure 19.2.8 are practically identical.

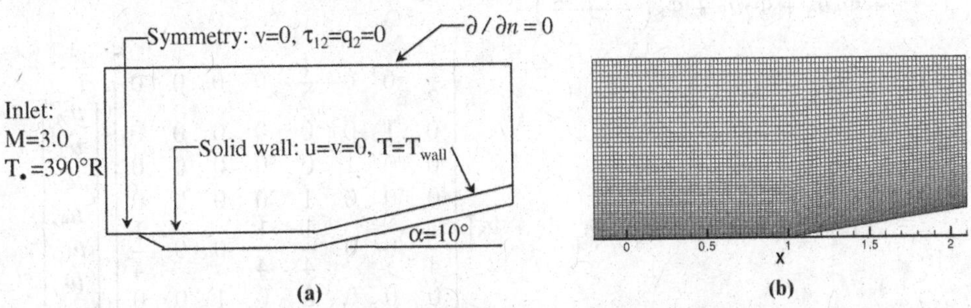

Figure 19.2.5 Geometries for adaptive mesh calculations for a compression corner supersonic flow using the flow field-dependent variation (FDV) parameters as error indicators. (a) Geometry and boundary conditions of compression corner. (b) Initial grid (4600 elements).

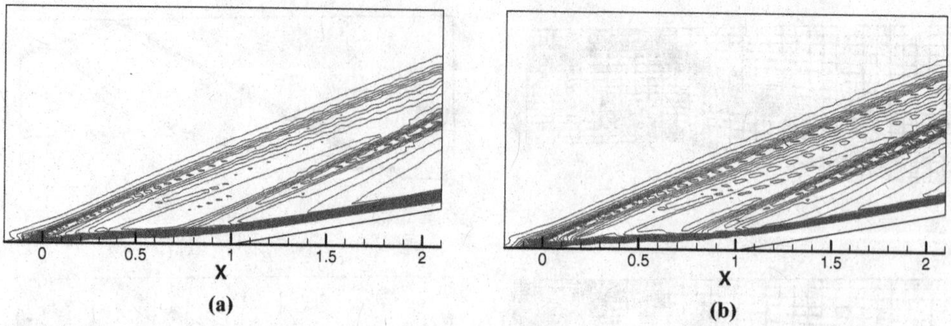

Figure 19.2.6 Contour plots of FDV parameters, resembling the flowfield for the compression corner flow of Figure 19.2.5. (**a**) First order convection FDV parameter (s_1). (**b**) First order diffusion FDV parameter (s_3).

19.2.1.3 Three-Dimensional Hexahedral Element

The refinement of three-dimensional hexahedral elements results in each hexahedral being divided into eight hexahedral elements, as shown in Figure 19.2.9a. During this refinement process, irregular, or hanging, nodes arise, similar to those produced during the refinement of two-dimensional quadrilateral elements. As demonstrated in the two-dimensional refinement process, only one irregular node is permitted along the side of an unrefined element. During this refinement process, elements may contain three, five, or six irregular nodes. The elimination process for these irregular nodes is demonstrated below.

Three Irregular Nodes

Figure 19.2.9b shows the results of a simple one-level refinement of a hexahedral element. In the refinement process, three irregular or hanging nodes arise (nodes 1, 5, and 8). In order for the element I to remain linear, nodes 1, 5, and 8 must be eliminated as follows:

$$u_1 = \frac{1}{2}(u_4 + u_A), \qquad u_5 = \frac{1}{4}(u_A + u_4 + u_B + u_C), \qquad u_8 = \frac{1}{2}(u_4 + u_B)$$

Then, any flow property, u, is calculated from

$$\begin{aligned}
u &= \Phi_1 u_1 + \Phi_2 u_2 + \Phi_3 u_3 + \Phi_4 u_4 + \Phi_5 u_5 + \Phi_6 u_6 + \Phi_7 u_7 + \Phi_8 u_8 \\
&= \Phi_1\left(\frac{u_4 + u_A}{2}\right) + \Phi_2 u_2 + \Phi_3 u_3 + \Phi_4 u_4 + \Phi_5\left(\frac{u_A + u_4 + u_B + u_C}{4}\right) \\
&\quad + \Phi_6 u_6 + \Phi_7 u_7 + \Phi_8\left(\frac{u_4 + u_B}{2}\right) \\
&= [\Phi_1 \quad \Phi_2 \quad \Phi_3 \quad \Phi_4 \quad \Phi_5 \quad \Phi_6 \quad \Phi_7 \quad \Phi_8]
\begin{bmatrix}
\frac{1}{2} & 0 & 0 & \frac{1}{2} & 0 & 0 & 0 & 0 \\
0 & 1 & 0 & 0 & 0 & 0 & 0 & 0 \\
0 & 0 & 1 & 0 & 0 & 0 & 0 & 0 \\
0 & 0 & 0 & 1 & 0 & 0 & 0 & 0 \\
\frac{1}{4} & 0 & 0 & \frac{1}{4} & \frac{1}{4} & 0 & 0 & \frac{1}{4} \\
0 & 0 & 0 & 0 & 0 & 1 & 0 & 0 \\
0 & 0 & 0 & 0 & 0 & 0 & 1 & 0 \\
0 & 0 & 0 & \frac{1}{2} & 0 & 0 & 0 & \frac{1}{2}
\end{bmatrix}
\begin{bmatrix} u_A \\ u_2 \\ u_3 \\ u_4 \\ u_C \\ u_6 \\ u_7 \\ u_B \end{bmatrix}
\end{aligned}$$

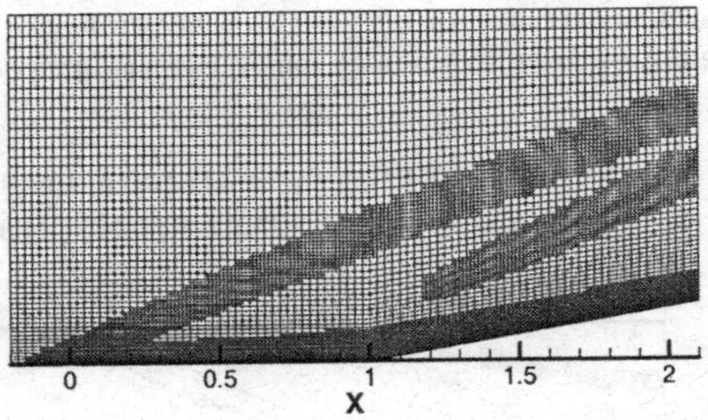

(i) Adaptive mesh based on FDV parameters (9466 elements)

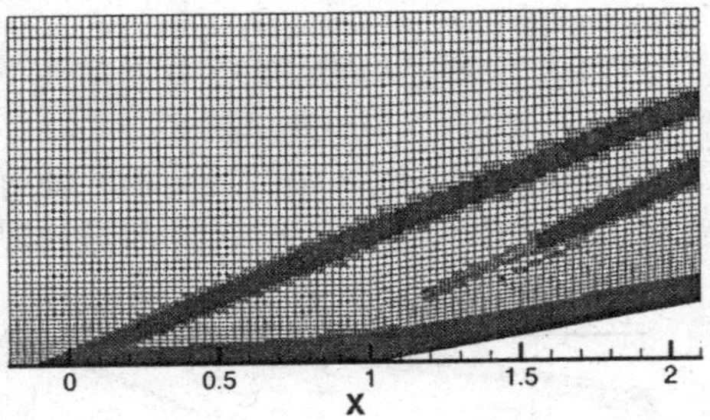

(i) Adaptive mesh based on FDV parameters (20836 elements)

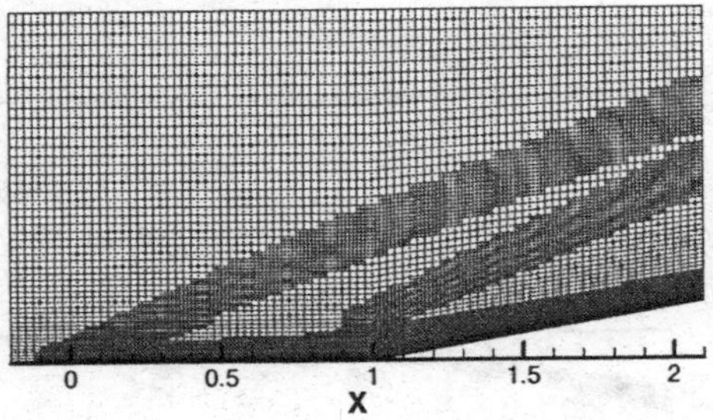

(ii) Adaptive mesh based on primitive variable error indicators (9988 elements)

(a)

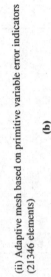

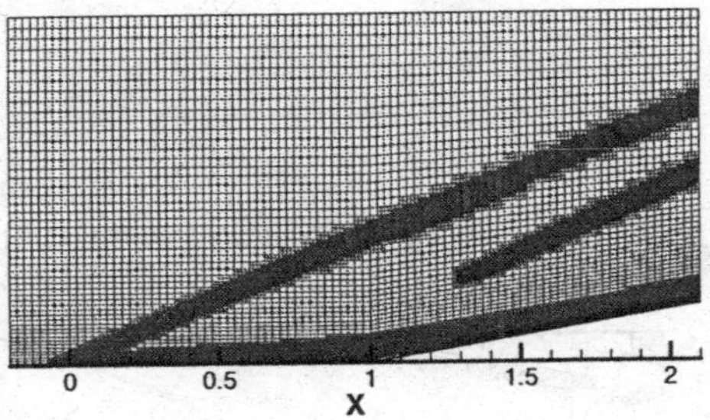

(ii) Adaptive mesh based on primitive variable error indicators (21346 elements)

(b)

Figure 19.2.7 Adaptive meshes for the first and second levels of refinement for the compression corner flow of Figure 19.2.5. **(a)** First grid refinement. **(b)** Second grid refinement.

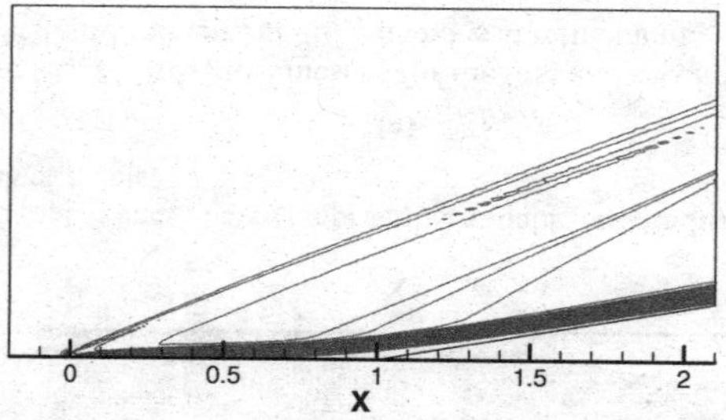

(i) Mach number contours calculated from FDV parameters

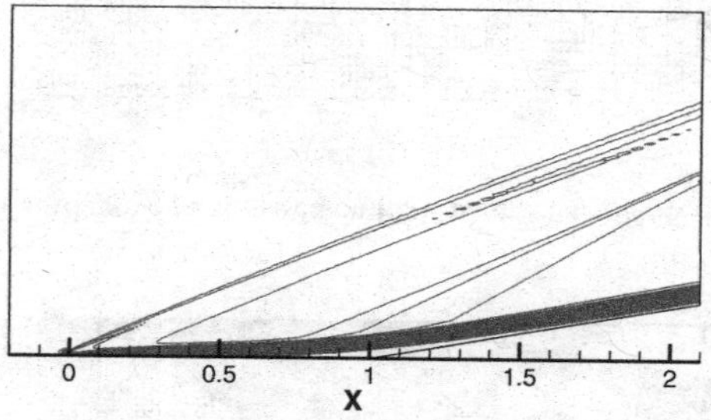

(ii) Mach number contours calculated from primitive variable error indicators

(a)

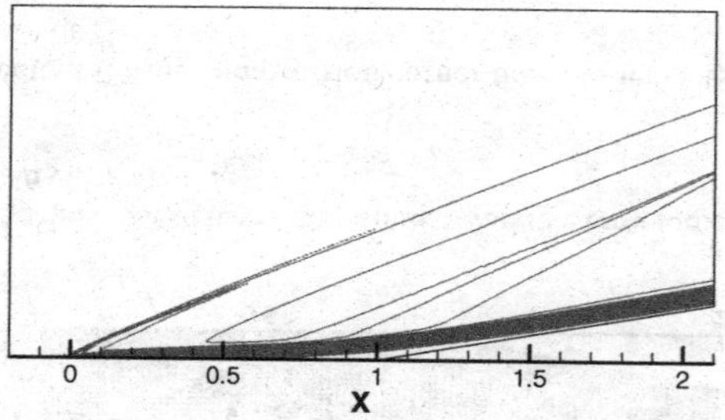

(i) Mach number contours calculated from FDV parameters

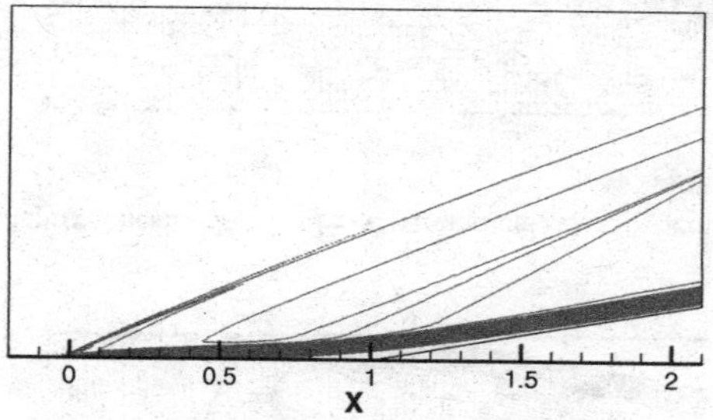

(ii) Mach number contours calculated from primitive variable error indicators

(b)

Figure 19.2.8 Mach number contours calculated from adaptive meshes, first and second levels of refinement using the error indicators based on the FDV parameters and primitive variable error indicators. **(a)** Mach number contours (first grid refinement). **(b)** Mach number contours (second refinement).

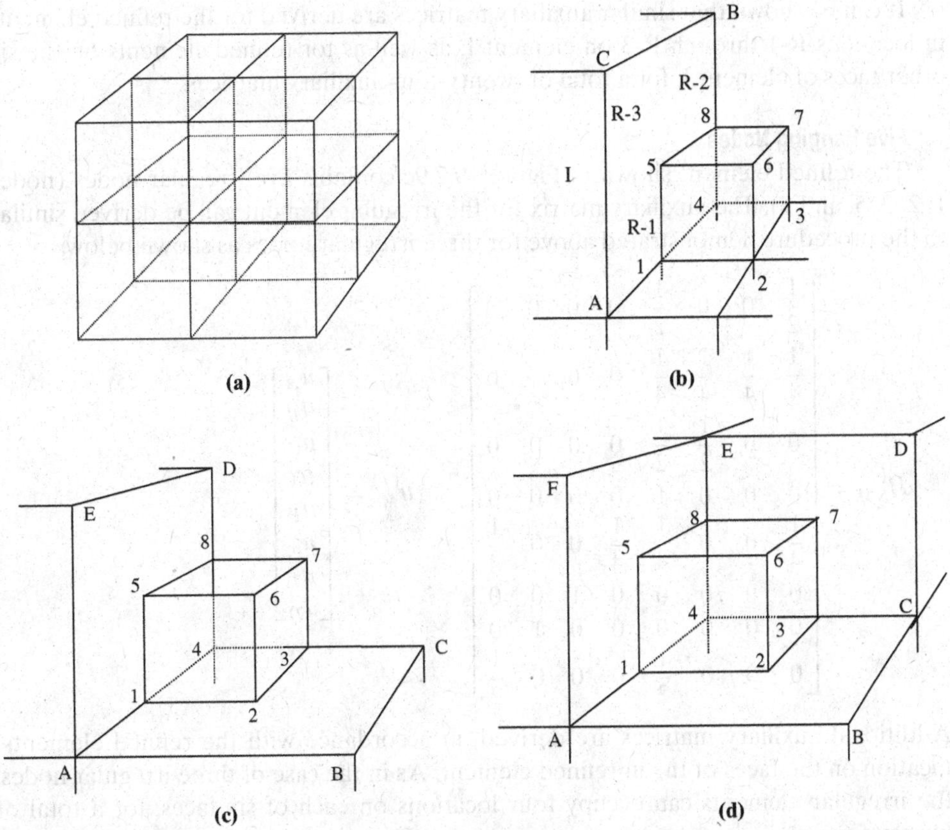

Figure 19.2.9 Mesh refinement of hexahedral element with hanging nodes. **(a)** Hexahedral elements. **(b)** Three hanging nodes. **(c)** Five hanging nodes. **(d)** Six hanging nodes.

or

$$u^{(T)} = \Phi_N H_{NM}^{(T)} u_M^{(T)}$$

with $\Phi_M^{(T)} = \Phi_N H_{NM}^{(T)}$ and $H_{NM}^{(T)}$ being the interpolation and auxiliarymatrix for the irregular node, respectively.

$$H_{NM}^{(T)} = \begin{bmatrix} \frac{1}{2} & 0 & 0 & \frac{1}{2} & 0 & 0 & 0 & 0 \\ 0 & 1 & 0 & 0 & 0 & 0 & 0 & 0 \\ 0 & 0 & 1 & 0 & 0 & 0 & 0 & 0 \\ 0 & 0 & 0 & 1 & 0 & 0 & 0 & 0 \\ \frac{1}{4} & 0 & 0 & \frac{1}{4} & \frac{1}{4} & 0 & 0 & \frac{1}{4} \\ 0 & 0 & 0 & 0 & 0 & 1 & 0 & 0 \\ 0 & 0 & 0 & 0 & 0 & 0 & 1 & 0 \\ 0 & 0 & 0 & \frac{1}{2} & 0 & 0 & 0 & \frac{1}{2} \end{bmatrix} \qquad u_M^{(T)} = \begin{bmatrix} u_A \\ u_2 \\ u_3 \\ u_4 \\ u_C \\ u_6 \\ u_7 \\ u_B \end{bmatrix}$$

$$\Phi_N^{(T)} = \left[\frac{\Phi_1}{2} + \frac{\Phi_5}{4} \quad \Phi_2 \quad \Phi_3 \quad \frac{\Phi_1}{2} + \Phi_4 + \frac{\Phi_5}{4} + \frac{\Phi_8}{2} \quad \frac{\Phi_5}{4} \quad \Phi_6 \quad \Phi_7 \quad \frac{\Phi_5}{4} + \frac{\Phi_8}{2} \right]$$

It can be shown that similar auxiliary matrices are derived for the refined elements in locations R-1 through R-3 on element I, as well as for refined elements on the six other faces of element I, for a total of twenty-four auxiliary matrices.

Five Hanging Nodes

The refined element shown in Figure 19.2.9c contains five irregular nodes (nodes 1, 2, 3, 5, and 8). The auxiliary matrix for the irregular element can be derived similar to the procedure demonstrated above for three irregular nodes as shown below.

$$H_{NM}^{(T)} = \begin{bmatrix} \frac{1}{2} & 0 & 0 & \frac{1}{2} & 0 & 0 & 0 & 0 \\ \frac{1}{4} & \frac{1}{4} & \frac{1}{4} & \frac{1}{4} & 0 & 0 & 0 & 0 \\ 0 & 0 & \frac{1}{2} & \frac{1}{2} & 0 & 0 & 0 & 0 \\ 0 & 0 & 0 & 1 & 0 & 0 & 0 & 0 \\ \frac{1}{4} & 0 & 0 & \frac{1}{4} & \frac{1}{4} & 0 & 0 & \frac{1}{4} \\ 0 & 0 & 0 & 0 & 0 & 1 & 0 & 0 \\ 0 & 0 & 0 & 0 & 0 & 0 & 1 & 0 \\ 0 & 0 & 0 & \frac{1}{2} & 0 & 0 & 0 & \frac{1}{2} \end{bmatrix} \qquad u_M^{(T)} = \begin{bmatrix} u_A \\ u_B \\ u_C \\ u_4 \\ u_E \\ u_6 \\ u_7 \\ u_D \end{bmatrix}$$

Additional auxiliary matrices are derived, in accordance with the refined elements' location on the faces of the unrefined element. As in the case of three irregular nodes, the irregular elements can occupy four locations on each of six faces, for a total of twenty-four auxiliary matrices.

Six Hanging Nodes

Similarly, the refined element in Figure 19.2.9d contains six irregular, or hanging, nodes (nodes 1, 2, 3, 5, 7, and 8). The auxiliary transition matrix for the irregular element shown in Figure 19.2.9 is

$$H_{NM}^{(T)} = \begin{bmatrix} \frac{1}{2} & 0 & 0 & \frac{1}{2} & 0 & 0 & 0 & 0 \\ \frac{1}{4} & \frac{1}{4} & \frac{1}{4} & \frac{1}{4} & 0 & 0 & 0 & 0 \\ 0 & 0 & \frac{1}{2} & \frac{1}{2} & 0 & 0 & 0 & 0 \\ 0 & 0 & 0 & 1 & 0 & 0 & 0 & 0 \\ \frac{1}{4} & 0 & 0 & \frac{1}{4} & \frac{1}{4} & 0 & 0 & \frac{1}{4} \\ 0 & 0 & 0 & 0 & 0 & 1 & 0 & 0 \\ 0 & 0 & \frac{1}{4} & \frac{1}{4} & 0 & 0 & \frac{1}{4} & \frac{1}{4} \\ 0 & 0 & 0 & \frac{1}{2} & 0 & 0 & 0 & \frac{1}{2} \end{bmatrix} \qquad u_M^{(T)} = \begin{bmatrix} u_A \\ u_B \\ u_C \\ u_4 \\ u_F \\ u_6 \\ u_D \\ u_E \end{bmatrix}$$

There are also twenty-four possible auxiliary matrices for elements with six irregular nodes. However, in practice, it is not necessary to store all possible auxiliary matrices in

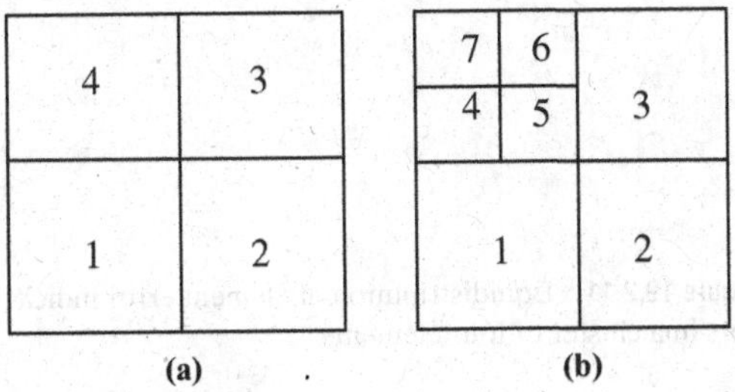

Figure 19.2.10 Illustration of refinement/unrefinement process. (a) Initial mesh. (b) Adapted refined mesh.

computer memory. Note that the rows of the auxiliary matrix for an irregular element can be determined simply by noting the number of nodes required for interpolation at the irregular node, and the node numbers used for interpolation. The nonzero entries are placed in the columns corresponding to the node numbers used for interpolation. Those nonzero entries are either $\frac{1}{2}$ or $\frac{1}{4}$, depending on whether the value is interpolated between two or four nodes.

Mesh Unrefinement

For time dependent problems, in which a discontinuity may be migrating over the grid during a given time interval, an unrefinement procedure is needed, as well as a refinement procedure. In a similar way as elements are refined if the error is greater than a specified number, elements are unrefined if the error is less than a specified number. One method of unrefinement is to require that only groups of elements that were refined before can be unrefined.

To ensure that the unrefinement occurs in the same way as the refinement, an array is added to the data structure to reflect the refinement history of the mesh [Devloo, Oden, and Pattani, 1988]. Toward this end, we introduce an array, NELGRP, in which NELGRP(I,IGR) is the Ith element of group IGR. If NELGRP(I,IGR) > 0, it refers to an element; if NELGRP(I,IGR) < 0, it refers to a group.

Using this array, we build a data structure of groups that contain elements or groups. When an element is refined, it is transformed into a group which contains four (or eight, for 3-D hexahedral elements) new subelements. All references to the original element are changed to the new group. For example, consider the simple grid of four elements in Figure 19.2.10a. If element 4 is refined as shown in Figure 19.2.10b, the NELGRP array for group 4 becomes:

$$\text{NELGRP}(1,4) = 4, \quad \text{NELGRP}(2,4) = 5, \quad \text{NELGRP}(3,4) = 6, \text{ and}$$
$$\text{NELGRP}(4,4) = 7$$

If element 5 were further refined, then NELGRP(2,4) would be changed to -5, referring to group 5, and the entries for NELGRP(,5) would be filled in with the new element numbers. Using this data structure, it becomes a simple matter of unrefining a group to recover the previous elements, when the error indicates unrefinement is needed.

19.2.2 MESH MOVEMENT METHODS (*r*-METHODS)

Instead of refining elements, grid points can be moved around (mesh redistribution) to provide clustering in certain regions by means of error indicators, known as the position

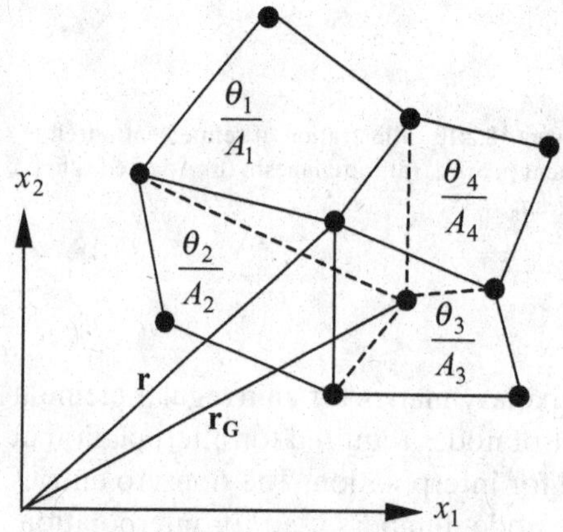

Figure 19.2.11 Equidistribution of element error indicators in a cluster of four elements.

vector approach [Oden, 1988]. This is in contrast to the clustering obtained by means of control functions in structured grids. Mesh movements may be combined with mesh refinements, known as the combined mesh refinement and mesh movement methods (*hr*-methods).

If the number of grid points is fixed, then it is desirable to relocate the nodes so that errors are equi-distributed over the entire mesh, in other words, the error over each element is the same. This method is often called the "moving mesh method." The procedure is as follows:

(1) Generate an initial mesh and obtain a trial solution.
(2) Compute the error indicator, θ.
(3) Determine the area weighted quantity θ_i/A_i for the element i (Figure 19.2.11).
(4) Let $\mathbf{r}_o$ be a position vector from the origin of a global coordinate system to the controlled element e_i of group G. The controlled element of the error of group G is

$$\mathbf{r}_G = \frac{\sum_{i=1}^{n} \mathbf{r}_i \theta_i / A_i}{\sum_{i=1}^{n} \theta_i / A_i} \tag{19.2.15}$$

where n denotes the number of elements surrounding a global node.
(5) Relocate the node at the centroid of group G to lie at the vertex r_G.
(6) Continue this process over each group G of n elements until the new location of each node remains within a prescribed tolerance.

An example of the r-method is shown in Figures 19.2.12, with applications demonstrated in Figure 19.2.13. The same approach can be used for triangular elements, and any number of surrounding elements can be accommodated at a global node under consideration.

19.2.3 COMBINED MESH REFINEMENT AND MESH MOVEMENT METHODS (*hr*-METHODS)

In this approach, the mesh is simultaneously refined and moved around as dictated by the error indicator. This adaptive process is implemented most conveniently in conjunction

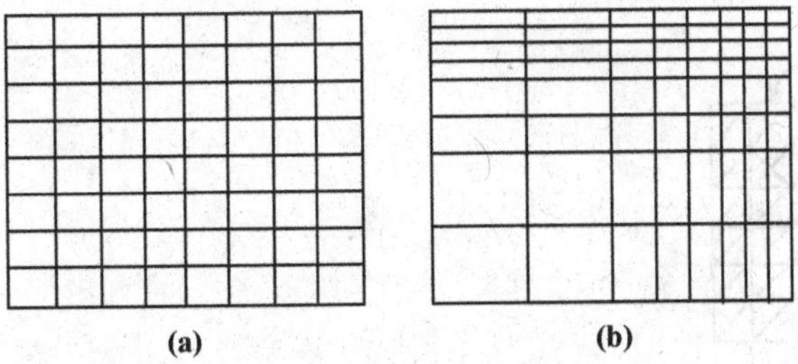

Figure 19.2.12 Example of r-methods. (a) Initial mesh. (b) Mesh after r-adaptivity.

with AFM or DVM mesh generators. There are two types of approaches in hr-methods: (1) mesh stretching and (2) local remeshing, described below.

Mesh Stretching

In this method, the measure of the error of each element is expressed in one dimension as

$$h_e^2 \left| \frac{d^2\rho}{dx^2} \right|_e = const \tag{19.2.16a}$$

With the second derivative of density evaluated at each node P or the current mesh, the new mesh may be generated with local spacing δ_P such that (Figure 19.2.14):

$$\delta_P^2 \left| \frac{d^2\rho}{dx^2} \right|_P = const \tag{19.2.16b}$$

For two dimensions with the local principal direction x_1 (major) and x_2 (minor),

$$\delta_{(1)P}^2 \left| \frac{d^2\rho}{dx_1^2} \right|_P = \delta_{(2)P}^2 \left| \frac{d^2\rho}{dx_2^2} \right|_P = \delta_{\min}^2 \left| \frac{d^2\rho}{dx_1^2} \right| = const \tag{19.2.17}$$

with $\partial^2\rho/\partial x_1^2 > \partial^2\rho/\partial x_2^2$ and $\delta_{(1)P}$ and $\delta_{(2)P}$ denoting node spacings in the x_1 and x_2 directions, respectively. Here $|\partial^2\rho/\partial x_1^2|_{\max}$ is the maximum value of $|\partial^2\rho/\partial x_1^2|_P$ over each node in the current mesh and $\delta_{\min}$ is a user-specified minimum value for δ in the new mesh. Thus, the local stretching parameter S_P is defined as

$$S_P = \sqrt{\left| \frac{d^2\rho}{dx_1^2} \right|_P \Big/ \left| \frac{d^2\rho}{dx_2^2} \right|_P} \tag{19.2.18}$$

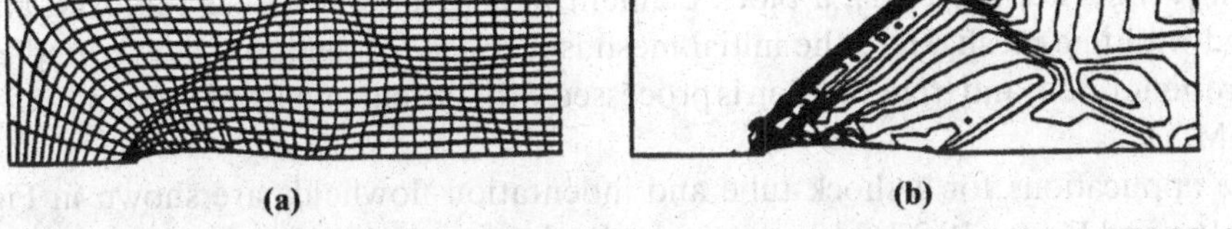

Figure 19.2.13 Example of an r-method for NACA 0012 airfoil in supersonic wind tunnel. (a) Mesh redistributions (10 applications). (b) Density contours.

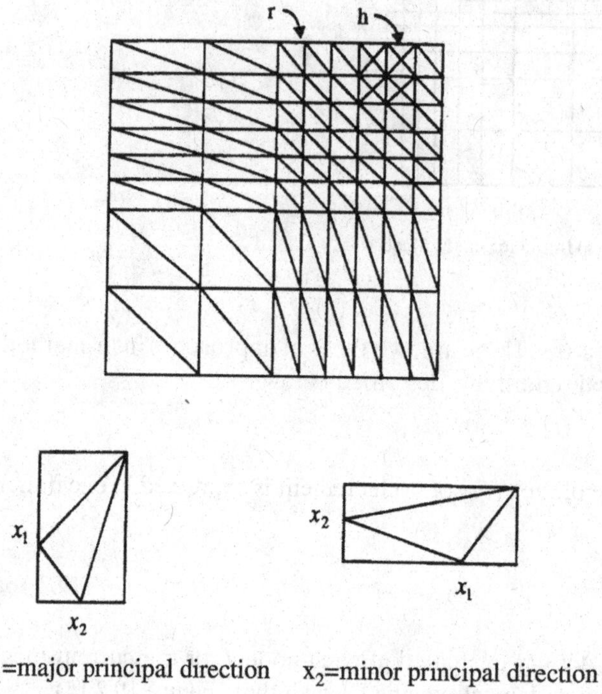

Figure 19.2.14 Example of mesh stretching scheme of *h*-method.

If δ_P computed from (19.2.17) is larger than the user-specified value δ_{max}, then we set $\delta_P = \delta_{max}$. Similarly, the node spacing will be controlled such that $\delta_P = \delta_{max}$ (user-specified maximum allowable spacing). It is thus expected from (19.2.18) that high stretching occurs only in the vicinity of one-dimensional flow features with low curvature.

In this manner, the mesh is regenerated in accordance with computed distribution of the mesh parameters and the solution of the problem recomputed on the new mesh. Obviously, the δ_{min} chosen governs the number of elements in the new mesh. This process continues until an acceptable quality of solution is achieved.

An example of a regular shock reflection at a wall with the sequence of remeshing is shown in Figure 19.2.15 [Peraire et al., 1987]. This method is prone to an excessive stretching, which is often an undesirable consequence.

Local Remeshing

To circumvent the excessive stretching, local remeshing may be employed. In this approach [Probert et al., 1991], a block element having large errors is removed and remeshed with fine mesh. Here the initial mesh is marked for deletion, new boundary points are generated, and triangulation is processed with the current front in conjunction with AFM.

Some applications for a shock tube and indentation flowfields are shown in Figure 19.2.16a and Figure 19.2.16b, respectively [Probert et al., 1991].

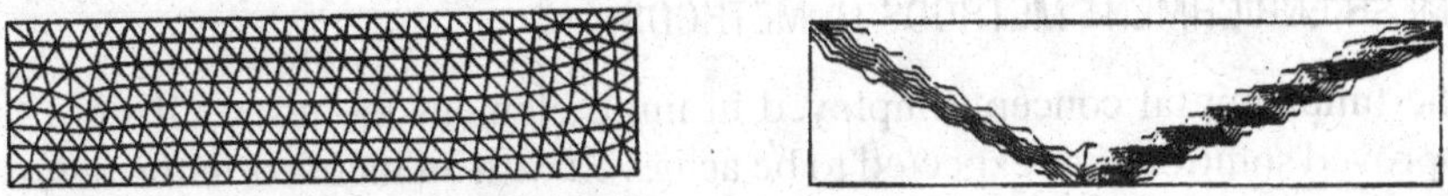

478 elements, 279 nodes, $\delta_{min}=0.13$, $\delta_{max}=0.13$, $S_{max}=1.0$

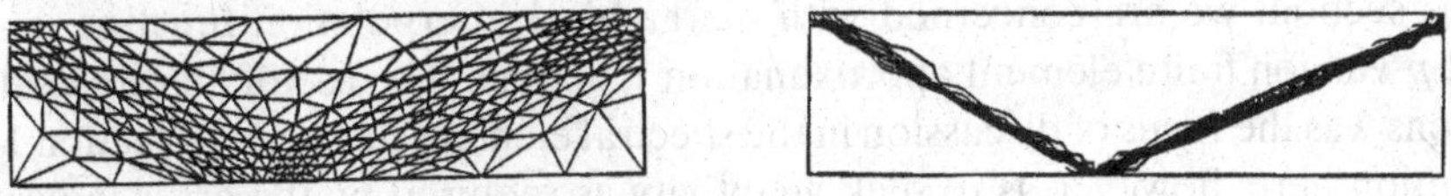

479 elements, 265 nodes, $\delta_{min}=0.05$, $\delta_{max}=0.60$, $S_{max}=3.0$

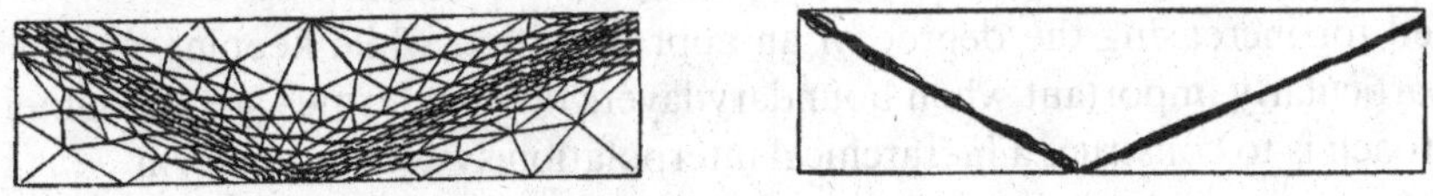

528 elements, 285 nodes, $\delta_{min}=0.02$, $\delta_{max}=0.75$, $S_{max}=6.0$

Figure 19.2.15 Local remeshing process for regular shock reflection at a wall and corresponding flowfields [Peraire et al., 1987].

(1) Fine mesh, $N_{element}=11200$, $N_{point}=6015$

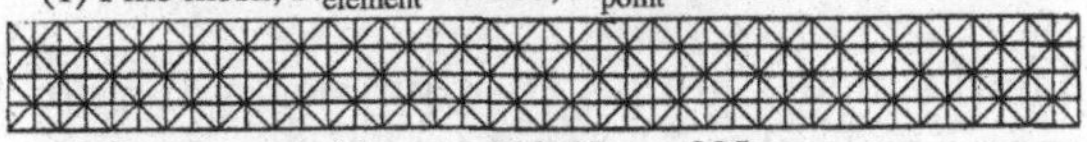

(2) Coarse mesh, $N_{element}=320$, $N_{point}=205$

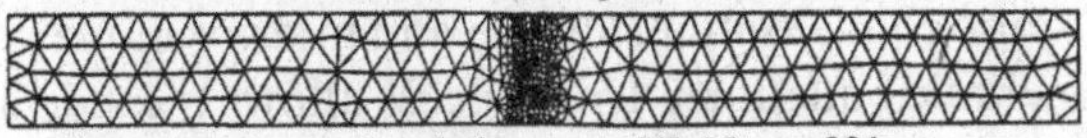

(3) Initial adaptive mesh, $N_{element}=663$, $N_{point}=391$

(4) Adaptive mesh at time t=2, $N_{element}=2112$, $N_{point}=1168$

(a)

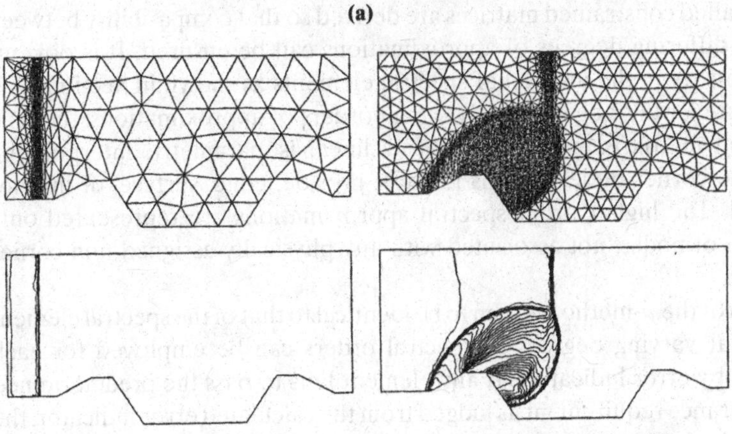

(1) NE=1228, NP=650 (2) NE=9226, NP=4693

(b)

Figure 19.2.16 Local remeshing with AFM [Probert et al., 1991]. **(a)** Propagation of a planar shock. **(b)** Computation of the flow field produced by a strong shock passing over an indentation showing the mesh and corresponding density contours at four different times.

19.2.4 MESH ENRICHMENT METHODS (p-METHODS)

This is the fundamental concept employed in finite element methods. Given a fixed mesh, improved solutions are expected to be achieved with an increase in the degree of the polynomials, or higher order approximations.

In this section, we are concerned with hierarchical interpolation function or the so-called p-version finite element approximation functions. The use of hierarchical interpolations was the focus of discussion in the spectral element methods in Section 14.1. Our attention here, however, is to seek adaptivity as required by the error indicator, resulting in various degrees of polynomials for different elements.

A need for increasing the degree of an approximation while keeping mesh sizes fixed is particularly important when boundary layers or singularities are encountered. One approach is to construct a hierarchical interpolation system in the form

$$\mathbf{U} = \Phi_\alpha \mathbf{U}_\alpha + \Phi_r^{(E)} \hat{\mathbf{U}}_r + \Phi_{rs}^{(F)} \hat{\mathbf{U}}_{rs} + \Phi_{rst}^{(I)} \hat{\mathbf{U}}_{rst} \tag{19.2.19}$$

for 3-D domain, similarly as in (14.1.16) with each function representing the tensor products of chosen polynomials (Chebyshev, Legendre, Lagrange, etc.). The degree p will be raised as required when the user-specified error indicator tolerance is exceeded. The hierarchical interpolation system (19.2.19) was detailed in Section 14.1.2 for the spectral element methods.

Recall that no side or interior nodes are installed physically (Figure 14.1.1), but higher order modes corresponding to the sides and interior are combined with the corner nodes. By means of static condensation, all side and interior mode variables are squeezed out of the final algebraic equations. This process allows the side and interior mode variables acting as the source terms, which are explicitly calculated.

In order to treat adjacent elements in which degrees of approximations are different as a result of adaptivity, special procedures are developed between the constrained and unconstrained nodes in the approach of Oden and co-workers [1989]. In such a procedure, the so-called constrained matrices are derived so that compatibility between two elements with differing degrees of approximations can be ensured. It is obvious that this is not necessary in the method of spectral elements as shown in Section 14.1. This is because whatever the Legendre polynomial orders of approximations, the final form of the element matrix is transformed into a linear isoparametric interpolation in terms of only the corner nodes. In this process, no side, edge, surface, of interior nodes are required. The higher order spectral approximations are represented only through summation of nodes, not associated with any physically assigned non-corner nodes.

Implementation of the p-method is seen to be identical to that of the spectral element methods, except that varying degrees of spectral orders can be employed for each element as dictated by error indicators. If any element fails to pass the predetermined (user-specified) tolerance requirement as judged from the calculated error indicator, the spectral order for this element must be raised. Then, along the boundaries (sides, edges, faces) of adjacent elements, there exist differences in degrees of freedom. In this case, we set the higher order element to dictate the degrees of freedom along the adjoining boundary. Other than the adaptive procedure, details of formulations for p-methods are identical to the SEM of Section 14.1.

19.2.5 COMBINED MESH REFINEMENT AND MESH ENRICHMENT METHODS (*hp*-METHODS)

If shock waves are interacting with (turbulent) boundary layers, the *p*-method alone is not adequate. Shock wave discontinuities can best be resolved through mesh refinements, and it is thus necessary that mesh enrichments which are efficient for boundary layers be combined with mesh refinements.

The simplest approach in this case is that the *h*-method is applied with only corner nodes of isoparametric elements until the shock waves are captured. Then we employ the *p*-version process with Legendre polynomials for boundary layer resolutions. This combined operation is to continue until all error indicator criteria are satisfied, with density and velocity gradients, respectively, being used for the *h*-version (shock waves) and *p*-version (boundary layers).The *hp* methods have been studied extensively by Babuska and his co-workers [1986–1998] and Oden and his co-workers [1986–1998].

In the process of adaptation, as dictated by the error indicator, a decision has to be made at any stage, whether *h*-refinements or *p*-enrichments are to be performed. One approach is to begin with low order polynomials and continue until *h*-refinements reach a certain level (for example, shock discontinuities have been resolved), followed by *p*-enrichments which are designed for resolving turbulence microscales such as in wall boundary layers or free shear layers. Another option is to rely on an optimization process in which an automatic decision is made as to whether *h*-refinements or *p*-enrichments are more desirable at any given stage of adaptation.

In the *hp* adaptivity, the error estimates and error indicators discussed in the *h*-version and *p*-version are combined. For a particular mesh and *p*-distribution, however, it is not possible to predict the accuracy a priori. Thus, we must rely on a posteriori error estimates using the finite element solutions.

To this end, we consider any function $u \in H^r(k)$ and a sequence of interpolations w^{hp} such that for any $0 \le s \le r$, and polynomial of degree $\le P_k$

$$\|u - w^{hp}\|_{s,k} \le \frac{c\, h_k^{\mu-s}}{P_k^{r-s}} \|u\|_{r,k}, \quad P_k = 1,\ 2,\ \ldots \tag{19.2.20}$$

with

$$\mu = \min(P_k + 1,\ r) \tag{19.2.21}$$

This is the error estimate applicable for the *hp* process [Babuska and Suri, 1990; Oden et al., 1995], with the error indicator given by

$$\theta = \frac{h_k}{P_k} |u|_k, \quad r = 2 < P + 1 \tag{19.2.22}$$

In practice the error indicator can be determined using the element residual technique. The fine mesh is obtained by raising the order of approximation by one for each node uniformly throughout the mesh. Then for each element *k*, the added shape function is interpolated in the sense of *hp* interpolation using the old shape functions. By subtracting the interpolates from each of the added shape functions, we effectively construct a basis for the element space of bubble function (Legendre polynomials, Chebyshev polynomials, Lagrange polynomials, etc.). The constrained approximation is fully taken into account. Next, the local problems are formulated and solved and the element error indicators are calculated using the gradients of variables as shown in (19.2.1) through (19.2.9).

A typical adaptive *hp*-method based on the error estimate proceeds as follows:

(1) Input initial data, global tolerance E_G, and local tolerance $E_L < E_G$.
(2) Solve the problem on the current finite element mesh.
(3) For each element k in the mesh, calculate the error indicator θ_k, if $\theta_k > E_L$, then refine the element.
(4) Calculate the global estimate

$$\theta_G = \sqrt{\sum_k \theta_k^2} \tag{19.2.23}$$

If $\theta_G > E_G$ then decrease the local tolerance $E_L = 90\%\ E_G$, go to (2).

In order to estimate the local quality of an error estimate, we introduce the local effectivity index γ_k:

$$\gamma_k = \frac{\theta_k}{\|e\|_k} \tag{19.2.24}$$

Introducing a discrete measure (weight) w_k

$$w_k = \frac{\|e\|_k^2}{\|e\|^2} \tag{19.2.25}$$

we obtain

$$\gamma^2 = \sum_k \gamma_k^2 w_k \tag{19.2.26}$$

Thus, the global effectivity index (squared) can be interpreted as the average of the local indices (square) weighted with respect to the discrete measure; more emphasis is placed upon elements with large errors and less on elements for which the error is small.

We may utilize the notion of standard deviation σ as a quantity estimating the discrepancy of the local effectivity indices.

$$\sigma^2 = \sum_k \left(\gamma_k^2 - \gamma^2\right)^2 w_k \tag{19.2.27}$$

This can be normalized to

$$\overline{\sigma}^2 = \sum_k \left(\overline{\gamma}_k^2 - 1\right)^2 w_k \tag{19.2.28}$$

with

$$\overline{\gamma}_k = \frac{\theta_k}{\|e\|\gamma - 1} \tag{19.2.29}$$

Equation (19.2.28) may be used as a criterion to compare the quality of various error estimates.

Our objective in the *hp*-method is to optimize the distribution of mesh size h and polynomial degree p over a finite element. For given h-refinements, the p-distributions may vary from element to element, as shown in Figure 14.1.2. Notice that boundaries between the higher and lower p's are dictated by the higher degrees polynomial with irregular nodes and elements treated as discussed in Section 19.2.1.

Toward this end, we examine the global error indicator Θ_k for element k which depends on h_k and p_k,

$$\Theta_k = \int_\Omega \theta_k(h, p)\, d\Omega \tag{19.2.30}$$

where $\theta_k(h, p)$ is the local error density. Thus, the total error indicator is expressed as

$$\Theta = \sum_k \Theta_k \tag{19.2.31}$$

Similarly, the total number of degrees of freedom is

$$N = \sum_k N_k = \int_\Omega n_k(h, p)\, d\Omega \tag{19.2.32}$$

where $n_k(h, p)$ denotes a degree of freedom density. Assume that the optimal mesh arises at $n = n_0$. Thus, the optimality condition can be achieved by constructing the Lagrange multiplier constraint

$$\lambda(n - n_o) = 0 \tag{19.2.33}$$

so that the functional

$$f = \theta(h, p) - \lambda(n - n_o) \tag{19.2.34}$$

achieves an optimality at

$$\delta f = \frac{\partial f}{\partial h}\delta h + \frac{\partial f}{\partial p}\delta p = 0 \tag{19.2.35}$$

Since δh and δp are arbitrary, we must have

$$\frac{\partial f}{\partial h} = \frac{\partial \theta}{\partial h} - \lambda \frac{\partial n}{\partial h} = 0 \tag{19.2.36}$$

$$\frac{\partial f}{\partial p} = \frac{\partial \theta}{\partial p} - \lambda \frac{\partial n}{\partial p} = 0 \tag{19.2.37}$$

These conditions lead to the optimal *hp* distribution,

$$\left.\frac{\partial \theta}{\partial n}\right|_{p=\text{constant}} = \lambda|_p \tag{19.2.38}$$

$$\left.\frac{\partial \theta}{\partial n}\right|_{h=\text{constant}} = \lambda|_h \tag{19.2.39}$$

The derivatives in (19.2.38) and (19.2.39) may be approximated by $\Delta\theta/\Delta n$, with $\Delta\theta$ denoting the change in error due to a change in number of degrees of freedom Δn. The process to reduce the error as much as possible would make the change in error per change in number of degrees of freedom as large as possible. Thus, the larger of the two quantities,

$$\lambda|_p = \left.\frac{\Delta\theta}{\Delta n}\right|_p = \text{constant} \tag{19.2.40}$$

or

$$\lambda|_h = \left.\frac{\Delta\theta}{\Delta n}\right|_h = \text{constant} \tag{19.2.41}$$

should be used as the result of optimization.

Notice that to modify a trial mesh, one refines those elements with $|\Delta^+\theta_k|$ below λ and unrefines those for which $|\Delta^+\theta_k|$ is above λ. For optimality, we refine elements for which the anticipated decrease of the error per unit new degrees of freedom is the largest.

For two-dimensional problems, refinements are not restricted in one element. This is because the approximation inside two neighboring elements is affected by the p-enrichment and h-refinement causing subdivision of neighboring elements. However, it is possible to extrapolate the one dimensional strategy to perform refinements for which the anticipated decreases of the error per new degree of freedom are as large as possible.

It may be argued that raising p gives a larger decrease in error than subdividing the element for some problems, but the mesh is achieved when geometrically well graded toward singularity with low p. The general procedure for the hp process is as follows:

(1) Compute the anticipated degrees of errors for all elements in an initial mesh.
(2) Evaluate $\frac{\Delta\theta}{\Delta n}|_p$ and $\frac{\Delta\theta}{\Delta n}|_h$ for every element.
(3) Identify $(\frac{\Delta\theta}{\Delta n})_{\max} = A$.
(4) Identify those elements for which $\frac{\Delta\theta}{\Delta n} \geq \alpha A$ where α is a predetermined number for refinement.
(5) Perform refinements based on Steps (2) and (4) and solve the problem on the new mesh.
(6) Calculate the global error $\Theta = \sum_k \Theta_k$. If $\Theta \leq \beta$ where β is a predetermined error tolerance, then stop; otherwise go to (1).

In the process of hp refinements, it is frequently required that adjacent elements have larger or smaller degrees of polynomial approximations than the element under consideration. This will result in irregular elements with irregular nodes. In this case, the adjoining boundaries are dictated by the higher order approximations of either element.

Oden et al. [1995] reports numerical results for the incompressible flow Navier-Stokes solution using the three-step hp methods in which the following three steps are implemented:

(1) Estimate the error indicator (19.2.2) on the initial mesh
(2) Compute n_k in (19.2.32) to construct a second mesh
(3) Calculate the distribution of polynomial degrees p_k to construct a third mesh.

An application of the above procedure to a back-step channel problem [Oden et al., 1995] is presented in Figure 19.2.17 and Table 19.2.1. The geometry features of the

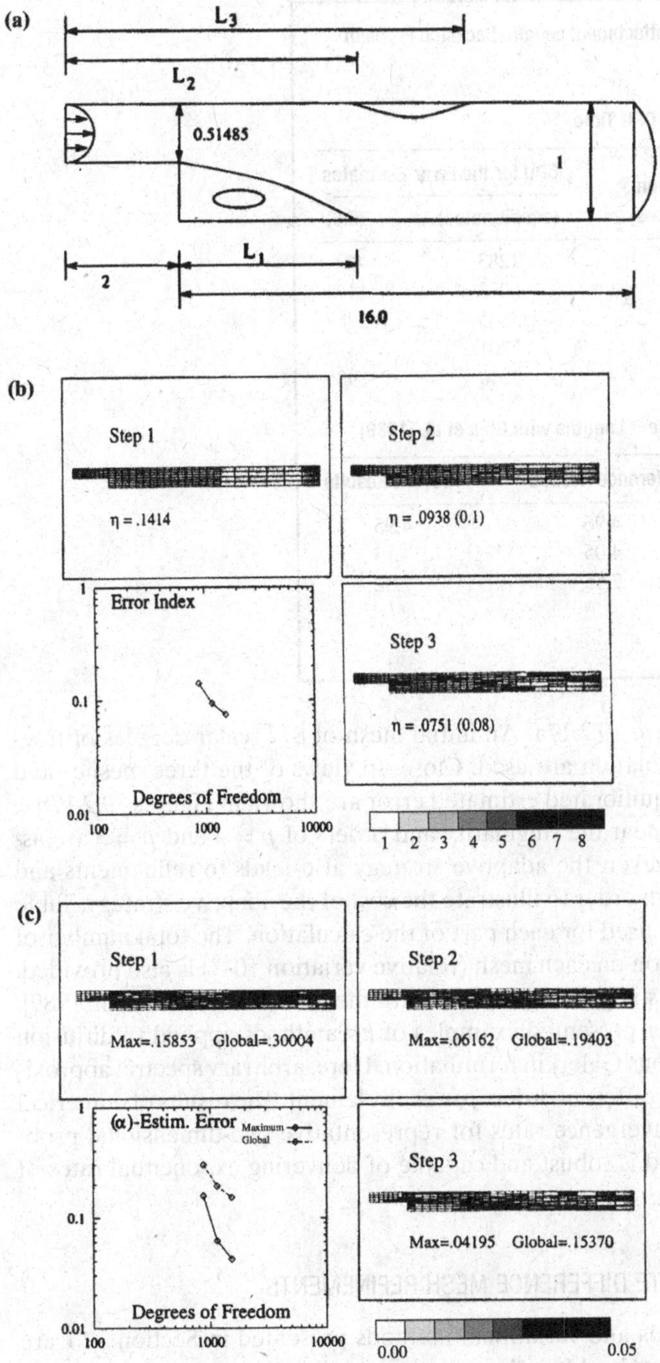

Figure 19.2.17 Analysis of a backstep channel problem with *hp* adaptive method (Re = 300) [Oden et al., 1995]. **(a)** Geometry for the backstep problem. **(b)** Close-up view of the three adaptive meshes. **(c)** Equilibrated estimated error.

Table 19.2.1 CPU Time and Reattachment Length, Backstep Problem of Figure 19.2.17

(a) CPU Time

Mesh	CPU for the Solution (number of iterations)	CPU for the Error Estimates (equilibriated)	(0.5)
1	12246(21)	1283	866
2	3333(4)	2073	1171
3	9264(5)	3845	2787
Total	24843	7201	4824
	100%	28%	19%

(b) Comparison of Reattachment Lengths with Ghia et al. [1989]*

Reattachment Lengths	Reference Results*	Present Results
L_1	4.96	4.95
L_2	4.05	4.13
L_3	7.55	7.32

Sources: [Oden et al., 1995]

problem are defined in Figure 19.2.17a. An initial mesh of 877 scalar degrees of freedom and a quadratic interpolation are used. Close-up views of the three meshes and error index evolution and equilibrated estimated error are shown in Figures 19.2.17b,c. The elements are h-refined near the singularity and orders of $p = 4$ and $p = 3$ are assigned near this point. However, the adaptive strategy also leads to refinements and enrichments in other areas. In order to illustrate the cost of the adaptive strategy, Table 19.2.1a shows the CPU time used for each part of the calculation. The total number of iterations to reach the solution on each mesh (relative variation 10^{-9}) is also provided. Table 19.2.1b presents results in good agreement with the literature [Ghia et al., 1989].

Oden et al. [1998] further presented examples of hp methods applied to diffusion problems using a discontinuous Galerkin formulation. Here, arbitrary spectral approximations are constructed with different orders p in each element. The results of numerical experiments on h and p-convergence rates for representative two-dimensional problems suggest that the method is robust and capable of delivering exponential rates of convergence.

19.2.6 UNSTRUCTURED FINITE DIFFERENCE MESH REFINEMENTS

The control function methods and variational methods presented in Section 19.1 are suitable for structured grids only. After the adaptive process, the entire mesh still remains structured. In the mesh refinement methods, it is desirable that such restriction be removed even for the FDM formulation. We examine this possibility for FDM.

The simplest case of mesh refinement may be illustrated for finite difference formulations as demonstrated by Altas and Stephenson [1991]. Consider a square S given by

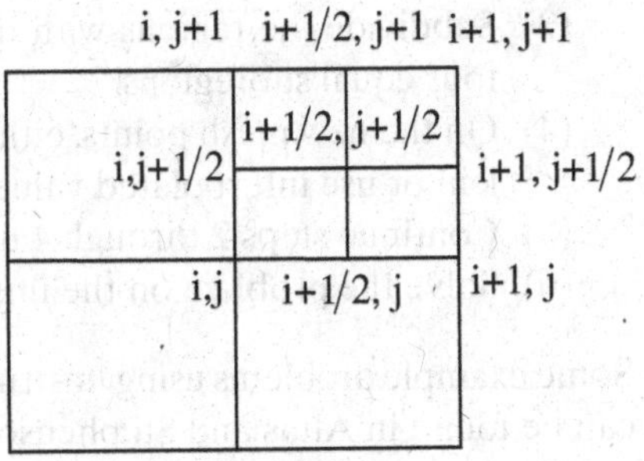

Figure 19.2.18 Comparison of errors between a square and subsquares.

(i, j), $(i+1, j)$, $(i+1, j+1)$, and $(i, j+1)$ and its subsquares, as shown in Figure 19.2.18. The computational error between the square and subsquares may be characterized as

$$e = \left|\left[\int\!\!\int u(x, y)ds - e_1\right] - \left[\int\!\!\int u(x, y)ds - e_2\right]\right| \tag{19.2.42}$$

where

$$e_1 = \frac{1}{4}(x_{i+1} - x_i)(y_{i+1} - y_i)[u(x_i, y_j) + u(x_{i+1}, y_j) + u(x_i, y_{j+1}) + u(x_{i+1}, y_{j+1})]$$

$$e_2 = \frac{1}{16}(x_{i+1} - x_i)(y_{i+1} - y_i)\{u(x_i, y_j) + u(x_i, y_{j+1}) + u(x_{i+1}, y_{j+1}) + 2[u(x_{i+\frac{1}{2}}, y_j)$$
$$+ u(x_{i+1}, y_{j+\frac{1}{2}}) + u(x_{i+\frac{1}{2}}, y_{j+1}) + u(x_i, y_{j+\frac{1}{2}})] + 4u(x_{i+\frac{1}{2}}, y_{j+\frac{1}{2}})\}$$

$$e = |e_1 - e_2|$$
$$= \frac{1}{16}(x_{i+1} - x_i)(y_{i+1} - y_i)\,|\{3[u(x_i, y_j) + u(x_{i+1}, y_j) + u(x_i, y_{j+1})$$
$$+ u(x_{i+1}, y_{j+1})] - 2[u(x_{i+\frac{1}{2}}, y_j) + u(x_{i+1}, y_{j+\frac{1}{2}}) + u(x_{i+\frac{1}{2}}, y_{j+1})$$
$$+ u(x_i, y_{j+\frac{1}{2}})] - 4u(x_{i+\frac{1}{2}}, y_{j+\frac{1}{2}})\}| \tag{19.2.43}$$

It can be shown using Taylor series expansions of the functions about the center point $(x_{i+\frac{1}{2}}, y_{j+\frac{1}{2}})$ of S that

$$e = \frac{1}{16}(x_{i+1} - x_i)(y_{i+1} - y_i)\,|\,2(x_{i+1} - x_i)^2 u_{xx}(x_{i+\frac{1}{2}}, y_{j+\frac{1}{2}})$$
$$+ 2(y_{i+1} - y_i)^2 u_{yy}(x_{i+\frac{1}{2}}, y_{j+\frac{1}{2}}) + R\,| \tag{19.2.44}$$

where R denote the remainder terms in Taylor expansions.

Here u is known only at vertices (Figure 19.2.18). Thus we construct a linear interpolation for side nodes and interior nodes. An adaptive mesh is created for all squares for which

$$e \geq E$$

where E is the user-defined tolerance.

(1) Start by using the subregions with a uniform mesh.
(2) Evaluate E using (19.2.44) on each subregion.

(3) Subdivide the regions with the quantity E larger than a given tolerance $\in$ into four equal subregions.
(4) On the new mesh points, either obtain a new approximate solution to the problem or use interpolated values of the previously obtained solution.
(5) Continue steps 2 through 4 until the largest value of E is less than $\in$.
(6) Solve the problem on the final mesh.

Some example problems using unstructured adaptive finite difference mesh refinements can be found in Altas and Stephenson [1991].

19.3 SUMMARY

Adaptive mesh methods were developed in structured grids using control functions and variational functions for FDM formulations. Obviously, in geometrical configurations not suitable for structured grids, control functions or variational functions are difficult to apply.

Unstructured adaptive methods have been extensively developed for FEM applications. Mesh refinement methods (h-methods) with error estimates and error indicators, mesh movement methods (r-methods), combined mesh refinement and mesh movement methods (hr-methods), mesh enrichment methods (p-methods), and combined mesh refinement and mesh enrichment methods (hp methods) were introduced in this chapter.

It is shown in Section 19.2.6 that adaptive unstructured mesh refinements can be performed by finite differences, although severely limited in utility and flexibility. Much greater efficiency can be provided with finite elements. For the last two decades, Oden and his co-workers and Babska and his co-workers have made significant contributions in FEM adaptive mesh methods. Developments of adaptive mesh methods in unstructured grids constitute one of the great achievements in the FEM research.

REFERENCES

Altas, I. and Stephenson, J. W. [1991]. A two-dimensional adaptive mesh generation method. *J. Comp. Phys.*, 94, 201–24.

Babuska, I. and Suri, M. [1990]. The p- and h-p versions of the finite element method. An overview. *Comp. Meth. Appl. Mech. Eng.*, 80, 5–26.

Babuska, I., Zienkiewicz, O. C., Gago, J., and Oliveira, E. R. A. (eds.) [1986]. *Accuracy Estimates and Adaptive Refinements in Finite Element Computations*. Chichester: Wiley.

Brackbill, J. U. [1982]. Coordinate System Control: Adaptive Meshes, Numerical Geneneration, Proceedings of a Symposium on the Numerical Generation of Curvilinear Coordinate Systems and their Use in the Numerical Solution of Partial Differential Equations (J. F. Thompson, ed.), New York: Elsevier, 277–94.

Brackbill, J. U. and Saltzman, J. S. [1982]. Adaptive zoning for singular problems in two dimensions. *J. Comp. Phys.*, 46, 342.

Chung, T. J. [1996]. *Applied Continuum Mechanics*. New York: Cambridge University Press.

Devloo, P., Oden, J. T., and Pattani, P. [1988]. An adaptive h-p finite element method for complex compressible viscous flows. *Comp. Meth. Appl. Mech. Eng.*, 70, 203–35.

Dwyer, H. A., Smooke, D. Mitchell, and Kee, Robert, J. [1982]. Adaptive gridding for finite difference solutions to heat and mass transfer problems. In J. F. Thompson (ed.). *Numerical Grid Generation*, New York: North-Holland, 339.

Eiseman, P. R. [1985]. Alternating direction adaptive grid generation. *AIAA J.*, 23, 551–60.
———. [1987]. Adaptive grid generation. *Comp. Meth. Appl. Mech. Eng.*, 64, 321–76.
Ghia, K. N., Osswald, G. A., and Ghia, U. [1989]. Analysis of incompressible massively separated viscous flows using unsteady Navier-Stokes equations. *Int. J. Num. Meth. Fl.*, 9, 1025–50.
Gnoffo, P. A. [1980]. Complete supersonic flowfields over blunt bodies in a generalized orthogonal coordinate system. NASA TM 81784.
Heard, G. A. and Chung, T. J. [2000]. Numerical simulation of 3-D hypersonic flow using flowfield-dependent variation theory combined with an *h*-refinement adaptive mesh. Presented at FEF2000, The University of Texas/Austin.
Kim, H. J. and Thompson, Joe F. [1990]. Three-dimensional adaptive grid generation on a composite-block grid. *AIAA J.*, 28, no. 3, 470–77.
Lohner, R. and Baum, J. D. [1990]. Numerical simulation of shock interaction with complex geometry three-dimensional structures using a new adaptive *h*-refinement scheme on unstructured grids. 28th Aerospace Sciences Meeting, January 8–11, 1990, Reno, Nevada, AIAA 90-0700.
Nakamura, S. [1982]. Marching grid generation using parabolic partial differential equations. In J. F. Thompson (ed.). *Numerical Grid Generation*, New York: North-Holland, 775.
Oden, J. T. [1988]. Adaptive FEM in complex flow problems. In J. R.Whiteman (ed.). *The Mathematics of Finite Elements with Applications*, Vol. 6, London: Academic Press, Lt., 1–29.
———. [1989]. Progress in adaptive methods in computational fluid dynamics. In J. Flaherty, et al. (ed.). *Adaptive Methods for Partial Differential Equations*, Philadelphia: SIAM Publications.
Oden, J. T., Babuska, I., and Baumann, C. E. [1998]. A discontinuous *hp* finite element method for diffusion problems. *J. Comp. Phys.*, 146, 491–519.
Oden, J. T., Strouboulis, T., and Devloo, P. [1986]. Adaptive finite element methods for the analysis of inviscid compressible flow: I. Fast refinement/unrefinement and moving mesh methods for unstructured meshes. *Comp. Meth. Appl. Mech. Eng.*, 59, no. 3, 327–62.
Oden, J. T., Wu, W., and Legat, V. [1995]. An *hp* adaptive strategy for finite element approximations of the Navier-Stokes equations. *Int. J. Num. Meth. Fl.*, 20, 831–51.
Peraire, J., Vahdati, M., Morgan, K., and Zienkiewicz, O. C. [1987]. Adaptive remeshing for compressible flow computations. *J. Comp. Phys.*, 72, no. 2, 449–66.
Probert, J., Hassan, O., Peraire, J., and Morgan, K. [1991]. An adaptive finite element method for transient compressible flows. *Int. J. Num. Meth. Eng.*, 32, 1145–59.
Yoon, W. S. and Chung, T. J. [1991]. Liquid propellant combustion waves. Washington, D.C.: AIAA paper AIAA-91-2088.

CHAPTER TWENTY

Computing Techniques

In Part Two and Part Three, various numerical schemes in CFD including FDM, FEM, and FVM have been discussed. We have presented methods of grid generation and adaptive meshing in both structured and unstructured grids in Part Four. Equation solvers for both linear and nonlinear algebraic equations resulting from FDM, FEM, and FVM have also been discussed in appropriate chapters. We are now at the stage of embarking on extensive CFD calculations in large-scale industrial problems, which will be presented in Part Five. To this end, it is informative to examine computational aspects associated with supercomputer applications and multi-processors. Among them are the domain decomposition methods (DDM), multigrid methods (MGM), and parallel processing. In DDM the domain of study is partitioned into substructures to make solvers perform more efficiently with reduction of storage requirements, whereas in MGM the solution convergence is accelerated with low-frequency errors being removed through coarse mesh configurations and with high-frequency errors removed through fine mesh configurations. These two methods lend themselves to parallel processing to speed up and reduce computer time. Development of parallel programs and both static and dynamic load balancing will be presented. The topics in this chapter are designed toward more robust computational strategies in dealing with geometrically complicated, large-scale CFD problems. Some selected example problems are also included.

20.1 DOMAIN DECOMPOSITION METHODS

In dealing with geometrically large, complicated systems, it is natural to seek an approach to split the domain into small pieces, known as domain decomposition methods (DDM). This is one of many possible applications to parallel processing to be discussed in Section 20.3. The basic idea of DDM was originated from the concept of linear algebra in solving the partial differential equations iteratively in subdomains, known as the Schwarz method [Schwarz, 1869]; and subsequently implemented in applications [Lions, 1988; Glowinski and Wheeler, 1988, among others]. The main advantages of DDM include efficiency of solvers, savings in computational storage conducive to parallel processing, and applications of different differential equations in different subdomains (representing viscous flow in one subdomain and inviscid flow in another subdomain, for example).

There are two approaches in the Schwarz method: (1) Multiplicative procedure which resembles the block Gauss-Seidel iteration, and (2) Additive procedure analogous to a block Jacobi iteration. We elaborate these procedures in the following sections.

20.1.1 MULTIPLICATIVE SCHWARZ PROCEDURE

In a typical domain decomposition approach, we divide the domain Ω into subdomains Ω_i such that

$$\Omega = \bigcup_{i=1}^{n} \Omega_i \tag{20.1.1}$$

an example of which is shown in Figure 20.1.1 In this example, there are three interior domains, $\Omega_1(1-12)$, $\Omega_2(13-21)$, $\Omega_3(22-27)$, and three boundary interfaces, $\Gamma_{1,2}, \Gamma_{1,3}, \Gamma_{2,3}(28-36)$. Here, for simplicity, boundary interface nodes are labeled last.

Let us consider the Poisson equation and the resulting matrix equations from FDM, FEM, or FVM formulations for this geometry in the form,

$$\begin{bmatrix} \underset{27\times 27}{K_{aa}} & \underset{27\times 9}{K_{ab}} \\ \underset{9\times 27}{K_{ba}} & \underset{9\times 9}{K_{bb}} \end{bmatrix} \begin{bmatrix} \underset{27\times 1}{U_a} \\ \underset{9\times 1}{U_b} \end{bmatrix} = \begin{bmatrix} \underset{27\times 1}{F_a} \\ \underset{9\times 1}{F_b} \end{bmatrix} \tag{20.1.2}$$

where the subscripts a, b denote the interior subdomains and interfaces, respectively, as related to the global stiffness matrix K_{aa} (27×27) with the subdomain stiffness matrices, $K_1(12 \times 12)$, $K_2(9 \times 9)$, $K_3(6 \times 6)$ for $\Omega_1, \Omega_2, \Omega_3$, respectively, and the boundary interface stiffness matrix, $K_{bb}(9 \times 9)$ together with the interface-subdomain interaction stiffness matrices $K_{ab}(27 \times 9)$ and $K_{ba}(9 \times 27)$ as shown in Figure 20.1.2. From the subdomain equations, we obtain

$$U_a = K_{aa}^{-1}(F_a - K_{ab}U_b) \tag{20.1.3}$$

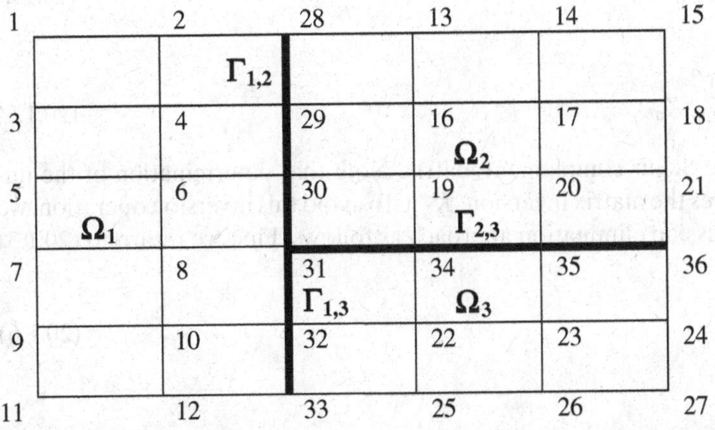

Figure 20.1.1 Decomposed domain (subdomains): Interior nodes (1–27), subdomain Ω_1 (1–12), subdomain Ω_2 (13–21), subdomain Ω_3 (22–27), Interfaces Γ_{12}, Γ_{13}, Γ_{23} (28–36).

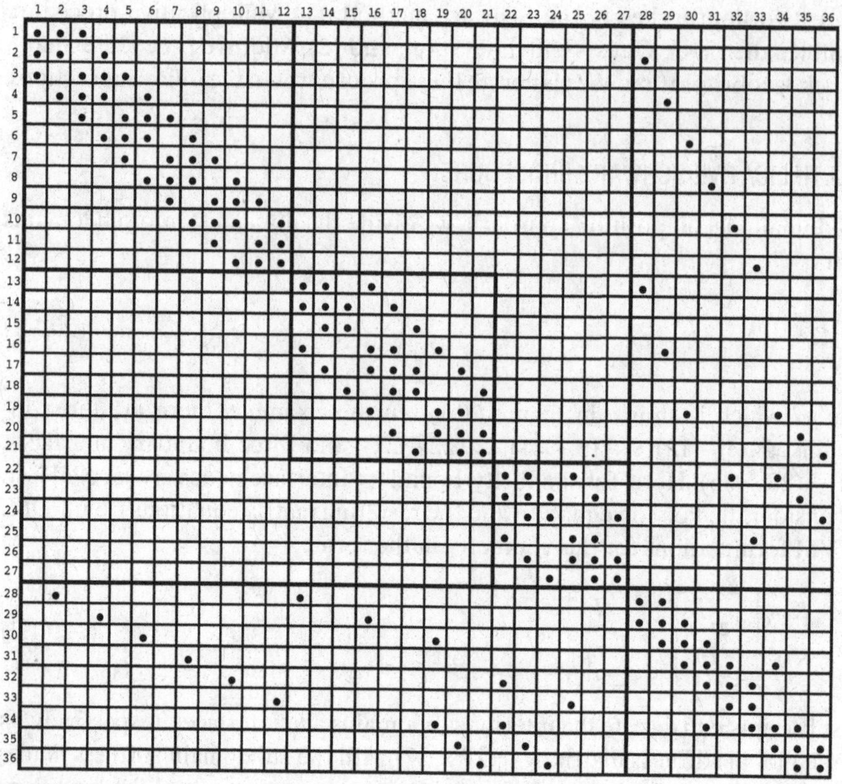

Figure 20.1.2 Global stiffness matrix, $K_{aa}(27 \times 27)$ for Figure 20.1.1 with the subdomain stiffness matrices $K_1(12 \times 12)$, $K_2(9 \times 9)$, $K_3(6 \times 6)$, for Ω_1, Ω_2, Ω_3, respectively, and the boundary interface stiffness matrix, $K_{bb}(9 \times 9)$ together with the interface-subdomain interaction stiffness matrices $K_{ab}(27 \times 9)$ and $K_{ba}(9 \times 27)$.

Substituting (20.1.3) into the interface equations leads to

$$S_{bb}U_b = F_b - K_{ba}K_{aa}^{-1}F_a \tag{20.1.4}$$

with

$$S_{bb} = K_{bb} - K_{ba}K_{aa}^{-1}K_{ab} \tag{20.1.5}$$

which is known as the Schur complement matrix. Note that determination of the unknowns U_a, U_b requires the matrix inversion, K_{aa}^{-1}. To avoid this inversion operation, we employ the block Gaussian elimination approach as follows: First we return to (20.1.3) and write in the form

$$U_a = F_a^* - K_{ab}^*U_b \tag{20.1.6}$$

with

$$F_a^* = K_{aa}^{-1}F_a \tag{20.1.7}$$

$$K_{ab}^* = K_{aa}^{-1}K_{ab} \tag{20.1.8}$$

Premultiplying F_a^* by K_{aa}, and K_{ab}^* by K_{aa}, we obtain, respectively,

$$K_{aa}F_a^* = K_{aa}K_{aa}^{-1}F_a = F_a \tag{20.1.9}$$

$$K_{aa}K_{ab}^* = K_{aa}K_{aa}^{-1}K_{ab} = K_{ab} \tag{20.1.10}$$

Now, any standard equation solver may be used to solve F_a^* and K_{ab}^* from (20.1.9) and (20.1.10), respectively. We then compute

$$F_b^* = F_b - K_{ba}F_a^* \tag{20.1.11}$$

and the Schur complement matrix in the form

$$S_{bb} = K_{bb} - K_{ba}K_{ab}^* \tag{20.1.12}$$

Finally, we solve the interface unknowns U_b using (20.1.11) and (20.1.12) from

$$S_{bb}U_b = F_b^* \tag{20.1.13}$$

and the interior subdomain unknowns using (20.1.9) and (20.1.10) from (20.1.3)

$$U_a = F_a^* - K_{ab}^*U_b \tag{20.1.14}$$

It is well known that any system of equations may be altered in such a manner that conditioning of the equations (eigenvalues) can be improved in order to assure accuracy. To this end, let us examine the global equation of the form

$$\underset{n\times n}{K}\;\underset{n\times 1}{U} = \underset{n\times 1}{F} \tag{20.1.15}$$

The preconditioned system of (20.1.15) may be written as

$$M^{-1}KU = M^{-1}F \tag{20.1.16}$$

where M is the preconditioning matrix and M^{-1} is the preconditioning operator. This is called the multiplicative Schwarz procedure which is equivalent to a block Gauss-Seidel iteration. In order to derive this preconditioning operator, we seek the restriction operator R_i and the prolongation operator (transpose of the restriction operator) with the subscript i denoting the number of subdomains such that

$$\underset{(n_i\times n_i)}{K_i} = \underset{(n_i\times n)}{R_i}\;\underset{(n\times n)}{K}\;\underset{(n\times n_i)}{R_i^T} \tag{20.1.17}$$

or

$$K^{-1} = R_i^T K_i^{-1} R_i \tag{20.1.18}$$

where the n_i refers to the total number of nodes for each subdomain and its boundary interface. Note that the subscript i here is not a tensorial index. For example, for the geometry represented by Figure 20.1.1, we have $n = 36$ and n_i for Ω_1, Ω_2, Ω_3 are 18, 16, 12, respectively, leading to the global stiffness matrix K shown in Figure 20.1.2. Here, the restriction matrices R_i consist of ones at associated nodes and zeros elsewhere (Figure 20.1.3), resulting in subdomain stiffness matrices as shown in Figure 20.1.4.

Let us assume that at each iterative solution step there is an error given by the error vector d,

$$d = U^* - U \tag{20.1.19}$$

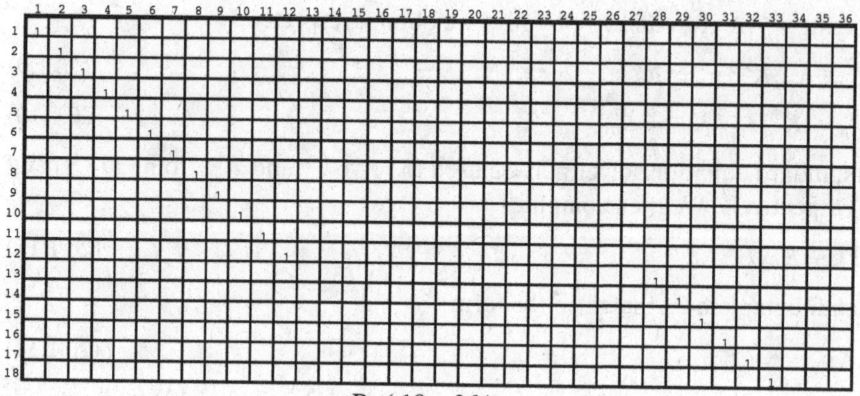

R_1 (18×36)

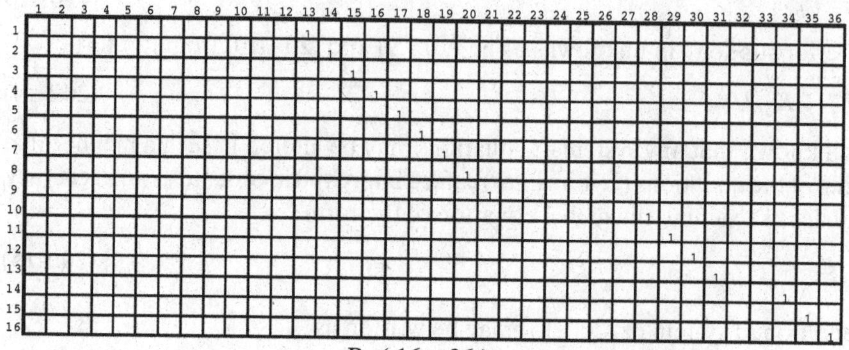

R_2 (16×36)

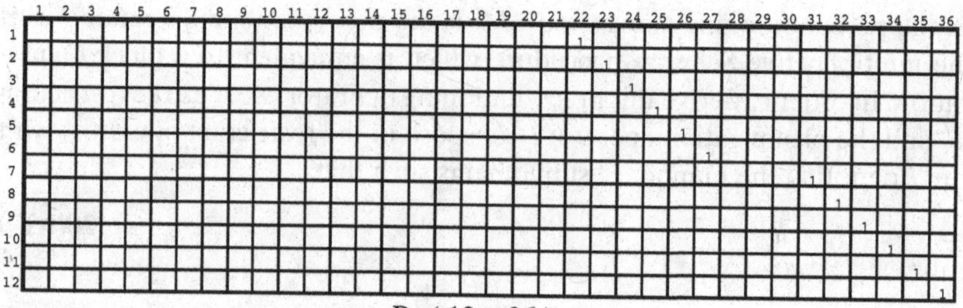

R_3 (12×36)

Figure 20.1.3 Restriction operators for subdomains given in Figure 20.1.1.

where U^* is the solution at the current step with U being the previous step. Then, we have

$$F - KU = Kd = K(U^* - U) \tag{20.1.20}$$

It follows from the above relations that

$$d = K^{-1}(F - KU) \tag{20.1.21}$$

$$U^* = U + R_i^T K_i^{-1} R_i(F - KU) \tag{20.1.22}$$

	1	2	3	4	5	6	7	8	9	10	11	12	13	14	15	16	17	18
1	$K_{1,1}$	$K_{1,2}$	$K_{1,3}$	0	0	0	0	0	0	0	0	0	0	0	0	0	0	0
2	$K_{2,1}$	$K_{2,2}$	0	$K_{2,4}$	0	0	0	0	0	0	0	0	$K_{2,28}$	0	0	0	0	0
3	$K_{3,1}$	0	$K_{3,3}$	$K_{3,4}$	$K_{3,5}$	0	0	0	0	0	0	0	0	0	0	0	0	0
4	0	$K_{4,2}$	$K_{4,3}$	$K_{4,4}$	0	$K_{4,6}$	0	0	0	0	0	0	0	$K_{4,29}$	0	0	0	0
5	0	0	$K_{5,3}$	0	$K_{5,5}$	$K_{5,6}$	$K_{5,7}$	0	0	0	0	0	0	0	0	0	0	0
6	0	0	0	$K_{6,4}$	$K_{6,5}$	$K_{6,6}$	0	$K_{6,8}$	0	0	0	0	0	0	$K_{6,30}$	0	0	0
7	0	0	0	0	$K_{7,5}$	0	$K_{7,7}$	$K_{7,8}$	$K_{7,9}$	0	0	0	0	0	0	0	0	0
8	0	0	0	0	0	$K_{8,6}$	$K_{8,7}$	$K_{8,8}$	0	$K_{8,10}$	0	0	0	0	0	$K_{8,31}$	0	0
9	0	0	0	0	0	0	$K_{9,7}$	0	$K_{9,9}$	$K_{9,10}$	$K_{9,11}$	0	0	0	0	0	0	0
10	0	0	0	0	0	0	0	$K_{10,8}$	$K_{10,9}$	$K_{10,10}$	0	$K_{10,12}$	0	0	0	0	$K_{10,32}$	0
11	0	0	0	0	0	0	0	0	$k_{11,9}$	0	$k_{11,11}$	$K_{11,12}$	0	0	0	0	0	0
12	0	0	0	0	0	0	0	0	0	$k_{12,10}$	$k_{12,11}$	$K_{12,12}$	0	0	0	0	0	$k_{12,33}$
13	0	$k_{13,2}$	0	0	0	0	0	0	0	0	0	0	$k_{13,28}$	$K_{13,29}$	0	0	0	0
14	0	0	0	$k_{14,4}$	0	0	0	0	0	0	0	0	$k_{14,28}$	$K_{14,29}$	$k_{14,30}$	0	0	0
15	0	0	0	0	0	$k_{15,6}$	0	0	0	0	0	0	0	$K_{15,29}$	$k_{15,30}$	$k_{15,31}$	0	0
16	0	0	0	0	0	0	0	$k_{16,8}$	0	0	0	0	0	0	$k_{16,30}$	$k_{16,31}$	$k_{16,32}$	0
17	0	0	0	0	0	0	0	0	0	$k_{17,10}$	0	0	0	0	0	$k_{17,31}$	$k_{17,32}$	$k_{17,33}$
18	0	0	0	0	0	0	0	0	0	0	0	$k_{18,12}$	0	0	0	0	$k_{18,32}$	$k_{18,33}$

	1	2	3	4	5	6	7	8	9	10	11	12	13	14	15	16
1	$K_{1,13}$	$K_{1,14}$	0	$K_{1,16}$	0	0	0	0	0	$K_{1,28}$	0	0	0	0	0	0
2	$K_{2,13}$	$K_{2,14}$	$K_{2,15}$	0	$K_{2,17}$	0	0	0	0	0	0	0	0	0	0	0
3	0	$K_{3,14}$	$K_{3,15}$	0	0	$K_{3,18}$	0	0	0	0	0	0	0	0	0	0
4	$K_{4,13}$	0	0	$K_{4,16}$	$K_{4,17}$	0	$K_{4,19}$	0	0	0	$K_{4,29}$	0	0	0	0	0
5	0	$K_{5,14}$	0	$K_{5,16}$	$K_{5,17}$	$K_{5,18}$	0	$K_{5,20}$	0	0	0	0	0	0	0	0
6	0	0	$K_{6,15}$	0	$K_{6,17}$	$K_{6,18}$	0	0	$K_{6,21}$	0	0	0	0	0	0	0
7	0	0	0	$K_{7,16}$	0	0	$K_{7,19}$	$K_{7,20}$	0	0	0	$K_{7,30}$	0	$K_{7,34}$	0	0
8	0	0	0	0	$K_{8,17}$	0	$K_{8,19}$	$K_{8,20}$	$K_{8,21}$	0	0	0	0	0	$K_{8,35}$	0
9	0	0	0	0	0	$K_{9,18}$	0	$K_{9,20}$	$K_{9,21}$	0	0	0	0	0	0	$K_{9,36}$
10	$K_{10,13}$	0	0	0	0	0	0	0	0	$K_{10,28}$	$K_{10,29}$	0	0	0	0	0
11	0	0	0	$k_{11,16}$	0	0	0	0	0	$k_{11,28}$	$k_{11,29}$	$k_{11,30}$	0	0	0	0
12	0	0	0	0	0	0	$k_{12,19}$	0	0	0	$k_{12,29}$	$k_{12,30}$	$k_{12,31}$	0	0	0
13	0	0	0	0	0	0	0	0	0	0	0	$k_{13,30}$	$k_{13,31}$	$k_{13,34}$	0	0
14	0	0	0	0	0	0	$k_{14,19}$	0	0	0	0	0	$k_{14,31}$	$k_{14,34}$	$k_{14,35}$	0
15	0	0	0	0	0	0	0	$k_{15,20}$	0	0	0	0	0	$k_{15,34}$	$k_{15,35}$	$k_{15,36}$
16	0	0	0	0	0	0	0	0	$k_{16,21}$	0	0	0	0	0	$k_{16,35}$	$k_{16,36}$

	1	2	3	4	5	6	7	8	9	10	11	12
1	$K_{1,22}$	$K_{1,23}$	0	$K_{1,25}$	0	0	0	$K_{1,32}$	0	$K_{1,34}$	0	0
2	$K_{2,22}$	$K_{2,23}$	$K_{2,24}$	0	$K_{2,26}$	0	0	0	0	0	$K_{2,35}$	0
3	0	$K_{3,23}$	$K_{3,24}$	0	0	$K_{3,27}$	0	0	0	0	0	$K_{3,36}$
4	$K_{4,22}$	0	0	$K_{4,25}$	$K_{4,26}$	0	0	0	$K_{4,33}$	0	0	0
5	0	$K_{5,23}$	0	$K_{5,25}$	$K_{5,26}$	$K_{5,27}$	0	0	0	0	0	0
6	0	0	$K_{6,24}$	0	$K_{6,26}$	$K_{6,27}$	0	0	0	0	0	0
7	0	0	0	0	0	0	$K_{7,31}$	$K_{7,32}$	0	$K_{7,34}$	0	0
8	$K_{8,22}$	0	0	0	0	0	$K_{8,31}$	$K_{8,32}$	$K_{8,33}$	0	0	0
9	0	0	0	$K_{9,25}$	0	0	0	$K_{9,32}$	$K_{9,33}$	0	0	0
10	$K_{10,22}$	0	0	0	0	0	$K_{10,31}$	0	0	$K_{10,34}$	$K_{10,35}$	0
11	0	$k_{11,23}$	0	0	0	0	0	0	0	$K_{11,34}$	$k_{11,35}$	$k_{11,36}$
12	0	0	$k_{12,24}$	0	0	0	0	0	0	0	$k_{12,35}$	$k_{12,36}$

Figure 20.1.4 Final forms of stiffness matrices.

Define the error e^* to be the difference between the right-hand side and the left-hand side of (20.1.22),

$$e^* = e - R_i^T K_i^{-1} R_i K(U^* - U) \tag{20.1.23}$$

which may be rewritten for subiteration steps i and $i-1$ as

$$e_i = e_{i-1} - R_i^T K_i^{-1} R_i K e_{i-1} \tag{20.1.24}$$

with $i = 1, \ldots s$, s being the total number of subdomains. This gives

$$e_i = (I - P_i)e_{i-1} \tag{20.1.25}$$

where P_i is known as the projector,

$$P_i = R_i^T K_i^{-1} R_i K \tag{20.1.26}$$

For the error at step s, we have

$$e_s = (I - P_s)(I - P_{s-1}) \ldots (I - P_1)e_0 \tag{20.1.27}$$

or

$$e_s = Q_s e_0 \tag{20.1.28}$$

with

$$Q_s = (I - P_s)(I - P_{s-1}) \ldots (I - P_1)$$

The multiplicative Schwarz procedure described above may be extended to overlapping subdomains, which will be elaborated in Section 20.4.1 together with parallel processing.

20.1.2 ADDITIVE SCHWARZ PROCEDURE

In contrast to the multiplicative Schwarz procedure, which is similar to the block Gauss-Seidel iteration, the additive Schwarz procedure consists of updating all the new block components from the same residual, analogous to a block Jacobi iteration, and thus the components in each subdomain are not updated until a whole cycle of updates through all domains is completed.

It follows from (20.1.22) and (20.1.26) that

$$U^* = \left(I - \sum_{i=1}^{s} P_i\right) U + \sum_{i=1}^{s} T_i F \tag{20.1.29}$$

with

$$T_i = P_i K^{-1} = R_i^T K_i^{-1} R_i \tag{20.1.30}$$

Note that, upon convergence, $U^* = U$, the solution (20.1.29) becomes

$$\sum_{i=1}^{s} P_i U = \sum_{i=1}^{s} T_i F \tag{20.1.31}$$

Comparing (20.1.16) and (20.1.31), we find that

$$\begin{aligned} \sum_{i=1}^{s} P_i &= M^{-1} K \\ \sum_{i=1}^{s} T_i &= \sum_{i=1}^{s} P_i K^{-1} = M^{-1} \end{aligned} \tag{20.1.32}$$

which identifies the preconditioning as given by (20.1.16),

$$M^{-1} K U = M^{-1} F$$

It is seen that the preconditioned iterative solution (20.1.29) has multiple benefits. Here, only the restricted and prolongated subdomain matrices are involved, the solution is more accurate due to preconditioning, convergence is faster, and computational storage requirements are less with domain decomposition.

The domain decomposition may be carried out in unstructured grids. The basic algebra for the structured grids presented above can be applied equally well to the unstructured grids. Furthermore, the domain decomposition lends itself to parallel processing which will be presented in Section 20.3. Examples of both overlapping and nonoverlapping subdomains together with parallel processing will be presented in Section 20.3.4.

20.2 MULTIGRID METHODS

20.2.1 GENERAL

The basic idea of multigrid methods (MGM), as originally pioneered by Brandt [1972, 1977, 1992], is to accelerate the convergence of iterative solvers. The low-frequency or large wavelength components of error on a fine mesh become high frequency or small wavelength components on a coarser mesh. Thus, it is preferable to use coarse grids to remove low-frequency errors, with accuracy ensured by means of fine grids. Two or more levels of solutions from fine to coarse grids (restriction process) and from coarse to fine grids (prolongation process) may be repeated until convergence is reached. In general, MGM is regarded as the most efficient technique to accelerate convergence among the iterative methods in solving the linear and nonlinear algebraic equations.

In multigrid operations, asymptotic behavior of the error (or of the residual) is dominated by the eigenvalues of the amplification matrix close to one in absolute value. The error components situated in the low-frequency range of the spectrum of the space-discretization are the slowest to be damped in the iterative process. The higher frequencies are the first to be reduced and a large part of the high-frequency error components will be damped, thus acting as a smoother of the error.

The simplest case of a multigrid procedure consists of nested structured grid in which a fine grid is coarsened by eliminating every other node in all directions so that all nodes in the coarse mesh appear in the fine mesh. In contrast, unstructured grids are in general unnested. We present the general procedure of nested structured multigrid methods in Section 20.2.2, followed by unnested unstructured multigrid methods in Section 20.2.3.

20.2.2 MULTIGRID SOLUTION PROCEDURE ON STRUCTURED GRIDS

For structured grid FDM computations, we may begin with the finest grid and coarsen the mesh by eliminating every other node, resulting in nested grids. An example for the three-level nested multigrid system is shown in Figure 20.2.1. In practice, several levels of multigrid discretization are desirable. The simplest descriptions of multigrid methods may be given as follows:

Restriction Process

Do n iterations (two or three relaxation sweeps) on the fine grid using any iterative solution method such as the Gauss-Seidel scheme. Interpolate the residual R onto the

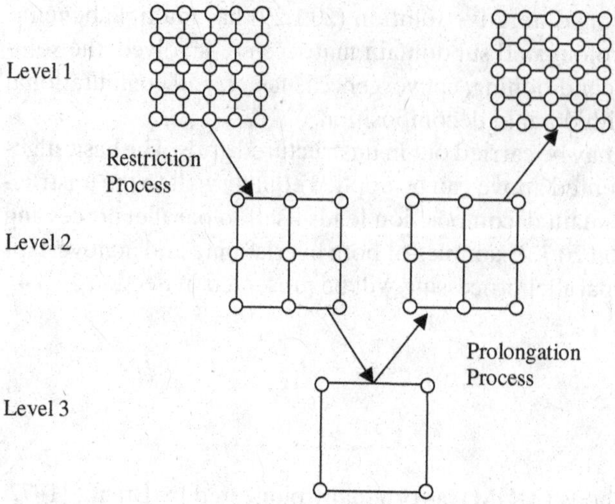

Figure 20.2.1 Three-level multigrid discretization.

coarse grid, with the residuals not updated during the iterations because we want the computed values to represent corrections to the fine-grid solution. This is known as the restriction process (restricted to the next coarsest grid), represented by the following steps:

Step 1 (finest grid)	$LU_1 = R_1$
Step 2 (level 2 coarse grid)	$L\Delta U_2 = -I_1^2 R_1$
Step 3 (level 3 coarse grid)	$L\Delta U_3 = -I_2^3 R_2 = -I_2^3(L\Delta U_2 + I_1^2 R_1)$
.	
.	

where L is the differential or difference operator and the subscript represents the previous step with the superscript being the destination. For illustration, the restriction process of two-level one-dimensional Poisson problems is shown in Figure 20.2.2(a) with the restriction operator I_1^2 given by

$$I_1^2 = \frac{1}{4}\begin{bmatrix} 1 & 2 & 1 & & \\ & & 1 & 2 & 1 \end{bmatrix}$$

Prolongation Process

With the corrections now available on the coarsest grid, they are to be prolongated or interpolated onto the next finer grid by adding them to the previous corrections obtained at the restriction process, until the finest grid is reached. These steps are given as follows:

Step 4 (level 3 fine grid)	$L\Delta U_3 = -I_3^2 R_3$	(updated from Step 3)
Step 5 (level 2 fine grid)	$L\Delta U_2 = -I_2^1 R_2$	(updated from Step 2)
Step 6 (level 1 fine grid)	$L\Delta U_1 = R_1$	(updated from Step 1)

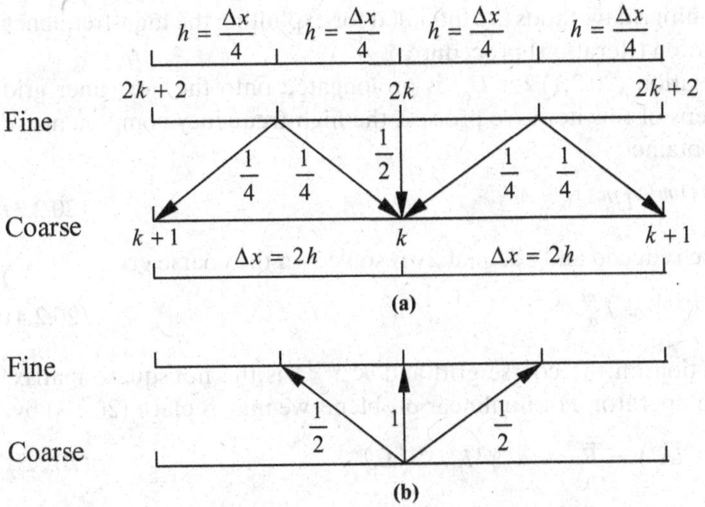

Figure 20.2.2 Restriction and prolongation. (a) Restriction (weighting). (b) Prolongation (interpolation).

where the subscript denotes the previous step with the superscript being the destination. Here, the prolongation operator for one-dimensional Poisson solver assumes the form,

$$I_2^1 = \frac{1}{2}\begin{bmatrix} 1 & \\ 2 & \\ 1 & 1 \\ & 2 \\ & 1 \end{bmatrix}$$

The above procedure is known as the V-cycle [Figure 20.2.3(a)]. The V-cycle may be repeated as many times as required until the desired accuracy is obtained, resulting in the so-called W-cycles [Figure 20.2.3(b)].

To illustrate the MGM procedure using the FEM notations, let us assume that we have a sequence of grids m and $m+1$ to solve the FEM equations of the form,

$$K_{\alpha\beta}^{m,m} U_\beta^m = F_\alpha^m \qquad \text{on coarse grid} \tag{20.2.1}$$

$$K_{\alpha\beta}^{m+1,m+1} U_\beta^{m+1} = F_\alpha^{m+1} \qquad \text{on fine grid} \tag{20.2.2}$$

with α, β denoting the number of global nodes for either coarse grid or fine grid. A low-frequency component on a fine grid becomes a high-frequency component on a

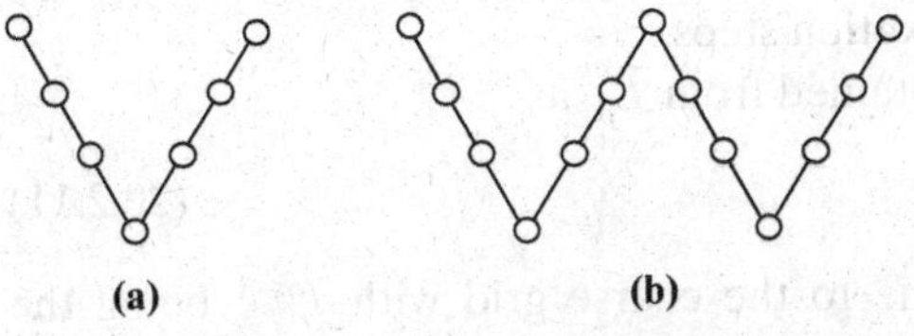

Figure 20.2.3 (a) V-cycle multigrid. (b) W-cycle multigrid.

coarse grid. Thus, the multigrid methods are intended for exploiting the high-frequency smoothing of the relaxation (iteration) procedure.

The coarse grid equation (20.2.1) for U_α^m is prolongated onto the next finer grid (20.2.2). After a few steps of this iterative process, the high-frequency components of the residual E^{m+1} are obtained

$$E_\alpha^{m+1} = F_\alpha^{m+1} - K_{\alpha\beta}^{m+1,m+1} U_\beta^{m+1} \tag{20.2.3}$$

The residual can then be reduced and adequately resolved on the coarse grid:

$$K_{\alpha\beta}^{m,m} \overline{U}_\beta^m = K_{\alpha\beta}^{m,m+1} E_\beta^{m+1} = \overline{F}_\alpha^m \tag{20.2.4}$$

where $\overline{U}_\beta^m$ is the correction on the coarse grid and $K_{\alpha\beta}^{m,m+1}$ is the nonsquare matrix, known as the restriction operator. For nonlinear problems we may replace (20.2.3) by

$$K_{\alpha\gamma}^{m,m}\left(K_{\gamma\beta}^{m,m+1} U_\beta^{m+1} + \overline{U}_\gamma^m\right) = \overline{F}_\alpha^m + K_{\alpha\gamma}^{m,m} K_{\gamma\beta}^{m,m+1} U_\beta^{m+1} \tag{20.2.5}$$

or

$$K_{\alpha\beta}^{m,m} \overline{U}_\beta^m = \overline{F}_\alpha^m \tag{20.2.6}$$

The solution of either (20.2.4) for linear problems or (20.2.6) for nonlinear problems enables U_β^{m+1} to be updated by adding to it the prolongation of $\overline{U}_\beta^m$ onto the finer grid so that U_β^{m+1} as calculated from (20.1.2) is updated to $\overline{U}_\beta^{m+1}$ as

$$\overline{U}_\alpha^{m+1} = U_\alpha^{m+1} + K_{\alpha\gamma}^{m+1,m}\left(U_\gamma^m - K_{\gamma\beta}^{m,m+1} U_\beta^{m+1}\right) \tag{20.2.7}$$

where $K_{\alpha\gamma}^{m+1,m}$ is the nonsquare matrix, known as the prolongation operator. The procedure described above will be repeated until the converged solution of (20.2.2) is obtained.

If FDM discretizations are employed, the restriction and prolongation operators can be replaced by appropriate finite difference formulas. To identify these operators, let us begin with the FDM formulations using the FEM notations.

$$K_{\alpha\beta}^{m+1} \Delta U_\beta^{m+1} = F_\alpha^{m+1} - K_{\alpha\beta}^{m+1} U_\beta^m = E_\alpha^{m+1} \tag{20.2.8}$$

with

$$U_\alpha^{m+1} = U_\alpha^m + \Delta U_\alpha^{m+1} \tag{20.2.9}$$

The residual E_α^m upon a few relaxation steps on the $(m+1)$th grid to smooth the high-frequency components is of the form

$$\overline{E}_\alpha^m = E_\alpha^m - K_{\alpha\beta}^m \Delta \overline{U}_\beta^m \tag{20.2.10}$$

where $\Delta \overline{U}_\beta^m$ is obtained through a few relaxation steps.

The residual E_α^{m-1} on the mth grid is obtained from $\overline{E}_\alpha^m$ as

$$E_\alpha^{m-1} = I_r^{m-1} E_{\alpha r}^m \tag{20.2.11}$$

which represents the transfer from the fine to the coarse grid with I_r^{m-1} being the restriction operator similar to $K_{\alpha\beta}^{m,m+1}$ in (20.2.4). This operator shows how the mesh values on the coarse grid are derived from the surrounding fine mesh values. This is a contrast to the nonsquare matrix $K_{\alpha\beta}^{m,m+1}$ relating the stiffness matrix of the coarse grid to the fine grid.

These relaxation processes (20.2.10) and restriction processes (20.2.11) continue until the coarsest grid $m = 1$ is reached. The correction $\Delta\overline{U}_\alpha^{m+1}$ is obtained from $\Delta\overline{U}_\alpha^{m}$ by prolongation factor I_r^{m+1} as we move from a coarse grid to a fine grid,

$$\Delta U_\alpha^{m+1} = I_r^{m+1}\Delta U_{\alpha r}^{m} \tag{20.2.12}$$

in which a few relaxation steps are to be made on a fine grid.

20.2.3 MULTIGRID SOLUTION PROCEDURE ON UNSTRUCTURED GRIDS

Although the multigrid methods can be applied to the structured grids for both FEM and FDM with nested grids in which all coarse grid nodes appear in the fine grid nodes, such nested structured grids may not be possible in general or impractical for complicated geometries. Multigrid methods for unstructured and unnested grids have been studied by many investigators in the past [Löhner and Morgan, 1987; Mavriplis and Jameson, 1990, among others]. To this end, it is convenient to use FEM or FVM in which any arbitrary geometries may easily be accommodated.

Let us consider an unstructured grid system as shown in Figure 20.2.4 in which the fine mesh A, B, C, D contains the coarse grid node 1 and the coarse grid element 1, 2, 3, 4 contains a fine grid element H, I, J, K. Our task is to determine how the flow variables, residuals, and corrections are to be transformed between fine and coarse grids. This can be achieved as follows:

(1) **Flow variables** A linear interpolation is used to transfer from the fine grid nodes A, B, C, D to the coarse grid node 1.
(2) **Residuals** (Restriction process) The fine grid residual at H is linearly distributed to the coarse grid nodes 1, 2, 3, 4 enclosing H. It is convenient to use the finite element trial functions for this purpose.
(3) **Corrections** (Prolongation process) The corrections from the coarse grid back to the fine grid E can be transferred using a simple linear interpolation of the

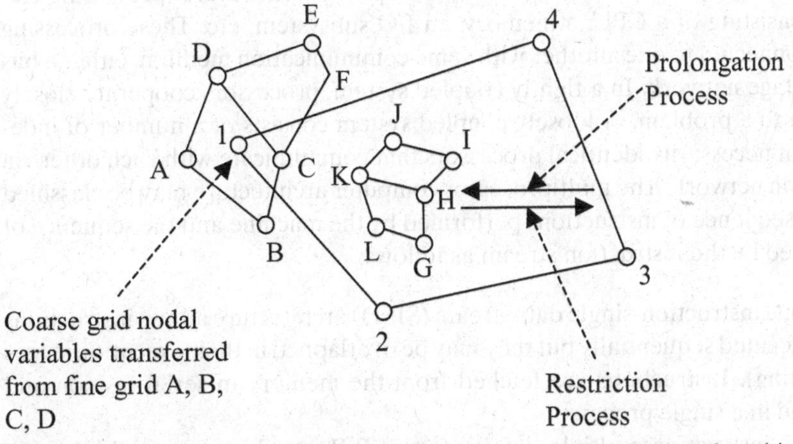

Figure 20.2.4 Grid transfers from coarse to fine and vice versa in unstructured quadrilateral multigrids.

coarse nodes 1, 2, 3, 4. An efficient strategy such as tree search algorithm may be employed to locate the coarse grid cell enclosing a particular fine grid node. In this algorithm, it requires information about the neighbors of each node or cell and a series of tests are carried out to determine if the coarse grid cell encloses the fine grid node.

As was indicated in Section 20.1 for domain decomposition, the parallel processing can be applied to multigrid methods also to obtain speedup in computer time. We shall discuss the subject of parallel processing in Section 20.3.

20.3 PARALLEL PROCESSING

20.3.1 GENERAL

Computational procedures in CFD in general as well as the adaptive mesh (Chapter 19), domain decomposition (Section 20.1), and multigrid methods (Section 20.2) discussed earlier will benefit from parallel processing, in which significant computational efficiency can be achieved. There are different forms of parallelism: multiple functional units, pipelining, vector processing, multiple vector pipelines, multiprocessing, and distributed computing.

In multiple functional units, we multiply the number of functional units such as adders and multipliers together. Here, the control units and the registers are shared by the functional units.

The concept of pipelining resembles an automobile assembly line. Let us assume that n number of operations takes s stages to complete in time t. The speedup factor S in this case can be given by the ratio, $S = nst/[(n+s-1)t]$. It is seen that for a large number of operations, the speedup factor is approximately equal to the number of stages.

Vector computers are equipped with vector pipelines such as a pipeline floating point adder or multiplier. Also, vector pipe lines can be duplicated to take advantage of any fine grain parallelism available in loops.

A multiprocessor system is a set of several computers with several processing elements, each consisting of a CPU, a memory, an I/O subsystem, etc. These processing elements are connected to one another with some communication medium, either a bus or some multistage network. In a tightly coupled system, processors cooperate closely on the solution to a problem. A loosely coupled system consists of a number of independent and not necessarily identical processors that communicate with each other via a communication network. The multiprocessor computer architecture may be classified in terms of the sequence of instructions performed by the machine and the sequence of data manipulated by the instruction stream as follows:

(1) The single instruction-single data stream (SISD) architecture allows instructions to be executed sequentially but they may be overlapped in their execution stages (pipelining). Instructions are fetched from the memory in serial fashion and executed in a single processor.
(2) In single instruction-multiple data stream (SIMD) architecture multiple processing elements are all supervised by the same control unit. All processors

receive the same instructions broadcast from the control unit, but operate on different data sets from distinct data streams.

(3) With multiple instruction-multiple data stream (MIMD), each processor has its own control unit and the processors execute independently. The processors interact with each other either through shared memory or by using message passing to execute an application.

Distributed computing is a more general form of multiprocessing, linked by some local area network such as the parallel virtual machine (PVM) and the message passing interface (MPI). This system is cost effective for large applications with high volume of computation performed before more data is to be exchanged. In distributed multiprocessors, each processor has a private or local memory but there is no global shared memory in the system. The processors are connected using an interconnection network, and they communicate with each other only by passing messages over the network.

Multiprocessors rely on distributed memory in which processing nodes have access only to their local memory, and access to remote data is accomplished by request and reply messages. Numerous designs on how to interconnect the processing nodes and memory modules include Intel Paragon, N-Cube, and IBM's SP systems. As compared to shared memory systems, distributed (or message passing) systems can accommodate a larger number of computing nodes.

Although parallel processing systems, particularly those based on the message passing (or distributed memory) model, have led to several large-scale computing systems and specialized supercomputers, their use has been limited for very specialized applications. This is because message passing is difficult when a sequential version of the program as well as the message passing version is to be maintained. Thus, the new trend is that the programmers approach the two versions completely independently and that programming on a shared memory multiprocessor system (SMP) is considered easier. In shared memory paradigm, all processors or threads of computation share the same logical address space and access directly any part of the data structure in a parallel computation. A single address space enhances the programmability of a parallel machine by reducing the problems of data partitioning, migration, and local balancing. The shared memory also improves the ability of parallelizing compilers, standard operating systems, resource management, and incremental performance.

In the following sections, we discuss the development of parallel algorithms, parallel solution of linear systems on SIMD and MIMD machines, and applications of parallel processing in domain decomposition and multigrid methods, new trends in parallel processing, and some selected CFD problems.

20.3.2 DEVELOPMENT OF PARALLEL ALGORITHMS

SIMD and MIMD Structures

In numerical methods such as CFD, the basis for development of parallel algorithms is the evaluation of arithmetic expressions. The evaluation can be represented by graphs or trees. To this end, let us consider the problem of mapping a given arithmetic expression E into an equivalent expression $\tilde{E}$ that can be performed parallel on SIMD or MIMD computers by means of commutative, distributive, or associative laws of linear

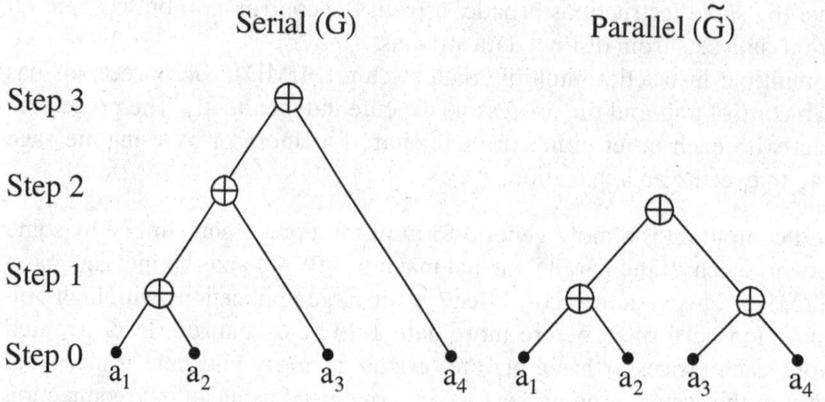

Figure 20.3.1 SIMD structure.

algebra. For example, two additions can be made parallel as follows:

$$E = a_4 + [a_3 + (a_2 + a_1)] \tag{20.3.1}$$

This can be transformed by the associativity of addition into

$$\tilde{E} = (a_4 + a_3) + (a_2 + a_1) \tag{20.3.2}$$

A typical SIMD structure is characterized by

$$E = a_1 + a_2 + a_3 + a_4 \tag{20.3.3}$$

By using the associative property of addition, we obtain

$$\tilde{E} = (a_1 + a_2) + (a_3 + a_4) \tag{20.3.4}$$

as schematically shown in Figure 20.3.1 in which G and $\tilde{G}$ denote the serial tree and parallel tree, respectively.

In MIMD structure, if we wish to compute

$$E = a_1 + a_2 \times a_3 + a_4 \tag{20.3.5}$$

it should be noted that the serial tree G is not a unique tree, and no tree height reduction can be obtained by applying the associative law. Instead, we apply the commutative property of addition with E being transformed into

$$\tilde{E} = (a_1 + a_4) + a_2a_3 \tag{20.3.6}$$

with the tree height reduced by one step as shown in Figure 20.3.2.

The speedup of a parallel algorithm is given by

$$S_p = T_1/T_p \tag{20.3.7}$$

where T_p is the execution time using p processors. The efficiency is defined by

$$E_p = S_p/p \tag{20.3.8}$$

Thus, for the case shown in Figure 20.3.2, we obtain $T_2 = 2$, $S_2 = T_1/T_2 = 3/2$, $E_2 = S_2/2 = 3/4$. In parallel processing, we must determine how many tree height reductions

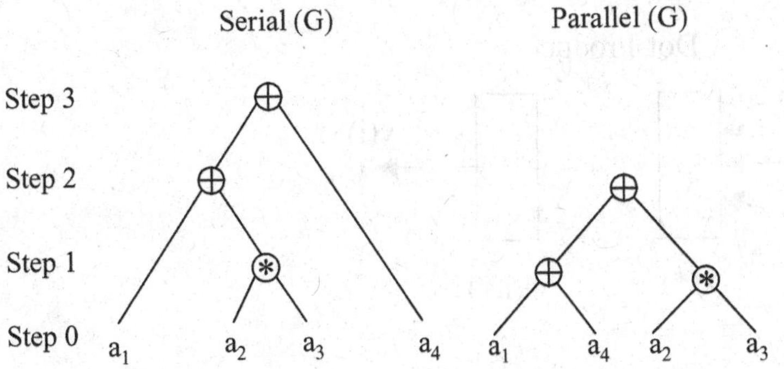

Figure 20.3.2 MIMD structure.

can be achieved for a given arithmetic expression and how many processors are needed for optimality.

Matrix-by-Vector Products in Parallel Processing

Matrix-by-vector multiplications are easy to implement on high-performance computers. Consider the matrix-by-vector product $y = Ax$. One of the most general schemes for storing matrices is the compressed sparse row (CSR) format. Here, the data structure consists of three arrays: a real array $A(1\!:\!nnz)$ to store the column positions of the elements row-wise, an integer array $JA(1\!:\!nnz)$ to store the column positions of the elements in the real array A, and finally, a pointer array $IA(1\!:\!n+1)$, the ith entry of which points to the beginning of the ith row in the arrays A and JA. Here, we note that each component of the resulting vector y can be computed independently as the dot product of the ith row of the matrix with the vector x. The algorithm for CSR format-dot product form may be given as follows:

1. $Do\ i = 1, n$
2. $k1 = ia(i)$
3. $k2 = ia(i+1) - 1$
4. $y(i) = dot\ product(a(k1 : k2), x(ja(k1 : k2)))$
5. $EndDo$

Note that the outer loop can be performed in parallel on any parallel platform. On some shared memory machines, the synchronization of this outer loop is inexpensive and the performance of the above program can be effective. On distributed memory machines, the outer loop can be split in a number of steps to be executed on each processor. It is possible to assign a certain number of rows (often contiguous) to each processor and to also assign the component of each of the vectors similarly. When performing a matrix-by-vector product, interprocessor communication will be necessary to get the needed components of the vector x that do not reside in a given processor.

The indirect addressing involved in the second vector in the dot product is called a *gather* operation. The vector $x(ja(k1 : k2))$ is first "*gathered*" from memory into a vector of contiguous elements. The dot product is then carried out as standard dot-product operation between two dense vectors, as illustrated in Figure 20.3.3.

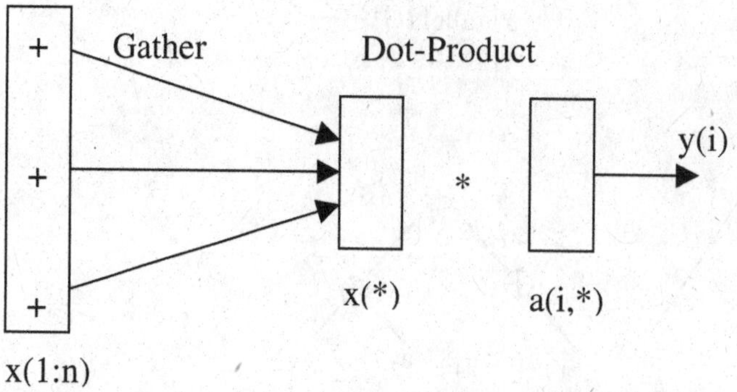

Figure 20.3.3 Row-oriented matrix-by-vector multiplication – Gather operation.

Let us now assume that the matrix is stored by compressed sparse column (CSC) storage format. The matrix-by-vector product can be performed by the following algorithm:

1. $y(1:n) = 0.0$
2. $Do\, i = 1, n$
3. $k1 = ia(i)$
4. $k2 = ia(i+1) - 1$
5. $y(ja(k1:k2)) = y(ja(k1:k2)) + x(j)^{*}a(k1:k2)$
6. *EndDo*

The above code initializes y to zero and then adds the vectors $x(j) \times a(1{:}n, j)$ for $j = 1, \ldots, n$ to it. It can also be used to compute the product of the transpose of a matrix by a vector, when the matrix is stored (row-wise) in the CSR format. Normally the vector $y(ja(k1{:}k2))$ is gathered and the vector update operation is performed in vector mode. Then the resulting vector is "*scattered*" back into the positions $ja(*)$ by what is called a *scatter* operation, as illustrated in Figure 20.3.4.

Notice that the inner loop involves writing back results of the right-hand side into memory positions that are determined by the indirect address function *ja*. To be correct, $y(ja(1))$ must be copied first, followed by $y(ja(2))$, etc. However, if it is known that the mapping $ja(i)$ is one-to-one, then the order of the assignments no longer matters. Since

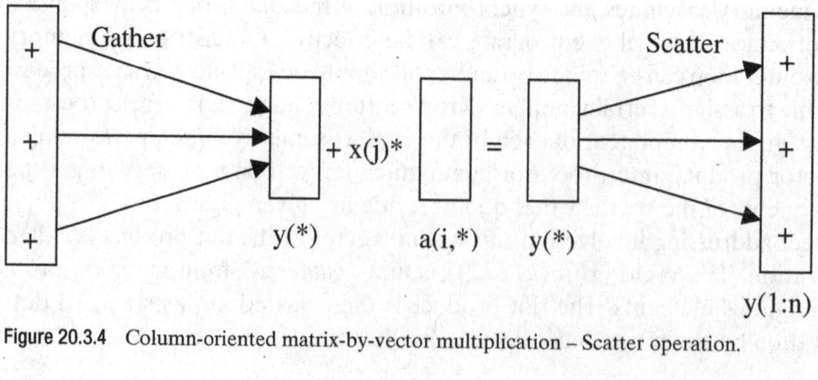

Figure 20.3.4 Column-oriented matrix-by-vector multiplication – Scatter operation.

compilers are not capable of deciding whether this is the case, a compiler directive from the user is necessary for the scatter to be invoked.

20.3.3 PARALLEL PROCESSING WITH DOMAIN DECOMPOSITION AND MULTIGRID METHODS

Although it is difficult to characterize multiprocessors in a simple manner, we may assume that they are individual processors and memory modules that are interconnected in some way. This interconnection can occur in a number of ways, but in general, processor memory modules communicate with one another directly or through a common shared memory. The processing unit in the model can be a simple bit processor, a scalar processor, or a vector processor. The memory unit in the module can be a few registers or a cache memory. Because of nonlinearity in fluid mechanics, it is important that the interaction between the computer modules in a multiprocessing system be controlled by a single operating system.

There are two forms of multiprocessors: the loosely coupled or distributed memory multiprocessors and the tightly coupled or shared memory multiprocessors. In a loosely coupled system, each computer module has a relatively large local memory where it accesses most of the instructions and data. Because there is no shared memory, processes executing on different computer modules communicate by exchanging messages through an interconnection network. In fact, the communication topology of this interconnection network is the crucial factor of these systems. Thus, loosely coupled systems are usually efficient when the interaction between computational tasks is minimal.

Tightly coupled multiprocessor systems communicate through a globally shared memory. Hence, the rate at which data can communicate from one computer module to the other is of the order of the bandwidth of the memory. Because of the complete connectivity between the computer modules and memory, the performance may tend to degrade due to memory contentions.

Ideal numerical models for multiprocessors are those that can be broken down into algebraic tasks, each of which can be executed independently on a computer module without ever having to obtain or pass data between the modules during the course of the execution. This framework allows a mechanism for analyzing the movement of data within a multiprocessing system. The basic idea is to regard the computational tasks being performed by the individual computer modules as numerical solutions of individual boundary value problems. In this way numerical data being obtained or transmitted between computer modules are the initial and boundary data of the differential equations. The solution of the overall mathematical model is then provided by "piecing" together each of the subproblems.

For the domain decomposition methods presented in Section 20.1, the domain $\Omega(t)$ is expressed as a union of subdomains (such as in Figure 20.1.1)

$$\Omega(t) = \bigcup_{j=1}^{k(t)} \Omega_j(t) \tag{20.3.9}$$

Each processor then assumes the task of solving one or more of the partial differential equations over a prescribed time interval Δt. At the end of this time interval, a new substructuring of the domain is performed:

$$\Omega(t+\Delta t) = \bigcup_{j=1}^{k(t+\Delta t)} \Omega_j(t, \Delta t) \tag{20.3.10}$$

and the process is repeated. The numerical mathematical relationship between the computed subdomain solutions and the solution of the global problem is delicate and is a function of the partial differential equation being solved. However, it is precisely this relationship that determines the efficiency of the computation on a multiprocessing system.

New Trends in Parallel Processing

It appears that the use of small clusters of SMP systems, often interconnected to address the needs of complex problems requiring the use of large numbers of processing nodes, is gaining popularity [Kavi, 1999]. Even when working with networked resources, programmers are relying on messaging standards such as MPI and PVM or relying on systems software to automatically generate message passing code from user-defined shared memory programs. The reliance on software support to provide a shared memory programming model (i.e., distributed shared memory systems) can be viewed as a logical evolution in parallel processing. Distributed shared memory (DSM) systems aim to unify parallel processing systems that rely on message passing with the shared memory systems. The use of distributed memory systems as shared memory systems addresses the major limitation of SMPs, namely scalability.

The growing interest in multithreading programming and the availability of systems supporting multithreading (Pthreads, NT-threads, Linux threads, Java) further emphasizes the trend toward shared memory programming model [Nichol, Buttlar, and Farrell, 1996]. The so-called OpenMP Fortran is designed for the development of portable parallel programs on shared memory parallel computer systems. One effect of the OpenMP standard will be to increase the shift of complex scientific and engineering software development from the supercomputer world to high-end desktop workstations.

Distributed shared memory systems (DSM) attempt to unify the message passing and shared memory programming models. Since DSMs span both physically shared and physically distributed memory systems, DSMs are also concerned with the interconnection networks that provide the data to the requesting processor in an efficient and timely fashion. Both the bandwidth (amount of data that can be supplied in a unit time) and latency (the time it takes to receive the first piece of requested data from the time the request is issued) are important to the design of DSM. It should be noted that because of the generally longer latencies encountered in large-scale DSMs, multithreading has received considerable attention in order to tolerate (or mask) memory latencies.

The management of large logical memory space involves moving data dynamically across the memory layers of a distributed system. This includes the mapping of the user data to the various memory modules. The data may be uniquely mapped to a physical address as done in cache coherent systems, or replicating the data to several physical addresses as done in reflective memory systems and, to some extent, in cache-only systems. Even in uniquely mapped systems, data may be replicated in lower levels of memories (i.e., cache). Replication, in turn, requires means for maintaining consistency. Directories are often the key to tracking the multiple copies and play a key role in coherency of replicated data. Hardware or software can maintain the coherency.

The granularity of the data that is shared and moved across memory hierarchies is another design consideration. The granularity can be based on objects without semantic meaning, based purely on a sequence of bytes (e.g., a memory word, a cache block, or a page) or it can be based on objects with semantic basis (e.g., variable, data structures or objects in the sense of object-oriented programming model). Hardware solutions often use finer grained objects (often without semantic meaning) while software implementations rely on coarser grained objects.

Synchronization and coordination among concurrent computations (i.e., processes or threads) in shared memory programming rely on the use of mutual exclusions, critical sections, and barriers. Implementation of the mechanisms needed for mutual exclusion (often based on the use of locks) and barriers explores the design space spanning hardware instructions, spin locks, tree barriers, etc. These solutions explore trade-offs between the performance (in terms of latency, serialization, network traffic) and scalability.

In addition to the issues mentioned above, resource management – particularly in terms of minimizing data migration, thread migration, and messages exchanged – are all significant in achieving cost-performance ratios in making DSM systems as contenders.

In summary, threads are used in multitasking to better utilize processor capabilities, resulting in better throughput and response time. For example, we can assign a separate task/thread to process user command inputs, while other tasks/threads can do the actual processing. This eliminates the need for "freezing" keyboard while processing a previous request. Thus, threads represent "concurrency." The complex flow without threads and simple flow with threads are illustrated in Figure 20.3.5a and Figure 20.3.5b, respectively. It is seen that a complex program can be viewed as concurrent activities and the flow of control becomes cleaner (Figure 20.3.5b). Without threads, we can still simulate the concurrent activities by interleaving the execution. Such structures cannot utilize multiple CPUs or kernel threads. With threads, we can use multiple processors or kernel threads. If multiple processors are available, we can increase the performance by using

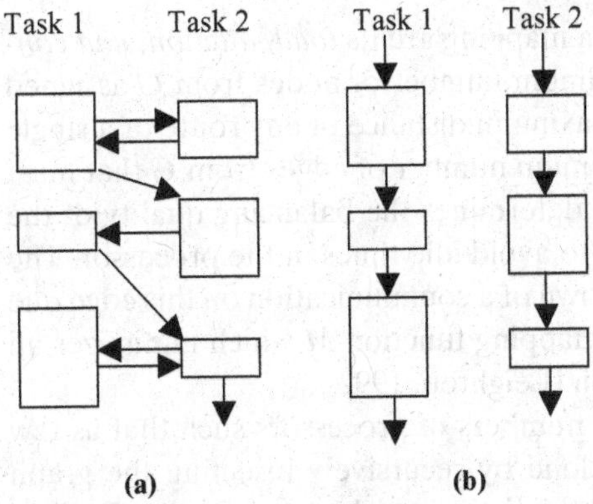

Figure 20.3.5 Threads and concurrency. (a) Complex flow without threads or concurrency. (b) Simple flow with threads or concurrency.

multiple concurrent activities. Multitasking or other concurrent programming methods utilize the multiple processing units. Multithreaded programs can be executed either on a single processor system or on an SMP with minimum changes. This is in contrast to traditional (old) parallel programming which requires careful and tedious changes to the program structure to utilize the multiple-processing unit. As each workstation is becoming more powerful and cheaper, the trend has been to use a network of such systems instead of supercomputers or massively parallel systems.

20.3.4 LOAD BALANCING

An important consideration in CFD is the problem of distributing the mesh across the memory of the machine at runtime so that the calculated load is evenly balanced and the amount of interprocessor communication is minimized. Load balancing is difficult in large distributed systems. Algorithms must minimize both load balance and communication overhead of the application. These algorithms should balance the load with as little overhead as possible, and they should be scalable.

We consider a parallel system as with P processors as a graph $H = (U, F)$ with nodes $U = \{0, \ldots, P-1\}$ and edges $F \subseteq U \times U$. Similarly, a parallel application is modeled as graph $G = (V, E, \rho, \sigma)$ with nodes $V = \{0, \ldots, N-1\}$, edges $E \subseteq V \times V$, node weights $\rho : V \to \tilde{R}$, and edge weights $\sigma : E \to \tilde{R}$.

We may view the load balancing as a graph embedding problem. Our task is to find a mapping $M : G \to H$ of the application graph to the processor graph minimizing a cost function. The processor graph H is usually static (constant during the runtime), whereas the parallel application graph G may be static or dynamic, that is, the computational load of the application may or may not change during runtime.

The Static Load Balancing

In the static load balancing, neither the structure nor the weights of the application graph G change during runtime. It is assumed that G is completely known prior to the start of the application such as in nonadaptive methods for numerical simulation. The static load balancing problem calculates a good mapping of the application graph $G = (V, E)$ onto the processor graph $H = (U, F)$.

Cost functions determining the quality of a mapping are its *load, dilation, and congestion*. The load of a mapping M is the maximum number of nodes from G assigned to any single node of H. The dilation is the maximum distance of any route of a single edge from G in H. The congestion is the maximum number of edges from G that must be routed via any single edge in H. The load determines the balancing quality of the mapping. It should be kept as low as possible to avoid idle times of the processor. The dilation of and edge of G determine the slowdown of a communication on this edge due to routing latency in H. The goal is to find a mapping function M which minimizes all three measures – load, dilation, and congestion [Leighton, 1992].

A graph is split into as many as there are numbers of processors such that as few as possible edges are external. This can be done by recursively bisecting the graph into two pieces. There are efficient solution heuristics which approximate the best value in terms of numbers of external edges. Some of the examples are (1) global methods partitioning the nodes into two subsets of equal size [Jones and Plassmann,

1994; Kaddoura, Ou, and Ranka, 1995]; (2) local methods where local heuristics determine equally sized sets of nodes which can be exchanged between parts such that the size of the cut decreases [Kerninghan and Lin, 1970; Fiduccia and Mattheyses, 1982; Hendrickson and Leland, 1993]; (3) multilevel hybrid methods in which a large graph is shrunk to a smaller one with similar characteristics, efficiently partitioned, and extrapolated to the original graph [Karypis and Kumar, 1995; Hendrickson and Leland, 1993].

Dynamic Load Balancing

The application graph $G = (V, E, \rho, \sigma)$ of problems in this class is dynamic; that is, nodes and edges are generated or deleted during runtime. Here, operations are carried out in phases. Changes to G do not occur at arbitrary, nonpredictable times but in synchronized manner. The mesh is usually refined based on error estimates of the current solution [Bornemann, Erdmann, and Kornhuber, 1993]. In general, we split the task of load balancing into two steps. First, we calculate how much load is to be shifted between processors, and second we determine which load is to be moved [Diekmann, Meyer, and Monien, 1997; Lüling and Monien, 1993].

Lin and Keller [1987] proposed a gradient model in which they assign a status of high, medium, or low to processors depending on their load. The algorithm then pushes the load from high to low. Lüling and Monien [1992] make processors balance their load with a fixed set of neighbors if the load difference between them increases above a certain threshold. Rudolph, Slivkin-Allouf, and Upfal [1991] showed that if processor j initiates a balancing action with a randomly chosen other processor with probability $(c/load\, j)$, then the expected load of j is at most c times the average load plus a constant.

The first step of the load balancing is to calculate how much load has to be transferred across each edge of H in order to achieve a globally balanced system. There are many approaches to this task: (1) Token distribution. This is the synchronized setting of the re-embedding problem in which a number of independent tokens on a network of processors are evenly distributed [Meyer et al., 1996]. (2) Random matchings. Ghosh et al. [1995] show that the load deviation halves in a minimal number of steps if a random matching of H's edges is chosen and some load is sent via these edges when the corresponding processors are not balanced. However, this approach is impractical in general situations. (3) Diffusion. A simple diffusive distributed load balancing strategy in which each processor balances its load with all its neighbors in each round was suggested by Cybenko [1989] and Boillat [1990]. These rounds are iterated until the load is completely balanced.

In addition to determining how much load is to be transferred, it is also important to choose load items which can be migrated in order to fulfill the flow requirements. For example, global iterative methods for solving linear systems such as multigrid or conjugate gradient computations can be parallelized by choosing load items so that the communication demands are minimized. Here, we must take into account the total length of subdomain boundaries, communication characteristics of the parallel system, etc. An example of recursive graph bisection for airfoils as demonstrated by Diekmann et al. [1997] is shown in Figure 20.3.6a. An aspect ratio optimization may be applied as shown in Figure 20.3.6b.

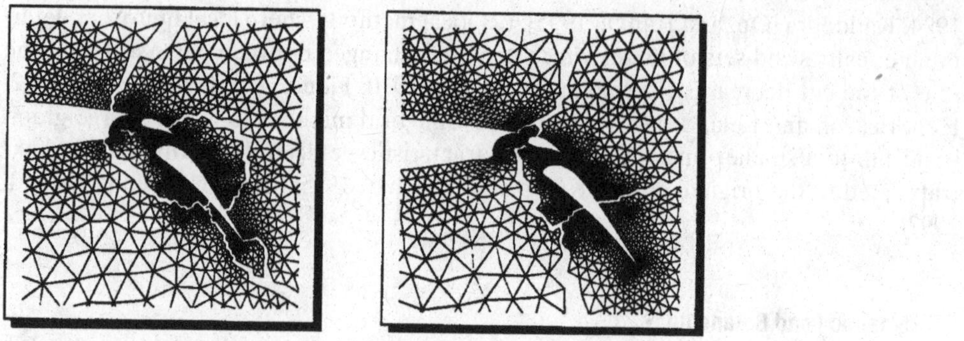

(a) Spectral bisection (b) Aspect ratio optimization

Figure 20.3.6 Dynamic load balancing of airfoil grid generation [Diekmann et al., 1997].

20.4 EXAMPLE PROBLEMS

In this section, two examples of parallel processing with domain decomposition are presented. Solutions of Poisson equation and Navier-Stokes system of equations will be discussed.

20.4.1 SOLUTION OF POISSON EQUATION WITH DOMAIN DECOMPOSITION PARALLEL PROCESSING

Domain decompositions methods are used effectively in parallel processing. Subdomains may be nonoverlapping, or overlapping. First, let us consider a nonoverlapping case (Figure 20.4.1a) and construct the matrix equations of the form,

$$Lu = \begin{bmatrix} K_{11} & 0 & K_{13} \\ 0 & K_{22} & K_{23} \\ K_{31} & K_{32} & K_{33} \end{bmatrix} \begin{bmatrix} u_1 \\ u_2 \\ u_3 \end{bmatrix} = \begin{bmatrix} f_1 \\ f_2 \\ f_3 \end{bmatrix} = f \tag{20.4.1}$$

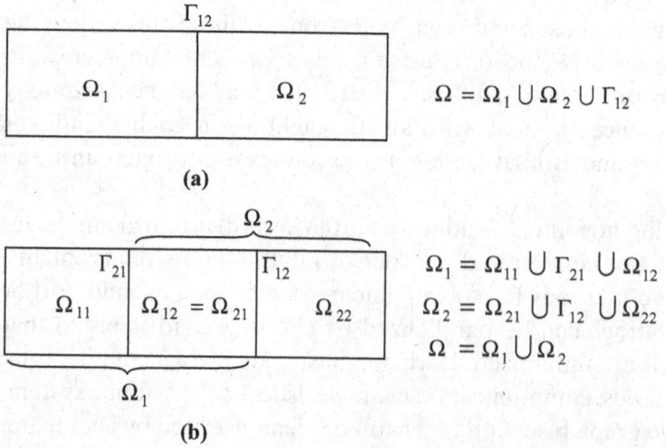

Figure 20.4.1 Domain decomposition. (a) Nonoverlapping subdomains. (b) Overlapping subdomains.

which is similar to (20.1.2). Here, the first two rows indicate subdomains Ω_1 and Ω_2, with the third row representing the boundary interface Γ_{12}. The subdomain variables u_1 and u_2 are calculated as

$$u_1 = K_{11}^{-1}(f_1 - K_{13}u_3)$$
$$u_2 = K_{22}^{-1}(f_2 - K_{23}u_3) \tag{20.4.2}$$

where the boundary interface variables u_3 are determined from

$$(K_{33} - K_{31}K_{11}^{-1}K_{13} - K_{32}K_{22}^{-1}K_{23})u_3 = f_3 - K_{31}K_{11}^{-1}f_1 - K_{32}K_{22}^{-1}f_2 \tag{20.4.3}$$

The above unknowns can be solved using two MIMD parallel processors. Here, we may utilize the preconditioning operator as described in Section 20.1.1.

The two subdomains used in the above example may be overlapped as shown in Figure 20.4.1b. In this case, the matrix equations take the form

$$Lu = \begin{bmatrix} K_{11} & K_{12} & 0 & 0 & 0 \\ K_{21} & K_{22} & K_{23} & 0 & 0 \\ 0 & K_{32} & K_{33} & K_{34} & 0 \\ 0 & 0 & K_{43} & K_{44} & K_{45} \\ 0 & 0 & 0 & K_{54} & K_{55} \end{bmatrix} \begin{bmatrix} u_1 \\ u_2 \\ u_3 \\ u_4 \\ u_5 \end{bmatrix} = \begin{bmatrix} f_1 \\ f_2 \\ f_3 \\ f_4 \\ f_5 \end{bmatrix} = f \tag{20.4.4}$$

which is partitioned into two systems, Ω_{11} and Ω_{22} such that

$$L_1 v_1 = \begin{bmatrix} K_{11} & K_{12} & 0 \\ K_{21} & K_{22} & K_{23} \\ 0 & K_{32} & K_{33} \end{bmatrix} \begin{bmatrix} v_{11} \\ v_{12} \\ v_{13} \end{bmatrix} = \begin{bmatrix} g_{11} \\ g_{12} \\ g_{13} \end{bmatrix} = F_1 \tag{20.4.5}$$

$$L_2 v_2 = \begin{bmatrix} K_{33} & K_{34} & 0 \\ K_{43} & K_{44} & K_{45} \\ 0 & K_{54} & K_{55} \end{bmatrix} \begin{bmatrix} u_3 \\ u_4 \\ u_5 \end{bmatrix} = \begin{bmatrix} f_3 \\ f_4 \\ f_5 \end{bmatrix} = F_2 \tag{20.4.6}$$

with

$$F_1 = \begin{bmatrix} f_1 \\ f_2 \\ f_3 \end{bmatrix} - \begin{bmatrix} 0 & 0 & 0 \\ 0 & 0 & 0 \\ 0 & K_{34} & 0 \end{bmatrix} \begin{bmatrix} u_3 \\ u_4 \\ u_5 \end{bmatrix} = F_1' - G_2 v_2 \tag{20.4.7}$$

$$F_2 = \begin{bmatrix} f_3 \\ f_4 \\ f_5 \end{bmatrix} - \begin{bmatrix} 0 & K_{32} & 0 \\ 0 & 0 & 0 \\ 0 & 0 & 0 \end{bmatrix} \begin{bmatrix} u_1 \\ u_2 \\ u_3 \end{bmatrix} = F_2'' - G_1 v_1 \tag{20.4.8}$$

The above process results in the system of equations in the form

$$\begin{bmatrix} L_1 & G_2 \\ G_1 & L_2 \end{bmatrix} \begin{bmatrix} v_1 \\ v_2 \end{bmatrix} = \begin{bmatrix} F_1 \\ F_2 \end{bmatrix} \tag{20.4.9}$$

This can be solved using the block Jacobi scheme:

$$\begin{bmatrix} L_1 & 0 \\ 0 & L_2 \end{bmatrix} \begin{bmatrix} v_1 \\ v_2 \end{bmatrix}^{k+1} = \begin{bmatrix} F_1 \\ F_2 \end{bmatrix} - \begin{bmatrix} 0 & G_2 \\ G_1 & 0 \end{bmatrix} \begin{bmatrix} v_1 \\ v_2 \end{bmatrix}^k \tag{20.4.10}$$

This system suggests that we can utilize two processors on a MIMD machine, forming a global and inner parallelism of the algorithm.

20.4.2 SOLUTION OF NAVIER-STOKES SYSTEM OF EQUATIONS WITH MULTITHREADING

Multithreaded programming is utilized to take advantage of multiple computational elements on the host computer [Schunk et al., 1999]. Typically, a multithreaded process will spawn multiple threads which are allocated by the operating system to the available computational elements (or processors) within the system. If more than one processor is available, the threads may execute in parallel, resulting in a significant reduction in execution time. If more threads are spawned than available processors, the threads appear to execute concurrently as the operating system decides which threads execute while the others wait. One unique advantage of multithreaded programming on shared memory multiprocessor systems is the ability to share global memory. This alleviates the need for data exchange or message passing between threads as all global memory allocated by the parent process is available to each thread. However, precautions must be taken to prevent deadlock or race conditions resulting from multiple threads trying to simultaneously write to the same data.

Threads are implemented by linking an application to a shared library and making calls to the routines within that library. Two popular implementations are widely used: the Pthreads library [Nichol et al., 1996] (and its derivatives) that are available on most Unix operating systems and the NTthreads library that is available under Windows NT. There are differences between the two implementations, but applications can be ported from one to the other with moderate ease and many of the basic functions are similar albeit with different names and syntax.

Domain decomposition methods (Section 20.3.1) can be used in conjunction with multithreaded programming to create an efficient parallel application. The subdomains resulting from the decomposition provide a convenient division of labor for the processing elements within the host computer. The additive Schwarz domain decomposition method discussed in Section 20.1.2 is utilized. The method is illustrated below (Figure 20.4.2.1) for a two-dimensional square mesh that is decomposed into four subdomains. The nodes belonging to each of the four subdomains are denoted with geometric symbols while boundary nodes are identified with bold crosses. The desire is to solve for each node implicitly within a single subdomain. For nodes on the edge of each subdomain, this is accomplished by treating the adjacent node in the neighboring subdomain as a boundary. The overlapping of neighboring nodes between subdomains is illustrated in Figure 20.4.2.2. Higher degrees of overlapping, which may improve convergence at the expense of computation time, are also used.

In a parallel application, load balancing between processors is critical to achieving optimum performance. Ideally, if a domain could be decomposed into regions requiring an identical amount of computation, it would be a simple matter to divide the problem between processing elements as shown in Figure 20.4.2.3 for four threads executing on an equal number of processors.

Unfortunately, in a "real world" application the domain may not be decomposed such that the computation for each processor is balanced, resulting in lost efficiency. If the execution time required for each subdomain is not identical, the CPUs will become idle for portions of time as shown in Figure 20.4.2.4.

One approach to load balancing, as implemented in this application, is to decompose the domain into more subdomains than available processors and use threads to perform

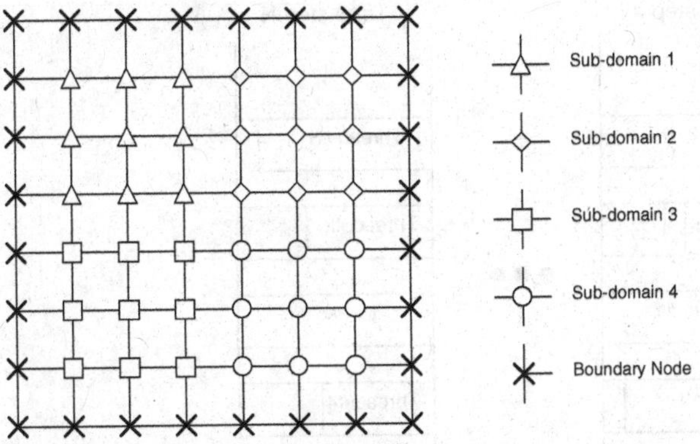

Figure 20.4.2.1 Multiple subdomains.

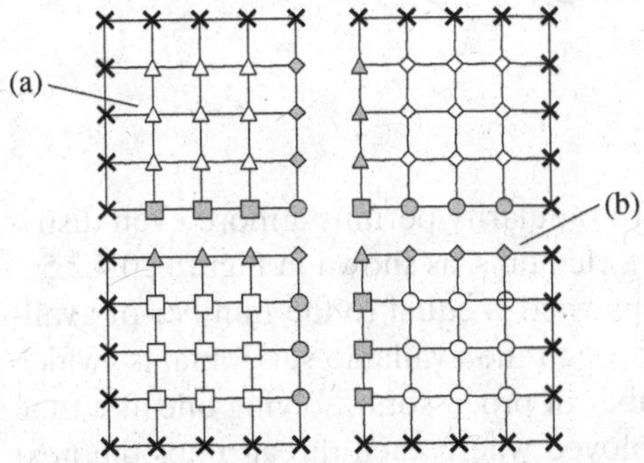

Figure 20.4.2.2 Domain decomposition. **(a)** Nodes shaded in white are solved implicitly within each subdomain. **(b)** Nodes from neighboring subdomains (shaded) are treated as boundary nodes and allowed to lag one time step.

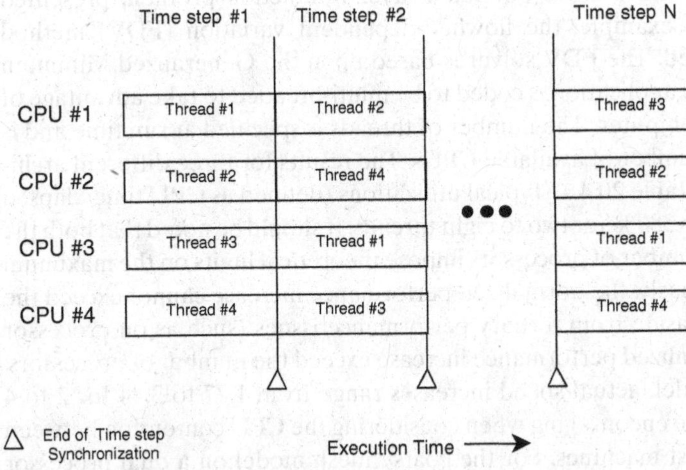

Figure 20.4.2.3 Ideal load balancing.

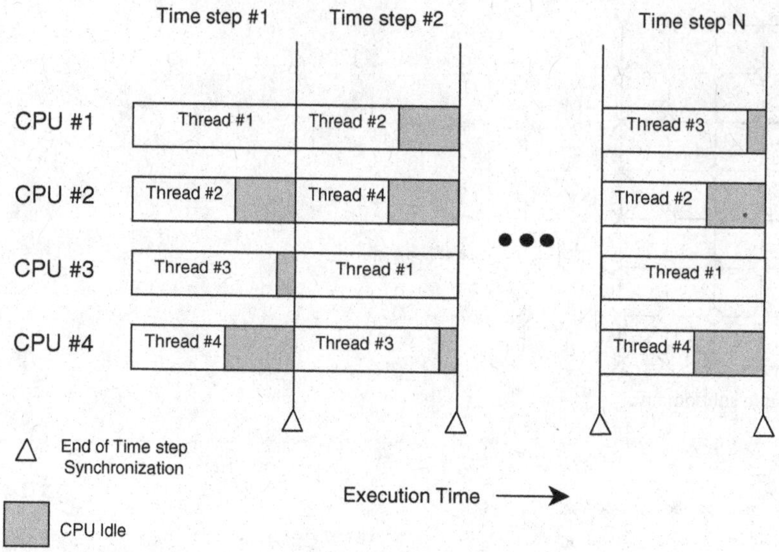

Figure 20.4.2.4 "Real world" load balancing.

the computations within each block. The finer granularity permits a more even distribution of work among the available processing elements, as shown in Figure 20.4.2.5.

In this approach, the number of threads spawned is equal to the number of available processors, with each thread marching through the available subdomains (which preferably number at least two times the number of processors), solving one at a time in an "assembly-line" fashion. A stack is employed where each thread pops the next subdomain to be solved off of the top of the stack. Mutual exclusion locks are employed to protect the stack pointer in the event two or more threads access the stack simultaneously. Each thread remains busy until the number of subdomains is exhausted. If the number of subdomains is large enough, the degree of parallelism will be high, although decomposing a problem into too many subdomains may adversely affect convergence. This approach is illustrated in Figures 20.4.2.6 and 20.4.2.7.

We revisit the triple shock wave boundary layer interaction problem presented in Section 6.7.5. In this example, the flowfied-dependent variation (FDV) method (Section 6.5) FDM is used. The FDV solver is based upon the Generalized Minimum Residual (GMRES). The application is coded to be multithreaded to take advantage of parallelism in the host computer. The number of threads is specified at run time and is based on the expected number of available CPUs. The results for three different architectures are provided in Table 20.4.1. Typical utilizations (defined as CPU time/elapsed time) range from 180% to 380% for two to eight threads. It should be noted that both the number of threads and number of processors impose theoretical limits on the maximum performance gain. Obviously, the normalized performance increase cannot exceed the number of threads and, aside from tertiary performance issues (such as on processor cache), nor can the normalized performance increase exceed the number of processors. For the coarse mesh model, actual speed increases range from 1.77 to 3.44 for 2 to 4 processors. The results are encouraging when considering the CPU contention between multiple users on the host machines. For the coarse mesh model on a dual processor Pentium II workstation (with no other users) a CPU utilization of 196% is observed

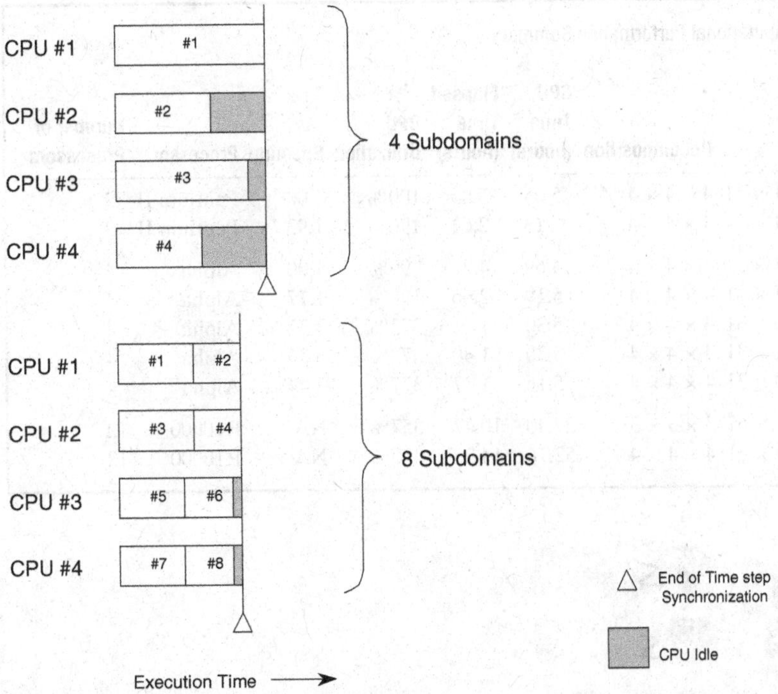

Figure 20.4.2.5 Domain decomposition improves parallelism.

with a real-time speedup of 1.92. The four-processor machine did exhibit a significant amount of overhead when moving beyond a single thread. Utilizing four threads for the fine mesh model resulted in CPU utilizations of 357% and 370% for a domain decomposed into 27 and 64 regions, respectively. The fine mesh model was not run with a single processor or thread so that no relative speedup data are available. The CPU utilization is encouraging considering the high CPU contention on the twelve-processor machine.

Density contours for the inviscid shock interaction (x-z plane, as viewed from above the wind tunnel model) are shown in Figure 20.4.2.8. The 15° fins produce inviscid shocks that are predicted to intersect and reflect approximately 92 mm from the ramp entrance. The reflected shock does not intersect with the exit corner of the ramp as expected. Two

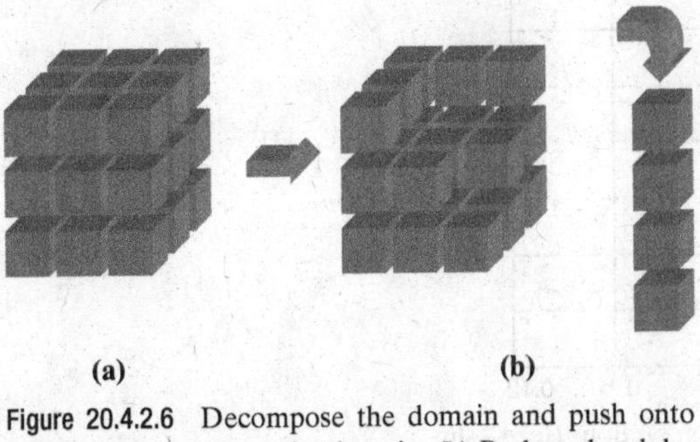

(a) (b)

Figure 20.4.2.6 Decompose the domain and push onto stack. **(a)** Decompose the domain. **(b)** Push each subdomain onto a software stack.

Table 20.4.1 Computational Performance Summary

Threads	Grid	Decomposition	CPU Time (hours)	Elapsed Time (hours)	CPU Utilization	Speedup	Processor	Number of Processors
1	55 × 41 × 31	4 × 4 × 4	5.05	5.05	100%	1.00	Pentium II	2
2	55 × 41 × 31	4 × 4 × 4	5.13	2.62	196%	1.93	Pentium II	2
1	55 × 41 × 31	4 × 4 × 4	4.69	4.72	99%	1.00	Alpha	4
2	55 × 41 × 31	4 × 4 × 4	5.19	2.66	195%	1.77	Alpha	4
4	55 × 41 × 31	4 × 4 × 4	5.30	1.42	373%	3.32	Alpha	4
6	55 × 41 × 31	4 × 4 × 4	5.30	1.40	378%	3.36	Alpha	4
8	55 × 41 × 31	4 × 4 × 4	5.16	1.37	377%	3.44	Alpha	4
4	109 × 81 × 61	3 × 3 × 3	37.40	10.47	357%	NA	R10000	12
4	109 × 81 × 61	4 × 4 × 4	52.76	14.25	370%	NA	R10000	12

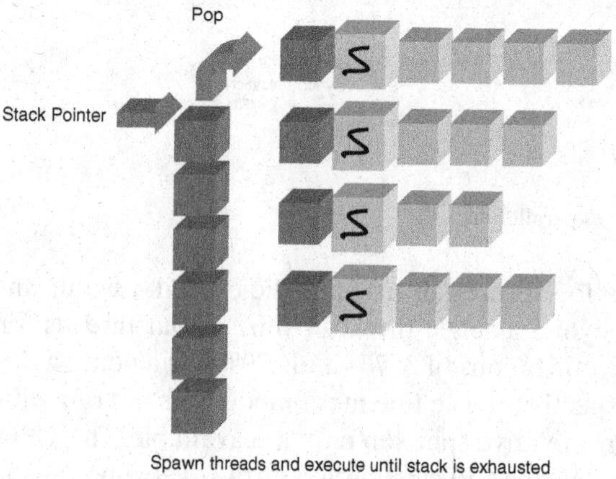

Figure 20.4.2.7 Allow threads to process each subdomain.

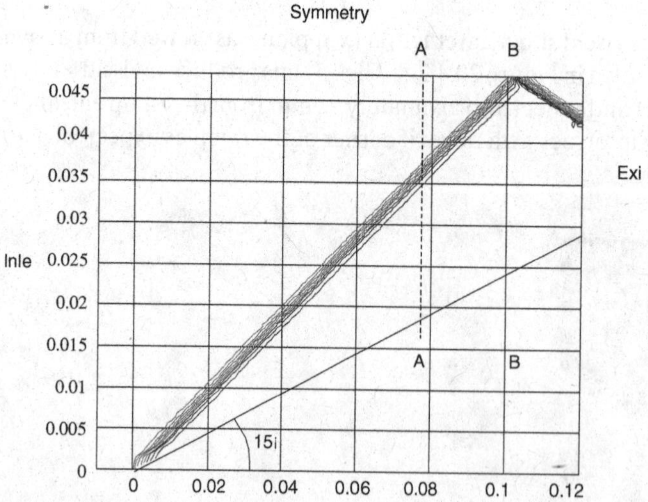

Figure 20.4.2.8 Density contours for X-Z cross section (top), slip boundary.

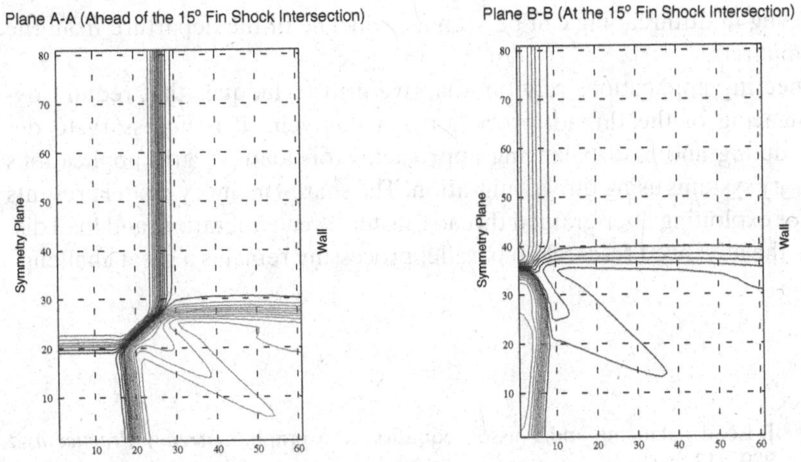

Figure 20.4.2.9 Density contours for Y-Z cross section, slip boundary.

cross sections, located at 67 mm and 92 mm, respectively, from the entrance are noted on the plot.

Density contours for the flow in x-y planes located 67 mm (upstream of the inviscid shock intersection) and 92 mm (coincident with the inviscid shock intersection) from the combined fin/ramp entrance are shown in Figure 20.4.2.9. It appears that the upstream predictions correlate well with the experimental images. The inviscid ramp and fin shocks, as well as the corner reflection, are easily discernible in the upstream figure (see left). Interestingly, it appears that the triangular-shaped slip lines are present in the numerical results of the upstream plane. Since the sliplines divide constant pressure regions with differing velocities, this feature is not visible in the static pressure plots. As in the experimental imagery, the inviscid fin shocks merge together in the symmetry plane at the point where the inviscid shocks intersect (see right). No curvature of the inviscid fin shock intersection is observed in the numerical predictions. The reflection of the corner shock about the symmetry plane is observed, but the ramp embedded shock is lower relative to the height of the fin than in the experimental results.

20.5 SUMMARY

Three of the most important computing techniques have been discussed: domain decomposition, multigrids, and parallel processing. For large geometrical configurations, domain decomposition provides efficiency in data managements. The number of resulting algebraic equations can still be very large, and the multigrid method of solutions of the large algebraic system of equations is considered a most effective approach.

The trends in parallel processing have been leaning toward the use of small clusters of Symmetric Multiprocessors (SMP), often interconnected to address the needs of complex problems requiring a large number of processing nodes. In the past, programming based on message passing paradigms on massively parallel computers or specialized supercomputers has been used. These systems are becoming less popular (or available) and distributed networks of SMP clusters are becoming the preferred choice for engineering. The growing interest in multithreaded programming and the availability of systems supporting multithreading can be seen as evidence of the departure from the use of supercomputers.

Many engineering applications rely on adaptive grid techniques that require dynamic load balancing of the threads/processors. In this vein, it is necessary to develop new scheduling and load balancing approaches for adaptive grid applications on shared memory systems using thread migration. The shared memory model presents opportunities for exploiting finer-grained threads, faster thread migration, and load distribution. Thus, the advanced research in parallel processing remains a great challenge in the future.

REFERENCES

Boillat, J. E. [1990]. Load balancing and Poisson equation in a graph. *Currency Practice and Experience*, 2, 4, 289–313.

Bornemann, F., Erdmann, B., and Kornhuber, R. [1993]. Adaptive multilevel methods in three space dimensions. *Int. J. Num. Meth. Eng.*, 36, 3187–3203.

Brandt, A. [1972]. Multilevel adaptive technique (MLAT) for fast numerical solutions to boundary value problems. *Lecture Notes in Physics 18*, Berlin: Springer-Verlag, 82–89.

——— [1977]. Multilevel adaptive solutions to boundary value problems. *Math. Comp.* 31, 333–90.

——— [1992]. On multigrid solution of high Reynolds incompressible entering flows. *J. Com. Phys.*, 1101, 151–64.

Cybenko, G. [1989]. Load balancing for distributed memory multiprocessors. *J. Par. Distr. Comp.*, 7, 279–301.

Diekmann, R., Meyer, D., and Monien, B. [1997]. Parallel decomposition of unstructured FEM-meshes. *Proc. IRREGULAR 95*, Springer LNCS, 199–215.

Fiduccia, C. M. and Mattheyses, R. M. [1982]. A linear-time heuristic for improving network partitions. Proc. 19th IEEE Design Automation Conference, 175–81.

Ghosh, B., Leighton, F. T., Maggs, B. M., and Muthukrishnan, S. [1995]. Tight analyses of two local load balancing algorithms. Proc. 27th ACM Symp. in Theory of Computing (STOC, 95), 548–58.

Glowinski, R. and Wheeler, M. F. [1987]. Domain decomposition and mixed finite element methods for elliptic problems. In R. Glowinski et al., (ed.). *Domain Decomposition Methods for Partial Differential Equations*. SIAM Publications, 144–72.

Hendrickson, B. and Leland, R. [1993]. A multilevel algorithm for partitioning graphs. Technical Report SAND93-1301, Sandia National Lab., Sandia.

Jones, M. T. and Plassmann, P. E. [1994]. Parallel algorithms for the adaptive refinement and partitioning of unstructured meshes. Proc. Scalable High Performance Computing Conf., IEEE Computing Conf., IEEE Computer Society Press, 478–85.

Kaddoura, M., Ou, C. W., and Ranka, S. [1995]. Mapping unstructured computational graphs for adaptive and nonuniform computational environments. *IEEE Par. and Dir. Technology*.

Karypis, G. and and Kumar, V. [1995]. A fast and high quality multilevel scheme for partitioning irregular graphs. Tech. Report. 95-035, CD-Dept, University of Minnesota.

Kavi, K. M. [1999]. Multithreaded system implementations. *J. Microcomp. App.*, 18, 2.

Kerninghan, B. W. and Lin, S. [1970]. An effective heuristic procedure for partitioning graphs. The Bell Systems Tech. J., 291–308.

Leighton, F. T. [1992]. *Introduction to Parallel Algorithms and Architectures*. Morgan Kaufmann Publishers.

Lin, F. C. H. and Keller, R. M. [1987]. The gradient model load balancing methods. IEEE Trans. on Software Engineering. 13, 32–38.

Lions, P. L. [1988]. On the Schwarz alternating method. In R. Glowinski et al. (eds.). *Domain Decomposition Methods for Partial Differential Equations*. Philadelphia: SIAM Publications, 1–42.

Löhner, R. and Morgan, K. [1987]. An unstructured multigrid method for elliptic problems, *Int. J. Num. Eng.*, 24, 101–15.

Lüling, R. and Monien, B. [1992]. Load balancing for distributed branch and bound algorithms. Proc. 6th Int. Parallel Processing Symp. (IPPS, 92), 543–49.

——— [1993]. A dynamic distributed load balancing algorithm with provable good performance. Proc. 5th Annual ACM Symp. on Parallel Algorithms and Architectures (SPPS, 92), 543–49.

Mavriplis, D. J. and Jameson, A. [1990]. Multigrid solution of the Navier-Stokes equations on triangular meshes. *AIAA J.*, 28, 8, 1415–25.

Meyer, F., Heide, A. D., Oesterdiekhoff, B., and Wanka, R. [1996]. Strongly adaptive token distribution. *Algorithmica.*, 15, 413–27.

Nichol, B., Buttlar, D., and Farrell, J. [1996]. *Pthreads Programming*. Paris: O'Reilly and Associates.

Rudoph, L., Slivkin-Allouf, M., and Upfal, E. [1991]. A simple load balancing scheme for task allocation in parallel machines. Proc. 3rd Annual ACM Symp. On Parallel Algorithms and Architectures (APAA, 91), 237–45.

Schunk, R. G., Canabal, F., Heard, G., and Chung, T. J. [1999]. Unified CFD methods via flowfield-dependent variation theory. AIAA paper 99–3715.

——— [2000]. Airbreathing propulsion system analysis using multithreaded parallel processing. AIAA paper, AIAA-2000-3467.

Schwarz, H. A. [1869]. Uber einige abbildungsaufgauben. *J. fur Die Reine und Angewandte Mathematik*, 70, 1005–20.

PART FIVE

APPLICATIONS

Having studied various computational methods in Parts Two and Three and automatic grid generation, adaptive methods, and computing techniques in Part Four, we are now prepared to re-examine these methods and test our knowledge on some selected engineering problems of application. For the past four decades, many applications have been accumulated to such a great extent that it is impossible to review them all in this text. Rather, we limit our scope of study to the following areas: turbulence (Chapter 21), chemically reactive flows and combustion (Chapter 22), acoustics (Chapter 23), combined mode radiative heat transfer (Chapter 24), multiphase flows (Chapter 25), electromagnetic flows (Chapter 26), and relativistic astrophysical flows (Chapter 27).

The selection of computational methods depends on many factors such as types of flows, ranges of speeds, dimensions of domain, etc. A decision as to the choice of FDM, FEM, or FVM is now a matter of preference and judgments of the analyst in view of the information presented in the previous chapters.

In the following chapters, example problems and computational methods are chosen randomly, depending on availability of sources. Some of them are drawn from the student works at the University of Alabama in Huntsville, and others are from those available in the open literature. In each of the applications, the corresponding governing equations and associated physics are first introduced. This is then followed by the computational methods used, numerical results and evaluations, each example being self-contained as much as possible.

It is hoped that these examples serve as a reasonable guidance for the uninitiated reader toward his or her direction and destination in CFD research. Some examples are elementary, and others represent the research results which are highly specialized. Thus, the reader may wish to explore subject areas selectively.

CHAPTER TWENTY-ONE

Applications to Turbulence

21.1 GENERAL

Turbulence is a natural phenomenon in fluids that occurs when velocity gradients are high, resulting in disturbances in the flow domain as a function of space and time. Examples include smoke in the air, condensation of air on a wall, flows in a combustion chamber, ocean waves, stormy weather, atmospheres of planets, and interaction of the solar wind with magnetosphere, among others.

Although turbulence has been the subject of intensive study for the past century, it appears that many difficulties still remain unresolved, particularly in flows with high Mach numbers and high Reynolds numbers. Turbulent flows arise in contact with walls or in between two neighboring layers of different velocities. They result from unstable waves generated from laminar flows as the Reynolds number increases downstream. With velocity gradients increasing, the flow becomes rotational, leading to a vigorous stretching of vortex lines, which cannot be supported in two dimensions. Thus, turbulent flows are always physically three-dimensional, typical of random fluctuations. This makes 2-D simplifications unacceptable in most of the numerical simulation.

In turbulent flows, large and small scales of continuous energy spectrum, which are proportional to the size of eddy motions, are mixed. Here, eddies are overlapping in space, with large ones carrying small ones. In this process, the turbulent kinetic energy transfers from larger eddies to smaller ones, with the smallest eddies eventually dissipating into heat through molecular viscosity. In direct numerical simulation (DNS), a refined mesh is used so that all of these scales, large and small, are resolved. This is known as the deterministic method. Although some simple problems have been solved using DNS, it is not possible to undertake industrial problems of practical interest due to the prohibitive computer cost.

Since turbulence is characterized by random fluctuations, statistical methods rather than deterministic methods have been studied extensively in the past. In this approach, time averaging of variables is carried out in order to separate the mean quantities from fluctuations. This results in new unknown variable(s) appearing in the governing equations. Thus, additional equation(s) are introduced to close the system, the process known as turbulence modeling or Reynolds averaged Navier-Stokes (RANS) methods. In this approach, all large and small scales of turbulence are modeled so that mesh refinements needed for DNS are not required. We discuss this topic in Sections 21.3 and 21.7.1.

A compromise between DNS and RANS is the large eddy simulation (LES) which has become very popular in recent years. Here, large-scale eddies are computed and small scales are modeled. Small-scale eddies are associated with the dissipation range of isotropic turbulence, in which modeling is simpler than in RANS. Since the large-scale turbulence is to be computed, the mesh refinements are required much more than in RANS, but not as much as in DNS because the small-scale turbulence is modeled. Governing equations and examples for LES are presented in Section 21.4 and Section 21.7.2.

Finally, we examine the physical aspects associated with DNS in Section 21.5, followed by numerical examples in Section 21.7.3.

21.2 GOVERNING EQUATIONS

Turbulent flowfields can be calculated with the Navier-Stokes system of equations averaged over space or time. When this averaging is performed, the equations describing the mean flowfield contain the averages of products of fluctuating velocities. In general, this will result in more unknowns than the number of equations available. Such difficulty can be resolved by turbulence modeling with additional equations being provided to match the number of unknowns. Such models are designed to approximate the physical behavior of turbulence. There are numerous ways of averaging flow variables: time averages, ensemble averages, spatial averages, and mass averages.

Time Averages

Any variable f is assumed to be the sum of its mean quantity $\overline{f}$ and its fluctuation part f',

$$f(x,t) = \overline{f}(x,t) + f'(x,t) \tag{21.2.1}$$

where $\overline{f}$ is the time average of f,

$$\overline{f}(x,t) = \frac{1}{\Delta t}\int_t^{t+\Delta t} f(x,t)dt \tag{21.2.2}$$

with

$$\overline{f'} = \frac{1}{\Delta t}\int_t^{t+\Delta t} f' dt = 0 \tag{21.2.3a}$$

The time average of the product of fluctuation parts of two different variables f' and g' is given by

$$\overline{f'g'} = \frac{1}{\Delta t}\int_t^{t+\Delta t} f'g' dt \neq 0 \tag{21.2.3b}$$

Here, the time interval Δt is chosen compatible with the time scale of the turbulent fluctuations, not only for the variable f but also for other variables within the physical domain.

Ensemble Averages

In terms of measurements of N identical experiments, $f(x,t) = f_n(x,t)$, we may determine the average,

$$\overline{f}(x,t) = \lim_{N\to\infty} \frac{1}{N}\sum_{n=1}^{N} f_n(x,t) \tag{21.2.4}$$

Spatial Averages

When the flow variable is uniform on the average such as in homogeneous turbulence, we may choose to use a spatial average defined as

$$\overline{f}(t) = \lim_{\Omega\to\infty} \frac{1}{\Omega}\int_{\Omega} f(x,t)d\Omega \tag{21.2.5}$$

Mass (Favre) Averages

For compressible flows, it is often more convenient to use mass (Favre) averages instead of time averages,

$$f = \tilde{f} + f'' \tag{21.2.6}$$

where the mean quantity $\tilde{f}$ is defined as

$$\tilde{f} = \frac{\overline{\rho f}}{\overline{\rho}} = \overline{f} + \frac{\overline{\rho' f'}}{\overline{\rho}} \tag{21.2.7}$$

and the fluctuation f'' has the property

$$\overline{\rho f''} = 0 \tag{21.2.8a}$$

whereas

$$\overline{f''} = -\overline{\rho' f'}/\overline{\rho} \neq 0 \tag{21.2.8b}$$

for the case of a time average. It is clear that the correlation of density fluctuations, ρ', with the fluctuating quantity, f', gives rise to a nonzero mean Favre fluctuation field, $\overline{f''}$. Thus, it is seen that the Favre average makes the turbulent compressible flow equations simpler with their form resembling those of incompressible flows. Despite these simplifications, however, the density fluctuations or compressibility effects must still be resolved; only the mathematical simplifications are achieved through Favre averages.

With time averages for incompressible flows and mass averages for compressible flows, the conservation equations can be derived as follows:

Time-Averaged Incompressible Flows

Continuity

$$\overline{v}_{i,i} = 0 \tag{21.2.9a}$$

Momentum

$$\rho \frac{\partial \overline{v}_j}{\partial t} + \rho \overline{v}_{j,i}\overline{v}_i = -\overline{p}_{,j} + (\overline{\tau}_{ij} + \tau^*_{ij})_{,i} \tag{21.2.9b}$$

with

$$\overline{\tau}_{ij} = 2\mu \overline{d}_{ij}, \qquad \overline{d}_{ij} = \frac{1}{2}(\overline{v}_{i,j} + \overline{v}_{j,i}), \qquad \tau^*_{ij} = -\rho \overline{v'_i v'_j}$$

Energy

$$\frac{\partial \overline{T}}{\partial t} + \overline{v}_i \overline{T}_{,i} = -(\overline{q}_i - q^*_i)_{,i} \tag{21.2.9c}$$

with

$$\overline{q}_i = -\alpha \overline{T}_{,i} \qquad q^*_i = -\overline{v'_i T'}$$

Mass (Favre)-Averaged Compressible Flows

Continuity

$$\frac{\partial \overline{\rho}}{\partial t} + (\overline{\rho}\tilde{v}_i)_{,i} = 0 \tag{21.2.10a}$$

Momentum

$$\frac{\partial}{\partial t}(\overline{\rho}\tilde{v}_j) + (\overline{\rho}\tilde{v}_i\,\tilde{v}_j)_{,i} = -\overline{p}_{,j} + (\overline{\tau}_{ij} + \tilde{\tau}^*_{ij})_{,i} \tag{21.2.10b}$$

with

$$\overline{\tau}_{ij} = 2\mu\left(\overline{d}_{ij} - \frac{1}{3}\overline{d}_{kk}\delta_{ij}\right), \qquad \tau^*_{ij} = -\rho\overline{v''_i v''_j}$$

Energy

$$\frac{\partial}{\partial t}(\overline{\rho}E) + [\overline{\rho}\tilde{v}_i H]_{,i} = -\left(\overline{q}_i + q^*_i - \overline{\tau_{ij} v''_j} + \frac{1}{2}\overline{\rho v''_i v''_j v''_j}\right)_{,i} + [(\overline{\tau}_{ij} + \tau^*_{ij})\tilde{v}_j]_{,i} \tag{21.2.10c}$$

with

$$E = \tilde{\varepsilon} + \frac{1}{2}\tilde{v}_i\tilde{v}_i, \qquad H = \tilde{H} + \frac{1}{2}\tilde{v}_i\tilde{v}_i, \qquad q^*_i = -\rho\overline{v''_i H''},$$

For time averaged incompressible flows, $-\rho\overline{v'_i v'_j}$ in (21.2.9b) and $-\overline{v'_i T'}$ in (21.2.9c) are identified as the Reynolds (turbulent) stress and Reynolds (turbulent) heat flux, respectively. The counterparts for mass-averaged compressible flows are $-\rho\overline{v''_i v''_j}$ in (21.2.10b) and $-\rho\overline{v''_i H''}$ in (21.2.10c), respectively. If time averages are used for compressible flows, the Reynolds stress components would be much more complicated. For this reason, mass averages are preferred for compressible flows.

These Reynolds stress tensors and Reynolds heat flux vectors are additional unknown variables. Therefore, additional governing equations other than those given in (21.2.9) and (21.2.10) matching the same number of unknowns must be provided. This is the process known as the turbulence closure or turbulence modeling. We discuss this subject in the next section.

21.3 TURBULENCE MODELS

There are many options in providing the closure process: zero-equation (algebraic) models, one-equation models, two-equation models, second order closure (Reynolds stress) models, and algebraic stress models as applied to incompressible flows. They are presented in Sections 21.3.1 through 21.3.4 with the effects of compressibility in Section 21.3.5.

21.3.1 ZERO-EQUATION MODELS

The purpose of zero-equation models is to close the system without providing extra differential equations. This may be achieved by the classical method of Prandtl mixing length [Prandtl, 1925]. Recent and more popular models are those advanced by Cebeci and Smith [1974] or Baldwin and Lomax [1978]. These models provide the Reynolds (turbulent) stress in terms of eddy (turbulent) viscosity μ_T,

$$\tau^*_{ij} = -\rho\overline{v'_i v'_j} = 2\mu_T d_{ij} = \mu_T(\overline{v}_{i,j} + \overline{v}_{j,i}) \tag{21.3.1}$$

where μ_T is computed by various approaches as described below.

Prandtl's Mixing Length Model

Historically, this is the earliest model proposed by Prandtl [1925] which applies to 2-D boundary layer problems:

$$\mu_T = \rho\ell^2\left|\frac{d\overline{u}}{dy}\right| \tag{21.3.2}$$

where the Prandtl mixing length ℓ is given by

$$\ell = \kappa y$$

with κ being the von Karman constant ($\kappa = 0.41$).

The turbulent shear stress for the incompressible boundary layer flow is given by

$$\tau^* = \mu_T\frac{d\overline{u}}{dy} = \rho\ell^2\left(\frac{d\overline{u}}{dy}\right)^2 \tag{21.3.3}$$

Upon integration of the above expression and using the empirical constant of integration from experiments, it can be shown that

$$u^+ = \frac{1}{\kappa}\text{In } y^+ + 5.5 \tag{21.3.4}$$

with $u^+ = \overline{u}/u_*$ and $y^+ = yu_*/\nu$ being the nondimensional relative velocity and nondimensional relative distance, respectively. A part of the turbulent velocity profile, called the law of the wall as given by (21.3.4) is valid only to the relative distance of approximately $y^+ = 30$; below this is the buffer zone and viscous sublayer as shown in Figure 21.3.1. From experiments, the viscous sublayer is identified by the range where y^+ is approximately equal to u^+. A smooth curve connects between the points $y^+ = 5$ and $y^+ = 30$. For flows such as in pipes or flat plates, the log layer deviates (defect layer) significantly at $y^+ \cong 500$ and above.

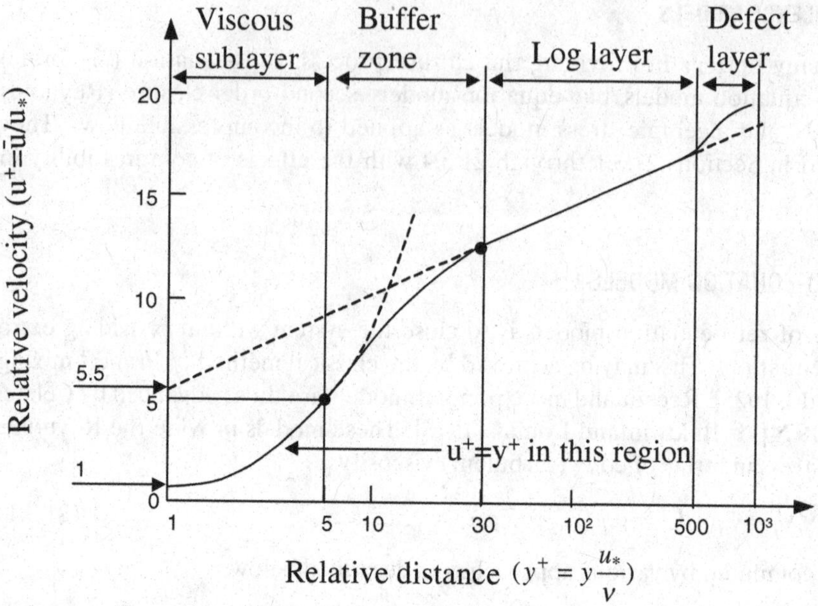

Figure 21.3.1 Two-dimensional turbulent boundary layer velocity profile showing various layers in terms of shear velocity $u_* = \sqrt{\tau/\rho}$.

Cebeci-Smith Model

The Cebeci-Smith model [1974] is designed for boundary layers consisting of inner layer region (Figure 21.3.2) with $\nu_T = \nu_T^{(I)}$ for $y < y_c$ and outer region with $\nu_T = \nu_T^{(o)}$ for $y > y_c$ where y_c is the critical height at which $\nu_T^{(I)}$ and$\nu_T^{(o)}$ coincide,

$$\nu_T^{(I)} = \rho\ell^2|\nabla \times \bar{\mathrm{v}}| \tag{21.3.5a}$$

$$\nu_T^{(o)} = 0.0168\rho\bar{u}_e\,\delta^* F \tag{21.3.5b}$$

Here ℓ is the Van Driest [1956] correction of the Prandtl mixing length,

$$\ell = \kappa y\,\lfloor 1 - \exp(-y^+/A^+)\rfloor \tag{21.3.6}$$

with

$$y^+ = y\frac{u_*}{\nu}, \qquad u_* = \sqrt{\frac{\tau_0}{\rho}}, \qquad A^+ = 26$$

τ_o = wall shear stress, $\bar{u}_e$ = external velocity at the boundary layer

δ^* = boundary layer displacement thickness.

$$F = \frac{1}{1 + 5.5\,(y/\delta)^6}$$

Here F denotes the intermittence at the edge of the boundary layer and δ is the boundary layer thickness.

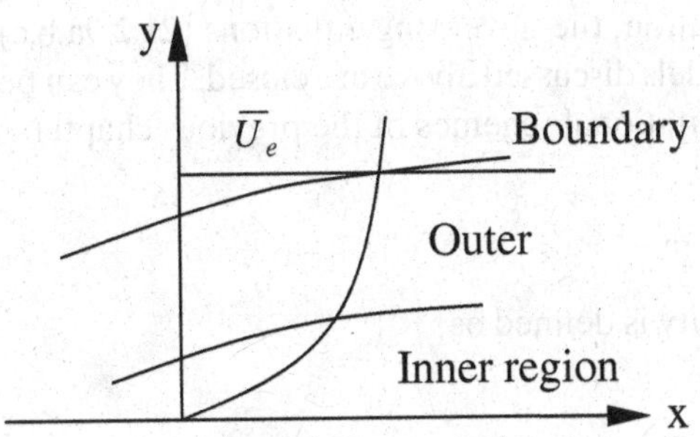

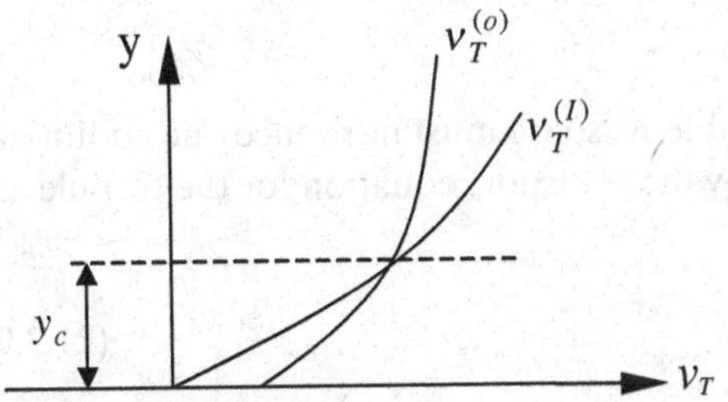

Figure 21.3.2 One-equation model [Cebeci-Smith, 1974; Baldwin-Lomax, 1978].

Baldwin-Lomax Model

The model given by (21.3.5) often encounters difficulties due to an uncertainty of the external velocity at the boundary layer $\overline{u}_e$ in (21.3.5b). To rectify this situation, Baldwin and Lomax [1978] proposed that the outer eddy viscosity be defined as

$$\nu_T^{(o)} = 0.0168\beta F y_{\max}\Gamma_{\max} \tag{21.3.7}$$

$$F = \frac{1}{1 + 5.5\,(\alpha y/y_{\max})^6}$$

$$\Gamma_{\max} = y\,[1 - \exp\,(-y^+/A)]\,|\nabla \times \overline{\mathrm{v}}|$$

$$\alpha = 0.3, \qquad \beta = 1.6$$

For shear layer applications, only the outer eddy viscosity will apply. In general, the zero-equation models fail to perform well in the region of recirculation and separated flows.

Turbulent Heat Flux Vector

The unknown quantity in (21.2.9c) is the turbulent heat flux $q^*_i = -\overline{\mathrm{v}_i' T'}$. This may be modeled as

$$q^*_i = \frac{\mu_T c_p}{\Pr_T}\overline{T}_{,i} \tag{21.3.8}$$

where $\Pr_T$ is the turbulent Prandtl number.

In the absence of thermoviscous dissipation, the governing equations (21.2.9a,b,c) together with any one of the turbulence models discussed above are closed. They can be solved simultaneously using suitable computational schemes of the previous chapters.

21.3.2 ONE-EQUATION MODELS

In the one-equation model, the eddy viscosity is defined as

$$\nu_T = c_\mu \ell \sqrt{K}, \quad c_\mu = 0.09$$

where K is the turbulent kinetic energy,

$$K = \frac{1}{2}\overline{v'_i v'_i}$$

Note that we have introduced one new variable K, so we must introduce one additional governing equation. This can be provided by the transport equation for the turbulence kinetic energy K,

$$\frac{DK}{Dt} = (\nu_k K_{,i})_{,i} + (\overline{\tau}_{ij}\overline{v}_i)_{,j} \tag{21.3.9}$$

with

$$\nu_k = \nu + \nu_T$$

This turbulent kinetic energy transport equation (21.3.9) is added to the Navier-Stokes system of equations for simultaneous solution, with ν_T calculated as shown in Section 21.3.1.

21.3.3 TWO-EQUATION MODELS

K–ε Model

There are many two-equation models used in practice today. Among them is the K–ε model, which has been used most frequently for low-speed incompressible flows in isotropic turbulence. In this model, the turbulent stress tensor is given

$$\tau^*_{ij} = 2\mu_T \overline{d}_{ij} - \frac{2}{3}\overline{\rho} K \delta_{ij} \tag{21.3.10}$$

where the turbulent (eddy) viscosity μ_T is defined as

$$\mu_T = \overline{\rho} c_\mu \frac{K^2}{\varepsilon} \tag{21.3.11a}$$

with ε being the turbulent kinetic energy dissipation rate,

$$\varepsilon = \nu \overline{v'_{i,j} v'_{i,j}} \tag{21.3.11b}$$

Thus, the turbulent viscosity in (21.3.11a) contains two unknown variables, K and ε. It is therefore necessary that transport equations for K and ε be provided, which can be derived from the momentum equations. To obtain the turbulent kinetic energy transport equation, we take a time average of the product of the fluctuation component of the velocity with the turbulent flow momentum equations. After some algebra, we arrive at

$$\overline{\rho}\frac{\partial K}{\partial t} + \overline{\rho}\overline{v}_i K_{,i} = A_{(k)} + B_{(k)} + C_{(k)} \tag{21.3.12a}$$

with $A_{(k)}$, $B_{(k)}$, $C_{(k)}$ denoting the production, dissipation, and diffusion transport, respectively,

$$A_{(k)} = \overline{\tau}_{ij}\overline{v}_{j,i}$$
$$B_{(k)} = -\rho\varepsilon$$
$$C_{(k)} = \left(\mu K_{,i} - \frac{1}{2}\overline{\rho}\overline{v'_i v'_j v'_j} - p' v'_i\right)_{,i}$$

where the first, second, and third terms of C_k represent the molecular diffusion, turbulent diffusion, and pressure diffusion, respectively.

Similarly, the dissipation energy transport equation can be derived by taking a time average of the product of $2\nu v'_{i,j}$ with the derivative of momentum equations, resulting in

$$\overline{\rho}\frac{\partial \varepsilon}{\partial t} + \overline{\rho}\overline{v}_i\varepsilon_{,i} = A_{(\varepsilon)} + B_{(\varepsilon)} + C_{(\varepsilon)} \tag{21.3.12b}$$

with

$$A_{(\varepsilon)} = -2\mu(\overline{v'_{i,k}v'_{j,k} + v'_{k,i}v'_{k,j}})\overline{v}_{i,j}$$
$$B_{(\varepsilon)} = -2\mu\overline{v'_k v'_{i,j}}\overline{v}_{i,jk} - 2\mu\overline{v'_{i,k}v'_{i,j}v'_{k,j}} - 2\mu\nu\overline{v'_{i,kj}v'_{i,kj}}$$
$$C_{(\varepsilon)} = (\mu\varepsilon_{,j} - \mu\overline{v'_j v'_{i,k}v'_{i,k}} - 2\mu p'_{,i}v'_{j,i})_{,j}$$

which represent production of dissipation, dissipation of dissipation, and dissipation transport terms, respectively. Here, the first, second, and third terms of $C_{(\varepsilon)}$ indicate molecular dissipation, turbulent dissipation, and pressure dissipation, respectively.

As a consequence of (21.3.12a,b), we are now confronted with more unknowns than we originally started in (21.3.11a,b). To avoid such additional unknowns, Launder and Spalding [1972] proposed the so-called K–ε model in which the turbulent kinetic energy and dissipation energy transport equations can be written as follows:

$$\frac{\partial}{\partial t}(\overline{\rho}K) + (\overline{\rho}K\overline{v}_i)_{,i} = (\overline{\tau}_{ij}\overline{v}_j)_{,i} - \overline{\rho}\varepsilon + (\mu_k K_{,i})_{,i} \tag{21.3.13a}$$

$$\frac{\partial\overline{\rho}\varepsilon}{\partial t} + (\overline{\rho}\varepsilon\overline{v}_i)_{,i} = c_{\varepsilon 1}(\overline{\tau}_{ij}\overline{v}_j)_{,i} - c_{\varepsilon 2}\overline{\rho}\frac{\varepsilon^2}{K} + (\mu_\varepsilon\varepsilon_{,i})_{,i} \tag{21.3.13b}$$

with

$$\mu_\kappa = \mu + \frac{\mu_T}{\sigma_\kappa}, \qquad \mu_\varepsilon = \mu + \frac{\mu_T}{\sigma_\varepsilon}$$
$$c_\mu = 0.09, \quad c_{\varepsilon 1} = 1.45 \sim 1.55, \quad c_{\varepsilon 2} = 1.92 \sim 2.00, \quad \sigma_\kappa = 1, \quad \sigma_\varepsilon = 1.3 \tag{21.3.14}$$

Notice that the first, second, and third terms on the right-hand side of (21.3.13a,b) correspond to the production, dissipation, and transport terms, respectively, as defined in (21.3.12a,b). The closure constants given in (21.3.14) are obtained from the experimental data. They may also be correlated (calibrated) by direct numerical simulation discussed in Section 21.5. It is seen that no new variables other than K and ε are contained in (21.3.13a,b). These two equations can now be combined in the solution of the Navier-Stokes system of equations.

Nonlinear (anisotropic) *K*–ε Model

An improved version of the K–ε model was proposed by Speziale [1987] in which the turbulent stress tensor includes the frame indifferent Oldroyd derivative.

$$\tau_{ij}^* = 2\mu_T \overline{d}_{ij} - \frac{2}{3}\overline{\rho} K\delta_{ij} + \hat{\tau}_{ij} \tag{21.3.15}$$

where $\hat{\tau}_{ij}$ represents the nonlinear anisotropic turbulence,

$$\hat{\tau}_{ij} = 4\overline{\rho} C_D c_\mu^2 \frac{K^3}{\varepsilon^2}\left(\hat{d}_{ij} - \frac{1}{3}\hat{d}_{kk}\delta_{ij} + \overline{d}_{ik}\overline{d}_{kj} - \frac{1}{3}\overline{d}_{ki}\overline{d}_{kj}\right)$$

$$\hat{d}_{ij} = \frac{\partial \overline{d}_{ij}}{\partial t} + \overline{v}_k \overline{d}_{ij,k} - d_{kj}\overline{v}_{i,k} - d_{ki}\overline{v}_{j,k}$$

with $C_D = 1.68$ as calibrated from the experimental data.

K–ω Model

The basic idea of the K–ω model was originated by Kolmogorov [1942] with turbulence associated with vorticity, ω, being proportional to K^2/ℓ,

$$\omega = c\frac{K^2}{\ell} \tag{21.3.16a}$$

where c is a constant. Thus, the eddy viscosity may be written as

$$\mu_T = \rho K/\omega \tag{21.3.16b}$$

The transport equations for k and ω [Wilcox, 1988] may be written as

$$\frac{\partial}{\partial t}(\overline{\rho} K) + (\overline{\rho} K\overline{v}_{i,i}) = (\mu_k K_{,i})_{,i} + (\overline{\tau}_{ij}\overline{v}_j)_{,i} - \beta^*\overline{\rho} K\omega \tag{21.3.17a}$$

$$\frac{\partial}{\partial t}(\overline{\rho}\omega) + (\overline{\rho}\omega\overline{v}_i)_{,i} = (\mu_\varepsilon \omega_{,i})_{,i} + \alpha\frac{\omega}{K}(\overline{\tau}_{ij}\overline{v}_j)_{,i} - \beta\overline{\rho}\omega^2 \tag{21.3.17b}$$

with the closure constants,

$$\alpha = 5/9, \qquad \beta = 3/40, \qquad \beta^* = 9/100, \qquad \sigma = 1/2, \qquad \sigma^* = 1/2$$

Wall Functions

At the wall boundary, the velocity gradients are high, requiring excessive mesh refinements. In order to alleviate such excessive mesh refinements, the so-called wall function [Launder and Spalding, 1972] is needed. To this end, the boundary conditions for K and ε in the near wall regions may be specified as

$$K = \frac{|\tau_w|}{\sqrt{c_\mu}}, \qquad \varepsilon = \frac{|\tau_w|}{a\delta}$$

where the wall shear stress τ_w is given by

$$|\tau_w| = \frac{a|u_*|\left(c_\mu^{0.5}K\right)^{0.5}}{\ell_n(E\delta^+)} \tag{21.3.18}$$

with the turbulent kinetic energy K computed iteratively at a distance $\delta^+ \geq 12$, $a = 0.419$, $\varepsilon = 9.793$, and

$$\delta^+ = \text{Re}\ \delta\left(c_\mu^{0.5}K\right)^{0.5}$$

For $\delta^+ < 12$ the laminar stress is given by

$$|\tau_w| = \frac{|u_*|}{\text{Re}\ \delta} \tag{21.3.19}$$

where the viscosity in the near wall regions is estimated as

$$\mu_* = \text{Re}\ \delta\frac{|\tau_w|}{|u_*|}$$

If the flow velocity increases, however, it has been observed that the role of the wall function becomes unrealistic and the K–ε model is considered unreliable. The K–ε model described here is based on isotropic turbulence and is referred to as standard K–ε model.

The following boundary conditions are typically imposed for a wall-bound turbulent flow:

(a) *Inflow*: specify $\overline{u}$, K, and ε
(b) *Outflow*: specify $\overline{v}$ by extrapolation, $\overline{u}$ by mass balance; $\overline{p}$, K, and ε by extrapolation
(c) Wall boundaries
   (i) Standard two-layer form of the law of the wall

$$u^+ = \frac{1}{\kappa}\ln y^+ + 5, \qquad \frac{K}{u_*^2} = c_\mu^{-\frac{1}{2}}, \qquad \varepsilon = c_\mu^{\frac{1}{2}}\frac{K^{\frac{3}{2}}}{\kappa y} \tag{21.3.20}$$

   These conditions are applied at the first grid point y away from the wall if $y^+ \equiv yu_*/\nu \geq 11.6$ with $u^+ = \overline{u}/u_*$. If $y^+ < 11.6$, then $\overline{u}$, K, and ε are interpolated to the wall values based on viscous sublayer constraints.
   (ii) Three-layer form of the law of the wall

$$u^+ = \begin{cases} y^+ & \text{for } \leq 5 \\ -3.05 + 5\ln y^+ & \text{for } 5 < y^+ \leq 30 \\ 5.5 + \dfrac{1}{\kappa}\ln y^+ & \text{for } y^+ > 30 \end{cases} \tag{21.3.21}$$

For the K–ω model, Wilcox [1989] proposes the wall function for ω in the form,

$$\omega = \frac{K^{1/2}}{c_\mu^{1/4}} \tag{21.3.22a}$$

and further argued that the pressure gradient must be included for high-pressure gradient flows.

$$\omega = \frac{u_*}{0.41y\sqrt{c_\mu}}\left(1 - 0.32\frac{y}{\rho\sqrt{u_*}}\frac{dp}{dx}\right) \qquad (21.3.22b)$$

21.3.4 SECOND ORDER CLOSURE MODELS (REYNOLDS STRESS MODELS)

Effects of streamline curvature, sudden changes in strain rate, secondary motion, etc. can not be accommodated in the two equation models presented in Section 21.3.3. The second order closure models or Reynold stress models are designed to handle these features. The stress tensor is given by

$$\tau_{ij} = \overline{\tau}_{ij} + \tau^*_{ij}$$

with τ^*_{ij} being the Reynold stress

$$\tau^*_{ij} = -\rho\overline{v'_i v'_j}$$

The Reynolds stress transport equation is of the form

$$\frac{\partial \tau^*_{ij}}{\partial t} + (\overline{v}_k \tau^*_{ij}),_k = A_{ij} + B_{ij} + C_{ij} + D_{ij} \qquad (21.3.23)$$

where A_{ij}, B_{ij}, C_{ij}, and D_{ij}, denote production, dissipation (destruction), diffusion, and pressure strain, respectively.

$$A_{ij} = -\tau^*_{ik}\overline{v}_{j,k} - \tau^*_{jk}\overline{v}_{i,k} \qquad (21.3.24)$$

$$B_{ij} = -2\mu\overline{v'_{i,k}v'_{j,k}} \qquad (21.3.25)$$

$$C_{ij} = \lfloor -(\rho\overline{v'_i v'_j v'_k} + \overline{p'v'_i}\delta_{jk} + \overline{p'v'_j}\delta_{ik}) + \mu\tau^*_{ij,k}\rfloor,_k \qquad (21.3.26)$$

$$D_{ij} = \overline{p'(v'_{i,j} + v'_{j,i})} \qquad (21.3.27)$$

Note that new variables are introduced in C_{ij} and D_{ij}, whereas A_{ij} and B_{ij} contain no new variables. Thus, we must model the diffusion transport and pressure-strain tensors. Although dissipation occurs at the smallest scales and one can use the Kolmogorov hypothesis of local isotropy, it may become anisotropic close to the wall, and thus modeling is needed. We discuss below some of the well-known second order closure models.

Dissipation Tensor

Since ε is the dissipation rate, this may be treated similarly as in the K–ε model. However, Hanjalic and Launder [1976] propose to add an extra term representing anisotropy close to the wall.

$$B_{ij} = -\frac{2}{3}\rho\varepsilon\delta_{ij} - 2f\rho\varepsilon b_{ij} \qquad (21.3.28)$$

where f is a damping function and b_{ij} denotes the dimensionless anisotropy tensor, respectively,

$$f = (1 + 0.1\,\mathrm{Re}^*)^{-1}, \quad \mathrm{Re}^* = K^2/(\varepsilon\nu)$$

$$b_{ij} = -\left(\frac{\tau^*_{ij} - \frac{2}{3}\rho K\delta_{ij}}{2\rho K}\right)$$

Diffusion Transport Tensor

The turbulence transport is characterized by the diffusion tensor C_{ijk}. Launder, Reece, and Rodi [1975] proposed that this tensor be modeled as

$$C_{ijk} = -\frac{2}{3}c\frac{K^2}{\varepsilon}(\tau^*_{ij,k} + \tau^*_{ik,j} + \tau^*_{jk,i}) + \mu\tau^*_{ij,k} \tag{21.3.29a}$$

with $c \cong 0.11$. They also postulated a more general form,

$$C_{ijk} = -c'\frac{K}{\rho\varepsilon}(\tau^*_{ij,m}\tau^*_{mk} + \tau^*_{ik,m}\tau^*_{mj} + \tau^*_{jk,m}\tau^*_{mi}) + \mu\tau^*_{ij,k} \tag{21.3.29b}$$

with $c' \cong 0.25$.

Pressure-Strain Correlation Tensor

This is an important contribution in turbulence since the terms involved in the pressure-strain tensor are of the same order of magnitude as the production terms. Pressure can be obtained by solving the pressure Poisson equation in which the forcing functions consist of slow and rapid fluctuations. To see this, we examine the pressure Poisson equation in the form,

$$p_{,ii} = -\rho(\mathrm{v}_{i,j}\mathrm{v}_j)_{,i} = -\rho(\mathrm{v}_{i,ji}\mathrm{v}_j + \mathrm{v}_{i,j}\mathrm{v}_{j,i})$$

In terms of mean and fluctuating components, we obtain

$$p'_{,ii} = -\rho(f_s + f_r) \tag{21.3.30}$$

where the slow forcing function f_s and rapid forcing function f_r are given by

$$f_s = \left(\mathrm{v}'_i\mathrm{v}'_j - \overline{\mathrm{v}'_i\mathrm{v}'_j}\right)_{,ij} \tag{21.3.31a}$$

$$f_r = 2\overline{\mathrm{v}}_{i,j}\mathrm{v}'_{j,i} \tag{21.3.31b}$$

The solution of (21.3.30) via Green functions results in integral forms corresponding to (21.3.31a) and (21.3.31b) such that the pressure-strain tensor can be written as

$$D_{ij} = E_{ij} + F_{ijkm}\overline{\mathrm{v}}_{k,m} \tag{21.3.32}$$

where E_{ij} and $F_{ijkm}\overline{\mathrm{v}}_{k,m}$ denote the slow pressure strain and rapid pressure strain, respectively. For inhomogeneous turbulence, the mean velocity present in the rapid pressure strain (21.3.31b) implies the process is not localized, leading to the argument that the single-point correlation may not be adequate. This would require that the products of fluctuating properties be correlated at two separate physical locations (two-point correlation). This task is difficult, and the so-called *locally homogeneous approximation* may be adopted as described below.

Rotta [1951] postulated that the slow pressure strain is of the form

$$E_{ij} = c_1 \frac{\varepsilon}{K}\left(\tau^*_{ij} + \frac{2}{3}\rho K\delta_{ij}\right) \quad 1.4 \le c_1 \le 1.8 \tag{21.3.33}$$

whereas Launder, Reece, and Rodi [1975] (known as LRR method) proposed that the rapid pressure-strain for homogeneous turbulence may be correlated by

$$F_{ijkm}\overline{\mathrm{v}}_{k,m} = \alpha\left(A_{ij} - \frac{1}{3}A_{kk}\delta_{ij}\right) - \beta\left(G_{ij} - \frac{1}{3}G_{kk}\delta_{ij}\right) - \gamma\rho K d_{ij} \tag{21.3.34}$$

with

$$D_{ij} = \tau^*_{im}\overline{\mathrm{v}}_{m,j} + \tau^*_{jm}\overline{\mathrm{v}}_{m,i} \tag{21.3.35}$$

$$\alpha = \frac{8+c_2}{11}, \qquad \beta = \frac{8c_2-2}{11}, \qquad \gamma = \frac{60c_2-4}{55}, \qquad 0.4 \le c_2 \le 0.6 \tag{21.3.36}$$

There are many other schemes for second order closure models. Among them are the tensor invariant method [Lumley, 1978], multi-scale method [Wilcox, 1988], nonlinear stress method [Speziale, Sarker, and Gatski, 1991], and modified LRR method [Launder, 1992].

21.3.5 ALGEBRAIC REYNOLDS STRESS MODELS

The purpose of algebraic Reynolds stress models is to avoid the solution of differential equations such as (21.3.23), and to obtain the Reynolds stress components directly from algebraic relationships. If mean strain rates are ignored in the Reynolds stress transport equations (21.3.23), it follows from the strain-dependent generalization of nonlinear constitutive relation that the turbulent stress tensor may be written as [Rodi, 1976; Gatski and Speziale, 1992],

$$\tau^*_{ij} = \frac{K}{\varepsilon}(D_{ij} + B_{ij}) \tag{21.3.37}$$

with

$$D_{ij} = c_1 \frac{\varepsilon}{K}\left(\tau^*_{ij} + \frac{2}{3}\rho K\delta_{ij}\right) \tag{21.3.38}$$

$$B_{ij} = -\frac{2}{3}\rho\varepsilon\delta_{ij} \tag{21.3.39}$$

Thus, if the mean strain rate vanishes, then we have

$$\tau^*_{ij} = -\frac{2}{3}\rho K\delta_{ij} \tag{21.3.40}$$

This suggests that the algebraic stress model is confined to isotropic turbulence. Thus, the algebraic stress model fails to properly account for sudden changes in the mean strain rate. If this algebraic Reynolds stress model is combined with the K–ε model, however, it may be possible to obtain satisfactory results for secondary motions as reported by So and Mellor [1978] and Dumuren [1991].

A fully explicit, self-consistent algebraic expression for the Reynolds stress, which is the exact solution to the Reynolds stress transport equation in the weak equilibrium limit can be derived as shown by Girimaji [1995]. Preliminary tests indicate that the model performs adequately, even for three-dimensional mean flow cases.

21.3.6 COMPRESSIBILITY EFFECTS

The turbulent models discussed above are applicable to incompressible flows with time averages. For compressible flows, however, it is more convenient to use Favre averages than time averages as mentioned in Section 21.2. The Favre-averaged unknowns in (21.2.10) are modeled as follows:

Favre-averaged turbulent stress tensor

$$\tau_{ij}^{*} = -\overline{\rho \mathrm{v}_i'' \mathrm{v}_j''} = 2\mu_T\left(d_{ij} - \frac{1}{3}d_{kk}\delta_{ij}\right) - \frac{2}{3}\bar{\rho} K\delta_{ij} \tag{21.3.41}$$

Favre-averaged turbulent heat flux vector

$$q_i^{*} = \overline{\rho \mathrm{v}_i'' H''} = -\frac{\mu_T c_p}{\mathrm{Pr}_T}\tilde{T}_{,i} = -\frac{\mu_T}{\mathrm{Pr}_T}\tilde{H}_{,i} \tag{21.3.42}$$

Favre-averaged turbulent molecular diffusion and turbulent transport

$$\overline{\tau_{ij}\mathrm{v}_j''} - \frac{1}{2}\overline{\rho \mathrm{v}_i''\mathrm{v}_j''\mathrm{v}_j''} = \left(\mu + \frac{\mu_T}{\sigma_k}\right)K_{,i} \tag{21.3.43}$$

The kinetic energy transport equations and Reynolds stress transport equations for compressible turbulent flows are written as follows:

Compressible turbulent kinetic energy transport equation

$$\frac{\partial \bar{\rho} K}{\partial t} + (\bar{\rho}\tilde{\mathrm{v}}_i K)_{,i} = A_{(k)} + B_{(k)} + C_{(k)} + D_{(k)} \tag{21.3.44}$$

with

$$A_{(k)} = \tau_{ij}^{*}\tilde{\mathrm{v}}_{j,i}$$

$$B_{(k)} = -\bar{\rho}\varepsilon$$

$$C_{(k)} = \left(\overline{\tau_{ij}\mathrm{v}_j''} - \frac{1}{2}\overline{\rho \mathrm{v}_i''\mathrm{v}_j''\mathrm{v}_j''} - \overline{p'\mathrm{v}_i''}\right)_{,i}$$

$$D_{(k)} = -\overline{\mathrm{v}_i''}\bar{p}_{,i} + \overline{p'\mathrm{v}_{i,i}''}$$

The first three terms on the right-hand side of (21.3.44) are similar to the case of incompressible flows with extra terms in $D_{(k)}$ representing the pressure work and pressure dilatation due to density and pressure fluctuations.

Compressible Reynolds stress transport equation

$$\frac{\partial \tau_{ij}^*}{\partial t} + (\tilde{v}_k \tau_{ij}^*),_k = A_{ij} + B_{ij} + C_{ij} + D_{ij} + \hat{D}_{ij} \tag{21.3.45}$$

with

$$A_{ij} = -\tau_{ij}^* \tilde{v}_{j,k} - \tau_{jk}^* \tilde{v}_{i,k}$$

$$B_{ij} = \overline{-\tau_{jk}^* v''_{i,k} - \tau_{ik}^* v''_{j,k}}$$

$$C_{ij} = \left[\overline{\rho v''_i v''_j v''_k} + \overline{p' v''_i}\delta_{jk} + \overline{p' v''_j}\delta_{ik} - \left(\overline{\tau_{jk} v''_i + \tau_{ik} v''_j}\right)\right]_{,k}$$

$$D_{ij} = \overline{-p'(v''_{i,j} + v''_{j,i})}$$

$$\hat{D}_{ij} = \overline{v}''_i \overline{p}_{,j} + \overline{v}''_j \overline{p}_{,i}$$

Here again the first four terms on the right-hand side of (21.3.45) have analogs for the incompressible flow with the last terms in $\hat{D}_{ij}$ for nonvanishing pressure gradients.

With additional new unknowns appearing in (21.3.44) and (21.3.45), we are faced with the difficult task of modeling them. Modeling in compressible turbulent flows for Reynolds averaged Navier-Stokes (RANS) system of equations has not been developed to a satisfactory extent. This is because the large-scale motions are difficult to model particularly in compressible flows. One way to resolve this problem is to use the large eddy simulation (LES) in which only subgrid (small) scales need be modeled. This will be discussed in Section 21.4.3.

Modifications From Incompressible Flows

Although the K–ε model has been applied to an incompressible flow with reasonable success, its performance in high-speed compressible flows met with difficulties. Sarkar et al. [1989] and Zeman [1990] independently proposed schemes which take into account the compressibility corrections by providing the so-called dilatational component ε_d in addition to the solenoidal component ε of the turbulence kinetic energy dissipation rate for the source term of the turbulence kinetic energy transport equation. Thus, (21.3.13a) is modified as

$$\frac{\partial}{\partial t}(\overline{\rho} K) + (\overline{\rho} K \overline{v}_i)_{,i} = (\mu_k K_{,i})_{,i} + (\overline{\tau}_{ij}\overline{v}_j)_{,i} - \overline{\rho}(\varepsilon + \varepsilon_d) \tag{21.3.46}$$

where

$$\varepsilon_d = \xi^* F(M_t) t$$

Sarkar Model

$$\xi^* = 1$$

$$F(M_t) = M_t^2$$

$$M_t = \frac{2K}{a^2} \quad \text{(Turbulent Mach Number)} \tag{21.3.47}$$

Zeman Model

$$\xi^* = \frac{3}{4}$$

$$F(M_t) = 1 - \exp\left[-\tfrac{1}{2}(\gamma+1)(M_t - M_{to})^2/\lambda^2\right] H(M_t - M_{to})$$

$H =$ Heavy side step function

$$\left.\begin{aligned} M_{to} &= 0.1\sqrt{2/(\gamma+1)} \\ \lambda &= 0.6 \end{aligned}\right\} \text{ free shear flows}$$

$$\left.\begin{aligned} M_{to} &= 0.25\sqrt{2/(\gamma+1)} \\ \lambda &= 0.66 \end{aligned}\right\} \text{ wall boundary layers}$$

Wilcox [1992] suggests that the Sarkar model can be improved by using

$$\xi^* = \frac{3}{2}$$

$$F(M_t) = \left(M_t^2 - M_{to}^2\right) H\left(M_t - M_{to}\right)$$

$$M_{to} = \frac{1}{4}$$

The K–ω model with compressibility effects may be given by [Wilcox, 1992]

$$\frac{\partial}{\partial t}(\overline{\rho} K) + (\overline{\rho} K \overline{v}_i)_{,i} = [(\mu + \sigma^* \mu_T) K_{,i}]_{,i} + (\overline{\tau}_{ij}\overline{v}_i)_{,i} - \beta^* \rho \omega K \tag{21.3.48}$$

$$\frac{\partial}{\partial t}(\overline{\rho}\omega) + (\overline{\rho}\omega\overline{v}_i)_{,i} = [(\mu + \sigma^* \mu_T)\,\omega_{,i}]_{,i} + \alpha\frac{\omega}{K}(\overline{\tau}_{ij}\overline{v}_j)_{,i} - \beta\rho\omega\left[\omega + \xi\sqrt{|2\Omega_{mn}\Omega_{mn}|}\right] \tag{21.3.49}$$

with

$$\omega = \frac{\varepsilon}{\beta^* K}, \qquad \beta^* = \beta_o^*[1 + \xi^* F(M_t)]$$

$$\mu_T = \frac{\rho K}{\omega}, \qquad \Omega_{mn} = \frac{1}{2}(\overline{v}_{i,j} - \overline{v}_{j,i})$$

$$\beta = \beta_o - \beta_o^* \xi^* F(M_t)$$

where β_o^* and β_o are the corresponding incompressible values of β^* and β as given in (21.3.16).

Hanine and Kourta [1991] reported comparisons of the performance of various turbulence models to predict the near wall compressible flows and emphasized the importance of compressibility corrections. Wilcox [1992a] also studied the supersonic turbulent boundary layer flows. He showed that neither the Sarkar nor the Zeman compressibility term is completely satisfactory for both the compressible mixing layer and wall-bounded flows [Wilcox, 1992b]. The compressibility corrections cause a decrease in the effective von Karman constant, which yields the unwanted decrease in skin friction. However, for the K–ε model, the constant in the law of the wall varies with density ratio in a nontrivial manner. Wilcox [1992] then combines Sarkar's simple functional dependence of dilatational dissipation on turbulence Mach number with Zeman's lag effect to produce a compressibility term that yields reasonably accurate predictions. Subsequently, Huang, Bradshaw, and Coarley [1992] reexamined the independent studies of Wilcox, Zeman, and Sarkar and concluded that the extension of incompressible turbulence models to compressible flow requires density corrections to the closure coefficients to satisfy the law of the wall. They further suggest that the K–ε model is more attractive than the K–ε model at high Mach numbers, because the coefficients of the unwanted density gradient terms are smaller. In view of these observations, the compressibility corrections which were originally developed for incompressible flows should be used with caution for applications into high-speed compressible turbulent flows.

The various turbulence models discussed in Section 21.3 represent a brief summary of historical developments for the period of nearly half a century. In Section 21.7, we present some limited numerical applications for the K–ε models. It appears, however, that the current interest in turbulence research is directed toward large eddy simulation and direct numerical simulation. We discuss these subjects in the following sections.

21.4 LARGE EDDY SIMULATION

Despite a great deal of effort and advancement in turbulence modeling for the past century, difficulties still remain in geometrically and physically complicated flowfields. The large eddy simulation (LES) is an alternative approach toward achieving our goal for more efficient turbulent flow calculations. Here, by using more refined meshes than usually required for Reynolds averaged Navier-Stokes (RANS) system of equations discussed in Section 21.3, large eddies are calculated (resolved) whereas small eddies are modeled. The rigor of LES in terms of performance and ability is somewhere between RANS of Section 21.3 and the direct numerical simulation (DNS) to be discussed in Section 22.5. There are two major steps involved in the LES analysis: filtering and subgrid scale modeling. Traditionally, filtering is carried out using the box function, Gaussian function, or Fourier cutoff function. Subgrid modeling includes eddy viscosity model, structure function model, dynamic model, scale similarity model, and mixed model, among others. These and other topics are presented below.

21.4.1 FILTERING, SUBGRID SCALE STRESSES, AND ENERGY SPECTRA

In order to define a velocity field containing only the large-scale components of the total field, it is necessary to filter the variables of the Navier-Stokes system of equations, resulting in the local average of the total field. To this end, using one-dimensional notation for simplicity, the filtered variable f may be written as

$$\overline{f} = \int G(x,\xi) f(\xi) d\xi \tag{21.4.1}$$

$$\text{with} \quad \int G(x,\xi) d\xi = 1$$

where $G(x, \xi)$ is the filter function which is large only when x and ξ are close together. They include box (tophat) function, Gaussian function, and Fourier cutoff function.

Box

$$G(x) = \begin{cases} 1/\Delta & \text{if } |x| \leq \Delta/2 \\ 0 & \text{otherwise} \end{cases} \tag{21.4.2}$$

Gaussian

$$G(x) = \sqrt{\frac{6}{\pi \Delta^2}} \exp\left(-\frac{6x^2}{\Delta^2}\right) \tag{21.4.3}$$

Fourier cutoff

$$\hat{G}(k) = \begin{cases} 1 & \text{if } k \leq \pi/2 \\ 0 & \text{otherwise} \end{cases} \tag{21.4.4}$$

The filtered momentum equation takes the form

$$\frac{\partial \overline{v}_j}{\partial t} + (\overline{v_i v_j})_{,i} = -\frac{1}{\rho}\overline{p}_{,j} + \overline{\tau}_{ij,i} \tag{21.4.5}$$

with

$$\begin{aligned} \overline{v_i v_j} &= \overline{(\overline{v}_i + v'_i)(\overline{v}_j + v'_j)} = \overline{\overline{v}_i \overline{v}_j} + \overline{v'_i \overline{v}_j} + \overline{\overline{v}_i v'_j} + \overline{v'_i v'_j} \\ &= \overline{\overline{v}_i \overline{v}_j} + \overline{v}_i \overline{v}_j - \overline{v}_i \overline{v}_j + \overline{v'_i \overline{v}_j} + \overline{\overline{v}_i v'_j} + \overline{v'_i v'_j} \\ &= \overline{v}_i \overline{v}_j - \tau^*_{ij} \end{aligned} \tag{21.4.6}$$

Substituting (21.4.6) into (21.4.5) yields

$$\frac{\partial \overline{v}_j}{\partial t} + (\overline{v}_i \overline{v}_j)_{,i} = -\frac{1}{\rho}\overline{p}_{,j} + \overline{\tau}_{ij,i} + \tau^*_{ij,i}$$

with the subgrid stress tensor τ^*_{ij} identified from (22.4.6) as

$$-\tau^*_{ij} = L_{ij} + C_{ij} + R_{ij} = \overline{v_i v_j} - \overline{v}_i \overline{v}_j \tag{21.4.7}$$

where L_{ij}, C_{ij}, and R_{ij} are known as the Leonard stress tensor, cross stress tensor, and subgrid scale Reynolds stress tensor, respectively.

$$\begin{aligned} L_{ij} &= \overline{\overline{v}_i \overline{v}_j} - \overline{v}_i \overline{v}_j \\ C_{ij} &= \overline{v'_i \overline{v}_j} + \overline{\overline{v}_i v'_j} \\ R_{ij} &= \overline{v'_i v'_j} \end{aligned} \tag{21.4.8}$$

Here, the Leonard stress represents the interaction between resolved scales, transferring energy to small scales (known as outscatter). The Leonard stress can be computed explicitly from the filtered velocity field. The cross stress represents the interaction between resolved and unresolved scales, transferring energy to either large or small scales. The subgrid scale Reynolds stress represents the interaction of two small scales, producing energy from small scales to large scales (known as backscatter).

The cross stress tensor may be simplified in terms of resolved scales using the so-called Galilean scale similarity model [Bardina et al., 1980],

$$C_{ij} = \overline{\mathrm{v}'_i \overline{\mathrm{v}}_j} + \overline{\overline{\mathrm{v}}_i \mathrm{v}'_j} = \overline{\mathrm{v}}_i \overline{\mathrm{v}}_j - \overline{\overline{\mathrm{v}}}_i \overline{\overline{\mathrm{v}}}_j \tag{21.4.9a}$$

Summing (22.4.8a) and (22.4.9a) leads to

$$K_{ij} = L_{ij} + C_{ij} = \overline{\overline{\mathrm{v}}_i \overline{\mathrm{v}}_j} - \overline{\overline{\mathrm{v}}}_i \overline{\overline{\mathrm{v}}}_i \tag{21.4.9b}$$

It is seen that the sum of the Leonard and cross stresses can be calculated from the resolved scales and thus only the subgrid scale Reynolds stress need be modeled. Thus, the turbulent stress tensor to be modeled is given by (21.4.6) or (21.4.7) as

$$\tau^*_{ij} = -(\overline{\mathrm{v}_i \mathrm{v}_j} - \overline{\mathrm{v}}_i \overline{\mathrm{v}}_j) \tag{21.4.10}$$

Before we discuss subgrid scale models, it is informative to examine the physical significance of the filtering in terms of the Kolmogorov' "−5/3 law" for the energy spectrum [Kolmogorov, 1941]. The energy spectrum $E(\kappa)$ is related by the turbulent kinetic energy,

$$K = \frac{1}{2}\overline{\mathrm{v}'_i \mathrm{v}'_i} = \int_0^\infty E(\kappa)d\kappa \tag{21.4.11}$$

The distribution of energy spectrum $E(\kappa)$ vs wave number κ is divided into three regions as shown in Figure 21.4.1: the region of energy containing large eddies, followed by the inertial subrange and energy dissipation range, between the wave numbers identified by the reciprocals of the energy bearing length scale ℓ (integral scale) and the Kolmogorov microscale η,

$$\eta = (\nu^3/\varepsilon)^{1/4} \tag{21.4.12}$$

Note that the inertial subrange is characterized by a straight line, known as the Kolmogorov's "−5/3 law,"

$$E(\kappa) = \alpha\varepsilon^{2/3}\kappa^{-5/3} \tag{21.4.13}$$

where α is a constant. In this range, eddies are small and dissipation becomes important at smallest scales. Thus, the filtering process is designed to identify this range with a

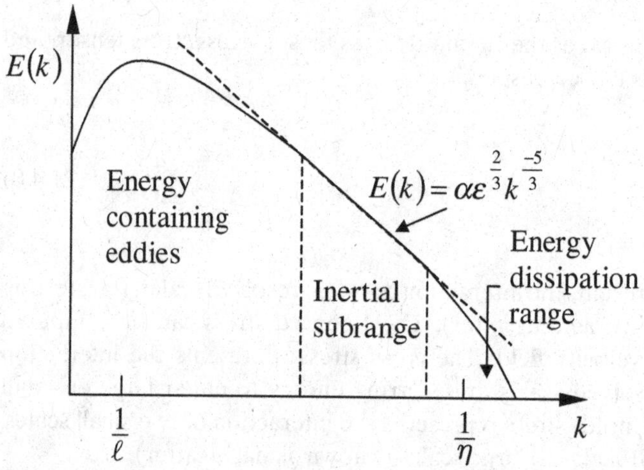

Figure 21.4.1 Energy spectrum vs. wave number space (log-log scales).

suitable filter width. In what follows, our discussion will be based on filtering by the box function.

21.4.2 THE LES GOVERNING EQUATIONS FOR COMPRESSIBLE FLOWS

The Navier-Stokes system of equations for LES may be written in terms of Favre averages using the filtering process presented in Section 22.4.1. The filtered continuity, momentum, and energy equations for compressible flows are described below.

Construction of turbulent closure models for high Mach numbers and high Reynolds numbers in hypersonic flows is difficult, particularly for large turbulence scales. For this reason, one may wish to explore the possibility of LES in the hope that the subgrid scale (SGS) modeling is still feasible. To this end, we rewrite the Favre-filtered compressible flow governing equations as follows:

$$\frac{\partial \overline{\rho}}{\partial t} + (\overline{\rho}\tilde{\mathrm{v}}_i)_{,i} = 0 \tag{21.4.14a}$$

$$\frac{\partial}{\partial t}(\overline{\rho}\tilde{\mathrm{v}}_j) + (\overline{\rho}\tilde{\mathrm{v}}_i\tilde{\mathrm{v}}_j)_{,i} + \overline{p}_{,j} - (\overline{\tau}_{ij} + \tau^*_{ij})_{,i} = \mathbf{0} \tag{21.4.14b}$$

$$\frac{\partial}{\partial t}(\overline{\rho}\tilde{E}) + [(\overline{\rho}\tilde{E} + \overline{p})\tilde{\mathrm{v}}_i - \tilde{\tau}_{ij}\tilde{\mathrm{v}}_j + \tilde{q}_i]_{,i} + (q_i^{(H)} + q_i^{(T)} + q_i^{(\mathrm{v})})_{,i} = 0 \tag{21.4.14c}$$

where the SGS variables are the turbulent stress τ^*_{ij}, turbulent heat flux $q_i^{(H)}$, turbulent diffusion $q_i^{(T)}$, and turbulent viscous diffusion $q_i^{(V)}$. They are expressed as

$$\begin{aligned}
\tau^*_{ij} &= -\overline{\rho}\,(\widetilde{\mathrm{v}_i\mathrm{v}_j} - \tilde{\mathrm{v}}_i\tilde{\mathrm{v}}_j)\\
q_i^{(H)} &= \overline{\rho}\tilde{c}_p(\widetilde{\mathrm{v}_i T} - \tilde{\mathrm{v}}_i\tilde{T})\\
q_i^{(T)} &= \frac{1}{2}\overline{\rho}\,(\widetilde{\mathrm{v}_i\mathrm{v}_j\mathrm{v}_j} - \tilde{\mathrm{v}}_i\tilde{\mathrm{v}}_j\tilde{\mathrm{v}}_j)\\
q_i^{(\mathrm{v})} &= -\,(\overline{\tau_{ij}\mathrm{v}_j} - \tilde{\tau}_{ij}\tilde{\mathrm{v}}_j)
\end{aligned} \tag{21.4.15a,b,c,d}$$

These unknown variables may be modeled by several different ways. Among them are (1) eddy viscosity model, (2) scale similarity model, and (3) mixed model. We describe these methods in the next section.

21.4.3 SUBGRID SCALE MODELING

The solution of the filtered Navier-Stokes system of equations enables only the large eddies to be resolved, leaving the small eddies still unresolved. Since these small eddies are more or less isotropic, the modeling is much easier than in the case of RANS. However, for compressible flows, particularly for supersonic and hypersonic flows in which turbulent heat flux, turbulent diffusion, and viscous diffusion may become significant, the SGS modeling process is far from satisfactory.

There are three different approaches for developing the SGS turbulent stress models. The eddy viscosity model is most widely used in which the global effect of SGS terms is taken into account, neglecting the local energy events associated with convection and diffusion [Smagorinsky, 1963; Yoshizawa, 1986; Moin et al., 1991; Gao and O'Brien, 1993].

The scale similarity model assumed that the most active subgrid scales are those close to the cutoff wave number and uses the smallest resolved SGS stresses. This approach does account for the local energy events, but tends to underestimate the dissipation [Bardina et al., 1980].

To compensate the drawbacks of the eddy viscosity model and scale similarity model, Erlebacher et al. [1992] proposed the mixed model in which the dissipation is adequately provided to the scale similarity model.

Germano [1992] proposed that the closure constants involved in the SGS turbulent stress tensors be calculated dynamically (flowfield dependent), known as the dynamic model. The advantage of the dynamic model has been demonstrated by many investigators.

Attempts have been made to provide SGS modeling for turbulent diffusion and viscous diffusion in the energy equation by some investigators. Among them are Normand and Lesieur [1992], Meneveau and Lund [1997], and Knight et al. [1998].

In what follows, we introduce some of the well-known models of SGS turbulent eddy viscosity, turbulent heat flux, and turbulent diffusion.

SGS Eddy Viscosity Model for Stress Tensor with Time Averages

In this model, the traditional gradient-diffusion approach (molecular motion) is used so that the turbulent stress tensor for compressible flows is written as

$$\tau_{ij}^* = 2\mu_T\left(\overline{d}_{ij} - \frac{1}{3}\overline{d}_{kk}\delta_{ij}\right) - \frac{2}{3}\overline{K}\delta_{ij} \tag{21.4.16}$$

$$\mu_T = \overline{\rho}(C_s\Delta)^2|\overline{d}|, \qquad \Delta \cong \ell, \qquad \overline{d}_{ij} = \frac{1}{2}(\overline{v}_{i,j} + \overline{v}_{j,i}), \qquad |\overline{d}| = (2\overline{d}_{ij}\overline{d}_{ij})^{1/2}$$

where C_s is the Smagorinsky constant and $\overline{K}$ is the subgrid scale turbulent kinetic energy. This constant can be evaluated by assuming the existence of an inertial range spectrum given in Figure 21.4.1. To this end, it has been suggested in [Lilly, 1966] that

$$|\overline{d}|^2 \cong 2\int_0^{\pi/\Delta} \kappa^2 E(\kappa)d\kappa = 2C_k\varepsilon^{2/3}\int_0^{\pi/\Delta} \kappa^{1/3}d\kappa = \frac{3}{2}C_k\varepsilon^{2/3}\left(\frac{\pi}{\Delta}\right)^{4/3} \tag{21.4.17}$$

where $C_k = 1.41$ is the Kolmogorov constant. Thus, we arrive at

$$C_s \cong \frac{1}{\pi}\left(\frac{2}{3\alpha}\right)^{3/4} = 0.18 \tag{21.4.18}$$

The isotropic parts, $\overline{K}$ and $\overline{d}_{kk}$ terms, on the right-hand side of (22.4.16) may be neglected for incompressible flows. For further details on the subgrid scale modeling for the isotropic parts in compressible flows, see Squires [1991], Erlebacher et al. [1992], and Vreman, Geurts, and Kuerten [1995].

SGS Eddy Viscosity Model for Stress Tensor with Favre Averages

The subgrid scale stress tensor as given by (21.4.15a) may now be written for the compressible flow Favre averages as

$$\tau_{ij}^* = -\overline{\rho}(\widetilde{v_i v_i} - \tilde{v}_i\tilde{v}_i) = 2\mu_T\left(\tilde{d}_{ij} - \frac{1}{3}\tilde{d}_{kk}\delta_{ij}\right) - \frac{2}{3}\tilde{K}\delta_{ij} \tag{21.4.19}$$

with

$$\mu_T = \overline{\rho}(C_s\Delta)^2|\tilde{d}|$$
$$\tilde{K} = \overline{\rho}C_I\Delta^2|\overline{d}|^2 \tag{21.4.20}$$

with $C_s = 0.16$ and $C_I = 0.09$.

Moin et al. [1991] extended the Germano's dynamic model [Germano, 1992] for Favre averages. The Favre averaged mixed model was developed by Speziale, Zang, and Hussaini [1988] and used by Erlebacher et al. [1992].

SGS Structure Function Model

Metais and Lesieur [1992] proposed the structure function model in the form

$$\nu_T = 0.105C_k^{-3/2}\Delta x[F(\mathbf{x}, \Delta x)]^{1/2} \tag{21.4.21}$$

where F is calculated as

$$F(\mathbf{x}, \Delta\Omega) = \frac{1}{6}\sum_{i=1}^{3}[\|u(\mathbf{x}) - u(\mathbf{x}+\Delta x_i\mathbf{i}_i)\|^2 + \|u(\mathbf{x}) - u(\mathbf{x}-\Delta x_i\mathbf{i}_i)\|^2]\left(\frac{\Delta\Omega}{\Delta\Omega_i}\right)^{2/3} \tag{21.4.22}$$

with $\Delta\Omega = (\Delta x_1 \Delta x_2 \Delta x_3)^{1/3}$. In the limit of $\Delta x \to 0$, Comte [1994] suggested that

$$\nu_T \cong 0.777(C_s\Delta x)^2\sqrt{2\overline{d}_{ij}\overline{d}_{ij} + \overline{\omega}_i\overline{\omega}_i} \tag{21.4.23}$$

where C_s is the Smagoronsky's constant and $\overline{\omega}_i$ is the vorticity of the filtered field.

Dynamic SGS Eddy Viscosity Model with Time Averages

It has been shown in the literature that superior results may be obtained by updating the model coefficients based on the current flowfields, known as the dynamic model [Germano, Piomelli, Moin, and Cabot, 1991]. Here, in addition to the subgrid scale filtering, a test filter is introduced with the test filter width Δ_t larger than the grid filter width Δ (usually $\Delta_t = 2\Delta$ is used) in order to obtain information from the resolved flowfield. Based on this model, Lilly [1992] suggested that

$$\mu_T = C_d\overline{\rho}\Delta^2|\overline{d}| \tag{21.4.24}$$

with

$$C_d = \frac{A_{ij}M_{ij}}{M_{km}M_{km}} \tag{21.4.25}$$

$$A_{ij} = \langle\overline{\rho}\overline{\mathrm{v}}_i\overline{\mathrm{v}}_j\rangle - \frac{\langle\overline{\rho}\overline{\mathrm{v}}_i\rangle\langle\overline{\rho}\overline{\mathrm{v}}_j\rangle}{\langle\overline{\rho}\rangle}$$

$$M_{ij} = -2\Delta_t^2\langle\overline{\rho}\rangle\langle|\overline{d}|\rangle\left\langle\overline{d}_{ij} - \frac{1}{3}\overline{d}_{kk}\delta_{ij}\right\rangle + 2\Delta^2\left\langle\overline{\rho}|\overline{d}|\left(\overline{d}_{ij} - \frac{1}{3}\overline{d}_{kk}\delta_{ij}\right)\right\rangle$$

where $\langle\ \rangle$ implies a test filtered quantity.

The test filter operation can be performed as

$$\langle f(x,t)\rangle = \int \langle G(x,\xi)\rangle f(\xi,t)d\xi \tag{21.4.26}$$

If the box function is used, we have

$$\langle G(x-\xi)\rangle = \begin{cases} \dfrac{1}{\Delta_t} & \text{if } x_i - \Delta_t/2 \le \xi_i \le x_i + \Delta_t/2 \\ 0 & \text{otherwise} \end{cases} \tag{21.4.27}$$

The test filter can be calculated using the trapezoidal rule, Simpson's rule, or interpolation function methods. For example, the one-dimensional filtering operation with the trapezoidal rule assumes the form,

$$\langle f_i\rangle = \frac{1}{\Delta_t}\int_{x-\Delta_t/2}^{x+\Delta_t/2} f(\xi)d\xi = \frac{1}{4}(f_{i-1} + 2f_i + f_{i+1}) \tag{21.4.28}$$

We then apply this one-dimensional approximation successively in each coordinate direction for multidimensional problems.

SGS Heat Flux Closure with Favre Averages

The subgrid scale modeling for the energy equation has not received much attention. This is because, for low Mach number flows, the effect of turbulence modeling is negligible. For high Mach number flows, we may use the standard gradient diffusion model (eddy viscosity)

$$q_i^{(H)} = \overline{\rho}\tilde{c}_p(\mathrm{v}_i T - \tilde{\mathrm{v}}_i \tilde{T}) = \frac{\overline{\rho}\tilde{c}_p \nu_T}{\mathrm{Pr}_T}\tilde{T}_{,i} \tag{21.4.29}$$

with

$$\nu_T = C\Delta^2|\tilde{d}| \tag{21.4.30}$$

The eddy viscosity in (21.4.30) may be expressed dynamically as shown in (21.4.25).

SGS Turbulent Diffusion and Viscous Diffusion Closures

Vreman et al. [1995] shows further details of subgrid modeling for the energy equation using the Fabre averaged variables. The SGS turbulent diffusion closure has been proposed by Knight et al. [1998] and the SGS viscous diffusion closure model studied by Meneveau and Lund [1997]. The scale similarity approach was applied in both cases. Future developments in these areas are needed to substantiate the accuracy of models, particularly for high Reynolds number and high Mach number hypersonic flows.

As a result of the LES solution of the Navier-Stokes system of equations, we obtain the flow variables which contain not only the mean quantities but also the fluctuations. We then compute the mean flowfield values by various schemes of averaging or filtering methods (time averages, spatial averages, or filtered Favre averages, etc.). The difference between the LES solution and the averages will lead to the turbulence fluctuations. From these fluctuations, detailed turbulence statistics can be computed. Among them are the turbulent intensities, distributions of energy spectra with respect to wave numbers, production, dissipation, and diffusion of turbulent kinetic energy and Reynolds stresses, compressibility effects as reflected by dilatation, high-speed flow heat transfer, details of shock wave turbulent boundary layer interactions through transition to full turbulence, and physics of relaminarization. Some examples on LES computations will be presented in Section 22.8, Applications.

21.5 DIRECT NUMERICAL SIMULATION

21.5.1 GENERAL

As we have seen in the previous chapters, turbulence modeling is not an easy task. Even in large eddy simulation, in which we only need to model small scales of isotropic motions, the process becomes complicated in dealing with energy equation for high-speed compressible flows. Thus, our final resort may seem to be a direct numerical solution in which no turbulence modeling is needed. However, we require excessive mesh refinements and higher order accurate numerical schemes. The computational cost for DNS particularly in high-speed compressible flows will be prohibitive.

In direct numerical simulations (DNS), the Navier-Stokes system of equations is solved directly with refined meshes capable of resolving all turbulence length scales including the Kolmogorov microscale,

$$\eta = (\nu^3/\varepsilon)^{1/4} \tag{21.5.1}$$

All turbulence scales ranging from the large energy-containing eddies to the dissipation scales,

$$0.1 \leq k\eta \leq 1$$

with k being the wave number must be resolved (see Figure 21.4.1). To meet this requirement, the number of grid points required is proportional to $L/\eta \approx \mathrm{Re}^{3/4}$ where L is the characteristic length and Re is the Reynolds number referenced to the integral scale of the flow. This leads to the number of grid points in 3-D to be proportional to

$$N = \mathrm{Re}^{9/4} \tag{21.5.2}$$

The number of grid points required for a channel flow may be estimated in terms of turbulence Reynolds number Re_T [Moser and Moin, 1984; Kim, Moin, and Moser, 1987] as

$$N = (3\mathrm{Re}_T)^{9/4} \tag{21.5.3}$$

with

$$\mathrm{Re}_T = \frac{u_T H}{2\nu} \tag{21.5.4}$$

where u_T is the shear velocity (approximately 5% of the mean average velocity) and H is the channel height.

Similarly, the time step is limited [Kim et al., 1987] by the Kolmogorov time scale, $\tau = (\nu/\varepsilon)^{1/2}$, as

$$\Delta t \cong \frac{0.003H}{u_T\sqrt{\mathrm{Re}_T}} \tag{21.5.5}$$

These restrictions are clearly too severe for DNS to be a practical design tool in industry in view of currently available computer capacity.

21.5.2 VARIOUS APPROACHES TO DNS

The DNS applications have been carried out most successfully using spectral methods in simple geometries. Fourier series are applied to the streamwise and spanwise directions whereas Chebyshev polynomials or B-splines are used for the wall-normal direction. However, the spectral methods are not suitable for practical industrial problems with complex geometries and boundary conditions. The use of FDM, FEM, or FVM, although not as accurate as spectral methods, is more flexible in handing arbitrary geometries and boundary conditions. In view of the fact that turbulence is three-dimensional in nature and DNS requires excessive grid refinements, FDM calculations with uniform structured grids have been used predominantly in the past. DNS in unstructured arbitrary practical geometries and boundary conditions at high Reynolds number flows are severely limited by available computer resources.

Applications of DNS in incompressible or subsonic flows and compressible or supersonic flows are distinguished by several factors: (1) For incompressible simulations, the viscous terms are treated usually implicitly, allowing the viscous stability limit to be relaxed, whereas for compressible flows the time discretization is explicit and the allowable time step is limited by the viscous stability limit rather than by the convection condition; (2) Toward transition to turbulence, instability growth rates are slower in compressible flows than in incompressible flows. This will require longer time integration; (3) High-speed transitional disturbance modes have high gradients for compressible flows requiring much more mesh refinements and higher order accuracy in spatial approximations than for incompressible flows.

In DNS, we may use either the temporal or spatial simulation approach. The temporally evolving simulation is usually limited to periodic inflow and outflow boundary conditions and a parallel flow without the consideration of the boundary layer growth. The spatially evolving approach is more general and practical in which nonperiodic inflow and outflow boundary conditions are used and the evolution of nonparallel boundary layer is accounted for. Some recent advancements for both temporally and spatially evolving simulations are reported in Guo, Kleiser, and Adams [1996].

The earlier works on transition and turbulence in boundary layer flows using DNS include Kim, Moin, and Moser [1985], Spalart and Yang [1987], Fasel, Rist, and Konzelmann [1990], Rai and Moin [1993], among others.

The DNS solution of the Navier-Stokes system of equations provides the flow variables which contain not only the mean quantities but also the fluctuations similarly as in LES discussed in Section 21.4. The objective of DNS is to obtain more accurate results for turbulence statistics than in LES at the expense of computing costs. Since the disadvantages resulting from possible inadequate subgrid scale modeling are eliminated in DNS, it is anticipated that the DNS results may be used as a guidance of improving any or all modeling processes for turbulence presented in the previous sections. Details of applications in DNS will be presented in Section 21.7.

21.6 SOLUTION METHODS AND INITIAL AND BOUNDARY CONDITIONS

Although explicit methods may be used in turbulent flows in general, it is often necessary to employ implicit methods in order to handle viscosity in wall-bounded turbulence. Various numerical schemes such as Runge-Kutta, Crank-Nicolson, Adams-Brashforth, among others, have been used in RANS, LES, and DNS calculations using FDM and FVM via FVM. For FEM formulations, the FEM equations may be solved using conjugate gradient or GMRES.

Initial and boundary conditions in turbulent flows are more sensitive to the solution as compared with laminar flows. This is because a small change in the initial state of turbulent flow is amplified exponentially in time. Since this is physical rather than numerical, it is difficult to assess the numerical error if one changes the numerical methods to improve the numerical methods or refine the mesh to obtain more accurate results. So, the question is: how do we know if we have a good solution? This question can be answered with reference to Figure 21.4.1. If the energy spectrum in the smallest scales with the wave number larger than the inertial subrange is much smaller than the peak in the smaller wave number region, then we may assume that the solution is satisfactory.

For inflow initial and boundary conditions, periodic boundary conditions are convenient to use (particularly suitable for spectral methods) if flows do not vary in a given direction. Otherwise, the initial and boundary conditions may be obtained from other simulations, adopted from isotropic turbulence.

For outflow boundaries, one may use the extrapolation conditions, requiring the derivatives of all variables normal to the surface set equal to zero,

$$(\rho u)_{,i} n_i = 0 \tag{21.6.1}$$

If the flow is unsteady, then it appears that time-dependent boundary conditions be implemented by enforcing the time-dependent mass flux conservation at the outflow boundary,

$$\Delta(\rho u) = -\Delta t u_0 (\rho u)_{,i} n_i \tag{21.6.2}$$

with u_0 being the average velocity of the outflow boundary. This tends to keep the reflected pressure waves from moving back to the domain.

On the solid boundary, the standard no-slip condition can be applied. Because of turbulence microscales close to the wall leading to complicated turbulent structures including separated flows, one must use highly refined meshes adjacent to the wall. Furthermore, in this region, turbulence may remain unsteady even when the flow away from the wall has reached a steady state.

In DNS and LES, the resolved flow may become unsymmetric even if the geometry and the flow boundary conditions are symmetric. Thus, the symmetry condition should not be used in the simulation of turbulence using DNS or LES.

As we have seen in multigrid methods (Section 20.2) in which low frequency (small wave number) errors are eliminated in coarse mesh, large-scale turbulence can be resolved quickly in the coarse mesh so that computational efficiency can be realized if the solution is then performed on the fine mesh subsequently. This suggests that multigrid methods are particularly useful in DNS and LES.

21.7 APPLICATIONS

21.7.1 TURBULENCE MODELS FOR REYNOLDS AVERAGED NAVIER-STOKES (RANS)

Exhaustive numerical demonstrations for turbulence model applications are not attempted in this section. Instead, we focus on some representative incompressible flow applications for RANS. In this illustration, we introduce the work of Thangam and Speziale [1992] which shows the comparison of various types of K–ε models as applied to the backward-facing step shown in Figure 21.7.1.1a. The finite volume method via FDM [Thangam and Hur, 1991] is employed with a computational grid of 200×100 mesh (a coarser version is shown in Figure 21.7.1.1b) and $Re = 1.32 \times 10^5$. Computed results for the standard K–ε model with the wall boundary conditions of the two-layer case are shown in Figure 21.7.1.2a.

As compared with the experimental data of Kim, Kline, and Johnston [1980], it is seen that reattachment length for the two-layer model ($X_r = 6.0$) is about 15% underestimated (experimental value, $X_r = 7.1$, from Kim et al [1980]). Despite this discrepancy, the mean velocity profiles appear to be in good agreement (Figure 21.7.1.2b), although the turbulent intensity profiles (Figure 21.7.1.2c) and shear stress profiles (Figure 21.7.1.2d) show some deviations from the experimental data.

For the three layer model, the reattachment length is $X_r = 6.25$ (Figure 21.7.1.3a), about 5% improvement from the two-layer case. Mean velocity profiles (Figure 21.7.1.3b), turbulent intensity profiles (Figure 21.7.1.3c), and shear stress profiles (Figure 21.7.1.3d) appear to be the same as in the two-layer model.

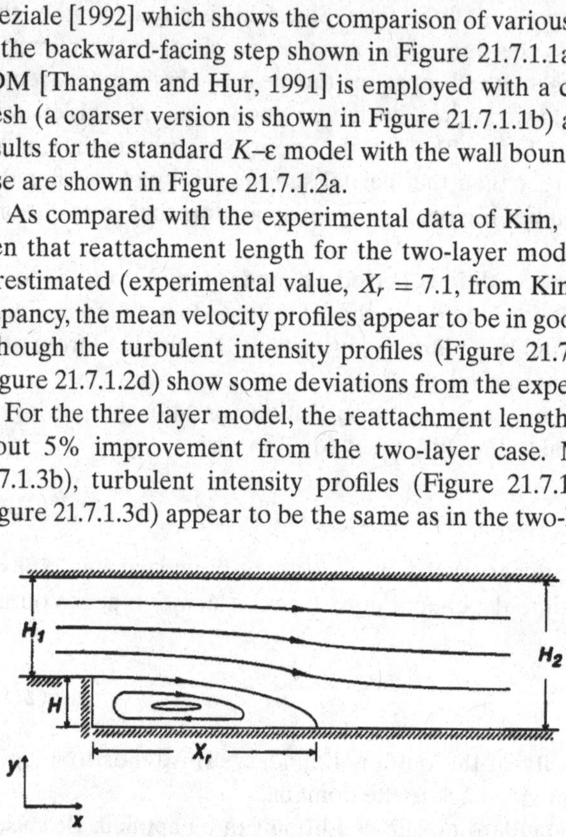

(a) Physical configuration and coordinate system

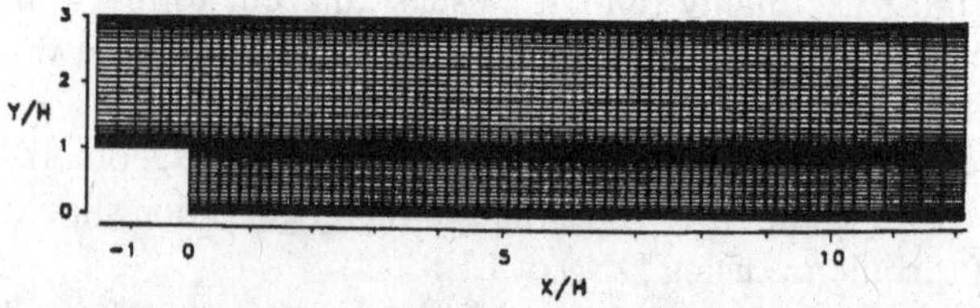

(b) Computational grid: Y/H=3, Re=1.32×10^5, 200 × 100 mesh

Figure 21.7.1.1 Incompressible turbulent flow backward facing step, 2-D geometry for K–ε model analysis, $C_\mu = 0.09$, $C_{\varepsilon 1} = 1.44$, $C_{\varepsilon 2} = 1.92$, $\sigma_k = 1.92$, $\sigma_\varepsilon = 1.0$, $CD = 1.68$ [Thangam and Speziale, 1988].

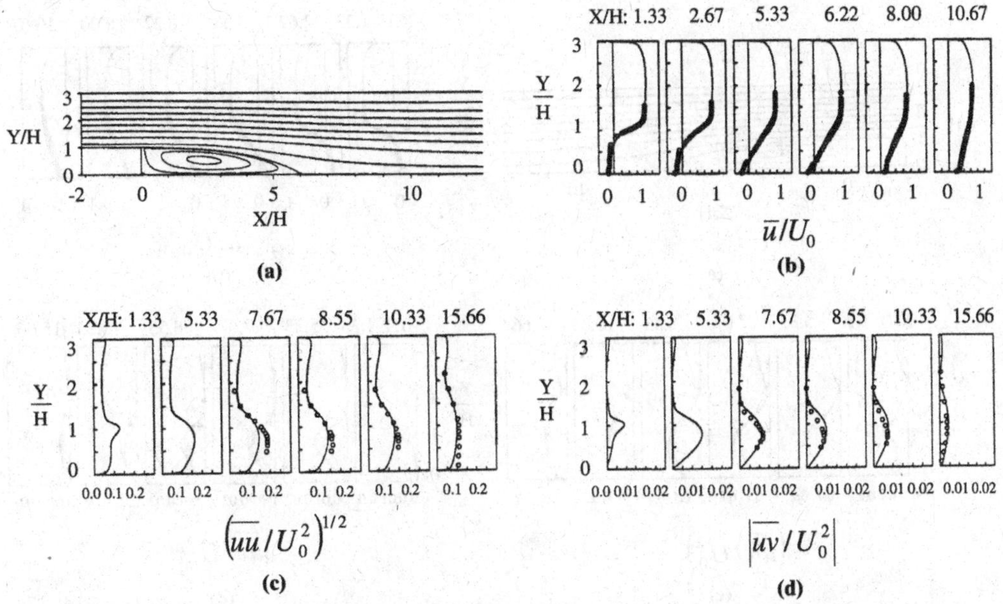

Figure 21.7.1.2 Results with the standard K–ε two-layer model [Thangam and Speziale 1988], compared with Kim et al. [1980]. **(a)** Contours of mean streamlines. **(b)** Mean velocity profiles at selected locations, compared with experiments [Kim et al., 1980]. **(c)** Turbulence intensity profiles. **(d)** Turbulence shear stress profiles.

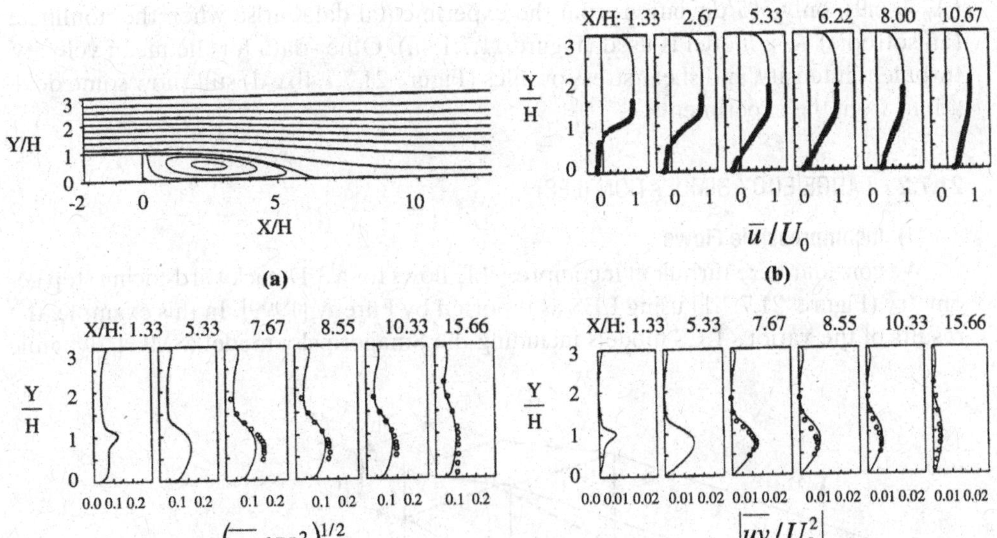

Figure 21.7.1.3 Results with the standard K–ε three-layer model [Thangam and Speziale, 1988]. **(a)** Contours of mean streamlines. **(b)** Mean velocity profiles at selected locations, compared with experiments [Kim et al., 1980]. **(c)** Turbulence intensity profiles. **(d)** Turbulence shear stress profiles.

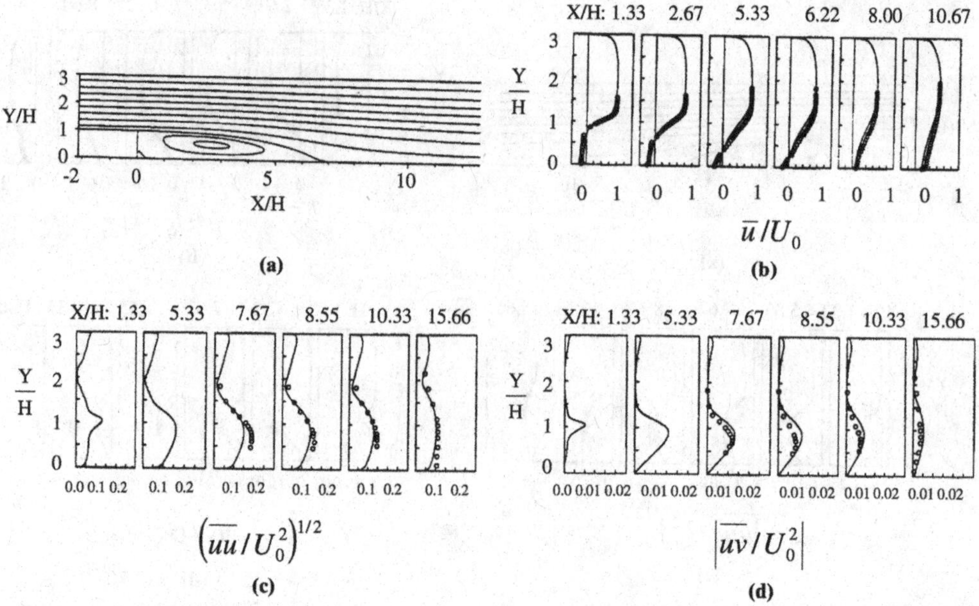

Figure 21.7.1.4 Results with the nonlinear (anisotropic) K–ε three-layer model [Thangam and Speziale, 1988]. **(a)** Contours of mean streamlines. **(b)** Mean velocity profiles at selected locations, compared with experiments [Kim et al., 1980]. **(c)** Turbulence intensity profiles. **(d)** Turbulence shear stress profiles.

It is interesting to note that significant improvements for the reattachment length ($X_r = 6.9$), only 3% deviation from the experimental data, arise when the nonlinear (anisotropic) K–ε model is used (Figure 21.7.1.4a). Other data for the mean velocity, turbulent intensity, and shear stress profiles (Figure 21.7.1.4b,c,d) still show some deviations from the experiments.

21.7.2 LARGE EDDY SIMULATION (LES)

(1) Incompressible Flows

We consider here turbulent incompressible flows for a 3-D backward-facing step geometry (Figure 21.7.2.1) using LES as reported by Fureby [1999]. In this example, the results of the various LES models including the Smagorinsky model (SMG), dynamic

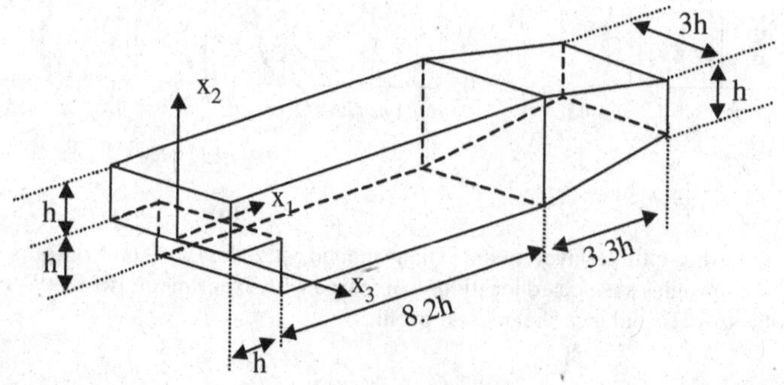

Figure 21.7.2.1 Backward-facing step 3-D geometry for LES analysis [Fureby, 1999].

Table 21.7.2.1 Overview of Simulations, Grids, and Global Quantities

Case/run	Re, $\times 10^4$	SGS mode	Grid; resolution	λ/h	$\partial\delta_\omega/\partial x_1$	Sr, $x_1/h=1$	Sr, $x_1/h=6$	Ξ, $x_1/h=3$
A1	1.5	OEEVM	87,104;2Δ	6.8	0.25	0.20	0.07	0.13
A2	1.5	OEEVM	204,460;3Δ/2	6.6	0.26	0.19	0.07	0.10
B1	2.2	OEEVM	366,750;3Δ/2	7.1	0.27	0.23	0.06	0.11
B2	2.2	OEEVM	170,400;2Δ	7.1	0.25	0.23	0.06	0.16
B3	2.2	SMG	170,400;2Δ	7.2	0.24	0.22	0.07	0.17
B4	2.2	DSMG	170,400;2Δ	7.1	0.27	0.24	0.07	0.17
B5	2.2	MILES	170,400;2Δ	7.4	0.25	0.23	0.05	0.15
B6	2.2	OEEVM	1,152,600; Δ	7.0	0.28	0.23	0.06	0.06
C1	3.7	OEEVM	366,750;2Δ	6.9	0.26	0.25	0.06	0.17
Exp*	1.5	—	—	6.5	0.28	—	—	—
Exp*	2.2	—	—	7.0	0.28	—	—	—
Exp*	3.7	—	—	6.8	0.28	—	—	—

*Pitz and Daily [1981].

Smagorinsky model (DSM), one-equation eddy viscosity model (OEEVM) [Lesieur and Metais, 1996], and monotonically integrated large eddy simulation (MILES) [Fureby, 1999] are compared with those of the experimental results of Pitz and Daily [1981]. In MILES, the Navier-Stokes system of equations are solved using the monotonic integration with flux limiters in which high-resolution monotone methods with embedded nonlinear filters providing implicit closure models so that explicit SGS models need not be used. Various test cases are summarized in Table 21.7.2.1.

Contours of streamwise instantaneous velocity as shown in Figure 21.7.2.2a indicate the free shear layer terminating at approximately $x_1/h \cong 7$. Figure 21.7.2.2b shows the

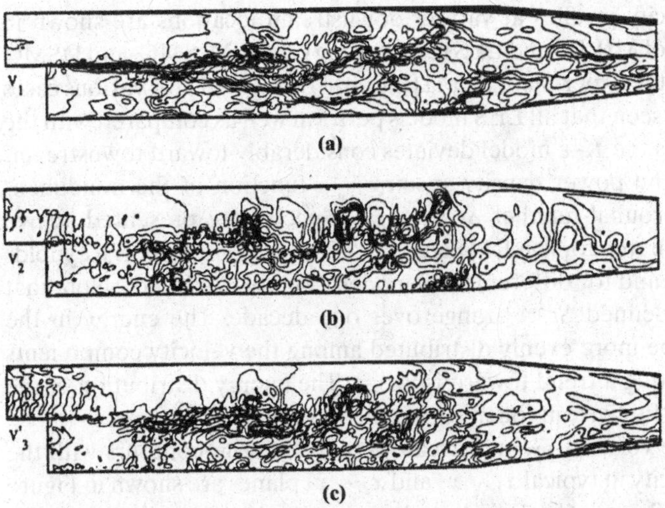

Figure 21.7.2.2 Instantaneous velocity and velocity fluctuation contours in the centerplane [Fureby, 1999]. (a) Streamwise velocity component. (b) Vertical velocity component. (c) Spanwise velocity component.

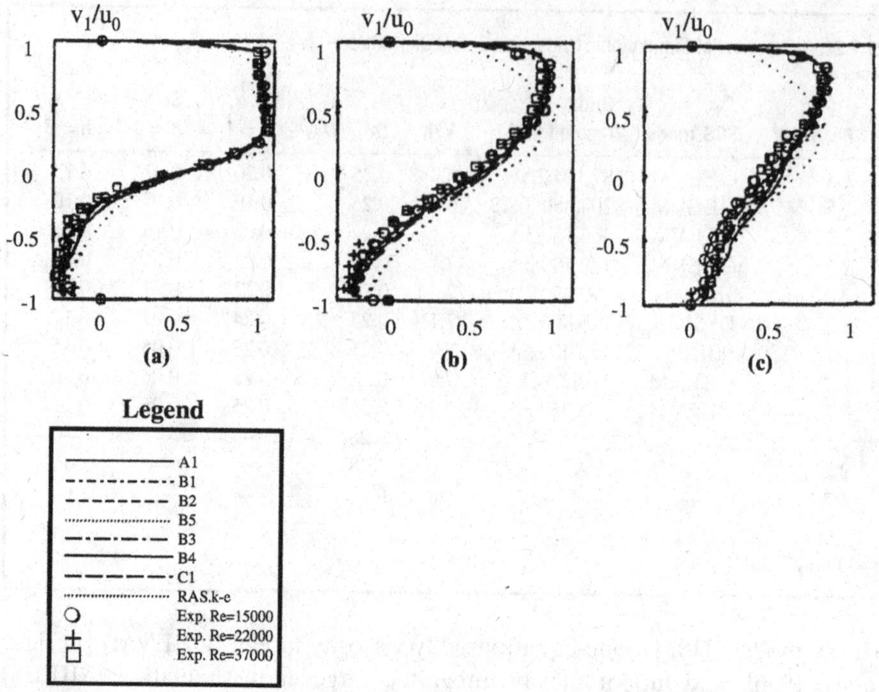

Figure 21.7.2.3 Streamwise mean velocity profiles $< \overline{v}_1 >$ downstream of the step at **(a)** $x_1/H = 2$, $Re = 15 \times 10^3$ **(b)** $x_1/H = 5$, $Re = 22 \times 10^3$, and **(c)** $x_1/H = 7$, $Re = 37 \times 10^3$ [Fureby, 1999].

vertical flow patches, with alternating positive and negative $\overline{v}_2$ regions of spanwise Kelvin-Helmoltz vortices. Spanwise velocity fluctuations are shown in Figure 21.7.2.2c, with peak values reaching as high as 0.5 u_0 near reattachment. The near wall region appears laminar-like in the simulation as well as in the experiment [Pitz and Daily, 1981].

Streamwise mean velocity profiles at various downstream locations are shown in Figure 21.7.2.3. The results of MILES and LES results using OEEVM, SMG, and DSMG models are compared with the experimental data [Pitz and Daily, 1981] for various cases given in Table 21.7.2.1. It is seen that all LES models perform well as compared with the experimental data, whereas the K–ε model deviates considerably toward townstream.

Figure 21.7.2.4 shows the power density spectra as a function of the nondimensional frequency or the Strouhal number $Sr = fh/\overline{v}_1$. Spectra are presented at two locations downstream of the step for run B1 (Figure 21.7.2.4a,b), for different Reynolds numbers (Figure 21.7.2.4c) and for different SGS models (Figure 21.7.2.4d). Note that all spectra exhibit a well-defined $Sr^{-5/3}$ range over one decade. The energy in the smaller scales is found to be more evenly distributed among the velocity components (Figure 21.7.2.4a,b), indicating a trend toward isotropy. The energy distribution in the larger scales is anisotropic, the $\overline{v}_1$ component being the most energetic.

Instantaneous spanwise vorticity $\overline{\omega}_3$ and streamwise vorticity $\overline{\omega}_1$ contours with the step height and inflow velocity in typical $x_1 - x_2$ and $x_2 - x_3$ planes are shown in Figure 21.7.2.5 for runs $A2$, $B1$, $B2$, and C_1. The shear layer separating from the step rolls up into coherent $\overline{\omega}_3$ vortices due to the shear layer instability. They undergo helical

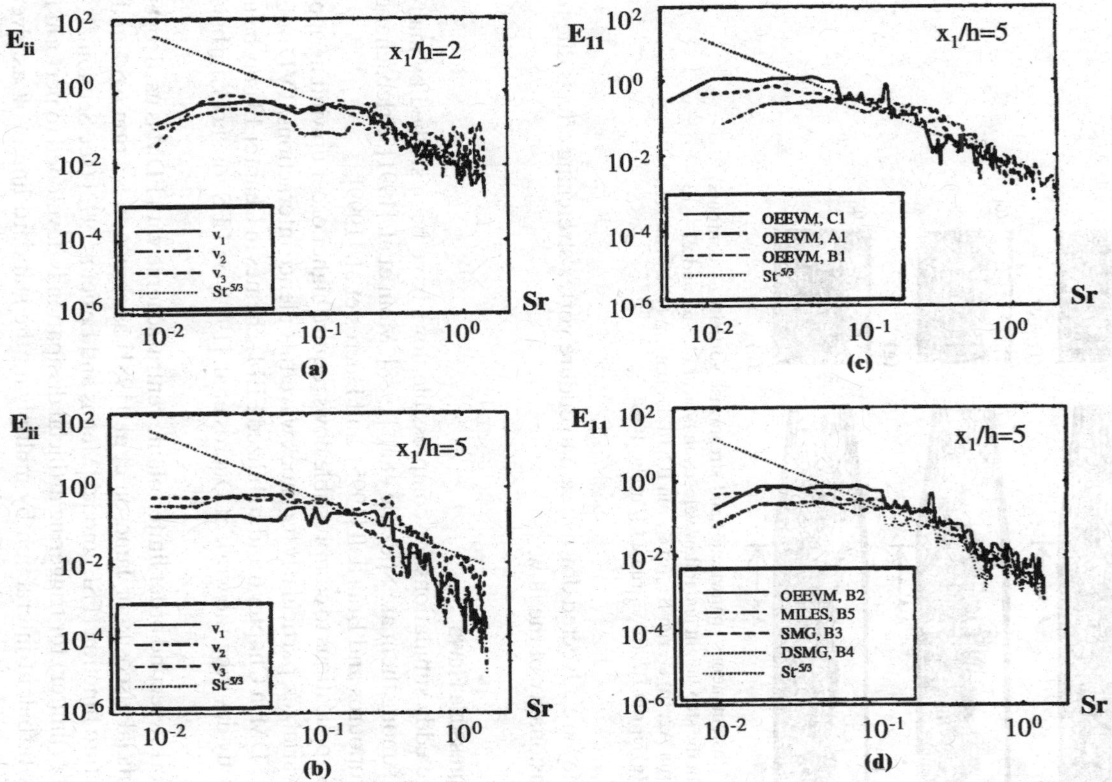

Figure 21.7.2.4 Energy spectra downstream of the step [Fureby, 1999]. **(a)** Component-based spectra for case B2 at $x_1/h = 2$. **(b)** Component-based spectra for case B2 at $x_1/h = 5$. **(c)** $\overline{v}_1$-based spectra for different Reynolds numbers at $x_1/h = 5$. **(d)** $\overline{v}_1$-based spectra for different LES models at $x_1/h = 5$.

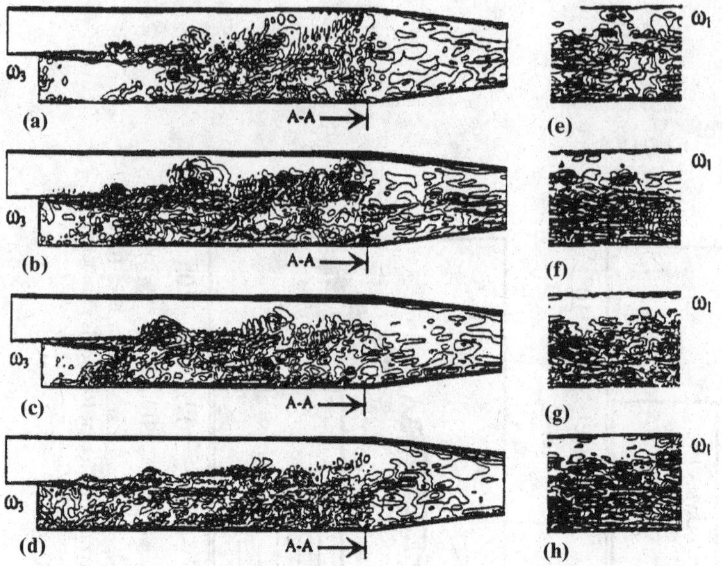

Figure 21.7.2.5 Instantaneous spanwise $\overline{\omega}_3$ and streamwise vorticity $\overline{\omega}_1$ contours normalized with the step height and inflow velocity in typical $x1 - x2$ and $x2 - x3$ planes: (a) $\overline{\omega}_3$ run A2, (b) $\overline{\omega}_3$ run B1, (c) $\overline{\omega}_3$ run B2, (d) $\overline{\omega}_3$ run C1, (e) $\overline{\omega}_1$ run A2, (f) $\overline{\omega}_1$ run B1, (g) $\overline{\omega}_1$ run B2, (h) $\overline{\omega}_1$ run C1 [Fureby, 1999].

pairing as they are advected downstream to endure vortex stretching effects due to the three-dimensionality of the flow.

(2) Compressible Flows

The large eddy simulations for compressible flows have been studied by many investigators. Among them are Speziale et al. [1988], Moin et al. [1991], Erlebacher et al. [1992], Spyropouros and Braisdell [1995], and Ducros et al. [1999].

The LES applications to compressible flows require higher order upwinding approximations of variables, particularly for shock wave/turbulence interactions. TVD or ENO schemes for FDM in Chapter 6 and stabilized FEM schemes in Chapter 13 may be used. Here we briefly describe the work of Ducros et al. [1999] on LES for shock/turbulence interactions.

In this work, a second order finite volume central scheme with FDM is used together with the artificial viscosity of Jameson et al. [1981] as shown in Section 6.6.1 and the filtered structure function (FSF) model [Metais and Lesieur, 1992] (see Section 21.4.3). Here a sensor (limiter) for triggering artificial dissipation is developed to perform LES of the shock/turbulent interaction by multiplying the standard limiter Ψ as given in (6.6.2) by the local function Φ defined as

$$\Phi = \frac{(\nabla \cdot \mathbf{v})^2}{(\nabla \cdot \mathbf{v})^2 + \omega^2 + \varepsilon} \tag{21.7.2.1}$$

where ω is the resolved vorticity and $\varepsilon = 10^{-30}$ is a positive small number so that this function varies between 0 for weakly compressible regions to about 1 in shock regions.

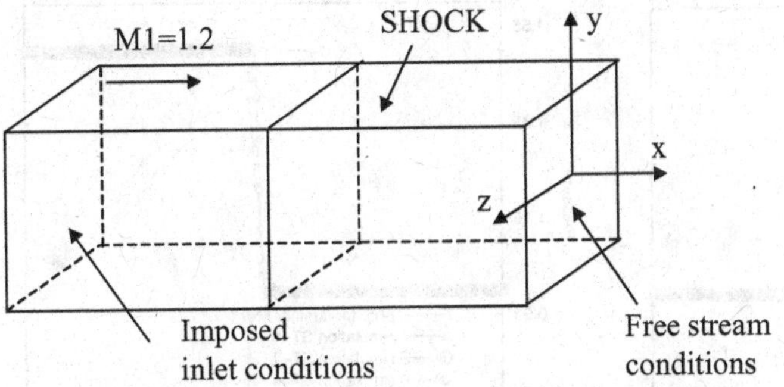

Figure 21.7.2.6 Schematic diagram of the computational domain for simulations ST-1 to ST-5. Periodic conditions are applied in the y- and z-directions [Ducros et al., 1999].

Thus, the artificial viscosity takes the form modified from that in (6.6.1) as

$$\varepsilon_{i+1/2}^{(2)} = k^{(2)} R_{i+1/2} \Psi_{i+1/2} \Phi_{i+1/2} \tag{21.7.2.2}$$

with

$$\Psi_{i+1/2}\Phi_{i+1/2} = \max(\Psi_i \Phi_i, \Psi_{i+1}\Phi_{i+1}) \tag{21.7.2.3}$$

The geometric configuration for the analysis is shown in Figure 21.7.2.6. The mean flow is in the x-direction, with the periodic boundary conditions applied in the y- and z-directions. Table 21.7.2.2 shows the various test cases, ST-1 through ST-5, with Ψ and $\Phi\Psi$ indicating the unmodified and modified versions, respectively. Figure 21.7.2.7 shows the distributions of the mean streamwise velocity, pressure, and Mach number. Note that the refined mesh gives a closer Rankine-Hugoniot jump condition.

Figure 21.7.2.8a,b shows the evolution of the normalized turbulent kinetic energy and turbulent Mach number for some simulations of Table 21.7.2.2. It is interesting to note that only the modified limiter $\Phi\Psi$ predicts a correct decay of turbulent kinetic energy for the preshock region, whereas the standard limiter [Jameson et al., 1981] exhibits a spurious dissipation (ST-1 and ST-3). As observed in Lee et al. [1993] and Lee et al. [1997], the isotropic flow becomes axisymmetric through the shock. This is

Table 21.7.2.2 Parameters of Simulations for the Three-Dimensional Shock/Turbulence Interaction

Simulation	(n_x, n_y, n_z)	Grid	K^2	k^3	Limiter
ST-1	64 × 32 × 32	Isotropic	1.5	0.02	ψ
ST-2	64 × 32 × 32	Isotropic	1.5	0	$\phi\psi$
ST-3	262 × 32 × 32	Locally refined	1.5	0.02	ψ
ST-4	262 × 32 × 32	Locally refined	1.5	0	$\phi\psi$
ST-5	156 × 32 × 32	Locally refined	1.5	0	$\phi\psi$

Note: The resolutions are referred to as resolution 1 (respectively, 2, 3) for 64 × 32 × 32 (respectively, 262 × 32 × 32 and 156 × 32 × 32).

Source: [Ducros et al., 1999]. Reprinted with permission from Academic Press.

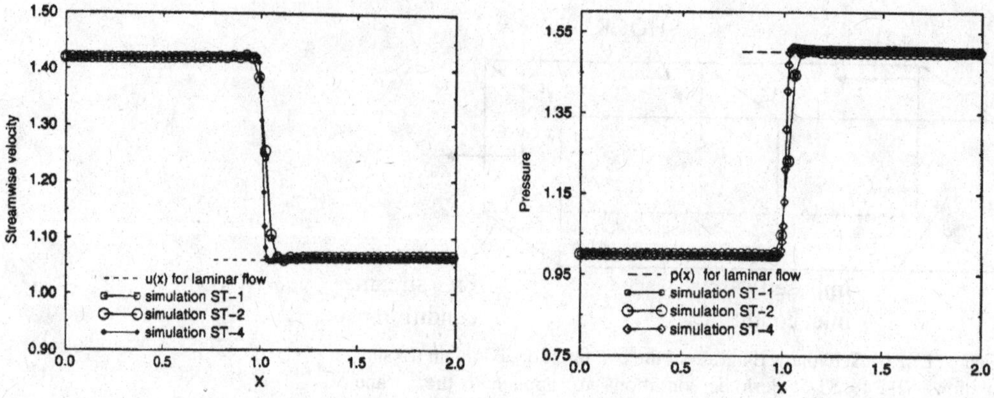

(a)

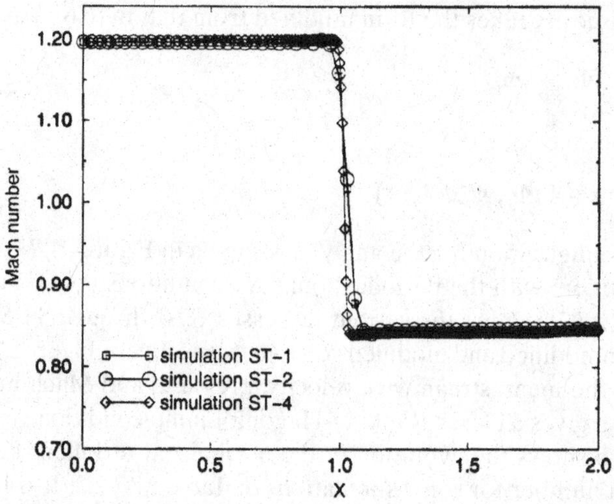

(b)

Figure 21.7.2.7 The x distribution of mean streamwise velocity, pressure, and Mach number [Ducros et al., 1999]. **(a)** X distribution of mean steamwise velocity $\overline{u}$ (top) and mean pressure $\overline{p}$ (bottom) accross the shock wave for simulations ST-1, ST-2, and ST-4; dashed lines denote the laminar values satisfying Rankine-Hugoniot jump conditions. **(b)** The x distribution of mean Mach number for simulations ST-1, ST-2, and ST-4 with the same legend as the previous figure.

shown in Figure 21.7.2.8c by the streamwise distribution of the Reynolds stresses (ST-2 and ST-4).

The streamwise and spanwise distributions of normalized vorticity fluctuations are displayed in Figure 21.7.2.9. Note that the cases of standard limiter (ST-1 and ST-3) leads to a spurious decay of vorticity, whereas this non-physical behavior is corrected by means of the modified limiter (ST-2, ST-4, ST-5).

Figure 21.7.2.10a shows a cut of instantaneous streamwise and spanwise components of vorticity for ST-1. No change in size and intensity of the scales for both components is visible, although the size of the smallest scales is larger than the width of the shock. The same variables for ST-4 are shown in Figure 21.7.2.10b. Here, the x-component

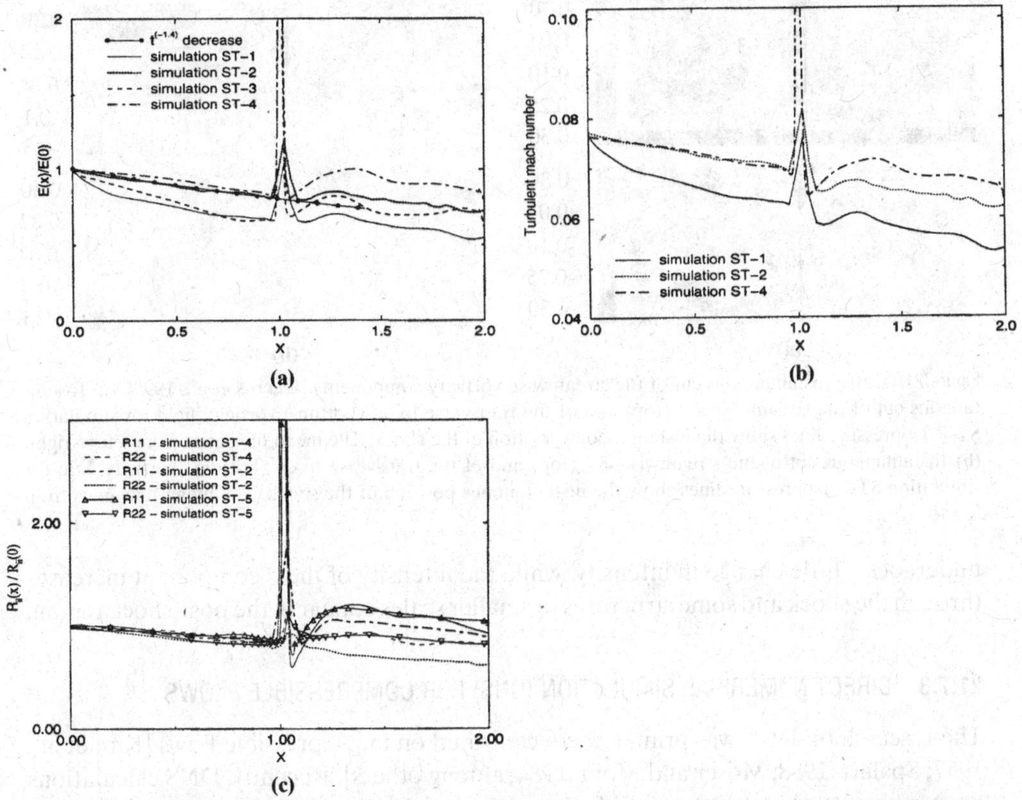

Figure 21.7.2.8 The x distribution of normalized turbulence kinetic energy, turbulent kinetic Mach number, and normalized Reynolds stresses [Ducros et al., 1999]. **(a)** The x distribution of normalized turbulence kinetic energy $\overline{E(x)}/\overline{E(0)}$ for simulations ST-1-4. **(b)** The x distribution of turbulence Mach number $\overline{M}_t$ for simulations ST-1, 2, 4. **(c)** The x distribution of normalized Reynolds stress $R_{ii}(x)/R_{ii}(0)$ for stimulations ST-2, 4, 5.

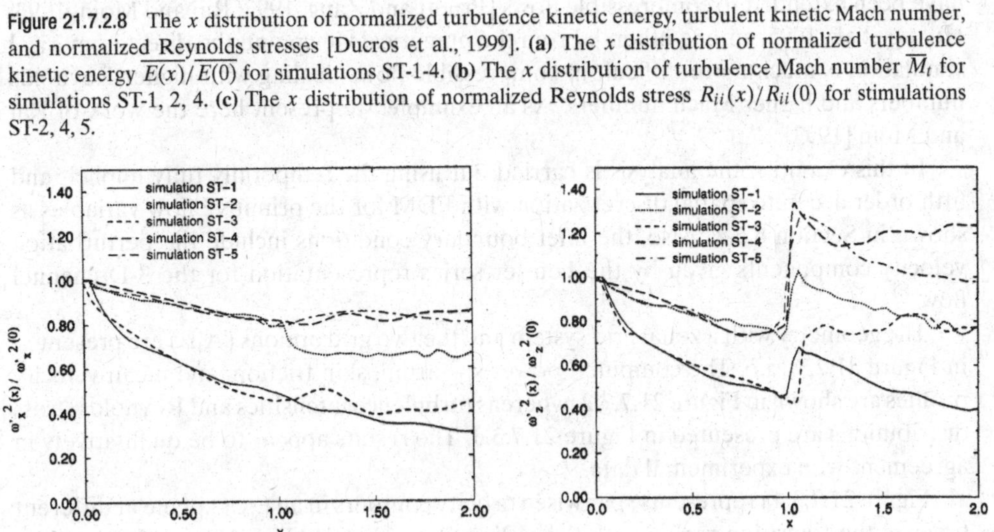

Figure 21.7.2.9 The x distribution of normalized fluctuation vorticity components [Ducros et al., 1999]. **(a)** The x distribution of normalized fluctuations vorticity component $\overline{\omega}_x^2(x)/\overline{\omega}_x^2(0)$ for simulations ST-1-5. **(b)** The x distribution of normalized fluctuations vorticity component $\overline{\omega}_z^2(x)/\overline{\omega}_z^2(0)$ for simulations ST-1-5.

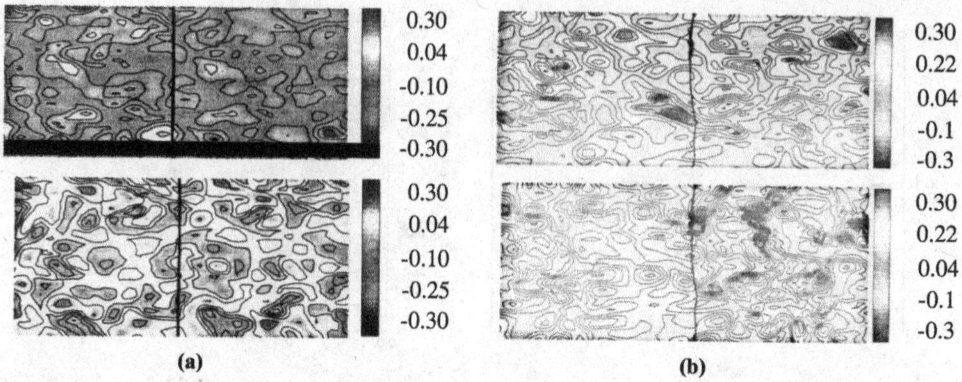

Figure 21.7.2.10 Instantaneous cut of the streamwise vorticity components [Ducros et al., 1999]. **(a)** Instantaneous cut of the streamwise ω_x, (top) and of the transverse to ω_z (bottom) vorticity field for simulation ST-1. Isopressure lines show the instantaneous position of the shock. The mean flow goes from left to right. **(b)** Instantaneous cut of the streamwise ω_x, (top) and of the transverse to ω_z (bottom) vorticity field for simulation ST-4. Isopressure lines show the instantaneous position of the shock. The mean flow goes from left to right.

undergoes a little change in intensity, while the intensity of the z-component increases through the shock and some structures of smaller scales appear in the post-shock region.

21.7.3 DIRECT NUMERICAL SIMULATION (DNS) FOR COMPRESSIBLE FLOWS

The research on DNS was primarily concentrated on incompressible flows [Kim et al., 1987; Spalart, 1988; Moser and Moin, 1984, among others]. Recently, DNS calculations have been extended to compressible flows [Pruett and Zang, 1992; Rai and Moin, 1993; Huang et al., 1995, among others]. From the numerical viewpoint, the direct numerical simulation is much more difficult in compressible flows dealing with higher Reynolds numbers and higher Mach numbers. As an example, we present here the work of Rai and Moin [1993].

In this example, the analysis is carried out using the temporally fully implicit and fifth order accurate spatial discretization with FDM for the primitive flow variables as shown in Section 6.6.2. Also, the inlet boundary conditions include the perturbation velocity components given by the Fourier series representation for the 3-D channel flow.

The geometry with a zonal grid system and the two grid options (A,B) are presented in Figure 21.7.3.1a,b. The computed power spectrum, skin friction, and mean velocity profiles are shown in Figure 21.7.3.2, whereas turbulence intensities and Reynolds stress distributions are presented in Figure 21.7.3.3. The results appear to be qualitatively in agreement with experimental data.

Figure 21.7.3.4a represents spanwise vorticity contours in an (x, y) plane at different times in the transition region with the y-direction expanded by a factor of 10 and the letter "d" on the ordinate indicating the laminar boundary layer thickness at $\text{Re}_x = 2.5 \times 10^5$. This figure shows the rollup of its tip into a spanwise vortex. Streamwise vorticity contours at $y^+ = 34.5$ are presented in Figure 21.7.3.4b. The letter "s" on the ordinate denotes the dimension of the computational region in the z-direction. Here.

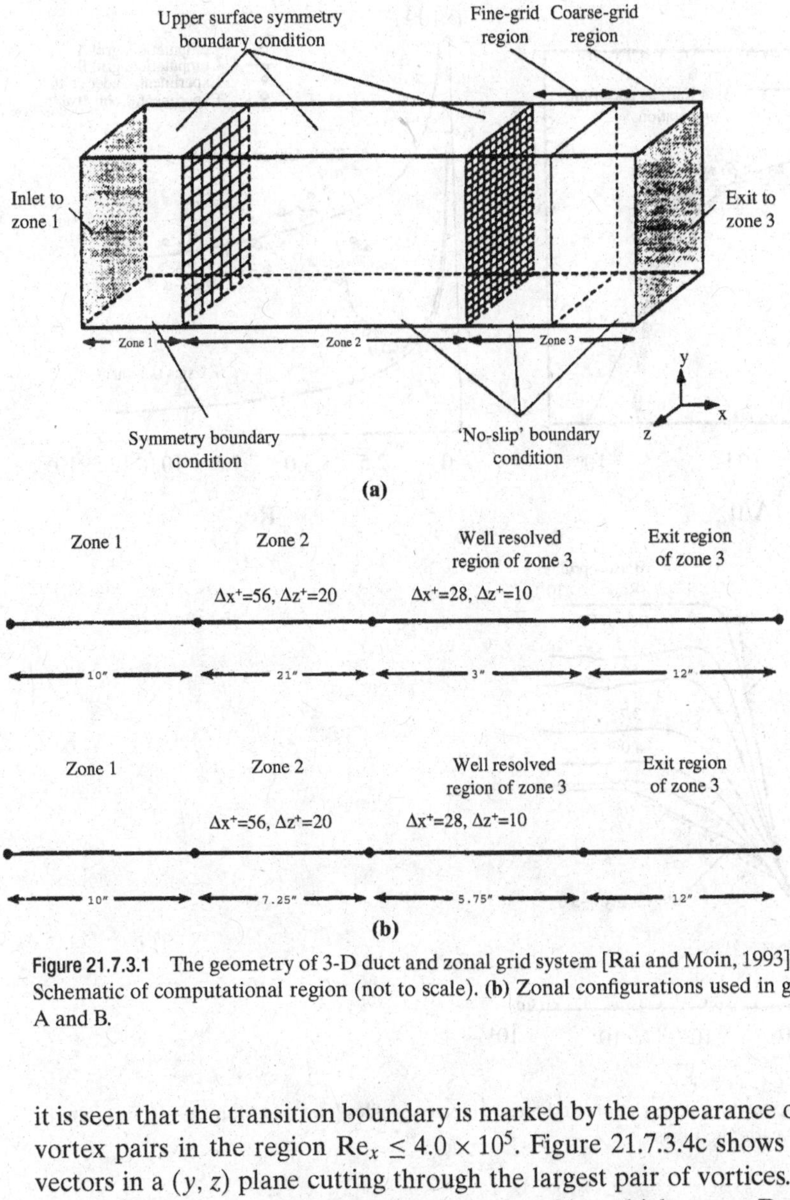

Figure 21.7.3.1 The geometry of 3-D duct and zonal grid system [Rai and Moin, 1993]. **(a)** Schematic of computational region (not to scale). **(b)** Zonal configurations used in grids A and B.

it is seen that the transition boundary is marked by the appearance of counter-rotating vortex pairs in the region $Re_x \leq 4.0 \times 10^5$. Figure 21.7.3.4c shows crossflow velocity vectors in a (y, z) plane cutting through the largest pair of vortices. The letter "d" on the ordinate represents the laminar boundary layer thickness at $Re_x = 4.0 \times 10^5$. The cross sectional structure of this pair of vortices is clearly seen in this figure. Further details are given in Rai and Moin [1993].

Some recent contributions in DNS include Pointsot and Lele [1992], Pruett and Zang [1992], Choi et al. [1993], Lee et al. [1993], Huser and Biringen [1993], Huang et al [1995]. Pruett et al [1995], Mittal and Balachandar [1996], and Guo et al. [1996], among others. In all cases, the main features in DNS are that higher order accurate computational methods must be used with refined mesh, and thus the computer cost will be very excessive.

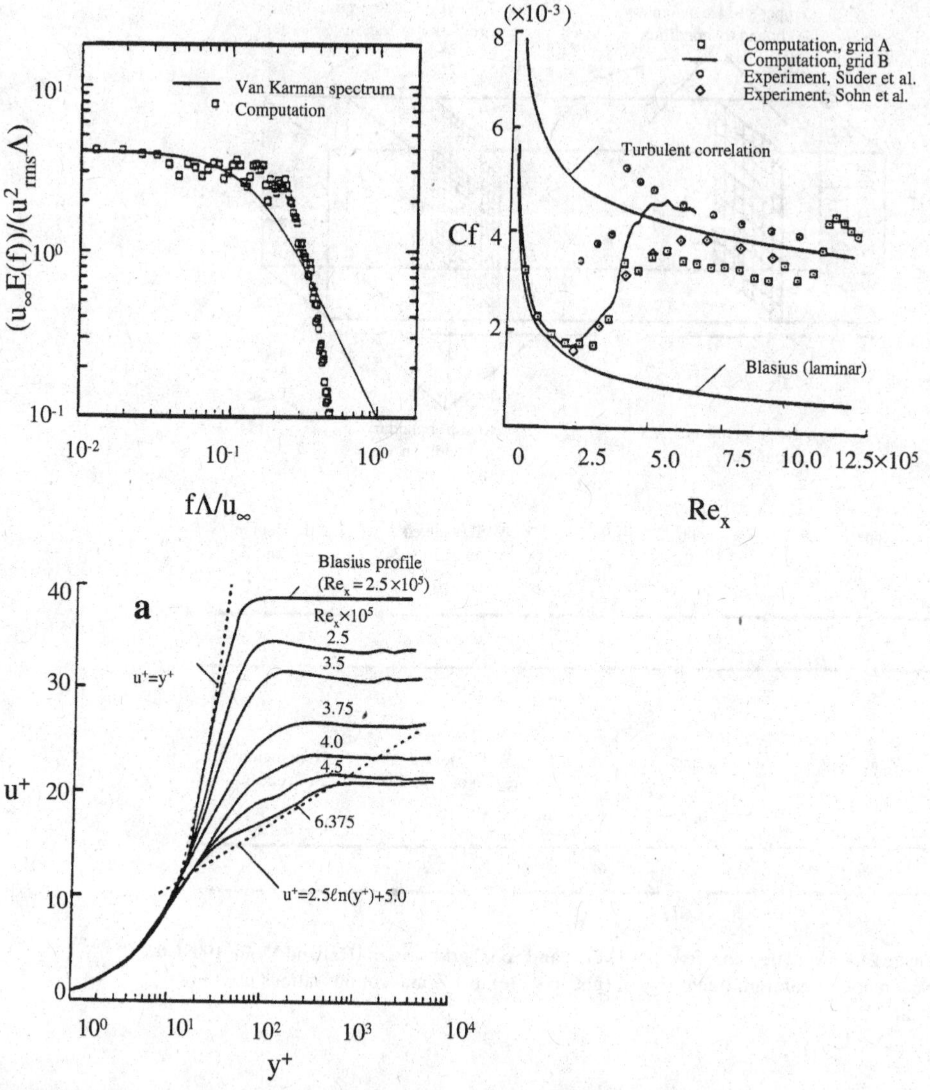

Figure 21.7.3.2 Power spectrum, skin friction, and mean velocity profiles [Rai and Moin, 1993]. Reprinted with permission from Academic Press.

21.8 SUMMARY

In this chapter, we have provided a brief review of the current state of the art on turbulence, including not only the theory of turbulence but also the examples of computations. Turbulence models with Reynolds averaged Navier-Stokes equations (RANS), large eddy simulation (LES), and direct numerical simulation (DNS) are covered.

Turbulence models include zero-equation models, one-equation models, two-equation models, second order closure models (Reynolds stress models), algebraic Reynolds stress models, and models with compressibility effects. Their advantages and disadvantages are noted.

Although the turbulence model approaches are still used in practice, there is a trend toward favoring LES for more accuracy, in which large scales are calculated and only the

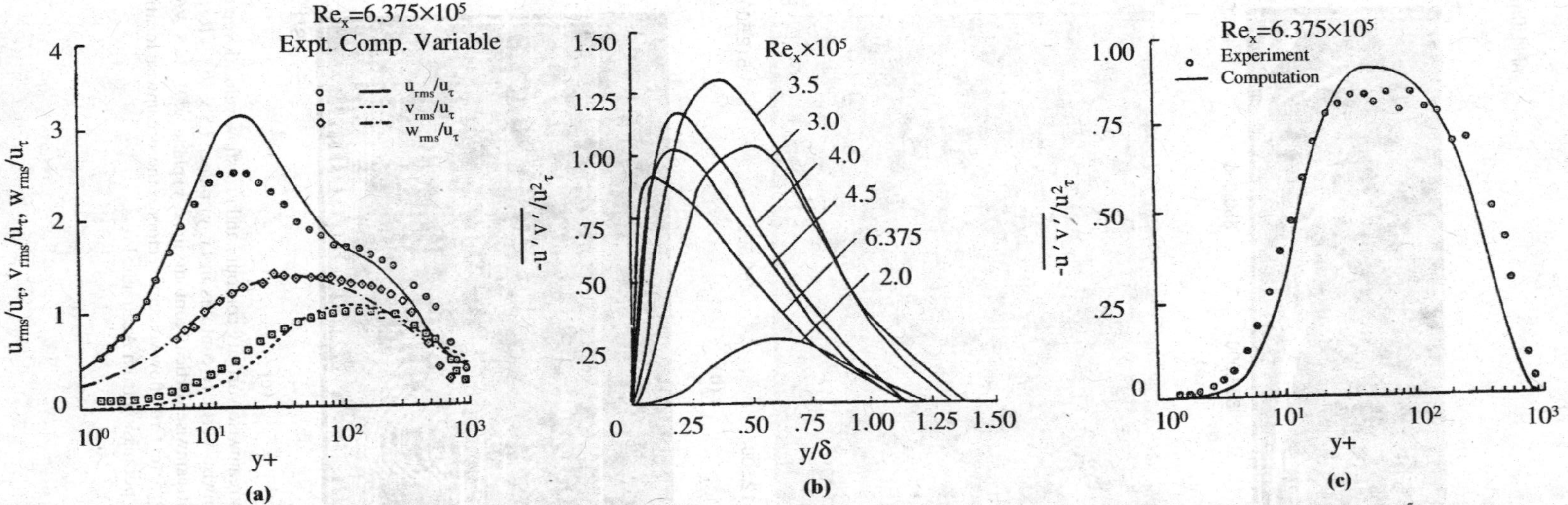

Figure 21.7.3.3 Turbulence intensities and Reynolds stress distributions [Rai and Moin, 1993]. **(a)** Turbulence intensities at streamwise location $\mathrm{Re}_x = 6.375 \times 10^5$, normalized by wall-shear velocity and plotted in wall coordinates. **(b)** Reynolds shear-stress distributions at various streamwise locations, normalized by the square of the wall-shear velocity. **(c)** Reynolds shear-stress distributions at the streamwise location $\mathrm{Re}_x = 6.375 \times 10^5$, normalized by the square of the wall-shear velocity and plotted in wall coordinates. Reprinted with permission from Academic Press.

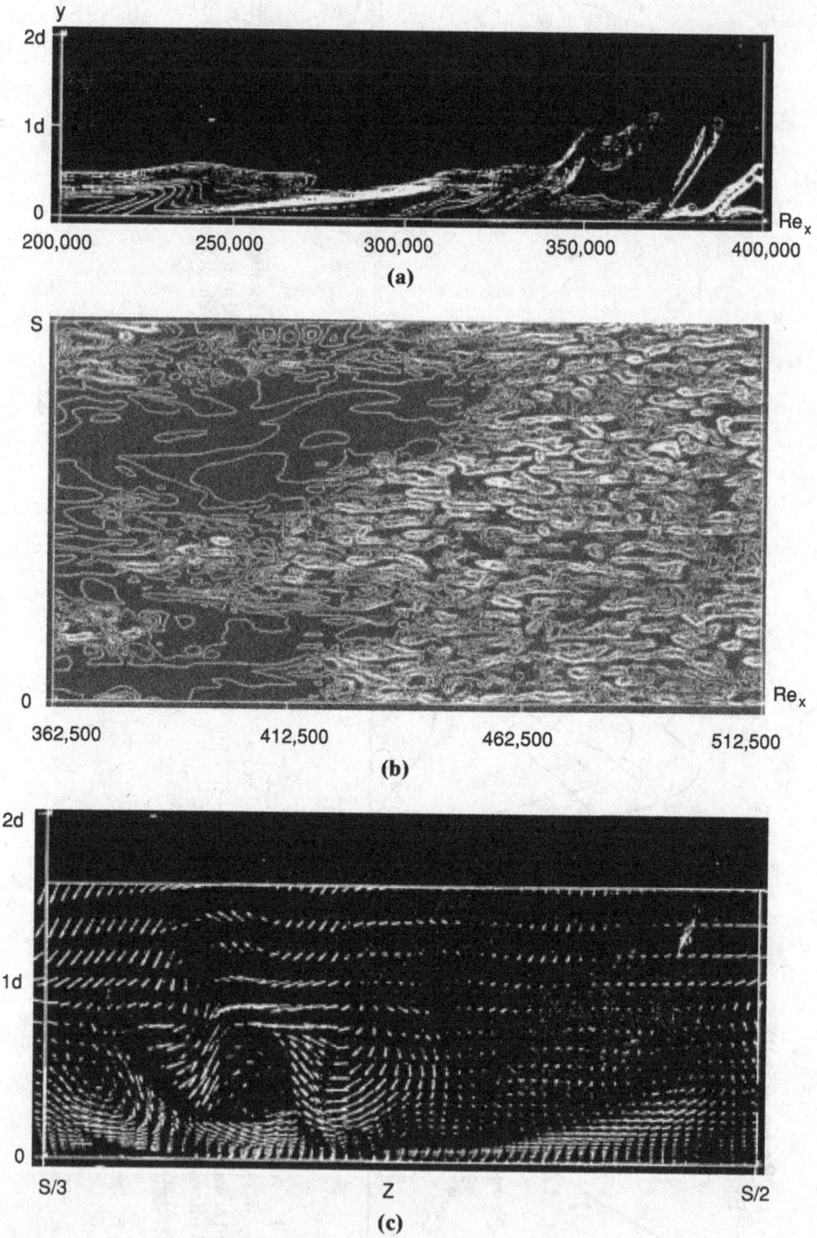

Figure 21.7.3.4 Spanwise and streamwise vorticity contours and crossflow velocity vectors [Rai and Moin, 1993]. **(a)** Spanwise vorticity contours in (x, y) plane, $2.5 \times 10^5 \leq \mathrm{Re}_x \leq 4.0 \times 10^5, t = 51.25\delta^*/u_\infty$. **(b)** Streamwise vorticity contours in (x, y) plane, $y^+ = 34.5, 3.6 \times 10^5 \leq \mathrm{Re}_x \leq 5.1 \times 10^5, 0 \leq Z \leq 5$. **(c)** Crossflow velocity vectors at the streamwise location $\mathrm{Re}_x = 384{,}375$. Reprinted with permission from Academic Press.

small scales are modeled. However, the small scale modeling is still in need of further research for high-speed compressible flows and reactive flows.

Our ultimate goal is then the DNS in which no modeling is required. Unfortunately, the state of the art on DNS is far from practical applications due to demands in unavailable computer resources. If and when DNS becomes a reality, then our concern is the most accurate numerical simulation approaches from those introduced in Parts Two and Three. This will be the focus of our research in the future. In this vein, the FDV theory introduced in Sections 6.5 and 13.6 will be particularly useful in resolving turbulence microscales as accurately as possible. Some examples of FDV applications with K–ε turbulence models for combustion are presented in Section 22.6.2.

REFERENCES

Baldwin, B. S. and Lomax, H. [1978]. Thin-layer approximation and algebraic model for separated turbulent flows. AIAA paper 78-257.

Bardina, J., Ferziger, J. H., and Reynolds, W. C. [1980]. Improved subgrid-scale models for large eddy simulation. AIAA paper, 80-1357.

Cebeci, T. and Smith, A. M. O. [1974]: Analysis of turbulent boundary layer. In *Appl. Math. Mech., 15*, Academic Press.

Choi, H., Moin, P., and Kim, J. [1993]. Direct numerical simulation of turbulent flow over rivets. *J. Fluid Mech.* 255, 503–39.

Comte, P. [1994]. Structure-function based models for compressible transitional shear flows. *ERCOFTAC Bull.*, 22, 9–14.

Comte, P. and Lesieur, M. [1989]. Coherent structure of mixing layers in large eddy-simulation in topological fluid dynamics. In H. K. Moffatt (ed.). *Topological Fluid Dynamics.* New York: Cambridge University Press, 360–80.

Ducros, F., Ferrand, V., Nicoud, F., Weber, C., Darracq, D., Gacherieu, C., and Poinsot, T. [1999]. Large-eddy simulation of the shock/turbulence interaction. *J. Comp. Phys.*, 152, 517–49.

Dumuren, A. O. [1991]. Calculation of turbulent-driven secondary motion in ducts with arbitrary cross section. *AIAA J.*, 29, 4, 531–37.

Erlebacher, G., Hussaini, M. Y., Speziale, C. G., and Zang, T. A. [1992]. Towards the large eddy simulation of compressible turbulent flows. *J. Fl. Mech.*, 238, 155–85.

Fasel, H. F., Rist, U., and Konzelmann, U. [1990]. Numerical investigation of the three-dimensional development in boundary layer transition. *AIAA J.*, 28, 1, 29–37.

Fureby, C. [1999]. Large eddy simulation of rearward-facing step flow. *AIAA J.*, 37, 11, 1401–10.

Gao, F. and O'Bien, E. E. [1991]. Direct numerical simulation of reacting flows in homogeneous turbulence. *AIChE J.*, 37, 1459–70.

Gatski, T. B. and Speziale, C. G. [1992]. On explicit algebraic stress models for complex turbulent flows. ICASE Report No. 92-58, Univ. Space Research Assoc., Hampton, VA.

Germano, M. [1992]. Turbulence: the filtering approach. *J. Fl. Mech.*, 238, 325–36.

Germano, M., Piomelli, U., Moin, P., and Cabot, W. H. [1991]. A dynamic subgrid-scale eddy viscosity model. *Phys. Fl.*, A, 3, 1760–65.

Girimaji, S. S. [1995]. Fully explicit and self-consistent algebraic Reynolds stress model, ICASE Report No. 95-82, NASA Langley Research Center.

Guo, Y. Kleiser, L., and Adams, N. A. [1996]. Comparison of temporal and spatial direct numerical simulation of compressible boundary layer transition. *AIAA J.*, 34, 4, 683–90.

Hanine, F. and Kourta, A. [1991]. Performance of turbulence models to predict supersonic boundary layer flows. *Comp. Meth. Appl. Mech. Eng.*, 89, 221–35.

Huang, P. G., Bradshaw. P., and Coakley. T. J. [1992]. Assessment of Closure Coefficients for Compressible-Flow Turbulence Models, NASA TM-103882.

Huang, P. G., Coleman, G. N., and Bradshaw, P. [1995]. Compressible turbulent channel flows: DNS results and modeling. *J. Fluid Mech.*, 305, 185–218.
Huser, A. and Biringen, S. [1993]. Direct numerical simulation of turbulent flow in a square duct. *J. Fluid Mech.*, 257, 65–95.
Jameson, A., Schmidt, W., Turkel, E. [1981]. Numerical solutions of the Euler equations by finite volume methods using Runge-Kutta time stepping. AIAA paper, 81-1250.
Kim, J., Kline, S. J., and Johnston, J. P. [1980]. Investigation of a reattaching turbulent shear layer: Flow over a backward-facing step. *ASME J. Fl. Eng.*, 102, 302–8.
Kim, J., Moin, P., and Moser, R. [1987]. Turbulence statistics in fully developed channel flow at low Reynolds number. *J. Fl. Mech.*, 177, 133–66.
Knight, D., Zhou, G., Okong'o, N., and Shukla, V. [1998]. Compressible large eddy simulation using unstructured grids. AIAA paper, 98-0535.
Kolmogorov, A. N. [1941]. Local structure of turbulence in incompressible viscous fluid for very large Reynolds number. *Doklady AN. SSR*, 30, 299–303.
———[1942]: Equations of turbulent motion of an incompressible fluid. *Izvestia Academy of Sciences, USSR, Physics*, 6, 1, 56–58.
Launder, B. E. (ed.). [1992]. *Fifth Biennial Colloquium on Computational Fluid Dynamics*. Manchester Institute of Science and Technology, England.
Launder B. E., Reece G. J., and Rodi W. [1975]. Progress in the development of Reynolds stress turbulent closure. *J. Fl. Mech.*, 68, 537–66.
Launder, B. E. and Spalding, B. [1972]. *Mathematical Models of Turbulence*. New York: Academic Press.
Lee, S., Lele, S. K., and Moin, P. [1993]. Direct numerical simulation of isotropic turbulence interacting with a weak shock wave. *J. Fl. Mech.*, 251, 533–62.
Lesieur, M. [1997]. *Turbulence in Fluids*. London: Kluwer Academic Publishers.
Lesieur, M and Metais, O. [1996]. New trends in large eddy simulation of turbulence. *Ann. Rev. Fl. Mech.*, 28, 45–63.
Lilly, D. K. [1966]. On the application of the eddy viscosity concept in the inertial subrange of turbulence. NCAR-123, National Center for Atmospheric Research, Boulder, CO.
———[1992]. A proposed modification of the Germano subgrid-scale closure methods. *Phys. Fl., A*, 4, 633–35.
Lumley, J. L. [1978]. Computational modeling of turbulent flows, *Adv. Appl. Mech.*, 18, 123–76.
Meneveau, C. and Lund, T. S. [1997]. The dynamic Smagorinsky model and scalar-dependent coefficients in the viscous range of turbulence. *Phys. Fl.*, 9, 3932–34.
Metais, O. and Lesieur, M. [1992]. Spectral large eddy simulations of isotropic and stably stratified turbulence. *J. Fl. Mech.*, 239, 157–94.
Mittal, R. and Balachandar, S. [1996]. Direct numerical simulation of flow past elliptic cylinders. *J. Comp. Phys.*, 124, 351–67.
Moin, P., Squires, K., Cabot, W., and Lee, S. [1991]. A dynamic subgrid-scale model for compressible turbulence and scalar transport. *Phys. Fl.*, 3, 2746–57.
Moser, R. D. and Moin, P. [1984]. Direct numerical simulation of curved turbulent channel flow. NASA TM-85974.
Normand, X. and Lesieur, M. [1992]. Direct and large-eddy simulation of laminar break-down in high-speed axisymmetric boundary layers. *Theor. Comp. Fl. Dyn.*, 3, 231–52.
Pierce, C. D. and Moin, P. [1999]. A dynamic model for subgrid-scale variance and dissipation rate of a conserved scalar. *Phys. Fl.*, 10, 3041–44.
Pitz, R. W. and Daily, J. W. [1981]. Experimental study of combustion: the turbulent structure of a reacting shear layer formed at a rearward facing step. NASA CR 165427.
Poinsot, T. J. and Lele, S. K. [1992]. Boundary conditions for direct simulations of compressible viscous flows. *J. Comp. Phys.*, 101, 104–29.
Prandtl, L. [1925]. Uber die ausgebildete turbulenz. *Z. Angew. Math. Mech.*, 136–39.

Pruett, C. D. and Zang, T. A. [1992]. Direct numerical simulation of laminar breakdown in high-speed, axisymmetric boundary layers. *Tjeoret. Comp. Fl. Dyn.*, 3, 345–67.

Pruett, C. D., Zang, T. A., Chang, C. L., and Carpenter, M. H. [1995]. Spatial direct numerical simulation of high-speed boundary layer flows. Part I: Algorithmic considerations and validation. *Theor. Comp. Fl. Dyn.* 7, 49–76.

Rai, M. M. and Moin, P. [1993]. Direct numerical simulation of transition and turbulence in a spatially evolving boundary layer. *J. Comp. Phys.*, 109, 169–92.

Rodi, W. [1976]. A new algebraic relation for calculating Reynolds stresses. *ZAMM*, 56, 219.

Rotta, J. C. [1951]. Statisische theorie nichthomogener turbulenz. *Zeitschrift fur Physik*, 129, 547–72.

Sarker, S., Erlebacher, G., Hussaini, M. Y., and Kreiss, H. O. [1989]. The analysis and modeling of dilatational terms in compressible turbulence, ICAS Report 89-79. Hampton VA: Univ. Space Research Assoc.

Smagorinsky, J. [1963]. General circulation experiments with the primitive equations, I. The basic experiment. *Mon. Weather Rev.*, 91, 99–164.

So, R.M.C. and Melloe, G. L. [1978]. Turbulent boundary layers with large streamline curvature effects. *ZAMP*, 29, 54–74.

Spalart, P. R. [1988]. Direct simulation of a turbulent boundary layer up to Re = 1410. *J. Fl. Mech.*, 187, 61–98.

Spalart, P. R. and Yang, K. S. [1987]. *J. Fl. Mech.*, 178, 345–58.

Spalding, D. B. [1972]. A novel finite difference formulation for differential equations involving both first and second derivatives. *Int. J. Num. Meth. Eng.*, 4, 551–59.

Speziale, C. G. [1987]. On non-linear K–ℓ and K–ε model of turbulence. *J. Fl. Mech.*, 178, 459–75.

Speziale, C. G., Erlebacher, G., Zang, T. A., and Hussaini, M. Y. [1988]. The subgrid-scale modeling of compressible turbulence. *Phys. Fl.*, 31, 940.

Speziale, C. G., Sarker, S., and Gatski, T. B. [1991]. Modeling of the pressure-strain correlation of turbulence, *J. Fl. Mech.*, 227, 245–72.

Speziale, C. G., Zang, T. A., and Hussaini, M. Y. [1988]. The subgrid scale modeling of compressible turbulence. *Phys. Fl.*, 31, 940–42.

Spyropoulos, E. T. and Blaisdell, G. A. [1995]. Evaluation of the dynamic subgrid-scale model for large eddy simulation of compressible turbulent flows. AIAA paper, 95-0355.

Squires, K. D. [1991]. Dynamic subgrid-scale modeling of compressible turbulence. *Annual Research Briefs*, Center for Turbulence Research, Stanford University, 207–23.

Thangam, S. and Hur, N. [1991]. A highly resolved numerical study of turbulent separated flow past a backward-facing step. *Int. J. Eng. Sci.*, 29, 5, 607–15.

Thangam, S. and Speziale, C. G. [1992]. Turbulent flow past a backward-facing step: a critical evaluation of two-equation models. *AIAA J.*, 30, 5, 1314–20.

Van Driest, E. R. [1956]. On turbulent flow near a wall. *J. Aero. Sci.*, 23, 1007–11.

Vreman, B., Geurts, B., and Kuerten, H. [1995]. Subgrid-modeling in LES of compressible flows. *Appl. Sci. Res.*, 54, 191–203.

Wilcox, D. C. [1988]. Multiscale model for turbulent Flows. *AIAA J.*, 26, 11, 1311–20.

———[1989]. Wall matching, a rational alternative to wall functions. AIAA paper, 89-611.

———[1992a]. Dilatation-dissipation corrections for advanced turbulence models, *AIAA J.*, 30, 11, 2639–46.

———[1992b]. *Turbulence Modeling for CFD*, DCW Industries, Inc., La Canada, CA.

Yoshzawa, A. [1986]. Statistical theory for compressible turbulent shear flows with the application to subgrid modeling. *Phys. Fl. A*, 29, 2152–64.

Zang, Y., Street, R. L., and Koseff, J. R. [1993]. A dynamic mixed subgrid-scale and its application to turbulenct recirculating flows. *Phys. Fl.* A, 5, 3186–96.

Zeman, O. [1990]. Dilatation dissipation: The concept and application in modeling compressible mixing layers. *Phys. Fl.* A, 2, no. 2, 178–88.

CHAPTER TWENTY-TWO

Applications to Chemically Reactive Flows and Combustion

22.1 GENERAL

In this chapter, we examine computations for reactive flows in general with computational combustion in particular. In reactive flows, the conservation equations for chemical species are added to the Navier-Stokes system of equations. This addition also requires a modification of the energy equation. Furthermore, the sensible enthalpy is coupled with the chemical species, which contributes to the heat source and diffusion of species interacting with temperature. Chemical reactions in high-speed turbulent flows with high temperatures are of practical interest. They are involved in hypersonic aircraft and reentry vehicles. In this case, it is necessary that the vibrational and electronic energies be taken into account, in which the ionization of chemical species may be important. Thus, the chemically reactive flows and combustion require significant modifications of not only the governing equations but also the existing computational methods discussed in previous chapters.

In general, we are concerned with characterizing ordinary flame and detonation by different time scales. These scales range over many orders of magnitude. When reaction phenomena are modeled such that characteristic times of variation are shorter than the time step used, the equations describing such physical phenomena become numerically *stiff* with respect to convection and diffusion.

Another type of difficulty is the disparity in spatial scales occurring in combustion. To model the steep gradients at a flame front, an extremely small grid spacing is required. In addition, complex phenomena such as turbulence, which occur on intermediate spatial scales, lead to difficult modeling problems.

The third set of obstacles arises because of the geometric complexity associated with real systems. Most of the detailed models developed to date have been one-dimensional. Thus, they give a very limited picture of how the energy release affects the hydrodynamics. Even though many processes in a combustion system can be modeled in one dimension, there are others, such as boundary layer growth or the formation of vortices and flow separation, which clearly require at least two-dimensional hydrodynamics. Combustion in the presence of shock wave turbulent boundary layer interactions demands a complete three-dimensional analysis.

The final consideration is the physical complexity. Combustion systems usually have many interacting species. These are represented by sets of many coupled equations which must be solved simultaneously. Complicated ordinary differential equations describing the chemical reactions or large matrices describing the molecular differential equations are costly and increase calculation time by orders of magnitude over idealized or empirical models. The fundamental processes in combustion include chemical kinetics, laminar and turbulent hydrodynamics, thermal conductivity, viscosity, molecular diffusion, thermochemistry, radiation, nucleation, surface effects, evaporation, condensation, etc. Before a model of a whole combustion system can be assembled, each individual process must be identified. These submodels are either incorporated into the larger model directly or, if the time and spatial scales are too disparate, they must be incorporated phenomenologically. In this process, microscopic details of chemistry and physics are not considered. Instead, we take a macroscopic or continuum view of the domain under study.

It is quite common that reactive flows and combustion occur in turbulent environments. The subject of turbulence, discussed in Chapter 21, then plays a new role in reactive flows and combustion. Both spatial and temporal scales must be reevaluated. Reynolds numbers and Damköhler numbers affect suitable selections of numerical schemes. For high-speed flows, the situation is even more complex. High Mach numbers associated with shock waves must be compromised in determining both spatial and temporal scales. Thus, the reactive flows and combustion in shock wave turbulent boundary layer interactions represent extremely difficult physical phenomena for a numerical simulation. Most likely, in this case, temperature gradients are high and the role of Peclet numbers is crucial as well. The reaction rates for many common chemical reactions are affected by turbulent flow. Thus, much of the data on file for reaction rates is also altered.

With these basic items of consideration in mind, our focus then will be the computational strategies in solving the governing equations involved in reactive flows in general with combustion in particular. These governing equations are summarized in Section 22.2, followed by computation of chemical equilibrium in Section 22.3, chemistry-turbulence interaction models in Section 22.4, and hypersonic reactive flows in Section 22.5. Finally, we examine some applications in Section 22.6. These examples include supersonic inviscid reactive flows (premixed hydrogen-air), turbulent reactive flow analysis with RANS models, PDF models for turbulent diffusion combustion, spectral element methods for spatially developing mixing layer analysis, spray combustion for turbulent reactive flows, LES and DNS analyses for turbulent reactive flows, and hypersonic nonequilibrium reactive flows with vibrational and electronic energies taken into account.

22.2 GOVERNING EQUATIONS IN REACTIVE FLOWS

22.2.1 CONSERVATION OF MASS FOR MIXTURE AND CHEMICAL SPECIES

Before we discuss the reactive flow governing equations, let us summarize definitions of variables involved in reactive flows.

Mass Concentration, ρ_k

The mass of species k per unit volume of the mixture.

Molar Concentration, $C_k = \rho_k/W_k$

The number of moles of species k per unit volume with W_k being the molecular weight.

Mass Fraction, $Y_k = \rho_k/\rho$

The ratio of mass concentration of species k to the total mass density of the mixture, $\sum_{k=1}^{N} Y_k = 1$, with N being the total number of species.

Mole Fraction, $X_k = C_k/C$

The ratio of molar concentration of species k to the total molar density C of the mixture, with $\sum_{k=1}^{N} X_k = 1$.

Number Density, $N_k = C_k/\rho = Y_k/W_k$

Actual number of moles of species k.

Partial Pressure for a Mixture

$$p = \sum_{k=1}^{N} p_k, \quad with \quad X_k = \frac{p_k}{p}$$

Equation of State

$$p = \rho R^0 T \sum_{k=1}^{N} \frac{Y_k}{W_k}$$

with R^0 being the universal gas constant (8.3143 J/g-mol K).

Stoichiometric Condition

This is the most stable condition of chemical reactions, defined by the equivalence ratio

$$\phi = \frac{F/O}{(F/O)_{st}} = 1$$

with F = mass of fuel, O = mass of oxidant, and the subscript st denoting the stoichiometric condition (most stable condition).

Mixture Fraction

$$f = \frac{\xi_M - \xi_A}{\xi_F - \xi_A}$$

where ξ denotes any extensive property (total energy, mass, etc.), $\xi = Y_F - (F/O)_{st} Y_o$, with subscripts F, A, and M representing fuel, air, and mixture, respectively.

The Law of Mass Action

Chemical reactions are characterized by the chemical reaction equations of the form

$$\sum_{k=1}^{N} \nu'_{ki} M_k \underset{k_b}{\overset{k_f}{\rightleftarrows}} \sum_{k=1}^{N} \nu''_{ki} M_k \quad (i = 1, \ldots M) \tag{22.2.1}$$

in which ν_{ki} is the stoichiometric coefficient of the species k for the reaction step i, with the prime and double primes representing the reactant and product, respectively. M_k is the chemical symbol for the species k, and k_f and k_b denote the specific reaction rate constants for the forward and backward reactions, respectively. These reactions are governed by the so-called law of mass action related by the reaction rate ω_k,

$$\omega_k = W_k \sum_{i=1}^{M} (\nu''_{ki} - \nu'_{ki}) \left[k_{fi} \prod_{j=1}^{N} C_j^{\nu'_{ji}} - k_{bi} \prod_{j=1}^{N} C_j^{\nu''_{ji}} \right] \tag{22.2.2}$$

where C_j is the molar concentration. Using the Arrhenius law, the specific reaction rate constants of species k are in the form

$$k_{fi} = A_i T^{\alpha_i} \exp\left(-\frac{E_i}{R^0 T}\right), \qquad k_{bi} = \frac{k_{fi}}{K_c}, \qquad K_c = \prod_{j=1}^{N} C_{j,e}^{(\nu''_{ji} - \nu'_{ji})} \tag{22.2.3}$$

Here, A_i is the frequency factor, α_i is the constant, E_i is the activation energy, and R^0 is the universal gas constant, K_c denotes the equilibrium constant, and $C_{j,e}$ refers to the molar concentration at thermodynamic equilibrium.

The law of mass action, as confirmed by numerous experimental observations, states that the rate of disappearance of a chemical species is proportional to the products of the concentrations of the reacting chemical species, each concentration being raised to the power equal to the corresponding stoichiometric coefficients. Thus, it follows from (22.2.1) and (22.2.3) that the forward reaction can be given by

$$\omega_k = W_k \sum_{i=1}^{M} (\nu''_{ki} - \nu'_{ki}) A_i T^{\alpha_i} \exp\left(-\frac{E_i}{R^0 T}\right) \prod_{j=1}^{N} \left(\frac{X_j p}{R^0 T}\right)^{\nu'_{ji}} \tag{22.2.4}$$

where the pressure p is related by the partial pressure p_j and mole fraction X_j as

$$p = \sum_j p_j, \qquad p_j = X_j p$$

Chemical kinetics and thermodynamic models and constants for various chemical reactions are available in the literature [Gardiner (ed), 1984; Westbrook and Dryer, 1984].

Mixture Conservation Equations

Let us now consider the continuity equation for component A in a binary mixture with a chemical reaction at a rate ω_A (kg m^{-3} sec^{-1}), known as the mass rate of production of species A,

$$\frac{\partial \rho_A}{\partial t} + \nabla \cdot (\rho_A \mathbf{v}_A) = \omega_A \tag{22.2.5a}$$

Similarly, the equation of continuity for component B is

$$\frac{\partial \rho_B}{\partial t} + \nabla \cdot (\rho_B \mathbf{v}_B) = \omega_B \tag{22.2.5b}$$

Adding (22.2.5a) and (22.2.5b) gives

$$\frac{\partial \rho}{\partial t} + \nabla \cdot (\rho \mathbf{v}) = 0 \tag{22.2.6}$$

The above equation results from the law of conservation of mass

$$\omega_A + \omega_B = 0, \qquad \rho_A + \rho_B = \rho, \qquad \rho_A \mathbf{v}_A + \rho_B \mathbf{v}_B = \rho \mathbf{v}$$

where the mixture velocity $\mathbf{v}$ is related by the diffusion velocity $\mathbf{V}_k$ and the species velocity $\mathbf{v}_k$

$$\mathbf{V}_k = \mathbf{v}_k - \mathbf{v} \tag{22.2.7a}$$

with

$$\mathbf{v} = \sum_{\mathbf{k}} \rho_{\mathbf{k}} \mathbf{v}_{\mathbf{k}} \Big/ \sum_{\mathbf{k}} \rho_{\mathbf{k}} \tag{22.2.7b}$$

which leads to

$$\sum_k \rho_k \mathbf{V}_k = \sum_k \rho Y_k \mathbf{V}_k = 0, \qquad \sum_k Y_k \mathbf{V}_k = 0 \tag{22.2.7c}$$

In terms of the molar units, the continuity equation takes the form

$$\frac{\partial C_A}{\partial t} + \nabla \cdot (C_A \mathbf{v}_A) = \overline{\omega}_A \tag{22.2.8}$$

where $\overline{\omega}_A$ is the molar rate of production of A per unit volume.

The species mass flow may be written in terms of the Fick's first law of diffusion,

$$\rho_A \mathbf{V}_A = -\rho D_{AB} \nabla Y_A \tag{22.2.9}$$

where D_{AB} is the diffusion constant for rigid spheres of two unequal mass (m_A, m_B) and diameter (d_A, d_B) [Hirschfelder, Curtis, and Bird, 1954; Gardiner, 1984]

$$D_{AB} = \frac{2}{3}\left(\frac{k^3}{\pi^3}\right)^{\frac{1}{2}} \left(\frac{1}{2m_A} + \frac{1}{2m_B}\right)^{\frac{1}{2}} \frac{T^{\frac{1}{2}}}{p\left(\frac{d_A + d_B}{2}\right)^2}$$

with k being the Boltzmann constant. More elaborate forms of diffusion constant will appear in Section 22.5.

It follows from (22.2.5)–(22.2.9) that

$$\frac{\partial \rho_A}{\partial t} + \nabla \cdot (\rho_A \mathbf{v}) = \nabla \cdot (\rho D_{AB} \nabla Y_A) + \omega_A \tag{22.2.10}$$

Similarly, we have, from (22.2.8) and (22.2.7),

$$\frac{\partial C_A}{\partial t} + \nabla \cdot (C_A \mathbf{v}) = \nabla \cdot (C D_{AB} \nabla X_A) + \overline{\omega}_A \tag{22.2.11}$$

where X_A denotes the molar fraction for the species A.

Notice that if chemical reactions are absent and all velocities vanish, then

$$\frac{\partial C_A}{\partial t} = D_{AB} \nabla^2 C_A \tag{22.2.12}$$

which is called Fick's second law of diffusion, and is valid in solids or stationary non-reacting fluids.

In view of (22.2.7) and (22.2.5), the continuity equation for a multicomponent system for species becomes

$$\frac{\partial}{\partial t}(\rho Y_k) + \nabla \cdot [\rho Y_k(\mathbf{v} + \mathbf{V}_k)] = \omega_k, \quad (k = 1, 2, \ldots, N) \tag{22.2.13}$$

where we have used the relation $\rho_k = \rho Y_k$. Carrying out differentiation in (22.2.13) and satisfying (22.2.7), we obtain

$$\rho \frac{\partial Y_k}{\partial t} + \rho(\mathbf{v} \cdot \nabla) Y_k + \nabla \cdot (\rho Y_k \mathbf{V}_k) = \omega_k \tag{22.2.14}$$

Using the Fick's first law of diffusion in (22.2.14), we obtain the conservation of mass equation for Y_k in the form

$$\rho \frac{\partial Y_k}{\partial t} + \rho(\mathbf{v} \cdot \nabla) Y_k - \nabla \cdot (\rho D_{km} \nabla Y_k) = \omega_k \tag{22.2.15}$$

which indicates the existence of N species equations. Thus, (22.2.6) and (22.2.15) constitute the conservation of mass for the mixture and individual species. It is now obvious that any one of these N equations may be replaced by the continuity equation for the mixture in any given problem, indicating that only $N-1$ equations of the Y_k species are independent.

22.2.2 CONSERVATION OF MOMENTUM

For reacting fluids with a mixture of species k, the body force, $\rho\mathbf{F}$, acting on species k will contribute to the rate of change of the momentum.

$$\rho\mathbf{F} = \rho \sum_{k=1}^{N} Y_k \mathbf{f}_k$$

in which $\mathbf{f}_k$ is the external force per unit mass on species k. Thus, the momentum equation takes the form

$$\rho \frac{\partial \mathbf{v}}{\partial t} + \rho(\mathbf{v} \cdot \nabla)\mathbf{v} = -\nabla p + \frac{\partial \tau_{ij}}{\partial x_i}\mathbf{i}_j + \rho \sum_{k=1}^{N} Y_k \mathbf{f}_k \tag{22.2.16}$$

where τ_{ij} is the viscous stress tensor,

$$\tau_{ij} = \mu\left(\frac{\partial v_i}{\partial x_j} + \frac{\partial v_j}{\partial x_i} - \frac{2}{3}\frac{\partial v_k}{\partial x_k}\delta_{ij}\right) \tag{22.2.17}$$

with μ being the viscosity and δ_{ij} is the Kronecker delta. Substituting (22.2.17) in (22.2.16) we obtain:

$$\rho \frac{\partial \mathbf{v}}{\partial \mathbf{t}} + \rho(\mathbf{v} \cdot \nabla)\mathbf{v} = -\nabla p + \mu\left[\nabla^2 \mathbf{v} + \frac{1}{3}\nabla(\nabla \cdot \mathbf{v})\right] + \rho \sum_{k=1}^{N} Y_k \mathbf{f}_k \tag{22.2.18a}$$

The momentum equation may be written in the conservation form

$$\frac{\partial}{\partial t}(\rho v_j) + \frac{\partial}{\partial x_i}(\rho v_i v_j) + \frac{\partial}{\partial x_i}\left[p\delta_{ij} - \mu\left(\frac{\partial v_j}{\partial x_i} + \frac{1}{3}\frac{\partial v_i}{\partial x_j}\right)\right] = \rho\sum_{k=1}^{N} Y_k f_{kj} \qquad (22.2.18b)$$

in which both continuity (22.2.3) and momentum (22.2.18a) equations are satisfied. Throughout this chapter, the subscripts for species and indices for tensors are interchangeably used, so the reader should distinguish them from the physical aspect of each case.

22.2.3 CONSERVATION OF ENERGY

The reactive flow energy equation may be written in various forms. Let us define the stagnation energy E as

$$E = \varepsilon + \frac{1}{2}\mathbf{v}\cdot\mathbf{v} \qquad (22.2.19a)$$

where ε is the specific internal energy density

$$\varepsilon = \sum_{k=1}^{N} Y_k H_k - \frac{p}{\rho} \qquad (22.2.19b)$$

with H_k being the enthalpy given by

$$H_k = H_k^o + \overline{H}_k \qquad (22.2.20a)$$

where $\overline{H}_k$ is the sensible enthalpy above the zero-point enthalpy H_k^0,

$$\overline{H}_k = \int_{T_o}^{T} c_{pk} dT \qquad (22.2.20b)$$

so that the static enthalpy is of the form

$$H = \sum_k Y_k H_k = \sum_{k=1}^{N} Y_k\left(H_k^o + \int_{T_o}^{T} c_{pk} dT\right) = \sum_k Y_k H_k^0 + \overline{H} \qquad (22.2.21)$$

with $\overline{H} = \sum_k Y_k \overline{H}_k$. A general form of the specific heat for k species (thermodynamic model) is given by

$$c_{pk} = A_k + B_k T + C_k T^2 + D_k T^3 + E_k T^4 \qquad (22.2.22a)$$

If we consider a linear form (first two terms on RHS above), then the integral in (22.2.21) becomes

$$\int_{T_o}^{T} c_{pk} dT = A_k T + \frac{1}{2} B_k T^2 \qquad (22.2.22b)$$

The coefficient of these polynomials are available from the general data bank in the JANNAF Tables, or Hirschfelder et al. [1954].

The nonconservation form of the energy equation may be written as (2.2.9c)

$$\rho\frac{D\varepsilon}{Dt} = -\nabla\cdot\mathbf{q} - p\nabla\cdot\mathbf{v} + \tau_{ij} v_{j,i} \qquad (22.2.23)$$

with

$$\mathbf{q} = \mathbf{q}^{(C)} + \mathbf{q}^{(D)}$$

where $\mathbf{q}^{(C)}$ and $\mathbf{q}^{(D)}$ are the heat fluxes due to conduction and chemical species diffusion, respectively,

$$\mathbf{q}^{(C)} = -k\nabla T \tag{22.2.24}$$

$$\mathbf{q}^{(D)} = \rho \sum_{k=1}^{N} H_k Y_k V_k \tag{22.2.25}$$

Using the Fick's first law of diffusion, we have

$$\mathbf{q}^{(D)} = -\rho \sum_{k=1}^{N} H_k D_{km} \nabla Y_k \tag{22.2.26}$$

The additional heat fluxes from the Dufour effect (influence of species gradient on temperature), and Soret effect (influence of temperature gradient on species diffusion), and radiative heat transfer may be added as necessary [Hirschfelder, Curtis, and Bird, 1954].

It follows from (22.2.23) that the energy equation takes the form

$$\rho \frac{DH}{Dt} = \frac{Dp}{Dt} + \nabla \cdot (k\nabla T) + \nabla \cdot \left(\sum_k \rho H_k D_{km} \nabla Y_k \right) + \tau_{ij} \mathrm{v}_{j,i} \tag{22.2.27}$$

Take a substantial derivative of (22.2.21) in the form,

$$\frac{DH}{Dt} = \frac{D\overline{H}}{Dt} + \sum_k H_k^0 \frac{DY_k}{Dt} \tag{22.2.28}$$

Inserting (22.2.14) into (22.2.28), we obtain

$$\rho \frac{DH}{Dt} = \rho \frac{D\overline{H}}{Dt} + \sum_k H_k^0 (-\nabla \cdot \rho Y_k \mathbf{V}_k + \omega_k) \tag{22.2.29}$$

Equating (22.2.27) and (22.2.29) and using the Fick's first law of diffusion lead to the nonconservation form of the energy equation,

$$\rho \frac{D\overline{H}}{Dt} - \frac{Dp}{Dt} - \nabla \cdot (k\nabla T) - \nabla \cdot \left(\sum_k \rho \overline{H}_k D_{km} \nabla Y_k \right) - \tau_{ij} \mathrm{v}_{j,i} = -\sum_k H_k^0 \omega_k \tag{22.2.30}$$

Using the relation (22.2.20), we may write (22.2.30) in the conservation form as

$$\frac{\partial}{\partial t}(\rho E) + \frac{\partial}{\partial x_i}(\rho E \mathrm{v}_i + p\mathrm{v}_i) - \frac{\partial}{\partial x_i}\left(kT_{,i} + \sum_k \rho H D_{km} Y_{k,i} + \tau_{ij} \mathrm{v}_j \right)$$
$$= S - \sum_k H_k^0 \omega_k \tag{22.2.31}$$

Here, the total energy E is given by

$$E = \overline{H} + \sum_k H_k^0 Y_k - \frac{p}{\rho} + \frac{1}{2}v_i v_i \tag{22.2.32a}$$

and the energy due to the body force is of the form

$$S = \rho \sum_{k=1}^{N} Y_k \mathbf{f}_k \cdot \mathbf{v} \tag{22.2.32b}$$

Equation (22.2.31) is the most general expression of the energy equation for reacting flows. By carrying out differentiation as implied in (22.2.31) and having satisfied conservation of mass (22.2.6), momentum (22.2.18a), and species (22.2.14), the remaining terms represent the nonconservation form of energy equation, given by (22.2.30) or

$$\rho c_p\left[\frac{\partial T}{\partial t} + (\mathbf{v}\cdot\nabla)T\right] - \frac{\partial p}{\partial t} - (\mathbf{v}\cdot\nabla)p - \tau_{ij}v_{j,i} - k\nabla^2 T - \sum_k \rho c_{pk} D_{km}(\nabla Y_k \cdot \nabla)T$$
$$= -\sum_k H_k^0 \omega_k \tag{22.2.33}$$

in which substitutions

$$\overline{H} = \sum_k Y_k \int_{T_0}^{T} c_{pk} dT = c_p T \quad \text{and} \quad \overline{H}_k = \int_{T_0}^{T} c_{pk} dT = c_{pk} T$$

are made for the zero-point enthalpy.

It should be noted that the energy equation (22.2.27) does not include coupling with species equations through (22.2.28). The direct influence of the reaction rate appearing on the right-hand side of (22.2.31) is important if chemical reactions dominate the diffusion process. The chemical reaction as represented by the energy equation in (22.2.31) can be either exothermic or endothermic if the relative enthalpy change (ratio of the enthalpy change to the total energy) is positive (heat release) or negative (heat absorption), respectively. Thus, combustion is the exothermic process.

22.2.4 CONSERVATION FORM OF NAVIER-STOKES SYSTEM OF EQUATIONS IN REACTIVE FLOWS

Grouping all governing equations for continuity, momentum, energy, and species, the conservation form of the Navier-Stokes system of equations in reactive flows is written as follows:

$$\frac{\partial \mathbf{U}}{\partial t} + \frac{\partial \mathbf{F}_i}{\partial x_i} + \frac{\partial \mathbf{G}_i}{\partial x_i} = \mathbf{B} \tag{22.2.34}$$

where $\mathbf{U}$, $\mathbf{F}_i$, and $\mathbf{G}_i$ are the conservation variables,

$$\mathbf{U} = \begin{bmatrix} \rho \\ \rho v_j \\ \rho E \\ \rho Y_k \end{bmatrix}, \quad \mathbf{F}_i = \begin{bmatrix} \rho v_i \\ \rho v_i v_j + p\delta_{ij} \\ \rho E v_i + p v_i \\ \rho Y_k v_i \end{bmatrix}, \quad \mathbf{G}_i = \begin{bmatrix} 0 \\ -\tau_{ij} \\ -\tau_{ij} v_j - kT_{,i} - \sum_{k=1}^{N} \rho \overline{H} D_{km} Y_{k,i} \\ -\rho D_{km} Y_{k,i} \end{bmatrix},$$

$$\mathbf{B} = \begin{bmatrix} 0 \\ \rho \sum_{k=1}^{N} Y_k f_{kj} \\ S - 2\sum_{k=1}^{N} H_k^0 \omega_k \\ \omega_k \end{bmatrix}$$

To prove that (22.2.34) is indeed the correct *conservation form*, we perform the differentiation implied in (22.2.34) and recover the nonconservation forms of the equations for momentum (22.2.16), energy (22.2.33), and species (22.2.15). If integrated, however, conservation properties across discrete boundaries are guaranteed through physical discontinuities such as shock waves as observed in Section 2.2.

Relationships between chemical reactions and flowfield phenomena which control the mixing process are characterized by Damköhler numbers. Each term in the species equation and energy equation influences such relationships, with temperature changes closely linked to the chemical reactions. Thus, the Damköhler number, *Da*, is defined in many different ways (see Table 22.2.1): as the ratio of the mass source to

Table 22.2.1 Various Definitions of Peclet and Damköhler Numbers

$$\underbrace{\nabla\cdot(\rho E + p)\mathbf{v}}_{A} - \underbrace{k\nabla^2 T}_{B} - \underbrace{\nabla\cdot(\rho H_k D_{km}\nabla Y_k)}_{C} = \underbrace{-H_k^0\omega_k}_{D}$$

$$\underbrace{\nabla\cdot(\rho Y_k \mathbf{v})}_{E} - \underbrace{\nabla\cdot(\rho D_{km}\nabla Y_k)}_{F} = \underbrace{\omega_k}_{G}$$

Peclet number, I	Pe_I	$\frac{uL}{\alpha}$	$\frac{A}{B} = \frac{\text{convective heat transfer}}{\text{conductive heat transfer}}$
Peclet number, II	Pe_{II}	$\frac{uL}{D_{km}}$	$\frac{E}{F} = \frac{\text{convective mass transfer}}{\text{diffusive mass transfer}}$
Damköhler number, I	Da_I	$\frac{\omega_k L}{\rho u Y_k}$	$\frac{G}{E} = \frac{\text{mass source}}{\text{convective mass transfer}}$
Damköhler number, II	Da_{II}	$\frac{\omega_k L^2}{\rho D_{km} Y_k}$	$\frac{G}{F} = \frac{\text{mass source}}{\text{diffusive mass transfer}}$
Damköhler number, III	Da_{III}	$\frac{\omega_k L}{\rho u}$	$\frac{D}{A} = \frac{\text{heat source}}{\text{convective heat transfer}}$
Damköhler number, IV	Da_{IV}	$\frac{H_k \omega_k L^2}{kT}$	$\frac{D}{B} = \frac{\text{heat source}}{\text{conductive heat transfer}}$
Damköhler number, V	Da_V	$\frac{H_k^0 \omega_k L^2}{\rho H D_{km} Y_k}$	$\frac{D}{C} = \frac{\text{heat source}}{\text{diffusive heat transfer}}$

the species convection(Da_I) or to the species diffusion(Da_{II}) of the species equation. Similarly, Damköhler numbers are defined also as the ratio of the heat source to the heat convection(Da_{III}), to the heat conduction(Da_{IV}), or to the heat diffusion(Da_V) of the energy equation. For example, the Damköhler number Da_I is defined as

$$Da = Da_I = \frac{\omega_k L}{\rho u Y_k} = \frac{\tau_d}{\tau_r} \tag{22.2.35}$$

where L is the characteristic length, τ_d and τ_r are the characteristic diffusion time and characteristic reaction time, respectively. If the reaction is very fast, $\tau_r \ll \tau_d$ or $Da \to \infty$, this is known as the *equilibrium chemistry*. The so-called *frozen chemistry* results if $\tau_r \gg \tau_d$ or $Da \to 0$. The *finite rate chemistry* prevails for $0 < Da < \infty$. It will be shown later that the Damköhler numbers are instrumental in determining appropriate numerical schemes as dictated by the dominance of each of the terms in both the energy and species equations, indicative of stiffness or time and length scales (Section 13.6 for FDV methods).

In correspondence with the above definitions, the equilibrium chemistry may be represented by the last equation of (22.2.34) with

$$\frac{\partial}{\partial x_i}\left(\rho Y_k \mathrm{v}_i - \rho D_{km}\frac{\partial Y_k}{\partial x_i}\right) = 0$$

which leads to

$$\frac{\partial}{\partial t}(\rho Y_k) = \omega_k \tag{22.2.36a}$$

or

$$\omega_k = \frac{d}{dt}(\rho Y_k) = \frac{d\rho_k}{dt} = W_k \sum_{i=1}^{M} (\nu''_{ki} - \nu'_{ki}) A_i T^{\alpha_i} \exp\left(-\frac{E_i}{R^0 T}\right) \prod_{j=1}^{N}\left(\frac{X_j p}{R^0 T}\right)^{\nu'_{ji}} \tag{22.2.36b}$$

The frozen chemistry occurs for $\omega_k = 0$ so that the species equation takes the form

$$\frac{\partial}{\partial t}(\rho Y_k) + \frac{\partial}{\partial x_i}\left(\rho Y_k \mathrm{v}_i - \rho D_{km}\frac{\partial Y_k}{\partial x_i}\right) = 0 \tag{22.2.37}$$

Our objective is to solve simultaneously the Navier-Stokes system of equations for the compressible reacting flow given by (22.2.34). The main variable solution vector is the conservation flow variables $\boldsymbol{U}$. Once the solution is obtained, it is necessary to convert (decode) the conservation variables into the primitive variables. Although the process is trivial for nonreacting equations, this is not the case for the reacting flows. In order to calculate temperature, we utilize the Lagrange interpolation polynomials for the total enthalpy as follows.

To begin, we equate the total enthalpy for the chemical species to the total flowfield static enthalpy.

$$\sum_{k=1}^{N} Y_k H_k = E + RT - \frac{1}{2}\mathrm{v}_i \mathrm{v}_i \tag{22.2.38}$$

Consider the j discrete temperature abscissa, $j = 1, \ldots, J$, such that

$$H_k(T_j) = H_{k,j} \quad \text{for } T_j = (j-1)\Delta T \quad \text{with} \quad T_1 \le T \le T_J$$

Applying the Lagrange interpolation polynomials to $H_{k,j}$ leads to

$$H_{k,j} = \left[\frac{(T-T_j)(T-T_{j+1})}{(T_{j-1}-T_j)(T_{j-1}-T_{j+1})}\right] H_{k,j-1} + \left[\frac{(T-T_{j-1})(T-T_{j+1})}{(T_j-T_{j-1})(T_j-T_{j+1})}\right] H_{k,j} + \left[\frac{(T-T_{j-1})(T-T_j)}{(T_{j+1}-T_{j-1})(T_{j+1}-T_j)}\right] H_{k,j+1} \tag{22.2.39}$$

with

$$|T - T_j| \le \Delta T \quad (j = 2, \ldots, J-1)$$

Assuming that ΔT is constant, we have

$$\begin{aligned} H_{k,j} &= \frac{1}{2}\left[\frac{T}{\Delta T} - (j-1)\right]\left[\frac{T}{\Delta T} - j\right] H_{k,j-1} - \left[\frac{T}{\Delta T} - (j-2)\right]\left[\frac{T}{\Delta T} - j\right] H_{k,j} \\ &\quad + \frac{1}{2}\left[\frac{T}{\Delta T} - (j-2)\right]\left[\frac{T}{\Delta T} - (j-1)\right] H_{k,j+1} \\ &= A\alpha^2 + B\alpha + C \end{aligned} \tag{22.2.40}$$

with

$$A = \left[\frac{1}{2} H_{k,j-1} - H_{k,j} + \frac{1}{2} H_{k,j+1}\right]$$

$$B = \left[-\frac{1}{2}(2j-1) H_{k,j-1} - (2j-2) H_{k,j} - \frac{1}{2}(2j-3) H_{k,j+1}\right]$$

$$C = \left[\frac{1}{2} j(j-1) H_{k,j-1} - j(j-2) H_{k,j} + \frac{1}{2}(j-1)(j-2) H_{k,j+1}\right]$$

$$\alpha = \frac{T}{\Delta T}$$

Substituting (22.2.40) to (22.2.38) leads to

$$a\alpha^2 + b\alpha + c = 0$$

or

$$\alpha = \frac{b + \sqrt{b^2 - 4ac}}{2a} \tag{22.2.41}$$

where

$$a = \sum_{k=1}^{N} Y_k A$$

$$b = \sum_{k=1}^{N} Y_k \left(B + \frac{R^0 \Delta T}{W_k}\right)$$

$$c = \sum_{k=1}^{N} Y_k \left(C - E + \frac{1}{2} \mathrm{v}_i \mathrm{v}_i\right)$$

Thus, the temperature and pressure are determined as

$$T = \alpha \Delta T$$
$$p = \rho R^0 T \sum_{k=1}^{N} \frac{Y_k}{W_k} \tag{22.2.42}$$

The solution of the Navier-Stokes system of equations in conservation form is desirable for high speed compressible flows. However, as indicated earlier (Section 6.4), the preconditioning of the time dependent term is an important choice particularly for low speed incompressible flows, in which the solution vector is altered already in terms of primitive variables and thus the cumbersome process of conversion of the conservation flow variables to primitive variables as shown above can be eliminated.

22.2.5 TWO-PHASE REACTIVE FLOWS (SPRAY COMBUSTION)

The combustion of liquid fuel sprays has numerous important applications in diesel engines, gas turbines, and space shuttle main engines. The prediction of the flow properties of spray flames requires the consideration of two phases in the flowfield. Various approaches [Faeth, 1979; Sirignano, 1993; Sirignano, 1999, among others] have been suggested to model the coupling of the discontinuous gas-liquid phase. There are three approaches to spray combustion modeling: Eulerian-Eulerian formulation, Eulerian-Lagrangian formulation, and probabilistic formulation.

The Eulerian-Eulerian approach treats both gaseous and liquid phases as continuum. In the Eulerian-Lagrangian approach, the gas field is described in Eulerian coordinates and the liquid droplet field is described in the Lagrangian formulation. This approach employs computational particles to represent a collection of physical particles having the same attributes such as spatial location, velocity, mass, temperature etc. The motion of the droplet is simulated using a Lagrangian formulation to predict the droplet behavior under the gas phase. The influence of the liquid phase on the gas phase is treated by inclusion of coupling source terms arising due to the gas and liquid phase interaction. In the probabilistic formulation, we define a droplet number density function or, in other words, a droplet number probability density function (PDF). This function $f(x, t, R, , \mathrm{v}, e_\ell)$ depends upon spatial position x, time t, droplet radius R, droplet velocity v, and droplet thermal energy e_ℓ. An excellent discussion of spray combustion and other related topics may be found in Sirignano [1999]. We introduce below a portion of the governing equations on Eulerian-Lagrangian formulation whose applications will be presented in Section 22.6.2.

If all external effects except the drag force are neglected, the equations of motion for the droplet can be expressed as

$$\frac{dx_{ik}}{dt} = U_{ik} \tag{22.2.43}$$

$$\frac{dU_{ik}}{dt} = \frac{3C_D \mu \,\mathrm{Re}_k}{16\rho_k r_k^2}(U_i - U_{ik}) \tag{22.2.44}$$

with

$$\mathrm{Re}_k = \frac{2r_k\rho}{\mu}|U_i - U_{ik}| \tag{22.2.45}$$

$$C_D = \frac{24}{\mathrm{Re}_k}\left(1 + \frac{\mathrm{Re}_k^{2/3}}{6}\right) \tag{22.2.46}$$

where x_{ik} is the displacement of the droplet characteristics k in the coordinate direction i, U_{ik} is the corresponding droplet velocity component, U_i is the gas velocity component, Re_k is the relative Reynolds number, and C_D is the drag coefficient.

The droplet evaporation rate and heat balance equation are given as

$$\begin{aligned} \dot{m}_k &= C_a C_b \\ \frac{dT_k}{dt} &= Q_L/m_k c_{pk} \end{aligned} \tag{22.2.47}$$

where C_a and C_b denote the evaporation coefficient and the correction factor for the convection effect, respectively; T_k is the droplet surface temperature, Q_L is the heat transferred into the droplet interior, m_k is the droplet mass, and c_{pk} is the droplet specific heat at constant volume. The parameters involved in (22.2.47) have been proposed by various investigators [Lefebvre, 1989; Abramzon and Sirignano, 1988, among others].

Chin and Lefebvre [1983] proposed that

$$\begin{aligned} C_a &= 4\pi r_k \lambda_g / c_{pg} \ln(1 + B_m) \\ C_b &= 1 + 0.276\,\mathrm{Re}_k^{1/2}\,\mathrm{Pr}^{1/3} \\ Q_L &= \dot{m}\left[\frac{c_{pg}(T - T_k)}{B_m} - H_v\right] \end{aligned} \tag{22.2.48}$$

where B_m is the mass transfer number,

$$B_m = \frac{Y_{fs} - Y_{f\infty}}{1 - Y_{fs}}$$

$$Y_{fs} = \left[1 + \left(\frac{p}{p_{fs}} - 1\right)\frac{W_a}{W_f}\right]^{-1}$$

Here, Y_{fs} and p_{fs} are the mass fraction and the fuel vapor pressure at the droplet surface, p and $Y_{f\infty}$ are the ambient pressure and the fuel mass fraction at the outer boundary of the film, and W_a and W_f are the molecular weights of air and fuel, respectively.

Another approach proposed by Abramzon and Sirignano [1988] is given by

$$\begin{aligned} C_a &= 4\pi r_k \rho_g D_g \ln(1 + B_m) \\ C_b &= 1 + (Sh_o/2 - 1)/F(B_m) \\ Q_L &= \dot{m}\left[\frac{c_{pf}(T - T_k)}{B_m} - H_v\right] \end{aligned} \tag{22.2.49}$$

with

$$F(B) = (1 + B)^{0.7}\frac{\ln(1 + B)}{B}$$

$$Sh_o = 1 + (1 + \mathrm{Re}_k\, Sc)^{1/3} f(\mathrm{Re}_k)$$

$$f(\mathrm{Re}_k) = \begin{cases} 1 & \text{for } \mathrm{Re}_k \leq 1 \\ \mathrm{Re}^{0.077} & \text{for } 1 \leq \mathrm{Re}_k \leq 4_k 00 \end{cases}$$

$$B_t = (1 + B_m)^{\phi} - 1, \qquad \phi = \frac{c_{pf}}{c_{pg}}\frac{Sh^*}{Nu^*}\frac{1}{L_e}, \qquad Sh^* = C_b$$

$$Le = \lambda_g / \rho_g D_g c_{pg}$$

$$Nu^* = 1 + (Nu_o/2 - 1)/F(B)$$

$$Nu_o = 1 + (1 + \mathrm{Re}_k \mathrm{Pr})^{1/3} f(\mathrm{Re}_k)$$

$$H_{v,eff} = \frac{c_{pf}(T - T_k)}{B_t}$$

where $F(B)$ is the film thickness correctin factor and D_g is the diffusion coefficient in the film.

Recent advances in two-phase reactive flows or spray combustion may be found in Sirignano [1999]. Further discussions on reactive turbulent flows in fluid-particle mixtures will be presented in Section 25.3.3.

22.2.6 BOUNDARY AND INITIAL CONDITIONS

Boundary and initial conditions for reacting flows are similar to those for nonreacting flows except that inflow boundary conditions must include chemical species based on reactant species being either premixed or nonpremixed. For the nonpremixed case, reactants are specified at separate inflow boundaries, whereas they are specified together at the inflow boundaries for the premixed case.

The Neumann boundary conditions are applied on Γ_N at the wall and outflow boundaries as

$$(\mathbf{F}_i + \mathbf{G}_i) n_i = \mathbf{N} \tag{22.2.50}$$

Mixture Mass Flux

$$\rho \mathrm{v}_i n_i = A \tag{22.2.51a}$$

Momentum Flux

$$(\rho \mathrm{v}_i \mathrm{v}_j + p\delta_{ij} - \tau_{ij}) n_i = B_j \tag{22.2.51b}$$

Energy Flux

$$\left(\rho E \mathrm{v}_i + p\mathrm{v}_i - \tau_{ij}\mathrm{v}_j - kT_{,i} + \sum_{k=1}^{N} \rho_k H V_{ki}\right) n_i = C \tag{22.2.51c}$$

Species Mass Flux

$$(\rho Y_k \mathrm{v}_i + \rho Y_k V_{ki}) n_i = D_k \tag{22.2.51d}$$

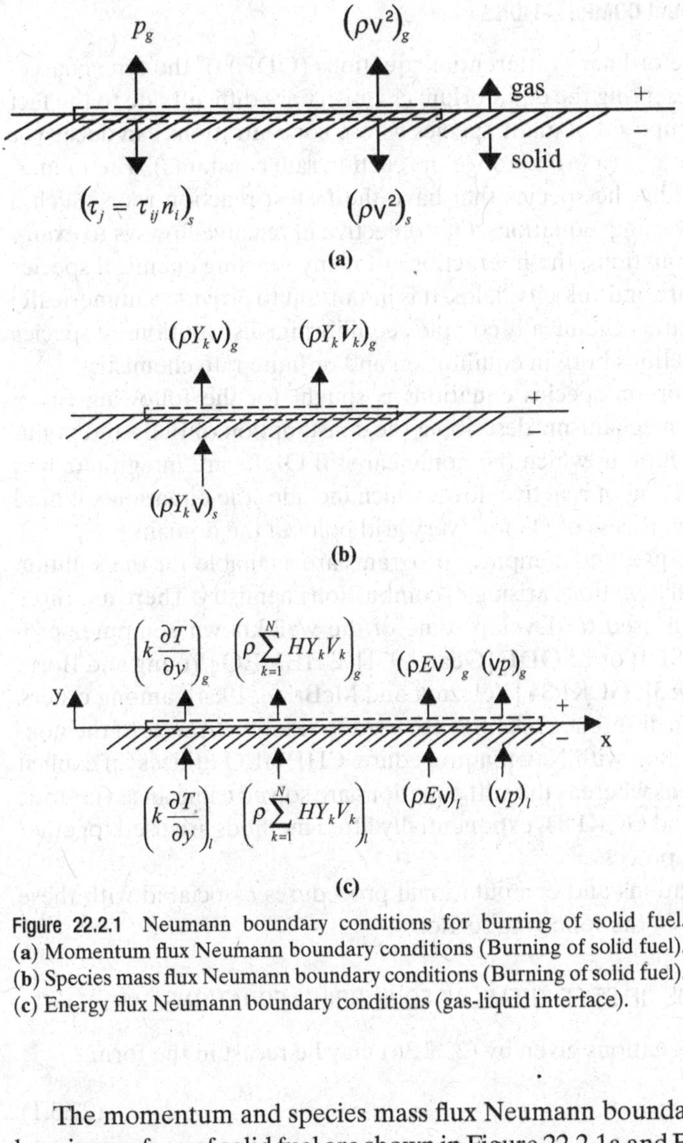

Figure 22.2.1 Neumann boundary conditions for burning of solid fuel.
(a) Momentum flux Neumann boundary conditions (Burning of solid fuel).
(b) Species mass flux Neumann boundary conditions (Burning of solid fuel).
(c) Energy flux Neumann boundary conditions (gas-liquid interface).

The momentum and species mass flux Neumann boundary conditions for a typical burning surface of solid fuel are shown in Figure 22.2.1a and Figure 22.2.1b, respectively. Similarly, the energy flux Neumann boundary conditions for a liquid fuel burning surface are depicted in Figure 22.2.1c.

If Dirichlet boundary conditions are specified on Γ_D, then the Neumann boundary conditions need not be specified. The Dirichlet boundary conditions are not to be specified for the case of Neumann boundary conditions vanishing along the walls and outflow boundaries.

Initial conditions for all chemical kinetics, equation of state, molecular, and thermal transport data should be provided at the beginning of the calculation, rather than appended as constraints at each time step. Care should be exercised, as these data may be the cause for large errors.

22.3 CHEMICAL EQUILIBRIUM COMPUTATIONS

Numerical solutions of the ordinary differential equations (ODE) of the type characterized by (22.2.36b) representing the equilibrium chemistry are difficult due to the fact that a kinetic system is composed of many species whose concentrations can decay (or grow) at widely disparate rates (a broad range of reaction rate constants). The numerical solution is dominated by the species that have the fastest reaction rates. Such a system constitutes *stiff* governing equations. Our objective in reactive flows is to examine, by solving such stiff equations, the interactions of many reacting chemical species with fluctuating temperature and velocity fields. It is important to provide a numerically efficient scheme for calculating chemically complex equilibrium distributions of species mole numbers or mass fractions both in equilibrium and/or finite rate chemistry.

The solution of equilibrium species equations is sought for the following cases: (1) we model the reaction mechanisms describing the consumption of fuels and pollutant formation and destruction in which the nonlinear stiff ODEs are integrated, and (2) multidimensional modeling of reactive flows, which includes the equations of fluid motion, thus repeating the process of (1) for every grid point in the domain.

Many numerical techniques and computer programs are available for the solution of stiff ordinary differential equations arising in combustion chemistry. There are three approaches that have been used to develop some of the well-known computer programs. They include DIFFSUB or LSODE [Gear, 1971]; CHEMEQ [Young and Boris, 1977]; CREK1D [Pratt, 1983]; GCKP84 [Zeleznik and McBride, 1984], among others. In LSODE, backward finite difference schemes are used to resolve stiffness of the nonlinear equations in conjunction with Newton procedure. CHEMEQ utilizes an explicit method for regular equations whereas the stiff equations are solved using an asymptotic integration. In CREK1D and GCKP84, exponentially fitted methods are used together with the Newton-Raphson process.

Some of the basic equations and computational procedures associated with these programs will be discussed in the following section.

22.3.1 SOLUTION METHODS OF STIFF CHEMICAL EQUILIBRIUM EQUATIONS

The ordinary differential equations given by (22.2.36) may be recast in the form

$$\frac{dY_k}{dt} = f_k(N_i, T) \quad k, i = 1, n \tag{22.3.1}$$

$$f_k = \frac{1}{\rho}\sum_{i}^{m}(\nu''_{ki} - \nu'_{ki})(R_{fi} - R_{bi}) \tag{22.3.2}$$

with the initial conditions $Y_k(t=0)$ and $T(t=0)$ given. Here, R_{fi} and R_{bi} are the forward and backward molar reaction rates per unit volume, respectively, with

$$R_{fi} = k_{fi}\prod_{i=1}^{n}(\rho Y_k)\nu'_{ji}, \qquad R_{bi} = k_{bi}\prod_{i=1}^{n}(\rho Y_k)\nu''_{ji} \tag{22.3.3}$$

$$k_{fi} = A_{fi}T^{\alpha_{fi}}\exp\left(\frac{-T_{fi}}{T}\right), \qquad T_{fi} = \frac{E_{fi}}{R^0} \tag{22.3.4a}$$

$$k_{bi} = A_{bi}T^{\alpha_{bi}}\exp\left(\frac{-T_{bi}}{T}\right), \qquad T_{bi} = \frac{E_{bi}}{R^0} \tag{22.3.4b}$$

where k_{fi} and k_{bi} are the forward reaction rate constant and backward reaction rate constant, respectively, and

$$k_{bi} = k_{fi}(R^0T)^{\Delta N_i} \exp\left(\frac{1}{R^0T}\sum_{k=1}^{n}(\nu'_{ki} - \nu''_{ki})g_k^0\right) \tag{22.3.5}$$

$$T_{bi} = T_{fi} + \frac{1}{R^0}\sum_{k=1}^{n}(\nu''_{ki} - \nu'_{ki})H_k \tag{22.3.6}$$

where g_k^0 is the 1 atm molar-specific Gibbs function of species k, H_k is the molar-specific enthalpy of species k, and ΔN_i is given by

$$\Delta N_i = \sum_{k=1}^{n}(\nu''_{ki} - \nu'_{ki}) \tag{22.3.7}$$

Equating the temperature exponents in (22.3.4b) and (22.3.5) for $\hat{k}_i$, we obtain

$$\hat{\alpha}_i = \alpha_i + \Delta N_i \tag{22.3.8}$$

To solve (22.3.1), we require the enthalpy constraint condition given by

$$\sum_{k=1}^{n} Y_k H_k = H^0 = \text{constant} \tag{22.3.9}$$

Differentiating (22.3.9) with respect to time and using (23.3.1) leads to

$$\sum_{k=1}^{n} Y_k c_{pk}\frac{dT}{dt} + \sum_{k=1}^{n} f_k H_k = 0 \tag{22.3.10}$$

To implement the constraint condition (22.3.10) for the nonlinear equation solvers such as in the Newton-Raphson method, it is necessary to have derivatives of quantities dT/dt and f_k in (22.3.10) with respect to temperature and the mass fraction as follows:

$$\frac{\partial}{\partial T}\left(\frac{dT}{dt}\right) = -\frac{\sum_{k=1}^{n}\frac{\partial f_k}{\partial T}H_k + \sum_{k=1}^{n} f_k c_{pk} + \left(\frac{dT}{dt}\right)\sum_{k=1}^{n} Y_k\frac{dc_{pk}}{dT}}{\sum_{k=1}^{N} Y_k c_{pk}}$$

$$\frac{\partial}{\partial Y_j}\left(\frac{dT}{dt}\right) = -\frac{\sum_{k=1}^{n}\frac{\partial f_k}{\partial Y_j}H_k + c_{pj}\left(\frac{dT}{dt}\right)}{\sum_{k=1}^{n} Y_k c_{pk}} \tag{22.3.11}$$

$$\frac{\partial f_k}{\partial T} = \frac{f_k}{T} + \frac{1}{\rho T}\sum_{i=1}^{m}(\nu''_{ki} - \nu'_{ki}) \times \left[R_{fi}\left(\alpha_i + \frac{T_i}{T} - \sum_{k=1}^{n}\nu'_{ki}\right) - R_{bi}\left(\hat{\alpha}_i + \frac{T_i}{T} - \sum_{i=1}^{n}\nu''_{ki}\right)\right]$$

In addition to these constraint derivatives, we must have the derivatives of f_k with respect to the mass fraction Y_k.

$$\frac{\partial f_k}{\partial Y_j} = (\rho Y_k)^{-1} \sum_{i=1}^{m} (v''_{ki} - v'_{ki})(v'_{ji} R_{fi} - v''_{ji} R_{bi}) + \frac{f_k}{\sum_{k=1}^{n} Y_k} - \left(\rho \sum_{i=1}^{n} Y_k\right)^{-1}$$

$$\times \sum_{i=1}^{n} (v''_{ki} - v'_{ki})(v'_{ji} R_{fi} - v''_{ji} R_{bi}) \tag{22.3.12}$$

The derivatives given above constitute the Jacobian matrix **J** for the Newton-Raphson solution of (22.3.1) (see Section 11.5.1 for Newton-Raphson methods) such that

$$\mathbf{J} \Delta \mathbf{y}_k = -\mathbf{R}_k \tag{22.3.13}$$

where $\mathbf{R}_k$ represents the residual of (22.3.1).

In CREK1D and GCKP84, the Gibbs function is minimized in order to achieve equilibrium. Chemical equilibrium is reached when the Gibbs function G and the Helmholtz free energy Φ are minimum. The partial molar Gibbs function for a species k is given by

$$g_k = h_k - TS_k = \varepsilon_k + pV_k - TS_k = \Phi_k + pV_k \tag{22.3.14}$$

where Φ_k is the Helmholtz free energy,

$$\Phi_k = \varepsilon_k - TS_k$$

Minimization of (22.3.14) gives

$$dg_k = d\Phi_k + d(pV_k) = d\Phi_k + RT\frac{dV_k}{V_k} + RT\frac{dp}{p} \tag{22.3.15}$$

Setting

$$d\Phi_k = 0, \quad \frac{V_k}{V_k^0} = \frac{\rho_k^0}{\rho} = \frac{N_k}{N}, \quad \text{and} \quad g_k^0 = h_k^0 - TS_k^0, \tag{22.3.16}$$

and integrating (22.3.15), we obtain

$$g_k = g_k^0 + RT \ln \frac{N_k}{N} + RT \ln \frac{p}{p_0} \tag{22.3.17}$$

The mass specific Gibbs function for the mixture is given by

$$G = \sum_{k=1}^{n} g_k N_k \tag{22.3.18}$$

subject to the conservation of atomic species,

$$\sum_{k=1}^{n} \sum_{i=1}^{m} (a_{ik} N_k - b_i) = 0, \quad k = 1, n, \quad i = 1, m \tag{22.3.19}$$

where a_{ik} represents the number of atoms of element i per mole of species k, b_i is the atom number of element i in the mixture, and m is the number of reaction equations for atomic species.

Multiplying (22.3.19) by the Lagrange multiplier λ_i, adding it to (22.3.18), and minimizing the sum with respect to N_k, we obtain

$$\sum_{k=1}^{n}\left(g_k + \sum_{k=1}^{n}\sum_{i=1}^{m}\lambda_i a_{ik}\right)dN_k = 0 \tag{22.3.20}$$

which leads to the equilibrium equation,

$$f_k = g_k + \sum_{i=1}^{m}\lambda_i a_{ik} = 0, \qquad N_k(k=1,n), \quad \lambda_i(i=1,m) \tag{22.3.21}$$

The minimization process of the mass specific Gibbs function consists of the following:

(a) Minimize (22.3.17) and (22.3.8),

$$\begin{aligned} dG &= \sum_{j=1}^{n}\frac{\partial}{\partial N_j}\left\{\sum_{k=1}^{n}N_k\left[g_k^0 + RT\ln N_k - RT\ln\sum_{i=1}^{n}N_i + RT\ln\frac{p}{p_0}\right]\right\}dN_j \\ &= \sum_{j=1}^{n}\left\{\sum_{k=1}^{n}\left[N_k\left(\frac{RT}{N_k}\delta_{jk} - \frac{RT}{N}\right) + g_k\delta_{jk}\right]\right\}dN_j = 0 \end{aligned} \tag{22.3.22}$$

or

$$dG = \sum_{k=1}^{n} g_k dN_k = 0 \tag{22.3.23}$$

(b) Multiply (22.3.19) by the Lagrange multiplier λ_i and minimize,

$$\lambda_i \sum_{k=1}^{n} a_{ik} dN_k = 0 \tag{22.3.24}$$

(c) Add (22.3.23) to (22.3.24),

$$\sum_{k=1}^{n}\left(g_k + \sum_{i=1}^{m}\lambda_i a_{ik}\right)dN_k = 0 \tag{22.3.25}$$

The final form (22.3.25) leads to

$$g_k + \sum_{i=1}^{m}\lambda_i a_{ik} = 0 \quad k = 1, n \quad i = 1, m \tag{22.3.26}$$

This is in addition to the m constraint equations given by (22.3.19).

We now have $(n+m)$ equations for the $(n+m)$ unknowns, $(N_k(k=1,n))$ and $(\lambda_i(i=1,m))$, which is a greater number than the n equations and unknowns required in the equilibrium constant formulation. However, it is possible to reduce the system to an m-dimensional system of equations and unknowns. It follows from (22.3.26) and (22.3.19) that

$$f_k = \frac{g_k}{RT} - \sum_{i=1}^{m} B_i a_{ik} \quad k = 1, n \tag{22.3.27}$$

$$f_i = b_i - b_i^* \quad i = 1, m \tag{22.3.28}$$

with $B_i = -\lambda_i/RT$, $b_i = a_{ik}N_k$.

These functions must vanish at equilibrium. To this end, the Newton-Raphson process for f_i with unknowns (correction variables) x_j can be carried out as follows.

$$\sum_{j=1}^{n} \frac{\partial f_k}{\partial x_j} \Delta x_j = -\Delta f_k \quad k = 1, n \tag{22.3.29}$$

Here, the appropriate equations are expanded in Taylor series with all terms containing derivatives higher than the first omitted. In view of the Gibbs function being given in the logarithmic quantities (23.3.17), the derivatives (Jacobians) are carried out with respect to the log function of N_j, N, and T,

$$\sum_{j=1}^{n} \frac{\partial f_k}{\partial \ln N_j} \Delta \ln N_j + \frac{\partial f_k}{\partial \ln N} \Delta \ln N + \frac{\partial f_k}{\partial \ln T} \Delta \ln T = -\Delta f_k \quad k = 1, n \tag{22.3.30}$$

Similar derivatives are required for f_i (22.3.28) and the enthalpy function (22.3.9). This process leads to the determination of corrections to the initial estimates of compositions N_j, Lagrange multipliers λ_i, mole number N, and temperature T. See further details in Gordon and McBride [1971] or Pratt and Wormeck [1976]. Comparisons of performance of various codes are presented in Radhakrishnan [1984].

22.3.2 APPLICATIONS TO CHEMICAL KINETICS CALCULATIONS

In order to solve (22.3.1) and (22.3.10), it is necessary to have information on elementary chemical reaction rates for a given reaction mechanism. To illustrate, let us consider the global system of hydrogen and oxygen:

$$H_2 + O_2 = OH + OH$$

for which the reaction rate is calculated from

$$k = 10^B T^s \exp\left(-\frac{E}{RT}\right)$$

with $B = 13$, $s = 0$, $E = 43$ kcal/mole. Such information is available from the existing literature. For example, the reaction rates data for hydrocarbon combustion chemistry are provided by Westbrook and Dryer [1984] and the C-H-O system by Warnats [1984]. In most of the combustion calculations, there are several hundred reactions that can be considered. However, due to limited computational resources, it is customary to select only important reaction mechanisms, neglecting those that are less important. For the purpose of illustration, some computed results for the H-N-O systems reported by Radhakrishnan [1984] using LSODE [Gear, 1971] are presented in Figure 22.3.1a for the reactions given in Table 22.3.1a and in Figure 22.3.1b for the reactions given in Table 22.3.1b. Notice that the mole fractions for all species appear to have reached equilibrium at approximately $t \cong 10^{-3}$ seconds for both cases.

For the nonequilibrium finite rate chemistry, it is necessary that the complete Navier-Stokes system of equations (22.2.34) be solved, in which the convection and diffusion terms are included in the species equations. Modifications in (22.2.34) will result in various types of simplified reactive flows. As in nonreactive flows, CFD calculations may be divided into reactive inviscid flows, reactive laminar flows, and reactive turbulent

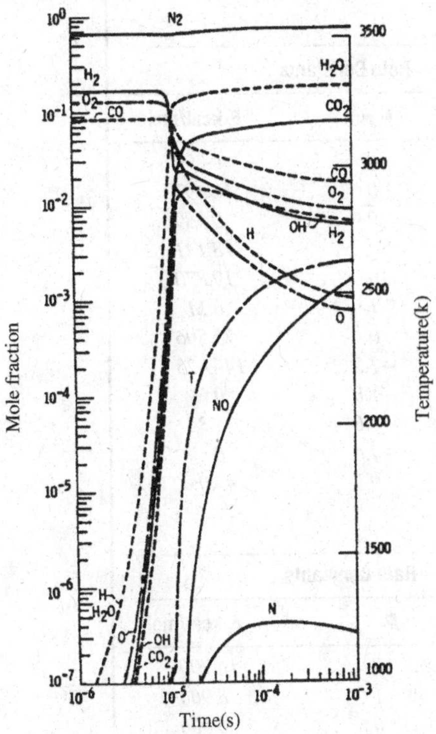

(a) Data from Table 22.3.1a

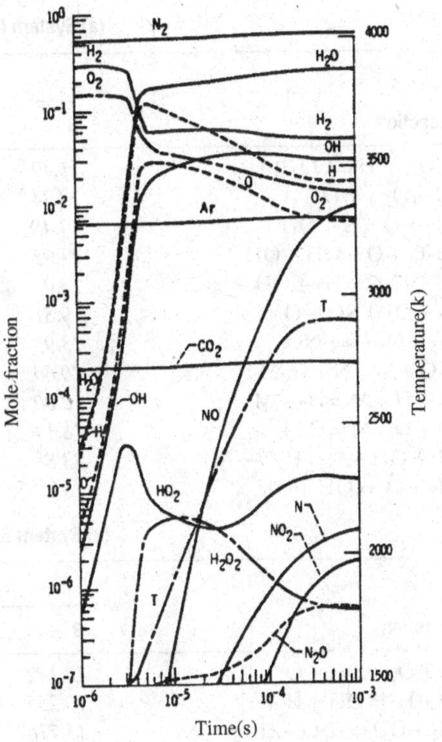

(b) Data from Table 22.3.1b.

Figure 22.3.1 Variation with time of species mole fractions and temperature.

flows. Computational schemes dealing with these topics have been introduced in earlier chapters. However, the method of the probability density function (PDF) as applied to reactive turbulent flows has not been covered in Chapter 21. This subject is introduced in the next section.

22.4 CHEMISTRY-TURBULENCE INTERACTION MODELS

Although many different turbulence models for RANS have been used extensively for nonreacting flows, detailed studies of applicability of such models in reacting flows are incomplete. Among the various alternatives, probability density function (PDF) methods have been found very favorable in applications to reacting flows. In the following subsections, we summarize some of the representative probability density function approaches used along with two-equation models.

22.4.1 FAVRE-AVERAGED DIFFUSION FLAMES

In reacting flows, the temperature of the products is higher than that of the reactants since the chemical reactions are exothermic. This trend is more prominent in turbulent flows due to the possibility of more enhanced mixing, leading to inhomogeneous density

Table 22.3.1 Reaction Mechanisms and Rate Constants for H-N-O

(a) System A

Reaction	Rate Constants		
	B	*N*	*E*, kcal/mole
$CO+OH=CO_2+H$	11.49	0	0.596
$H+O_2+OH$	14.34	0	16.492
$H_2+O=H+OH$	13.48	0	9.339
$H_2O+O=OH+OH$	13.92	0	18.121
$H+H_2O=H_2+OH$	14.0	0	19.870
$N+O_2=NO+O$	9.81	1.0	6.M
$N_2+O=N+NO$	13.95	0	75.506
$NO+M=N+O+M$	20.60	−1.5	149.025
$H+H+M=H_2+M$	18.00	−1.0	0
$O+O+M=O_2+M$	18.14	−1.0	.34
$H+OH+M=H_2O+M$	23.88	−7.6	0
$H_2+O_2=OH+OH$	13.00	0	43.0

(b) System B

Reaction	Rate Constants		
	B	*N*	*E*, kcal/mole
$H+O_2=OH+O$	14.342	0	16.790
$O+H_2=OH+H$	10.255	1.0	8.900
$H_2+OH=H_2O+H$	13.716	0	6.500
$OH+OH=O+H_2O$	12.799	0	1.093
$H+O_2+M=HO_2+M$	15.176	0	−1.000
$O+O+M=O_2+M$	13.756	0	−1.788
$H+H+M=H_2+M$	17.919	−1.0	0
$H+OH+M=H_2O+M$	21.924	−2.0	0
$H_2+HO_2=H_2O+OH$	11.857	0	18.700
$H_2O_2+M=OH+OH+M$	17.068	0	45.500
$H_2+O_2=OH+OH$	13.000	0	43.000
$H+HO_2=OH+OH$	14.398	0	1.900
$O+HO_2=OH+O_2$	13.699	0	1.000
$OH+HO_2=H_2O+O_2$	13.699	0	1.000
$HO_2+HO_2=H_2O_2+O_2$	12.255	0	0
$OH+H_2O_2=H_2O+HO_2$	13.000	0	1.800
$O+HM=OH+HO_2$	13.903	0	1.000
$H+H_2O_2=H_2O+OH$	14.505	0	9.000
$HO_2+NO=NO_2+OH$	13.079	0	2.390
$O+NO_2=NO+O_2$	13.000	0	.596
$NO+O+M=NO_2+M$	15.750	0	−1.160
$NO_2+H=NO+OH$	14.462	0	.795
$N+O_2=NO+O$	9.806	1.0	6.250
$O+N2=NO+N$	14.255	0	76.250
$N+OH=NO+H$	13.602	0	0
$N2O+M=N_2+O+M$	14.152	0	51.280
$O+N2O=N_2+O_2$	13.794	0	24.520
$O+N2O=NO+NO$	13.491	0	21.8W
$N+NO_2=NO+NO$	12.556	0	0
$OH+N2=N2O+H$	12.505	0	80.280

distributions. Thus, the mass average (often known as Favre average) is particularly useful in turbulent reacting flows.

For completeness, we record the summary of the mass-average process below:

$$\tilde{v}_i(x) = \frac{\overline{\rho v_i(x)}}{\overline{\rho}} \tag{22.4.1}$$

in which the bar indicates the conventional time average, whereas the tilde denotes a mass-averaged quantity. Thus, the velocity v_i consists of

$$v_i(x) = \tilde{v}_i(x) + v_i''(x,t) \tag{22.4.2}$$

where the double prime denotes the fluctuations about the mass-averaged mean.

The mass-averaged conservation equations are given by

Continuity Equation

$$\frac{\partial \rho}{\partial t} + (\overline{\rho}\tilde{v}_i)_{,i} = 0 \tag{22.4.3}$$

Momentum Equation

$$\frac{\partial \overline{\rho v}_j}{\partial t} + \overline{\rho}\tilde{v}_{j,i}\tilde{v}_i + \overline{p}_{,j} + (\overline{\tau}_{ji} - \overline{\rho v_i'' v_j''})_{,i} = 0 \tag{22.4.4}$$

Energy Equation

$$\begin{aligned}\frac{\partial \overline{\rho}\tilde{h}}{\partial t} &+ (\overline{\rho}\tilde{h}\tilde{v}_i)_{,i} - \frac{\partial \overline{p}}{\partial t} - \tilde{v}_i\overline{p}_{,i} + \overline{v_i'' p_{,i}} - \overline{\sigma}_{ij}\tilde{v}_{j,i} - \overline{\sigma_{ij}v_{j,i}''} + (\overline{q}_i + \overline{\rho h'' v_i''})_{,i} \\ &-(\overline{\rho} c_p D\tilde{T}\tilde{Y}_{k,i})_{,i} - (\overline{\rho c_p D T'' Y_{k,i}''})_{,i} = -\sum_{k=1}^{N} \overline{h\rho_k\omega_k}\end{aligned} \tag{22.4.5}$$

Species

$$\frac{\partial}{\partial t}(\overline{\rho}\tilde{Y}_k) + (\overline{\rho}\tilde{v}_i\tilde{Y}_k)_{,i} - (\overline{\rho D Y_{k,i}} - \overline{\rho v_i'' Y_k})_{,i} = \overline{\omega}_k \tag{22.4.6}$$

Equation of State

$$\overline{p} = R_o \sum_{i=1}^{N} (\overline{\rho}\tilde{T}\tilde{Y}_k + \overline{\rho T'' Y_k})\frac{1}{W_k} \tag{22.4.7}$$

where $\tilde{h}$ and W_k are the static enthalpy and molecular weight, respectively.

A variety of closure models for reacting turbulent flows have been proposed. The most widely used approaches are the K–ε model and the Reynolds stress model (second order closure) written in terms of the Favre average as follows:

$$\overline{\rho}\frac{\partial K}{\partial t} + \overline{\rho}\tilde{v}_i\frac{\partial K}{\partial x_i} = \frac{\partial}{\partial x_i}\left[\left(\frac{\mu_t}{\sigma_K} + \mu\right)\frac{\partial K}{\partial x_i}\right] - \overline{\rho v_i'' v_j''}\frac{\partial \tilde{v}_i}{\partial x_j} + \frac{\mu_t}{\overline{\rho}^2}\frac{\partial \overline{\rho}}{\partial x_i}\frac{\partial \overline{p}}{\partial x_i} - \overline{\rho}\varepsilon \tag{22.4.8}$$

$$\overline{\rho}\frac{\partial \varepsilon}{\partial t} + \overline{\rho}\tilde{v}_i\frac{\partial \varepsilon}{\partial x_i} = \frac{\partial}{\partial x_i}\left[\left(\frac{\mu_t}{\sigma_\varepsilon} + \mu\right)\frac{\partial \varepsilon}{\partial x_i}\right] - c_1\frac{\varepsilon}{K}\left(\overline{\rho v_i'' v_j''}\frac{\partial \tilde{v}_i}{\partial x_j} + \frac{\mu_t}{\overline{\rho}^2}\frac{\partial \overline{\rho}}{\partial x_i}\frac{\partial \overline{p}}{\partial x_i}\right) - c_2\overline{\rho}\frac{\varepsilon^2}{K} \tag{22.4.9}$$

with

$$\bar{\rho}\overline{v_i''v_j''} = \frac{2}{3}\delta_{ij}\left(\rho K + \mu_t \frac{\partial \tilde{v}_i}{\partial x_i}\right) - \mu_t\left(\frac{\partial \tilde{v}_i}{\partial x_j} + \frac{\partial \tilde{v}_j}{\partial x_i}\right) \tag{22.4.10}$$

$$\bar{\rho}\overline{v_i''\phi''} = \frac{\mu_t}{\sigma_t}\frac{\partial \tilde{\phi}}{\partial x_i}$$

$$K = \frac{\overline{v_i''v_i''}}{2}, \qquad \varepsilon = \nu\overline{v_{i,j}''v_{i,j}''}, \qquad \mu_t = \frac{c_\mu \bar{\rho} K^2}{\varepsilon} \tag{22.4.11}$$

The Reynolds stress model calls for the following transport equations:

$$\begin{aligned}
\bar{\rho}\frac{\partial}{\partial t}(\overline{v_i''v_j''}) + \bar{\rho}\tilde{v}_k\frac{\partial}{\partial x_k}(\overline{v_i''v_j''}) = & -\frac{\partial}{\partial x_k}(\bar{\rho}\overline{v_i''v_j''v_k''}) - \overline{v_i''}\frac{\partial \bar{p}}{\partial x_j} - \overline{v_j''}\frac{\partial \bar{p}}{\partial x_i} \\
& -\left(\overline{v_j'\frac{\partial p'}{\partial x_i}} + \overline{v_j''\frac{\partial p'}{\partial x_j}}\right) - \bar{\rho}\overline{v_i''v_k''}\frac{\partial \tilde{v}_j}{\partial x_k} - \bar{\rho}\overline{v_i''v_k''}\frac{\partial \tilde{v}_j}{\partial x_k} \\
& -\bar{\rho}\overline{v_j''v_k''}\frac{\partial \tilde{v}_i}{\partial x_k} - \left(\overline{\tau_{ki}'\frac{\partial v_j''}{\partial x_k}} + \overline{\tau_{kj}'\frac{\partial v_i''}{\partial x_k}}\right)
\end{aligned} \tag{22.4.12}$$

$$\begin{aligned}
\bar{\rho}\frac{\partial}{\partial t}(\overline{v_i''\phi''}) + \bar{\rho}\tilde{v}_j\frac{\partial}{\partial x_j}(\overline{v_i''\phi''}) = & -\frac{\partial}{\partial x_j}(\bar{\rho}\overline{v_i''v_j''\phi''}) - \overline{\phi''}\frac{\partial \bar{p}}{\partial x_i} - \overline{\phi''\frac{\partial p'}{\partial x_i}} - \bar{\rho}\overline{v_j''\phi''}\frac{\partial \tilde{v}_i}{\partial x_j} \\
& -\bar{\rho}\overline{v_i''v_j''}\frac{\partial \tilde{\phi}}{\partial x_j} - \left(\overline{\tau_{ij}\frac{\partial \phi''}{\partial x_j}} + \overline{\phi_k'\frac{\partial v_i''}{\partial x_k}}\right) + \bar{\rho}\overline{v_i''}Q(\phi)
\end{aligned} \tag{22.4.13}$$

Here $Q(\phi)$ is the source or sink term and ϕ denotes any variable other than pressure ($\phi = T$, or $\phi = Y_k$, etc.).

22.4.2 PROBABILITY DENSITY FUNCTIONS

The mean reaction rate cannot be expressed in terms of mean concentrations. For diffusion type flames, it is convenient to assume fast reactions and an appropriate shape for the probability density distributions of a conserved scalar, known as the probability density function (PDF). This can be taken to be the mixture fraction defined as the mass fraction of fuel in both burned and unburned forms.

The PDF, $P(f, x_i)$, is usually described in terms of two parameters $\tilde{f}$ (mixture fraction) and $\tilde{g}$ (square of fluctuations of mixture fraction, $\tilde{f}''^2$),

$$\tilde{f} = \int_0^1 fP(f, x_i)df \tag{22.4.14}$$

$$\tilde{g} = \tilde{f}''^2 = \int_0^1 (f - \tilde{f})^2 P(f, x_i)df \tag{22.4.15}$$

which may be obtained by solving the partial differential equations

$$\rho\tilde{v}_j \frac{\partial \tilde{f}}{\partial x_j} = \frac{\partial}{\partial x_j}\left(\frac{\mu_t}{\sigma_t}\frac{\partial \tilde{f}}{\partial x_j}\right) \tag{22.4.16}$$

$$\rho\tilde{v}_j \frac{\partial \tilde{g}}{\partial x_j} = \frac{\partial}{\partial x_j}\left(\frac{\mu_t}{\sigma_t}\frac{\partial \tilde{g}}{\partial x_j}\right) + 2\frac{\mu_t}{\sigma_t}\left(\frac{\partial \tilde{f}}{\partial x_j}\frac{\partial \tilde{f}}{\partial x_j}\right) - C_D\frac{\overline{\rho}\varepsilon}{K}\tilde{g} \tag{22.4.17}$$

Various forms of probability density function [Bilger, 1980] include (1) double-delta or rectangular-wave variation of mixture fraction with time, (2) clipped Gaussian distribution, (3) intermittency function, (4) beta probability density function, and (5) joint PDF for mixture function and reaction progress variable.

The PDF is inapplicable to phenomena such as ignition and extinction where direct kinetic effects are important. Furthermore, the definition of the mixture fraction, f, is not suitable for premixed flames. For premixed flames, therefore, the mean reaction rate must be evaluated.

In physically controlled diffusion flames, it is assumed that the chemistry is sufficiently fast and intermediate species do not play a significant role. The reaction takes place in an irreversible, single step as follows:

Oxidizer + Fuel = Product

For fast chemistry and the one step irreversible reaction, there will be no oxidant present for mixtures richer than stoichiometric and no fuel present when the mixture is weaker than stoichiometric. Both will be zero when the mixture is stoichiometric.

For the physically controlled diffusion flames, the mixture composition can be related to one conserved scalar quantity. In a two-feed system, the mixture fraction is conserved under chemical reactions and is defined by

$$\xi = \frac{(sY_{fu} - Y_{ox}) + Y_{ox,A}}{sY_{fu,F} + Y_{ox,A}} \tag{22.4.18}$$

Here, Y_{fu} and Y_{ox} denote the mass fractions of fuel and oxidizer, respectively; s, the stoichiometric oxidant required to burn 1 kg fuel; the subscripts A and F, the air and fuel stream conditions at the inlet. At the location where $Y_{ox} = sY_{fu}$, combustion is complete and the mixture fraction is in stoichiometric condition.

$$\xi_{st} = \frac{Y_{ox,A}}{sY_{fu,F} + Y_{ox,A}} \tag{22.4.19}$$

The corresponding location is called the flame sheet. The assumption of chemical equilibrium is now made so that

$$0 \le \xi \le \xi_{st}, \qquad Y_{fu} = 0, \qquad Y_{ox} = Y_{ox,A}\frac{\xi_{st} - \xi}{\xi_{st}} \tag{22.4.20}$$

$$\xi_{st} \le \xi \le 1, \qquad Y_{ox} = 0, \qquad Y_{fu} = Y_{fu,F}\frac{\xi - \xi_{st}}{\xi_{st}} \tag{22.4.21}$$

The mass fraction of the products can be obtained by the mass conservation

$$Y_{pr} = 1.0 - (Y_{ox} + Y_{fu}) \tag{22.4.22}$$

For adiabatic operation of gaseous flames, the enthalpy is a conserved scalar, and for unit Lewis number, the instantaneous enthalpy (h) and thermochemical properties are related to the instantaneous value of the mixture fraction

$$h(\xi) = \xi h_F + (1-\xi)h_A \tag{22.4.23}$$

$$\int_0^T \overline{c}_p dT = h(\xi) - Y_{fu}H_{fu} \tag{22.4.24}$$

$$\overline{c}_p(\xi) = \sum_i Y_i(\xi)c_{pi}(\xi) \tag{22.4.25}$$

$$\rho(\xi) = \frac{M(\xi)P}{RT(\xi)} \tag{22.4.26}$$

$$\frac{1}{M(\xi)} = \frac{Y_{fu}(\xi)}{M_{fu}} + \frac{Y_{ox}(\xi)}{M_{ox}} + \frac{Y_{pr}(\xi)}{M_{pr}} \tag{22.4.27}$$

The density-weighted mean values (ϕ) of any property are evaluated by convoluting the property functions with a probability density function:

$$\tilde{\phi} = \int_0^1 \phi(\xi)P(\xi, x_i)d\xi \tag{22.4.28}$$

Let us consider, for example, two of many possible probability density function approaches: (1) double-delta PDF and (2) beta PDF, as described below.

Double-Delta PDF

$$P(\xi, x_i) = a\delta(\xi_-) + (1-a)\delta(\xi_+) \tag{22.4.29}$$

$$\xi_+ = f + \sqrt{g}, \qquad \xi_- = f - \sqrt{g} \tag{22.4.30}$$

$$\delta(\xi_-)|_{\xi_-<0} = \delta(0), \qquad \delta(\xi_+)|_{\xi_+>0} = \delta(1) \tag{22.4.31}$$

$$a = \begin{cases} 0.5 & \text{for } 0 < \xi_\pm < 1 \\ (1-f)/1 - f + g/(1-f) & \text{for } \xi_+ > 1 \\ g/[f(fg/f)] & \text{for } \xi_- < 0 \end{cases} \tag{22.4.32}$$

where $\delta(\xi)$ is the Dirac delta function.

Beta PDF

$$P(\xi, x_i) = \frac{\xi^{a-1}(1-\xi)^{b-1}}{\int_0^1 \xi^{a-1}(1-\xi)^{b-1}d\xi} \tag{22.4.33}$$

$$a = f\left[\frac{f(1-f)}{g} - 1\right] \tag{22.4.34}$$

$$b = (1-f)\left[\frac{f(1-f)}{g} - 1\right] \tag{22.4.35}$$

The fluctuations g must satisfy the following conditions

$$0 < g \leq f(1 \leq f) \tag{22.4.36}$$

The constraints of (22.4.36) imply

$$a \geq 0 \quad \text{and} \quad b \geq 0 \tag{22.4.37}$$

The integration of (22.4.28) with β-PDF can be performed using a standard procedure, but singularities $\xi - 0$ or 1 must be analytically removed before weighting any property with β-PDF.

There are many other options to the PDF methods such as the rectangular-wave variation of mixture fraction with time [Spalding, 1971; Khalil, Spalding, and Whitelaw, 1975], clipped Gaussian distribution [Lockwood and Naguib, 1975], and joint PDF for mixture fraction and reaction-progress variable [Janicka and Kollmann, 1980]. An excellent account of PDF approaches can be found in Pope [1985] and a review paper by Kollmann [1990].

Boundary Treatments and Numerical Solutions

The general boundary conditions for axisymmetric cylindrical coordinates are u or τ_x and v or τ_r, specified as

$$\tau_x = n_r\left(\frac{\partial \mathrm{v}}{\partial x} + \frac{\partial u}{\partial r}\right)\mu_t + n_x\left(2\frac{\partial u}{\partial x} - \frac{2}{3}\nabla\cdot\mathbf{v}\right)\mu_t \tag{22.4.38}$$

$$\tau_r = n_r\left(2\frac{\partial \mathrm{v}}{\partial r} - \frac{2}{3}\nabla\cdot\mathbf{v}\right)\mu_t + n_x\left(\frac{\partial \mathrm{v}}{\partial x} + \frac{\partial u}{\partial r}\right)\mu_t \tag{22.4.39}$$

where n_x and n_r are the direction cosines of the outward normal to the boundary Γ. For other scalar variables ϕ (i.e., K, ε, f, and g), general boundary conditions are simply ϕ or $\partial\phi/\partial n$ specified on Γ.

For the inlet boundaries of a coaxial jet, all variables (u, v, K, ε, f, g) are specified. The turbulent kinetic energy is specified by experimental data or reasonable profiles. Since no measurements are available for the length scale, the following expression is used for the calculation of the dissipation rate:

$$\varepsilon = \frac{c_\mu K^{\frac{3}{2}}}{0.03 D_h} \tag{22.4.40}$$

where D_h is the hydraulic diameter. The mixture fraction at the inlet stream is, by definition, $f_A = 0.0$, $f_F = 1.0$ and thus, the fluctuations (g) of the mixture fraction are, by definition, zero for the inlet of the oxidizer and fuel side. At outlet boundaries, traction-free boundary conditions ($\tau_x = \tau_r = 0$) or ($\tau_x = \mathrm{v} = 0$) are used with $\partial\phi/\partial n = 0$. At symmetry, the normal gradients of all scalar variables ($\partial\phi/\partial n$) are zero, and the radial velocity component (v) and tangential surface traction (τ_r) are zero.

The wall regions present several flow characteristics that distinguish them from the other regions of the flow, such as steep gradients and a relatively low level of turbulence. To account for flow phenomena in wall regions, the wall function method is commonly employed. In the context of finite elements, the wall function method can be implemented by assuming a constant shear stress up to a distance δ within the near-wall region of the flow. With this assumption, the shear stress is calculated by the modified log law

$$|\tau_w| = \frac{\rho \kappa u_\tau c_\mu^{\frac{1}{4}} K^{\frac{1}{2}}}{\ln E\delta^+} \tag{22.4.41}$$

$$\delta^+ = \frac{\rho \delta c_\mu^{\frac{1}{4}} K^{\frac{1}{2}}}{\mu} \tag{22.4.42}$$

in which u_τ and K are the potential values computed at the previous time step. Once the near-wall values of the shear stresses are evaluated, near-wall values of K and ε can be calculated from

$$K = \frac{|\tau_w/\rho|}{c_\mu^{\frac{1}{2}}} \tag{22.4.43}$$

$$\varepsilon = \frac{|\tau_w/\rho|^{\frac{3}{2}}}{\kappa\delta} \tag{22.4.44}$$

In finite element formulation, τ_w is used as a Neumann boundary condition for calculating the new tangential velocity component with the normal component being zero. The surface integral form for the wall function can be written as

$$\int_\Gamma \overset{*}{\Phi}_\alpha r \tau_w n_r d\Gamma = \int_\Gamma \overset{*}{\Phi}_\alpha r \frac{\rho \kappa u_\tau c_\mu^{\frac{1}{4}} K^{\frac{1}{2}}}{\ln E\delta^+} n_r d\Gamma \tag{22.4.45}$$

The near-wall heat flux is determined by

$$\dot{q}_w = \begin{cases} \dfrac{u c_p (T - T_w)}{\Pr \delta} & \text{for } \delta^+ < 11.6 \\ \dfrac{(T - T_w)\rho c_p c_\mu^{\frac{1}{4}} K^{\frac{1}{2}}}{\Pr_t \left[\dfrac{1}{\kappa} \ln(E\delta^+) + \dfrac{P(\Pr)}{\Pr} \right]} & \text{for } \delta^+ \geq 11.6 \end{cases} \tag{22.4.46}$$

where the function $P(\Pr)$ is of the form [Launder and Spalding, 1974],

$$\frac{P(\Pr)}{\Pr_T} = 9.24 \left[\left(\frac{\Pr}{\Pr_T} \right)^{\frac{3}{4}} - 1 \right] \left[1 + 0.28 \exp\left(-0.007 \frac{\Pr}{\Pr_T} \right) \right] \tag{22.4.47}$$

Here τ_w and $\dot{q}_w$ are specified as Neumann boundary conditions in the momentum and energy equations, respectively.

In turbulent reacting flows, the strong coupling between the velocity and pressure fields and the nonlinear stiff source terms in the turbulence equations have a dominant influence on the solution strategy. In treating the continuity and momentum equations, a coupled velocity-pressure formulation leads to an improvement of the solution convergence. Such a coupled solution eliminates the need for the transformation of the continuity equation into a pressure or pressure-correction equation as required in the sequential solution method. The coupled solution is relatively insensitive to Reynolds numbers, grid density, and grid aspect ratio. Other scalar transport equations (K, ε, f, and g) are solved sequentially. The stiff source terms in the K–ε turbulence equations are treated implicitly for numerical stability.

The overall solution procedure is outlined below:

(1) Guess the values of all variables.
(2) Calculate auxiliary variables such as temperature, density, etc., from the associated combustion model.
(3) Solve the coupled continuity and momentum equations.
(4) Solve the transport equations for other variables (K, ε, f, and g).

Treat the new values of the variables as improved guesses and return to Step 2 and repeat the process until convergence.

For solutions with the conservation form of the Navier-Stokes system of equations, it is necessary to obtain appropriate modeling for the energy and species equations. These topics are discussed in the next section.

22.4.3 MODELING FOR ENERGY AND SPECIES EQUATIONS IN REACTIVE FLOWS

Favre Averages

Additional governing equations for reacting turbulent flows include the energy equation and species equations in terms of Favre averages:

$$\frac{\partial \bar{\rho}\tilde{E}}{\partial t} + (\bar{\rho}\tilde{E}\tilde{v}_i)_{,i} = \left[\left(\frac{\bar{\mu}}{\mathrm{Pr}} + \frac{\bar{\mu}_T}{\mathrm{Pr}_T}\right)\tilde{H}_{,i} + \left(\bar{\mu} + \frac{\bar{\mu}_T}{\sigma_k}\right)\tilde{K}_{,i}\right]_{,i} + [(\bar{\tau}_{ij} + \tau^*_{ij} - \bar{p}\delta_{ij})\tilde{v}_j]_{,i} \tag{22.4.48}$$

$$\frac{\partial \bar{\rho}\tilde{Y}_k}{\partial t} + (\bar{\rho}\tilde{Y}_k\tilde{v}_i)_{,i} - \left[\left(\frac{\bar{\mu}}{Sc} + \frac{\bar{\mu}_T}{Sc_T}\right)\tilde{Y}_{k,i}\right]_{,i} = \bar{\omega}_k \tag{22.4.49}$$

in which the standard K–ε model is used. Additionally, we must model the reaction rate ω_k. To this end, we return to the law of mass action given by (22.2.2). Here, the forward reaction rate constant in (22.2.3) is modified to

$$k_{fi} = A_i(\tilde{T} + T'')^{\alpha_i} \exp\left(-\frac{\hat{T}_i}{\tilde{T} + T''}\right) \tag{22.4.50}$$

where $\hat{T}_i$ is the species activation temperature. Assuming that $\frac{T''}{\tilde{T}} < 1$ and expanding the sum $\tilde{T} + T''$ in series, we obtain the Favre averaged reaction rate constant,

$$\tilde{k}_{fi} = (1 + s)A_i\tilde{T}^{\alpha_i} \exp\left(-\frac{\hat{T}_i}{\tilde{T}}\right) \tag{22.4.51}$$

with

$$s = \left[(\alpha_i - 1)\left(\frac{\alpha_i}{2} + \frac{\hat{T}_i}{\tilde{T}}\right) + \frac{1}{2}\left(\frac{\hat{T}_i}{\tilde{T}}\right)^2\right]\frac{\widetilde{T''T''}}{\tilde{T}^2} \tag{22.4.52}$$

where the terms higher than second order in $T''/\tilde{T}$ are neglected.

Direct Stress Model

An alternative approach is to use the direct stress method in which we introduce the transport equation for the Reynolds (turbulent) heat flux in the form,

$$\bar{\rho}\frac{D}{Dt}(\widetilde{v''_i T''}) = A_i + B_i + C_i + D_i + \hat{D}_i \tag{22.4.53}$$

where A_i, B_i, C_i, D_i, and $\hat{D}_i$ denote production, dissipation (destruction), diffusion, pressure strain, and nonvanishing pressure gradient, respectively [Launder, Reece, and Rodi, 1975]:

$$\begin{aligned}
A_i &= -\bar{\rho}(\mathrm{v}''_m T''\bar{\mathrm{v}}_{i,m} + \widetilde{\mathrm{v}''_m \mathrm{v}''_i T''_{,m}}) \\
B_i &= \widetilde{\mathrm{v}''_i q''}/c_p = -C_q \frac{K}{\varepsilon} \widetilde{\mathrm{v}''_i \mathrm{v}''_m} q_{T,m} \\
C_i &= \left[C_T \bar{\rho} \frac{K}{\varepsilon} \widetilde{\mathrm{v}''_m \mathrm{v}''_n} (\widetilde{\mathrm{v}''_i T''})_{,n} \right]_{,m} \\
D_i &= \overline{p T''_{,i}} = -\overline{p T''}_{,i} - C_{1T} \bar{\rho} \frac{\varepsilon}{K} \widetilde{\mathrm{v}''_i T''} + C_{2T} \bar{\rho}\, \widetilde{\mathrm{v}''_m T''} \mathrm{v}''_{i,m} \\
\hat{D}_i &= -\overline{T''}\bar{p}_{,i}
\end{aligned} \qquad (22.4.54a,b,c,d,e)$$

with $C_q = 1$, $C_T = 0.15$, $C_{1T} = 3.0$, $C_{2T} = 3.3$ [Gibson and Launder, 1978; Launder et al., 1975]. The Favre averaged temperature is modeled as

$$\tilde{T}'' = \frac{\overline{\rho T''}}{\bar{\rho}} = \frac{\overline{(\rho T'') T''}}{\bar{\rho}\bar{T}} = \frac{\tilde{T}''^2}{\bar{T}} \qquad (22.4.55)$$

where the Favre mean temperature fluctuation can be determined from the transport equation,

$$\bar{\rho} \frac{D\tilde{T}''^2}{Dt} = \left[C_T \bar{\rho} \frac{K}{\varepsilon} \widetilde{\mathrm{v}''_m \mathrm{v}''_n} (\tilde{T}''^2)_{,n} \right]_{,m} - 2\bar{\rho}\, \widetilde{\mathrm{v}''_m T''}\, \bar{T}_{,m} + 2\widetilde{T'' q''}/c_p - C_L \frac{\varepsilon}{K} \tilde{T}''^2 \qquad (22.4.56)$$

with $C_L = 2$. It should be noted that all unknowns have been defined (correlated) except for q_T in (22.4.54b) and $\widetilde{T'' q''}$ in (22.4.56). They can be correlated with the laminar flamelet model and thermochemical approach [Bray, 1979; Bradley et al., 1990; Al-Masseeh et al., 1990] as follows:

$$q_T = erf\left(\frac{s_q}{\bar{s}\pi^{0.5}} \right) \int_0^1 q_l(\theta) P(\theta) d\theta \qquad (22.4.57)$$

$$\widetilde{T'' q''} = erf\left(\frac{s_q}{\bar{s}\pi^{0.5}} \right) (T_b - T_u) \int_0^1 (\bar{\theta} - \theta) q_l(\theta) P(\theta) d\theta \qquad (22.4.58)$$

Here, s_q is the critical flame quenching value, $\theta = (T - T_u)/(T_b - T_u)$ is the dimensionless reaction progress variable with the subscripts b and u implying fully burned and unburned gaseous temperatures, $q_l(\theta)$ is the heat release rate for a one-dimensional laminar flame, $\bar{s}$ is the mean strain rate acting on the flamelets, and $P(\theta)$ is the Gaussian PDF of the reacting progress variable. Further details are given in Al-Masseeh et al. [1990].

22.4.4 SGS COMBUSTION MODELS FOR LES

For applications of LES in combustion, we may consider two approaches: the conserved scalar method discussed in Section 22.4.2 and the direct closure method [Bilger, 1980].

Here, we consider an exothermic, single-step, irreversible chemical reaction of the type $A + rB \rightarrow (1+r)P$ where r represents the stoichiometric ratio of oxidizer to fuel mass.

Derivation of the direct closure models begins with the reaction rate for the kth species, ω_k, appearing in (22.2.14). The spatially filtered reaction rate is of the form.

$$\overline{\omega}_k = \overline{\omega_k(\rho, T, Y_1, Y_2, \ldots Y_n)} \tag{22.4.59}$$

which may be decomposed in two different ways.

$$\begin{aligned} \overline{\omega}_{k1} &= \overline{\omega_k(\overline{\rho}, \tilde{T}, \tilde{Y}_1, \tilde{Y}_2, \ldots \tilde{Y}_n)} + \omega_{SGS1} \\ \overline{\omega}_{k2} &= \omega_k(\overline{\rho}, \tilde{T}, \tilde{Y}_1, \tilde{Y}_2, \ldots \tilde{Y}_n) + \omega_{SGS2} \end{aligned} \tag{22.4.60}$$

with

$$\begin{aligned} \omega_{SGS1} &= \overline{\omega_k(\rho, T, Y_1, Y_2, \ldots Y_n)} - \overline{\omega_k(\overline{\rho}, \tilde{T}, \tilde{Y}_1, \tilde{Y}_2, \ldots \tilde{Y}_n)} \\ \omega_{SGS2} &= \overline{\omega_k(\rho, T, Y_1, Y_2, \ldots Y_n)} - \omega_k(\overline{\rho}, \tilde{T}, \tilde{Y}_1, \tilde{Y}_2, \ldots \tilde{Y}_n) \end{aligned} \tag{22.4.61}$$

The first decomposition breaks the filtered reaction rate into filtered large-scale and SGS contributions, whereas the second decomposition leads to resolved large-scale and SGS contributions, with ω_{SGS1} and ω_{SGS2} representing the contribution of SGS fluctuations, but requiring models. To this end, these terms are filtered again at the same filter level, resulting in

$$\begin{aligned} \overline{\overline{\omega}}_{k1} &= \overline{\overline{\omega_k(\overline{\rho}, \tilde{T}, \tilde{Y}_1, \tilde{Y}_2, \ldots \tilde{Y}_n)}} + \overline{\omega}_{SGS1} \\ \overline{\overline{\omega}}_{k2} &= \overline{\omega_k(\overline{\rho}, \tilde{T}, \tilde{Y}_1, \tilde{Y}_2, \ldots \tilde{Y}_n)} + \overline{\omega}_{SGS2} \end{aligned} \tag{22.4.62}$$

which may be expressed in terms of large-scale and SGS contributions to the twice-filtered reaction rate, using the same decomposition strategies as in (22.4.60) as follows:

$$\begin{aligned} \overline{\overline{\omega}}_{k1} &= \overline{\overline{\omega_k(\tilde{\tilde{\rho}}, \tilde{\tilde{T}}, \tilde{\tilde{Y}}_1, \tilde{\tilde{Y}}_2, \ldots \tilde{\tilde{Y}}_n)}} + \hat{\omega}_1 + \overline{\omega}_{SGS1} \\ \overline{\overline{\omega}}_{k2} &= \omega_k(\tilde{\tilde{\rho}}, \tilde{\tilde{T}}, \tilde{\tilde{Y}}_1, \tilde{\tilde{Y}}_2, \ldots \tilde{\tilde{Y}}_n) + \hat{\omega}_2 + \overline{\omega}_{SGS2} \end{aligned} \tag{22.4.63}$$

where

$$\begin{aligned} \hat{\omega}_1 &= \overline{\overline{\omega_k(\overline{\rho}, \tilde{T}, \tilde{Y}_1, \tilde{Y}_2, \ldots \tilde{Y}_n)}} - \overline{\overline{\omega_k(\overline{\overline{\rho}}, \tilde{\tilde{T}}, \tilde{\tilde{Y}}_1, \tilde{\tilde{Y}}_2, \ldots \tilde{\tilde{Y}}_n)}} \\ \hat{\omega}_2 &= \overline{\omega_k(\overline{\rho}, \tilde{T}, \tilde{Y}_1, \tilde{Y}_2, \ldots \tilde{Y}_n)} - \omega_k(\overline{\overline{\rho}}, \tilde{\tilde{T}}, \tilde{\tilde{Y}}_1, \tilde{\tilde{Y}}_2, \ldots \tilde{\tilde{Y}}_n) \end{aligned} \tag{22.4.64}$$

with $\tilde{\tilde{a}} = \overline{\overline{\rho}\tilde{a}}/\overline{\rho}$ for any variable a. Invoking scale similarity, we may express $\omega_{SGS1} = K_1\hat{\omega}_1$, $\omega_{SGS2} = K_2\hat{\omega}_2$ with K_1, K_2 as model coefficients. Thus, returning to (22.4.60), the so-called similarity filtered reaction rate model (SFRRM) and scale similarity resolved reaction rate model (SSRRRM) are given by, respectively,

$$\begin{aligned} (\overline{\omega}_k)_{SFRRM} &= \overline{\omega_k(\overline{\rho}, \tilde{T}, \tilde{Y}_1, \tilde{Y}_2, \ldots\ldots \tilde{Y}_n)} + K_1\hat{\omega}_{k1} \\ (\overline{\omega}_k)_{SSRRRM} &= \omega_k(\overline{\rho}, \tilde{T}, \tilde{Y}_1, \tilde{Y}_2, \ldots\ldots \tilde{Y}_n) + K_2\hat{\omega}_2 \end{aligned} \tag{22.4.65}$$

There are other options for SGS reaction rate modeling such as in Pope [1990], Moller, Lundgren, and Fureby [1996], Norris and Edward [1997], among others. Applications of these models will be demonstrated in Section 22.6.6.

22.5 HYPERSONIC REACTIVE FLOWS

22.5.1 GENERAL

Computations in hypersonic flows present new challenges. The reason for this is that, when the Mach number is higher than about 5, most or all variable gradients increase significantly close to the wall. Typical cases of external and internal flows are depicted in Figure 22.5.1. We are concerned with high pressure gradients, high entropy gradients, high velocity gradients, and high temperature gradients.

For the external flow (Figure 22.5.1a), high pressure gradients will result in thin shock layers on sharp nose and highly curved shock layers on a blunt nose. A possible merging with the viscous boundary layer will complicate calculations for high Mach numbers coupled with low Reynolds numbers. Across the shock wave, entropy increases sharply particularly at the nose, thus forming the entropy layer which flows downstream. In this process, strong vortical flows are generated, contributing to turbulence. In the vicinity of the wall, high velocity gradients are prevalent. This will cause turbulence microscale motions, resulting in high pressure and high skin friction on the wall. The

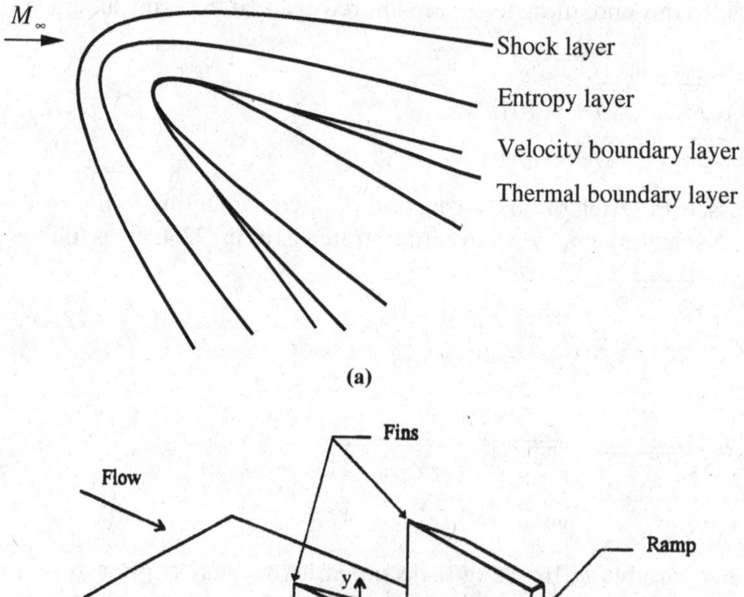

(a)

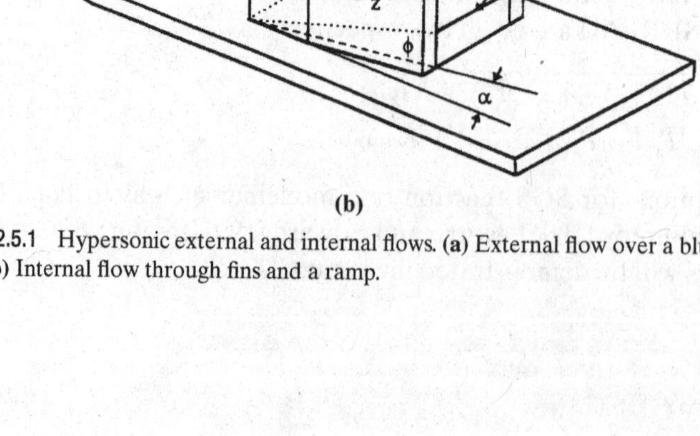

(b)

Figure 22.5.1 Hypersonic external and internal flows. (a) External flow over a blunt body. (b) Internal flow through fins and a ramp.

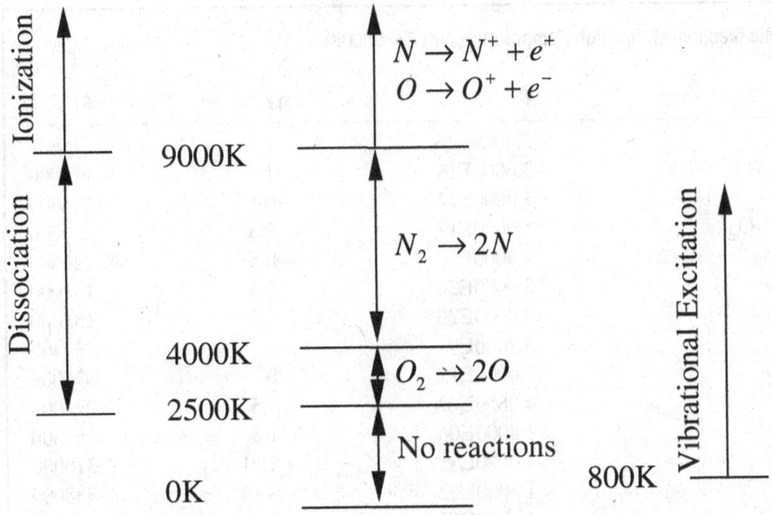

Figure 22.5.2 Ranges of vibrational excitation, dissociation, and ionization for air at 1 atm.

viscous boundary layer due to the high velocity gradient will grow as the Mach number increases. As the boundary layer moves closer to the entropy layer and shock layer, the so-called viscous interaction with inviscid regions leads to difficulties in obtaining accurate computational solutions.

For the internal flow (Figure 22.5.1b), high temperature gradients close to the wall lead to the rise of temperature due to viscous dissipation of energy. Most of the currently available CFD methods encounter difficulties in predicting the correct heat flux. Triple shock waves are formed with two fin shocks interacting with the ramp shock. In the vicinity of triple shock interactions, complex boundary layer separations and reattachments also cause numerical difficulties in predicting turbulence microscale behavior.

In case of a reentry vehicle, the kinetic energy of a high speed, hypersonic flow is dissipated due to friction, resulting in a thermal boundary layer with extremely high temperatures (Figure 22.5.2). This will excite vibrational energy within molecules and possibly cause dissociation and even ionization within the gas, leading to a chemically reacting boundary layer. For air at 1 atm, O_2 dissociation ($O_2 \rightarrow 2O$) begins at about 2000 K and the molecular oxygen is essentially entirely dissociated at 4000 K. At this temperature, N_2 dissociation ($N_2 \rightarrow 2N$) begins and is essentially totally dissociated at 9000 K. Above 9000 K, ionization takes place ($N \rightarrow N^+ + e^-$, $O \rightarrow O^+ + e^-$) and the gas becomes a partially ionized plasma. These high temperature gases are known as real gases. If the vibrational excitation and chemical reactions take place very rapidly in comparison with the flow diffusion velocity, then this is referred to as the equilibrium flow. If the opposite is true, then we have nonequilibrium flow, which is much more difficult in computations. High temperature chemically reacting flows influence lift, drag, and moments for a hypersonic aircraft and if the shock-layer temperature is very high, then heat transfer may be dominated by radiation. When ionization takes place, the free electrons absorb radio frequency waves, causing the communication blackout. Examples of chemical reaction equations are shown in Table 22.5.1.

Table 22.5.1 Kinetic Mechanism for High-Temperature Air ($T > 9000$K)

Reaction	C_f	η_f	k_f
$O_2 + N = 2O + N$	3.6000E18	−1	118800
$O_2 + NO = 2O + NO$	3.6000EI8	−1	118800
$N_2 + O = 2N + O$	1.9000EI7	−0.5	226000
$N_2 + NO = 2N + NO$	1.9000EI7	−0.5	226000
$N_2 + O_2 = 2N + O_2$	1.9000EI7	−0.5	226000
$NO + O_2 = N + O + O_2$	3.9000E20	−1.5	151000
$NO + N_2 = N + O + N_2$	3.9000E20	−1.5	151000
$O + NO = N + O_2$	3.2000E9	1	39400
$O + N_2 = N + NO$	7.0000EI3	0	76000
$N + N_2 = 2N + N$	4.0850E22	−1.5	226000
$O + N = NO^+ + e^-$	1.4000E06	1.5	63800
$O + e^- = O^+ + 2e^-$	3.6000E31	−2.91	316000
$N + e^- = N^+ + 2e^-$	1.1000E32	−3.14	338000
$O + O = O^+ + e^-$	1.6000EI7	−0.98	161600
$O + O_2^+ = O_2 + O^+$	2.9200EI8	−1.11	56000
$N_2 + N^+ = N + N_2^+$	2.0200E11	0.81	26000
$N + N = N_2^+ + e^-$	1.4000E13	0	135600
$O + NO^+ = NO + O^+$	3.6300EI5	−0.6	101600
$N_2 + O^+ = O + N_2^+$	3.4000EI9	−2	46000
$N + NO^+ = NO + N^+$	I.0000E19	−0.93	122000
$O_2 + NO^+ = NO + O_2^+$	1.8000EI5	0.17	66000
$O + NO^+ = O_2 + N^+$	1.3400EI3	0.31	154540
$O_2 + O = 2O + O$	9.0000EI9	−1	119000
$O_2 + O_2 = 2O + O_2$	3.2400E19	−1	119000
$O_2 + N_2 = 2O + N_2$	7.2000EI8	−1	119000
$N_2 + N_2 = 2N + N_2$	4.7000EI7	−0.5	226000
$NO + O = N + 2O$	7.8000E20	−1.5	151000
$NO + N = O + 2N$	7.8000E20	−1.5	151000
$NO + NO = N + O + NO$	7.8000E20	−1.5	151000
$O_2 + N_2 = NO + NO^+ + e^-$	1.3800E20	−1.84	282000
$NO + N_2 = NO^+ + e^- + N_2$	2.2000EI5	−0.35	216000

For high-altitude flights, about 150 km or above, the Knudsen number K_N is $K_N > 1$, where the continuum theory (Euler and Navier-Stokes system of equations) fails, and we must resort to the kinetic theory of gas (or free molecular flow theory). A hypersonic vehicle entering the atmosphere from space will experience the full range of these low-density effects.

22.5.2 VIBRATIONAL AND ELECTRONIC ENERGY IN NONEQUILIBRIUM

The statistical thermodynamics and kinetic theory of gases are used in derivations of the governing equations for hypersonic flows. The basic foundations are well established in the literature [Wilke, 1950; Hirschfelder et al., 1954; Brokaw, 1958; Lee, 1985; Park, 1990]. The Navier-Stokes system of equations governing the hypersonic flows includes not only the conservation of mass, momentum, and species, but also the conservation of vibrational energy and electronic energy. Thus, the conservation form of the Navier-Stokes system of equations is written as

$$\frac{\partial \mathbf{U}}{\partial t} + \frac{\partial \mathbf{F}_i}{\partial x_i} + \frac{\partial \mathbf{G}_i}{\partial x_i} = \mathbf{B} \tag{22.5.1}$$

$$\mathbf{U} = \begin{bmatrix} \rho \\ \rho \mathrm{v}_i \\ \rho E \\ \rho Y_k \\ \rho E_v \\ \rho E_e \end{bmatrix} \qquad \mathbf{F}_i = \begin{bmatrix} \rho \mathrm{v}_i \\ \rho \mathrm{v}_i \mathrm{v}_j + p\delta_{ij} \\ (\rho E + p)\mathrm{v}_i \\ \rho Y_k \mathrm{v}_i \\ \rho E_v \mathrm{v}_i \\ (\rho E_e + p_e)\mathrm{v}_i \end{bmatrix}$$

$$\mathbf{G}_i = \begin{bmatrix} 0 \\ -\tau_{ij} \\ -\tau_{ij}\mathrm{v}_j + q_i + \hat{q}_i \\ -\rho D_{km} Y_{ki} \\ q_{\mathrm{v}i} \\ q_{ei} \end{bmatrix} \qquad \mathbf{B} = \begin{bmatrix} 0 \\ 0 \\ -\sum_{k=1}^{N} H_k^0 \omega_k \\ \omega_k \\ \dot{E}_{\mathrm{v}} \\ \dot{E}_e \end{bmatrix} \tag{22.5.2}$$

with

$$q_i = k_h T_{,i} + k_e T_{e,i}$$

$$\hat{q}_i = \left[-\sum_k (\varepsilon_{rk} + \varepsilon_{\mathrm{v}k} + \varepsilon_{ek} + H_k) \sum_j N_j \mathbf{V}_j \right]_i$$

$$q_{\mathrm{v}i} = \left[-\sum_k \varepsilon_{\mathrm{v}k} \sum_j N_j \mathbf{V}_j \right]_i$$

$$q_{ei} = -f_e k_e T_{e,i} - \left[\sum_k \varepsilon_{ek} \sum_j N_j V_j \right]_i$$

$$E_{\mathrm{v}} = \sum_m \varepsilon_{\mathrm{v}m} N_m, \qquad E_e = \frac{3}{2} N_e k T_e + \sum_k \varepsilon_{ek}(T_e) N_k$$

$$E = E_1 + E_2 + E_3 + E_4 + E_5 + E_6 + E_7$$

$$\begin{aligned}
E_1 &= \frac{3}{2} kT \sum N_k && \textit{Translation (heavy particle)} \\
E_2 &= \frac{3}{2} kT_e N_e && \textit{Electron translation} \\
E_3 &= kT \sum N_k && \textit{Rotation (molecule)} \\
E_4 &= \sum \varepsilon_{vi}(T_v) N_k && \textit{Vibration (molecule)} \\
E_5 &= \sum \varepsilon_{ek}(T_e) N_k && \textit{Electronic excitation} \\
E_6 &= \sum H_k N_k && \textit{Chemical} \\
E_7 &= \frac{1}{2} \mathrm{v}_i \mathrm{v}_i && \textit{Kinetic}
\end{aligned} \tag{22.5.3}$$

where k is the Boltzmann constant, with the subscripts ν and e indicating the vibration and electronic energy, respectively, and i denoting the number of heavy particles or molecules as well as the coordinates x_i (for simplicity of notation). The rates of change of the vibrational and electronic energy in the source terms are given by

$$\dot{E}_v = \dot{E}_{v1} + \dot{E}_{v2} + \dot{E}_{v3}$$

$$\dot{E}_e = \dot{E}_{e1} + \dot{E}_{e2} + \dot{E}_{e3} + \dot{E}_{e4} + \dot{E}_{e5} + \dot{E}_{e6} + \dot{E}_{e7}$$

Vibrational relaxation energy rate, $\dot{E}_{v1} = \sum_{k=m} N_k f_v \left(\frac{\varepsilon_{vE} - \varepsilon_v}{\tau_L} \right)_k$, ε_{vE} = *equilibrium internal energy*

Vibrational N_2 *energy exchange rate,* $\dot{E}_{v2} = (N_2) \frac{\varepsilon_{vE}(T_e) - \varepsilon_v}{\tau_e}$, τ_e = *relaxation time*

Vibrational molecule energy exchange rate, $\dot{E}_{v3} = \sum_{k=m} \bar{\varepsilon}_{vk} \frac{\partial N_k}{\partial t}$, $\bar{\varepsilon}_{vk}$ = *average removed energy*

Electronic ionization energy exchange rate, $\dot{E}_{e1} = -\sum_{k=ion} E_{\infty k} \left(\frac{\partial N_k}{\partial t} \right)_+$ $E_{\infty k}$ = *ionization potential*

Electronic impact dissociation energy rate, $\dot{E}_{e2} = D(N_2) \left(\frac{\partial N_e}{\partial t} \right)_{de}$, D = *dissociation energy*

Electronic energy gain rate, $\dot{E}_{e3} = 2N_e \sum_{k=all} \nu_k \frac{m_e}{m_k} \frac{3}{2} k(T - T_e)$, ν_k = *collision frequency*

Electron–vibration energy exchange rate, $\dot{E}_{e4} = -(N_2) \frac{\varepsilon_{vE}(T_e) - \varepsilon_v}{\tau_e}$

Electronic excitation energy rate, $\dot{E}_{e5} = \sum_{k=all} \varepsilon_{ek} \frac{\partial N_k}{\partial t}$

Electronic associative ionization energy, $\dot{E}_{e6} = \sum_{kl} \varepsilon_k \frac{\partial N_k}{\partial t}$

Electronic radiative energy rate, $\dot{E}_{e7} = -Q_R$

The diffusion velocity $\mathbf{V}_i$ in (22.5.3) may be obtained by solving the multicomponent diffusion equation of the form

$$\nabla X_k = \sum_{j=1}^{n} \frac{X_k X_j}{D_{kj}} (\mathbf{V}_j - \mathbf{V}_k) + (Y_k - X_k) \frac{\nabla p}{p} + \frac{\rho}{p} \sum_{j=1}^{n} Y_k Y_j (\mathbf{f}_k - \mathbf{f}_j)$$
$$+ \sum_{j=1}^{n} \frac{X_k X_j}{\rho D_{kj}} \left(\frac{D_j}{Y_j} - \frac{D_k}{Y_k} \right) \frac{\nabla T}{T} \tag{22.5.4}$$

Thus, it can be shown that

$$\mathbf{V}_i = \frac{N}{\rho X_k} \sum_j m_j D_{kj} \left[\nabla X_j + \left(X_j - \frac{\rho_j}{\rho} \right) \frac{1}{p} \nabla p - \frac{N_j}{p} Z_j e E \right] - \frac{1}{\rho_k T} D_k \nabla T \tag{22.5.5}$$

where D_{kj} is the binary diffusion coefficient, D_k is the thermal diffusion, Z_j is the number of electrostatic charge ($= 0$ for neutral species, $= 1$ for positive ions, and $= -1$ for electrons), e is the electronic charge, and E is the electrostatic field intensity. A simplification of the diffusion velocity given in (22.5.5) leads to the Fick's first law of diffusion (22.2.9).

Following Vos [1963], the diffusion coefficients D_{kj} may be written as

$$D_{kj} = \frac{kT}{p\Delta_{kj}^{(\ell,s)}} \tag{22.5.6}$$

where

$$\Delta_{kj}^{(\ell,s)} = \frac{8}{3}\sqrt{\frac{2m_k m_j}{\pi kT(m_k + m_j)}}\pi\Omega_{kj}^{(\ell,s)}, \quad (\ell, s = 1) \tag{22.5.7a}$$

$$\pi\Omega_{kj}^{(\ell,s)} = \frac{\int_0^\infty \int_0^\pi e^{-\gamma^2}\gamma^{2s+3}(1 - \cos^\ell \chi)4\pi\sigma_{kj}\, d\chi\, d\gamma}{\int_0^\infty \int^\pi e^{-\gamma^2}\gamma^{2s+3}(1 - \cos^\ell \chi)\sin\chi\, d\chi\, d\gamma} \tag{22.5.7b}$$

with σ_{kj} and χ being the differential cross section and scattering angle, respectively, and

$$\gamma = \sqrt{\frac{m_k m_j}{2(m_k + m_j)kT}}g \tag{22.5.8}$$

where g is the relative velocity of the colliding particles.

The thermal conductivity k_h for heavy particles and k_e for electron energy are defined as

$$k_h = \frac{15}{4}k\sum_k \frac{X_k}{\sum_{j=1}^n \alpha_{kj} X_j \Delta_{kj}^{(\ell,s)}}, \quad (\ell, s = 2) \tag{22.5.9}$$

$$k_h = \frac{15}{4}k\sum_k \frac{X_e}{\sum_{j=1}^n \alpha_{ej} X_j \Delta_{kj}^{(\ell,s)}(T_e)}, \quad (\ell, s = 2) \tag{22.5.10}$$

with

$$\alpha_{kj} = 1 + \frac{(1 - m_k/m_j)(0.45 - 2.54 m_k/m_j)}{(1 + m_k/m_j)^2} \tag{22.5.11}$$

The viscosity constant μ associated with the stress tensor is given by Wilke [1950] as

$$\mu = \sum_k \frac{m_k X_k}{\sum_{j \neq k} X_j \Delta_{kj}^{(\ell,s)}} \tag{22.5.12}$$

with

$$\Delta_{kj}^{(\ell,s)} = \frac{16}{5}\sqrt{\frac{2m_k m_j}{\pi kT(m_k + m_j)}}\pi\Omega_{kj}^{(\ell,s)}, \quad (\ell, s = 2) \tag{22.5.13}$$

It is apparent that the inclusion of vibrational and electronic energy components will be computationally intensive.

The chemical species Y_k in (22.5.2) consist of

$$Y_k = [O, N, O_2, N_2, NO, O^+, N^+, O_2^+, N_2^+, NO^+, e^-]^T$$

with typical chemical reactions in air occurring in six different ways.

(1) Thermal dissociation of O_2

$$O_2 + M \Leftrightarrow O + O + M$$

(2) Dissociation of N_2

$$N_2 + e^- \Leftrightarrow N + N + e^-$$

(3) Exchange reactions of NO (known as Zeldovich reactions)

$$O + N_2 \Leftrightarrow NO + N$$

$$NO + O \Leftrightarrow O_2 + N$$

(4) Associative ionization, dissociative recombination

$$N + O \Leftrightarrow NO^+ + e^-$$

$$O + O \Leftrightarrow O_2^+ + e^-$$

$$N + N \Leftrightarrow N_2^+ + e^-$$

(5) Ionization of O

$$O + e^- \Leftrightarrow O^+ + e^- + e^- - e^-$$

(6) Exchange reactions

$$NO^+ + O \Leftrightarrow N^+ + O_2$$

With all ingredients that enter the most general form of the governing equation (22.5.1), the solution undergoes a laborious process. For applications to numerical simulations, we must provide adequate thermochemical models. They include the vibrational model, electronic excitation model, and chemical reaction model. Note that there are six different temperatures corresponding to six different energies shown in (22.5.3) with the kinetic energy excluded. Candler [1989] shows an illustration of effects of these temperatures upon the computational results for all other variables. Park and Yoon [1991] demonstrate the validity of using two temperatures (corresponding to translational and vibrational energies only). The thermochemical model in Park [1990] is described below.

Neglecting the ionizing phenomena, only five neutral species, $\rho_1 = O_2$, $\rho_2 = N$, $\rho_3 = NO$, $\rho_4 = O_2$, and $\rho_5 = N_2$, are considered. We further note that O_2 and N_2 can be expressed as a linear combination of other species from the elemental conservation condition. Thus, only the first three species can be treated as the species variables.

Vibrational Model

The vibrational energy is then given by

$$E_v = \sum_k n_k \varepsilon_{vk}, \quad (\mathrm{J/m^2}) \tag{22.5.14}$$

where

$$\varepsilon_{vk} = 8.314 \frac{\theta_k}{\exp[(\theta_k T_v - 1)]}, \quad \text{(J/mole)} \tag{22.5.15}$$

with θ_k being the characteristic vibrational temperature of the molecules, $\theta_k = 2740$, 2273, 3393 K for $k = 3, 4$, and 5, respectively.

The rate of change of the average vibrational energy of the molecules k by collisiosn with species j is of the form

$$\omega_{kj} = \frac{\varepsilon_{vE} - \varepsilon_{vk}}{\tau_{Lkj} + \tau_c} \left| \frac{T_s - T_v}{T_s - T_{vs}} \right|^{s-1} \quad \text{(J/mole·s)} \tag{22.5.16}$$

where ε_{vE} is the average vibrational energy of the species k per mole evaluated as the translational temperature. The quantity τ_{Lkj} is the vibrational relaxation time of the Landau-Teller model [Millikan and White, 1963],

$$\tau_{Lkj} = \exp(A_{kj} T^{-1/3} - B_{kj})/p_c \tag{22.5.17}$$

where p_c is the partial pressure of the colliding particles in atm. The quantity τ_c is the average collision time,

$$\tau_c = (c\bar{n}\sigma_v)^{-1}$$

where c is the average molecular speed $c = \sqrt{8KT/\pi m}$, $\bar{n}$ is the total number density of the mixture, and σ_v is the limiting cross section, $\sigma_v = 10^{-21}(50{,}000/T)^2$. T_s and T_{vs} are the translational temperature-rotational and vibrational-electronic temperatures immediately behind the shock wave, respectively. The exponent s is given by

$$s = 3.5 \exp(-T_s 5000)$$

The parameters A_{kj} and B_{kj} are adjusted for conformity with experimental data [Park and Yoon, 1990].

The rate of change of vibrational energy per unit volume of the flow in J/m^3 · s takes the form

$$\dot{E}_v = \sum_k \left[\sum_j \left(n_k \omega_{kj} - \bar{\varepsilon}_k \frac{\omega_k}{W_k} \right) \right] \tag{22.5.18}$$

where $\bar{\varepsilon}_k$ is the average vibrational energy removed in the dissociation of molecule k, approximately 80% of the dissociation of molecule k. In a rapidly expanding flow or in a boundary layer, this model may not be valid.

Electronic Excitation Model

The electronic excitation energy of the species is given by

$$E_e(n_k, T_v) = \sum_k n_k \varepsilon_{ek} \quad \text{(J/m}^3\text{)} \tag{22.5.19}$$

where the expression for the electronic energy ε_{ek} is given in Lee [1985]. The rate of change of electronic excitation energy of the flow is

$$\dot{E}_e = \sum_k \varepsilon_{ek} \frac{\omega_k}{W_k} \tag{22.5.20}$$

Chemical Reaction Model

With the vibrational and electronic energies calculated as described above, the translational-rotational temperature can be determined by the equation

$$E = \left(\sum_k c_{vk} n_k \right) T + E_v + E_e + \sum_k H_k^0 n_k + \frac{\rho}{2} v_i v_i \quad (\mathrm{J/m^3}) \tag{22.5.21}$$

where c_{vk} is the frozen specific heat at constant volume for species k for translational and rotational energies ($c_{v1} = c_{v2} = 12.47$ and $c_{v3} = c_{v4} = c_{v5} = 20.79$ J/mole). The quantity H_k^0 is the energy of formation of species k ($H_k^0 = 246.81, 470.70, 89.79, 0, 0$).

The average temperature [Park, 1990] is given by

$$T_a = \sqrt{T_v T} \tag{22.5.22}$$

The forward reaction rate coefficient for reaction j with the third body k is

$$k_{fjk} = C_{jk} T_a^{n_{jk}} \exp\left(-\frac{T_{djk}}{T_a} \right) \quad (\mathrm{mole/m^3{\cdot}s}) \tag{22.5.23}$$

where C_{jk} and n are the rate parameters (Table 22.5.2). The backward reaction rate coefficient is given by

$$k_{bjk} = k_{fjk} / K_{ej}(T_a) \tag{22.5.24}$$

The equilibrium constants K_{ej} are calculated using partition functions from the atomic and molecular constants [Park, 1990].

$$K_{ej} = \exp\left[a_1 z + a_2 + a_3 \ln\left(\frac{1}{z} \right) + \frac{a_4}{z} + \frac{a_5}{z^2} \right] \tag{22.5.25}$$

with $z = T_a/10{,}000$ and the coefficients a_i given in Table 22.5.3.

Table 22.5.2 Reaction Rate Parameters C_{jk}, n_{kj} and T_{jk} in (22.5.23)

j	k	Reaction	C_j m³/moles	n_j	T_{dj}, K
1	1	$O_2+O=O+O+O$	1.0×10^{16}	−1.5	59,500
2	2	$O_2+N=O+O+N$	1.0×10^{16}	−1.5	59,500
3	4	$O_2+NO=O+O+NO$	2.0×10^{15}	−1.5	59,500
4	4	$O_2+O_2=O+O+O_2$	1.0×10^{16}	−1.5	59,500
5	5	$O_2+N_2=O+O+N_2$	2.0×10^{15}	−1.5	59,500
6		$N_2+O=NO+N$	1.8×10^{8}	0.0	76,000
7		$NO+O=O_2+N$	2.2×10^{3}	1.0	19,500
8	1	$N_2+O=N+N+O$	3.0×10^{16}	−1.6	113,200
9	2	$N_2+N=N+N+N$	3.0×10^{16}	−1.6	113,200
10	3	$NO_2+NO=N+N+NO$	7.0×10^{15}	−1.6	113,200
11	4	$N_2+O_2=N+N+O_2$	7.0×10^{15}	−1.6	113,200
12	5	$N_2+N_2=N+N+N_2$	7.0×10^{15}	−1.6	113,200

Table 22.5.3 Coefficients a_j in (22.5.25)

j	a_1	a_2	a_3	a_4	a_5
1-5	0.55388	16.27551	1.77630	−6.5720	0.03144
6	0.97646	0.89043	0.74572	−3.9642	0.00712
7	0.004815	−1.7443	−1.2227	−0.95824	−0.045545
8-12	1.53510	15.4216	1.2993	−11.4940	−0.00698

In the following section, some selected example problems for various topics in reactive flows and combustion are presented.

22.6 EXAMPLE PROBLEMS

22.6.1 SUPERSONIC INVISCID REACTIVE FLOWS (PREMIXED HYDROGEN-AIR)

(1) Global Two-Step Model (Quasi-1-D and 2-D Analysis), Rapid Expansion Diffuser

Examples of combustion with hydrogen-air reactions are numerous. Among them are Janicka and Kollmann [1979], Evans and Schexnayder [1980], Rogers and Schexnayder [1981], Rogers and Chinitz [1983], Drummond, Hussaini, and Zang [1985], Kim [1987], and Chung, Kim, and Sohn [1987]. To illustrate the simplest cases of hydrogen-air combustion, we begin with a two-step global model of Rogers and Chinitz [1983],

$$H_2 + O_2 \underset{k_{b1}}{\overset{k_{f1}}{\rightleftarrows}} 2OH \tag{22.6.1a}$$

$$2OH + H_2 \underset{k_{b2}}{\overset{k_{f2}}{\rightleftarrows}} 2H_2O \tag{22.6.1b}$$

with

$$k_{f,k} = A_k(\phi)T^{N_i}\exp\left(-\frac{E_k}{R^\circ T}\right)$$

$$A_1(\phi) = (8.917\phi + 31.433/\phi - 28.95)\times 10^{47}\,(\text{cm}^3/\text{mole}\cdot\text{s})$$

$$E_1 = 4865\ \text{cal/mole}$$

$$N_1 = -10$$

$$A_2(\phi) = (2 + 1.333/\phi - 0.833\phi)\times 10^{64}\,(\text{cm}^6/\text{mole}^2\cdot\text{s})$$

$$E_2 = 42{,}500\ \text{cal/mole}$$

$$N_2 = -13$$

These data are for initial temperature of 1000–2000 K and equivalent ratio, $0.2 \leq \phi \leq 2$.

Using the law of mass action (22.2.2), we can construct from (22.6.1a,b), four nonlinear simultaneous ordinary differential equations of the form

$$\begin{aligned}
\omega_1/W_1 &= \frac{dC_1}{dt} = -aC_1C_2 + bC_3^2 - cC_1C_3^2 + dC_4^2 \\
\omega_2/W_2 &= \frac{dC_2}{dt} = -aC_1C_2 + bC_3^2 \\
\omega_3/W_3 &= \frac{dC_3}{dt} = 2aC_1C_2 - 2bC_3^2 - 2cC_1C_3^2 + 2dC_4^2 \\
\omega_4/W_4 &= \frac{dC_4}{dt} = 2cC_1C_3^2 - 2dC_4^2
\end{aligned} \qquad (22.6.2a,b,c,d)$$

with

$$C_1 = C_{H_2}, \quad C_2 = C_{O_2}, \quad C_3 = C_{OH}, \quad C_4 = C_{H_2O},$$
$$a = k_{f1}, \quad b = k_{b1}, \quad c = k_{f2}, \quad d = k_{b2}.$$

More complete models have been proposed by various investigators. For example, an eighteen-step model of Rogers and Schexnayder [1980] is shown in Table 22.6.1.1.

In what follows, we demonstrate quasi–one-dimensional calculations for the supersonic inviscid reactive flows in a diffuser, as shown in Figure 22.6.1.1a with the two-step global model (22.6.1a,b). The various approaches used in this analysis include: (1) Implicit Adams-Moulton finite differences [Drummond et al., 1985], (2) Spatial Chebyshev spectral method with the temporal Runge-Kutta iterations

Table 22.6.1.1 Combustion Mechanism for Eighteen-Step Hydrogen-Air

Reaction	*A* (moles)	*N* (cm³)	*E* (cal/gm-mole)
(1) $O_2 + H_2 = OH + OH$	1.70×10^{13}	0	48150
(2) $O_2 + H = OH + O$	1.42×10^{14}	0	16400
(3) $H_2 + OH = H_2O + H$	3.16×10^{7}	1.8	13750
(4) $H_2 + O = OH + H$	2.07×10^{14}	0	13750
(5) $OH + OH = H_2O + O$	5.50×10^{13}	0	7000
(6) $H + OH + M = H_2O + M$	2.21×10^{22}	−2.0	0
(7) $H + H + M = H_2 + M$	6.53×10^{17}	−1.0	0
(8) $H + O_2 + M = HO_2 + M$	3.20×10^{18}	−1.0	0
(9) $OH + HO_2 = O_2 + H_2O$	5.0×10^{13}	0	1000
(10) $H + HO_2 = H_2 + O_2$	2.53×10^{13}	0	700
(11) $H + HO_2 = OH + OH$	1.99×10^{14}	0	1800
(12) $O + HO_2 = O_2 + OH$	5.0×10^{13}	0	1000
(13) $HO_2 + HO_2 = O_2 + H_2O_2$	1.99×10^{12}	0	0
(14) $H_2 + HO_2 = H + H_2O_2$	3.01×10^{11}	0	18700
(15) $OH + H_2O_2 = H_2O + HO_2$	1.02×10^{13}	0	1900
(16) $H + H_2O_2 = H_2 + HO_2$	5.0×10^{14}	0	10000
(17) $O + H_2O_2 = OH + HO_2$	1.99×10^{13}	0	5900
(18) $H_2O_2 + M = OH + OH + M$	1.21×10^{17}	0	45500

Source: [Rogers and Schexnayder, 1981].

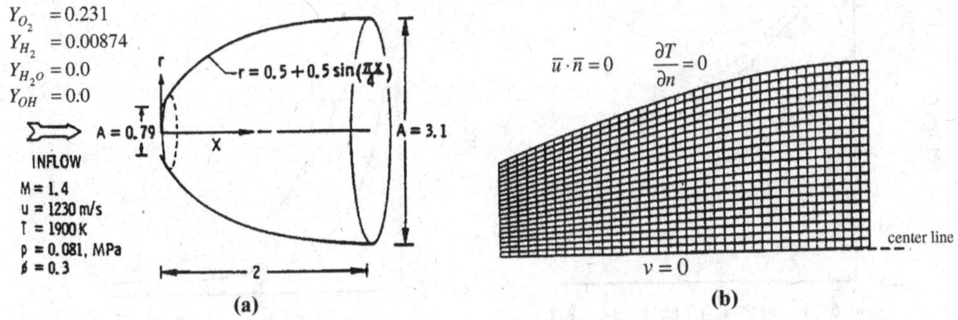

Figure 22.6.1.1 Global 2-step chemical reactions (H_2-air), rapid-expansion diffuser. (a) Rapid expansion supersonic diffuser for quasi–1-D analysis. (b) Upper half of (a) for 2-D analysis.

[Drummond et al., 1985], and (3) Operator splitting/point implicit Taylor-Galerkin method (Section 13.2.2) [Chung and Karr, 1980; Kim, 1987; Chung et al., 1987].

The governing equations are of the form

$$\frac{\partial \mathbf{U}}{\partial t} + \frac{\partial \mathbf{F}}{\partial x} = \mathbf{B} \tag{22.6.3}$$

$$\mathbf{U} = \begin{bmatrix} \rho A \\ \rho u A \\ \rho E A \\ \rho Y_k A \end{bmatrix} \qquad \mathbf{F} = \begin{bmatrix} \rho u A \\ \rho u^2 A + pA \\ \rho u H A \\ \rho u Y_k A \end{bmatrix} \qquad \mathbf{B} = \begin{bmatrix} 0 \\ p\dfrac{dA}{dx} \\ 0 \\ \omega_k A \end{bmatrix} \tag{22.6.4}$$

where A is the cross-sectional area as defined in Figure 22.6.1.1a with initial and boundary conditions. The thermodynamic model for the specific heat and the total enthalpy is as given in (22.2.22).

To compare the results of the quasi–one-dimensional analysis with those of two-dimensional analysis, we show the analysis using the operator splitting/point implicit Taylor-Galerkin method (see Section 13.2.2) with the discretization as shown in Figure 22.6.1.1b [C. S. Yoon, 1992]. In this case, we use the conservation form of the full Navier-Stokes system of equations (22.2.34) without the diffusion terms.

Although not shown, normal shocks are formed at the inlet, contrary to the nonreactive flows of a similar case shown in Figure 13.7.2. Due to chemical reactions, inlet normal shocks and high gradients of temperature, pressure, and mass fractions of all reactants and products are clearly evident in Figure 22.6.1.2. Approximately 2,000 iterations are required before convergence to the steady state. This is contrary to 1,000 iterations for the case of non-reacting flows demonstrated in Figure 13.7.2.

Our intention here is to compare the effect of quasi–one-dimensional analysis with the two-dimensional calculations and also to compare the results of the finite rate chemistry with those of equilibrium chemistry. The steady state quasi–1-D results of Drummond et al., [1985] and Kim [1987] with the finite rate chemistry are identical, both shown by the solid lines, whereas the 2-D results (along the center line) of Yoon [1992] (dash-dot-dash lines) show considerable differences. Both temperature and pressure are higher for the 2-D analysis, indicating the significant convection effects which

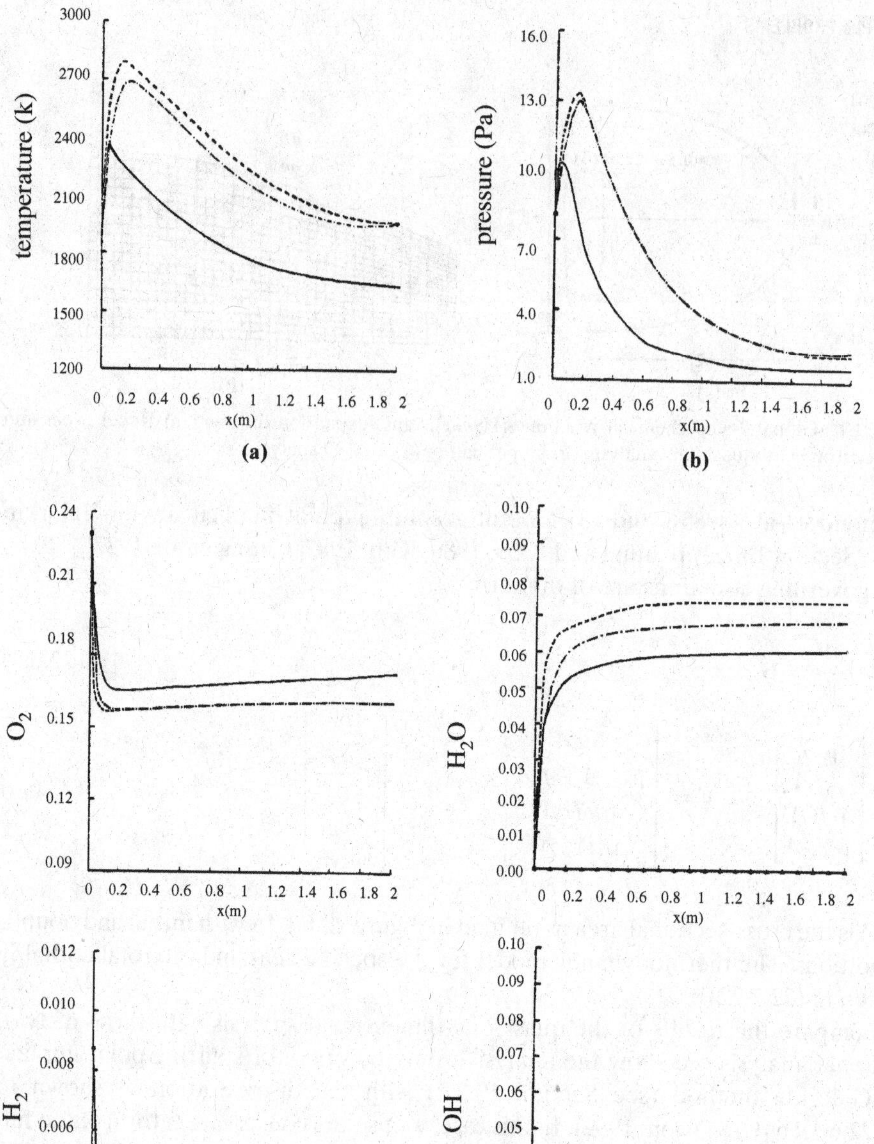

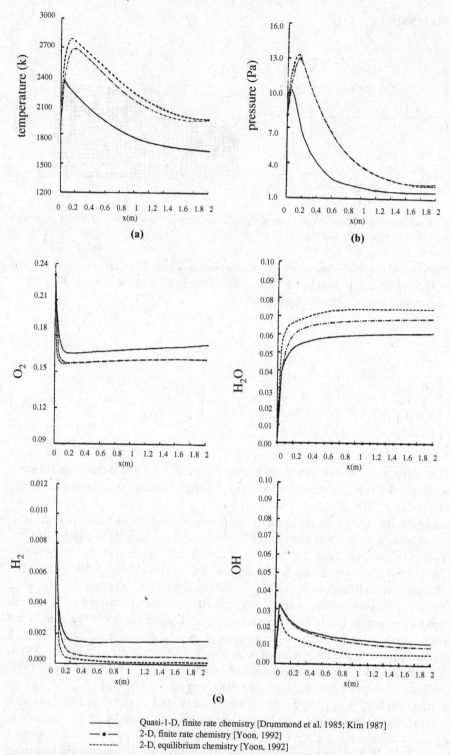

Figure 22.6.1.2 Hydrogen-air reactive supersonic inviscid flow, comparison between quasi–1-D and 2-D analyses, and comparison of finite rate chemistry with equilibrium chemistry and frozen chemistry for 2-D calculations [Drummond, 1985; Kim, 1987; Yoon, 1992]. **(a)** Axial temperature profile. **(b)** Axial pressure profile. **(c)** Axial mass fraction distributions.

promote the reaction process. This leads to a more rapid consumption of reactants (H_2, O_2), causing the product (H_2O) to be produced in a larger amount with (OH) remaining about the same as in the quasi–1-D simulation. The equilibrium solution shows that temperature is higher with an increase of H_2 consumption and H_2O production, resulting in a decrease in the radical (OH) dissociation. This trend shows the inadequacy of an equilibrium model in which the effect of convection and diffusion is absent.

(2) Comparison of Global Two-Step Model with Eighteen-Step Model, Ramjet Combustion
To investigate the effect of different reaction models, we examine in this example the comparison of the global two-step model with the eighteen-step model (Table 22.6.1.1) using the ramjet combustor (15° ramp) shown in Figure 22.6.1.3a [Yoon, 1992]. The supersonic inflow and outflow and adiabatic wall conditions are assumed.

Because of the inlet temperature of 900 K, which is less than the ignition temperature of 1,000 K, there should not be any reaction until the corner shock raises the temperature beyond this limit. Contour lines for temperature and mass fractions of various species clearly indicating corner shocks for the eighteen-step model are shown in

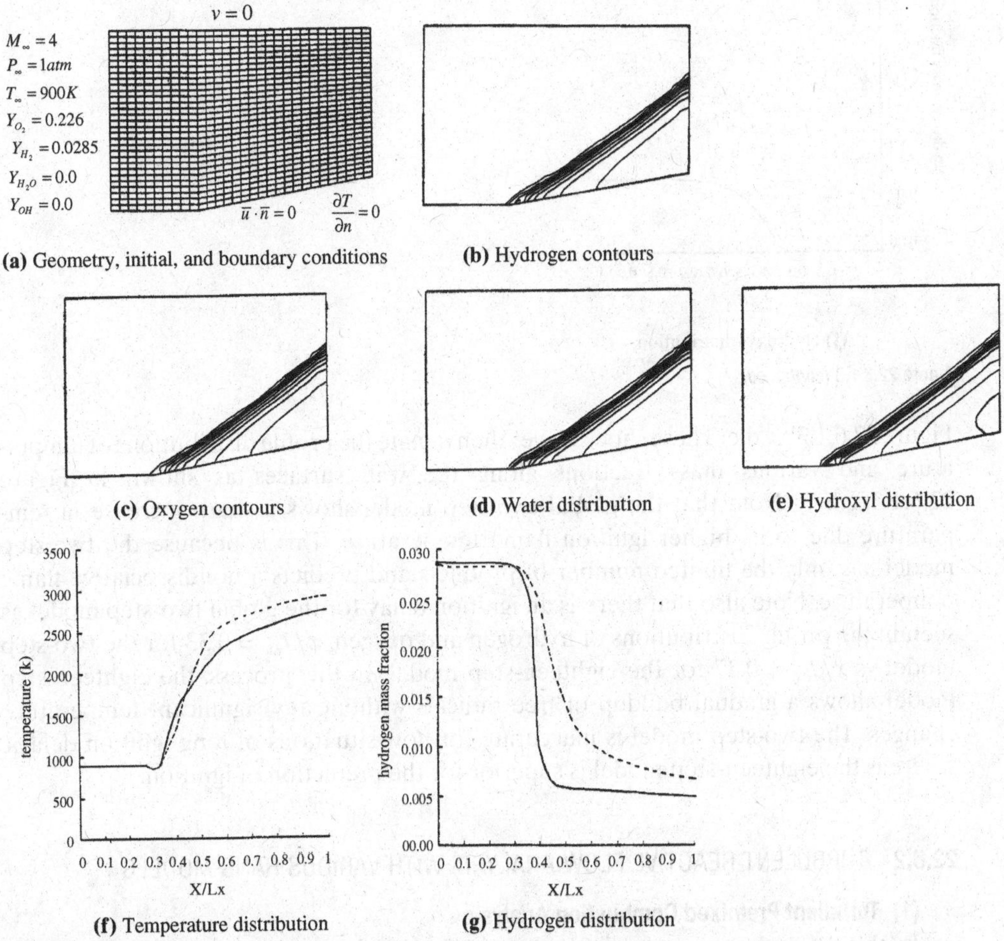

(a) Geometry, initial, and boundary conditions **(b)** Hydrogen contours

(c) Oxygen contours **(d)** Water distribution **(e)** Hydroxyl distribution

(f) Temperature distribution **(g)** Hydrogen distribution

Figure 22.6.1.3 Ramjet combustion (hydrogen-air reactions), comparison of the result of 18-step with 2-step reactions [Yoon, 1992], —— 18-step, ----- global 2-step.

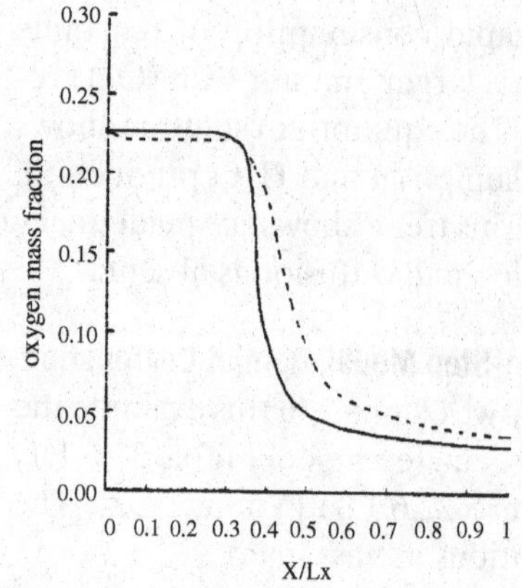

(h) Oxygen distribution

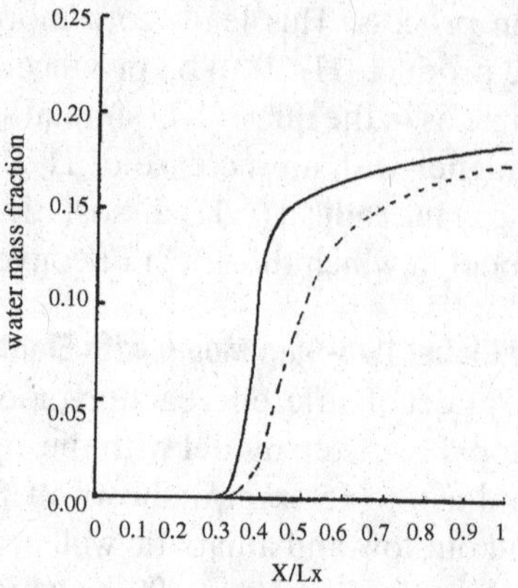

(i) Water distribution

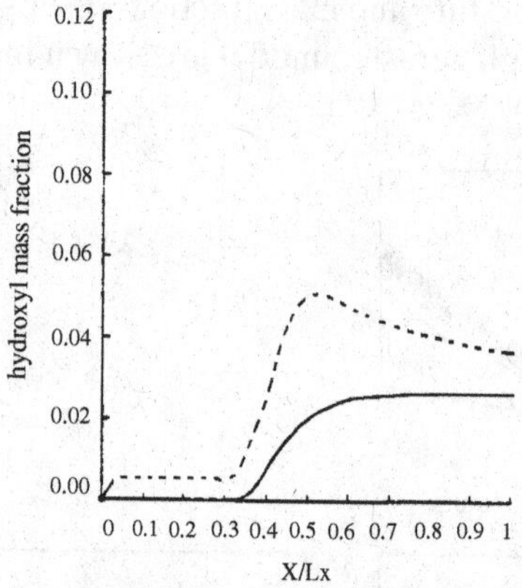

(j) Hydroxyl distribution

Figure 22.6.1.3 ***(continued).***

Figure 22.6.1.3b,c,d,e. These shock waves then dictate the profile distributions of temperature and various mass fractions along the wall surfaces as shown in Figure 22.6.1.3f,g,h,i,j. Note that the global two-step model shows a sharp increase in temperature due to its higher ignition flame temperature. This is because the two-step model has only the limited number of products and predicts a nondissociative flame temperature. Note also that there is an ignition delay for the global two-step model as seen in the profile distributions of hydrogen and oxygen, $x/L_x = 0.33$ for the two-step model vs $x/L_x = 0.27$ for the eighteen-step model. In this process, the eighteen-step model allows a gradual buildup of free radicals without any significant temperature changes. The two-step model is inaccurate for flow situations of long ignition delays, whereas the eighteen-step model is superior for the prediction of ignition.

22.6.2 TURBULENT REACTIVE FLOW ANALYSIS WITH VARIOUS RANS MODELS

(1) Turbulent Premixed Combustion Analysis

In this example, we examine the work of Al-Masseeh et al., [1990] in which the K–ε model and the direct stress model (Reynolds stress model) are compared with the experimental data for turbulent reactive flows (premixed CH_4^- air). The turbulent (Reynolds) heat flux transport equation and its related equations as shown in (22.4.53–22.4.58) in addition to the standard Reynolds stress transport equation are used. The geometry, initial, and boundary conditions are shown in Figure 22.6.2.1a. Using the SIMPLE algorithm (Section 5.3.1), the following results are obtained. Both nonswirling and swirling cases are included.

The temperature contours of nonswirling gases with an annular axial velocity of 60 m/s are shown in Figure 22.6.2.1b. Both $K-\varepsilon$ model and the direct stress model predict a conelike turbulent flame and the velocity vectors similar to the experimental data. The flame length temperature of 1700 K occurs at approximately 80 mm for all cases. However, the predicted temperature contours differ considerably from those of measured data with the predicted flame thickness being much thinner.

Figure 22.6.2.1c shows the temperature distributions for the lower inlet velocity of 30 m/s. Here, instead of the conelike flames, the direct stress model exhibits an annular jet flame as expected and confirmed in the experiment. This is not the case for the $K-\varepsilon$ model, which predicts a rather different field and a threshold velocity of about 24 m/s.

In the case of swirling flows (Figure 22.6.2.1d) (swirling numbers, $S = 0.53$ and $S = 0.69$) with the inlet velocity of 30 m/s, both sets of prediction are in better agreement with the experimental data. It is seen that the contours are substantially thicker and shorter for the swirling flames. Note that the $K-\varepsilon$ model overpredicts the flame thickness due to the higher turbulence dissipation rate in the $K-\varepsilon$ model solution and consequent increased strain rate, causing error function in the heat release rate expression in (22.4.57) to be less than that with the direct stress model.

(2) Turbulent Scramjet Flame Holder Combustion Analysis

The purpose of this example [W. S. Yoon, 1992; Yoon and Chung, 1991, 1992; Chung, 1993a,b] is to compare the results of the turbulent scramjet flame holder combustion with $K-\varepsilon$ model with those of laminar and inviscid flame. Calculations are carried out using the flowfield-dependent variation (FDV) method (Section 13.6). In this example all FDV parameters (s_1, s_2, s_3, s_4) are made independent of the flowfield (Mach number and Reynolds number) and set equal to 0.5. The geometry (10° ramp) and finite element discretization is shown in Figure 22.6.2.2a. The inlet initial and boundary conditions are:

$$\rho = 0.4437\ \text{kg/m}^3, \qquad p = 0.119\ \text{MPa}, \qquad T = 900\ \text{K}, \qquad M = 4,$$
$$Y_{O_2} = 0.2356, \qquad Y_{H_2} = 0.0029, \qquad Y_{OH} = Y_{H_2O} = 0,$$
$$\phi = 0.1, \qquad M = 4.0, \qquad Re = 106.$$

The calculated contours of the various variables are plotted in Figure 22.6.2.2b for the turbulent flow. The temperature and various species mass fraction distributions along the vertical direction at different axial locations are shown in Figure 22.6.2.2c,d. At an upstream position ($x = 0.4$), the inviscid flame remains constant along the vertical plane, whereas the laminar flame oscillates slightly in the vicinity of both upper and lower walls with H_2 and O_2 remaining still constant. For turbulence, the temperature and products close to the boundary edges rise sharply due to mixing. Somewhere downstream ($x = 2.5$) the trend rapidly changes for the inviscid flame. Temperature rises sharply toward the lower wall due to the shock wave interactions, causing chemical reactions predominantly at the lower wall. For the laminar flame, the viscous effects

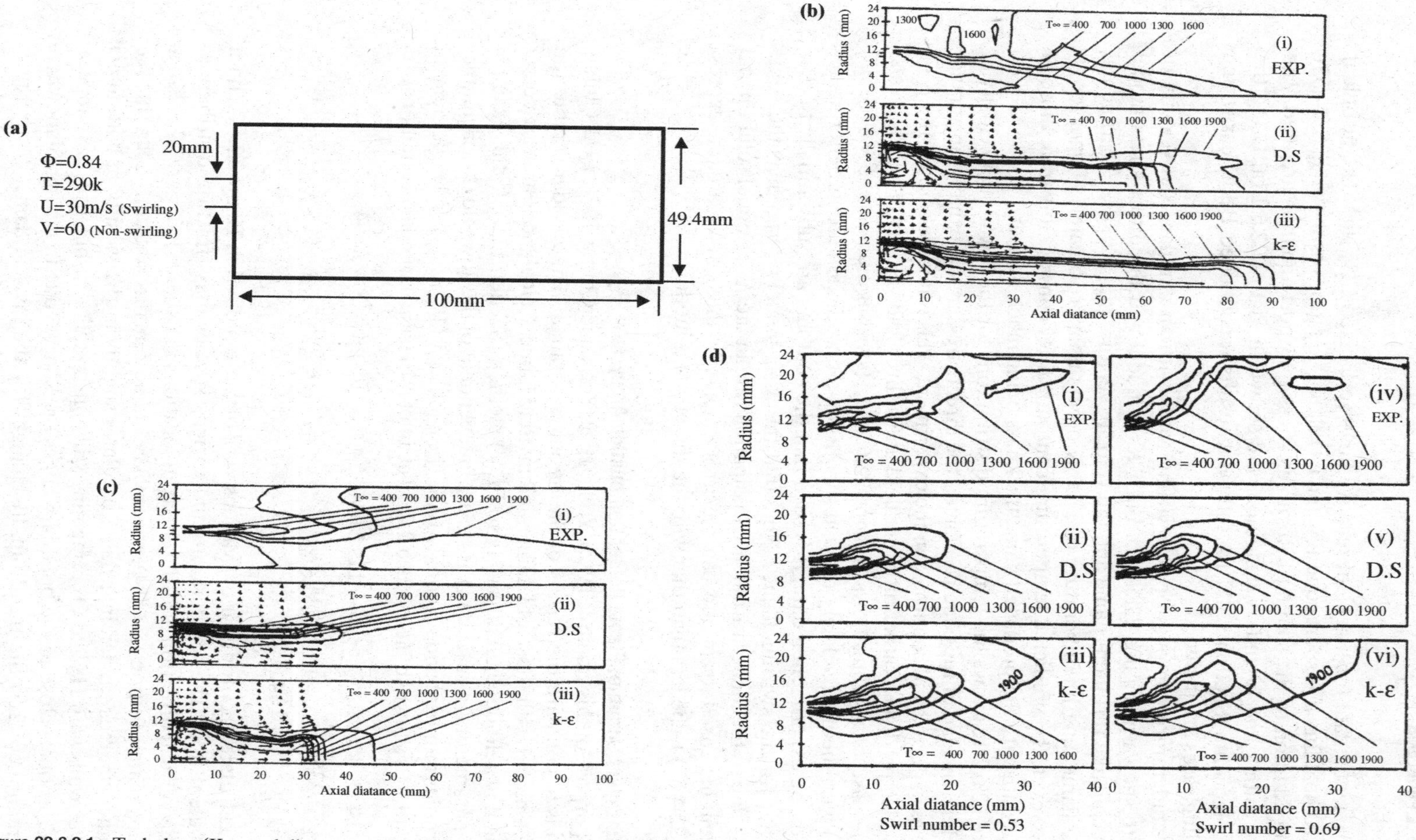

Figure 22.6.2.1 Turbulent (K–ε and direct stress model) premixed nonswirling and swirling combustion (CH_4-air) with strained flamelet model [Al-Masseeh et al., 1990]. **(a)** Geometry, initial, and boundary conditions (not to scale). **(b)** Temperature contours: (i) measured. (ii) direct stress and (iii) K–ε models. All for $S = 0$ and $U = 60$ m/s. **(c)** Temperature contours: (i) measured. (ii) direct stress and (iii) k–ε models. All for $S = 0$ and $U = 30$ m/s. **(d)** Temperature contours: (i) measured. (ii) direct stress and (iii) k–ε models. All for $S = 0.53$ and $U = 30$ m/s; (iv) measured. (v) direct stress and (vi) k–ε models. All for $S = 0.69$ and $U = 30$ m/s.

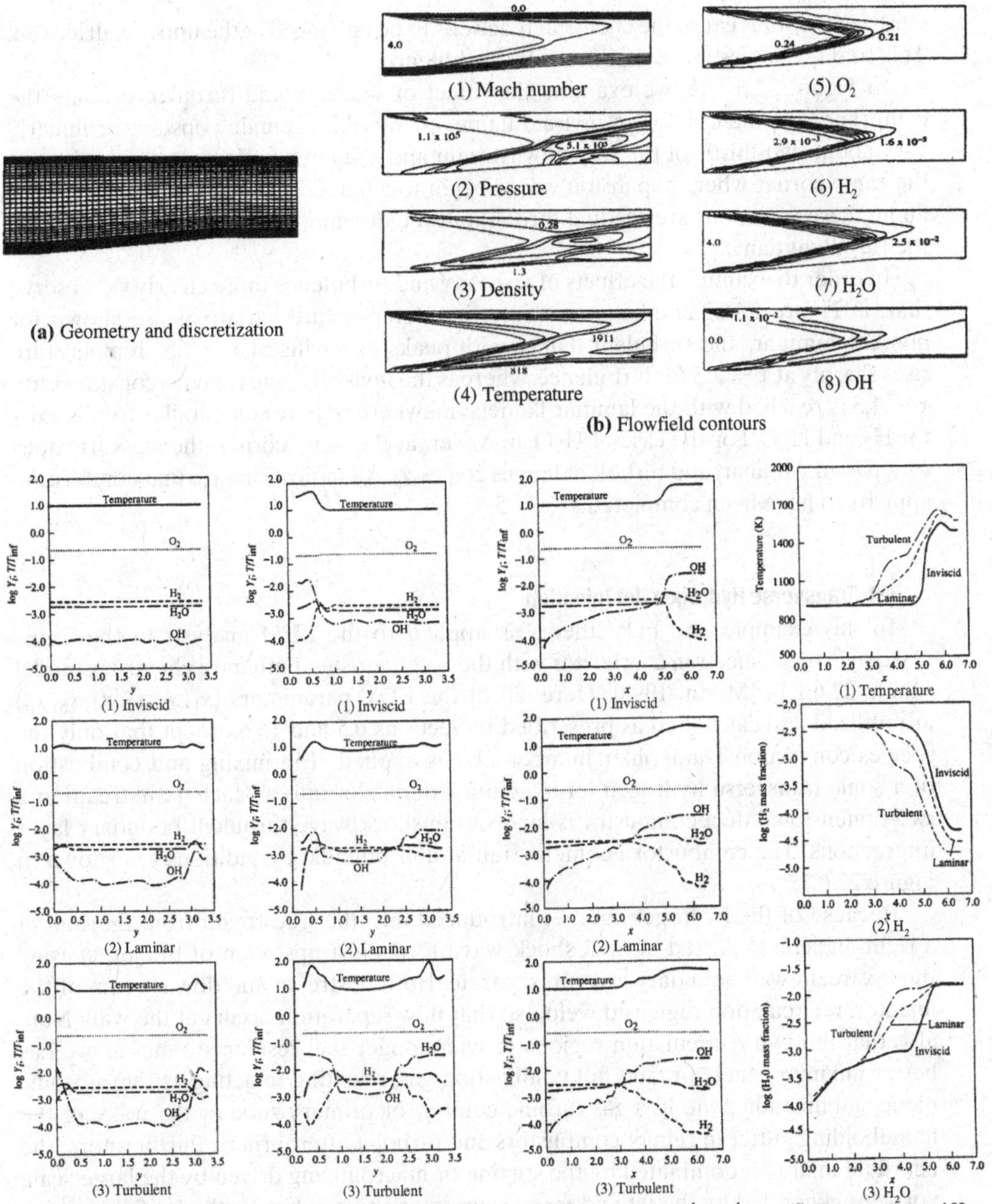

(c) Temperature and species mass fraction plots along the vertical direction (x=0.4)

(d) Temperature and species mass fraction plots along the vertical direction (x=2.5)

(e) Temperature and species mass fraction plots along the center line

(f) Temperature and H_2 and H_2O mass fraction plots along the center line for inviscid, laminar, and turbulent flows

Figure 22.6.2.2 Turbulent (k–ε model) scramjet flame holder combustion, comparison with inviscid and laminar flames [Yoon, 1992].

closer to the walls cause the chemical reactions to be enhanced at the upper wall as well. This trend becomes more significant for turbulence.

In Figure 22.6.2.2e, we examine the effect of viscosity and turbulence along the centerline. For the case of an inviscid flame, all variables remain constant or linearly vary about two-thirds of the way downstream and suddenly undergo perturbations at the ramp corner where expansion waves begin to emerge. In contrast, for the laminar flame these variations are gradual throughout the domain. This trend also prevails for the turbulent flame.

In order to examine the effects of viscosity and turbulence more clearly we observe that, in Figure 22.6.2.2f, along the centerline, temperature variations are shown for inviscid, laminar, and turbulent flames with peaks occurring at $x = 5.5$. Temperature rises sharply at $x = 2.5$ for turbulence, whereas the inviscid flame remains constant until $x = 4.5$ is reached with the laminar flame somewhere in between. Similar trends exist for H_2 and H_2O. For the case of H_2O, however, at the ramp corner, the mass fractions for inviscid, laminar, and turbulent flames coalesce. All indications are that combustion appears to have been completed at $x = 5.5$.

(3) Transverse Hydrogen Jet Injection

In this example, the FDV theory is applied to the FEM analysis to the transverse hydrogen injection combustor with the eighteen-step finite rate chemistry model (Table 22.6.1.1) [Moon, 1998]. Here, all of the FDV parameters $(s_1, s_2, s_3, s_4, s_5, s_6)$ are utilized and calculated as prescribed in Sections 6.5 and 13.6 except that only the species convection Damkohler number Da_I is applied. The mixing and combustion of a sonic transverse hydrogen jet injection from a slot into a Mach 4 airstream in a two-dimensional duct combustor is involved in shock wave turbulent boundary layer interactions. The combustor geometry, initial and boundary conditions are shown in Figure 22.6.2.3.

Because of the hydrogen fuel jet introduced into the freestream from the wall at a right angle, a detached normal shock wave forms just upstream of the jet, causing the upstream wall boundary layer to separate. Both upstream and downstream of the injector, recirculation regions develop so that flow separation occurs at the wall. Note also that the two recirculation regions provide longer fuel residence times as well as better mixing of fuel, air, and hot combustion gas, resulting in acting as the subsonic flame stabilization zone in a gas turbine combustor primary zone or the wake of the flameholding gutter in ramjet combustors and turbojet afterburners. Furthermore, the nearfield mixing is dominated by the stirring or macromixing driven by the large-scale vorticies generated by the jet and freestream interaction, whereas the far-field mixing depends on the small-scale turbulence within the plume and mixing layer.

The static pressure contours are presented in Figure 22.6.2.3b. The leading edge shock and the incidence and reflection of the bow shock to and from the symmetric plane of the duct can be seen. Velocity distributions in the vicinity of the injector are shown in Figure 22.6.2.3c. For clarity of presentation the velocity components for every other grid point are shown. Both recirculation zones and mixing layers can be identified. The mass fraction contours of H_2 and H_2O are shown in Figure 22.6.2.3d,e. It is noted that high reaction rate regions spread downstream along the mixing layer.

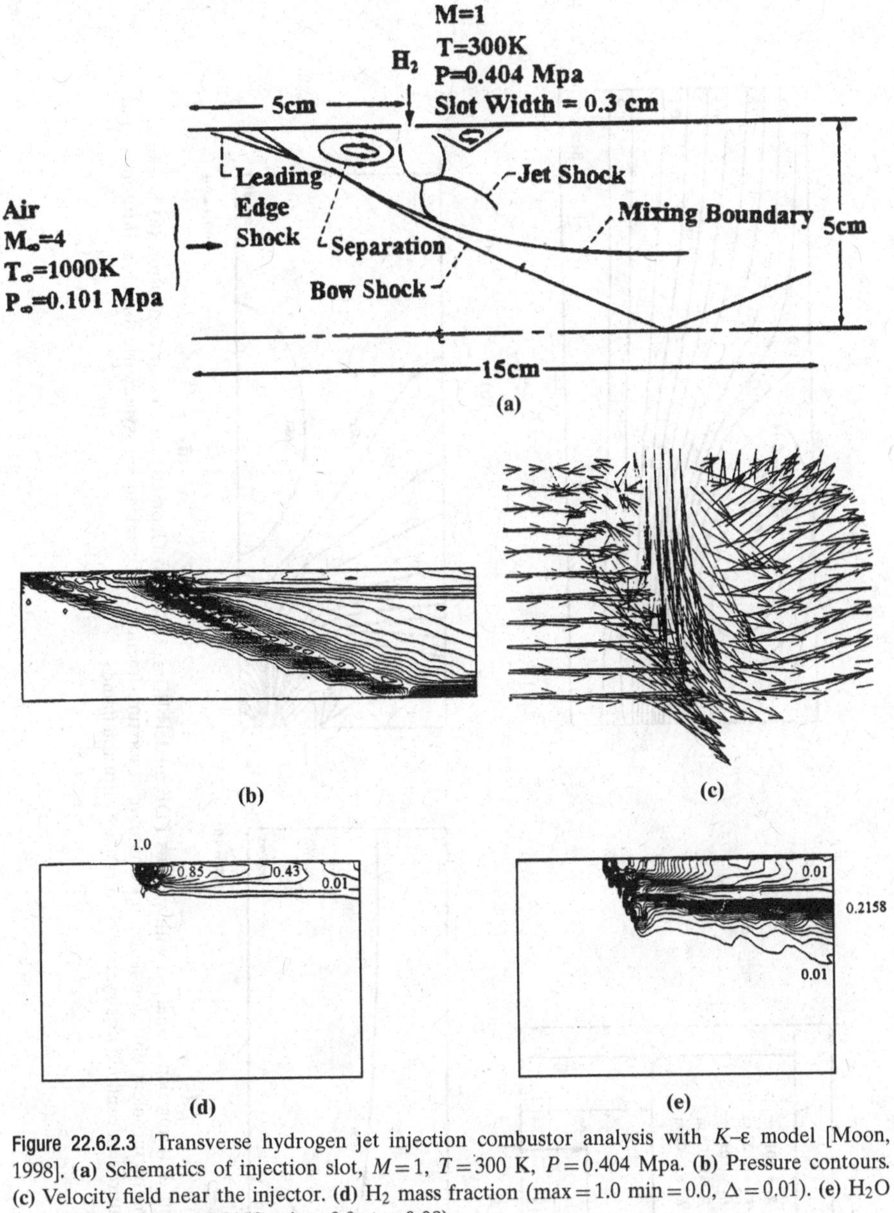

Figure 22.6.2.3 Transverse hydrogen jet injection combustor analysis with K–ε model [Moon, 1998]. **(a)** Schematics of injection slot, $M=1$, $T=300$ K, $P=0.404$ Mpa. **(b)** Pressure contours. **(c)** Velocity field near the injector. **(d)** H_2 mass fraction (max $=1.0$ min $=0.0$, $\Delta=0.01$). **(e)** H_2O mass fraction (max $=0.2158$, min $=0.0$, $\Delta=0.08$).

22.6.3 PDF MODELS FOR TURBULENT DIFFUSION COMBUSTION ANALYSIS

The use of PDF approach in combustion is widespread. In PDF applications, we employ the assumed-PDF approach. On the form of the assumed PDF, however, various choices are available such as the modeling of scalar mixing with mapping closure methods [Pope, 1985; Girimaji, 1991; Frolov et al., 1997], among others.

In this example, we demonstrate the PDF approach presented in Section 22.4.2 using the K–ε model with GPG-FEM [Kim, 1987]. The geometry of a coaxial combustor is shown in Figure 22.6.3.1a. The fuel properties and inlet conditions are: Stoichiometric

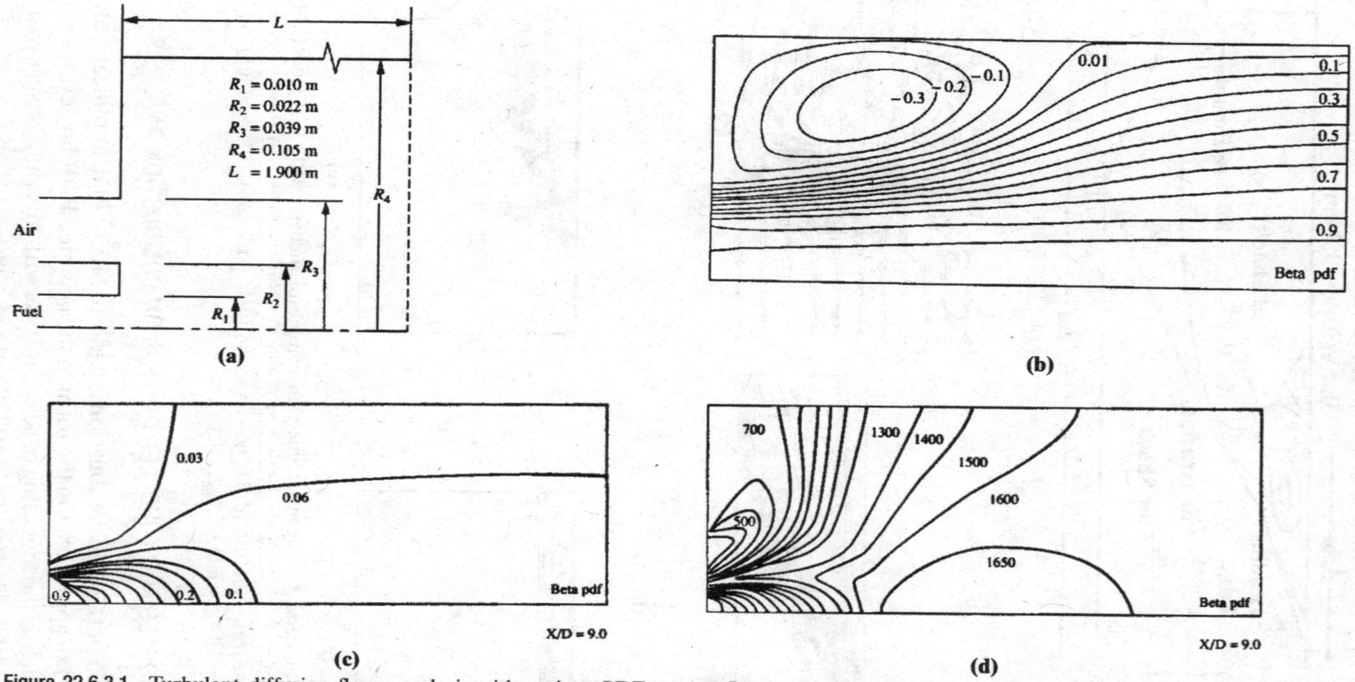

Figure 22.6.3.1 Turbulent diffusion flame analysis with various PDF models [Kim, 1987]. **(a)** Geometry of coaxial combustor. **(b)** Streamlines in a confined axis-symmetric turbulent diffusion flame. **(c)** Contours of mixture fraction in a confined axis-symmetric turbulent diffusion flame. **(d)** Contours of temperature in a confined axis-symmetric turbulent diffusion flame.

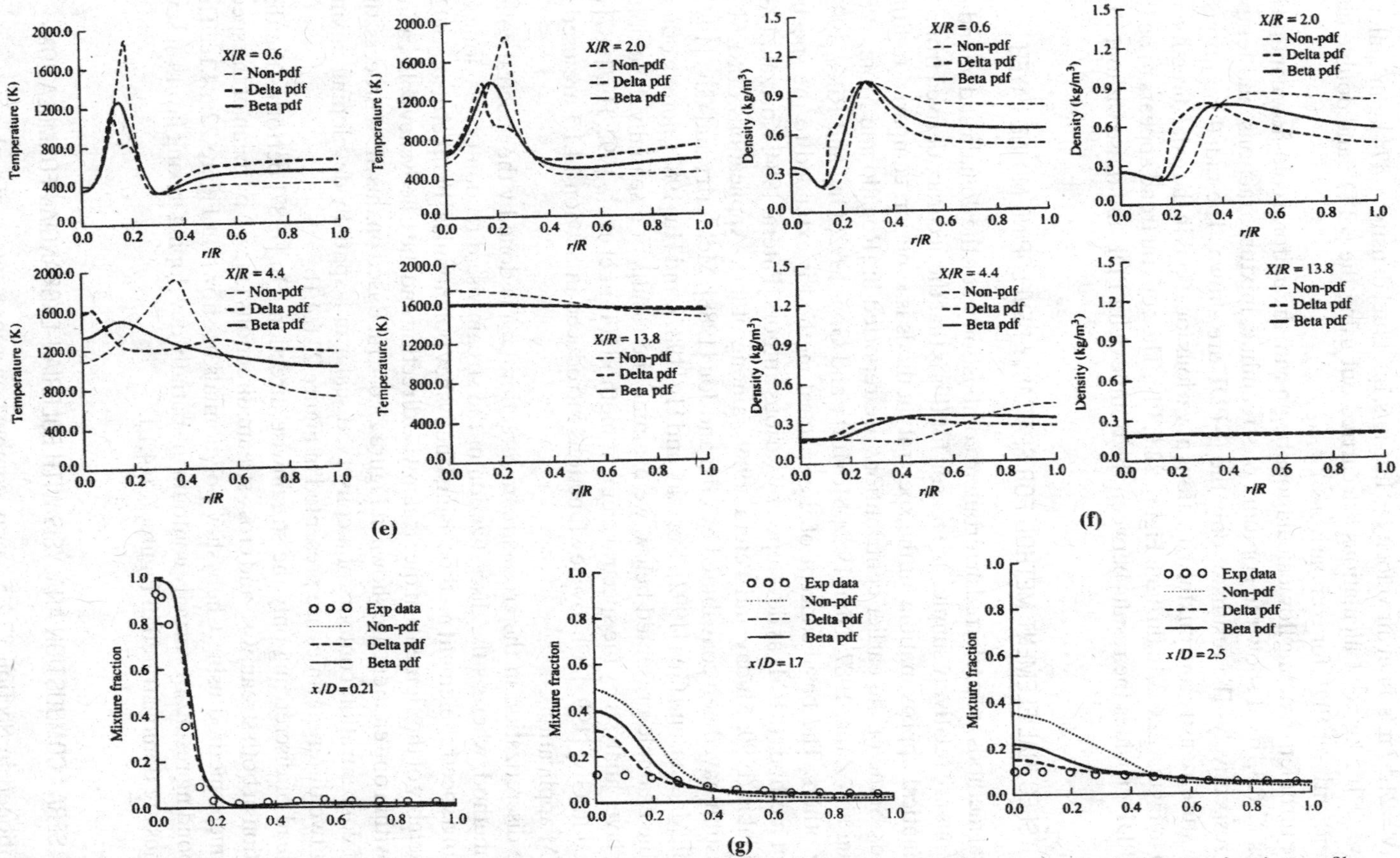

Figure 22.6.3.1 (*continued*) (e) Radial profiles of predicted mean temperature. (f) Radial profiles of predicted mean density. (g) Mixture fraction profiles at various x/D locations.

A/F ratio = 10.6, heat of reaction = 2.63×10^7 (J/kg), inlet A/F ratio = 15.75, inlet fuel velocity = 21.57 m/s, inlet air velocity = 1.46 m/s, inlet fuel density = .474 kg/m^3, inlet air density = 1.165 kg/m^3. Calculations are carried out using the β-PDF and double delta PDF and without PDF for comparisons.

The turbulent reacting flow calculations begin with the uniform cold-flow conditions. Figures 22.6.3.1b,c,d show the contours of streamline, mixture fractions, and temperature, respectively. The results of only the β-PDF are shown. The radial variations of temperature, density, and mixture fractions at various locations in the axial direction are shown in Figure 22.6.3.1e through Figure 22.6.3.1g. The general trend appears to be that the β-PDF provides the results between the double delta PDF and those without PDF.

22.6.4 SPECTRAL ELEMENT METHOD FOR SPATIALLY DEVELOPING MIXING LAYER

Spectral methods are preferred in turbulent combustion when the domain and boundary conditions are relatively simple. The reason for this is that the accuracy derived from the mathematical approximations in the spectral methods is superior, compared to other methods. Some of the earlier contributions are reported in [Rogallo and Moin, 1984; Hussaini and Zang, 1987; Givi, 1989; McMurty and Givi, 1992; Givi and Riley, 1992], among others. The basic concept of the spectral method is extended to the spectral element methods (SEM) as developed by various authors [Patera, 1984; Korczak, 1985; Karniadakis, 1990; Maday and Petera, 1989], among others. Applications of SEM to combustion have been contributed by Givi and Jou [1988], McMurtry and Givi [1992], Frankel, Madina, and Givi [1992], Korczak and Hu [1987], and Hu [1987], among others.

In the example presented below, we examine the results of a spatially developing mixing layer analysis by the spectral element method [Frankel et al., 1992; Hu, 1987] as reported in Givi [1993]. Chebyshev functions introduced in Section 14.1.1 are used in the SEM applications.

The discretization in the cross-stream direction (x_2) is done by the spectral collocation method, whereas the discretization in the streamwise direction (x_1) is done by means of a spectral-element method using Chebyshev polynomials [Frankel et al., 1992]. The assembly of the elements in the streamwise direction and the Chebyshev collocation points within one element are shown in Figure 22.6.4.1a. Based on this SEM process, the plots of concentration contours of a conserved scalar in a spatially developing mixing layer at two different times are presented in Figure 22.6.4.1b.

Instead of discretizing only the streamwise direction, Hu [1987] performs the discretization in both streamwise and cross-stream directions (x_1, x_2) by mean of the spectral element method using Chebyshev polynomials as shown in Figure 22.6.4.1c. The corresponding results of vorticity contours in spatially developing mixing layers at several times are demonstrated in Figure 22.6.4.1d.

22.6.5 SPRAY COMBUSTION ANALYSIS WITH EULERIAN-LAGRANGIAN FORMULATION

As mentioned in Section 22.2.5, spray combustion represents a two-phase flow and may be analyzed by either one of the three approaches: Eulerian-Eulerian, Eulerian-Lagrangian, and probabilistic formulation. From the computational point of view, the

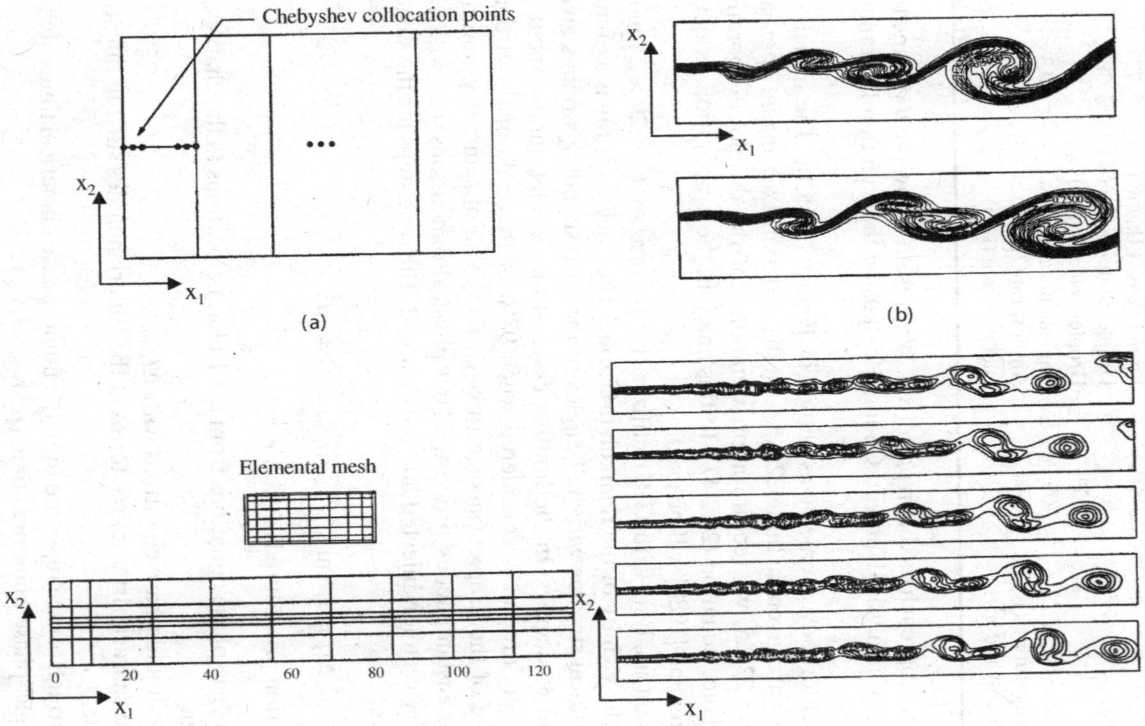

Figure 22.6.4.1 Spatially developing mixing layer analysis with spectral method [Givi, 1993]. **(a)** The assembly of the elements in the streamwise direction and the Chebyshev collocation points within one element [Frankel et al., 1992]. **(b)** Plots of concentration contours of a conserved scalar in a spatially developing mixing layer at two different times. **(c)** The assembly of the elements and the Chebyshev collocation points within each element [Hu, 1987]. **(d)** Plots of vorticity contours in spatially developing mixing layers at several times. The discretization in both streamwise and cross-stream directions (x_1, x_2) is done by means of the spectral-element method using Chebyshev polynomials [Hu, 1987].

Table 22.6.5.1 Initial Conditions Used in the Prediction

Gas-Phase Boundary Conditions		Liquid-Phase Initial Conditions	
Velocity (m/s)	30	Fuel	n-decane
Temperature (K)	1000	Liquid density (kg/m^3)	773
Pressure (atm)	10	Droplet temperature (K)	300
Density (kg/m^3)	3.399	Droplet velocity (m/s)	20
Duct wall temperature (K)	700	Equivalent ratio	0.3
Centerbody wall temperature (K)	1000	Air flow rate (kg/s)	0.9
Turbulent kinetic energy (m^2/s^2)	$0.01u^2$	Fuel flow rate (kg/s)	0.018

Eulerain-Lagrangian approach has been preferred. We present below the n-decane fuel centerbody combustor analysis [Kim and Chung, 1990] using the Eulerian Lagrangian formulation.

Consider the centerbody geometry as shown in Figure 22.6.5.1a. The initial and boundary conditions are given in Table 22.6.5.1. In the Eulerian-Lagrangian approach described in Section 22.2.5, we require approximations for the droplet evaporation rate in the heat balance equation (22.2.47). In this analysis, the evaporation model of Abramzon and Sirignano [1988] will be used.

The finite element analysis with GPG utilizes the discretization of 29 × 24 mesh with finer mesh in the vicinity of the recirculation zone. The injected spray is assumed to comprise four conical streams with half-angles of the corresponding streams given by $\theta = 5$, 15, 25, and 35 degrees. In the limiting cases of the droplet impingement on the chamber walls, the droplet is considered when 97% of the mass of the droplet is vaporized. In case of the droplet passage through the plane of symmetry, another droplet with similar instantaneous properties and physical dimensions, but with the mirror image velocity vector, is injected into the flowfield. The time steps for the steady state calculations are:

$$\Delta t_{inj} = 1.6 \text{ m/s}, \qquad \Delta t_g = 1.6 \text{ m/s}, \qquad \Delta t_{\ell,m} = 0.04 \text{ m/s}$$

The overall solution procedure is as follows:

(a) Integrate the gas-phase equations from the Eulerian locations to the characteristic location.
(b) Integrate the liquid-phase equations with $\Delta t_{\ell,m}$.
(c) Evaluate the characteristic source terms at the Eulerian nodes surrounding the characteristic.
(d) Steps (a) through (c) are repeated until the liquid-phase numerical time catches up with the gas-phase numerical time ($n\Delta t_{\ell,m} = \Delta t_g$)
(e) Solve the gas-phase equations.
(f) Steps (a) through (e) are repeated until the iteration converges before advancing to the next step for unsteady calculations.

Figure 22.6.5.1b shows the droplet trajectories and vaporization process. The four droplet groups are identified by the volume of the droplet and the characteristic location. It is seen that the droplet motion is initially governed by the droplet inertia force

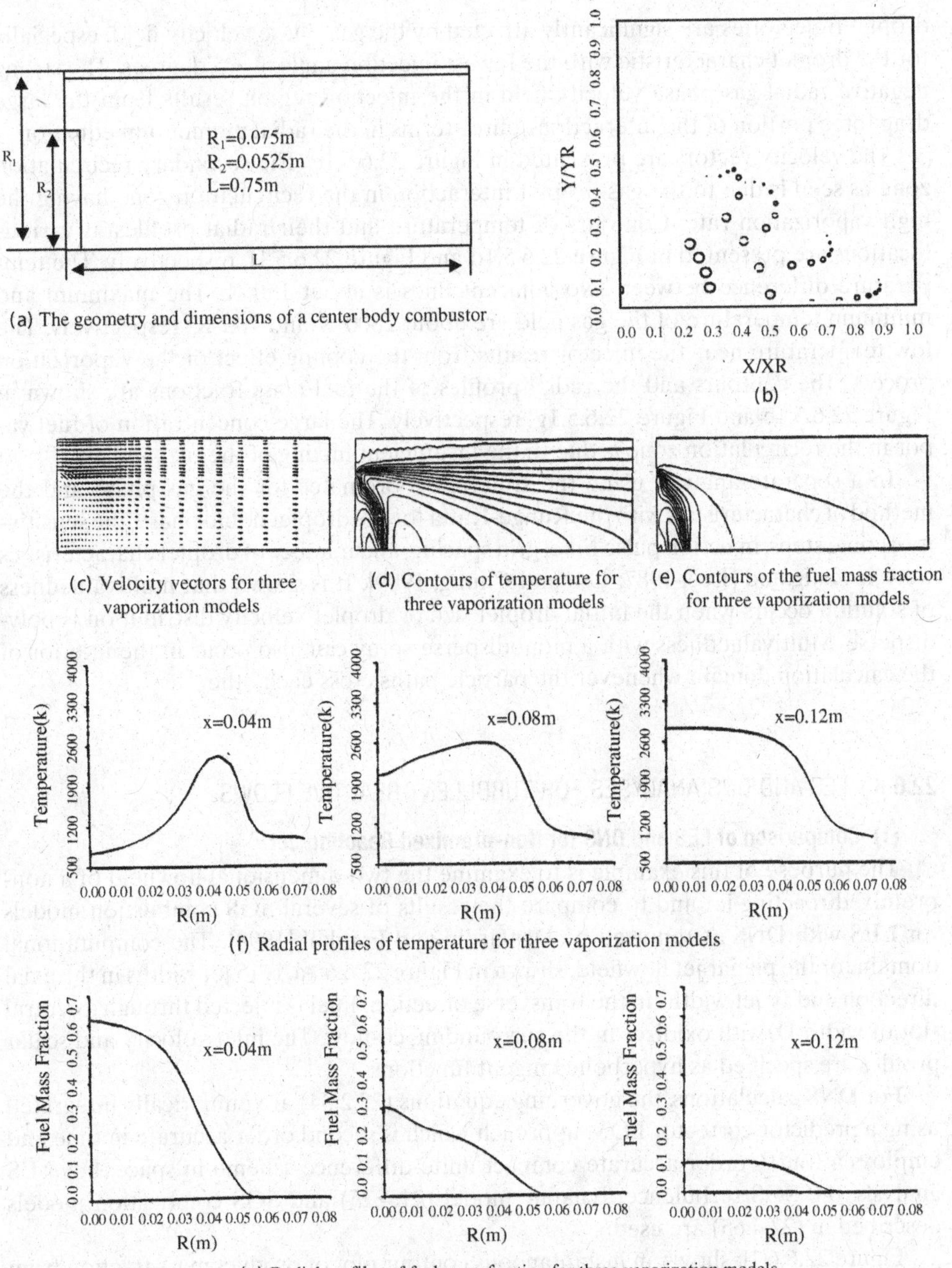

Figure 22.6.5.1 Spray combustion of center body combustor [Kim and Chung, 1990].

before the inertia force causes the droplets to decelerate and the droplet path is eventually determined by the gas-phase flowfield. Most of the vaporization occurs within the recirculation zone because the smaller droplets are unable to penetrate downstream. Because of the strong negative radial gas-phase velocity field near the injector, the droplet trajectories are significantly affected by the gas-phase velocity field, especially for the droplet characteristic with the lowest injection angle, $\theta = 5$ degrees. The strong negative radial gas-phase velocity field in the injection region results from the large drag force portion of the interaction source terms in the radial momentum equation.

The velocity vectors are presented in Figure 22.6.5.1c. The secondary recirculation zone as seen is due to the gas-droplet interaction in the recirculation zone having the high vaporization rate. Contours of temperature and their radial profiles at various locations are presented in Figure 22.6.5.1d and Figure 22.6.5.1f, respectively. The temperature difference between two adjacent lines is about 150°K. The maximum and minimum temperature of the gas field are about 2800°K and 700°K, respectively. The low temperature near the injector results from the cooling effect of the vaporization process. The contours and the radial profiles of the fuel mass fractions are shown in Figure 22.6.5.1e and Figure 22.6.5.1g, respectively. The large concentration of fuel vapor in the recirculation zone is due to the insufficient mixing of the fuel and air.

In a separate analysis using the Eulerian coordinates for the gas phase and the method of characteristics with the Runge-Kutta for the droplet liquid phase, the sensitivity of time steps, injection pulse time, grid spacing, and number of droplet characteristics were investigated [Lee, 1987; Lee and Chung, 1989]. It is shown that multivaluedness of solution occurs when the initial droplet size or droplet velocity distribution is polydisperse. Multivaluedness with a monodisperse spray can also occur in the interior of the calculation domain whenever the particle paths cross each other.

22.6.6 LES AND DNS ANALYSES FOR TURBULENT REACTIVE FLOWS

(1) Comparison of LES and DNS for Non-premixed Reacting Jet

The purpose of this example is to examine the two-dimensional flowfield of a non-premixed reacting jet and to compare the results of several SGS combustion models for LES with DNS as reported by DesJardin and Frankel [1998]. The computational domain for the planar jet flowfield, shown in Figure 22.6.6.1a, is 15 jet widths in the axial direction and 10 jet widths in the transverse direction. Fuel is injected through a central slot of width D, with oxidizer in the surrounding co-flow. The inlet velocity and scalar profiles are specified as hyperbolic tangent functions.

For DNS calculations, the governing equations (22.2.34) are numerically integrated using a predictor-corrector FDM approach which is second order accurate in time and employs a fourth order accurate compact finite-difference scheme in space. For LES analysis, the SGS turbulence dynamic model (21.4.25) and SGS combustion models described in (22.4.65) are used.

Figure 22.6.6.1b shows an instantaneous contour plot of product mass fraction from the LES with the SSFRRM [see (22.4.65a)], which is qualitatively (not at same times) compared with the counterpart calculated from DNS as shown in Figure 22.6.6.1c. The difference in appearance is attributed to the effects of the SGS model.

In Figure 22.6.6.1d,e, LES predictions of mean and rms product mass fraction are compared to DNS results. Here, DNS_c denotes a coarser grid used. The notations FRRM and RRRM refer to SSFRRM and SSRRRM with the model coefficients K_1, K_2 set equal to zero, respectively, in (22.4.65). Also, SLFDM and SLFBM are the strained

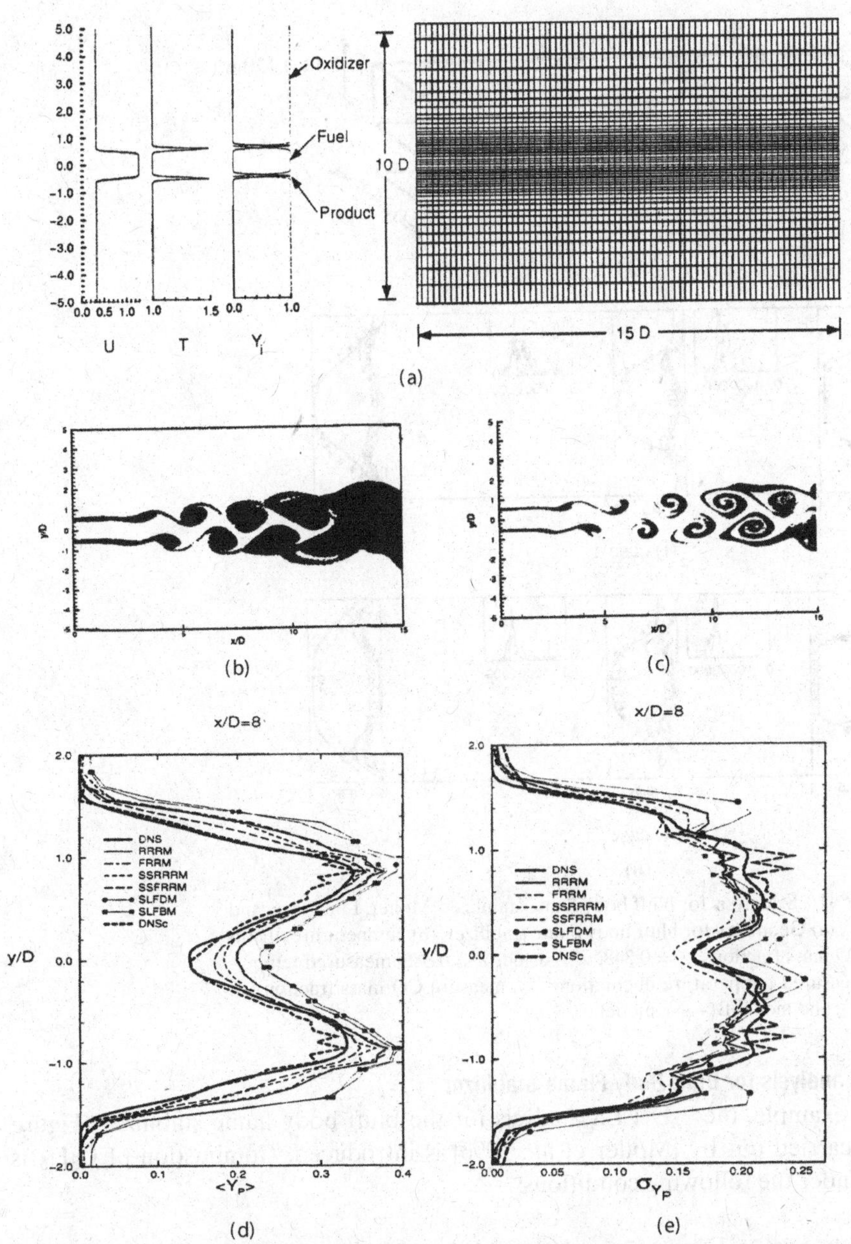

Figure 22.6.6.1 LES and DNS analysis of non-premixed reacting jet [DesJardin and Frankel, 1998]. **(a)** Schematic of computational domain: LES grid and inflow conditions. **(b)** Product mass fraction, LES. **(c)** Product mass fraction, DNS. **(d)** Transverse mean mass product. **(e)** Transverse rms mass product.

laminar flamelet delta model and strained laminar flamelet beta model, respectively, as related to the double delta and beta PDF models discussed in Section 22.4.2 [Cook, Riley, and Kosary, 1997]. As seen in Figure 22.6.6.1d,e, LES model predictions appear to be in agreement with DNS better than the flamelet models.

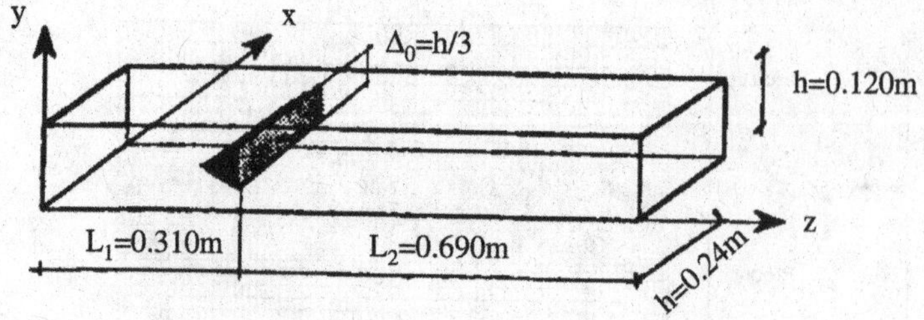

(a)

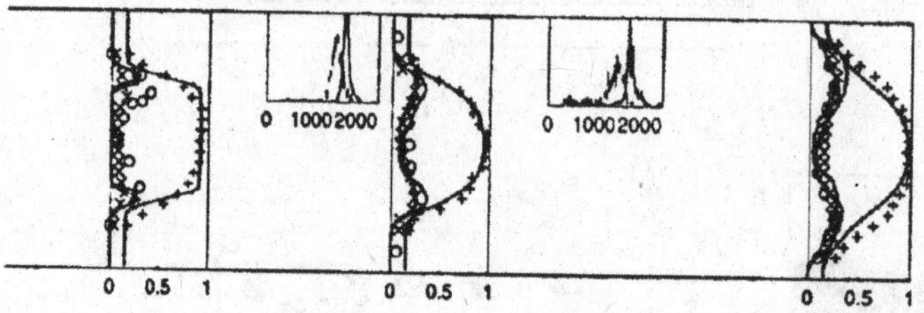

(i) case 1

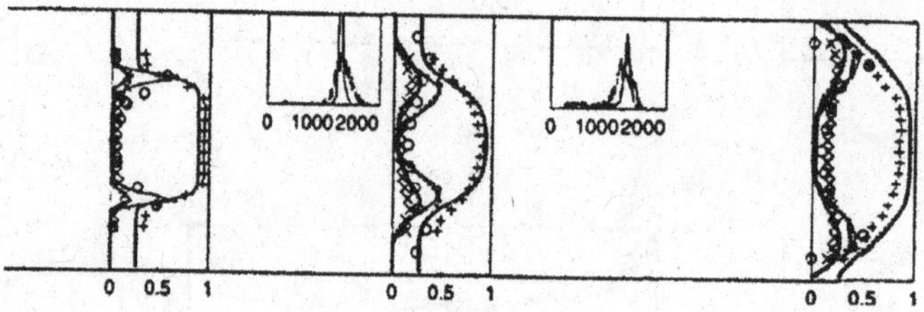

(ii) case 2

(b)

Figure 22.6.6.2 LES analysis for bluff body flame stabilizer [Moller, Lundgren, and Fureby, 1996]. **(a)** Geometry for bluff body flame stabilizer. **(b)** Temperature fluctuations and CO mass fractions, $Z = 0.348$, $Z = 0.460$, $Z = 0.686$, measured temperature (+), measured temperature fluctuations (x), measure CO mass fraction (o) model A(—), and model B(-----), model C(.....).

(2) LES analysis for Bluff Body Flame Stabilizer

In this example, the 3-D LES analysis for the bluff body flame stabilizer (Figure 22.6.6.2a) carried out by [Moller et al., 1996] is introduced. Combustion of C_3H_5 is modeled under the following conditions:

$$\text{Case1: } \phi = 0.62, \quad Re = 47.5 \times 10^3, \quad M = 0.056, \quad u = 17 \text{ m/s}, \quad T = 288 \text{ K}$$
$$\text{Case 2: } \phi = 0.62, \quad Re = 31.6 \times 10^3, \quad M = 0.113, \quad u = 34 \text{ m/s}, \quad T = 600 \text{ K}$$

There are three cases for combustion modeling. Model A: the eddy viscosity model of Fureby and Moller [1995], Model B: PDF reaction rate modeling of Dopazo and O'Brien [1973], and Model C: MILES model by Fureby [1996]. Computed results are compared with their own measured experimental data.

The domain is discretized with $40 \times 80 \times 340$ mesh. The governing equations (22.2.34) are solved using FVM with the third order accurate upwinding for convection,

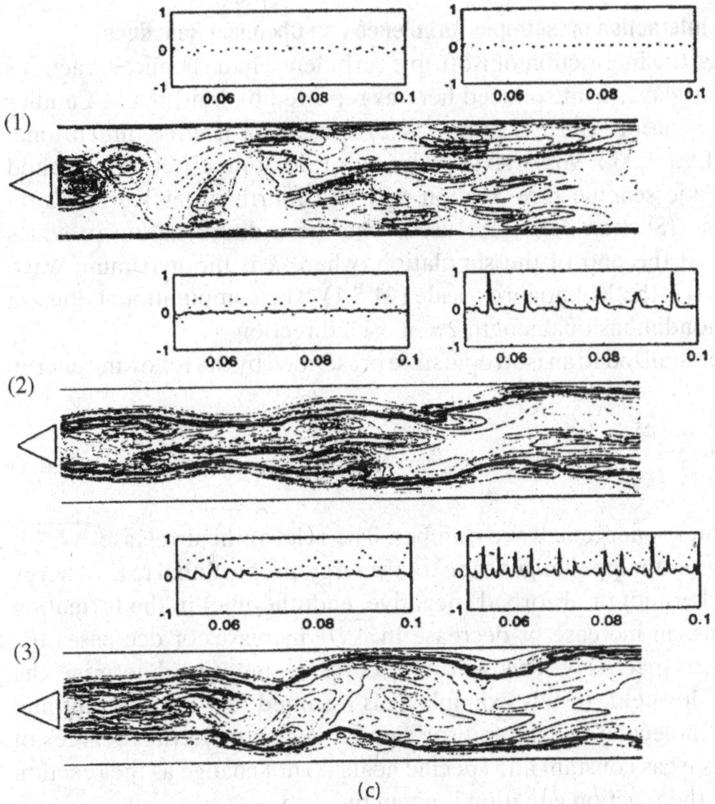

Figure 22.6.6.2 (*continued*) (c) Instantaneous isocontours at $x = 0.12, 0.31 \leq Z \leq 0.87$ (case 1) (2) and (3): flame surface for model B, superimposed on contours of the spanwise vorticity at the same section for cases 1 and 2; normalized pressure (...), normalized Rayleigh parameter are also shown.

fourth order accurate finite differencing for diffusion, and Crank-Nicolson for temporal approximations.

In Figure 22.6.6.2b, the time-averaged temperature and its rms fluctuations together with the time-averaged CO mass fraction and the flame front dynamics in terms of temperature PDFs are shown for reacting cases, along with experimental and simulated profiles. The formation of CO is restricted to the reaction zone along the flame front, which coincides with regions of high temperature fluctuations. Note that the amount of CO in Case 3 is larger than in Case 2. The increase of reaction rate for conversion of C_3H_5 to CO in the preheated case is larger than the increase of rate of formation of CO_2 from CO. Consequently, more CO is accumulated in the reaction zone in the preheated case.

The instantaneous isocontours of the spanwise vorticity in a section between $z = 0.31$ and $z = 0.87$ at $x = 0.12$ are shown in Figure 22.6.6.2c. For Cases 2 and 3, the flame surface is superimposed. An important effect of the energy release on the macroscopic features of the flow is that the vorticity is less structured in Cases 2 and 3 compared with Case 1 and that multiple local extremes occur. Another effect of the heat release is to decrease the magnitude of the vorticity at the center of the vortex structures.

(3) DNS Analysis for Interaction of Isotropic Turbulence and Chemical Reactions

In this DNS analysis, the interaction of isotropic turbulence and chemical reactions in a hypersonic boundary layer is introduced here as reported by Martin and Candler [1998]. A simplified version of (22.2.34) and (22.2.2) is used for a three-dimensional domain with a mesh of $96 \times 96 \times 96$. A sixth order accurate finite difference method based on a compact Pade scheme (see Section 6.7) and fourth order Runge-Kutta time integration scheme (Section 4.4.3) are used. The mesh discretization provides a resolution of $k\eta = 1$ at the end of the simulation, where k is the maximum wave number resolvable and η is the Kolmogorov scale (21.5.1). The computational domain is a periodic box with nondimensional length 2π in each direction.

The velocity field is initialized to an isotropic state prescribed by the following energy spectrum:

$$E(k) \approx k^4 \exp\left(-2\left(\frac{k}{k_0}\right)^2\right) \tag{22.6.5}$$

where k_0 denotes the most energetic wave number. The relative heat release $\overline{\Delta H^0}$ is defined as the ratio of the enthalpy change to the total energy, proportional to the energy released (positive, exothermic) or absorbed (negative, endothermic) in the formation of product species. Thus, an increase or decrease in $\overline{\Delta H^0}$ increases or decreases the energy in the flowfield, respectively. This will be used as an input to determine the various features of the flowfield. In this example, it is assumed that the reactant and product have the same molecular weight and the same number of internal degrees of freedom; thus the mixture gas constant and specific heats do not change as the reaction progresses. In this case, the reaction equation is given by

$$S1 + M \Leftrightarrow S2 + M$$

In Figure 22.6.6.3a, we notice a large increase in the rms magnitude of the temperature when the heat release is increased. A positive temperature fluctuation causes an exponential increase in the reaction rate. However, because of the turbulent motion, the heated fluid may move to a different location before the reaction progresses further, reducing or eliminating the feedback process. Thus, the interaction between the chemical heat release and the turbulent motion should depend on the amount of heat released.

The energy spectrum as defined in (21.4.11–21.4.130) along with (22.6.5) may be decomposed into its incompressible and compressible components at several different times during the simulations. Figure 22.6.6.3b shows these spectra for the nonreacting solution. Note that the compressible modes are about two orders of magnitude less energetic than the incompressible modes at all but the smallest scales. It is seen that there seems to be aliasing errors near $k\eta \geq 1$, indicating that small scales are not resolved. As time evolves, the compressible energy spectrum decays slightly at all scales, whereas the incompressible modes decrease at the large scales, and increase at the small scales. Figure 22.6.6.3c plots the endothermic case and the trend is similar to the nonreacting case. For the case of exothermic reaction (Figure 22.6.6.3d), however, the energy spectrum rises about two orders of magnitude larger than in the case of the endothermic reaction, closer to the incompressible counterpart. It is interesting to note that, in Figure 22.6.6.3e, the compressible mode becomes more energetic, whereas the incompressible mode is not affected by the increase of the heat release.

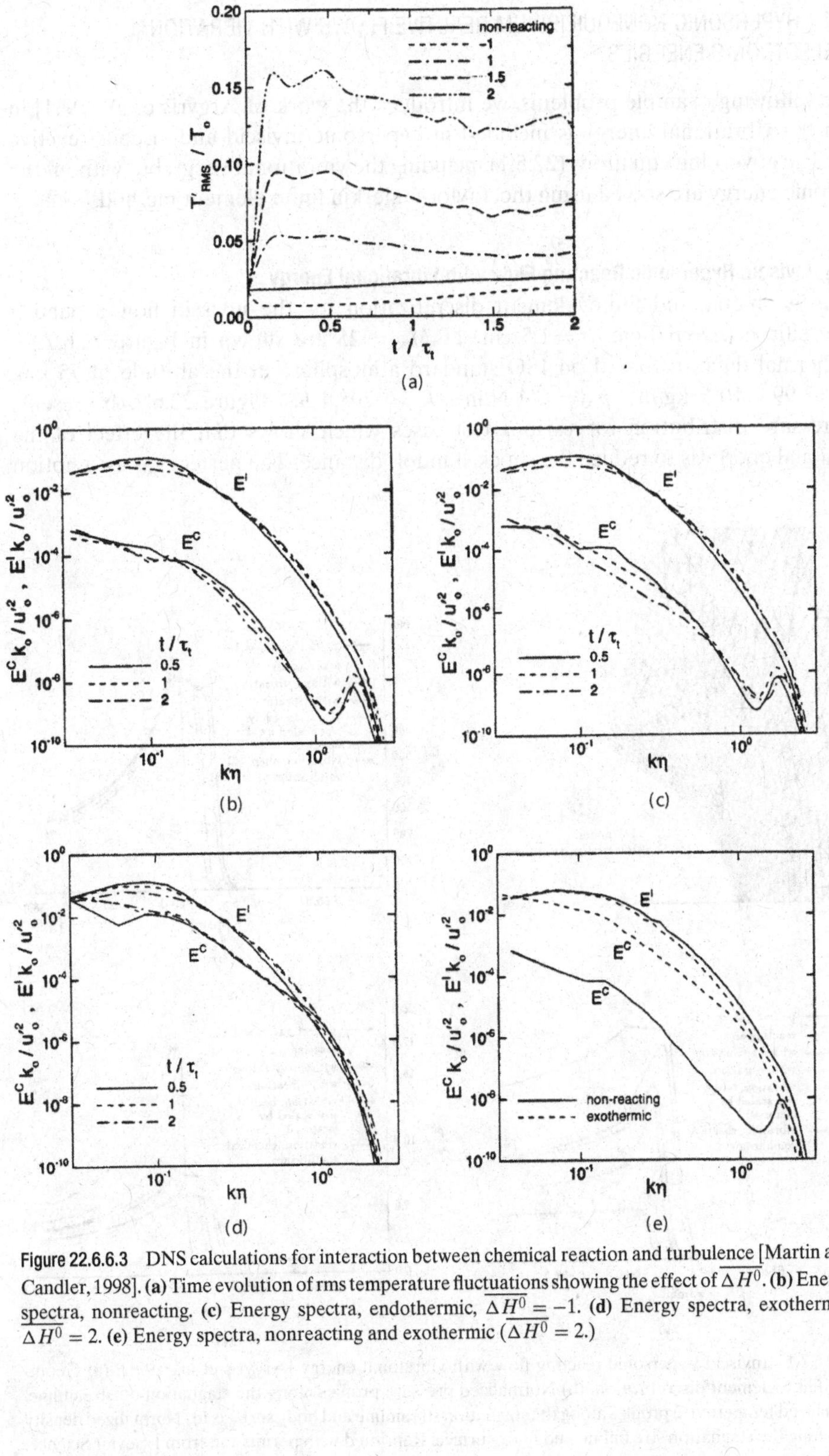

Figure 22.6.6.3 DNS calculations for interaction between chemical reaction and turbulence [Martin and Candler, 1998]. **(a)** Time evolution of rms temperature fluctuations showing the effect of $\overline{\Delta H^0}$. **(b)** Energy spectra, nonreacting. **(c)** Energy spectra, endothermic, $\overline{\Delta H^0} = -1$. **(d)** Energy spectra, exothermic, $\overline{\Delta H^0} = 2$. **(e)** Energy spectra, nonreacting and exothermic ($\overline{\Delta H^0} = 2$.)

22.6.7 HYPERSONIC NONEQUILIBRIUM REACTIVE FLOWS WITH VIBRATIONAL AND ELECTRONIC ENERGIES

In the following example problems, we introduce the work of Argyris et al. [1991] in which the vibrational energy is included in hypersonic inviscid and viscous reactive flows. The governing equations (22.5.1) including the vibrational enegy, but without the electronic energy are solved using the Taylor-Galerkin finite element method.

(1) Inviscid Hypersonic Reacting Flow with Vibrational Energy

The geometry and finite element discretization for the inviscid flow around a simple ellipse ($a = 6.0$ cm, $b = 1.5$ cm) at $M_\infty = 25$ are shown in Figure 22.6.7.1a. The thermal data are based on ISO standard atmosphere at the altitude of 75 km. ($\rho_\infty = 3.99 \times 10^{-5}$ kg/m^3, $p_\infty = 2.4$ N/m^2, $T_\infty = 208.4$ K.) Figure 22.6.7.1b presents the pressure distribution for various test cases which shows that the effect of the vibrational energy is to reduce the shock standoff distance. The perfect gas assumption

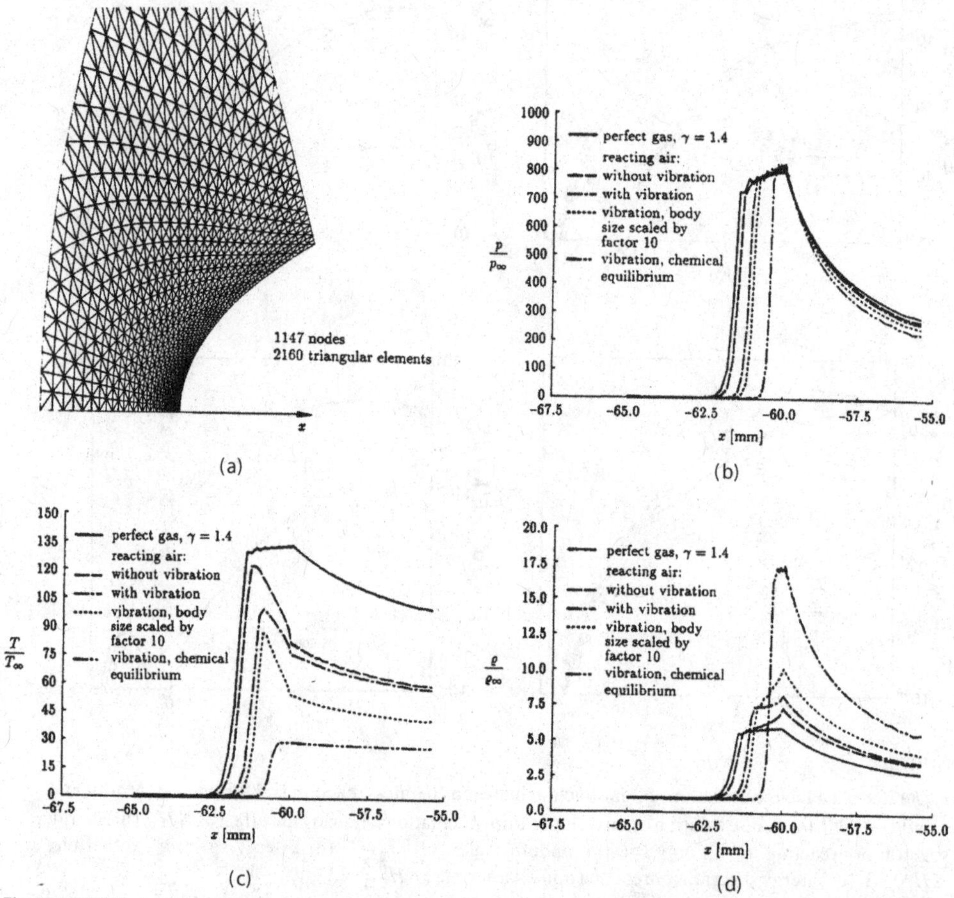

Figure 22.6.7.1 Inviscid hypersonic reacting flow with vibratonal energy [Argyris et al., 1991]. **(a)** Geometry and finite element discretization. **(b)** Normalized pressure profiles along the stagnation on streamline. **(c)** Normalized temperature profiles along the stagnation streamline and body surface. **(d)** Normalized density profiles along the stagnation streamline and body surface. Reprinted with permission from Elsevier Science.

without the vibrational energy provides the largest standoff distance. In Figure 22.6.7.1c, it is clearly shown that the inclusion of vibrational energy causes the temperature to decrease significantly as compared to the case of perfect gas without vibration. This results in an increase in density as shown in Figure 22.6.7.1d.

(2) Viscous Hypersonic Reacting Flow with Vibrational Energy

In this example, we examine the viscous hypersonic reactive flow with vibrational energy. The geometry and finite element discretization and schematics of the shock wave and boundary layer are presented in Figure 22.6.7.2a,b. The effects of various conditions including the frozen flow, equilibrium flow, and finite rate chemistry with and without mass diffusion on the Stanton number, $St = \dot{q}_w / \lfloor \rho_\infty c_{p\infty} u_\infty (T_{0\infty} - T_w) \rfloor$ are shown in Figure 22.6.7.2c. In this case, the vibrational energy is not included. Note that the finite rate chemistry without mass diffusion provides the lowest wall heat flux with the frozen chemistry giving the largest magnitude. There is an indication that the thin boundary layer is not sufficiently resolved for the case of frozen chemistry, as seen from the fact that the peak value of Stanton number fails to occur at the stagnation point as it should. The results with vibrational energy are shown in Figure 22.6.7.2d. We observe that the Stanton number increases for some distance downstream of the stagnation point before it decreases further downstream. The effect of mass diffusion is clearly evident, causing the heat flux to be reduced. In Figure 22.6.7.2e, the influence of vibration and mass diffusion on the species distribution is shown. Note that the mass fraction of atomic oxygen (Y_O) is reduced significantly due to vibration and mass diffusion. The excitation of vibrational energy reduces the flow temperature and subsequently decreases the dissociation process.

(3) Thermochemical Nonequilibrium Hypersonic Flows with Two-Temperature Model

In this example, the work of Park and Yoon [1991] is introduced to illustrate an implementation of vibrational, electronic excitation, and chemical reaction models described in (22.5.14–22.5.24) for thermochemical nonequilibrium flows at suborbital flight speeds. Here the nonequilibrium vibrational and electronic excitation and dissociation are taken into account without ionization. The steady-state of the resulting system of equations is carried out by using lower-upper factorization and symmetric Gauss-Seidel sweeping technique through Newton-Raphson iteration, together with the Roe's upwinding scheme.

Sample calculations are made for flows over a circular cylinder of 1-inch diameter with its axis perpendicular to the flow direction, placed in the test section of a shock tunnel as used by Hornung [1972] for interferometry experiments. The diameter of the cylinder is 2 inches. The freestream conditions are: nitrogen density $= 5.349 \times 10^{-3}$ kg/m^3, velocity $= 5.59$ km/s, nitrogen atom mass fraction $= 0.073$, and temperature $= 1833$ K. The flow Mach number is 6.13, and the Reynolds number based on the body diameter is 24,000. The nitrogen flow is calculated using the 5-species model (Table 22.5.3) by setting the mole fraction of oxygen to be 10^{-6}.

To compare with the experimental interferometry results of Hornung [1972], the interferometric fringes are computed from

$$\Delta\rho = \frac{4160 F \lambda}{\ell(1 + 0.28 Y_N)} \ (\text{kg/m}^3)$$

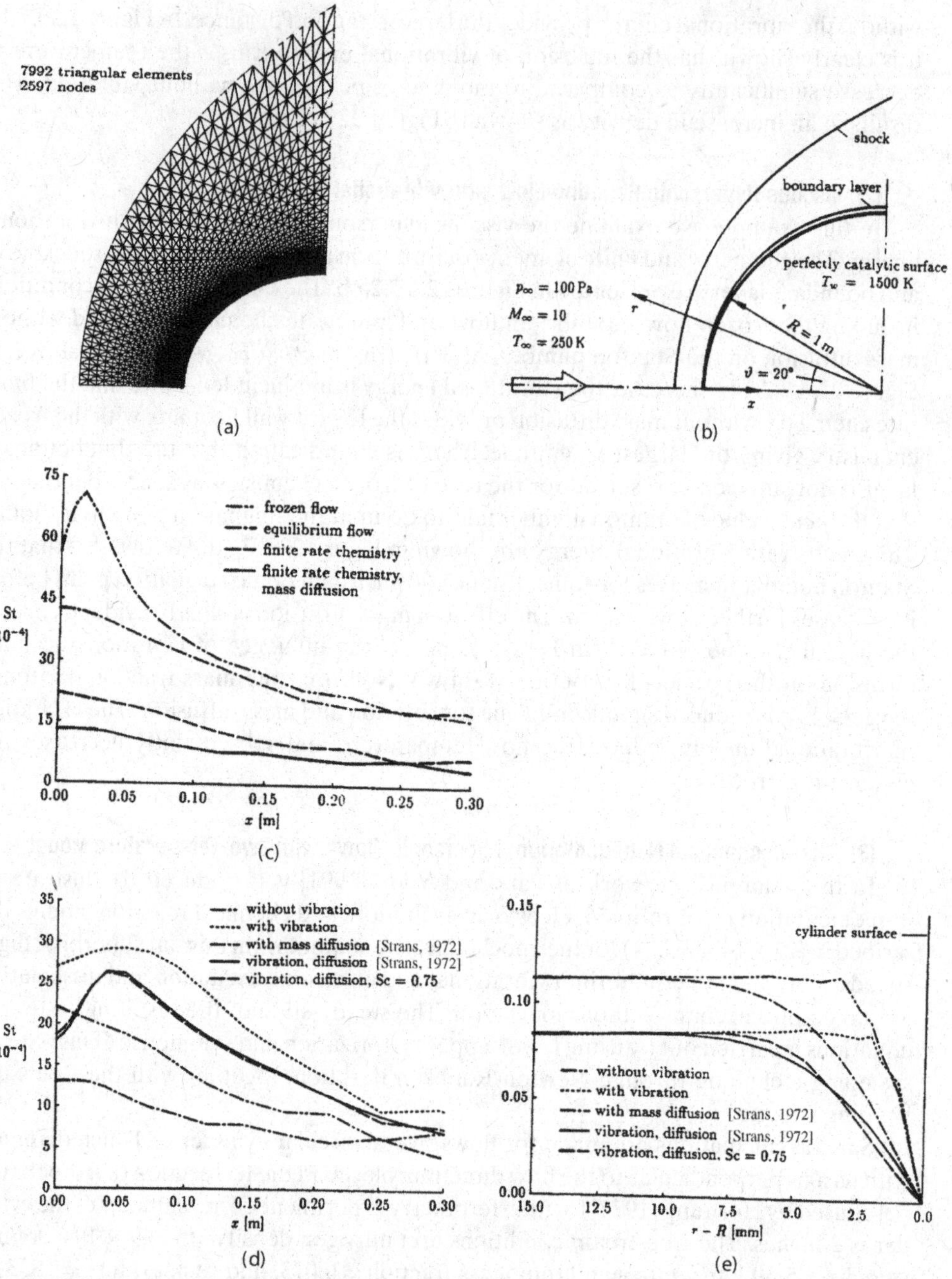

Figure 22.6.7.2 Viscous hypersonic reacting flow with vibrational energy [Argyris et al., 1991]. **(a)** Geometry and finite element discretization. **(b)** Schematics of shock wave and boundary layer. **(c)** Stanton number (St) on the cylinder surface at $0° \leq \theta \leq 45°$ for different chemical models. **(d)** Stanton number (St) on the cylinder at $0° \leq \theta \leq 45°$ for different internal degress of freedom. **(e)** Profiles of mass fraction of atomic oxygen normal to the cylinder surface at an angle of $\theta = 20°$. Reprinted with permission from Elsevier Science.

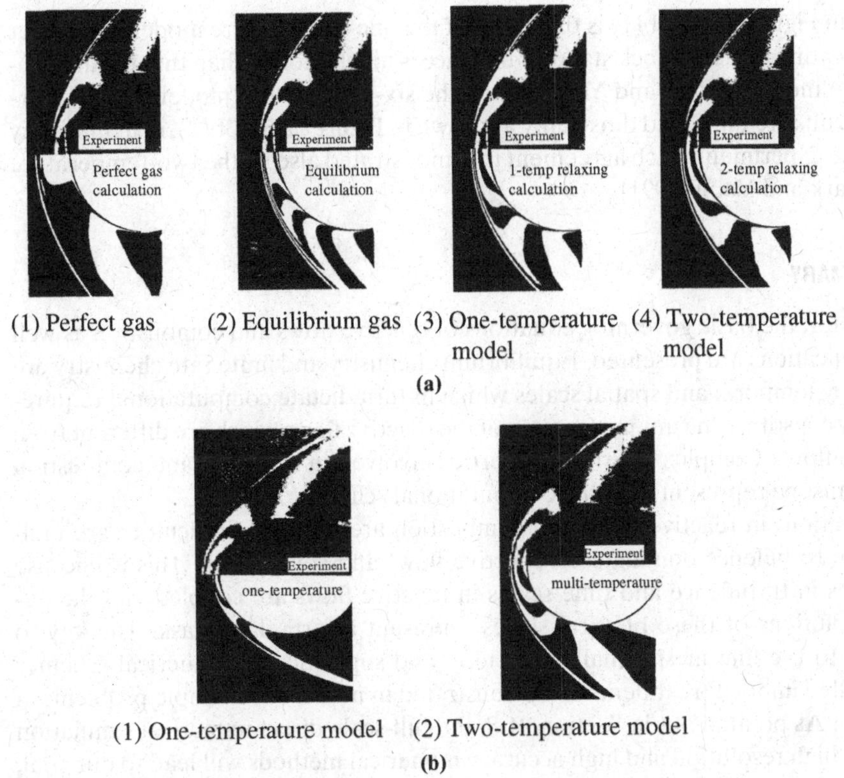

Figure 22.6.7.3 Comparison of hypersonic flows over a cylinder [Park and Yoon, 1991; Candler, 1989]. **(a)** Hypersonic flow analysis, 2-inch diameter cylinder. **(b)** Hypersonic flow analysis, 2 inch diameter cylinder.

where F is the fringe number, λ is the wavelength, and ℓ is the experiment's geometric path.

In Figure 22.6.7.3a(1), the calculated shock standoff distance for a perfect gas is very much larger than the measured value [Hornung, 1972]. The calculated fringes have no resemblance to the experimental fringes. It is interesting to note that as shown in Figure 22.6.7.3a(2), the shock standoff distance for the equilibrium gas is shorter than the measured value with the appearance of fringes still quite different from those of the experiments.

In contrast, the results of the one-temperature model [Figure 22.6.7.3a(3)] become closer to the experiment. With the two-temperature model, the shock standoff distance and fringes match very well with the experiment as shown in Figure 22.6.7.3a(4).

(4) Thermomechanical Nonequilibrium Hypersonic Flows with Multi-Temperature Model

In this example, we compare the two-temperature model of Park and Yoon [1991] with the multi-temperature model of Candler [1989]. Here, we consider seven species (N_2, O_2, NO, NO^+, N, O, e^-) and six temperatures, with all other data being equal to those of Park and Yoon. However, the vibrational temperatures of different molecular species are calculated independently and the electron temperature is calculated separately.

Shown in Figure 22.6.7.3b(1) is the result of the one-temperature model of Candler [1989], indicating that the shock standoff distance is much shorter than that for the one-temperature model of Park and Yoon. When the six-temperature model is used, however, the results are improved drastically as shown in Figure 22.6.7.3b(2), matching very well with the experiment. Such agreement is demonstrated also by the two-temperature model of Park and Yoon [1991].

22.7 SUMMARY

In this chapter, the basic governing equations of reactive flows and combustion as well as their applications are presented. Equilibrium chemistry and finite rate chemistry are controlled by temporal and spatial scales which in turn dictate computational requirements. They constitute the unique features of the reactive flows which are different from nonreactive flows. Complex physical properties involved in reactions and combustion processes must be represented in the computational schemes.

Computations in reactive flows and combustion are difficult. Difficulties are multiplied when turbulence dominates in reactive flows and combustion. This is because spatial scales in turbulence and time scales in reactive flows are coupled and the numerical resolutions of these physical scales represent a formidable task. The key to the issue is to use fine mesh, small time steps, and sophisticated numerical schemes with controlled implicit treatments as demonstrated in numerous example problems in Section 22.6. As pointed out in Section 21.8, the full-scale direct numerical simulation (DNS) with high resolution and high accuracy numerical methods will lead to our goal, hopefully when computer resources become available.

As mentioned in Chapter 21, the role of FDV theory with various variation parameters, particularly in terms of the Damköhler numbers, should be investigated. Only with the most accurate numerical schemes will DNS be fully effective.

REFERENCES

Abramzon, B. and Sirignano, W. A. [1988]. Droplet vaporization model for spray combustion. AIAA paper, 88-0636.

Al-Masseeh, W. A., Bradley, D., Gaskell, P. H., and Lau, A. K. C. [1990]. Turbulent premixed, swirling combustion: direct stress, strained flamelet modeling and experimental investigation. Twenty-Third Symposium (International) on Combustion/The Combustion Institute, 825–33.

Argyris, J. A., Doltsinis, I. S., Fritz, H., Urban, J. [1991]. An exploration of chemically reacting viscous hypersonic flow. *Comp. Meth. Appl. Mech. Eng.*, 89, 85–128.

Bilger, R. W. [1980]. Turbulent flows with non-premixed Reactants. In P. A. Libby and F. A. Williams (eds.). *Turbulent Reacting Flows*. Berlin: Springer-Verlag, 65–114.

Bradley, D. and Law, A. K. C. [1990]. *Pure and Applied Chemistry*, 62, 803.

Bray, K. N. C. [1979]. *Seventeenth Symposium (International), The Combustion Institute*, Pittsburgh: 223.

Brokaw, R. S. [1958]. Approximate formulas for the viscosity and thermal conductivity of gas mixtures. *J. Chem. Phys.*, 29, 391–397.

Candler, G. [1989]. On the computation of shock shapes in nonequilibrium hypersonic flows. AIAA Paper, 89-0312.

Chin, J. S. and Lefebvre, A. H. [1983]. The role of the heat-up period in fuel droplet evaporation. AIAA Paper, 83-0068.

Chung, T. J. and Karr, G. R. [1980]. Analysis of nonlinear chemically reactive flow characteristic of high energy laser systems. *Int. J. Num. Meth. Eng.*, 16, 1–12.

Chung, T. J., Kim, Y. M., and Sohn, J. L. [1987]. Finite element analysis in combustion phenomena. *Int. J. Num. Meth. Fl.*, 7, 989–1012.

Chung, T. J. [1993a]. Recent advances in finite element analysis for laminar reacting flows in combustion. In T. J. Chung, Taylor, and Francis (eds.). *Numerical Modeling in Combustion*. Moscow: 133–76.

———. [1993b]. Finite element methods in turbulent combustion. In T. J. Chung, Taylor, and Francis (eds.). *Numerical Modeling in Combustion*. Moscow: 375–98.

Chung, T. J. [1997]. A new computational approach with flowfield dependent variation algorithm for applications to supersonic combustion. In G. Roy, S. Frolov, and P. Givi (eds.). *Advanced Computation and Analysis of Combustion*. Moscow: ENAS Publishers, 466–89.

Cook, A. W., Riley, J. J., and Kosary, G. [1997]. A laminar flamelet approach to subgrid-scale chemistry in turbulent flows. *Comb. Flame*, 109, 332–45.

DesJardin, P. E. and Frankel, S. H. [1998]. Large eddy simulation of a nonpremixed reacting jet: Application and assessment of subgrid-scale combustion models. *Phys. Fl.*, 10, 9, 2298–2314.

Dopazo, C. and O'Brien, E. E. [1973]. Isochoric turbulent mixing of two rapidly reacting chemical species with chemical heat release. *Phys. Fl.*, 16, 2075–87.

Drummond, J. P., Hussaini, M. Y., and Zang, T. A. [1985]. Spectral methods for modeling supersonic chemically reacting flow fields. AIAAS Paper, 85-0302.

Evans, J. S. and Schexnayder C. J. [1980]. Influence of chemical kinetics and unmixedness on burning in supersonic hydrogen flames. *AIAA J.*, 18, 2, 188–93.

Faeth, G. M. [1977]. Current status of droplet and liquid combustion. *Prog. Energy Comb. Sci.*, 3, 191–224.

Frankel, S. H., Madina, C. K., and Givi, P. [1992]. Modeling of the reactant conversion rate in a turbulent shear flow. *Chem. Eng. Comm.*, 113, 197–209.

Frolov, S. M., Basevich, V. A., Neuhaus, M. G., and Tatchl, R. [1997]. A joint velocity-scalar PDF method for modeling premixed and nonpremixed combustion. In G. Roy, S. Frolov, and P. Givi (eds.). *Advanced Computation and Analysis of Combustion*. Moscow: ENAS Publishers, 537–61.

Fureby, C. [1996]. On subgrid scale modeling in large eddy simulations of compressible fluid flow. *Phys. Fl.*, 8, 1301.

Fureby, C. and Moller, S. I. [1995]. Large eddy simulation of reacting flows applied to bluff body stabilized flames. *AIAA J.*, 33, 12, 2339–2347.

Gardiner, W. C. (ed.). [1984]. *Combustion Chemistry*. Springer-Verlag.

Gear, C. W. [1971]. *Numerical Initial Value Problems in Ordinary Differential Equations*. Englewood Cliffs, NJ: Prentice-Hall.

Gibson, M. M. and Launder, B. E. [1978]. Group effects on pressure fluctuations in the atmospheric boundary layer. *J. Fluid Mech.*, 86, 491–511.

Girimaji, S. S. [1991]. Assumed β-PDF model for turbulent mixing: validation and extension to multiple scalar mixing, *Comb. Sci. Tech.*, 78, 177–96.

Givi, P. [1993]. Spectral methods in combustion. In T. J. Chung, Taylor, and Francis (eds.). *Numerical Modeling in Combustion*, 409–52.

Givi, P. and Jou, W. H. [1988]. Mixing and chemical reaction in a spatially developing mixing layer. *J. Nonequil. Thermod.*, 13, 4, 355–72.

Givi, P. and Riley, J. J. [1992]. Some current issues in the analysis of reacting shear layers: Computational challenges. In R. Voit (ed.). *Major Research Topics in Combustion*, New York: 558–650.

Gordon, S. and McBride, J. [1971]. Computer program for calculation of complex chemical equilibrium compositions, rocket performance, incident and reflected shocks, and Chapman-Jouguet detonations. NASA SP-273, 1971.

Hirschfelder, J. O., Curtis, C. F., and Bird, R. [1954]. *Molecular Theory of Gases and Liquids*. New York: Wiley.

Hornung, H. G. [1972]. Non-equilibrium dissociating nitrogen flow over spheres and cylinders. *J. Fl. Mech.*, 53, Part 1, 149–76.

Hu, D. [1987]. Direct numerical simulation of plane mixing layer flows with the isoparametric spectral element method. M.S. thesis, Case Western Reserve University.

Hussaini, M. Y. and Zang, T. A. [1987]. Spectral methods in fluid dynamics. *Ann. Rev. Fl. Mech.*, 19, 339–67.

Janicka, J. and Kollmann, W. [1979]. A two-variables formalism for the treatment of chemical reactions in turbulent H_2-air diffusion flames. *Seventeenth Symposium (International) on Combustion*, The Combustion Institute, 421–29.

Janicka, J. and Kollmann [1980]. A prediction model for turbulent diffusion flames including NO formation. *AGARD Proc.* No. 275.

Khalil, E. E., Spalding, D. B., and Whitelaw, J. H. [1975]. The calculation of local flow properties in two-dimensional furnaces. *Int. J. Heat Mass Trans.*, 18, 775–84.

Kim, Y. M. [1987]. Finite element methods in turbulent combustion. Ph.D. diss. The University of Alabama, Huntsville.

Kim, Y. M. and Chung, T. J. [1989]. Finite element analysis of turbulent diffusion flames. *AIAA J.*, 27, 3, 330–39.

———. [1990]. Finite element model for turbulent spray combustion. AIAA paper, 90-0359.

———. [1991]. Turbulent combustion analysis with various probability density functions. AIAA paper, 89-1991.

Kollmann, W. [1990]. The PDF approach to turbulent flow. *Theor. Comp. Fl. Dyn.*, 1, 249–85.

Korczak, K. Z. and Hu, D. [1987]. Turbulent mixing layers – direct spectral element simulation. AIAA paper, 87-0113.

Launder, B. E. and Spalding, D. B. [1974]. The numerical computation of turbulent flows. *Comp. Meth. Appl. Meth. Eng.*, 3, 239–53.

Launder, B. E., Reece, G. J., and Rodi, W. [1975]. Progress in the development of a Reynolds stress turbulence closure. *J. Fluid Mech.*, 86, 537–66.

Lee, J. H. [1985]. Basic governing equations for the flight regimes of aeroassisted orbital transfer vehicles, *Progress in Astronautics and Aeronautics*, ed. H. F. Nelson, AIAA, 96, 3–53.

Lee, S. K. [1987]. Numerical modeling of spray vaporization using finite elements. Ph.D. diss. The University of Alabama, Huntsville.

Lee, S. K. and Chung, T. J. [1989]. Axisymmetric unsteady droplet vaporization and gas temperature distribution. *AIAA J.*, 111, 5, 487–94.

Lefebvre, A. H. [1989]. *Atomization and Spray*. Washington, DC: Hemisphere.

Lockwood, F. C. and Naguib, A. S. [1975]. The prediction of the fluctuations in the properties of free, round-jet turbulent, diffusion flames. *Comb. Flame*, 24, 109–24.

Martin, M. P. and Candler, G. V. [1998]. Effect of chemical reactions on decaying isotropic turbulence. *Phys. Fl.*, 10, 7, 1715–24.

McMurtry, P. A. and Givi, P. [1991]. Spectral simulations of reacting turbulent flows. In E. S. Oran and J. P. Boris (eds.). *Numerical Approaches to Combustion Modeling*, New York: AIAA, 257–303.

Millikan, R. C. and White, D. R. [1963]. Systematics of vibrational relaxation. *J. Chem. Phys.*, 139, 3209–13.

Moller, S. I., Lundgren, E., and Fureby, C. [1996]. Large eddy simulation of unsteady combustion. *Twenty-Sixth Symposium (International) on Combustion*. The Combustion Institute, 241–48.

Moon, S. Y. [1998]. Applications of FDMEI to chemically reacting shock wave boundary layer interactions. Ph.D. diss. The University of Alabama, Huntsville.

Moon, S. Y., Yoon, K. T., and Chung, T. J. [1996]. Numerical simulation of heat transfer in chemically reacting shock wave-turbulent boundary layer interactions. *Num. Heat Trans.*, Part A, 30, 55–72.

Norris, J. W. and Edwards, J. R. [1997]. Large-eddy simulations of high-speed, turbulent diffusion flames with detailed chemistry. AIAA paper 97-0370.

Park, C. [1990]. *Nonequilibrium Hypersonic Aerothermodynamics*. New York: Wiley.

Park, C. and Yoon, S. [1991]. Fully coupled implicit method for thermochemical nonequilibrium air at suborbital flight speeds. *J. Spacecraft*, 28, 1, 31–39.

Pope, S. B. [1985]. PDF methods for turbulent reactive flows. *Prog. Energy Comb. Sci.*, 11, 119–92.

Pope, S. B. [1990]. Computations of turbulent combustion: progress and challenges, Proc. 23rd Symposium (Internatial) on Combustion, Pittsburg, Combustion Institute, 591–612.

Pratt, D. T. [1983]. CREK-1D: A computer code for transient, gas phase combustion kinetics. Spring meeting of the Western States of the Combustion Institute, WSCI 83–21.

Pratt, D. T. and Wormeck, J. J. [1976]. A computer program for calculation of turbulent flow WSA-ME-TEL-76-1. Washington State University.

Radhakrishnan, K. [1984]. Comparison of numerical techniques for integration of stiff ordinary differential equations arising in combustion chemistry, NASA Technical Paper 2372.

Rogallo, R. S. and Moin, P. [1984]. Numerical simulation of turbulent flows. *Ann. Rev. Fl. Mech.*, 16, 99–137.

Rogers, R. C. and Chinitz, W. [1983]. Using a global hydrogen-air combustion model in turbulent reacting flow calculations. *AIAA J.*, 21, 4, 586–92.

Rogers, R. C. and Schexnayder, C. J. [1981]. Chemical kinetic analysis of hydrogen-air ignition and reaction times. NASA TP-1856.

Sirignano, W. A. [1993]. Computational spray combustion. In T. J. Chung (ed.). *Numerical Modeling in Combustion*. Washington DC: Hemisphere.

———. [1999]. *Fluid Dynamics and Transport of Droplets and Sprays*. UK: Cambridge University Press.

Spalding, D. B. [1971]. Mixing and chemical reaction in steady confined turbulent flames. *Thirteenth Symposium (International) on Combustion*. The Combustion Institute, 649–58.

Warnats, J. [1984]. Chemistry of high temperature combustion of alkanes up to octane. *Twentieth Symposium (International) on Combustion*. The Combustion Institute, 845–56.

Westbrook, C. K. and Dryer, F. L. [1984]. Chemical kinetic modeling of hydrocarbon combustion. *Prog. Energy Comb. Sci.*, 10, 1–57.

Wilke, C. R. [1950]. A viscosity equation for gas mixtures. *J. Chem. Phys.*, 18, 517–19.

Yoon, C. S. [1992]. Finite element analysis for supersonic combustion with finite rate chemistry. Ph.D. diss., The University of Alabama, Huntsville.

Yoon, W. S. [1992]. Analysis of turbulence and shock wave interactions and wave instabilities in combustion. Ph.D. diss. The University of Alabama, Huntsville.

Yoon, W. S. and Chung, T. J. [1991]. Liquid propellant combustion waves. AIAA paper, 91-2088.

———. [1992]. Numerical studies on supersonic and hypersonic combustion. AIAA paper, 92-0094.

———. [1993a]. Numerical simulation of airbreathing combustion at all speed regimes. AIAA paper, 93-1972.

———. [1993b]. Finite rate chemical reactions in subsonic, supersonic, and hypersonic turbulent flows. AIAA paper, 93-2993.

Young, T. R. and Boris, J. P. [1977]. A numerical technique of solving stiff ordinary differential equations associated with the chemical kinetics of reactive flow problems. *J. Phys. Chem.*, 81, 2424–27.

Zeleznik, F. J. and McBride, B. J. [1984]. Modeling the internal combustion engine. NASA RP-1094.

CHAPTER TWENTY-THREE

Applications to Acoustics

23.1 INTRODUCTION

Acoustics is the science that deals with sound waves such as in a combustion chamber, jet noise, oceanography, meteorology, architectural acoustics, and environmental acoustics. Sound waves may occur in the quiescent air even with extremely small pressure disturbances. This could lead to a noise audible to the human ear. In this case, changes in all flow variables other than the pressure remain constant. On the other hand, the noise level can be extremely high (thunder or explosion), but still with fluctuations of all variables other than the pressure remaining more or less constant. This phenomenon may be referred to as the *pressure mode acoustics*. When fluids undergo circulations causing significant velocity gradients, vortical waves are generated, which then produce pressure disturbances. The noise coming from this action (vorticity) may be categorized as the *vorticity mode acoustics*. In many instances in nature or in engineering, we encounter rapid changes in temperature such as in hypersonic flows over a spacecraft creating an entropy boundary layer between the shock layer and velocity boundary layer, subsequently leading to pressure fluctuations. Entropy waves are predominant in this case. We may identify the noise generated by entropy waves as the *entropy mode acoustics*.

The categorization suggested above was actually originated by Kovasznay [1953]. It is our intention to follow his suggestion in this chapter. However, it appears that the research in the acoustics community in general has been centered around acoustic waveforms (linear and nonlinear–N-waves), sound emission (radiation), and sound absorption (viscous dissipation), under which a large number of subdivided disciplines can be identified. Selection of example problems under such vast subject areas is difficult for the purpose of this chapter, which is concerned only with an introduction of computational acoustics. Thus, instead, in adopting the suggested categorization by pressure mode acoustics, vorticity mode acoustics, and entropy mode acoustics, it is necessary that appropriate governing equations be identified. For example, we may select suitable topics for the pressure mode acoustics in which the Helmholtz equation or its variant such as the Kirchhoff's formula is used. For the vorticity mode acoustics, standard vorticity transport equation(s), Lighthill's acoustic analogy, or Ffowcs Williams-Hawkings equation may be invoked. Pressure disturbances arising from the solution of these equations will contribute to the vorticity mode acoustics. Using the first and second laws of thermodynamics, the energy equation can be derived in terms of entropy in a variety of forms. In this process, the pressure disturbances and the entropy noise from this calculation is identified as the entropy mode acoustics. In all cases, it is useful to obtain analytical expressions for the pressure fluctuations from the governing equations used for each of the modes. Here, the concept of Green's function will play a major role. In general, however, solutions of the Navier-Stokes system of equations can provide the most useful information. With proper filtering process (or time averages), the fluctuation components of all variables including the pressure fluctuations and the root mean squared pressure (p_{rms}) are calculated to determine the noise level.

It is quite possible that the noise level calculated may actually be the combination of all three modes in a given physical situation, regardless of the equations being used for the solution. Thus, the quantitative determination of the magnitude of the noise level from the dominant mode and from the possible contributions of other less dominant modes in the system would be of interest. This is not attempted in this chapter. Instead, our focus will be to select suitable example problems under the suggested categorization, discuss the governing equations and computational methods, and evaluate the results.

Some basic definitions used in acoustics are introduced below. The time-averaged value of a fluid property, say f, is defined as

$$\overline{f} = \langle f \rangle = \frac{1}{\Delta t}\int_t^{t+\Delta t} f\,dt \tag{23.1.1}$$

where the symbols '‾' and '⟨ ⟩' imply time averages to be used interchangeably in what follows. From this result, the acoustic intensity **I** (Watt/m^2) is defined as, with $f = p$

$$\mathbf{I} = \langle p\mathbf{v}\rangle = \frac{1}{\Delta t}\int_t^{t+\Delta t} p\mathbf{v}\,dt \tag{23.1.2}$$

from which the acoustic power π can be calculated:

$$\pi = \int_\Gamma \mathbf{I}\cdot\mathbf{n}\,d\Gamma = \rho_0 a_0 \langle u^2\rangle \Gamma \tag{23.1.3}$$

where Γ denotes the surface area. The noise level is then determined either by the acoustic intensity level (IL) or by the sound pressure level (SPL).

$$IL = 10\log_{10}\frac{I}{I_{ref}} \quad \textit{in dB (decibel)} \tag{23.1.4}$$

$$SPL = 20\log_{10}\frac{p_{rms}}{p_{ref}} \quad \textit{in dB (decibel)} \tag{23.1.5}$$

where $I_{ref} = 10^{-12}$ Watt/m^2 at 1000 Hz (barely audible sound to human ear) and I is the scalar acoustic intensity normal to the surface as determined from (23.1.2). The reference pressure ($p_{ref} = 2.04 \times 10^{-5}$ N/m^2) corresponds almost to I_{ref} in a plane wave, and p_{rms} is the root-mean square pressure.

In many engineering problems, we are concerned with unstable waves rather than the noise generation such as occur in combustion instability. They are undesirable physical phenomena in view of efficiency of the engineering performance. In this case, the acoustic energy tends to grow without bound in resonance, leading to an inefficient combustion and/or severe vibrations of the system. In this chapter, we shall address this subject as well as the acoustic noise generation.

Acoustic problems are generally classified as a function of the generating source. For each class of problems there are standard techniques used to solve them. In the following sections, the most commonly used equations are described along with the solution methodology.

23.2 PRESSURE MODE ACOUSTICS

23.2.1 BASIC EQUATIONS

The most commonly used equation in acoustics is the wave equation. It is derived from the continuity and momentum equations (in the absence of body forces and sources and sinks of mass) by writing the variables as the sum of the freestream and a fluctuation,

$$p = p_0 + p'$$

and linearizing the equations around the freestream state. This leads to

$$\frac{1}{a_0^2}\frac{\partial^2 p'}{\partial t^2} - \nabla^2 p' = 0 \tag{23.2.1}$$

where a_0 is the speed of sound. Equation (23.2.1) assumes a zero convection velocity (i.e., $\mathbf{v} \equiv \mathbf{0}$). For a nonzero $\mathbf{v}$, the convected wave equation is given by

$$\frac{1}{a_0^2}\left(\frac{\partial}{\partial t} + \mathbf{v}\cdot\nabla\right)^2 p - \nabla^2 p = 0 \tag{23.2.2}$$

where the prime in (23.2.1) is neglected for simplicity. Multiplying (23.2.1) by $e^{i\omega t}$ and integrating by parts over an appropriate time interval results in the well-known Helmholtz equation

$$\left[\nabla^2 + \left(\frac{\omega}{a_0}\right)^2\right] p = 0 \tag{23.2.3}$$

where $p = \hat{p}e^{i\omega t}$ and ω is the circular frequency. Equation (23.2.2) can also be written in the form of (23.2.3) using a change of reference frame defined by $\mathbf{x_0} = \mathbf{x} - \mathbf{v}\tau$ and $\tau = t$.

In order to study the linear acoustic wave propagation generated by a known source (say a vibrating sphere), one can either use the various CFD methods presented in the previous chapters or, if possible, find a close form solution. For instance, depending upon the complexity of the source, a time domain Green's function solution can be found for (23.2.1) and (23.2.2) and a frequency domain one for (23.2.3) [Howe, 1998]. For acoustic waves generated by large amplitude pressure disturbances, the nonlinear Euler equations should be used to capture the nonlinear wave propagation phenomena. In more complex problems involving shocks, boundary layers, and jets, the Navier-Stokes system of equations should be used (see Section 2.2.11).

23.2.2 KIRCHHOFF'S METHOD WITH STATIONARY SURFACES

Kirchhoff's formula is used in the theory of diffraction of light and in other electromagnetic problems. It also has many applications to problems of wave propagation in acoustics [Pierce, 1981]. The idea of Kirchhoff's formula is to surround the region of a nonlinear flowfield and acoustic sources by a closed surface. In the domain inside the surface, a nonlinear aerodynamic computation is carried out, which provides the pressure distribution on the surface as well as its time history. Outside this surface the acoustic disturbance satisfies the stationary wave equation (23.2.1). To determine $p(\mathbf{x}, t)$, consider the homogeneous Helmholtz equation given by (23.2.3) whose solution is the Green's function $G(\mathbf{y}, \mathbf{x}; \omega)$. It can be shown that

$$p(x, \omega) = \oint_S \left\{ G(\mathbf{x}, \mathbf{y}; \omega)\frac{\partial p}{\partial y_j}(\mathbf{y}, \omega) - p(\mathbf{y}, \omega)\frac{\partial G}{\partial y_j}(\mathbf{x}, \mathbf{y}; \omega) \right\} n_j dS(\mathbf{y}) \tag{23.2.4}$$

where $\mathbf{n}$ is the unit normal on S directed into the fluid. Making use of the convolution theorem, the time domain solution can be written as

$$p(x, t) = \oint_S \left\{ -G(\mathbf{x}, \mathbf{y}; t-\tau)\frac{\partial p}{\partial y_j}(\mathbf{y}, \tau) + p(\mathbf{y}, \tau)\frac{\partial G}{\partial y_j}(\mathbf{x}, \mathbf{y}; t-\tau) \right\} n_j dS(\mathbf{y}) d\tau \tag{23.2.5}$$

where the retarded time integration is taken over $(-\infty, \infty)$. The linearized momentum equation gives $\rho_0 \frac{\partial \mathrm{v}_j}{\partial \tau} = -\frac{\partial p}{\partial y_j}$ in the absence of the body forces, which leads to

$$p(x, t) = \oint_S \left\{ G(\mathbf{x}, \mathbf{y}; t-\tau)\rho_0 \frac{\partial \mathrm{v}_j}{\partial \tau}(\mathbf{y}, \tau) + p(\mathbf{y}, \tau)\frac{\partial G}{\partial y_j}(\mathbf{x}, \mathbf{y}; t-\tau) \right\} n_j dS(\mathbf{y}) d\tau. \tag{23.2.6}$$

Using the free space Green function given by

$$G(\mathbf{x}, \mathbf{y}; t-\tau) = \frac{1}{4\pi|\mathbf{x}-\mathbf{y}|}\delta(t-\tau-|\mathbf{x}-\mathbf{y}|/a_0)$$

equation (23.2.6) becomes

$$\begin{aligned} p(x, t) = {} & \frac{\rho_0}{4\pi}\frac{\partial}{\partial t}\oint_S \frac{\mathrm{v}_n(\mathbf{y}, t-|\mathbf{x}-\mathbf{y}|/a_0)}{|\mathbf{x}-\mathbf{y}|} dS(\mathbf{y}) \\ & - \frac{1}{4\pi}\frac{\partial}{\partial x_j}\oint_S \frac{p(\mathbf{y}, t-|\mathbf{x}-\mathbf{y}|/a_0)}{|\mathbf{x}-\mathbf{y}|} n_j dS(\mathbf{y}). \end{aligned} \tag{23.2.7}$$

Equation (23.2.7) can be reduced to the following form:

$$4\pi p(\mathbf{x}, t) = \int_S \left[\frac{p}{r^2}\frac{\partial r}{\partial \mathbf{n}} - \frac{1}{r}\frac{\partial p}{\partial \mathbf{n}} + \frac{1}{a_0 r}\frac{\partial r}{\partial \mathbf{n}}\frac{\partial p}{\partial \tau} \right] dS \tag{23.2.8}$$

where $|\mathbf{x}-\mathbf{y}| = r$ is the distance between the observer and the source, $\partial r/\partial \mathbf{n} = \cos\theta$ where θ is the angle between the normal vector and the radial direction and $\mathbf{n}$ the outward normal vector. In (23.2.8), the notation "[]" is used to denote the retarded time.

23.2.3 KIRCHHOFF'S METHOD WITH SUBSONIC SURFACES

Hawkings [1977] proposed using the formula for predicting the noise of high-speed propellers and helicopter rotors. His idea consisted of surrounding the rotating blades by a closed surface, which moves at the forward speed of the helicopter. A nonlinear computation is carried out inside the closed surface, which gives the pressure distribution on the surface and its time history. Outside this surface, (23.2.6) is modified to account for the motion and is written as follows:

$$4\pi p(\mathbf{x},t) = \int_{S_1} \left[\frac{p}{r_1^2} \frac{\partial r_1}{\partial \mathbf{n}_1} - \frac{1}{r_1} \frac{\partial p}{\partial \mathbf{n}_1} + \frac{1}{a_0 r_1 \beta^2} \frac{\partial p}{\partial \tau} \left(\frac{\partial r_1}{\partial \mathbf{n}_1} - M_0 \frac{\partial x_1}{\partial \mathbf{n}_1} \right) \right] dS_1 \tag{23.2.9}$$

where the subscript '1' denotes the transformed coordinates. The transformation used is that given by Prandtl-Glauret:

$$x_1 = x, \qquad y_1 = \beta y \quad \text{and} \quad z_1 = \beta z$$

In (23.2.9), we have

$$r_1 = \left[(x - x')^2 + \beta[(y - y')^2 + (z - z')^2] \right]^{1/2}, \quad \beta = (1 - M_0^2)^{1/2}$$

and the retarded time becomes

$$\tau = \frac{[r_1 - M_0(x - x')]}{a_0 \beta^2}$$

where M_0 is the freestream Mach number. The position of the source is given by (x', y', z'). For a zero freestream velocity, (23.2.9) reduces to (23.2.8), as can be easily shown.

23.2.4 KIRCHHOFF'S METHOD WITH SUPERSONIC SURFACES

The convective wave equation (23.2.2) is still the governing equation; however, for a supersonically moving surface the time delay τ is not uniquely defined. It is given by

$$\tau^{\pm} = [\pm r_1 - M_0(x - x')]/a B^2 \quad \text{where} \quad B = \left(M_0^2 - 1 \right)^{0.5}.$$

The radiated pressure field takes the form

$$4\pi p(\mathbf{x},t) = \int_{S_1} \left[\frac{p}{r_1^2} \frac{\partial r_1}{\partial \mathbf{n}_1} - \frac{1}{r_1} \frac{\partial p}{\partial \mathbf{n}_1} + \frac{1}{a_0 r_1 B^2} \frac{\partial p}{\partial \tau} \left(\pm \frac{\partial r_1}{\partial \mathbf{n}_1} - M_0 \frac{\partial x_1}{\partial \mathbf{n}_1} \right) \right]_{\tau^{\pm}} dS_1 \tag{23.2.10}$$

where $\pm$ notation indicates evaluation for both retarded times τ^+ and τ^-. This equation, however, still presents a singularity at $M_0 = 1$. In order to overcome this difficulty, Farassat [1996] and Farassat and Farris [1999] recently developed a Kirchhoff formula applicable across the whole speed range, but particularly useful for supersonic surfaces.

The theories presented here allow the computation of farfield sound given the detailed flow field in the vicinity of the source. The choice of theory to be used is problem dependent. In Section 23.5.1, several examples are presented and solved.

23.3 VORTICITY MODE ACOUSTICS

23.3.1 LIGHTHILL'S ACOUSTIC ANALOGY

The sound generated by vorticity in an unbounded fluid is generally referred to as *aerodynamic sound* [Lighthill, 1952, 1954]. Most fluid flows of engineering interest are unsteady in nature, of high Reynolds number and turbulent. These flows are known to generate noise; that is, turbulent boundary layers, jets, and shear layers. Though the acoustic radiation is a very small by-product of the fluid motion, which creates a numerical challenge, it is becoming an important part of the flow solution.

The theory of aerodynamic sound was developed by Lighthill [1952], who rewrote the Navier-Stokes equations into an exact, inhomogeneous wave equation whose source terms are important only within the turbulent region. Furthermore, at low Mach numbers, the sound generation and subsequent propagation can be decoupled from the fluid motion. The momentum equation for an ideal, stationary fluid of density ρ_0 and sound speed a_0 subject to the externally applied stress T_{ij} is

$$\frac{\partial(\rho v_i)}{\partial t} + \frac{\partial\left(a_0^2(\rho - \rho_0)\right)}{\partial x_i} = -\frac{\partial T_{ij}}{\partial x_j}. \tag{23.3.1}$$

Using the continuity equation to eliminate (ρv_i) results in the well-known Lighthill acoustic analogy equation

$$\left(\frac{1}{a_0^2}\frac{\partial^2}{\partial t^2} - \nabla^2\right)\left[a_0^2(\rho - \rho_0)\right] = \frac{\partial^2 T_{ij}}{\partial x_i \partial x_j}. \tag{23.3.2}$$

In the derivation of (23.3.1) an ideal, linear fluid is assumed. In such a fluid, the momentum transfer is produced solely by the pressure. In (23.3.1) and (23.3.2), T_{ij} is the Lighthill stress tensor given by

$$T_{ij} = \rho v_i v_j + \left((p - p_0) - a_0^2(\rho - \rho_0)\right)\delta_{ij} - \tau_{ij}. \tag{23.3.3}$$

Solution of (23.3.2) requires an accurate determination of the Lighthill stress tensor given by (23.3.3). When the mean density and sound speed are uniform, the variation in ρ produced by low Mach number, high Reynolds number velocity fluctuations are of order $\rho_0 M^2$, and $\rho v_i v_j \approx \rho_0 v_i v_j$ with a relative error $\sim O(M^2) \ll 1$. Similarly, we have

$$p - p_0 - a_0^2(\rho - \rho_0) \approx (p - p_0)(1 - a_0^2/a^2) \sim O(\rho_0 v^2 M^2).$$

Therefore, $T_{ij} \approx \rho_0 v_i v_j$, when viscous stresses are neglected, the solution to Lighthill equation can be written as

$$\begin{aligned} p(\mathbf{x}, t) &\approx \frac{\partial^2}{\partial x_i \partial x_j}\int \frac{\rho_0 v_i v_j(\mathbf{y}, t - |\mathbf{x} - \mathbf{y}|/a_0)}{4\pi|\mathbf{x} - \mathbf{y}|} d^3\mathbf{y} \\ &\approx \frac{x_i x_j}{4\pi a_0^2 |\mathbf{x}|^3}\frac{\partial^2}{\partial t^2}\int \rho_0 v_i v_j(\mathbf{y}, t - |\mathbf{x} - \mathbf{y}|/a_0) d^3\mathbf{y}, \quad |\mathbf{x}| \to \infty \end{aligned} \tag{23.3.4}$$

where $p(\mathbf{x}, t) = a_0^2(\rho - \rho_0)$ is the perturbation pressure in the far field. In general, in order to compute farfield noise from a jet, a shear layer or turbulent boundary layer; it is necessary to carry out an accurate CFD computation in the near field to determine the Reynolds stresses and then use (23.3.4) for the farfield computations. Examples of accurate CFD computations include Direct Numerical Simulation (DNS), Large Eddy Simulation (LES), or even a well-resolved Unsteady Reynolds Averaged Navier-Stokes (URANS).

23.3.2 FFOWCS WILLIAMS-HAWKINGS EQUATION

When Lighthill's acoustic analogy is used in flows with moving boundaries, moving sources or in turbulent shear layers separating a quiescent medium from a high-speed flow, it is necessary to introduce control surfaces. These surfaces can coincide with existing physical surfaces or correspond to a convenient interface between fluid regions of widely differing mean properties. Suitable boundary conditions are applied on these surfaces.

Let $f(\mathbf{x}, t)$ be an indicator function that vanishes on the surface S and satisfies $f(\mathbf{x}, t) > 0$ in the fluid where Lighthill's equation is to be solved, and $f(\mathbf{x}, t) < 0$ elsewhere. Multiply (23.3.1) by $H(f)$ and rearrange into the form

$$\frac{\partial}{\partial t}(\rho \mathbf{v}_i H(f)) + \frac{\partial}{\partial x_i}(H(f)a_0^2(\rho - \rho_0))$$
$$= -\frac{\partial}{\partial x_j}(H(f)T_{ij}) + (\rho \mathbf{v}_i(\mathbf{v}_j - \overline{\mathbf{v}}_j) + (p - p_0)\delta_{ij} - \tau_{ij})\frac{\partial H}{\partial x_j}(f). \tag{23.3.5}$$

A similar process can be applied to the continuity equation to obtain

$$\frac{\partial}{\partial t}(H(f)(\rho - \rho_0)) + \frac{\partial}{\partial x_i}(H(f)\rho \mathbf{v}_i) = (\rho(\mathbf{v}_i - \overline{\mathbf{v}}_i) + \rho_0 \overline{\mathbf{v}}_i)\frac{\partial H}{\partial x_i}(f). \tag{23.3.6}$$

Elimination of $H\rho\mathbf{v}_i$ between the two equations above leads to the well-known Ffowcs Williams-Hawkings equation

$$\left(\frac{1}{a_0^2}\frac{\partial^2}{\partial t^2} - \nabla^2\right)[H(f)a_0^2(\rho - \rho_0)]$$
$$= \frac{\partial^2(H(f)T_{ij})}{\partial x_i \partial x_j} - \frac{\partial}{\partial x_i}\left([\rho \mathbf{v}_i(\mathbf{v}_j - \overline{\mathbf{v}}_j) + (p - p_0)\delta_{ij} - \tau_{ij}]\frac{\partial H}{\partial x_j}(f)\right)$$
$$+ \frac{\partial}{\partial t}\left([\rho(\mathbf{v}_j - \overline{\mathbf{v}}_j) + \rho_0\overline{\mathbf{v}}_j]\frac{\partial H}{\partial x_j}(f)\right). \tag{23.3.7}$$

This equation is valid throughout the whole space. Using Green's function, one can write down a formal outgoing wave solution. Written in an integral form, the Ffowcs Williams-Hawkings equation [1969] is

$$H(f)a_0^2(\rho - \rho_0) = \frac{\partial^2}{\partial x_i \partial x_j}\int_{V(\tau)}[T_{ij}]\frac{d^3\mathbf{y}}{4\pi|\mathbf{x} - \mathbf{y}|}$$
$$- \frac{\partial}{\partial x_i}\oint_{S(\tau)}[\rho \mathbf{v}_i(\mathbf{v}_j - \overline{\mathbf{v}}_j) + p'_{ij}]\frac{dS_j(\mathbf{y})}{4\pi|\mathbf{x} - \mathbf{y}|}$$
$$+ \frac{\partial}{\partial t}\oint_{S(\tau)}[\rho(\mathbf{v}_j - \overline{\mathbf{v}}_j) + \rho_0\overline{\mathbf{v}}_j]\frac{dS_j(\mathbf{y})}{4\pi|\mathbf{x} - \mathbf{y}|}, \tag{23.3.8}$$

where $p'_{ij} = (p - p_0)\delta_{ij} - \tau_{ij}$, and the square bracket "[]" denotes the retarded time ($\tau = t - |\mathbf{x} - \mathbf{y}|/a_0$). The surface integrals indicates a monopole and a dipole source contribution from the surface while the volume integral indicates a quadripole source. When $\overline{v}$ is zero in (23.3.5) (i.e., stationary surface), the generalized Kirchhoff formula is recovered.

23.4 ENTROPY MODE ACOUSTICS

23.4.1 ENTROPY ENERGY GOVERNING EQUATIONS

If temperature gradients are high, the fluctuation components of temperature can be very large, leading to entropy waves. In this case, we invoke the first and second laws of thermodynamics. At equilibrium in a continuous flowfield domain, the combined first and second laws of thermodynamics and Maxwell's relations lead to [Chung, 1996]

$$T\frac{DS}{Dt} = \frac{D\varepsilon}{Dt} + p\frac{D}{Dt}\left(\frac{1}{\rho}\right) \tag{23.4.1}$$

$$T\frac{DS}{Dt} = \frac{DH}{Dt} - \frac{1}{\rho}\frac{Dp}{Dt} \tag{23.4.2}$$

$$T\frac{DS}{Dt} = c_p\frac{DT}{Dt} - \frac{\beta T}{\rho}\frac{Dp}{Dt} \tag{23.4.3}$$

where S is the specific entropy, β is the thermal expansion coefficient,

$$\beta = -\frac{1}{\rho}\frac{\partial \rho}{\partial T}$$

and other variables are defined in Section 2.2 and Section 22.2. It can be shown that the equations of momentum, energy, continuity, and vorticity transport for 3-D compressible flows are of the form

Momentum

$$\frac{\partial \mathbf{v}}{\partial t} + \nabla\hat{H} - \mathbf{v}\times\omega = T\nabla S + \nu\left(\nabla^2\mathbf{v} + \frac{1}{3}\nabla(\nabla\cdot\mathbf{v})\right) \tag{23.4.4}$$

Energy

$$\rho c_p\frac{DT}{Dt} - \beta T\frac{Dp}{Dt} - \tau_{ij}v_{j,i} - k\nabla^2 T = 0 \quad \text{for sound emission} \tag{23.4.5a}$$

$$\rho T\frac{DS}{Dt} = \tau_{ij}v_{j,i} - k\nabla^2 T = 0 \quad \text{for sound absorption} \tag{23.4.5b}$$

Continuity

$$\frac{1}{\rho a^2}\frac{Dp}{Dt} + \nabla\cdot\mathbf{v} = \frac{\beta T}{c_p}\frac{DS}{Dt} \tag{23.4.6}$$

Vorticity Transport

$$\frac{\partial\omega}{\partial t} + (\mathbf{v}\cdot\nabla)\omega + \omega\nabla\cdot\mathbf{v} - (\omega\cdot\nabla)\mathbf{v} = \nabla T\times\nabla S + \nu\nabla^2\omega \tag{23.4.7}$$

where $\hat{H}$ denotes the total enthalpy. Taking a time derivative of (23.4.5) and combining with (23.4.4), we obtain the acoustic analogy equation which may be used for determining unstable entropy waves,

$$\frac{\partial}{\partial t}\left(\frac{1}{\rho a^2}\frac{\partial p}{\partial t}\right) - \nabla\cdot\left(\frac{1}{\rho}\nabla p\right) = (v_i v_j)_{,ij} + \frac{\partial}{\partial t}\left(\frac{\beta T}{c_p}\frac{DS}{Dt}\right) \tag{23.4.8}$$

Although unstable entropy waves can be calculated from (23.4.8), it is more convenient to use a form in which the entropy term is replaced by thermodynamic relationships. This approach, known as the entropy-controlled instability (ECI) method, is intended to include pressure and vorticity modes as well as the entropy mode. Following Yoon and Chung [1994], the mathematical formulation is described below.

23.4.2 ENTROPY CONTROLLED INSTABILITY (ECI) ANALYSIS

In this approach, the energy equation is first written in conservation form. Upon differentiation of the convective terms, we isolate the derivative of total energy in terms of pressure gradients and subsequently in terms of entropy gradients. All variables are replaced by the sum of their mean and fluctuating parts. Furthermore, the logarithmic form of entropy changes is replaced by truncated infinite series to retain highly nonlinear physical aspects of the system. The energy equation is then integrated by parts spatially, resulting in both domain and boundary surface integrals. These surface integral terms constitute the nonlinear, nonisentropic acoustic intensity acting on the solid boundaries. These terms are driven by the fluctuation of the enclosed fluid. The next step is to take time averages of all terms of the time-dependent domain integrals and time-independent surface integrals. These processes convert the partial differential equation of energy into a nonlinear ordinary differential equation, characterizing the stability or instability of wave motions with the energy growth factor as the main dependent variable.

For a nonisentropic flow, the pressure gradient is written as

$$p_{,i} = a^2\rho_{,i} + (\rho a^2/c_p)S_{,i} \tag{23.4.9}$$

The spatial derivative of the stagnation energy E is given by

$$\begin{aligned} E_{,i} &= \left(c_p T - \frac{p}{\rho} + \frac{1}{2}v_j v_j\right)_{,i} \\ &= \frac{c_v}{R\rho}P_{,i} - \frac{c_v}{R}\frac{p}{\rho^2}\rho_{,i} + v_j v_{j,i} \end{aligned} \tag{23.4.10}$$

where R is the specific gas constant. Substituting (23.4.9) into (23.4.10) yields

$$\rho E_{,i} = \frac{p}{\rho}\rho_{,i} + \frac{p}{R}S_{,i} + \rho v_j v_{j,i} \tag{23.4.11}$$

Consider now the energy equation written in the conservation form

$$\frac{\partial}{\partial t}(\rho E) + (\rho E v_i - \sigma_{ij}v_j + q_i)_{,i} = 0 \tag{23.4.12}$$

Substituting (23.4.11) into (23.4.12) yields

$$\frac{\partial}{\partial t}(\rho E) + (\rho E v_i)_{,i} + v_i\left[\frac{p}{\rho}\rho_{,i} + \frac{p}{R}S_{,i} + \rho v_j v_{j,i}\right] - \rho v_i E_{ji} - (\sigma_{ij} v_j)_{,i} + q_{i,i} = 0 \tag{23.4.13}$$

It is interesting to note that for large entropy gradients, (23.4.13) will be dominated by the term $v_i(p/R)S_{,i}$, instrumental in nonlinear, nonisentropic wave oscillations.

The most general form of the nonlinear, nonisentropic wave equation may be obtained by integrating (23.4.13) by parts spatially and taking a time average of the resulting equation.

$$\left\langle \int_\Omega \frac{\partial}{\partial t}(\rho E)\, d\Omega \right\rangle - \left\langle \int_\Omega \left[E_{,i}\rho v_i + \rho\left(v_i \frac{p}{\rho}\right)_{,i} + \left(v_i \frac{p}{R}\right)_{,i} S + (\rho v_i v_j)_{,i} v_j \right] d\Omega \right\rangle$$
$$+ \left\langle \int_\Gamma \left[\rho E v_i n_i + v_i\left(\frac{p}{\rho}\rho n_i + \frac{p}{R} S n_i + \rho v_j v_j n_i\right) - \sigma_{ij} v_j n_i - k T_{,i} n_i \right] d\Gamma \right\rangle = 0 \tag{23.4.14}$$

where Ω and Γ represent the domain and boundary surface, respectively, $\langle\cdot\rangle$ implies time averages and n_i denotes the ith component of the outward normal vector to the surface. Physically, the time derivative term (first term) and spatial derivative terms (next five terms) represent the temporal growth and spatial growth of waves, respectively. The last six terms of boundary integrals imply the so-called acoustic intensity on the solid boundaries.

From thermodynamic relations for an ideal gas, we may write the entropy difference in the form

$$S - S_o = R \ln\left[\left(1 + \frac{p'}{\overline{p}}\right)^{\frac{1}{(\gamma-1)}} \left(1 + \frac{\rho'}{\overline{\rho}}\right)^{\frac{-\gamma}{(\gamma-1)}}\right] \tag{23.4.15}$$

Expanding the right-hand side of (23.4.15) in infinite series we obtain

$$S = R[S_{(1)} + S_{(2)} + S_{(3)} + S_{(4)} + \cdots] + S_O \tag{23.4.16}$$

with

$$S_{(1)} = \frac{1}{(\gamma-1)}\frac{p'}{\overline{p}} - \frac{\gamma}{(\gamma-1)}\frac{\rho'}{\overline{\rho}}$$

$$S_{(2)} = -\frac{1}{2}\left[\frac{1}{(\gamma-1)}\left(\frac{p'}{\overline{p}}\right)^2 - \frac{\gamma}{(\gamma-1)}\left(\frac{\rho'}{\overline{\rho}}\right)^2\right]$$

$$S_{(3)} = -\frac{1}{3}\left[\frac{1}{(\gamma-1)}\left(\frac{p'}{\overline{p}}\right)^3 - \frac{\gamma}{(\gamma-1)}\left(\frac{\rho'}{\overline{\rho}}\right)^3\right]$$

$$S_{(4)} = \frac{1}{4}\left[\frac{1}{(\gamma-1)}\left(\frac{p'}{\overline{p}}\right)^4 - \frac{\gamma}{(\gamma-1)}\left(\frac{\rho'}{\overline{\rho}}\right)^4\right]$$

Note that the higher order terms (fifth order or higher) are neglected as they are small in comparison with lower order terms.

23.4.3 UNSTABLE ENTROPY WAVES

At this point, all variables may be written as

$$p = \overline{p} + \varepsilon p', \qquad \mathrm{v}_i = \overline{\mathrm{v}}_i + \varepsilon \mathrm{v}_i', \qquad \rho = \overline{\rho} + \varepsilon\rho', \qquad T = \overline{T} + \varepsilon T' \tag{23.4.17}$$

where the symbols, bar and prime, denote the spatial or temporal mean and fluctuating parts, respectively; and ε signifies the energy growth factor ($0 \leq \varepsilon \leq \infty$) which is temporally dependent but spatially independent. Notice that $\varepsilon = 0$ indicates vanishing of the fluctuating parts whereas $\varepsilon = \infty$ implies an unbounded growth of fluctuations as a function of time.

Substituting (23.4.16) and (23.4.17) into (23.4.14) yields

$$\frac{\partial}{\partial t}(\varepsilon^2 E_1 + \varepsilon^3 E_2 + \varepsilon^4 E_3) - \varepsilon^2 I_1 - \varepsilon^3 I_2 - \varepsilon^4 I_3 = 0 \tag{23.4.18}$$

where

$$E_1 = \left\langle \int_\Omega a^{(1)} d\Omega \right\rangle, \qquad E_2 = \left\langle \int_\Omega a^{(2)} d\Omega \right\rangle, \qquad E_3 = \left\langle \int_\Omega a^{(3)} d\Omega \right\rangle \tag{23.4.19a–c}$$

$$I_1 = \left\langle \int_\Omega b^{(1)} d\Omega \right\rangle - \left\langle \int_\Gamma c_i^{(1)} n_i d\Gamma \right\rangle \tag{23.4.20a}$$

$$I_2 = \left\langle \int_\Omega b^{(2)} d\Omega \right\rangle - \left\langle \int_\Gamma c_i^{(2)} n_i d\Gamma \right\rangle \tag{23.4.20b}$$

$$I_3 = \left\langle \int_\Omega b^{(3)} d\Omega \right\rangle - \left\langle \int_\Gamma c_i^{(3)} n_i d\Gamma \right\rangle \tag{23.4.20c}$$

where $a^{(i)}$, $b^{(i)}$, and $c^{(i)}$ ($i = 1, 2, 3$) consist of mean and fluctuating parts of variables.

A glance at (23.4.18) indicates that the zeroth-order terms in ε are canceled and first order terms vanish due to time averages, and $E_{(i)}$ is no longer an explicit function of time because of its time averages. Thus, the partial derivative with respect to time in (23.4.18) involves only ε, not $E_{(i)}$, so that

$$\frac{d\varepsilon}{dt} = \frac{\varepsilon^2 I_1 + \varepsilon^3 I_2 + \varepsilon^4 I_3}{2\varepsilon E_1 + 3\varepsilon^2 E_2 + 4\varepsilon^3 E_3} \tag{23.4.21}$$

or

$$\frac{d\varepsilon}{dt} = (\varepsilon I_1 + \varepsilon^2 I_2 + \varepsilon^3 I_3)\frac{1}{2E_1}\left[1 - \varepsilon\frac{3E_2}{2E_1} + \varepsilon^2\left(\frac{9E_2}{4E_1} - \frac{2E_3}{E_1}\right)\right] \tag{23.4.22}$$

where higher-order terms and those terms much smaller than unity have been neglected. With some algebra we arrive at the nonlinear, ordinary differential equation, known as the stability equation, of the form

$$\frac{d\varepsilon}{dt} - \alpha_1\varepsilon - \alpha_2\varepsilon^2 - \alpha_3\varepsilon^3 = 0 \tag{23.4.23}$$

with the energy growth rate parameters α_i $(i = 1, 2, 3)$ defined as

$$\alpha_1 = \frac{1}{2E_1} I_1, \qquad \alpha_2 = \frac{1}{2E_1}\left(I_2 - \frac{3E_2}{2E_1} I_1\right)$$

$$\alpha_3 = \frac{1}{2E_1}\left\{I_3 - \frac{3E_2}{2E_1} I_2 + \left[\frac{9}{4}\left(\frac{E_2}{E_1}\right)^2 - \frac{2E_3}{E_1}\right] I_1\right\}$$

Notice that (23.4.23) is identical in form to Flandro [1985], although the basic approach to the formulation and the solution procedures differ, resulting in the energy growth rate parameters $\alpha_{(i)}$ entirely different from those in [Flandro, 1985]. For the case of linear and isentropic acoustic behavior, (23.4.23) is reduced to the results of Cantrell and Hart [1964] as demonstrated in Chung and Yoon, 1991 and Yoon and Chung, 1994.

The energy growth factor ε and energy growth rate parameters α_i are so named because they both represent energy growth. However, ε is devised such that it changes only as a function of time, which determines the stability of the entire domain as solved from the ordinary differential equation, (23.4.23). It does not change from point to point in the domain. On the other hand, the energy growth rate parameters α_i are spatially dependent, calculated through numerical integrations of quantities $a^{(i)}$, $b^{(i)}$, and $c^{(i)}$ in (23.4.19) and (23.4.20) as a result of the time-dependent flowfield solutions of Navier-Stokes system of equations. Through these combined processes, both ε and α_i can now be considered to depend on time and space, because (23.4.23) cannot be solved without the updated spatially dependent α_i.

It is seen that for linear stability $(\alpha_2 = \alpha_3 = 0)$, we have

$$\frac{d\varepsilon}{dt} - \alpha_1 \varepsilon = 0 \tag{23.4.24}$$

This is a special case of (23.4.23) for linear and isentropic waves which Cantrell and Hart [1964] obtained from the integral method. Since the solution of (23.4.24) is $\varepsilon = e^{\alpha_1 t}$, given the initial condition, $\varepsilon(0) = 1$, this condition is satisfied as follows:

$$\text{stable when } \varepsilon = 0, \quad \text{for } \alpha_1 = -\infty \tag{23.4.25a}$$

$$\text{neutrally stable when } \varepsilon = 1, \quad \text{for } \alpha_1 = 0 \text{ or } t = 0 \tag{23.4.25b}$$

$$\text{unstable when } \varepsilon = \infty, \quad \text{for } \varepsilon = 0, \text{ for } \alpha_1 = \infty \tag{23.4.25c}$$

From (23.4.25a) and (23.4.25b) we have

$$0 \le \varepsilon < l \quad \text{with} \quad -\infty \le \alpha_1 < 0 \tag{23.4.25d}$$

and from (23.4.25b) and (23.4.25c), we obtain

$$l < \varepsilon \le \infty \quad \text{with} \quad 0 < \alpha_1 \le \infty \tag{23.4.25e}$$

Stability is assured with $-\infty \le \alpha_1 < 0$ which implies that $0 \le \varepsilon < 1$. This criterion applies only to linear and isentropic waves.

The linear stability criteria discussed above do not hold true for nonlinear and nonisentropic waves with $\alpha_2 \ne 0$ and $\alpha_3 \ne 0$. The initial condition for ε for the case of $\alpha_2 \ne 0$ and $\alpha_3 \ne 0$ is unknown at $t = 0$. The magnitude of ε is no longer the measure of stability. Instead, the time rate of change of ε determines the state of instability regardless of initial conditions for ε at $t = 0$. Thus the stability criteria for nonlinear and nonisentropic waves are determined by the time rate of change of ε (energy growth or decay) as follows:

$$\text{stable when } \frac{d\varepsilon}{dt} < 0 \tag{23.4.26a}$$

$$\text{neutrally stable when } \frac{d\varepsilon}{dt} = 0 \tag{23.4.26b}$$

$$\text{unstable when } \frac{d\varepsilon}{dt} > 0 \tag{23.4.26c}$$

The variables involved in the energy growth parameters α_1, α_2, and α_3 are obtained from the numerical solution of Navier-Stokes system of equations. For this purpose the energy growth factor ε becomes a scaling factor at a neutral state ($\varepsilon = 1$) which corresponds to the standard definition such that, for any variable f, we write for $\varepsilon = 1$

$$f = \overline{f} + f' \tag{23.4.27}$$

This is in contrast to (23.4.7) in which fluctuations of all variables were scaled by the energy growth factor ε. Once the Navier-Stokes solution for f and its time average $\overline{f}$ is obtained, then the difference between the two represents the fluctuating part f'. The scaling of f' (or $\varepsilon f'$) is then required in the sense of (23.4.7) so as to determine whether the fluctuations may have a tendency to grow or decay. Thus $f' = f - \overline{f}$ for $\varepsilon = 1$, initially, allows us to provide data to compute α_1, α_2, and α_3 so that (23.4.23) can be solved. The initial values of ε at $t = 0$ are independent of the stability criteria of (23.4.26a), (23.4.26b), and (23.4.26c). A numerical example using this concept of entropy-controlled instability (ECI) method will be presented in Section 23.5.3.

23.5 EXAMPLE PROBLEMS

23.5.1 PRESSURE MODE ACOUSTICS

(1) Fluid-Structure-Acoustic Interaction

Kirchhoff's equations are widely used in the theory of light diffraction and other electromagnetic problems, boundary elements [Morino and Tseng, 1990], as well as in problems of wave propagation in acoustics [Pierce, 1981]. The example presented in this section deals with the special case of acoustic radiation from a vibrating structure [Frendi, Mastrello, and Ting, 1995]. The goal of the study is to establish the efficiency of Kirchhoff's equation for the computation of acoustic radiation from vibrating structures. In order to do this, two models are used: one in which the acoustic field is fully coupled to the structural vibration, and in the other, the two are decoupled. In the fully coupled model, the nonlinear Euler equations are used to describe the acoustic wave propagation and the nonlinear plate equations are used to describe the structural vibrations. These vibrations are excited using a harmonic plane wave acoustic source placed on the opposite side of the propagation domain.

In addition to (23.2.4) and (23.2.5), the ideal gas law given by $p = \rho RT$ is used, where p is the pressure, R the gas constant, and T the temperature. The initial conditions in the acoustic fluid are: $u = v = w = 0$, $p = p_0$ and $\rho = \rho_0$. The boundary conditions on the rigid and flexible structure are: $v(x,0,z,t)=0$ on the rigid surface and $v(x,0,z,t)=\eta_t(x,z,t)$ on the flexible part with $\eta_t(x,z,t)$ being the transverse velocity of the surface. The motion of the flexible surface is described by a system of three-dimensional, nonlinear partial differential equations, given by

$$D\nabla^4\eta+\rho_p h\eta_{tt}+\gamma\eta_t=\Delta p+\frac{Eh}{1-\nu^2}\left[\left(u_x^0+\frac{1}{2}\eta_x^2\right)(\eta_{xx}+\nu\eta_{zz})+\left(w_z^0+\frac{1}{2}\eta_z^2\right)\right.$$
$$\left.\times(\eta_{zz}+\nu\eta_{xx})+(1-\nu)\eta_{xz}(u_z^0+w_x^0+\eta_x\eta_z)\right],$$

$$\begin{aligned} u_{xx}^0+d_1u_{zz}^0+d_2w_{xx}^0&=-\eta_x(\eta_{xx}+d_1\eta_{zz})-d_2\eta_z\eta_{xz},\\ w_{zz}^0+d_1w_{xx}^0+d_2u_{xz}^0&=-\eta_z(\eta_{zz}+d_1\eta_{xx})-d_2\eta_x\eta_{xz}\end{aligned} \tag{23.5.1.1}$$

where $\nabla^4\eta=\eta_{xxxx}+2\eta_{xxzz}+\eta_{zzzz}$, $d_1=(1-\nu)/2$, $d_2=(1+\nu)/2$ and $D=Eh^3/12(1-\nu^2)$. In (23.5.1.1) u^0 and w^0 are the in-plane displacements, and η is the transverse displacement. The physical constants appearing in (23.5.1.1) are the stiffness, D, the density, ρ_p, the thickness, h, the physical damping, γ, the modulus of elasticity, E, and the Poisson ratio, ν. Equations (23.5.1.1) are solved subject to the initial and boundary conditions,

$$\begin{aligned} t&=0, & u^0&=w^0=\eta=\eta_t=0,\\ x&=x_0,x_0+L, & u^0&=w^0=\eta=\eta_x=0,\\ z&=z_0,z_0+W, & u^0&=w^0=\eta=\eta_z=0.\end{aligned} \tag{23.5.1.2}$$

In (23.5.1.1), the load Δp is defined by $\Delta p=2p^{(t)}-[p^{(i)}+p^{(r)}]$, where $p^{(t)}$, $p^{(i)}$ and $p^{(r)}$ are the transmitted, incident and reflected pressures, respectively. The sum $[p^{(i)}+p^{(r)}]$ represents the loading and is chosen to be of the form $p^{(i)}+p^{(r)}=[\varepsilon\sin(\omega t)+\varepsilon^*\sin(2\omega t)]H(t)$, where ε, ε^* and ω are the amplitudes and frequency of the wave, and $H(t)$ denotes the Heaviside unit step function. For small values of ε, $\varepsilon^*=0$ (linear), while for large values of ε, $\varepsilon^*=\varepsilon/2$ (nonlinear).

In the uncoupled model, the velocity potential $\phi(t,x,y,z)$ is governed by the simple wave equation given in (23.2.1) and the acoustic pressure and velocity are related to the potential by $p=-\rho_0\frac{\partial\phi}{\partial t}$ and $v=\frac{\partial\phi}{\partial y}$. To produce the incident excitation given above, the incident potential should be of the form

$$\phi^{(i)}(t,x,y,z)=\frac{\varepsilon}{2\rho_0\omega}\left[\cos\left[\omega\left(t+\frac{y}{a_0}\right)\right]+2\frac{\varepsilon^*}{\varepsilon}\cos\left[2\omega\left(t+\frac{y}{a_0}\right)\right]\right]H\left(t+\frac{y}{a_0}\right) \tag{23.5.1.3}$$

which is a solution to (23.2.1). In the case of a rigid surface $\phi^{(r)}(t,x,y,z)=\phi^{(i)}(t,x,-y,z)$, the total potential on the incident side is $\phi=\phi^{(i)}+\phi^{(r)}+\phi^{(s)}$ and that on the transmitted side is $\phi=\phi^{(t)}$, with both $\phi^{(s)}$ and $\phi^{(t)}$ satisfying (23.2.1). Since the ambient fluid above and below the flexible structure is the same, the velocity potential induced by the structural vibration is anti-symmetric in y, in other words, $\phi^{(t)}(t,x,y,z)=-\phi^{(s)}(t,x,-y,z)$; this leads to a pressure difference across the structure of

$$\Delta p=2\rho_0\left[\frac{\partial\phi^{(i)}}{\partial t}+\frac{\partial\phi^{(s)}}{\partial t}\right].$$

Solving the wave equation for $\phi^{(s)}$ subject to the initial and boundary conditions given above we obtain Kirchhoff's equation

$$\phi^{(s)}(t, x, y > 0, z) = -\frac{1}{2\pi}\int\int \frac{[\eta_t(t, x', z')]}{R} dx'dz' \tag{23.5.1.4}$$

where the $R = [(x - x')^2 + y^2 + (z - z')^2]^{1/2}$ denotes the distance from a point $(x, y > 0, z)$ to a source located at $(x', 0, z')$ and $[\cdot]$ denotes the retarded value as defined in earlier sections. Since the time derivative of $\phi^{(s)}$ is needed in the pressure difference equation, one can differentiate the above equation to obtain

$$\phi_t^{(s)}(t, x, y > 0, z) = -\frac{1}{2\pi}\int\int \frac{[\eta_{tt}(t, x', z')]}{R} dx'dz'. \tag{23.5.1.5}$$

Equations (23.5.1.1) and (23.5.1.5) are solved together to obtain the structural response and acoustic radiation. These equations form the decoupled model.

The nonlinear Euler equations are solved using an FDM scheme developed by Gottlieb and Turkel [1976], which is a modified version of the McCormack scheme, while the plate equations are solved by an FEM method [Robinson, 1990]. The integral given by (23.5.1.5) is computed by a combination of Simpson's and the trapezoidal rule. Results are obtained using both models for an excitation frequency of 751 Hz which corresponds to a natural frequency of the structure. At high levels of excitation, a harmonic is used (1502 Hz) in addition to the fundamental (751 Hz), in order to simulate an experimental study. The structural parameters are considered to be uniform and are given by: density $\rho_p = 4450.15$ kg/m^3, modulus of elasticity $E = 1.013 \times 10^5$ N/m^2, Poisson ratio $\nu = 0.33$, and a damping ratio of 0.01 is used. The acoustic fluid properties are: temperature $T_0 = 288.33$ K, density $\rho_0 = 1.23$ kg/m^3, pressure $p_0 = 1.013 \times 10^5$ N/m^2, sound speed $a_0 = 340$ m/s, specific heat at constant volume $c_v = 1.004$ kJ/(kg K), and the ratio of specific heats is $\gamma = c_p/c_v = 1.4$.

For a small amplitude excitation, 130 dB or 6.8×10^{-4} atm, the structural response is linear as shown by Figures 23.5.1.1 and 23.5.1.2. The nondimensional time histories of the displacement and near-field radiated pressure are shown in addition to their respective power spectra. The displacement is nondimensionalized with respect to the thickness, while the pressure is nondimensionalized with respect to $\rho_0 a_0^2$. Both the power spectra and the time histories show the presence of a single frequency indicative of a linear behavior. When the excitation level is increased to 174 dB, 0.22 atm, the response of the flexible structure becomes nonlinear as shown by both the time history and the power spectra (Figure 23.5.1.3). The response predictions obtained using both models are in agreement. In addition, both the near-field and far-field acoustic pressure results obtained using both models compare well (Figures 23.5.1.4 and 23.5.1.5). This is an important result since the cost, in terms of CPU time, is an order of magnitude lower in the uncoupled case. One can use the uncoupled model to predict, with reasonable accuracy, the structural response and the resulting acoustic radiation.

(2) Blade-Vortex-Interaction Hover Noise

Xue and Lyrintzis [1994] carried out three-dimensional computations of the noise generated by the blade-vortex-interaction (BVI) in a transonic regime. The near field

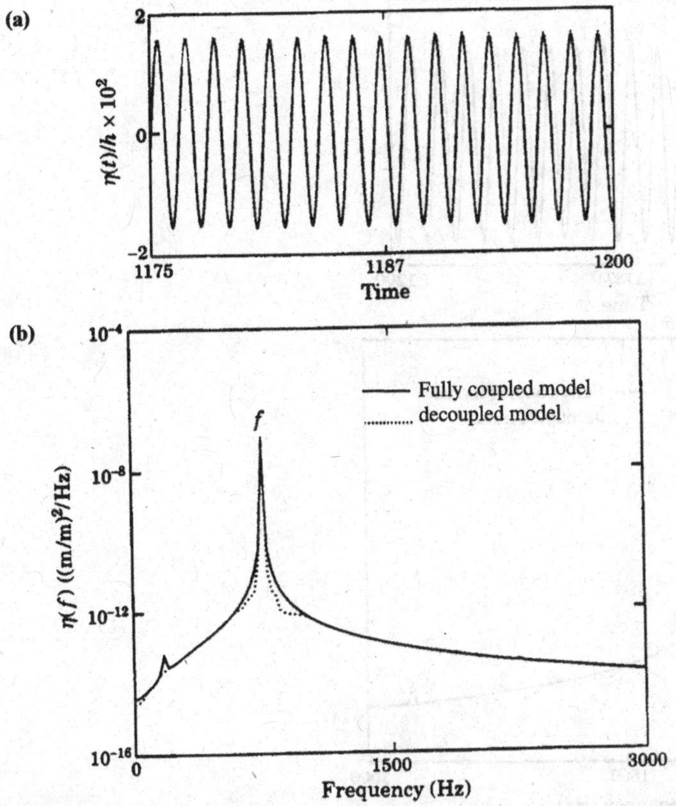

Figure 23.5.1.1 (a) The time history and (b) power spectral density of the panel center displacement for an excitation amplitude 130 dB (linear). [Frendi et al., 1995]. Reprinted with permission from Academic Press.

is assumed inviscid and irrotational, which allows a potential flow formulation. These assumptions are reasonable since the flow is dominated by the inviscid effects and the shocks are not strong enough to generate sizable vorticity. The potential equations are solved using FDM. These near-field results are then coupled with a rotating Kirchhoff method to obtain the far-field noise.

The full potential equations written in a strong conservation form and in generalized coordinates are

$$\frac{\partial}{\partial \tau}\left(\frac{\rho}{J}\right)+\frac{\partial}{\partial \xi}\left(\frac{\rho U}{J}\right)+\frac{\partial}{\partial \eta}\left(\frac{\rho V}{J}\right)+\frac{\partial}{\partial \zeta}\left(\frac{\rho W}{J}\right)=0 \quad (23.5.1.6)$$

with the density given by

$$\rho=\left\{1-\frac{\gamma-1}{2}[2\Phi_\tau+(U+\xi_t)\Phi_\xi+(V+n_t)\Phi_\eta+(W+\varsigma_t)\Phi_\varsigma]\right\}^{\frac{1}{\gamma-1}} \quad (23.5.1.7)$$

and U, V, and W are contravariant velocities along the ξ, η, and ς directions,

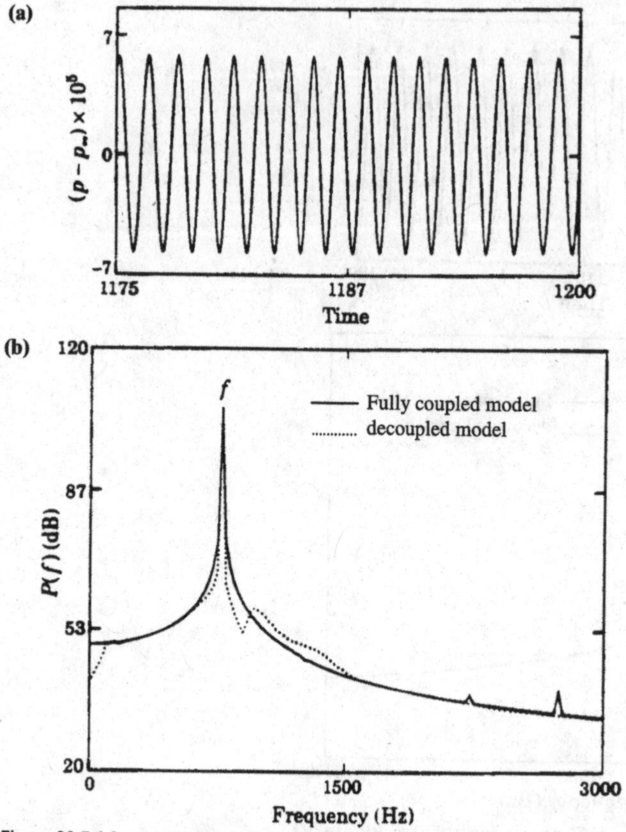

Figure 23.5.1.2 (a) The time history and (b) power spectral density of the transmitted near-field pressure, 2.54 cm from the panel center, for an excitation amplitude 130 dB (linear) [Frendi et al., 1995]. Reprinted with permission from Academic Press.

respectively, and are given by

$$\begin{aligned} U &= \xi_t + A_1\Phi_\xi + A_4\Phi_\eta + A_5\Phi_\varsigma \\ V &= \eta_t + A_4\Phi_\xi + A_2\Phi_\eta + A_6\Phi_\varsigma \\ W &= \varsigma_t + A_5\Phi_\xi + A_6\Phi_\eta + A_3\Phi_\varsigma \end{aligned} \tag{23.5.1.8}$$

and

$$\begin{aligned} A_1 &= \xi_x^2 + \xi_y^2 + \xi_z^2 \\ A_2 &= \eta_x^2 + \eta_y^2 + \eta_z^2 \\ A_3 &= \varsigma_x^2 + \varsigma_y^2 + \varsigma_z^2 \\ A_4 &= \xi_x\eta_x + \xi_y\eta_y + \xi_z\eta_z \\ A_5 &= \xi_x\varsigma_x + \xi_y\varsigma_y + \xi_z\varsigma_z \\ A_6 &= \eta_x\varsigma_x + \eta_y\varsigma_y + \eta_z\varsigma_z \\ J &= \xi_x(\eta_y\varsigma_z - \eta_z\varsigma_y) + \eta_x(\xi_z\varsigma_y - \xi_y\varsigma_z) + \varsigma_x(\xi_y\eta_z - \xi_z\eta_y). \end{aligned} \tag{23.5.1.9}$$

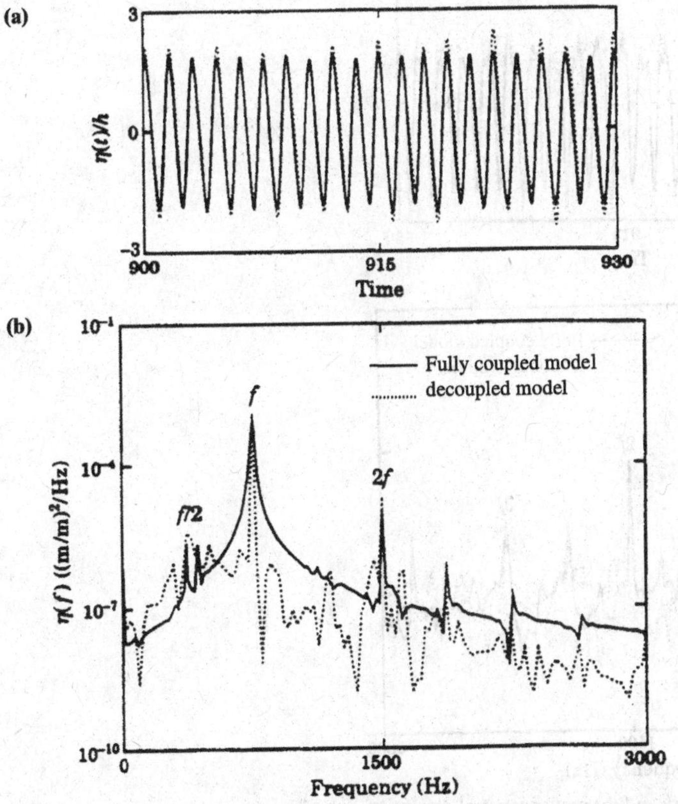

Figure 23.5.1.3 (a) The time history and (b) power spectral density of the panel center displacement for an excitation amplitude 174 dB (nonlinear) [Frendi et al., 1995]. Reprinted with permission from Academic Press.

Here, A_1 through A_6 are metric quantities, and J is the Jacobian of the transformation $|\partial(\xi, \eta, \varsigma)/\partial(x, y, z)|$. A tangency boundary condition is used on the blade surface and a nonreflecting boundary condition at the outer radial position. In order to carry out the Kirchhoff integration to the far field, a rigid control surface wrapped around the rotating blade is used. Therefore, the control surface rotates with the blade.

Using Morino's rotating Kirchhoff formulation, we obtain the pressure of an observer in the rotating coordinate system as

$$p(\mathbf{x}_*, t_*) = \frac{1}{4\pi} \iint_S \left[\frac{-1}{\hat{\rho}} \frac{\partial p}{\partial \mathbf{n}} - \frac{\partial}{\partial \hat{\mathbf{n}}} \left(\frac{-1}{\hat{\rho}} \right) p + \frac{-1}{\hat{\rho}} \frac{\partial p}{\partial t} \left(\frac{\partial \theta}{\partial \hat{\mathbf{n}}} + 2 \frac{\mathbf{M} \cdot \mathbf{n}}{a_0} \right) \right]_{\tau_*} dS \tag{23.5.1.10}$$

with

$$\frac{\partial}{\partial \hat{\mathbf{n}}} = -\mathbf{n} \cdot \nabla_{x_*} + \mathbf{M} \cdot \mathbf{n} \mathbf{M} \cdot \nabla_{x_*} \tag{23.5.1.11}$$

The above equations are valid for any arbitrary moving surface. For the case of a rotating

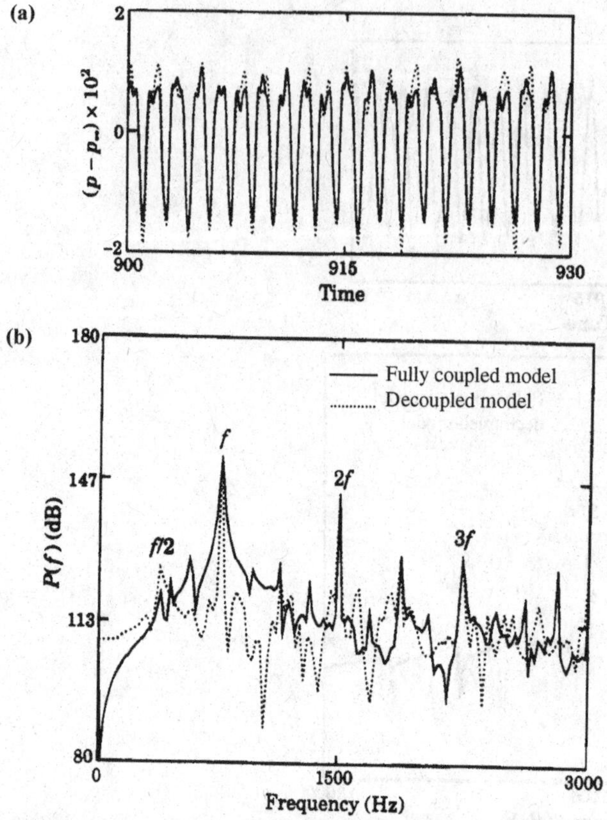

Figure 23.5.1.4 (a) The time history and (b) power spectral density of the transmitted near-field pressure, 2.54 cm from the panel center, excitation amplitude 174 dB (nonlinear) [Frendi et al., 1995]. Reprinted with permission from Academic Press.

surface (i.e., hover)

$$\mathbf{M} = \frac{1}{a_0}(\omega \times \mathbf{X}), \qquad \mathbf{M}_* = \frac{1}{a_0}(\omega \times \mathbf{X}_*)\; \hat{\rho} = r(1 - M_r). \tag{23.5.1.12}$$

The location having the subscript '*' represents the observer's location, the other represents the source. The symbol θ in (23.5.1.10) is given by $\theta = \tau^* - t_*$, with τ^* being the source emission time and is the solution to

$$\tau - t_* + \frac{|X_*(t_*) - X(\tau)|}{a_0} = 0 \tag{23.5.1.13}$$

where τ represents the source time.

Farassat and Myers [1988] derived a rotating Kirchhoff formula for a stationary observer and a rotating control surface. Their formulation is mathematically identical to that of Morino (above) after making a coordinate transformation. The advantage of Farassat and Myers' formula is that it allows direct comparison with experiments.

The model and numerical techniques were tested using a problem with a known analytical solution. The results showed good agreement between all the solutions. Results were then obtained for a nonlifting rotor with the NACA 0012 airfoil section shown

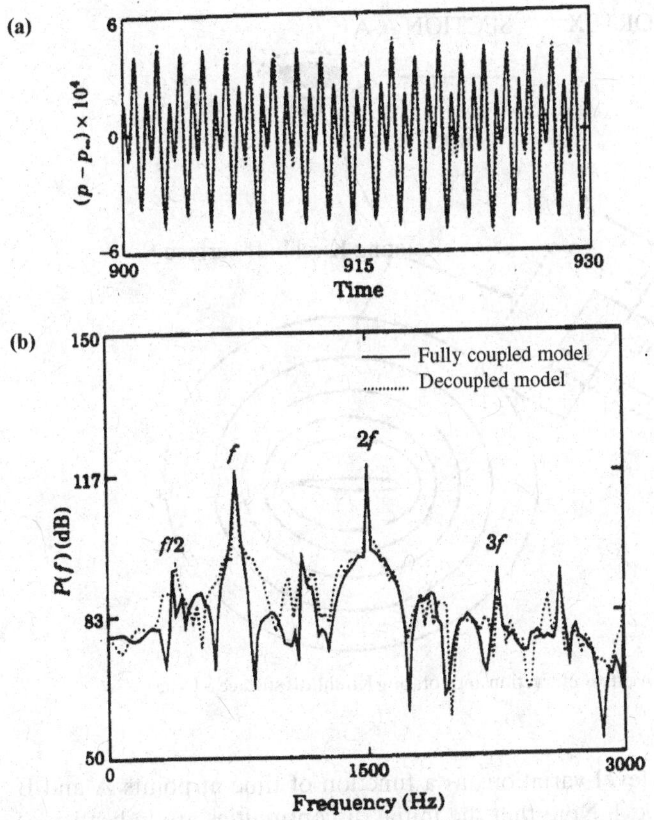

Figure 23.5.1.5 (a) The time history and (b) power spectral density of the transmitted far-field pressure, 203.2 cm from the panel center, excitation amplitude 174 dB (nonlinear) [Frendi et al., 1995]. Reprinted with permission from Academic Press.

in Figure 23.5.1.6. An $80 \times 25 \times 25$ grid was used, and 600 time-steps were taken. Each time step corresponds to a 0.3-deg rotor azimuth per time step. It should be noted that to get better results for lift coefficient versus blade rotation Ψ, 0.125 deg should be used for $\Delta\Psi$. The observer position is defined using distance d and angles α and β shown in Figure 23.5.1.7. Figure 23.5.1.8 shows the noise signal for $\alpha = 60$ and $\beta = -30$. It can be seen that with increased tip Mach number the two disturbances increase proportionally. Figure 23.5.1.9 shows the effect of vortex strength and location. Both parameters have a significant effect in the resulting noise as shown in Figure 23.5.1.9. A higher vortex strength increases both disturbances, while a higher distance decreases both of them and especially the second one. A low vortex strength and a higher distance are desirable for low BVI noise.

(3) Noise Level in Rocket Nozzle Exhaust

To illustrate, let us consider the rocket nozzle investigated by Carofano [1984] using FDM/TVD and subsequently by Chung and Yoon [1993] with FEM/FDV. The geometry, initial and boundary conditions shown in Figure 23.5.1.10a, along with the shock positions of both computations at $t = 0.0012$ sec favorably compared in

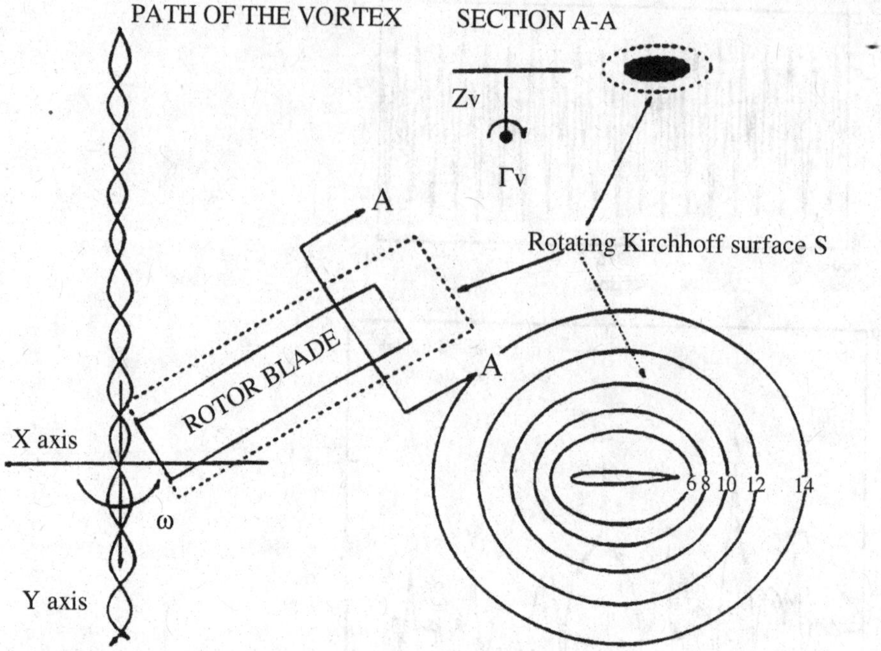

Figure 23.5.1.6 Hover parallel blade-vortex interaction and rotating Kirchhoff surface S [Xue and Lyrintzis, 1994].

Figure 23.5.1.10b. The noise level variations as a function of time at points A and B are shown in Figure 23.5.1.10c,d. Note that the initial discontinuities are indicative of the secondary shock and adjacent vorticity formation. The noise level reaches the constant value of approximately 193 dB at $t = 0.00045$ sec for A, with some delay (0.0005 sec) at point B.

Figure 23.5.1.11a shows the density distribution at $t = 0.0012$ sec in which the primary shock wave dominates the front, followed by slip (contact) surface and by the secondary shock. Vorticity develops at the upstream region. Expansion waves start at

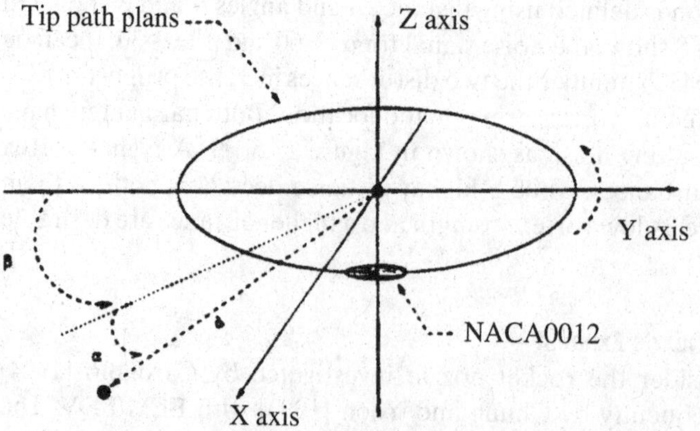

Figure 23.5.1.7 Observer's position for a rotating blade [Xue and Lyrintzis, 1994].

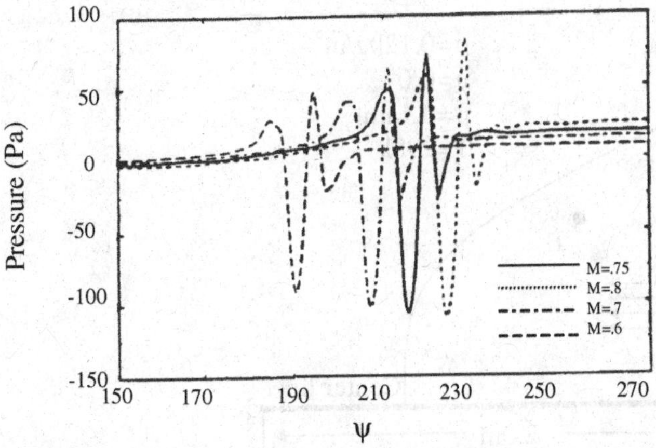

Figure 23.5.1.8 Different Mach numbers for hover parallel BVI, $M = 0.8, 0.75, 0.7$, and 0.6, $\Gamma_v = 0.2$, $Z_v = -0.26$ at position $d = 3R$, $\alpha = 60$, $\beta = -30$ [Xue and Lyrintzis, 1994].

the exit separation points, which are connected to the secondary and slip surface. These physical phenomena are consistent with a typical plume flowfield. The corresponding pressure, temperature, and Mach number distribution are shown in Figures 23.5.1.11b–d. In Figure 23.5.1.11e, the two-dimensional distributions of noise level at $t = 0.0012$ sec are shown. Note that beyond regions of vorticity and second shock, the noise level once again becomes constant, reflecting the flowfield. Similar plots are shown in Figure 23.5.12a–e for $t = 0.0025$ sec, demonstrating the progress of flowfield and noise level.

(4) Noise Control in Perforated Muzzle Brake

The geometry of flow through a vent hole in a perforated muzzle brake is shown in Figure 23.5.1.13a initially investigated by Carofano [1987] using FDM/TVD and

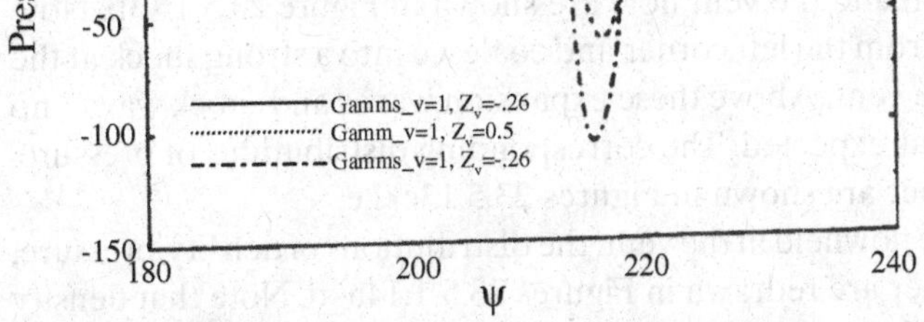

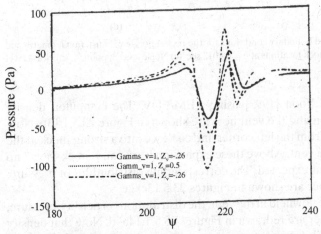

Figure 23.5.1.9 Different Mach numbers for hover parallel BVI, $M = 0.75$ at position $d = 3R$, $\alpha = 60$, $\beta = -30$ [Xue and Lyrintzis, 1994].

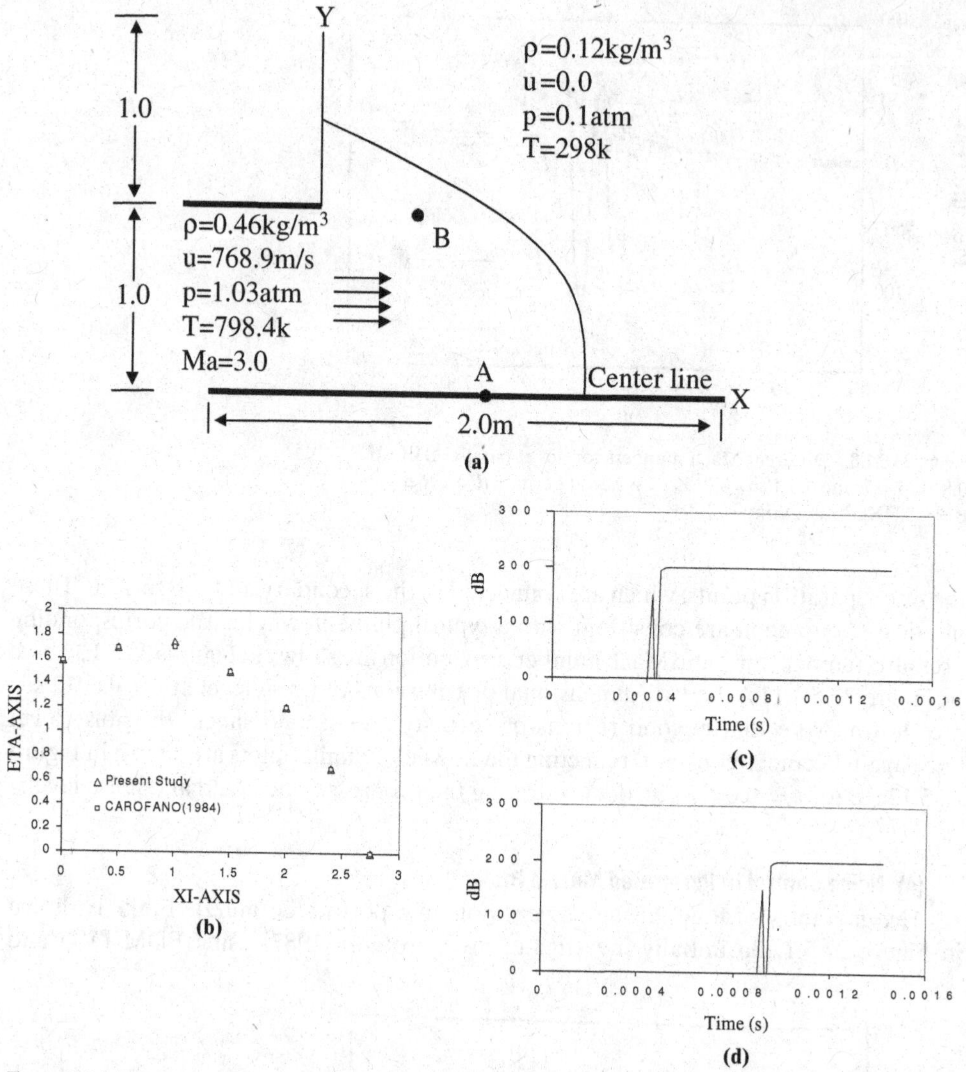

Figure 23.5.1.10 Geometry, initial and boundary conditions for the rocket nozzle exhaust. **(a)** Geometry and initial and boundary condition. **(b)** Shock locations at $t = 0.0012$ sec. **(c)** Noise level at point A. **(d)** Noise level at point B.

subsequently by Chung and Yoon [1993] using FEM/FDV. The computed density distributions in the test section and the vent hole are shown in Figure 23.5.1.13b. Note that expansion waves fan out from the left corner and coalesce into a strong shock at the right corner extending into the vent. Above these expansion waves and shock waves, no significant flowfield develops as expected. The corresponding distributions of pressure, temperature, and Mach number are shown in Figures 23.5.13c,d,e.

To accentuate details of the flowfield in the vent, the distributions of density, pressure, temperature, and Mach number are redrawn in Figures 23.5.1.14a–d. Note that density and pressure distributions exhibit sharp changes in gradients toward downstream of the vent. This indicates that the role of the vent is to reduce the noise at the expense of

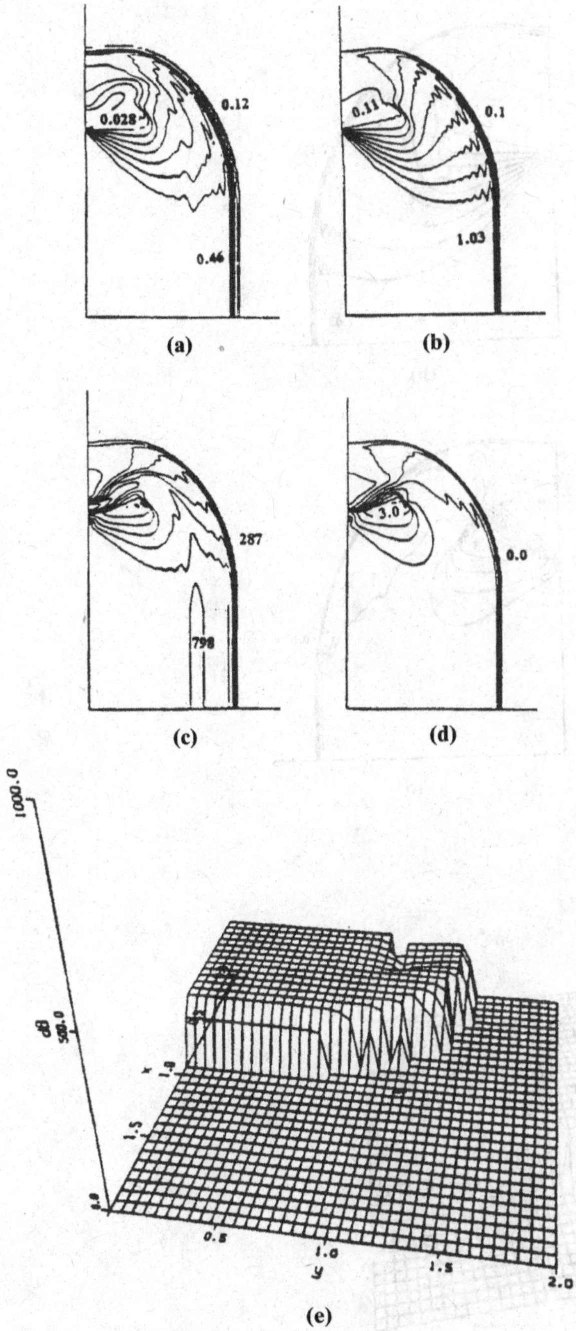

Figure 23.5.1.11 Noise level (dB) distribution and corresponding flowfields at $t = 0.0012$ sec. **(a)** Density distribution. **(b)** Pressure distribution. **(c)** Temperature distribution. **(d)** Mach number distribution. **(e)** Noise level (dB).

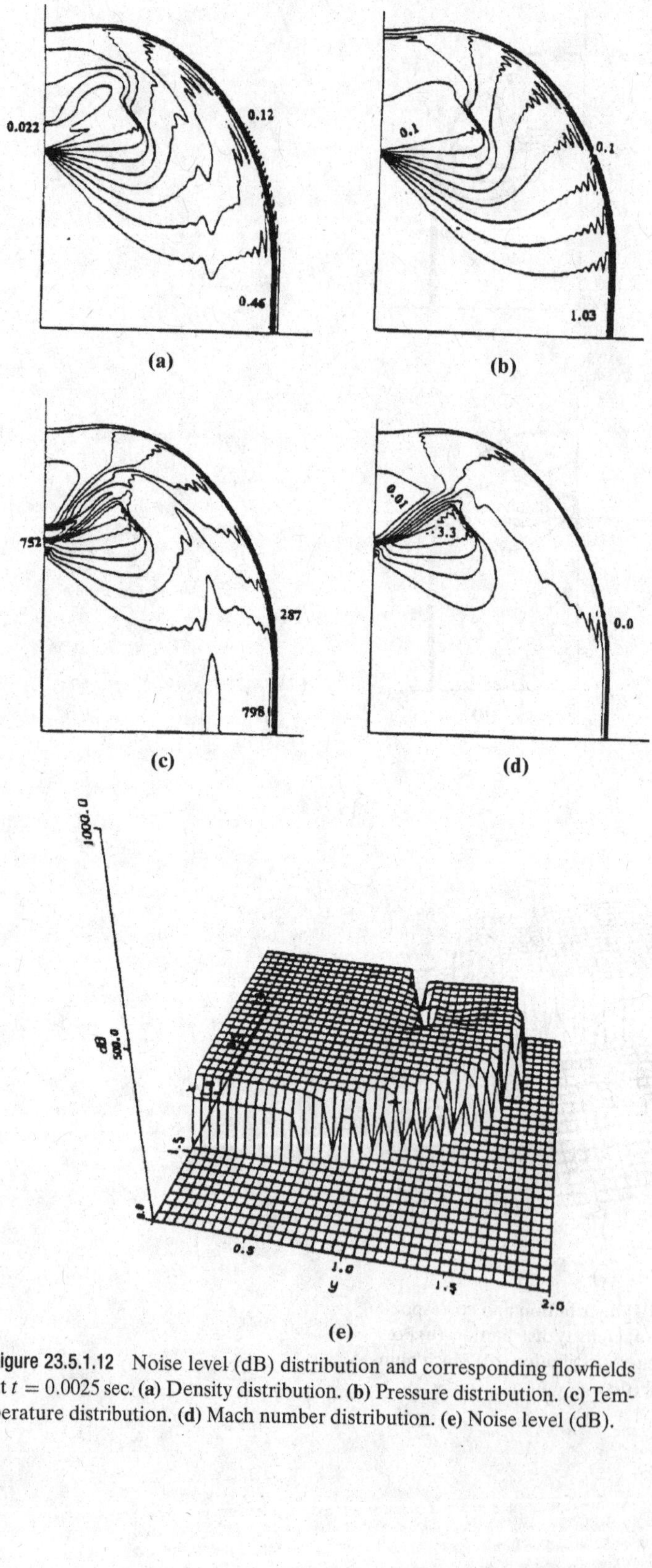

Figure 23.5.1.12 Noise level (dB) distribution and corresponding flowfields at $t = 0.0025$ sec. **(a)** Density distribution. **(b)** Pressure distribution. **(c)** Temperature distribution. **(d)** Mach number distribution. **(e)** Noise level (dB).

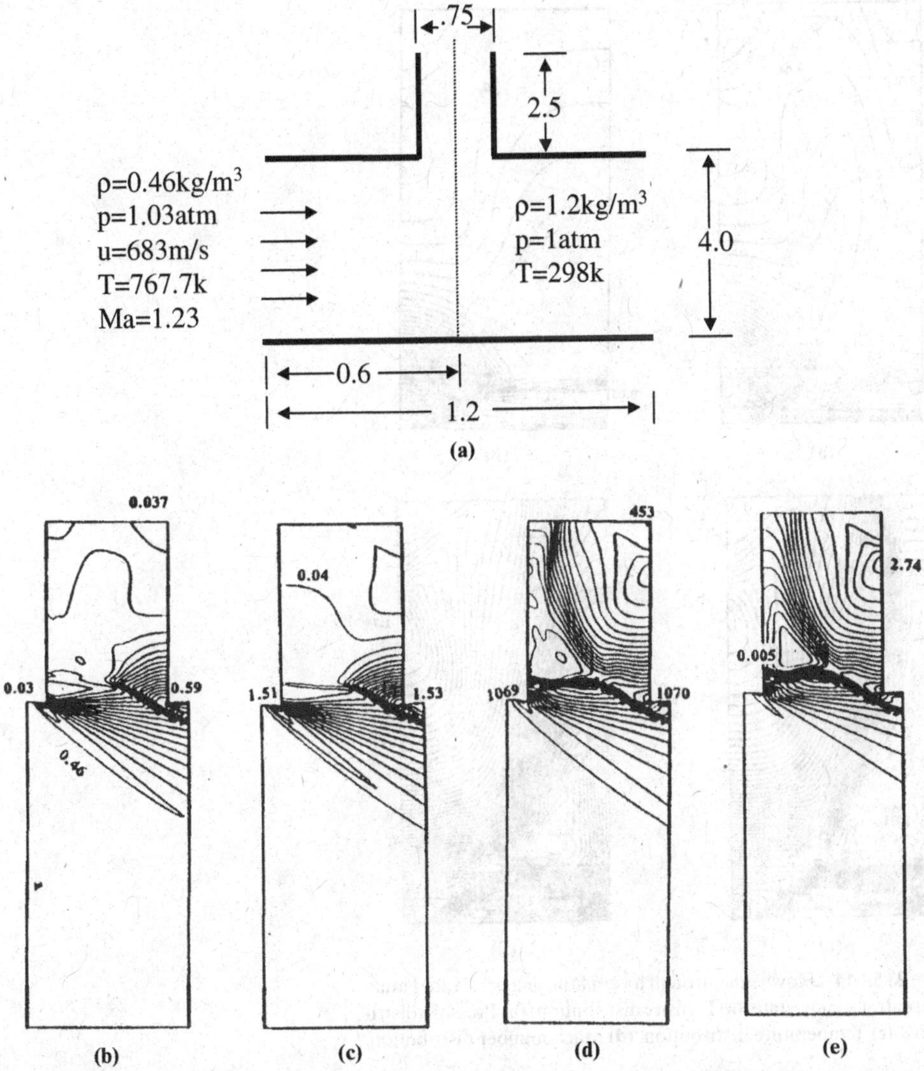

Figure 23.5.1.13 Geometry and flowfields in a perforated muzzle break at steady-state. **(a)** Geometry and initial boundary conditions. **(b)** Density. **(c)** Pressure. **(d)** Temperature. **(e)** Mach number.

heat exhaust. Thus, the proper sizes and geometries of the vent can serve as an excellent noise control device.

For the purpose of comparison of these results with the experimental data of Carofano, we examine the pressure ratios at various locations (Figure 23.5.1.15a). At $x/D = 0.0$ the pressure ratio (p/p_o) vs. y/D are shown in Figure 23.5.1.15b. The results of FEM/FDV (solid line) are identical to the experimental results (solid circle) at $y/D < 0.3$. Some deviations of the Carofano calculations (dotted line) from the FEM/FDV solution are noted. For the rest of the locations (Figure 23.5.1.15c–g) the results of the FEM/FDV are compared favorably with the experiments. It should be noted

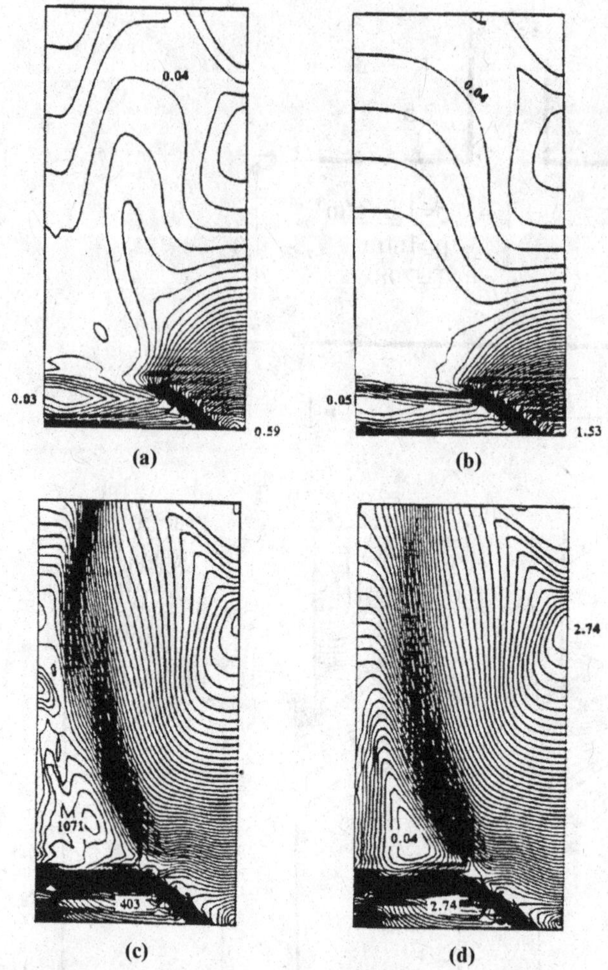

Figure 23.5.1.14 Flowfields through a vent hole in a perforated muzzle break at steady-state. (a) Density distribution. (b) Pressure distribution. (c) Temperature distribution. (d) Mach number distribution.

that Carofano used the low-resolution TVD scheme. Had the high-resolution scheme been used, however, it is anticipated that a closer agreement with the experiments and FEM/FDV would have been achieved.

23.5.2 VORTICITY MODE ACOUSTICS

(1) Isotropic Turbulence

Applications of the Lighthill acoustic analogy are numerous. Recently, Sarkar and Hussaini [1993] developed a Hybrid Direct Numerical Simulation for the computation of sound radiated from isotropic turbulence. This method consisted of using DNS to resolve the turbulent flow together with the Lighthill acoustic analogy for the farfield sound. They suggested that using the first form of (23.3.1.4) would be more advantageous

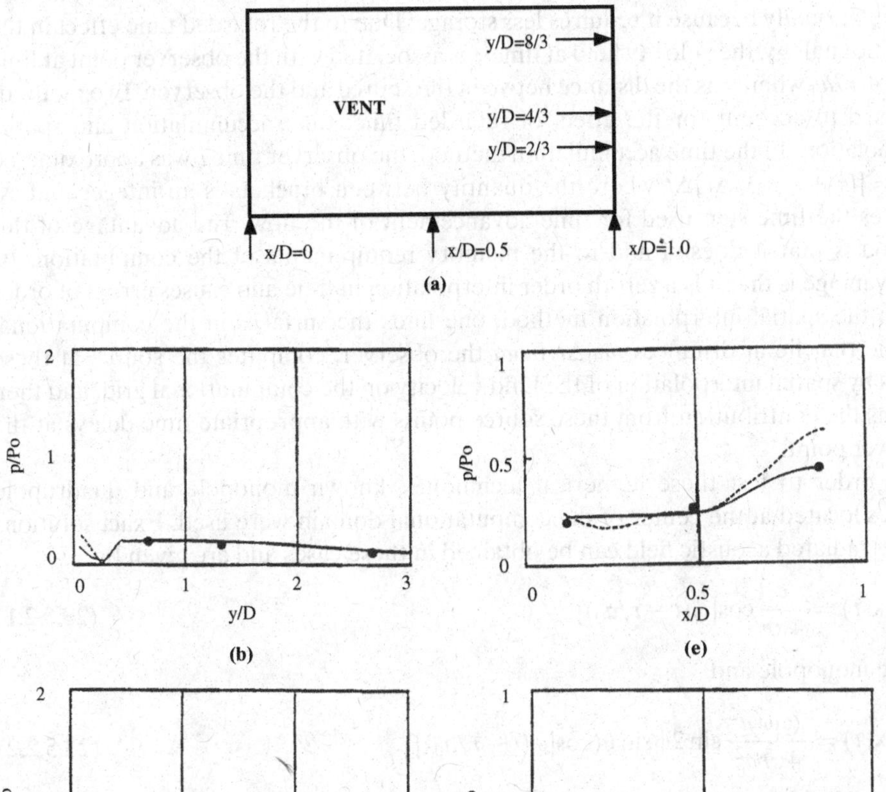

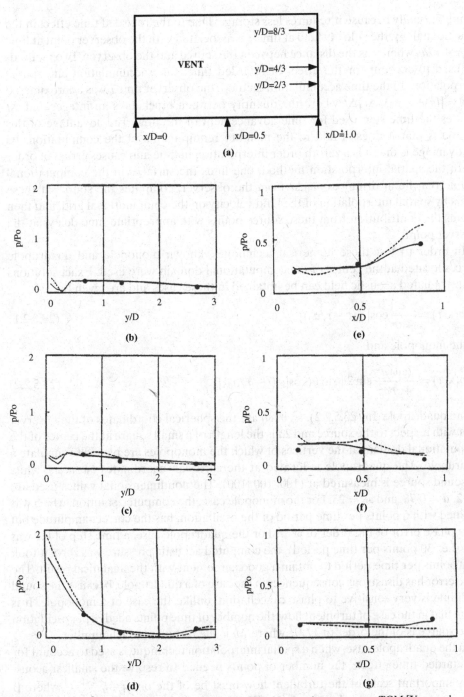

Figure 23.5.1.15 Comparison of pressure rations (P/P_O) with experiments, —— TGM [Yoon and Chung, 1994], ------ TVD [Carofano, 1987], • • Experiment [Carofano, 1987]. **(a)** Locations of data. **(b)** $X/D = 0.0$. **(c)** $X/D = 0.5$. **(d)** $X/D = 1.0$. **(e)** $Y/D = 2/3$. **(f)** $Y/D = 4/3$. **(g)** $Y/D = 8/3$.

computationally because it requires less storage. Due to the retarded time effect in the acoustic analogy, the velocity field at time τ is associated with the observer point at time $t = \tau + r/a_0$ where r is the distance between the source and the observer. Two methods are used to account for the effect of retarded time: time accumulation and spatial interpolation. In the time accumulation method, the observer time t was approximated by $t = [(\tau + r/a_0)/\Delta t]\Delta t$ where the quantity between brackets is an integer and Δt denotes the time step used for time advancement of the flow. The advantage of this method is that it doesn't add to the memory requirements of the computation. Its disadvantage is that it is a zeroth order interpolation in time and causes errors of order Δt. In the spatial interpolation method, one finds the surfaces in the computational volume that lie at distances $na_0\Delta t$ from the observer, computes the source at these points by spatial interpolation of the fluid velocity on the computational grid, and then obtains the contribution from these source points with appropriate time delays at the observer point.

In order to test these numerical techniques, known monopole and quadrupole sources located at the center of the computational domain were used. Exact solutions for the radiated acoustic field can be obtained in these cases and are given by

$$p(\mathbf{x}, t) = \frac{c}{4\pi r} \cos[\omega(t - r/a_0)] \tag{23.5.2.1}$$

for the monopole and

$$p(\mathbf{x}, t) = \frac{2c\omega^2 d^2}{4\pi r a_0^2} \sin 2\phi \sin\theta(\cos[\omega(t - r/a_0)]) \tag{23.5.2.2}$$

for the quadrupole. In (23.5.2.2), (r, θ, ϕ) are the spherical coordinates of the observer point with respect to the source and $2d$ is the length of a small square at the center of the computational domain at the vertices of which the monopoles are placed to simulate a quadrupole. The quadrupole is located at the cartesian coordinates (π, π, π), while the sound source is measured at (100, 100, 100). The nondimensional values used are $\omega = 2$, $d = 0.78$, and $a_0 = 25$. For the monopole case, the computed solution, which was obtained with 6 points per time period of the oscillation, has the correct amplitude but had a phase error of the order of $\omega\Delta t$. For the quadrupole case, a time step of 0.1 was used (i.e., 30 points per time period), the computed acoustic pressure was zero; it took 3,000 points per time period to obtain a good agreement with the analytical result. The phase error has disastrous consequences in the case of a quadrupole, because the sound amplitude is very sensitive to phase cancellation, unlike the case of a monopole. It is shown that in the case of turbulent flow, the number of time points needed per oscillation of the source is of the order of $1/M_t^2$ where M_t is the turbulent Mach number.

In the quadrupole case, when a spatial interpolation technique is used to account for the retarded time effects, the number of points needed to resolve the smallest acoustically important scale of the turbulent flow must be of the order $M_t^{-2/(\beta+1)}$ where β is the order of the interpolation scheme. This implies that a high order interpolation scheme is needed for low Mach number turbulence (third order or higher). Because of the severe time step restrictions in low Mach number turbulence, the second form of (23.3.1.4) is used in the following example.

Once these initial tests were completed, the actual problem of sound radiation from an isotropic turbulence is addressed. The turbulent flow inside a cubical domain is computed by solving the Navier-Stokes equations numerically. Since the turbulence is homogeneous, periodic boundary conditions are used in all three directions. A Fourier collocation method for the spatial discretization of the governing equation is used together with a third order low storage Runge-Kutta scheme for time advancement. Initial conditions are needed for v_i', ρ, p, and T. The initial velocity field is divided into two components: solenoidal and compressible velocity fields. The solenoidal velocity field which satisfies $\nabla \cdot v^{I'} = 0$ is chosen to be a random Gaussian field with the power spectrum $E(k) = k^4 \exp(-2k^2/k_m^2)$ where k_m corresponds to the peak of the power spectrum. The compressible velocity field which satisfies $\nabla \times v^{C'} = 0$ is also chosen to be a random Gaussian field satisfying the same power spectrum. The power spectra of the two velocity components are scaled so as to obtain a prescribed v_{rms} and a prescribed $\chi = v_{rms}^C / v_{rms}$ which is the compressible fraction of kinetic energy. The pressure associated with the incompressible velocity is evaluated using the Poisson equation $\nabla^2 p^{I'} = -\overline{\rho} v_{i,j}^{I'} v_{j,i}^{I'}$. The mean density $\overline{\rho}$ is chosen to be unity, $\overline{p}$ is chosen so as to obtain a prescribed Mach number characterizing the turbulence ($\approx v_{rms}/\sqrt{\gamma \overline{p}/\overline{\rho}}$). The fluctuating density and compressible pressure are chosen as random fields with the same power spectrum as that given above.

The results are obtained for a Lighthill tensor of the form $T_{ij} = \overline{\rho} v_i v_j$. Two computations are carried out using two different time steps, the second case having half the time step of the first, that is, $\Delta t \varepsilon_0 / K_0 = 0.00375$. The spatial discretization used is a uniform 64^3 grid, while the initial parameters are: viscosity $\nu_0 = 1/225$, the turbulent Mach number $M_{t,0} = 0.05$, and the Taylor microscale Reynolds number $R_{\lambda,0} = 38$ ($R_{\lambda,0} = q\lambda/\nu$, $q = \sqrt{v_i' v_i'}$, $\lambda = q/\sqrt{\omega_i' \omega_i'}$ while $M_t = q/\overline{a}$ where $\overline{a}$ is mean speed of sound). Figure (23.5.2.1a) shows the evolution of the acoustic pressure at a far-field point as a function of eddy turnover time. The acoustic pressure decreases with time

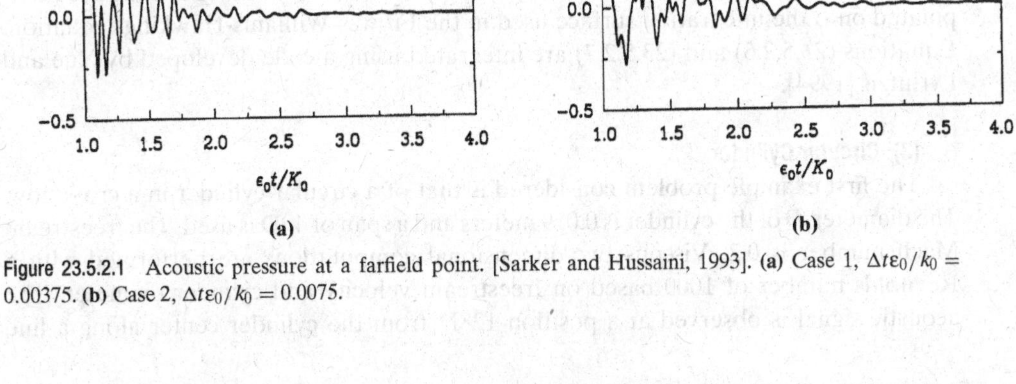

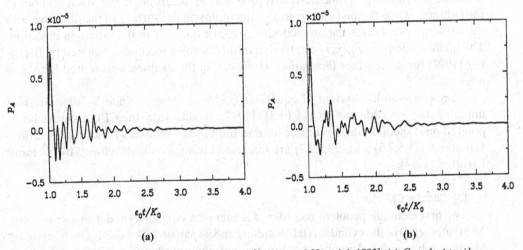

Figure 23.5.2.1 Acoustic pressure at a farfield point. [Sarker and Hussaini, 1993]. (a) Case 1, $\Delta t \varepsilon_0 / k_0 = 0.00375$. (b) Case 2, $\Delta t \varepsilon_0 / k_0 = 0.0075$.

because of the decay in turbulence. Reducing the time step by half leads to the far-field acoustic pressure shown on Figure 23.5.2.1b. A small difference in the details of the time history can be seen.

(2) The Ffowcs Williams-Hawkings Equation

Singer et al. [1999] examined the difficulties involving the use of a hybrid scheme coupling a CFD flow computation with the Ffowcs Williams-Hawkings equation to predict noise generated by vortices passing over a sharp edge. Three sound radiation model problems are studied: a circular cylinder in a cross flow, a two-dimensional vortex filament moving around the edge of a half plane, and vortices convecting past the trailing edge of an airfoil. Following Brentner and Farassat [1998], the differential form of the Ffowcs Williams-Hawkings equation (23.3.7) is used. An integral representation of this equation can be written directly by utilizing formulation 1A of Farassat and Myers [1988] and Brentner [1986]:

$$p(\mathbf{x},t) = p_T(\mathbf{x},t) + p_L(\mathbf{x},t) + p_Q(\mathbf{x},t) \tag{23.5.2.3}$$

where

$$4\pi p_T(\mathbf{x},t) = \int_{f=0}\left[\frac{\rho_0(\dot{U}_n + U_{\dot{n}})}{r(1-M_r)^2}\right]dS + \int_{f=0}\left[\frac{\rho_0 U_n(r\dot{M}_r + a_0(M_r - M^2))}{r^2(1-M_r)^3}\right]dS \tag{23.5.2.4}$$

and

$$4\pi p_L(\mathbf{x},t) = \frac{1}{a_0}\int_{f=0}\left[\frac{\dot{L}_r}{r(1-M_r)^2}\right]dS + \frac{1}{a_0}\int_{f=0}\left[\frac{L_r(r\dot{M}_r + a_0(M_r - M^2))}{r^2(1-M_r)^3}\right]dS + \int_{f=0}\left[\frac{L_r - L_M}{r^2(1-M_r)^2}\right]dS. \tag{23.5.2.5}$$

Here the dot indicates a time derivative, $L_M = L_i M_i$ where M_i is the Mach number in the i-direction, r is the distance from a source point on the surface to the observer, and the subscript r indicates the projection of a vector quantity in the radiation direction. The quadrupole term $p_Q(\mathbf{x},t)$ can be determined using a method developed by Brentner [1997] (or some other technique). However, in the examples presented here, it is neglected.

The Navier-Stokes system of equations is solved using a Finite Volume Method implemented in a code known as CFL3D [1997]. Results from the CFD code are interpolated onto the integration surface used in the Ffowcs Williams-Hawkings equation. Equations (23.5.2.6) and (23.5.2.7) are integrated using a code developed by Xue and Lyrintzis [1994].

(3) Circular Cylinder

The first example problem considered is that of a circular cylinder in a cross flow. The diameter D of the cylinder is 0.019 meters and a span of 40D is used. The freestream Mach number is 0.2. Viscous two-dimensional computations are performed with a Reynolds number of 1000 based on freestream velocity and cylinder diameter. The acoustic signal is observed at a position 128D from the cylinder center along a line

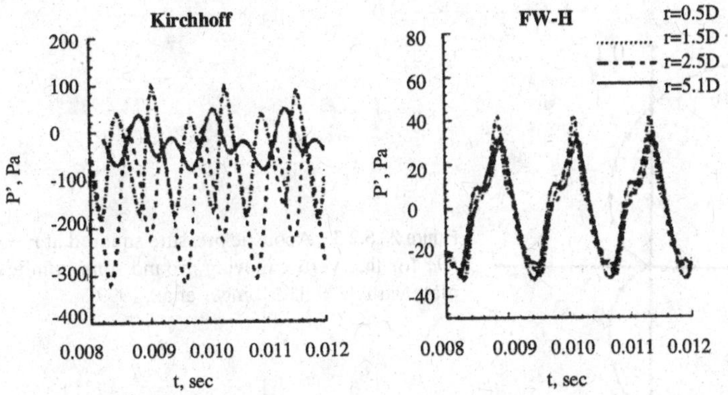

Figure 23.5.2.2 Acoustic signals computed for various integration surfaces [Singer et al., 1999].

mutually perpendicular to the freestream direction and the cylinder axis. The CFD grid about the cylinder is stretched both radially and circumferentially such that more points are concentrated in the cylinder wake than upstream of the cylinder. The Ffowcs Williams-Hawkings integration surfaces are taken on the cylinder itself and at distances 1.5D, 2.5D, and 5.1D.

Figure 23.5.2.2 shows the computed pressure signals at the observer point for the different integration surfaces and using both Kirchhoff and the Ffowcs Williams-Hawkings formulations. The computed acoustic signal is not expected to vary with the integration surface location because the contribution of the quadrupole sources are negligible [Cox, 1997]. The use of integration surfaces that cut through the cylinder wake does not appear to adversely affect the results with the Ffowcs Williams-Hawkings formulation. However, the figure shows that widely varying results are obtained from a standard Kirchhoff formulation for the same problem with the same integration surfaces.

(4) Vortex Line

The next model problem to be considered is that of a line vortex moving around the edge of a half plane. An analytical solution was derived by Creighton [1972] for low Mach numbers. The velocity potential, ϕ, is given by

$$\phi(x, y, t) = \frac{4^{3/4}}{[(r-t)^2+4]^{1/4}} \frac{M^{1/2}\sin(\theta/2)}{r^{1/2}} \tag{23.5.2.6}$$

where r is the nondimensional distance from the edge and θ is the angle measured relative to the positive x axis. To test the ability of the Ffowcs Williams-Hawking equation to propagate a scattered acoustic field, the exact acoustic solution to the problem, (23.5.2.8), with $M = 0.01$ is used to establish flow conditions on an integration surface defined by a rectangular prism with $-200a < x < 2a$, $-2a < y < 2a$ and $-500a < z < 500a$ where z is the spanwise coordinate.

Figure 23.5.2.3 shows a comparison between the exact solution derived from (23.5.2.6) and the numerically propagated solution for observers at a radius $50a$ from the edge. To achieve these results, 250 cells are used in the x-direction, 50 cells are used in the y-direction, and 2500 cells are used in the z-direction. Reduction of the number

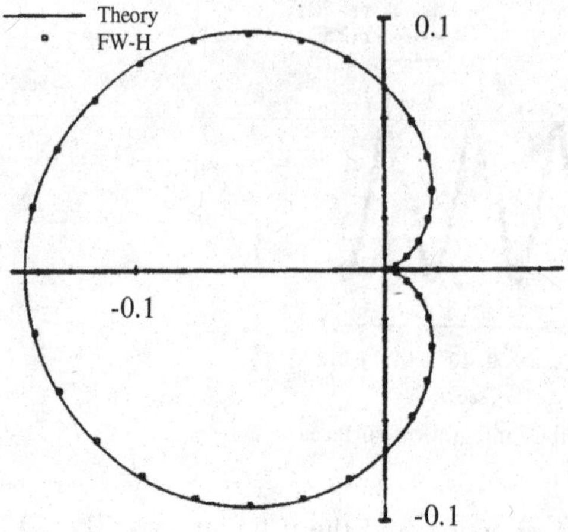

Figure 23.5.2.3 Acoustic pressure squared at $r = 50a$ for line vertex moving around semi-infinite plate with $M = 0.01$ [Singer et al., 1999].

of cells or a significant shortening of the domain will result in a degradation of the results. The good agreement between the numerical and analytical solutions shown by Figure 23.5.2.3 is obtained using a large z-direction domain. This is due to the no-penetration boundary condition used which makes much of the plate appear to radiate. Therefore, the domain must be made large enough to contain all the acoustic sources.

(5) Airfoil Trailing Edge

The next model problem studied is that of vortices convecting past the trailing edge of an airfoil. In order to avoid the Reynolds number variation for different Mach numbers, inviscid computations are carried out. The small amount of numerical dissipation was thought to be enough to produce vortex roll-up shortly downstream of the vortex generator plate. A 2.6% thickness NACA 00 series airfoil was used. The chord C is chosen to be 1 meter. For purposes of the acoustic calculation, the span of the airfoil is twice the chord. The flat plate (or vortex generator) is introduced at 98% of the chord and extends from 0.0015C to 0.0025C above the airfoil chordline. In the presence of flow, vortices roll up just downstream of the flat plate, alternately near the plate's top and bottom edges. The dominant shedding frequency f_s is related to the plate height via the Strouhal number $St = Lf_s/U_0$ where U_0 is the freestream velocity.

Figure 23.5.2.4 shows the seven-zone patched grid computational domain used. This partition is chosen so as to capture the relevant physics. The computational domain extends 2C upstream of the leading edge, 2C downstream of the trailing edge, and 2C above and below the airfoil chordline. The boundary conditions used are freestream conditions upstream, farfield conditions above and below the airfoil, and extrapolation at the downstream boundary. The effect of time and grid resolution on the CFD results is shown on Figure 23.5.2.5 where the spectra of the pressure coefficient at a location directly under the vortex generator are shown. Increasing the resolution makes the spectrum fuller and shifts the dominant frequencies to slightly lower values. In order to calculate the acoustic field using the Ffowcs Williams-Hawking equation, two integration surfaces are considered, one on the airfoil and the other 1% of the chord off

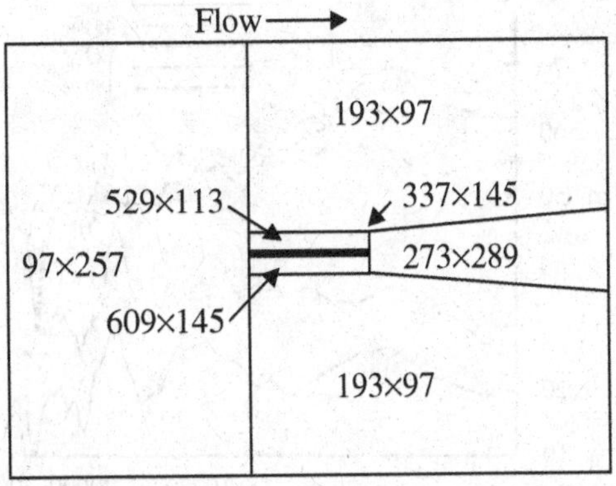

Figure 23.5.2.4 Grid resolution in the various computational zones [Singer et al., 1999].

the airfoil. Far-field acoustic signals are obtained at several locations 10C away from the trailing edge of the airfoil. Figure 23.5.2.6 shows spectra of the acoustic signal for several observer positions. The angular measurements are increasing conterclockwise from the 0 degrees position pointing directly upstream of the trailing edge. The figure shows much reduced noise levels directly upstream and downstream from the trailing edge.

23.5.3 ENTROPY MODE ACOUSTICS

(1) Entropy Controlled Instability (ECI) Analysis Rocket Motor Combustion

We have shown that the thermodynamic formulation of unstable wave phenomena characterized by the pressure, vorticity, and entropy modes leads to the nonlinear ordinary differential equation in Section 23.4. However, to determine the energy growth rate parameters α_1, α_2, and α_3, it is necessary that the mean and fluctuating parts of all

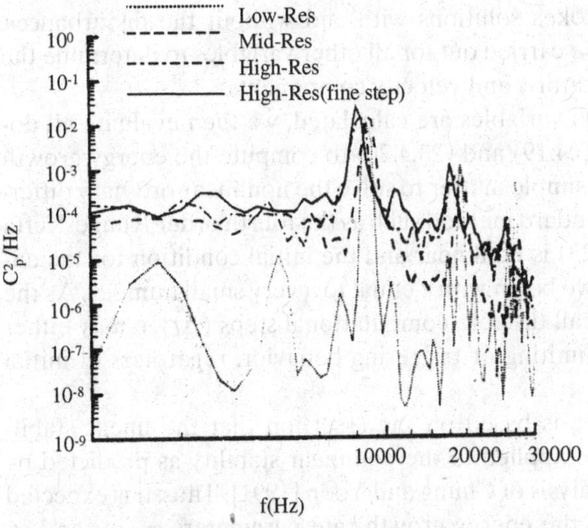

Figure 23.5.2.5 Spectra of pressure coefficient under vortex-generator plate for different grid resolutions $M = 0.2$ [Singer et al., 1999].

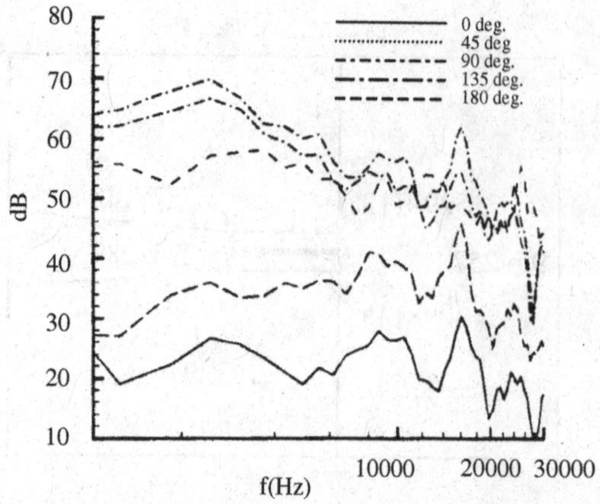

Figure 23.5.2.6 Spectra of acoustic signals (referenced to 20 μPa) for various observers all located 10C from trailing edge of airfoil. On-airfoil-body integration surface used, $M = 0.2$ [Singer et al., 1999].

variables be calculated. Toward this end, we must first solve the Navier-Stokes system of equations using any of the CFD methods discussed in Parts Two and Three. In this analysis, the finite element Taylor-Galerkin scheme is used. The initial time-dependent oscillatory pressure (initial condition at boundaries) is assumed to be of the form

$$p = \overline{p}(1 + d \sin \omega t) \tag{23.5.3.1}$$

where $\overline{p}$ is the mean pressure, d is the percent disturbance from the mean pressure, and ω is the fundamental driving frequency. The fluctuating part is obtained from

$$p' = p - \overline{p} \tag{23.5.3.2}$$

where p and $\overline{p}$ are the Navier-Stokes solutions with and without the disturbances, respectively. Similar calculations are carried out for all other variables to determine the fluctuating parts of density, temperature, and velocity components.

Once the fluctuation parts of all variables are calculated, we then evaluate all domain and boundary integrals of (23.4.19) and (23.4.20) to compute the energy growth parameters in (23.4.23). Now it is a simple matter to solve the nonlinear ordinary differential equation (23.4.23) using a standard method such as the fourth order Runge-Kutta scheme. At this point, since (23.4.23) is nonlinear and the initial condition for the energy growth factor, ε, is unknown, we begin with ε equal to a very small number. As the solution process continues with small discrete computational steps (Δt), ε may either increase or decrease, resulting in limiting or triggering behavior, regardless of initial values of ε used for the analysis.

The procedure described above is based on the assertion that the linear stability criteria for (23.4.24) may not be applied to the nonlinear stability as predicted by (23.4.23), contrary to the earlier analysis of Chung and Yoon [1991]. Thus, it is expected that, with certain combinations of the energy growth rate parameters α_1, α_2, and α_3 which are dictated by the solution of the full Navier-Stokes system of equations, two prominent stability phenomena can be identified: limit cycles and triggering toward an

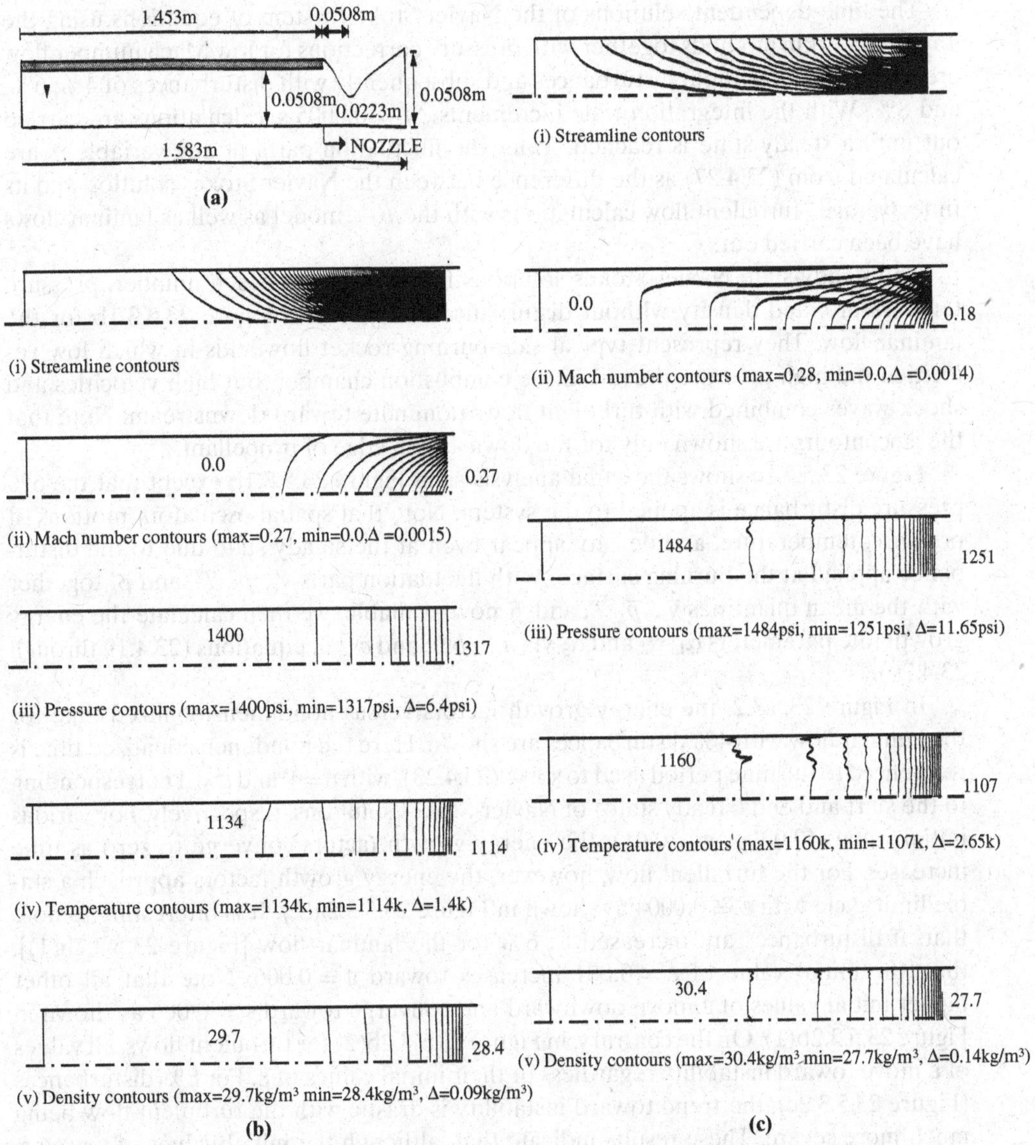

Figure 23.5.3.1 State flowfield for side-burning rocket motor, burning surface boundary conditions: $\rho = 29.7\ \text{kg/m}^3$, $V = 1.22$ m/s, $T = 1134$ K, $P = 1400$ psi, Mach no. $= 0.0018$, laminar calculations [Yoon and Chung, 1994]. (a) Geometry. (b) 0% disturbance. (c) 6% disturbance.

unbounded instability as discussed by Flandro [1985]. We shall examine these physical aspects through oscillations of a typical rocket combustion chamber, based on Yoon and Chung [1994].

Consider an axisymmetric typical side-burning rocket shown in Figure 23.5.3.1a. The bottom face represents the axisymmetric centerline. The initial and boundary conditions are: On the burning surface the tangential and axial velocity components are zero, with $M = 0.0018$, $P = 1400$ psi, $\rho = 29.6742\ \text{kg/m}^3$, $T = 1095$ K, and $\gamma = 1.4$.

The time-dependent solutions of the Navier-Stokes system of equations using the Taylor-Galerkin methods together with pressure corrections for low Mach number flow are obtained first without disturbances and subsequently with disturbances of 4%, 6%, and 8%. With the integration time increments $\Delta t = 0.0005$ s, calculations are carried out until a steady state is reached. Thus, the fluctuation parts of any variable f are calculated from (23.4.27) as the difference between the Navier-Stokes solution and its time average. Turbulent flow calculations with the K–ε model as well as laminar flows have been carried out.

The steady-state Navier-Stokes solutions for streamlines, Mach number, pressure temperature, and density without disturbance are shown in Figure 23.5.3.1b for the laminar flow. They represent typical side-burning rocket flowfields in which low velocities and high pressure prevail in the combustion chamber, but high velocities and shock waves combined with turbulent flows dominate toward downstream. Note that the isocontours are shown only for the downstream edge of propellant.

Figure 23.5.3.1c shows the same analysis as in Figure 23.5.3.1b except that the 6% pressure disturbance is applied to the system. Note that spatial oscillatory motions of pressure, temperature, and density appear even at the steady state due to the disturbance applied at the burning surface. With fluctuation parts v_i', p', T', and ρ' together with the mean quantities $\overline{v}_i$, $\overline{p}$, $\overline{T}$, and $\overline{\rho}$ now available, we then calculate the energy growth rate parameters α_1, α_2 and α_3 via $a^{(i)}$, $b^{(i)}$, and $c^{(i)}$ in equations (23.4.19 through 23.4.20).

In Figure 23.5.3.2, the energy growth factors versus nondimensionalized time for the laminar flow with 4% disturbances are shown. Here the nondimensionalized time is referenced to the time period used to solve (23.4.23), with $t = 0$ and $t = 1$ corresponding to the start and end (steady state) of Navier-Stokes solutions, respectively. For various initial values ($0.001 < \varepsilon < 0.01$), the energy growth factors converge to zero as time increases. For the turbulent flow, however, the energy growth factors approach a stable limit cycle with $\varepsilon \cong 0.0004$ as shown in Figure 23.5.3.2a(2). It is interesting to note that, if disturbances are increased to 6% for the laminar flow [Figure 23.5.3.2b(1)], then the initial value of $\varepsilon = 0.001$ increases toward $\varepsilon = 0.006$. Note that all other higher initial values of ε move downward and converge toward $\varepsilon = 0.006$ as shown in Figure 23.5.3.2b(1). On the contrary, in Figure 23.5.3.2b(2) for turbulent flows, all values of ε move toward instability regardless of their initial values of ε. For 8% disturbances (Figure 23.5.3.2c), the trend toward instability is drastic with the turbulent flow being much more severe. These results indicate that, although the initial values of ε are arbitrarily chosen, the stable limit cycles prevail for smaller disturbances, but they are triggered into an unbound instability for larger disturbances.

In Figure 23.5.3.3, the time rates of change of ε, $(d\varepsilon/dt)$ vs. ε itself are shown for both laminar and turbulent flows at various disturbance percentages. These results confirm the trend observed in the previous figures. It is seen that turbulence induces instability otherwise stable in laminar flows, and that the increase of percent disturbances drives the system toward instability.

In order to demonstrate the role of nonlinearity in (23.4.23) with $\alpha_2 \neq 0$, $\alpha_3 \neq 0$ in comparison with the linear analysis (23.4.24) with $\alpha_2 = 0$, $\alpha_3 = 0$, various cases of $d\varepsilon/dt$ vs. energy growth factors are shown in Figure 23.5.3.4. It should be noted that most of the nonisentropic properties are associated with α_2 and α_3 and that they are zero

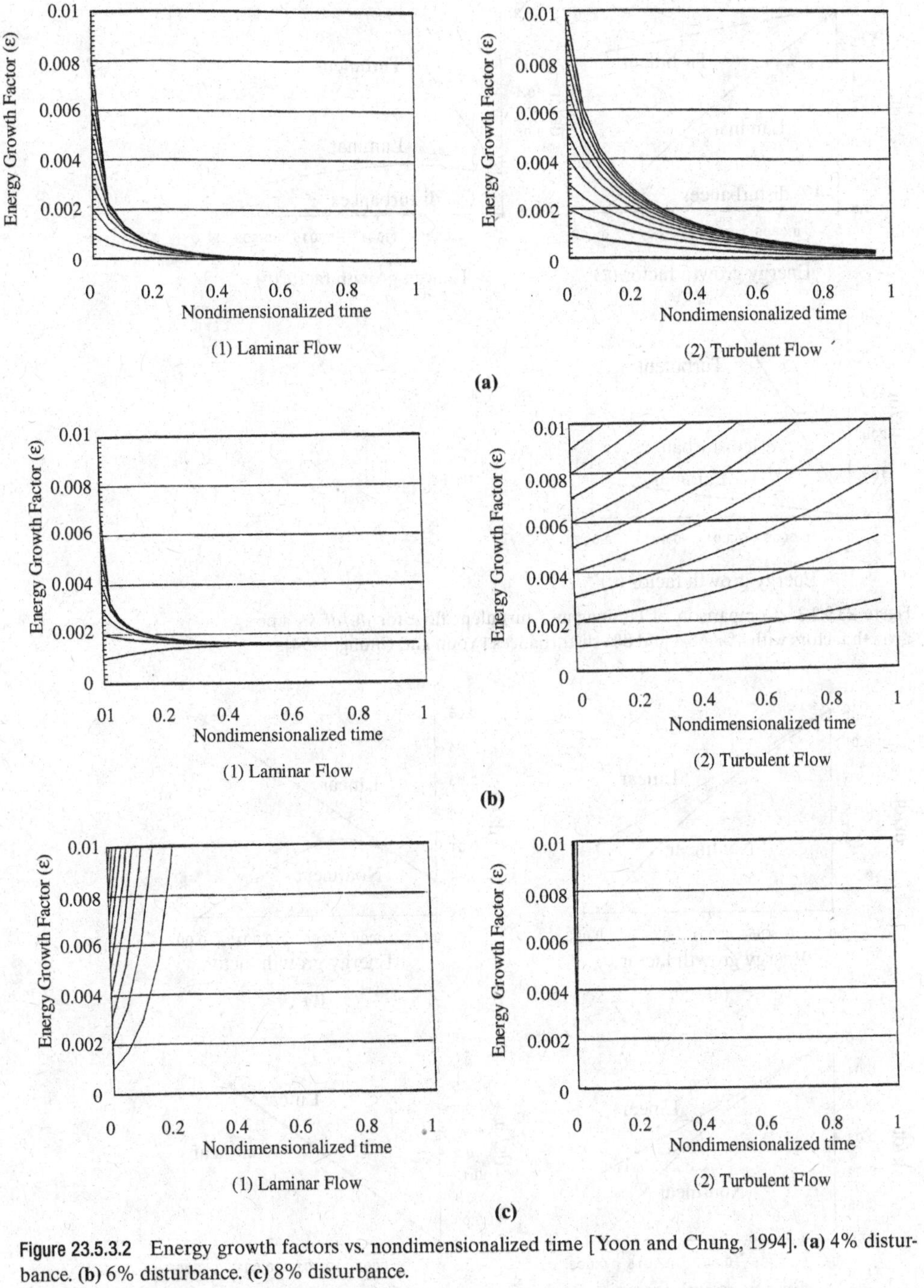

Figure 23.5.3.2 Energy growth factors vs. nondimensionalized time [Yoon and Chung, 1994]. **(a)** 4% disturbance. **(b)** 6% disturbance. **(c)** 8% disturbance.

for the linear analysis. Non-isentropy is involved in energy dissipation and in nonlinear waves. Solutions of the linear equation (23.4.24) are, therefore, expected to overestimate the instability behavior. This prediction is clearly evident in Figure 23.5.3.4. As the % disturbances increase and the laminar flow changes to turbulence, the difference

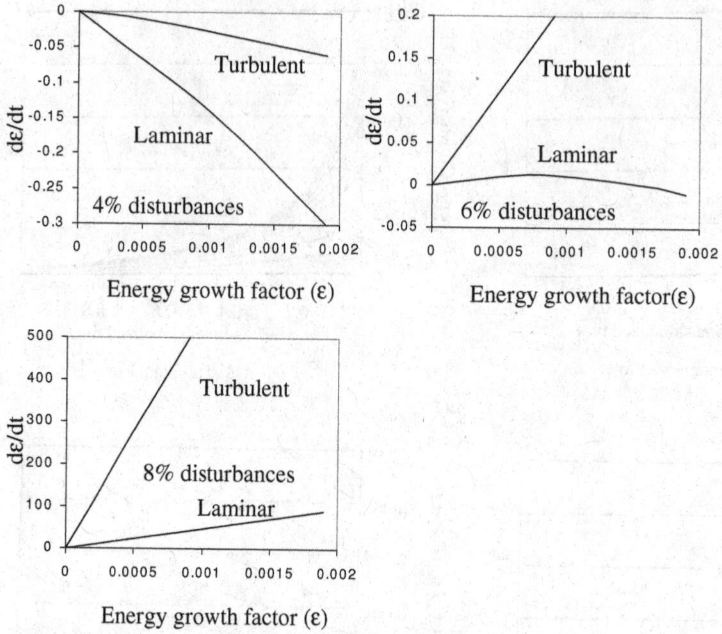

Figure 23.5.3.3 Comparison of laminar and turbulent flow for $d\varepsilon/dt$ vs. energy growth factors with 4%, 6%, and 8% disturbances [Yoon and Chung, 1994].

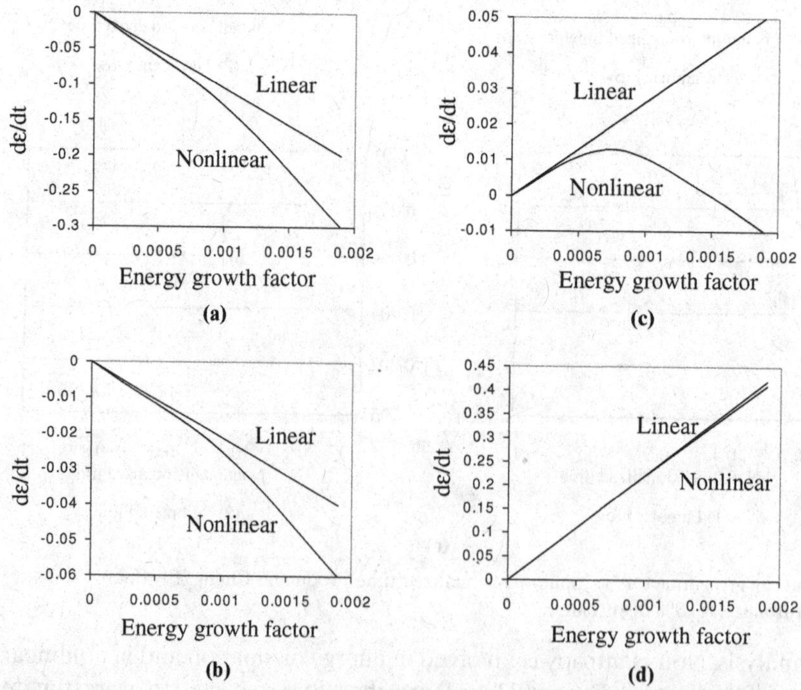

Figure 23.5.3.4 Comparison of linear and nonlinear analysis for $d\varepsilon/dt$ vs. energy growth factors [Yoon and Chung, 1994]. **(a)** Laminar with 4% disturbance. **(b)** Turbulence with 4% disturbance. **(c)** Laminar with 6% disturbance. **(d)** Turbulence with 6% disturbance.

between the linear and nonlinear analyses becomes smaller. The nonlinear analysis with nonisentropic properties becomes more critical when the % disturbances are small.

Overall, the roles of acoustic, vortical, and entropy modes are clearly exhibited in Figures 23.5.3.3 and 23.5.3.4. For low Mach number flows such as at the head end, no pressure discontinuities (no shocks) occur. The pressure fluctuations are linear, sinusoidal, and isentropic. For high Mach number flows such as prevalent at the throat and nozzle, the acoustic mode changes into entropy mode due to shock waves, irreversibility, high temperature, or low density. For regions of recirculation or vortical motions such as at converging and diverging sections, the vortical mode dictates wave motions. The wave instability determined by (23.4.23) is associated with the wave motions due to the combination of all modes of acoustics. This is particularly true for a side-burning rocket motor such as examined here. However, because of high speeds downstream, the entropy mode eventually dominates. This is evident from the comparison of the results of linear and isentropic analysis with those of nonlinear and nonisentropic analysis as shown in Figure 23.5.3.4. It is seen that the linear and isentropic analysis greatly overestimates the instability. As to the trends of the vortical mode evidenced in Figure 23.5.3.3 by turbulent flows (large vortical mode effect) versus laminar flows (small vortical mode effect), it is seen that the effect of turbulence or vortical mode is reflected by higher energy growth rate (greater instability). The effect of acoustic mode is embedded in both Figures 23.5.3.3 and 23.5.3.4 and to the cases of laminar flows and linear waves as well.

(2) Unstable Waves of Flame Propagation in a Closed Tube

In this analysis, an acoustic instability problem associated with the premixed flame propagation in a closed tube is described as reported by Gonzales [1996]. It is shown that the flame front displays a cellular structure that has a close connection with the acoustic waves. First pressure oscillations are triggered as the flames reach the lateral walls and suddenly decelerate. Then, a slender cusp appears which subsequently collapses due to the periodic acceleration driving the flames to display a cellular pattern. This eventually leads to the total heat release (temperature fluctuations) oscillating in phase with pressure, causing a violent instability. For this reason, this example is identified as the entropy mode acoustics.

The Navier-Stokes system of equations for reactive flows given by (22.2.34) is used with a single species reactant (Y, hydrocarbon) with all variables given by the nondimensional quantities. The standard FVM with predictor-corrector semi-implicit scheme is employed. Calculations cover a plane tube with the aspect ratio of 6. The reaction rate is given by

$$\omega = -Y\left[Y - \left(1 - \frac{1}{\phi}\right)\right]\exp\left[-T_A\left(\frac{1}{T} - \frac{1}{T_b}\right)\right]$$

with ϕ (*equivalence ratio*) $= 0.97$, T_A (*activation temprature*) $= 33$, T_b (*burned gas-temp*) $= 7.67$.

Here, the Damköhler number is set equal to 1.1×10^4 with $\text{Re} = 25$ and $M = 3 \times 10^{-3}$ which ensures that the acoustic time has the same order of magnitude as the transit time of the flame. The time step (Δt) satisfying the CFL condition is used.

Figure 23.5.3.5a shows the various stages of flame propagation. The first stage is identified by the initially curved, flat, and cusped [Figure 23.5.3.5a(i) through (vi)] flame

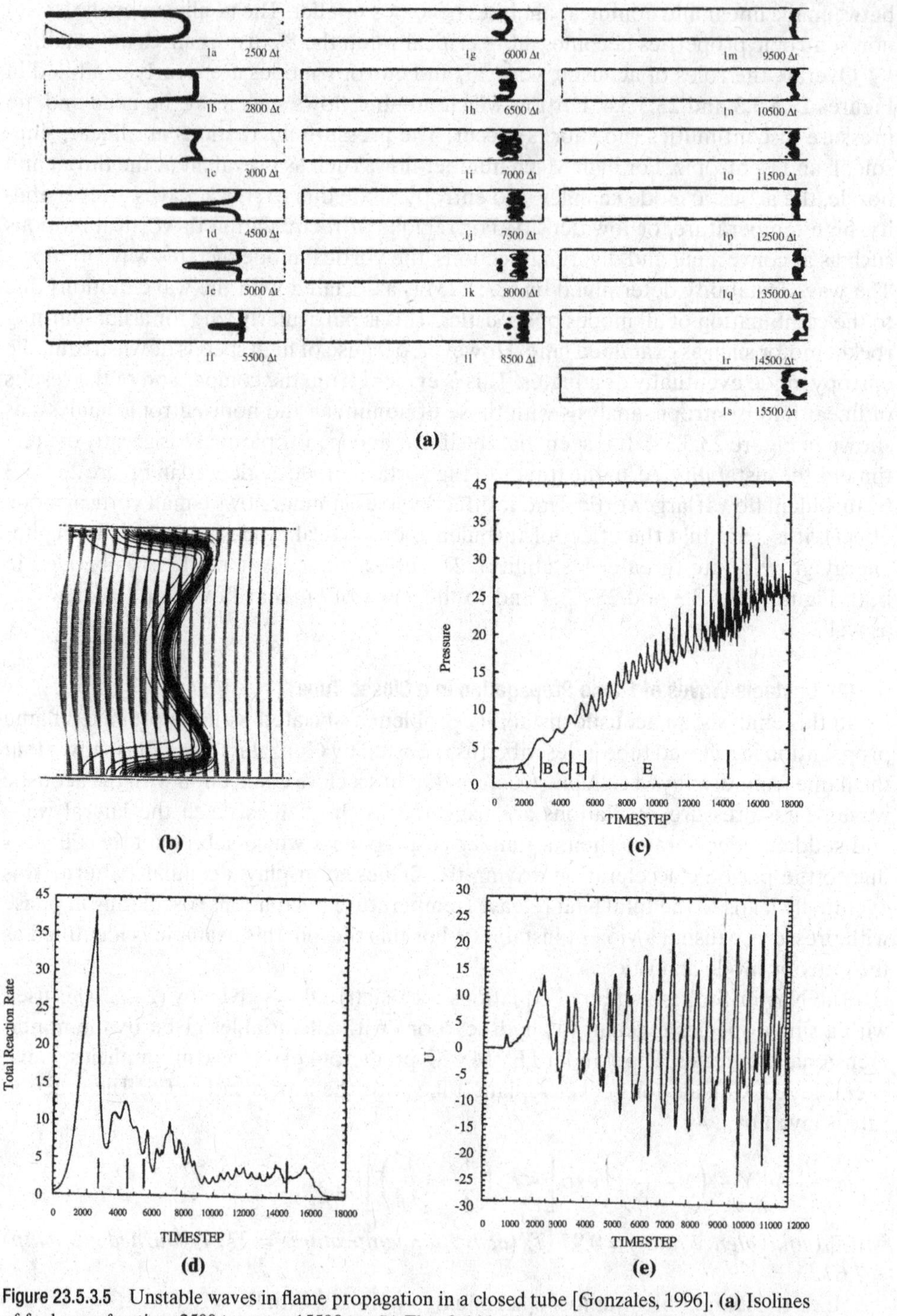

Figure 23.5.3.5 Unstable waves in flame propagation in a closed tube [Gonzales, 1996]. **(a)** Isolines of fuel mass fraction, $2500\Delta t \leq t \leq 15500\Delta t$. **(b)** Flowfield in the vicinity of the flame $t = 3000\Delta t$. **(c)** Temporal evolution of pressure at the unburned end of the tube. **(d)** Temporal evolution of total reaction rate. **(e)** Temporal evolution of longitudinal velocity on the plane of symmetry at $x/L = 0.9$.

shapes which are clearly evident as verified in experiments. A snapshot of a cusp at $t = 3000\Delta t$ is shown in Figure 23.5.3.5b. The second stage begins with the collapse of the primary cusp and the formation of the cellular flame (Figure 23.5.3.5c). This is then followed by the third stage where the cellular shapes are severely distorted before vanishing at the end.

The temporal evolution of pressure and total reaction rate are presented in Figures 23.5.3.5c,d. Unstable pressure oscillations and diminishing reaction rates are evident in these figures. Finally, Figure 23.5.3.5e displays the time evolution of the logitudinal component of the gas velocity in the unburned medium at a point located on the plane of symmetry, close to the bottom wall. It is shown that velocity oscillations are triggered as the flames reach the lateral walls, around $t = 3000\Delta t$. Corresponding to the pressure variations, velocity oscillations become sharper and increase in amplitude as the flame approaches the end wall.

(3) Unstable Waves in Combustion Dynamics

Wave instabilities in combustion dynamics were investigated using perturbation expansions of the conservation equations with all excited frequencies calculated by the eigenvalue analyses [Kim, 1985; Chung and Kim, 1985]. Unsteady oscillatory combustion waves were examined for the high-frequency responses across the long flame such as in the double-base propellants [Park and Chung, 1987; Park, 1988]. Another aspect of the combustion dynamics is the unstable waves due to the coupling of pressure and vortical modes. The Orr-Sommerfeld equation was solved to determine the wave numbers and unsteady stream functions from which vortically coupled acoustic instability growth constants were calculated [Sohn, 1986; Chung and Sohn, 1986]. It is found that stability boundaries for coupled pressure and vorticity mode oscillations are similar to the classical hydrodynamics stability boundaries, but they occur in the form of multiple islands [Chung and Sohn, 1986].

23.6 SUMMARY

In this chapter, it is shown that the subject of acoustics may be categorized into three areas: the pressure mode acoustics, the vorticity mode acoustics, and the entropy mode acoustics. The reason for this categorization is that the acoustic fields can be computed by the wave equation with the Kirchhoff's formula, the momentum equations with the Lighthill's stress tensor, or by the entropy energy equation with the first and second laws of thermodynamics.

The pressure mode acoustics includes the Kirchhoff's method with stationary surfaces, subsonic surfaces, and supersonic surfaces. It is shown that the basic idea of the Kirchhoff's formula is to surround the region of a nonlinear flowfield and acoustic sources by a closed surface. In the domain inside the surface, a nonlinear aerodynamic computation is carried out, which provides the pressure distribution on the surface as well as its time history. Hawkings surface pressure modifications are used for subsonic and supersonic flows.

The vorticity mode acoustics is based on the aerodynamic sound theory of Lighthill's acoustic anology as applied to the turbulent jet and shear boundary layers. The governing equations are shown to be derived from the Navier-Stokes system of equations, resulting in an exact, nonhomogeneous wave equation with Ffowcs Williams-Hawkings modifications for moving boundaries.

For high temperatures and high temperature gradients coupled with pressure mode and vorticity mode acoustics, unstable waves are likely to dominate the flowfield. In this case, it is shown that implementation of the first and second laws of thermodynamics into the energy equation leads to the nonlinear, nonisentropic wave equation in terms of entropy. Integration of this equation results in a total of six terms of the acoustic intensity on solid surface boundaries, representing the sources of acoustic intensity due to total energy, pressure oscillations, entropy changes, vortical oscillations, viscous dissipation, and temperature changes.

Representative examples of numerical calculations for the pressure mode acoustics, vorticity mode acoustics, and entropy mode acoustics have been demonstrated. A coverage of the entire field of acoustics is beyond the scope of this chapter, but the suggested categorization of acoustics into the three areas has been adequately represented by a limited number of examples.

REFERENCES

Brentner, K. S. [1986]. Prediction of helicopter discrete frequency rotor noise – A computer program incorporating realistic blade motions and advanced formulation. *NASA-TM* 87724.

———. [1997]. An efficient and robust method for predicting helicopter rotor high-speed impulsive noise. *J. Sound Vibr.*, 203, no. 7, 87–100.

Brentner, K. S. and Farassat, F. [1998]. An analytical comparison of the acoustic analogy and Kirchhoff formulation for moving surfaces. *AIAA J.*, 36, no. 8, 1379–86.

Cantrell, R. H. and Hart, R. W. [1964]. Interaction between sound and flow in acoustic cavities: mass, momentum, and energy considerations. *J. Acous. Soc. Am.*, 36, 697–706.

Carofano, G. C. [1984]. Blast computation using Harten's total variation diminishing schemes, ARLCB-TR-84029.

———. [1987]. An experimental and numerical study of the flow through a vent hole in a perforated muzzle brake. ARCCB-TR-87016, June.

Chung, T. J. [1996]. *Applied Continuum Mechanics*. London: Cambridge University Press.

Chung, T. J. and Kim, P. K. [1985]. Unsteady combustion of solid propellants. In R. Glowinski, B. Larrouturou, and R. Temam (eds.). *Numerical Simulation of Combustion Phenomena, Lecture Notes in Physics*. Berlin: Springer-Verlag.

Chung, T. J. and Sohn, J. L. [1986]. Interactions of coupled acoustic and vortical instability. *AIAA J.*, 24, 10, 1582–95.

Chung, T. J. and Yoon, W. S. [1991]. Wave instability in combustion. *Comp. Meth. Appl. Mech. Eng.*, 26, 95–106.

———. [1993]. Flowfield simulation and acoustic control. Final Report, DAAH01-92-R002, U.S. Army Missile Command.

Cox, J. S. [1997]. Computation of vortex shedding and radiated sound for a circular cylinder: Subcritical and transcritical Reynolds numbers. Master's thesis, The George Washington University.

Creighton, D. G. [1972]. Radiation from vortex filament motion near a half-plane. *J. F. Mech.*, 182, 357–62.

Farassat, F. [1996]. The Kirchhoff formula for moving surfaces in aeroacoustics – the subsonic and supersonic cases. *NASA-TM* 110285.

Farassat, F. and Farris, M. [1999]. The mean curvature of the influence surface of wave equation with sources on a moving surface. *Math. Meth. Appl. Sci.*, 22, 1485–1503.

Farassat, F. and Myers, M. K. [1988]. Extension of Kirchhoff formula to radiation from moving surfaces. *J. Sound Vib.*, 123, no. 3, 451–61.

Farassat, F. and Succi, G. P. [1988]. The prediction of helicopter discrete frequency noise. *Vertica*, 7, no. 4, 309–20.

Ffowcs Williams, J. E. and Hawkings, D. L. [1969]. Sound generated by turbulence and surfaces in arbitrary motion. *Phil. Trans. Roy. Soc.*, A264, 324–42.

Flandro, G. A. [1985]. Energy balance analysis of nonlinear combustion instability. *AIAA J. Propul.*, 1, 3, 210–21.

Frendi, A., Maestrello, L., and Ting, L. [1995]. An efficient model for coupling structural vibration with acoustic radiation. *J. Sound Vib.*, 182, no. 5, 741–57.

Gonzalez, M. [1996]. Acoustic instability of a premixed flame propagating in a tube. *Comb. Flame*, 107, 245–59.

Gottlieb, D. and Turkel E. [1976]. Dissipative two-four methods for time dependent problems. *Math. Compu.*, 30, 703–23.

Hawkings, D. L. [1979]. Noise generation by transonic open rotors. Westland Research paper 599.

———. [1989]. Comments on the extension of Kirchhoff's formula to radiation from moving surfaces. *J. Sound Vib*., 132, 1, 160–79.

Howe, M. S. [1998]. *Acoustics of Fluid-Structure Interactions*. New York: Cambridge University Press.

Kim, P. K. [1985]. Unsteady flame zone combustion response of solid propellant rocket motors. Ph.D. diss. The University of Alabama, Huntsville.

Kovasznay, L. S. G. [1953]. Turbulence in supersonic flow. *J. Aero. Sci.*, 20, 657–82.

Lighthill, M. J. [1952]. On sound generated aerodynamically, Part I: general theory. *Proc. Roy. Soc. London*, A241, 564–87.

———. [1954]. On sound generated aerodynamically, Part II: Turbulence as a source of sound. *Proc. Roy. Soc. London*, A222, 1–32.

Morino, L. and Tseng, K. [1990]. A general theory of unsteady compressible potential flows with applications to airplanes and rotors. In P. K. Benerjee and L. Morino (eds.). *Developments in Boundary Element Methods*, Vol. 6, Barking, UK: Elsevier Applied Science Publisher, pp. 183–245.

Park, O. Y. [1988]. Nonlinear combustion dynamics analysis of solid propellants. Ph.D. diss. The University of Alabama, Huntsville.

Park, O. Y. and Chung T. J. [1987]. Two-dimensional solid propellant combustion modeling by finite elements. AIAA paper, 87–0566.

Pierce, A. D. [1981]. *Acoustics: An Introduction to Its Physical Principles and Applications*. New York: McGraw-Hill.

Robinson, J. H. [1990]. Finite element formulation and numerical simulation of the random response of composite plates. Master's thesis, Mechanical Engineering, Old Dominion University, Norfolk, VA.

Rumsey, C., Biedron, R., and Thomas, J. [1997]. CFL3D: Its history and some recent applications. NASA-TM 112861.

Sarkar, S. and Hussaini, M. Y. [1993]. A hybrid direct numerical simulation of sound radiated from isotropic turbulence. FED-Vol. 147, Computational Aero- and Hydro-acoustics, ASME 1993.

Singer, B. A., Brentner, K. S., Lockard, D. P., and Lilley, G. M. [1999]. Simulation of acoustic scattering from a trailing edge. AIAA paper 99-0231.

Smagorinski, J. [1963]. General circulation experiments with the primitive equations. I. The basic experiment. *Monthly Weather Rev.*, 91, 99–164.

Sohn, J. L. [1986]. Interaction of couples acoustic and vortical instabilities in rocket combustion chambers. Ph.D. diss. The University of Alabama, Huntsville.

Strawn, R. C. and Caradonna, F. X. [1987]. Conservative full potential model for unsteady transonic rotor flows. *AIAA J.*, 25, no. 2, 193–98.

Xue, Y. and Lyrintzis, A. S. [1994]. Rotating Kirchhoff method for three-dimensional transonic blade-vortex interaction hover noise. *AIAA J.*, 32, no. 7, 1350–59.

Yoon, W. S. [1992]. Analysis of turbulence and shock wave interactions and wave instabilities in combustion. Ph.D. diss. The University of Alabama, Huntsville.

Yoon, W. S. and Chung, T. J. [1994]. Nonlinearly unstable waves dominated by entropy mode. *J. Acoust. Soc. Am.*, 96(2), 1096–1103.

CHAPTER TWENTY-FOUR

Applications to Combined Mode Radiative Heat Transfer

24.1 GENERAL

In heat transfer, there are three different modes – conduction, convection, and radiation. We have included conduction and convection in the Navier-Stokes system of equations discussed in the previous chapters. Radiative heat transfer is another mode of heat transfer to be examined in this chapter. Heat transfer by radiation occurs in many engineering applications of nonparticipating and participating media. In this chapter, we study this subject as a separate mode of heat transfer first and then as a combined mode integrated into other modes.

In nonparticipating media, conduction and convection are absent. Here we are concerned with view factors, radiative boundary conditions, and radiative heat transfer in absorbing, emitting, and scattering media. Radiative heat transfer is associated with the radiative heat flux which involves integrals with respect to the wavelength, solid angle, and optical depth. The governing equation for radiative heat transfer, then, takes the form of integrodifferential equations. This aspect of the radiative heat transfer is unique and requires a special computational treatment.

Participating media combines the radiative heat transfer with conduction and/or convection. The most significant feature in the combined mode heat transfer is the fact that the radiative heat flux is always three-dimensional, even if the computational domain is chosen to be one- or two-dimensional. For this reason, special mathematical formulations and computational schemes must be developed. We discuss this subject in Sections 24.2.4 and 24.3.3.

For the sake of completeness and future reference, some basic definitions and formulas in radiative heat transfer are summarized below.

Planck's Law

Monochromatic (spectral) emissive power of black body is given by

$$e_{b\nu}(T) = \frac{2\pi\nu^3 n^2}{c_0^2\left[\exp\left(\dfrac{h\nu}{KT}\right) - 1\right]} \tag{24.1.1}$$

with $\nu =$ frequency of radiation, c_0 (*speed of light in vacuum*) $= 2.998 \times 10^{10}$ cm/s, $n =$ index of refraction ($n = 1$ for vacuum), $T =$ absolute temperature, K (Boltzmann's constant) = 1.38×10^{-16} erg/°K, h (Planck's constant) = 6.625×10^{-27} cm/s. If n is independent of frequency ν or wavelength λ, then $\nu = c_0/n\lambda$, $d\nu = -c_0 d\lambda/n\lambda^2$. This gives $e_{b\lambda} = -e_{b\nu} d\nu$, and

$$e_{b\lambda}(T) = \frac{C_1}{n^2\lambda_0^5\left[\exp\left(\frac{C_2}{KT}\right) - 1\right]} \tag{24.1.2}$$

where $e_{b\lambda}(T)$ is known as the Planck's function and $C_1 = 2\pi c_0^2 h = 3.74 \times 10^{-5}$ erg cm²/s and $C_2 = hc_0/K = 1.4387$ cm °K.

Stefan-Boltzmann Law

The black-body emissive power per unit time and area over all frequencies is given by

$$e_b(T) = \int_0^\infty e_{b\nu}(T) d\nu = n^2 \sigma T^4 \tag{24.1.3}$$

where the integral is evaluated using (24.1.1) and σ is the Stefan-Boltzmann constant,

$$\sigma = \frac{2\pi^5 K^4}{15 c_0^2 h^3} = 5.668 \times 10^{-5} \text{ erg/s cm}^2 \, ^\circ\text{K}^4$$

Intensity of Radiation

The amount of energy passing in a given direction is described in terms of the intensity of radiation i_b as shown in Figure 24.1.1a,

$$i_b = \frac{d\Phi}{d\omega \cos\theta} \tag{24.1.4}$$

where Φ is the radiant energy per unit time and unit area leaving a given surface in the direction θ (polar angle) from the normal and contained within a solid angle $d\omega$. The energy flux passing from the surface into the hemispherical space above the surface is then

$$\Phi = \int i_b \cos\theta \, d\omega$$

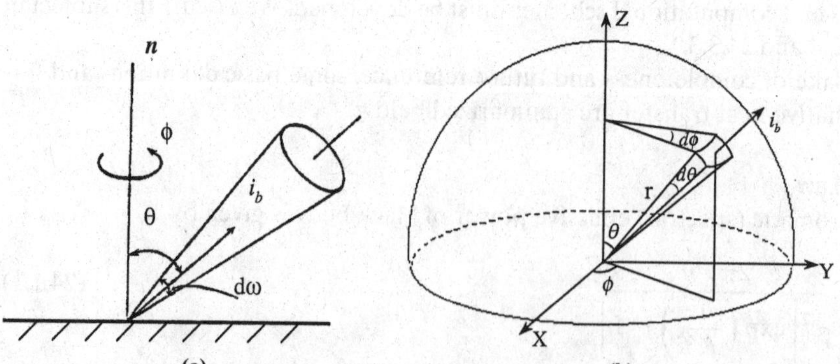

Figure 24.1.1 Basic geometry for radiation. (a) The intensity of radiation. (b) Integration of intensity over solid angle.

The solid angle is defined as the surface element on the hemisphere divided by the square of the radius

$$d\omega = \sin\theta d\theta d\phi \tag{24.1.5}$$

where ϕ is the azimuthal angle as shown in Figure 24.1.1b. Thus,

$$\Phi = \int_0^{2\pi}\int_0^{\pi/2} i_b \cos\theta \sin\theta d\theta d\phi \tag{24.1.6}$$

If the intensity of radiation is independent of direction, then,

$$\Phi = \pi i_b \tag{24.1.7}$$

Note that this definition is limited to the case of radiation leaving a surface. For radiation through absorbing, emitting, and scattering media, the net rate at which energy is locally transferred within the medium must be considered.

Absorption and Scattering

Let a_λ be the monochromatic absorption coefficient for radiation of wave length λ and intensity I_λ. The local monochromatic absorption per unit time and unit volume within the isotropic medium due to an incident beam is

$$Q_a = a_\lambda \int_{4\pi} I_\lambda d\omega \tag{24.1.8}$$

Similarly, the monochromatic energy that is scattered per unit time, per unit area normal to the pencil of rays, per unit solid angle, and per unit volume is

$$Q_s = \gamma_\lambda \int_{4\pi} I_\lambda d\omega \tag{24.1.9}$$

where γ_λ is the monochromatic scattering coefficient. We define β_λ as the monochromatic extinction coefficient

$$\beta_\lambda = a_\lambda + \gamma_\lambda \tag{24.1.10}$$

which is related to the mean free path λ_p for photons of wavelength λ as

$$\lambda_p = \frac{1}{\beta_\lambda} \tag{24.1.11a}$$

For nonscattering media, we have

$$\lambda_p = \frac{1}{a_\lambda} \tag{24.1.11b}$$

Emission

The local monochromatic emission of radiant energy J_λ is expressed as (Kirchhoff's law)

$$J_\lambda = a_\lambda I_\lambda = \frac{a_\lambda}{\pi} e_{b\lambda}(T) \tag{24.1.12a}$$

The total monochromatic emission per unit time and per unit volume is obtained by multiplying (24.1.12a) by the total solid angle 4π,

$$\hat{J}_\lambda = 4a_\lambda e_{b\lambda}(T) \tag{24.1.12b}$$

If a'_λ denotes the absorption coefficient characterizing true absorption including induced emission, then we have

$$J'_\lambda = \frac{a'_\nu}{\pi}(1 - e^{-h\nu/KT})e_{b\lambda}(T) \tag{24.1.12c}$$

The total emission of radiation locally within a medium is obtained through integration of (24.1.12b) over all wavelengths,

$$J = 4a_p e_b(T)$$

where

$$a_p = \frac{\int_0^\infty a_\lambda e_{b\lambda}(T)d\lambda}{e_b(T)} \tag{24.1.13}$$

which is known as the Planck mean absorption coefficient, a property of the medium in local thermodynamic equilibrium.

Similarly, in view of (24.1.8) and (24.1.9), the total absorption and scattering coefficients are defined as, respectively,

$$a = \frac{\int_0^\infty a_\lambda \int_{4\pi} I_\lambda d\omega d\lambda}{\int_{4\pi} I d\omega} \tag{24.1.14}$$

$$\gamma = \frac{\int_0^\infty \gamma_\lambda \int_{4\pi} I_\lambda d\omega d\lambda}{\int_{4\pi} I d\lambda} \tag{24.1.15}$$

Note that a and γ are not equilibrium properties because I_λ is a function of the medium and the surfaces.

Surface Radiation

In general, the monochromatic hemispherical emittance ε_λ is a function of both wavelength and temperature,

$$\varepsilon_\lambda = \frac{e_\lambda}{e_{b\lambda}} \tag{24.1.16}$$

The monochromatic hemispherical absorptance α_λ is given by

$$\alpha_\lambda = \frac{H_{\lambda,a}}{H_{\lambda,i}} \tag{24.1.17}$$

where $H_{\lambda,a}$ is the energy absorbed and $H_{\lambda,i}$ is the spectral energy density of the radiation incident per unit area and time.

A portion of the incident radiation, $H_{\lambda,r}$, may be reflected back into the hemispherical space, characterized by the monochromatic hemispherical reflectance ρ_λ,

$$\rho_\lambda = \frac{H_{\lambda,r}}{H_{\lambda,i}} \tag{24.1.18}$$

which satisfies the relation for an opaque material,

$$\alpha_\lambda + \rho_\lambda = 1 \tag{24.1.19}$$

On diffuse surfaces, Kirchhoff's law states that

$$\alpha_\lambda = \varepsilon_\lambda \tag{24.1.20}$$

The total hemispherical emittance ε is given by

$$\varepsilon = \frac{e}{e_b} = \frac{\int_0^\infty \varepsilon_\lambda e_{b\lambda} d\lambda}{\int_0^\infty e_{b\lambda} d\lambda} = \frac{\int_0^\infty \varepsilon_\lambda e_{b\lambda} d\lambda}{\sigma T^4} \tag{24.1.21}$$

This gives

$$e = \varepsilon \sigma T^4 \tag{24.1.22}$$

The total hemispherical absorption α is defined as

$$\alpha = \frac{\int_0^\infty \alpha_\lambda H_{\lambda,i} d\lambda}{\int_0^\infty H_{\lambda,i} d\lambda} \tag{24.1.23}$$

The total hemispherical reflectance ρ is given by

$$\rho = \frac{\int_0^\infty \rho_\lambda H_{\lambda,i} d\lambda}{\int_0^\infty H_{\lambda,i} d\lambda} \tag{24.1.24}$$

The following relationships hold for an opaque material and gray material, respectively,

$$\rho = 1 - \alpha, \qquad \rho = 1 - \varepsilon \tag{24.1.25a,b}$$

With these definitions, the governing equations and computational procedures involved in radiative heat transfer of nonparticipating media will be presented in Section 24.2 and the combined mode heat transfer of participating media in Section 24.3. Numerical solutions can be carried out using a variety of methods (FDM, FEM, and FVM) presented in Parts Two and Three. For simplicity, however, finite element and finite volume formulations will be used to demonstrate numerical aspects of radiative heat transfer. In Section 24.4, we include and discuss example problems solved with various numerical schemes using FDM, FEM, and FVM. Although Monte-Carlo methods and discrete ordinate methods have been used extensively in radiative heat transfer in the past, they are not included in this chapter since they are unrelated to the CFD methods discussed in this book.

24.2 RADIATIVE HEAT TRANSFER

24.2.1 DIFFUSE INTERCHANGE IN AN ENCLOSURE

Nonparticipating media include most monatomic and diatomic gases as well as air and vacuum. Consider a region within which there is a black-body radiation on an enclosed space whose walls have a uniform temperature T_e. In view of (24.1.22), if there is a body with surface area A at temperature T within the enclosure, the net rate of radiant outflow Q can be written as

$$\frac{Q}{A} = \varepsilon\sigma T^4 - \alpha\sigma T_e^4 \tag{24.2.1}$$

Note that the first and second terms of the right-hand side denote, respectively, the radiant energy emitted per unit time and unit area by the body and the corresponding radiant flux. For the gray-body condition ($\varepsilon = \alpha$),

$$\frac{Q}{A} = \varepsilon\sigma\left(T^4 - T_e^4\right) \tag{24.2.2}$$

The view factor provides information on the fraction of the diffusely distributed radiant energy leaving one surface, A_i, that arrives at a second surface, A_j, denoted by $F_{A_i-A_j}$, having the following relationships:

$$A_i F_{A_i-A_j} = A_j F_{A_j-A_i} \tag{24.2.3}$$

$$\sum_{j=1}^{N} F_{A_i-A_j} = 1 \tag{24.2.4}$$

where $F_{A_i-A_i} = 0$ for convex surface, $F_{A_i-A_i} \neq 0$ for concave surface.

Let us consider the surfaces A_i and A_j of the enclosure in Figure 24.2.1. The net radiative heat flux q_i at the surface A_i is equal to the difference between the leaving (B_i) and incident (H_i) fluxes

$$q_i = B_i - H_i$$

where B_i, known as the radiosity, is given by the sum of radiative flux emitted at T_i and incident radiative flux reflected by the surface,

$$B_i = \varepsilon_i \sigma T_i^4 + \rho_i H_i$$

From the relation (24.2.3) the total radiative energy leaving all the zones of the enclosure and incident upon the surface A_i

$$H_i = \frac{1}{A_i}\sum_{j=1}^{N} B_j A_j F_{i-j}$$

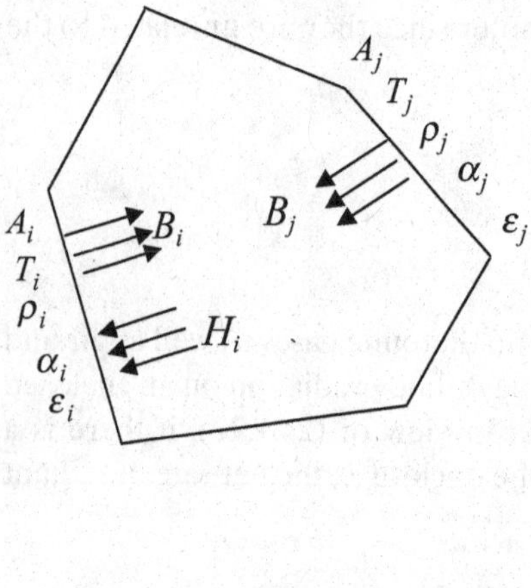

Figure 24.2.1 Enclosure filled with nonparticipating medium.

thus

$$q_i = B_i - \sum_{j=1}^{N} B_j F_{i-j} \tag{24.2.5a}$$

$$B_i = \varepsilon_i \sigma T_i^4 + \rho_i \sum_{j=1}^{N} B_j F_{i-j} \tag{24.2.5b}$$

Substituting (24.2.5b) into (24.2.5a) yields

$$q_i = \frac{\varepsilon_i \sigma T_i^4 - (1-\rho_i) B_i}{\rho_i} = \frac{\varepsilon_i}{\rho_i}(\sigma T_i^4 - B_i) \tag{24.2.6a}$$

The radiosities B_i may be calculated by rearranging (24.2.5a–24.2.6a)

$$M_{\alpha\beta} R_\beta = D_\alpha \quad (1 \le \alpha, \beta \le N) \tag{24.2.6b}$$

with N being the number of surfaces of the enclosure, and

$$M_{\alpha\beta} = \delta_{\alpha\beta} + \delta_{\alpha\gamma} A_\gamma \frac{\varepsilon_\beta}{\rho_{(\beta)}} - A_\alpha F_{(\alpha)-\beta}$$

$$D_\alpha = A_\alpha \frac{\varepsilon_{(\alpha)}}{\rho_{(\alpha)}} \sigma T_{(\alpha)}^4$$

where the subscript within the parenthethesis is not an index, not subject to summation. Solving the radiosities from (24.2.6b) and substituting into (24.2.6a) yield the heat flux q_i at the surface A_i. The unknowns in this process are the view factors (Table 24.2.1) which are described next.

Table 24.2.1 View Factors F_{A-B} for Two Square Planes, Two-Point Gaussian Quadrature

Solution Schemes		Two Parallel Planes	Two Intersecting Planes 30°	60°	90°	120°	150°
Analytic Solution		0.19983	0.62020	0.37120	0.20004	0.08700	0.02151
Finite Elements	3 × 3	0.19980	1.53905	0.51115	0.23359	0.09541	0.02299
			1.09247	0.45421			
	5 × 5	0.19982	1.17043	0.45474	0.22015	0.09196	0.02236
	8 × 8	0.19982	0.96347	0.42319	0.21561	0.08998	0.02199
			0.79660	0.40214			
	20 × 20	—	0.75673	0.39177	0.20506	0.08797	0.02160
	30 × 30	—	0.71082	0.38481	0.20339	0.08751	0.02152
	40 × 40	—	0.68786	0.38133	0.20255	0.08729	0.02147

24.2.2 VIEW FACTORS

The view factors between various surfaces (Figure 24.2.2a) are given as follows:

$$F_{dA_1-A_2} = \int_{A_2} \frac{\cos\theta_1 \cos\theta_2}{\pi L^2}\, dA_2 \tag{24.2.7a}$$

$$dF_{A_2-dA_1} = \frac{dA_1}{A_2}\int_{A_2} \frac{\cos\theta_1 \cos\theta_2}{\pi L^2}\, dA_2 \tag{24.2.7b}$$

$$F_{A_1-A_2} = \frac{1}{A_1}\int_{A_1}\int_{A_2} \frac{\cos\theta_1 \cos\theta_2}{\pi L^2}\, dA_2\, dA_1 \tag{24.2.7c}$$

$$F_{A_2-A_1} = \frac{1}{A_2}\int_{A_2}\int_{A_1} \frac{\cos\theta_1 \cos\theta_2}{\pi L^2}\, dA_1\, dA_2 \tag{24.2.7d}$$

The net radiant interchange between the two surfaces is

$$Q = \int_{A_1}\int_{A_2} \frac{\cos\theta_1 \cos\theta_2\, \sigma\left(T_1^4 - T_2^4\right)}{\pi L^2}\, dA_1\, dA_2 \tag{24.2.8}$$

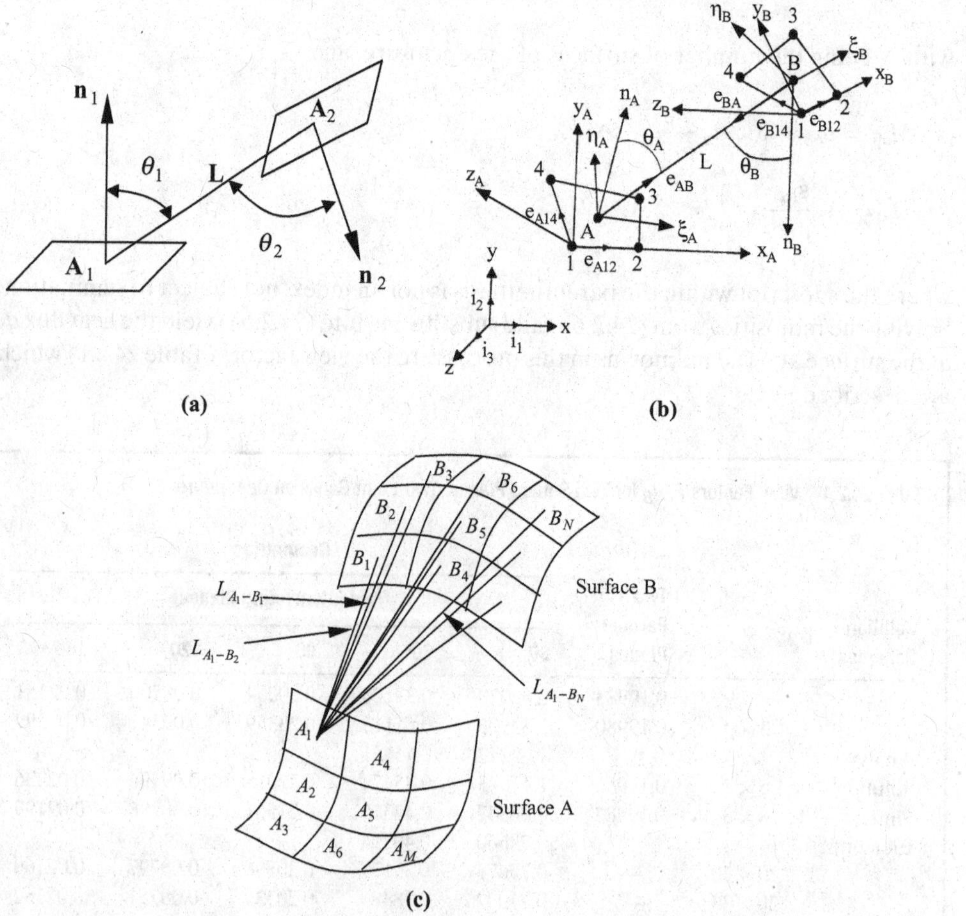

Figure 24.2.2 Geometries for view factor calculations. **(a)** Coordinates for view factor. **(b)** Normal and coordinate transformations between local and global coordinates. **(c)** Global view factor – assembly of local contributions.

where θ_1 and θ_2 are the angles between the normal and the line L separating the two surfaces dA_1 and dA_2.

Consider two diffusely reflecting, small gray bodies. The radiant interchange between the two bodies is

$$Q = AF\sigma\left(T_1^4 - T_2^4\right) \tag{24.2.9a}$$

where the reciprocity between the surfaces is given by

$$AF = A_1F_{1-2} = A_2F_{2-1}$$

For radiation interchange between large parallel gray plates, all reflected radiation is returned to the emitter. Here we note that the view factor is unity. Thus

$$Q = \frac{1}{\dfrac{1}{\varepsilon_1} + \dfrac{1}{\varepsilon_2} - 1} A\sigma\left(T_1^4 - T_2^4\right) \tag{24.2.9b}$$

View factors for simple geometries can be analytically integrated. For complicated and arbitrary geometries, however, numerical integrations are required. There are numerous numerical methods available, as detailed in the open literature. One such method is the finite element calculations, described below.

Evaluation of integrals involved in (24.2.7) can be carried out via Gaussian quadrature (Chapter 9). To this end, we first establish the coordinate systems as depicted in Figure 24.2.2b for isoparametric coordinates (ξ, η), local three-dimensional cartesian coordinates (x, y, z) with the origin at the node 1, the x-axis along the nodes 1-2, and x-y plane on the surface element 1-2-3-4, and the global coordinates (X, Y, Z). Let us now consider the normal vectors on the surface to calculate the angles θ_A and θ_B and coordinate transformations between the local and global coordinates (Figure 24.2.2b),

Surface A

The unit vector $\mathbf{e}_{A12}$ in the direction from node 1 to node 2 is of the form

$$\mathbf{e}_{A12} = \lambda_{Ai}\,\mathbf{i}_i \tag{24.2.10}$$

where

$$\lambda_{A1} = \frac{X_{A21}}{L_{A12}}, \qquad \lambda_{A2} = \frac{Y_{A21}}{L_{A12}}, \qquad \lambda_{A3} = \frac{Z_{A21}}{L_{A12}}$$

$$L_{A12} = \sqrt{(X_{A2} - X_{A1})^2 + (Y_{A2} - Y_{A1})^2 + (Z_{A2} - Z_{A1})^2}$$

$$X_{A21} = X_{A2} - X_{A1}, \text{ etc.}$$

Similarly, the unit vector in the direction from node 1 to node 4 is written as

$$\mathbf{e}_{A14} = \mu_{Ai}\mathbf{i}_i \tag{24.2.11}$$

where

$$\mu_{A1} = \frac{X_{A41}}{L_{A14}}, \qquad \mu_{A2} = \frac{Y_{A41}}{L_{A14}}, \qquad \mu_{A3} = \frac{Z_{A41}}{L_{A14}}$$

$$L_{A14} = \sqrt{(X_{A4} - X_{A1})^2 + (Y_{A4} - Y_{A1})^2 + (Z_{A4} - Z_{A1})^2}$$

$$X_{A41} = X_{A4} - X_{A1}, \text{ etc.}$$

The unit normal vector out of the surface A is

$$\mathbf{n}_A = \mathbf{e}_{A12} \times \mathbf{e}_{A14} = \nu_{Ai}\mathbf{i}_i$$

where

$$\nu_{A1} = \lambda_{A2}\mu_{A3} - \lambda_{A3}\mu_{A2}$$

$$\nu_{A2} = \lambda_{A3}\mu_{A1} - \lambda_{A1}\mu_{A3}$$

$$\nu_{A3} = \lambda_{A1}\mu_{A2} - \lambda_{A2}\mu_{A1}$$

Surface B

The unit vectors $\mathbf{e}_{B12}$, $\mathbf{e}_{B14}$, and $\mathbf{n}_B$ for the Surface B can be derived similarly as the Surface A.

$$\mathbf{e}_{B12} = \lambda_{Bi}\mathbf{i}_i \tag{24.2.12}$$

$$\mathbf{e}_{B14} = \mu_{Bi}\mathbf{i}_i \tag{24.2.13}$$

$$\mathbf{n}_B = \mathbf{e}_{B12} \times \mathbf{e}_{B14} = \nu_{Bi}\mathbf{i}_i \tag{24.2.14}$$

Angles θ_A and θ_B

Let L be the length of the line connecting the two surfaces A and B at arbitrary points. The angles θ_A and θ_B are measured from their normals to the surfaces of the line L. The unit vector along this line is

$$\mathbf{e}_{AB} = \kappa_{ABi}\mathbf{i}_i \tag{24.2.15}$$

where

$$\kappa_{AB1} = \frac{X^{BA}}{L}, \qquad \kappa_{AB2} = \frac{Y^{BA}}{L}, \qquad \kappa_{AB3} = \frac{Z^{BA}}{L}$$

$$L = \sqrt{(X^B - X^A)^2 + (Y^B - Y^A)^2 + (Z^B - Z^A)^2}$$

$$X^{BA} = X^B - X^A, \text{ etc.}$$

The angles θ_A and θ_B can be determined from the relationships

$$\cos\theta_A = \frac{\mathbf{n}_A \cdot \mathbf{e}_{AB}}{|\mathbf{n}_A||\mathbf{e}_{AB}|} \tag{24.2.16a}$$

and

$$\cos\theta_B = \frac{\mathbf{n}_B \cdot \mathbf{e}_{BA}}{|\mathbf{n}_B||\mathbf{e}_{BA}|} \tag{24.2.16b}$$

The local and global coordinates for the Surface A are related by

$$x_i^A = a_{ij}^A X_j^A \tag{24.2.17}$$

where

$$a_{11}^A = \lambda_{A1}, \qquad a_{12}^A = \lambda_{A2}, \qquad a_{13}^A = \lambda_{A3}$$

The unit vector in the direction of y on the Surface A is obtained by

$$\mathbf{e}_y^A = \mathbf{n}_A \times \mathbf{e}_{A12} = \gamma_i^A \mathbf{i}_i \tag{24.2.18}$$

Thus

$$a_{21}^A = \gamma_1^A, \qquad a_{22}^A = \gamma_2^A, \qquad a_{23}^A = \gamma_3^A$$

and

$$a_{31}^A = \nu_{A1}, \qquad a_{32}^A = \nu_{A2}, \qquad a_{33}^A = \nu_{A3}$$

Similarly for the surface B,

$$x_i^B = a_{ij}^B X_j^B \tag{24.2.19}$$

$$a_{11}^B = \lambda_{B1}, \qquad a_{12}^B = \lambda_{B2}, \qquad a_{12}^B = \lambda_{B3}$$

$$\mathbf{e}_y^B = \mathbf{n}_B \times \mathbf{e}_{B12} = \gamma_i^B \mathbf{i}_i \tag{24.2.20}$$

$$a_{21}^B = \gamma_1^B, \qquad a_{22}^B = \gamma_2^B, \qquad a_{23}^B = \gamma_3^B$$

$$a_{31}^B = \nu_{B1}, \qquad a_{32}^B = \nu_{B2}, \qquad a_{33}^B = \nu_{B3}$$

The transformation between the isoparametric coordinates and the local cartesian coordinates is related by

$$dxdy = |J| d\xi d\eta \tag{24.2.21}$$

where

$$|J| = \begin{vmatrix} \dfrac{\partial x}{\partial \xi} & \dfrac{\partial y}{\partial \xi} \\ \dfrac{\partial x}{\partial \eta} & \dfrac{\partial y}{\partial \eta} \end{vmatrix} \tag{24.2.22}$$

At this point, we introduce isoparametric finite element functions to relate the variation of the global coordinates with nodal values,

$$X_i^A = \Phi_N^A(\xi_A, \eta_A) X_{Ni}^A \tag{24.2.23a}$$

or

$$X_i^B = \Phi_N^B(\xi_B, \eta_B) X_{Ni}^B \tag{24.2.23b}$$

where Φ_N^A and Φ_N^B may be chosen as linear isoparametric interpolation functions. It follows from (24.2.17) that

$$x^A = a_{11}^A X^A + a_{12}^A Y^A + a_{13}^A Z^A$$

$$y^A = a_{21}^A X^A + a_{22}^A Y^A + a_{23}^A Z^A$$

and from (24.2.23),

$$\frac{\partial x^A}{\partial \xi_A} = a_{11}^A \frac{\partial \Phi_N^A}{\partial \xi_A} X_N^A + a_{12}^A \frac{\partial \Phi_N^A}{\partial \xi_A} Y_N^A + a_{13}^A \frac{\partial \Phi_N^A}{\partial \xi_A} Z_N^A$$

$$\frac{\partial y^A}{\partial \xi_A} = a_{21}^A \frac{\partial \Phi_N^A}{\partial \xi_A} X_N^A + a_{22}^A \frac{\partial \Phi_N^A}{\partial \xi_A} Y_N^A + a_{23}^A \frac{\partial \Phi_N^A}{\partial \xi_A} Z_N^A$$

etc., and similarly for the surface B.

From (24.2.21) for the surfaces A and B, we have

$$dA_A = dx_A dy_A = |J|_A d\xi_A d\eta_A \tag{24.2.24a}$$

$$dA_B = dx_B dy_B = |J|_B d\xi_B d\eta_B \tag{24.2.24b}$$

Let us consider two finite surfaces A and B, each discretized into M and N numbers of isoparametric elements, respectively (Figure 24.2.2c). The view factor F_{A-B} is obtained as

$$F_{A-B} = \sum_{\alpha=1}^{M}\sum_{\beta=1}^{N} F_{A\alpha-B\beta} = \frac{1}{A_A}\sum_{\alpha=1}^{M}\sum_{\beta=1}^{N} \hat{F}_{A\alpha-B\beta} \tag{24.2.25a}$$

where

$$A_A = \sum_{\alpha=1}^{M} A_{A\alpha} \tag{24.2.25b}$$

$$\hat{F}_{A\alpha-B\beta} = \int_{A_{A\alpha}}\int_{B_{B\beta}} \frac{\cos\theta_{A\alpha}\cos\theta_{B\beta}}{\pi L^2_{A\alpha-B\beta}}\, dA_{A\alpha}\, dA_{B\beta} \tag{24.2.25c}$$

To utilize the Gaussian quadrature integration, we write

$$\hat{F}_{A\alpha-B\beta} = \int_{-1}^{1}\int_{-1}^{1}\int_{-1}^{1}\int_{-1}^{1} C\,|J|_{A\alpha}|J|_{B\beta}\, d\xi_{A\alpha} d\eta_{A\alpha} d\xi_{B\beta} d\eta_{B\beta}$$

where

$$C = \frac{\cos\theta_{A\alpha}\cos\theta_{B\beta}}{\pi L^2_{A\alpha-B\beta}}$$

which is determined by combining (24.2.23) with (24.2.15) and (24.2.16). Thus

$$F_{A-B} = \frac{\sum_{\alpha=1}^{M}\sum_{\beta=1}^{N}\sum_i\sum_j\sum_k\sum_l W_i W_j W_k W_l\, f(\xi_i^{A\alpha}, \eta_j^{A\alpha}, \xi_k^{B\beta}, \eta_l^{B\beta})}{\sum_{\alpha=1}^{M} A_{A\alpha}} \tag{24.2.26}$$

The function f contains the integrand $C\,|J|_{A\alpha}|J|_{B\beta}$ and the routine Gaussian quadrature integration may be carried out [Chung and Kim, 1982]. An example problem for view factor calculations are shown in Section 24.4.1(1).

Radiation Boundary Conditions with View Factors

Radiation boundary conditions in nonparticipating media can be implemented in the energy equation of the form,

$$\frac{\partial}{\partial t}(\rho c_p T) + \frac{\partial}{\partial x_i}(\rho c_p T\,\overline{v}_i - kT_{,ii}) = 0 \tag{24.2.27}$$

Here the convection velocity $\overline{v}_i$ is taken as a constant, but should be treated as a variable when the energy equation (24.2.27) is solved simultaneously with the equations of continuity and momentum. For simplicity of discussion, let us consider the Galerkin finite element formulation of (24.2.27b) in the form

$$\int_\Omega \Phi_\alpha \left[\frac{\partial}{\partial t}(\rho c_p T) + \frac{\partial}{\partial x_i}(\rho c_p T\,\overline{v}_i - kT_{,ii}) \right] d\Omega = 0 \qquad (24.2.28)$$

Integrating (24.2.28) by parts,

$$\left(\int_\Omega \rho c_p \Phi_\alpha \Phi_\beta \, d\Omega \right) \frac{\partial T_\beta}{\partial t} + \int_\Gamma \overset{*}{\Phi}_\alpha (\rho c_p T\,\overline{v}_i n_i - kT_{,i} n_i)\, d\Gamma - \left(\int_\Omega \rho c_p \overline{v}_i \Phi_{\alpha,i} \Phi_\beta \, d\Omega \right) T_\beta$$
$$+ \left(\int_\Omega k \Phi_{\alpha,i} \Phi_{\beta,i} \, d\Omega \right) T_\beta = 0$$

or

$$\int_\xi \hat{W}(\xi)\,(A_{\alpha\beta}\dot{T}_\beta + B_{\alpha\beta}T_\beta + K_{\alpha\beta}T_\beta - G_\alpha)\, d\xi = 0 \qquad (24.2.29)$$

where

Heat capacity matrix $\quad A_{\alpha\beta} = \int_\Omega \rho c_p \Phi_\alpha \Phi_\beta \, d\Omega$

Heat convection matrix $\quad B_{\alpha\beta} = -\int_\Omega k \Phi_{d,i} \Phi_{\beta,i} \, d\Omega$

Heat conduction matrix $\quad K_{\alpha\beta} = \int_\Omega \rho c_p \overline{v}_i \Phi_{\alpha,i} \Phi_\beta \, d\Omega$

The Neumann boundary vector is contributed by

$$\rho c_p T\,\overline{v}_i n_i - kT_{,i} n_i = g \quad \text{on } \Gamma_N \qquad (24.2.30)$$

The heat flux normal to the boundary surface takes the form

$$-kT_{,i} n_i = \overline{q}^{(CD)} + \overline{q}^{(CV)} + \overline{q}^{(R)} \qquad (24.2.31)$$

where the superscripts (CD), (CV), and (R) denote conduction, convection, and radiation, respectively. Here $\overline{q}^{(CD)}$ represents the conduction heat flux applied on the boundary surface, $\overline{q}^{(CV)}$ is the convection heat flux

$$\overline{q}^{(CV)} = \rho c_p T\,\overline{v}_i n_i = \overline{\alpha}(T - T') \quad \text{on } \Gamma_N^{(CV)} \qquad (24.2.32)$$

with $\overline{\alpha}$ and T' being the heat transfer coefficient and ambient temperature, respectively.

The radiation boundary heat flux $\overline{q}^{(R)}$ is of the form

$$\overline{q}^{(R)} = F\sigma\varepsilon(T^4 - T_r^4) \quad \text{on } \Gamma_N^{(R)} \qquad (24.2.33)$$

Here F is the view factor and T_r denotes the radiation boundary temperature of a separate body to which radiation exchange occurs.

It should be noted that the Neumann boundary conditions contain the variable T in (24.2.32) and (24.2.33). This implies that temperature is unknown and must be computed. Thus, the surface integral containing the temperature variable consists of the boundary surface convection matrix $\overset{*}{C}_{\alpha\beta}$ and the boundary surface radiation vector $\overset{*}{R}_\alpha$,

$$\overset{*}{C}_{\alpha\beta} = \int_\Gamma \overline{\alpha}\,\overset{*}{\Phi}_\alpha\,\overset{*}{\Phi}_\beta\, d\Gamma \tag{24.2.34}$$

$$\overset{*}{R}_\alpha = \int_\Gamma F\sigma\,\varepsilon\,\overset{*}{\Phi}_\alpha\,\overset{*}{\Phi}_\beta\,\overset{*}{\Phi}_\gamma\,\overset{*}{\Phi}_\delta\,\overset{*}{\Phi}_\lambda\, d\Gamma\, T_\alpha T_\beta T_\gamma T_\delta T_\lambda \tag{24.2.35}$$

It follows from (24.2.34), (24.2.35), and (24.2.31) that the Neumann boundary condition G_α is the mixed boundary condition (sometimes called the Cauchy boundary condition),

$$G_\alpha = \overset{*}{G}_\alpha - \overset{*}{C}_{\alpha\beta} T_\beta - \overset{*}{R}_\alpha \tag{24.2.36}$$

where $\overset{*}{G}_\alpha$ denotes the known data on the boundary surface

$$\overset{*}{G}_\alpha = \int_\Gamma \overset{*}{\Phi}_\alpha \left(\overline{q}^{(CD)} + \overline{\alpha}\, T' + F\sigma\varepsilon T_r^4\right) d\Gamma \tag{24.2.37}$$

Substituting (24.2.36) into (24.2.29)

$$\int_\xi \hat{W}(\xi)\left(A_{\alpha\beta}\dot{T}_\beta + B_{\alpha\beta} T_\beta + K_{\alpha\beta} T_\beta + \overset{*}{C}_{\alpha\beta} T_\beta + \overset{*}{R}_\alpha - \overset{*}{G}_\alpha\right) d\xi = 0 \tag{24.2.38}$$

As discussed in Chapters 10 and 11, the transient problem in (24.2.38) may be recast in a time-marching scheme. In terms of the temporal parameter η, we write

$$\begin{aligned}\left[A_{\alpha\beta} + \eta\Delta t\left(B_{\alpha\beta} + K_{\alpha\beta} + \overset{*}{C}_{\alpha\beta}\right)\right] T_\beta^{n+1} &= \left[A_{\alpha\beta} - (1-\eta)\Delta t\left(B_{\alpha\beta} + K_{\alpha\beta} + \overset{*}{C}_{\alpha\beta}\right)\right] T_\beta^n \\ &\quad + \Delta t\left(\overset{*}{G}_\alpha - \overset{*}{R}_\alpha\right)\end{aligned} \tag{24.2.39}$$

The algebraic equations resulting from (24.2.37) are nonlinear because of the radiation boundary term: $\overset{*}{R}\alpha$, and the standard Newton-Raphson iteration method should be used as described in Section 11.5.1. This will lead to

$$J_{\alpha\beta}\Delta T_\beta^{n+1,r+1} = -E_\alpha^{n+1,r} \tag{24.2.40}$$

where

$$J_{\alpha\beta} = A_{\alpha\beta} + \eta\Delta t\left(B_{\alpha\beta} + K_{\alpha\beta} + \overset{*}{C}_{\alpha\beta}\right) + \Delta t\, S_{\alpha\beta}$$

$$\begin{aligned}S_{\alpha\beta} &= S_{\alpha\beta\gamma\delta}(T_\gamma T_\delta T_\lambda)^{n+1,r} + S_{\alpha\eta\beta\delta\lambda}(T_\eta T_\delta T_\lambda)^{n+1,r} + S_{\alpha\eta\gamma\beta\lambda}(T_\eta T_\gamma T_\lambda)^{n+1,r} \\ &\quad + S_{\alpha\eta\gamma\delta\beta}(T_\eta T_\gamma T_\delta)^{n+1,r}\end{aligned}$$

$$\begin{aligned}E_\alpha^{n+1,r} &= \left[A_{\alpha\beta} + \eta\Delta t\left(B_{\alpha\beta} + K_{\alpha\beta} + \overset{*}{C}_{\alpha\beta}\right)\right] T_\beta^{n+1,r} \\ &\quad - \left[A_{\alpha\beta} - (1-\eta)\Delta t\left(B_{\alpha\beta} + K_{\alpha\beta} + \overset{*}{C}_{\alpha\beta}\right)\right] T_\beta^{n,r} + \Delta t\left(\overset{*}{G}_\alpha - \overset{*}{R}{}_\alpha^{n,r}\right)\end{aligned}$$

An alternative approach for $\overset{*}{R}_\alpha$ is to assume a linear variation of T_4 within a small element

$$T^4 = \Phi_\alpha T_\alpha^4 \tag{24.2.41}$$

where T_α^4 is calculated from initial and/or boundary conditions and subsequently from previous values during the time-marching process described in Chapters 10 and 11.

Numerical examples for the implementation of radiation boundary conditions are shown in Section 24.4.1(2) for both steady-state and transient heat transfer problems.

24.2.3 RADIATIVE HEAT FLUX AND RADIATIVE TRANSFER EQUATION

The governing equations for radiative heat transfer in participating media have been well established [Sparrow and Cess, 1970; Siegel and Howell, 1992, among many others]. Optical thicknesses measured in terms of absorption properties or the photon mean free path are involved in the radiative heat transfer. We define the monochromatic optical thickness of the medium as

$$\tau_{o\lambda} = a_\lambda L$$

or

$$\tau_{o\lambda} = \frac{L}{\lambda_p}$$

where L is a characteristic length, a_λ is the absorption coefficient independent of temperature, and λ_p is the photon mean free path. It is seen that the optical thickness $\tau_{o\lambda}$ is a reciprocal photon Knudsen number. We define $\tau_{o\lambda} \ll 1$ as optically thin and $\tau_{o\lambda} \gg 1$ as optically thick. The limiting case $a_\lambda = 0$ would represent a nonparticipating medium (transparent) where the radiative flux vector $\mathbf{q}_R$ is constant ($\nabla \cdot \mathbf{q}_R = 0$), whereas $a_\lambda = \infty$ corresponds to an opaque medium in which $\mathbf{q}_R = 0$.

Consider two surfaces depicted in Figure 24.2.3a for one-dimensional radiative transfer with the intensity of radiation directed at an angle θ from the normal, denoted

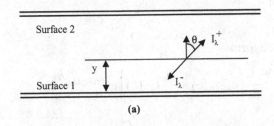

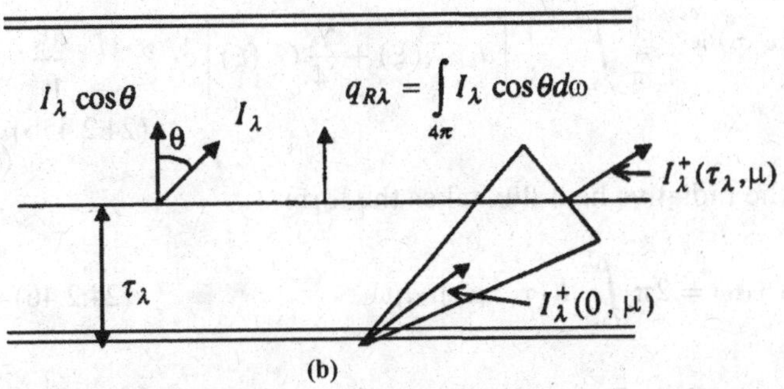

Figure 24.2.3 Geometries for one-dimensional radiative heat transfer. **(a)** Coordinate system for one-dimensional radiative transfer. **(b)** The radiation heat flux.

as $I^+(y, \theta)$ and $I^-(y, \theta)$. The differential equations for these intensities assume the form

$$\cos\theta \frac{dI_\lambda^+}{dy} + \beta_\lambda I_\lambda^+ = \frac{a_\lambda}{\pi} e_{b\lambda}(y) + \frac{\gamma_\lambda}{4\pi} G_\lambda(y) \tag{24.2.42a}$$

$$\cos\theta \frac{dI_\lambda^-}{dy} + \beta_\lambda I_\lambda^- = \frac{a_\lambda}{\pi} e_{b\lambda}(y) + \frac{\gamma_\lambda}{4\pi} G_\lambda(y) \tag{24.2.42b}$$

where $G_\lambda(y)$ is the radiation function given by

$$G_\lambda(y) = 2\pi \int_0^\pi I_\lambda(y, \theta') \sin\theta' \, d\theta' \tag{24.2.43}$$

We may recast these equations as follows:

$$\mu \frac{dI_\lambda^+}{d\tau_\lambda} + I_\lambda^+ = \frac{1}{\pi\beta_\lambda}\left[a_\lambda e_{b\lambda}(\tau_\lambda) + \frac{\gamma_\lambda}{4} G_\lambda(\tau_\lambda)\right] \tag{24.2.44a}$$

$$\mu \frac{dI_\lambda^-}{d\tau_\lambda} + I_\lambda^- = \frac{1}{\pi\beta_\lambda}\left[a_\lambda e_{b\lambda}(\tau_\lambda) + \frac{\gamma_\lambda}{4} G_\lambda(\tau_\lambda)\right] \tag{24.2.44b}$$

where

$$\tau_\lambda = \int_0^y \beta_\lambda dy, \qquad \tau_{o\lambda} = \int_0^L \beta_\lambda dy, \qquad \mu = \cos\theta$$

With the boundary conditions of the form,

$$I_\lambda^+(\tau_\lambda, \mu) = I_\lambda^+(0, \mu), \quad \tau_\lambda = 0$$

$$I_\lambda^-(\tau_\lambda, \mu) = I_\lambda^-(\tau_{o\lambda}, \mu), \quad \tau_\lambda = \tau_{o\lambda}$$

we obtain the solution to (24.2.44) in the form

$$I_\lambda^+(\tau_\lambda, \mu) = I_\lambda^+(0, \mu)\, e^{-\tau_\lambda/\mu} - \frac{1}{\pi}\int_0^{\tau_\lambda} \frac{1}{\beta_\lambda}\left[a_\lambda e_{b\lambda}(\xi) + \frac{\gamma_\lambda}{4} G_\lambda(\xi)\right] e^{-(\tau_\lambda - \xi)/\mu} \frac{d\xi}{\mu} \tag{24.2.45a}$$

$$I_\lambda^-(\tau_\lambda, \mu) = I_\lambda^-(\tau_{o\lambda}, \mu)\, e^{(\tau_{o\lambda} - \tau_\lambda)/\mu} + \frac{1}{\pi}\int_{\tau_\lambda}^{\tau_{o\lambda}} \frac{1}{\beta_\lambda}\left[a_\lambda e_{b\lambda}(\xi) + \frac{\gamma_\lambda}{4} G_\lambda(\xi)\right] e^{-(\tau_\lambda - \xi)/\mu} \frac{d\xi}{\mu} \tag{24.2.45b}$$

As defined in Figure 24.2.3b, the radiative heat flux takes the form

$$q_{R\lambda}(\tau_\lambda) = \int_{4\pi} I_\lambda(\tau_\lambda, \mu) \cos\theta \, d\omega = 2\pi \int_{-1}^{1} I_\lambda(\tau_\lambda, \mu)\, \mu \, d\mu \tag{24.2.46}$$

or

$$q_{R\lambda}(\tau_\lambda) = 2\pi \int_0^1 I_\lambda^+ \mu \, d\mu - 2\pi \int_0^{-1} I_\lambda^- \mu \, d\mu$$

In view of (24.2.45), we obtain

$$q_{R\lambda}(\tau_\lambda) = 2\pi \int_0^1 I_\lambda^+(\tau_\lambda, \mu)\, e^{-\tau_\lambda/\mu} \mu\, d\mu - 2\pi \int_0^{-1} I_\lambda^-(\tau_{o\lambda}, -\mu)\, e^{-(\tau_{o\lambda}-\tau_\lambda)/\mu} \mu\, d\mu$$
$$+ 2 \int_0^{\tau_\lambda} \frac{1}{\beta_\lambda}\left[a_\lambda e_{b\lambda}(\xi) + \frac{\gamma_\lambda}{4} G_\lambda(\xi)\right] E_2(\tau_\lambda - \xi)\, d\xi$$
$$- 2 \int_0^{\tau_{o\lambda}} \frac{1}{\beta_\lambda}\left[a_\lambda e_{b\lambda}(\xi) + \frac{\gamma_\lambda}{4} G_\lambda(\xi)\right] E_2(\xi - \tau_\lambda)\, d\xi \tag{24.2.47}$$

where the $E_n(\xi)$ is given by

$$E_n(\xi) = \int_0^1 \mu^{n-2} e^{-\xi/\mu} d\mu$$

This integral for $n = 1, 2, 3$ is evaluated as

$$E_1(\xi) = -a - \ln\xi + \xi - \frac{\xi^2}{2\cdot 2!} + \frac{\xi^3}{3\cdot 3!} + \cdots \tag{24.2.48a}$$

$$E_2(\xi) = 1 + (a - 1 + \ln\xi)\xi - \frac{\xi^2}{1\cdot 2!} + \frac{\xi^3}{2\cdot 3!} + \cdots \tag{24.2.48b}$$

$$E_3(\xi) = \frac{1}{2} - \xi + \frac{1}{2}\left(-a + \frac{3}{2} - \ln\xi\right)\xi^2 + \frac{\xi^3}{1\cdot 3!} + \cdots \tag{24.2.48c}$$

with $a = 0.5772$ (Euler's constant). The total radiation flux is

$$q_R(y) = \int_0^\infty q_{R\lambda}(\tau_\lambda)\, d\lambda$$

The divergence of the radiation flux vector takes the form

$$\nabla \cdot \mathbf{q}_R = \frac{dq_R}{dy} = \int_0^\infty \frac{dq_{R\lambda}}{dy} d\lambda = \int_0^\infty \beta_\lambda \frac{dq_{R\lambda}}{d\tau_\lambda} d\lambda \tag{24.2.49}$$

where

$$\frac{dq_{R\lambda}}{d\tau_\lambda} = -G_\lambda(\tau) + \frac{4a_\lambda}{\beta_\lambda} e_{b\lambda}(\tau_\lambda) + \frac{\gamma_\lambda}{\beta_\lambda} G_\lambda(\tau_\lambda) \tag{24.2.50}$$

where $G_\lambda(\tau_\lambda)$ is the incidence radiation function given by

$$G_\lambda(\tau_\lambda) = 2\pi \int_0^1 I_\lambda^+(\tau_\lambda, \mu')\, d\mu' - 2\pi \int_0^{-1} I_\lambda^-(\tau_\lambda, \mu')\, d\mu'$$

or

$$G_\lambda(\tau_\lambda) = 2\pi \int_0^1 I_\lambda^+(0, \mu)\, e^{-\tau_\lambda/\mu}\, d\mu + 2\pi \int_0^1 I_\lambda^-(\tau_{o\lambda}, -\mu)\, e^{-(\tau_{o\lambda}-\tau_\lambda)/\mu} d\mu$$
$$+ 2 \int_0^{\tau_{o\lambda}} \frac{1}{\beta_\lambda}\left[a_\lambda e_{b\lambda}(\xi) + \frac{\gamma_\lambda}{4} G_\lambda(\xi)\right] E_2(|\tau_\lambda - \xi|)\, d\xi \tag{24.2.51}$$

Here we note that Planck's function eb_λ is temperature dependent and thus equations (24.2.50) and (24.2.51) constitute nonlinear integrodifferential equations.

The gray-medium assumption leads to simplification of (24.2.50) and (24.2.51) in which the absorption and scattering coefficients are assumed to be independent of wavelength. Denoting that

$$\int_0^\infty e_{b\lambda} d\lambda = \sigma T^4, \qquad \int_0^\infty I_\lambda d\lambda = I, \qquad \int_0^\infty G_\lambda d\lambda = G$$

we rewrite (24.2.47) as

$$\begin{aligned} q_R(\tau) = {} & 2\pi \int_0^1 I_\lambda^+(0, \mu) e^{-\tau/\mu} \mu \, d\mu - 2\pi \int_0^1 I_\lambda^-(\tau_o, -\mu) e^{-(\tau_o-\tau)/\mu} \mu \, d\mu \\ & + 2 \int_0^\tau \frac{1}{\beta} \left[a\sigma T^4(\xi) + \frac{\gamma}{4} G(\xi) \right] E_2(\tau - \xi) \, d\xi \\ & - 2 \int_0^{\tau_o} \frac{1}{\beta} \left[a T^4(\xi) + \frac{\gamma}{4} G(\xi) \right] E_2(\tau - \xi) \, d\xi \end{aligned} \qquad (24.2.52)$$

This gives

$$\frac{dq_R(\tau)}{d\tau} = -G(\tau) + 4\frac{a\sigma}{\beta} T^4(\tau) + \frac{\gamma}{\beta} G(\tau) \qquad (24.2.53)$$

$$\begin{aligned} G(\tau) = {} & 2\pi \int_0^1 I^+(0, \mu) e^{-\tau/\mu} \, d\mu + 2\pi \int_0^1 I^-(\tau_o, -\mu) e^{-(\tau_o-\tau)/\mu} \, d\mu \\ & + 2 \int_0^{\tau_o} \frac{1}{\beta_\lambda} \left[a\sigma T^4(\xi) + \frac{\gamma}{4} G(\xi) \right] E_1(|\tau - \xi|) \, d\xi \end{aligned} \qquad (24.2.54)$$

For radiative equilibrium, in which radiation is the predominant mode of heat transfer and the system is in steady state, we have $\nabla \cdot \mathbf{q}_R = 0$ or here $dq_R/dy = 0$. Then from (24.2.53) we obtain, for $\gamma = 0$,

$$G(\tau) = 4\sigma T^4(\tau) \qquad (24.2.55)$$

with the conservation of energy given by

$$\begin{aligned} 2\sigma T^4(\tau) = {} & \pi \int_0^1 I^+(0, \mu) e^{-\tau/\mu} \, d\mu + \pi \int_0^1 I^-(\tau_o, -\mu) e^{-(\tau_o-\tau)/\mu} \, d\mu \\ & + \sigma \int_0^{\tau_o} T^4(\xi) E_1(|\tau - \xi|) \, d\xi \end{aligned} \qquad (24.2.56)$$

The nonscattering media can be represented by (24.2.47) and (24.2.50) with $\gamma_\lambda = 0$, thus, (24.2.51) is no longer necessary. For pure scattering ($a_\lambda = 0$), (24.2.50) becomes

$$\frac{dq_{R\lambda}}{d\tau_\lambda} = 0$$

which implies that the energy equation is uncoupled from the radiation transfer process. Thus, the governing equations for scattering can be obtained by (24.2.50) and (24.2.51) with $a_\lambda = 0$.

For a diffuse surface, $I_\lambda^+(0, \mu)$ and $I_\lambda^-(\tau_{o\lambda}, -\mu)$ are independent of direction, that is, independent of μ. Thus, we set

$$I_\lambda^+(0, \mu) = \frac{B_{1\lambda}}{\pi}, \qquad I_\lambda^-(\tau_{o\lambda}, -\mu) = \frac{B_{2\lambda}}{\pi}$$

where $B_{1\lambda}$ and $B_{2\lambda}$ are the surface radiosities. We then have

$$2\pi \int_0^1 I^+(0,\mu)\, e^{-\tau/\mu}\, \mu\, d\mu = 2B_{1\lambda} E_3(\tau_\lambda) \tag{24.2.57a}$$

$$2\pi \int_0^1 I_\lambda^-(\tau_{o\lambda}, -\mu)\, e^{-(\tau_{o\lambda}-\tau_\lambda)/\mu}\, \mu d\mu = 2B_{2\lambda} E_3(\tau_{o\lambda}-\tau_\lambda) \tag{24.2.57b}$$

$$2\pi \int_0^1 I_\lambda^+(0,\mu)\, e^{-\tau_\lambda \mu} d\mu = 2B_{1\lambda} E_2(\tau_\lambda) \tag{24.2.57c}$$

$$2\pi \int_0^1 I_\lambda^-(\tau_{o\lambda}, -\mu)\, e^{-(\tau_{o\lambda}-\tau_\lambda)/\mu}\, d\mu = 2B_{2\lambda} E_3(\tau_{o\lambda}-\tau_\lambda) \tag{24.2.57d}$$

To determine $B_{1\lambda}$ and $B_{2\lambda}$, we proceed as follows: Substitute (24.2.57a,b) into (24.2.47) with $\tau_\lambda = 0$ for Surface 1. Then

$$q_{R\lambda}(0) = B_{1\lambda} - H_{1\lambda} \tag{24.2.58}$$

$$H_{1\lambda} = 2B_{2\lambda} E_3(\tau_{o\lambda}) + 2\int_0^{\tau_{o\lambda}} \frac{1}{\beta_\lambda}\left[a_\lambda e_{b\lambda}(\xi) + \frac{\gamma_\lambda}{4} G_\lambda(\xi)\right] E_2(\xi)\, d\xi$$

Here $B_{1\lambda}$ is the radiant energy leaving Surface 1 and $H_{1\lambda}$ is the incident energy. We may define the surface radiosity as

$$B_{1\lambda} = \varepsilon_{1\lambda} e_{b1\lambda} + (1-\varepsilon_{1\lambda}) H_{1\lambda}$$

This leads to

$$B_{1\lambda} = \varepsilon_{1\lambda} e_{b1\lambda} + 2(1-\varepsilon_{1\lambda})\left\{B_{2\lambda} E_3(\tau_{o\lambda}) + 2\int_0^{\tau_{o\lambda}} \frac{1}{\beta_\lambda}\left[a_\lambda e_{b\lambda}(\xi) + \frac{\gamma_\lambda}{4} G_\lambda(\xi)\right] E_2(\xi) d\xi\right\} \tag{24.2.59a}$$

Likewise, for Surface 2

$$B_{2\lambda} = \varepsilon_{2\lambda} e_{b2\lambda} + 2(1-\varepsilon_{2\lambda}) \times\left\{B_{1\lambda} E_3(\tau_{o\lambda}) + 2\int_0^{\tau_{o\lambda}} \frac{1}{\beta_\lambda}\left[a_\lambda e_{b\lambda}(\xi) + \frac{\gamma_\lambda}{4} G_\lambda(\xi)\right] E_2(\tau_{o\lambda}-\xi) d\xi\right\} \tag{24.2.59b}$$

For black surfaces, $B_{1\lambda} = e_{b1\lambda}$ and $B_{2\lambda} = e_{b2\lambda}$. For nonblack surfaces, the radiosities $B_{1\lambda}$ and $B_{2\lambda}$ can be determined by solving (24.2.59a,b) simultaneously.

Optically Thin Limit

Using (24.2.48a), $E_2(\xi) = 1 + O(\xi)$ and $E_3(\xi) = \frac{1}{2} - \xi + O(\xi^2)$, we express the monochromatic radiation flux as

$$q_{R\lambda}(\tau_\lambda) = B_{1\lambda}(1-2\tau_\lambda) - B_{2\lambda}(1-2\tau_{o\lambda}+2\tau_\lambda) + 2\int_0^{\tau_{o\lambda}} \frac{1}{\beta_\lambda}\left[a_\lambda e_{b\lambda}(\xi) + \frac{\gamma_\lambda}{4} G_\lambda(\xi)\right] d\xi - 2\int_0^{\tau_{o\lambda}} \frac{1}{\beta_\lambda}\left[a_\lambda e_{b\lambda}(\xi) + \frac{\gamma_\lambda}{4} G_\lambda(\xi)\right] d\xi \tag{24.2.60}$$

For the optically thin case, $\tau_{o\lambda} \ll 1$, a further simplification can be made,

$$q_{R\lambda} = B_{1\lambda} - B_{2\lambda} \tag{24.2.61}$$

With similar simplifications from (24.2.59), we obtain

$$B_{1\lambda} = \frac{\varepsilon_{1\lambda} e_{b1\lambda} + (1-\varepsilon_{1\lambda})\varepsilon_{2\lambda} e_{b2\lambda}}{1-(1-\varepsilon_{1\lambda})(1-\varepsilon_{2\lambda})} \tag{24.2.62a}$$

$$B_{2\lambda} = \frac{\varepsilon_{2\lambda} e_{b2\lambda} + (1-\varepsilon_{2\lambda})\varepsilon_{1\lambda} e_{b1\lambda}}{1-(1-\varepsilon_{1\lambda})(1-\varepsilon_{2\lambda})} \tag{24.2.62b}$$

These expressions correspond to radiation transfer through nonparticipating media. Differentiation of (24.2.60) gives

$$\frac{dq_{R\lambda}}{d\tau_\lambda} = -2B_{1\lambda} - 2B_{2\lambda} + \frac{4}{\beta_\lambda}\left[a_\lambda e_{b\lambda}(\tau_\lambda) - \frac{\gamma_\lambda}{4} G_\lambda(\tau_\lambda)\right] \tag{24.2.63}$$

with

$$G_\lambda(\tau_\lambda) = 2B_{1\lambda} + 2B_{2\lambda}$$

Thus,

$$\frac{dq_{R\lambda}}{d\tau_\lambda} = -\frac{2a_\lambda}{\beta_\lambda}[B_{1\lambda} + B_{2\lambda} - 2e_{b\lambda}(\tau_\lambda)] \tag{24.2.64}$$

or

$$\frac{dq_{R\lambda}}{dy} = -2a_\lambda[B_{1\lambda} + B_{2\lambda} - 2e_{b\lambda}(y)] \tag{24.2.65}$$

This indicates that the radiation transfer to or from a volume element is independent of the scattering coefficient. It is realized that the optically thin conditions are free from integral equations.

Integration of (24.2.65) over all wavelengths leads to

$$\frac{dq_R}{dy} = 2a_m(T,\, T_1)B_1 + 2a_m(T,\, T_2)B_2 - 4a_p(T)\sigma T^4(y) \tag{24.2.66}$$

with

$$a_m(T,\, T_1) = \frac{\int_0^\infty a_\lambda(T) B_{1\lambda} d\lambda}{B_1} \tag{24.2.67a}$$

$$a_m(T,\, T_2) = \frac{\int_0^\infty a_\lambda(T) B_{2\lambda} d\lambda}{B_2} \tag{24.2.67b}$$

For black surfaces with the monochromatic absorption coefficient independent of temperature, we obtain

$$a_m(T,\, T_1) = \frac{\int_0^\infty a_\lambda e_{b\lambda}(T_1) d\lambda}{e_b(T_1)} = a_p(T_1) \tag{24.2.68a}$$

$$a_m(T,\, T_2) = \frac{\int_0^\infty a_\lambda e_{b\lambda}(T_2) d\lambda}{e_b(T_2)} = a_p(T_2) \tag{24.2.68b}$$

Thus, from (24.2.66)

$$-\frac{dq_R}{dy} = 2\sigma\left[a_p(T_1)T_1^4 + a_p(T_2)T_2^4 - 2a_p(T)T^4(y)\right] \tag{24.2.69}$$

For a gray medium ($a = a_p$), equation (24.2.57) yields

$$\frac{dq_R}{dy} = -2\sigma a_p(T)\left[T_1^4 + T_2^4 - 2T^4(y)\right] \tag{24.2.70}$$

Optically Thick Limit

Let us define

$$S_\lambda(\tau_\lambda) = \frac{1}{\beta_\lambda}\left[a_\lambda e_{b\lambda}(\tau_\lambda) + \frac{\gamma_\lambda}{4}G_\lambda(\tau_\lambda)\right]$$

Expanding $S_\lambda(\xi)$ in a Taylor series about $\xi = \tau_\lambda$,

$$S_\lambda(\xi) = S_\lambda(\tau_\lambda) + \frac{dS_\lambda}{d\tau_\lambda}(\xi - \tau_\lambda) + \frac{1}{2}\frac{d^2S_\lambda}{d\tau_\lambda^2}(\xi - \tau_\lambda)^2 + \cdots \tag{24.2.71}$$

Let $z = \tau_\lambda - \xi$ and $z' = \xi - \tau_\lambda$ and substitute (24.2.71) into (24.2.47) with $\tau_{o\lambda} \to 0$ or $\tau_\lambda \to \infty$ and $\tau_{o\lambda} - \tau_\lambda \to \infty$. Then we obtain

$$q_{R\lambda} = -4\frac{dS_\lambda}{d\tau_\lambda}\int_0^\infty zE_2(z)\,dz$$

or

$$q_{R\lambda} = -\frac{4}{3}\frac{d}{d\tau_\lambda}\left\{\frac{1}{\beta_\lambda}\left[a_\lambda e_{b\lambda}(\tau_\lambda) + \frac{\gamma_\lambda}{4}G_\lambda(\tau_\lambda)\right]\right\} \tag{24.2.72}$$

Similarly, we obtain

$$G_\lambda(\tau_\lambda) = 4S_\lambda(\tau_\lambda)\int_0^\infty E_1(z)dz = 4S_\lambda(\tau_\lambda) = 4e_{b\lambda}(\tau) \tag{24.2.73}$$

which leads to

$$q_R = -\frac{4}{3}\frac{de_{b\lambda}}{d\tau_\lambda} = -\frac{4}{3\beta_\lambda}\frac{de_{b\lambda}}{dy} \tag{24.2.74}$$

The total radiation heat flux is then

$$q_R = -\int_0^\infty \frac{4}{3\beta_\lambda}\frac{de_{b\lambda}}{dy}d\lambda = -\frac{4}{3\beta_R}\frac{de_b}{dy} \tag{24.2.75}$$

where β_R is the Rosseland mean extinction coefficient defined by

$$\frac{1}{\beta_R} = \int_0^\infty \frac{1}{\beta_\lambda}\frac{de_{b\lambda}}{de_b}d\lambda \tag{24.2.76}$$

For nonscattering media ($\gamma_\lambda = 0$), this reduces to the Rosseland mean absorption coefficient

$$\frac{1}{a_R} = \int_0^\infty \frac{1}{a_\lambda}\frac{de_{b\lambda}}{de_b}d\lambda \tag{24.2.77}$$

with $a_R = a_p = a$ for a gray medium, $a_p > a_R$, otherwise.

Radiative Equilibrium

For gray and diffuse bounding surfaces, the radiation flux is given

$$q_R = 2B_1E_3(\tau) - 2B_2E_3(\tau_o - \tau) + 2\sigma\int_0^{\tau} T^4(\xi)E_2(\tau - \xi)\,d\xi - 2\sigma\int_0^{\tau_o} T^4(\xi)E_2(\xi - \tau)\,d\xi \tag{24.2.78}$$

Upon differentiating (24.2.78), we obtain

$$2\sigma T^4(\xi) = B_1E_3(\tau) + B_2E_3(\tau_o - \tau) + \sigma\int_0^{\tau_o} T^4(\xi)E_1(|\tau - \xi|)\,d\xi \tag{24.2.79}$$

Introducing dimensionless quantities,

$$\phi(\tau) = \frac{\sigma T^4(\tau) - B_2}{B_1 - B_2}, \qquad Q = \frac{q_R}{B_1 - B_2}$$

we rewrite (24.2.78) and (24.2.79), respectively,

$$Q = 2E_3(\tau) + 2\int_0^{\tau}\phi(\xi)E_2(\tau - \xi)\,d\xi - 2\int_0^{\tau_o}\phi(\xi)E_2(\xi - \tau)\,d\xi \tag{24.2.80a}$$

$$2\phi(\xi) = E_2(\tau) + \int_0^{\tau_o}\phi(\xi)E_1(|\tau - \xi|)\,d\xi \tag{24.2.80b}$$

Since $q_{R1} = -q_{R2} = q_R$, we obtain

$$q_R = \frac{\varepsilon_1}{1 - \varepsilon_1}\left(\sigma T_1^4 - B_1\right) = -\frac{\varepsilon_2}{1 - \varepsilon_2}\left(\sigma T_2^4 - B_2\right)$$

and

$$\frac{B_1 - B_2}{\sigma\left(T_1^4 - T_2^4\right)} = \frac{1}{1 + \left(\dfrac{1}{\varepsilon_1} + \dfrac{1}{\varepsilon_2} - 2\right)Q} \tag{24.2.81}$$

This indicates that $B_1 - B_2$ can be found if Q is known as a function of τ_o.

Let us now consider the more realistic case of a nongray medium. For black surfaces with $\gamma_\lambda = 0$, equation (24.2.47) becomes

$$q_{R\lambda}(\tau_\lambda) = 2e_{b1\lambda}e_3(\tau_\lambda) - 2e_{b2\lambda}E_3(\tau_{o\lambda} - \tau_\lambda) + 2\int_0^{\tau_\lambda} e_{b\lambda}(\xi)E_2(\tau_\lambda - \xi)\,d\xi - 2\int_0^{\tau_{o\lambda}} e_{b\lambda}(\xi)E_2(\xi - \tau_\lambda)\,d\xi \tag{24.2.82a}$$

and

$$q_R = \int_0^{\infty} q_{R\lambda}\,d\lambda = \text{constant} \tag{24.2.82b}$$

Note that the previous problem of a gray medium was linear in T^4, but the present case is nonlinear with $e_{b\lambda}(T)$ being a different function of T for every value of λ.

24.2.4 SOLUTION METHODS FOR INTEGRODIFFERENTIAL RADIATIVE HEAT TRANSFER EQUATION

It follows from (24.2.44) that the governing integrodifferential equation for radiative heat transfer in participating media takes the form

$$\frac{dI(\mathbf{r},\mathbf{s})}{ds} = -[a(\mathbf{r}) + \gamma_s(\mathbf{r})]I(\mathbf{r},\mathbf{s}) + S(\mathbf{r},\mathbf{s}) = -\beta(\mathbf{r})I(\mathbf{r},\mathbf{s}) + S(\mathbf{r},\mathbf{s}) \tag{24.2.83}$$

where S is the source function,

$$S = a(\mathbf{r})I_b(\mathbf{r}) + \frac{\gamma_s(\mathbf{r})}{4\pi}\int_{4\pi} I(\mathbf{r},\mathbf{s}')\Phi(\mathbf{s}',\mathbf{s})\,d\omega' \tag{24.2.84}$$

with $\mathbf{r}$ denoting the position vector, $\mathbf{s}$ the unit vector in the direction of the ray, and Φ the scattering phase function. The boundary conditions for gray-diffuse surfaces are of the form,

$$I_w = \frac{\varepsilon\sigma T_w^4}{\pi} + \frac{1-\varepsilon}{\pi}\int_{\mathbf{n}\cdot\mathbf{s}>0} I(\mathbf{s})\mathbf{s}\cdot\mathbf{n}\,d\omega \tag{24.2.85}$$

where $\mathbf{n}$ is the unit vector outward normal to the boundary surface. If FVM is used, for example as in Figure 24.2.4a, this is one of the boundary surfaces for the control volume A, with $\mathbf{n}$ being normal to this surface. One such location is point 8 (Figure 24.2.4b,c) at the center of the boundary surface where the product $\mathbf{s}\cdot\mathbf{n}$ is to be calculated.

The finite volume formulation of (24.2.83) with respect to the volume and solid angle leads to

$$\int_\omega\int_\Omega\left(\frac{dI}{ds} + \beta I - S\right)d\Omega d\omega = 0 \tag{24.2.86}$$

Integrating, we obtain

$$\int_\omega\int_\Omega(\beta I - S)d\Omega d\omega + \int_\omega\int_\Gamma Is_i n_i d\Gamma d\omega = 0 \tag{24.2.87}$$

For the finite volume method via finite elements as described in Chapter 15, we notice that integration over the solid angle is combined with the domain integral [first term in (24.2.87)] and with the boundary surface integral [second term in (24.2.87)] as shown in Figure 24.2.4b,c. For example, it is shown that the integral of control angle coordinates for solid angles along $\mathbf{s}$ are fixed at node 8 for simultaneous numerical integration with the boundary surface integral along $\mathbf{n}$. The integral equation (24.2.87) is transformed into a discrete finite volume summing process,

$$\left[\sum_{CV}(\beta + \hat{\Phi})\omega_n\Delta\Omega - \sum_{CS}\hat{s}_i n_i\Delta\Gamma\right]I_n = \sum_{CV} aI_b\omega_n\Delta\Omega \tag{24.2.88}$$

with

$$\hat{\Phi} = \frac{\gamma_s}{4\pi}\sum_m I_m\overline{\Phi}(m,n), \quad m,n = \textit{index for solid angle}$$

$$\overline{\Phi}(m,n) = \frac{1}{\omega'_m\omega_n}\int_{\omega_m}\int_{\omega_n}\Phi(\mathbf{s}',\mathbf{s})\,d\omega' d\omega \tag{24.2.89a,b,c}$$

$$\hat{s}_i = \int_{\omega_n} s_i d\omega$$

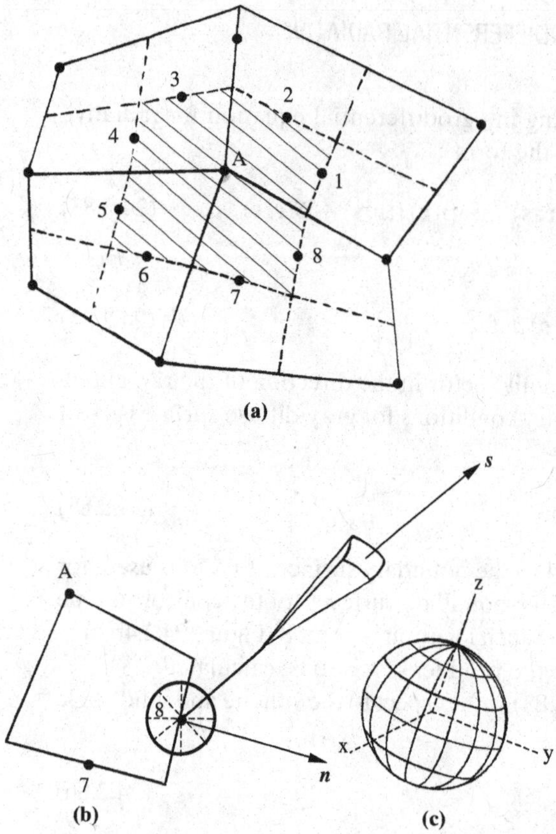

Figure 24.2.4 Finite volume representation via finite elements for control boundary surfaces and control angle directions. (a) Finite volume representation at A with boundary nodes, 1-8. (b) Designation of **n** and **s** at boundary node 8. (c) Subdivision of directional solid angles at boundary node 8.

where (24.2.89c) can be integrated analytically, using (24.1.5). By carrying out the summation process indicated above, one obtains a system of algebraic equations to determine nodal values of the radiative intensity. The procedure described here can be extended to hexahedral elements for 3-D applications. The geometrical configurations are detailed in Chapter 7.

The numerical solution of (24.2.83) has been studied extensively in the past, using a variety of methods such as Monte Carlo methods, discrete ordinate methods as well as FDM, FEM, and FVM. We present some of the FVM examples for the solution of (24.2.83) in Section 24.4.2.

24.3 RADIATIVE HEAT TRANSFER IN COMBINED MODES

24.3.1 COMBINED CONDUCTION AND RADIATION

If the medium conducts heat as well as absorbs, emits, and scatters thermal energy, the total heat flux vector is the sum of the contributions of conduction heat flux $\mathbf{q}_c$ and the radiative heat flux $\mathbf{q}_R$. Thus, the energy equation becomes

$$\nabla \cdot (\mathbf{q}_c + \mathbf{q}_R) = 0 \tag{24.3.1}$$

with

$$-k\nabla^2 T + \nabla \cdot \mathbf{q}_R = 0 \tag{24.3.2}$$

For a one-dimensional case, we have

$$k\frac{d^2T}{dy^2} = \frac{dq_R}{dy} \tag{24.3.3}$$

For a gray and diffuse parallel bounding surface, the radiation equations follow from (24.2.47), (24.2.50), (24.2.51), (24.2.57), and (24.2.59a,b):

$$\begin{aligned} q_R(\tau) &= 2B_1E_3(\tau) - 2B_2E_3(\tau_o - \tau) + \frac{2a}{\beta}\int_0^\tau \left[\sigma T^4(\xi) + \frac{\gamma}{4a}G(\xi)\right]E_2(\tau - \xi)\,d\xi \\ &\quad - \frac{2a}{\beta}\int_0^{\tau_o}\left[\sigma T^4(\xi) + \frac{\gamma}{4a}G(\xi)\right]E_2(\xi - \tau)\,d\xi \end{aligned} \tag{24.3.4}$$

$$\begin{aligned} \frac{dq_R}{d\tau} &= -2B_1E_2(\tau) + 2B_2E_2(\tau_o - \tau) \\ &\quad - \frac{2a}{\beta}\int_0^{\tau_o}\left[\sigma T^4(\xi) + \frac{\gamma}{4a}G(\tau)\right]E_1(|\tau - \xi|)d\xi + \frac{4a\sigma}{\beta}T^4(\tau) - \frac{\gamma}{4a}G(\tau) \end{aligned} \tag{24.3.5}$$

$$G(\tau) = 2B_1E_2(\tau) + 2B_2E_2(\tau_o - \tau) + \frac{2a}{\beta}\int_0^{\tau_o}\left[\sigma T^4(\xi) + \frac{\gamma}{4a}G(\tau)\right]E_1(|\tau - \xi|)\,d\xi \tag{24.3.6}$$

Denoting $\tau = \beta y$, $\tau_o = \beta L$, equation (24.3.3) becomes

$$\frac{d^2T}{d\tau^2} = \frac{(1 - \gamma/\beta)}{k\beta}[4\sigma T^4(\tau) - G(\tau)] \tag{24.3.7}$$

Introducing nondimensional quantities,

$$\theta(\tau) = \frac{T(\tau)}{T_1}, \qquad \theta_2 = \frac{T_2}{T_1}, \qquad \eta(\tau) = \frac{G(\tau)}{\sigma T_1^4}, \qquad N = \frac{k\beta}{4\sigma T_1^3}, \qquad \omega_o = \frac{\gamma}{\beta}, \qquad X = \frac{B}{\sigma T_1^4}$$

Equations (24.3.7) and (24.3.6) are reduced to

$$N\frac{d^2\theta}{d\tau^2} = (1 - \omega_o)\left[\theta^4(\tau) - \frac{1}{4}\eta(\tau)\right] \tag{24.3.8}$$

$$\begin{aligned} \eta(\tau) &= -2X_1E_2(\tau) + 2X_2E_2(\tau_o - \tau) \\ &\quad + 2\int_0^{\tau_o}\left[(1 - \omega_o)\theta^4(\xi) + \frac{\omega_o}{4}\eta(\xi)\right]E_1(|\tau - \xi|)\,d\xi \end{aligned} \tag{24.3.9}$$

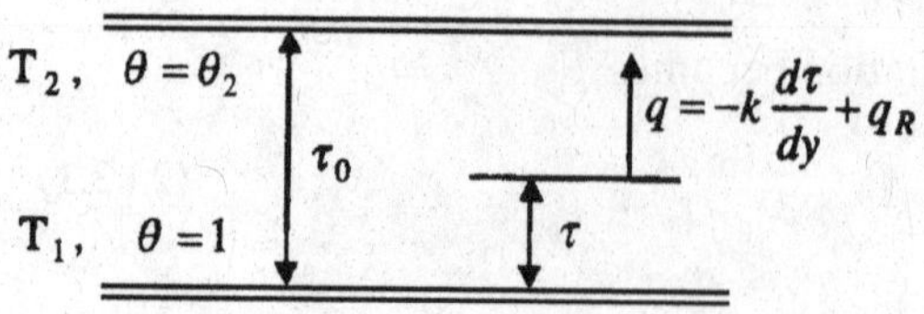

Figure 24.3.1 One-dimensional combined conduction and radiation.

where

$$X_1 = \varepsilon_1 + 2(1-\varepsilon_1)\left\{X_2 E_3(\tau_o) + \int_0^{\tau_o}\left[(1-\omega_o)\,\theta^4(\xi) + \frac{\omega_o}{4}\eta(\xi)\right]E_2(\xi)\,d\xi\right\} \tag{24.2.10a}$$

$$X_2 = \varepsilon_2\theta_2^4 + 2(1-\varepsilon_2)\left\{X_1 E_3(\tau_o) + \int_0^{\tau_o}\left[(1-\omega_o)\,\theta^4(\xi) + \frac{\omega_o}{4}\eta(\xi)\right]E_2(\tau_o-\xi)\,d\xi\right\} \tag{24.2.10b}$$

The boundary conditions are $\theta(0) = 1$, $(\tau_o) = \theta_2$, as shown in Figure 24.3.1.

Note that $N = \infty$ and $N = 0$ indicate the conditions of solely conduction and radiation, respectively. The parameter $\omega_o = \gamma/\beta$ is known as the albedo of scattering and represents the fraction of attenuated energy due to scattering. It is seen that $\omega_o = 0$ implies a nonscattering medium and equations (24.3.8) and (24.3.9) are reduced to a single equation,

$$N\frac{d^2\theta}{d\tau^2} = \theta^4(\tau) - \frac{X_1}{2}E_2(\tau) + \frac{X_2}{2}E_2(\tau_o - \tau) + \frac{1}{2}\int_0^{\tau_o}\theta^4(\xi)E_1(|\tau-\xi|)\,d\xi \tag{24.3.11}$$

For the case of an optically thin medium, it follows from (24.3.61) and (24.3.62) that

$$q_R = \sigma(T_1^4 - T_2^4)\left(\frac{1}{\varepsilon_1} + \frac{1}{\varepsilon_2} - 1\right)^{-1} \tag{24.3.12}$$

or

$$q = \frac{k}{L}(T_1 - T_2) + q_R$$

which indicates a nonparticipating medium.

The radiation flux for an optically thick medium is given by (24.3.74) as

$$q_R = \frac{4}{3\beta}\frac{d(\sigma T^4)}{dy} = -\frac{16\sigma T^3}{3\beta}\frac{dT}{dy} \tag{24.3.13}$$

or

$$q = \frac{k}{L}(T_1 - T_2) + \frac{4\sigma}{3\beta L}(T_1^4 - T_2^4) \tag{24.3.14}$$

or in nondimensional form,

$$\frac{q}{\sigma T_1^4}\tau_o - 4N(1-\theta_2) + \frac{3}{4}(1-\theta_2^4) \tag{24.3.15}$$

For pure scattering ($\omega_o = 1$), the energy equation is uncoupled from the radiation transfer process with $\tau_o = \gamma L$ and thus

$$q_R = (B_1 - B_2)Q(\tau_o) \tag{24.3.16}$$

Formulation of Finite Element Equations

For a gray and diffuse parallel bounding surface, the governing equations are given by (24.3.8) and (24.3.9),

$$N\frac{d^2\theta}{dT^2} = (1-\omega_o)\left[\theta^4(\tau) - \frac{1}{4}\eta(\tau)\right] \tag{24.3.17}$$

$$\eta(\tau) = 2X_1E_2(\tau) + 2X_2E_2(\tau_o - \tau) + 2IE_1(|\tau - \xi|) \tag{24.3.18}$$

where

$$I[E_i(\xi)] = \int_0^{\tau_o}\left[(1-\omega_o)\theta^4(\xi) + \frac{\omega_o}{4}\eta(\xi)\right]E_i(\xi)\,d\xi \tag{23.3.19}$$

$$\begin{aligned}X_1 = \varepsilon_1 + \frac{1}{D}\{&2\varepsilon_2(1-\varepsilon_1)E_3(\tau_o)\theta_2^4 + 4\varepsilon_1(1-\varepsilon_1)(1-\varepsilon_2)E_3^2(\tau_o)\\ &+8(1-\varepsilon_1)^2(1-\varepsilon_2)E_3^2(\tau_o)I[E_2(\xi)] + 4(1-\varepsilon_1)(1-\varepsilon_2)E_3(\tau_o)I[E_2(\tau_o-\xi)]\}\\ &+2(1-\varepsilon_2)I[E_2(\xi)]\end{aligned} \tag{23.3.20a}$$

$$\begin{aligned}X_2 = \frac{1}{D}\{&\varepsilon_2\theta_2^4 + 2(1-\varepsilon_2)E_3(\tau_o)\varepsilon_1 + 4(1-\varepsilon_1)(1-\varepsilon_2)E_3(\tau_o)I[E_2(\xi)]\\ &+2(1-\varepsilon_2)I[E_2(\tau_o-\xi)]\}\end{aligned} \tag{23.3.20b}$$

$$D = 1 - 4(1-\varepsilon_1)(1-\varepsilon_2)E_3^2(\tau_o) \tag{23.3.21}$$

Substituting (24.3.20) and (24.3.21) into (24.3.19) yields

$$\eta(\tau) = f_1E_1(\tau) + f_2E_2(\tau_o - \tau)f_3 \tag{24.3.22}$$

where

$$\begin{aligned}
f_1 &= g_1 + g_2I[E_2(\xi)] + g_3I[E_2(\tau_o-\xi)]\\
f_2 &= g_4 + g_5I[E_2(\xi)] + g_6I[E_2(\tau_o-\xi)]\\
f_3 &= 2I[E_1(|\tau-\xi|)]\\
g_1 &= 2\varepsilon_1 + \frac{2}{D}[2\varepsilon_2(1-\varepsilon_1)E_3(\tau_o)\theta_2^4 + 4\varepsilon_1(1-\varepsilon_1)(1-\varepsilon_2)E_3^2(\tau_o)]\\
g_2 &= \frac{16}{D}(1-\varepsilon_1)^2(1-\varepsilon_2)E_3^2(\tau_o) + 4(1-\varepsilon_1)\\
g_3 &= \frac{8}{D}(1-\varepsilon_1)(1-\varepsilon_2)E_3(\tau_o)\\
g_4 &= \frac{2}{D}[\varepsilon_2\theta_2^4 + 2(1-\varepsilon_2)E_3(\tau_o)\varepsilon_1]\\
g_5 &= \frac{8}{D}(1-\varepsilon_1)(1-\varepsilon_2)E_3(\tau_o)\\
g_6 &= \frac{4}{D}(1-\varepsilon_2)
\end{aligned} \tag{24.3.23}$$

For applications of finite elements into the governing equations (24.3.18) and (24.3.19), we introduce the global functions Φ_α, such that

$$\xi = \Phi_\alpha^{(\xi)}\xi_\alpha, \qquad \theta = \Phi_\alpha^{(\theta)}\theta_\alpha, \qquad \theta^4 = \Phi_\alpha^{(\theta^4)}\theta_\alpha^4, \qquad \eta = \Phi_\alpha^{(\eta)}\eta_\alpha \tag{24.3.24}$$

For simplicity, it is reasonable to assume linear functions for all of these variables. The Galerkin finite element representation for (24.3.18) and (24.3.19) takes the form

$$\int_0^{\tau_o} \Phi_\alpha \left[\frac{d^2\theta}{d\tau^2} - \frac{(1-\omega_o)}{N}\left(\theta^4 - \frac{1}{4}\eta\right)\right] d\tau = 0 \tag{24.3.25a}$$

$$\int_0^{\tau_o} \Phi_\alpha[\eta - f_1 E_2(\tau) - f_2 E_2(\tau_o - \tau) - f_3]\, d\tau = 0 \tag{24.3.25b}$$

Here, α denotes the global nodes. Integrating (24.3.25) by parts, we obtain

$$A_{\alpha\beta}\theta_\beta + B_{\alpha\beta}\eta_\beta = F_\alpha + G_\alpha \tag{24.3.26}$$

$$C_{\alpha\beta}\eta_\beta = H_\alpha \tag{24.3.27}$$

Note that the calculations of f_1, f_2, and f_3 in (24.3.23) can be carried out by use of Gaussian quadrature integration. To this end, we choose to use linear isoparametric coordinates such that

$$\Phi_1 = \frac{1}{2}(1-\zeta), \qquad \Phi_2 = \frac{1}{2}(1+\zeta)$$

$$\Phi_\alpha = \bigcup_{e=1}^{E} \Phi_N \Delta_{N\alpha}$$

where $\Delta_{N\alpha}$ is the Boolean matrix, and E is the total number of elements. For example, consider

$$I[E_2(\xi)] = \int_0^{\tau_o} \left[(1-\omega_o)\theta^4(\xi) + \frac{\omega_o}{4}\eta(\xi)\right] E_2(\xi)\, d\xi \tag{24.3.28}$$

where from (24.3.48b)

$$E_2(\xi) \cong 1 - 0.4228\,\xi + \xi \ln \xi - \frac{\xi^2}{2} + \frac{\xi^3}{12} \tag{24.3.29}$$

Let h be the length of an element. Then

$$\zeta = \frac{\xi}{h}, \qquad d\xi = \frac{\partial \xi}{\partial \zeta} d\zeta = h\, d\zeta$$

Thus

$$I[E_2(\xi)] = \bigcup_{e=1}^{E} h \int_{-1}^{1} \left[(1-\omega_o)\Phi_N^{(e)}\Delta_{N\alpha}\theta_\alpha^4 + \frac{\omega_o}{4}\Phi_N \Delta N\alpha \eta_\alpha\right]\left[1 - 0.4228\,\Phi_N\Delta_{N\alpha}\xi_\alpha \right.$$
$$\left. + \Phi_N\Delta_{N\alpha}\xi_\alpha \ln(\Phi_N\Delta_{N\alpha}\xi_\alpha) - \frac{1}{2}\Phi_N\Delta_{N\alpha}\xi_\alpha^2 + \frac{1}{12}\Phi_N\Delta_{N\alpha}\xi_\alpha^2 + \cdots\right] d\zeta \tag{24.3.30}$$

The Gaussian quadrature integration can now be applied to (24.3.29)

$$I[E_2(\xi)] = \bigcup_{e=1}^{E} \sum_{i=1}^{N} w_i\, f(\eta_i) \tag{24.3.31}$$

Returning to (24.3.28) and (24.3.29), denoting $\nu = \tau/h$, $d\tau = hd\nu$, we write the explicit forms of the matrices,

$$A_{\alpha\beta} = \bigcup_{e=1}^{E} \int_0^h \frac{d\Phi_N}{d\tau}\frac{d\Phi_M}{d\tau} d\tau \Delta_{N\alpha}\Delta_{N\beta} = \bigcup_{e=1}^{E} \frac{1}{h}\int_{-1}^{1} \frac{d\Phi_N}{d\nu}\frac{d\Phi_M}{d\nu} \Delta_{N\alpha}\Delta_{N\beta} d\nu \tag{24.3.32}$$

$$B_{\alpha\beta} = -\bigcup_{e=1} \frac{(1-\omega_o)h}{4N}\int_{-1}^{1} \Phi_N\Phi_M \Delta_{N\alpha}\Delta_{N\beta}\, d\nu \tag{24.3.33}$$

$$C_{\alpha\beta} = \bigcup_{e=1}^{E} \frac{1}{h}\int_{-1}^{1} \Phi_N\Phi_M \Delta_{N\alpha}\Delta_{N\beta} d\nu \tag{24.3.34}$$

$$F_\alpha = \bigcup_{e=1}^{E} \frac{d\theta}{d\tau} \overset{*}{\Phi}_M \Delta_{N\alpha}\Big|_0^h \tag{24.3.35}$$

$$G_\alpha = -\bigcup_{e=1}^{E} \frac{(1-\omega_o)h}{N}\int_{-1}^{1} \Phi_N\Phi_M \Delta_{N\alpha}\Delta_{N\beta}\theta_\beta^4 d\nu \tag{24.3.36}$$

$$\begin{aligned} H_\alpha = \bigcup_{e=1}^{E} h\int_{-1}^{1} &\{f_1[\Phi_N\Delta_{N\alpha} - 0.4228\Phi_N\Phi_M\Delta_{N\alpha}\Delta_{M\beta}\tau_\beta \\ &+ \Phi_N\Phi_M\Delta_{N\alpha}\Delta_{M\beta}\ln(\Phi_p\Delta_{p\gamma}\tau_\gamma) + \cdots] + f_2[0.5772\Phi_N\Delta_{N\alpha} \\ &+ 0.4228\Phi_N\Phi_M\Delta_{N\alpha}\Delta_{M\beta}\tau_\beta + \Phi_N\Delta_{N\alpha}\ln(1-\Phi_M\Delta_{M\beta}\tau_\beta) \\ &- \Phi_N\Phi_M\Delta_{N\alpha}\Delta_{M\beta}\ln(1-\Phi_p\Delta_{p\gamma}\tau_\gamma) + \cdots] + f_3\Phi_N\Delta_{N\alpha}\}d\nu \end{aligned} \tag{24.3.37}$$

where the last term for H_α is given by

$$\begin{aligned} \bigcup_{e=1}^{E} h\int_{-1}^{1} f_3\Phi_N\Delta_{N\alpha}d\nu = \bigcup_{e=1}^{E} h\int_{-1}^{1}\int_{-1}^{1} &\Phi_N\Delta_{N\alpha}\left[(1-\omega_o)\Phi_M\Delta_{M\beta}\theta_\beta^4 + \frac{\omega_o}{4}\Phi_M\Delta_{M\beta}\eta_\beta\right] \\ &\times[-0.5772 - \ln(\Phi_p\Delta_{p\gamma}\tau_\gamma - \Phi_p\Delta_{p\gamma}\xi_\gamma) \\ &+ \Phi_p\Delta_{p\gamma}\tau_\gamma - \Phi_p\Delta_{p\gamma}\xi_\gamma + \cdots]\, d\zeta\, d\nu \end{aligned}$$

The boundary conditions are given by (Figure 24.3.1)

$$\theta = 1 \quad \text{at } \tau = 0$$

$$\theta = \theta_2 \quad \text{at } \tau = \tau_o$$

Since the Dirichlet boundary conditions are provided at both ends, the Neumann boundary conditions F_α (24.3.36) need not be prescribed.

All matrices in (24.3.27) and (24.3.28) are integrated using the Gaussian quadrature. Here, θ^4 can be calculated initially from the Dirichlet boundary conditions and placed in G_α and H_α. The solution of (24.3.28) and (24.3.29) is obtained iteratively by updating G_α and H_α until convergence, in which the nodal values θ_α^4 and η_α on the right-hand sides of (24.3.28) and (24.3.29) are replaced by those values of the previous iterative cycle.

Emitting and Absorbing Media

With $\omega_o = 0$ which represents nonscattering media, the expressions (24.3.18) and (24.3.19) are reduced to a single equation (24.3.53). The finite element analog takes the form

$$A_{\alpha\beta}\theta_\beta = F_\alpha + \hat{G}_\alpha + \hat{H}_\alpha \tag{24.3.38}$$

where G_α and H_α are the same as in (24.3.37) and (24.3.38), respectively, with the value of ω_o set equal to zero.

The radiation flux for optically thin media is given by (24.3.13) which is a constant and indicates that the medium is nonparticipating.

With the radiation flux given by (24.3.14), the Galerkin finite element equation for the combined conduction and radiation takes the form (with $T = \Phi_\alpha T_\alpha$)

$$\int_0^L \Phi_\alpha \left(\frac{d^2T}{dy^2} + g(T)\frac{dT}{dy} \right) dy = 0 \tag{24.3.39}$$

where

$$g(T) = \frac{48\sigma T^2}{3\beta k + 16\sigma T^3} \tag{24.3.40}$$

Integrating (24.3.40) by parts, we obtain

$$A_{\alpha\beta} T_\beta + B_{\alpha\beta}(g) T_\beta = F_\alpha \tag{24.3.41}$$

where

$$A_{\alpha\beta} = \int_0^L \frac{d\Phi_\alpha}{dy}\frac{d\Phi_\beta}{dy} dy = \sum_{e=1}^{E} \int_0^h \frac{d\Phi_N}{dy}\frac{d\Phi_M}{dy} \Delta_{N\alpha}\Delta_{N\beta} dy \tag{24.3.42}$$

$$B_{\alpha\beta}(g) = -\bigcup_{e=1}^{E} \int_o^h \Phi_N \frac{\Phi_M}{dy} g(T)\Delta_{N\alpha}\Delta_{N\beta} dy \tag{24.3.43}$$

$$F_\alpha = \frac{dT}{dy} \overset{*}{\Phi}_\alpha \Big|_0^L \tag{24.3.44}$$

Since $B_{\alpha\beta}(g)$ is nonlinear, the Newton-Raphson iterations may be used to solve (24.3.42). To this end, it is particularly advantageous to use isoparametric coordinates and Gaussian quadrature for easy integration. An alternative method would be to write

$$g = \Phi_\alpha g_\alpha \tag{24.3.45}$$

and

$$B_{\alpha\beta}(g) = -\bigcup_{e=1}^{(e)} \int_o^h \Phi_N \frac{d\Phi_M}{dy} \Phi_p \Delta_{N\alpha}\Delta_{M\beta}\Delta_{p\gamma} g_\gamma dy \tag{24.3.46}$$

This will require updating of g_γ at each iterative cycle until convergence.

Example problems for the combined mode conduction and radiation are presented in Section 24.4.3.

24.3.2 COMBINED CONDUCTION, CONVECTION, AND RADIATION

The most general form of the energy equation is written as

$$\rho c_v \frac{DT}{Dt} = -\nabla \cdot q - p\nabla \cdot \mathbf{v} + \mu\Phi \tag{24.3.47}$$

where

$$-p\nabla \cdot \mathbf{v} + \mu\Phi = \sigma_{ij} d_{ij}$$
$$q = -k\nabla T + q_R$$

An alternative form is

$$\rho c_p \frac{DT}{Dt} = \nabla \cdot k\nabla T + bT\frac{Dp}{Dt} + \mu\Phi - \nabla \cdot q_R \tag{24.3.48}$$

where b is the thermal expansion coefficient of the fluid.

Boundary Layer Energy Equation

Let us consider a boundary layer form of (24.3.48) for a two-dimensional flow.

$$u\frac{\partial T}{\partial x} + \mathrm{v}\frac{\partial T}{\partial y} = \alpha\frac{\partial^2 T}{\partial y^2} + \frac{bTu}{\rho c_p}\frac{dP}{dx} + \frac{\mu}{\rho c_p}\left(\frac{\partial u}{\partial y}\right)^2 - \frac{1}{\rho c_p}\nabla \cdot \mathbf{q}_R \tag{24.3.49}$$

where u and v denote x and y components of velocity, respectively, with x being the streamwise coordinate and y the transverse coordinate, and where α is the thermal diffusivity of the fluid. The simplification here is based on the large Peclet number

$$Pe = \frac{u_\infty L}{\alpha} \gg 1$$

where L is a characteristic dimension. For small Eckert number

$$E_c = \frac{u_\infty^2}{c_p \Delta T} \ll 1$$

we may also neglect the second and third terms on the right-hand side of (24.3.49). Furthermore, if

$$u\frac{\partial T}{\partial x} \gg \frac{1}{\rho c_p}\frac{\partial q_{Rx}}{\partial x}$$

then it is reasonable to neglect the radiation in the x-direction. These simplifications lead to

$$u\frac{\partial T}{\partial x} + \mathrm{v}\frac{\partial T}{\partial y} = \alpha\frac{\partial^2 T}{\partial y^2} - \frac{1}{\rho c_p}\frac{\partial q_R}{\partial y} \tag{24.3.50}$$

Flow through Ducts

For a fully developed flow through ducts with $E_c \ll 1$, we may neglect $\mathrm{v}\partial T/\partial y$ in (24.3.50)

$$u\frac{\partial T}{\partial x} = \alpha\frac{\partial^2 T}{\partial y^2} - \frac{1}{\rho c_p}\frac{\partial q_R}{\partial y} \tag{24.3.51}$$

where

$$u = 6u_m\left(\frac{y}{L} - \frac{y^2}{L^2}\right) \tag{24.3.52}$$

and, for a nonscattering gray fluid and black plate surfaces,

$$-\frac{dq_R}{d\tau} = 2\sigma T_w^4 E_2(\tau) + 2\sigma T_w E_2(\tau_o - \tau) + 2\sigma\int_0^{\tau_o} T^4(\xi)E_1(|\tau - \xi|)\,d\xi - 4\sigma T^4(\tau) \tag{24.3.53}$$

The axial temperature gradient may be assumed as

$$\frac{\partial T}{\partial x} = \left(\frac{T_w - T}{T_w - T_b}\right)\frac{q_w}{\rho c_p u_m L} \tag{24.3.54}$$

where T_b is the local bulk-fluid temperature and q_w is the total wall-heat flux,

$$q_w = -k\left(\frac{\partial T}{\partial y}\right)_{y=0} + 2\sigma T_w^4[1 - E_3(\tau_o)] - 2\sigma\int_0^{\tau_o} T^4(\xi)E_2(\xi)\,d\xi \tag{24.3.55}$$

Upon combining (24.3.51) through (24.3.55), we now have

$$2N\frac{d^2\theta}{d\tau^2} - \frac{6\psi}{\tau_o}\left(\frac{\tau}{\tau_o} - \frac{\tau^2}{\tau_o^2}\right)\left(\frac{1-\theta}{1-\theta_b}\right)$$
$$= -E_2(\tau) - E_2(\tau_o - \tau) - \int_0^{\tau_o}\theta^4(\xi)E_1(|\tau - \xi|)\,d\xi - 2\theta^4(\tau) \tag{24.3.56}$$

with

$$N = \frac{ka}{4\sigma T_w^3}, \qquad \psi = \frac{q_w}{\sigma T_w^3}, \qquad \theta = \frac{T}{T_w}, \qquad \theta_b = \frac{T_b}{T_w}$$

$$\psi = -4N\left(\frac{d\theta}{d\tau}\right)_{\tau=0} + 2 - 2E_2(\tau_o) - 2\int_0^{\tau_o}\theta^4(\xi)E_2(\xi)\,d\xi \tag{24.3.57}$$

We have boundary conditions for (24.3.56) in the form

$$\theta(0) = \theta(\tau_o) = 1$$

Nonviscous, Nonconducting Flow over a Flat Plate

In this case, we have $u = u_\infty$, $v = 0$, and $dP/dx = 0$. The governing equation becomes

$$u_\infty\frac{\partial T}{\partial x} = -\frac{1}{\rho c_p}\frac{\partial q_R}{\partial y} \tag{24.3.58}$$

Setting $\omega_o = 0$ and $\varepsilon_w = 1$, $\tau_o = \infty$, we get

$$q_R = 2\sigma T_w^4 E_3(\tau) + 2\sigma\int_0^{\tau} T^4(x,t)E_2(\tau - \xi)d\xi - 2\sigma\int_{\tau}^{\infty} T^4(x,t)E_2(\xi - \tau)\,d\xi \tag{24.3.59}$$

$$-\frac{dq_R}{d\tau} = 2\sigma T_w^4 E_2(\tau) + 2\sigma\int_0^{\infty} T^4(x,\xi)E_1(|\tau - \xi|)\,d\xi - 4\sigma T^4(x,\tau) \tag{24.3.60}$$

Furthermore, we may assume that temperature differences within the flow are sufficiently small such that T^4 can be expressed as a linear function of temperature. Toward this end, we expand T^4 in a Taylor series about T_∞ neglecting higher order terms. Thus,

$$T^4 \cong 4T_\infty^3 T - 3T_\infty^4 \tag{24.3.61}$$

Introducing nondimensional quantities,

$$\theta = \frac{T - T_w}{T_\infty - T_w}, \qquad \zeta = \frac{2\sigma a T_\infty^3 x}{\rho c_p u_\infty}$$

we obtain

$$\frac{\partial \theta}{\partial \zeta} = 4\int_0^\infty \theta(\zeta, \xi) E_1(|\tau - \xi|)\, d\xi - 8\theta(\zeta, \tau) \tag{24.3.62}$$

with the boundary condition

$$\theta(0, \tau) = 1 \tag{24.3.63}$$

In view of (24.3.61), (24.3.62), and (24.3.59) with $\tau = 0$ we have

$$\frac{q_{Rw}}{\sigma(T_w^4 - T_\infty^4)} = 2\int_0^\infty \theta(\zeta, \xi) E_2(\xi)\, d\xi \tag{24.3.64}$$

with

$$T_w^4 - T_\infty^4 \cong 4T_\infty^3 (T_w - T_\infty) \tag{24.3.65}$$

Optically Thin Boundary Layer

We consider effects of both viscosity and heat conduction in a laminar flow of a constant property gray fluid over a black isothermal plate governed by

$$u\frac{\partial T}{\partial x} + v\frac{\partial T}{\partial y} = \alpha\frac{\partial^2 T}{\partial y^2} + \frac{2\sigma a}{\rho c_p}\left[T_w^4 E_2(\tau) + \int_0^\infty T^4(x, \xi) E_1(|\tau - \xi|)\, d\xi - 2\sigma T^4(x, \tau)\right] \tag{24.3.66}$$

In the outer region the velocity in the x direction has the free stream u_o and neglecting heat conduction, we have

$$u_o\frac{\partial T}{\partial x} = \frac{2\sigma a}{\rho c_p}\left[T_w^4 E_2(\tau) + \int_0^\infty T^4(x, \xi) E_1(|\tau - \xi|)\, d\xi - 2T^4(x, \tau)\right] \tag{24.3.67}$$

For an approximate solution, we substitute the incoming free stream temperature T_o for the temperature on the right side as first approximation and then carry out the integral to obtain a second approximation. This yields ($T = T_o$ at $x = 0$),

$$T(x, \tau) = T_o + \sigma\left(T_w^4 - T_o^4\right) E_2(\tau)\frac{2ax}{\rho c_p u_o} + \cdots \tag{24.3.68}$$

At the edge of the thermal layer $\tau = ay = a\delta$ where δ is the inner boundary layer thickness, which is small so that $E_2(a\delta) \cong E_2(0) = 1$. Thus, (24.3.68) becomes

$$T(x,\delta) = T_o + \sigma\left(T_w^4 - T_o^4\right)\frac{2ax}{\rho c_p u_o} + \cdots \tag{24.3.69}$$

Now the last integral in (24.3.67) is carried out in two steps, one from $\tau = 0$ to $a\delta$ and one from $a\delta$ to ∞. The first portion, neglected as the thermal layer, is optically thin and the second is evaluated via (24.3.68). This leads to

$$u\frac{\partial T}{\partial x} + \mathrm{v}\frac{\partial T}{\partial y} = \alpha\frac{\partial^2 T}{\partial y^2} + \frac{2\sigma a}{\rho c_p}\left(T_w^4 + T_o^4 - 2T^4\right) \tag{24.3.70}$$

The boundary conditions consist of (24.3.69) at $y = \delta$ and $T = T_w$ at $y = 0$.

Another approach in terms of dimensionless forms may be used. Denoting

$$u = u_\infty f', \qquad \mathrm{v} = \frac{1}{2}\sqrt{\frac{\nu u_\infty}{x}}(\eta f' - f) \tag{24.3.71}$$

where $f(\eta)$ is the dimensionless Blasius stream function, and

$$\eta = y\sqrt{\frac{u_\infty}{\mathrm{n}x}}$$

Let $\tau = ay$ and

$$\zeta = \frac{2\sigma a T_\infty^3 x}{\rho c_p u_\infty}$$

we obtain

$$f'\frac{\partial T}{\partial \zeta} + \frac{1}{2}\sqrt{\frac{2\,\mathrm{Pr}\,N}{\xi}}(\eta f' - f)\frac{\partial T}{\partial \tau}$$
$$= 2N\frac{\partial^2 T}{\partial \tau^2} + \frac{1}{T_\infty^3}\left[T_w^4 E_2(\tau) + \int_0^\infty T^4(\zeta,\xi)E_1(|\tau - \xi|)\,d\xi - 2T^4(\zeta,\tau)\right] \tag{24.3.72}$$

where

$$N\frac{\kappa a}{4\sigma T_\infty^3}, \qquad \eta = \frac{\tau}{\sqrt{2\,\mathrm{Pr}\,N\zeta}}, \qquad \mathrm{Pr} = \frac{\nu}{\alpha}$$

Setting $N = 0(f'(\eta) < f'(\infty) = 1)$ in (24.3.72) leads to

$$\frac{\partial T_o}{\partial \zeta} = \frac{1}{T_\infty^3}\left[T_w^4 E_2(\tau) + \int_0^\infty T_o^4(\zeta,\xi)E_1(|T - t|)\,d\xi - 2T_o^4(\zeta,\tau)\right] \tag{24.3.73}$$

Note that this equation fails to satisfy temperature continuity near the surface. Assuming $\mathrm{Pr} = 0(1)$ and introducing

$$\overline{\tau} = \frac{\tau}{\sqrt{2\,\mathrm{Pr}\,N}}$$

it follows from (24.3.72) that

$$f'\frac{\partial T}{\partial \zeta} + \frac{1}{2\sqrt{\zeta}}(\eta f' - f)\frac{\partial T}{\partial \overline{\tau}} = \frac{1}{\Pr}\frac{\partial^2 T}{\partial \overline{\tau}^2} + \frac{1}{T_\infty^3}\left[T_w^4 + \int_0^\infty T^4(\zeta,\xi)E_1(\xi)\,d\xi - 2T^4(\zeta,\overline{\tau})\right] \tag{24.3.74}$$

Note further that if $T(\zeta,\tau) = T_o(\zeta,\tau)$ for $>\tau_\delta$, then we get

$$\int_0^\infty T^4(\zeta,\xi)E_1(\xi)\,d\xi = \int_0^\infty T_o^4(\zeta,\xi)E_1(\xi)\,d\xi - \int_0^{\tau_\delta} T_o^4(\zeta,\xi)E_1(\xi)\,d\xi + \int_0^{\tau_\delta} T_o^4(\zeta,\xi)E_1(\xi)\,d\xi + \int_0^{\tau_\delta} T^4(\zeta,\xi)E_1(\xi)\,d\xi$$

Here the second and third integrals on the right side are of the order $\sqrt{2\Pr N}$ and may be neglected, and

$$\int_0^\infty T_0^4(\zeta,\xi)E_1(\xi)\,d\xi = T_\infty^3\frac{dT_o(\zeta,0)}{d\zeta} + 2T_o^4(\zeta,0) - T_w^4$$

Thus,

$$f'\frac{\partial T}{\partial \zeta} + \frac{1}{2\sqrt{\zeta}}(\eta f' - f)\frac{\partial T}{\partial \overline{\tau}} = \frac{1}{\Pr}\frac{\partial^2 T}{\partial \overline{\tau}^2} + \frac{dT_o(\zeta,0)}{d\zeta} + \frac{2}{T_\infty^3}\left[T_o^4(\zeta,0) - T^4(\zeta,\overline{\tau})\right] \tag{24.3.75}$$

The boundary conditions are

$$T = T_w, \qquad \overline{\tau} = 0$$

$$T \to T_o(\zeta, 0), \qquad \overline{\tau} \to \infty$$

with

$$T(\zeta,\infty) = T_o(\zeta,0) \quad \text{for } N \ll 1$$

Optically Thick Boundary Layer

Although, in general, the optically thin boundary layer is a physically realistic model, an optically thick model may be used if the thermal layer has become very thick or the medium is highly absorbing. Again for small Eckert numbers, it follows from (24.3.50) and (24.3.14) that, with $\beta = a$,

$$q_R = -\frac{4\sigma}{3a}\frac{\partial T^4}{\partial y}$$

we obtain

$$u\frac{\partial T}{\partial x} + v\frac{\partial T}{\partial y} = \alpha\frac{\partial^2}{\partial y^2}\left(T + \frac{4\sigma}{3ka}T^4\right) \tag{24.3.76}$$

Using the similarity transformation

$$T(\eta) = T(x, y)$$

(24.3.76) becomes

$$\frac{1}{\Pr}\frac{\partial^2}{\partial\eta^2}\left(T+\frac{T^4}{3T_\infty^3 N}\right)+\frac{1}{2}f\frac{\partial T}{\partial\eta}=0 \tag{24.3.77}$$

with the boundary conditions

$$\theta=\frac{T-T_w}{T_\infty-T_w}$$

and using (24.3.61), equation (24.3.77) reduces to

$$\frac{1}{\Pr}\left(1+\frac{4}{3N}\right)\frac{\partial^2\theta}{\partial\eta^2}+\frac{1}{2}f\frac{\partial\theta}{\partial\eta}=0 \tag{24.3.78}$$

with

$$\theta(0)=0,\qquad \theta(\infty)=1$$

Combined Conduction-Convection-Radiation in Scattering Medium

When scattering is considered in the absorbing and emitting medium for a parallel-plate channel with the velocity and temperature profiles fully developed, we have the governing equation

$$u\frac{\partial T}{\partial x}=\alpha\frac{\partial^2 T}{\partial y^2}+\frac{\mu}{\rho c_p}\frac{dP}{dx}+\frac{\mu}{\rho c_p}\left(\frac{\partial u}{\partial y}\right)^2-\frac{1}{\rho c_p}\frac{\partial q_R}{\partial y}$$

It can easily be shown that, from (24.3.49)

$$\frac{d^2\theta}{d\tau^2}=\frac{1}{N}(1-\omega_o)\left[\theta^4(\tau)-\frac{1}{4}\eta(\tau)\right]-\frac{3\Psi}{N\tau_o}\left(\frac{\tau}{\tau_o}-\frac{\tau^2}{\tau_o^2}\right)\left(\frac{1-\theta}{1-\theta_b}\right)$$
$$+\frac{36Ec\Pr}{\tau_o^2}\left(1-\frac{2\tau}{\tau_o}\right)^2=0 \tag{24.3.79}$$

where $\theta=T/T_1$, $\theta_b=T_b/T_w$, $Ec=u_b^2/c_pT_w^2$ (Eckert number) and $\Pr=c_p\mu/\kappa$ (Prandl number) and η is defined in (24.3.9). The total heat flux Ψ at $y=0$ is given by

$$\Psi=-4N\left(\frac{d\theta}{d\tau}\right)_{\tau=0}+\Phi \tag{24.3.80}$$

where

$$\Phi=2[X_1E_3(0)-X_2E_3(\tau_o)]-\int_0^{\tau_o}\left[(1-\omega_o)\theta^4(\xi)-\frac{1}{4}\eta(\xi)\right]E_2(\xi)\,d\xi$$

Note that the last two terms on the right-hand side of (24.3.79) represent scattering and viscous dissipation. If they are dropped, then we recover (24.3.8) for the case of nonscattering.

Flow through Ducts

For a fully developed flow through ducts given by (24.3.51) and the nondimensional form of (24.3.56), the finite element equations may be derived in the form similar to (24.3.27). We proceed with

$$\theta=\Phi_\alpha\theta_\alpha,\qquad \theta^4=\Phi_\alpha\theta_\alpha^4,\qquad \tau=\Phi_\alpha\tau_\alpha$$

and obtain

$$A_{\alpha\beta}\theta_\beta = F_\alpha + G_\alpha \tag{24.3.81}$$

where $A_{\alpha\beta}$ and F_α are the same as (24.3.33) and (24.3.36), respectively, and G_α represents the rest of the terms in (24.3.56).

Although the boundary layer equations are simplified from the general form, the solution by numerical methods often leads to instability unless extremely small elements are provided in the boundary layer region. We consider finite element equations for the optically thin and thick cases.

For optically thin boundary layers let us consider the nondimensional form (24.3.75) which results from (24.3.66). It is seen that the outer solution to (ζ,τ) is required a priori. For linearized radiation, we assume

$$\hat{\theta}(\zeta,\tau) = \frac{T_o(\zeta,\tau) - T_w}{T_\infty - T_w} \tag{24.3.82a}$$

$$\theta = \frac{T(\zeta,\tau) - T_w}{T_\infty - T_w} \tag{24.3.82b}$$

Thus, the linearized equation takes the form

$$f'\frac{\partial\theta}{\partial\zeta} + \frac{1}{2\sqrt{\zeta}}(\eta f' - f)\frac{\partial\theta}{\partial\overline{\tau}} = \frac{1}{\Pr}\frac{\partial^2\theta}{\partial\overline{\tau}^2} + \frac{d\hat{\theta}(\zeta,0)}{d\zeta} + 8\,\hat{\theta}(\zeta,0) - 8\,\theta(\zeta,\overline{\tau}) \tag{24.3.83}$$

The required boundary conditions are

$$\theta \to 0 \qquad \text{at } \overline{\tau} = 0$$
$$\theta \to \hat{\theta}(\zeta,0) \quad \text{for } \overline{\tau} \to \infty$$

The finite element representation of (24.3.83) is

$$A_{\alpha\beta}\theta_\beta + B_{\alpha\beta}\theta_\beta = F_\alpha + G_\alpha + H_\alpha \tag{24.3.84}$$

in which

$$A_{\alpha\beta} = \frac{1}{\Pr}\int_\Omega \frac{\partial\Phi_\alpha}{\partial\overline{\tau}}\frac{\partial\Phi_\beta}{\partial\overline{\tau}}d\Omega \tag{24.3.85a}$$

$$B_{\alpha\beta} = \int_\Omega 8\Phi_\alpha\Phi_\beta d\Omega \tag{24.3.85b}$$

$$F_\alpha = \frac{1}{\Pr}\frac{\partial\theta}{\partial\overline{\tau}}\overset{*}{\Phi}_\alpha\Big|_0^{\overline{\tau}=\infty} \tag{24.3.85c}$$

$$G_\alpha = \int_\Omega\left(\Phi_\alpha\frac{\partial\Phi_\beta}{\partial\zeta} + 8\Phi_\alpha\Phi_\beta\right)\hat{\theta}_\beta d\Omega \tag{24.3.85d}$$

$$H_\alpha = -\int_\Omega\left(\Phi_\alpha\frac{\partial\Phi_\beta}{\partial\zeta}\theta_\beta\Phi_\gamma f'_\gamma + \Phi_\alpha\frac{\partial\Phi_\beta}{\partial\overline{\tau}}\theta_\beta\Phi_\gamma a_\gamma\right)d\Omega \tag{24.3.85e}$$

where

$$a_\gamma = \left[\frac{1}{2\sqrt{\zeta}}(\eta f' - f)\right]_\gamma \tag{24.3.85f}$$

To carry out calculations initially, the velocity field must be given and boundary layer thickness assumed. This will require once again an iterative procedure until convergence.

For more rigorous analysis, the continuity and momentum equations can be solved simultaneously with the energy equation. In the presence of thermal buoyancy with natural convection, we have

$$\frac{\partial u}{\partial x} + \frac{\partial \mathrm{v}}{\partial y} = 0 \tag{24.3.86a}$$

$$u\frac{\partial u}{\partial x} + \mathrm{v}\frac{\partial u}{\partial y} = \nu\frac{\partial^2 u}{\partial y^2} + gb(T - T_\infty) \tag{24.3.86b}$$

$$u\frac{\partial T}{\partial x} + \mathrm{v}\frac{\partial T}{\partial y} = \alpha\frac{\partial^2 T}{\partial y^2} + \frac{1}{\rho c_p}[2a\sigma(T_w^4 - T_\infty^4 - 2T^4)] \tag{24.3.86c}$$

Written in nondimensional form, we have

$$\frac{\partial U}{\partial X} + \frac{\partial V}{\partial Y} = 0 \tag{24.3.87a}$$

$$U\frac{\partial U}{\partial X} + V\frac{\partial U}{\partial Y} = \frac{1}{\mathrm{Re}}\frac{\partial^2 U}{\partial Y^2} + \frac{R_a\theta}{\mathrm{Re}^2\,\mathrm{Pr}} \tag{24.3.87b}$$

$$U\frac{\partial \theta}{\partial X} + V\frac{\partial \theta}{\partial Y} = \frac{1}{\mathrm{Re}\,\mathrm{Pr}}\frac{\partial^2 \theta}{\partial Y^2} + \frac{\zeta_L}{(m-1)}\{1 + m^4 - 2[1 + (m-1)\,\theta]\} \tag{24.3.87c}$$

where

$$U = \frac{u}{u_\infty}, \qquad V = \frac{\mathrm{v}}{u_\infty}, \qquad X = \frac{x}{L}, \qquad Y = \frac{y}{L}$$

$$\theta = \frac{T - T_\infty}{T_w - T_\infty}, \qquad m = \frac{T_w}{T_\infty}, \qquad \mathrm{Re} = \frac{u_\infty L}{\nu}$$

$$\zeta_L = \frac{2\sigma a T_\infty^3 L}{\rho c_p u_\infty}, \qquad R_a = \frac{gb(T_w - T_\infty)L^3}{\nu k}, \qquad \mathrm{Pr} = \frac{\nu}{\alpha}$$

Boundary conditions are

$$\theta = 1, \qquad U = V = 0 \quad \text{at } X, Y = 0 \tag{24.3.88a}$$

$$T(x,\delta) \cong T_\infty + \frac{2a\sigma x}{\rho c_p u_\infty}(T_w^4 - T_\infty^4), \quad y = \delta \tag{24.3.88b}$$

$$Y = \frac{\delta_x}{L}, \qquad \theta = \zeta(m+1)(m^2+1) \tag{24.3.88c}$$

where boundary layer thickness is given by

$$\delta_x = \frac{5x}{\sqrt{\mathrm{Re}}} \tag{24.3.88d}$$

To solve (24.3.87) we may use FDM, FEM, or FVM developed in Parts Two and Three. However, treatments of convective terms and the fourth order nonlinearity of temperature must be resolved. In particular, the optimal control methods (OCM) presented in Section 14.2 and also in Kim, 1982, Chung and Kim, 1984, Utreja, 1982, and Utreja and Chung, 1989 are found to be efficient in the solution of (24.3.87). In this method the formulation begins with construction of the cost function,

$$J = \frac{1}{2}\left[\int_\Omega \left(\frac{\partial U}{\partial X} + \frac{\partial V}{\partial Y}\right)^2 d\Omega + \int_\Omega \left(U\frac{\partial U}{\partial X} + V\frac{\partial U}{\partial Y} - \frac{1}{\text{Re}}\frac{\partial^2 U}{\partial Y^2} - \frac{R_a\theta}{\text{Re}^2\,\text{Pr}}\right)^2 d\Omega \right.$$
$$+ \int_\Omega \left(U\frac{\partial \theta}{\partial X} + V\frac{\partial \theta}{\partial Y} - \frac{1}{\text{Re}\,\text{Pr}}\frac{\partial S}{\partial Y} + 2\zeta_L\theta - a\right)^2 d\Omega$$
$$\left. + \lambda_1\int_\Omega \left(R - \frac{\partial U}{\partial Y}\right)^2 d\Omega + \lambda_2 \int_\Omega \left(S - \frac{\partial \theta}{\partial Y}\right)^2 d\Omega\right] \tag{24.3.89}$$

where λ_1 and λ_2 are the penalty constants, with R and S related by

$$R = \frac{\partial U}{\partial Y} \tag{24.3.90}$$

$$S = \frac{\partial \theta}{\partial Y} \tag{24.3.91}$$

$$\frac{\partial^2 U}{\partial Y^2} = \frac{\partial R}{\partial Y} \tag{24.3.92}$$

$$\frac{\partial^2 \theta}{\partial Y^2} = \frac{\partial S}{\partial Y} \tag{24.3.93}$$

Minimizing (24.3.89) with respect to all nodal variables, we obtain

$$\delta J = \frac{\partial J}{\partial U_\alpha}\delta U_\alpha + \frac{\partial J}{\partial V_\alpha}\delta V_\alpha + \frac{\partial J}{\partial \theta_\alpha}\delta\theta_\alpha + \frac{\partial J}{\partial R_\alpha}\delta R_\alpha + \frac{\partial J}{\partial S_\alpha}\delta S_\alpha = 0 \tag{24.3.94}$$

since $\delta U_\alpha, \delta V_\alpha, \delta\theta_\alpha, \delta R_\alpha$ and δS_α , are arbitrary, and for (24.3.94) to be valid with respect to every nodal value of these infinitesimal quantities, we must have

$$\frac{\partial J}{\partial U_\alpha} = 0, \qquad \frac{\partial J}{\partial V_\alpha} = 0, \qquad \frac{\partial J}{\partial \theta_\alpha} = 0, \qquad \frac{\partial J}{\partial R_\alpha} = 0, \qquad \frac{\partial J}{\partial S_\alpha} = 0 \tag{24.3.95}$$

which provide simultaneous algebraic equations. The boundary conditions given by (24.3.88) can be imposed easily in (24.3.95).

For optically thick boundary layer flow, the procedure for a finite element analysis is now routine, either via Galerkin or optimal control methods.

If the Galerkin approach is used for (24.3.76), then we get

$$A_{\alpha\beta}T_\beta + B_{\alpha\beta}T_\beta = F_\alpha + G_\alpha \tag{24.3.96}$$

where $A_{\alpha\beta}$ and F_α are the same as (24.3.42) and (24.3.44), respectively, and

$$B_{\alpha\beta} = \int_\Omega \left(\frac{1}{\alpha}u\,\Phi_\alpha\frac{\partial \Phi_\beta}{\partial x} + \frac{1}{\alpha}\text{v}\,\Phi_\alpha\frac{\partial \Phi_\beta}{\partial x}\right) d\Omega \tag{24.3.97}$$

$$G_\alpha = \int_\Omega m\,\Phi_\alpha \frac{\partial \Phi_\beta}{\partial y} T_\beta^4\, d\Omega \tag{24.3.98}$$

with

$$m = \frac{4\sigma}{3ka\,\alpha} \tag{24.3.99}$$

It is seen that simplification using the nondimensional form (24.3.78) leads to

$$A_{\alpha\beta}\theta_\beta + B_{\alpha\beta}\theta_\beta = F_\alpha \tag{24.3.100}$$

where

$$A_{\alpha\beta} = \int_\Omega \frac{\partial \Phi_\alpha}{\partial \eta}\frac{\partial \Phi_\beta}{\partial \eta} d\Omega \tag{24.3.101}$$

$$B_{\alpha\beta} = \int_\Omega g\,\Phi_\alpha \frac{\partial \Phi_\beta}{\partial \eta} d\Omega \tag{24.3.102}$$

$$F_\alpha = \frac{\partial \theta}{\partial \eta}\overset{*}{\Phi}_\alpha\bigg|_0^1 \tag{24.3.103}$$

$$g = \frac{3\,\mathrm{Pr}\,Nf}{2(3N+4)} \tag{24.3.104}$$

The corresponding boundary conditions are

$$\theta(0) = 0 \tag{24.3.105}$$

$$\theta(\infty) = 1 \tag{24.3.106}$$

The formulation using OCM is self-explanatory and requires no further elaboration.

Scattering Medium

The combined conduction-convection-radiation heat transfer in a scattering medium governed by (24.3.79) and (24.3.80) can be solved most effectively by the optimal control penalty finite element method. This is because the Galerkin method is likely to suffer instability due to the convective term or nonself-adjointness of the differential equation.

Following the OCM procedure, we write the cost function of the type (24.3.89) in a more general form

$$J = \frac{1}{2}\int_\Omega \left(R_n R_n + \lambda_{(m)} S_m^{(i)} S_m^{(i)}\right) d\Omega \tag{24.3.107}$$

where R_n $(n = 1, 2, \ldots)$ denotes the residual of the nth governing equation, $\lambda_{(m)}$ refers to the mth penalty function corresponding to the mth auxiliary constraint equation introduced for a reduction of the second order derivative into first order such that

$$S_m^{(i)} = G_m^{(i)} - Y_{m,i} \tag{24.3.108}$$

with Y_m the mth unknown and the comma denotes a partial derivative with respect to the coordinate x_i. Thus, for the problem at hand, we write

$$R_1 = \frac{dG}{d\tau} - \frac{1-\omega_o}{N}\left[\theta^4(\tau) - \frac{1}{4}\eta(\tau)\right] - \frac{3\Psi}{N\tau_o}\left(\frac{\tau}{\tau_o} - \frac{\tau^2}{\tau_o^2}\right)\left(\frac{1-\theta}{1-\theta_b}\right)$$
$$+ \frac{36Ec\,\mathrm{Pr}}{\tau_o^2}\left(1 - \frac{2\tau}{\tau_o}\right)^2 = 0 \tag{24.3.109}$$

$$R_2 = \eta(\tau) - 2\left\{ X_1 E_2(\tau) - X_2 E_2(\tau_o - \tau) + \int_0^{\tau_o} \left[(1-\omega_o)\theta^4(\xi) + \frac{\omega_o}{4}\eta(\xi) \right] E_1(|\tau - \xi|)\, d\xi \right\} = 0 \tag{24.3.110}$$

$$S_1 = G - \frac{d\theta}{d\tau} = 0 \tag{24.3.111}$$

It is implied that the unknowns to be calculated consist of the temperature $\theta = Y_1$, the radiation function $\eta = Y_2$, and the temperature gradient $G = G_1^{(1)}$. It is noted that the ranges of indices n, m, and i in (24.3.107) are $n = 1, 2, 3$; $m = 1$, $i = 1$. At this point we require that all variables, Y_1, Y_2, $G_1^{(1)}$, be interpolated for finite elements,

$$Y_m = \Phi_\alpha Y_{m\alpha} \rightarrow \begin{cases} \theta = \Phi_\alpha \theta_\alpha \\ \eta = \Phi_\alpha \eta_\alpha \end{cases} \tag{24.3.112}$$

$$G_m^{(i)} = \Phi_\alpha G_{m\alpha}^{(i)} \rightarrow G = \Phi_\alpha G_\alpha \tag{24.3.113}$$

where $Y_{m\alpha}$ and $G_{m\alpha}^{(i)}$ are the values of Y_m and $G_m^{(i)}$ at a global node α, and Φ_α denotes the global interpolation function [Razzaque, Klein, and Howell, 1982].

Finally, we minimize the cost function such that

$$\delta J = \frac{\partial J}{\partial \theta_\alpha}\delta\theta_\alpha + \frac{\partial J}{\partial \eta_\alpha}\delta\eta_\alpha + \frac{\partial J}{\partial G_\alpha}\delta G_\alpha = 0 \tag{24.3.114}$$

Since $\delta\theta_\alpha$, $\delta\eta_\alpha$, and δG_α are arbitrary, we must have

$$\int_\Omega \left(R_1 \frac{\partial R_1}{\partial \theta_\alpha} + R_2 \frac{\partial R_2}{\partial \theta_\alpha} + \lambda S_1 \frac{\partial S_1}{\partial \theta_\alpha} \right) d\Omega = 0 \tag{24.3.115a}$$

$$\int_\Omega \left(R_1 \frac{\partial R_1}{\partial \eta_\alpha} + R_2 \frac{\partial R_2}{\partial \eta_\alpha} + \lambda S_1 \frac{\partial S_1}{\partial \eta_\alpha} \right) d\Omega = 0 \tag{24.3.115b}$$

$$\int_\Omega \left(R_1 \frac{\partial R_1}{\partial G_\alpha} + R_2 \frac{\partial R_2}{\partial G_\alpha} + \lambda S_1 \frac{\partial S_1}{\partial G_\alpha} \right) d\Omega = 0 \tag{24.3.115c}$$

Combining these equations, we obtain

$$A_{ij}(X) X_j = f_i \tag{24.3.116}$$

with

$$X_j = (\theta_\alpha, \eta_\alpha, G_\alpha) \tag{24.3.117}$$

Note that integrations are required along optical depth as indicated by (24.3.26, 27) plus the finite element domain of (24.3.115). However, the integration limits for both the optical depth and finite element domain are identical since all radiation and flow variables along the axial direction are constant. The significant feature of (24.3.115) is that the resulting nonlinear algebraic equations are symmetric, positive definite, and well conditioned with a proper choice of the penalty constant λ. The solution of the nonlinear equations of (24.3.115) may best be carried out by the Newton-Raphson technique in the form

$$J_{ij}\Delta X_j^{(n+1)} = -f_i^{(n)} \tag{24.3.118}$$

$$J_{ij} = \frac{\partial f_i^{(n)}}{\partial X_j} \tag{24.3.119}$$

$$\Delta X_j^{(n+1)} = X_j^{(n+1)} - X_j^{(n)} \tag{24.3.120}$$

with $f_i^{(n)}$ denoting the ith finite element equation in (24.3.116). Note that, in (24.3.120), an inversion of the Jacobian matrix is avoided. The solution involves calculation of $\Delta X_j^{(n+1)}$, and the unknowns $X_j^{(n+1)}$ are then determined from (24.3.120).

It should be noted that the gradient boundary conditions can be specified simultaneously at any given boundary node, which is not possible nor permitted in the Galerkin finite element method. This fact can be considered an advantage in the optimal control penalty finite elements, but at the same time deterioration of the solution may result due to non-specification of such boundary conditions. It is further reminded that the optimal control penalty finite elements offer no advantage over the standard Galerkin approach if the governing equation is self-adjoint, in which no convective terms are present.

Example problems for the two-dimensional analysis of combined mode conduction, convection, and radiation are presented in Section 24.4.4.

24.3.3 THREE-DIMENSIONAL RADIATIVE HEAT FLUX INTEGRAL FORMULATION

Energy transfer in absorbing, emitting, and scattering media is an important consideration in rocket propulsion, plasma generators for nuclear fusion, ablating systems, hypersonic shock layers, nuclear explosions, etc. Equations governing such energy transfer may represent a combined mode heat transfer by conduction, convection, and radiation.

Our objective is to compute the radiation function in terms of surface and volume integrals through arbitrary optical coordinates. In this approach, no limitations on optical thickness are imposed. Numerical solutions of the governing equations are implemented through the Galerkin finite elements. It is shown that use of isoparametric elements facilitates numerical integration via Gaussian quadrature, unlimited by optical depths of the participating media.

Example problems to demonstrate efficiency of the solution procedure include two-dimensional diverging and converging channels. Effects of combined mode (conduction-convection-radiation), albedo, optical thickness, etc., are investigated [Chung and Kim, 1984].

In general, it is convenient to have the energy equation in terms of the fluid temperature and heat capacity rather than internal energy. It is also assumed that the radiant energy and the radiation stresses are much smaller than the corresponding molecular quantities and can therefore be neglected even at very high temperatures. Thus, if the radiating fluid is an ideal gas, we have, for a steady state

$$\rho c_p u_i \frac{\partial T}{\partial x_i} - k\frac{\partial^2 T}{\partial x_i \partial x_i} + \frac{\partial q_i^R}{\partial x_i} = 0 \tag{24.3.121}$$

Consider a spectral optical volume, V_λ, and a spectral surface area, A_λ, as defined in Goulard [1962]

$$dV_\lambda = \tau_\lambda^2 d\tau_\lambda d\omega \tag{24.3.122a}$$

$$dA_\lambda = \tau_\lambda^2 d\omega/\cos\phi \tag{24.3.122b}$$

where ϕ is the angle between the normal vector to the surface and the direction ω as shown in Figure 24.3.2a. Note that V_λ and A_λ are the functions of the spectral optical length, τ_λ measured from the point M at r in the direction ω. Expressing the spectral optical space defined in equations (24.3.122a,b), we have

$$q_{\lambda i}^R(r) = \int_{A_\lambda} I_\lambda(r_w)\frac{e^{-\tau_{\lambda w}}}{\tau_{\lambda w}'^2}\ell_i \cos\phi\, dA_\lambda + \int_{V_\lambda} S_\lambda(r')\frac{e^{-\tau_\lambda'}}{\tau_\lambda'^2}\ell_i\, dV_\lambda \tag{24.3.123}$$

$$q_{\lambda i,i}^R(r) = 4\pi a_\lambda(r)B_\lambda(r) - a_\lambda(r)\int_{A_\lambda} I_\lambda(r_w)\frac{e^{-\tau_{\lambda w}}}{\tau_{\lambda w}'^2}\cos\phi\, dA_\lambda + a_\lambda(r)\int_{V_\lambda} S_\lambda(r')\frac{e^{-\tau_\lambda'}}{\tau_\lambda'^2}\, dV_\lambda \tag{24.3.124}$$

For a gray medium with gray bounding surfaces, it can be shown that

$$q_i^R(r) = \int_A \frac{\sigma T_w^4}{\pi}\frac{e^{-\tau_w}}{\tau^2}\ell_i \cos\phi\, dA + \int_V S(r')\frac{e^{-\tau'}}{\tau'^2}\ell_i\, dV \tag{24.3.125}$$

and

$$\frac{\partial q^R(r)}{\partial x_i} = 4\beta\left(1 - \frac{\gamma}{\beta}\right)\left(\sigma T^4(r) - \frac{1}{4}\overline{H}\right) \tag{24.3.126}$$

where T_w represents the surface temperature and H is an integral defined by

$$\overline{H} = \int_A \frac{\sigma T_w^4}{\pi}\frac{e^{-\tau_w}}{\tau_w^2}\cos\phi\, dA + \int_V S(r')\frac{e^{-\tau'}}{\tau'^2}\, dV \tag{24.3.127}$$

with

$$S(r') = \left(1 - \frac{\gamma}{\beta}\right)\frac{\sigma T^4}{\pi} + \frac{\omega_o}{4\pi}\overline{H} \tag{24.3.128}$$

Assume that all the radiative physical properties (i.e., β, a, and γ) are constant throughout the flow domain and define dimensionless quantities

$$\mathrm{v}_i = \frac{u_i}{U_o}, \qquad \theta = \frac{T}{T_o}, \qquad X_i = \beta x_i, \qquad \tau_o = \beta L$$

where U_o, T_o, and L are, respectively, the reference velocity, temperature, and length. Then the energy equation can be written in dimensionless form as

$$\frac{\partial^2\theta}{\partial x_i \partial x_i} - \frac{\mathrm{Re\,Pr}}{\tau_o}\mathrm{v}_i\frac{\partial\theta}{\partial x_i} - \frac{1-\omega_o}{N}\left(\theta^4 - \frac{1}{4\pi}H\right) = 0 \tag{24.3.129}$$

Here H is the radiation function given by

$$H = \int_V \eta_s \frac{e^{-\tau}}{\tau^2}\, dV + \int_A \theta_w^4 \frac{e^{-\tau_w}}{\tau_w^2}\cos\phi\, dA \tag{24.3.130}$$

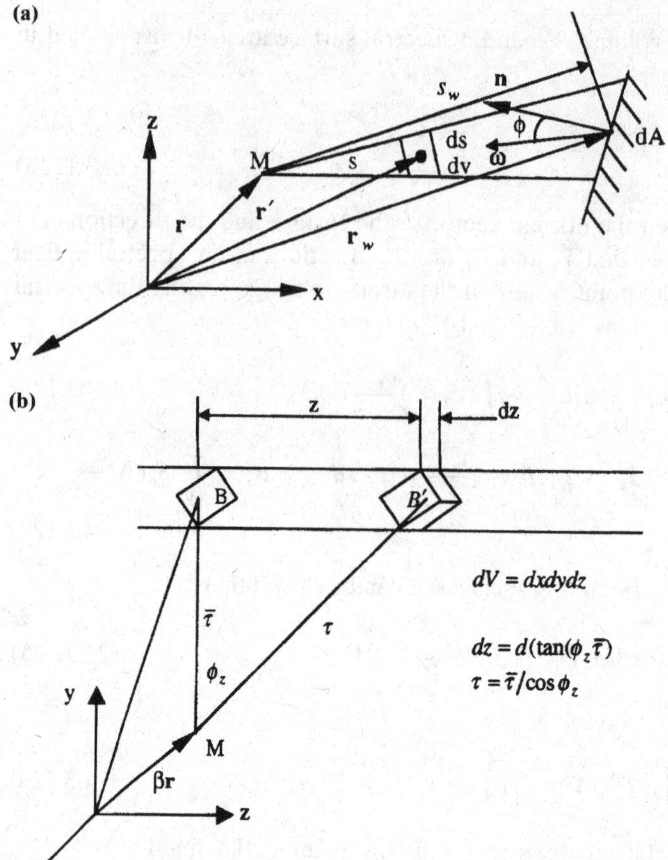

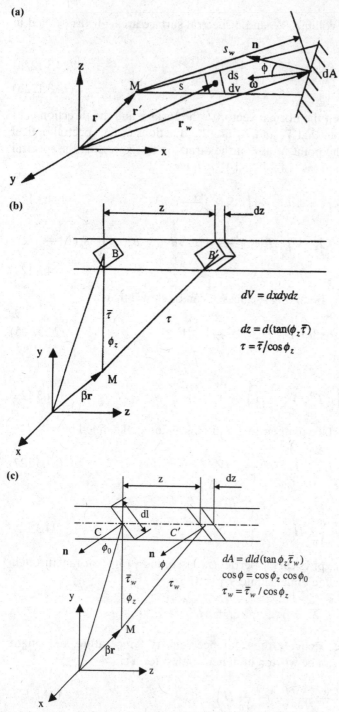

Figure 24.3.2 Three-dimensional heat flux integral formulation. (a) Radiation contribution of a surface element, dA, and a volume element. dv, to a point M in a direction ω. (b) Geometry for volume integral. (c) Geometry for surface integral.

where η_s is the dimensionless source function defined by $\eta_s = \pi S/\sigma T_o^4$. That is

$$\eta_s(r) = (1-\omega_o)\left(\theta^4 + \frac{\omega_o}{4\pi}H\right) \tag{24.3.131}$$

To determine H, let us now consider an optical volume element, dV, centered at B as shown in Figure 24.3.2b, in which all distances are measured by the optical coordinates X, Y, and Z. Let ϕ_z be the angle at which the point, M, sees the volume element, dV, with respect to a point B at $Z=0$, and let $\bar{\tau}$ and τ be the optical distances of $\overline{MB}$ and $\overline{MB'}$, respectively. The volume integral of H is then evaluated as follows:

$$\begin{aligned}\int_V \eta_s \frac{e^{-\tau}}{\tau^2}\,dV &= \iiint_{-\infty}^{\infty} \eta_s(B)\frac{e^{-\tau}}{\tau^2}\,dZ\,dX\,dY\\ &= \iiint_{-\infty}^{\infty} \eta_s(B)\frac{e^{-\bar{\tau}/\cos\phi_z}}{(\bar{\tau}^2/\cos\phi_z)^2}\,d(\tan\phi_z\bar{\tau})\,dX\,dY\\ &= \iint \eta_s F_v(\bar{\tau})\,dX\,dY\end{aligned} \tag{24.3.132}$$

where B is a point in the two-dimensional domain of the X- and Y-coordinates, and $F_v(\bar{\tau})$ is a geometric function of the optical distance, $\bar{\tau}$, measured from the point, M, involved in the volume integral,

$$F_v(\bar{\tau}) = 2\int_0^{\pi/2} \frac{e^{-\bar{\tau}/\cos\phi_z}}{\bar{\tau}}\,d\phi_z \tag{24.3.133}$$

To determine the surface integral of H, we consider a surface element $dA = d\ell dZ$, in which $d\ell$ is a boundary segment of the two-dimensional domain (Figure 24.3.2). Let ϕ be the angle between the normal vectors to the surface element centered at C' and the line $\overline{MC'}$. Also, let ϕ_o be the angle at $Z=0$.

Noting $\cos\phi = \cos\phi_z \cos\phi_o$ where ϕ_z is the angle between $\overline{MC}$ and $\overline{MC'}$, it follows that

$$\int_A \theta_w^4 \frac{e^{-\tau_w}}{\tau_w^2}\cos\phi_z dA = \iint_{-\frac{\pi}{2}}^{\frac{\pi}{2}} \theta_w^4 \frac{e^{-\bar{\tau}/\cos\phi_z}}{(\bar{\tau}_w/\cos\phi_z)}\cos\phi_z\cos\phi_o(\tan\phi_z\bar{\tau})\,d\ell \tag{24.3.134}$$

Thus, the integral, H, at the point, M, at r in the domain can be evaluated with η_s solved simultaneously by means of equation (24.3.131).

Galerkin finite elements to solve problems such as equations (24.3.129, 24.3.130) are straightforward. We consider that the temperature, θ, and radiation function, H, are approximated as

$$\theta = \Phi_\alpha \theta_\alpha, \qquad H = \Phi_\alpha H_\alpha \tag{24.3.135}$$

where Φ_α denotes the four-node isoparametric interpolation function, with α representing the global nodes. Substituting equation (24.3.135) into equation (24.3.129) and equation (24.3.130), we arrive at the Galerkin finite element equations in the form

$$A_{\alpha\beta}\theta_\beta + B_{\alpha\beta}\theta_\beta + C_{\alpha\beta}H_\beta = F_\alpha + G_\alpha \tag{24.3.136a}$$

$$D_{\alpha\beta}H_\beta = L_\alpha + M_\alpha \tag{24.3.136b}$$

where

$$A_{\alpha\beta} = \int_\Omega \frac{\partial \Phi_\alpha}{\partial x_i}\frac{\partial \Phi_\beta}{\partial x_i} d\Omega \qquad B_{\alpha\beta} = \frac{\mathrm{Re\,Pr}}{\tau_o}\int_\Omega v_i\,\Phi_\alpha \frac{\partial \Phi_\beta}{\partial x_i} d\Omega$$

$$C_{\alpha\beta} = \frac{1-\omega_o}{4\pi N}\int_\Omega \Phi_\alpha \Phi_\beta\, d\Omega \qquad F_\alpha = \int_\Gamma \frac{\partial \theta}{\partial x_i} n_i \overset{*}{\Phi}_\alpha\, d\Gamma$$

$$G_\alpha = \frac{1-\omega_o}{N}\left[\int_\Omega \Phi_\alpha \Phi_\beta\, d\Omega\right]\theta_\beta^4 \qquad D_{\alpha\beta} = \int_\Omega \Phi_\alpha \Phi_\beta\, d\Omega$$

$$L_\alpha = -\int_\Omega \left(\int_V \eta_s \frac{e^{-\tau_w}}{\tau^2} dV\right)\Phi_\alpha\, d\Omega \qquad M_\alpha = -\int_\Gamma \left(\int_A \theta_w^4 \frac{e^{-\tau_w}}{\tau_w^2}\cos\phi\, dA\right)\overset{*}{\Phi}_\alpha\, d\Gamma$$

Here, $\overset{*}{\Phi}_\alpha$ represents boundary interpolation functions.

Combining equations (24.3.136a) and (24.3.136b), we write the resulting equations in the form

$$K_{ij} X_j = f_i \tag{24.3.137}$$

with

$$X_j = \begin{bmatrix} \theta_\alpha \\ H_\alpha \end{bmatrix}$$

The solution of nonlinear equations (24.3.137) may best be carried out by the Newton-Raphson technique in the form

$$J_{ij}\,\Delta X_j^{(n+1)} = -f_i^{(n)} \tag{24.3.138}$$

where

$$J_{ij} = \frac{\partial f_i^{(n)}}{\partial X_j}$$

$$\Delta X_j^{(n+1)} = X_j^{(n+1)} - X_j^{(n)} \tag{24.3.139}$$

with $f_i^{(n)}$ denoting the ith finite element equation in (24.3.137). Note that, in (24.3.137), an inversion of the Jacobian matrix is avoided. The solution involves calculation of $\Delta X_j^{(n+1)}$, and the unknowns $X_j^{(n+1)}$ are then determined from equation (24.3.139).

Example problems for the combined mode conduction, convection, and radiation with three-dimensional flux integration are presented in Section 24.4.5.

24.4 EXAMPLE PROBLEMS

24.4.1 NONPARTICIPATING MEDIA

(1) View Factors

Calculate view factors for (1) two parallel 1 × 1 square planes, one unit apart and (Figure 24.4.1.1a) and (2) two intersecting 1× 1 square planes at angles 30°, 60°, 90°, 120°,

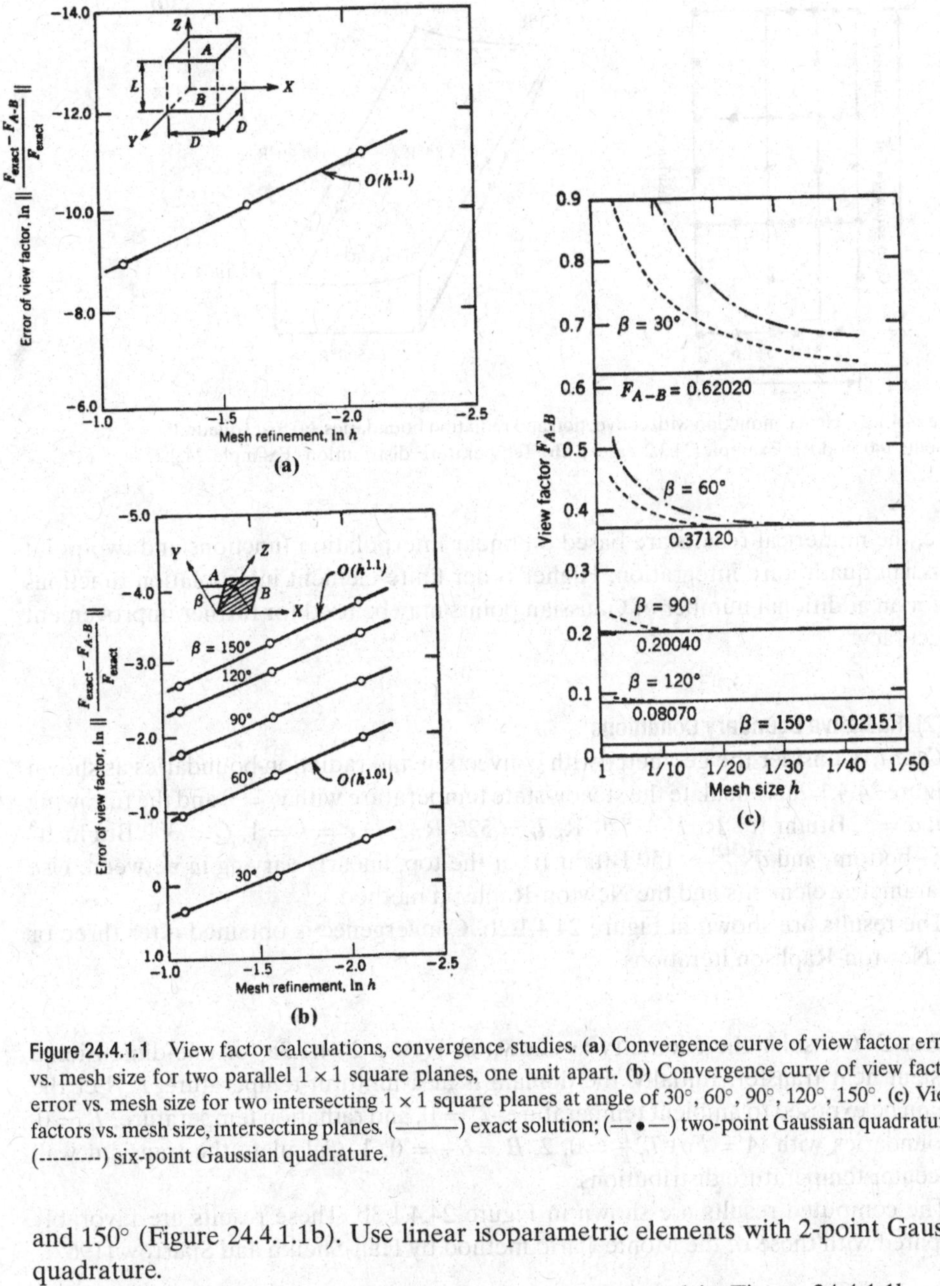

Figure 24.4.1.1 View factor calculations, convergence studies. (a) Convergence curve of view factor error vs. mesh size for two parallel 1 × 1 square planes, one unit apart. (b) Convergence curve of view factor error vs. mesh size for two intersecting 1 × 1 square planes at angle of 30°, 60°, 90°, 120°, 150°. (c) View factors vs. mesh size, intersecting planes. (———) exact solution; (—•—) two-point Gaussian quadrature; (---------) six-point Gaussian quadrature.

and 150° (Figure 24.4.1.1b). Use linear isoparametric elements with 2-point Gaussian quadrature.

The convergence curves for the results are presented in Figure 24.4.1.1b,c. It is shown that the most accurate results are obtained for parallel surfaces. In the case of intersecting surfaces with smaller angles, more refined grids and an additional number of Gaussian points are required for convergence. It is reminded that the power of the finite element method is its capability to handle irregular geometries other than a simple case as shown in this example. If convergence is guaranteed from the basic mathematical viewpoint, the accuracy of the solution for irregular geometries can be guaranteed.

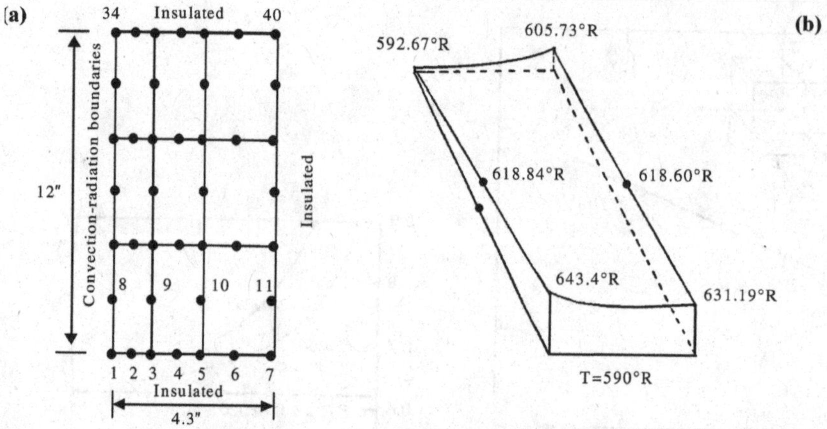

Figure 24.4.1.2 Heat conduction with convection and radiation boundaries. **(a)** 3 × 3 quadratic elements (40 nodes), Example 24.2.2 case 1. **(b)** Temperature distribution, Example 24.2.2 case 1.

Here, the numerical results are based on linear interpolation functions and two-point Gaussian quadrature integration. Higher order finite element interpolation functions and/or an additional number of Gaussian points may be used for further improvement in accuracy.

(2) Radiative Boundary Conditions

Case 1. Consider the geometry with convection and radiation boundaries as shown in Figure 24.4.1.2a. Calculate the steady-state temperature with $\overline{v}_i = 0$ and the following data: $\overline{\alpha} = 1$, Btu/hr ft^2 °R. $T' = 520$°R, $T_r = 520$°R, $F = \varepsilon = k = 1$, $Q = 300$ Btu/hr ft^3 at the bottom, and $\overline{q}^{(CD)} = 150$ Btu/hr ft^2 at the top, linearly varying in between. Use isoparametric elements and the Newton-Raphson method.

The results are shown in Figure 24.4.1.2b. Convergence is obtained after three or four Newton-Raphson iterations.

Case 2. Consider the geometry as shown in Figure 24.4.1.3a for one-dimensional transient heat transfer. Initially, the domain is at a uniform temperature T_o. Let the domain be exposed to ambient temperature, $T' = 0$, and radiation temperature, $T_r = 0$, at boundaries with $A = F\sigma\varepsilon T_0^3 \frac{L}{k} = 0, 2$, $B = \overline{\alpha}\frac{L}{k} = 0, 1$. Calculate the transient wall and center temperature distributions.

The computed results are shown in Figure 24.4.1.3b. These results are favorably compared with those of the Monte Carlo method by Haji-Sheikh and Sparrow [1967].

24.4.2 SOLUTION OF RADIATIVE HEAT TRANSFER EQUATION IN NONPARTICIPATING MEDIA

Two examples problems using FVM/FEM are presented in this section. It should be noted that Chapter 7 discusses the finite volume methods via finite difference

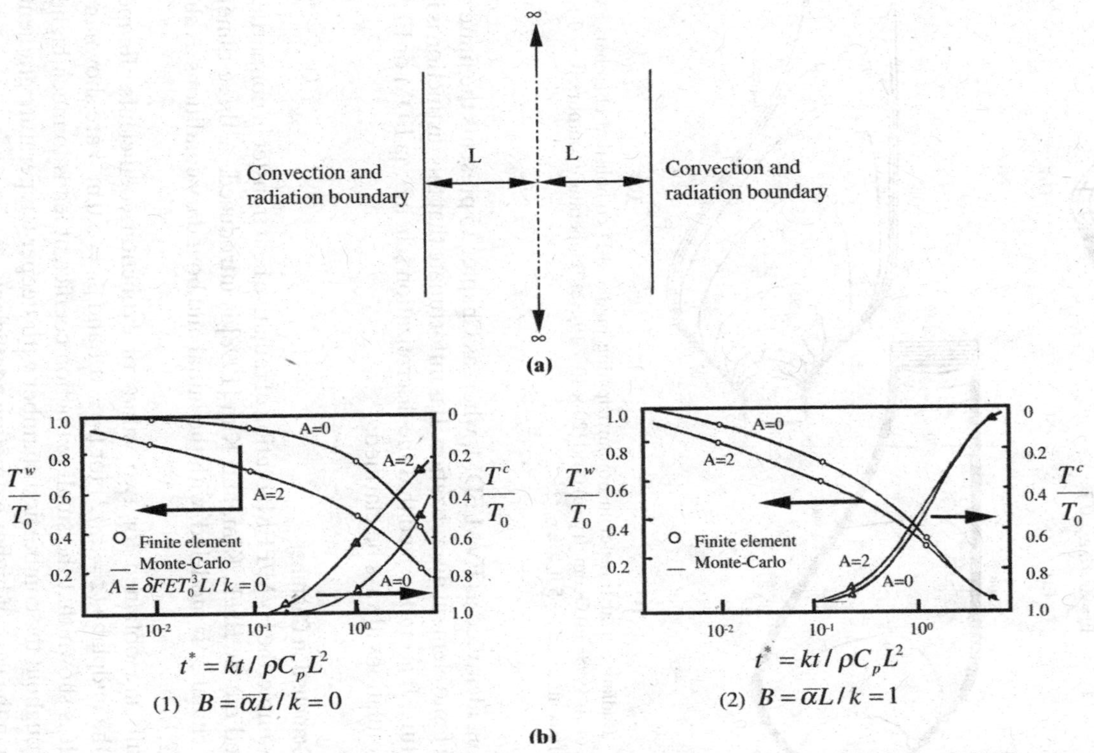

Figure 24.4.1.3 1-D transient radiation-convection. **(a)** One-dimensional analysis, Example 24.2.2, case 2. **(b)** Temperature distribution, Example 24.2.2, case 2.

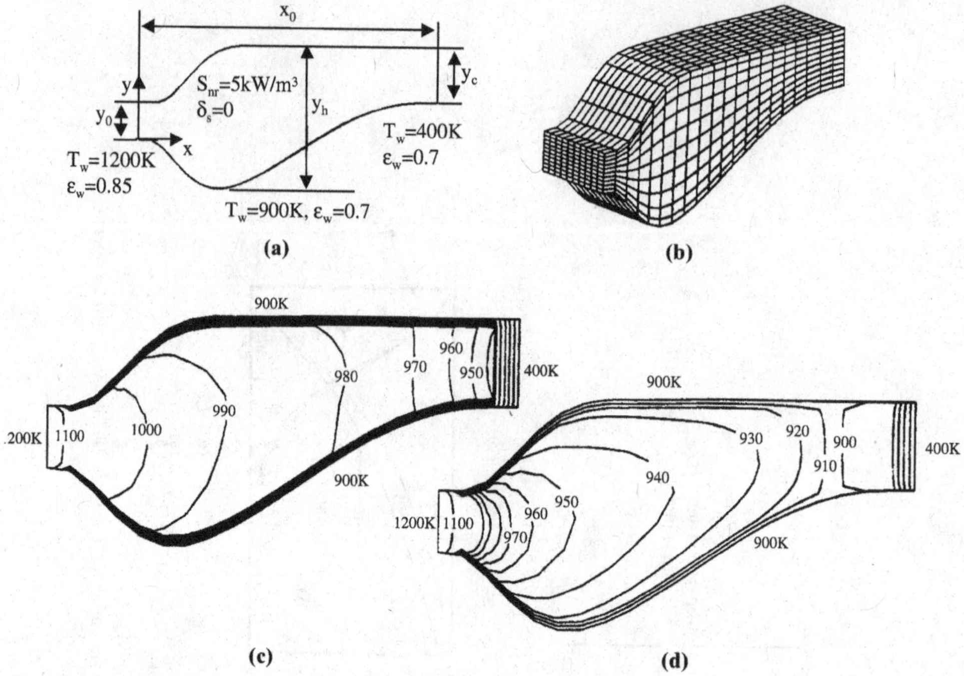

Figure 24.4.2.1 Radiative heat transfer in nonparticipating media, 3-D combustion chamber, FVM/FEM solution [Baek et al., 1998]. **(a)** Geometry. **(b)** Grid (20 × 10 × 10). **(c)** Temperature contours, $\beta_0 = 0.1, z = z_0/2$. **(d)** Temperature contours, $\beta_0 = 5.0, z = z_0/2$.

discretization, identified as FVM/FDM, whereas Chapter 15 presents the finite volume methods via finite element discretization. It is unfortunate that this distinction is ignored in the literature. It is emphasized that the discretization scheme via FDM or FEM used in FVM formulations always be clarified.

(1) 3-D Combustion Chamber

In this example, the FVM/FEM solution of radiative heat transfer in nonparticipating media carried out by Baek, Kim, and Kim [1998] is introduced. A three-dimensional combustion chamber is modeled, with the initial and boundary conditions as shown in Figure 24.4.2.1a,b.

The results are obtained for two values of extinction coefficients. Temperature contours at the midplane ($z = z_0/2$) for $\beta_0 = 0.1$ and $\beta_0 = 5.0\ \mathrm{m}^{-1}$ are shown in Figure 24.4.2.1c,d. It is shown that the small extinction coefficient leads to much higher temperature throughout the combustion chamber with steeper temperature gradient at the side walls than the case of higher extinction coefficient.

(2) 3-D Enclosure and Reflective Walls

The FVM/FEM analysis by Raithby [1999] for the 3-D enclosure and reflective walls is presented here. The geometry and solid angle discretization are shown in Figure 24.4.2.2a,b. The section of the wall $x = 0, 0 \le y \le 12, 0 \le z \le 17$ is black and held at 1000°C. All other walls are diffuse and fully reflective (adiabatic to radiation). The

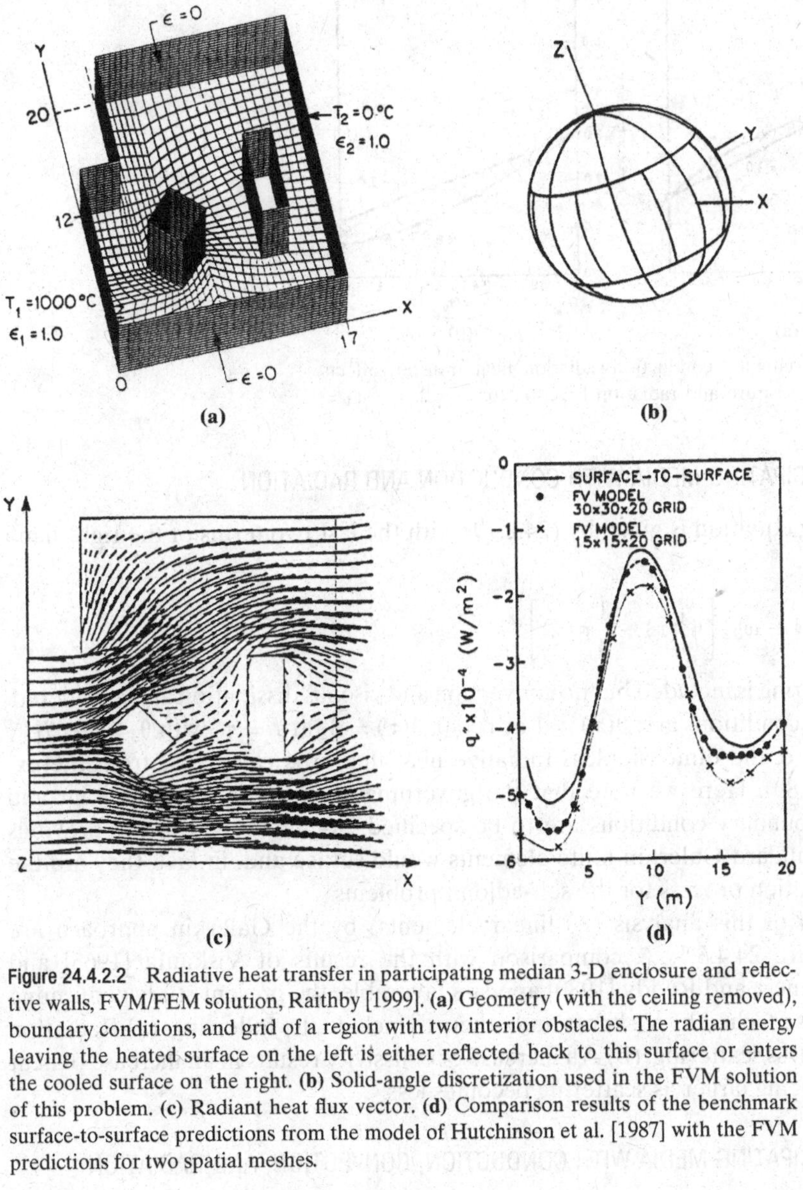

Figure 24.4.2.2 Radiative heat transfer in participating median 3-D enclosure and reflective walls, FVM/FEM solution, Raithby [1999]. **(a)** Geometry (with the ceiling removed), boundary conditions, and grid of a region with two interior obstacles. The radian energy leaving the heated surface on the left is either reflected back to this surface or enters the cooled surface on the right. **(b)** Solid-angle discretization used in the FVM solution of this problem. **(c)** Radiant heat flux vector. **(d)** Comparison results of the benchmark surface-to-surface predictions from the model of Hutchinson et al. [1987] with the FVM predictions for two spatial meshes.

angular grid shown in Figure 24.4.2.2b has $L = 32$ solid angles. There are 8 solid angles in the range of polar angle $0 \leq \phi \leq 60°$, 8 in the range $120° \leq \phi \leq 180°$, and 16 in the range $60° \leq \phi \leq 120°$. All the solid angles are equal in size.

The radiation heat flux distributions are shown in Figure 24.4.2.2c on the plane $z = 7.05$ m. It is seen that the reflection of the radiation by the two interior walls and by the exterior walls transport the radiation to the cold walls. The effect of mesh discretization on the net heat flux is shown in Figure 24.4.2.2d, with the finer mesh approaching closer to the benchmark solution of Hutchinson, Stefurak, and Gerber [1996].

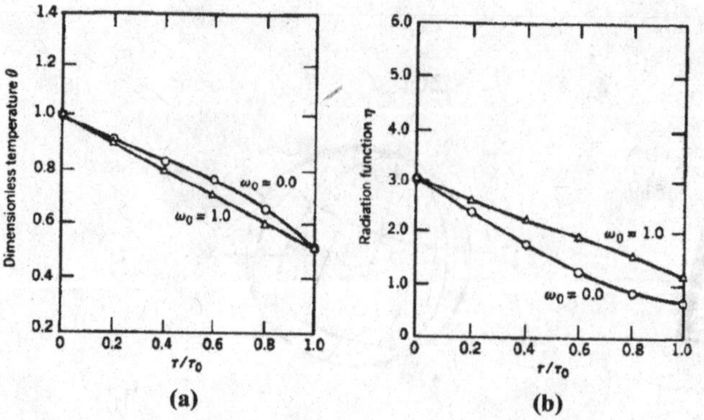

Figure 24.4.3.1 Combined conduction-radiation heat transfer, effects of ω_0 on the temperature and radiation function for $N = 1, \tau_0 = \varepsilon_1 = \varepsilon_2 = 1.0$.

24.4.3 PARTICIPATING MEDIA WITH CONDUCTION AND RADIATION

The governing equation is given by (24.2.79) with the last two terms of the right-hand side neglected,

$$\frac{d^2\theta}{d\tau^2} = \frac{1}{N}(1 - \omega_0)\left[\theta^4(\tau) - \frac{1}{4}\eta(\tau)\right]$$

in which scattering is included but no convection and viscous dissipation are considered. The boundary conditions are: $\theta(\tau) = 1$ at $\tau = 0$, $\theta(\tau) = \theta_2$ at $\tau = \tau_0$ with $\theta_2 = T_2/T_1$.

Furthermore, the dimensionless radiative heat flux Ψ is calculated from the expression (24.3.80). Here we note that the governing equation is of self-adjoint and no gradient boundary conditions are to be specified. As mentioned in the previous section, the standard Galerkin finite elements would suffice and, in fact, they are the best approximation process for the self-adjoint problems.

The results of this analysis (20 linear elements) by the Galerkin approach are shown in Figure 24.4.3.1. A comparison with the results of Viskanta [1965] and Fernandes, Francis, and Reddy [1980] appears favorable, the present study confirming the conclusions reached by Viskanta and others to include the following: (a) Radiation increases with less scattering, (b) An increase of emissivity results in an increase of heat flux, this rate being larger as scattering becomes less.

24.4.4 PARTICIPATING MEDIA WITH CONDUCTION, CONVECTION, AND RADIATION

(1) Combined Mode Heat Transfer Without and With Scattering and Viscous Dissipation

The equation governing this subject is obtained from (24.3.79) by setting $\omega_o = 0$ and $EcPr = 0$:

$$\frac{d^2\theta}{d\tau^2} = \frac{1}{N}(1 - \omega_o)\left[\theta^4(\tau) - \frac{1}{4}\eta(\tau)\right] + \frac{3\Psi}{N\tau_o}\left(\frac{\tau}{\tau_o} - \frac{\tau^2}{\tau_o^2}\right)\frac{1-\theta}{1-\theta_b}$$

Because of the presence of the convective term here, the standard Galerkin finite elements fail and the optimal control penalty finite elements are shown to be effective.

The boundary conditions for this problem (Figure 24.4.4.1a) are: $\theta = \theta_w = 1$ at $\tau = 0$ and $\tau = \tau_o$. Due to symmetry about the center line, we may also use: $\theta = \theta_w$ at $\tau = 0$,

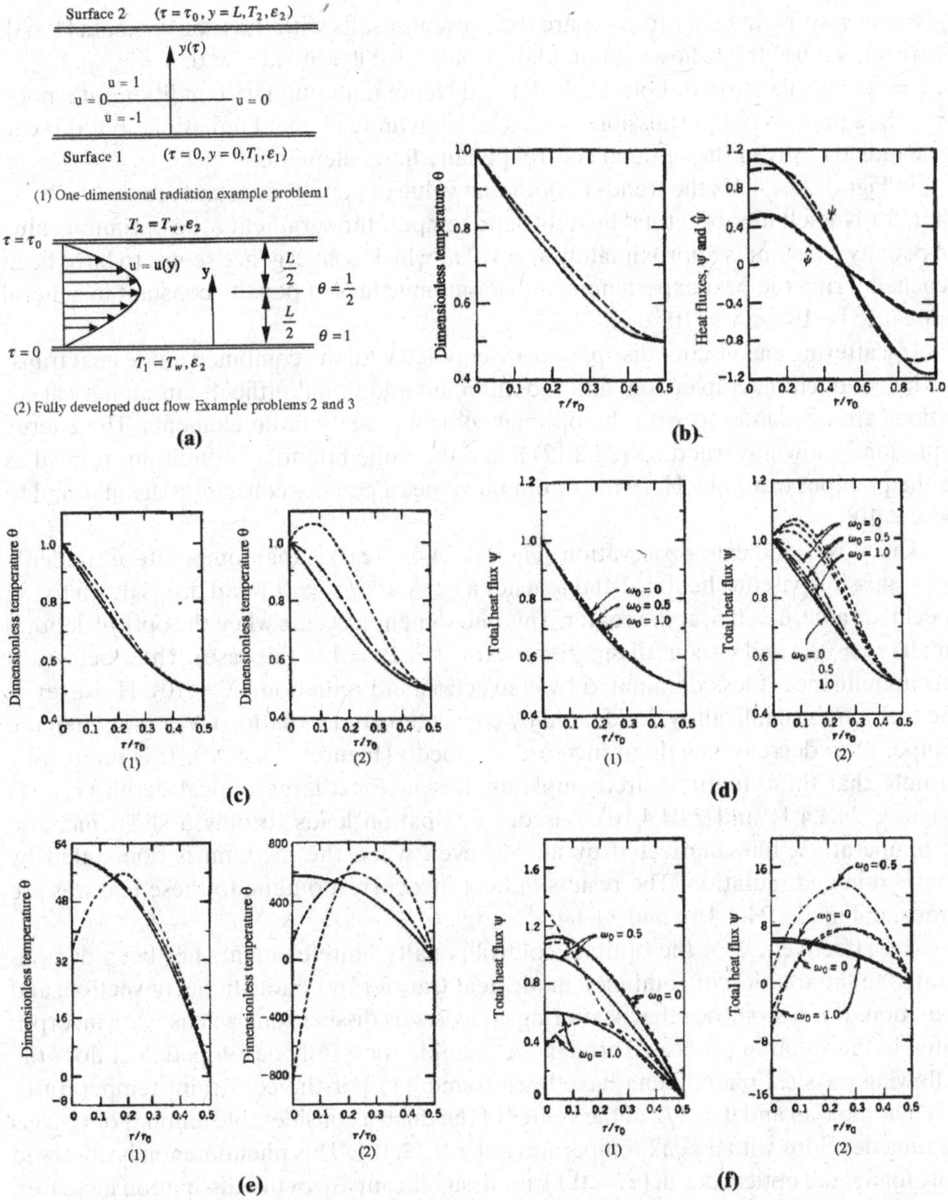

Figure 24.4.4.1 Combined mode conduction, convection, and radiation with and without scattering and viscous dissipation [Chung and Kim, 1984]. **(a)** Example problem. (1) One-dimensional radiation, example problem 1; (2) fully developed duct flow, example problems 2 and 3. **(b)** Combined conduction-convection-radiation heat transfer without scattering and viscous dissipation. $N = 0.1$; $\omega_0 = 0$; $Ec\,Pr = 0$; $\tau_0 = \varepsilon_1 = \varepsilon_2 = 1$. **(c)** Combined conduction-convection-radiation heat transfer with scattering and viscous dissipation. $N = 0.1$; $\varepsilon_1 = \varepsilon_2 = 1$. (1) $\tau_0 = 1.0$; (2) $\tau_0 = 0.1$. (——) $Ec\,Pr = 0$, $\omega_0 = 1.0$; (—•—) $Ec\,Pr = 0.5$, $\omega_0 = 0.5$; (------) $Ec\,Pr = 1.0$, $\omega_0 = 0$. **(d)** Combined conduction-convection-radiation heat transfer, with scattering and viscous dissipation. $N = 0.1$; $\varepsilon_1 = \varepsilon_2 = 1$. (1) $\tau_0 = 1.0$; $\tau_0 = 0.1$. (——) $Ec\,Pr = 0$; (—•—) $Ec\,Pr = 0.5$; (------) $Ec\,Pr = 1.0$. **(e)** Combined conduction-convection-radiation heat transfer (temperature distributions) with scattering and viscous dissipation. $N = 0.1$; $\varepsilon_1 = \varepsilon_2 = 1$. (1) $\tau_0 = 1.0$; $\tau_0 = 0.1$. (——) $Ec\,Pr = 0$, $\omega_0 = 1.0$; (—•—) $Ec\,Pr = 0.5$, $\omega_0 = 0.5$; (------) $Ec\,Pr = 1.0$, $\omega_0 = 0$. **(f)** Combined conduction-convection-radiation heat transfer (heat flux distributions) with scattering and viscous dissipation. $N = 0.1$; $\varepsilon_1 = \varepsilon_2 = 1$. (1) $\tau_0 = 1.0$; (2) $\tau_0 = 0.1$. (——) $Ec\,Pr = 0$; (------) $Ec\,Pr = 1.0$.

$\frac{d\theta}{d\tau} = 0$ at $\tau = \frac{\tau_o}{2}$. In order to compare the present results with those of Viskanta [1963], however, we use the following boundary conditions: $\theta = \theta_w$ at $\tau = 0$, $\theta = \theta_o$ and $\frac{d\theta}{d\tau} = 0$, $\tau = \frac{\tau_o}{2}$. Specification of both Dirichlet and Neumann boundary conditions at a node ($\tau = \tau_o/2$ here) is not permissible in the Galerkin finite element equations, but this can be handled easily in the optimal control penalty finite elements.

In Figure 24.4.4.1b, the trends of optimum value of penalty constant are shown. Note that, for Nusselt number, total heat flux and temperature gradients, the optimum value of penalty constant is approximately $\lambda = 10^5$ at which convergence seems to have been reached. From the past experiences, the optimum values of penalty constant in general appear to be $10^4 < \lambda < 10^{10}$.

If scattering and viscous dissipation are coupled with the combined mode heat transfer by conduction, convection, and radiation, no additional difficulty in numerical solutions are encountered with the optimal control penalty finite elements. The energy equation is now governed by (24.2.121), and the same boundary conditions prevail as in the previous example. Here the optimum value of penalty constant is again found to be $\lambda = 10^5$.

The most significant observation (Figures 24.4.4.1c,d) is that temperature suddenly rises (possible viscous heating) and reaches a peak at $\tau/\tau_o = 0.1$ and drops down to the specified value, $\theta = 0.5$, at the center. This phenomenon occurs when the optical depth is small ($\tau_o = 0.1$) and viscous dissipation is large (Ec Pr = 1 in this case). The albedo (ω_o) has no influence if less dominated by convection and radiation ($N = 10$). However, if the medium is significantly dominated by convection and radiation ($N = 0.1$), then the temperature decreases with an increase of albedo (Figure 24.4.4.1d). It is interesting to note that these features are completely absent for a large optical depth ($\tau_o = 1$) (Figures 24.4.4.1c and 24.4.4.1d). Viscous dissipation leads to only a slight increase in temperature, but unaffected by albedo even when the medium is dominated by convection and radiation. The results of heat flux corresponding to these features are shown in Figures 24.4.4.1e and 24.4.4.1f.

The effectiveness of the optimal control penalty finite elements has been demonstrated in the solution of combined mode heat transfer by conduction, convection, and radiation. It is also shown that scattering and viscous dissipation can easily be incorporated in the solution process. Through an example for a fully developed duct flow, the following physical phenomena have been found: (1) For the constraint temperatures $\theta = 1$ at the wall and $\theta = 1/2$ at the center of the duct, a considerable amount of viscous heating develops with a peak temperature at $\tau/\tau_o \cong 0.1$. This phenomenon is observed only for a small optical depth ($\tau_o - 0.1$), more significant as viscous dissipation increases. Note also that heat flux distributions are such that negative Nusselt numbers appear in the region where temperature jumps occur. (2) An increase of emissivity leads to a large Nusselt number when scattering is absent in the convection-radiation dominated medium with a large optical depth. If the medium is dominated by convection and radiation ($N = 0.1$), temperature decreases with an increase of albedo of scattering. This influence disappears as the medium begins to be dominated by conduction ($N = 10$). For large optical depths ($\tau_o = 1$), the effect of viscous dissipation diminishes and the temperature field is not affected by the albedo even when the medium is dominated by convection and radiation.

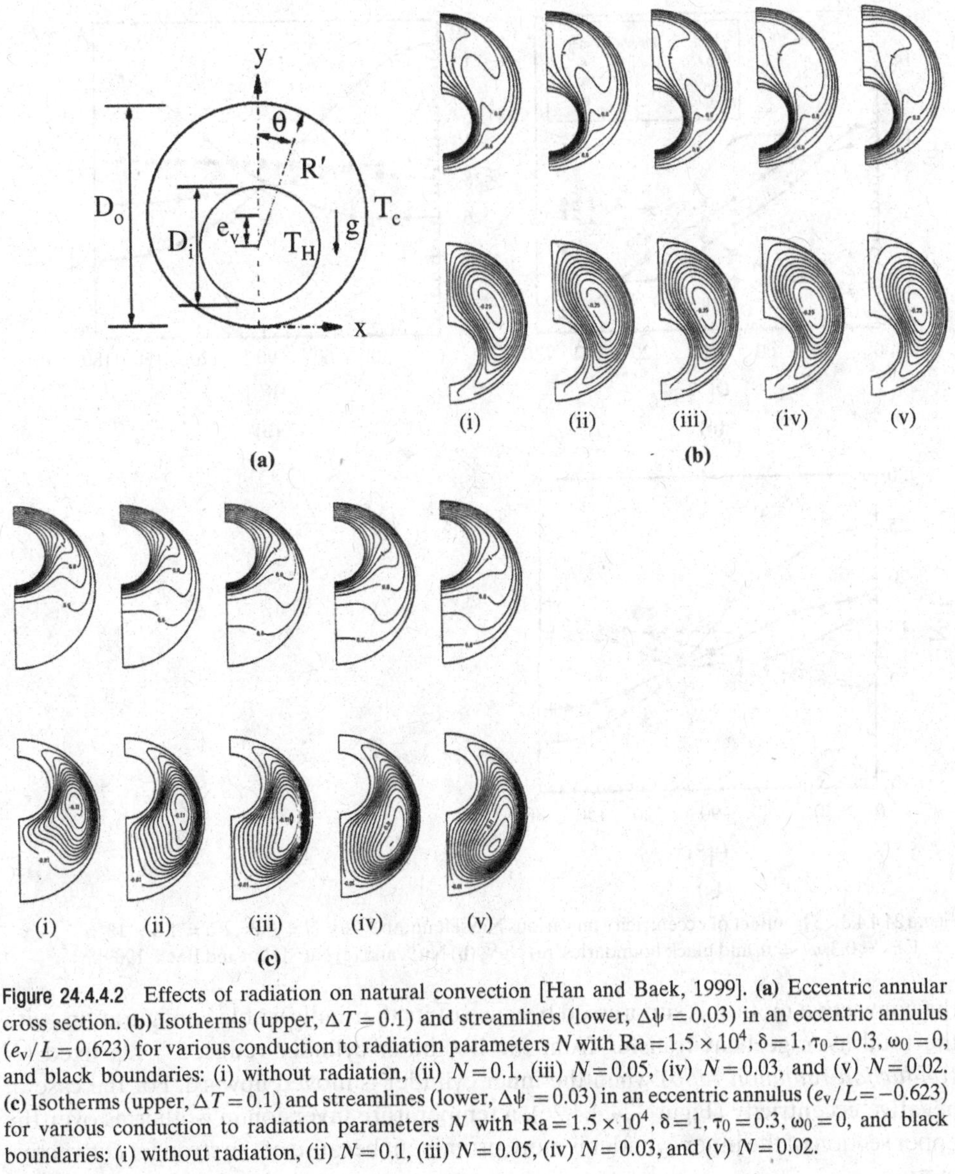

Figure 24.4.4.2 Effects of radiation on natural convection [Han and Baek, 1999]. **(a)** Eccentric annular cross section. **(b)** Isotherms (upper, $\Delta T = 0.1$) and streamlines (lower, $\Delta\psi = 0.03$) in an eccentric annulus ($e_v/L = 0.623$) for various conduction to radiation parameters N with $\mathrm{Ra} = 1.5 \times 10^4$, $\delta = 1$, $\tau_0 = 0.3$, $\omega_0 = 0$, and black boundaries: (i) without radiation, (ii) $N = 0.1$, (iii) $N = 0.05$, (iv) $N = 0.03$, and (v) $N = 0.02$. **(c)** Isotherms (upper, $\Delta T = 0.1$) and streamlines (lower, $\Delta\psi = 0.03$) in an eccentric annulus ($e_v/L = -0.623$) for various conduction to radiation parameters N with $\mathrm{Ra} = 1.5 \times 10^4$, $\delta = 1$, $\tau_0 = 0.3$, $\omega_0 = 0$, and black boundaries: (i) without radiation, (ii) $N = 0.1$, (iii) $N = 0.05$, (iv) $N = 0.03$, and (v) $N = 0.02$.

(2) Effect of Radiation on Natural Convection

This example presents the analysis investigating the effect of radiation on natural convection in the eccentrically positioned cylindrical annulus (Figure 24.4.4.2a) as reported by Han and Baek [1999]. The flowfield equation is solved using the compressible SIMPLER [Karki and Patankar, 1989] (similar to equation (5.3.20) for diagonal dominance), whereas the radiative transfer equation is solved using FVM/FEM. The spatial and angular domains are discretized into 41×63 nonuniform spatial control volumes and 2×24 control angles with uniform $\Delta\phi$ and $\Delta\theta$, respectively. Isotherms and streamline contours for the positive eccentricity are shown in Figure 24.4.4.2b. Here,

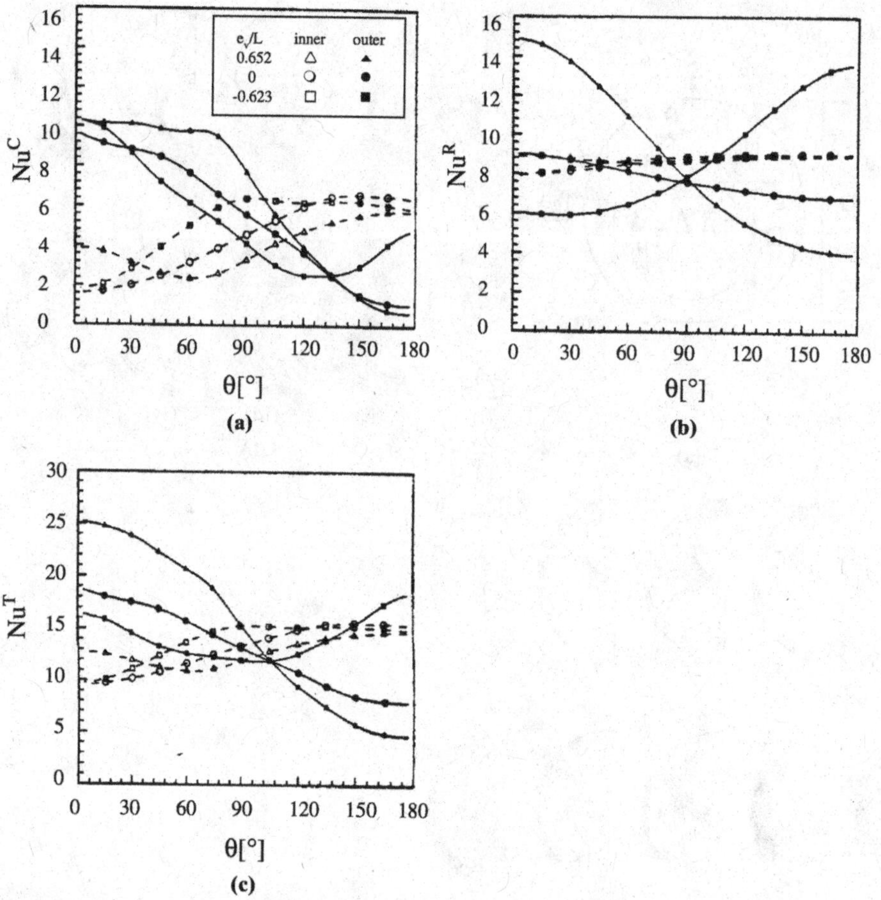

Figure 24.4.4.3 The effect of eccentricity on various Nusselt numbers for $N = 0.05$, $Ra = 1.5 \times 10^4$, $\delta = 1$, $\tau_0 = 0.3$, $\omega_0 = 0$, and black boundaries: **(a)** Nu^C, **(b)** Nu^R, and **(c)** Nu^T [Han and Baek, 1999].

the convective motion is suppressed because of its thermally stable configuration, with the total average heat transfer from the hot inner cylinder reduced for a fixed (*N conduction-radiation ratio*) when the inner cylinder is moved upward. For the case of negative eccentricity (Figure 24.4.4.2c), a temperature inversion in isotherms over the upper section of the inner cylinder is more prominent because of the stronger convective motion.

In Figure 24.4.4.3, the conductive, radiative, and total Nusselt number variations around the inner and outer cylinder walls are shown. It is seen that the conductive wall heat flux directed into the outer cylinder is dominant over the upper region, especially for the positive eccentricity.

24.4.5 THREE-DIMENSIONAL RADIATIVE HEAT FLUX INTEGRATION FORMULATION

Consider a divergent or convergent channel flow through two infinitely wide plates with an angle, δ, having different wall temperatures, as shown in Figure 24.4.5.1a [Kim, 1982]. It is assumed that the velocity profile of the channel flow is fully developed, laminar,

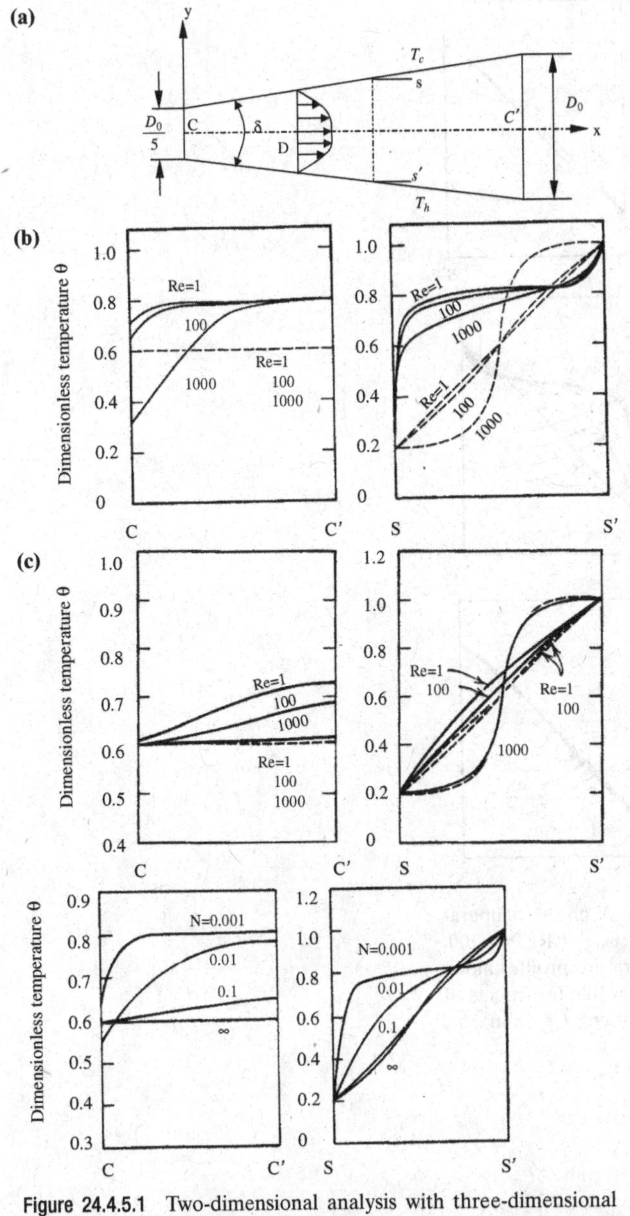

Figure 24.4.5.1 Two-dimensional analysis with three-dimensional heat flux formulation in combined conduction, convection, and radiation. **(a)** Geometry of two-dimensional radiation problem. **(b)** Effects of Reynolds number on the temperature profiles along $\overline{CC'}$ and $\overline{SS'}$. $\tau_0 = 1.0$; Pr = 1.0. (——) ω_0; (------) $\omega_0 = 1$. **(c)** Effects of Reynolds number on the temperature profiles along $\overline{CC'}$ and $\overline{SS'}$. $\tau_0 = 0.1$; $N = 0.001$; Pr = 1.0. (——) $\omega_0 = 0$; (---) $\omega_0 = 1$. **(d)** Effects of N on the temperature profiles along $\overline{CC'}$ and $\overline{SS'}$. $\tau_0 = 0.1$; $\omega_0 = 0$; Re Pr = 100.

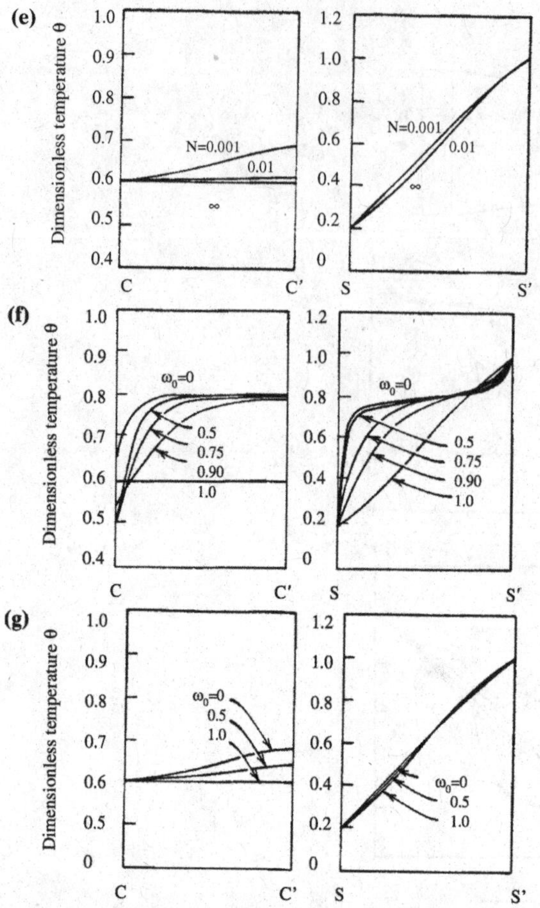

Figure 24.4.5.1 (*continued.*) (e) Effects of N on the temperature profiles along $\overline{CC}'$ and $\overline{SS}'$. $\tau_0 = 0.1$; $\omega_0 = 0$; Re Pr = 100. (f) Effects of albedo ω_0 on the temperature profiles along $\overline{CC}'$ and $\overline{SS}'$. $\tau_0 = 1.0$; $N = 0.00$; Re Pr = 100. (g) Effects of albedo ω_0 on the temperature profiles along $\overline{CC}'$ and $\overline{SS}'$. $\tau_0 = 0.1$; $N = 0.00$; Re Pr = 100.

and approximately given by

$$u(x, y) = \frac{3}{2} u_m(x) \left[1 - 4 \left(\frac{y}{D(x)} \right)^2 \right], \qquad u_m(x) = \frac{D_o}{D(x)} u_m(x - L)$$

where $u_m(x)$ and $D(x)$ are, respectively, the mean flow velocity and the channel width at a section, and D_o denotes the width at the channel exit. The upper and lower surfaces are also assumed to have uniform temperature T_c and T_h, respectively, and they are assumed to be black for simplicity. Furthermore, the inlet and outlet sections of the channel are assumed to be imaginary porous black surfaces, through which the flowing medium passes without any restrictions. The outlet mean velocity and channel width, D_o, are used as reference velocity and length, respectively, and the lower plate temperature, T_h, is assumed to be the reference temperature.

With the foregoing assumptions, the energy equation (24.3.129) and the equation (24.3.130) are solved simultaneously using the Galerkin finite elements. The boundary conditions are: $\theta = \theta_h$ on the lower surface, $\theta = \theta_c$ on the upper surface where $\theta_h = T_h/T_o$ and $\theta_c = T_c/T_o$. Note also that normal temperature gradients at the entrance and exit are set equal to zero.

In this example, it is assumed that $\theta_h = 1.0$, $\theta_c = 0.2$, and $\delta = 15$ deg. For given Reynolds and Prandtl numbers, the temperature distributions along the center line $\overline{CC'}$ and the middle section $\overline{SS'}$, are investigated for selected values of conduction-radiation ratio, N, optical thickness, τ_o, and albedo, ω_o.

A total of 72 linear two-dimensional isoparametric elements with 91 nodes are used in this example. An average of 6 iterations for the Newton-Raphson process was required for convergence with 0.1% error.

For the Prandtl number of unity, Figures 24.4.5.1b and 24.4.5.1c show the effects of the Reynolds number, Re, and the optical thickness, τ_o, on the temperature profiles for a small $N = 0.001$. It is noted that for pure scattering ($\omega_o = 1$), the temperature profile at the center line $\overline{CC'}$ is independent of Re and τ_o, while the profile at the middle section $\overline{SS'}$ is strongly dependent on the Reynolds number but not the optical thickness. If the medium does not scatter but only absorbs radiation ($\omega_o = 0$), the center line temperatures become close to $\theta = 0.8$ for a lower Reynolds number when it is optically thick ($\tau_o = 1.0$). However, they become close to $\theta = 0.6$ for a higher Re when optically thin ($\tau_o = 0.1$). It is also indicated that the middle section temperature profile becomes closer to a straight line as Re increases for $\tau_o = 1.0$ (optically thick) and $\omega_o = 0$, whereas the opposite is true in the case of $\omega_o = 1.0$.

On the other hand, if conduction energy transfer dominates over radiation (large N), there are very little effects of τ_o on the temperature profiles for $\omega_o = 0$, as noted in Figures 24.4.5.1d and 24.4.5.1e. Also, the figures indicate that the profiles strongly depend on N if it is optically thick, but the dependence on N is moderate for a small optical depth ($\tau_o = 0.1$). The same trend also appears in Figures 24.4.5.1f and 24.4.5.1g, which show the dependence of the temperature distributions on albedo, ω_o, for a low N. However, for a higher N the profiles converge to those for pure scattering.

It should also be noted that, for pure scattering, the temperature profile at the middle section, $\overline{SS'}$, is strongly dependent on the Reynolds number, but along the centerline, $\overline{CC'}$, it remains independent of Reynolds numbers since the temperatures for upper and lower boundaries are kept constant.

The finite element solution for the two-dimensional radiation flux combined with convection and conduction has been obtained. The following conclusions are reached: (1) Isoparametric finite elements offer advantages of easy integration for the two-dimensional radiation function involving the specular volume and specular surface elements through Gaussian quadrature. (2) For the diverging channel, for pure scattering, the temperature profile at the center line is independent of Reynolds number and optical thickness. In the absence of scattering, however, the middle section temperature profile becomes linear as the Reynolds number increases for large optical thickness. (3) If conduction energy transfer dominates over radiation, there are very little effects of optical thickness on the temperature profile in the absence of scattering. (4) For the converging channel, the radiation effect on the temperature profiles is small even when conduction and convection are small. (5) Standard Galerkin finite elements may be used if the convection domination is relatively small (Re Pr < 1000). However, for large Reynolds numbers (Re Pr >1000), it is concluded that the nonsymmetric form and ill-conditioning of the matrix from the convective terms would cause the solution to deteriorate. In this case, the optimal control penalty finite elements can be used to overcome such difficulties.

24.5 SUMMARY

The subject of radiative heat transfer in general is divided into the nonparticipating and participating media. Often, the term "radiation transfer" refers to the radiative heat transfer for the nonparticipating media. Both nonparticipating and participating media have been treated in this chapter.

The highlights of the radiative heat transfer in the nonparticipating media are the view factor calculations and the solution of radiative heat transfer equations. For the participating media, detailed formulations are presented for the combined conduction and radiation problems and combined conduction, convection, and radiation problems. A special feature in this chapter is the integral formulation of three-dimensional heat flux with respect to the optical thickness and spatial volumetric domain. Although the coverage is not complete, it is intended that the subject of the combined mode radiative heat transfer be self-contained.

Based on the theoretical foundations and numerical schemes presented in Sections 2 and 3, various example problems are demonstrated, including view factor calculations, solutions of radiative heat transfer equations, participating media with conduction and radiation, participating media with conduction, convection, and radiation, and the solution of the energy equation for the combined mode radiative heat transfer with the three-dimensional radiative heat flux integral equation. In these examples, FDM, FEM, and FVM are employed selectively.

REFERENCES

Baek, S. W., Kim, M. Y., and Kim, J. S. [1998]. Nonorthogonal finite volume solutions of radiative heat transfer in a three-dimensional enclosure. *Num. Heat Trans. Part B*, 34, 419–37.

Chung, T. J. [1988]. Integral and integrodifferential systems. In W. J. Minkowycz and E. M. Sparrow (eds.). *Handbook of Numerical Heat Transfer*, 579–624.

Chung, T. J. and Kim, J. Y. [1982]. Radiation view factors by finite elements. *J. Heat Trans.*, 104, 793–95.

———. [1984]. Two-dimensional, combined-mode heat transfer by conduction, convection and radiation in emitting, absorbing and scattering Media – solution by finite elements. *J. Heat Trans.*, 106, 448–52.

Fernandes, R. and Francis, J. [1982]. Combined conductive and radiative heat transfer in an absorbing, emitting, and scattering cylindrical medium. *ASME J. Heat Trans.*, 104, 594–601.

Fernandes, R., Francis, J., and Reddy, J. N. [1980]. A finite element approach to combined conductive and radiative heat transfer in a planar medium. AIAA 15th Thermophysics Conf., July 14–16, Snowmass, CO.

Goulard, R. [1962]. Fundamental equations of radiation gas dynamics, Purdue University School of Aero. and Eng. Sci. Report 62.

Haji-Sheikh, A. and Sparrow, E. M. [1967]. The solution of heat conduction problems by probability methods. *J. Heat Trans.*, 121–31.

Han, C. Y. and Baek, S. W. [1999]. Natural convection phenomena affected by radiation in concentric and eccentric horizontal cylindrical annuli. *Num. Heat Trans., Part A*, 36, 473–88.

Hutchinson, B. R., Stefurak, G., and Gerber, A. [1996]. Using the hemi-cube method to simulate automotive heating and cooling including solar and thermal radiation. Paper 960691, SAE Int. Congress & Exposition, Michigan.

Karki, K. C. and Patankar, S. V. [1989]. Pressure-based calculation procedure for viscous flows at all speeds in arbitrary configurations. *AIAA J.*, 27, 9, 1167–74.

Kim, J. Y. [1982]. New finite element applications in special problems in fluids and heat transfer. Ph.D. diss. The University of Alabama, Huntsville.

Raithby, G. D. [1999]. Discussion of the finite volume methods for radiation and its application using 3-D unstructured meshes. *Num. Heat Trans., Part B*, 35, 389–405.

Razzaque, M. M., Klein, D. E., and Howell, J. R [1982]. Finite element solution of radiative heat transfer in a two-dimensional rectangular enclosure with gray participating media. ASME Paper 82-WA/HT-51.

Siegel, R. and Howell, J. R. [1992]. *Thermal Heat Transfer.* Washington, D.C.: Hemisphere.

Sparrow, E. M. and Cess, R. D. [1970]. *Radiation Heat Transfer*. Belmont, CA: Brooks/Cole Publishing.

Utreja, L. R. [1982]. Radiative convective heat transfer in compressible boundary layers by optimal control penalty finite elements. Ph.D. diss. The University of Alabama, Huntsville.

Utreja, L. R. and Chung, T. J. [1989]. Combined convection-conduction-radiation boundary layer flows using optimal control penealty finite elements. *J. Heat Trans.*, 111, 433–37.

Viskanta, R. [1963]. Interaction of heat transfer by conduction, convection and radiation in a radiating fluid. *J. Heat Tran.*, 85, 318–29.

———. [1965]. Heat transfer by conduction and radiation in absorbing and scattering materials. *J. Heat Trans.*, 87, no. 1, 143–50.

Viskanta, R. and Grosh, R. J. [1962]. Heat transfer by simultaneous conduction and radiation in an absorbing medium. *Heat Trans.*, 84, 63–72.

CHAPTER TWENTY-FIVE

Applications to Multiphase Flows

25.1 GENERAL

Multiphase flow is a common observation such as occurs in evaporation or condensation in which a liquid particle is transformed into a gas or vice versa. Other examples include phase changes involved in the boiling of liquids, tracking of free surfaces between gas and liquid, or rocket solid propellants which, upon ignition, change into a liquid phase and subsequently into a gas phase. Furthermore, rigid body motions of solids in the presence of gases or liquids such as in sedimentation and fluidized beds, and reactive laminar and turbulent flows in fluid-particle mixtures, are the complicated physical phenomena in multiphase flows.

Interfaces are present as a specified initial condition or as a result of phase changes through evaporation, condensation, melting, solidification, merging (coalescence), or breakup. Surface tension plays an important role in these interfaces. Interface kinematics dealing with interface tracking in the free surface flows may be described and solved in many different ways. Among them are: volume tracking methods, front tracking methods, level set methods, phase field formulations, continuum advection schemes, boundary integral methods, particle-based methods, and moving mesh methods. A brief review of these methods is given below.

Volume tracking methods, often known as the volume of fluid (VOF) method, were originated by Nichols and Hirt [1975] and Noh and Woodward [1976], and further extended by Hirt and Nichols [1981]. Since then, the VOF method has been improved significantly over the years [Rudman, 1997; Rider and Kothe, 1998]. The VOF method is based on the conservation of the volume fraction function F with respect to time and space, expressed as

$$\frac{\partial F}{\partial t} + (\mathbf{v} \cdot \nabla) F = 0 \tag{25.1.1}$$

This method will be further elaborated in Section 25.2.

The basic idea of the front tracking methods resides in the original marker and cell (MAC) formulation [Harlow and Welch, 1966; Daly, 1967]. The interface is represented discretely by Lagrangian markers connected to form a front which lies within and moves through a stationary Eulerian mesh. As the front moves and deforms, interface points are added, deleted, and reconnected as necessary. Further details may be found in Glimm and McBryan [1985], Churn et al. [1986], and Tryggvason et al. [1998].

Level set methods have been successful as evidenced in the literature [Osher and Sethian, 1988; Sethian, 1996]. The basic premise of the level set method is to embed the propagating interface $\Gamma(t)$ as the zero level set of a higher dimensional function ϕ, defined as $\phi(x, t = 0) = \pm d$, where d is the distance from x to $\Gamma(t = 0)$, chosen to be positive (negative) if x is outside (inside) the initial $\Gamma(t = 0) = \phi(x, t = 0) = 0$, then a dynamical equation for $\phi(x, t)$ that contains the embedded motion for $\Gamma(t)$ as the level set $\phi = 0$ can be derived similarly as in the volume of fluid conservation equation (25.1.1).

Phase field formulations are applied predominantly to crystal growth problems and Hele-Shaw flows [Caginalp, 1989; Kobayashi, 1993; Wang et al., 1993; Wheeler, Murray, and Schaefer, 1993]. Applications to the Navier-Stokes system of equations have also been made recently [Antanovskii, 1995; Jacqumin, 1996]. In these formulations, interfacial forces are modeled as continuum forces by smoothing interface discontinuities and forces over thin but numerically resolvable layers. This smoothing allows conventional numerical approximations of interface kinematics on fixed grids.

In continuum advection schemes, the solution of (25.1.1) is carried out with schemes normally required for an hyperbolic system, as an integral part of the conventional fluid dynamics problems [Rider and Kothe, 1995; Chan, Pericleous, and Cross, 1991; Pericleous, Chan, and Cross, 1995]. High order approximations for the volume fraction function F will be the main factor for success.

Boundary integral methods are designed to track the interface explicitly, as in front tracking methods, although the flow solution in the entire domain is deduced solely from information possessed by discrete points along the interface [Geller, Lee, and Leal, 1986; Hou, 1995; Rallison and Acrivos, 1978]. The advantage of these methods is the reduction of the flow problem by one dimension involving quantities of the interface only.

Particle-based methods use discrete "particles" to represent macroscopic fluid parcels [Monaghan, 1985]. Here, Lagrangian coordinates are used to solve the Navier-Stokes system of equations on "particles" having properties such as mass, momentum, and energy. The nonlinear convection term is modeled simply as particle motion and by knowing the identity and position of each particle, material interfaces are automatically tracked. By using particle motion to approximated the convection terms, numerical diffusion across interfaces (where particles change identity) is virtually zero; hence interface widths are well defined. Particle-based methods may be categorized in two groups: (1) a scheme similar to particle-in-cell (PIC) methods [Harlow, 1988] and (2) meshless method [Belytschko et al., 1996] such as the smooth particle hydrodynamics (SPH) methods [Gingold and Monaghan, 1977; Monaghan, 1992].

In moving mesh methods, the position history of discrete points x_i lying on the interface is tracked for all time by integrating the evolution equation, forward in time.

$$\frac{dx_i}{dt} = \mathrm{v}_i \tag{25.1.2}$$

A moving mesh is Lagrangian if every point is moved, and mixed (Lagrangian-Eulerian) if grid points in a subset of the domain are moved. Mixed methods are used for mold filling simulations [Noh, 1964], where the mold computational domain can be held stationary and the molten liquid is followed with a Lagrangian mesh [Lewis, Navti, and Taylor, 1997; Muttin et al., 1993].

In other cases of multiphase problems such as rigid body motions in fluids or gases, the flowfield depends on the moments and products of inertia of the solid and torque acting on the surface of the solid. Hu, Joseph, and Crochet [1992] developed a numerical scheme using finite elements for simulating solid-liquid mixture motions of a few sedimenting circular and elliptic cylinders confined in a channel. This work was then extended to simulate a large number of solid particles with moving unstructured grids in the arbitrary Lagrangian-Eulerian (ALE) coordinates [Hu, 1995]. Glowinski et al. [1999] studied a distributed Lagrange multiplier/fictitious domain method for suspended solid particles in fluids. A finite element discretization in space and an operator-splitting technique of discretization in time were used in this analysis. Subsequently, Maury [1999] developed direct simulations of 2-D fluid-particle flows in biperiodic domains using the ALE finite elements. It was shown that this method provides long-time simulations of many-body motions (up to 5,000 particles).

Direct numerical simulation (DNS) of particle-turbulence was demonstrated by Pedinotti, Martiotti, and Banerjee [1992], Pan and Banerjee [1996, 1997], and Li, Mosyak, and Hetsroni [1999]. In these studies, the fluid flow in a horizontal channel is solved using DNS, whereas a Lagrangian approach is used for the particle motion.

If the liquid-gas mixture or solid-gas mixture system is reactive, then computations become complicated. The turbulent spray combustion discussed in Section 22.2.5 is the liquid-gas mixture flow [Kim and Chung, 1990]. In this chapter, we examine more broad and general approaches to the liquid-gas mixture and liquid-solid mixture flows [Smirnov, Nikitin, and Legros, 1997; Mashayek, Taulbee, and Givi, 1997; Udaykumar et al., 1997].

In Section 25.2, we discuss the volume of fluid (VOF) formulations with emphasis on surface tension using the continuum surface force (CSF). Laminar flows for the fluid-particle mixture with rigid boy motions of solids are presented in Section 25.3. Also included in this section are the turbulent flows and reactive turbulent flows in fluid-particle mixtures. Selected example problems are presented in Section 25.4.

25.2 VOLUME OF FLUID FORMULATION WITH CONTINUUM SURFACE FORCE

25.2.1 NAVIER-STOKES SYSTEM OF EQUATIONS

One of the most widely used approaches for simulating the liquid-gas phase interfaces subjected to surface tension is the volume of fluid (VOF) concept developed originally by Nichols and Hirt [1975] and others and subsequently extended by Blackbill, Kothe, and Zemach [1992] for implementation of continuum surface force (CSF) model. The VOF with CSF may be embedded into the conservation form of the Navier-Stokes system of equations as

$$\frac{\partial \mathbf{U}}{\partial t} + \frac{\partial \mathbf{F}_i}{\partial x_i} + \frac{\partial \mathbf{G}_i}{\partial x_i} = \mathbf{B} \tag{25.2.1}$$

with

$$\mathbf{U} = \begin{bmatrix} \rho \\ \rho F \\ \rho v_j \\ \rho E \end{bmatrix}, \quad \mathbf{F}_i = \begin{bmatrix} \rho \hat{v}_i \\ \rho \hat{v}_i F \\ \rho \hat{v}_i v_j + p\delta_{ij} \\ \rho \hat{v}_i E + p v_i \end{bmatrix}, \quad \mathbf{G}_i = \begin{bmatrix} 0 \\ 0 \\ -\tau_{ij} \\ -\tau_{ij} v_j + q_i \end{bmatrix}, \quad \mathbf{B} = \begin{bmatrix} 0 \\ 0 \\ \rho f_j \\ \rho f_j v_j \end{bmatrix}$$

with the second equation denoting the space conservation for the volume fraction F:

$$\frac{\partial}{\partial t} \int_\Omega \rho F d\Omega + \frac{\partial}{\partial x_i} \int_\Omega \rho F \hat{v}_i d\Omega = 0 \tag{25.2.2}$$

which represents the volume of fluid (VOF) [Nichols, Hirt, and Hotchkiss, 1980]. Here, the total velocity (partial velocity) v_i is the sum of the convection (Eulerian) velocity $\hat{v}_i$ and grid (Lagrangian) velocity $\overset{*}{v}_i$,

$$v_i = \hat{v}_i + \overset{*}{v}_i \tag{25.2.3}$$

This leads to the Lagrangian description for $\overset{*}{v}_i = v_i$ and the Eulerian description for $\overset{*}{v}_i = 0$. Thus, we have an arbitrary (mixed) Lagrangian-Eulerian (ALE) description for $\overset{*}{v}_i \neq 0$ and $v_i \neq \overset{*}{v}_i$. Initially, we set $\overset{*}{v}_i = 0$ and calculate v_i^{n+1} from (25.2.1). Then, calculate the grid displacement

$$u_i^{n+1} = u_i^n + \Delta t \left(v_i^{n+1} - v_i^n \right) \tag{25.2.4}$$

This provides the basis for the computation of $\overset{*}{v}{}_i^{n+1}$ as

$$\overset{*}{v}{}_i^{n+1} = \overset{*}{v}{}_i^n + \frac{u_i^{n+1} - u_i^n}{\Delta t} \tag{25.2.5}$$

The body force ρf_j consists of the gravity ρg_j and the volume force $Q_j(\Omega)$:

$$\rho f_j = \rho g_j + Q_j(\Omega) \tag{25.2.6}$$

Derivations of the volume force $Q_j(\Omega)$, which represents the contribution of surface tension between the liquid and gas interfaces, will be presented later in this section.

On the droplet-free surface we define the volume fraction F as

$$F = \frac{v^{(\ell)}}{v^{(\ell)} + v^{(g)}} \tag{25.2.7}$$

with the subscripts (g) and (ℓ) indicating gas and liquid, respectively. The total density ρ is given by

$$\rho = \rho^{(\ell)} F + \rho^{(g)} (1 - F) \tag{25.2.8}$$

In the formulation presented above, there are two options: The first is to remesh the grid network as dictated by the grid velocity and the grid displacement. The second option is to maintain the Eulerian coordinates and redefine the original element based on the volume fraction so that the element properties are updated for the subsequent time steps.

Multiphase interactions are usually studied in low-speed incompressible flows, although the Navier-Stokes system of equations (25.2.1) written in conservation form for generality, is capable of handling both incompressible and compressible flows. This is convenient for applications of the flowfield-dependent variation (FDV) method which will be discussed in Section 25.3.1.

25.2.2 SURFACE TENSION

The pressure p is the sum of the pressure due to surface tension (p_s) and the vapor pressure (p_v). This gives

$$p_s = p - p_v = \sigma\kappa \tag{25.2.9}$$

where σ is the surface tension and κ is the curvature,

$$\kappa = \frac{1}{R_1} + \frac{1}{R_2}$$

with R_1 and R_2 being the radii of curvature for a doubly curved surface.

The rate of change of force due to surface tension in a unit dimension x_j is given by

$$\int_\Gamma \frac{\partial \sigma}{\partial x_j} d\Gamma = \int_\Omega \frac{\partial}{\partial x_j}(p^{(g)} - p^{(\ell)} + \sigma\kappa)\, d\Omega - \int_\Omega \frac{\partial}{\partial x_i}\left(\tau_{ij}^{(g)} - \tau_{ij}^{(\ell)}\right) d\Omega \tag{25.2.10}$$

Integrating the right-hand side, we obtain

$$\int_\Gamma \frac{\partial \sigma}{\partial x_j} d\Gamma = \int_\Gamma (p^{(g)} - p^{(\ell)} + \sigma\kappa)\, n_j\, d\Gamma - \int_\Gamma \left(\tau_{ij}^{(g)} - \tau_{ij}^{(\ell)}\right) n_i\, d\Gamma \tag{25.2.11}$$

where n_j is the component of the unit normal vector on the surface. This leads to

$$\frac{\partial \sigma}{\partial x_j} = (p^{(g)} - p^{(\ell)} + \sigma\kappa) n_j - \left(\tau_{ij}^{(g)} - \tau_{ij}^{(\ell)}\right) n_i \tag{25.2.12}$$

Thus, for $p^{(g)} = p^{(\ell)}$ and $\tau_{ij}^{(g)} = \tau_{ij}^{(\ell)}$, we must have

$$\frac{\partial \sigma}{\partial x_j} = \sigma\kappa n_j \tag{25.2.13}$$

Let us now introduce the tangential vector component t_i so that (see Figure 25.2.1a),

$$\frac{\partial \sigma}{\partial x_i} t_i t_j = \frac{\partial \sigma}{\partial x_i}(\delta_{ij} - n_i n_j) \tag{25.2.14a}$$

or

$$\nabla_T \sigma = \nabla \sigma - \nabla_N \sigma \tag{25.2.14b}$$

with

$$\nabla_N \sigma = \mathbf{n}(\mathbf{n} \cdot \nabla)\sigma$$

$$\nabla_T \sigma = \mathbf{t}(\mathbf{t} \cdot \nabla)\sigma$$

Thus, the normal and tangential projections of (25.2.11) lead to the scalar pressure boundary conditions at the interface. Along the unit normal and tangential directions, the boundary conditions can be obtained by projecting (25.2.10) onto the normal

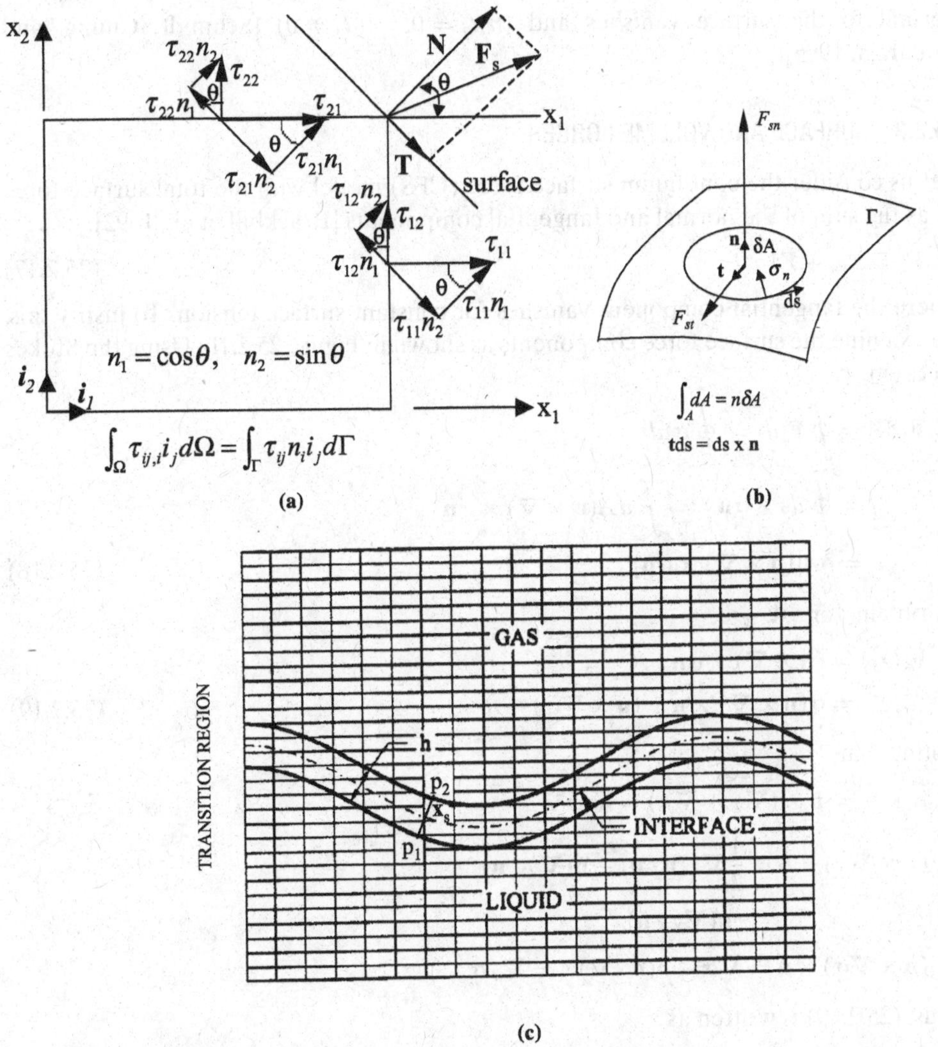

Figure 25.2.1 Surface tension interfaces. **(a)** Normal and tangential components of stress. **(b)** Normal and tangential surface components. **(c)** Transition region.

direction n_j and tangential direction t_j as follows:

Normal Direction

$$p^{(g)} - p^{(\ell)} + \sigma\kappa = \left(\tau_{ij}^{(g)} - \tau_{ij}^{(\ell)}\right)n_i n_j \tag{25.2.15}$$

Tangential Direction

$$\frac{\partial\sigma}{\partial x_j}t_j = \left(\tau_{ij}^{(\ell)} - \tau_{ij}^{(g)}\right)n_i t_j \tag{25.2.16}$$

with

$$\tau_{ij}^{(\alpha)} = \mu^{(\alpha)}\left(\frac{\partial \mathbf{v}_i^{(\alpha)}}{\partial x_j} + \frac{\partial \mathbf{v}_j^{(\alpha)}}{\partial x_i}\right)$$

These results are based on the fact that the spatial rate of change of surface tension normal to the surface vanishes and ($n_j t_j = 0, \quad n_i t_j \neq 0$) [Schmidt, Chung, and Nadaraja, 1995].

25.2.3 SURFACE AND VOLUME FORCES

Let us consider the continuum surface force (CFS) model with the total surface force $\mathbf{F}_s$ as the sum of the normal and tangential components [Brackbill et al., 1992],

$$\mathbf{F}_s = \mathbf{F}_{sn} + \mathbf{F}_{st} \tag{25.2.17}$$

where the tangential component vanishes for constant surface tension. To justify this, we examine the surface force components as shown in Figure 25.2.1b. Using the Stokes theorem,

$$\begin{aligned} \mathbf{F}_s \delta A &= \oint \mathbf{F}_s ds = \oint \sigma \mathbf{t} ds \\ &= \oint d\mathbf{s} \times \sigma \mathbf{n} = \iint dA(\mathbf{n} \times \nabla) \times \sigma \mathbf{n} \\ &= \delta A[(\mathbf{n} \times \nabla) \times \sigma \mathbf{n}] \end{aligned} \tag{25.2.18}$$

we obtain, for $\delta A \to 0$,

$$\begin{aligned} \mathbf{F}_s(x_s) &= (\mathbf{n} \times \nabla) \times \sigma \mathbf{n} \\ &= \sigma(\mathbf{n} \times \nabla) \times \mathbf{n} + (\mathbf{n} \times \nabla\sigma) \times \mathbf{n} \end{aligned} \tag{25.2.19}$$

Noting that

$$\mathbf{n} \times \nabla = \mathbf{n} \times (\nabla_T + \nabla_N) = \mathbf{n} \times \nabla_T$$

$$\begin{aligned} (\mathbf{n} \times \nabla_T) \times \mathbf{n} &= \frac{1}{2}\nabla_T(\mathbf{n} \cdot \mathbf{n}) - \mathbf{n}(\nabla_T \cdot \mathbf{n}) \\ &= -\mathbf{n}(\nabla_T \cdot \mathbf{n}) \end{aligned}$$

$$(\mathbf{n} \times \nabla\sigma) \times \mathbf{n} = \nabla\sigma - \mathbf{n}(\mathbf{n} \cdot \nabla)\sigma = \nabla_T \sigma$$

Thus, (25.2.19) is written as

$$\mathbf{F}_s(x_s) = -\sigma \mathbf{n}(\nabla_T \cdot \mathbf{n}) + \nabla_T \sigma \tag{25.2.20}$$

where we identify the normal and tangential components of the surface force as

$$\mathbf{F}_{sn} = -\sigma \mathbf{n}(\nabla_T \cdot \mathbf{n}) = -\sigma \kappa \mathbf{n}$$

$$\mathbf{F}_{st} = \nabla_T \sigma$$

where

$$\kappa = -\nabla_T \cdot \mathbf{n} = -\nabla \cdot \mathbf{n} \tag{25.2.21}$$

with the negative sign implying that the center of curvature is in the gas phase. The sign change will occur if it is in the liquid phase.

For a constant surface tension, the tangential component of the surface force vanishes, and we have

$$\mathbf{F}_s(x_s) = -\sigma \kappa \mathbf{n} \tag{25.2.22}$$

where x_s denotes the interface (Figure 25.2.1c).

To examine the volume force, let us first consider the density phase function $\rho(\mathbf{x})$ (characteristic function or color function) as follows (see Figure 25.2.1c):

$$\rho(x) = \begin{cases} \rho^{(g)} & \text{if in gas phase} \\ \rho^{(\ell)} & \text{if in liquid phase} \\ \langle\rho\rangle & \text{at the interface} \end{cases}$$

with

$$\langle\rho\rangle = \frac{1}{2}(\rho^{(g)} + \rho^{(\ell)})$$

The volume force $\mathbf{F}_v(\mathbf{x})$ is defined as

$$\lim_{h\to 0} \int_\Omega \mathbf{F}_v(\mathbf{x})\, d\Omega = \int_\Gamma \mathbf{F}_s(x_s)\, d\Gamma \tag{25.2.23}$$

with

$$\mathbf{F}_v(\mathbf{x}) = 0 \quad \text{for } |\mathbf{n}(x_s)\cdot(\mathbf{x}-\mathbf{x}_s)| \geq h$$

Consider the mollified density phase function $\tilde{\rho}(\mathbf{x})$ given by

$$\tilde{\rho}(x) = \frac{1}{h^3}\int_\Omega \rho(x')S(x'-x)d\Omega' \tag{25.2.24}$$

with

$$h^3 = \int_\Omega S(x)d\Omega$$

$$S(x) = 0 \quad \text{for } |x| \geq \frac{h}{2}$$

$$\lim_{h\to 0}\tilde{\rho}(x) = \rho(x)$$

where S is an interpolation function. Taking the gradient of $\tilde{\rho}(\mathbf{x})$ and denoting $[\rho] = \rho^{(\ell)} - g^{(g)}$, we have

$$\begin{aligned} \nabla\tilde{\rho}(\mathbf{x}) &= \frac{1}{h^3}\int_\Omega \rho(\mathbf{x}')\nabla\mathbf{S}(\mathbf{x}-\mathbf{x}')\,d\Omega' \\ &= \frac{[\rho]}{h^3}\int_\Gamma \mathbf{n}(\mathbf{x}_s)\,\mathbf{S}(\mathbf{x}-\mathbf{x}_s)\,d\Gamma \\ &\cong \frac{[\rho]}{h^3}\mathbf{n}(\mathbf{x}_{so})\int_\Gamma \mathbf{S}(\mathbf{x}-\mathbf{x}_s)\,d\Gamma + O\left(\frac{h}{R}\right)^2 \end{aligned} \tag{25.2.25}$$

where R is the radius of the curvature at $\mathbf{x}_{so}$ (surface point closest to $\mathbf{x}$) so that

$$\frac{1}{h^2}\int_\Gamma \mathbf{S}(\mathbf{x}-\mathbf{x}_s)\,d\Gamma \leq \mathbf{S}(\mathbf{x}-\mathbf{x}_{so}) \tag{25.2.26}$$

where the funtion $S(\mathbf{x}-\mathbf{x_s})$ plays the role of a delta function such that it is zero everywhere except at $\mathbf{x} = \mathbf{x_s}$. Thus, we have

$$\lim_{h\to 0}\int_\Gamma \mathbf{n}(\mathbf{x}_{so})\cdot\nabla\tilde{\rho}(\mathbf{x})d\Gamma = [\rho] \tag{25.2.27}$$

$$\lim_{h\to 0}\nabla\tilde{\rho}(\mathbf{x}) = \mathbf{n}[\rho]\,\delta\,[\mathbf{n}\cdot(\mathbf{x}-\mathbf{x}_s)] = \nabla\rho(\mathbf{x}) \tag{25.2.28}$$

where $\delta[\cdot]$ is the delta function. This gives

$$\int_\Gamma \mathbf{F}(\mathbf{x}_s)d\Gamma = \int_\Omega \mathbf{F}_s(\mathbf{x})\,\delta[\mathbf{n}(\mathbf{x}_s)\cdot(\mathbf{x}-\mathbf{x}_s)]d\Omega = \int_\Omega \sigma\kappa(\mathbf{x})\mathbf{n}(\mathbf{x})\,\delta[\mathbf{n}(\mathbf{x}_s)\cdot(\mathbf{x}-\mathbf{x}_s)]d\Omega \tag{25.2.29}$$

Substituting (25.2.28) into (25.2.29) yields

$$\int_\Gamma \mathbf{F}(\mathbf{x}_s)d\Gamma = \lim_{h\to 0}\int_\Omega \sigma\kappa(\mathbf{x})\frac{\nabla\tilde{\rho}(\mathbf{x})}{[\rho]}d\Omega \tag{25.2.30}$$

It follows from (25.2.23) and (25.2.30) that

$$\mathbf{F}_v(\mathbf{x}) = \sigma\kappa(\mathbf{x})\frac{\nabla\tilde{\rho}(\mathbf{x})}{[\rho]} \tag{25.2.31}$$

and

$$\int_{P_1}^{P_2} F_v(x)\,d\,(n\cdot x) = \int_{\rho^{(\ell)}}^{\rho^{(g)}} \sigma\kappa(x)\,\mathbf{n}(x)\frac{d\tilde{\rho}(x)}{[\rho]} \tag{25.2.32}$$

$$\cong \sigma\kappa(x_s)\hat{\mathbf{n}}(x_s) \quad \text{for } h>0$$

$$\lim_{h\to 0}\mathbf{F}_v(\mathbf{x}) = \mathbf{F}_s\,\delta\,[\mathbf{n}(\mathbf{x}_s)\cdot(\mathbf{x}-\mathbf{x}_s)] \tag{25.2.33}$$

Note that this is equivalent to the conventional definition

$$p_s = p^{(\ell)} - p^{(g)} = \sigma\kappa$$

25.2.4 IMPLEMENTATION OF VOLUME FORCE

The body force ρf_j consists of the gravity ρg_i and the volume force $Q_j(\Omega)$ as defined in (25.2.6). It follows from (25.2.30) and (25.2.31) that

$$\mathbf{Q}(\Omega) = \mathbf{F}_v(\mathbf{x}) = \sigma\kappa(\mathbf{x})\frac{\nabla\tilde{\rho}(\mathbf{x})}{[\rho]}\frac{\tilde{\rho}(\mathbf{x})}{\langle\rho\rangle} \tag{25.2.34}$$

where $\tilde{\rho}(\mathbf{x})/\langle\rho\rangle$ with $\langle\rho\rangle = (\rho^{(\ell)}+\rho^{(g)})/2$ is multiplied to the right-hand side of (25.2.34) to signify the process $h\to 0$, coinciding (25.2.31) at the interface

$$(\tilde{\rho}(\mathbf{x})/\langle\rho\rangle = 1).$$

Note that when the acceleration due to surface tension is independent of the density, neighboring contours in the transition region tend to remain a constant distance apart under the action of surface tension. Denser fluid elements in the transition region experience the same acceleration as lighter fluid elements when $\rho(\mathbf{x})/\langle\rho\rangle$ is included in $\mathbf{F}_v(\mathbf{x})$. Otherwise, the interface tends to thicken when $\mathbf{F}_v(\mathbf{x})$ is directed toward the fluid having the smaller density, and too thin when $\mathbf{F}_v(\mathbf{x})$ is directed toward the fluid having the lower density.

The unit normal vector $\mathbf{n}$ can be determined from the gradient of $\tilde{\rho}(\mathbf{x})$ as

$$\mathbf{n} = \frac{\nabla\tilde{\rho}(\mathbf{x})}{|\nabla\tilde{\rho}(\mathbf{x})|} \tag{25.2.35}$$

This definition with (25.2.21) leads to

$$\kappa \nabla \tilde{\rho} = -\nabla \tilde{\rho}(x)(\nabla \cdot \mathbf{n}) \tag{25.2.36}$$

where $\nabla \tilde{\rho}$ is nonzero only in the transition region and thus, the volume force is nonzero only in the transition region.

The curvature κ is calculated from

$$\kappa = -n_{i,i} = \tilde{\rho}_{,i} \frac{|\tilde{\rho}_{,i}|_{,i}}{|\tilde{\rho}_{,i}|^2} - \frac{\tilde{\rho}_{,ii}}{|\tilde{\rho}_{,i}|} \tag{25.2.37}$$

Since the volume fraction F as defined in (25.2.7) is indeed the mollified density phase function $\tilde{\rho}(\mathbf{x})$, we now set

$$\tilde{\rho}_{,j} = F_{,j}$$

with

$$[\rho] = 1 \quad \text{for} \quad \tilde{\rho} = F$$

and

$$\lim_{h \to 0} \int_\Omega Q_j(\Omega) d\Omega = \int_\Gamma Q_j(\Gamma) d\Gamma$$

with

$$Q_j(\Gamma) = \sigma \kappa n_j \tag{25.2.38a}$$

$$Q_j(\Omega) = \sigma \kappa \frac{\tilde{\rho}_{,j}}{[\rho]} \tag{25.2.38b}$$

Thus, the volume forces $Q_j(\Gamma)$ and $Q_j(\Omega)$ are given by

$$Q_j(\Gamma) = \sigma \left(\frac{F_{,ii}}{|F_{,k}|} - F_{,i} \frac{|F_{,k}|_{,i}}{|F_{,k}|^2} \right) n_j \tag{25.2.39a}$$

$$Q_j(\Omega) = \sigma \left(\frac{F_{,ii}}{|F_{,k}|} - F_{,i} \frac{|F_{,k}|_{,i}}{|F_{,k}|^2} \right) \frac{\tilde{\rho}_{,j}}{[\rho]} \tag{25.2.39b}$$

or

$$Q_j(\Omega) = \sigma \left(\frac{F_{,ii}}{|F_{,k}|} - F_{,i} \frac{|F_{,k}|_{,i}}{|F_{,k}|^2} \right) F_{,j} \tag{25.2.39c}$$

with

$$[\rho] = 1 \quad \text{for } \tilde{\rho} = F$$

The expression (25.2.39c) is now inserted into (25.2.6), which will then complete the Navier-Stokes system of equations (25.2.1).

25.2.5 COMPUTATIONAL STRATEGIES

We consider that the convection and diffusion process, as well as the distribution of body forces in two-phase flows, are very much flowfield dependent. To this end, we follow the

FDV formulation as described in Section 13.6. To derive convection, diffusion, diffusion gradient, and source term Jacobians we define the various conservation flow variables as

$$\rho F = \ell, \qquad \rho u = m_1, \qquad \rho \mathrm{v} = m_2, \qquad \rho \hat{u} = \hat{m}_1$$
$$\rho \overset{*}{u} = \overset{*}{m}_1, \qquad \rho \hat{\mathrm{v}} = \hat{m}_2, \qquad \rho \overset{*}{\mathrm{v}} = \overset{*}{m}_2, \qquad \rho E = e \tag{25.2.40}$$

These definitions lead to the conservation variables, convection flux variables, diffusion flux variables, and the source terms,

$$\mathbf{U} = \begin{bmatrix} \rho \\ \ell \\ m_1 \\ m_2 \\ e \end{bmatrix} \qquad \mathbf{F}_1 = \begin{bmatrix} \hat{m}_1 \\ \dfrac{\ell \hat{m}_1}{\rho} \\ \dfrac{\hat{m}_1 m_1}{\rho} + p \\ \dfrac{\hat{m}_1 m_2}{\rho} \\ \dfrac{e\hat{m}_1 + p m_1}{\rho} \end{bmatrix} \qquad \mathbf{F}_2 = \begin{bmatrix} \hat{m}_2 \\ \dfrac{\ell \hat{m}_2}{\rho} \\ \dfrac{\hat{m}_2 m_1}{\rho} \\ \dfrac{\hat{m}_2 m_2}{\rho} + p \\ \dfrac{e\hat{m}_2 + p m_2}{\rho} \end{bmatrix}$$

$$\mathbf{G}_1 = \begin{bmatrix} 0 \\ 0 \\ -\tau_{11} \\ -\tau_{12} \\ -\tau_{11}\mathrm{v}_1 - \tau_{12}\mathrm{v}_2 + q_1 \end{bmatrix} \qquad \mathbf{G}_2 = \begin{bmatrix} 0 \\ 0 \\ -\tau_{21} \\ -\tau_{22} \\ -\tau_{21}\mathrm{v}_1 - \tau_{22}\mathrm{v}_2 + q_2 \end{bmatrix}$$

$$\mathbf{B} = \begin{bmatrix} 0 \\ 0 \\ \rho f_1 \\ \rho f_2 \\ \rho f_1 \mathrm{v}_1 + \rho f_2 \mathrm{v}_2 \end{bmatrix}$$

with

$$\rho f_j = \rho g_j + Q_j \qquad Q_j = \sigma\left(\frac{F_{,jj}}{|F_{,k}|} - F_{,j}\frac{|F_{,k}|_{,j}}{|F_{,k}|^2}\right)F_{,j}$$

$$\begin{aligned} \tau_{ij} &= \mu\left(\frac{\partial \mathrm{v}_i}{\partial x_j} + \frac{\partial \mathrm{v}_j}{\partial x_i}\right) - \frac{2}{3}\mu\delta_{ij}\frac{\partial \mathrm{v}_k}{\partial x_k} \\ &= \tau_{ij}^{(\ell)} F + \tau_{ij}^{(g)}(1-F) \\ &= \left[\mu^{(\ell)}(\mathrm{v}_{i,j} + \mathrm{v}_{j,i}) - \frac{2}{3}\mu^{(\ell)}\mathrm{v}_{k,k}\delta_{ij}\right]F + \left[\mu^{(g)}(\mathrm{v}_{i,j} + \mathrm{v}_{j,i}) - \frac{2}{3}\mu^{(g)}\mathrm{v}_{k,k}\delta_{ij}\right](1-F) \end{aligned}$$

$$p = (\gamma - 1)\left[e - \frac{1}{2\rho}(m_1^2 + m_2^2)\right]$$

$$T = T^{(\ell)}F + T^{(g)}(1-F)$$

These governing equations can be solved using FDM, FEM, or FVM. In general, the following solution steps may be followed.

(1) Solve the Navier-Stokes system of equations given by (25.2.1) with initial and boundary conditions.
(2) Calculate u_i^{n+1} from (25.2.4).
(3) Calculate $\overset{*}{\mathrm{v}}_i^{n+1}$ from (25.2.5).
(4) Repeat Steps (2) through (4) with updated values.
(5) Repeat until steady state is reached.

Notice that in the above solution process, the treatment of the transition region and the interface are the most critical aspects of the two phase flow problems. To this end, the determination of the deformed surface curvature and subsequently the volume force is important.

The mixed Lagrangian-Eulerian treatments given by (25.2.3) through (25.2.5) allow the mesh movement for remeshing at each time step. However, it is possible to retain the Eulerian coordinates if so desired by setting $\mathrm{v}_i = \hat{\mathrm{v}}_i$. In this case, the value of the volume fraction F alone will determine the phase.

The formulations presented in the previous sections have been implemented in Brackbill et al., 1992 and Kothe and Mjolsness, 1992 for the solutions of jet-induced tank mixing and water rod collision using finite differences. In their solutions, pressure corrections processes through the pressure Poisson equations are implemented together with solutions of the incompressible momentum equations as described in Section 5.3.

Flowfield-Dependent Variation (FDV) Method

Difficulties in the analysis of two phase flows include determination of the effect of any one variable upon another. In the liquid-gas system, we are concerned with the process of how the effects of surface tension on temperature, pressure, density, velocity, and volume changes of the liquid and gas can be properly taken into account. In this regard, formulations required in the FDV theory (Section 6.5 and Section 13.6) address these concerns through the FDV parameters as well as the various Jacobians of convection, diffusion, and source terms. The most critical ones in two-phase flows are the source term Jacobians.

$$\mathbf{d} = \frac{\partial \mathbf{B}}{\partial \mathbf{U}}$$

Explicit forms of the source term Jacobians are presented in Appendix C.

25.3 FLUID-PARTICLE MIXTURE FLOWS

25.3.1 LAMINAR FLOWS IN FLUID-PARTICLE MIXTURE WITH RIGID BODY MOTIONS OF SOLIDS

There are many practical applications of fluid-particle mixture flows in which rigid body motions of solid particles are important, such as in sedimentation or fluidized beds. Here, the surface tension is considered insignificant. Instead, rigid-body motions of solids require the moments and products of inertia and torque to be taken into account.

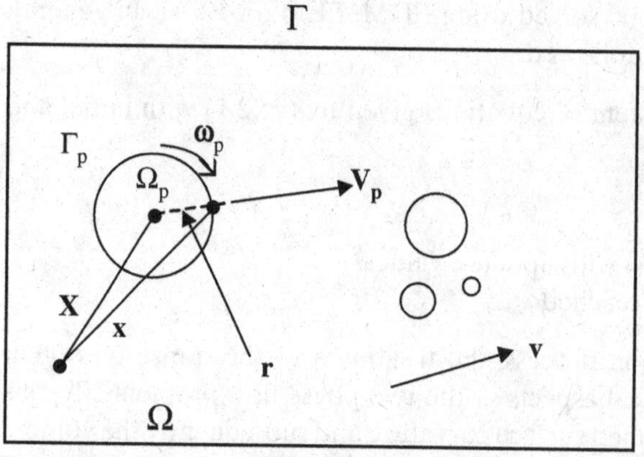

Figure 25.3.1 Fluid-particle mixture flow with solid particles undergoing rigid-body motions.

In addition to the governing equations for fluids, the momentum equations for particles and their kinematic equations are needed. Let v_{pi} be the translational velocity of the pth particle in the direction i (see Figure 25.3.1). The equations for momentum and torque are given by, respectively,

$$M_{ji}\frac{dv_{pi}}{dt} = F_{pj} + G_{pj} \tag{25.3.1}$$

$$I_{ji}\frac{d\omega_{pi}}{dt} + (\omega \times I_{ji}\omega_i)_p = T_{pj} \tag{25.3.2}$$

where M_{ji} is the 3×3 diagonal mass matrix, F_{pj} is the force imposed on the particle by the fluid (hydrodynamic force), G_{pj} is the body force such as gravity, I_{ji} is the 3×3 matrix of moments and products of inertia of the particle, and T_{pj} is the torque imposed on the particle by the fluid. The translational velocity is related by the generalized position X_{pi} as

$$v_{pi} = \frac{dX_{pi}}{dt} \tag{25.3.3}$$

The fluid velocity is expressed as the sum of the velocity components due to translation and rotation through an angular velocity ω, for a particle at $\mathbf{r} = \mathbf{x} - \mathbf{X}$,

$$\mathbf{v} = \mathbf{v}_p + \omega \times \mathbf{r}$$

or written as a component form

$$v_i = v_{pi} + (\varepsilon_{ijm}\omega_j r_m)_p \quad \text{on } \Gamma_p \quad (p = 1, 2, \ldots, N) \tag{25.3.4}$$

with N denoting the number of particles and

$$\omega_{pi} = \frac{d\Theta_{pi}}{dt} \tag{25.3.5}$$

where Θ_{pi} is the angular orientation of the pth particle.

The hydrodynamic forces and torques are given by

$$F_{pj} = \int_{\Gamma_p} \sigma_{ij} n_i d\Gamma_p \tag{25.3.6}$$

$$T_{pj} = \int_{\Gamma_p} (\mathbf{r} \times \sigma_{im} n_i \mathbf{i}_m)_i d\Gamma_p \tag{25.3.7}$$

with

$$\sigma_{ij} n_i \mathbf{i}_j = \mathbf{s} \quad \text{on} \quad \Gamma_p (p = 1, 2, \ldots, N)$$

It should be noted that for 2-D the second term on the left-hand side in (25.3.2) vanishes.

In order to utilize the above equations of momentum for the particle motion, it is useful to examine the nonconservation form of the momentum equations for fluids of incompressible flow together with the continuity equation in the ALE coordinates.

$$\rho_f \left(\frac{\partial \mathbf{v}}{\partial t} + (\hat{\mathbf{v}} \cdot \nabla)\mathbf{v} \right) = \sigma_{ij,i} \mathbf{i}_j + \rho_f \mathbf{g} \tag{25.3.8a}$$

$$\nabla \cdot \hat{\mathbf{v}} = \mathbf{0} \tag{25.3.8b}$$

The finite element formulation of (25.3.8) and integration by parts lead to the traction boundary conditions as the sum of the contribution of both domain wall surfaces and particle surfaces,

$$\int_{\Gamma} \sigma_{ij} n_i d\Gamma = \int_{\Gamma_w} \sigma_{ij} n_i d\Gamma_w + \int_{\Gamma_p} \sigma_{ij} n_i d\Gamma_p \tag{25.3.9}$$

Notice that the second term on the right-hand side of (25.3.9) is identical to the sum of the hydrodynamic force and torque given in (25.3.6). Thus, when the finite element equations for (25.3.8a,b) and the momentum and torque equations (25.3.1, 25.3.2) are constructed and combined, it is seen that the surface traction force acting on the particle surface and the hydrodynamic force of the particle are cancelled out.

It should be realized that during the computation, particles may collide with each other or with the wall. In order to prevent this, there are a number of options that have been reported in the literature. Among them are: Collision repulsive model [Glowinski et al., 1999], which adds a fictitious force to the particle momentum equations to prevent collision; inelastic restitution model [Johnson and Yezduyar, 1996], which monitors the conservation of linear momentum at the contact surface; thin liquid film gap model [Hu, 1996], which provides a thin layer of mesh around the particle that is fixed to the particle surface, moving together with the particle; and coupled variational formulation [Maury, 1999], which leads to a symmetric linear system.

In general, the computational procedure can be described as follows:

(1) Introduce initial mesh.
(2) Initialize $\mathbf{v}(\mathbf{x_0}, 0)$, $p(\mathbf{x},0)$, $\mathbf{X_p}(0)$, $\mathbf{v}_p(0)$ for $p = 1, 2 \ldots, N$.
(3) Select time step Δt^{n+1} and solve the fluid momentum and continuity equations without particles.
(4) Introduce the particles into the flowfield.

(5) Solve the particle momentum and torque equations at $t^{n+1} = t^n + \Delta t^{n+1}$ from Step (3).
(6) Update particle position, $\mathbf{X}^{n+1} = \mathbf{X}^n + \Delta t^{n+1}\mathbf{v}^{\mathbf{n}}_{pi}$.
(7) Update mesh nodes, $\mathbf{x}^{n+1} = \mathbf{x}^n + \Delta t^{n+1}\mathbf{v}^{\mathbf{n}}$
(8) Remesh and project (if the mesh distortion is severe).
    (a) Generate a new mesh.
    (b) Project the flowfield data onto the new mesh.
(9) Return to the fluid momentum and continuity equations and repeat the process until convergence.

Some selected numerical examples are presented in Section 25.4.2.

25.3.2 TURBULENT FLOWS IN FLUID-PARTICLE MIXTURE

We have seen the complexity of turbulence in Chapter 21. Thus, it is of interest to examine interactions of turbulence with particle-laden flows. Experimental data indicate that the addition of particles may increase or decrease the turbulent kinetic energy of the carrier fluid. The presence of small particles in isotropic turbulence reduces the turbulent kinetic energy, whereas the opposite is true for larger particles [Hetsroni and Sokolov, 1971; Parthasarathy and Faeth, 1987]. However, for anisotropic turbulence, Vinberg, Zaichick, and Pershukov [1991] showed that the addition of small particles can enhance turbulence. For coarse particles, the level of fluctuations is determined by vortex shedding and turbulent kinetic energy depends on the drag coefficient [Yarin and Hetsroni, 1993].

Turbulence is also affected by the one-way coupling or two-way coupling. In one-way coupling, the fluid moves the particles, but there is no feedback from the particles on the fluid motion. Pedinotti et al. [1992] carried out DNS analysis, assuming that the particle concentration is low enough to allow the use of one-way coupling. However, some turbulence mechanisms may be significantly influenced by both particle-particle and particles-wall interactions so that the two-way coupling must be considered [Hetsroni and Rozenblit, 1994; Li, et al., 1999].

If the effect of surface tension and rotational force (torque) is negligible, the equation of motion of a particle is simpler than in the cases examined in the previous sections. In this case, the approximate form of the equation for the motion of a single particle is

$$\frac{dv_{pi}}{dt} = \frac{f}{\tau_p}(\hat{v}_i - v_{pi}) \tag{25.3.10}$$

where τ_p is the particle time constant for Stokesian drag of a spherical particle.

$$\tau_p = \frac{\rho_p d_p^2}{18\mu} \tag{25.3.11}$$

where d_p is the particle diameter. The function f is an empirical correction to Stokesian drag for large particle Reynolds number,

$$\mathrm{Re}_p = \frac{\hat{\rho}_f d_p^2 |\hat{v}_i - v_{pi}|}{\mu} \tag{25.3.12a}$$

$$f = 1 + 0.15\mathrm{Re}_p^{0.687} \tag{25.3.12b}$$

where the symbol ^ denotes the values of the fluid variables at the particle location. The particle Reynolds number valid in (25.3.12) and (25.3.13) is $\mathrm{Re}_p \leq 1000$.

Taking into account the effects of particles moving parallel and perpendicular to a wall [Kim and Karrila, 1991], the particle equations of motion may be written in the form

$$m_p \frac{d\mathbf{v}}{dt} = \frac{3}{4} C_D \frac{\rho}{\rho_p} \frac{1}{d_p} |\hat{\mathbf{v}} - \mathbf{v}_p| h(H) + m_f \frac{D\hat{\mathbf{v}}}{Dt} + \frac{1}{2} m_f \frac{d}{dt}(\hat{\mathbf{v}} - \mathbf{v}_p) + (m_p - m_f)\mathbf{g} \tag{25.3.13}$$

with

$$C_D = \frac{24}{\mathrm{Re}_p} + \frac{6}{1 + \mathrm{Re}_p^{0.5}} + 0.4, \qquad 0 \leq \mathrm{Re}_p \leq 2 \cdot 10^5, \qquad \mathrm{Re}_p = \frac{|\hat{\mathbf{v}} - \mathbf{v}_p| d_p}{\nu}$$

$$h(H) = \frac{1}{1 - \frac{9}{16}\frac{d_p}{2H} + \frac{1}{8}\left(\frac{d_p}{2H}\right)^3} \quad \text{for a particle moving parallel to a wall}$$

$$h(H) = \frac{1}{1 - \frac{9}{8}\frac{d_p}{2H} + \frac{1}{8}\left(\frac{d_p}{2H}\right)^3} \quad \text{for a particle moving perpendicular to a wall}$$

where H is the height of the wall.

The corresponding momentum equations for the fluid are the same as (25.3.8a) except that we subtract on the right-hand side of (25.3.8a) the momentum source term, W_j, representing the effects of the particle drag which is calculated by volume averaging the contributions from all of the individual particles within the cell volume.

$$\rho_f \left[\frac{\partial \mathrm{v}_j}{\partial t} + (\mathrm{v}_i \mathrm{v}_j)_{,i} \right] = -p_{,j} + \tau_{ij,i} + \rho_f g_j - W_j \tag{25.3.14}$$

with

$$W_j = \frac{1}{\rho_f V} \sum_{p=1}^{N_p} \left(\frac{f m_p}{\tau_p} (\hat{\mathrm{v}}_j - \mathrm{v}_{pj}) \right) \tag{25.3.15}$$

where V is the cell volume, the particle mass is $m_p = \pi\rho_p d_p^3/6$, and N_p is the number of particles within the cell volume.

Based on the governing equations above, a number of investigators studied DNS solutions [Mashayek et al., 1997; Li et al., 1999] and LES [Hansell, Kennedy, and Kollman, 1992], among others. Mashayek et al. [1997] developed algebraic Reynolds stress models for two phase flows. Comparisons with DNS calculations show reasonable agreements. The procedure of development of algebraic models for two-phase flows is similar to the derivation of algebraic stress model of a single-phase flow reported by Taulbee [1992].

25.3.3 REACTIVE TURBULENT FLOWS IN FLUID-PARTICLE MIXTURE

The laminar flows in fluid-particle mixture discussed in the previous section can be extended to include chemical reactions such as in Smirnov [1988], van der Wel et al.

[1993], Eckhoff [1994], and Smirnov et al. [1997], among others. Again, the Eulerian frame for the gas and Lagrangian coordinates for particles are the preferred approach. In most practical problems in engineering, the fluid is air and the particle is the condensed phase consisting of either liquid droplet or minute solid particles such as dust. The spray combustion presented in Section 22.2.5 is an example of reactive turbulent flows in fluid-particle mixture in which liquid fuel droplets are considered as the condensed phase.

As in the laminar flow, there are two types of models for particle-laden gas flows: one-way coupling and two-way coupling. In turbulent flows, the two-way coupling becomes significant, with the high rates of mass and energy fluxes from the particles in combustion process which may cause major changes in the flowfield. This is particularly important in dust explosion phenomena.

The conventional RANS models proved to be satisfactory for a homogeneous system, but unsuitable for heterogeneous polydispersed phases due to uncertainties involved in modeling process and instability in the nature of turbulent flows. Thus, it is advantageous to use deterministic methods such as DNS in order to examine adequacy of RANS models.

Governing Equations for the Gas Phase

Let the volume fraction of gas be given by α such that the volume fraction of particles is $(1-\alpha)$. The governing equations using the K–ε model are written as

$$\frac{\partial \mathbf{U}}{\partial t}+\frac{\partial \mathbf{F}_i}{\partial x_i}+\frac{\partial \mathbf{G}_i}{\partial x_i}=\mathbf{B} \tag{25.3.16}$$

with

$$\mathrm{U}=\begin{bmatrix}\overline{\alpha\rho}\\ \overline{\alpha\rho}\tilde{\mathrm{v}}_j\\ \overline{\alpha\rho}\tilde{E}\\ \overline{\alpha\rho}K\\ \overline{\alpha\rho}\varepsilon\\ \overline{\alpha\rho}\tilde{Y}_k\end{bmatrix}\qquad \mathrm{F}_i=\begin{bmatrix}\overline{\alpha\rho}\tilde{\mathrm{v}}_i\\ \overline{\alpha\rho}\tilde{\mathrm{v}}_i\tilde{\mathrm{v}}_j+\overline{\alpha}\overline{p}\delta_{ij}\\ \overline{\alpha\rho}\tilde{\mathrm{v}}_i\tilde{E}+\overline{\alpha}\overline{p}\tilde{\mathrm{v}}_i\\ \overline{\alpha\rho}\tilde{\mathrm{v}}_iK\\ \overline{\alpha\rho}\tilde{\mathrm{v}}_i\varepsilon\\ \overline{\alpha\rho}\tilde{\mathrm{v}}_i\tilde{Y}_k\end{bmatrix}$$

$$\mathrm{G}_i=\begin{bmatrix}0\\ -\tau_{ij}-\tau_{ij}^*\\ -\overline{\tau}_{ij}\overline{\mathrm{v}}_j-\tau_{ij}^*\tilde{\mathrm{v}}_j+\overline{q}_i+q_i^*\\ \overline{\alpha\rho}(\nu+\frac{\nu^*}{\sigma_K})K_{,i}+\tau_{ij}^*\tilde{\mathrm{v}}_j\\ \overline{\alpha\rho}(\nu+\frac{\nu^*}{\sigma_\varepsilon})K_{,i}+\frac{\varepsilon}{K}C_{1\varepsilon}\tau_{ij}^*\tilde{\mathrm{v}}_j\\ \overline{\alpha\rho}\tilde{Y}_{K,i}+\overline{\alpha\rho}\mathrm{v}_i''Y_{k,i}''\end{bmatrix}\qquad \mathbf{B}=\begin{bmatrix}\overline{\overline{M}}\\ \overline{\alpha\rho}g_j+\overline{\overline{N}}_j\\ \overline{\alpha\rho}g_j\mathrm{v}_j+\overline{\overline{Q}}\\ -\overline{\alpha\rho}\varepsilon\\ -C_{2\varepsilon}\overline{\alpha\rho}\varepsilon\\ \overline{\omega}_k+\overline{\overline{M}}\end{bmatrix}$$

$$\tau_{ij}''=-\overline{\rho}\widetilde{\mathrm{v}_i''\mathrm{v}_j''}=\overline{\rho}\nu^*\left(\tilde{\mathrm{v}}_{i,j}+\tilde{\mathrm{v}}_{j,i}-\frac{2}{3}\tilde{\mathrm{v}}_{k,k}\delta_{ij}\right)-\frac{2}{3}\overline{\rho}K\delta_{ij}$$

$$\tilde{E}=\sum_k Y_k\big(c_{vk}\tilde{T}+H_k^0\big)+\frac{1}{2}\tilde{\mathrm{v}}_i\tilde{\mathrm{v}}_i$$

$$q_i^*=\overline{\alpha\rho}\sum_k c_{pk}\tilde{Y}_k\widetilde{\mathrm{v}_i''T''}+\sum_k\big(c_{pk}\tilde{T}+H_k^0\big)\overline{\alpha\rho}\widetilde{\mathrm{v}_i''Y_k}$$

where $\overline{\overline{M}}$ is the mass flux per unit volume from the other phases, $\overline{\overline{N}}_j$ is the momentum flux from other components and phases, and $\overline{\overline{Q}}$ is the energy flux from other phases.

Mathematical Model for Particle Phase

A stochastic approach may be used to describe the motion of polyhedral particles with a group of representative variables such as the mass and velocities of model particles involved. The equations of motion and energy balance of kth particles are

$$m_k \frac{d\mathbf{V}_k}{dt} = m_k \mathbf{g} + \mathbf{F_k} - \frac{m_k}{\rho_k} \nabla p, \quad \frac{d\mathbf{r}}{dt} = \mathbf{V_k} \tag{25.3.17}$$

$$m_k \frac{de_k}{dt} = q_k + \hat{Q}_k \tag{25.3.18}$$

where e_k is the specific internal energy and $\hat{Q}_k$ is the heat release or absorption on the particle surface due to chemistry or phase transition.

Particle mass depletion is determined by

$$\frac{dm_k}{dt} = \sum_k \dot{m}_{kj} \tag{25.3.19}$$

The variations of particle radius and volume are calculated from the depletion of the skeleton component instead of the total particle mass in terms of suitable probability density functions for particle radius distributions. As a result, the extraction of volatiles can cause the decay of particle mean density, resulting in longer flotation in the atmosphere.

Mathematical Modeling of Phase Interactions

The two-way coupling can be modeled by determining the mass, momentum, and energy fluxes between model particles and the surrounding gas. Subsequently, total fluxes from the particle phase to the gas on the basis of the statistical processing can be evaluated.

Mass exchange processes between particles and gas occur as a result of phase transition from evaporation or condensation on the surface of liquid droplets, devolatization of dust particles, chemical reactions on the interface, etc. In addition to volatiles extraction, an overall reaction is assumed to take place on a particle's surface:

$$C + \frac{1}{2} O_2 \rightarrow CO \tag{25.3.20}$$

Thus, two components from the gas phase (O_2 and CO) take part in the reaction and the generalized volatiles component can be extracted. Finally, fluxes $\overline{\overline{M}}_k$, $\overline{\overline{M}}$, $\overline{\overline{N}}_j$, $\overline{\overline{Q}}$ to gas phase as well as the volume fraction of particles $(1-\alpha)$ are calculated by evaluating the corresponding fluxes from the volume of model particles. It is assumed that the volatiles L extracted from organic dust consist of

$$L = (O_2, CO, CO_2, H_2O, N_2, CH_4, H_2, NH_3)$$

There are two overall reactions assumed to control chemical transformation:

$$L + \nu_{O_2} O_2 \rightarrow \nu_{CO} CO + \nu_{H_2O} H_2O + \nu_{CO} CO_2 + \nu_{N_2} N_2$$

$$2CO + O_2 \Leftrightarrow 2CO_2$$

The values of stoichiometric coefficients in the volatile oxidation reaction are calculated according to the concentrations of components in volatiles. Further details can be found in Smirnov [1988] and Smirnov et al. [1997].

25.4 EXAMPLE PROBLEMS

25.4.1 LAMINAR FLOWS IN FLUID-PARTICLE MIXTURE

In this example, we discuss finite element calculations for laminar flow in fluid-particle mixture with effects of rigid body motions of solid particles upon fluid flows as reported by Maury [1999]. Both translational forces and torques acting on the surface of particles are taken into account. Variational finite element equations are derived from governing equations given in Section 25.3.1; nonuniform biperiodic unstructured meshes of domains with holes are generated.

The main feature in this formulation is the average behavior of a large number of particles. An extra term is added to the pressure, representing the Lagrange multiplier associated with the verticle volume conservation constraint.

Following the ALE approach and the method of characteristics, the time-discretization turns out to be a generalized Stokes problem. A suitable variational formulation leads to a symmetric system involving all the unknowns which are then solved, using the conjugate gradient Uzawa algorithm [Elman and Golub, 1994].

Figure 25.4.1a shows the boundary of the periodic window containing 1,000 2-D elliptical particles of various sizes. The biperiodic unstructured triangular mesh corresponding to the selected zone representing the left side block is shown in

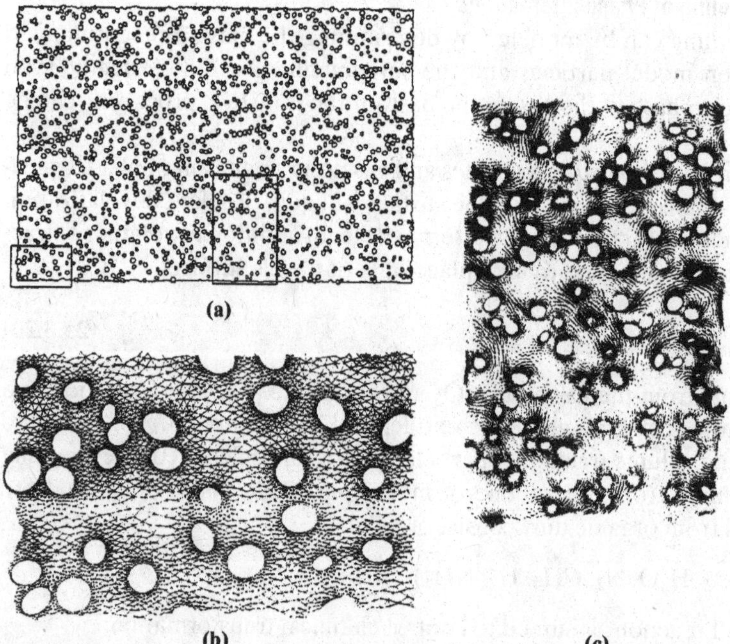

Figure 25.4.1 Laminar flow in fluid-particle mixture [Maury, 1999]. (a) Mesh boundary. (b) Mesh detail, left block in (a). (c) Velocity field, right block in (a).

Figure 25.4.1b. The computed velocity field for the zone designated by the right side block is demonstrated in Figure 25.4.1c. Further details are shown in Maury [1999].

25.4.2 TURBULENT FLOWS IN FLUID-PARTICLE MIXTURE

We present in this example the results of direct numerical simulation (DNS) of interactions between solid particles and near-wall turbulence as reported by Li et al. [1999]. The flowfield of horizontal channel is solved using a Lagrangian approach for the particle motion. Two-way coupling is used to account for the effect of the particles on the structure of the near-wall turbulence, and on the mainstream.

In this analysis, the vorticity transport equations (12.2.10) and the curl of the vorticity transport equations given by (12.2.11) are used for fluid motions. The rigid-body particle motion includes translational force without rotational torque. However, the particle equations of motion as given in (25.3.10) are modified to include the effects of height of the channel H with respect to the particle size and additional temporal rate of changes of flowfield [Kim and Karrila, 1991]. Computational procedures are as follows:

(1) The flowfield is calculated to the steady state without particles using the spectral method. Equations (12.2.10) and (12.2.11) are used with the solution expanded to the finite Fourier series in the x_1, x_2 directions and Chebyshev polynomials to the normal direction x_3 (see Chapter 14).
(2) Introduce the particles into this flowfield and calculate their motions with one-way coupling using (25.3.13). The particles are considered as points at this time and allowed to reach stationary distribution. However, the particles are relatively large and each covers a number of collocation points. The fluid velocity at various locations in the particle is averaged with a three-dimensional cubic spline interpolation scheme and applied on the particle through (25.3.13).
(3) In order to implement the two-way coupling, all the velocities in the collocation points occupied by the particle are set equal using (25.3.14).
(4) Iterations between Step (2) and Step (3) continue until convergence.

Figure 25.4.2a shows the geometry for this example. The calculations are carried out in a computational domin of $1074 \times 537 \times 171$ wall units in the x_1, x_2, and x_3 directions with a resolution of $128 \times 128 \times 129$. The density of particle is 1050 kg. The turbulent Reynolds number $Re^* = 2hv^*/\nu = 85.4$, the corresponding bulk Reynolds number $Re = 2hU/\nu = 2600$ with $h = 37$ mm are used.

Figure 25.4.2b shows the distribution of particles (dimensionless diameter $d^+ = 8.5$) in the x_1, x_2 plane, compared with the experimental data (Figure 25.4.2c). It is seen that the tendency of particles to agglomerate into the streaks depends on the particle size and flow conditions. In Figure 25.4.2d, distributions of particles in the x_1, x_3 plane and x_2, x_3 plane are shown. Although not shown, the coarser particles affect the velocity fluctuations of the carrier fluid significantly. Turbulent intensity and Reynolds stresses are increased considerably as the particle size increases. Further details are presented in Li et al. [1999].

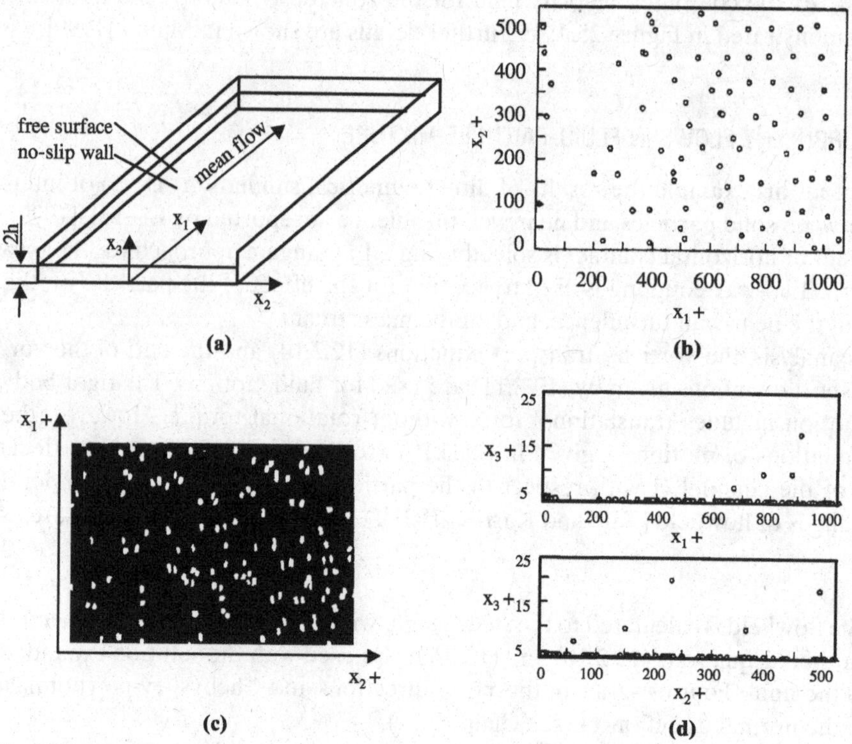

Figure 25.4.2 Turbulent flow in fluid-particle mixture [Li et al., 1999]. **(a)** Flow geometry. **(b)** Distribution of particles ($d+ = 8.5$), numerical simulation, x_1-x_2 plane. **(c)** Distribution of particles, experimental data, x_1-x_2 plane. **(d)** Distribution of particles ($d+ = 8.5$), numerical simulation, x_1-x_3 and x_2-x_3 planes.

25.4.3 REACTIVE TURBULENT FLOWS IN FLUID-PARTICLE MIXTURE

This example shows thermogravitational instability in large-scale combustion of dispersed dust-air mixtures and its contribution to turbulence as studied by Smirnov et al. [1997]. Two-way coupling effects in gas-particle interactions and combination of both deterministic and stochastic approaches are demonstrated.

The $K - \varepsilon$ approximations are used to calculate the gas phase flow with the account for mass, momentum, and energy fluxes from the particle's phase. The equations of motion for particles take into consideration those turbulent fluctuations in the gas flow. The models for phase transitions and chemical reactions accommodate thermal destruction of dust particles, volatilization, chemical reactions in the gas phase, and heterogeneous oxidation of particles. The influence of inert and chemically reacting particles on the flowfield induced by heating from below and by sedimentation is adequately resolved. The related equations for this analysis are presented in Section 25.3.3.

The computational domain is 2000 m in the horizontal direction and 1000 m in the vertical direction. Initial and boundary conditions are: $U = 10\,\text{m/s}$, $p_0 = 1.013 \cdot 10^5$ Pa,

$$T_0 = 280\text{K}, \qquad Q = 1.67 \cdot 10^5\ \text{W/m}^3, \qquad \alpha_{s0} = 0.0005,$$
$$K_0 = 0.1\ \text{m}^2/\text{s}^2, \qquad \varepsilon_0\, 0.01\ \text{m}^2/\text{s}^3.$$

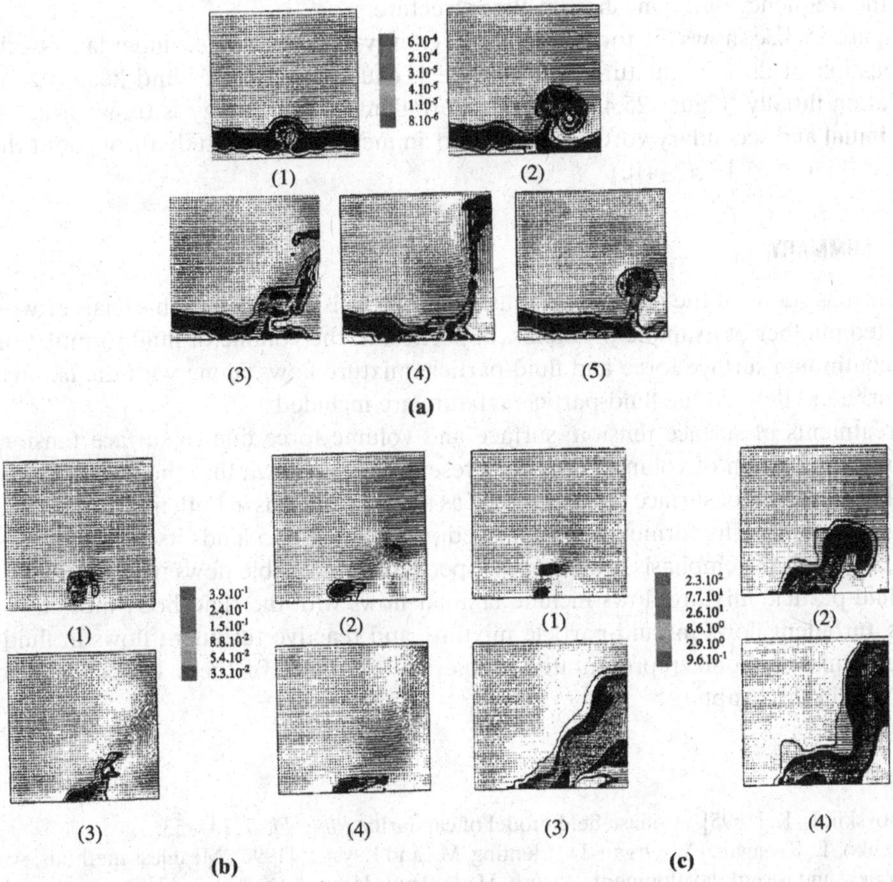

Figure 25.4.3 Reactive turbulent flow in fluid-particle mixture [Smirnov et al., 1997]. **(a)** Dust volume share distribution in combustion of air-dust mixtures over the heat source. (1) t = 21s, (2) 52, (3) 88, (4) 137, (5) 542. **(b)** Volatiles oxidation intensity in combustion of air-dust mixtures over the heat source. (1) t = 21s, (2) 52, (3) 88, (4) 137. **(c)** Turbulent kinetic energy distribution in the gas phase under large scale combustion of air-dust mixtures. (1) t = 21s, (2) 52, (3) 88, (4) 137.

Figure 25.4.3a shows the dust volume share distribution in combustion of air-dust mixtures over the heat source. Here the particles are pushed aside by the upgoing blob of hot air above the zone of heat release [Figure 25.4.3a(1)]. Particles are then lifted up by the vortices created and form a well-known mushroom-type structure that is deformed due to the wind [Figure 25.4.3a(2)]. The new ignition and combustion of particles delivered into the heated zone causes the formation of another upgoing blob of heated dust-air mixture [Figure 25.4.3a(3)]. The second blob is then sucked into the primary vortices and rises further above [Figure 25.4.3a(4)]. Meanwhile, a new ignition of fresh dust-air mixture begins and the process is repeated [Figure 25.4.3a(5)].

Volatiles oxidation intensity in combustion of air-dust mixtures over heat source is presented in Figure 25.4.3b. It is seen that the initial ignition [Figure 25.4.3b(1)] is followed by several hot spots transported by vortices [Figure 25.4.3b(2)]. The main reaction zone oscillates due to periodic contractions and expansions of the heat zone

[Figure 25.4.3b(3,4)]. The characteristic frequency of these oscillations is much lower than the frequency of the mushroom-type structure.

Figure 25.4.3c shows the turbulent kinetic energy in the gas phase under large-scale combustion of dust-air mixtures. Turbulence is caused by vortices and heated zone oscillation intially [Figure 25.4.3c(1)]. The turbulent kinetic energy is transported by both initial and secondary vortices, leading to an increased magnitude throughout the process [Figure 25.4.3c(2–4)].

25.5 SUMMARY

The current status of the research in multiphase flows is reviewed in this chapter with a limited number of example problems. In particular, the volume of fluid formulation with continuum surface force and fluid-particle mixture flows, along with the laminar and turbulent flows in the fluid-particle mixture are included.

Treatments of surface tension, surface and volume force due to surface tension, and implementation of volume force are presented. It is shown that the volume force calculated in terms of surface tension is used as the source terms in both momentum and energy equations. The formulation suggested in Section 25.2.5 lends itself to all speed flows, although the emphasis is on the low-speed incompressible flows in general.

Fluid-particle mixture flows include laminar flows with the rigid-body motions of solids, turbulent flows in fluid-particle mixture, and reactive turbulent flows in fluid-particle mixture. Some representative numerical examples of these topics are also examined in this chapter.

REFERENCES

Antanovskii, L. K. [1995]. A phase field model of capilarity. *Phys. Fl.*, 7, 747–53.

Belytschko, T., Krongauz, Y., Organ, D., Fleming, M., and Ksysl, P. [1996]. Meshless methods: An overview and recent developments. *Comp. Meth. Appl. Mech. Eng.*, 139, 3–47.

Blackbill, J. U., Kothe, D. B., and Zemach, C. [1992]. A continuum method for modeling surface tension. *J. Comp. Phys.*, 100, 335–54.

Caginalp, G. [1989]. Stefan and Hele-Shaw models as asymptonic limits of the phase-field equations. *Phys. Rev. A.*, 39, 5887.

Chan, K. S., Pericleous, K. A., and Cross, M. [1991]. Numerical simulation of flows encountered during mold-fitting. *Appl. Math. Mod.*, 15, 624–31.

Churn, I. L., Glimm, J., McBtyan, O., Plohr, B., and Yanic, S. [1986]. Front tracking for gas dynamics. *J Comp. Phys.*, 62, 83–110.

Daly, B. J. [1967]. Numerical study of two fluid Rayleigh-Taylor instabilities. *Phys. Fl.*, 10, 297–307.

Eckhoff, R. K. [1994]. Prevention and mitigation of dust explosions in process industries: A survey of recent research and development. Proc. 6th Colloquium (International) on Dust Explosions, Shenyan, 5–34.

Elman, H. C. and Golub, G. [1994]. Inexact and preconditioned Uzawa algorithms for saddle point problems. *SIAA J. Num. Anal.*, 31, 6, 1645–52.

Geller, A. S., Lee, S. H., and Leal, L. G. [1986]. Motion of a particle normal to a deformable surface. *J. Fl. Mech.*, 169, 27–69.

Gingold, R. A. and Monaghan, J. J. [1977]. Smoothed particle hydrodynamics: Theory and application to non-spherical stars. *Monthly Notices of Royal Astron. Soc.*, 181, 375–35.

Glimm, J. and McBryan, O. A. [1985]. A computational model for interfaces. *Adv. Appl. Math.*, 6, 422–35.

Glowinski, R., Pan, T. W., Hesla, T. I., and Joseph, D. D. [1999]. A distributed Lagrange multiplier/fictitious domain method for particulate flows. *Int. J. Multiphase Flow.*, 25, 755–94.

Hansell, D., Kennedy, I. M., and Kollmann, W. [1992]. On the Lagrangian simulation of turbulent jet. *Phys. Fl.*, 3, 11, 2571–79.

Harlow, F. H. [1988]. PIC and its progeny. *Comp. Phys. Commun.* 48, 1–11.

Harlow, F. H. and Welch, J. E. [1966]. Numerical study of large-amplitude free-surface motion. *Phys. Fl.*, 9, 842–51.

Hetsroni, G. and Rozenblit, R. [1994]. Heat transfer to a liquid-solid mixture in a flume. *Int. J. Multiphase Flow.*, 20, 671–89.

Hetsroni, G. and Sokolov, M. [1971]. Distribution of mass, velocity, and intensity of turbulence in two-phase jet. *J. Appl. Mech.*, 38, 315–27.

Hirt, C. W. and Nichols, B. D. [1981]. Volume of fluid (VOF) method for the dynamics of free boundaries. *J. Comp. Phys.*, 39, 201–25.

Hou, T. Y. [1995]. Numerical solutions to free boundary problems. *Acta Num.*, 335–415.

Hu, H. H. [1995]. Direct simulation of flows of solid-liquid mixtures. *Int. J. Multiphase Flows.*, 22, 2, 335–52.

Hu, H. H., Joseph, D. D., and Crochet, M. J. [1992]. Direct simulation of fluid particle motions. *Theor. Comp. Fl. Dyn.*, 3, 285–306.

Jacqumin, D. [1996]. An energy approach to the continuum surface method. AIAA paper, 96-0858.

Johnson, A. A. and Yezduyar, T. [1996]. Simulation of multiple spheres falling in a liquid-filled tube. *Comp. Meth. Appl. Mech. Eng.*, 134, 351–73.

Kim, Y. M. and Chung, T. J. [1990]. Finite element model for turbulent spray combustion. AIAA paper, 90-0359.

Kim, S. and Karrila, S. [1991]. *Microhydrodynamics*. London: Butterworth.

Kobayashi, R. [1993]. Modeling and numerical simulations of dentric crystal growth. *Physica D.* 63, 410–23.

Kothe, D. B. and Mjosness, R. C. [1992]. RIPPLE: A new model for incompressible flows with free surface. *AIAA J.*, 30, 2694–2700.

Lewis, R. W., Navti, S. E., and Taylor C. [1997]. A mixed Lagrangian-Eulerian approach to modeling fluid flow during mould filling. *Int. J. Num. Meth. Fl.*, 25, 931–52.

Li, C., Mosyak, A., and Hetsroni, G. [1999]. Direct numerical simulation of particle-turbulence interaction. *Int. J. Multiphase Flow.*, 25, 187–200.

Mashayek, F., Taulbee, D. B., and Givi, P. [1997]. Statistical modeling and direct simulation of two-phase turbulent flows. In G. D. Roy, S. M. Frolov, and P. Givi (ed.) Advanced Computation & Analysis of Combustion, Moscow: ENAS Publishers, 161–74.

Maury, B. [1999]. Direct simulations of 2D fluid-particle flows in biperiodic domains. *J. Comp. Phys.*, 156, 325–51.

Monaghan, J. J. [1985]. Particles methods for hydrodynamics. *Comp. Phys. Rep.*, 3, 71–124.

———. [1992]. Smoothed particle hydrodynamics. *Ann. Rev. Astron. Astrophys.*, 30, 543–74.

Muttin, F., Coupez, T., Bellet, M., and Chenot, J. L. C. [1993]. Lagrangian finite element analysis of time dependent viscous free surface flow using an automatic remeshing technique. *Int. J. Num. Meth. Eng.*, 36, 2001–15.

Nichols, B. D. and Hirt, C. W. [1975]. Methods for calculating multi-dimensional, transient free surface flows past bodies. Technical Report LA-UR-75-1932, Los Alamos National Laboratory, NM.

Nichols, B. D., Hirt, C. W., and Hotchkiss, R. S. [1980]. SOLA-VOF: A solution algorithm for transient fluid flow with multiple free boundaries. Los Alamos National La., LA-8355, NM.

Noh, W. F. [1964]. CEL: A time-dependent, two-space-dependent, coupled Euler-Lagrange code. *Meth. Comp. Phys.*, 3, 117–79.

Noh, W. F. and Woodward, P. R. [1976]. SLIC (simple line interface methods). In A. I. Vande Voore and P. J. Zandbergen (ed.), *Lecture Notes in Physics*, 59, 330–40.

Osher, S. and Sethian, J. A. [1988]. Fronts propagating with curvature-dependent speed: Algorithms based on Hamilton-Jacobi formulation. *J. Comp. Phys.*, 79, 12–49.

Pan, Y. and Banerjee, S. [1996]. Numerical simulation of particle interactions with turbulence. *Phys. Fl.*, 8, 2733–35.

———. [1997]. Numerical investigation of the effects of large particles on wall-turbulence. *Phys. Fl.*, 9, 3786–07.

Parthasarathy, R. N. and Faeth, G. M. [1987]. Structure of a turbulent particle-laden water jet in still water. *Int. J. Multiphase Flow.*, 13, 699–714.

Pedinotti, S., Martiotti, G., and Banerjee, S. [1992]. Direct numerical simulation of particle behavior in the wall region of turbulent flows in horizontal channels. *Int. J. Multiphase Flow.*, 18, 927–41.

Pericleous, K. A., Chan, K. S., and Cross, M. [1995]. Free surface flow and heat transfer in cavities: The SEA algorithm. *Num. Heat Trans.,* Part B, 27, 487–507.

Rallison, J. M. and Acrivos, A. [1978]. A numerical study of the deformation and burst of a viscous drop in an extensional flow. *J. Fl. Mech.*, 89, 191–200.

Rider, W. J. and Kothe, D. B. [1995]. A marker particle method for interface tracking . In H. A. Dwyer (ed.), *Proceedings of the Sixth International Synposium on Computational Dynamics*, Davis, CA, 976–81.

Rider, W. J. and Kothe, D. B. [1998]. Reconstructing volume tracking. *J. Comp. Phys.*, 141, 112–52.

Rudman, M. [1997]. Volume tracking methods for interfacial flow calculations. *Int. J. Num. Meth. Fl.*, 24, 671–91.

Schmidt, G. R., Chung, T. J., and Nadaraja, A. [1995]. Thermocapillary flow with evaporation and condensation at low gravity. Part I, Non-deforming surface, Part II, deformable surface. *J. Fl. Mech.*, 294, 323–66.

Sethian, J. A. [1996]. *Level Set Methods*. London: Cambridge University Press.

Smirnov, N. N. [1988]. Combustion and detonation in multiphase media. Initiation of detonation in dispersed systems behind a shock wave. *Int. J. Heat`Mass Trans.*, 31, 4, 779–93.

Smirnov, N. N., Nikitin, V. F., and Legros, J. C. [1997]. Turbulent combustion of multiphase gas-particles mixtures. Thermodynamical instability. In G. D. Roy, S. M. Frolov, and P. Givi (eds.), *Advanced Computation & Analysis of Combustion*, Moscow: ENAS Publishers, 136–60.

Taulbee, D. B. [1992]. Improved algebraic Reynolds stress model and corresponding nonlinear stress model. *Phys. Fl. A.*, 4, 2555–61.

Tryggvason, G., Bunner, B., Ebrat, O., and Tauber, W. [1998]. Computations of multi-phase flow by a finite difference/front tracking method. I. Multi-fluid flows. In *Lecture Notes for the 29th Computational Fluid Dynamics Lecture Series*, Karman Institute for Fluid Mechanics, Belgium.

Udaykumar, H. S., Shyy, W., Segal, S., and Pai S. [1997]. Phase change characteristics of energetic solid fuels in turbulent reacting flows. In G. D. Roy, S. M. Frolov, and P. Givi (eds.), *Advanced Computation & Combustion*, Moscow: ENAS Publishers, 195–207.

Van der Wel, P. G. J., Lemkowitz, S. M., Timmers, P., and Scarlett, B. [1993]. The role of turbulence on the propagation mechanisms and behavior or dust explosions. Proc. 5th Coloquium (International) on Dust Explosions, Pultusk, 199–209.

Vinberg, A. A., Zaichick, L. I., and Pershukov, V. A. [1991]. Computational model for turbulent gas-particles jet streams. *J. Eng. Phys.*, 61, 554–63.

Wang, S. L., Sekerka, R. F., Wheeler, A. A., Murray, B. T., Coriell, S. R., and McFadden, G. B. [1993]. Thermodynamically consistent phase-field model for solidification. *Physica D.*, 69, 189–200.

Wheeler, A. A., Murray, B. T., and Schaefer R. J. [1993]. Computation of dendrites using a phasefield model. *Physica. D.*, 66, 243–62.

Yarin, L. P. and Hetsroni, G. [1993]. Turbulent intensity in dilute two-phase flow. *Int. J. Multiphase Flow.*, 20, 27–44.

CHAPTER TWENTY-SIX

Applications to Electromagnetic Flows

In this chapter, computations involved in electromagnetism are discussed, including magnetohydrodynamics, rarefied gas dynamics, and plasma dynamics. To deal with these physical phenomena, Maxwell equations and Boltzmann equations are introduced. It is shown how these equations are solved separately and together with the standard fluid dynamics equations.

Section 26.1 introduces all governing equations involved in electromagnetism, followed by solutions of Boltzmann equation using the BGK model discussed in Section 26.2. We discuss in Section 26.3 semiconductor plasma processing, including charged particle kinetics in plasma discharge, discharge modeling with moment equations, and reactor model for chemical vapor deposition. In Section 26.4, some applications are presented, including magnetohydrodynamic flows in coronal mass ejection and various aspects of plasma processing in semiconductors.

26.1 MAGNETOHYDRODYNAMICS

Magnetohydrodynamics (MHD) deals with the motion of a highly conducting fluid in the presence of a magnetic field. Such a motion generates electric currents which change the magnetic field, and the disturbed field in turn gives rise to mechanical forces which affect the flowfield. This coupling between the electromagnetic and mechanical forces then characterizes hydromagnetic phenomena. Celestial bodies which contain large conducting masses are known to exhibit pronounced hydrodynamic phenomena.

The electromagnetic field is produced by a distribution of electric current and charge. The motion of charge constitutes a current that is determined by the magnitude of the charge and velocity. The current density at a point is defined as the vector **J** by the equation

$$\mathbf{J} = \rho \mathbf{v} \tag{26.1.1}$$

where ρ is the charge density and $\mathbf{v}$ the velocity vector. It follows that in metals and valves, where the electricity is carried by electrons that are negatively charged, the direction of the current density vector is opposite to that of the moving electrons.

The current I across a surface is defined to be the rate at which a charge crosses that surface. Since a charge can cross S only by virtue of its velocity normal to S, we have

$$I = \mathbf{J} \cdot \mathbf{n} dS \tag{26.1.2}$$

where $\mathbf{n}$ is a unit vector normal to S.

Consider now two isolated charges e and e_1 moving in free space. The charge e is acted on by certain electrical forces due to e_1. If e is at rest, the electrical force is $e\mathbf{E}$. The vector $\mathbf{E}$ is called the electric intensity. If e is moving with velocity $\mathbf{v}$, there is an additional force $e\mathbf{v} \times \mathbf{B}$ where the vector $\mathbf{B}$ is called the magnetic flux density. Two other vectors play a role in specifying the electromagnetic field, and they are related to the lines of force which emanate from charge and currents. The vector $\mathbf{D}$, which is called the electric flux density, effectively measures the number of lines of force which originate from a charge. The vector $\mathbf{H}$, which is called the magnetic intensity, is such that its value on a closed curve effectively measures the current which passes through the curve.

We shall assume that the vectors $\mathbf{E}$, $\mathbf{B}$, $\mathbf{D}$, and $\mathbf{H}$ are continuous and possess continuous derivatives at ordinary points at which Maxwell's equations

$$\boldsymbol{\nabla} \times \mathbf{E} + \frac{\partial \mathbf{B}}{\partial t} = 0 \tag{26.1.3a}$$

$$\boldsymbol{\nabla} \times \mathbf{H} - \frac{\partial \mathbf{D}}{\partial t} = \mathbf{J} \tag{26.1.3b}$$

$$\boldsymbol{\nabla} \cdot \mathbf{B} = 0 \tag{26.1.3c}$$

$$\boldsymbol{\nabla} \cdot \mathbf{D} = \rho \tag{26.1.3d}$$

are satisfied. Since the divergence of the curl of any vector vanishes identically, we obtain, by taking the divergence of (26.1.3b)

$$\boldsymbol{\nabla} \cdot \mathbf{J} = -\boldsymbol{\nabla} \cdot \frac{\partial \mathbf{D}}{\partial t} = -\frac{\partial}{\partial t}(\boldsymbol{\nabla} \cdot \mathbf{D}) \tag{26.1.4}$$

Substitution of (26.1.3b) into (26.1.4) gives

$$\frac{\partial \rho}{\partial t} + \boldsymbol{\nabla} \cdot \mathbf{J} = 0 \tag{26.1.5}$$

By analogy with a corresponding equation in hydrodynamics, (26.1.5) is called the equation of continuity.

In a field of infinite electrical conductivity, the fluid particles are tied to the lines of force of the magnetic field so that the lines of force may be thought of as possessing inertia, the mass per unit length equal to the density of the fluid ρ.

To describe the magnetohydrodynamic behavior, we must have: (1) the mechanical equations embodying the effect of the electromagnetic forces as well as other forces on the motion, (2) the equation at continuity, (3) the equation of heat transport, and (4) the equation of state as well as the Maxwell equations given in (26.1.3).

Consider a viscous fluid in motion in which the only body forces are gravity and electromagnetic forces. The equation of motion can be written as

$$\rho \frac{\partial \mathbf{v}}{\partial t} + \rho(\mathbf{v} \cdot \boldsymbol{\nabla})\mathbf{v} = -\boldsymbol{\nabla} p + \rho \mathbf{g} + \mathbf{J} \times \mathbf{B} + e\mathbf{E} + \mu \left[\boldsymbol{\nabla}^2 \mathbf{v} + \frac{1}{3} \boldsymbol{\nabla}(\boldsymbol{\nabla} \cdot \mathbf{v}) \right] \tag{26.1.6}$$

in which

$$\mathbf{J} \times \mathbf{B} = (\nabla \times \mathbf{B}) \times \mathbf{B} \tag{26.1.7}$$

The equations of continuity, energy, and state are the same as given in Chapter 2 except that the product of velocity with the magnetic and electric forces must be added to the energy equation. The condition to be satisfied at a fluid-fluid boundary or fluid-vacuum boundary can be obtained by integration of the relevant equation across a thin stratum coinciding with the surface.

The new variables introduced in the Maxwell's equations and the momentum equations may be combined to form the electromagnetic Navier-Stokes system of equations as follows:

$$\frac{\partial \mathbf{U}}{\partial t} + \frac{\partial \mathbf{F}_i}{\partial x_i} + \frac{\partial \mathbf{G}_i}{\partial x_i} = \mathbf{S} \tag{26.1.8}$$

with

$$\mathbf{U} = \begin{bmatrix} \rho \\ \rho \mathrm{v}_j \\ \rho \hat{E} \\ B_j \end{bmatrix} \quad \mathbf{F}_i = \begin{bmatrix} \rho \mathrm{v}_i \\ \rho \mathrm{v}_i \mathrm{v}_j + p^* \delta_{ij} \\ (\rho \hat{E} + p^*)\mathrm{v}_i \\ \mathrm{v} B_j - \mathrm{v}_j B_i \end{bmatrix} \quad \mathbf{G}_i = \begin{bmatrix} 0 \\ -\tau^*{}_{ij} \\ -\tau^*{}_{ij}\mathrm{v}_j + q_i \\ \frac{1}{\mu_0 \sigma} B_{j,i} \end{bmatrix} \quad \mathbf{S} = \begin{bmatrix} 0 \\ \rho F_j \\ \rho F_i \mathrm{v}_i + E_i J_i \\ 0 \end{bmatrix} \tag{26.1.9}$$

$$P^* = P + \frac{1}{2\mu_0} B_k B_k \tag{26.1.10a}$$

$$\tau_{ij}^* = \tau_{ij} + \tau_{ij}^{(m)} \tag{26.1.10b}$$

$$\tau_{ij} = 2\mu\left(d_{ij} - \frac{1}{3} d_{kk}\right) \tag{26.1.10c}$$

$$\tau_{ij}^{(m)} = \frac{1}{\mu_0} B_i B_j \tag{26.1.10d}$$

$$d_{ij} = \frac{1}{2}(\mathrm{v}_{i,j} + \mathrm{v}_{j,i}) \tag{26.1.10e}$$

$$\hat{E} = \frac{3}{2} N K_B T = \frac{3}{2} p, \quad (K_B = \text{Boltzmann's constant})$$

There are four conservation variables to be solved: ρ, $\rho \mathrm{v}_j$, $\rho \hat{E}$, B_j, and subsequently the primitive variables are calculated from the constraint conditions. Note that the electromagnetic forces may be written in conservation forms using gradients of squares and products of the magnetic flux density in (26.1.9). This may be desired for computational efficiency in dealing with discontinuities and/or fluctuations.

The motion of ionized gas belongs to the regime of plasma dynamics. The charged particles in a magnetic field are of interest in many physical phenomena such as occur in astrophysics, semiconductor, etc. The characteristic feature of the motion of a charged particle in a magnetic field is its tendency to spiral around the magnetic lines of force: on this is superposed a slow drift normal to the magnetic field if this is not uniform. This drift will be in opposite sense for oppositely charged particles in a gravitational field or a field of force other than an electrical field. But in the case of crossed electrical and magnetic fields, the drift will be the same for the charges of opposite sign, irrespective of their masses and charges.

Computations involved in magnetohydrodynamic flows such as in coronal mass ejection may be carried out using (26.1.8). Computational difficulties or solution convergence can occur due to physical discontinuities arising from the relationship between the Lundquist number S and the magnetic resistivity η.

$$S = \tau_d/\tau_A \tag{26.1.11}$$

$$\eta = v_A L/S \tag{26.1.12}$$

where τ_d and τ_A denote the magnetic diffusion time and Alfve'nic time, respectively, with v_A and L being the local Alfve'nic speed and scale height of the solar atmosphere, respectively,

$$L = kT_0/mg \tag{26.1.13}$$

where k is the Boltzmann constant and m is the proton mass. It is interesting to note that the Alfve'nic time and Lundquist number resemble the chemical reaction time and Damköler number in Newtonian reactive flows, respectively, whereas the magnetic resistivity is anagolous to viscosity. Thus, it is expected that the governing equations given by (26.1.8) may become stiff, resulting in difficulties of convergence to accurate solutions. To this end, it is worth investigating the merit of the flowfield-dependent variation (FDV) approach presented in Sections 6.5 and 13.6.

Plasma reactors used for semiconductor manufacturing can be described by a continuum CFD model coupling plasma transport, neutral species dynamics, gas flow, heat transfer and power coupling from an external source. Such a multicomponent, multitemperature system is simulated by the mass conservation for each species, momentum conservation of the mixture, and energy transport of electrons and neutrals [Bose et al., 1999]. The mass-averaged flow velocity is given by (26.1.6). The mass fraction for each of N species is described by

$$\frac{\partial \rho_s}{\partial t} + \nabla \cdot \mathbf{v}\rho_s = -\nabla \cdot J_s + \sum_{r=1}^{N} R_{sr} \tag{26.1.14}$$

where ρ_s is the species density (the product of number density n and molecular mass ms), $\mathbf{J}_s$ is a mass flux due to gradients of density, pressure, and electrostatic forces. The flux $\mathbf{J}_s$ can be written in a form ensuring the mass conservation [Bose et al., 1999]. The source R_{sr} denotes the mass rate production or consumption of species s from reaction r. The rates of electron-induced reactions in the plasma are functions of electron distribution function (EDF) which can be found as a solution of the Boltzmann equation.

The Boltzmann equation is a continuity equation in a six-dimensional space (three coordinates in domain space and three in velocity space)

$$\frac{\partial f}{\partial t} + (\mathbf{v} \cdot \nabla) f + (\mathbf{F} \cdot \nabla_v) f = \left(\frac{\partial f_e}{\partial t}\right)_c \tag{26.1.15}$$

where ∇_v stands for the gradient vector operator in the velocity space. Here $(\partial f_e/\partial t)_c$ represents the collisional force equal to the rate of change by encounters in the number of the class $\mathbf{v}$, $d\mathbf{v}$, in a fixed element of volume $d\mathbf{r}$ at $\mathbf{r}$, t. Each charged particle of a mass

m is acted upon by a force $m\mathbf{F}$ given by

$$\mathbf{F} = \frac{e}{m}(\mathbf{E} + \mathbf{v} \times \mathbf{B}) \tag{26.1.16}$$

The Boltzmann equation for charged particle of mass m and carrying a charge e is therefore

$$\frac{\partial f}{\partial t} + (\mathbf{v} \cdot \boldsymbol{\nabla}) f + \left[\frac{e}{m}(\mathbf{E} + \mathbf{v} \times \mathbf{B}) \cdot \boldsymbol{\nabla}_{\mathbf{v}}\right] f = \left(\frac{\partial f_e}{\partial t}\right)_c \tag{26.1.17}$$

Numerical implementations of (26.1.6) and applications to plasma instability are shown in Section 7.3 [Chung, 1978]. Numerical solutions of the magnetohydrodynamic Navier-Stokes system of equations (26.1.8) as applied to coronal magnetic field have been reported by Wu and his co-workers [Wu and Wang, 1987; Wu et al., 2000], among others. Some of their results will be presented in Section 26.4.1. Applications of Boltzmann equation in the form given in (26.1.17) will be demonstrated in plasma glow discharge processing for semiconductors in Section 26.4.2.

26.2 RAREFIED GAS DYNAMICS

26.2.1 BASIC EQUATIONS

Let us consider a rarefied gas flowing in a horizontal duct with irregular cross section with z being the coordinate parallel to the flow and x, y the coordinates of the cross section normal to z. The Boltzmann equation, linearized in the manner of Bhatnagar, Gross, and Krook [1954], known as the BGK model, may be written in the form

$$\frac{\partial f}{\partial t} + (\mathbf{c} \cdot \boldsymbol{\nabla}) f = \frac{\beta_0}{\sigma}(f_{eq} - f) \tag{26.2.1}$$

where $\mathbf{c}$ is the dimensionless velocity defined by

$$\mathbf{c} = \mathbf{v}\beta_0 \qquad \beta_0 = \sqrt{\frac{m}{2kT_0}} \tag{26.2.2}$$

f is the single local dimensionless distribution function, $\mathbf{v}$ the molecular velocity, σ the collision frequency, and k the Boltzmann constant. Applying the Chapman and Enskog method of successive approximation, we write

$$\begin{aligned} f &= f_0(1 + \Phi) \\ n &= n_0(1 + \nu) \\ T &= T_0(1 + \tau) \end{aligned} \tag{26.2.3}$$

with

$$f_0 = n\beta_0^3 \pi^{-3/2} \exp(-c^2) \tag{26.2.4}$$

and

$$f_{eq} = n\beta_0^3 \pi^{-3/2} \exp\left[\frac{1}{(1+\tau)}(\mathbf{c} - \mathbf{q}^*)^2\right] \tag{26.2.5}$$

where $\mathbf{q}^*$ is the dimensionless flow velocity which is defined as $\mathbf{q}^* = \mathbf{q}\,\beta_0$ and n is the number density. For the problem we consider here, the flow velocity can be expressed as

$$q_x^* = q_y^* = 0 \tag{26.2.6}$$

and

$$q_z^* = \frac{1}{n\beta_0^3}\int_{-\infty}^{\infty}\int_{-\infty}^{\infty} f c_z \Phi dc_x dc_y dc_z \tag{26.2.7}$$

For small Mach number flow where $|q_z| \ll 1, \nu \ll 1\, \tau \ll 1$, and $\Phi \ll 1$, the equilibrium local distribution function can be linearized as

$$f_{eq} = f_0\left[1 + 2c_z q_z^* + \left(c^2 - \frac{3}{2}\right)\tau\right] \tag{26.2.8}$$

For simplicity, we consider a pure shear flow without heat transfer, namely, an isothermal flow at T_0. Thus, the temperature change due to compression will be ignored; that is, $\tau = 0$. This implies that the gas flows are resulting from a density gradient along the z direction, which in turn is caused by a pressure gradient. Therefore, the Boltzmann equation is linearized as

$$\frac{\partial \Phi}{\partial t} + c_x\frac{\partial \Phi}{\partial x} + c_y\frac{\partial \Phi}{\partial y} + c_z\overline{K} = \frac{1}{\theta}(2c_z q_z^* - \Phi) \tag{26.2.9}$$

where the nonlinear terms $(\Phi/n)(dn/dz)$ are neglected, and furthermore, we have restricted our attention to a fully developed flow, with $\partial\Phi/\partial z$ and $(d/dz)(1/p)(dp/dz)$ being zero, leading to $\overline{K} = (1/p)(dp/dz)$, and $\theta = \sigma/\beta_0$.

On the walls of the duct, we shall assume that it reflects diffusely the molecules impinged on it. Thus, the boundary conditions for the distribution function will be characterized by Maxwellian; namely, on the boundary (F_0)

$$f(-\text{sgn}\, c_x; \mathbf{r}_0, \mathbf{c}) = f(-\text{sgn}\, c_y; \mathbf{r}_0, \mathbf{c}) = f_0 \tag{26.2.10}$$

with

$$\text{sgn}\, c_x = +1 \quad \text{for } c_x > 0 \text{ and } c_y > 0$$

$$\text{sgn}\, c_y = -1 \quad \text{for } c_x < 0 \text{ and } c_y < 0$$

the boundary condition for Φ then becomes

$$\Phi(-\text{sgn}\, c_x; \mathbf{r}_0, \mathbf{c}) = \Phi(-\text{sgn}\, c_y; \mathbf{r}_0, \mathbf{c}) = 0 \tag{26.2.11}$$

Introducing the dimensionless variables of the form

$$x^* = \frac{x}{r_0}, \qquad y^* = \frac{y}{r_0}, \qquad \Phi^* = \frac{\Phi}{Kr_0}, \qquad \delta = \frac{r_0}{\theta} \tag{26.2.12}$$

and also assuming that

$$\Phi^* = c_z\psi(x^*, y^*, c_x, c_y) \tag{26.2.13}$$

we can rewrite (26.2.9) as

$$\frac{\partial\psi}{\partial t} + c_z\frac{\partial\psi}{\partial x} + c_y\frac{\partial\psi}{\partial y} + \overline{K} = \frac{1}{\theta}\left[\frac{1}{\pi}\int_{-\infty}^{\infty}\int_{-\infty}^{\infty}\psi\exp(-c_x^2 - c_y^2)dc_x dc_y - \psi\right] \tag{26.2.14}$$

where the superscript * is deleted for simplicity.

Our objective is to determine ψ, called the perturbation function, and subsequently Φ from which we can calculate flow velocity by (26.2.7). In order to obtain the values for ψ, we shall apply the half-range method [Gross, Jackson, and Ziering, 1957] by dividing the ψ into four parts: namely,

$$\psi = \psi^{\pm\pm}(c_x, c_y) \tag{26.2.15}$$

and that $\psi^{\pm\pm}$ is only defined for $c_x > 0$ and $c_y > 0$, ψ^{+-} for $c_x > 0$ and $c_y < 0$, ψ^{-+} for $c_x < 0$ and $c_y < 0$. The integral in (6.2.14) may be written as

$$\begin{aligned}
&\int_{-\infty}^{\infty}\int_{-\infty}^{\infty} \psi \exp(-c_x^2 - c_y^2)dc_x dc_y \\
&= \int_{0}^{\infty}\int_{0}^{\infty} \psi^{++} \exp(-c_x^2 - c_y^2)dc_x dc_y + \int_{0}^{\infty}\int_{-\infty}^{0} \psi^{-+} \exp(-c_x^2 - c_y^2)dc_x dc_y \\
&\quad + \int_{-\infty}^{0}\int_{0}^{\infty} \psi^{+-} \exp(-c_x^2 - c_y^2)dc_x dc_y + \int_{-\infty}^{0}\int_{-\infty}^{0} \psi^{--} \exp(-c_x^2 - c_y^2)dc_x dc_y
\end{aligned} \tag{26.2.16}$$

Finally, we calculate the volume flow rate Q_z

$$Q_z = \iint q_z(x, y)dxdy \tag{26.2.17}$$

In the following section, we demonstrate how these calculations are performed using the finite element technique.

26.2.2 FINITE ELEMENT SOLUTION OF BOLTZMANN EQUATION

In the final form of the Boltzmann equation with BGK collision model, we assume that the perturbation function $\psi^{\pm\pm}(x, y, c_x c_y)$ is given by [Chung, Oden, and Wu, 1974; Chung, 1978]

$$\psi^{\pm\pm}(x, y, c_x, c_y) = \sum_{m=0}^{\infty}\sum_{n=0} \phi_{mn}(x, y) H_{mn}^{\pm\pm}(c_x, c_y) \tag{26.2.18}$$

wherein $H_{mn}^{\pm\pm} = h_m^{\pm}(c_x)h_n^{\pm}(c_y)$ and $h_m^{\pm}(c_x)$ and $h_n^{\pm}(c_y)$ are the Hermite polynomials of order m. Substituting (26.2.18) into (26.2.14), we obtain the residual function

$$\begin{aligned}
&R^{\pm\pm}(x, y, c_x, c_y) \\
&= \sum_{m=0}^{\infty}\sum_{n=0}^{\infty} \frac{\partial \phi_{mn}}{\partial t} H_{mn}^{\pm\pm} + \sum_{m=0}^{\infty}\sum_{n=0}^{\infty} c_x \frac{\partial \phi_{mn}}{\partial x} H_{mn}^{\pm\pm} + \sum_{m=0}^{\infty}\sum_{n=0}^{\infty} c_y \frac{\partial \phi_{mn}}{\partial y} H_{mn}^{\pm\pm} \\
&\quad + \frac{1}{\theta}\left(\sum_{m=0}^{\infty}\sum_{n=0}^{\infty} \phi_{mn} H_{mn}^{\pm\pm} - \frac{1}{\pi}\int_{-\infty}^{\infty}\int_{-\infty}^{\infty} \sum_{m=0}^{\infty}\sum_{n=0}^{\infty} \phi_{mn} H_{mn}^{\pm\pm} \exp(-c_x^2 - c_y^2)dc_x dc_y \right)
\end{aligned} \tag{26.2.19}$$

In the approximation (26.2.18), we choose ϕ_{mn} in such a manner that the averages of R

with respect to $H_{ij}^{\pm\pm}$

$$\overline{R}_{ij}(x, y) = \int_{-\infty}^{\infty}\int_{-\infty}^{\infty} RH_{ij}^{\pm\pm} dc_x dc_y \tag{26.2.20}$$

vanish in velocity domain so as to obtain systems of partial differential equations in ϕ_{mn} of the form

$$\sum_{m=0}^{\infty}\sum_{m=0}^{\infty} \frac{\partial \phi_{mn}}{\partial t} W_{mnij} + \sum_{m=0}^{\infty}\sum_{m=0}^{\infty} \frac{\partial \phi_{mn}}{\partial x} A_{mnij} + \sum_{m=0}^{\infty}\sum_{m=0}^{\infty} \frac{\partial \phi_{mn}}{\partial y} B_{mnij} + \overline{K} E_{ij} + \frac{1}{\theta}\left(\sum_{m=0}^{\infty}\sum_{m=0}^{\infty} \phi_{mn} C_{mnij} - \sum_{m=0}^{\infty}\sum_{m=0}^{\infty} \phi_{mn} D_{mn} E_{ij}\right) = 0 \tag{26.2.21}$$

where

$$W_{mnij} = \int_{-\infty}^{\infty}\int_{-\infty}^{\infty} H_{mn}^{\pm\pm}(c_x, c_y) H_{ij}^{\pm\pm}(c_x, c_y) dc_x dc_y$$

$$A_{mnij} = \int_{-\infty}^{\infty}\int_{-\infty}^{\infty} c_x H_{mn}^{\pm\pm}(c_x, c_y) H_{ij}^{\pm\pm}(c_x, c_y) dc_x dc_y$$

$$B_{mnij} = \int_{-\infty}^{\infty}\int_{-\infty}^{\infty} c_y H_{mn}^{\pm\pm}(c_x, c_y) H_{ij}^{\pm\pm}(c_x, c_y) dc_x dc_y$$

$$C_{mnij} = \int_{-\infty}^{\infty}\int_{-\infty}^{\infty} c_x H_{mn}^{\pm\pm} H_{ij}^{\pm\pm} dc_x dc_y$$

$$E_{ij} = \int_{-\infty}^{\infty}\int_{-\infty}^{\infty} c_x H_{mn}^{\pm\pm} dc_x dc_y$$

Thus, we have reduced the problem of solving (26.2.1) to that of solving an infinite system of partial differential equations (26.2.21) in the function $\phi_{mn} = \phi_{mn}(x, y)$. We shall proceed to obtain approximate solution of a truncated version of (26.2.21) by the finite element method.

Introduce the functional relationship in the form,

$$\phi_{mn}(x, y) = S_N(x, y)\phi_{Nmn} \tag{26.2.22}$$

with N being the local node. Substituting (26.2.22) into (26.2.21), we obtain the new local residual

$$\hat{R}_{ij}(x, y) = \sum_{m=0}^{\infty}\sum_{n=0}^{\infty}\left\{ W_{mnij} S_N \frac{\partial \phi_{Nmn}}{\partial t} + A_{mnij} \frac{\partial S_N}{\partial x} \phi_{Nmn} + B_{mnij} \frac{\partial S_N}{\partial y} \phi_{Nmn} + \frac{1}{\theta}(C_{mnij} S_N \phi_{Nmn} - D_{mn} E_{ij} S_N \phi_{Nmn})\right\} + \overline{K} E_{ij} \tag{26.2.23}$$

We now choose the local nodal values of ϕ_{Nmn} in such a manner that the local residual $\hat{R}_{ij}(x, y)$ is orthogonal to the subspace spanned by the functions $S_N(x, y)$ for each finite element; that is

$$\iint S_N \hat{R}_{ij} dx dy = 0 \tag{26.2.24}$$

This is basically the Galerkin method used throughout this book. Thus, the local finite element equations are of the form

$$\sum_{m=0}^{\infty}\sum_{n=0}^{\infty}\left\{W_{mnij}w_{NM}\frac{\partial\phi_{Mmn}}{\partial t}+\left[A_{mnij}a_{NM}+B_{mnij}b_{NM}\right.\right.$$
$$\left.\left.+\frac{1}{\theta}(C_{mnij}-D_{mn}E_{ij})c_{NM}\right]\phi_{Mmn}\right\}+E_{ij}\overline{K}d_N=0 \tag{26.2.25}$$

where

$$\begin{aligned}
w_{NM} &= \iint S_N S_M dxdy \\
a_{NM} &= \iint \frac{\partial S_N}{\partial x} S_M dxdy \\
b_{NM} &= \iint \frac{\partial S_N}{\partial y} S_M dxdy \\
c_{NM} &= w_{NM} \\
d_N &= \iint S_N dxdy
\end{aligned} \tag{26.2.26}$$

Equation (26.2.25) represents the general local finite element model of (26.2.21) which will then be assembled into a global form. Boundary conditions amount to simply prescribing nodal values of $\phi_{mn}(x, y)$ at boundary nodes. In specific applications, appropriate forms of the interpolation function $S_N(x, y)$ must be chosen and only a finite number of terms of the series in (26.2.23) can be used.

Having completed all integrations in (26.2.26), we obtain the finite element equations of the form

$$J_{mnij}^{NM}\dot{\phi}_{mn}^{M}+K_{mnij}^{NM}\phi_{mn}^{M}=F_{ij}^{N} \tag{26.2.27}$$

where

$$\begin{aligned}
J_{mnij}^{NM} &= \sum_{m=0}^{\infty}\sum_{n=0}^{\infty} W_{mnij}w_{NM} \\
K_{mnij}^{NM} &= \sum_{m=0}^{\infty}\sum_{n=0}^{\infty}\left(A_{mnij}a_{NM}+B_{mnij}b_{NM}+\frac{1}{\theta}(C_{mnij}-D_{mn}E_{ij})c_{NM}\right) \\
F_{ij}^{N} &= E_{ij}\overline{K}d_N
\end{aligned}$$

The number of equations generated in (26.2.27) depends on the order of Hermite polynomial approximations and the number of nodes in an element. Consider the mth Hermite polynomial defined by

$$H_m(\varsigma)=(-1)^m\exp(\varsigma^2)\frac{d^m}{d\varsigma^m}\exp(-\varsigma^2) \tag{26.2.28}$$

The number of local finite element equations is determined by

$$r=(1+m)^2 N$$

where N is the number of nodes in an element: for example, if we choose $m = 3$ and $N = 4$, then the number of local finite element equations becomes 64 with 16 equations at each node. The total number of equations for the entire cross section is 16 time the total number of nodes.

Numerical solutions of (26.2.27) for a square duct were carried out [Chung et al., 1974; Chung, 1978]. It was shown that the orthogonal projection in the Euclidean space dealing with both spatial domain and velocity dimension leads to an effective approach to the solution of the Boltzmann equation.

26.3 SEMICONDUCTOR PLASMA PROCESSING

26.3.1 INTRODUCTION

Plasma dynamics describing the motion of ionized gas studied in Sections 26.1 may be extended to charged particle kinetics combined with reactive flows of Chapter 22 for applications to integrated circuits (IC) in semiconductor. Physics and chemistry of plasma-enhanced chemical vapor deposition (PECVD) are important processes for semiconductor device fabrication. Low temperature, partially ionized discharges used in IC manufacturing are characterized by a number of interacting effects: plasma generation of active species; plasma power deposition and loss mechanisms; surface processes proceeding at the wafer, reactor walls, and fixtures; particulate generation; and gas flow and heat transfer patterns, etc. Here, the role of CFD will be extremely important in resolving plasma process simulation and discharge modeling [Meyyappan, 1995].

A plasma is a collection of charged particles where the long-range electromagnetic fields set up collectively by the charged particles have an important effect on the particles' behavior. In the case of a semiconductor, the fields the plasma sets up will be mostly electric fields. This electrical field is created because electrons in the plasma tend to move much faster than ions. The fast-moving electrons hit the wall and charge up negatively, and this negative charge pushes other electrons away at the same time as attracting positive ions, with the rates of arrival of electrons and positive ions made about equal at the steady state.

The negative space charge at dielectric walls repels plasma electrons from the wall, exposing the positive charge in a region close to the wall known as a *sheath*. The electric fields in the positively charged sheath region are strong, whereas those in the plasma interior are weak. The strong electrical field in the sheath accelerates the ions in a direction normal to the surface. The high energy ions moving toward the surface represent the vital aspect of the plasma processing of materials. The ions, among other things, sputter (knock) material off the surface, damage the surface, provide heat to the surface, or implant in the surface. This process is known as *etching*. Because the sheath electric fields cause ions to arrive at near normal incidence, they tend to hit the bottom of a trench rather than a vertical sidewall of a trench. This is crucial for the faithful transfer of a pattern during etching. Plasma ions striking a semiconductor through a hole in a mask will more likely dig the hole straight down. The electrons ionize neutrals, break molecules apart to form radicals, create molecules in excited states, and heat the surface. The radicals are frequently responsible for etching the surface. Radicals may also polymerize on the surface and form a layer that protects the surface. Chemical etching by itself is isotropic and has high selectivity but is not desirable because there is no preferred direction, whereas etching by ions provides anisotropy but often lacks selectivity. Thus, a commonly used example of the combination of chemical and physical etching is etching in the presence of a gas that will polymerize on the surface. Anisotropy is achieved because ions only clean the polymer off the bottom of a trench and not off the side walls. Selectivity might depend on the chemical etch of the bare surface only being effective for (say) Si or SiO_2. In etching using CF_4 gas, a polymer is formed on the surface of Si but not SiO_2. The SiO_2 gives up oxygen to form CO_2 from the radicals, which contain carbon. This prevents the formation of the polymer on SiO_2, and so SiO_2 will be etched while Si is protected from etching by the polymer [Kirmse et al., 1996].

The electric and magnetic fields in the plasma are set up self-consistently by the plasma, and the plasma is controlled by those fields. The conditions in which the electric field is strongly affected by the plasma may be described by Poisson's equation,

$$\nabla \cdot \mathbf{E} = \frac{\rho}{\varepsilon} \tag{26.3.1a}$$

or

$$\nabla^2 \Phi = -\frac{\rho}{\varepsilon} \tag{26.3.1b}$$

where **E** is the electric intensity, ρ is the charge density, ε is the surface energy, and Φ is the electrostatic potential in the main chamber. One-dimensional analysis of (26.3.1b) shows that the plasma number density in a plasma reactor has to be in excess of 10^6 cm^{-3} which will affect the electrostatic potential. This is a very low density compared to densities used in plasma processing. The usual situation in a plasma reactor is that the density of ions and electrons is high enough to shield out a typical applied voltage in a very short distance, so that the plasma interior can be nearly free from strong electric fields, and the electrons can provide a charge density that nearly neutralizes the plasma in the interior. Further details of the basic principles of plasma processing have been well documented in the literature [Chapman, 1980; Boenig, 1982; Manos and Flamm, 1989; and Hitchon, 1999; among others].

In plasma processing reactors, a self-sustaining glow discharge is made available from a direct current (DC), radio frequency (RF), or microwave power source. The electrons gain energy from the applied electrical field but do not lose energy significantly from the numerous elastic collisions with the gas due to their small mass compared to the gas atoms (molecules). As a result, the electrons attain a very high temperature, while the background gas and heavy ions are relatively cold. Hence, these discharges, which are only partially ionized, are cold plasmas. A variety of electron-impact reactions result in inelastic collisions with the gas molecules in the discharge that are responsible for the creation of reactive and nonreactive fragments from the parent gas. Inelastic collision between heavy particles leads to recombination and other chemical reaction. The net result from all of this chemical activity is a partially ionized discharge consisting of electrons, positive ions, negative ions, atoms, radicals, other neutral fragments, and the parent gas(es). This system is not in thermodynamic equilibrium. It is indeed the deviation from thermodynamic equilibrium that is responsible for the effectiveness of a discharge in material processing and permits low-temperature processing. In addition,

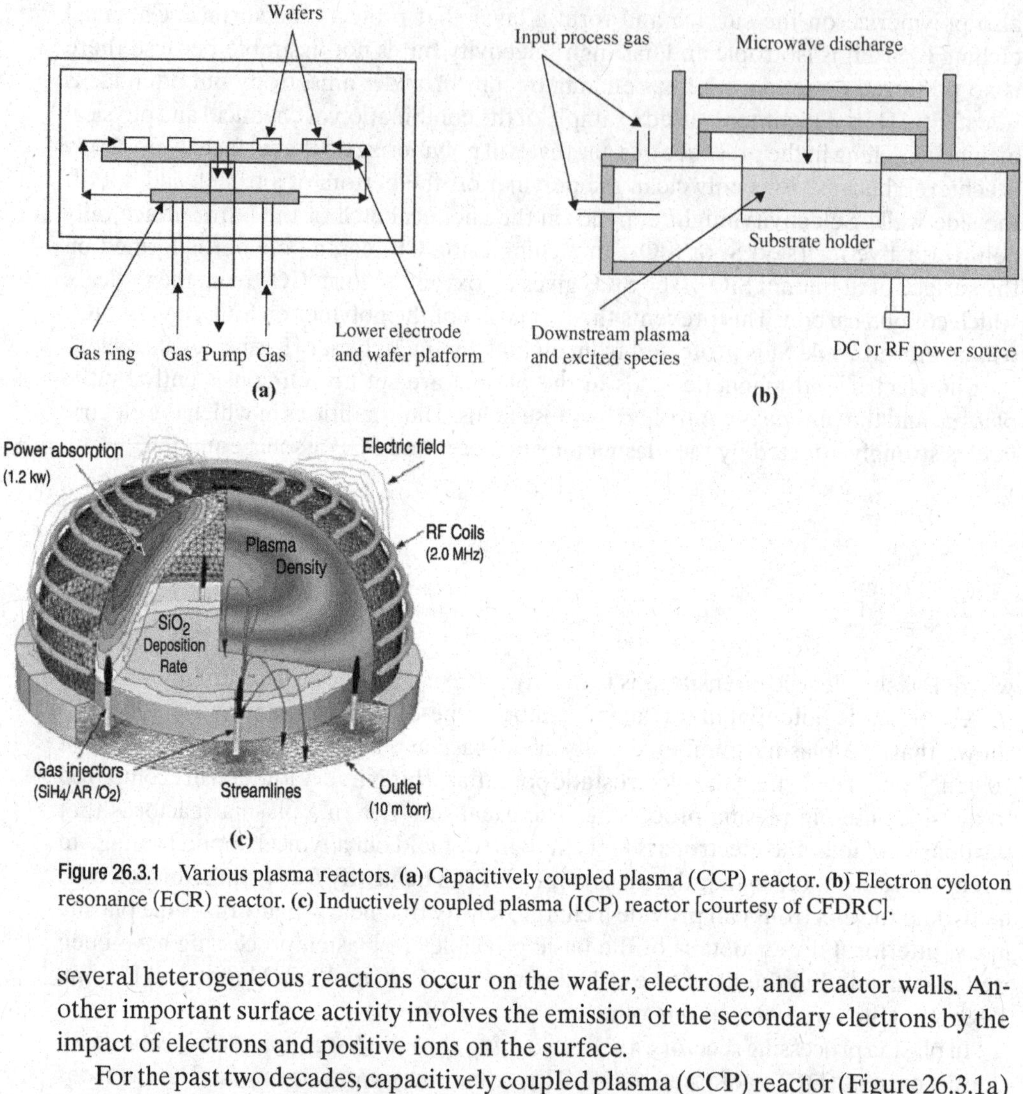

Figure 26.3.1 Various plasma reactors. **(a)** Capacitively coupled plasma (CCP) reactor. **(b)** Electron cycloton resonance (ECR) reactor. **(c)** Inductively coupled plasma (ICP) reactor [courtesy of CFDRC].

several heterogeneous reactions occur on the wafer, electrode, and reactor walls. Another important surface activity involves the emission of the secondary electrons by the impact of electrons and positive ions on the surface.

For the past two decades, capacitively coupled plasma (CCP) reactor (Figure 26.3.1a) has been employed using typically 13.56-MHz of RF power source. Recently, efforts are being made to develop new plasma sources capable of high processing rates and uniformity over large wafers (larger than 200 mm) with minimum wafer damage, achieving a structure scale size of 150 nm or less. The electron cycloton resonance (ECR) reactor is currently receiving much attention for both deposition and etching (Figure 26.3.1b). Here, microwave energy at 2.45 GHz is coupled to the natural resonance frequency of the electron gas in the presence of a strong magnetic field (875 gauss). Another new plasma source is an inductively coupled plasma (ICP) reactor in which the plasma is driven inductively, with a power source that operates at the standard RF of 13.56 MHz. ICP reactors use a variety of different coil designs as illustrated in Figure 26.3.1c.

In the past, computer calculations were carried out with simplifications, using either plasma discharge model (PDM) or gas flow model (GFM) in chemical vapor deposition (CVD). In PDM, we consider only the discharge physics aspects, setting aside gas flow, gas and wafer heating, and reactor issues. Here, it is assumed that the loss of charged particles (electrons and ions) due to gas convective flow is negligible. In this approach, we are primarily concerned with the physical features of the discharge such as electrical characteristics, utilization of applied power, and variations of densities, plasma potential, sheath thickness, ion flux, etc., as a function of process input parameters. On the other hand, in GFM, we disregard the discharge aspects and focus on an idealized reactor model. In this approach, we assume an electron density distribution in the reactor. Reaction rate constants are either taken from experiments or evaluated using some assumed form for electron energy distribution function (EEDF).

Obviously, the most desirable approach is to combine both PDM and GFM at the expense of computational resources, known as the complete process model (CPM). The governing equations for PDM are those presented in Section 26.1, whereas the reactive flow equations of Chapter 22 are applied to GFM. Traditionally, Monte Carlo methods and particle-in-cell methods have been used for PDM, with FDM, FEM, and FVM favored for GFM. For the sake of completeness, the governing equations for CPM as a combination of PDM and GFM are summarized below.

26.3.2 CHARGED PARTICLE KINETICS IN PLASMA DISCHARGE

The charged particle kinetics for PECVD may be governed by the Boltzmann equation (26.1.12) as

$$\frac{\partial f}{\partial t} + (\mathbf{v} \cdot \nabla) f + (\mathbf{F} \cdot \nabla_{\mathbf{v}}) f = \left(\frac{\partial f}{\partial t} \right)_c \tag{26.3.2}$$

with $\mathbf{F}$ being the electromagnetic force,

$$\mathbf{F} = \frac{e}{m}(\mathbf{E} + \mathbf{v} \times \mathbf{B}) \tag{26.3.3}$$

and the velocity distribution function (VDF) f may be given by the two-term approximation in the form,

$$f(\mathbf{r}, \mathbf{v}, t) = f_0(\mathbf{r}, \mathbf{v}, t) + \frac{\mathbf{v}}{\mathrm{v}} \cdot \mathbf{f}_1(\mathbf{r}, \mathbf{v}, t) \tag{26.3.4}$$

where f_0 and f_1 denote the isotropic part and anisotropic part, respectively, given by Shkarofsky et al. [1966] as

$$\frac{\partial f_0}{\partial t} + \frac{1}{3}(\mathbf{v} \cdot \nabla)\mathbf{f}_1 - \frac{1}{\mathrm{v}^2}\frac{\partial}{\partial \mathrm{v}}\left[\frac{e\mathrm{v}^2}{3\mathrm{m}}(\mathbf{E} \cdot \mathbf{f}_1) + \frac{1}{2}\mathrm{v}^3 \delta \nu f_0\right] + S_e$$
$$= -\nu^*(\mathrm{v}) f_0 + \frac{\mathrm{v}'}{\mathrm{v}}\nu^*(\mathrm{v}') f_0(\mathbf{r}, \mathrm{v}' t) + I_e \tag{26.3.5}$$

$$\frac{\partial \mathbf{f}_1}{\partial t} + (\mathbf{v} \cdot \nabla) f_0 - \frac{e\mathbf{E}}{m}\frac{\partial f_0}{\partial \mathrm{v}} = -\nu \mathbf{f}_1 \tag{26.3.6}$$

where ν and ν^* are the moment transfer frequency and the total frequency of inelastic collisions, $\mathrm{v}' = (\mathrm{v}^2 + 2\varepsilon^*/m)^{1/2}$, S_e is the electron-electron Coulomb collision operator, and I_e is the source of newly born electrons.

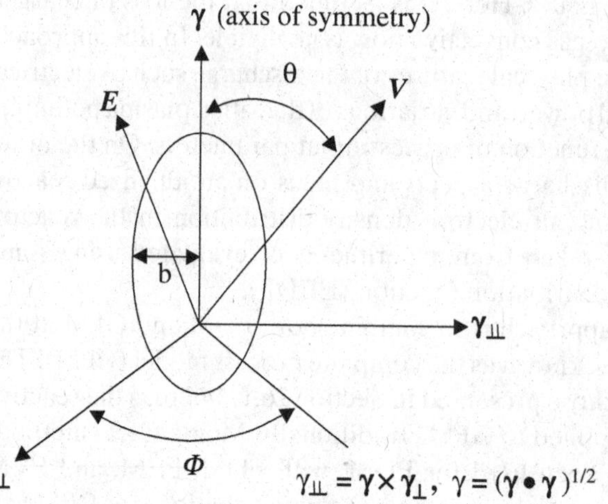

Figure 26.3.2 The surface of an ellipsoid of revolution describes the distribution function in three dimensions. The electric field is shown for the general case in which it is not aligned with the axis of symmetry ($0 < \gamma < 1$).

Recently, Richley [1999] proposed an elliptic representation of the Boltzmann equation with validity for all degrees of anisotropy. By choosing an ellipsoid of revolution to describe the angular dependence of the velocity distribution, the Boltzmann equation can be reduced to a set of two equations which may be applicable to a wide range of conditions. These equations are reduced to the two-term representation of (26.3.4) for nearly isotropic cases.

To this end, Richley considers an ellipsoid of revolution as shown in Figure 26.3.2 in which the magnitude of the distribution function is taken to be the length of a line extending from one focus to a point on the surface such that

$$f = \frac{b\sqrt{1-\gamma^2}}{1-\gamma\cdot\dfrac{\mathbf{v}}{\mathrm{v}}} \cong b\left(1+\gamma\cdot\frac{\mathbf{v}}{\mathrm{v}}\right) \tag{26.3.7}$$

where γ is the vector in the direction of the axis of symmetry, with magnitude equal to the eccentricity of the ellipsoid and $0 \leq \gamma \leq 1$. It can be shown that the elliptic representation is equivalent to a three-term spherical harmonic expansion,

$$f(\mathbf{r},\mathbf{v}) = f_0(\mathbf{r},\mathrm{v}) + \frac{\mathbf{v}}{\mathrm{v}}\cdot\mathbf{f}_1 + \tilde{f}_2 : \frac{\mathbf{vv}}{\mathrm{v}^2} \tag{26.3.8}$$

with the colon denoting a tensor product. Here, $\tilde{f}_2$ and $\mathbf{vv}$ represent second order tensors the product of which results in a scalar. This is analogous to the product of stress tensor and velocity gradients representing thermoviscous dissipation in the fluid mechanics energy equation. To express the nature of anisotropy, it is convenient to use new quantities formed by integration over all solid angles ω as defined in (24.1.5), leading to

$$n(\mathbf{r},\mathrm{v}) = \int_\omega f d\omega \tag{26.3.9}$$

$$\mathbf{W} = \int_\omega \frac{\mathbf{v}}{\mathrm{v}} f d\omega \tag{26.3.10}$$

It can be shown that substitution of (26.3.9) and (26.3.10) into (26.3.2) results in two moment equations of the form

$$\frac{\partial n}{\partial t}+\nabla\cdot(\mathrm{v}\mathbf{W})-\frac{q_0}{m}\frac{1}{\mathrm{v}^2}\frac{\partial}{\partial \mathrm{v}}(\mathrm{v}^2\mathbf{E}\cdot\mathbf{W})=\left(\frac{\partial n}{\partial t}\right)_c \tag{26.3.11a}$$

$$\frac{\partial \mathbf{W}}{\partial \mathbf{t}}+\nabla\cdot\left[\frac{n\mathrm{v}}{2}\left(\frac{3X}{\gamma}-1\right)\gamma\gamma\right]+\nabla\left[\frac{n\mathrm{v}}{2}\left(1-\frac{X}{\gamma}\right)\right]-\frac{q_0}{m}\frac{1}{\mathrm{v}^3}\frac{\partial}{\partial \mathrm{v}}\left[\frac{n\mathrm{v}^3}{2}\left(\frac{3X}{\gamma}-1\right)\mathbf{E}\cdot\gamma\gamma\right]$$
$$-\frac{q_0}{m}\mathbf{E}\left\{\frac{\partial}{\partial \mathrm{v}}\left[\frac{n}{2}\left(1-\frac{X}{\gamma}\right)\right]+\frac{n}{2\mathrm{v}}\left(1-\frac{3X}{\gamma}\right)\right\}=\left(\frac{\partial \mathbf{W}}{\partial \mathbf{t}}\right)_c \tag{26.3.11b}$$

with γ being the unit vector along the axis of symmetry, and

$$X=|\mathbf{W}|/n \tag{26.3.12}$$

which may be called the "anisotropy parameter" ($X \ll 1$, small anisotropy; $X=1$, strong anisotropy). It should be noted that $\gamma\gamma$ represents the second order tensor which when dotted with a vector results in another vector.

Let us now assume that

$$n=4\pi f_0 \tag{26.3.13}$$

$$\mathbf{W}=(4\pi/3)\mathbf{f}_1 \tag{26.3.14}$$

$$\tilde{f}_2=\frac{5}{4}\left(3\frac{X}{\gamma}-1\right)(3\gamma\gamma-\tilde{\mathbf{I}})f_0 \tag{26.3.15}$$

with $\tilde{\mathbf{I}}$ being the unity tensor. If we substitute these to (26.3.11a) and (26.3.11b), the two-term spherical harmonic expansion can be obtained as demonstrated by Richley [1999]. Thus, the elliptic representation can be thought of as being identically the two-term spherical harmonic expansion, with closure of the hierarchy according to (26.3.15).

Returning to (26.3.8), the isotropic part f_0 defines the scalar characteristics of particles such as density, mean energy, etc. The vector part $\mathbf{f}_1$ denotes vector quantities such as current. The tensor part $\tilde{\mathbf{f}}_2$ contributes to things like directed energy. Based on the elliptic representation of the Boltzmann equation described above, it is possible to obtain special cases of (26.3.5) and (26.3.6), valid for strongly or weakly anisotropic VDF.

(1) Strongly Anisotropic VDF

The Boltzmann equations for strongly anisotropic VDF may be written as

$$\frac{\partial f_0}{\partial t}+\frac{1}{\mathrm{v}}\nabla\cdot(\mathrm{v}^2\mathbf{f}_1)=S_0 \tag{26.3.16}$$

$$\frac{\partial \mathbf{f}_1}{\partial t}+\frac{1}{\mathrm{v}}\nabla\cdot\mathrm{v}\tilde{\mathbf{Y}}+\mathrm{v}\nabla Y_1+\frac{q\mathbf{E}f}{2m\mathrm{v}}\left(\frac{3X}{\gamma}-1\right)(\mathbf{I}-\gamma\gamma)=\mathbf{S}_1 \tag{26.3.17}$$

where

$$\tilde{\mathbf{Y}}=\frac{f_0\mathrm{v}^2}{2}\left(\frac{3X}{\gamma}-1\right)\gamma\gamma \tag{26.3.18}$$

$$Y_1=\frac{f_0}{2}\left(1-\frac{X}{\gamma}\right) \tag{26.3.19}$$

with S_0 and $\mathbf{S}_1$ being the collision terms given in Richley [1999].

Equations (26.3.16) and (26.3.17) are useful for problems with strong anisotropy of the VDF such as the fast electrons in the cathode region of glow discharges and positive ions in gaseous plasmas. The elliptic representation is an efficient alternative to statistical methods especially attractive for multidimensional problems with substantial anisotropy of the VDF where statistical methods are computationally expensive.

(2) VDF with Small Anisotropy

The expressions of (26.3.18) and (26.3.19) may be simplified for small anisotropy [Kortchagen, Buch, and Tsendin, 1996] as $\mathbf{Y} = \mathbf{0}$ and $Y_1 = f_0/3$ with $\gamma = 0$. Furthermore, we set

$$\mathbf{f}_1 = \mathbf{f} + \bar{\mathbf{f}}\, e^{i\omega t} \tag{26.3.20}$$

with

$$\mathbf{f} = -\frac{\mathrm{v}}{3\nu_m}\nabla f_0 \tag{26.3.21}$$

$$\mathbf{f} = \frac{q\mathbf{E}\mathrm{v}}{(\nu_m + i\omega)}\frac{\partial f_0}{\partial \varepsilon} \tag{26.3.22}$$

where ω is the angular field frequency, ν_m is the transport collision frequency. Substituting these expressions into (23.3.16) leads to

$$\frac{\partial f_0}{\partial t} + \frac{1}{\mathrm{v}}\nabla \cdot (\mathrm{v} D_r \nabla f_0) + \frac{1}{\mathrm{v}}\frac{\partial}{\partial \varepsilon}\left(\mathrm{v} D_\varepsilon \frac{\partial f_0}{\partial \varepsilon}\right) = S_0 \tag{26.3.23}$$

where $D_r = \mathrm{v}^2/3\nu_m$ is the diffusion coefficient in configuration space,

$$D_\varepsilon = D_r \frac{\tilde{E}^2}{2}\frac{\nu_m^2}{(\nu_m^2 + \omega^2)} \tag{26.3.24}$$

is a diffusion coefficient along the energy axis.

(3) Weakly Collisional Discharges with Hot Plasma Effects

If the particle mean free path becomes comparable to or larger than the characteristic size of the system, then specific kinetic effects appear in the weakly collisional operating regimes. For low-pressure RF discharges, these effects are due to thermal electron motion and include collisionless electron heating and anomalous skin effect [Kolobov and Economou, 1997; Lieberman and Godyak, 1998]. The collisionless power absorption dominates at pressures below 10 mTorr and can exceed the collisional power absorption by an order of magnitude at the lowest gas pressure. In the weakly collisional regimes, the oscillating part of the electron distribution function (EDF) can be written as an integral along the electron trajectories [Kolobov, 1998]:

$$f(\mathbf{r},\mathrm{v},t) = e\int ds e^{-\mathrm{v}s}\mathbf{v}(\mathrm{t} - \mathrm{s}) \cdot \mathbf{E}(\mathbf{r}(t - s), t - s)\frac{\partial f_0}{\partial \varepsilon} \tag{26.3.25}$$

Consequently, the electron current at a given point depends on the field values in other points, leading to a variety of hot plasma effects [Godyak et al., 1999]. In this regime the energy relaxation length of electrons is large compared to the discharge dimensions. Thus, the isotropic part of the EDF is given by [Aliev, Kaganovich, and Schluter, 1997]

$$\frac{\partial f_o}{\partial t} + \frac{1}{\bar{v}}\frac{\partial \bar{v}}{\partial \varepsilon}\left(D_\varepsilon \frac{\partial f_0}{\partial \varepsilon} + V_\varepsilon f_0\right) = S \tag{26.3.26}$$

where $\bar{v}$ denotes the electron velocity averaged over discharge volume accessible to electrons with total energy ε. It should be noted that in spite of simplicity (26.3.26) contains enough information about electron kinetics in the spatially inhomogeneous plasma. The focal point is the energy diffusion coefficient D_ε which describes the peculiarities of electron heating [Godyak and Kolobov, 1998].

The solution of the Boltzmann equation has been carried out using particle-in-cell/Monte Carlo collision (PIC/MCC) methods [Surendra and Graves, 1991a,b]. Since these methods are extremely time consuming, research into deterministic methods such as finite volume methods will be highly desirable.

26.3.3 DISCHARGE MODELING WITH MOMENT EQUATIONS

When should we use the Boltzmann equation, and when is the continuum model sufficient? Figure 26.3.3 illustrates a hierarchy of transport descriptions of electrons in weakly ionized gaseous and solid-state plasmas [Bringuier, 1999]. This hierarchy is based on peculiarities of electron collision dynamics resulting in a great distinction of momentum and energy relaxation rates. When electrons move through a background of

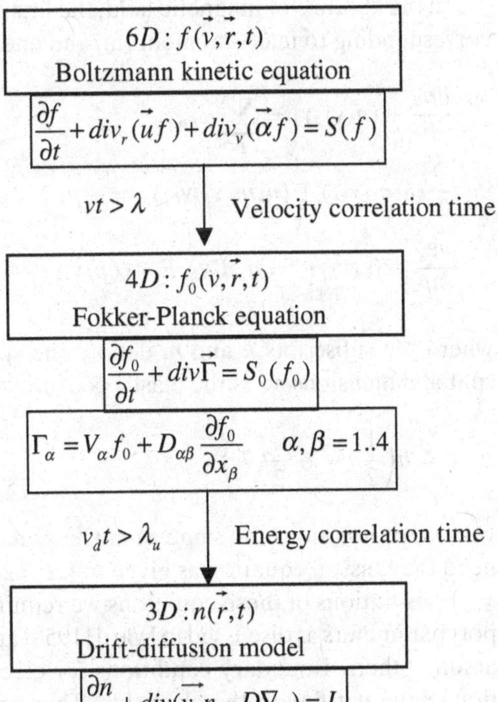

Figure 26.3.3 A hierarchy of transport descriptions. The left-hand side of each transport equation contains a divergence of flux in a relevant space. The flux consists of drift or drift-diffusion. Arrows connect two levels of descriptions of decreasing complexity. As space and time scales increase, a lesser description suffices. Over time scales longer than the velocity-correlation time, the energy and position suffice to specify the particle state. Over time scales exceeding the energy correlation time, a macroscopic description in coordinate space is sufficient.

neutral gas atoms or in a solid lattice they significantly change the direction of their motion, but only slightly change their energy in elastic collisions with heavy particles. Thus, the energy relaxation length λ_u (and time τ_u) considerably exceeds the velocity relaxation length λ and collision time τ. Over time scales shorter than the energy correlation time, and for spatial scales smaller than the energy relaxation length, the continuum (drift-diffusion) approximation in position space is not sufficient and kinetic analysis is necessary (low arrow in Figure 26.3.3). However, owing to the great distinction of momentum and energy relaxation rates, for the time scales exceeding collision time and for spatial scales exceeding the mean free path λ, the six-dimensional Boltzmann kinetic equation can be reduced to a much simpler Fokker-Planck equation (26.3.23) in a four-dimensional energy-position manifold (upper arrow in Figure 26.3.3).

It is difficult to solve spatially inhomogeneous Boltzmann and Fokker-Planck equations. As an alternative, continuum approach has been widely used for modeling gas discharges. This approach uses the moments of the distribution to obtain the macroscopic properties of the discharge, assuming a certain form of the distribution function or solving local BE to calculate transport coefficients and rates of chemical reactions. An infinite chain of coupled moment equations would be equivalent to the Boltzmann equation. In practice, only a finite set of moment equations can be used. The most practical number of moment equations would be three, representing mass, momentum, and energy. Higher moments require more closure relations, resulting in a difficult compromise associated with new unknowns as discussed in Chapman and Cowling [1952] and Gogolides and Swain [1992], among others.

In the absence of magnetic field, the first three moments of the Boltzmann equation corresponding to mass, momentum, and energy are of the form

$$\frac{\partial n_k}{\partial t} + (n_k v_{ki})_{,i} = \sum_m \omega_{km} \tag{26.3.27}$$

$$\frac{\partial}{\partial t}(n_k m_k v_{kj}) + (n_k m_k v_{ki} v_{kj})_{,i} = -p_{k,j} + n_k q_k E_j - n_k m_k \nu_k \tag{26.3.28}$$

$$\frac{\partial \varepsilon_k}{\partial t} + (\varepsilon_k v_{ki})_{,i} = q_k n_k v_{ki} E_i - (p_k v_{ki})_{,i} + (k_k T_{k,i})_{,i} - \sum_m \omega_{km} H_m \tag{26.3.29}$$

where the subscripts k and m denote the species, tensorial indices i and j represent spatial dimensions, ν_k is the elastic collision frequency, and

$$\varepsilon = n\left(\frac{1}{2}mv^2 + \frac{3}{2}kT\right) \tag{26.3.30}$$

with other notations the same as in Chapter 22. In addition to the above equations we need the Poisson equation as given in (26.3.2).

For solutions of these equations we require appropriate rate expressions and transport parameters as discussed in Ward [1958] and Richards, Thompson, and Swain [1987], among others. Boundary conditions for electrons in RF discharges require specification of the net flux at the electrode. They are given by the sum of electrons lost due to recombination and electrons generated by secondary electron emission. These and other conditions are detailed in Chantry [1987], Graves and Jensen [1986], and Barnes,

Colter, and Elter [1987], among others. Some example problems will be presented in Section 26.3.5.

26.3.4 REACTOR MODEL FOR CHEMICAL VAPOR DEPOSITION (CVD) GAS FLOW

The basic governing equations and assumptions in chemical vapor deposition gas flow are similar to those in reactive flows presented in Chapter 22. In addition, listed below are assumptions made specifically for chemical vapor deposition gas flow:

(1) The mixture of feed gas and generated species is treated as continuum in which the mean free path of the gas molecules must be much smaller than the reactor dimensions. Thus, the collisions in the gas phase would be more dominant than collisions on the walls.
(2) The gases in the plasma reactor are assumed to be ideal; ideal gas law and Newton's law of viscosity can be applied.
(3) The Reynolds number is small enough that the low is laminar.
(4) The Mach number is low so that the effects of pressure variation on the density of the gas mixture may be neglected.
(5) Gas is weakly ionized.

Here we consider an ideal reactor model that ignores the details of the glow discharge. This is done by assuming an electron density distribution in the reactor. The rate constants are either taken from experiments or evaluated using some assumed form for EEDF.

The conservation equations for mass, momentum, energy, and species are the same as given in (22.2.34). For the sake of completeness we repeat here,

$$\frac{\partial \mathbf{U}}{\partial t} + \frac{\partial \mathbf{F}_i}{\partial x_i} + \frac{\partial \mathbf{G}_i}{\partial x_i} = \mathbf{B} \tag{26.3.31}$$

with

$$\mathbf{U} = \begin{bmatrix} \rho \\ \rho v_i \\ \rho E \\ \rho Y_k \end{bmatrix}, \quad \mathbf{F}_i = \begin{bmatrix} \rho v_j \\ \rho v_i v_j + p\delta_{ij} \\ \rho E v_i + p v_i \\ \rho Y_k v_i \end{bmatrix}, \quad G_i = \begin{bmatrix} 0 \\ -\tau_{ij} \\ -\tau_{ij} v_j - k T_{,i} - \sum_{k=1}^{N} \rho H_k D_{km} Y_{k,i} \\ -\rho D_{km} Y_{k,i} \end{bmatrix},$$

$$B = \begin{bmatrix} 0 \\ \rho \sum_{k=1}^{N} Y_k f_{ki} \\ S - \sum_{k=1}^{N} H_k^0 \omega_k \\ \omega_k \end{bmatrix}$$

Instead of solving the entire equations required for CFD, many options for simplifications have been shown in the literature. For example, a fully developed flow may be assumed such as in Chen [1983], Meyyappan and Buggeln [1990], and Venkatesan, Trachtenburg, and Edgar [1992]. Another example is a plug flow model in which it is assumed that there is no velocity gradient in the axial direction [Lii et al., 1990]. If the inner-electrode spacing is small compared to the length in the flow direction, concentration is more uniform in the axial direction compared to the radial direction. In this case axially average one-dimensional species equations can be solved to study polymer etching in an oxygen discharge [Economou and Alkier, 1988] and silicon etching in NF_4 discharge [Stenger et al., 1987].

26.4 APPLICATIONS

26.4.1 APPLICATIONS TO MAGNETOHYDRODYNAMIC FLOWS IN CORONA MASS EJECTION

Wu and his co-workers [Wu and Wang, 1987; Wu et al., 2000] developed the fully implicit continuous Eulerian (FICE) method using finite difference discretizations of SIMPLE algorithm (Section 5.3.1) and boundary condition implementations for compatibility and characteristic properties and nonreflecting boundaries as discussed in Section 6.7.1. They used this method in solving specialized cases of (26.1.8) as applied to magnetohydrodynamic flows in coronal mass ejection. Here, the governing equations are based on resistive MHD theory in which the field topology is changed due to the magnetic reconnection process. They consist of conservation laws of mass, momentum, energy, and induction equations describing the dynamical interaction between plasma flow and magnetic field. The viscous dissipation is neglected as it is two orders smaller than the magnetic dissipation.

Wu et al. [2000] simulated recent observations at the 1996 solar minimum obtained by the Large Angle Spectrometric Coronagraph (LASCO) on the Solar and Heliospheric Observatory revealing the motion of density enhancements in the coronal streamer belt, known as the plasma blobs. Figure 26.4.1.1 shows the initial condition used for the analysis. It represents the physical parameters of the magnetic resistive MHD simulation

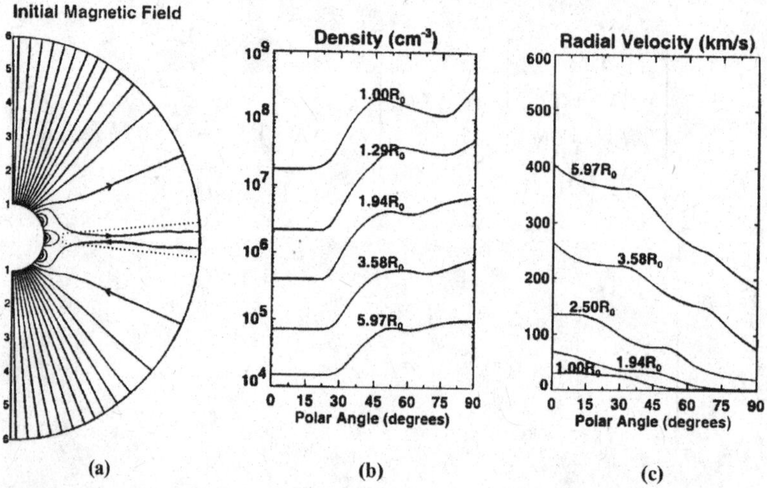

Figure 26.4.1.1 Physical parameters of MHD simulation model: **(a)** the magnetic field configuration, the dotted line region indicates the region of reconnection, **(b)** Plasma density as a function of latitude at various solar radii; **(c)** solar wind radial velocity as a function of latitude at various solar radii [Wu et al., 2000].

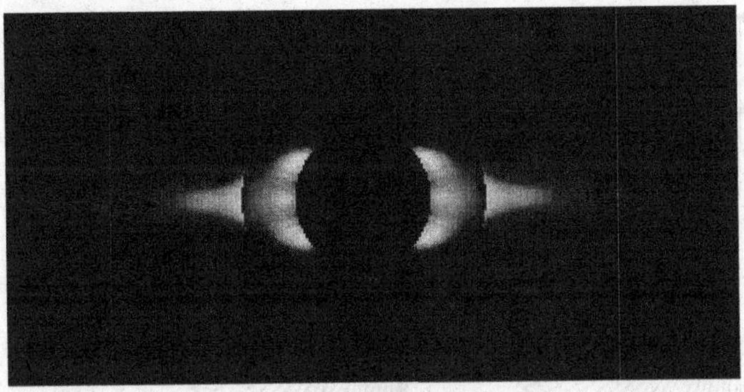

(a)

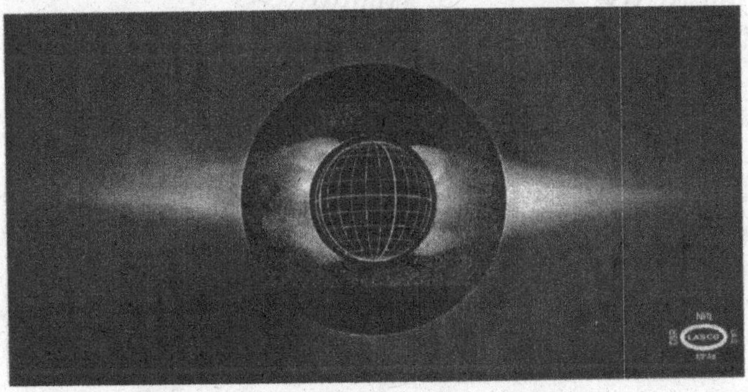

(b)

Figure 26.4.1.2 Calculated and observed corona [Wu et al., 2001]. (**a**) Numerical MHD simulated images. (**b**) Observed typical corona.

model for a global coronal magnetic field to investigate the formation and propagation of the observed plasma blobs. This choice is important to create the appropriate dynamics for the formation of the plasma blobs by magnetic reconnection. The solution changes and evolves due to the introduction of finite magnitude resistivity in the induction equations.

The calculated results show the quiescent corona as shown in Figure 26.4.1.2a as compared to the observed typical corona at solar minimum (Figure 26.4.1.2b). Figure 26.4.1.3 shows the evolution of the magnetic field topology. It is seen that the change of magnetic field topology caused by magnetic reconnection occurs in four stages, with reconnection taking place at five points marked by O_1 through O_5. A radial pressure gradient is created between the inner corona and outer corona, as shown in Figure 26.4.1.4. It is also shown that after the second stage reconnection, some of the magnetic flux feeds into two neighboring loops and pushes the central loop outward together with outward pressure gradient. Further details are provided in Wu et al. [2000].

26.4.2 APPLICATIONS TO PLASMA PROCESSING IN SEMICONDUCTORS

(1) Capacitively Coupled RF Glow Discharge

In this analysis [Gogolides and Swain, 1992], a plasma is simulated using a two-moment model in the study of on an electropositive (Ar) and an electronegative

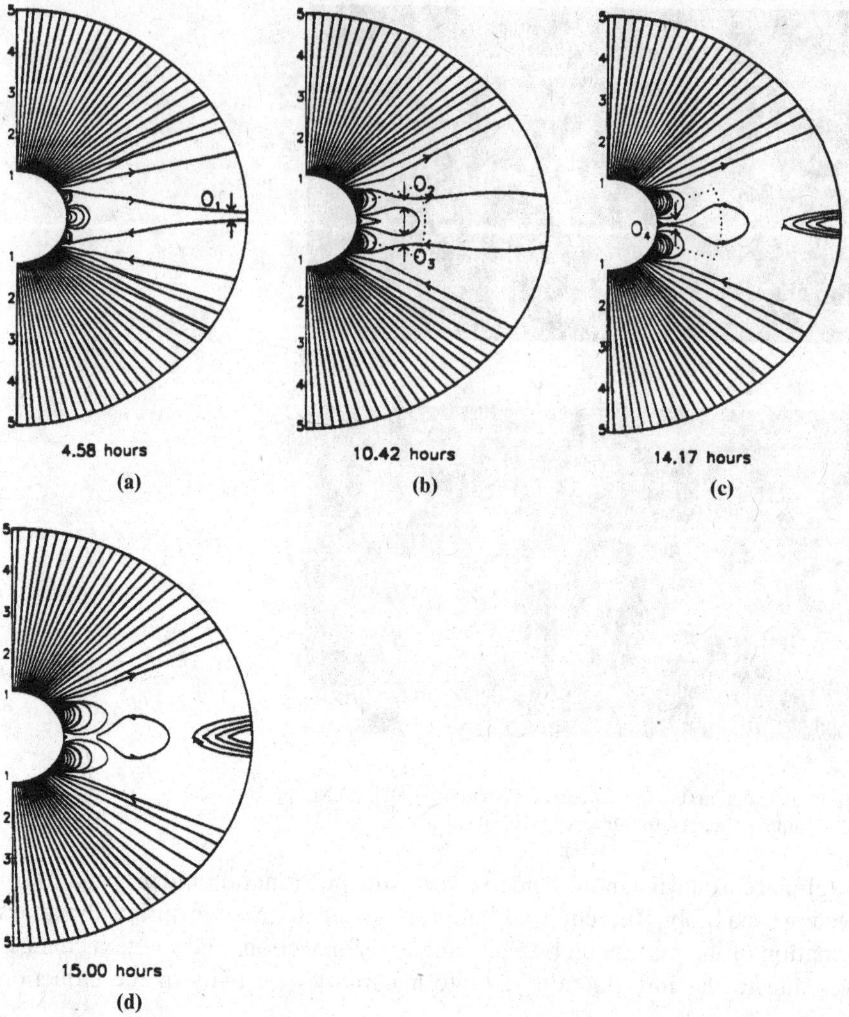

Figure 26.4.1.3 Evolution of magnetic field configuration during the magnetic reconnection where $OI(I = 1\text{-}5)$ represents the location of magnetic reconnection occurring at times of 4.58 hours, 10.42 hours, 14.17 hours, and 16.25 hours, respectively, after introduction of the magnetic resistivity for $S = 100$ [Wu et al., 2000].

(SF_6) discharge. The effects of pressure and current amplitude variation and the effect of ion mobility variation with electric field are examined. The predicted emission of the 750.4 nm argon line (Figure 26.4.2.1a) is compared with experiments (Figure 26.4.2.1b). Given conditions are: 1 torr, 2-cm spacing, 13.5 MHz, and 0.14 W/cm^2, and 0.14 W/cm^2.

(2) Two-dimensional Capacitively Coupled RF Glow Discharge

Young and Wu [1993] modeled electrons with a three-moment nonequilibrium model and ions are modeled with a nonequilibrium single-moment model which includes an ionic effective electric field. The nonuniform plasma density profiles and the

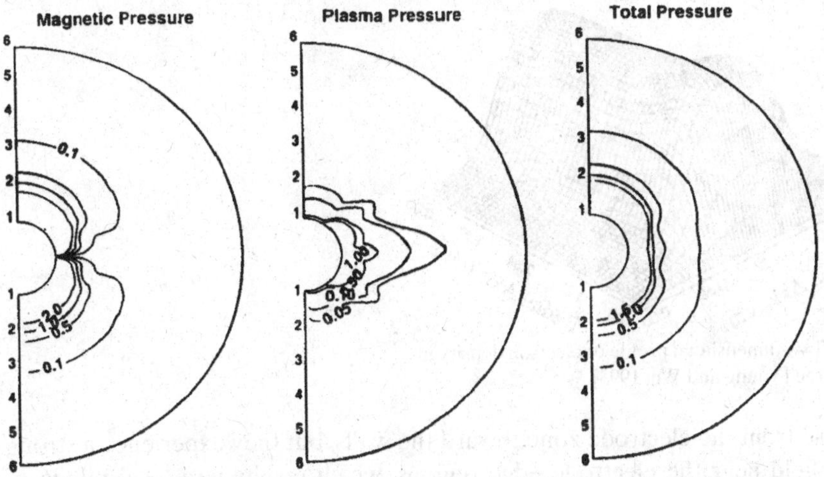

Figure 26.4.1.4 Pressure distributions in the meridianal plane at $t = 12.5$ hours after introduction of magnetic resistivity for S = 100 [Wu et al., 2000].

radial sheath width variation with various gas pressures are investigated. The electron density profiles from a two-dimensional simulation for argon are shown in Figure 26.4.2.2. The ambipolar structure in the axial direction is evident. The two-dimensional results for plasma variables in the center of the discharge are only 5% different from the corresponding one-dimensional results. It is near the edge that the density is different from that in the center. Indeed, the density near the walls is higher. The electrons

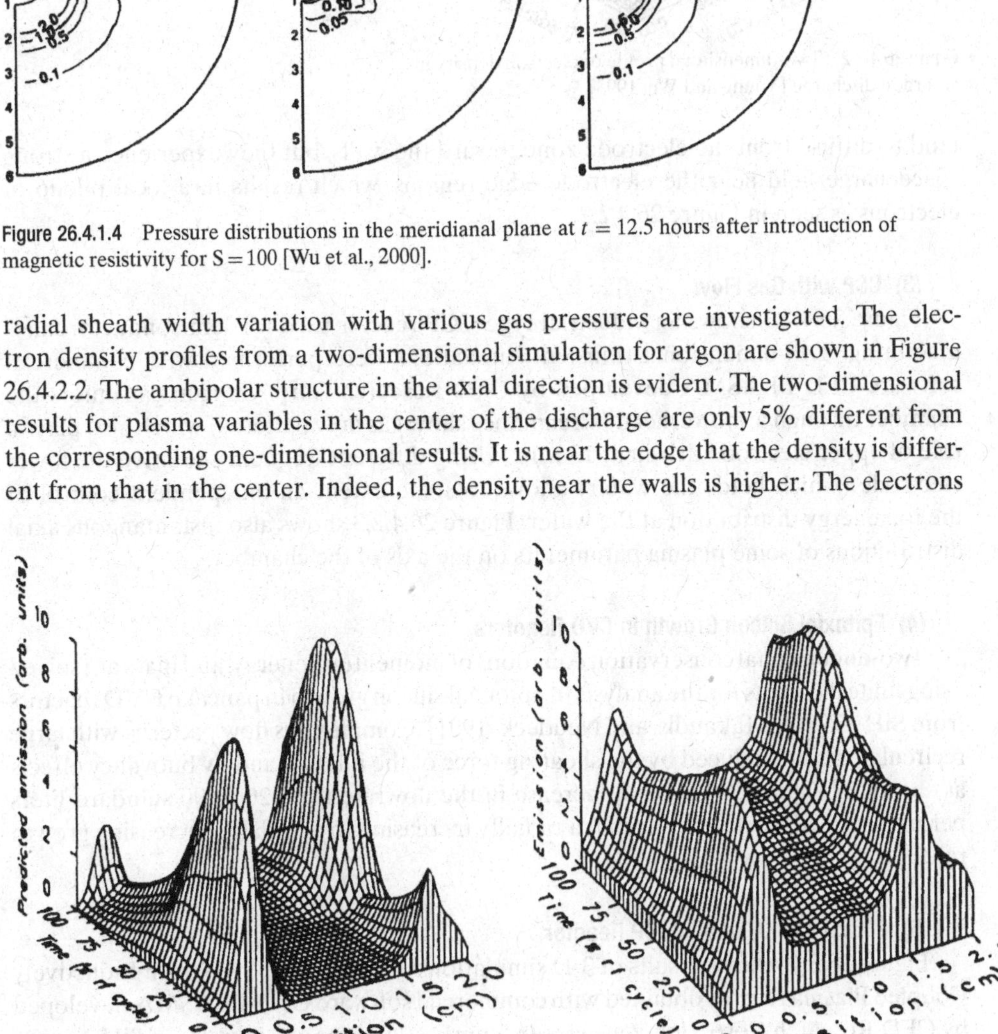

Figure 26.4.2.1 RF glow discharge [Gogolides and Swain, 1992]. **(a)** Predictions from a two-moment model for emission of the 750.4-nm argon line (1 torr, 2 cm spacing, 13.5 Mhz, and 0.14 W/cm^2). **(b)** Experimentally measured emission.

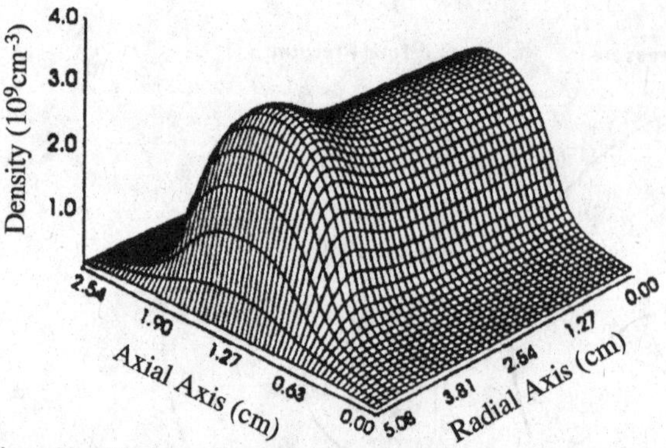

Figure 26.4.2.2 Two-dimensional profile of electron density in an argon discharge [Young and Wu, 1993].

tend to diffuse from the electrode zone toward the walls, but they experience a strong spacecharge field near the electrode edge regions, which results in a local pileup of electrons as seen in Figure 26.4.2.2.

(3) CCP with Gas Flow

Figure 26.4.2.3 shows an industrial type CCP reactor used for semiconductor manufacturing. Two-dimensional transient simulations were performed with commercial software CFD-PLASMA developed by CFD Research Corp. Oxygen gas enters the chamber through a showerhead (streamlines show gas flow patterns) which also plays a role of upper electrode driven at 13 Mhz. Using different frequency to drive the lower electrode (which holds the wafer) allows to some extent an independent control of the ion energy distribution at the wafer. Figure 26.4.2.3 shows also instantaneous axial distributions of some plasma parameters on the axis of the chamber.

(4) Epitaxial Silicon Growth in CVD Reactors

Two-dimensional conservation equations of momentum, energy, and mass are solved using finite elements for the analysis of epitaxial silicon growth in pancake CVD reactors from SiH_2Cl_2 [Oh, Takaudis, and Neudeck, 1991]. Complex gas flow patterns with large recirculations are induced by the shearing force of the inlet jet and by buoyancy effects as shown in Figure 26.4.2.4. An increase in the flowrate from 20 to 90 standard liters per minute leads to a reverse from a radially increasing to a radially decreasing growth rate.

(5) PECVD of SiO_2 in a 3D ICP Reactor

Figure 26.3.1c shows results of 3-D simulations of PECVD of SiO_2 in an Inductively Coupled Plasma reactor obtained with commercial software CFD-PLASMA developed by CFD Research Corp. (*http://www.cfdrc.com/~cfdplasma*). A mixture of $SiH_4/Ar/O_2$ is injected into the gas chamber through vertical injectors and exits through an outlet at the bottom of the chamber (streamlines show the gas flow patterns). The total gas pressure in the reactor is maintained at the 10 mTorr level. RF current in coils wrapped around a dome chamber induces an azimuthal electric field which maintains the plasma.

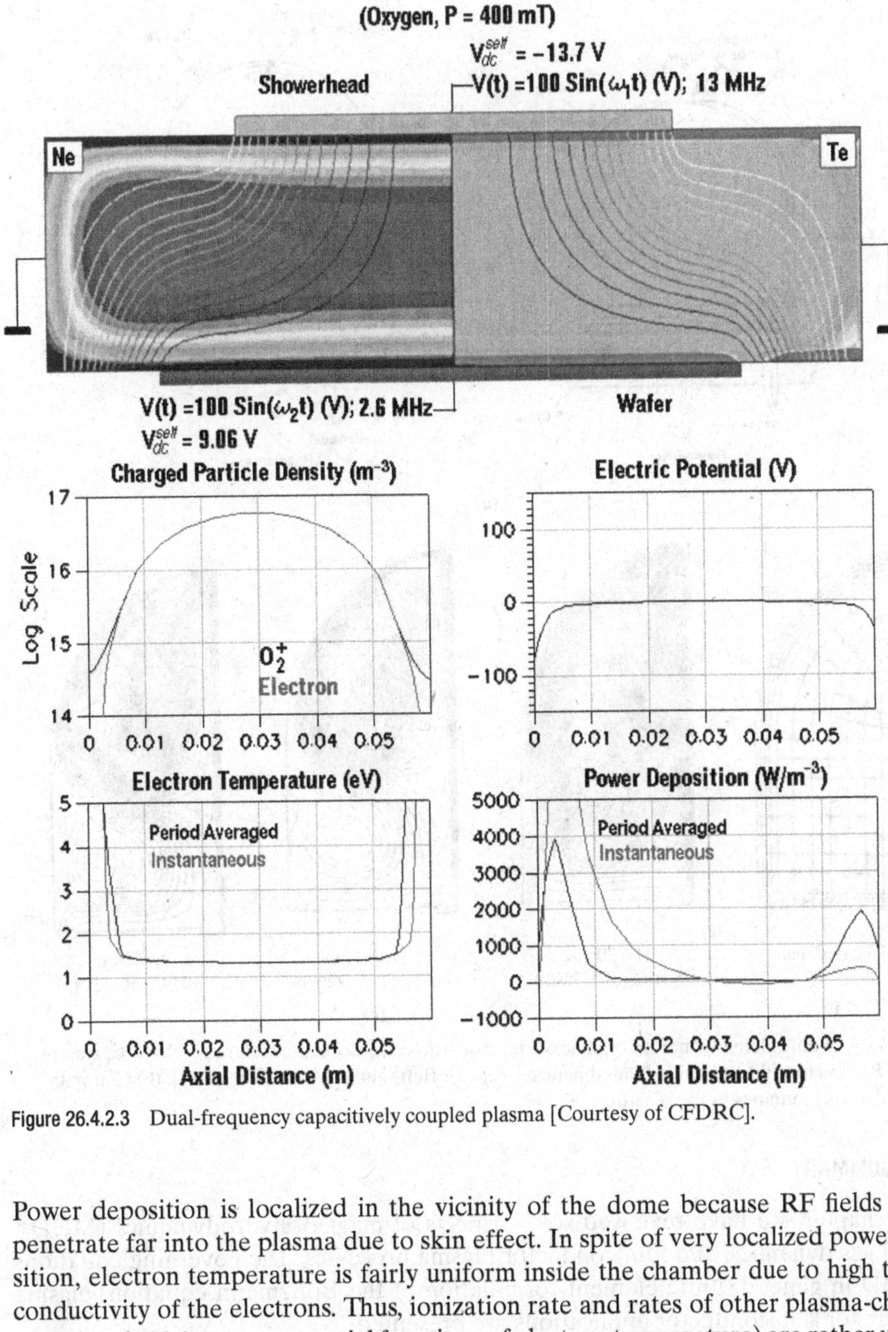

Figure 26.4.2.3 Dual-frequency capacitively coupled plasma [Courtesy of CFDRC].

Power deposition is localized in the vicinity of the dome because RF fields cannot penetrate far into the plasma due to skin effect. In spite of very localized power deposition, electron temperature is fairly uniform inside the chamber due to high thermal conductivity of the electrons. Thus, ionization rate and rates of other plasma-chemical reactions (which are exponential functions of electron temperature) are rather uniform within a volume and the plasma density has a maximum in the chamber center because of recombination loss of charged particles at the walls of the chamber. A processing wafer is placed on top of an electrostatic chuck (a pedestal in Figure 26.3.1c). Contours of calculated SiO_2 deposition rate are shown in Figure 26.3.1c.

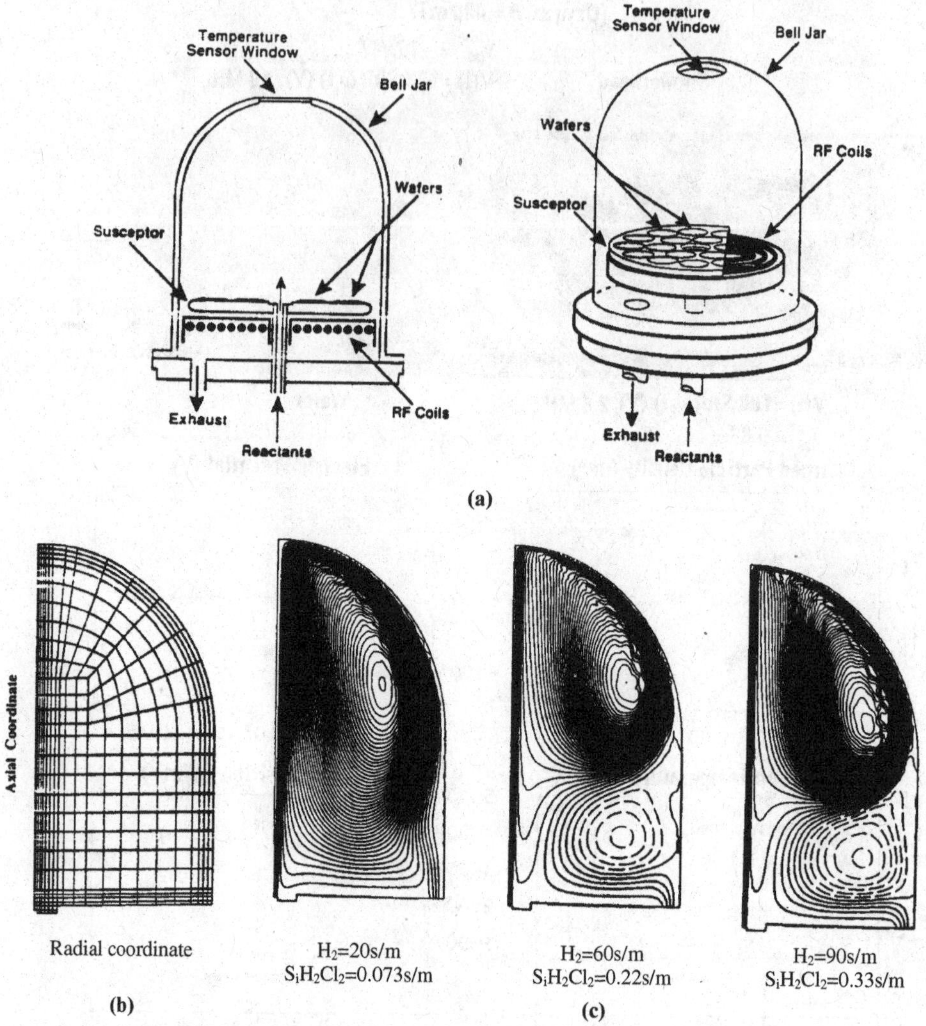

Figure 26.4.2.4 Finite element analysis of pancake reactor. The susceptor temperature $T_s = 950°C$, system pressure $P = 150$ torr. **(a)** Induction heated pancake reactor (left) and cross-section (right). **(b)** Computational grid. **(c)** Streamlines in the reactor.

26.5 SUMMARY

In this chapter, we have reviewed some aspects of magnetohydrodynamics (MHD), rarefied gas dynamics, and semiconductor plasma processes. The governing equations for MHD in general, finite element formulation of the Boltzmann equation, plasma discharge for semiconductor applications are presented.

Some of the example problems in coronal mass ejection using the fully implicit continuous Eulerian method [Wu et al., 2000] are reviewed. Here the governing equations are based on the resistive MHD theory in which the field topology is changed due to the magnetic reconnection process. They consist of conservation laws of mass, momentum, energy, and induction equations describing the dynamical interaction between plasma flow and magnetic field.

Example problems for plasma processing in semiconductors include capacitively coupled RF glow discharge and radial flow in an RF glow discharge. A brief discussion of epitaxial silicon growth in chemical vapor deposition reactors is also presented. In addition, the results of 3-D simulations of PECVD of SiO_2 in inductively coupled plasma rector are demonstrated, representing the current state of the art.

REFERENCES

Aliev, Y. M., Kaganovich, I. D., and Schluter, H. [1997]. Quasilinear theory of collisionless electron heating in rf gas discharges. *Phys. Plasmas.*, 4, 2413–21.

Barnes, M. S., Colter, T. J., Elter, M. E. [1987]. Large-signal time-domain modeling of low pressure RF glow discharges. *J. Appl. Phys.*, 61, 1, 81–89.

Bhatnagar, E. P., Gross, E. O., and Krook, M. [1954]. A model for collision processes in gases; I. Small amplitude processes in charged and neutral one-component systems. *Phy. Rev.*, 94, 511.

Boenig, H. V. [1982]. *Plasma Science and Technology*. Ithaca, NY: Cornell University Press.

Bose, D., Govindan, T. R., and Meyyappan, M. [1999]. A continuum model for the inductively coupled plasma reactor in semiconductor processing. *J. Electrochem. Soc.*, 146, 7, 2705–11.

Bringuier, E. [1999]. The current equation in strong electric fields. *Philosoph. Mag. B*, 79, 10, 1659–72.

Chantry, P. J. [1987]. A simple formula for diffusion calculations involving wall reflection and low density. *J. Appl. Phys.*, 62, 4, 1141–48.

Chapman, B. [1980]. *Glow Discharge Processes*. New York: Wiley.

Chapman, B. and Cowling, T. G. [1952]. *The Mathematical Theory of Non-uniform Gases*. New York: Cambridge University Press.

Chen, I. [1983]. Mass transfer analysis of the plasma deposition process. *Thin Solid Fils.*, 101, 41–53.

Chung, T. J. [1978]. *Finite Element Analysis in Fluid Dynamics*. New York: McGraw-Hill.

Chung, T. J., Oden, J. T., and Wu, S. T. [1974]. The finite element analysis of transient rarefied gas flow. *Proceedings of Int. Symposium in Finite Element Methods in Flow Problems*. Swansea, England.

Economou, D. J. and Alkier, R. C. [1988]. A mathematical model for a parallel plate plasma etching reactor. *J. Electrochem. Soc.*, 135, 11, 2787–94.

Godyak, V. A. and Kolobov, V. I. [1998]. Effect of collisionless heating on electron energy distribution in inductively coupled plasma. *Phys. Rev. Letters.*, 81, 369–72.

Godyak, V. A., Piejak, R. B., Alexandrovich, B. M., and Kolobov, V. I. [1999]. Hot plasma and nonlinear effects in inductive discharges. *Phys. Plasmas*, 6, 1804–12.

Gogolides, E. and Swain, H. H. [1992]. Continuum modeling of radiofrequency glow discharges I: Theory and results for electropositive and electronegative gases. *J. Appl. Phys.*, 72, 3971–87, II: Parametric studies and sensitivity analysis. 3988–4002.

Graves, D. B. and Jensen, K. F. [1986]. A continuum model of DC and RF discharges. *IEEE Trans. Plasma Sci.*, PS-14, 78–89.

Gross, E. P., Jackson, E. A., and Ziering, S. [1957]. Boundary value problems in kinetic theory of gases. *Annals Phys.*, 1, 141–67.

Hitchon, W. N. G. [1999]. *Plasma Processing for Semiconductor Fabrication*. London: Cambridge University Press.

Kirmse, K. H. R., Wendt, A. E., Disch, S. B., Wu, J. Z., Abraham, I. C., Meyer, J. A., Breun, R. A., and Woods, R. C. [1996]. *J. Vacuum Sci. Tech.*, B 14, 710.

Kolobov, V. I. [1998]. Anomalous skin effect in bounded systems. In U. Kortschagen and L. D. Tsendin (eds.) *Electron Kinetics and Applications of Glow Discharges*. New York: Plenum Press.

Kolobov, V. I. and Economou, D. J. [1997]. Anomalous skin effect in gas discharge plasmas (Review). *Plasma Sources Sci. Tech.*, 6, R1.

Kolobov, V. I. and Godyak, V. A. [1995]. Non-local electron kinetics in collisional plasmas (Review), *IEEE Trans. Plasma Sci.*, 23, 503.

Kortchagen, U., Buch, C. and Tsendin, L. D. [1996]. On simplifying approaches to the solution of the Boltzmann equation in spatially inhomogeneous plasma (Review), *Plasma Source Sci. Tech.*, 5, 1.

Lieberman, M. A. and Godyak, A. [1998]. From Fermi acceleration to stochastic electron heating in gas discharges. *IEE Trans. Plasma Sci.*, 26, 955–86.

Lii, Y., Jorne, J, Cadien, K. C., and Schoenholtz, Jr. [1990]. Plasma etching of silicon in SF_6: experimental and reactor modeling. *J. Electrochem. Soc.*, 137, 11, 3633–39.

Manos, D. M. and Flamm, D. L. [1989]. *Plasma Etching: An Introduction*. San Diego: Academic.

Meyyappan, M. (ed.) [1995]. *Computational Modeling in Semiconductor Processing*. Boston: Artech House.

Meyyappan, M. and Buggeln, R. [1990]. A process model for reactive ion and etching and study of the effects of magnetron enhancement. Materials Research Society Symposium Proceedings. Vol. 158, Material Research Society, Pittsburgh, 395–400.

Oh, I. H., Takaudis, C. G., and Neudeck, G. W. [1991]. Mathematical modeling of epitaxial silicon growth in pancake chemical vapor deposition reactors. *J. Electrochem. Soc.*, 138, 2, 554–67.

Richards, A. D., Thompson, B. E., and Swain, H. H. [1987]. Continuum modeling of argon radio frequency glow discharges. *Applied Phy. Lett.*, 50, 6, 2782–92.

Richley, E. A. [1999]. Elliptic representation of the Boltzman equation with validity for all degree of anisotropy. *Phys. Rev. E*, 59, 4533–41.

Shkarofsky, I. P., Johnston, T. W., and Vachynski, M. P. [1966]. *The Particle Kinetics of Plasmas*. Reading, MA: Addison-Wesley.

Stenger, Jr., H. G., Caran, H. S., Sullivan, C. F., and Russo, W. M. [1987]. Reaction kinetics and reactor modeling of silicon. *AIChE J.*, 33, 7, 1187–96.

Surendra, M. and Graves, D. B. [1991a]. Particle simulation of radio-frequency glow discharges. *IEEE Trans. Plasma Sci.*, 19, 2, 144–57.

———. [1991b]. Capacitively coupled glow discharges at frequencies above 13.56 MHz, *Appl. Phys. Lett.*, 17, 2091–93.

Venkatesan, S. P., Trachtenburg, and Edgar, T. F. [1992]. Effect of flow direction on etch uniformity in parallel–plate (radial flow) isothermal plasma reactors. *J. Electrochem. Soc.*, 134, 12, 3193–97.

Ward, A. L. [1958]. Effect of space charge in cold-cathode gas discharges. 112, 6, 1852–57.

Wu, S. T. and Wang, J. F. [1987]. Numerical tests of a modified full implicit continuous Eulerian (FICE) scheme with projected normal characteristic boundary conditions for MHD flows. *Comp. Meth. Appl. Mech. Eng.*, 64, 267–82.

Wu, S. T., Wang, A. H., Plunkett, S. P., and Michels, D. J. [2001]. Evolution of global coronal magnetic field due to magnetic reconnection: The formation of the observed blob motion in the coronal streamer belt. *Ap. J.* 545, 1101–15.

Young, F. F. and Wu, C. H. [1993]. Radial flow effects in a multidimensional, three-moment fluid model of a radio frequency glow discharge. *Appl. Phys. Lett.*, 62, 5, 473–75.

CHAPTER TWENTY-SEVEN

Applications to Relativistic Astrophysical Flows

27.1 GENERAL

Relativistic theory is divided into two categories: special relativity and general relativity. In special relativity, we follow Einstein's postulate establishing the universality of the speed of light, *c*, relative to any unaccelerated observer, regardless of the motion of the light's source from the observer. General relativity arises as an extension to special relativity to describe the motion of particles evolving under the presence of gravitational fields. In order to take into account the effect of gravitation, however, we must abandon the Eulerian coordinates used in Newtonian fluid dynamics. Instead, it is necessary to invoke a curvilinear four-dimensional manifold (the spacetime) to represent particle's trajectories.

Many of the problems encountered in astrophysics are involved in the numerical solution of special or general relativistic fluid dynamics equations. Active research in this subject area has been in progress for the past 40 years. Earlier studies include structure and evolution of stars [Chandrasekhar, 1942; Aller and McLaughlin, 1965, among others]. Black hole accretion flows have been studied extensively as evident from numerous publications [Paczynski and Wiita, 1980; Katz, 1980; Eggum et al., 1988; Hawley et al., 1984a,b; Clarke et al., 1985; Stella and Vietri, 1997; Bromley et al., 1998; Font et al., 1998a,b; Koide et al., 1999; Font et al., 1999]. Some of the recent activities include Gamma ray bursts [Meszaros and Rees, 1993; Sari and Piran, 1998; Fishman and Meegan, 1995; and Panattescu and Meszros, 1998], explosive and jet phenomena [Norman, 1997], and astrophysical turbulence flows and instability [Bulbus and Hawley, 1998].

Despite these developments in "computational astrophysical fluid dynamics," many difficulties remain unresolved. Among them are the rapidly rotating stars, detailed accretion disk structure and evolution, evolving and interacting binaries, etc. First of all, boundary conditions are unknown in many instances. It is difficult to predict the geometry for the solution until the problem is actually solved. The shape of the outer surface of a rotating star or accretion disk can be part of the solution. In order to resolve such difficulties, arbitrary assumptions such as spherical symmetry or slowly rotating perturbation problems have been suggested [Kippenhahn and Thomas, 1970]. Uncertainties of boundary conditions may be resolved using iterative processes usually accommodated in FDM, FEM, or FVM.

The main computational issues involved in relativistic hydrodynamics stem from discontinuities associated with shock waves. High-resolution shock-capturing techniques developed in Newtonian computational fluid dynamics (CFD) have been proven successful also in dealing with relativistic astrophysical flows. Balsara [1994] studied an extension of the Riemann solver to resolve shock waves using a secant method and Newton method for iterative solutions. In this work, an exact treatment of transverse velocities across general, oblique shocks was enforced and, as a result, the equivalence to the nonrelativistic limit was demonstrated.

Another approach to the Riemann solver with various reconstruction schemes in three-dimensions was reported by Aloy et al. [1999] as an extension of earlier studies [Marti et al., 1991; Marti et al., 1995]. The results including spherical shock reflection with the Lorentz factor of 700 and larger are shown to be satisfactory.

Gravitational effects in symmetric spherical coordinates with pseudo-Newtonian approximations have been studied by Nobuta and Hanawa [1999] using the total variation diminishing (TVD) scheme [Roe, 1981] in their investigation of time-dependent inviscid hydrodynamical accretion flows onto a black hole. In this work, accretion that consists of hot tenuous gas with low specific angular momentum and cold dense gas with high specific angular momentum was considered, resulting in the hot gas accreting continuously and the cold gas intermittently as blobs. It is shown that the high specific angular momentum gas blobs bounce at the centrifugal barrier and create shock waves.

Meier [1999] examined finite element methods (FEM) for applications to multidimensional astrophysical structural and dynamical analysis to study rotating stars, interacting binaries, thick advecting accretion disks, and four-dimensional spacetime problems in general. In this approach, the complex differential equations on the arbitrary curvilinear grid are generated automatically by the FEM integrals.

In this chapter, we explore applications of the existing CFD technologies to relativistic astrophysical fluid dynamics. We first review the governing equations in Section 27.2. This is followed by some selected example problems presented in Section 27.3 [Richardson et al., 1999; Richardson, 2000; Richardson and Chung, 2002].

27.2 GOVERNING EQUATIONS IN RELATIVISTIC FLUID DYNAMICS

27.2.1 RELATIVISTIC HYDRODYNAMICS EQUATIONS IN IDEAL FLOWS

The equations describing the evolution of a relativistic fluid are local conservation laws of the stress-energy, $T^{\mu\nu}$, and the matter current density, J^{μ}, given by covariant derivatives as follows:

$$T^{\mu\nu}_{;\mu} = 0 \quad (\mu, \nu = 0, 1, 2, 3) \tag{27.2.1}$$

$$J^{\mu}_{;\mu} = 0 \tag{27.2.2}$$

with

$$J^{\mu} = \rho u^{\mu} \tag{27.2.3}$$

where ρ is the rest mass density and u^μ the 4-velocity of the fluid. Note also that the semicolon implies the covariant derivative with respect to the 4-metric of the underlying spacetime.

For simplicity, we consider a perfect fluid (viscosity and thermal conduction are neglected). In this case the energy-momentum tensor can be written as

$$T^{\mu\nu} = \rho h u^\mu u^\nu + p g^{\mu\nu} \tag{27.2.4}$$

with $g^{\mu\nu}$ being the 4-metric describing the spacetime and h the specific enthalpy defined as

$$h = 1 + \varepsilon + \frac{p}{\rho} \tag{27.2.5}$$

where ε is the specific internal energy and p the isotropic pressure. The above system of equations can be closed by the normalization condition for the four-velocity,

$$g_{\mu\nu} u^\mu u^\nu = -1 \tag{27.2.6}$$

and the equation of state,

$$p = p(\rho, \varepsilon) \tag{27.2.7}$$

Dynamics of the gravitational field in general relativity theory is described by the Einstein' field equation

$$G_{\mu\nu} = 8\pi T_{\mu\nu} \tag{27.2.8}$$

where $G_{\mu\nu}$ is the Einstein tensor associated with the ten metric components $g_{\mu\nu} = g_{\nu\mu}$ of the spacetime and the stress energy tensor $T_{\mu\nu}$. There are various forms of the Einstein's equations suitable for numerical analyses such as suggested by Arnowitt, Deser, and Misner [1962], known as the ADM or $(3+1)$ formulation, and Bona et al. [1995], known as the BM hyperbolic formulation.

In the ADM formulation, the spacetime is considered to be foliated into a set of non-interacting spacelike hypersurfaces. There are two kinematic variables which describe the evolution between these surfaces: the lapse function $\alpha = (-g^{tt})^{-\frac{1}{2}}$, which describes the rate of advance of time along a timelike unit vector n^μ normal to a surface, and the spacelike shift vector β^i that describes the motion of coordinates. These parameters are related to the line element,

$$ds^2 = -(\alpha^2 - \beta_i \beta^i) dt^2 + 2\beta_i dx^i dt + \eta_{ij} dx^i dx^j \tag{27.2.9}$$

with η_{ij} being the three-dimensional metric tensor. The ADM formulation, then, casts the Einstein's equations into a first order (in time) quasi-linear system of equations.

In the BM hyperbolic formulation, the evolution equations are written as a first order balance law with the same mathematical structure as the hydrodynamic equations usually employed in traditional CFD approaches.

Instead of using the covariant derivatives, it is convenient to formulate the numerical process via a coordinate system $(x^0 = t, x^1, x^2, x^3)$ and express (27.2.1) and (27.2.2) in terms of coordinate derivatives. To this end, we write the governing conservation equations in the form [Hawley et al., 1984; Banyuls et al., 1997; Font et al., 1998],

$$\frac{\partial \mathbf{U}}{\partial t} + \frac{\partial \mathbf{F}^i}{\partial x^i} = \mathbf{B} \tag{27.2.10}$$

with the conservation variables **U** written as

$$\mathbf{U} = \begin{bmatrix} D \\ S_j \\ \tau \end{bmatrix} = \begin{bmatrix} \sqrt{\eta} W \rho \\ \sqrt{\eta} \rho h W^2 v_j \\ \sqrt{\eta}(\rho h W^2 - p - W\rho) \end{bmatrix} \tag{27.2.11}$$

where η is the determinant of η_{ij}, v_j is the fluid 3-velocity, and W is the Lorentz factor,

$$W = (1 - \eta_{ij} v^i v^j)^{-1/2} \tag{27.2.12}$$

where we invoke the natural unit,

$$G = c = 1$$

and the spatial components of the 4-velocity u^i are related to the 3-velocity v_i by

$$u^i = W(v^i - \beta^i/\alpha) \tag{27.2.13}$$

The convection flux vector $\mathbf{F}^i$ and the source vector **B** are given by

$$\mathbf{F}^i = \begin{bmatrix} \alpha\left(v^i - \dfrac{\beta^i}{\alpha}\right) D \\ \alpha\left(\left(v^i - \dfrac{\beta^i}{\alpha}\right) S_j + \sqrt{\eta} p \delta^i_j\right) \\ \alpha\left(\left(v^i - \dfrac{\beta^i}{\alpha}\right)\tau + \sqrt{\eta} v^i p\right) \end{bmatrix} \tag{27.2.14}$$

$$\mathbf{B} = \begin{bmatrix} 0 \\ \alpha\sqrt{\eta} T^{\mu\nu} g_{\nu\sigma} \Gamma^{\sigma}_{\mu j} \\ \alpha\sqrt{\eta}(T^{\mu 0}\alpha_{,\mu} - \alpha T^{\mu\nu}\Gamma^{0}_{\mu\nu}) \end{bmatrix} \tag{27.2.15}$$

where D, S_j, and τ are defined in (27.2.11) and $\Gamma^{\sigma}_{\mu\nu}$ is the 4-Christoffel symbol,

$$\Gamma^{\sigma}_{\mu\nu} = \frac{1}{2} g^{\alpha\beta}(g_{\nu\beta,\mu} + g_{\mu\beta,\nu} - g_{\mu\nu,\beta}) \tag{27.2.16}$$

The numerical solution of (27.2.10) can be carried out using any one of the schemes developed in Chapter 6 for FDM, Chapter 13 for FEM, or Chapters 7 and 15 for FVM. Some applications of (27.2.10) without the source term are presented in Section 27.3.4.

27.2.2 RELATIVISTIC HYDRODYNAMICS EQUATIONS IN NONIDEAL FLOWS

In special relativity, the effect of gravitational fields is neglected. Viscosity and heat conduction may be included in both general and special relativistic flows. However, one of the least investigated physical phenomena in the relativistic flows is the effect of viscosity and heat conduction upon the flow field. The primary reason for this lack of studies is the difficulty involved in computation, as is usually the case for shock wave turbulent boundary layer interactions in the Newtonian flows. Anticipating that this will be a future research topic in the relativistic hydrodydamics, some of the basic governing equations are presented in this section.

The metric tensor components in special relativity [Misner, Thorne, and Wheeler, 1973] are defined by the Minkowski geometry, $g_{\alpha\beta}$. The Kerr black hole geometry is commonly used in general relativity since it describes a black hole that has angular momentum and since it easily reduces to the Schwarzschild metric for non-rotating black holes. In general, the components of the metric tensor are derived from the line element ds^2 given by

$$ds^2 = g_{\alpha\beta} dx^\alpha dx^\beta$$

For the special relativistic line element, we have

$$g_{\alpha\beta} = \begin{bmatrix} -1 & 0 & 0 & 0 \\ 0 & 1 & 0 & 0 \\ 0 & 0 & 1 & 0 \\ 0 & 0 & 0 & 1 \end{bmatrix}$$

$$ds^2 = -dt^2 + dx^2 + dy^2 + dz^2$$

The Kerr line element for angular momentum a per unit mass M is written as [Misner et al., 1973]

$$ds^2 = -\frac{\Delta - a^2 \sin^2\theta}{\omega^2} dt^2 - 2a \frac{2Mr \sin^2\theta}{\omega^2} dt\, d\phi + \frac{(r^2+a^2)^2 - a^2 \Delta \sin^2\theta}{\omega^2} \sin^2\theta\, d\phi^2 + \frac{\omega^2}{\Delta} dr^2 + \omega^2 d\theta^2$$

with

$$\Delta \equiv r^2 - 2Mr + a^2$$

$$\omega^2 \equiv r^2 + a^2 \cos^2\theta$$

The general relativistic Kerr metric is of the form,

$$g_{\alpha\beta} = \begin{bmatrix} -\dfrac{\Delta + a^2 \sin^2\theta}{\omega^2} & 0 & 0 & \dfrac{a \sin^2\theta(\Delta - (r^2+a^2))}{\omega^2} \\ 0 & \dfrac{\omega^2}{\Delta} & 0 & 0 \\ 0 & 0 & \omega^2 & 0 \\ \dfrac{a \sin^2\theta(\Delta - (r^2+a^2))}{\omega^2} & 0 & 0 & -\dfrac{a^2 \sin^4\theta\, \Delta + \sin^2\theta (r^2+a^2)^2}{\omega^2} \end{bmatrix}$$

Further details of metrics and Christoffel symbols are presented in Appendix D-1.

The fluid four-acceleration is defined as

$$A_\alpha = v_{\alpha;\beta} v^\beta \tag{27.2.17}$$

with

$$\mathrm{v}_{\alpha;\beta} = \mathrm{v}_{\alpha;\gamma} P^{\gamma}_{\beta} - c^{-2} A_\alpha \mathrm{v}_\beta \tag{27.2.18}$$

where P^{γ}_{β} is the projection tensor defined as

$$P^{\gamma}_{\beta} = \delta^{\gamma}_{\beta} - c^{-2} \mathrm{v}^{\gamma} \mathrm{v}_\beta \tag{27.2.19}$$

Defining the symmetric rotation tensor $\Omega_{\alpha\beta}$ as

$$\Omega_{\alpha\beta} = \frac{1}{2}\left(\mathrm{v}_{\alpha,\gamma} P^{\gamma}_{\beta} - \mathrm{v}_{\beta,\gamma} P^{\gamma}_{\alpha}\right) \tag{27.2.20}$$

and the symmetric shear tensor $E_{\alpha\beta}$ as

$$E_{\alpha\beta} = \frac{1}{2}\left(\mathrm{v}_{\alpha,\gamma} P^{\gamma}_{\beta} + \mathrm{v}_{\beta,\gamma} P^{\gamma}_{\alpha}\right) \tag{27.2.21}$$

we may express the velocity gradient in the form,

$$\mathrm{v}_{\alpha;\beta} = D_{\alpha\beta} + \Omega_{\alpha\beta} + \frac{1}{3}\theta P_{\alpha\beta} - c^{-2} A_\alpha \mathrm{v}_\beta \tag{27.2.22}$$

where

$$D_{\alpha\beta} = E_{\alpha\beta} - \frac{1}{3}\theta P_{\alpha\beta} \tag{27.2.23}$$

$$\theta = \mathrm{v}^{\alpha}_{;\alpha} \tag{27.2.24}$$

Introducing the shear viscosity μ and the bulk viscosity ς, and using the general form for speed of light, the energy stress tensor can be written as

$$M_{\alpha\beta} = \rho_{00} \mathrm{v}_\alpha \mathrm{v}_\beta + p P_{\alpha\beta} - 2\mu D_{\alpha\beta} - \varsigma\,\theta P_{\alpha\beta} + c^{-2}(Q_\alpha \mathrm{v}_\beta + Q_\beta \mathrm{v}_\alpha) \tag{27.2.25}$$

where ρ_{00} is the total mass density

$$\rho_{00} = \rho_0(1 + \varepsilon/c^2) \tag{27.2.26}$$

and Q_α is the four-vector generalization of the heat flux,

$$Q_\alpha = -kP_{\alpha\beta}\left(T_{,\beta} + T\mathrm{v}_{\alpha;\beta}\mathrm{v}^{\beta}\right)$$

To derive the equations of motion, we take a covariant derivative of the energy stress tensor,

$$\begin{aligned} M^{\alpha\beta}_{;\beta} = &\left\{\frac{D\rho_{00}}{D\tau} + \theta[\rho_{00} + c^{-2}(p - \varsigma\theta)]\right\}\mathrm{v}^{\alpha} + [\rho_{00} + c^{-2}C]A^{\alpha} + P^{\alpha\beta}(p - \varsigma\theta)_{,\beta} \\ &-2\mu D^{\alpha\beta}_{;\beta} + c^{-2}\left[\frac{DQ^{\alpha}}{D\tau} + \frac{4}{3}\theta Q^{\alpha} + \mathrm{v}^{\alpha} Q^{\beta}_{;\beta} + Q_\beta(D^{\alpha\beta} + \Omega^{\alpha\beta})\right] = F^{\alpha} \end{aligned} \tag{27.2.27}$$

where τ is the relativistic proper time, related to the Newtonian time t as

$$\frac{D}{D\tau} = \left(\frac{dt}{d\tau}\right)\frac{\partial}{\partial t} + \left(\frac{dx^i}{d\tau}\right)\frac{\partial}{\partial x^i} = \frac{\mathrm{v}^0}{c}\frac{\partial}{\partial t} + \mathrm{v}^i\frac{\partial}{\partial x^i}$$

It follows from (27.2.27) that

$$F_\alpha = P_{\alpha\gamma} F^\gamma \tag{27.2.28}$$

Substituting (27.2.27) into (27.2.28), we obtain the equations of motion

$$\begin{aligned}\left[\rho_{00} + c^{-2}(p - \theta\varsigma)\right]\frac{Dv_\alpha}{D\tau} &= F_\alpha - P^\beta_\alpha(p - \theta\varsigma)_{,\beta} + 2\mu P_{\alpha\gamma} D^{\gamma\beta}_{;\beta} \\ &\quad - c^{-2}\left[P_{\alpha\gamma}\frac{DQ^\gamma}{D\tau} + \frac{4}{3}\theta Q_\alpha + P_{\alpha\gamma} Q_\beta (D^{\gamma\beta} + \Omega^{\gamma\beta})\right]\end{aligned} \tag{27.2.29}$$

Note that the speed of sound used in (27.2.25) through (27.2.29) will be set equal to unity.

It is interesting to note that, in cartesian coordinates, we obtain

$$\frac{D}{D\tau} \to \frac{D}{Dt}$$

$$P_{ij} \to \delta_{ij}$$

with (27.2.29) being transformed into the Newtonian momentum equations.

The relativistic energy equation can be derived first by multiplying (27.2.27) with the relativistic velocity v_α, by invoking the first and second laws of thermodynamics, together with the entropy S, and following the procedure similar to the Newtonian counter part [Chung, 1996, 175–77]. This process will lead to

$$\rho_0 T\frac{DS}{D\tau} = \rho_0\left[\frac{D\varepsilon}{D\tau} + p\frac{D}{D\tau}\left(\frac{1}{\rho_0}\right)\right] = 2\mu D^{\alpha\beta} D_{\alpha\beta} + \varsigma\theta^2 - \left(Q^\beta_{;\beta} + c^{-2} Q^\alpha A_\alpha\right) \tag{27.2.30}$$

The solution of (27.2.42) in terms of entropy is cumbersome. Thus, the energy equation is cast in the form

$$\rho_0\left[\frac{D\varepsilon}{D\tau} + \frac{p}{\rho_0} v^\alpha_{;\alpha}\right] = 2\mu D^{\alpha\beta} D_{\alpha\beta} + \varsigma\theta^2 - \left(Q^\beta_{;\beta} + c^{-2} Q^\alpha A_\alpha\right) \tag{27.2.31}$$

Additional details are available in Eckert, 1940; Weinberg, 1972; Misner et al., 1973; and Miharas and Miharas, 1984, among others.

In order to achieve convergence to accurate solutions for the problems with shock discontinuities, the governing equations must be written in conservation form. To this end, the nonconservation forms of the momentum and energy equations given above, together with the continuity equation, are transformed into a conservation form. The acceptable conservation form should be capable of recovering the nonconservation form.

$$\frac{\partial \mathbf{U}}{\partial t} + \frac{\partial \mathbf{F}_i}{\partial x_i} + \frac{\partial \mathbf{G}_i}{\partial x_i} = \mathbf{B} \tag{27.2.32}$$

$$\mathbf{U} = \begin{bmatrix} \sqrt{-g}\rho\alpha W \\ \sqrt{-g}(\rho h\alpha^2 W^2 v^1) \\ \sqrt{-g}(\rho h\alpha^2 W^2 v^2) \\ \sqrt{-g}(\rho h\alpha^2 W^2 v^3 + Pg^{03}) \\ \sqrt{-g}(\rho h\alpha^2 W^2 + Pg^{00}) \end{bmatrix} \tag{27.2.33a}$$

$$\mathbf{F}^i = \begin{bmatrix} \sqrt{-g}\rho\alpha W v^i \\ \sqrt{-g}(\rho h\alpha^2 W^2 v^i v^1 + Pg^{11}\delta_1^i) \\ \sqrt{-g}(\rho h\alpha^2 W^2 v^i v^2 + Pg^{22}\delta_2^i) \\ \sqrt{-g}(\rho h\alpha^2 W^2 v^i v^3 + Pg^{33}\delta_3^i) \\ \sqrt{-g}(\rho h\alpha^2 W^2 v^i + Pg^{30} v^i) \end{bmatrix} \tag{27.2.33b}$$

$$\mathbf{G}^i = \begin{bmatrix} 0 \\ \sqrt{-g}\left(-\varsigma\,\theta P^{i1} - 2\eta\left[\frac{1}{2}(v^i_{;\gamma} P^{\gamma 1} + v^1_{;\gamma} P^{\gamma i}) - \frac{1}{3} v^\gamma_{;\gamma} P^{i1}\right] + Q^i v^1 + Q^1 v^i\right) \\ \sqrt{-g}\left(-\varsigma\,\theta P^{i2} - 2\eta\left[\frac{1}{2}(v^i_{;\gamma} P^{\gamma 2} + v^2_{;\gamma} P^{\gamma i}) - \frac{1}{3} v^\gamma_{;\gamma} P^{i2}\right] + Q^i v^2 + Q^2 v^i\right) \\ \sqrt{-g}\left(-\varsigma\,\theta P^{i3} - 2\eta\left[\frac{1}{2}(v^i_{;\gamma} P^{\gamma 3} + v^3_{;\gamma} P^{\gamma i}) - \frac{1}{3} v^\gamma_{;\gamma} P^{i3}\right] + Q^i v^3 + Q^3 v^i\right) \\ \sqrt{-g}\left(-\varsigma\,\theta P^{i0} - 2\eta\left[\frac{1}{2}(v^0_{;\gamma} P^{\gamma i} + v^i_{;\gamma} P^{\gamma 0}) - \frac{1}{3} v^\gamma_{;\gamma} P^{i0}\right] + Q^i W\right) \end{bmatrix} \tag{27.2.33c}$$

$$\mathbf{B} = \begin{bmatrix} 0 \\ \sqrt{-g}M^{\mu\alpha}\Gamma^1_{\alpha\mu} \\ \sqrt{-g}M^{\mu\alpha}\Gamma^2_{\alpha\mu} \\ \sqrt{-g}M^{\mu\alpha}\Gamma^3_{\alpha\mu} \\ \sqrt{-g}M^{\mu\alpha}\Gamma^0_{\alpha\mu} \end{bmatrix} \tag{27.2.33d}$$

where the Christoffel symbols are evaluated in terms of metric tensors (Appendix D).

Once the conservation variables, $\mathbf{U}$, $\mathbf{F}_i$, $\mathbf{G}_i$, and $\mathbf{B}$, are solved, it is necessary to covert these conservation variables into the primitive variables such as ρ, v_i, p, h, T, and W. It can be shown that these relations lead to a quartic equation. The real roots for the quartic equation can be found using the polynomical solution techniques [Richardson, 2000].

Simplification with Minkowski Coordinate Transformation

Neglecting the lapse function and the shift vector ($\alpha = 1, \beta_i = 0$) and using the Minkowski coordinate transformation [Miharas and Miharas, 1984], the conservation form of the special relativistic hydrodynamic equations in cartesian coordinates is written as

$$\frac{\partial \mathbf{U}}{\partial t} + \frac{\partial \mathbf{F}_i}{\partial x_i} + \frac{\partial \mathbf{G}_i}{\partial x_i} = 0 \tag{27.2.34}$$

with

$$\mathbf{U} = \begin{bmatrix} \rho W \\ \rho h W^2 \mathrm{v}_j \\ \rho h W^2 - p \end{bmatrix}, \quad \mathbf{F}_i = \begin{bmatrix} \rho W \mathrm{v}_i \\ \rho h W^2 \mathrm{v}_i \mathrm{v}_j + p\delta_{ij} \\ \rho h W^2 \mathrm{v}_i + p\mathrm{v}_i \end{bmatrix}, \quad \mathbf{G}_i = \begin{bmatrix} 0 \\ -\tau_{ij} \\ -\tau_{ij}\mathrm{v}_j + q_i \end{bmatrix} \tag{27.2.35}$$

where τ_{ij} and q_i denote the stress tensor and heat flux vector, respectively,

$$\tau_{ij} = 2\mu \hat{d}_{ij} \tag{27.2.36}$$

$$q_i = kP_{ij}(T,_j + TA_j) \tag{27.2.37}$$

with

$$\hat{d}_{ij} = \frac{1}{2}(\mathrm{v}_{i,k}P_{kj} + \mathrm{v}_{j,k}P_{ki}) - \frac{1}{3}\mathrm{v}_{k,k}P_{ij} \tag{27.2.38}$$

$$P_{ij} = \delta_{ij} + \mathrm{v}_i \mathrm{v}_j \tag{27.2.39}$$

$$A_j = \frac{d\mathrm{v}_j}{d\tau} = W^2 \frac{d\mathrm{v}_j}{dt} \tag{27.2.40}$$

Here, the quantity $\mathrm{v}_i \mathrm{v}_j$ in (27.2.39) is dimensionless (divided by c^2) in the natural unit, τ is the independent variable, known as proper time, such that $d\tau = Wdt$. It should also be noted that $\sqrt{\eta} = 1$ in (27.2.11). All other variables are as defined in Section 27.2.1.

27.2.3 PSEUDO-NEWTONIAN APPROXIMATIONS WITH GRAVITATIONAL EFFECTS

The general relativistic flows may be simplified using the Pueudo-Newtonian approximations [Paczynski and Wiita, 1980; Nobuta and Hanawa, 1999] associated with the black hole accretion. Here, the gravitational effect is introduced in terms of the Schwarzschild radius r_s

$$r_s = \frac{2GM}{c^2} \tag{27.2.41}$$

where G and M denote the gravitational constant and black hole mass. The pseudo-Newtonian potential is given by

$$\Phi(r) = -\frac{GM}{r - r_s} \tag{27.2.42}$$

It is now possible to write the conservation variables, convection flux, diffusion flux, and source terms with gravity similarly as in the Newtonian fluid dynamics except that the pressure and energy must be expressed in terms of the Boltzmann constant instead of the standard gas constant [Miharas and Miharas, 1984]. Thus, we have

$$\mathbf{U} = \begin{bmatrix} \rho \\ \rho \mathrm{v}_j \\ \rho E \end{bmatrix}, \quad \mathbf{F}_i = \begin{bmatrix} \rho \mathrm{v}_i \\ \rho \mathrm{v}_i \mathrm{v}_j + p\delta_{ij} \\ (\rho E + p)\mathrm{v}_i \end{bmatrix}, \quad \mathbf{G}_i = \begin{bmatrix} 0 \\ -\tau_{ij} \\ -\tau_{ij}\mathrm{v}_j - kT_{,i} + q_R \end{bmatrix},$$

$$\mathbf{B} = \begin{bmatrix} 0 \\ (\rho\Phi)_{,j} \\ (\rho\Phi)_{,i}\mathrm{v}_i \end{bmatrix} \tag{27.2.43}$$

with

$$E = \varepsilon + \frac{1}{2}\mathrm{v}_i\mathrm{v}_i$$

$$p = \frac{\rho KT}{w_m m_H} + \frac{\sigma T^4}{3} \tag{27.2.44}$$

$$\varepsilon = \frac{1}{\gamma - 1}\frac{KT}{w_m m_H} + \frac{\sigma T^4}{\rho} \tag{27.2.45}$$

where K is the Boltzmann constant, w_m is the mean molecular weight, m_H is the mass of the hydrogen atom, and q_R is the radiative heat flux (see Chapter 24). For spherical coordinates, all partial derivatives will be converted to covariant derivatives.

27.3 EXAMPLE PROBLEMS

27.3.1 RELATIVISTIC SHOCK TUBE

The nonrelativistic (Newtonian) shock tube problem shown in Figure 6.5.1 may be transformed to the case of relativistic flow using the jump condition as deduced from (27.2.17) in the form,

$$[\rho W u] = 0 \tag{27.3.1.a}$$

$$[\rho h W^2 u^2 + p] = 0 \tag{27.3.1.b}$$

$$[\rho h W^2 u + pu] = 0 \tag{27.3.1.c}$$

Closed form solutions of (27.3.1) are of the form [Hawley et al., 1984a]

$$\frac{\rho_2}{\rho_1} = \frac{\gamma + 1}{\gamma - 1} + \frac{\gamma}{\gamma - 1}(W_m - 1) \tag{27.3.2a}$$

$$u_s = \left\{\frac{1 + \gamma/(\gamma - 1)(W_m u_m)^2}{\rho_1 + [\gamma/(\gamma - 1)]W_m p_m}\right\}\frac{p_m}{W_m u_m} \tag{27.3.2b}$$

where the subscript m denotes the intermediate stage. The finite difference solution of

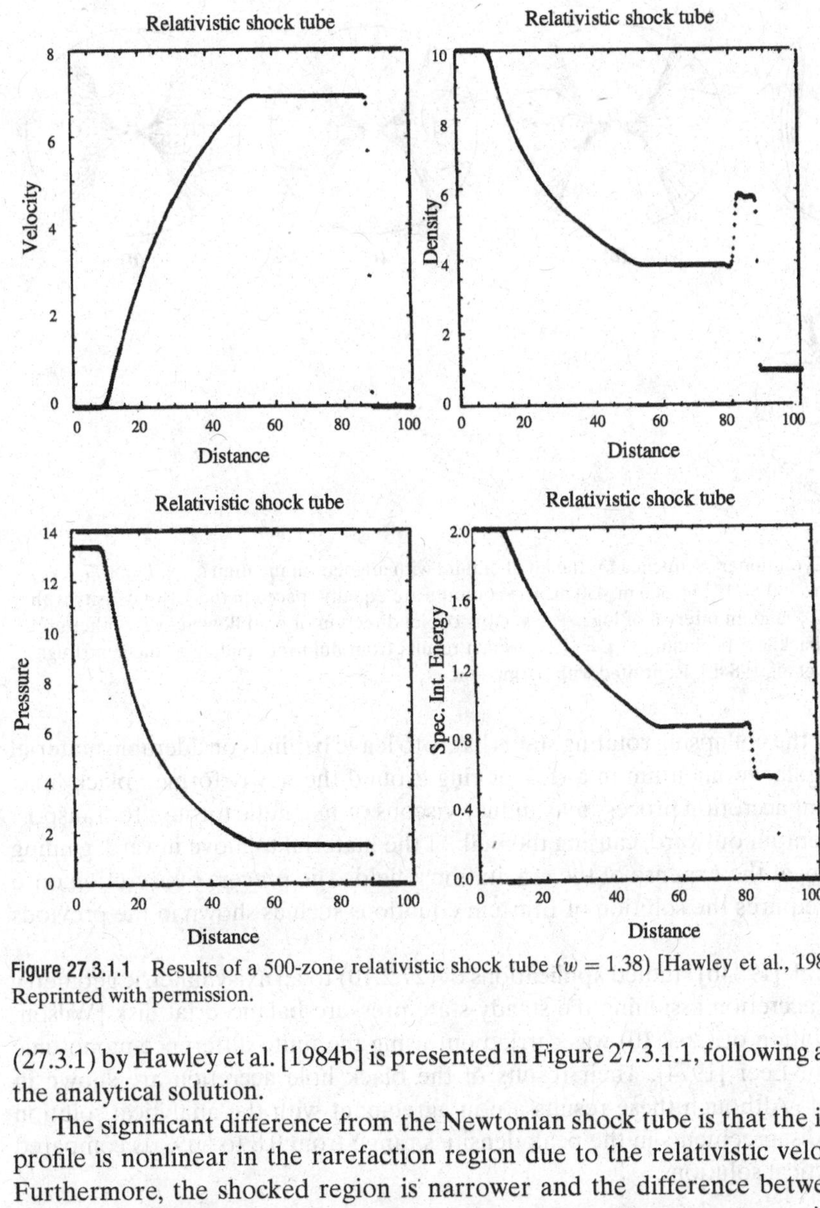

Figure 27.3.1.1 Results of a 500-zone relativistic shock tube ($w = 1.38$) [Hawley et al., 1984b]. Reprinted with permission.

(27.3.1) by Hawley et al. [1984b] is presented in Figure 27.3.1.1, following almost exactly the analytical solution.

The significant difference from the Newtonian shock tube is that the initial velocity profile is nonlinear in the rarefaction region due to the relativistic velocity addition. Furthermore, the shocked region is narrower and the difference between the shock velocity and intermediate velocity is smaller because of Lorentz contraction, approaching the velocity of light. These effects lead to the density change across the shock becoming larger than the Newtonian flow. Note also that temperature and pressure jumps are not as prominent as in the case of nonrelativistic flow.

27.3.2 BLACK HOLE ACCRETION

It is well known that accretion is the origin of X-ray and gamma-ray emission as well as jets emerging from some active black holes. Examples of black hole accretion include the stellar collapse to a black hole, a black hole in a binary system, and a supermassive black hole in an active galactic nuclei, with the accreting matter gaining angular momentum.

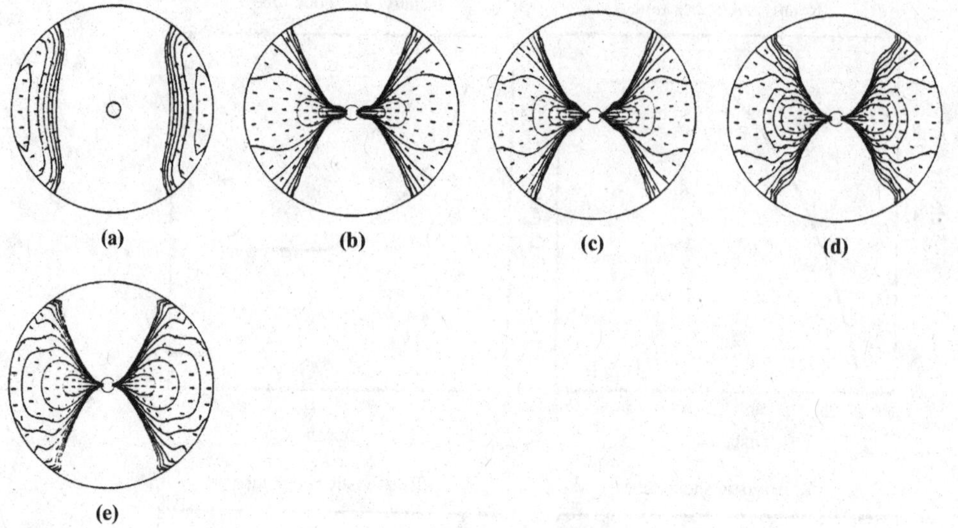

Figure 27.3.2.1 Evolutionary sequence for the infall of fluid with angular momentum $l_{ms} < l = 3.77 < l_{ms}$ at times, $t = 43$ m, 130 m, 172 m, 300 m, 430 m. The contours are equally spaced in the log of density with a minimum of 10^{-6} and an interval of log2. The vectors depict direction of fluid flow at every other grid point. Note the backflow beginning at $t = 172$ m which results from fluid rebounding at the centrifugal barrier. [Hawley et al., 1984b]. Reprinted with permission.

In these cases, the collapsing rotating star is likely to leave behind considerable material with large angular momentum in a disk or ring around the newly formed black hole. The subsequent accretion process may induce viscous or magnetic torques to transport angular momentum outward, causing the bulk of the material to move inward, gaining internal energy at the expense of the gravitational field. The process of accretion onto a balck hole requires the solution of Einstein equations such as shown in the previous section.

Hawley et al. [1984b] studied applications of (27.2.10) to an axisymmetric geometry of black hole accretion assuming the steady-state pressure-balanced fat disk [Wilson, 1972]. The solution of (26.2.10) was carried out using the finite difference monotonic scheme of Van Leer [1974]. Their results of the black hole accretion are shown in Figure 27.3.2.1. Although these results are in agreement with the analytical solution qualitatively, disagreements in the peak densities range from 0.8 to 50% as compared with the analytical solution.

27.3.3 THREE-DIMENSIONAL RELATIVISTIC HYDRODYNAMICS

The Riemann solvers as used in Newtonian fluids may equally be efficient in relativistic hydrodynamics. Aloy et al. [1999] solved three-dimensional relativistic equations (27.2.10) without source terms, using FDM Riemann solvers [LeVeque, 1991].

In their work of the three-dimensional simulation of a relativistic jet propagating through an homogeneous atmosphere, Aloy et al. [1999] reports snapshots of the proper rest-mass density distribution, pressure, specific internal energy, and Lorentz factor of the relativistic jet model as shown in Figure 27.3.3.1. The input data include: the beam

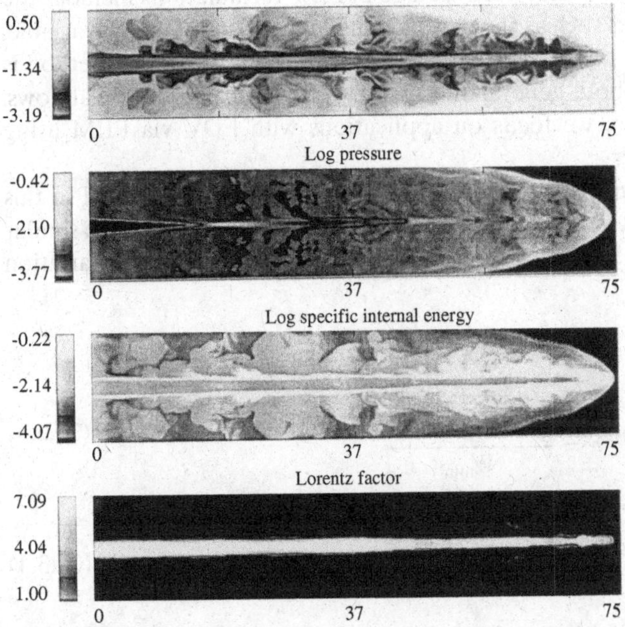

Figure 27.3.3.1 Snapshots (top to bottom) of the proper rest-mass density distribution, pressure, specific internal energy (all on a logarithmic scale), and Lorentz factor of the relativistic jet model discussed in the text ($v = 0.99c$, $M_b = 6.0$, $\eta = 0.01$, $\gamma = 5/3$) after 160 units of time. The resolution is four zones/R_b [Aloy et al., 1999]. Reprinted with permission.

flow velocity $u_b = 0.99c$, the beam Mach number $M_b = 6.0$, and the ratio of the rest mass density of the beam and the ambient medium $\eta = 0.01$. The ambient medium consists of $15R_b \times 15R_b \times 75R_b$ with R_b being the beam radius. The jet is injected at $z = 0$ in the direction of the positive z-axis through a circular nozzle ($x^2 + y^2 \leq R_b^2$) and is in pressure equilibrium with the ambient medium ($\gamma = 1.4$). Simple outflow boundary conditions are imposed except at the injection region.

The gross morphological and dynamical properties of highly supersonic relativistic jets as shown in Figure 27.3.3.1 is qualitatively similar to those obtained in two-dimensional analysis [Marti et al., 1997].

27.3.4 FLOWFIELD DEPENDENT VARIATION (FDV) METHOD FOR RELATIVISTIC ASTROPHYSICAL FLOWS

The most important aspect of the relativistic hydrodynamics is how to resolve shock waves as seen in the previous two sections. Turbulent boundary layers are even more difficult problems, particularly when the shock waves are interacting with turbulent boundary layers. Although there are many options available for the solution approaches, we shall examine, in this section, the FDV theory introduced in Sections 6.5 and 13.6 as applied to the general and special relativistic flows. Any one of the governing equations presented in this section, (27.2.10), (27.2.17), or the conservation equations with (27.2.26) can be accommodated in the FDV formulation.

As shown in Sections 6.5 and 13.6, the advantage of FDV formulations includes the ability to capture discontinuities such as shock waves and high gradients of any variable, and to deal with disparity and nonlinearity of source terms. Thus, the FDV method is considered to be effective for both general and special relativistic astrophysical flows. For the purpose of illustration, we focus on applications with FDV via FEM using (27.2.32).

The variation parameters needed for the scope of the problems discussed in this section are those of convection, viscosity, and source term. These parameters are similar to the ones shown previously in Sections 6.5 and 13.6. The first order convection variation parameters are of the form:

$$s_1 = \begin{cases} \min(r,1) & r > \alpha \\ 0 & r < \alpha, \qquad r = \dfrac{\sqrt{W_{max}^2 - W_{min}^2}}{W_{min}} \\ 1 & W_{min} = 1 \end{cases} \tag{27.3.3}$$

$$s_2 = \frac{1}{2}(1 + s_1^\eta) \quad 0 < \eta < 1 \tag{27.3.4}$$

where α is small number ($\alpha \approx 0.01$) and W is the Lorentz factor, replacing the Mach number used in the Newtonian flows.

The viscous variation parameters are defined by the relativistic Reynolds number similarly as in Newtonian flows (Sections 6.5 and 13.6). They are of the form:

$$s_3 = \begin{cases} \min(s,1) & s > \alpha \\ 0 & s < \alpha, \qquad s = \dfrac{\sqrt{Re_{max}^2 - Re_{min}^2}}{Re_{min}} \\ 1 & Re_{min} = 0 \end{cases} \tag{27.3.5}$$

$$s_4 = \frac{1}{2}(1 + s_3^\eta) \quad 0 < \eta < 1 \tag{27.3.6}$$

Recall that Damköler number was used for the source term variation parameters in chemically reacting flows (Section 13.6). For gravitational effects, however, we must employ the source term variation parameters associated with gravitation in terms of relativistic Froude number Fr.

$$s_5 = \begin{cases} \min(t,1) & t > \alpha \\ 0 & t < \alpha, \qquad t = \dfrac{\sqrt{Fr_{max}^2 - Fr_{min}^2}}{Fr_{min}} \\ 1 & Fr_{min} = 0 \end{cases} \tag{27.3.7}$$

$$s_6 = \frac{1}{2}(1 + s_5^\eta) \quad 0 < \eta < 1 \tag{27.3.8}$$

with

$$\mathrm{Fr} = \frac{\mathrm{v}_i^2}{\mathrm{GL}} = \frac{\mathrm{v}_i^2}{\mathrm{L}}$$

Using these variation parameters, the flux and source term Jacobians are defined similarly as in Newtonian flows:

$$a_i = \frac{\partial \mathbf{F}_i}{\partial U}, \qquad b_i = \frac{\partial \mathbf{G}_i}{\partial U}, \qquad c_{ik} = \frac{\partial \mathbf{G}_i}{\partial U_{,k}}, \qquad d = \frac{\partial \mathbf{B}}{\partial U} \tag{27.3.9}$$

Explicit forms of convection, diffusion, and source term Jacobians are presented in Appendix D-2. The diffusion gradient Jacobians c_{ik} are calculated numerically.

The conservation form of the relativistic hydrodynamic equations given by (27.2.17) is written in terms of the FDV formulation,

$$\left(\mathbf{A} + \mathbf{E}_i^n \frac{\partial}{\partial x_i} + \mathbf{E}_{ij}^n \frac{\partial^2}{\partial x_i \partial x_j}\right) \Delta \mathbf{U}^{n+1} = -\mathbf{Q}^n \tag{27.3.10}$$

and its FEM applications lead to

$$(A_{\alpha\beta}\,\delta_{rs} + B_{\alpha\beta rs})\Delta U_{\beta s}^{n+1} = H_{\alpha r}^n + N_{\alpha r}^n \tag{27.3.11}$$

with details of the algebra for each term in (27.3.11) carried out similarly as in (13.6.21) through (13.6.24).

Computations of the convection and diffusion Jacobians follow the same procedure as in the Navier-Stokes system of equations such that all convection and diffusion flux terms are differentiated with respect to each of the conservation variables. Unlike the nonrelativistic flows, extraction of primitive variables $(\rho, \mathrm{v}_i, p, T)$ from the conservation variables requires the solution of quartic equations or iterative processes through integration time steps [Richardson, 2000].

Two test problems are used to evaluate the FDV theory in the relativistic regime. The first is the special relativistic shock tube to test the shock capturing scheme. This is followed by the general relativistic hydrodynamic equations, examining the "dust infall" problem.

Special Relativistic Shock Tube

The special relativistic shock tube and how it differs from the Newtonian shock tube was briefly discussed in Section 27.3.1, and we present here the special relativistic shock tube solved using FDV theory. The example utilizes the Minkowski geometry used in Font et al. [1998b].

The initial conditions have the left-hand parameters given by $P_l = 13.3$ and $\rho_1 = 10.0$, and the initial right-hand parameters by $P_r = 6.67\text{e-}7$ and $\rho_1 = 1.0$. The initial velocity is zero along the entire length of the tube. These parameters carry normalized units as discussed in Section 27.2.1. The equation of state is given by $P = (\Gamma - 1)\rho\varepsilon$ where the adiabatic exponent, Γ, is equal to 5/3. The shock tube is one unit in length $(x = -0.5, 0.5)$ with the initial pressure boundary located at $x = 0$. The tube is divided into 400 nodes such that $\Delta x = 1/400$, and a CFL number of 0.18 is used. The FDV variation parameter constants from (27.3.3) and (27.3.4) are $\alpha = 0.001$ and $\eta = 0.25$.

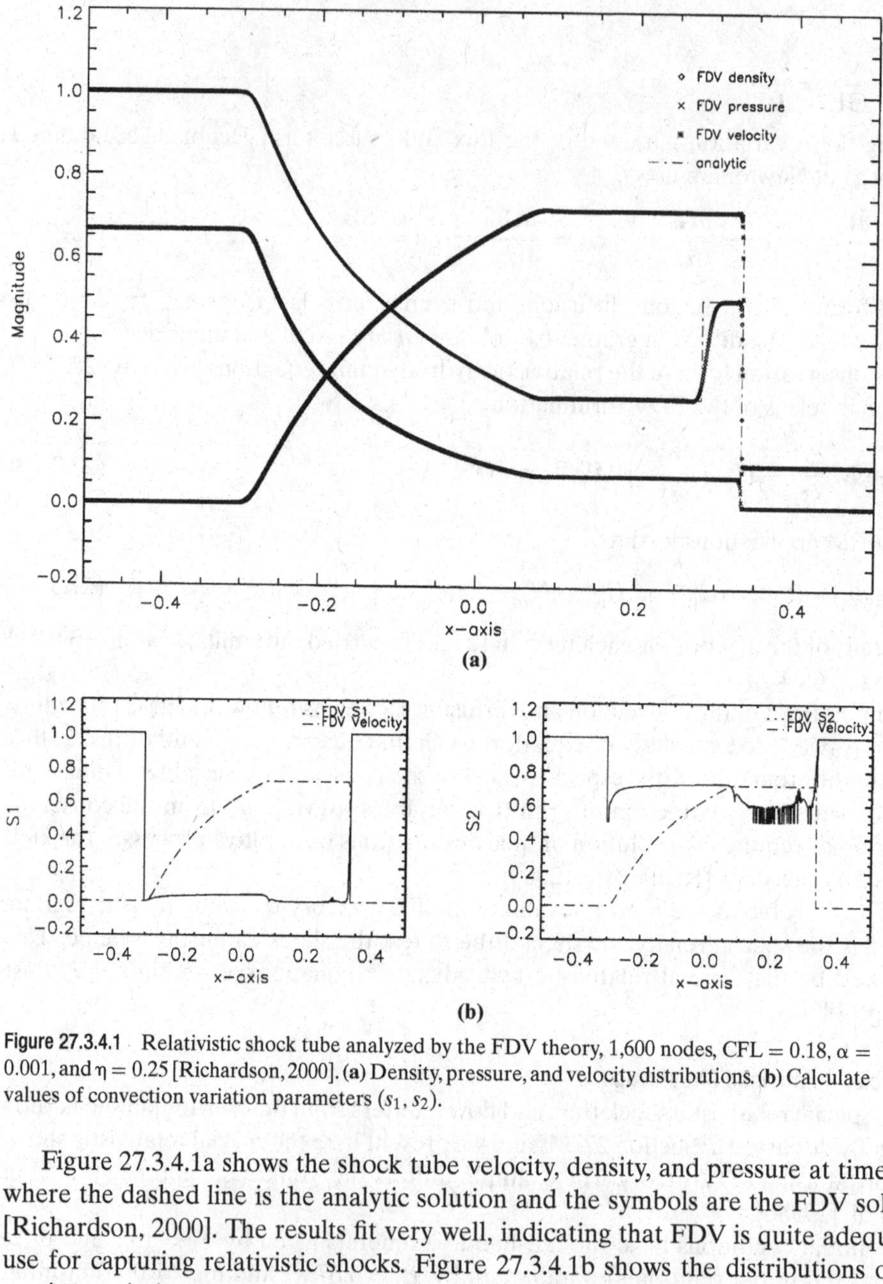

Figure 27.3.4.1 Relativistic shock tube analyzed by the FDV theory, 1,600 nodes, CFL = 0.18, α = 0.001, and η = 0.25 [Richardson, 2000]. **(a)** Density, pressure, and velocity distributions. **(b)** Calculate values of convection variation parameters (s_1, s_2).

Figure 27.3.4.1a shows the shock tube velocity, density, and pressure at time = 0.4, where the dashed line is the analytic solution and the symbols are the FDV solutions [Richardson, 2000]. The results fit very well, indicating that FDV is quite adequate to use for capturing relativistic shocks. Figure 27.3.4.1b shows the distributions of convection variation parameters (s_1 and s_2). It is interesting to note that discontinuities of shock waves follow precisely those of the FDV variation parameters as have been demonstrated in Newtonian fluids in Sections 6.5 and 13.6.

Dust Infall

The shock tube alone is not an adequate test for demonstrating the abilities of FDV for astrophysical problems since it uses a flat space and does not include a general

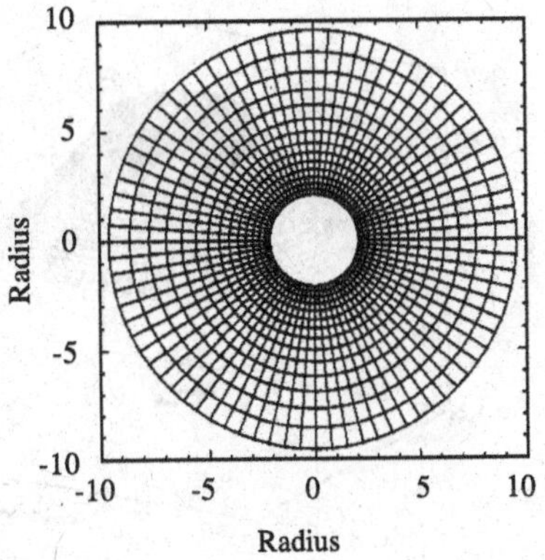

Figure 27.3.4.2 Grid geometry.

relativistic metric. The 'dust infall' problem is used to demonstrate FDV's ability to accurately solve the general relativistic hydrodynamic equations since it utilizes the Kerr metric. This example and the analytic solution are presented in detail in Hawley et al. [1984b]. The FDV results for this problem carried out by Richardson [2000] are presented below.

In this example problem, the geometry of Figure 27.3.4.2 is used. Here, particles are released at a specified distance from the black hole and the 'free-fall' velocity of the particles into the black hole due to the gravitational pull and the density distribution are calculated along the radial components as the matter accretes into the black hole. Although the two dimensional geometry is used for simulation (Figure 27.3.4.2), basically this is a one-dimensional steady-state problem in which pressure is ignored since particles in free-fall are noninteracting. The outer boundary condition is that of the analytic solution which corresponds to a velocity of 0.296 and a density of 0.0093 at a radial distance of 18.1. Initially, no inner boundary condition is applied. The lack of an applied inner boundary condition is an important concept since the boundary conditions for most astrophysical problems are not known. The fixed boundary condition at the outer boundary is the same as the initial condition. Elsewhere, the initial conditions, velocity and density, are set to negligibly small values. Once the density starts to accumulate, the proportional flux is applied, using the FDV flux boundary conditions specified close to the event horizon.

Figure 27.3.4.3a shows the dust infall velocity profile where the coordinates (r, ϕ) are plotted in the plane and velocity is along the z-axis with the black hole in the center [Richardson, 2000]. Matter is equally distributed around the outer boundary nodes. The matter flows toward the black hole with increasing velocity until a point where the velocity suddenly drops to zero near the surface relative to the outside observer. The results are symmetric indicating the correct application of the multidimensionality. Figure 27.3.4.3b represents the density profile. Particles flow into the system and build up at the surface. Some of the particles are passed through the boundary into the black hole as discussed above; otherwise the density would continue to build rather than reaching a steady-state solution. The no-boundary condition case was applied to obtain these results.

The density and velocity curves for the case where no inner boundary condition was applied is shown in Figure 27.3.4.4. A CFL number of 0.25 was used for this case, and

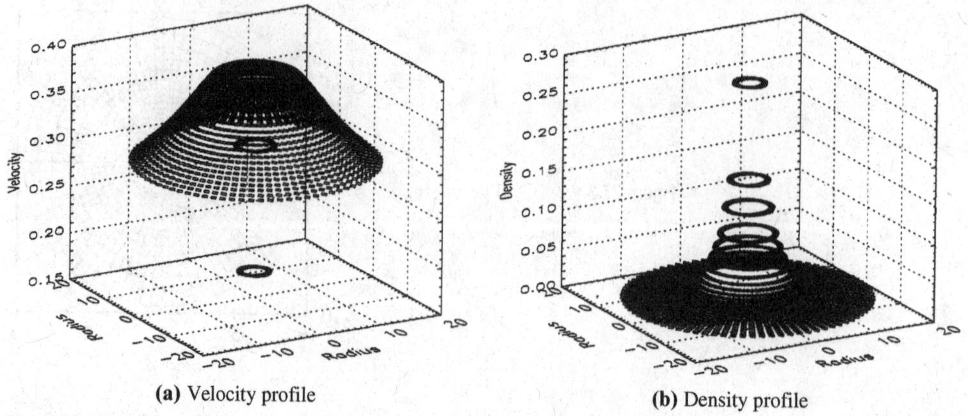

Figure 27.3.4.3 Dust infall velocity and density profiles [Richardson, 2000].

the steady-state solution was found after approximately 550 iterations. The dashed line shows the analytic solution. Since the FEM flux boundary conditions are imbedded, not applying them with specific fluxes is not the same as having no flux at the surface. This is shown, otherwise a steady-state solution would not have been found. The solution is very stable, and the profiles are very similar to the exact solution. The error from the exact solution was found to be 11.0% for the velocity and 6.0% for the density. The profiles of the variation parameters s_5 and s_6 are interesting since the sharp V-shape feature corresponds to the point where the velocity profile begins to decrease rapidly to zero. The s_1 and s_2 profiles are very similar to the density profile.

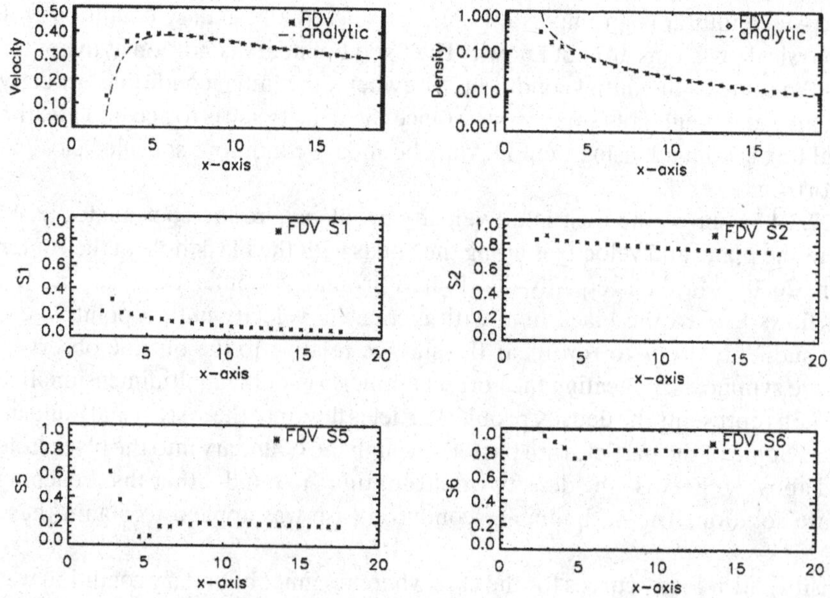

Figure 27.3.4.4 Dust infall with no inner boundary condition [Richardson, 2000].

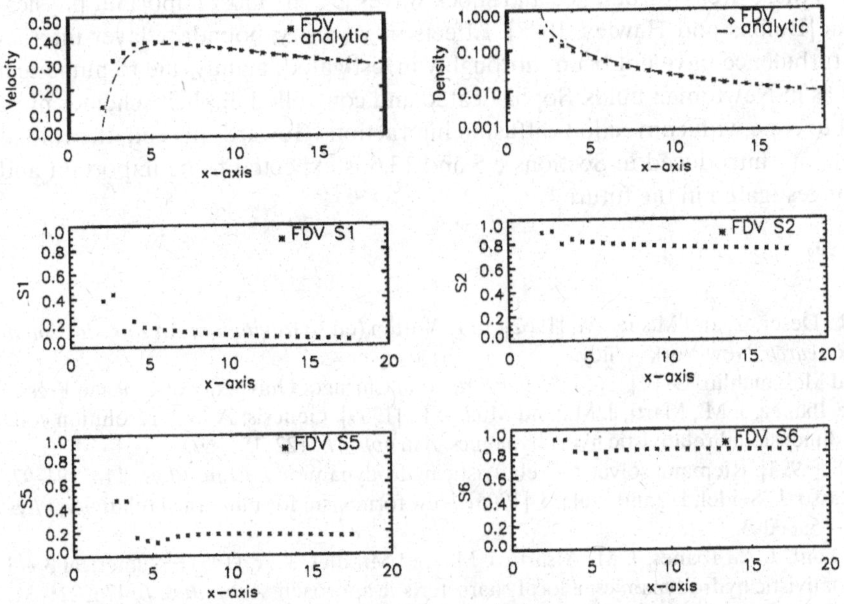

Figure 27.3.4.5 Dust infall with proportional inner boundary condition of 0.1 [Richardson, 2000].

Figure 27.3.4.5 shows the density and velocity where a flux equal to 0.1 times the final density was applied at the inner boundary. A CFL of 0.2 was used in this case. This solution is not as stable as the one shown in Figure 27.3.4.4 where no inner boundary condition was applied. The errors for this case were 11.0% for the velocity and 6.3% for the density [Richardson, 2000].

27.4 SUMMARY

Active research toward computational relativistic hydrodynamics has been in progress for the past three decades. Efforts are being made to provide reliable computer codes such as in Font et al. [1998b, 1999], in which the FDM with Roe schemes and flux vector splitting techniques have been examined to resolve relativistic shock wave problems. Recently, FEM applications have been reported in Meir [1999]. Richardson and Chung [2000] and Richardson [2000] examined the flowfield-dependent variation theory via FEM.

It is shown in the work reported by Font et al. [1998b] that sources of error depend on the initial data being evolved in spacetime or hydrodynamical evolution. For the shock tube problem, only the hydrodynamical evolution was relevant since the evolution took place on a flat background metric. For an evolution along a coordinate axis, the Roe scheme was superior to the flux vector splitting. For an evolution where the shock front is along the diagonal, the flux vector splitting was slightly more accurate. The BM system tends to be more accurate than the ADM system.

The large amount of observation data involving general relativistic phenomena requires the integration of numerical relativity with the traditional tools of astrophysics such as hydrodynamics, magnetohydrodynamics, nuclear astrophysics, and radiation transport. In these areas, turbulence and shock waves are the most important physical phenomena [Bulbus and Hawley, 1998]. Effects of viscosity, boundary layer interactions, and turbulence have not been thoroughly investigated, mainly due to numerical difficulties as in Newtonian fluids. Sophisticated and controlled implicit schemes must be devised to cope with convection-diffusion interactions. Toward this end, the role of the FDV theory introduced in Sections 6.5 and 13.6 is expected to be important and should be investigated in the future.

REFERENCES

Arnowitt, R., Deser, S., and Misner, W. [1962]. In L. Witten (ed.) *Gravitation: An Introduction to Current Research*. New York: Wiley.

Aller, L. and McLaughlin, D. B. [1965]. *Stellar Structure*. Chicago: University of Chicago Press.

Aloy, M. A., Ibanez, J. M., Marti, J. M., and Müller, E. [1999]. Genesis: A high-resolution code for three-dimensional relativistic hydrodynamics. *Astrophy. J.*, 122, 151–66.

Balsara, D. S. [1994]. Riemann solver for relativistic hydrodynamics. *J. Com. Phys.*, 114, 284–97.

Bona, C., Masso, J., Seidel, E., and Stela, J. [1995]. New formalism for numerical relativity. *Phys. Rev. Lett.*, 75, 600–3.

Banyuls, F., Font, J. A., Ibanes, J. M., Msarti, J. M., and Miralles, J. A. [1997]. Numerical 3 + 1 general relativistic hydrodynamics: a local characteristic approach. *Astrophys. J.*, 476, 221–31.

Bromley, B. C., Miller, W. A., and Pariev, V. I. [1998]. The inner edge of the accretion disk around a supermassive black hole N&V. *Nature*, 391, 54–55.

Bulbus, S. A. and Hawley, J. F. [1998]. Instability, turbulence, and enhanced transport in accretion disks. *Rev. Mod. Phys.*, 70, 1–53.

Chandrasekhar, S. [1942]. *Principles of Stellar Dynamics*. Chicago: University of Chicago Press.

Chung, T. J. [1996]. *Applied Continuum Mechanics*. London: Cambridge University Press.

Chung, T. J. [1999]. Transitions and interactions of inviscid/viscous, compressible/incompressible, and laminar/turbulent flows. *Int. J. Num. Meth. Fl.*, 31, 223–46.

Clarke, D., Karoik, S., and Henrikssen, R. N. [1985]. Numerical simulation of the growth of thick accretion disks. *Astrophys. J.*, 58, 81–106.

Dave, R., Dubinski, D. R., and Hernquist, L. [1997]. Parallel Tree SPH. *New Astron.*, 2, 277–97.

Eckert, C. [1940]. The thermodynamics of irreversible processes. *Phys. Rev.*, 58, 919.

Eggum, G. E., Coronitti, F. V., and Katz, J. I. [1980]. Radiation hydrodynamic calculation of super Eddington accretion disks. *Astrophys. J.*, 330, 142–67.

Fishman, G. and Meegan, C. A. [1995]. Gamma-ray bursts. *Annu. Rev. Astrophys.*, 33, 415–58.

Font, J. A., Ibanez, J. M., and Papadopoulos, P. [1998a]. A horizon adapted approach to the study of relativistic accretion flows onto rotating holes. *Ap. J.* 507, L67–L70.

Font, J. A., Miller, M., Suen, W. M., and Tobias, M. [1998b]. Three-dimensional numerical general relativistic hydrodynamics I: Formulations, methods, and code tests. *Phys. Rev. D*, 61, 044011.

Font, J. A., Ibanez, J. M., Papadopoulos, P. [1999]. Non-axisymmetric relativistic Bondi-Holyle accretion on to a Kerr black hole. *Mon. Not. R. Astro. Soc.*, 305, 920–36.

Hawley, J. R., Smarr, L. L., and Wilson, J. R. [1984a]. A numerical study of nonspherical black hole accretion. 1. Equations and test problems. *Ap. J.*, 277, 296–311.

———. [1984b]. A numerical study of nonspherical black hole accretion. II. Finite differencing and code calibration. *Astrophys. J.*, 55, 211–78.

Katz, J. [1980]. Acceleration, radiation, and recession in SS 433. *Astrophys. J. Lett.*, 236, L127–L130.

Kippenhahn, R. and Thomas, H. C. [1970]. *Stellar Rotation*, D. Slettebak (ed.), Dordrecht: Reidel.

Koide, S., Shibuta, K., and Kudoh, T. [1999]. Relativistic jet formation from black hole magnetized accretion disks: method, tests, and applications of a general relativistic magetohydrodynamic numerical code. *Ap. J.* 522, 727–752.

LeVeque, R. J. [1991]. *Numerical Methods for Conservation Law*. Basel: Birkhauser.

Marti, J. M., Ibanez, J. M., and Miralles, J. A. [1991]. Numerical relativistic hydrodynamics. *Phys. Rev.*, D43, 3794–3801.

Marti, J. M., Muller, E., Font, J. A., Ibanez, J. M., and Marquina, A. [1995]. Morphology and dynamics of highly supersonic relativistic jets. *Astrophys. J. Lett.*, 448, L105–L108.

———. [1997]. *Ap. J.* 479, 151–63.

Meir, D. L. [1999]. Multi-dimensional astrophysical strucural and dynamical analysis. I. Development of a non linear finite element approach. *Astrophys. J.*, 518, 788–813.

Meszaros, P. and Rees, M. J. [1993]. Relativistic fireballs and their impact on external matter models for cosmological gamma-ray bursts. *Ap. J.*, 405, 278–84.

Miharas, D. and Miharas, B. W. [1984]. *Foundations of Radiation Hydrodynamics*. New York: Oxford University Press.

Misner, C. W., Thorne, K. S., and Wheeler, J. A. [1973]. *Gravitation*. San Francisco: Freeman.

Nobuta, K. and Hanawa, T. [1999]. Jets from time-dependent accretion flows onto a black hole. *Astrophys. J.*, 510, 614–30.

Norman, M. L. [1997]. *Computational Astrophysics*. D. A. Clark and M. J. West (eds.) ASP: San Francisco.

Paczynski, B. and Wiita, P. J. [1980]. Thick accretion disks and supercritical luminosities. *Astron. Astrophys.*, 88, 23–31.

Penattescu, A. and Meszaros, P. [1998]. Radiative regimes in gamma-ray bursts and afterglows. *Ap. J.*, 501, 772–79.

Richardson, G. A. [2000]. The development and application of the finite element general relativistic astrophysical flow and shock solver. Ph.D. dissertation, The University of Alabama in Huntsville, AL.

Richardson, G. A., Cassibly, J. T., Chung, T. J., and Wu, S. T. [2010]. Finite element form of FDV for widely varying flowfilds. *J. of Com. Physics*, 229, 149–167.

Richardson, G. A. and Chung, T. J. [2002]. Computational relativistic astrophysics using the flowfield-dependent variation theory. *Astrophys. J. Suppl. Series*, 139, 539–563.

Richardson, G. A., Chung, T. J., Karr, G. R., and Pendleton, G. N. [1999]. Flowfield dependent variation method for complex relativistic fluids. AIP (American Institute of Physics) Conference Proceedings. 526, 494–498, Melville, NY.

Roe, P. [1981]. Approximate Riemann solvers, parameter vectors, and difference schemes. *J. Comp. Phys.*, 43, 357–72.

Sari, R. and Piran, T. [1998]. Spectra and light curves of gamma-ray burst after glows. *Astrophys. J.*, 497, L17–L20.

Stella, L. and Vietri, M. [1997]. Lense-Thirring precession and quasi-periodic oscillations in low mass x-ray binaries. *Astrophys. J.*, 492, L59–L62.

Van Leer, B. [1974]. Towards the ultimate conservative difference scheme. II. Monotonicity and conservation combined in a second order scheme. *J. Comp. Phys.*, 14, 361–70.

Weinberg, S. [1972]. *Gravitation and Cosmology: Principles and Applications of the General Theory of Relativity*. New York: Wiley.

Wilson, J. R. [1972]. Numerical study of fluid flow in a kerr space. *Astrophys. J.*, 173, 431–38.

APPENDIXES

APPENDIX A

Three-Dimensional Flux Jacobians

For three-dimensional flows, the vector of conservation variables, **U**, can be defined in terms of a new set of variables, $l = \rho u$, $m = \rho v$, $n = \rho w$, and $e = \rho E$

$$\mathbf{U} = \begin{bmatrix} U_1 \\ U_2 \\ U_3 \\ U_4 \\ U_5 \end{bmatrix} = \begin{bmatrix} \rho \\ \rho u \\ \rho v \\ \rho w \\ \rho E \end{bmatrix} = \begin{bmatrix} \rho \\ l \\ m \\ n \\ e \end{bmatrix}$$

In terms of these variables, the convection and diffusion flux variables are written as

$$\mathbf{F}_1 = \begin{bmatrix} l \\ \dfrac{l^2}{\rho} + p \\ \dfrac{lm}{\rho} \\ \dfrac{ln}{\rho} \\ \dfrac{l}{\rho}(e+p) \end{bmatrix} \qquad \mathbf{F}_2 = \begin{bmatrix} m \\ \dfrac{lm}{\rho} \\ \dfrac{m^2}{\rho} + p \\ \dfrac{mn}{\rho} \\ \dfrac{m}{\rho}(e+p) \end{bmatrix} \qquad \mathbf{F}_3 = \begin{bmatrix} n \\ \dfrac{ln}{\rho} \\ \dfrac{mn}{\rho} \\ \dfrac{n^2}{\rho} + p \\ \dfrac{n}{\rho}(e+p) \end{bmatrix}$$

$$\mathbf{G}_1 = \begin{bmatrix} 0 \\ -\tau_{11} \\ -\tau_{12} \\ -\tau_{13} \\ -\dfrac{l}{\rho}\tau_{11} - \dfrac{m}{\rho}\tau_{12} - \dfrac{n}{\rho}\tau_{13} - k\dfrac{\partial T}{\partial x} \end{bmatrix} \qquad \mathbf{G}_2 = \begin{bmatrix} 0 \\ -\tau_{21} \\ -\tau_{22} \\ -\tau_{23} \\ -\dfrac{l}{\rho}\tau_{21} - \dfrac{m}{\rho}\tau_{22} - \dfrac{n}{\rho}\tau_{23} - k\dfrac{\partial T}{\partial y} \end{bmatrix}$$

$$\mathbf{G}_3 = \begin{bmatrix} 0 \\ -\tau_{31} \\ -\tau_{32} \\ -\tau_{33} \\ -\dfrac{l}{\rho}\tau_{31} - \dfrac{m}{\rho}\tau_{32} - \dfrac{n}{\rho}\tau_{33} - k\dfrac{\partial T}{\partial z} \end{bmatrix}$$

$$p = (\gamma - 1)\left[e - \frac{1}{2\rho}(l^2 + m^2 + n^2)\right] \qquad T = \frac{1}{\rho C_v}\left[e - \frac{1}{2\rho}(l^2 + m^2 + n^2)\right]$$

Explicit forms of Jacobians for convection, diffusion, and diffusion gradients are derived as follows:

$$\mathbf{a}_i = \frac{\partial \mathbf{F}_i}{\partial \mathbf{U}} = \begin{bmatrix} \frac{\partial \mathbf{F}_i^1}{\partial \mathbf{U}_1} & \frac{\partial \mathbf{F}_i^1}{\partial \mathbf{U}_2} & \frac{\partial \mathbf{F}_i^1}{\partial \mathbf{U}_3} & \frac{\partial \mathbf{F}_i^1}{\partial \mathbf{U}_4} & \frac{\partial \mathbf{F}_i^1}{\partial \mathbf{U}_5} \\ \frac{\partial \mathbf{F}_i^2}{\partial \mathbf{U}_1} & \frac{\partial \mathbf{F}_i^2}{\partial \mathbf{U}_2} & \frac{\partial \mathbf{F}_i^2}{\partial \mathbf{U}_3} & \frac{\partial \mathbf{F}_i^2}{\partial \mathbf{U}_4} & \frac{\partial \mathbf{F}_i^2}{\partial \mathbf{U}_5} \\ \frac{\partial \mathbf{F}_i^3}{\partial \mathbf{U}_1} & \frac{\partial \mathbf{F}_i^3}{\partial \mathbf{U}_2} & \frac{\partial \mathbf{F}_i^3}{\partial \mathbf{U}_3} & \frac{\partial \mathbf{F}_i^3}{\partial \mathbf{U}_4} & \frac{\partial \mathbf{F}_i^3}{\partial \mathbf{U}_5} \\ \frac{\partial \mathbf{F}_i^4}{\partial \mathbf{U}_1} & \frac{\partial \mathbf{F}_i^4}{\partial \mathbf{U}_2} & \frac{\partial \mathbf{F}_i^4}{\partial \mathbf{U}_3} & \frac{\partial \mathbf{F}_i^4}{\partial \mathbf{U}_4} & \frac{\partial \mathbf{F}_i^4}{\partial \mathbf{U}_5} \\ \frac{\partial \mathbf{F}_i^5}{\partial \mathbf{U}_1} & \frac{\partial \mathbf{F}_i^5}{\partial \mathbf{U}_2} & \frac{\partial \mathbf{F}_i^5}{\partial \mathbf{U}_3} & \frac{\partial \mathbf{F}_i^5}{\partial \mathbf{U}_4} & \frac{\partial \mathbf{F}_i^5}{\partial \mathbf{U}_5} \end{bmatrix}$$

$$\mathbf{a}_1 = \begin{bmatrix} 0 & 1 & 0 & 0 & 0 \\ \frac{\gamma-3}{2}u^2 + \frac{\gamma-1}{2}(v^2+w^2) & (3-\gamma)u & (1-\gamma)v & (1-\gamma)w & \gamma-1 \\ -uv & v & u & 0 & 0 \\ -uw & w & 0 & u & 0 \\ -\gamma Eu + (\gamma-1)u(u^2+v^2+w^2) & \gamma E + \frac{1-\gamma}{2}(3u^2+v^2+w^2) & (1-\gamma)uv & (1-\gamma)uw & \gamma u \end{bmatrix}$$

$$\mathbf{a}_2 = \begin{bmatrix} 0 & 0 & 1 & 0 & 0 \\ -uv & v & u & 0 & 0 \\ \frac{\gamma-3}{2}v^2 + \frac{\gamma-1}{2}(u^2+w^2) & (1-\gamma)u & (3-\gamma)v & (1-\gamma)w & \gamma-1 \\ -vw & 0 & w & v & 0 \\ -\gamma Ev + (\gamma-1)v(u^2+v^2+w^2) & (1-\gamma)uv & \gamma E + \frac{1-\gamma}{2}(u^2+3v^2+w^2) & (1-\gamma)vw & \gamma v \end{bmatrix}$$

$$\mathbf{a}_3 = \begin{bmatrix} 0 & 0 & 0 & 1 & 0 \\ -uw & w & 0 & u & 0 \\ -vw & 0 & w & v & 0 \\ \frac{\gamma-3}{2}w^2 + \frac{\gamma-1}{2}(u^2+v^2) & (1-\gamma)u & (1-\gamma)v & (3-\gamma)w & \gamma-1 \\ -\gamma Ew + (\gamma-1)w(u^2+v^2+w^2) & (1-\gamma)uw & (1-\gamma)vw & \gamma E + \frac{1-\gamma}{2}(u^2+v^2+3w^2) & \gamma w \end{bmatrix}$$

$$\tau_{11} = \mu\left(\frac{4}{3}\frac{\partial u}{\partial x} - \frac{2}{3}\left(\frac{\partial v}{\partial y} + \frac{\partial w}{\partial z}\right)\right) = \mu\left[\frac{4}{3}\frac{\partial}{\partial x}\left(\frac{l}{\rho}\right) - \frac{2}{3}\left[\frac{\partial}{\partial y}\left(\frac{m}{\rho}\right) + \frac{\partial}{\partial z}\left(\frac{n}{\rho}\right)\right]\right]$$

$$\frac{\partial}{\partial x}\left(\frac{l}{\rho}\right) = \frac{1}{\rho}l_{,1} - \frac{l}{\rho^2}\rho_{,1}, \qquad \mu_R = 2\mu + \lambda \text{ and } \lambda = -\frac{2}{3}\mu,$$

$$\mathbf{b}_1 = \frac{\partial \mathbf{G}_1}{\partial \mathbf{U}} = \begin{bmatrix} 0 & 0 & 0 & 0 & 0 \\ b^1_{21} & b^1_{22} & b^1_{23} & b^1_{24} & 0 \\ b^1_{31} & b^1_{32} & b^1_{33} & b^1_{34} & 0 \\ b^1_{41} & b^1_{42} & b^1_{43} & b^1_{44} & 0 \\ b^1_{51} & b^1_{52} & b^1_{53} & b^1_{54} & b^1_{55} \end{bmatrix} \qquad \mathbf{b}_2 = \frac{\partial \mathbf{G}_2}{\partial \mathbf{U}} = \begin{bmatrix} 0 & 0 & 0 & 0 & 0 \\ b^2_{21} & b^2_{22} & b^2_{23} & b^2_{24} & 0 \\ b^2_{31} & b^2_{32} & b^2_{33} & b^2_{34} & 0 \\ b^2_{41} & b^2_{42} & b^2_{43} & b^2_{44} & 0 \\ b^2_{51} & b^2_{52} & b^2_{53} & b^2_{54} & b^2_{55} \end{bmatrix}$$

$$b^1_{21} = \frac{1}{\rho^2}[\mu_R l_{,1} + \lambda(m_{,2} + n_{,3}) - \mu_R(2u\rho_{,1} - \mathrm{v}\rho_{,2} - w\rho_{,3})] \qquad b^1_{22} = \frac{\mu_R}{\rho^2}\rho_{,1}$$

$$b^1_{23} = \frac{\lambda}{\rho^2}\rho_{,2} \qquad b^1_{24} = \frac{\lambda}{\rho^2}\rho_{,3} \qquad b^1_{31} = \frac{\mu}{\rho^2}[l_{,2} + m_{,1} - 2u\rho_{,2} - 2\mathrm{v}\rho_{,1}]$$

$$b^1_{32} = \frac{\mu}{\rho^2}\rho_{,2} \qquad b^1_{33} = \frac{\mu}{\rho^2}\rho_{,1} \qquad b^1_{34} = 0$$

$$b^1_{41} = \frac{\mu}{\rho^2}(l_{,3} + n_{,1} - 2u\rho_{,3} - 2w\rho_{,1}) \qquad b^1_{42} = \frac{\mu}{\rho^2}\rho_{,3} \quad b^1_{43} = 0 \qquad b^1_{44} = \frac{\mu}{\rho^2}\rho_{,1}$$

$$b^1_{51} = \frac{1}{\rho}(u\tau_{11} + \mathrm{v}\tau_{12} + w\tau_{13}) + ub^1_{21} + \mathrm{v}b^1_{31} + wb^1_{41}$$
$$- \frac{k}{\rho^2 C_\mathrm{v}}[-(\rho E)_{,1} + (2E - 3u^2 - 3\mathrm{v}^2 - 3w^2)\rho_{,1} + 2ul_{,1} + 2\mathrm{v}m_{,1} + 2wn_{,1}]$$

$$b^1_{52} = ub^1_{22} + \mathrm{v}b^1_{32} + wb^1_{42} - \frac{\tau_{11}}{\rho} - \frac{k}{\rho^2 C_\mathrm{v}}[2u\rho_{,1} - l_{,1}]$$

$$b^1_{53} = ub^1_{23} + \mathrm{v}b^1_{33} + wb^1_{43} - \frac{\tau_{12}}{\rho} - \frac{k}{\rho^2 C_\mathrm{v}}(-m_{,1} + 2\mathrm{v}\rho_{,1})$$

$$b^1_{54} = ub^1_{24} + \mathrm{v}b^1_{34} + wb^1_{44} - \frac{\tau_{13}}{\rho} - \frac{k}{\rho^2 C_\mathrm{v}}(-n_{,1} + 2w\rho_{,1}) \qquad b^1_{55} = \frac{k}{\rho^2 C_\mathrm{v}}\rho_{,1}$$

$$b^2_{21} = b^1_{31} \qquad b^2_{22} = b^1_{32} \qquad b^2_{23} = b^1_{33} \qquad b^2_{24} = 0$$

$$b^2_{31} = \frac{1}{\rho^2}[\lambda(l_{,1} + n_{,3}) + \mu_R(m_{,2} + u\rho_{,1} - 2\mathrm{v}\rho_{,2} + w\rho_{,3})] \qquad b^2_{32} = \frac{\lambda}{\rho^2}\rho_{,1}$$

$$b^2_{33} = \frac{\mu_R}{\rho^2}\rho_{,2} \qquad b^2_{34} = \frac{\lambda}{\rho^2}\rho_{,3}$$

$$b^2_{41} = \frac{\mu}{\rho^2}(m_{,3} + n_{,2} - 2\mathrm{v}\rho_{,3} - 2w\rho_{,2}) \qquad b^2_{42} = 0 \qquad b^2_{43} = \frac{\mu}{\rho^2}\rho_{,3}$$

$$b^2_{44} = \frac{\mu}{\rho^2}\rho_{,2}$$

$$b^2_{51} = \frac{1}{\rho}(u\tau_{21} + \mathrm{v}\tau_{22} + w\tau_{23}) + ub^2_{21} + \mathrm{v}b^2_{31} + wb^2_{41}$$
$$- \frac{k}{\rho^2 C_\mathrm{v}}[-(\rho E)_{,2} + (2E - 3u^2 - 3\mathrm{v}^2 - 3w^2)\rho_{,2} + 2ul_{,2} + 2\mathrm{v}m_{,2} + 2wn_{,2}]$$

$$b_{52}^2 = -\frac{\tau_{21}}{\rho} + ub_{22}^2 + \text{v}b_{32}^2 + wb_{42}^2 - \frac{k}{\rho^2 C_\text{v}}[-l_{,2} + 2u\rho_{,2}]$$

$$b_{53}^2 = ub_{23}^2 + \text{v}b_{33}^2 + wb_{43}^2 - \frac{\tau_{22}}{\rho} - \frac{k}{\rho^2 C_\text{v}}[2\text{v}\rho_{,2} - m_{,2}]$$

$$b_{54}^2 = ub_{24}^2 + \text{v}b_{34}^2 + wb_{44}^2 - \frac{\tau_{23}}{\rho} - \frac{k}{\rho^2 C_\text{v}}[2w\rho_{,2} - n_{,2}] \qquad b_{55}^2 = \frac{k}{\rho^2 C_\text{v}}\rho_{,2}$$

$$b_{21}^3 = b_{41}^1 \qquad b_{22}^3 = b_{42}^1 \qquad b_{23}^3 = 0 \qquad b_{24}^3 = b_{44}^1$$

$$b_{31}^3 = b_{41}^2 \qquad b_{32}^3 = 0 \qquad b_{33}^3 = b_{43}^2 \qquad b_{34}^3 = b_{44}^2$$

$$b_{41}^3 = \frac{1}{\rho^2}[\lambda(l_{,1} + m_{,2}) + \mu_R(n_{,3} + u\rho_{,1} + \text{v}\rho_{,2} - 2w\rho_{,3})] \qquad b_{42}^3 = \frac{\lambda}{\rho^2}\rho_{,1}$$

$$b_{43}^3 = \frac{\lambda}{\rho^2}\rho_{,2} \qquad b_{44}^3 = \frac{\mu_R}{\rho^2}\rho_{,3}$$

$$b_{51}^3 = \frac{1}{\rho}(u\tau_{31} + \text{v}\tau_{32} + w\tau_{33}) + ub_{21}^3 + \text{v}b_{31}^3 + wb_{41}^3$$
$$-\frac{k}{\rho^2 C_\text{v}}[-(\rho E)_{,3} + (2E - 3u^2 - 3\text{v}^2 - 3w^2)\rho_{,3} + 2ul_{,3} + 2\text{v}m_{,3} + 2wn_{,3}]$$

$$b_{52}^3 = -\frac{\tau_{31}}{\rho} + ub_{22}^3 + \text{v}b_{32}^3 + wb_{42}^3 - \frac{k}{\rho^2 C_\text{v}}[-l_{,3} + 2u\rho_{,3}]$$

$$b_{53}^3 = ub_{23}^3 + \text{v}b_{33}^3 + wb_{43}^3 - \frac{\tau_{32}}{\rho} - \frac{k}{\rho^2 C_\text{v}}[2\text{v}\rho_{,3} - m_{,3}]$$

$$b_{54}^3 = ub_{24}^3 + \text{v}b_{34}^3 + wb_{44}^3 - \frac{\tau_{33}}{\rho} - \frac{k}{\rho^2 C_\text{v}}[2w\rho_{,3} - n_{,3}] \qquad b_{55}^3 = \frac{k}{\rho^2 C_\text{v}}\rho_{,3}$$

$$\mathbf{c}_{ij} = \frac{\partial \mathbf{G}_i}{\partial \mathbf{U}_{,j}}$$

$$\mathbf{c}_{11} = \begin{bmatrix} 0 & 0 & 0 & 0 & 0 \\ c_{21}^{11} & c_{22}^{11} & 0 & 0 & 0 \\ c_{31}^{11} & 0 & c_{33}^{11} & 0 & 0 \\ c_{41}^{11} & 0 & 0 & c_{44}^{11} & 0 \\ c_{51}^{11} & c_{52}^{11} & c_{53}^{11} & c_{54}^{11} & c_{55}^{11} \end{bmatrix}, \qquad \mathbf{c}_{12} = \begin{bmatrix} 0 & 0 & 0 & 0 & 0 \\ c_{21}^{12} & 0 & c_{23}^{12} & 0 & 0 \\ c_{31}^{12} & c_{32}^{12} & 0 & 0 & 0 \\ 0 & 0 & 0 & 0 & 0 \\ c_{51}^{12} & c_{52}^{12} & c_{53}^{12} & 0 & 0 \end{bmatrix},$$

$$\mathbf{c}_{13} = \begin{bmatrix} 0 & 0 & 0 & 0 & 0 \\ c_{21}^{13} & 0 & 0 & c_{24}^{13} & 0 \\ 0 & 0 & 0 & 0 & 0 \\ c_{41}^{13} & c_{42}^{13} & 0 & 0 & 0 \\ c_{51}^{13} & c_{52}^{13} & 0 & c_{54}^{13} & 0 \end{bmatrix}$$

$$\mathbf{c}_{21} = \begin{bmatrix} 0 & 0 & 0 & 0 & 0 \\ c_{21}^{21} & 0 & c_{23}^{21} & 0 & 0 \\ c_{31}^{21} & c_{32}^{21} & 0 & 0 & 0 \\ 0 & 0 & 0 & 0 & 0 \\ c_{51}^{21} & c_{52}^{21} & c_{53}^{21} & 0 & 0 \end{bmatrix}, \quad \mathbf{c}_{22} = \begin{bmatrix} 0 & 0 & 0 & 0 & 0 \\ c_{21}^{22} & c_{22}^{22} & 0 & 0 & 0 \\ c_{31}^{22} & 0 & c_{33}^{22} & 0 & 0 \\ c_{41}^{22} & 0 & 0 & c_{44}^{22} & 0 \\ c_{51}^{22} & c_{52}^{22} & c_{53}^{22} & c_{54}^{22} & c_{55}^{22} \end{bmatrix},$$

$$\mathbf{c}_{23} = \begin{bmatrix} 0 & 0 & 0 & 0 & 0 \\ 0 & 0 & 0 & 0 & 0 \\ c_{31}^{23} & 0 & 0 & c_{34}^{23} & 0 \\ c_{41}^{23} & 0 & c_{43}^{23} & 0 & 0 \\ c_{51}^{23} & 0 & c_{53}^{23} & c_{54}^{23} & 0 \end{bmatrix}$$

$$\mathbf{c}_{31} = \begin{bmatrix} 0 & 0 & 0 & 0 & 0 \\ c_{21}^{31} & 0 & 0 & c_{24}^{31} & 0 \\ 0 & 0 & 0 & 0 & 0 \\ c_{41}^{31} & c_{42}^{31} & 0 & 0 & 0 \\ c_{51}^{31} & c_{52}^{31} & 0 & c_{54}^{31} & 0 \end{bmatrix}, \quad \mathbf{c}_{32} = \begin{bmatrix} 0 & 0 & 0 & 0 & 0 \\ 0 & 0 & 0 & 0 & 0 \\ c_{31}^{32} & 0 & 0 & c_{34}^{32} & 0 \\ c_{41}^{32} & 0 & c_{43}^{32} & 0 & 0 \\ c_{51}^{32} & 0 & c_{53}^{32} & c_{54}^{32} & 0 \end{bmatrix},$$

$$\mathbf{c}_{33} = \begin{bmatrix} 0 & 0 & 0 & 0 & 0 \\ c_{21}^{33} & c_{22}^{33} & 0 & 0 & 0 \\ c_{31}^{33} & 0 & c_{33}^{33} & 0 & 0 \\ c_{41}^{33} & 0 & 0 & c_{44}^{33} & 0 \\ c_{51}^{33} & c_{52}^{33} & c_{53}^{33} & c_{54}^{33} & c_{55}^{33} \end{bmatrix}$$

$$c_{21}^{11} = \mu_R \frac{l}{\rho^2} \qquad c_{22}^{11} = -\frac{\mu_R}{\rho} \qquad c_{31}^{11} = \mu \frac{m}{\rho^2} \qquad c_{33}^{11} = -\frac{\mu}{\rho}$$

$$c_{41}^{11} = \mu \frac{n}{\rho^2} \qquad c_{44}^{11} = -\frac{\mu}{\rho}$$

$$c_{51}^{11} = uc_{21}^{11} + \mathrm{v}c_{31}^{11} + wc_{41}^{11} - \frac{k}{\rho^2 C_\mathrm{v}}\left(-e + \frac{1}{\rho}(l^2 + m^2 + n^2)\right) \qquad c_{52}^{11} = uc_{22}^{11} + \frac{k}{\rho^2 C_\mathrm{v}} l$$

$$c_{53}^{11} = \mathrm{v}c_{33}^{11} + \frac{k}{\rho^2 C_\mathrm{v}} m \qquad c_{54}^{11} = wc_{44}^{11} + \frac{k}{\rho^2 C_\mathrm{v}} n \qquad c_{55}^{11} = -\frac{k}{\rho C_\mathrm{v}}$$

$$c_{21}^{12} = \lambda \frac{m}{\rho^2} \qquad c_{23}^{12} = -\frac{\lambda}{\rho} \qquad c_{31}^{12} = \mu \frac{l}{\rho^2} \qquad c_{32}^{12} = -\frac{\mu}{\rho}$$

$$c_{51}^{12} = uc_{21}^{12} + \mathrm{v}c_{31}^{12} \qquad c_{52}^{12} = \mathrm{v}c_{32}^{12} \qquad c_{53}^{12} = uc_{23}^{12}$$

$$c_{21}^{13} = \lambda \frac{n}{\rho^2} \qquad c_{24}^{13} = -\frac{\lambda}{\rho} \qquad c_{41}^{13} = \mu \frac{l}{\rho^2} \qquad c_{42}^{13} = -\frac{\mu}{\rho}$$

$$c_{51}^{13} = uc_{21}^{13} + wc_{41}^{13} \qquad c_{52}^{13} = wc_{42}^{13} \qquad c_{54}^{13} = uc_{24}^{13}$$

$$c_{21}^{21} = c_{31}^{11} \qquad c_{23}^{21} = c_{33}^{11} \qquad c_{31}^{21} = \lambda \frac{l}{\rho^2} \qquad c_{32}^{21} = -\frac{\lambda}{\rho}$$

$$c_{51}^{21} = uc_{21}^{21} + \mathrm{v}c_{31}^{21} \qquad c_{52}^{21} = \mathrm{v}c_{32}^{21} \qquad c_{53}^{21} = uc_{23}^{21}$$

$$c_{21}^{22} = c_{31}^{12} \qquad c_{22}^{22} = c_{32}^{12} \qquad c_{31}^{22} = \mu_R \frac{m}{\rho^2} \qquad c_{33}^{22} = -\frac{\mu_R}{\rho}$$

$$c_{41}^{22} = \mu \frac{n}{\rho^2} \qquad c_{44}^{22} = -\frac{\mu}{\rho}$$

$$c_{51}^{22} = uc_{21}^{22} + \mathrm{v}c_{31}^{22} + wc_{41}^{22} - \frac{k}{\rho^2 C_\mathrm{v}}\left[-e + \frac{1}{\rho}(l^2 + m^2 + n^2)\right]$$

$$c_{52}^{22} = uc_{22}^{22} + \frac{k}{\rho^2 C_\mathrm{v}} l \qquad c_{53}^{22} = \mathrm{v}c_{33}^{22} + \frac{k}{\rho^2 C_\mathrm{v}} m \qquad c_{54}^{22} = wc_{44}^{22} + \frac{k}{\rho^2 C_\mathrm{v}} n$$

$$c_{55}^{22} = -\frac{k}{\rho C_\mathrm{v}} \qquad c_{31}^{23} = \lambda \frac{n}{\rho^2} \qquad c_{34}^{23} = -\frac{\lambda}{\rho} \qquad c_{41}^{23} = \mu \frac{m}{\rho^2}$$

$$c_{43}^{23} = -\frac{\mu}{\rho} \qquad c_{51}^{23} = vc_{31}^{23} + wc_{41}^{23} \qquad c_{53}^{23} = wc_{43}^{23} \qquad c_{54}^{23} = vc_{34}^{23}$$

$$c_{21}^{31} = c_{41}^{11} \qquad c_{24}^{31} = c_{44}^{11} \qquad c_{41}^{31} = \lambda \frac{l}{\rho^2} \qquad c_{42}^{31} = -\frac{\lambda}{\rho}$$

$$c_{51}^{31} = uc_{21}^{31} + wc_{41}^{31} \qquad c_{52}^{31} = wc_{42}^{31} \qquad c_{54}^{31} = uc_{24}^{31}$$

$$c_{31}^{32} = c_{41}^{22} \qquad c_{34}^{32} = c_{44}^{22} \qquad c_{41}^{32} = \lambda \frac{m}{\rho^2} \qquad c_{43}^{32} = -\frac{\lambda}{\rho}$$

$$c_{51}^{32} = \mathrm{v}c_{31}^{32} + wc_{41}^{32} \qquad c_{53}^{32} = wc_{43}^{32} \qquad c_{54}^{32} = \mathrm{v}c_{34}^{32}$$

$$c_{21}^{33} = c_{41}^{13} \qquad c_{22}^{33} = c_{42}^{13} \qquad c_{31}^{33} = c_{41}^{23} \qquad c_{33}^{33} = c_{43}^{23}$$

$$c_{41}^{33} = \mu_R \frac{n}{\rho^2} \qquad c_{44}^{33} = -\frac{\mu_R}{\rho}$$

$$c_{51}^{33} = uc_{21}^{33} + \mathrm{v}c_{31}^{33} + wc_{41}^{33} - \frac{k}{\rho^2 C_\mathrm{v}}\left[-e + \frac{1}{\rho}(l^2 + m^2 + n^2)\right] \qquad c_{52}^{33} = uc_{22}^{33} + \frac{k}{\rho^2 C_\mathrm{v}} l$$

$$c_{53}^{33} = \mathrm{v}c_{33}^{33} + \frac{k}{\rho^2 C_\mathrm{v}} m \qquad c_{54}^{33} = wc_{44}^{33} + \frac{k}{\rho^2 C_\mathrm{v}} n \qquad c_{55}^{33} = -\frac{k}{\rho C_\mathrm{v}}$$

Gaussian Quadrature

Gaussian quadrature is one of the most accurate numerical integration methods. In general, the points of subdivision may not necessarily be equidistant, but they must be symmetrically placed with respect to the midpoint of the interval of integration.

Consider the integral under the curve $u = f(x)$ between the interval a and b depicted in Figure B.1a. The endpoints may be replaced by the nondimensional quantities – 1 and 1, as seen in Figure B.1b for $u = f(\xi)$.

The integral for Figure B.1a is

$$I(x) = \int_{a}^{b} f(x)\,dx \tag{B.1}$$

where x may be written in terms of ξ as

$$x = \frac{a+b}{2} + \left(\frac{b-a}{2}\right)\xi \tag{B.2}$$

and

$$dx = \left(\frac{b-a}{2}\right)d\xi \tag{B.3}$$

Thus,

$$u = f(x) = f\left[x = \frac{a+b}{2} + \left(\frac{b-a}{2}\right)\xi\right] \to f(\xi)$$

where

$$I(x) = \frac{b-a}{2}\int_{-1}^{1} f(\xi)\,d\xi = \frac{b-a}{2}I(\xi) \tag{B.4}$$

$$I(\xi) = \int_{-1}^{1} f(\xi)\,d\xi \tag{B.5}$$

It is possible to write (B.5) in the form

$$I(\xi) = w_1 f(\xi_1) + w_2 f(\xi_2) + \cdots + w_n f(\xi_n) \tag{B.6}$$

in which w_i and $f(\xi_i)$, with $i = 1, \ldots, n$, are the weight coefficient and abscissae, respectively. This implies that $I(\xi)$ contains $2n$ unknowns and requires $2n$ equations

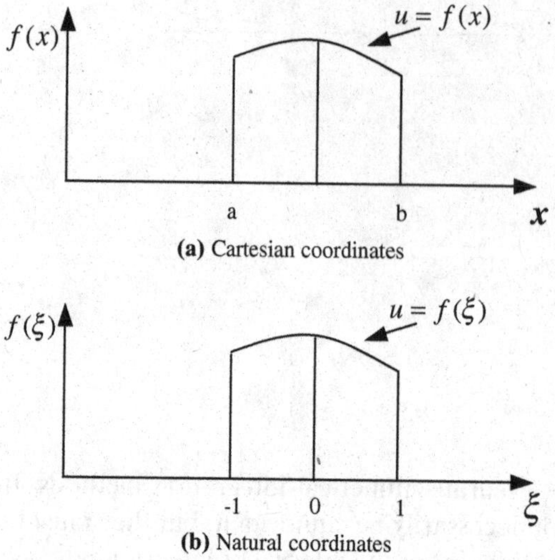

Figure B.1 Integrals under functions $u = f(x)$ and $u = f(\xi)$.

to uniquely define these unknowns. Let $f(\xi)$ be written as

$$f(\xi) = c_1 + c_2\xi + c_3\xi^2 + \cdots + c_m\xi^{m-1} \tag{B.7}$$

with $m = 2n$. Substituting (B.7) into (B.5) gives

$$I(\xi) = \int_{-1}^{1} f(\xi)\, d\xi = 2c_1 + c_3\frac{2}{3} + c_3\frac{2}{5} + \cdots \tag{B.8}$$

Writing (B.7) at each point of subdivision yields

$$\begin{aligned} f(\xi_1) &= c_1 + c_2\xi_1 + \cdots + c_m\xi_1^{m-1} \\ f(\xi_2) &= c_1 + c_2\xi_2 + \cdots + c_m\xi_2^{m-1} \\ &\vdots \\ f(\xi_n) &= c_1 + c_2\xi_n + \cdots + c_m\xi_n^{m-1} \end{aligned}$$

Substituting these into (B.6) leads to

$$\begin{aligned} I(\xi) = {} & w_1\left(c_1 + c_2\xi_1 + \cdots + c_m\xi_1^{m-1}\right) \\ & + w_2\left(c_1 + c_2\xi_2 + \cdots + c_m\xi_2^{m-1}\right) \\ & \vdots \\ & + w_n\left(c_1 + c_2\xi_n + \cdots + c_m\xi_n^{m-1}\right) \end{aligned}$$

or

$$\begin{aligned} I(\xi) = {} & c_1(w_1 + w_2 + \cdots + w_n) \\ & + c_2(w_1\xi_1 + w_2\xi_2 + \cdots + w_n\xi_n) \\ & + c_3\left(w_1\xi_1^2 + w_2\xi_2^2 + \cdots + w_n\xi_n^2\right) \\ & \vdots \\ & + c_m\left(w_1\xi_1^{m-1} + w_2\xi_2^{m-1} + \cdots + w_n\xi_n^{m-1}\right) \end{aligned} \tag{B.9}$$

Equating (B.8) and (B.9) yields

$$\begin{aligned} &w_1 + w_2 + \cdots + w_n = 2 \\ &w_1\xi_1 + w_2\xi_2 + \cdots + w_n\xi_n = 0 \\ &w_1\xi_1^2 + w_2\xi_2^2 + \cdots + w_n\xi_n^2 = \frac{2}{3} \\ &\vdots \end{aligned} \tag{B.10}$$

Writing $2n$ of these equations and solving them simultaneously would make it possible to yield the values of $2n$ quantities of $\xi_1, \xi_2, \ldots, \xi_n, w_1, w_2, \ldots, w_n$.

The numerical integration performed in the manner described above is called Gaussian quadrature. The reader should consult the standard book on Gaussian quadrature for tabulated results for the weight coefficients w_i and abscissae $f(\xi_i)$. For $n = 2$ and 3, these values are

n	w_i	$\pm\xi_i$
2	1.0000000000	0.5773502691
3	$\frac{5}{9} = 0.5555555555$	$\sqrt{\frac{3}{5}} = 0.7745966692$
	$\frac{8}{9} = 0.8888888888$	0.0000000000

Here the weight coefficients are symmetric about $\xi = 0$ for the abscissae being antisymmetric about $\xi = 0$. For example, for $n = 2$, there is $\xi_i = \pm 0.5773502691$ for which $w_i = 1$. Similarly, for $n = 3$, there is $\xi_i = \pm\, 0.7745966692$ for which $w_i = 0.5555555555$.

Note that the solution of simultaneous equations (B.10) is laborious. To avoid this difficulty, various polynomials of standard form (Legendre, Hermite, Chebyshev polynomials, etc.) may be utilized. It is known that the Legendre polynomials are considered most efficient for this purpose.

Gaussian Quadrature by Legendre Polynomials

We consider the integral (B.5) in the form

$$\int_{-1}^{1} f(\xi)\, d\xi = \sum_{k=1}^{n} w_k f(\xi_k) \tag{B.11}$$

Our problem is to determine the $2n$ constants, $w_1, w_2, \ldots, w_n, \xi_1, \xi_2, \ldots, \xi_n$, and it is noted that the integral of (B.11) is exact if the integrand $f(\xi)$ is a polynomial of degree $2n$ or less.

The associated points ($k = 1, 2, \ldots n$) are equal to the values of the roots of a Legendre polynomial $\phi_n(\xi)$. Let us arbitrarily take a polynomial $g_n(\xi)$ of degree n such that

$$g_n(\xi) = \alpha_0\phi_0(\xi) + \alpha_1\phi_1(\xi) + \cdots + \alpha_n\phi_n(\xi) \tag{B.12}$$

where $\phi_0(\xi), \phi_2(\xi), \ldots, \phi_n(\xi)$ may be found in a standard text. As an example, let us suppose that, for $n = 3$,

$$g_3(\xi) = \alpha_0 + \alpha_1\xi + \alpha_2\frac{1}{2}(3\xi^2 - 1) + \alpha_3\frac{1}{2}(5\xi^3 - 3\xi) \tag{B.13}$$

Comparing (B.13) with

$$g_n(\xi) = 1 + 3\xi + 4\xi^2 - 7\xi^3 \tag{B.14}$$

we obtain

$$\alpha_0 = \frac{7}{3}, \qquad \alpha_1 = -\frac{6}{3}, \qquad \alpha_2 = \frac{8}{3}, \qquad \alpha_3 = -\frac{14}{5}$$

which yields

$$g_3(\xi) = \frac{7}{3}\phi_0(\xi) - \frac{6}{5}\phi_1(\xi) + \frac{8}{3}\phi_2(\xi) - \frac{14}{5}\phi_3(\xi)$$

This simple example serves to show that any polynomial $g_n(\xi)$ can be written in terms of the Legendre polynomials.

From the orthogonality property of the Legendre polynomials,

$$\int_{-1}^{1} \phi_m(\xi)\phi_n(\xi)\,d\xi = \begin{cases} 0, & (m \neq n) \\ \dfrac{2}{2n+1} & (m = n) \end{cases}$$

we have

$$\int_{-1}^{1} g_n(\xi)\phi_n(\xi)d\xi = \int_{-1}^{1} \alpha_0\phi_0(\xi)\phi_n(\xi)d\xi + \int_{-1}^{1} \alpha_1\phi_1(\xi)\phi_n(\xi)d\xi$$
$$+ \cdots + \int_{-1}^{1} \alpha_n\phi_n(\xi)\phi_n(\xi)d\xi = 0 \tag{B.15}$$

Comparing (B.15) with (B.11) and noting that $g_n(\xi)\phi_n(\xi)$ is the integrand, we have

$$w_1 g_n(\xi_1)\phi_n(\xi_1) + w_2 g_n(\xi_2)\phi_n(\xi_2) + \cdots + w_n g_n(\xi_n)\phi_n(\xi_n) = 0 \tag{B.16}$$

Since $g_n(\xi)$ is an arbitrarily chosen polynomial, the only way that condition (B.16) may be satisfied is by

$$\phi_n(\xi_1) = \phi_n(\xi_2) = \cdots = \phi_n(\xi_n) = 0$$

In other words, the associated points $\xi_1, \xi_2, \ldots, \xi_n$ are the roots and the Legendre polynomial $\phi_n(\xi) = 0$. For $n = 3$, the roots of $\phi_n(\xi) = 0$ are

$$\phi_n(\xi) = \phi_3(\xi) = \frac{1}{2}(5\xi^3 - 3\xi) = 0$$

$$\xi_k = \left(-\sqrt{\frac{3}{5}}, \quad 0, \quad \sqrt{\frac{3}{5}}\right)$$

Now we turn to the determination of the values of the weighting functions $w_k (k = 1, 2, \ldots, n)$. By definition of the Lagrange polynomial, any polynomial $\phi_n(\xi)$ of degree

n passing through $\xi_k (k = 1, 2, \ldots, n)$ points may be expressed in the form

$$\phi_n(\xi) = \sum_{k=1}^{n} \phi(\xi_k) L_k(\xi) \tag{B.17}$$

Hence

$$\int_{-1}^{1} \phi_n(\xi)\, d\xi = \int_{-1}^{1} \sum_{k=1}^{n} \phi(\xi_k) L_k(\xi)\, d\xi = \sum_{k=1}^{n} \phi_n(\xi_k) \int_{-1}^{1} L_k(\xi)\, d\xi \tag{B.18}$$

Comparing (B.18) with (B.11) yields

$$w_k = \int_{-1}^{1} L_k(\xi)\, d\xi \quad (k = 1, 2, \ldots n) \tag{B.19}$$

By virtue of (B.11), we can rewrite (B.19) as

$$w_k = \frac{1}{\phi'_n(\xi_k)} \int_{-1}^{1} \frac{\phi_n(\xi)}{\xi - \xi_k} d\xi \tag{B.20}$$

For $n = 3$, we have

$$\phi_n(\xi) = \phi_3(\xi) = \frac{1}{2}(5\xi^3 - 3\xi)$$

$$\phi'_n(\xi) = \phi'_3(\xi) = \frac{3}{2}(5\xi^2 - 1)$$

Thus, using $\xi_1 = -\sqrt{\frac{3}{5}},\ \xi_2 = 0,\ \xi_3 = \sqrt{\frac{3}{5}}$

$$w_1 = \frac{1}{\frac{3}{2}\left[5\left(\frac{3}{5}\right) - 1\right]} \int_{-1}^{1} \frac{\frac{1}{2}(5\xi^3 - 3\xi)}{\xi + \sqrt{\frac{3}{5}}} d\xi = \frac{5}{9}$$

$$w_2 = \frac{1}{\frac{3}{2}[5(0) - 1]} \int_{-1}^{1} \frac{\frac{1}{2}(5\xi^3 - 3\xi)}{\xi + 0} d\xi = \frac{8}{9}$$

$$w_3 = \frac{1}{\frac{3}{2}\left[5\left(\frac{3}{5}\right) - 1\right]} \int_{-1}^{1} \frac{\frac{1}{2}(5\xi^3 - 3\xi)}{\xi - \sqrt{\frac{3}{5}}} d\xi = \frac{5}{9}$$

It can be shown that the general form of the integral (B.20) is

$$w_k = \frac{2}{[\phi'_n(\xi_k)]^2 (1 - \xi_k^2)} = \frac{2(1 - \xi_k^2)}{(n+1)^2 [\phi_{n+1}(\xi_k)]^2} \tag{B.21}$$

The values of weighting functions for $n = 3$ can be directly obtained by using (B.21):

$$w_k = \frac{2(1-\xi_k^2)}{(3+1)^2\left[\frac{1}{8}(35\xi_k^4 - 30\xi^2 + 3)\right]^2}$$

$$w_1 = \frac{2\left(1-\frac{3}{5}\right)}{16\left[\frac{1}{8}\left(35\frac{9}{25} - 30\frac{3}{5} + 3\right)\right]^2} = \frac{5}{9}$$

$$w_2 = \frac{2(1-0)}{16\left[\frac{1}{8}(35(0) - 30(0) + 3)\right]^2} = \frac{8}{9}$$

$$w_3 = \frac{2\left(1-\frac{3}{5}\right)}{16\left[\frac{1}{8}\left(35\frac{9}{25} - 30\frac{3}{5} + 3\right)\right]^2} = \frac{5}{9}$$

The abscissae and weight coefficients of the Gaussian quadrature formula calculated in this manner are tabulated in Table 9.3.1 for the range $n = 2$ through $n = 10$.

Example 1

Using the three Gaussian points $n = 3$, integrate the following with Gaussian quadrature:

$$I = \int_0^3 x^2 \cos x dx$$

For this problem,

$$x = \frac{a+b}{2} + \frac{b-a}{2}x = \frac{0+3}{2} + \left(\frac{3-0}{2}\right)\xi = \frac{3}{2} + \frac{3}{2}\xi$$

$$dx = \frac{3}{2}d\xi$$

thus,

$$I = \int_0^3 x^2 \cos x dx = \frac{3}{2}\int_{-1}^{1} f(\xi)\, d\xi$$

$$= \frac{3}{2}\int_{-1}^{1}\left[\frac{3}{2}(1+\xi)\right]^2 \cos\left[\frac{3}{2}(1+\xi)\right] d\xi$$

$$= \frac{3}{2}[w_1 f(\xi_1) + w_2 f(\xi_2) + w_3 f(\xi_3)]$$

$$= \frac{3}{2}\left\{\frac{5}{9}\left[\frac{3}{2}\left(1-\sqrt{\frac{3}{5}}\right)\right]^2 \cos\left[\frac{3}{2}\left(1-\sqrt{\frac{3}{5}}\right)\right] + \frac{8}{9}\left[\frac{3}{2}(1-0)\right]^2 \cos\left[\frac{3}{2}(1-0)\right]\right.$$

$$\left. + \frac{5}{9}\left[\frac{3}{2}\left(1+\sqrt{\frac{3}{5}}\right)\right]^2 \cos\left[\frac{3}{2}\left(1+\sqrt{\frac{3}{5}}\right)\right]\right\} = -4.936$$

Since the exact solution for this problem is $I = -4.9522$, the error in this case is 0.327%.

Example 2

For 2-D and 3-D problems, the Gaussian quadrature formulas are, respectively:

$$\iint f(x,y)dxdy = \int_{-1}^{1}\int_{-1}^{1} f(\xi,\eta)\,d\xi \quad d\eta = \sum_{i=1}^{n}\sum_{j=1}^{n} w_i w_j f(\xi_i,\eta_j)$$

$$\iiint f(x,y,z)dxdydz = \int_{-1}^{1}\int_{-1}^{1}\int_{-1}^{1} f(\xi,\eta,\zeta)\,d\xi$$

$$d\eta d\zeta = \sum_{i=1}^{n}\sum_{j=1}^{n}\sum_{k=1}^{n} w_i w_j w_k f(\xi_i,\eta_j,\zeta_k)$$

Note that the abscissae values for η_j, ζ_k are the same as for ξ_i.

Example 3

Consider the integral for the isoparametric element:

$$I(x,y) = \iint \frac{\partial}{\partial x_i} f(\xi,\eta)dx_1dx_2$$

$$I(\xi,\eta) = \sum_{i=1}^{n}\sum_{j=1}^{n} w_i w_j f(\xi_i,\eta_j)$$

Let $N = M = 1$, $n = 3$, and assume, for simplicity of illustration, that

$$f_{11}(\xi_i,\eta_j) = \frac{1+\xi^2+\eta^2+\eta\xi}{1+\xi+\eta}$$

The Gaussian quadrature integration becomes

$$\begin{aligned}\int_{-1}^{1}\int_{-1}^{1} f_{11}(\xi_i,\eta_j)\,d\xi d\eta = \sum_{i=1}^{3}\sum_{j=1}^{3} w_i w_j f_{11}(\xi_i,\eta_j) \\ = w_1w_1 f_{11}(\xi_1,\eta_1) + w_1w_2 f_{11}(\xi_1,\eta_2) + w_1w_3 f_{11}(\xi_1,\eta_3) \\ + w_2w_1 f_{11}(\xi_2,\eta_1) + w_2w_2 f_{11}(\xi_2,\eta_2) + w_2w_3 f_{11}(\xi_2,\eta_3) \\ + w_3w_1 f_{11}(\xi_3,\eta_1) + w_3w_2 f_{11}(\xi_3,\eta_2) + w_3w_3 f_{11}(\xi_3,\eta_3)\end{aligned}$$

$$
\begin{aligned}
&= (0.555)^2\left[\frac{1+(-0.774)^2+(-0.774)^2+(0.774)^2}{1-0.774-0.774}\right] \\
&\quad + (0.555)(0.888)\left[\frac{1+(-0.774)^2+0+0}{1-0.774-0} \right] \\
&\quad + (0.555)^2\left[\frac{1+(-0.774)^2+(-0.774)^2+(0.774)(-0.774)}{1-0.774+0.774}\right] + \cdots \\
&= 8.4444
\end{aligned}
$$

For the finite element analysis using two- or three-dimensional isoparametric elements in general, one may obtain reasonably accurate results with several Gaussian points in each direction.

APPENDIX C

Two Phase Flow – Source Term Jacobians for Surface Tension

$$\mathbf{U} = \begin{bmatrix} U_1 \\ U_2 \\ U_3 \\ U_4 \\ U_5 \end{bmatrix} = \begin{bmatrix} \rho \\ \rho F \\ \rho v_1 \\ \rho v_1 \\ \rho E \end{bmatrix} = \begin{bmatrix} \rho \\ \ell \\ m_1 \\ m_2 \\ e \end{bmatrix} \qquad \mathbf{B} = \begin{bmatrix} 0 \\ 0 \\ \rho f_1 \\ \rho f_2 \\ \rho f_1 v_1 + \rho f_2 v_2 \end{bmatrix}$$

$$\mathbf{d} = \frac{\partial \mathbf{B}}{\partial \mathbf{U}} = \begin{bmatrix} 0 & 0 & 0 & 0 & 0 \\ 0 & 0 & 0 & 0 & 0 \\ d_{31} & d_{32} & 0 & 0 & 0 \\ d_{41} & d_{42} & 0 & 0 & 0 \\ d_{51} & d_{52} & d_{53} & d_{54} & 0 \end{bmatrix}$$

where

$$\rho f_i = \rho g_i + Q_i \qquad Q_i = \sigma\left(\frac{F_{,jj}}{|F_{,k}|} - F_{,j}\frac{|F_{,j}|_{,j}}{|F_{,k}|^2}\right)F_{,i}$$

Dimensionless form:

$$\rho f_i = \frac{\rho^*}{\mathrm{Fr}_i} + \frac{1}{\mathrm{We}}\left(\frac{F_{,jj}}{|F_{,k}|} - F_{,j}\frac{|F_{,j}|_{,j}}{|F_{,k}|^2}\right)F_{,i}$$

where Froude and Weber numbers are used to make the relationship dimensionless:

$$\mathrm{Fr} = \frac{v_\infty^2}{g_i L} \qquad \mathrm{We} = \frac{\rho_\infty v_\infty^2 L}{\sigma}$$

Since $\rho F = \ell$ replace $F = \dfrac{\ell}{\rho}$ in terms of the independent variables

$$|F_{,k}| = (F_{,k}F_{,k})^{\frac{1}{2}} = \left[\left(\frac{\ell}{\rho}\right)_{,k}\left(\frac{\ell}{\rho}\right)_{,k}\right]^{\frac{1}{2}}, \qquad |F_{,k}|^2 = (F_{,k}F_{,k}) = \left[\left(\frac{\ell}{\rho}\right)_{,k}\left(\frac{\ell}{\rho}\right)_{,k}\right] \text{ and}$$

$$|F_{,k}|_{,j} = \left[(F_{,k}F_{,k})^{\frac{1}{2}}\right]_{,j} = \left[\left\{\left(\frac{\ell}{\rho}\right)_{,k}\left(\frac{\ell}{\rho}\right)_{,k}\right\}^{\frac{1}{2}}\right]_{,j}$$

$$Q_i = \sigma\left(\frac{\left(\frac{\ell}{\rho}\right)_{,jj}}{\left[\left(\frac{\ell}{\rho}\right)_{,k}\left(\frac{\ell}{\rho}\right)_{,k}\right]^{\frac{1}{2}}} - \left(\frac{\ell}{\rho}\right)_{,j}\frac{\left[\left\{\left(\frac{\ell}{\rho}\right)_{,k}\left(\frac{\ell}{\rho}\right)_{,k}\right\}^{\frac{1}{2}}\right]_{,j}}{\left[\left(\frac{\ell}{\rho}\right)_{,k}\left(\frac{\ell}{\rho}\right)_{,k}\right]}\right)\left(\frac{\ell}{\rho}\right)_{,j}$$

let $A = \frac{F_{,jj}}{|F_{,k}|} = \frac{\left(\frac{\ell}{\rho}\right)_{,jj}}{\left[\left(\frac{\ell}{\rho}\right)_{,k}\left(\frac{\ell}{\rho}\right)_{,k}\right]^{\frac{1}{2}}}$ and

$$B = F_{,j}\frac{|F_{,k}|_{,j}}{|F_{,k}|^2} = \left(\frac{\ell}{\rho}\right)_{,j}\frac{\left[\left\{\left(\frac{\ell}{\rho}\right)_{,k}\left(\frac{\ell}{\rho}\right)_{,k}\right\}^{\frac{1}{2}}\right]_{,j}}{\left[\left(\frac{\ell}{\rho}\right)_{,k}\left(\frac{\ell}{\rho}\right)_{,k}\right]}$$

$$Q = \sigma(A - B)\left(\frac{\ell}{\rho}\right)_{j}$$

$$\frac{\partial \mathbf{B}}{\partial \mathbf{U}_1} = \begin{bmatrix} d_{11} \\ d_{21} \\ d_{31} \\ d_{41} \\ d_{51} \end{bmatrix} = \frac{\partial}{\partial \rho}\begin{bmatrix} 0 \\ 0 \\ \rho f_1 \\ \rho f_2 \\ \rho f_1 + \rho f_2 \end{bmatrix} = \frac{\partial}{\partial \rho}\begin{bmatrix} 0 \\ 0 \\ \rho g_1 + Q_1 \\ \rho g_2 + Q_2 \\ \rho g_1 \mathrm{v}_1 + \rho g_2 \mathrm{v}_2 + \mathrm{v}_1 Q_1 + \mathrm{v}_2 Q_2 \end{bmatrix}$$

$d_{31} = \frac{\partial}{\partial \rho}(\rho g_1 + Q_1) = g_1 + \frac{\partial Q_1}{\partial \rho}$, $\quad d_{41} = \frac{\partial}{\partial \rho}(\rho g_2 + Q_2) = g_2 + \frac{\partial Q_2}{\partial \rho}$ and

$$d_{51} = \frac{\partial}{\partial \rho}(\rho g_1 \mathrm{v}_1 + \rho g_2 \mathrm{v}_2 + \mathrm{v}_1 Q_1 + \mathrm{v}_2 Q_2) = g_1 \mathrm{v}_1 + g_2 \mathrm{v}_2 + \mathrm{v}_1 \frac{\partial Q_1}{\partial \rho} + \mathrm{v}_2 \frac{\partial Q_2}{\partial \rho}$$

where

$$\frac{\partial Q_i}{\partial \rho} = \frac{\partial}{\partial \rho}\left[\sigma(A - B)\left(\frac{\ell}{\rho}\right)_{,j}\right] = \sigma\left[\left(\frac{\partial A}{\partial \rho} - \frac{\partial B}{\partial \rho}\right)\left(\frac{\ell}{\rho}\right)_{,j} + (A - B)\left(-\frac{\ell}{\rho^2}\right)_{,j}\right]$$

$$\frac{\partial A}{\partial \rho} = \frac{\partial}{\partial \rho}\left\{\frac{\left(\frac{\ell}{\rho}\right)_{,jj}}{\left[\left(\frac{\ell}{\rho}\right)_{,k}\left(\frac{\ell}{\rho}\right)_{,k}\right]^{\frac{1}{2}}}\right\} = \frac{\partial}{\partial \rho}\left[\left(\frac{\ell}{\rho}\right)_{,jj}\right]\left[\left(\frac{\ell}{\rho}\right)_{,k}\left(\frac{\ell}{\rho}\right)_{,k}\right]^{-\frac{1}{2}}$$

$$+ \frac{\partial}{\partial \rho}\left\{\left[\left(\frac{\ell}{\rho}\right)_{,k}\left(\frac{\ell}{\rho}\right)_{,k}\right]^{-\frac{1}{2}}\right\}\left(\frac{\ell}{\rho}\right)_{,jj} = \frac{\left(-\frac{\ell}{\rho^2}\right)_{,jj}}{\left[\left(\frac{\ell}{\rho}\right)_{,k}\left(\frac{\ell}{\rho}\right)_{,k}\right]^{\frac{1}{2}}}$$

$$-\left(\frac{\ell}{\rho}\right)_{,jj}\frac{\left[\left(-\frac{\ell}{\rho^2}\right)_{,k}\left(\frac{\ell}{\rho}\right)_{,k}\right]}{\left[\left(\frac{\ell}{\rho}\right)_{,k}\left(\frac{\ell}{\rho}\right)_{,k}\right]^{\frac{3}{2}}} = -\frac{\left(-\frac{F}{\rho}\right)_{,jj}}{|F_{,k}|} + F_{,jj}\frac{\left[\left(\frac{F}{\rho}\right)_{,k} F_{;k}\right]}{|F_{,k}|^3}$$

where

$$\left(\frac{F}{\rho}\right)_{,j} = \frac{1}{\rho}F_{,j} - \frac{1}{\rho^2}F\rho_{,j} \text{ and } \left(\frac{F}{\rho}\right)_{,jj} = -\frac{2}{\rho^2}\rho_{,j}F_{,j} + \frac{2}{\rho^3}F(\rho_{,j})^2 + \frac{1}{\rho}F_{,jj} - \frac{1}{\rho^2}F\rho_{,jj}$$

$$\frac{\partial B}{\partial \rho} = \frac{\partial}{\partial \rho}\left\{\left(\frac{\ell}{\rho}\right)_{,j}\frac{\left[\left\{\left(\frac{\ell}{\rho}\right)_{,k}\left(\frac{\ell}{\rho}\right)_{,k}\right\}^{\frac{1}{2}}\right]_{,j}}{\left[\left(\frac{\ell}{\rho}\right)_{,k}\left(\frac{\ell}{\rho}\right)_{,k}\right]}\right\} = \frac{\partial}{\partial \rho}\left[\left(\frac{\ell}{\rho}\right)_{,j}\right]\frac{\left[\left\{\left(\frac{\ell}{\rho}\right)_{,k}\left(\frac{\ell}{\rho}\right)_{,k}\right\}^{\frac{1}{2}}\right]_{,j}}{\left[\left(\frac{\ell}{\rho}\right)_{,k}\left(\frac{\ell}{\rho}\right)_{,k}\right]}$$

$$+\frac{\partial}{\partial \rho}\left\{\frac{\left[\left\{\left(\frac{\ell}{\rho}\right)_{,k}\left(\frac{\ell}{\rho}\right)_{,k}\right\}^{\frac{1}{2}}\right]_{,j}}{\left[\left(\frac{\ell}{\rho}\right)_{,k}\left(\frac{\ell}{\rho}\right)_{,k}\right]}\right\}\left(\frac{\ell}{\rho}\right)_{,j}$$

expand second $\frac{\partial}{\partial \rho}$ term

$$\frac{\partial}{\partial \rho}\left\{\frac{\left[\left\{\left(\frac{\ell}{\rho}\right)_{,k}\left(\frac{\ell}{\rho}\right)_{,k}\right\}^{\frac{1}{2}}\right]_{,j}}{\left[\left(\frac{\ell}{\rho}\right)_{,k}\left(\frac{\ell}{\rho}\right)_{,k}\right]}\right\} = \frac{\frac{\partial}{\partial \rho}\left(\left[\left\{\left(\frac{\ell}{\rho}\right)_{,k}\left(\frac{\ell}{\rho}\right)_{,k}\right\}^{\frac{1}{2}}\right]_{,j}\right)}{\left[\left(\frac{\ell}{\rho}\right)_{,k}\left(\frac{\ell}{\rho}\right)_{,k}\right]}$$

$$+\frac{\partial}{\partial \rho}\left\{\left[\left(\frac{\ell}{\rho}\right)_{,k}\left(\frac{\ell}{\rho}\right)_{,k}\right]^{-1}\right\}\left\{\left[\left(\frac{\ell}{\rho}\right)_{,k}\left(\frac{\ell}{\rho}\right)_{,k}\right]^{\frac{1}{2}}\right\}_{,j}$$

$$= \frac{\left[\frac{1}{2}\left\{\left(\frac{\ell}{\rho}\right)_{,k}\left(\frac{\ell}{\rho}\right)_{,k}\right\}^{-\frac{1}{2}}\left\{2\left(-\frac{\ell}{\rho^2}\right)_{,k}\left(\frac{\ell}{\rho}\right)_{,k}\right\}\right]_{,j}}{\left[\left(\frac{\ell}{\rho}\right)_{,k}\left(\frac{\ell}{\rho}\right)_{,k}\right]}$$

$$-\left[\left(\frac{\ell}{\rho}\right)_{,k}\left(\frac{\ell}{\rho}\right)_{,k}\right]^{-2} 2\left[\left(-\frac{\ell}{\rho^2}\right)_{,k}\left(\frac{\ell}{\rho}\right)_{,k}\right]$$

$$\times\left[\left\{\left(\frac{\ell}{\rho}\right)_{,k}\left(\frac{\ell}{\rho}\right)_{,k}\right\}^{\frac{1}{2}}\right]_{,j}$$

$$\frac{\partial B}{\partial \rho} = \left(-\frac{\ell}{\rho^2}\right)_{,j}\frac{\left[\left\{\left(\frac{\ell}{\rho}\right)_{,k}\left(\frac{\ell}{\rho}\right)_{,k}\right\}^{\frac{1}{2}}\right]_{,j}}{\left[\left(\frac{\ell}{\rho}\right)_{,k}\left(\frac{\ell}{\rho}\right)_{,k}\right]}$$

$$+\left(\frac{\ell}{\rho}\right)_{,j}\left[\frac{\left[\left\{\left(\frac{\ell}{\rho}\right)_{,k}\left(\frac{\ell}{\rho}\right)_{,k}\right\}^{-\frac{1}{2}}\left\{\left(-\frac{\ell}{\rho^2}\right)_{,k}\left(\frac{\ell}{\rho}\right)_{,k}\right\}\right]_{,j}}{\left[\left(\frac{\ell}{\rho}\right)_{,k}\left(\frac{\ell}{\rho}\right)_{,k}\right]}\right.$$

$$- \frac{2\left[\left(-\frac{\ell}{\rho^2}\right)_{,k}\left(\frac{\ell}{\rho}\right)_{,k}\right]\left[\left\{\left(\frac{\ell}{\rho}\right)_{,k}\left(\frac{\ell}{\rho}\right)_{,k}\right\}^{\frac{1}{2}}\right]_{,j}}{\left[\left(\frac{\ell}{\rho}\right)_{,k}\left(\frac{\ell}{\rho}\right)_{,k}\right]^2}\Bigg]$$

$$= \left(\frac{F}{\rho}\right)_{,j}\frac{|F_{,k}|_{,j}}{|F_{,k}|^2} - F_{,j}\left[\frac{\left[\frac{\left\{\left(\frac{F}{\rho}\right)_{,k}F_{,k}\right\}}{|F_{,k}|}\right]_{,j}}{|F_{,k}|^2} - \frac{2\left[\left(\frac{F}{\rho}\right)_{,k}F_{,k}\right]|F_{,k}|_{,j}}{|F_{,k}|^4}\right]$$

where

$$\left(\frac{F}{\rho}\right)_{,j} = \frac{1}{\rho}F_{,j} - \frac{1}{\rho^2}F\rho_{,j}, \qquad |F_{,k}|_{,j} = \left[(F_{,k}F_{,k})^{\frac{1}{2}}\right]_{,j} = \frac{1}{2}(F_{,k}F_{,k})^{-\frac{1}{2}}(F_{,k}F_{,k})_{,j}$$

$$= \frac{1}{2}\frac{(F_{,jk}F_{,k} + F_{,k}F_{,jk})}{|F_{,k}|} = \frac{(F_{,jk}F_{,k})}{|F_{,k}|}$$

$$\left[\frac{\left\{\left(\frac{F}{\rho}\right)_{,k}F_{,k}\right\}}{|F_{,k}|}\right]_{,j} = \frac{\left\{\left(\frac{F}{\rho}\right)_{,k}F_{,k}\right\}_{j}}{|F_{,k}|} - \frac{1}{2}\frac{(F_{,k}F_{,k})_{,j}}{|F_{,k}|^3}\left\{\left(\frac{F}{\rho}\right)_{,k}F_{,k}\right\}$$

and

$$\left\{\left(\frac{F}{\rho}\right)_{,k}F_{,k}\right\}_{,j} = \left\{\left(\frac{F}{\rho}\right)_{,k}\right\}_{,j}F_{,k} + \left(\frac{F}{\rho}\right)_{,k}F_{,jk} = \left(\frac{F_{,k}}{\rho} - \frac{F\rho_{,k}}{\rho^2}\right)_{,j}F_{,k} + \left(\frac{F}{\rho}\right)_{,k}F_{,jk}$$

$$\left(\frac{F_{,k}}{\rho} - \frac{F\rho_{,k}}{\rho^2}\right)_{,j} = \left(\frac{1}{\rho}\right)_{,j}F_{,k} + \frac{1}{\rho}F_{,jk} - \left(-\frac{1}{\rho^2}F\right)_{,j}\rho_{,k} - \frac{1}{\rho^2}F\rho_{,jk}$$

$$= -\frac{1}{\rho^2}\rho_{,j}F_{,k} + \frac{1}{\rho}F_{,jk} + \left(\frac{2}{\rho^3}\rho_{,j}F - \frac{1}{\rho^2}F_{,j}\right)\rho_{,k} - \frac{1}{\rho^2}F\rho_{,jk}$$

$$(F_{,k}F_{,k})_{,j} = 2F_{,jk}F_{k}$$

$$\frac{\partial \mathbf{B}}{\partial \mathbf{U}_2} = \begin{bmatrix} d_{12} \\ d_{22} \\ d_{32} \\ d_{42} \\ d_{52} \end{bmatrix} = \frac{\partial}{\partial \ell}\begin{bmatrix} 0 \\ 0 \\ \rho f_1 \\ \rho f_2 \\ \rho f_1 + \rho f_2 \end{bmatrix} = \frac{\partial}{\partial \ell}\begin{bmatrix} 0 \\ 0 \\ \rho g_1 + Q_1 \\ \rho g_2 + Q_2 \\ \rho g_1 v_1 + \rho g_2 v_2 + v_1 Q_1 + v_2 Q_2 \end{bmatrix}$$

$$d_{32} = \frac{\partial}{\partial \ell}(\rho g_1 + Q_1) = \frac{\partial Q_1}{\partial \ell},$$

$$d_{42} = \frac{\partial}{\partial \ell}(\rho g_2 + Q_2) = \frac{\partial Q_2}{\partial \ell}, \text{ and}$$

$$d_{52} = \frac{\partial}{\partial \ell}(\rho g_1 v_1 + \rho g_2 v_2 + v_1 Q_1 + v_2 Q_2) = v_1 \frac{\partial Q_1}{\partial \ell} + v_2 \frac{\partial Q_2}{\partial \ell}$$

$$\frac{\partial Q_i}{\partial \ell} = \frac{\partial}{\partial \ell}\left[\sigma(A-B)\left(\frac{\ell}{\rho}\right)_{,j}\right] = \sigma\left[\left(\frac{\partial A}{\partial \ell} - \frac{\partial B}{\partial \ell}\right)\left(\frac{\ell}{\rho}\right)_{,j} + (A-B)\left(\frac{1}{\rho}\right)_{,j}\right]$$

$$\frac{\partial A}{\partial \ell} = \frac{\partial}{\partial \ell}\left\{\frac{\left(\frac{\ell}{\rho}\right)_{,jj}}{\left[\left(\frac{\ell}{\rho}\right)_{,k}\left(\frac{\ell}{\rho}\right)_{,k}\right]^{\frac{1}{2}}}\right\} = \frac{\left(\frac{1}{\rho}\right)_{,jj}}{\left[\left(\frac{\ell}{\rho}\right)_{,k}\left(\frac{\ell}{\rho}\right)_{,k}\right]^{\frac{1}{2}}} - \left(\frac{\ell}{\rho}\right)_{,jj}\frac{\left[\left(\frac{1}{\rho}\right)_{,k}\left(\frac{\ell}{\rho}\right)_{,k}\right]}{\left[\left(\frac{\ell}{\rho}\right)_{,k}\left(\frac{\ell}{\rho}\right)_{,k}\right]^{\frac{3}{2}}}$$

$$= \frac{\left(-\frac{1}{\rho}\right)_{,jj}}{|F_{,k}|} + F_{,jj}\frac{\left[\left(\frac{1}{\rho}\right)_{,k} F_{,k}\right]}{|F_{,k}|^3}$$

$$\frac{\partial B}{\partial \ell} = \frac{\partial}{\partial \ell}\left\{\left(\frac{\ell}{\rho}\right)_{,j}\frac{\left[\left\{\left(\frac{\ell}{\rho}\right)_{,k}\left(\frac{\ell}{\rho}\right)_{,k}\right\}^{\frac{1}{2}}\right]_{,j}}{\left[\left(\frac{\ell}{\rho}\right)_{,k}\left(\frac{\ell}{\rho}\right)_{,k}\right]}\right\}$$

$$= \left(\frac{1}{\rho^2}\right)_{,j}\frac{\left[\left\{\left(\frac{\ell}{\rho}\right)_{,k}\left(\frac{\ell}{\rho}\right)_{,k}\right\}^{\frac{1}{2}}\right]_{,j}}{\left[\left(\frac{\ell}{\rho}\right)_{,k}\left(\frac{\ell}{\rho}\right)_{,k}\right]} + \left(\frac{\ell}{\rho}\right)_{,j}\left\{\frac{\left[\left\{\left(\frac{\ell}{\rho}\right)_{,k}\left(\frac{\ell}{\rho}\right)_{,k}\right\}^{-\frac{1}{2}}\left\{\left(\frac{\ell}{\rho}\right)_{,k}\left(\frac{\ell}{\rho}\right)_{,k}\right\}\right]_{,j}}{\left[\left(\frac{\ell}{\rho}\right)_{,k}\left(\frac{\ell}{\rho}\right)_{,k}\right]}\right.$$

$$\left. - \frac{2\left[\left(\frac{1}{\rho}\right)_{,k}\left(\frac{\ell}{\rho}\right)_{,k}\right]\left[\left\{\left(\frac{\ell}{\rho}\right)_{,k}\left(\frac{\ell}{\rho}\right)_{,k}\right\}^{\frac{1}{2}}\right]_{,j}}{\left[\left(\frac{\ell}{\rho}\right)_{,k}\left(\frac{\ell}{\rho}\right)_{,k}\right]^2}\right\}$$

$$\frac{\partial B}{\partial \ell} = \left(\frac{1}{\rho^2}\right)_{,j}\frac{|F_{,k}|_{,j}}{|F_{,k}|} + F_{,j}[\,]\left\{\frac{\left[\left\{\left(\frac{\ell}{\rho}\right)_{,k}\left(\frac{\ell}{\rho}\right)_{,k}\right\}^{-\frac{1}{2}}\left\{\left(\frac{\ell}{\rho}\right)_{,k}\left(\frac{\ell}{\rho}\right)_{,k}\right\}\right]_{,j}}{\left[\left(\frac{\ell}{\rho}\right)_{,k}\left(\frac{\ell}{\rho}\right)_{,k}\right]}\right.$$

$$\left. - \frac{2\left[\left(\frac{1}{\rho}\right)_{,k}\left(\frac{\ell}{\rho}\right)_{,k}\right]\left[\left\{\left(\frac{\ell}{\rho}\right)_{,k}\left(\frac{\ell}{\rho}\right)_{,k}\right\}^{\frac{1}{2}}\right]_{,j}}{\left[\left(\frac{\ell}{\rho}\right)_{,k}\left(\frac{\ell}{\rho}\right)_{,k}\right]^2}\right\}$$

$$\frac{\partial B}{\partial \ell} = \left(\frac{1}{\rho^2}\right)_{,j} \frac{|F_{,k}|_{,j}}{|F_{,k}|} + F_{,j} \left[\frac{\left[\dfrac{\left\{\left(\frac{1}{\rho}\right)_{,k} F_{,k}\right\}}{|F_{,k}|}\right]_{,j}}{|F_{,k}|^2} - \frac{2\left[\left(\frac{1}{\rho}\right)_{,k} F_{,k}\right]|F_{,k}|_{,j}}{|F_{,k}|^4} \right]$$

where

$$\left[\frac{\left\{\left(\frac{1}{\rho}\right)_{,k} F_{,k}\right\}}{|F_{,k}|}\right]_{,j} = -\frac{(F_{,jk}F_{,k})}{|F_{,k}|^3}\left\{\left(\frac{1}{\rho}\right)_{,k} F_{,k}\right\} + \frac{\left\{\left(\frac{1}{\rho}\right)_{,k} F_{,k}\right\}_{,j}}{|F_{,k}|} \text{ and}$$

$$\begin{aligned}\left\{\left(\frac{1}{\rho}\right)_{,k} F_{,k}\right\}_{,j} &= \left\{\left(\frac{1}{\rho}\right)_{,k}\right\}_{,j} F_{,k} + \left(\frac{1}{\rho}\right)_{,k} F_{,jk} \\ &= \left(\frac{2}{\rho^3}\rho_{,j}\rho_{,k} - \frac{1}{\rho^2}\rho_{,jk}\right)F_{,k} - \frac{1}{\rho^2}\rho_{,k}F_{,jk}\end{aligned}$$

$$\frac{\partial \mathbf{B}}{\partial \mathbf{U}_3} = \begin{bmatrix} d_{13} \\ d_{23} \\ d_{33} \\ d_{43} \\ d_{53} \end{bmatrix} = \frac{\partial}{\partial(\rho v_1)} \begin{bmatrix} 0 \\ 0 \\ \rho f_1 \\ \rho f_2 \\ \rho f_1 + \rho f_2 \end{bmatrix} = \begin{bmatrix} 0 \\ 0 \\ 0 \\ 0 \\ \rho g_1 + Q_1 \end{bmatrix}$$

$$\frac{\partial \mathbf{B}}{\partial \mathbf{U}_4} = \begin{bmatrix} d_{14} \\ d_{24} \\ d_{34} \\ d_{44} \\ d_{54} \end{bmatrix} = \frac{\partial}{\partial(\rho v_2)} \begin{bmatrix} 0 \\ 0 \\ \rho f_1 \\ \rho f_2 \\ \rho f_1 + \rho f_2 \end{bmatrix} = \begin{bmatrix} 0 \\ 0 \\ 0 \\ 0 \\ \rho g_2 + Q_2 \end{bmatrix}$$

$$\frac{\partial \mathbf{B}}{\partial \mathbf{U}_5} = \begin{bmatrix} d_{15} \\ d_{25} \\ d_{35} \\ d_{45} \\ d_{55} \end{bmatrix} = \frac{\partial}{\partial(\rho E)} \begin{bmatrix} 0 \\ 0 \\ \rho f_1 \\ \rho f_2 \\ \rho f_1 + \rho f_2 \end{bmatrix} = \begin{bmatrix} 0 \\ 0 \\ 0 \\ 0 \\ 0 \end{bmatrix}$$

APPENDIX D

Relativistic Astrophysical Flow Metrics, Christoffel Symbols, and FDV Flux and Source Term Jacobians

D.1 METRICS AND CHRISTOFFEL SYMBOLS

The metric tensor $g_{\alpha\beta}$ is related by the square of the line element as

$$ds^2 = g_{\alpha\beta} dx^\alpha dx^\beta$$

with the Minkowski form,

$$g_{\alpha\beta} = \begin{bmatrix} -1 & 0 & 0 & 0 \\ 0 & 1 & 0 & 0 \\ 0 & 0 & 1 & 0 \\ 0 & 0 & 0 & 1 \end{bmatrix}$$

$$ds^2 = -dt^2 + dx^2 + dy^2 + dz^2$$

The Kerr line element is of the form

$$ds^2 = -\frac{\Delta - a^2 \sin^2\theta}{\omega^2} dt^2 - 2a \frac{2Mr \sin^2\theta}{\omega^2} dt d\phi$$
$$+ \frac{(r^2 + a^2)^2 - a^2 \Delta \sin^2\theta}{\omega^2} \sin^2\theta d\phi^2 + \frac{\omega^2}{\Delta} dr^2 + \omega^2 d\theta^2$$

with

$$\Delta \equiv r^2 - 2Mr + a^2 \qquad \omega^2 \equiv r^2 + a^2 \cos^2\theta$$

The general relativistic Kerr metric with the angular momentum a per unit unit mass M,

$$g_{\alpha\beta} = \begin{bmatrix} -\dfrac{\Delta + a^2 \sin^2\theta}{\omega^2} & 0 & 0 & \dfrac{a \sin^2\theta(\Delta - (r^2 + a^2))}{\omega^2} \\ 0 & \dfrac{\omega^2}{\Delta} & 0 & 0 \\ 0 & 0 & \omega^2 & 0 \\ \dfrac{a \sin^2\theta(\Delta - (r^2 + a^2))}{\omega^2} & 0 & 0 & -\dfrac{a^2 \sin^4\theta \Delta + \sin^2\theta(r^2 + a^2)^2}{\omega^2} \end{bmatrix}$$

The Christoffel symbols for the Kerr metric are calculated from the metric terms,

$$g_{00,1} = \frac{2\left(-Mr^2 + Ma^2\cos^2\theta + a^2 r\right)}{\omega^4}$$

$$g_{00,2} = \frac{-4a^2\sin\theta\cos\theta(r^2 + a^2 - Mr)}{\omega^4}$$

$$g_{03,1} = g_{30,1} = \frac{2aM\sin^2\theta(r^2 - a^2\cos^2\theta)}{\omega^4}$$

$$g_{03,2} = g_{30,2} = \frac{-4aMr\sin\theta\cos\theta(r^2 + a^2)}{\omega^4}$$

$$g_{11,1} = \frac{2(-Mr^2 + a^2\cos^2\theta(M - r) + a^2 r)}{\omega^4}$$

$$g_{11,2} = \frac{-2a^2\sin\theta\cos\theta}{\omega^4}$$

$$g_{22,1} = 2r \qquad g_{22,2} = -2a^2\cos\theta\sin\theta$$

$$g_{33,1} = \frac{-2a^2\sin^2\theta\begin{pmatrix}a^2\sin^2\theta\cos^2\theta(r - M) + r\sin^2\theta(Mr - a^2)\\ +2a^2 r\cos^2\theta(r^2 + a^2) + r^5 - a^4 r\end{pmatrix}}{\omega^4}$$

$$g_{33,2} = \frac{-2\sin\theta\cos\theta\begin{pmatrix}\sin^2\theta(3a^2r^4 - 4Mr^3a^2) + r^2(r^3 + a^4 + 2a^2r^2) + 4a^4r^2\\ +2a^6 + \cos^2\theta 2r^4a^2 + 4a^4\sin^2\theta\cos^2\theta\Delta + 2a^4\sin^4\theta\Delta\end{pmatrix}}{\omega^4}$$

$$g_{00,0} = g_{00,3} = g_{03,0} = g_{03,3} = g_{11,0} = g_{11,3} = g_{22,0} = g_{22,3}$$
$$= g_{30,0} = g_{30,3} = g_{33,0} = g_{33,3} = 0$$

$$g^{00} = \frac{g_{33}}{g_{00}g_{33} - g_{03}g_{30}} \qquad g^{03} = g^{30} = \frac{-g_{30}}{g_{00}g_{33} - g_{03}g_{30}}$$

$$g^{11} = \frac{1}{g_{11}} \qquad g^{22} = \frac{1}{g_{22}} \qquad g^{33} = \frac{g_{00}}{g_{00}g_{33} - g_{03}g_{30}}$$

$$\Gamma^0{}_{00} = \Gamma^0{}_{03} = \Gamma^0{}_{30} = \Gamma^0{}_{11} = \Gamma^0{}_{12} = \Gamma^0{}_{21} = \Gamma^0{}_{22} = \Gamma^0{}_{33} = 0$$

$$\Gamma^1_{01} = \Gamma^1_{10} = \Gamma^1_{02} = \Gamma^1_{20} = \Gamma^1_{13} = \Gamma^1_{31} = \Gamma^1_{23} = \Gamma^1_{32} = 0$$

$$\Gamma^2_{01} = \Gamma^2_{10} = \Gamma^2_{02} = \Gamma^2_{20} = \Gamma^2_{13} = \Gamma^2_{31} = \Gamma^2_{23} = \Gamma^2_{32} = 0$$

$$\Gamma^3_{00} = \Gamma^3_{03} = \Gamma^3_{30} = \Gamma^3_{11} = \Gamma^3_{12} = \Gamma^3_{21} = \Gamma^3_{22} = \Gamma^3_{33} = 0$$

$$\Gamma^0{}_{01} = \Gamma^0{}_{10} = \frac{1}{2}(g^{00}g_{00,1} + g^{03}g_{30,1}) \quad \Gamma^0{}_{02} = \Gamma^0{}_{20} = \frac{1}{2}(g^{00}g_{00,2} + g^{03}g_{30,2})$$

$$\Gamma^0{}_{13} = \Gamma^0{}_{31} = \frac{1}{2}(g^{00}g_{03,1} + g^{03}g_{33,1}) \quad \Gamma^0{}_{23} = \Gamma^0{}_{32} = \frac{1}{2}(g^{00}g_{03,2} + g^{03}g_{33,2})$$

$$\Gamma^1_{00} = -\frac{1}{2}g^{11}g_{00,1} \quad \Gamma^1_{03} = \Gamma^1_{30} = -\frac{1}{2}g^{11}g_{03,1} \quad \Gamma^1_{11} = \frac{1}{2}g^{11}g_{11,1}$$

$$\Gamma^1_{12} = \Gamma^1_{21} = \frac{1}{2}g^{11}g_{11,2} \quad \Gamma^1_{22} = -\frac{1}{2}g^{11}g_{22,1} \quad \Gamma^1_{33} = -\frac{1}{2}g^{11}g_{33,1}$$

$$\Gamma^2_{00} = -\frac{1}{2}g^{22}g_{00,2} \quad \Gamma^2_{03} = \Gamma^2_{30} = -\frac{1}{2}g^{22}g_{03,2} \quad \Gamma^2_{11} = -\frac{1}{2}g^{22}g_{11,2}$$

$$\Gamma^2_{12} = \Gamma^2_{21} = \frac{1}{2}g^{22}g_{22,1} \quad \Gamma^2_{22} = \frac{1}{2}g^{22}g_{22,2} \quad \Gamma^2_{33} = -\frac{1}{2}g^{22}g_{33,2}$$

$$\Gamma^3_{01} = \Gamma^3_{10} = \frac{1}{2}(g^{30}g_{00,1} + g^{33}g_{30,1}) \quad \Gamma^3_{02} = \Gamma^3_{20} = \frac{1}{2}(g^{30}g_{00,2} + g^{33}g_{30,2})$$

$$\Gamma^3_{13} = \Gamma^3_{31} = \frac{1}{2}(g^{30}g_{03,1} + g^{33}g_{33,1}) \quad \Gamma^3_{23} = \Gamma^3_{32} = \frac{1}{2}(g^{30}g_{03,2} + g^{33}g_{33,2})$$

D.2 FDV Flux and Source Term Jacobians

The three-dimensional Jacobians for ideal fluids for the general relativistic Kerr metric are derived similarly as in Newtonian flows. The special relativistic Jacobians can be deduced from these by setting $g^{00} = -1$, $g^{11} = g^{22} = g^{33} = 1$, and $g^{03} = g^{30} = 0$. It is important to note that the Christoffel symbols are not dependent on the conservation variables. This allows the source term Jacobians to be written as a combination of the other Jacobians.

Convection Flux Jacobian

$$a_1 = \begin{bmatrix} 0 & \frac{1}{hW} & 0 & 0 & 0 \\ -hW\left(V_1^2 + \frac{g^{11}}{g^{00}}\right) & 2V_1 & 0 & 0 & \frac{g^{11}}{g^{00}} \\ (-hWV_1V_2) & V_2 & V_1 & 0 & 0 \\ \begin{bmatrix} -hWV_1V_3+ \\ V_1\frac{g^{30}}{g^{00}}\left(hW + \frac{P}{\rho W}g^{00}\right) \end{bmatrix} & V_3 & 0 & V_1 & -V_1\frac{g^{30}}{g^{00}} \\ 0 & 1 & 0 & 0 & 0 \end{bmatrix}$$

$$a_2 = \begin{bmatrix} 0 & 0 & \frac{1}{hW} & 0 & 0 \\ (-hWV_1V_2) & V_2 & V_1 & 0 & 0 \\ -hW\left(V_2^2 + \frac{g^{22}}{g^{00}}\right) & 0 & 2V_2 & 0 & \frac{g^{22}}{g^{00}} \\ \begin{bmatrix} -hWV_2V_3+ \\ V_1\frac{g^{30}}{g^{00}}\left(hW + \frac{P}{\rho W}g^{00}\right) \end{bmatrix} & 0 & V_3 & V_2 & -V_2\frac{g^{30}}{g^{00}} \\ 0 & 0 & 1 & 0 & 0 \end{bmatrix}$$

$$a_3 = \begin{bmatrix} \frac{g^{30}}{g^{00}} & 0 & 0 & \frac{1}{hW} & -\frac{1}{hW}\frac{g^{30}}{g^{00}} \\ hWV_1\left(\frac{g^{30}}{g^{00}} - V_3\right) & V_3 & 0 & V_1 & -V_1\frac{g^{30}}{g^{00}} \\ hWV_2\left(\frac{g^{30}}{g^{00}} - V_3\right) & 0 & V_3 & V_2 & -V_2\frac{g^{30}}{g^{00}} \\ hW\left(2V_3\frac{g^{30}}{g^{00}} - V_3^2 - \frac{g^{33}}{g^{00}}\right) & 0 & 0 & 2V_3 & -2V_3 + \frac{g^{33}}{g^{00}} \\ 0 & 0 & 0 & 1 & 0 \end{bmatrix}$$

Diffusion Flux Jacobians

$$b_i = \begin{bmatrix} 0 & 0 & 0 & 0 & 0 \\ b_i[5] & b_i[6] & b_i[7] & b_i[8] & b_i[9] \\ b_i[10] & b_i[11] & b_i[12] & b_i[13] & b_i[14] \\ b_i[15] & b_i[16] & b_i[17] & b_i[18] & b_i[19] \\ b_i[20] & b_i[21] & b_i[22] & b_i[23] & b_i[24] \end{bmatrix}$$

$$b1[5] = -W\omega\left(C_6A_2 + 2A_3\frac{V^1}{\alpha}V^1W^2\right) + 2\xi\left(\begin{matrix} W^3\left(\frac{V^1}{\alpha}V^1_{,t} + V_1\frac{V^1_{,t}}{\alpha}\right) + W\left(\left(\frac{V^1}{\alpha}\right)_{,1} + \left(\frac{V^1}{\alpha}\right)_{,2} + \left(\frac{V^1}{\alpha}\right)_{,3}\right) \\ +W^3D_4\left(2\frac{V^1}{\alpha}V^1 + \frac{V^1}{\alpha}V^2 + \frac{V^2}{\alpha}V^1 + \frac{V^1}{\alpha}V^3 + \frac{V^3}{\alpha}V^1\right) \\ \times(g^{11} + W^2A_{10}) \end{matrix}\right) + 2W\left(q_{1,\alpha}V^1 - q_1\frac{V^1}{\alpha}\right)$$

$$b1[6] = W\omega(B_{,1}A_2 + 2A_3BV^1W^2) - 2\xi\left(\begin{matrix} W^3(BV^1{}_{,t} + V_1B_{,t}) + W(B_{,1} + B_{,2} + B_{,3})(g^{11} + W^2A_{10}) \\ +W^3A_{26}(2BV^1 + BV^2 + BV^3) \end{matrix}\right) + 2W(q_{1,\varepsilon}V^1 - q_1B)$$

$$b1[7] = W\omega(B_{,2}A_2) - 2\xi(W^3A_{26}BV_1) + 2W(q_{1,\eta}V^1), \quad b1[8] = W\omega(B_{,3}A_2) - 2\xi(W^3A_{26}BV_1) + 2W(q_{1,\upsilon}V^1), \quad b1[9] = 0$$

$$b1[10] = -W^3\omega\left(A_{19}V^1V^2 + 2A_3\frac{V^1}{\alpha}V^2\right) + W\left(q_{1,\alpha}V^2 - \frac{q_1}{\alpha}V^2 + q_{2,\alpha}V^1 - \frac{q_2}{\alpha}V^1\right) + \xi W^3\left(\frac{V^2}{\alpha}(V^1{}_{,t} + A_6) + V^2\left(\left(\frac{V^1{}_{,t}}{\alpha}\right) + A_{23} + A_5\right) + \frac{V^1}{\alpha}(V^2{}_{,t} + A_7) + V^1\left(\left(\frac{V^2{}_{,t}}{\alpha}\right) + A_9 + A_{18}\right)\right)$$

$$b1[11] = W^3\omega(B_{,1}V^1V^2 + A_3BV^2) - \xi W^3(B(V^2{}_{,t} + A_7) + V^2(B_{,t} + BV^1{}_{,1} + A_1)$$
$$+ V^1V^2{}_{,1}B) + W(q_{1,\varepsilon}V^2 + q_{2,\varepsilon}V^1 + q_2B)$$
$$b1[12] = W^3\omega(B_{,2}V^1V^2 + A_3BV^1) - \xi W^3(B(V^1{}_{,t} + A_6) + V^1(B_{,t} + BV^2{}_{,2} + A_1)$$
$$+V^2V^1{}_{,2}B) + W(q_{1,\eta}V^2 + q_{2,\eta}V^1 + q_1B)$$
$$b1[13] = W^3\omega(B_{,3}V^1V^2) - \xi W^3(BV^2V^1{}_{,3} + V^1V^2{}_{,3}B) + W(q_{1,\upsilon}V^2 + q_{2,\upsilon}V^1),$$
$$b1[14] = 0$$
$$b1[15] = -W^3\omega\left(A_{19}V^1V^3 + 2A_3\frac{V^1}{\alpha}V^3\right) + W\left(q_{1,\alpha}V^3 - \frac{q_1}{\alpha}V^3 + q_{3,\alpha}V^1 - \frac{q_3}{\alpha}V^1\right)$$
$$+ \xi W^3\left(\frac{V^3}{\alpha}(V^1{}_{,t} + A_6) + V^3\left(\left(\frac{V^1{}_{,t}}{\alpha}\right) + A_{23} + A_5\right) + \frac{V^1}{\alpha}(V^3{}_{,t} + A_8)\right.$$
$$\left. +V^1\left(\left(\frac{V^3{}_{,t}}{\alpha}\right) + A_4 + A_{24}\right)\right)$$
$$b1[16] = W^3\omega(B_{,1}V^1V^3 + A_3BV^3) - \xi W^3(B(V^3{}_{,t} + A_8) + V^3(B_{,t} + BV^1{}_{,1} + A_1)$$
$$+ V^1V^3{}_{,1}B) + W(q_{1,\varepsilon}V^3 + q_{3,\varepsilon}V^1 + q_3B)$$
$$b1[17] = W^3\omega(B_{,2}V^1V^3) - \xi W^3(BV^3V^1{}_{,2} + V^1V^3{}_{,2}B) + W(q_{1,\eta}V^3 + q_{2,\eta}V^1)$$
$$b1[18] = W^3\omega(B_{,3}V^1V^3 + A_3BV^1) - \xi W^3(B(V^1{}_{,t} + A_6) + V^1(B_{,t} + BV^3{}_{,3} + A_1)$$
$$+ V^3V^1{}_{,3}B) + W(q_{1,\upsilon}V^3 + q_{3,\upsilon}V^1 + q_1B)$$
$$b1[19] = 0,$$
$$b1[20] = -W^3\omega\left(A_{19}V^1 + A_3\frac{V^1}{\alpha}\right) + \xi\left(W^3(CA_{23} + A_5)\right.$$
$$\left. + W\frac{V^1{}_{,t}}{\alpha}(W^2 + g^{00} + 2W^2V^3 + 2g^{30})\right) + Wq_{1,\alpha}$$
$$b1[21] = W^3\omega(B_{,1}V^1 + A_3B) - \xi(W^3(BV^1{}_{,1} + A_1)$$
$$+WB_{,t}(W^2 + g^{00} + 2W^2V^3 + 2g^{30})) + Wq_{1,\varepsilon}$$
$$b1[22] = W^3\omega(B_{,2}V^1) - \xi(W^3BV^1{}_{,2}) + Wq_{1,\eta},$$
$$b1[23] = W^3\omega(B_{,3}V^1) - \xi(W^3BV^1{}_{,3}) + Wq_{1,\upsilon},$$
$$b1[24] = 0$$

where

$$A_1 = V^1B_{,1} + V^2B_{,2} + V^3B_{,3}, \qquad A_2 = g^{11} + (V^1V^1W^2),$$
$$A_3 = V^1{}_{,1} + V^2{}_{,2} + V^3{}_{,3}$$
$$A_4 = V^3{}_{,1}\frac{V^1}{\alpha} + V^3{}_{,2}\frac{V^2}{\alpha} + V^3{}_{,3}\frac{V^3}{\alpha},$$

$$A_5 = V^1\left(\frac{V^1}{\alpha}\right)_{,1} + V^2\left(\frac{V^1}{\alpha}\right)_{,2} + V^3\left(\frac{V^1}{\alpha}\right)_{,3}, \qquad A_6 = V^1V^1{}_{,1} + V^2V^1{}_{,2} + V^3V^1{}_{,3}$$

$$A_7 = V^1V^2{}_{,1} + V^2V^2{}_{,2} + V^3V^2{}_{,3}, \qquad A_8 = V^1V^3{}_{,1} + V^2V^3{}_{,2} + V^3V^3{}_{,3},$$

$$A_9 = V^2{}_{,1}\frac{V^1}{\alpha} + V^2{}_{,2}\frac{V^2}{\alpha} + V^2{}_{,3}\frac{V^3}{\alpha} \qquad A_{10} = V^1V^1 + V^1V^2 + V^1V^3,$$

$$A_{11} = T_{,1} + T_{,2} + T_{,3}, \qquad A_{12} = V^1{}_{,t} + V^2{}_{,t} + V^3{}_{,t}$$

$$A_{13} = V^2V^1 + V^2V^2 + V^2V^3, \qquad A_{14} = V^3V^1 + V^3V^2 + V^3V^3,$$

$$A_{15} = g^{22} + (V^2V^2W^2) \qquad A_{16} = g^{33} + (V^3V^3W^2),$$

$$A_{17} = V^1\frac{V^2}{\alpha} + V^2\frac{V^1}{\alpha} + 2V^2\frac{V^2}{\alpha} + V^3\frac{V^2}{\alpha}V^2\frac{V^3}{\alpha}$$

$$A_{18} = V^1\left(\frac{V^2}{\alpha}\right)_{,1} + V^2\left(\frac{V^2}{\alpha}\right)_{,2} + V^3\left(\frac{V^2}{\alpha}\right)_{,3},$$

$$A_{19} = \left(\frac{V^1}{\alpha}\right)_{,1} + \left(\frac{V^2}{\alpha}\right)_{,2} + \left(\frac{V^3}{\alpha}\right)_{,3}$$

$$A_{20} = V^1{}_{,1} + V^2{}_{,1} + V^3{}_{,1}, \qquad A_{21} = V^1{}_{,2} + V^2{}_{,2} + V^3{}_{,2},$$

$$A_{22} = V^1{}_{,3} + V^2{}_{,3} + V^3{}_{,3} \; A_{23} = V^1{}_{,1}\frac{V^1}{\alpha} + V^1{}_{,2}\frac{V^2}{\alpha} + V^1{}_{,3}\frac{V^3}{\alpha},$$

$$A_{24} = V^1\left(\frac{V^3}{\alpha}\right)_{,1} + V^2\left(\frac{V^3}{\alpha}\right)_{,2} + V^3\left(\frac{V^3}{\alpha}\right)_{,3}$$

$$\Omega = A_{11} + \tau W^2(A_{12} + A_6 + A_7 + A_8),$$

$$B = \frac{1}{\rho h W^2}, \qquad \mu = \frac{2\xi}{3}, \qquad \omega = \mu - \chi, \qquad \alpha = \rho W$$

$$q_1 = -k\Omega(g^{11} + W^2A_{10}), \qquad q_2 = -k\Omega(g^{22} + W^2A_{13}),$$

$$q_3 = -k\Omega(g^{33} + W^2A_{14})$$

$$q_{1,\alpha} = k\begin{pmatrix} W^2A_{17}\Omega + \\ (g^{11} + W^2A_{10})\tau W^2\begin{pmatrix}\dfrac{V^1{}_{,t}}{\alpha} + \dfrac{V^2{}_{,t}}{\alpha} + \dfrac{V^3{}_{,t}}{\alpha} \\ + A_{23} + A_5 + A_9 + A_{18} + A_4 + A_{24}\end{pmatrix}\end{pmatrix}$$

$$q_{1,\varepsilon} = -k\Big(BW^2\Omega(2V^1 + V^2 + V^3) + (g^{11} + W^2A_{10})\tau W^2(B_{,t} + A_1 + BA_{20})\Big)$$

$$q_{1,\eta} = -k(BW^2V^1\Omega + (g^{11} + W^2A_{10})\tau W^2(B_{,t} + A_1 + BA_{21})),$$

$$q_{1,\upsilon} = -k(BW^2V^1\Omega + (g^{11} + W^2A_{10})\tau W^2(B_{,t} + A_1 + BA_{22}))$$

$$d = \begin{bmatrix} 0 & 0 & 0 & 0 & 0 \\ d[5] & d[6] & d[7] & d[8] & d[9] \\ d[10] & d[11] & d[12] & d[13] & d[14] \\ d[15] & d[16] & d[17] & d[18] & d[19] \\ d[20] & d[21] & d[22] & d[23] & d[24] \end{bmatrix}$$

$$d[5] = (a3[20] + b3[20])\Gamma^1{}_{03} + (a1[5] + b1[5])\Gamma^1{}_{11} + (a2[5] + b2[5])\Gamma^1{}_{12}$$
$$+(a1[10] + b1[10])\Gamma^1{}_{21} + (a3[15] + b3[15])\Gamma^1{}_{33} + (a1[10] + b1[10])\Gamma^1{}_{22}$$

$$d[6] = \Gamma^1{}_{30} + (a3[21] + b3[21])\Gamma^1{}_{03} + (a1[6] + b1[6])\Gamma^1{}_{11} + (a2[6] + b2[6])\Gamma^1{}_{12}$$
$$+(a1[11] + b1[11])\Gamma^1{}_{21} + (a3[16] + b3[16])\Gamma^1{}_{33} + (a1[11] + b1[11])\Gamma^1{}_{22}$$

$$d[7] = \Gamma^1{}_{30} + (a3[22] + b3[22])\Gamma^1{}_{03} + (a1[7] + b1[7])\Gamma^1{}_{11} + (a2[7] + b2[7])\Gamma^1{}_{12}$$
$$+(a1[12] + b1[12])\Gamma^1{}_{21} + (a3[17] + b3[17])\Gamma^1{}_{33} + (a1[12] + b1[12])\Gamma^1{}_{22}$$

$$d[8] = \Gamma^1{}_{30} + (a3[23] + b3[23])\Gamma^1{}_{03} + (a1[8] + b1[8])\Gamma^1{}_{11} + (a2[8] + b2[8])\Gamma^1{}_{12}$$
$$+(a1[13] + b1[13])\Gamma^1{}_{21} + (a3[18] + b3[18])\Gamma^1{}_{33} + (a1[13] + b1[13])\Gamma^1{}_{22}$$

$$d[9] = \Gamma^1{}_{00} + (a3[24] + b3[24])\Gamma^1{}_{03} + (a1[9] + b1[9])\Gamma^1{}_{11} + (a2[9] + b2[9])\Gamma^1{}_{12}$$
$$+(a1[14] + b1[14])\Gamma^1{}_{21} + (a3[19] + b3[19])\Gamma^1{}_{33} + (a1[14] + b1[14])\Gamma^1{}_{22}$$

$$d[10] = (a3[20] + b3[20])\Gamma^2_{03} + (a1[5] + b1[5])\Gamma^2_{11} + (a2[5] + b2[5])\Gamma^2_{12}$$
$$+(a1[10] + b1[10])\Gamma^2_{21} + (a3[15] + b3[15])\Gamma^2_{33} + (a1[10] + b1[10])\Gamma^2_{22}$$

$$d[11] = \Gamma^2_{30} + (a3[21] + b3[21])\Gamma^2_{03} + (a1[6] + b1[6])\Gamma^2_{11} + (a2[6] + b2[6])\Gamma^2_{12}$$
$$+(a1[11] + b1[11])\Gamma^2_{21} + (a3[16] + b3[16])\Gamma^2_{33} + (a1[11] + b1[11])\Gamma^2_{22}$$

$$d[12] = \Gamma^2_{30} + (a3[22] + b3[22])\Gamma^2_{03} + (a1[7] + b1[7])\Gamma^2_{11} + (a2[7] + b2[7])\Gamma^2_{12}$$
$$+(a1[12] + b1[12])\Gamma^2_{21} + (a3[17] + b3[17])\Gamma^2_{33} + (a1[12] + b1[12])\Gamma^2_{22}$$

$$d[13] = \Gamma^2_{30} + (a3[23] + b3[23])\Gamma^2_{03} + (a1[8] + b1[8])\Gamma^2_{11} + (a2[8] + b2[8])\Gamma^2_{12}$$
$$+(a1[13] + b1[13])\Gamma^2_{21} + (a3[18] + b3[18])\Gamma^2_{33} + (a1[13] + b1[13])\Gamma^2_{22}$$

$$d[14] = \Gamma^2_{00} + (a3[24] + b3[24])\Gamma^2_{03} + (a1[9] + b1[9])\Gamma^2_{11} + (a2[9] + b2[9])\Gamma^2_{12}$$
$$+(a1[14] + b1[14])\Gamma^2_{21} + (a3[19] + b3[19])\Gamma^2_{33} + (a1[14] + b1[14])\Gamma^2_{22}$$

$$d[15] = (a1[20] + b1[20])\Gamma^3_{01} + 2(a2[20] + b2[20])\Gamma^3_{02} + (a3[5] + b3[5])\Gamma^3_{13}$$
$$+(a1[15] + b1[15])\Gamma^3_{31} + (a3[10] + b3[10])\Gamma^3_{23} + (a2[15] + b2[15])\Gamma^3_{32}$$

$$d[16] = \Gamma^3_{10} + \Gamma^3_{20} + 2(a1[21] + b1[21])\Gamma^3_{01} + (a2[21] + b2[21])\Gamma^3_{02}$$
$$+(a3[6] + b3[6])\Gamma^3_{13} + (a1[16] + b1[16])\Gamma^3_{31} + (a3[11]$$
$$+b3[11])\Gamma^3_{23} + (a2[16] + b2[16])\Gamma^3_{32}$$

$$d[17]=\Gamma^3_{10}+\Gamma^3_{20}2(a1[22]+b1[22])\Gamma^3_{01}+(a2[22]+b2[22])\Gamma^3_{02}+(a3[7]+b3[7])\Gamma^3_{13}$$
$$+(a1[17]+b1[17])\Gamma^3_{31}+(a3[12]+b3[12])\Gamma^3_{23}+(a2[17]+b2[17])\Gamma^3_{32}$$
$$d[18]=\Gamma^3_{10}+\Gamma^3_{20}+(a1[23]+b1[23])\Gamma^3_{01}+2(a2[23]+b2[23])\Gamma^3_{02}$$
$$+(a3[8]+b3[8])\Gamma^3_{13}+(a1[18]+b1[18])\Gamma^3_{31}+(a3[13]$$
$$+b3[13])\Gamma^3_{23}+(a2[18]+b2[18])\Gamma^3_{32}$$
$$d[19]=(a1[24]+b1[24])\Gamma^3_{01}+(a2[24]+b2[24])\Gamma^3_{02}+(a3[9]+b3[9])\Gamma^3_{13}$$
$$+(a1[19]+b1[19])\Gamma^3_{31}+(a3[14]+b3[14])\Gamma^3_{23}+(a2[19]+b2[19])\Gamma^3_{32}$$
$$d[20]=(a1[20]+b1[20])\Gamma^0_{01}+(a2[20]+b2[20])\Gamma^0_{02}+(a3[5]+b3[5])\Gamma^0_{13}$$
$$+(a1[15]+b1[15])\Gamma^0_{31}+(a3[10]+b3[10])\Gamma^0_{23}+(a2[15]+b2[15])\Gamma^0_{32}$$
$$d[21]=\Gamma^0_{10}+\Gamma^0_{20}+(a1[21]+b1[21])\Gamma^0_{01}+(a2[21]+b2[21])\Gamma^0_{02}$$
$$+(a3[6]+b3[6])\Gamma^0_{13}+(a1[16]+b1[16])\Gamma^0_{31}+(a3[11]$$
$$+b3[11])\Gamma^0_{23}+(a2[16]+b2[16])\Gamma^0_{32}$$
$$d[22]=\Gamma^0_{10}+\Gamma^0_{20}(a1[22]+b1[22])\Gamma^0_{01}+(a2[22]+b2[22])\Gamma^0_{02}+(a3[7]+b3[7])\Gamma^0_{13}$$
$$+(a1[17]+b1[17])\Gamma^0_{31}+(a3[12]+b3[12])\Gamma^0_{23}+(a2[17]+b2[17])\Gamma^0_{32}$$
$$d[23]=\Gamma^0_{10}+\Gamma^0_{20}+(a1[23]+b1[23])\Gamma^0_{01}+(a2[23]+b2[23])\Gamma^0_{02}$$
$$+(a3[8]+b3[8])\Gamma^0_{13}+(a1[18]+b1[18])\Gamma^0_{31}+(a3[13]$$
$$+b3[13])\Gamma^0_{23}+(a2[18]+b2[18])\Gamma^0_{32}$$
$$d[24]=(a1[24]+b1[24])\Gamma^0_{01}+(a2[24]+b2[24])\Gamma^0_{02}+(a3[9]+b3[9])\Gamma^0_{13}$$
$$+(a1[19]+b1[19])\Gamma^0_{31}+(a3[14]+b3[14])\Gamma^0_{23}+(a2[19]+b2[19])\Gamma^0_{32}$$

Homework Problems

The following homework problems are prepared assuming that this book can be divided into three semester courses with three credit hours each: CFD I (Chapters 1 through 4 and 8 through 11), CFD II (Chapters 5 through 7 and 12 through 16), and CFD III (Chapters 17 through 27). Instead of providing homework assignments at the end of each chapter, some selected problems are given in this appendix. An emphasis is placed on comparisons between FDM, FEM, and FVM. Through these exercises, it is hoped that the reader gain appreciation for studying all available methods without prejudices so that, at the end, advantages and disadvantages of each method can be identified. This will be beneficial in making decisions on the most suitable choices for your problems at hand. A sample computer program can be found at http://www.uah.edu/cfd as detailed at the end of this appendix.

Homework problems for CFD I

1. One-dimensional problems

1.1 Given the differential equation

$$\frac{d^2u}{dx^2} - 2u = f(x) \quad 0 < x < 1, \quad f(x) = 4x^2 - 2x - 4$$

Boundary conditions:

$$\text{(A) } u(0) = 0, \quad \text{(B) } u(1) = -1, \quad \text{(C) } \frac{du(0)}{dx} = 1, \quad \text{(D) } \frac{du(1)}{dx} = -3$$

Develop a computer program to solve the above differential equation by FDM, FEM, FVM via FDM, and FVM via FEM, using 4 elements, 8 elements, and 16 elements. Draw the solution curves using computer graphics for the following boundary conditions:

(1) (A) and (B), (2) (A) and (D), (3) (B) and (C)

Compare with the exact solution and provide comments on your results.

1.2 Given the differential equation

$$\frac{d^2u}{dx^2} - u = f(x), \quad f(x) = 2 - 2\cos x - x^2, \quad 0 < x < 1$$

Exact solution: $u = x^2 + \cos x$

Boundary conditions:

$$(A)\ u(0) = 1, \quad (B)\ u(1) = 1.54,$$

$$(C)\ \frac{du}{dx}(0) = 0, \quad (D)\ \frac{du}{dx}(1) = 1.16$$

Develop a computer program to solve by FDM, FEM, FVM via FDM, FVM via FEM, using 8 elements, 16 elements, and 32 elements for the following boundary conditions:

(1) (A) and (B), (2) (A) and (D), (3) (B) and (C)

Compare with the exact solution and provide comments on your results.

2. Two-dimensional elliptic partial differential equation

Consider the two-dimensional heat conduction equation:

$$\frac{\partial^2 T}{\partial x^2} + \frac{\partial^2 T}{\partial y^2} = 0$$

in a rectangular plate ($L = 2$ m, $H = 1$ m) with the boundary conditions as shown. Use $T = 0$ at all interior nodes as an initial guess. Develop a computer program to solve using the 40 × 20 mesh. Double and triple the mesh sizes to compare with the exact solution:

$$T = T_0 \left[2 \sum_{n=1}^{N} \frac{1 - (-1)^n}{n\pi} \frac{\sinh (n\pi (H - y)/L)}{\sinh (n\pi H/L)} \sin \frac{n\pi x}{L} \right] \quad \text{(Try } N = 100 \text{ and } 500\text{)}$$

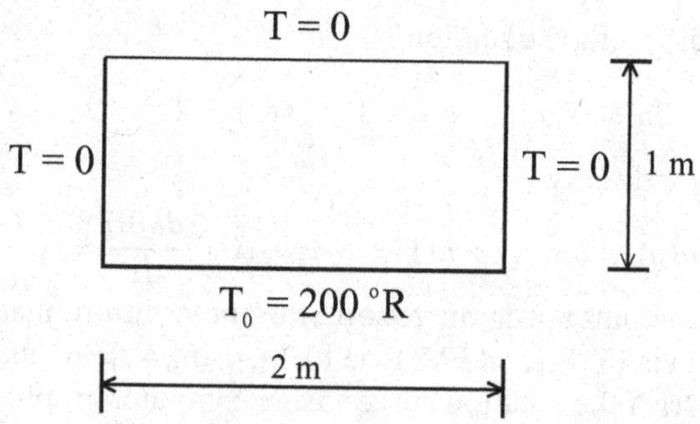

Use FDM (direct method, Jacobi, Point-Gauss–Seidel, PSOR, LSOR, and ADI).

3. One-dimensional parabolic partial differential equation for Couette flow

$$\frac{\partial u}{\partial t} - \nu \frac{\partial^2 u}{\partial y^2} = 0$$

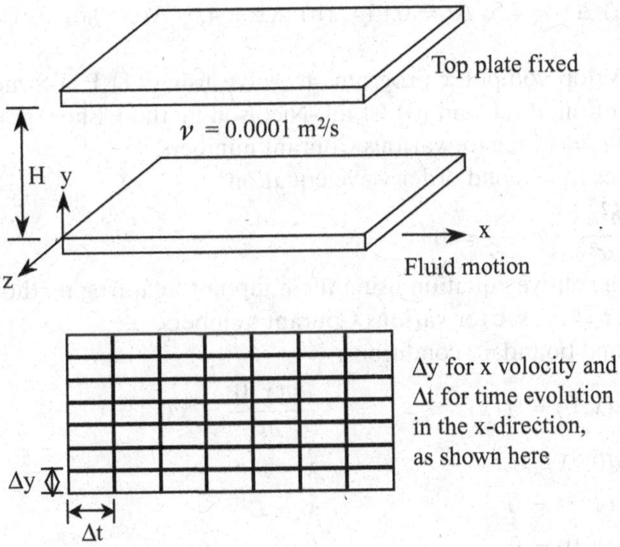

$u_0 = 20$ m/s	Initial conditions	$t = 0$	$u = u_0, y = 0$
H = 30 mm			$u = 0, 0 < y < H$
$\Delta y = 2$ mm	Boundary conditions	$t > 0$	$u = u_0, y = 0$
			$u = 0, y = H$

Develop computer programs using the following methods, show the results graphically, and provide comments.

FTCS Explicit Method:

(1) $\Delta t = 0.02$; (2) $\Delta t = 0.0205$

Crank-Nicolson Method:

(1) $\Delta t = 0.02$; (2) $\Delta t = 0.0205$

4. One-dimensional hyperbolic partial differential equation

4.1 Consider the first order wave equation:

$$\frac{\partial u}{\partial t} + a\frac{\partial u}{\partial x} = 0$$

with $a = 330$ m/s.

Initial and boundary conditions:

$$u(0,t) = 0 \qquad x = 0$$
$$u(L,t) = 0 \qquad x = L$$
$$u(x,0) = 0 \qquad 0 \le x \le 10$$
$$u(x,0) = 15\sin\left(\frac{\pi(x-10)}{30}\right) \qquad 10 \le x \le 80$$
$$u(x,0) = 0 \qquad 80 \le x \le 200$$

with (a) $\Delta x = 4.5$, $\Delta t = 0.014$; (b) $\Delta x = 4.5$, $\Delta t = 0.01$; (c) $\Delta x = 4.5$, $\Delta t = 0.006$.

Develop computer programs to solve using (1) FTBS method, (2) Lax-Wendroff method, and (3) Crank-Nicolson method. Show graphically the results at $t = 0.5$ sec for various Courant numbers.

4.2 Consider the second order wave equation:

$$\frac{\partial^2 u}{\partial t^2} - a^2 \frac{\partial^2 u}{\partial x^2} = 0$$

Solve the above equation using the midpoint Leapfrog method. Show the results at $t = 0.3$ sec for various Courant numbers.

Initial and boundary conditions:

$$u(x,0) = f(x), \qquad \frac{\partial u(x,0)}{\partial t} = g(x) = 0$$

$$u(0,t) = 0 \qquad x = 0$$

$$u(L,t) = 0 \qquad x = L$$

$$u(x,0) = 0 \qquad 0 \le x \le 10$$

$$u(x,0) = 60 \sin\left(\frac{\pi(x-10)}{30}\right) \qquad 10 \le x \le 20$$

$$u(x,0) = 0 \qquad 20 \le x \le 30$$

4.3 Consider the nonlinear convection equation:

$$\frac{\partial u}{\partial t} + u\frac{\partial u}{\partial x} = 0$$

or

$$\frac{\partial u}{\partial t} + \frac{\partial F}{\partial x} = 0, \quad F = \frac{1}{2}u^2$$

with

$$u(x,0) = 2 \qquad 0 \le x \le 3$$

$$u(x,0) = 0.5 \qquad 3 \le x \le 5$$

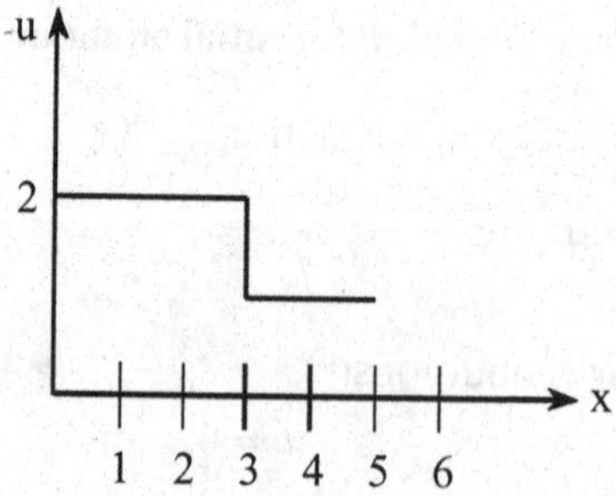

Solve using (1) Lax method, (2) Lax–Wendroff method, and (3) MacCormack method, with $\Delta t/\Delta x = 1$ and $\Delta t/\Delta x = 0.5$. Show the results graphically at $t = 0, 0.5, 1.0, 1.5,$ and 2.0 sec.

5. One-dimensional Burgers' equation

Consider the following:

Nondimensional form:

$$\frac{\partial u}{\partial t} + u\frac{\partial u}{\partial x} - \nu\frac{\partial^2 u}{\partial x^2} = 0$$

Conservation form:

$$\frac{\partial u}{\partial t} + \frac{\partial F}{\partial x} - \nu\frac{\partial^2 u}{\partial x^2} = 0, \qquad F = \frac{1}{2}u^2$$

Alternate form:

$$\frac{\partial u}{\partial t} + A\frac{\partial u}{\partial x} = \nu\frac{\partial^2 u}{\partial x^2}, \qquad A = \frac{\partial F}{\partial u}$$

Solve using (1) FTCS explicit method, (2) MacCormack explicit method, and (c) BTCS implicit method.

Boundary conditions:

$u = 2$ at $x = -9$ and $u = -2$ at $x = 9$

Exact solution for these boundary conditions, $\nu = 1$:

$$u = -\frac{2\sinh x}{\cosh x - e^{-t}}$$

Use $\Delta x = 0.2$, $\Delta t = 0.01$. Compute at $t = 0.1, 0.4, 0.8, 1.0$ sec with (a) $\Delta x = 0.2$, $\Delta t = 0.02$, (b) $\Delta x = 0.2$, $\Delta t = 0.05$, (c) $\Delta x = 0.5$, $\Delta t = 0.01$, and (d) $\Delta x = 0.5$, $\Delta t = 0.05$.

6. Repeat Problem 2 using FEM (GGM, TGM, and GPG)

7. Repeat Problem 3 using FEM (GGM, TGM, and GPG)

8. Repeat Problem 4 using FEM (GGM, TGM, and GPG)

9. Solve the two-dimensional Poisson equation using FEM (GGM, TGM, GPG) and FDM

$$\frac{\partial^2 u}{\partial x^2} + \frac{\partial^2 u}{\partial y^2} + f(x, y) = 0, \quad f(x, y) = -2y$$

Exact solution: $u = x^2 y$

Boundary conditions and initial conditions are to be specified (using the exact solution) as shown in the figure below, with Neumann boundary conditions to be specified at nodes with letter N, and Dirichlet elsewhere. Begin with all interior nodes specified as $u = 0$. Compare the results of coarse, intermediate, and fine grids. For FEM use both triangular elements and quadrilateral isoparametric elements for comparisons. Use the five-point scheme for FDM.

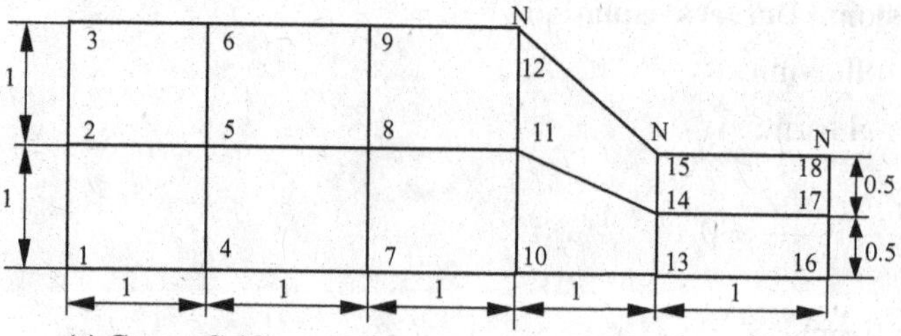

(a) Coarse Grid

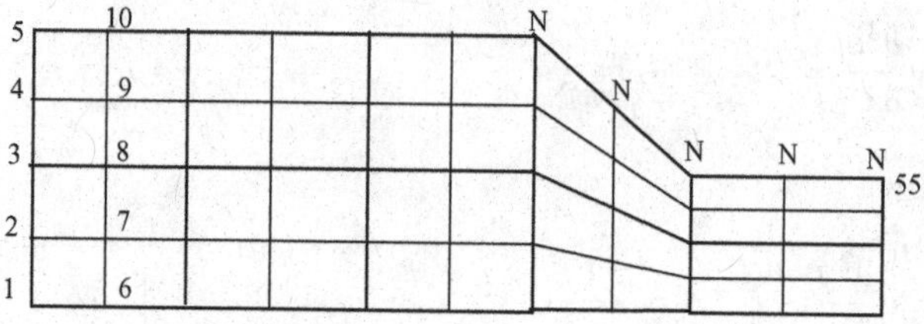

(b) Intermediate Grid (Halved from the Coarse Grid)

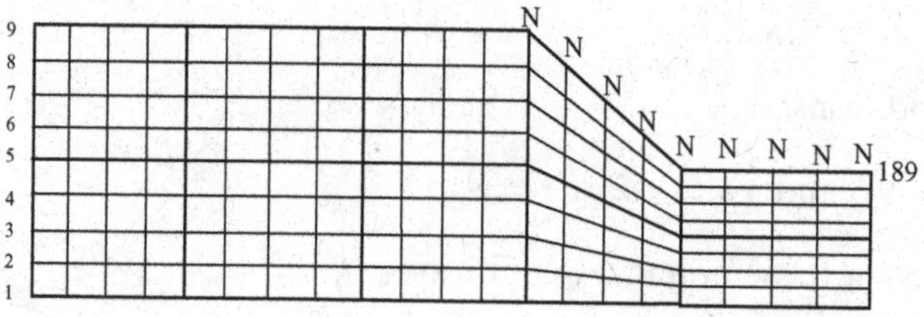

(c) Fine Grid (Halved from the Intermediate Grid)

10. Repeat Problem 9 for the two-dimensional transient problem

$$\frac{\partial u}{\partial t} - \frac{\partial^2 u}{\partial x^2} - \frac{\partial^2 u}{\partial y^2} - f(x, y) = 0, \quad f(x, y) = -\frac{1}{(1+t)^2} - 2y, \quad u = \frac{1}{1+t} + x^2 y$$

11. Repeat Problem 9 for the two-dimensional transient convection-diffusion equations

$$\frac{\partial u}{\partial t} + u\frac{\partial u}{\partial x} + v\frac{\partial u}{\partial y} - \nu\left(\frac{\partial^2 u}{\partial x^2} + \frac{\partial^2 u}{\partial y^2}\right) - f_x = 0,$$

$$f_x = \frac{1}{1+t}\left(x^2 + 2xy - \frac{1}{1+t}\right) + 3x^3y^2 - 2\nu y$$

$$\frac{\partial v}{\partial t} + u\frac{\partial v}{\partial x} + v\frac{\partial v}{\partial y} - \nu\left(\frac{\partial^2 v}{\partial x^2} + \frac{\partial^2 v}{\partial y^2}\right) - f_y = 0,$$

$$f_y = \frac{1}{1+t}\left(y^2 + 2xy - \frac{1}{1+t}\right) + 3x^2y^3 - 2\nu x$$

$$\nu = 10^{-3}, 1, 10^3$$

Exact solution:

$$u = \frac{1}{1+t} + x^2 y, \quad v = \frac{1}{1+t} + xy^2$$

Homework Problems for CFD II

1. Lid-driven cavity incompressible flow

Use FDM (ACM, SIMPLE, SIMPLER, SIMPLEC, and PISO). Develop a computer program and draw streamline distributions for Re $= 10, 10^2, 10^3, 10^4$, $v = 1$. Boundary conditions: $u_0 = 1$ and $v_0 = 0$ at the top, and $u = v = 0$ at walls.

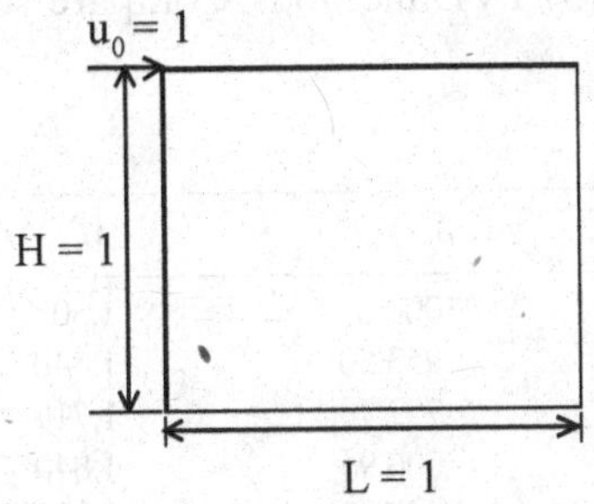

2. Repeat Problem 1 for a backstep geometry as shown with $u_{max} = 1$ at inlet

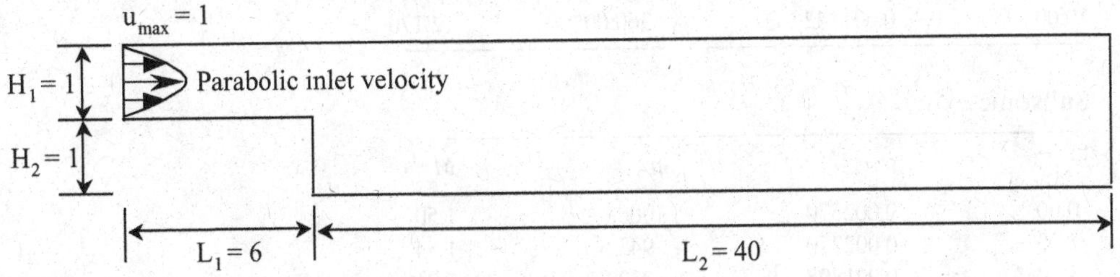

3. Repeat Problems 1 and 2 using the vortex method

4. Consider the Euler equation (compressible flow)

$$\frac{\partial \mathbf{U}}{\partial t} + \frac{\partial \mathbf{F}}{\partial x} - \mathbf{H} = 0$$

$$\mathbf{U} = \begin{bmatrix} \rho \\ \rho u \\ \rho E \end{bmatrix}, \quad \mathbf{F} = A \begin{bmatrix} \rho u \\ \rho u^2 + p \\ (\rho E + p)u \end{bmatrix}, \quad \mathbf{H} = \frac{dA}{dx} \begin{bmatrix} 0 \\ p \\ 0 \end{bmatrix}$$

These equations represent the flow of a compressible gas inside a diverging nozzle (10 ft long) with cross section given by $A(x) = 1398 + 0.347 \tanh(0.8x - 4)\text{ft}^2$ with $\gamma = 1.4$, $R = 1716 \frac{\text{ft}^2}{\text{sec}^2\ {}^\circ R}$

Inlet:

$$M = 1.5, \qquad p = 1000\ \text{lbf/ft}^2, \qquad \rho = 0.00237\ \text{slug/ft}^3$$

$$\rho u = 2.7323\ \text{slug/(ft}^2\ \text{sec)}, \qquad \rho E = 4075\ \text{slug/(ft sec}^2)$$

Exit:

Either supersonic or subsonic with $u = 390.75$ ft/sec

Supersonic exit:

$\rho = 0.00237, \quad \rho u = 2.7323, \quad \rho E = 4075 \quad \text{for } 0 \leq x \leq 10$

Subsonic exit:

$\rho = 0.00237, \quad \rho u = 2.7323, \quad \rho E = 4075 \quad \text{for } 0 \leq x \leq 2.8$

$\rho = 0.00237, \quad \rho u = 0.92608, \quad \rho E = 2680.93 \quad \text{for } 2.8 \leq x \leq 10$

Solve with FDM using (1) Lax-Wendroff explicit, (2) explicit MacCormack, (3) Flux splitting, (4) MUSCL, and (5) TVD methods. Compare your results with the analytical solution.

Supersonic exit:

x	ρ	p	M
0.00	0.002370	1000.0	1.50
3.00	0.002270	953.50	1.540
4.20	0.001898	732.70	1.710
4.80	0.001650	600.95	1.840
5.60	0.001335	447.90	2.030
6.80	0.001170	371.15	2.150
8.00	0.001148	362.60	2.165
10.00	0.001142	360.00	2.170

Subsonic exit:

x	ρ	p	M
0.00	0.002370	1000.0	1.50
3.00	0.002270	943.5	1.54
4.20	0.001898	732.7	1.71
5.00	0.00153	539.6	1.91
5.01	0.00386	2206.6	0.59
6.00	0.00410	2400.7	0.47
7.00	0.00415	2445.1	0.44
10.00	0.00417	2459.6	0.43

5. Solve the Euler equations for two-dimensional domain shown below

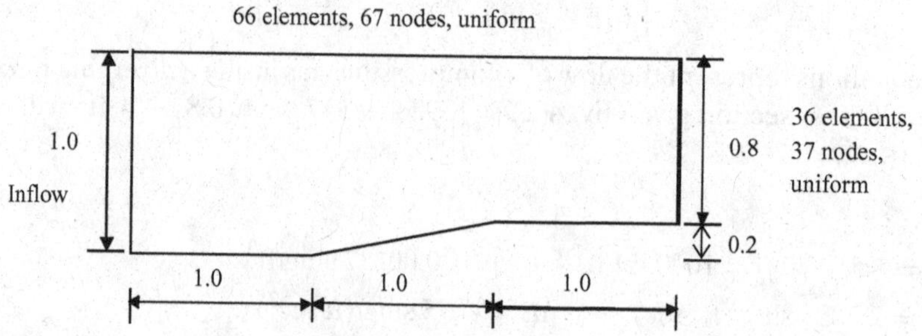

$$\frac{\partial \mathbf{U}}{\partial t} + \frac{\partial \mathbf{F}_i}{\partial x_i} = 0, \quad \mathbf{U} = \begin{bmatrix} \rho \\ \rho V_i \\ \rho E \end{bmatrix}, \quad \mathbf{F}_i = \begin{bmatrix} \rho V_i \\ \rho V_i V_j + p\delta_{ij} \\ \rho E V_i + p V_i \end{bmatrix}$$

Inlet:

$$M = 2, \quad \gamma = 1.4, \quad R = 1716 \text{ ft}^2/\text{sec}^2/\text{sec}^2\,°R, \quad T = 519\,°R$$

$$a = \sqrt{\gamma RT} = 117 \text{ ft/sec}, \quad \rho = 0.002378 \text{ slugs/ft}^3$$

$$u = 2 \times 1117 = 2234 \text{ ft/sec}, \quad v = 0, \quad p = 2116 \text{ lbf/ft}^2$$

$$\rho E = \frac{p}{\gamma - 1} + \frac{1}{2}\rho(u^2 + v^2) = 11224 \text{ lbf/ft}^2$$

Initial conditions: Use inlet conditions as initial conditions for all nodes.
Boundary conditions: Supersonic inlet, supersonic exit, slip wall conditions.
Solve using MacCormack, Lax-Wendroff, flux vector splitting, MUSCL, TVD methods.

6. Repeat Problems 4 and 5 using FVM via FDM
7. Repeat Problem 1 using FEM (TGM, GPG, and FDV)
8. Repeat Problem 2 using FEM (TGM, GPG, and FDV)
9. Repeat Problem 3 using FEM (TGM and GPG)
10. Repeat Problem 4 using FEM (TGM, GPG, and FDV)
11. Repeat Problem 6 using FVM via FEM
12. Develop programs to solve the Navier-Stokes system of equations for Problem 5 using all methods required for Problems 5 through 11. Repeat these programs for a geometry in three-dimensional, with the depth of x_3-direction given as 1 in the figure of Problem 5.

$$\frac{\partial \mathbf{U}}{\partial t} + \frac{\partial \mathbf{F}_i}{\partial x_i} + \frac{\partial \mathbf{G}_i}{\partial x_i} = 0$$

$$\mathbf{U} = \begin{bmatrix} \rho \\ \rho V_j \\ \rho E \end{bmatrix}, \quad \mathbf{F}_i = \begin{bmatrix} \rho V_i \\ \rho V_i V_j + p\delta_{ij} \\ \rho E V_i + p V_i \end{bmatrix}, \quad \mathbf{G}_i = \begin{bmatrix} 0 \\ -\tau_{ij} \\ -\tau_{ij} V_j + q_i \end{bmatrix}$$

Homework Problems for CFD III

1. Develop computer programs to reproduce grids as shown in

(a) Fig. E17.1.1, physical and transformed geometries
(b) Fig. E17.1.2, quadratic Lagrange polynomials
(c) Fig. E17.1.3, three-dimensional grids
(d) Fig. E17.1.4, clustering of mesh lines
(e) Fig. E17.1.5, conical body
(f) Example 17.1.6

2. Develop computer programs to reproduce grids as shown in

(a) Fig. E17.2.1, elliptic grid generation, TFI
(b) Fig. E17.3.4, surface grid generation, Bezier curve

3. Develop computer programs to reproduce grids as shown in

(a) Fig. 18.1.7, Delaunay-Voronoi, Watson algorithm
(b) Fig. 18.1.9, Delaunay-Voronoi, Bowyer algorithm

4. Develop computer programs to reproduce grids as shown in

(a) Fig. 18.2.3, advancing front method (AFM)
(b) Fig. 18.2.4, AFM smoothing
(c) Fig. 18.4.1, tetrahedral elements, NACA0012 airfoil

5. Develop computer programs to reproduce grids as shown in

(a) Fig. 19.2.4, adaptive mesh refinement (*h*-method), GPG
(b) Fig. 19.2.5, adaptive mesh refinement (*h*-method), FDV

6. Develop a computer program for an example of domain decomposition
7. Develop a computer program for an example of multigrid methods
8. Develop a computer program for an example of parallel processing
9. Special term projects: One or two chapters in Part V may be used for special term projects so that the automatic mesh generation studied in Part IV can be utilized, leading to a complete CFD project.

Note: Implementations of boundary conditions and methods of solutions for algebraic equations vary considerably, depending on flow conditions, geometries, and types of equations. They have been discussed in various chapters and sections as summarized below.

Boundary conditions: 1.6.1, 1.6.2, 2.3, 6.7.1, 6.7.2, 10.1.2, 11.1.1, 11.1.2, 13.6.6
Equation solvers: 4.2.7, 4.4.2, 4.5.1, 10.3.1, 11.5.1, 11.5.2, 11.5.3

A Computer Program (Fortran 90) for the Solution of Navier-Stokes System of Equations Using the Flowfield-Dependent Variation (FDV) Method with Finite Elements

Note: Computed results and source code available at http://www.uah.edu/cfd.

This is a computer program for the solution of Navier-Stokes system of equations in which all features of flows are included to accommodate a wide variety of Mach numbers and Reynolds numbers (compressible, incompressible, inviscid, and viscous flows). The governing equations are of the form (conservation form of the Navier-Stokes system of equations):

$$\frac{\partial \mathbf{U}}{\partial t} + \frac{\partial \mathbf{F}_i}{\partial x_i} + \frac{\partial \mathbf{G}_i}{\partial x_i} = \mathbf{B}$$

The solution is carried out using the flowfield-dependent variation (FDV) method with element-by-element (EBE) assembly via generalized minimal residual (GMRES) solution scheme using finite element discretizations with isoparametric elements and Gaussian quadrature integrations.

The advantages of FDV method are as follows:

(1) The first-order FDV parameters (s_1, s_3) as calculated from the current flowfield variables (Mach numbers and Reynolds numbers) assure the accuracy of solution. They alter the roles of each term in the governing equations in different positions of the domain, reflecting the incompressible behavior very close to the wall and compressible behavior or shock wave discontinuities away from the wall automatically. This can be demonstrated by contour plots of the FDV parameters themselves resembling the actual flowfields. The FDV scheme provides accurate solutions in turbulence with DNS mesh configurations and in supersonic combustion through FDV Jacobians.
(2) The second-order FDV parameters (s_2, s_4) assure the stability of solution process.
(3) A single program based on the FDV theory is capable of accommodating all different flow physics, high speed or low speed, compressible or incompressible, viscous or inviscid, in one-, two-, and three-dimensional geometries, reflecting the interactions between various physical phenomena.
(4) The FDV method can be applied to both FDM and FEM geometries.

Example problems include:

I. Incompressible viscous flow
   A. Lid-driven cavity flow (two-dimensional, three-dimensional)
   B. Backstep flow (two-dimensional, three-dimensional)
II. Compressible (inviscid or viscous) flow
   A. Shock tube (one-dimensional)
   B. Transonic flow (variable cross sections with one-dimensional formulation)
   C. Flat plate flow (two-dimensional, three-dimensional)
   D. Compression corner flow (two-dimensional, three-dimensional)
   E. Supersonic combustion chamber fin-inlet flow (three-dimensional)

Index

Printed in the United States
By Bookmasters